Contents

TRADEMARKS

Algon®	Algonquin Industries, Inc.
Apple II™ **IIc™** **IIe™** **II Plus™**	Apple Computer, Inc.
Cyber™	Control Data Corp.
Cray X-MP™	Cray Research, Inc.
DEC™ **PDP™** **Unibus™**	Digital Equipment Corp.
Nichrome™	Driver Harris Co.
Hypalor® **Kevlar®** **Mylar®** **Teflon™** **Tefzel™**	E.I. DuPont de Nemours and Co., Inc.
Alumel™ **Chromel™**	Hoskins Manufacturing Co.
Multibus™	Intel Corporation
IBM™ **IBM PC™** **PC AT™** **PC XT™**	International Business Machines Corp.
HEXFET™	International Rectifier, Semiconductor Div.
Moog®	Moog Inc., Industrial Div.
TMOS™ **VMEbus™**	Motorola Semiconductor Products Sector.
Teletype®	Teletype Corporation
Telex™	Telex Communications, Inc.
Remington Rand™	Unisys Corp.
Ethernet® **Xerox®**	Xerox Corp.

Preface

The *Encyclopedia of Electronics—2nd Edition* is intended as a general reference in the rapidly expanding field of electronics. Where so comprehensive a subject as electronics is concerned it is inevitable that there should be an overlap with physics, mathematics, chemistry, and computer science. The introduction of new concepts and products into the field is being reflected in the language. Electronics-related words, phrases, acronyms, jargon, and even "buzz" words have kept pace with technology over the past five years, since the publication of the first edition.

Many words and phrases have migrated from the more forbidding lexicon of pure and applied science. People with little or no formal education in science and technology now regularly discuss products and concepts in terms that were heard only in the laboratory or seen only in professional journals a few years ago. High-technology terminology is entering the mainstream through the popular media—television, radio, newspapers, and general interest magazines. Some comprehension of advanced technology is needed to make sense of the copy in the ads for the latest TVs, VCRs, stereo systems, automobiles, and appliances—to say nothing of personal computers, cordless telephones, and facsimile machines.

Even those who profess to have no special interest in science and technology are now surrounded by products that contain such recent examples of high technology as LEDs, liquid crystals, integrated circuits, and microcontrollers. Products range from watches, calculators, and cameras to TV sets, stereos, VCRs, and microwave ovens. Traditional home appliances from telephones to washing machines have been improved with electronics. Many games and toys are now sophisticated electronics products. There may not be a personal computer in every home, but the numbers are rising.

Outside the home the impact of electronics is also conspicuous. Automobiles are now packed with electronic controls, safety devices, and entertainment products. The banks have computer-based automated tellers; service stations have computer-based test equipment; and the instruments and apparatus at the doctor's office have been updated with electronics. The new technology has been absorbed by business, industry, telecommunications, aviation, and even recreational boating. However, the most significant trend discernible over the past five years has been the ongoing merger between computers and communications.

Examples of words drawn from the electronics lexicon and not widely heard or seen just five years ago include *compact disks* (CDs), *cellular radio*, *communications satellites*, *dynamic memory* (D-RAM or DRAM), *facsimile* (FAX), *fiberoptic cable*, and *laser printers*. Those engaged in occupations within or related to the electronics field are sure to be aware of *ASIC*, *CMOS*, *VMEbus*, *Multibus*, *cache*, *emulation*, *coprocessor*, and *Ethernet*. There are frequent references to *CISC*, *RISC*, *SRAM*, *file servers*, *LANs*, *PLCCs*, and *PLDs*.

Despite the influx of new words and phrases, the classical vocabulary of electronics (largely adopted from electrical engineering) is still in wide circulation. These words are heard and written in the classroom, laboratory, repair shop, and factory. Some terminology has become obsolete and has disappeared along with the technology. There is little discussion these days of *triodes*, *pentodes*, *selenium rectifiers*, or even *germanium transistors*.

This encyclopedia was edited to fill a gap in reference sources between the dictionary (or dictionary of electronics) and formal textbooks or handbooks for the electronics engineering professional. It assumes that the average reader may want to know more about a subject than is given in a brief definition, yet may not be prepared to obtain that information from a professional-level text or handbook.

Even the reader with a formal background might not wish to take the time and make the effort to research the formal academic papers, journals, or texts just to find a simple, clear explanation of a topic in the field.

The articles in this encyclopedia are descriptive. They make fewer demands on the reader's educational background in electronics than do more academically rigorous references. Mathematics is used sparingly, and then only when necessary to explain a topic. Encyclopedias of science and technology cover many of the same subjects, but in ten or more volumes—and they may not be up to date on leading-edge subjects.

This encyclopedia does not call for minimum educational level, although a basic knowledge of high school physics and chemistry or practical electronics would be helpful. The single volume is intended to satisfy the reader's "need to know" with more than just a few sentences, any alternate meanings, and the correct spelling of the word or phrase. It is written to be a useful "stepping stone" or refresher for the reader wishing to delve deeper into specific subjects discussed in more advanced texts and papers.

This second edition has introduced many new articles and illustrations not found in the first edition. Some of the topics covered in the first edition that are considered obsolete have been deleted to make room for newer topics believed to be of more interest to a wider group of readers. The result is a book of about the same length as the first edition with more emphasis on coverage of the latest developments in electronics and computer hardware.

Because of its ongoing infiltration into all walks of human endeavor, even a fundamental knowledge of electronics is essential for an educated person. An otherwise well-informed person who has been "out of touch" with business and commerce for the past five years might have difficulty in reading and understanding some of today's popular articles without current knowledge. Articles on business, finance, and world economics now regularly cover the international electronics and computer industries, their products, and the politics of world trade.

Even persons with college degrees are finding it necessary to take formal courses on such recent subjects as word processing, computer graphics, computer-aided design, and desktop publishing because of the complex nature of those subjects.

This encyclopedia takes the view that computer science is a separate subject from, although it overlaps with, electronics. Thus the volume is neither a comprehensive encyclopedia of computers nor computer software; however, it contains many articles on topics in those fields. Selection was made on the basis of their relative importance to the wider field of electronics.

Standard Symbols for Use in Schematic Diagrams

Ammeter
AND Gate
Antenna, Balanced
Antenna, General etc. . . .
Antenna, Loop, Shielded
Antenna, Loop, Unshielded
Attenuator, Fixed
Attenuator, Variable
Battery + -
Capacitor, Feedthrough
Capacitor, Fixed, Nonpolarized
Capacitor, Variable, Ganged
Capacitor, Variable, Single
Capacitor, Variable, Split-Stator
Cathode, Indirectly Heated
Cavity Resonator
Cell Electrochemical + -
Cell, Photoconductive
Cell, Solar
Circuit Breaker
Coaxial Cable
Coupler, Directional
Coupler, Optoelectronic

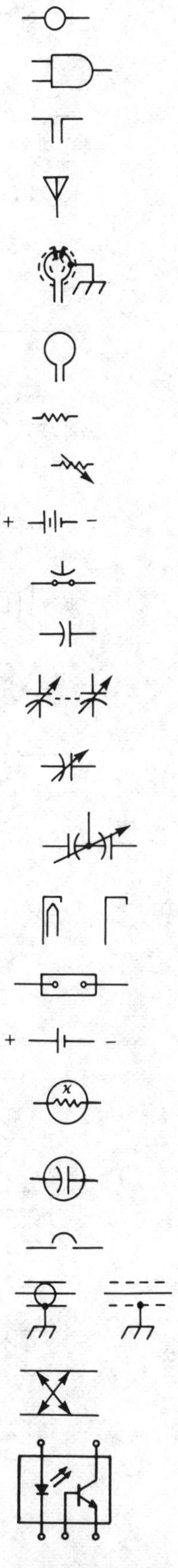

Crystal, Piezoelectric
Delay Line
Diac
Diode, General
Diode, Gunn
Diode, Light-Emitting
Diode, Photosensitive
Diode, Photovoltaic
Diode, Pin
Diode, Varactor
Diode, Zener
Exclusive-OR Gate
Female Contact, General
Fuse
Galvanometer G
Ground, Chassis
Ground, Earth
Headphone, Double
Inductor, Air-Core
Inductor, Iron-Core
Inductor, Tapped
Inductor, Variable
Integrated Circuit
Inverter or Inverting Amplifier
Jack, Coaxial

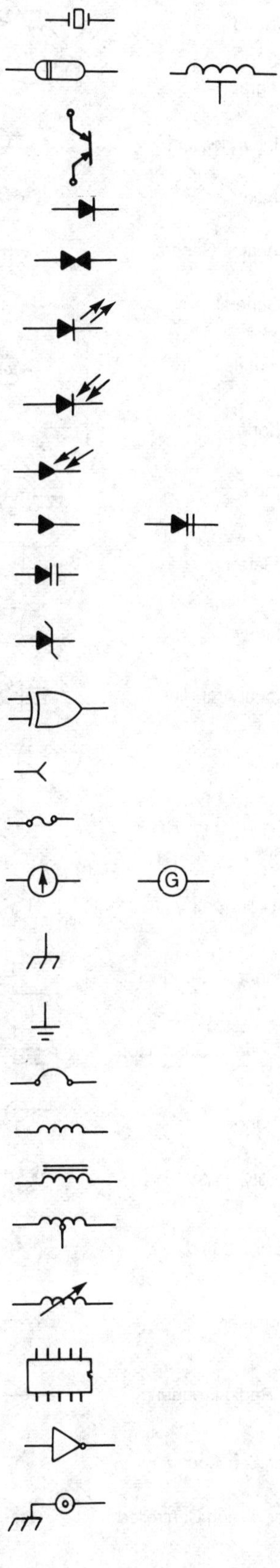

Jack, Phone, 2-Conductor

Jack, Phone, 2-Conductor Interrupting

Jack, Phone, 3-Conductor

Jack, Phono

Key, Telegraph

Lamp, Incandescent

Lamp, Neon

Male Contact, General

Meter, General

Microammeter

Microphone

Milliammeter

NAND Gate

NOR Gate

Operational Amplifier

OR Gate

Outlet, Utility, 117-V

Outlet, Utility, 234-V

Plug, Phone, 2-Conductor

Plug, Phone, 3-Conductor

Plug, Phono

Plug, Utility, 117-V

Plug, Utility, 234-V

Potentiometer

Probe, Radio-Frequency

Rectifier, Semiconductor

Rectifier, Silicon-Controlled

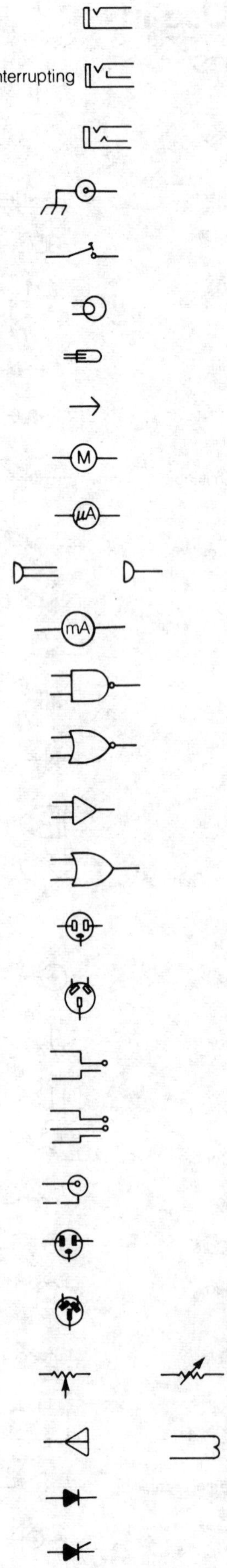

Relay, DPDT

Relay, DPST

Relay, SPDT

Relay, SPST

Resistor

Resonator

Rheostat

Saturable Reactor

Shielding

Signal Generator

Speaker

Switch, DPDT

Switch, DPST

Switch, Momentary-Contact

Switch, Rotary

Switch, SPDT

Switch, SPST

Thermistor

Thermocouple

Transformer, Air-Core

Transformer, Iron-Core

Transformer, Tapped Primary

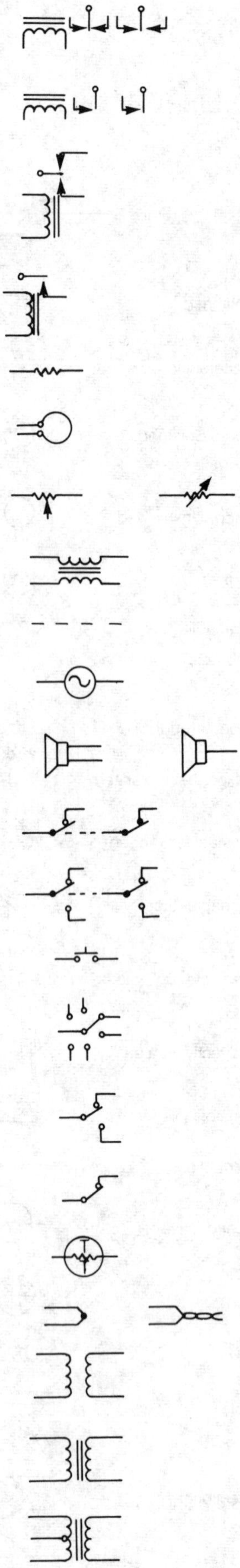

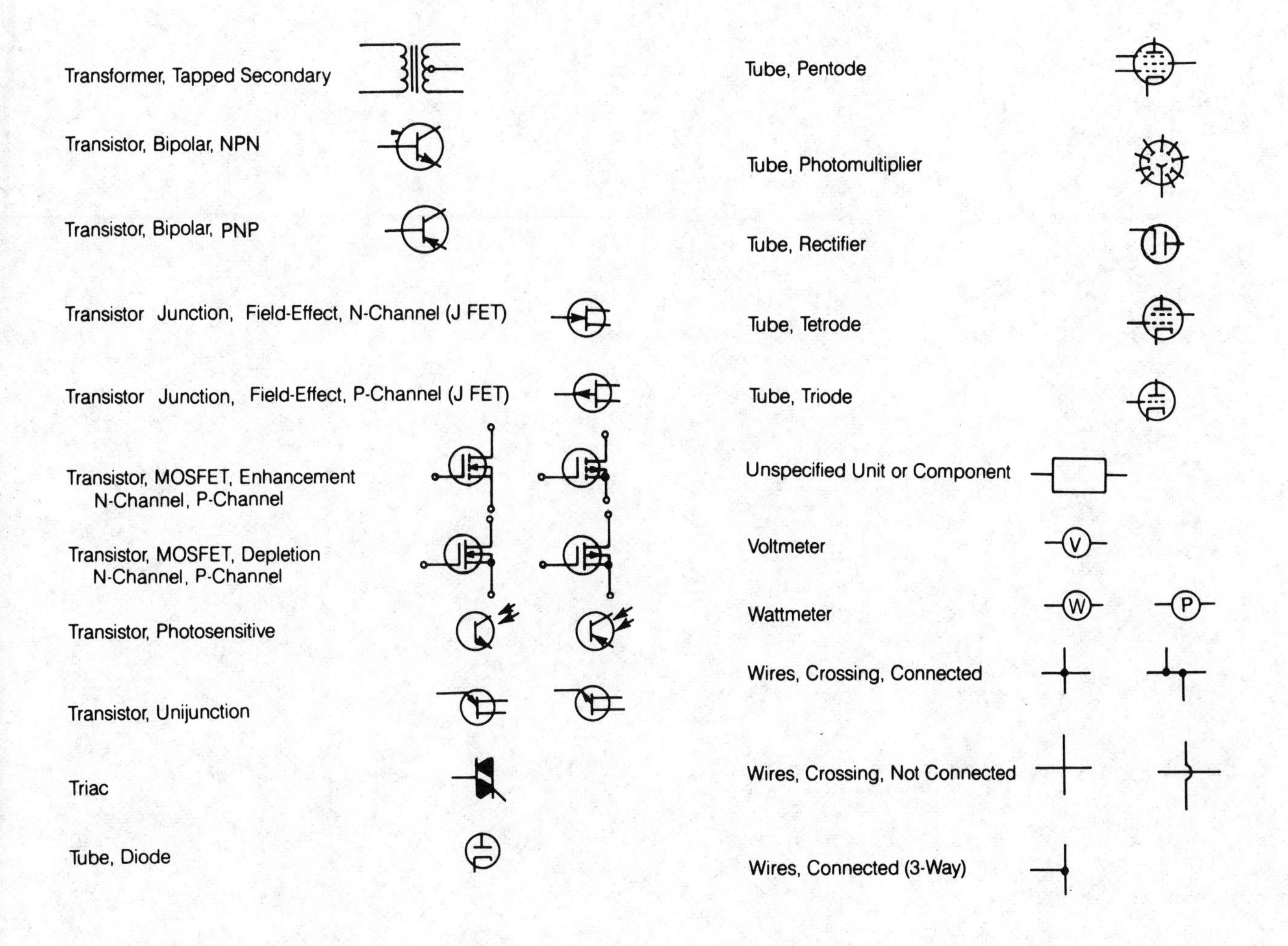
Transformer, Tapped Secondary
Transistor, Bipolar, NPN
Transistor, Bipolar, PNP
Transistor Junction, Field-Effect, N-Channel (J FET)
Transistor Junction, Field-Effect, P-Channel (J FET)
Transistor, MOSFET, Enhancement
N-Channel, P-Channel
Transistor, MOSFET, Depletion
N-Channel, P-Channel
Transistor, Photosensitive
Transistor, Unijunction
Triac
Tube, Diode
Tube, Pentode
Tube, Photomultiplier
Tube, Rectifier
Tube, Tetrode
Tube, Triode
Unspecified Unit or Component
Voltmeter
V
Wattmeter
W
P
Wires, Crossing, Connected
Wires, Crossing, Not Connected
Wires, Connected (3-Way)

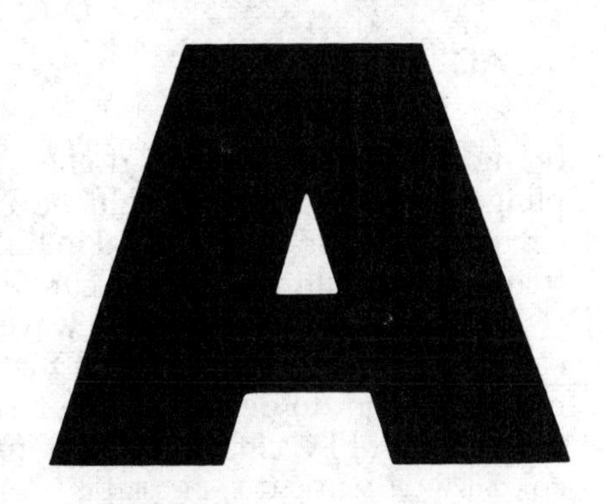

ABSOLUTE TEMPERATURE SCALE

See KELVIN TEMPERATURE SCALE.

ABSORPTANCE

When visible light strikes a substance, some of that light might pass through the substance or be reflected. The remainder of the energy is absorbed and converted to some other form of energy, usually heat. For transparent or semitransparent materials, absorptance is expressed in decibels per unit length (*see* DECIBEL). For reflective materials, absorptance is simply expressed in decibels. If P_1 is the impinging power and P_2 is the power following absorption so that P_2 is less than P_1, then (see the illustration):

$$\text{absorptance } (\textit{decibels}) = 10 \log_{10} (P_1/P_2)$$

Absorptance can be defined for any electromagnetic wavelength and any material substance. The resulting quantity is dependent on the wavelength of the electromagnetic energy as well as on the substance. Certain ultrapure glass can have absorptance levels as low as 2 decibels per kilometer. A perfect vacuum is considered to have an absorptance of 0 decibels per kilometer at all wavelengths. *See also* ABSORPTION, TRANSMITTANCE.

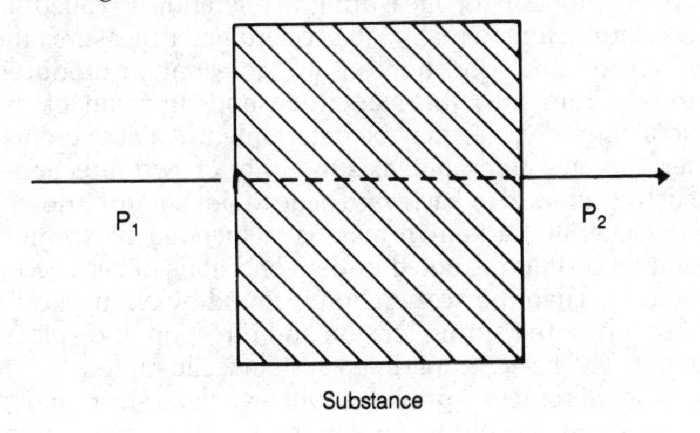

ABSORPTANCE: When light, radio, or sound waves enter a substance, interaction and dissipation occur. The output power P_2 is always less than the input power P_1.

ABSORPTION

When any energy or current is converted to some other form of energy and some energy is lost or dissipated, absorption is said to have taken place. Light might be converted to heat, for example, when it strikes a dark surface. Generally, absorption refers to an undesired conversion of energy.

Radio signals encountering the ionosphere undergo absorption, as well as refraction. The amount of absorption depends on the wavelength, the layer of the ionosphere, the time of day, the time of year, and the level of sunspot activity. Sometimes most of the radio signal is converted to heat in the ionosphere; sometimes very little is absorbed and most of it is refracted or allowed to continue on into space. (See A, B, and C in the illustration.)

The amount of absorption in a given situation is called *absorptance*, and is expressed in decibels or decibels per unit length. *See also* ABSORPTANCE.

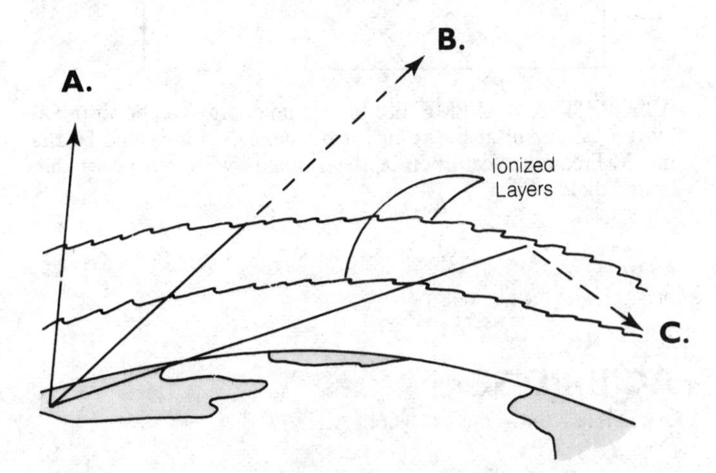

ABSORPTION: Radio waves directed into space might pass through the ionosphere unaffected (A), be partially absorbed before passing through (B), or be partially absorbed and reflected back to earth (C).

ABSORPTION WAVEMETER

An absorption wavemeter is a circuit for measuring radio frequencies. It consists of a tuned LC (inductance-capacitance) circuit, loosely coupled to the source to be measured. If the resonant frequency of the tuned circuit for various capacitor or inductor settings is known (by means of a calibrated dial), the LC circuit can be adjusted until maximum energy transfer occurs. This condition is indicated by a peaking of rf (radio frequency) voltage as shown by a meter. See the illustration.

When using an absorption wavemeter, it is important that it not be too tightly coupled to the circuit under test. An rf probe, or short wire pickup, is connected to the LC circuit of the absorption wavemeter and the probe brought near the source of rf energy. If too much coupling occurs, reactance might be introduced into the circuit under test, which might in turn change the circuit frequency, resulting in an inaccurate measurement.

A special kind of absorption wavemeter has a built-in oscillator. Known as a grid-dip meter, this instrument makes it easier to determine the resonant frequency or frequencies of LC circuits and antenna systems. *See also* COAXIAL WAVE METER, GRID-DIP METER.

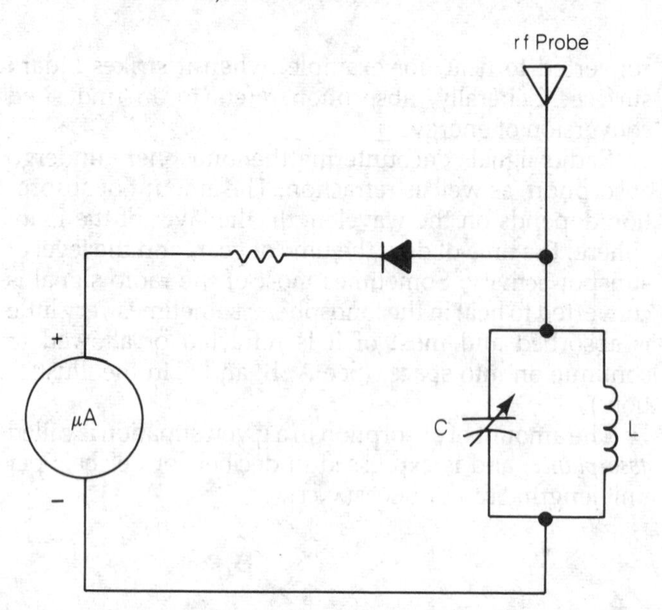

ABSORPTION WAVEMETER: Maximum energy will be absorbed by the LC circuit at the resonant frequency, as indicated by the meter. Frequency can then be determined by reference to a calibrated dial or chart.

AC

See ALTERNATING CURRENT.

AC BRIDGE

See ALTERNATING-CURRENT BRIDGE.

AC GENERATOR

See ALTERNATING-CURRENT GENERATOR.

AC NETWORK

See ALTERNATING-CURRENT NETWORK.

AC RELAY

See RELAY.

AC RIPPLE

See ALTERNATING-CURRENT RIPPLE.

AC VOLTAGE

See ALTERNATING-CURRENT VOLTAGE.

ACCELERATION

Acceleration is the rate of change in velocity of an object. Velocity is determined in a given direction and a given speed. When an object continues in a particular direction at a constant speed, the acceleration is zero. If either the speed or direction (or both) are changing, then the object is accelerating.

Velocity and acceleration are vector quantities, because both have a definite direction and intensity. When a car moves forward at increasing speed, its direction of acceleration is forward, in the same direction as the velocity. When the brakes are applied, the acceleration is directed backward, opposite the velocity. At a constant cruising speed along a curved road, the direction of the acceleration is toward the center of curvature, at roughly right angles to the velocity.

Electrons and other charged particles can be accelerated by electromagnetic fields, making possible radio-wave propagation. The acceleration might be back and forth, along a straight wire (linear acceleration). Acceleration might also take the form of circular motion.

ACCELERATION SWITCH

A switch that opens or closes when it is accelerated is known as an acceleration switch. Acceleration always produces a force opposite the direction of acceleration. If the force reaches a threshold intensity, the acceleration switch is activated.

A simple acceleration switch typically consists of a weighted spring with an electrical contact, which the weight will trip when the acceleration force is sufficient. Acceleration switches are frequently used in aerospace systems.

ACCELEROMETER

An instrument for measuring acceleration is called an accelerometer. Actually, an accelerometer measures the intensity and direction of an acceleration-produced force. From this, the magnitude and direction of the acceleration vector may be determined. An accelerometer operates along one axis in either of two directions, such as forward or backward, side to side, or up or down.

Generally, a known mass is suspended by springs, and its position is noted under conditions of zero acceleration. Then the acceleration is found by electronically measuring the spring tension and direction of displacement. The basic spring-mass system is illustrated.

Accelerometers provide input signals to steer guided missiles. A computer can determine course errors from the readings and produce instructions for corrective maneuvers.

Two linear (two-direction) accelerometers can be combined at right angles to form a two-dimensional accelerometer; three linear accelerometers can be combined orthogonally to form a three-dimensional accelerometer. The vector components are determined and then added. (*See* VECTOR, VECTOR ADDITION.)

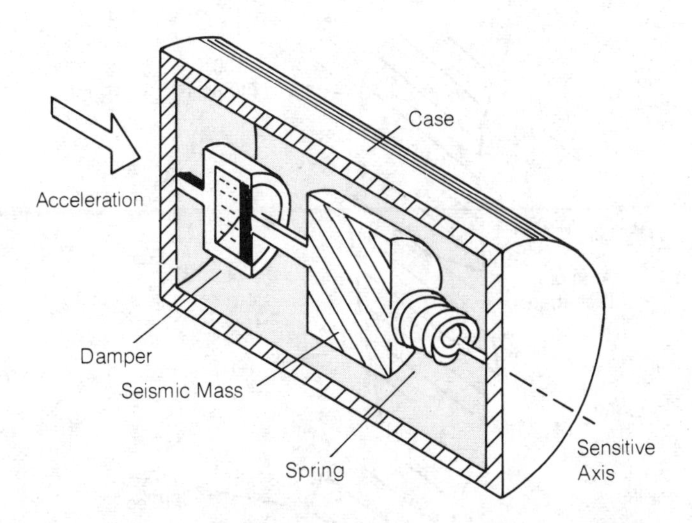

ACCELEROMETER: Acceleration is measured along the sensitive axis as the seismic mass shifts left with respect to the case. Motion is damped by the damper and the spring restores equilibrium.

ACCESS TIME

Access time is the time elapsed between a request for data from a memory storage device and the instant that data is received. Access time is a waiting time specified as minimum, maximum, or average.

Individual memory devices have specified access times. These vary depending on the size of the memory device and the technology used in its manufacture. Semiconductor memories have maximum access times from about 10 to 500 nanoseconds. For example, 1 megabit CMOS (complementary metal oxide semiconductors) dynamic RAMs (random access memories) are available with maximum access time between 85 and 120 nanoseconds. Bipolar memories have a speed advantage over MOS memories. By contrast, charge-coupled device memories have slower access times in the range of approximately 5 to 200 microseconds, and MBMs (magnetic bubble memories) have average access times of 10 to 40 microseconds.

Rigid magnetic disk drives have average access times, also called seek times, of 10 to 85 microseconds, and flexible or floppy disk drives have average access times of 75 to 100 microseconds. Seek time must be distinguished from latency time, which is the time for a single revolution of the disk. Latency time is a function of disk speed in revolutions per minute and is measured in milliseconds. *See* CHARGE-COUPLED DEVICE, DISK DRIVE, SEMICONDUCTOR MEMORY.

ACCUMULATOR

The accumulator of a computer is the register and associated equipment used to accumulate the results of arithmetic or logic operations. It usually stores one quantity and, on receipt of another, forms the sum and temporarily stores the result. In some computers the accumulator cannot be referenced directly, but in others, several addressable accumulators are available. *See* COMPUTER, MICROPROCESSOR.

ACOUSTIC COUPLER

An acoustic coupler is a device that provides an interface between a computer and a standard commercial dial-up telephone line through a standard telephone headset. After a receiving number is dialed, the telephone headset is placed in cushioned muffs on a desktop stand designed to cradle the headset. The serial data from the sending computer is then converted by a modem for transmission. Similarly, the computer can receive data over the phone line from the modem. Acoustic couplers are being replaced by direct-wired connections between the computer modem and the commercial telephone line. *See* MODEM.

ACOUSTIC FEEDBACK

Acoustic feedback is the positive feedback of acoustic energy in a system. If it reaches sufficient magnitude in an amplifier, it will cause oscillation. In a public-address system, acoustic feedback might occur not only electrically, between the input and output component wiring, but between the speaker(s) and microphone. The result is a loud audible rumble, tone, or squeal. It might take almost any frequency, and totally disables the public-address system for its intended use. This kind of feedback can also occur between a radio receiver and transmitter, if both are voice-modulated and operated in close proximity.

Another form of acoustic feedback occurs in voice-operated communications systems (*see* VOX). While receiving signals through a speaker, the sound might reach sufficient amplitude to actuate the transmitter switching circuits. This results in intermittent, unintended transmissions, and makes reception impossible. Compensating circuits in some radio transceivers equipped with VOX reduce the tendency toward this kind of acoustic feedback.

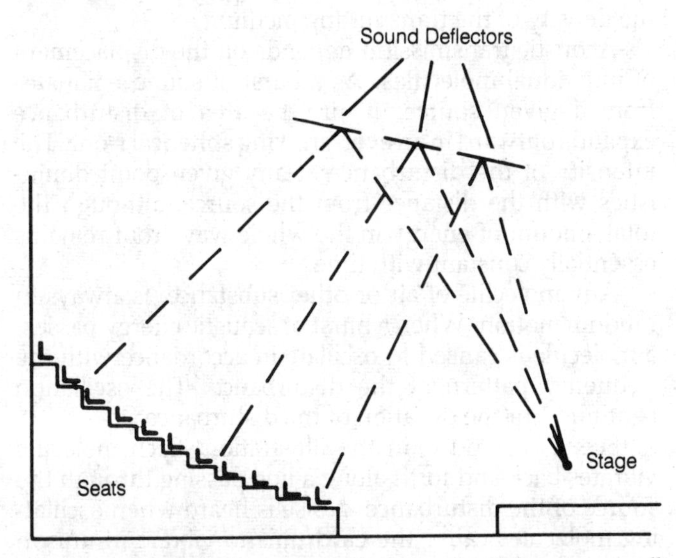

ACOUSTICS: Sound deflectors enhance audibility in a well-designed auditorium.

ACOUSTICS

The science of sound engineering is known as *acoustics*.

The excellent sound-distributing qualities of a symphony concert hall are the result of careful design. (See the illustration.) If the room is badly designed acoustically, there will be areas where certain sound frequencies are amplified or attenuated out of proportion. Violins might be heard but not the cello. Sound from parts of the stage might not be amplified enough to be heard in all parts of the hall.

Acoustics are important in high-fidelity sound recording. The higher frequencies must be carefully balanced with the low and midrange frequencies. The channel balance in stereo recordings must be accurate. Also, relative loudness must be faithfully retrieved. *See also* HIGH FIDELITY.

ACOUSTIC TRANSDUCER

A sensor that converts sound energy to another form of energy, or some form of energy to sound energy, is an acoustic transducer. The most common acoustic transducers are the familiar speaker and microphone. Buzzers, bells, and sound or vibration detectors, however, are also acoustic transducers. *See also* MICROPHONE, SPEAKER, ULTRASONIC TRANSDUCER.

ACOUSTIC TRANSMISSION

Acoustic transmission is the transfer of energy in the form of regular mechanical vibration through a solid, liquid, or gaseous medium. The most familiar form of acoustic transmission is the sound wave. In air, sound waves travel at approximately 1100 feet per second. In water, sound waves travel considerably faster; in a metallic solid such as steel, sound travels even faster. In general, the speed of acoustic transmission is roughly proportional to the density of the transmitting medium.

Acoustic transmission depends on the displacement of individual molecules. As a burst of sound emanates from a given source in air, the area of disturbance expands outward in an ever-growing spherical front. The intensity of the disturbance at any given point diminishes with the distance from the source, although the total amount of energy in the whole wavefront remains essentially constant with time.

Any molecule of air or other substance, is always in random motion. When a burst of acoustic energy passes, a molecule is caused to oscillate in accordance with the frequency pattern of the disturbance. The oscillation continues for the duration of the disturbance.

See A, B, and C in the illustration. Each molecule vibrates back and forth along a line passing through the source of the disturbance. Noise is heard when oscillating molecules cause the eardrums to vibrate in unison with them. A transducer (such as a microphone) converts acoustic disturbances into electrical impulses. Acoustic transmission cannot take place in a vacuum, because there are no molecules to provide a medium for sound waves.

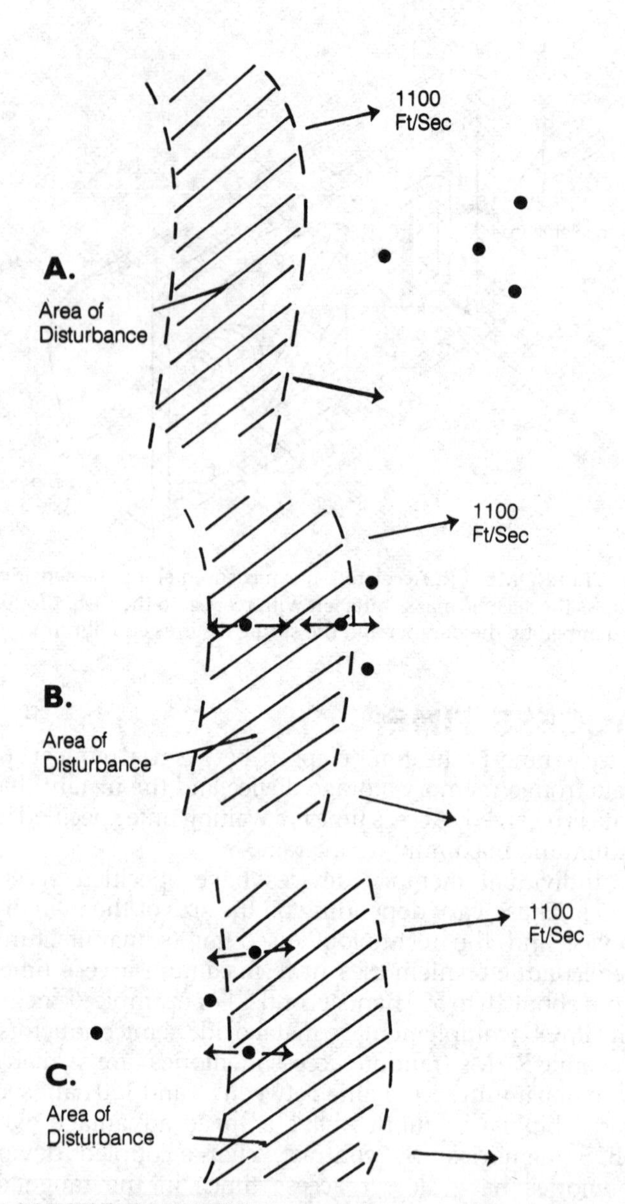

ACOUSTIC TRANSMISSION: Sound waves in air approach molecules in random motion (A), causing them to oscillate in response to the frequency of the longitudinal vibrations (B), before they return to random motion (C).

ACTIVE ANALOG FILTER

An *active analog filter* includes an operational amplifier and external resistors and capacitors. These devices act as filters to eliminate the need for inductors. They can amplify the signal or provide gain, but they require a power source. The values of the resistors and capacitors must be precise for active filtering. Fully integrated active resistor-capacitor filters have not been successful because neither resistors nor capacitors can be fabricated precisely enough on a silicon chip. Moreover, integrated resistors take up too much space on the chip.

In contrast to passive and active analog filters, integrated filters called switched-capacitor filters (SFC) have variables such as gain, Q factor, center frequency, and bandwidth that can be closely controlled. The precision of these filters depends on the ratios of the on-chip capacitors, not on their absolute values. Ratios are easy to control on MOS (metal-oxide silicon) ICs (integrated circuits). Because SCFs are monolithic, their variables are very stable over time and do not respond to changes in temperature. Their stability is 10 to 20 times better than active filters. *See* SWITCH-CAPACITOR FILTER.

ACTIVE COMPONENT

In electronics, an active component is one that requires a source of power to function. Active components are used for gain, oscillation, switching action, or rectification. If components draw power from an external source such as a battery or power supply, they are called active components. Active components include integrated circuits, transistors, vacuum tubes, and diodes.

A passive component, by contrast, works with no outside source of power. Passive components include resistors, capacitors, inductors, and some diodes.

AC-TO-DC CONVERTER

An ac-to-dc (alternating current to direct current) converter, more commonly called a dc power supply, consists of rectifier and filter circuits. The rectifier is typically a diode bridge, and the filter circuit is a network containing capacitors and maybe resistors or chokes. *See also* POWER SUPPLY, RECTIFICATION, RECTIFIER.

ADAPTIVE CONTROL

Adaptive control is a method of achieving near optimum performance of an industrial process. Adaptive control continuously and automatically makes adjustments for errors or deviations in tooling or the work in response to sensed changes in process variables. Adaptive control is typically accomplished by means of sensors or transducers that measure changes in the working conditions and initiate corrections. Factors such as wear in tooling, variations in the size or shape of the work, alterations in tool-work orientation, or changes in the environment are sensed in a feedback loop closed around a computer. A program routine in the computer processes sensor information and sends appropriate signals to correct the operation of the end effector.

Adaptive control on a machine tool, for example, might use a control loop to measure tool wear or misalignment and send signals through the computer back to the machine tool to reposition the tool holder. Another example of adaptive control can be found in robotic welding. Sensors in a closed computer control loop initiate the repositioning of an arc welding torch to compensate for changes in the position of the workpiece that have occurred since the robot was taught the job during setup. Adaptive controls also permit the torch to compensate for variations in the size or shape of the workpiece and the length of the welding rod. *See* ROBOT.

ADCOCK ANTENNA

See ANTENNA DIRECTORY.

ADDER

An adder or binary adder is a logic circuit that receives data from two sources and provides a sum and a carry. Adder circuits are fundamental computational building blocks in digital computers. Computers work with binary numbers from 8 to 32 bits wide. There must be as many adder circuits as bit width in a computer.

An add operation can be performed with a circuit known as a half adder as shown in Fig. 1A. This can be formed from AND, OR, and NOT gates or can be simplified to an exclusive OR and an AND gate as illustrated by logic symbols. By the rules of binary arithmetic, the sum of any two input numbers, 1 or 0, is a single number 1 or 0.

The half adder has two inputs—A and B—and two outputs—sum (S) and carry (C), as shown. The truth table for a half adder (*see* TRUTH TABLE) shows that if both inputs are 0, the sum and carry are 0; if either of the inputs is 1 or 0, the sum will be 1. But, if both inputs are 1, the sum is 1 and the carry is 1.

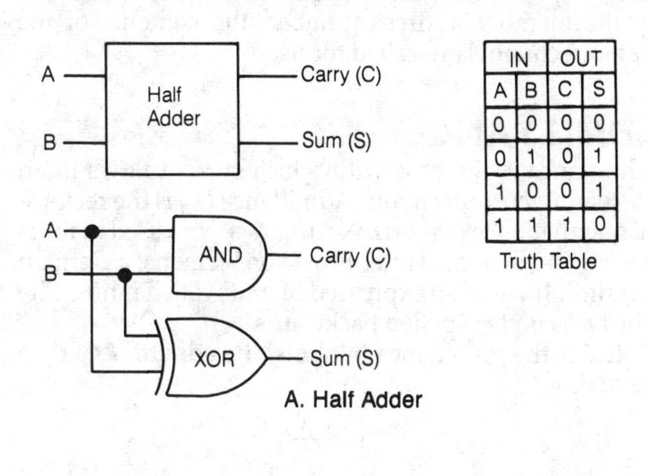

IN		OUT	
A	B	C	S
0	0	0	0
0	1	0	1
1	0	0	1
1	1	1	0

Truth Table

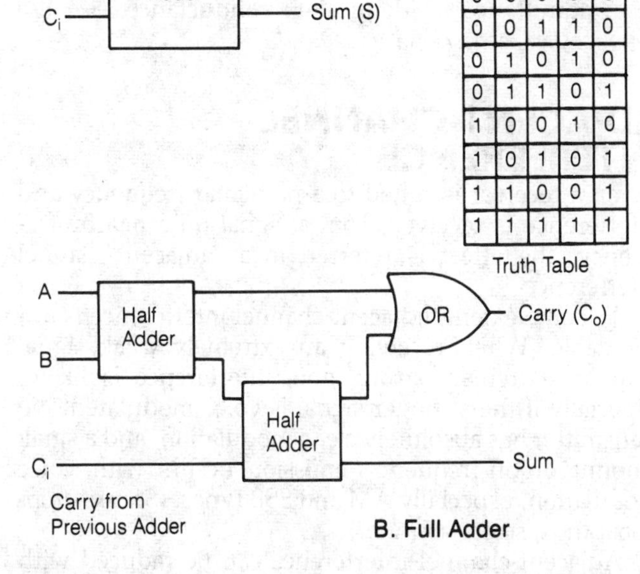

IN			OUT	
A	B	C_i	S	C_o
0	0	0	0	0
0	0	1	1	0
0	1	0	1	0
0	1	1	0	1
1	0	0	1	0
1	0	1	0	1
1	1	0	0	1
1	1	1	1	1

Truth Table

ADDER: A half adder symbol, logic circuit and truth table are shown at (A) and a full adder symbol, logic circuit, and truth table are shown at (B).

The full adder circuit shown in section B of the illustration provides carries. It is formed from two half adder circuits as shown. The truth table for the full adder shows all the possible conditions for the three inputs—A, B, and C_i, (the input carry). The output part of the table shows the binary signals at sum and carry output (C_o) for these conditions. For example, if both A and B are 1 and a carry is present, the sum will be 1 and the carry at C_o will be 1. *See* AND GATE, EXCLUSIVE OR GATE, INVERTER, OR GATE.

ADDRESS

Computer memory is stored in discrete packages for easy access. Each memory location bears a designator, usually a number, called the address. By selecting a particular address by number, the corresponding set of memory data is made available for use.

A digital computer can have several different memory channels for storing numbers. Each channel is itself designated by a number, for example, 1 through 8. By actuating a memory-address function control, followed by the memory address number, the contents of the memory channel are called for use.

ADMITTANCE

Admittance is the ease with which current flows in an alternating-current circuit. Admittance (Y) is the reciprocal of impedance (*see* IMPEDANCE), and is expressed in units of siemens (S) named after Ernst von Siemens, a German inventor. It was also expressed in units called mhos. The word *mho* is *ohm* spelled backwards.

If Z is the impedance in ohms, the admittance, Y, in siemens is:

$$Y = \frac{1}{Z}$$

Resistors, inductors, and capacitors all contribute to the admittance in an ac circuit. In a dc circuit, only the resistance determines the admittance. In this specialized case, admittance is identical to conductance. *See also* CONDUCTANCE, RESISTANCE.

ADJACENT-CHANNEL INTERFERENCE

When a receiver is tuned to a particular frequency and interference is received from a signal on a nearby frequency, the effect is referred to as adjacent-channel interference.

To some extent, adjacent-channel interference is unavoidable. When receiving an extremely weak signal near an extremely strong one, interference is likely, especially if the stronger signal is voice modulated. No transmitter has absolutely clean modulation, and a small amount of off-frequency emission occurs with voice modulation, especially AM and SSB types. (*See* AMPLITUDE MODULATION, SINGLE SIDEBAND.)

Adjacent-channel interference can be reduced with proper engineering techniques in transmitters and receivers. Transmitter audio amplifiers, modulators, and rf amplifiers should produce as little distortion as the state of the art will permit. Receivers should employ selective filters of the proper bandwidth for the signals to be received and the adjacent-channel response should be as low as possible. A flat response in the passband (*see* PASSBAND) and a steep drop-off in sensitivity outside the passband, are characteristics of good receiver design. See A and B in the illustration.

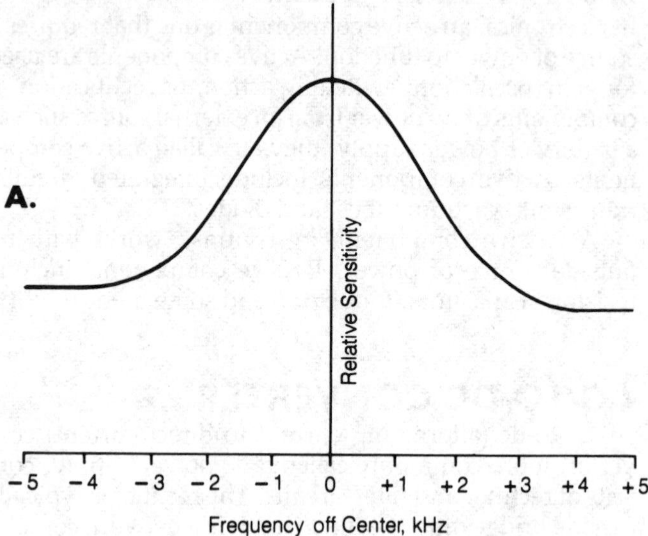

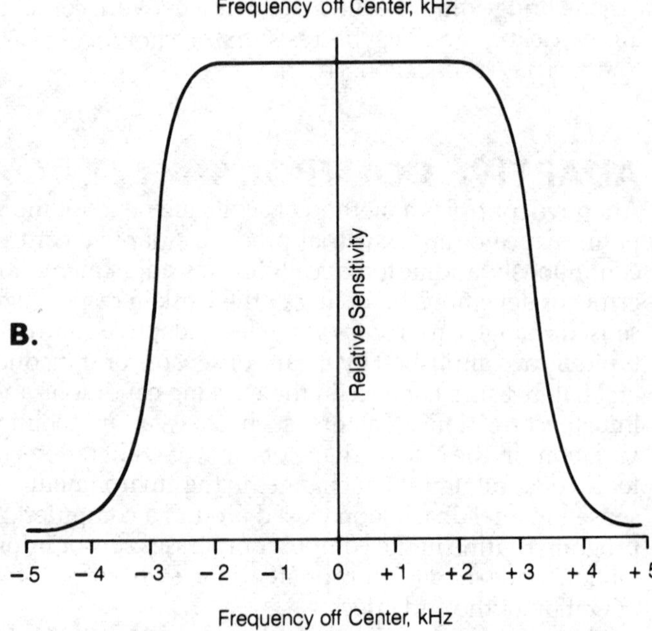

ADJACENT-CHANNEL INTERFERENCE: The plot of a poor bandpass response in a receiver (A), and a more selective bandpass response that reduces adjacent channel interference (B).

AFC

See AUTOMATIC FREQUENCY CONTROL.

AGC

See AUTOMATIC GAIN CONTROL.

AGING

All electronic components deteriorate with time. This process is called aging. Some components age more rapidly than others. Others, such as wire and carbon-composition resistors, are almost immune to aging.

Aging can be regarded as either mechanical or electrical. Examples of mechanical aging include the deterioration of control shaft bearings and frequently used screw threads, physical scratching and denting of cabinets, and similar wear and tear. Examples of electrical aging are the dielectric breakdown of capacitors (especially electrolytics), contamination of coaxial cables, drying and cracking of insulation, and other age-related changes that directly affect the electrical specifications of equipment.

AIR COOLING

Components that generate excessive heat—such as vacuum tubes, transistor power amplifiers, and some resistors—must be provided with some means for cooling them or damage may result. These components may be air cooled or conduction cooled (*see* CONDUCTION COOLING). Air cooling can take place as heat radiation, or as convection.

In high-powered, vacuum-tube transmitters, a fan is usually provided to force air over the tubes or through special cooling fins. By using fans greater heat dissipation is possible than would be possible without them, and this allows higher input and output power levels.

Low-powered transistor amplifiers use small heatsinks to conduct heat away from the body of the transistor (*see* HEATSINK). The heatsink can then radiate the heat into the atmosphere as infrared energy, or the heat can be dissipated into a large, massive object such as a block of metal. Ultimately, however, some of the heat from conduction-cooled equipment is dissipated in the air as radiant heat.

AIR CORE

The term *air core* is usually applied to inductors or transformers. At higher radio frequencies, air-core coils are used because the required inductance is small. Powdered-iron and ferrite cores greatly increase the inductance of a coil as compared to an air core (*see* FERRITE CORE). This occurs because these materials cause a concentration of magnetic flux within the coil. The magnitude of this concentration is referred to as permeability (*see* PERMEABILITY); air is given, by convention, a permeability of 1 at sea level.

Air-core inductors and transformers can be identified in schematic diagrams by the absence of parallel straight lines near the turns. In a ferrite or powdered-iron core, two parallel straight lines indicate the presence of a permeability-increasing substance in the core (see A and B in the illustration). Coils are sometimes wound on forms made of dielectric material such as glass or plastic. Because these materials have essentially the same permeability as air (with minor differences), they are considered air-core inductors in schematic representations.

AIR DIELECTRIC

Air is a dielectric material of variable quality (*see* DIELECTRIC), depending on the amount of dust, water vapor, and other impurities present. Even with considerable dust and humidity, however, the loss caused by air as a dielectric material is small (*see* DIELECTRIC LOSS).

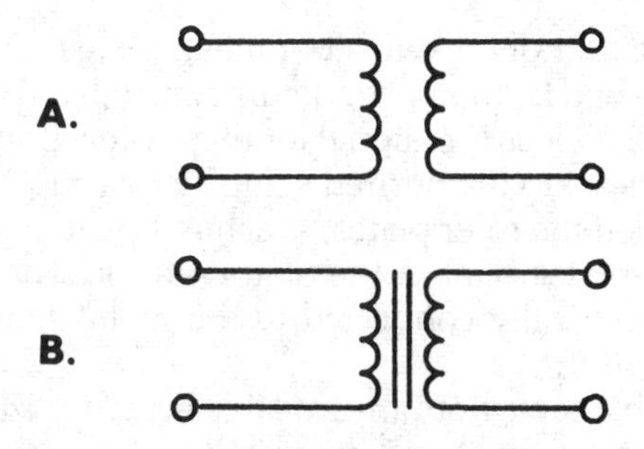

AIR CORE: The schematic symbol for air-core transformers (A), and the schematic symbol for powdered-iron or ferrite-core transformers (B).

Small-value variable capacitors use air as their dielectric material (*see* AIR-VARIABLE CAPACITOR). For low-power and receiving applications, the plate spacing is small, but at high power levels the plate spacing may be quite large.

Coaxial and parallel-wire transmission lines sometimes have an air dielectric. Some prefabricated lines are sealed and filled with a dry, inert gas such as helium or nitrogen. These air-dielectric lines have relatively low loss per-unit-length. Dry, pure air is given, by convention, a dielectric constant of 1 at sea level (*see* DIELECTRIC CONSTANT). Air dielectric has a high tolerance for potential differences although at sufficient voltages it will ionize and conduct. A typical and frequent example of the dielectric breakdown of air is a lightning stroke. (*See* DIELECTRIC BREAKDOWN.) The ionization of air is more likely to occur when the relative humidity or dust particle concentration is high. Dielectric breakdown is not as likely when the air is pure and dry.

AIR GAP

An air gap is formed by two pointed contacts, spaced a given distance apart, aiding discharge through the air when the voltage between the contacts reaches a certain value. A common application of the air gap is in a transmission-line lightning arrestor. When sufficient potential builds up between the contacts, the air ionizes and forms a conductive path for the flow of electrons to allow discharge. The greater the spacing between the pointed contacts, the greater the voltage required to cause dielectric breakdown of the air between the contacts.

Pointed contacts are generally used in air gaps because electrons discharge more easily from a sharp point than from a blunt object. The metal in the air-gap contacts must have a fairly high melting temperature to keep the points from being dulled by repeated arcing. *See also* ARCBACK, ARC RESISTANCE.

AIR-VARIABLE CAPACITOR

An air-variable capacitor is a component whose capacitance is adjustable, usually by means of a rotating shaft.

One rotating and one fixed set of metal plates are positioned in meshed fashion with rigidly controlled spacing. Air forms the dielectric material for these capacitors. The capacitance is set to the desired value by rotating one set of plates, called the rotor plates, to achieve overlap with a fixed set of plates, called the stator plates. The rotor plates are electrically connected to the metal shaft and frame of the unit.

Air-variable capacitors are available in many sizes and shapes. For receiving, and low-power rf transmitting applications, the plate spacing can be as small as a fraction of a millimeter. At high levels of rf power, the plates can be spaced an inch or more apart. The capacitance of an air variable capacitor can range from a few picofarads (pF) to 1000 pF. The maximum capacitance depends on the size and number of plates used, and on their spacing.

Because the dielectric material in air-variable capacitors is air, they are efficient capacitors as long as they are not subjected to excessive voltages resulting in flashover. The vacuum-variable is a capacitor in an evacuated enclosure that is even more efficient than an air variable capacitor.

Air-variable capacitors are frequently found in tuned circuits, rf power-amplifier output networks, and antenna matching systems. *See also* ANTENNA MATCHING, CAPACITOR.

ALARM SYSTEM

Electronic burglar and fire alarm systems are widely used in homes and businesses.

Burglar alarm systems, interconnected with the telephone lines, inform a central station when a break-in or suspected break-in occurs. Ultrasonic devices can be used to detect motion; when suspicious changes in the ultrasonic interference pattern take place, both silent and audible alarms are triggered. A silent alarm activates the telephone line to the central station; an audible alarm sets off a loud bell or siren.

Another means of detecting unauthorized personnel is the infrared (heat) sensor. This device reacts to the body heat of an intruder but it is immune to mechanical vibrations. Unless an alarm is properly disabled when someone enters the building, the police may be summoned by the central station.

Conducting metallic strips are often placed around windows. When the window glass is broken, the metal strip fractures, triggering the alarm. Some alarms, not connected to the telephone lines, set off a loud siren or other audible device. They also might turn on strategically positioned lights in an effort to scare a potential intruder away.

Fire alarms operate by detecting smoke or heat. The common battery-operated smoke detector emits a loud buzz or squeal when the concentration of smoke in the air reaches a threshold level. Fire alarms might have power supplies such as batteries, independent of the ac line, because commercial power is often cut off before the smoke or heat level is sufficient to activate the device. *See also* SMOKE DETECTOR, ULTRASONIC TRANSDUCER.

ALBEDO

The reflectivity of an object or substance is its albedo. Usually, albedo refers to the visible-light reflectivity, but it can also be specified for infrared, ultraviolet, X rays, radio waves, or even sound waves.

Albedo can be expressed as a ratio or as a percentage. As a percentage, if A represents the albedo, e_i the impinging energy, and e_r the reflected energy, then:

$$A = 100\ (e_r/e_i)$$

Thus, albedo can range from 0 to 100 percent.

An object with a zero albedo at all electromagnetic wavelengths is called a black body (*see* BLACK BODY). It absorbs all the energy it receives. With respect to visible light, freshly fallen snow has an albedo of nearly 100 percent. At radar wavelengths, objects with high albedo show up better than objects with low albedo (*see* RADAR); the stealth aircraft, designed to have an extremely low albedo at radar frequencies, is difficult to detect and track.

The energy not reflected by an object is either absorbed or transmitted through the object. *See also* ABSORPTANCE, TRANSMITTANCE.

ALEXANDERSON ANTENNA

See ANTENNA DIRECTORY.

ALIASING

In communications systems, aliasing is the blending of high-frequency signals into the baseband because a signal was sampled at too low a rate. (*See* NYQUIST RATE, SAMPLING THEOREM.) As a result of aliasing, high-frequency components appear as low frequencies. This problem is solved by the use of an antialiasing filter that removes high frequencies so that they cannot appear at a lower frequency. (*See* FILTER.)

In video displays, antialiasing signal conditioning is used at the edges between two displayed areas to minimize the effect of "jaggies," the steplike discontinuities that appear on diagonal lines and curves on cathode ray tube monitors as a result of the finite number of scanning lines in the display raster.

ALGEBRA

Algebra is the branch of mathematics dealing with the determination of unknown quantities. One or more variables, usually denoted by lowercase letters *a* through *z*, are determined from one or more equations.

Algebraic equations are given an order of magnitude, corresponding to the highest exponential value attached to the variables. An equation of order 2, such as $x^2 + 2x + 1 = 0$, is referred to as a quadratic equation, and an equation of order 3, such as $x^3 + 3 = 0$, is called a cubic equation. In general, the higher the order of a single equation, the more difficult it is to solve. Some equations, such as $x^2 = -1$, have no real-number solutions, although they may have solutions in the set of complex numbers (*see* COMPLEX NUMBER, J OPERATOR).

Linear algebra deals with the solutions to sets of first-order, or linear, equations. There can be hundreds of equations and hundreds of variables in such a system. Computers must often be used to arrange the variables and coefficients into a pattern called a matrix. The set of equations is then solved by an algorithm (*see* ALGORITHM).

ALGORITHM

An algorithm is any problem-solving procedure. The term is usually applied to the solution-finding process for a mathematical problem. However, any procedure that can be broken down into discrete steps is an algorithm. The number of steps in an algorithm must be countable, or finite. One example of a rather complicated algorithm for a simple process is the tying of a shoelace. (Try writing down the procedure, step by step, without illustrations!)

A mathematical example of an algorithm is procedure of extracting the square root of a number. By repeating, or iterating, this algorithm many times, any desired number of significant digits can be determined.

ALLOY TRANSISTOR

An alloy transistor is manufactured by alloying impurity elements into the opposing sides of a chip or die of single crystal germanium. The original wafer can be either N-type or P-type germanium. Acceptor elements can then be alloyed into the opposing sides of the N-type material to form a PNP transistor. Donor elements also can be alloyed into the opposing sides of the P-type die to form an NPN transistor. The construction of a PNP junction transistor made by the alloying process is shown in the figure.

Two indium pellets are attached to the opposing sides of a germanium wafer, and the combined structure is fired in a furnace to melt the indium pellets. The molten indium dissolves the germanium. When cooled, the germanium recrystallizes, and its indium content forms a P region on each side of the die. The collector is made as the larger of the two indium disks. This construction permits efficient collection of carriers from the emitter, thus achieving high current gain.

In the NPN junction transistor, the wafer is P-type germanium, perhaps doped with indium. The two N-regions are formed by alloying antimony pellets into the opposite sides of the P-type die.

Neither the alloy nor the point-contact processes are used any longer for the commercial manufacture of transistors because the technology is obsolete and no longer economical. They have been replaced by diffusion and epitaxial growth processes. As a result, both are of interest primarily for historical reasons. *See* TRANSISTOR.

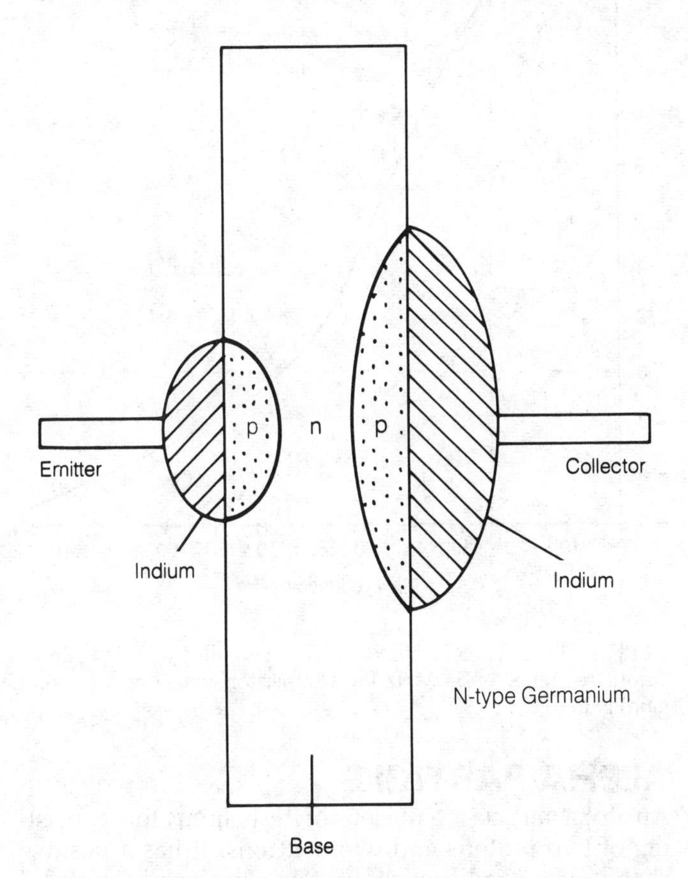

ALLOY TRANSISTOR: A cross section of a germanium NPN junction transistor made by the alloy process.

ALKALINE CELL

See BATTERY.

ALLOY-DIFFUSED SEMICONDUCTOR

See ALLOY TRANSISTOR.

ALL-PASS FILTER

An all-pass filter is a device or network designed to have constant attenuation at all frequencies of alternating current. However, a phase shift can be introduced, and this phase shift is also constant for all alternating-current

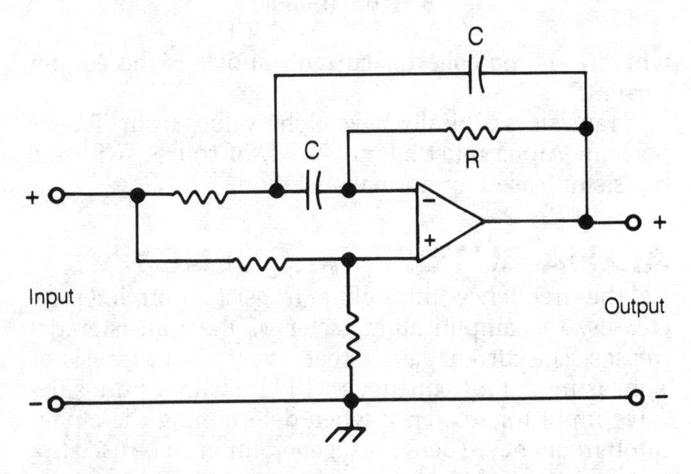

ALL-PASS FILTER: Phase delay is determined by the values of R and C in the operational amplifier feedback loops.

frequencies. In practice, the attenuation is usually as small as possible (*see* ATTENUATION).

All-pass filters are generally made with noninverting operational amplifiers. An example of this is shown in the figure. The amount of phase delay is determined by the values of resistor R and capacitors C. The amount of attenuation is regulated by the values of the other resistors. *See also* OPERATIONAL AMPLIFIER.

ALNICO

Alnico is a term for a family of permanent-magnet alloys containing aluminum, nickel, cobalt, and iron. Alnico magnets provide very high flux density but low coercive force, the property of a magnet that permits it to retain its magnetic field. This quality makes Alnico magnets relatively easy to demagnetize. To achieve the highest magnetic force, Alnico magnets must be magnetized lengthwise. Their high flux density makes them suitable as permanent magnets for dc motors, magnetrons, and other microwave power tubes.

Many permanent-magnet materials have relatively large values of coercive force. Ferrite or ceramic magnets, for example, have low flux density but high coercive force. This quality permits them to be satisfactorily magnetized in the widthwise direction. Rare earth or samarium-cobalt magnets, however, exhibit both high magnetic flux density and high coercive force. *See* MAGNETRON, PERMANENT MAGNET.

ALPHA

In a transistor, the ratio between a change in collector current and a change in emitter current is known as the alpha. Alpha is represented by the first letter of the Greek alphabet, lowercase (α) and is determined in the grounded-base arrangement.

The collector current in a transistor is always smaller than the emitter current because the base draws some current from the emitter-collector path when the transistor is forward biased. Generally, the alpha of a transistor is given as a percentage:

$$\alpha = 100\ (I_C/I_E),$$

where I_C is the collector current, and I_E is the emitter current.

Transistors typically have alpha values from 95 to 99 percent. Alpha must be measured, of course, with the transistor biased for normal operation.

ALPHA-CUTOFF FREQUENCY

As the frequency through a transistor amplifier increases, the amplification factor of the transistor decreases. The current gain, or beta (*see* BETA) of a transistor is measured at a frequency of 1 kHz, with a pure sine-wave input for reference when determining the alpha-cutoff frequency. Then, a test generator must be used (*see* SIGNAL GENERATOR) which has a constant output amplitude over a wide range of frequencies. The frequency to the amplifier input is increased until the current gain in the common-base arrangement decreases by 3 dB (decibels) with respect to its value at 1 kHz. A decrease in current gain of 3 dB represents a drop to 0.707 of its previous magnitude. The frequency at which the beta is 3 dB below the beta at 1 kHz is called the alpha-cutoff frequency for the transistor.

Depending on the type of transistor involved, the alpha-cutoff frequency might be only a few MHz, or perhaps hundreds of MHz. The alpha-cutoff frequency is an important specification in the design of an amplifier. An alpha-cutoff frequency that is too low for a given amplifier requirement will result in poor gain characteristics. If the alpha-cutoff frequency is unnecessarily high, money is wasted. Under these conditions there is a greater tendency toward unwanted vhf parasitic oscillation (*see* PARASITIC OSCILLATION).

As the input frequency is increased past the alpha-cutoff frequency, the gain of the transistor continues to decrease until it reaches unity, or zero dB. At still higher frequencies, the gain becomes smaller than unity. The drawing illustrates a sample gain-*vs*-frequency curve for a hypothetical transistor, showing the alpha-cutoff and unity-gain frequencies. *See also* DECIBEL, GAIN.

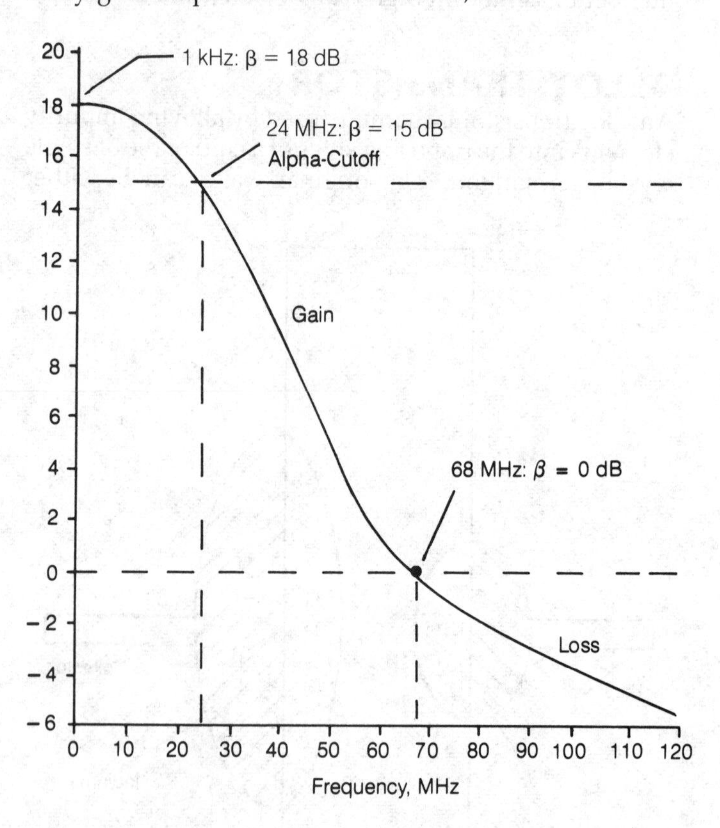

ALPHA-CUTOFF FREQUENCY: This plot illustrates an alpha-cutoff frequency of 24 MHz for a transistor whose beta (current gain) is 15 dB.

ALPHA PARTICLE

An alpha particle is a nucleus of the helium atom consisting of two protons and two neutrons. It has a positive charge. Many radioactive substances emit alpha particles as they decay.

An alpha particle has twice the positive charge of a single proton, and four times the atomic mass. When alpha particles are accelerated to high speeds, they are capable of breaking up or modifying the nuclei of heavier atoms. Alpha particles, when numerous and traveling at great speed, are called alpha rays. They are the least energetic of radioactive emissions. Some alpha particles strike the atmosphere of the earth as they arrive from distant stars and galaxies. Many of these alpha rays are absorbed by the atmosphere before they reach the ground. *See also* COSMIC RADIATION.

ALPHANUMERIC

An expression containing letters of the alphabet and numerals 0 through 9 is called an alphanumeric expression. Alphanumeric characters are arranged in the order ABCDEFGHIJKLMNOPQRSTUVWXYZ0123456789. When placing an expression in alphabetic-numeric sequence, the first character is evaluated first, then the second, and so on until a disagreement occurs between that expression and its neighbors. The digits 0 through 9 are usually treated as letters of the alphabet following Z. Alphanumeric expressions are commonly used as designators for variables in computer programs.

ALTERNATE ROUTING

Alternate routing is the backup system used when the primary system of communications between two points breaks down. Alternate routing can also be used in power transmission, in case of interruption of a major power line, to prevent prolonged and widespread blackouts. Power from other plants is routed to cities affected by the failure of one particular generating plant or transmission line.

As an example, the primary communications link for a particular system might be with a geostationary satellite, as shown in A. If the satellite fails, another satellite can be used in its place if one is available (B). This is alternate routing. If the second satellite ceases to function or is not available, further backup systems might be used, such as a high-frequency shortwave link (C) or telephone connection (D). Alternate routing systems should be set up and planned in advance, before the primary system goes down, so that communications may be maintained with a minimum of delay.

ALTERNATING CURRENT

Whenever electrons in a conductor flow in both directions, alternating current exists. If the electrons always move in the same direction, even if it is at variable speed, the current is considered direct (*see* DIRECT CURRENT).

The most familiar example of alternating current is the 60-Hz line current used in homes and offices. Radio-frequency energy in an antenna or transmission line is also alternating current, as is the audio-frequency energy in a stereo speaker system. There are many different kinds of alternating current.

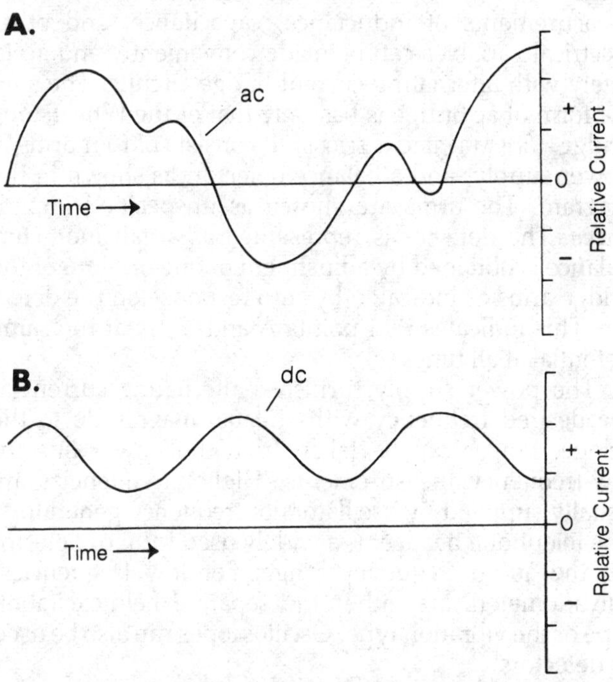

ALTERNATING CURRENT: The polarity of the current must be reversed to be an alternating current (A). A current that varies but does not reverse is variable direct current (B).

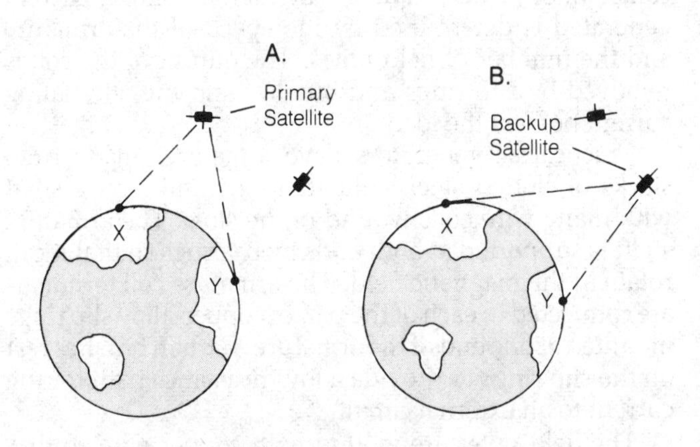

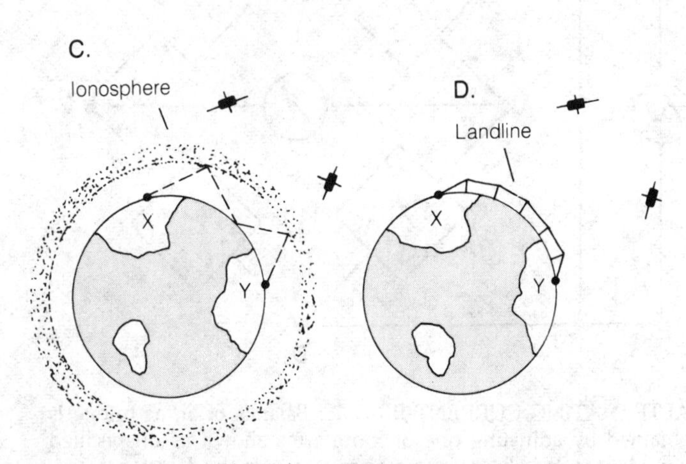

ALTERNATE ROUTING: Alternate routing is used when a primary communications satellite fails (A). These systems include a backup satellite (B), ionospheric communications (C), and terrestrial cable or landline (D).

An ac waveform might be quite simple, such as a sine wave, square wave or sawtooth wave. Or, it might be irregular in shape and very complex, like a voice pattern. The prime criterion for a current to be regarded as ac is that the electrons repeatedly reverse their direction (see A and B in the illustration). Alternating current can also be converted to direct current with a rectifier and filter.

Alternating current has an advantage over direct current for power transmission in the ease of voltage transformation (*see* TRANSFORMER). Direct current generates less electromagnetic noise than alternating current and is more efficiently transferred over long distances, but the transformation of voltage is more difficult. *See also* SAWTOOTH WAVE, SINE WAVE, SQUARE WAVE.

ALTERNATING-CURRENT BRIDGE

Measurements of inductance, capacitance, and other electrical variables can be made conveniently and accurately with alternating-current bridge circuits. The simple form of ac bridge is basically that of the Wheatstone bridge. (*See* WHEATSTONE BRIDGE). It consists of four arms, a power supply, and a balance detector, as shown in the diagram. The arms are shown as unspecified impedances. The detector is represented as a galvanometer. Balance is obtained by adjustment of one or more of the bridge arms as indicated by zero response on the detector. This indicates that points A and C are at the same potential at all times.

The power supply furnishes alternating current at the desired frequency with suitable magnitude to the bridge. Power can be taken from the power line for low-frequency measurements. Higher frequencies are usually supplied by oscillators or frequency generators. The telephone headset is a widely used form of detector for the audio frequency range. For low frequencies, galvanometers are either the separate field-excitation type or the vibration type. Oscilloscopes can also be used as detectors.

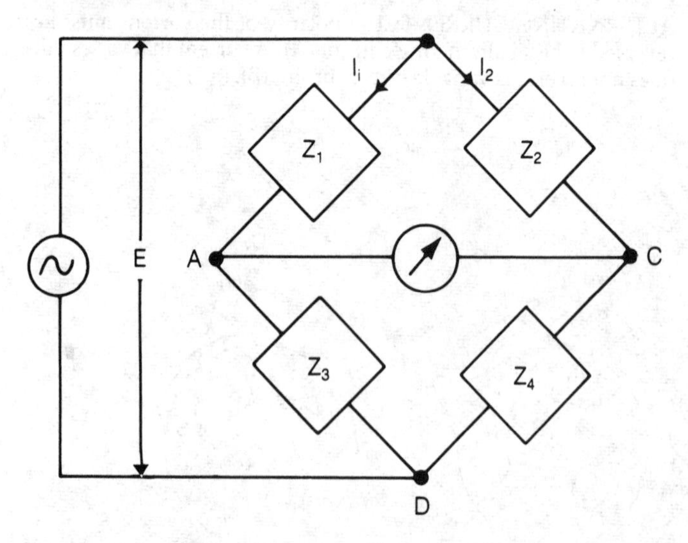

ALTERNATING-CURRENT BRIDGE: Balance of an ac bridge is obtained by adjusting one or more arms shown as unspecified impedances and observing a zero response on the detector.

Alternating-current bridges are used for the measurement of capacitance and inductance. The best known ac bridge is the Maxwell bridge for the measurement of inductance by capacitance. (*See* MAXWELL BRIDGE.) Others include the Hay and Owen bridges, both modifications of the Maxwell bridge for the measurement of inductance. (*See* HAY BRIDGE, OWEN BRIDGE.) All three bridges simplify to a pure capacitance in the arm opposite the pure inductance, and the other two arms become pure resistances. The differences among the three bridges are in the methods used to balance the resistance component of the coil.

The Schering bridge is one of the most important ac bridges for the measurement of capacitance; it has generally replaced the Wien bridge. (*See* SCHERING BRIDGE.) Other ac inductance bridges include the Anderson, Stroud, and Oats bridges. (*See* ANDERSON BRIDGE.) There are also direct-current bridges for use in electrical measurement.

ALTERNATING-CURRENT GENERATOR

An alternating-current (ac) generator, or alternator, is a machine capable of generating alternating current power when driven by an external force. The operation of the ac generator is based on the principles of electromagnetism and electromagnetic induction for the creation of electrical energy from mechanical energy. The same principles are used by the motor to create mechanical energy from electrical energy. The ac generator can be run as a motor by changing the electrical connections.

A basic ac generator is illustrated. An armature coil is rotated by any external mechanical means—from hand cranking to internal combustion engine or steam, water, or air turbine. The closed-wire loop forms one coil of an armature, which is free to rotate about an axis parallel with the ends of the magnet pole pieces. The magnetic field is typically formed by an electric field as shown.

The rotating coil cuts the magnetic flux, and electric currents are induced to flow in the coil. The direction of electron flow is determined by the direction of conductor movement in relation to the magnetic flux field. Current will flow in opposite directions in the opposite sides of the coil and reverse direction during the next half cycle of coil rotation. The frequency of the alternating current generated is determined by the speed of the armature and the number of field poles. The output of the coil is removed by slip rings and brushes and the alternating current powers the load.

Practical ac generators have armatures made from stacks of slotted steel laminations mounted on a shaft with many wire coils wound in the slots. The armature shaft is supported at both ends by bearings so that it can rotate in the magnetic field. The armature coil terminals are connected to each of the pair of copper-alloy slip rings mounted at one end of the armature. Carbon brushes rest on the slip rings to provide a low-resistance path for the current to an external circuit.

The field poles are wound with copper wire so that the magnetic field strength will be increased when an external direct current from an outside source is supplied to the field windings. A variable resistor, called a rheostat, can control the field strength in the circuit. *See* RHEOSTAT.

The power output of a generator is determined by the number of lines of force cut by the armature conductors per unit of time. The rate of cutting flux lines can be increased or decreased by altering the following:

- Strength of the magnetic field
- Speed of armature rotation
- Number of field poles
- Number of turns per armature coil, number of armature coils, or both.

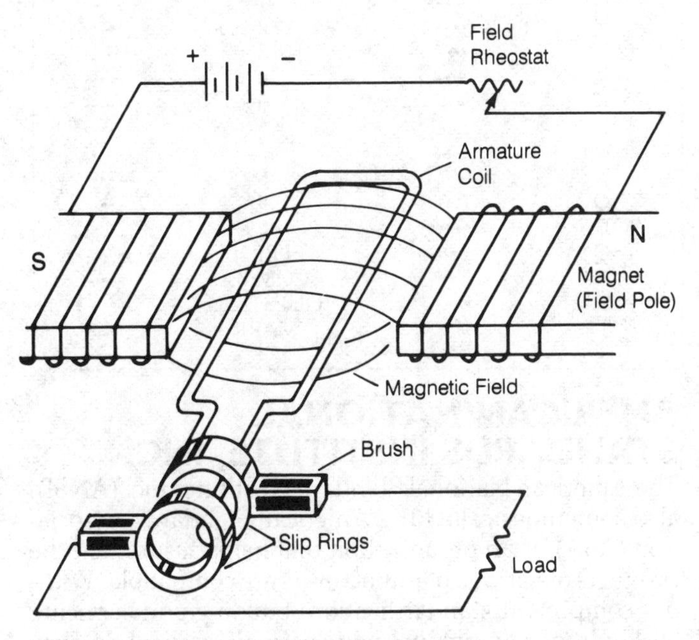

ALTERNATING-CURRENT GENERATOR: A basic ac generator is an armature coil rotated by an external drive within a magnetic field. Slip rings deliver the ac power to the load.

ALTERNATING-CURRENT NETWORK

An alternating-current network is a circuit containing resistance and reactance. It differs from a direct-current network that only contains resistance. Resistance is provided by all electrical conductors, including resistors, coils, and lamps. When a current passes through a resistance, heat is generated. Pure reactances, however, do not convert electrical current into heat. Instead, they store the energy and release it later. In storing energy, a reactance offers opposition to the flow of alternating current. Inductors and capacitors are the most elementary sources of reactances. Reactance is always either positive (inductive) or negative (capacitance). Certain semiconductor circuits can be made to act like coils or capacitors by showing reactance at a particular ac frequency. Shorted or open lengths of transmission line also behave like reactances at some frequencies.

While simple resistance is a one-dimensional quantity, and reactance is also one-dimensional (though it can be positive or negative), their combination in an ac network is two-dimensional. Resistance ranges from zero to infinite values; reactance ranges from zero to infinite positive or negative values. Their combination in an alternating-current circuit is called impedance. Any ac network has a net impedance at a given frequency. The impedance generally changes as the frequency changes, unless it is a pure resistance. The impedance is expressed as a single point on a resistance-reactance half-plane. Reactance is multiplied by the imaginary number j, called the j factor, for mathematical convenience. This quantity is defined as the square root of −1, also sometimes denoted by i. *See also* IMPEDANCE, J OPERATOR, REACTANCE.

ALTERNATING-CURRENT RIPPLE

Alternating-current ripple, usually referred to simply as ripple, is undesired modulation of a signal or power source. The most common form of ripple is 60- or 120-Hz ripple originating from ac-operated power supplies.

In theory, the output of any power supply will contain some ripple when the supply delivers current. This ripple can, and should, be minimized in practice because it will cause undesirable performance of equipment. Sufficient filtering, if used, will ensure that the ripple will not appear in the output of the circuit.

Alternating current ripple is virtually eliminated by using large inductors in series with the output of a power supply, and by connecting large capacitors in parallel with the supply output. The more current the supply is required to deliver, the more inductance and capacitance will be required in the filter stage. *See also* POWER SUPPLY.

ALTERNATING-CURRENT VOLTAGE

There are several different ways to measure voltage in an alternating-current circuit: peak; peak-to-peak; and root-mean-square (rms) methods.

An ac waveform is not necessarily a simple sine wave. It can be square, sawtooth, or irregular in shape. But whatever the shape of an ac waveform, the peak voltage is definable as the largest instantaneous value the waveform reaches. See A in the illustration. The peak-to-peak voltage is the difference between the largest instantaneous values the waveform reaches to either side of zero. See B. Usually, the peak-to-peak voltage is exactly twice the peak voltage. However, if the waveform is not symmetrical, the peak value can differ in the negative direction from the peak value in the positive direction, and the peak-to-peak voltage might not be twice the positive-peak or negative-peak voltage.

The rms voltage is the most commonly specified property of an ac voltage. Root-mean-square voltage is defined as the dc voltage needed to cause the same amount of heat dissipation in a simple, nonreactive resistor as a given alternating-current voltage. For symmetrical, sinusoidal waveforms, the rms voltage is 0.707 times the peak voltage and 0.354 times the peak-to-peak voltage. *See also* ROOT MEAN SQUARE.

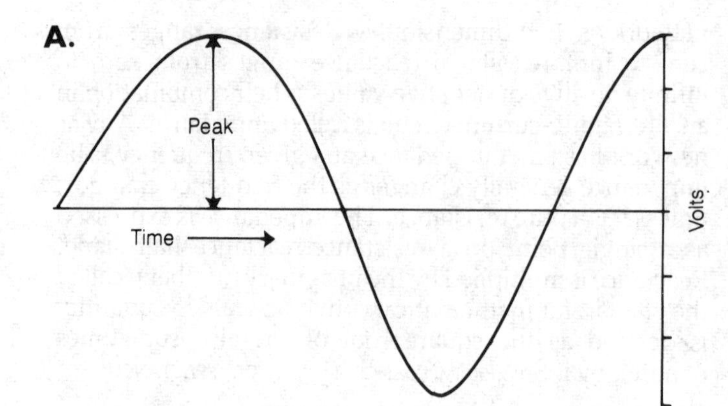

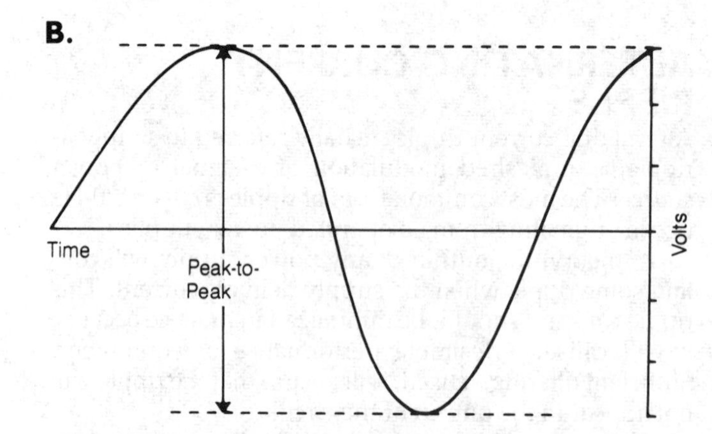

ALTERNATING-CURRENT VOLTAGE: The positive peak voltage of a sinusoidal ac waveform (A) and the peak-to-peak voltage (B).

ALTERNATOR

See ALTERNATING-CURRENT GENERATOR.

ALUMINUM

Aluminum is a dull, light, somewhat brittle metal with an atomic number of 13, and an atomic weight of 27. Aluminum is commonly used as a conductor for electricity. Aluminum is very strong in proportion to its weight and has replaced much heavier metals such as steel and copper in many applications.

Aluminum, like other metallic elements, is found in the earth's crust. It occurs in a rock called bauxite. Recent advances in mining and refining of bauxite have made aluminum one of the most widely used and inexpensive industrial metals.

Aluminum is resistant to corrosion and is an excellent choice in the construction of communications antennas.

AMERICAN MORSE CODE

The American Morse Code is a system of dot and dash symbols, first used by Samuel Morse in telegraph communications. The American Morse Code is not widely used today. It has been largely replaced by the International Morse Code, but some telegraph operators still use American Morse.

The American Morse Code differs from the International Morse Code. The table shows the American Morse symbols, sometimes called Railroad Morse. Some letters contain internal spaces. This causes confusion for operators familiar with International Morse Code. Some letters are also entirely different in the two codes. *See also* INTERNATIONAL MORSE CODE.

AMERICAN MORSE CODE: American Morse Code symbols (sometimes called Railroad Morse).

Character	Symbol	Character	Symbol
A	•–	U	••–
B	–•••	V	•••–
C	•• •	W	•––
D	–••	X	•–••
E	•	Y	•• ••
F	•–•	Z	••• •
G	––•	1	•––•
H	••••	2	••–••
I	••	3	•••–•
J	–•–•	4	••••–
K	–•–	5	–––
L	——	6	••••••
M	––	7	––••
N	–•	8	–••••
O	• •	9	–••–
P	•••••	0	———
Q	••–•	PERIOD	••––••
R	• ••	COMMA	•–•–
S	•••	QUESTION MARK	–••–•
T	–		

AMERICAN NATIONAL STANDARDS INSTITUTE, INC.

The American National Standards Institute, Inc. (ANSI), also sometimes called the American Standards Association (ASA), is an organization that helps assure that the products of various manufacturers are compatible. Without component standardization, building and servicing all kinds of equipment, nonelectronic as well as electronic, would be much more difficult.

AMERICAN WIRE GAUGE

Metal wire is available in many different sizes or diameters. Wire is classified according to diameter by giving it a number. The designator for a given wire is known as the American Wire Gauge (AWG). In England, a slightly different system is used (*see* BRITISH STANDARD WIRE GAUGE). The numbers in the American Wire Gauge system range from 1 to 40, although larger and smaller gauges exist. The higher the AWG number, the thinner the wire.

The table shows the diameter and corresponding AWG designator for AWG 1 through 40. The higher the AWG number for a given conductor metal, the smaller the current-carrying capacity becomes. The AWG designator does not include any coatings on the wire such as enamel, rubber, or plastic insulation. Only the metal part of the wire is taken into account.

AMMETER

Ammeters are instruments for measuring current flow. Ammeters are placed in series with the line carrying the current to be measured. There are two general classes of

AMERICAN WIRE GAUGE:
American Wire Gauge equivalents in millimeters.

AWG	Diameter, millimeters	AWG	Diameter, millimeters
1	7.35	21	0.723
2	6.54	22	0.644
3	5.83	23	0.573
4	5.19	24	0.511
5	4.62	25	0.455
6	4.12	26	0.405
7	3.67	27	0.361
8	3.26	28	0.321
9	2.91	29	0.286
10	2.59	30	0.255
11	2.31	31	0.227
12	2.05	32	0.202
13	1.83	33	0.180
14	1.63	34	0.160
15	1.45	35	0.143
16	1.29	36	0.127
17	1.15	37	0.113
18	1.02	38	0.101
19	0.912	39	0.090
20	0.812	40	0.080

ammeters: alternating current (ac) and direct current (dc). The dc ammeter is a galvanometer with a low-resistance shunt placed in parallel with the coil circuit as shown in the figure. The shunt diverts all but a fraction of the current flowing in the circuit from the sensitive meter movement. The galvanometer is a sensitive permanent-magnet D'Arsonval meter movement that works on the principles of the electric motor. The meter coil can withstand only about 50 mA without being damaged. *See* D'ARSONVAL MOVEMENT, GALVANOMETER.

By adjusting the ratio of current flow between the moving coil and the shunt, the meter scale can be calibrated in microamperes, milliamperes, or amperes. This adjustment can be made by using the proper resistance relationship between the shunt and the moving coil. The illustration shows a moving-coil dc ammeter.

If the current is to be divided so that 1/10 goes through the moving coil of the galvanometer and 9/10 goes through the shunt, the shunt must carry nine times as much current as the moving coil. The resistance of the shunt must be 1/9 of the resistance of the moving coil of the galvanometer. Similarly, if the resistance of the shunt is 1/99, then 1/100 of the current will pass through the galvanometer.

The moving-coil meter will not operate in an alternating current circuit because the magnitude and direction of flow of ac is always changing. The meter attempts to follow those changes. To obtain a steady reading, the current can be rectified. For example, a Wheatstone bridge with diode rectifiers in each arm converts the ac into pulsating current and the rectified current energizes the ammeter. *See* WHEATSTONE BRIDGE.

Moving-vane and electrodynamometer instruments can also be used for measuring alternating currents and voltages. *See* ELECTRODYNAMOMETER. Analog multimeters have internal circuits that can be switched to make either ac or dc measurements. Both ac and dc current measurements can also be made with semiconductor circuits in digital multimeters. *See* DIGITAL MULTIMETER.

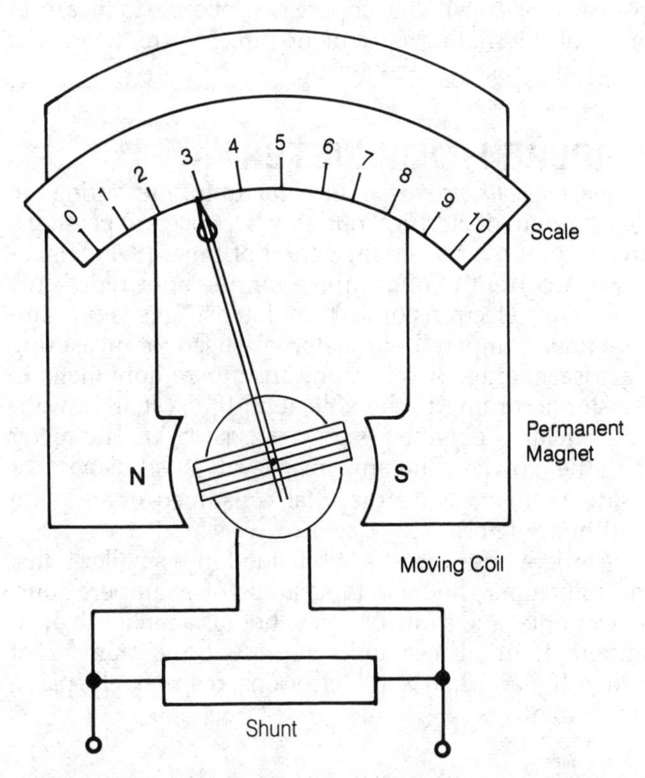

AMMETER: An ammeter is a D'Arsonval galvanometer with most of the current bypassed by a shunt or parallel resistor.

AMPERE

The *ampere* is the unit of electric current. A flow of one coulomb per second, or 6.28×10^{18} electrons per second, past a given fixed point in an electrical conductor, is a current of one ampere (see figure).

Various units smaller than the ampere are often used to measure electric current. A milliampere (mA) is one thousandth of an ampere, or a flow of 6.28×10^{15} electrons per second past a given fixed point. A microampere (μA) is one millionth of an ampere, or a flow of 6.28×10^{12} electrons per second. A nanoampere (nA) is a billionth of an ampere; it is the smallest unit of electric current you are likely to use. It represents a flow of 6.28×10^{9} electrons per second past a given fixed point.

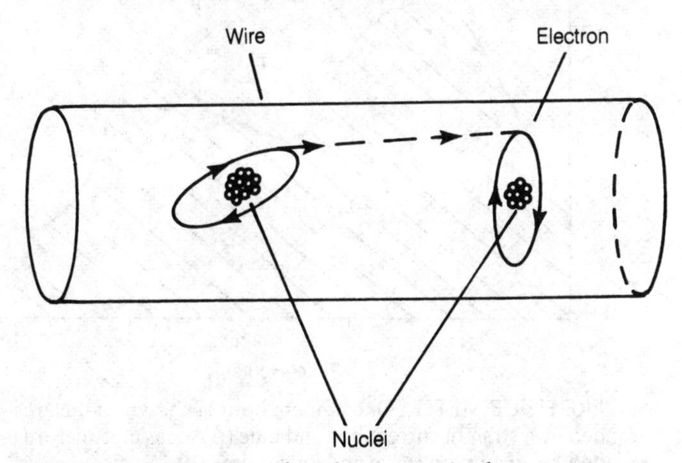

AMPERE: An illustration of an electron moving between atoms.

A current of one ampere is produced by a voltage of one volt across a resistance of one ohm. This is Ohm's law (*see* OHM'S LAW). The ampere is applicable to measurement of alternating current or direct current. *See also* ALTERNATING CURRENT, DIRECT CURRENT.

AMPERE-HOUR METER

An *ampere-hour meter* is an instrument for measuring the total amount of electrical quantity (*see* COULOMB) passing a given point over a certain period of time. (See illustration.) A current of one ampere for one hour represents 2.26×10^{22} electrons or 3600 coulombs. This is one ampere hour. Ampere-hour meters are used for measuring electrical energy. By calibrating an ampere-hour meter to register according to the voltage in the system, a watt-hour meter is obtained (*see* WATT-HOUR METER). The meter that the power company installs at any business or residence to measure the total consumed energy is a watt-hour meter.

Ampere hours can be subdivided into smaller units; the milliampere hour is a thousandth of an ampere hour, for example, and a microampere hour is a millionth of an ampere hour. These units represent the transfer of 2.26×10^{19} and 2.26×10^{16} electrons, respectively, past a given point.

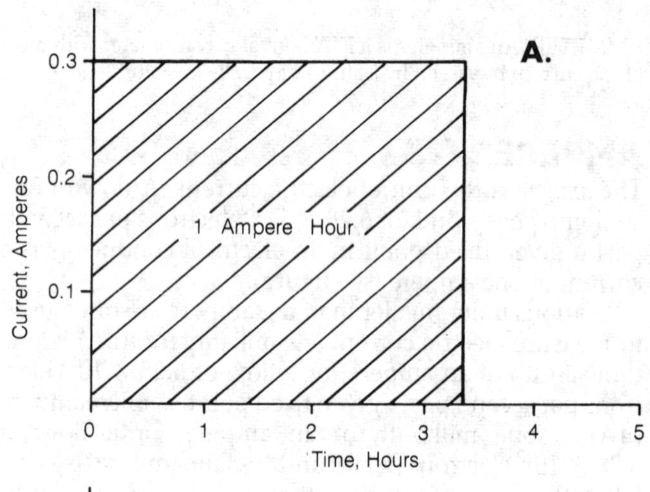

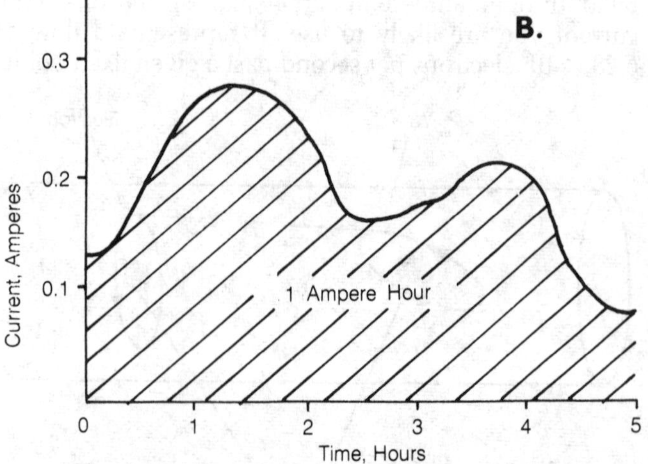

AMPERE-HOUR METER: One ampere hour is shown as the area bounded by a straight current line and time (A), or as an equal area bounded by an ampere curve and longer time (B).

AMPERE'S LAW

The direction of an electric current is generally regarded as the direction of positive charge transfer. This is opposite to the direction of the movement of the electrons because electrons carry a negative charge. By convention, current is considered to move from the positive to the negative terminal of a battery or power supply.

According to Ampere's law, the magnetic field or flux lines generated by a current in a wire travel counterclockwise when the current is directed toward the observer (see illustration). This rule is also called the right-hand rule for magnetic-flux generation. A more universal rule for magnetic flux, applying to motors and generators, is called Fleming's Rule (*see* FLEMING'S RULES). As the right hand is held with the thumb pointed outward and the fingers curled, a current in the direction pointed by the thumb will generate a magnetic field in the circular direction pointed by the fingers.

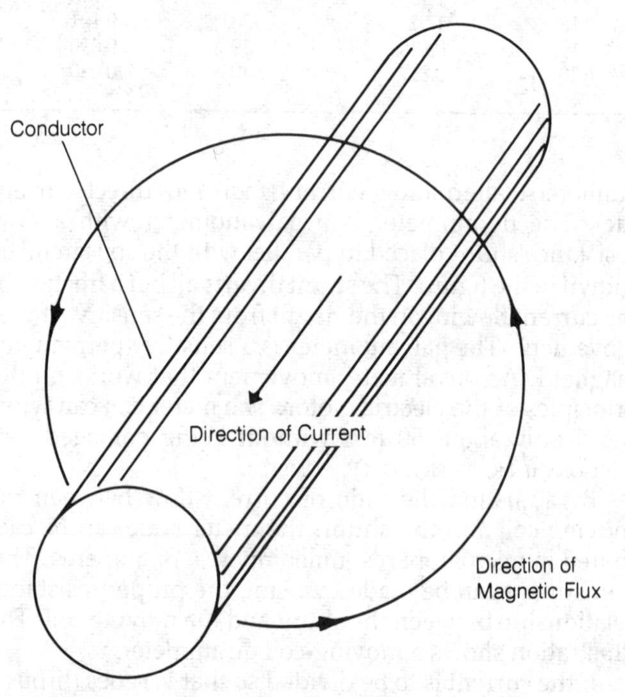

AMPERE'S LAW: Magnetic flux is considered to circulate clockwise when current is directed toward the observer in accordance with Fleming's right hand rule.

AMPERE TURN

The ampere turn is a measure of magnetomotive force. One ampere turn is developed when a current of one ampere flows through a coil of one turn, or, in general, when a current of $1/n$ amperes flows through a coil of n turns.

One ampere turn is equal to 1.26 gilberts. The gilbert is the conventional unit of magnetomotive force. *See also* GILBERT, MAGNETOMOTIVE FORCE.

AMPLIDYNE DRIVE

The amplidyne drive is a dc (direct-current) motor drive using a special dc generator called an amplidyne in place of a conventional dc generator. The amplidyne drive acts

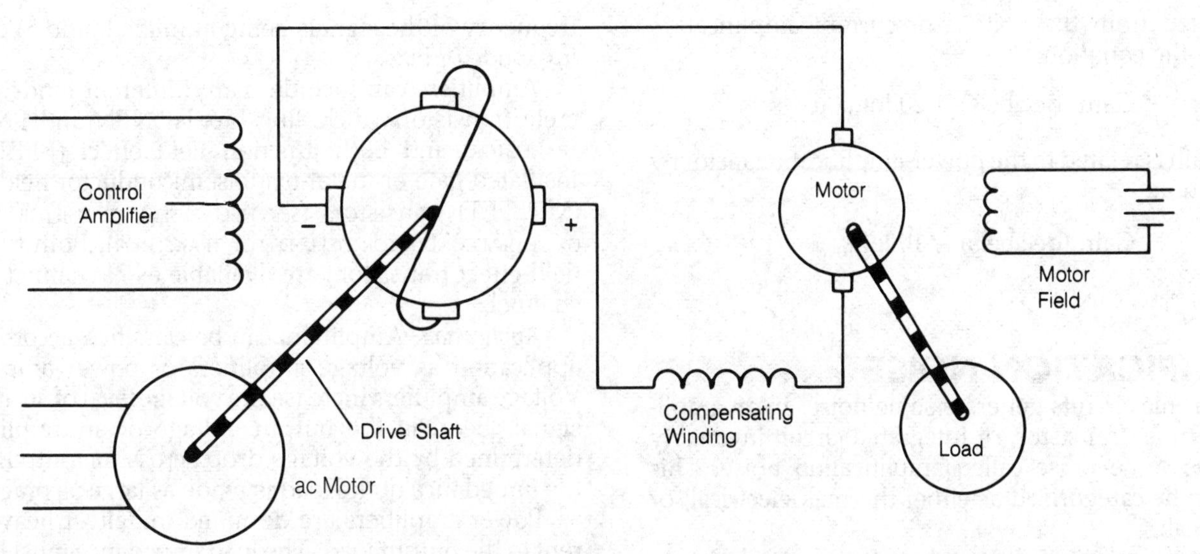

AMPLIDYNE: The amplidyne is a dc generator that functions as a power amplifier in servo systems. A second set of brushes in the dc generator permits rapid field excitation changes and high amplification.

as an electromechanical power amplifier because the amplidyne field requires only a fraction of the control power of a dc generator to produce the same output power.

The amplidyne drive consists of the basic system as shown in the diagram. The symbol for the amplidyne generator is similar to that of a conventional dc generator, except that it has an extra set of brushes connected by a curved shorting bar. The amplidyne generator is ordinarily driven by an ac motor because ac motors provide essentially a constant speed. The control field is shown as a split winding because it is common to supply the fields with a control amplifier having separate outputs for each polarity of the applied error signal. The field of the dc motor can be supplied by a rectifier or a pair of permanent magnets. *See* DC GENERATOR, SERVO SYSTEM.

AMPLIFICATION

Amplification is an increase in the magnitude of a current, voltage, or wattage. Amplification makes it possible to transmit radio signals of high power, sometimes over a million watts. Amplification also makes it possible to receive signals that are extremely weak. It allows the operation of such diverse instruments as light meters, public-address systems, and television receivers.

Usually, amplification includes increasing the magnitude of a change in a certain quantity. For example, a change from −1 to +1 V, or 2 V peak-to-peak, might be amplified so that the range becomes 0 to +10 V, or five times greater. Alternating currents, when amplified, produce effective voltage gain (if the impedance is correct) and power gain.

Direct-current amplification is usually done to increase the sensitivity of a meter or other measuring instrument. Alternating-current amplification is employed primarily in audio-frequency and radio-frequency applications for the purpose of receiving or transmitting a signal. *See also* AMPLIFIER.

AMPLIFICATION FACTOR

Amplification factor is the ratio of the output amplitude in an amplifier to the input amplitude. The quantity is expressed for current, voltage or power, and is abbreviated by the Greek lowercase letter mu (μ).

Usually, the amplification factor is determined from the peak-to-peak voltage or current, as shown. If the voltage amplification factor for an amplifier is a certain value, the current amplification factor need not be, and probably will not be, the same. Voltage or current gain is

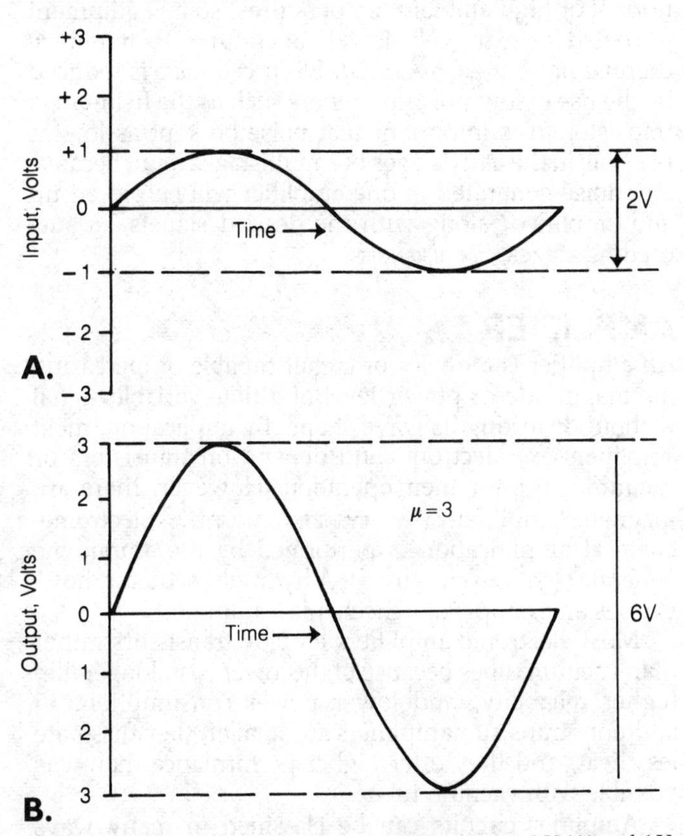

AMPLIFICATION FACTOR: Peak-to-peak voltage is 2 V (A) and 6 V at (B), so the voltage amplification factor is 3.

determined from the voltage or current amplification factor by the equation:

$$\text{Gain (decibels)} = 20 \log_{10} \mu$$

Power gain is related to the power-amplification factor by the equation:

$$\text{Gain (decibels)} = 10 \log_{10} \mu$$

See also DECIBEL, GAIN.

AMPLIFICATION NOISE

All electronic circuits generate some noise. In an amplifier, the tube, transistor, or integrated circuit invariably generates some noise called amplification noise. This noise can be categorized as either thermal, electrical, or mechanical.

The molecules of all metals and nonmetals in an electronic circuit, are in constant random motion. The higher the temperature, the more active the molecules. This generates thermal noise in any amplifier.

As the electrons in a circuit move from atom to atom, or impact against the metal anode of a vacuum tube, electrical noise is generated. The larger the amount of current flowing in a circuit, in general, the more electrical noise there will be.

Mechanical noise is produced by the vibration of the circuit components in an amplifier. Sturdy construction and, if needed, shock-absorbing devices, reduce the mechanical noise.

Although nothing can be done about thermal noise at normal or high ambient temperatures, some equipment is cooled to extremely low temperatures to minimize thermal noise (*see* CRYOGENICS). Electrical noise is reduced by the use of low-noise amplifiers such as the field-effect transistor. It is important that noise be kept as low as possible in the early stages of a multistage circuit because any noise generated in one amplifier will be picked up and amplified, along with the desired signals, in succeeding stages. *See also* NOISE.

AMPLIFIER

An amplifier is a device or circuit capable of increasing the magnitude or power level of a time-variable signal without distorting its wave shape. By implication, most amplifiers are electronic and depend on transistors or vacuum tubes for their operation. However, there are magnetic amplifiers (*see* MAGNETIC AMPLIFIER). Electromechanical amplification is performed by the amplidyne generator (*see* AMPLIDYNE DRIVE). A hydraulic actuator, however, is an example of a mechanical amplifier.

Most electronic amplifiers employ transistors rather than vacuum tubes because of the lower cost, longer life, higher reliability, and lower power consumption. In addition, transistor amplifiers are smaller, they dissipate less heat, and they offer higher performance than was possible with vacuum tubes.

Amplifier circuits can be classified in many ways including: 1) application, 2) circuit configuration, 3) coupling (if more than one stage is used), 4) bandwidth and frequency of the signals being amplified, and 5) operating mode or bias.

Amplifiers can include many different kinds of discrete transistors, including bipolar NPN and PNP (*see* TRANSISTOR) and both junction field effect (J-FET) and insulated gate or metal-oxide semiconductor field-effect (MOSFET) transistors (*see* FIELD-EFFECT TRANSISTOR, METAL-OXIDE SEMICONDUCTOR FIELD-EFFECT TRANSISTOR). Both types of field-effect transistors are available as N channel and P channel.

Application. Amplifiers can be classified according to application as voltage amplifiers or power amplifiers. Voltage amplifiers increase the voltage level of an applied signal. Because the output voltage of an amplifier is determined by the voltage drop across the output load, the impedance of the load is made as large as practical.

Power amplifiers are designed to deliver heavy current to the output load. The load impedance must be low enough to allow a high-current output, but not so low that the signal is distorted excessively. These amplifiers are also called current amplifiers. *See* POWER AMPLIFIER. Other amplifiers include the buffer amplifier, (*see* BUFFER STAGE), square-wave amplifier (*see* SQUARE WAVE) and frequency doubler.

Circuit Configuration. Transistor amplifiers can be classified according to the way the principal elements are returned to ground. In amplifier circuits with electron tubes, there are grounded-cathode, grounded-grid, and grounded-plate amplifiers. Similarly, in transistor amplifier circuits, there are grounded-base, grounded-emitter, and grounded-collector amplifiers. They are also known as common-base (CB), common-emitter (CE), and common-collector (CC) amplifiers.

The element does not have to be connected directly to ground but can be connected to ground through a resistor or capacitor. The potential of the grounded element is at ac ground, the zero-potential reference point for measuring all signal voltages.

Amplifiers also can use these transistors in integrated-circuit form. The metal-oxide field-effect transistor (MOSFET) is easier to fabricate and occupies less space on a chip than the bipolar transistor. The control electrode or gate draws practically no current, and this results in high input impedance. MOSFETs introduce less noise than bipolar transistors. The recent availability of high-current MOSFETs makes possible the replacement of bipolar transistors in some power amplifiers.

The basic electron tube amplifier consists of a vacuum triode tube and associated circuitry. *See* TRIODE TUBE. The signal voltage to be amplified is applied to the grid circuit, and the amplified signal appears across the load resistor in the plate circuit. A gain in voltage and power is provided. By selection of tubes and associated circuitry, these gains can be maximized.

Coupling. In applications where more than a single amplifier stage is used, the amplifiers are often classified according to the way they are coupled: 1) resistance capacitance (RC), 2) impedance coupling, 3) transformer coupling, and 4) direct coupling.

The output load of an RC-coupled amplifier is usually a high resistance value. RC-coupled amplifiers have good frequency response characteristics over a relatively wide frequency range. However, gain falls off above and below this range. The decrease in gain at the lower frequencies is caused by the increased reactance of the coupling capacitor. At higher frequencies, the decrease in gain is caused by the decreased reactance of the interelectrode capacitances of the stages, as well as the stray capacitance of the wiring between stages. *See* RESISTANCE-CAPACITANCE CIRCUIT.

In transformer coupling, the output of one circuit is coupled to the input of the next circuit by a transformer. Additional amplification can be obtained if the transformer has a step-up turns ratio. Transformer coupling is widely used in tuned radio-frequency and intermediate-frequency amplifiers. *See* TRANSFORMER COUPLING.

In direct coupling, the output of the first stage is applied directly to the input of the second stage. The dc amplifier can amplify both dc and ac signals. Special circuitry either in the coupling network or amplifier stage eliminates the need for a coupling capacitor. Direct-current amplifiers are widely used for amplifying low-frequency signals.

Bandwidth. Amplifiers can be classified by bandwidth as tuned or untuned. A tuned amplifier has a tuned (resonant) circuit in its input or output circuit, or both, that passes a relatively narrow band of frequencies. The center of this frequency is the resonant frequency of the tuned circuit. The width of the band depends on the Q of the tuned circuit. Tuned amplifiers are important in radio frequency and intermediate frequency sections of radio and television receivers. (*See* TUNED CIRCUIT, TUNING.) The receiver is tuned to the carrier frequency of the desired signal, and the tuned amplifiers amplify that signal. The resonant frequency of some tuned amplifiers can be set by varying either the inductance or capacitance in the circuit. Gain is normally maximum at the resonant frequency and decreases for frequencies on either side of the resonant frequency.

An untuned amplifier is one that is not tuned to any specific band of frequencies. The range of frequencies that it can amplify is limited by the circuit components and stray capacitances. However, an untuned amplifier can amplify a wider range of frequencies than a tuned amplifier.

Frequency. Amplifiers may also be classified according to frequency as: 1) direct current (dc), 2) audio frequency (af), 3) intermediate frequency (i-f), 4) radio frequency (rf) and, 5) video frequency. Direct-current amplifiers amplify dc signals, and audio amplifiers amplify signals in the audio-frequency range from about 20 to 20,000 Hz. Video amplifiers amplify signals as high as 200 MHz. I-f and rf amplifiers are not defined by specific frequency range. They are generally tuned amplifiers and therefore amplify a relatively small band of frequencies. *See* DC AMPLIFIER, AUDIO AMPLIFIER.

Operating Method. Amplifiers also can be classified according to their operating method or biasing conditions as class A, class B, class AB, or class C. The distinction among these classes is determined for a sinusoidal signal voltage applied to the input. The class of operation is determined by the quiescent point set by the bias of the input signal and by the amount of ac signal.

Class A amplifiers are biased in the center of the operating curve. This biasing allows the output current to flow during the entire cycle or the full 360 degrees of the input signal, without any part of the signal being cut off. This biasing methods also results in minimum distortion of the output signal, but Class A is the most inefficient configuration. Class A amplifiers are widely used in audio systems where it is important that distortion be low. *See* CLASS A AMPLIFIER.

A Class B amplifier is biased at cutoff so that output current flows for one-half or approximately 180 degrees of its input-signal voltage cycle. When no input signal is present, no output current flows. Thus a Class B amplifier cuts off one half of the ac input signal waveform. *See* CLASS B AMPLIFIER.

A Class AB amplifier is biased so that output current flows for more than one half of the input cycle, but for less than the entire cycle. In effect, the Class AB amplifier offers a compromise between the low distortion of Class A amplifiers and the high efficiency of Class B amplifiers. *See* CLASS AB AMPLIFIER.

A Class C amplifier is biased beyond cutoff so that its output current flows only during the positive-going peak of the input cycle. Class C amplifiers are capable of high power output, but they also introduce high distortion, preventing their use in audio applications. *See* CLASS C AMPLIFIER.

A single audio amplifier must be Class A but two audio amplifiers can operate Class B in a push-pull circuit where each supplies opposite halves of the signal. Class C operation is used for tuned radio frequency amplification because an LC (inductive-capacitive) circuit can reconstruct full sine waves at the output.

AMPLITUDE

The strength of a signal is called its amplitude. Amplitude can be defined in terms of current, voltage, or power for any given signal.

Knowing the root-mean-square (*see* ROOT MEAN SQUARE) current, *I*, and the root-mean-square voltage, *E*, for a particular ac signal, the power amplitude in watts is given by:

$$P = EI$$

If the circuit impedance, *Z*, and either the current or voltage is known, power amplitude, *P*, is:

$$P = I^2Z = E^2/Z$$

Amplitude is usually described in reference to the strength of a radio-frequency signal, either at some intermediate point in a receiver or transmitter circuit, or at the output of a transmitter. Amplitude is measured with a wattmeter, oscilloscope, or spectrum analyzer (*see* SPECTRUM ANALYZER). On an oscilloscope, signals of increasing amplitude (see A, B, and C in the illustration) appear as waveforms of greater and greater height, but of the same wavelength, assuming the frequency remains constant.

For weak signals at the antenna terminals of a receiver, the term strength is usually employed. These signals are measured in microvolts. *See also* SENSITIVITY.

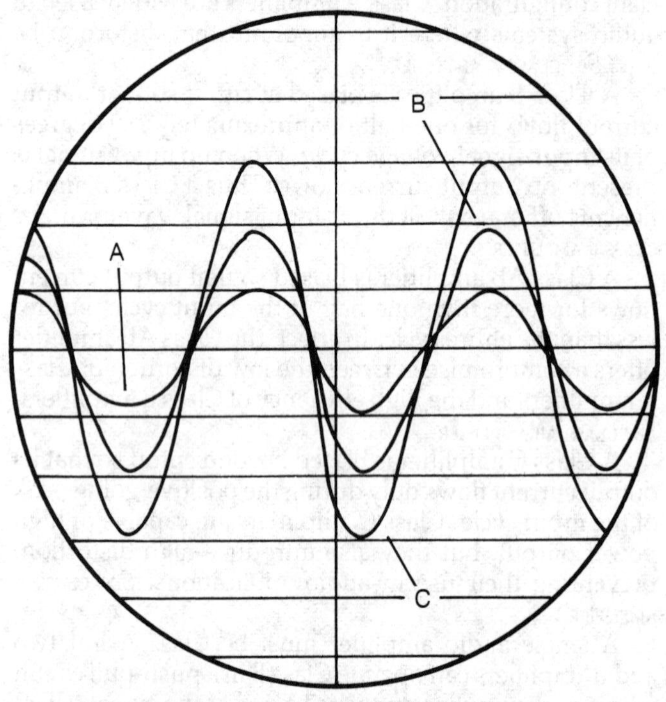

AMPLITUDE: An oscilloscope display of waves with equal frequency and phase but different amplitudes. The reference wave (A), is doubled (B) and tripled (C) in amplitude.

AMPLITUDE MODULATION

Amplitude modulation (AM) is the process of impressing information on a radio-frequency signal by varying its amplitude. The simplest example of amplitude modulation is Morse-code transmission where the amplitude switches from zero to maximum.

Generally, amplitude modulation is used for relaying messages by voice, television, facsimile, or other modes that are relatively sophisticated. The process is always the same: audio or low frequencies are impressed upon a carrier wave of much higher frequency. In the figure, A, B, and C show the amplitude modulation of a carrier wave by a sine-wave or sinusoidal audio tone. The amplitude of the carrier is greatest on positive peaks of the sinusoidal tone, and smallest on negative peaks.

The modulation of an AM signal can be strong and it can be very small. The intensity or degrees of modulation is expressed as a percentage. This percentage can vary from zero to more than 100. An unmodulated carrier, as shown in the figure, has zero percent modulation by definition. If the negative peaks drop to zero amplitude, the signal is defined to have 100 percent modulation. At C, a signal with modulation is about 75 percent. If the modulation percentage exceeds 100, the negative peaks drop to zero amplitude and remain there for a part of the audio cycle. This is undesirable, because it causes distortion of the information reproduced by the receiver.

When a given radio-frequency signal is amplitude modulated, mixing occurs (*see* MIXER) between the modulation frequencies (f_M) and the carrier frequency (f_C). For sine-wave modulation such as shown in the figure, this mixing results in new radio-frequency signals, f_{LSB} and f_{USB}, given by:

$$f_{LSB} = f_C - f_M$$
$$f_{USB} = f_C + f_M$$

These new frequencies are called sidebands. They are referred to as the upper sideband (USB) and lower sideband (LSB).

A special kind of amplitude modulation, commonly called single sideband (SSB), but properly named single-sideband, suppressed-carrier (SSSC), eliminates the carrier frequency (f_C) and also one of the sideband frequencies at the transmitter. Only one sideband is left at the output. This sideband is combined with a local oscillator signal at the receiver, resulting in a perfect reproduction of the modulating signal, provided the receiver frequency is correctly set. *See also* SINGLE SIDEBAND.

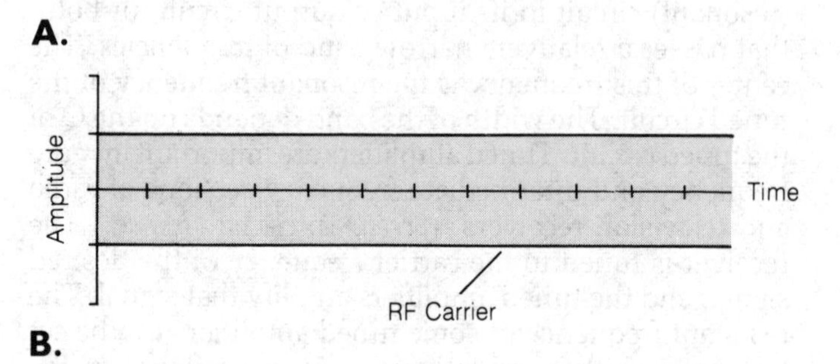

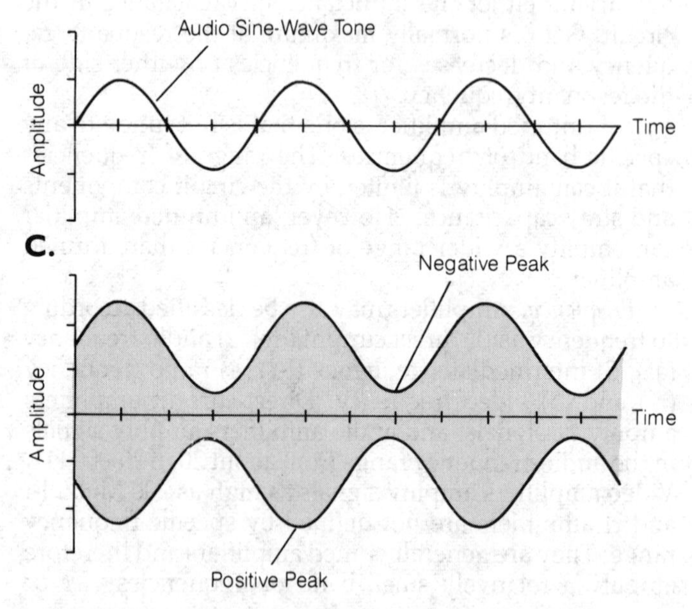

AMPLITUDE MODULATION: An unmodulated carrier (A), is modulated by a sine wave (B), and the result is shown (C).

ANALOG

Quantities or representations that are variable over a continuous range are referred to as analog. In electronics, analog quantities are differentiated from digital quantities by the fact that analog variables can take an infinite number of values, but digital variables are limited to defined states.

ANALOG COMPUTER

An analog computer carries out computations in the form of electrical analogs of numbers or physical variables. The computer receives input signals representing variable physical quantities and combines them to produce continuously varying outputs that relate to the input and processing methods. An analog computer can simulate a model of a system being studied. The outputs can be displayed graphically or used as control signals in a related system.

The most important components of an electronic analog computer are dc (direct command) or operational amplifiers. (*See* OPERATIONAL AMPLIFIER.) These amplifiers perform addition and integration when combined with active and passive feedback components or networks and appropriate input impedances. Analog computer inputs can be obtained from sensors or transducers that produce time-varying output signals.

An analog computer requires a method for setting up coefficients in a problem. The coefficients are set up either by the use of precision potentiometers or by adjusting the ratio of feedback and input impedances applied to the operational amplifiers. Analog computers can also have sets of controls for starting and stopping computation.

A control station is established that is capable of holding the problem solution at any point in the calculation. Control operations are performed by a series of relays. The control system typically has provision for the automatic application of the initial conditions of the problem while the computer controls are in the reset condition. Operator knowledge of the actual operation of the operate-reset relays is important for the solution of complex problems.

General-purpose analog computers must have components capable of multiplying variable quantities if they are to be useful in solving problems other than those involving linear differential equations with constant coefficients. For the solution of even more complex problems, the computer must have circuitry capable of generating functions that are not easily represented mathematically. Similarly, more complex problems often require the representation of nonlinear phenomena. This representation can be done with diodes or relays in the computing system. Suitable recorders are also valuable peripherals for general-purpose analog computers. Results are plotted graphically as continuous functions of variable quantities.

ANALOG-TO-DIGITAL CONVERTER

An analog-to-digital (A/D) converter is a circuit that converts continuously variable analog data—usually voltage—into an equivalent digital form. These converters can be built as monolithic integrated circuits, high-performance hybrid circuits, or discrete component modules. A/D converters are included in digital panel meters (DPM), digital multimeters (DMM), and other instruments. The most important characteristics of A/D converters include absolute and relative accuracy, linearity, no missing codes, resolution, conversion speed, and stability. Other characteristics considered are: input ranges, digital output codes, interfacing techniques, multiplexing, signal conditioning, and memory circuitry within the device.

The most popular conversion technique for communication and computer applications is successive approximations. Successive approximation is popular because it offers an excellent compromise between speed and accuracy. The dual-ramp and voltage-to-frequency integrating conversion methods are used in electronic measurement instrumentation and industrial process control. However, the faster, high-speed flash techniques are widely used for converting video signals.

Voltage-to-frequency converters provide high-resolution conversion and can also provide such special features as long-term integration (from seconds to years), frequency modulation, voltage isolation, and arbitrary frequency division or multiplication. Synchro-, and resolver-to-digital converters are used where angular or linear position must be measured precisely with high resolution and converted into digital form.

A/D converters can be assembled from D/A converters and comparator circuits can be combined with a microprocessor to perform tracking or successive-approximation conversion.

The time required for a complete measurement by an analog-to-digital converter is called conversion time. For

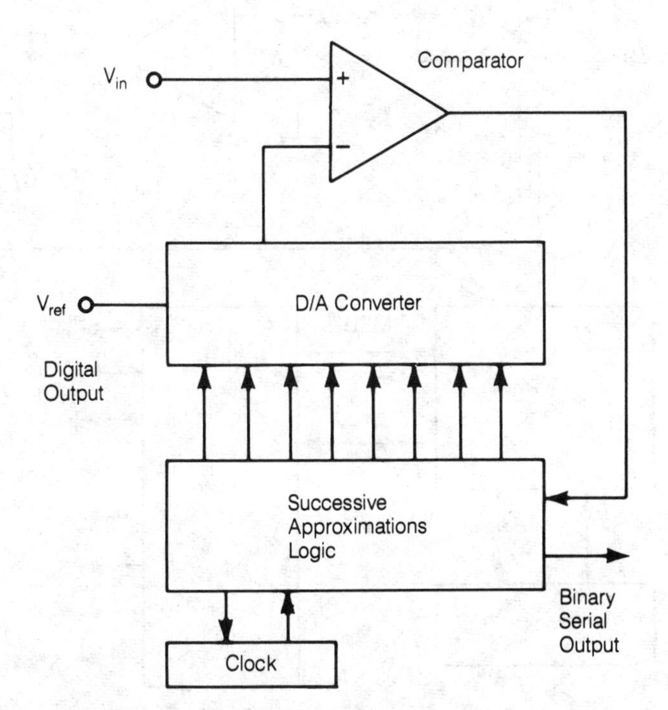

ANALOG-TO-DIGITAL CONVERTER: Fig. 1. A successive approximation A/D converter includes a shift register, control logic, and output register.

most converters, conversion time is essentially identical to the inverse of conversion rate.

The successive approximation A/D converter, as shown in the block diagram, Fig. 1, compares an unknown against a group of weighted references. Successive-approximation conversion is generally similar to the orderly weighting of an unknown quantity on a precision balance. A set of weights in descending order of value is tried, starting with the largest. Any weight that tips the scale is removed. At the end of the process when balance is achieved, the sum of the weights remaining on the scale represents the unknown value. Successive approximation A/D conversion is very fast. For example, 10-bit SA converters can make 10-bit conversions in less than 20 microseconds.

Most V/F (voltage-to-frequency) converters use a charge-balancing circuit. A block diagram is shown in Fig. 2. Conversion begins when a capacitor is charged from a current source that is proportional to the input voltage. The capacitor is then discharged with a precise current each time the charge on the capacitor reaches a preset level. V/F converters are not effective in measuring low input voltages because of offset voltage errors. In addition, the slew rate and settling time of the amplifier limits the upper frequency. In typical V/F converters, the comparator output pulses are fed to a counter for a fixed time. The accumulated count is proportional to the input voltage.

A dual-slope integrating A/D converter as shown in the block diagram, Fig. 3, converts the unknown signal to a proportional time interval which can then be measured digitally. The unknown signal at the input is integrated for a predetermined length of time. Then a reference input is switched to the integrator that integrates down from the level of the unknown until a zero level is reached. A counter counts the time for the second integration process, which is proportional to the average of the unknown signal over a predetermined integrating period. The counter then provides the digital readout of

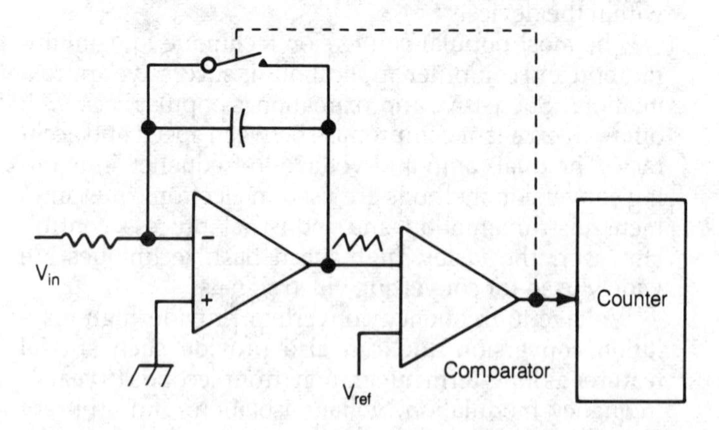

ANALOG-TO-DIGITAL CONVERTER: Fig. 2. A reference capacitor in a voltage-to-frequency converter is charged from a current source to a voltage proportional to the input voltage. The capacitor is discharged with a precise current level each time the charge on the capacitor reaches a preset level.

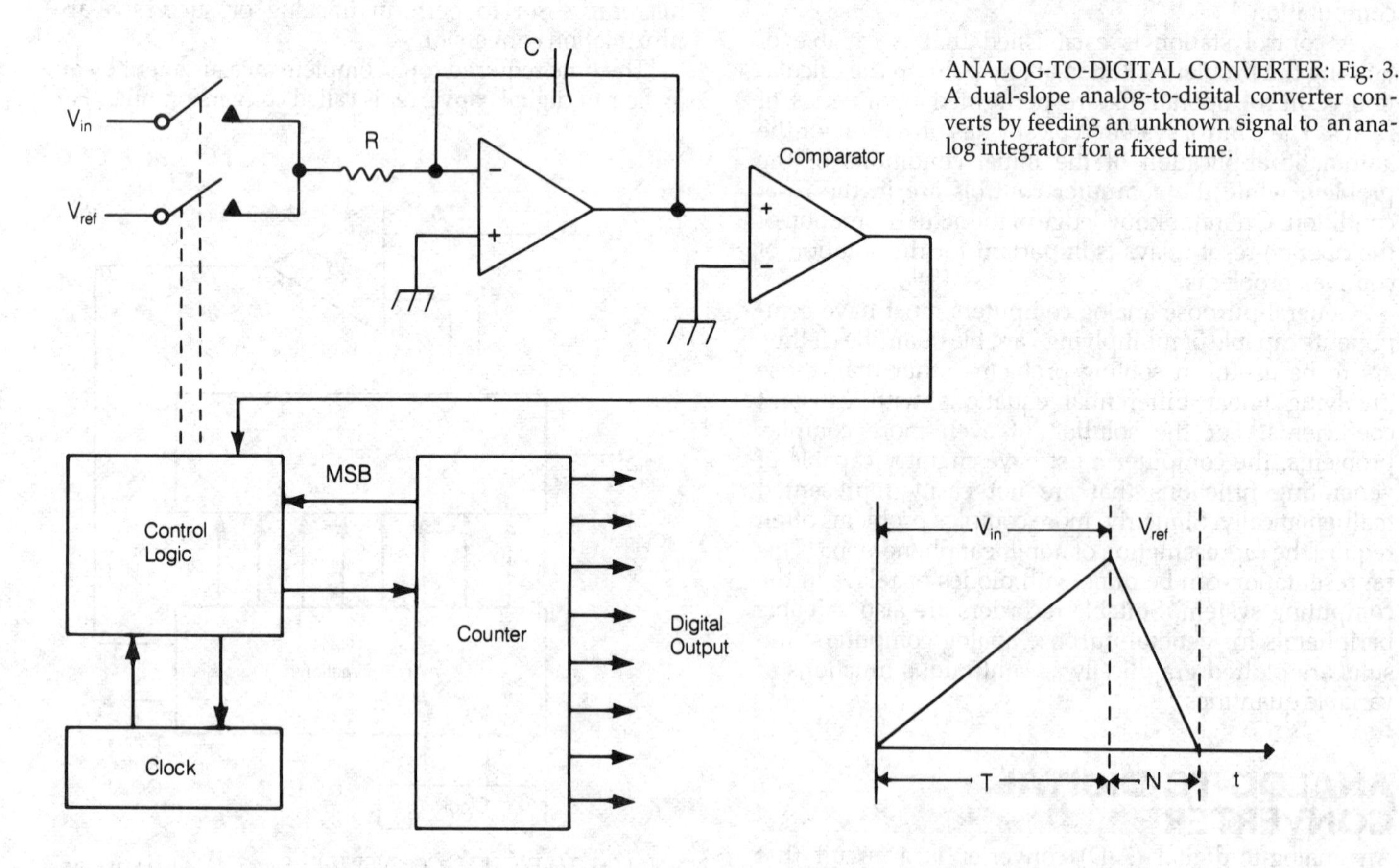

ANALOG-TO-DIGITAL CONVERTER: Fig. 3. A dual-slope analog-to-digital converter converts by feeding an unknown signal to an analog integrator for a fixed time.

time that represents the average value of the signal for the conversion period.

A flash converter is an A/D converter for high-speed, low-resolution applications. A simplified block diagram is shown in Fig. 4. It performs the A/D conversion function by using one comparator for each possible level and feeding the input signal to all comparators. Flash converters require $(2^n - 1)$ comparators for an n-bit binary word. The comparator outputs are encoded into the appropriate binary words.

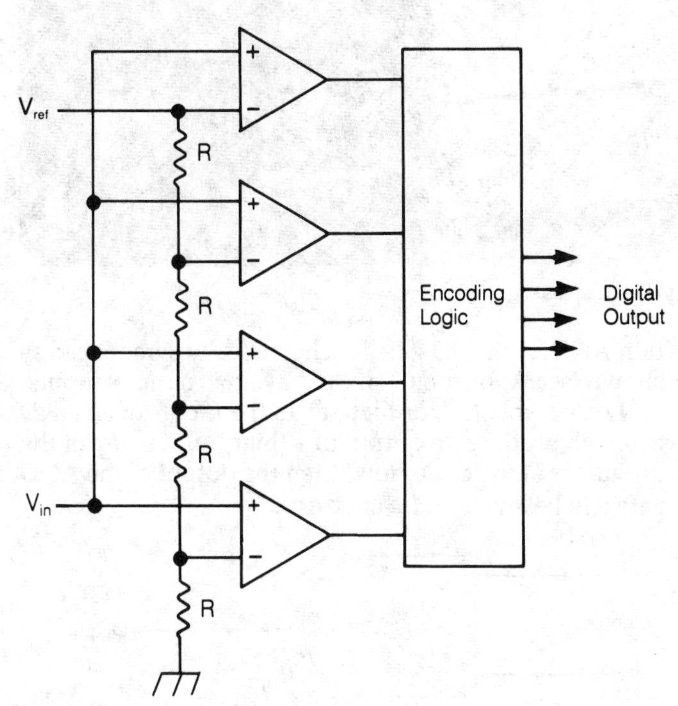

ANALOG-TO-DIGITAL CONVERTER: Fig. 4. A high-speed flash converter uses multiple parallel converters. The comparator outputs are encoded into the appropriate binary word.

ANALOG FUNCTION CIRCUIT

Analog function circuits or analog-to-analog converters are computational circuits and special-purpose devices for conditioning analog signals. These circuits can be used to relieve a computer central processor of the burden of conditioning analog signals and saving the additional programming required.

Among the more popular analog function circuits are those that perform multiplication, taking ratios, raising to powers, taking roots and performing special-purpose nonlinear functions such as linearizing transducers. Analog function circuits can also make rms measurements, compute trigonometric functions and vector sums, integrate and differentiate, and transform current to voltage or voltage to current. Some functional circuits can be purchased as fabricated multiplier/dividers or log/antilog amplifiers.

ANALOG MULTIPLEXER

An analog multiplexer is a circuit that serially switches a number of different analog input signals onto a single line or channel. An analog multiplexer can be used if data from many analog signal sources must be processed by the same computer or communications channel with a single A/D (analog-to-digital) converter. The analog multiplexer couples the input signals into the A/D converter in some preset or random sequence. Any individual input channel normally is accessed through a digital address code applied to some digital inputs. The logic address input determines which data source is to be coupled to the converter at any time.

Multiplexers can also be used in reverse, as distributors, or demultiplexers. If a converter must distribute analog information to many different channels, the multiplexer, fed by a high-speed D/A (digital-to-analog) converter, can continually refresh the various output channels with updated information. In practice, each channel must have analog storage to retain its information until the next update.

Most commercial analog multiplexers are built as monolithic integrated circuits, CMOS (complementary metal-oxide semiconductor) technology is most widely used because of the superior switching capabilities of FET (field-effect transistor) switches. Time-division multiplexing can be visualized as a rotating commutator that momentarily and sequentially connects each of several inputs to a single common output. Analog multiplexers are used in data logging systems, data acquisition systems, automatic test equipment, and control systems.

ANALOG PANEL METER

An analog panel meter (APM) is an instrument that provides a readout of changing variables in an analog format such as a moving pointer or bar. Commercial APMs are based on the D'Arsonval moving-coil meter movement or electronically generated moving bars. APMs provide a readout of measurements as well as an indication of trend in those measurements.

The traditional APM is based on the EM (electromechanical) D'Arsonval meter movement. A pointer attached to a moving coil is supported by bearings between two poles in a permanent magnetic field. The moving-coil meter operates according to the motor rule. Direct current that is proportional to the input signal in the coil creates a field that interacts with the permanent magnet field causing the coil movement within mechanical restraints. (*See* D'ARSONVAL MOVEMENT.) The pointer moves across a graduated scale that has been calibrated in units of measurement such as voltage, current, velocity, or distance. EM APMs are packaged in standard cases for mounting in panel cutouts. These instruments are adapted to measuring variables with series resistors, shunts or rectifiers. Because they get their power from the input signal, they do not require additional power.

Standard EM APMs have rated accuracies of ± 0.2 to 5 percent of full scale. However, custom graduated and calibrated meters can provide ± 0.1 percent accuracy. Typical meters have round faces with diameters of 2 inches as well as rectangular faces measuring 4 inches by 4 inches. These instruments can display dc (direct cur-

ANALOG PANEL METER: Analog panel meters with liquid crystal bar graph displays.

rent) electrical values or rectified ac (alternating current) signals.

Electronic APMs, like DPMs digital panel meters, use the analog-to-digital converters to convert input signals to values that can be displayed as moving bar. (*See* ANALOG-TO-DIGITAL CONVERTER, DIGITAL PANEL METER.) APM bar graphs provide information on rapidly changing variables that can be seen many feet from the meter. The resolution of the electronic APM depends on the number of incremental elements in the display. As many as 100 rectangular LEDs (light-emitting diodes) can be stacked or arrayed to form a moving bar. (*See* LIGHT-EMITTING DIODE.) Alternatively, bar increments can be formed as electrodes on an LCD (liquid crystal display). See Figure. (*See* LIQUID CRYSTAL DISPLAY.) Some DPMs include bar graphs on their displays to supplement the digital readout, especially if the input is subject to rapid change.

ANALOG SWITCH

An analog switch is a semiconductor switch that short circuits or opens paths for transmitting or blocking an analog signal between two circuit positions. The shorted or open switch position is usually digitally controlled. Analog switches are made with CMOS (complementary metal-oxide semiconductor) technology because of the superiority of FET (field-effect transistor) switches. Commercial analog switches are available in many styles and switch configurations. These include single, dual, and quad single-pole, single-throw (SPST), normally open (NO) and normally closed (NC). Also included are single and dual single-pole, double-throw (SPDT) and double-pole, single-throw (DPST) monolithic devices. Analog switches are used in instruments, automatic test equipment, communication systems, telephone equipment, process control, and telemetry. *See* SWITCH.

AND GATE

An AND gate is a circuit that performs the logical operation AND. An AND gate is schematically symbolized as shown (see A, B, and C). It can have two or more inputs.

Logic symbols 1, or high, at all the inputs of an AND gate will produce an output of 1 (high). But if any of the inputs are at logic 0, or low, then the output of the AND gate will be low. *See also* LOGIC GATE.

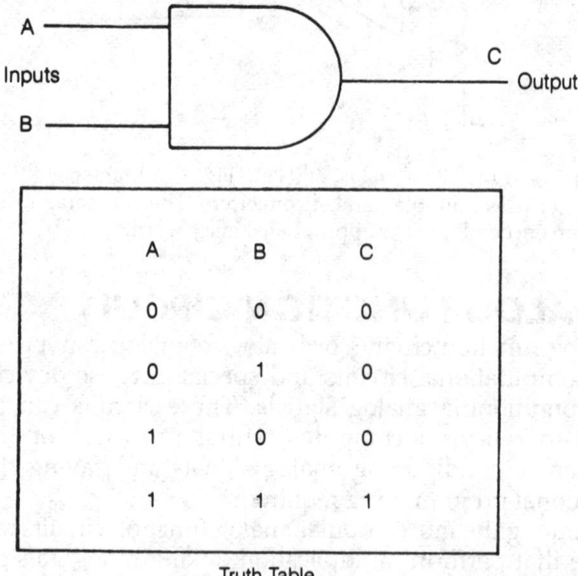

A	B	C
0	0	0
0	1	0
1	0	0
1	1	1

Truth Table

AND GATE: The symbol and truth table for an AND gate.

ANDERSON BRIDGE

An Anderson bridge is a circuit for determining unknown capacitances or inductances. The schematic diagram shows an Anderson bridge designed to measure inductances.

For the proper operation of an Anderson bridge, it is necessary to have a frequency standard, a way to balance this standard with the known reactance and the unknown reactance, and an inductor to show when balance has been achieved. A galvanometer (*see* GALVANOMETER) can be used as the indicator. It shows both positive and negative deflections from zero (null).

The Anderson bridge actually measures reactance, from which the inductance or capacitance is determined. Inductance bridges are calibrated in millihenrys or microhenrys. Capacitance bridges are calibrated in microfarads or picofarads. Some bridges are capable of measuring either inductance or capacitance. *See also* REACTANCE.

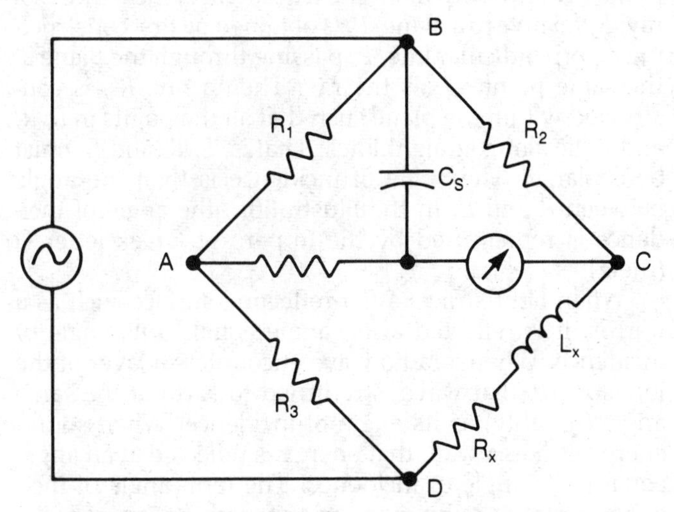

ANDERSON BRIDGE: This bridge is a modification of the Maxwell bridge for measuring inductance.

ANGLE OF DEFLECTION

In a cathode-ray tube (*see* CATHODE-RAY TUBE), a narrow beam of electrons is sent through a set of electrically charged deflection plates for writing on the screen. The angle of deflection of the beam is the number of degrees the beam is diverted from a straight path (see illustration). If the beam of electrons continues through the deflection plates in a perfectly straight line, the angle of deflection is zero.

In general, the greater the amplitude of the input signal to an oscilloscope, the greater the angle of deflection of the electron beam. The angle of deflection is directly proportional to the input voltage. Therefore, if an ac voltage of 2 volts peak-to-peak causes an angle of deflection of ± 10 degrees, an ac voltage of 4 volts peak-to-peak will result in an angle of deflection of ± 20 degrees. The angle of deflection in an oscilloscope is always quite small, so the displacement on the screen is essentially proportional to the angle of deflection. Some oscilloscopes are calibrated so that the angle of deflection increases in logarithmic proportion, rather than in direct proportion, to the input signal voltage. *See also* OSCILLOSCOPE.

ANGLE OF DEPARTURE

The term *angle of departure* refers to the angle, relative to the horizon, with which a radio signal leaves a transmitting antenna.

At high frequencies, or about 3 to 30 MHz, the angle of departure from a horizontal antenna depends on the height of the antenna above effective ground. (*See* EFFECTIVE GROUND.) Sometimes a high angle of departure is desirable, such as when local communication is attempted at long wavelengths. At other times, a very low angle of departure is needed, such as for working over great distances. The closer an antenna is to the level of effective ground, the higher the angle of departure at a particular wavelength. Any height of ¼ wavelength or less results in an angle of departure of 90 degrees, or directly upward.

The angle of departure is related to the distance at which a radio signal is returned to earth via the ionosphere. For E-layer propagation when the ionized layer is at an average height of 65 miles, the single-hop return distance as a function of the angle of departure is shown in A. For F-layer propagation when the ionized layer is at an average altitude of 200 miles, the single-hop return distance as a function of the angle of departure is shown in B. Multi-hop propagation allows communication over distances much greater than those possible via single-hop paths. *See also* E LAYER, F LAYER, PROPAGATION CHARACTERISTICS.

ANGLE OF DIVERGENCE

The spread of an electron beam in an oscilloscope, or the spread of a light beam from a collimating device or laser, is called the angle of divergence. A perfectly straight, parallel beam is impossible to realize in either case, but the angle of divergence should be as small as possible. A perfectly straight beam would have an angle of divergence of zero degrees.

In practice, the electron beam in a good oscilloscope has an angle of divergence of 2 degrees or less. The

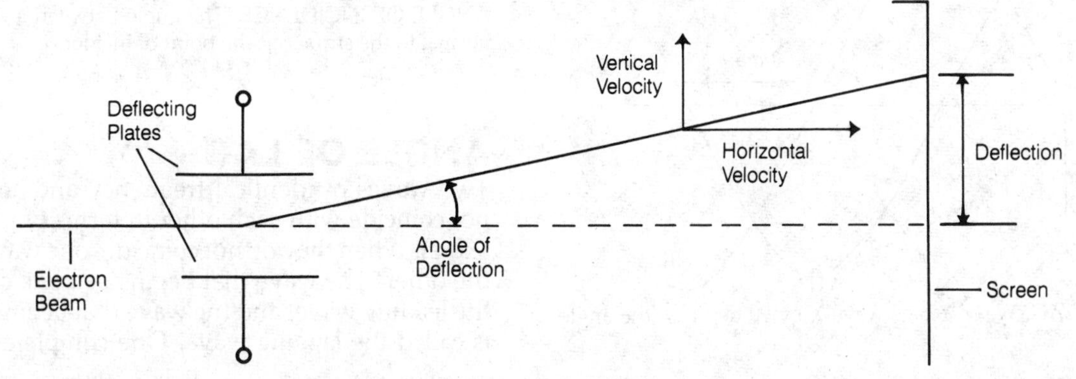

ANGLE OF DEFLECTION: The angle of deflection in a cathode ray tube is the angle of direction of the electron beam after passing between the deflecting plates.

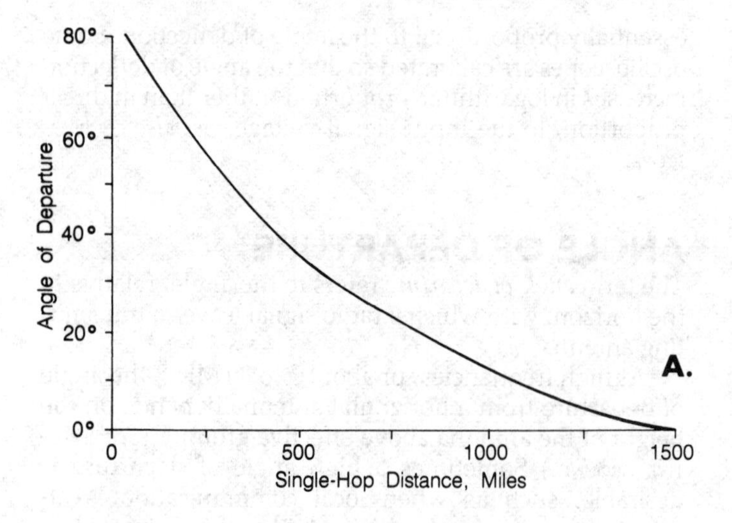

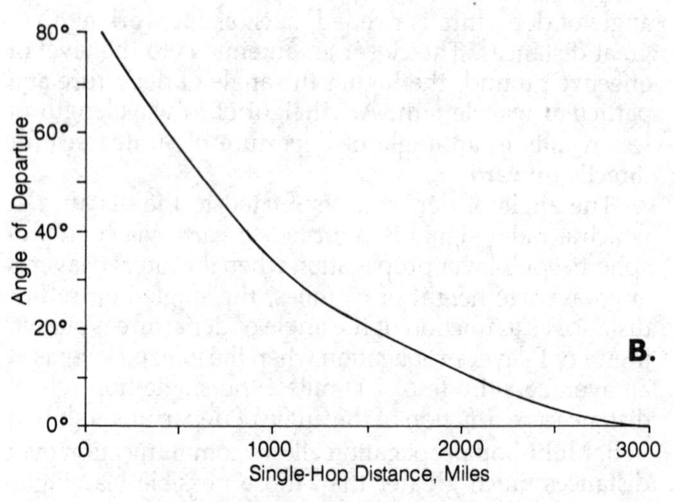

ANGLE OF DEPARTURE: Single-hop distances of radio waves reflected from layers of the ionosphere as functions of the departure angle are shown for the E layer (A) and the F layer (B).

greater the angle of divergence, the thicker the oscilloscope trace line will appear on the screen (*see* OSCILLOSCOPE).

Laser devices can be designed to produce an almost perfectly parallel beam of light, with an angle of divergence of essentially zero (*see* LASER).

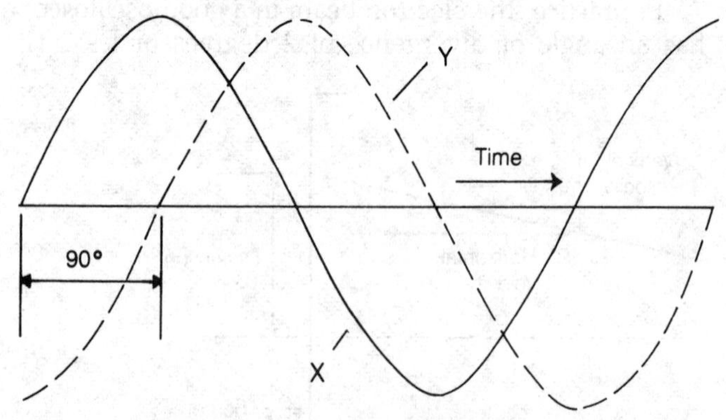

ANGLE OF LAG: Wave Y lags wave X by 90 degrees, the angle of lag.

ANGLE OF INCIDENCE

A ray of energy impinging on a locally flat surface or region is said to have a certain angle of incidence. The angle of incidence is measured between the normal to the boundary and the ray, and thus can vary from zero to 90 degrees (see figure). When determining the angle of incidence of a ray, some confusion is possible, since the measuring device must be correctly oriented with respect to the ray and the plane. The true angle of incidence of a ray, *R*, relative to a plane, *P*, is obtained by first constructing a perpendicular line, *L*, passing through the plane at the same point, *Q*, as the ray. Then a ray, *R'*, is constructed within the plane such that all the points in *L*, *R*, and *R'* lie along straight lines. That is, *L*, *R*, and *R'* must be coplanar. The angle of incidence is then the angle between *R* and *L*. In the illustration, the angle of incidence is represented by the uppercase Greek letter Θ (theta).

When light strikes a flat reflecting surface such as a mirror, it is reflected at an angle equal to its angle of incidence. When a radio wave encounters a layer of the ionosphere, that wave is returned to earth at the same angle (roughly) as its angle of incidence. When sound energy strikes a wall, that energy is reflected at an angle equal to its angle of incidence. The term angle of incidence refers to impinging, or approaching, energy. *See also* ANGLE OF REFLECTION, ANGLE OF REFRACTION.

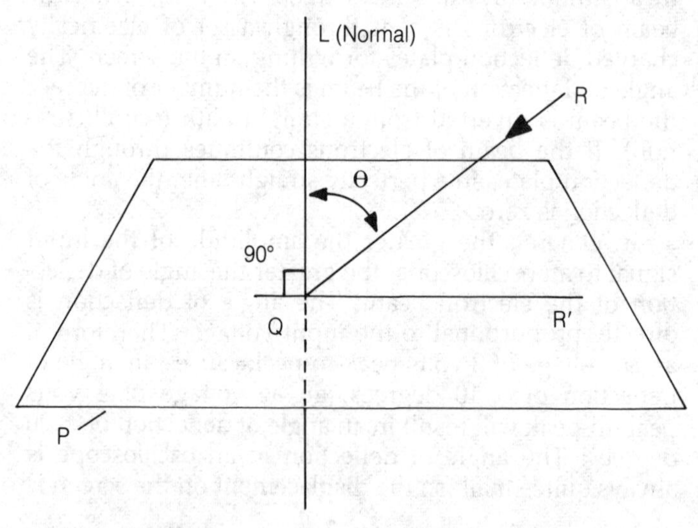

ANGLE OF INCIDENCE: The angle between a light ray and the normal to the surface at the point of incidence.

ANGLE OF LAG

Two waves of identical frequency and amplitude need not coincide with each other in terms of phase (*see* PHASE ANGLE). When they do not coincide, one wave is said to lag the other. The wave that begins its cycle earlier is called the leading wave, and the wave that begins its cycle later is called the lagging wave. One complete wave cycle is represented by 360 degrees. One-half cycle is 180 degrees. One wave can lag another by any angle from zero to 180 degrees as shown. If one wave lags another by more than a half cycle, consider that wave to be leading the other by some angle less than 180 degrees. When the angle of lag between two waves is precisely 180 degrees, the waves are said to be in phase opposition. When the angle of lag is 90 degrees, the waves are in phase quadrature.

Angle of lag is an important quantity in alternating-current circuit theory. In a circuit containing resistance and no reactance, the voltage and current waves are exactly in phase. In an inductive reactance, the current cycle lags the voltage cycle by 90 degrees. In a capacitive reactance, the voltage lags the current by 90 degrees. In a circuit containing some resistance and some reactance, the waves are separated by some value between zero and 90 degrees. This occurs because reactances do not simply dissipate energy, as do resistances, but instead they store energy and release it later in the cycle. *See also* ANGLE OF LEAD, REACTANCE.

ANGLE OF LEAD

When two waves have the same frequency but are not in phase, one wave leads the other by a certain number of degrees. In the illustration in ANGLE OF LAG, wave Y lags wave X (*see* ANGLE OF LAG). However, wave X also leads wave Y. This means exactly the same thing.

The angle of lead is specified as some value between zero and 180 degrees, as is the angle of lag. The leading wave is identified by the fact that its cycle begins earlier than that of the lagging wave, by an amount less than a half cycle. In most amplitude-vs-time illustrations, including that in ANGLE OF LAG, the lagging wave is displaced to the right and the leading wave is on the left.

If a wave leads another by some angle, ϕ, of more than 180 degrees, we consider it to lag the other by $360 - \phi$ degrees. This keeps confusion to a minimum. *See also* PHASE ANGLE, REACTANCE.

ANGLE OF RADIATION

See ANGLE OF DEPARTURE.

ANGLE OF REFLECTION

When a ray of energy strikes a flat object or barrier and is reflected, we speak of its departure angle as its angle of reflection. While this angle is also sometimes called the angle of departure, the latter term is generally reserved for energy generated at or near the flat surface or barrier, rather than energy reflected from it (*see* ANGLE OF DEPARTURE).

The angle of reflection is always the same as the angle of incidence for a locally flat, smooth surface (see figure). Further, the plane containing the incident and reflected rays is always perpendicular to the plane of the barrier at the point of reflection. *See also* ANGLE OF INCIDENCE.

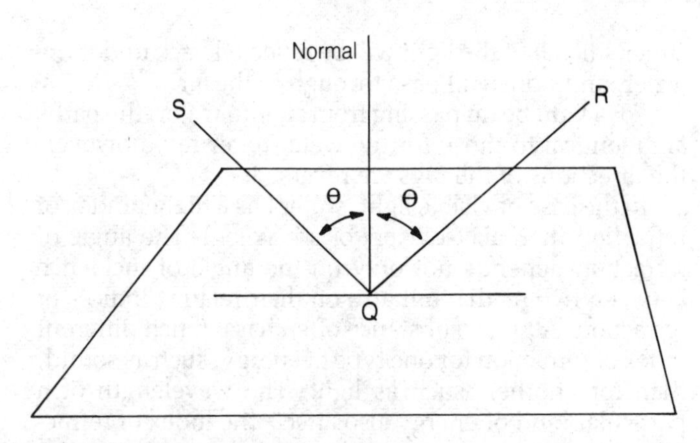

ANGLE OF REFLECTION: The angle between a light ray and the normal to the surface at the point of reflection. The angle of incidence equals the angle of reflection.

ANGLE OF REFRACTION

Energy passing from one medium to another is often refracted at the barrier (*see* REFRACTION). A good example is a visible light ray passing into or out of a pool of water as shown. Refraction can take place with sound waves, radio waves, infrared energy, ultraviolet energy, and X rays, as well as with visible light.

In the illustration, a ray of light, *R*, leaves a pool of water. At the surface, the ray changes its direction by a certain angle, ϕ. The angle of incidence of *R* with respect to the water surface is given by θ. Because of the refraction, ray *R* leaves the water surface with a new angle, θ', such that

$$\theta' = \theta + \phi$$

The angle of refraction is defined as θ'. The angle of direction change is ϕ.

When the angle of incidence θ is 0 degrees, the measure of ϕ will be zero. As θ is made larger, ϕ increases. At a certain point, the takeoff angle will become zero; then the angle of refraction is 90 degrees. That is, the ray will follow the water surface. Then if θ is made

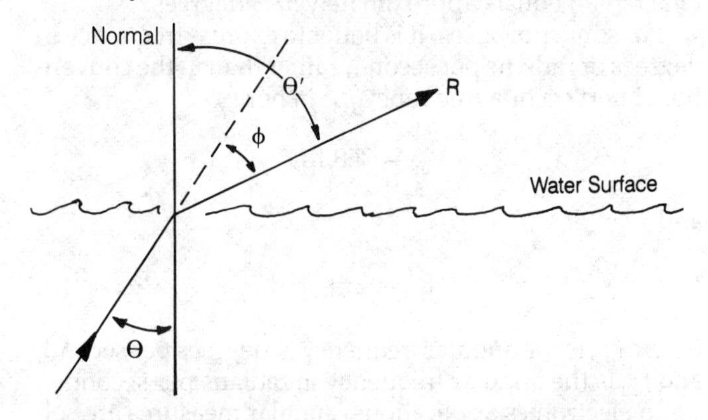

ANGLE OF REFRACTION: A light ray entering another medium obliquely will be bent toward the normal when the velocity is reduced, and it will be bent away from the normal when the velocity is increased. In this example, the angle of refraction exceeds the angle of incidence.

larger still, all of the light will be reflected back under the water, and none will pass through to the air.

For a light beam passing from air into water, the paths are identical to those for the water-to-air ray; however, the directions of the rays are reversed.

In the case of visible light, water has a higher index of refraction than air (*see* INDEX OF REFRACTION). The angle of refraction depends not only on the angle of incidence between two media, but also on their relative indices of refraction. A given substance often has a much different index of refraction for one type of energy, such as sound, than for another, such as light. The wavelength of a particular kind of energy also affects the index of refraction.

The ionosphere acts to bend shortwave radio signals because its index of refraction is greatly different, at high frequencies, than the index of refraction of air. *See also* PROPAGATION, PROPAGATION CHARACTERISTICS.

ANGSTROM UNIT

The angstrom unit (Å or AU) has been used for the measurement of optical wavelengths.

$$1 \text{ Å } 10^{-10} \text{ meters} = 0.0001 \text{ micron} = 3.937 \times 10^{-9} \text{ inch}$$

Formerly used only in physics or optics, the angstrom unit is now a unit of measurement for determining the thickness of epitaxial layers of elements and compounds grown on semiconductor substrates. Heterojunction layers of elements and compounds are being grown in layers measurable in angstrom units on substrates of silicon, gallium arsenide, and other semiconductor materials.

ANGULAR FREQUENCY

Frequency is expressed in cycles per second, or hertz (*see* HERTZ). One complete cycle passes through 360 degrees, representing one revolution around a circle. This same revolution can be expressed as 2π radians (*see* RADIAN). One radian equals approximately 57.3 degrees.

For some purposes, it is better to express frequency in degrees or radians per second, rather than in the conventional hertz. For a frequency f_{Hz} in hertz,

$$f_{d/s} = 360 \, f_{Hz}$$

and

$$f_{r/s} = 2\pi f_{Hz}$$

where $f_{d/s}$ is the angular frequency in degrees per second, and $f_{r/s}$ is the angular frequency in radians per second.

In electronics applications, angular measures are seldom used to express frequency. *See also* ANGULAR MOTION.

ANGULAR MOTION

Angular motion is motion in a circle, also called rotational motion.

For a given object, X, orbiting a central point, C, as shown in Fig. 1, the angular displacement from t_1 to time t_2 is given by the angle θ traversed by the line segment CX. The angular velocity is measured in degrees per second or radians per second (*see* RADIAN). The angular

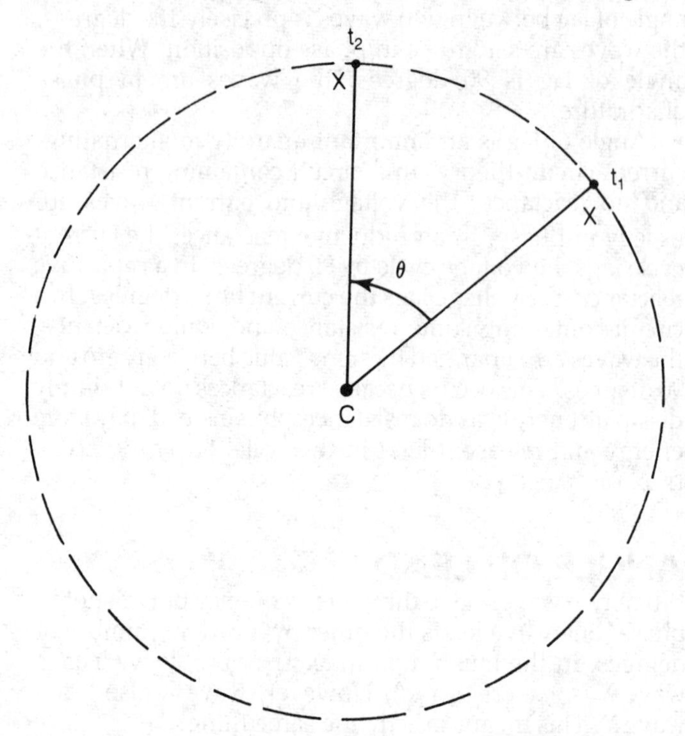

ANGULAR MOTION: Fig. 1. The angular or circular displacement of X from time t_1 to t_2 (measured in degrees or radians) is 0.

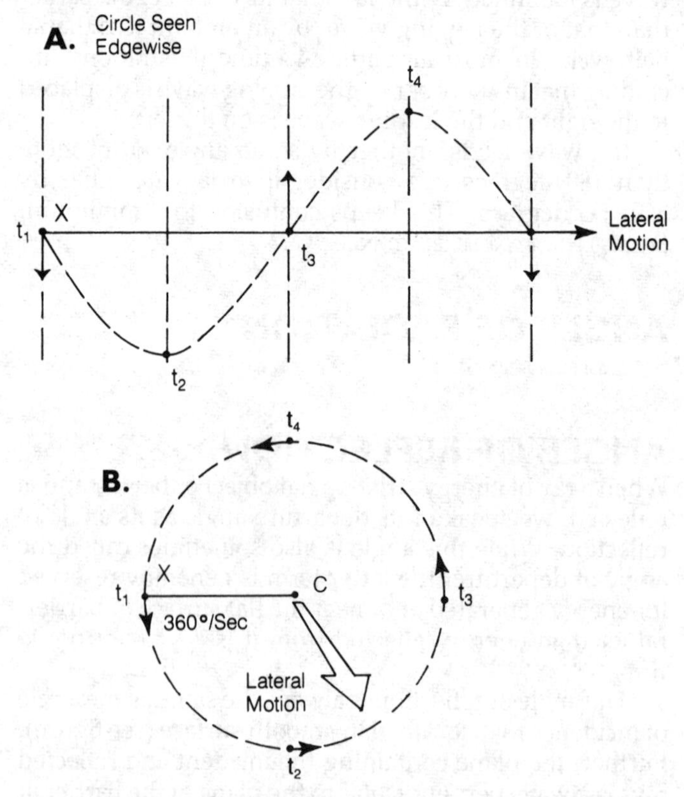

ANGULAR MOTION: Fig. 2. Angular motion at a constant velocity generates a sine wave (A), and the generation process is shown at (B).

acceleration is the rate at which the angular velocity changes. It is expressed in degrees or radians-per-second-per-second.

Angular motion at constant velocity forms an exact mathematical representation of a sine wave. If the object X in the first illustration rotates with constant speed, completing one revolution per second around the central point C, its angular velocity is 360 degrees or 2π radians per second. If the circle is viewed edge-on, a back-and-forth oscillation of object X is observed. If the circle is then moved laterally (perpendicular to the plane in which it lies) as shown in Fig. 2 (A and B), this back-and-forth oscillation becomes a sine wave with a frequency of 1 Hz. This mathematical representation expresses how a sine wave is represented in terms of degrees: One cycle is given by one rotation of a point in a circle, or 360 degrees. A frequency of 1 Hz thus is equal to 360 degrees per second, or 2π radians per second. A half cycle is 180 degrees or π radians. A quarter cycle is 90 degrees or $\pi/2$ radians. *See also* SINE WAVE.

ANODE

In a vacuum tube or semiconductor diode, the anode is the electrode toward which the electrons flow. The anode of a vacuum tube is also known as the plate. The anode is always positively charged relative to the cathode (*see* CATHODE) under conditions of forward bias, and negatively charged relative to the cathode under conditions of reverse bias. Current therefore flows only when there is forward bias. (A small amount of leakage current flows with reverse bias, but it is usually negligible.) If the reverse voltage becomes excessive, there might be a sudden increase in current in the reverse direction (*see* ARCBACK) in a diode vacuum tube. A and B illustrate the normal conditions in a tube.

The term anode is sometimes used in reference to the positive terminal or electrode in a cell or battery. *See also* DIODE.

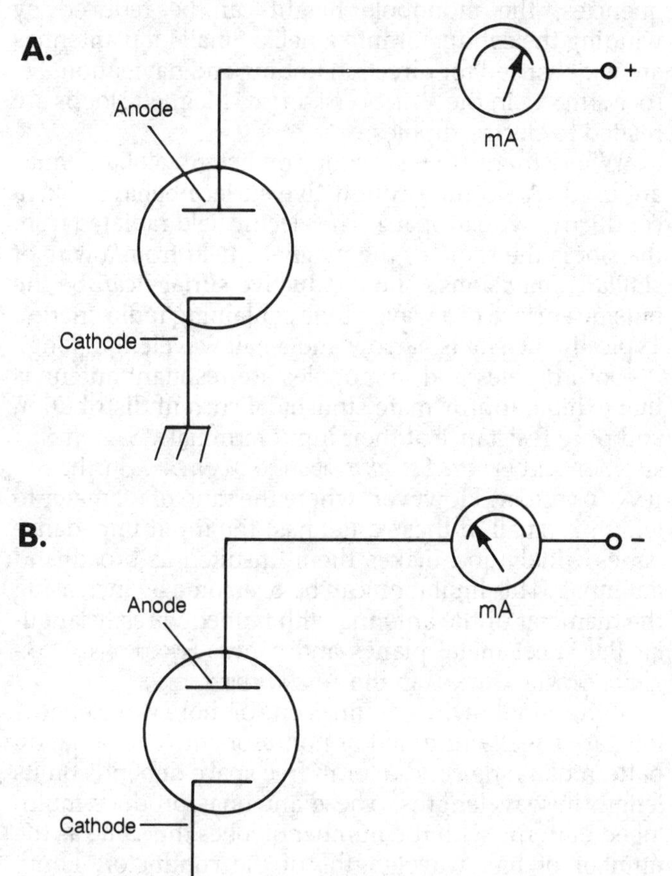

ANODE: Current flows from the cathode to the anode of a tube when the anode is positive (A), and current is cut off when the anode is negative (B).

ANTENNA

An antenna is the component in a radio system that either couples electromagnetic energy to or from free space for transmission or reception. At the radio transmitter, the antenna radiates the energy into free space in the form of radio waves. At the radio receiver, the antenna intercepts some of the radio energy and couples it into the receiver circuitry, which amplifies it to a useful level and then recovers the intelligence carried by the signal. Antennas have the property of reciprocity: the same antenna can usually be used either for transmitting or receiving. In most descriptions of antennas, radiation is the only function described; it is understood that reception is reciprocal.

Antennas are usually classified by use, but they can also be classified by operating frequency. An antenna for low radio frequencies might be a mile long, but one for microwaves might be measured in feet or inches. In practice, antennas are measured in units of wavelength of their operating frequency rather than in standard measurement units. Antennas are measured that way so that structures differing significantly in size can be described in similar terms. In this encyclopedia, antenna measurements are given in terms of their principal wavelength. All references to specific antenna types listed in this section are given in alphabetical order in the ANTENNA DIRECTORY.

An antenna can be a single conductor or an array of conductors that radiate or intercept energy in the form of electromagnetic waves. The simplest antenna, or aerial, is a length of metal wire. When the output of a transmitter is applied to an antenna, current flows back and forth along its length. However, because the antenna is not a closed circuit, the current creates an uneven distribution of electrons. In a simple, center-fed, half-wavelength antenna, the current distribution curve is out of phase with the voltage or charge distribution curve. At the ends of the antenna, the charge is maximum, and the current is zero. At the center, the charge is zero, and the current is maximum. Both current and charge buildup along the antenna vary sinusoidally with the input and produce fields in space around the antenna.

The antenna current produces a magnetic field, and the charge produces an electric field. These fields are 90 degrees out of phase with each other. If the frequency of the fields is high enough, a portion of both the magnetic and electric fields around the antenna is detached and moves outward in space. A moving electric field (E) creates a magnetic field (H) and a moving magnetic field creates an electric field. These fields are in phase with and have a direction perpendicular to the fields that created them. The E and H fields add together vectorially in space to produce a single sinusoidally varying electromagnetic field called the radio wave. The laws governing this radiation are described by Maxwell's equations.

The field strength of a radio wave is maximum in the immediate vicinity of the antenna and decreases inversely with distance from the antenna. The radiation pattern or polar diagram of an antenna shows how the field strength varies with distance and direction from the antenna.

The simplest wire radiator or antenna is the elementary dipole or doublet. The height of the antenna above ground, the conductivity of the earth below it, and the shape and dimensions of an antenna all affect the radiated field pattern in space. In most applications, antenna radiation is directed between specified angles in both the horizontal or vertical planes.

The Hertz antenna is a simple dipole one-half wavelength or any even or odd multiple of a half wavelength long. Hertz antennas are installed above ground and can be mounted either vertically or horizontally; they need not be connected conductively to the ground. *See* HERTZ ANTENNA, VERTICAL ANTENNA, and VERTICAL DIPOLE ANTENNA in the ANTENNA DIRECTORY following this section.

The Marconi antenna is a grounded antenna one quarter wavelength long that operates as a half-wavelength antenna. The transmitter can be connected between the bottom of the antenna and ground. A reflection of current and voltage distribution set up from the antenna is provided by the ground. The wave emitted from the antenna-ground combination is the same as that emitted by a Hertz antenna operated at the same frequency. *See* GROUND-PLANE ANTENNA, MARCONI ANTENNA, and VERTICAL ANTENNA in the ANTENNA DIRECTORY.

For maximum transfer of power from the transmitter to the antenna, the output impedance of the transmitter must be matched to the antenna impedance. Antenna input impedance determines the antenna current at the feed point for any given rf (radio frequency) voltage at that point. In a half-wave antenna, the effective current is maximum at the center and minimum or zero at the ends; similarly the effective voltage is a maximum at the ends and minimum at the center. Therefore, the impedance varies along the antenna and is minimum at the center and maximum at the ends. If energy is fed to a half-wave antenna at its center, the antenna is said to be center fed. If the energy is fed at the ends, it is said to be end fed. The impedance at the center of a half-wave Hertz antenna is approximately 73 Ω and at the ends it is approximately 2,500 Ω. The intermediate points have intermediate values of impedance.

The antenna at the end of the transmission line is equivalent to a resistance that absorbs a certain amount of energy from the generator. Neglecting the losses that occur in the antenna, this is the energy that is radiated into space. The value of resistance that would dissipate the same power that the antenna dissipates is called radiation resistance. *See* RADIATION RESISTANCE.

The position of a simple antenna in space determines the polarization of the emitted wave. Polarization defines the orientation of the electric field component of the wave with respect to ground. Antennas can be polarized either horizontally or vertically. The transmitting and receiving antennas in a system must both be polarized in the same direction for efficient reception. *See* ANTENNA POLARIZATION. The variation of signal strength around an antenna can be shown graphically by polar diagrams. *See* ANTENNA PATTERN.

Bandwidth of an antenna is related to its input impedance. It can be limited by pattern shape, polarization, and impedance characteristics. If the amount of reactive energy stored in the antenna is large with respect to radiated resistive energy, the bandwidth will be narrow. Where wide bandwidth is important, antenna designs are selected for wide inherent bandwidth.

Antennas whose dimensions are short compared to their operating wavelengths are called electrically small antennas. They typically exhibit low radiation resistance and high reactance resulting in high Q and narrow bandwidth. Some examples of electrically small antennas include the kinds of end-fed monopole antennas commonly used at low frequencies for long-range communication, commercial broadcasting, and mobile applications. Where height is a limiting factor at higher frequencies, the monopole height can be reduced by winding the conductor into a helix. Small loop antennas are widely used for direction finding and navigation. *See* LOOP ANTENNA in the ANTENNA DIRECTORY. Magnetic loops are related to electric dipoles.

Where there are restrictions on height, slot antennas are used. A slot in a conductive surface behaves like a conductive wire in space. The electric field radiated from the slot is the same as the magnetic field from a wire of similar dimensions. The conductive surface can be the outside surface of a wave guide containing radio energy. Typically the slot is narrow and a half wavelength long.

Both dipoles and monopoles are resonant antennas that exhibit approximate sinusoidal current distribution and pure resistance at their input terminals. *See* ANTENNA RESONANT FREQUENCY. *See also* MULTIBAND ANTENNA in the ANTENNA DIRECTORY. However, where the ratio of diameter to length is small in these antennas, the input impedance varies widely and makes them unsuited as broadband antennas. This limitation can be overcome by increasing the diameter of the antenna with fanned wires, triangular flat sheet metal planes and cones. *See* BICONICAL ANTENNA, BOWTIE ANTENNA in the ANTENNA DIRECTORY.

Long, single-wire antennas one or more wavelengths long are usually untuned or nonresonant. The radiation pattern of a long conductor in free space depends on its length in wavelengths. These antennas produce multilobed patterns with the number of lobes the same as the number of half wavelengths of the conductor. Long, single-conductor antennas that are terminated with a resistor radiate one major lobe in the direction of wave travel down the line, but the radiation pattern of the unterminated line is more symmetrical. *See* LONGWIRE ANTENNA in the ANTENNA DIRECTORY.

Two horizontal nonresonant, long-conductor antennas can be formed into a horizontal V array and terminated with resistors to provide an improved radiation pattern. The angle between the V array is determined by the length of the conductors and the conditions necessary for major lobe addition in phase. Even more improvement in horizontal radiation pattern can be achieved with the rhombic array in which the two long-conductor antennas are bent to form a diamond pattern and given a common terminating resistor. Major lobes of the four legs add in phase to form a major lobe. *See* RHOMBIC ANTENNA in the ANTENNA DIRECTORY.

Antennas can be made more directional by concentrating the transmitted energy. Directional antennas consist of a number of separate elements that function together to provide improved directivity. Multielement arrangements are called antenna arrays. Their characteristics are determined by the number and types of elements they use. Three commonly used elements for antenna arrays are dipoles, reflectors, and directors. A dipole is typically a single conductor antenna fed at the center and usually operated half wave. Reflectors and directors are directive elements that alter the radiation pattern of the dipole. *See* DIRECTOR. Reflectors and directors are called parasitic elements and the antenna systems that use them are called parasitic arrays. *See* PARASITIC ARRAY.

Driven arrays consist of two or more elements, usually half-wave dipoles, with each element being driven by the output of the transmitter. They are used for high power applications. *See* DRIVEN ELEMENT. They can be divided into three basic types, broadside, end-fire and collinear arrays.

For a discussion of other array antennas designed for maximum radiation or sensitivity in one direction *see* LOG-PERIODIC ANTENNA, MULTIELEMENT ANTENNA, PHASED-ARRAY ANTENNA in the ANTENNA DIRECTORY.

Dielectric materials formed as solid rods or cylinders can function as waveguides and antennas. The wavelength inside a large, solid dielectric rod is less than the free-space wavelength. If the rod diameter is large compared to the contained wavelength, most of the energy in the wave travels inside the dielectric. But if the diameter is reduced below a half wavelength in a gradual taper, the wave continues beyond the end of the rod and is propagated into free space. The major lobe is in the direction of the rod.

Direct aperture antennas for use at short wavelengths can be built as horns, mirrors, or lenses. These antennas use conductors and dielectrics as surfaces or solids in contrast to the lower frequency antennas that use conductors as discrete lines. Direct aperture horns intended to produce high-gain beams over broad bands include the conical horn and the pyramidal horn. *See* HORN ANTENNA in the ANTENNA DIRECTORY.

The Luneberg lens is a dielectric sphere that can be used as an antenna because of its ability to focus radio energy for increased gain. It has an index of refraction that varies with the distance from the center of the sphere. Energy is both fed into the lens at the focal point for transmission and removed from that point for receiving. The pencil beam formed by the lens may be steered by changing the position of the feed point.

Three-dimensional reflectors improve gain, modify patterns, and eliminate backward radiation at higher frequencies. Primary apertures such as low-gain dipoles, slots, or horns radiate toward larger reflectors called secondary apertures. The large reflector further shapes the radiated wave to produce the desired pattern. These include plane-sheet reflectors, corner reflectors, and parabolic reflectors. *See* CORNER REFLECTOR, PLANE REFLECTOR.

The horn parabolic reflector, a portion of the paraboloid, was designed to eliminate the presence of the feed horn in the path of the reflected wave. These reflectors are commonly used in point-to-point microwave systems and satellite communications ground stations because of their broad band and very low noise characteristics.

Directional electromagnetic energy beams can be formed in limited spaces with two reflector systems. The most common two-reflector antenna is the Cassegrain system. *See* CASSEGRAIN FEED.

ANTENNA DIRECTORY

Following are brief summaries of various antennas, arranged in alphabetical order.

Adcock Antenna An Adcock antenna is a special type of phased array (*see* PHASED-ARRAY ANTENNA) for achieving a bidirectional, or figure-eight, radiation pattern in the horizontal plane.

A pair of vertical antennas separated by ½ wavelength (or occasionally less) is connected in such a way

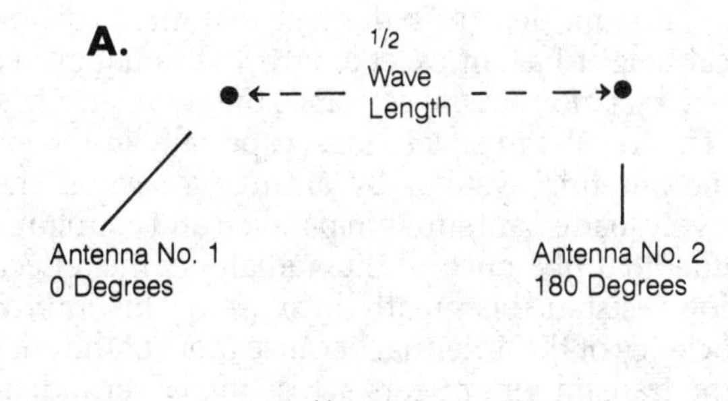

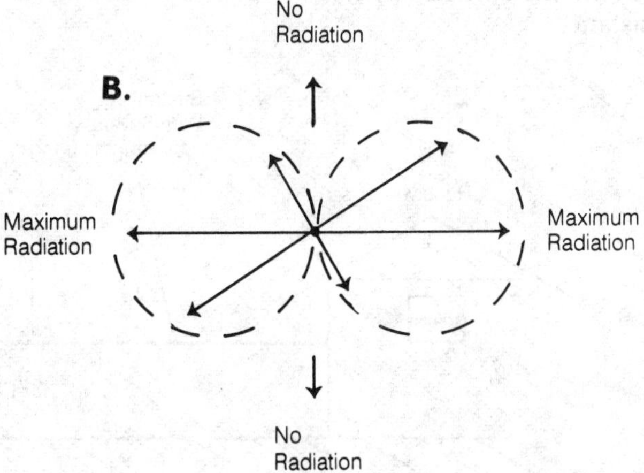

ADCOCK ANTENNA: The Adcock antenna (A) has two vertical radiators spaced ½ wavelength apart and driven 180 degrees out of phase. The radiation pattern is shown at (B).

that the signal reaching one antenna is 180 degrees out of phase with the signal reaching the other antenna. This is usually accomplished by making one transmission-line section ½ wavelength electrically longer than the other. This antenna combination radiates well in horizontal directions that line up with the two antennas, but poorly in directions perpendicular to the line connecting the antennas. See A and B. The signals from the antennas cancel in the latter directions. In intermediate directions, the antenna system radiates, but not as well as in the most favored directions.

Adcock antennas can be used for receiving as well as transmitting. An interfering signal may be attenuated by orienting the Adcock antenna so that the undesired signal falls into the notch in the figure-eight pattern. Some AM (amplitude modulation) broadcast stations must use directional transmitting antennas, such as the Adcock antenna, to minimize interference with each other. A nearby station on the same channel, or an adjacent channel, as a given station might otherwise cause interference. Arrays of Adcock antennas are often used as direction-finding systems.

The Adcock antenna displays a different impedance than a simple vertical antenna, and matching systems are therefore required for proper transmitter and feed line operation. *See also* ANTENNA MATCHING.

Alexanderson Antenna An antenna for use at low or very low frequencies, the Alexanderson antenna consists of several base-loaded vertical radiators connected together at the top and fed at the bottom of one radiator. The figure illustrates the concept of the antenna.

At low frequencies, the principal problem with transmitting antenna design is the fact that any radiator of practical height has an exceedingly low radiation resistance (*see* RADIATION RESISTANCE), since the wavelength is so large. This results in severe loss, especially in the earth near the antenna system. By arranging several short, inductively loaded antennas in parallel, and coupling the feed line to only one of the radiators, the effective radiation resistance is greatly increased. This improves the efficiency of the antenna, because more of the energy from the transmitter appears across the larger radiation resistance.

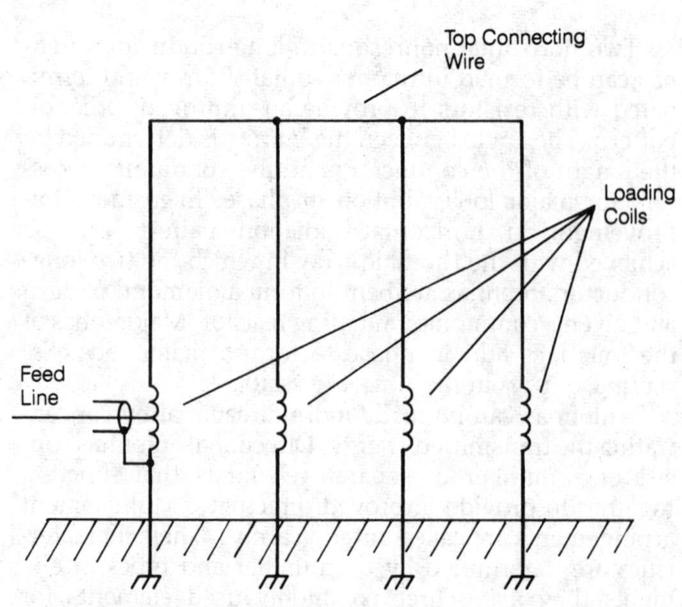

ALEXANDERSON ANTENNA: This antenna consists of loaded vertical radiators arranged in parallel.

The Alexanderson antenna has not been extensively used at frequencies above the standard AM broadcast band. But where available ground space limits the practical height of an antenna and prohibits the installation of a large system of ground radials, the Alexanderson antenna could be a good choice at frequencies as high as perhaps 5 MHz. The Alexanderson antenna requires a far less elaborate system of ground radials than a single-radiator vertical antenna worked against ground. The radiation resistance of an Alexanderson array, as compared to a single radiator of a given height, increases according to the square of the number of elements. *See also* INDUCTIVE LOADING.

Beverage Antenna A Beverage antenna is a form of traveling-wave antenna used for receiving at medium and high frequencies.

It consists of a long, straight wire of at least several wavelengths run close to the ground as shown in the illustration. A 300 to 600 Ω noninductive resistor is generally connected between the far end of the antenna and a ground rod. This antenna responds well to signals

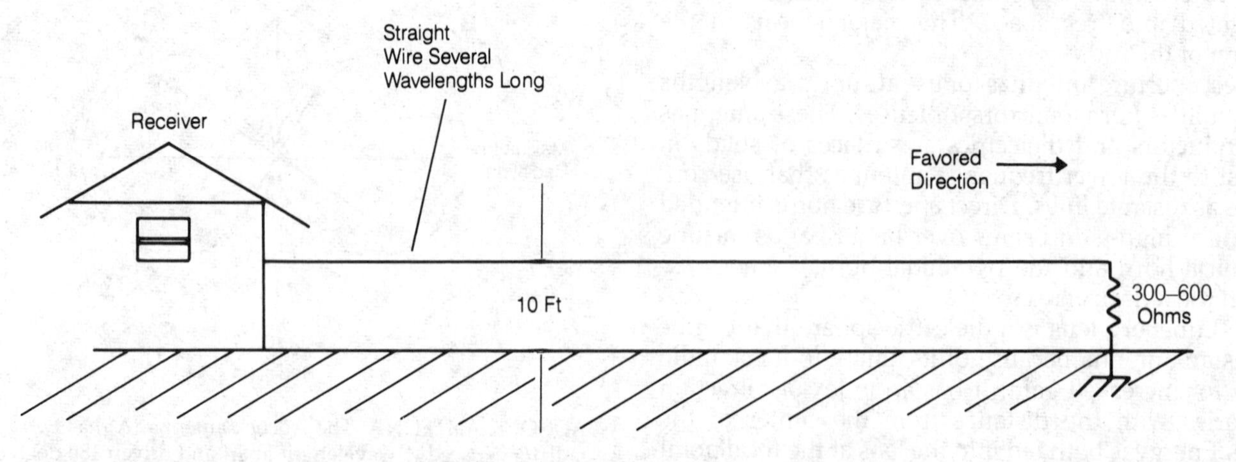

BEVERAGE ANTENNA: This unidirectional receiving antenna is used at medium and high frequencies.

arriving from the direction in which the wire is pointed, but it responds poorly in all other directions. Traveling waves are set up along the wire as electromagnetic fields arrive from the favored direction.

A Beverage antenna is a form of terminated longwire (*see* LONGWIRE ANTENNA). Because of its proximity to the ground, the Beverage is a poor antenna for transmitting. Its highly directional receiving characteristics, along with its relatively minimal noise pickup, make it a favorite among radio amateurs. The Beverage is frequently used by radio amateurs for listening on their 160-meter (1.8 MHz) and 80-meter (3.5 MHz) bands.

The resistor at the far end of the Beverage antenna should, ideally, be matched to the characteristic impedance of the wire for unidirectional operation. The resistor may be eliminated or short-circuited, resulting in bidirectional operation if desired. Without the terminating resistor, reception remains excellent in the favored direction of the illustration, but the response is also excellent in the opposite direction, making the antenna bidirectional.

Biconical Antenna A biconical antenna is a balanced broadband antenna that consists of two metal cones arranged so that they meet at or near the vertices, as shown in the drawing. The biconical antenna is fed at the point where the vertices meet. The exact feed-point impedance of a biconical antenna depends on the flare angle of the cones and the separation between their vertices.

A binconical antenna displays resonant properties at frequencies above that at which the height h of the cones is ¼ wavelength in free space. The highest operating frequency is several times the lowest operating frequency. A biconical antenna oriented vertically, as shown, emits and receives vertically polarized electromagnetic waves.

The biconical antenna is often used as vhf, but its size becomes prohibitively large at lower frequencies. However, one of the cones can be replaced by a disk or ground plane to reduce the physical dimensions of the antenna while retaining the broadband characteristics. If the top cone is replaced by a disk, the antenna becomes a discone. If the lower cone is replaced by a ground plane, the antenna becomes a conical monopole. Both the discone and the conical monopole are practical for use at frequencies as low as about 2 MHz. *See also* CONICAL MONOPOLE ANTENNA, DISCONE ANTENNA.

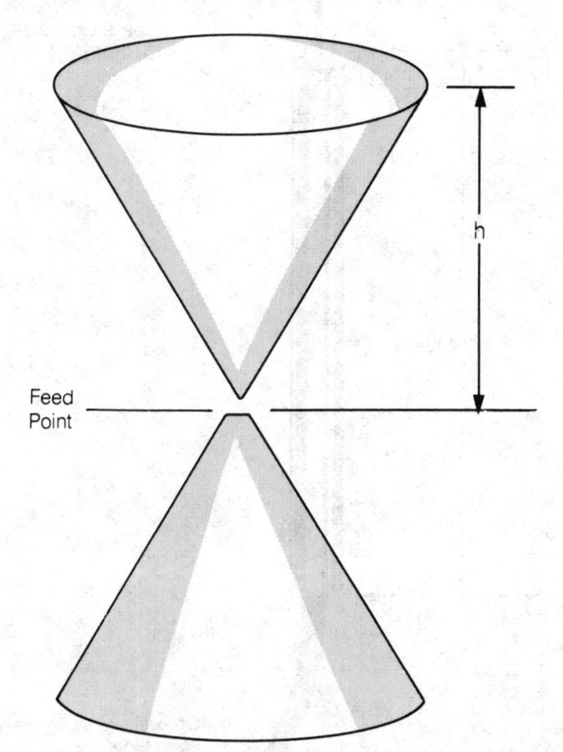

BICONICAL ANTENNA: This antenna functions up to several octaves above the frequency where the cone height, h, is ¼ wavelength.

Bowtie Antenna The bowtie antenna, a broadbanded antenna used at vhf and uhf (very high frequencies and ultrahigh frequencies), consists of two triangular pieces of stiff wire, or two triangular flat metal plates, as shown in the drawing. The feed point is at the gap between the apexes of the triangles. There may be a reflecting screen to provide unidirectional operation.

The bowtie antenna is a two-dimensional form of the biconical antenna (*see* BICONICAL ANTENNA). It obtains its broadband characteristics from the same principles as the biconical antenna. The feed-point impedance depends on the apex angles of the two triangles. The polarization of the bowtie antenna is along a line running through the feed point and the centers of the triangle bases.

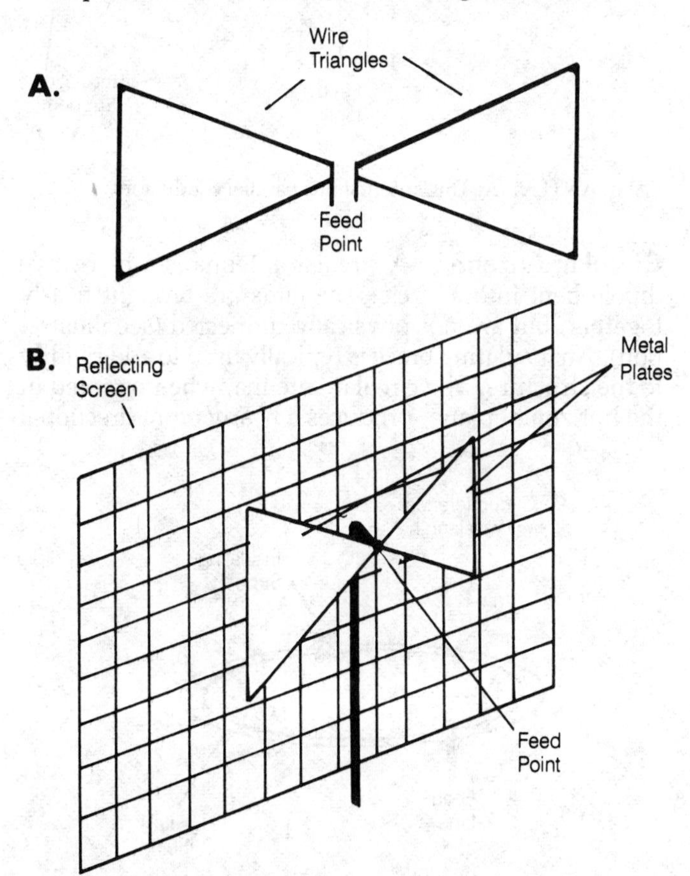

BOWTIE ANTENNA: A wire bowtie antenna (A) and a bowtie antenna consisting of two triangular metal plates with a reflecting screen (B).

Bowtie antennas are occasionally used for transmitting and receiving at frequencies as low as about 20 MHz. Below that frequency, the size of the bowtie antenna becomes prohibitive.

Cage Antenna A cage antenna is similar to a dipole antenna. Several half-wavelength, center-fed conductors are connected together at the feed point, as shown in the illustration. The cage antenna is so named because it resembles a cage. The multiplicity of conductors increases the effective bandwidth of the antenna and reduces losses caused by the resistance of the wire conductors.

A cage antenna with wires or rods arranged in the form of two cones, with apexes that meet at the feed point, is called a biconical antenna. This antenna offers a good impedance match for 50- or 75-Ω feed lines, over a frequency range of several octaves. *See also* BICONICAL ANTENNA.

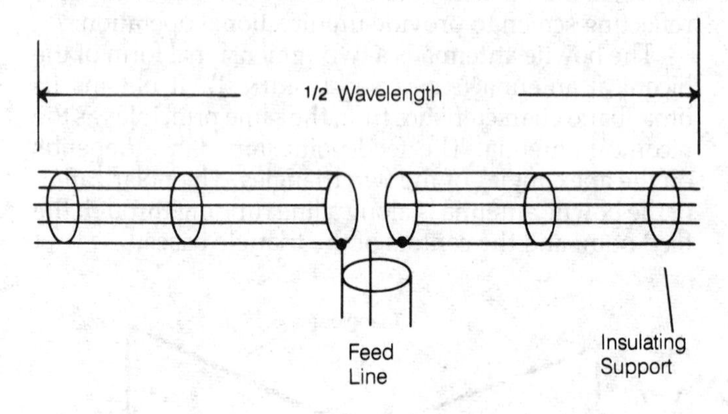

CAGE ANTENNA: This antenna has parallel conductors.

Circular Antenna A circular antenna is a half-wave dipole bent into a circle. The ends are brought nearly together, but are not physically connected (see illustration). An insulating brace is typically used to add rigidity to the structure. The circular antenna, when oriented in the horizontal plane, produces a nearly omnidirectional radiation pattern in all azimuth directions. Sometimes the circular antenna is called a halo.

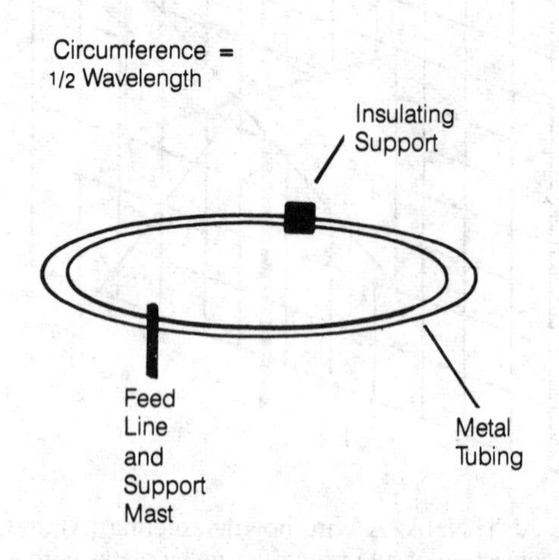

CIRCULAR ANTENNA: This antenna is a horizontal half-wave dipole formed in a loop with insulated ends.

Circular antennas, constructed from metal tubing, may be used in mobile installations at frequencies above approximately 50 MHz. The horizontal polarization of the halo results in less fading or "picket fencing" than does vertical polarization. Several circular antennas may be stacked at ½-wavelength intervals to produce omnidirectional gain in the horizontal plane.

Circular antennas can be operated at odd multiples of the fundamental frequency, and a reasonably good impedance match will result when 50- or 75-Ω feed lines are employed.

Coaxial Antenna A coaxial antenna is a half-wave vertical dipole fed through one radiating element with coaxial cable. At the feed point, the cable center conductor is extended ¼ wavelength directly upward; the shield is folded back along the cable for ¼ wavelength. The illustration shows the construction of a coaxial antenna.

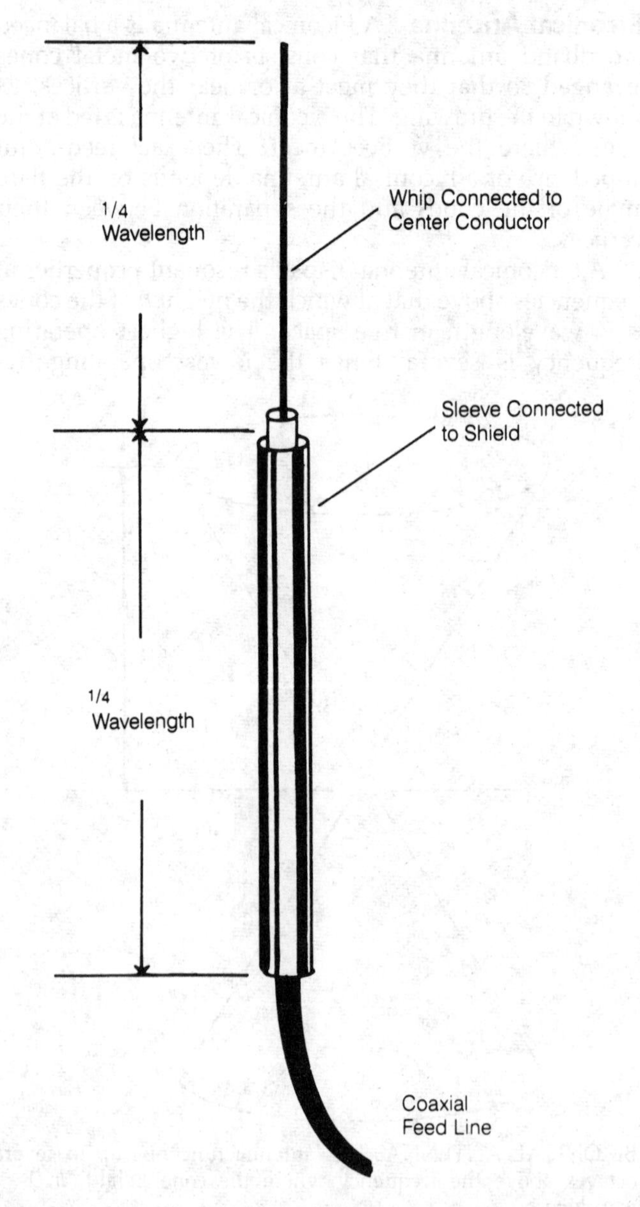

COAXIAL ANTENNA: The structure of a coaxial antenna.

Coaxial antennas are frequently used on the 27-MHz Citizen's Band. These antennas are about 18 feet high—and the antenna provides a low angle of radiation. The radiation is vertically polarized.

Induced currents on the feed line can be a problem with coaxial antennas. These currents can be choked off by winding the coaxial transmission line into a tight coil at the base of the antenna. Coaxial antennas are practical at frequencies above about 5 MHz where a vertical half-wave structure can be supported. *See also* VERTICAL DIPOLE ANTENNA.

Collinear Antenna A set of half-wave radiators fed in phase and positioned so that all the driven elements lie along one straight line is called a collinear antenna. These antennas may be either horizontal or vertical, as shown in the illustration. A set of stacked collinear antennas is sometimes called a collinear array.

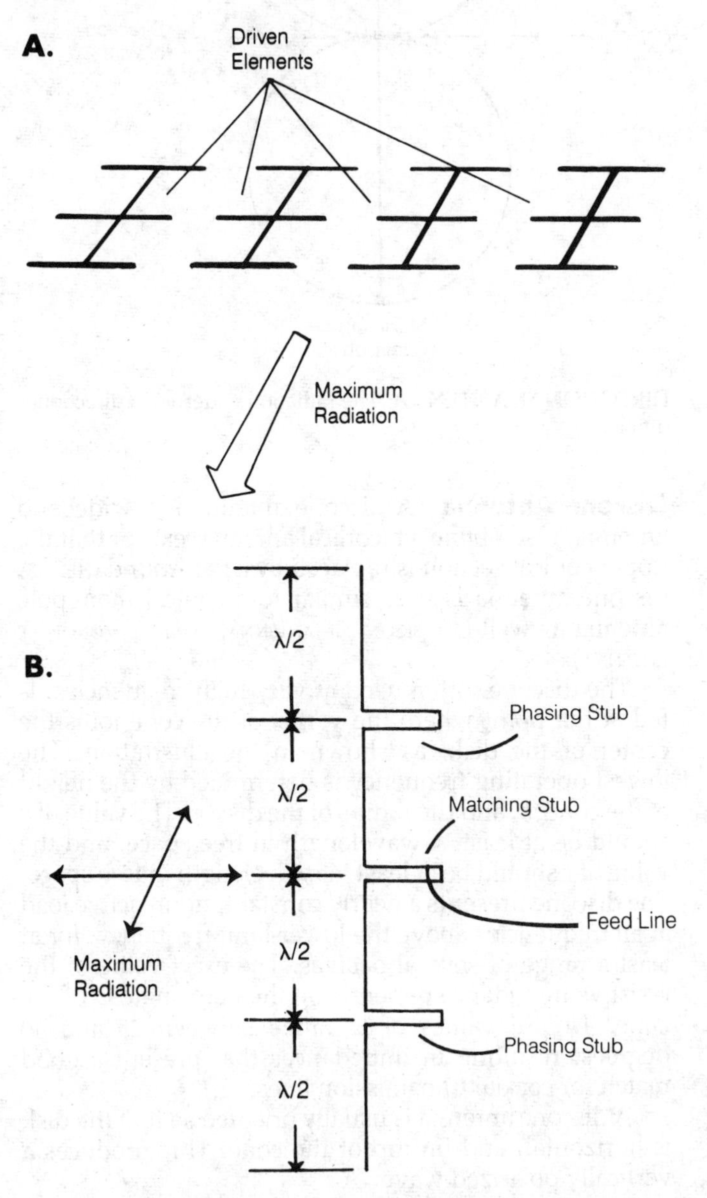

COLLINEAR ANTENNA: A collinear antenna formed from Yagi antennas (A), and a vertical version of the antenna for ominidirectional radiation (B).

Vertical collinear antennas are used in mobile and base installations at very-high and ultrahigh frequencies, to obtain omnidirectional gain in the azimuth plane. A two-element collinear vertical provides about 3 dB power gain over a single dipole; a four-element collinear vertical gives about 6 dB gain over a vertical dipole.

When Yagi antennas are combined in collinear fashion, the assembly exhibits far greater gain than one Yagi alone. Sometimes twenty or more Yagi antennas are oriented in a matrix to form a large collinear array. These arrays can produce forward gain in excess of 20 dB with respect to a half-wave dipole.

Conical Monopole Antenna A conical monopole antenna is a form of biconical antenna in which the lower cone has been replaced by a ground plane. The upper cone is usually bent inward at the top.

A conical monopole displays a wideband frequency response. The lowest operating frequency is determined by the size of the cone. At all frequencies up to several octaves above that at which the slant height (see illustration) of the cone is ¼ wavelength, the conical monopole provides a reasonably good match to a 50-Ω feed line.

Conical monopole antennas have vertical polarization and are often constructed in the form of a wire cage as shown. This closely approximates a solid metal cone.

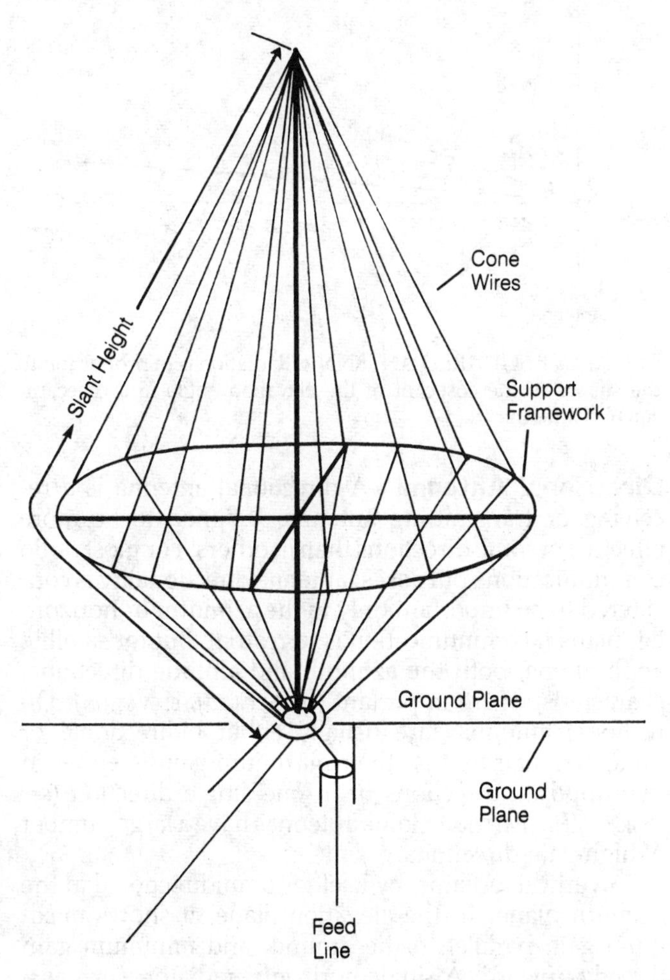

CONICAL MONOPOLE ANTENNA: The structure of a conical monopole.

A good ground system must be furnished for the conical monopole to work well. The conical monopole is an unbalanced antenna, and coaxial cable is best for the feed system. *See also* BICONICAL ANTENNA, DISCONE ANTENNA.

Cosecant-Squared Antenna A cosecant-squared antenna is a microwave radar antenna designed to give echoes of the same intensity from targets at all distances. It is named from the shape of its vertical-plane radiation pattern. The radiation intensity varies according to the square of the cosecant of the elevation angle.

The least beam intensity is radiated directly upward, at an elevation angle of 90 degrees with respect to the horizon. The greatest beam intensity is radiated horizontally at an elevation angle of zero degrees. The drawing illustrates this pattern. The intensity of radiation from a cosecant-squared antenna varies in that the path attenuation is essentially the same over all signal paths to and from points in a given horizontal plane. *See also* RADAR.

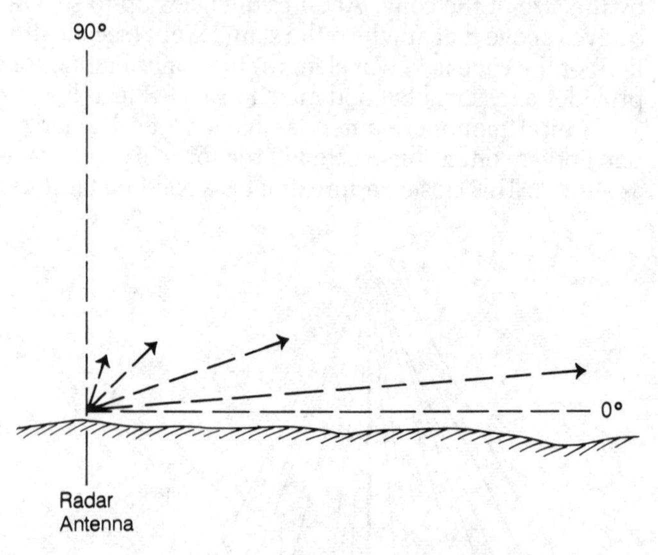

COSECANT-SQUARED ANTENNA: Radiation is in proportion to the square of the cosecant of the elevation angle in a cosecant-squared antenna.

Directional Antenna A directional antenna is a receiving or transmitting antenna designed to be more effective in some directions than in others. For most radio communications purposes, antenna directionality is considered to be important only in the azimuth, or horizontal, plane if communication is terrestrial. But for satellite applications, both the azimuth and altitude directional characteristics are important (*see* AZIMUTH, ELEVATION). Directional antennas are usually either bidirectional or unidirectional; that is, their maximum gain is either in two opposite directions or in one single direction (*see* BIDIRECTIONAL PATTERN). Some antennas have a large number of high-gain directions.

A vertical radiator, by itself, is omnidirectional in the azimuth plane. In the elevation plane, it shows maximum gain parallel to the ground, and minimum gain directly upward. A single horizontal radiator, such as a dipole antenna, produces more gain off the sides than off the ends, and therefore it is directional in the azimuth plane, as shown in the illustration. A dipole is considered a directional antenna, since it shows a bidirectional pattern.

Sophisticated types of directional antennas provide large amounts of signal gain in their favored directions. This gain and directionality can be obtained in a variety of ways.

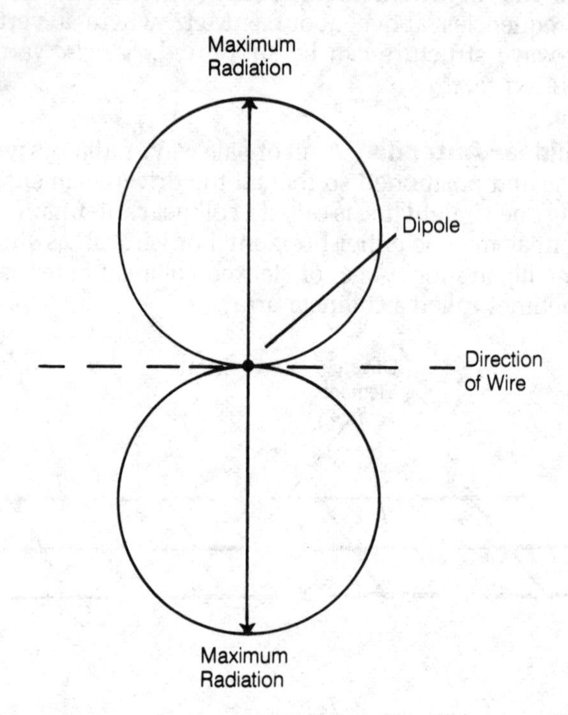

DIRECTIONAL ANTENNA: The radiation pattern of a directional dipole.

Discone Antenna A discone antenna is a wideband antenna, resembling a biconical antenna, except that the upper conical section is replaced by a flat, round disk. A discone antenna is very similar to a conical monopole antenna as well (*see* BICONICAL ANTENNA, CONICAL MONOPOLE ANTENNA).

The discone, often used at very high frequencies, is fed at the point where the vertex of the cone joins the center of the disk, as shown in the illustration. The lowest operating frequency is determined by the height of the cone, h, and the radius of the disk, r. The value of h should be at least 1/4 wavelength in free space, and the value of r should be at least 1/10 wavelength in free space. The discone presents a nearly constant, nonreactive load at all frequencies above the lower-limit frequency, for at least a range of several octaves. The exact value of the resistive impedance depends on the flare angle Θ of the cone. Typical values of Θ range between 25 and 40 degrees, resulting in impedances that present a good match for coaxial transmission lines.

A discone antenna is usually oriented so that the disk is horizontal, and on top of the cone. This produces a vertically polarized wave.

Dish Antenna A dish antenna is a high-gain antenna for transmission and reception of ultrahigh frequency

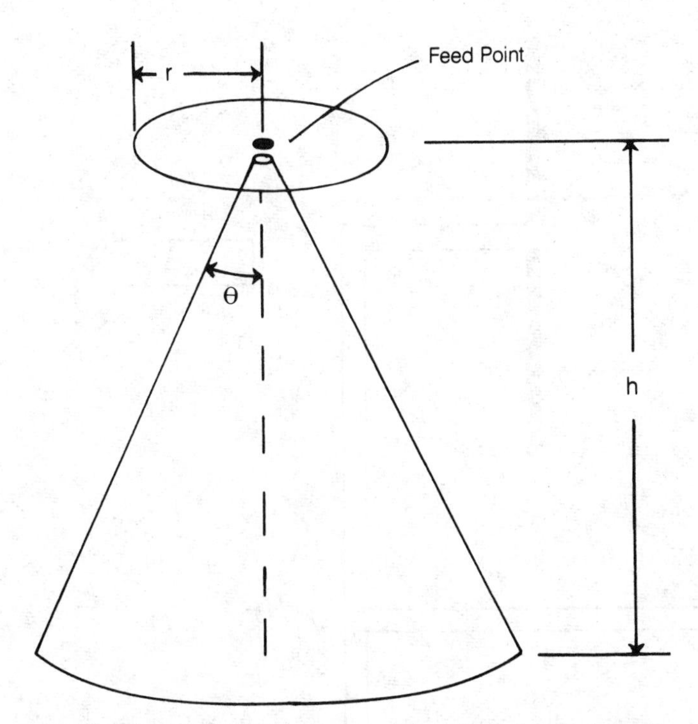

DISCONE ANTENNA: This antenna is usually fed at the intersection of the cone and the disk with the disk in the horizontal plane.

and microwave signals. The dish antenna consists of a driven element or other form of radiating device, and a large spherical or parabolic reflector, as shown in the illustration. The driven element is placed at the focal point of the reflector.

Signals arriving from a great distance, in parallel wavefronts, are reflected off the dish and brought together at the focus. Energy radiated by the driven element is reflected by the dish and sent out as parallel waves. The principle is exactly the same as that of a flashlight or latern reflector, except that radio waves are involved instead of visible light.

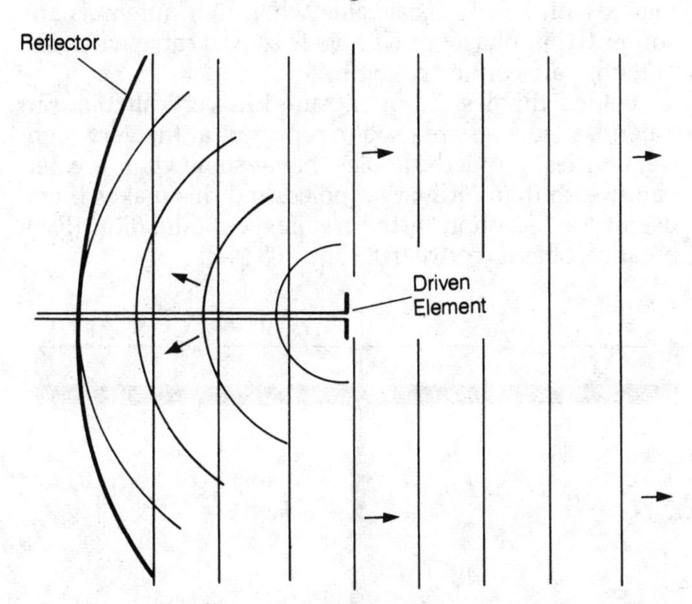

DISH ANTENNA: A section view of a dish antenna showing how the reflector forms parallel wave fronts from the transmitted signal and also focuses received parallel wave fronts back to the driven element.

A dish antenna must be at least several wavelengths in diameter for effective operation. Otherwise, the waves will be diffracted around the edges of the dish reflector. The dish is thus an impractical choice of antenna, in most cases, for frequencies below the ultrahigh range. The reflecting element of a dish antenna can be made of sheet metal, or it can be fabricated from a screen or wire mesh. In the latter case, the spacing between screen or mesh conductors must be a very small fraction of a wavelength in free space.

Dish antennas typically show very high gain. The larger a dish with respect to a wavelength, the greater the gain of the antenna. It is essential that a dish antenna be correctly shaped, and that the driven element be located at the focal point. Dish antennas are used in radar, and in satellite communications systems. Some television receiving antennas employ this configuration as well. *See* ANTENNA POWER GAIN, PARABOLOID ANTENNA.

Double-V Antenna A double-V, or fan, antenna is a form of dipole antenna. In any antenna, the bandwidth increases when the diameter of the radiating element is made larger. A double-V takes advantage of this by using two elements rather than one, and geometrically separating them so that they behave as a single, very broad radiator. Two dipoles are connected in parallel and positioned at an angle, as in the illustration.

The maximum radiation from a double-V antenna, and the maximum gain for receiving, occur in all directions perpendicular to a line bisecting the angle between the dipoles. In the horizontal plane, this is broadside to the plane containing the elements. The polarization is parallel with the line bisecting the apex angles of each pair of jointed conductors. The impedance at the feed point is somewhat higher than that of an ordinary dipole antenna; it is generally on the order of 200 to 300 Ω. This provides a good match for the common types of prefabricated twin-lead line.

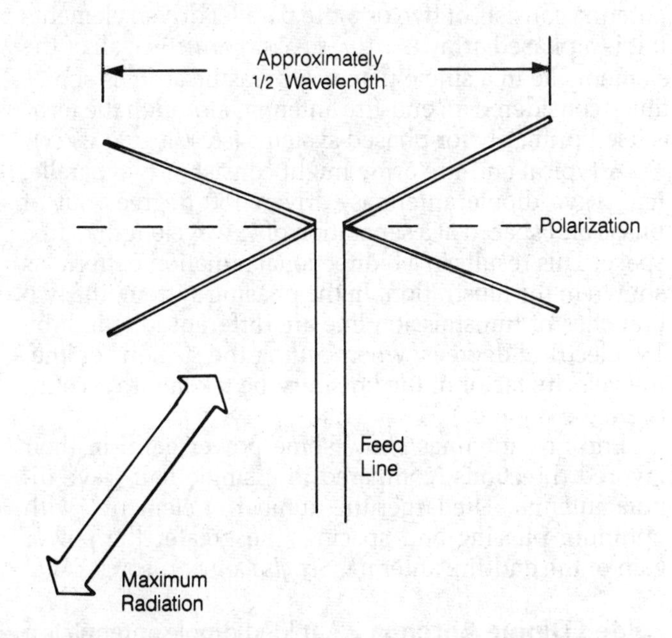

DOUBLE-V ANTENNA: This antenna is basically two half-wave dipoles connected in parallel.

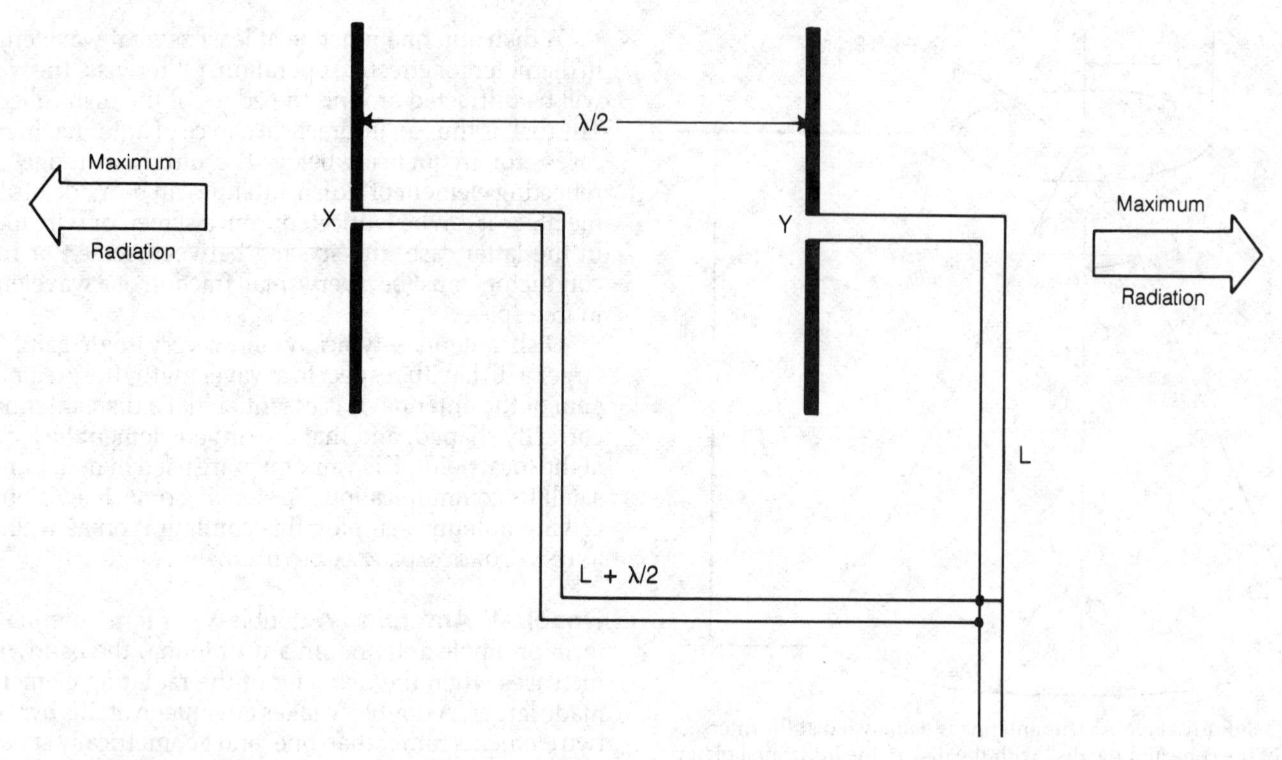

END-FIRE ANTENNA: One section of the phasing system of this antenna is a half wavelength longer than the other and the two elements are fed in opposing phase.

Double-V antennas can be used for either transmitting or receiving, and are effective at any frequency on which a dipole antenna is useful. They are quite often used as receiving antennas for the standard frequency-modulation (FM) broadcast band at 88 to 108 MHz, because of their broad frequency-response characteristics.

End-Fire Antenna An end-fire antenna is a bidirectional or unidirectional antenna in which the greatest amount of radiation takes place off the ends. An end-fire antenna consists of two or more parallel driven elements if it is a phased array (*see* PHASED-ARRAY ANTENNA); all of the elements lie in a single plane. A parasitic array is sometimes considered an end-fire antenna, although the term is used primarily for phased systems (*see* PARASITIC ARRAY).

A typical end-fire array might consist of two parallel half-wave dipole antennas, driven 180 degrees out of phase and spaced at a separation of ½ wavelength in free space. This results in a bidirectional radiation pattern, as shown in the illustration. In the phasing system, the two branches of transmission line are different in length by 180 electrical degrees; when cutting the sections of line, the velocity factor of the line must be taken into account (*see* VELOCITY FACTOR).

End-fire antennas show some power gain, in their favored directions, compared to a single half-wave dipole antenna. The larger the number of elements, with optimum phasing and spacing, the greater the power gain of the end-fire antenna. *See also* ANTENNA POWER GAIN.

Folded Dipole Antenna A folded dipole antenna is a half-wavelength, center-fed antenna constructed of parallel wires in which the outer ends are connected together. The folded-dipole antenna may be thought of as a squashed full-wave loop (see illustration).

The folded dipole has exactly the same gain and radiation pattern, in free space, as a dipole antenna. However, the feed-point impedance of the folded dipole is four times that of the ordinary dipole. Instead of approximately 73 Ω, the folded dipole presents a resistive impedance of almost 300 Ω. This makes the folded dipole desirable for use with high-impedance, parallel-wire transmission lines. It also can be used to obtain a good match with 75-Ω coaxial cable when four antennas are connected in phase, or with 50-Ω coaxial cable when six antennas are connected in phase.

Folded dipoles are often found in vertical collinear antennas, such as are used in repeaters at the very high frequencies. Folded dipoles have somewhat greater bandwidth than ordinary dipoles, and this makes them useful for reception in the frequency-modulation (FM) broadcast band, between 88 and 108 MHz.

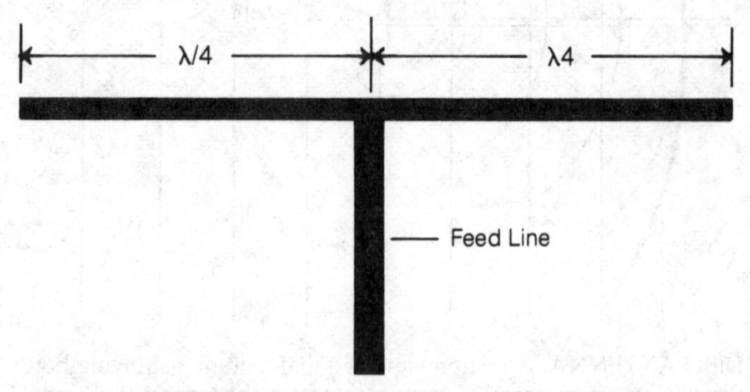

FOLDED-DIPOLE ANTENNA: This antenna has the same radiation pattern as a conventional dipole, but its feed point impedance is four times as large.

Ground-Plane Antenna A ground-plane antenna is a vertical radiator operated against a system of quarter-wave radials and elevated at least a quarter wavelength above the effective ground. The radiator itself may be any length, but should be tuned to resonance at the desired operating frequency.

When a ground plane is elevated at least 90 electrical degrees above the effective ground surface, only three or four radials are necessary in order to obtain an almost lossless system. The radials are usually run outward from the base of the antenna at an angle that may vary from 0 degrees to 45 degrees with respect to the horizon. The drawing illustrates a typical ground-plane antenna.

A ground-plane antenna is an unbalanced system, and should be fed with coaxial cable. A balun can be used, however, to allow the use of a balanced feed line. The feed-point impedance of a ground-plane antenna having a quarter-wave radiator is about 37 Ω if the radials are horizontal; this impedance increases as the radials are drooped, reaching about 50 Ω at an angle of 45 degrees. The radials may be run directly downward, in the form of a quarter-wave tube concentric with the feed line. Then the feed-point impedance is approximately 73 Ω. This configuration is known as a coaxial antenna.

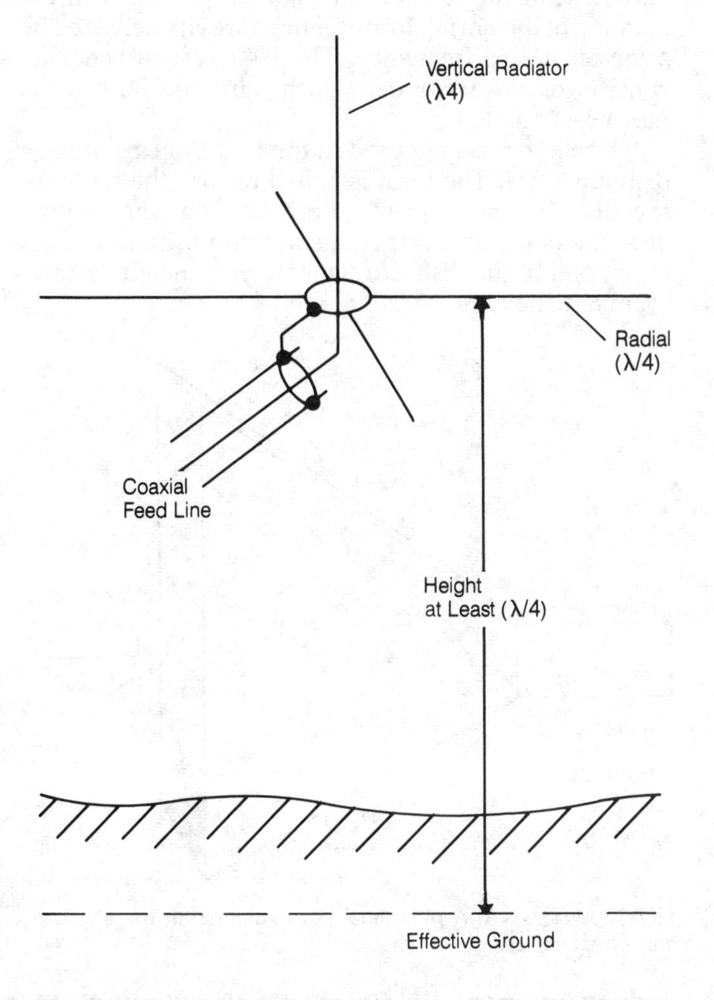

GROUND-PLANE ANTENNA: This antenna has three or four quarter-wave radials, and the feed point is at least a quarter wavelength above the ground.

Half-Wave Antenna A half-wave antenna is a radiating element that measures an electrical half wavelength in free space. This antenna can have a physical length anywhere from practically zero to almost the physical dimensions of a half wavelength in free space. A dipole antenna is the simplest example of a half-wave antenna.

In theory, a half wavelength in free space is given in feet according to the equation:

$$L = \frac{492}{f}$$

where *L* is the linear distance and *f* is the frequency in megahertz. A half wavelength in meters is given by:

$$L = \frac{150}{f}$$

In practice, an additional factor must be added to the above equations, because electromagnetic fields travel somewhat more slowly along the conductors of an antenna than they do in free space. For ordinary wire, the results as obtained above are multiplied by about 0.95. For tubing or large-diameter conductors, the factor is slightly smaller, and may range down to about 0.90 (*see* VELOCITY FACTOR).

A half-wave antenna may be made much shorter than a physical half wavelength. This is accomplished by inserting inductances in series with the radiator. The antenna may be made much longer than the physical half wavelength by inserting capacitances in series with the radiator. *See also* CAPACITIVE LOADING, ELECTRICAL WAVELENGTH, INDUCTIVE LOADING.

Halo Antenna A halo antenna is a special form of horizontal half-wave antenna. Basically, the halo consists of a dipole whose elements have been bent into a circle, so that the circumference of the circle is ½ electrical wavelength. The ends of the dipole are insulated from each other at the opposite side of the circle from the feed point (see illustration).

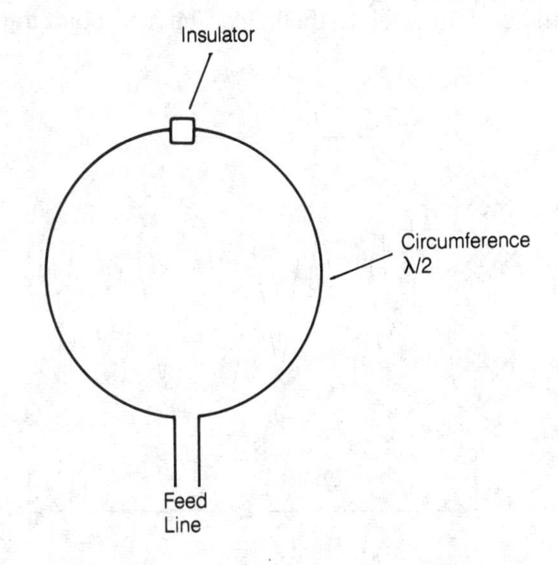

HALO ANTENNA: This antenna is a half-wave dipole formed in a loop.

The halo antenna is used mostly in the very high frequency part of the radio spectrum. At 30 MHz, for example, the circumference of a halo antenna is only about 15 feet, so that the diameter is less than 5 feet. Copper or aluminum tubing is practical for the construction of a halo in the very high frequency range.

Halo antennas exhibit a nearly omnidirectional radiation pattern in the horizontal plane. The polarization is horizontal. Halo antennas are often vertically stacked to obtain omnidirectional gain in the horizontal plane.

Helical Antenna A helical antenna is a form of circularly polarized, high-gain antenna used mostly in the ultrahigh-frequency part of the radio spectrum. Circular polarization offers certain advantages at these frequencies (*see* CIRCULAR POLARIZATION).

The drawing illustrates a typical helical antenna. The reflecting device may consist of sheet metal or screen, in a disk configuration with a diameter *d* of at least 0.8 wavelength at the lowest operating frequency. The radius *r* of the helix should be approximately 0.15 wavelength at the center of the intended operating frequency range. The longitudinal spacing between turns of the helix, given by *s*, should be approximately ¼ wavelength in the center of the operating frequency range. The overall length of the helix, shown by *L*, may vary, but should be at least 1 wavelength at the lowest operating frequency. The longer the helix, the greater the forward power gain. Gain figures in excess of 15 decibels can be realized with a single, moderate-sized helical antenna; when several such antennas are phased, the gain increases accordingly. Bays of two or four helical antennas are quite common.

The helical antenna illustrated will show a useful operating bandwidth equal to about half the value of the center frequency. An antenna centered at 400 MHz will function between approximately 300 and 500 MHz, for example. The helical antenna is normally fed with coaxial cable. The outer conductor should be connected to the reflecting screen or sheet, and the center conductor should be connected to the helix. The feed-point impedance is about 100 to 150 ohms throughout the useful operating frequency range.

Helical antennas are ideally suited to satellite communications, since the circular polarization of the transmitted and received signals reduces the amount of fading as the satellite orientation changes. The sense of the circular polarization may be made either clockwise or counterclockwise, depending on the sense of the helix.

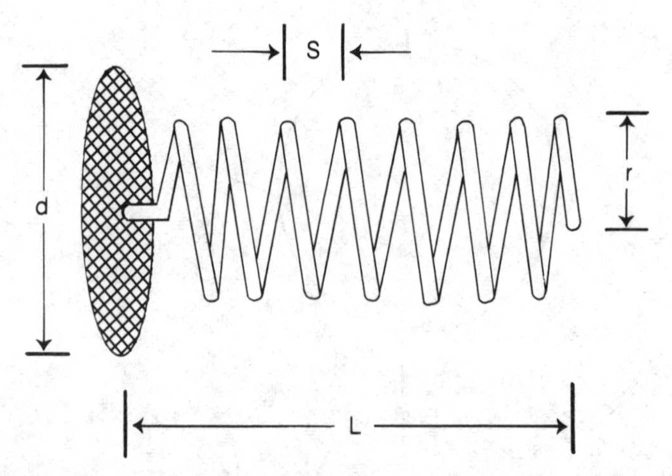

HELICAL ANTENNA: This antenna produces high gain at short wavelengths.

Hertz Antenna A Hertz antenna is any horizontal, half-wavelength antenna. The feed point can be at the center, at either end, or at some intermediate point. The Hertz antenna operates independently of the ground and is therefore a balanced antenna.

An example of the Hertz antenna is the dipole antenna. The driven element of a Yagi antenna may be the Hertz configuration.

Horn Antenna A horn antenna is a waveguide termination for the transmission and reception of signals at ultrahigh and microwave frequencies. There are several different configurations of the horn antenna, but they all look similar. The illustration is a drawing of a commonly used pyramidal horn antenna.

The horn antenna provides a unidirectional radiation pattern with the favored direction coincident with the opening of the horn. Horn antennas are characterized by a lowest usable frequency. The feed system generally consists of a waveguide, which joins the horn at its narrowest point.

Horn antennas are used in the feed systems of large dish antennas. The horn is aimed toward the center of the dish, in the opposite direction from the favored direction of the dish. When the horn is positioned at the focal point of the dish extremely high gain and narrow-beam radiation are realized.

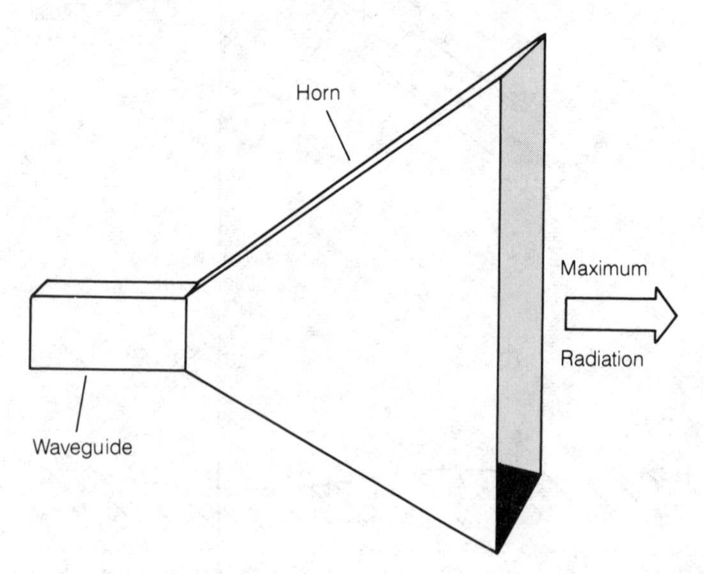

HORN ANTENNA: A pyramidal horn antenna for use at microwave frequencies.

Indoor Antenna For radio and television reception or transmission an outdoor antenna is always preferable to an indoor antenna. However, in certain situations it is not possible to install an outdoor antenna.

Indoor antennas are always a compromise, especially at the very low, low, medium, and high frequencies. Electrical wiring interferes with wave propagation at these frequencies. In a concrete-and-steel structure, the shielding effect is even more pronounced than in a wood frame building. Indoor antennas are generally more subject to manmade interference from electrical appliances. If an indoor antenna is used for transmitting purposes, the chances of electromagnetic interference are greatly increased, compared with the use of an outdoor antenna. The efficiency of an indoor transmitting antenna is generally lower than that of an identical outdoor antenna.

An indoor antenna has some advantages over an outdoor antenna. It is less susceptible to induced voltages resulting from nearby lightning strikes. It is not subject to corrosion from exposure to weather, and maintenance is simpler. At very high and ultrahigh frequencies, an indoor antenna can perform very well if it is located high above the ground.

Isotropic Antenna An isotropic antenna is an antenna that radiates electromagnetic energy equally well in all directions. This antenna is a theoretical construct, and does not actually exist. However, the isotropic-antenna concept is occasionally used for antenna-gain comparisons.

The power gain of an isotropic antenna is about −2.15 decibels with respect to a half-wave dipole in free space. That is, the field strength from a half-wave dipole antenna, in its favored direction, is approximately 2.15 decibels greater than the field strength from an isotropic antenna at the same distance and at the same frequency.

The radiation pattern of the isotropic antenna, in three dimensions, appears as a perfect sphere, since the device works equally well in all directions. In any given plane, the radiation pattern of the isotropic antenna is a perfect circle, centered at the antenna. *See also* ANTENNA PATTERN, ANTENNA POWER GAIN, dBd, dBi.

Kooman Antenna A Kooman antenna is a high-gain, unidirectional antenna used at ultrahigh and microwave frequencies. The antenna uses a reflector in conjunction with collinear and broadside characteristics (see illustration).

Several full-wave, center-fed conductors are oriented horizontally and stacked above each other. They are separated by intervals of ½ electrical wavelength (as measured along the phasing line). The phasing line consists of two parallel conductors. The conductors are transposed at each succeeding driven element. Thus, all of the driven elements are fed in the same phase. The reflecting network may consist of a wire mesh or another set of conductors. *See also* BROADSIDE ARRAY, COLLINEAR ANTENNA.

Loaded Antenna The natural resonant frequency of an antenna can be changed by placing reactance in series with the radiating element. An antenna that has such a reactance in its radiating element is called a loaded antenna.

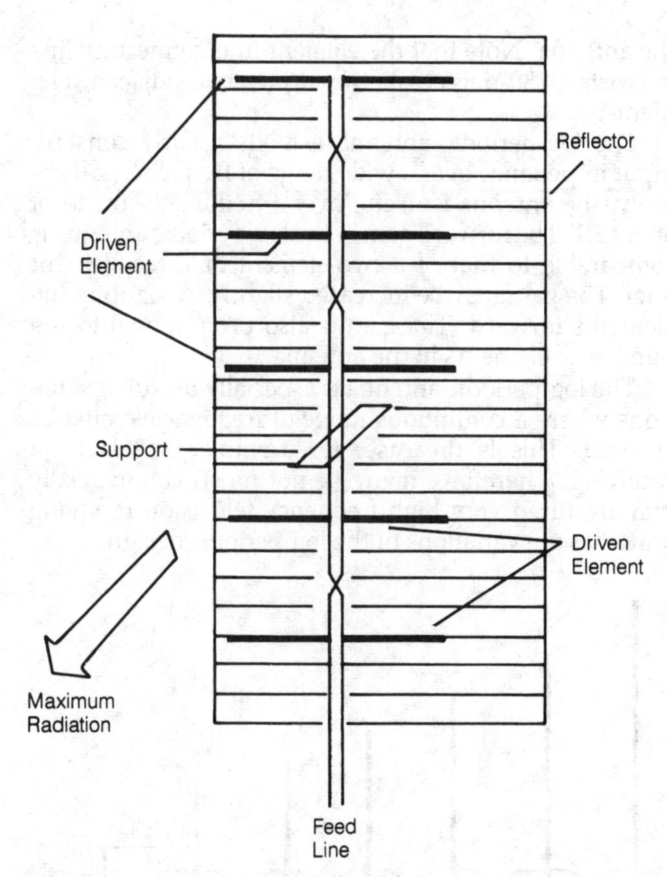

KOOMAN ANTENNA: This antenna has a stacked array of full-wave radiating elements and a reflector.

The most common type of loaded antenna has a resonant frequency that has been lowered by the series installation of an inductor. This type of loaded antenna is often used for mobile operation at medium and high frequencies. A capacitance hat, or loading disk, can be placed at the end or ends of such an antenna to increase the bandwidth.

The series connection of capacitors raises the resonant frequency of an antenna element. This kind of loaded antenna is sometimes used at very-high and ultrahigh frequencies. *See also* CAPACITIVE LOADING, INDUCTIVE LOADING.

Log-Periodic Antenna A log-periodic antenna, also known as a log-periodic dipole array (LPDA), is a special form of unidirectional, broadband antenna. The log-periodic antenna is sometimes used in the high frequency and very high frequency parts of the radio spectrum, for transmitting or receiving.

The log-periodic antenna consists of a special arrangement of driven dipoles, connected to a common transmission line. The illustration is a schematic diagram of the general form of a log-periodic antenna. The design parameters are beyond the scope of this discussion, since they vary depending on the gain and bandwidth desired. In general, the elements become shorter nearer the feed point (forward direction) and longer toward the back of the antenna. Note that the element-interconnecting line is twisted 180 degrees between any two adjacent elements.

The log-periodic antenna exhibits a fairly constant input impedance over a wide range of frequencies. Typically, the antenna is useful over a frequency spread of about 2:1. The forward gain of the log-periodic antenna is comparable to that of a two-element or three-element Yagi. The gain may be increased slightly by slanting the elements forward. The gain is also proportional to the number of elements in the antenna.

The log-periodic antenna is especially useful in situations where a continuous range of frequencies must be covered. This is the case, for example, in television receiving. Therefore, many (if not most) commercially manufactured very high frequency television receiving antennas are variations of the log-periodic design.

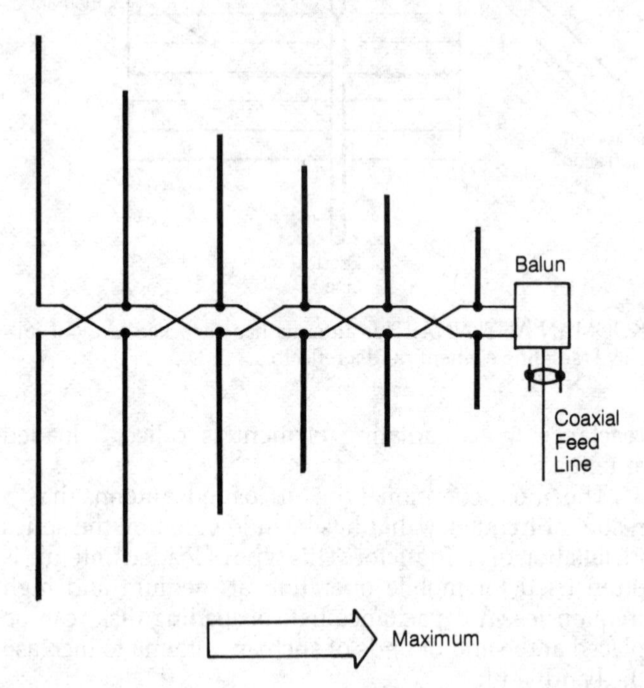

LOG-PERIODIC ANTENNA: The layout of a log-periodic dipole array.

Longwire Antenna A wire antenna measuring 1 wavelength or more, and fed at a current loop or at one end, is called a longwire antenna. Longwire antennas are sometimes used for receiving and transmitting at medium and high frequencies.

Longwire antennas offer some power gain over the half-wave dipole antenna. The longer the wire, the greater the power gain. For an unterminated longwire antenna, measuring several wavelengths, the directional pattern resembles A in the illustration. As the wire is made longer, the main lobes get more nearly in line with the antenna, and their amplitudes increase. As the wire is made shorter, the main lobes get farther from the axis of the antenna, and their amplitudes decrease. If the longwire is terminated at the far end (opposite the feed point), half of the pattern disappears, as at B.

Longwire antennas have certain advantages. They offer considerable gain and low-angle radiation, provided they are made long enough. The graph (C) shows the theoretical power gain, with respect to a dipole, that can be realized with a longwire antenna. The gain is a function of the length of the antenna. Longwire antennas are inexpensive and easy to install, provided there is sufficient real estate. The longwire antenna must be as straight as possible for proper operation.

There are two main disadvantages to the longwire antenna. First, it cannot conveniently be rotated to change the direction in which maximum gain occurs. Second, a great deal of space is needed, especially in the medium-frequency spectrum and the longer high-frequency wavelengths. For example, a 10-wavelength longwire antenna measures about 1,340 feet at 7 MHz. This is more than a quarter of a mile.

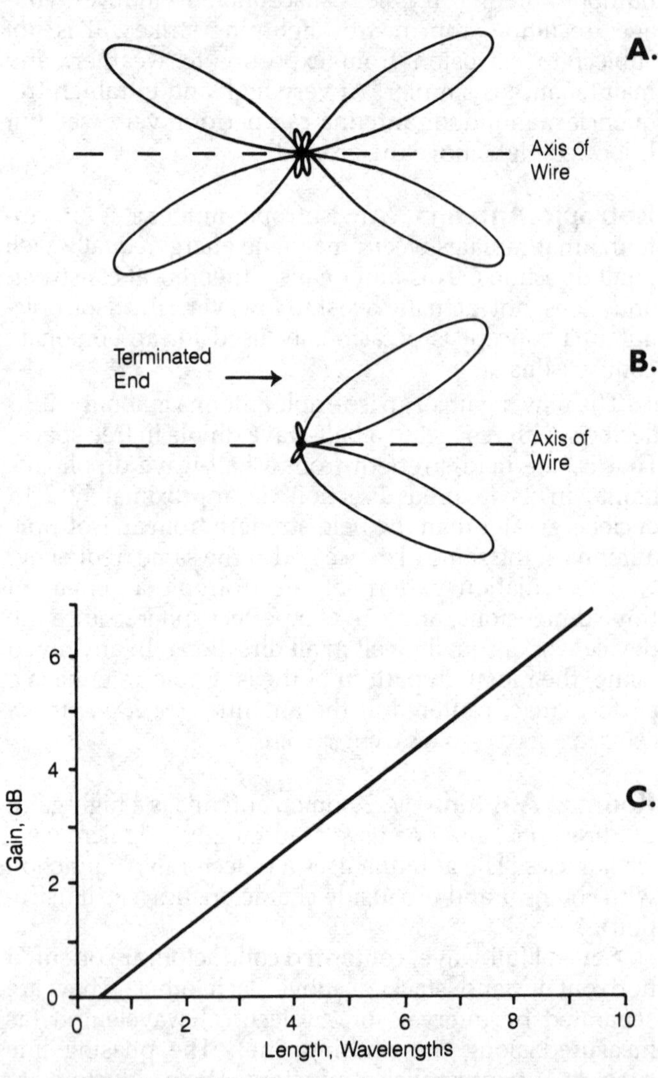

LONGWIRE ANTENNA: The directional pattern for an unterminated longwire antenna (A), the directional pattern for a terminated longwire (B), and the plot of gain vs. wavelength for these antennas (C).

Loop Antenna Any receiving or transmitting antenna, consisting of one or more turns of wire forming a direct-current short circuit, is called a loop antenna. Loop antennas can be categorized as either small or large.

Small loops have a circumference of less than 0.1 wavelength at the highest operating frequency. Such antennas are suitable for receiving, and exhibit a sharp null along the loop axis. The small loop can contain just one turn of wire, or it may contain many turns. The loop can be electrostatically shielded to improve the directional characteristics. Figure 1 shows a multiturn unshielded loop antenna (A). A single-turn shielded loop antenna is shown at B. The ferrite-rod antenna is a form of small loop in which a ferromagnetic core is used to enhance the signal pickup. Small loops are used for direction finding and for eliminating manmade noise or strong local interfering signals. Small loops are not suitable for transmitting, because of their extremely low radiation resistance.

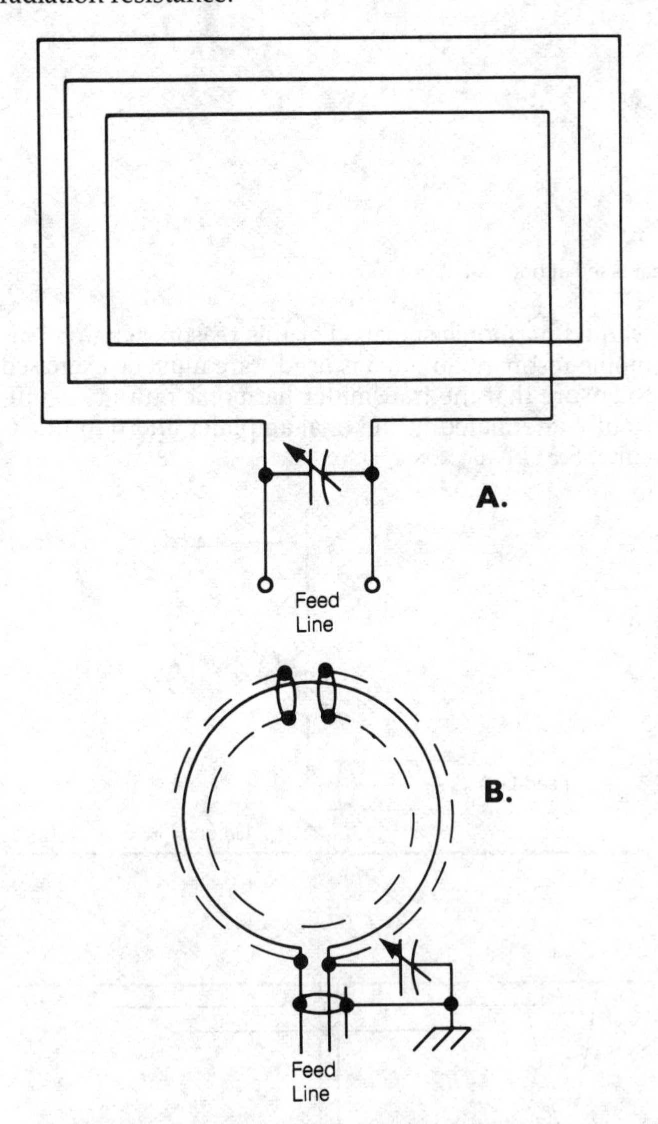

LOOP ANTENNA: Fig. 1. An unshielded loop antenna with several turns (A) and a shielded loop antenna with a single turn (B).

Large loops have a circumference of either 0.5 wavelength or 1 wavelength at the operating frequency. The half-wavelength loop (A in Fig. 2) presents a high impedance at the feed point, and the maximum radiation occurs in the plane of the loop. The full-wavelength loop (B) presents an impedance of about 50 ohms at the feed point, and the maximum radiation occurs perpendicular to the plane of the loop. A large loop can be used either for transmitting or receiving. The half-wavelength loop exhibits a slight power loss relative to a dipole, but the full-wavelength loop shows a gain of about 2 dBd (decibels). The full-wavelength loop forms the driven element for the popular quad antenna.

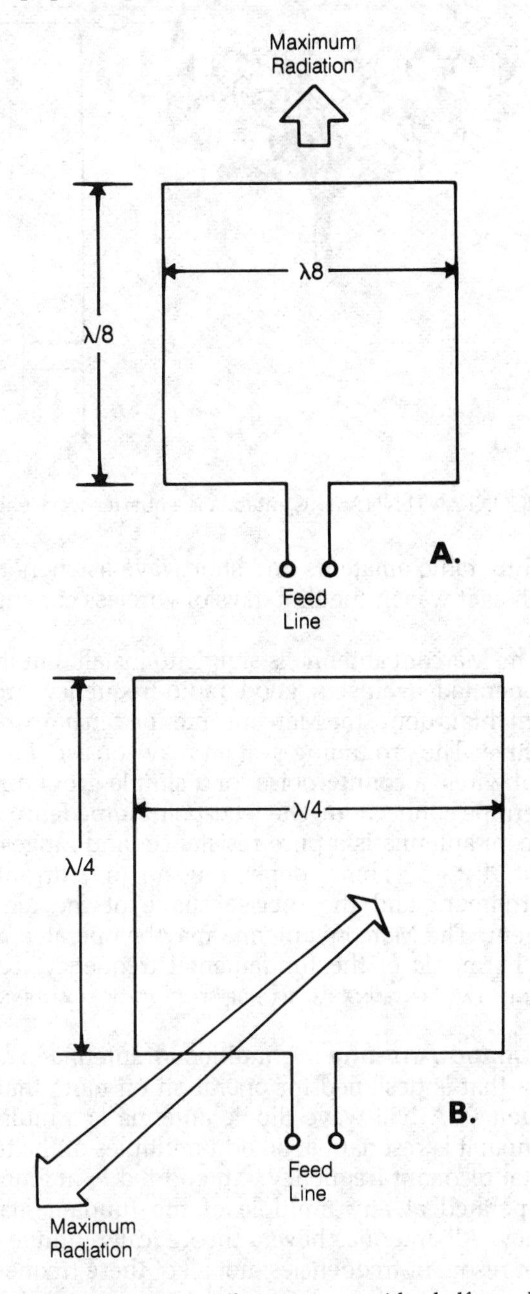

LOOP ANTENNA: Fig. 2. A loop antenna with a half wavelength circumference (A) and a wavelength circumference (B).

Marconi Antenna Any antenna measuring ¼ electrical wavelength at the operating frequency, and fed at one end, is called a Marconi antenna. One of the earliest communications antennas was of the Marconi type, used at low frequencies, and consisting of an end-fed wire with a short vertical section and a long horizontal span (see illustration). The most common modern form of Marconi antenna is the ground-mounted, ¼ wavelength vertical antenna. However, a simple length of wire, connected directly to a transmitter, is also sometimes

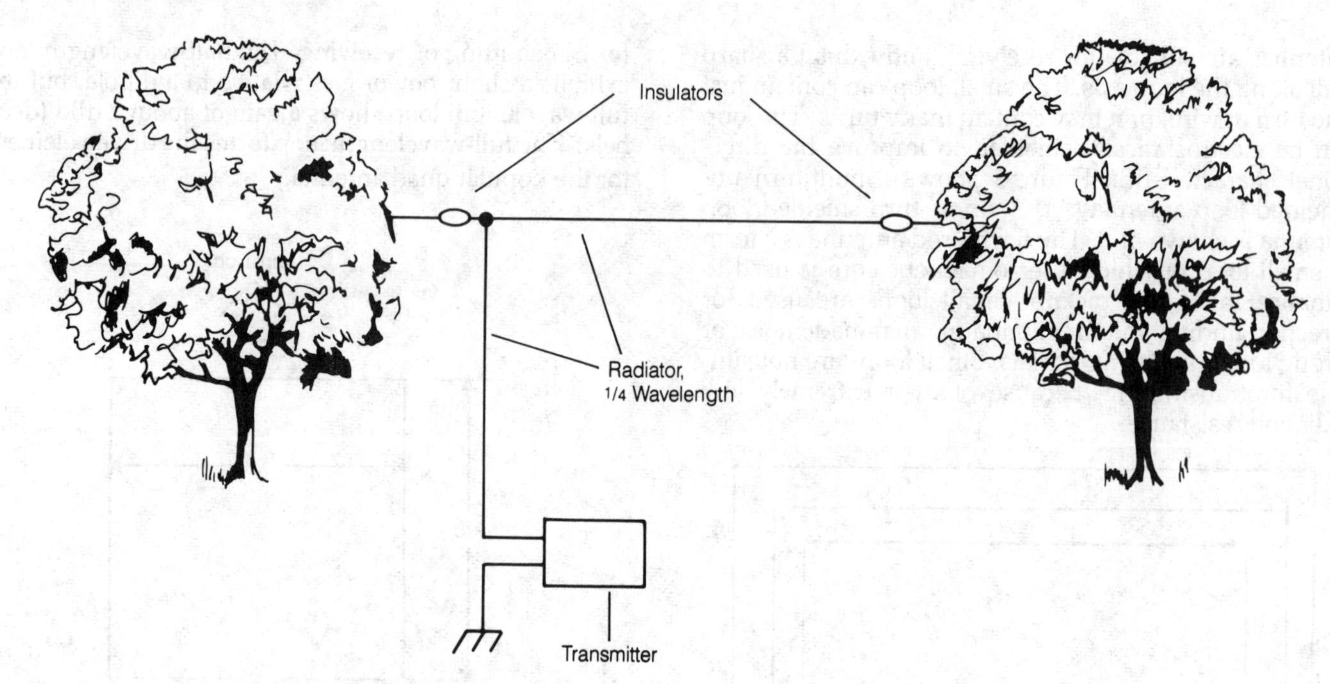

MARCONI ANTENNA: This antenna is a quarter-wavelength radiator that is fed at one end.

used by radio amateurs and shortwave listeners today, much as it was in the first days of wireless communication.

The Marconi antenna is simple to install, but it must be operated against a good radio-frequency ground. When this is done, the Marconi antenna can be extremely effective. The grounding system may consist of a set of radial wires, a counterpoise, or a simple ground-rod or waterpipe connection. The feed-point impedance of the Marconi antenna is a pure resistance, and ranges from about 20 to 50 ohms, depending on the surrounding environment and the precise shape of the radiating element. The Marconi antenna may be operated at any odd harmonic of the fundamental frequency. *See also* GROUND-PLANE ANTENNA, MARCONI EFFECT, VERTICAL ANTENNA.

Multiband Antenna A multiband antenna is an antenna that is designed for operation on more than one frequency. A half-wave dipole antenna is a multiband antenna; it is resonant at all odd multiples of the fundamental resonant frequency. An end-fed ½ antenna can be operated at any multiple of the fundamental frequency. All antennas have a theoretically infinite number of resonant frequencies; not all of these frequencies, however, are useful.

Multiband antennas can be designed deliberately for operation on specific frequencies. Traps are commonly used for achieving multiband operation (*see* TRAP). A variable inductor can be used for changing the resonant frequency of an antenna, as shown at A in the illustration (*see* INDUCTIVE LOADING). A tuning network may be used to adjust the resonant frequency of an antenna/feeder system (*see* TUNED FEEDERS). Several different antennas can be connected in parallel to a single feed system to achieve multiband operation as at B.

Multiband antennas offer convenience; it is simple to switch from one frequency band to another. However, harmonic-resonant multiband antennas can radiate unwanted harmonic signals. For this reason, when a harmonic-resonant antenna is used, care must be exercised to ensure that the transmitter harmonic output is sufficiently attenuated in the final-amplifier and output circuits. *See also* HARMONIC, RESONANCE.

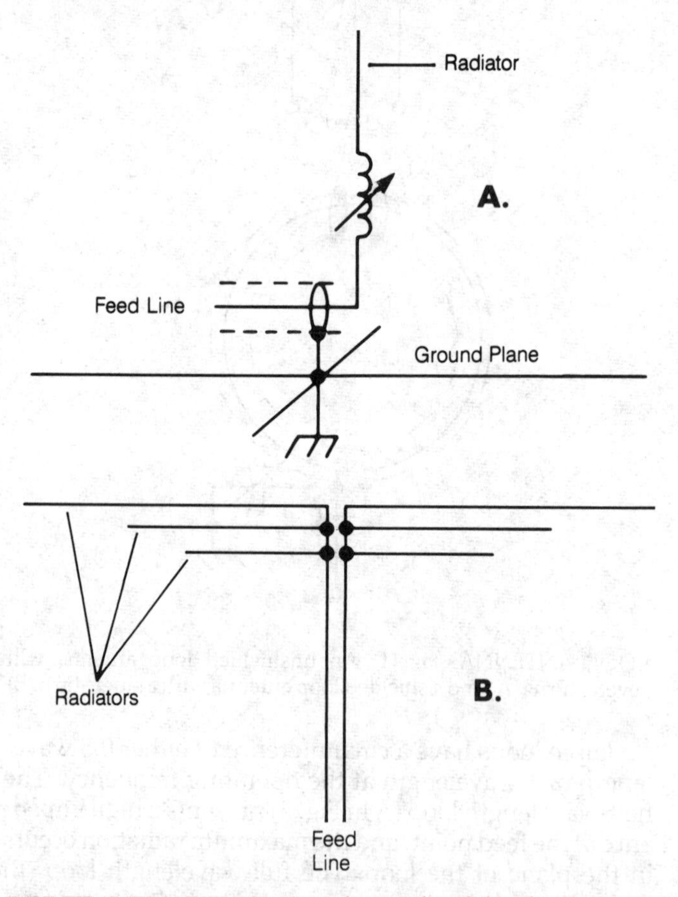

MULTIBAND ANTENNA: Two forms of multiband antenna are the short radiator adjusted for resonance with a variable inductor (A) and several half-wave antennas connected in parallel at a common feed point (B).

Multielement Antenna Some antennas employ multiple elements for improved directional response. An antenna element consists of a length of conductor. A conductor may be directly connected to the feed line; then it is called an active or driven element. A conductor can be physically separate from the feed line; then it is called a passive or parasitic element.

Multielement antennas can be broadly classified as either parasitic or phased arrays. In the parasitic array, passive elements are placed near a single driven element (*see* DIRECTOR, DRIVEN ELEMENT, PARASITIC ARRAY). In the phased array, two or more elements are driven together (*see* PHASED-ARRAY ANTENNA).

Multielement antennas are used mostly at high frequencies, although some large broadcast installations make use of phased vertical arrays at low and medium frequencies. The main advantages of a multielement antenna are power gain and directivity, which enhance both the transmitting and receiving capability of a communications station. The design and construction of a multielement antenna is, however, more critical than for a single-element antenna.

Paraboloid Antenna A paraboloid antenna is a form of dish antenna with a reflecting surface that is a geometric paraboloid. A paraboloid is the surface of rotation of a parabola about its axis (*see* PARABOLA). The paraboloid antenna has a focal point where rays arriving parallel to the antenna axis converge. This distinguishes the paraboloid antenna from the spherical antenna in which the rays converge but not to an exact point. The distance of the focal point from the center of the surface depends on the degree of curvature of the paraboloid.

Paraboloid reflectors can be constructed from wire mesh, or they can be spun from sheet metal. Paraboloid antennas are generally used at ultrahigh frequencies. Wire mesh is satisfactory at the longer wavelengths, but at very short wavelengths, sheet metal is preferable. The driven element is located at the focal point of the paraboloid. (*See also* DISH ANTENNA.)

Phased-Array Antenna A phased-array antenna is an antenna with two or more driven elements. The elements are fed with a certain relative phase, and they are spaced at a certain distance, resulting in a directivity pattern that exhibits gain in some directions and little or no radiation in other directions.

Phased arrays can be very simple, consisting of only two elements. Two examples of simple pairs of phased dipoles are shown in the illustration. At A, the two dipoles are spaced ¼ wavelength apart in free space, and they are fed 90 degrees out of phase. The result is that the signals from the two antennas add in phase in one direction, and cancel in the opposite direction, as shown by the arrows. In this particular case, the radiation pattern is unidirectional. However, phased arrays might have directivity patterns with two, three, or even four different optimum directions. A bidirectional pattern can be obtained, for example, by spacing the dipoles at one wavelength, and feeding them in phase, as shown at B.

More complicated phased arrays are used by radio transmitting stations. Several vertical radiators, arranged in a specified pattern and fed with signals of specified phase, produce a designated directional pattern. This is done to avoid interference with other broadcast stations on the same channel.

Phased arrays might have fixed directional patterns, or they might have rotatable or steerable patterns. The pair of phased dipoles (A) may, if the wavelength is short enough to allow construction from metal tubing, be mounted on a rotator for 360-degree directional adjustability. With phased vertical antennas, the relative signal phase can be varied, and the directional pattern can be adjusted. *See also* PHASE, RADAR.

Quad Antenna A quad antenna is a form of parasitic array that operates according to the same principles as the Yagi antenna except that full-wavelength loops are used instead of ½-wavelength straight elements.

A full-wavelength loop has approximately 2 decibels gain, in terms of effective radiated power, compared to a ½-wavelength dipole. This implies that a quad antenna should have 2 decibels gain over a Yagi with the same number of elements. Experiments have shown that this is true.

A two-element quad antenna may consist of a driven element and a director, or a driven element and a reflector. A three-element quad has one driven element, one director, and one reflector. The director has a circumference of approximately 0.97 electrical wavelength; the driven element measures exactly 1 wavelength around; the reflector measures about 1.03 wavelength in circum-

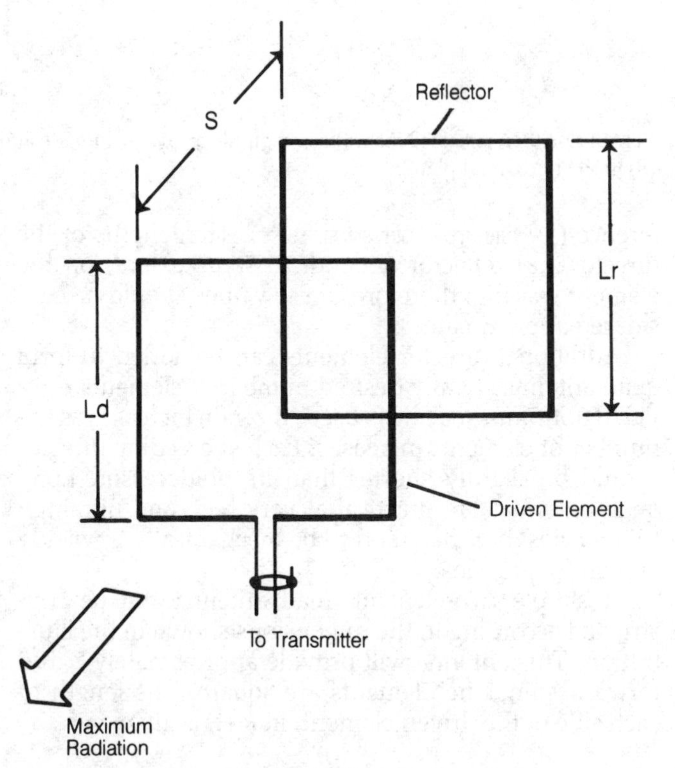

QUAD ANTENNA: A quad antenna is formed as a driven loop with a reflecting loop.

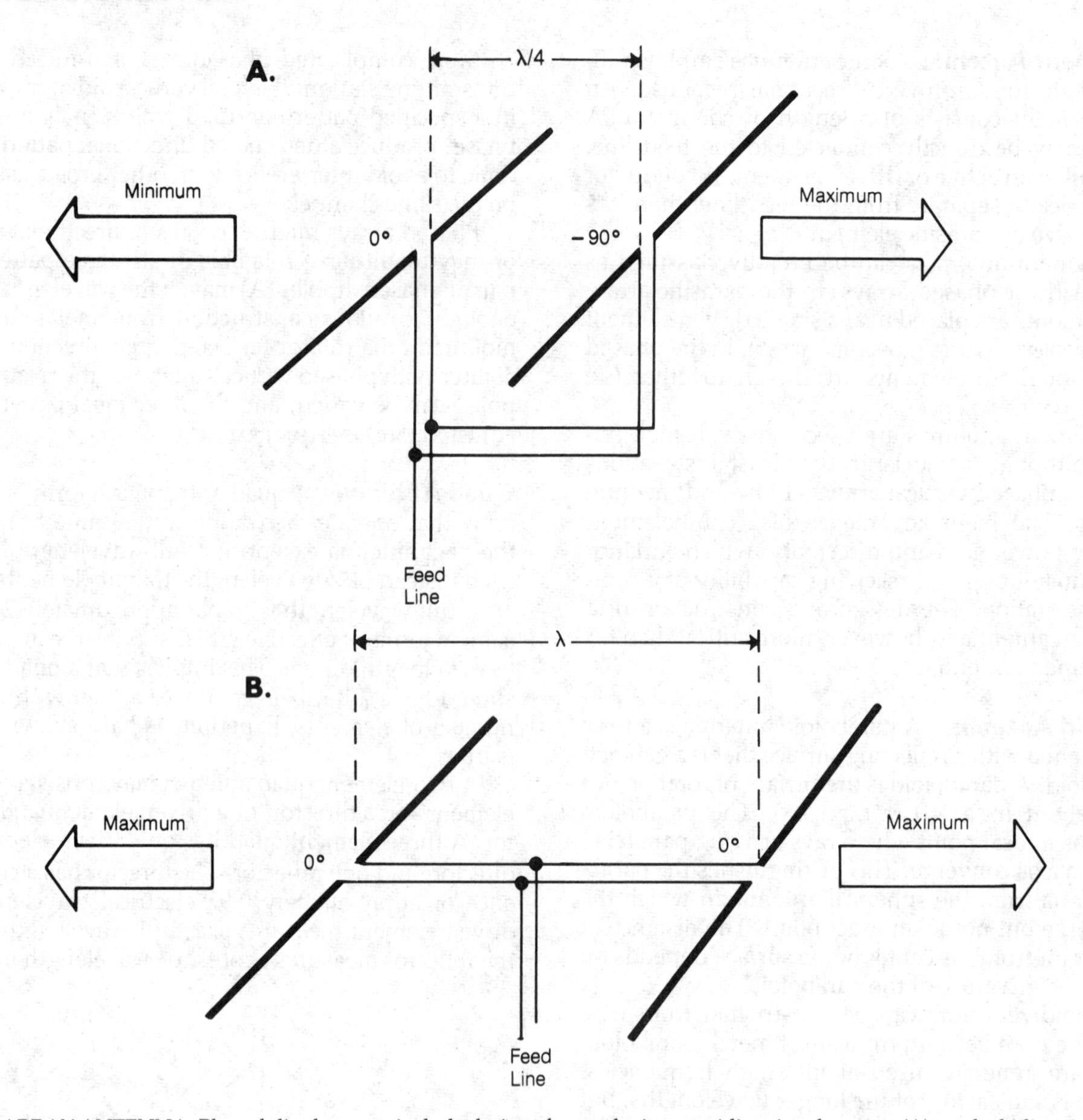

PHASED-ARRAY ANTENNA: Phased dipole arrays include designs for producing a unidirectional pattern (A), and a bidirectional pattern (B) in the horizontal plane.

ference (*see* DIRECTOR, DRIVEN ELEMENT). The lengths of the director and reflector depend, to some extent, on the element spacing; therefore, these values should be considered approximate.

Additional director elements can be added to form quad antennas of any desired number of elements. Provided optimum spacing is used, the gain increases as the number of elements increases. Each succeeding director should be slightly shorter than its predecessor. Long quad antennas are practical at very high and ultrahigh frequencies, but they tend to be mechanically unwieldy at high frequencies.

A simple two-element quad antenna can be constructed according to the dimensions shown in the illustration. This antenna will provide approximately 7 dBd forward gain. The elements are square. The length of each side of the driven element, in feet, is given by:

$$\text{Ld} = 251/\text{f}$$

where *f* is the frequency in megahertz. The length of each side of the reflector is given by:

$$\text{Lr} = \frac{258}{\text{f}}$$

The element spacing, in feet, is given by:

$$\text{s} = \frac{200}{\text{f}}$$

Geometrically, the quad antenna shown has an almost perfect cube shape. For this reason, two-element quad antennas are often called cubical quads. *See also* PARASITIC ARRAY, YAGI ANTENNA.

Reference Antenna For antenna power gain to have meaning, a reference antenna is needed. Antenna power gain is generally measured in decibels relative to a reference dipole. Sometimes the reference antenna is an isotropic radiator (*see* ANTENNA POWER GAIN, dBd, dBi, ISOTROPIC ANTENNA).

For the measurement of transmitted-signal gain, the reference antenna and the test antenna are set up side by

RHOMBIC ANTENNA: The rhombic antenna provides a choice of radiation patterns.

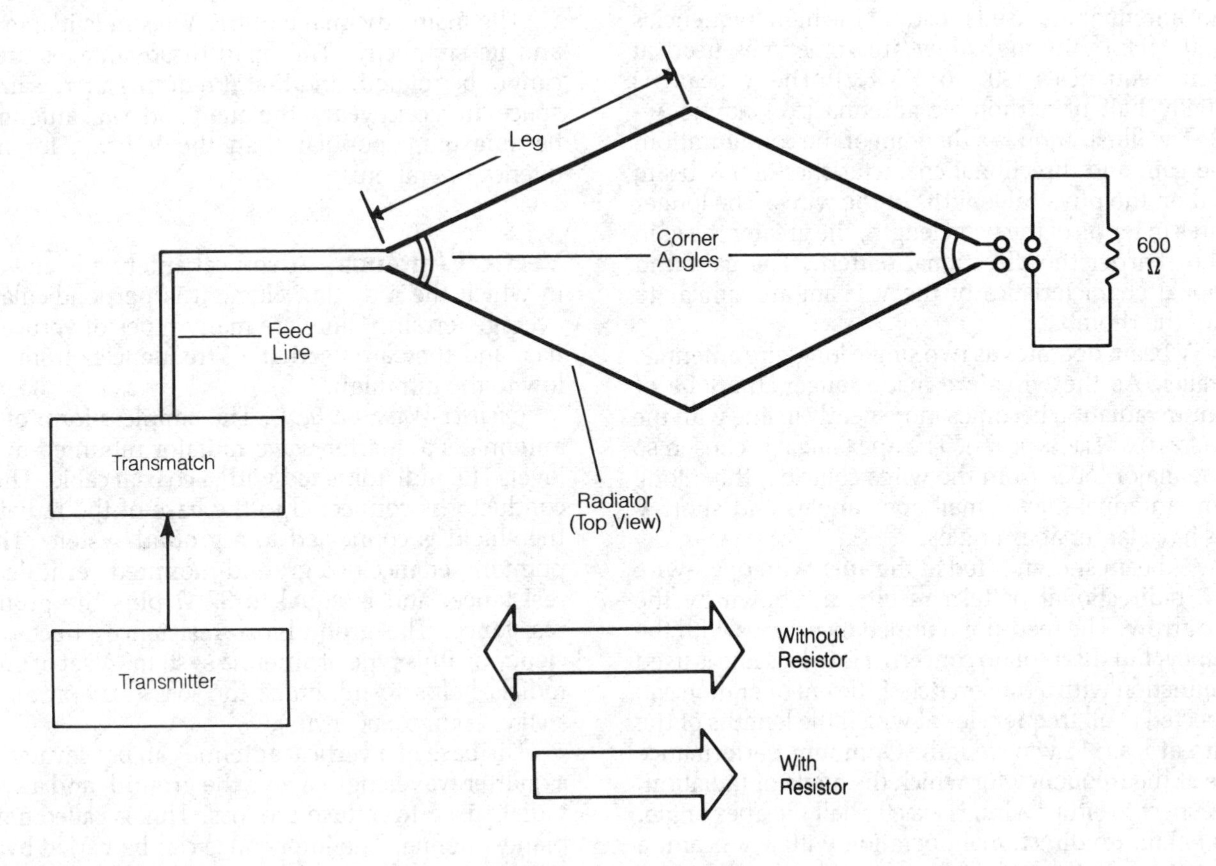

side. An accurately calibrated field-strength meter is positioned at a distance of at least several wavelengths from both antennas. The reference and test antennas are oriented so that the field-strength meter is located in the favored direction (major lobe) of both antennas. A signal of P watts is applied to the reference antenna, and the field-strength meter is set to show a relative indication of 0 decibels. Then the same signal of P watts is applied to the test antenna, and the gain is read from the field-strength meter. *See* TEST RANGE.

To measure received-signal gain, a similar arrangement is used. The reference and test antennas are placed side by side with their major lobes oriented toward a signal generator/antenna several wavelengths distant. A field-strength meter or receiver, equipped with an accurate S meter, is connected to the reference antenna, and the receiver gain is adjusted until a relative indication of 0 dB is obtained. Then the receiver is connected to the test antenna. A perfect impedance match must be maintained for both antennas. The gain of the test antenna can then be read fro the receiver S meter. *See* S METER.

Rhombic Antenna A rhombic antenna is a form of longwire antenna that exhibits gain in one or two fixed directions. Rhombic antennas are typical at the high frequencies (3 to 30 MHz), and are constructed of wire. The rhombic antenna is so named because it is shaped like a rhombus (see illustration).

The gain of a rhombic antenna depends on the physical size in wavelengths, and also on its corner angles. The larger the rhombic antenna in terms of wavelength, the more elongated the rhombus must be order to realize optimum gain. A rhombic antenna with legs measuring ½ wavelength can produce approximately 2 dBd decibels of power gain; this figure increases to 5 dBd decibels for legs of 1 wavelength, 10 dBd decibels for legs of 3 wavelengths, and 12 dBd decibels for legs of 5 wavelengths, provided the corner angles are optimized.

The rhombic antenna shown can be fed at either corner at which the angle between the wires is less than 90 degrees. The wires at the opposite corner are left free, resulting in the bidirectional pattern of radiation and reception (shown by the double arrow). The feed-point impedance varies with the frequency, but open-wire line can be used in conjunction with a transmatch for efficient operation at all frequencies at which the length of each leg of the rhombus in ½ wavelength or greater.

A 600-Ω, noninductive, high-power resistor can be connected at the far end to obtain a unidirectional pattern (shown by the single arrow). The addition of the terminating resistor results in a nearly constant feed-point impedance of 600 Ω at all frequencies at which the leg length is ½ wavelength or more.

The main advantages of the rhombic antenna are its power gain, and the fact that it can produce this gain over a wide range of frequencies. The principal disadvantages are that it cannot be rotated, and that it covers a large amount of space. In recent years, the quad and yagi antennas have become much more common than the rhombic. *See also* QUAD ANTENNA, YAGI ANTENNA.

V-Beam Antenna A V-beam is a form of longwire antenna that exhibits gain in one or two fixed directions. V-beam antennas are used mostly at the high frequencies (3 to 30 MHz), although they are sometimes used at medium frequencies (300 to 3 MHz). The V beam is essentially half of a rhombic antenna (*see* RHOMBIC ANTENNA). The illustration is a diagram of the configuration.

The gain and directional characteristics of a V beam depend on the physical lengths of the wires. The longer the wires in terms of the wavelength, the greater the gain and the sharper the directional pattern. The gain and directional characteristics of the V beam are similar to those of the rhombic.

The V beam operates as two single longwire antennas in parallel. As the wires are made longer, the lobe of maximum radiation becomes more nearly in line with the wires (*see* LONGWIRE ANTENNA). The apex angle is chosen so that the major lobes from the wires coincide; thus, long V-beam antennas have small apex angles and short V beams have larger apex angles.

The V beam shown is fed at the apex with open-wire line. A bidirectional pattern results, as shown by the double arrow. The feed-point impedance varies with the frequency, but this is of no concern if low-loss line is used in conjunction with a transmatch. Efficient operation can be expected at all frequencies at which the lengths of the legs are at least ½ wavelength. Optimum performance will be at the frequency for which the angle of radiation, with respect to either wire, is exactly half the apex angle.

To obtain unidirectional operation with a V beam, a 600-Ω, noninductive, high-power resistor can be connected between the far end of each wire and ground. This eliminates half of the pattern, as shown by the single arrow. The addition of the terminating resistors results in a nearly constant feed-point impedance of 300 to 450 Ω at all frequencies for which the wires measure ½ wavelength or longer.

The main advantage of the V beam is its power gain and its simplicity. The main disadvantages are that it cannot be rotated, and that it requires a large amount of space. In recent years, the quad and Yagi antennas have become more popular than the V beam for high-frequency operation.

Vertical Antenna A vertical antenna is any antenna in which the radiating element is perpendicular to the average terrain. There are many types of vertical antennas, and they are used at all frequencies from the very low to the ultrahigh.

Quarter-Wave Verticals. The simplest form of vertical antenna is a quarter-wave radiator mounted at ground level. The radiator is fed with a coaxial cable. The center conductor is connected to the base of the radiator, and the shield is connected to a ground system. The feed-point impedance of a ground-mounted vertical is a pure resistance, and is equal to 37 Ω plus the ground-loss resistance. The ground-loss resistance affects the efficiency of athis type of antenna system. A set of grounded radials helps to minimize the loss (*see* ANTENNA GROUND SYSTEM, EARTH CONDUCTIVITY).

The base of a vertical antenna can be elevated at least a quarter wavelength above the ground, and a system of radials used to reduce the loss. This is called a ground-plane antenna. The impedance can be varied by adjusting the angle of the radials with respect to the horizontal.

At some frequencies, the height of a quarter-wave vertical is unmanageable unless inductive loading is used to reduce the physical length of the radiator. This technique is used mostly at very low, low, and medium

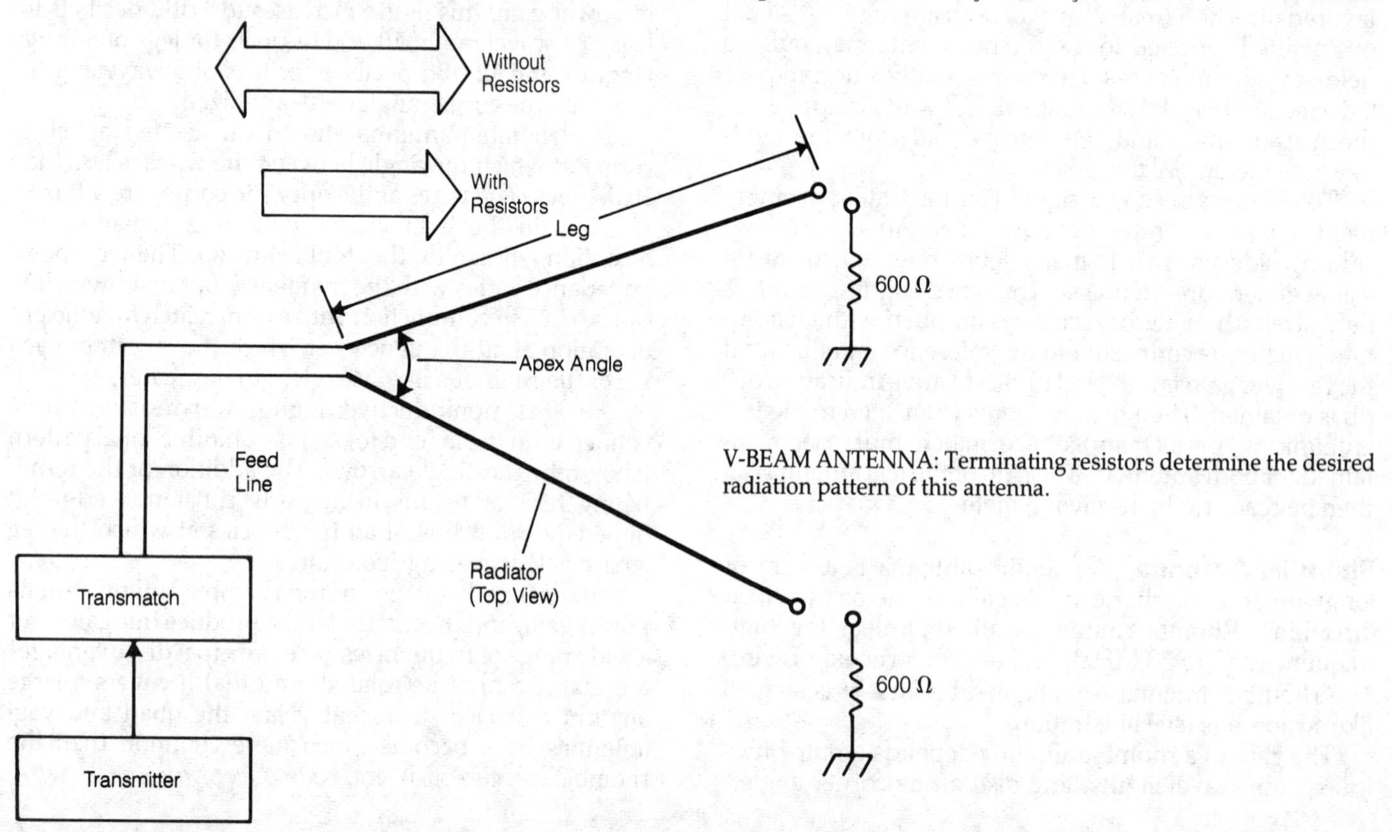

V-BEAM ANTENNA: Terminating resistors determine the desired radiation pattern of this antenna.

frequencies, and to some extent at high frequencies (*see* INDUCTIVE LOADING).

A vertical antenna can be made resonant on several frequencies by the use of multiple loading coils, or by inserting traps at specific points along the radiator.

Half-Wave Verticals. A half-wave radiator can be fed at the base by using an impedance transformer. This provides more efficient operation for a ground-mounted antenna, as compared with a quarter-wavelength radiator, because the radiation resistance is much higher. A quarter-wave matching transformer can be attached to a half-wave vertical to form a J antenna. A half-wave radiator can be fed at the center with coaxial cable to form a vertical dipole (*see* VERTICAL DIPOLE ANTENNA). If the feed line is run through one element of a vertical dipole, the configuration is called a coaxial antenna (*see* COAXIAL ANTENNA). A half-wave element can even be fed at the top with open-wire line.

Multiple Verticals. Two or more vertical radiators can be combined in various ways to obtain gain and/or directivity. Parasitic elements can be placed near vertical radiators to achieve the same results (*see* PARASITIC ARRAY, PARASITIC ELEMENT). Two or more vertical antennas can be stacked in collinear fashion to obtain an omnidirectional gain pattern (*see* COLLINEAR ANTENNA). A Yagi can be oriented so that all of its elements are vertical (*see* YAGI ANTENNA).

Advantages and Disadvantages of Vertical Antennas. Vertical antennas radiate, and respond to, electromagnetic fields having vertical polarization (*see* VERTICAL POLARIZATION). This is an asset at very low, low, and medium frequencies, because it facilitates efficient surface-wave propagation (*see* SURFACE WAVE). Vertical antennas provide good low-angle radiation at high frequencies, an attribute for long-distance ionospheric communication. At medium and high frequencies, vertical antennas are often more convenient to install than horizontal antennas because of limited available space. Self-supporting vertical antennas require essentially no real estate. At very high and ultrahigh frequencies, some communications systems use vertical polarization; a vertical antenna is obviously preferable to a horizontal antenna in such cases.

One of the chief disadvantages of a vertical antenna, especially at high frequencies, is its susceptibility to reception of man-made noise; most artificial noise sources emit vertically polarized electromagnetic fields. Ground-loss problems present another disadvantage of vertical antennas at high frequencies. A ground-mounted vertical antenna usually requires an extensive radial system if the efficiency is to be reasonable.

Vertical-Dipole Antenna A half-wave vertical radiator, fed at the center, is called a vertical dipole antenna. A vertical dipole exhibits excellent low-angle radiation characteristics, good efficiency without the need for a radial system, and relative simplicity.

The illustration shows a vertical dipole for use at high frequencies. The design is very practical at frequencies of about 10 MHz or more, and is conceivable at frequencies considerably lower than that, provided adequate guying is used. The radiating element is constructed from aluminum tubing. The feed line should be run away from the radiator at a right angle for a distance of at least a quarter wavelength. The feed-point impedance is approximately 73 Ω at the resonant frequency of the antenna.

A vertical dipole can be fed with coaxial line through one of the halves of the radiating element. This type of arrangement is called a coaxial antenna. *See also* COAXIAL ANTENNA.

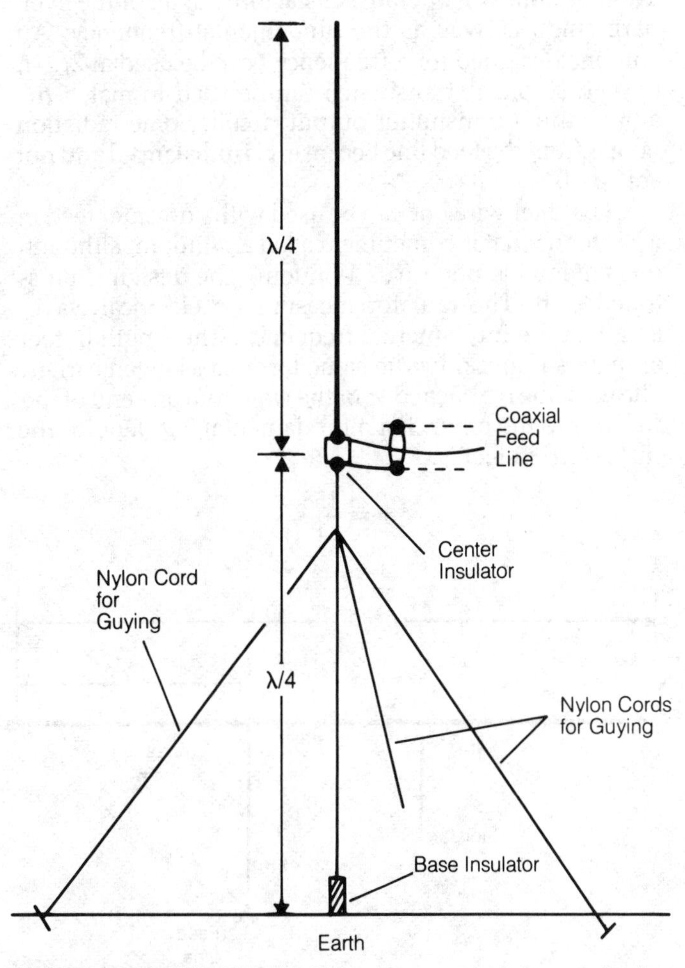

VERTICAL-DIPOLE ANTENNA: A diagram of a balanced dipole antenna.

Windom Antenna A Windom antenna is a multiband wire antenna that uses a single-wire feed line. The antenna is named for its inventor. The Windom is a ½ wavelength horizontal antenna, fed slightly off center (see illustration A).

For a given fundamental frequency *f* in metahertz, the length *d* of a windom, in feet, is:

$$d = \frac{468}{f}$$

If *d* is expressed in meters, then:

$$d = \frac{143}{f}$$

The feed line is attached at a point 36 percent of the way

from one end of the radiator to the other end. For a given frequency *f* in megahertz, the distance *r*, in feet, from the near end of the radiator to the line is:

$$r = \frac{168}{f}$$

If *r* is expressed in meters,

$$r = \frac{51}{f}$$

The Windom will operate satisfactorily at all of the even harmonics, as well as the fundamental frequency. An antenna designed for a frequency *f* can be used at 2*f*, 4*f*, 6*f*, and so on. A transmatch can be used to match the antenna to a transmitter output circuit. Some radiation occurs from the feed line because it is unbalanced and not shielded.

A parallel-wire line can be used with off-center feed in an antenna that is sometimes called a Windom, although this antenna is not a true Windom. The design is illustrated at B. The radiator measures ½ electrical wavelength at the fundamental frequency. The length in feet or meters if found by the same formula as given earlier. The feed line is attached ⅓ of the way from one end of the radiator to the other. For a fundamental frequency *f*, the distance *r*, in feet is:

$$r = \frac{156}{f}$$

WINDOM ANTENNA: The basic Windom antenna (A) and a variation of that antenna (B).

and in meters:

$$r = \frac{48}{f}$$

This configuration results in a feed-point impedance of about 300 Ω, at frequencies *f*, 2*f*, 3*f*, 4*f*, 5*f*, and so on. Some radiation occurs from the line because the system is not perfectly balanced. A 4:1 balun or a transmatch can be used at the transmitter end of the feed line to obtain a more or less purely resistive impedance of approximately 75 Ω.

Yagi Antenna A Yagi antenna is a form of parasitic array that operates by electromagnetic coupling between a driven element and one or more separate, parallel conductors called parasitic elements (*see* PARASITIC ARRAY, PARASITIC ELEMENT). The Yagi antenna is named for the Japanese engineer who discovered the properties of parasitic elements. Yagi antennas are used at high, very high, and ultrahigh frequencies for receiving and transmitting radio signals.

In the Yagi, the power is applied to a radiator called the driven element. The parasitic elements act to produce power gain and directional characteristics (*see* ANTENNA POWER GAIN, FRONT-TO-BACK RATIO). The driven element is usually an electrical half wavelength. The parasitic elements are somewhat shorter or longer. In general, a reflector is slightly longer than the driven element, and a director is somewhat shorter (*see* DIRECTOR, DRIVEN ELEMENT).

A Yagi antenna may have one or two reflectors, and one or more directors. There is usually only one drive element, although two or more can be phased to provide some gain. A common configuration at high frequencies is one driven element, one director, and one reflector. The director measures approximately 0.48 electrical

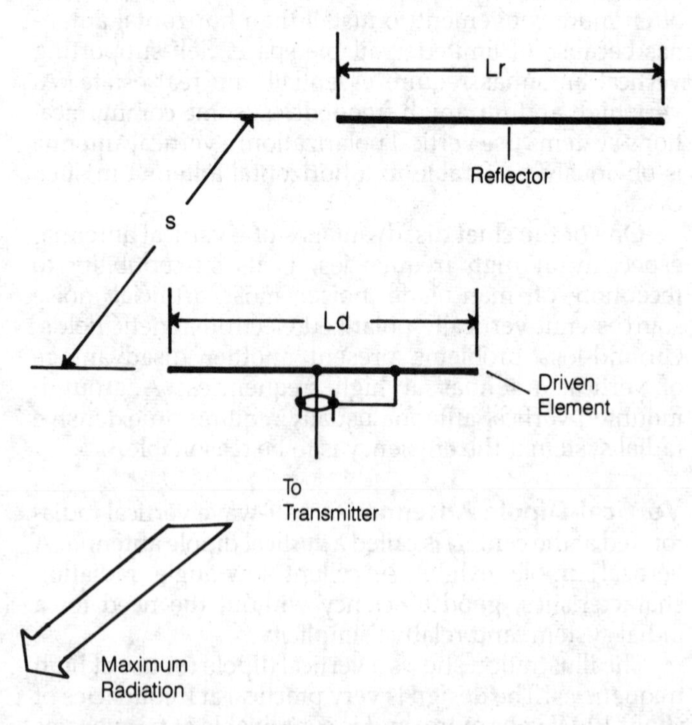

YAGI ANTENNA: A diagram of a two-element Yagi antenna.

wavelength, and the reflector near 0.52 wavelength. Because the exact lengths depend on the element spacing, these figures are approximate.

Additional director elements can be added to obtain enhanced forward gain and front-to-back ratio. Provided optimum element spacing is used, the gain increases as the number of elements increases. Each succeeding director should be slightly shorter than its predecessor. Long Yagi antennas are practical at very-high and ultra-high frequencies, but they are difficult to construct at high frequencies because of their large size. A simple two-element Yagi antenna can be built according to the dimensions shown in the illustration. This antenna will provide about 5 decibels of forward gain. The elements are straight. The length L_d of the driven element, in feet, is given in terms of the frequency *f*, in megahertz, by the equation:

$$L_d = \frac{460}{f}$$

The length of the reflector L_r, in feet, is given by:

$$L_r = \frac{480}{f}$$

(These equations are based on the assumption that the element diameter is very small compared with the element length. If heavy tubing is used, the elements will be slightly shorter.) The element spacing *s*, in feet, is:

$$s = \frac{200}{f}$$

Some adjustment of L_d, L_r, and *s* are required to obtain optimum forward gain.

Resonant loops can be used in a configuration similar to that of the Yagi antenna. This type of antenna is called a quad. *See also* QUAD ANTENNA.

ANTENNA EFFICIENCY

Not all the electromagnetic field received by an antenna from its feed line is ultimately radiated into space. Some power is dissipated in the ground near the antenna in buildings and trees, in the earth itself, and in the conducting material of the antenna. If *P* represents the total amount of available power at a transmitting antenna, P_R the amount of power eventually radiated into space, and P_L the power lost in surrounding objects and the antenna conductors, then:

$$P_L + P_R = P, \text{ and}$$
$$\text{Efficiency (percent)} = 100\ (P_R/P)$$

The efficiency of a nonresonant antenna is difficult to determine in practice, but at resonance (*see* RESONANCE), when the antenna impedance, *Z*, is a known pure resistance, the efficiency can be found by the equation:

$$\text{Efficiency (percent)} = 100\ (R/Z),$$

where *R* is the theoretical radiation resistance of the antenna at the operating frequency (*see* RADIATION RESISTANCE). The actual resistance, *Z*, is always larger than the theoretical radiation resistance *R*. The difference is the loss resistance, and there is always some loss.

Antenna efficiency is optimized by making the loss resistance as small as possible. Some means of reducing the loss resistance are the illustration of a good rf-ground system (if the antenna needs a good rf ground), the use of low-loss components in tuning networks (if they are used), and locating the antenna itself as high above the ground, and as far from obstructions, as possible.

ANTENNA GROUND SYSTEM

Some antennas must operate against a ground system, or rf (radio frequency) reference potential. Others do not need an rf ground. In general, unbalanced or asymmetrical antennas need a good rf ground, while balanced or symmetrical antennas do not. The ground-plane antenna requires an excellent rf ground in order to function efficiently. However, the half-wave dipole antenna does not need an rf ground. (*See* GROUND-PLANE ANTENNA in the ANTENNA DIRECTORY.

A good dc (direct-current) ground does not necessarily constitute a good rf ground. An elevated ground-plane antenna has a very effective rf ground that does not have to be connected physically to the earth. A single thin wire hundreds of feet long can be terminated at a ground rod or the grounded side of a utility outlet and work very well as a dc ground, but it will not work well at radio frequencies.

The earth affects the characteristics of any antenna. The overall radiation resistance and impedance are affected by the height above ground. However, an rf ground system does not necessarily depend on the height of the antenna. Capacitive coupling to ground is usually sufficient, in the form of a counterpoise (*see* COUNTERPOISE).

A ground system for lightning protection must be a direct earth connection. Some antennas are grounded through inductors that do not conduct rf, but serve to discharge static buildup before a lightning stroke hits.

ANTENNA IMPEDANCE

Any antenna displays a defined impedance at its feed point at a particular frequency. Usually, this impedance changes as the frequency changes. Impedance at the feed point of an antenna consists of radiation resistance (*see* RADIATION RESISTANCE) and either capacitive or inductive reactance (*see* REACTANCE). Both the radiation resistance and the reactance are defined in ohms.

An antenna is said to be resonant at a particular frequency when the reactance is zero. Then, the impedance is equal to the radiation resistance in ohms. The frequency at which resonance occurs is called the resonant frequency of the antenna. Some antennas have only one resonant frequency. Others have many. The radiation resistance at the resonant frequency of an antenna depends on several factors, including the height above ground and the harmonic order, and whether the antenna is inductively or capacitively tuned. The radiation resistance can be as low as a fraction of an ohm, or as high as several thousand ohms.

When the operating frequency is made higher than the resonant frequency of an antenna, inductive reactance appears at the feed point. When the operating frequency is below the resonant point, capacitive reactance appears. To get a pure resistance, a reactance of the opposite kind from the type present must be connected in series with the antenna. *See also* ANTENNA RESONANT FREQUENCY, IMPEDANCE.

ANTENNA MATCHING

For optimum operation of an antenna and feed-line combination, the system should be at resonance. This is usually done by eliminating the reactance at the feed point, where the feed line joins the antenna radiator. In other words, the antenna itself is made resonant. The remaining radiation resistance is then transformed to a value that closely matches the characteristic impedance of the feed line.

If the reactance at the antenna feed point is inductive, series capacitors are added to cancel out the inductive reactance. If the reactance is capacitive, series coils are employed. An example of this is shown at A in the figure. The inductances are adjusted until the antenna is resonant. Both coils should have identical inductances to keep the system balanced (since this is a balanced antenna).

Once the antenna is resonant, only resistance remains. This value may not be equal to the characteristic impedance of the line. Generally, coaxial lines are designed to have a characteristic impedance, or Z_O, of 50 to 75 Ω, which closely approximates the radiation resistance of a half-wave dipole in free space. But, if the radiation resistance of the antenna is much different from the Z_O of the line, a transformer should be used to match the two parameters, as shown in B. This results in the greatest efficiency for the feed line. Without the transformer, standing waves on the line will cause some loss of signal. The amount of loss caused by standing waves is sometimes inconsequential, but sometimes it is large (*see* STANDING WAVE, STANDING-WAVE RATIO).

In some antenna systems, no attempt is made to obtain an impedance match at the feed point. Instead, a matching system is used between the transmitter or receiver and the feed line. This allows operating convenience when the frequency is changed often. However, it does nothing to reduce the loss on the line caused by standing waves. *See also* TUNED FEEDERS.

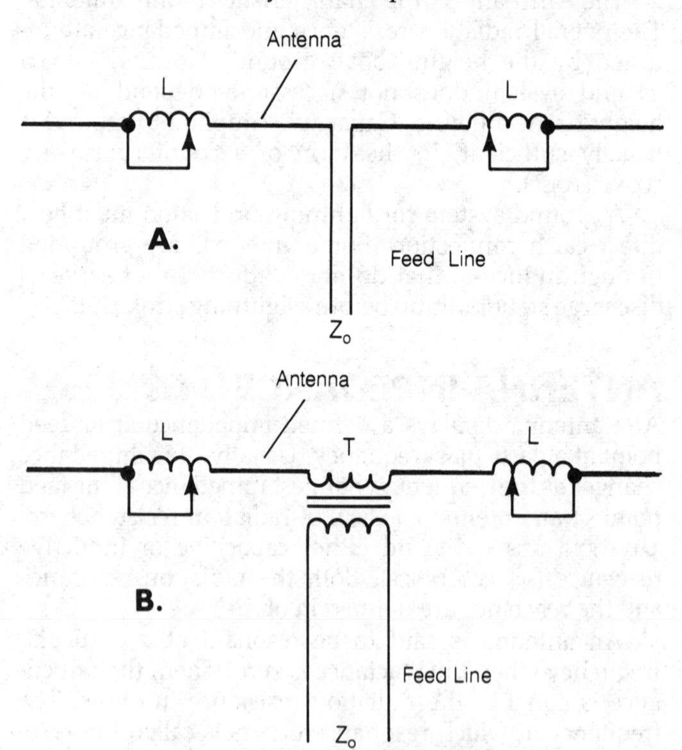

ANTENNA MATCHING: The dipole is tuned to resonance with inductors (A), and a transformer is used to match the radiation resistance of the dipole to the characteristic impedance of the feed line (B).

ANTENNA PATTERN

The directional characteristics of any transmitting or receiving antenna, when graphed on a polar coordinate system, are called the antenna pattern. The simplest

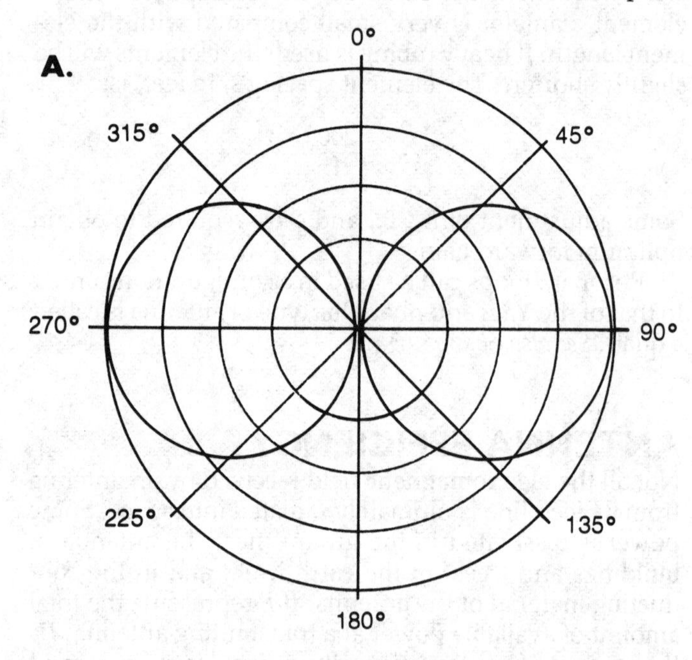

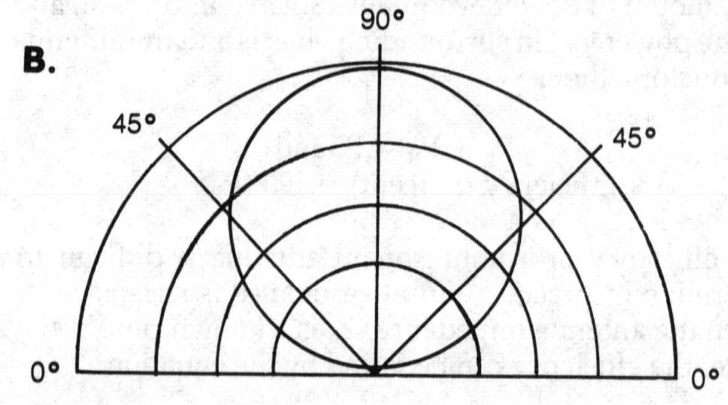

ANTENNA PATTERN: A horizontal-plane pattern for a ½ wavelength dipole (A) and its vertical plane pattern as viewed from the end (B) at a height of ¼ wavelength.

possible antenna pattern occurs when an isotropic antenna is used (*see* ISOTROPIC ANTENNA in the ANTENNA DIRECTORY), although this is a theoretical ideal. It radiates equally well in all directions in three-dimensional space.

Antenna patterns are represented by diagrams such as those in A and B. The location of the antenna is assumed to be at the center of the coordinate system. The greater the radiation or reception capability of the antenna in a certain direction, the farther from the center the points on the chart are plotted. A dipole antenna, oriented horizontally so that its conductor runs in a north-south direction, has a horizontal-plane (H-plane) pattern similar to that in Fig. A. The elevation-plane (E-plane) pattern depends on the height of the antenna above effective ground at the viewing angle. With the dipole oriented so that its conductor runs perpendicular to the page, and the antenna ¼ wavelength above effective ground, the E-plane antenna pattern will resemble B.

The patterns in A and B are quite simple. Many antennas have patterns that are very complicated. For all antenna pattern graphs, the relative power gain (*see* ANTENNA POWER GAIN) relative to a dipole is plotted on the radial axis. The values thus range from 0 to 1 on a linear scale. Sometimes a logarithmic scale is used. If the antenna has directional gain, the pattern radius will exceed 1 in some directions. Examples of antennas with directional gain are the log periodic, longwire, quad, and Yagi. Some vertical antennas have gain in all horizontal directions. This occurs at the expense of gain in the E plane.

ANTENNA POLARIZATION

The polarization of an antenna is determined by the orientation of the electric lines of force in the electromagnetic field radiated or received by the antenna. Polarization may be linear, or it may be rotating (circular). Linear polarization can be vertical, horizontal, or somewhere in between. In circular polarization, the rotation can be either counterclockwise or clockwise (*see* CIRCULAR POLARIZATION).

For antennas with linear polarization, the orientation of the electric lines of flux is parallel with the radiating element, as shown in A and B. Therefore, a vertical element produces signals with vertical polarization, and a horizontal element produces horizontally polarized fields in directions broadside to the element.

In receiving applications, the polarization of an antenna is determined according to the same factors involved in transmitting. Thus, if an antenna is vertically polarized for transmission of electromagnetic waves, it is also vertically polarized for reception.

In free space, with no nearby reflecting objects to create phase interference, the circuit attenuation between a vertically polarized antenna and a horizontally polarized antenna, or between any two linearly polarized antennas at right angles, is approximately 30 decibels compared to the attenuation between two antennas having the same polarization.

Polarization affects the propagation of electromagnetic energy to some extent. A vertical antenna works much better for transmission and reception of surface-wave fields (*see* SURFACE WAVE) than a horizontal antenna. For sky-wave propagation (*see* SKY WAVE), the polarization is not particularly important, since the ionosphere causes the polarization to be randomized at the receiving end of a circuit.

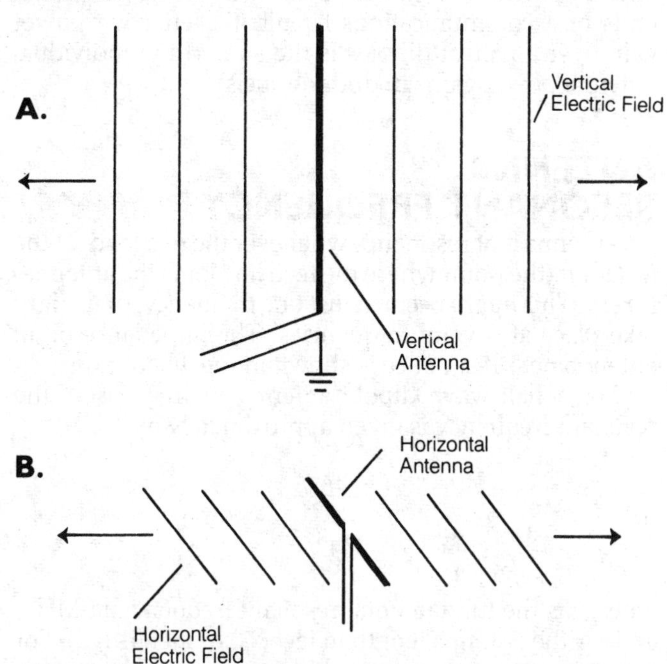

ANTENNA POLARIZATION: A diagram showing vertical polarization of an antenna (A) and horizontal polarization (B).

ANTENNA POWER GAIN

The power gain of an antenna is the ratio of the effective radiated power (*see* EFFECTIVE RADIATED POWER) to the actual rf (radio frequency) power applied to the feed point. Power gain may also be expressed in decibels. If the effective radiated power is P_{ERP} watts and the applied power is P watts, then the power gain in decibels (dB) is:

$$\text{Power Gain (dB)} = 10 \log_{10} (P_{ERP}/P)$$

Power gain is always measured in the favored direction of an antenna. The favored direction is the azimuth direction in which the antenna performs the best. For power gain to be defined, a reference antenna must be chosen with a gain assumed to be unity, or 0 decibels. This reference antenna is usually a half-wave dipole in free space. Power gain figures taken with respect to a dipole are expressed in dBd (decibels). The reference antenna for power-gain measurements may also be an isotropic radiator, in which case the units of power gain are called dBi (decibels). *See* ISOTROPIC ANTENNA in the ANTENNA DIRECTORY. For any given antenna, the power gains in dBd and dBi are different by approximately 2.15 decibels.

$$\text{Power Gain (dBi)} = 2.15 + \text{Power Gain (dBd)}$$

Directional transmitting antennas can have power gains in excess of 20 dBd. At microwave frequencies, large dish antennas (*see* DISH ANTENNA, PARABOLOID ANTENNA in the ANTENNA DIRECTORY) can be built with power gains of 30 dBd (decibels) or more.

Power gain is the same for reception, with a particular antenna, as for transmission of signals. Therefore, when antennas with directional power gain are used at both ends of a communications circuit, the effective power gain over a pair of dipoles is the sum of the individual antenna power gains in dBd (decibels).

ANTENNA RESONANT FREQUENCY

An antenna is at resonance whenever the reactance at the *feed point* (the point where the feed line joins the antenna) is zero. This might occur at just one frequency, or it might take place at several frequencies. The impedance of an antenna near resonance is shown in the illustration.

For a half-wave dipole antenna in free space, the resonant frequency is given approximately by:

$$f = 468/s_{ft}$$

or

$$f = 143/s_m,$$

where f is the fundamental resonant frequency in MHz, and s is the antenna length in feet (s_{ft}) or meters (s_m). For a quarter-wave vertical antenna operating against a perfect ground plane:

$$f = 234/h_{ft},$$

or

$$f = 71/h_m,$$

where h is the antenna height in feet (h_{ft}) or meters (h_m).

The dipole antenna and quarter-wave vertical antenna display resonant conditions at all harmonics of their fundamental frequencies. Therefore, if a dipole or quarter-wave vertical is resonant at a particular frequency f, it will also be resonant at $2f$, $3f$, $4f$, and so on. The impedance is not necessarily the same, however, at harmonics as it is at the fundamental. At frequencies corresponding to odd harmonics of the fundamental, the impedance is nearly the same as at the fundamental. At even harmonics, the impedance is much higher.

An antenna operating at resonance, where the radiation resistance is almost the same as the characteristic impedance, or Z_O, of the feed line, will perform with good efficiency, provided the ground system (if a ground system is needed) is efficient. *See also* CHARACTERISTIC IMPEDANCE, RADIATION RESISTANCE.

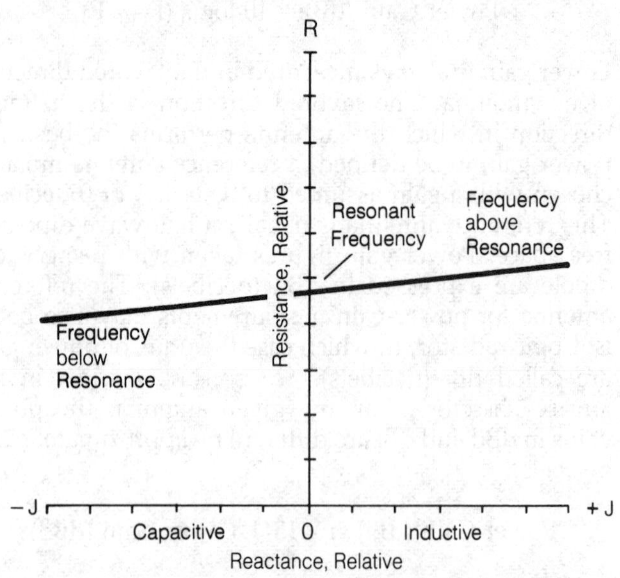

ANTENNA RESONANT FREQUENCY: The plot of antenna resonance as a function of resistance and reactance.

ANTENNA TUNING

Antenna tuning is the process of adjusting the resonant frequency of an antenna or antenna system (*see* ANTENNA RESONANT FREQUENCY). This is usually done by means of a tapped or variable inductor at the antenna feed point, or somewhere along the antenna radiator (see figure). It is also sometimes done with a transmatch at the transmitter, so that the feed line and antenna together form a resonant system (*see* TUNED FEEDERS). Antenna tuning can be done with any sort of antenna.

An antenna made of telescoping sections of tubing is tuned exactly to the desired frequency by changing the amount of overlap at the tubing joints, thus changing the physical length of the radiating or parasitic elements. In a Yagi array, the director and reflector elements must be precisely tuned to obtain the greatest amount of forward power gain and front-to-back ratio. Phased arrays must be tuned to give the desired directional response. In some antennas, tuning is not critical, while in others it must be done precisely to obtain the rated specifications. In general, the higher the frequency, the more exacting are the tuning requirements.

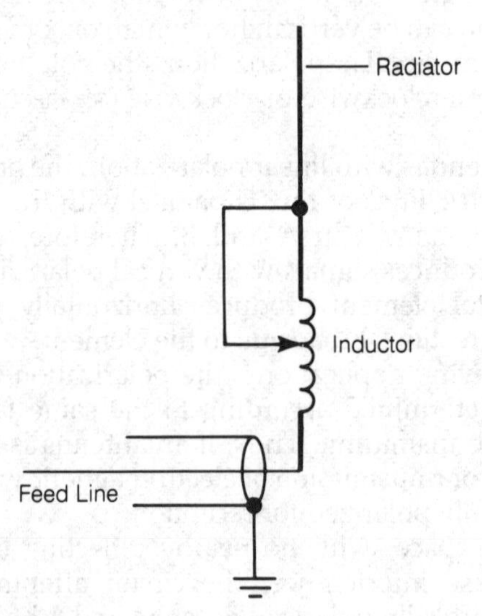

ANTENNA TUNING: A short vertical antenna can be tuned with a variable inductor.

APPARENT POWER

In an alternating-current circuit containing reactance, the voltage and current reach their peaks at different times.

That is, they are not exactly in phase. This complicates the determination of power. In a nonreactive circuit, consider:

$$P = EI$$

where P is the power in watts, E is the rms (*see* ROOT MEAN SQUARE) voltage in volts, and I is the rms current in amperes. In a circuit with reactance, this expression is referred to as the apparent power. It is called *apparent* because it differs from the true power (*see* TRUE POWER) that would be dissipated in a resistor or resistive load. Only when the reactance is zero is the apparent power identical to the true power.

In a nonresonant or improperly matched antenna system, a wattmeter placed in the feed line will give an exaggerated reading. The wattmeter reads apparent power, which is related to the sum of the true transmitter output power and the reactive or reflected power (*see* REACTIVE POWER). To determine the true power, the reflected reading of a directional wattmeter is subtracted from the forward reading. The more severe the antenna mismatch, the greater the difference between the apparent and true power. As the antenna approaches a state of pure reactance, the apparent power approaches twice the true power in the feed line.

In the extreme, all of the apparent power in a circuit is reactive. This occurs when an alternating-current circuit consists of a pure reactance, such as a coil, capacitor, or short-circuited length of transmission line. *See also* REACTANCE.

APPLICATIONS PROGRAM

An applications program is a computer program written to solve a specific problem or to be used in a general application. It can be as simple as a 10-statement program in BASIC for the repetitive solution to a simple mathematical equation or it could consist of thousands of statements for the design of an aircraft wing. There are two basic types of application programs: custom and package.

Custom programs are written to solve problems or perform tasks that are unique to a particular user. Some custom programs are developed by the user, particularly if they involve proprietary products or processes. Others might be developed under contract by an outside software company. An example is a program for operating an automatic pick-and-place component location or insertion machine for loading electronic circuit boards.

Package programs are prewritten by commercial software houses for common applications and sold competitively on the open market. These programs are written for use with specific operating systems. (*See* OPERATING SYSTEM). Most can be used by many different persons with little or no modification. They are sold through retail stores or by direct mail. Examples of package programs are those for playing video games (including simulation of automobile and aircraft controls), word processing, computer graphics, desktop publishing, business and accounting spreadsheets, and income tax preparation.

APPLICATION-SPECIFIC DEVICE

The term *application specific integrated circuit* (ASIC) applies to custom and semicustom integrated circuits as opposed to standard or off-the-shelf products. ASICs can be custom-designed by manual methods, designed with the aid of computers from information on logic gates and complex gate functions (called macrocells) stored in a computer memory, or simply completed with the aid of computer-drawn masks.

In the changing technology of integrated circuit design, even hand-crafted custom IC (integrated circuit) designs can be computer assisted with the manual effort confined to optimizing specific parts of the layout. Standard cells are original designs dependent on the compilation of logic gates and macrocells stored in computer memory (silicon compilation). However, gate arrays are prefabricated arrays of transistors that are organized to perform specific functions by interconnects and masks made by computer.

There is no universally accepted definition of ASIC, although there is agreement that they are ICs designed and fabricated or finished (dedicated) to perform in unique or restricted applications. Some have extended the term to include field- as well as factory-programmed logic ICs such as programmable logic devices (PLDs). *See* GATE ARRAY, PROGRAMMABLE LOGIC DEVICE, STANDARD CELL.

ARC

An arc occurs when electricity flows through space. Lightning is a good example of an arc. When the potential difference between two objects becomes sufficiently large, the air (or other gas) ionizes between the objects, creating a path of relatively low resistance through which current flows.

An arc can be undesirable and destructive, such as a flashover across the contacts of a wafer switch. Or, an arc can be put to constructive use. A carbon-arc lamp is an extremely bright source of light, and is sometimes seen in large spotlights or searchlights where other kinds of lamps would be too expensive for the illumination needed.

Undesirable or destructive arcing is prevented by keeping the voltage between two points below the value that will cause a flashover. An antenna lightning arrestor allows built-up static potential to discharge across a small gap (*see* AIR GAP) before it gets so great that arcing occurs between components of the circuit and ground.

ARCBACK

Arcback is a flow of current in the reverse direction in a vacuum-tube rectifier. Ordinarily, electrons should flow from the cathode to the anode, but not vice-versa. This is how the tube rectifies alternating current.

If the reverse voltage becomes too large, electrons will start to flow from the anode to the cathode, and the tube will no longer rectify. The maximum reverse voltage that a tube can tolerate without arc back is called the peak inverse voltage (PIV), or peak reverse voltage (PRV) (*see* PEAK INVERSE VOLTAGE). In a semiconductor diode, the peak inverse voltage is sometimes called the avalanche voltage (*see* AVALANCHE VOLTAGE).

ARC RESISTANCE

Arc resistance is a measure of the durability of an insulating or dielectric material against arcing. When an arc occurs next to a dielectric material, a conductive path along the surface of the material will eventually be formed. The greater the arc resistance of the material, the longer it will be before this conductive path is formed. Such a conductive path, of course, ruins the insulating properties of the substance.

Arc resistance also refers to the electrical resistance of an arc, for example in an arc lamp. With a certain interelectrode voltage, E, and current, I, the arc resistance, R, is given by Ohm's law (*see* OHM'S LAW) as:

$$R = E/I$$

For alternating current, the voltage and current must be rms values (*see* ROOT MEAN SQUARE) for this determination to be accurate.

AREA REDISTRIBUTION

An irregularly shaped pulse is difficult to measure in terms of duration; it might be difficult to pinpoint the moments of its beginning and ending. Area redistribution is a means of measuring the effective duration of an irregular pulse.

In A, an irregular pulse is shown as it might appear on an oscilloscope. The shaded region has a definite area, measurable in terms of amplitude and time (such as millivolt-seconds). It is virtually impossible to define the beginning and ending moments of this irregular pulse, and therefore its duration is vague. However, a rectangular pulse may be constructed containing the same peak amplitude and the same area as the irregular pulse (B). In a circuit, this pulse is constructed by using a rectangular pulse generator. The duration of the rectangular pulse that delivers the same amount of energy is considered to be the effective duration of the irregular pulse.

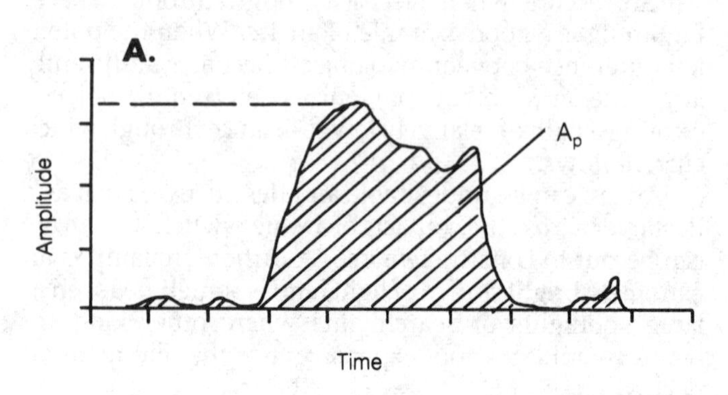

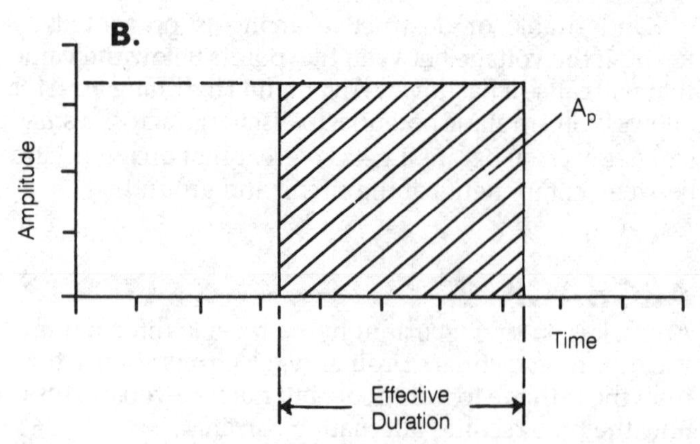

AREA REDISTRIBUTION: An irregular pulse waveform (A) is transposed into an equivalent area with the same amplitude (B) for determining effective duration time.

ARGUMENT

An argument is a variable in a function. The value of any function depends on the values of its arguments. In computer terminology the term argument is used in the same context as the term variable in algebra.

A function, f, of arguments x, y, and z is denoted by $f(x,y,z)$. Suppose that a function of three variables is specified by:

$$f(x,y,z) = 3x - 2y + z$$

Then the values for f are computed for various combinations of argument values.

The set of possible values for the arguments of a function is called the domain of the function. We might have different domains for each argument. For example, the domain for x might be the set of integers, the domain for y might be all real numbers between, but not including, 0 and 1, and the domain for z might be the set of rational numbers. The range of a function is the set of all possible values resulting from combinations of the arguments. Functions can have only one argument, or hundreds of arguments. *See also* FUNCTION.

ARITHMETIC LOGIC UNIT

The arithmetic logic unit (ALU) of a computer is one of the three basic functional sections of the central processing unit (CPU) of a computer and a microprocessor as shown in the illustration. (*See* COMPUTER, MICROPROCESSOR.) The other functions of the CPU are the control and timing unit and memory registers. The ALU performs arithmetic and logical operations under the direction of the control and timing unit. This includes addition, subtraction and logical operations as well as trigonometric operations such as sine, cosine, and tangent with AND, OR, and NOT gates. A typical logical operation involves comparing two numbers, then selecting one of two (or more) program paths depending on the result of the comparison. Acting in harmony with the control and timing unit, the ALU can test two numbers and cause the computer to branch to one of two possible program paths.

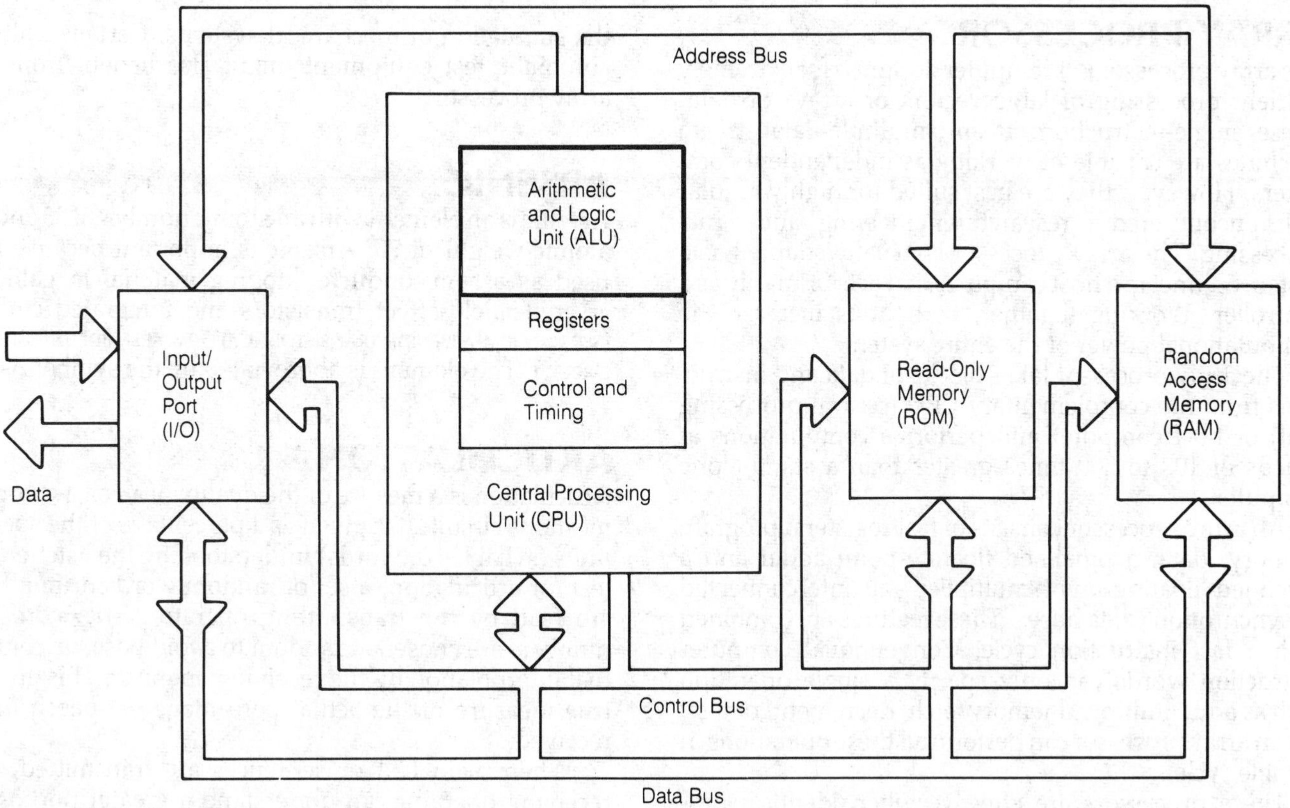

ARITHMETIC LOGIC UNIT: Arithmetic and logic is performed by the ALU that is part of the central processing unit of computers.

ARITHMETIC MEAN

The arithmetic mean of a set of numbers:

$$\{x_1, x_2, x_3, \ldots, x_n\}$$

is a special function of those numbers, defined as:

$$a(x_1, x_2, x_3, \ldots, x_n) = \frac{x_1 + x_2 + x_3 + \ldots + x_n}{n}$$

This function is also called the average of the set of numbers. The arithmetic mean is always larger than or equal to the smallest number in a set, and smaller than or equal to the largest number in a set. The arithmetic mean is not necessarily midway between the smallest and largest numbers in a given set. If x_{min} is the smallest number and x_{max} is the largest number in a set, then:

$$m(x_1, x_2, x_3, \ldots, x_n) = \frac{x_{min} + x_{max}}{2}$$

is called the median of the set:

$$\{x_1, x_2, x_3, \ldots, x_n\}.$$

Occasionally, numbers will be averaged by a different function than a $(x_1, x_2, x_3, \ldots, x_n)$, called the geometric mean. This does not usually result in the same value as the arithmetic mean. *See also* GEOMETRIC MEAN, MEDIAN.

ARMSTRONG OSCILLATOR

See OSCILLATOR CIRCUITS.

ARRAY

The term *array* can apply to many different subjects in electronics. In antennas it refers to an antenna with a group of active dipoles, radiators and directors mounted together in an arrangement that permits the radiated energy to be added in phase for greater output power. The arrangement also permits the transmitting and receiving beams to be directed electronically. Array antennas provide directional characteristics for the received and transmitted radio frequency energy. *See* PHASED-ARRAY ANTENNA, YAGI ANTENNA in the ANTENNA DIRECTORY.

In semiconductor technology, the term array relates to repetitive patterns of transistors or functional circuits called gates on an integrated circuit chip. For gate arrays, they are patterns of uncommitted logic gates on a common substrate that can be dedicated by final metal masking steps. *See* GATE ARRAY, PROGRAMMABLE LOGIC DEVICE. ROMs (programmable read-only memories) are integrated circuit arrays of uncommitted memory cells that can be factory or field programmed for specific applications. *See* SEMICONDUCTOR MEMORY. Arrays of electrode gates are present on charged-coupled devices. *See* CHARGE-COUPLED DEVICE. In LED (light-emitting diode) displays, arrays of LED dies form the alphanumeric characters. *See* OPTOELECTRONICS. Computer programming data can be arranged in arrays of data.

ARRAY PROCESSOR

The array processor is a computer designed for extremely efficient processing of large vectors or arrays of data. These single-instruction stream, multiple-data stream machines are capable of working as independent computers. However, they are best suited for highly regular tasks encountered in research, engineering, and signal processing. The array processor is usually adapted as a system adjunct to a host computer that acts primarily as a controller. Working together, both units increase the computational power of the entire system.

The array processor takes blocks of data and instructions from the complementary CPU (central processing unit) or host computer and performs computations at speeds of 100 to 200 times greater than a stand-alone computer.

An array processor consists of fast registers, program memory, data, a pipelined floating-point adder and a pipelined floating-point multiplier—all interconnected by synchronous data buses. These features are combined with a fast instruction cycle. Conventional computer instruction words can only specify a single operation such as add, multiply, memory fetch, decrement, or test, but an array processor can perform all these operations in a single cycle.

These processors are effective when identical processing must be done on many items of data. All processors receive the same instructions. The array processor is limited in that individual computations must depend only on the data in a particular element and its immediate neighbors.

Array processing is well suited for a large class of scientific and economic data that is very consistent in format and that must be processed in some systematic manner. The actual numerical operations are very repetitive. For example, the program might dictate that eight words of data be brought out of memory and processed by some systematic algorithm. The resultant four words of memory are then to be restored to memory. The pattern of mathematical operations will be repeated over and over again. These are called vector or array processes.

This processing contrasts with the more typical data processing, which calls for random combinations of logical and arithmetic operations as well as conventional branching.

All computers might, at some time, carry out array or vector processes (that is, one array of data is processed to produce another array of data). However, the general-purpose computer is not designed to carry out these highly patterned and organized kinds of calculation efficiently.

Array processors are well suited for Fourier analysis. They are also used for the calculation, reconstruction, and enhancement of CAT (computer-aided tomography) and MR (magnetic resonance) scanners, satellite, sonar, radar, and seismic images. They can also be used for the conversion of speech signals into compressed digital data and the subsequent resynthesis of that data. Other applications include the composition of images in radio astronomy, the statistical analysis of economic data, and the simulation of mechanical systems. Certain realtime automatic test equipment might also benefit from the array processor.

ARSENIC

Arsenic is an element with an atomic number of 33 and an atomic weight of 75. Arsenic is important because it is used as a semiconductor doping material in gallium-arsenide field-effect transistors and integrated circuits (*see* GALLIUM-ARSENIDE TRANSISTOR, GALLIUM ARSENIDE INTEGRATED CIRCUIT). The element is abbreviated by the symbol *As*.

ARTICULATION

Articulation is a mesure of the quality of a voice-communications circuit. It is given as a percentage of the speech units (syllables or words) understood by the listener. To test for articulation, a set of random words or numbers are read by the transmitting operator. The words or numbers are chosen at random to avoid possible contextual interpolation by the receiving operator. This gives a true measure of the actual percentage of speech units received.

When plain text or sentences are transmitted, the receiving operator can understand a greater portion of the information because he or she can figure out some of the missing words or syllables by guesswork. The percentage of speech units received with plain-text transmission is called intelligibility (*see* INTELLIGIBILITY).

Articulation and intelligibility differ from fidelity. Perfect reproduction of the transmitted voice is not as important in a communications system as the accurate transfer of information. The best articulation generally occurs when the voice frequency components are restricted to approximately the range of 200 to 3000 Hz. Articulation can also be enhanced at times by the use of speech compressors or rf clipping (*see* SPEECH CLIPPING, SPEECH COMPRESSION).

ARTIFICIAL GROUND

An artificial ground is an rf (radio-frequency) ground not directly connected to the earth. A good example of an artificial ground is the system of quarter-wave radials in an elevated ground-plane antenna (*see* GROUND-PLANE ANTENNA in the ANTENNA DIRECTORY). A counterpoise (*see* COUNTERPOISE) is also a form of artificial ground.

In some situations, it is difficult or impossible to obtain a good earth ground for an antenna system. A piece of wire ¼ wavelength, or any odd multiple of ¼ wavelength, at the operating frequency can operate as an artificial ground in such a case. This arrangement does not form an ideal ground; the wire will radiate some energy, and thus is actually a part of the antenna. But an artificial ground is much better than no ground at all. A dc (direct-current) ground should be used in addition to the rf ground to minimize the danger of electrical shock from built-up static on the antenna and from possible short circuits in the transmitting or receiving equipment. *See also* DC GROUND.

ARTIFICIAL INTELLIGENCE

Artificial intelligence (AI) is the study of methods for designing computer programs to perform cognitive tasks. The subject includes image understanding, natural-language processing and understanding, mathematical theorem proving, and expert knowledge such as medical diagnosis and economic analysis. However, the content of artificial intelligence is not precisely defined.

AI methods include search techniques and knowledge-based techniques. Search techniques involve description and analysis without reference to a specific problem, and knowledge-based techniques depend on the knowledge accumulated in solving a wide variety of problems.

An example of a search technique is a program written for the recognition of objects by means of computer-aided vision. (*See* COMPUTER VISION.) Its objectives include the design of computer programs that perform at high levels of competence in cognitive tasks. AI is not dependent on methods that mimic the internal structure of human behavior.

An example of a knowledge-based technique is the writing of a computer program that performs decision functions that are modeled on the processing of information by a human expert. Once those hypotheses are documented, a computer can predict their consequences for hypothetical situations.

A-SCAN

See RADAR.

ASCII

American National Standard Code for Information Interchange (ASCII) is a seven-unit digital code for the transmission of teleprinter data. Letters, numerals, symbols, and control operations are represented. ASCII is designed primarily for computer applications, but is also used in some teletypewriter systems.

Each unit is either 0 or 1. In the binary number system, there are 2^7, or 128, possible representations. Table 1 gives the ASCII code symbols for the 128 characters.

The other commonly used teletype code is the Baudot code (*see* BAUDOT CODE). The speed of transmission of ASCII or Baudot is called the baud rate (*see* BAUD RATE). If one unit pulse is *s* seconds in length, then the baud rate is defined as l/*s*. For example, a baud rate of 100 represents a pulse length of 0.01 second, or 10 ms. The speed of ASCII transmission in words per minute, or WPM (*see* WORDS PER MINUTE) is the same as the baud rate. Commonly used ASCII data rates range from 110 to 19,200 baud, as shown in Table 2.

ASPECT RATIO

The aspect ratio of a rectangular image is the ratio of its width to its height. For television in the United States, the aspect ratio of a picture is 4:3. That is, the picture is 1.33 times as wide as it is high. This ratio must be maintained in a television receiver or distortion of the picture will result. *See also* TELEVISION.

ASCII Table 1: Symbols for ASCII teleprinter code.

First Four Signals	Last Three Signals							
	000	001	010	011	100	101	110	111
0000	NUL	DLE	SPC	0		P	/	p
0001	SOH	DC1	!	1	A	Q	a	q
0010	STX	DC2	"	2	B	R	b	r
0011	ETX	DC3	#	3	C	S	c	s
0100	EOT	DC4	$	4	D	T	d	t
0101	ENQ	NAK	%	5	E	U	e	u
0110	ACK	SYN	&	6	F	V	f	v
0111	BEL	ETB	'	7	G	W	g	w
1000	BS	CAN	(	8	H	X	h	x
1001	HT	EM	)	9	I	Y	i	y
1010	LF	SUB	*	:	J	Z	j	z
1011	VT	ESC	+	;	K	[	k	{
1100	FF	FS	,	<	L	/	l	/
1101	CR	GS	—	=	M	]	m	}
1110	SO	RS	.	>	N		n	~
1111	SI	US	|	?	O	—	o	DEL

ACK: Acknowledge
BEL: Bell
BS: Back space
CAN: Cancel
CR: Carriage return
DC1: Device control no. 1
DC2: Device control no. 2
DC3: Device control no. 3
DC4: Device control no. 4
DEL: Delete
DLE: Data link escape
ENQ: Enquiry
EM: End of medium
EOT: End of transmission
ESC: Escape
ETB: End of transmission block
ETX: End of text
FF: Form feed
FS: File separator
GS: Group separator
HT: Horizontal tab
LF: Line feed
NAK: Do not acknowledge
NUL: Null
RS: Record separator
SI: Shift in
SO: Shift out
SOH: Start of heading
SPC: Space
STX: Start of text
SUB: Substitute
SYN: Synchronous idle
US: Unit separator
VT: Vertical tab

ASCII Table 2: Speed rates for the ASCII code.

Baud Rate	Length of Pulse, milliseconds	WPM
110	9.09	110
150	6.67	150
300	3.33	300
600	1.67	600
1200	0.833	1200
1800	0.556	1800
2400	0.417	2400
4800	0.208	4800
9600	0.104	9600
19,200	0.052	19,200

ASSEMBLER and ASSEMBLY LANGUAGE

An assembler is a computer program designed to produce a machine-language program (*see* MACHINE LANGUAGE). The high-level programming languages generally used by a computer operator, such as BASIC, FORTRAN, and COBOL, are not directly used by the computer; the computer understands only machine language. It is extremely difficult and tedious to compose an entire computer program in machine language. The assembler translates the higher-order operator language into machine language. This makes computer programming much easier and faster.

An assembler program must be written for translating a given higher-order language into machine language. Assembly language is used to compose this program. Various higher-order languages require different assembler programs; BASIC is not the same, for example, as FORTRAN. Therefore the assembly language for each kind of higher-order language is unique. *See also* COMPUTER, HIGHER-ORDER LANGUAGE.

ASSOCIATIVE STORAGE

Data can be stored in memory by assigning it an address (*see* ADDRESS) and recalling the information by calling the appropriate address. Alternatively, data can be stored according to part or all of its actual contents. This latter method of storing information is called associative storage.

Suppose, for example, that a schedule of events, by month, is stored according to the associative method. Then, to get a printout of the schedule for January, the input command might read "JAN," or "JANUARY," or "JAN92." Without associative storage, each month would require a separate address, such as A01 through A12 for a given year.

Associative storage has obvious advantages for situations in which a computer is operated by personnel with little or no training. Associative storage makes a computer more friendly.

ASTABLE MULTIVIBRATOR

An astable circuit is a form of oscillator. (The word *astable* means unstable). An astable multivibrator consists of two tubes or transistors arranged so that the output of one is fed directly to the input of the other. Two identical resistance-capacitance networks determine the frequency at which oscillation will occur. The amplifiers are connected in a common-cathode or common-emitter configuration as shown in the diagram.

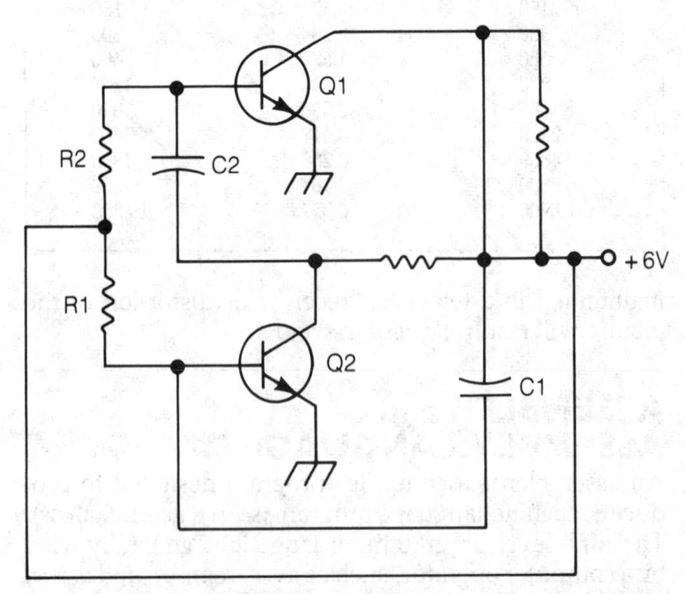

ASTABLE MULTIVIBRATOR: This two-transistor circuit functions as an audio oscillator.

In the common-cathode or common-emitter circuit, the output of each tube or transistor is 180 degrees out of phase with the input. An oscillating pulse may begin, for example, at the base of Q1 in the illustration. It is inverted at the collector of Q1, and goes to the base of Q2. It is again inverted at the collector of Q2, and therefore returns to the base of Q1 in its original phase. This produces positive feedback, resulting in sustained oscillation.

The astable multivibrator is frequently used as an audio oscillator, but is not often seen in rf applications because its output is extremely rich in harmonic products. *See also* MULTIVIBRATOR, OSCILLATOR.

ASTIGMATISM

Astigmatism is the condition of an image in focus in one direction and out of focus in another direction. In optics, this applies to lenses. Astigmatism can occur in the human eye. In electronics, it can apply to the electron beams in a cathode-ray tube (*see* CATHODE-RAY TUBE), or to dish and paraboloid antennas (*see* DISH ANTENNA, PARABOLOID ANTENNA in the ANTENNA DIRECTORY).

With no horizontal or vertical signal input, the electron beam in an oscilloscope cathode-ray tube should focus as a sharp round spot (A). It should remain that way when the beam is deflected to any part of the screen. If the beam is out of focus, the spot will appear blurred and larger than normal. A focusing control is provided in most oscilloscopes to ensure that the beam is maintained in the sharpest possible focus.

If the vertical and horizontal focal points differ, however, astigmatism is produced and the beam will strike a spot with distorted shape (B). Focusing then becomes impossible; when the correct adjustment is reached in one plane, it will be incorrect in the other plane. Astigmatism degrades the accuracy of oscilloscope measurements. It causes hazy areas on a television picture screen, and blurred regions on a cathode-ray tube alphanumeric display. Astigmatism can be corrected by careful alignment of the peripheral circuitry of a cathode-ray tube.

In a dish or paraboloid antenna, astigmatism widens the antenna pattern in one plane. This results in less effective concentration of the radiated energy, and therefore less power gain, than the rated values. This condition is caused by physical distortion of the antenna reflector. *See also* ANTENNA POWER GAIN.

ASYMMETRICAL CONDUCTIVITY

Ordinarily, a conductor carries direct current equally well throughout its cross section. Electrons flow just as easily on one side of a wire or length of tubing as on the opposite side. This is not generally true, however, for the earth. Current often follows an irregular path between

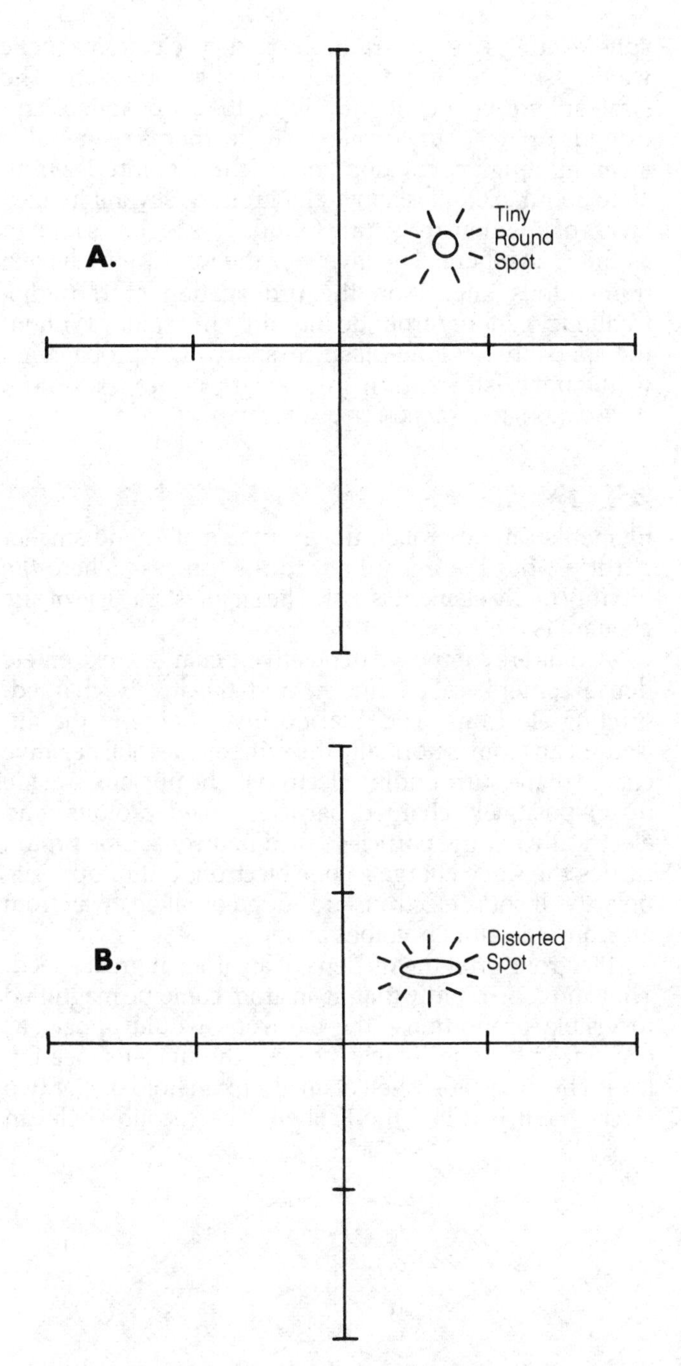

ASTIGMATISM: A properly adjusted oscilloscope displays a round spot (A), but if astigmatism is present the spot is elongated (B).

two ground electrodes with a potential difference. This is asymmetrical conductivity of the ground.

When an alternating current passes through a conductor such as metal wire or tubing, the flow of current should ideally be symmetrical about the center of the wire or tubing. Although skin effect (*see* SKIN EFFECT) will cause the conductivity at high frequencies to be better at the periphery than at the center, the nonuniformity is symmetrical about the center. If the conductivity is not symmetrical, the resistance will increase. In a carbon-composition resistor, asymmetrical conductivity is especially undesirable because it causes overheating of some parts of the resistor, even when the component is carrying less than its maximum rated current.

ASYMMETRICAL DISTORTION

In a binary system of modulation, the high and low conditions (or 1 and 0 states) have defined lengths for each bit of information. The modulation is said to be distorted when these bits are not set to the proper duration. If the output bits of one state are too long or too short, compared with the signal input bits, the distortion is said to be asymmetrical.

A simple example of asymmetrical distortion is found in a Morse-code signal. Morse code is a binary modulation system with bit lengths corresponding to the duration of one dot. Ideally, a string of dots (such as the letter *H* in A and B) has high and low states of precisely equal length. An electronic keyer can produce signals of this nature, as illustrated in A. However, because of the shaping network in a continuous wave (CW) transmitter, the high state is often effectively prolonged, as shown in B, because the decay time is lengthened. While the rise or attack time is made slower by the shaping network, the change is much greater on the decay side in most cases. This creates asymmetrical distortion because the dot-to-space ratio, 1:1 at the input, is greater than 1:1 at the transmitter output.

Asymmetrical distortion makes reception of a binary signal difficult and less accurate than would be the case for an undistorted signal. The effect may not be objectionable if it is small; in CW communications, a certain amount of shaping makes a signal more pleasant to the ear. Excessive asymmetrical distortion should be avoided. In teleprinter communications, asymmetrical distortion causes frequent printing errors. An oscilloscope can be used to check for proper adjustment of the high-to-low ratio.

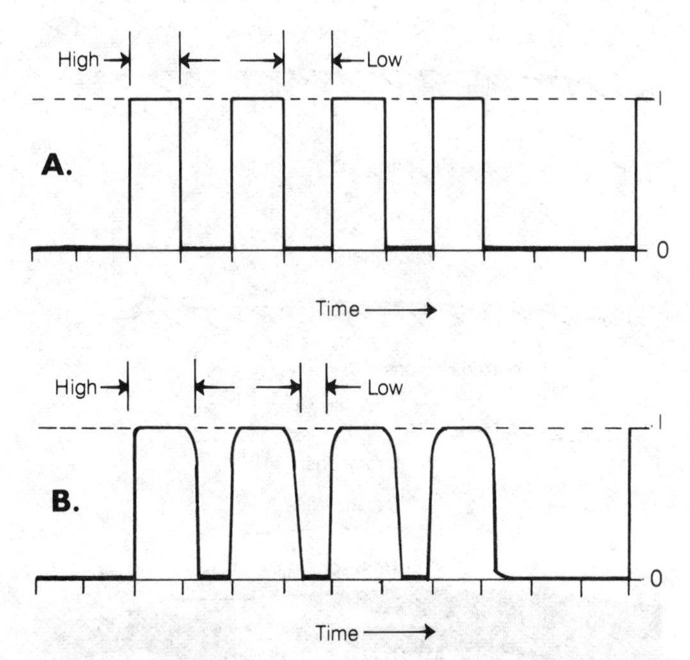

ASYMMETRICAL DISTORTION: An undistorted Morse letter H as it appears on an oscilloscope (A). The same waveform is shown demonstrating asymmetrical distortion (B).

ASYMMETRICAL SIDEBAND

See VESTIGIAL SIDEBAND.

ASYNCHRONOUS DATA

Asynchronous data is information not based on a defined time scale. An example of asynchronous data is manually sent Morse code, or CW. Machine-sent Morse code, in contrast, is synchronous. Synchronous transmission offers a better signal-to-noise ratio in communications systems among machines than asynchronous transmission. Nevertheless, the simple combination of a hand key and human ear for CW—the most primitive communications system—is commonly used when more sophisticated systems will not work.

Asynchronous data need not be as simple as hand-sent Morse code. A manually operated teletypewriter station and a voice system are other examples of asynchronous data transfer. *See also* SYNCHRONOUS DATA.

ATMOSPHERE

The atmosphere is the shroud of gases that surrounds the earth. Other planets also have atmospheres.

Our atmosphere exerts an average pressure of 14.7 pounds per square inch at sea level. As the elevation above sea level increases, the pressure of the atmosphere drops, until it is practically zero at an altitude of 100 miles. Effects of the atmosphere, however, extend to altitudes of several hundred miles.

The atmosphere is defined in terms of three layers, as shown. The lowest layer, the troposphere, is where all weather disturbances take place. It extends to a height of approximately 8 to 10 miles above sea level. The troposphere affects certain radio-frequency electromagnetic waves (*see* DUCT EFFECT, TROPOSPHERIC PROPAGATION). The stratosphere begins at the top of the troposphere and extends up to about 40 miles. No weather is seen in this layer, although circulation does occur. At altitudes from 40 to about 250 miles above the ground, several ionized layers of low-density gas are found. This region is known as the ionosphere. The layers of the ionosphere have a tremendous impact on the propagation of rf (radio-frequency) energy from dc into the vhf region. Without the ionosphere, long-distance shortwave propagation would not exist. *See also* D LAYER, E LAYER, F LAYER, IONOSPHERE, PROPAGATION, PROPAGATION CHARACTERISTICS.

250 Miles
Ionized Regions
Ionosphere
Ionized Regions
40 Miles
Stratosphere
8 to 10 Miles
Troposphere

ATMOSPHERE: The earth's atmosphere extends to an altitude of about 250 miles and is divided into the layers shown.

ATOM

Elements can be broken down into smaller and smaller particles, but there is a limit to this process where the identity of the element is lost. The smallest particle of any element is the atom.

Atoms are composed of positively charged, extremely dense centers, called nuclei, and negatively charged, orbiting electrons. The total positive charge in the nucleus of an atom is normally the same as the total negative charge of the surrounding electrons. The nucleus is made up of positively charged particles called protons, and electrically neutral particles called neutrons. One proton carries the same charge as one electron, but of opposite polarity. If some electrons are added or taken away from an atom, the atom becomes an ion.

Electrons orbit the nucleus of an atom at great speed. They move so rapidly that, if an atom could be magnified to visible proportions, the electrons would appear as concentric spherical shells around the nucleus (see figure). The innermost shell of an atom can hold one or two electrons; it is called the K shell. The second shell can

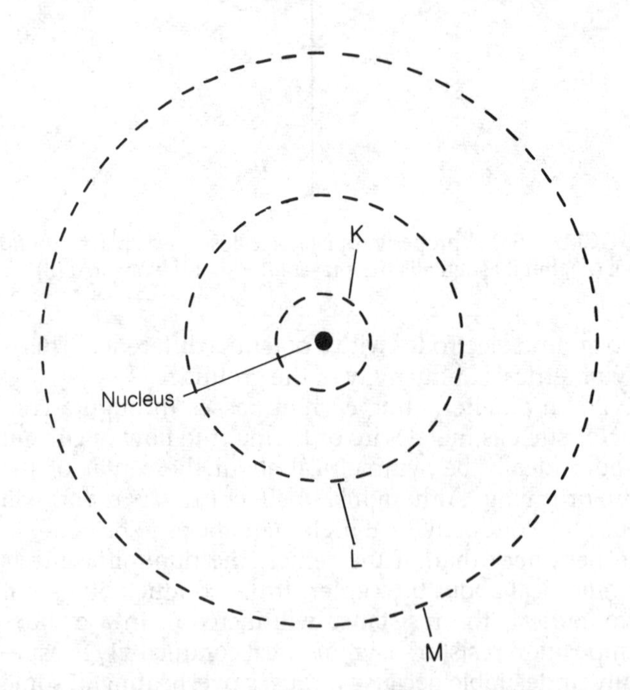

ATOM: A diagram of an atom showing the nucleus and the first three spherical electron shells, K, L, and M.

hold up to eight electrons, and is called the L shell. The third shell can hold up to 18 electrons, and is called the M shell. Heavy atoms may have more than 100 protons and 100 electrons, with several concentric shells. In general, the nth shell of electrons in an atom can have as few as 1 or as many as $2n^2$ electrons.

The negatively charged electron shells of adjacent atoms repel each other. This is why a chair, for example, does not merge with or sink into the floor. The repulsive force among atoms keeps objects from gravitationally collapsing upon themselves. *See also* ELECTRON, NEUTRON, PROTON.

ATOMIC CHARGE

When an atom contains more or less electrons than normal, it is called an ion. The atomic charge of a normal atom is zero. The atomic charge of an ion is positive if there is a shortage of electrons, and negative if there is a surplus of electrons. Some atoms ionize quite easily; others do not readily ionize.

The unit of atomic charge is the amount of electric charge carried by a single electron or proton; they carry equal but opposite charges. This is called an electron unit. One coulomb is a charge of 6.28×10^{18} electron units. Therefore, an electron unit is 1.59×10^{-19} coulomb (*see* COULOMB). An ion might have an atomic charge of $+3$ electron units or -2 electron units. The first case would indicate a deficiency of three electrons; the second case, an excess of two electrons. In general, the atomic charge number is always an integer, since the electron unit is the smallest possible quantity of charge. *See also* ELECTRON.

ATOMIC CLOCK

Time standards throughout the world today are based on atomic clocks. These clocks also function as frequency standards. In the United States, time and frequency are broadcast by WWV in Fort Collins, Colorado, and by WWVH on the island of Kauai, Hawaii, by the National Bureau of Standards (*see* NATIONAL BUREAU OF STANDARDS, WWV/WWVH).

The element cesium is a frequently used atomic standard. The clock functions according to the resonant frequency of the vibration of this element. These clocks are accurate to three parts in 10^{12}, or 3×10^{-10} percent, per year. Other atomic time-standard sources include the hydrogen maser and the rubidium gas cell.

Atomic clocks are generally set according to the mean solar day. This day is divided into 24 equal hours. Corrections are made periodically to compensate for irregularities in the rotational speed of the earth. *See also* COORDINATED UNIVERSAL TIME.

ATOMIC MASS UNIT

An atomic mass unit, or AMU, is defined as precisely 1/16 the mass of a neutral atom of O16, the most common isotope of oxygen (*see* ISOTOPE). One AMU is 1.66×10^{-24} gram, or approximately the mass of a single neutron or proton.

Atomic weights are expressed in atomic-mass units. The lightest element, hydrogen, has an atomic weight of 1 AMU. Lawrencium has an atomic weight of 257 AMU. *See also* ATOMIC WEIGHT.

ATOMIC NUMBER

The atomic number of an element is the number of protons in the nucleus of a single atom of that element. Hydrogen has a nucleus consisting of one proton alone, and thus has an atomic number of 1. Helium has a nucleus made up of two protons and two neutrons; its atomic number is 2. There are elements corresponding to every possible atomic number up to over 100. The atom of uranium, with 92 protons, has the largest atomic number of any naturally occurring element. Man-made elements have greater atomic numbers.

The table lists atomic numbers of elements from 1 to 103. Atomic weights (*see* ATOMIC WEIGHT) are also given.

ATOMIC WEIGHT

The atomic weight of an element is the mass of that element in atomic mass units (*see* ATOMIC MASS UNIT). The atomic weight is approximately equal to the total number of protons and neutrons in the nucleus of an atom. An element of a given atomic number can have several different possible atomic weights, depending on the number of neutrons. These variations in the structure of the nucleus of an element are known as isotopes. For example, an atom of carbon (atomic number 6) usually has six neutrons in its nucleus, giving it an atomic weight of 12. However, some atoms of carbon have eight neutrons instead of six; this isotope is called carbon 14.

As the atomic number increases, the atomic weight generally increases as well. However, there are some instances in which atoms of higher atomic number have smaller atomic weight.

An electron has almost no mass—about 5×10^{-4} AMU—and therefore electrons affect the atomic weight very little. The table for ATOMIC NUMBER shows the atomic weights of the most common isotopes of the elements having atomic numbers from 1 to 103. The atomic weights in the table have been rounded off to the nearest whole number.

ATTACK

The rise time for a pulse is sometimes called the attack or attack time. In music, the attack time of a note is the time required for the note to rise from zero amplitude to full loudness. The attack time for a control system, such as an automatic gain control (*see* AUTOMATIC GAIN CONTROL, AUTOMATIC LEVEL CONTROL), is the time needed for that system to fully compensate for a change in input parameters. The graphics show the attack time for a musical tone (A) and a dc pulse (B).

The attack time of a musical note affects its sound quality. A fast rise time sounds hard, and a slow rise time sounds soft. With automatic gain or level control, an attack time that is too fast may cause overcompensation,

ATOMIC NUMBER AND ATOMIC WEIGHT: Alphabetical list of elements with atomic numbers and atomic weights. (Atomic weights are rounded to nearest whole number.)

Element Name	Atomic Number	Atomic Weight	Element Name	Atomic Number	Atomic Weight
Actinium	89	227	Molybdenum	42	96
Aluminum	13	27	Neodymium	60	144
Americium	95	243	Neon	10	20
Antimony	51	122	Neptunium	93	237
Argon	18	40	Nickel	28	59
Arsenic	33	75	Niobium	41	93
Astatine	85	210	Nitrogen	7	14
Barium	56	137	Nobelium	102	254
Berkelium	97	247	Osmium	76	190
Beryllium	4	9	Oxygen	8	16
Bismuth	83	209	Palladium	46	106
Boron	5	11	Phosphorus	15	31
Bromine	35	80	Platinum	78	195
Cadmium	48	112	Plutonium	94	242
Calcium	20	40	Polonium	84	210
Californium	98	251	Potassium	19	39
Carbon	6	12	Praseodymium	59	141
Cerium	58	140	Promethium	61	145
Cesium	55	133	Proactinium	91	231
Chlorine	17	35	Radium	88	226
Chromium	24	52	Radon	86	222
Cobalt	27	59	Rhenium	75	186
Copper	29	64	Rhodium	45	103
Curium	96	247	Rubidium	37	85
Dysprosium	66	163	Ruthenium	44	101
Einsteinium	99	254	Samarium	62	150
Erbium	68	167	Scandium	21	45
Europium	63	152	Selenium	34	79
Fermium	10	257	Silicon	14	28
Fluorine	9	19	Silver	47	108
Francium	87	223	Sodium	11	23
Gadolinium	64	157	Strontium	38	88
Gallium	31	70	Sulfur	16	32
Germanium	32	73	Tantalum	73	181
Gold	79	197	Technetium	43	99
Hafnium	72	178	Tellurium	52	128
Helium	2	4	Terbium	65	159
Holmium	67	165	Thallium	81	204
Hydrogen	1	1	Thorium	90	232
Indium	49	115	Thulium	69	169
Iron	26	56	Tin	50	119
Krypton	36	84	Titanium	22	48
Lanthanum	57	139	Tungsten	74	184
Lawrencium	103	257	Uranium	92	238
Lead	82	207	Vanadium	23	51
Lithium	3	7	Xenon	54	131
Lutetium	71	175	Ytterbium	70	173
Magnesium	12	24	Yttrium	39	89
Manganese	25	55	Zinc	30	65
Mendelevium	101	256	Zirconium	40	91
Mercury	80	201			

while an attack time that is too slow will cause a loud popping sound at the beginning of each pulse or syllable.

The time required for a note or pulse to drop from full intensity back to zero amplitude is called the decay or release time. *See also* DECAY, DECAY TIME.

ATTENUATION

Attenuation is the decrease in amplitude of a signal between any two points in a circuit. It is usually expressed in decibels (*see* DECIBEL). Attenuation is the opposite of amplification (*see* AMPLIFICATION), and can be defined for voltage, current, or power. Occasionally the attenuation in a particular circuit is expressed as a ratio. For example, the reduction of a signal in amplitude from ±5 volts peak to ± 1 volt peak is an attenuation factor of 5.

In general, the attenuation in voltage for an input E_{IN} and an output E_{OUT} is:

$$\text{Attenuation (dB)} = 20 \log_{10} (E_{IN}/E_{OUT})$$

Similarly, the current attenuation for an input of I_{IN} and an output of I_{OUT} is:

$$\text{Attenuation (dB)} = 20 \log_{10} (I_{IN}/I_{OUT})$$

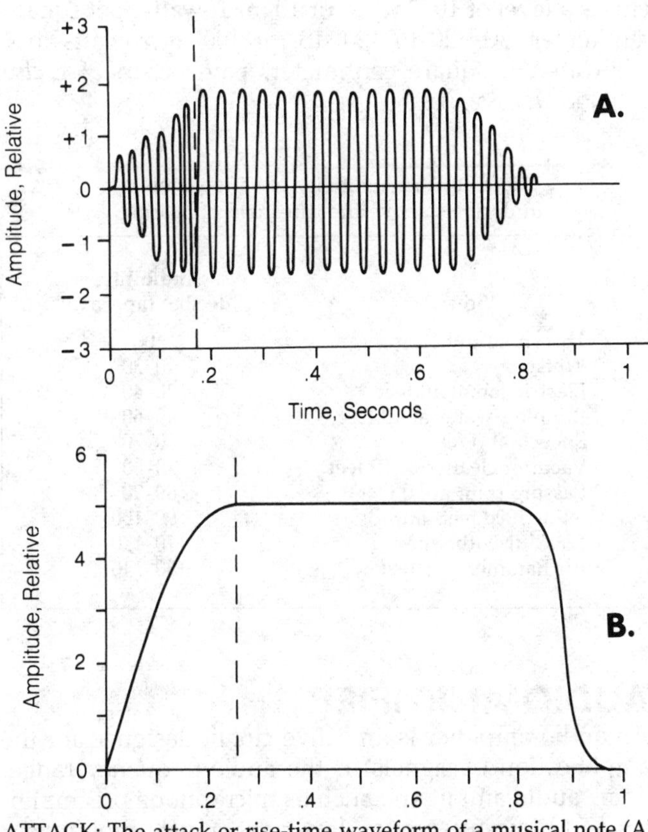

ATTACK: The attack or rise-time waveform of a musical note (A) and the attack waveform of a direct-current pulse (B).

For power, given an input of P_{IN} watts and an output of P_{OUT} watts, the attenuation in decibels is given by:

$$\text{Attenuation (dB)} = 10 \log_{10} (P_{IN}/P_{OUT})$$

If the amplification factor (*see* AMPLIFICATION FACTOR) is X dB, then the attenuation is −X dB. That is, positive attenuation is the same as negative amplification, and negative attenuation is the same as positive amplification. *See also* ATTENUATOR.

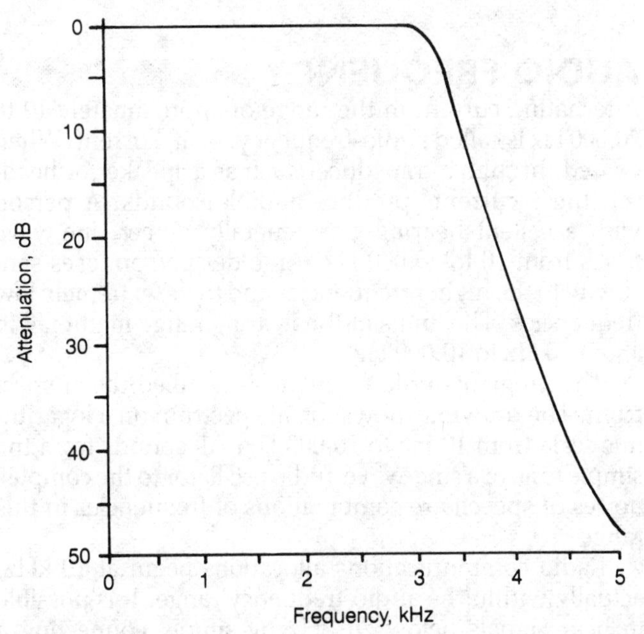

ATTENUATION: The plot of attenuation versus frequency for a lowpass filter.

ATTENUATION DISTORTION

Attenuation distortion is an undesirable attenuation characteristic over a particular range of frequencies (*see* ATTENUATION VS FREQUENCY CHARACTERISTIC). This can occur in radio-frequency as well as audio-frequency applications. The low-pass response of the illustration in ATTENUATION is an advantage for a voice communications circuit because most of the frequencies in the human voice fall below 3 kHz. However, for the transmission of music, the curve in the illustration would represent a circuit with objectionable attenuation distortion, because music contains audio frequencies as high as 20 kHz or more.

In sound systems, attenuation distortion depends to some extent on individual preferences. An equalizer (*see* EQUALIZER) is sometimes used to eliminate the attenuation distortion for a particular listener or application.

Attenuation distortion in a sound system can be the result of poor circuit design, use of improper speakers or headsets, use of poor-quality tape for recording, or excessively slow tape speed, and many other factors. *See also* HIGH FIDELITY.

ATTENUATION EQUALIZER

See EQUALIZER.

ATTENUATION vs FREQUENCY CHARACTERISTIC

The attenuation-vs-frequency characteristic of a circuit is the amount of loss through the circuit as a function of frequency. This function is generally shown with the amplitude, in decibels relative to a certain reference level, on the vertical scale and the frequency on the horizontal scale. The illustration in ATTENUATION shows the attenuation-versus-frequency characteristic curve for an audio low pass filter. This filter is designed for use as an intelligibility enhancer for a voice-communications circuit.

Various circuits are used for precise adjustment of the attenuation-vs-frequency characteristic in different situations. In high-fidelity recording equipment, the familiar equalizer or graphic equalizer (*see* EQUALIZER) is often used in place of simple bass and treble controls for obtaining exactly the desired audio response. In some rf (radio-frequency) amplifiers, it is advantageous to introduce a lowpass, highpass, or bandpass response to prevent oscillation or excessive radiation of unwanted energy. *See also* BANDPASS FILTER, BAND-REJECTION FILTER, HIGHPASS FILTER, LOWPASS FILTER.

ATTENUATOR

An attenuator is a network, usually passive rather than active, designed to cause a reduction in the amplitude of a signal. These circuits are useful for sensitivity measurements and other calibration purposes. In a well-designed attenuator, the amount of attenuation is constant over the entire range of frequencies in the system to be checked. An attenuator introduces no reactance, and therefore no phase shift. Attenuators are made for a wide variety of input and output impedances. It is important that the impedances be properly matched, or the attenuator will not function properly.

Two simple passive attenuators made from noninductive resistors are shown in A and B. The circuit at A is called a pi-network attenuator, and the circuit at B is called a T-network attenuator. These circuits will function from the audio-frequency range well into the vhf (very high frequency) spectrum. At uhf and above, however, the resistors begin to show inductive reactance because the wavelength is so short that the leads are quite long electrically. Then the attenuators will no longer perform their intended function.

Attenuators for uhf must be designed especially to suit the short wavelengths at those frequencies. Such circuits must be physically small.

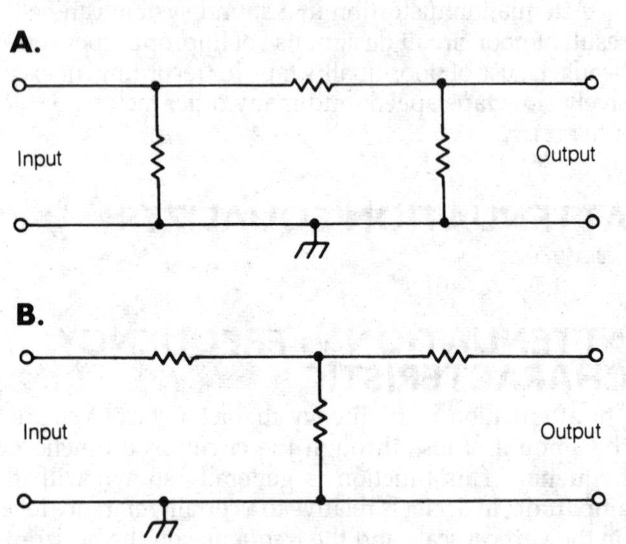

ATTENUATOR: Attenuators made with noninductive resistors are a pi network (A), and the T network (B).

AUDIBILITY

Audibility is a measure of the intensity of sound. It is expressed in decibels relative to the threshold of hearing—the weakest sound level that can be heard by the human ear. In a soundproof chamber, the audibility is 0 decibels. A certain amount of background noise is created by the motion of blood through the capillaries of the ear. Some noise is also caused by the random motion of air molecules around the eardrum.

The table shows the audibility levels, in decibels, for some familiar sounds. The threshold of hearing for an average person is considered to be 10^{-16} watt per square centimeter. The audibility is given as:

$$\text{Audibility (dB)} = 10 \log_{10} (X/10^{-16}) = 160 + 10 \log_{10} X,$$

where X is the intensity of the given sound in watts per square centimeter. At 110 dB, tingling is felt in the ear. This is a level of 10^{-5} watt, or 10 microwatts, per square centimeter. At 120 to 130 dB, or 100 microwatts to 1 milliwatt per square centimeter, pain occurs. *See also* LOUDNESS, VOLUME, VOLUME UNIT.

AUDIBILITY: Audibility of various noises in decibels above the threshold of hearing.

Sound	Audibility, decibel (approx.)
Threshold of hearing	0
Whisper	10–20
Electric fan at 10 feet	30–40
Running water at 10 feet	40–60
Speech at 5 feet	60–70
Vacuum cleaner at 10 feet	70–80
Passing train at 50 feet	80–90
Jet at 1000 feet altitude	90–100
Loud discotheque	110–120
Air hammer at 5 feet	130–140

AUDIO AMPLIFIER

An audio amplifier is an active circuit designed for the amplification of signals in the audio-frequency range. Some audio amplifiers, such as microphone preamplifiers and radio-receiver audio stages, must amplify very small signal input, but they need not generate much output power. Other audio amplifiers, such as those found in high-fidelity or public-address systems, must develop large amounts of power output, sometimes thousands of watts.

In communications systems, the frequency response of an audio amplifier is restricted to a relatively narrow range. But in a high-fidelity system, a flat response is desired from perhaps 10 Hz to well above the range of human hearing. *See also* AUDIO POWER, AUDIO RESPONSE, HIGH FIDELITY, OPERATIONAL AMPLIFIER.

AUDIO FREQUENCY

Alternating current in the range of approximately 10 to 20,000 Hz is called audio-frequency, or af, current. When passed through a transducer such as a speaker or headset, these currents produce audible sounds. A person with excellent hearing can generally detect sine-wave tones from 10 to 20,000 Hz. An older person loses sensitivity to the higher frequencies and to the extremely low frequencies. The limits of the hearing range might fall to about 40 Hz to 10,000 Hz.

The range of audio frequencies is called the af spectrum. The drawing shows the af spectrum on a logarithmic scale from 10 Hz to 20,000 Hz. All sounds from the simple tone of a sine-wave audio oscillator to the complex noises of speech are combinations of frequencies in this range.

Radio communications allocations begin at 10 kHz, actually within the audio-frequency range. It is possible to hear signals below 20 kHz by simply connecting a sensitive audio amplifier to an antenna.

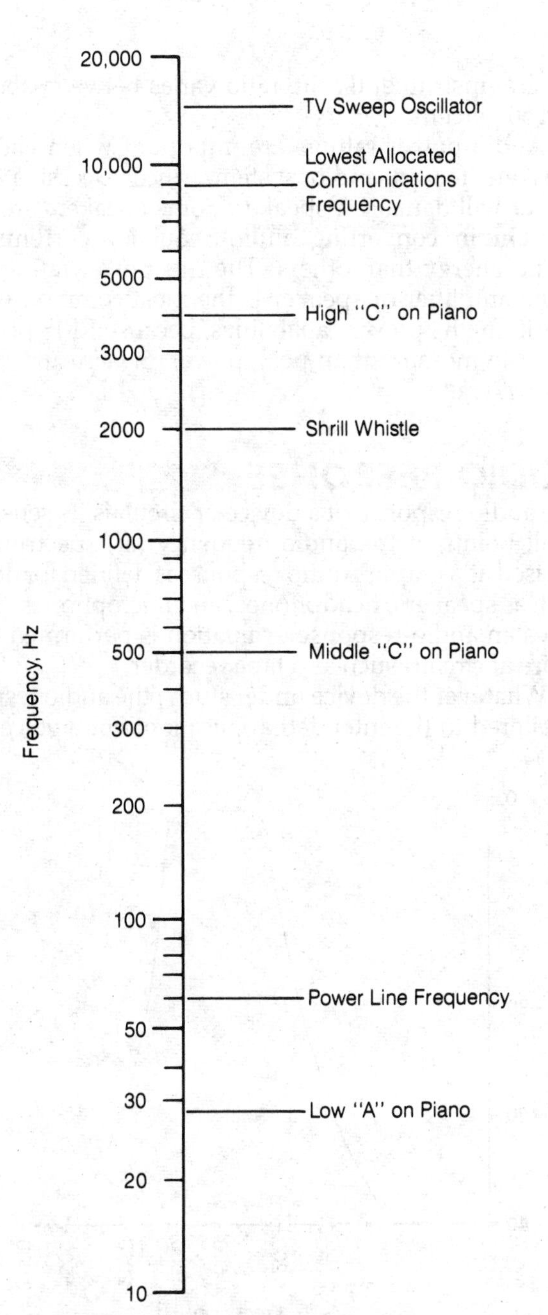

AUDIO FREQUENCY: The audio frequency band covers the range of 10 Hz to 20 kHz.

AUDIO-FREQUENCY TRANSFORMER

An audio-frequency transformer is used for matching impedances at audio frequencies. The output of an audio amplifier might have an impedance of 200 Ω, and the speaker or headset an impedance of only 8 Ω. The af (audio frequency) transformer provides the proper termination for the amplifier. This assures the most efficient possible transfer of power.

The impedance-matching ratio of a transformer is proportional to the square of the turns ratio. Thus, if the primary winding has N_{PRI} turns and the secondary winding has N_{SEC} turns, the ratio of the primary impedance to the secondary impedance, Z_{PRI} to Z_{SEC}, is:

$$\frac{Z_{PRI}}{Z_{SEC}} = \left(\frac{N_{PRI}}{N_{SEC}}\right)^2$$

Also:

$$\frac{N_{PRI}}{N_{SEC}} = \sqrt{\frac{Z_{PRI}}{Z_{SEC}}}$$

Audio-frequency transformers are available in various power ratings and impedance-matching ratios. There are also different kinds of audio-frequency transformers to meet various frequency-response requirements. These transformers are similar to ordinary alternating-current power transformers. They are wound on laminated or powdered-iron cores. *See also* TRANSFORMER.

AUDIO IMAGE

An audio image is a sound that comes from, or appears to come from, a certain point in space. Audio images are created by high-fidelity stereo systems because of the relative intensity and phase of the sound in the channels. By proper synthesis of amplitude and phase in the left-hand and right-hand channels, such sounds as a busy street, a passing train, or a concert orchestra are faithfully reproduced in a pair of speakers or stereo headphones.

Audio images always have an apparent distance and an apparent direction. Sometimes there is difficulty in determining the distance to a sound image; a loud sound that is far away sounds very similar to a weaker sound nearby. There might be ambiguity in direction if two sounds from different places arrive at the ears with the same relative phase and amplitude. It is very hard, for example, to distinguish (with the eyes closed) among voices directly in front, behind, above, and below. All these directions are equidistant from both ears, and so the sound arrives in the same phase and at the same amplitude in all cases. *See also* STEREOPHONICS.

AUDIO LIMITER

An audio limiter is a circuit that prevents the amplitude of an audio-frequency signal from exceeding a limiting value. All audio-frequency voltages below the limiting value are not affected. A limiter can be either active or passive.

The illustration shows a simple diode limiter. If germanium diodes are employed, the limiting voltage is approximately ±0.3 volt peak. If silicon diodes are used, the limiting voltage is about ±0.6 volt peak. Larger limiting voltages are obtained by connecting diodes in series.

Diode limiters, as well as most other kinds of audio limiters, cause considerable distortion at volume levels above the limiting voltage. If fidelity is important, an automatic gain or level control is preferable to audio limiting (*see* AUTOMATIC GAIN CONTROL, AUTOMATIC LEVEL CONTROL). In radio communications transmitters, audio limiters are sometimes used in conjunction with amplifiers and lowpass filters to increase the average talk power. The signal is first amplified, and then the limiter is employed to prevent overmodulation. Finally, a lowpass filter prevents excessive distortion products from being transmitted over the air. This is called speech clipping (*see* SPEECH CLIPPING). Under poor conditions, speech clipping can result in improved intelligibility and articulation (*see* ARTICULATION, INTELLIGIBILITY).

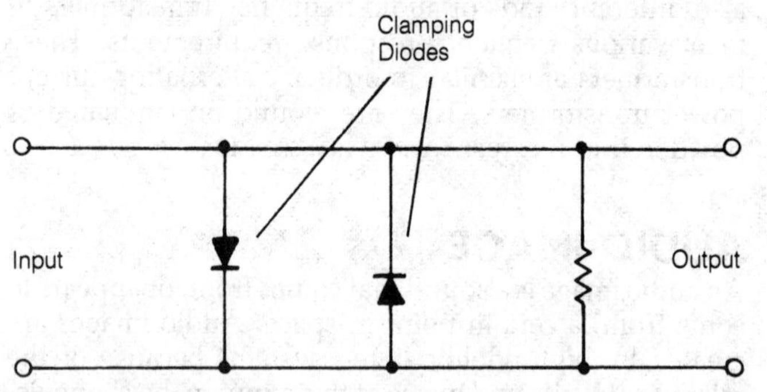

AUDIO LIMITER: Two semiconductor diodes in reverse parallel perform audio limiting.

AUDIO MIXER

An audio mixer is a circuit that combines two or more audio signals. It has an input terminal, or connection, for each input source. It also has an output signal terminal. Level controls are provided independently for each input, and also for the output.

Combining audio signals is not a simple matter of joining wires. Unless the proper steps are taken to avoid interaction among the input signal sources, distortion and poor efficiency will result. The impedances at the input terminals of an audio mixer must remain constant, and at the required values, as the level controls are adjusted. The impedances must also remain constant at all frequencies and under conditions of varying signal amplitude. The output impedance must also remain constant. This requires careful design. Nonlinearity of the components must be minimized. The frequency response, or attenuation-vs-frequency characteristics, must be as flat as possible. Phase distortion must also be avoided. *See also* ATTENUATION VS FREQUENCY CHARACTERISTIC.

AUDIO POWER

Audio power is the power, in watts, dissipated in the output load of an audio amplifier. If the speakers or headphones of an audio system were replaced by nonreactive resistors of an identical impedance, all the audio power would be spent as heat. It could then be easily and accurately measured. Voltmeters are generally used in audio systems to measure power. They are calibrated in watts according to the impedance of the output.

Audio power can be expressed either as root-mean-square, or rms, watts (*see* ROOT MEAN SQUARE) or as peak watts. For a pure sine-wave tone, the rms power is 50 percent of the peak power. For a human voice, the rms power is generally 15 to 25 percent of the peak power. For musical instruments, the ratio varies between about 20 and 50 percent.

Audio power ratings are important when choosing speakers for an audio system, since excessive audio power will damage a speaker. Some speakers are more efficient in converting audio-frequency currents into sound energy than others. The rms power rating of an audio amplifier or speaker is the most common way of specifying its power capabilities, because RMS power is easier to measure than peak power. *See also* AUDIO AMPLIFIER, SPEAKER.

AUDIO RESPONSE

The audio response of a device or circuit is its sensitivity at all points in the audio-frequency (af) spectrum, expressed as a graph. Audio response is defined for devices such as speakers, headphones, and microphones. Sometimes an audio-response evaluation is performed on an entire af circuit, such as a tape recorder.

Whatever the device under study, the audio response is tailored to the intended application. For a voice com-

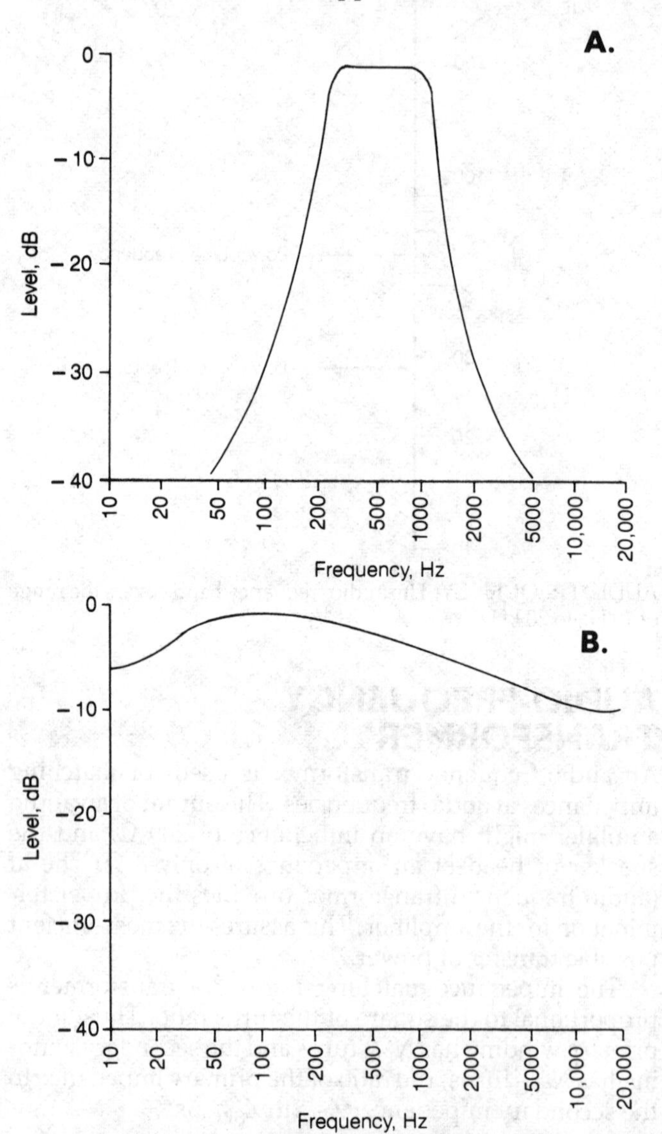

AUDIO RESPONSE: A response curve suitable for voice communications (A), and one suitable for music reproduction (B).

munications system, the audio response of a microphone or speaker should be similar to that shown in A. In a high-fidelity application, however, this response would be unsatisfactory. The more nearly flat response of B is essential for the faithful reproduction of music.

In filtering networks with no gain, the frequency response is sometimes called the attenuation-vs-frequency characteristic (*see* ATTENUATION VS FREQUENCY CHARACTERISTIC).

The audio response of a communications system is usually fixed. In a high-fidelity system, however, it is variable. A bass-treble, or tone, control may consist of one or two potentiometers to vary the low-frequency and high-frequency amplification. A more precise adjustment of the audio response in a high-fidelity system is accomplished by a circuit called an equalizer. This circuit allows control of several audio-frequency ranges independently. *See also* BASS, EQUALIZER, TONE CONTROL, TREBLE.

AUDIO SIGNAL GENERATOR

An audio signal generator, sometimes also called an audio function generator, is used in the testing, troubleshooting, and adjustment of audio-frequency circuits.

Modern audio signal generators can produce a variety of waveforms. Sine and square waves are the most common, but sawtooth and triangular waves can also be obtained with many of the units (*see* SAWTOOTH WAVE, SINE WAVE, SQUARE WAVE, TRIANGULAR WAVE). The frequency range may extend well above and below the human hearing range. A typical audio signal generator costs a few hundred dollars.

By using an audio signal generator with output voltage independent of the frequency, audio-frequency response tests can be performed (*see* AUDIO RESPONSE). A special voltmeter for audio-frequency amplitude measurement, or an oscilloscope, is used to check the output. As the frequency is varied, the output amplitude can be plotted on a graph. The resulting curve clearly shows the audio response of the device under test.

A specialized form of audio signal generator/voltmeter measures the amount of distortion produced in an audio amplifier. A pure sine-wave tone is supplied at the input of the amplifier and is notched out at the output, leaving only the harmonic components. The combined amplitude of these components is then measured. This kind of test instrument is called a distortion analyzer (*see* DISTORTION ANALYZER).

The most sophisticated instruments can generate many different waveforms, simulating various musical instruments. The Moog synthesizer is an example of this kind of generator (*see* MOOG SYNTHESIZER). This audio signal generator is not normally used as a test instrument, but for making electronic music.

AUDIO TAPE

See RECORDING TAPE.

AUDIO TAPER

The human ear does not perceive sound intensity linearly. Instead, the apparent intensity, or audibility, of a sound increases according to the logarithm of the actual amplitude in watts per square centimeter. An increase of approximately 26 percent is the smallest detectable change for the human ear, and is called the *decibel*, abbreviated dB (*see* AUDIBILITY, DECIBEL). A threefold increase in sound power is 5 dB; a tenfold increase is 10 dB, and a hundredfold increase represents 20 dB.

The audio-taper volume control compensates for the nonlinearity of human hearing. A special kind of potentiometer is used to give the impression of a linear increase in audibility as the volume is raised. Most radio receivers, tape players, and high-fidelity systems use audio-taper potentiometers in their af amplifier circuits. This potentiometer is sometimes also called a logarithmic-taper, or log-taper, component.

For linear applications, a linear-taper potentiometer is needed. Use of a linear-taper potentiometer is undesirable in a nonlinear situation. The converse is also true; an audio-taper device does not function well in a linear circuit. *See also* LINEAR TAPER.

AUDIO-VISUAL COMMUNICATION

An audio-visual communication system uses both sight and sound to convey information. The most common example of this is television.

In an audio-visual communications system, it is important that the visible image by synchronized with the sound. This usually presents no problem unless the information is recorded.

A form of audio-visual communication requiring very little spectrum space is the slow-scan television and single-sideband (*see* SINGLE SIDEBAND) system. The picture is sent by one sideband, at the rate of one still frame every seven seconds; the voice is sent over the other sideband. This requires only about 6 kHz of spectrum, or the space taken by an average AM (amplitude modulation) voice broadcast signal.

Audio-visual telephones are already available. However, a wide audio spectrum is required for transmission of a detailed motion picture.

AURORA

A sudden eruption on the sun, called a solar flare, causes high-speed atomic particles to be ejected far into space from the surface of the sun. Some of these particles are drawn to the upper atmosphere over the north and south magnetic poles of the earth. This results in dramatic ionization of the thin atmosphere over the Arctic and Antarctic. On a moonless night, the glow of the ionized gas can be seen as the Aurora Borealis (Northern Lights) and Aurora Australis (Southern Lights).

The aurora are important in radio communication because they affect the propagation of electromagnetic waves in the high-frequency and very high frequency ranges. Signals might be returned to earth at much higher frequencies than usual when the aurora are active. The unusual solar activity that causes the aurora, however, also can degrade normal ionospheric propagation. Sometimes the disturbance is so pronounced that power lines are disrupted by the changing magnetic fields. *See also* AURORAL PROPAGATION, IONOSPHERE, PROPAGATION, PROPAGATION CHARACTERISTICS, SUNSPOT, SUNSPOT CYCLE.

AURORAL PROPAGATION

In the presence of unusual solar activity, the aurora (*see* AURORA) often return electromagnetic energy to earth. This is called auroral propagation. See illustration.

The aurora occur in the ionosphere, at altitudes of about 40 to 250 miles above the surface of the earth (*see* IONOSPHERE). Theoretically, auroral propagation is possible, when the aurora are active, between any two points on the ground from which the same part of the aurora can be seen. Auroral propagation seldom occurs when one end of the circuit is at a latitude less than 35 degrees north or south of the equator.

Auroral propagation is characterized by a rapid flutter, which makes voice signals such as AM, FM, or SSB unreadable most of the time. Morse code is generally most effective in auroral propagation, and even this mode may sometimes be heard but not understood because of the extreme distortion. The flutter is the result of severe selective fading and multipath distortion (*see* MULTIPATH FADING, SELECTIVE FADING). Auroral propagation can take place well into the very high frequency region, and is usually accompanied by moderate to severe deterioration in ionospheric conditions of the E and F layers. *See also* E LAYER, F LAYER, PROPAGATION, PROPAGATION CHARACTERISTICS.

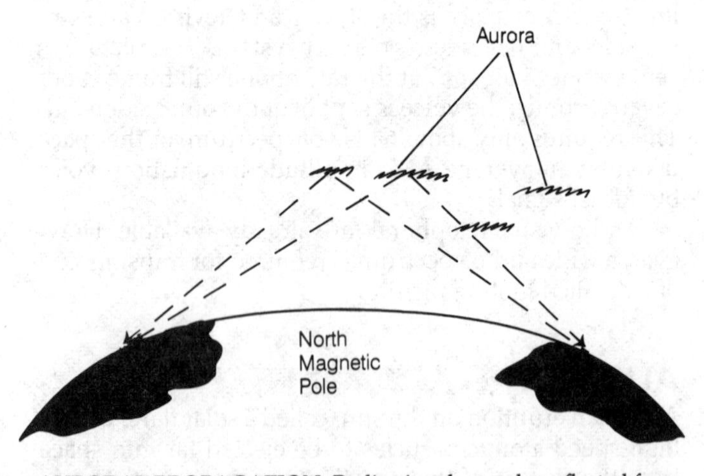

AURORAL PROPAGATION: Radio signals may be reflected from the ionized auroral "curtains" over both north and south poles of the earth.

AUTOALARM

An autoalarm is a receiving device tuned to 500 kHz, the international distress frequency. Whenever a signal is received on this frequency, the autoalarm is activated. The alarm produces an antenna signal, and then the receiver is switched to a speaker so that the signal can be heard.

AUTODYNE

An autodyne is a combined oscillator and mixer using a field-effect transistor. The oscillator can be crystal-controlled or variable in frequency. See the simple autodyne circuit diagram.

The source, gate, and drain form the oscillator, in this case a Hartley type. The signal to be mixed with the oscillator frequency is applied to gate 2.

The autodyne is often used as a detector in a direct-conversion receiver, or regenerative receiver (*see* DIRECT-CONVERSION RECEIVER, REGENERATIVE DETECTOR). The incoming signal beats with the oscillator signal, forming audio-frequency mixing products. This af is then amplified using ordinary audio amplifiers. The autodyne is sometimes used as a mixer when circuit simplicity is important. The autodyne replaces both the local oscillator and mixer stages in a receiver.

AUTOMATIC BIAS

Automatic bias is a method of obtaining the proper bias for a tube or transistor with a resistor, usually in the cathode or emitter circuit. A capacitor can also be used in parallel with the resistor, to stabilize the bias voltage. The illustration shows an example of automatic bias. In this case, reverse bias is produced at the emitter-base junction of the transmitter, facilitating the operation of a Class C amplifier.

Other means of developing bias include resistive voltage-dividing networks and special batteries or power supplies. *See also* BIAS, BIAS STABILIZATION, VOLTAGE DIVIDER.

AUTOMATIC BRIGHTNESS CONTROL

An automatic-brightness-control circuit is used in a television receiver to keep the relative brightness of the picture within certain limits. Excessively large changes in picture brightness are objectionable to the eye. Such variations also affect the amount of contrast in the picture. Some combinations of brightness and contrast appear unnatural.

An automatic-brightness control operates in the same manner as automatic-gain and level controls (*see* AUTOMATIC GAIN CONTROL, AUTOMATIC LEVEL CONTROL). An increase in the brilliance of the scene causes the automatic brightness control to dim the picture; a decrease in brilliance causes the automatic brightness control to lighten the picture. The amount of compensation is adjusted so that some variation in picture brightness occurs, but to a lesser extent than the changes in the actual scene. Improperly adjusted automatic brightness control will result in unnatural television pictures. *See also* BRIGHTNESS.

AUTOMATIC CONTRAST CONTROL

Automatic contrast control maintains the proper amount of contrast in a television picture receiver. The contrast of

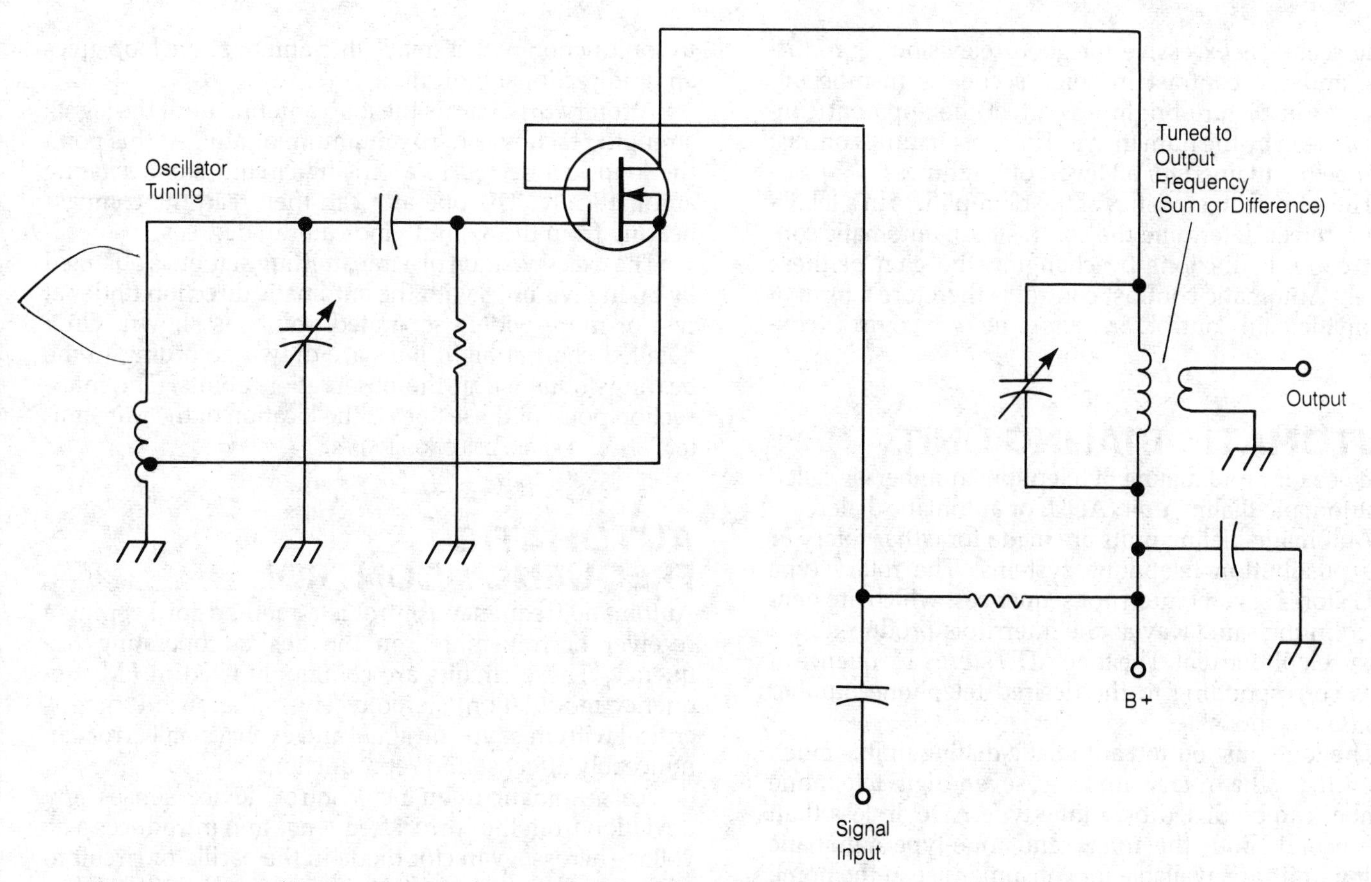

AUTODYNE: An oscillator and mixer in a single stage form an autodyne circuit.

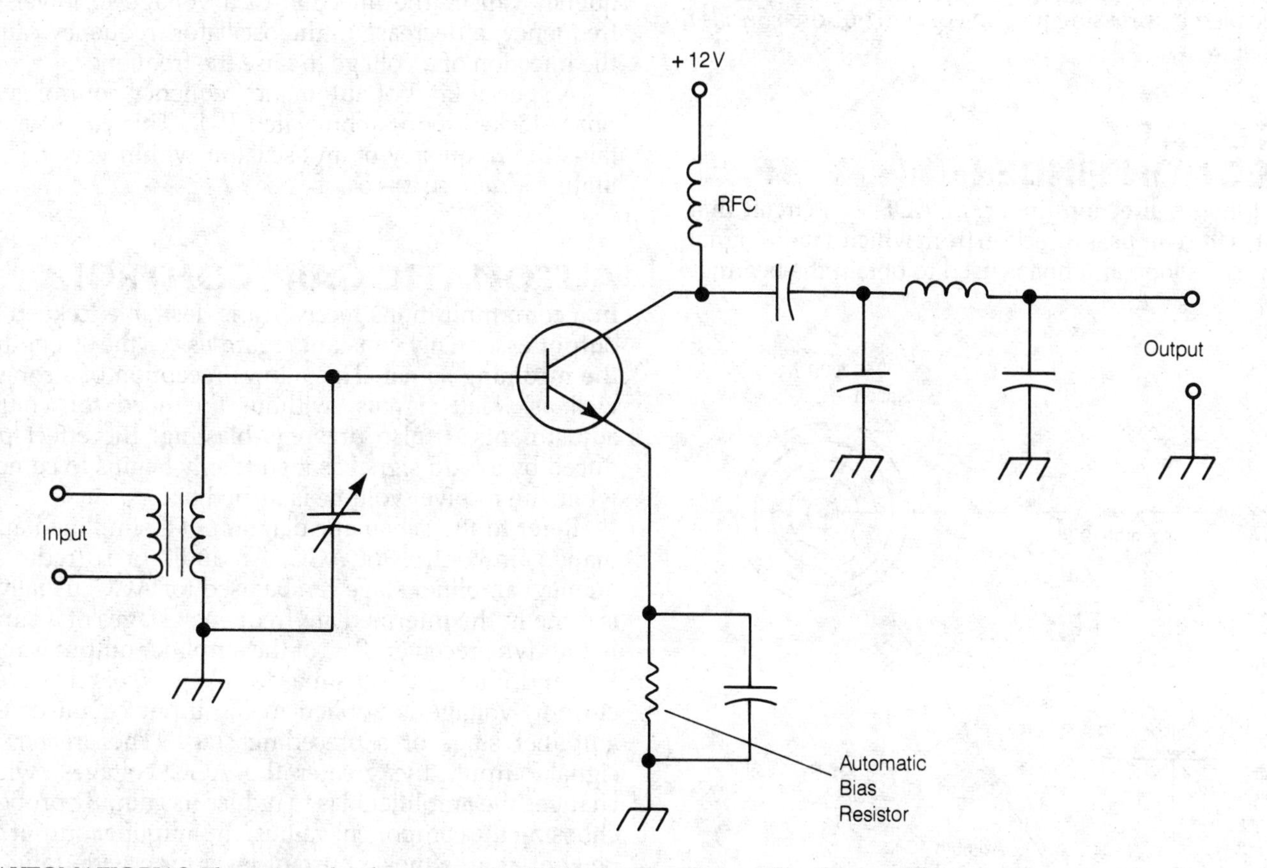

AUTOMATIC BIAS: The resistor develops an automatic bias for Class C operation of an amplifier.

some scenes is excessive for good television reproduction, and the contrast in some scenes is insufficient. Changes in picture brightness affect the apparent contrast as seen by the human eye. The most natural contrast must be maintained for all levels of brightness.

The gain characteristics of the rf amplifiers in a television receiver determine the contrast. An automatic contrast control functions by changing the gain of these stages. Automatic-contrast control is therefore a form of automatic gain control. *See* AUTOMATIC GAIN CONTROL, CONTRAST.

AUTOMATIC-DIALING UNIT

A device for rapid dialing of telephone numbers is called an automatic-dialing unit (ADU) or automatic dialer.

Automatic-dialing units are made for either rotary or tone pushbutton telephone systems. The rotary type ADU stores several interrupt sequences, which are generated in the same way as the interrupts produced by a rotary telephone dial. The tone ADU stores a sequence of tones corresponding to the desired telephone number (*see also* TOUCHTONE®).

The tone pushbutton automatic-dialing unit is much faster than the rotary unit. A seven-digit telephone number can be dialed by a tone-type ADU in less than one second. Both the rotary and tone-type automatic dialing units are available for consumer use in the home or business. Numbers are programmed into the ADU and recalled by pressing just one or two buttons. *See also* TELEPHONE SYSTEM.

AUTOMATIC DIRECTION FINDER

An automatic direction finder, or ADF, is a circuit that indicates the compass direction from which a radio signal is coming. A loop antenna is used to obtain the bearing.

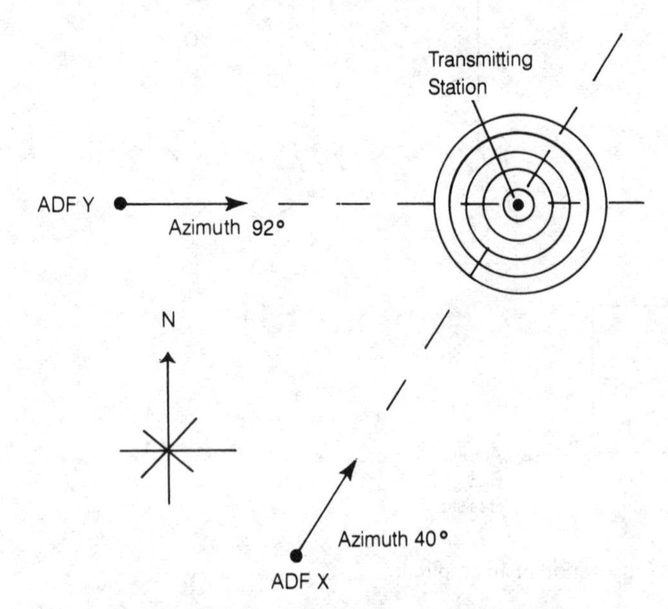

AUTOMATIC DIRECTION FINDER: Bearings taken by two automatic direction finders can locate a radio transmitter or beacon.

In conjunction with a small whip antenna, the loop gives an unidirectional indication.

A rotary drive turns the loop antenna until the signal strength reaches a sharp minimum, or null. At this point the loop, connected to a sensing circuit, stops rotating automatically. The operator can then read the compass bearing from the azimuth indicator in degrees.

The exact position of a transmitting station is obtained by taking readings with the automatic direction finder at two or more widely separated points as shown. On a detailed chart straight lines are drawn according to the bearings obtained at the observation points. The intersection point of these lines is the location of the transmitter. *See also* RADIO DIRECTION FINDER.

AUTOMATIC FREQUENCY CONTROL

Automatic frequency control is a method for keeping a receiver or transmitter on the desired operating frequency. These circuits are commonly used in FM (frequency modulation) stereo receivers, because tuning is critical with this type of signal and even a small error can noticeably affect sound reproduction.

An automatic-frequency-control device senses any deviation from the correct frequency and introduces a dc voltage across a varactor diode in the oscillator circuit to compensate for the drift. An increase in oscillator frequency causes the injection of a voltage to lower the frequency; a decrease in the oscillator frequency causes the injection of a voltage to raise the frequency.

A special kind of automatic frequency control is the phase-locked loop, abbreviated PLL. This circuit maintains the frequency of an oscillator within very narrow limits. *See also* PHASE-LOCKED LOOP.

AUTOMATIC GAIN CONTROL

In a communications receiver it is desirable to keep the output essentially constant regardless of the strength of the incoming signal. This allows reception of strong as well as weak signals, without the need for volume adjustments. It also prevents blasting, the effect produced by a loud signal as it suddenly begins to come in while the receiver volume is turned up.

Refer to the schematic diagram of a simplified automatic gain control, or AGC. An audio or rf (radio frequency) amplifier stage can be used for AGC; usually, it is done in the intermediate-frequency stages of a superheterodyne receiver. Part of the amplifier output is rectified and filtered to obtain a dc voltage. This dc (direct-current) voltage is applied to the input circuit of that amplifier stage or a preceding stage. The greater the signal output, the greater the AGC voltages, which changes the amplifier bias to reduce its gain. By properly choosing the component values, the amplifier output can be kept at an almost constant level for a wide range of signal-input amplitudes. Automatic gain control is sometimes also called automatic level control or automatic volume control (*see* AUTOMATIC LEVEL CONTROL), although the term automatic level control is generally used to define a broader scope of applications.

The attack time (*see* ATTACK) of an AGC system must be rapid, and the decay time (*see* DECAY) slower. Some receivers have AGC with variable decay or release time. The decay time is easily set to the desired value by suitably choosing the value of the filter capacitor in the dc circuit. The time constant of this decay circuit should be slow enough to prevent background noise from appearing between elements of the received signal, but rapid enough to keep up with fading effects. *See also* RECEIVER.

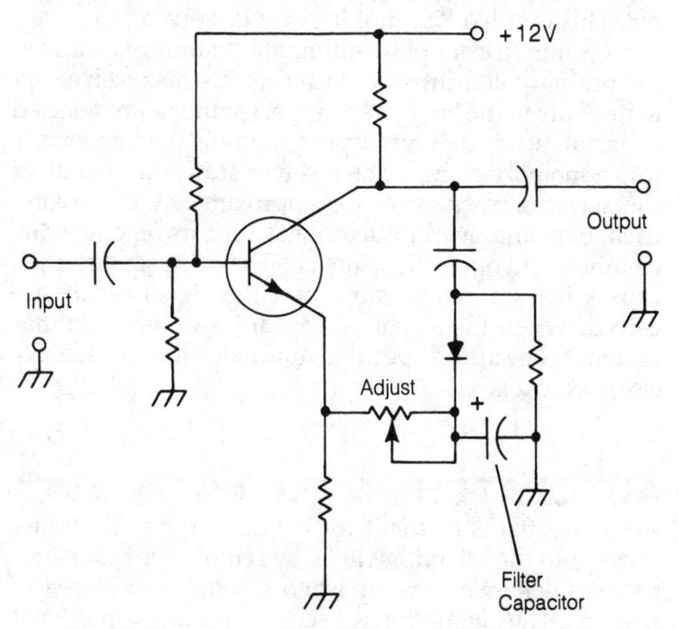

AUTOMATIC GAIN CONTROL: Rectified output applied to the emitter of a transistor amplifier stage reduces gain in the presence of strong signals.

AUTOMATIC INTERCEPT

An automatic intercept circuit receives incoming messages and records them for later handling. A common example of this circuit is a telephone answering machine. These machines are available to consumers in various designs from simple to very sophisticated for use in the home and office.

The automatic intercept answers the telephone after it rings and plays a recorded announcement stating that the called party is not presently available. The caller can then wait for a tone, whereupon the automatic intercept machine records a message of specified length, usually 30 to 45 seconds. Some automatic intercepts can record dozens of messages in this way.

AUTOMATIC LEVEL CONTROL

Automatic level control (ALC) is a generalized form of automatic gain control, or AGC (*see* AUTOMATIC GAIN CONTROL). While AGC is employed in communications receivers, it is also in a wide variety of other equipment from tape recorders to AM, SSB, and FM transmitters. Both AGC and ALC function in the same way: Part of the output is rectified and filtered, and the resulting dc voltage is fed back to a preceding stage to control the amplification. This keeps the output essentially constant, even when the input amplitude changes greatly.

Automatic level control is frequently used in voice transmitters to prevent flat topping of the modulated waveform (*see* FLAT TOPPING). Automatic-level control can also be used to boost the low-level modulation because the amplifier can be made more sensitive without fear of overmodulation. A circuit designed to increase the talk power in a voice transmitter is called a speech compressor or speech processor (*see* SPEECH COMPRESSION) and is a form of ALC in conjunction with an audio amplifier. Automatic-level control should not be confused with limiting or clipping (*see* AUDIO LIMITER, CLIPPER, SPEECH CLIPPING).

In a transmitter modulation circuit, ALC is sometimes called automatic-modulation control (*see* AUTOMATIC MODULATION CONTROL).

AUTOMATIC MESSAGE HANDLING

Messages are routed to the proper destination rapidly and efficiently by a system called automatic message handling. A message in this system is categorized by an electronic code such as a combination of tone bursts or a subaudible tone. This eliminates the need for manual processing.

The placing of a direct-dial, long-distance telephone call is an example of automatic message handling. The first digit, 1, is represented by a combination of two audio tones or a single pulse. It indicates to a computer that the caller intends to use the direct-dial system. The next three digits, also represented by either tones or pulses, define the geographical region to which the call is directed. The next three digits select the number prefix, further narrowing the possibilities. The final four digits direct the call to a particular business or residence.

Automatic message handling is done today by digital computers. This saves the time and cost of operator assistance.

AUTOMATIC MODULATION CONTROL

Automatic modulation control is a form of automatic-level control (*see* AUTOMATIC LEVEL CONTROL). In an AM (amplitude modulation), SSB (single sideband), or FM (frequency modulation) transmitter, automatic modulation control prevents excessive modulation while allowing ample microphone gain.

In an AM or SSB transmitter, overmodulation causes a phenomenon known as splatter, which greatly increases the signal bandwidth and can cause interference to stations on nearby channels or frequencies. In an FM transmitter, overmodulation results in overdeviation. This makes the signal appear distorted in the receiver, and can also cause interference to stations on adjacent channels. The output of a sine-wave-modulated FM transmitter is shown both without (at A) and with (at B) automatic modulation control. In this particular system, the proper deviation is ±5 kHz. In operation, while observing a deviation monitor, an operator would find it impossible to cause overdeviation, even by loud shouting into the microphone. The dotted lines in A and B show the proper modulation limits.

Without automatic modulation control, proper adjustment of a voice transmitter would be exceedingly critical. Therefore, nearly all voice transmitters use some form of automatic modulation control. *See also* AMPLITUDE MODULATION, FREQUENCY MODULATION, SINGLE SIDEBAND.

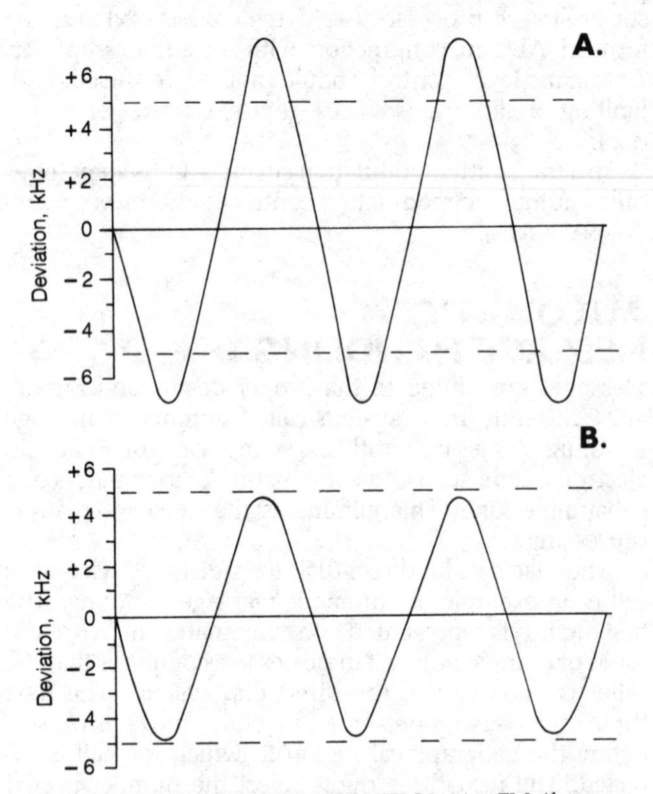

AUTOMATIC MODULATION CONTROL: An FM (frequency-modulation) signal without modulation control shows excessive deviation (A). With automatic modulation control, it appears as (B).

AUTOMATIC NOISE LIMITER

An automatic noise limiter, abbreviated ANL, is a circuit that clips impulse and static noise peaks while not affecting the desired signal. An automatic noise limiter sets the level of limiting, or clipping, according to the strength of the incoming signal. The AGC voltage (*see* AUTOMATIC GAIN CONTROL) in the receiver usually forms the reference for the limiter. The stronger the signal, the greater the limiting threshold. A manual noise limiter (*see* NOISE LIMITER) differs from the automatic noise limiter; the manual limiter must be set by the operator, and will cause distortion of received signals if not properly adjusted.

A noise limiter prevents noise from becoming stronger than the signals within the receiver passband, but the signal-to-noise ratio remains marginal in the presence of strong impulse noise. For improved reception with severe impulse noise, the noise blanker has been developed. Noise blankers are extremely effective against many kinds of manmade noise, but noise limiters are generally better for atmospheric static.

AUTOMATIC SCANNING RECEIVER

Some receivers have the ability to search for an occupied channel automatically, without the need for an operator. A channelized receiver with automatic-scanning capability can be programmed to constantly sweep through a given set of frequencies until a busy channel is found. This is called the busy-scan mode. Alternatively, the scanner can be set to search for an empty channel. This is called the vacant-scan mode. In some cases, a subaudible tone of a certain frequency may be necessary to cause the automatic scanning receiver to stop on a particular channel. This is called Private-Line®, or PL, operation.

A common example of automatic scanning is found in the ordinary vhf/uhf scanner receiver. This receiver operates only in the busy-scan mode. Channels are selected by installing crystals, or by programming the frequencies into a memory circuit. The receiver scans through all of the selected frequencies for approximately 0.1 second each, over and over, until a signal appears on one of the channels and opens the squelch. The open squelch then causes the scanner to stop, and the signal can be received. When the signal disappears, scanning resumes automatically after a specified time delay has elapsed. *See also* SCAN FUNCTION, SCANNING.

AUTOPATCH

An autopatch is a circuit for connecting a radio transceiver into the telephone lines by remote control. Autopatch is generally accomplished through repeaters (*see* REPEATER). The illustration is a schematic representation of an autopatch system.

The repeater is first accessed by transmitting a sequence of tone-coded digits. This is done with a telephone type keypad connected to the microphone input of the transmitter (*see* TOUCHTONE®). The autopatch system is then activated by another set of digits. At this point, the operator may dial the desired telephone number.

Autopatch telephone conversations are not always the same as an ordinary phone hookup. Interruption of the radio operator is impossible if his receiver is disabled while he speaks. A simple form of autopatch system that works over a limited range, allowing duplex operation, is the increasingly popular cordless telephone. *See also* CORDLESS TELEPHONE.

AUTOTRANSFORMER

The autotransformer is a special type of step-by-step or step-down transformer with only one winding. Usually, a transformer has two separate windings called the primary and secondary electrically isolated from each other, as shown in A. The autotransformer, however, uses a single tapped winding.

For step-down purposes, the input to the autotransformer is connected across the entire winding, and the output is taken from only part of the winding as shown in B. For step-up purposes, the situation is reversed: the

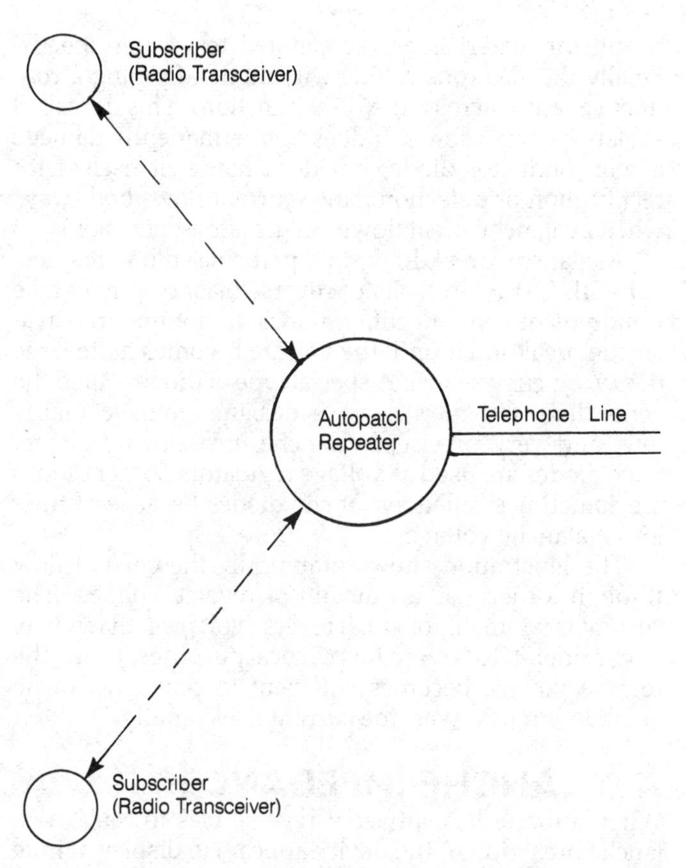

AUTOPATCH: An autopatch repeater allows radio transceivers to be coupled to the commercial telephone system.

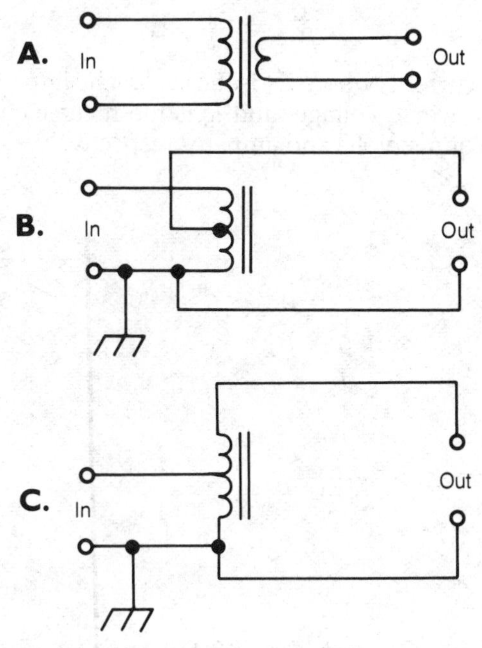

AUTOTRANSFORMER: The schematic symbols for a conventional transformer (A), a step-down autotransformer (B), and a step-up autotransformer (C).

input is applied across part of the winding, and the output appears across the entire coil (C).

The autotransformer obviously contains less wire than a transformer with two separate windings. If electrical (direct-current) isolation is needed between the input and output, however, the autotransformer cannot be used.

The step-up or step-down voltage ratio of an autotransformer is determined according to the same formula as for an ordinary transformer (*see* TRANSFORMER). If N_{PRI} is the number of turns in the primary winding, and N_{SEC} is the number of turns in the secondary winding, then the voltage transformation ratio is:

$$\frac{E_{PRI}}{E_{SEC}} = \frac{N_{PRI}}{N_{SEC}}$$

where E_{PRI} and E_{SEC} represent the ac voltages across the primary and secondary windings, respectively. The impedance-transfer ratio is:

$$\frac{Z_{PRI}}{Z_{SEC}} = \left(\frac{N_{PRI}}{N_{SEC}}\right)^2$$

where Z_{PRI} and Z_{SEC} are the impedances across the primary and secondary windings, respectively.

AVAILABLE POWER

When a transmission line is connected to an antenna, or a source of energy is connected to a load, power is dissipated in the load. This dissipation can occur as heat in a resistor, or as heat and visible light in a light bulb. It can also take the form of electromagnetic radiation, as in a radio antenna, or sound, as in a speaker or headset. There are many forms of energy dissipation. The amount of power dissipated in a load is measured in watts, and represents the rate of energy transfer.

If the load impedance is a conjugate match to the source or input impedance, then the transfer of energy from the source to the load occurs with 100 percent efficiency. However, if the source impedance and load impedance are not conjugate-matched, some of the energy from the source is not absorbed by the load. Instead, it is returned to the source as reactive power (*see* IMPEDANCE, IMPEDANCE MATCHING, REACTIVE POWER). The amount of power dissipated in the load is therefore not as great as it could be.

The available power from a generator, or source, is the amount of power dissipated in a conjugate-matched load. With a given generator, the true power dissipated by the load decreases as the match gets worse and worse. The available power remains the same no matter what the match. *See also* CONJUGATE IMPEDANCE, TRUE POWER.

AVAILABLE POWER GAIN

The available power gain of a circuit is the ratio of the available output power to the available input power. When impedances at the input and output of a circuit are properly matched (as they should be for most efficient operation), the available power gain is the same as the true power gain (*see* AVAILABLE POWER, TRUE POWER, TRUE POWER GAIN).

Available power gain is usually expressed in decibels, or dB (*see* DECIBEL). If P_{IN} is the available power input in watts and P_{OUT} is the available power output in watts, then:

Available power gain (dB) = 10 $\log_{10}$ (P_{OUT}/P_{IN})

By comparing the available power gain with the true power gain of an amplifier, it is possible to tell whether or not a given circuit is performing as well as it should. The gain of an amplifier can be seriously degraded simply because impedances are not matched, making the true power gain much smaller than it could be if proper matching were arranged.

AVALANCHE

Avalanche occurs when an electric field becomes so strong that ionization occurs in a gas. One electron, accelerated by the field, collides with molecules of the gas and ionizes them. This sets many electrons free, each of which is accelerated by the field to strike, and ionize, more molecules. The eventual result is a conductive path through the gas.

The avalanche effect also occurs at semiconductor junctions, but in a somewhat different way. Ordinarily, a junction of N-type and P-type material (*see* N-TYPE SEMICONDUCTOR, P-TYPE SEMICONDUCTOR) will not conduct when reverse biased. However, if the reverse-bias voltage is sufficiently strong, conduction will take place. This is called avalanche breakdown. *See also* AVALANCHE BREAKDOWN.

AVALANCHE BREAKDOWN

When reverse voltage is applied to a P-N semiconductor junction, very little current will flow as long as the voltage is relatively small. But as the voltage is increased, the holes and electrons in the P-type and N-type semiconductor materials are accelerated to greater speeds. Finally the electrons collide with atoms, forming a conductive path across the P-N junction. This is called avalanche breakdown. It does not permanently damage a semiconductor diode, but does render it useless for rectification or detection, since current flows both ways when avalanche breakdown occurs (*see* RECTIFICATION).

Avalanche breakdown in a particular diode may occur with just a few volts of reverse bias, or it may take hundreds of volts. Rectifier diodes do not undergo avalanche breakdown until the voltage becomes quite large (*see* AVALANCHE VOLTAGE). A special type of diode, called the zener diode (*see* ZENER DIODE), is designed to have a fairly low, and very precise, avalanche breakdown voltage. Such diodes are used as voltage regulators. Zener diodes are sometimes called avalanche diodes because of their low avalanche voltages.

The illustration shows graphically the current flow through a diode as a function of reverse voltage. The current is so small, for small reverse voltages, that it may be considered to be zero for practical purposes. When the reverse voltage becomes sufficient to cause avalanche breakdown, however, the current rises rapidly.

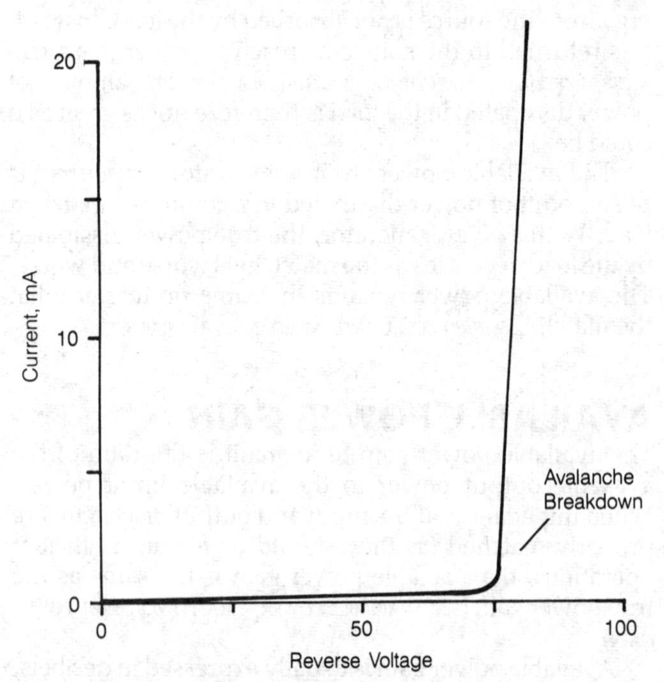

AVALANCHE BREAKDOWN: Avalanche breakdown occurs in a reverse-biased diode when the current rises sharply.

AVALANCHE IMPEDANCE

When a diode has sufficient reverse bias to cause avalanche breakdown, the device appears to display a finite resistance. This resistance fluctuates with the amount of reverse voltage, and is called the avalanche resistance or avalanche impedance. It is given by:

$$Z_A = \frac{E_R}{I_R}$$

for direct current, where Z_A is the avalanche impedance, E_R is the reverse voltage, and I_R is the reverse current. Units are ohms, volts and amperes respectively.

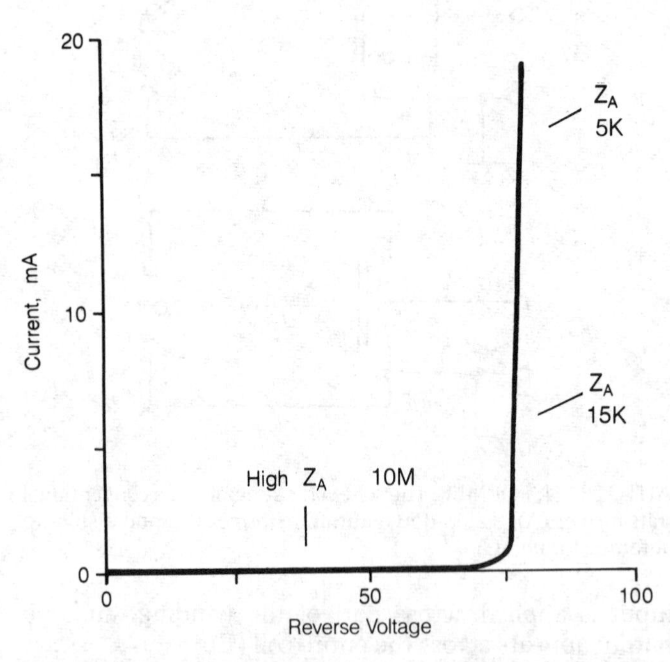

AVALANCHE IMPEDANCE: Avalanche impedance in a reverse-biased diode is a function of reverse voltage; it drops sharply beyond the avalanche breakdown value.

When E_R is smaller than the avalanche voltage (*see* AVALANCHE VOLTAGE), the value of Z_A is extremely large, because the current is small, as shown in the figure. When E_R exceeds the avalanche voltage, the value of Z_A drops sharply. As E_R is increased further and further, the value of Z_A continues to decrease.

The magnitude of Z_A for alternating current is practically infinite in the reverse direction, as long as the peak ac voltage never exceeds the avalanche voltage. When the peak ac reverse voltage rises to a value greater than the avalanche voltage, the impedance drops. The maximum voltage that a semiconductor diode can tolerate, for rectification purposes, without avalanche breakdown is called the peak inverse voltage (*see* PEAK INVERSE VOLTAGE).

The exact value of the avalanche impedance for alternating current is the value of resistor that would be necessary to allow the same flow of reverse current. The average Z_A for alternating current differs from the instantaneous value, which fluctuates as the voltage rises and falls. Avalanche is usually undesirable in alternating-current applications. *See also* INSTANTANEOUS EFFECT.

AVALANCHE PHOTODIODE

An avalanche photodiode operates in the avalanche breakdown region. As shown in the drawing, photons with energy that is high enough are able to generate electron-hole pairs in the depletion region. The positively charged hole moves toward the negative side of the battery, and the negatively charged electron moves toward the positive side. Electrons moving in a high electric field are accelerated and collide with other bound electrons. The high velocity of impact when electrons collide creates additional hole-electron pairs. These pairs, when accelerated, produce other electron-hole pairs and cause an avalanche multiplication process. One photon can produce as many as 100 electrons in an avalanche photodiode. *See* DIODE, PHOTODETECTOR.

AVALANCHE TRANSISTOR

An avalanche transistor is an NPN or PNP transistor designed to operate with a high level of reverse bias at the emitter-base junction. Normally, transistors are forward biased at the emitter-base junction except when cutoff conditions are desired when there is no signal input.

Avalanche transistors are seen in some switching applications. The emitter-base junction is reverse biased almost to the point where avalanche breakdown takes place. A small additional reverse voltage, supplied by the input signal, triggers avalanche breakdown of the junction and resultant conduction. Therefore, the entire transistor conducts, switching a large amount of current in a very short time. The extremely sharp knee of the reverse-voltage-versus-current curve facilitates this switching capability (see the illustrations in AVALANCHE BREAKDOWN and AVALANCHE IMPEDANCE). A small amount of input voltage can thus cause the switching of large values of current.

AVALANCHE VOLTAGE

The avalanche voltage of a P-N semiconductor junction is the amount of reverse voltage required to cause avalanche breakdown (*see* AVALANCHE, AVALANCHE BREAKDOWN). Normally, the N-type semiconductor is negative with respect to the P-type in forward bias, and the N-type semiconductor is positive with respect to the P-type in reverse bias.

In some diodes, the avalanche voltage is very low, as

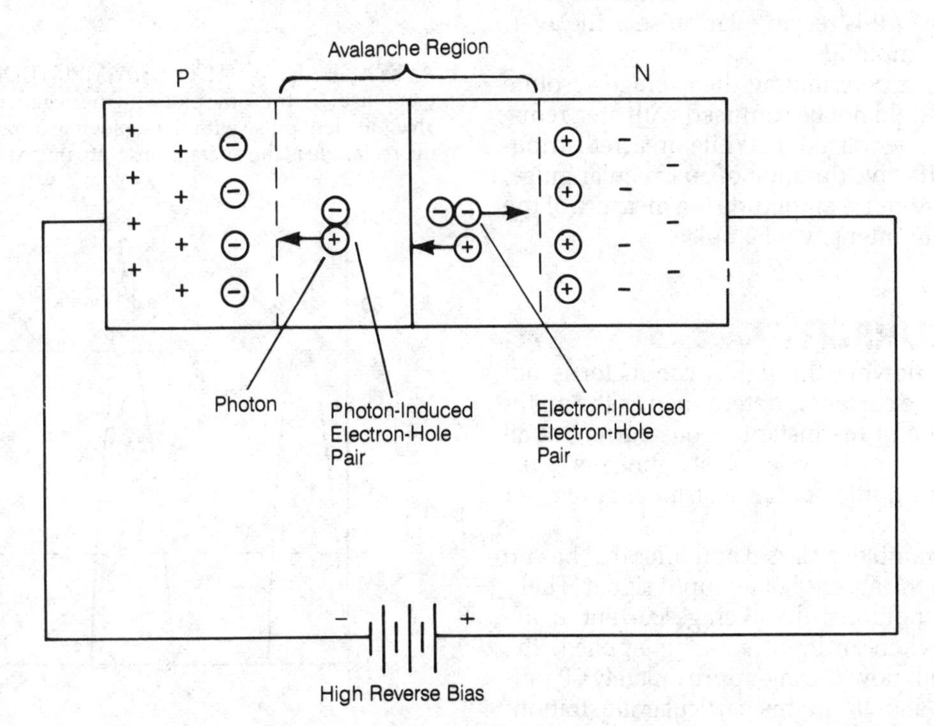

AVALANCHE PHOTODIODE: This photodiode is sensitive to light because it operates in the avalanche breakdown region.

small as 6 V. In other diodes, it might be hundreds of volts. When the avalanche voltage is reached, the current abruptly rises from near zero to a value that depends on the reverse bias (see the illustrations in AVALANCHE BREAKDOWN and AVALANCHE IMPEDANCE).

Diodes with high avalanche voltage ratings are used as rectifiers in dc power supplies. The avalanche voltage for rectifier diodes is called the peak inverse voltage or peak reverse voltage (*see* PEAK INVERSE VOLTAGE). In the design of a dc (direct-current) power supply, diodes with sufficiently high peak-inverse-voltage ratings must be chosen, so that avalanche breakdown does not occur.

Zener diodes are deliberately designed so that an effect similar to avalanche breakdown occurs at a relatively low, and well defined voltage. (*see* ZENER DIODE). They are used as voltage-regulating devices in low-voltage dc power supplies. *See also* POWER SUPPLY.

AVERAGE ABSOLUTE PULSE AMPLITUDE

A pulse of voltage, current, or power is often irregularly shaped. Sometimes the polarity of the pulse reverses one or more times. A rather complicated pulse is shown at A in the illustration.

The average absolute pulse amplitude is a measure of the pulse intensity. In determining the average absolute pulse amplitude, the beginning and ending times, t_0 and t_1, must be known. Then, the absolute value of the pulse polarity is taken by inverting the negative part or parts of the pulse, forming a positive mirror image (B). Next, the total area under the pulse curve is found. It is shown by the shaded region. Finally, a rectangle is constructed, having the same beginning and ending times, t_0 and t_1, as the original pulse, and also the same area under the curve (C). The amplitude of this rectangular pulse is the average absolute pulse amplitude.

The procedure for determining the average absolute pulse amplitude should not be confused with area redistribution (*see* AREA REDISTRIBUTION). While area redistribution involves the effective duration of an irregular pulse, the average absolute-pulse amplitude is a measure of the effective strength, or intensity, of a pulse.

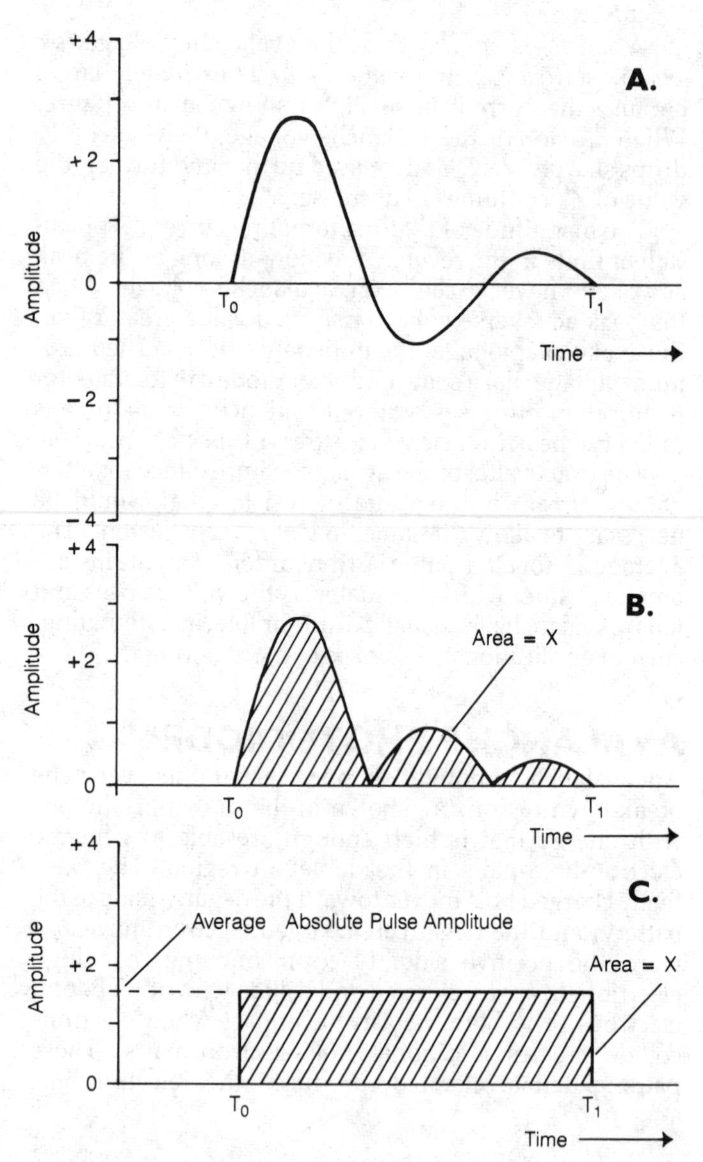

AVERAGE ABSOLUTE PULSE AMPLITUDE: The plot of an irregular pulse (A), the same plot with the negative part inverted (B), and an equivalent pulse with the same time duration (C). The height of the rectangle is the average absolute pulse amplitude.

AVERAGE CURRENT

When the current flowing through a conductor is not constant, the average current is determined as the mathematical mean value of the instantaneous current at all points during one complete cycle. Most ammeters register average current. Some special instruments register peak current.

Consider, for example, a class B amplifier that has no collector current in the absence of an input signal. Then, under no-signal conditions, the average current at the collector is zero. When an input signal is applied, the collector current will flow during approximately 50 percent of the cycle as shown. In this particular illustration, the peak collector current (*see* PEAK CURRENT) is 70 mA. The average current is smaller—about 22 mA.

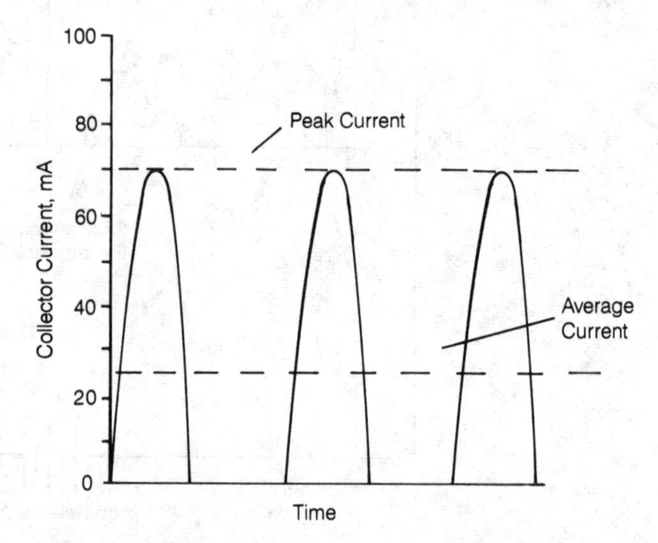

AVERAGE CURRENT: Average and peak current in a Class B amplifier output.

For alternating current, the average current is usually zero, because the polarity is positive during half the cycle and negative during half the cycle, with peak values and waveforms identical in both the positive and negative directions. Of course, the effects of alternating current are very different from the effects of zero current! Average current is given little importance in ac circuits; the root-mean-square, or rms, current is more often specified as the effective current in such instances (*see* ROOT MEAN SQUARE).

In a sine-wave half cycle, the average current is 0.637 times the peak current.

AVERAGE LIFE

After excess carriers (electrons or holes) in a semiconductor N-type or P-type piece of material have been injected, they eventually combine with carriers of the opposite polarity. Holes are filled in by electrons, and electrons vacate atoms to create holes, restoring the material to its original condition. The time it takes for this process to be completed is called the average life. It is also sometimes called the lifetime or recombination time.

Excess electrons are introduced into an N-type semiconductor by a negative charge. Excess holes are put into P-type material by a positive charge (*see* HOLE, N-TYPE SEMICONDUCTOR, P-TYPE SEMICONDUCTOR).

The recombination of excess carriers, as a function of time, is exponential. It occurs very rapidly at first, and then more and more slowly, in a manner similar to the discharging of a capacitor (see illustration). In practice, the recombination process is effectively complete when most of the excess carriers have disappeared. This takes very little time. In silicon, it may be as little as a few nanoseconds (billionths of a second). In high-speed switching or vhf/uhf (very high frequency/ultrahigh frequency) applications, it is important to have an extremely short average life. To make the recombination period as short as possible, impurity atoms are added to the semiconductor material. These impurity atoms, such as gold, act as recombination catalysts. *See also* DOPING.

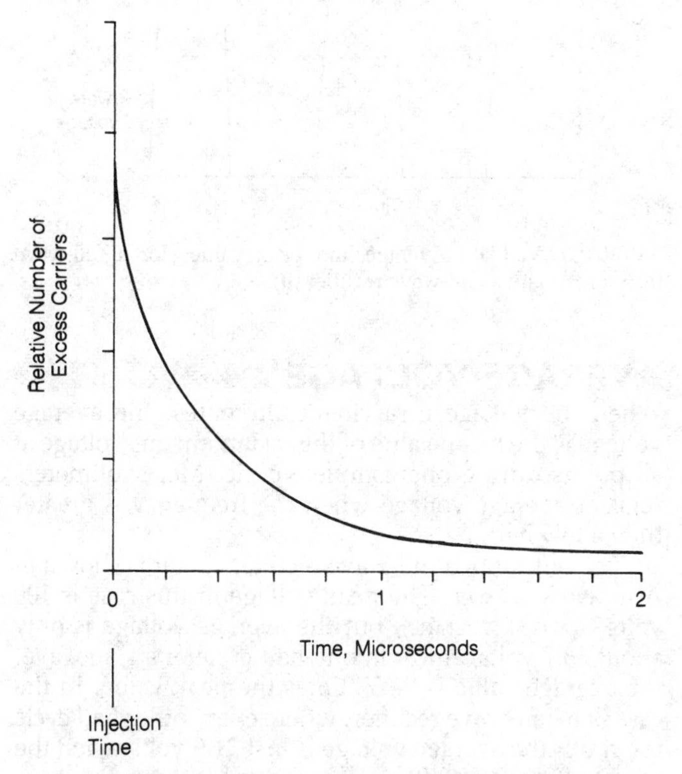

AVERAGE LIFE: A plot of the time required for excess injected carriers to recombine across a semiconductor junction after a polarity change.

AVERAGE NOISE FIGURE

The average noise figure of a device is its noise figure (*see* NOISE FIGURE) summed over all possible frequencies. This is theoretically dc to infinity. The illustration shows the noise input and output of a device from 1 Hz to 10^{12} Hz, which is a range of zero to infinity for most practical purposes. The input noise is attributable only to thermal agitation (*see* THERMAL NOISE), at a temperature of 290 Kelvin, or about 63° Fahrenheit, for determination of the average noise figure.

The total output-noise power in the figure is represented by the area under the high curve. The total input thermal-noise power is represented by the area under the lower curve. Noise figures are generally expressed in decibels (*see* DECIBEL). The best possible noise figure is 0 decibels, where the output-noise power is the same as the input-noise power. This is a theoretical ideal; it is never achieved in practice.

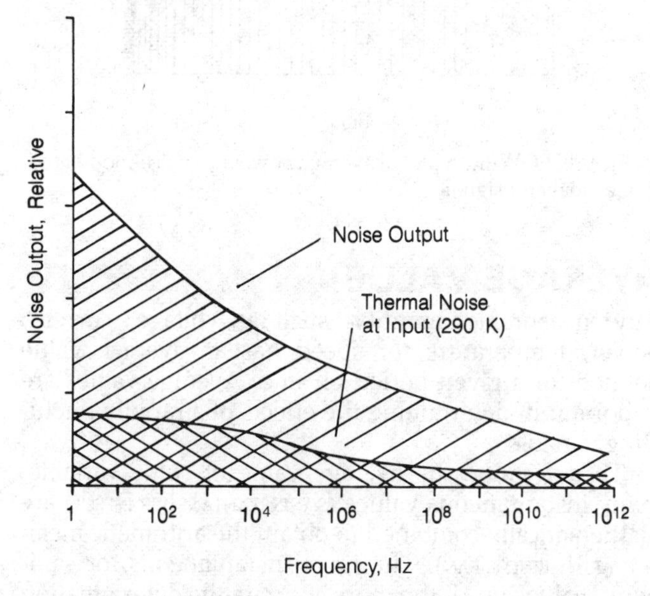

AVERAGE NOISE FIGURE: A plot of the ratio of thermal noise at 290 K to noise output versus frequency for circuits across the entire practical frequency range.

AVERAGE POWER

When the level of power changes rapidly as in the modulation envelope of a single-side band transmitter (see the graph), the signal power may be defined in terms of peak power or average power. An ordinary wattmeter reads out in average power; special meters are needed to determine peak power.

In a given period of time, the area under the curve in the figure, if geometrically rearranged to form a perfect rectangle, will have a height corresponding to the average power. Between instants t_0 and t_1, chosen arbitrarily, the average power can be evaluated by considering the envelope as an irregular pulse (*see* AVERAGE ABSOLUTE PULSE AMPLITUDE).

The average power output of a circuit is never greater than the peak power output. If the power level is constant, as with an unmodulated carrier or frequency-modulated signal, the average and peak power are the same. If the amplitude changes, then the average power is less than the peak power. The output of a single-sideband signal modulated by a voice will have an average power level of about half the peak power level, although the exact ratio depends on the characteristics of the voice. *See also* PEAK ENVELOPE POWER, PEAK POWER.

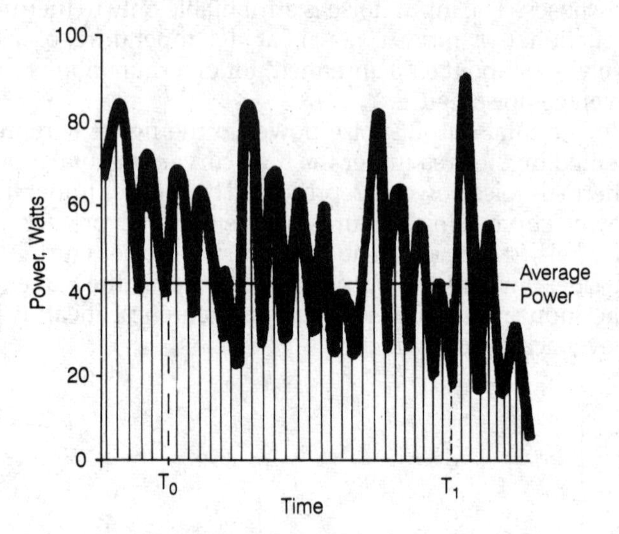

AVERAGE POWER: A plot of average power in a voice-modulated, single-sideband signal.

AVERAGE VALUE

Any measurable quantity, such as voltage, current, power, temperature, or speed has an average value defined for a given period of time. Average values are important in determining the effects of a rapidly fluctuating variable.

To determine the average value of some variable, many instantaneous values (*see* INSTANTANEOUS EFFECT) are mathematically combined to obtain the arithmetic mean (*see* ARITHMETIC MEAN). The more instantaneous, or sampling, values used, the more accurate the determination of the average value. For a sine-wave half cycle (such as one of the pulses in A or B), having a duration of 10 milliseconds (0.01 second) at an ac frequency of 50 Hz, instantaneous readings might be taken at intervals of 1 millisecond, then 100 microseconds, 10 microseconds, 1 microsecond, and so on. This would yield first 10, then 100, then 1000, and finally 10,000 sampling values. The average value determined from many sampling values is more accurate than the average value determined from just a few sampling values.

There is no limit to the number of sampling values that may be averaged in this way. The true average value, however, is defined as the arithmetic mean of all the instantaneous values in a given time interval. While the mathematical construction of this is rather complicated, a true average value always exists for a quantity evaluated over a specific period of time.

An equivalent method of evaluating the average value for a variable quantity calls for the construction of a rectangle, with a total area equal to the area under the curve, for a certain time interval. *See also* AVERAGE ABSOLUTE PULSE AMPLITUDE.

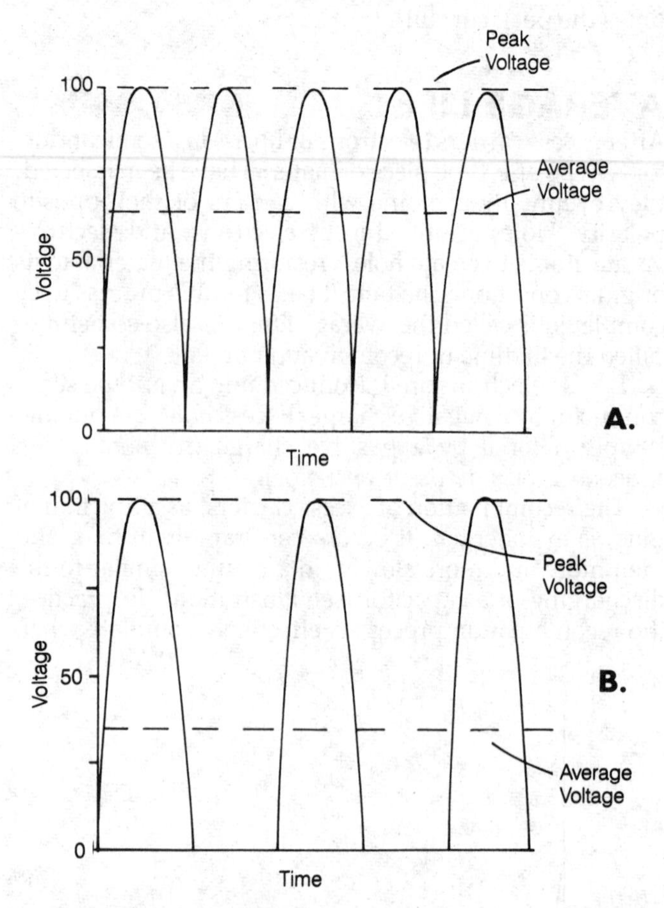

AVERAGE VALUE: Average and peak values for a full-wave rectifier (A) and a half-wave rectifier (B).

AVERAGE VOLTAGE

When the voltage in a circuit alternates, the average voltage is the mean value of the instantaneous voltage at all points during one complete cycle. Most voltmeters register average voltage when the frequency is greater than a few hertz.

The output of a full-wave rectifier circuit is shown in A in AVERAGE VALUE. The peak voltage in this case is 100 volts (*see* PEAK VOLTAGE), but the average voltage is only about 63.7 volts, since, in one-half cycle of a sine wave, the average value is 0.637 times the peak value. In the case of a half-wave rectifier, where every other half cycle is cut off, the average voltage is just 31.9 volts when the peak voltage is 100 (B in AVERAGE VALUE).

For a source of alternating current, the average voltage is usually zero over a long period of time, since the polarity is positive during half the cycle and negative during the other half, and the peak values are identical in both directions. An ac voltage has, of course, very different effects from zero voltage, and therefore the average voltage is given little importance in ac circuits. The root-mean square, or rms, voltage is more often specified as the equivalent dc voltage. *See also* ROOT MEAN SQUARE.

AVOGADRO CONSTANT

The Avogadro constant, often called Avogadro's number, is approximately 6.022169×10^{23}, or 602,216,900, 000,000,000,000,000. It is abbreviated *N*. This number of atoms or molecules is called one mole. The atomic weight (*see* ATOMIC WEIGHT) of a given element is the weight, in grams, of one mole of atoms of that element. One gram is therefore 6.022169×10^{23} atomic-mass units. Avogadro's number is primarily of interest to chemists and physicists.

AWG

See AMERICAN WIRE GAUGE.

AXIAL LEADS

A component with axial leads has leads projecting from both ends along a common, linear axis. Many resistors, semiconductor diodes, capacitors, and inductors have axial leads.

Axial leads give a component extra mechanical rigidity on a circuit board because the component can be mounted flush with the board. Axial leads have less mutual inductance than radial leads because of their greater separation and their collinear orientation.

Axial leads are sometimes inconvenient when a component must be mounted within a small space on a printed-circuit board. In these situations, surface mounted components are preferred (*see* SURFACE-MOUNT TECHNOLOGY).

AXIS

The axis of an object is a straight line around which the object revolves or rotates, or could be revolved or rotated. For example, the axis of earth is a straight line passing through the geographic poles.

An axis may also be defined as a straight line about which an object is symmetrical. The leads of an axial resistor for example, are called axial leads because the body of the component is symmetrical about the line containing the leads.

When one object orbits in a circle around another, the axis of revolution is a straight line passing through the center of the orbit and perpendicular to the plane containing the orbit.

The term *axis* can be applied to the scales of a graph. The horizontal axis is called the abscissa, representing the independent variable; the vertical axis is called the ordinate, representing the dependent variable.

AYRTON SHUNT

The sensitivity of a galvanometer is reduced by means of a shunt resistor. The addition of variable shunt resistors, however, sometimes affects the meter movement, or damping (*see* GALVANOMETER). A special type of shunt, called an Ayrton shunt, allows adjustment of the range of a galvanometer without affecting the damping. This shunt consists of a series combination of resistors. Several different galvanometer ranges can be selected by connecting the input current across variable portions of the total resistance (see illustration).

Because the resistance across the galvanometer coil is the same regardless of the range selected, the damping of the meter movement does not depend on the range chosen. The effective input resistance to the meter, however, does change. The more sensitive the meter range, the greater the resistance at the meter input terminals. A microammeter, for example, with a range of 0 to 100 μA, can be used as a milliammeter or ammeter without changing the damping when a suitable Ayrton shunt system is provided (*see* AMMETER). The common volt-ohm-milliammeter often uses an Ayrton shunt to facilitate operation in its various modes. *See also* VOLT-OHM-MILLIAMMETER.

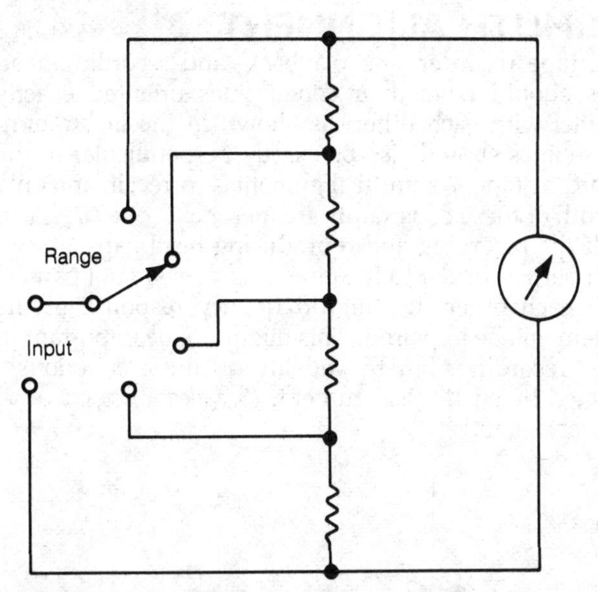

AYRTON SHUNT: This resistive circuit permits high meter sensitivity without changing the total resistance across the meter.

AZIMUTH

The term *azimuth* refers to the horizontal direction or compass bearing of an object or radio signal. True north is azimuth zero. The azimuth bearing is defined in degrees, measured clockwise from true north, as shown in the illustration. Thus, east is azimuth 90 degrees, south is azimuth 180 degrees, and west is azimuth 270 degrees. Provided the elevation is less than 90 degrees, as is usually defined by convention, these azimuth bearings hold for any object in space.

Azimuth bearings are always defined as being at least zero, but less than 360 degrees, avoiding ambiguity. Measures smaller than 0 degrees, and greater than or equal to 360 degrees, are not defined.

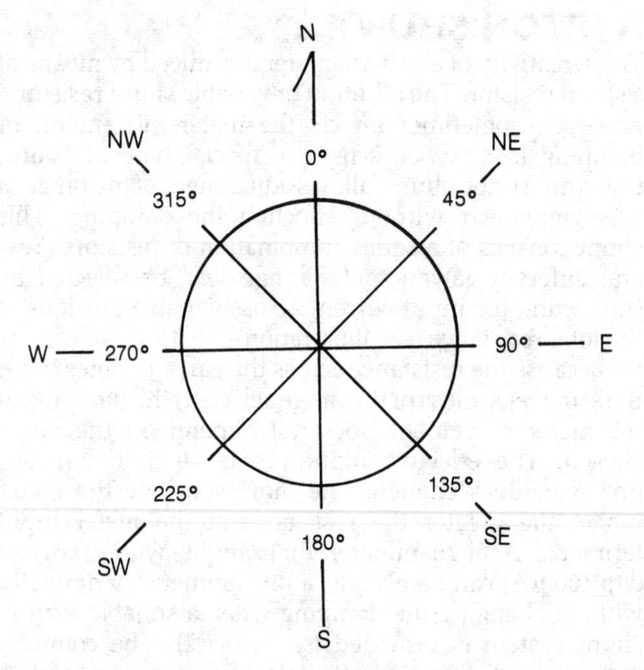

AZIMUTH: Azimuth is direction from true north measured clockwise in degrees in the horizontal plane.

AZIMUTH ALIGNMENT

In a tape recorder, the playback- and recording-head gaps should have their center lines oriented exactly parallel with each other, as shown in the illustration. These lines should also be exactly perpendicular to the recording tape. Azimuth alignment is correct if, and only if, both of these constraints are met.

If the recording and reproducing head gaps are not both perpendicular to the tape, or if they are not parallel with each other, the high-frequency response of the system will be impaired. This might not be important in voice recording, but the fidelity of music is seriously affected by such misalignment. (*See also* AUDIO RESPONSE, RECORDING HEAD.)

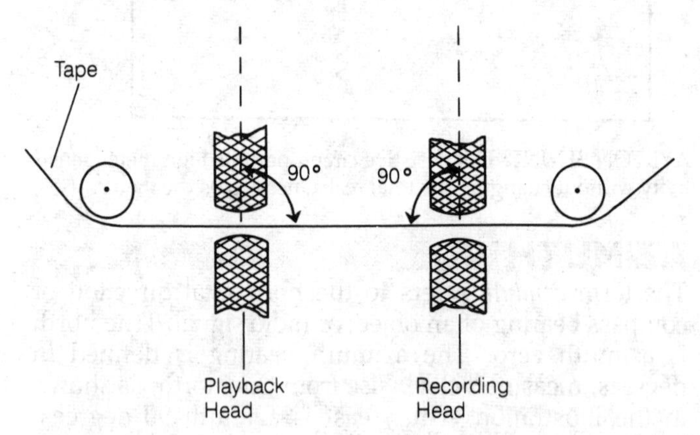

AZIMUTH ALIGNMENT: The heads of a tape recorder are oriented perpendicular to the tape in azimuth alignment.

AZIMUTH RESOLUTION

Azimuth resolution is the ability of a radar system to distinguish between objects at the same distance from the antenna but with slightly different bearings.

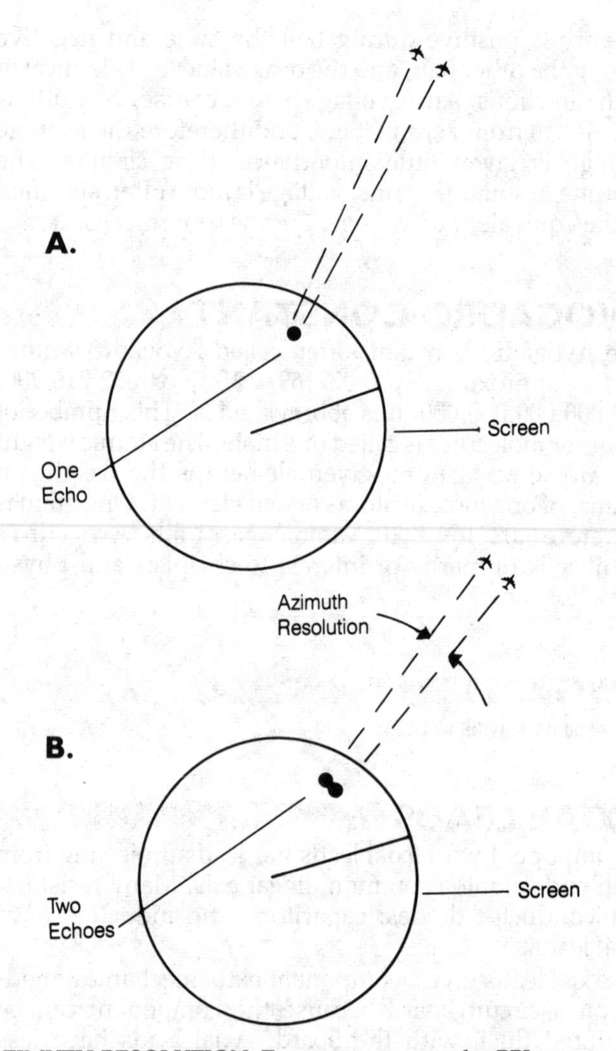

AZIMUTH RESOLUTION: Two targets on a radar PPI screen are too close for azimuth resolution so they appear as a single echo (A). With azimuth resolution, they appear as two echoes (B).

A and B illustrate the determination of azimuth resolution for a radar system. At A, the two aircraft are too close together to show up as separate echoes on the radar screen. But at B, the aircraft are sufficiently far apart for the radar to "see" them as two, rather than one.

There is a certain minimum angle that the objects must subtend with respect to the radar antenna for the radar to distinguish them as two separate objects rather than a single echo. This angle is the azimuth resolution of the system. It depends on many variables, including the radar wavelength, the size of the antenna, the pulse frequency, the speed at which the antenna rotates, and the resolution of the cathode-ray display tube.

Azimuth resolution is specified as an angle. Therefore, the farther away the objects are from the radar antenna, the greater the space between the objects must be in order for them to show up as separate echoes. For example, as two planes fly next to each other at constant spacing, directly away from the radar antenna, their echoes may appear distinct until they reach a specific distance. When they pass this distance, their echoes will appear to merge together because the angle between them has become smaller than the azimuth resolution of the system. *See also* AZIMUTH, RADAR, RESOLUTION.

B

BACK BIAS

Back bias is a voltage taken from one part of a multistage circuit and applied to a previous stage. This voltage is obtained by rectifying a portion of the signal output from an amplifier circuit. Back bias can be either regenerative or degenerative, or it can be used for control purposes.

A typical application of back bias is the automatic level control (*see* AUTOMATIC GAIN CONTROL, AUTOMATIC LEVEL CONTROL). A degenerative voltage is applied to either the input of the stage from which it is derived or to a previous stage. This prevents significant changes in output amplitude under conditions of fluctuating input signal strength.

Back bias can be applied to either the anode or cathode of an active device. A negative voltage, for example, might be applied to the plate of a tube or the collector of an NPN transistor to reduce its gain. Alternatively, a positive voltage could be applied to the cathode of a tube or the emitter of an NPN transistor, if the cathode or emitter is elevated above dc ground. Still another form of degenerative back bias is the application of a negative voltage to the grid of a tube or the base of an NPN transistor.

Sometimes the term *back bias* is used to define reverse bias. This condition occurs when the anode of a diode, transistor, or tube is negative with respect to the cathode. *See also* REVERSE BIAS.

BACK DIODE

A back diode is a special kind of tunnel diode, operated in the reverse-bias mode. Tunnel diodes are used as oscillators and amplifiers in the uhf and microwave part of the electromagnetic spectrum (*see* TUNNEL DIODE).

The back diode has a negative-resistance region at low levels of reverse bias. This means that, within a certain range of voltage, an increase in reverse voltage will cause a decrease in the flow of current.

An ordinary semiconductor diode has a voltage-versus-current curve similar to that shown at A in the illustration. The back diode has a voltage-versus-current characteristic similar to that shown at B. The back diode is normally operated in the region indicated, at small values of reverse voltage (E_R). This results in oscillation at microwave frequencies.

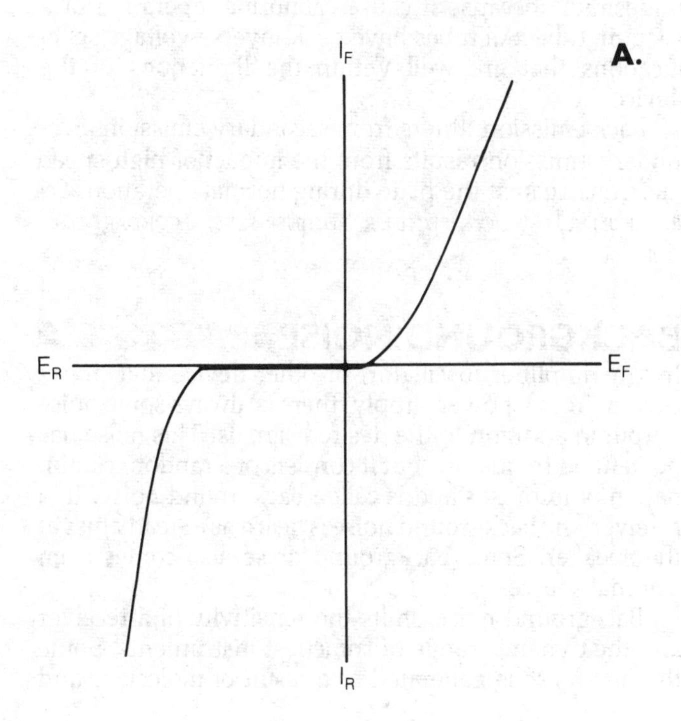

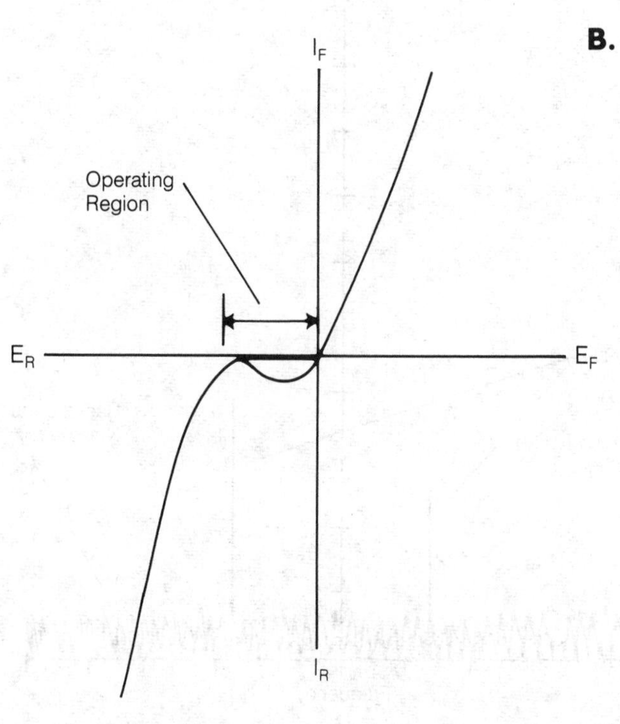

BACK DIODE: A voltage-versus-current curve for a typical diode (A), and for a back diode (B).

BACK EMISSION

When a diode, triode, tetrode, or pentrode tube is forward biased, electrons are emitted by the cathode and flow to the anode. Under conditions of reverse bias, when the anode or plate is negative with respect to the cathode, very little current flows. There is some emission of electrons by the plate under conditions of reverse bias. This is called back emission.

Back emission is generally very small, resulting in negligible current flow. If the reverse bias voltage is increased sufficiently, a sudden large increase in back emission occurs. This is called arcback (*see* ARCBACK). It is undesirable because it causes abnormal operation of a vacuum tube. All tubes have peak-inverse-voltage specifications that are well within the limitations of the device.

Back emission differs from secondary emission. Secondary emission results from the impact of high-speed electrons against the plate during normal operation. *See also* PRIMARY EMISSION, RECTIFIER, RECTIFIER TUBE, SECONDARY EMISSION.

BACKGROUND NOISE

In any amplifier, oscillator, or other device that draws current from a power supply, there is always some noise output in addition to the desired signals. This noise has no defined frequency, but it consists of a random combination of impulses and is called background noise. In a receiver, the background noise is heard as a steady hiss at the speaker. Some background noise also comes from external sources.

Background noise limits the sensitivity of a receiver and the dynamic range of some test instruments. Some thermal noise is generated as a result of molecular and atomic motion in all substances. Some noise is generated by the motion of electrons through conductors and semiconductors, from the impact of electrons against the plate and grids of vacuum tubes, and similar effects. A certain amount of background noise originates in outer space, emitted from distant stars and galaxies. Our own sun is a significant source of background noise at some frequencies.

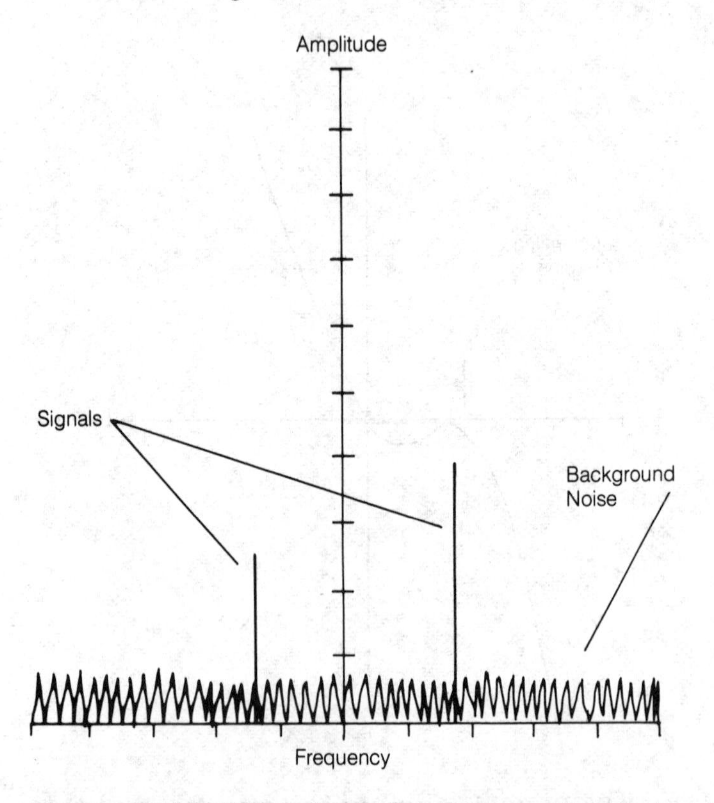

BACKGROUND NOISE: Background noise appears as "grass" on a spectrum analyzer.

On a spectrum analyzer, background noise and signals can be clearly identified (see illustration). Any signal that is weaker than the average level of the background noise will be undetectable. *See also* NOISE, SENSITIVITY, SIGNAL-TO-NOISE RATIO.

BACKLASH

Any rotary or continuously adjustable control will have some mechanical imprecision. The accuracy of adjustment is limited by the ability of an operator to interpolate readings. (A good example is the ordinary receiver or transmitter frequency dial.) Some inaccuracy, however, is a consequence of the mechanical shortcomings of parts including gears, planetary drives, and slide-contact devices. This is called backlash.

If a certain setting for a control is desired, it might depend on whether the control is turned clockwise or counterclockwise. For example, suppose it is necessary to tune precisely to the frequency of radio station WWV at 15.0000 MHz. This is usually done in receivers by zero beating (*see* ZERO BEAT). When tuning upward from some frequency below 15 MHz, the dial might indicate 15.0002 MHz at zero beat. When tuning downward from some frequency above 15 MHz, the dial might read 14.9997 at zero beat. The difference, 15.0002 − 14.9997 or 0.0005 MHz, (0.5kHz), is the dial backlash.

Because backlash is a purely mechanical phenomenon, and not an electrical effect, it can be eliminated by the use of digital displays. The precision of the adjustment is limited only by the resolution of the display; backlash and human error are overcome. *See also* DIGITAL CONTROL.

BACKPLANE

A backplane, also known as a backpanel, is a circuit board that functions as an interconnection between two or more circuit boards within an enclosure of an electronic product or system. It is a rigid circuit board on which printed circuit board connectors have been assembled in uniform rows, as shown in the illustration. Backplanes are often referred to as mother boards, and all of the other circuit boards that are interconnected to it are termed daughter boards.

Backplanes are usually custom fabricated with conventional printed wiring or circuit board techniques. A typical backplane is made of glass fiber-filled epoxy laminate, FR-4. Thickness may vary from 0.060 to 0.250 inch. Circuitry might be formed on one or both sides of the backpanel, or the backpanel might be a multilayered unit with up to 12 layers. Multilayer backplanes can provide better electrical properties than two-sided backplanes. Capacitance, voltage drop, and ground loops are minimized. Heavy internal copper layers up to 0.030 inch thick provide the necessary high current carrying capability for the use of backplanes with high-speed logic and ICs (integrated circuits).

Military specification backplanes are made from sheet aluminum with insulation provided by press-in insulated bushings. The bushings are pressed into a series of very accurately positioned holes in the panel. Backplanes or backpanels laminated from aluminum sheets provide greater package strength than conventional backpanels and provide radio frequency shielding. Sheet vinyl is often used between the aluminum sheets in the lamination. One aluminum panel is the ground plane, and the others are voltage planes. These panels can be designed for the desired capacitance and impedance characteristics and to provide a rigid support structure for the cardedge connectors.

Backplanes can be substrates for either one-piece edge connectors or two-piece, box-style PC connectors. Connector pins are terminated with solder eyelets or wiring posts. The spacing between connectors determines the spacing between adjacent circuit boards. Where backplanes are fitted with two-piece PC board connectors, the connector half with the pins (male side) is mounted on the backpanel, and the connector half of the daughter PC board contains the sockets (female side). The case or enclosure provides protection against accidental misalignment of the rows of pins.

Backplanes, like PC boards or cards, are fabricated to commercial or military specifications. If the electronic system is subject to changes during its operating life, the wire connections can be made by wire wrapping. This technique is also employed if the equipment is dedicated to a specific application at the point of installation. Wire-wrapped panels may have the necessary impedance characteristics to permit effective use of emitter-coupled logic (ECL).

Press-fit connectors are used where it is necessary to mount the connectors in prepared holes in the backplane. The press-fit concept depends on pressing connector contacts into pre-formed, plated-through holes. These precision holes typically have inside diameters of 0.040 to 0.043 inch. The press-fit interface is both mechanically secure and electrically stable. It permits a gas-tight electrical connection to be made between the side walls of the plated-through holes and the contact.

The contacts have square cross sections that deform the sidewalls with enough pressure to cut into the plating to seal out oxides and contamination. The shanks of some contacts are made so they compress on entry into the plated-through hole, but expand to provide more surface contact between the pin and the inner diameter of the hole than rigid shanks.

After insertion of the pins, an insulator housing is seated over the contacts to form the connector receptacle. Some systems allow individual contact replacement and accessibility to the printed circuit conductors for change or repair.

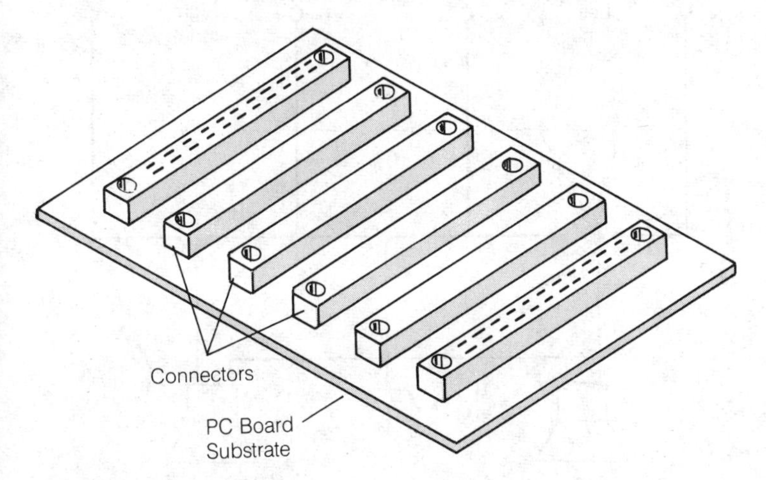

BACKPLANE: A backplane or backpanel can have from 2 to 10 layers. Complient contacts on the connectors are press fit into the panel for interconnection purposes.

BACKSCATTER

Backscatter is a form of ionospheric propagation via the E and F layers (*see* E LAYER, F LAYER).

Generally, when a high-frequency electromagnetic field encounters an ionized layer in the upper atmosphere, the angle of return is roughly equal to the angle of incidence (*see* ANGLE OF INCIDENCE). However, a small amount of the field energy is scattered in all directions, as shown in the drawing. Some energy is scattered back toward the transmitting station, and into the skip zone if there is a skip zone (*see* SKIP ZONE). A receiving station within the skip zone, not normally able to hear the transmitting station, may hear this scattered energy if it is sufficiently strong. This is called backscatter.

Backscatter signals are often, if not usually, too weak to be heard. Under the right conditions, however, communication is possible at high frequencies with backscatter over distances that are normally too short for E-layer or F-layer propagation. Backscatter signals are characterized by rapid fluttering and fading. This makes reception of amplitude-modulated or single-sideband signals almost impossible. The multipath nature of backscatter propagation (*see* MULTIPATH FADING) often makes even CW (continuous wave) reception very poor. *See also* PROPAGATION, PROPAGATION CHARACTERISTICS.

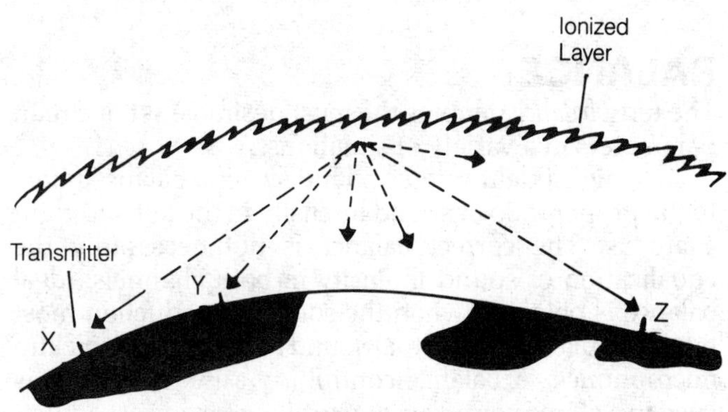

BACKSCATTER: Backscatter from the ionosphere permits radio communication between separated skip zones.

BAFFLE

A baffle is a sound-shielding or sound-reflecting object. Baffles are commonly used in speakers to prevent the backward-radiated sound waves from bouncing off the rear of the speaker enclosure and interfering with the desired forward radiation of sound energy (see illustration). Under certain conditions, a baffle is deliberately not used, or a hole is cut so that a portion of the reflected sound energy is radiated in a forward direction. An example of this is the bass-reflex enclosure (*see* BASS-REFLEX ENCLOSURE).

Baffles are sometimes used in concert halls to direct sound from the stage to the entire audience in a uniform pattern. These baffles may be hung in random configurations from the ceiling of the auditorium.

In an optical communications system, light baffles are sometimes used to reduce the level of interfering background light picked up by the receiving device. Excessive ambient light can degrade the sensitivity and efficiency of such systems. *See also* OPTICAL TRANSMISSION.

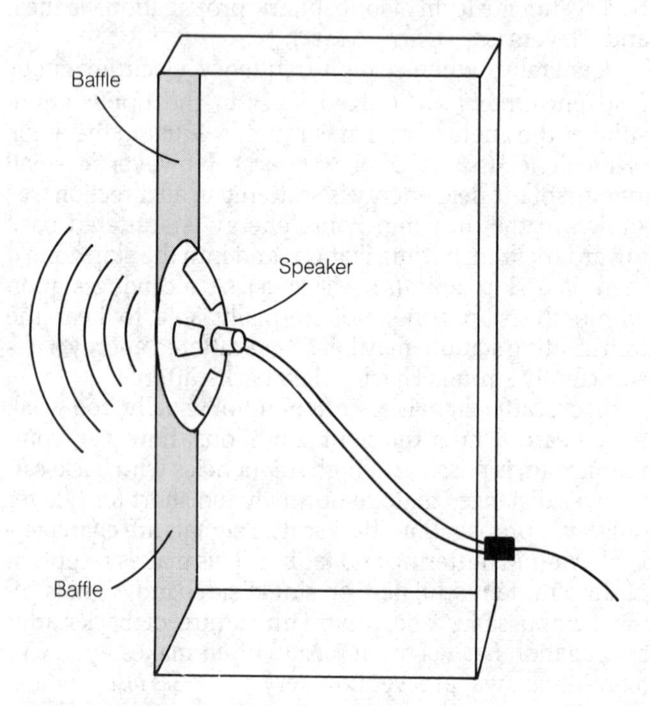

BAFFLE: A baffle prevents sound reflections within the speaker enclosure from altering the quality of the sound output.

BALANCE

The term *balance* denotes the most desirable set of circuit parameters in a variety of situations.

In a high-fidelity stereo sound system, balance refers to the proper ratio of sound intensity in the left and right channels. The correct balance is not necessarily the equalization of sound intensity in both channels; ideal balance is obtained when the sound reproduction most nearly duplicates the actual sound arriving at the pickup microphones. A balance control (*see* BALANCE CONTROL) is used in a stereo sound system to allow for correct adjustment of the balance.

In a radio-frequency transmission line consisting of two parallel conductors, balance is the condition in which the currents in the conductors flow with equal magnitude, but in opposite directions, everywhere along the line. A properly operating parallel-wire transmission line should always be kept in balance (*see* BALANCED TRANSMISSION LINE).

When two or more vacuum tubes or transistors are operated in push-pull or in parallel, it is desirable to balance the devices so that they both have essentially the same gain characteristics. Supposedly identical tubes or transistors often display slightly different amounts of gain under the same conditions. If one tube or transistor draws more current than the other, a phenomenon known as current hogging can take place (*see* CURRENT HOGGING). Differences in input impedances, even if very small, can disrupt the efficiency and linearity of a circuit containing devices in push-pull or in parallel.

In a single-sideband transmitter, carrier balance refers to the most nearly complete suppression of the carrier signal, leaving only the sidebands. An adjustment is usually provided in a balanced modulator to facilitate proper carrier balance (*see* BALANCED MODULATOR).

BALANCE CONTROL

A balance control is usually made up of one or two potentiometers in a high-fidelity stereo sound system for adjusting the balance between the left and right channels (*see* BALANCE). The schematic shows an example of a single stereo balance control consisting of two potentiometers in tandem. The amplifier gain is held constant in this example; the entire output of the left and right channel amplifiers appears across the resistances. This system may be used in low-power configurations. In stereophonic systems with high-power amplifiers, the amplifier input is usually adjusted to control the gain.

The degree of balance necessary to achieve faithful reproduction of the original sound is usually, but not always, obtained when the left channel gain is the same as the right channel gain. The positioning of the speakers, and the characteristics of the room in which the stereo system is used, can affect the relative gain between

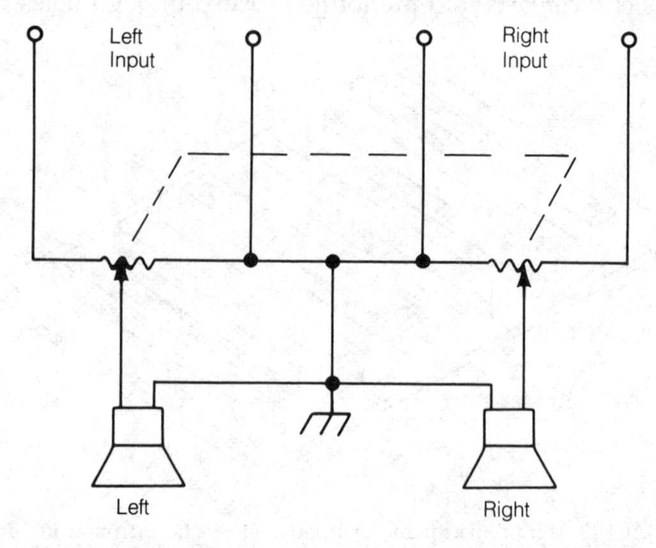

BALANCE CONTROL: Two potentiometers are connected in tandem for sound-balance control.

the channels that results in the best balance. *See also* STEREOPHONICS.

BALANCED CIRCUIT

When an amplifier, oscillator, or modulator contains two elements operating on opposite-phase components of the signal cycle, with equal gain, the circuit is said to be balanced. A push-pull amplifier (see illustration) is one example of a balanced circuit. Each transistor operates on exactly half of the input cycle. Each transistor has the same gain as the other. The combination of their outputs therefore forms a complete, and undistorted, cycle.

Other examples of balanced circuits include the balanced modulator, push-pull amplifier/doubler, astable multivibrator, and parallel-transistor or parallel-tube amplifier. (*See* ASTABLE MULTIVIBRATOR, BALANCED MODULATOR, PARALLEL TRANSISTORS, PARALLEL TUBES.)

If the active components of a circuit, which is intended to be balanced, have unequal gain characteristics or current drain, unbalance will be introduced. This results in degraded circuit performance. It can also lead to current hogging, and possible destruction of one or both of the amplifying or oscillating devices (*see* CURRENT HOGGING).

The direct input and output of a balanced circuit are indicated by the fact that neither lead is grounded. A transformer, however, allows an unbalanced input or load to be used with a balanced circuit.

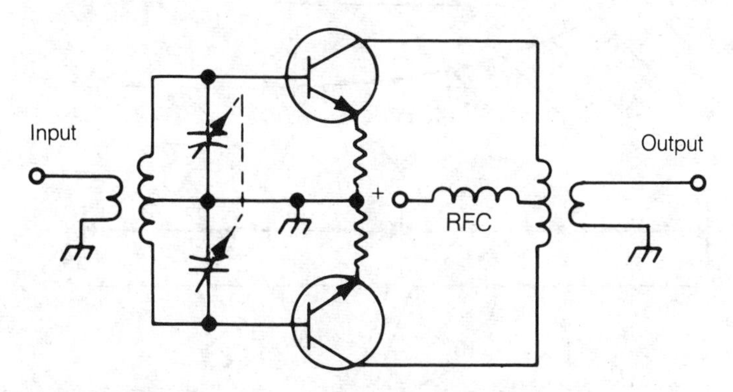

BALANCED CIRCUIT: The push-pull amplifier is an example of a balanced circuit.

BALANCED DETECTOR

A balanced detector is a special kind of demodulator for reception of frequency-modulated (FM) signals. Any FM detector should, if possible, be sensitive only to variations in frequency, and not variations in amplitude, of the received signals. A simple circuit that accomplishes this, called a balanced detector, is shown in the schematic.

The tuned circuits, consisting of L1/C1 and L2/C2, are set to slightly different frequencies. By adjusting C1 so that the resonant frequency of L1/C1 is slightly above the center frequency of the signal, and by adjusting C2 so that the resonant frequency of L2/C2 is slightly below the center frequency of the signal, the circuit becomes a sort of frequency comparator. Whenever the signal frequency is at the center of the channel, midway between the resonant frequencies of L1/C1 and L2/C2, the transistors Q1 and Q2 produce equal outputs, and the output of the balanced detector is zero. When the signal frequency rises, the output of Q1 increases and the output of Q2 decreases. When the signal frequency falls, the output of Q2 increases and the output of Q1 decreases. The filtering capacitors, C_F, smooth out the rf component of the signals, leaving only audio frequencies at the output.

An amplitude-modulated (AM) signal will produce no output from the balanced detector. The outputs of Q1 and Q2 are always exactly equal for an AM signal at the center of the channel, although they may both change as the signal amplitude changes. The gain characteristics of Q1 and Q2, if properly chosen, will minimize the response of the balanced detector to AM signals not at the center of the channel. *See also* DISCRIMINATOR, FREQUENCY MODULATION, RATIO DETECTOR, SLOPE DETECTION.

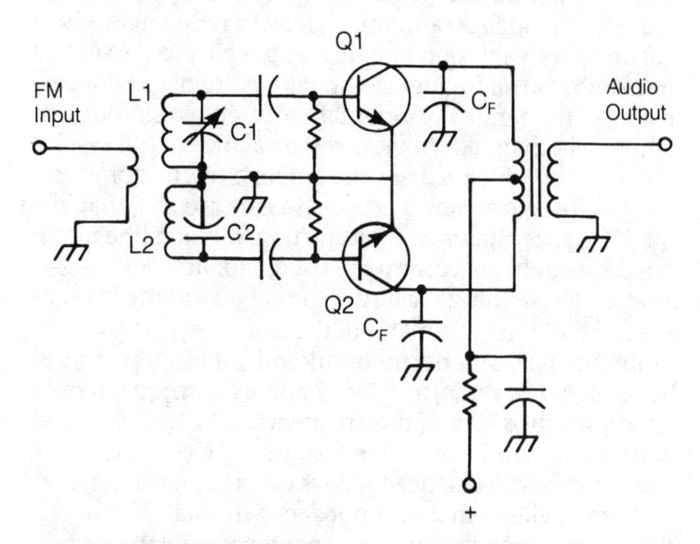

BALANCED DETECTOR: Signal frequency changes cause amplitude changes at the output of the balanced detector.

BALANCED LINE

Any electrical line consisting of two conductors, each of which displays an equal impedance with respect to ground, is called a balanced line if the currents in the conductors are equal in magnitude and opposite in direction at all points along the line.

The advantage of a balanced line is that it effectively prevents external fields from affecting the signals it carries. The equal and opposite nature of the currents in a balanced line results in the cancellation of the electromagnetic fields set up by the currents everywhere in space except in the immediate vicinity of the line conductors.

Precautions must be taken to ensure that balance is maintained in a line that is supposed to be balanced. Otherwise, the line loss increases, and the line becomes susceptible to interference from external fields (in receiving applications) or radiation (in transmitting applications).

Some balanced lines use four, instead of two, parallel conductors arranged at the corners of a geometric square in the transverse plane. Diagonally opposite wires are connected together. This is called a four-wire line. It displays more stable balance than a two-wire line. *See also* FOUR-WIRE TRANSMISSION LINE, OPEN-WIRE LINE, TRANSMISSION LINE, TWIN LEAD.

BALANCED LOAD

A balanced load is a load that presents the same impedance, with respect to ground, at both ends or terminals. This means that the resistance, capacitance and inductance are identical at both terminals. A balanced load may be reversed—that is, the terminals transposed—without affecting circuit performance. A balanced load is required at the termination of a balanced transmission line, to ensure that the currents in the line will be equal and opposite (*see* BALANCED LINE, BALANCED TRANSMISSION LINE).

A good example of a balanced load is a center-fed length of conductor in free space, as shown at A in the illustration. Such an antenna, if fed at a right angle with a balanced transmission line, presents a balanced load to a transmitter at all frequencies in the electromagnetic spectrum. If the length of orientation of either side of this antenna is changed, however, the balance of the system will be upset. One side of the antenna might, for example, be closer to some object (such as the ground) than the other side, as shown at B, or the transmission line might not be brought away from the conductor at a right angle, as at C. These things will ruin the balance of the load, as seen at the transmitter end of the feed line.

In practice, a perfectly balanced load can never truly be achieved in radio-frequency applications; it can only be approached. At audio frequencies, where the load does not radiate appreciable amounts of electromagnetic energy, a nearly balanced load is often achievable.

Some believe that an impedance mismatch between the transmission line and the antenna upsets the system balance. Actually, a perfect impedance match between a line and the load is not essential for balance to exist. While an impedance mismatch does not affect system balance, poor balance can create an impedance mismatch where none would exist if the balance were maintained. *See also* IMPEDANCE MATCHING.

BALANCED MODULATOR

To generate single-sideband signals, a balanced modulator is needed to suppress the carrier energy (*see* SINGLE SIDEBAND). One circuit is shown in simplified form in the drawing.

Transistors Q1 and Q2 act as ordinary amplitude modulators. While the audio input to Q1 is in phase with the audio input to Q2, however, their rf (radio frequency) carrier inputs are out of phase. The collectors are connected in parallel and are thus in phase. The carrier is therefore cancelled in the output circuit, leaving only the sideband energy. The resulting signal is called a double-sideband, suppressed-carrier signal.

There are several ways to build balanced modulators using both active devices (as shown here) and passive devices, such as diodes. All balanced modulators are

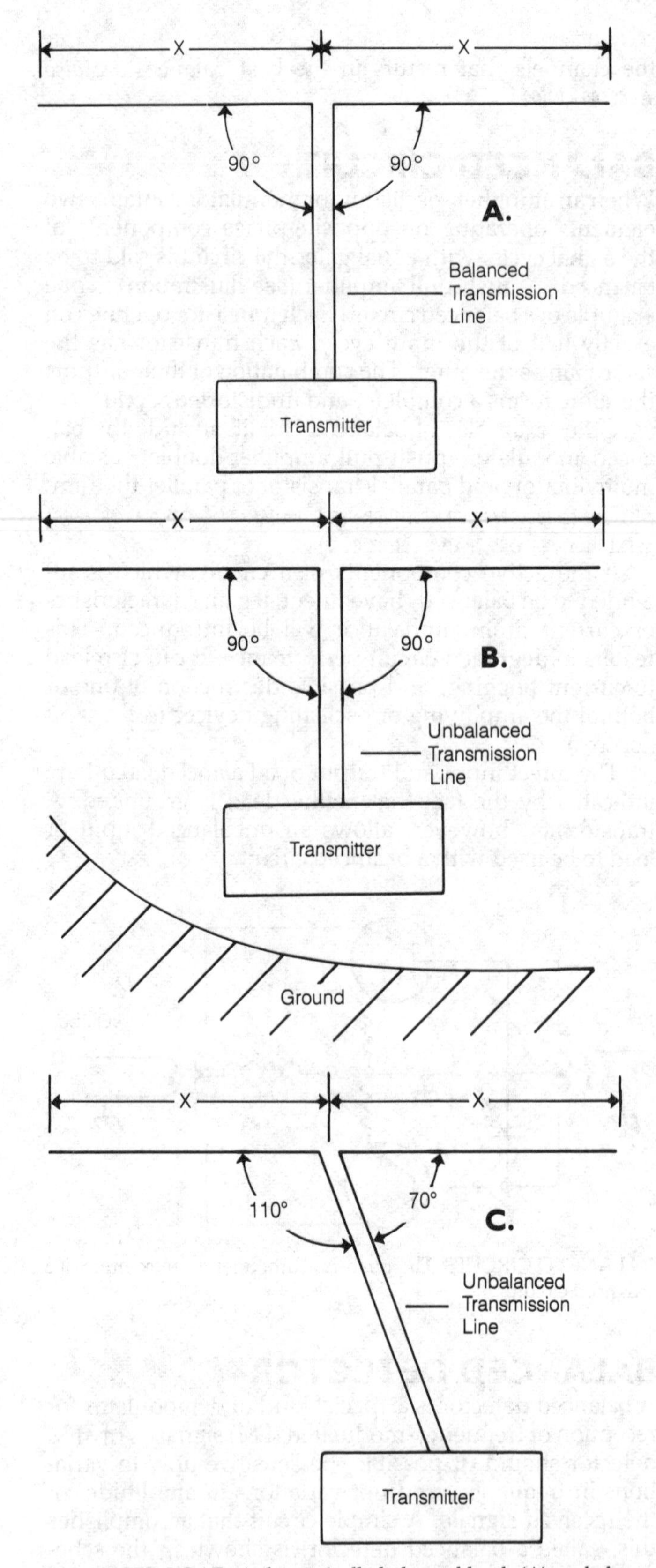

BALANCED LOAD: A theoretically balanced load. (A), unbalance caused by uneven ground (B), and unbalance caused by a nonsymmetrical feed (C).

designed to produce the double-sideband, suppressed-carrier type of signal. To get single-sideband emission, one of the sidebands from the balanced modulator is eliminated either by filtering or by phase cancellation.

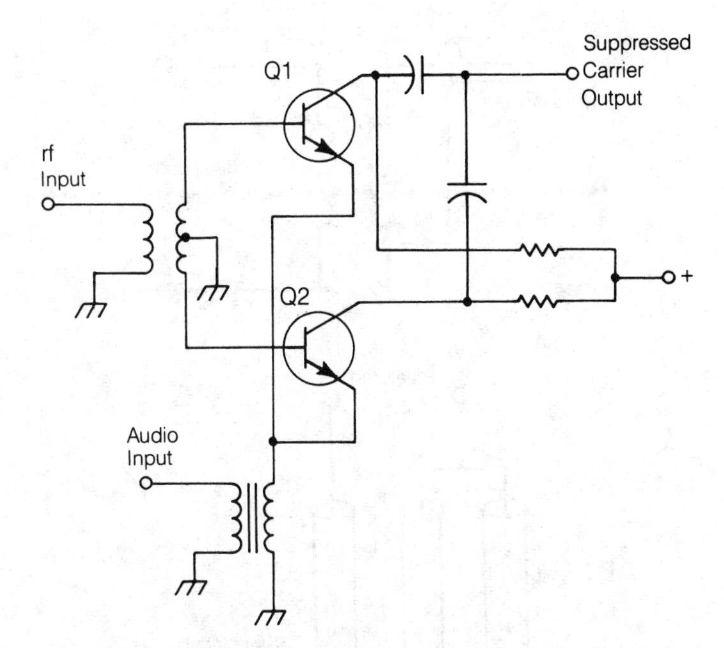

BALANCED MODULATOR: The carrier signal is canceled out, but the sidebands remain in a balanced modulator.

BALANCED OSCILLATOR

See OSCILLATOR CIRCUITS.

BALANCED OUTPUT

A balanced-output circuit is designed to be used with a balanced load and a balanced line. There are two output terminals in this circuit; each terminal displays the same impedance with respect to ground. The phase of the output is opposite at either terminal with respect to the other (see illustration). The peak amplitudes at both terminals are identical.

A balanced output can be used as an unbalanced output (*see* UNBALANCED SYSTEM) by grounding one of the terminals. However, an unbalanced output cannot be used as a balanced output. A special transformer is needed for this purpose. In radio-frequency applications, such a transformer is called a balun, short for *balanced/unbalanced*. *See also* BALUN.

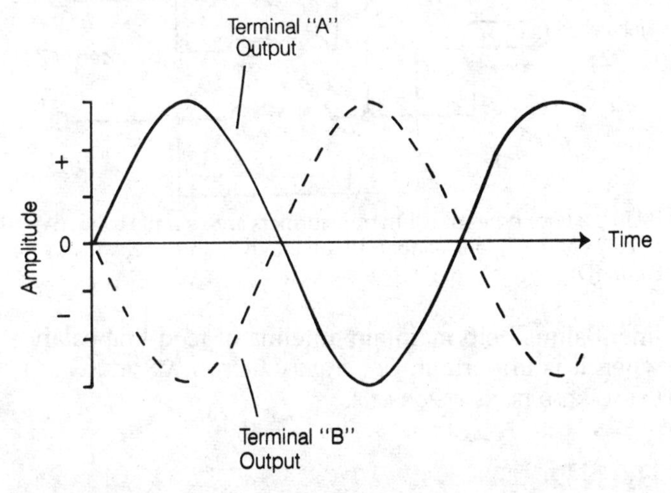

BALANCED OUTPUT: Signals at the two terminals of a balanced output are equal in strength but opposite in phase.

BALANCED TRANSMISSION LINE

A balanced transmission line is a form of balanced line used in radio-frequency antenna transmitting and receiving systems. This line may have two or four parallel wires. A two-wire balanced transmission line operates as illustrated.

The currents in the parallel wires are equal in magnitude but opposite in direction (heavy arrows). These currents set up electric, or E, fields, shown by the dashed lines, and magnetic, or M, fields, shown by the solid circles. The E and M fields are perpendicular to each other everywhere in space, but they are of appreciable strength only in the immediate vicinity of the conductors.

Because the E and M fields are mutually perpendicular, an electromagnetic field is produced. This field propagates in a direction perpendicular to the E and M fields, or right along the conductors of the transmission line. This happens with a speed of almost 300,000,000 meters per second, or the velocity of light.

With a balanced transmission line such as that shown, it is important that the currents in the conductors be equal in amplitude and opposite in direction at all points along the line. This keeps the electromagnetic energy traveling along the line, and prevents it from being radiated into space. If the current balance is upset for any reason, some electromagnetic energy will be radiated away from the line. If an improperly balanced two-wire line is used for receiving, some signal pickup will occur along the line length. This adversely affects the performance of an antenna system, especially a directional one.

Balanced transmission lines usually have lower signal loss than unbalanced lines. In a balanced line, some of the dielectric is air, which has low loss, but in an unbalanced line, the dielectric often is entirely polyethylene, which has somewhat higher loss. Low loss is the chief

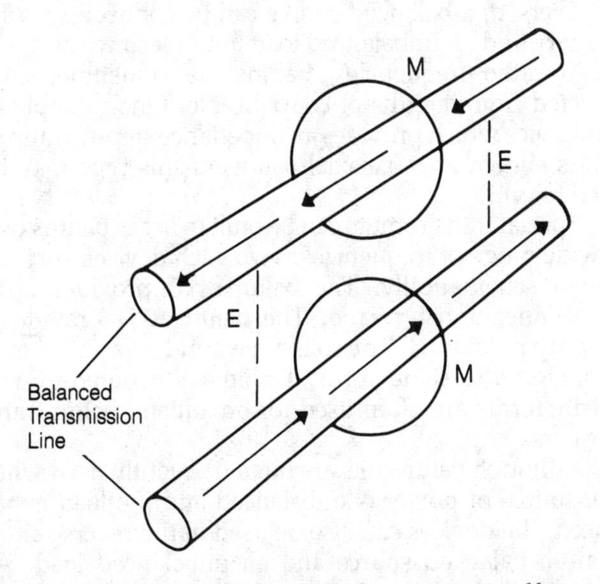

BALANCED TRANSMISSION LINE: Currents are equal but opposite everywhere along the conductors of a balanced transmission line.

advantage of open-wire balanced lines. Unbalanced lines are more convenient to install because they are less susceptible to the effects of nearly metallic objects. *See also* FOUR-WIRE TRANSMISSION LINE, OPEN-WIRE LINE, TRANSMISSION LINE, TWIN LEAD, UNBALANCED TRANSMISSION LINE.

BALLAST

The term *ballast* refers to current regulation or control under conditions of changing voltage. Without ballast, a sudden, large increase in the voltage supplied to a circuit can result in malfunction or damage. In most electronic circuits today, voltage regulation in power supplies makes ballast devices unnecessary (*see* POWER SUPPLY). Ballast devices are sometimes used with portable ac generators, which have somewhat variable output voltages.

A ballast lamp is a bulb that exhibits increased resistance with voltage. This keeps the dissipated power nearly constant. If there is a sudden surge in the voltage, a ballast lamp will not burn out. If the voltage drops somewhat, the ballast lamp will maintain its brilliance.

A ballast tube maintains constant current flow in a circuit when connected in series under unstable voltage conditions. A ballast resistor operates on the same principle. The greater the input voltage, the larger the voltage drop across the resistor, and the flow of current remains nearly constant.

BALUN

When a balanced load is connected to an unbalanced source of power, a balun may be used. The word *balun* is a contraction of *bal*anced/*un*balanced. A balun has an unbalanced input and a balanced output.

The simplest form of balun is an ordinary transformer. The isolation between the primary and secondary windings allows an unbalanced source of power to be connected to one end, and a balanced load to the other, as shown at A in the illustration. This condition can also be reversed; a balanced source can be connected to the primary and an unbalanced load to the secondary.

At radio frequencies, baluns are sometimes constructed from lengths of coaxial cable. One example of this device which provides an impedance step-up ratio of 1:4, is shown at B. Coaxial baluns of this type may be used for vhf.

Special transformers can be built to act as baluns over a wide range of frequencies. Two such devices are illustrated schematically. The balun at C provides a 1:1 impedance-transfer ratio. The balun at D provides a step-up ratio of 1:4. The coils are wound adjacent to each other on the same form. Toroidal powdered-iron or ferrite forms are often used for broadband balun transformers.

Although balun coils are quite frequently used when the source of power is unbalanced and the load is balanced, the devices can also be used in the reverse situation—a balanced source and an unbalanced load. An example of the first type of application is a dipole or Yagi fed with coaxial cable. An example of the latter application is a vertical ground-plane antenna fed with balanced

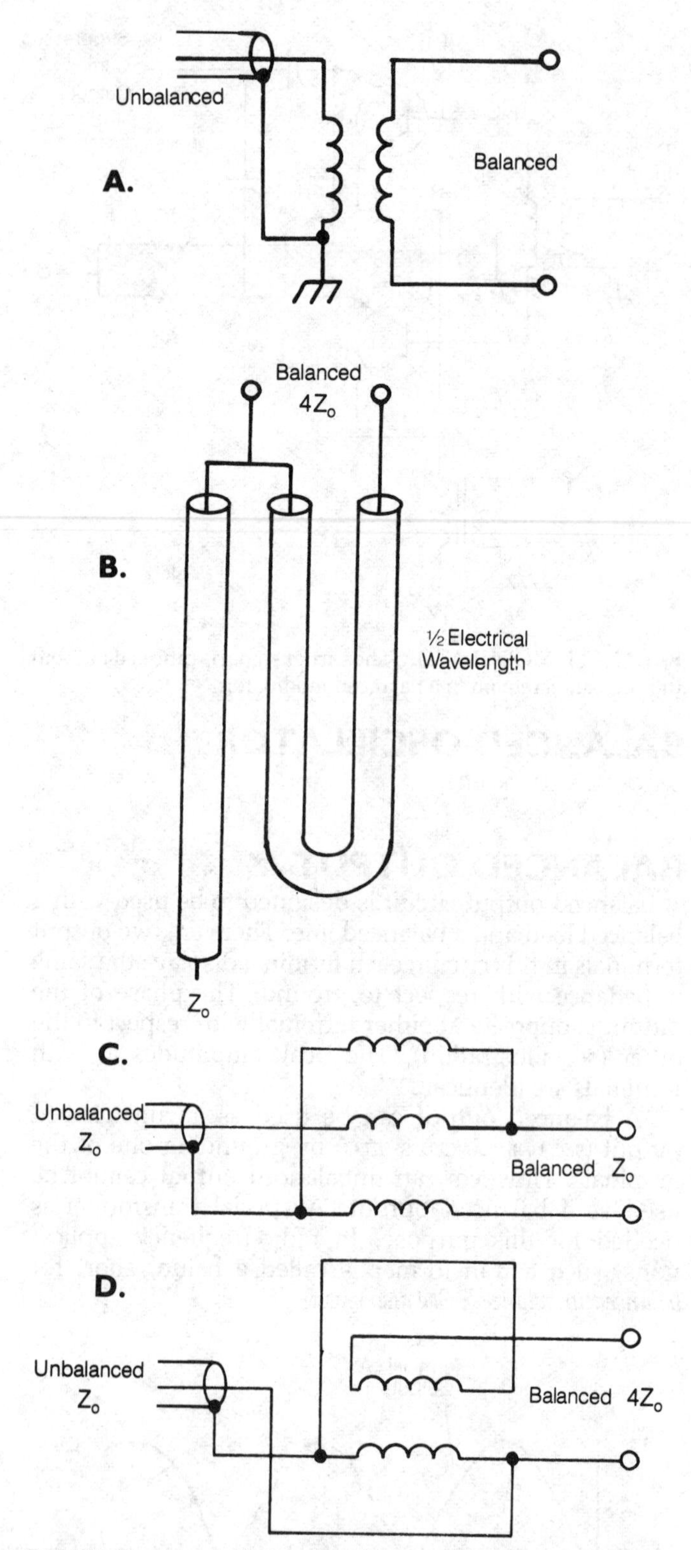

BALUN: Four types of balun transformers are a simple balun (A), a coaxial balun (B), a broadband 1:1 balun (C), and a broadband 1:4 balun (D).

line. Baluns help maintain antenna or feed-line balance when it is important. *See also* BALANCED TRANSMISSION LINE, UNBALANCED TRANSMISSION LINE.

BAND

Any range of electromagnetic frequencies, marked by lower and upper limits for boundary purposes, is called a band. The AM (amplitude modulation) broadcast band is from 535 to 1605 kHz. The FM (frequency modulation) broadcast band extends from 88 to 108 MHz. The 40-meter amateur band has a lower limit of 7.000 MHz and an upper limit of 7.300 MHz. There is no limit to how narrow or wide a band can be; it must only cover a certain range.

In radio-communication, the electromagnetic spectrum is subdivided into bands according to frequency, as shown in the table. The very low frequency (vlf) band is only 21 kHz wide, but it covers more than a threefold range of frequencies. All of the higher bands cover a tenfold range of frequencies. For a more detailed breakdown of the bands in the radio-frequency spectrum, *see* FREQUENCY ALLOCATIONS, SUBBAND.

BAND: Radio-frequency bands.

Designation	Frequency	Wavelength
Very low (vlf)	9 kHz–30 kHz	33 km–10km
Low (lf)	30 kHz–300 kHz	10 km–1 km
Medium (mf)	300 kHz–3 MHz	1 km–100 m
High (hf)	3 MHz–30 MHz	100 m–10 m
Very high (vhf)	30 MHz–300 MHz	10 m–1 m
Ultrahigh (uhf)	300 MHz–3 GHz	1 m–100 mm
Super high (shf)	3 GHz–30 GHz	100 mm–10 mm
Extremely high (ehf)	30 GHz–300 GHz	10 mm–1 mm

BAND GAP

An atom has certain allowed amounts of energy called energy levels. These levels differ because of differences in atomic structure, dynamics and temperature. The energy level of an electron in an atom is related to its orbit; the larger the orbit, the greater the energy level of the electron. A given electron can have only certain allowed energy levels. Consequently, all other energy levels are forbidden, and the electron will only be found in the allowed energy levels as shown in Fig. 1.

When atoms combine to form a crystal, variations in energy level of individual atoms are introduced because of the proximity of adjacent atoms. As a result, the precise levels are expanded into bands. Within a crystal, electrons can be found only in an allowed energy band. The bands where electrons are forbidden are called forbidden energy gaps or band gaps.

Each atom has a filled valence band in which electrons participate in bonding the crystal, and each atom has a conduction or empty band located at a higher energy level. The two bands are separated by the forbidden energy gap, also called the band gap. The width of the band gap is equal to the amount of energy a valence electron must gain to break its covalent bond and become a conduction band electron. The band gap is measured in electron volts (eV). (An electron volt is the amount of energy one electron acquires in traveling through a potential difference of one volt.)

Figure 2 shows typical energy level bands (a) in an insulator, (b) in a semiconductor, and (c) in a metal. The energy bands of an insulator can be 7 eV or more. However, silicon (S) has a band gap 1.1 eV, and gallium arsenide (GaAs) has a band gap of 1.4 eV. (Both materials are semiconductors.) In a metal, the energy bands of conduction and valence overlap, aiding in the passage of electrons.

At absolute zero temperature, an intrinsic semiconductor, essentially a pure crystal, has a filled valence band and an empty conduction band. As can be seen in Fig. 2, the empty band acts like an insulator, but it has a narrower band gap. However, as the temperature of an intrinsic semiconductor rises, some electrons jump to the

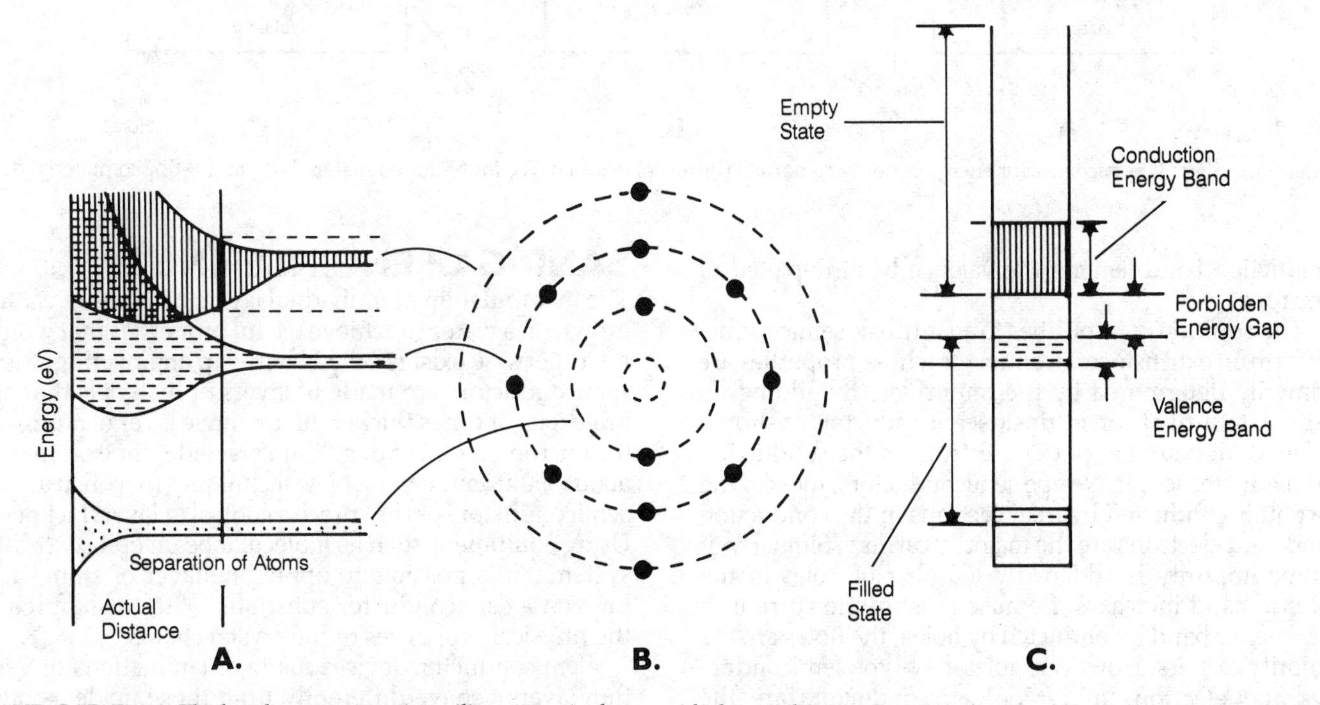

BAND GAP: Fig. 1. The band-gap concept in a silicon atom is shown with the overlapping or *smear* of energy bands shown at (A), the Bohr concept of atomic structure (B), and the energy band levels (C).

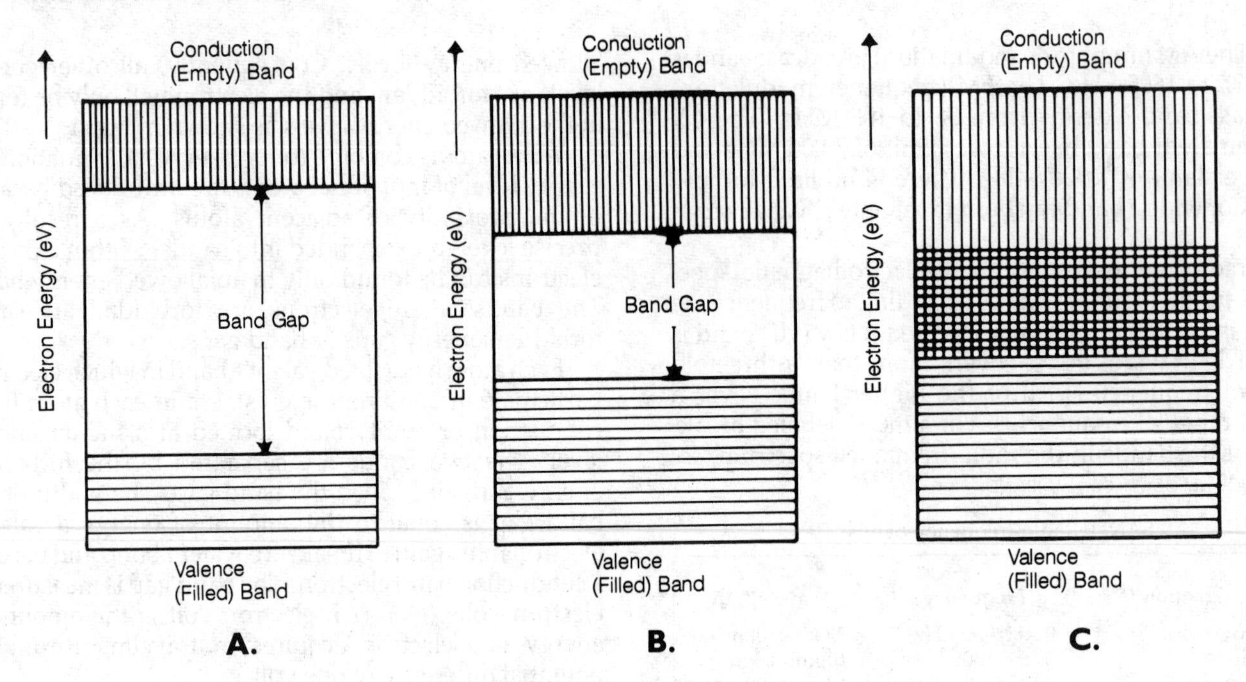

BAND GAP: Fig. 2. Typical energy bands for an insulator (A), a semiconductor (B), and a metal (C).

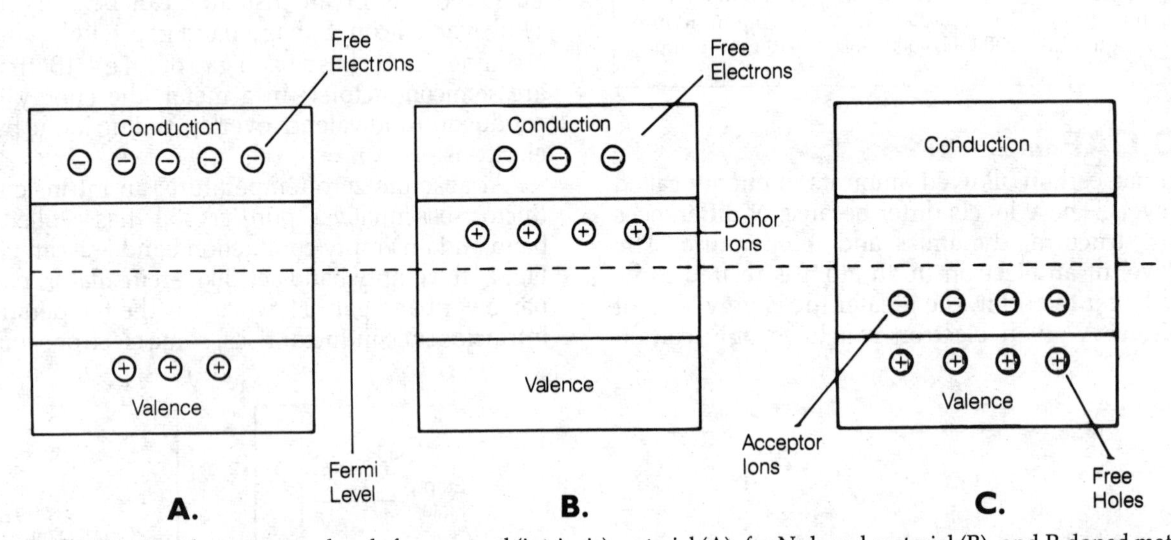

BAND GAP: Fig. 3. Semiconductor energy bands for a normal (intrinsic) material (A), for N-doped material (B), and P-doped material (C).

conduction band leaving the valence band depleted of electrons.

The addition of impurities to an intrinsic semiconductor form an extrinsic semiconductor whose properties are primarily determined by the impurities. By adding an N-type impurity to an intrinsic semiconductor, as shown in Fig. 3, the number of free electrons in the conduction band is increased. In N-type semiconductors, most of the current is conducted by free electrons in the conduction band, and electrons are the majority carriers. Similarly, if P-type impurity is added, the number of holes in the valence band increases. Because most of the current in the valence band is conducted by holes, the holes are the majority carriers. However, holes in N-type semiconductors and electrons in P-type semiconductors are the minority carriers.

BAND-GAP ENGINEERING

The manipulation of individual layers of atoms in crystal grown on a wafer to achieve useful properties that would not otherwise exist is called band-gap engineering. Most semiconductors are made of layers of materials that are hundreds of times thicker than a single layer of atoms. A 0.1 micrometer layer of gallium arsenide, for example, is about 350 atomic layers. New techniques for constructing semiconductors permit precise control of layer thickness. Using equipment such as molecular beam epitaxy (MBE) systems, it is possible to apply one layer of atoms at a time on a semiconductor substrate. With this approach the physical properties of the crystal change.

Semiconductor devices made of laminations of very thin layers behave differently from those made as bulk crystals. It is now possible to make devices such as diode lasers and transistors with electronic or optical properties unobtainable with conventional techniques of diffusion: epitaxy and ion implantation. The superthin layers have a direct effect on the band gap of semiconductor atoms, or the difference in energy between the valance and conduction bands of those atoms. (*See* BAND GAP.) The velocity at which electrons or holes may travel through a crystal lattice of atoms can be raised or lowered. The wavelength of light emitted by electrons confined to planes only nanometers thick can be tuned.

Band-gap engineering can be done by: 1) alloying to change a semiconductor chemical composition, 2) forming layers of different (heterogeneous) materials together to form heterojunctions, and 3) introducing mechanical strain between crystal layers. So far, band-gap engineering has been most successful when applied to gallium arsenide and other compounds of elements from Groups III and V of the periodic table. Practical photonic devices have been made from alternating layers of gallium arsenide and aluminum gallium arsenide. *See* COMPOUND SEMICONDUCTOR, EPITAXY.

BANDPASS FILTER

Any resonant circuit, or combination of resonant circuits, designed to discriminate against all frequencies except a frequency f_0, or a band of frequencies between two limiting frequencies of f_0 and f_1 is called a bandpass filter. In a parallel circuit, a bandpass filter shows a high impedance at the desired frequency or frequencies, and a low impedance at unwanted frequencies. In a series configuration, the filter has a low impedance at the desired frequency or frequencies, and a high impedance at unwanted frequencies. Three types of bandpass filter circuits are illustrated.

Some bandpass filters are built with components other than true coils and capacitors, but all these filters operate on the same principle. The crystal-lattice filter uses piezoelectric materials, usually quartz, to obtain a bandpass response (*see* CRYSTAL-LATTICE FILTER). A mechanical filter uses vibration resonances of certain substances. A resonant antenna is itself a form of bandpass filter, since it allows efficient radiation at one frequency and neighboring frequencies, but discriminates against others. In optics, a simple color filter, discriminating against all light wavelengths except within a certain range, is a form of bandpass filter.

Bandpass filters are designed to have a very sharp, defined, resonant frequency. Sometimes the resonance is spread out over a fairly wide range. *See also* BANDPASS RESPONSE.

BANDPASS RESPONSE

The attenuation-versus-frequency characteristics of a bandpass filter is called the bandpass response (*see* ATTENUATION-VS-FREQUENCY CHARACTERISTIC). A bandpass filter may have a single, well-defined resonant frequency (A in the illustration), denoted by f_0; or, the response might be rectangular, having two well-defined limit frequencies f_0 and f_1, as shown at B. The bandwidth (*see* BANDWIDTH) might be just a few hertz, such as with an audio filter designed for reception of Morse code. Or, the bandwidth might be several MHz, as in a helical filter designed for the front end of a vhf receiver.

A bandpass response is always characterized by high attenuation at all frequencies except within a particular range. The actual attenuation at desired frequencies is called the insertion loss (*see* INSERTION LOSS).

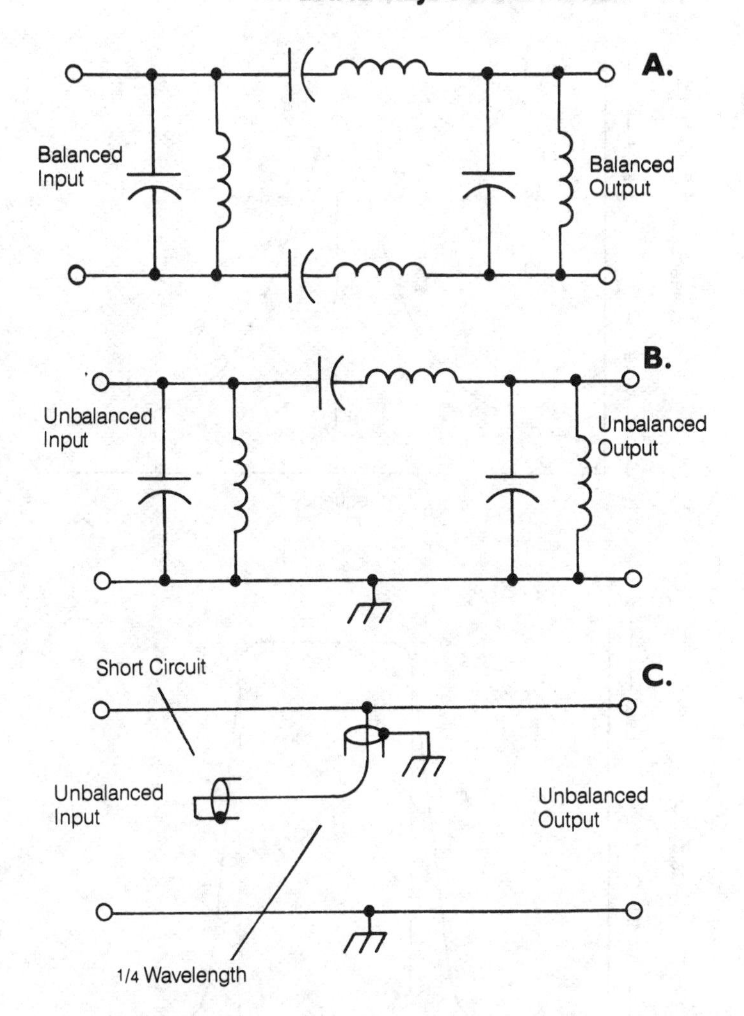

BANDPASS FILTER: A balanced LC (inductive-capacitive) filter (A), an unbalanced LC filter (B), and an unbalanced filter made from a length of coaxial cable (C).

BAND-REJECTION FILTER

A band-rejection filter is a resonant circuit designed to pass energy at all frequencies, except within a certain range. The attenuation is greatest at the resonant frequency, f_0, or between two limiting frequencies, f_0 and f_1. Three examples of band-rejection filters are shown in the illustration. Note the similarity between the band-rejection and bandpass filters (*see* BANDPASS FILTER). The fundamental difference is that the band-rejection filter consists of parallel LC circuits connected in series with the signal path, or series LC circuits in parallel with the signal path; in bandpass filters, series-resonant circuits are connected in series, and parallel-resonance circuits in parallel.

Band-rejection filters need not necessarily be made up of actual coils and capacitors, but they usually are.

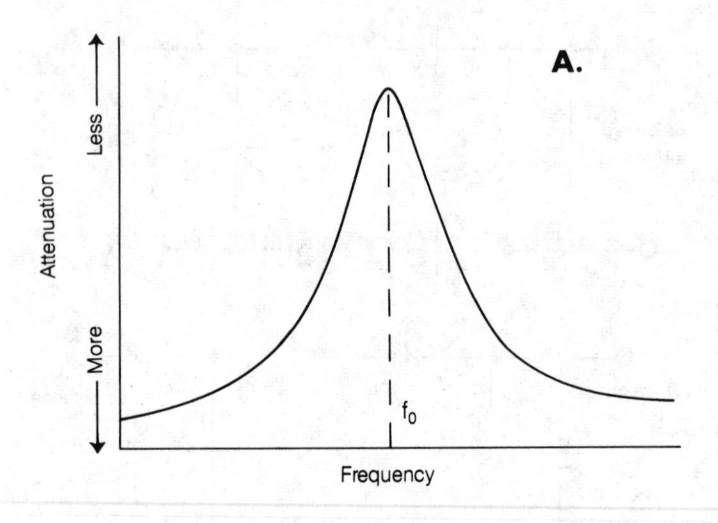

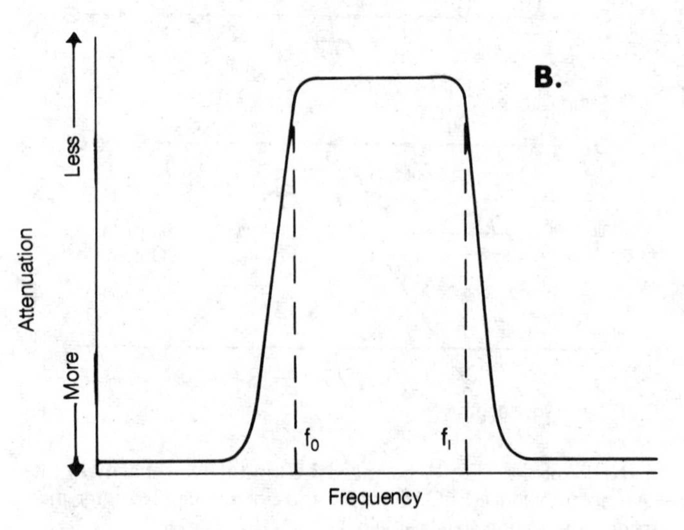

BANDPASS RESPONSE: A peaked single-frequency bandpass response (A), and a wide-band or flat-topped bandpass response (B).

Quartz crystals are sometimes used as band-rejection filters. Lengths of transmission line, either short-circuited or open, act as band-rejection filters for certain frequencies. A common example of a band-rejection filter is the notch filter found in some of the more advanced communications receivers. Another example is the antenna trap. Still another is the parasitic suppressor generally seen in the plate lead of a high-power rf (radio frequency) amplifier. *See also* NOTCH FILTER, PARASITIC SUPPRESSOR, TRAP.

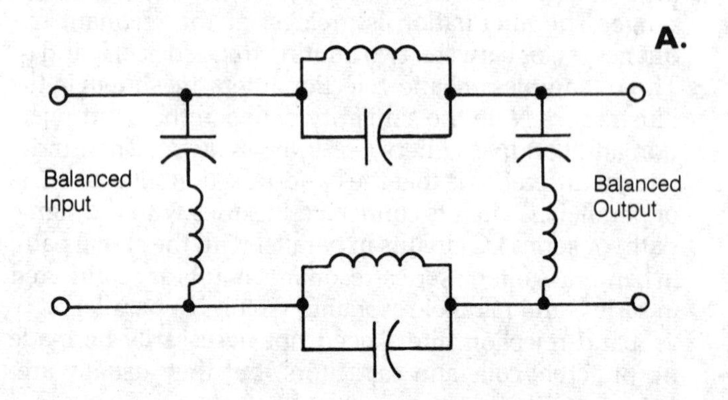

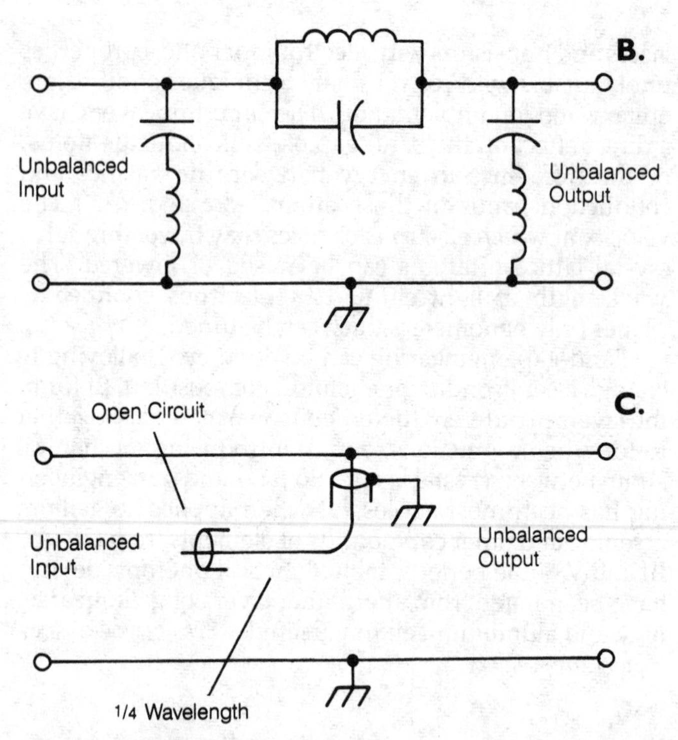

BAND-REJECTION FILTER: A balanced LC (inductive-capacitive) filter (A), an unbalanced LC filter (B), and an unbalanced filter made from a length of coaxial cable (C).

BAND-REJECTION RESPONSE

All band-rejection filters show an attenuation-vs-frequencies characteristic (*see* ATTENUATION-VS-FREQUENCY CHARACTERISTIC) marked by low loss at all frequencies except within a prescribed range. The illustration shows two types of band-rejection response. A sharp response occurs at or near a single resonant frequency f_0. A rectangular response is characterized by low attenuation below a limit f_0 and above a limit f_1, and high attenuation between these limiting frequencies.

Most band-rejection filters have a relatively sharp response. This is true of antenna traps and notch filters. Parasitic suppressors, used in high-frequency power amplifiers to prevent vhf parasitic oscillation, have a more broadband response. *See also* NOTCH FILTER, PARASITIC SUPPRESSOR, TRAP.

BANDWIDTH

Bandwidth is a term used to define the frequency occupied by a signal and required for effective transfer of the information to be carried by that signal. The term is also sometimes used in reference to the form of a bandpass or band-rejection filter response (*see* BANDPASS RESPONSE, BAND-REJECTION RESPONSE).

The bandwidths of several common types of signals are given by the table. The receivers for these signals must have a bandpass response at least as great as the signal bandwidth. If the receiver bandpass response is too narrow, the signal cannot be readily understood. In the extreme, where the receiver bandpass response is very narrow compared to the signal bandwidth, no information can be conveyed.

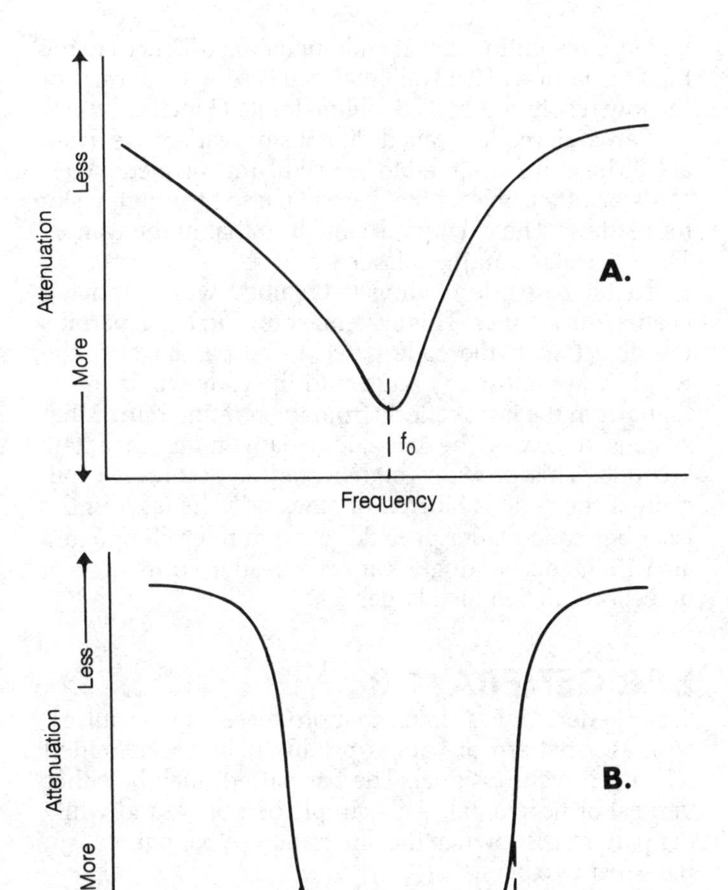

BAND-REJECTION RESPONSE: A null at a single frequency, band-rejection response (A), and a wideband band-rejection response (B).

While an excessively narrow bandpass is obviously undesirable for a signal having a given bandwidth, an unnecessarily wide receiver response is wasteful. If the response is much wider than the bandwidth of the signal, a lot of noise enters the channel along with the desired signal. Other signals on nearby frequencies may also enter the channel and disrupt communications. Therefore, the receiver bandpass response should always be tailored to the bandwidth of the signal received.

Bandwidth is sometimes used as a specification for the sharpness of a bandpass filter response. Two frequencies, f_0 and f_1, are found, at which the power attenuation is 3 decibels with respect to the level at the center frequency of the filter. The bandwidth is the difference between these two frequencies. In general, if a signal occupies a certain bandwidth, the receiver filter bandwidth (as determined from the 3-decibel power attenuation points) should be the same for best reception.

BANDWIDTH: Bandwidths of some common signals.

Emission Type	Typical Bandwidth
Morse code (CW)	10 Hz–250 Hz
Radioteletype (RTTY)	200 Hz–800 Hz
Single sideband (SSB)	3 kHz
Slow-scan television (SSTV)	3 kHz
Amplitude modulation, voice (AM)	6 kHz
Amplitude modulation, music (AM)	10 kHz
Frequency modulation, voice (FM)	10 kHz
Frequency modulation, music (FM)	100 kHz
Television (TV)	3 MHz–10 MHz

BAR-CODE READER

Bar-code readers convert the universal bar-code formats now printed on retail items into digital signals that can be interpreted by a computer for registering the sale, providing current price data, and maintaining the inventory. The bar code (a stamp-sized patch of alternating thick and thin stripes with different spacing on a white background) is printed or affixed to retail merchandise. The narrowest bars printed have a nominal width of 0.3 millimeter (0.012 inch). Each bar translates into a number 1 or 0.

The digital bar code is a highly effective alternative to keyboard data entry. Bar-code scanning is faster and more accurate than key entry and permits faster merchandise checkout. In addition, bar-code scanning has a higher first-read rate and greater data accuracy than optical character recognition. When compared to mag-

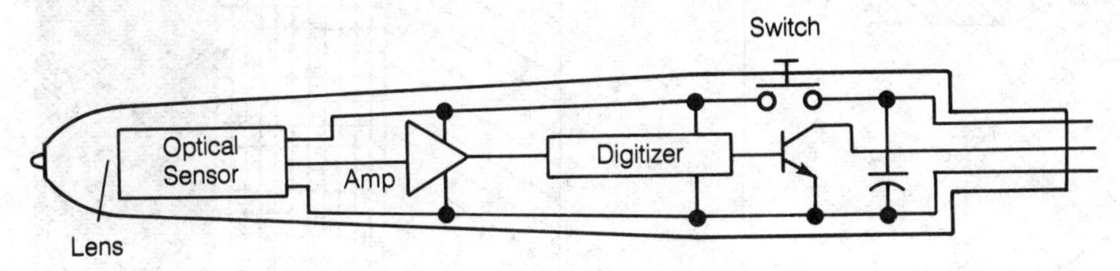

BAR-CODE READER: Fig. 1. Handheld bar-code reader (called a wand) contains an optical detector with an infrared or light emitter and detector combination to read printed bar codes.

netic stripe encoding, bar code offers significant advantages in flexibility of media, symbol placement, and immunity to electromagnetic fields.

Almost all products that are widely distributed for sale in retail stores are now marked with this product code. Retail stores have a central computer programed to translate the code to the product name as well as quantity and price information as desired. The computer keeps the name and current price of the item in memory and sends that information back to the point-of-sale terminal for display and printout on the customer's receipt. Then the quantity of the item purchased is subtracted from the quantity in inventory maintained by the computer.

Bar-code reading is also used in other applications, including remote data collection, ticket identification, security checkpoint verification, work-in-process tracking, material handling, inventory control, library circulation control, medical file folder tracking, magazine, book or general publication distribution, and identification of work in progress for manufacturing, repair, or testing.

The handheld bar-code reader (called a wand) is illustrated in Fig. 1. The wands contain an optical LED sensor sensitive to 655 nanometer (nm) visible red light, 700 nm visible red light or an 820 nm infrared energy, a photodetector IC (integrated circuit), and precision aspheric optics. Internal signal conditioning circuitry converts the optical data into a logic level pulse width representation of the bars and spaces.

In addition to the optical sensor, the wand contains an analog amplifier, a digitizing circuit, and an output transistor. These elements provide a digital logic compatible output from a single voltage supply in the 3 to 6 V range. A nonreflecting black bar results in a logic high (1) level, and a reflecting white space will cause a logic low (0). The spot sizes of wands can vary from the general purpose resolution of 0.19 millimeter (0.0075 inch) to the high resolution of 0.13 millimeter (0.005 inch) spot size or the low resolution of 0.38 millimeter (0.13 inch).

Bar-code readers called digital slot readers are available. These are designed to be used with encoded paper, cards or other objects that can be passed through a slot for reading. The circuitry is similar to that of the wands. There are also handheld laser scanners.

Faster bar-code reading is obtained with a moving beam laser scanner. This system, shown in Fig 2, permits the object with the code patch to be passed over the window at almost any angle and the code will be read. Light from the laser reflects from the rotating mirror that sweeps it across the bar code many times in a few seconds. This provides higher reading accuracy, especially if the code is blurred or smeared. The laser in the laser bar-code reader provides a much finer illuminated area than the handheld bar-code reader so its area of detection can be much larger.

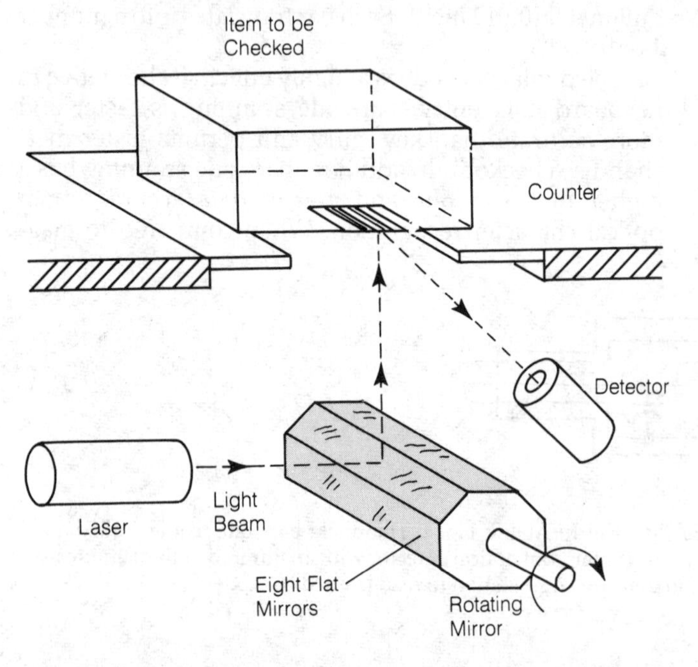

BAR-CODE READER: Fig. 2. Higher accuracy and reliability is obtained because the laser in the laser-scanning bar-code reader does multiple scanning as long as the coded item is held over the aperture.

BAR GENERATOR

A bar generator is a circuit that produces regular pulses, so that a pattern of bars is produced in a transmitted television picture signal. The bar pattern may be either vertical or horizontal. An example of a black-and-white bar pattern is shown in the illustration. A bar pattern can have just two bars.

The pattern produced by a bar generator is used to check for vertical or horizontal linearity in a television transitter and receiver. Vertical bars facilitate alignment of horizontal circuits. Horizontal bars allow adjustment for vertical linearity. In a color television system, colored bars are used to align the circuits for accurate color reproduction. In a black-and-white system, the various shades of gray are used to adjust the relative brightness and contrast. The lines of demarcation between the bars allows adjustment of the focus.

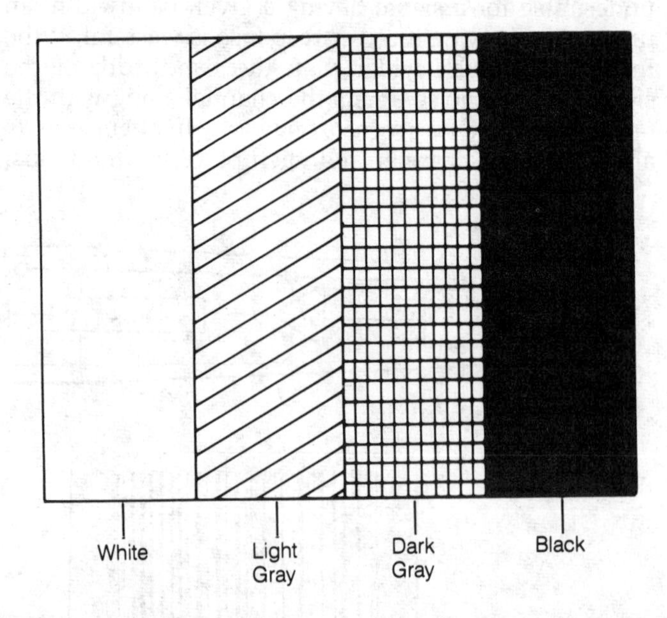

BAR GENERATOR: A black-and-white bar-generator pattern for adjusting horizontal linearity, focus, contrast, and brightness on a television receiver.

A bar pattern can be used for complete alignment of a television transmitting and receiving system. *See also* TELEVISION.

BARIUM

Barium is an element with an atomic number of 56 and an atomic weight of 137. This element is oxidized and used in the cathodes of vacuum tubes. A coating of barium oxide prolongs the life, and enhances the electron-emitting properties, of the cathode.

A substance called barium titanate, a compound of barium and titanium, can be used in place of quartz in piezoelectric crystals. *See also* CRYSTAL, PIEZOELECTRIC EFFECT, QUARTZ, SONAR, SUPERCONDUCTIVITY.

BARKHAUSEN EFFECT

When the grid of an amplifier tube is sufficiently capacitively coupled to the plate, oscillation at very-high or ultrahigh frequencies may take place. This oscillation is called vhf or uhf parasitic oscillation (*see* PARASITIC OSCILLATION) because it is usually undesirable.

Parasitic oscillation may occur without amplifier drive. In this case it can be detected by the presence of unusual levels of plate current while the exciter is off (*see* EXCITATION). Parasitic oscillation can also take place when there is sufficient drive to cause the grid to go positive during part of the cycle. In this case, a spectrum analyzer is usually necessary to ascertain the existence of vhf (very high frequency) or uhf (ultrahigh frequency) parasitics.

Barkhausen effect is deliberately created in a circuit called a Barkhausen-Kurz oscillator. *See also* BARKHAUSEN-KURZ OSCILLATOR.

BARKHAUSEN-KURZ OSCILLATOR

A Barkhausen-Kurz oscillator is a special kind of oscillator for generating ultrahigh frequency energy. The tube used has a grid voltage that is positive with respect to all the other electrodes, including the plate. Variations in the electric field between the grid and the plate of the tube cause electrons to oscillate in the interelectrode space. The frequency of oscillation is controlled by external tuned circuits.

Barkhausen oscillation can occur in a tube even when it is not wanted. This form of oscillation is called parasitic oscillation. Parasitic oscillation is detrimental to the performance of an amplifier because it produces outputs on frequencies not intended. However, in the Barkhausen-Kurz oscillator, this form of oscillation is used constructively. *See also* OSCILLATION, OSCILLATOR, PARASITIC OSCILLATION.

BARNETT EFFECT

When a cylinder of iron or other magnetic material is rotated about its axis, as shown in the drawing, a small amount of magnetization occurs. This is called the Barnett effect.

Magnetized metals differ from ordinary metals in that the atoms are aligned along a certain axis to a greater extent than would occur by probability. This happens only in metals containing iron and/or nickel. Magnetization causes attractive or repulsive forces. The Barnett effect results in permanent magnetization. *See also* MAGNETIC MATERIAL, MAGNETIZATION, PERMANENT MAGNET.

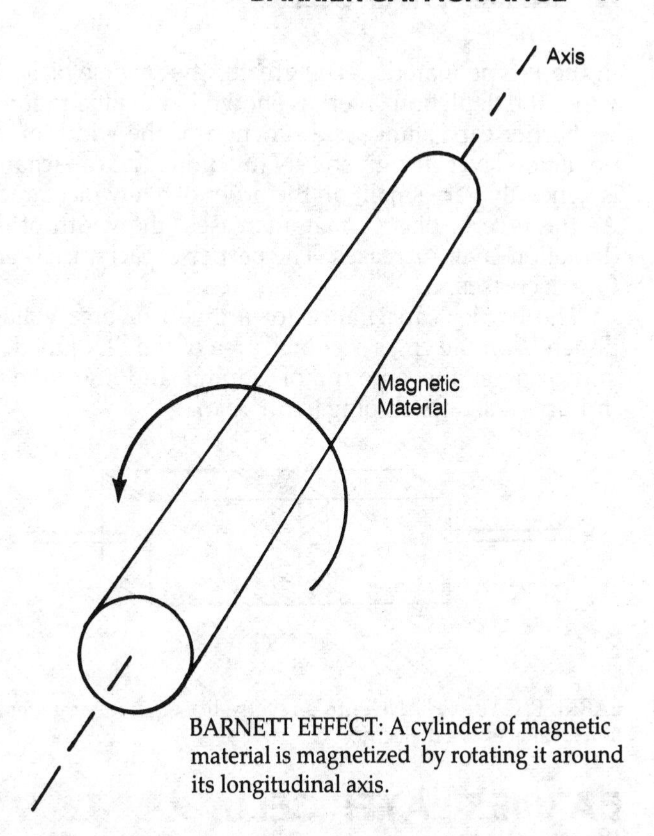

BARNETT EFFECT: A cylinder of magnetic material is magnetized by rotating it around its longitudinal axis.

BARRETTER

A barretter is a form of ballast devise (*see* BALLAST). A barretter tube consists of a helically wound length of iron wire inside a hydrogen-filled container. When connected in series at the output of a power supply or at the input to a circuit, the barretter tube keeps the current flow at a constant value for a wide range of voltages. Thus it acts as a current regulator for the circuit.

A resistor with a positive temperature coefficient (*see* TEMPERATURE COEFFICIENT), deliberately designed for a certain amount of resistance increase for each degree of temperature rise, is sometimes called a barretter or barretter resistor. This resistor should not be confused with a thermistor, which is a semiconductor device. *See also* THERMISTOR.

BARRIER CAPACITANCE

When a semiconductor P-N junction is reverse biased so that it does not conduct, capacitance exists between the P-type and N-type semiconductor materials, called barrier capacitance. It is also sometimes called the depletion-layer or junction capacitance.

Under conditions of reverse bias, a depletion region, or potential barrier, forms between the P-type and N-type materials (*see* DEPLETION LAYER). Positive ions are created in the N-type material, and negative ions are created in the P-type material. The greater the reverse bias, the wider the depletion layer, as shown in the illustration.

Barrier capacitance is a function of the width of the depletion layer in a given P-N junction. The capacitance is generally very small, on the order of a few picofarads. As the reverse-bias voltage increases, the width of the depletion layer increases. The barrier capacitance therefore decreases.

The barrier capacitance for a given reverse voltage depends on the cross sectional area of the P-N junction, and also on the amount of doping and the kind of impurity used for doping (*see* DOPING).

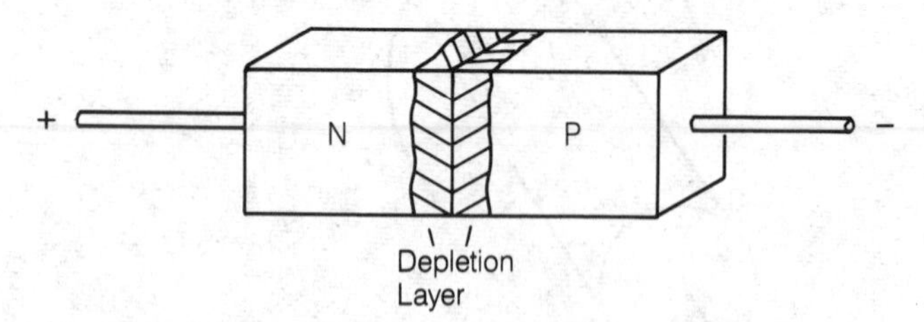

BARRIER CAPACITANCE: In a reverse-biased diode, the depletion layer acts as a dielectric.

BARRIER-LAYER CELL

See PHOTODETECTOR.

BARRIER VOLTAGE

Barrier voltage is the forward bias necessary to cause conduction across a junction of two unlike materials. In a semiconductor P-N junction, the barrier voltage is approximately 0.3 V for germanium and 0.6 V for silicon.

The forward-voltage-vs-current characteristic for a semiconductor diode looks like that shown in the graph.

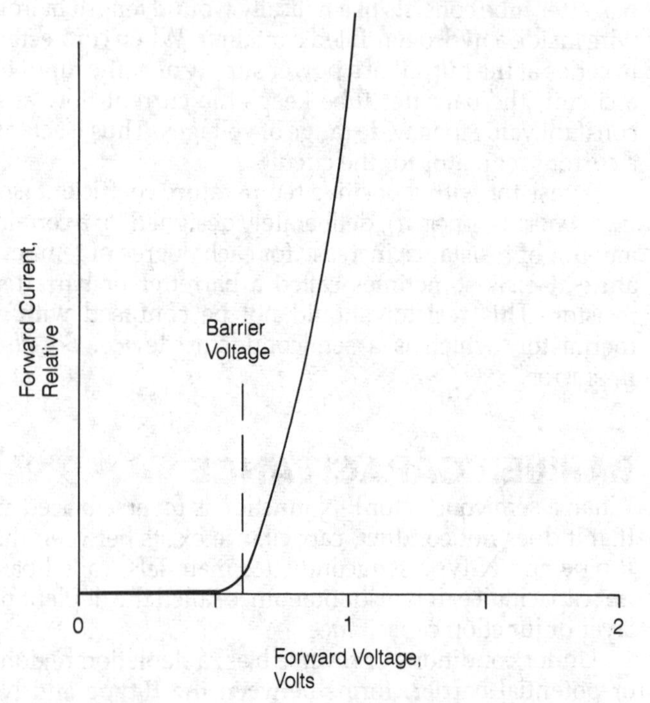

BARRIER VOLTAGE: The barrier voltage is the forward voltage at which conduction begins in a semiconductor diode.

With small values of forward voltage, very little current flows. But when the voltage reaches the barrier voltage, a sharp rise in the flow of current is observed.

Barrier voltage is not the same thing as avalanche voltage, which takes place under conditions of reverse bias and with much higher voltages. *See also* AVALANCHE VOLTAGE.

BASE

A semiconductor transistor consists of one of two configurations, called PNP, illustrated at A, and NPN, as at B. In the PNP device, a wafer of N-type semiconductor material is sandwiched in between two wafers of P-type material. In the NPN transistor, a wafer of P-type semiconductor material is sandwiched in between two wafers of N-type material (*see* N-TYPE SEMICONDUCTOR, P-TYPE SEMICONDUCTOR). The center wafer is called the *base* of the transistor, since the device is built geometrically around it.

The base of a transistor acts very much like the grid of a vacuum tube. However, the base of a transistor operates on a different principle. The input signal in a transistor amplifier is applied between the base and the emitter. The output is taken between the emitter and collector, or between the collector and ground. *See also* COLLECTOR, EMITTER, TRANSISTOR.

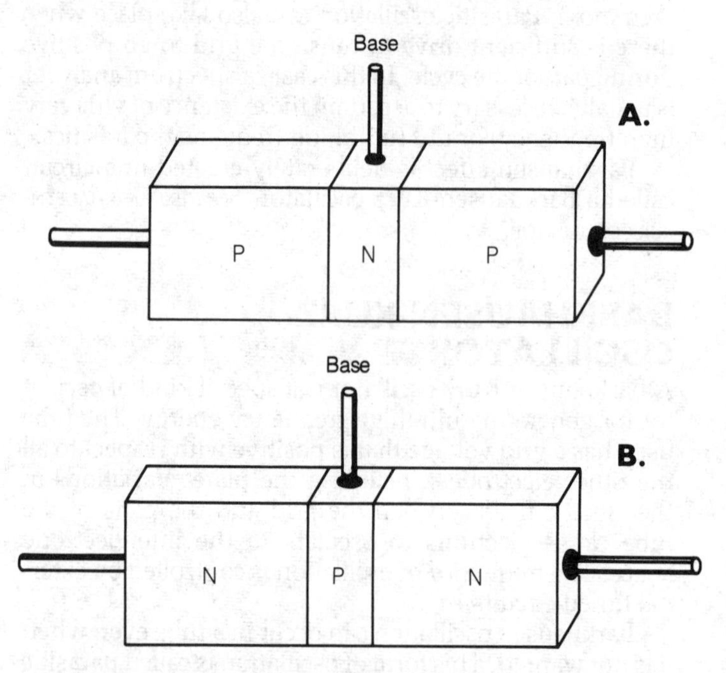

BASE: The base is the central part of the structure of a PNP transistor (A) or of an NPN transistor (B).

BASEBAND

When a signal is amplitude-modulated or frequency-modulated, the range of modulating frequencies is called the baseband. For a human voice, intelligible transmissions are possible with baseband frequencies from about 200 to 3000 Hz. Experiments have been conducted to determine the minimum baseband needed for effective voice communication, and with analog methods, efficiency is impaired if the baseband is restricted to a range much smaller than 200 to 3000 Hz. One method of compressing baseband signals for voice communication is called narrow-band voice modulation (*see* NARROW-BAND VOICE MODULATION).

For good reproduction of music, the baseband should cover at least the range of frequencies from 20 to 5000 Hz, and preferably up to 15 kHz or more. The baseband in a standard AM (amplitude modulation) broadcast signal is restricted to the range zero to 5 kHz, since the channel spacing is 10 kHz and a larger baseband would result in interference to stations on adjacent channels. In FM (frequency modulation) broadcasting, the baseband covers the entire range of normal human hearing. *See also* AUDIO FREQUENCY, HIGH FIDELITY.

BASE BIAS

The base bias of a transistor is the level of the base voltage with respect to the emitter or ground. Base bias can be derived from a battery, a resistive voltage-dividing network, or a resistor-capacitor combination. When the base is at the same voltage as the emitter of a transistor, the condition is called zero bias. Various kinds of base bias are shown in the illustration.

A transistor is normally cut off unless the forward bias at the emitter-base PN junction is greater than the barrier voltage (*see* BARRIER VOLTAGE). In a PNP transistor, forward base bias occurs when the base is negative with respect to the emitter. In an NPN device, forward bias means that the base is more positive than the emitter. To cause current to flow, this voltage difference must be at least 0.3 for a germanium transistor, and 0.6 volts for a silicon transistor.

For various amplification purposes, the base may be biased at class A, AB, B, or C. Class A operation involves forward bias, as does class AB operation. Class B usually results with zero bias. Class C operation can be achieved either with zero bias or reverse bias, depending on the particular transistor. *See also* CLASS A AMPLIFIER, CLASS AB AMPLIFIER, CLASS B AMPLIFIER, CLASS C AMPLIFIER.

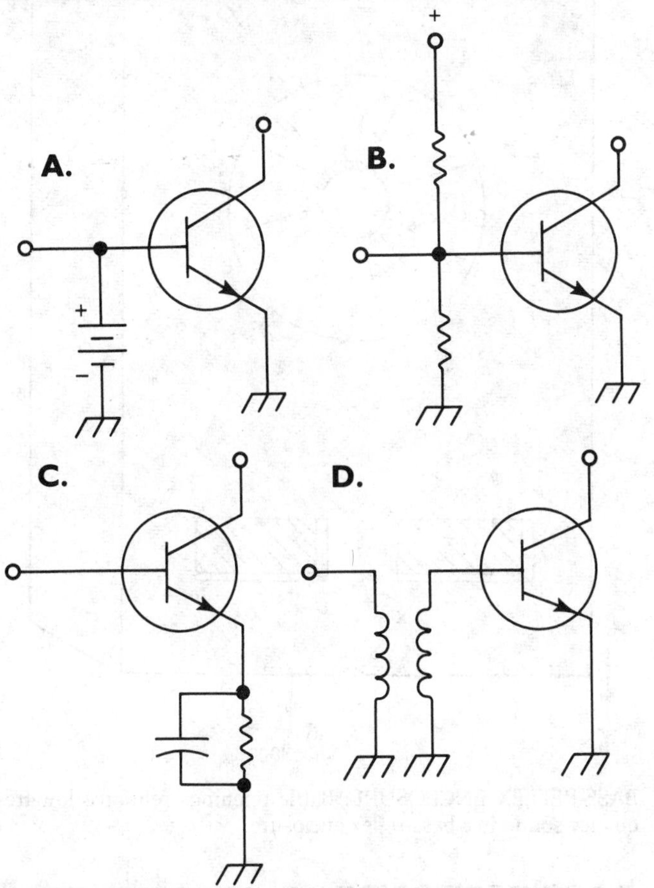

BASE BIAS: Four methods of achieving base bias are with a battery (A), resistor voltage division (B), RC combination in the emitter circuit (C), and transformer coupling for zero bias (D).

BASE LOADING

Base loading is a term used to denote the connection of a series inductance or capacitance at the bottom of a vertical antenna radiator, for changing the resonant frequency. Usually, a base-loading system consists of a coil which lowers the resonant frequency of the radiator for quarter-wave operation. Base loading is also frequently used to bring a ½-wave or ⅝-wave antenna to a matched condition. Figure 1 shows a base-loading coil, housed in protective plastic, for a ⅝-wave vhf mobile antenna.

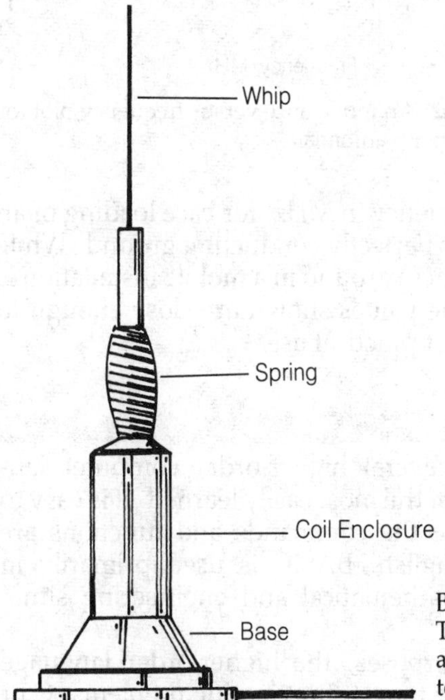

BASE LOADING: Fig. 1. The loading network is at the base of the antenna.

For quarter-wave resonant operation with a radiator less than ¼ wavelength in height, some form of inductive loading is required to eliminate the capacitive reactance at the feed point. When the coil is placed at the antenna base, the scheme is called base loading. When the coil is placed above the base of the radiating element, it is center loading (*see* CENTER LOADING). Center loading requires more inductance, for the same frequency and radiator length, than base loading.

An 8-foot mobile whip antenna can be brought to quarter-wave resonance by base loading at all frequencies below its natural quarter-wave resonant frequency, which is approximately 29 MHz. Figure 2 gives the value of loading inductance, *L*, in microhenrys (μH), as a

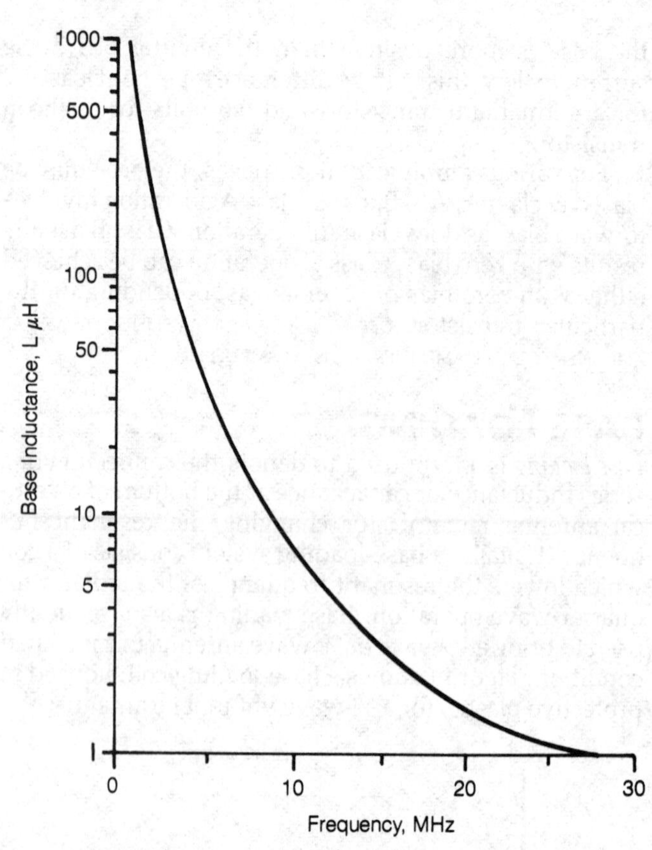

BASE LOADING: Fig. 2. An inductance versus frequency plot for base loading an 8-foot whip antenna.

function of the frequency in MHz, for base loading of an 8-foot radiator over perfectly conducting ground. While the rf (radio frequency) ground in a mobile installation is far from perfect, the values shown are close enough to actual values to be of practical use.

BASIC

BASIC is one of several higher-order computer languages, and is one of the most easily learned. It is easy to understand because the commands and functions are similar to plain English. BASIC is used primarily in teaching and in mathematical and engineering situations.

For business purposes, the higher-order language called COBOL is preferred. In scientific problems, C or FORTRAN is more efficient than BASIC. *See also* COBOL, FORTRAN, HIGHER-ORDER LANGUAGE.

BASS

The term *bass,* pronounced *base,* refers to low-frequency sound energy. On a piano or musical staff, any note below middle C is in the bass clef. All sounds above middle C are in the treble clef (*see* TREBLE). Middle C represents a frequency of approximately 261.6 Hz.

A tone control in a high-fidelity recording or receiving sound system can have separate bass and treble adjustments, or it may consist of a single knob or slide potentiometer. The amount of bass, compared to the amplitudes of other audio frequencies, affects the presence of the sound. Too little bass content causes a thin diluted sound. Too much bass causes sound to seem muffled and heavy, and also creates annoying vibration. *See also* AUDIO RESPONSE, BASS RESPONSE, TONE CONTROL.

BASS-REFLEX ENCLOSURE

A type of speaker cabinet that reinforces the bass frequencies is called a bass-reflex enclosure. By providing an opening at the front of the speaker in the baffle (*see* BAFFLE), and by suitably choosing the internal dimensions of the speaker cabinet, bass frequencies are reflected from the rear of the enclosure and are reinforced in phase as they emerge through the opening or openings (see illustration).

A bass-reflex enclosure improves the bass response of physically small speakers. Generally, bass reproduction requires the use of large speaker cones that can follow the relatively slow rate of vibration of bass audio. A bass-reflex enclosure takes advantage of sound resonance, so that a smaller speaker can achieve good bass quality in a high-fidelity system. *See also* BASS RESPONSE, WOOFER.

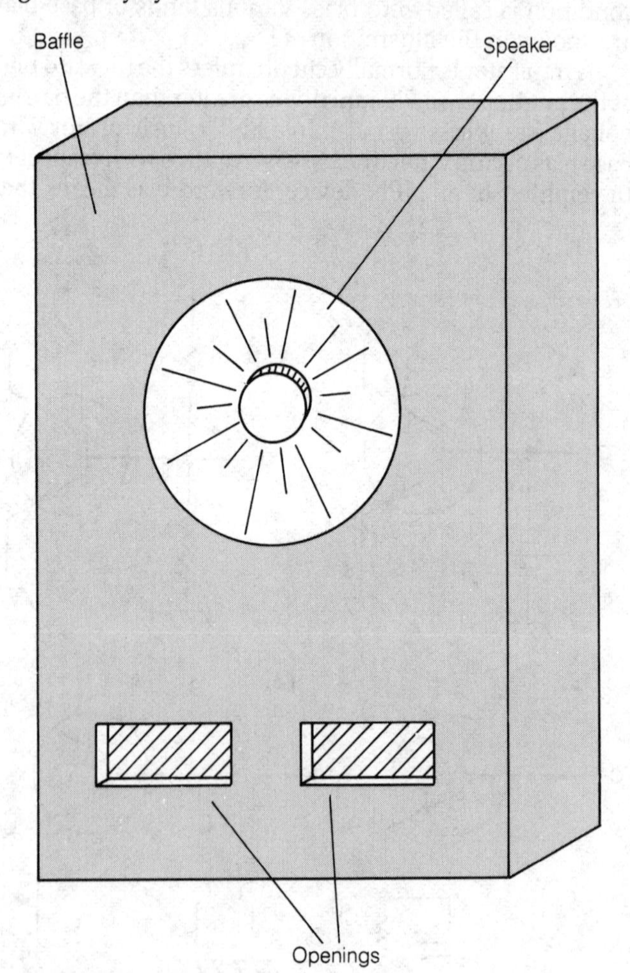

BASS-REFLEX ENCLOSURE: Baffle openings reinforce low-frequency sound in a bass-reflex enclosure.

BASS RESPONSE

Bass response is the ability of a sound amplification or reproduction system to respond to bass audio-frequency energy. Bass response can be defined for microphones, speakers, radio receivers, record players, and recording tape—anything that involves the transmission of audio energy. In most high-fidelity sound systems, the bass response is adjustable by means of a tone control or a bass gain control (*see* TONE CONTROL).

Good bass response involves more than the simple ability of a sound system to reproduce bass audio. The sound will not be accurately reproduced if the audio response (*see* AUDIO RESPONSE) curve is not optimized. For example, a sound system might have excellent gain at low frequencies, with low distortion, but the gain must also be fairly uniform over the bass range of frequencies (A in the illustration). A nonuniform bass response will result in poor sound, as shown at B.

A device called an equalizer (*see* EQUALIZER) allows compensation for differences in sound systems. The bass and treble response, as well as the response at midrange and high frequencies, can be tailored for the best possible sound reproduction by using such a device.

BATCH PROCESSING

When a series of programs is run sequentially by a computer, it is in the batch-processing mode. This is usually done by large computers at high speed. Programs arrive from many different sources, or subscribers, in the form of electronic data. Batch processing is often the most economical way to run a computer program, although it is inconvenient for program debugging, since the results are not known until the information is relayed back.

In contrast to batch processing, programs are sometimes run simultaneously. This is called multiprogramming or time sharing (*see* TIME SHARING).

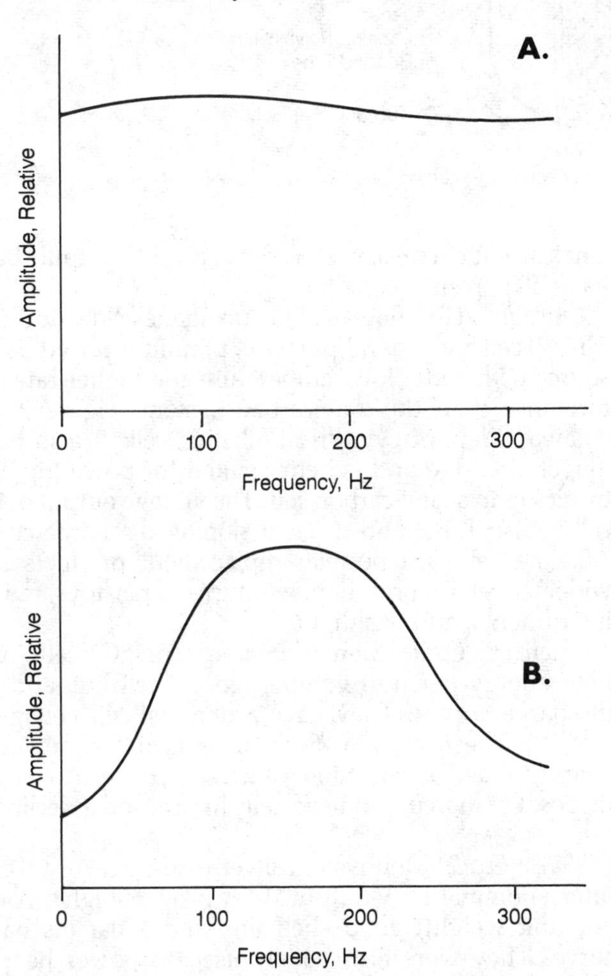

BASS RESPONSE: The plot of a good bass response for music reproduction (A) and a poor response (B).

BATTERY

A battery is an assembly of two or more electrochemical power cells that transform chemical energy into electrical energy. If the cells are connected in series, the output voltage is the sum of individual cell voltages. If they are connected in parallel, the output current is the sum of individual cell currents, but the voltage remains that of a single cell.

In common usage, the term *battery* is also applied to a single packaged power cell. A power cell is a closed case containing two dissimilar metals in contact with an electrolyte. When the electrodes are externally connected, the electrochemical reaction produces electrons at the anode which flow in the external circuit. The electron flow in the external load creates power.

If the cell electrolyte is a paste or powder it is called a dry cell, but if it is liquid it is called a wet cell. There are also semiliquid gelled electrolyte cells.

Batteries are important power sources for electronics and they have made possible many cordless electronic products including portable radios, tape recorders, TV sets, cellular telephones and test instruments. Miniature button cells are available for powering watches, hearing aids, cameras, and heart pacemakers. Cylindrical cells and round, flat coin provide standby power on CMOS (complementary metal-oxide semiconductor) memory circuit boards to preserve the data stored in the memory whenever power is shut down—accidentally or routinely. CMOS memory backed up by cells functions as ROM (read-only memory).

The discussion of batteries here is limited to those having direct application to electronics. Many different electrochemistries are employed in the cells and batteries for powering electronics. The measure of performance of a power cell is determined by its energy density in terms of weight and volume. Comparisons can be made between systems based on watt-hours per pound (WH/lb) or watt-hours per cubic inch (WH/in^3). In most electronic applications where batteries are small, WH/in^3 is a more meaningful indicator. But comparisons must be made only between cells with the same size and volume. Rated capacity, nominal cell voltage, typical operating temperature and shelf life are other important specifications characteristics to be compared.

Power cells and batteries can be classed as primary (nonrechargeable) or secondary (chargeable). Primary batteries are capable of one continuous or intermittent discharge, but secondary batteries can be recharged following partial or complete discharge. Direct current is passed through them in a direction opposite to the current flow or discharge. *See* BATTERY CHARGING.

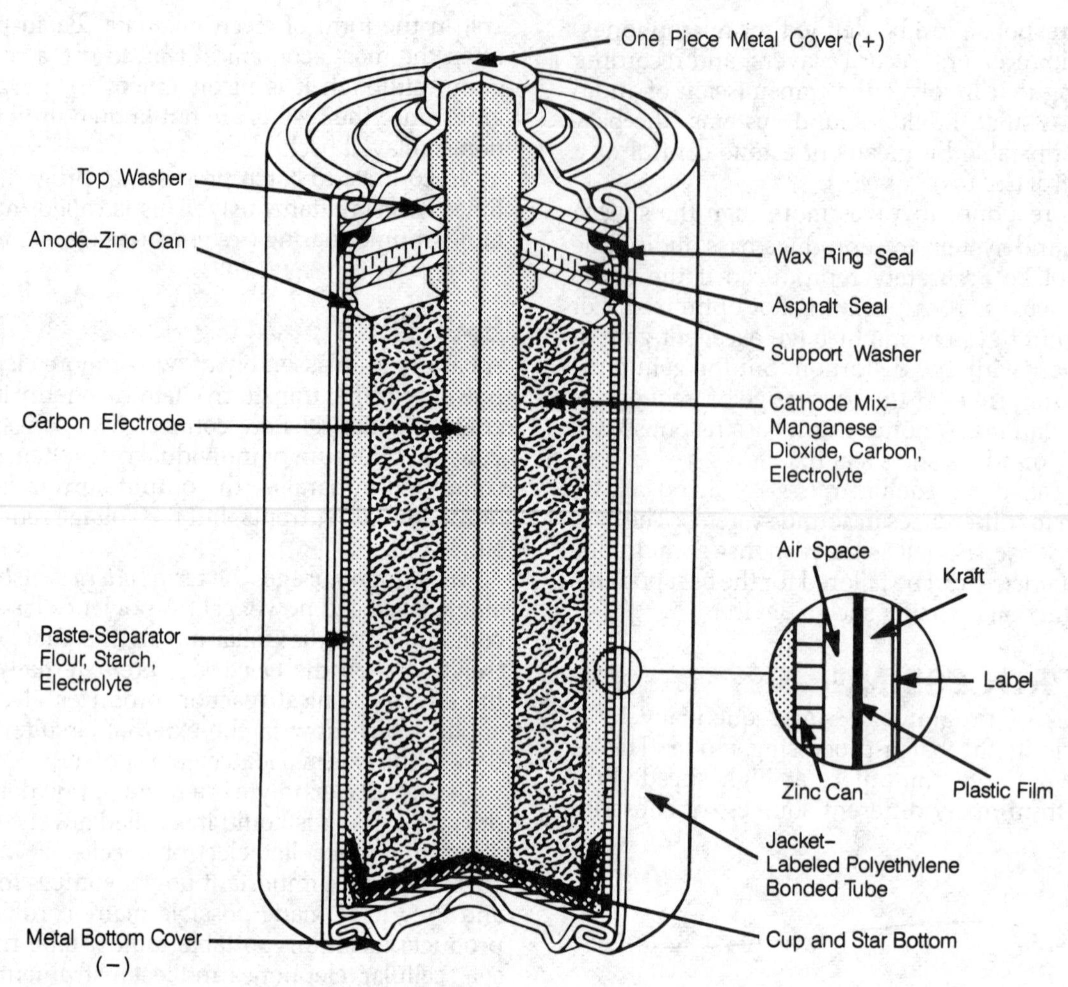

BATTERY: Fig. 1. Cutaway view of a general purpose cylindrical zinc-carbon (Leclanche) cell. (Courtesy Eveready Battery Company, Inc.)

For electronics applications, the primary cell and battery systems in greatest use are alkaline and lithium. Among the secondary systems, nickel cadmium is favored for handheld portable products. Where weight and volume are of less importance, sealed lead-acid cells and batteries are preferred. Lead-acid cells and batteries for electronics have sealed cases to prevent the escape of corrosive sulfuric acid, the liquid electrolyte that could damage electronic hardware. Secondary batteries with lithium as the negative electrode are being developed. One example is lithium-molybdenum disulfide.

Primary Systems The most widely used primary or nonrechargeable cells for powering electronics are zinc-alkaline manganese cells. However, lithium cells with output voltages of about 1.5 V are becoming more popular for powering consumer products and they are replacing mercuric-oxide and silver-oxide cells.

Zinc-Carbon. The zinc-carbon (Zn-C) cell is a popular, low-cost power source with a moderate shelf life. Figure 1 is a cutaway view of a cylindrical general purpose zinc-carbon Leclanche cell. It has a nominal cell voltage of 1.5 V and exhibits a declining discharge characteristic. Zinc-carbon cells and batteries are widely used as general purpose power sources for flashlights and lamps. The zinc-chloride heavy duty system is an improvement on zinc-carbon chemistry. It also has a 1.5 V output, but it has a 50 percent longer life.

Alkaline. The zinc-alkaline manganese dioxide (Zn/MnO2) cell is a general-purpose, premium-priced power source with better low-temperature and higher-rate performance than the zinc-carbon system. Figure 2 is a cutaway view of cylindrical alkaline cell. It also has a longer shelf life and is better suited for powering electronics than a zinc-carbon cell. The voltage output of this cell is also 1.5 V and it has a sloping discharge curve. Alkaline cells are popular replacement products sold widely in retail stores for powering cameras, toys, radios, instruments, and flashlights.

Mercuric Oxide. Mercuric oxide (ZnHgO) cells offer high energy per unit volume, good shelf life, and flat discharge curves. They have a nominal cell voltage of 1.35 V. These cells were widely used for powering hearing aids, pagers, and cameras. Concern over the disposal of mercury in used cells has caused a decline in sales.

Silver Oxide. Monovalent silver-oxide (Zn-Ag_2O) cells offer a nominal 1.5 V output. They provide high capacity per unit weight, good shelf life, and a flat discharge curve. They were also widely used to power hearing aids, watches, and cameras. The higher price of silver has caused a decline in sales of these cells.

Lithium Cells. The lithium electrochemistries have become more popular for powering electronic products because of their high energy (watt-hours per unit volume). Moreover, some lithium systems offer twice the voltage output of the popular primary cells such as alkaline, and they can be assembled to form smaller batteries. Widely favored in military applications there is, however, commercial production of about six different lithium battery systems in a range of standard packages. The nominal output voltage of a lithium cell varies from 1.5 to 3.6 V (see the table).

BATTERY: Commercial battery system energy density comparison.

Electrochemical Systems		Nominal Voltage	Watt-hours/ Cubic Inch (Typical)
Carbon-zinc	(Zn/C)	1.5	2.0
Alkaline-Manganese	(Zn/MnO_2)	1.5	2.0 to 3.0
Silver oxide (monovalent)	(Zn/Ag_2O)	1.5	8.0
Mercuric oxide	(Zn/HgO)	1.35	8.0
Silver oxide (divalent)	(Zn/AgO)	1.55	9.0
Zinc-air	(Zn/O_2)	1.4	14.0
Nickel-cadmium	($Cd/Ni(OH_2)$)	1.2	1.2
Lead acid (sealed)	(Pb/PbO_2)	2.0	1.1
Lithium Electrochemical Systems			
Sulfur dioxide	($Li\text{-}SO_2$)	3.0	7.4
Solid state (iodine)	($Li\text{-}I_2$)	2.8	9.8
Manganese dioxide	($Li\text{-}MnO_2$)	3.0	8.5
Carbon monofluoride	(Li-CFx)	3.0	7.5
Thionyl chloride	($Li\text{-}SOCl_2$)	3.6	11.0
Copper oxide	(Li-CuO)	1.5	8.2

Lithium cells are classified by reference to their electrolytes: organic liquids, inorganic liquids, or solid. Lithium technology is identified by the chemical abbreviation for lithium (Li) followed by the abbreviation for the electrolyte. The most popular lithium chemistries are:

Sulfur Dioxide	(Li/SO_2)
Thionyl Chloride	($Li/SOCl_2$)
Solid state (iodine)	(Li/I_2)
Manganese Dioxide	(Li/MnO_2)
Carbon Monofluoride	(Li/CF_X)
Copper Oxide	(Li-CuO)

Some lithium systems have energy output ratings (WH/in^3) that are three times those of alkaline cells. Their higher cost is justified because a smaller number of cells is needed for equivalent energy. For example, one lithium cell might replace two alkaline cells in the same case size in certain applications. The lithium-iodine cell is the principal power source for heart pacemakers.

The lithium sulfur-dioxide (Li/SO_2) system offers excellent high-rate and low-temperature performance. Pressurized and hermetically sealed SO_2 cells offer excellent shelf life and a flat discharge characteristic. These cells have a nominal cell voltage of 3.0 V. They are being used in both military and special industrial applications where their high capacity, high rate, and ability to operate in extreme temperatures is valued. Typical applications include powering appliances, atmospheric telemetry, computers, clocks, CMOS memory, radio transceivers, emergency lights, medical instruments, security systems, and sonobuoys.

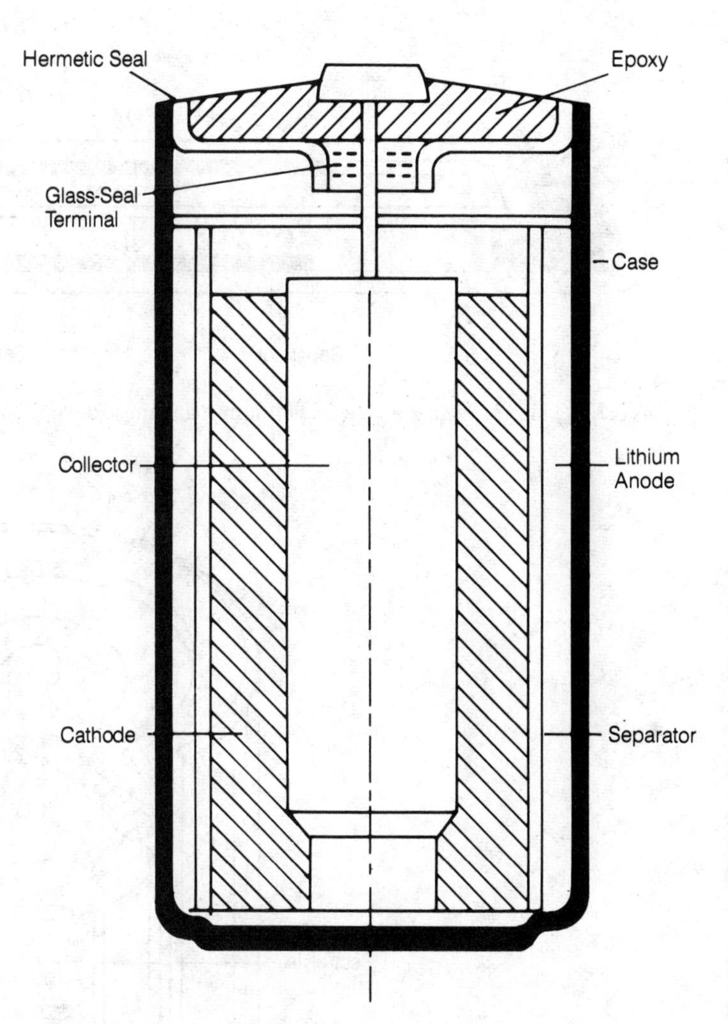

BATTERY: Fig. 2. Sectional view of a cylindrical lithium-thionyl chloride cell.

Lithium-thionyl chloride ($SOCl_2$) cells and batteries can also be used for many general purpose applications as well as CMOS memory backup. Figure 2 is a section view of a cylindrical lithium-thionyl chloride cell. These cells have a nominal output voltage of 3.4 V and a flat discharge curve. Packaged in hermetically sealed cases, they will operate satisfactorily at a temperature as high as 100°C.

Lithium-manganese dioxide (Li/MnO_2) cells also offer good rate capability, low-temperature performance, excellent shelf life and a flat discharge curve. Figure 3 is a section view of a lithium-manganese dioxide coin cell. These cells have a voltage of 3.0 V. They are used to power cameras, watches, calculators, CMOS memories, and instruments.

Lithium-carbon monofluoride (Li-CFX) cells have nominal voltages of 2.8 V. They are widely used as coin cells for CMOS memory backup. The cells are not hermetically sealed.

Zinc-Air. The zinc-air (Zn/O_2) cell uses oxygen from

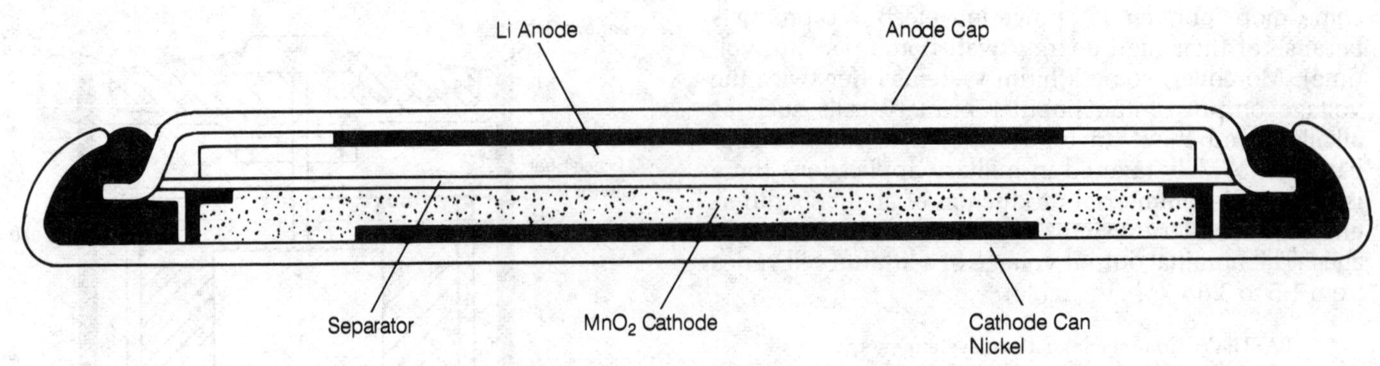

BATTERY: Fig. 3. Sectional view of a lithium-manganese dioxide coin cell for computer-memory backup. (Courtesy Eveready Battery Company, Inc.)

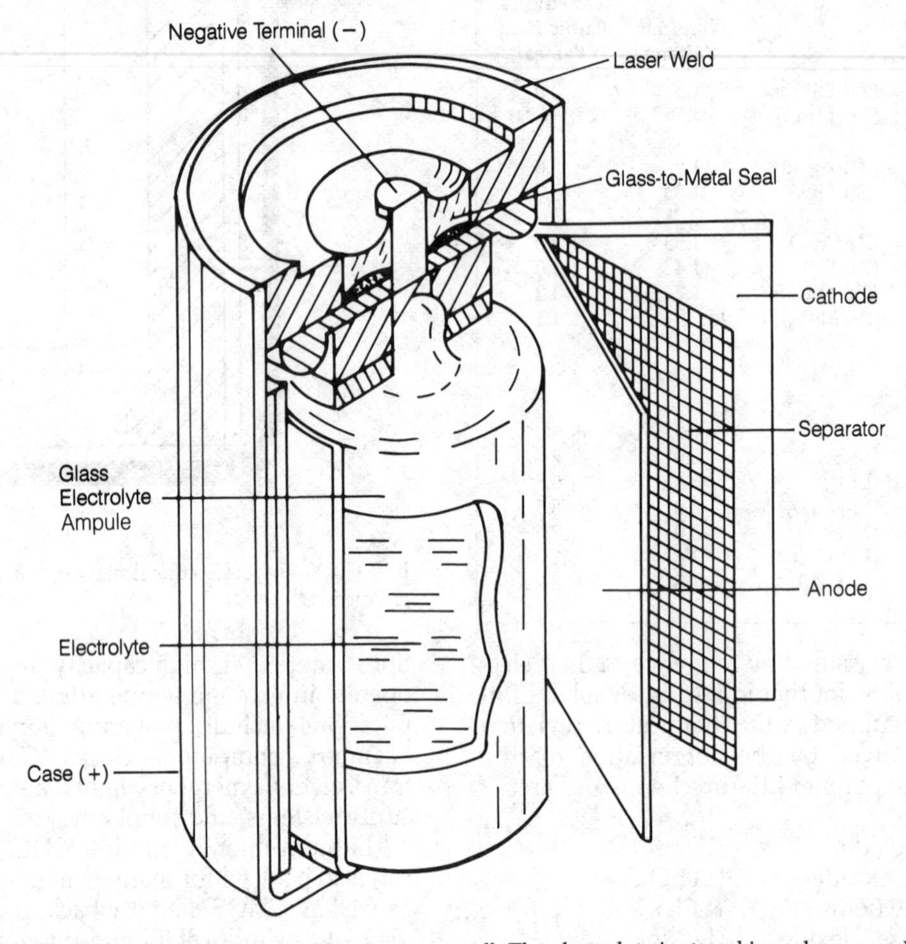

BATTERY: Fig. 4. Cutaway view of a lithium-thionyl chloride reserve cell. The electrolyte is stored in a glass ampule.

the air as the active cathode material. The air cathode has only a tenth of the internal volume of a cell so the anode can be larger than for other primary cells. The zinc-air system offers the highest energy density per unit volume and weight of any primary cell, but nominal cell voltage is only 1.2 V.

The anode is an amalgam of powdered zinc mixed with a gelling agent. The electrolyte is a water-based solution containing potassium hydroxide and zinc oxide. The air cathode assembly is a mixture of carbon, Teflon, and manganese dioxide impressed onto a nickel-plated screen. A semipermeable membrane of Teflon separates the electrodes and prevents moisture from entering and leaving the cell. The cell is activated when the seal is removed and oxygen from the air provides the active cathode material. Zinc-air cells are packaged in button cell cases. Typical applications are in hearing aids, medical instruments, and pagers.

Secondary Batteries

Nickel-Cadmium. A nickel-cadmium cell consists of a negative electrode of cadmium, a positive electrode of nickel hydroxide, and an aqueous solution of potassium hydrochloride as the electrolyte. Its nominal open-circuit voltage is 1.2 V. Typical applications are in handheld videotape cameras, laptop computers, portable television sets, and power tools.

Lead-Acid. The sealed lead-acid battery uses a lead negative electrode, a lead dioxide positive electrode, and a solution of sulfuric acid for the electrolyte. The electrochemistry is the same as that of the standard automobile battery. The nominal voltage of a lead-acid cell is 2.0 V.

There are two types of sealed lead-acid cells: gelled electrolyte and starved electrolyte. The electrolyte in the gelled type cell is a polymeric gell, and the cell is packaged in a rectangular cases. The starved-electrolyte cell contains very little free fluid electrolyte. It has a spiral-wound construction and is packaged in a cylindrical case. Both types of lead-acid cells are used in handheld videotape cameras, laptop computers, portable television sets, and power tools.

Battery Packaging Standard commercial case sizes were developed for zinc-carbon and alkaline cells and batteries. These include the cylindrical AAA, AA, C, D, and N flashlight battery cases and the rectangular 9 V radio battery cases. The radio battery case has snap-on terminals for positive contact connections. Standard rectangular lantern batteries in several sizes have been developed for higher output power. Miniature alkaline, lithium, mercuric-oxide, silver-oxide, and zinc-air button cells have been developed for hearing aids, cameras, calculators, watches, and clocks. Lithium systems have been adapted to some standard as well as a wide range of nonstandard cylindrical cases. They are also offered in coin cell packages.

Commercial- or consumer-grade nickel-cadmium cells are packaged in the standard AAA, AA, C, and D cylindrical as well as the 9 V radio battery cases as replacements for primary cells. Because of their lower voltage ratings, more nickel-cadmium cells are needed to provide equivalent energy.

Gelled lead-acid batteries are available in many different rectangular packages, but some starved-electrolyte cells are packaged in D-size cases, which may'be assembled to form multicell batteries.

Reserve Batteries Many military applications require reliable primary reserve batteries capable of power output after years of storage. These reserve sources must achieve full cell voltage within a short time after activation. They then function as conventional cells and batteries. Until recently, the most widely procured reserve batteries were based on the zinc-silver oxide (Zn/AgO) cell. However, the lithium-thionyl chloride reserve cell as shown in Fig. 4 is replacing them. The liquid electrolyte is held in a glass ampule. After activation by a mechanical or explosive impulse that fractures the glass ampule, the cell voltage rises rapidly to its nominal 3.65 V. These cells and batteries made from them are used to power military survival radios and beacons.

Other military applications require primary reserve batteries able to produce high electrical power output for periods up to an hour within a fraction of a second of activation. These power sources must also be reliably

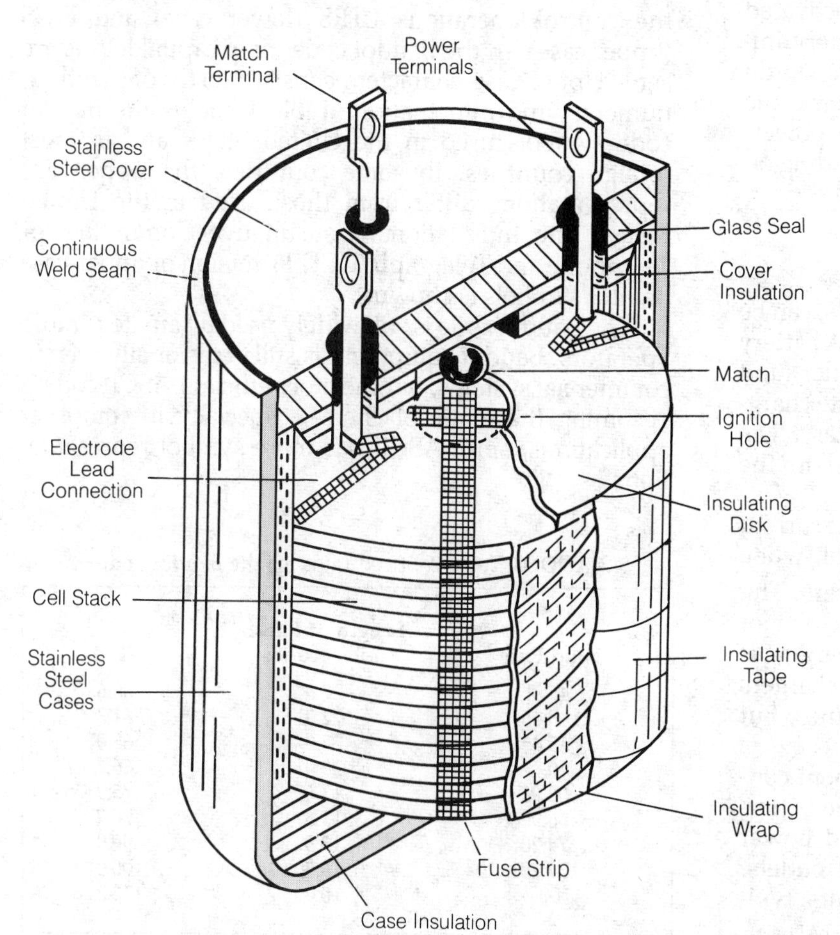

BATTERY: Fig. 5. Cutaway view of a lithium-iron disulfide thermal battery.

activated after 20 years of storage, but they are only required to provide a one-shot burst of power. These requirements are being met by lithium-thermal batteries as shown in Fig. 5. Designed and built to withstand extreme centrifugal and linear acceleration forces, they are being used to power electronics in missiles, torpedoes, and submarine decoys.

One electrochemistry suitable for this application is lithium alloy/iron disulfide. A solid electrolyte separates the lithium alloy anode and iron disulfide cathode of each cell in the battery while it is on standby. When activated electrically, pyrotechnic sources within the battery initiate a chain reaction that raises the temperature within the case to the 300° to 450°C necessary to melt the electrolyte in each cell and produce electrical power. Output voltages can be from 1.5 to 100 V, depending on the battery size and internal connections.

BATTERY CHARGING

Certain types of batteries, called storage batteries, can be recharged after their energy has been used up. A battery charger, consisting of a transformer and rectifier (and sometimes a milliammeter or ammeter to indicate charging current), is used to recharge a storage battery. The positive terminal of the charger is connected to the positive terminal of the battery for charging.

While a battery is charging, it draws the most current initially, and the level drops continuously until, when the battery is fully charged, the current is near zero. The total number of ampere hours indicated by a plot of current versus time would correspond, when charging is complete, to the capacity of the battery. Some chargers operate quickly, within a period of a few minutes, but charging usually takes several hours.

Portable appliances and some radio equipment contain rechargeable nickel-cadmium batteries. The batteries are charged with a small current-regulated power supply that fits standard household electrical outlets. Nickel-cadmium batteries require four to six hours, typically, to reach a fully charged condition. *See also* BATTERY.

BAUDOT CODE

Baudot is a 5-unit digital code for the transmission of teleprinter data. Letters, numerals, symbols, and a few control operations are represented. Baudot was one of the first codes used with mechanical printing devices. Sometimes this code is called the Murray code. In recent years, Baudot has been replaced in some applications by the ASCII code (*see* ASCII).

There are 2^5, or 32, possible combinations of binary pulses in the Baudot code, but this number is doubled by the control operations LTRS (lower case) and FIGS (upper case). In the Baudot code, only capital letters are sent. Upper-case characters consist mostly of symbols, numerals and punctuation. Table 1 shows the Baudot code symbols used in the United States and in most foreign countries. In some countries, the upper-case representations differ from those used in the United States. The International Consultative Committee for Telephone and Telegraph (CCITT) version of upper-case Baudot symbols is shown.

The Baudot code is still widely used by amateur radio operators. Baudot equipment is still occasionally seen in commercial systems, but the more efficient ASCII code is becoming the mode of choice, especially in computer applications, since ASCII has more symbol representations.

BAUDOT Table 1: Symbols for Baudot teleprinter code.

Digits	Ltrs	US Figs	CCITT Figs	Digits	Ltrs	US Figs	CCITT Figs
00000	BLANK	BLANK	BLANK	10000	T	5	5
00001	E	3	3	10001	Z	″	+
00010	LF	LF	LF	10010	L	)	)
00011	A	—	—	10011	W	2	2
00100	SPACE	SPACE	SPACE	10100	H	#	£
00101	I	BELL	′	10101	Y	6	6
00110	S	8	8	10110	P	0	0
00111	U	7	7	10111	Q	1	1
01000	CR	CR	CR	11000	0	9	9
01001	D	$	WRU	11001	B	?	?
01010	R	4	4	11010	G	&	&
01011	J	′	BELL	11011	FIGS	FIGS	FIGS
01100	N	,	,	11100	M	●	●
01101	F	;	;	11101	X	/	/
01110	C	:	:	11110	V	;	=
01111	K	(	(	11111	LTRS	LTRS	LTRS

BAUDOT Table 2: Speed rates for the Baudot code.

Baud Rate	Length of Pulse, milliseconds	WPM
45.45	22.0	60.6
45.45	22.0	61.3
45.45	22.0	65.0
50	20.0	66.7
56.92	17.6	75.9
56.92	17.6	76.7
74.20	13.5	99.0
74.20	13.5	100.0
100	10.0	133.3

The most common Baudot data speeds range from 45.45 to 100 bauds, or about 60 to 133 words per minute (WPM), as shown in Table 2. *See also* BAUD RATE, WORDS PER MINUTE.

BAUD RATE

Baud rate is a measure of the speed of transmission of a digital code. One baud consists of one element or pulse. The baud rate is simply the number of code elements transmitted per second. Sometimes the baud rate is stated simply as *baud*.

Baud rates for teleprinter codes range from 45.45, the slowest Baudot speed, to 19,200 for the fastest ASCII speed. *See also* ASCII, BAUDOT CODE.

BAZOOKA BALUN

A bazooka, or bazooka balun, is a means of decoupling a coaxial transmission line from an antenna.

When coaxial feed lines are used with antenna radiators, the unbalanced nature of the line allows rf (radio frequency) currents to flow along the outer conductor of the cable. This can create problems such as changes in the antenna directional pattern, and possibly radiation from the line itself. One method of choking off such currents is shown in the illustration. A metal cylinder or braided sleeve, measuring ¼ wavelength, is placed around the coaxial cable as shown. The end of the sleeve near the antenna is left free, and the other end is connected to the shield of the cable. For a given frequency, *f*, in MHz, the length of the bazooka in feet is given by

$$\text{Length (feet)} = 234/f$$

The bazooka is not a transformer, and consequently it cannot correct impedance mismatches. The bazooka, however, is useful for preventing radiation from a coaxial transmission line at high and very high frequencies. *See also* BALUN.

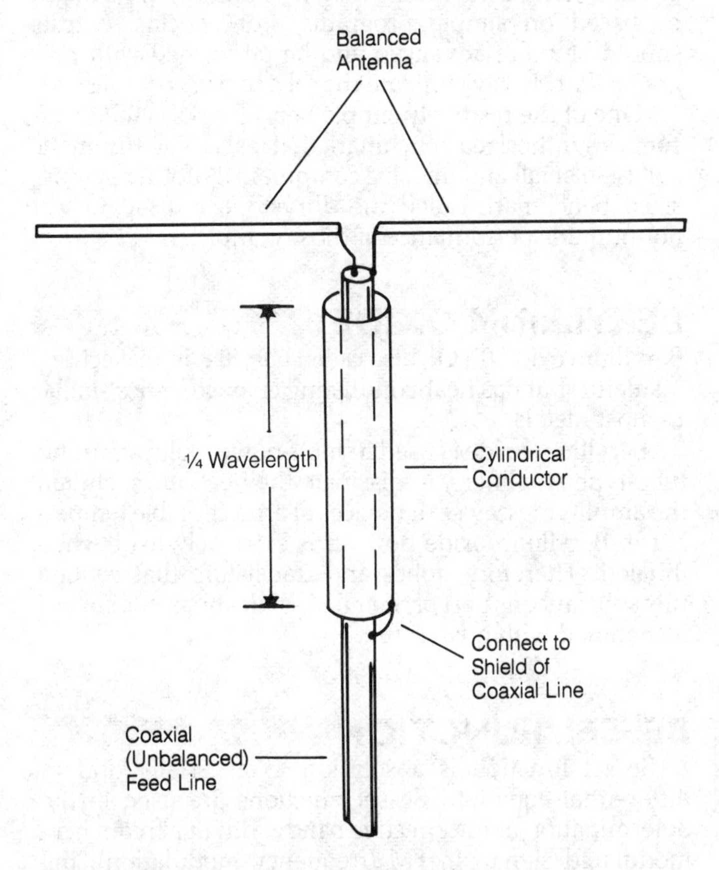

BAZOOKA BALUN: This assembly chokes off antenna currents on a feed line.

B DISPLAY

See RADAR.

BCD

See BINARY-CODED DECIMAL.

BEACON

A beacon is a transmitting station, usually with low rf (radio frequency) power output, designed for aiding in the monitoring of propagation conditions. The beacon sends a steady signal with frequent identification by call sign and geographical location. By listening on the beacon frequency, propagation conditions between the beacon and a receiving station are easily checked.

Beacons also are used for radio location and for easy spotting of certain objects on radar equipment. Beacons are a valuable aid in radar tracking, especially if an object is difficult to see or identify. Some radio beacons are called transponders. *See also* PROPAGATION, RADAR.

BEAMWIDTH

The beamwidth of a directional antenna or transducer is a measure of its response or output concentration. Antenna beamwidth is usually specified in horizontal direction, or azimuth. However, beamwidth may also be specified in the vertical plane.

To determine the beamwidth, the favored direction of the antenna (the direction in which it radiates the greatest amount of power) must be found, as in the illustration. Normally, less power is radiated to the left or right of, or above and below, this favored direction. As a field strength meter is moved back and forth at a distance from the antenna, the two directions are found at which the power output from the antenna is 3 decibels below the output in the favored direction. A power drop of 3 decibels is a 50-percent reduction (*see* DECIBEL). Hence, these two directions, or bearings, are called the half-power points.

The beamwidth is the angle in degrees, in a specified plane (vertical or horizontal), between the half-power points as shown. Generally, the greater the power gain of an antenna, the narrower its bandwidth. A two-element quad or Yagi antenna might have a beamwidth of 60 to 80 degrees in the horizontal plane. A large parabolic dish antenna, at microwave frequencies, can have a beamwidth of less than one degree.

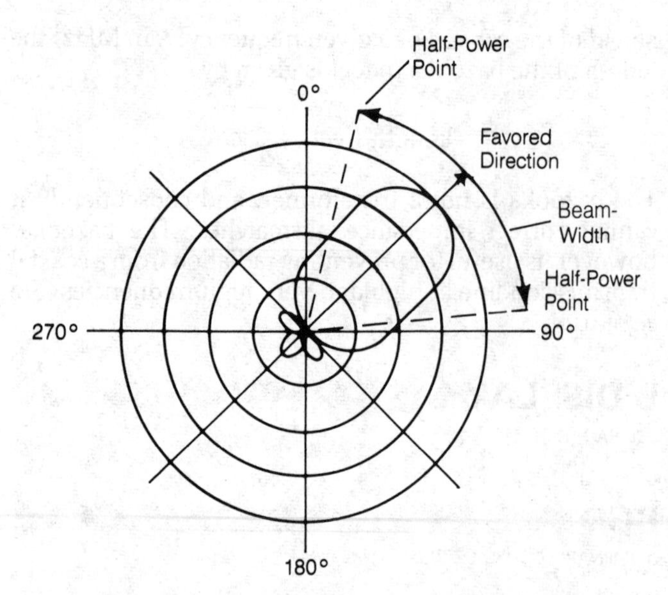

BEAMWIDTH: The beamwidth of an antenna is measured in degrees between its half-power points in the azimuth plane.

BEARING

The term *bearing* refers to horizontal direction and is normally given in degrees. The compass bearing is the azimuth based on the north magnetic pole. A signal might be received from great distance, and its bearing found to be 90 degrees. This means the signal arrives from the east.

Bearings in terrestrial propagation define great circles. Charts of great circle bearings to various parts of the world, centered on certain locations, are available for communications purposes.

The term also means a support for a rotating shaft. Bearings are found in motors, variable capacitors and potentiometers.

BEAT

When two dissimilar waves are combined or superimposed, a beat frequency occurs. Waves beat together to create the appearance of either a change in amplitude (if the frequencies differ by just a few hertz) or new frequencies, called beat frequencies or heterodynes (if the original frequencies are far apart). Beat frequencies are the sum and difference of the original frequencies.

Sound waves often beat together when two musical notes are combined. Two tuning forks, having frequencies of, say, middle C (261.6 Hz) and D (293.7Hz) will beat together to create sound waves at 32.1 Hz and 555.3 Hz, the difference and sum frequencies, respectively. Beat frequencies are an important part of the sound of music, even though listeners are not usually aware of them.

Beat frequencies are important in the operation of mixers and superheterodyne receivers and transmitters. Heterodynes are responsible for the undesirable effects of cross modulation and intermodulation distortion. Heterodynes occur in any nonlinear circuit where two or more frequencies exist. *See also* CROSS MODULATION, HETERODYNE, INTERMODULATION, MIXER, SUPERHETERODYNE RADIO RECEIVER.

BEAT-FREQUENCY OSCILLATOR

See OSCILLATOR CIRCUITS.

BENCHMARKING

Benchmarking is a technique for rating computer performance by running a set of well-known programs on the computer to compare its performance with that of other machines. The term is derived from the use by surveyors of permanent markers established at specified locations with precisely established lattitude, longitude, and elevation. The markers serve as reference points for surveyers in plotting property lines, maps, and charts. The term is also used by artisans who make personal marks on their workbenches to determine if the work in progress is machined within established measurement tolerances.

Computer benchmarking permits measuring the capabilities of a new computer system either in absolute terms (will the computer do what is required?) or relative terms (will computer A do a better job than computer B in any given application?). Computer designers use benchmark tests to help them improve system architectures; prospective customers use benchmarking for guidance in purchasing a system that will provide optimum execution of their applications programs.

Benchmarks are important in evaluating computer performance, but the tests must be appropriate to the intended tasks. The benchmark should reflect the nature of the work to be accomplished. Preliminary tests might be based on simple programs, but conclusive tests should be more advanced and be performed with programs that closely simulate the jobs to be accomplished.

One of the relatively simple benchmarks is the whetsone, a synthesized benchmark that tests basic arithmetic ability in small and midsize computers. Another synthesized benchmark called the dhrystone is used to test nonnumeric performance of those computers.

BERYLLIUM OXIDE

Beryllium oxide (BeO), also called beryllia, is an electrical insulator, but has heat-conducting characteristics similar to most metals.

Beryllium oxide is used in high-power solid state and tube-type amplifiers, to dissipate the heat and maintain the amplifying device or devices at an acceptable temperature. Beryllium-oxide powder is extremely toxic when inhaled. Therefore, tubes and transistors that contain this substance as part of their heat-sink apparatus should be handled with great care.

BESSEL FUNCTION

A Bessel function is a solution to a certain kind of differential equation. Bessel functions are used in the determination of the effective bandwidth of a frequency-modulated signal. In FM (frequency modulation), the distribution of sidebands above and below the carrier frequency depends on the deviation-to-modulating-frequency ratio, or modulation index (*see* MODULATION INDEX). This quantity affects the sideband amplitudes at various distances from the center frequency. Theoretically, FM sidebands extend infinitely to either side of the carrier. In practice, their amplitudes are significant only within a certain range, given by a Bessel function.

Bessel functions are used by engineers in the design of bandpass filters. The optimum shape of the bandpass response can be determined with Bessel functions (*see* ARGUMENT) where n is a non-negative integer called the order of the function. When graphed, Bessel functions look like damped sine waves. The drawing shows examples of the Bessel functions $J_0(x)$ and $J_1(x)$ for values of x between 0 and 12.

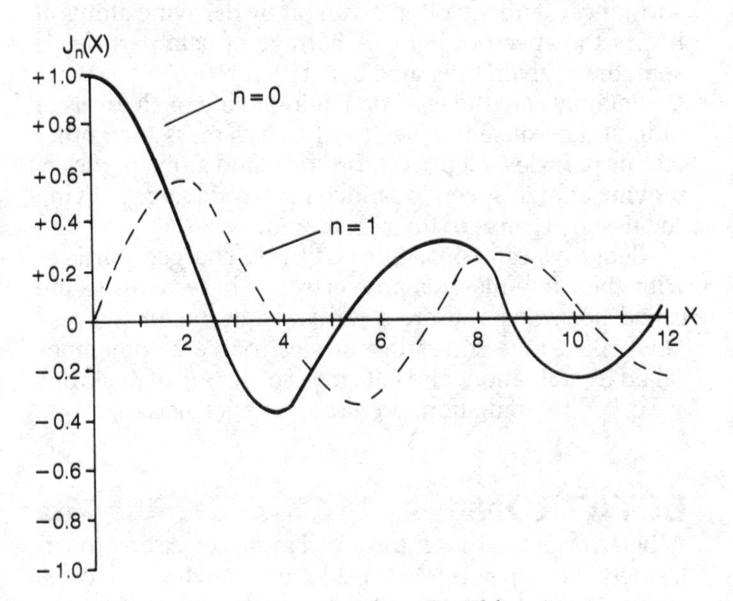

BESSEL FUNCTION: The graphs of two Bessel functions.

BETA

The term *beta* is used to define the current gain of a bipolar transistor when connected in a grounded-emitter circuit. A small change in the base current, I_B, causes a change in the collector current, I_C. The ratio of the change in I_C to the change in I_B is called the beta, and is abbreviated by the Greek letter β:

$$\beta = \frac{\Delta I_C}{\Delta I_B}$$

The collector voltage is kept constant when measuring the beta of a transistor.

The accompanying graph shows a typical I_C-vs-I_B curve for a transistor in a grounded-emitter configuration with constant collector voltage. The beta of the transistor is the slope of the curve at the selected operating point. A line drawn tangent to the curve at the no-signal operating point, as shown, has a definite ratio ($\Delta I_C/\Delta I_B$) which is easily determined by inspection. *See also* ALPHA, TRANSISTOR.

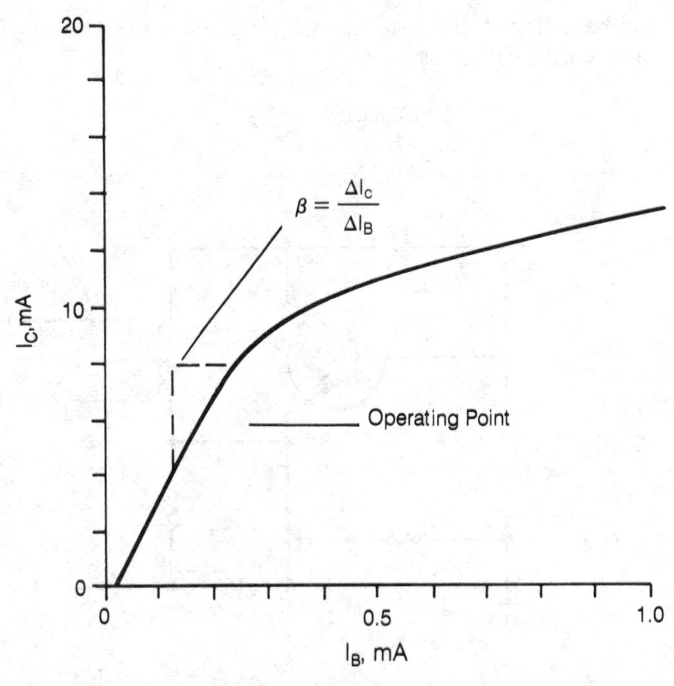

BETA: Beta is a measure of amplification in a transistor.

BETA-ALPHA RELATION

The beta of a transistor is the base-to-collector current gain with the emitter at ground potential (*see* BETA). The alpha is the emitter-to-collector current gain with a grounded-base circuit (*see* ALPHA). The alpha of a transistor is always less than 1, and the beta is always greater than 1.

The beta of a transistor can be mathematically determined if the alpha is known according to the equation:

$$\beta = \frac{\alpha}{1 - \alpha}$$

where the Greek letters α and β represent the alpha and beta, respectively. The following ratios also hold:

$$\frac{\alpha}{\beta} = 1 - \alpha = \frac{1}{1 + \beta}$$

$$\frac{\beta}{\alpha} = \frac{1}{1 - \alpha} = 1 + \beta$$

The alpha is mathematically derived from the beta by the equation:

$$\beta = \alpha\,(1 + \beta)$$

A typical bipolar transistor with an alpha of 0.95 has a beta of 19. As the beta approaches infinity, the alpha approaches 1, and vice versa.

BETA CIRCUIT

Some amplifiers use feedback, either negative, for stabilization, or positive for enhanced gain and selectivity. The part of the amplifier circuit responsible for the feedback is called a beta circuit. The schematic illustrates an amplifier with feedback, showing the beta circuit. In this particular case, the feedback is negative, intended to prevent oscillation.

In rf (radio frequency) amplifiers, neutralization is a form of feedback, and the neutralizing capacitor forms the beta circuit. *See also* FEEDBACK AMPLIFIER, NEUTRALIZATION, NEUTRALIZING CAPACITOR.

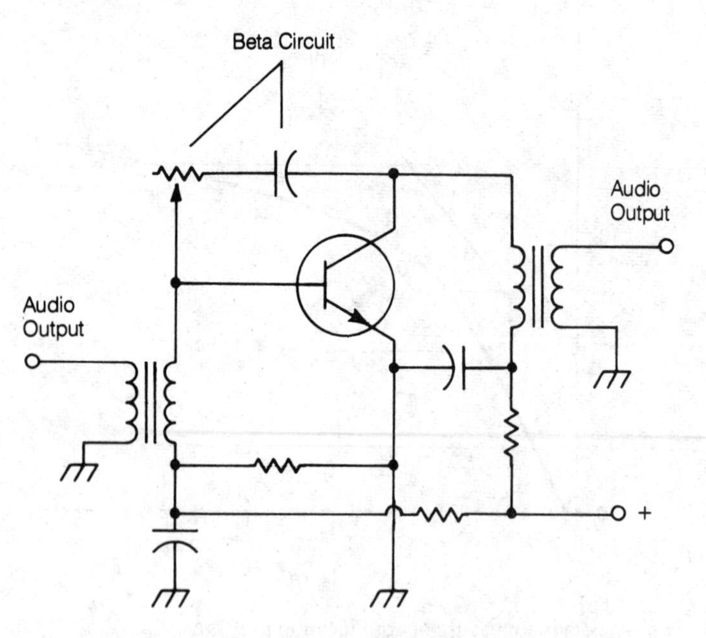

BETA CIRCUIT: The beta circuit is a negative feedback circuit.

BETA-CUTOFF FREQUENCY

As the operating frequency of an amplifier transistor is increased, the current amplification, or beta, decreases. In a common-emitter amplifier, the frequency at which the beta drops 3 decibels (dB) compared to its value at 1 kHz is called the beta-cutoff frequency. The beta-cutoff frequency is determined in the same manner as the alpha-cutoff frequency (*see* ALPHA-CUTOFF FREQUENCY). The only difference is that the emitter, rather than the base, is at ground potential.

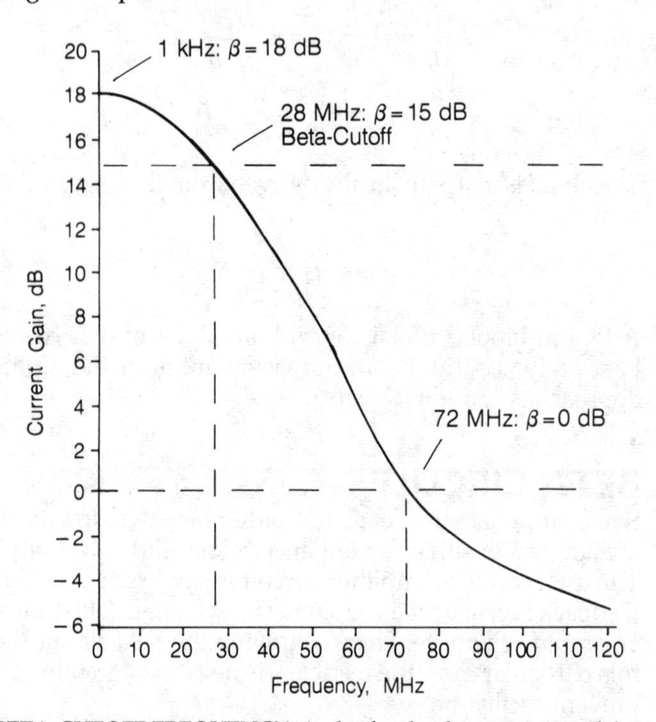

BETA-CUTOFF FREQUENCY: A plot for the determination of the beta-cutoff frequency for a transistor.

Depending on the type of transistor, the beta-cutoff frequency might be a few MHz or several hundred MHz. The beta-cutoff frequency differs slightly from the alpha-cutoff frequency, since a common-emitter configuration is used instead of a common-base configuration. Both specifications are useful, however, in amplifier design with bipolar transistors at high and very high radio frequencies. The graph illustrates the determination of beta-cutoff frequency.

BETA PARTICLE

A beta particle is a high-speed, electrically charged particle with the same mass and charge quantity as an electron. Beta particles are emitted by many radioactive substances, and are often given off by decaying atoms at nearly the speed of light. A barrage of beta particles is sometimes given the name beta rays.

Beta rays are the least damaging to life of all forms of radiation because the electron has less mass than other atomic particles. Neutrons, protons, and alpha particles, moving at high speed, do much greater damage to living cell tissue because of their greater mass.

Beta rays can consist of positively charged particles with the same mass as an electron. These particles are called positrons and are a form of anti-matter (*see* POSITRON). Beta rays consisting of electrons are sometimes called $\beta-$ radiation, and beta rays consisting of positrons are called $\beta+$ radiation. *See also* RADIATION HARDNESS.

BETATRON

A betatron is a system for accelerating electrons to extremely high speeds. When electrons move at velocities approaching the speed of light, they become beta particles, a form of radioactive energy (*see* BETA PARTICLE). When

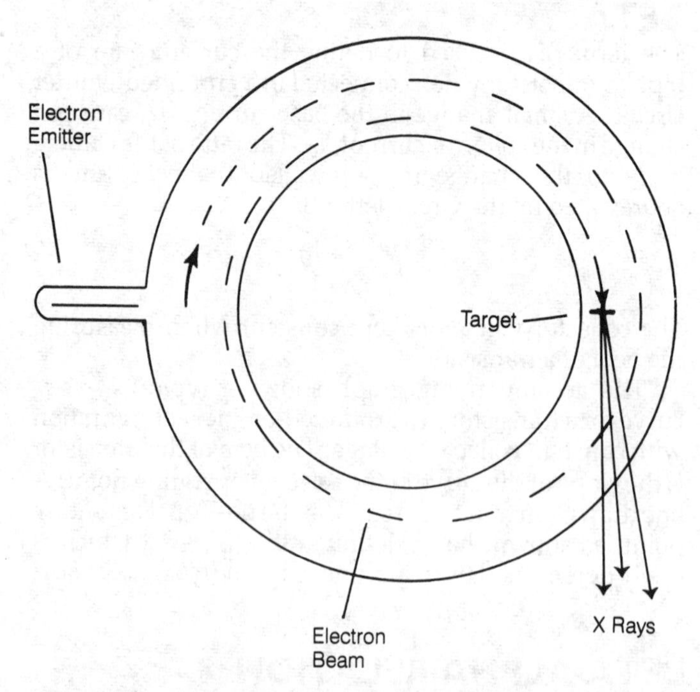

BETATRON: A betatron is an accelerator for producing X rays by electron bombardment.

the electrons strike a metal target, X rays for use in medical and industrial applications are produced.

The illustration is a diagram of a betatron. Electrons are injected into a toroidal, or doughnut-shaped, evacuated chamber. Powerful magnetic fields accelerate the electrons around many times to impact against the target. These magnetic fields are carefully controlled to direct the electrons to the target with the maximum attainable speed.

BEVERAGE ANTENNA

See ANTENNA DIRECTORY.

BFO

See OSCILLATOR CIRCUITS.

B-H CURVE

A B-H curve is a graph that illustrates the magnetic properties of a material. The magnetic field flux density, *B*, is plotted on the vertical scale as a function of the magnetizing force, *H*. Flux density is measured in units called gauss, and magnetizing force is measured in units called oersteds (*see* GAUSS, OERSTED). B-H curves are of importance in determining the suitability of a particular magnetic material for use in the cores of inductors and transformers. The graph shows an example of a B-H curve, also called a hysteresis loop.

When the magnetizing force is increasing, the flux density increases also. While the magnetizing force decreases, flux density follows. The flux density does not change exactly in step with the magnetizing force in most core materials. This sluggishness of the magnetic response is called hysteresis. *See also* HYSTERESIS, HYSTERESIS LOOP, HYSTERESIS LOSS.

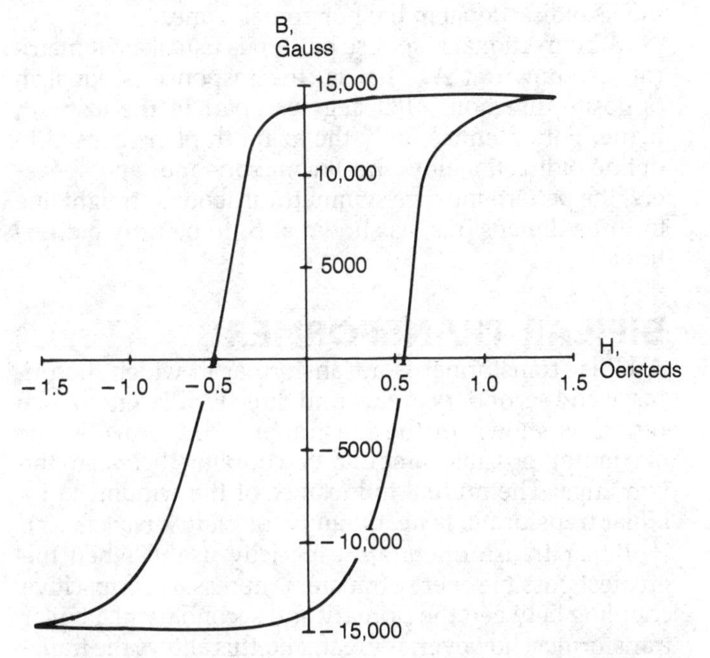

B-H CURVE: A B-H curve shows the lag between magnetizing force and flux density in a magnetic core of an alternating-current coil.

BIAS

Bias refers to a potential difference applied deliberately between two points for controlling a circuit. In a vacuum tube, the bias is the voltage between the cathode and control grid. In a bipolar transistor, the bias is the voltage between the emitter and the base, or the emitter and the collector. In a field-effect transistor, the bias is the voltage between the source and the gate, or the source and the drain.

Certain bias conditions are used for specified purposes. In forward bias, the cathode of a vacuum tube is negative with respect to the grid or plate. In reverse bias, the opposite is true: The cathode is positive with respect to the grid or plate. In a semiconductor P-N junction, forward bias occurs when the P-type material is positive with respect to the N-type materials; in reverse bias, the P-type material is negative with respect to the N-type material. When two electrodes are at the same potential, they are said to be at zero bias. *See also* FORWARD BIAS, REVERSE BIAS, ZERO BIAS.

BIAS COMPENSATION

Bias compensation is a means of balancing the pickup arm of a record player. Ordinarily, the pickup arm has a tendency to pull inward toward the center of the record as the record rotates. If the stylus is too large for the grooves of the record (*see* STYLUS), it may actually slide inward across the surface.

By providing an outward-directed tension on the pickup arm, the tendency of the arm to slide toward the center can be reduced or eliminated. This bias compensation is properly adjusted according to the speed of the record and the weight of the pickup arm. Bias compensation prevents the possibility of damage to a record from a sliding stylus, or from pressure of the stylus against the grooves of the record. This prolongs the life of the record and the stylus.

BIAS CURRENT

Bias current is the current between the emitter and the base of a bipolar transistor under conditions of no input signal. When the emitter-base junction is forward-biased, the bias current is normally a few microamperes or milliamperes. When the junction is at zero bias, the bias current is zero. When the junction is reverse-biased, the bias current is negligibly small.

A typical transistor has an optimum bias-current operating point. When the bias voltage is properly chosen, the bias current is such that the distortion is minimal but the amplification adequate. Bias-current specifications apply to Class A and Class AB amplifiers. In Class B and Class C amplifiers, the bias current is normally zero. *See also* BIAS, CLASS A AMPLIFIER, CLASS AB AMPLIFIER, FORWARD BIAS.

BIAS DISTORTION

When a tube, transistor, or other amplifying device is operated with a bias resulting in nonlinearity, the distor-

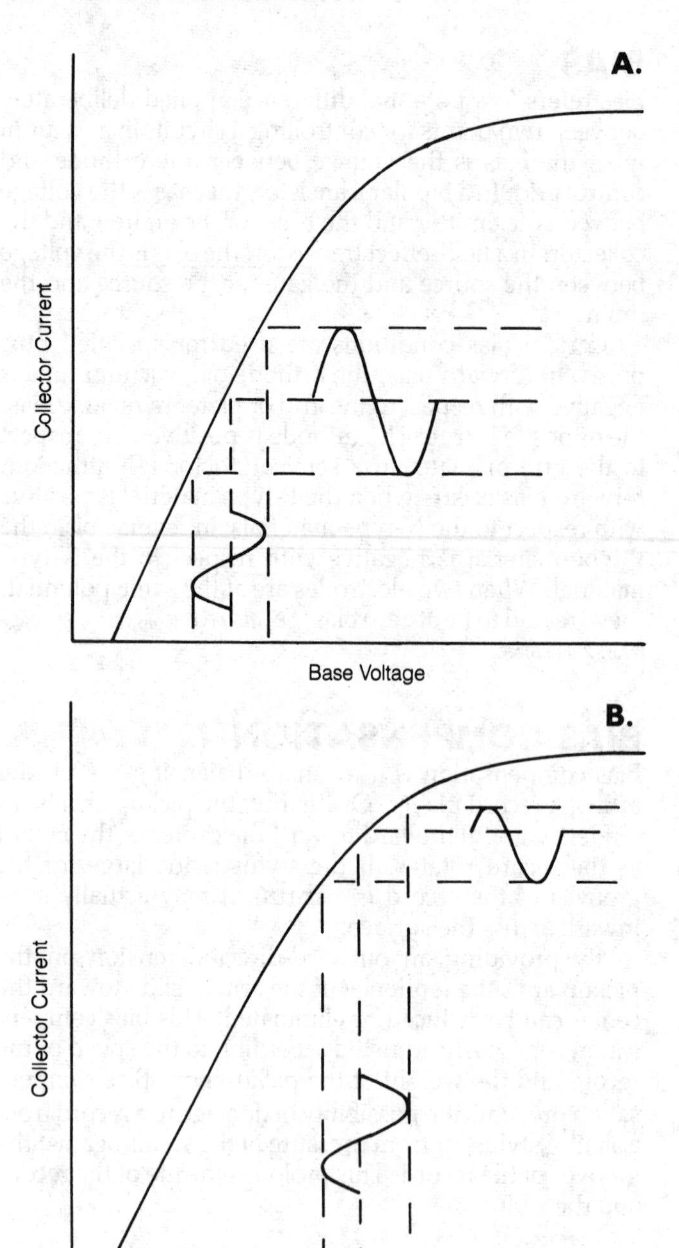

BIAS DISTORTION: Bias distortion occurs if the operating point of a transistor is improperly set. No distortion occurs at (A), but bias distortion or unsymmetrical output occurs at (B).

tion in the output signal is called bias distortion. Bias distortion is unimportant, in rf (radio frequency) Class B and Class C amplifiers. In high-fidelity audio equipment, however, the bias must be chosen carefully to avoid this distortion.

In the accompanying graphs, two operating points are shown. At A, the operating point is selected for linear operation, and there is essentially no distortion. At B, the operating point has been improperly selected; the bias current is too large. Therefore, the amplification factor is lower on the positive part of the input cycle than on the negative part. At the output, the waveform is lopsided. This is bias distortion.

Bias distortion in an amplifier is not the same as the distortion from excessive input-signal amplitude. While the two forms of distortion may produce similar output waveforms, excessive input amplitude, or overdrive, causes distortion no matter how the bias is set. The drive level must be regulated to avoid distortion. *See also* BIAS, CLASS A AMPLIFIER.

BIAS STABILIZATION

Bias stabilization is a method of ensuring that the bias in a circuit will remain constant. In a transistor circuit, a common means of bias stabilization is the resistive voltage divider. By choosing the appropriate ratio of resistances and the correct magnitude of resistance values, the base voltage can be maintained within a precise range.

Without bias stabilization, the base of a transistor can become improperly biased because of temperature changes or because of variations in the strength of an input signal. In some cases, changes in bias can lead to thermal runaway, resulting in component overheating, loss of gain and linearity, and possibly even destruction of a transistor. *See also* BIAS, THERMAL RUNAWAY.

BICONICAL ANTENNA

See ANTENNA DIRECTORY.

BIDIRECTIONAL PATTERN

Any transducer that is sensitive or emits in two different directions is said to be bidirectional. It can be an antenna, microphone, speaker, or other device designed for converting one form of energy to another. A in the illustration shows a bidirectional response pattern in the azimuth (horizontal) plane.

Bidirectional patterns are often preferred over unidirectional patterns. Bidirectional antennas are quite common; the ordinary half-wave dipole antenna, for example, is bidirectional in the horizontal plane.

A bidirectional response pattern is usually symmetrical, as shown at A. That is, the response is equal in opposite directions, 180 degrees apart in the azimuth plane. For antennas, only the azimuth plane is used to define bidirectionality. But for microphones and speakers, the pattern must be symmetrical about a straight line in three dimensions, as shown at B, to be truly bidirectional.

BIFILAR TRANSFORMER

A bifilar transformer is a transformer in which the primary and secondary are wound directly adjacent to each other, as shown in the illustration. This provides the maximum possible amount of coupling between the windings. The mutual inductance of the windings of a bifilar transformer is nearly unity (*see* MUTUAL INDUCTANCE).

Bifilar transformers are especially useful when the greatest possible energy transfer is necessary. Capacitive coupling between the primary and secondary of a bifilar transformer, however, is great, and this allows the transfer of harmonic energy with essentially no attenuation. *See also* TRANSFORMER.

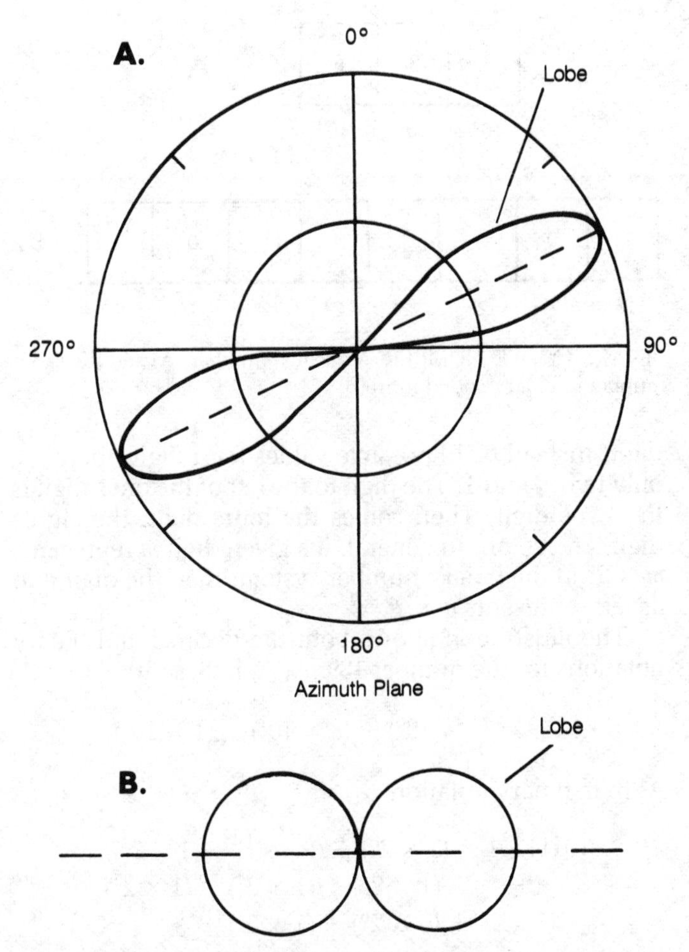

BIDIRECTIONAL PATTERN: A bidirectional antenna pattern in the azimuth plane (A) and the same pattern when viewed in cross section (B).

BIFILAR WINDING

A wirewound resistor is made noninductive by a technique called bifilar winding. The resistance wire is first bent double. Then the wire is wound on the resistor form, beginning at the joined end of the double wire and continuing until all the wire has been put on the form. This results in cancellation of most of the inductance, because the currents in the adjacent windings flow in opposite directions. The field from one half of the winding thus cancels the field from the other half.

In general, a bifilar winding is any winding that consists of two wires wound adjacent to each other on the same form. A bifilar transformer (*see* BIFILAR TRANSFORMER) uses bifilar windings for maximum energy transfer. Some broadband balun transformers (*see* BALUN) use bifilar, trifilar, or even quadrifilar windings. *See also* QUADRIFILAR WINDING, TRIFILAR WINDING, RESISTOR.

BILATERAL NETWORK

Any network or circuit with a voltage-versus-current curve that is symmetrical with respect to the origin is called a bilateral network. The voltage-versus-current curve will remain symmetrical with respect to the origin even if the polarity of the voltage is reversed. Moreover, the current will remain the same.

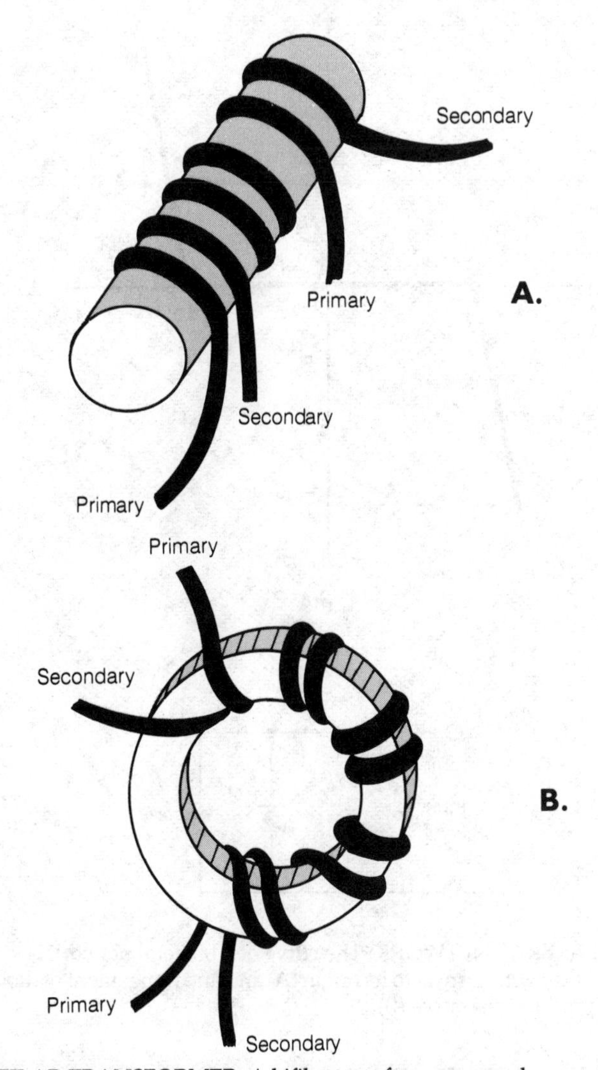

BIFILAR TRANSFORMER: A bifilar transformer wound as a solenoid (A) and as a toroid (B).

Any network of resistors is a bilateral network. Certain combinations of other components are also bilateral circuits. The illustration shows a symmetrical voltage-versus-current curve, A, characteristic of a bilateral network. At B, a schematic diagram of one circuit, which gives the curve at A, is shown. There are many other circuits that will provide this same response.

Most tube and transistor circuits are not bilateral because current ordinarily flows through such devices in only one direction.

BINARY-CODED DECIMAL

The binary-coded decimal (BCD) system of writing numbers assigns a 4-bit binary code to each numeral 0 through 9 in the base-10 system, as shown in the table. The 4-bit BCD code for a given digit is its representation in binary notation. Four places are always assigned.

Numbers larger than 9, with two or more digits, are expressed digit by digit in the BCD notation. For example, the number 189 in base-10 form is written as 0001 1000 1001 in BCD notation. The BCD representation of a number is not the same as its binary representation; in binary form, for example, 189 is written 10111101. *See also* BINARY-CODED NUMBER.

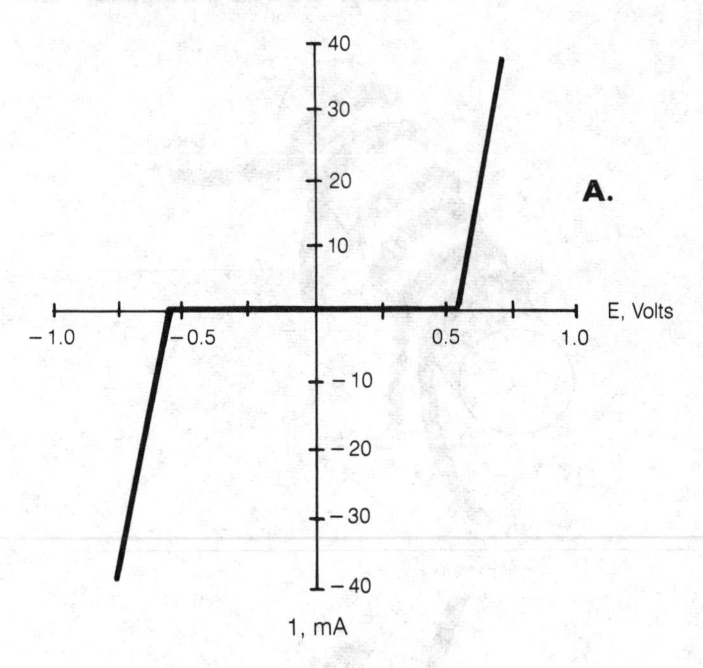

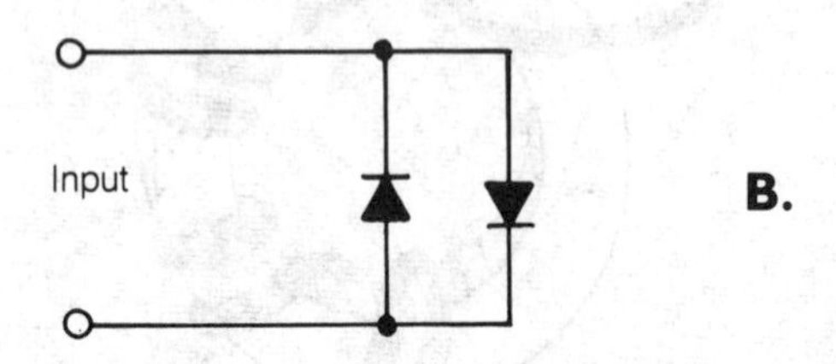

BILATERAL NETWORK: The curve of a bilateral network is symmetrical with respect to its origin (A) and an arrangement of diodes to produce this curve (B).

BINARY-CODED DECIMAL: Binary-coded decimal representations of the digits 0 through 9.

Digit in Base 10	BCD Notation
0	0000
1	0001
2	0010
3	0011
4	0100
5	0101
6	0110
7	0111
8	1000
9	1001

BINARY-CODED NUMBER

A binary-coded number is a number expressed in the binary system. The binary system of numbers is sometimes called base 2. Normally, numbers are written in base 10, called the decimal system.

In the base-10 system, the digit farthest to the right is the ones digit, and the digit immediately to its left is the tens digit. In general, if a given digit m represents $m \times 10^p$ in the decimal system, then the digit n, to the immediate left of m, represents $n \times 10^{p+1}$.

In base 2, the digit farthest to the right is the ones digit. Instead of 10 possible values for a digit, there are only two: 0 and 1. The digit to the left of the ones digit is the twos digit. Then comes the fours digit, the eights digit, and so on. In general, if a given digit m represents $m \times 2^p$ in the binary number system, then the digit n to its left represents $n \times 2^{p+1}$.

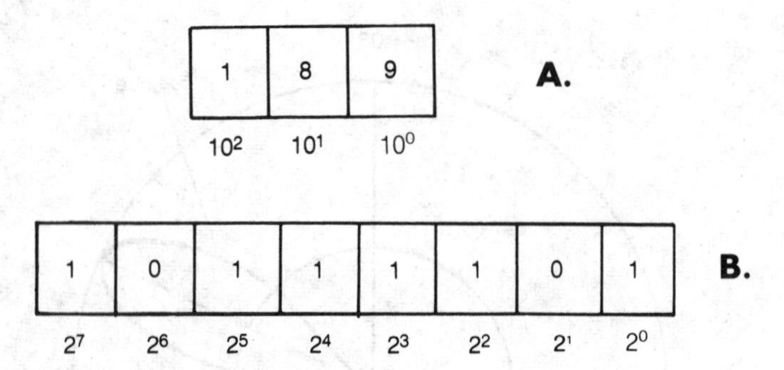

BINARY-CODED NUMBER: A decimal number (A) and the same number in binary-coded form (B).

The illustration shows both the decimal and binary notations for the number 189. At A, in base 10:

$$189 = (9 \times 10^0) + (8 \times 10^1) + (1 \times 10^2)$$

At B, in binary notation:

$$\begin{aligned} 10111101 = {} & (1 \times 2^0) + (0 \times 2^1) + (1 \times 2^2), \\ & + (1 \times 2^3) + (1 \times 2^4) + (1 \times 2^5) \\ & + (0 \times 2^6) + (1 \times 2^7), \end{aligned}$$

which is simply another way of saying that:

$$189 = 1 + 4 + 8 + 16 + 32 + 128.$$

Binary-coded numbers are used by calculators and computers, because the 0 and 1 digits are easily handled as off and on conditions. Although a binary-coded number has more digits than its decimal counterpart, this presents no problem for a computer with capability of handling thousands or millions of bits. *See also* COMPUTER.

BINARY COUNTER

A binary counter is a device that counts pulses in the binary-coded number system. Each time a pulse arrives, the binary code stored by the counter increases by 1. A simple algorithm (*see* ALGORITHM) ensures that the counting procedure is correct.

Binary counters form the basis for most frequency counters (*see* FREQUENCY COUNTER). These circuits actually count the number of cycles in a specified time period, such as 0.1 second or 1 second. These high-speed counters must work very fast to measure frequencies approaching 1 GHz which is 10^9 Hz.

A simple divide-by-two circuit is sometimes called a binary counter or binary scaler. This circuit produces one output pulse for every two input pulses. A T flip-flop (*see* FLIP-FLOP) is sometimes called a binary counter. Several divide-by-two circuits, when placed in series, permit digital division by any power of 2. *See also* DIGITAL CIRCUITRY.

BIOELECTRONICS

The application of electronics to biological problems is called the field of bioelectronics. An example is the development of electronic circuits to perform or regulate certain body functions. Probably the most common example of a bioelectronic device is the hearing aid. This consists of a microphone, audio amplifier, and earphone. Another frequently used bioelectronic device is the heart pacemaker.

BIONICS

Bionics is a scientific study and development of replacement body organs with bioelectronic and biomechanical devices (*see* BIOELECTRONICS). While the development of a complete android, or bionic man, is probably impossible, certain living organs can be replaced, at least temporarily, by bionic substitutes. One example of this is the kidney dialysis machine.

Bionics is closely related to robotics (*see* ROBOT). Robot arms and legs have been developed, and as the bionic technology advances, it might someday be possible to build complete bionic arms and legs to replace injured limbs. It may also be possible to build bionic organs that actually perform better than their living counterparts.

BIPOLAR TRANSISTOR

See TRANSISTOR.

BIRMINGHAM WIRE GAUGE

The Birmingham Wire Gauge is a standard for measurement of the size, or diameter, of wire. Usually it is used for steel wire. Wire is classified according to diameter by giving it a number. The Birmingham Wire Gauge designators differ from the American and British Standard designators (*see* AMERICAN WIRE GAUGE, BRITISH STANDARD WIRE GAUGE), but the sizes are nearly the same. The higher the designator number, the thinner the wire.

The table shows the diameter versus Birmingham Wire Gauge number for designators 1 through 20. The Birmingham Wire Gauge designator does not include any coatings that might be on the wire, such as enamel, rubber, or plastic insulation. Only the metal part of the wire is taken into account.

BIRMINGHAM WIRE GAUGE: Birmingham Wire Gauge equivalents in millimeters.

BWG	Diameter, millimeter	BWG	Diameter millimeter
1	7.62	11	3.05
2	7.21	12	2.77
3	6.58	13	2.41
4	6.05	14	2.11
5	5.59	15	1.83
6	5.16	16	1.65
7	4.57	17	1.47
8	4.19	18	1.25
9	3.76	19	1.07
10	3.40	20	0.889

BISTABLE CIRCUIT

Any circuit that can attain either of two conditions, and maintain the same conditions until a change command is received, is called a bistable circuit. The flip-flop (*see* FLIP-FLOP) is the most common bistable circuit.

A very simple kind of bistable circuit is the pushbutton switch, used with any device such as an electric light. When the button is pushed once, the lamp or appliance is switched on; pressing the button again turns the circuit off. These switches can be either mechanical or electronic. The bistable circuit always maintains the same condition indefinitely unless a change-of-state command is received.

BISTABLE MULTIVIBRATOR

See FLIP-FLOP.

BIT

Bit is an acronym for *binary digit*. Each numeral place in a binary number represents one bit; for example, the binary number 11101 has five bits. A bit can be either 0 or 1. In the binary-coded decimal, or BCD, notation, each decimal digit is represented by four bits (*see* BINARY-CODED DECIMAL).

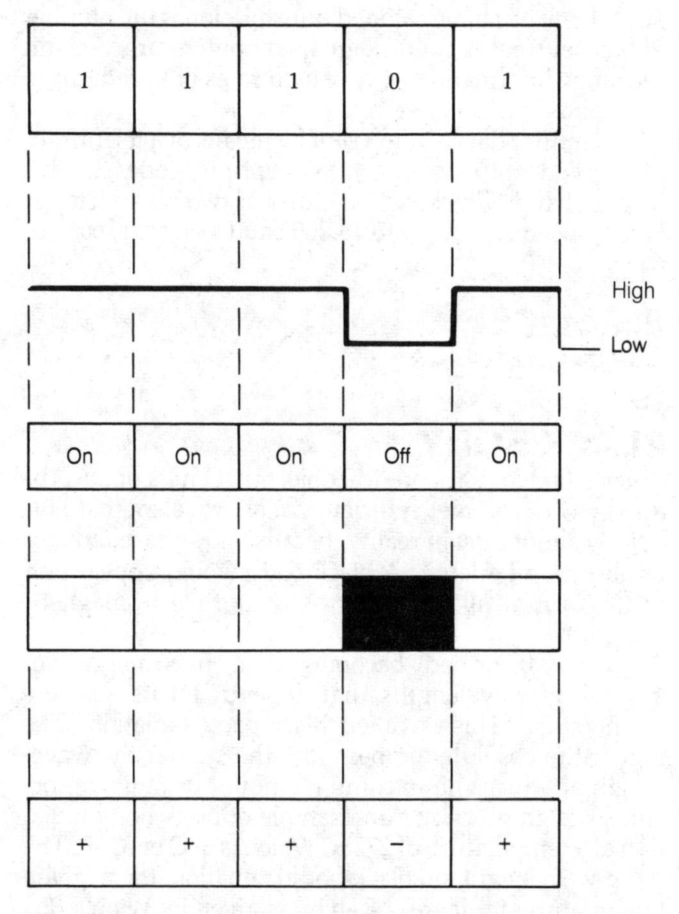

BIT: A bit can exist either in the high (on) state or low (off) state.

The 0 or 1 state of a bit can be denoted in various ways, such as low and high, off and on, black and white, or negative and positive (see illustration).

A group of bits is called a byte (*see* BYTE). Usually, a byte consists of eight bits, although the number varies depending upon the application.

BIT MAPPING

Bit mapping is a technique for storing data in computer memory for the generation of a computer graphics image. The data bits reside at addresses in memory that match pixel locations on the screen and are read sequentially to generate an image. With bit mapping, whole raster lines can be read directly out of memory by parallel accessing of a sequence of bits. The lines are transferred to high-speed shift registers, and the bits are shifted out serially to form the display. Data values stored in a bit-map memory are converted into display signals by a digital-to-analog converter. (*See* DIGITAL-TO-ANALOG CONVERTER.)

Bit mapping makes it easier to present document graphics and text on the same screen. It also simplifies windowing, or the display of smaller graphics or text displays within the context of the larger screen display. Bit mapping also aids in presentation of variable text fonts. *See* COLOR TV TUBE, COMPUTER GRAPHICS, RASTER.

BIT RATE

The bit rate of a binary-coded transmission is the number of bits sent per second. Digital computers process and exchange information at very high rates of speed, up to thousands of bits per second.

Computer data is often sent by means of a teleprinter code called ASCII (*see* ASCII). In teleprinter codes, a bit is called a baud, which can be either of two binary states. ASCII speeds range up to 19,200 baud. *See also* BAUD RATE.

BIT SLICE

See MICROPROCESSOR.

BLACK BODY

A black body is a theoretical object that emits or absorbs energy with complete efficiency at all wavelengths. This object cannot exist in reality, because no surface radiates or absorbs all energy. A black body would appear perfectly dark at all wavelengths of the electromagnetic spectrum.

When a black body becomes hot, it emits energy over a range of wavelengths that depends on its absolute temperature. This is called black-body radiation. The higher the absolute temperature, the shorter the wavelength at which the maximum amount of radiation occurs. The graph shows an example of black-body radiation at a temperature of 273 K, which is 0° C or 32° F. The peak wavelength of black-body radiation for a given temperature, in degrees Kelvin, is given by Wien's displacement formula as:

$$\lambda = 2898 / T$$

where λ is the wavelength in microns (millionths of a meter) and T is the temperature in degrees Kelvin. *See also* KELVIN TEMPERATURE SCALE.

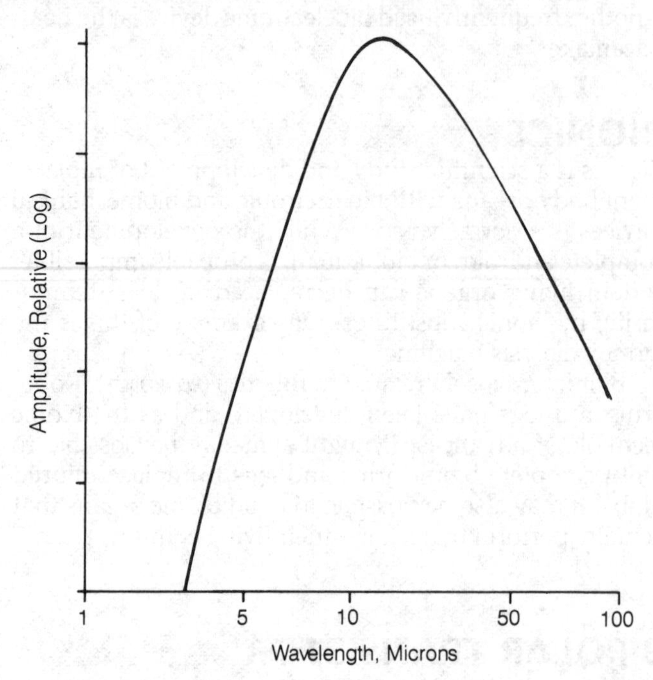

BLACK BODY: A plot of black-body radiation for a temperature of 273 K or 0°C.

BLACK BOX

Any circuit in which the internal details are unknown, or irrelevant is called a black box. This circuit becomes, in its intended application, a component itself. An example of a black box is an integrated circuit. While the internal details of various integrated circuits are complex and different, this device appears as an empty box in a circuit diagram.

Two black boxes that behave in identical fashion under a certain set of circumstances may be used interchangeably even though they are internally different. An engineer does not have to be concerned with differences that are inconsequential in practice.

BLACK-LIGHT LAMP

See ULTRAVIOLET RADIATION.

BLACK TRANSMISSION

Black transmission is a form of amplitude-modulated facsimile signal (*see* FACSIMILE) in which the greatest copy density, or darkest shade, corresponds to the maximum amplitude of the signal. Black transmission is the opposite of white transmission, in which the brightest shade corresponds to the maximum signal amplitude (*see* WHITE TRANSMISSION).

In a frequency-modulated facsimile system, black transmission means that the darkest copy corresponds to the lowest transmitted frequency.

BLANKETING

Blanketing is a form of interference or jamming, in which a desired signal is obliterated by a more powerful undesired signal. A high-powered, frequency-modulated transmitter is capable of sending several kHz of signal bandwidth, rendering a sizable portion of the spectrum useless. Generally a complex combination of modulating tones is used. The multiplicity of FM (frequency modulation) sidebands extends more or less uniformly across a specified range of frequencies. *See also* JAMMING.

BLANKING SIGNAL

In television picture transmission, a blanking signal is a pulse that cuts off the receiver picture-tube during return traces. The blanking signal prevents the return trace from showing up on the screen, where it would interfere with the picture. This pulse is a square wave with rise and decay times that are very short.

Blanking signals are sometimes used in radar equipment to shut off the display for a certain length of time with each antenna rotation. This makes the radar blind in certain directions, eliminating unwanted echoes. *See also* RADAR.

BLEEDER RESISTOR

A bleeder resistor is a high-value, low-wattage resistor connected across the output terminals of a power supply. Bleeder resistors serve two purposes: first to aid in voltage regulation, and second to reduce the shock hazard in high-voltage systems after power has been shut off. The schematic illustrates the connection of a bleeder resistor in a power supply.

The bleeder resistor prevents excessive voltages from developing across the power-supply filter capacitors when there is no load. Without this current-draining device, the voltage at the supply output might build up to values substantially larger than the rated value.

In a high-voltage power supply, the filter capacitors will remain charged for some time after the power has been shut off unless some means of discharging them is provided. Dangerous electrical shocks, from seemingly safe equipment, are made less likely by a bleeder resistor. However, the high-voltage terminal of the supply should always be shorted to ground before servicing the equipment. This will ensure that no voltage is present. A metal rod with a well-insulated handle should be used to discharge any residual voltages.

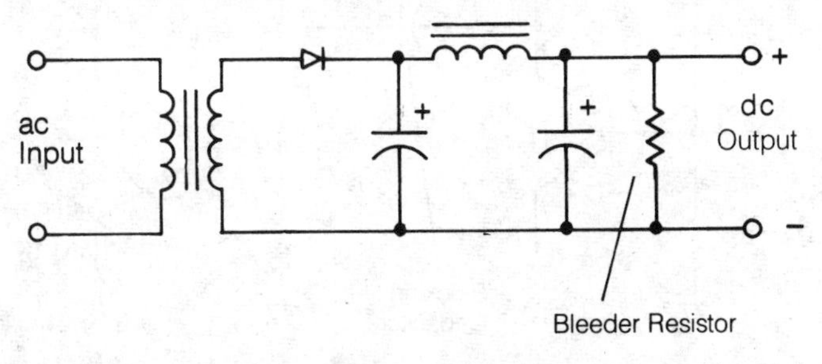

BLEEDER RESISTOR: A bleeder resistor in a power supply regulates the voltage under no-load conditions.

BLIND ZONE

In radar, the term *blind zone* refers to a direction from which echoes cannot be received. The usual reason for the existence of blind zones is an object that gets in the way of the radar signals (see drawing). This obstruction shows up as a large, bright area on a radar display screen, unless azimuth blanking is used to eliminate it (*see* RADAR).

In any communications system, a blind zone is an area or direction from which signals cannot be received, or to which messages cannot be transmitted. When communicating through the ionosphere, reception is often impossible within a certain distance called the skip zone (*see* SKIP ZONE). At vhf (very high frequency) and above, a hill or building can create a blind zone. As the wavelength decreases, smaller and smaller obstructions cause blind zones.

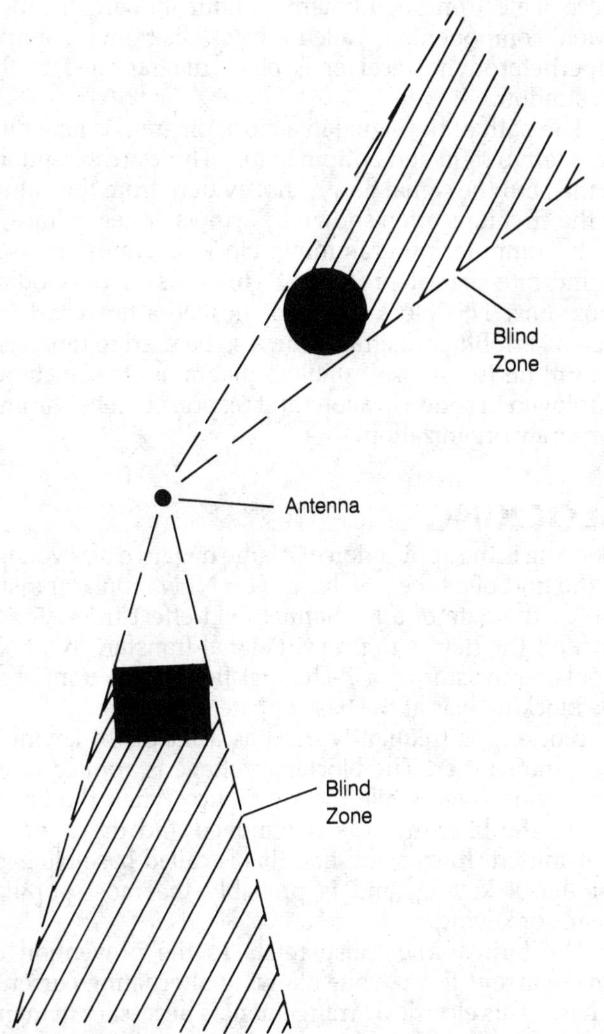

BLIND ZONE: Obstacles near radar can interfere with echoes in some directions.

BLOCK DIAGRAM: A block diagram shows discrete stages, not individual components.

BLOCK DIAGRAM

A circuit diagram showing the general configuration of a piece of electronic equipment without showing the individual components is called a block diagram. A simple superheterodyne receiver is block diagrammed in the illustration.

The path of the signal in a block diagram is generally from left to right and bottom to top. There are exceptions to this, but the signal flow, if not evident from the nature of the circuit, is often shown by arrows or heavy lines.

In computer programming, block diagrams are used to indicate logical processes. This aids in developing programs. This block diagram is called a flowchart (*see* FLOWCHART). Block diagrams can also be used to represent algorithms (*see* ALGORITHM). Block diagrams are sometimes employed to show divisions and responsibilities within a corporate organization.

BLOCKING

Blocking is the application of a large negative bias voltage to the grid of a tube, the base of an NPN bipolar transistor, or the gate of a N-channel field-effect transistor to turn off the device to prevent signal transfer. In a PNP bipolar transistor or a P-channel field-effect transistor, the blocking bias at the base or gate is positive.

Blocking is frequently used as a means of keying a code transmitter. The blocking voltage is applied to an amplifying stage while the key is up. When the key is down, the blocking bias is removed and the signal is transmitted. In transmitters, this is called base-block or gate-block keying, and is probably the most popular means of keying.

The term *blocking* also refers to the prevention of direct-current flow, while allowing alternating currents to pass. This electrical arrangement is necessary in many types of oscillators, and in capacitive coupling networks between amplifier stages. *See also* BLOCKING CAPACITOR.

BLOCKING CAPACITOR

When a capacitor is used for blocking the flow of direct current, but allowing the passage of alternating current, it is called a blocking capacitor. Blocking capacitors permit the application of different dc bias voltages to two points in a circuit. This situation is common in multistage amplifiers. The schematic illustrates the connection of a blocking capacitor between two stages of a transistorized amplifier.

Blocking capacitors are usually fixed rather than variable and should be selected so that attenuation does not occur at any point in the operating-frequency range. Generally, the blocking capacitor should have a value that allows ample transfer of signal at the lowest operating frequency, but the value should not be larger than the minimum to accomplish this. At vhf (very high frequency), the value of a blocking capacitor in a high-impedance circuit may be only a fraction of one pico-

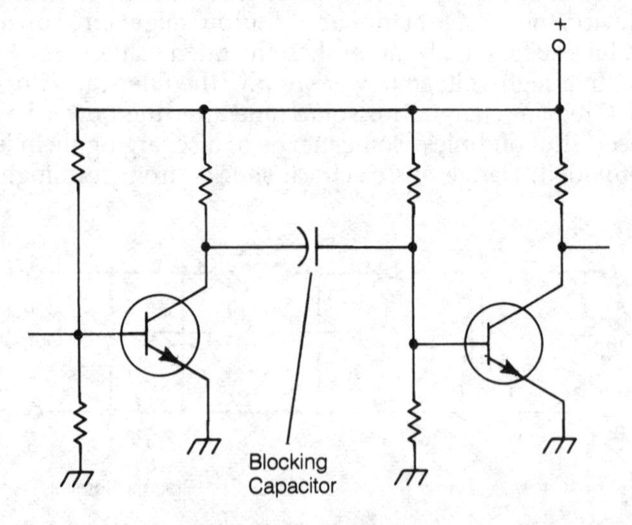

BLOCKING CAPACITOR: A blocking capacitor between two amplifier stages blocks the flow of direct current.

farad. At audio frequencies in low impedance circuits, values may range up to about 100 microfarads.

Blocking capacitors are used in the feedback circuits of some kinds of oscillators. The value of the feedback capacitor in an oscillator should be the smallest that will allow stable operation. *See also* CAPACITIVE COUPLING.

BLOCKING OSCILLATOR

See OSCILLATOR CIRCUITS.

BNC CONNECTOR

See CONNECTOR.

BODY CAPACITANCE

When the human body is brought near an electronic circuit, some capacitance is introduced between the circuit wiring and the ground unless the circuit is completely shielded. Usually, the effects of this body capacitance are not noticeable, because the value of capacitance is never more than a few picofarads. In certain tuned circuits such as ferrite-rod antennas or loop antennas, however, body capacitance can have a pronounced effect on the tuning, especially when the shunt capacitor is set for a small value. Usually, the hand is brought very near the shunt tuning capacitor as adjustments are made.

In general, the smaller the capacitances in a circuit, the more likely it is that body capacitance will be noticed. Shielding becomes very important at high frequencies.

BOHR ATOM

In 1913, Niels Bohr theorized that the atoms of all substances include negatively charged particles, called electrons, in orbit about positively charged particles, called nuclei. The inward force, keeping the electrons from flying off into space, is described as the electrical attraction between the nucleus and the electron, because they are of opposite polarity. The drawing illustrates the Bohr model of the atom.

According to the Bohr theory, electrons can exist in different levels of orbit. A higher orbit, farther from the nucleus, represents a higher energy state than a lower orbit close to the nucleus. If an electron gains energy, it moves into a higher orbit; a loss of energy results in a lower orbit. The modern theory of the atoms is similar to the Bohr model. *See also* ATOM, ELECTRON.

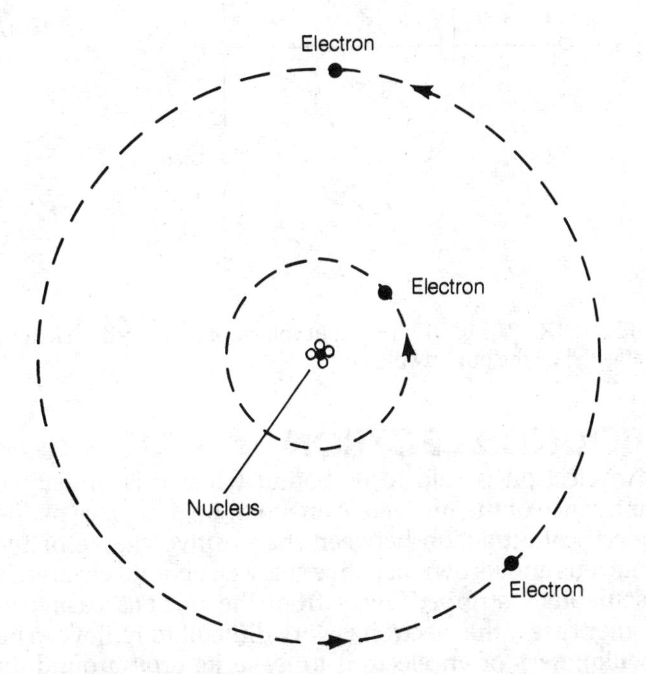

BOHR ATOM: A simplified diagram of an atom according to Bohr.

BOLOMETER

A bolometer is a sensor that changes its resistance in the presence of radiant energy. The most common example of a bolometer is a thermistor (*see* THERMISTOR).

The bolometer converts incident radiation into heat energy. The temperature change is then determined, and the amount of temperature change can be converted by known relation into a measure of the radiation intensity. The thermistor performs this function by itself; radiation intensity can be shown by a meter reading that depends on the resistance of the device with a constant supply voltage.

BOLTZMANN CONSTANT

The Boltzmann constant is a coefficient that defines the relationship between temperature and electron energy. The relation is linear, and is determined by this constant, which is abbreviated k. The higher the temperature, the greater the amount of energy in an electron.

The Boltzmann constant can be specified in many different units. The two most common expressions are:

$$k = 1.38 \times 10^{-23} \text{ J/K}$$
$$k = 8.61 \times 10^{-5} \text{ eV/K}$$

where J/K represents joules per degree Kelvin, and eV/K represents electron volts per degree Kelvin. *See also* ELECTRON VOLT, JOULE, KELVIN TEMPERATURE SCALE.

BOMBARDMENT

In electronics, bombardment occurs when an electrode is subjected to a stream of high-speed impinging electrons. With sufficient bombardment, secondary emissions occur (*see* SECONDARY EMISSION). If a massive metal electrode made of tungsten is bombarded by electrons with great velocity, X rays are produced.

In general, bombardment can refer to any situation in which a material is exposed to high-speed atomic particles. By bombarding certain elements with accelerated protons or alpha particles, the nucleus structure is changed, creating new elements or isotopes.

The earth is always under bombardment from space by high-speed particles. Fortunately, most of these particles collide with atoms in the atmosphere before they can reach the ground. In this way, our atmosphere protects life from the deadly effects of this cosmic radiation. *See also* ALPHA PARTICLE, BETA PARTICLE, COSMIC RADIATION.

BOOLEAN ALGEBRA

Boolean algebra is a system of mathematical logic, using the functions AND, NOT, and OR. In the Boolean system, AND is represented by multiplication, NOT by complementation, and OR by addition. Thus X AND Y is written XY or $X \cdot Y$; NOT X is written X' or $\overline{X}$; and X OR Y is written $X + Y$. Table 1 shows the values of these functions, where true is represented by 1 and false by 0. Boolean functions are used in the design of digital logic circuits.

Using the Boolean representations for logical functions, some of the mathematical properties of multiplication, addition and complementation can be applied to form equations. The logical combinations on either side of the equation are equivalent. In some cases, the properties of the logical functions are not identical to their mathematical counterparts. As an example, multiplication is distributive with respect to addition in ordinary arithmetic, and it is true also in Boolean algebra. Thus:

$$X(Y + Z) = XY + XZ,$$

which means that the logical statements:

X AND (Y OR Z)

and:

(X AND Y) OR (X AND Z)

are equivalent. The statement:

$$XX = X$$

however, is quite alien to the ordinary kind of mathematical algebra.

Table 2 lists some of the most common theorems of Boolean algebra. *See also* LOGIC EQUATION, LOGIC FUNCTION.

BOOLEAN ALGEBRA Table 1: Boolean truth tables. At A, the AND function (multiplication); at B, the NOT function (complementation); at C, the OR function (addition).

		A	B	C
X	Y	XY	X′	X+Y
0	0	0	1	0
0	1	0	1	1
1	0	0	0	1
1	1	1	0	1

BOOLEAN ALGEBRA Table 2: Some theorems in Boolean algebra.

1. X+0 = X (additive identity)
2. X1 = X (multiplicative identity)
3. X+1 = 1
4. X0 = 0
5. X+X = X
6. XX = X
7. (X′) ′ = X (double negation)
8. X+X′ = 1
9. X′X = 0
10. X+Y = Y+X (commutativity of addition)
11. XY = YX (commutativity of multiplication)
12. X+XY = X
13. XY′+Y = X+Y
14. X+Y+Z = (X+Y)+Z = X+(Y+Z) (associativity of addition)
15. XYZ = (XY)Z = X(YZ) (associativity of multiplication)
16. X(Y+Z) = XY+XZ (distributivity)
17. (X+W) (Y+Z) = XY+XZ+WY+WZ (distributivity)

BOOTSTRAP CIRCUIT

A circuit in which the output is taken from the source or emitter is called a bootstrap circuit (see illustration), because the output voltage directly affects the bias. In this amplifier negative output pulses cause an increase in the negative voltage at the input, and positive output causes a reduction in the negative voltage at the input.

The input of a bootstrap circuit is applied between the source and gate of an FET (field-effect transistor), or between the emitter and base of a transistor. This differentiates the bootstrap circuit from an ordinary grounded drain or grounded collector amplifier in which the input is applied between the gate and ground, or between the base and ground.

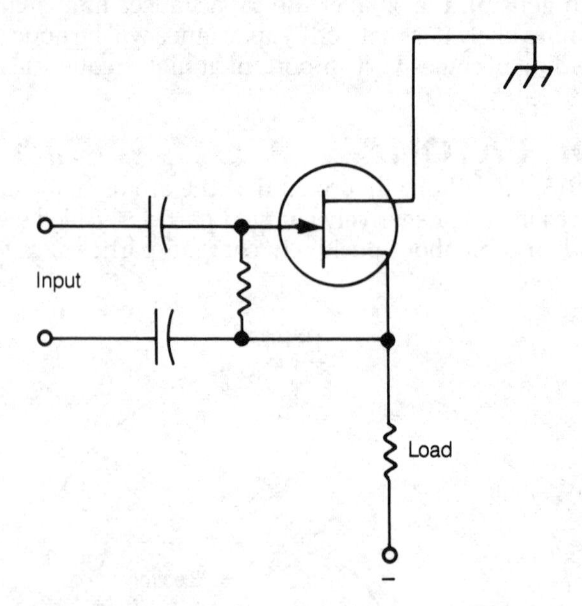

BOOTSTRAP CIRCUIT: The input voltage of the bootstrap circuit is affected by the output voltage.

BOUND ELECTRON

An electron is said to be bound when it is under the influence of the nucleus of an atom, held in place by the electrical attraction between the positive charge of the nucleus and its own negative charge. A bound electron is sometimes stripped away from the nucleus easily; in other cases, the electron is very difficult to remove. The willingness of an electron to leave its orbit around an atomic nucleus depends on the position of the electron, and the number of other electrons in that orbit. Electrons that are stripped away in ordinary reactions among elements are called valence electrons.

When electrons are not under the influence of atomic nuclei, they are said to be free electrons. A free electron becomes bound when a nucleus captures it. Bound electrons are sometimes shared among more than one nucleus. *See also* FREE ELECTRON, VALENCE ELECTRON.

BOWTIE ANTENNA

See ANTENNA DIRECTORY.

BREADBOARD TECHNIQUE

The breadboard technique is a method of constructing experimental circuits. A circuit board, usually made of phenolic or similar material, is supplied with a grid of holes. Components are mounted in the holes and wired together temporarily, either using hookup wire, or by copper plating on the underside of the board. Connections are easily removed or changed. The term comes from the early radio experimenter's practice of mounting components on a wooden board of the kind used to mix bread dough. The term *brassboard* is used in essentially the same way to describe a handmade prototype circuit.

BREAK

When a receiving operator takes control of a communications circuit, a break is said to occur. The sending operator might say "break" to give the receiving operator a chance to ask a question or make a comment. The sending operator then pauses, or breaks, awaiting a possible reply.

When a third party with an important message interrupts a contact between two stations, the interruption is called a break. The third station indicates its intention to interrupt by saying "break" during a pause in the transmission of one of the other stations.

The term *break* is also used to refer to opening contacts in a circuit that is alternately opened and closed. For example, in Morse-code transmission, the transition between key-down and key-up conditions is called the break. The change from key-up to key-down condition is called *make*. *See also* MAKE.

BREAKDOWN

When the voltage across a space becomes sufficient to cause arcing, a breakdown occurs. In a gas, breakdown takes place as the result of ionization which produces a conductive path. In a solid, breakdown actually damages the material permanently, and the resistance becomes much lower than usual.

In a semiconductor diode, the term breakdown is generally used in reference to a reverse-bias condition in which current flow occurs (*see* AVALANCHE BREAKDOWN). Ordinarily, the reverse resistance of a diode is extremely high, and the current flow is essentially zero. When breakdown occurs, the resistance abruptly drops and the current flow increases to a large large value.

Dielectric materials should be chosen so that breakdown will not take place under ordinary operating conditions. A sufficient margin of safety should be provided to allow for the possibility of a sudden change in the voltage across the dielectric. Rectifier diodes should have sufficient peak inverse-voltage ratings so that avalanche breakdown does not take place. Some semiconductor diodes, however, are deliberately designed to take advantage of their breakdown characteristics. *See also* ZENER DIODE.

BREAKDOWN VOLTAGE

The breakdown voltage of a dielectric material is the voltage at which the dielectric becomes a conductor because of arcing. In a solid dielectric, this usually results in permanent damage to the material. In a gaseous or liquid dielectric, the arcing causes ionization, but the breakdown damage is not permanent because of the fluid nature of the material.

In a reverse-biased diode, the breakdown voltage is the voltage at which the flow of current begins to rapidly increase (see illustration). Ordinarily, the current flow is very small in the reverse direction through a semiconductor diode; when the reverse voltage is smaller than the breakdown voltage, virtually no current flows. When the breakdown voltage is reached, a fairly large current flows. This is called avalanche breakdown. The transition between low and high current flow is abrupt. *See also* AVALANCHE BREAKDOWN.

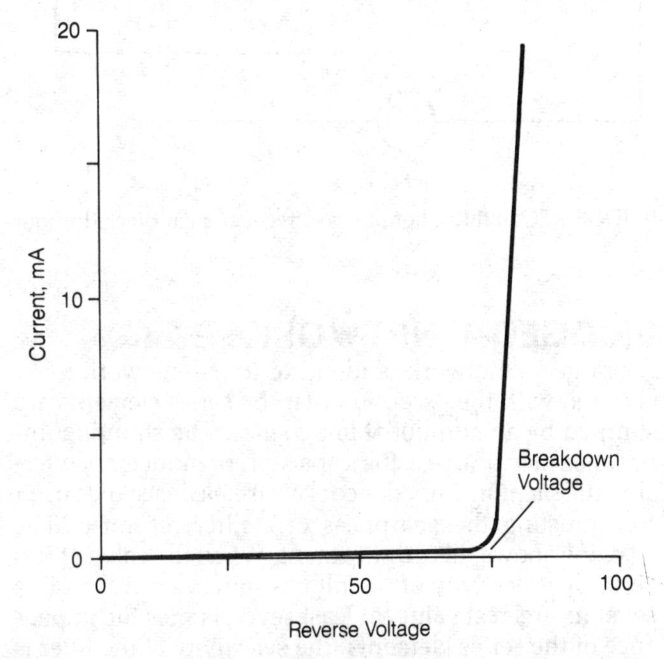

BREAKDOWN VOLTAGE: Significant conduction in a diode occurs in the reverse direction at the breakdown voltage.

BRIDGE RECTIFIER

A bridge rectifier is a form of full-wave rectifier circuit consisting of four diodes arranged as shown in the illustration. The diodes may be either tubes or semiconductors.

During that part of the alternating-current cycle in which point *X* is negative and point *Y* is positive, electrons flow from point *X* through D3, then through the load *R*, and back through D2 to point *Y*. When point *X* is positive and point *Y* is negative, electrons flow from point *Y* through D4, then through the load *R*, and back through D1 to point *X*. With either polarity at *X* and *Y*, the electrons flow through *R* in the same direction.

Bridge rectifier circuits are commonly used in modern solid-state power supplies. Some integrated circuits are built especially for use as bridge rectifiers; they contain four semiconductor diodes in a bridge configuration, encased in a single package. Sometimes, circuits such as the one shown are used as detectors in radio receiving equipment as well as in mixers. *See* RECTIFIER BRIDGE.

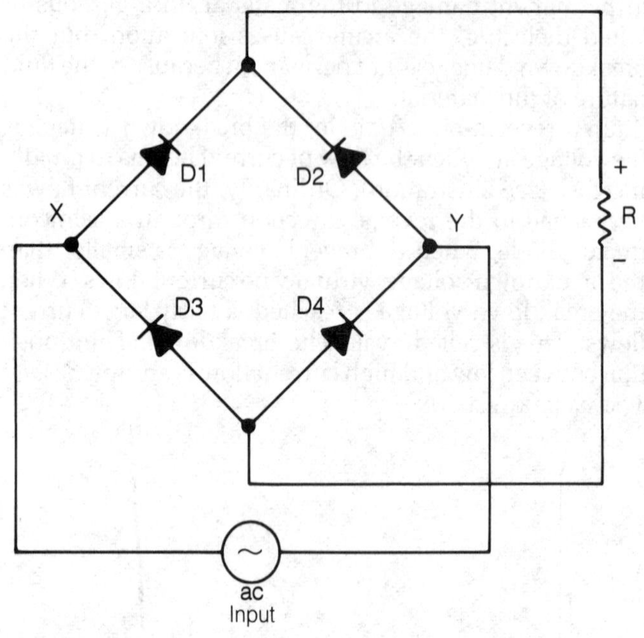

BRIDGE RECTIFIER: The full-wave rectifier circuit includes four diodes.

BRIDGED-T NETWORK

A bridged-T network is identical to a T network (*see* T NETWORK) with the exception that the series elements are shunted by an additional impedance. The shunting impedance may be a resistor, capacitor, or inductor. Generally, the *shunting* impedance is a variable resistance used for adjusting the sharpness of a filter response. The drawing shows this arrangement. When the value of R is zero, the selectivity of the filter is minimum. When R is set at its greatest value, at least several times the impedance of the series elements, the selectivity of the filter is maximum.

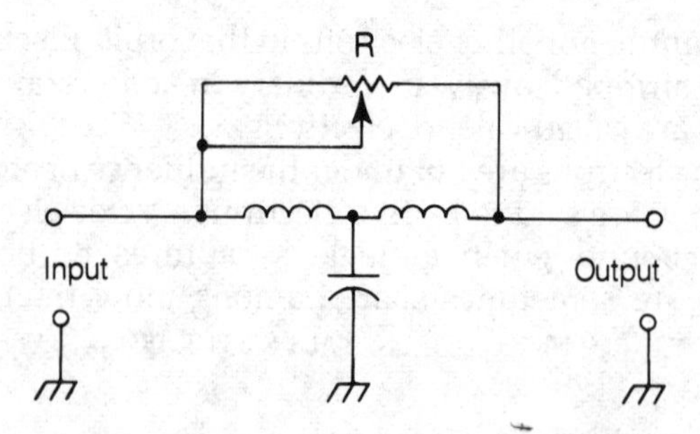

BRIDGED-T NETWORK: Resistor R changes the selectivity of the LC (inductive-capacitive) network of a bridged-T network lowpass filter.

BRIDGING CONNECTION

When a high-impedance circuit is connected across a component of measurement, the addition of the high-impedance circuit should not noticeably affect the operation of the device under test. This connection is called a bridging connection. The ordinary voltmeter is used in this way, as are the test oscilloscope and the spectrum analyzer. Any loss caused by a bridging connection is called bridging loss. Some bridging loss is unavoidable, but the greater the impedance of the bridging circuit with respect to the circuit under test, the smaller the bridging loss.

The term bridging connection is used interchangeably with the term shunt connection. *See also* SHUNT.

BRIGHTNESS

Brightness is the perceived intensity of visible-light radiation or reflection from an object. In a cathode-ray tube, the brightness of the image depends on the intensity of the electron beam striking the phosphor material.

All television receivers and computer monitors have a brightness control, which regulates the flow of electrons in the picture tube. Brightness and contrast are set with separate controls on computer monitors. An improperly adjusted brightness control creates unnatural pictures on a television receiver. A bar pattern can be used to set the brightness control manually on a television receiver.

The brightness of a television picture is also interrelated with the contrast. To keep the brightness and contrast of a television picture in proper proportion, automatic brightness and contrast controls are now employed in television receivers. *See also* AUTOMATIC BRIGHTNESS CONTROL, AUTOMATIC CONTRAST CONTROL, CONTRAST.

BRITISH STANDARD WIRE GAUGE

Metal wire is available in many different sizes or diameters. Wire is classified according to diameter by giving it a number. The designator most commonly used in the United States is called the American Wire Gauge, abbreviated AWG (*see* AMERICAN WIRE GAUGE). In some other countries, the British Standard Wire Gauge is used. The higher the number, the thinner the wire. The British Standard Wire Gauge sizes for designators 1 through 40 are shown in the table.

The larger the designator number for a given conductor metal, the smaller the current-carrying capacity becomes. The British Standard Wire Gauge designator does not take into account any coatings on the wire, such as

BRITISH STANDARD WIRE GAUGE: National British Standard Wire Gauge (NBS SWG) equivalents in Inches.

NBS SWG	Diameter, inches	NBS SWG	Diameter, inches
1	0.300	21	0.032
2	0.276	22	0.028
3	0.252	23	0.024
4	0.232	24	0.022
5	0.212	25	0.020
6	0.192	26	0.018
7	0.176	27	0.0164
8	0.160	28	0.0148
9	0.144	29	0.0136
10	0.128	30	0.0124
11	0.116	31	0.0116
12	0.104	32	0.0108
13	0.092	33	0.0100
14	0.080	34	0.0092
15	0.072	35	0.0084
16	0.064	36	0.0076
17	0.056	37	0.0068
18	0.048	38	0.0060
19	0.040	39	0.0052
20	0.036	40	0.0048

enamel, rubber, or plastic insulation. Only the diameter of the metal itself is included in the measurement.

BRITISH THERMAL UNIT

The British Thermal Unit is a measure of energy transfer. It is abbreviated Btu. One Btu of energy will raise the temperature of one pound of pure water by one degree Fahrenheit. There are many other units of energy measure. The British Thermal Unit is commonly used to specify the cooling capacity of an air conditioner. *See also* ENERGY CONVERSION, ERG, FOOT-POUND, JOULE, KILOWATT HOUR, WATTHOUR.

BROADBAND MODULATION

Broadband modulation is a relatively new concept in radiocommunications. Sometimes it is called wideband modulation.

The transfer of information by a radio signal requires a certain minimum amount of spectrum space. This minimum depends on the rate at which the information is conveyed. A very slow CW (continuous wave) signal requires just 10 to 20 Hz of bandwidth. An ordinary voice signal requires about 2 to 3 kHz. High-speed teletype may require up to several kilohertz. The complex signal of a television picture needs several megahertz of spectrum space. If two signals overlap at the receiving end of a communications circuit there will be interference. Thus, a given amount of spectrum space has room for a limited number of signals.

In the past, attempts have been made to minimize the required bandwidth of radio signals to make the most possible room in the available spectrum. However, by spreading the channel of a signal over a frequency range many times greater than its normal bandwidth, the density of power is made very small in all channels (see drawing).

A band might be filled to capacity by ordinary signals. But by spreading the energy of one more signal uniformly across the entire band, the interference to and from any fixed channel is negligible.

Various methods of obtaining broadband modulation are being tested. One common way of achieving this kind of modulation is to constantly vary the frequency of an ordinary transmitter in a predetermined manner over a wide range. The receiver then simply follows the transmitter according to a certain algorithm.

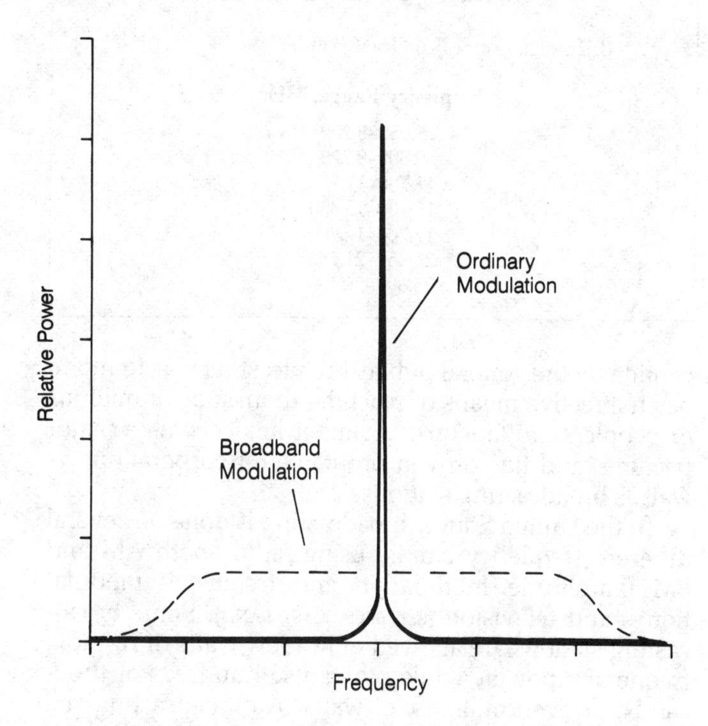

BROADBAND MODULATION: The low average concentration of energy across the band in broadband modulation is compared with the peak in conventional modulation.

BROADCAST BAND

In the United States, the standard AM (amplitude modulation) broadcast band is probably the most well known, extending from 535 to 1605 kHz. The FM broadcast band occupies the space from 88 to 108 MHz. There are four television broadcast bands. The lower three of these are collectively called the vhf (very high frequency) TV broadcast band and they contain channels 2 through 13. The fourth band, called the uhf TV broadcast band, contains channels 14 through 69. All of these bands, along with the AM and FM (frequency modulation) radio broadcast bands, are listed in the table.

Several shortwave bands are allocated for international broadcasting because of the nature of propagation at high frequencies. These are listed also in the table. These bands are sometimes identified according to the approximate wavelength of the signals in meters. For example, the band 15.100 to 15.450 MHz is known as the 19-meter band. *See also* FREQUENCY ALLOCATIONS.

BROADCASTING

Broadcasting is one-way transmission, intended for re-

BROADCAST BAND: U.S. and international broadcast bands.

A. United States Broadcast Bands

Frequency Range, MHz	Allocation
0.535-1.605	Standard AM
88.0-108.0	Standard FM
54.0-72.0	vhf Television
76.0-88.0	vhf Television
174.0-216.0	vhf Television
470.0-806.0	uhf Television

B. International Broadcast Bands

Frequency Range, MHz
5.950-6.200
9.500-9.775
11.700-11.975
15.100-15.450
17.700-17.900
21.450-21.750
25.600-26.100

ception by the general public. Broadcasting was found to be an effective means of reaching thousands or millions of people simultaneously. Thus it has become a major pastime, and has proven profitable for corporations as well as broadcasting stations.

In the United States, broadcasting is done on several different frequency bands, using radio—both AM and FM (amplitude modulation and frequency modulation)—and television (*see* BROADCAST BAND). Some broadcasting stations are allowed only a few watts of rf (radio frequency) power, while others use hundreds of thousands, or even millions, of watts. All broadcasting stations in the United States must be licensed by the Federal Communications Commission.

BROADSIDE ARRAY

A broadside array is a phased array of antennas (*see* ANTENNA DIRECTORY), arranged so that the maximum radiation occurs in directions perpendicular to the plane containing the driven elements. This requires that all of the antennas be fed in phase, and special phasing harnesses are required to accomplish this.

At frequencies as low as about 10 MHz, a broadside array may be constructed from as few as two driven antennas. At vhf and uhf, there might be several antennas. The antennas themselves may have just a single element, or they can consist of Yagi antennas, loops, or other systems with individual directive properties. If a reflecting screen is placed behind the array of dipoles in the drawing, the system becomes a billboard antenna.

The directional properties of a broadside array depend on the number of elements, the gain of the elements and on the spacing among the elements.

BROWN AND SHARP GAUGE

See AMERICAN WIRE GAUGE.

BUBBLE MEMORY

A bubble memory or magnetic bubble memory (MBM) is part of an alterable, nonvolatile serial memory system. This high-density data storage device can be used as magnetic mass storage for computers. Capable of performing large block transfers, bubble memories are similar to semiconductor shift registers. Figure 1 is a simplified block diagram of a magnetic bubble-memory system. The MBM stores data in the form of tiny cylindrical magnetic bubbles or domains that are about three microns in diameter. The domains are formed and arranged vertically in a thin, magnetic garnet layer generally grown by liquid-phase epitaxy over a nonmagnetic substrate.

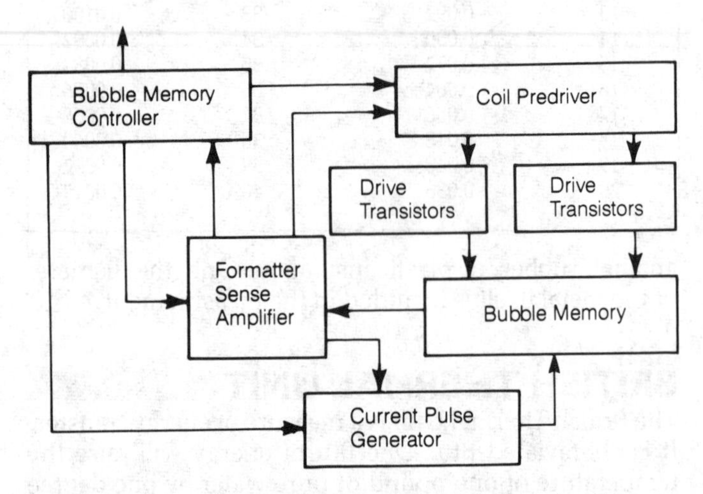

BUBBLE MEMORY: Fig. 1. The basic functional blocks of a magnetic bubble memory (MBM) system.

The domains are created from electrical signals by a bubble generator within the memory device. They are herded in the desired direction by external, rotating magnetic driving fields. They are herded to permalloy magnetized metal patterns deposited on the substrate. The domains are stored and even destroyed when no longer needed. The domains are reconverted to electrical signals by an internal detector. The presence of a bubble is interpreted as binary one, and the absence of a bubble is interpreted as a zero. Serial data is shifted at a relatively slow kilohertz range.

MBMs are intended to perform the same kinds of memory functions as magnetic disk drives, but MBMs have no moving parts that can be damaged by shock or vibration. They are classed as nonvolatile read/write memories. Advanced bubble memory units have a capacity of 1 megabit or 128 k bytes. MBMs are interfaced to a computer with an external bubble memory controller, and their operation requires a set of semiconductor IC (integrated circuit) support devices.

The bubble memory substrate is made from a wafer of nonmagnetic crystalline material called gadolinium-gallium garnet (GGG) on which a thin film of garnet is grown epitaxially. Figure 2 is a crosssectional view of part of the wafer showing nonmagnetic conductors, bubble steering patterns of magnetic metal, insulation, passivation, and bonding pads. These components are depos-

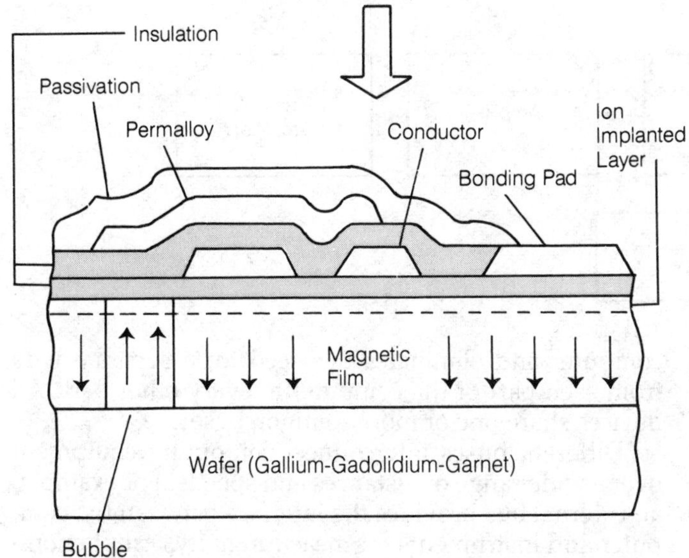

BUBBLE MEMORY: Fig. 2. Cross section of magnetic bubble memory substrate shows magnetic thin film grown on gallium-gadolinium-garnet substrate.

ited on the wafer by methods similar to those used in semiconductor IC fabrication.

The bubbles, once formed, are oriented perpendicular to the film. When a magnetic bias field is applied, the domains form the tiny cylinders, and they become magnetized in opposition to the applied field. A magnetic bubble will move from a region of low magnetic field strength to one of higher magnetic field strength.

In an MBM system, the substrate is mounted between two permanent magnets which create a continuous bias field as shown in Fig. 3. Currents in two electromagnetic coils wrapped around the substrate at right angles to each other create the rotating magnetic field that propels the bubbles through the film. The coil currents are generated by electronic circuits external to the bubble substrate.

The substrate has a large number of endless storage loops for the circulation of blocks of data. These are in parallel with and coupled to an input track and an output track. The input track writes the data in the storage loops, and an output track reads the data under the direction of a controller. Exchange or replication of data between the tracks and loops occurs in all loops simultaneously.

If power fails, the rotating field disappears, and the bubbles stop moving. However, because the bias field created by the permanent magnets is not affected, the bubbles and data they represent are held in the film. The data becomes accessible again when power is restored.

BUFFER

A buffer is a circuit that temporarily stores data while the data is being transmitted from one part of a system to another. Buffers are used in computer and communication equipment to compensate for a difference in the flow of data or time of occurrence of events. A buffer can be a register or memory with no limit on capacity.

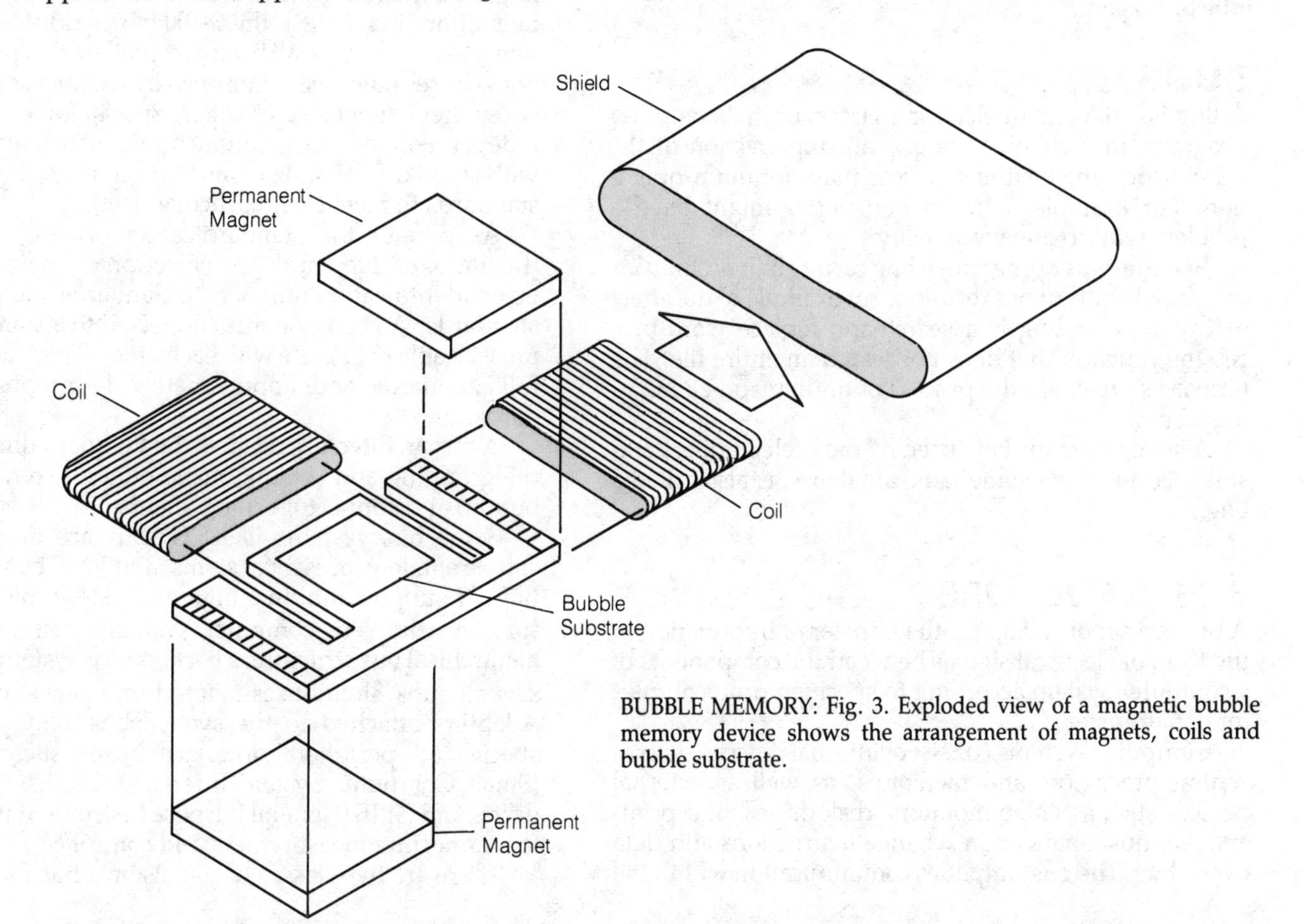

BUBBLE MEMORY: Fig. 3. Exploded view of a magnetic bubble memory device shows the arrangement of magnets, coils and bubble substrate.

BUFFER STAGE: A buffer stage follows the oscillator in this continuous-wave transmitter.

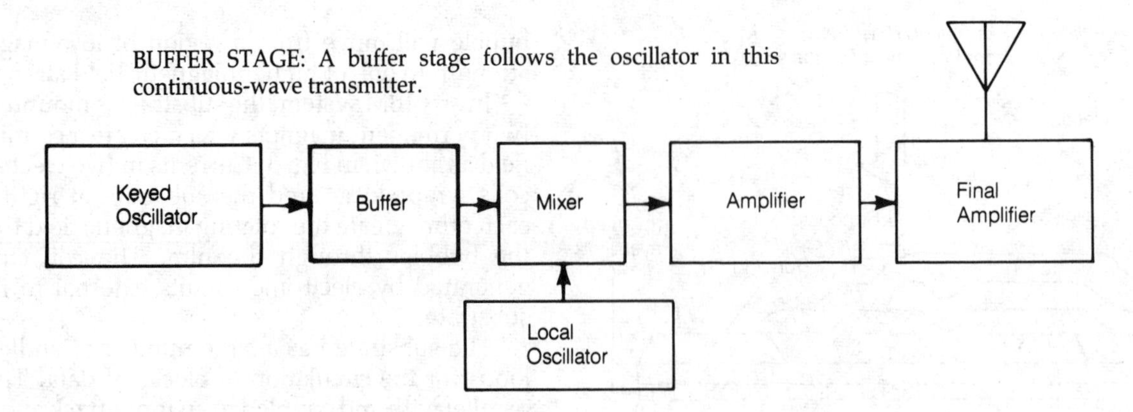

BUFFER STAGE

A buffer stage is a single-tube or single-transistor stage used to provide isolation between two other stages in a radio circuit. Buffer stages are commonly used following oscillators, especially keyed oscillators, to present a constant load impedance to the oscillator. As the amplifiers following an oscillator are tuned, their input impedances change, and this can affect the frequency of the oscillator unless a buffer state is used. When an oscillator is keyed, the input impedance of the following amplifier stage may change; or, when an amplifier is keyed, its impedance can change. The buffer stage serves to isolate this impedance change from the oscillator, and stabilize the oscillator frequency.

The block diagram illustrates the location of a buffer stage in a CW (continuous wave) transmitter. In this case, the oscillator is keyed. A buffer circuit generally has little or no gain.

BUG

A bug is a flaw in an electronic circuit or in a computer program. In a circuit, a bug is an imperfection in the general design, resulting in less-than-optimum operation. For example, a radio transmitter might have a problem with frequency stability.

In a computer program, a bug results in inaccurate or incomplete output or execution. An example of the effect of this type of bug is a search-and-replace word-processing function that does not search an entire file. The term *debug* refers to the process of finding and eliminating a bug.

A semiautomatic key, used by radiotelegraphers and still used today by some radio amateurs, is also called a bug.

BUS STRUCTURE

A bus is a set of wiring for the transfer of information in the form of electrical signals between the components of a computer system according to specified protocol rules for data transfer.

Computer systems consist of internal devices such as central processors and memories, as well as external devices such as video monitors, disk drives, and printers. All must share and exchange instructions and data over a bus. The bus simplifies communication within the computer and eliminates the need for a separate wire from each part of the computer to every other part. All devices share one or more common buses.

Different buses interconnect computing equipment over a wide range of distances and speeds. For example, an external bus provides the interface between the computer and instruments for measuring physical phenomena in real-time data acquisition and control systems.

There are many different bus standards. No single bus is suitable for all applications. Bus structures become standards when a particular manufacturer's bus is accepted by many users and equipment manufacturers who make equipment that is compatible with it. Examples are Digital Equipment Corporation's open Unibus and Qbus, Motorola's VME bus and IBM's PC bus. Some developers of proprietary bus structures have submitted their buses for endorsement by national and international standards groups. The General Purpose Instrumentation Bus (GPIB), the S-100 bus and Intel Corporation's Multibus have all been accepted by this procedure. Some buses have been improved by technical review and subsequent suggested changes. Special interest groups, independent of any manufacturer, also have worked with standards organizations to design and develop bus standards to meet their particular needs.

As part of the standardization process, the IEEE (Institute of Electrical and Electronic Engineers) has given identification numbers to standards and proposed standards. All components connected to a standard bus must be able to operate with each other. There are now 12 full standards and approximately 14 proposed standards.

A bus architecture dictates the PC board dimensions, the grouping and cooling of components, power distribution, and connector arrangements.

Some bus systems like STD bus are intended for simple single-processor systems, and VME bus and Multibus II support multiple high-end, 32-bit microprocessors. A complete computer system often includes a hierarchical bus structure: a backbone or system bus, and several subsystem buses tailored for specific functions. Adapters attached to the system bus allow access to specialized peripheral interface buses such as SCSI (Small Computer System Interface) for disk and tape drives and GPIB (General Purpose Instrumentation Bus), for connecting measurement and control equipment.

There are five classes of signals on a bus for transferring data: address, data, control, response, and timing. Together they form a transaction bus. Address and data signals can flow sequentially over the same conductors and this multiplexing reduces the size of the bus. An arbitration bus assures orderly access to the bus among competing devices, and the interrupt bus accepts signal requests for attention from devices.

BUTTERFLY CAPACITOR

At very high and ultrahigh frequency (vhf and uhf) ranges, a tuning device called a butterfly capacitor is used in place of the conventional coil-and-capacitor tank circuit. A butterfly capacitor includes inductance as well as capacitance. The device gets its name from the fact that its plates resemble the opened wings of a butterfly.

The butterfly capacitor has a very high Q factor (*see* Q FACTOR). That is, it displays excellent selective characteristics although the cavity resonator is better at vhf and uhf (*see* CAVITY RESONATOR). The drawing shows a butterfly capacitor.

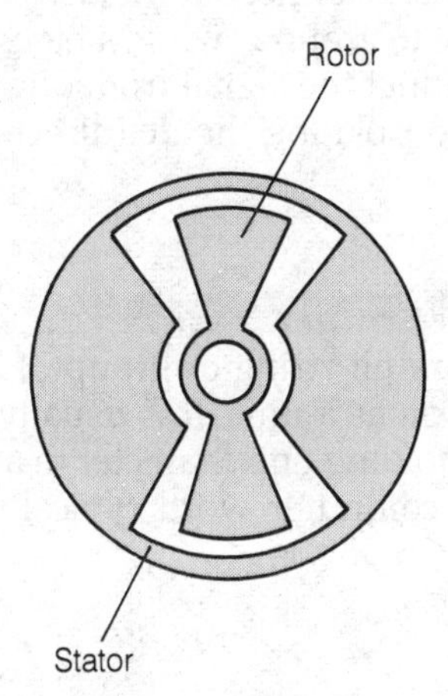

BUTTERFLY CAPACITOR: The butterfly capacitor is used to tune high-frequency circuits.

BUTTERWORTH FILTER

A Butterworth filter is a special type of selective filter designed to have a flat response in its passband and a uniform roll-off characteristic. A Butterworth filter can be designed for a lowpass, highpass, bandpass, or band-rejection response.

Figure 1 shows some ideal Butterworth responses for lowpass, highpass, and bandpass filters (A, B, and C). Note the absence of peaks in the passband. Schematic diagrams of sample lowpass (A), highpass (B), and bandpass (C) Butterworth filters are shown in Fig. 2. The input and load impedances must be correctly chosen for proper operation; the values of the filter resistors, inductors and capacitors depend on the input and load impedances.

A similar type of selective filter, called the Chebyshev filter, is also commonly used for lowpass, highpass, bandpass, and band-rejection applications. *See also* BANDPASS FILTER, BAND-REJECTION FILTER, CHEBYSHEV FILTER, FILTER, HIGHPASS FILTER, LOWPASS FILTER.

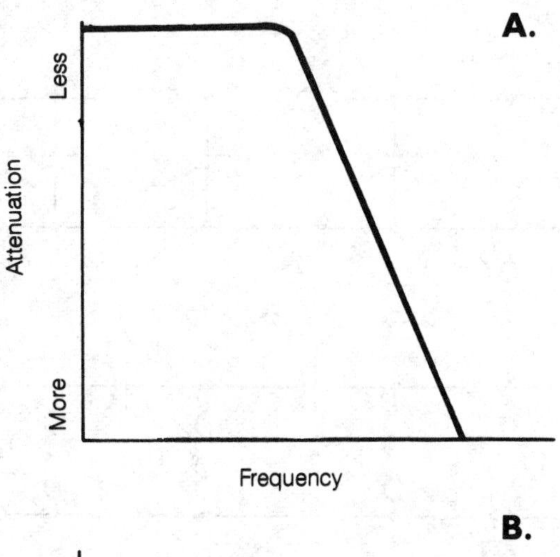

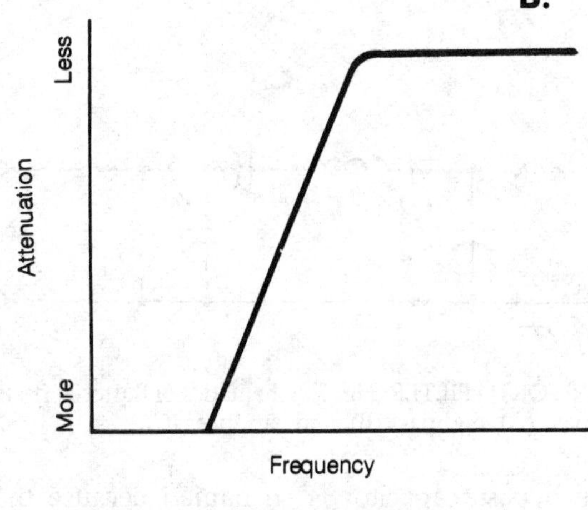

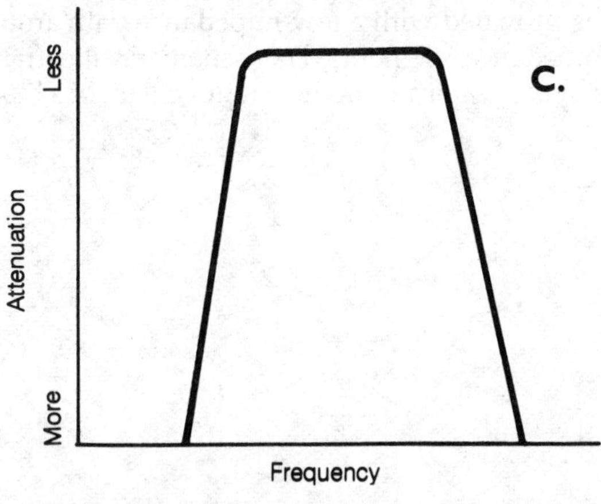

BUTTERWORTH FILTER: Fig. 1. The ideal Butterworth filter responses include lowpass (A), highpass (B), and bandpass (C).

BYPASS CAPACITOR

A capacitor that provides a low impedance for a signal while allowing dc (direct current) bias, is called a bypass capacitor. Bypass capacitors are frequently used in the emitter circuits of transistor amplifiers to provide stabilization. They are also used in the source circuits of field-effect transistors and in the cathode circuits of tubes. The bypass capacitor places a circuit element at signal ground potential, although the dc potential may be several hundred volts.

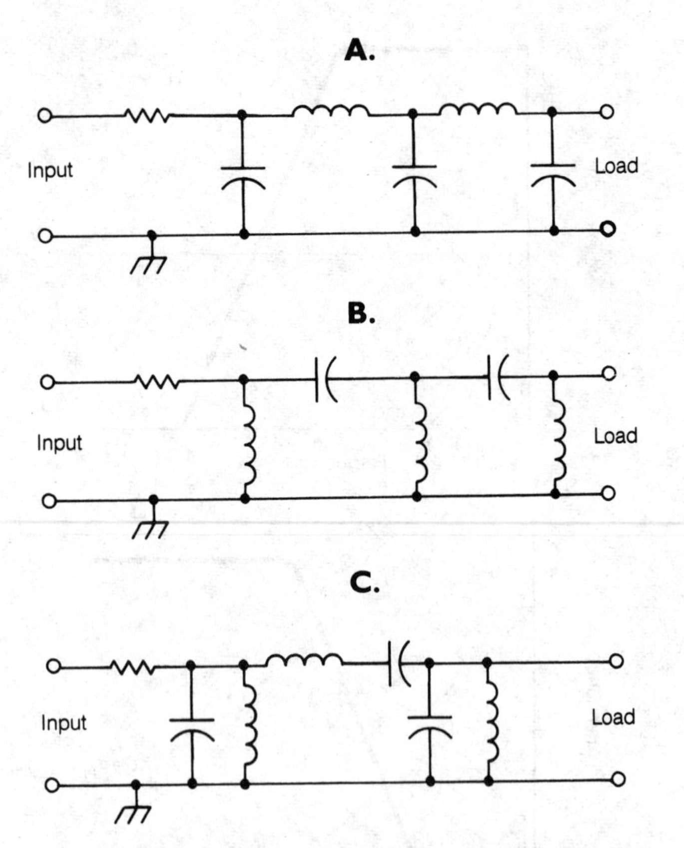

BUTTERWORTH FILTER: Fig. 2. Schematics of Butterworth filters for lowpass (A), highpass (B), and bandpass (C).

The bypass capacitor is so named because the ac signal is provided with a low-impedance path around a high-impedance element. The schematic illustrates a typical bypass capacitor in the emitter circuit of a bipolar-

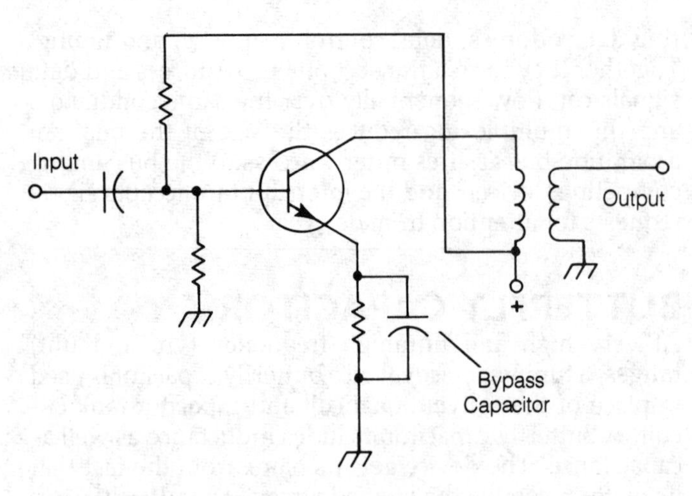

BYPASS CAPACITOR: The bypass capacitor provides stabilization and emitter bias in this amplifier circuit.

transistor amplifier. A bypass capacitor is different from a blocking capacitor (*see* BLOCKING CAPACITOR); a bypass capacitor is usually connected in such a manner that it shorts the signal to ground, while a blocking capacitor is intended to conduct the signal from one part of a circuit to another while isolating the dc (direct current) potentials.

BYTE

A byte is a binary bit string or group of adjacent binary digits operated on as a unit. It is usually eight bits long and capable of holding one character in a character set of a computer. A computer word typically consists of at least two bytes. *See also* BIT.

CABLE

Cable consists of two or more insulated conductors in a common jacket. Cables can be divided into two categories: 1) those with parallel insulated conductors and 2) those with insulated conductors that are bundled and twisted together.

There are four kinds of parallel cable: 1) uninsulated conductors covered with a common insulation such as 50 and 95 Ω transmission line, 300 Ω twin lead, and antenna rotator cable; 2) uninsulated conductors laminated between two layers of insulating materials (known as flat cable or flat flexible cable); 3) individually insulated wires bonded or woven together (ribbon cable); and 4) individually insulated wires enclosed in an extruded overall jacket.

Common Insulation Cable. Common insulation cable is produced by drawing the desired number of uninsulated wires simultaneously through the extruder to form an integral plastic covering over them as shown in Figs. 1A and 1B. The size and shape of the extrusion die determines wall thickness and the shape of the finished cross section. The conductors can be distinguished and polarity can be identified by the following methods:

- Contrasting color stripe along one edge of the insulation.
- Tinned and bare copper conductors.
- Ridges or fins along one edge.

Flat Cable. Flat cable—also known as flat-conductor cable (FCC), flat flexible cable, and laminated cable—refers to cable with two or more parallel conductors sandwiched between layers of insulation as shown in Fig. 2A. The conductors are placed between two layers of a sheet insulation which may have an adhesive applied. Under heat and pressure, the insulation layers bond to the conductors and to each other.

The conductors for flat cable can be bare or coated flat copper ribbon or round wire that is solid or stranded. The edges of flat conductors are usually rolled to remove burrs and sharp edges, and the conductors are annealed to relieve mechanical stresses. Many kinds of insulation are used in making flat cable including vinyls (polyvinyl chloride—PVC), polyester composites, cross-linked polyolefin, Teflon, and polyimide. The vinyls are used for commercial products because of their low cost. Polyester composites exhibit good heat conductive, shrink, stretch, and cold flow characteristics as well as dielectric strength. Polyester also has high flex life, good tensile strength, and excellent abrasion resistance.

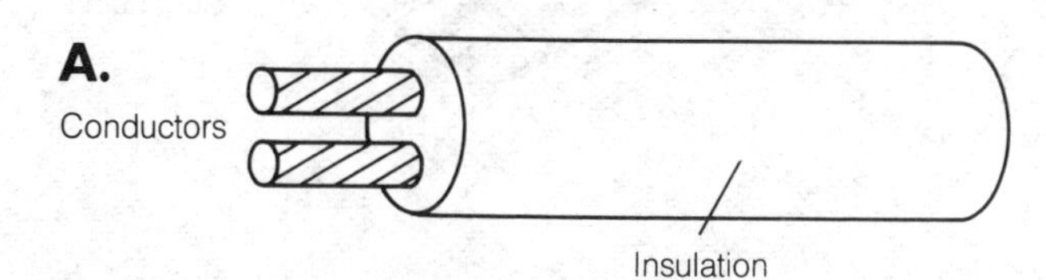

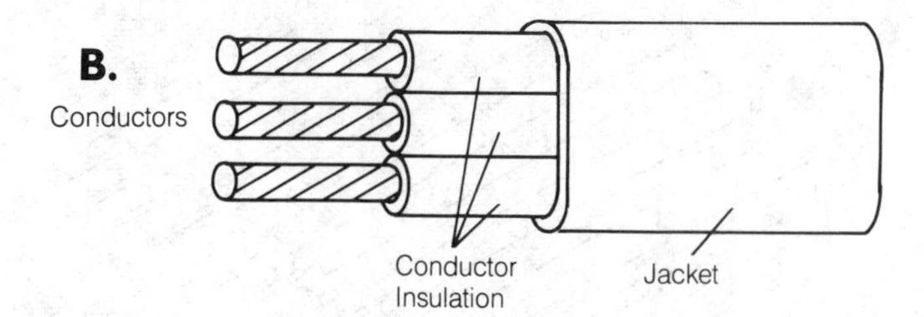

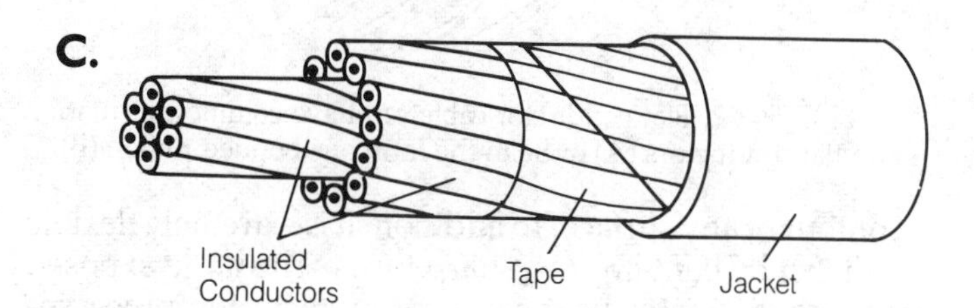

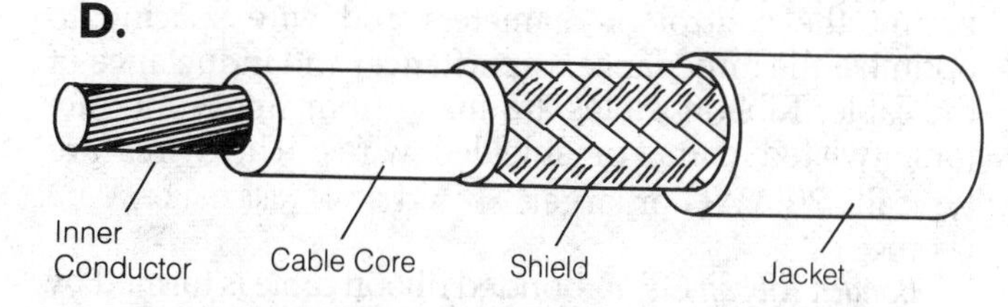

CABLE: Fig. 1. Multiconductor cable includes simple two- (A) and three-conductor (B) cables, multiconductor cables (C), and coaxial cable (D).

Polyolefin has excellent electrical properties, a 150°C temperature rating, and good resistances to hot solder and chemicals. Both TFE and FEP Teflon have excellent electrical properties and remain chemically stable when subjected to a wide range of environmental conditions. TFE is an effective insulation for temperatures up to 260°C, and FEP is useful to 200°C. Polyimide is usually combined with Teflon FEP as a bonding agent. It has excellent dimensional stability, abrasion resistance, and electrical properties.

Ribbon Cables. Ribbon cables are flat multiconductors with individual insulated conductors that can easily be separated. This cable is widely used where space limitations make the installation of conventional cables difficult. It saves space and weight and will follow the

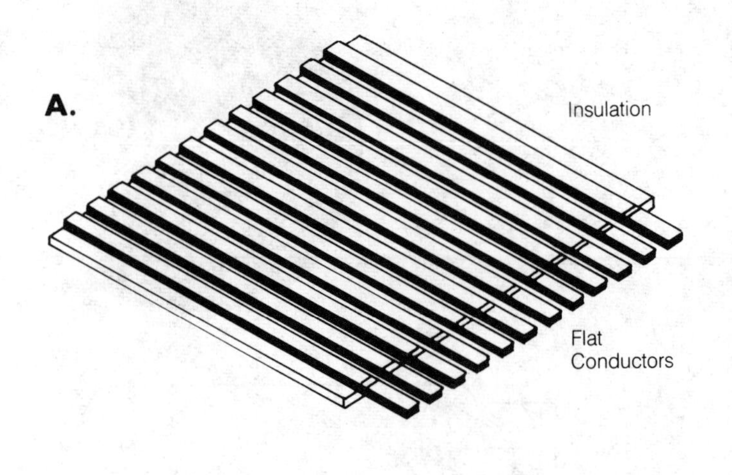

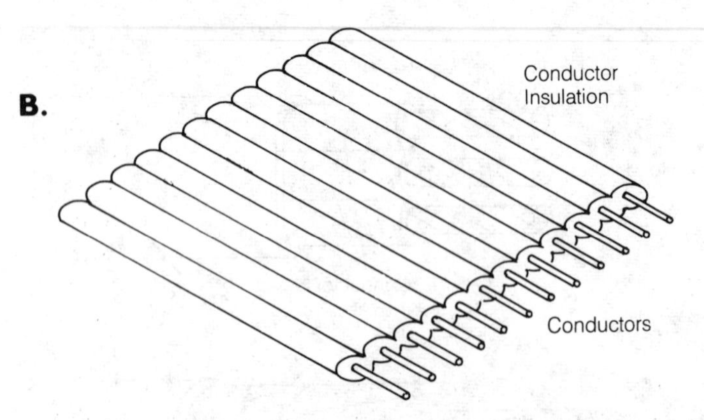

CABLE: Fig. 2. Flat conductor cable can have common insulation and flat conductors (A) or be in the form of a bonded ribbon (B).

contour of any surface. In addition, it is extremely flexible and can be bent around sharp turns. The large exposed surface dissipates heat easily. Manufacturing processes permit the control of diameters and wire spacing to optimize the impedance, capacitance, and inductance of the cable. Ribbon cables are made from single conductors, twisted pairs, or shielded wires. The wires are typically 30 AWG or larger. *See* AMERICAN WIRE GAUGE, CONDUCTOR.

Bonded Ribbon Cable. Bonded ribbon cable is formed by heat sealing or bonding wires with vinyl (PVC) insulation. Bonded ribbon cable as shown in Fig. 2B provides the most weight and space savings of any flat cable. Each insulated conductor can easily be separated without damaging its insulation. This cable can be terminated with mass-termination insulation displacement (MTIDC) connectors. *See* CONNECTOR.

Woven-Ribbon Cable. Woven-ribbon cable permits the assembly of wires that do not have vinyl insulation that can be heat bonded or combinations of vinyl and other kinds of insulation. The cable is formed by weaving. Weaving results in a very flexible assembly of wire with inherent strain relief, useful in preventing conductor flexing fatigue.

Shielded Flat Cable. Shielded flat cable is used where lengths of flat ribbon cable must be run outside of a shielded enclosure (as between two separate cabinets of a system). The cable is shielded with copper braid and jacketed with a flat vinyl jacket. Alternatively the cable can be in a round-to-flat configuration with the flat cable beyond the connectors rolled, shielded with copper braid, and enclosed in a round PVC jacket. These cables can be terminated on either end with conventional post-and-box PC (printed circuit) connectors or D-type subminiature connectors. *See* CONNECTOR.

Twisted-Component Cables. Cables made from twisted wires as shown in Fig. 1C generally are more flexible and easier to handle than those made from parallel conductors. The desired electrical and physical properties of the cable can be obtained more easily by varying the type, grouping, and positioning of individual conductors. These cables can be classed as bunched and concentric. Bunched cables consist of a number of conductors twisted together. They can be identical single wires or twisted pairs. Concentric cables consist of a central component (filler, single wire, or cable group) surrounded by one or more layers of helically laid wires.

Coaxial Cables. Coaxial cables are intended for transmitting radio frequency signals with frequencies that extend into the microwave region above 1 GHz. Standard coaxial cable has a single inner conductor and an outer shield of copper braid over a dielectric core. It is jacketed with vinyl (PVC) for protection as shown in Fig. 1D. However, there are also multiconductor versions of coaxial cable: those with two conductors are called twinax, and those with three conductors are called triax. Semirigid coaxial cable with an external metal sheath is used for the transmission of microwave and near-microwave frequency signals. It minimizes rf (radio frequency) losses because the bending radius is larger than that of conventional coaxial cable. *See* COAXIAL CABLE.

Fiberoptic Cable. Fiberoptic cable consists of single and multiple strands of glass fiber enclosed in protective jackets to permit them to be handled and installed without damage to the glass fibers. Single-mode fibers are widely used for long-distance telecommunications, and multimode fibers are used in short-hauls systems under one kilometer in length. The insulating and jacketing materials used in the manufacture of fiber optic cable are similar to those used in the fabrication of conventional wire cables. *See* FIBEROPTIC CABLE AND CONNECTORS.

Cable Manufacture. In conventional cable manufacture, individual copper conductors within cables are insulated prior to their assembly into cables. The insulation can be color coded to assist in making interconnections and later circuit diagnosis. Large bundles of small diameter insulated conductors form a generally round cross section, but small numbers of large diameter conductors form a fluted cross section.

Fillers. Fillers are included in cables made from a relatively small number of large-diameter wires or cables to round out the depressions that would otherwise appear under the outer jacket. Filling is usually done to improve appearance of the cable, but it can also provide proper conductor separation in low-loss transmission lines. Fillers act as cushioning in heavy-duty cables that are subject to flexing and impact. The most commonly used fillers are lightweight, nonconducting cotton, jute, vinyl, polyethylene, and twisted polyethylene monofilaments.

Binders. Binders are threads wound spirally over individual groups of insulated wire to hold the assembly together for subsequent processing. Colored binders can be used to separate and identify identical groups of conductors. Nylon is widely used for electronic cables, but polyester or polypropylene are more commonly used for telephone cables.

Tapes. Tapes are often wound around insulated wire bundles as an added protection against mechanical abuse and to prevent damage to the insulation. For example, tapes can be placed between shields and adjacent conductors. They make it easier to strip a jacket and ensure a smooth sheath surface. Polyester is the most frequently used tape material, but paper tapes are also popular. They can be applied either spirally or longitudinally.

Jackets. Jackets—also referred to as sheaths—cover and protect enclosed wires of a cable against abrasion, mechanical damage, spilled chemicals, and fire. Jackets can cover single conductors or they may cover the entire cable. Nylon inner jackets, where used, are typically 0.002 to 0.006 inch thick. Outer jackets, most frequently vinyl, polyethylene, neoprene, or polyurethane, can account for 10 percent of the cable core diameter.

Jackets are usually pulled on over the core so they adhere rather loosely and are easy to strip. However, neoprene jackets are pressure extruded to fill all the voids and convolutions in the cable core. They have a smooth firm surface and are more difficult to strip unless the underlying conductors are insulated with a different type of plastic or covered with a separator or barrier.

Vinyl (PVC) jackets, inherently flame and abrasion resistant, are used for indoor and general-purpose applications. Neoprene is used where the cable is to be exposed to abuse and severe handling. It will not stiffen in subzero temperatures, and it is oil, ozone, and weather resistant. Cables with neoprene jackets can be buried or placed in conduits, trays, racks, or ducts. Hypalon has most neoprene properties, but it has superior resistance to ozone, oxidation, and heat.

Many electronic product manufacturers prefer premanufactured cable assemblies for external interconnects. In applications where the cable is subject to physical or environmental abuse, molded assemblies are used. Molding with an appropriate sealant keeps moisture and dust out of the critical spaces under the cable jacket where it enters the terminating connectors because the molding bonds the jacket to the connector.

Other jacketing materials include ethylene-propylene rubber (EPR), polyethylene, Kynar, polyurethane, thermoplastic elastomer (TPE), Teflon FEP, and Tefzel.

CABLE TELEVISION

In locations far from any television broadcast stations or where line-of-sight obstructions exist, cable television is popular. In recent years, cable television has replaced the conventional TV antenna in many cities. Cable television offers better picture quality than the individual antenna and cable is also less susceptible to outside interference from automobile engines, airplanes, and other sources.

Most cities in the United States and Canada have only two or three television broadcast stations, but with cable TV, all 12 vhf (very high frequency) channels can be filled in every city in the United States. For example, a viewer in Miami might watch stations from Atlanta, New York, and Orlando, all with cable.

When television signals are transmitted by cable, the signals can be heterodyned, or converted, to lower frequencies for more efficient transmission. All cable systems have increased loss as the frequency is increased. For long-distance cable transmission, uhf (ultrahigh frequency) channels are generally heterodyned down to vhf. This is why a cable TV viewer might switch to channel 3 on a TV set and find it occupied by channel 17. Often, vhf channels are converted to lower vhf channels for the same reason.

CACHE MEMORY

A cache memory is a small amount of high-speed memory that buffers or holds a copy of the data or instructions currently being used by the central processing unit (CPU). These data and instructions are copies of those in the slower, main memory. Thus, cache memory improves the computer's speed by providing rapid access to currently used data and instructions. Cache memory also may be contained in one or more integrated circuit memory devices. *See also* SCRATCH-PAD MEMORY.

CADMIUM

Cadmium is an element with an atomic number of 48 and an atomic weight 112. It is a metallic substance often used as a plating material for steel to prevent corrosion. Cadmium adheres well to solder, and it is therefore a good choice for plating of chassis for the construction of electronic equipment.

Cadmium is used in the manufacture of photocells, devices that change resistance according to the amount of light that strikes them. Cadmium is also used in nickel-cadmium cells and batteries. *See also* BATTERY.

CADMIUM PHOTOCELL

See PHOTOCONDUCTIVE CELL.

CAGE ANTENNA

See ANTENNA DIRECTORY.

CALCULATOR

An electronic calculator is an instrument that performs calculations using semiconductor devices. Some calculations that would require several minutes to perform on paper can be performed in just a few seconds on an electronic calculator. Calculators are now small enough to fit in a shirt pocket. Some calculators are made for desktop or office use. They are capable of printing the calculations on a narrow strip of paper.

The calculator is less complex than the computer.

Computers can solve complicated algebraic equations that might require years of time to solve with a calculator. However, the line of distinction between the calculator and computer is not well defined. Some calculators can be programmed to solve problems, just like computers. *See also* COMPUTER.

CALCULUS

Calculus is a mathematical technique for determining the rate of change of a quantity precisely. Calculus is also used to determine the net average effect of a quantity over a period of time.

Differential calculus is concerned with the instantaneous rate of change of a variable (*see* INSTANTANEOUS EFFECT). The illustration shows a graph of collector current as a function of the base voltage for an NPN bipolar transistor. If a fairly accurate mathematical representation of this function can be found, a precise determination of the ratio of collector-current change to base-voltage change can be derived for any operating point.

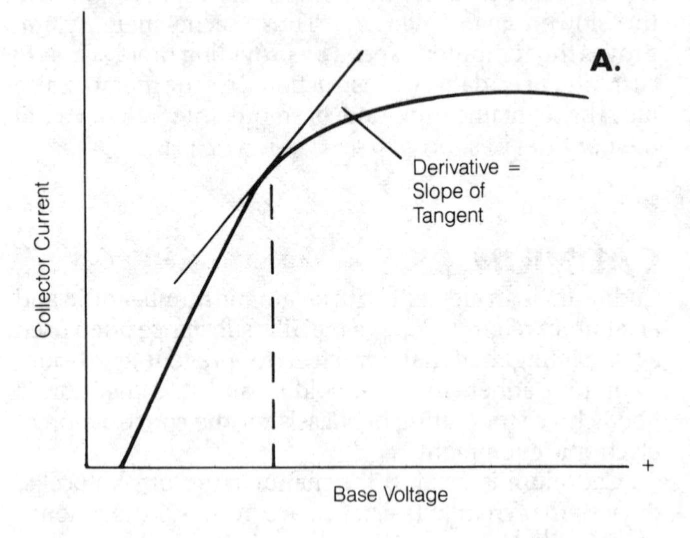

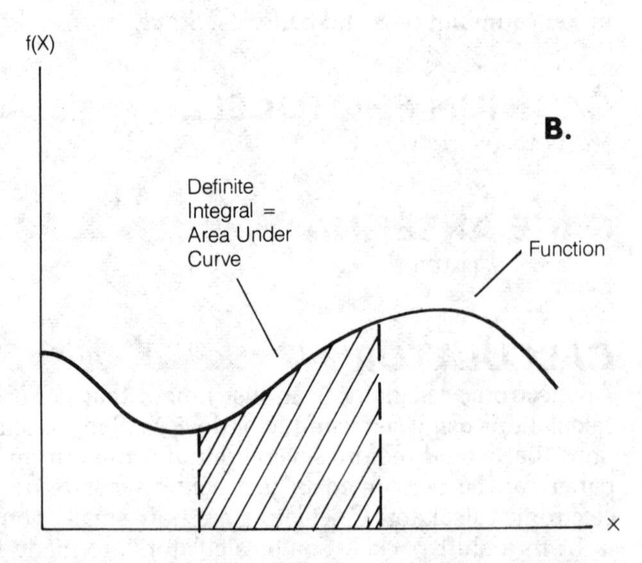

CALCULUS: The slope of a tangent line in differential calculus (A) and the area under a curve between two points determined by integral calculus (B).

This ratio is the slope of a line tangent to the curve at the operating point. The rate of change of any variable is called the derivative.

In integral calculus, the area under the curve of a function is determined between two points on the abscissa or independent-variable axis. An example of this is illustrated at B. Regions above the abscissa (horizontal axis in the illustration) are considered to have positive area; regions below the abscissa have negative area. Integral calculus may be used to compute volumes in three dimensions, as well as areas. The area under a curve between two specified points is called a definite integral. The area function itself is called the indefinite integral.

Integration and differentiation are, in a certain sense, opposites. Usually, given a function *f*, the integral of the derivative yields *f*, and the derivative of the integral also yields *f*.

Special circuits, called differentiators and integrators produce an output waveform that is the derivative or integral of the input waveform function. *See also* DIFFERENTIATOR CIRCUIT, INTEGRATOR CIRCUIT.

CALIBRATION

Calibration is the adjustment of a measuring instrument for an accurate indication. The frequency dials of radio receivers and transmitters, as well as the dials and meters on test instruments, must be calibrated periodically to ensure their accuracy.

The precision of calibration is usually measured as a percentage, plus or minus the actual value. For example, the frequency calibration of a radio receiver might be specified as plus or minus 0.0001 percent, written as: ±0.0001 percent. At a frequency of 1 MHz, this is a maximum error of one part in a million, or 1 Hz. At 10 MHz, it represents a possible error of 10 Hz either side of the indicated frequency.

Calibration requires the use of reference standards. Time and frequency are broadcast by radio stations WWV and WWVH in the United States (*see* WWV/WWVH), with extreme accuracy. Standard values for all units are determined by the National Bureau of Standards (*see* NATIONAL BUREAU OF STANDARDS).

CALORIMETER

A calorimeter is an instrument for measuring heat. The calorimeter is used in some radio-frequency wattmeters. A special resistive load which dissipates all the power applied to it as heat, is connected across the transmission line. The resistance of the load is precisely matched to the characteristic impedance of the transmission line, and the wattmeter serves as a dummy load. All the power from the transmitter is dissipated as heat in the load; the amount of heat is measured by the calorimeter, and this power is read on a meter as wattage. *See also* WATTMETER.

CAMERA TUBE

A camera tube is a component that focuses a visible

image onto a small screen and scans this image with an electron beam to develop a composite video signal. The current in the electron beam varies, depending on the brightness of the image in each part of the screen. The response must be very rapid because scanning is done at a high rate of speed. The position of the electron beam, relative to the image, is synchronized between the camera tube and the receiver picture tube.

The camera tube must often used in television broadcast stations is called the image orthicon. A simpler and more compact camera tube, used primarily in industry and space applications, is called the vidicon. *See also* COMPOSITE VIDEO SIGNAL, IMAGE ORTHICON, VIDICON.

CANADIAN STANDARDS ASSOCIATION

The Canadian Standards Association (CSA) is an agency that prepares specifications for use by Canadian industry and government. Its equivalent in the United States is the Underwriters' Laboratories (UL) for electricity and electronics.

CANDELA

The candela (cd) is the standard unit of light-source intensity. It was formerly called the candle. One candela is equal to one lumen per steradian (lm/sr).

One candela is defined as the luminous intensity of 1/600,000 of one square meter of projected area of a black-body radiator operating at the temperature of solidification of platinum at a pressure of 101, 325 newtons per square meter. A 40-W electric bulb has a luminous intensity of about 3000 candela. An ordinary candle has a luminous intensity of about 1 candela. *See also* LIGHT INTENSITY, LUMEN, STERADIAN.

CANDLE

See CANDELA.

CAPACITANCE

Capacitance is the ability of an object to store an electric charge. Capacitance, represented by the capital letter C in equations, is measured in units called farads. One farad is equal to one coulomb of stored electric charge per volt of applied potential. Thus:

$$C \text{ (farads)} = Q/E$$

where Q is the quantity of charge in coulombs, and E is the applied voltage in volts.

The farad is an extremely large unit of capacitance. Generally, capacitance is specified in microfarads, abbreviated μF, or in picofarads, abbreviated pF. These units are, respectively, a millionth and a trillionth of a farad:

$$1\ \mu F = 10^{-6}\ F$$
$$1\ pF = 10^{-12}\ F$$

Any two electrically isolated conductors will introduce a certain amount of capacitance. When two conductive materials are deliberately placed near each other to produce capacitance, the device is called a *capacitor* (*see* CAPACITOR).

Capacitances connected in parallel add together. For n capacitances $C_1, C_2, \ldots, C_n$ in parallel, then, the total capacitance C is:

$$C = C_1 + C_2 + \ldots + C_n$$

In series, capacitances add as follows:

$$C = \frac{1}{1/C_1 + 1/C_2 + \ldots + 1/C_n}$$

See also CAPACITIVE REACTANCE.

CAPACITIVE COUPLING

Capacitive coupling is the electrostatic connection between the stages of amplification in a radio circuit. The output of one stage is coupled to the input of the succeeding stage by a capacitor, as illustrated in the schematic diagram. Capacitor C is the coupling capacitor.

Capacitive coupling allows the transfer of an alternating-current signal, but prevents the short-circuiting of a direct-current potential difference. Thus, the desired bias may be maintained on either side of the capacitor. A coupling capacitor is sometimes called a blocking capacitor (*see* BLOCKING CAPACITOR).

The chief disadvantage of capacitive coupling is the difficulty it presents in making a good impedance match between stages. This makes power transfer inefficient. Also, capacitive coupling allows the transfer of all harmonic energy, as well as the fundamental or desired frequency, from stage to stage. Tuned circuits are necessary to reduce the transfer of this unwanted energy. Sometimes, the use of coupling transformers is better than the use of capacitors for interstage connection. *See also* TRANSFORMER COUPLING.

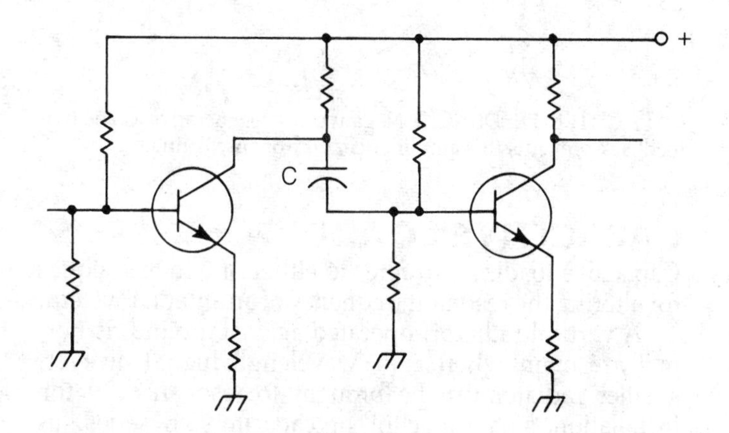

CAPACITIVE COUPLING: The capacitor passes the alternating-current signal but blocks the direct current.

CAPACITIVE DIODE

See VARACTOR DIODE.

CAPACITIVE FEEDBACK

Capacitive feedback is a means of returning a portion of the output signal of a circuit back to the input with a capacitor (see illustration). In most amplifier circuits, the output is 180 degrees out of phase with the input, so connecting a capacitor between the output and the input reduces the gain. If the output is fed back in phase, the gain of an amplifier increases, but the probability of oscillation is much greater.

Capacitive feedback may occur as a result of coupling between the input and output wiring of a circuit. This can produce undesirable effects such as amplifier instability or parasitic oscillation. This kind of capacitive feedback is also possible within a tube or transistor, because of interelectrode capacitance. It is nullified by a process called neutralization (*see* NEUTRALIZATION).

Capacitive feedback is commonly employed in oscillator circuits including the multivibrator, Hartley, Pierce, and Colpitts types. The Armstrong oscillator uses inductive feedback through coupling between coils. *See also* INDUCTIVE FEEDBACK.

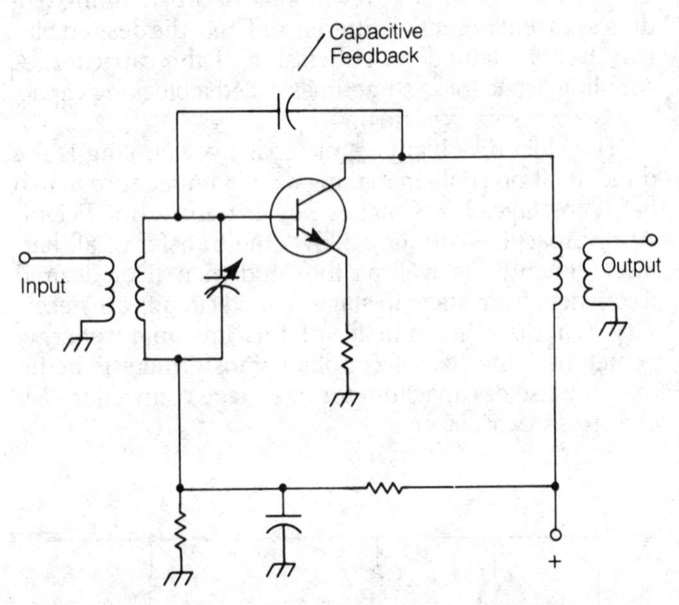

CAPACITIVE FEEDBACK: Negative or degenerative capacitive feedback prevents this amplifier circuit from oscillating.

CAPACITIVE LOADING

Capacitive loading can refer to either of two techniques for altering the resonant frequency of an antenna system.

A vertical radiator, operated against ground, is normally resonant when it is ¼ wavelength high. However, a taller radiator can be brought to resonance by the installation of a capacitor or capacitors in series, as illustrated in the diagram (A). This is commonly called capacitive loading or tuning.

A vertical radiator shorter than ¼ wavelength can be resonated by the use of a component called a capacitance hat (B in the illustration). A capacitance hat is often used in conjunction with a loading coil. The capacitance hat results in an increase in the useful bandwidth of the antenna, as well as a shorter physical length at a given resonant frequency.

Both of the methods of capacitive loading described here are also used with dipole antennas. They can also be employed in the driven elements of Yagi antennas and in various kinds of phased arrangements.

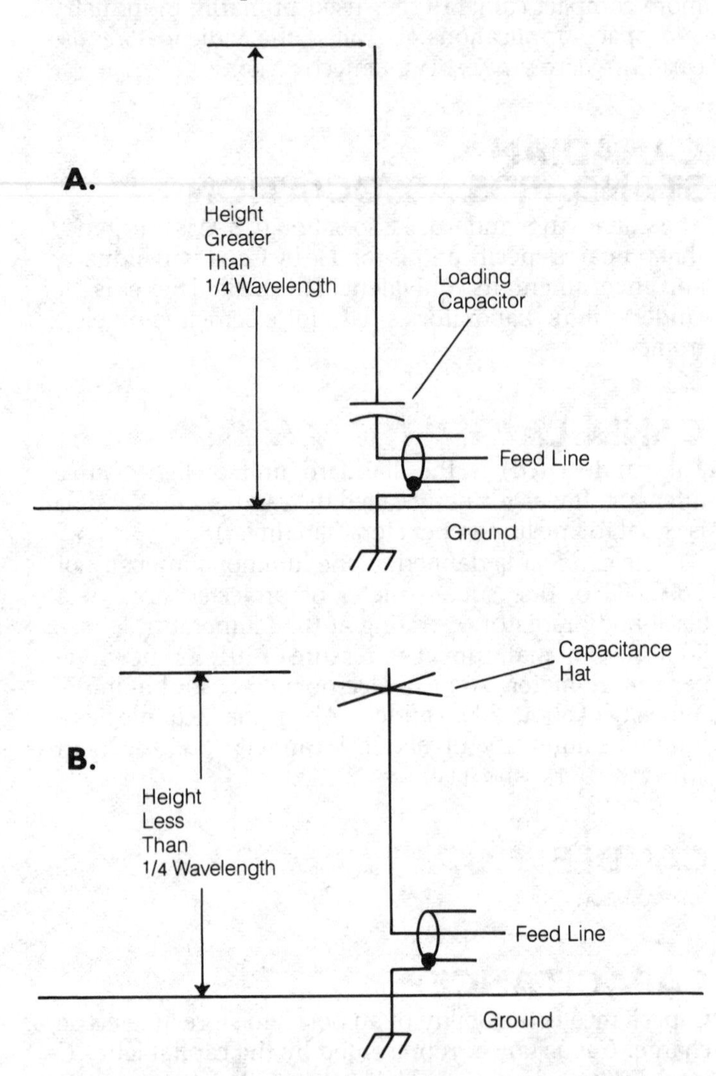

CAPACITIVE LOADING: Two types of capacitive loading for antennas are a capacitor in series with the antenna to raise its resonant frequency (A) and a network of wires forming a capacitance hat to increase the bandwidth and lower the resonant frequency (B).

CAPACITIVE REACTANCE

Reactance is the opposition that a component offers to alternating current. The reactance of a component can be either positive or negative, and it varies in magnitude depending on the frequency of the alternating current. Capacitive reactance is negative, and inductive reactance is positive. These choices of positive and negative are arbitrary, having been chosen for convenience.

The reactance of a capacitor of C farads, at a frequency

of *f* hertz, is given by:

$$X_c = \frac{1}{2\pi fC}$$

This formula also holds for values of C in microfarads, and frequencies in megahertz. (When using farads, it is necessary to use hertz; when using microfarads, it is necessary to use megahertz. Units cannot be mixed.)

The magnitude of capacitive reactance, for a given component, approaches zero as the frequency is raised. It becomes larger negatively as the frquency is lowered. This effect is illustrated by the graphs. At A, the reactance of a 0.01-μF capacitor is shown as a function of the frequency in MHz. At B, the reactance of various values of capacitance are shown for a frequency of 1 MHz.

Reactances, like dc resistances, are expressed in ohms. In a complex resistance-reactance circuit, the reactance is multiplied by the imaginary number $\sqrt{-1}$, which is mathematically represented by the letter j in engineering notation. Thus a capacitive reactance of − 20 ohms is specified as − 20j; a combination of 10 ohms resistance and − 20 ohms reactance is an impedance of 10 − 20j. *See also* IMPEDANCE, J OPERATOR.

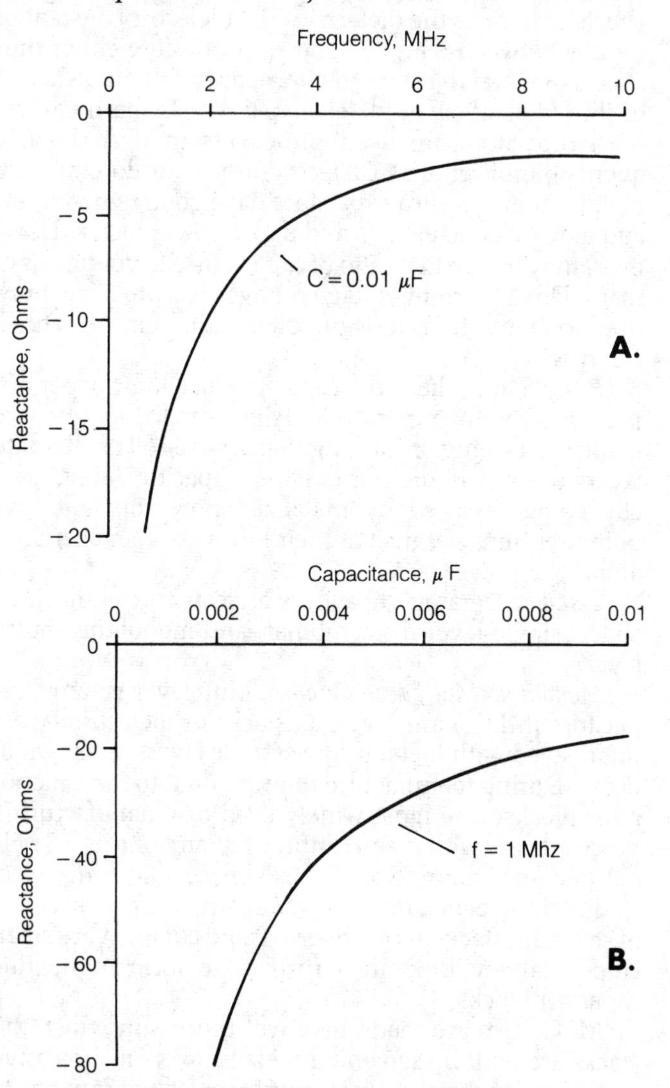

CAPACITIVE REACTANCE: The plot of reactance versus frequency for a capacitor (A) and the plot of reactance as a function of frequency at a specified frequency (B).

CAPACITOR

A capacitor is an electronic component capable of storing electrical energy. A capacitor consists of two metal plates insulated from each other by a dielectric. The capacitance value is determined by the following formula:

$$C = \frac{KA}{t}$$

where *C* = coulombs/volt (farad)
K = relative dielectric constant of the insulator
A = area of overlay of the plates
t = thickness of the dielectric

Capacitors used in electronics applications can be classified by basic type as electrostatic or electrolytic. Electrostatic capacitors have solid material or air dielectrics and are further categorized as air, film (plastic), paper, ceramic, glass, and mica (see the table). Electrolytic capacitors are further classified according to the metal used to form the oxide dielectric by electrochemical reaction: aluminum or tantalum. Electrolytic capacitors can have foil or solid-slug anodes made from those metals.

Capacitors can also be classified as either fixed or variable. The capacitance of fixed capacitors remains essentially unchanged except for small variations caused by temperature variations. By contrast, the capacitance of variable capacitors can be set to any value within a preset range of values.

Variable Capacitors. Some variable capacitors for use in tuning radios and instruments are made to be adjusted continuously by means of knob and shaft between their minimum and maximum limits. *See* AIR-VARIABLE CAPACITOR.

Trimmer Capacitors. Trimmer capacitors are miniature variable capacitors for use at radio frequencies. There are three major styles: multiturn, single-turn, and compression. Multiturn trimmer capacitors can have glass, quartz, sapphire, plastic, ceramic, or air dielectrics and are typically used in military rf (radio frequency) equipment. Single-turn trimmer capacitors, used primarily in commercial rf equipment, have ceramic, plastic, or air dielectrics. Compression trimmer capacitors use a mica dielectic. Useful into the microwave range, trimmer capacitors are infrequently adjusted.

Electrostatic Capacitors. Electrostatic capacitors with ceramic, mica, glass, or plastic-film dielectrics have plates that can be metal foil, evaporated metal thin-films, or screened-on and fired metallic inks. Electrostatic capacitors are made by various methods including rolling flexible metal foil and dielectric sheets or metalized dielectric sheets and stacking metalized dielectric sheets.

Plastic Film. Plastic-film capacitors are made with dielectric films of such plastics as polyester, polypropylene, polystyrene, polycarbonate, and polysulfone. Dielectric thickness can range from 0.06 mil (1.5 microns) to over 0.8 mil (20 microns). Plastic-film capacitors most commonly used in electronics have capacitance values of

CAPACITOR: Common types of capacitors and their applications.

Capacitor Type	Approximate Frequency Range	Voltage Range
Air variable	lf, mf, hf, vhf, uhf	Med to high
Ceramic	lf, mf, hf, vhf	Med. to high
Electrolytic	af, vlf	Low to med.
Mica	lf, mf, hf, vhf	Low to med.
Mylar	vlf, lf, mf, hf	Low to med.
Paper	vlf, lf, mf, hf	Low to med.
Polystyrene	af, vlf, lf, hf	Low
Tantalum	af, vlf	Low
Trimmer	mf, hf, vhf, uhf	Low to med.

Frequency Abbreviations

af: Audio frequency (0 to 20 kHz)
vlf: Very low frequency (10 to 30 KHz)
lf: Low frequency (30 to 300 kHz)
mf: Medium frequency (300 kHz to 3 MHz)
hf: High frequency (3 to 30 MHz)
vhf: Very high frequency (30 to 300 MHz)
uhf: Ultra high frequency (300 MHz to 3 GHz)

0.001 to 10 microfarads (μF) although they are available in values of 50 picofarads (pF) to 500 μF. Working voltages range from 50 Vdc to 1,600 Vdc. Capacitance tolerance is from ± 1 to ± 20 percent.

Film capacitors are manufactured as film and foil or metallized film. In film and foil construction, tin or aluminum foil with a thickness of about 0.00025 inch is interleaved with the film dielectric. In metallized film construction, the plates are formed by vacuum depositing very thin layers of aluminum or zinc (200 to 500 angstroms) onto the film. Metallized-film construction permits size and weight reduction for comparable capacitance-voltage ratings.

Film capacitors are made by rolling the foil and film or metallized foil on a mandrel or by cutting and stacking the metallized foil. Metallized-film capacitors are self healing because, after a dielectric breakdown occurs, the metallization around the fault is evaporated and isolated without shorting adjacent plates.

Both film and foil and metallized units are available with axial and radial leads in many case styles: hermetically sealed in tubular or rectangular metal cases, molded plastic tubular or box styles, wrap and fill, preformed plastic box, and conformally coated with resin.

Polyester film (tradenamed Mylar) is the most popular general-purpose film dielectric. Capacitors made from this film are smaller than those made from other films, and they exhibit low leakage, a moderate temperature coefficient over the temperature range of − 55° to 85°C, and moderate dissipation factor. Capacitance tolerance is typically ± 10 percent. Foil and film capacitors are used in consumer electronics and instruments, and metallized units are used for general blocking, coupling, decoupling, bypass and filtering.

Polypropylene film provides capacitor characteristics that are superior to those of polyester, and capacitors made from this film are used in both low-frequency and high-frequency applications. Polypropylene has properties that are similar to polystyrene, but it has a higher ac (alternating current) current rating. Polypropylene capacitors can operate at 105°C and their volumetric efficiency is better than that of polyester. Foil and polypropylene capacitors are used in CRT (cathode ray tube) deflection, pulse-forming circuits, and radio-frequency circuits. Its capacitance tolerance is ± 5 percent, and its temperature coefficient is linear.

Polystyrene film has characteristics similar to polypropylene, but it is less popular because capacitors made from this film are larger than comparably rated polypropylene units. These capacitors are used in timing, integrating, and tuned circuits, but their temperature limit is only 85°C.

Polycarbonate film permits operating temperatures of − 55° to 125°C with tolerances of ± 5 percent. Capacitors made from this film have high insulation resistance and capacitance stability. They are widely used in military applications.

Polysulfone is an experimental dielectric not used in commercial capacitor production. Other dielectrics with limited applications are Kapton and Teflon.

Mica. Mica dielectric capacitors use thin rectangular sheets of mica as the dielectric. The dielectric constant of mica is between 6 and 8. The electrodes are either thin sheets of metal foil stacked alternately with mica sheets, or thin films of silver ink screened directly on the mica. Mica capacitors are used principally in rf (radio frequency) applications. Dielectric losses are low at very high frequency. Mica capacitors have good temperature and aging characteristics and a low power factor. However, they have a low ratio of capacitance to volume or to mass. Direct-current voltage ratings are from a few hundred to many thousands of volts, and rf current ratings can reach 50 A.

Paper. Paper dielectric capacitors use kraft paper impregnated with mineral oil. Paper capacitors also are made by stacking or forming them into a roll of alternate layers of foil and dielectric. Paper capacitors are gradually being replaced by metalized polypropylene and polyester films because of their lower cost, smaller size, and lower power factor.

Ceramic. Ceramic capacitors are constructed in three styles: single-layer disk, tubular, and monolithic multilayer.

Monolithic Multilayer Ceramic. Multilayer ceramic capacitors (MLCs) are made as stacks of dielectric layers interleaved with metallized electrode layers as shown in Fig. 1. During manufacture they are fired to form monolithic blocks. The most widely used dry manufacturing process starts with green (unfired) ceramic strips 0.8 mils (20 microns) thick. Rows of silver-palladium ink electrodes are screened on the ceramic strips. Up to 40 layers of strips are stacked, compressed, and cut into very small chips that are fired in a furnace to form monolithic capacitor blocks.

MLCs also are made by a wet process in which the stacks are built by screening metallic inks on successive layers of wet ceramic slurry until the desired number of layers is obtained. This stack is then also compressed, cut, and fired.

MLC end terminals are made by plating successive

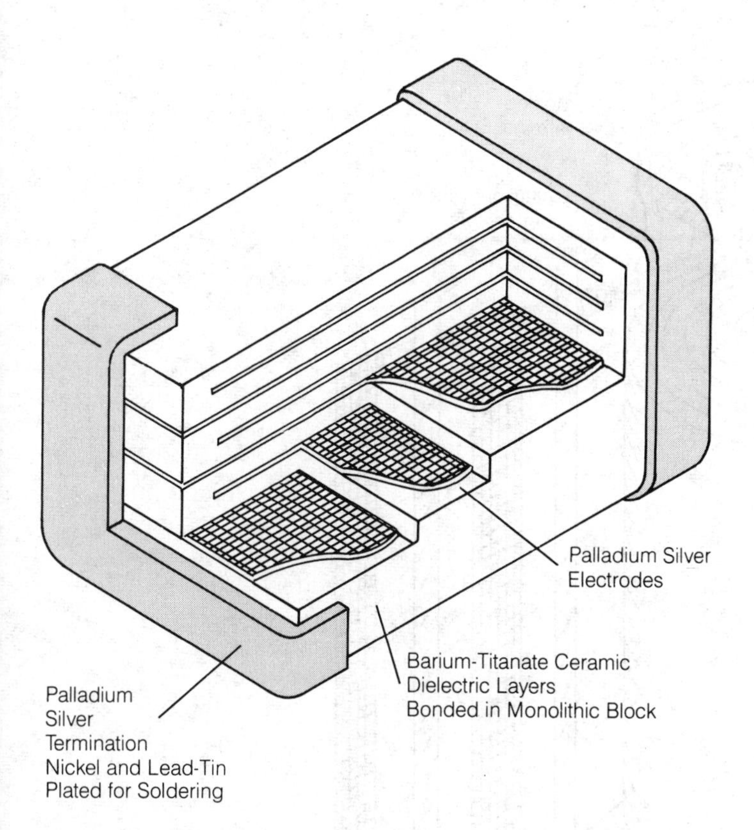

CAPACITOR: Fig. 1. Cutaway view of a ceramic multilayer capacitor formed by depositing layers of ceramic dielectric. Deposited noble metal electrodes are pressed and fired to form a monolithic chip.

layers of silver-palladium, nickel and tin, or lead-tin. Variations in termination method depend on whether the MLC is to be leaded and coated or to remain a bare chip for surface mounting. MLCs are used to bypass noise to ground on integrated circuit boards. They are also employed in timing and frequency selection applications. *See* SURFACE MOUNT TECHNOLOGY.

MLCs offer low inductance and low resistance, a wide range of capacitance values in a given volume, and a wide range of temperature coefficients. MLCs also exhibit lower inductance and resistance than tantalum capacitors with similar ratings.

Ceramic dielectric materials are classified by dielectric constant, K, as Class I, Class II, and Class III. Class I dielectrics exhibit low K values but have excellent temperature stability. Class II dielectrics have generally high K values and volumetric efficiency but lower temperature stability.

Class I dielectrics include NPO (negative positive zero), also designated COG and BY. These ceramics are made by combining magnesium titanate (with a positive temperature coefficient) and calcium titanate (with a negative coefficient) to form a dielectric with excellent temperature stability. Their properties are essentially independent of frequency. They have ultrastable temperature coefficients of 0 ± 30 ppm/C over the temperature range of − 55° to 125°C. These dielectrics show a flat response to both ac and dc (alternating or direct current) voltage changes. Low-K ceramics are used in MLCs intended for resonant circuit and filter applications.

Class II dielectrics are high-K ceramics called ferroelectrics and are based on barium titanate. With the addition of barium stanate, barium zirconate, or magnesium titanate, the dielectric constant can be lowered from values as high as 8000. These compounds stabilize the capacitor over a wider temperature range. Class II dielectrics include the general purpose X7R (BX) and Z5U (BZ). X7R has a stable characteristic, but capacitance can vary ± 15 percent over the temperature range of − 55° to 125°C. Capacitance value decreases with dc voltage but increases with ac voltage. Z5U compositions exhibit maximum temperature-capacity changes of + 22 and − 56 percent for the range of 10° to 85°C.

Class III dielectrics were developed primarily for ceramic-disk capacitors. These dielectrics give the capacitor high volumetric efficiency but with high leakage resistance and dissipation factor. Capacitors made with Class III dielectrics have low working voltages.

Nonleaded-chip MLCs are widely used in the fabrication of hybrid circuits and for surface-mount assembly. These capacitors will withstand the 450°F (230°C) reflow solder temperatures and the 540°F (280°C) wave-solder temperatures. MLC chips are now available in standard sizes. Examples include the 0.08 by 0.05 inch MLC, designated 0805; the 0.125 by 0.063 inch MLC, designated 1206; and the 0.05 by 0.225 inch MLC, designated 2225. *See also* HYBRID CIRCUIT.

Disk Ceramic. Disk ceramic capacitors are made by metalizing surfaces on both sides of a ceramic disk with silver ink to form electrodes and attaching a lead to each electrode. The capacitor is usually dipped in an encapsulant of phenolic resin or epoxy. Disk capacitors are widely used in tuning-circuit applications. They are being replaced by tubular-ceramic capacitors to conserve PC (printed circuit) board space and permit automatic PC board mounting. Inner and outer surfaces of the ceramic tube are painted with silver ink, and the tube acts as a single layer. The capacitors are then leaded and coated.

Electrolytic Capacitors. Electrolytic capacitors are made by electrochemical processes in which the dielectrics are formed on aluminum or tantalum foil or on sintered slugs formed from tantalum powder. The metal foils are acid etched to increase their porosity and effective area from 6 to 20 times. The thin oxide layers formed on the plates give electrolytic capacitors their high capacitance-to-volume ratios.

Aluminum-Electrolytic Capacitors Aluminum-electrolytic capacitors are constructed by interleaving two strips of etched aluminum foil between separators soaked in electrolyte as shown in Fig. 2. The stacked materials are wound in jelly-roll fashion and enclosed in a tubular aluminum case. External connections are made from the electrodes to the outside terminals on the case. After electric current is passed through the system, a layer of oxide forms on the anode, and the layer becomes the thin aluminum oxide dielectric medium. A dc (direct) current must flow in one direction during capacitor formation to maintain the deposition of the oxide layer. A positive terminal is marked externally. An electrolyte in contact with the metal foil acts as a cathode.

Aluminum-electrolytic capacitors are polarized, and the polarity must be observed when using the capacitor. Nonpolarized electrolytic capacitors can be made for use

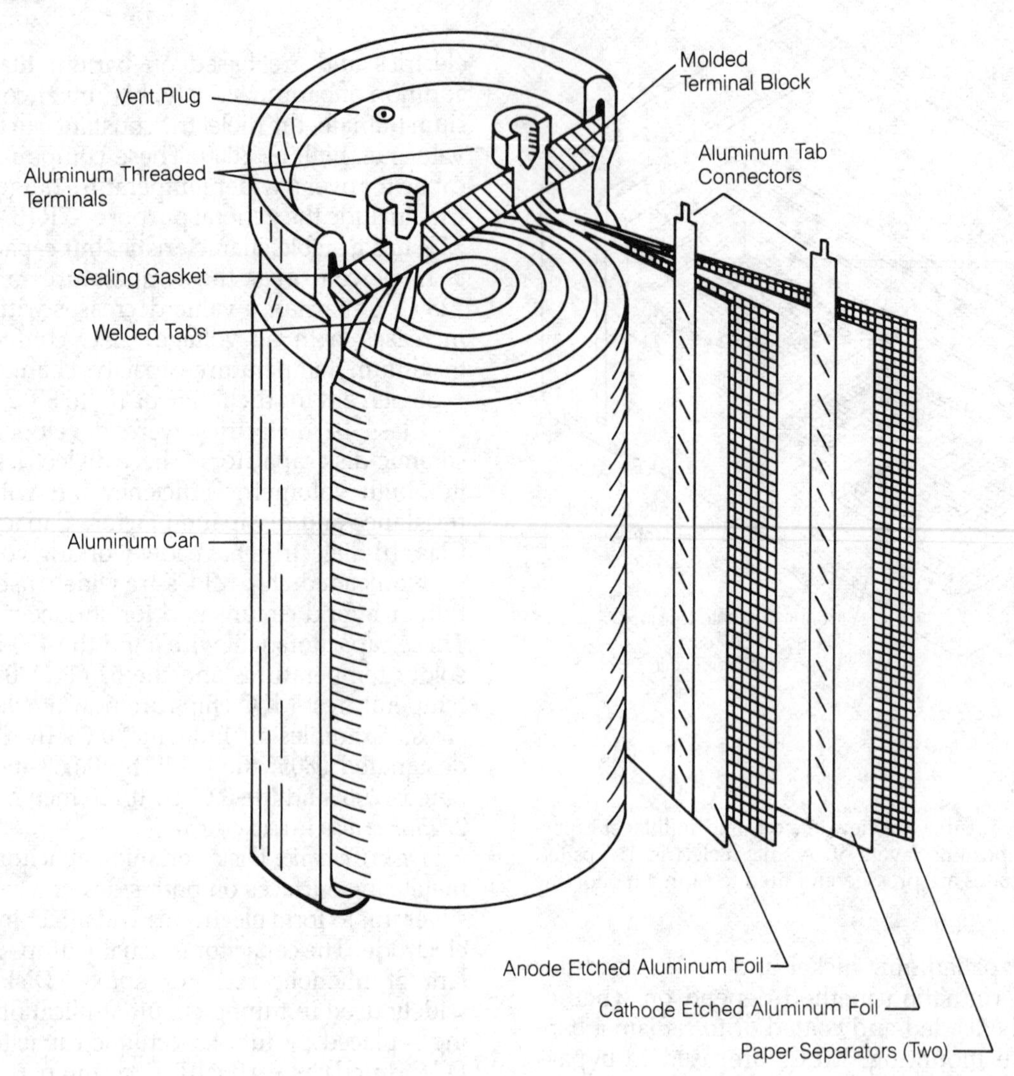

CAPACITOR: Fig. 2. Cutaway view of an aluminum electrolytic capacitor made by rolling aluminum foil anodes and cathodes separated by electrolyte-filled paper.

in ac (alternating current) circuits by placing two polarized capacitors in series with their cathode terminals connected. The anode terminals are used for external circuit connection while the cathode terminals are isolated from the circuit by an insulator.

Tantalum. There are three types of tantalum electrolytic capacitors: wet-foil, wet-slug, and dry-slug or solid. Wet-foil tantalum capacitors are made by a process similar to that used to make aluminum electrolytic capacitors. They are able to withstand voltages as high as 300 Vdc and are most frequently used in military applications. Tantalum capacitors as a group have higher CV ratings than comparably sized aluminum electrolytic capacitors. The dielectric is tantalum oxide (Ta_2O_5). All tantalum capacitors are inherently polar units.

However, both the wet- and dry- or solid-slug tantalum capacitors are based on a porous pellet of tantalum. The anode slug is made by pressing fine tantalum powder in a mold and firing it in a vacuum furnace at about 2000°C to form the porous structure. The fired slug, like the etched foil, provides a large effective surface area. Figure 3 illustrates a wet slug capacitor.

Both wet-slug tantalum and tantalum-foil capacitors use a highly conductive liquid (most often sulfuric acid), sealed in a silver or tantalum case, as the second electrode. Wet-slug capacitors are widely used in timing circuits and in other critical applications where leakage currents must be extremely low. They are considered to be the most reliable capacitors.

Solid-tantalum capacitors have a thin, solid film of manganese dioxide (MnO_2) chemically deposited on the tantalum oxide dielectric to act as a second electrode. They can be sealed in metal cases, or they can be dipped or molded in resin as shown in Fig. 4. These capacitors are available with radial or axial leads. A chip form of the dry-tantalum capacitor, shown in Fig. 5, is intended for use in surface-mount circuit assembly.

Electrolytic capacitors have higher leakage currents than electrostatic capacitors because of impurities in the foil and electrolyte. With an increase in temperature, leakage current increases while voltage breakdown decreases. Also, power-factor losses are higher than in electrostatic capacitors. These losses are defined in terms of equivalent series resistance (ESR).

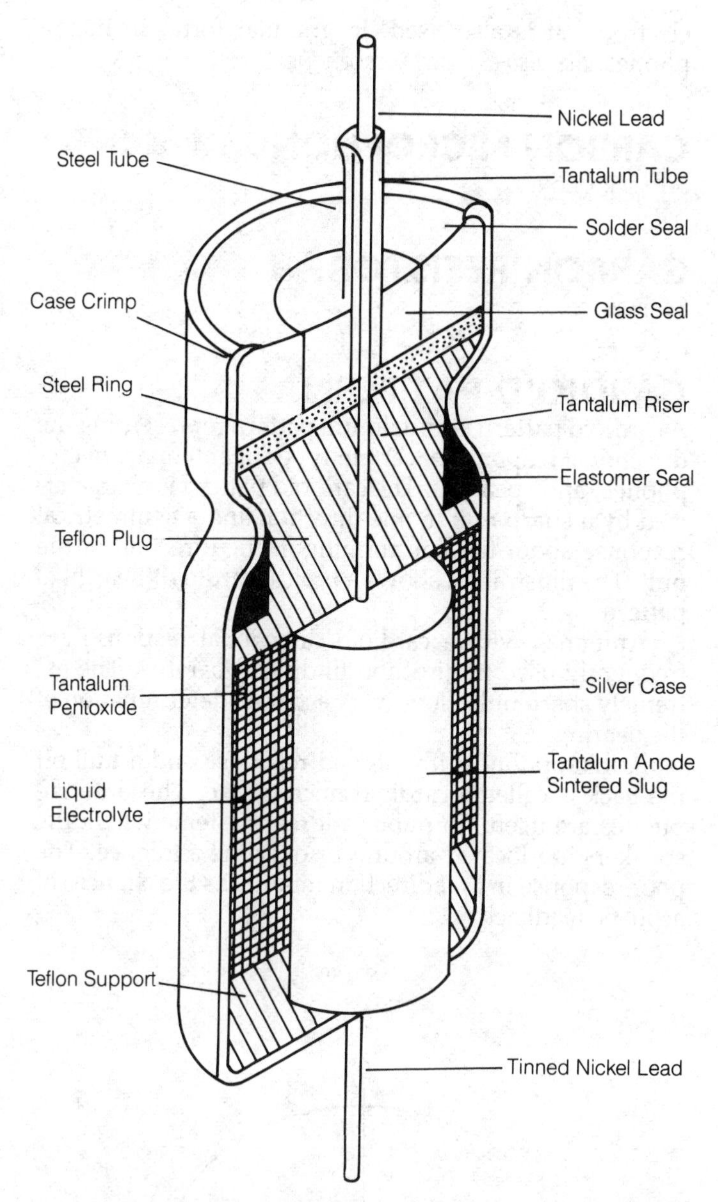

CAPACITOR: Fig. 3. Cutaway view of a wet-slug tantalum capacitor for high reliability applications.

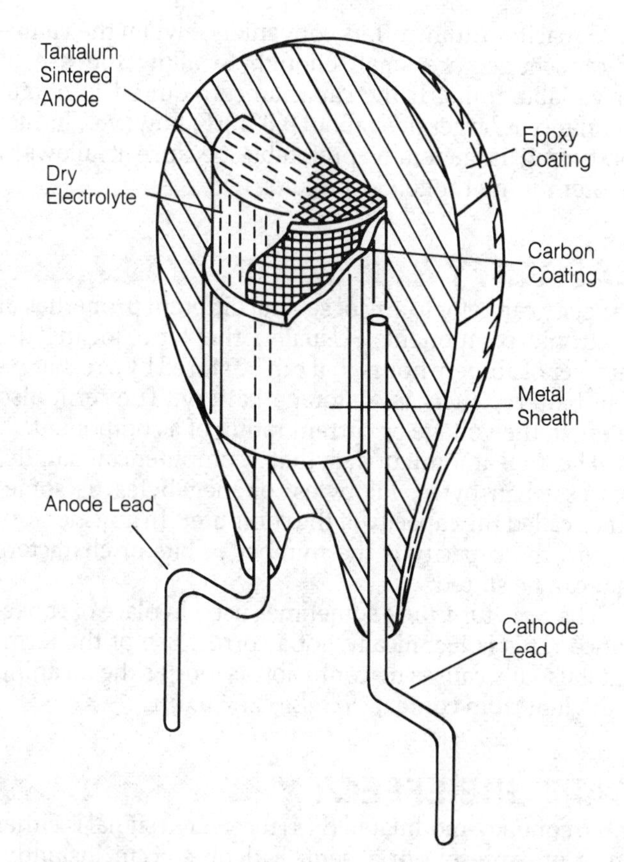

CAPACITOR: Fig. 4. Cutaway view of an epoxy-dipped solid tantalum capacitor with radial leads for PC-board mounting.

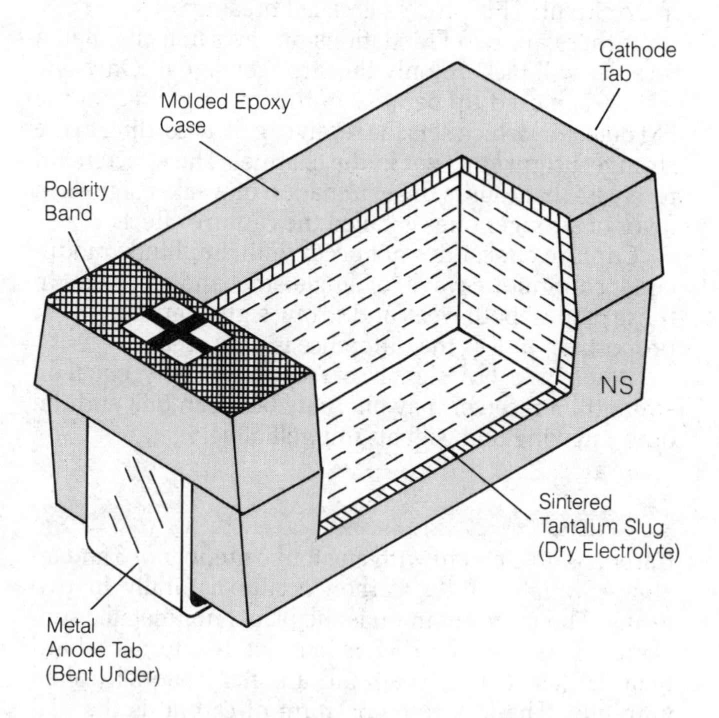

CAPACITOR: Fig. 5. Cutaway view of a solid tantalum chip capacitor for surface mounting.

CAPACITOR MICROPHONE

See MICROPHONE.

CAPACITOR TUNING

Capacitor tuning is a means of adjusting the resonant frequency of a tuned circuit by varying the capacitance while leaving the inductance constant. When this is done, the resonant frequency varies according to the inverse of the square root of the capacitance. This means that if the capacitance is doubled, the resonant frequency drops to 0.707 of its previous value. If the capacitance is cut to one-quarter, the resonant frequency is doubled. Mathematically, the resonant frequency of a tuned inductance-capacitance circuit is given by the formula:

$$f = \frac{1}{2\pi\sqrt{LC}}$$

where f is the frequency in hertz, L is the inductance in henrys, and C is the capacitance in farads. Alternatively, the units may be given as megahertz, microhenrys, and microfarads.

Capacitor tuning offers convenience when the values of capacitance are small enough to allow the use of air-variable units. If the capacitance required to obtain resonance is larger than about 0.001 μF, however, inductor tuning is generally preferable because it allows a greater tuning range. *See* INDUCTOR TUNING.

CAPACITY

Capacity can refer to any of several different properties of electronic components. Usually, the term means the number of ampere hours that can be stored by a rechargeable battery, such as a storage battery. The term also refers to the voltage or current rating of a component.

The total amount of data that a computer can handle, expressed in bytes, kilobytes, or megabytes, is sometimes called the capacity of the computer. In any memory circuit, the capacity is the number of bits or characters that can be stored.

The term *capacity* is sometimes used in place of capacitance. This is technically not a correct use of the term, but it usually causes no confusion as long as the meaning is evident from context. *See also* CAPACITANCE.

CAPTURE EFFECT

In a frequency-modulated (FM) receiver, a signal is either on or off—present or absent. Fadeouts occur instantly, rather than gradually as they do in other modes. When an FM signal is too weak to be received, it becomes intermittent. This process is called breakup.

If there are two FM stations on the same channel, a receiver will pick up only the stronger signal. Only one station is picked up because of the limiting effect of the FM detector, which sets the receiver gain according to the strongest signal present in the channel. The effect in FM receivers that causes the elimination of weaker signals in favor of stronger ones is called the capture effect.

Capture effect does not occur with amplitude modulated, continuous-wave, or single-sideband receivers. In those modes, both weak and strong signals are audible in proportion, unless the difference is very great.

When two FM signals are approximately equal in strength, a receiver may alternate between one and the other, making both signals unintelligible.

CARBON

Carbon is an element with an atomic number of 6 and an atomic weight of 12. Carbon occurs naturally in two forms. The more common is the black, nonmetallic substance. This type of carbon is found in two forms: a hard material (as seen in coal) and a softer material (called graphite). The less common form of carbon is the diamond.

The resistance of carbon is much greater than that of metal, and thus graphite is often used as the conducting material in the fabrication of noninductive resistors. Other substances are added to the graphite to vary the conductivity.

Carbon is used in some dry cells as the positive electrode. It is also used, in granular form, in microphones. *See also* BATTERY, MICROPHONE.

CARBON MICROPHONE

See MICROPHONE.

CARBON RESISTOR

See RESISTOR.

CARDIOID PATTERN

A cardioid pattern is an azimuth pattern representing the directional response of certain types of antennas, microphones, and speakers. The cardioid pattern is characterized by a sharp null in one direction and a symmetrical response about the line running in the direction of the null. The illustration shows a typical cardioid (heartlike) pattern.

Antennas with a cardioid directional response are frequently used in direction-finding apparatus. The extremely sharp null allows very accurate determination of the bearing.

A microphone with a broad response and a null off the back is called a cardioid microphone. These microphones are useful in public-address systems where the speakers are located around or over the audience. The poor response in that direction minimizes the chances of acoustic feedback.

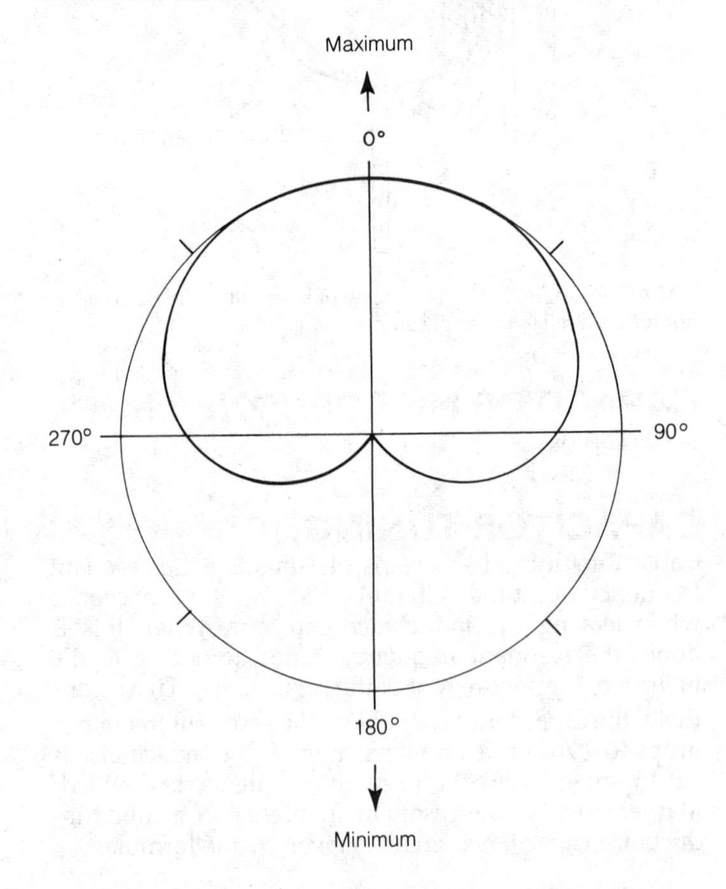

CARDIOID PATTERN: A cardioid pattern with a sharp null in one direction.

CARRIER

A carrier is an alternating-current wave of constant frequency, phase, and amplitude. By varying the frequency, phase, or amplitude of a carrier wave, information is transmitted. The term is generally used in reference to the electromagnetic wave in a radio communications system; it may also refer to a current transmitted along a wire. An unmodulated carrier has a theoretical bandwidth of zero; all of the power is concentrated at a single frequency in the electromagnetic spectrum (see illustration). A modulated carrier has a nonzero, finite bandwidth (*see* BANDWIDTH).

The term *carrier* is also used for electrons in N-type semiconductor material, or for holes in P-type semiconductor material. In these cases, carriers are the presence or absence of extra electrons in atoms of the substance. *See also* ELECTRON, HOLE, N-TYPE SEMICONDUCTOR, P-TYPE SEMICONDUCTOR.

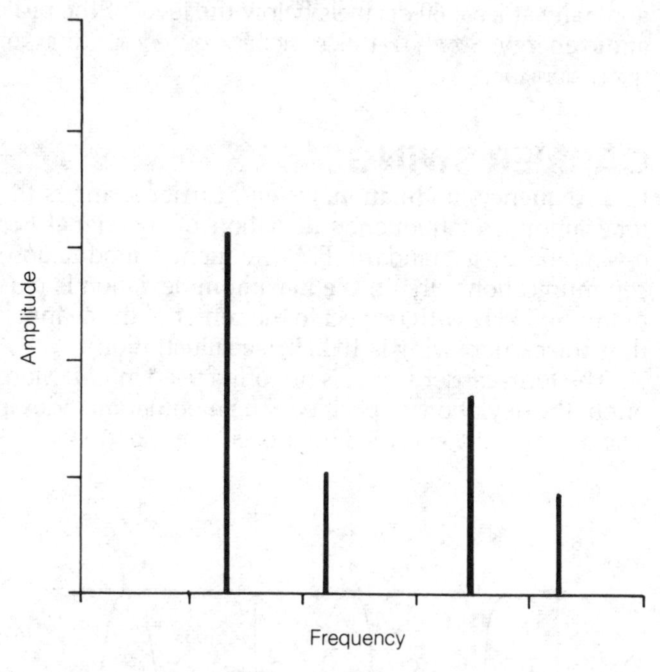

CARRIER: The plot of four unmodulated carriers with single frequencies and theoretically zero bandwidths.

CARRIER CURRENT

Carrier current is the magnitude of radio-frequency alternating currents in a transmitter carrier. Carrier current is generally measured along a transmission line. It may be expressed as a peak value, a peak-to-peak value, or a root-mean-square (rms) value. The magnitude of the carrier current depends on the transmitter output power or carrier power, and on the impedance of the transmission line or antenna system. It also depends on the transmitter output power or carrier power, and on the impedance of the transmission line or antenna system. It also depends on the standing-wave ratio along the transmission line.

Carrier current can be measured with a radio-frequency ammeter. If the standing-wave ratio is 1 and the feed-line impedance, Z_O, is known, the rms current, I, can be calculated from a wattmeter reading P watts according to the formula:

$$I = \sqrt{\frac{P}{Z_O}}$$

under conditions of zero transmitter modulation. *See also* CARRIER POWER, ROOT MEAN SQUARE.

CARRIER FREQUENCY

The carrier frequency of a signal is the average frequency of its carrier wave (*see* CARRIER). For continuous-wave (CW) and amplitude-modulated (AM) signals, the carrier frequency should be as constant as the state of the art will permit. The same is true for the suppressed-carrier frequency of a single-sideband (SSB) signal. For frequency-modulated (FM) signals, the carrier frequency is determined under conditions of no modulation. This frequency should always remain as stable as possible.

In frequency-shift keying (FSK), there are two signal frequencies, separated by a certain value called the carrier shift. The shift frequency is usually 170 Hz for amateur communications and 425 Hz or 850 Hz for commercial circuits. The space-signal frequency is generally considered to be the carrier frequency in a frequency-shift-keyed circuit.

When specifying the frequency of any signal, the carrier frequency is always given.

CARRIER MOBILITY

In a semiconductor material, the electrons or holes are called carriers (*see* ELECTRON, HOLE). The average speed of carrier movement, per unit of electric-field intensity, is called the carrier mobility of the semiconductor material. Electric-field units are specified in volts per centimeter or volts per meter. The mobility of holes is usually not the same as the mobility of electrons in a given material. However, the mobility depends on the amount of doping (*see* DOPING). At very high doping levels, the mobility decreases because of scattering of the carriers. As the electric field is made very intense, the mobility also decreases as the speed of carriers approaches a maximum, or limiting, velocity. This limiting velocity is approximately 10^4 to 10^5 meters per second, or 10^6 to 10^7 centimeters per second, in most materials.

CARRIER POWER

Carrier power is a measure of the output power of a transmitter for continuous wave (CW), amplitude modulation (AM), frequency-shift keying (FSK), or frequency modulation (FM). Carrier power is measured by an rf wattmeter under conditions of zero modulation, and with the transmitter connected to a load equivalent to its normal rated operating load.

The output power of a CW transmitter is determined under key-down conditions. With FSK or FM, the output power does not change with modulation, but remains constant at all times. With AM, the output power varies somewhat with modulation.

In single-sideband (SSB) or double-sideband, suppressed-carrier operation, there is no carrier and hence no carrier power. Instead, the output power of the transmitter is measured as the peak-envelope power. *See also* PEAK ENVELOPE POWER, RF POWER.

CARRIER SHIFT

Carrier shift is a method of transmitting a radioteletype signal. The two carrier conditions are called mark and space. While no information is being sent, a steady carrier is transmitted at the space frequency. When a letter, word, or message is sent, the carrier frequency is shifted in pulses by a certain prescribed number of hertz. In amateur communications, the standard carrier shift is 170 Hz; in commercial and military communications, values of 425 Hz or 850 Hz are generally used, although nonstandard carrier-shift values are sometimes employed.

The transmitter output power remains constant under both mark and space conditions. *See also* FREQUENCY-SHIFT KEYING.

CARRIER SUPPRESSION

In a single-sideband (SSB) transmitter, the carrier is eliminated along with one of the sidebands. The carrier is also eliminated in the double-sideband, suppressed-carrier (DSB) mode. The carrier is usually eliminated by phase cancelation in a special modulator called a balanced modulator.

The illustration shows spectral displays of typical amplitude-modulated AM, DSB, and SSB signals. The carrier can never be entirely eliminated, although it is generally at least 60 decibels below the level of the peak audio energy. *See also* BALANCED MODULATOR, DOUBLE SIDEBAND, SINGLE SIDEBAND.

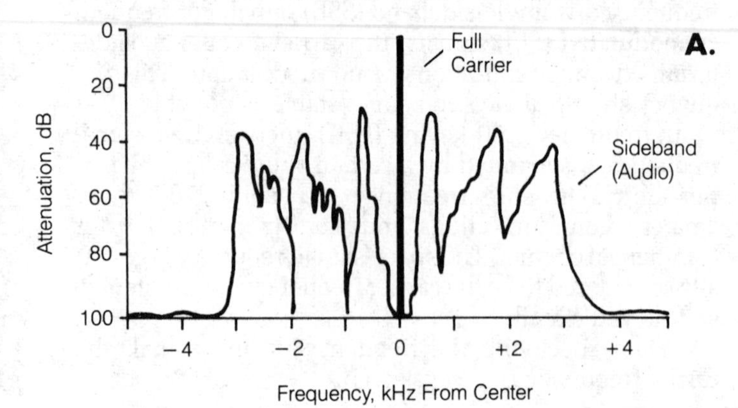

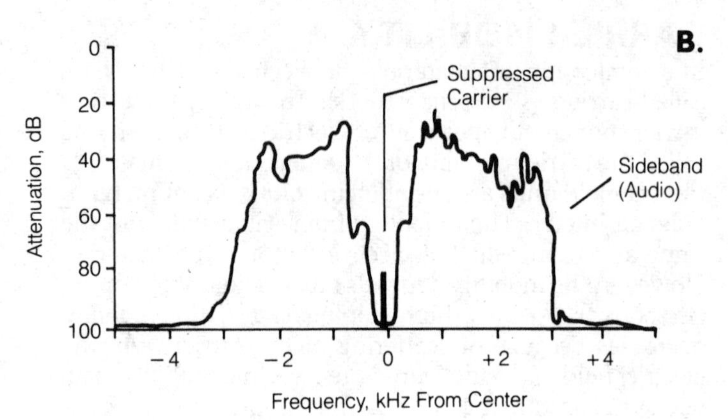

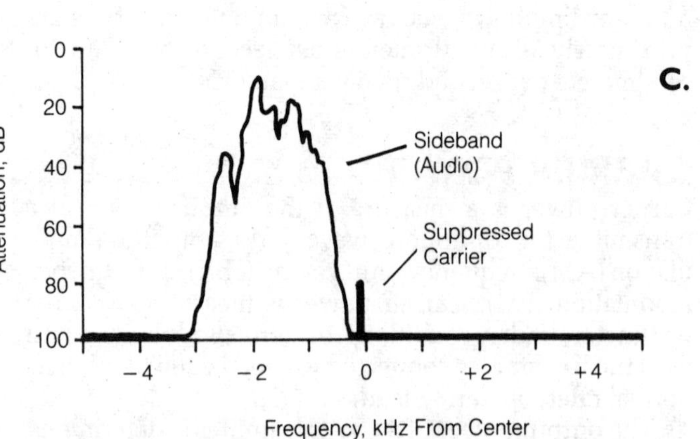

CARRIER SUPPRESSION: An amplitude-modulated signal on a spectrum analyzer (A), a double-sideband suppressed carrier signal (B), and a single-sideband signal (C).

CARRIER SWING

In a frequency-modulation system, carrier swing is the total amount of frequency deviation of the signal (*see* DEVIATION). In a standard FM (frequency modulation) communications circuit, the maximum deviation is plus or minus 5 kHz with respect to the center of the channel; thus the carrier swing is 10 kHz (see illustration).

The term carrier swing is not often used in FM. More often, the deviation is specified. The modulation index is also occasionally specified (*see* MODULATION INDEX).

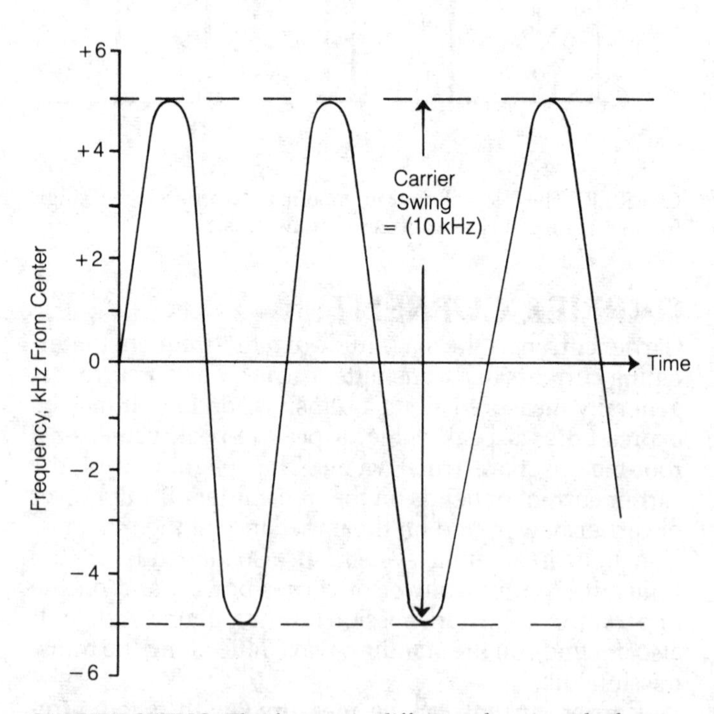

CARRIER SWING: The frequency difference between the lowest and highest instantaneous frequencies of a frequency-modulated signal is the carrier swing.

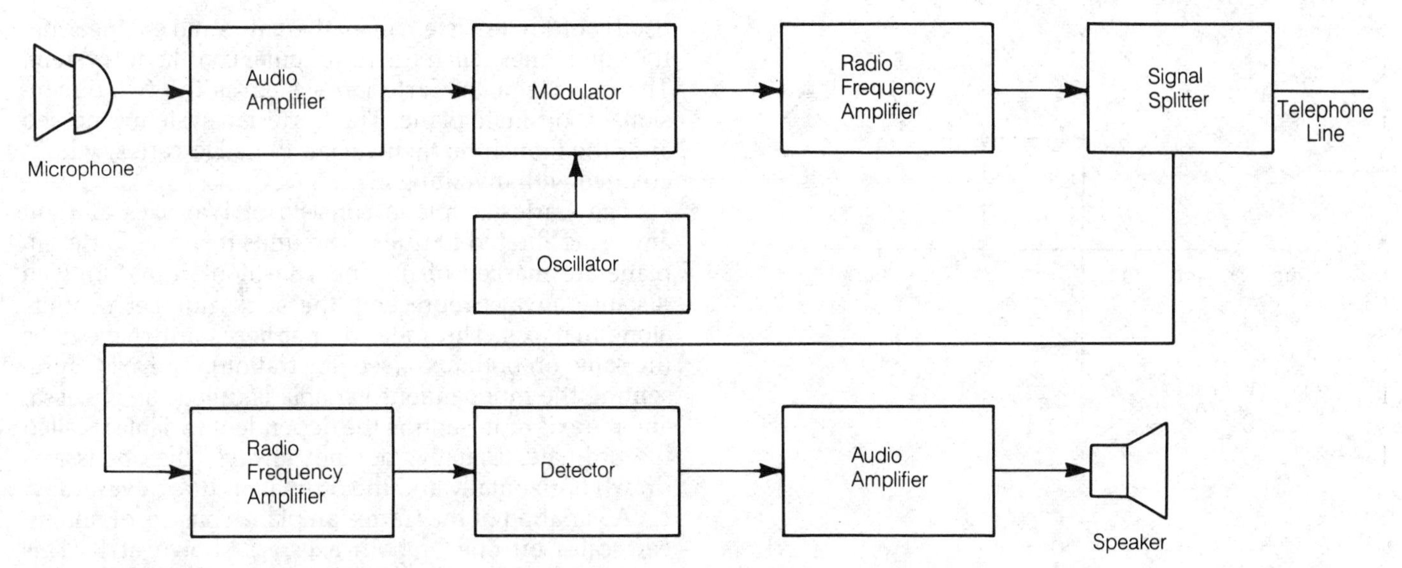

CARRIER TERMINAL: A block diagram of a carrier terminal.

CARRIER TERMINAL

A carrier terminal is a transmitter and receiver circuit used in a carrier-current commuication system (*see* CARRIER CURRENT). The carrier terminal is actually a low-power, radio-frequency transmitter or set of transmitters, and a radio-frequency receiver or set of receivers. The drawing shows a block diagram of a single-frequency carrier terminal.

The transmitter and receiver frequencies are usually different to allow duplex operation. At the far end of the circuit, the transmitting and receiving frequencies are transposed. A carrier terminal is sometimes designed for single-sideband operation; in this case, transmission may be by one sideband and reception by the other sideband. *See also* DUPLEX OPERATION.

CARRIER VOLTAGE

Carrier voltage is the alternating-current voltage of a transmitter radio-frequency carrier. It may be expressed as a peak voltage, a peak-to-peak voltage, or a root-mean-square (rms) voltage; the magnitude of the carrier voltage depends on the transmitter output power (carrier power) and on the impedance of the feed line and antenna system. It also depends on the standing-wave ratio along the transmission line.

Carrier voltage can be measured with an rf (radio frequency) voltmeter. If the standing-wave ratio is 1 and the feed-line impedance, Z_O is known, the rms carrier voltage, *E*, can be calculated from an rf wattmeter reading *P* watts according to the formula:

$$E = \sqrt{PZ_O}$$

under conditions of zero transmitter modulation. *See also* CARRIER POWER, ROOT MEAN SQUARE.

CARRY

A carry is a mathematical operation in the addition of numbers. When the sum of a column of single digits exceeds 9 in the decimal system, or 1 in the binary system, some digits must be carried. In a digital computer, the carry operation is done electronically.

In subtraction, a carry is sometimes called a borrow. It is done when the difference of two digits is less than zero. *See also* COMPUTER.

CARRYING CAPACITY

Carrying capacity is the ability of a component to handle a given amount of current or power. Generally, the term is used to define the current-handling capability of wire. The maximum amount of current that can be safely carried by a wire depends on the type and diameter of the metal used. The table gives the carrying capacity for various sizes of copper wire in the American Wire Gauge (AWG) system, for continuous duty in open air.

When the carrying capacity of wire is exceeded, there is a possibility of fire because of overheating. If the carrying capacity of a wire is greatly exceeded, the wire becomes soft and subject to stretching or breakage. It might even melt. This sometimes happens when there is a short circuit and no fuse or circuit breaker. It often happens when lightning strikes a conductor.

CARRYING CAPACITY: Continuous current ratings, in open air, of some AWG sizes of copper wire.

Size, AWG	Current, Amperes
8	73
10	55
12	41
14	32
16	22
18	16
20	11

CARTESIAN COORDINATES

The Cartesian coordinate system is the most commonly

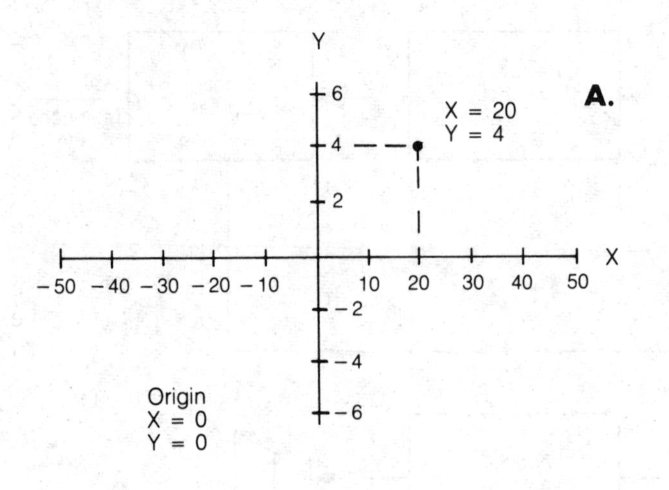

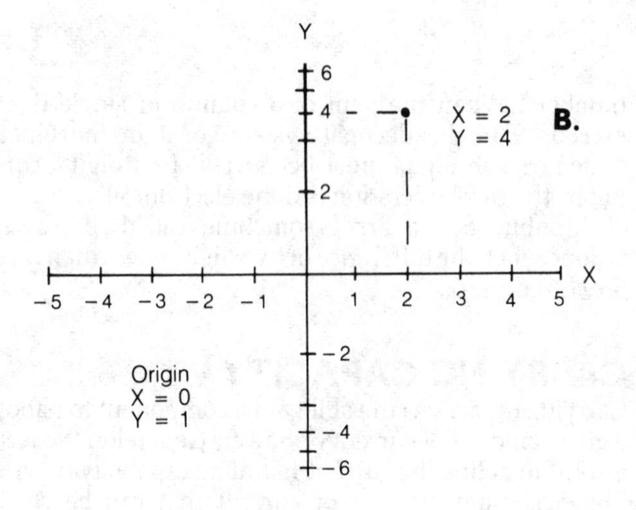

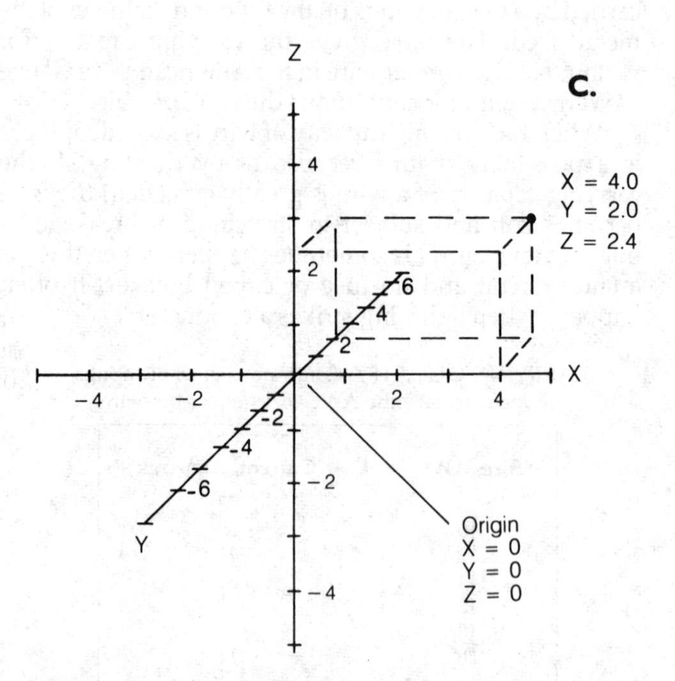

CARTESIAN COORDINATES: A two-dimensional Cartesian coordinate system with points plotted on linear scales (A), a version with y values plotted on a logarithmic scale (B), and the three-dimensional Cartesian coordinate system (C).

used coordinate system in mathematics and engineering. It is sometimes called the rectangular coordinate system. The most familiar Cartesian system is the two-dimensional coordinate plane. The Cartesian system is named after the French mathematician René Descartes, who is credited with inventing it.

The Cartesian plane consists of two axes at right angles, calibrated in units. The units in a true Cartesian plane are marked off in linear fashion so that a given distance always represents the same number of units along that axis. The scales do not necessarily have to be the same on both axes (see illustration). The axis representing the independent variable is called the abscissa, and the axis representing the dependent variable is called the ordinate. Usually, but not always, the abscissa is drawn horizontally and the ordinate is drawn vertically.

A variation of the Cartesian plane consists of nonlinear scales on one or both axes, as shown at B. This nonlinearity is usually in the form of a logarithmic scale.

Cartesian coordinate systems may be constructed in three or more dimensions. A typical three-dimensional Cartesian system is shown at C. All three axes are mutually perpendicular at their point of intersection. In four or more dimensions, the Cartesian system becomes impossible to visualize, although it can be represented mathematically.

Other types of coordinate systems exist. The most popular of these is the polar system. Less common are curvilinear systems such as the Smith chart. *See also* CURVILINEAR COORDINATES, POLAR COORDINATES.

CARTRIDGE

A cartridge is an electromechanical transducer in a phonograph. It picks up the vibrations of the stylus, and converts them to electrical impulses. These are then amplified by the audio circuits in the phonograph. The illustration is a pictorial diagram of a typical cartridge.

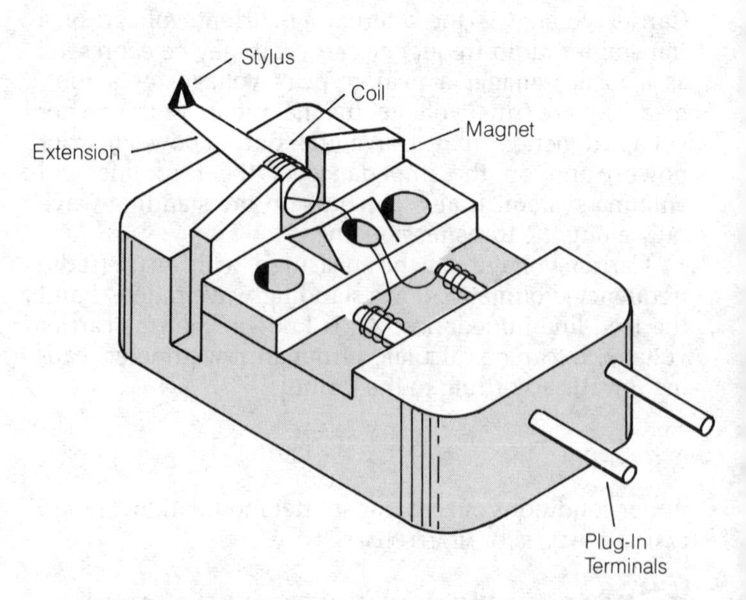

CARTRIDGE: A phonograph cartridge picks up vibrations from a record groove and transforms them into electrical signals.

CASCADE

When two or more amplifying stages are connected in series, the arrangement is called a cascade circuit. Several amplifying stages can be cascaded to obtain more signal gain than is possible with only one amplifying stage.

There is a practical limit to the number of amplifying stages that can be cascaded. The noise generated by the first stage, in addition to any noise present at the input of the first stage, will be amplified by succeeding stages along with the desired signal. When many stages are connected in cascade, the probability of feedback and resulting oscillation increases greatly.

Tuned circuits and stabilizing networks are used to maximize the number of amplifiers that can be connected in cascade. By shielding the circuitry of each stage, and by making sure that the coupling is not too tight, the chances of oscillation are minimized. Proper impedance matching ensures the best transfer of signals, with the least amount of noise. When several stages of amplification are connected in cascade, the entire unit is called a cascade amplifier.

CASCADE VOLTAGE DOUBLER

The cascade voltage doubler is a circuit for obtaining high dc (direct current) voltages with an alternating-current input. It is a form of power supply used in tube circuits with low to medium current requirements. A schematic diagram of a cascade voltage doubler is illustrated.

Capacitor C1 charges through diode D2 to the peak ac input potential during half of the cycle. During the other

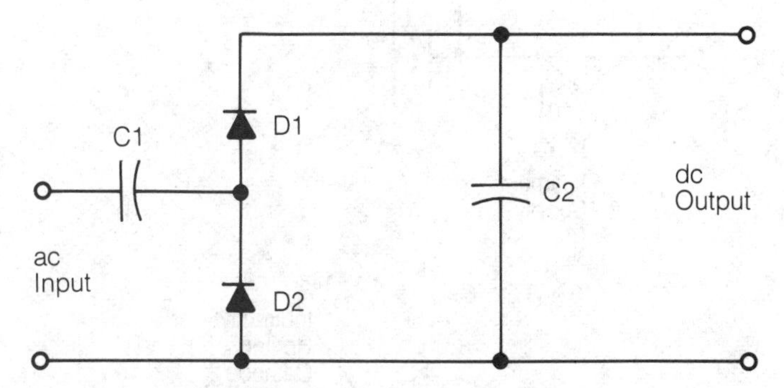

CASCADE VOLTAGE DOUBLER: This circuit doubles the direct-current output from an alternating-current input.

half of the cycle, it discharges through D1 to charge C2 to twice the peak ac input potential. Capacitor C2 also serves as a filtering capacitor to smooth out the ripple in the supply output.

Cascade voltage-doubler circuits have the advantage of a common input and output terminal, allowing unbalanced operation. *See also* VOLTAGE DOUBLER.

CASCODE

A circuit capable of obtaining high gain with tubes or transistors and of providing excellent impedance match between two amplifying stages is called a cascode amplifier.

The input stage is a grounded-cathode or grounded-emitter circuit. The plate or collector of the first stage is fed directly to the cathode or emitter of the second stage, which is a grounded-grid or grounded-base amplifier. The illustration shows a schematic diagram of a cascode amplifier using a pair of NPN bipolar transistors.

Because of the low-noise characteristics of the cascode circuit, it is often used in preamplifiers for improving the sensitivity of hf and vhf (high frequency and very high frequency) receivers. *See also* PREAMPLIFIER.

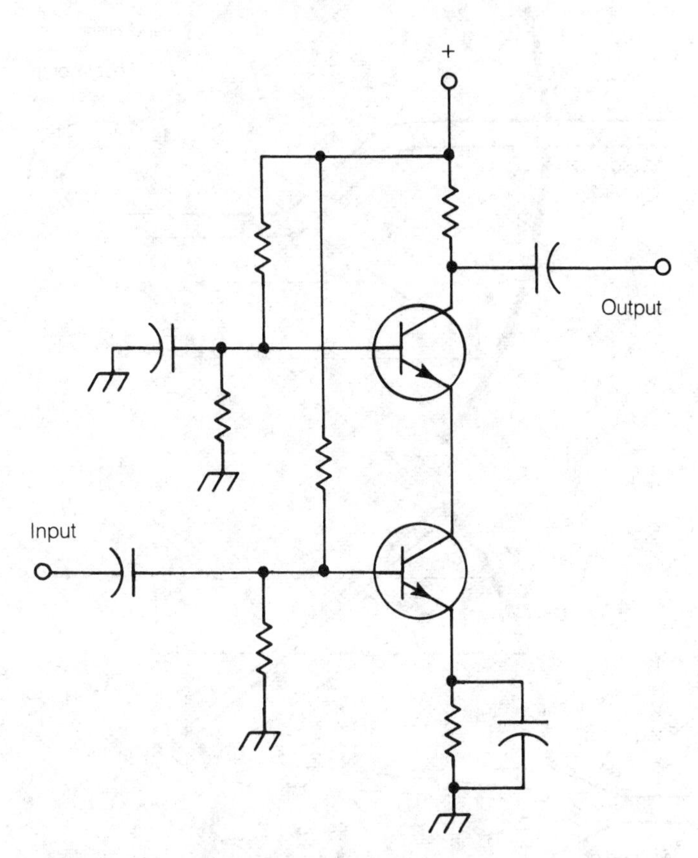

CASCODE: A cascode circuit with NPN transistors.

CASSEGRAIN FEED

The Cassegrain configuration is a method of feeding a parabolic or spherical dish antenna. It eliminates the need for mounting the feed apparatus at the focal point of the dish. In the conventional dish feed system, the waveguide terminates at the position where the rf energy emerges from the dish in a parallel beam (A in the illustration).

In the Cassegrain transmission system the waveguide terminates at the base of the dish which is more convenient from the standpoint of installation. The waveguide output is beamed to a small convex reflector at the focus of the dish, and the convex reflector directs the energy back to the main dish, shown at B. The beam emerges from the antenna in parallel rays. In reception, the rays of electromagnetic energy follow the same path, but in the opposite direction.

Dish antennas are used at uhf and microwave frequencies where the diameter of the antenna and the focal length, are much larger than the wavelength of the transmitted and received energy. *See also* DISH ANTENNA, PARABOLOID ANTENNA in the ANTENNA DIRECTORY.

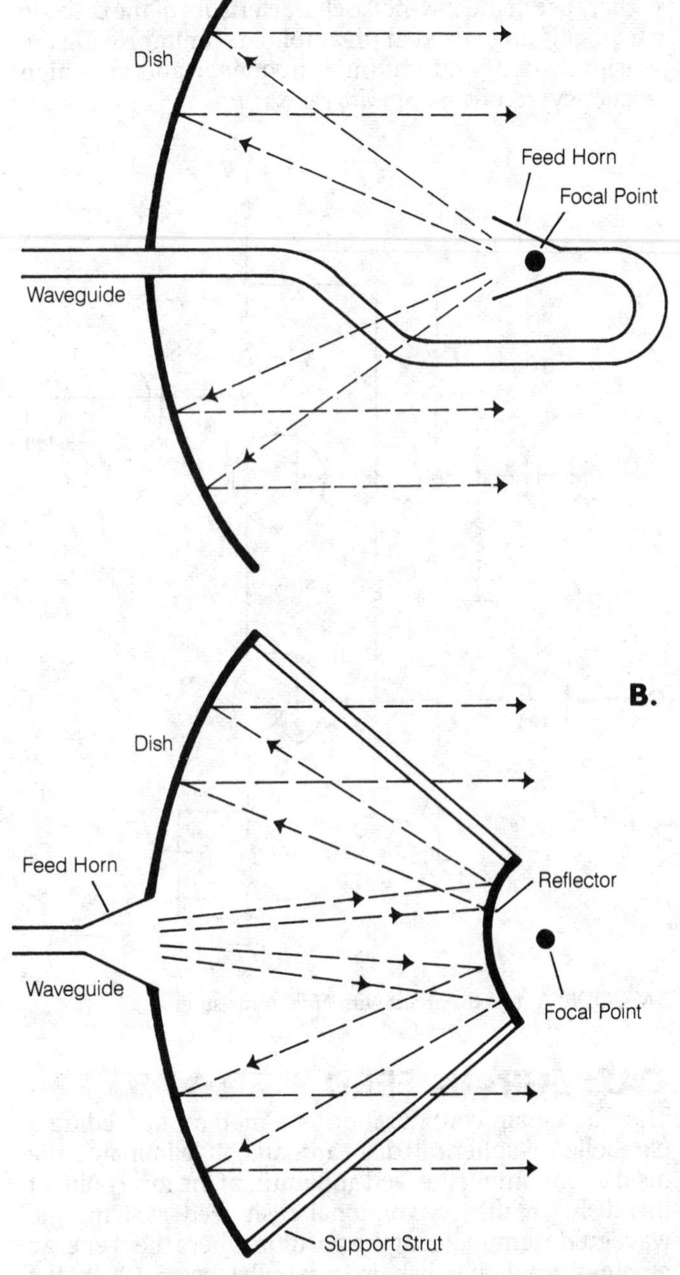

CASSEGRAIN FEED: Conventional dish-antenna feed (A) and a Cassegrain feed (B).

CASSETTE

A package that holds recording tape is called a cassette. Cassettes are made in a variety of sizes and shapes for various applications. A typical cassette, used in a portable tape recorder, plays for 30 minutes on each side; longer playing cassettes allow recording for as much as 60 minutes per side. The internal construction of a cassette is shown.

Some cassettes, such as those used in stereo tape players, have endless loops of tape. Four recording paths, each with two channels for the left and the right half of the sound track, provide a total of eight recording tracks.

In recent years, cassette recorders have been more widely used in video applications (*see* VIDEO TAPE RECORDING).

CATHODE

The cathode is the electron-emitting electrode in a vacuum tube. To enhance is emission properties, the cathode of a tube is heated to a high temperature by a wire called the heater (*see also* FILAMENT).

There are two types of cathode in common use. The directly heated cathode consists of a specially treated filament, which itself emits electrons when connected to a potential that is negative with respect to the anode or plate of the tube. The indirectly heated cathode is a cylinder of metal, separated from the heater but physically close to it.

Directly heated cathodes require a direct current for heating in amplifier and oscillator applications, because

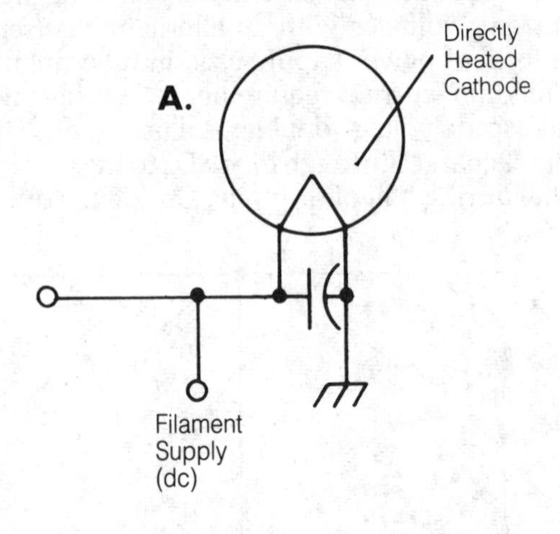

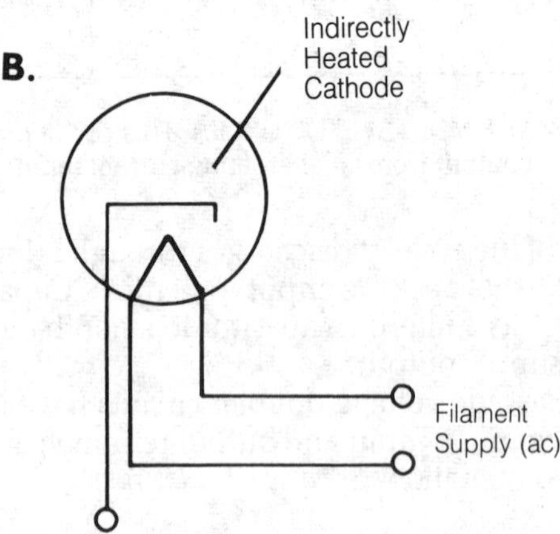

CATHODE: The schematic symbol for a directly heated cathode (A) and the symbol for an indirectly heated cathode (B).

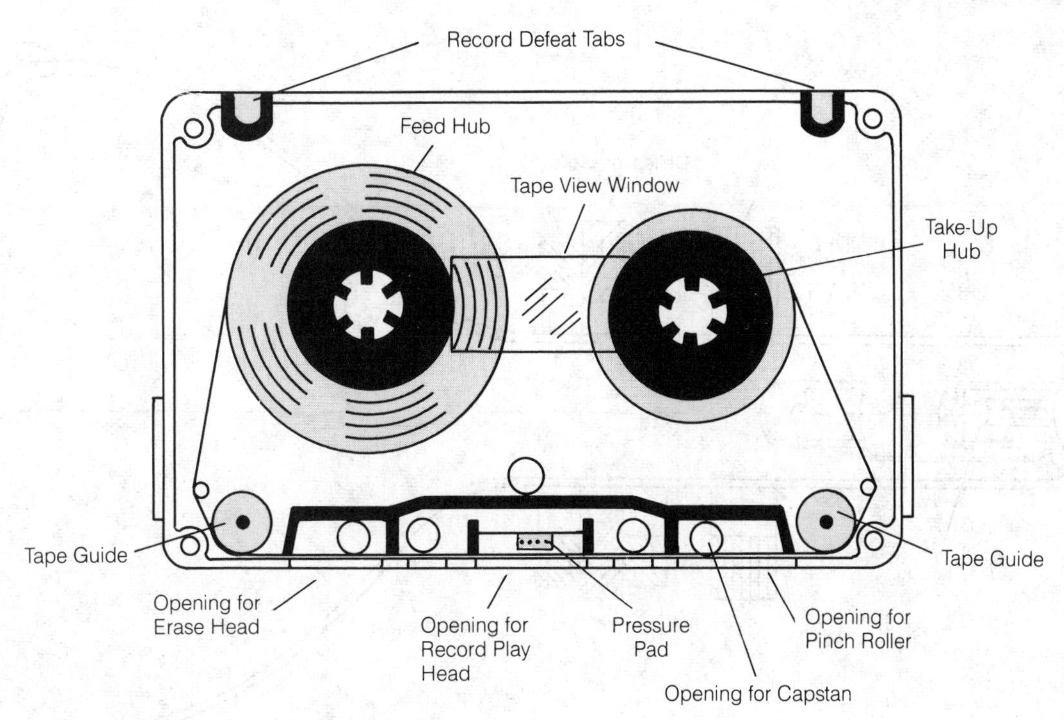

CASSETTE: The internal details of a standard audio cassette.

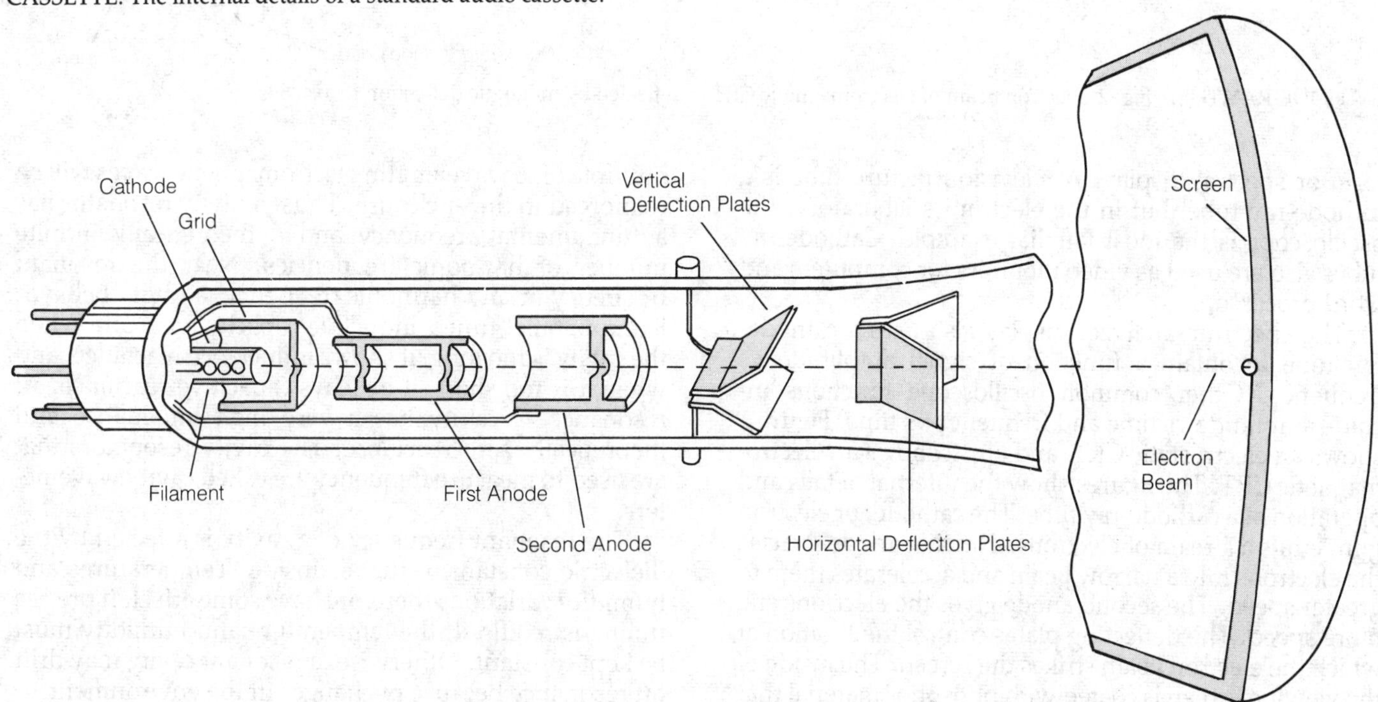

CATHODE-RAY TUBE: Fig. 1. Electron beam of electrostatic CRT is deflected by horizontal and vertical deflection plates.

alternating current would produce ripple in the output signal. The indirectly heated cathode employs an alternating-current filament supply; the cathode acts as a shield against the ripple emission from the filament. The drawing shows schematic representations of both types of cathode.

In a bipolar transistor, the equivalent of the cathode is the emitter. In a field-effect transistor, the equivalent of the cathode is the source. *See also* EMITTER, SOURCE.

CATHODE-RAY OSCILLOSCOPE

See OSCILLOSCOPE.

CATHODE-RAY TUBE

A cathode-ray tube is a component for obtaining a graphic display of an electronic function such as a wave-

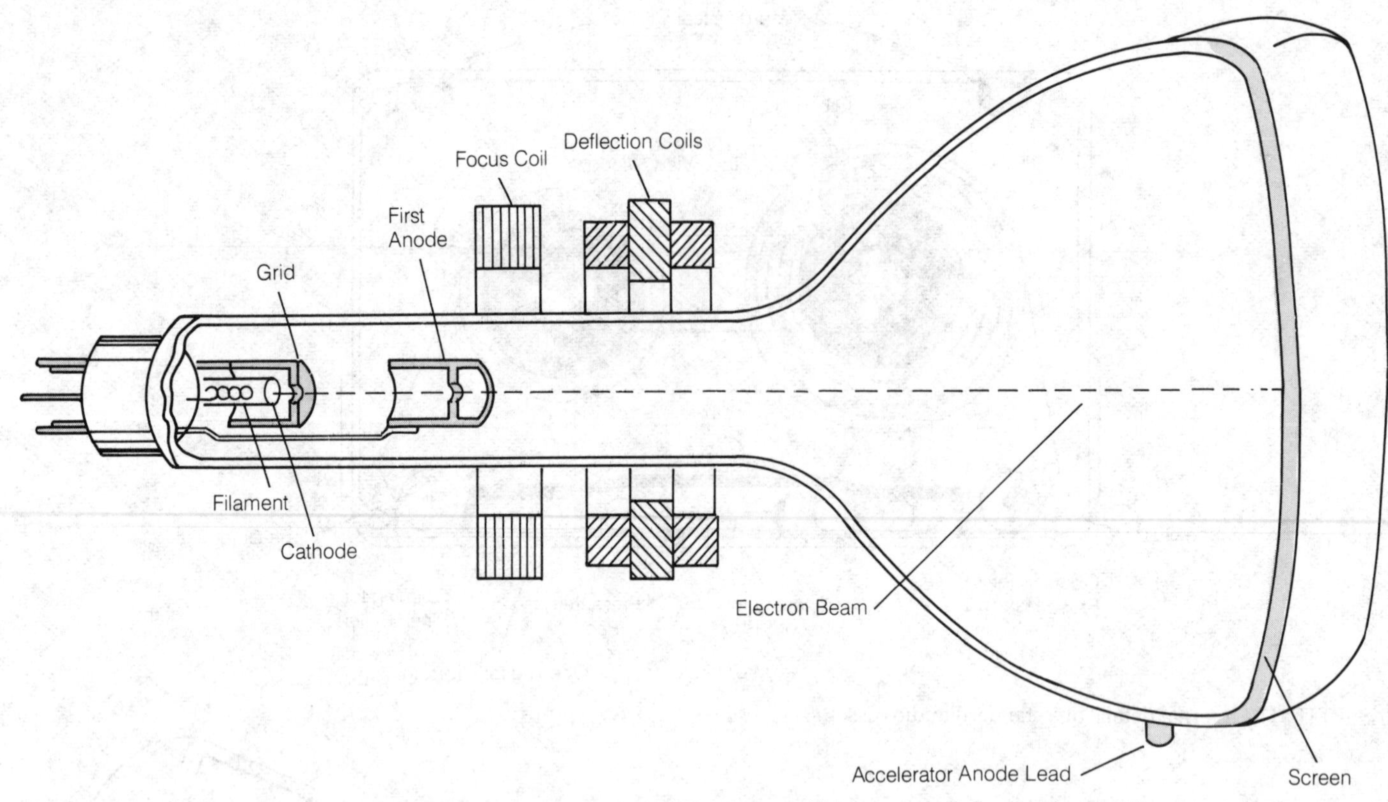

CATHODE-RAY TUBE: Fig. 2. Electron beam of electromagnetic CRT is deflected by magnetic deflection focus coils.

form or spectral display. A television picture tube is a cathode-ray tube, but in the electronics laboratory, the oscilloscope is the most familiar example. Cathode-ray tubes also are used as video monitors for computers and word processors.

The spectrum-analyzer display also uses a cathode-ray tube to obtain a function of signal amplitude vs frequency. Other common oscilloscope functions include amplitude vs time and frequency vs time. Figure 1 shows an electrostatic CRT, and Fig. 2 shows an electromagnetic CRT. The figures show the internal details and operation of a cathode-ray tube. The cathode, or electron gun, emits a stream of electrons. The first anode focuses the electrons into a narrow beam and accelerates them to greater speed. The second anode gives the electrons still more speed. The deflecting plates control the location at which the electron beam strikes the screen. The inside of the viewing screen is coated with phosphor material that glows when the electrons hit it.

The cathode-ray tube is capable of deflecting an electron beam at an extreme rate of speed. Some cathode-ray tubes can show waveforms at frequencies of hundreds of megahertz. *See also* OSCILLOSCOPE.

CAVITY RESONATOR

A cavity resonator is a metal enclosure, usually shaped like a cylinder or rectangular prism (see illustration). Cavity resonators operate as tuned circuits, and are practical for use at frequencies above about 200 MHz.

A cavity has an infinite number of resonant frequencies. When the length of the cavity is equal to any integral multiple of ½ wavelength, electromagnetic waves will be reinforced in the enclosure. Thus, a cavity resonator has a fundamental frequency and a theoretically infinite number of harmonic frequencies. Near the resonant frequency or any harmonic frequency, a cavity behaves like a parallel-tuned inductive-capacitive circuit. When the cavity is too long, it shows inductance reactance, and when it is too short, it displays capacitive reactance. At resonance, a cavity has a very high impedance and theoretically zero reactance. The cavity resonators that are used to measure frequency are called cavity wavemeters.

The resonant frequency of a cavity is affected by the dielectric constant of the air inside. Temperatures and humidity variations therefore have some effect. If precise tuning is required, the temperature and humidity must be kept constant. Otherwise, a resonant cavity may drift off resonance because of changes in the environment.

A length of coaxial cable, short-circuited at both ends, is sometimes used as a cavity resonator. Such resonators can be used at lower frequencies than can rigid metal cavities, since the cable does not have to be straight. The velocity factor of the cable must be taken into account when designing cavities of this kind. *See also* VELOCITY FACTOR.

CAVITY WAVEMETER

See CAVITY RESONATOR.

CELL

See BATTERY.

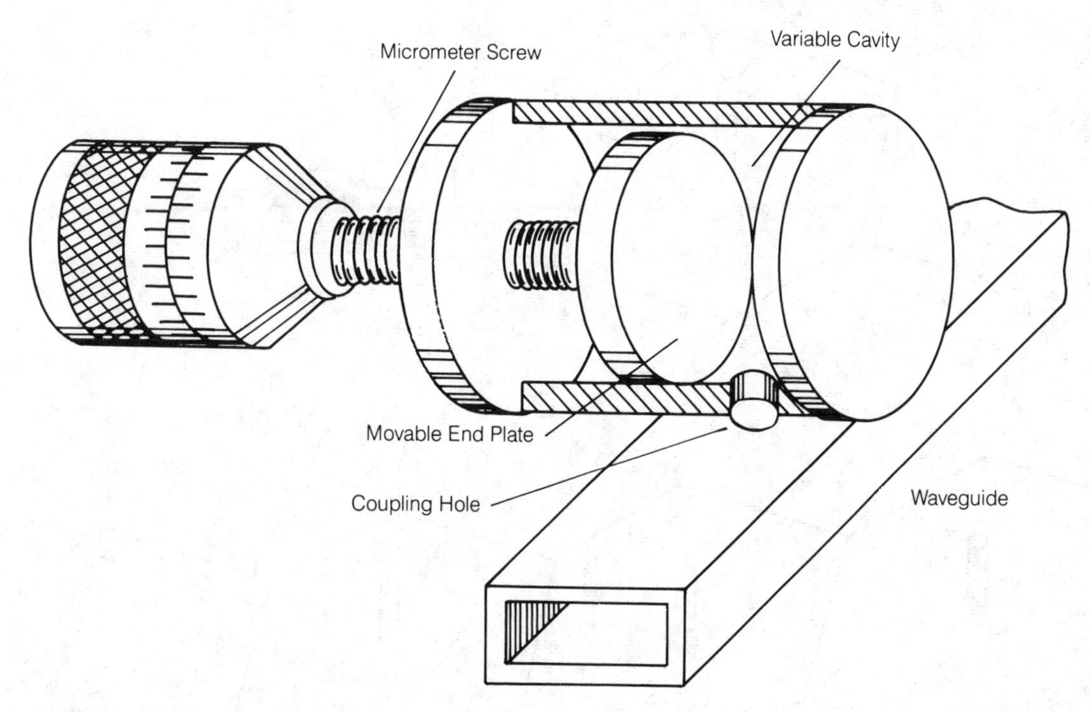

CAVITY RESONATOR: Micrometer screw adjusts the movable plate to change size and frequency of the resonant cavity. This wavemeter is coupled to a waveguide by an interconnecting hole.

CELLULAR MOBILE RADIO TELEPHONE

Cellular mobile radio, also known as cellular mobile radio telephone, is a terrestrial mobile telephone technology that increases the number of available channels by dividing a geographic region into small hexagonal cells, each with a base telephone office repeater. The figure shows a cluster of hexagonal cells, each containing a base station indicated by the tower. The base stations are interconnected with one another and to one or more central office switching systems by wires or microwave links. *See* CENTRAL OFFICE SWITCHING SYSTEM, REPEATER.

Subscribers to these systems have mobile radio telephones in their vehicles. As the subscriber drives through a region, the mobile telephone operation is automatically switched from repeater to repeater as shown in the illustration. *See* MOBILE TELEPHONE, TELEPHONE SYSTEM.

A low-power transmitter at a base station can reach any user within its cell with minimal signal fading. The frequency of the radio transmissions in one cell can be used in another if it is sufficiently far from the first cell, but adjacent cells may not use the same frequency.

To make a telephone call from one mobile unit in one cell to another mobile unit in another cell, the handset is taken off hook. This starts transmission of digital source-and-destination radio data signals that are just strong enough to reach the nearest base station.

All of the base stations are connected to a mobile telephone switching office. It has the dual role of a conventional telephone switching office with connections to the regular land-based telephone service and that of a control station for the cellular radio system. In narrowband systems, cells are assigned their own frequency bands to avoid interference from adjacent cells. The bands can be assigned in a number of patterns.

The switching office circulates a paging signal from one base station to another until it finds the mobile unit whose number has been dialed. The located mobile unit then responds by transmitting an acknowledgment to its local base station. In this way the switching office knows that both parties are ready to call and the cells each party is in.

In this preliminary procedure both mobile units, as well as the switching office, have been using special setup channels to communicate the digital data for the call-initiation. These setup channels are shared by all the users of a given cell. The switching center assigns a channel pair to each of the mobile units, and the units use these channels for voice links as long as the mobile units stay in their original cells. These channel pairs are not necessarily the same pair for the two cells.

When a mobile unit moves from one cell to another during a call, an automatic handoff procedure is initiated. This calls for further data exchange over the setup channels. The message must be rerouted by the base station of the new cell. When the call is terminated, on-hook signals are exchanged between the mobile unit and its base station, and the channels are cleared for further use.

An outstanding advantage of cellular mobile telephone is the large number of users that can be handled in the entire service area by reducing the size of the cells. Power levels can be reduced for smaller cells, with a corresponding reduction in the interference range around each cell.

The most common form of modulation in land mobile communications today is narrowband FM (frequency modulation). Each message is assigned a carrier with a frequency that is unique in the local cell cluster. The signal frequency modulates that carrier with a deviation

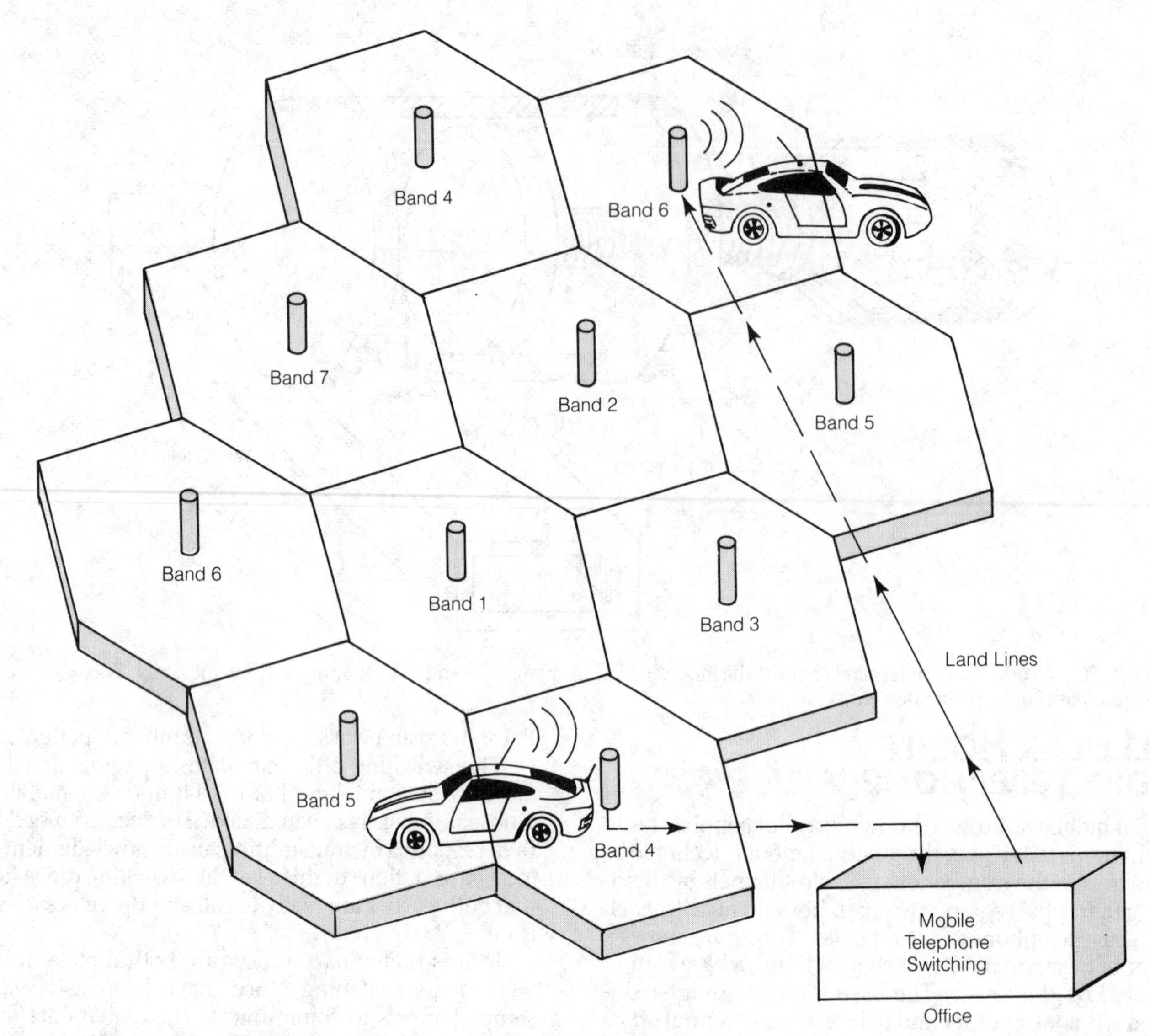

CELLULAR MOBILE RADIO TELEPHONE: Area of coverage is divided into adjacent cells, each containing a base station. Mobile units communicate by radio to the base stations, which are connected to a telephone switching office.

that can range from 5 to 12 kHz. Carrier frequencies may be separated by 25 or 30 kHz.

One mobile telephone service now in service is operated at frequencies between 825 and 890 MHz. Each radio channel connecting the mobile phones with the cell stations consists of a pair of one-way channels separated by 45 MHz. A channel width of 30 kHz is used.

Each cell station is connected by wire to a central mobile telephone switching office. This office also determines when a mobile phone should be handed off to another cell during phone conversation. By monitoring the strength of signals received from the mobile unit by adjacent cells, it hands off the call to the cell receiving the strongest signal. The switching office than signals over the voice channels for the mobile phone to switch to a new reception channel. This procedure causes only about a 0.05-second interruption of conversation.

CELSIUS TEMPERATURE SCALE

The Celsius temperature scale is a scale in which the freezing point of pure water at one atmosphere is assigned the value zero degrees, and the boiling point of pure water at one atmosphere is assigned the value 100 degrees. The Celsius scale was formerly called the Centigrade scale. The word *Celsius* is generally abbreviated by the capital letter C.

Temperatures in Celsius and Fahrenheit are related by the equations:

$$C = \frac{5}{9}(F - 32)$$

$$F = \frac{9}{5}C + 32$$

where C represents the Celsius temperature and F represents the Fahrenheit temperature. The nomograph aids in quick conversion between temperature readings in the two scales.

Celsius temperature is related to Kelvin, or absolute, temperature by the equation:

$$K = C + 273$$

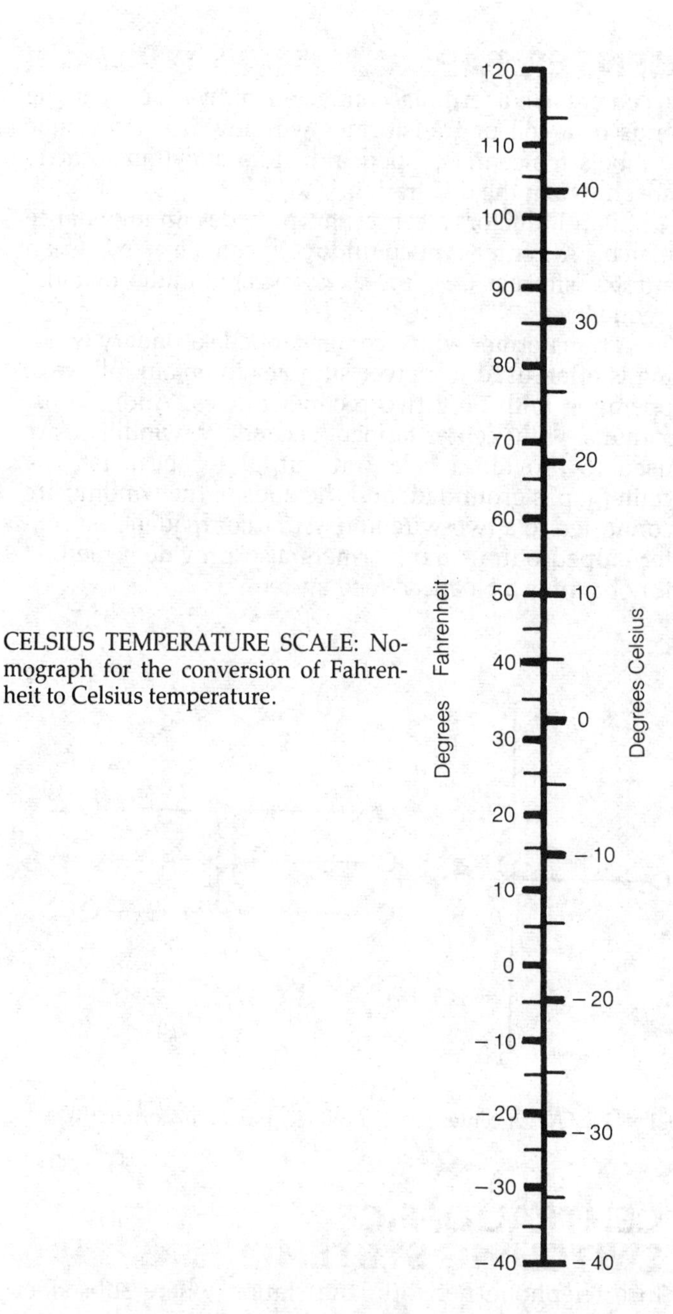

CELSIUS TEMPERATURE SCALE: Nomograph for the conversion of Fahrenheit to Celsius temperature.

where K represents the temperature in ° Kelvin. A temperature of −273° Celsius is called absolute zero, the coldest possible temperature. *See also* FAHRENHEIT TEMPERATURE SCALE, KELVIN TEMPERATURE SCALE.

CENTER FEED

When an antenna element, resonant or nonresonant, is fed at its center, the antenna is said to have center feed. Usually, this antenna is a half-wave dipole, the driven element of a Yagi, or one of the elements of a phased array.

Center feed provides good electrical balance when a two-wire transmission line is used, (see drawing), whether the element is ½ wavelength or any other length, provided the two halves of the antenna are at nearly equal distances from surrounding objects such as trees, utility wires, and the ground.

For antenna elements measuring an odd multiple of ½ wavelength, center feed results in a purely resistive load impedance between approximately 50 and 100 Ω. For antenna elements measuring an even number of half wavelengths, center feed results in a purely resistive load impedance ranging from several hundred to several thousand ohms. If the element is not an integral multiple of ½ wavelength, reactance is present at the load in addition to resistance.

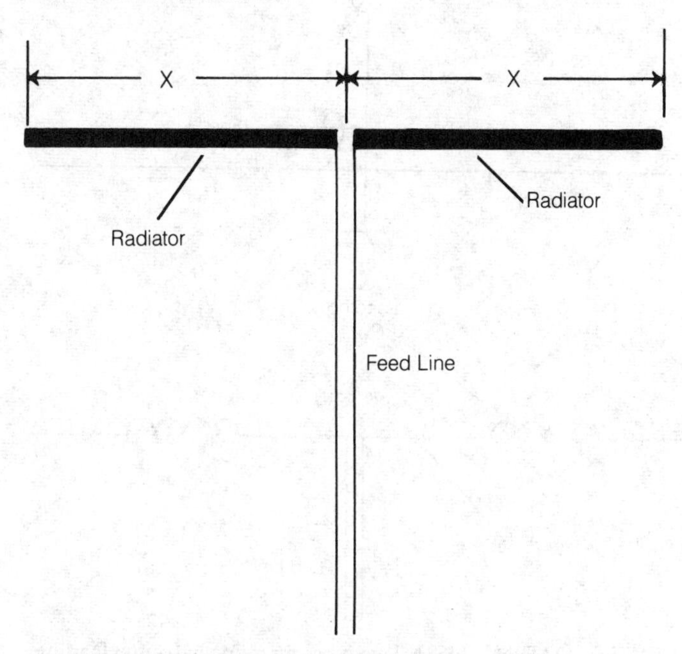

CENTER FEED: A center-fed antenna is symmetrical.

CENTER LOADING

Center loading is a method of altering the resonant frequency of an antenna radiator. An inductance or capacitance is placed along the physical length of the radiator, approximately halfway between the feed point and the end. Figure 1 shows center loading of a vertical radiator fed against ground (A) and a balanced, horizontal radiator (B).

Inductances lower the resonant frequency of a radiator having a given physical length. Generally, for quarter-wave resonant operation with a radiator less than ¼ wavelength in height, some inductive loading is necessary to eliminate the capacitive reactance at the feed point. For quarter-wave resonant operation with a radiator between ¼ and ½ wavelength in height, a capacitor must be used to eliminate the inductive reactance at the feed point.

An 8-foot mobile whip antenna can be tuned to quarter-wave resonance with inductive center loading at all frequencies below its natural quarter-wave resonant frequency, which is about 29 MHz. Figure 2 gives the value of loading inductance L, in microhenrys (μH), as a function of the frequency in megahertz (MHz), for center loading of an 8-foot radiator over perfectly conducting ground. While the rf ground in a mobile installation is far from perfect, the values given are close enough to be of practical use.

When the inductor or capacitor in an antenna loading scheme is placed at the feed point, the system is called based loading. *See also* BASE LOADING.

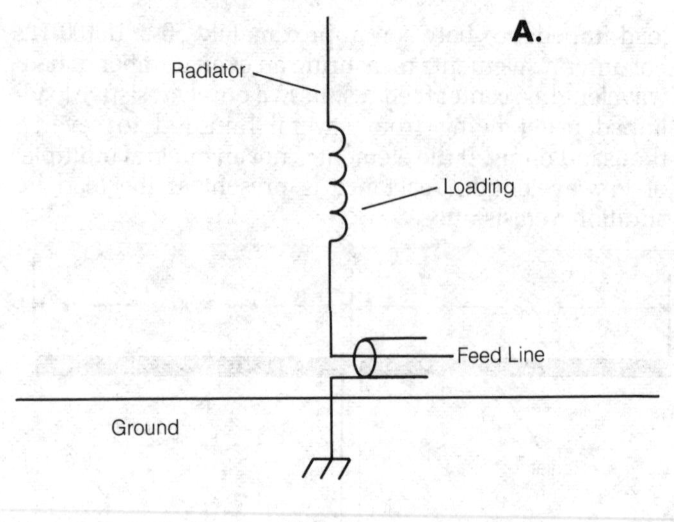

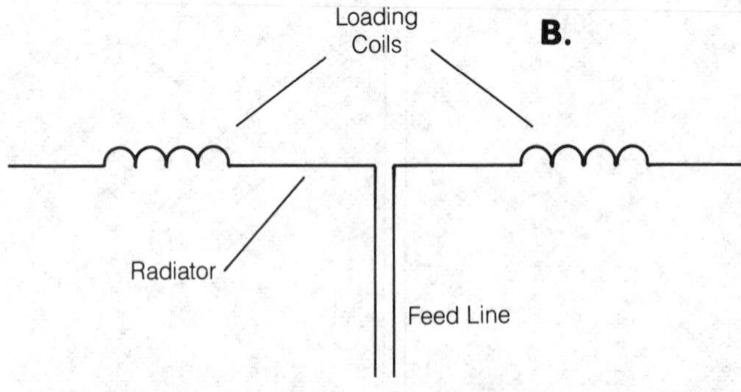

CENTER LOADING: Fig. 1. Two types for inductive center loading an antenna include one for resonating a radiator less than a quarter wavelength long (A) and one for resonating an antenna less than a half wavelength long.

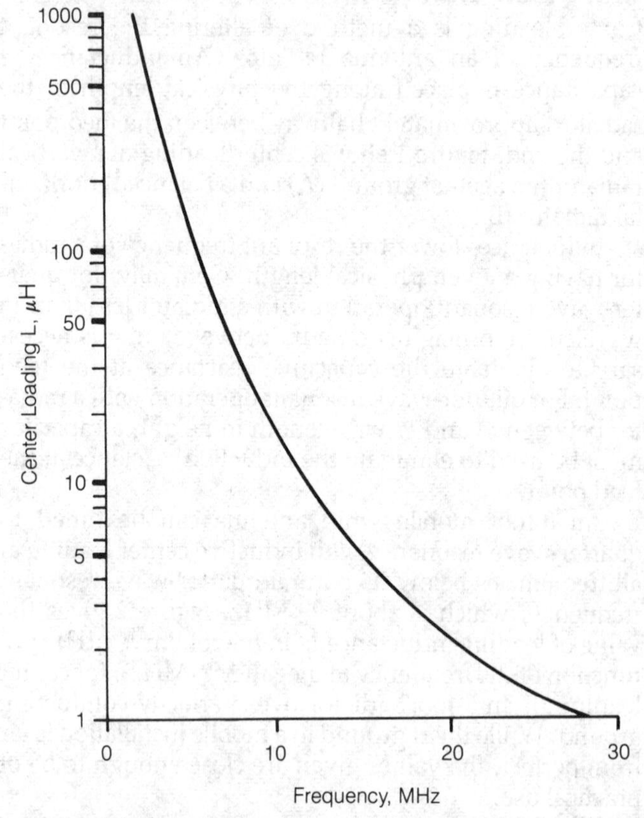

CENTER LOADING: Fig. 2. A plot of center loading inductance versus frequency for an 8-foot vertical whip antenna.

CENTER TAP

A center tap is a terminal connected midway between the ends of a coil or transformer winding. The schematic symbols for center-tapped inductors and transformers are shown in the illustration.

In an inductor, a center tap provides an impedance match. A center-tapped inductor can be used as an autotransformer (*see* AUTOTRANSFORMER) at audio or radio frequencies.

A transformer with a center-tapped secondary winding is often used in power supplies to obtain full-wave operation with only two rectifier diodes. Audio transformers with center-tapped secondary windings are used to provide a balanced output to speakers. The center tap is grounded, and the ends of the winding are connected to a two-wire line. At radio frequencies, center-tapped output transformers also provide a method for obtaining a balanced feed system.

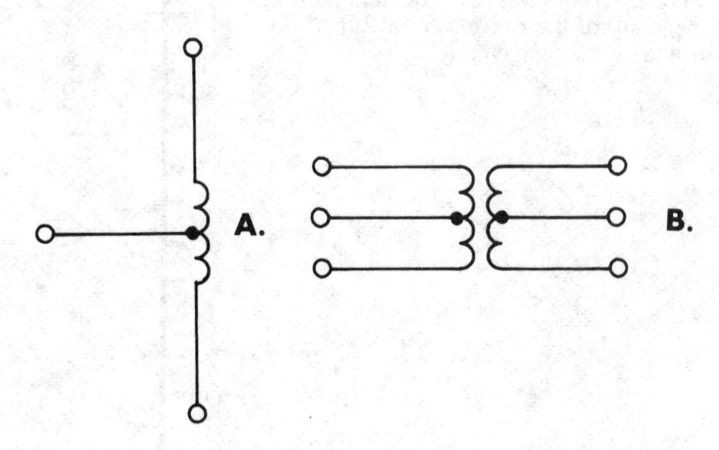

CENTER TAP: A center tap in a coil (A) and a transformer (B).

CENTRAL OFFICE SWITCHING SYSTEM

In a telephone network, the station where subscriber lines converge is called the central office. There are many different possible interconnections; a large central office requires extensive switching apparatus to enable any customer to call any other at any time. Modern central-office switching systems are computer-controlled.

Each local area typically has its own central office. For long-distance communication, one central office is connected to another with a trunk line (*see* TRUNK LINE.) A call originates on one subscriber line, goes to a central office, is routed along a trunk line to another central office (if the called line is connected to a different central office), and finally arrives at the distant subscriber line (see illustration). This entire process can be carried out in seconds without an aid of a human operator. *See also* TELEPHONE SYSTEM.

CENTRAL PROCESSING UNIT

See COMPUTER (COMPUTER ORGANIZATION).

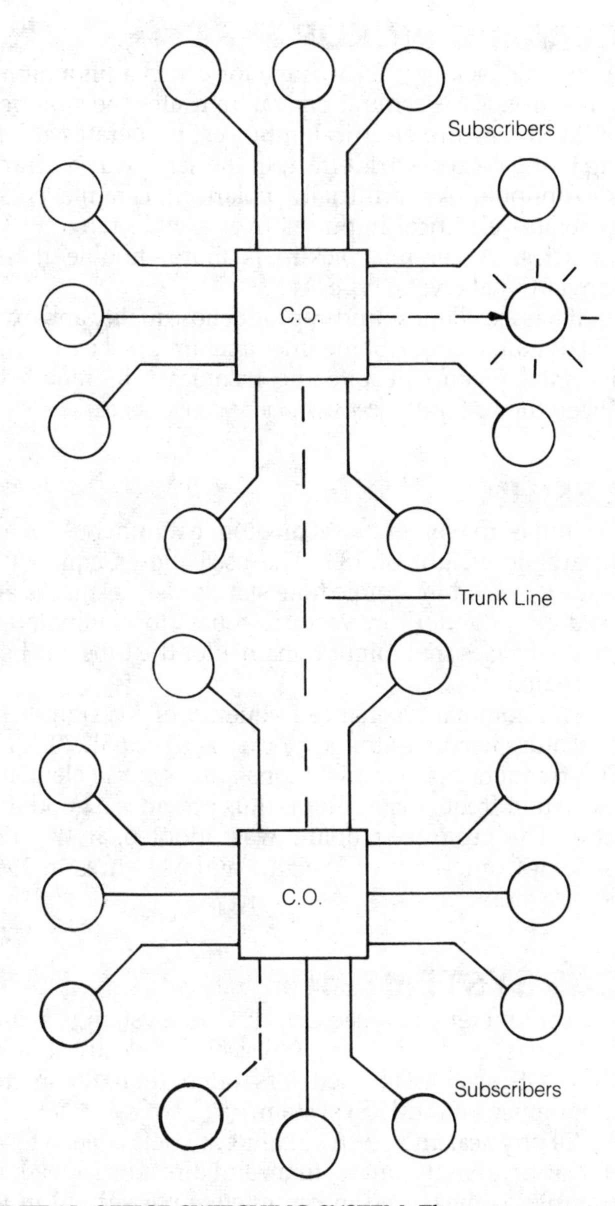

CENTRAL OFFICE SWITCHING SYSTEM: The system connects individual telephone subscribers to the trunk line.

CENTRIFUGAL FORCE

When an object is revolving around a central point, a force is introduced in a direction opposite to the direction of the acceleration. The force is called centrifugal force. In circular motion, the acceleration is directed inward, and thus the centrifugal force occurs outward, away from the center of revolution (see drawing). The greater the magnitude of the acceleration, the greater the centrifugal force.

The force, *F*, in kilogram-meters per-second-per-second (kgm/s^2), is calculated according to the formula:

$$F = ma$$

where *m* is the mass of the object in kilograms, and *a* is the acceleration in meters per-second-per-second. The acceleration of a revolving body is determined by the formula:

$$a = 4\pi r^2/T^2$$

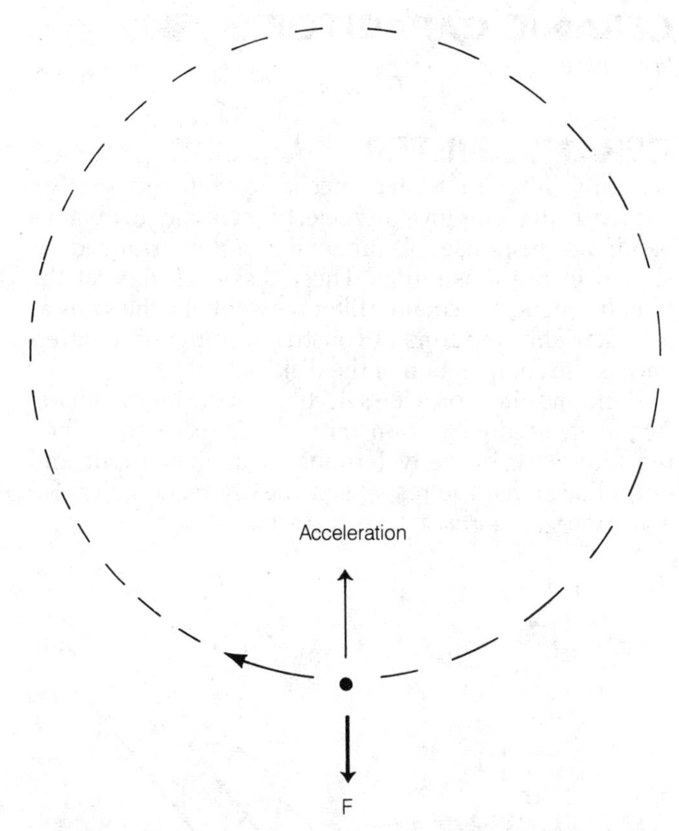

CENTRIFUGAL FORCE: Centrifugal force opposes acceleration in circular motion.

where *r* is the radius of revolution in meters, and *T* is the time in seconds for one revolution at constant speed.

CENTRIFUGAL SWITCH

A centrifugal switch is a switch actuated by centrifugal force. An acceleration switch is used for this purpose because centrifugal force is actually acceleration force. When the acceleration reaches a specified threshold intensity, the switch changes state. *See also* ACCELERATION SWITCH.

CERAMIC

Ceramic is a manufactured compound consisting of aluminum oxide, magnesium oxide, or other similar materials. Materials such as steatite and barium titanate are other examples of ceramics. The most familiar example of a ceramic material for use in electronics is alumina, or aluminum oxide.

Ceramics are used in a wide variety of electronic applications. Some capacitors employ a ceramic material as the dielectric. Certain inductors are wound on ceramic forms, because ceramic is an excellent insulator and is physically strong. Ceramics are used in the manufacture of microphones, phonographic cartridges, and transducers for sonar and depth finders. Some bandpass filters, intended for use at radio frequencies, have resonant ceramic crystals or disks. Ceramic materials are used in the manufacture of superconductors. *See also* CAPACITOR, CERAMIC FILTER, CERAMIC MICROPHONE, CERAMIC PICKUP, SUPERCONDUCTIVITY.

CERAMIC CAPACITOR

See CAPACITOR.

CERAMIC FILTER

A ceramic filter is a form of mechanical filter (*see* MECHANICAL FILTER) that employs piezoelectric ceramics to obtain a bandpass response. Ceramic disks are arranged as shown in the illustration. These disks resonate at the filter frequency. A ceramic filter is essentially the same as a crystal filter in terms of construction; the only difference is the composition of the disk material.

Ceramic filters provide selectivity in the intermediate-frequency sections of transmitters and receivers. When the filters are properly terminated at their input and output sections, the response is nearly rectangular. *See also* BANDPASS RESPONSE, CRYSTAL-LATTICE FILTER.

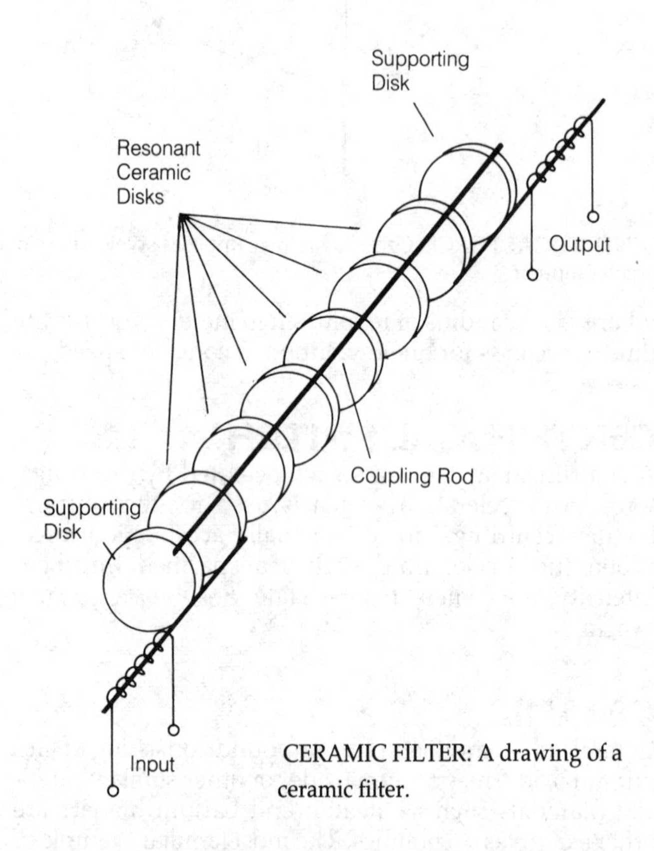

CERAMIC FILTER: A drawing of a ceramic filter.

CERAMIC MICROPHONE

A ceramic microphone uses a ceramic cartridge to transform sound energy into electrical impulses. Its construction is similar to that of a crystal microphone (*see* MICROPHONE). When subjected to the stresses of mechanical vibration, certain ceramic materials generate electrical impulses.

Ceramic and crystal microphones must be handled with care, since they are easily damaged by impact. Ceramic microphones display a high output impedance and excellent audio-frequency response. *See also* MICROPHONE.

CERAMIC PICKUP

A ceramic pickup is a cartridge for use in a phonograph (*see* CARTRIDGE). A ceramic crystal translates the vibrations of the stylus into electrical impulses. It operates according to the piezoelectric effect, in the same way a ceramic microphone. An artificially polarized ceramic crystal generates electrical impulses over a wide range of frequencies. A ceramic pickup is more durable than a conventional crystal type.

It has excellent sound-reproduction quality and excellent dynamic range. Sometimes a ceramic pickup is called a crystal pickup despite the technical differences between them. *See also* CERAMIC, CRYSTAL TRANSDUCER.

CESIUM

Cesium is an element with an atomic number of 55, and an atomic weight of 133. The oscillation frequency of cesium is used in atomic time standards. Cesium is also used as a "getter" in vacuum tubes, to eliminate any residual gases that might remain after the tube has been evacuated.

The nominal resonance frequency of a cesium-beam oscillator, used in atomic clocks, is 9192.631770 MHz. This frequency is extremely constant, as are all elemental oscillation frequencies, and it thus provides a good time base. The cesium standard was adopted at the 12th General Conference of Weights and Measures in 1964. *See also* ATOMIC CLOCK.

CGS SYSTEM

The centimeter-gram-second, or CGS, system is a standard basis for determining physical and electrical units. The CGS system is used less often than the meter-kilogram-second (MKS) system.

All physical and electrical units are definable in terms of quantity, length, mass, time, and direction. Speed, for example, is measured in centimeters per second in the CGS system. Current is defined as coulombs (electrical quantity) per second. Voltage is defined in terms of mass, length, time, and quantity. Resistance is defined on the basis of current and voltage.

In the CGS system, electrical units are called ab units; *for example,* the abampere, abvolt, and abohm. Electrostatic units in the CGS system are called stat units. *See also* CURRENT, MKS SYSTEM, VOLTAGE.

CHANNEL

A channel is a particular band of frequencies to be occupied by one signal, or one two-way conversation in a given mode. The term is also used to refer to the current path between the source and drain of a field-effect transistor (*see* FIELD-EFFECT TRANSISTOR).

The amplitude-modulation (AM) broadcast band and the frequency modulation (FM) broadcast band are allocated in channels by legal authority of the Federal Communications Commission in the United States. Some frequency bands, such as the Amateur-Radio FM band at

144 MHz, are allocated into channels by common agreement among the users. Some bands, such as the high-frequency Amateur bands, are not allocated into channels. Operation in these bands is done with variable-frequency oscillators.

Signals require bandwidth for efficient transfer of information. This bandwidth is called the channel, or channel width, of the signal. A typical AM broadcast signal is 10 kHz wide, or 5 kHz above and below the carrier frequency at the center of the channel (see illustration).

Some signal channels are as narrow as 3 to 5 kHz; some are several megahertz wide. *See also* BANDWIDTH, FREQUENCY ALLOCATIONS.

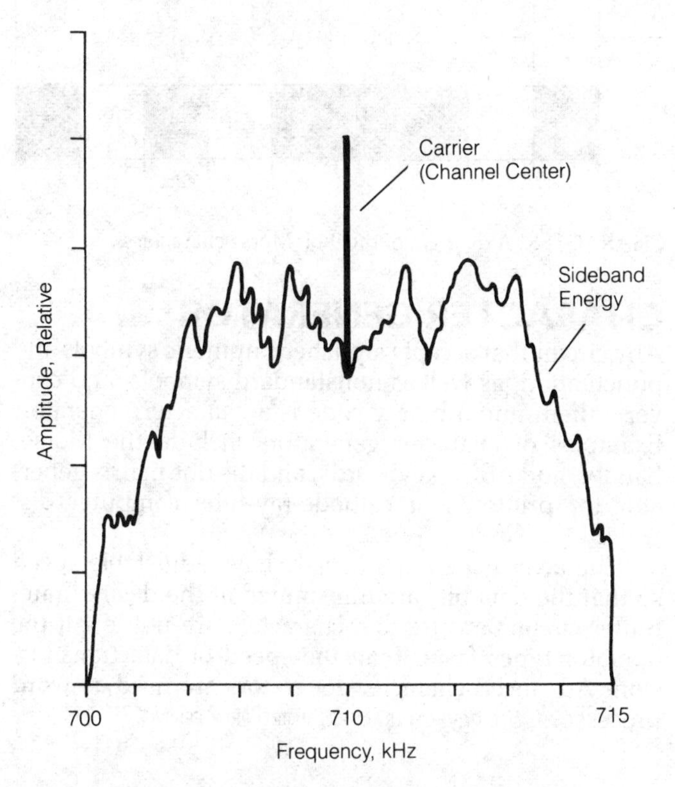

CHANNEL: The amplitude-versus-frequency plot of a broadcast band channel for an amplitude-modulated signal.

CHANNEL ANALYSIS

When a signal is checked to ensure that all its components are within the proper assigned channel, the procedure is called channel analysis. Channel analysis requires a spectrum analyzer to obtain a visual display of signal amplitude as a function of frequency (*see* SPECTRUM ANALYZER).

The illustration shows an amplitude-modulated (AM) signal as it would appear on a spectrum-analyzer display. The normal bandwidth of an AM broadcast signal is plus or minus 5 kHz relative to the channel center. (A communications signal is often narrower than this, about plus or minus 3 kHz.) At A, a properly operated AM transmitter produces energy entirely within the channel limits. At B, overmodulation causes excessive bandwidth. At C, an off-frequency signal results in out-of-band emission. Channel analysis can reveal almost any problem with a modulated signal.

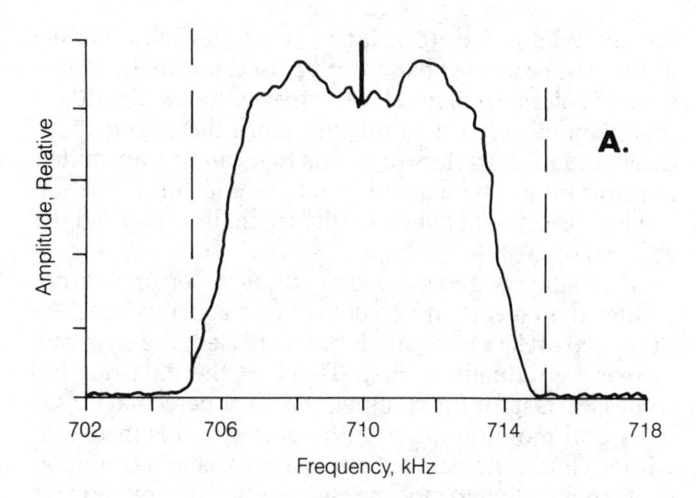

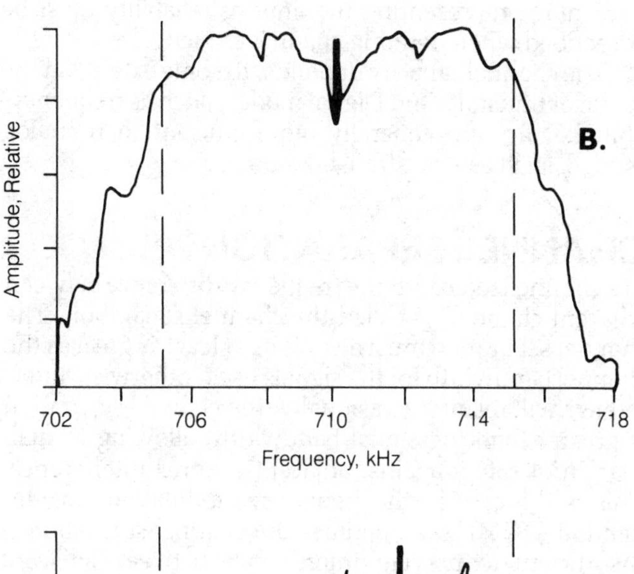

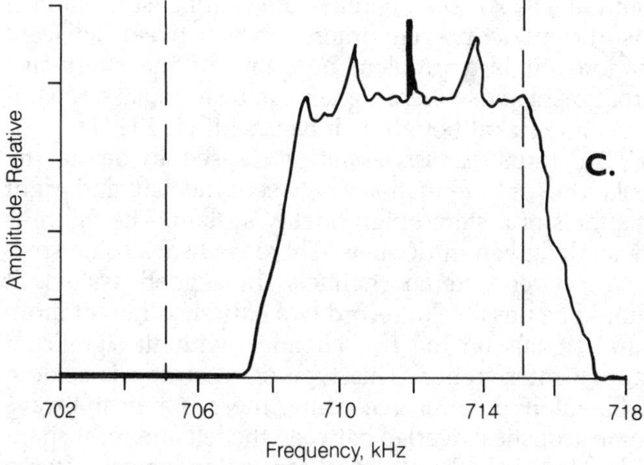

CHANNEL ANALYSIS: Waveforms of a properly operating amplitude-modulated transmitter (A), one with excessive bandwidth (B), and one with an off-frequency signal (C).

CHANNEL CAPACITY

Any channel is capable of carrying some information at a certain rate of speed. A channel just a few hertz wide can accommodate a code (continuous wave) signal; a channel in the AM broadcast band can be used for the transmission of voices. In general, the greater the bandwidth of the channel (*see* BANDWIDTH), the more the information, measured in characters or words that can be transmitted per unit time. The maximum rate at which information can be reliably sent over a particular channel is called the channel capacity.

Reliability is generally defined as an error rate not greater than a certain percentage of characters sent. As the speed of data transmission is increased, the error rate increases gradually at first. Then, as the data rate becomes too fast for the channel, the error percentage rises more and more rapidly. As the data speed is increased without limit, the data reception approaches a condition nearly equivalent to random characters. The precise error percentage representing the limit of reliability must be prescribed when defining channel capacity.

The channel capacity depends, to some extent, on the mode of transmission. Digital modes, such as frequency-shift keying, are generally more efficient than analog modes, such as voice transmission.

CHANNEL SEPARATION

In a channelized band, the frequency difference between adjacent channels is called the channel separation. The channel separation must always be at least as great as the channel bandwidth for the signals used; otherwise, interference will result. Occasionally, the channel separation is greater than the signal bandwidth, allowing a small margin of safety against adjacent-channel interference. This is the case in the frequency-modulation amateur band at 144 MHz. Sometimes the channel separation is insufficient to prevent interference between adjacent stations. It is a problem on some of the shortwave broadcast bands where signals can be found at a separation of 5 kHz, although their bandwidth is 10 kHz.

The term *channel separation* is used to define the isolation, or attenuation, between the left and right channels of a stereo high-fidelity system. The figure is generally given in decibels. There is always some spillover between stereo channels. In a good system, it should be possible to record two entirely different monaural signals on the two channels without significant mutual interference. This type of recording is seldom deliberately done; in recording, there is almost always some acoustical overlap between the left and right channels. However, the electrical separation between stereo channels should be as great as the state of the art will permit. *See also* STEREOPHONICS.

CHARACTER

A character is an elementary symbol or unit of data. It may consist of a binary digit (bit), a decimal digit (0 through 9), a letter of the alphabet, a punctuation mark, or a nonstandard symbol. A space is technically considered a character as well. A character, in most codes, is represented by several bits. For example, in the International Morse code, the letter *L* is represented by the combination of bits 101110101, as shown in the illustration, where 0 represents the key-up condition, 1 represents the key-down condition, and the length of one dot is defined as one bit. (The space following the letter is not included as part of the letter.)

Words are made up of combinations of characters. In the English language, the length of a word may vary considerably. In some computer languages, a word is always the same length; in others, it varies. *See also* BIT, WORD.

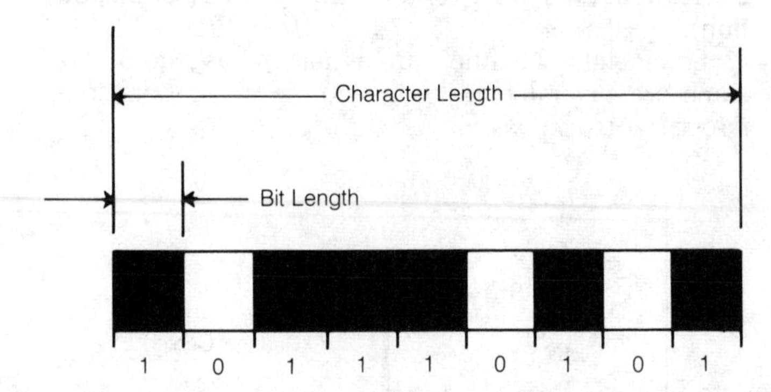

CHARACTER: A diagram of the 9-bit Morse character *L*.

CHARACTER GENERATOR

Any circuit that accepts alphabetic numeric symbols and punctuation, as well as nonstandard symbols, and converts them into a binary code is a character generator. Examples of character generators include the Morse, Baudot, and ASCII keyboards, and the dot-matrix generator for printouts or cathode-ray-tube computer displays.

Character generators usually have adjustable speed so that the data bits are transmitted at the desired rate. Buffer circuits ensure that characters are not lost if the operator types faster than the speed of data transmission. Advanced character generators are used as word processors. *See also* CHARACTER, WORD PROCESSING.

CHARACTERISTIC

Any specific property of a material or device is called a characteristic. A characteristic is definable and measurable, and can be expressed in electrical, electromagnetic, hydraulic, magnetic, mechanical, nuclear, or thermal terms.

Examples of characteristics include hardness, conductivity, temperature coefficients, and radiation resistance.

In logarithms, the characteristic is the portion of the logarithm to the left of the decimal point. For positive logarithms, the characteristic is the integral portion. For negative logarithms, the characteristic is the integral portion minus 1. For example, in the equation:

$$\log_{10} 253 = 2.403$$

the characteristic is 2. In the equation:

$$\log_{10} 0.253 = -0.5969$$

the characteristic is −1. *See also* LOGARITHM.

CHARACTERISTIC CURVE

A function or relation defining the interdependence of two quantities is called a *characteristic curve.* In electronics, characteristic curves are generally mentioned in reference to semiconductors or vacuum tubes.

A common example of a characteristic curve is illustrated. This curve defines the relation between the gate voltage (E_G) and the drain current (I_D) for an N-channel field-effect transistor, given a drain voltage of 3 V.

Characteristic curves are used to find the best operating bias for an oscillator or amplifier. Different kinds of amplifiers operate at different points on the characteristic curve for a particular device. A linear amplifier should be biased at a point where the characteristic curve is nearly straight. A Class C amplifier is biased beyond the cutoff or pinchoff point. *See also* FIELD-EFFECT TRANSISTOR, TRANSISTOR.

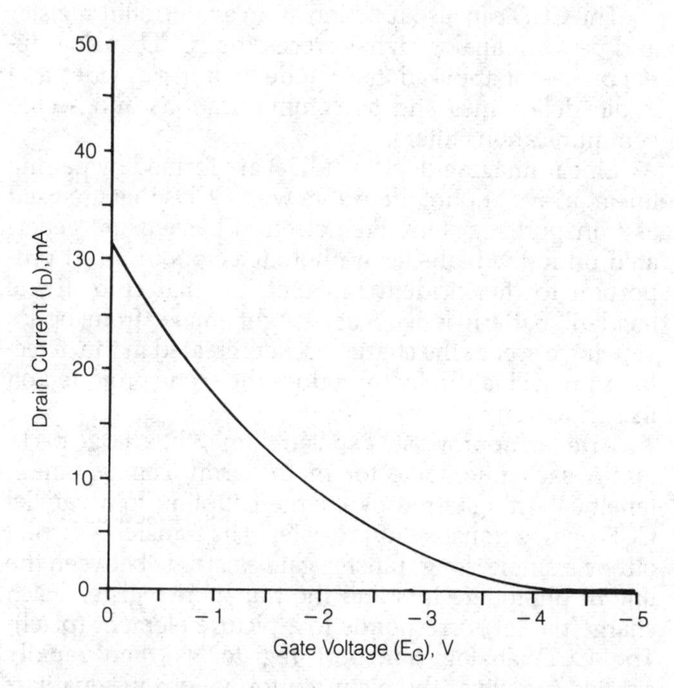

CHARACTERISTIC CURVE: A characteristic curve for a field-effect transistor (FET).

CHARACTERISTIC DISTORTION

Characteristic distortion is a fluctuation in the characteristic curve of a semiconductor device or vacuum tube. The characteristic curve is affected by any change of bias. A change of bias can be caused by the signal itself. Normally, the operation of an amplifier is fairly predictable from the direct-current bias and the characteristic curve of the device. The addition of an alternating-current signal, especially if it is high in voltage, can alter the bias. To obtain an accurate prediction of the operation of an amplifier in which there is high drive voltage, characteristic distortion must be considered. *See also* CHARACTERISTIC CURVE.

CHARACTERISTIC IMPEDANCE

The ratio of the signal voltage (*E*) to the signal current (*I*) in a transmission line depends on several things. Ideally, there should be no variations in current and voltage at different places along the line. When the load consists of a noninductive resistor of the correct ohmic value, the voltage-to-current (*E*/*I*) ratio Z_O will be constant all along the line, so that:

$$Z_O = E/I$$

The value Z_O under these circumstances is called the characteristic, or surge, impedance of the transmission line.

The characteristic impedance of a given line is a function of the diameter and spacing of the conductors used. It is also a function of the type of dielectric material. For air-dielectric coaxial line, in which the inside diameter of the outer conductor is *D*, and the outside diameter of the inner conductor is *d* (see A in the illustration), the characteristic impedance is:

$$Z_O = 138 \log_{10} (D/d)$$

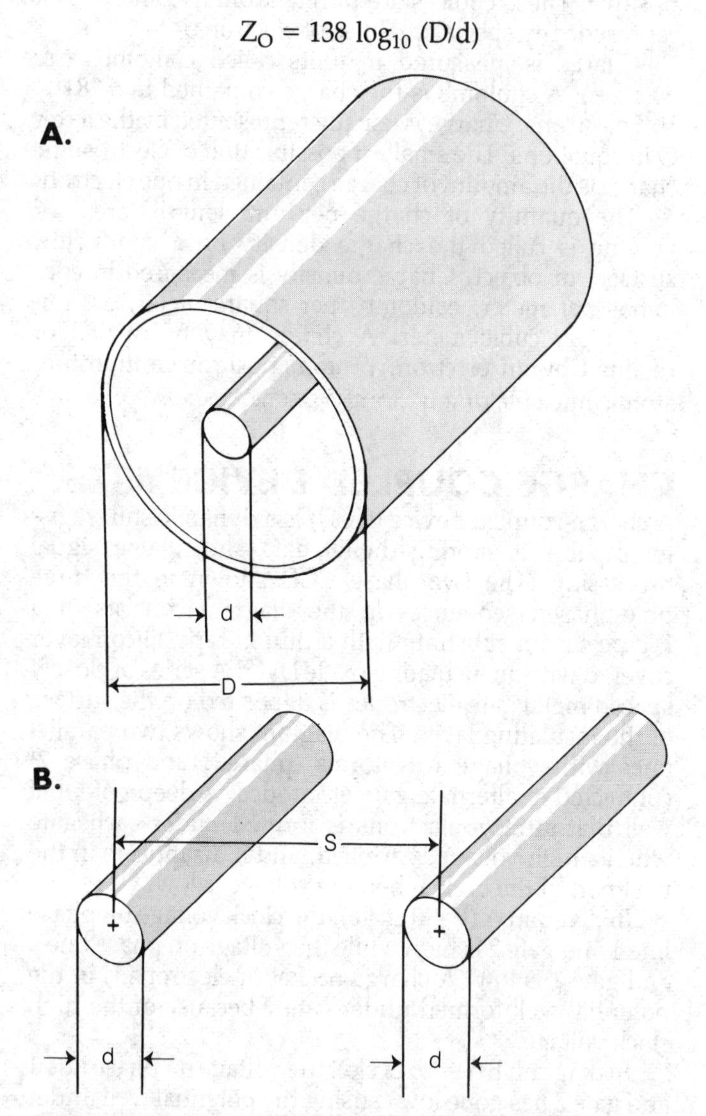

CHARACTERISTIC IMPEDANCE: Dimensions for determining the characteristic impedance of a coaxial cable (A) and a two-wire transmission line.

For air-spaced two-wire line, in which the conductor diameter is d and the center-to-center conductor spacing is s (B in the illustration,) the formula is:

$$Z_O = 276 \log_{10} (2s/d)$$

The units for D, d, and s must, of course, be uniform in calculation, but any units can be used.

Generally, transmission lines do not have air dielectrics. Solid dielectrics, such as polyethylene, are often used, and this lowers the characteristic impedance for a given conductor size and spacing. For optimum operation of a transmission line, the load impedance should consist of a pure resistance R at the operating frequency, such that $R = Z_O$. *See also* TRANSMISSION LINE.

CHARGE

Charge is an electrostatic quantity, measured as a surplus or deficiency of electrons on a given object. When there is an excess of electrons, the charge is called negative. When there is a shortage of electrons, the charge is called positive. These choices are purely arbitrary, and do not represent any special qualities of electrons.

Charge is measured in units called coulombs (*see* COULOMB). A coulomb is the charge contained in 6.281×10^{18} electrons. Charge is usually represented by the letter Q in equations. The smallest possible unit of electrostatic charge is the amount of charge contained in one electron.

The quantity of charge per unit length, area, or volume is called the charge density on a conductor, surface, or object. Charge density is measured in coulombs per meter, coulombs per square meter, or coulombs per cubic meter. A charge may be carried or retained by an electron, proton, positron, antiproton, atomic nucleus, or ion. *See also* ELECTRON, IONIZATION.

CHARGE-COUPLED DEVICE

A charge-coupled device (CCD) is a dynamic shift register capable of storing digital data and analog signal processing. The two-phase CCD shown in the three time-phased sequences in the diagram consists of a P-type silicon substrate with a thin N-type silicon layer covered with an insulating oxide layer. A series of closely spaced metal gate electrodes is deposited on the surface of the insulating layer. The diagram shows two parallel lines with voltage waveforms (phase 1 and phase 2) connected to alternate gate electrodes. A deep potential well that attracts electrons is formed under each gate when a high voltage is applied, and it disappears in the next part of the cycle when the voltage is low.

In diagram A (t = 0 cycle), the clock voltage on phase line 1 and gate 2 is high while the voltage on phase line 2 and gate 1 is low. A charge packet of electrons is in the potential well formed under gate 2 because of the high clock voltage.

In diagram B (t = ½ cycle), the voltage on phase line 1 and gate 2 has gone low causing the potential well under gate 2 to collapse. At the same time, the voltage on phase line 2 and gate 3 is high. This has created a new potential well under gate 3 and has shifted the electron charge packet to the right.

In diagram C (t = 1 cycle), the clock voltage on phase line 1 and gate 4 has gone high again so the potential well under electrode 3 has collapsed and a new one is formed under gate 4. The electron packet now steps to the right again to the new well under electrode 4. As a result of alternately raising and lowering the voltage on the gates, the charge packets beneath them can be passed along more than 1000 times with no substantial loss. Each charge packet can represent an analog value, or the CCD can function as a digital shift register with each charge packet representing a serial code. Diode gate structures put information in and take it out of the CCD register.

CCDs can act as digital memory because they are basically shift registers. They are classed as dynamic memories because they require periodic refreshing, and they are volatile. CCD digital memory is block-access oriented rather than random-bit accessed because of its shift-register characteristics.

The CCD can also function as an analog shift register and perform analog signal processing. CCD analog signal processor applications include their use as video and audio delay lines and as communications and secure communications filters.

Linear imaging devices (LIDs) are formed by pairing linear arrays of photodetectors with CCDs that are used as transport registers. Free electron packets are generated on individual silicon photodetectors in direct proportion to the incident radiation on that chip. If the incident pattern is a focused light image from optics viewing a scene, the charge packets created in the detector array will faithfully reproduce the scene projected on its surface.

After an appropriate exposure time, the charge packets in each photodetector of the array can be simultaneously transferred by charge coupling to a parallel CCD analog transfer shift register. The transfer is carried out by a single, long transfer-gate electrode between the line of photodetectors and the transport register. Each charge packet corresponds to a picture element (pixel). The CCD analog transport register is then rapidly clocked to deliver the picture information, in serial format, to the device output circuit. LIDs sense and deliver information a line at a time and are electronically scanned in one dimension. They are also called line-scan devices.

LIDs are commercially available with as many as 1728 elements. LID applications include their use in high-speed, high-resolution facsimile machines for copying text, maps, drawings, and photographs. They are also used in bar-code reading and the sorting of mail, currency, and merchandise.

Area imaging devices (AIDs) are two-dimensional X–Y arrays of photodetecting elements capable of sensing an area image. They have both vertical and horizontal transfer gates and transport registers. An entire field of video information can be delivered after each exposure period as a series of lines of video signal. Typical sizes of AIDs for television cameras are arrays of 400 by 400 or 488 by 380 elements.

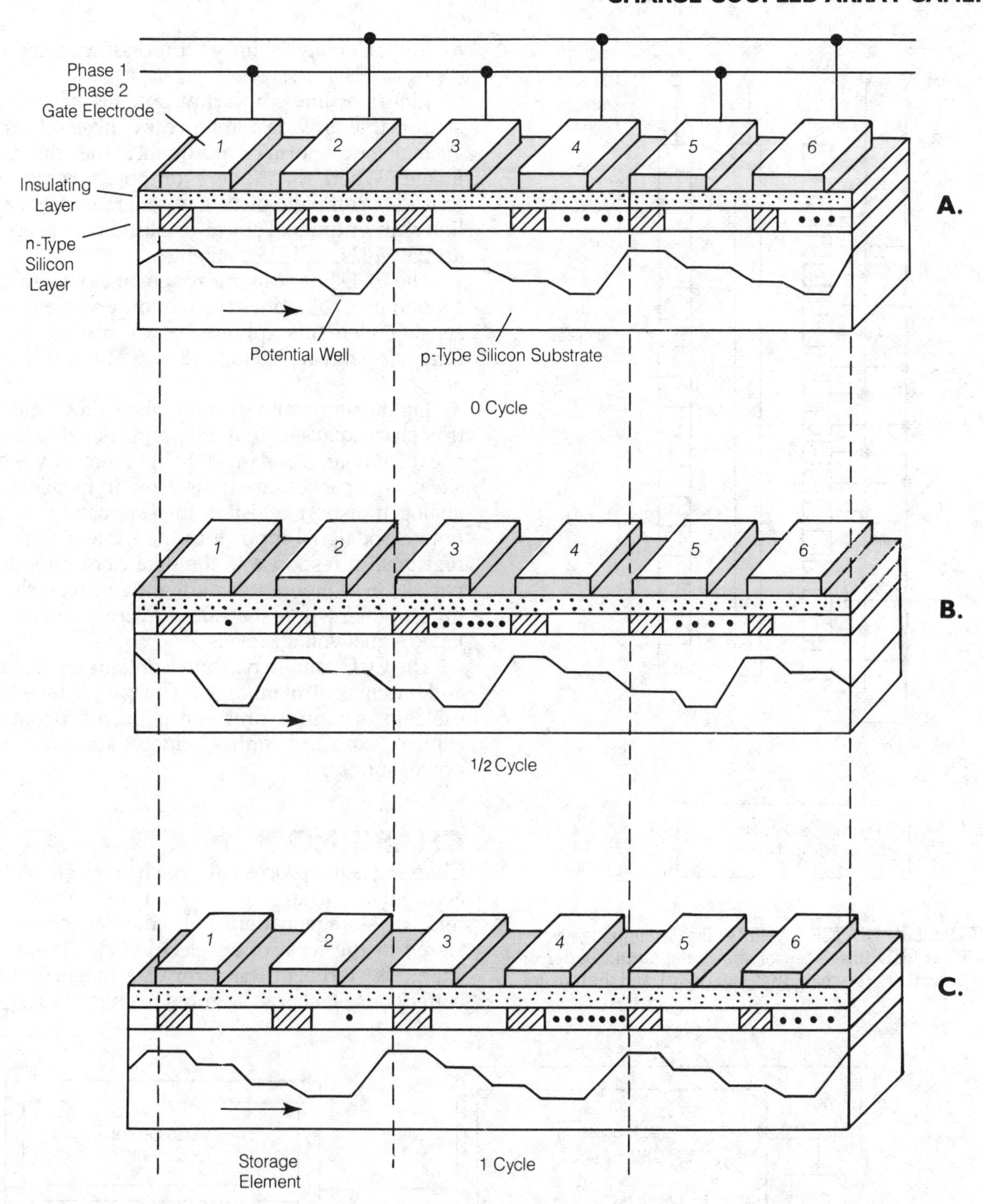

CHARGE-COUPLED DEVICE: Three two-phase CCDs are shown in a sequence of three time cycles. The internal clock voltage on the electrodes cause the electrons to step to the right in packets through the N-type semiconductor layer.

CHARGE-COUPLED ARRAY CAMERA

Charge-coupled device (CCD) cameras contain arrays of evenly spaced, solid-state photosensitive elements as shown in Fig. 1. Light falling on these elements creates a charge that is scanned out sequentially cell by cell. The camera output is a sequence of voltage levels representing individual cell brightness. CCD cameras are available in resolutions ranging from 40 by 40 to 512 by 512 and higher. CCD cameras also are available as line-scan versions with even higher numbers of line sensors. Figure 2 is a subsystem block diagram showing the major functional units of a CCD camera. The output can be interlaced for compatibility with conventional video equipment.

CCD cameras are used in precision, noncontact industrial optical inspection of objects and optical data acquisition. The application includes measurement of position, size, and shape of objects. The cameras can also be used for detection and categorization of defects in objects as well as the sorting of objects for size, shape, and color.

Some CCD television cameras provide standard television output signals for display of images on monitors or

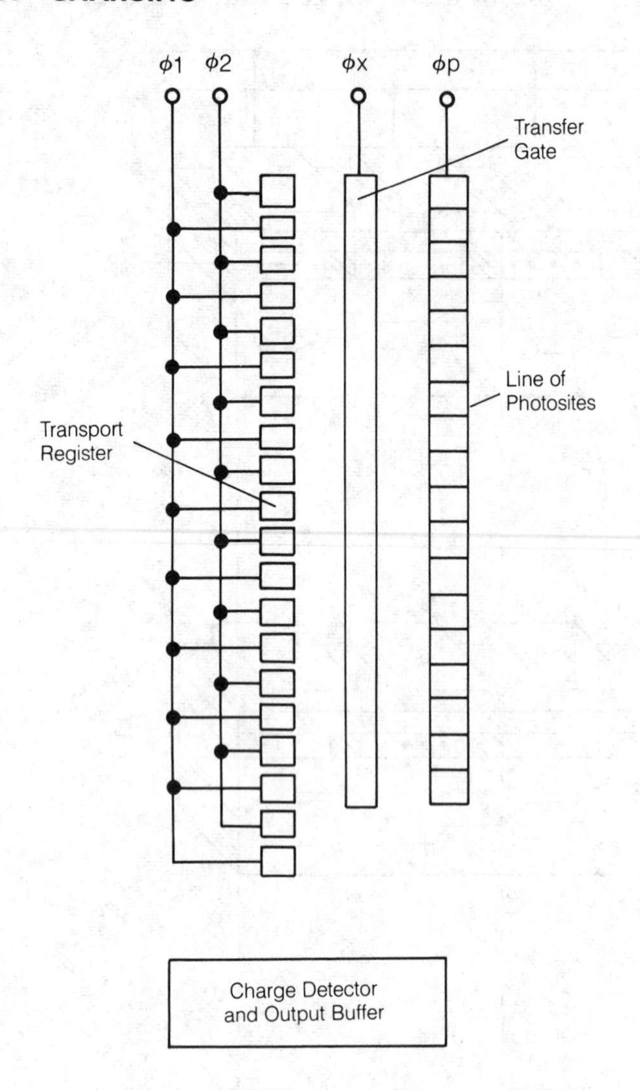

CHARGE-COUPLED DEVICE CAMERA: Fig. 1. Simplified schematic of a linear CCD image sensor shows the output buffer or transport register (left), the charge detector (right), and the transfer gate (center).

for digital analysis image processing equipment. *See* COMPUTER VISION, TELEVISION.

Linear or line-scan array cameras have CCD linear sensors that provide a single row of pixels. They produce an analog waveform proportional to the brightness of the image. When the camera or object being sensed is moved, a complete picture is generated from a series of line-scan outputs. Typical resolutions of these cameras are 256, 1024, or 1728 elements.

The CCD line scan image sensor as shown in Fig. 2 is a monolithic component containing a single row of image sensing elements (photosites or pixels), two analog transport shift registers, and two output sense amplifiers.

Light energy falling on the photosites generates electron charge packets that are proportional to the product of exposure time and incident light intensity. The photosite charge packets are transferred in parallel to the two analog transport registers in response to an exposure time clock signal input into the camera. The transport registers, in response to the data clock rate, deliver the packets in sequence to an integrated circuit charge sensing amplifier where they are converted into proportional video signal voltage levels.

The CCD camera typically contains signal processing and timing control modules. The camera is connected by cable to a control unit consisting of a video output control, exposure control, video data rate control and power supply.

CHARGING

Charging is the process by which an electrostatic charge accumulates. Charging can occur with capacitors, inductors, or storage batteries. In the capacitor, charge is stored in the form of an electric field. In an inductor, charge is stored in the form of a magnetic field. In a battery or cell, charge is stored in chemical form.

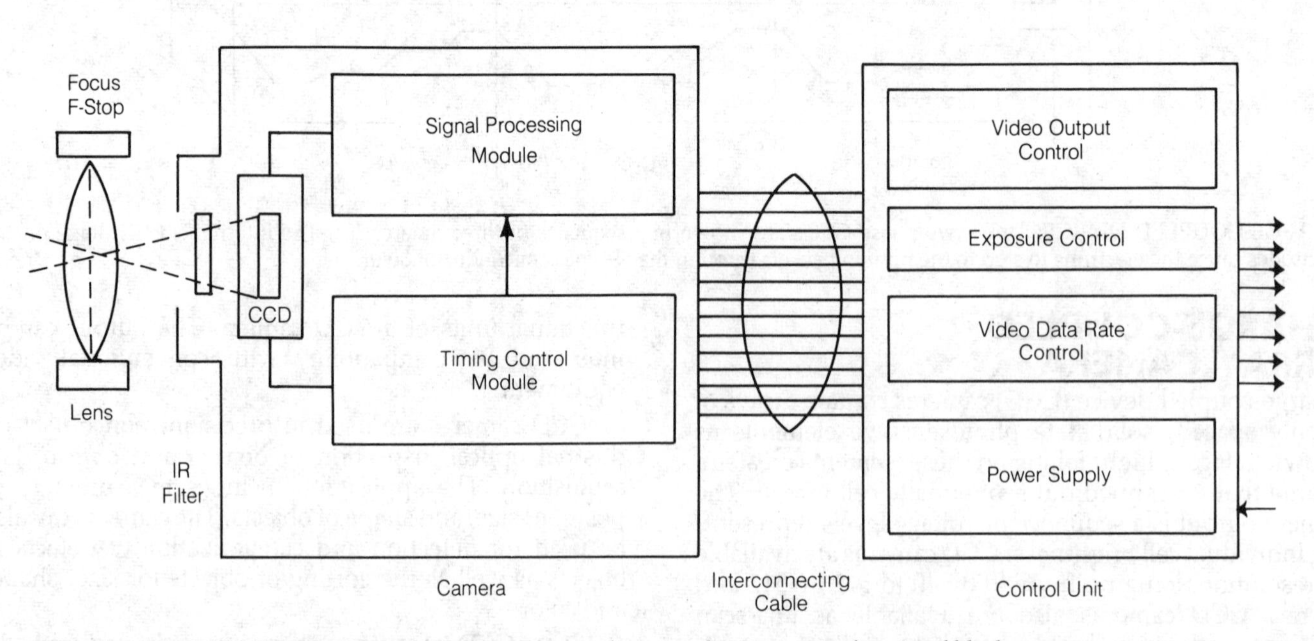

CHARGE-COUPLED DEVICE CAMERA: Fig. 2. CCD camera subsystem showing major functional blocks.

The rate of charging is measured in coulombs per second, which represents a certain current in amperes. This charging current can be measured with an ammeter. In a storage battery, charging occurs rapidly at first, and then the rate slows as the storage capacity is reached. The same is true of capacitors and inductors. The charging current of a battery must be maintained at the proper levels for periods as long as hours to ensure optimum charging. *See also* BATTERY.

CHATTER

When the voltage to a relay alters or is not quite sufficient to keep the contacts closed continuously, the contacts may open and close intermittently. This can occur at a high rate of speed, producing a characteristic sound that is called chatter.

Chatter can be deliberately produced by connecting the relay contacts in series with the coil power supply so that oscillation is produced. As soon as the coil receives voltage, the contacts open and the voltage is interrupted. A spring then returns the contacts to the closed condition, and the cycle repeats itself. A resistance-capacitance combination determines the frequency of the oscillation. By varying the oscillation weight and speed, a simple buzzer can be constructed.

Generally, chatter is an undesirable effect, and can totally disable a piece of electronic equipment. Solid-state switching is preferable to relay switching in many modern applications for this reason. Transistors and diodes do not chatter. *See also* RELAY.

CHEBYSHEV FILTER

A Chebyshev (also spelled *Tschebyscheff, Tschebyshev*) filter is a special type of selective filter with a nearly flat response within its passband, nearly complete attenuation outside the passband, and a sharp cutoff response. The extremely steep skirt selectivity of the Chebyshev filter is its primary advantage over other filters. The Chebyshev filter, similar to the Butterworth filter (*see* BUTTERWORTH FILTER) can be designed for lowpass, highpass, and band-rejection applications, as well as for bandpass use.

Figure 1 shows the Chebyshev responses for lowpass, highpass, and bandpass filters. The ripple shown is usually of little or no consequence. Figure 2 shows schematic diagrams for simple Chebyshev lowpass, highpass, and bandpass filters. The input and load impedances must be properly chosen for the best filter response. The values of the inductors and capacitors depend on the input and load impedances, as well as the frequency response desired. *See also* BANDPASS FILTER, BAND-REJECTION FILTER, HIGHPASS FILTER, LOWPASS FILTER.

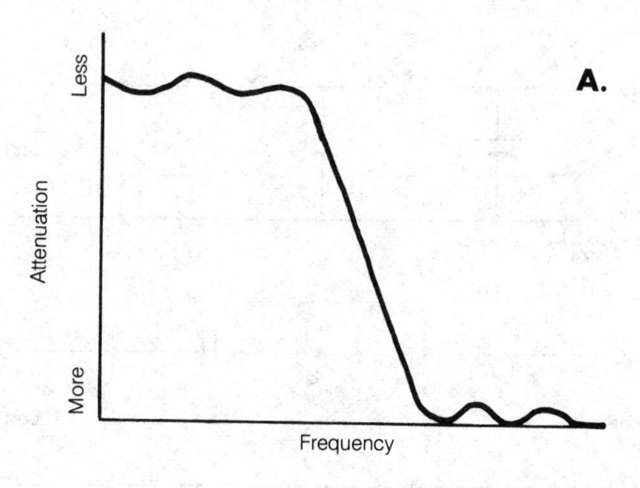

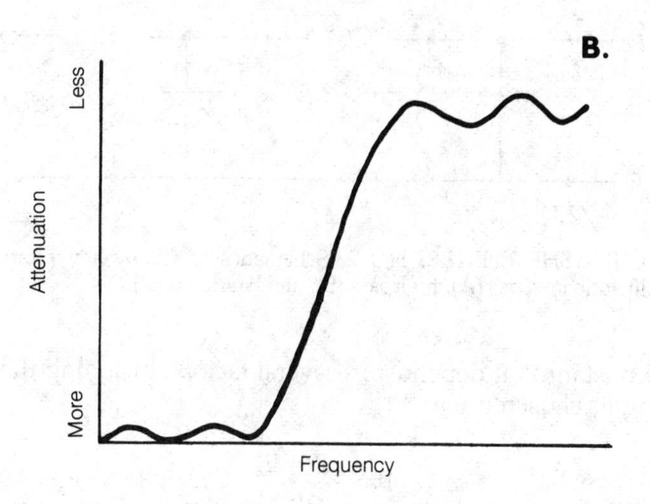

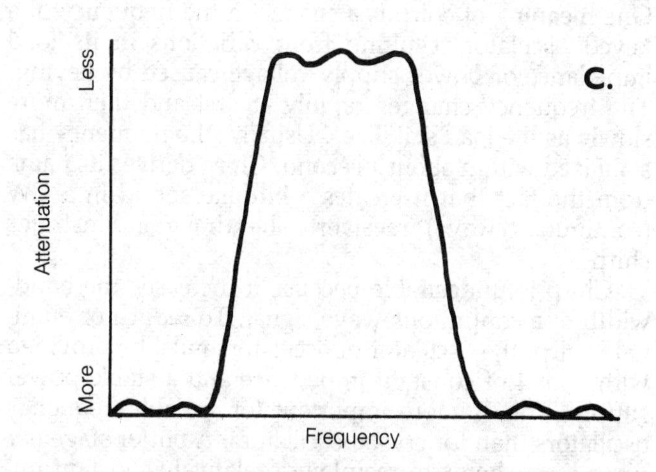

CHEBYSHEV FILTER: Fig. 1. Curves for responses of Chebyshev filters include lowpass (A), highpass (B), and bandpass (C).

CHILD'S LAW

In a diode vacuum tube, the current varies with the 3/2 power (the square root of the cube) of the voltage, and inversely with the square of the distance between the electrodes. This relationship is called Child's Law. Mathematically, if E is the voltage between the cathode and the plate, I is the current through the device, d is the distance between the cathode and the plate, and k is a

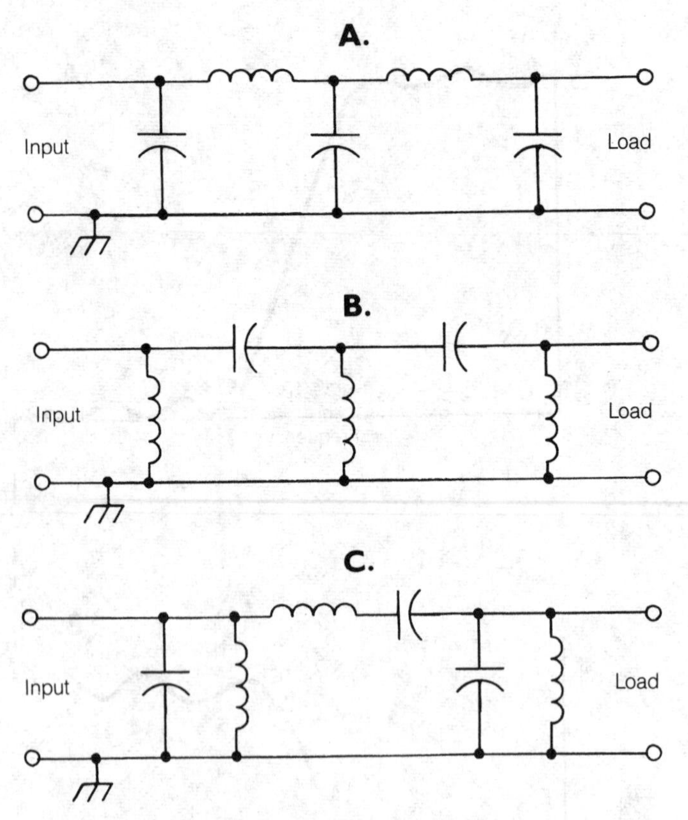

CHEBYSHEV FILTER: Fig. 2. Schematics of Chebyshev filters include lowpass (A), highpass (B), and bandpass (C).

constant that depends on several factors, including the units chosen, then:

$$I = \frac{kZ\sqrt{E^3}}{d^2}$$

See also TUBE.

CHIRP

One meaning of chirp is a change in the frequency of a keyed oscillator resulting from variations in its load impedance or power supply voltage caused by keying. The frequency changes rapidly at first and then more slowly as the load stabilizes. Usually, the frequency has stabilized within about 1 second. Chirp derives its name from the fact that it creates a birdlike sound in a CW (continuous wave) receiver. The drawing illustrates chirp.

Chirp is undesirable because it increases the bandwidth of a continuous-wave signal. To reduce or eliminate chirp, the oscillator or oscillators must be provided with a load of constant impedance and a stable power supply. This is more important for variable-frequency oscillators than for crystal oscillators. A buffer stage (*see* BUFFER STAGE) helps to maintain a relatively constant impedance at the output of an oscillator. Another means of reducing chirp is to key an amplification stage of a transmitter, rather than the oscillator itself.

The term *chirp* is used in radar to refer to the widening or narrowing of pulses for the purpose of improving the signal-to-noise ratio. Narrow pulses are deliberately expanded at the transmitter, and reduced again as the signal is received. This results in a greater duty cycle for the transmitter and greater average power output. It does not affect the range or resolution of the radar, but it improves the ability of the system to detect small targets. *See also* RADAR.

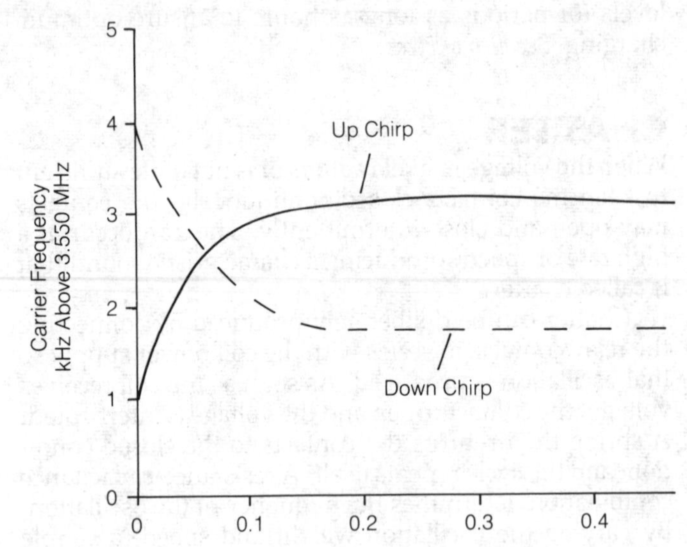

CHIRP: Chirps are diagrammed as fast changes in signal frequency.

CHOKE

An inductor, used for passing direct current while blocking an ac (alternating-current) signal, is called a choke. Typically, an inductor shows no reactance at dc (direct current) and increasing reactance at progressively higher ac frequencies. (See graph.) Some chokes are designed for cutting off only radio frequencies; others are designed to cut off audio as well as radio frequencies. Some chokes are designed to cut off essentially all alternating currents, even the 60-Hz line current, and are used as power-supply filtering components. Together with large capacitances, these chokes make extremely effective power-supply filters.

Chokes can have air cores if they are intended for radio frequencies. Larger chokes use cores of powdered

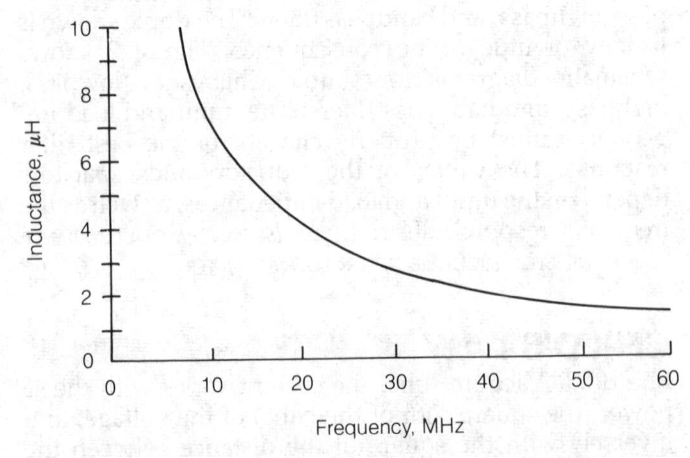

CHOKE: An inductance-versus-frequency plot for a 500-Ω choke.

iron or ferrite. This increases the inductance of the winding for a given number of turns. Some chokes are wound on toroidal, or doughnut-shaped, cores to provide a large increase in the inductance. Others are wound on cores shaped like the core of an ac power transformer so they can handle large amounts of current.

Chokes are useful when it is necessary to maintain a certain dc bias without short-circuiting the desires signal. *See also* INDUCTOR.

CHOKE-INPUT FILTER

A power-supply filter eliminates ripple from the direct-current output of a rectifier circuit. A capacitor may be the only filtering component. However, the addition of a large value of inductance, connected in series with the ungrounded side of the supply output, improves the quality of the output by further reducing the ripple (*see* RIPPLE). The drawing illustrates a choke-input power-supply filter.

The choke, *L*, is placed on the rectifier side of the filter capacitors. Capacitor C1 provides ripple suppression, and is a large-value electrolytic unit. A smaller Mylar or disk-ceramic capacitor, C2, effectively prevents radio-frequency energy from entering the supply through the power leads.

Choke-input filters are bulkier and heavier than simple capacitor filters, especially when the supply must deliver a large amount of current at a high voltage. However, the choke-input filter improves the supply regulation and reduces the amount of ripple reaching the load. This is particularly desirable when the load draws a large amount of current, or when the most ripple-free direct current is needed. Sometimes, two or even three stages of choke/capacitor filtering are used. *See also* CHOKE, POWER SUPPLY.

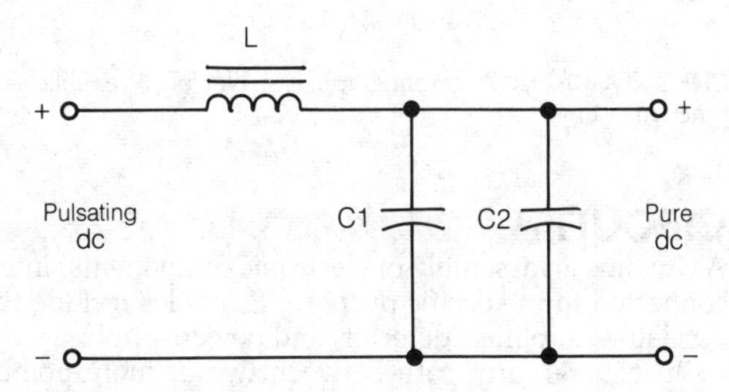

CHOKE-INPUT FILTER: The choke-input filter provides better voltage regulation than a single capacitor.

CHOPPER

A chopper is a circuit that interrupts a direct current, producing pulses of constant amplitude and frequency. A chopper consists of a low-frequency oscillator, which opens and closes, a high-current switching transistor. An example of a chopper circuit is shown in the illustration.

A direct-current voltage can be passed through a chopper. The resulting waveform can then be amplified or attenuated, and finally filtered to obtain a direct-current voltage that differs from the input. Choppers can also be used as rectifiers; they can eliminate or invert one half of an alternating-current cycle.

A rotating, perforated disk or shutter, used to modulate a beam of light or a stream of atomic particles, is also called a chopper.

A chopper can be used to obtain a high-voltage source of alternating current from a battery. This allows the operation of some household appliances from an automotive electrical system. *See also* CHOPPER POWER SUPPLY, VIBRATOR.

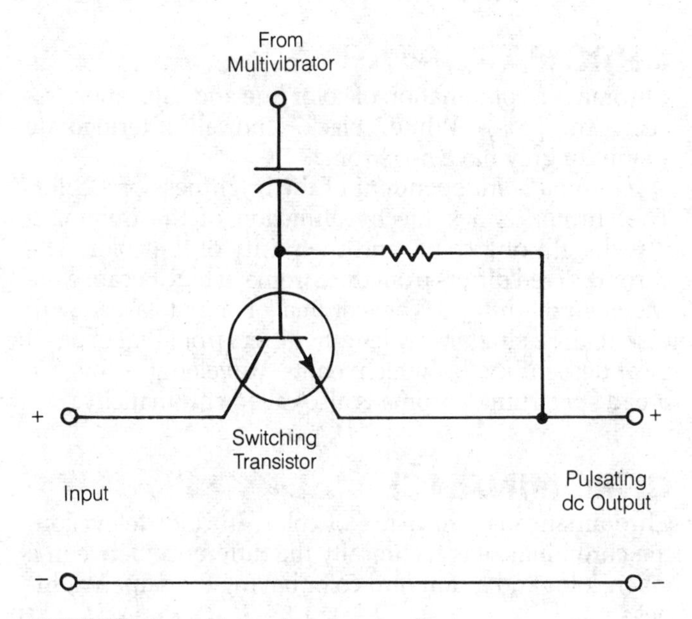

CHOPPER: A transistor chopping circuit.

CHOPPER POWER SUPPLY

A circuit for obtaining a high-voltage source of alternating current from a direct-current source is called a chopper power supply. These devices were once used in the operation of mobile or portable equipment containing vacuum tubes. Chopper power supplies also permit the use of certain standard household appliances in boats and automobiles.

A block diagram of a chopper power supply is shown in the illustration. The chopper interrupts the direct-current power source at regular intervals, and the amplifier then increases the pulse amplitude. The transformer converts the pulsating direct current to alternating current. This supply produces power similar to that found in household utility outlets.

The output of a chopper power supply is not a good sine wave. For this reason, many chopper power supplies are unsuitable for use with electric clocks and other appliances that need a nearly perfect, 60 Hz sine-wave supply.

If the transformer shown is replaced by a filtering network, a direct-current transformer or amplifier can be obtained. This transformer can provide hundreds of volts from a simple 12 V, direct-current electrical system. *See also* CHOPPER, DC-TO DC CONVERTER.

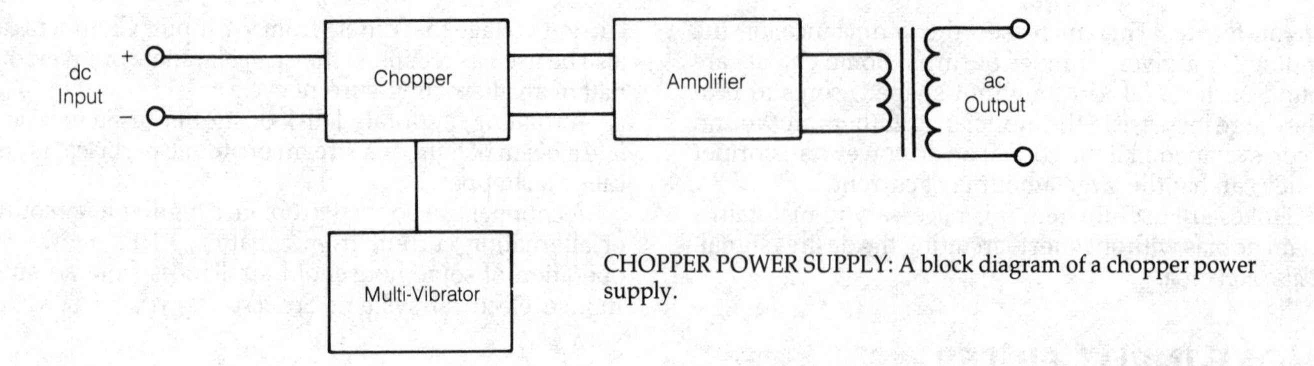

CHOPPER POWER SUPPLY: A block diagram of a chopper power supply.

CHROMA

Chroma is a combination of color hue and saturation (*see* HUE, SATURATION). White, black, and all intermediate shades of gray have no chroma.

Chroma is independent of the brightness of a color. The chroma is a subjective function of the dominant wavelength of a color, and the purity of the color. The chroma of red differs from the chroma of blue because the wavelength differs. The chroma of a red laser, with essentially a single wavelength, differs from that of a red incandescent bulb, which emits wavelengths over a broad spectrum. Chroma is also called chromaticity.

CHROMINANCE

Chrominance is a measure of color. In color television, the chrominance is technically the difference between a given color and a standard color having the same brightness.

Chrominance primaries are the three colors that, in various combinations, can produce any color. The three chrominance primaries are red, green, and blue. Red and green combine to form yellow; red and blue combine to form yellow; red and blue combine to form magenta; green and blue combine to form cyan, or blue-green; and red, green, and blue in equal proportions combine to form white.

The red, green and blue chrominance components of a color television signal are transmitted independently. When they are combined at the receiver, a full-color picture results. A crystal-controlled oscillator generates a subcarrier at 3.57945 MHz, plus or minus 10 Hz, for the chrominance detector. *See also* COLOR TELEVISION.

CHRONOGRAPH

A chronograph is a plot of any quantity as a function of time. A machine that records a quantity as a function of time is called a chronograph or chronograph recorder.

Chronographs are frequently used in electronics. The familiar sine-wave display is an example of a chronograph. Plots of temperature-versus-time, amplitude-versus-time, and other functions are frequently used. The drawing illustrates a chronograph of amplitude versus time for a sine-wave amplitude-modulated signal.

A chronograph is characterized by an accurate, and usually linear, time display along the horizontal axis. This allows precise determination of the period of a function. In the illustration, the period of the modulating waveform is 1 millisecond.

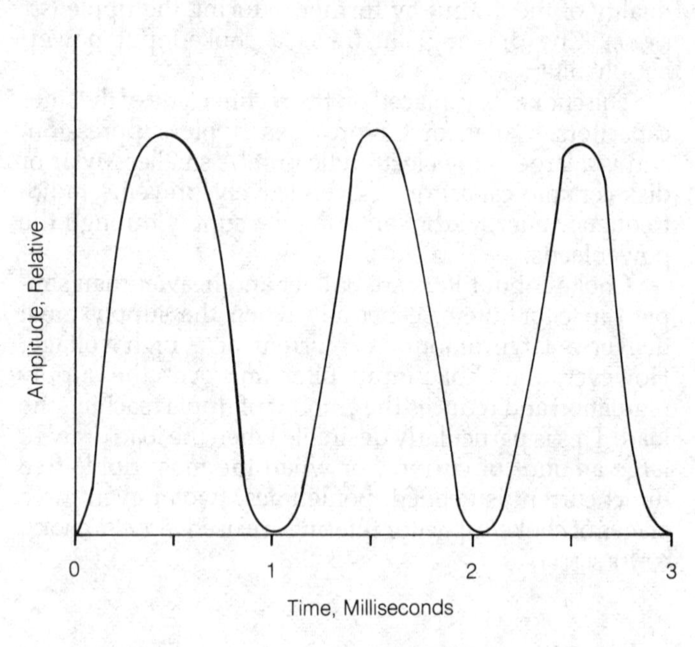

CHRONOGRAPH: A chronograph is a plot of a variable as a function of time.

CIRCUIT

A circuit is an assembly of electronic components, interconnected for a specific purpose. Examples include the oscillator, amplifier, detector, and power supply.

In general, any path between two or more points, capable of signal transfer, is a circuit. A telephone line, and the terminals at either end, is a circuit; so is any electromagnetic path in a communications system.

Any electronic circuit can be represented by a schematic or block diagram. *See also* BLOCK DIAGRAM, SCHEMATIC DIAGRAM.

CIRCUIT BOARD OR CARD

Circuit boards or cards are substrates, typically made from phenolic-impregnated paper or epoxy-impregnated glass cloth, for the insertion of components to make functional circuit modules. The most widely used circuit boards are printed circuit boards (PCBs) or cards, also known as printed wiring (PW) boards or cards. Both of these expressions are misnomers because only conductors, not circuits, are formed on the substrate by photolithography and plating rather than printing conductive ink. (See illustration.)

Most PC (PCB) boards or cards are manufactured today by the subtractive process—selectively removing copper from foil laminated to the substrate to leave the desired patterns of conductive paths and pads. Alternatives to the PCB include automatically wired panels, wire wrapping panels, and backplanes or backpanels. PC boards are classified by the number of signal layers. Single-sided PCBs have conductors on only one side; double-sided PCBs have conductors on both sides. Multilayer PCBs have internal conductors (which can be power or ground planes) as well as conductors on both sides. Regardless of the number of layers, PCBs are about 0.0625 inch thick.

Automatically wired circuit boards are made by a proprietary process for connecting specified points on a bare laminated board with fine insulated wires using a computer-controlled X-Y wire placement machine. The wires, which are permitted to cross, are bonded to surface of the laminate with adhesives.

Wire-wrapping panels are laminated boards with arrays of vertical posts inserted at specific points to permit interconnection by wrapping wire around them manually or by machine. These panels are used in equipment subject to field wiring changes, in circuit development or in small-lot prototype production.

PCB Fabrication. Most PCBs today are made by the subtractive process in which copper foil laminated to the substrate is etched away leaving the desired conductor pattern. There are many variations in the fabrication process depending conductor width and complexity. Typically, a pattern or mask for the removal of the copper is produced by photographic reduction of a large-scale drawing (perhaps enlarged 10 times the original size). Simple masks can be drawn or prepared manually, but complex masks are produced on a computer-aided design workstation and an X-Y plotter. The actual mask, like a photographic negative, has opaque and transparent areas outlining the conductor pattern.

A light-sensitive liquid or dry film called photoresist is applied to the copper laminate, and the mask is placed over it and exposed to ultraviolet light. Light passing through the transparent regions of the mask hardens the photoresist so the unexposed photoresist can be removed by a chemical solvent. The exposed bare copper is then etched away to leave the bare substrate. The hardened photoresist over the conductors is then removed in another process, leaving the copper conductors exposed. These surfaces can be lead-tin plated or dipped in lead-tin solder to improve their solderability. Multilayer boards are made by completing individual layers and then stacking and bonding them together under heat and pressure.

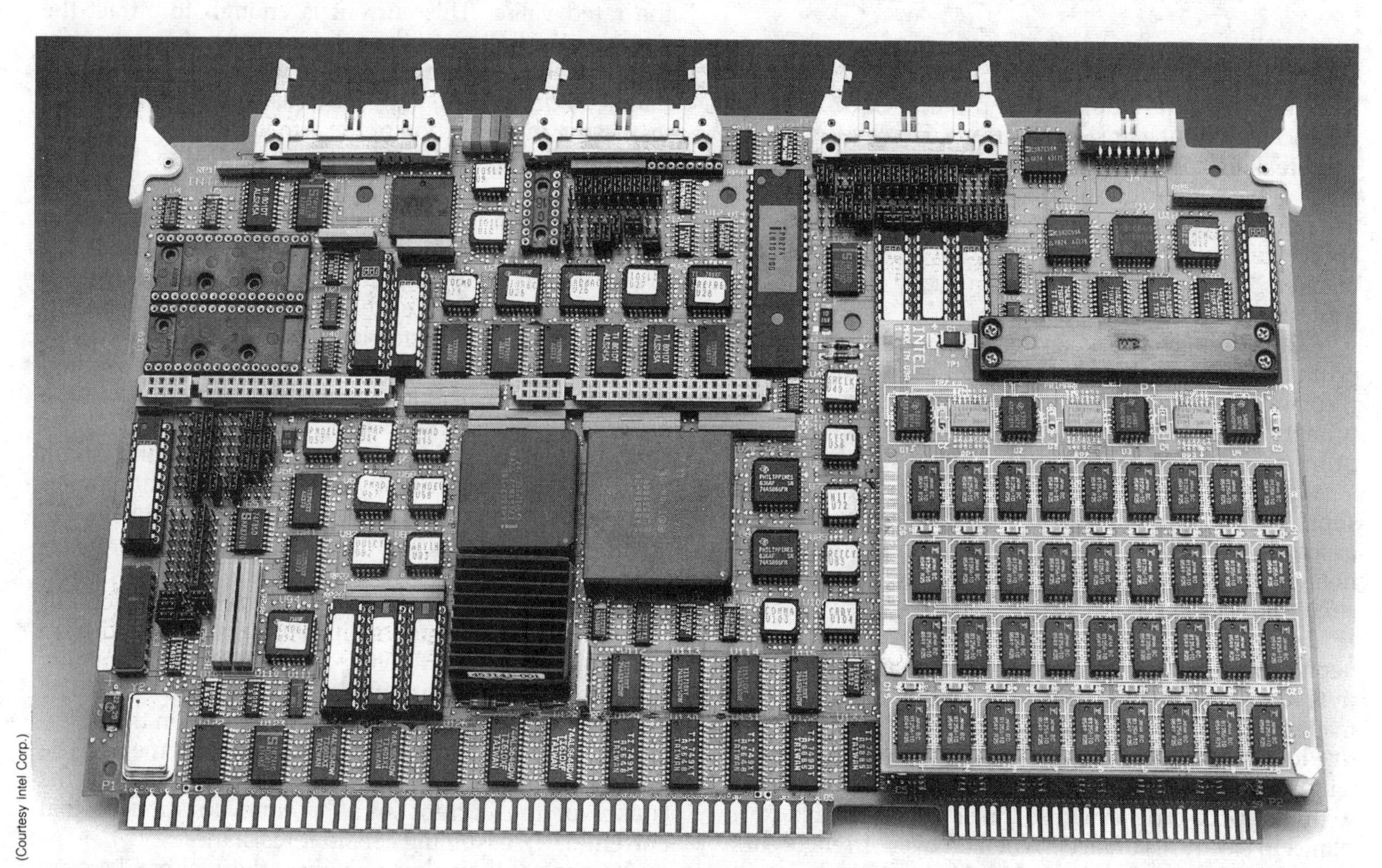

(Courtesy Intel Corp.)

CIRCUIT BOARD: A single-board computer is an example of an advanced circuit board.

In the alternative additive process, the conductive lines and pads are plated on the bare surface of the substrate that has been chemically treated to accept copper. In later steps, the thickness of the initial deposition can be increased with copper plating. As in the subtractive process, the conductors can be lead-tin plated or solder-dipped to improve solderability.

Plated-through Holes. Plated-through holes were an important advancement in the fabrication of double-sided and multilayer PCBs. They eliminate the need to insert metal eylets for vertical conductor interconnection—a costly, time-consuming, and unreliable process. The plated-through holes can connect internal conductors to external conductors in any combination.

Holes are formed or drilled at various points through the PCB where component leads are to be inserted. The inside walls of the holes are coated with a chemical, which causes copper to adhere to it. The thin initial copper sleeves are reinforced with more copper plating before they are lead-tin plated. Component leads are inserted in the plated-through holes which act as sockets. During wave soldering, molten solder is drawn up around the leads by capillary action to form a secure bond.

The surface-mount assembly process and leadless components have eliminated the need for costly plated-through holes. The leadless components can be soldered to the surface conductors and pads of the holeless surface-mount PCB. *See* SURFACE-MOUNT TECHNOLOGY.

CIRCUIT BREAKER

A circuit breaker is a device that protects electronic and electrical circuits against overcurrent. However, the circuit breakers specified for electronics are smaller, lower rated devices than the primary ac (alternating current) line circuit breakers for building service equipment and appliance protection. As a result, the secondary circuit breakers for electronics use are usually called circuit protectors.

Circuit protectors for electronics can be in series with primary circuit breakers in the power lines, but they can respond faster to lower overcurrent in the line than the primary circuit breakers. However, circuit protectors are primarily intended to respond to local conditions within a cabinet or enclosure such as a faulty component or short circuit. The current limiting protectors that are based on circuit breaker designs are instantaneous-trip or time-delay magnetic devices. There are also thermal circuit protectors.

Instantaneous Magnetic Circuit Protectors. Instantaneous magnetic circuit protectors are designed so that their contacts open in the presence of the increased magnetic field caused by an overcurrent. They are sensitive to current, not temperature. The protective element within an electromagnetic circuit protector is a solenoid with a clapper-type armature linked by collapsible couplings to electrical contacts. When an overcurrent occurs (due to a short circuit or component failure), the solenoid is actuated, opening the contacts. The increased magnetic field attracts a hinged armature similar to that found on an electromagnetic relay. *See* RELAY.

Unlike the relay, the protector armature does not directly open the electrical contacts. Instead it collapses a train of linkage within the protector case, permitting fast-acting springs to snap open (in as fast as 2 milliseconds) the contacts to avoid arcing or sticking. When the overcurrent condition is cleared, the linkage can be reset manually with a toggle or lever. These protective devices can be dual-purpose components combining the functions of a toggle switch and a protective device. In most products, the contacts will remain open when there is an overload or fault, even if the toggle is manually held in the on position.

Time-delay Magnetic-hydraulic Circuit Protectors. Time-delay protectors include a hydraulic time-delay dashpot as shown in the figure. They are otherwise similar to instantaneous magnetic circuit breakers. The current flowing through the windings of a coil creates a flux, attracting an armature to the cylinder cap and snapping open the protector contacts. The protector coil is wound around a hollow dashpot cylinder that houses a piston. The dashpot slows the response of the protector by allowing the contacts to remain closed during nondestructive overcurrents to minimize nuisance tripping. The time delay of the dashpot can be set to meet specific circuit protection requirements.

For example, the time delay might be initiated only when the current is between 100 and 125 percent above the rated value. This current is enough to attract the piston at the bottom of the cylinder and pull it upwards. As the piston moves up, it adds to the permeability of the magnetic circuit, causing an increase in flux density and increasing the attraction of the armature.

The time it takes for the piston to travel to the top of the cylinder is determined in part by the viscosity of a silicone fluid in the sealed cylinder. When the cylinder reaches the top, the armature will be tripped. If the overcurrent is removed before the piston reaches the top of the cylinder, the breaker will not trip, and a return spring will push the piston back to the bottom. However, if the current surge instantly exceeds six times the rating, the time delay dashpot is overridden, and the contacts are opened instantaneously.

For all practical purposes, magnetic-hydraulic circuit protectors are unaffected by changes in ambient temperature, although changes in viscosity of the fluid in the dashpot could have a slight effect on the time delay.

Thermal Protectors. Thermal circuit-breaking protective devices, like fuses, are heat-sensitive devices. A bimetallic blade contact opens as a result of head from excess load current flowing through it. The opening releases a latch, causing the contact to snap open. The bending of the bimetallic element depends on the difference in coefficient of expansion of the contact metals. The heat produced is directly related to the product of the square of the current and the time duration of that current, (I^2t), assuming constant resistance.

Because of their slow response, thermal breakers are generally used to protect wiring from overheating and

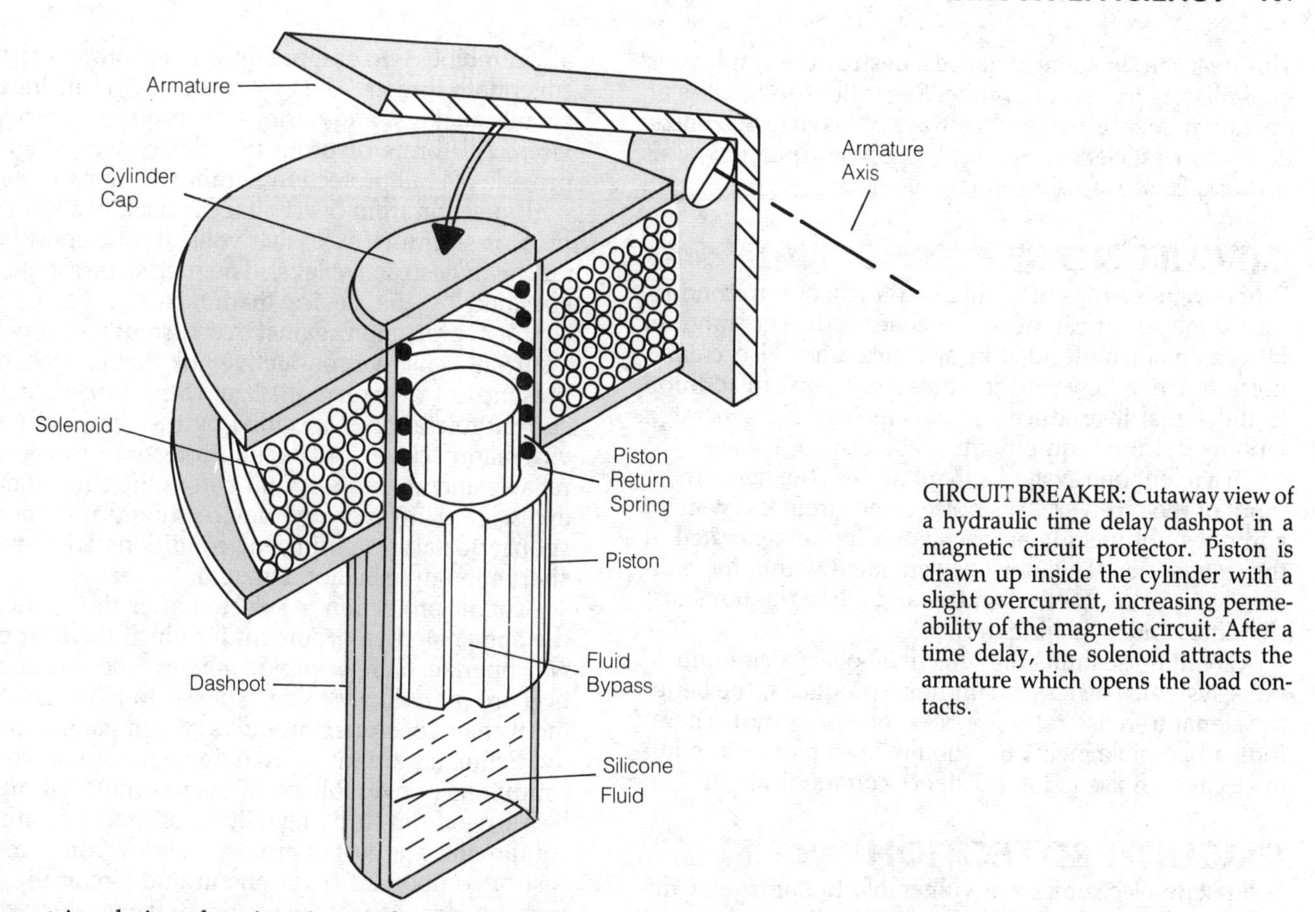

CIRCUIT BREAKER: Cutaway view of a hydraulic time delay dashpot in a magnetic circuit protector. Piston is drawn up inside the cylinder with a slight overcurrent, increasing permeability of the magnetic circuit. After a time delay, the solenoid attracts the armature which opens the load contacts.

subsequent insulation deterioration and not to protect semiconductor devices. But some thermal types, like all the magnetic-hydraulic types, can be used as power switches, thus saving the expense of a separate switch.

There are two kinds of thermal circuit protectors: standard and positive pressure. The bimetallic element of the standard unit bends upward, causing the pressure between the contact surfaces to decrease, until the blade finally snaps open and separates the contacts. With this design, the heat rise due to overcurrent can be excessive before blade opening due to a decrease in contact pressure. This could lead to early failure or even contact welding.

The positive pressure blade is designed so that contact pressure increases with overcurrent heat. When the critical point is reached, the blade snaps open instantly, and the arc is extinguished. Thermal circuit protectors, like magnetic-hydraulic protectors, offer circuit resistance that helps to limit circuit fault current.

Some thermal protectors, like magnetic-hydraulic protectors, can be manually reset. However, the element cannot be reset until the bimetallic element cools sufficiently to be reset. The thermal protector has a slower response time than the magnetic protector and there is no provision for a time delay. Thus it is more prone to false triggering. Moreover, its established set point can change with changes in ambient temperature.

CIRCUIT CAPACITY

In a communications system, the number of channels that can be accommodated simultaneously without overload or mutual interference is called the circuit capacity. For example, in a 100 kHz radio-frequency band, with amplitude-modulated signals that occupy 10 kHz of spectrum space apiece, the circuit capacity is 10 channels. It is impossible to increase the number of channels without sacrificing efficiency.

In an electrical system, the circuit capacity is usually specified as the number of amperes that the power supply can deliver without overloading. A typical household branch circuit has a capacity of 10 to 30 A. A fuse or circuit breaker prevents overload. *See also* CIRCUIT BREAKER, FUSE.

CIRCUIT DIAGRAM

See SCHEMATIC DIAGRAM.

CIRCUIT EFFICIENCY

Circuit efficiency is the proportion of the power in a circuit that does the work required of that circuit. Efficiency is expressed as a percentage. In an amplifier or oscillator, the efficiency is the ratio of the output power to the input power. For example, if an amplifier has a power input of 100 watts (W) and a power output of 50 W, its efficiency is 50/100, or 50 percent.

The circuit efficiency of an amplifier depends on the class of operation. Some Class A amplifiers have a circuit efficiency as low as 20 to 30 percent. Class B amplifiers usually have an efficiency of about 50 to 60 percent. Class C amplifiers often have an efficiency rating of more than 80 percent.

An inefficient circuit generates more heat for the same amount of input power compared to an efficient one.

This heat can, in some instances, destroy the amplifying or oscillating transistor or tube. Regardless of the class of operation, measures should always be taken to maximize the circuit efficiency. *See also* CLASS A AMPLIFIER, CLASS AB AMPLIFIER, CLASS B AMPLIFIER, CLASS C AMPLIFIER.

CIRCUIT NOISE

Whenever electrons move in a conductor or semiconductor, some electrical noise is generated. The random movement of molecules in any substance also creates noise. A circuit always generates some noise in addition to the signal it produces or transfers. Noise generated within electronic equipment is called circuit noise.

In a telephone system, circuit noise is the noise at the input of the receiver. This noise comes from the system, and does not include any acoustical noise generated at the transmitter. This noise is generated within the electrical circuits of the transmitter and along the transmission lines and switching networks.

Circuit noise limits the sensitivity of any communications system. The less circuit noise produced, the better the signal-to-noise ratio (*see* SIGNAL-TO-NOISE RATIO). Therefore, all possible measures should be employed to minimize circuit noise in long-distance communication.

CIRCUIT PROTECTION

Solid-state electronics are vulnerable to damage or destruction from voltages or currents that exceed the specified ratings of the individual components. Semiconductor devices, particularly transistors and integrated circuits, are also vulnerable to excess temperature beyond the specified limits. Excessive heat causes a breakdown within the transistors.

Individual semiconductors can suffer from insulation breakdown or punch through from overvoltage spikes. Internal conductive paths can literally be burned out by excessive current and the resulting heating effect. Overvoltage can have many different causes: the use of an unregulated or poorly regulated power supply, the failure of one or more components within the circuit changing current or voltage distribution, natural causes such as lightning-induced over-voltage and overcurrent on the power line, or electrostatic discharge (ESD). *See* ELECTROSTATIC DISCHARGE.

Semiconductor devices can also be destroyed by overheating due to inadequate heat radiators or inadequate circulating air that inhibits radiation or convection cooling. The circuitry might be exposed accidentally to environmental extremes beyond the design limits. Inadequate cooling can be corrected by forced-air fan or blower cooling, larger heat sinks, liquid cooling, or other measures. Even sensitive passive components such as capacitors or resistor networks can fail because of overheating, voltage spikes, or current surges.

A semiconductor device can be irreparably damaged in picoseconds with trillionths of a second exposure to overvoltage. Damage from overcurrent typically takes longer: milliseconds or thousandths of a second. Practical circuit-protection plans for sensitive electronics usually combines two or more different protective devices to cover both threats. The protective devices include fuses, circuit breakers, and various overvoltage protective devices. All forms of protective device work together to provide overall protection without mutual interference.

Protection from overvoltage is usually a form of clipping or shunting excessive voltage to ground before it reaches a destructive level. The precise threshold level is specified by the device manufacturer. *See* CLIPPER. In contrast, protection against overcurrent takes the form of a current-sensitive conductive link that is opened in the presence of overcurrent. Again, the threshold level of current tolerance is specified by the semiconductor device manufacturer. All circuit protective devices must be reliable and able to respond before the current becomes excessive. However, the device should not open in response to false or transient conditions with durations short enough to be nondamaging.

Circuit protection is selected after the evaluation of the anticipated environment in which the host product will operate. For example, military-specification products are more likely to be exposed to a hostile environment than consumer products of comparable complexity. Estimates are made by judgement of the worst-case conditions of overvoltage or overcurrent, but much can be learned from studying failures of similar products and equipment. The cost of protection is low compared to the cost of component replacement and circuit repair. The consequences of a loss of service of the product while it is being repaired is another factor. Although many circuits are underprotected, some are overprotected.

Overvoltage Protection. There are three general classes of products designed for protecting semiconductors against overvoltages: (1) silicon transient voltage suppressors (TVS's), (2) metal oxide varistors (MOVs), and surge voltage protectors (SVPs).

Transient Voltage Suppressors. The silicon TVS is a specialized zener avalanche PN junction that protects because of reverse bias voltage clamping. The TVS diode breaks down and becomes a short circuit when the applied voltage exceeds its rated avalanche level. (*See* AVALANCHE.) When the applied voltage (reverse bias) falls below breakdown level, current is restored to its normal level.

Silicon TVSs are optimized for circuit protection. They have better surge handling capabilities than conventional zener diodes, low series-resistance values, and response times measurable in picoseconds. (*See* ZENER DIODE.) While zener diodes are not suitable for circuit protection, TVSs can be used for circuit regulation and clamping as zener diode replacements.

The TVS is useful in dc (direct current) circuits, but protection in ac (alternating current) circuits can be obtained by placing two TVSs together back-to-back. Dual devices are available in a single package. Most recently, surface-mount versions of these varistors have become available.

The three most important characteristics of the TVS are 1) pulse power (peak pulse power multiplied by the clamping voltage), 2) standoff voltage, and 3) maximum clamping voltage. Standoff voltages range from 5 to 170 V, and clamping voltages range from 7 to 210 V.

Metal Oxide Varistors (MOVs). MOVs are variable resistors (or varistors) intended primarily for protection against ac voltage transients because they behave like back-to-back TVSs. MOVs are nonlinear resistors whose resistance value changes are a function of the applied voltage. As a result of their symmetrical, bilateral characteristics, MOVs can provide clamping during both positive and negative swings of the ac waveform. When the voltage exceeds the MOV rating, its resistance drops sharply, and it becomes a short circuit. While the voltage transient is bypassed, the body of the devices is able to absorb any current flowing at the time of the transient without destroying the device. *See* METAL OXIDE VARISTOR.

Spark Gap Voltage Discharge Tubes (SVPS). The SVP is a circuit protective device capable of handling higher voltages than the solid-state suppressors, TVS and MOV. The SVP provides a low-resistance path for successive voltage transient when its voltage ratings are exceeded. Gas within the metal or ceramic tube ionizes during an overvoltage, causing the SVP to change from a nonconducting to a conducting state. The arc formed shorts out the SVP, grounding any high currents as well. After the transient has passed, the gas de-ionizes, and the SVP is reset. The response time of SVPs is generally slower than either TVSs or MOVs.

Overcurrent Protection. Electronic circuits can be protected from overcurrent by appropriate fuses. They can also be protected by light-duty secondary circuit breakers called circuit protectors. These devices are distinguished from the heavy-duty primary circuit protectors used in electrical power circuits by their smaller size and lower current ratings. Circuit protectors have contacts that open in the presence of overcurrent surges. Many of them used for electronics protection can be manually reset. Intended as fuse replacements, there are two fundamentally different forms of circuit protector: electromagnetic and thermal.

Electromagnetic Circuit Protectors. Electromagnetic protectors are designed so that their contacts will open in the presence of the electrical field caused by an overcurrent. One form of circuit protector is the electromagnetic circuit protector. The protective mechanism is a solenoid with a clapper-type armature linked by collapsible couplings to electrical contacts. When an overcurrent occurs (caused by a short circuit or component failure), the solenoid is actuated. The resulting magnetic field attracts a hinged armature similar to that of an electromagnetic relay.

However, unlike the relay, the protector armature closure does not directly open the electrical contacts. Instead it collapses a train of linkage within the protector case, permitting fast-acting springs to snap open the contacts to avoid arcing or sticking. When the overcurrent condition is cleared, the linkage can be reset manually with a toggle or lever. These protective devices can be dual-purpose components able to perform a switch as well as protection.

Electromagnetic protectors for electronics are likely to include an hydraulic time-delay dashpot. The dashpot slows the response of the protector by allowing the contacts to remain closed during nondestructive overcurrents to minimize nuisance tripping.

The time delay of the dashpot can be set to meet specific circuit protection requirements. For example, the time delay may be initiated when the current is 25 percent above the rated value. However, if the current surge exceeds six times the rating, the time delay dashpot will be overridden, and the contacts will be opened instantaneously.

For all practical purposes, the electromagnetic-hydraulic circuit protectors are unaffected by changes in ambient temperature, although changes in viscosity of the oil in the dashpot could have a slight effect on the time delay. *See* CIRCUIT BREAKER.

Thermal Protectors. Thermal circuit-breaking protective devices are based on the bending of a bimetallic contact element in the presence of heat caused by an overcurrent. The bimetallic content bends to open the circuit. The amount of bending of the bimetallic element depends on the difference in coefficient of expansion of the contact metals. The heat is directly related to the product of the square of the current and the duration of that current (I^2t), assuming constant resistance.

Some thermal protectors, like magnetic-hydraulic protectors, can be manually reset. However, the element can not be reset until the bimetallic element cools sufficiently to be reset. The thermal protector has a slower response time than the magnetic protector, and there is no provision for a time delay. Thus it is more prone to false triggering. Moreover, its established set point is changed with changes in ambient temperature. *See* CIRCUIT BREAKER.

Fuses. Fuses provide the lowest-cost, but slowest-acting overcurrent protection. The fuse, like the circuit protector, is placed in series with the electronic circuit it is protecting. When the current exceeds the fuse rated value, the conductive element in the fuse melts or blows, opening the circuit.

Because the response time of fuses is measurable in milliseconds, they offer no protection against high-speed voltage transients. They are fail-safe, disposable components that responds to the same I^2t product as thermal protectors. Four classes of fuses are specified for electronics applications: very fast acting, non time delay, time delay; and dual-element time delay. For more detailed information on fuses, *see* FUSE.

CIRCULAR ANTENNA

See ANTENNA DIRECTORY.

CIRCULAR CONNECTOR

Circular connectors, also called cylindrical connectors, are designed to terminate multiconductor cables. They all have three or more pin-and-socket contacts within mating plugs or receptacles with cylindrical shells. Figure 1 illustrates a plug (left) and receptacle (right) for a circular connector. The plug is the part of the connector that is attached to the free or moving end of the cable. The receptacle, is the part of the connector that is attached to a wall, chassis, or enclosure.

CIRCULAR CONNECTOR: Fig. 1. Connector consists of the mating plug (left) and receptacle (right) of a MIL-C-38999 Series V connector.

Originally developed more than 45 years ago for military applications, cylindrical connectors can simultaneously terminate as many as 128 conductors. The first of these connectors were designed for external coupling between separate enclosed military system components. The Department of Defense is still the largest user of these connectors.

Circular connectors are included in both military and commercial avionics systems for interconnecting modules of communications, radar, and navigation systems. They are also used on military aircraft and missiles for electronics warfare, countermeasures, and guidance. In addition, the connectors are specified for many applications on ships and submarines for communications, electronic warfare, countermeasures, fire control, navigational aid, radar, sonar, surveillance, and test equipment.

Circular connectors are also used in tanks and other mobile weapons systems for fire control, radar, and electronic countermeasures. They also have a place in mobile vans and fixed ground stations for communications, radar, surveillance, and aircraft ground control. Commercial circular connectors are used in radio and television broadcasting, industrial robots, and process control systems.

Figure 2 is a sectional view of an important and popular military style cylindrical connector—the MIL-C-38999, Series IV. An important feature of all circular connectors is the rapidity with which they can be aligned and coupled or uncoupled without the aid of tools. The metal shells of the receptacles and plugs provide protection against abrasion and external forces that might crush the connectors. They also protect the mating contacts from airborne contaminants, dust, moisture, and salt spray. Metal shells shield against both transmitted and received radio frequency energy if the mating cables are shielded.

The integral locking mechanisms, either complete or partially threaded shells (breech lock) or bayonet-style (pin-and-curved slot), provide a self-supporting interconnection. Neither internal nor external screws are required for securing the mating parts. Cylindrical connectors can also withstand the high shock and vibration forces encountered in aircraft, ships, and vehicles.

A knurled outer ring is twisted to unlock the mating shells. In addition, built in keys assure rapid and correct alignment or polarization of the mating contacts. Circular connectors are precision made products designed to assure maximum conductivity between the circuits being connected.

Bulkhead connectors serve as transitions between two cables located on opposite sides of a metal case or bulkhead in aircraft or ships. These permit watertight or pressurized integrity to be maintained at the bulkhead.

Circular connector pins are machined from brass or nickel-silver, and sockets are formed from nonferrous materials such as beryllium-copper, phosphor-bronze, or nickel-silver. Sockets include flexible innerleaf spring contact surfaces to grip the pins with sufficient force to obtain high electrical conductivity even after many engagements and disengagements.

Gold is preferred for plating mating contacts because it permits nondestructive sliding contact and resists corrosion, oxidation, and other contamination that blocks low-level dry signals. A minimum gold-plating thickness of 100 microinches is specified for military specification connectors, but 15 to 50 microinches is acceptable for many commercial avionics and industrial applications.

Pins and sockets are placed in aluminum or stainless steel shells by inserting them in multihole spacers that establish contact distribution and spacing. The pins must be resilient to prevent damage to either pins or sockets caused by misalignment when the parts are coupled. Permissible center-to-center spacing of pins and sockets is determined by the voltage, current, and frequency of the signals to be transmitted.

Military-style connectors have removable contacts that are crimped to the individual wires of the cable. The wires with crimped, poke-home contacts are inserted into the shells of the connector with a special hand tool that compresses the springs, locking the contacts into position within the shell. This design permits the pins and sockets to be removed easily for inspection. Field changes or repairs can be performed on individual wire terminations without disturbing adjacent wires.

Special coaxial cables with pins and sockets are available for cylindrical connectors. Fiberoptic cables also have been adapted to multipin circular connectors. Fiberoptic contacts have been designed to fit the space allowed by these connectors and they may be intermixed with lower frequency contacts.

CIRCULAR CONNECTOR: Fig. 2. Cross section view of a mated MIL-C-38999 Series IV connector.

Hermetically sealed connectors are made with contacts rigidly fixed within a spacer of glass or ceramic. Because the contacts are not removable, wires must be attached individually by welding or soldering. Care must be taken in the use of these connectors to prevent cracking the hermetic seals and destroying the hermetic integrity.

Military standard MIL-STD-1353A lists the preferred connectors for military applications. There are three size classifications: standard, miniature, and subminiature. Construction and testing of all three size classifications is similar. Standard size circular connectors are most likely to be specified for large shipboard and ground-based systems, although there has been a trend toward specifying miniature connectors in these applications. Subminiature connectors are favored for avionics because of their lighter weight and smaller size.

Five military connector families are being procured in the largest volumes today: MIL-C-38999, MIL-C-26482, MIL-C-5015, MIL-C-83723, and MIL-C-22992. The four versions of MIL-C-38999 are: Series I, II, III, and IV. Each has mechanical differences indicating modifications by the originating military service working with one of the connector manufacturers.

Quick-release, or breakaway, cylindrical connectors permit rapid disengagement as missiles are launched from aircraft, ships, submarines, ground silos, or vehicles. Braided-metal lanyards are fastened to the outer shell. As the missile blasts off, it applies tension to the lanyard, which pulls out threaded coupling segments for instant release.

Commercial versions of these military specification connector families are available; they do not require the same costly full qualification testing and traceability documentation as the military products.

CIRCULAR POLARIZATION

The polarization of an electromagnetic wave is the orientation of its electric-field lines of flux. Polarization can be horizontal, vertical, or at a slant (*see* HORIZONTAL POLARIZATION, VERTICAL POLARIZATION). The polarization can also be rotating, either clockwise or counterclockwise. Uniformly rotating polarization is called circular polarization. The orientation of the electrical field lines of flux completes one rotation for every cycle of the wave, with constant angular speed.

Antennas for circular polarization are not turned to produce the rotating electromagnetic field; the rotation is easily accomplished by electrical means. The illustration shows a typical antenna for generating waves with circular polarization. The antennas are fed 90 degrees out of phase by making feed-line stub X a quarter wavelength longer than stub Y. The signals from the two antennas thus add vectorially to create a rotating field. The direction, or sense, of the rotation can be reversed by adding ½ wavelength to either stub X or stub Y (but not both).

Circular polarization is compatible with a 3-decibel power loss, with either horizontal or vertical polarization, or with slanted linear polarization. When communicating with another station also using circular polarization, the senses must be in agreement. If a circularly polarized signal arrives with opposite sense from that of the receiving antenna, the attenuation is about 30 decibels compared with matched rotational sense.

In uniform circular polarization, the vertical and horizontal signal components may have equal magnitude, but this is not always the case. A more general form of rotating polarization, in which the components may have different magnitude, is called elliptical polarization. *See also* ELLIPTICAL POLARIZATION.

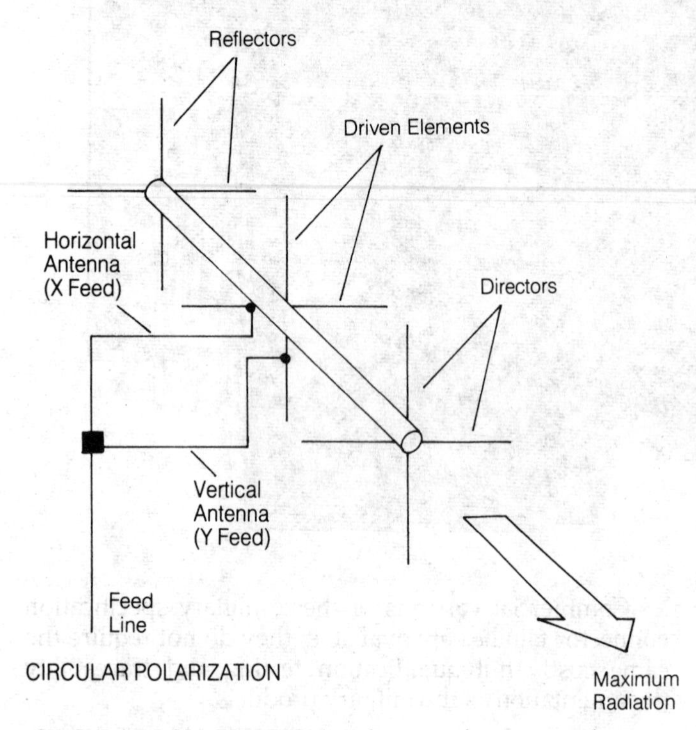

CIRCULAR POLARIZATION: A diagram of an antenna for transmitting circularly polarized signals.

CIRCULAR SCANNING

See RADAR.

CIRCULAR SWEEP

A circular sweep, or circular trace, is an oscilloscope trace that describes a circle. This sweep is obtained by applying two sine-wave signals of equal strength and frequency to the horizontal and vertical inputs 90 degrees out of phase.

CLAMP CIRCUIT

A clamp circuit is one that sets, or holds, the operating level of a component or another circuit. One type of clamp circuit called a dc (direct current) restorer, is used to restore the dc component to an ac (alternating current) waveform after the ac has been amplified. Early television receivers used one or more diodes to restore the dc level to the video information before it was applied to the cathode ray tube. This caused the black (or white) level of the picture to remain constant from one side of the screen to the other. Modern receivers have multipurpose integrated circuits that perform this feature automatically.

Waveform clippers or peak limiters using diodes have been called clamp circuits although this terminology is not strictly correct. Their purpose is to limit the positive or negative excursions of an ac waveform. *See also* CLIPPER.

CLASS A AMPLIFIER

A Class A amplifier is a linear amplifier in which the plate, collector, or drain current flows for 100 percent of the input cycle. The tube or transistor is never driven to the cutoff point, and the input signal occurs over the linear part of the characteristic curve. The tube or transistor is biased at the middle of the linear part of the characteristic curve for Class A operation, as shown in the illustration.

Class A amplifiers are often used in audio-frequency applications where a minimum amount of waveform distortion is important. A Class A amplifier may be either single-ended or push-pull (*see* PUSH-PULL AMPLIFIER). The efficiency of a Class A amplifier is low, approximately 20 percent. Class A amplifiers draw essentially no power from the input source; they are recommended for use as receiver preamplifiers and front-end amplifiers. Radio-frequency power amplifiers are usually operated in Class AB, B, or C. *See also* AMPLIFIER, CHARACTERISTIC CURVE, CIRCUIT EFFICIENCY.

CLASS AB AMPLIFIER

A Class AB amplifier is an amplifier operated at bias conditions between the Class A and Class B amplifiers *see* CLASS A AMPLIFIER, CLASS B AMPLIFIER). Plate, collector, or drain current flows for all or most of the input signal cycle. However, the tube or transistor is driven into the nonlinear portion of the characteristic curve, and therefore some waveform distortion occurs in the output.

There are two kinds of Class AB amplifier. The Class AB1 amplifier is not driven to cutoff, although the input signal drives the tube or transistor into the nonlinear region near cutoff. In a Class AB2 amplifier, the tube or transistor is cut off during a small part of the cycle. See the illustration for approximate bias points for Class AB1 and Class AB2 operation of an NPN bipolar transistor.

The efficiency of a Class AB amplifier is slightly higher than that of a Class A amplifier, but it is not as good as that of a Class B amplifier. A Class AB1 amplifier draws very little power from the signal input source, but a Class AB2 circuit draws significant power. Some distortion of the signal waveform occurs in both cases; thus Class AB amplifiers are not used for high-fidelity applications. Class AB circuits are sometimes used as radio-frequency power amplifiers. *See also* AMPLIFIER, CHARACTERISTIC CURVE.

CLASS B AMPLIFIER

A Class B amplifier is an amplifier operated at or near the cutoff point of the characteristic curve of a tube or transistor. Plate, collector, or drain current flows for approximately 50 percent of the signal input cycle. During the other part of the cycle, the device is cut off. The

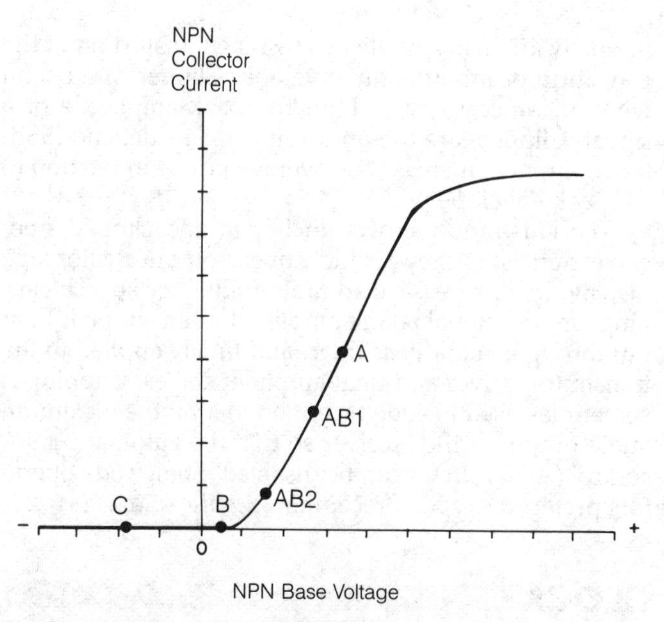

CLASS A, AB, B, AND C AMPLIFIERS: The operating points for the various amplifier classes plotted on an NPN transistor characteristic curve.

illustration shows the direct-current bias point for Class B operation of an NPN bipolar transistor.

Class B amplifiers are used as radio-frequency amplifiers. In the push-pull configuration, Class B circuits offer low distortion of the waveform, and good efficiency at audio frequencies. The efficiency of a Class B amplifier is usually less than 50 percent. Class B amplifiers draw considerable power from the source in the single-ended configuration. *See also* AMPLIFIER, CHARACTERISTIC CURVE, CIRCUIT EFFICIENCY.

CLASS C AMPLIFIER

A Class C amplifier is an amplifier operated beyond the cutoff point of the characteristic curve of a tube or transistor. The plate, collector, or drain current flows for less than half of the signal input cycle. During the remainder of the cycle, the device is cut off. The direct-current base bias for Class C operation of an NPN bipolar transistor is shown in the illustration.

Class C amplifiers are unsuitable for audio-frequency applications because the output waveform is severely distorted. Class C amplifiers are also unsuitable for weak signal use, and for linear applications. However, Class C amplifiers are often used in continuous-wave or frequency-modulated transmitters, where there is no variation of carrier amplitude. The efficiency of the Class C amplifier is high. In some cases it can approach 80 percent. Class C amplifiers require a large amount of driving power to overcome the cutoff bias. *See also* AMPLIFIER, CHARACTERISTIC CURVE, CIRCUIT EFFICIENCY.

CLEAR

The term *clear* refers to the resetting or reinitialization of a circuit. Active memory contents of a computer are erased by the clear operation. Auxiliary memory is retained when active circuits are cleared.

All electronic calculators have a clear function button. When this button is actuated, the calculation is discontinued and the display reverts to zero. By switching a calculator off and then back on, the clear function is done automatically.

CLEAR CHANNEL

A clear channel is an amplitude-modulation (AM) broadcast-band channel that renders service over a large area and is protected within that area against interference. The maximum amount of power allowed a clear-channel station is 50 kW. This is the greatest amount of power permissible on the standard AM broadcast band in the United States. Some stations in other countries use considerably more power.

Clear-channel status is allocated to comparatively few stations. Much more common are the regional-channel and local-channel stations, with lower maximum power limits. *See also* LOCAL CHANNEL.

CLICK FILTER

When a switch, relay, or key is opened and closed, a pulse of radio-frequency energy is emitted, especially when the device carries a large amount of current. A capacitor connected across the device slows down the decay time from the closed to the open condition, where the click is most likely to occur. The illustration shows this arrangement at A. A small resistor in series, combined with the capacitor in parallel, slows down the

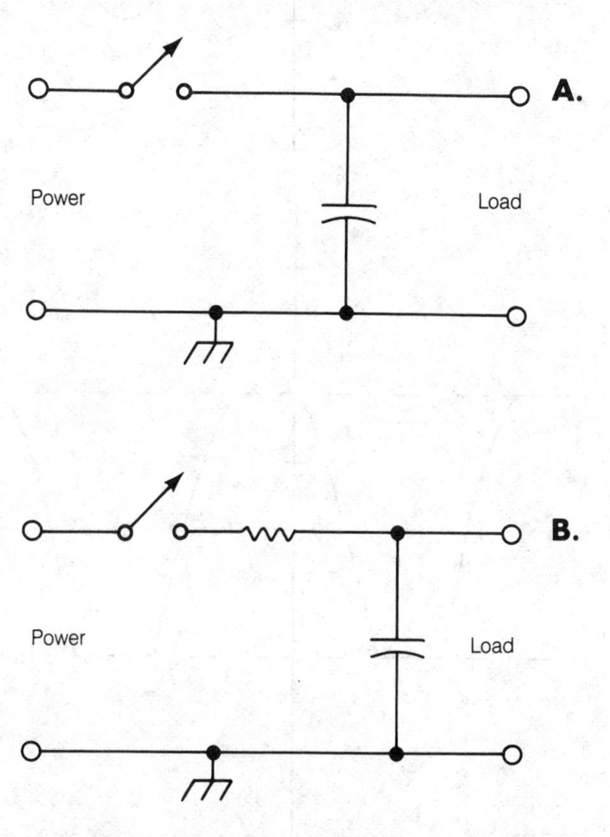

CLICK FILTER: The capacitor in a click filter smooths out decay in a code signal envelope (A), and the added series resistor smooths out both rise and decay (B).

make time from the open to the closed state (at B). These devices are called click filters. Sometimes a choke is used in place of the resistor for circuits that draw high current.

In a code transmitter, a click filter is used to regulate the rise and decay times of the signal. Without this filter, the rapid rise and decay of a signal can cause wideband pulses to be radiated at frequencies well above and below that of the carrier itself. This can result in serious interference to other stations.

CLIPPER

A clipper is a circuit that limits the peak amplitude of a signal at some value smaller than the peak value it normally attains. This always results in distortion of the waveform or modulation envelope. Clippers are useful when it is necessary to limit the peak amplitude of a signal. Clippers are often used in single-sideband (SSB) transmitters to increase the average power in relation to the peak-envelope power.

The illustration shows unclipped and clipped sine-wave signals as they would appear on an oscilloscope display. In a process called radio-frequency speech clipping, an SSB signal is first amplified, then clipped, then put through a bandpass filter, and finally applied to the transmitter driver and final-amplifier stages. Clipping is sometimes used in code reception to limit the maximum audio output of the receiver so that the automatic-gain-control (AGC) circuit can be disabled. Some code operators prefer this type of reception. *See also* SPEECH CLIPPING.

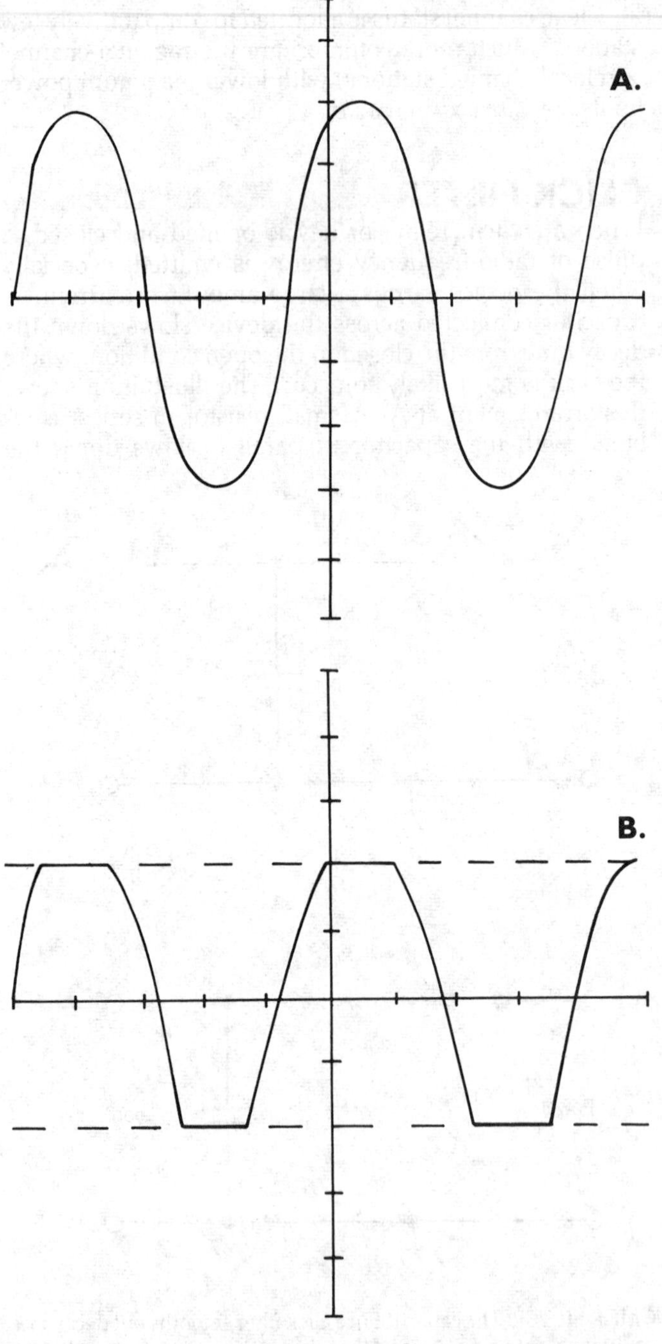

CLIPPER: An unclipped waveform (A) and a clipped waveform (B).

CLOCK

A clock is a pulse generator that serves as a time-synchronizing standard for digital circuits. The clock sets the speed of operation of a microprocessor, microcomputer, or computer.

The clock produces a stream of electrical pulses with extreme regularity. Some clocks are synchronized with time standards. The speed may also be controlled by a resistance-capacitance network or by a piezoelectric crystal. The clock frequency is generally specified in pulses per second, or hertz.

CLOSED CIRCUIT

Any complete circuit that allows the flow of current is called a closed circuit. All operating circuits are closed.

A transmission sent over a wire, cable, or fiberoptics medium and not broadcast for general reception is called a closed-circuit transmission. A telephone operates as closed circuit (except when used as a radio telephone). Some closed-circuit radio and television systems function as intercoms or security monitoring devices. *See also* CARRIER CURRENT.

COAXIAL ANTENNA

See ANTENNA DIRECTORY.

COAXIAL CABLE

Coaxial cable is designed to transmit signals efficiently between 1 kHz and 4000 MHz with minimum loss and little or no distortion. This frequency range extends from radio through the higher frequencies of TV well into the microwave band. A coaxial cable is made of a central signal conductor covered with an insulating material (the dielectric core) which, in turn, is covered by an outer tubular conductor (the return path). The cable is called coaxial because the two conductors are separated by the dielectric core. All of these parts of the cable are covered with a protective jacket. Figure A shows the construction of a typical coaxial cable.

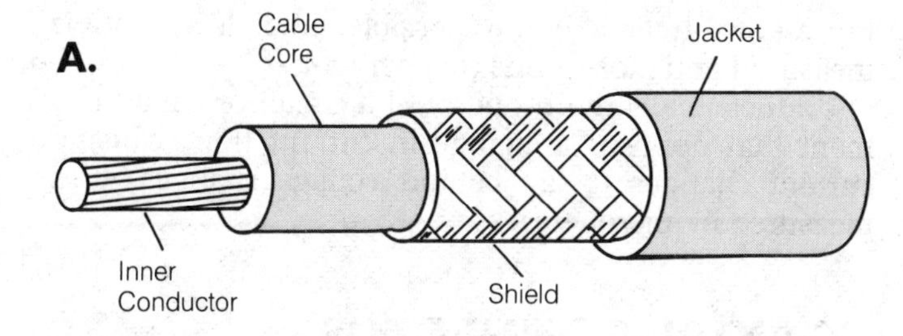

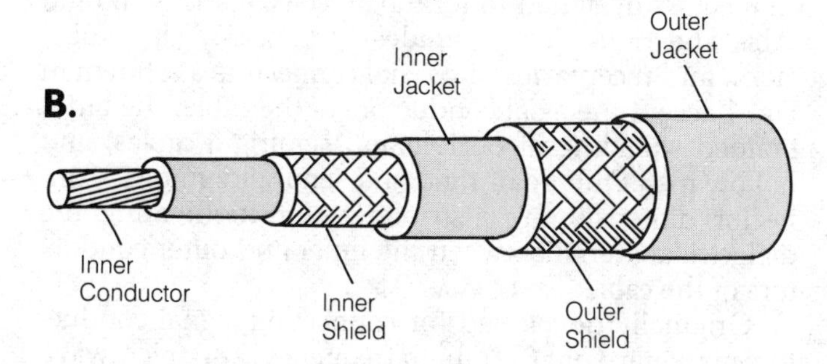

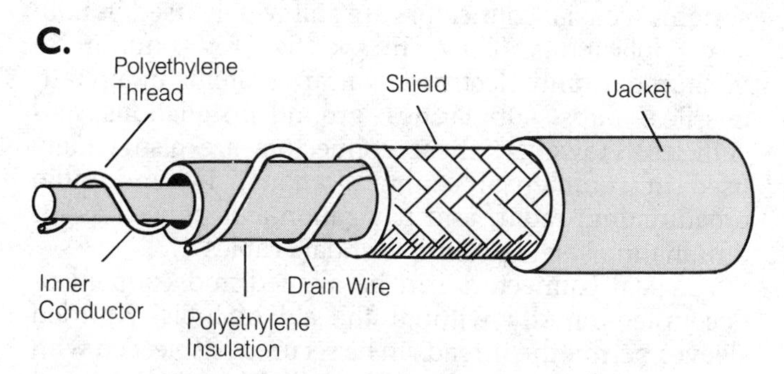

COAXIAL CABLE: Coaxial cables can be conventional single shield and single jacket (A), double-shielded, double-jacketed triaxial (B), and an air-spaced coaxial (C).

The inner conductor is typically solid or stranded annealed copper wire that can be bare, tinned, or silver coated. The attenuation is lower with solid conductors, but cable flexibility is greater with stranded conductors. Annealed copper wire is generally preferred because of its excellent electrical properties. However, for mechanical strength, copper-covered steel or copper alloy conductors can be used.

Five different core dielectric materials are now used in coaxial cable. Four of these are based on polyethylene, and the fifth is Teflon FEP. Extruded polyethylene is the most widely used core dielectric. It has a low cost and has very good electrical properties, but it is not generally suitable for applications where the ambient temperature will exceed 80°C. Polyethylene is flammable, but the shield and PVC (polyvinyl chloride) cable jacket reduce this hazard. It has the highest dielectric constant of the five materials: 2.27.

Cross-linked (irradiated) polyethylene can be used in ambient temperatures up to 125°C. With approximately the same electrical characteristics as conventional polyethylene, its additional toughness is obtained from the irradiation process that transforms it into a thermosetting material. This dielectric has a greater resistance to abra-

COAXIAL CABLE: Characteristics of prefabricated coaxial transmission lines.

Type	Characteristic Impedance, Ω	Velocity Factor	Outside Dia., inches	Picofarads per Foot
RG-8/U	52	0.66	0.41	29.5
RG-9/U	51	0.66	0.42	30.0
RG-11/U	75	0.66	0.41	20.6
RG-17/U	52	0.66	0.87	29.5
RG-58/U	54	0.66	0.20	28.5
RG-59/U	73	0.66	0.24	21.0
RG-174/U	50	0.66	0.10	30.8
Hard line				
1/2-inch	50	0.81	0.50	25.0
	75	0.81	0.50	16.7
Hard line				
3/4-inch	50	0.81	0.75	25.0
	75	0.81	0.75	16.7

sion, ozone, solvents, heat from soldering, and cracking due to stress. The dielectric constant is 2.45.

Cellular (foamed) polyethylene is used where there is a requirement for lower dielectric constants. Dielectric constants as low as 1.5 have been achieved. Cellular polyethylene can also be cross linked for greater strength and toughness. Flame-retardant polyethylene has additives that impart flame retardance to polyethylene, but this benefit is at the cost of an increase in dielectric constant to 2.5.

Teflon FEP has excellent electrical properties, and it can be used in ambient temperatures up to 200°C. Its low dielectric constant of 2.15 makes it suitable for use in miniature coaxial cables.

The dielectric constant (K) of a core material is important because it determines the diameter and weight of the finished cable. In cables where the electrical properties of attenuation and impedance are to be kept the same, as K decreases, the distance between inner and outer conductor is reduced. This permits the manufacture of smaller and lighter cable. Core materials with the lowest dielectric constants are recommended for airborne applications where space and weight reductions are of prime importance.

The dielectric constant can be lowered with air-spaced (semisolid) cores. These are made by wrapping a low-loss filament around the inner conductor as shown in C. A tube of the same material is then extruded over the spiral leaving air space. The effective K will vary with the air-to-plastic ratio.

Shield (the outer conductor) for coaxial cables consists of groups of small-diameter bare, tinned, or silver-coated copper wires braided together. Double braids are sometimes used to provide more effective shielding. In applications where crosstalk and noise must be minimized, a double-shielded, double-jacketed triaxial cable as shown in Fig. 1B can be specified.

Military standard MIL-C-17 mandates braided shields. However, there are other, lower-cost shields for coaxial cables: spiral shields, aluminum/Mylar shields, and conductive plastics. These alternate shields are accepted for nonmilitary commercial use, but they are not acceptable for RG cables.

Jacket. The jacket provides overall protection against environmental hazards for the coaxial cables. Both polyethylene and vinyl are used as jacket material. Conventional vinyl with good weathering, abrasion and moisture resistance characteristics is accepted as the MIL-C-17 Type 1 jacket. However, the plasticizers used in this formulation can migrate into the cable-core dielectric and alter its electrical properties so that attenuation increases.

Military specification MIL-C-17 Type IIa jacket has all the features of Type I jackets but does not have the migrating plasticizers. The vinyl will withstand low temperatures and is suitable for direct burial in the ground.

COAXIAL CABLE ELECTRIC CHARACTERISTICS

Characteristic impedance (Z_O) for coaxial cables can be defined as the total opposition to the flow of current in the cable. It can also be defined in terms of its direct relation to the ratio of the sizes of the inner and outer conductors, and its inverse relation to the dielectric constant of the core material. Thus characteristic impedance is the resistance that a cable terminated by its own characteristic impedance offers to a transmitter.

Maximum power can be transferred in a coaxial cable only when the characteristic impedances of transmitter, rf (radio frequency) line, and receiver (or antenna) are equal. If the match is exact, losses are due only to resistance in the line (attenuation). If there is a mismatch, there will be reflection losses. The characteristic impedance of a cable, unlike conductor resistance, does not vary with length. Coaxial cable are generally designed to match 50, 75, or 95 Ω impedances. Characteristic impedance is measured in ohms (Ω).

Capacitance in coaxial cables is that property of the combination of conductors and dielectrics that permits electrical energy to be stored when the conductors are at different potentials. Capacitance, like impedance, is dependent upon the inner and outer conductor sizes and the dielectric constant of the core, but it is a reciprocal relationship. Thus capacitance increases as impedance decreases in a cable with the same dielectric constant. Capacitance in coaxial cables is usually expressed in picofarads per foot.

Attenuation in coaxial cable is defined as the loss of electrical power in a length of cable. Losses occur in the conductor and dielectric as well as from radiation. An increase in conductor size will reduce attenuation because electrical loss is decreased. It is possible to increase conductor size while keeping the cable dimensions the same by using a dielectric material with a lower dielectric constant. Attenuation is measured in decibels per 100 feet.

Velocity of propagation is the speed of transmission of electrical energy in a cable as compared with its speed in air considered to represent 100 percent. Velocity is inversely proportional to the dielectric constant, so a lower constant causes an increase in velocity.

Time delay is the elapsed time between the initial transmission of a signal from one point to its appearance or detection at another point. The time-delay calculation represents the maximum possible time delay and is measured in nanoseconds (ns) per foot.

Inductance is the property of a circuit or circuit element that opposes a change in current flow, causing current changes to lag behind voltage changes. It is measured in microhenries.

COAXIAL CONNECTOR

Coaxial connectors are a class of cylindrical or circular connectors designed to terminate coaxial and semirigid cable and rigid circular guides. (*See* CABLE.) The plugs, jacks, and receptacles of coaxial connectors as shown in Fig. 1 accept the axial conductors of the cable, the outer braided shielding of flexible and semirigid cables, and hollow metal tubing of rigid and semirigid guides. Connector dimensioning also allows for terminating the dielectric material between the inner and outer conductors of the cable. *See* COAXIAL CABLE.

Originally developed for connecting signal conductors in military uhf (ultrahigh frequency) and microwave systems, coaxial connectors are still widely used in military applications. They are specified for communications, radar, and electronic warfare systems in aircraft, missiles, ships, submarines, ground installations, and vehicles. However, these connectors are also widely used in commercial communications, TV and radio broadcasting, radar, and navigation aids as well as for terminating shielded computer data cables.

Coaxial connectors can be aligned and coupled or decoupled rapidly without the aid of tools. Knurled sleeves permit the threads to be secured or released with thumb and finger pressure. The cylindrical shells protect against abrasion and crushing forces. Some connectors have bayonet-style (pin-and-curved slot) locking mechanisms as shown in Fig. 2. The shells seal against moisture, salt spray, dust, airborne contaminants, and emitted or received rf energy.

Coaxial connectors are grouped in five product categories: standard (C, N, twinax, triax and uhf); miniature (BNC and TNC); subminiature (SMA, SMB, and SMC); microminiature (SSMA, SSMB, and SSMC); and precision (APC, APC-3.5, and APC-7. The 2.4, 3.5, and 7 refer to the mating cable dimensions in millimeters.)

The uhf connectors are designed to operate at frequencies up to 300 MHz. With peak voltage ratings of 500 V, they are available in two sizes—N and C. N connectors are medium-size, weatherproof units with threaded coupling for use into the microwave frequencies; they are impedance matched to either 50 or 70 Ω cables. C connectors are available in two styles: standard, with peak voltage ratings of 1500 V, useful to 10 GHz; and high voltage with ratings to 4000 V, useful to 2 GHz.

BNC coaxial connectors are miniature, bayonet-locked units, designed to operate at frequencies up to 11 GHz. TNC connectors, with the same electrical specifications, are weatherproof units with threaded couplings.

SMA connectors are semiprecision, subminiature units specified to operate at frequencies up to 18 GHz on semirigid cable and up to 12.4 GHz on flexible cable. Slightly smaller SMB connectors are specified for use

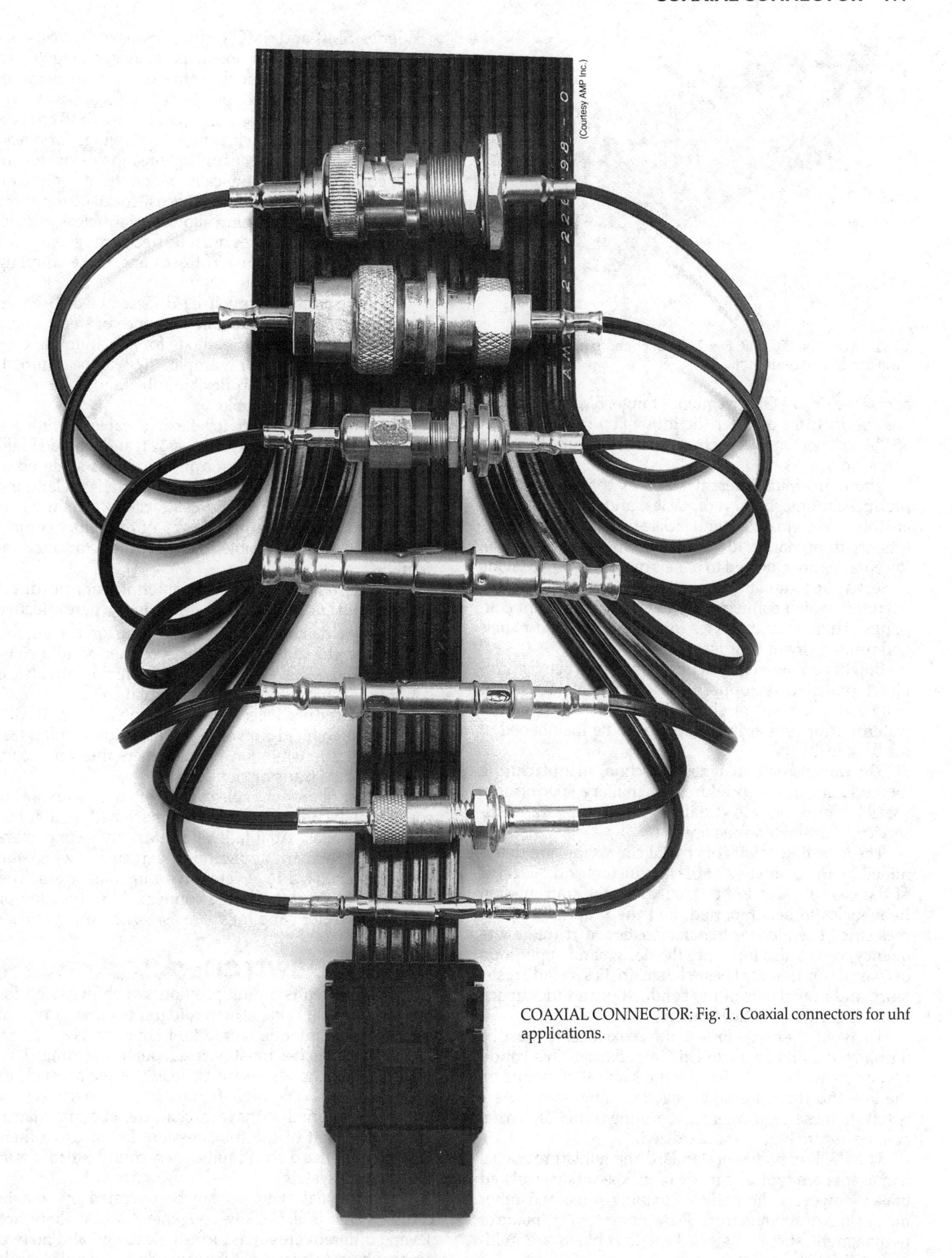

COAXIAL CONNECTOR: Fig. 1. Coaxial connectors for uhf applications.

COAXIAL CONNECTOR: Fig. 2. Bayonet-and-pin locking mechanism of a coaxial connector.

across the dc to 4 GHz frequency range. SMC connectors are specified for a wider (dc to 6 GHz) range. Smaller SSMA connectors have characteristics similar to SMA connectors.

There are four different kinds of coaxial connector products: plugs, jacks, receptacles, and adaptors. Plugs are the male connectors, and jacks are the female connectors used primarily to terminate unsupported cable. Receptacles are attached to panels or chassis and provide a means for terminating cable at that panel or chassis. Adaptors match connector elements with different couplings. There also are styles for isolated ground panels and printed circuit boards.

Bulkhead connectors act as transitions between cables terminated on opposite sides of a metal case or bulkhead in aircraft or ships. These connectors permit watertight or pressurized integrity to be maintained at the bulkhead.

The dimensions, materials selection, manufacturing methods and test procedures for military specification coaxial connectors are dictated by MIL-C-39012, a tri-service, coordinated specification.

The key dimensions of coaxial connectors are determined by the diameters of the conductors and dielectric of the coaxial cable to be terminated, the transmission frequencies to be supported, and the properties of the dielectric. The electrical characteristics of rf (radio frequency) cables also influence the design and application of coaxial connectors. These characteristics include resistance and inductance of the conductors and the capacitance and leakage between them.

To avoid rf energy losses, the coaxial cable must be terminated in its characteristic impedance. This condition exists when the voltage is the same at all points on the line and there are no voltage standing waves. As a result of these requirments, the dimensions of coaxial connectors are highly standardized.

The shells or bodies of standard and miniature coaxial connectors are typically made of nickel- or silver-plated brass. However, the bodies, coupling nuts, and other metal parts of subminiature SMA connectors are made of nonmagnetic stainless steel. Female contacts are gold-plated beryllium copper.

Series SMB and SMC connectors have bodies, coupling nuts, and other metal parts made of gold- or silver-plated brass. Female contacts are also made of gold-plated beryllium copper. Series SMC connectors are mated with threaded couplings, and series SMB connectors are spring mated with snap-fit couplings for quick-connect/quick disconnect applications. Insulators for coaxial connectors are typically made from polytetrafluorethylene (PTFE). Some low-cost coaxial connectors are cast from die-cast zinc alloy. Nevertheless, all the machined or cast surfaces must be free of rough surfaces or burrs that would cause rf losses and make coupling and decoupling difficult.

Coaxial connectors must be disassembled to be attached to the coaxial cable in a sequence of steps. There are at least ten different methods for attaching the connectors to the cables. For example, they can be clamped, crimped, or soldered on flexible cable and soldered or clamped on semirigid cable.

The standard clamp for flexible cable includes a threaded nut, flat compression gasket, and tapered braid clamp. The center conductor of the cable is soldered to the center contact of the connector. Solderless clamping is obtained with a threaded cable clamp, flat washer, and threaded body clamp. The tip of the center contact is threaded over the cables' center conductor after assembly.

Coaxial connectors are used for terminating dc (direct current) and audio-frequency signal lines where effective shielding is desired. The inner conductor carries the signal, and the outer conductor is connected to a common ground. The external braided shield provides a continuous shield to protect the signal conductor from externally produced electromagnetic interference (EMI). Similarly the shield prevents the radiation of high-frequency data transmission signals where the connectors terminate data transmission lines.

The first fiberoptic cable connectors were standard rf coaxial cable connectors specially machined to adapt to fiberoptic cables. Adaptations of SMA style connectors are popular for many applications, particularly where the system is exposed to severe environmental stress. The fiberoptic connectors can be connected by threaded or pin-and-bayonet couplings. *See* FIBEROPTIC CABLE AND CONNECTORS.

COAXIAL SWITCH

A coaxial switch is a multiposition switch designed for use with coaxial cable. The photograph shows a typical coaxial switch intended for radio-frequency use.

Coaxial switches must have adequate shielding. This requires that the enclosure be made of metal, such as aluminum. At very high frequencies a coaxial switch must be designed to have a characteristic impedance identical to that of the transmission line in use; otherwise, impedance discontinuities may contribute to loss in the antenna system.

Some coaxial switches can be operated by remote control. This is especially convenient when there are several different antennas on a single tower, and no two of them have to be used at the same time. A single length

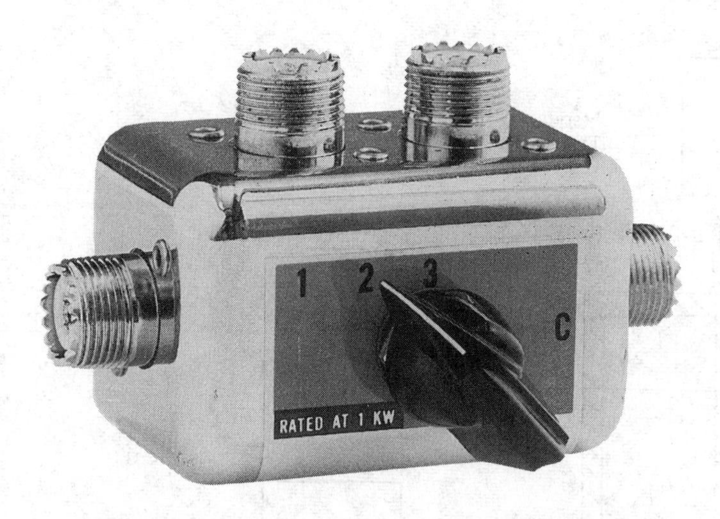

COAXIAL SWITCH: A coaxial switch for radio frequency transmission.

of cable may then be used as the main feed line, and each antenna can be connected to a separate branch by the switch.

COAXIAL TANK CIRCUIT

A coaxial cable, cut to any multiple of ¼ electrical wavelength, can be used in place of an inductance-capacitance tuned circuit. If the length of the cable is an even multiple of ¼ wavelength, a low impedance is obtained by short-circuiting the far end, or a high impedance is obtained by opening the far end (A and B in the illustration). If the length of the cable is an odd multiple of ¼ wavelength, a high impedance is obtained by short-circuiting the far end and a low impedance is obtained by opening the far end (C and D).

Coaxial tank circuits are used mostly at very high and ultrahigh frequencies where ¼ or ½ wavelength is a short length. Coaxial tank circuits have excellent selectivity. Cavity resonators are also used as tuned circuits at very high and ultrahigh frequencies. *See also* CAVITY RESONATOR, COAXIAL WAVEMETER.

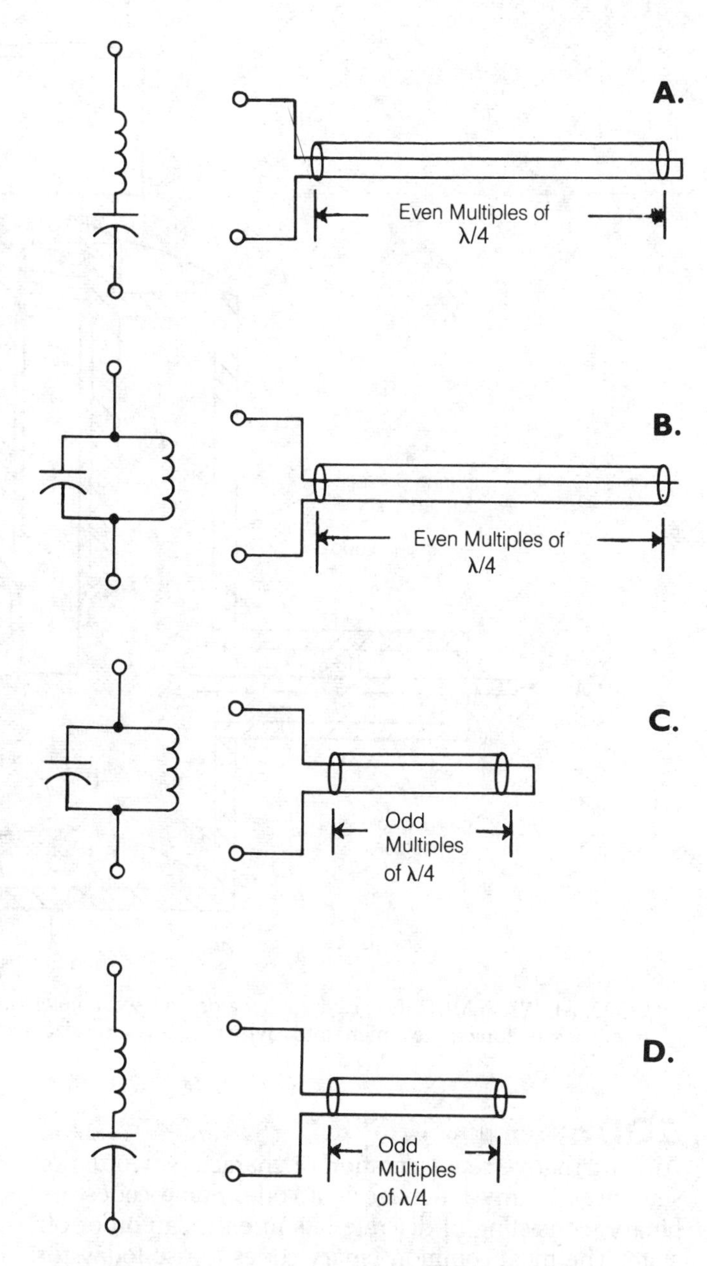

COAXIAL TANK CIRCUIT: Schematics and drawings for quarter-wavelength coaxial tank circuits include even multiples (A and B) and odd multiples (C and D).

COAXIAL WAVEMETER

For measuring very high, ultrahigh, and microwave frequencies, a coaxial wavemeter can be employed. This device consists of a rigid metal cylinder with an inner conductor along its central axis, and a sliding disk that shorts the cylinder and the inner conductor. The coaxial wavemeter is a variable-frequency coaxial tank circuit (*See* CAVITY RESONATOR.)

By adjusting the position of the shorting disk, resonance can be obtained. Resonance is indicated by a dip or peak in an rf voltmeter or ammeter. The length of the resonant section is easily measured; this allows determination of the wavelength of the applied signal. The frequency is determined from the wavelength by the formula:

$$f = 300k/\lambda$$

where f is the frequency in megahertz; λ (Greek lambda) is the wavelength in meters; and k is the velocity factor of the cable tank circuit, typically 0.95 for air dielectric. *See also* COAXIAL TANK CIRCUIT, VELOCITY FACTOR.

COBOL

COBOL is a computer language. The acronym stands for Common Business-Oriented Language. Commands and functions are expressed as words in the English language. COBOL is primarily intended for business use; it is an internationally standardized computer language.

For problems in mathematics and physics of the type encountered in scientific research the languages BASIC and FORTRAN are generally preferred. *See also* BASIC, FORTRAN.

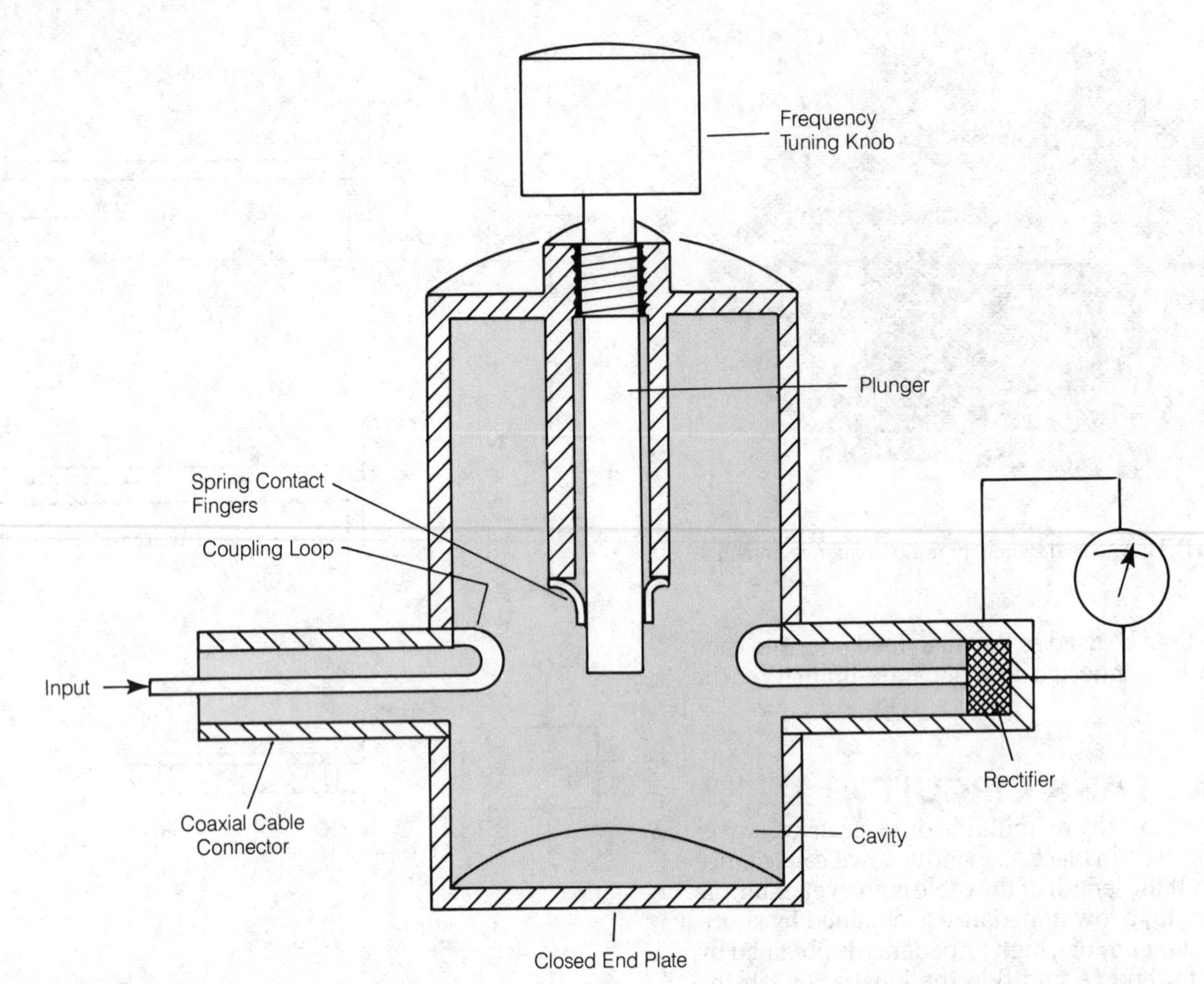

COAXIAL WAVEMETER: A cavity resonator, this wavemeter can make transmission or absorption measurements at microwave frequencies. Plunger movement into cavity reduces cavity size increasing resonant frequency; retraction decreases frequency.

CODE

Any alternative representation of characters, words, or sentences in any language is a code. Some codes are binary, consisting of discrete bits in either an on or off state. The most common binary codes in use today for communications are ASCII, Baudot, and the International Morse code (*see* ASCII, BAUDOT CODE, INTERNATIONAL MORSE CODE). The Q and 10 signals, which are abbreviations for various statements, are codes. Words in computer languages are a form of code.

Binary codes allow accurate and rapid transfer of information, since digital states provide a better signal-to-noise ratio than analog forms of modulation. The oldest telecommunications mode a combination of the Morse code and the human ear, is still used today when all other modes fail.

CODE TRANSMITTER

A code transmitter is the simplest kind of radio-frequency transmitter. It consists of an oscillator and one or more stages of amplification. One of the amplifiers is keyed to turn the carrier on and off. The block diagram shows a simple code transmitter.

Sophisticated code transmitters caused mixers for multiband operation. Many amplitude-modulated, frequency-modulated, or single-sideband transmitters can function as code transmitters. An unmodulated carrier is simply keyed through the amplifying stages.

Ideally, the output of a code transmitter is a pure, unmodulated sine wave at the operating frequency. Changes in amplitude under key-down conditions are undesirable. The rise and decay times of the carrier, as the transmitter is keyed, must be regulated to prevent key clicks. The frequency should be stable to prevent chirp. *See also* CHIRP.

CODING

The process of formulating a code is called coding. When preparing a code language, it is necessary to decide whether the smallest code element will present a character, a word, or a sentence. The ASCII, Baudot, and Morse codes (*see* ASCII, BAUDOT CODE, INTERNATIONAL MORSE CODE) represent each character by a combination of digital pulses. Computer languages use digital words to perform specific functions. Communications codes use a group of characters, such as ORX or 10-4, to represent an entire thought or sentence.

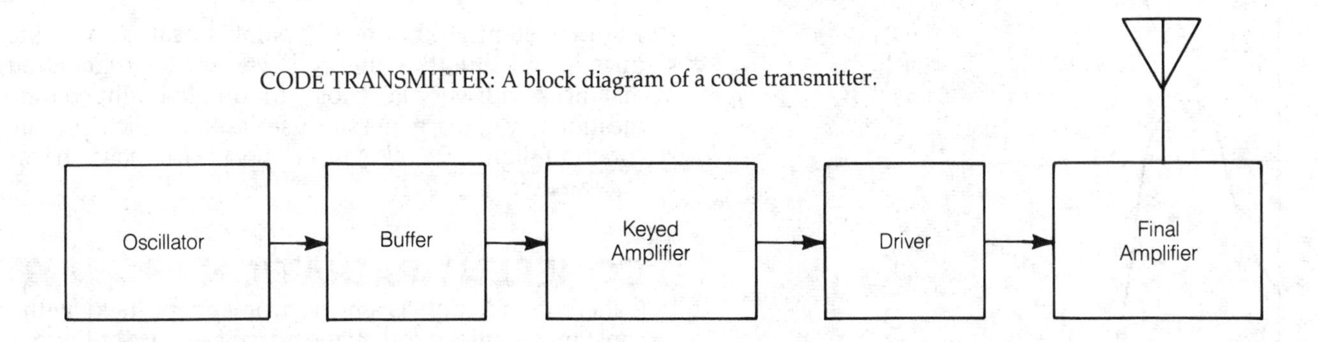

CODE TRANSMITTER: A block diagram of a code transmitter.

When a language is translated into code, the process is called encoding. When a code is deciphered back into ordinary language, the process is called decoding. These functions may be done either manually or by machine. *See also* DECODING, ENCODING.

COEFFICIENT OF COUPLING

Two circuits can interact to a greater or lesser extent. The degree of interaction, or coupling, between two alternating-current circuits is expressed as a quantity called the coefficient of coupling, abbreviated in equations by the letter k. Usually, the coefficient of coupling is used in reference to inductors.

The coefficient of coupling, k, is related to the mutual inductance M and the values of two coils L_1 and L_2 according to the formula:

$$k = \frac{M}{\sqrt{\frac{L_1}{L_2}}}$$

where the inductances are specified in henrys.

For impedances Z_1 and Z_2 in general, where they are of the same kind (predominantly capacitive or inductive):

$$k = \frac{M}{\sqrt{\frac{Z_1}{Z_2}}}$$

where M is the mutual impedance. *See also* MUTUAL IMPEDANCE, MUTUAL INDUCTANCE.

COERCIVE FORCE

A magnetic material with some residual magnetism may be demagnetized by the application of a magnetic force opposing that of the existing field. For example, a permanent magnet, placed inside a coil carrying a direct current, can be demagnetized by that current if the field created by the current is opposite to that of the magnet. The amount of magnetizing force H needed to demagnetize a certain object is called coercive force.

Specific kinds of magnetic material can be magnetized only to a set limit called the saturation induction. The amount of magnetizing force, H, required to completely magnetize a given material, is called the coercive force. The ease with which a material is magnetized and demagnetized is called the coercivity of that material. *See also* COERCIVITY.

COERCIVITY

Some magnetic materials are fairly easy to magnetize and demagnetize while others are difficult to magnetize and demagnetize. The coercivity of a given material is an expression of the readiness with which it is magnetized and demagnetized.

In some applications, it is desirable to have a high degree of coercivity—that is, the material should be difficult to magnetize. Examples include the permanent magnet in a speaker and a recording tape intended for the prolonged storage of information.

In some applications, a small amount of coercivity is desirable—that is, the material should be easy to magnetize and demagnetize. The core of an electromagnet has low coercivity. Some kinds of magnetic tape, intended for frequent erasing and re-recording, have low coercivity.

In general, the more coervice force required to demagnetize a material, the larger its coercivity. *See also* COERCIVE FORCE.

COHERENT LIGHT

Coherent light is light with a single frequency and phase. Most light, even if it appears to be monochromatic, consists of a certain range of wavelengths, and has random phase combinations. White light of nearly equal radiation intensity at all visible frequencies; red light consists primarily of radiation at long visible wavelengths; green light is composed mostly of light in the middle of the visible frequency range. Figure 1 shows a spectral graph of sunlight passed through a red color filter (A). At B, the electromagnetic waves transmitted through the red color filter are in random phase, and variable frequency combinations.

The light transmitted by a helium-neon laser appears red, just as does sunlight through a red color filter. However, the laser light is emitted at just one wavelength (A in Fig. 2) and all the waves coming from the laser are in perfect phase alignment (B). Thus, the helium-neon laser emits coherent red light, while the red color filter transmits incoherent light.

Coherent light travels with greater efficiency—that is, lower attenuation per kilometer—than incoherent light.

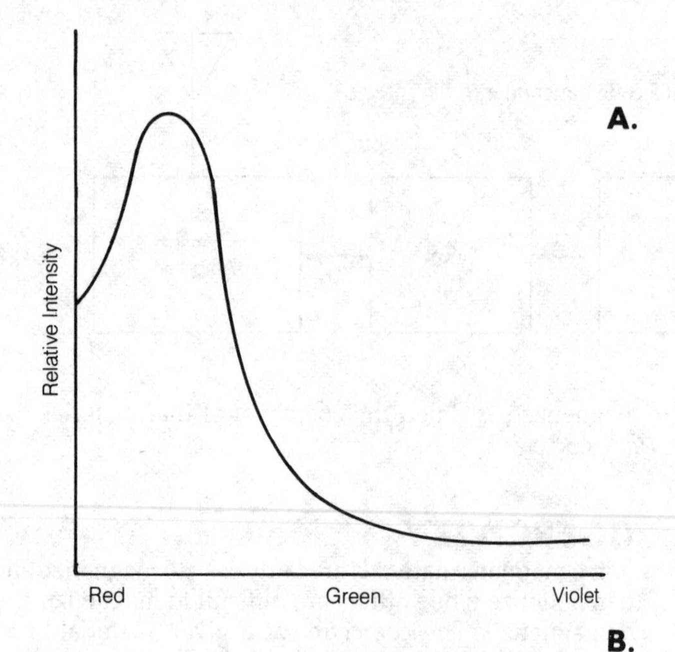

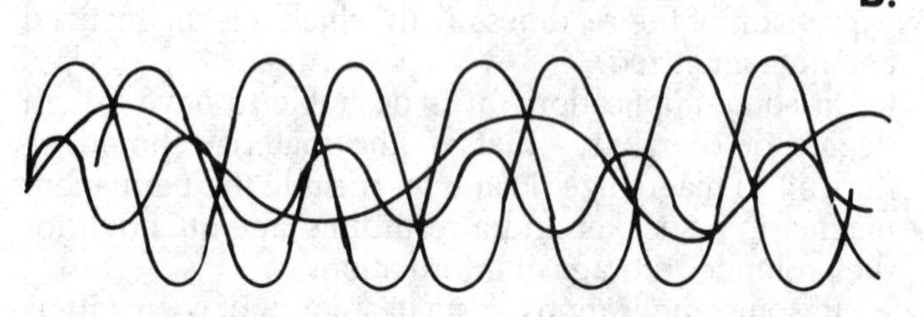

COHERENT LIGHT: Fig. 1. A spectral plot of incoherent light (A) and its wavelength and phase relationships (B).

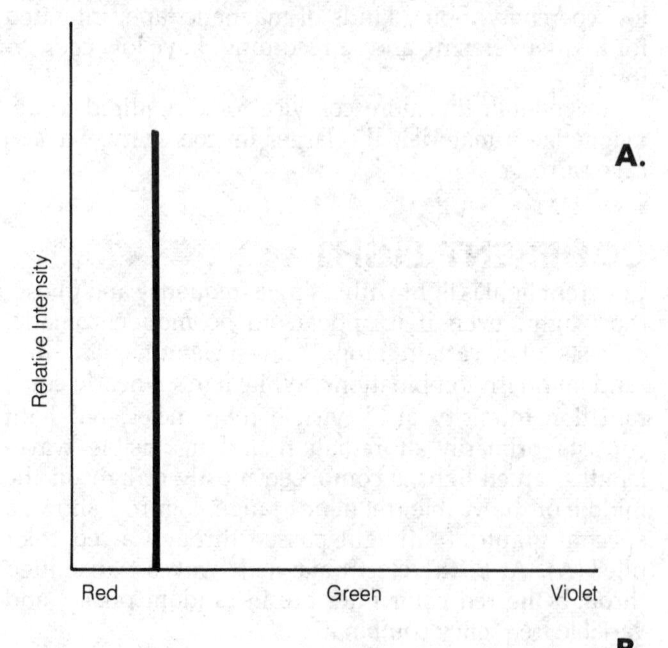

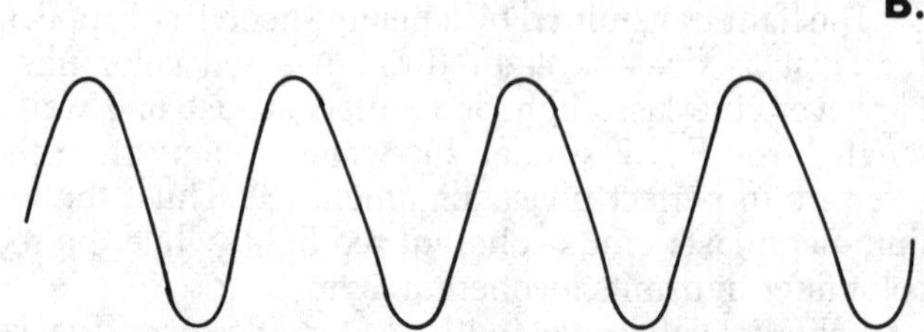

COHERENT LIGHT: Fig. 2. A spectral plot of coherent light (A) and a single waveform of a coherent light generator (B).

With coherent light, a nearly parallel beam can be produced, and thus the energy is carried for tremendous distances with very little loss. Modulated-light communications systems generally use lasers, which produce coherent light. *See also* LASER, MODULATED LIGHT, OPTICAL TRANSMISSION.

COHERENT RADIATION

Coherent radiation is an electromagnetic field with a constant, single frequency and phase. A continuous-wave radio-frequency signal is an example of coherent radiation. The static, or sferics, produced by a thunderstorm is an example of incoherent electromagnetic radio emission.

Energy is transferred more efficiently by coherent radiation than by incoherent radiation. The laser is an example of a device that produces coherent light. *See also* COHERENT LIGHT, LASER.

COIL

A coil is a helical winding of wire usually intended to provide inductive reactance. The most common form of wire coil is the solenoidal winding (A in the drawing). The wire may be wound on an air core or a core with magnetic permeability to increase the inductance for a given number of turns. Some coils are toroidally wound, as shown at B.

Coils are used in speakers, earphones, microphones, relays, and buzzers to set up or respond to a magnetic field. They are employed in transformers for stepping a voltage up or down, or for impedance matching. A coil wound on a ferrite rod can act as a receiving antenna at low, medium, and high frequencies. In electronic circuits, coils are generally used to provide inductance. *See also* COIL WINDING, INDUCTANCE, INDUCTOR.

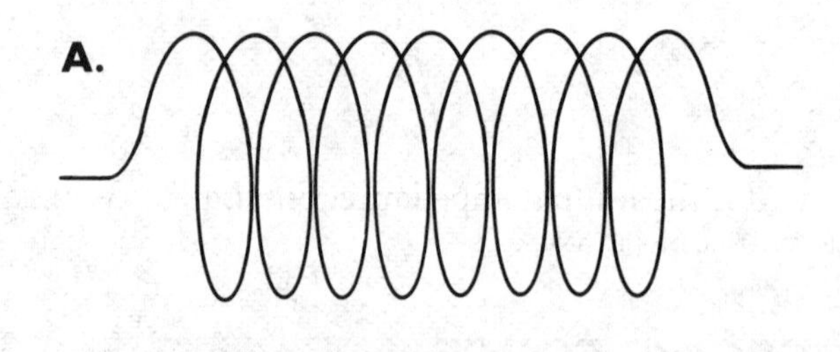

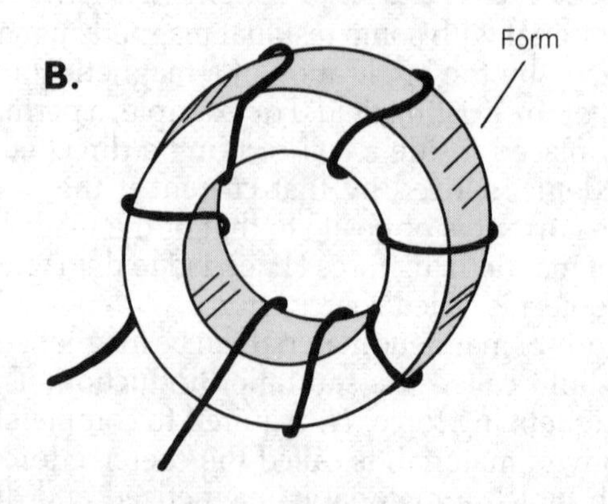

COIL: A solenoidal winding (A) and a toroidal winding (B).

COIL WINDING

When winding a coil to obtain a specific value of inductance, the dimensions of the coil, the number of turns, the type of core material, and the shape of the coil must be specified.

Usually, if a powdered-iron or ferrite core material is used for coil winding, data is furnished with the core as a guide for obtaining the desired value of inductance. For air-core solenoidal coils with only one layer of turns, the inductance L in microhenrys is given by the formula:

$$L = \frac{r^2N^2}{9r + 10m}$$

where r is the coil radius in inches, N is the number of turns, and m is the length of the winding in inches (see illustration). The inductance of a single-layer air-core solenoid increases with the square of the number of turns, and directly with the coil radius. For a coil with a given radius and number of turns, the greatest inductance is obtained when the length m is made as small as possible. *See also* INDUCTANCE, INDUCTOR.

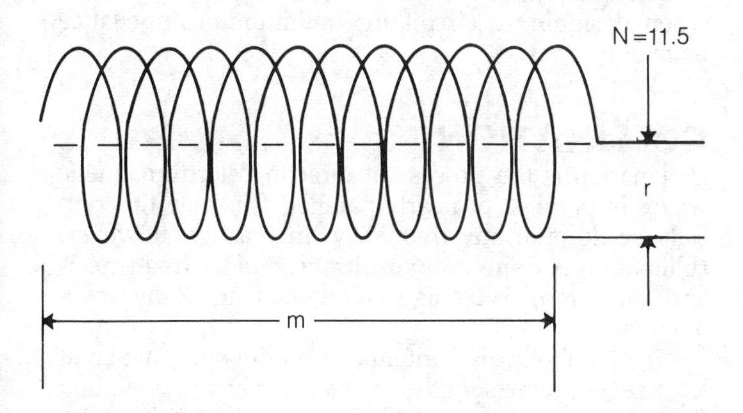

COIL WINDING: A drawing of a single-layer, solenoidal coil showing its key dimensions.

COINCIDENCE CIRCUIT

A coincidence circuit is any digital circuit that requires a certain combination of input pulses to generate an output pulse. The input pulses must arrive within a designated period of time.

The most common form of coincidence circuit is a combination of AND gates (*see* AND GATE). For an output pulse to occur all the inputs of an AND gate must be in the high state. Any complex logic circuit can be considered a coincidence circuit, but the term is generally only used with reference to the logical operation AND.

COLD CATHODE

When a tube is operated without a filament, the cathode is said to be cold. Certain tubes are designed to work with cold cathodes; these include photoelectric tubes, mercury-vapor rectifiers, and some voltage-regulator tubes. In a cold-cathode tube, the electrons are literally pulled from the cathode by a high positive anode potential.

The schematic symbol for a cold-cathode diode tube is shown in the drawing. *See also* PHOTOTUBE.

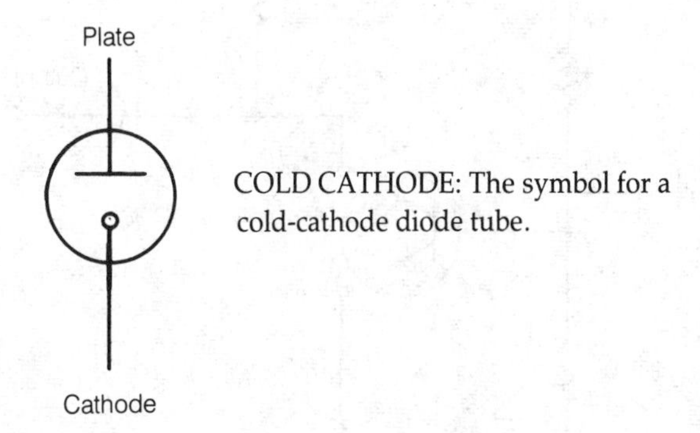

COLD CATHODE: The symbol for a cold-cathode diode tube.

COLLECTOR

The collector is the part of a semiconductor bipolar transistor into which carriers flow from the base under normal operating conditions. The base-collector junction is reverse-biased; in a PNP transistor the base is positive with respect to the collector. In an NPN transistor, the base is negative with respect to the collector.

The output from a transistor oscillator or amplifier is usually taken from the collector. The collector can be placed at ground potential in some situations, but it is usually biased with a direct-current power supply. The amount of power dissipated in the base-collector junction of a transistor must not be allowed to exceed the rated value, or the transistor will be destroyed. Resistors are often used to limit the current through the collector; these resistors are placed in series with either the emitter or collector lead. In some transistors, the collector is bonded to the outer case to facilitate heat conduction away from the base-collector junction.

COLLECTOR CURRENT

In a bipolar transistor, the collector current is the average value of the direct current that flows in the collector lead. When there is no signal input, the collector current is a pure, constant direct current determined by the bias at the base, the series resistance, and the collector voltage. The collector current for proper operation of a transistor varies considerably, depending on the application.

When a signal is applied to the base or emitter circuit of a transistor amplifier, the collector current alternates. But its average value, as indicated by an ammeter in the collector circuit (see drawing), may change only slightly. The collector current is the difference between the emitter current and the base current.

COLLECTOR RESISTANCE

The internal resistance of the base-collector junction of a bipolar transistor is called the collector resistance. This resistance may be specified either for direct current or for alternating current.

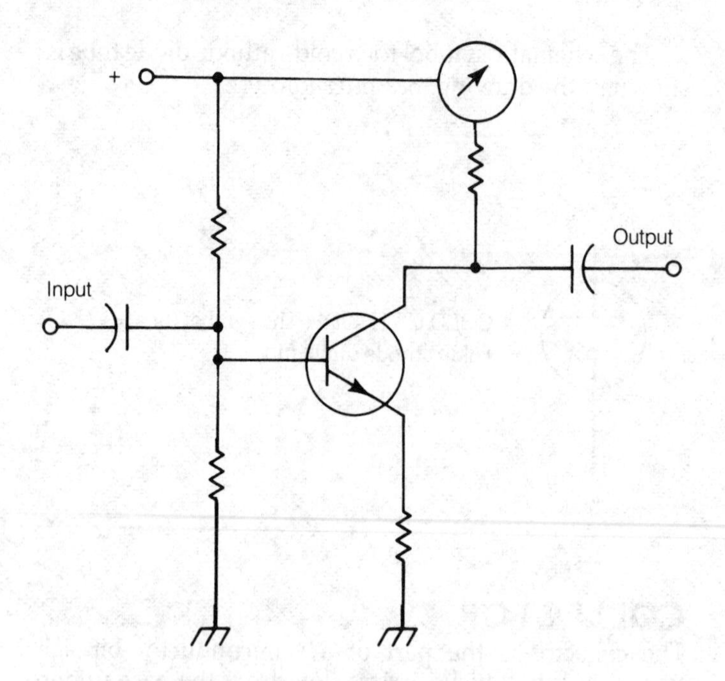

COLLECTOR CURRENT: A circuit for measuring collector current in a transistor amplifier.

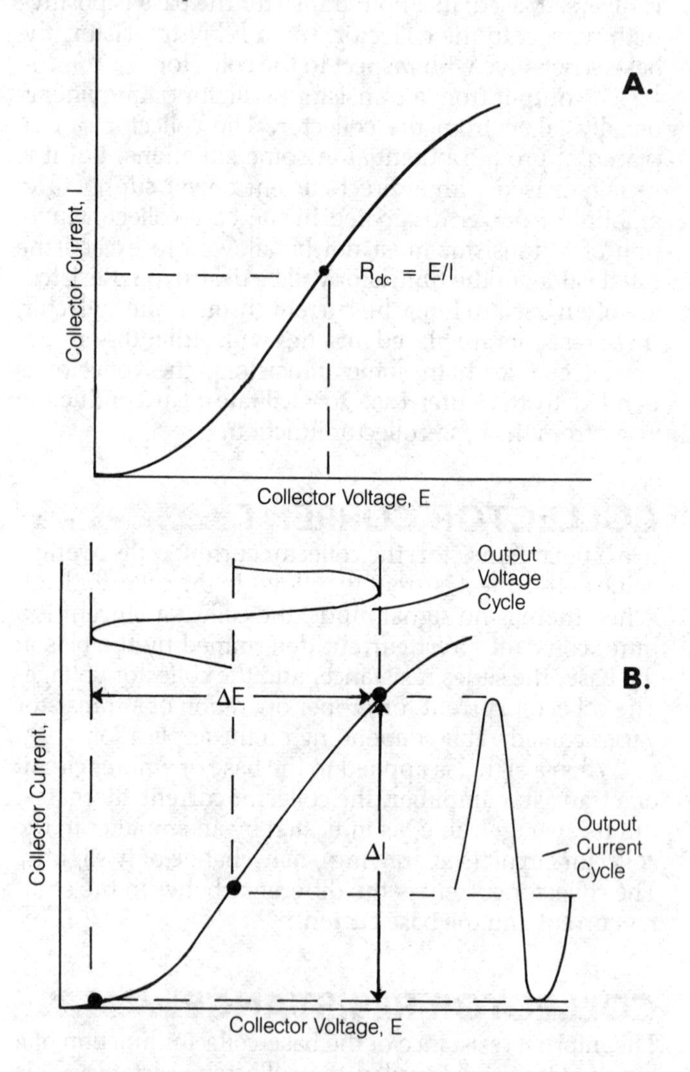

COLLECTOR RESISTANCE: Curves for determining collector resistance for direct current (A) and alternating current (B).

The direct-current collector resistance, R_{dc}, is given by:

$$R_{dc} = E/I$$

where E is the collector voltage and I is the collector current (see A in the illustration). This resistance varies with the supply voltage, the base bias, and any resistances in series with the emitter or collector.

The alternating-current resistance, R_{ac}, is given approximately by:

$$R_{ac} = \Delta E/\Delta I$$

where ΔE and ΔI are the ranges of maximum-to-minimum collector voltage and current, as the fluctuating output current goes through its cycle. The illustration shows a method of approximately determining this dynamic resistance (B). The value of R_{ac} is affected by the same factors that influence R_{dc}. In addition, the class of operation has an effect, as does the magnitude of the input signal.

The alternating-current collector resistance is useful when designing a circuit for optimumum impedance matching.

COLLIMATION

Collimation is the process of directing electromagnetic waves in parallel, or nearly parallel. This may theoretically be done at any frequency. In practice, however, collimation is done only at ultrahigh radio frequencies and at infrared, visible-light, ultraviolet, and X ray wavelengths.

The parabolic dish antenna (A in drawing), when at least several wavelengths in diameter at the operating frequency, is a common example of a collimating device (*see* DISH ANTENNA, PARABOLOID ANTENNA in the ANTENNA DIRECTORY). A spherical dish will work almost as well. Waves are emitted by a small horn at the focal point of the reflector. The waves then bounce off the reflector and are emitted from the assembly as a parallel beam. A parabolic or spherical mirror acts as a collimating device for visible light. Most flashlights and lanterns use this principle.

At infrared, visible-light, and ultraviolet wavelengths, collimation can be accomplished with a refracting lens, as shown at B. In a cathode-ray tube, the electron beam is collimated by the fields between sets of focusing electrodes. Electric fields may also be employed to collimate a beam of protons, alpha particles, or other charged atomic particles.

COLLINEAR ANTENNA

See ANTENNA DIRECTORY.

COLOR-BAR GENERATOR

A color-bar generator is an instrument for testing and adjustment of a color television receiver. It operates in a

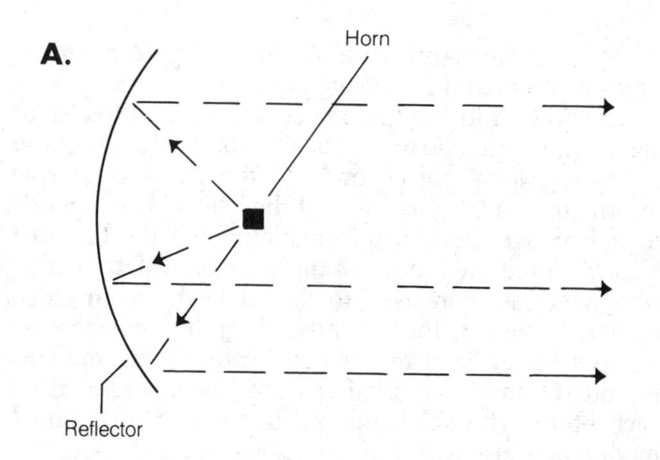

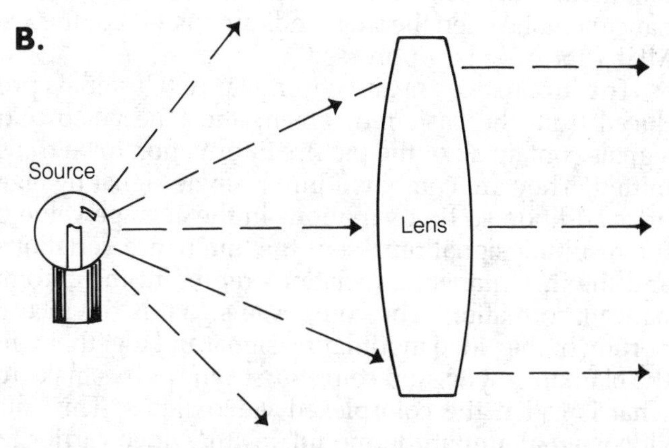

COLLIMATION: Collimation is achieved for radio energy with a parabolic reflector (A) and for light with a lens (B).

manner similar to a black-and-white bar generator (*See* BAR GENERATOR).

The bar pattern can be either vertical or horizontal, and in various color combinations. Color reproductions, as well as brightness, contrast, horizontal and vertical linearity, and focus can be adjusted. *See also* COLOR TELEVISION, TELEVISION.

COLOR CODE

A color code represents component values and characteristics with colors. This scheme is used with almost all resistors (*see* RESISTOR COLOR CODE). Color codes are sometimes used on capacitors, inductors, transformers, and transistors.

When a cable has multiple conductors, the individual wires are usually color-coded as a means of identification of conductors at opposite ends of the cable. In direct-current power leads, the color black might signify ground or negative polarity, and red signifies the hot or positive lead. In house wiring, color coding can vary.

COLOR FIDELITY

The faithfulness or accuracy of the color in a color television is called color fidelity. Color fidelity depends on the proper intensity and linearity of the three primary colors—red, blue, and green—in the color signal from the transmitter. It also depends on proper alignment of the receiving equipment. Color fidelity can be degraded by interference.

The color fidelity of a television receiver is adjusted with a color-bar generator (*see* COLOR-BAR GENERATOR). Any scene will then appear nearly life-like. Fine adjustment of the color-intensity and color-tint functions are provided as front-panel controls on most television receivers. *See also* COLOR TELEVISION.

COLORIMETRY

The process of measuring color is called colorimetry. This is done with an instrument called a colorimeter, which compares any color to a reference standard. The three primary colors—red, blue, and green—are synthesized with constant chroma, but the luminance (brightness) of each color can be adjusted. By properly combining the three primary colors, the color under test can be matched. The relative components of the color are then indicated very precisely. Matching is done by visual comparison. *See also* CHROMA, HUE, SATURATION.

COLOR PICTURE SIGNAL

A color picture signal is a modulated radio-frequency signal that contains all the information needed to reproduce accurately a scene in full color. The channel width of a color-television picture signal is typically 6 MHz.

The illustration shows the modulation envelope of a single horizontal line of the video information in a color picture signal. The equalizing pulses, horizontal synchronization pulses, and vertical synchronization pulses are not shown. The horizontal blanking pulse turns off the picture-tube electron beam as it retraces from the end of one line to the beginning of the next line. This pulse is followed by a color-burst signal, which consists of eight or nine cycles at 3.579545 MHz.

The color (chroma) is encoded as a phase- and amplitude-modulated (relative to the color burst) 3.58 Mhz signal that is superimposed on the intensity (black-and-white) level signal. The video information is then set as the electron beam scans from left to right. There are 525 or 625 horizontal lines per frame.

The illustration at B shows the equalizing, horizontal synchronization, and vertical synchronization pulses of the picture signal. These pulses are sent after each complete frame. *See also* TELEVISION.

COLOR TELEVISION

The transmitted images of color television are reproduced at the TV receiver in colors closely matching those of the original scene. The color information is transmitted by the color television signal. Color television signals are compatible with both black-and-white and color television receivers. This means that a monocolor receiver obtains a black-and-white picture from a color signal, but a color receiver can receive both black-and-white and color pictures.

Both the color and black-and-white signal consist of an FM (frequency modulation) sound carrier and an AM

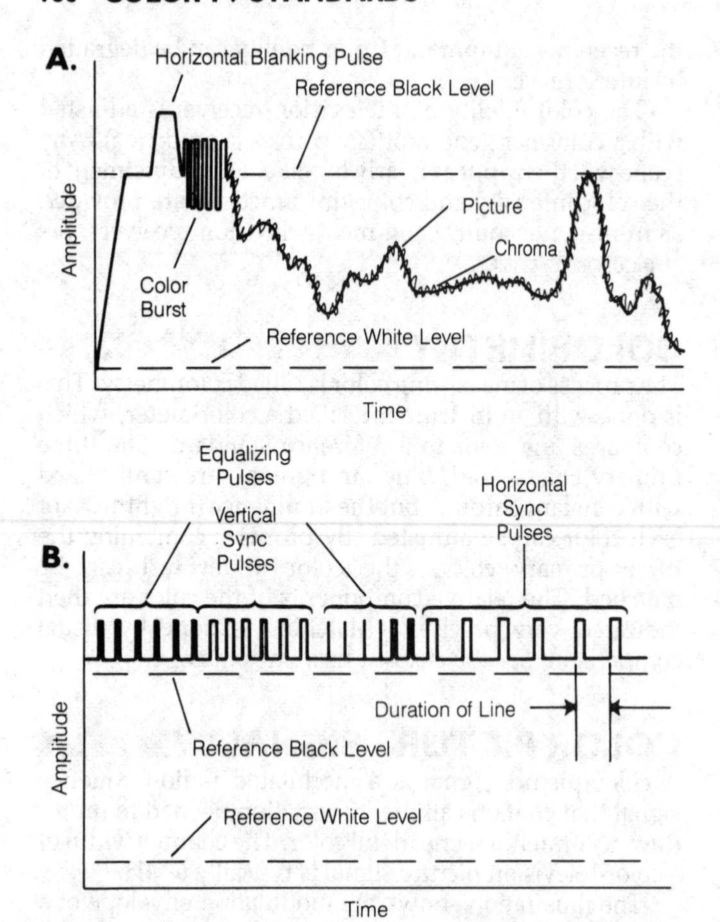

COLOR PICTURE SIGNAL: Waveforms of the video information contained in a single horizontal sweep (A) and the control pulses of a television receiver (B).

(amplitude modulation) video carrier within a 6-MHz channel. The video portion of the color signal, like that of the black-and-white signal consists of horizontal lines of picture information, with the individual lines followed by sync (synchronous) and blanking pulses.

The part of the black-and-white signal that carries the picture information is the carrier amplitude, modulated to represent the brightness or darkness of the original image. However, the part of the color video signal that carries picture information is a composite of color information and the amplitude variations. In addition, the color signal has an additional feature, the color-sync burst, immediately after the horizontal sync pulses. Each horizontal line of the video portion of the color signal consists of picture information, horizontal blanking, and sync pulses, and a color sync burst.

All of the picture information in a color television signal is derived from the red, green, and blue video signals obtained by scanning the scene to be transmitted. The amplitude of these voltage signals tracks the variations in the color content of the scene being scanned. The other colors, including white, can be obtained by mixing the primary colors in the proper ratios.

The three primary colors are mixed or combined to produce the luminance, or *Y*, signal corresponding to the light and dark variations of the televised scene. This waveform is similar to the modulating signal used for black-and-white television. The proportions of red, green, and blue in this white signal are 30 percent red, 59 percent green, and 11 percent blue.

In addition to the luminance signal, the three color signals are combined to produce two other signals: *Q* and *I*. The Q signal corresponds to the green or purple information in the picture, and the I signal corresponds to the orange or cyan information. Both the Q and I signals amplitude modulate the same 3.58 MHz subcarrier, which is then used to modulate the main video carrier. However, the subcarrier is shifted in phase 90 degrees before it is modulated by the Q signal. This permits Q and I modulation to be distinguished from each other. The sidebands produced by the Q and I modulation are added vectorially to form the chrominance signal. When the sidebands are combined, the 3.58 MHz subcarrier is suppressed.

The luminance (Y) and chrominance (C) signals produced from the basic red, green, and blue video color signals contain all of the picture information to be transmitted. They are combined into a single signal by algebraic addition so that variations in the average value of the resulting signal represent the luminance variations, and the instantaneous variations represent the chrominance information. The composite signal is the picture portion of the video modulating signal and together with the blanking, sync, and color-burst sync pulses make up what is called the colorplexed video signal. This total video signal amplitude modulates the video carrier for transmission to the receiver.

Each color-sync burst consists of a few cycles of the unmodulated, 3.58 MHz subcarrier used to produce the Q and I signals. However, the subcarrier was suppressed after the Q and I signals were generated so it must be reinserted in the receiver for detection of the Q and I signals. The receiver uses the color-sync burst to synchronize the phase of the reinserted subcarrier with that of the original subcarrier at the transmitter.

Both the luminance and chrominance signals are detected in color receivers. Together with blanking, sync, and color-burst sync pulses, the Y and C signals produce the color picture. The chrominance signal provides the color variations for the picture, and the luminance signal provides the variations in intensity or brightness of the colors.

COLOR TV STANDARDS

Three different analog color TV broadcast standards are used in the world and they are incompatible. The NTSC standard devised by the National Television System Committee was approved by the Federal Communications Commission (FCC) in 1953 was the first worldwide standard. It is now in use in the United States, Canada, Japan, Mexico, and some South American and Asian countries. The second of the three main standards is PAL (for phase-alternation line) adopted in Western Europe, Australia, and some South American and African countries. The third is the Secam standard (for sequential color with memory) adopted by France, the USSR, various Eastern Bloc countries, and a few African countries.

The standards were intended to make color television

fully compatible with monochrome (black-and-white) television in countries where it existed. The dual objectives were the reception of high quality black-and-white images on monochrome receivers from color signals and the reception of high quality black-and-white images from monochrome signals on color receivers. There are many variations on the three major worldwide standards because of local differences in black-and-white television. Most of the variants differ in the number of lines scanned, the bandwidth of the various carriers, and other characteristics.

All NTSC systems scan 525 lines per frame at a rate of 30 frames (60 fields) per second. By contrast, most PAL and Secam systems scan 625 lines per frame at a rate of 25 frames (50 fields) per second.

Three characteristics are needed for a receiver to reconstruct a faithful color image transmitted by the television camera: (1) brightness or intensity, (2) hue, and (3) saturation. Hue is the spectral wavelength that defines each color in a scene, and saturation is the vividness of the color with respect to a neutral gray of the same brightness.

The luminescence signal occupies a wide bandwidth within the transmitted television channel of all standards. When received on a conventional monochrome television receiver, the luminance signal produces a black-and-white version of the color picture. To a color television receiver, the luminance signal provides the brightness information. The NTSC, PAL, and Secam standards all produce the luminance signal by similar means, but each occupies a different bandwidth.

A second signal, the chrominance, occupies a narrower bandwidth and conveys all the chromatic information about hue and saturation. Each component of the chrominance signal is carried as a color-difference signal in the standard systems. The signal is expressed as a difference between each of the basic red, green, or blue (R, G, or B) color signals and luminance signal (Y): R-Y, G-Y, and B-Y.

Color-difference signals modulate a subcarrier, which is transmitted together with the luminance signal as a composite video signal. The receiver decodes the color-difference signals, separates them, and adds them individually to the luminance signaling three separate paths (red, green, and blue) to recreate the color image.

In the NTSC standard, the chrominance information is carried by simultaneously amplitude modulating and phase modulating a subcarrier contained within the high frequency portion of the luminance signal. The frequency of the subcarrier is equal to an odd multiple of half the horizontal line-scanning rate of the image. Thus the chrominance and luminance information are interleaved across the video spectrum somewhat like the teeth of a comb. The hue is determined by the instantaneous phase of the subcarrier. The saturation is determined by the ratio of the instantaneous amplitude of the luminance signal.

The color signals R, G, and B are converted at the transmitter by appropriate matrix circuity into two other chrominance components designated I and Q (for in-phase and quadrature). These I and Q components make up the NTSC chrominance signal. This chrominance signal is added to the luminance signal along with appropriate horizontal and vertical blanking synchronization signal to create the composite video signal of the NTSC standard.

COLOR TV TUBE

The color television tube is a modification of the basic black-and-white electromagnetic cathode ray tube (CRT). These tubes use a metal shadow mask for the selection of red, blue, and green components to create the desired TV picture. The red, green, and blue (RGB) colors are generated from phosphors that are excited by electrons in beams emitted from three different electron guns. There is one gun for each color.

Figure 1 shows the relationship between the shadow mask, the three electron guns and the phosphors. Each beam converges at the shadow mask as shown in Fig. 2A, and each beam approaches the holes at a slightly different angle. Because of this angular relationship, the red beam only strikes the red phosphor, the blue beam strikes only the blue phosphor, and the green beam strikes only the green phosphor. But each beam illuminates more than one hole as shown in Fig. 2B. With separate grid modulation of the three electron gun beams, the three basic colors can be mixed in different proportions to yield a wide range of colors.

There are four different architectures for color TV and monitor CRTs (cathode ray tubes). A delta gun arrangement with a dot mask was popular in TV sets until the mid-1970s. The metal mask had evenly spaced holes and the RGB phosphors were clustered on the glass faceplate in groups of three or triads as shown in Fig. 2. But this style lost favor because convergence (the ability of the three beams to meet at the shadow mask at all positions) was difficult to obtain.

The slotted mask with in-line guns was devised in the TV industry. The metal mask has a series of vertical slots in place of the holes and the phosphors on the glass faceplate are uniformly spaced vertical RGB bars rather than triads. Sony Corporation invented the Trinitron in 1968, similar in design except that it has a three-beam electron gun, and the metal mask is a series of metal strips. The in-line gun design, as shown in Fig. 2, simplified convergence with properly designed yokes. Increased brightness is achieved with this design.

TV images are usually scenes where the objects are generally blobs of intensity and color. Transitions from one object to the next are fuzzy because of the limited bandwidth of the video signal. By contrast, the image in computer displays is composed of lines with sharp transitions of luminance. The in-line/strip-mask CRT provides excellent TV pictures, but the in-line gun/dot-mask architecture gives better results for the display of text and graphics.

COLPITTS OSCILLATOR

See OSCILLATOR CIRCUITS.

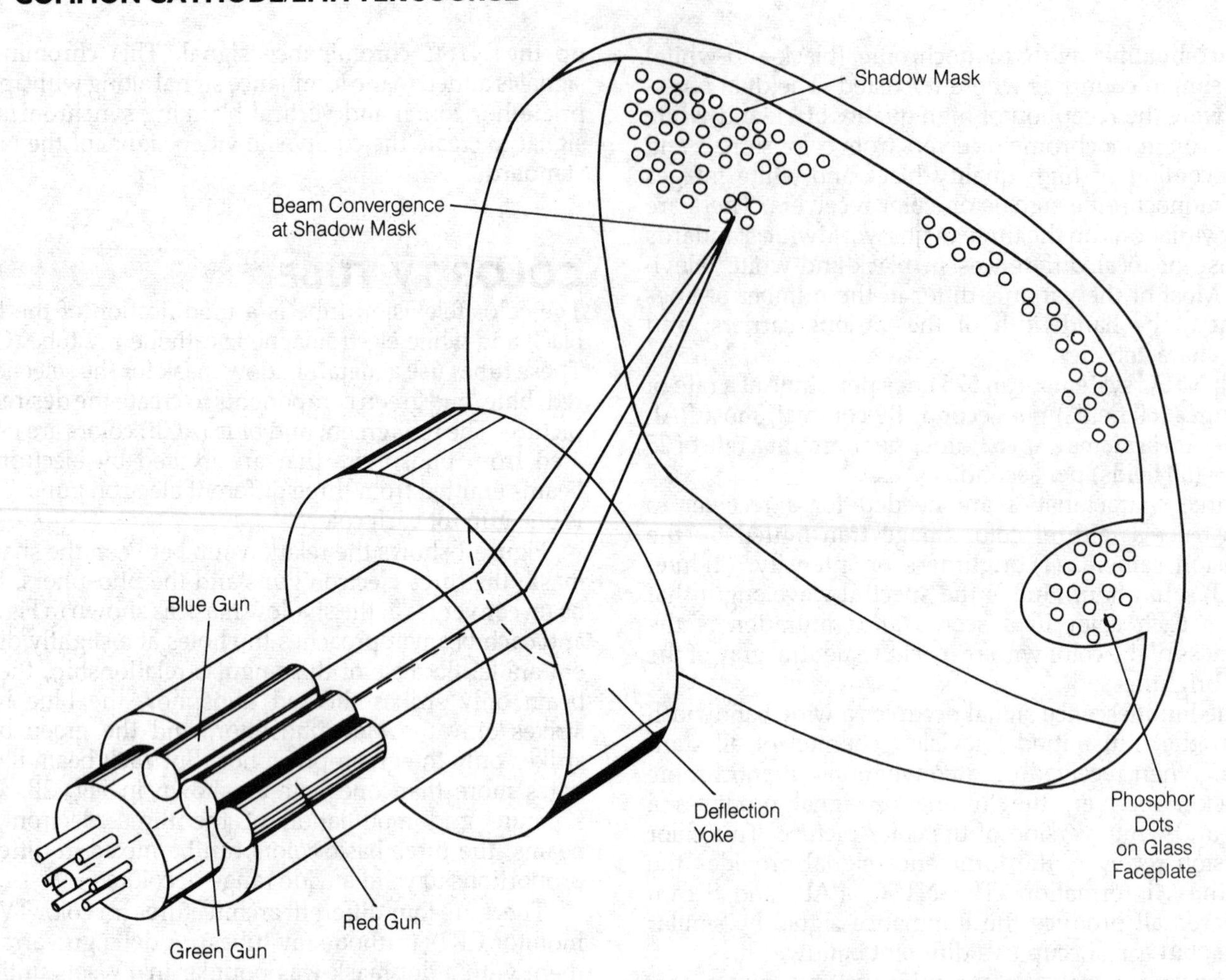

COLOR TV TUBE: Fig. 1. Cathode ray tube with in-line electron guns, metal dot mask and phosphor dot triads on glass faceplate.

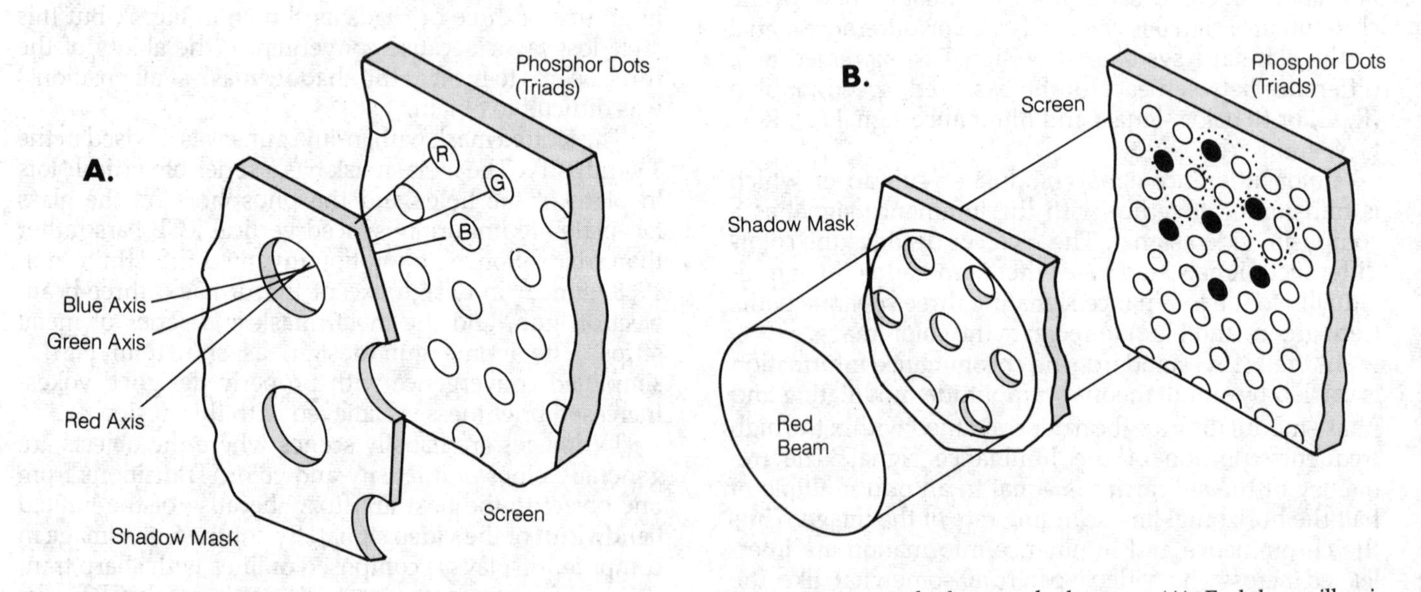

COLOR TV TUBE: Fig. 2. Convergence of blue, green and red electron beams at shadow mask shown at (A). Each beam illuminates more than one phosphor dot triad (B).

COMMON CATHODE/EMITTER/SOURCE

The common-cathode, common-emitter, and common-source circuits are probably the most frequently used amplifier arrangement with tubes, transistors, and field-effect transistors. The schematics show simple common-cathode, common-emitter, and common-source amplifier circuits. The cathode, emitter, or source is always operated at ground potential with respect to the signal; it need not be at ground potential for direct current.

The common-cathode and common-source circuits have high input and output impedances. The common-emitter circuit is characterized by moderately high input impedance and high output impedance. In all three circuits, the input and output signals are 180 degrees out of phase. *See also* FIELD-EFFECT TRANSISTOR, TRANSISTOR.

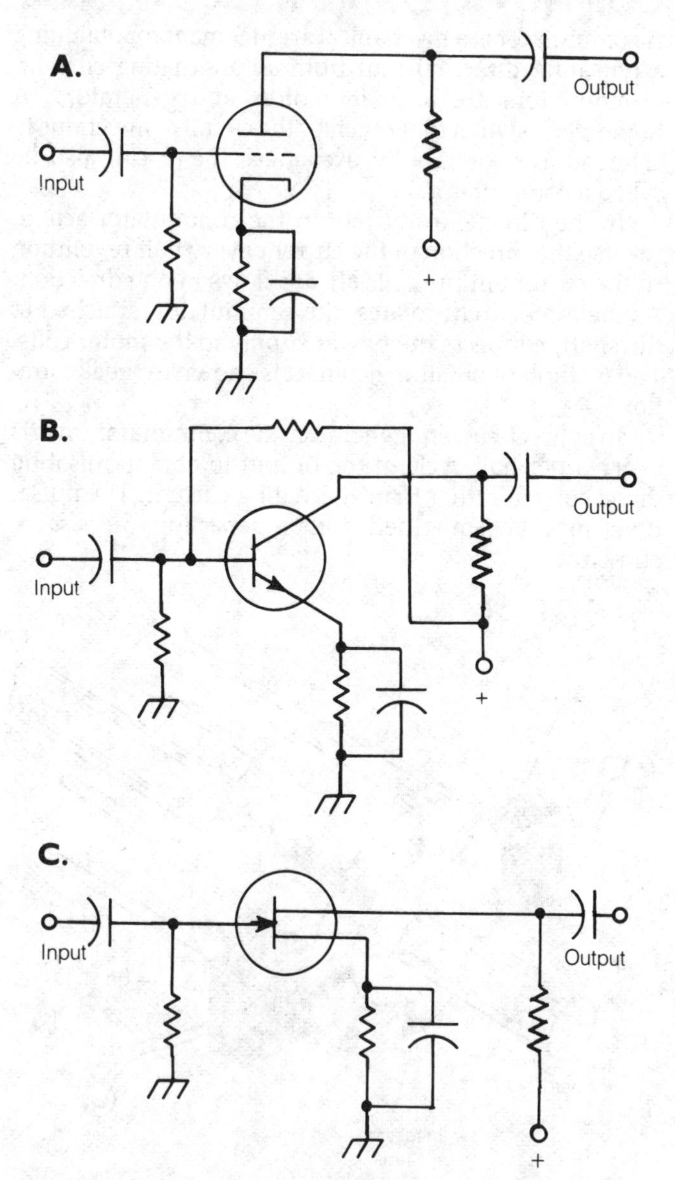

COMMON CATHODE/EMITTER/SOURCE: Schematics of amplifiers—tube with common cathode (A), transistor with common emitter (B), and field-effect transistor with common source (C).

COMMON GRID/BASE/GATE

The common-grid, common-base, and common-gate circuits are amplifier or oscillator arrangements with tubes, transistors, and field-effect transistors respectively. These circuits have excellent stability as amplifiers. They are less likely to break into unwanted oscillation than the common-cathode, common-emitter, and common-source circuits (*see* COMMON CATHODE/EMITTER/SOURCE).

The grid, base, or gate is usually connected directly to ground; occasionally a direct-current bias may be applied and the grid, base, or gate shunted to signal ground with a bypass capacitor. The drawings illustrate typical common-grid, common-base, and common-gate amplifiers.

The common-grid, common-base, and common-gate circuits display low input impedance. They require considerable driving power. The output impedance is high; the input and output waveforms are in phase. This amplifier is often used as a power amplifier at radio frequencies. *See also* FIELD-EFFECT TRANSISTOR, TRANSISTOR.

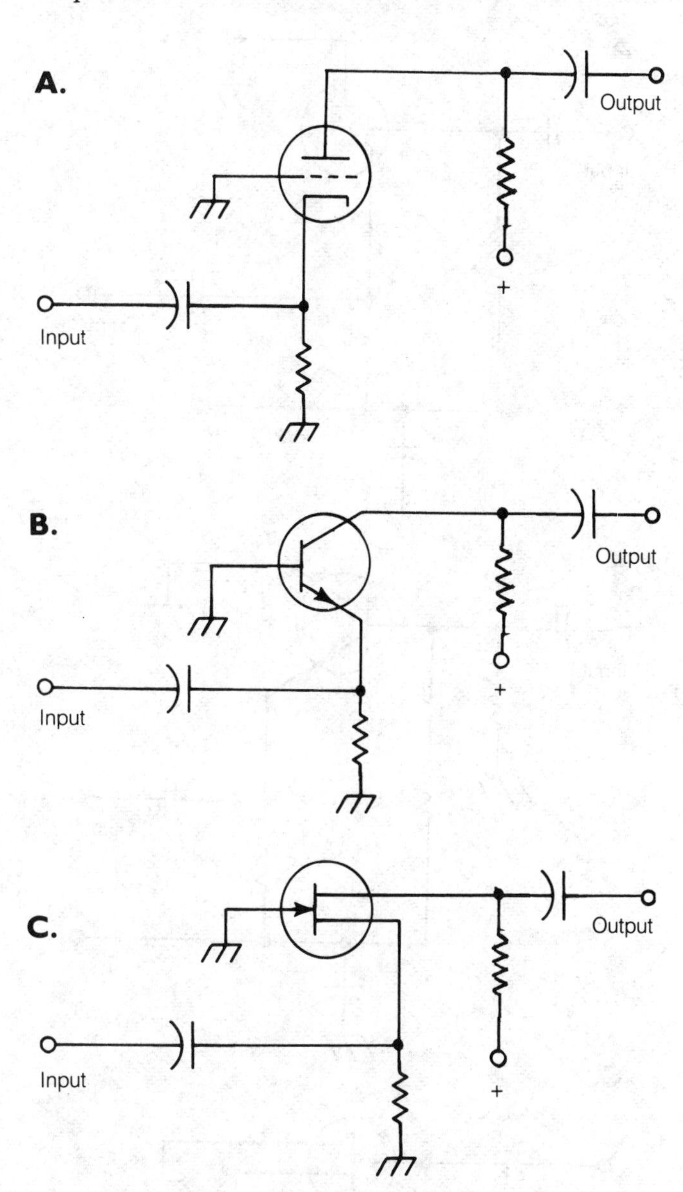

COMMON GRID/BASE GATE: Schematics of amplifiers—tube with common grid (A), transistor with common base (B), and field-effect transistor with common gate (C).

COMMON PLATE/ COLLECTOR/DRAIN

The common-plate, common-collector and common-drain circuits are generally used in applications where a high-impedance generator must be matched to a low-impedance load. The input impedances of the common-plate, common-collector, and common-drain circuits are high; the output impedances are low. The schematics show typical common-plate, common-collector, and common-drain circuits. They are sometimes called cathode-follower, emitter-follower, and source-follower circuits. The gain is always less than unity (*see* EMITTER FOLLOWER, SOURCE FOLLOWER).

The plate, collector, or drain can be grounded directly. Biasing can be accomplished in a manner identical to that of the common-cathode, common-emitter, and common-source circuits (*see* COMMON CATHODE/EMITTER/SOURCE); the plate, collector, or drain is then placed at signal ground with a bypass capacitor, and the output is taken across a cathode, emitter, or source resistor or transformer. *See also* FIELD-EFFECT TRANSISTOR, TRANSISTOR.

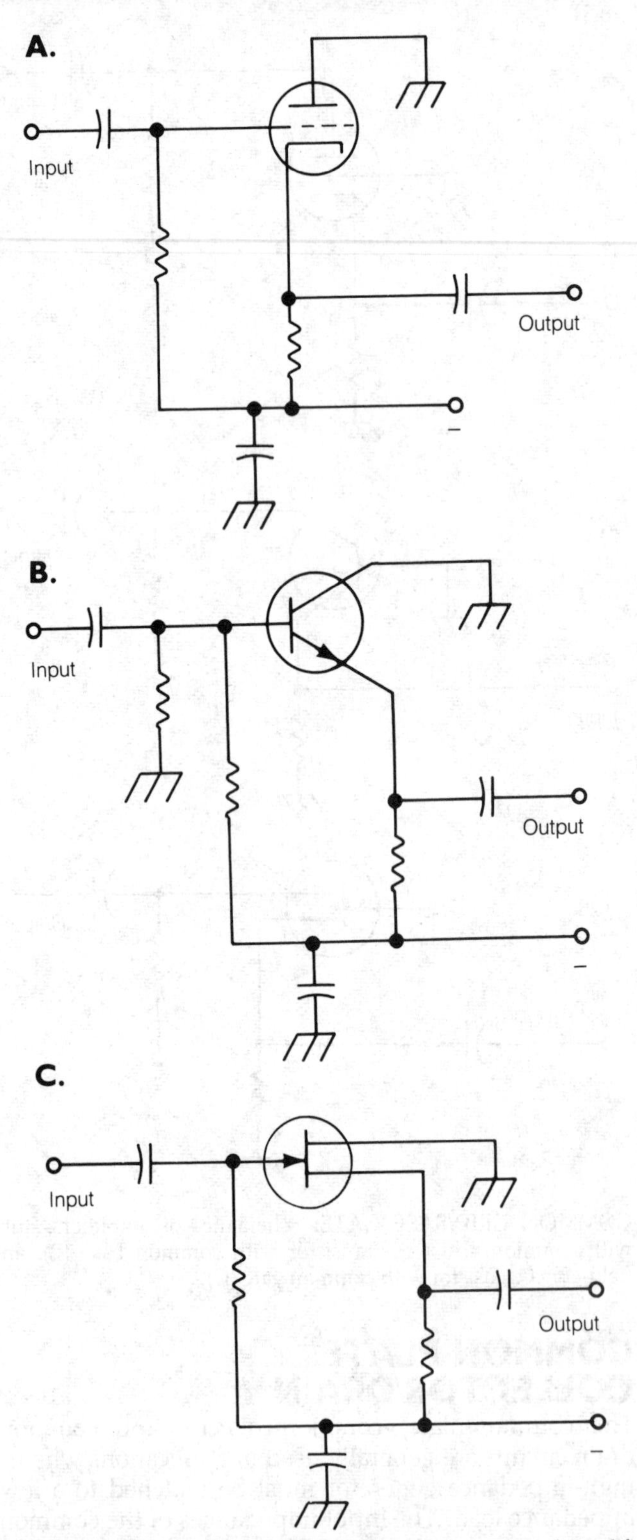

COMMON PLATE/COLLECTOR/DRAIN: Schematics of amplifiers—tube with common plate (A), transistor with common collector (B), and transistor with common drain (C).

COMMUTATOR

A commutator is a mechanical arrangement for obtaining a pulsating direct current from an alternating current. Commutators are used in motors and generators. A high-speed switch that reverses the circuit connections to a transducer, or rapidly exchanges them, can also be called a commutator.

In the direct-current motor, the commutator acts to reverse the direction of the current every half revolution so the current in the coils always flows in one direction. As the motor shaft rotates, the commutator, attached to the shaft, connects the power supply to the motor coils. The method of obtaining contact is shown in the illustration.

In a direct-current generator, the commutator inverts every other half cycle of the output to obtain pulsating direct current rather than alternating current. The pulsations may be smoothed with a capacitor. *See also* DC GENERATOR.

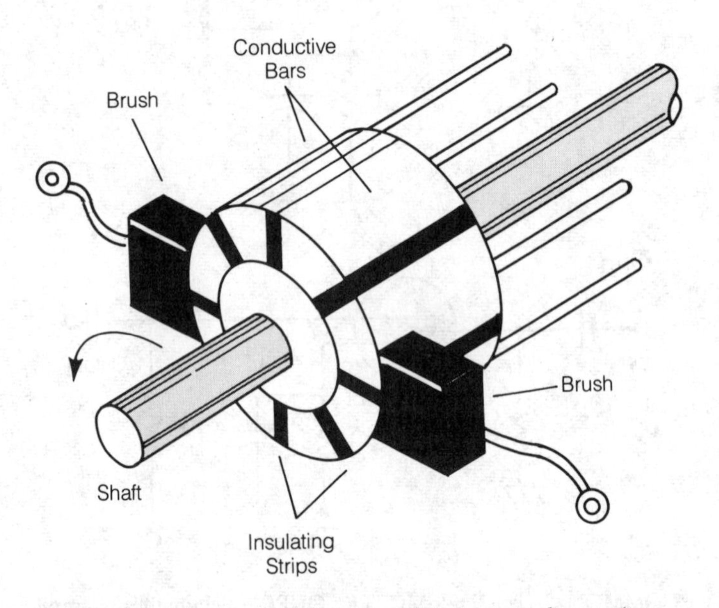

COMMUTATOR: A commutator and brushes make continuous electrical contact with the rotating armature of a direct-current motor or generator.

COMPACT DISK

A compact disk (CD) is an alternate to magnetic-disk and tape recording that has become a successful competitor to both vinyl records and tape recordings for the storage of speech and music. A variation of the CD is used to store video information. CDs provide the sonic detail and realism of far more costly professional turntable/tonearm/cartridge systems. Between selections or during musical pauses, there is total silence without annoying ticks or pops. A small section of a CD is illustrated and

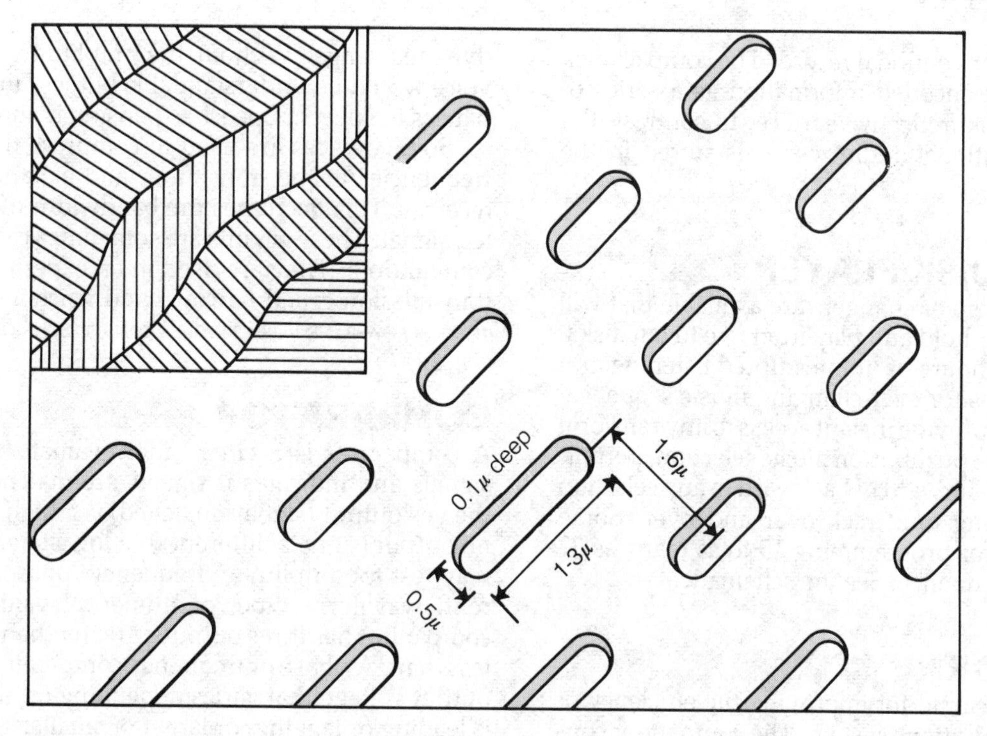

COMPACT DISK: Music is stored digitally on a compact disk as a series of microscopic pits. Inset at upper left illustrates the far larger grooves of a conventional disk.

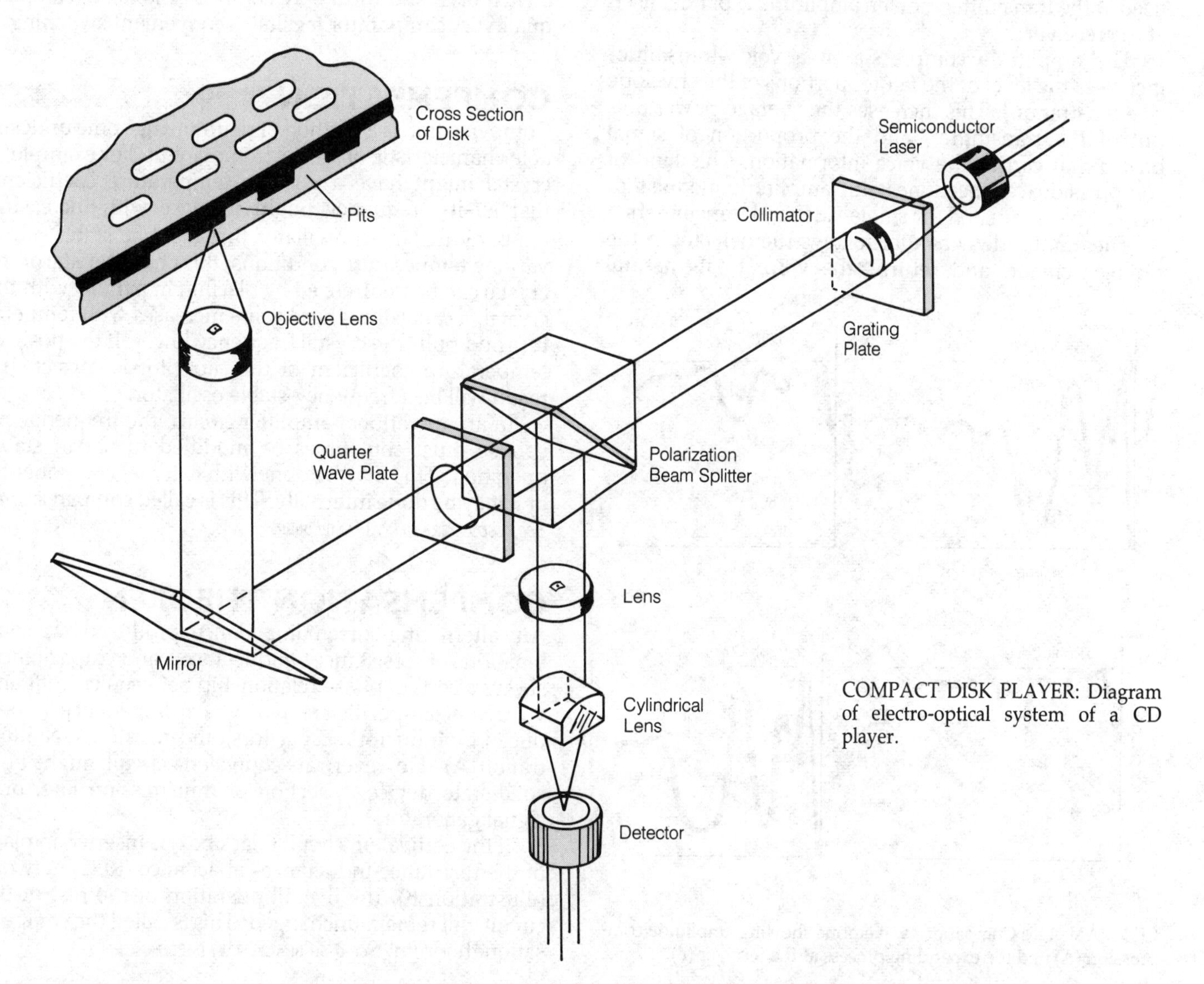

COMPACT DISK PLAYER: Diagram of electro-optical system of a CD player.

compared with a conventional record. The compact disk stores the digitally encoded information as a series of microscopic pits on a reflective surface. In contrast, the music of a conventional audio record is stored in the grooves.

COMPACT DISK PLAYER

Compact disk players or changers are available that will play a single disk or hold and play from five to ten disks. Changers provide hours of uninterrupted listening and random-play modes for ever-changing music sequence. Most CD changers provide instant access to any music on the disk. Many have pushbutton music selection, permitting the user to skip forward or backward to any selection or to keep repeating one track over and over. Some include provision for programming 15 to as many as 32 selections in any sequence. See the schematic.

COMPANDOR

A compandor is a circuit for improving the efficiency of an analog communications system. The compandor consists of two separate circuits: an amplitude compressor used at the transmitter, and an amplitude expander used at the receiver.

The amplitude compressor in a voice transmitter increases the level of the fainter portions of the envelope (see illustration). This increases the average power output of the transmitter, and the proportion of signal power that carries the voice information. This kind of compression can be done with amplitude-modulated, frequency-modulated, or single-sideband transmitters.

The amplitude expander follows the detector in the receiver circuit, and returns the voice to its natural dynamic range. Without the amplitude expander, the voice would be understandable but less intelligible after pauses in speech. Speech expansion is shown at B.

Some devices are used to compress the modulating frequencies at the transmitter and expand them at the receiver. This decreases the bandwidth of the transmitted signal. These circuits are sometimes called frequency compandors. They are used in an experimental form of transmission called narrow-band voice modulation. *See also* NARROW-BAND VOICE MODULATION, SPEECH COMPRESSION.

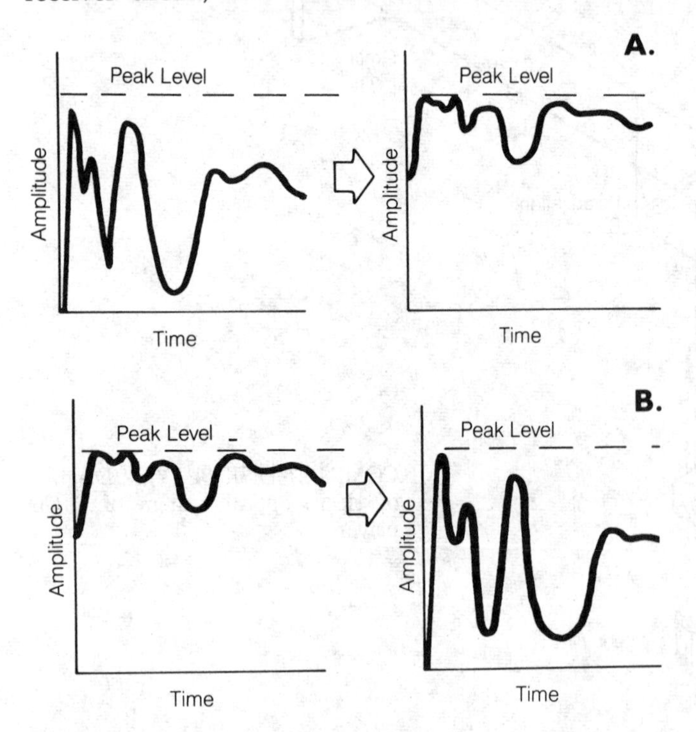

COMPANDOR: Compandor waveforms showing amplitude compression (A) and the expanding process at the receiver (B).

COMPARATOR

A comparator is a circuit that evaluates two or more signals and indicates if signals are matched. Typically, the yes output (signals matched) is a high state, and the no output (signals different) is a low state. Comparators may test for amplitude, frequency, phase, voltage, current, waveform type, or numerical value. A number comparator has three outputs: greater than, equal to, and less than. A phase or frequency comparator may have an output voltage that varies depending on which quantity is leading or lagging, or larger or smaller.

In high-fidelity audio testing, it might be desirable to switch back and forth between two systems to compare quality. A comparator facilitates convenient switching.

COMPENSATION

Compensation is a method of neutralizing some undesirable characteristic of an electronic circuit. For example, a crystal might have a positive temperature coefficient; that is, its frequency might increase with increasing temperature. In an oscillator that must be stable under varying temperature conditions, this characteristic of the crystal can be neutralized by placing, in parallel with the crystal, a capacitor whose value increases with temperature and pulls the crystal frequency lower. If the positive temperature coefficient of the capacitor is correct, the result will be a frequency-stable oscillator.

In an operational-amplifier circuit, the frequency response must sometimes be modified to obtain stable operation. This may be done with external components, or it may be done internally. This is called compensation. *See also* OPERATIONAL AMPLIFIER.

COMPENSATION THEOREM

Any alternating-current impedance, produced by a combination of resistance, inductance, and capacitance, shows a certain phase relationship between current and voltage at a specific frequency. An impedance is produced by inductors, capacitors, and resistors (see illustration A). However, an equivalent circuit might be a solid-state device, a section of transmission line, or a signal generator.

If the equivalent circuit (black box) is inserted in place of the resistance-inductance-capacitance (RLC) network (illustration B), the overall operation of the rest of the circuit will remain unchanged. This is called the compensation theorem. *See also* BLACK BOX, IMPEDANCE.

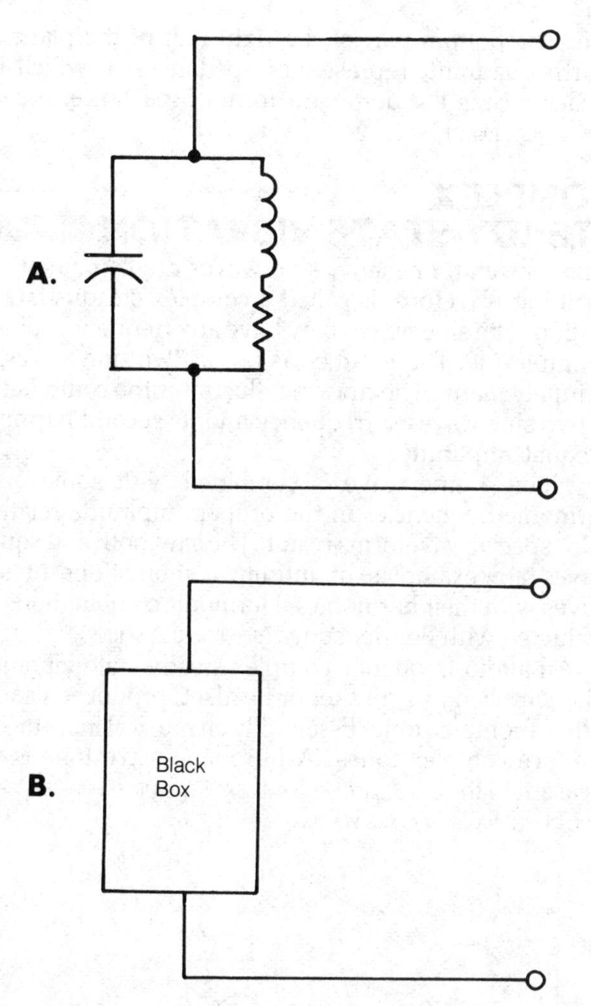

COMPENSATION THEOREM: A schematic of a resistance-inductance-capacitance (RLC) circuit (A) and a block diagram of its equivalent circuit (black box) (B).

COMPILER

In a digital computer, the compiler is the program that converts the higher-order language—such as ALGOL, APL BASIC, COBOL, FORTRAN, Pascal, and PL/1—into machine language. The operator understands and uses the higher-order language; the computer operates in machine language. A compiler is thus an electronic translator.

The compiler must be programmed to translate a higher-order language into machine language, and vice versa. This program is called the assembler. Assembler programs are written in assembly language. *See also* ASSEMBLER AND ASSEMBLY LANGUAGE, HIGHER-ORDER LANGUAGE, MACHINE LANGUAGE.

COMPLEMENT

The complement is a function of numbers—in particular, binary numbers. In the base-2 system, where the only possible digits are 0 and 1, the complement of 0 is 1 and the complement of 1 is 0. In logic, the complement function is the same as negation.

When there are several digits in a binary number, the complement is obtained simply by reversing each digit. For example, the complement of 10101 is 01010. *See also* BINARY-CODED NUMBER.

COMPLEMENTARY COLORS

Two colors are called complementary when they combine to form white as seen by the human eye. Examples of primary colors are red and cyan, green and magenta, and yellow and blue. The two colors must be combined with the proper relative luminosity to appear white. Also, the levels of saturation of the two colors must be in the correct proportions.

When two colored pencil marks combine to form black, the pencils are said to have complementary pigments. Pigment differs from color. The term color refers to radiant energy seen directly from a source, or reflected from a white object. Pigment is a reflecting characteristic of a surface. *See also* PRIMARY COLORS.

COMPLEMENTARY METAL-OXIDE SEMICONDUCTOR

Complementary metal-oxide semiconductor, abbreviated CMOS and pronounced *seamoss*, is the name for a semiconductor technology. Both N-channel and P-channel field-effect transistors are employed. A CMOS integrated circuit is fabricated on a silicon substrate as shown in the figure.

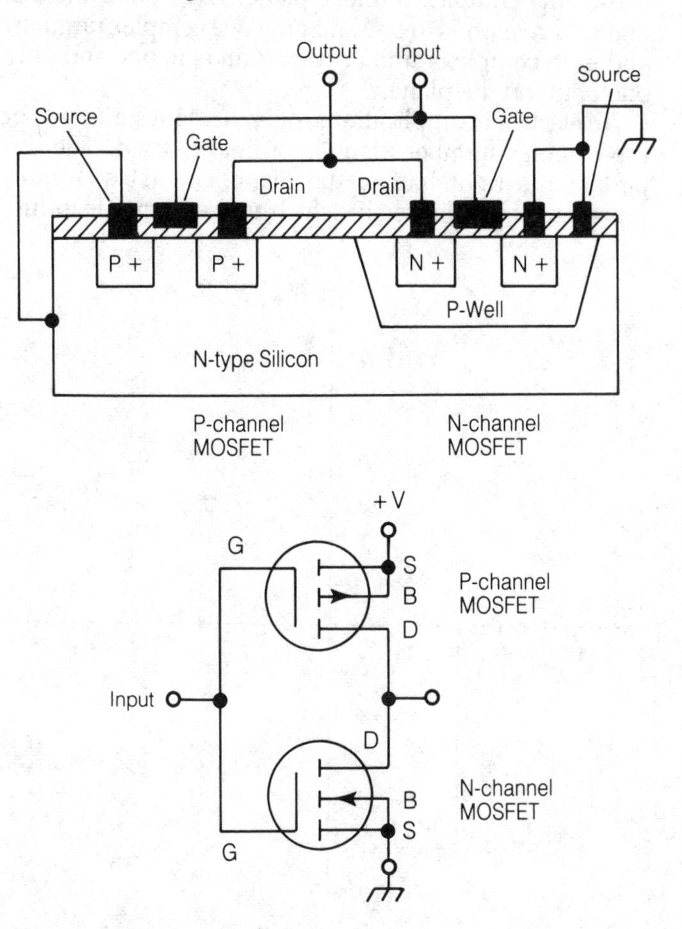

COMPLEMENTARY MOS DEVICE: Basic CMOS gate contains coupled PMOS and CMOS transistors.

The chief advantage of CMOS is its extremely low current consumption. The main disadvantage of CMOS is its susceptibility to damage by static electricity. A CMOS integrated circuit should be stored with its pins embedded in a conducting foam material, available for this purpose. When building or servicing equipment containing CMOS devices, adequate measures must be taken to prevent static buildup. *See also* METAL-OXIDE SEMICONDUCTOR FIELD-EFFECT TRANSISTOR.

COMPLEX NUMBER

A complex number is a quantity of the form $a + bi$, where a and b are real numbers, and $i = \sqrt{-1}$. The number a is called the real part of the complex quantity, and the number bi is called the imaginary part. In electronics, the number i is usually called j, and the complex number is written in the form $R \times jX$ (*see* J OPERATOR). Complex numbers are used by electrical engineers to represent impedances: the real part is the resistance and the imaginary part is the reactance. This representation is used because the mathematical properties of complex numbers are well suited to definition of impedance.

Complex numbers are represented on a two-dimensional Cartesian plane (*see* CARTESIAN COORDINATES). The real part, a, is assigned to the horizontal axis; the imaginary part, bi, is assigned to the vertical axis. The drawing shows the complex-number plane. Each point on this plane corresponds to one and only one complex number, and each complex number corresponds to one and only one point on the plane.

In electronics applications, only the right-hand side of the complex-number plane is ordinarily used. The top part of the right half of the plane, or first quadrant, represents impedances in which the reactance is inductive. The bottom part of the right half of the plane, or fourth quadrant, represents impedances in which the capacitance is the dominant form of reactance. *See also* IMPEDANCE, REACTANCE.

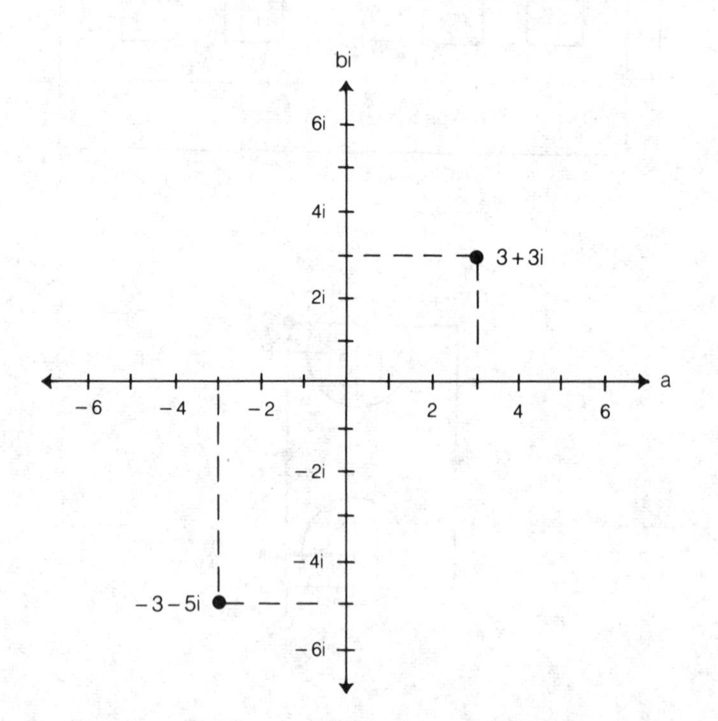

COMPLEX NUMBER: A plot of the complex coordinate plane.

COMPLEX STEADY-STATE VIBRATION

When several, or many, sine waves are combined, the resulting waveform is called a complex steady-state vibration. The sine waves may have any frequency, phase, or amplitude. There can be as few as two sine waves, or infinitely many. The drawing illustrates the combination of two sine waves: a frequency and its second harmonic at equal amplitude.

When a sine wave is combined with some of its harmonic frequencies in the proper amplitude relationship, special waveforms result. The sawtooth and square waves are examples of infinite combinations of sine waves with their harmonics. Harmonic combinations are evaluated with Fourier series (*see* FOURIER SERIES).

An audio-frequency complex steady-state vibration, when applied to a speaker or headset, produces a sound called a complex tone. Essentially all musical instruments produce complex tones. A pure sine-wave tone is unpleasant to the ear. *See also* COMPLEX WAVEFORM, FOURIER SERIES, SAWTOOTH WAVE, SQUARE WAVE.

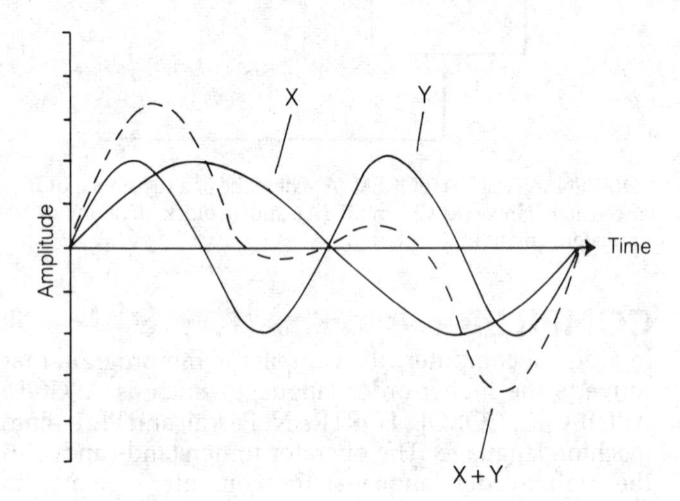

COMPLEX STEADY-STATE VIBRATION: The two sine waves, X and Y, add to produce the complex steady-state waveform X + Y.

COMPLEX WAVEFORM

Any nonsinusoidal waveform is a complex waveform. The human voice produces a complex waveform. Almost all sounds have complex waveforms. The most complex of all waveforms is called white noise, a random combination of sound frequencies and phases. If a complex waveform displays periodic properties, it is called a complex steady-state vibration (*see* COMPLEX STEADY-STATE VIBRATION).

All complex waveforms consist of combinations of sine waves. Even white noise can be resolved to sine-wave tones. A pure sine-wave tone has a rather irritating quality, especially at the midrange and high audio frequencies. *See also* SINE WAVE.

COMPLIANCE

Compliance is a measure of how well a phonograph stylus responds to vibration. Compliance is measured in terms of displacement per unit force, usually in centimeters or millimeters per dyne. In a speaker, compliance is the flexibility of the cone suspension. Compliance can be measured for either static or dynamic forces. Static compliance is the measure of displacement for a simple, direct force; dynamic compliance is the ease of displacement for an alternating force.

The greater the compliance of a stylus, the better the low-frequency response. The same is true of speakers; woofers should have a high degree of compliance for good reproduction of bass sounds.

COMPONENT

In any electrical or electronic system, a component is a basic functional unit. Examples of components include resistors, capacitors, inductors, switches, integrated circuits, and relays.

COMPONENT LAYOUT

The arrangement of electronic components on a circuit board is called the component layout. The component layout is an important part of the design of a circuit. In modern printed-circuit boards, the placement of conductors dictates the component layout. Poor circuit layout can increase the chances of unwanted feedback, either negative or positive. It can also cause inconvenience in servicing and adjustment of the equipment.

Component layout becomes more important as the frequency increases. In some very high frequency and ultrahigh-frequency circuits, the component layout alone can make the difference between a circuit that works and a circuit that does not work.

COMPOSITE VIDEO SIGNAL

A composite video signal is the modulating waveform of a television signal. It contains video, sync, and blanking information. When the composite video signal is used to modulate a radio-frequency carrier, the video information can be transmitted over long distances by electromagnetic propagation. By itself, the composite video signal is similar to the audio-frequency component of a voice signal, except that the bandwidth is greater.

Most video monitors used with computers and communications terminals operate directly from the composite video signal. Some monitors require a modulated, very high frequency picture signal, usually on one of the standard television broadcast channels. *See also* COLOR PICTURE SIGNAL.

COMPOUND

When two or more atoms combine by chemical reaction, the result is a compound. A compound consists of two atomic nuclei sharing electrons. For example, hydrogen and oxygen are elements; two atoms of hydrogen can be joined with one atom of oxygen to form the familiar compound H_2O—water.

Some combinations of elements join readily to form compounds. Some do not. Atoms whose outer shells are filled with electrons, such as helium and neon, are called non-reactive or inert elements because they do not ordinarily form compounds.

A compound differs from a mixture. In a mixture, there may be several elements and/or compounds, but no chemical reaction takes place among them. The atmosphere is a mixture of nitrogen, oxygen, carbon dioxide, and traces of many other gases. Some man-made pollutant gases combine to form compounds in the atmosphere.

Compounds are widely used in electronics as heat conductors, insulators, and dielectric materials. Compounds are also used in the manufacture of semiconductor devices, such as the metal-oxide semiconductor field-effect transistor (MOSFET). *See also* COMPOUND SEMICONDUCTOR.

COMPOUND CONNECTION

When two tubes or transistors are connected in parallel, or in any other configuration, the arrangement can be called a compound connection. Examples of compound-connected devices include the Darlington pair, the push-pull configuration, certain rectifier circuits, and many different digital-logic circuits.

Compound connections are used in audio-frequency and radio-frequency circuits for improving the efficiency or amplification factor. Compound connections may also be used to provide an impedance match.

COMPOUND MODULATION

When a modulated signal is itself used to modulate another carrier, the process is called compound modulation. An example of compound modulation is the impression of an amplitude-modulated signal of 100 kHz onto a microwave carrier. The main carrier may have many secondary signals impressed on it. The secondary signals do not all have to employ the same kind of modulation. For example, a code signal, an amplitude-modulated signal, a frequency-modulated signal and a single-sideband signal may all be impressed on a microwave carrier at the same time. The illustration shows compound modulation.

With compound modulation, the main signal frequency must be at least several times the frequency of the modulating signals. Otherwise efficiency is poor and the signal-to-noise ratio is degraded.

Compound modulation is useful in telephone trunk lines where a large number of conversations must be carried over a single circuit. Compound modulation is used in the transmission of signals satellites. It is also used in fiberoptics systems where many thousands of separate signals can be impressed on a single beam of light. *See also* FIBEROPTIC COMMUNICATION, MODULATED LIGHT, TRUNK LINE.

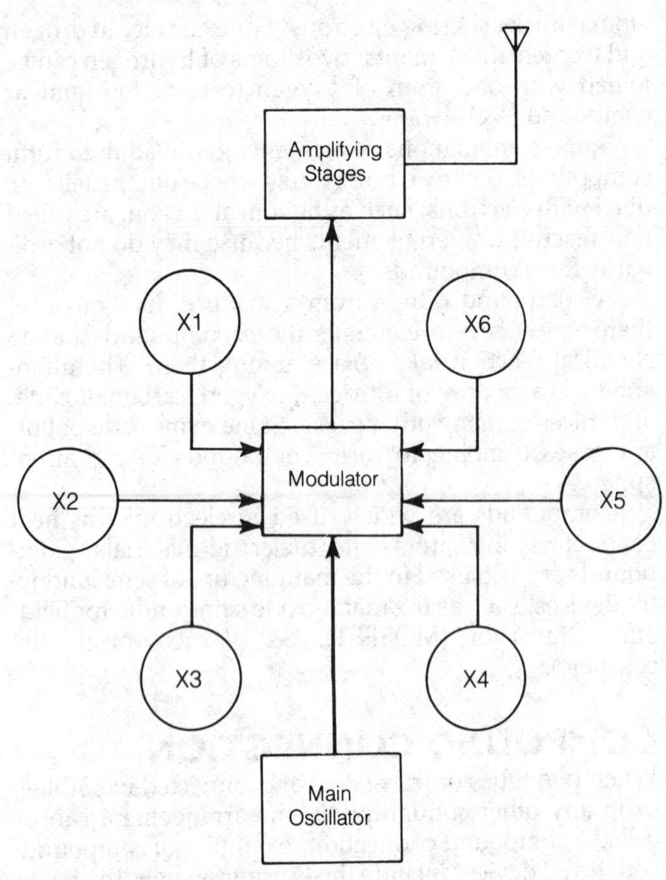

COMPOUND MODULATION: Two or more signals are impressed on the carrier.

COMPOUND SEMICONDUCTOR

Compound semiconductors exhibit semiconductor properties similar to those of silicon and germanium. Both silicon and germanium have four valence electrons, and the binary compounds of interest for semiconductor applications are the Group III-V compounds with an average of four valence electrons. The most popular of these is gallium arsenide (GaAs) with a fixed atomic ratio of 1:1. Gallium has three valence electrons and arsenic has five valence electrons giving the compound an average of four valence electrons. Other semiconductor compounds can be formed from Group II and Group VI elements. An example is cadmium sulfide (CdS).

The principal binary semiconductor compounds in addition to gallium arsenide are gallium phosphide (GaP) and indium phosphide (InP). These compounds have properties that differ from those of silicon and germanium, particularly in carrier mobility and energy gaps.

The preparation of N-type and P-type compound semiconductors is similar to that for elemental semiconductors. The doping of III-V compounds is accomplished by adding a Group VI element, such as tellurium, as a donor impurity to produce N-type material or by adding a Group II element, such as zinc.

Some semiconductor compounds form solid solutions. This characteristic can be used to obtain a material with a specific energy band gap value. For example, if one compound with an energy gap of 0.03 eV (electron volt) forms a solid solution with another compound with an energy band gap of 1.2 eV, a solution can be prepared by proportioning various amounts of each compound to obtain an energy band gap between 0.3 and 1.2 eV. This technology is referred to as band-gap engineering. *See* BAND-GAP ENGINEERING.

Gallium arsenide is an important binary compound used in the fabrication of semiconductor lasers, microwave diodes, microwave transistors and oscillators, infrared light-emitting diodes (LEDs), and microwave, high-frequency analog and digital integrated circuits. It has an energy band gap of 1.43 eV, making it useful for the fabrication of devices operating at temperatures up to 350°C. It also has an electron mobility of 4,300 centimeter^2/volt-s and a dielectric constant of 12.93.

Indium phosphide (InP) has an energy gap of 1.25 eV, an electron mobility of 4,000 centimeter^2/volt-s and a dielectric constant of 10.8.

Antimony is a Group V element, and it combines with Group III elements to form compound semiconductors. The most important of these are indium antimonide (InSb), gallium antimonide (GaSb), and aluminum antimonide (AlSb).

COMPRESSION

Compression is the process of modifying the modulation envelope of a signal. This is done in the audio stages, before the modulator. The weaker components of the audio signal are amplified to a greater extent than the stronger components. The illustration shows a compression function with the output amplitude dependent on the input signal amplitude. The peak, or maximum amplitude of the compressed signal is the same as the

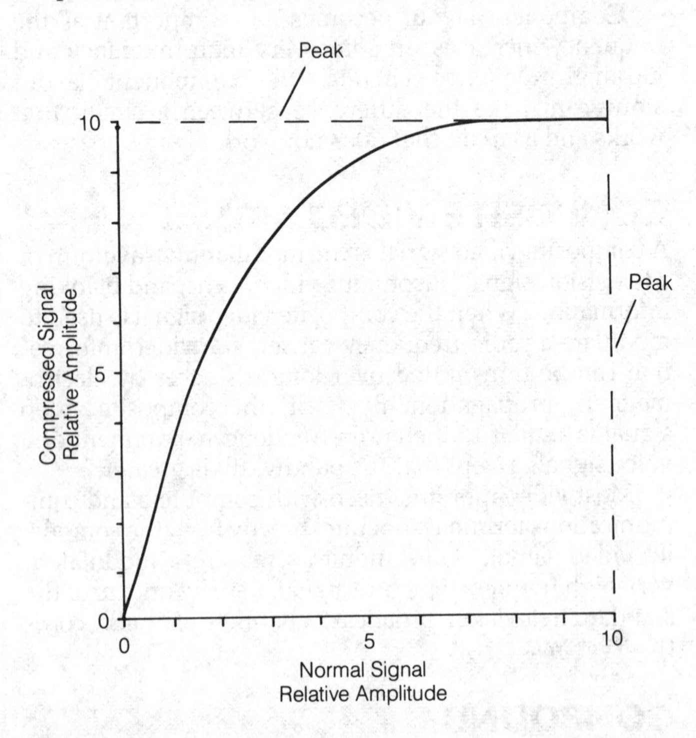

COMPRESSION: A curve showing a compression function with the output amplitude dependent on the input signal amplitude.

amplitude of the normal signal. The weaker the instantaneous amplitude of the signal, the greater the amplification factor.

Compression is often used in communications systems to improve intelligibility under poor conditions. *See also* COMPRESSION CIRCUIT, SPEECH COMPRESSION.

COMPRESSION CIRCUIT

A compression circuit is an amplifier that displays variable gain, depending on the amplitude of the input signal. The lower the input signal level, the greater the amplification factor. A compression circuit operates in a manner similar to an automatic-level-control circuit (*see* AUTOMATIC LEVEL CONTROL).

One method of obtaining compression is to apply a rectified portion of the signal to the input circuit of the amplifying device, changing the bias as the signal amplitude changes. This rectified bias should reduce the gain as the signal level increases. The time constant must be adjusted for the least amount of distortion. An example is shown in the diagram.

There is a practical limit to the effectiveness of compression. Too much compression will cause the system to emphasize acoustical noise. In a voice communications transmitter, several decibels of effective gain can be realized using audio compression. *See also* SPEECH COMPRESSION.

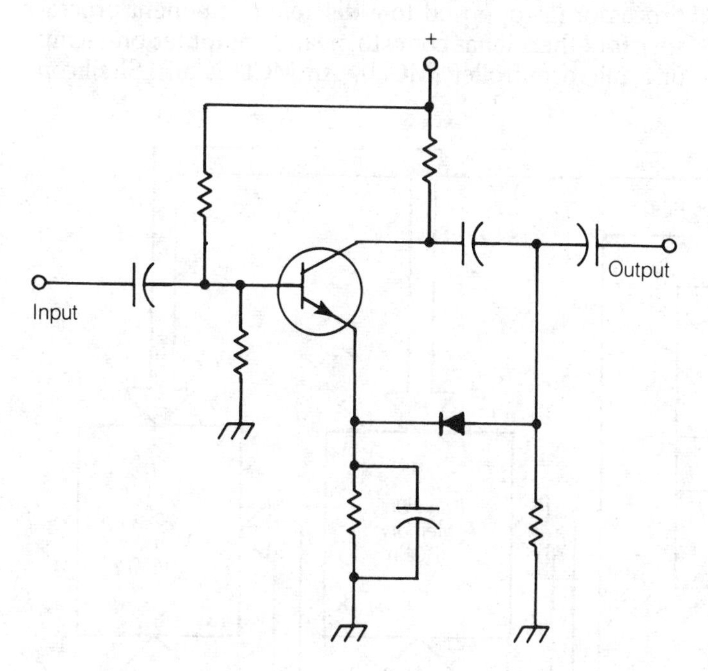

COMPRESSION CIRCUIT: A schematic for a compression circuit.

COMPTON EFFECT

When photons (particles of radiant energy) strike electrons, the wavelength of the photons changes because some of the photon energy is transferred to the electrons. The change in wavelength is called the Compton effect. The amount of change in the wavelength is a function of the scattering angle. Compton effect is generally observed with X rays.

At a grazing angle of collision, very little energy is transferred to the electron and the wavelength of the photon therefore increases only slightly (A in the illustration). At a sharper angle of collision (B), the photon loses more energy, and the change in wavelength is greater. At a nearly direct angle of collision, the wavelength change is the greatest, as shown at C.

Compton effect causes the wavelength of X rays to be spread out when the radiation passes through an obstruction. *See also* X RAY.

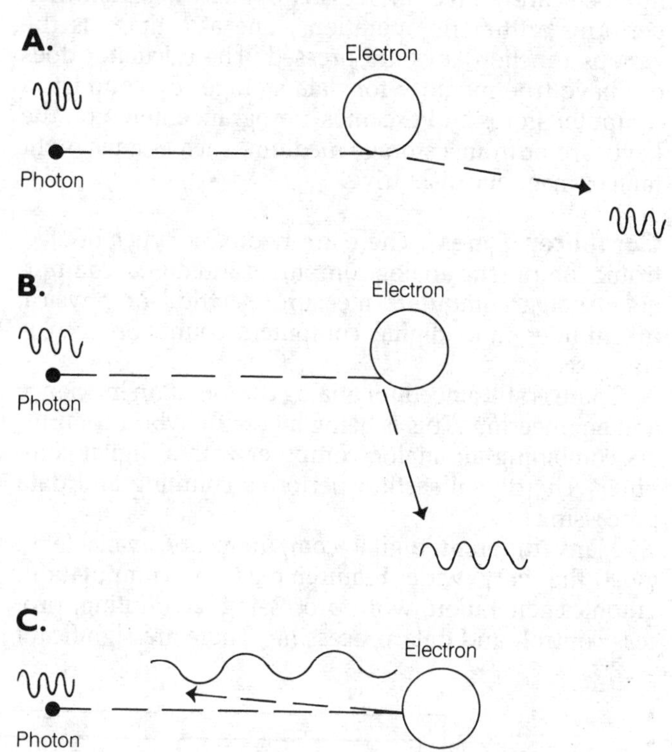

COMPTON EFFECT: Changes in emitted wavelength from photon-electron collisions (Compton effect) are diagrammed. At a shallow angle the effect is negligible (A), at a sharper angle the effect increases (B), and a direct hit produces the maximum effect.

COMPUTER

Any device that aids in computation, from an abacus or slide rule to a mechanical adding machine or an electronic calculator, can be called a computer. Some electromechanical machines based on gears, motors, clutches, and other mechanisms use in the performance of specialized calculations have also been referred to as computers. In the 1940s and 1950s, electronic equipment, based on potentiometers and vacuum-tube operational amplifiers, capable of performing a wide range of analog computations were referred to as computers. However, today, by common usage, the word *computer* has come to mean (and is used in this book) to mean a stored-program electronic digital computer. Figure 1 is a basic block diagram of a digital computer.

An electronic digital computer is distinguished from other computing devices by its speed, internal memory, and automatic execution of a program stored in its memory. The speed of an electronic computer is obtained with integrated circuit logic and memory devices.

The internal memory of an electronic stored-program computer stores both data and instructions. A sequence of instructions for input, processing and output is performed automatically, without human intervention. By contrast, an electronic calculator requires human direction from a keyboard at each step of the computation.

The difference between an electronic calculator and a computer is significant. Electronic calculators accept numbers that are entered on a keyboard, and these numbers are stored in registers. The calculator then performs arithmetic operations, one at a time, as the various function keys are pressed. The calculator does not have true memory for data storage. By contrast, a computer stores and executes a program entered on the keyboard or from a storage medium, such as a magnetic tape or magnetic disk drive.

Computer Types There are two basic types of electronic computers: analog computers and digital computers. Analog computers measure electrical or physical magnitudes, and digital computers count. *See* ANALOG COMPUTER.

There is still a need for analog computation in science and engineering. This is being met with hybrid computers combining an analog computer with a digital computer. The digital section performs counting and data processing.

Many different digital computers are available to meet the very wide requirements for computation, graphics generation, word processing, accounting, process control, and data processing. There are significant differences in capabilities, performance, size, form factor, and price in available digital computers today. New computers are becoming available with significant differences from the traditional or Von Neumann architecture. *See* COMPUTER ARCHITECTURE.

There are also differences in input/output (I/O) peripheral devices in use. Printers and plotters are used to produce hard copy printed pages. The CRT displays transactions and also is an interactive device. Other interactive devices include keyboards, light pens, the mouse, and data entry tablets.

At the low end of the computer hierarchy are the simple, low-cost microcomputers for playing electronic games on home television sets or carrying out routine process control duties. At the next higher level there are the personal computers that offer a wide range of options and performance. At an even higher performance level are the minicomputers and the computer-aided design (CAD) graphics workstations. At the highest levels are the high-speed supercomputers used in making extensive scientific and engineering calculations. Figure 2 shows a personal computer.

For many years, digital computers were simply classed as microcomputers (because of their use of microprocessors—MPUs—as central processing units or CPUs), minicomputers, and mainframe computers. The term microcomputer has a dual meaning today. For some it means any computer with a microprocessor as a central processor (as opposed to a discrete component processor); for others it has come to mean a computer-on-a-chip or a microcontroller (MCU). An MCU is an LSI silicon

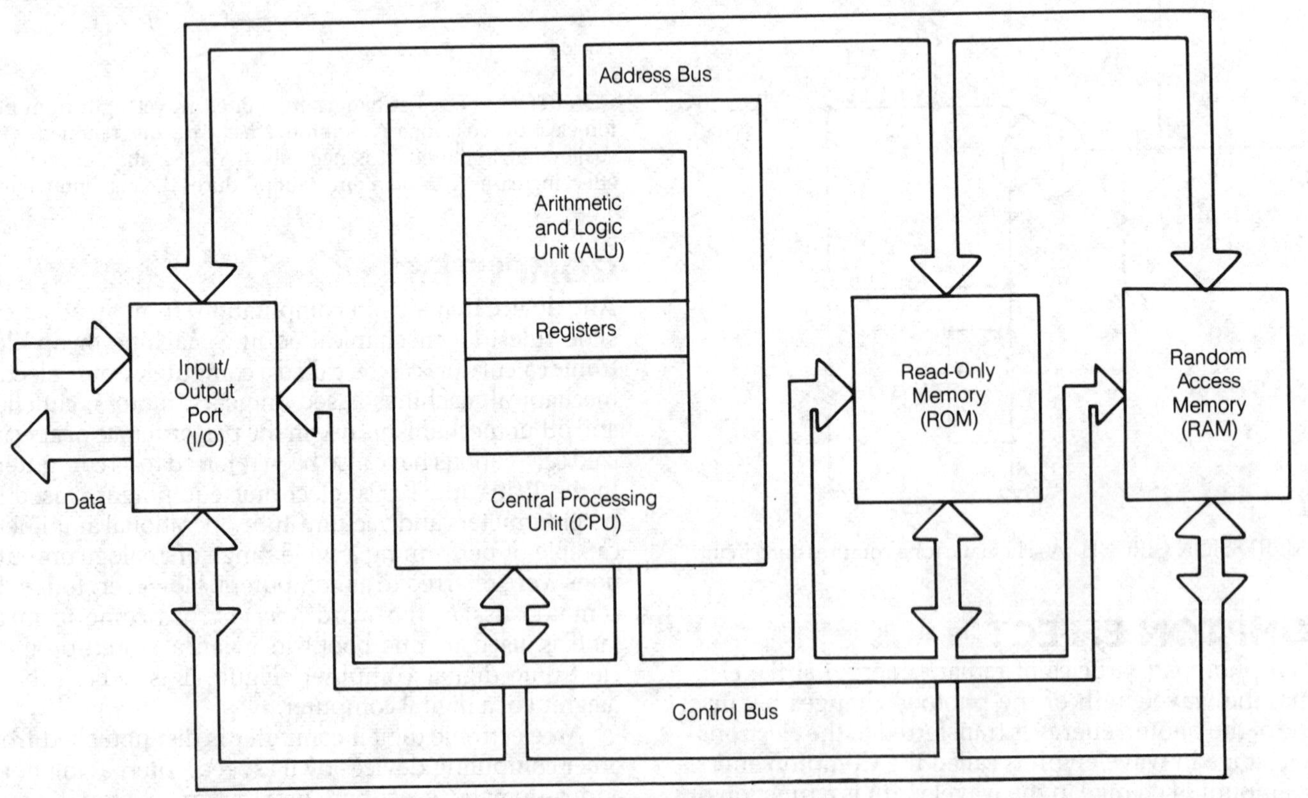

COMPUTER: Fig. 1. Block diagram of the organization of a digital computer.

COMPUTER: Fig. 2. Personal computer based on a 32-bit microprocessor equipped with keyboard and mouse.

chip containing a CPU, limited random access memory (RAM) and read-only memory (ROM), and on-chip (I/O) functions. *See* MICROPROCESSOR.

Today the term *personal computer* (or PC) is understood to mean a general purpose desktop (or laptop) unit for computation, process control, word processing and even desktop publishing. It can also function as a smart video data terminal or VDT (*See* VIDEO DISPLAY TERMINAL). Most personal computers in use are based on 8-bit and 16-bit microprocessors, but designs on 32-bit units are available.

Workstations are a new classification of computers specialized for computer-aided design (CAD) and computer-aided engineering (CAE) with heavy emphasis on graphics capability. (*See* COMPUTER-AIDED DESIGN AND COMPUTER-AIDED ENGINEERING.) Current generation workstations are based on 16-bit and 32-bit microprocessors.

The term *minicomputer* is still in use, but it is now defined more in terms of word length capability than its former industrial control specialty. Most minicomputers are based on 32-bit, custom-designed microprocessors. The term *mainframe* computer today is ambiguous. It has been replaced by such terms as supercomputer, super minicomputer, and mini supercomputer.

Computer Organization All digital computers, regardless of size or complexity—from personal computers to supercomputers—have the same functional components as shown in Fig. 1. This arrangement is named sequential machine architecture and it dates back to the World War II period of 1942 to 1945. From this diagram it can be seen that the computer reads or inputs data from sources such as keyboards, modems, or secondary memory devices such as disk or tape drives. It writes or outputs to the CRT monitor and other devices such as printers.

The computer has two kinds of memory to store program instructions as well as the data that are being processed. The computer is under the direction of a program and it has a CPU (central processing unit) that interprets the program instructions and supervises their execution.

The computer also has a section that performs addition, division, and other arithmetic operations called the

ALU (arithmetic logic unit). The ALU can also perform logical operations such as comparing the magnitude of two numbers. The flow of data in Fig. 1 is on the data bus and the flow of control information is on the control bus.

Central Processing Unit (CPU). The arithmetic and logic unit (ALU), the control and timing unit, and the registers are integrated into a single physical unit called the central processing unit or CPU. The CPU is the brain or nerve center of a computer. It controls the operation of the computer through the control and timing unit and performs arithmetic and logical operations through the arithmetic and logic unit. *See* ARITHMETIC LOGIC UNIT.

Control and Timing Unit. The control and timing unit supervises all operations of the computer under direction of a stored program. First the control and timing unit determines which instruction is to be executed next by the computer. The control and timing unit then fetches the instruction from the main memory and interprets the instruction. The instruction is then executed by other computer units, under the direction of the control unit.

The Arithmetic Logic Unit (ALU). The ALU performs computations and data manipulations such as addition, subtraction, comparison, and logical operations. A typical logical operation involves comparing two numbers, and then selecting one of two (or more) program paths depending on the result of the comparison. *See* ARITHMETIC LOGIC UNIT.

Acting in harmony with the control and timing unit, the ALU can test numbers and cause the computer to branch to one of two possible program paths. This ability to test or compare two numbers and to branch to one of several paths depending on the results of the comparison gives the computer great power and flexibility. It is a major reason why the digital computer is so useful in many different applications.

Main Memory. A computer must have a memory for storage of data and instructions. This memory can be in three levels: primary, secondary, and tertiary. The computer main memory consists of arrays of semiconductor memory devices. Data and instructions are stored in areas called locations. Each location in main memory has an address so that data can be located. The capacity of main memory is determined by the size and application of the computer.

The computer central processing unit (CPU) can only operate on the data from the main memory, and only instructions from the main memory can control the computer. The basic building block of a computer main memory is the integrated circuit semiconductor memory. Each memory IC (integrated circuit) contains thousands of transistors that function as switches in representing the binary digits zero and one. (*See* SEMICONDUCTOR MEMORY.) In most computers, main memory is divided into sections. For example, there is often a division between random access memory (RAM) and read-only memory (ROM) as shown in the figure.

Main memory is usually supplemented by secondary storage mass memory which includes floppy-disk drives, rigid-disk drives (typically Winchester type). These memory devices are random access memories that can read and write data rapidly, making them virtual main memories. *See* DISK DRIVE.

Demand-Paged. Demand-paged memory systems divide both disk and RAM (virtual and physical) memories into fixed-sized pages. In moving from disk to RAM, blocks of information are switched into the same number of pages. Demand-paged virtual memory permits multiple users and multiprocessing. Segmented memory systems impose partitions on RAM. This segmentation must accommodate the longest program or data construct needed.

Many computer systems also have tertiary memory that serve as backup, long-term, or archival data and program storage. The most commonly used tertiary memories are magnetic tape drives with the memory media in the form of tape cassettes or tape reels (streaming tape). Tape drives are serial-access memories with relatively long access times. Write-once, read-many (WORM) optical disk drives and compact disk (CD) ROMs are now also being used for data storage. *See* COMPACT DISK, OPTICAL DISK DRIVE.

COMPUTER-AIDED DESIGN and COMPUTER-AIDED ENGINEERING

Computers are now widely used in the design and manufacture of products and systems. Computer-aided design (CAD) and computer-aided engineering (CAE) refer to the use of computers to aid the designer from product conception through the preparation of engineering drawings, specifications, and parts lists. Computers can develop elevation views and three-dimensional drawings of engineering structures, analyze those structures, and prepare manufacturing drawings. They can also perform interference checking, tool-path definition, parts listing, and numerical control (NC) tape preparation.

Computer-aided manufacturing (CAM) refers to the use of computers in inventory and quality control as well as the supervision and direct control of manufacturing machines and processes. Ideally design concepts originated at the CAD/CAE level are transferred directly into instructions for the manufacture of products because they share a common database. The same database can be used for testing, quality control, inventory control and production scheduling.

Computer-aided design has evolved from automatic drafting. The computer-based video workstation terminal displayed the end product as a two-dimensional representation of a conventional engineering drawing. Information on the third dimension was obtained from orthogonal views.

However, in the design of circuit boards and integrated circuits, the third dimension could be introduced by displaying layers positioned and registered on one another. Design work can be done on individual layers and a composite presentation can distinguish each layer by means of color coding. By following sets of rules programmed into the computer, wiring runs and interconnects can be made without overlap or interference.

Computer solids modeling was developed for the design of more complex objects such as machines, ships, and aircraft. The object can be represented on the screen as a three-dimensional line drawing that can be moved about in space. These presentations suggested isometric drawings. With further developments these wire-frame images could be filled in to achieve solids modeling and surface modeling in shades of gray or colors.

Finite element analysis (FEA) was a CAD development that enabled a complex solid structure is broken down into a large number of simple elements so each simple element could be analyzed. For example, the forces on the structure or loading of building, ship, or aircraft structures could be analyzed.

Computer-aided engineering is an outgrowth of computer-aided design, and it becomes increasingly difficult to distinguish the difference between them. Engineering software has been developed to permit the solution of problems other than stress analysis. For example, electrical and electronic circuit problems can now be solved with programs that analyze the specific circuits displayed in schematic form. Data and design rules on the physical and electrical characteristics of components are stored in memory to be called up as needed.

CAE systems permit circuit simulation at the abstract symbolic level. Substitution of components can be made in the models to determine their effect on circuit performance. This simulation saves on the cost of completing drawings and building hardware prototypes. CAE seeks to analyze and optimize designs in many fields of engineering.

Computer-aided testing (CAT) is another outgrowth of computer-aided design and engineering. The electrical characteristics of devices, components, and subsystems can be stored in computer memory for comparison with actual measured characteristics with the computer-aided test equipment. CAT systems speed up the testing of complex integrated circuits in wafer form or as packaged devices. They can also test complete circuit boards and subsystems. CAT systems can provide printed records of product testing and can even direct the marking and ejection of failures from production lines.

Computer-aided manufacturing is an outgrowth of the numerical control (NC) of machines. NC machines were designed for digital control of the machine by a program introduced on a punched and coded paper tape. Computers can now control banks of NC machining centers, milling machines, or lathes in a hierarchical organization known as computer numerical control (CNC). They can also manage tool wear and replacement, scheduling, parts resupply, productivity, and other administrative details.

Robots are example machines that are directly controlled by computers. General-purpose robots are modified by the installation of special tooling to perform a range of specific tasks such as arc welding, spot welding, painting, gluing and assembly. The robot can be taught by manually leading it through the task so that the robot memory will store a record of the movements of each robot axis in three dimensions. The robot controller then replays the sequence on command when the work is to be done automatically. Some robots can also be programed off line by writing programs based on assembling standardized sequences without tying up the robot for teaching. *See* ROBOT.

Many kinds of specialized machines that do not qualify as robots such as automatic pick-and-place machines for inserting or positioning electronic component on circuit boards may also be controlled by a computer. Separate programs are prepared giving the location of each leaded or surface mount component and this information is stored in a computer memory library as in the case of robots.

Computer vision (CV) systems can be programed to recognize specific objects in a field of similar objects for sorting. They can also be used to inspect parts for quality or completeness by comparing their images against a master image stored in memory. *See* COMPUTER VISION.

Computer integrated manufacturing (CIM) seeks to combine and standardize communications among different kinds of workstations and different software to permit a totally integrated factory. Basic design information can be developed off line at a laboratory or factory and be transmitted over compatible communications lines and bus structures to permit the organization of factories from a centralized hierarchical computer.

Under CIM, all phases of design engineering, planning, manufacture, and inventory control can be under centralized control. This activity is in its early stages. Adherents of the concept are encountering the problems of communication between disparate systems and software as well as the unwillingness of many suppliers of embedded computers and software to subordinate their products to an overall supervisory control system.

COMPUTER-AIDED MEDICAL IMAGING

Computer-aided medical imaging refers to a number of different techniques that permit physicians and surgeons to view the inside of the human body without intrusive surgery. Signals obtained from X rays, nuclear magnetic resonance, ultrasonic waves, and radioactive isotopes, for example, are organized by computers and applications software to present views of living human beings unobtainable with earlier X ray and sonogram methods. The views are typically presented on a cathode ray tube (CRT). (*See* CATHODE RAY TUBE.) These scans permit doctors to watch vital organs at work, identify blockages and growths, detect disease, and even detect warning signs of diseases not yet present—all without exploratory surgery.

Five different computerized body scanning techniques are discussed in this section. All are based on the ability of the computer to enhance or construct more detailed images than those that could be derived directly from the basic technology. The techniques are:

1. Computer tomography (CT) or computer-aided tomography (CAT) is computerized body scanning based on X rays and the construction of detailed cross section slice views with the aid of computer programming. Two- and three-dimensional images can be formed for analysis and diagnosis.
2. Magnetic-resonance imaging (MRI), formerly called nuclear magnetic resonance imaging (NMRI) is imaging of the human body based on the interaction of atoms with strong magnetic fields and radio waves. It uses a computer to form two- or three-dimensional images for analysis and diagnosis.
3. Digital subtraction angiography (DSA) is a specialized form of imaging for diagnosis of cardiovascular problems. DSA uses opaque dyes in the blood system and X rays. A computer to construct images of blood vessels on a television monitor free of nonessential background.
4. Positron-emission tomography (PET) is a form of radioisotope scanning. It depends on the radioactive emissions from substances with short half-lives to locate disease and anomalies in the human body.
5. Sonography is based on ultra sound and is a kind of specialized sonar or depth finder adapted to viewing and analysis within the human body. Images are formed on a monitor from short-range ultrasonic echos.

Computed Tomography Computer tomography converts signals generated by X ray beams into video images on a cathode ray tube. X rays are absorbed (shadowed) by dense structures like bones or metallic foreign objects within the body as the X rays pass through the body. However, softer tissue such as muscles, organs, and skin are more easily penetrated by X rays (*See* X RAY). Conventional X ray radiograms represent the amount of X ray penetration by different grey (intensity) levels in a photographic image.

Conventional X ray radiograms, which view the body from only one angle, can be difficult even for experts to interpret. Shadows of bones, muscles, and organs are superimposed on one another, partially masking whatever lies behind them. Conventional X ray images can be digitized and computer-enhanced to permit clearer distinction between bones, tissues and organs.

For computer tomography an X ray source is mounted on a circular frame permitting it to revolve in a plane through 360 degrees. The patient is placed within the frame as shown in Fig. 1. The X ray source produces a thin, fan-shaped beam as it rotates around the patient's body, exposing all sides within the region to X rays. Sensitive detectors on the opposite side of circular frame convert the changing patterns of radiation received from the scanner into digital signals.

A computer program processes the data from the detectors and compares the views from many source positions to form a single video image. The digital data is displayed on a television monitor. The monitor presents views of the patient's body at any point along its length as cross sections or thin slices. With rotary scanning, all sides of internal bones, organs, and tissues are exposed to X rays, thus avoiding the masking of conventional X ray images.

Magnetic-Resonance Imaging Magnetic-resonance imaging (MRI) is based on the phenomenon of nuclear resonance exhibited when atoms are constrained by a magnetic field and subjected to radio-frequency pulses. Atoms within the human body align themselves when subjected to a strong magnetic field, which acts as an elastic restraint. Radio frequency energy at selected wavelengths is focused on these atoms, forcing their nuclei out of alignment. When the radio energy is turned off, the atomic nuclei realign themselves to the magnetic field, and they resonate like springs or tuned circuits, generating weak radio signals. *See* RESONANCE, TUNED CIRCUIT.

MRI is effective because the body is composed primarily of hydrogen atoms that can generate an image with their returning radio signals. Parts of the body containing higher water density generate a stronger return signal than those parts containing little water (like teeth and bones). The signal strengths are shown as intensities or colors in the displayed image. The image generated differs from that of the X ray CT scanner because of its ability to show tissue and bone marrow clearly, even when surrounded by bone. MRI can distinguish between white matter and water-rich gray matter in the brain better than is possible with X ray CT scan.

MRI equipment consists of a huge, superconducting electromagnet that is cooled by liquid helium. It is constructed as a large tube with an inside diameter of about 1 meter to admit the patient. (See Fig. 2.) The electromagnet of a typical MRI system weighs more than 20 tons and is capable of producing a uniform magnetic field of 1.5 Tesla within its bore. (This is approximately 30,000 times the strength of the earth's natural magnetic field.) *See* TESLA. Additional coils adjust the magnetic field gradient.

The MRI system also includes a radio-frequency generator with coils. Signals are produced either by surface coils or by cylindrical head or larger whole-body coils housed within the bore of the magnet. The system also includes a computer for image formation. The complete MRI system must be shielded from external radio frequencies with an appropriate metallic enclosure.

Protons within hydrogen atoms normally spin like tops and point in random directions. However, within the strong electromagnetic field of an MRI scanner, the protons align themselves in the direction of the electromagnet poles. However, the field does not hold them in rigid alignment, and they wobble or process at a known frequency or rate. The frequency is proportional to the strength of the magnetic field.

The MRI scanner excites these protons with a radio

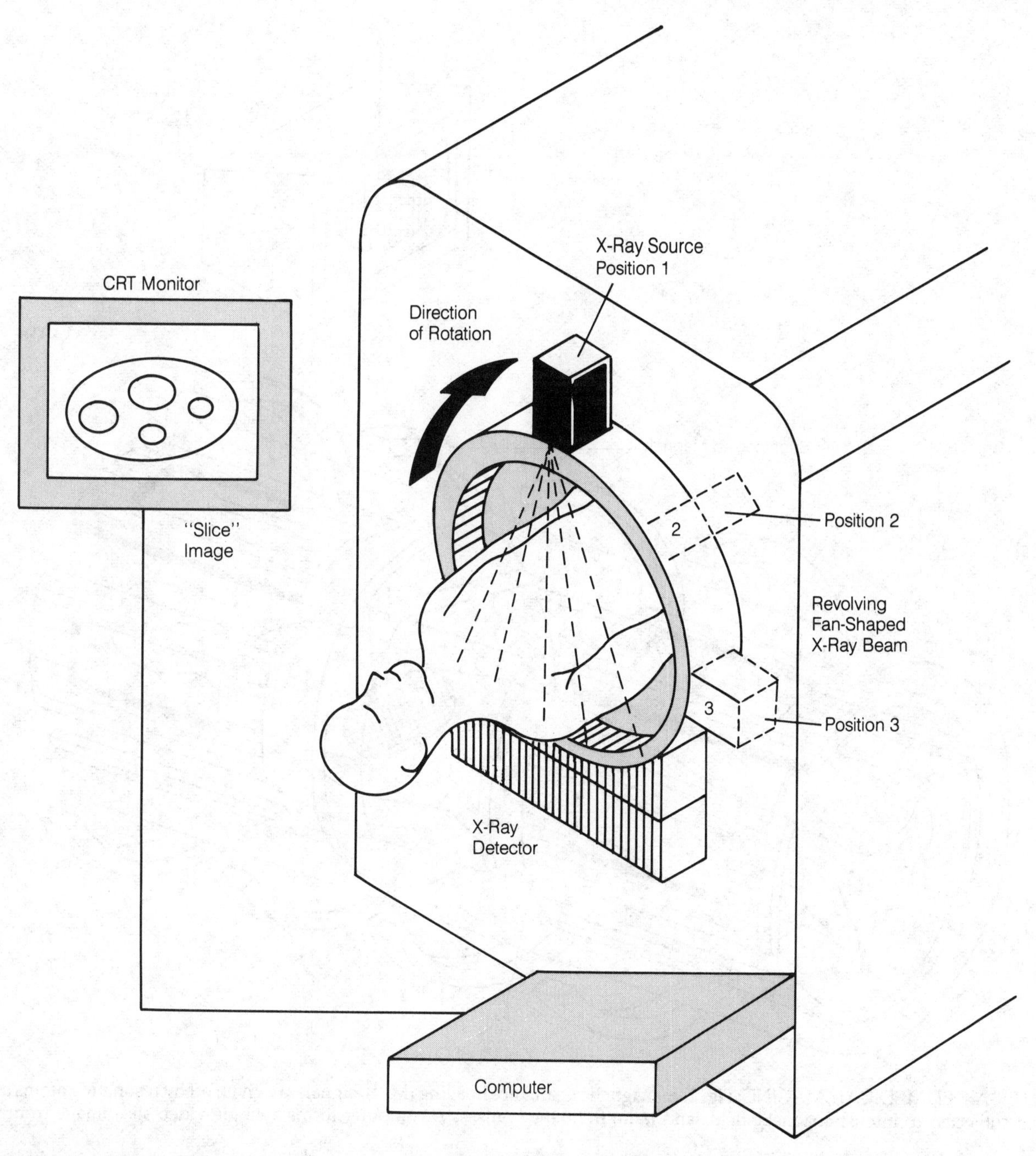

COMPUTER-AIDED MEDICAL IMAGING: Fig. 1. In computed tomography (CT) imaging, an X ray source revolves around the patient and sensitive detectors on the opposite side of the ring record the X-ray beam penetration. A computer compares the many views to form a single video slice image of the body.

pulses synchronized with the proton precession frequency, destabilizing them and forcing them out of alignment. When the radio pulse ends, the protons resonate for a few milliseconds while they realign with the magnetic field. This resonance emits faint, characteristic radio-frequency signals.

A computer converts these faint signals into an image of the area scanned as shown in Fig. 2. The image shows the densities of the hydrogen atoms and their interaction with surrounding tissues in a crosssection of the body. Because the presence of hydrogen is proportional to water content, it is possible to distinguish between tissues and organs.

The computer establishes a grid of bones in three dimensions—X, Y, and Z—called *voxels* (volume elements). First the magnetic field is varied in the Z direction to define a plane of interest where the body will be scanned. Radio frequency coils within the magnet emit a pulse at the frequency needed to cause the nuclei of the selected atoms to resonate and produce faint radio signals.

Before the protons realign themselves and the resonance is damped out, the gradient coils are used to vary the magnetic strength in the Y direction. This causes the protons to precess at different rates from the top of the plane to the bottom. The computer is able to locate voxels in the Y direction after detecting hundreds of resonance cycles.

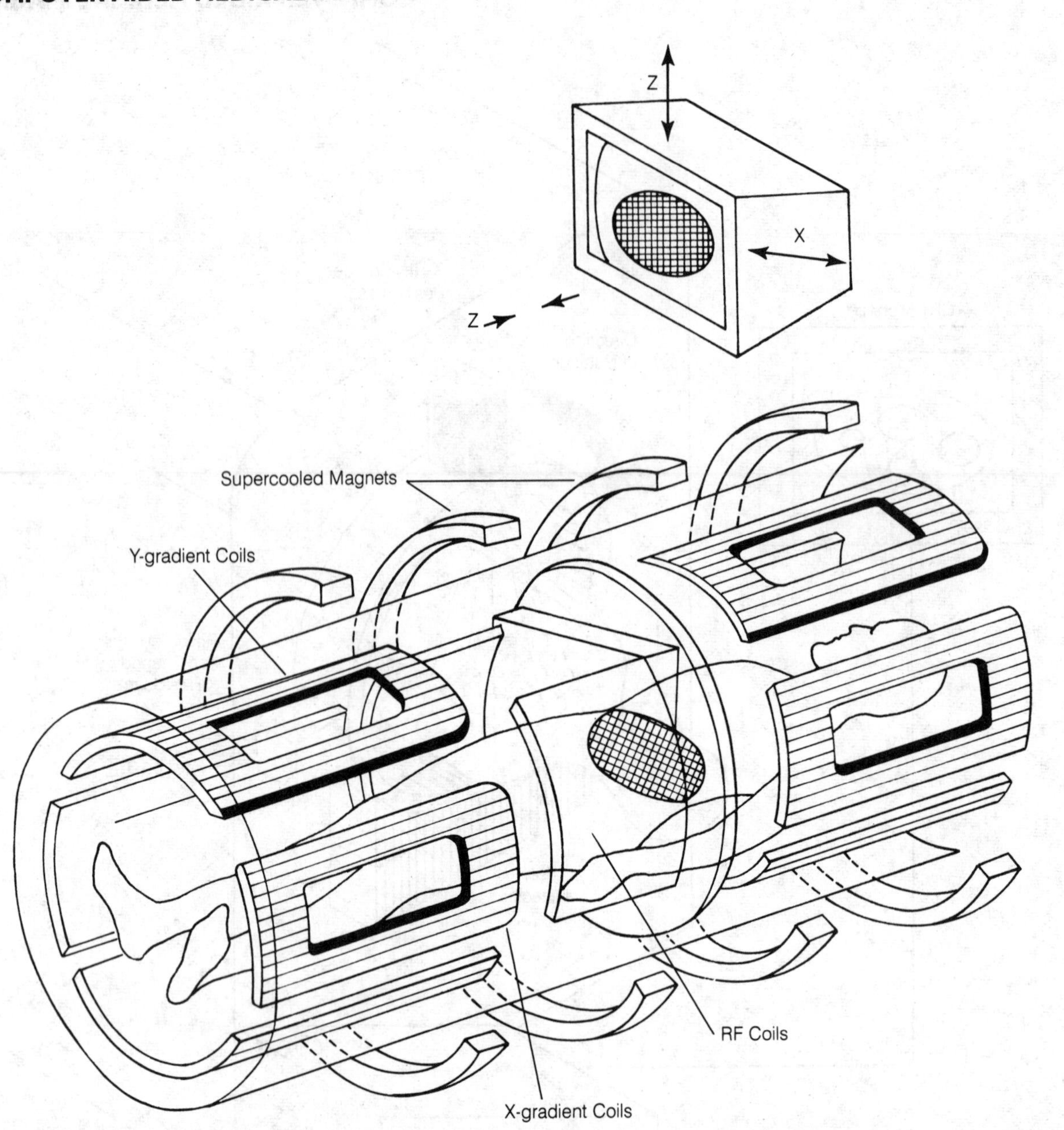

COMPUTER-AIDED MEDICAL IMAGING: Fig. 2. In magnetic resonance imaging (MRI), signals are given off by resonating atoms after they have been subjected to intense magnetic fields and radio frequency pulses. A computer forms a single video slice image from the data collected.

The gradient coils then vary the magnetic field in the X direction, causing protons to resonate at different frequencies as they stabilize themselves. After each voxel is located in the X, Y, and Z directions, the computer displays the voxel on the screen of the CRT monitor as a pixel. The brightness of the pixel is determined by the number of hydrogen protons within the voxel and the magnetic properties of the tissue. The pixels form a readable image when raster scanned on the video monitor screen.

For medical diagnostics, hydrogen is used as the basis for MRI scanning. However, other elements such as iron, sodium, and phosphorous will resonate, and their presence or absence can provide early warning signs of strokes or heart attacks. MR scanners with magnetic fields of 4 Teslas—80,000 times the strength of the earth's magnetic field—have been built. Higher magnetic fields provide more detailed images.

The use of the MRI equipment in chemical analysis is called magnetic-resonance spectroscopy. In spectroscopic imaging, low-resolution spectra maps showing the distribution and concentration of key metabolites (biomolecules involved in metabolism) in a given area of the body can be produced. The map consists of an array of voxels, each containing a spectrum indicating the chemical makeup of that particular region.

Sonography Sonography forms images of the interior of the human body by beaming high-frequency sound waves into the human body. A computer translates the echoes that are returned into an image, similar to that of an active sonar system or depth finder. (*See* DEPTH SOUNDER,

SONAR). Sonography is the only computer-aided scanning technique recommended for pregnant women. It is also suited for the examination of other body organs such as heart, liver, and gall bladder.

The transducer in the sonographic system is a piezoelectric crystal that produces sound waves, which penetrate the body. The sound waves are reflected back to the crystal, which converts them back into electrical signals. When these signals are processed and displayed, an image of internal organs is formed because of the differences in densities and reflective properties of those organs.

Digital-Subtraction Angiography Digital subtraction angiography (DSA) is a computed X ray analysis of the cardiovascular system that uses fluids opaque to X rays (contrast agents) for image enhancement. DSA removes everything from the image except the specific veins or arteries under examination.

An X ray picture is first made by the digital X ray scanner to provide a reference. Then a contrast agent (opaque to X rays) is injected through a catheter into the arteries or veins. A second X ray image is made showing the agent moving through the blood vessels. A computer subtracts the first image from the second, leaving only the difference image: blood vessels containing the agent.

By highlighting dynamic aspects of the human body, such as the passage of blood through the heart, DSA can be used to study disorders and predict possible occurrence of disease or heart attack.

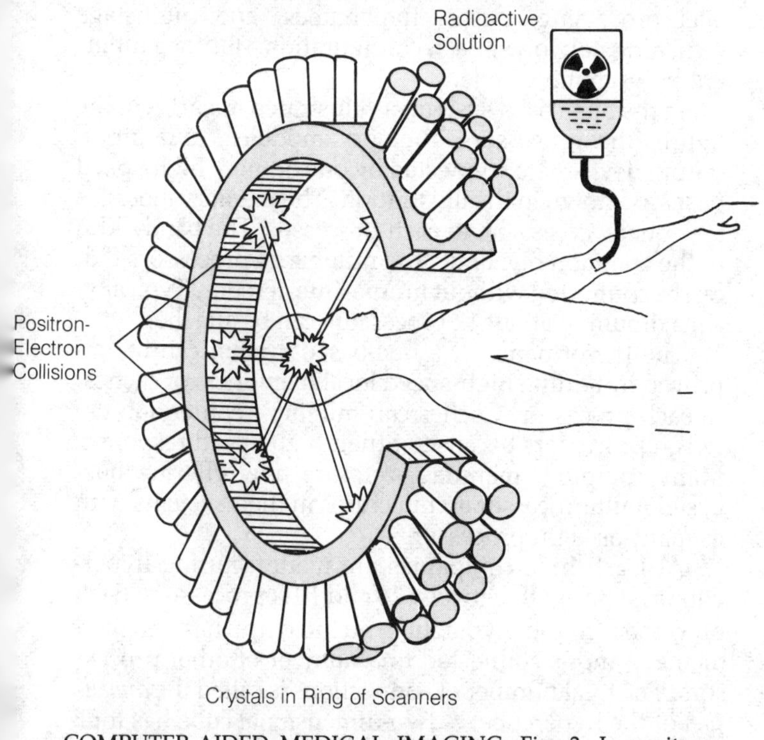

COMPUTER-AIDED MEDICAL IMAGING: Fig. 3. In positron emission tomography (PET) imaging, radioactive tracers injected into the body emit positrons which collide with electrons to form gamma rays. Crystal detectors around the ring detect the gamma rays and the computer records emission and plots the source of radiation to form an image.

Positron-Emission Tomography Positron emission computed tomography (PET) uses trace amounts of radioisotopes to form an image. It is well suited for studying epilepsy, schizophrenia, Parkinson's disease, and stroke.

A PET scanner is shown in Fig 3. It consists of a ring of radiation detection sensors around the patient's body. A small, low-energy cyclotron prepares isotopes with short half-lives for the PET scan. These substances can lose half their radioactivity within minutes or hours of creation. (*See* ISOTOPE.) When injected into the body, the radioactive solution emits positrons that are detected when they generate gamma rays.

When in the human body, the positrons collide with electrons and the two annihilate each other, releasing two gamma rays. (*See* GAMMA RAY, POSITRON.) The emitted rays move in opposite directions, leave the body, and strike the ring of radiation detectors. The detectors respond to the incidence of gamma rays by emitting a flash of energy (scintillation) that is then converted into electronic signals. A computer records the location of each energy flash and plots the source of radiation within the patient's body by comparing flashes and looking for pairs of flashes that arise from the same positron-electron annihilation. It then translates that data into a PET scan image. The PET monitor displays the concentration of isotopes in various colors indicating level of activity.

COMPUTER ARCHITECTURE

Computer architecture refers to the functional arrangements of the processing elements in a digital computer. Most computers are based on the Von Neumann sequential processor architecture. These computers do one operation at a time, using a single central processor. They also use a sequential centralized control unit; sequential machine language; and linearly addressed, fixed-width memory. To achieve higher speeds with a sequential machine, the individual components must run faster. *See* COMPUTER GENERATIONS.

Some computers achieve higher speed by using parallel or concurrent methods with architectures that permit the machine to perform many operations simultaneously. Array processors are examples of computer circuits that are capable of concurrent computation. They have single-instruction, multiple-data (SIMD) path architectures. Array processors can be integrated with conventional, Von Neumann programming methods using sequential instruction streams. *See* ARRAY PROCESSOR.

SIMD machines are well suited for problems where the data is primarily structured in regular dense arrays. Examples occur in image processing and physical simulations. Because they are not general-purpose computers, they are usually used as peripheral processors for host Von Neumann computers.

A common method to achieve higher computational speed is by connecting two or more processors with high-speed buses, packet communications networks, or other circuit-switching designs. This method allows parallel processing on some tasks. *See* DATA COMMUNICATIONS.

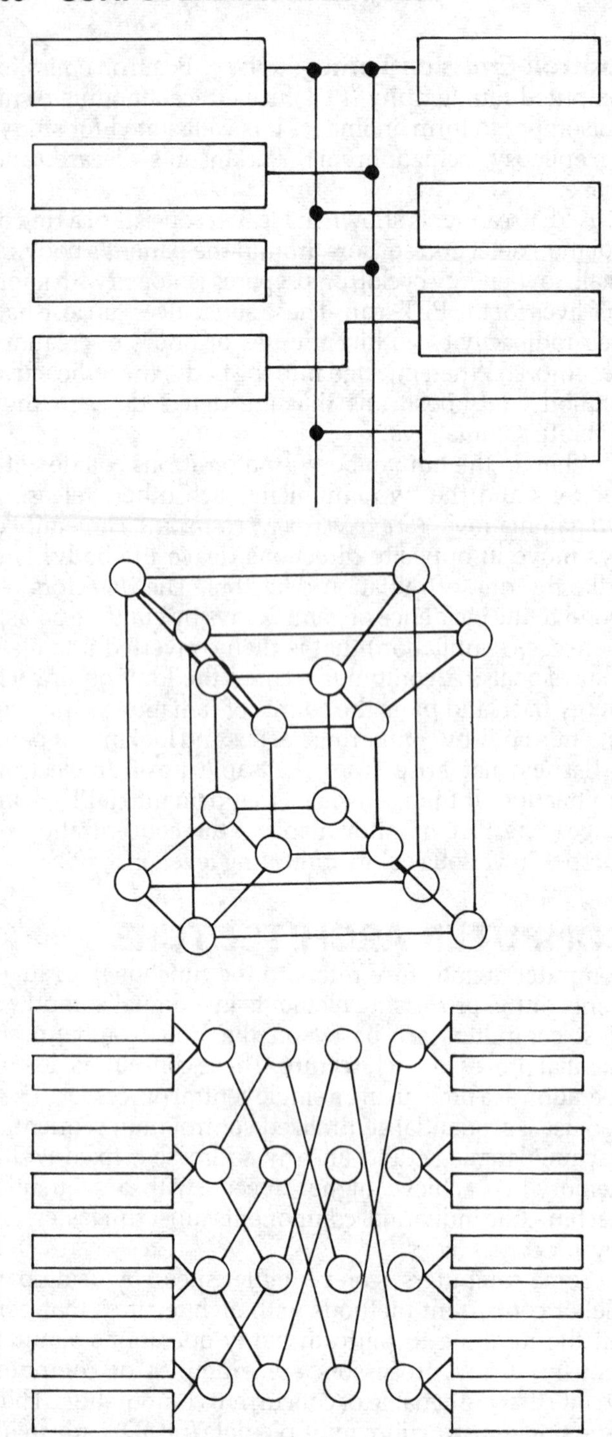

COMPUTER ARCHITECTURE: Interconnection architectures for parallel processors are buses (A), four-dimensional hypercube (B), and multistage switching network (C).

In the design of a concurrent or parallel computer, a decision must be made on control, the size of the basic processor, communications, topology, programming languages, and task allocation. Control can either be centralized or decentralized.

There are three main approaches to parallel computer architecture: control flow, demand flow, and data flow.

Control-flow machines use a single central processor to schedule and allocate tasks to a large number of subsidiary processors. Instructions are passed down from a single centralized source of instructions to many processors dealing with parallel streams of data. As in Von Neumann designs, instructions are executed when the control program orders them. Program counters can be used and the central processor sends out instructions to all processors simultaneously.

Demand-flow computers are organized in a hierarchy. Tasks arriving at a group of processors are systematically divided into subtasks and distributed among the lower-ranking processors. These processors carry out instructions when results are needed for other calculations. They send the results back to higher-level processors to be combined into a final result.

Data-flow computers are highly decentralized; they are made up of a group of processors of equal rank that work cooperatively on the same level rather than being subordinated or organized into a hierarchy. The program dictates the flow of data and the destination of the results. Each processor carries out instructions and calculates its results at its own pace as data becomes available. It then sends the results on to another designated processor. Data-flow computers must be programmed with data-flow languages that differ from conventional programming languages.

Pipelined architectures are a variation of data flow architecture, in which all processors process data at the same rate. Data flow through a sequence of pipelined processors at a constant speed, similar to the way products being manufactured flow along an assembly line and each processor performs different operations.

Three interconnection architectures dominate in parallel processing: buses, hypercubes, and multistage switching networks. Each configuration calls for a different programming style.

In the bus-based system, as illustrated in part A of the figure, the processors, memory modules, and input/output devices are connected by one or more high-speed buses as shown in the illustration. The memory modules are equally accessible to each processor. The bandwidth of the common bus limits the number of processors that can be connected without mutual interference. Typically, a maximum of about 12 processors can be attached.

The performance of a bus-based system can be improved by adding high-speed local memories, or caches, to each processor. Caches can minimize contention between processors because of their high-speed response. Many computer manufacturers are now offering bus-based multiprocessor architectures in their systems with as many as eight processors.

At least three companies are manufacturing hypercube systems with between 16 and 1024 processors. Each processor in the hypercube architecture has its own memory and is connected to a number of other processors. The total number of connections is called the *dimension* of the hypercube. A two-dimensional cube has four processors, each connected with two others; a three-dimensional cube has eight processors, each with three connections; and a four-dimensional organization has 16 processors, each with four connections as shown in part B.

There is no common bus to limit the total number of

processors in hypercube architecture. Hypercubes are best suited to message-passing programming techniques. These match the point-to-point organization of the physical communications links. It is not possible to send a message directly between two processors that are not directly connected. The message must be forwarded through intermediary processors until it reaches its destination.

Multistage switching network architecture connects processors and memory modules through a specialized switching network. See C in the illustration. As in bus-based architecture, the entire memory can be accessed by any processor, permitting effective use of shared memory programming. However, multistage switch network architecture can be expanded to 200 processors or more. As more processors are added, the switching network expands, and the switching bandwidth increases. The architecture permits many processors simultaneous access to many memories because of the multiple paths throughout the network.

COMPUTER BOARD-LEVEL MODULE

Computer board-level modules are pretested, ready-to-plug-in assembled circuit boards available from many commercial sources. They permit manufacturers to assemble their own proprietary computer systems, embed computers in industrial equipment or scientific instruments, or add capability to their existing computer systems. These modules permit the user to avoid the expense of owning or leasing plant facilities and hiring personnel for circuit-board assembly.

Board-level products are being offered by semiconductor manufacturers, particularly by those offering microprocessors and microcontrollers whose sales they would like to increase. However, there are also many independent manufacturers unaffiliated with semiconductor firms that purchase all of the needed components on the open market for the assembly and resale of these products. Many are specialists in certain classes of board products such as memory, input/output, printer, or disk-drive controllers.

The purchase of a board-level products permits original equipment manufacturers to concentrate on their prime specialties, which can be as varied as process controls, robots, or scientific instruments. Building electronic circuit boards in-house would divert financial and manufacturing resources. However, some organizations that employ experienced electronics engineers might wish to concentrate that resource on software, systems integration, and applications, rather than production.

(Courtesy Intel Corp.)

COMPUTER BOARD-LEVEL MODULE: Single-board computer with a 32-bit microprocessor and 2 Mbytes of memory.

Even computer manufacturers have found the economics favorable for purchasing completed circuit boards from outside vendors. In addition, the board-level products serve a large add-on aftermarket for an existing computer systems.

Both proprietary and universal board-level products are available, many as catalog items in stock. Manufacturers often provide multiple selections in each category. The general categories are as follows:

1. Microcomputer modules consisting of a microprocessor and a selection of peripheral devices. These could include parallel and serial I/O (input/output) ports and memory, sockets for additional memory, and options for a numeric coprocessor. *See* COPROCESSOR.
2. Memory modules: boards with arrays of memory ICs (integrated circuits) to augment existing computer system main memory. These can be DRAMs, SRAMs, EPROMs, EEPROMs, NVRAMs, ROMs or combinations of these memory devices. *See* SEMICONDUCTOR MEMORY.
3. Mass storage controllers for controlling floppy-disk drives, hard-disk drives, and tape drives.
4. Digital I/O and timer modules, to increase system I/O capability.
5. Analog I/O modules for instrumentation and process control.
6. Communications controllers for RS-232, IEEE-488 and other interface standards.
7. Interface modules for functions such as interactive graphics and electronic speech synthesis.

Many of the board-level products are intended for use with popular microcomputers and minicomputers. These range from IBM's PC, PC XT, and PC AT personal computers to Digital Equipment Corporation's LSI-11, VAX-11. In addition, many of the board-level products are offered by semiconductor manufacturers who want more of their microprocessor systems included (embedded) in systems ranging from medical and electronic test instruments to graphics workstations.

The compatibility of a board-level module with a specific brand of computer implies compatibility with the bus systems architecture employed in that computer. For example, in the case of IBM personal computers, there is an IBM PC bus; in the case of DEC computers there is a Q bus, both proprietary organizations. There are also other prominent bus structures including Multibus I and Multibus II from Intel Corporation, for its 16-bit and 32-bit microprocessors and VMEbus from Motorola suitable for its 32-bit microprocessors. *See* BUS STRUCTURE.

The independent fabricators of computer board-level products often mix and match various microprocessors with generally accepted bus standards. By contrast, semiconductor companies usually offer only their own microprocessor sets on boards compatible with their proprietary bus structures.

COMPUTER GENERATIONS

Practical digital computers were first developed in the 1940s. The earliest models were custom-made experimental machines that were cumbersome and slow. Nevertheless, they were able to solve practical problems such as the plotting of trajectories for use in World War II gunnery tables. Among the earliest computers were the MARK I, ENIAC, EDVAC, and EDSAC.

The MARK I was developed at Harvard University by Howard Aiken during the period from 1937 to 1944. It was basically a very large electromechanical calculator. It was based on electromechanical relay switching that was operated from a punched paper tape program that looked like a player piano roll. This machine qualified as an automatic electrical digital computer.

The ENIAC (for electronic numerical integrator and calculator) was developed by Eckert and Mauchly at the University of Pennsylvania between 1943 and 1946. The first all-electronic computer to employ electronic tubes instead of relays, ENIAC stored its programs on wired boards similar to those used in telephone switching. Although it did not have an internally stored program, ENIAC is recognized as the first electronic digital computer.

EDVAC (for electronic digital vacuum-tube automatic computer) was developed at the University of Pennsylvania and EDSAC (for electronic delay storage automatic computer) was developed at Cambridge University in England in 1949. Both were the first true electronic computers with internally stored programs.

The UNIVAC I, produced by the UNIVAC division of Remington Rand (now UNISYS Corporation) was the first commercially available computer. UNIVAC I had self-checking circuitry and a high-performance magnetic tape system for input and output of data, but it was very slow. The first UNIVAC I was purchased by the U.S. Bureau of the Census in 1951 for computing national census data. The first computer sold for business data processing was a UNIVAC I delivered to the General Electric Company in Louisville in 1954.

The International Business Machines Corporation (IBM) installed its first commercial computer, the IBM 650, in 1954. However, IBM did not become a dominant force in the data processing industry until the 1960s, when it introduced the 1401. IBM also introduced an even more popular 360 series of computers during the 1960s. This was the first series of computers designed and built as a compatible family. The 360 is recognized as the first of the third generation of computers, which features integrated circuits and magnetic disk storage.

The organization of computers into generations is somewhat arbitrary, but there are five major points to consider in the classifications:

1. Principal type of electronic circuit hardware.
2. Principal secondary data storage.
3. Computer software.
4. Telecommunications technology available.
5. Computer performance.

First Generation First generation computers are considered to be those that were built between 1946 to 1956. These computers used vacuum tubes as the principal switching devices in their processor circuitry. Magnetic drums were used for memory, and cathode-ray tubes (CRTs) were used to display input and output data. Punched cards were the primary medium for storing data files and the input of data to the computer.

Computer languages consisted primarily of machine language in the earliest computers, and later assembly language was introduced. Computer operating systems were primitive and processing was done sequentially under manual supervision of an operator. Teletype terminals could be used for data entry and output.

ENIAC, EDVAC, UNIVAC I and the IBM 650 represent this first generation. Computer performance during that period rose to 2-kilobyte memories and 10 kiloinstructions per second (KIPS).

First-generation computers demonstrated the usefulness of computers in data-processing applications. However, they were expensive, slow, and unreliable. They required large amounts of power for their inefficient and space-consuming vacuum-tube circuitry. As a result, the computers occupied large rooms and required extensive air conditioning to dissipate the waste heat generated by the tubes.

Second Generation Second-generation computers featured discrete semiconductor transistor circuitry to replace vacuum tube circuitry. Computers built in the years of 1957 to 1963 are considered to be second generation. Transistors were far smaller in size, consumed less power, and were more reliable than vacuum tubes. Magnetic core memories were introduced to this generation, and they resulted in shrinking the space requirements for memory. Consequently, overall computer size and power was reduced while performance significantly increased. Housing space and building air conditioning requirements were lower.

Examples of second-generation computers include the National Cash Register NCR 501, the IBM 7094 and the Control Data Corporation CDC-6600.

Magnetic tape became the dominant form of secondary storage, permitting larger amounts of stored data and faster data inputs than did punched cards. High-level languages such as ALGOL, COBOL, and FORTRAN were introduced to this generation. Batch operating systems permitted rapid processing of files on magnetic tape. Data could be transmitted between second-generation computers by digital transmission lines. The performance milestones of the generation were 32-kilobyte memories and 200 kiloinstructions per second (KIPS).

Third Generation Third-generation computers were characterized by the introduction of silicon-integrated circuitry. Computers built from 1964 to 1981 are considered third generation. Before about 1970, small-scale integrated (SSI) circuits replaced thousands of discrete transistors in central processing and input/output functions. Thousands of transistors and hundreds of logic gates were fabricated on a single SSI chip.

After 1970, silicon integrated circuit memories were introduced to replace magnetic core memories, and both medium-scale integrated circuits (MSI) and large-scale integrated circuits (LSI) became available. Computers were built around semiconductor LSI microprocessors and magnetic disk drives were also introduced on a large scale.

This resulted in another round of size and power reductions in computers while speed, computing capability, and reliability improved dramatically. Examples of third-generation computers include the IBM 360 and 370, Digital Equipment Corporation's (DEC) PDP-11, the Honeywell 200, the Cray 1, the ILLIAC-IV and the Cyber 205.

Computers in this generation were able to communicate with one another by means of satellite links, microwave terrestrial links, networking, and optical fibers. Packet switching made intercomputer communications more efficient.

Very high level languages were introduced along with Pascal operating systems and structured programming. Magnetic disk storage became more popular than magnetic drum storage, permitting interactive time sharing. Computer-aided design workstations and computer-graphics systems were introduced in the 1970s. The first program for artificial intelligence (AI) called LISP was developed during this period. Computers in this era featured 2-megabyte memory and were capable of 5 megainstructions per second (MIPS).

Fourth Generation Fourth-generation computers are characterized by the first use of very large integrated circuitry (VLSI) in both logic and memory functions. All computers delivered between 1981 and the end of the 1980's are considered fourth-generation computers. Circuit densities with several thousand gates per chip are now in use and high-density CMOS (complementary metal-oxide semiconductors) circuitry has largely replaced bipolar TTL (transistor-transistor logic) in logic and memories.

New mass-storage devices, including bubble memories and magnetic-optical memory, have been introduced into this generation. The basic IC (integrated circuit) memory device in fourth-generation computers, the dynamic random access memory (DRAM), has reached the 250-kilobit level, and 1 megabit DRAMs are now being designed into advanced fourth-generation computers.

Some examples of fourth-generation computers include the Cray XMP, the IBM 308 and the Amdahl 580. There are now complete hierarchies of computers within this generation using many different internal architectures. These include the minicomputers and mainframe computers originally developed during the third generation as well as the supercomputers, minisupercomputers, superminicomputers and personal computers of this generation.

Significant advancements have been made in combining computer and communications technology in the fourth generation. Along with larger, faster internal memories with a significantly lower cost per bit for data processing, there have been advances in the capacities and cost per bit of rotating disk drive memories.

New user-oriented and applications-oriented languages have been introduced. The ADA language was developed for military applications along with different versions of operating systems such as DOS and UNIX. Virtual memory and reduced instruction set computers (RISC) are being developed. Large strides have been made in programming languages, and rudimentary artificial intelligence (AI) programs are now used in computers whose architecture is suited for these data-processing requirements.

The milestones of this generation are 8-megabyte memory and 30 MIPS.

Fifth Generation Fifth-generation computers are still under development. These computers will be marked by advances in software, distributed computing, and more extensive use of artificial intelligence. This generation represents the next step in the ongoing convergence of computer and communications technologies.

Earlier computer generations could be characterized by significant advancements in hardware: 1) vacuum tube, 2) transistor, 3) small-scale integrated circuit (SSI), and 4) large-scale integrated circuits and very large scale ICs (VLSIs). The hardware technology advancements of the fourth generation computers is being transferred to the fifth generation.

Among the hardware developments expected to influence the fifth computer generation are advanced packaging and interconnection techniques including surface mounting; ultralarge, perhaps even wafer-scale integration; three-dimensional IC (integrated circuit) design; the introduction of gallium arsenide IC technology; and more electro-optical components. Josephson junction technology might be re-activated, and superconducting materials could play an important part in computer design.

The software developments anticipated include concurrent languages, functional programming, and symbolic processing. Programming will be more user friendly through the use of natural language, vision, and speech as input.

COMPUTER GRAPHICS

Computer graphics (CG) or computer-aided graphics (CAG) is that part of computer science concerned with the presentation of graphical information such as sketches, drawings, instrument plots, histograms, maps, vectors, schematics, graphs, and projections on a computer graphics display. Computer graphics are used in a wide range of activities, from computer-aided design (CAD) and analysis of scientific and economic analysis to three-dimensional visual simulations. CG is also used in the preparation of art, animations, and special effects for motion pictures.

Vast quantities of data representing abstract mathematical and scientific concepts can be presented in a graphical form in computer graphics. Presentations also can be made of such physical, chemical, and biological structures as atoms, molecules, organic chains, and DNA.

The figure shows the development of a three-dimensional (3-D) solid object using computer graphics. The presentation starts with the drawing of a "wire-frame," which is then "filled in" by "painting." In the final step, 3-D simulation is obtained with appropriate shading to define contours.

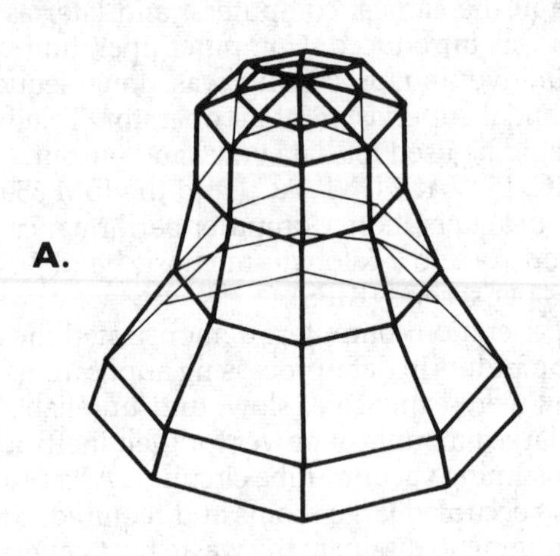

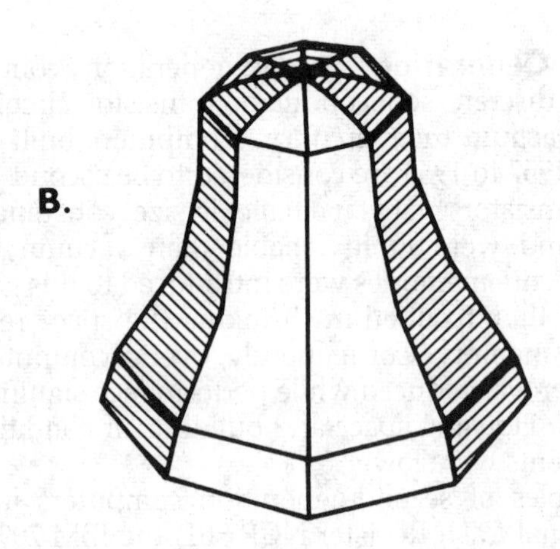

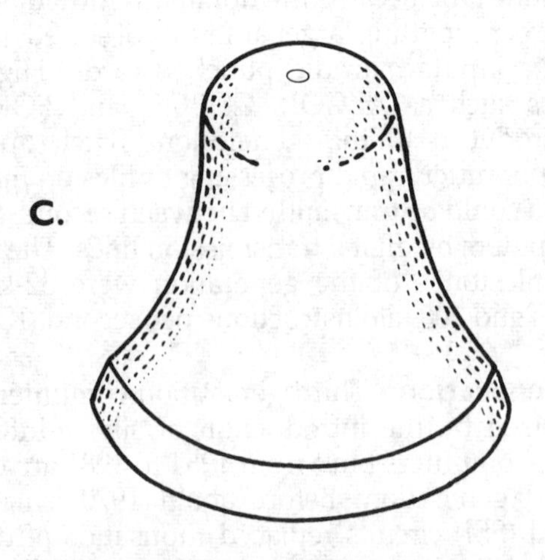

COMPUTER GRAPHICS: The development of a solid object in computer graphics starts with the wire frame (A) that is filled in by "painting" to provide a three-dimensional effect (B) and then contoured with shading (C).

Solids-modeling programs permit the building of databases that can be used by other programs to calculate masses and mass properties, to prepare a model for mathematical analysis, to check for interferences, and even to manufacture the part. *See* COMPUTER-AIDED DESIGN AND COMPUTER-AIDED ENGINEERING.

Cathode ray tube (CRT) displays and X-Y plotters are used as output devices in computer graphics. X-Y plotters provide permanent printed graphics (*see* CATHODE RAY TUBE). Two kinds of CRT writing techniques are used in computer graphics: cursive writing and raster scan.

In cursive writing, the CRT electron beam is moved sequentially so that it traces out lines (vectors) and points in a manner simulating manual drawing. For example, line-generation circuits may move the cursor from either designated end point or between both end points. This method is limited largely to line drawing.

Raster scanning in CG provides the highest graphics resolution. In raster scanning, deflection circuits trace a fixed pattern of parallel lines down the face of the screen from top to bottom, usually from left to right. The electron beam sweeps the entire screen, typically in a repetitive noninterlaced pattern. It passes over the entire CRT screen at some time during the raster scan. The raster is visible as an illuminated screen, without the presence of an image. Color CRTs monitors are used to enhance the graphics presentations. (*See* COLOR TV TUBE). In many computer graphics systems, the operator can interact with the screen to add, delete, or modify the presentation with a light pen, mouse, or X-Y input tablet. *See* COMPUTER MONITOR.

COMPUTER MONITOR

Computer monitors are cathode ray tube (CRT) displays designed and built specifically for computer graphics systems and computer-aided design workstations. The screens are typically 14 to 19 inches, measured diagonally. Most computer graphics monitors today present graphics in color, although some present graphics in a gray scale.

Most computer monitors employ raster scanning similar to that used in a television set. However, computer graphics monitors typically employ noninterlaced scanning that is typically refreshed 40 or more times per second (40 Hz) to eliminate perceptible flicker. In conventional TV raster scanning, interlacing is used and two passes must be made to form a picture (*See* COLOR TV TUBE.) Computer graphics monitors permit the operator to interact with the screen to add, delete, or modify the presentation with a light pen, mouse, or X-Y tablet.

Resolution is very important in computer monitors to provide high definition or a "photographlike" presentation. Resolution depends on the size of the screen, the size and spacing of the phosphor dots (triads), and the ability of the drive electronics to provide sufficient intensity during the raster scan. The quality of the color in the presentation is also considered to be a factor in the definition of resolution.

The *intensity* of the color of the image is a function of the electron beam intensity or voltage applied to a specific pixel. The intensity and colors change as the electron beam scans in the raster pattern across the screen, typically from left to right and top to bottom.

Because the raster pattern of a computer monitor addresses every pixel location on the screen during every refresh cycle, monitors can display lines, dots, and also solid or continuously varying areas of gray scale. The greater the number of phosphor dot triads for a given monitor size, the clearer and sharper are the alphanumeric characters. In addition, the small steps, or *jaggies,* visible on diagonal lines and curves are more subdued. Some 19-inch monitors might have more than 1000 phosphor dot triads in the horizontal dimension.

Character- or table-driven and point-by-point (pixel-by-pixel) methods are employed to address the host system memory. Character-driven graphics address characters in small (typically 7 by 9) pixel blocks. Pixel-by-pixel, or *bit mapping,* systems address individual pixels. *See* BIT MAPPING.

Pixel-by-pixel addressing at the full refresh rate is important in raster-scan CRT displays. It allows alphanumeric and graphic, vector-graphic, or photographic images—either static or animated—to be combined on a single display. Dynamic, interactive, or real-time graphics require that hundreds of thousands of pixels be written in every frame. If the images are to be displayed in color or in a range of gray-scale intensities, even more storage is needed for each pixel. Data values stored in a bit-map memory are converted into display signals within the time limits of the scanning process.

COMPUTER PROGRAMMING

To perform their functions, all computers must be programmed. Generally, a computer requires two different programs. The higher-order program allows the computer to interface, or communicate, with the operator. There are several different higher-order programming languages such as BASIC, COBOL, and FORTRAN, (*see* BASIC, COBOL, FORTRAN, HIGHER-ORDER LANGUAGE). The assembler program converts the higher-order language into binary machine language, and vice versa. The computer operates in machine language (*see* ASSEMBLER AND ASSEMBLY LANGUAGE, MACHINE LANGUAGE).

Preparation of assembler and higher-order programs is usually done manually. Some computers can generate their own programs for certain purposes. The program is a logical sequence of statements, telling the computer how to proceed.

COMPUTER VISION

Computer vision (CV), also known as machine vision, uses computers to acquire, interpret, and process visual information. In most systems, cameras based on vidicon tubes or charge-coupled devices (CCD) are used to sense light from a scene. The resulting signal is digitized and processed by computer. CV can be used to recognize and classify, position, or orient objects in the field of view of the camera. Noncontact measurements can also be made of object dimensions. In industrial applications, CV sys-

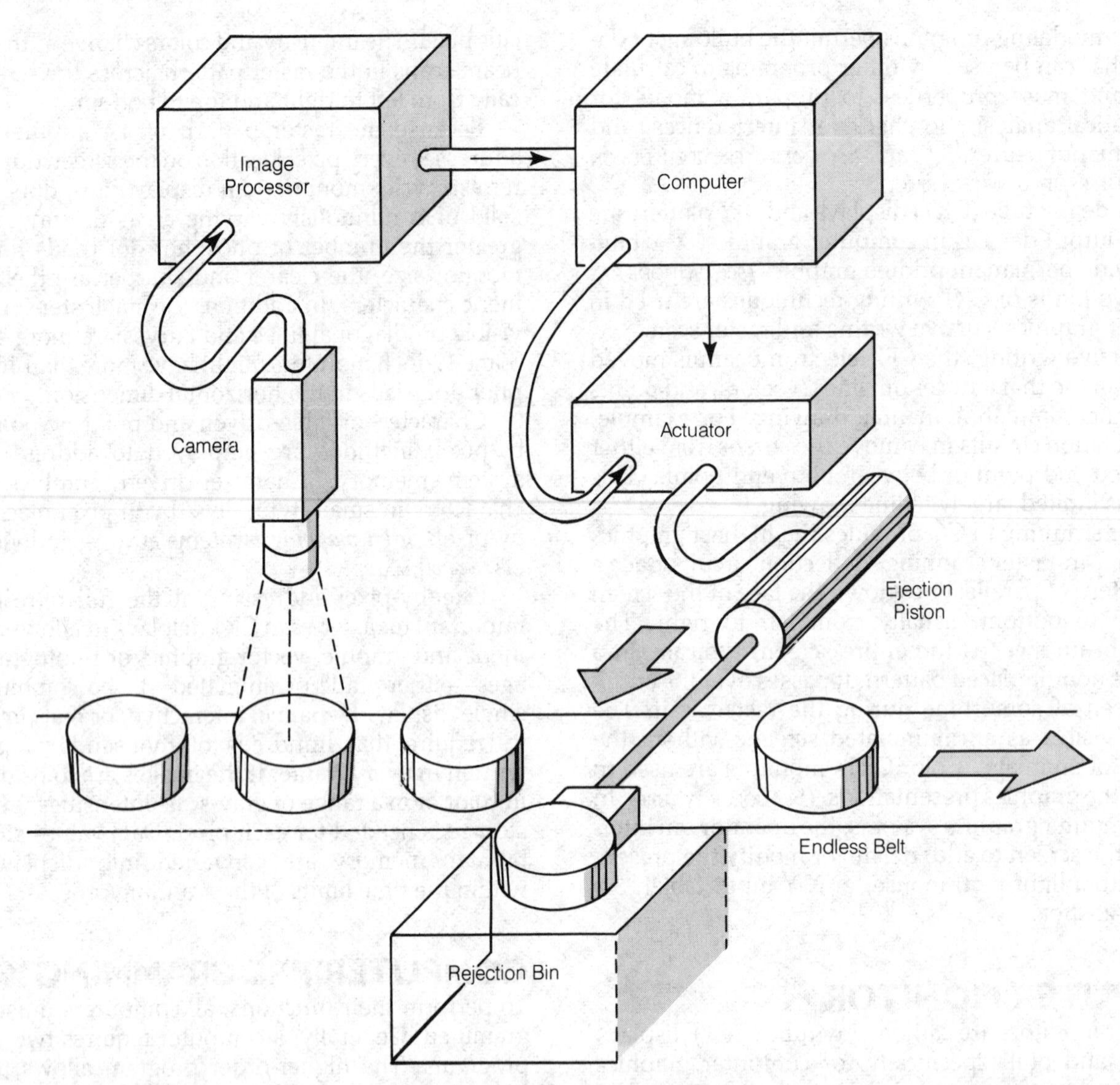

COMPUTER VISION: Objects on a conveyor are viewed by a TV camera which sends data to an image processor. A computer compares the features of the image against those of a stored reference. If the images do not match, a signal is sent to eject the object from the moving belt.

tems are used for inspection, location, counting, measuring, and control.

Computer-vision systems are widely used for object inspection. An example of an inspection system is shown in the illustration. This kind of a system is able to sort good parts from bad parts. Generally, the computer system contains the visual characteristics of different objects in memory so that it can classify an unknown object in its field of view by comparing it with these prototype objects.

Usually the prototypes are stored as a collection of feature values that represent numerical quantities or measurements. Those feature values that are independent of position and orientation can be used for object comparison.

Computer vision systems can also provide position and orientation information about an object in its field of view. Generally, the system must be given the zero position information about each object so it can report displacement from that position. Vision systems can also measure parts. They do this by finding one or more dimensions of the part. The accuracy of visual measurement depends on the resolution of the image, the placement of the camera, lighting, and other factors.

Attention must be given to the camera, lighting, optics, and interfaces to the computer when the workplace is being set up for CV. Objects of interest should be as prominent in the picture as possible to enhance the computer ability of the computer to extract information about them from the image. It must be able to separate the objects of interest from other objects that enter the camera view.

The camera lens determines the field of view. A lens with a short focal length gives a wide-angle view, and a long focal length gives a telephoto view. A camera lens with a long focal length must be placed farther away from the area to be viewed than one with a short focal length.

The camera is connected to a special computer that performs the image processing. This computer has hardware to digitize the camera signal. The digitizing circuit can be a thresholding circuit for a binary system or an analog-to-digital (A-D) converter for a gray-scale system.

A thresholding circuit compares the analog camera signal against a threshold level and delivers a 1 if the input is above the threshold and a 0 if the input is below the threshold. Systems that use binary images can recognize well-illuminated and non-overlapping parts in specific three-dimensional orientation, but they are not suited for factory applications where illumination is not constant and parts might overlap.

Six or eight bits of gray scale are used in most applications. The amount of information a system must process is determined by the number of pixels assigned to each image, and the number of possible intensities for each pixel. For example, an 8-bit format will have 256 different levels of intensity per pixel. Gray-scale processing reduces the need to physically segregate parts because the system can distinguish between overlapping parts by differences in the light intensities of the image.

Image-processing algorithms extract object features such as edges and orientation, while minimizing the effect of lighting changes. Edge detecting or finding is a technique used to locate the boundaries between regions in an image. Edge finding consists of locating picture elements in an image that are likely to be on an edge, and linking these possible edge points together into a coherent edge.

The improved performance of gray-scale processing requires far more data to be processed. A 512- by-512-pixel display with 256 gray levels contains over 2 million bits of information (256 bytes) for each image frame, as opposed to about 262,000 bits for a binary image.

An intermediate approach combines binary processing of the image and gray-scale image acquisition. The gray-scale image is processed to form a binary image that is more accurate than the threshold binary image. The approach offers better accuracy than binary systems and faster execution than gray-scale processing. Another method to reduce computing is to limit processing to areas where relevant data is likely to be present, notably the part boundaries. The background and majority of internal features of a part are ignored.

The model-based approach uses a digital memory representation of a part to reduce the amount of data it must process. To generate a model of the part to be inspected or recognized, a sample of a good part is placed under the camera so the system can capture its image. Then part features such as boundary length, curve, type, and change in light intensity between boundaries are measured and stored in the image processor's memory as a *template*, or model, of the part. Once the image processing system has all the part template, it can search for similar objects in other images.

The model approach lets the system deduce part features that the camera cannot see directly, such as overlapped features or features out of the field of view. The model-based approach speeds up part recognition by allowing the system to stop its search whenever it has identified a unique set of part features.

Computer vision systems are used in robotic materials-handling and assembly work. The system locates the part, identifies it, and might carry out different programmed responses. For example, the system may direct the robot gripper to a suitable grasping position, pick up the good parts, and take them to an assembly area. Both the camera and robot have unique coordinate systems. However, if the robot is used for grasping, it must know where the object lies in reference to its own coordinate system. This requires the transfer of information about the object location from the coordinate system of the camera to that of the robot.

CONDUCTANCE

Conductance is the mathematical reciprocal of resistance. The unit of conductance (G) is the siemens (S), named after Ernest von Siemens, a German inventor. It was also called the *mho—ohm* spelled backwards. Mathematically, if *R* is the resistance in ohms, the conductance *G* in siemens is:

$$G = \frac{1}{R}$$

Conductance is occasionally specified instead of resistance when the value of resistance is very low. *See also* ADMITTANCE, RESISTANCE.

CONDUCTION COOLING

Conduction cooling is a method of increasing the dissipation capability of a tube or transistor. Heat is carried from the device by a thermally conductive material such as beryllium oxide or a silicone compound, to a large metal surface. The metal, in turn, employs cooling fins to let the heat escape into space. The illustration shows a typical conduction-cooling arrangement for a transistor.

Conduction cooling is quieter than forced-air cooling, the other most common method of cooling. Blowers are

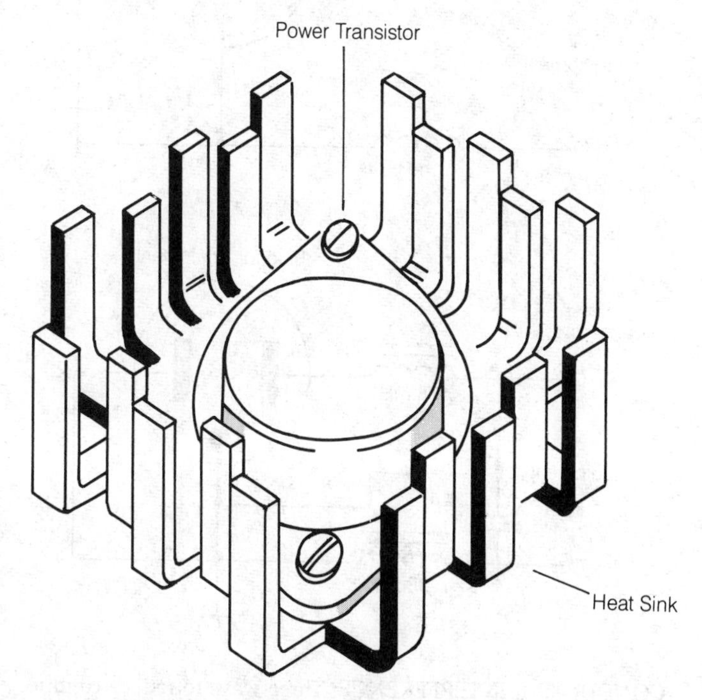

CONDUCTION COOLING: Conduction cooling of a bipolar transistor often is improved with a heat sink.

not necessary for conduction cooling. However, care must be exercised to ensure that a good thermal bond is maintained between the cooling device and the tube or transistor. *See also* AIR COOLING, HEATSINK.

CONDUCTIVE MATERIAL

A conductive material is a substance that conducts electricity very well. Most metals are good conductors; silver is the best. Strong liquid electrolytes are also good conductors.

Conduction occurs as a transfer of orbital electrons from atom to atom. When one atom accepts an extra electron, it usually gives up an electron, passing it on to a nearby atom. The best conductors do this readily. In a highly conductive material, a large current flows with the application of a relatively small voltage.

The most frequently used conductive material in electrical wiring is copper. Steel and aluminum are also employed. *See also* CONDUCTIVITY.

CONDUCTIVE INTERFERENCE

Interference to electronic equipment may originate in the power lines supplying that equipment. Interference can be radiated by the lines and picked up by a radio antenna, or it may be conducted to the equipment and coupled by the power-supply transformer. The latter mode is called conductive interference.

Conductive interference originates in many different kinds of electrical appliances such as razors, fluorescent lights, and electric blankets. Light dimmers are a well-known cause. Faulty power transformers can transmit interference on utility lines. At radio frequencies, these impulses are coupled by the capacitance between the primary and secondary windings of the supply transformer. At audio frequencies, the interference may pass through the transformer inductively.

Conductive interference can be very difficult to eliminate. The illustration shows some common alternating-current line filters for minimizing the effects of conductive interference at audio and radio frequencies.

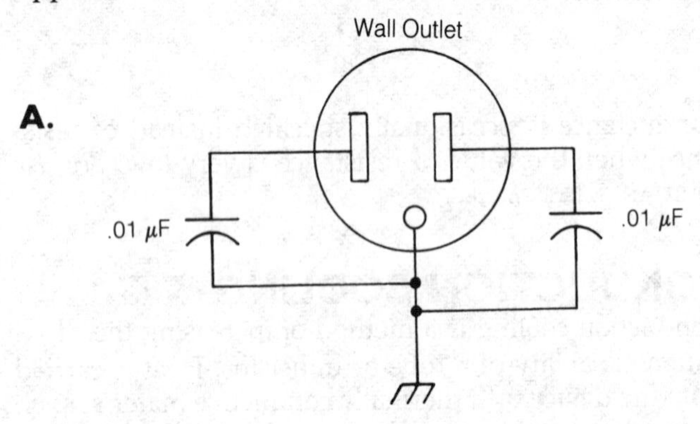

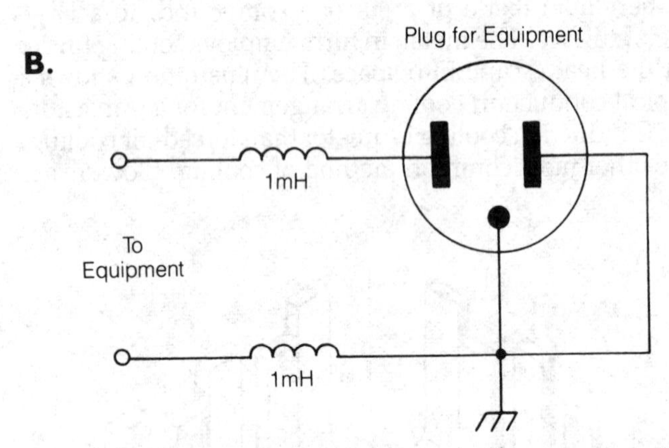

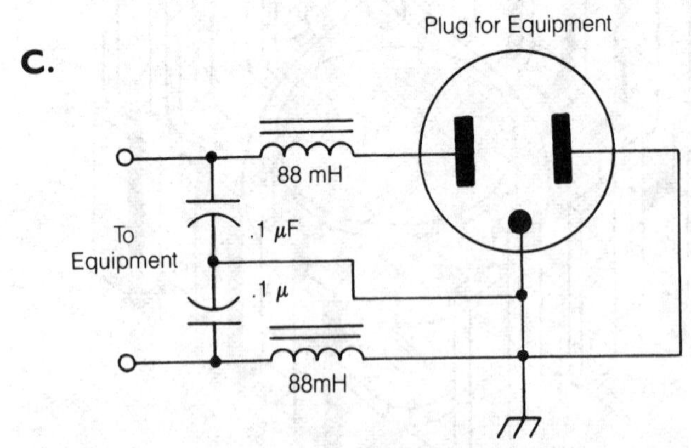

CONDUCTIVE INTERFERENCE: Three ways to reduce conductive interference are the use of bypass capacitors (A), series chokes (B), and combination chokes and bypass capacitors (C).

CONDUCTIVITY

Conductivity is a measure of the ease with which a wire or material carries an electric current. Conductivity is expressed in mhos, millimhos, or micromhos per unit length. The table shows the conductivity of copper wire for American Wire Gauges 2 through 34.

Conductivity varies with temperature. The values in the table are for room temperature. Some materials become more conductive as the temperature rises; most get less conductive as the temperature rises.

The better the conductivity of a material, the more current will flow when a specific voltage is applied. The

CONDUCTIVITY: Conductivity, in millimhos per meter, of various gauges of copper wire.

American Wire Gauge	Millimhos per Meter
2	1,913,375
4	1,203,319
6	756,705
8	475,879
10	299,411
12	188,265
14	118,369
16	74,451
18	46,820
20	29,449
22	18,518
24	11,647
26	7,323
28	4,606
30	2,897
32	1,822
34	1,146

greater the conductivity, the smaller the voltage required to produce a given current through a length of conductor. Conductivity is the opposite of resistivity. *See also* RESISTIVITY.

CONDUCTOR

Conductors for electronics include metal wires, metal strips and plates, or metallized surfaces on insulating substrates. They transmit and receive electrical signals and power. A wide selection of metals and alloys can function satisfactorily as bare or insulated signal and power conductors in electronics. Copper, copper-covered steel, high-strength copper alloys, and aluminum are widely used conductors. However, there are special applications for pure nickel, pure silver, copper-covered aluminum, and many other metals.

Copper remains the most widely used metal for electronics conductors. Among the desirable physical properties of copper are high electrical and thermal conductivity; ductility; malleability; solderability; high melting point; high resistance to corrosion, wear, and fatigue; and low cost. The grade of copper most widely used for electronics is classified as ETP copper (for electrolytic tough pitch). It contains minute amounts of copper oxide. Additional refining produces a copper with practically no oxide called OFHC (oxygen free, high conductivity).

Copper conductors can be solid wire, stranded wire with multiple strands, and flexible wires with many fine strands (typically 20 or more). The temper or degree of hardness of commercial copper conductor wire ranges from soft-drawn annealed (SD) and medium-hard drawn (MHD) to hard drawn (HD). The tensile strength of hard- and medium-hard drawn copper exceeds that of all soft-drawn copper with the same wire gauge.

Electronics wiring is rarely used in sizes greater than #10 American Wire Gauge (AWG). (*See* AMERICAN WIRE GAUGE.) The American Wire Gauge is used in the United States to identify the sizes of wire up to and including #4/0. The next size larger than #4/0 AWG is measured in circular mils (CM). It has a value of 250,000 CM or 250 MCM.

Stranded Conductors. Stranded wires are specified because they are more flexible than equivalent solid wires. Moreover, they can withstand more vibration and bending before fracture. Surface damage to a stranded conductor will have less effect than similar damage to a solid wire because the breakage of a single strand will have little effect on the current carrying capability of the wire.

Commonly used conductors in electronics have 7, 10, 16, 19, 26, or more strands, with 7 and 19 strands the most widely accepted. For any given wire size, the greater the number of strands, with corresponding decrease in individual strand diameter, the more flexible the wire and the more expensive the conductor. Conductor weight and resistance is proportional to circular mil area which, in stranded wires, depends on the number and diameter of component strands. However, the circular mil area of stranded wires only approximates that of their solid-wire equivalent.

Conductor Coatings. For copper wire, tin is the most widely applied coating to prevent oxidation and improve solderability. However, silver and nickel are also applied in certain applications. Occasionally, other coatings such as silver/nickel combinations can be specified. Tin plating of copper is acceptable for conductors that are not exposed to temperatures of more than 150°C. The drawback to tin coatings is that it increases the resistance of the finished wire.

Silver plating of soft-drawn copper wire can increase its maximum operating temperature from 140° to 200°C. It is useful in high-frequency applications where the higher conductivity of silver is desirable. Nickel plating can increase the operating temperature of copper wire from 140° to 260°C. Nickel plating is recommended for Teflon TFE hook-up wire operating for prolonged periods at temperatures from 200° to 260°C.

Conductor Insulation. Thermoplastics, thermosets, and elastomers are all used as primary insulation applied directly over solid and stranded conductors. These materials are also used as jackets over primary insulation, braids, shields, and cables. In general, these materials can be classified as extrudable or tape insulations.

The extrudable primary insulation materials include polyvinyl chloride (also known as vinyl or PVC), the polyolefins (polyethylene and polypropylene), and the fluorocarbons (principally Teflon TFE, FEP, and PFA). The tape insulations include polyester by itself, or laminated with materials such as Kapton. *See* CABLE, COAXIAL CABLE.

Printed Circuit-Board (PC) Conductors. PC conductors are formed on printed wiring boards by subtractive or additive methods. In the subtractive PC board and card manufacturing method, the process begins with the insulating substrate material, typically epoxy-filled glass fiber (FR-4) laminated on one or both sides with copper foil. In a succession of operations that includes masking, plating, and chemical etching, the unwanted copper is removed to leave thin conductors securely bonded to the substrate. These are typically plated with lead-tin solder to permit easier solder bonding of components. *See* PRINTED CIRCUIT BOARD.

In the additive process, copper conductors are formed on the surface of a bare, chemically treated substrate, typically, FR-4 by masking and plating processes. The final copper conductors on the board surface can also be lead-tin solder plated to ease the solder bonding of components.

Plated-through holes are conductors widely used to interconnect conductive layers of two-sided and multilayer PC boards in conventional PC board manufacture. The internal laminations can be power or ground planes. Conducting surfaces on opposite sides of two-sided PC boards are correctly registered as part of the manufacturing and lamination process. The external and internal conductors are joined by these plated-through holes.

As a first step in the operation, holes are drilled or punched through the conductors on the board to be joined. The inner walls of the holes are then chemically treated to accept copper plating which forms a conducting copper sleeve within the hole. This sleeve or eyelet joins all conducting surfaces adjacent to the hole.

In other wiring board fabrication processes, fine insulated wire is located on the surface of the PC board between designated locations with numerically controlled (NC) machines. The insulated wire is then chemically bonded to the surface of the board. This technique permits conductors to cross over each other without causing short circuits.

All conventional conductors dissipate power in the form of heat. Where the resistivity is critical, the conductors can be cooled in liquid nitrogen or helium. A class of materials called superconductors has been developed. These materials exhibit sharply reduced resistance values when their temperatures are lowered by immersion in liquified gasses. *See* SUPERCONDUCTIVITY.

CONE

In electronics, a cone is a diaphragm that radiates sound from a speaker. It is named because it is conical in shape, as shown in the illustration. Speaker cones are usually made of thin cardboard. The oscillating vibration of the cone, caused by the action of the speaker coil in the magnetic field, produces sound waves.

The speaker cone must be designed properly for optimum sound reproduction. In communications systems, a high-frequency cone response (about 3000 Hz) is undesirable. In a high-fidelity sound system, the response must extend to the limit of the human hearing range, or about 16,000 to 20,000 Hz. Special speaker combinations are necessary to achieve this frequency range. *See also* SPEAKER, TWEETER, WOOFER.

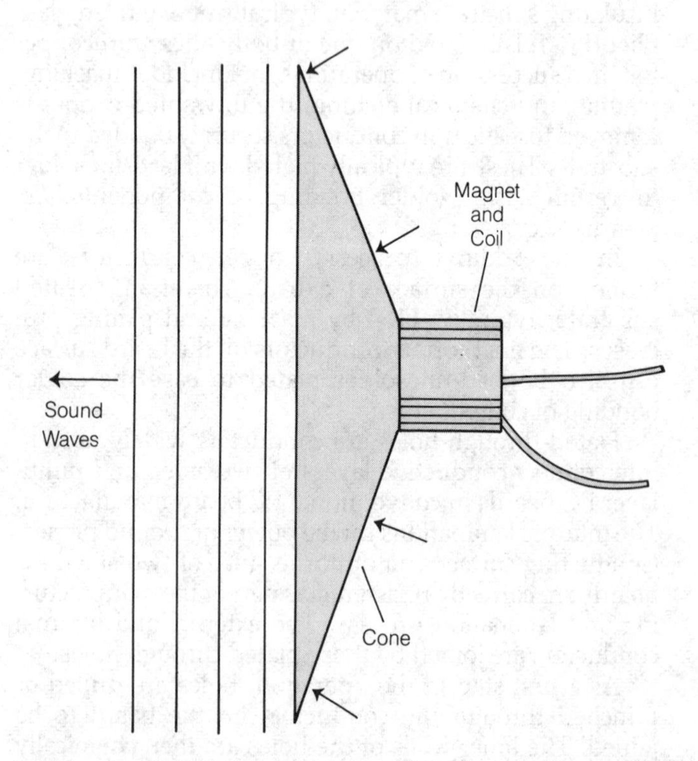

CONE: A speaker cone produces sound by compressing the air when it is vibrated.

CONE MARKER

A cone marker is a very high frequency or ultrahigh frequency beacon that radiates energy vertically, in a cone-shaped pattern. The vertex of the cone is located at the antenna, and the cone opens upward.

Cone markers are used by aircraft to locate their precise position. Knowing the exact location of two or more cone markers on the ground, and measuring the direction to each, the pilot of an aircraft can determine the the location of the plane.

CONE OF SILENCE

A low-frequency radiolocation beacon radiates energy mainly in the horizontal plane. An inverted, vertical cone over the antenna contains very little signal, as shown in the drawing. This zone is called the cone of silence.

Any omnidirectional antenna, designed for the radiation of signals primarily toward the horizon, has a cone of silence. The more the horizontal gain, the broader the apex of this cone. In some repeater installations using high-gain collinear vertical antennas located high above the level of average terrain, a cone of silence (poor repeater response) exists underneath the antenna. This zone can extend outward for several miles in all directions at ground level (B in the illustration). This effect often makes a nearby repeater useless. *See also* REPEATER.

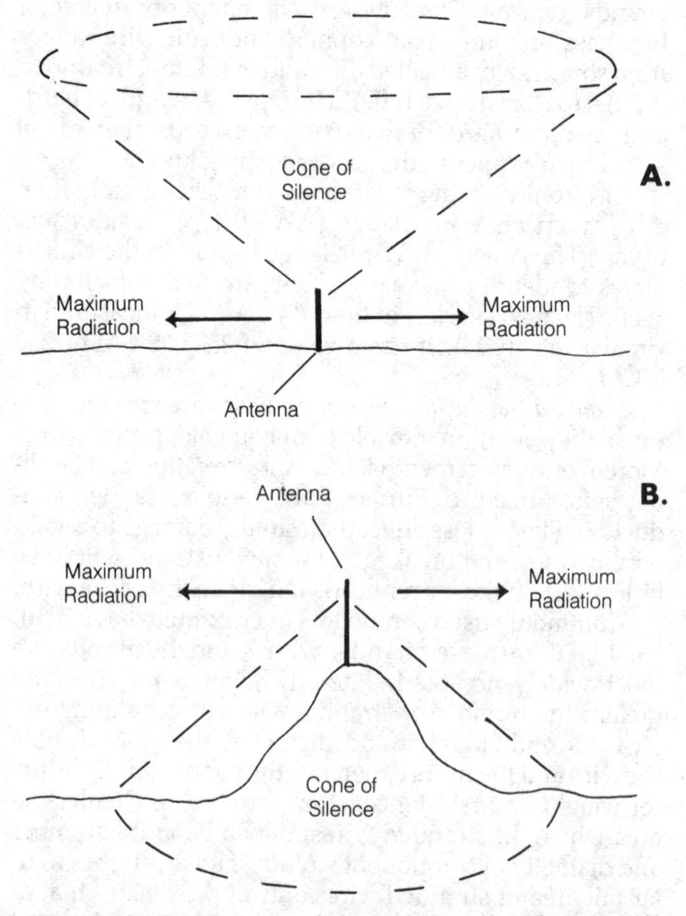

CONE OF SILENCE: A cone of silence above a vertical antenna (A), and one below a collinear repeater antenna (B).

CONFERENCE CALLING

When a telephone connection is made among three or more parties simultaneously, the mode is called conference calling. Each subscriber is then able to talk to any or all of the others. When one subscriber speaks, all the others can hear.

Conference calling requires special arrangements to match the impedance at the switching system. Several telephones, operating at the same time on one circuit, result in different loading conditions than a single telephone produces.

Conference calling is available for most home and business telephone subscribers in the United States today. Conference calling can be carried out over local or long-distance circuits. *See also* TELEPHONE SYSTEM.

CMOS

See COMPLEMENTARY METAL-OXIDE SEMICONDUCTOR.

CONICAL SCANNING

Conical scanning is a method of determining the direction of a target in a radar system. The elevation as well as the azimuth can be found by this method. The radiation pattern of the antenna is made slightly off center, and the antenna is electrically or mechanically rotated so the lobe of maximum radiation rotates about the boresight (axis) of the radar dish.

When the target is exactly aligned with the boresight, the returned signal is constant in amplitude (as shown at A in the illustration) because the target is always the same angular distance from the lobe of maximum radiation. When the target is not aligned with the boresight, as shown at B, the returned signal is modulated at the frequency of antenna rotation. In this way the antenna can be accurately pointed in the direction of the target. The conical-scanning method is very precise. *See also* RADAR.

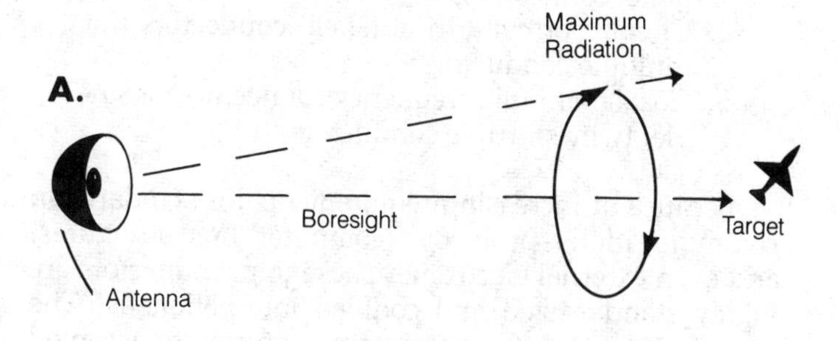

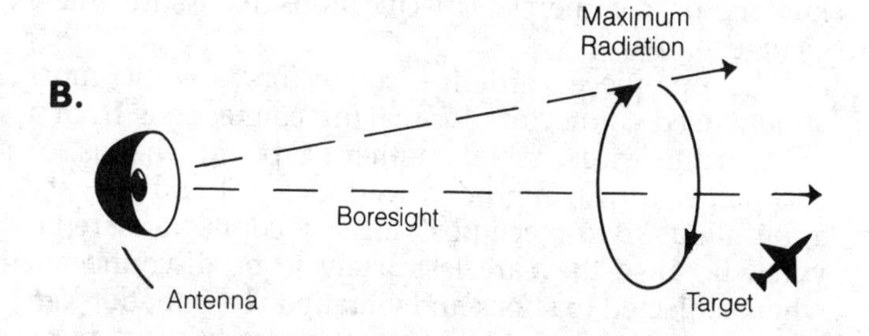

CONICAL SCANNING: In conical scanning radar the axis of maximum radiation is rotated conically around the antenna bore sight. When the antenna is on target the echo has a constant amplitude (A). When the target moves with respect to the cone of maximum radiation, the amplitude of the echo varies (B).

CONIC SECTION

A conic section is the geometric intersection of a cone and a plane. Conic sections include the circle, ellipse, parabola, and hyperbola (see drawing).

Conic sections can be generated, for illustrative purposes, using a bright flashlight and a large, flat surface such floor. When the flashlight is pointed straight down, the large area of relatively dim light (outside the bright central spot) is bounded by a circle. The flashlight must be oriented vertically, so that the bright central spot is directly below the bulb. In the Cartesian (x,y) plane, a circle is represented by the equation:

$$(x - a)^2 + (y - b)^2 = k$$

where a, b, and k are constants.

When the flashlight is pointed at an angle, in such a way that the perimeter of the area of dim light falls on the surface but is not a perfect circle, it forms an ellipse. In the Cartesian (x,y) plane, an ellipse is represented by the equation:

$$c(x - a)^2 + d(y - b)^2 = k$$

where a, b, c, d, and k are constants.

As the flashlight angle is raised, a point will be reached where the far edge of the perimeter of dim light no longer reaches the ground. At this angle, the outline of the dim region forms a parabola. In the Cartesian (x,y) plane, a parabola is represented by the formula:

$$(y - b)^2 = k(x - a)$$

where a, b, and k are constants.

If the angle of the flashlight is raised still further, the outline of the dim region becomes a hyperbola, until the dim region no longer reaches the ground at any point. In the Cartesian (x,y) plane, a hyperbola is represented by:

$$c(x - a)^2 - d(y - b)^2 = k$$

where a, b, c, d, and k are constants.

Conic sections are frequently seen as mathematical relations and functions. *See also* CARTESIAN COORDINATES.

CONJUGATE IMPEDANCE

An impedance contains resistance and reactance. The resistive component is denoted by a real number, a, and the reactive component by an imaginary number, jb (*see* COMPLEX NUMBER, IMPEDANCE, J OPERATOR, REACTANCE, RESISTANCE). The net impedance is denoted by $R + jX$. If X is positive, the reactance is inductive. If X is negative, the reactance is capacitive.

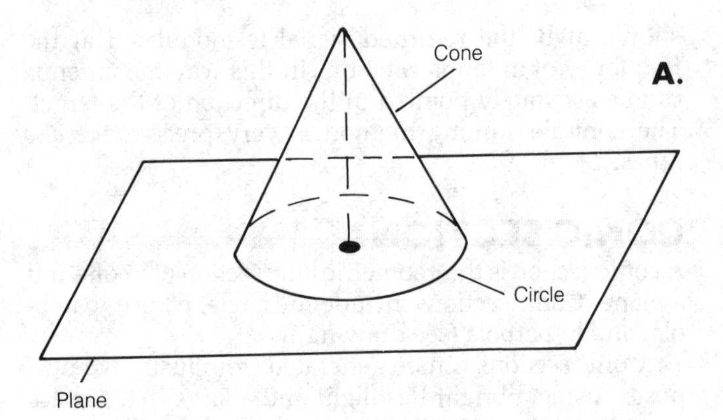

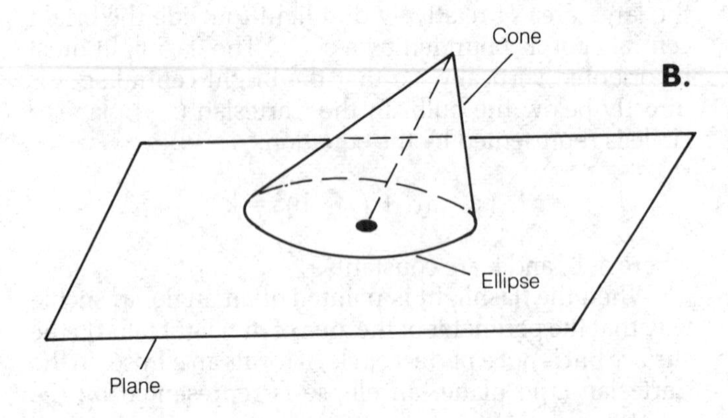

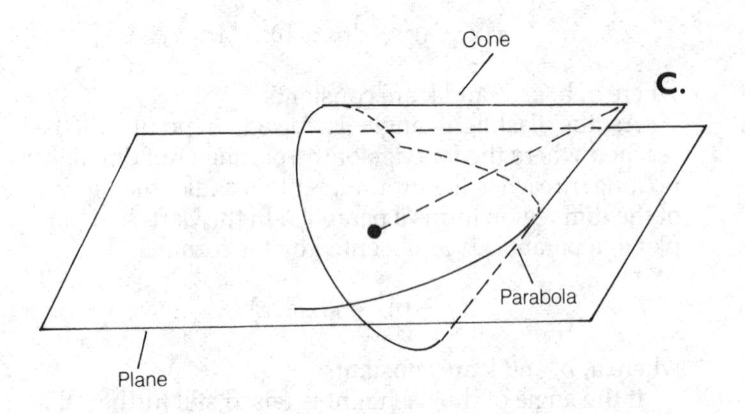

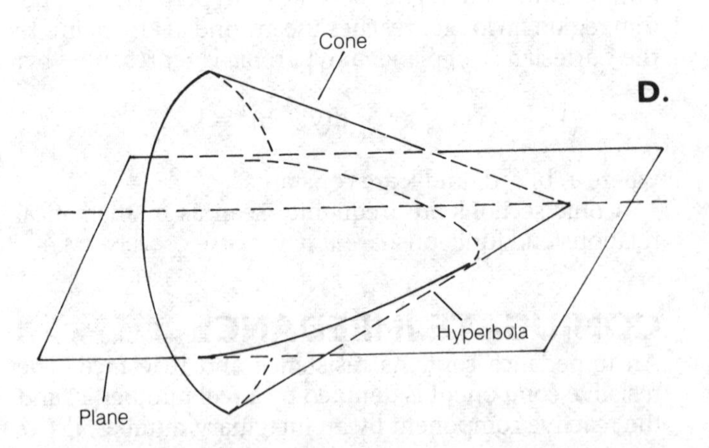

CONIC SECTION: Sections formed by a cone intersecting a plane: a circle (cone perpendicular to the plane, A), an ellipse (cone offset with respect to the plane, B), a parabola (cone surface parallel to the plane, C), and a hyperbola (cone axis parallel to the plane, D).

Given an impedance of $R + jX$, the conjugate impedance is $R - jX$. That is, the conjugate impedance contains the same value of resistance, but an equal and opposite reactance.

Conjugate impedances are important in the design of alternating-current circuits and in impedance matching between a generator and load.

CONNECTOR

In electronics, connectors are manufactured products designed to terminate conductors and cables between electronic circuits within a system, between systems, and between systems and external power sources and signal lines. Connectors interconnect circuits on circuit boards with backplanes or backpanels (*see* BACKPLANE) or wiring within an enclosure. They can also interconnect chassis and circuit boards in different enclosures.

Connectors can terminate the cables interconnecting the peripherals of a system, including their accessories and interfaces. A distinction is usually made between connectors and sockets, which are considered to be component-to-circuit board connectors. Sockets are available for mounting: 1) electron tubes, 2) relays, 3) dual-in-line (DIP) and single-in-line (SIP) semiconductor and component packages and, 4) leadless and leaded chip carriers (LCC) and pin-grid arrays (PGA). *See* SEMICONDUCTOR PACKAGE, SOCKET.

Electronic connectors today include:

1. Single-piece PC (printed circuit) or card-edge connectors.
2. Two-piece plug and receptacle PC board connectors.
3. Rectangular multipin connectors (rack-and-panel connectors).
4. Circular or cylindrical (shell) connectors (for multiple conductors).
5. Coaxial or radio-frequency connectors for single, twin, or triple conductors.

Because of increasing requirements for standards in communications protocols, computer bus structures, and even external electronics packaging, connectors are highly standardized and codified internationally. The manufacture of most connectors conforms to accepted military or commercial specifications to assure interchangeability.

The one-piece cardedge connectors are the most widely used connectors for making connections from a PC board to wires, cable, another PC board, chassis, or backplane. But in high-density industrial and military applications, two-piece post-and-box connectors are favored because they are less likely to be disconnected when subjected to shock and vibration. Termination can be made by solder, wire wrap, or press fit. Two-piece connectors, as shown in Fig. 1, are manufactured to conform to European as well as military specifications.

One-piece cardedge connectors have been adapted to accept flat ribbon cable with mass termination insulation displacement connection (MTIDC). The connectors are

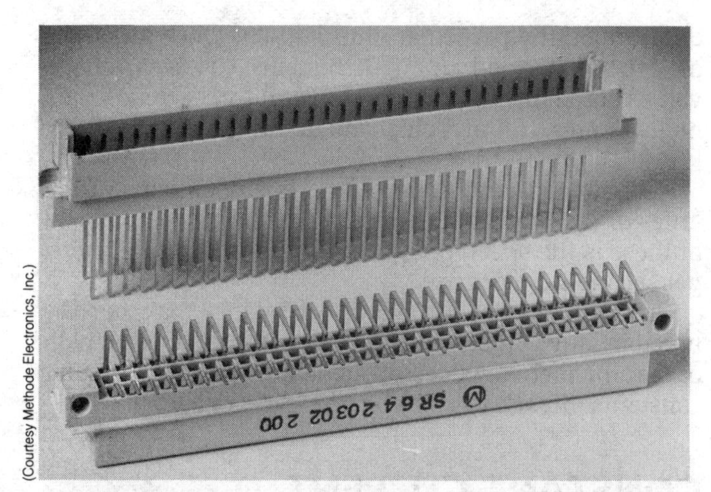

(Courtesy Methode Electronics, Inc.)

CONNECTOR: Fig. 1. Box-and-post two-piece PC-board connectors are used in computers and telecommunications equipment.

equipped with sets of movable teeth that align with individual wires in the flat cable. Simultaneous external pressure on the extended connector cusps forces them down in a single rank over the wires, shearing the insulation of all wires at the same time and forming multiple gas-tight connections.

Rectangular metal-shell plugs and receptacles are widely used to make connections from one internal chassis to another in the same case or cabinet. One example, the trapezoidal or D-type connector is illustrated in Fig. 2. There are many variations of these rectangular D-type connectors, also referred to as rack-and-panel connectors. Termination methods include crimp-removable contacts, solder and MTIDC where the connectors have been adapted to that method. In making external signal-level connections from one piece of communications equipment to another in a system, multipin

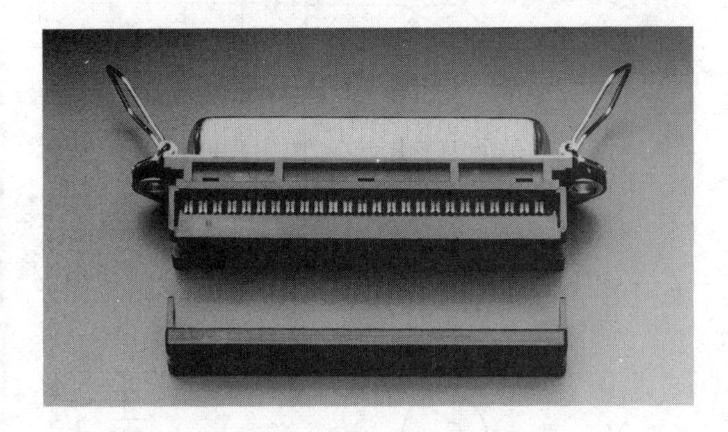

(Courtesy Fujitsu.)

CONNECTOR: Fig. 2. Shielded D-type subminiature connectors for computers and telecommunications equipment.

rectangular metal-shell connectors are preferred over shelless self-mounting connectors. Provision can be made for shielding these connectors from EMI and RFI (electromagnetic interference and radio frequency).

Other options for multiple conductor interconnection are the multipin circular or cylindrical connectors with breech or full threaded couplings (with or without environmental seals). Wire terminations are typically made by crimp-removable contacts. Circular connectors are available with up to 100 pins. These connectors can include internal contacts for radio-frequency coaxial cables. Most circular connectors are manufactured to exacting military specifications. Provisions can be made on circular connectors for shielding from EMI and RFI. *See* CIRCULAR CONNECTOR.

There are many types of coaxial or radio-frequency connectors available commercially for interconnecting high-speed data, video, and radio-frequency wires and cables. These connectors have provision for terminating an inner coaxial conductor (stranded or solid wire) and flexible metal shielding for two-conductor or shielded single-conductor service. *See* COAXIAL CABLE, COAXIAL CONNECTOR.

CONSERVATION OF ENERGY

The law of conservation of energy is an axiom of physics. When energy changes its form, the capacity for doing work remains the same. Energy cannot be created or destroyed within a system.

A battery might, for example, be used to run an electric motor or to illuminate a light bulb. If the battery is completely exhausted, the expended energy is the same in either case. With the motor, the chemical energy in the battery is converted mostly to mechanical energy. Some heat is produced as well because of losses in the conductors. With the light bulb, the chemical energy in the battery is converted to light and heat, which are forms of electromagnetic energy. *See also* ENERGY, WORK.

CONSONANCE

Consonance is a special manifestation of resonance. If two objects are near each other but not actually in physical contact, and both have identical or harmonically related resonant frequencies, then one might be activated by the other.

An example of acoustic consonance is shown by two tuning forks with identical fundamental frequencies. If one fork is struck and then brought near the other (see illustration at A), the second fork will begin vibrating. If the second fork has a fundamental frequency that is a harmonic of the frequency of the first fork, the same thing will happen; the second fork will then vibrate at its resonant frequency.

Electromagnetic consonance is produced when two antennas, both with the same resonant frequency, are brought in close proximity (B). If one antenna is fed with a radio-frequency signal at its resonant frequency, cur-

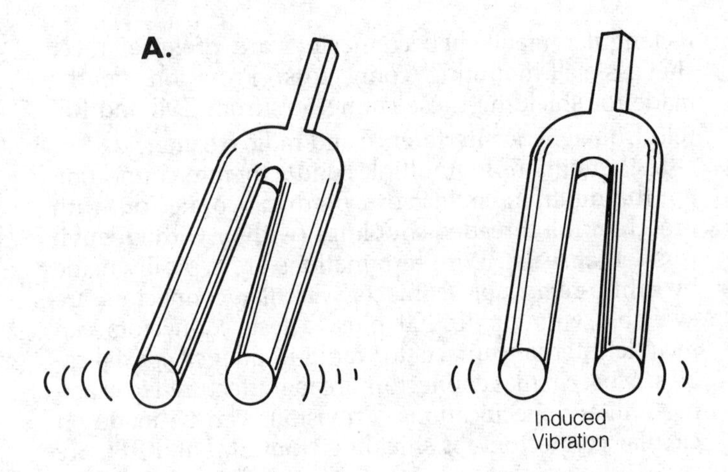

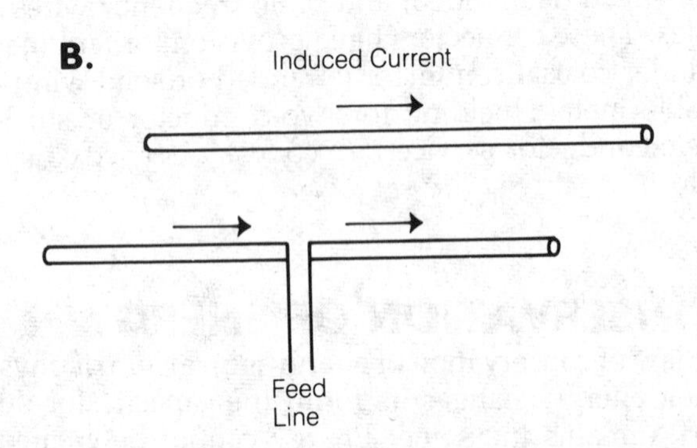

CONSONANCE: A vibrating tuning fork causes another fork to oscillate by acoustic consonance (A), and radio-frequency current flows in an isolated radiator due to electromagnetic consonance with an active antenna.

rents will be induced in the other antenna, and it, too, will radiate. Parasitic arrays operate on this principle. *See also* PARASITIC ARRAY, RESONANCE.

CONSTANT

In a mathematical equation, a constant is a fixed numerical value that does not change as the function is varied over its domain. In an equation with more than one variable, some of the variables may be held constant for certain mathematical evaluations. The familiar equation of Ohm's Law:

$$I = \frac{E}{R}$$

where *I* represents the current in amperes, *E* represents the voltage, and *R* represents the resistance in ohms, can be changed to the more general formula:

$$I^* = \frac{kE^*}{R^*}$$

where *I** is the current in amperes, milliamperes, or microamperes; *E** is the voltage in kilovolts, volts, millivolts, or microvolts; and *R** is the resistance in ohms, kilohms, or megohms. The value of the constant, k, then depends on the particular units chosen. Once the units have been decided upon, k remains constant for all values of the function. For example, *I** can be denoted in milliamperes, *E** in volts, and *R** in megohms; then $k = 0.001$, or 10^{-3}.

Certain constants are accepted as fundamental properties of the physical universe. Perhaps the best known of these is the speed of light, c, which is 3×10^8 meters per second in free space.

In mathematical equations, constants may be denoted by any symbol or letter. Letters from the first half of the alphabet, however, are generally used to represent constants. *See also* FUNCTION.

CONSTANT-CURRENT MODULATION

Constant-current modulation, also called Heising modulation, is a form of plate or collector amplitude modulation. The plates of collectors of the radio-frequency and audio-frequency amplifiers are directly connected through a radio-frequency choke, as shown in the diagram.

The choke allows the audio signal to modulate the supply voltage of the radio-frequency stage. This produces amplitude modulation in the radio-frequency amplifier. The plate or collector current is maintained at a constant level because of the extremely high impedance of the choke at the signal frequency. Constant-current modulation is sometimes called choke-coupled modulation, because of the radio-frequency choke. *See also* AMPLITUDE MODULATION.

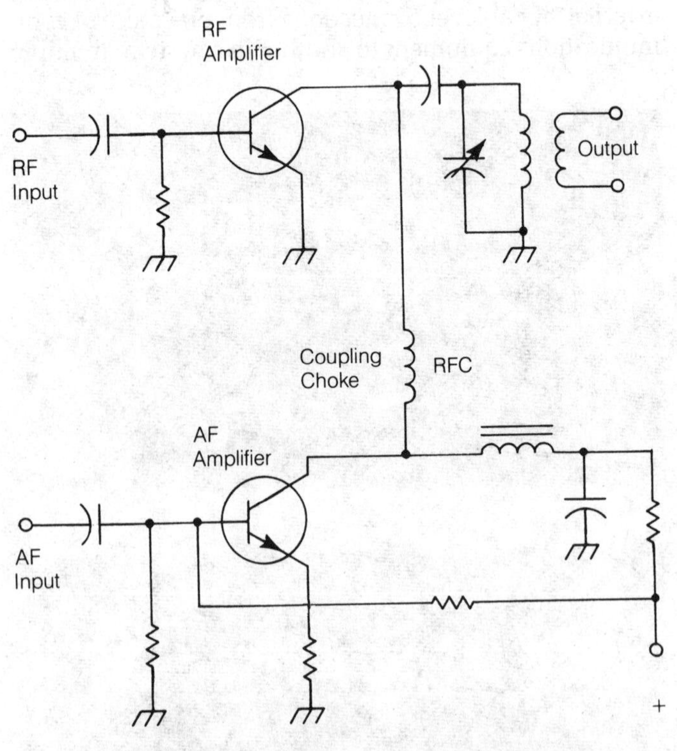

CONSTANT-CURRENT MODULATION: A choke feeds audio energy to the modulation stage of a radio frequency amplifier in constant-current modulation.

CONSTANT-CURRENT SOURCE

See POWER SUPPLY.

CONSTANT-K FILTER

A constant-k filter is a series of L networks (*see* L NETWORK). There might be only one L section or several. The L sections can be combined to form pi and T networks (*see* PI NETWORK, T NETWORK). The filter can be either a highpass or lowpass type (*see* HIGHPASS FILTER, LOWPASS FILTER). The diagrams illustrate various highpass and lowpass L-section filters.

The primary characteristic of a constant-k filter is the fact that the product of the series and parallel reactances is constant over a certain frequency range. Usually, only one kind of reactance is in series, and only one kind is in parallel. Capacitive reactance is given by the formula:

$$X_c = \frac{1}{2\pi fC}$$

where X_C is in ohms, f is in megahertz, and C is in microfarads. Inductive reactance is given by:

$$X_L = 2\pi fL$$

where X_L is in ohms, f is in megahertz, and L is in microhenries. The product of the series and parallel reactances in the illustration is thus:

$$k = X_L X_C = \frac{2\pi fL}{2\pi fC} = L/C$$

This value is a constant, and does not depend on the frequency. In practice, the inherent resistance of any circuit limits the frequency range over which it is useful.

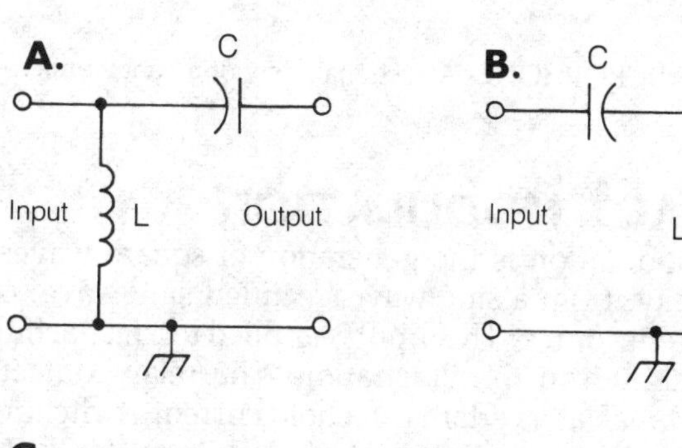

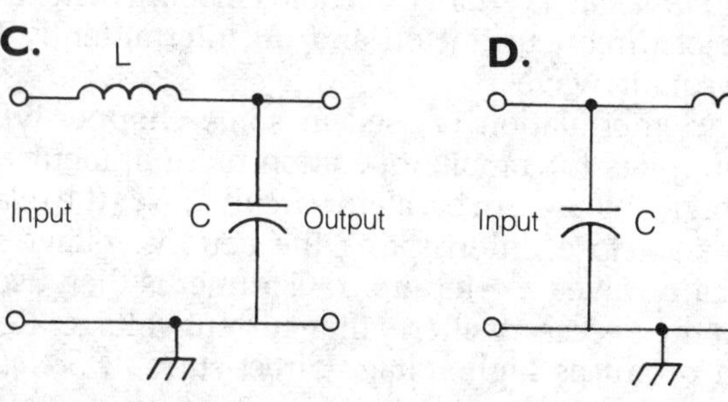

CONSTANT-K FILTER: Four types of constant-K filter are: the high pass with choke input (A), the highpass with capacitor input (B), the lowpass with choke input (C), and the lowpass with capacitor input (D).

CONSTANT-VELOCITY RECORDING

Constant-velocity recording is a method of recording phonograph disks. The amount of cutter movement is a function of the recording frequency. The higher the recording frequency, the smaller the excursion of the cutter.

When playing back a disk with a constant-velocity characteristic, the audio-frequency response of the equipment must compensate for the lower frequencies that cause greater vibration amplitude of the stylus. The reproducing characteristic is shown in the illustration. The reference frequency is 1000 Hz. The higher the frequency, the greater the gain of the playback equipment. The sound reproduced by the phonograph sounds natural.

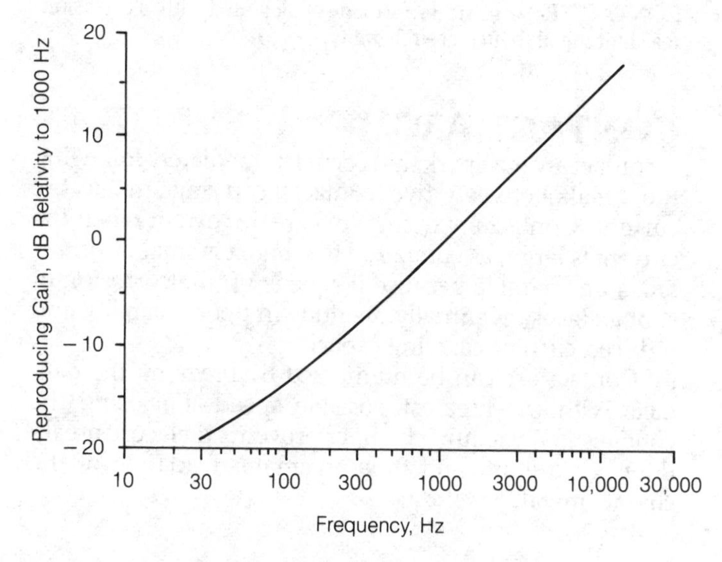

CONSTANT-VELOCITY RECORDING: A plot of gain versus frequency for constant-velocity recording.

CONSTANT-VOLTAGE POWER SUPPLY

See POWER SUPPLY.

CONTACT

A *contact* is a metal disk, prong, or cylinder, intended to make and break an electric circuit. Contacts are found in electric switches, relays, and keying devices. A typical pair of relay contacts is shown in the drawing.

The surface area of the contact determines the amount of current that it can safely carry. Generally, convex contacts cannot carry as much current as flat contacts because contact area is smaller when the surfaces are curved. Contacts may be made and broken either by a sliding motion or by direct movement. Contacts must be kept clean and free from corrosion to maintain their current-handling capacity. Metals that oxidize easily do not make good contacts. Gold and silver are often used where reliability must be high. Some relay contacts are housed in evacuated glass or metal envelopes to keep them from corroding. *See also* REED RELAY, RELAY, SWITCH.

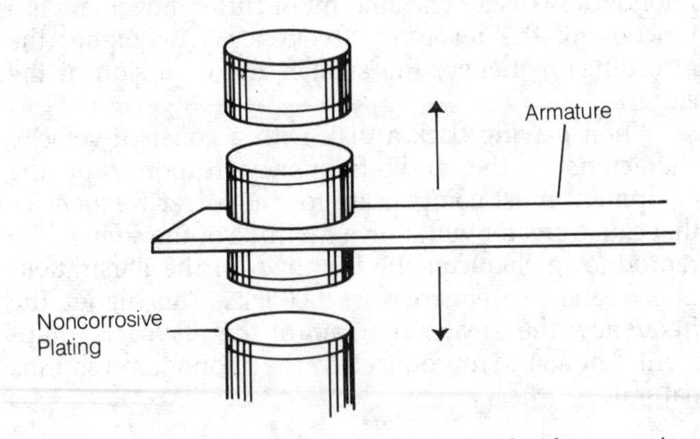

CONTACT: Relay contacts are made of, or plated with, a corrosion-resistant metal to increase reliability.

CONTACT ARC

A contact arc is a spark that occurs immediately following the break between two contacts carrying an electric current. Contact arc occurs to a greater extent when the current is large, as compared to when it is small. Contact arc is undesirable because it speeds up the corrosion of contact faces, eventually resulting in poor reliability and reduced current-carrying capacity.

Contact arc can be minimized by breaking the contacts with the greatest possible speed. Enclosing the contacts in a vacuum chamber prevents dielectric breakdown of the air, and thus eliminates oxidation of the contact metal. *See also* ARC.

CONTACT BOUNCE

Contact bounce is an intermittent make-and-break action of relay or switch contacts as they close. Contact bounce is caused by the elasticity of the metal in the contacts and armature. Contact bounce is undesirable because it often results in contact arc (*see* CONTACT ARC) as the contacts intermittently break. Contact bounce can sometimes cause a circuit to malfunction because of modulation in the flow of current.

The effects of contact bounce can be minimized by placing a small capacitance across the contacts. This helps to smooth out the modulating effect on direct current. The magnetizing current in a relay coil should not be much greater than the minimum needed to close the contacts reliably. Adjustment of the armature spring tension is also helpful in some cases. *See also* RELAY.

CONTACT MICROPHONE

A microphone that picks up conducted sound energy rather than sound propagated through the atmosphere is called a contact microphone. A contact microphone works in the same way as an ordinary microphone (*see* MICROPHONE). The diaphragm, instead of being vibrated by air molecules, is placed directly against the surface from which sound is to be received. The illustration shows a simplified cutaway view of a dynamic contact microphone. An electrically insulating layer prevents a metallic surface from affecting the operation of the microphone.

Contact microphones can be used to listen through walls. People speaking in an adjacent room cause the walls to vibrate, because the air molecules exert pressure on the walls. Contact microphones can also pick up minute vibrations in the ground caused by trucks, trains, and people walking. Some contact microphones are specially designed for underwater recording.

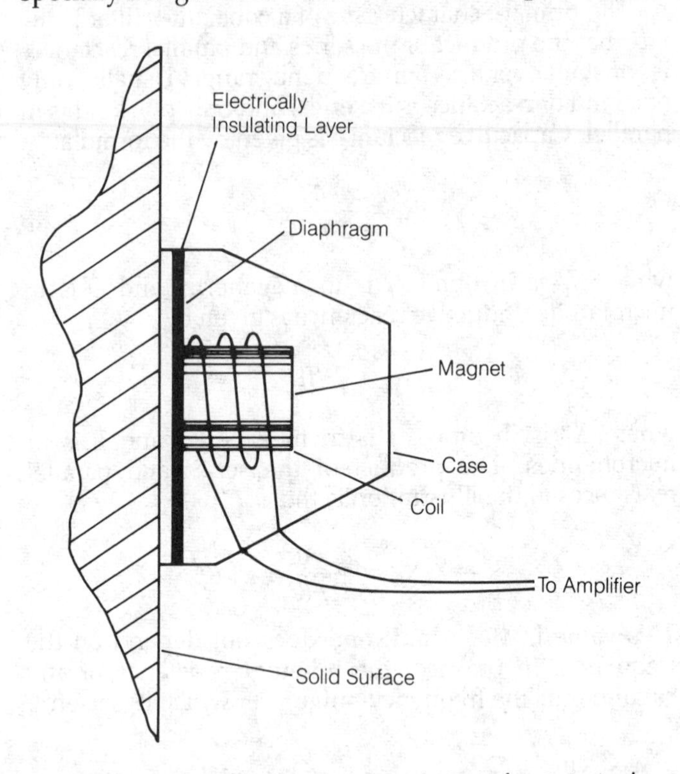

CONTACT MICROPHONE: A sectional view of a contact microphone.

CONTACT MODULATION

Contact modulation is the generation of square waves, using a relay, from a sine wave, rectified sine-wave, or direct-current source. The input is applied to a fast-acting relay, as shown in the illustration. The relay contacts make and break at a certain threshold current, switching a source of direct current on and off intermittently to produce square waves.

Contact modulation is used in some chopper-type power supplies for mobile operation of equipment requiring high voltage. An oscillator circuit causes the relay to open and close, interrupting the battery voltage to form square waves. A step-up transformer is then used to obtain the necessary alternating-current voltage. Rectification produces high-voltage direct current. *See also* CHOPPER POWER SUPPLY.

CONTACT RATING

The contact rating of a switch or relay is the amount of

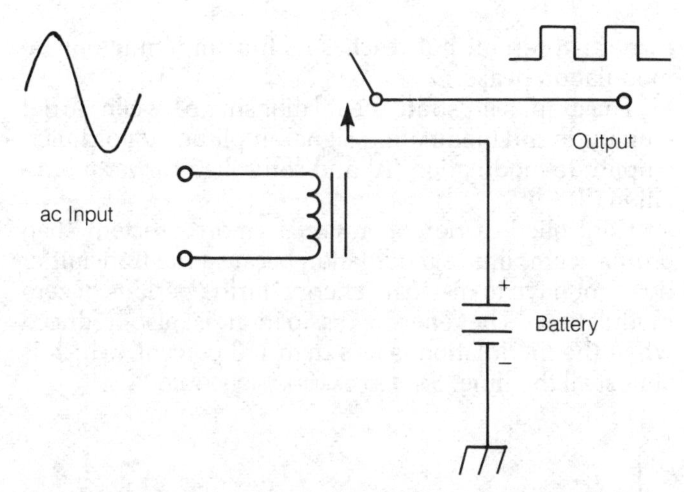

CONTACT MODULATION: The opening and closing of a relay contact converts alternating current to a square wave.

current that the contacts can handle. The contact rating depends on the physical size of the contacts and the contact area. Usually, the contact rating is greatest for direct current, and decreases with increasing alternating-current frequency.

A switch or relay should always be operated well within its contact ratings or arcing will occur and possibly even contact overheating with resulting damage. Contact ratings are usually specified in two ways—continuous and intermittent. The continuous rating is the maximum current that the contacts can handle indefinitely without interruption. The intermittent contact rating is the maximum current that the contacts can handle if they are closed half of the time. *See also* DUTY CYCLE.

CONTINUITY

In an electrical circuit, continuity is the existence of a closed circuit, allowing the flow of current. A simple instrument called a continuity tester is used to check for circuit continuity. The continuity tester consists of a power source and an indicating device such as a battery and lamp, as shown in the illustration. If the circuit under test is broken, or if the resistance is higher than it should be, the bulb will fail to light. For variable-resistance circuits, an ohmmeter can be used as a continuity tester (*see* OHMMETER).

CONTINUOUS-DUTY RATING

The continuous-duty rating of a device is the maximum amount of current, voltage, or power that it can handle or deliver with a 100-percent duty cycle (*see* DUTY CYCLE). This means that the device must operate constantly for an indefinite period of time.

A power supply, for example, might be specified as capable of delivering 6 A continuous duty. This means that if the necessary load is connected to the supply so that 6 A of current is drawn, the supply can be left in operation for an unlimited amount of time, and it will continue to deliver 6 amperes.

Continuous-duty ratings are often given in watts for

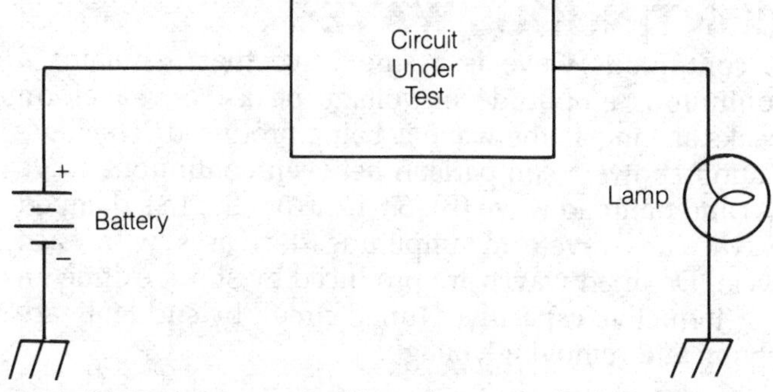

CONTINUITY: A battery and lamp form a continuity tester.

tubes, transistors, resistors, and other devices. Sometimes these devices can handle or deliver an amount of current, voltage, or power greater than the continuous-rated value for short periods of time. *See also* INTERMITTENT-DUTY RATING.

CONTINUOUS FUNCTION

A continuous function is a mathematical function in which the dependent variable or variables change in a continuous way over the domains of the independent variables (*see* DEPENDENT VARIABLE, FUNCTION, INDEPENDENT VARIABLE).

If a function is continuous at all points in its domain, it is called a continuous function. There are various mathematical ways to test a function to determine whether or not it is continuous. Examples of continuous functions are sine waves and transistor characteristic curves. Examples of discontinuous functions are square waves and saw-tooth waves.

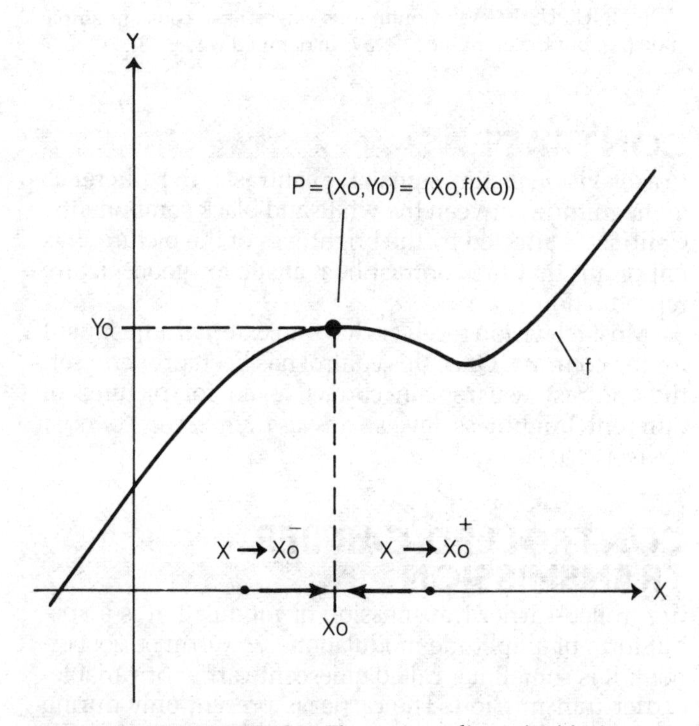

CONTINUOUS FUNCTION: There are no discontinuities or steps in a continuous function.

CONTINUOUS WAVE

A continuous wave is a sine wave that maintains a continuous amplitude of voltage peaks, or waveform peaks as long as the wave is being produced. The illustration shows a comparison between continuous wave (A) and damped wave (B). *See* DAMPED WAVE. In a damped wave, the waveform amplitude decreases with each cycle. Damped waves are produced by shock exiting an LC (inductive-capacitive) tuned circuit by suddenly applying and removing voltage.

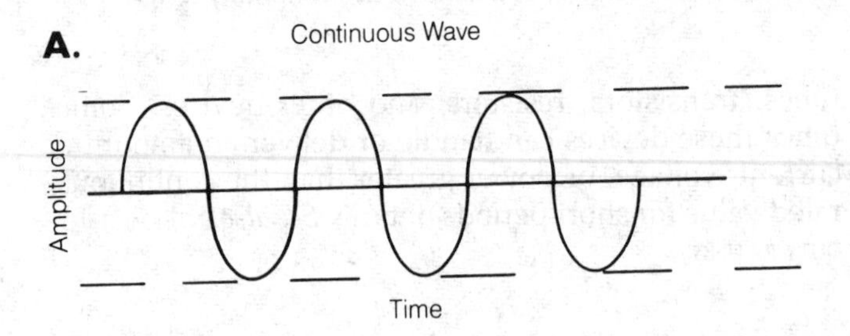

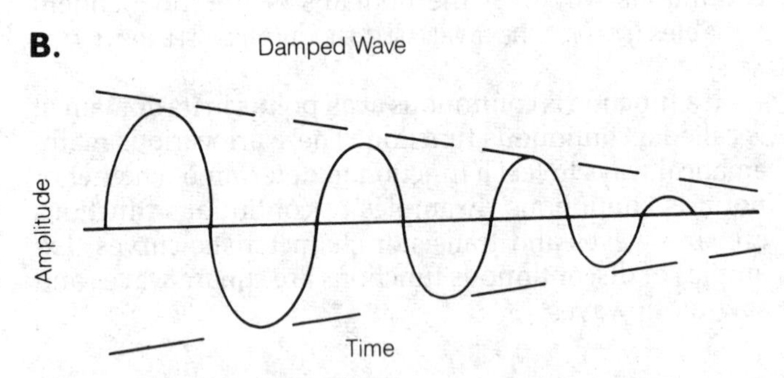

CONTINUOUS WAVE: Continuous waves have constant amplitude (A), but the amplitude decays in damped waves (B).

CONTRAST

In a television picture signal, the contrast is the difference in magnitude between the white and black components. Contrast is affected by the brightness of the picture. It is important that the contrast be realistic for good picture reproduction.

Most television receivers have an external adjustment for the contrast. Once this control has been properly set, the contrast will remain correct, even for pictures of different brightness levels. *See also* AUTOMATIC CONTRAST CONTROL, TELEVISION.

CONTROLLED-CARRIER TRANSMISSION

Controlled-carrier transmission or modulation is a special form of amplitude modulation (*see* AMPLITUDE MODULATION). It is sometimes called quiescent-carrier or variable-carrier transmission. The carrier is present only during modulation. In the absence of modulation, the carrier is suppressed. At intermediate levels of modulation, the carrier is present, but reaches its full amplitude only at modulation peaks.

The graphs illustrate the relationship between carrier amplitude and modulating-signal amplitude for ordinary amplitude modulation (A) and controlled-carrier modulation (B).

Controlled-carrier operation is more efficient than ordinary amplitude modulation because the transmitter does not have to dissipate energy during periods of zero modulation. The energy dissipation is also reduced when the modulation is less than 100 percent, which is almost all the time. *See also* CARRIER, MODULATION.

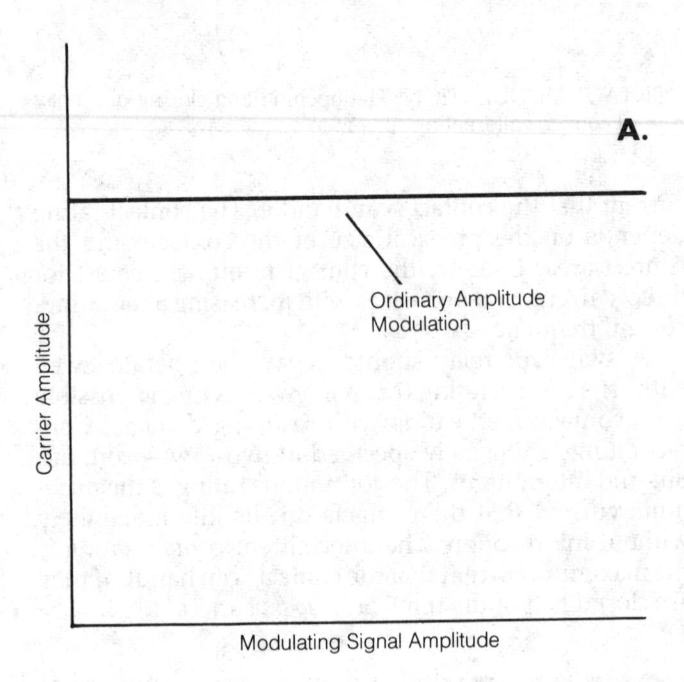

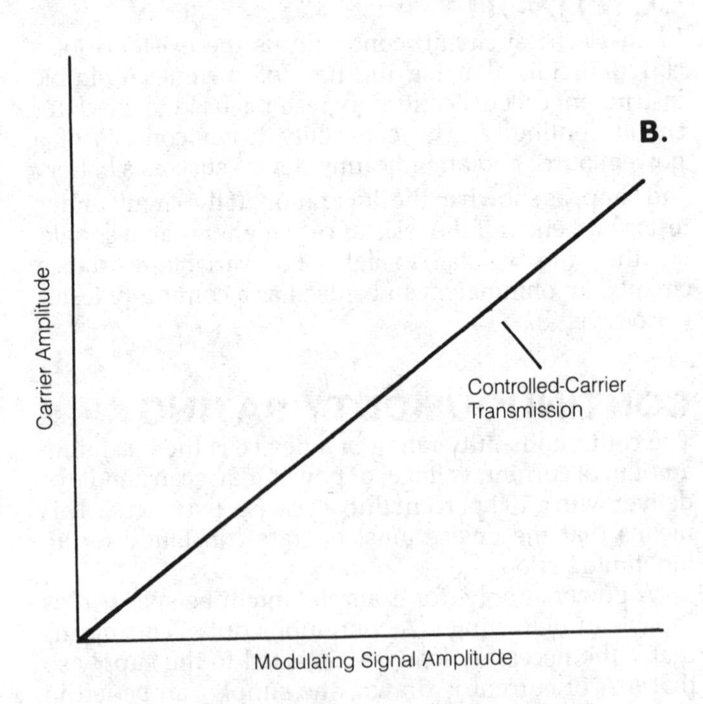

CONTROLLED-CARRIER TRANSMISSION: Conventional amplitude modulation is plotted (A), and controlled-carrier modulation is plotted (B).

CONTROLLED RECTIFIER

See SILICON-CONTROLLED RECTIFIER.

CONTROL REGISTER

A control register stores information for the regulation of digital-computer operation. The control register consists of an integrated circuit or a set of integrated circuits. While one program statement is executed, the control register stores the next statement. Once a statement has been carried out by the computer, the statement in the control register is executed, and a new statement enters the register. *See also* COMPUTER, REGISTER.

CONVECTION COOLING

Convection cooling is a form of air cooling. When a device such as a tube or transistor dissipates heat into the surrounding air, this well-known effect can be employed to cool the tube or transistor within an enclosure. Convection cooling can be enhanced by the addition of a fan at the bottom of the enclosure to blow air upward. The fan may also be located at the top of the chamber to pull air through.

To a certain extent, convection is part of the cooling system in almost all power supplies and amplifiers. Devices called heatsinks, with vertical fins, facilitate heat dissipation by convection in almost all high-power supplies. The heat is conducted to the heatsink which transfers it to the surrounding environment. *See also* AIR COOLING, CONDUCTION COOLING, HEATSINK.

CONVERGENCE

In a cathode-ray tube with more than one electron beam, convergence is the property of the electron beams to intersect a single point, on the phosphor screen of a cathode-ray tube. Convergence is important in color television receivers and monitors. Three electron guns are employed, one for each of the primary colors (red, blue, and green). If convergence is not obtained at all points on a color picture screen, the colors will not line up properly. This results in blurred colored borders around some objects in the picture. The drawing illustrates convergence of electron beams toward the phosphor screen of a cathode-ray tube.

As the electron beams scan across the tube, their point of convergence follows a two-dimensional surface in space, called the convergence surface. In a color TV tube or monitor the three color electron beams must not only converge but their convergence surface must correspond exactly with the phosphor surface of the picture tube. The controls in a color television receiver that provide precise alignment of the convergence surface are called the convergence and convergence-phase controls. *See also* COLOR TELEVISION, COLOR TV TUBE, MONITOR.

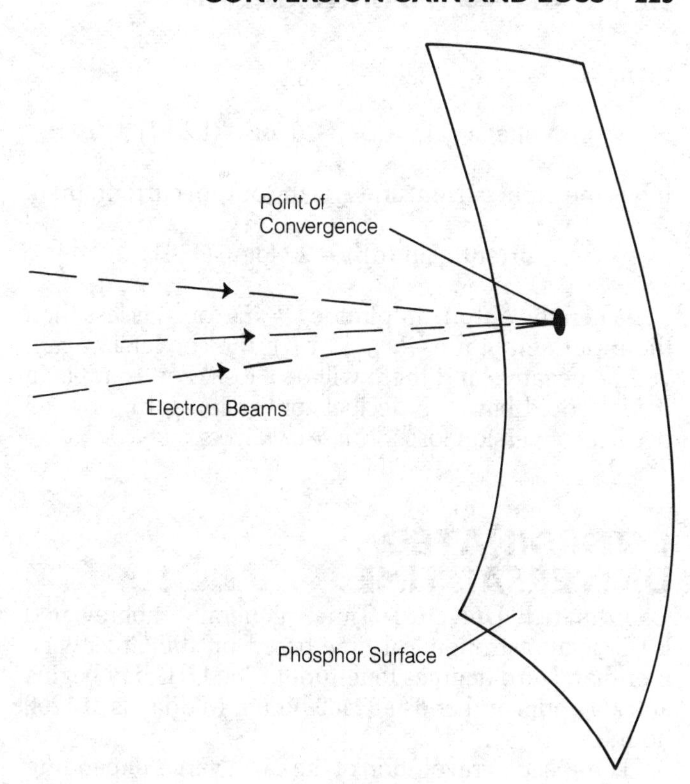

CONVERGENCE: Electron beams converge at a point on the phosphor coating in a cathode-ray tube.

CONVERSION EFFICIENCY

In a frequency converter, the ratio of the output power, voltage, or current to the input power, voltage, or current is called the conversion efficiency. For example, if a frequency converter is designed to change a 21 MHz signal to a 9 MHz signal, the conversion efficiency is the ratio of the 9 MHz signal power, voltage, or current to the 21 MHz signal power, voltage, or current. Conversion efficiency is always less than 100 percent.

In a rectifier, the conversion efficiency is defined as the ratio of the direct-current output power to the alternating-current input power. The conversion efficiency in a rectifier-and-filter power supply is always less than 100 percent because of losses in the rectifier devices and transformers. *See also* POWER SUPPLY, RECTIFIER.

CONVERSION GAIN AND LOSS

In a converter, the conversion gain or loss is the increase or decrease, respectively, of signal strength from input to output. This gain or loss can be determined in terms of power, voltage, or current. It is specified in decibels.

A passive converter, such as a diode mixer, always has some conversion loss. However, converters with amplifiers may produce conversion gain.

If P_1 is the input power to a converter and P_2 is the output power, then the conversion gain is given in decibels by:

$$\text{Power gain (dB)} = 10 \log_{10} (P2/P1)$$

If E_1 is the input voltage and E_2 is the output voltage, then:

$$\text{Voltage gain (dB)} = 20 \log_{10} (E2/E1)$$

If I_1 is the input current and I_2 is the output current, then:

$$\text{Current gain (dB)} = 20 \log_{10} (I2/I1)$$

When the output amplitude (P_2, E_2, or I_2) is less than the input amplitude (P_1 E_1, or I_1), the conversion gain will be negative and there will be a loss. For example, a device could have −3 decibel conversion gain, or a +3 decibel conversion loss. *See also* GAIN, LOSS.

COORDINATED UNIVERSAL TIME

Coordinated Universal Time, generally abbreviated UTC, is an astronomical time based on the Greenwich meridian (zero degrees longitude). The UTC day begins at 0000 hours and ends at 2400 hours. Midday is at 1200 hours.

The speed of revolution of the earth varies depending on the time of the year. Coordinated Universal Time is based on the mean, or average, period of the synodic (sun-based) rotation. The earth is slightly behind UTC near June 1, and is slightly ahead near October 1. This variation is the result of tidal effects caused by the gravitational interaction between the earth and the sun.

Coordinated Universal Time is five hours ahead of Eastern Standard Time in the United States. It is six hours ahead of Central Standard Time, seven hours ahead of Mountain Standard Time, and eight hours ahead of Pacific Standard Time. These differences are one hour less during the months of daylight-saving time.

Coordinated Universal Time is broadcast by several time-and-frequency-standard radio stations throughout the world. In the United States, the principal stations are WWV in Colorado and WWVH in Hawaii. These stations are operated by the National Bureau of Standards. *See also* TIME ZONE, WWV/WWVH.

COORDINATE SYSTEM

A coordinate system is a means of precisely locating a point on a line, on a plane, in three-dimensional space, or in a space of more than three dimensions. The simplest system of coordinates is the number line of one dimension. In two dimensions, two coordinate numbers are necessary to define a point uniquely. In three dimensions, three numbers are needed. Coordinate values may be represented either by distances or by angles. In some advanced mathematical systems, four, five, or more dimensions are used.

Coordinate systems are often used in electricity and electronics. The familiar graph showing one variable as a function of another is an example of the Cartesian coordinate plane. A radar system shows the locations of objects on a two-dimensional display corresponding to a Cartesian or polar coordinate plane, depending on the type of radar display. The orientation of an antenna system is often determined by means of a coordinate system. *See also* CARTESIAN COORDINATES, CURVILINEAR COORDINATES, POLAR COORDINATES, SPHERICAL COORDINATES.

COPPER

Copper is an element with an atomic number of 29 and an atomic weight of 64. In its pure form, copper appears as a dark, red color metal. It is flexible and malleable. Copper is used extensively in the manufacture of wire because it is an excellent conductor of electric current. It is also a good conductor of heat and is fairly resistant to corrosion. Copper oxide exhibits the properties of a semiconductor.

COPPER-CLAD WIRE

Copper-clad wire is wire with a core of hard metal such as iron or steel, and an outer coating of copper. Copper-clad wire is frequently used in radio-frequency antenna systems where both tensile strength and good electrical conductivity are essential. The central core provides mechanical strength to resist stretching and breakage and the copper coating carries most of the current. Copper-clad wire has higher direct-current resistance than pure copper wire of the same diameter because iron and steel are less effective electrical conductors than copper, but the resistance at radio frequencies is almost as low as that of pure copper wire because of the skin effect. At high alternating-current frequencies, most of the electron flow takes place near the outer part of a conductor (*see* SKIN EFFECT).

Copper-clad steel wire is more difficult to work than soft-drawn or hard-drawn pure copper wire. The advantages of greater tensile strength, and less tendency to stretch, however, make copper-clad steel wire preferable to pure copper wire in many applications. *See also* WIRE.

COPROCESSOR

In computer technology, a coprocessor, also called a math or numeric coprocessor, is a circuit capable of performing a large number of mathematical computations in a short time. The coprocessor, an integrated circuit or add-in board, acts as an extension of the host microprocessor. There are two additional types of math processors: the embedded math processor customized for a specific application and the general-purpose attached chip.

The standard coprocessor works in conjunction with the host processor to speed up floating-point mathematics and data processing. (Floating-point mathematics are calculations on data elements represented as a fixed-point component and an exponent.) The coprocessor receives both instructions and data from the data bus and is useful in accelerating computer-aided design and layout and filling out spread sheets. A typical coprocessor configuration is illustrated.

Dedicated numerics processors are customized for such applications as graphics workstations, digital signal

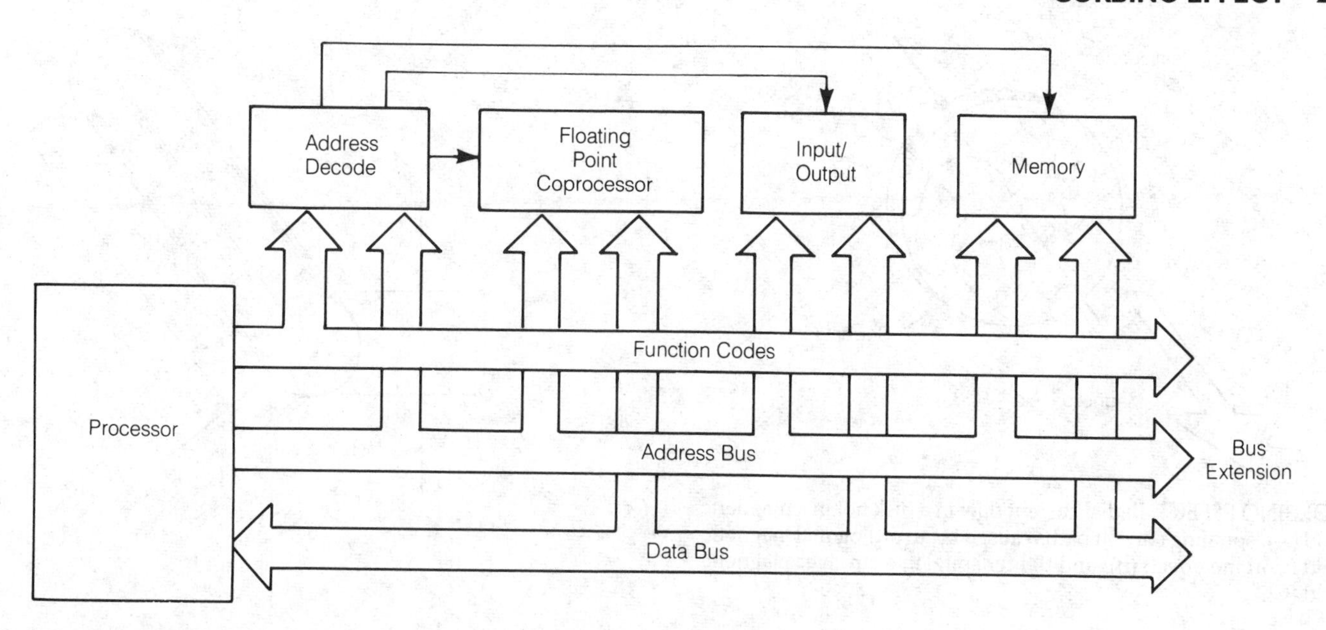

COPROCESSOR: A block diagram of a typical coprocessor configuration in a computer.

processors, and high-speed controllers. They can be 10 to 100 times faster than standard coprocessors. The general-purpose attached math processor combines the benefits of high-level language coding with the speed of dedicated processors. It receives data directly from memory and takes over control of math-intensive operations from the host microprocessor.

The speed of all these processors is measured in megaflops—millions of floating-point operations per second—rather than millions of instructions per second (MIPS). General-purpose computers typically perform 10 MIPS for each megaflop.

A standard coprocessor or may speed up the mathematical operations of a personal computer by 100 times. Standard coprocessors are proprietary products designed for use with specific microprocessors such as Motorola's 68030 or Intel's 80386. To the host processor, the coprocessor is an extension of its architecture. The microprocessor passes on protocols and instructions to the coprocessor. Standard coprocessors process at rates of 0.1 to 0.5 megaflops.

CORBINO EFFECT

When a disk carrying a radial current, either toward or away from the center, is subjected to a magnetic field whose lines of force run perpendicular to the disk, the flow of current is bent. The illustration shows this phenomenon, which is called the Corbino effect.

The Corbino effect allows a semiconductor disk in a variable magnetic field to act as a variable resistor. When no magnetic field is present, the current flows in straight lines between the center of the disk and the edge where one electrode is placed at the center and the other around the entire edge (see A). In the presence of a relatively weak magnetic field (B), the current path is bent slightly. Since the path is made longer from the center to the edge of the disk, the resistance increases. In an intense magnetic field (C), the current is forced to flow over a much longer path, since it is bent greatly. This increases the resistance by a large amount. A disk of semiconductor material, placed in a magnetic field to control its center-to-edge resistance, is called a Corbino disk.

The Corbino effect takes place because electrons moving perpendicular to a magnetic field tend to be deflected in a direction orthogonal to both the field orientation and the direction of electron motion. *See also* HALL EFFECT.

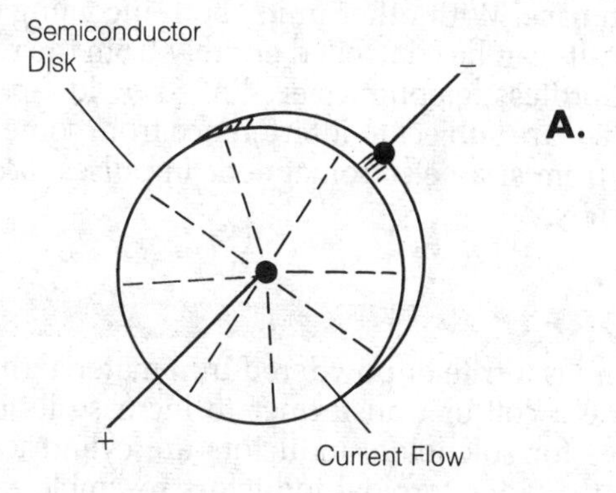

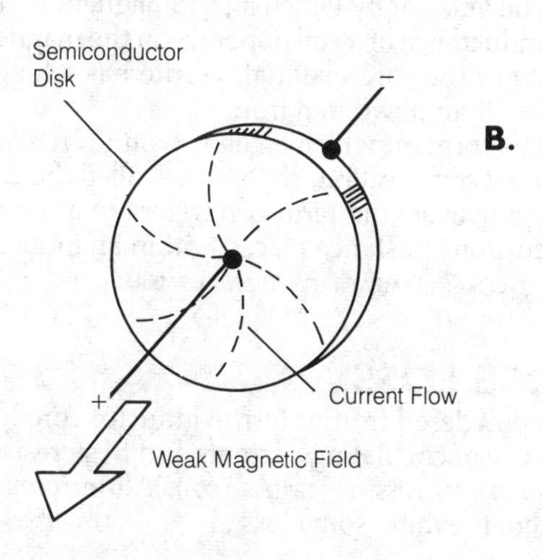

C.

Semiconductor Disk

Current Flow

Strong Magnetic Field

CORBINO EFFECT: Radial current flow in a disk not in a magnetic field (A), spiraling current paths caused by an orthogonal magnetic field (Corbino effect) (B), and tighter spiral in a stronger magnetic field (C).

CORDLESS TELEPHONE

A cordless telephone uses a radio connection between the receiver and the base rather than the usual cord. Two small antennas, one at the main unit and the other at the receiver, permit reception as far away as 600 to 800 feet under ideal conditions. This can be extremely convenient, allowing calls to be made or received anywhere around the house.

Cordless telephones are available in various configurations. With some units, calls can be received in a remote place, but outgoing calls must be dialed from the main base. With other units, both incoming and outgoing calls can be controlled entirely from the receiver.

Cordless telephones are linked by low-power radio, so they are subject to interference from some appliances and from stray electromagnetic impulses. *See also* MOBILE TELEPHONE.

CORE

A core is ferrite or powdered-iron material that is placed inside a coil or transformer to increase its inductance. Cores for solenoidal inductors are cylindrical in shape and those for toroidal inductors resemble a doughnut. The drawing illustrates these core configurations.

The amount by which a ferromagnetic core increases the inductance of a coil depends on the magnetic permeability of the core material. Ferrite has a higher permeability, than powdered iron.

The form on which a coil is wound, regardless of its material composition, can also be called the core. Generally, however, the term core refers to a ferrite or powdered-iron substance placed within an inductor. *See also* FERRITE CORE, PERMEABILITY, TRANSFORMER.

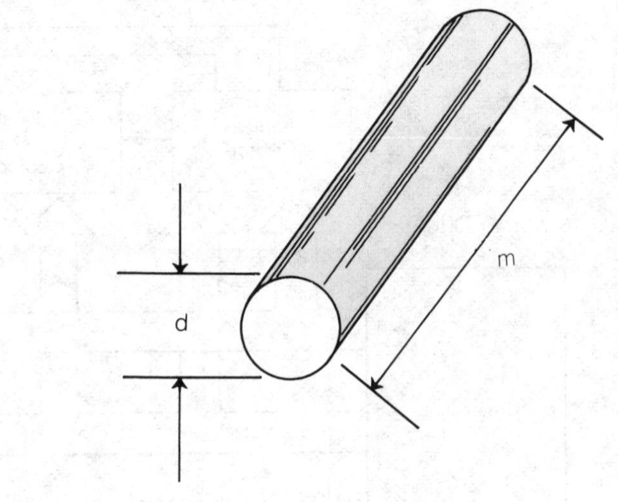

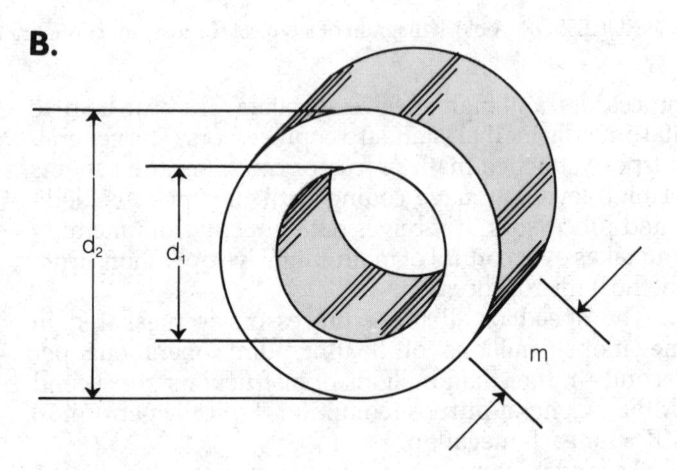

CORE: Drawings of ferrite inductor cores: solenoidal (A) and toroidal (B).

CORE LOSS

In a powdered-iron or ferrite inductor core losses occur because of circulating currents and hysteresis effects (*see* EDDY-CURRENT LOSS, HYSTERESIS LOSS). All ferromagnetic core materials exhibit some loss.

Generally, the loss in a core material increases as the frequency increases. Ferrite cores usually have a lower usable-frequency limit than powdered-iron cores. In general, the higher the magnetic permeability of a core material, the lower the maximum frequency at which it can be used without objectionable loss.

Core losses tend to reduce the Q factor of a coil (*see* Q FACTOR). In ferrite-rod antennas, core losses limit the frequency range for reception. In applications requiring large currents or voltages through or across an inductor, excessive core losses can result in over-heating of the core material. The core might actually break under these conditions. *See also* FERRITE CORE, PERMEABILITY, TRANSFORMER, TRANSFORMER EFFICIENCY.

CORE MEMORY

A core memory stores binary information in the form of magnetic fields. A group of small toroidal ferromagnetic cores comprises the memory. A magnetic field of a given polarity represents a digit 1; a field of the opposite polarity represents a digit 0. Wires are run through the cores to permit reversing the polarity of the magnetic flux in the cores. Core memory is retained even when power is removed from the equipment.

Core memory, once popular in computers, has been replaced by semiconductor memory in most applications. Some core memory is still used in military equipment because it is unaffected by nuclear radiation.

CORE SATURATION

As the magnetizing force applied to a ferrite or powdered-iron core is increased, the number of lines of flux through the core is also increased. However, there is a maximum limit to the number of flux lines a given core can accommodate. When the core has reached its limit of magnetization, it is said to be saturated. Some cores saturate more easily than others.

In inductors with ferrite or powdered-iron cores, saturation reduces the effective value of inductance for large currents, because the magnetic permeability of a saturated core is lessened. The change in inductance can have adverse effects on circuit performance in some situations. In a transformer, core saturation results in reduced efficiency and excessive heating of the core material. The current through an inductor or transformer should be kept below the level at which saturation occurs. *See also* CORE, FERRITE CORE.

CORNER EFFECT

The response of a bandpass filter should be in the shape of a perfect rectangle: zero attenuation within the passband, high attenuation outside the passband, and an instantaneous transition from zero attenuation to high attenuation at the limits of the passband. Lowpass, highpass, and band-rejection filters should have similar instantaneous transitions in the mathematically ideal case. The drawing illustrates this ideal response for a bandpass filter (A).

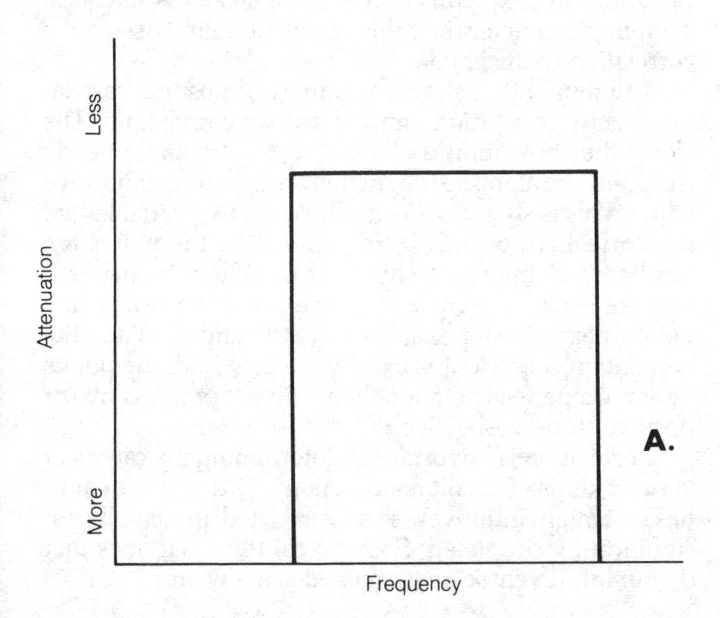

In practice, the rectangular response is never perfect. The corners of the rectangle—the transitions from zero to high attenuation—are not instantaneous, as shown at B. Adjacent to the passband, the corners appear rounded rather than sharp. This is called the corner effect, and it always occurs with selective filters. *See also* BANDPASS RESPONSE, BAND-REJECTION RESPONSE, HIGHPASS RESPONSE, LOWPASS RESPONSE.

CORNER REFLECTOR

A corner reflector is a reflector for producing antenna gain at very high frequencies. Two flat metal sheets or screens are attached together as shown in the illustration (A). Placed behind the radiating element of an antenna, the corner reflector produces forward gain. The beamwidth is quite broad (*see* BEAMWIDTH). The gain is the same for reception as for transmission at the same frequency. Corner-reflector antennas are often used for reception of television broadcast signals in the ultrahigh frequency band.

A corner reflector also can consist of three flat metal surfaces or screens, assembled at right angles to each other as shown in the drawing. This configuration should have dimensions at least several wavelengths across. It will always return electromagnetic energy in exactly the same direction from which it arrives. Because of this effect, corner reflectors of this type make excellent radar reflectors. These reflectors are sometimes called tri-corner reflectors.

CORONA

When the voltage on an electrical conductor, such as an antenna or high-voltage transmission line, exceeds a threshold value, the air around the conductor begins to ionize to form a blue or purple glow called corona. This glow can be seen at the end or ends of an antenna at

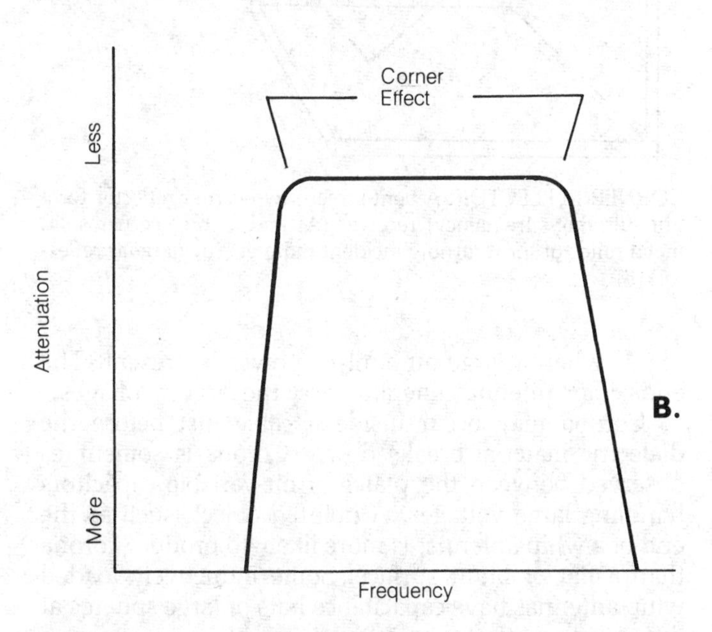

CORNER EFFECT: An ideal filter response curve (A) and one with rolloff at the corners (corner effect) (B).

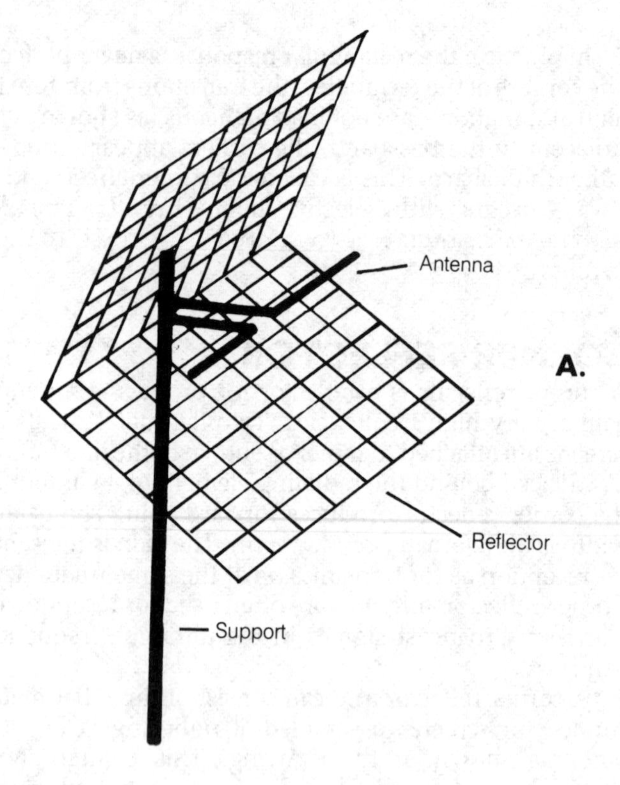

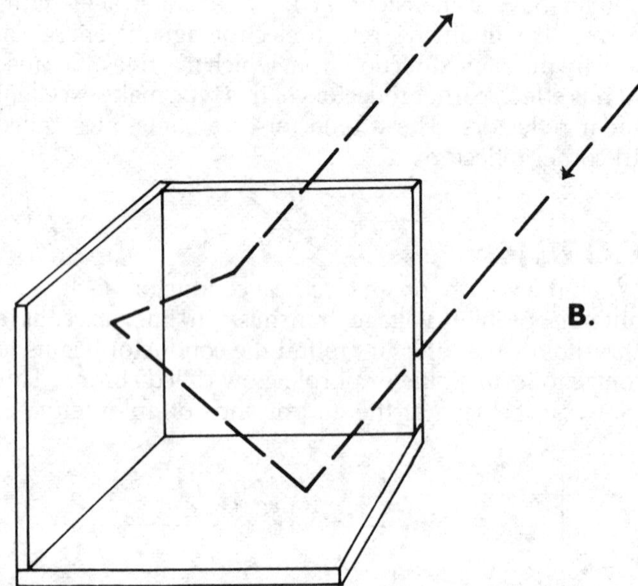

CORNER REFLECTOR: A bent-screen, two-corner reflector for a uhf (ultrahigh frequency) receiver (A) and a three-corner solid metal reflector for returning incident radio energy (a radar reflector) (B).

night, when a large amount of power is present. The ends of an antenna generally carry the largest voltages.

Corona may occur inside a cable just before the dielectric material breaks down. Corona is sometimes observed between the plates of air-variable capacitors handling large voltages. A pointed object, such as the end of a whip antenna, is more likely to produce corona than a flat or blunt surface. Some inductively loaded whip antennas have capacitance hats or large spheres at the ends to minimize corona.

Corona sometimes occurs as a result of high voltages caused by static electricity during thunderstorms. This display is occasionally seen at the tip of the mast of a sailing ship. Corona was first observed by sailors hundreds of years ago, and was called Saint Elmo's fire.

CORRECTION

In an approximation of the value of a quantity, a correction or correction factor is a small increment, added to or subtracted from one approximation to obtain a better approximation.

Corrections are generally used when external factors influence the reading or result obtained with an instrument. For example, a frequency counter may be based on a crystal intended for operation at 20°C. If the temperature is above or below 20°C, a correction factor may be necessary, such as 2 Hz for each degree Celsius.

Sometimes correction is performed by a circuit component especially designed for the purpose. In the example, a capacitor with a precisely chosen temperature coefficient may be placed in parallel with the crystal to correct the drift caused by temperature changes. This kind of correction is usually called compensation. *See also* COMPENSATION.

CORRELATION

Correlation is a mathematical expression for the relationship between two quantities. Correlation can be positive, zero, or negative. A positive correlation indicates that an increase in one parameter is accompanied by an increase in the other. A negative correlation indicates that an increase in one parameter occurs with a decrease in the other. Zero correlation means that variations in the two parameters are unrelated.

The coefficient of correlation between two variables is usually expressed as a number in the range between -1 and $+1$. The most negative correlation is given by -1, and the most positive by $+1$. The failure rate of a certain component (such as a transistor) may be correlated with the temperature. The higher the temperature, for example, the more frequently the component fails. A statistical sampling can determine this correlation, and assign it a correlation coefficient.

The illustration shows examples of positive correlation, zero correlation, and negative correlation. The closer the correlation coefficient to $+1$, the more nearly the points lie along a straight line with slope of a positive value. This is shown at A and B. When two variables are not correlated, or have zero correlation, the points are randomly scattered, as shown at C. When the parameters are negatively correlated, the points lie near a line with a negative slope, as shown at D and E. When the correlation coefficient is exactly -1 or $+1$, all the points lie along a perfectly straight line with either a negative or positive slope, respectively.

Correlation is important in determining the causes of various kinds of circuit malfunctions. Often, two quantities that might intuitively seem correlated are actually not significantly correlated. Sometimes, two quantities that do not intuitively seem correlated actually are.

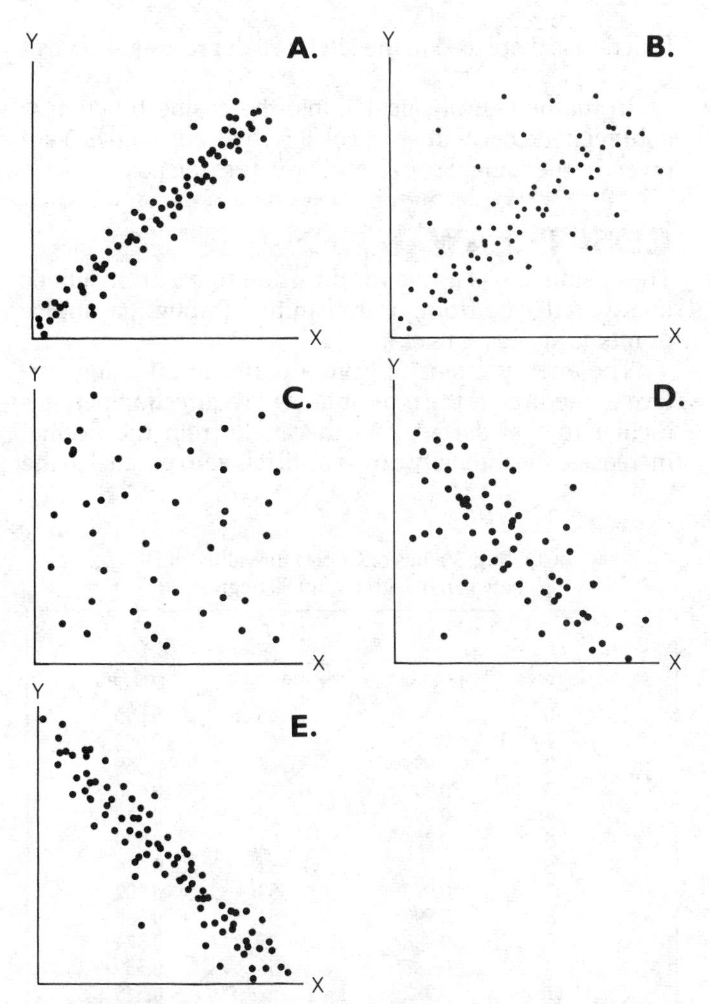

CORRELATION: Correlation is the measure of similarity between two groups of variables. Graph of strong positive correlation (A), weak positive (B), zero (C), weak negative (D), and strong negative (E).

CORRELATION DETECTION

Correlation detection is detection involving the comparison of an input signal with an internally generated reference signal. The output of a correlation detector varies depending on the similarity of the input signal to the internally generated signal; maximum output occurs when the two signals are identical.

A phase comparator is an example of a correlation detector. In-phase signals produce maximum output; if the signals are not perfectly in phase, the output is reduced. A frequency comparator is another example of correlation detection. Two signals of the same frequency produce maximum output. The greater the difference between the signal frequency and the internally generated signal frequency, the less the output. *See also* FREQUENCY COMPARATOR, PHASE COMPARATOR.

CORROSION

Corrosion is primarily a chemical reaction between a metal and the oxygen in the air. However, salt from the ocean, dissolved in water droplets in the air, can also cause corrosion. Chemicals created by manmade reactions, such as sulfur dioxide can corrode metals.

Corrosion is characterized by deterioration of the surface of a metal. Sometimes, the oxide of a metal can act to protect the metal against further corosion; an example of this is aluminum oxide. But the oxide of a metal increases the vulnerability of the metal to further corrosion by increasing the surface area exposed to the air. An example of this is iron oxide, or rust.

Strong electrical currents can accelerate corrosion. An example of this is the rapid corrosion of the surface of a ground rod placed in acid soil. Conducted currents, through the electrolyte soil cause rapid disintegration of the metal by electrolytic action. Reactions between certain chemicals placed in physical contact also can result in corrosion.

COSECANT

The cosecant function is a trigonometric function equal to the reciprocal of the sine function (*see* SINE). In a right triangle A in the illustration, the cosecant of an angle θ between zero and 90 degrees is equal to the length of the hypotenuse divided by the length of the side opposite the angle. The smaller the angle, the greater the value of the cosecant function for angles θ between 0 and 90

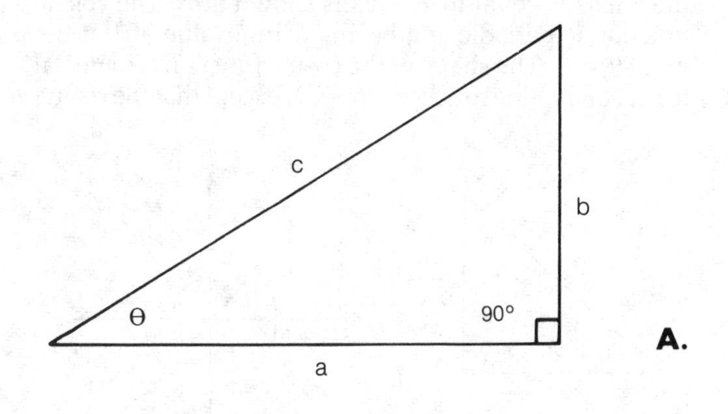

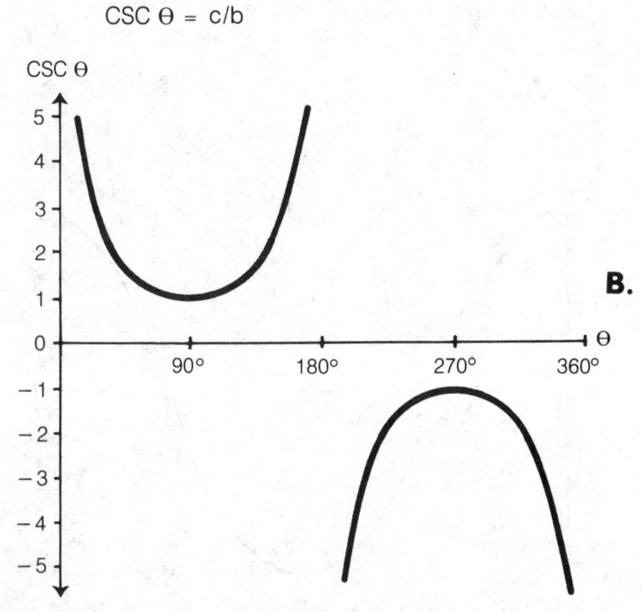

COSECANT: Triangle used to define the secant function (A) and a graph of the function between 0 and 360 degrees (B).

degrees. B illustrates the values of the cosecant function for angles ranging from zero to 360 degrees. The cosecant is undefined for $\theta = 0$ degrees and $\theta = 180$ degrees.

In mathematical calculations, the cosecant function is abbreviated csc, and is given by the formula:

$$\csc \theta = \frac{1}{\sin \theta}$$

where sin represents the sine function. *See also* TRIGONOMETRIC FUNCTION.

COSECANT-SQUARED ANTENNA

See ANTENNA DIRECTORY.

COSINE

The cosine function is a trigonometric function. In a right triangle, the cosine is equal to the length of the adjacent side, or base, divided by the length of the hypotenuse, as shown at A in the illustration. In the unit circle $x^2 + y^2 = 1$, plotted on the Cartesian (x,y) plane, the cosine of the angle θ measured counterclockwise from the x axis is equal to x. This is shown at B. The cosine function is periodic and begins with a value of 1 at the point $\theta = 0$. The shape of the cosine function is identical to that of the sine function (*see* SINE), except that the cosine function is displaced to the left by 90 degrees as shown at C.

In mathematical calculations, the cosine function is abbreviated cos. Values of cos θ for various angles θ are given in the table. *See also* TRIGONOMETRIC FUNCTION.

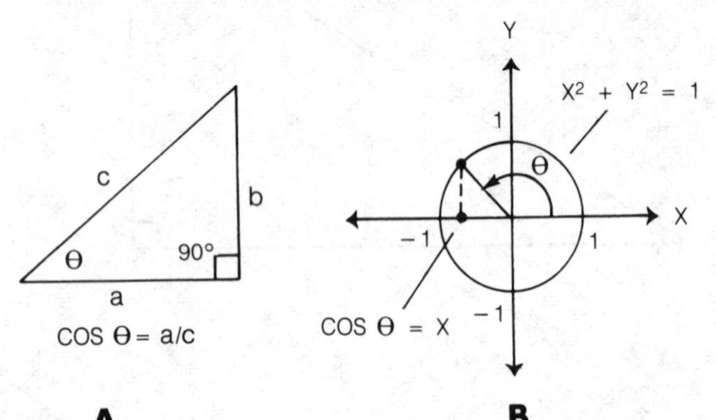

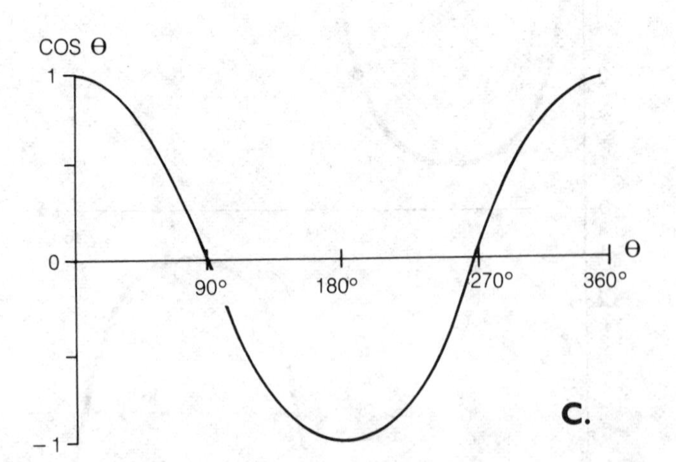

COSINE: A triangle used to define the cosine function (A), the unit circle model (B), and a graph of the function between 0 and 360 degrees (C).

COSINE LAW

The cosine law is a rule for diffusion of electromagnetic energy reflected from, or transmitted through, a surface or medium (*see* DIFFUSION).

The energy intensity from a perfectly diffusing surface or medium is the most intense in a direction perpendicular to that surface. As the angle from the normal increases, the intensity drops until it is zero parallel to the

COSINE: Values of COS Θ for values of Θ between 0 degrees and 90 degrees.

Θ, degrees	**cos Θ**	**Θ, degrees**	**cos Θ**
0	1.000	46	0.695
1	0.999	47	0.682
2	0.999	48	0.669
3	0.999	49	0.656
4	0.998	50	0.643
5	0.996	51	0.629
6	0.995	52	0.616
7	0.993	53	0.602
8	0.990	54	0.588
9	0.988	55	0.574
10	0.985	56	0.559
11	0.982	57	0.545
12	0.978	58	0.530
13	0.974	59	0.515
14	0.970	60	0.500
15	0.966	61	0.485
16	0.961	62	0.469
17	0.956	63	0.454
18	0.951	64	0.438
19	0.946	65	0.423
20	0.940	66	0.407
21	0.934	67	0.391
22	0.927	68	0.375
23	0.921	69	0.358
24	0.914	70	0.342
25	0.906	71	0.326
26	0.899	72	0.309
27	0.891	73	0.292
28	0.883	74	0.276
29	0.875	75	0.259
30	0.866	76	0.242
31	0.857	77	0.225
32	0.848	78	0.208
33	0.839	79	0.191
34	0.829	80	0.174
35	0.819	81	0.156
36	0.809	82	0.139
37	0.799	83	0.122
38	0.788	84	0.105
39	0.777	85	0.087
40	0.766	86	0.070
41	0.755	87	0.052
42	0.743	88	0.035
43	0.731	89	0.017
44	0.719	90	0.000
45	0.707		

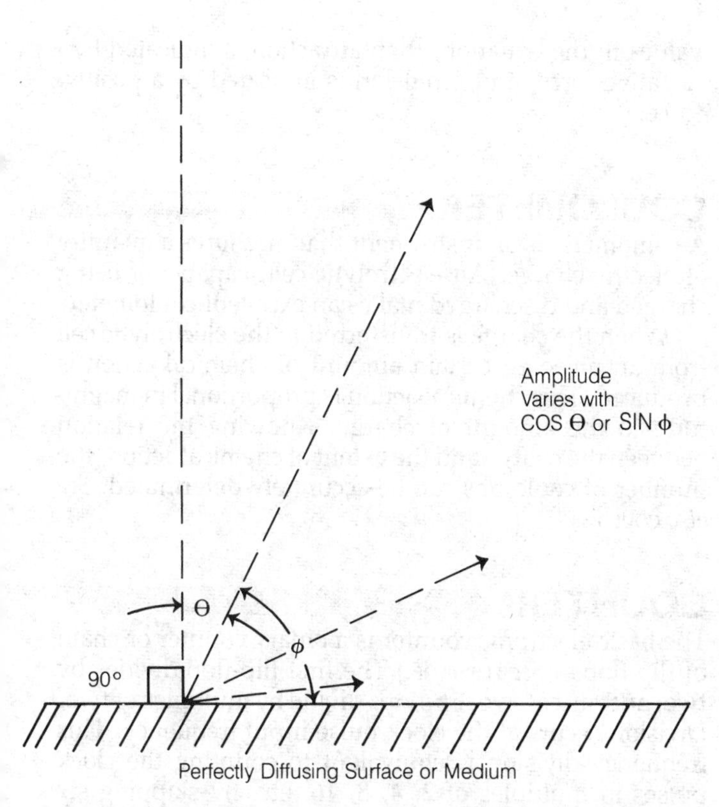

COSINE LAW: The intensity of transmitted or reflected energy from a surface is proportional to the cosine of its angle relative to the normal.

surface (see illustration). The intensity, according to the cosine law, varies with the cosine of the angle θ relative to the normal. The intensity also varies with the sine of the angle relative to the surface, where $\phi = 90° - \theta$.

COSMIC NOISE

Cosmic noise is electromagnetic energy arriving from distant planets, stars, galaxies, and other celestial objects. Cosmic noise occurs at all wavelengths, from the very low frequency radio band to the X ray band and above. At the lower frequencies, the ionosphere of the earth prevents the noise from reaching the surface. At some higher frequencies, atmospheric absorption prevents the noise from reaching us.

Cosmic noise limits the sensitivity obtainable with receiving equipment, because this noise cannot be eliminated. Radio astronomers deliberately listen to cosmic noise in an effort to gain better understanding of our universe. Manmade noise rather than cosmic noise limits the sensitivity of receiving equipment.

COSMIC RADIATION

Cosmic radiation is a barrage of high-speed atomic particles from outer space. This radiation is also called cosmic rays or cosmic bombardment. This radiation has high penetrating power, but the atmosphere of the earth absorbs much of it before it can reach the ground. Cosmic radiation may consist of alpha particles, beta particles, electrons, neutrons, protons, and other particles.

A Geiger counter or similar radiation detector can show the presence of cosmic radiation. A device called a cloud chamber, which renders the paths of atom particles visible, can also be used to prove the existance of cosmic radiation. Some cosmic rays originate in the sun, some come from relatively nearby stars and some originate in distant stars and galaxies in various stages of evolution. Atomic particles travel millions, or even billions, of light years. *See also* ALPHA PARTICLE, BETA PARTICLE, ELECTRON, NEUTRON, POSITRON, PROTON.

COTANGENT

The cotangent function is a trigonometric function, equal to the reciprocal of the tangent function (*see* TANGENT). In a right triangle, the cotangent of an angle θ between zero and 90 degrees is equal to the length of the adjacent side divided by the length of the opposite side (A in the illustration). The smaller the angle, the larger the ratio. The values of the cotangent function for angles ranging from zero to 360 degrees are shown at B. The cotangent function is undefined at $\theta = 0°$ and at $\theta = 180°$.

In mathematical calculations, the cotangent function is abbreviated cot, and is given by the formula:

$$\cot \theta = \frac{1}{\tan \theta}$$

where $\tan \theta$ represents the value of the tangent function at the angle θ. *See also* TRIGONOMETRIC FUNCTION.

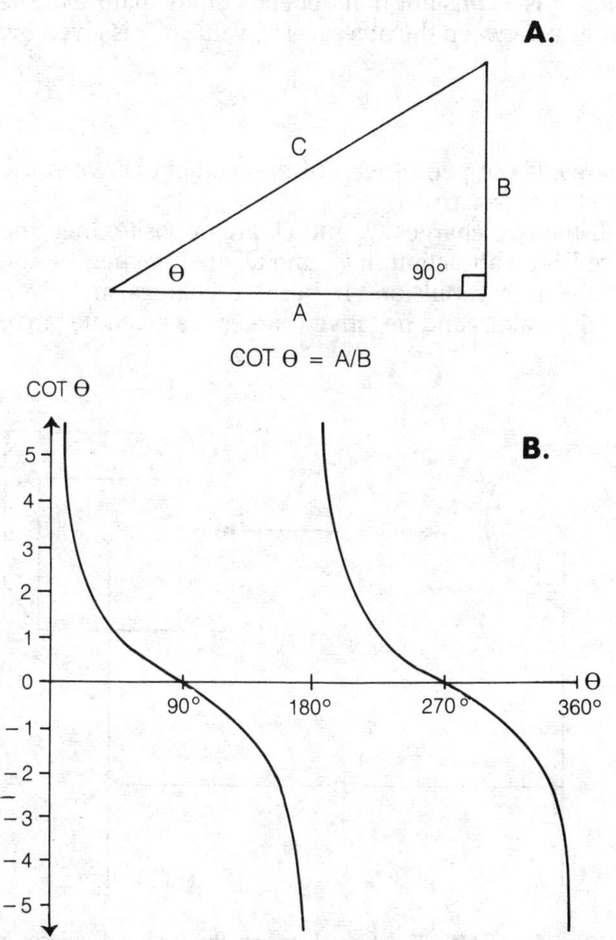

COTANGENT: A triangle used to define cotangent (A) and a graph of the cotangent function from 0 to 360 degrees (B).

COULOMB

The coulomb is the unit of electrical charge quantity. A coulomb of charge is contained in 6.28×10^{18} electrons. One electron thus carries 1.59×10^{-19} coulomb of negative electrical charge.

With a current of 1 ampere flowing in a conductor, exactly 1 coulomb of electrons (or other charge carriers) passes a fixed point in 1 second. The electron flow may take place in the form of electron transfer among atoms, or in the form of positive charge carriers called holes. A coulomb of positive charge indicates a deficiency of 6.28×10^{18} electrons on an object; a coulomb of negative charge indicates a surplus of 6.28×10^{18} electrons. *See also* ELECTRON, HOLE.

COULOMB'S LAW

The properties of electrostatic attraction and repulsion are given by a rule called Coulomb's law.

Given two charged objects X and Y (see illustration), separated by a charge-center to charge-center distance d containing charges Q_x and Q_y, Coulomb's law states that the force F between the objects, caused by the electrostatic field is:

$$F = \frac{kQ_XQ_Y}{d^2}$$

where k is a constant that depends on the nature of the medium between the objects. The value of k is given by:

$$k = \frac{1}{4\pi\epsilon}$$

where ϵ is the permittivity of the medium between the objects (*see* PERMITTIVITY).

If the two charges Q_x and Q_y are opposite, then the force F is an attraction. If Q_x and Q_y are like charges, the force F is a repulsion. If positive charges are given positive values and negative charges are given negative values in the equation, then attraction is indicated by a negative force, and repulsion is indicated by a positive force.

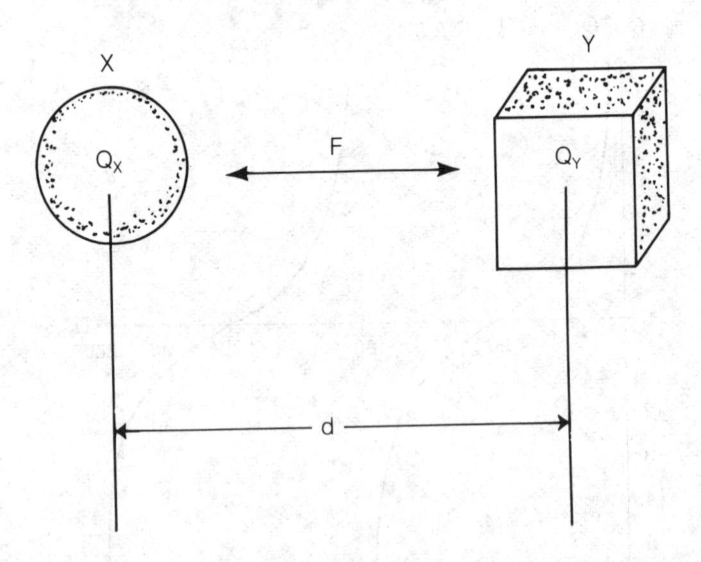

COULOMB'S LAW: The force between two charged objects is directly proportional to the product of their charges and inversely proportional the square of their separation distance.

COULOMETER

A coulometer is an instrument that measures a quantity of electric charge. An electrolytic cell, capable of being charged and discharged makes an excellent coulometer.

When the charge is transferred to the electrolytic cell from an object, a certain amount of chemical action is produced. This chemical action is proportional in magnitude to the amount of charge. Knowing the relation between the charge and the extent of chemical action, the number of coulombs can be accurately determined. *See also* COULOMB.

COUNTER

The basic electronic counter is a binary counter or chain of flip-flops. (*See* FLIP-FLOP.) The first flip-flop divides by two, and successive flip-flops divide by multiples of two. Division occurs at the clock pulse input frequency. This frequency division is equivalent to counting the clock pulses in multiples of 2, 4, 8, 16, etc. By skipping six counts, the count of 16 can be modified to 10 for a decade counter. The counter is also a register. (*See* REGISTER.) A counting register makes the count for each order of places. A shift register can shift the count between chains of flip-flops for each binary place in the count. *See* SHIFT REGISTER, CHARGE-COUPLED DEVICE.

A divide-by-eight counter is shown at A in the illustration. This is a ripple counter because the three flip-flops are cascaded. The J and K inputs of all the flip-flops are connected and brought to the +5 V supply. This input makes each flip-flop ready to toggle. When the set and reset are both high, the flip-flop will switch or toggle each time the clock input goes low.

The clock pulses are sent to the clock input of the first flip-flop, (FF-A). However, the clock input of FF-B is obtained from the Q output of FF-A. Similarly, the clock input of FF-C is obtained from the Q output of FF-B. The divide-by-eight output of the counter is taken from the Q output of flip-flop C, the final flip-flop.

This circuit is a binary counter that divides by two in three successive steps. For every eight changes of the clock, the Q output of FF-A changes four times, the Q output of FF-B changes twice, and the Q output of FF-C changes only once. As a result, the final output in the clock input divided by eight. A disadvantage of this cascade counter is its relatively slow speed. The slow speed is the result of each flip-flop waiting for the previous flip-flop to change state.

If a faster counter is desired, a synchronous counter is used. B shows a four-stage synchronous counter. It is synchronous because all four flip-flops are triggered at the same time by the same clock pulse. This counter counts 16 clock pulses before it returns to a starting state. To obtain a decade counter, four flip-flops are connected to obtain the maximum count of 16. However, a NAND gate and a NOR gate are added. The NAND gate is

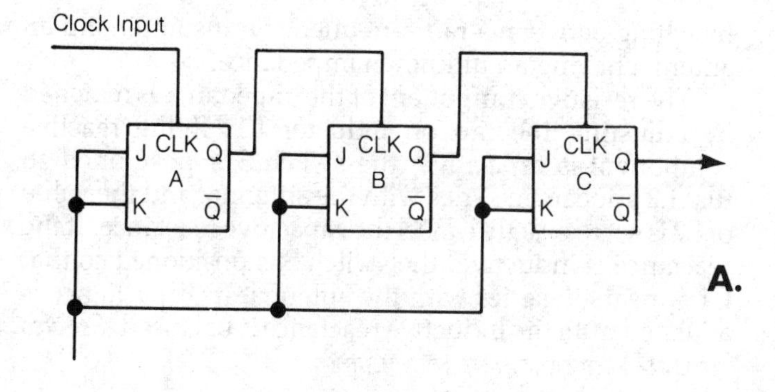

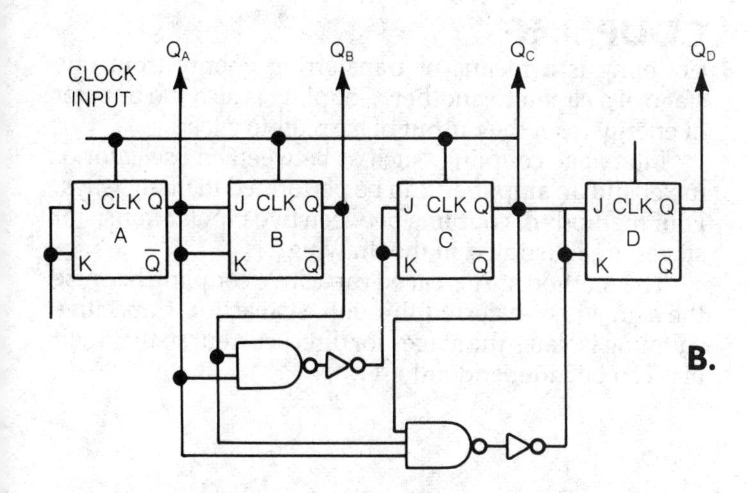

COUNTER: A divide-by-eight, ripple-type counter with three flip-flops (A) and a synchronous counter with four flip-flops (B).

connected so that six of the counts will not be used. The AND gate is used for steering the reset pulses for the flip-flops.

COUNTERMODULATION

Countermodulation is the bypassing of the cathode, emitter, or source resistor of the front end of a receiver, to eliminate cross modulation (*see* CROSS MODULATION) in the circuit. The capacitor value is chosen so that the radio-frequency signal is shunted to ground, but the audio frequencies are not. The result is that audio-frequency signals are canceled, or greatly reduced, by degenerative feedback. The desired radio-frequency signal is, however, easily passed through the amplifier.

The capacitor should have a reactance of less than one fifth the resistor value at frequencies below 20 kHz, and it should have a reactance of least five times the resistor value at the signal frequency. Therefore, the capacitance depends on the value of the resistor. Countermodulation becomes less effective at low and very low frequencies. The graph illustrates typical countermodulation bypass capacitor values as a function of the cathode, emitter, or source resistance.

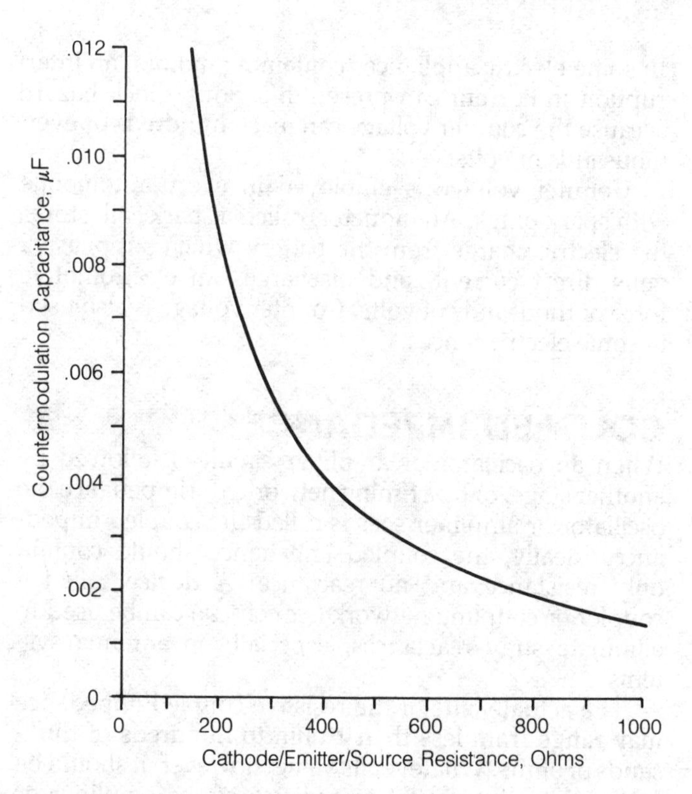

COUNTERMODULATION: Countermodulation capacitance plotted as a function of cathode, emitter, or source resistance.

COUNTERPOISE

A counterpoise is a means of obtaining a radio-frequency ground or ground plane without a direct earth-ground connection. A grid of wires is placed just above the actual surface to provide capacitive coupling to the ground. This greatly reduces ground loss at radio frequencies.

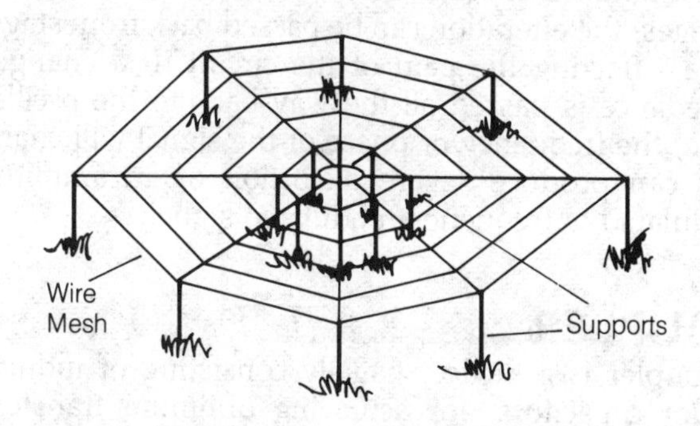

COUNTERPOISE: The capacitance coupling of a wire mesh counterpoise acts as a ground at radio frequencies.

A simple counterpoise is shown in the illustration. Ideally, the radius of a counterpoise should be at least ¼ wavelength at the lowest operating frequency for a given system. The counterpoise is especially useful at locations where the soil conductivity is poor, rendering a direct ground connection ineffective. A counterpoise can be used in conjunction with a direct ground connection. *See also* GROUND CONNECTION, GROUND PLANE.

COUNTER VOLTAGE

When the current through a conductor is cut off, a reverse voltage, called a counter voltage, appears across the coil. This voltage can be very high if the current through the coil is high, and if the coil inductance is large.

In some electric appliances containing motors, an interruption in current can present a serious shock hazard because the counter voltage can reach hundreds or even thousands of volts.

Counter voltage is employed in every automobile with spark plugs. An inductor, called a spark coil, stores the electric charge from the battery which supplies 12 volts direct current, and discharges an electromotive force of thousands of volts. Counter voltage is also used in some electric fences.

COUPLED IMPEDANCE

When an oscillator or amplifier circuit is followed by another stage, or by a tuning network, the impedance the oscillator or amplifier sees is called the coupled impedance. Ideally, the coupled impedance should contain only resistance and no reactance. A device called a coupler or coupling network (*see* COUPLER) can be used to eliminate stray reactances, especially in antenna systems.

The actual value of the resistive coupled impedance may range from less than 1 ohm to hundreds of thousands of ohms. Whatever its value, however, it should be matched to the output impedance of the amplifier or oscillator to which it is connected. The coupled impedance should also remain constant at all times. If it changes, the alteration can be passed back from stage to stage, affecting the gain of the circuit. If a change in impedance is passed all the way back to the oscillator stage, the frequency or phase of the signal will change. This can produce severe distortion of an amplitude-modulated or frequency-modulated signal.

COUPLER

A coupler is a device, usually consisting of inductors and/or capacitors, for achieving optimum transfer of power from an amplifier or oscillator to the next stage. A coupler also can be used between the output of a transmitter and an antenna. Some couplers have fixed components, and some are adjustable. A simple coupler circuit is illustrated in the diagram. It is intended for impedance matching between a radio-frequency transmitter and an antenna having an unknown impedance.

The resistive component of the impedance is matched by adjusting the tap on inductor L1. If the reactive component is capacitive, the switch S is positioned so that L2 appears in series with the antenna, and the value of L2 is set to exactly cancel the capacitive reactance. If the reactance is inductive, the switch S is positioned so that C1 appears in series with the antenna. Its capacitance is adjusted until the inductive reactance is balanced. *See also* ANTENNA MATCHING, ANTENNA TUNING.

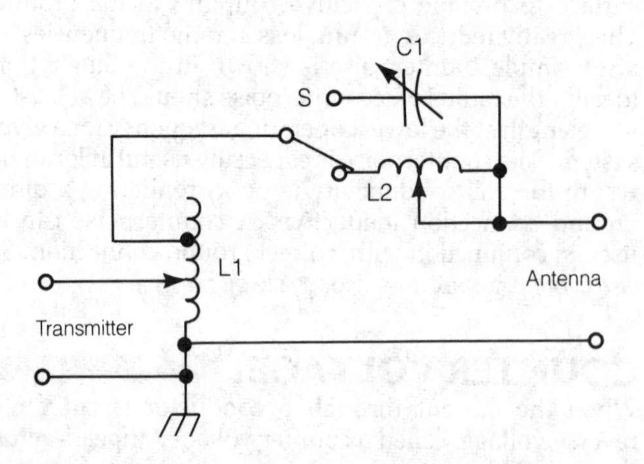

COUPLER: A schematic for an impedance-matching antenna coupler.

COUPLING

Coupling is a means of transferring energy from one stage of a circuit to another. Coupling is also the transfer of energy from the output of a circuit to a load.

Interstage coupling, such as between an oscillator or mixer and an amplifier, can be performed in many ways. Four methods of coupling between two bipolar transistor stages are illustrated in the drawing.

The method at A is called capacitive coupling because the signal is transferred through a capacitor. Capacitive coupling isolates the stages for direct current so that their bias can be independently set.

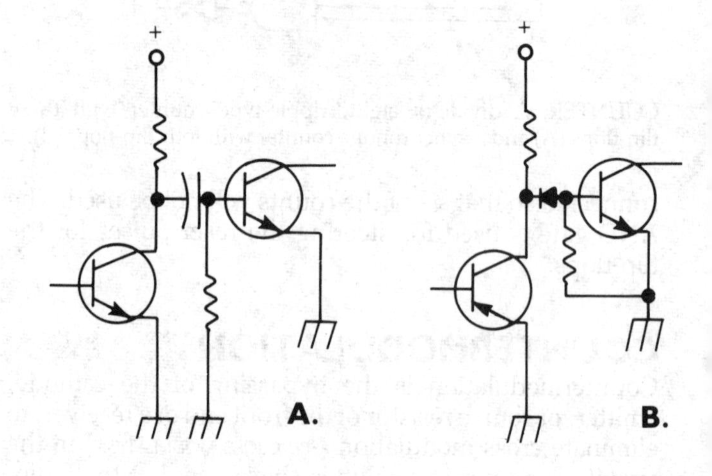

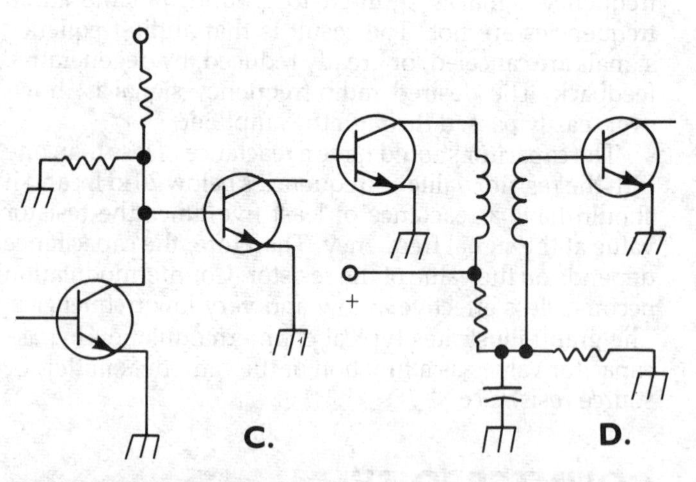

COUPLING: Three types of coupling are: capacitive (A), diode (B), direct (C), and transformer (D).

The coupling scheme at B is called diode coupling. The diode passes signal energy in one direction, but isolates the stages for direct current in that direction. (Note that the second stage uses a PNP transistor while the first stage uses an NPN transistor.) The second stage operates in Class B or Class C.

The coupling scheme at C is called direct coupling. The voltage at the collector of the first transistor is the same as the voltage at the base of the second. For this method to function, the collector voltage of the second NPN transistor must be much more positive than the collector voltage of the first transistor. Also, the base voltage of the first transistor must be carefully set to avoid saturation.

Transformer coupling is illustrated at D. Although this method is the most expensive of those shown, it is preferable because it allows precise impedance matching, and offers good harmonic attenuation. Transformer coupling isolates the two stages for direct current, and allows the use of tuned circuits for improved efficiency. The phase can be reversed if desired. With a well-designed transformer-coupled circuit, electrostatic coupling is kept to a minimum. This improves the stability of the circuit.

The four methods of coupling shown are only a sampling of the many different arrangements possible. The most common method of interstage coupling is the capacitive method.

Coupling between a radio-frequency transmitter and its antenna is accomplished by means of a circuit called a coupling network, or coupler. *See also* COUPLER.

COVALENT BOND

A covalent bond occurs when two or more atoms share electrons. Some atoms give up electrons easily; these elements are said to have negative valence numbers. Some atoms readily accept additional electrons; they have positive valence numbers (*see* VALENCE NUMBER).

Atoms contain electrons in discrete orbits, called shells. The innermost shell is called the K shell, and may contain at most two electrons. The next shell is called the L shell and can contain as many as eight electrons. The third shell, the M shell, can have up to 18 electrons. In general the nth shell of an atom can contain from zero to 2n electrons. When an atom has just one or two electrons in its outermost shell, it gives up the electrons easily. When an atom has a shortage of one or two electrons in the outer shell, it readily accepts one or two more.

A covalent bond occurs among atoms having an equal surplus and shortage of electrons. For example, oxygen with atomic number 8, has a deficiency of two electrons in its L shell. Hydrogen, with one electron, can either accept another to get two, or give one up to have none. Two hydrogen atoms can share their electrons with one atom of oxygen, creating the familiar compound H_2O. In this manner, all the atoms are satisfied.

Some elements have all their shells filled completely. These atoms do not usually produce covalent bonds. Helium and neon are two examples of these elements.

COVERAGE

Coverage refers to the frequency range of a receiver or transmitter. The term is also used to define the service area of a communications or broadcast station.

When specifying the frequency coverage of a transmitter or receiver, either the actual frequency range or the approximate wavelength range may be indicated. An amateur radio receiver might, for example, be specified to cover 80 through 10 meters. This usually means that it operates only on the amateur bands designated in this range. A general-coverage receiver might be specified to work over the range 535 kHz to 30 MHz. This implies continuous coverage.

The coverage area of a broadcast station is determined by the level of output power and the directional characteristics of the antenna system. The Federal Communications Commission limits the coverage allowed to broadcasting stations in the United States. This prevents mutual interference between different stations on the same frequency. Stations may be designated for local, regional, or national (clear-channel) coverage. *See also* CLEAR CHANNEL, LOCAL CHANNEL.

CPU

See COMPUTER (COMPUTER ORGANIZATION).

CREST FACTOR

The ratio of the peak amplitude to the root-mean-square amplitude of an alternating-current or pulsating direct-

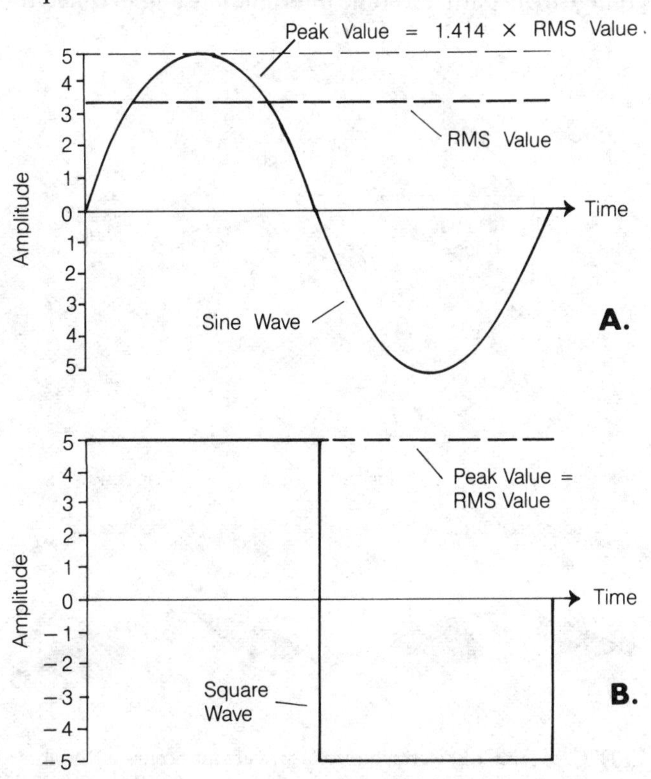

CREST FACTOR: The crest factor the ratio of peak amplitude to rms value. For a sine wave it is 1.414 (A), but for a square wave it is 1 because peak and rms values are equal (B).

current waveform is called the crest factor. Sometimes it is called the amplitude factor. The crest factor depends on the shape of the wave.

In the case of a sine wave, the crest factor is equal to $\sqrt{2}$, or approximately 1.414 (see A in illustration). In the case of a square wave, the peak and root-mean-square amplitudes are equal, and therefore the crest factor is equal to 1, as shown at B.

In a complicated waveform, the crest factor may vary considerably, and may change with time. It is never less than 1 because the root-mean-square (rms) voltage, current, or power is never greater than the peak voltage, current, or power. For a voice or music waveform, the crest factor is generally between 2 and 4. *See also* ROOT MEAN SQUARE.

CRITICAL ANGLE

When a beam of light or radio waves passes from one medium to another having a lower index of refraction, the energy may continue on into the second medium, as shown at A in the illustration, or it might be reflected off the boundary and remain within the original medium, as shown at B. Whether refraction or reflection occurs depends on the angle of incidence (*see* ANGLE OF INCIDENCE, INDEX OF REFRACTION).

If the angle of incidence is very large, reflection will take place. Then the energy remains in the region with the larger index of refraction. If the angle of incidence is 0 degrees, the energy passes into the medium with the lower index of refraction, generally, and there is no change in its path. At some intermediate angle, called the critical angle, reflection just begins to occur as the angle of incidence is made larger. The critical angle depends on the ratio of the indices of refraction of the two media, at the energy wavelength involved.

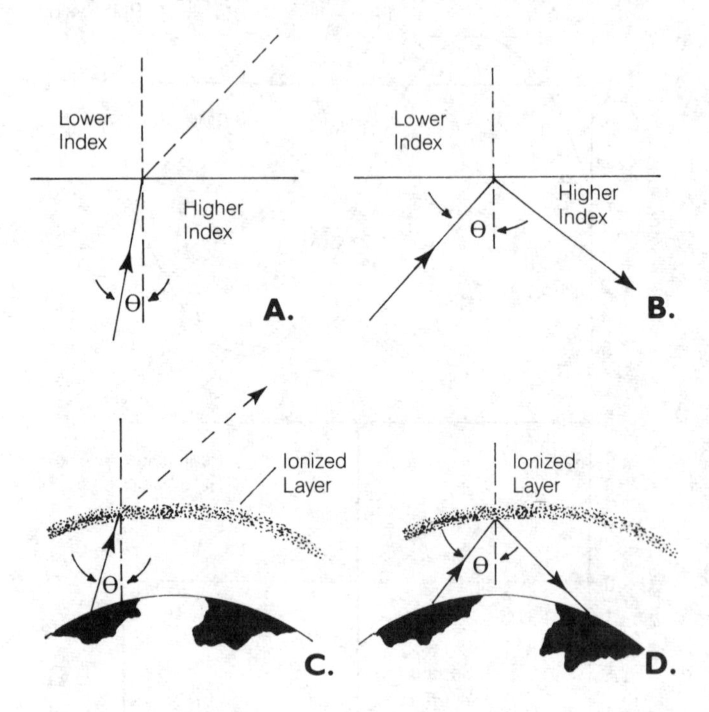

CRITICAL ANGLE: Refraction of a wavefront occurs at the interface between media with differing refractive indices (A). The critical angle is the angle of incidence beyond which all energy will be internally reflected (B). Radio energy propagated at greater than the critical angle passes through the ionosphere (C); at less than the critical angle it is reflected (D).

When radio waves encounter the E or F layers of the ionosphere (*see* E LAYER, F LAYER), the waves can be returned to the earth, or they can continue on into space (shown at C). The smallest angle of incidence, at which energy is consistently returned to the earth is called the critical angle (D). The critical angle for radio waves depends on the density of the ionosphere and the wavelength of the signal. However, some radio energy transmitted vertically may not be returned to earth so the critical angle is 0 degrees. In other cases, the radio energy may not be reflected to earth, no matter how great the angle of incidence. *See also* ANGLE OF REFRACTION, PROPAGATION CHARACTERISTICS.

CRITICAL COUPLING

When two circuits are coupled, the optimum value of coupling (for which the best transfer of power takes place) is called critical coupling. If the coupling is made tighter or looser than the critical value, the power transfer becomes less efficient.

The coefficient of coupling, *k*, for critical coupling is given by:

$$k = \frac{1}{\sqrt{Q_1 Q_2}}$$

where Q_1 is the Q factor of the primary circuit and Q_2 is the Q factor of the secondary circuit. *See also* COEFFICIENT OF COUPLING, Q FACTOR.

CRITICAL DAMPING

In an analog system, damping is the rapidity with which a mass reaches the quiescent state. The longer the time required for this, the greater the damping (*see* DAMPING). If the damping is insufficient, the mass may overshoot and oscillate back and forth before coming to rest. If the damping is excessive, the mass might not respond too fast to be useful for the desired purpose.

Critical damping is the smallest amount of damping that can be realized without overshoot.

CRITICAL FREQUENCY

At low and very low radio frequencies, all energy is returned to the earth by the ionosphere. This is true even if the angle of incidence of the radio signal with the ionized layer is 90 degrees (*see* ANGLE OF INCIDENCE).

As the frequency of a signal is raised, a point is reached where energy sent directly upward will escape into space. All signals impinging on the ionosphere at an angle of incidence smaller than 90 degrees will, however, still be returned to the earth. The frequency at which this occurs is called the critical frequency.

The critical frequency depends on the density of the ionized layers. This density changes with the time of day, the time of year, and the level of sunspot activity. The critical frequency for the ionospheric F layer is typically between about 3 and 5 MHz. *See also* PROPAGATION CHARACTERISTICS.

CROSSBAR SWITCH

A crossbar switch is a special of switch that provides a large number of different connection arrangements. A set of contacts is arranged in a matrix. The matrix may be two-dimensional or three-dimensional. An example of a crossbar switch is shown in the illustration. The matrix can be square or rectangular in two dimensions; it can be shaped like a cube or a rectanglar prism in three dimensions.

A shorting bar selects any adjacent pair of contacts that lie along a common axis. In this way, a large number of different combinations are possible with a relatively small number of switch contacts. The shorting bar is magnetically controlled.

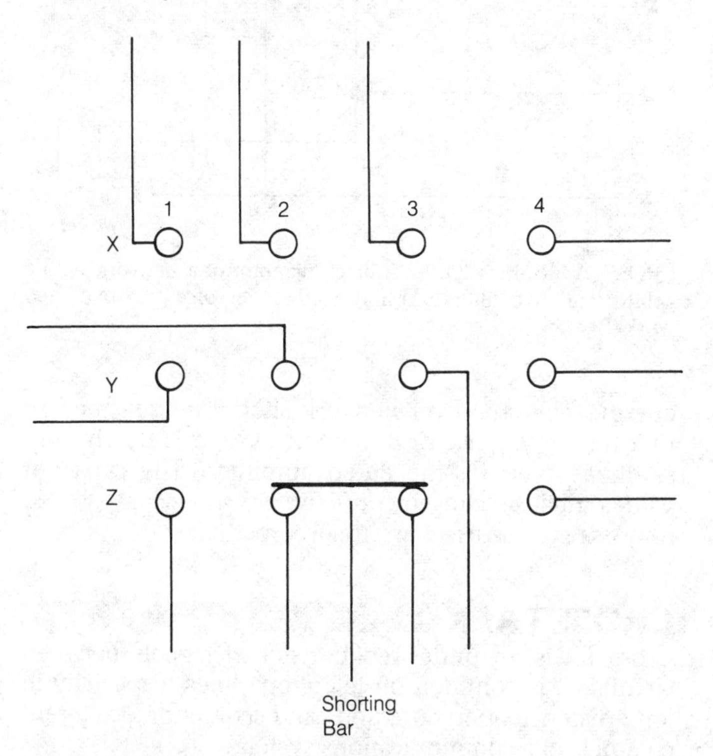

CROSSBAR SWITCH: A crossbar switch permits many different connections.

CROSS-CONNECTED NEUTRALIZATION

In a push-pull amplifier (*see* PUSH-PULL AMPLIFIER), instability may cause unwanted oscillation. This is prevented in all radio-frequency amplifiers by a procedure called neutralization (*see* NEUTRALIZATION). The neutralization method most often used in a push-pull amplifier is called cross-connected neutralization, or simply cross neutralization.

In cross neutralization, two feedback capacitors are used. The output of one half of the amplifier is connected to the input of the other half, as shown in the diagram. The fed-back signals are out of phase with the input signals for undesired oscillation energy. When the capacitor values are correctly set, the probability of oscillation is greatly reduced. The values of the two neutralizing capacitors must always be identical to maintain circuit balance.

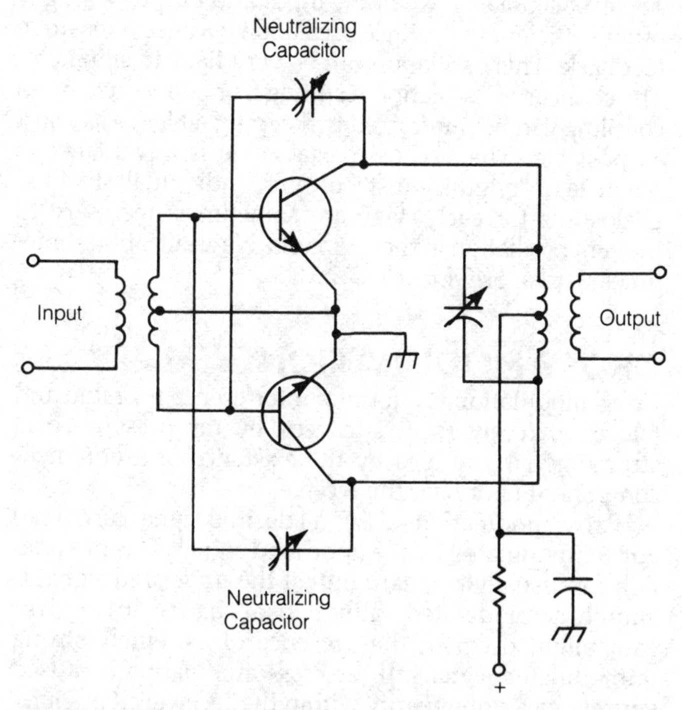

CROSS-CONNECTED NEUTRALIZATION: This neutralization method is used with push-pull amplifiers.

CROSS COUPLING

Cross coupling is a means of obtaining oscillation using two stages of amplification. Normally, an amplifying stage produces a 180-degree phase shift in a signal passing through. Some amplifiers produce no phase shift. In either case, the signal will have its original phase after passing through two amplifying stages. By coupling the output of the second stage to the input of the first stage with a capacitor, oscillation can be achieved. The

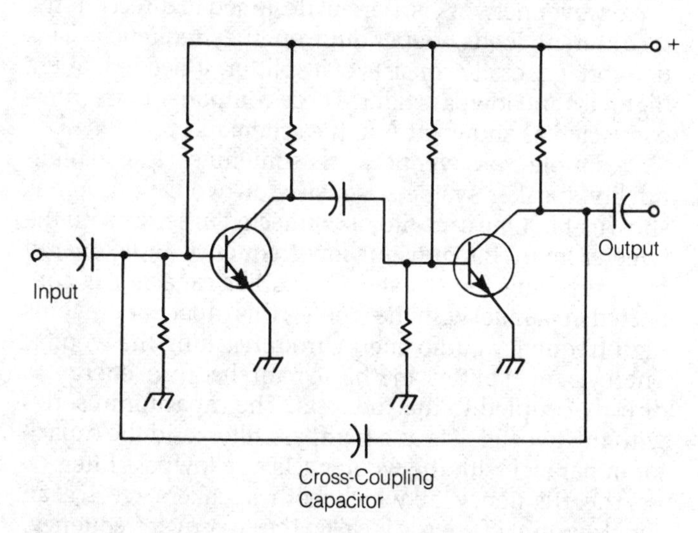

CROSS COUPLING: Cross coupling permits the oscillation of two amplifier stages.

diagram illustrates simple cross coupling in a two-stage transistor circuit.

Cross coupling frequently occurs when it is not wanted. This is likely in multistage, high-gain amplifiers. The capacitance between the input and output wiring is often sufficient to produce oscillation because of positive feedback. This oscillation can be very hard to eliminate. The chances of oscillation resulting from unwanted cross coupling can be minimized by keeping all leads as short as possible. The use of coaxial cable is advantageous when lead lengths must be long. Individual shielded enclosures for each stage are sometimes necessary to prevent oscillation in these circuits. Neutralization sometimes works. *See also* NEUTRALIZATION.

CROSS MODULATION

Cross modulation is a form of interference for radio and television receivers. It is caused by the presence of a strong signal and also by the existence of a nonlinear component in or near the receiver.

Cross modulation causes all desired signal carriers to appear modulated by the undesired signal. This modulation can usually be heard only if the undesired signal is amplitude-modulated, although a change in receiver gain might occur in the presence of extremely strong unmodulated signals. If the cross modulation is caused entirely by a nonlinearity within the receiver, it is sometimes called intermodulation (*see* INTERMODULATION).

Cross modulation can be prevented by attenuating the level of the undesired signal before it reaches nonlinear components, or drives normally linear components into nonlinear operation. If the nonlinearity is outside the receiver and antenna system, the cross modulation can be eliminated only by locating the nonlinear junction and eliminating it. Marginal electrical bonds between two wires, between water pipes, or even between parts of a metal fence, can be responsible.

CROSSOVER NETWORK

A crossover network is a circuit designed to direct energy to different loads, depending on the frequency. This network generally consists of a splitter, if needed, and a highpass and lowpass filter. These components are interconnected as shown at A in the schematic.

A simple crossover network, sometimes used in high-fidelity speaker systems, consists of two capacitors, as shown at B. One capacitor is connected in series with the tweeter line. This prevents low-frequency audio energy from reaching the tweeter. The other capacitor is connected in parallel with the woofer; this capacitor prevents high-frequency audio energy from reaching the woofer. There is no splitter in this circuit because energy is directly coupled to the speakers. The capacitor in series with the tweeter acts as a highpass filter, and the capacitor in parallel with the woofer acts as a lowpass filter.

The frequency at which both speakers receive an equal amount of energy is called the crossover frequency. Below the crossover point, the woofer gets more energy. Above the crossover frequency, the tweeter gets more energy. The crossover network should be designed so that the impedance of the whole system is nearly constant, as seen by the audio amplifier. The capacitor values must be chosen correctly. *See also* HIGHPASS FILTER, LOWPASS FILTER, SPLITTER, TWEETER, WOOFER.

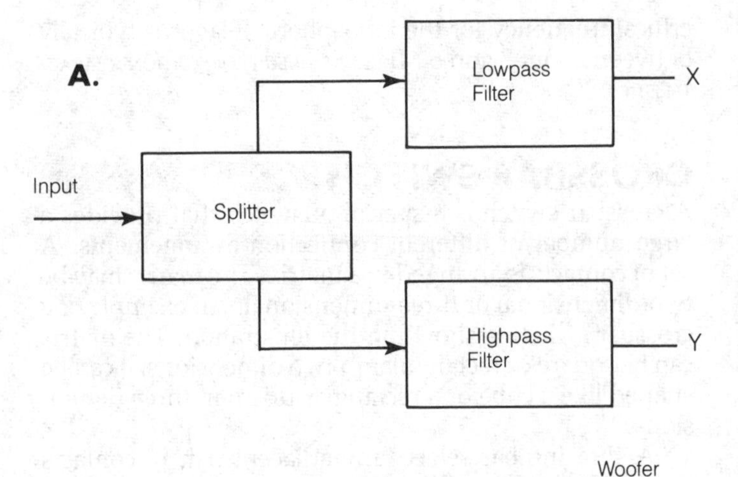

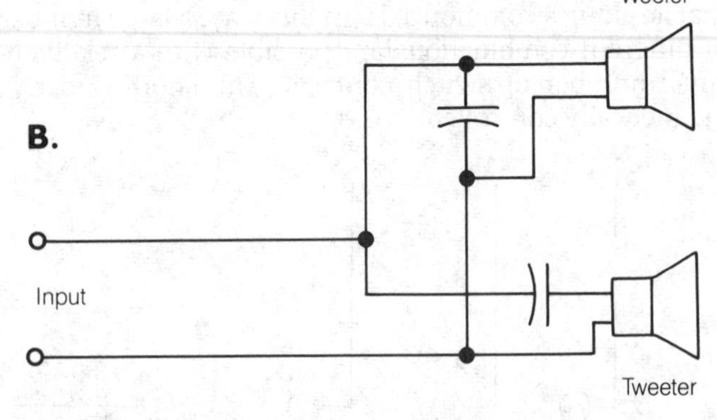

CROSSOVER NETWORK: A block diagram of a network with a splitter and two filters (A) and a schematic of a crossover network (B).

CROSSTALK

Crosstalk is an undesired transfer of signals between circuits. It is common on telephone lines, especially in long-distance operation, and can occur among different channels of a communications system.

Crosstalk rejection is the level of the crosstalk energy, in decibels, with respect to the desired signal. The crosstalk level is the amplitude of the crosstalk energy with respect to some reference value. Crosstalk loss is the effective degradation, in decibels, of the signal-to-noise ratio, caused by crosstalk in a communications circuit.

Crosstalk can be minimized by ensuring that a transmission line is properly shielded or balanced. In a carrier current communications link (*see* CARRIER CURRENT), crosstalk is minimized by linear networks and the maintenance of good electrical connections.

CRYOGENICS

Cryogenics is the science of the behavior of matter and energy at extremely low temperatures. The coldest possible temperature, called absolute zero, is the absence of all heat. This temperature is approximately −459.72°F or −273.16°C. The Kelvin temperature scale (*see* KELVIN TEMPERATURE SCALE) is based on absolute zero.

When the temperature of a conductor is brought to within a few degrees of absolute zero, the conductivity increases dramatically. If the temperature is cold enough, a current can be made to flow continuously in a closed loop of wire. This is called superconductivity. *See* SUPERCONDUCTIVITY.

CRYSTAL

A crystal is piezoelectric material for transforming mechanical vibrations into electrical impulses. Crystals are used in some microphones for this purpose (*see* MICROPHONE, CRYSTAL TRANSDUCER). Some crystal materials are used as detectors or mixers (*see* CRYSTAL DETECTOR, CRYSTAL SET).

Quartz crystals are widely used to generate radio-frequency energy. They are usually packaged in metal cases. Two wire leads extend from the base. These leads are internally connected to the faces of the crystal, which consists of a thin wafer of quartz.

The frequency at which a quartz crystal vibrates depends on the method for cutting the crystal and also on its thickness. The thinner the crystal, the higher the natural resonant frequency. The highest fundamental frequency of a common quartz crystal is about 15 to 20 MHz; above this frequency range, harmonics must be employed to obtain radio-frequency energy.

Quartz crystals have excellent frequency stability, their main advantage. Crystals are much better than coil-and-capacitor tuned circuits in this respect. A crystal, however, cannot be tuned over a wide range of frequencies. Some crystals are used as selective filters because of their high Q factors. *See also* CRYSTAL CONTROL, CRYSTAL LATTICE FILTER, Q FACTOR.

CRYSTAL CONTROL

Crystal control is a method of determining the frequency of an oscillator with a piezoelectric crystal. These crystals, usually made of quartz (*see* CRYSTAL), have excellent oscillating frequency stability. Crystal control is much more stable than the coil-and-capacitor method.

Crystal oscillators may operate either at the fundamental frequency of the crystal or at one of the harmonic frequencies. Crystals designed especially for operation at a harmonic frequency are called overtone crystals. Overtone crystals are almost always used at frequencies above 20 MHz because a fundamental-frequency crystal would be too thin at these wavelengths, and might easily crack.

The operating frequency of a piezoelectric crystal can be increased slightly by placing an inductor in parallel with the crystal leads. A capacitor across a crystal will reduce the oscillating frequency. Generally, the amount of frequency adjustment possible with these methods is very small—approximately ±0.1 percent of the fundamental operating frequency. *See also* OSCILLATOR.

CRYSTAL DETECTOR

In early radio receivers a crystal detector was used to demodulate an amplitude-modulated signal. A piece of mineral such as galena was placed in contact with a length of fine wire called a cat's whisker to obtain rectification. The cat's whisker usually had to be moved around on the galena to get satisfactory reception.

Modern receivers employ silicon semiconductor diodes for the detection of amplitude-modulated signals. A bipolar or field-effect transistor, biased for Class B operation, will also act as a detector. *See also* DETECTION.

CRYSTAL-LATTICE FILTER

A crystal-lattice filter is a selective filter, usually of the bandpass type. It is similar in construction to a ceramic filter (*see* BANDPASS FILTER, CERAMIC FILTER) when housed in one container. Some crystal-lattice filters consist of several separate piezoelectric crystals. The crystals are cut to slightly different resonant frequencies to obtain the desired bandwidth and selectivity characteristics.

Crystal-lattice filters are used in the intermediate stages of superheterodyne receivers. They are also used in the filtering stages of single-sideband transmitters. Properly adjusted crystal-lattice filters have an excellent rectangular response, with steep edges and high adjacent-channel attenuation. A simple crystal-lattice filter, using only two piezoelectric crystals at slightly different frequencies, is shown in the diagram.

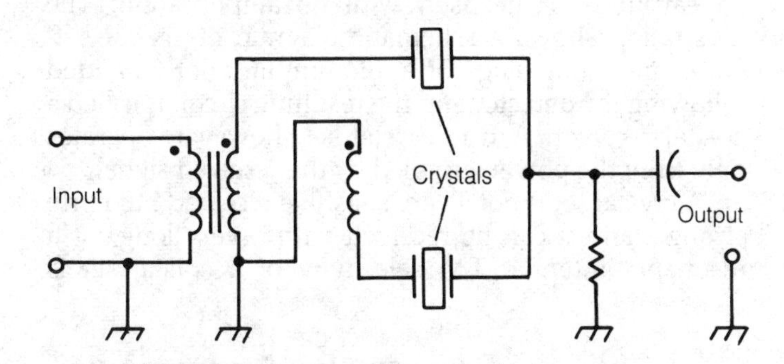

CRYSTAL-LATTICE FILTER: This filter contains crystals that are resonant at different frequencies to provide the desired bandwidth and selectivity characteristics.

CRYSTAL MICROPHONE

A crystal microphone is microphone that employs a piezoelectric crystal to convert sound vibrations into electrical impulses. The impulses may then be amplified for public-address or communications circuits.

In the crystal microphone, which operates in a manner similar to the ceramic microphone (*see* CERAMIC MICROPHONE), vibrating air molecules set a metal diaphragm in motion. The diaphragm, connected physically to the crystal, puts mechanical stress on the piezoelectric substance. This, in turn, results in small currents at the same frequency or frequencies as the sound. Crystal microphones have excellent fidelity characteristics. *See also* MICROPHONE.

CRYSTAL OSCILLATOR

See OSCILLATOR CIRCUITS.

CRYSTAL OVEN

A crystal oven is a heated chamber to keep the temperature of a crystal at a constant value. Most piezoelectric oscillator crystals shift slightly in frequency as the temperature changes. Some crystals increase in frequency with an increase in the temperature; this is called a positive temperature coefficient. Other crystals decrease in frequency when the temperature rises; this is called a negative temperature coefficient. A crystal oven houses crystals in circuits where extreme frequency accuracy is needed.

Crystal ovens employ thermostats and small heating elements like ordinary ovens. The temperature is kept at a level just above the temperature in the room where the circuit is operated. Several ovens may be used, one inside the other, to obtain even more precise temperature regulation. Frequency-standard oscillators can have this kind of crystal oven.

CRYSTAL SET

A crystal set is a simple amplitude-modulation receiver, consisting only of a tuned circuit, a passive detector, and an earphone or headset, with no amplification. The illustration shows a schematic diagram of crystal set. Sometimes, one stage of audio amplification is added following the detector and the resulting circuit is called a crystal receiver. A true crystal set, however, operates only from the power supplied by the received signal.

A crystal set is not a very sensitive receiver, but fairly strong stations can be received with a very long, or a resonant, antenna. The selectivity of a crystal set is generally rather poor. Crystal sets were originally used for the reception of spark-generated radio signals, in the early days of wireless communication.

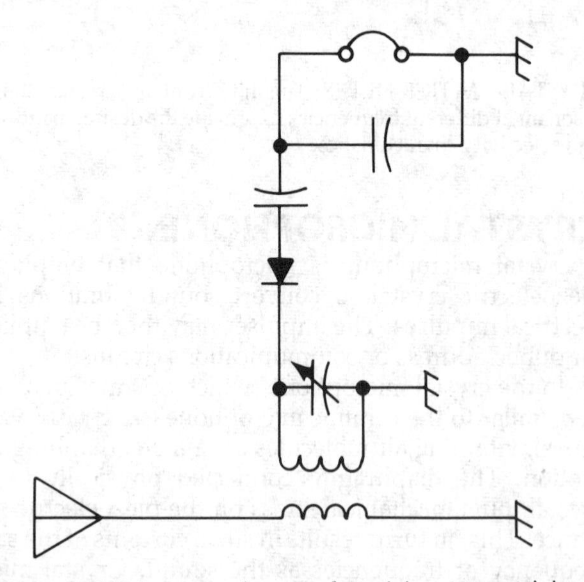

CRYSTAL SET: Schematic of a crystal receiving set containing a Galena detector.

CRYSTAL TEST CIRCUIT

A crystal test circuit is a circuit for testing piezoelectric crystals for proper operation. Most crystal test circuits only verify that the crystal will oscillate on the correct frequency under the specified conditions. This allows faulty crystals to be easily identified.

More advanced crystal testing circuits are used to determine the temperature coefficients, crystal current, and other operating variables. The properties of a crystal or ceramic transducer are checked by a special type of crystal tester.

CRYSTAL TRANSDUCER

Piezoelectric crystals convert mechanical vibrations into electrical impulses, and vice versa. This is called the piezoelectric effect (*see* PIEZOELECTRIC EFFECT). Crystals, either natural or synthetic, make excellent transducers for this reason. Piezoelectric crystals are used in microphones, phonograph cartridges, earphones, and buzzers. The tones now used on many devices with keyboards are generated by tiny piezoelectric crystals.

If alternating currents are applied to the crystal, it will vibrate, producing sound or ultrasound. Conversely, if sound or ultrasound impinges on the diaphragm, an alternating current will appear between the holder plates. Crystal transducers can operate at frequencies well above the range of human hearing. *See also* TRANSDUCER, SONAR.

CURRENT

Current is a flow of electric charge carriers past a point, or from one point to another. The charge carriers may be electrons, holes, or ions (*see* ELECTRON, HOLE). In some cases, atomic nuclei may carry charge.

Electric current is measured in units called amperes. A current of one ampere consists of the transfer of one coulomb of charge per second (*see* AMPERE, COULOMB). Current may be either alternating or direct. Current is symbolized by the letter I in most equations involving electrical quantities.

The direction of current flow is theoretically the direction of the positive charge transfer. Thus, in a circuit containing a dry cell and a light bulb, for example, the current flows from the positive terminal of the cell, through the interconnecting wires, the bulb, and finally to the negative terminal. This is a matter of convention. The electron movement is actually in opposition to the current flow.

CURRENT AMPLIFICATION

Current amplification is the increase in the flow of current between the input and output of a circuit. It is also called current gain. In a transistor, the current-amplification characteristic is called the beta (*see* BETA).

Some amplifier circuits are designed to amplify current and some are designed to amplify voltage. Others are designed to amplify power, which is the product of the current and the voltage. A current amplifier requires driving power to operate because this circuit draws current from its source. Current amplifiers generally have an output impedance that is lower than the input impedance. Therefore these circuits are often used in step-down matching applications. That is, they are used as impedance transformers.

Current amplification is measured in decibels. Mathematically, if I_{IN} is the input current and I_{OUT} is the output current, then:

$$\text{Current gain (dB)} = 20 \log_{10} \frac{I_{OUT}}{I_{IN}}$$

See also DECIBEL, GAIN.

CURRENT-CARRYING CAPACITY

See CARRYING CAPACITY.

CURRENT DRAIN

The amount of current that a circuit draws from a generator or other power supply is called the current drain. The amount of current drain determines the size of the power supply needed for proper operation of a circuit.

If the current drain is too great for a power supply, the voltage output of the supply will drop. Ripple might occur in supplies designed to convert alternating current to direct current. With battery power, the battery life is shortened, the voltage drops, and the battery may overheat dangerously.

Current drain is measured in three ways. The peak drain is the largest value of current drawn by a circuit in normal operation. The average current drain is measured over a long period. The total drain in ampere hours is divided by the operating time in hours. Standby current drain is the amount of current used by a circuit during standby periods. Power supplies should always be chosen to handle the peak current drain without malfunctioning. *See also* POWER SUPPLY.

CURRENT FEED

Current feed is a method of connecting a transmission line to an antenna at a point on the antenna where the current is maximum. This point is called a current loop (*see* CURRENT LOOP). In a half-wavelength radiator, the current maximum occurs at the center, and therefore current feed is the same as center feed (A in the illustration). In an antenna longer than ½ wavelength, current maxima exist at odd multiples of ¼ wavelength from either end of the radiator. There may be several different points on an antenna radiator that are suitable for current feed (as at B).

The impedance of a current-fed antenna is relatively low. The resistive component varies between about 70 and 200 ohms in most cases. Current feed results in good electrical balance in a two-wire transmission line, provided the current in the antenna is reasonably symmetrical. *See also* VOLTAGE FEED.

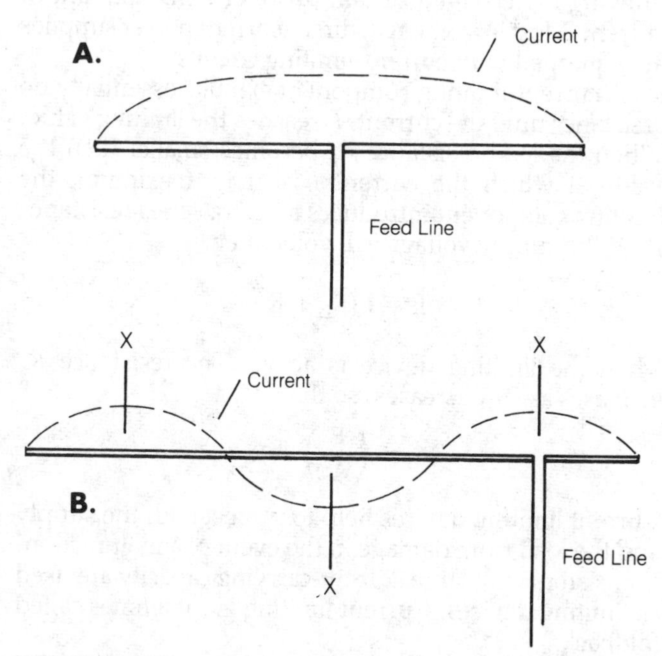

CURRENT FEED: A ½ wavelength dipole antenna is current fed at the center (A), but a 3⁄2 wavelength antenna can be current fed at any of the points marked X.

CURRENT HOGGING

When two active components are connected together in parallel or in push-pull configuration, one of them can draw most of the current. This situation, called current hogging, occurs because of improper balance between components. Current hogging can sometimes take place with poorly matched tubes or transistors connected in push-pull or parallel amplifiers.

Initially, one of the tubes or transistors exhibits a slightly lower resistance in the circuit. As a result, this component draws more current than its mate. If the resistive temperature coefficient of the device is negative, the tube or transistor carrying the larger current will show a lower resistance as it heats. The more the component heats up, the lower its resistance becomes, and the more current it draws. Ultimately, one of the tubes or transistors does all the work in the circuit. This might shorten its operating life. It also disturbs the linearity of a push-pull circuit and upsets the impedance match between the circuit and the load.

Current hogging may be prevented, or at least made unlikely, by placing small resistors in series with the emitter, source, or cathode leads of the amplifying devices. Careful selection of the devices, to ensure the most nearly identical operating characteristics, is also helpful.

CURRENT LIMITING

Current limiting is a process that prevents a circuit from

drawing more than a certain predetermined amount of current. Most low-voltage, direct-current power supplies are equipped with current-limiting circuits.

A current-limiting component exhibits essentially no resistance until the current, I, reaches the limiting value. When the load resistance R_L becomes smaller than the value at which the current I is at its maximum, the limiting component introduces an extra series resistance R_s. If the supply voltage is E volts, then:

$$E = I\,(R_L + R_s)$$

when the limiting device is active. The resistance R_s increases as R_L decreases, so that:

$$R_s = \left(\frac{E}{I}\right) - R_L$$

Current limiting devices help to protect both the supply and the load from damage in the event of a malfunction. Transistors with large current-carrying capacity are used as limiting devices. Current limiting is sometimes called foldback.

CURRENT LOOP

In an antenna radiating element, the current in the conductor depends on the location. At any free end, the current is negligible; the small capacitance allows only a tiny charging current to exist. At a distance of ¼ wavelength from a free end, the current reaches a maximum, called a current loop. A ½-wavelength radiator has a single current loop at the center. A full-wavelength radiator has two current loops. In general, the number of current loops in a longwire antenna radiator is the same as the number of half wavelengths. The drawing illustrates the current distribution in an antenna radiator of 3⁄2 wavelength, showing the locations of the current loops.

Current loops may occur along a transmission line not terminated in an impedance identical to its characteristic impedance. These loops occur at multiples of ½ wavelength from the resonant antenna feed point when the antenna impedance is smaller than the feed-line characteristic impedance. The loops exist at odd multiples of ¼ wavelength from the feed point when the resonant antenna impedance is larger than the feed-line characteristic impedance. Ideally, the current on a transmission line should be the same everywhere, equal to the voltage divided by the characteristic impedance. *See also* CURRENT NODE, STANDING WAVE.

CURRENT NODE

A current node is a current minimum in an antenna radiator or transmission line. The current in an antenna depends, to some extent, on the location of the radiator. Current nodes occur at free ends of a radiator, and at distances of multiples of ½ wavelength from a free end. The illustration shows the locations of current nodes along a 3⁄2-wavelength antenna radiator. The number of current nodes is equal to 1 plus the number of half wavelengths in a radiator.

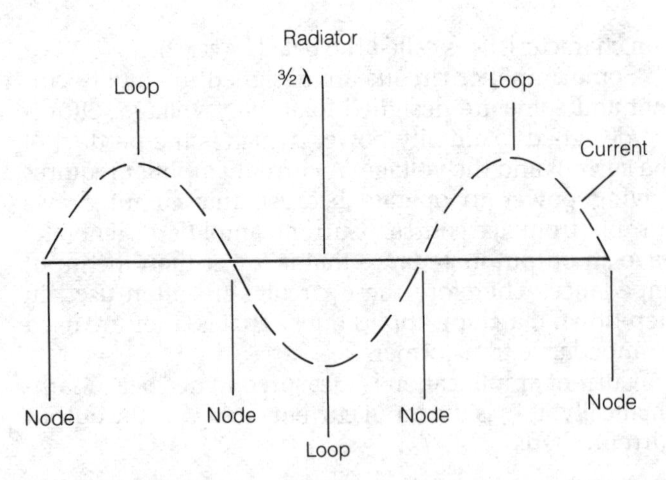

CURRENT LOOP AND NODE: Locations of current loops and nodes on a 3⁄2 wavelength antenna.

Current nodes may occur along a transmission line not terminated in an impedance equal to its characteristic impedance. These nodes occur at multiples of ½ wavelength from the resonant antenna feed point when the antenna impedance is larger than the feed-line characteristic impedance. They exist at odd multiples of ¼ wavelength from an antenna feed point when a resonant antenna impedance is smaller than the characteristic impedance of the line.

Current nodes are always spaced at intervals of ¼ wavelengths from current loops. Ideally, the current on a transmission line is the same everywhere, being equal to the voltage divided by the characteristic impedance. *See also* CURRENT LOOP, STANDING WAVE.

CURRENT REGULATION

Current regulation is the process of maintaining the current in a load at a constant value. This is done with a constant-current power supply (*see* POWER SUPPLY). A variable-resistance device is necessary to accomplish current regulation. When such a device is placed in series with the load, and the resistance increases in direct proportion to the supply voltage, the current remains constant.

CURRENT SATURATION

As the bias between the input points of a tube or semiconductor device is varied so that the output current increases, a point will eventually be reached at which the output current no longer increases. This condition is called current saturation. In a vacuum tube, the grid bias must be positive with respect to the cathode to obtain current saturation. In an NPN bipolar transistor, the base must be sufficiently positive with respect to the emitter (see illustration). In a PNP bipolar transistor, the base must be sufficiently negative with respect to the emitter. In a field-effect transistor, the parameters for current saturation are affected by the gate-source and drain-source voltages.

Saturation is usually not a desirable condition. It destroys the electrical ability of a tube or semiconductor

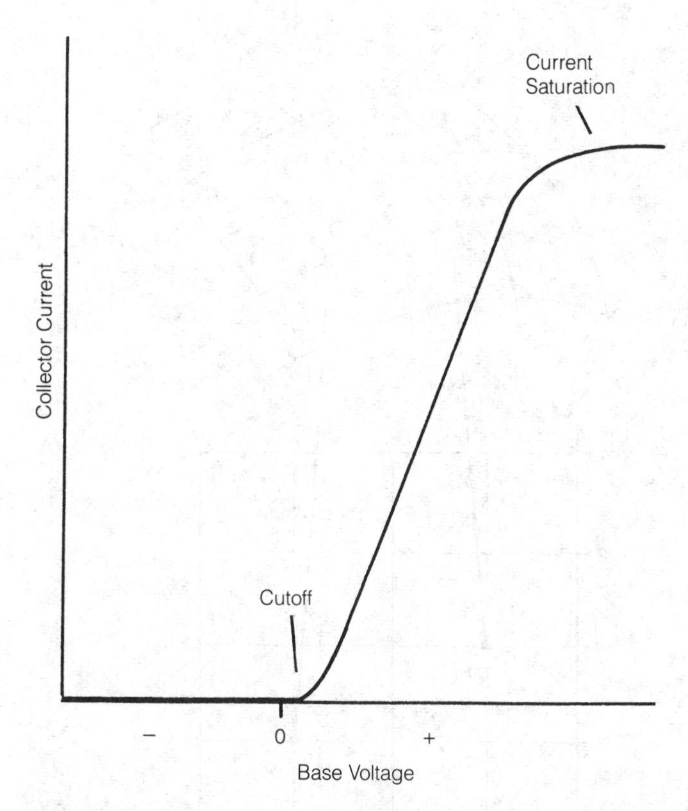

CURRENT SATURATION: A plot of collector current versus base voltage for an NPN transistor showing current leveling or saturation.

device to amplify. Saturation sometimes occurs in digital switching transistors where it may be induced deliberately in the high state to produce maximum conduction. *See also* SATURATION, SATURATION CURRENT, SATURATION CURVE.

CURRENT TRANSFORMER

A current transformer is a transformer for stepping current up or down. An ordinary voltage transformer functions as a current transformer, but in the opposite sense (*see* TRANSFORMER). The current step-up ratio of a transformer is the reciprocal of the voltage step-up ratio. If N_{PRI} is the number of turns in the primary winding and N_{SEC} is the number of turns in the secondary winding, then:

$$\frac{I_{SEC}}{I_{PRI}} = \frac{N_{PRI}}{N_{SEC}}$$

where I_{PRI} and I_{SEC} are the currents in the primary and secondary, respectively. The impedance of the primary and secondary, given by Z_{PRI} and Z_{SEC} are related to the currents by the equation:

$$\frac{Z_{PRI}}{Z_{SEC}} = \left(\frac{I_{SEC}}{I_{PRI}}\right)^2$$

These formulas assume a transformer efficiency of 100 percent. While this is an ideal theoretical condition which never actually occurs, the equations are accurate enough in practice. *See also* TRANSFORMER EFFICIENCY.

CURVE

A curve is a graphical illustration of a relation between two variables. In electronics, two-dimensional graphs are commonly used to show the characteristics of circuits and devices. Generally the Cartesian coordinate plane is used, but other coordinate systems can be employed (*see* CARTESIAN COORDINATES, CURVILINEAR COORDINATES, POLAR COORDINATES).

The illustration shows two examples of curves. At A, the relation between the grid voltage and the plate current of a typical vacuum tube is shown. At B, the

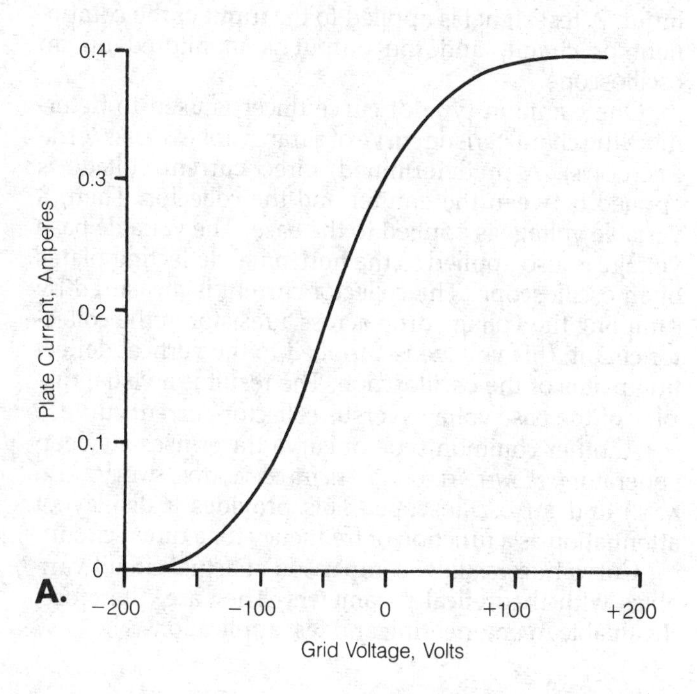

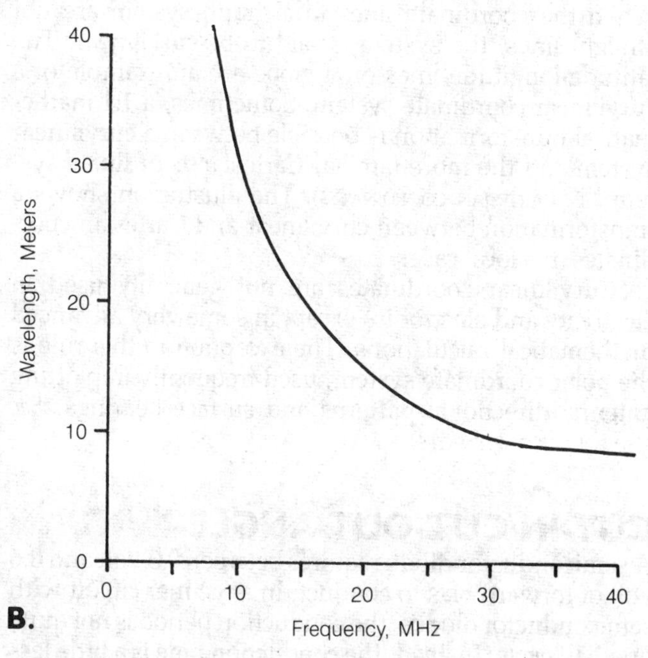

CURVE: Examples of curves are plate current versus grid voltage for a vacuum tube (A) and frequency versus wavelength for electromagnetic waves in free space (B).

relationship between frequency and wavelength for electromagnetic radiation in free space is illustrated. In both of these cases, the curves represent a special kind of relation called a function, because there is never more than one value on the vertical (dependent) axis for any value on the horizontal (independent) axis. *See also* FUNCTION.

CURVE TRACER

A curve tracer is a test circuit used to check the response of a component or circuit under conditions of variable input. A test signal is applied to the input of the component or circuit, and the output is monitored on an oscilloscope.

One common type of curve tracer is used to determine the characteristic curve of a transistor (*see* CHARACTERISTIC CURVE). A predetermined, direct-current voltage is applied between the emitter and the collector. Then, a variable voltage is applied to the base. The variable base voltage is also applied to the horizontal deflecting plates of an oscilloscope. The collector current is measured by sampling the voltage drop across a resistor in the collector circuit; this voltage is supplied to the vertical deflection plates of the oscilloscope. The result is a visual display of the base voltage versus collector-current curve.

Another common type of curve tracer uses a sweep generator (*see* SWEEP-FREQUENCY FILTER OSCILLATOR, SWEEP GENERATOR) and an oscilloscope. This provides a display of attenuation as a function of frequency for a tuned circuit.

Curve tracers allow comparison of actual circuit variables with theoretical parameters. They are, therefore, invaluable in engineering and test applications.

CURVILINEAR COORDINATES

When the coordinate lines in a graph system are not straight lines, the system is said to be curvilinear. The latitude-longitude lines on a globe are an example of a curvilinear coordinate system. Sometimes, a 1:1 mathematical transformation is possible between a curvilinear system and the more familiar Cartesian coordinate system (*see* CARTESIAN COORDINATES). The illustration shows a transformation between curvilinear and Cartesian coordinates in a local case.

Curvilinear coordinates are not generally used in electricity and electronics except in some very advanced mathematical calculations. The exception to this rule is the polar coordinate system, used frequently in plotting antenna directional patterns and surface bearings. *See also* POLAR COORDINATES.

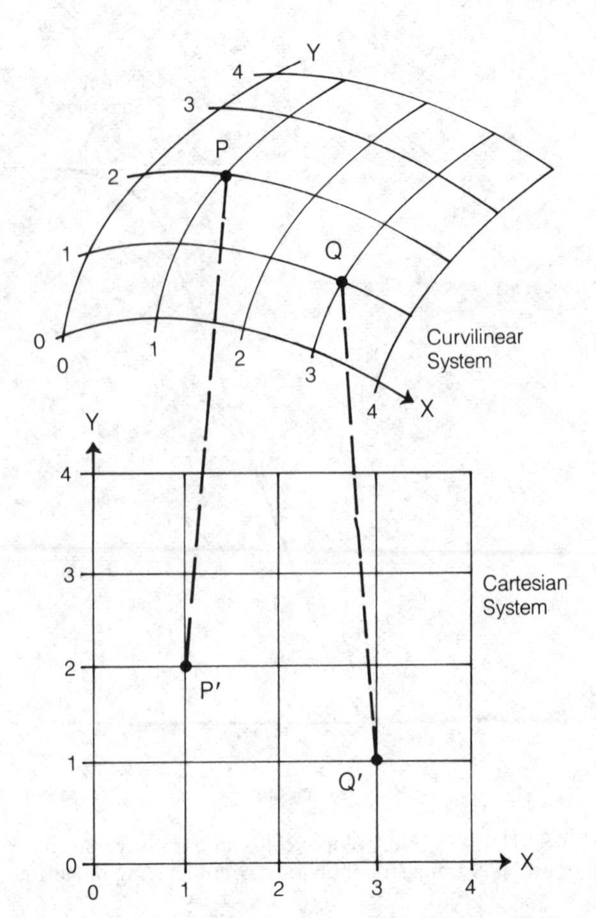

CURVILINEAR COORDINATES: The transformation of curvilinear coordinates to Cartesian coordinates.

CUT-IN/CUT-OUT ANGLE

A semiconductor diode requires between 0.3 volt and 0.6 volt of forward bias to conduct. In a rectifier circuit with semiconductor diodes, the conduction period is not quite one-half cycle. Instead, the conduction time is a little less than 180 degrees, as shown in the illustration.

The cut-in angle is the phase angle at which conduction begins. The cut-out angle is the phase angle at which conduction stops. In a 60-Hz rectifier circuit, if a phase angle of 0 degrees is represented by $t = 0$ second and a phase angle of 180 degrees is represented by $t = 1/120$ second or 8.33×10^{-3} second, then:

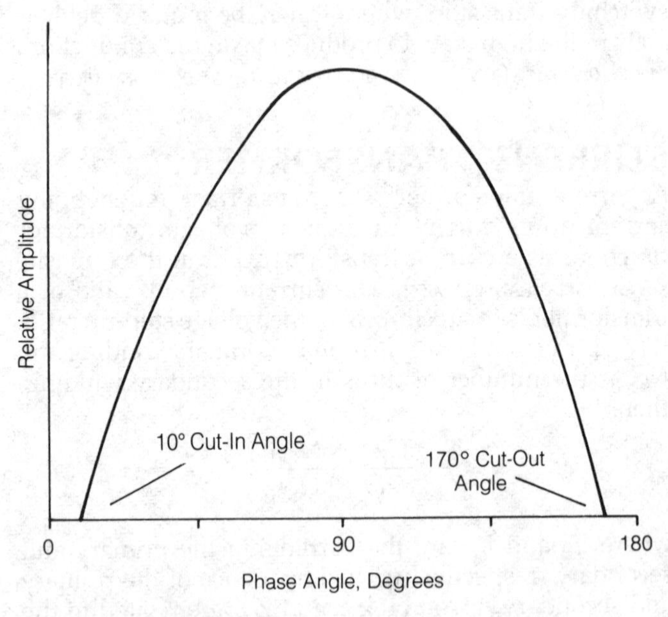

CUT-IN/CUT-OUT ANGLE: An illustration of cut-in and cut-out angles for a rectified wave.

$$\Theta_1 = 2.16 \times 10^4 \, t_1$$

and:

$$\Theta_2 = 2.16 \times 10^4 \, t_2$$

where Θ_1 and Θ_2 are the cut-in and cut-out angles, respectively, and t_1 and t_2 are the cut-in and cut-out times, respectively.

The cut-in and cut-out angles become closer to 0 and 180 degrees as the voltage of a sine-wave, alternating-current waveform increases. In a square-wave rectifier circuit, the angles Θ_1 and Θ_2 are essentially equal to 0 and 180 degrees. *See also* RECTIFICATION.

CUTOFF

Cutoff is a condition in a tube or bipolar transistor in which the grid or base voltage prevents current from flowing through the device. In a field-effect transistor, the condition of current cutoff is usually called pinchoff (*see* PINCHOFF).

In a vacuum tube, cutoff is achieved when the grid voltage is made sufficiently negative with respect to the cathode. In an NPN bipolar transistor, the base must generally be at either the same potential as the emitter, or more negative. In a PNP bipolar transistor, the base must usually be at either the same potential as the emitter, or more positive. In a field-effect transistor, pinchoff depends on the bias relationship among the source, gate, and drain.

The cutoff condition of an amplifier is often used to increase the efficiency when linearity is not important, or when waveform distortion is of no consequence. This is the case in the Class B and Class C amplifier circuits (*see* CLASS B AMPLIFIER, CLASS C AMPLIFIER). A cut-off tube or transistor may also be used as a rectifier or detector.

The term cutoff also refers to any point at which a certain parameter is exceeded in a circuit. For example, the cutoff frequency of a low-pass filter, or the alpha-cutoff frequency of a transistor.

CUTOFF ATTENUATOR

A waveguide has a minimum operating frequency below which it is not useful as a transmission line because it causes a large attenuation of a signal. The cutoff, or minimum usable, frequency of a waveguide depends on its cross-sectional dimensions (*see* WAVEGUIDE).

When a section of waveguide is deliberately inserted into a circuit and its cutoff frequency is higher than the operating frequency of the circuit, the waveguide becomes an attenuator. This device, used at very high and ultrahigh frequencies, is called a cutoff attenuator. The amount of attenuation depends on the difference between the operating frequency, f_o, and the cutoff frequency, f_c, of the waveguide. As f_c-f_o becomes larger, so does the attenuation. The amount of attenuation also depends on the length of the section of waveguide. The longer the lossy section of waveguide, the greater the attenuation. *See also* ATTENUATOR.

CUTOFF FREQUENCY

In any circuit or device, the term *cutoff frequency* can refer to either a maximum usable frequency or a minimum usable frequency.

In a transistor, the gain drops as the frequency is increased. The cutoff in this case is called the alpha-cutoff frequency or beta-cutoff frequency, depending on the application (*see* ALPHA-CUTOFF FREQUENCY, BETA-CUTOFF FREQUENCY).

As the frequency is raised in a lowpass filter, the frequency at which the voltage attenuation becomes 3 decibels, relative to the level in the operating range, is called the cutoff frequency. In a highpass filter, as the frequency is lowered, the frequency at which the voltage attenuation becomes 3 dB, relative to the level within the operating range, is called the cutoff frequency. This is illustrated in the graph. A bandpass or band-rejection filter has two cutoff frequencies. (*See* BANDPASS RESPONSE, BAND-REJECTION RESPONSE, HIGHPASS RESPONSE, LOWPASS RESPONSE.)

Many different circuits exhibit cutoff frequencies, either at a maximum, minimum, or both. A broadband antenna has a maximum and minimum cutoff frequency. All amplifiers have an upper cutoff frequency. Sections of coaxial or two-wire transmission line have an upper

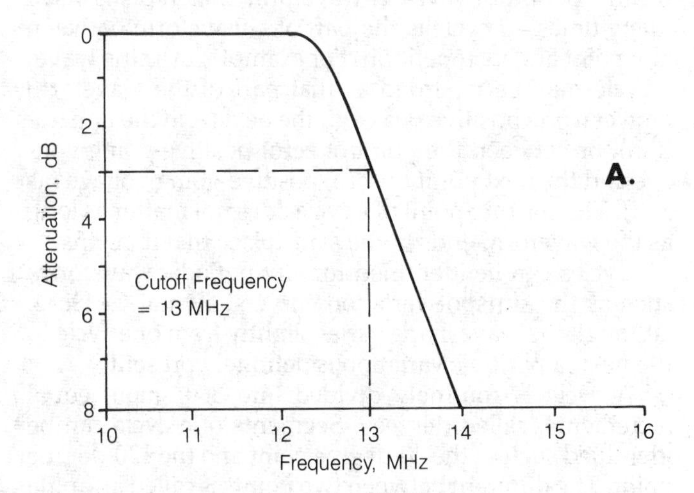

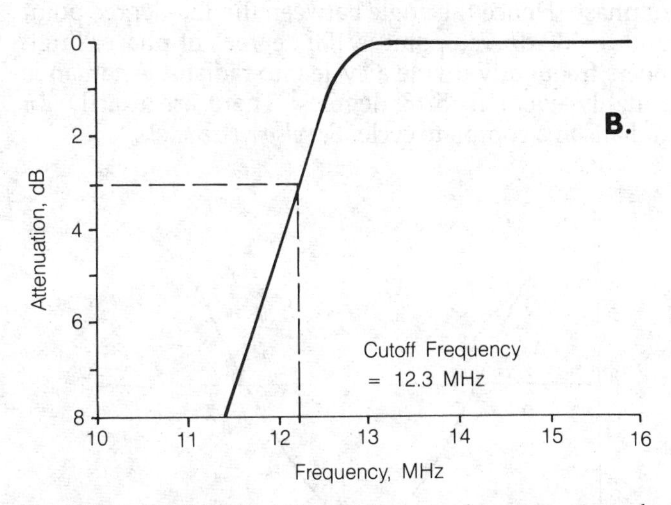

CUTOFF FREQUENCY: Attenuation-versus-frequency curves for a lowpass filter with a cutoff of 13.0 MHz (A) and a highpass filter with a cutoff of 12.3 MHz (B).

cutoff frequency; waveguides have a lower cutoff that is well defined. Usually, the specification for cutoff is 3 decibels, representing 70.7 percent of the current or voltage in the normal operating range. However, other attenuation levels are sometimes specified for special purposes. *See also* ATTENUATION.

CUTOFF VOLTAGE

The cutoff voltage of a vacuum tube or transistor is the level of grid, base, or gate voltage at which cutoff occurs. In a field-effect transistor, the cutoff voltage is usually called the pinchoff voltage.

A bipolar transistor is normally cut off. That is, when the base voltage is zero with respect to the emitter, the device does not conduct. Until approximately 0.3 to 0.6 V of base voltage is applied in the forward direction, the transistor remains cut-off. Above +0.3 to +0.6 V, an NPN transistor will begin to conduct; below −0.3 to −0.6 V a PNP transistor begins to conduct. This is based on the assumption that the collector is properly biased—positive for an NPN device and negative for a PNP device.

CYCLE

In any periodic wave—a waveform that repeats itself many times—a cycle is the part of a waveform between any point and its repetition. For example, in a sine wave, a cycle may be regarded as that part of the waveform between one positive peak and the next (A in the illustration), or between the point of zero, positive-going voltage and the next point of zero, positive-going voltage (as at B). The starting point of a cycle does not matter as long as the waveform ends at the same place that it begins.

Cycles can be identified for any periodic waveform, such as the sunspot-variation curve shown at C. Here, although the waveshape varies slightly from one cycle to the next, a periodic variation is definitely present.

A cycle is routinely divided into 360 small, equal increments, called degrees. Segments of a cycle can be identified such as the 30-degree point and the 130-degree point. The different between two points is called an angle of phase. Hence the angle between the 130-degree point and the 30-degree point is 100 degrees of phase. Engineers frequently divide a cycle into radians. A radian is roughly equal to 57.3 degrees. There are exactly 2π radians in a complete cycle. *See also* PHASE ANGLE.

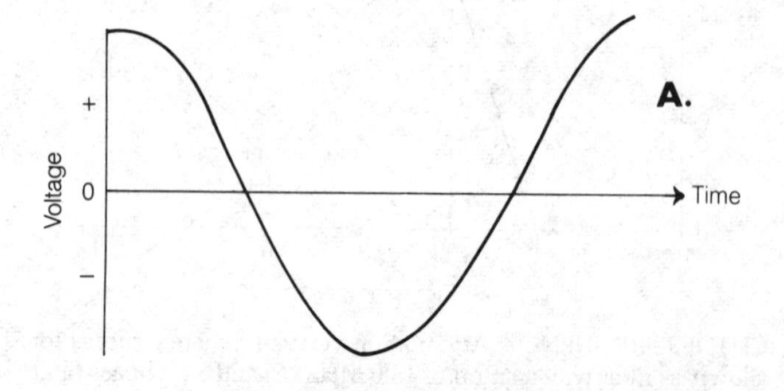

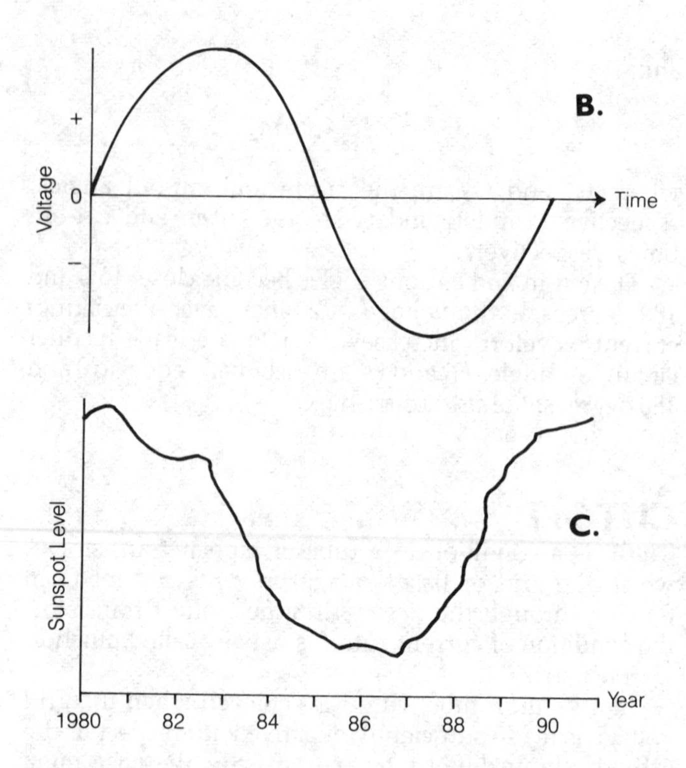

CYCLE: Plots of waveforms defining the cycle are a sine wave between positive peaks (A), a sine wave between two zero points (B), and a cycle of sunspot activity between positive peaks (C).

CYCLES PER SECOND

The term *cycles per second* is an obsolete expression for the frequency of a periodic wave. The commonly accepted electronic or electrical term today is hertz (*see* HERTZ). In older text and reference books, the frequency is expressed in cycles, kilocycles, or megacycles per second, abbreviated cps, kc, and Mc, respectively.

CYCLIC IONOSPHERIC VARIATION

The density of ionization in the upper atmosphere varies periodically with the time of day, the time of year, and the level of sunspot activity. Variations that occur on a regular basis are called cyclic ionospheric variations. These variations affect the properties of radiocommunication on the medium and high frequencies.

Generally, the density of ions is greater during the daylight hours than during the night because the ultraviolet radiation from the sun causes the atoms in the upper atmosphere to ionize, at heights ranging from about 40 to 250 miles. This produces a daily cycle, which reaches its peak sometime after midday and reaches its minimum shortly before sunrise.

During the summer months, the level of ionization is usually greater, on the average, than during the rest of the year. During the winter months, the level of ionization is the least (when it is winter in the northern hemisphere, it is summer in the southern hemisphere and vice versa). The sun remains above the horizon for the longest time in the summer, allowing more atoms to become ionized. Also, the ultraviolet radiation is somewhat more intense in the summer, especially at lower levels in the atmosphere. The daily cycle is impressed upon this annual cycle.

The level of sunspot activity varies over a period of about 11 years. The years of maxima for this era are 1958, 1969, 1980, 1991, and 2002 A.D. The years of minima are 1964, 1975, 1986, 1997, and 2008 A.D. Ionospheric density is, on the average, greatest during the sunspot maxima and least during the minima. The annual and daily cyclic variations are impressed on the 11-year cycle. *See also* PROPAGATION CHARACTERISTICS, SOLAR FLUX, SUNSPOT CYCLE.

CYCLIC SHIFT

A cyclic shift is a transfer of information in a storage register in one direction or the other, usually called the right or the left. In a cyclic shift toward the left, each digit or bit is moved one place toward the left, except for the extreme left-hand digit or bit, which replaces the one originally at the far right (A in the illustration). In a cyclic shift toward the right, each digit or bit is moved one position to the right, except for the extreme right-hand digit or bit, which replaces the one on the far left, as at B.

In an n-bit register, a succession of n cyclic shifts in the same direction results in the original information. Also, if m cyclic shifts are performed in one direction, followed by or combined with m cyclic shifts in the opposite direction, where m is any positive integer, the initial storage is obtained. *See also* SHIFT REGISTER.

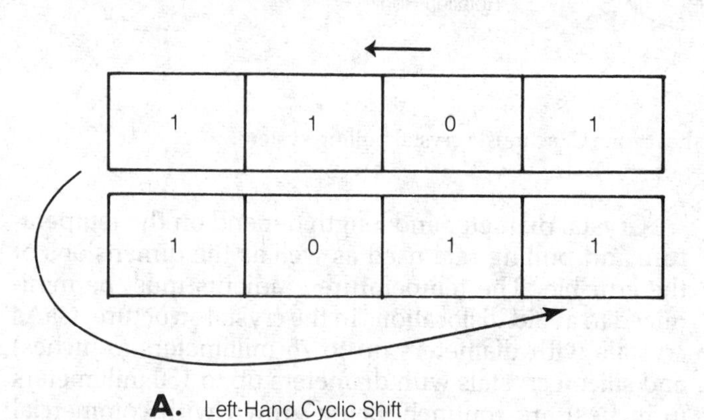

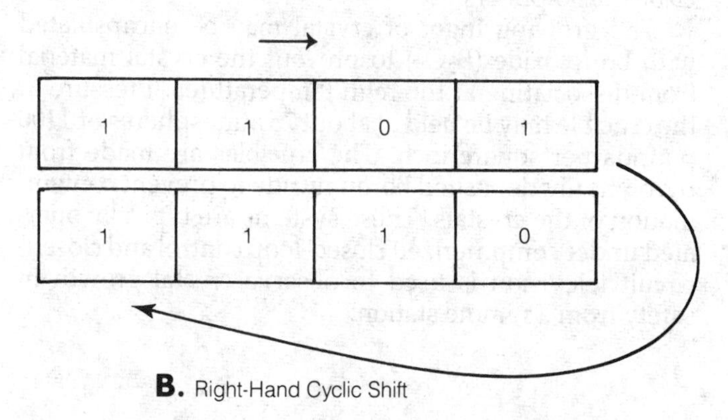

CYCLIC SHIFT: Diagrams of cyclic shifts in the binary number 1101—one unit to the left (A) and one unit to the right (B).

CYCLOTRON

A cyclotron is a particle accelerator used for atom smashing. This process can change one element to another. A unipolar magnetic field forces charged atomic particles, such as electrons, ions, protons, or nuclei of heavier atoms, into spiral orbits within two metal chambers. The drawing shows a simplified diagram of a cyclotron. Because of their shape, resembling the capital letter *D*, the chambers are called dees.

The charged particles are injected into the machine at the center. As the particles move outward in a spiral path that increases in radius, their angular frequency, or time to complete one revolution, remains constant or decreases only slightly. Therefore, their speed greatly increases. The beam of particles is ejected from the cyclotron at extreme speed, often a large fraction of the speed of light, so that relativisitic effects occur. These effects include changes in mass and spatial dimensions.

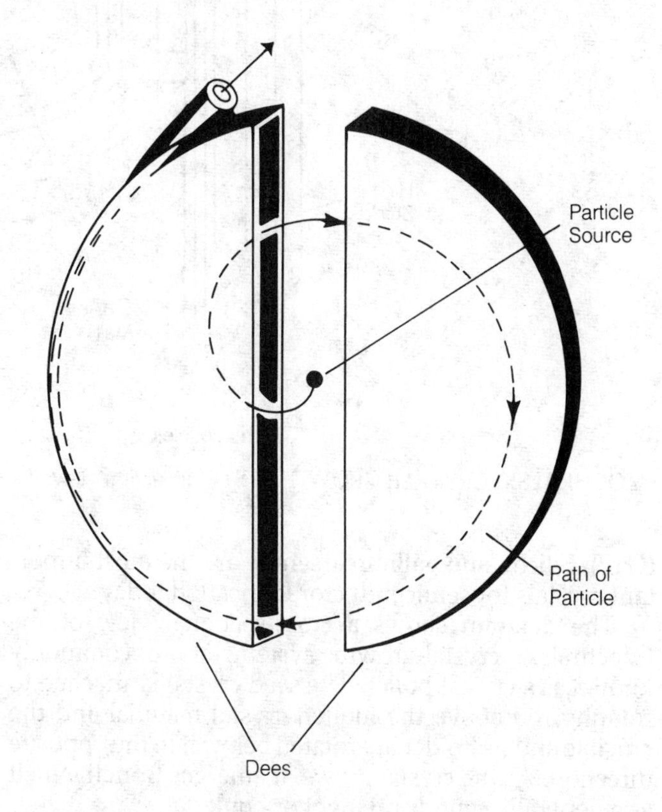

CYCLOTRON: Charged particles are forced into spiral paths and accelerated by the cyclotron.

CZOCHRALSKI CRYSTAL-GROWTH SYSTEM

The Czochralski crystal-growth method, also known as the high-pressure, liquid-encapsulated Czochralski (LEC) method, is used for growing large, single crystals. A seed crystal is immersed in molten crystal material, and the crystal is grown by a drawing or pulling action. This system can be used to grow Group IV materials such as silicon or synthetc rubies and Group III-V materials such as gallium arsenide (GaAs) and gallium phosphide

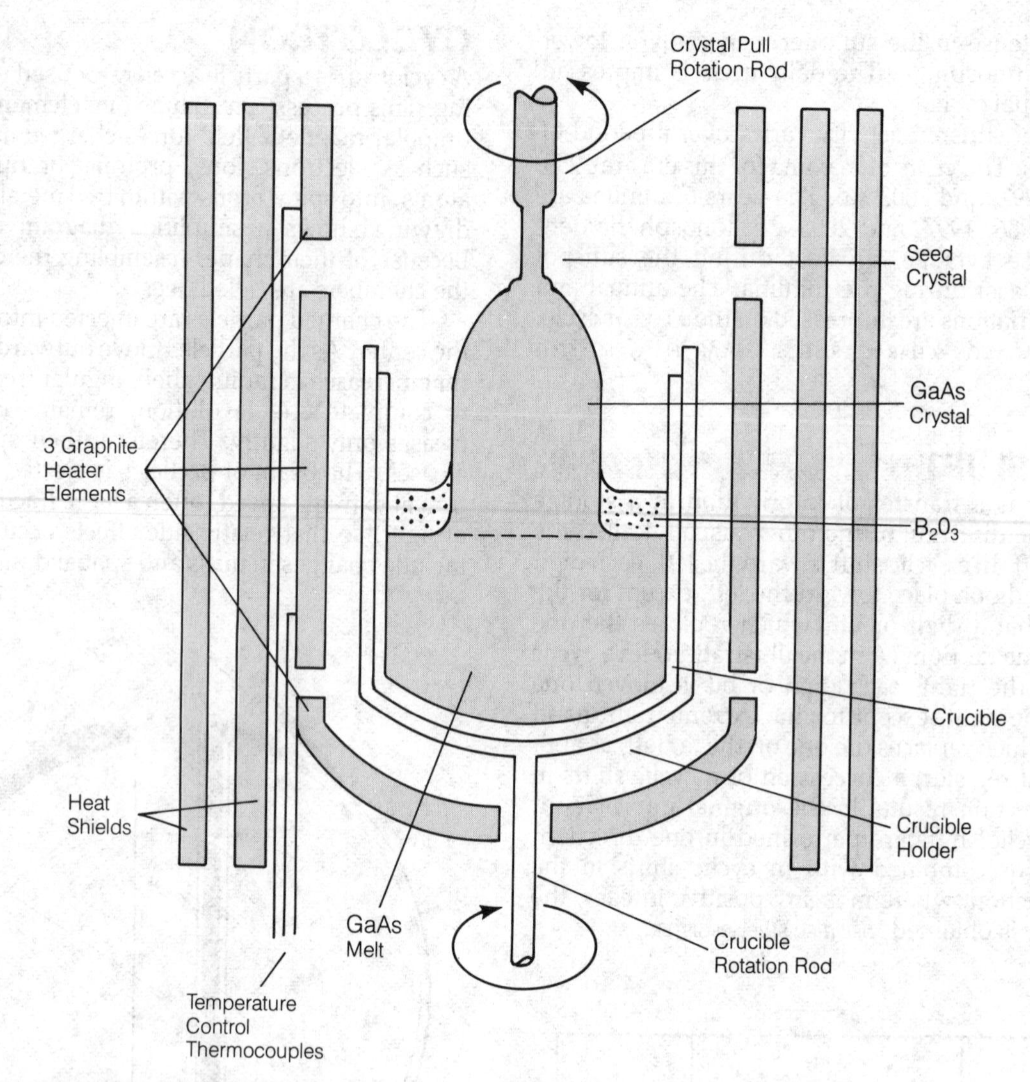

CZOCHRALSKI CRYSTAL-GROWTH SYSTEM: Section view of a three-heat zone Czochralski crystal pulling system.

(GaP). Silicon and gallium arsenide are the most important crystals for semiconductor fabrication today.

The diagram shows a cross-sectional view of the Czochralski crystal-growth system, more commonly known as a crystal puller. The seed crystal is attached to rotating rod above the molten crystal material and the crucible and its holder are rotated below it in the opposite direction. As the crystal grows, it emerges from the melt as a generally cylindrical ingot or boule.

The Czochralski method is performed under high pressure so that the crucible, melt, and seed crystal are all contained within a high-pressure chamber. Heating elements melt the raw materials prior to crystals growth. Rotation reduces radial temperature gradients, and slow withdrawal of the rotating seed results in uniform crystal growth. The conditions for optimum growth vary widely, but pulling rates can be as rapid as a few inches an hour.

Crystal diameter and length depend on the temperature and pulling rate used as well as the dimensions of the crucible. The temperature gradients must be minimized to avoid dislocations in the crystal structure. GaAs crystals with diameters up to 75 millimeters (3 inches) and silicon crystals with diameters up to 150 millimeters (6 inches) are routinely being grown with commercial Czochralski pullers.

The growing ingot of crystal may be encapsulated with boric oxide (B_2O_3) to prevent the crystal material from dissociating at the high temperatures. Pressure in the crucible may be held at about 75 atmospheres or 1100 pounds per square inch. The crucibles are made from pyrolytically deposited boron nitride to prevent contamination of the crystals. These systems are typically operated under computerized closed-loop control and closed-circuit television is used to observe crystal growth in safety from a remote station.

DAISY-WHEEL PRINTER

See PRINTER.

DAMPED WAVE

A damped wave is an oscillation whose amplitude decays with time, as shown in the figure. The damping might take place rapidly, in a few microseconds, to the point where the wave amplitude is essentially zero; it might also occur slowly, over a period of milliseconds or even several seconds. Damping can take place within the time of one cycle or less, or it can occur over a period of millions of cycles. Generally, the higher the Q factor in a circuit (*see* Q FACTOR), the more cycles occur before the signal amplitude decays essentially to zero. Decay in a damped wave occurs as a logarithmic function, called a logarithmic decrement. *See also* CONTINUOUS WAVE, LOGARITHM.

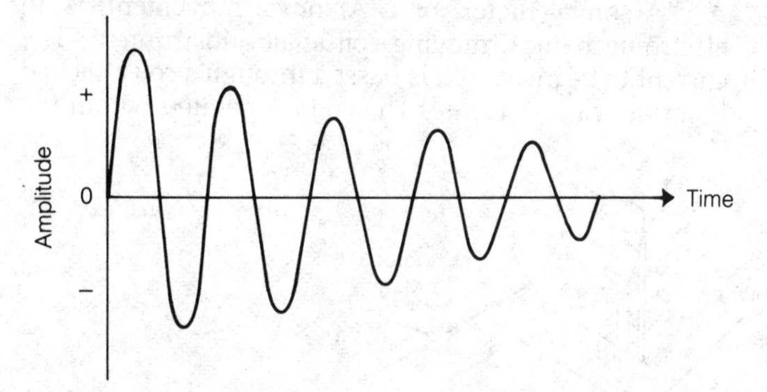

DAMPED WAVE: The form of a damped wave with decreasing amplitude.

DAMPING

Damping is the reduction of oscillatory energy in a mechanical or electrical system by absorption or radiation. Damping may be done to improve stability. There are three classifications of damping:

1. Critically damped—no overshoot or undershoot (optimum response).
2. Underdamped—overshoot occurs but oscillation is excessive.
3. Overdamped—no overshoot or undershoot and the response is too slow.

Damping can be applied to any closed-loop mechanical system as friction or braking.

Any absorption in a circuit, tending to reduce the amount of stored energy, is called damping. A resistor placed in a tuned circuit to lower the Q factor, for example, enhances damping. This tends to reduce the chances of oscillation in a high-gain, tuned amplifier circuit. Mechanical resistance can be built into a transducer, such as an earphone or microphone, to limit the frequency response. This also is called damping. *See also* Q FACTOR.

DAMPING FACTOR

The damping factor is the numerical quantity indicating the ability of an amplifier to operate a speaker satisfactorily. Values over 4 are considered to be satisfactory. In a high-fidelity sound system, the actual output impedance of the amplifier may be much smaller than the impedance of the speaker. The ratio of the speaker impedance, which is usually 4, 8, or 15 Ω, to the amplifier output impedance, which is often less than 1 Ω, is called the damping factor.

The effect of this difference in impedance is to minimize the effects of speaker acoustic resonances. The sound output should not be affected by such resonances in a high-fidelity system. The frequency response should be as flat as is practicable. The damping factor in a high-fidelity system is somewhat dependent on the frequency of the audio energy. It also is a function of the extent of the negative feedback in the audio amplifier circuit. Damping factors in excess of 60 are quite common.

In a damped oscillation, the quotient of the logarithmic decrement and the oscillation period is sometimes called the damping factor. In a damped-wave circuit, where the coil inductance is given by L and the radio-frequency resistance is given by R, the damping factor, a, is defined as:

$$a = \frac{R}{2L}$$

See also DAMPED WAVE, LOGARITHM.

DAMPING RESISTANCE

If the Q factor in a resonant circuit becomes too great (*see* Q FACTOR), an undesirable effect called ringing can occur. Ringing is especially objectionable in audio filters used in radioteletype demodulators and in code communications. To reduce the Q factor sufficiently, a damping resistor is placed across a parallel-resonant circuit. This resistor may also be placed in series with a series-resonant circuit. In a parallel-resonant circuit, the Q factor decreases as the shunt resistance decreases. In a series-resonant circuit, the Q factor decreases as the series resistance increases. The lower the Q factor, the less the tendency for the resonant circuit to ring.

Damping resistance is used to refer to a noninductive resistor placed across an analog meter to increase the damping. *See also* CRITICAL DAMPING, DAMPING.

DARAF

The daraf, the unit of elastance, is the reciprocal unit of the farad. The word *daraf* is *farad* spelled backwards. A value of 1 daraf is the reciprocal equivalent of 1 farad. The quantity 1/C, where C is capacitance, is called elastance (*see* ELASTANCE).

DARLINGTON AMPLIFIER

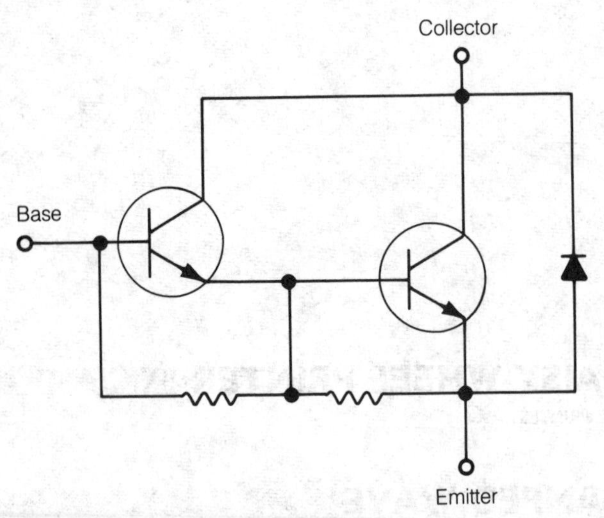

DARLINGTON AMPLIFIER: Two bipolar transistors with emitter-to-base connection form a Darlington amplifier.

A Darlington amplifier, or Darlington pair, is a compound connection between two transistors or tubes (see illustration). In the Darlington amplifier shown, the collectors of the transistors are connected. The input is supplied to the base of the first transistor. The emitter of the first transistor is connected directly to the base of the second transistor. The emitter of the second transistor serves as the emitter for the pair. The output is generally taken from both collectors.

The amplification of a Darlington pair is equal to the product of the amplification factors of the individual transistors as connected in the system. This does not necessarily mean that a Darlington amplifier will produce far more gain than a single bipolar transistor in the same circuit. The impedances must be properly matched at the input and output to ensure optimum gain. Some Darlington pairs are available in a single case. These devices are called Darlington transistors. The Darlington amplifier is sometimes called a double emitter follower or a beta multiplier. *See also* TRANSISTOR.

D'ARSONVAL MOVEMENT

A D'Arsonval meter, or D'Arsonval movement, is an electromechanical moving-coil analog instrument. The current to be measured is passed through a coil attached to an indicating needle. The coil is operated within the

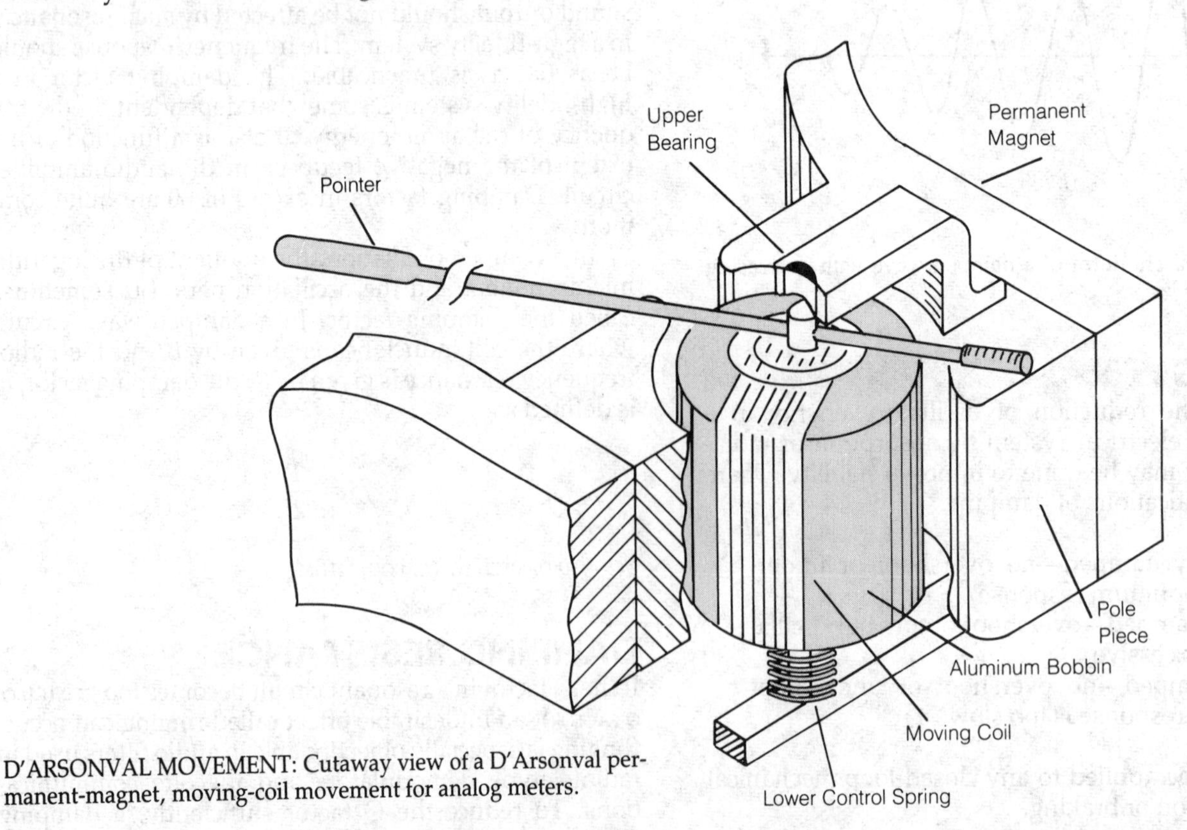

D'ARSONVAL MOVEMENT: Cutaway view of a D'Arsonval permanent-magnet, moving-coil movement for analog meters.

field of a permanent magnet. As a current passes through the coil, a magnetic field is set up around the coil. In accordance with the motor rule, torque appears between the permanent magnet field and the field of the coil. A spring allows the coil, and hence the pointer, to rotate over a restricted angular distance; the greater the current, the stronger the torque, and the farther the coil turns. A D'Arsonval movement is shown in the figure.

Generally, the coil in a D'Arsonval meter is mounted on jeweled bearings for maximum accuracy. D'Arsonval meters are widely employed as analog ammeters, milliammeters, and microammeters. With suitable peripheral circuitry, D'Arsonval meters are used as analog voltmeters and wattmeters as well, both for direct current and alternating current at all frequencies. *See also* AMMETER, ANALOG PANEL METER, MOTOR.

DATA

The term *data* is generally used in place of the word *information* in electronic and computer applications. The information stored and handled by a computer is called data.

Any alpha-numeric sequence representing some identifiable quantity is regarded as data. This can be, for example, the notation showing that the high temperature on August 5 was 85°F.

DATA ACQUISITION SYSTEM

A data-acquisition system is a system for monitoring, controlling, and logging or recording data from transducers or sensors. This type of system includes, at a minimum, one or more sensors or transducers, circuitry for interfacing that data with data conversion circuitry, and a means for displaying and/or recording the data. The simplest data-acquisition system is a single-channel digital panel meter (DPM) based system with a built-in circuit card for interfacing a specific sensor to the analog-to-digital (A/D) conversion circuitry of the DPM for display. The DPM might be dedicated to measuring a specific variable or it might be designed to accept a range of adaptor cards. Each card adapts to a different type of sensor for measuring temperature, force, pressure, flow, or level. The DPM might also contain circuitry for converting binary coded decimal (BCD) data into ASCII code

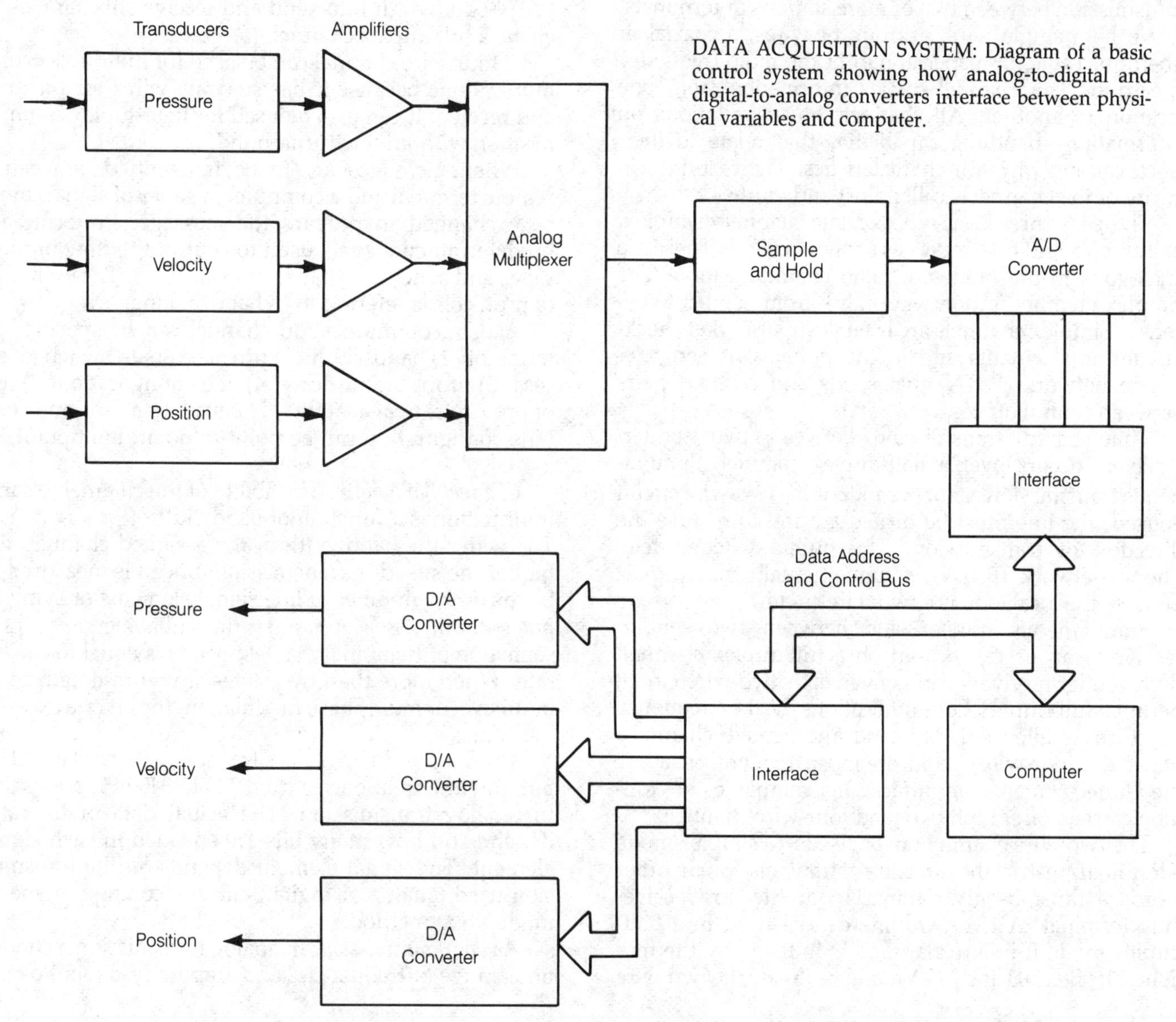

DATA ACQUISITION SYSTEM: Diagram of a basic control system showing how analog-to-digital and digital-to-analog converters interface between physical variables and computer.

for transmission to a computer-based data logger.

A more complex multisensor data-acquisition system is illustrated in the figure. Data from sensors for pressure, velocity, and temperature are multiplexed together, sampled, and converted for display and logging by a computer. The data can be transmitted to the computer over a transmission line in either analog or digital format, as appropriate to system requirements. There can be digital readout at the site of the sensor, at a central control station, or both.

Among the temperature sensors or transducers commonly used in data acquisition systems are thermoswitches, thermocouples, resistance temperature detectors (RTD), and thermistors. The force and pressure transducers include strain gauges and piezoelectric devices. Flow meters are commonly used to measure liquid or gas flow. *See* ANALOG-TO-DIGITAL CONVERTER, DIGITAL-TO-ANALOG CONVERTER, SAMPLE-HOLD AMPLIFIER.

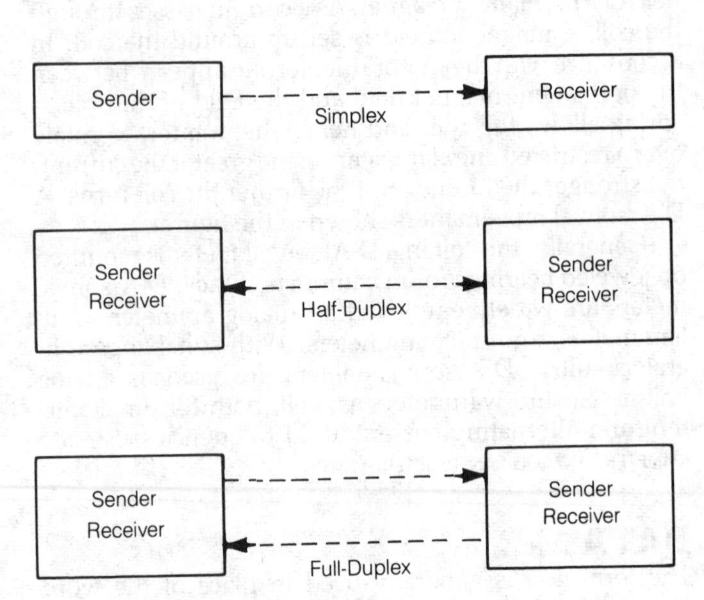

DATA CHANNEL: Fig. 1. Basic transmission methods.

DATA CHANNEL

A data channel, or communications link, is a path for transmission between two or more stations or terminals. It can be a single wire, a group of wires, a coaxial or fiberoptic cable, or a special part of the radio frequency spectrum. The channel carries information from one location to another. All channels have limitations on information—handling capabilities that relate to their electrical and physical characteristics. The carrying capacity of the channel is called the bandwidth.

There are three basic types of data channels: simplex, half-duplex and full-duplex as shown in Fig. 1. These are analogous to the modes of radio communication. The simplex channel is one way only—from sender to receiver. Simplex channels are used in supermarket checkout terminal circuits and in the public-switched telephone network (PSTN) that sends and receives pairs between central offices.

Time-shared transmission between two sender/receivers occurs over a half-duplex channel. Simultaneous transmission is not permitted. If a two-wire circuit is used, the line must be turned around to reverse the direction of transmission. The public-switched telephone network (PSTN) is fundamentally half-duplex because the local loop is a two-wire circuit.

Simultaneous transmission between two sender/receivers can be carried out on a full-duplex channel. Both sender/receivers can converse. A two-wire circuit permits full-duplex communications if the frequency spectrum is allocated into send and receive channels. However, four-wire circuits are most frequently used. In the United States, communications companies or common carriers offer both two- and four-wire channels.

The two-wire channel can be used in a simplex mode if terminals restrict the direction of transmission. In other words, a transmit-only terminal is connected to a receive-only terminal. A two-wire line can be used in a half-duplex mode if line turnaround is initiated by the modem. The Bell 103 and 212A modems divide the two-wire PSTN bandwidth into send and receive subchannels to create a full duplex channel. (*See* MODEM.)

A four-wire channel can be used for full-duplex communications because it has separate wires for transmit and receive. It can also be used for half-duplex communication without line turnaround.

When messages are to be transmitted between a remote terminal and a computer, a series of signals must be exchanged to prepare the message. Protocols are predetermined signals used to control the flow of messages and synchronize their transmission. The exchange of protocols is referred to as handshaking.

A data-communication channel can be specified in terms of: 1) bandwidth, 2) private versus switched access, 3) propagation delay, 4) line configuration, 5) use of protocols, 6) availability, 7) installation time and cost. Line configuration can be point-to-point, multipoint, or loop.

Channel Bandwidth. The ability of the channel to carry information is a function of bandwidth. (*See* BANDWIDTH.) The wider the bandwidth of the assigned channel, the higher the speed of transmission. Speed is measured in bauds or the number of line signal elements or symbols per second. (*See* BAUD RATE.) If the signal element represents one of two binary states, bauds is equal to the bit rate. When more than two states are represented, as in multilevel or multiphase modulation, the bit rate exceeds the bauds.

The bandwidth of a voice frequency channel is 5 kHz, but only the frequencies from 300 to 3400 Hz are usable for analog transmission. The actual data or bit rater depends on how many bits are encoded in each signal element. The signal element depends on the transmission used (analog or digital) and the coding scheme or modulation method.

Analog Transmission. In analog transmission a continuous range of frequencies and amplitudes is sent over a communications channel that includes both voice and data. Linear amplifiers, attenuators, filters and transformers are needed to maintain the signal quality. (*See* AMPLIFIER, ATTENUATOR, FILTER, TRANSFORMER.) However, amplifiers increase the noise as well as the information content of a signal so error rates are higher in analog transmission than digital transmission.

The capacitance, inductance, and resistance characteristics of the telephone line delay and attenuate signals with different frequencies. These alterations of the information content are called envelope delay and attenuation distortion. In some cases received signal strength might be unacceptable.

Analog transmission systems require linear amplifiers and filters at fixed distances to boost the signal and filter out noise. The spacing of this equipment depends on the end-to-end media. For example, amplifiers are spaced about every 6000 feet with telephone twisted-wire pairs.

Modems in the existing analog telephone network to convert digital signals to an analog format for transmission. They also reconvert those signals to a digital format at the receiver.

Digital Transmission. In digital transmission, pulses are sent over a communications channel at fixed data rates that are dependent on the digital carrier system. Digitized data, voice and compressed freeze-frame video can be transmitted in a digital form if sufficient bandwidth is available. Among the digital data networks available in the U.S. are AT&T's Dataphone Digital Service (DDS). There are also public digital data networks in Germany, Japan, Denmark, Finland, Norway, and Sweden.

Digital transmission systems use regenerative repeat-

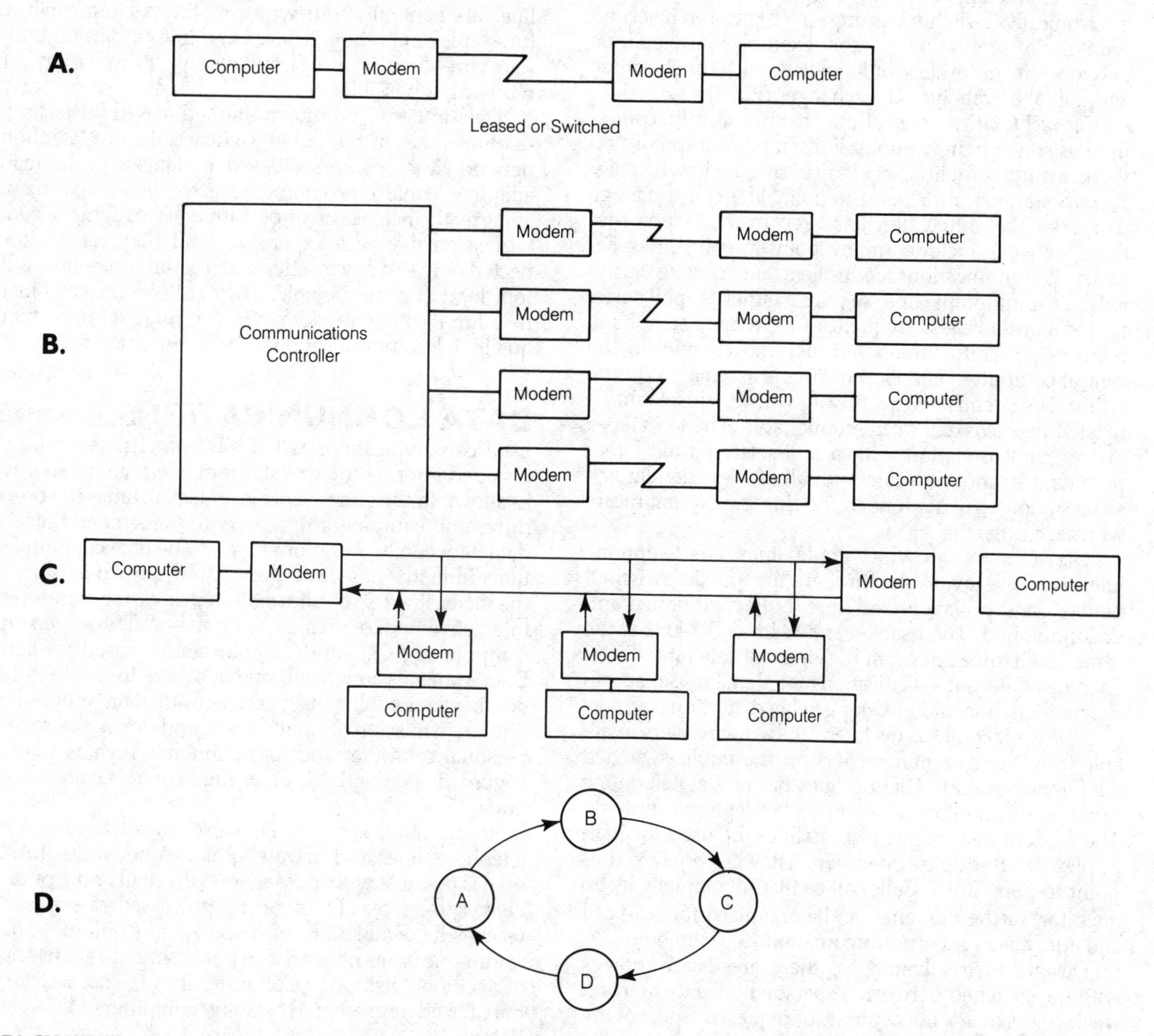

DATA CHANNEL: Fig. 2. Single point-to-point line (A), point-to-point network (B), multipoint line (C), and ring or loop topology (D).

ers to retime and reshape the digital pulses. These digital repeaters recreate the original waveform more reliably than the linear amplifiers in the analog systems and, as a result, there are fewer transmission errors per message. Error rates in digital transmission are typically 1 or 2 percent of those in analog transmission.

Different types of digitized information such as speech, computer data, word processing data, and facsimile can be mixed in the same digital system. In addition, digital transmission can be more effectively encrypted for security and it can be compressed to save bandwidth.

Digital systems can use time-division multiplexing rather than the frequency-division multiplexing used in analog systems. This can increase the information capacity of the transmission system. These systems are also better adapted for use with optical fiber transmission and satellite links.

The characteristics of digital signal are convenient for electronically switching groups of digits for insertion onto one of several data paths. Examples in telephone networks include electronic switches (channel banks) and digital private branch exchanges (PBXs).

Channel Configurations. Data channels can be configured as single point-to-point links, a point-to-point network, a multipoint line, or a loop or ring as shown in Fig. 2. Two stations on a point-to-point line can exchange data after the connection has been made. A point-to-point network includes many point-to-point links between communications controllers and remote terminals. A multipoint line requires either a poll-select protocol or dedicated frequencies for remote stations to regulate access to the shared inbound channel to the central computer. The loop or ring is assembled on the customer's premises with private wiring. This could be twisted pair, coaxial, or fiberoptic cable. The loop has a master control station with a poll-select protocol that permits it to communicate with all of the secondary or slave stations. All stations in the ring can communicate with each other.

Voice-Grade Lines. Voice-grade lines are telephone lines that are available through the public switched network and private leased lines, both conditioned and unconditioned. The usable bandwidth of 3.1 kHz is the same for all three lines, but the effective data rates differ. Each has different specifications for signal noise, amplitude attenuation and envelope delay distortion.

Dial-Up Lines. Dial-up lines are two-wire pairs available from the common carriers on the public-switched telephone network. Dial-up lines permit one telephone to reach any point on the world-wide telephone network. The modem determines if these lines are used for half-duplex or full-duplex operation. They are organized as point-to-point links. Calls can be placed manually by an operator, or the modems can be organized for auto call and auto answer, permitting unattended operation.

Despite their advantages, there are disadvantages with the switched network. A person is able to interpret weak voices over a noisy phone or request a repeat of the message or an alternate line if information is lost or not understood. However, computers or terminals without error-correcting features cannot make these decisions on the quality of digital information. Thus, they can lose or misinterpret data because of noise.

A second disadvantage relates to the delay distortion caused by the different speeds of transmission of the various frequency components of a signal being transmitted. This too can result in erroneous received data.

The time lost in the switched network due to connect, disconnect, and turnaround that limit the amount of data transmitted is a third disadvantage. A fourth disadvantage is the low reliability of telephone switching equipment.

Private leased lines avoid the problems associated with the switched dial-up network, but they are more costly. The private leased line provides ready availability and freedom from busy signals. It also provides point-to-point or multipoint operation and is conditioned for better data quality and higher transmission rates. Leased lines are generally four-wire circuits usable for half- or full-duplex operation. There is no line turnaround on a four-wire circuit, and simultaneous transmission and receiving is possible.

Data integrity on unconditioned leased lines can be ten times that of the public-switched dial-up telephone network. Microprocessor-based modems with automatic adaptive equalizers compensate for line impairments and greatly increase error performance at higher speeds.

Leased lines are expensive, and they can be connected only to a few locations. On a multipoint line, only four locations are possible. They are economically justified for users with demands for high-volume, high-quality telecommunications. *See* DIGITAL-DATA SERVICE.

DATA COMMUNICATIONS

Data communications is the electronic transmission of encoded information or data from one point to another, as shown in the diagram. The subject includes the procedures and equipment necessary to transmit and receive data between two or more points. The data communications industry has undergone rapid growth because of the increasing use of information processing computers, interactive video display terminals (VDTs), personal computers (PCs), and remote sensor-based systems. Data transmission from one location to another has become essentially inseparable from data processing. The transmission of data back and forth between a personal computer and its peripherals such as printer, keyboard and display is an example of data communications.

Information is now being exchanged faster by the telephone lines and private local area networks (LANs) than is possible with messengers, the mail or the priority delivery services. Data can be processed as soon as it reaches its destination for display and printout. A data communications network can link an organization locally or over long distances. Computer files can be electrically transferred and shared among computers, PCs, and VDTs.

The equipment used in data communications includes:

1. Intelligent VDTs able to provide a man-machine interface.
2. Cluster controllers that interface terminals or peripherals to communications lines.
3. Time-division multiplexers that combine two or more slow-speed lines into one high-speed line by means of time sharing.
4. Remote data concentrators that buffer messages from slow-speed lines and multiplex them over high-speed lines.
5. Modems that convert transmitted digital information into a modulated carrier and reconvert the carrier at the receiver.
6. Front-end processors that control communications lines for a host computer.
7. Communication test equipment that diagnoses failures in communication lines, modems, or terminals.
8. Voice/data private branch exchanges.
9. Data PBXs and port-contention units.

The five elements of the data communications process are:

1. Transmitter or source of information.
2. Message.
3. Binary serial interface.
4. Communication channel or link.
5. Receiver of transmitted information.

A data communications interface is often needed to make the binary serial data compatible with the communications channel. *See* DATA CHANNEL, DATA COMMUNICATIONS LINE SHARING EQUIPMENT, DATA SERIAL TRANSMISSION, DIGITAL SERIAL INTERFACE, LOCAL AREA NETWORK, MODEM.

DATA COMMUNICATIONS LINE SHARING EQUIPMENT

Data communications lines are shared to reduce costs, improve reliability, and simplify maintenance. Communications lines and modems can be shared with equipment called multiplexers, cluster controllers, remote data concentrators, modem-sharing units, port-sharing units, port selectors, and lineplexers.

A multiplexer or MUX is a circuit that can transmit two or more messages on a single communications channel by dividing the bandwidth into frequency or time slots. Typically, there is one MUX connected to the serial ports of a central computer and another MUX at the remote site connected to each of the communicating devices. The MUX should have no effect on data sent between the computer and the remote stations and it should have minimal effect on response time.

The basic types of multiplexers are:

Frequency-Division Multiplexer (FDM). The FDM divides the available transmission frequency bandwidth into smaller-segment narrower bands, each of which is used as a separate channel. FDM is used for lower-speed synchronous, full-duplex, leased-line transmission.

Time-Division Multiplexer (TDM). The TDM connects terminals one at a time, at regular intervals, to the entire communications bandwidth. TDM is usually used on T-1 carrier facilities at 1.544 megabits per second (*see* T-1 CARRIER) although it can be used for synchronous or

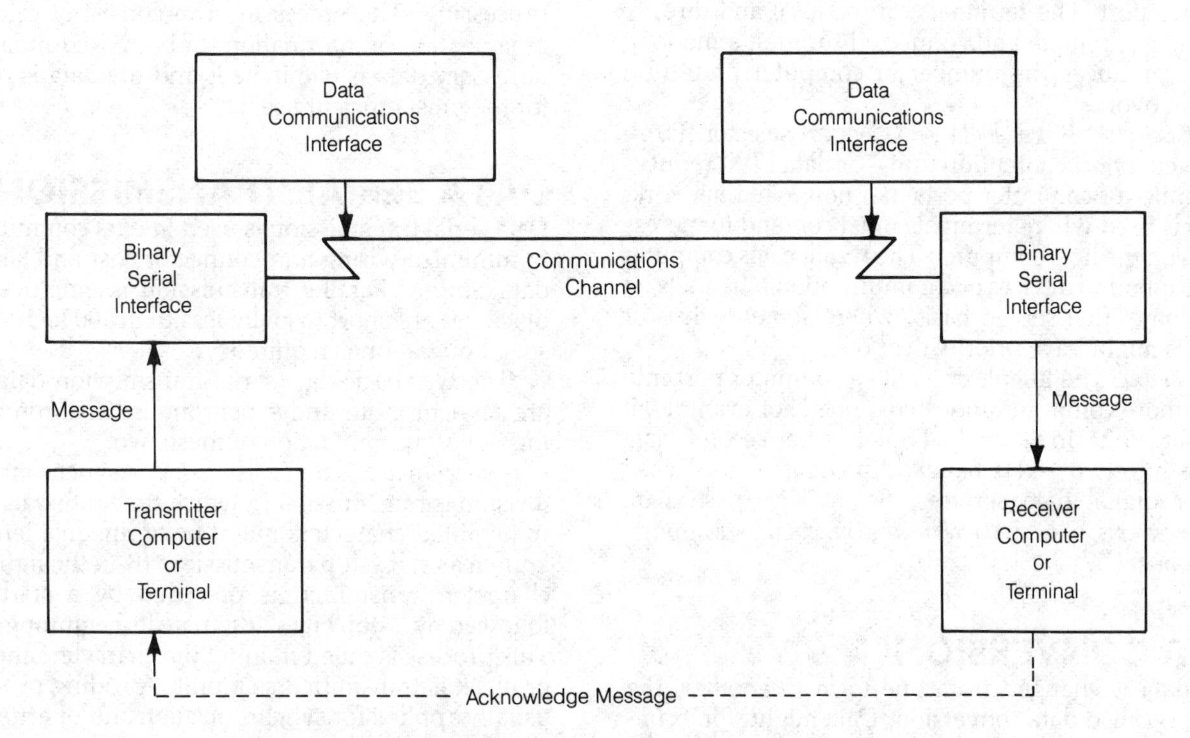

DATA COMMUNICATIONS: The data communications process.

asynchronous full-duplex leased line transmission.

Statistical multiplexers dynamically allocate time slots to attached devices based on activity. By making use of idle time, more data streams can share a common communications line. Typically, four to eight terminals or computer ports are connected to a statistical multiplexer, which uses synchronous full-duplex leased line transmission.

Cluster Controller. A cluster controller manages and directs messages to and from the connected remote devices when it receives poll and select commands from the central computer or front-end processor. The system devices include video display terminals (VDT), personal computers (PC), and printers. They are polled independently by the cluster controller for data to be sent to the central computer.

Remote Data Concentrator (RDC). A remote data concentrator is a communications processor like a front-end processor (FEP) that is positioned at a remote site. The RDC functions are similar to those of the statistical multiplexer, but it is not transparent to the data flow. Multiplexers can be used in any computer system, but RDCs are made by the computer manufacturers to be compatible with their product lines.

Modem-Sharing Units (MSU). A modem-sharing unit (MSU) is device that permits two to six synchronous terminals to share a synchronous modem. The terminals must be polled by the central computer so that the polled terminal recognizes its address and activates its request to send data. The MSU releases the modem to the first terminal that initiates a request to send signal; all other terminals are locked out.

Port-Sharing Unit (PSU). A port-sharing unit (PSU) is similar to an MSU, except that it has its own timing source for transmit and receive clocks. A PSU connects from two to six polled synchronous terminals to a single computer port. The terminals can be local and directly connected or remote and connected through a modem. The PSU reduces the number of computer ports in a polled network.

Port Selector (PS) or Data PBX. A port selector (PS)—also called a port contention unit—or data PBX, reduces the number of computer ports in a nonpolled network. The PS is used where terminals must contend for access to ports on the host computer. The PS allocates computer ports to inbound devices or communications channels on a first-come, first-served basis, where some devices or channels might have priority over others.

Lineplexer. The lineplexer splits a computer port into two or more communications channels. For example, it can split 19.2, 16.8, or 14.4 kilobits-per-second data streams into two 4 kHz bandwidth channels for leased lines or digital data services. *See* DATA CHANNEL, DATA COMMUNICATIONS, DATA SERIAL TRANSMISSION, DIGITAL SERIAL INTERFACE, MODEM.

DATA CONVERSION

When data is changed from one form to another, the process is called data conversion. Data might, for example, be sent in serial form, bit by bit, and then be converted into parallel form for use by a certain circuit (*see* PARALLEL DATA TRANSFER, SERIAL DATA TRANSFER). Data might initially be in analog form, such as a voice or television-picture signal, and then be converted to digital form for transmission (see illustration). In the receiving circuit, the digital data may then be converted back into analog data (*see* ANALOG, ANALOG-TO-DIGITAL CONVERTER, DIGITAL, DIGITAL-TO-ANALOG CONVERTER).

Data conversion is performed for the purpose of improving the efficiency and/or accuracy of data transmission. For example, the digital equivalent of a voice signal is propagated with a better signal-to-noise ratio, in general, than the actual analog signal. It therefore makes sense to convert the signal to digital form when accuracy is of prime importance.

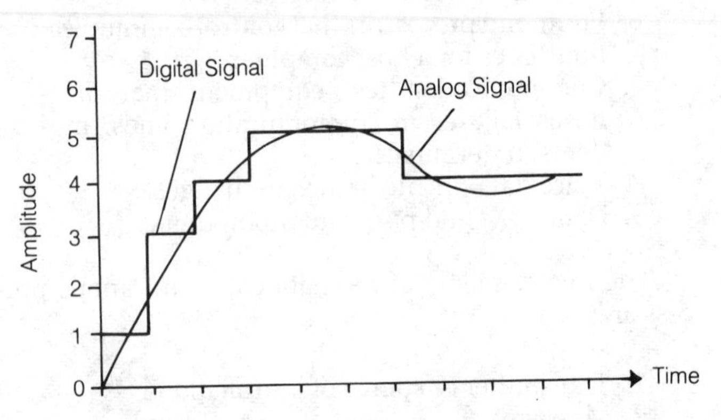

DATA CONVERSION: Step functions closely match the curve of an analog signal in data conversion.

DATA PROCESSING

The work actually performed on data, once the data has been acquired (*see* DATA ACQUISITION SYSTEM), is called data processing. The processing can consist of calculations, organization of information, or both. Redundant or unnecessary data is eliminated, and the data is organized for the most efficient use.

DATA SERIAL TRANSMISSION

Data serial transmission is used in data communications to minimize wiring and connector cost and simplifying data timing. Parallel transmission is employed where distances are short (usually less than 100 feet) and high-speed operation is required.

The two basic data serial transmission data formats are asynchronous and synchronous. Isochronous transmission is a combination of these two.

Asynchronous Data Transmission. Asynchronous transmission is transmission in which time intervals between transmitted characters might be of unequal length. It is known as start-stop transmission. (B in the figure). Each character transmitted is preceded by a start bit and followed by a stop bit to designate its beginning and end. This process is called framing the character, and a frame or block is transmitted as a unit. A coding procedure is usually applied for synchronization and/or error control. Characters can be sent at irregular intervals, making this

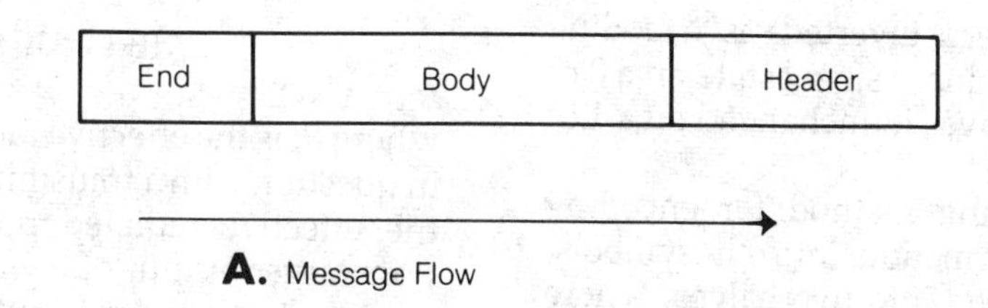

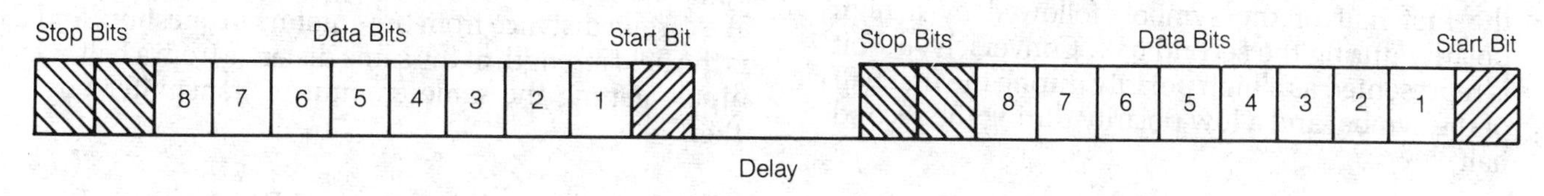

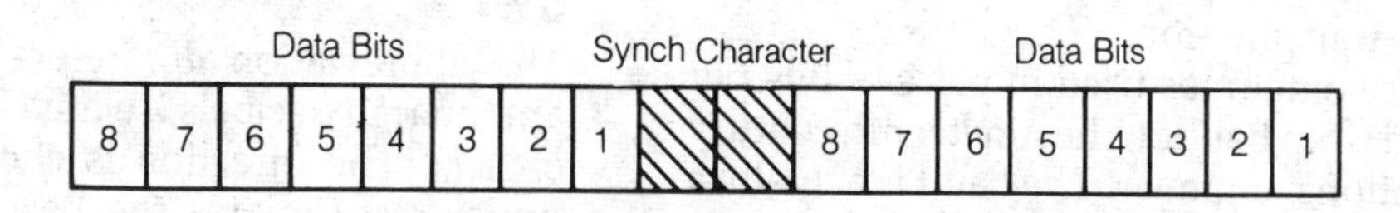

DATA SERIAL TRANSMISSION: Diagram of serial data flow (A), asynchronous transmission (B), and synchronous transmission (C).

transmission acceptable for low-speed keyboard terminals where data entry is intermittent.

The most commonly used code for data transmission is the ASCII code (American Standard Code for Information Interchange). This code uses seven bits per character, plus a parity bit. (An 8-bit ASCII code is also available.)

Synchronous Data Transmission. Synchronous transmission is data transmission in which the sending and receiving terminals are kept in constant synchronization by data sets (C in the figure). Start and stop bits are not required for each character. The sending and receiving devices are synchronized by exchanging a predetermined set of synchronization signals, either periodically or before the transmission of each message. The sending device transmits a long stream of characters without start/stop bits. The receiver counts off the first eight bits (assuming ASCII code), assumes this to be the character, and passes it on to the computer. It then counts off the next characters until the message is completed.

Asynchronous transmission is favored for slow-speed, manually operated terminals because it permits characters to be transmitted at irregular intervals. The disadvantage is that transmission is less efficient; each character must have start and stop bits. For example, with ASCII code, each 8-bit character of information requires a total of 11 bits in transmission. As a result, efficiency is 8/11 or about 73 percent.

Isochronous Transmission. Isochronous transmission combines the elements of both asynchronous and synchronous techniques, although it really is a special type of synchronous transmission. Each character has start and stop bits so that characters can be transmitted at irregular intervals. Synchronous modems can be used, yet there can be gaps in transmission because of the start and stop bits. The transmitter and receiver are synchronized during data transmission.

The isosynchronous mode has the advantage over the asynchronous mode because of its greater speed. Isosynchronous transmission permits speeds up to 9.6 kilobits per second without the need for large memory buffering of data, although synchronous transmission can even be faster. However, the asynchronous transmission is limited to about 1.8 kilobits per second.

Encoding for Synchronous and Isochronous Transmission. Synchronous data is often encoded to ensure that enough transitions exist in the data stream for the phase-locked loop (PLL) (*see* PHASE-LOCKED LOOP) in the model or terminal to extract the receive clock from the received data. Encoding embeds the transmit clock in the data, while decoding extracts the receive clock from the data.

The encoding schemes include:

- NRZ (nonreturn-to-zero) is a coding scheme in which 1 is a high-voltage level and 0 is a low-voltage level. The bit does not return to zero voltage in the middle of the bit cell.

- NRZI (Nonreturn to zero inverted) is a coding scheme that inverts the binary signal state on a 0 of the message data and leaves it unchanged on a 1 of the message data).
- Manchester is a signaling method for encoding clock and data bit information into bit symbols. Each bit symbol is divided into two halves, where the second half is the inverse polarity of the first half. A 0 bit is represented as a low polarity during the first half of the symbol, followed by a high polarity during the second half. Conversely, a 1 bit is represented as a high polarity during the first half of the symbol and a low polarity during the second half.

Data transmission rate is usually stated in bits per second (b/s or bps). This rate can be translated to characters per second by dividing it by the number of bits per character. For example, if a telephone line has a rated capacity of 4.8 kilobits per second, this is equivalent to 600 characters per second (4800/8) for an ASCII code using synchronous transmission.

The term *baud* is sometimes used in place of bits per second. It is derived from Baudot, the nineteenth century French communications engineer. The baud is actually a unit of telegraph signaling speed and is not synonymous with bits per second. However, the two do coincide for certain codes. *See* ASCII, BAUD RATE, CHANNEL, DATA COMMUNICATIONS, DIGITAL SERIAL INTERFACE.

dB

See DECIBEL.

dBa

The abbreviation *dBa* stands for adjusted decibels. Adjusted decibels are used to express relative levels of noise. First, a reference noise level is chosen, and assigned the value 0 dBa. All noise levels are then compared to this value. Noise levels lower than the reference level have negative values, such as −3 dBa. Noise levels greater than the reference level have positive values, such as +6 dBa.

The decibel is a means of expressing a ratio between two currents, power levels, or voltages. A reference level is therefore always necessary for the decibel to have meaning. *See also* DECIBEL.

dBd

The abbreviation *dBd* refers to the power gain of an antenna, in decibels, with respect to a half-wave dipole antenna. The dBd specification is the most common way of expressing antenna power gain (*see* DECIBEL).

The reference direction of the antenna under test is considered to be the direction in which it radiates the most power. The reference direction of the dipole is broadside to the antenna conductor.

Power gain in dBd is given by the formula:

$$\text{dBd} = 10 \log_{10} \left(\frac{P_a}{P_d}\right)$$

where P_a is the effective radiated power from the antenna in question with a transmitter output of P watts, and P_d is the effective radiated power from the dipole with a transmitter output of P watts.

An alternative method of measuring antenna power gain in dBd is possible using the actual field-strength values. If E_a is the field strength in microvolts per meter at a certain distance from the antenna in question, and E_d is the field strength at the same distance from a half-wave dipole getting the same amount of transmitter power, then

$$\text{dBd} = 20 \log_{10} \left(\frac{E_a}{E_d}\right)$$

See also ANTENNA POWER GAIN, dBi.

dBi

The abbreviation *dBi* refers to the power gain of an antenna, in decibels, relative to an isotropic antenna (*see* DECIBEL). The direction is chosen in which the antenna under test radiates the best. An isotropic antenna, in theory, radiates equally well in all directions (*see* ISOTROPIC ANTENNA in the ANTENNA DIRECTORY). The gain of any antenna in dBi is 2.15 dB greater than its gain in dBd; that is,

$$\text{dBi} = 2.15 + \text{dBd}$$

Power gain in dBi is given by the formula:

$$\text{dBi} = 10 \log_{10} \left(\frac{P_a}{P_i}\right)$$

Where P_a is the effective radiated power from the antenna in question with a transmitter output of P watts, and P_i is the effective radiated power from the isotropic antenna with a transmitter output of P watts.

An alternative method of measuring antenna power gain in dBi is possible using actual field-strength values. If E_a is the field strength in microvolts per meter at a certain distance from the tested antenna, and E_i is the field strength from an isotropic antenna getting the same amount of power from the transmitter, then:

$$\text{dBi} = 20 \log_{10} \left(\frac{E_a}{E_i}\right)$$

Actually, an isotropic antenna is not seen in practice. It is essentially impossible to construct a true isotropic antenna. Gain figures in dBi are sometimes used instead of dBd for various reasons. *See also* ANTENNA POWER GAIN, dBd.

dBm

The abbreviation *dBm* refers to the strength of a signal, in decibels, compared to 1 milliwatt, with a load impedance of 600 ohms. If the signal level is exactly 1 mW, its level is 0 dBm. In general:

$$\text{dBm} = 10 \log_{10} P$$

where P is the signal level in milliwatts.

With a 600-Ω load, 0 dBm represents 0.775 V, or 775 mV. With respect to voltage in a 600-Ω system, then:

$$\text{dBm} = 10 \log_{10} \left(\frac{E}{775}\right)$$

where E is the voltage in millivolts. A level of 0 dBm also represents a current of 1.29×10^{-3} A or 1.29 mA. With respect to current in a 600-Ω system:

$$\text{dBm} = 20 \log_{10} \left(\frac{I}{1.29}\right)$$

where I is the current in milliamperes (mA). *See also* DECIBEL.

DC

See DIRECT CURRENT.

DC AMPLIFIER

A direct current, or dc, amplifier is a circuit intended to increase the current, power, or voltage in a direct-current circuit. A common type of dc amplifier is used for the purpose of increasing the sensitivity of a meter or other indicating device. This amplifier can be simple, resembling an elementary alternating-current amplifier, as shown in the schematic.

Direct-current amplifiers can be used to amplify the voltage in an automatic-level-control circuit for speech compression. In this application, the time constant of the dc amplifier is critical. The microphone amplifiers are set to a high level of gain when there is no audio input; the greater the audio signal fed to the amplifiers, the more dc is supplied by the dc amplifiers acting on the automatic-level-control voltage. This dc voltage reduces the gain of the amplifying stages as the audio input increases. *See also* AMPLIFICATION, AMPLIFIER, AUTOMATIC LEVEL CONTROL, SPEECH COMPRESSION.

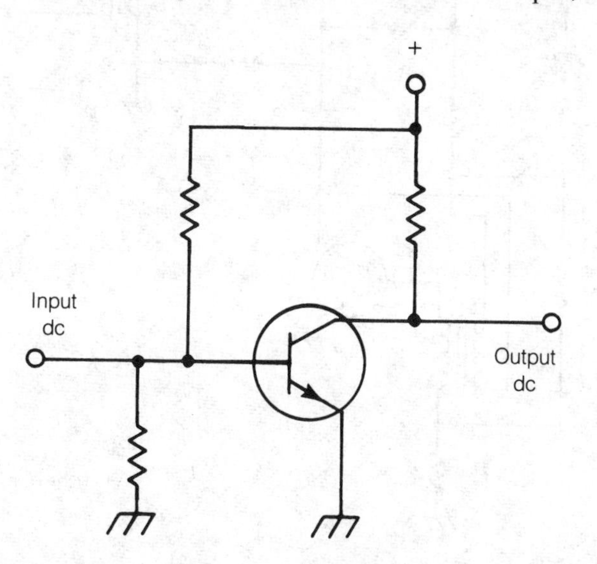

DC AMPLIFIER: A schematic for a direct-current amplifier.

DC COMPONENT

All waveforms have a direct-current, or dc, component and an alternating-current, or ac, component. Sometimes one component is zero. For example, in the output of a dry cell, the ac component is zero (see A in illustration). In a 60-Hz household outlet, the dc component is zero because the average voltage from this source is zero (B in illustration).

In a complex waveform, as illustrated at C, the dc component is the average value of the voltage. This average must be taken over a sufficient period of time. Some waveforms, such as the voltage at the collector of any amplifier circuit, have significant dc components.

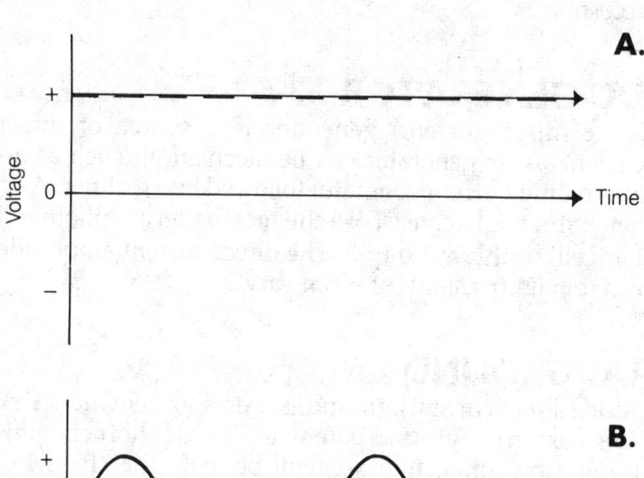

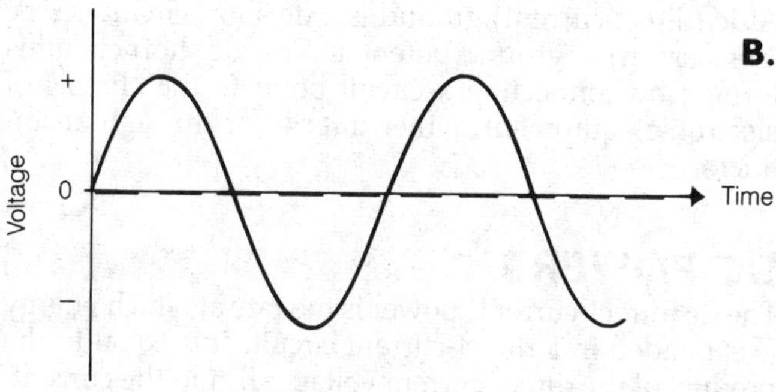

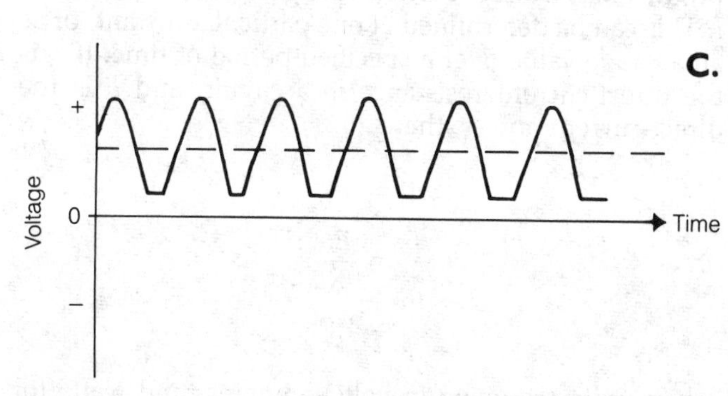

DC COMPONENT: A plot of pure dc voltage (A), alternating current with a zero direct-current component (B), and pulsating direct current with both alternating current and direct current components (C).

The dc component does not always change the practical characteristics of the signal, but the dc component must be eliminated to obtain satisfactory circuit operation in some situations. *See also* ALTERNATING CURRENT, DIRECT CURRENT.

DC ERASING HEAD

In a magnetic tape recorder, the dc erasing head is the head supplied with a pure direct current for the sole purpose of erasing all the magnetic impulses on the tape.

As the tape moves through the machine in the record mode, the dc erasing head is encountered first. All previously recorded information is removed prior to the tape's arrival at the recording head. This ensures that the tape will contain the lowest possible amount of background noise, and that the old information will not interfere with the new. It is essential that the erasing head be supplied with unmodulated direct current. The dc erasing head is, of course, disabled in the playback mode. *See also* MAGNETIC RECORDING, MAGNETIC TAPE, TAPE RECORDER.

DC GENERATOR

A dc (direct-current) generator is a source of direct current. A dc generator can be mechanical, such as an alternating-current generator followed by a rectifier. A dc generator might consist of a chemical battery, a photovoltaic cell, or thermocouple. The direct-current amplitude can remain constant, or it can vary.

DC GROUND

A dc (direct-current) ground is a dc short circuit (*see* DC SHORT CIRCUIT) to ground potential. This dc short circuit is formed by connecting a circuit point to the chassis of electronic equipment, either directly or through an inductor.

DC POWER

The dc (direct-current) power is the rate at which energy is expended in a direct-current circuit. It is equal to the product of the direct-current voltage, *E*, and the current, *I*. This can be determined at one particular instant, or as an average value over a specified period of time. If *R* is the direct-current resistance in a circuit, and *P* is the direct-current power, then:

$$P = EI = \frac{E^2}{R} = I^2R$$

when units are given in volts, amperes, and watts for voltage, current, and power, respectively.

Direct-current energy is the average direct-current power multiplied by the time period of measurement. The standard unit of energy is the watt hour, although it may also be specified in watt seconds, watt minutes, kilowatt hours, or other variations. If *W* is the amount of energy expended in watt hours, then:

$$W = Pt = EIt = \frac{E^2t}{R} = I^2Rt$$

where *t* is the time in hours. *See also* ENERGY, POWER.

DC POWER SUPPLY

See POWER SUPPLY.

DC SHORT CIRCUIT

A dc (direct-current) short circuit is a path that offers little or no resistance to direct current. The simplest example of a direct-current short circuit is a length of electrical conductor. However, an inductor also provides a path for direct current. But an inductor, unlike a plain length of conductor, offers reactance to alternating-current energy. (*See* INDUCTIVE REACTANCE.) Coils, or chokes, are often used in electronic circuits to provide a dc short circuit while offering high resistance to alternating-current signals. *See also* CHOKE, CHOKE-INPUT FILTER.

DC-TO-AC CONVERTER

A dc-to-ac (direct-current to alternating-current) converter is a form of power supply, often used to obtain

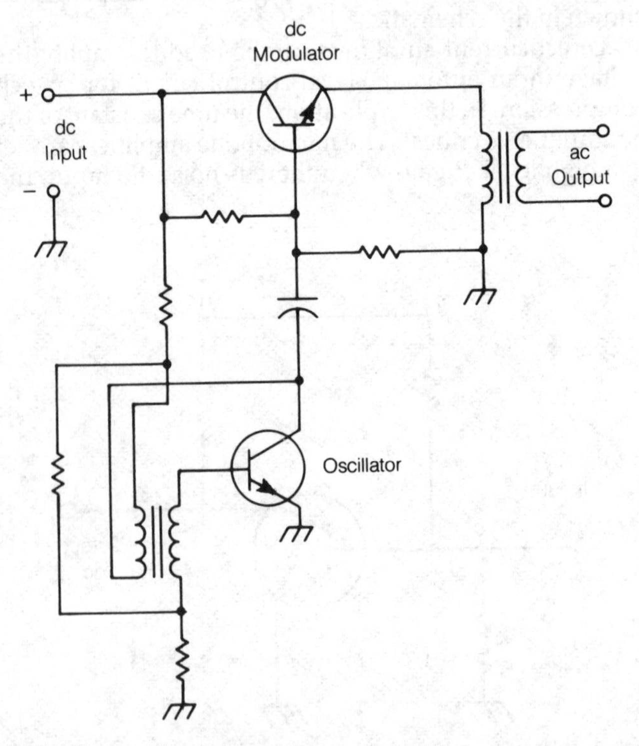

DC-TO-AC CONVERTER: A schematic for a dc-to-ac converter with an oscillator, modulator and output transformer.

equivalent alternating-current line power from a battery or other source of low-voltage direct current. The chopper power supply makes use of a dc-to-ac converter (*see* CHOPPER POWER SUPPLY).

A dc-to-ac converter operates by modulating, or interrupting, a source of direct current. A relay or oscillating circuit is used to accomplish this. The resulting modulated direct current is then passed through a transformer to eliminate the direct-current component, and to get the desired alternating-current voltage. A simple schematic diagram of a dc-to-ac converter is shown.

When a dc-to-ac converter is designed especially to produce 120-V, 60-Hz alternating current for the operation of household appliances, the device is called a power inverter. *See also* INVERTER.

DC-TO-DC CONVERTER

A dc-to-dc converter is a circuit that changes the voltage of a direct-current power supply. This circuit consists of a modulator, a transformer, a rectifier, and a filter, as shown in the illustration. It is similar to that of a dc-to-ac converter, except for the addition of the rectifier and filter (*see* DC-TO-AC CONVERTER).

A dc-to-dc converter may be used either to step-up or step-down voltage. Usually, such a circuit is used to obtain a high direct-current voltage from a comparatively low voltage. A common type of dc-to-dc converter is used as a power supply for electronic equipment when the only available source of power is a 12-V automotive battery or electrical system. The regulation of such a voltage step-up circuit depends on the ability of the battery or car alternator to handle large changes in the load current. A special regulator circuit is required if the voltage regulation must be precise.

Low-power dc-to-dc converters can be built into a small integrated-circuit package.

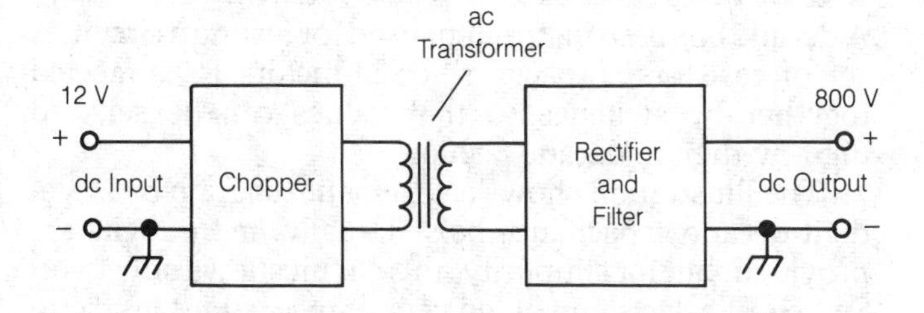

DC-TO-DC CONVERTER: A block diagram of a dc-to-dc converter with a rectifier and filter.

DC TRANSMISSION

When electric power is sent from one point to another as a direct current, the method of transfer is called direct-current, or dc, transmission. Most power-transmission lines today carry alternating current. However, direct-current transmission, especially for long lines, is still in use.

Alternating current can be stepped up or down at will by the use of transformers. When power must be sent over long distances, high voltages are used, since higher voltages result in lower losses. But at any voltage, direct current suffers less attenuation than alternating current because transmission-line losses always increase with increasing frequency. The primary obstacle to dc transmission is the difficulty in converting extremely high voltages from alternating to direct and vice versa. There is another advantage to dc transmission, however: in a direct-current line, the electric and magnetic fields do not alternate. This greatly reduces the emission of electromagnetic energy, which can create electromagnetic interference problems near high-voltage power lines. The only electromagnetic radiation from a line carrying pure direct current would be the result of occasional transient spikes or corona discharge.

DEAD BAND

When no signals are received within a certain frequency range in the electromagnetic spectrum, that band of frequencies is said to be dead. A dead band can result from geomagnetic activity that disrupts the ionosphere of the earth; in fact, the term *dead band* is used only on those frequency bands that are affected by ionospheric propagation. During a severe geomagnetic storm, propagation becomes virtually impossible through the ionosphere. The illustration shows a hypothetical situation in which propagation deteriorates, making the band sound dead (*see* GEOMAGNETIC FIELD, IONOSPHERE).

A dead band can be caused by the deterioration of the ionosphere with the setting of the sun, with low sunspot activity, and perhaps with coincidences of unknown origin. A band might sometimes appear dead simply because no one is transmitting on it at a particular time.

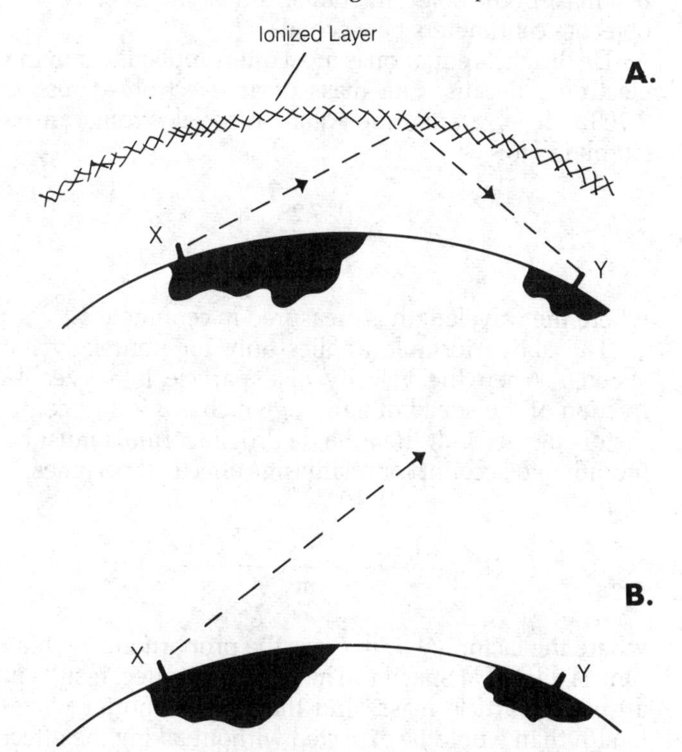

DEAD BAND: Ionospheric bounce normally permits radio communication between points X and Y (A), but with diminished ionization, signals are not received because of a dead band (B).

A band can go dead for just a few seconds or minutes, or it may remain unusable for hours or days. The range of frequencies can be as small as a few kilohertz, or it can extend for several megahertz. *See also* PROPAGATION CHARACTERISTICS.

The term *dead band* is also used to describe the lack of response of a servomotor system through a part of its range of operation. This lack of response can be caused by backlash in the gears or rotor of the servo, or by a lack of resolution of the position-sensing potentiometer (or other device) that feeds angular or position information to the servo system. The servo dead band can be expressed as degrees of arc, or as a percentage of total travel. Dead band is also hysteresis.

DE BROGLIE'S EQUATION

All moving particles display properties of wavelength. In 1923, a physicist named Louis V. de Broglie (pronounced *de Broily*) discovered that the wavelength of a moving particle is related to the mass of the particle and to its speed. The greater the mass, the greater the frequency of the wave action, and the shorter the wavelength. The wavelength also becomes shorter with increasing velocity.

De Broglie's equation is given by:

$$\lambda = \frac{h}{mv}$$

where λ is the wavelength in centimeters, h is Planck's constant of approximately 6.63×10^{-27} erg-sec, m is the rest mass of the object in grams, and v is the speed of the object in centimeters per second.

De Broglie's equation is most often applied to moving electrons. Because the mass of an electron at rest is 9.108×10^{-28} grams, the equation for electrons can be simplified to:

$$\lambda = \frac{7.28}{v}$$

where the wavelength is measured in centimeters.

The above formula applies only for nonrelativistic speeds. When the velocity of a particle is a sizeable fraction of the speed of light, c, which is 3×10^{11} centimeters per second, then the de Broglie formula must be modified to account for relativisitic effects. It becomes:

$$\lambda = \frac{6.63 \times 10^{-27} \sqrt{1 - v^2 / c^2}}{mv}$$

where the factor $\sqrt{1 - v^2 / c^2}$ is the proportion by which time is dilated at speed v. The relativistic effect results in a greater particle mass, and therefore a shorter wavelength than would be expected without taking the effect into account. De Broglie waves have been observed for electrons and other moving particles.

DEBUGGING

Debugging is the process of removing errors, or "bugs," from an electronic circuit or computer program. After a circuit design has been built as a prototype, it is tested. Problems may be detected and the methodical correction of those problems is debugging. Similarly, after a computer program is written, it is tested. Here debugging is the correction of all errors that prevent it from running as designed.

The process of debugging can be very simple; it might, for instance, involve only a small change in the value of a component. Sometimes, the debugging process requires that the entire design process be started all over. Occasionally, the bugs are hard to find, and do not appear until the product has been put into mass production or the program has been published and extensively used.

DECADE

A range of any variable in which the value at one end of the range is ten times the value at the other end, is called a decade. The radio-frequency bands are arbitrarily designated as decades: 30 to 300 kHz is called the low-frequency band, 0.3 to 3 MHz is called the medium-frequency band, 3 to 30 MHz is called the high-frequency band, and so on.

There are an infinite number of decades between any quantity and the zero value for that parameter. For example, there are frequency decades of 30 to 300 kHz, 3 to 30 kHz, 0.3 to 3 kHz, 30 to 300 Hz, etc.

The decade method of expressing quantities is often used by scientists and engineers because its logarithmic nature allows the evaluation of a larger range of quantities than with a simple linear system. *See also* ORDER OF MAGNITUDE, SCIENTIFIC NOTATION.

DECADE BOX

A decade box is an instrument used for testing a circuit. A set of resistors, capacitors, or inductors is connected together by switches so that values can be selected digit-by-digit in decade fashion.

The illustration shows a schematic diagram of a two-digit decade capacitance box. (Usually, more digits are provided, but for simplicity, this circuit shows only two.) Switch S1 selects any of ten capacitance values in microfarads: 0.00, 0.01, 0.02, 0.03, . . ., 0.09. Switch S2 also selects any of ten values of capacitance in microfarads, each ten times the values of capacitance in the circuit containing S1: 0.0, 0.1, 0.2, 0.3, . . ., 0.9. Therefore, there are 100 possible values of capacitance that can be selected by this system, ranging from 0.00 to 0.99 μF in increments of 0.01 μF.

Decade boxes are sometimes used to set the frequency of a digital radio receiver or transmitter. A signal generator and monitor may have a range of zero to 999.999999 MHz or higher, with frequencies selectable in increments as small as 1 Hz. This gives as many as 10^9

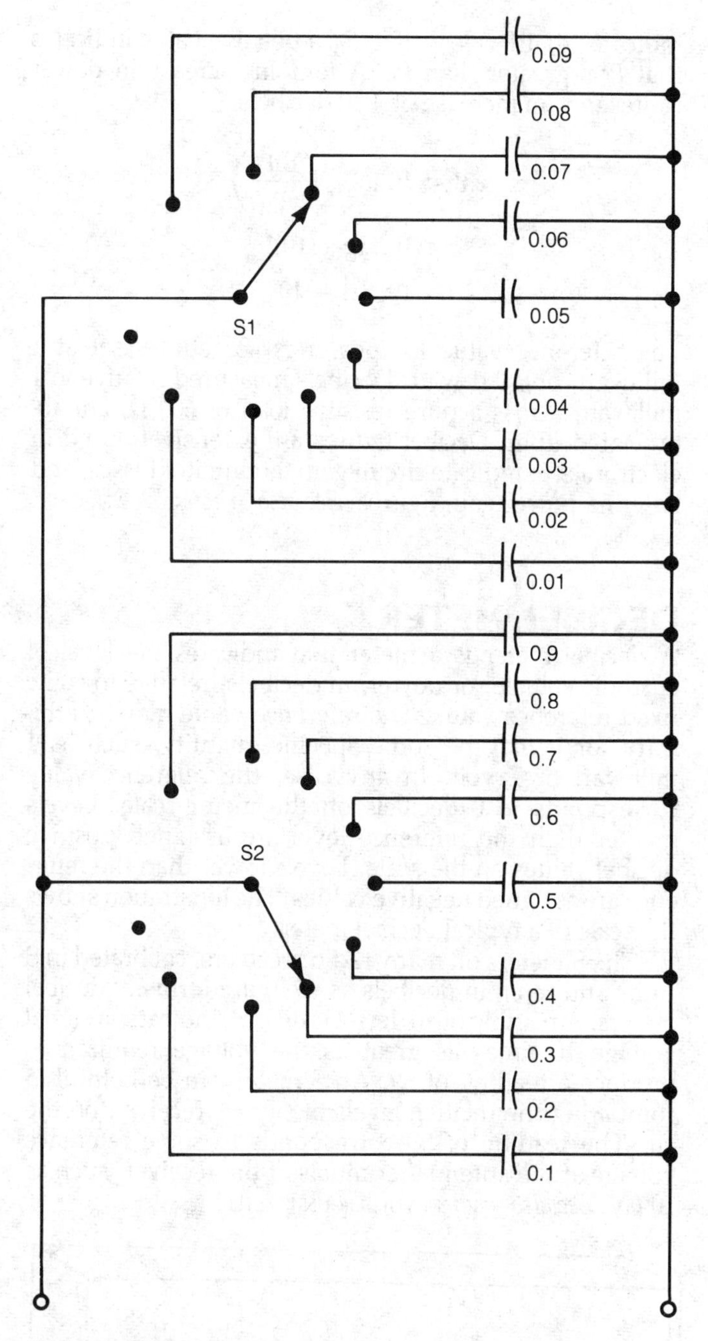

DECADE BOX: A schematic for a decade capacitance box.

possible frequencies, with only nine independent selector switches. *See also* DECADE.

DECADE COUNTER

A counter that proceeds in decimal fashion, beginning at zero and going through 9, then to 10 and up to 99, and so on, is called a decade counter. A decade counter operates in the familiar base-10 number system. Counting begins with the ones digit (10^0), then the tens digit (10^1), and so on up to the 10^n digit. In a counter going to the 10^n digit, there are $n+1$ possible digits.

The display on a decade counter often, if not usually, contains a decimal point. Decades can proceed downward toward zero, as well as upward to ever-increasing levels. However, fractions of a pulse or cycle are never actually counted. Rather, the pulses or cycles are counted for a longer time when greater accuracy is needed. For example, a counting time of 10 seconds gives one additional digit of accuracy, as compared to a counting time of 1 second. This allows decimal parts of a cycle to be determined. *See also* DECADE, FREQUENCY COUNTER.

DECAY

The decline in amplitude of a pulse or waveform is called its decay. The decay of a pulse or waveform, although appearing to be instantaneous, is not because a finite amount of time is always required for decay.

When a high-wattage incandescent bulb is switched off, for example, one can see its brightness decay. But the decay in brilliance of a neon bulb or light-emitting diode is too rapid to be seen. Nevertheless, even a light-emitting diode (LED) has a finite brightness-decay time.

The graph for decay time illustrates the decay of the waveform in a continuous-wave (CW) transmitter following the transition from key-down to key-up condition. The decay curve is a logarithmic function. *See also* DAMPED WAVE, DECAY TIME, LOGARITHM, RISE TIME.

DECAY TIME

The decay time of a pulse or waveform is the time required for the amplitude to decay to a specified percentage of the maximum amplitude. The time interval begins the instant the amplitude starts to fall (see graph) and ends when the determined percentage has been attained.

The decay of a pulse or waveform proceeds in a logarithmic manner (*see* LOGARITHM). Therefore, in theory, the amplitude never reaches zero. In practice, a point is always reached at which the pulse or wave amplitude can be considered zero. This point may be chosen for the determination of the decay time interval. In a capaci-

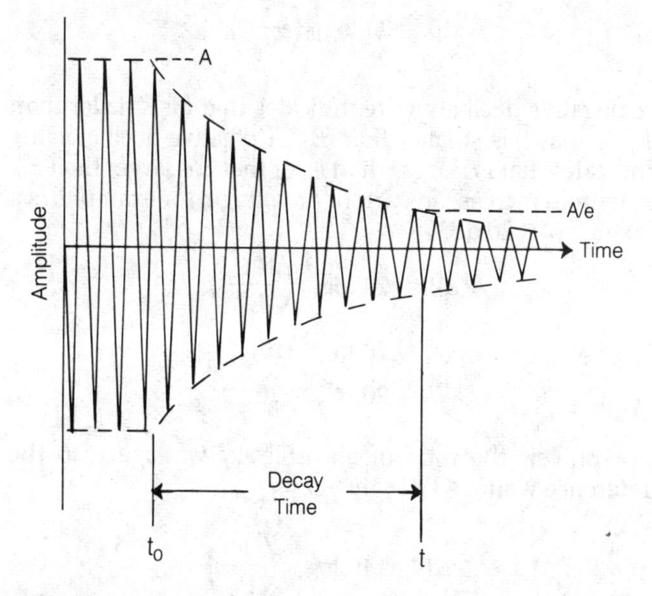

DECAY TIME: The time for a signal to decay from a maximum to a predetermined level shown as t_0 to t.

tance-resistance circuit, the decay time is considered to be the time required for the change voltage to drop to 37 percent of its maximum value. In this case, the final amplitude is equal to the initial amplitude divided by e, where e is approximately 2.718. *See also* DECAY, TIME CONSTANT.

DECCA

Decca is a British radio-navigation system. The system operates in the range of 70 to 130 kHz, or approximately the same frequency as the American radio-navigation system, Loran C. It is useful to about 200 nautical miles (nmi).

Decca is a phase-comparison system in which sets of hyperbolic lines of position are determined. Two or more fixed stations, with precisely coordinated phase, radiate the signals. An airplane pilot or ship's captain locates his position by comparing the phases from the two stations, using a special integrator circuit designed for the purpose. When additional transmitting stations are used, several different readings can be obtained, and they can be averaged to get a close approximation of position. *See also* LORAN.

DECIBEL

The decibel is a means of measuring relative levels of current, voltage, or power. A reference current I_0, or voltage E_0, or power P_0 must first be established. Then, the ratio (expressed as decibels, dB) of an arbitrary current I to the reference current I_0 is given by:

$$\text{dB} = 20 \log_{10}\left(\frac{I}{I_0}\right)$$

The ratio of an arbitrary voltage E to the reference voltage E_0 is given by:

$$\text{dB} = 20 \log_{10}\left(\frac{E}{E_0}\right)$$

A negative decibel figure indicates that I is smaller than I_0, or that E is smaller than E_0. A positive decibel value indicates that I is larger than I_0, or that E is larger than E_0. A tenfold increase in current or voltage, for example, is a change of +20 dB:

$$\begin{aligned}\text{dB} &= 20 \log_{10}\left(\frac{10E}{E_0}\right)\\ &= 20 \log_{10}(10)\\ &= 20 \times 1 = 20\end{aligned}$$

For power, the ratio of an arbitrary wattage P to the reference wattage P_0 is given by:

$$\text{dB} = 10 \log_{10}\left(\frac{P}{P_0}\right)$$

As with current and voltage, a negative decibel value indicates that P is less than P_0; a positive value indicates that P is greater than P_0. A tenfold increase in power represents an increase of +10 decibels:

$$\begin{aligned}\text{dB} &= 10 \log_{10}\left(\frac{10P}{P_0}\right)\\ &= 10 \log_{10}(10)\\ &= 10 \times 1 = 10\end{aligned}$$

The reference value for power is sometimes set at 1 milliwatt, or 0.001 watt. Decibels measured relative to 1 milliwatt, across a pure resistive load of 600 Ω, are abbreviated dBm. Decibel figures are extensively used in electronics to indicate circuit gain, attenuator losses, and antenna power gain figures. *See also* ANTENNA POWER GAIN, dBa, dBd, dBi, dBm.

DECIBEL METER

A decibel meter is a meter that indicates the level of current, voltage, or power, in decibels, relative to some fixed reference value. The reference value may be arbitrary, or it may be some specific quantity, such as 1 milliwatt or 1 volt. In any case, the reference value corresponds to 0 decibels on the meter scale. Levels greater than the reference level are assigned positive decibel values on the scale. Levels lower than the reference are assigned negative values. The illustration shows the scale of a typical decibel meter.

The S meters on many radio receivers, calibrated in S units and often in decibels as well, are forms of decibel meters. A reading of 20 decibels over S9 indicates a signal voltage 10 times as great as the voltage required to produce a reading of S9. An S meter can be helpful in comparing the relative levels of signals received on the air. The reading of S9 corresponds to some reference voltage at the antenna terminals of the receiver, such as 10 μV. *See also* S METER, VOLUME-UNIT METER.

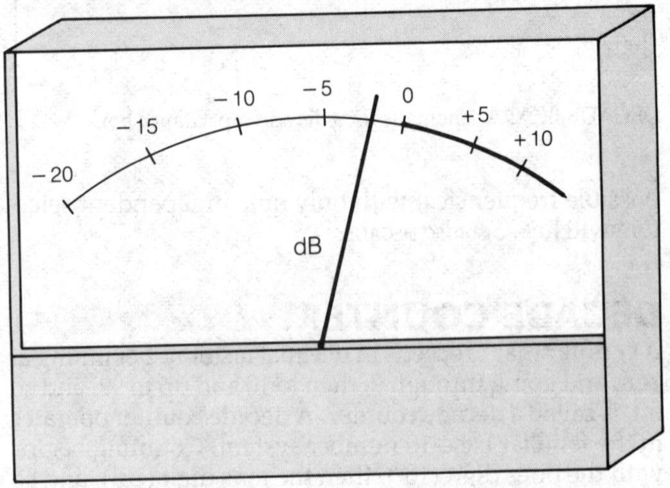

DECIBEL METER: A decibel meter gives a logarithmic readout of variables in decibels, relative to some fixed reference.

DECIMAL

The term *decimal* refers to a base-10 number system. In

this system, commonly used throughout the world, most familiar numbers are represented by combinations of ten different digits in various decimal places.

The digit farthest to the right, but to the left of the decimal point, is multiplied by 10^0, or 1; the digit next to the left is multiplied by 10^1, or 10; the digit to the left of this is multiplied by 10^2, or 100. With each move to the left, the base value of the digit increases by a factor of 10, so that the nth digit to the left of the decimal point is multiplied by 10^{n-1}.

The digit first to the right of the decimal point is multiplied by 10^{-1}, or 1/10; the digit next to the right is multiplied by 10^{-2}, or 1/100. This process continues, so that with each move to the right, the base value of the digit decreases by a factor of 10. Therefore, the nth digit to the right of the decimal point is multiplied by 10^{-n}.

Ultimately, the number represented by a decimal sequence is determined by adding the decimal values of all the digits. For example, the number 27.44 is equal to $(2 \times 10^1) + (7 \times 10^0) + (4 \times 10^{-1}) + (4 \times 10^{-2})$.

DECODING

Decoding is the process of converting a message, received in code, into plain language. This is generally done by a machine, although in the case of the Morse code, a human operator often acts as the decoding medium.

Messages can be coded either for the purpose of efficiency and accuracy, such as with the Morse code or other codes, or for the purpose of keeping a message secret, as with voice scrambling or special abbreviations. Both types of code can be used at the same time. In this case, decoding requires two steps: one to convert the scrambling code to the plain text, and the other to convert the code itself to English or another language.

The conversion of a digital signal to an analog signal can be called decoding. The opposite of the decoding process—the conversion of an analog signal to a digital signal, or the transformation of a plain-language message into coded form—is called encoding. Decoding is always done at the receiving end of a communications circuit. *See also* DIGITAL-TO-ANALOG CONVERTER, ENCODING.

DECOUPLING

When undesired coupling effects must be minimized, a technique called decoupling is employed. For example, a multistage amplifier circuit will often oscillate because of feedback among the stages. This oscillation usually takes place at a frequency different from the operating frequency of the amplifier. To reduce this oscillation, or eliminate it entirely, the interstage coupling should be made as loose as possible, consistent with proper operation at the desired frequency. This is called decoupling.

Another form of decoupling consists of the placement of chokes and/or capacitors in the power-supply leads to each stage of a multistage amplifier (see diagram). This minimizes the chances of unwanted interstage coupling through the power supply.

When several different loads are connected to a single transmission line, such as in a multiband antenna system, resonant circuits are sometimes employed to effectively decouple all undesired loads from the line at the various operating frequencies. The trap antenna decouples part of the radiator to obtain resonance on two or more different frequencies. *See also* DECOUPLING STUB, TRAP.

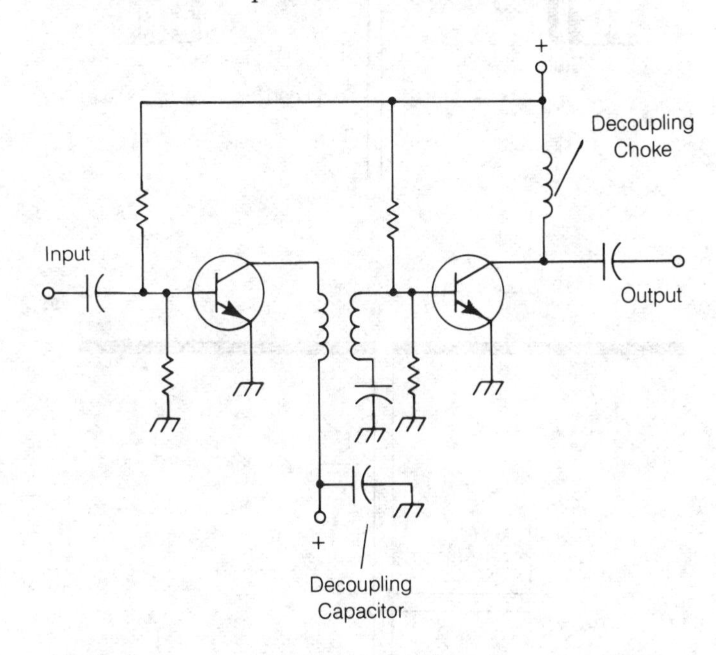

DECOUPLING: A decoupling capacitor and choke reduce the effects of feedback in a multistage amplifier.

DECOUPLING STUB

A decoupling stub is a length of transmission line that acts as a resonant circuit at a specific frequency, and is used in an antenna system in place of a trap (*see* TRAP). This stub usually consists of a ¼-wavelength section of transmission line, short-circuited at the far end. This arrangement acts as a parallel-resonant inductance-capacitance circuit.

At the resonant frequency of the stub, the impedance between the input terminals is extremely high. Therefore, this stub can decouple a circuit at the resonant frequency (*see* DECOUPLING). Decoupling might be desired in a multiband antenna system, or to aid in the rejection of an unwanted signal.

A ¼-wavelength section of transmission line, open at the far end, will act as a series-resonant inductance-capacitance circuit. At the resonant frequency, this section, also called a stub, has an extremely low impedance, essentially equivalent to a short circuit. This kind of stub can be extremely effective in rejecting signal energy at unwanted frequencies. By connecting a series-resonant stub across the antenna terminals, spurious responses or emissions are suppressed at the resonant frequency of the stub. The drawing illustrates the use of parallel-equivalent (A) and series-equivalent (B) stubs in antenna and feedline systems.

Some stubs are ½ wavelength long, rather than ¼ wavelength. A short-circuited ½-wavelength stub acts as a series-resonant circuit, and an open-circuited ½-wavelength stub acts as a parallel-resonant circuit. All

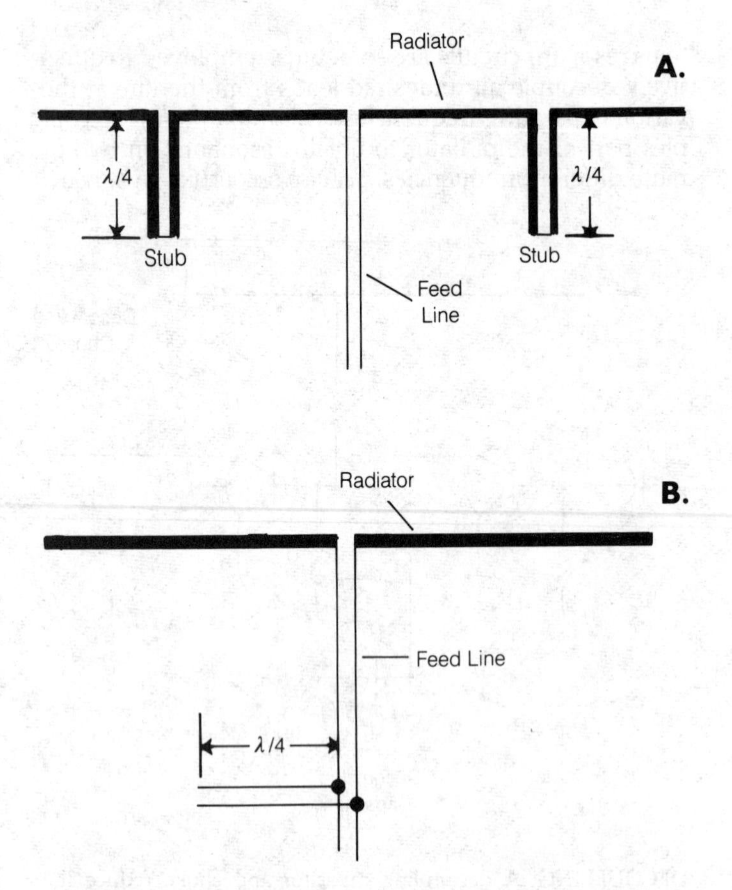

DECOUPLING STUB: Examples of decoupling stubs for dual-frequency operation (A) and attenuation of a specific frequency band (B).

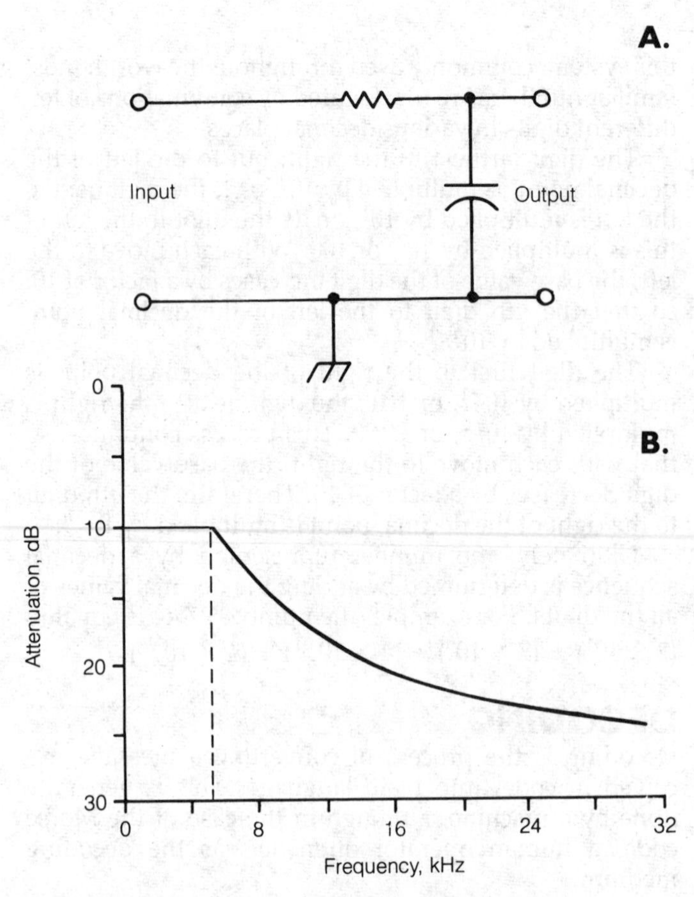

DEEMPHASIS: A simple deemphasis circuit (A) and a plot of attenuation versus frequency for deemphasis (B).

stubs show the same characteristics at odd harmonics of the fundamental frequency.

DEEMPHASIS

Deemphasis is the deliberate introduction of a lowpass network into the audio-frequency stages of a frequency-modulation receiver. This is done to offset the preemphasis introduced at the transmitter (*see* PREEMPHASIS). The schematic diagram shows a simple deemphasis network, A. The graph (B) illustrates, in a qualitative manner, the attenuation-versus-frequency characteristic of a typical deemphasis network.

By introducing preemphasis at the transmitter and deemphasis at the receiver in a frequency-modulation communications system, the signal-to-noise ratio at the upper end of the audio range is improved. As the transmitted-signal modulating frequency increases, the amplitude increases (because of preemphasis at the transmitter); when this amplitude is brought back to normal by deemphasis in the receiver, the noise is attenuated as well. *See also* FREQUENCY MODULATION, SIGNAL-TO-NOISE RATIO.

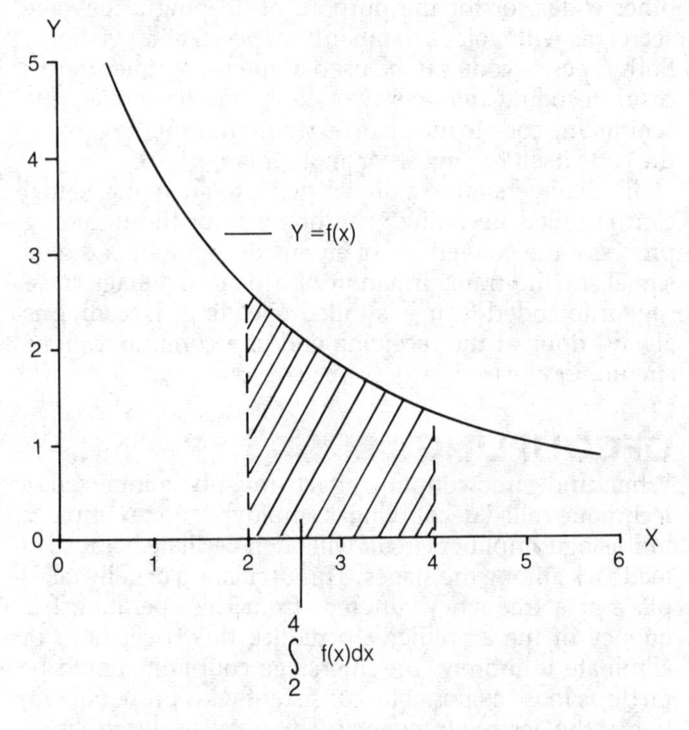

DEFINITE INTEGRAL: A definite integral shown as a shaded area under the curve between two values on the x axis.

DEFINITE INTEGRAL

A definite integral is an integral evaluated between two defined points in the domain of a function (*see* INDEFINITE INTEGRAL). Geometrically, the definite integral of a one-variable function, evaluated between two points, is equal to the area under the curve of the function between those two points. The illustration shows an example of a definite integral, according to this geometric interpretation.

A definite integral, evaluated between two points, *a* and *b*, on the x axis of a function f(*x*), is written:

$$\int_a^b f(x)\,dx$$

The expression *dx* is the differential of *x*, and is necessary to make the quantity meaningful. The details of integration are covered in calculus textbooks.

An area that lies above the abscissa, or x axis, in the definite integral is considered positive. An area below the abscissa is considered negative. In the illustration, the value of the integral is positive, since the shaded region lies entirely above the x axis. *See also* CALCULUS.

DEFLECTION

Deflection is a deliberately induced change in the direction of an energy beam. The beam can consist of sound waves, radio waves, infrared radiation, visible light, ultraviolet radiation, X rays, or atomic-particle radiation such as a stream of electrons.

In a cathode-ray tube, electron beams are deflected to focus and direct their energy to a certain spot on a phosphor screen. In loudspeaker enclosures, deflecting devices are used to get the best possible fidelity. Deflectors called baffles are used for acoustic purposes in concert halls and auditoriums (*see* ACOUSTICS). Deflecting mirrors are employed in optical telescopes. Heat deflectors are sometimes used to improve energy efficiency in homes and buildings. *See also* BAFFLE, CATHODE-RAY TUBE, DEFLECTOR.

DEFLECTOR

A deflector is an electrode used to deflect an energy beam. Generally, the term is used to describe the plates in a cathode-ray tube. These plates focus and direct the beam to the desired point on the phosphorescent screen.

In a loudspeaker, an attachment for spreading the sound waves over a broad angle is called a deflector. A loudspeaker deflector improves the fidelity of the sound by allowing the sound to be heard over a larger area.

DEGAUSSING

Degaussing is a procedure for demagnetizing an object. A circuit called a degausser or demagnetizer is used for this purpose. Sometimes the presence of a current in a nearby electrical conductor can cause an object to become magnetized. This effect may be undesirable. Examples of devices in which degaussing is sometimes required include television picture tubes, tape-recording heads, and relays. By applying an alternating current to produce an alternating magnetic field around the object, the magnetization can usually be eliminated. Another method of degaussing is the application of a steady magnetic field in opposition to the existing, unwanted field. *See also* DEMAGNETIZING FORCE.

DEGENERATION

See NEGATIVE FEEDBACK.

DEGREE

The degree is a unit of either temperature or angular measure. There are three common temperature scales: the Celsius, Fahrenheit, and Kelvin scales (*see* CELSIUS TEMPERATURE SCALE, FAHRENHEIT TEMPERATURE SCALE, KELVIN TEMPERATURE SCALE).

In the Fahrenheit scale, which is most generally used in the United States, the freezing point of pure water at one atmosphere pressure is assigned the value 32°. The boiling point of pure water at one atmosphere is 212° on the Fahrenheit scale. In the Celsius temperature scale, water freezes at 0° and boils at 100°. Thus, 1° in the Celsius scale is representative of a greater change in temperature than 1° in the Fahrenheit scale. In the Kelvin, or absolute, temperature scale, the degrees are the same size as in the Celsius scale, except that 0 Kelvin corresponds to absolute zero, or −273°C, the lowest temperature in the physical universe. In the respective temperature scales, readings are given in °F (Fahrenheit), °C (Celsius), or K (Kelvin).

In angular measure, a degree represents 1/360 of a complete circle. The number 360 was chosen in ancient times, when scientists and astronomers noticed that the solar cycle repeated itself approximately once in 360 days. One day thus corresponds to a degree in the circle of the year.

Phase shifts or differences are usually expressed in degrees, with one complete cycle represented by 360 degrees of phase. *See also* CYCLE, PHASE ANGLE, RADIAN.

DEHUMIDIFICATION

The operation of electronic circuits is affected by the temperature and relative humidity. If the amount of water vapor in the air is too high, corrosion is accelerated, and condensation might occur. In an electrical circuit, condensation can cause unwanted electrical conduction between parts that should be isolated. Condensation can also cause the malfunction of high-speed switches. Dehumidification is the process of removing excess moisture from the air.

There are various methods for dehumidification. The simplest method is to raise the temperature; for a given amount of water vapor in the air, the relative humidity decreases as the temperature rises. Dry crystals of calcium chloride or cobalt chloride, packed in a cloth sack, will absorb water vapor from the air, and will help to dehumidify an airtight chamber. Dehumidifying sprays are also available. *See also* HUMIDITY.

DEIONIZATION

When an ionized substance becomes neutral, the change is called deionization. A collection of positive ions, for example, may be neutralized by the introduction of electrons or by the variation of parameters such as the temperature or the intensity of the electric field.

Generally, gases tend to ionize at high temperatures and/or high electrical potentials. In an ionized gas, as the temperature is reduced, the atoms will eventually become neutral, assuming that the intensity of the electric field is not too great. The temperature at which this neutralization takes place is called the deionization temperature. Given a constant temperature, as the intensity of the electric field is reduced, an ionized substance will become neutral at a certain point, called the deionization potential. *See also* IONIZATION.

DELAY

A delay is a specific time interval between a cause and its effect, or between one effect and another related effect. A delay always occurs between two related events. This delay might be inconsequential and perhaps unimportant, such as the time required for the voice impulses to traverse a local telephone circuit during a casual conversation, or the time between the transmission of an amateur-radio message and its reception in a distant receiver. The delay may, however, be of great importance, or it might be deliberately introduced into a circuit. An example of this is the delay in broadcasting programming, usually about 7 seconds, between the actual event and its transmission over the air.

The shortest possible delay between two circuit points is determined by the speed at which electric impulses can flow in that circuit. In an alternating-current circuit, delay is sometimes expressed as a fraction of a cycle, or in degrees of phase. *See also* DELAY CIRCUIT, DELAY LINE, PHASE ANGLE.

DELAY CIRCUIT

A delay circuit is a set of electronic components designed deliberately for the purpose of introducing a time or phase delay. Such a circuit might be a passive combination of resistors, inductors, and/or capacitors. A delay circuit can consist of a simple length of transmission line. Or, the device can be an active set of integrated circuits and peripheral components.

Delay circuits are used in a wide variety of applications. Broadcast stations delay their transmissions by approximately 7 seconds. This allows the signal to be cut off, if necessary, before it is transmitted over the air. Phase-delay circuits are extremely common. Certain kinds of switches and circuit breakers have a built-in time delay. Sometimes, delay is undesirable and hinders the performance of a circuit. *See also* DELAY, DELAY DISTORTION, DELAYED AUTOMATIC GAIN CONTROL, DELAYED MAKE/BREAK, DELAYED REPEATER, DELAYED TRANSMISSION, DELAY TIME, DELAY TIMER.

DELAY DISTORTION

In some electronic circuits, the propagation time varies with the signal frequency (*see* PROPAGATION SPEED). When this happens, distortion takes place because signal components having different frequencies arrive at the receiving end of the circuit at different times. This is delay distortion, which can happen in a radio communications circuit, in a telephone system, or even within a single piece of electronic equipment.

Generally, higher frequencies are propagated at a lower rate of speed than lower frequencies. If the propagation time is extremely short, the delay distortion will be inconsequential. But the longer the propagation time, the greater the chances of delay distortion. Delay distortion can be minimized by making the percentage difference between the lowest and highest frequencies in a signal as small as possible. A baseband signal, with components as low as 100 Hz and as high as 3000 Hz, for example, is more subject to delay distortion than a single-sideband with the same audio characteristics and a suppressed-carrier frequency of 1 MHz. In the former case, the percentage difference between the lowest and highest frequencies is very large, but in the latter case, it is extremely small.

DELAYED AUTOMATIC GAIN CONTROL

A delayed automatic-gain-control circuit is a special form of automatic-gain-control, or AGC, circuit (*see* AUTOMATIC GAIN CONTROL). It is used in many communications receivers.

In a delayed AGC circuit, signals below a certain threshold level are passed through the receiver with maximum gain. Only when the signal strength exceeds this threshold amplitude, does the AGC become active. Then, as the signal strength continues to increase, the AGC provides greater and greater attenuation.

The delayed AGC circuit allows better weak-signal reception than does an ordinary automatic-gain-control circuit.

DELAYED MAKE/BREAK

When a circuit is closed or opened a short while after the actuating switch or relay is energized or deenergized, the condition is called delayed make or break. For example, the contacts of a relay may close several milliseconds, or even several seconds or minutes, following application of current to the circuit. The contacts of a relay or other switching device may not open until some time after current has been removed. The former device is a delayed-make circuit; the latter is a delayed-break circuit.

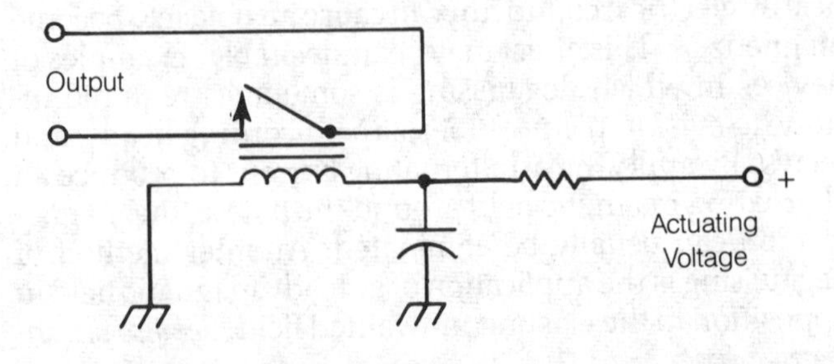

DELAYED MAKE/BREAK: The schematic of a relay for delayed make-and-break action with a capacitor to set the time delay.

Delayed-make and delayed-break devices are sometimes used in power supplies. For example, in a tube-type power amplifier, the filament voltage should be applied a few seconds or minutes before the plate voltage. A delayed-make circuit can be used in the plate supply, accomplishing this function automatically. A simple delayed-make/break relay arrangement, providing a delay of up to several seconds, is illustrated in the schematic diagram. The value of the capacitor determines the delay time. The larger the capacitance, the longer the delay. The resistor limits the current through the relay coil, and serves to lengthen the charging time of the capacitor. *See also* DELAY TIMER.

DELAYED REPEATER

A delayed repeater is a device that receives a signal and retransmits it later. The delay time between reception and retransmission may vary from several milliseconds to seconds or even minutes. Generally, a delayed repeater records the signal modulation envelope on magnetic tape, and plays the tape back into its transmitter.

A delayed repeater operates in essentially the same way as an ordinary repeater. The signal is received, demodulated, and retransmitted at a different frequency. An isolator circuit prevents interference between the receiver and the transmitter. The block diagram shows a simple delayed-repeater circuit. *See also* REPEATER.

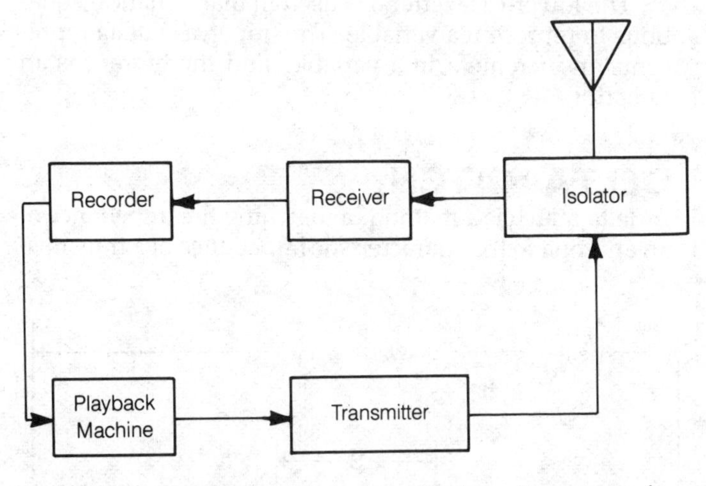

DELAYED REPEATER: A block diagram of a radio repeater for intercepting, demodulating, and recording a signal for later transmission on a different frequency.

DELAYED TRANSMISSION

When a signal is transmitted after the event actually takes place, the transmission is called a delayed transmission. A broadcast station usually delays the transmission of a program for about 7 seconds. This allows time to decide if the signals should be cut off before they are transmitted. This kind of delayed transmission is accomplished by recording the station program on magnetic tape, and playing it back into the transmitter after an elapsed time.

Another form of delayed transmission is the prerecorded program. An audio or audio-video tape recorder is used to record a program. Then the program is edited to meet the time and other requirements of the station, and broadcast at a later time.

DELAY LINE

A delay line is a circuit or length of transmission line that provides a delay for a pulse or signal traveling through it. All transmission lines carry energy with a finite speed. For example, electromagnetic fields travel along a solid polyethylene-dielectric coaxial cable at approximately 66 percent of the speed of light.

The illustration shows three different kinds of delay lines. At A, a length of coaxial cable is wound into a coil. Because the speed of propagation in the cable is about 2×10^8 meters per second, this line provides a delay of 5 nanoseconds (5×10^{-9} second) per meter. At B, a network of inductances and capacitances simulates a long length of transmission line. At C, an interior view of a helical delay line is illustrated. In this type of line, the center conductor is wound into a helix, like a spring, thus greatly increasing its length. This increases the propagation delay per unit length of the line. *See also* DELAY TIME.

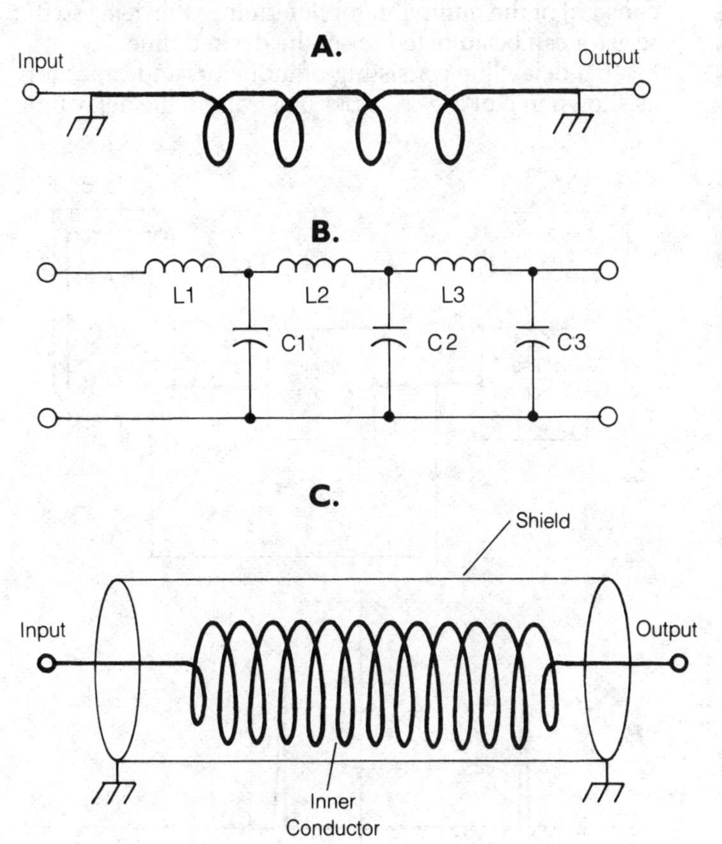

DELAY LINE: Three kinds of delay line are: a coiled length of coaxial cable (A), a ladder array of inductors and capacitors (B), and a coaxial line with an inner helical conductor (C).

DELAY TIME

The delay time is the length of time required for a pulse or signal to travel through a delay circuit or line, as compared to its travel over the same distance through free space. (*See* DELAY CIRCUIT, DELAY LINE.) Delay time is measured in seconds, milliseconds (10^{-3} second), microseconds (10^{-6} second), or nanoseconds (10^{-9} second).

Delay time is ordinarily measured with an oscilloscope. The illustration shows a simple arrangement for measuring the delay time through a circuit (A). A pulse or signal, supplied by the signal generator, is run through the delay circuit and also directly to the oscilloscope. This results in two pulses or waveforms on the oscilloscope screen. The frequency of the pulses or signal is varied, to be certain that the delay time observed is correct (it could be more than one cycle, misleading a technician if only one wavelength is used). The delay time is indicated by the separation of pulses or waves on the oscilloscope screen (shown at B).

Many types of electronic equipment perform a sequence of operations so it may be necessary either to delay the passage of a signal from one circuit to another or to delay the initiation of some operation. Circuits that do this are called time-delay circuits. An example is the monostable multivibrator (*see* MONOSTABLE MULTIVIBRATOR). This time interval between the application of the input pulse and the delivery of the output pulse is the time delay of the circuit. The resistor-capacitor (RC) time constant of the multivibrator determines the delay so the resistor can be adjusted to set the desired time.

In a delay line consisting of inductors and capacitors as shown in B of the DELAY LINE illustration, the delay time is the sum of individual sections. The charge of each capacitor is delayed by its associated inductor. The delay time of each inductor-capacitor (LC) section is given by:

$$t = \sqrt{LC}$$

where t is the delay time in seconds, and L and C are values of inductance and capacitance. Three of these sections would provide a delay of $3t$. *See also* PROPAGATION SPEED.

DELAY TIME: A block diagram of a delay timing circuit (A) and the waveform for measuring time delay on an oscilloscope (B).

DELAY TIMER

Any device that introduces a variable delay in the switching of a circuit is called a delay timer. A delay timer usually has a built-in, resettable clock. After the prescribed amount of time has elapsed, the switching is performed.

Delay timers are common in consumer electronics. They can be used to actuate lamps or alarms to deter possible intruders when a home is unoccupied, or they can be used to time the cooking of food in a microwave oven. *See also* RELAY.

DELTA

Delta is a letter of the Greek alphabet. The capital letter delta, Δ, looks like a triangle. The lowercase letter delta, δ, resembles the lowercase letter *d*.

The lower-case letter δ is used in mathematical equations to represent a variable. The uppercase delta represents an increment in a variable, and therefore it is an operator.

DELTA MATCH

A delta match is a method of matching the impedance of an antenna to the characteristic impedance of a transmis-

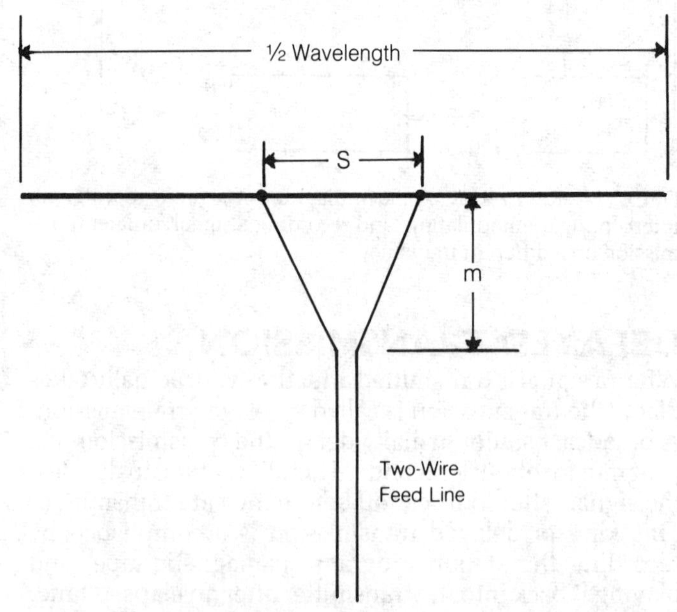

DELTA MATCH: A method for matching the characteristics of a two-wire transmission line to an antenna.

sion line. The delta-matching technique is used with balanced antennas and two-wire transmission lines.

The illustration shows a delta matching system. The length of the network, *m*, and the width or spacing between the connections, *s*, is adjusted until the standing-wave ratio on the feed line is at its lowest value. The length of the radiating element is ½ wavelength.

A variation of the delta match is called the T match. When the transmission line is unbalanced, such as is the case with a coaxial cable, a gamma match may be used for matching to a balanced radiating element. *See also* GAMMA MATCH, T MATCH.

DELTA MODULATION

Delta modulation is a form of pulse modulation (*see* PULSE MODULATION). In the delta-modulation scheme, the pulse spacing is constant, generated by a clock circuit. The pulse amplitude is also constant. But the pulse polarity can vary, being either positive or negative.

When the amplitude of the modulating waveform is increasing, positive unit pulses are sent. When the modulating waveform is decreasing in amplitude, negative unit pulses are sent. When the modulating waveform amplitude is not changing, the pulses alternate between positive and negative polarity (see illustration). Delta modulation gets its name from the fact that it follows the difference, or derivative, of the modulating waveform.

In the delta-modulation detector, the pulses are integrated. This results in a waveform that closely resembles the original modulating waveform. A filter eliminates most of the distortion caused by sampling effects. The integrator circuit can consist of a series resistor and a parallel capacitor; the filter is usually of the lowpass variety. *See also* INTEGRATION.

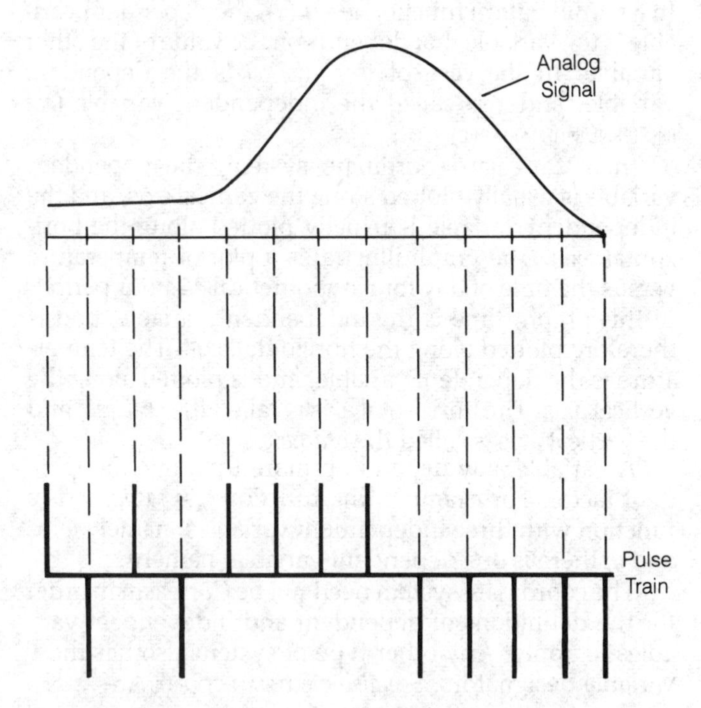

DELTA MODULATION: Delta modulation is a form of pulse modulation for converting analog signals to digital pulses.

DEMAGNETIZING FORCE

A demagnetizing force is any effect that reduces the magnetism of an object. A certain amount of magnetic force, in opposition to the field of a magnetized object, will reduce the magnetism of the object to zero. High temperatures can cause demagnetization; so can a hard physical blow.

Objects are demagnetized by a device called a degausser. The degausser generates either an alternating magnetic field, or a steady magnetic field in opposition to the field of the object to be demagnetized (*see* DEGAUSSING).

In a powdered-iron or ferrite inductor or transformer core, the demagnetizing force required to reduce the core flux to zero is a function of the hysteresis characteristics of the material. This in turn depends on the magnetic permeability of the core material, and on the frequency at which the core is used. *See also* HYSTERESIS, HYSTERESIS LOOP, PERMEABILITY.

DEMODULATION

See DETECTION.

DE MORGAN'S THEOREM

De Morgan's theorem, also called the de Morgan laws, involves sets of logically equivalent statements. Let the logical operation OR be represented by multiplication and the operation AND be represented by addition, as in Boolean algebra (*see* BOOLEAN ALGEBRA). Let the NOT operation be represented by complementation, indicated by an apostrophe ('). Then de Morgan's laws are stated as:

$$(X + Y)' = X'Y'$$
$$(XY)' = X' + Y'$$

The illustration shows these rules in schematic form, showing logic gates, at A and B.

From these rules, it can be proven that the same laws hold for any number of logical statements X_1, $X_2, \ldots, X_n$:

$$(X_1 + X_2 + \ldots + X_n)' = X_1'X_2' \ldots X_n'$$
$$(X_1X_2 \ldots X_n)' = X_1' + X_2' + \ldots + X_n'$$

De Morgan's theorem is useful in the design of digital circuits, since a given logical operation may be easier to obtain on one form than in the other.

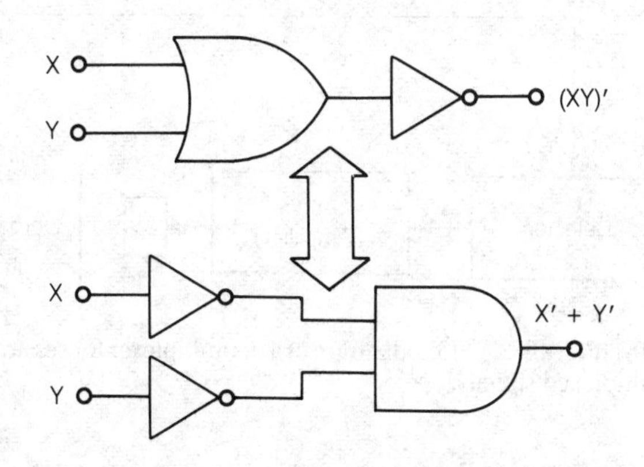

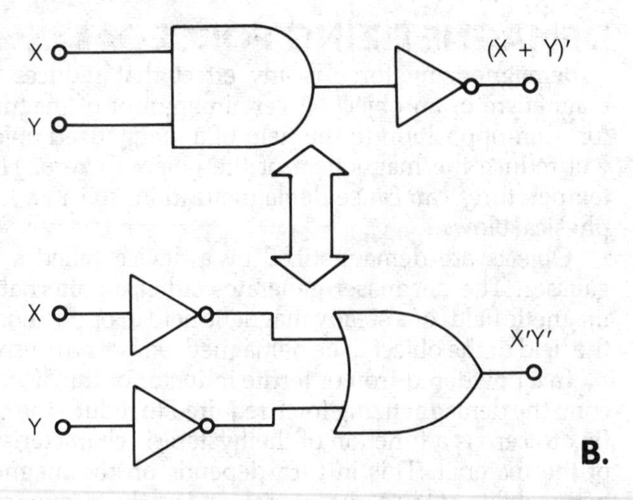

DE MORGAN'S THEOREM: De Morgan's theorem includes symbolic logic statements for showing the equivalence of the logic circuits shown at (A) and those shown at (B).

DEMULTIPLEXER

A demultiplexer is a circuit that separates multiplexed signals (*see* MULTIPLEX). Multiplexing is a method of transmitting several low-frequency signals on a single carrier having a higher frequency. Multiplexing can also be done on a time-sharing basis among many signals. A demultiplexer at the receiving end of a multiplex communications circuit allows interception of only the desired signal, and rejection of all the other signals.

The illustration shows a simplified block diagram of a demultiplexer. The main carrier is amplified and detected (*see* DETECTION), obtaining the subcarriers. A selective circuit then filters out all subcarriers except the desired one. A second detector demodulates the subcarrier, obtaining the desired signal. The main carrier modulation does not have to be the same as the subcarrier modulation. Thus, different types of detectors may be needed for the two carriers. For example, the main carrier might be amplitude-modulated, while the subcarrier is frequency-modulated. *See also* MODULATION.

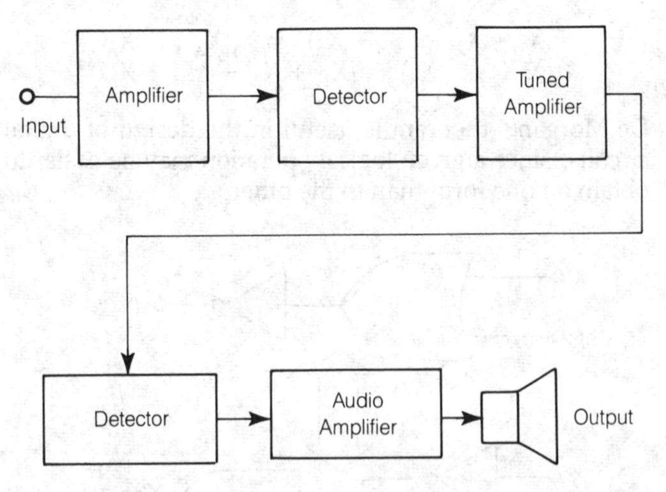

DEMULTIPLEXER: A block diagram of a demultiplexer for separating multiplexed signals.

DENSITY MODULATION

Density modulation is the modulation of the intensity of a beam of particles, by varying only the number of particles. An example of density modulation is shown in the drawing. This type of modulation can be applied to the electron stream from the electron gun of a cathode-ray tube. The greater the density of the electrons in the beam, the brighter the phosphorescent screen will glow when the electrons strike it. This is because, with increasing electron-beam density, more electrons strike the screen per unit time. The density is directly proportional to the number of particles passing any given point in a certain span of time.

In density modulation, all the particles of the beam move at constant speed, regardless of the intensity of the beam. When the speed of the electrons changes, the beam is said to be velocity-modulated. Density modulation may be combined with velocity modulation in some situations. *See also* VELOCITY MODULATION.

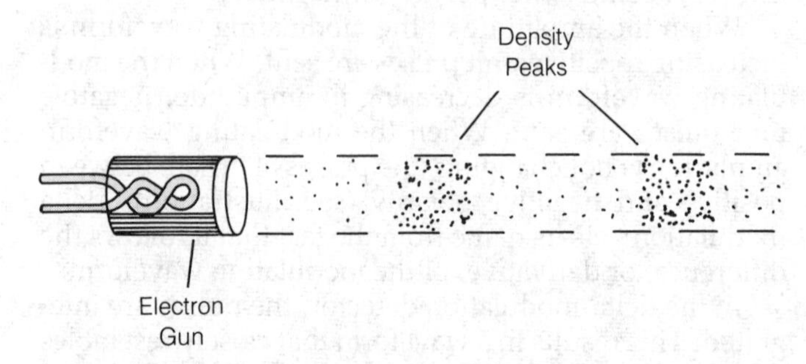

DENSITY MODULATION: In density modulation of an electron stream, the electron density is varied but velocity remains constant.

DEPENDENT VARIABLE

In a mathematical function $y = f(x)$, the dependent variable is the variable that depends on the value of the other variable. In the case of $y = f(x)$, y is the dependent variable, and x is called the independent variable (*see* INDEPENDENT VARIABLE).

In a Cartesian coordinate system, the dependent variable is usually plotted along the vertical axis, and the independent variable is usually plotted along the horizontal axis. The graph illustrates a plot of temperature versus the time of day for a hypothetical 24-hour period. In this graph, time is the independent variable, and is therefore plotted along the horizontal axis. The temperature is the dependent variable, and is plotted along the vertical axis. The horizontal axis is called the *abscissa*, and the vertical axis is called the *ordinate*.

A variable may depend on more than one independent factor. For example, the function $z = f(w,x,y)$ is a function with three independent variables, namely w, x, and y; there is one dependent variable, namely z.

The coordinate system need not be Cartesian in order for the definitions of dependent and independent variables to apply. Any other type of system also has these variable designators. *See also* CARTESIAN COORDINATES, FUNCTION.

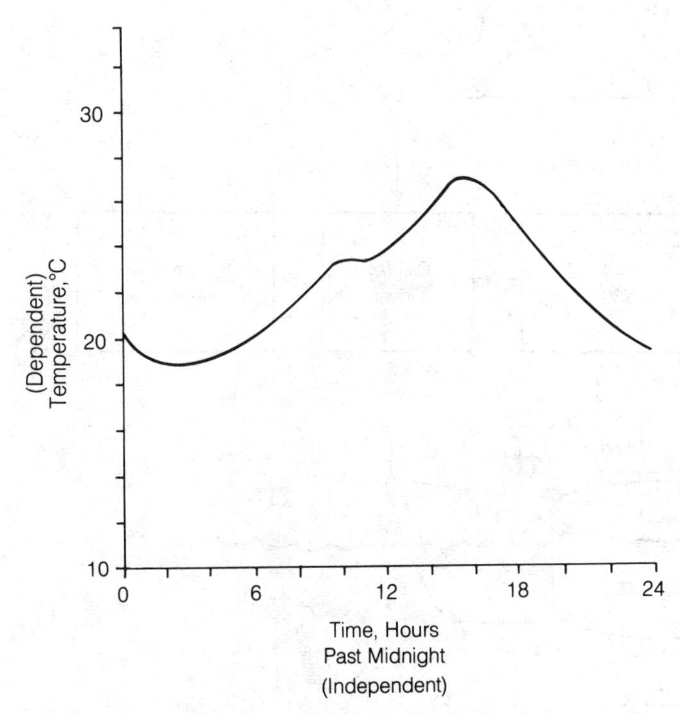

DEPENDENT VARIABLE: In this plot of temperature versus time, temperature is the dependent variable.

DEPLETION LAYER

A depletion layer forms at the junction of an N-type semiconductor and a P-type semiconductor whenever the junction is reverse-biased (*see* N-TYPE SEMICONDUCTOR, P-N JUNCTION, P-TYPE SEMICONDUCTOR).

A depletion layer is essentially devoid of charge carriers, and therefore no current can flow through the region. This is why a semiconductor diode will not conduct in the reverse direction. The depletion layer acts as a dielectric, and, in fact, a reverse-biased diode can be used as a capacitor.

DEPLETION MODE

See FIELD-EFFECT TRANSISTOR.

DEPOLARIZATION

Electrolyte material may form gases at, or leave deposits on, the electrodes in the presence of the flow of current. When this occurs, it is called polarization (*see* POLARIZATION). Polarization can prevent a dry cell or storage battery from operating properly. The deposits can act as electrical insulators, cutting off the flow of current by increasing the internal resistance.

Depolarization is the process of preventing, or at least minimizing, polarization in an electric cell. In dry cells, the compound manganese dioxide is used in the electrolyte material to retard polarization. *See* BATTERY.

DEPTH SOUNDER

Depth sounders are marine electronic instruments used for determining the depth of water beneath a ship. They are the modern replacements for the handheld lead line that has been used by sailors for centuries. Also known as depth finders or fathometers, they determine water depth by measuring the time for sound energy to travel from the transducer mounted under a ship hull to the seafloor and be reflected back to the transducer (see figure). The frequency of the sound waves used by most depth finders is about 200 kHz in the ultrasonic region. Because sound travels about 4800 feet per second in seawater, the round trip time of the pulse in 120 feet of water is about 1/20 second.

The principles of the depth sounder are closely related to those of sonar, and its major components are transmitter, oscillating transducer, receiver, and display. The transducer, typically a solid cylinder of piezoelectric ceramic, oscillates in much the same manner as a phonograph pickup or microphone. This vibration projects ultrasonic waves into the water in a relatively narrow beam directed at the bottom. The directional characteristic of the beam makes it necessary that the ship or boat remain relatively stable to obtain an accurate reading.

This ultrasonic signal is not in the range that humans can hear. It travels at 4800 feet per second (or 0.0002083 seconds per foot) in water. As the signal leaves the transducer, it triggers counting circuitry in a digital display or causes a neon lamp to flash at the zero-depth mark on a rotating display.

The time required for a signal to travel from the transducer to the bottom is the product of water depth (in feet) times the speed of sound in water (in seconds per foot). In 30 feet of water the signal will reach the bottom in 0.00625 seconds; it will be reflected back in 0.01250 seconds. Because the bottom seldom has a smooth, flat surface, the reflected signal is widely scattered. Therefore, only a small amount of the reflected signal actually reaches the transducer.

Many sounders have the transmitter, receiver, and display packaged together in a single case, and the transducer is connected to the central enclosure by coaxial cable. There are four kinds of displays on sounders: special rotating flasher indicators, analog (needle type), digital, and chart recorder. The rotating flasher indicator has a flashing lamp on the edge of a rotating disk mounted on a calibrated dial. The lamp flashes at the top (zero depth) when a pulse is transmitted and flashes again when an echo returns. Depth is read from the scale closest to the second lamp flash. Digital units might have LED (light-emitting diode), liquid crystal, or vacuum fluorescent displays. The analog meter display is a d'Arsonval meter movement.

Some sounders have recorders that plot a permanent record of the ocean floor. They may also show fish or various underwater obstructions. The recorder scrolls chemically treated paper past a heated stylus, which makes dry marks to trace out a depth profile on the paper. Some sounders with recorders also have an indicator for faster depth reading to aid in interpreting bottom conditions. Many sounders are also equipped with alarms that sound if the ship enters water that is shallower than the alarm setting. This is useful if the ship drags anchor with changing tide and wind.

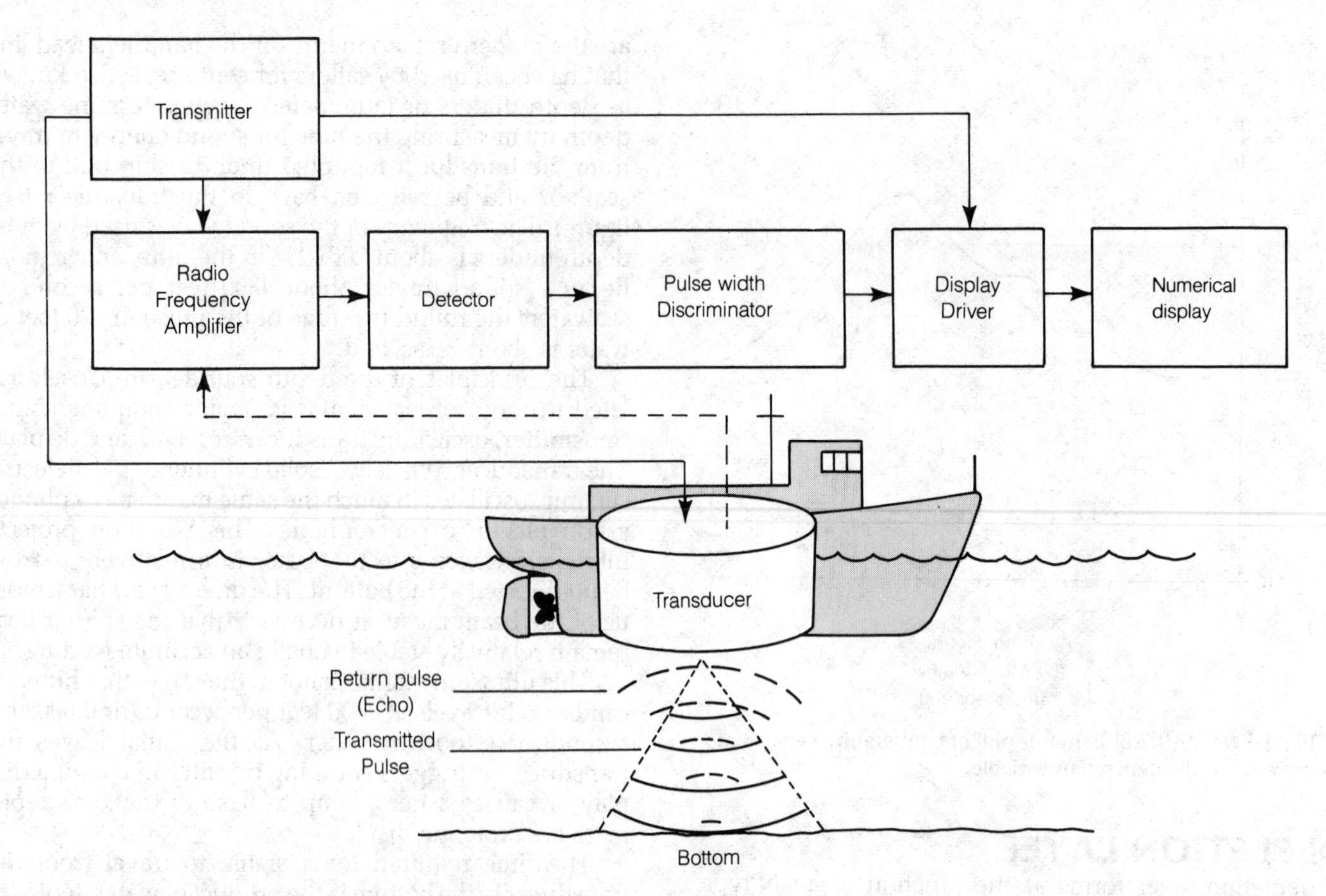

DEPTH SOUNDER: Diagram of a depth sounder or fathometer showing how water depth is determined by timing the round trip transit of sound waves and their echoes from the sea floor.

A depth sounder can be used as an aid in piloting at night or in the fog. A ship position can be estimated by comparing a sequence of depth readings with those shown in the chart for the area. However, corrections must be made for the tide if this information is to be reliable.

DERATING CURVE

A derating curve is a function that specifies the amount of dissipation a component can withstand safely. This function is plotted with time as the dependent variable, or ordinate, and dissipation or temperature as the independent variable, or abscissa (*see* DEPENDENT VARIABLE, INDEPENDENT VARIABLE).

DERIVATIVE

The derivative of a mathematical function is an expression of its rate of change, either at a certain instant, or in the general case. When the value of a function is constant, the rate of change, and therefore the derivative, is zero at all points. When the value of a function increases, the derivative is positive; the greater the rate of the increase, the larger the value of the derivative. When the value of a function decreases, the derivative is negative; greater rates of decrease result in derivatives that are more and more negative (*see* FUNCTION). The graph illustrates these three kinds of situations. The actual value of a derivative can be any real number at a specific point on

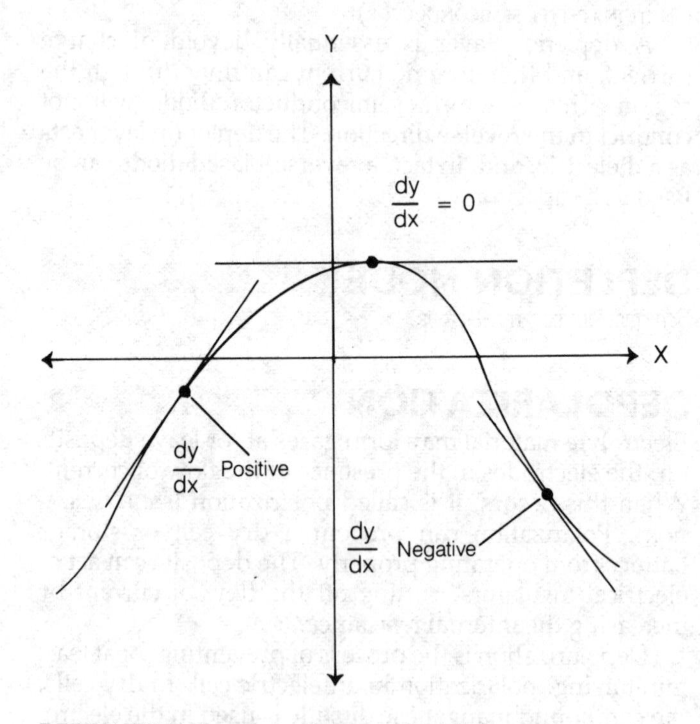

DERIVATIVE: The plots of three derivatives or tangents to the curve showing positive, zero, and negative slope.

a function. The derivative can, in some cases, be undefined at certain points of a function.

The derivative of a function may be the same at every point in the domain of a function, or it may change. The derivative of a function, in the general sense, is another function. For a given function $y = f(x)$, the derivative is denoted in a variety of ways, such as dy/dx, $df(x)/dx$, y', or $f'(x)$. The operation of finding a derivative for a function is called differentiation. Derivatives can be found in succession; such derivatives are called the second, third, fourth, and higher derivatives, generally written y'', y''', and so on, or $f''(x)$, $f'''(x)$, and so on.

The table shows the derivative functions for several of the most familiar mathematical functions. *See also* CALCULUS, DIFFERENTIATION, INDEFINITE INTEGRAL.

DERIVATIVE: Derivatives of some common mathematical functions. Constants are represented by the letters c and n. The value of e is approximately 2.7183.

f(x)	df(x)/dx
$y = c$	$y' = 0$
$y = cx$	$y' = c$
$y = cx^2$	$y' = 2cx$
$y = cx^n$	$y' = ncx^{n-1}$
$y = \log_e (x)$	$y' = 1/x$
$y = e^x$	$y' = e^x$
$y = \sin x$	$y' = \cos x$
$y = \cos x$	$y' = -\sin x$
$y = \tan x$	$y' = \sec^2 x$

DESENSITIZATION

Desensitization is a reduction in the gain of an amplifier, particularly when the amplifier is employed in the radio-frequency stages of a receiver. Desensitization can be introduced deliberately; the radio-frequency (rf) gain control is used to reduce the sensitivity of a receiver. Desensitization can occur when it is not wanted and may cause a loss in radio reception.

The most frequent cause on unwanted desensitization of a radio-frequency receiver is an extremely strong signal near the operating frequency. Such a signal, after passing through the front-end tuning network, can drive the first rf-amplifier stage to the point where its gain is radically diminished. When this happens, the desired signal appears to fade out. If the interfering signal is strong enough, reception may become impossible near its frequency.

Some receivers are more susceptible to unwanted desensitization, or desensing, than others. Excellent front-end selectivity, and the choice of amplifier designs that provide uniform gain for a wide variety of input signal levels, make this problem less likely. Sometimes, a tuned circuit in the receiver antenna feed line is helpful. *See also* FRONT END.

DESOLDERING

When replacing a faulty component in an electronic circuit, it is usually necessary to desolder some connections. The method of desoldering employed in a given situation is called the desoldering technique.

With most printed-circuit boards, the desoldering technique consists of the application of heat with a soldering iron, and the conduction of solder away from the connection by means of a braided wire called a wick. The connection and the wick must both be heated to a temperature sufficient to melt the solder. Excessive heat, however, should be avoided, so that only the desired connection is desoldered, and so that the circuit board and nearby components are not damaged.

Many sophisticated desoldering devices are available. One popular device employs an air-suction nozzle, which swallows the solder by vacuum action as a soldering iron heats the connection. This equipment is especially useful when a large number of connections must be desoldered rapidly. It is also useful in desoldering miniature connections are damaged easily if force is applied. *See also* SOLDER.

DETECTION

Detection is the extraction of the modulation from a radio-frequency signal. A circuit that performs this function is called a detector. Different forms of modulation require different detector circuits.

The simplest detector consists of a semiconductor diode, which passes current in only one direction. This detector is suitable for demodulation of amplitude-modulated (AM) signals. By cutting off one half of the signal, the modulation envelope is obtained. A Class AB, B, or C amplifier can be used to perform this function, and also provide some gain. This is called envelope detection (*see* CLASS AB AMPLIFIER, CLASS B AMPLIFIER, CLASS C AMPLIFIER, ENVELOPE DETECTOR).

There are several different ways of detecting a frequency-modulated (FM) signal. These methods also apply to phase modulation. An ordinary AM receiver can be employed to receive FM by means of a technique called slope detection (*see* SLOPE DETECTION). Circuits that actually sense the frequency or phase changes of a signal are called the discriminator and ratio detector (*see* DISCRIMINATOR, RATIO DETECTOR).

Single-sideband (SSB) and continuous-wave (CW) signals, as well as signals that employ frequency-shift keying (FSK), require a product detector for their demodulation. This circuit operates by mixing the received signal with the output of a local oscillator, resulting in an audio-frequency beat note.

The detector circuit in a radio receiver is generally placed after the radio-frequency or intermediate-frequency amplifying stages, and ahead of the audio-frequency amplifying stages. *See also* AMPLITUDE MODULATION, CONTINUOUS WAVE, FREQUENCY MODULATION, FREQUENCY-SHIFT KEYING, PHASE MODULATION, SINGLE SIDEBAND.

DETUNING

Detuning is a procedure in which a resonant circuit is deliberately set to a frequency other than the operating frequency of the equipment. Detuning can be employed in the intermediate-frequency stages of a receiver to reduce the chances of unwanted interstage oscillation. This is called stagger tuning (*see* STAGGER TUNING).

The preselector stage of a receiver can be tuned to a

higher or lower frequency than the intended one to reduce the gain in the presence of an extremely strong signal. A transmission line is usually made nonresonant, or detuned, to reduce its susceptibility to unwanted coupling with the antenna radiating element. *See also* PRESELECTOR.

DEUTERIUM

Deuterium is a form of the element hydrogen. It has an atomic number of 1 and an atomic weight of 2.

Ordinary hydrogen consists of a single electron in orbit around a single proton (*see* HYDROGEN). This atom is the most common in the universe. Hydrogen makes up about 55 percent of the interstellar material in the cosmos. When a neutron is added to the nucleus of the simple hydrogen atom, deuterium is formed. Deuterium occurs in the fusion process, where hydrogen is converted into helium, liberating energy.

Initially, in the hydrogen fusion reaction, two protons combine and throw off a positive charged particle of anti-matter, known as a positron (*see* POSITRON). This results in a nucleus of deuterium, containing one proton and one neutron. This nucleus joins with another proton to form a helium-3 nucleus, which contains two protons and one neutron. Two nuclei of helium-3 then combine, giving off two protons, to form a nucleus of helium-4. Helium-4 is the second most abundant element in the universe, comprising about 44 percent of the interstellar matter.

DEVIATION

In a frequency-modulated (FM) signal, deviation is the maximum amount by which the carrier frequency changes either side of the center frequency. The greater the amplitude of the modulating signal in an FM transmitter, the greater the deviation, up to a certain maximum. Deviation can be measured on a signal monitor, as shown in the photograph.

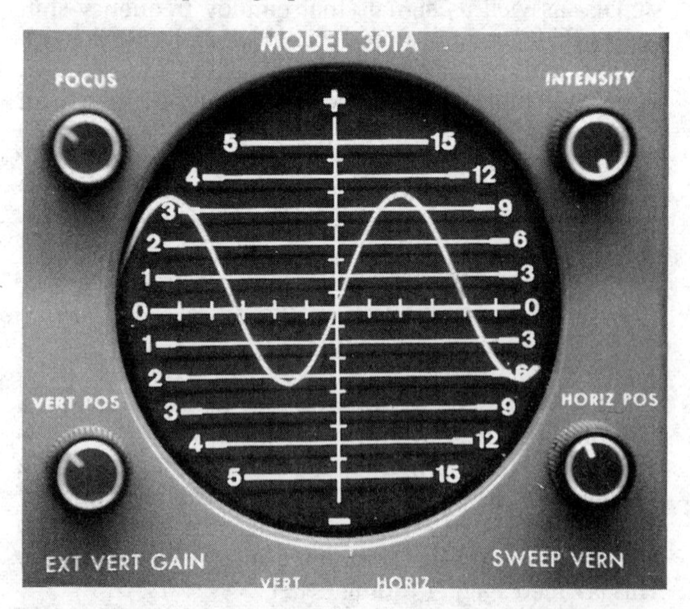

DEVIATION: A monitor display of deviation of an FM signal.

Deviation is generally expressed as a plus-or-minus (±) frequency figure. For example, the standard maximum deviation in a frequency-modulated communications system is ±5 kHz. When the deviation is greater than the allowed maximum, an FM transmitter is said to be over-deviating. This often causes the received signal to sound distorted. Sometimes, the downward deviation is not the same as the upward deviation. This is the equivalent of a shift in the signal carrier center frequency with modulation, and it, too, can result in a distorted signal at the receiver. *See also* CARRIER SWING, FREQUENCY MODULATION, MODULATION INDEX.

DIAC

A diac is a semiconductor device that resembles a PNP bipolar transistor, except that it has no base connection. A diac acts as a bidirectional switch. The illustration shows the construction of a diac at A. The schematic symbol is shown at B.

The diac can be used as a variable-resistance device for alternating current. The output voltage can be regulated independently of the load resistance. In conjunction with a triac, the diac can be employed as a motor-speed control, light dimmer, or other control. *See also* TRIAC.

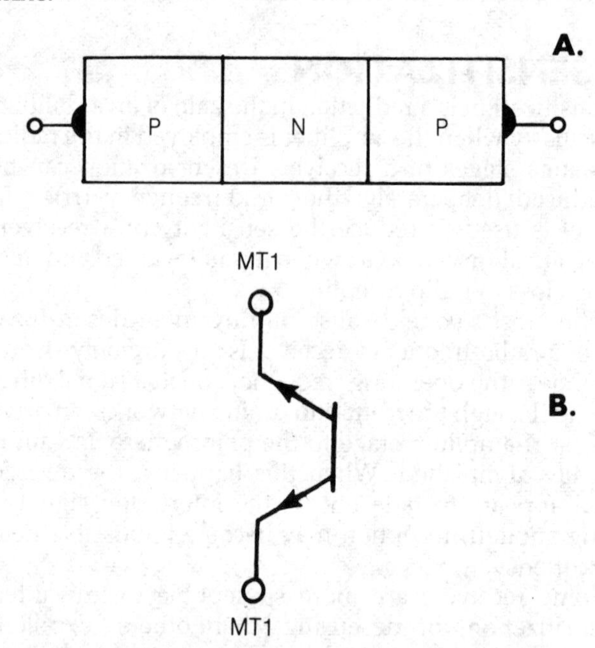

DIAC: A diagram of a diac (A) and its schematic symbol (B).

DIAGRAM

In electronics, a diagram is a drawing that depicts the organization of an electrical or electronic circuit. There are three basic kinds of diagrams: block, schematic, and wiring.

A block diagram shows only the major circuits comprising a system. Rectangles or circles are used to represent each circuit, and are labeled according to their function. The boxes or circles are interconnected by lines to show the general functional scheme. Arrows are used to show the path of a signal.

A schematic diagram shows all of the main components in a circuit and the way they are connected together. Component values are generally given, often with the reference designation. Wiring diagrams are still more complete; pin numbers for integrated circuits and tubes, color-code designators, and part numbers are shown for all components. This makes it easier for a technician to assemble the circuit.

In schematic and wiring diagrams, certain standard symbols are employed to represent the various components. *See also* BLOCK DIAGRAM, SCHEMATIC DIAGRAM, WIRING DIAGRAM.

DIALING

In a telephone system, dialing is a means of selecting the telephone subscriber to be called. A sequence of seven digits is used for this purpose in the United States. Long-distance calls require a three-digit area-code prefix in addition to the seven-digit local designators. This entire sequence might be preceded by another digit to indicate direct dialing or operator assistance.

Telephone dialing is performed with either of two mechanisms: the rotary dial, or the Touchtone® system. The rotary dial transmits a series of pulses, with one pulse representing the number 1, two pulses representing the number 2, and so on up to ten pulses, representing the number 0. The pulses are sent at the rate of approximately ten per second, although other speeds are sometimes used. The Touchtone® system uses a 12-key or 16-key pad, with pairs of audible tones representing the digits 0 through 9, along with the symbols * and #. In the 16-key pad, four extra functions are included, designated A, B, C, and D. The Touchtone® system permits much more rapid dialing than the rotary system. The tone system also makes it easy to interface the telephone with computer equipment. *See also* TELEPHONE SYSTEM, TOUCHTONE®.

DIAL SYSTEM

Dial systems are the mechanical devices used to set the frequency of a radio transmitter or receiver to the desired value. Dial systems are also widely employed in such controls as preselectors, amplifier tuning networks, and antenna impedance-matching circuits.

DIAMAGNETIC MATERIAL

Any material with a magnetic permeability less than 1 is called a diamagnetic material. The permeability of free space is defined as equal to 1.

Examples of diamagnetic materials include bismuth, paraffin wax, silver, and wood. In all these cases, the magnetic permeability is only a little less than 1. *See also* FERROMAGNETIC MATERIAL, PARAMAGNETIC MATERIAL, PERMEABILITY.

DIAPHRAGM

A diaphragm is a thin disk for converting sound waves into mechanical vibrations and vice versa. Diaphragms are used in speakers, headphones, and microphones.

In the speaker or headset, electrical impulses cause a coil to vibrate within a magnetic field. The diaphragm is attached to the coil, and moves along with the coil at the frequency of the electrical impulses. The result is sound waves, since the diaphragm imparts vibration to the air molecules immediately adjacent to it.

In a dynamic microphone, the reverse occurs. Vibrating air molecules set the diaphragm in motion at the same frequency as the impinging sound. This causes a coil, attached to the diaphragm, to move back and forth within a magnetic field supplied by a permanent magnet or electromagnet. In a crystal or ceramic microphone, vibration is transferred to a piezoelectric material, which generates electrical impulses. In any microphone, the electrical currents have the same frequency characteristics as the sound hitting the diaphragm.

Some devices can operate as either a microphone or a speaker/headphone. *See also* CRYSTAL TRANSDUCER, HEADPHONE, MICROPHONE, SPEAKER, TRANSDUCER.

DIELECTRIC

A dielectric is an electrical insulator used in the manufacture of cables, capacitors, and coil forms. Dry, pure air is an excellent dielectric. Other examples of dielectrics include wood, paper, glass, and various rubbers and plastics. Distilled water is also a good dielectric; water is a conductor only because it so often contains impurities that enhance its conductivity.

Dielectric materials are classified according to their ability to withstand electrical stress, and according to their ability to cause a charge to be retained when they are employed in a capacitor. Dielectric materials are also classified according to their loss characteristics. Generally, dielectric materials become lossier and less able to withstand electrical stress as the operating frequency is increased. *See also* DIELECTRIC ABSORPTION, DIELECTRIC BREAKDOWN, DIELECTRIC CONSTANT, DIELECTRIC CURRENT, DIELECTRIC HEATING, DIELECTRIC LENS, DIELECTRIC LOSS, DIELECTRIC POLARIZATION, DIELECTRIC RATING, DIELECTRIC TESTING.

DIELECTRIC ABSORPTION

After a dielectric material has been discharged, it may retain some of the electric charge originally placed across it. This effect is called dielectric absorption because the material seems to absorb some electric charge. In some capacitors, the dielectric absorption may make it necessary to discharge the component several times before the open-circuit voltage remains at zero.

DIELECTRIC BREAKDOWN

When the voltage across a dielectric material becomes sufficiently high, the dielectric, normally an insulator, will begin to conduct. When this occurs in an air-dielectric cable, capacitor, or feed line, it is called arcing (*see* ARC). The breakdown voltage of a dielectric substance is measured in volts or kilovolts per unit length. Some dielectric materials can withstand much greater electrical stress than others. The amount of voltage, per unit length, that a dielectric material can withstand is called its dielectric strength.

In most materials, dielectric breakdown is the result of ionization. At a pressure of one atmosphere, air breaks down at a potential of about 2 to 4 kilovolts per millimeter. The value depends on the relative humidity and the amount of dust and other matter in the air; the greater the humidity or the amount of dust, the lower the breakdown voltage. In a solid dielectric material such as polyethylene, which has a dielectric strength of about 1.4 kilovolts per millimeter, permanent damage can result from excessive voltage. *See also* DIELECTRIC, DIELECTRIC HEATING, DIELECTRIC RATING.

DIELECTRIC CONSTANT

The dielectric constant of a material, usually abbreviated by the lowercase letter k, is a measure of the ability of a dielectric material to hold a charge. The dielectric constant is generally defined in terms of the capability of a material to increase capacitance. If an air-dielectric capacitor has a value of C, then the same capacitor, with a dielectric substance of dielectric-constant value k, will have a capacitance of kC. The dielectric constant of air is thus defined as 1.

Various insulating materials have different dielectric constants. Some materials have a dielectric constant that changes considerably with frequency; usually the constant decreases as the frequency increases. The table shows the dielectric constants of ten materials at frequencies of 1 kHz, 1MHz, and 100 MHz, at approximately room temperature. *See also* DIELECTRIC, PERMITTIVITY.

DIELECTRIC CURRENT

Dielectric materials are electrical insulators but not perfect insulators. A small current flows through the best-quality dielectric substances. This current results in some heating, and therefore some loss (*see* DIELECTRIC HEATING, DIELECTRIC LOSS).

For direct current, the resistivity of a dielectric material is specified in ohm-millimeters or ohm-centimeters. This resistivity is always high, assuming the dielectric material is not contaminated; it ranges from about 10^{12} to 10^{20} ohm-centimeters at room temperature. The table shows the direct-current resistivity of several types of dielectric materials. The greater the resistivity, the lower the dielectric current for a given voltage, provided the voltage is not so great that breakdown occurs. In the event of dielectric breakdown, a conductive path forms through the material, and the current increases dramatically. The substance then loses its dielectric properties. *See also* DIELECTRIC BREAKDOWN.

DIELECTRIC HEATING

Dielectric heating is the result of energy losses in a dielectric material when it is subjected to an electric field (*see* DIELECTRIC LOSS). Dielectric materials heat up in direct proportion to the intensity of the electric field. The lossier a dielectric material, the hotter it will get in the presence of an electric field.

Dielectric heating is sometimes deliberately used for forming certain plastics. By subjecting the plastic to an intense radio-frequency electric field, the material becomes soft and pliable.

DIELECTRIC LENS

A dielectric lens focuses or collimates electromagnetic energy in the microwave frequency range. These lenses operate in a manner similar to an optical lens. Any dielectric material may be used, although the dielectric loss should be as low as possible for maximum transmission of energy.

The drawing illustrates the principle of dielectric lens operation. Waves striking the lens axially, that is, at the center, are not bent. When the waves do not strike the surface of the lens orthogonally, refraction occurs.

The focal length of a dielectric lens depends on the degree of curvature in the lens, and on the dielectric constant of the material from which the lens is made. The greater the curvature and/or the larger the dielectric constant, the shorter the focal length. *See also* FOCAL LENGTH, REFRACTION.

DIELECTRIC LOSS

Some dielectric materials transform very little of an electric field into heat, and others transform a large amount

DIELECTRIC CONSTANT, CURRENT, AND RATING:
Dielectric characteristics of various materials at room temperature (approximately 25°C).

Material	Dielectric Constant 1 kHz	1 MHz	100 MHz	Direct-Current Resistivity, Ω-Centimeter	Rating, kV/millimeter
Bakelite	4.7	4.4	4.0	10^{11}	0.1
Balsa wood	1.4	1.4	1.3		—
Epoxy resin	3.7	3.6	3.4	4×10^7	0.13
Fused quartz	3.8	3.8	3.8	10^{19}	0.1
Paper	3.3	3.0	2.8	—	0.07
Polyethylene	2.3	2.3	2.3	10^{17}	1.4
Polystyrene	2.6	2.6	2.6	10^{18}	0.2
Porcelain	5.4	5.1	5.0	—	—
Teflon	2.1	2.1	2.1	10^{17}	6
Water (pure)	78	78	78	10^6	—

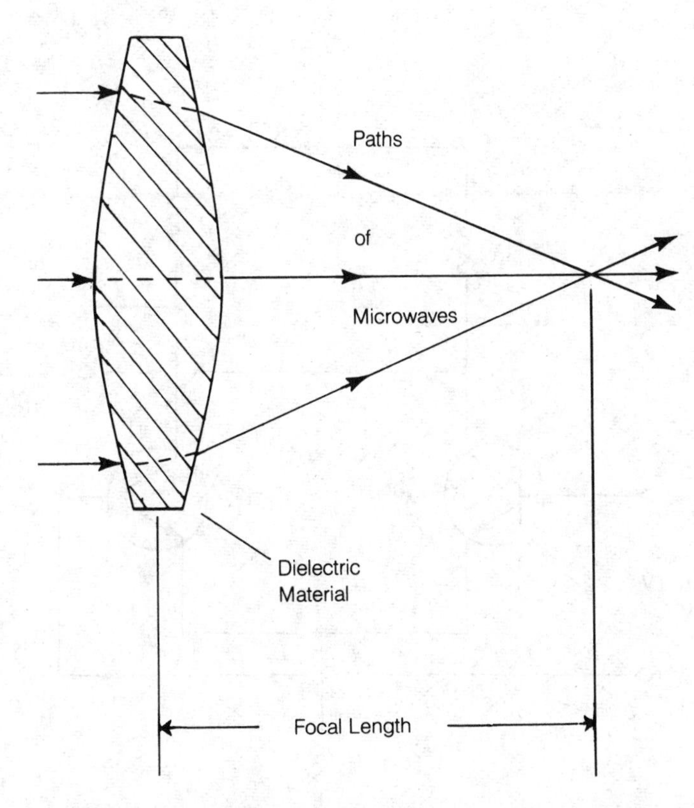

DIELECTRIC LENS: A dielectric lens can focus microwave energy as an optical lens focuses light.

of the field into heat. The dielectric loss in a material is related to its ability to become hot in the presence of an alternating electric field. Dielectric loss is expressed in terms of the dissipation factor (*see* DISSIPATION FACTOR).

The dielectric loss of an insulating substance usually increases as the frequency of the alternating electric field is raised. Some of the best dielectric materials, such as polystyrene, have very low loss levels. Other materials, such as nylon, show generally lower losses as the frequency increases. *See also* DIELECTRIC.

DIELECTRIC POLARIZATION

When an electric field is placed across a dielectric material, the location of the positive-charge center in each atom is slightly displaced relative to the negative-charge center. The greater the intensity of the surrounding electric field, the greater the charge displacement. This results in what is known as dielectric polarization (see illustration). At A, the atoms of the dielectric material are shown, with their charge-center locations, under conditions of no electric field. At B, the charge centers are illustrated under the influence of a weak electric field, with the positive part of the field toward the right in the diagram. At C, the atomic charge centers are shown under the influence of a strong electric field, again with the positive part of the field toward the right. When the polarity of the electric field is reversed, the dielectric polarization also reverses.

Dielectric polarization is caused by the forces of attraction between opposite charges, and repulsion between like charges. If the intensity of the electric field is sufficiently great, electrons are stripped from the atomic nuclei. This results in conduction, since the atoms then easily flow from one atom to the next, throughout the material. Sometimes this ionization causes permanent damage, such as in a solid dielectric. In the case of a gaseous dielectric material, such as air, the ionization does not cause permanent damage. *See also* DIELECTRIC, DIELECTRIC BREAKDOWN.

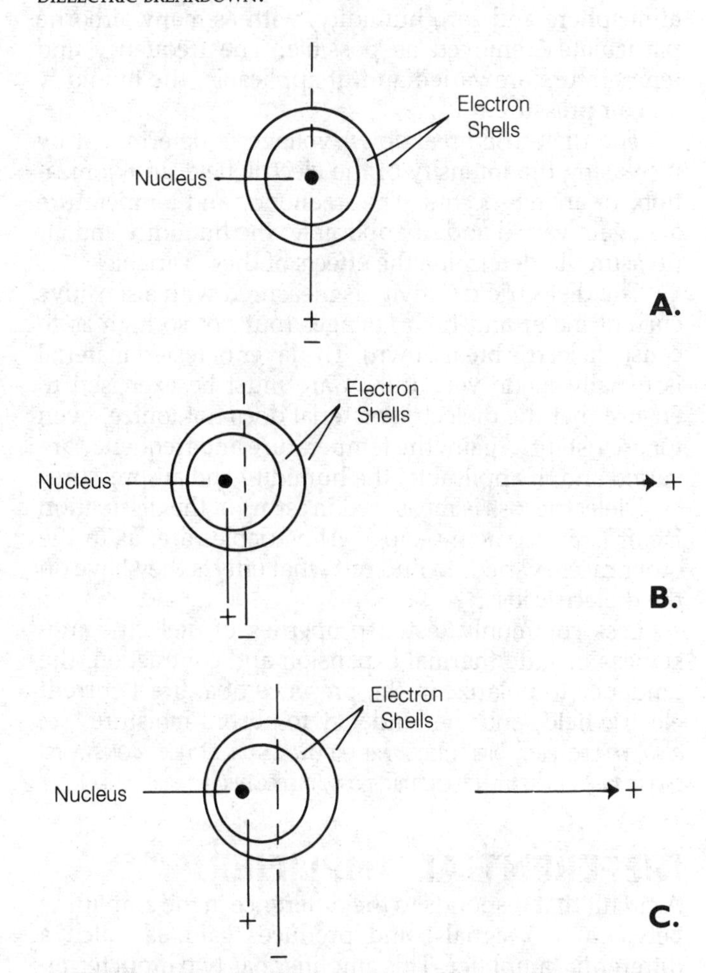

DIELECTRIC POLARIZATION: The positive charge is concentric within the negative charge of an atom in dielectric material with no electric field present (A), minor displacement of the positive charge in a weak field (B), and increased displacement in a strong field (C).

DIELECTRIC RATING

The dielectric rating of an insulating material is a measure of its ability to withstand electric fields without breaking down (*see* DIELECTRIC BREAKDOWN). The term can also be used to refer to the general characteristics of a dielectric material, such as the dielectric constant, the resistivity, and the loss. The dielectric characteristics of common substances are shown in the DIELECTRIC CONSTANT, CURRENT AND RATING table.

The dielectric rating of a capacitor is usually the breakdown voltage of the entire device. This rating depends on the thickness of the dielectric layer, as well as on the dielectric material used.

DIELECTRIC TESTING

Dielectric testing is a means of determining the dielectric properties of a material, such as the dielectric constant, breakdown voltage, or loss characteristics. These parameters often vary with the frequency, temperature, and sometimes with the relative humidity and atmospheric pressure.

To determine the dielectric constant of a material at a given electromagnetic frequency, the value of a capacitor is compared, using an air dielectric at a pressure of one atmosphere and zero humidity, with as many airborne particulates removed as possible. The frequency and temperature are varied and, if applicable, the humidity and air pressure.

The dielectric breakdown voltage is determined by increasing the intensity of the electric field until ionization, or arcing, occurs. The frequency and temperature are again varied and, if applicable, the humidity and air pressure, to determine the effects of these variables.

The dielectric resistivity is measured with a sensitive current meter and high voltages (but not so high as to cause dielectric breakdown). The layer of tested material is usually made very thin. Care must be exercised to ensure that the dielectric material does not ionize, even for an instant. Again, the temperature and frequency are varied and, if applicable, the humidity and air pressure.

Dielectric loss is measured in terms of the dissipation factor (*see* DISSIPATION FACTOR). All variables are, as in the other cases, varied, to find out what effects they have on the dielectric loss.

Less commonly tested properties of dielectric substances include thermal expansion and contraction, the tendency to polarize in the presence of a direct-current electric field, and the tendency to absorb moisture. *See also* DIELECTRIC, DIELECTRIC BREAKDOWN, DIELECTRIC CONSTANT, DIELECTRIC CURRENT, DIELECTRIC LOSS, DIELECTRIC POLARIZATION.

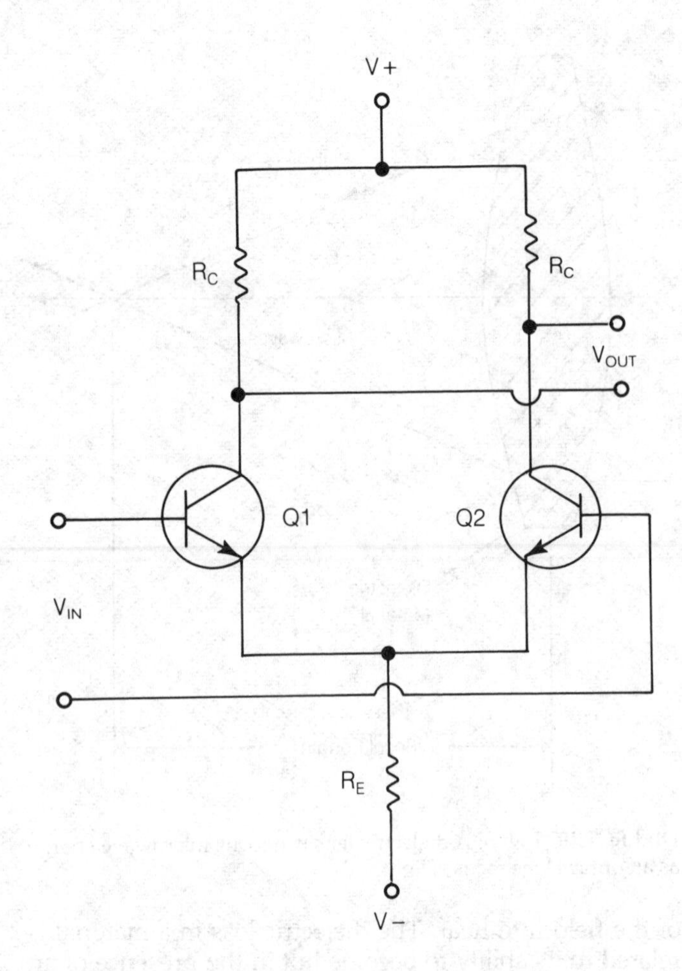

DIFFERENTIAL AMPLIFIER: A schematic of a differential amplifier that can provide differential output, differential input, single-ended input, and single-ended output.

DIFFERENTIAL AMPLIFIER

A circuit that responds to the difference in the amplitude between two signals, and produces gain, is called a differential amplifier. This amplifier has two input terminals and two output terminals. Differential amplifiers are now made as integrated circuits. The diagram illustrates a simple differential amplifier.

When two identical signals are applied to the input terminals of a differential amplifier, the output is zero. The greater the difference in the amplitudes of the signals, the greater the output amplitude. Differential amplifiers can be used as linear amplifiers and have a broad operating-frequency range. Differential amplifiers can also be used as mixers, detectors, modulators, and frequency multipliers. *See also* DIFFERENTIAL VOLTAGE GAIN.

DIFFERENTIAL CAPACITOR

A differential capacitor is a variable capacitor with two sets of stator plates, and a single rotor-plate set. As the capacitor rotor is turned, the rotor plates move into one set of stators and out of the other. There are two variable capacitors in a differential capacitor. The values of these two capacitors are inversely related; as the capacitance of one unit increases, the capacitance of the other unit decreases. When one capacitor is at minimum value, the other is at the maximum. The rotor plates of the two capacitors are connected together and form the common terminal. The stator sets are brought out separately. *See also* AIR-VARIABLE CAPACITOR.

DIFFERENTIAL EQUATION

A differential equation is a mathematical equation in which some or all of the variables are derivatives of differentials (*see* DERIVATIVE, DIFFERENTIATION). These equations can be categorized according to their form. Equations containing only first derivatives are called first-order differential equations. If the highest order derivative in a differential equation is the *n*th, then the equation is called an *n*th-order differential equation.

Differential equations can be fairly simple, or they can be quite complicated and hard to solve. The solution of a differential equation generally requires integration (*see* INDEFINITE INTEGRAL).

DIFFERENTIAL INSTRUMENT

A differential, or indicating, instrument is a meter that shows the difference between two input signals. Two identical coils are connected to two sets of input terminals. The coils carry currents in opposite directions. When the currents are equal, the meter is balanced and there is no deflection of the needle. When one current is greater in magnitude than the other, the net current flow is in one direction or the other. Differential instruments may use amplifiers for better sensitivity. *See also* DIFFERENTIAL AMPLIFIER.

DIFFERENTIAL KEYING

Differential keying is a form of amplifier/oscillator keying in a code transmitter. While oscillator keying has the advantage of allowing full break-in operation, since the transmitter is off during the key-up intervals, chirp often results because of loading effects of the amplifier stages immediately following the keyed oscillator (*see* CHIRP, CODE TRANSMITTER).

In differential keying, the amplifier is keyed along with the oscillator, but it is delayed. When the key is pressed, the oscillator comes on first. Then, a few milliseconds later, the amplifier is switched on (see drawing). This gives the oscillator time to chirp before the signal is actually transmitted over the air. When the key is lifted, both the oscillator and the amplifier are switched off together; alternatively, the amplifier may be switched off a few milliseconds before the oscillator.

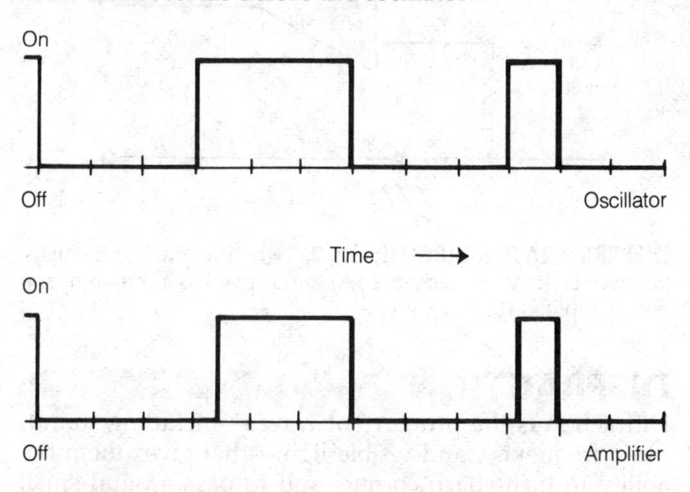

DIFFERENTIAL KEYING: The amplifier is switched on after the oscillator as a protective measure in differential keying.

DIFFERENTIAL TRANSDUCER

A differential transducer is a sensor device that has two input terminals and a single output terminal. The output is proportional to the difference between the input parameters. An example is a differential pressure transducer, which responds to the difference in mechanical pressure between two points. Any pair of transducers can be connected in a differential arrangement by using a differential amplifier (*see* DIFFERENTIAL AMPLIFIER).

When the two input parameters to a differential transducer are identical, the output is zero. The greater the difference between the two inputs, the greater the output will be. Most output occurs when one transducer section receives a large input and the other gets none. *See also* TRANSDUCER.

DIFFERENTIAL TRANSFORMER

A differential transformer is a transformer with one or two primary windings and two secondary windings, and an adjustable powdered-iron or ferrite core. The windings are generally placed on a solenoidal form. The primaries, if dual, are connected in series. The secondary windings are connected in phase-opposing series fashion (*see* TRANSFORMER). The schematic diagram shows a differential transformer. As the core is moved in and out of the solenoidal form, the coupling ratio between winding pair X and winding pair Y (as illustrated) varies. This affects the amplitude and phase of the transformer output. When the core is at the center, so that the coupling between the two winding pairs is equal, the output of the transformer is zero. The farther off center the core is positioned, the greater the output amplitude. However, the output phase is reversed with the core nearer winding pair X, as compared to when it is closer to pair Y.

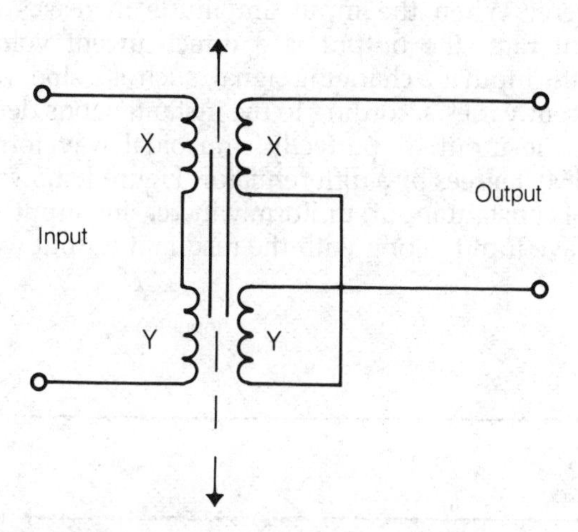

DIFFERENTIAL TRANSFORMER: Movement of the core in this transformer permits the adjustment of output voltage and phase.

DIFFERENTIAL VOLTAGE GAIN

The voltage gain of a differential amplifier is known as differential voltage gain. Any differential device has a differential voltage-gain figure that can be measured in decibels (*see* DECIBEL). However, a differential amplifier usually displays a signal gain of significant proportions; passive devices generally show a loss.

Differential voltage gain is the ratio, in decibels, between a change in output voltage and a change in input voltage applied to either input terminal. *See also* DIFFERENTIAL AMPLIFIER.

DIFFERENTIATION

The mathematical determination of the derivative of a function is called differentiation. Differentiation may be performed either at a single point on a function, or as a general operation involving the entire function (*see* CALCULUS, DERIVATIVE).

Certain electronic circuits differentiate the waveform supplied to their inputs. The output waveform of this circuit is the derivative of the input waveform, representing the rate of change of the input amplitude. If the amplitude is becoming more positive, then the output of the differentiator circuit is positive. If the amplitude is becoming more negative, then the output is negative. If the input amplitude to a differentiating circuit is constant, the output is zero.

Differentiation is, both electrically and mathematically, the opposite of integration. When an integrator and a differentiator circuit are connected in cascade, the output waveform is usually identical to the input waveform. *See also* DIFFERENTIATOR CIRCUIT, INTEGRATION, INTEGRATOR CIRCUIT.

DIFFERENTIATOR CIRCUIT

A differentiator circuit is an electronic circuit that generates the derivative, with respect to time, of a waveform (*see* DERIVATIVE). When the input amplitude to a differentiator is constant—that is, a direct current—the output is thus zero. When the input amplitude increases at a constant rate, the output is a direct-current voltage. When the input is a changing signal, such as a sine wave, the output varies according to the instantaneous derivative of the input. A perfectly sinusoidal waveform is shifted 90 degrees by a differentiator. Figure 1 shows the cases of constant input, uniformly increasing input, and sine-wave input, along with the resultant output waveforms.

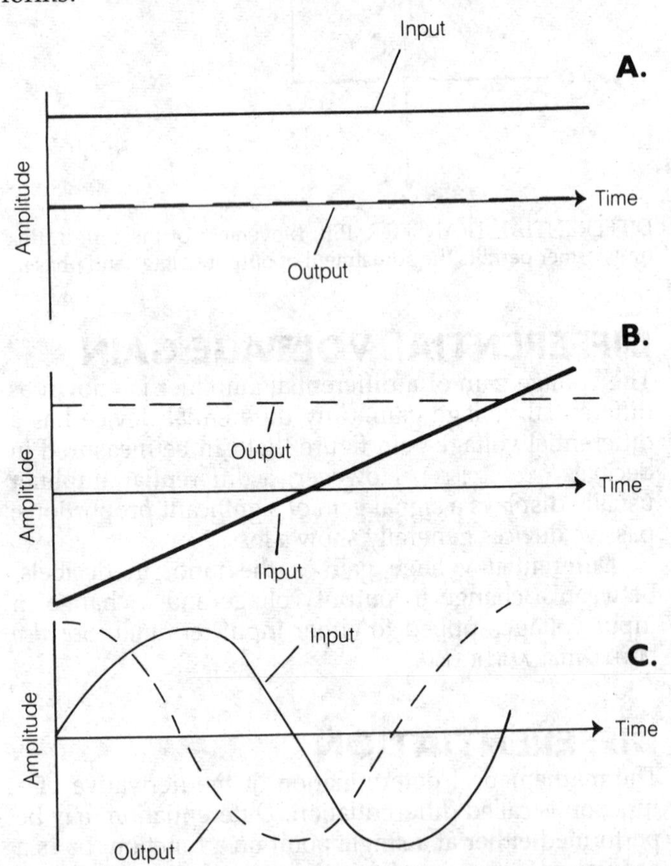

DIFFERENTIATOR CIRCUIT: Fig. 1. Differentiation of input waveforms produces zero output from dc (direct current) input (A), an increasing input yields a positive dc output (B), and a sine wave input yields a sine wave shifted by 90 degrees (C).

Figure 2 shows two examples of differentiating circuits. The circuit at A is a series resistance-capacitance network; at B, an operational amplifier is used.

The differentiator circuit acts in opposite manner to an integrator circuit. This does not mean, however, that cascading an integrator and a differentiator will always result in the same signal at the output and input. Such is usually the case, but certain waveforms cannot be duplicated in this way. *See also* INTEGRATION, INTEGRATOR CIRCUIT.

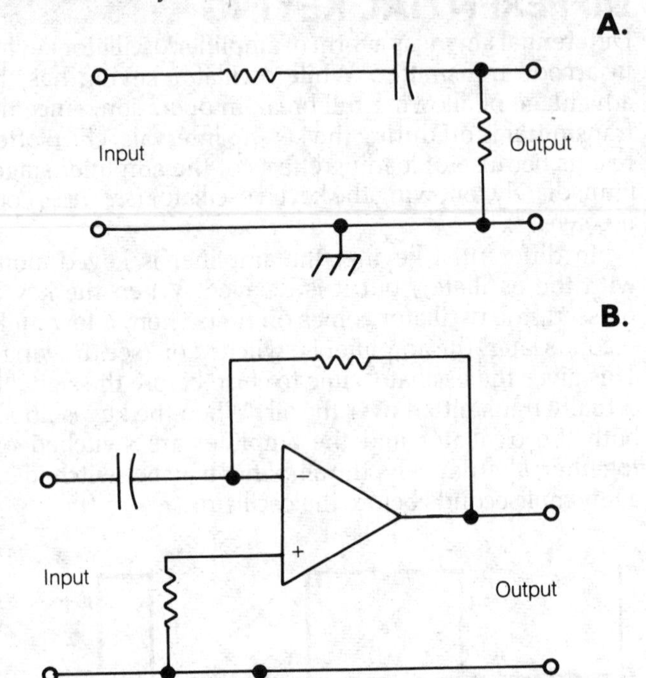

DIFFERENTIATOR CIRCUIT: Fig. 2. Two differentiator circuits—passive resistance-capacitance (A) and active based on an operational amplifier (B).

DIFFRACTION

Diffraction is the property of waves—including sound, radio frequency, and visible light—that gives them the ability to turn sharp corners and to pass around small obstructions. The wavelike nature of visible light was discovered by observing interference patterns caused by diffraction.

Diffraction allows speech to be heard from around the corner of an obstruction such as a building, even when there are no nearby objects to reflect the sound. This effect, called razor-edge diffraction, is shown at A in the illustration. The corner of the obstruction acts as a second source of wave action. Electromagnetic fields having extremely long wavelength, comparable to or greater than the diameter of an obstruction, are propagated around that obstruction with very little attenuation. For example, low-frequency radio waves are easily transmitted around a concrete-and-steel building that is small compared to a wavelength, at B. As the wavelength of the energy becomes shorter, however, the obstruction causes more and more attenuation. At very high frequencies, a concrete-and-steel building casts a definite shadow in the wave train of electromagnetic signals.

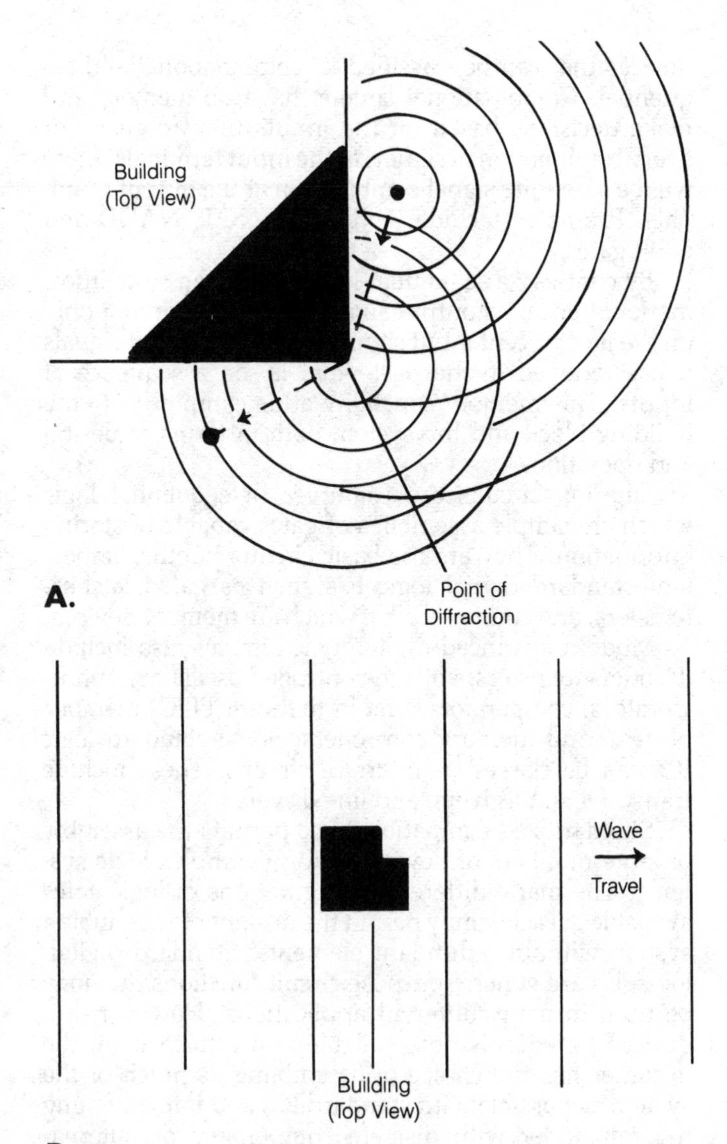

DIFFRACTION: Sound waves diffract around the corner of a building (A), and long radio waves diffract around obstructions that are small with respect to wavelength (B).

A piece of clear plastic, with thousands of opaque lines drawn on it, is called a diffraction grating. Light passing through a grating is split into its constituent wavelengths, in much the same way as light transmitted through a prism. This effect is the result of interference caused by diffraction through the many tiny slits. *See also* SPECTROSCOPE.

DIFFUSION

Diffusion is the process that occurs when one material spreads into, and permeates, another material. Some semiconductor devices are made by diffusing one kind of substance into another. N-type and P-type semiconductor materials, as well as metal alloys and oxides, can be diffused at high temperatures to form various devices with different properties. Integrated circuits are fabricated on a silicon wafer by means of diffusion techniques.

DIGITAL

Electronic circuits can be divided into two broad groups: analog and digital. In analog circuits, the voltage and current waveforms are similar to the signal variations. A digital signal, however, is a group of pulses with the same level but either on or off. This is also called a *binary signal* because it has just two bits of information. Digital circuits are used in numerical counting, computers, clocks, and test equipment.

Digital signals exhibit lower noise than analog signals. Digital transmission is more efficient and precise than analog transmission. Digital displays of variables are precise and easy to read. *See also* DIGITAL CIRCUITRY.

DIGITAL CIRCUITRY

Any circuit that operates in the digital mode is called a digital circuit (*see* DIGITAL). Examples of digital circuits include the flip-flop and the various forms of logic gates. Complicated digital devices are built up from simple fundamental units such as the inverter, AND gate, and OR gate. There is no limit, in theory, to the level of complexity attainable in digital circuit design. With increasingly sophisticated methods of miniaturization, more and more complex digital circuits are being packaged into smaller and smaller packages. Entire computers can now be installed on a PC board.

Digital circuitry is generally more efficient than older analog circuitry (*see* ANALOG) because of the finite number of possible digital states, compared to the infinite number of levels in an analog circuit. Many digital circuits can operate at greater speed than similar analog devices. *See also* AND GATE, FLIP-FLOP, INVERTER, OR GATE.

DIGITAL COMPUTER

See COMPUTER.

DIGITAL CONTROL

Digital control is a method of adjusting equipment by digital means. Variables frequently controlled by digital methods include the frequency of a radio receiver or transmitter, the programming of information into a memory channel, and the selection of a memory address.

Digital control involves the selection of discrete states or levels, rather than the adjustment of a parameter over a continuous range. When the adjustment is continuous, the device is analog controlled (*see* ANALOG). Digital controls are becoming increasingly common in all kinds of electronic devices.

DIGITAL DATA SERVICE

There are many of digital data services available for the transmission of computer data. Each of these services has a maximum data rate in bits per second (b/s). These services include:

- Dataphone digital service (DDS).

- Accunet T1.5, 1.544 megabits-per-second digital service.
- X.25 packet-switched public data network (PSPDN).

Dataphone Digital Service (DDS) was developed by AT&T Communications as a digital transmission network that would provide fewer errors, higher data rates and lower cost than conventional analog transmission facilities. DDS is now available in over 100 American cities. Full-duplex, private-line service is provided at synchronous data rates of 2.4, 4.8, 9.6, and 56 kilobits per second. DDS uses bipolar signaling (above and below zero volts) and pulse code modulation over an integrated combination of local distribution facilities, intermediate length lines (T-1 and T-2 carriers), and long-haul facilities. Regenerative repeaters are spaced about every 6000 feet to recreate the digital pulse train. The basic customer interfaces are a channel service unit (CSU) and a data service unit (DSU).

Accunet T1.5 is a service offered by AT&T Communications for 1.544 megabits-per-second point-to-point digital transmission that can be leased to customers. The internal digital network is based on the same T-1, T-1C, and T-2 carriers used in DDS (*see* T-1 CARRIER). The customer interfaces with this service with network channel-terminating equipment. AT&T claims 99.5 percent error-free seconds in any consecutive 24-hour period and 99.7 percent availability over a 12-month period.

X.25 Packet-Switched public data network (PSPDN) networks put user data into discrete units called packets and route them through internal switching nodes to their destination. High-speed links (9.6 kilobits per second to 1.544 megabits per second) are used internally to interconnect the packet-switching nodes. CCIT X.25 specifies a synchronous interface between the computer or terminal and the PSPDN.

DIGITAL-LOGIC INTEGRATED CIRCUIT

Standard digital-logic integrated circuits are building block ICs (integrated circuits) formed from logic gates in different semiconductor technologies. They might be simple gate-level logic, or they might be large-scale integrated circuits. Standard digital-logic ICs are manufactured for inventory and listed in catalogs. All products in a family built to a specific technology are compatible with each other and do not require special interfacing devices to work together in a system.

There are three basic types of logic gates—AND, OR and NOT—and two commonly used negative forms of these basic circuits—NAND and NOR. More complex circuits such as flip-flops are comprised of several gates. The digital circuit designer attempts to design a system with the minimum number of IC components to operate at the highest speeds (least average gate propagation delay) with the lowest average gate power dissipation. This objective has been the incentive for semiconductor manufacturers to strive for higher levels of integration and higher gate densities on a given size of silicon chip.

Standard commercial digital-logic IC families include devices that can be classified as combinational and sequential. Combinational circuits have no memory and make decisions based on the inputs they receive. For every combination of signals at the input terminals, there will be a definite signal combination at the output terminals. Examples include AND, OR, NOT, NAND and NOR gates.

By contrast, a sequential-logic device can store information; hence, its output signal might depend not only on the most recent input signal, but also on input signals it has received in the past—that is, on a sequence of inputs. This memory capability adds complexity to the building block and the system both in terms of design and operation.

Flip-flop circuits are examples of sequential logic which are simple assemblies of gates capable of storing information. They are the basic circuits in other important standard digital logic ICs such as gated latches, registers, and counters, all of which are memory devices.

Modern advanced digital logic families also include decoders, registers, voltage-controlled oscillators, multivibrators, comparators, first-in first-out (FIFO) memory buffers, and memory controllers. Some standard logic ICs can be classed as interface circuits. These include transceivers, receivers, and line drivers.

Standardized compatible logic permits the assembly of large numbers of devices to form complex logic systems. The many different configurations of logic gates available in each family permit the designer to assemble a system without redundant elements. Standard digital-logic ICs are general-purpose circuit functions that may be used in many different applications. However, if a desired function is not available as a catalog item, the designer has the choice of assembling as much of the system as possible with standard ICs and implementing the remainder with discrete components or, alternatively, resorting to custom or semicustom ICs.

In general, the more complex the IC, the less likely that it will be available as a standard product. The designer's decision on alternatives if a standard IC is not available will be influenced by the available circuit board space, power limitations, number of systems required, cost and availability considerations, and urgency in the completion of the design.

Even if a system has been fully developed from standard logic and completely tested, the designer might wish to consider other alternatives to reduce parts count, power, and circuit-board space requirements to reduce cost and increase reliability. The alternatives available today include field programmable logic devices (PLDs), factory-programmed gate arrays, standard cells, or full custom ICs. *See* APPLICATIONS-SPECIFIC DEVICE, GATE ARRAY, STANDARD CELL.

Standard digital-logic ICs available include small-scale integration (SSI) that is limited to about 12 gates, medium-scale integration (MSI) with up to 100 gates, and some large-scale integration (LSI) with up to 1000 gates. Many standard catalog LSI and very large scale integration (VLSI) digital logic ICs are dedicated to specific functions in automotive, telecommunications or consumer entertainment products.

The introduction of the microprocessor (MPU) and the microcontroller (MCU) have pre-empted many possible future uses for digital logic ICs. However, standard digital logic ICs are still in demand for linking or glueing systems that contain many LSI and VLSI devices.

DIGITAL-LOGIC TECHNOLOGY

There are many different digital logic families in wide use today. These families can be generally classified as bipolar or complementary metal-oxide silicon (CMOS). The largest number of families are bipolar, and the first families introduced were bipolar. These were resistor-transistor logic (RTL) and diode-transistor logic (DTL). Both families are now obsolete and are not used in new circuit designs.

The principal bipolar families in use today can be further subdivided into transistor-transistor logic (TTL, Fig. 1) and emitter-coupled logic (ECL). Many of these families have been available for 20 years or more, having replaced earlier gate-level RTL and DTL.

Five of the most prominent bipolar TTL families were originally developed by Texas Instruments and all are widely alternate-sourced by other manufacturers. The sixth family, FAST, is a derivative of TTL developed by Fairchild (now National Semiconductor). The bipolar families are:

1. Gold-doped TTL (Standard TTL).
2. Standard Schottky TTL (S-TTL).
3. Advanced Schottky TTL (AS-TTL).
4. Low-speed TTL (LS-TTL).
5. Advanced Schottky TTL.
6. Fairchild Advanced Schottky Technology (FAST).

Bipolar emitter-coupled logic predates TTL, and there are variations. Its prime merit lies in its high speed although it is at the cost of high power dissipation. The principal ECL families are 10k ECL and 100k ECL.

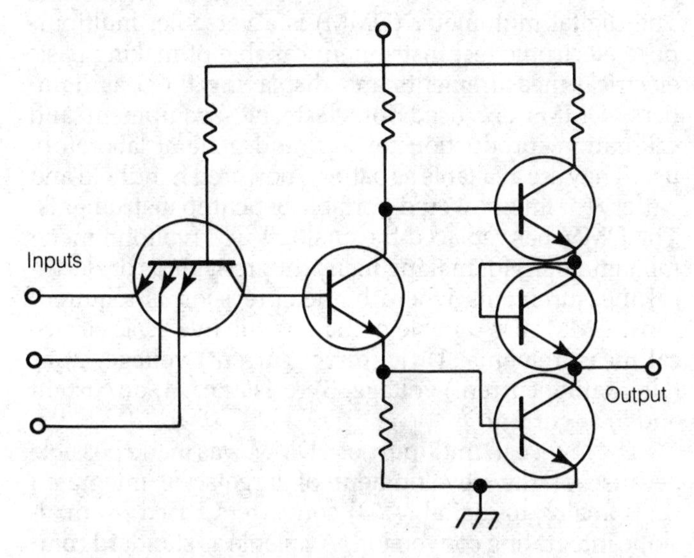

DIGITAL-LOGIC TECHNOLOGY: Fig. 1. transistor-transistor logic.

Complementary metal-oxide silicon is a relatively recent introduction to standard logic families. Originally developed by RCA (now Harris Corp.), it also is widely alternate sourced:

1. CMOS 4000 B (RCA developed).
2. High-speed CMOS (HC/HCT).
3. Improved high-speed CMOS (HCS/HCTS).
4. Advanced CMOS (ACL)(FACT).

Digital logic performance can be evaluated with a common digital logic configuration. The two-input NAND gate has been selected. Measurements are made of average gate propagation delay and average gate power dissipation for comparison purposes. Despite what might appear to be obvious choices in terms of speed and power, some logic families might not be selected because of other considerations such as anticipated extremes in environment or constraints on available power. These factors may force the selection of a less than optimum family. There is also price and availability to consider.

Switching speed is the speed at which a gate switches or changes its output from high to low or vice versa. It is stated in terms of average propagation delay and measured in nanoseconds (ns) or billionths of a second. The fastest switching speed is desired in all logic circuits so propagation delay should be as short as possible.

Power dissipation, specified as average gate-power dissipation, is a measure of the electrical energy that is converted to heat in powering the device. Measured in milliwatts (mW), this value should be as low as possible to conserve power and minimize cooling problems.

Speed-power product is the product of average gate propagation delay (ns) and average gate-power dissipation (mW). It is expressed in picojoules, a unit of energy.

Noise margin is a measure of how securely the digital logic IC transmits and receives information without errors in the presence of electrical noise. It is desirable to have as wide a noise margin as possible. The output voltage must be larger than the input voltage required to set the logic states correctly. Noise margin is measured in volts (V).

Fan-out is the number of inputs to other gates that can be driven successfully by the digital IC. High fan-out capability is important because it helps to reduce the number of ICs on the circuit board.

Density of circuitry is an indication of effective use of silicon real estate. It is important that gates, as active functions, occupy a minimum amount of space on the wafer as measured in square mils (thousandths of an inch). The number of transistors and gates that can be integrated on a chip depends on the technology as well as on such factors as required electrical isolation and heat dissipation ability.

Manufacturing cost for some digital logic IC technologies is inherently higher than others. Cost is influenced by the number of processing or masking steps required to produce the IC on the wafer, the test procedure specified, the size of the chip and the yield of the process.

The speed-power curve, Fig. 2, is a graphical presen-

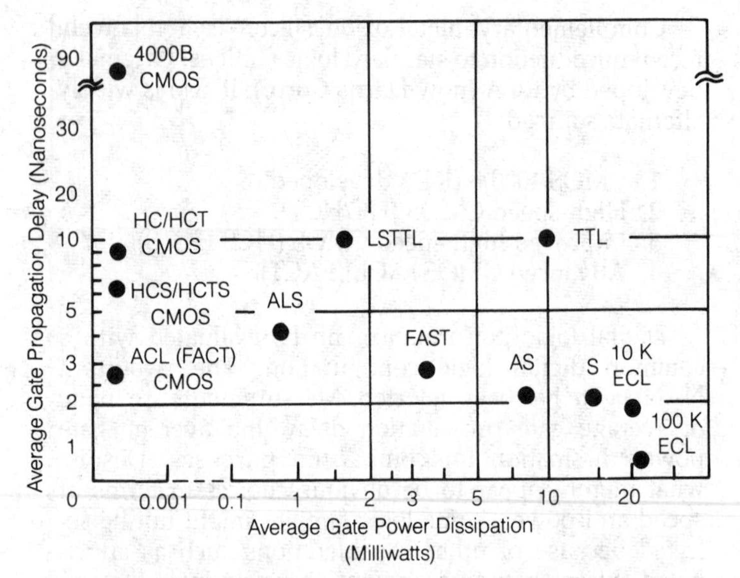

DIGITAL-LOGIC TECHNOLOGY: Fig. 2. Speed or average gate propagation is plotted against power or average gate power dissipation for families of digital logic.

tation that summarizes the relative merits of each of the digital logic families in two coordinates: average gate propagation delay speed and average gate power dissipation (power). The plot is based on averaged test data taken from sample two-input NAND gates made in each technology. It is further assumed that the same test procedures are applied to all of the gates under identical conditions or ambient temperature, packaging and other factors could influence the results.

Each point plotted on the speed-power curve is an arbitrary average point or centroid that lies within a closed performance envelope representing performance range. Not all manufacturers agree on the exact position for the points representing each technology although they claim their points lie within their speed-power envelopes. Plotted as irregularly shaped areas for each family, the envelopes could overlap, making interpretation of the plot more difficult. The plot portrays tradeoffs and relative merit.

The 10 nanosecond, 10 mW point on the plot representing standard gold-doped TTL is an important reference to indicate progress. The objective of all digital logic development over the past 20 years has been to move the point representing the logic family performance down and to the left toward the ideal represented by the origin (0 nanoseconds, 0 mW).

Improvements in Schottky TTL families such as ALS, TTL, and FAST have made significant strides toward power reductions and speed increases. But CMOS has made faster progress. An improved CMOS technology called ACL or its equivalent called FAST has come closer to the origin than any of the other families. It has an average gate speed of less than 3 nanoseconds and average gate power less than 0.001 mW.

DIGITAL MODULATION

Whenever a specific characteristic of a signal is varied for the purpose of conveying information, the process is called modulation (*see* MODULATION). If this is done by restricting the signal to discrete levels, or states, the process is known as digital modulation. Digital modulation differs from analog modulation (*see* ANALOG), in which a signal varies over a continuous range, and therefore has a theoretically infinite number of possible levels. The variable in a digital-modulation system can be any signal characteristic, such as amplitude, frequency, or phase. In addition, pulses can be digitally modulated (*see* PULSE MODULATION).

The figure illustrates a hypothetical digital pulse-amplitude modulation system. The analog waveform is shown by the solid line. A series of pulses, transmitted at uniform time intervals, approximates this waveform. The pulses can achieve any of eight different amplitude levels, as shown on the vertical scale. In this particular example, the amplitude of a given pulse corresponds to the instantaneous level nearest the amplitude of the analog signal.

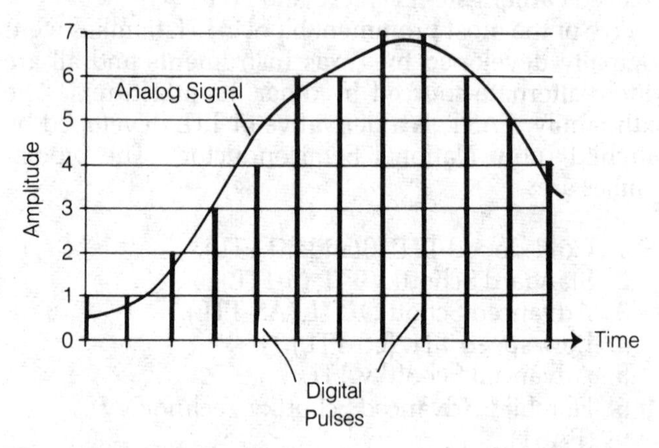

DIGITAL MODULATION: An analog waveform is converted into a series of digital pulses approximating the corresponding analog amplitude in digital modulation.

DIGITAL MULTIMETER

The digital multimeter (DMM) is a versatile, multipurpose electronic test instrument capable of making basic electrical measurements and displaying them as numbers. DMMs are used for electronics equipment and calibration, production testing, and general laboratory use. They are available as battery-powered handheld and battery- or line-powered portable benchtop instruments. The DMM has replaced the analog VOM (voltohmmeter milliammeter) for making many different kinds of electrical measurements where higher precision is required. Most DMMs are capable of making the five basic electrical measurements: 1) dc (direct current) voltage, 2) ac (alternating current) voltage, 3) resistance, 4) dc current and 5) ac current.

The low-cost, multipurpose DMM was made possible because of the development of large-scale integrated (LSI) analog-to-digital (A/D) converters based on dual-slope integrating conversion. Available as standard commercial single-chip devices, they are more efficient, they consume less power and they occupy less space than earlier proprietary discrete A/D converter circuits. The display is refreshed two to three times per second. *See* ANALOG-TO-DIGITAL CONVERTER.

The development of the DMM was further encouraged by the commercial availability of logic-compatible seven-segment digital displays, particularly light-emitting diode (LED) and liquid crystal diode (LCD). Early DMMs used neon displays, but they are no longer being offered because of the need for high-voltage conversion. The LCD made it practical to operate a five-to-eight-function handheld DMM with a large display (0.5-inch-high numerals) from a 9V disposable alkaline battery. *See* BATTERY, LIGHT-EMITTING DIODE DISPLAY, LIQUID CRYSTAL DISPLAY.

The most popular DMMs have 3½- and 4½-digit, seven-segment displays. A 3½-digit DMM has a resolution of 1 part in 1999 or ±0.05 percent; a 4½-digit DMM has resolution of 1 part in 19999 or ±0.005 percent. Accuracy in DMMs is stated as dc voltage measurement accuracy. (A typical value for a 3½-digit DMM is ±0.5 + 1 digit) or about one-tenth the value of the resolution. *See* DIGITAL PANEL METER.

A typical handheld DMM is housed in a case measuring about 7 inches long by 3½ inches wide by 2 inches high. It weighs about a pound. See Fig. 1. Three design concepts for switching handheld DMMs have evolved. Some instruments have a centrally located rotary multipurpose function and range switch. Others have one rotary multifunction switch and one rotary range switch. In a third design concept, a row of function switches is arranged along the left side of the case to be operated by the fingers of the right hand that is holding the instrument.

Benchtop DMMs are packaged in rectangular cases with the display and controls on the front of the instrument. The DMM has a large depth dimension for stability on the bench. The front face can be raised with a bail for easier reading. These units usually have more features

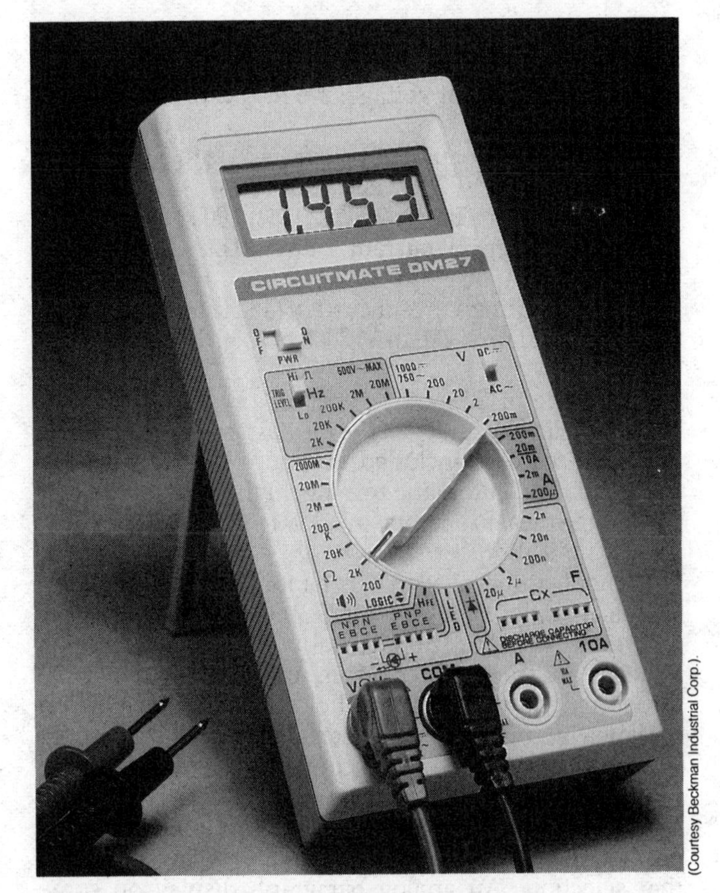

(Courtesy Beckman Industrial Corp.)

DIGITAL MULTIMETER (DMM): Fig. 1. Battery-powered, handheld digital multimeter (DMM) has a 3½-digit, 3200-count digital display augmented by a moving-bar analog display.

than handheld units. Line-powered bench DMMs typically have LED displays. Some are powered with rechargeable batteries for portability.

The following measurement ranges are typical for a

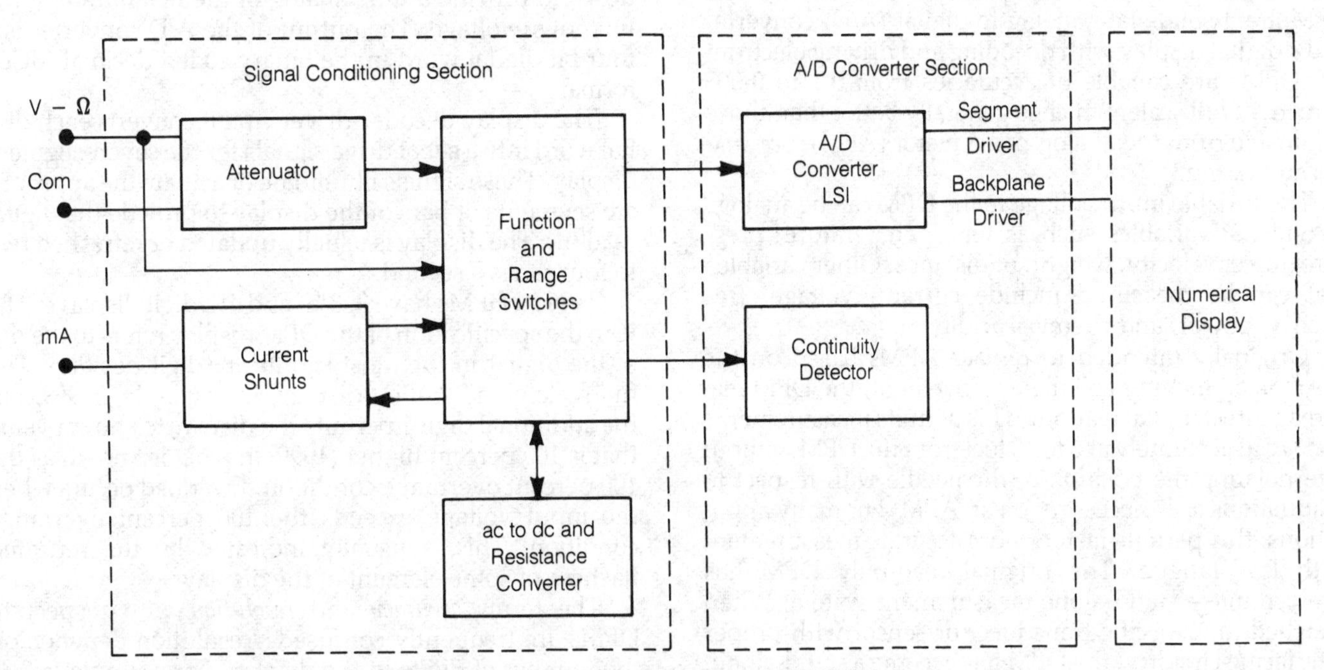

DIGITAL MULTIMETER: Fig. 2. Functional block diagram of a digital multimeter (DMM).

3½-digit DMM:

1. Direct-current voltage: 200 mV, 2 V, 200 V, or 1000 V.
2. Alternating-current voltage (45 Hz to 500 Hz): 200 mV, 2 V, 20 V, 200 V, or 750 V.
3. Resistance: 200 Ω, 2 kΩ, 20 kΩ, 200 kΩ, 2 MΩ.
4. Direct-current current: 2 mA, 20 mA, 200 mA, 2000 mA, 10 A.
5. Alternating-current current (45 Hz to 1 kHz): 2 mA, 20 mA, 200 mA, 2000 mA, or 10 A.

The price of DMMs is generally proportional to the resolution and accuracy of the display as well as features. Standard features included on many DMMs are automatic polarity, automatic zeroing, high-input impedance for voltage measurements, overload protection, and low-battery indication.

Other features available on handheld and bench-type DMMs include: 1) autoranging—the ability to select the range and position the decimal point automatically on voltages and resistance, 2) true root mean square (rms) measurement of ac, 3) peak-hold circuitry, 4) temperature measurement, 5) level detection, 6) diode testing for good-bad selection, and 7) continuity detection. Some DMMs sound audible tones to indicate continuity and relay closures. Microprocessors included in some DMMs provide automatic range selection and remind users of that selection. An analog bar graph display on some DMMs supplements the numerical display by permitting the observation of peaking or dipping trends.

DIGITAL PANEL METER

The digital panel meter, or DPM, is a compact electronic measuring system capable of measuring an analog variable and converting it into an accurate digital reading on the instrument display. The principal components of a DPM are a solid-state analog-to-digital (A/D) converter and a digital display with decoding and driving electronics. DPMs are capable of accuracies from 0.1 to 0.005 percent of full scale, which is generally better than those obtainable from the analog panel meter (APM). *See* ANALOG PANEL METER.

The variable input voltage to the DPM can be analogs of physical variables such as time, temperature, pressure, force, velocity, weight, or distance. Other variables that can be measured include current, voltage, frequency, power, and chemical acidity.

Originally intended to replace APMs based on the D'Arsonval moving-coil meter movement, the DPM features a direct digital readout. This permits measurements to be read accurately at 6 to 10 feet from the DPM without interpolating the position of the needle with respect to graduations as is necessary on an APM. For many applications, this permits faster, more accurate measurement with less fatigue. The original read-only DPM has evolved into a stand-alone measurement system. When matched to a specific transducer or sensor with proper interfacing circuitry, the DPM has become a stand-alone, self powered measuring system capable of powering the sensor, generating and transmitting data, and providing a local readout.

DPMs are available commercially as factory-manufactured products in a wide variety of sizes, form factors, choice of display technology, character height, number of digits, and power supply. The case sizes for most industrial DPMs are approximately 4 inches wide by 2 inches high by 4 inches deep. Most are line powered (117 or 220 V ac). Read-only DPMs intended as APM replacements are typically in smaller cases and are dc-logic level powered (5 V dc or battery).

The choices of seven-segment digital display available in commercial DPMs are: 1) light-emitting diode, 2) neon gas-discharge panel, 3) vacuum-tube fluorescent panel and, 4) liquid-crystal display (LCD) panel. The display may have from three to six digits with character heights from 0.43 to 0.80 inch. The first three displays are favored for line-powered industrial DPMs, and LCD displays are favored for logic-powered, display-only DPMs. (*See* LIGHT-EMITTING DIODE DISPLAY, LIQUID CRYSTAL DISPLAY, NEON GAS DISCHARGE DISPLAY.) Some DPMs with LED and LCD displays have additional analog moving-bar displays to supplement the digital readout and provide dynamic trend information.

The principal internal electronics in all DPMs is a monolithic dual-slope integrating analog-to-digital (A/D) converter. (*See* ANALOG-TO-DIGITAL CONVERTER.) This circuit provides high accuracy and resolution at low cost. During the measurement cycle, the unknown input voltage charges a capacitor in the conversion circuit while connected for about 100 milliseconds. In this interval (the first half of the dual slope) the capacitor reaches a value that is proportional to the unknown voltage.

The capacitor is then discharged during the second half of the dual slope at a constant rate by a known internal reference voltage of opposite polarity to the unknown input voltage. The time to discharge is counted down to provide a time analog of the magnitude of the unknown voltage. The output of the A/D converter is a four-bit digital word in the binary coded decimal (BCD) format.

The display decoder-driver circuit converts each digital word into a set of drive signals for the seven-segment display. These signals illuminate or actuate the appropriate segments or bars of the display to provide the digital readout. The display is usually updated or refreshed two to four times a second.

Typical DPMs have 3, 3½- and 4½-digit displays. The ½ in the specification of the DPM display refers to the use of the digit 1 in the most significant digit position. The full-scale on a 3½-digit display, for example, is 999, but the additional digit 1 permits the display to show a value that is 100 percent higher (1999) in what is known as the 100 percent overrange condition. Overload occurs when the input voltage exceeds the 100 percent overrange condition. This is usually indicated by the repeated flashing of some element of the display.

The terms *accuracy* and *resolution* with respect to DPMs are frequently confused. Resolution depends on the number of digits in the display. For example, a 3½-digit DPM can resolve one part in 2000 or 0.05 percent.

DIGITAL PANEL METER (DPM): Fig. 1. Line-powered 4½-digit DPM with LED display in a 1/8 DIN/NEMA metal case.

Similarly, a 4½-digit DPM can resolve one part in 20,000 or 0.005 percent. See Fig. 1. Accuracy is related to the quality of the instrument and it is closely tied to resolution. A quality 3½-digit DPM can be expected to have an accuracy of ±0.05 percent, ±1 count. See Fig. 2.

Because of the additional circuitry required, line-powered DPMs cost more than equivalent logic-powered DPMs. Pricing is also proportional to the number of digits in the display. Line-powered DPMs are widely specified in the process control industry because the cases are large enough to accommodate dedicated interface circuitry to personalize it for some sensor such as a thermocouple or resistance-temperature detector (RTD). The internal line-powered supply can power additional circuitry and the sensor as well as other features.

Latched BCD output is a feature offered on some DPMs. It is held between conversions and is continually updated. BCD can be used to drive a data logging printer, or it can be transmitted directly to a host computer. Alternatively, some DPMs have an internal microcontroller chip to convert BCD to standard ASCII code, thus relieving the host computer of this burden.

DIGITAL PANEL METER: Fig. 2. Battery-powered 3½- digit DPM with a liquid-crystal display.

Industrial grade and systems-oriented DPMs are now widely packaged in the Deutsche Industrie Normen-ausschuss/National Electrical Manufacturers Association (DIN/NEMA) standard plastic and metal case sizes. They are designed to fit in one of two popular cutout dimensions: 1.77 by 3.93 inches (45 by 99.8 millimeters) or 1.8 by 3.6 inches (45 by 92 millimeters).

DIGITAL SERIAL INTERFACE

Data terminals and computers exchange signals with modems over a standardized interface. In the U.S., this interface usually conforms to the standards established by the Electronic Industries Association (EIA). In the rest of the world, a similar set of standards has been defined by the Consultative Committee on International Telephone and Telegraphy (CCITT). The CCIT V series of recommendations applies to analog modems, the X series applies to public data networks (PDNs), and the I series applies to integrated services digital networks (ISDN). The International Organization for Standardization (ISO) defines mechanical standards for connections used in terminal/computer-to-modem interfaces.

Interfaces were initially intended for terminal- and computer-to-modem operation, but the interfaces can also be used for terminal/computer-to-terminal/computer and modem-to-modem operation if appropriate changes are made in the connectors by switching the signal leads. Synchronous transmission timing signals must also be supplied.

Four subjects covered in a physical interface standard are:

1. Function. Signals for each signal lead in the connecting cable between the terminal/computer-to-modem are defined.
2. Procedure. Organization of the signal wires for various line configurations and applications are defined.
3. Mechanical. Connector size and shape, pin spacing, assignment, and latching arrangements are defined.
4. Electrical. Voltage and current levels, noise margins, threshold levels, capacitive loading, slew rate, and signal transition times are defined.

EIA RS-232C Interface. The accepted standard for interfacing a terminal/computer with a modem for data rates up to 20 kilobits per second (kb/s), the accepted EIA standard is RS-232C. This standard applies to both asynchronous and synchronous serial, binary-data transmission. It is accepted for the public-switched telephone dial-up networks, private wire, or leased line in half-duplex or full-duplex mode.

RS-232C covers the functional, electrical, and mechanical characteristics of the interface. It defines 20 specific functions. The international equivalent of RS-232C is CCITT V.24 for function, V.28 for electrical, and ISO 2110 for mechanical. RS-232C signal leads are classified into data, control, and timing circuits for a primary and secondary channel. Twenty-one leads are defined in a 25-pin plug-in connector that has a male terminal for the computer and a female shell for the modem.

EIA RS-449 Interface. RS-449 became an EIA standard in 1977 for terminal/computer-to-modem connections. The standard specifies interface requirements for expanded transmission speeds up to 2 megabits per second (Mb/s), longer cable lengths, and ten additional functions. This standard applies to binary, serial, asynchronous, and synchronous communications in half- or full-duplex mode. The physical connection is made through a 37-contact connector. When service to secondary channel interchange circuits is required, a separate 9-contact connector is specified. Intended to overcome the shortcomings of RS-232C, the federal government adopted RS-449 in 1980. However, RS-449 has not been generally accepted.

RS-422A Balanced Digital Interface. RS is a standard that operates in conjunction with RS-449 and specifies the electrical characteristics for circuits with their own ground leads called balanced circuits. RS-422A characteristics include improved performance: up to 10 megabits per second in distances of meters and 100 kilobits per second at 1.2 kilometers.

DIGITAL SIGNAL PROCESSOR

A digital signal proccessor (DSP) is a microprocessor optimized to process sampled data at high rates. It performs operations such as accumulating the sum of multiple products at faster rates than a general purpose microprocessor. The architecture of the DSP is specifically designed to take advantage of the repetitive nature of signal processing by pipelining data flow or starting to execute another task before a preceding one is completed to achieve extra speed. *See* MICROPROCESSOR, PIPELINE PROCESSOR.

DSP circuits are widely used in telecommunications, particularly in the integrated services digital network (ISDN). (*See* INTEGRATED SERVICES DIGITAL NETWORK.) They also are used to process data for many applications including filtering (*see* ACTIVE ANALOG FILTER), computer-aided design and engineering (*see* COMPUTER-AIDED DESIGN and COMPUTER-AIDED ENGINEERING), speech recognition (*see* SPEECH RECOGNITION), speech compression (*see* SPEECH COMPRESSION), radar (*see* RADAR), and medical imaging (*see* COMPUTER-AIDED MEDICAL SCANNING).

Many DSP chips are available commercially, and they offer variations in throughput, address space, arithmetic precision and benchmark performance. The benchmark is a program that permits comparison of competing products for a given application. There are differences in engineering support and the quality of the hardware and software development tools available on these products.

DSP circuits can be classified as general purpose (for digital-signal processing) or applications specific. The figure shows a block diagram of a general purpose DSP. Applications-specific DSP ICs (integrated circuits) are designed to perform specified tasks more accurately, faster, or more cost effectively than general-purpose units. Examples are digital filters and Fourier transform chips. DSP single-chip ICs can also be grouped in terms of precision and arithmetic type. Fixed-point DSP circuits with 16, 24, and 32 bits of precision are available. Some have full floating-point arithmetic capabilities.

The fast array multiplier and accumulator distinguishes the DSP chip from the general-purpose microprocessor. It allows the multiplication of two numbers and the addition of the product to previously accumulated results in a single clock period. In a typical fixed-point multiplier-accumulator, two 16-bit numbers can be multiplied, and the 32-bit product can be added to a 32-bit accumulator register in a single instruction cycle. A typical microprocessor would require about 25 clock cycles to carry out this task.

Most DSP chips include cache memories to service program and data pipelines. A cache is a small, fast memory located between a larger, slower memory and a processor. The cache improves access to data and instructions. The DSP chips that carry out block floating-point arithmetic most efficiently have barrel shifters—circuits that allow arbitrary shifting of data as well as other circuitry to detect the presence of exponents.

A typical DSP chip has two data memories and two data buses. It can deliver two operands required for a single-cycle execution of the multiply-accumulate function. In contrast with general-purpose microprocessors that store both instructions and memory in the same memory, most DSP processors have separate program and data memories so that instructions and data can be retrieved simultaneously. However, some DSPs store certain kinds of static data in the program memory for transfer to the faster, smaller data memory when needed. Instruction cycle or multiply-accumulate times per operation range from 60 to 200 nanoseconds. Word lengths are 8 to 32 bits, with 16 bits being typical.

DSP chips have multiple memories and corresponding buses to enhance system throughput by increasing the rate of data access. Some devices have dual internal data memories and can access external data memory. Typical internal memories have 129 to 512 16-bit words. Another way to increase throughput depends on external program and data memories that are fast enough to supply operands and instructions in a single cycle. DSP applications, like image processing, require that DSP chips address at least 64,000 words of external data or program memory.

In the simplest applications, the DSP chip stands alone between analog-to-digital and digital-to-analog converters and external program and data memory. Analog input data is converted for processing and reconverted to an analog output. In other applications, the DSP chip can act as a slave or peripheral to a microprocessor with program and data memory.

DSP ICs are packaged in dual-in-line (DIP), plastic and ceramic leadless chip carrier (LCC and PLCC), and pin-grid array (PGA) packages. *See* SEMICONDUCTOR PACKAGE.

DIGITAL-TO-ANALOG CONVERTER

A digital-to-analog (D/A) converter (DAC) is a circuit

capable of converting a digital signal to a time-varying output signal. The classical approach to converting a digital signal to an analog signal has been the binary-weighted method. The circuit illustrated consists of a series of switches with an operational amplifier used as an adder. The switch for a particular bit is closed for logic 1 and opened for logic 0. The diagram shows the position of the switches for the binary number 0101.

To overcome the difficulty in matching resistors in ratios greater than about 20:1, the R-2R ladder network was developed. Only two different resistor values are needed in this circuit. Each switch has a single-pole, double-throw (SPDT) form that connects the 2R leg to either the reference voltage Vref or ground. The inverted R-2R ladder concept adapts the circuit to semiconductor design and fabrication. The DAC produces a voltage that is proportional to the product of the digital input and the reference voltage.

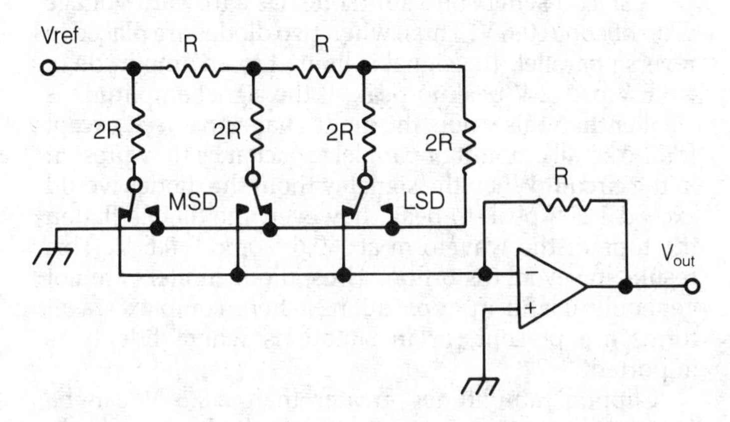

DIGITAL-TO-ANALOG CONVERTER: The inverted R-2R digital-to-analog converter is the most popular D to A circuit. MSD is the most significant digit, LSD is the least significant.

DIGITAL TRANSMISSION SYSTEM

Any system that transfers information by digital means is a digital transmission system. The simplest digital transmission system is a Morse code transmitter and receiver, along with the attendant operators. Computers communicate by digital transmission.

Analog signals, such as voice and picture waveforms, can be transmitted by digital methods. At the transmitting station, an analog-to-digital converter changes the signal to digital form. This signal is then transmitted, and the receiver employs a digital-to-analog converter to restore the original analog signal.

Digital transmission provides a better signal-to-noise ratio, over a given communications link, than analog transmission. This results in better efficiency. *See also* ANALOG-TO-DIGITAL CONVERTER, DIGITAL MODULATION, DIGITAL-TO-ANALOG CONVERTER.

DIMMER

A dimmer is a device that controls the voltage supplied to a set of lights. The circuit may be a simple potentiometer, although a true dimmer usually provides a voltage that does not depend on the resistance of the load.

Dimmers are used in household light switches. They can control the level of illumination by regulating the alternating-current voltage. Generally, the voltage can be set at any desired value from 0 V to 120 V. Dimmers can use diacs or triacs to provide a voltage drop. The dimmer is similar in design to a motor-speed control. *See also* DIAC, TRIAC.

DIODE

A diode is a tube or semiconductor device intended to pass current in only one direction. The semiconductor diode is far more common than the tube diode in modern electronic circuits. The drawing at A shows the construction of a typical semiconductor diode; it consists of N-type semiconductor material, usually silicon, and P-type material. Electrons flow into the N-type material and out of the P terminal. The schematic symbol for a semiconductor diode is shown at B. Positive current flows in the direction of the arrow. Electron movement is contrary to the arrow. The positive terminal of a diode is called the anode, and the negative terminal is called the cathode.

Semiconductor diodes are used for many different purposes in electronics. They can be used as amplifiers, frequency controllers, oscillators, voltage regulators, switches, and mixers. *See also* DIODE ACTION, DIODE CAPACITANCE, DIODE CLIPPING, DIODE DETECTOR, DIODE FIELD-STRENGTH METER, DIODE MIXER, DIODE-TRANSISTOR LOGIC, DIODE TUBE, DIODE TYPES, GUNN DIODE.

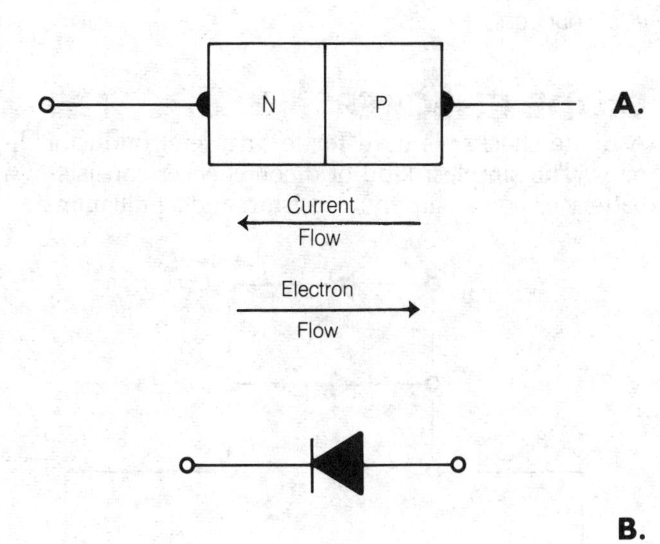

DIODE: A diagram of a diode (A) and its schematic symbol (B).

DIODE ACTION

Diode action is the property of an electronic component that permits it to pass current in only one direction. In a tube or semiconductor diode, the electrons can flow from the cathode to the anode, but not vice versa. Diode action occurs in all tubes and bipolar transistors, as well as in semiconductor diodes.

A voltage across the diode that allows current to flow through a diode is called forward bias. This occurs when the cathode is negative with respect to the anode. A voltage of the opposite polarity is called reverse bias. With most tubes and transistors, as well as with semiconductor diodes, a certain amount of forward bias voltage is necessary in order for current to flow.

DIODE CAPACITANCE

When a diode is reverse-biased, so that the anode is negative with respect to the cathode, the device will not conduct. Under these conditions in a semiconductor diode, a depletion layer forms at the P-N junction (*see* DEPLETION LAYER, P-N JUNCTION). The greater the reverse-bias voltage, the wider the depletion region.

The depletion region in a semiconductor diode has such a high resistance that it acts as a dielectric material (*see* DIELECTRIC). Because the P and N materials both conduct, the reverse-biased diode acts as a capacitor, assuming the bias remains reversed during all parts of the cycle.

Some diodes are used as variable capacitors. These devices are called varactors (*see* VARACTOR DIODE). The capacitance of a reverse-biased diode limits the frequency at which it can effectively be used as a detector, since at sufficiently high frequencies the diode capacitance allows significant signal transfer in the reverse direction. Diode capacitance, when undesirable, is minimized by making the P-N junction area as small as possible.

A tube type diode also displays capacitance when reverse bias is applied to it because of interelectrode effects. The capacitance of a reverse-biased tube diode does not depend directly on the value of the voltage. *See also* DIODE TUBE.

DIODE CHECKER

A diode checker is used for testing semiconductor diodes. The simplest kind of diode checker consists of a battery or power supply, a resistor, and a milliammeter, as shown in the figure. This instrument can be used to determine if current will flow in the forward direction (anode positive) and not in the reverse direction (cathode positive).

An ohmmeter can be a good diode checker. More advanced diode checkers show whether or not a particular diode is within its rated specifications. However, for most purposes, the circuit shown, or an ohmmeter, is adequate. When a diode fails, the failure is generally an open or short circuit. *See also* DIODE.

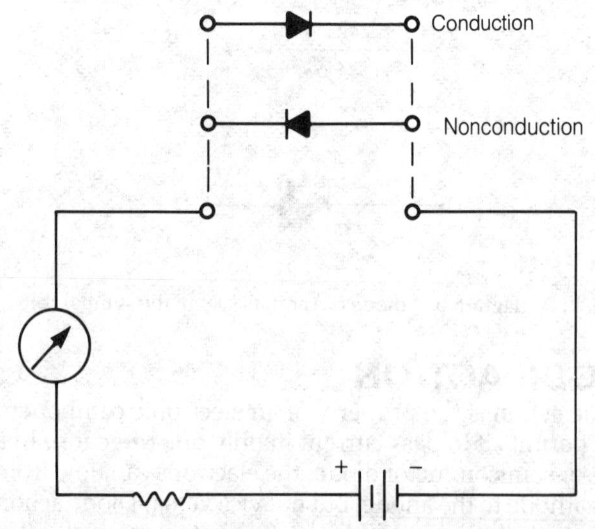

DIODE CHECKER: A diode checking circuit includes a battery, meter, and resistor.

DIODE CLIPPING

A diode clipper, or diode limiter, is a circuit that uses diodes for limiting the amplitude of a signal. This circuit generally consists of two semiconductor diodes connected in reverse parallel. One is shown at A in the illustration.

A silicon semiconductor diode has a forward voltage drop of about 0.6 V. Thus, when two diodes are placed in reverse parallel, the signal is limited to an amplitude of ±0.6 V or 1.2 V peak-to-peak. If the signal amplitude is smaller than this value, the diodes have no effect, except for the small amount of parallel capacitance they present in the circuit. When the signal without the diodes would exceed 1.2 V peak-to-peak, however, the diodes flatten the tops of the waveform at +0.6 V and −0.6 V. This results in severe distortion. Thus, diode limiters are not generally useful in applications where complex waveforms are present, or in situations where fidelity is important.

Clipping amplitudes greater than ±0.6 V can be obtained by connecting two or more diodes in series in each branch of the clipping circuit. The diode clipper shown at B, using silicon diodes, limits the signal ampli-

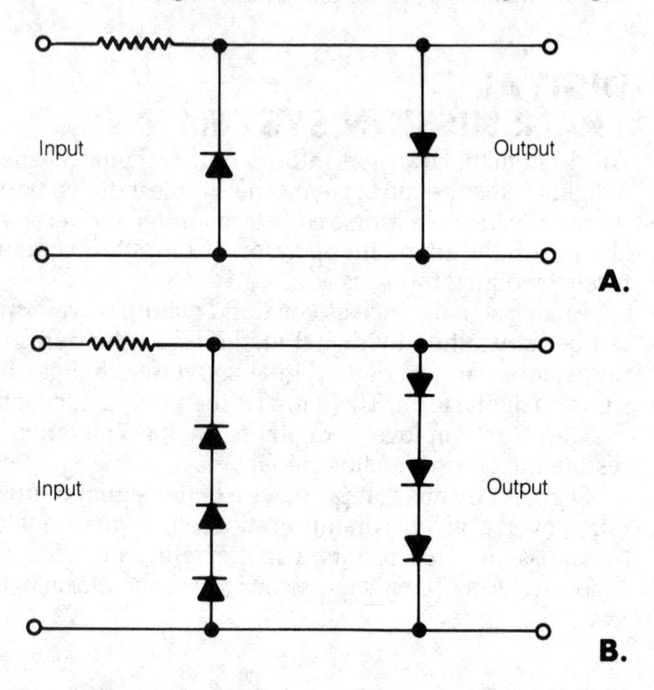

DIODE CLIPPING: Diode arrangements for clipping—two diodes in reverse parallel (A) and series diodes in reverse parallel for a higher clipping threshold (B).

tude to ±1.8 V, or 3.6 V peak-to-peak. If germanium diodes are used, the clipping level is ±0.9 V, or 1.8 V peak-to-peak.

The diode clipper circuit has a maximum effective operating frequency. As the signal frequency is raised, the diodes begin to show capacitance when they are reverse-biased. Because one diode or the other is always reverse-biased in the configurations shown in the drawings, performance is degraded at excessively high frequencies. The maximum usable frequency depends on the type of diodes used, and on the impedance of the circuit. *See also* DIODE.

DIODE DETECTOR

A diode detector is an envelope-detector circuit (*see* ENVELOPE DETECTOR). This detector is generally used for the demodulation of an amplitude-modulated signal.

When the alternating-current signal is passed through a semiconductor diode, half of the wave cycle is cut off. This results in a pulsating direct-current signal of variable peak amplitude. The rate of pulsation corresponds to the signal carrier frequency, and the amplitude fluctuations are the result of the effects of the modulating information. A capacitor is used to filter out the carrier pulsations, in a manner similar to the filter capacitor in a power supply. The remaining waveform is the audio-frequency modulation envelope of the signal. This waveform contains a fluctuating direct-current component, the result of the rectified and filtered carrier. A transformer or series capacitor can be used to eliminate this. The resulting output is then identical to the original audio waveform at the transmitter.

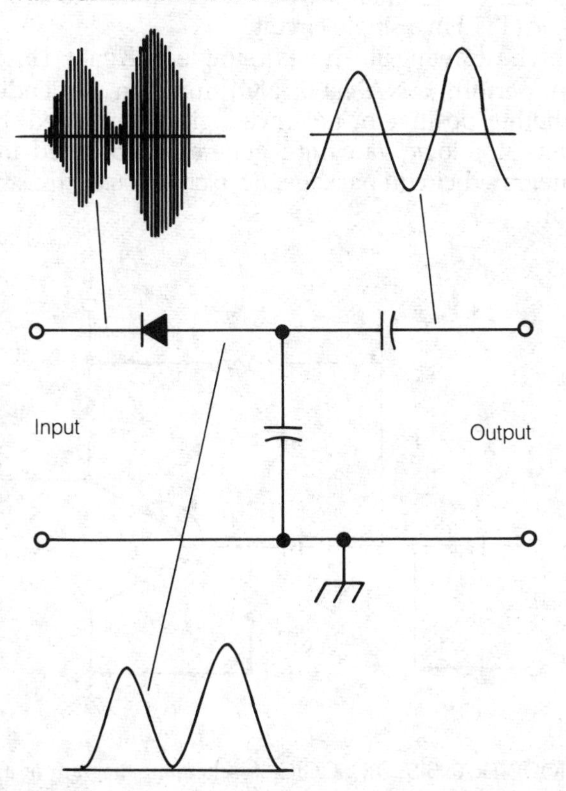

DIODE DETECTOR: A diagram of a simple diode detector circuit showing waveforms at three points.

The illustration is a simple schematic diagram of a diode detector. The waveforms at various points in the circuit are shown. *See also* DETECTION.

DIODE FIELD-STRENGTH METER

A field-strength meter that uses a semiconductor diode to obtain a direct current for driving a microammeter is called a diode field-strength meter. This field-strength meter, as shown in the schematic diagram, is the simplest meter possible for measuring relative levels of electromagnetic field strength, but it is not very sensitive. More advanced field-strength meters have amplifiers and tuned circuits built in and more nearly resemble radio receives than simple meters.

The diode field-strength meter is portable and can check a transmission line for proper shielding or balance. A handheld diode field-strength meter should show very little electromagnetic energy near a transmission line, but in the vicinity of the antenna radiator, a large reading is normally obtained. Some commercially made SWR (standing-wave-ratio) meters have a built-in diode field-strength meter, so that a small whip antenna can be used to monitor relative field strength. *See also* FIELD STRENGTH, FIELD-STRENGTH METER.

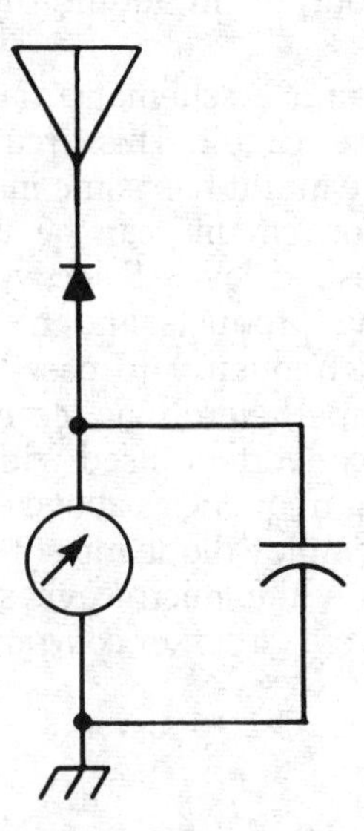

DIODE FIELD-STRENGTH METER: A diagram of a simple circuit for rectifying radio-frequency energy and measuring field strength.

DIODE IMPEDANCE

The impedance of a diode is the vector sum of the resistance and reactance of the device in a particular circuit (*see* IMPEDANCE, REACTANCE, RESISTANCE). Both the resistance and the capacitive reactance of a diode depend on the voltage across the device. The inductive reactance of a diode is essentially constant, and is primarily the result of inductance in the wire leads.

Generally, the resistance of a heavily forward-biased diode is extremely low, and the device in this case acts as a nearly perfect short circuit. When the forward bias is not strong, the resistance is higher, and the capacitive reactance is small. A reverse-biased diode has extremely high resistance. The capacitive reactance of a reverse-biased semiconductor diode increases as the reverse-bias voltage rises because the depletion region becomes wider (*see* DEPLETION LAYER, DIODE CAPACITANCE). The smaller the capacitance, the larger the capacitive reactance. *See also* DIODE, DIODE TUBE.

DIODE MATRIX

A diode matrix is a form of high-speed, digital switching circuit, using semiconductor diodes in a large array. Two sets of wires, one shown horizontally on a circuit diagram and the other shown vertically, are interconnected at various points by semiconductor diodes. Diode matrices can be small, as in a simple counter, or they can be large, as in a digital computer.

Diode matrices are employed as decoders, memory circuits, and rotary switching circuits. *See also* DECODING, SEMICONDUCTOR, MEMORY.

DIODE MIXER

A diode mixer is a circuit that uses the nonlinear characteristics of a diode for mixing signals (*see* MIXER). Whenever two signals with different frequencies are fed into a nonlinear circuit, the sum and difference frequencies are obtained at the output, in addition to the original frequencies.

The illustration is a schematic diagram of a typical, simple diode mixer circuit. This circuit has no gain, because it is passive and it has some insertion loss. However, amplification circuits can be used to boost the output to the desired level. Selective circuits reject all unwanted mixing products and harmonics, allowing only the desired frequency to pass. Diode mixers are often found in superheterodyne receivers and transmitters. Frequency converters, used with receivers to provide operation on frequencies above or below their normal range, may employ diode mixers. These circuits can function well into the microwave spectrum. *See also* FREQUENCY CONVERSION, FREQUENCY CONVERTER, MIXING PRODUCT.

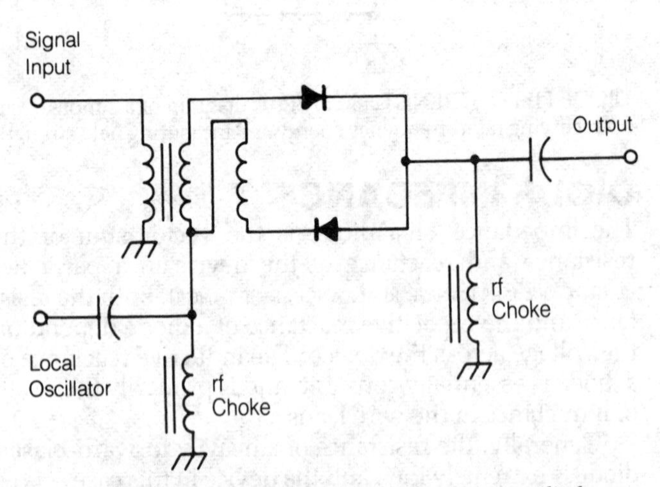

DIODE MIXER: The nonlinearity of a semiconductor diode produces the mixing products in this diode mixer circuit.

DIODE OSCILLATOR

See GUNN DIODE.

DIODE RATING

The rating of a diode refers to its ability to handle current, power, or voltage. Some semiconductor diodes are intended for small-signal applications, and can handle only a few microamperes or milliamperes of current. Other semiconductor diodes, called rectifiers, are capable of handling peak-inverse voltages of hundreds or even thousands of volts, and currents of several amperes. These rugged diodes are used in power-supply rectifier circuits.

Diode ratings are specified in terms of the peak inverse voltage, or PIV, and the maximum forward current. Zener diodes, used mostly for voltage regulation, are rated in breakdown or avalanche voltage and power-handling capacity. Other characteristics of diodes, that can be called ratings or specifications, are temperature effects, capacitance, voltage drop, and the current-voltage curve. *See also* AVALANCHE VOLTAGE, DIODE CAPACITANCE, DIODE IMPEDANCE, PEAK INVERSE VOLTAGE, ZENER DIODE.

DIODE-TRANSISTOR LOGIC

Diode-transistor logic (DTL) is a form of digital-logic design in which a diode and transistor act to amplify and invert a pulse. Diode-transistor logic has a slower switching rate than most other bipolar logic families. The power-dissipation rating is medium to low. Diode-transistor logic is sometimes mixed with transistor-transistor logic (TTL) in a single circuit.

The drawing illustrates a simple DTL gate. This circuit can perform a NAND or NOR function, depending on whether positive or negative logic is employed. Diode-transistor logic gates are generally fabricated into an integrated-circuit package. *See* DIGITAL-LOGIC INTEGRATED CIRCUIT.

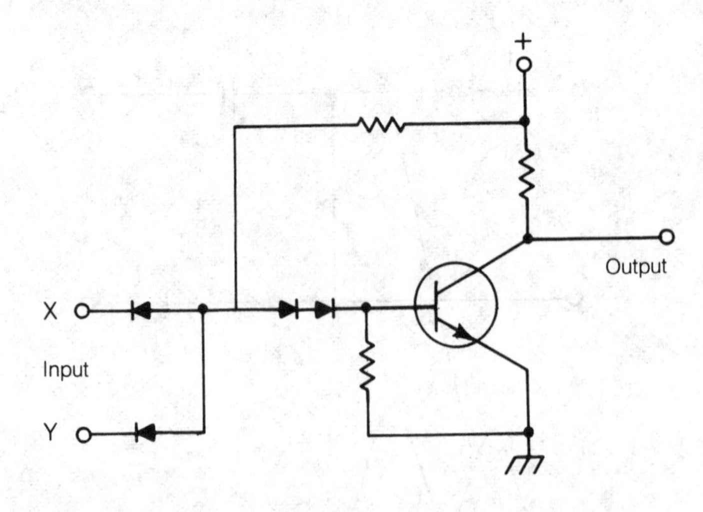

DIODE-TRANSISTOR LOGIC: A schematic diagram of a diode-transistor logic (DTL) gate.

DIODE TUBE

A diode tube is a vacuum tube with only two elements: a cathode and an anode. Diode tubes are used for the same purposes as semiconductor diodes; however, diode receiving tubes are obsolete in modern circuits.

The illustration shows the schematic symbols for diode tubes. At A, a directly heated-cathode tube is shown. At B, an indirectly heated-cathode tube is shown. The tubes at A and B require a power supply for the purpose of heating the cathode. At C, a cold-cathode tube is shown.

Magnetrons and neon gas discharge displays are examples of diode tubes used in modern electronics. *See* MAGNETRON.

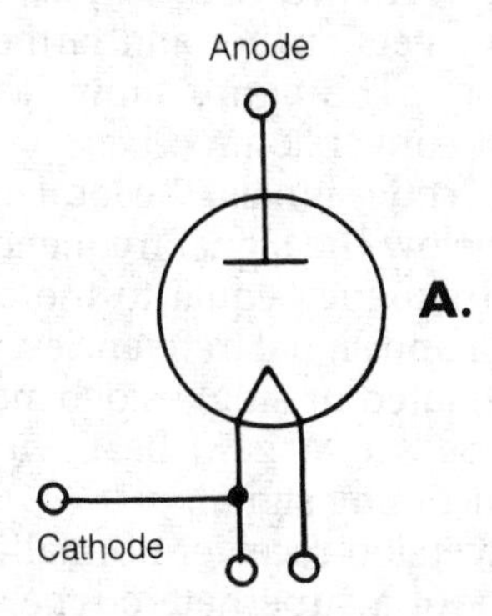

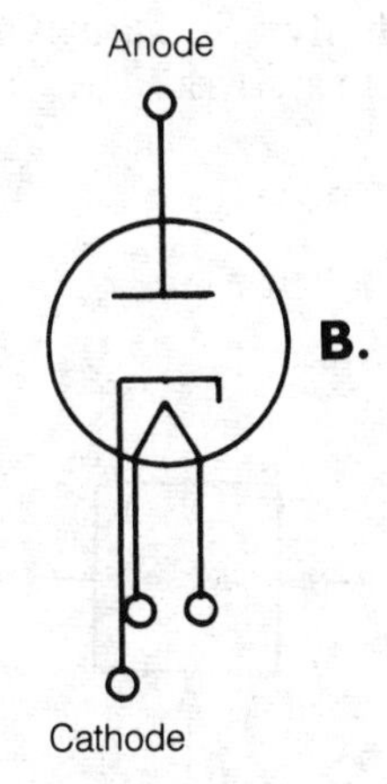

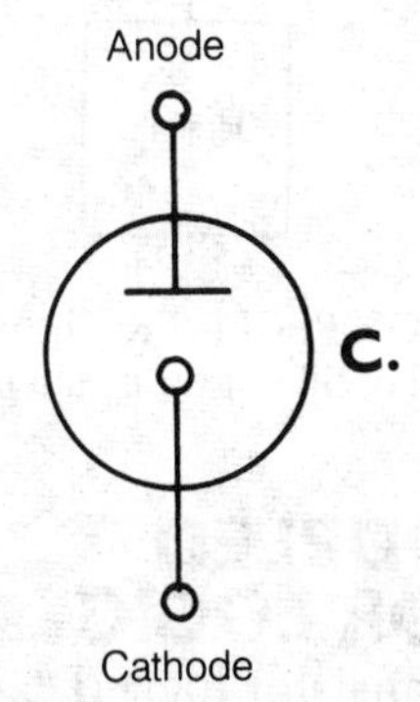

DIODE TUBE: Schematic symbols for diode tubes: directly heated cathode (A), indirectly heated cathode (B), and cold cathode (C).

DIODE TYPES

There are many different types of semiconductor diodes, each intended for a different purpose. The most common application for a diode is the conversion of alternating current to direct current. Detection and rectification use the ability of a diode to pass current in only one direction.

Light-emitting diodes, called LEDs, produce visible light when forward-biased. Solar-electric diodes generate direct current from visible light. Zener diodes are used as voltage regulators and limiters. Gunn diodes and tunnel diodes can be employed as oscillators at ultrahigh and microwave frequencies. Varactor diodes are used for amplifier tuning. A device called a PIN diode, which exhibits very low capacitance, is used as a high-speed switch at radio frequencies. Hot-carrier diodes are used as mixers and frequency multipliers. Frequency multiplication is also accomplished effectively using a step-recovery diode. The impact-avalanche-transit-time diode, or IMPATT diode, can act as an amplifying device.

Details of various diodes types and uses are discussed under the following headings: DIODE, DIODE ACTION, DIODE DETECTOR, DIODE FIELD-STRENGTH METER, DIODE MATRIX, DIODE-TRANSISTOR LOGIC, GUNN DIODE, IMPATT DIODE, LIGHT-EMITTING DIODE, PIN DIODE, P-N JUNCTION, SOLAR CELL, TUNNEL DIODE, VARACTOR DIODE, ZENER DIODE.

DIP

The dual-in-line package (DIP) is a familiar type of integrated circuit. *See also* SEMICONDUCTOR PACKAGE.

In electronics the term *dip* can also refer to the adjustment of a certain parameter for a minimum value. A common example is the dipping of the plate current in a tube type radio-frequency amplifier. The dip indicates that the output circuit is tuned to resonance, or optimum condition. Antenna tuning networks are adjusted for a dip in the standing-wave ratio. A dip is also sometimes called a null (*see* NULL).

DIPLEX

When more than one receiver or transmitter are connected to a single antenna, the system is called a diplex or multiplex circuit. The diplexer allows two transmitters or receivers to be operated with the same antenna at the same time.

The most familiar example of a diplexer is a television feed-line splitter, which allows two television receivers to be operated simultaneously using the same antenna. This device must have impedance-matching circuits to equalize the load for each receiver. Simply connecting two or more receivers together by splicing the feed lines will result in ghosting because of reflected electromagnetic waves along the lines. Diplexers for transmitters operate in a similar manner to those for receivers.

Multiplex transmission is sometimes called diplex transmission when two signals are sent over a single carrier. Each of the two signals in a diplex transmission consists of a low-frequency, modulated carrier called a subcarrier. The main carrier, much higher in frequency than the subcarriers, is modulated by the subcarriers. *See also* MULTIPLEX.

DIP METER

See GATE-DIP METER, GRID-DIP METER.

DIPOLE

When an atom or molecule has a negative and positive charge that are separated in space, the atom or molecule is known as an electric dipole. A molecule or atom with separate north and south magnetic poles is called a magnetic dipole. An electric dipole is surrounded by an electric field, often represented by lines of flux as shown in the illustration at A. A magnetic dipole is also surrounded by lines of flux representing the magnetic field as at B.

Depending on the electric or magnetic field in the vicinity of a dipole, the atom or molecule tends to orient itself in a certain direction because of the attraction between opposite charges or magnetic poles, and because of the repulsion between like charges or poles. An oscillating field will cause the dipoles to move back and forth at a rate corresponding to the frequency of the oscillation. An oscillating dipole, likewise, produces an oscillating field. The forces of electric and magnetic attraction and repulsion are responsible for the propagation of radio signals through space. *See also* ELECTROMAGNETIC FIELD.

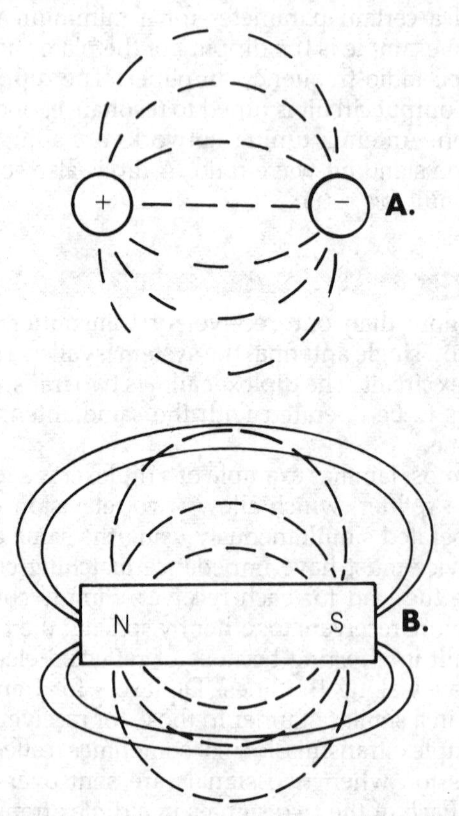

DIPOLE: Diagrams of dipoles with flux lines: electric (A) and magnetic (B).

DIPOLE ANTENNA

See HALF-WAVE ANTENNA IN THE ANTENNA DIRECTORY.

DIRECT-CONVERSION RECEIVER

A direct-conversion receiver is a receiver whose intermediate frequency is actually the audio signal heard in the speaker or headset. The received signal is fed into a mixer, along with the output of a variable-frequency local oscillator. As the oscillator is tuned across the frequency of an unmodulated carrier, a high-pitched audio beat note is heard, which becomes lower until it vanishes at the zero-beat point. Then it rises to pitch again as the oscillator frequency gets farther and farther away from the signal frequency. The figure shows a simple block diagram of a direct-conversion receiver.

For reception of code signals, the local oscillator is set slightly above or below the signal frequency. The audio tone will have a frequency equal to the difference between the oscillator and signal frequencies. For reception of amplitude-modulated or single-sideband signals, the oscillator should be set to zero beat with the carrier frequency of the incoming signal.

The direct-conversion receiver normally cannot provide the selectivity of a superheterodyne, since single-signal reception with the direct-conversion receiver is impossible. Audio filters are often used to provide some measure of selectivity. *See also* INTERMEDIATE FREQUENCY, SUPERHETERODYNE RADIO RECEIVER.

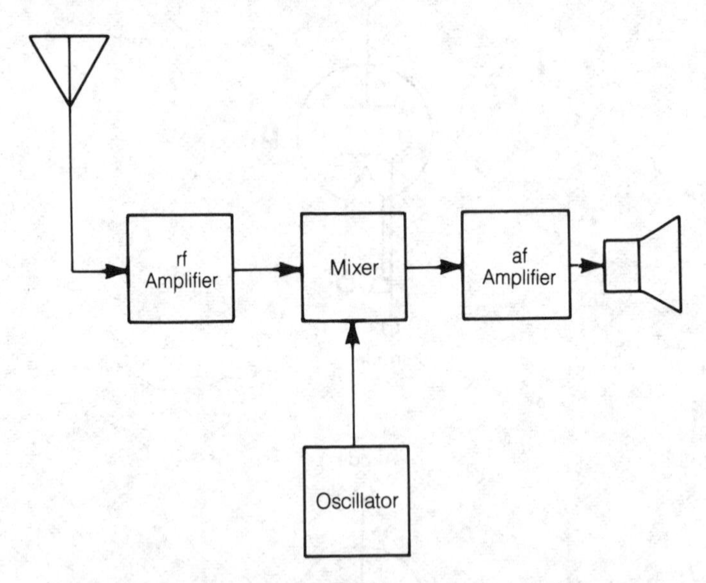

DIRECT-CONVERSION RECEIVER: The incoming signal is mixed directly with the local oscillator output in this receiver.

DIRECT-COUPLED TRANSISTOR LOGIC

Direct-coupled transistor logic (DCTL) is a bipolar logic family. It was the earliest logic design used in commercially manufactured integrated circuits. The DCTL logic scheme is fairly simple, as shown in the schematic.

Direct-coupled transistor logic has poor noise rejection characteristics. Current hogging can also cause some problems. Newer forms of direct-coupled transistor logic are available, and these designs have better operating characteristics than the original form. The signal voltages in DCTL are low. The switching speed and power-handling capabilities are about average, compared with other bipolar logic families. *See also* DIGITAL-LOGIC INTEGRATED CIRCUIT.

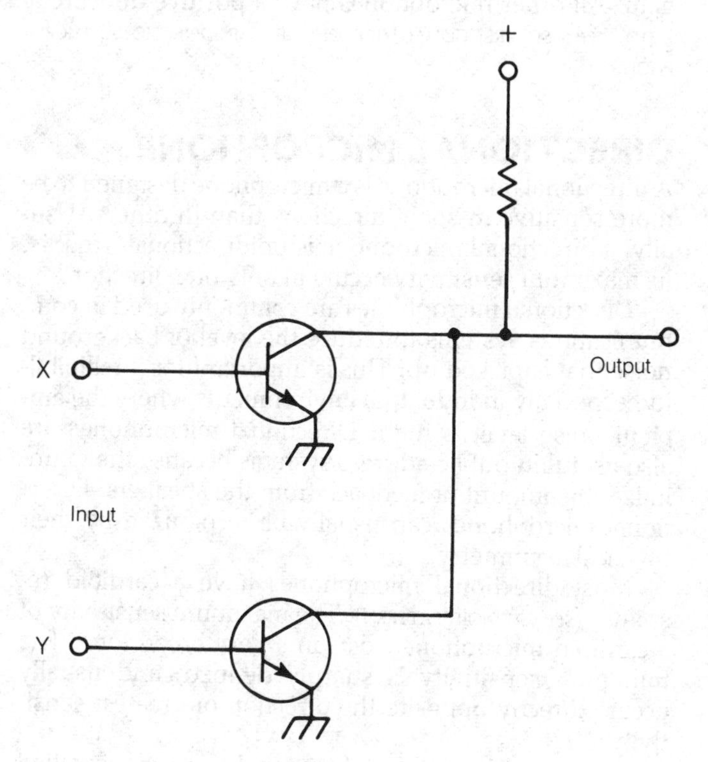

DIRECT-COUPLED TRANSISTOR LOGIC: Logic gates are directly connected transistors in this logic form.

DIRECT COUPLING

Direct coupling is a form of circuit coupling, usually employed between stages of an amplifier. In direct coupling, the output from one stage is connected, by a direct wire circuit, to the input of the next stage. This increases the gain of the pair over the gain of a single component, provided the bias voltages are correct. The Darlington amplifier is an example of direct coupling (*see* DARLINGTON AMPLIFIER). The schematic diagram showing direct-coupled transistor logic shows an example of direct coupling used to increase the sensitivity of a switching network.

Direct coupling is characterized by a wideband frequency response because there are no intervening reactive components, such as capacitors or inductors, to act as tuned circuits. Therefore, direct-coupled amplifiers are more subject to noise than tuned amplifiers. Direct coupling transmits the alternating-current and direct-current components of a signal. Current hogging can sometimes be a problem, especially if adequate attention is not given to the maintenance of proper bias. *See also* CURRENT HOGGING.

DIRECT CURRENT

A direct current is a current that always flows in the same direction; that is, the polarity never reverses.

Physicists consider the current in a circuit to flow from the positive pole to the negative pole. This is purely a convention. The movement of electrons in a direct-current circuit is opposite to the theoretical direction of the current. In a P-type semiconductor material, however, the motion of the positive charge carriers (holes) is the same as the direction of the current.

Typical sources of direct current include electronic power supplies, as well as batteries and cells. The intensity, or amplitude, of a direct current can vary with time, and this variation can be periodic.

DIRECT-DRIVE TUNING

When the tuning knob of a radio receiver or transmitter is mounted directly on the shaft of a variable capacitor, the tuning is said to be directly driven. A half turn of the tuning knob thus covers the entire range. Direct-drive tuning is found in small portable transistor radio receivers for the standard broadcast band. It is difficult to obtain precise tuning with a direct-drive control.

Most controls for the adjustments of frequency are indirectly driven. Thus, several turns of the control knob are needed to cover the entire range. This method of control is called vernier drive, and it may be employed in various other situations besides radio tuning. Cable-driven controls are also sometimes used to spread out the tuning range of a radio receiver or transmitter.

Most circuit-adjustment controls, such as volume and tone, are directly driven. However, vernier or cable drives may be used with any control to obtain precise adjustment. *See also* VERNIER.

DIRECTIONAL ANTENNA

See ANTENNA DIRECTORY.

DIRECTIONAL FILTER

A directional filter is a frequency-selective filter or set of filters in a carrier-current communications system. Generally, the filter consists of a lowpass and/or highpass filter (*see* CARRIER CURRENT, HIGHPASS FILTER, LOWPASS FILTER). The cutoff frequency of the filter is set in the middle of the band in which communication is carried out.

A directional-filter set allows communication in only one direction for the lower half of the band, and in only the other direction for the upper half of the band. The illustration is a block diagram of a carrier-current communications system, showing a directional-filter pair at each end of the circuit. At station X, the transmitter operates in the upper half of the band, at a higher frequency than the receiver, which is set for the lower half of the band. At station Y, this is exactly reversed: The transmitter operates in the lower part of the band, and the receiver in the upper part.

The directional filters prevent the signal components of a transmitter from overloading the station receiver. With sufficient selectivity, full duplex operation is possible over a single current-carrying conductor. *See also* DUPLEX OPERATION.

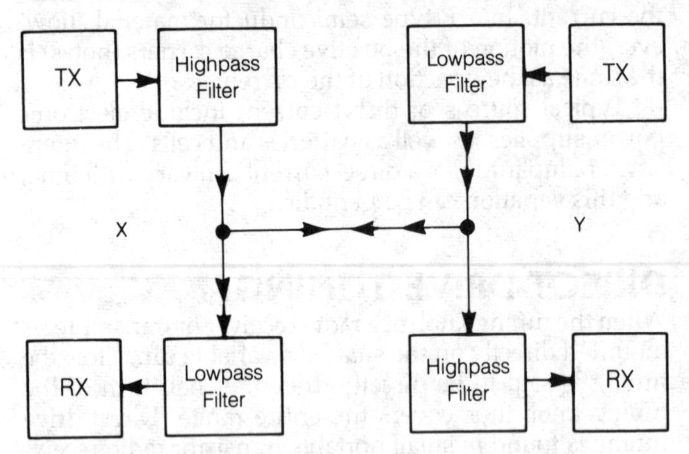

DIRECTIONAL FILTER: Duplex communication can be obtained over a single conductor with this filter.

DIRECTIONAL GAIN

Directional gain is a measurement for expressing the sound-radiating characteristics of a speaker. The term can also be applied to the acoustic-pickup characteristics of a microphone.

Generally, speakers do not radiate sound equally well in all directions. Instead, most of the sound energy is concentrated in a narrow cone, with its axis perpendicular to the speaker face. If a speaker were able to radiate equally well in all directions, its directional gain would be zero.

Consider a theoretical speaker, supplied with P watts of audio-frequency power, that produces a sound-intensity level of p_x dynes per square centimeter at a fixed distance of m meters, in all directions. The directional gain of this speaker is zero, because its sound radiation is the same in all directions. Imagine that this theoretical speaker is replaced by a real speaker, which radiates more effectively in some directions than in others. Now consider that the real speaker has the same efficiency as the hypothetical one, and receives the same amount of power. If the sound pressure at a distance of m meters from this speaker, along the axis of its maximum sound output, is p dynes per square centimeter, then the directional gain of the real speaker, in decibels (dB), is:

$$\text{Directional gain (dB)} = 10 \log_{10} \left(\frac{p}{p_x}\right)$$

All real speakers have positive directional gain. When sound radiation is enhanced in one direction, it must be sacrificed in other directions. The converse of this is also true: a reduction in sound in some directions results in an increase in other directions, assuming the same total power output from the speaker.

Directional gain is also expressed in decibels for microphones. But for a microphone, the directional gain is given in terms of pickup sensitivity, or the sound pressure required to produce a specified electrical output at the microphone terminals. For transducers in general, the directional gain is expressed as the directivity index (*see* DIRECTIVITY INDEX).

An omnidirectional microphone has zero directional gain. All other microphones have a positive directional gain. *See also* DIRECTIONAL MICROPHONE, OMNIDIRECTIONAL MICROPHONE.

DIRECTIONAL MICROPHONE

A directional microphone is a microphone designed to be more sensitive in some directions than in others. Usually, a directional microphone is unidirectional—that is, its maximum sensitivity occurs in only one direction.

Directional microphones are commonly used in communications systems to reduce the level of background noise that is picked up. This is important for intelligibility, especially in industrial environments where the ambient noise level is high. Directional microphones are also useful in public-address systems because they minimize the amount of feedback from the speakers. Directional microphones can usually be recognized by their physical asymmetry.

Most directional microphones have a cardioid response (*see* CARDIOID PATTERN). The maximum sensitivity of a cardioid microphone exists in a very broad lobe. The minimum sensitivity is sharply defined, and usually occurs directly opposite the direction of greatest sensitivity.

A microphone without directional properties is called an omnidirectional microphone. *See* OMNIDIRECTIONAL MICROPHONE.

DIRECTION FINDER

See RADIO DIRECTION FINDER.

DIRECTIVITY INDEX

The directivity index of a transducer is a measure of its directional properties. The directional index is similar to the directional gain of a speaker or microphone (*see* DIRECTIONAL GAIN); however, the mathematical definition is slightly different in terms of its expression.

For a sound-emitting transducer, let p_{av} be the average sound intensity in all directions from the device, at a constant radius m, assuming that the transducer receives an input power P. Then if p is the sound intensity on the acoustic axis, or favored direction, at a distance m from the device, the directivity index in decibels is given by:

$$\text{Directivity index (dB)} = 10 \log_{10} \left(\frac{p}{p_{av}}\right)$$

For a pickup transducer, the same mathematical concept applies. However, it is in the reverse sense. If a given sound source, at a distance m from the transducer and having a certain intensity, produces an average voltage E_{av} at the transducer terminals as its orientation is varied in all possible directions, and the same source provides a voltage E when at a distance m from the transducer on the acoustic axis, then:

$$\text{Directivity index (dB)} = 20 \log_{10}\left(\frac{E}{E_{av}}\right)$$

See also TRANSDUCER.

DIRECTIVITY PATTERN

See ANTENNA PATTERN.

DIRECT MEMORY ACCESS

In a digital computer, direct memory access (DMA) is a means of obtaining information from the memory circuits without the intercession of the central processing unit (CPU).

Direct memory access saves time and is more efficient than getting memory information by routing through the central processing unit. There are several different methods for obtaining direct memory access. The process varies among different computer models. Direct memory access is used for the purpose of transferring memory data from one location to another, when it is not necessary to perform any operations on it.

DIRECTOR

A director is a form of parasitic element in an antenna system designed to generate power gain in certain directions. This increases the efficiency of a communications system, both by maximizing the effective radiated power of a transmitter, and by reducing the interference from unwanted directions in a receiving system. *See* ANTENNA POWER GAIN, PARASITIC ARRAY.

An example of the operation of a director is found in all Yagi- or quad-type antennas. A half-wave dipole antenna has a gain of 0 dBd (*see* ANTENNA DIRECTORY, dBd DECIBELS) in free space. This means that its gain with respect to a dipole is zero. The figure illustrates the directive pattern of a half-wave dipole in free space at A.

If an unattached length of conductor measuring approximately ½ wavelength is brought near the half-wave dipole and parallel to it, the directivity pattern of the antenna changes radically. This was noticed by a Japanese engineer named Yagi. (The Yagi antenna is named after him.) When the free element is a certain distance from the dipole, gain is produced in the direction of the free element, as shown at B. The free element is called a director. At certain other separations, the free element causes the gain to occur in the opposite direction; then it is called a reflector.

The most power gain that can be obtained using a dipole antenna and a single director is, theoretically, about 6 dBd. In practice it is closer to 5 dBd because of ohmic losses in the antenna conductors. When two full-wavelength loops are brought in close proximity parallel to each other, with one loop driven (connected to the transmission line) and the other loop free, the same effect is observed. An antenna using loops in this manner is called a quad antenna.

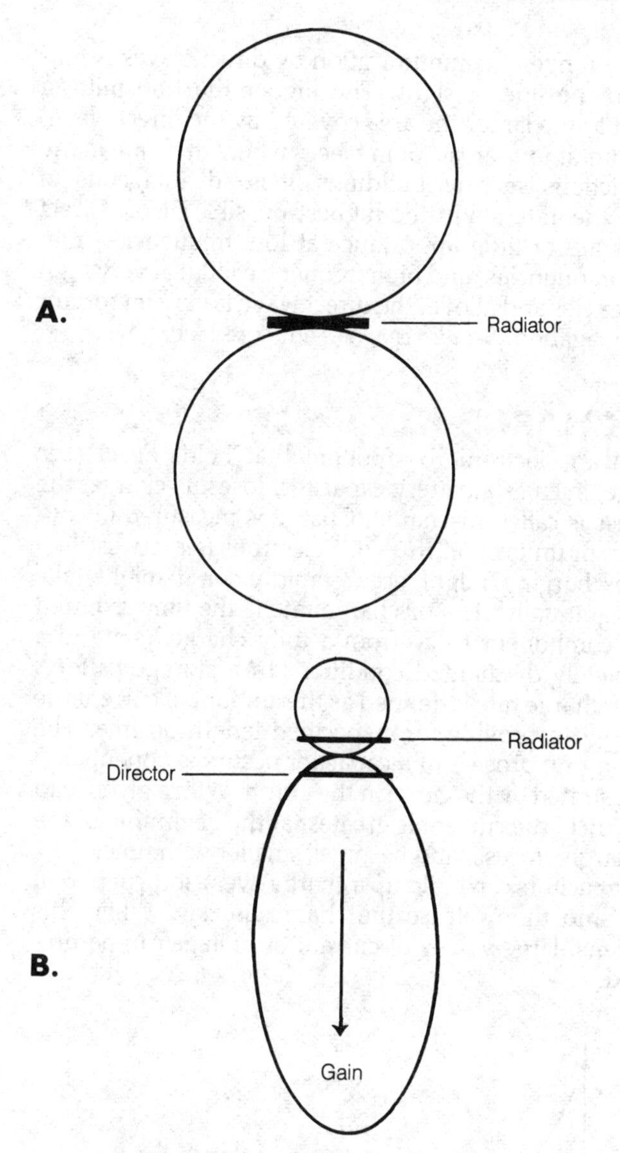

DIRECTOR: A radiation pattern of a dipole (A) and the change in pattern made by introducing a director (B).

DIRECT WAVE

In radio communications, the direct wave is the electromagnetic field that travels from the transmitting antenna to the receiving antenna along a straight line through space. Direct waves are responsible for part, but not all, of the signal propagation between two antennas when a line connecting the antennas lies entirely above the ground. The surface wave and the reflected wave also contribute to the overall signal at the receiving antenna in this case. The combination of the direct wave, the reflected wave, and the surface wave is sometimes called the ground wave (*see* GROUND WAVE, REFLECTED WAVE, SURFACE WAVE). Depending on the relative phases of the direct, reflected, and surface waves, the received signal over a line-of-sight path may be very strong or practically nonexistent.

The range of communication by direct waves is limited to the line of sight. The higher the transmitting antenna, the larger the area covered by the direct wave. In mountainous areas, or in places where there are many obstructions such as buildings, it is advantageous to locate the antenna in the highest possible place. Direct waves are of little importance at low frequencies, medium frequencies, and high frequencies. But at very high frequencies and above, the direct wave is very important in propagation. *See also* LINE-OF-SIGHT COMMUNICATION.

DISCHARGE

When an electronic component that holds an electric charge, such as a battery capacitor, loses its charge, the process is called discharge. Charge is measured in coulombs, or units of 6.218×10^{18} electrons (*see* CHARGE).

Discharging might occur rapidly, or it might take place gradually. The discharge rate is the time required for a component to go from a fully charged state to a completely discharged condition. In a storage battery, the discharge rate is defined as the amount of current the battery can provide for a specified length of time. The discharging process of a capacitor occurs exponentially, as illustrated by the curve in the graph. With a given load resistance, the current is greatest at the beginning of the discharge process, and becomes smaller with time.

Capacitors can build up a charge over a long period of time, and then release the charge quickly. When this happens, large values of current or voltage can be produced.

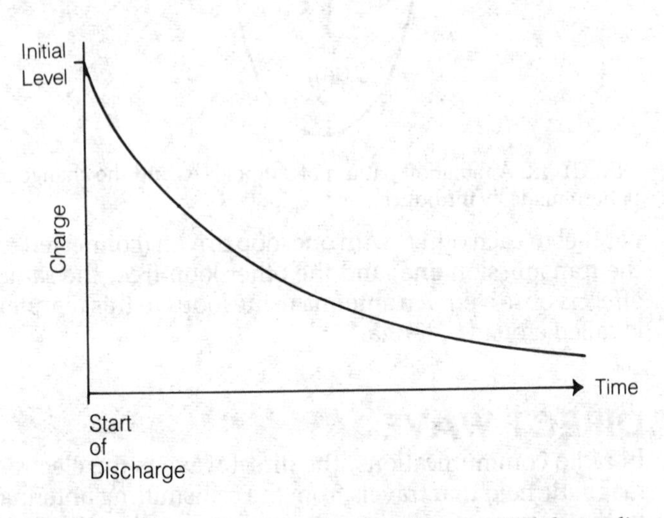

DISCHARGE: The plot of a generalized exponential-decay discharge curve.

DISCONE ANTENNA

See ANTENNA DIRECTORY.

DISCONTINUITY

A discontinuity in an electrical circuit is a break, or open circuit, that prevents current from flowing. Discontinuity may occur in the power-supply line to equipment, resulting in failure. Sometimes a discontinuity occurs because of a faulty solder connection or a break within a component. These circuit breaks can be extremely difficult to find. Sometimes the discontinuity is intermittent, or affects the operation of a circuit only in certain modes or under certain conditions.

In a transmission line, a discontinuity is an abrupt change in the characteristic impedance (*see* CHARACTERISTIC IMPEDANCE). This may occur because of damage or deterioration in the line, a short or open circuit, or a poor splice. A transmission-line discontinuity can be introduced deliberately by splicing two sections of line with different characteristic impedances. This technique is used for certain impedance-matching applications, and for distributing the power uniformly among two or more separate antennas.

In a discontinuous mathematical function, a discontinuity is a point in the domain of the function (along the axis of the independent variable) at which the function jumps from one value to another, or is undefined. Some functions have no discontinuities; these are called continuous functions. Some functions have many discontinuities, perhaps even an infinite number of them. *See also* CONTINUOUS FUNCTION, DISCONTINUOUS FUNCTION, FUNCTION.

DISCONTINUOUS FUNCTION

A discontinuous function is a function that is not continuous at every point in its domain (*see* CONTINUOUS FUNCTION, FUNCTION). On a graph, as the illustration, this appears as either a jump in value, or an undefined value. At A, the

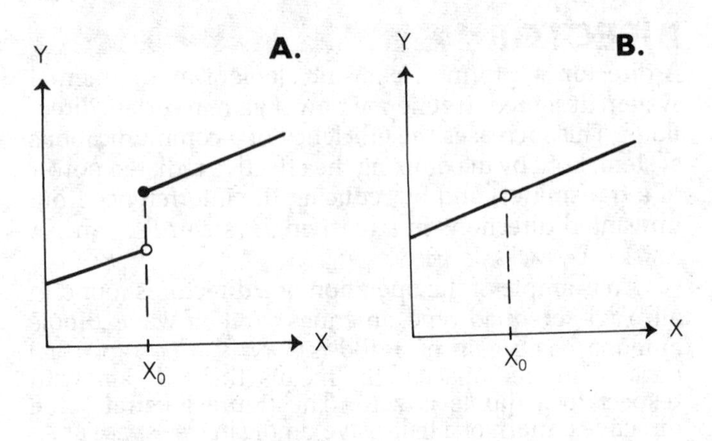

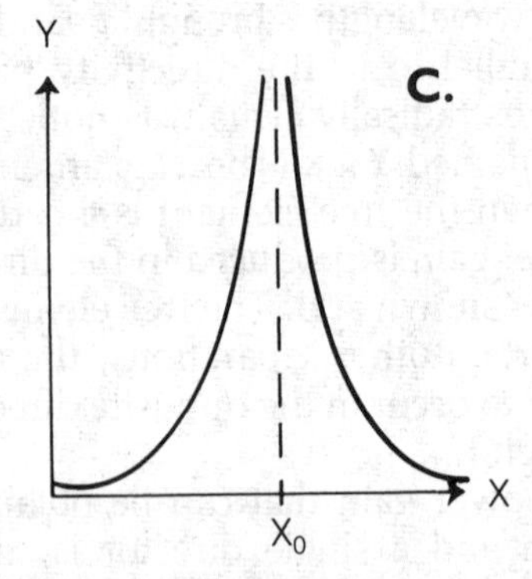

DISCONTINUOUS FUNCTION: Examples of discontinuous functions—a plot showing a jump (A) and plots with undefined points (B and C).

value of the function abruptly changes. At B and C, a point in the domain is not mapped into any value; that is, the value of the function is not defined.

Suppose x_0 is a point in the domain of the function $y = f(x)$, at which the function is not continuous. Then, the limiting value of the function as x approaches x_0 from the negative (left) side, and the limiting value of the function as x approaches x_0 from the positive (right) side, are not both equal to the value of the function $y_0 = f(x_0)$. This is expressed as an assertion that the following is not true:

$$\lim_{x \to x_0^-} f(x) = \lim_{x \to x_0^+} f(x) = y_0$$

Examples of discontinuous functions are the tangent, cotangent, secant, and cosecant functions. All of these functions have undefined points. Another common example of a discontinuous function is the step function. *See also* COSECANT, COTANGENT, SECANT, TANGENT.

DISCRETE COMPONENT

An electronic component such as a resistor, capacitor, inductor, or transistor is called a discrete component if it is packaged as one or two functional elements per package. However, resistors, capacitors, or transistors in an integrated circuit are not discrete; they are manufactured with the whole package, which may contain thousands of individual components.

In the early years of electronics, all circuits were built from discrete components. Only after the development of solid-state circuit integration did other designs emerge. At first, discrete components were assembled and sealed in a package called a compound circuit. But modern technology has provided the means for fabricating thousands of individual components on the surface of a semiconductor wafer.

Although discrete components are used less often in a circuit than they were only a few years ago, there will always be a place for them. Some devices such as fuses, circuit breakers, and power transistors must remain discrete. However, there will probably be even fewer discrete components in electronic circuits in the coming years, as digital-control techniques become more accepted. *See also* DIGITAL CONTROL, INTEGRATED CIRCUIT.

DISCRIMINATOR

A detector used in frequency-modulation receivers is called a discriminator (*see* FREQUENCY MODULATION). The discriminator circuit produces an output voltage that depends on the frequency of the incoming signal. In this way, the circuit detects the frequency-modulated waveform.

When a signal is at the center of the passband of the discriminator, the voltage at the output of the circuit is zero. If the signal frequency drops below the channel center, the output voltage becomes positive. The greater the deviation of the signal frequency below the channel center, the greater the positive voltage at the output of the discriminator. If the signal frequency rises above the channel center, the discriminator output voltage becomes negative; and, again, the voltage is proportional to the deviation of the signal frequency. The amplitude of the voltage at the output of the discriminator is linear, in proportion to the frequency of the signal. This ensures that the output is not distorted.

The illustration is a schematic diagram of a simple discriminator circuit suitable for use in a frequency-modulation receiver. A shift in the input signal frequency causes a phase shift in the voltages on either side of the transformer. When the signal is at the center of the channel, the voltages are equal and opposite, so that the net output is zero.

A discriminator circuit is somewhat sensitive to amplitude variations in the signal, as well as to changes in the frequency. Therefore, a limiter circuit is usually necessary when the discriminator is used in a frequency-modulation receiver. A circuit called a ratio detector, developed by RCA, is not sensitive to amplitude variations in the incoming signal. Thus, it acts as its own limiter. Immunity to amplitude variations is important in frequency-modulation reception, because it enhances the signal-to-noise ratio. *See also* LIMITER, RATIO DETECTOR.

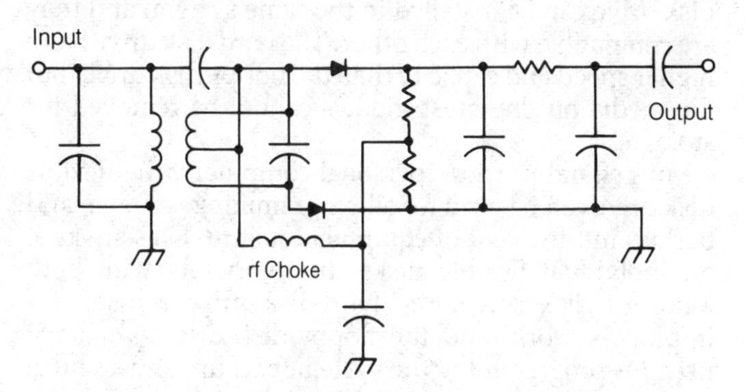

DISCRIMINATOR: A schematic of a circuit for frequency modulation detection.

DISH ANTENNA

See ANTENNA DIRECTORY.

DISK CAPACITOR

See CAPACITOR.

DISK DRIVE

A disk drive is an electromechanical subsystem that stores and retrieves digital data on one or more surfaces of disk-shaped magnetic media. The disks rotate under electromagnetic heads. A disk drive is capable of taking digital data from memory and writing it on disks or reading it from disks and restoring it to memory. Read-write heads on all disk drives are capable of altering the magnetic domains of the surface media to form small regions that represent ones or zeros.

The read-write heads can move in two directions over the spinning media: radially and vertically. All disk drives can be accessed randomly. Disk drives are secondary or mass memory for computers ranging in size and capability from personal computers to supercomputers.

The term *disk drive* generally applies to two different electromechanical configurations: 1) hard-disk drives and 2) flexible- or floppy-disk drives. The operating principles and magnetic storage techniques of both drives are similar. Both use the same methods for organizing data in tracks and sectors. Hard-disk drives use rigid media, typically aluminum disks or platters coated with magnetic material. Depending on the drive model selected, the disks might or might not be routinely removable from the drive. By contrast, a flexible or floppy-disk drive uses flexible, removable media, typically a sheet-plastic disk coated with magnetic material. The disk is supported and protected with an external protective jacket. The flexible disk rotates within a rigid or semirigid protective jacket and is called a floppy disk or diskette.

The floppy- or flexible-disk drive is standard equipment for most small computers or microcomputers (personal computers, desktop workstations, and desktop word processors) as well as many minicomputers. Hard-disk drives are optional mass-storage peripherals for microcomputers but are standard equipment for most minicomputers and mainframe computers. Both forms of disk drive can be installed in the same system, and they are compatible with each other. The hard disk drive has a higher speed and capacity than the floppy disk drive, but the media on the latest models cannot be removed for storage.

In general-purpose personal computers, the flexible disk drive can be used for all programming, storage, and backup in the event of equipment failure. Files are kept on individual flexible disks. In computers with both kinds of disk drive, the hard-disk drive is used for in-process work, and the floppy-disk drive is usually used for programming the system, backup, and archival storage. A magnetic tape drive might also be used as a peripheral for archival storage of data. *See* TAPE DRIVE.

The name *disk drive* has taken on the meaning of hard-disk drive with magnetic media. Although it is also a disk drive, the complete name *floppy-disk drive* is usually used to identify that machine. Both of these drives should be distinguished from disk drives that use a compact disc (CD) optical disk as in CD ROM, a write-once, read-many (WORM) optical disk, and a magnetic erasable optical disk. Lasers are used to write digital data on all three of these disks. However, the CD and WORM disk drives are only capable of reading previously stored data by means of internal lasers. But the laser system of the erasable optical-magnetic disk drive can write and erase unwanted data by overwriting. *See* OPTICAL DISK DRIVE.

The most common form of hard-disk drive in use is the Winchester disk drive, characterized by media sealed within an environmentally conditioned case. However, hard disk drives with removable media are still in use. Hard disk drive media is a set of one or more rigid aluminum disks or platters coated with a magnetic material. This material can be an iron or cobalt oxide layer similar to that used on video or audio recording tape, or it can be the more recently introduced magnetic media that is a deposited, thin-film metal alloy. The alloy coating is evaporated on the surface of the disk in a vacuum chamber. It has a smoother surface than the oxide coating, which helps to reduce errors. A single Winchester disk drive can use from four to nine disks or platters on a common motor-drive spindle.

Floppy or flexible disks are thin sheets of plastic coated with ferromagnetic iron or cobalt oxide similar to that used on audio and video recording tape. The 5¼-inch and 8-inch floppy disks are enclosed in a semirigid plastic jacket, but the 3½-inch floppy disks are enclosed in rigid cartridges.

Disk drives write and read only digital information on the disks. They work on the same basic magnetic recording principles as magnetic tape recorders except that the media spins under magnetic read-write heads, which can be moved from track to track. This contrasts with tape drives, in which the flexible plastic tape moves past the fixed read-write heads. Digital data tape drives are used with computers as peripherals where rapid access of data is not a requirement. *See* TAPE DRIVE, MAGNETIC RECORDING.

The principal advantage of any disk drive over a tape drive is faster accessibility to the stored data because the data can be accessed randomly. It is not necessary to search through a long tape to find the desired data as is necessary with serially accessed tape.

Information is stored on the disks by forcing a change in the state of magnetic domains on the surface of the magnetic material. Magnetism is binary in that it forms two poles: north and south. The magnetic state in one direction can be considered positive (or binary 1), and in the other direction it can be considered negative (or binary zero).

All disk drives have one or more spinning magnetic disks or platters and one or more read-write heads. The heads are able to swing radially across the face of the spinning disk so that all active surfaces of the disk will pass under a head during a finite amount of time. The head is capable of altering the magnetic state of a well-defined region of each disk as it passes under (or over) the head. See Fig. 1.

Read-write heads are in contact with the disk in flexible disk drives, but they fly a few microinches above the disks on a thin film of air in the hard-disk drives. A pulse of current passed through the head creates a magnetic field. This field induces a polarity change in the domains of the magnetic material passing under it. In effect, a bar magnet is formed in a small region. If the current flow is reversed, the polarity of the domain will be reversed.

Flexible-Disk Drives The flexible- or floppy-disk drive was invented to overcome the high cost of hard disk drives for personal computers, dedicated word processors, industrial controls, and robots based on microprocessors. Initially the cost of the hard-disk drive exceeded that of the small computer. The two forms of disk drive are compatible with each other, and many computers are equipped with both. The floppy-disk drive has replaced punched tape, and the tape reader as the primary means for entering programs into computer memory. However, the decreasing cost of hard-disk

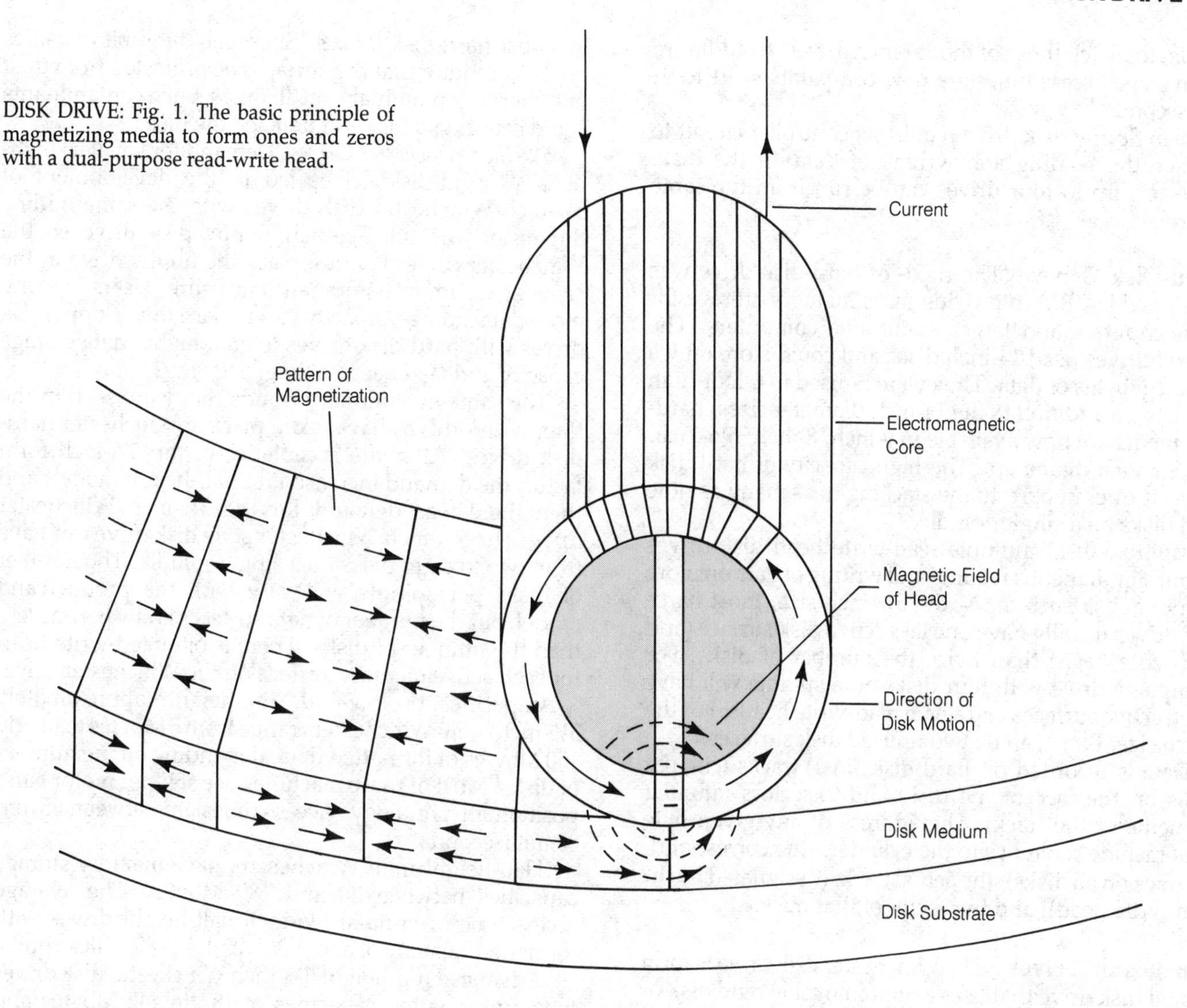

DISK DRIVE: Fig. 1. The basic principle of magnetizing media to form ones and zeros with a dual-purpose read-write head.

drives combined with increasing demand for more data storage in small computer systems has limited the applications for floppy-disk drives.

The thin plastic disk coated with ferromagnetic iron or cobalt oxide is enclosed in a semirigid plastic protective jacket in the 5¼- and 8-inch sizes, but it is in a rigid cartridge in the 3½-inch size. The drive spins the plastic disk within the protective envelope, and the read-write head accesses the disk through a radial slot in the envelope.

The floppy disk remains popular for the storage and distribution of programs as well as data because it is removable from the drive. The disk can be easily distributed, mailed, or stored. It will withstand considerable handling without data loss if reasonable precautions are taken to protect it against environmental extremes and abuse.

Floppy-disk data capacity depends on the diameter of the disk, and this diameter determines the physical dimensions of the disk drive. Floppy-disk drives have only one disk per drive. The disk is rotated at 300 revolutions per minute (rpm) by a spindle motor, and the read-write head is moved by a stepper motor across the face of the disk. Single-sided disk drives have only one read-write head; double-sided disk drives have two.

The early 8-inch disks had far higher capacity than the 5¼-inch disks. However, by improving formatting and read-write heads, there were progressive increases in the capacity of the smaller disks. The amount of data that can now be stored on 5¼-inch disk surpasses what could be stored on earlier 8-inch disks. As the capacity of smaller disk drives was increased, demand for them increased.

The 5¼-inch minifloppy drive evolved from a unit with a single read-write head and a one-sided disk; it had a capacity of 125 kilobytes (kbytes). The latest models have two read-write heads and double-sided disks with a capacity up to 1.6 megabytes (Mbytes).

The next step in the evolution was height reduction. Half-height versions of 8-inch and 5¼-inch drives were developed so that two drives could be located in the same volume formerly occupied by a single unit. Half-height 5¼-inch drives are now available in unformatted capacities of 360 kbytes to 1.6 Mbytes, with average access times between 90 and 96 milliseconds (ms).

The 5¼-inch disk and drive are being supplanted by the 3½-inch microfloppy disk and drive as the data capacity of the smaller disk increases. The 3½-inch rigid envelope is easier to store and ship than the 5¼-inch floppy disk. The 3½-inch drives are now available with unformatted capacities of 0.5 Mbyte, 1 Mbyte, and 2

Mbyte, to match those of the 5¼-inch drives. In addition, the average access times are now comparable—91 to 94 milliseconds.

Each floppy disk drive requires a controller circuit to manage the reading and writing of data on the disk. However, up to four drives can be run from the same controller.

Hard-Disk Drives The rigid- or hard-disk drive was developed by IBM to provide mass memory storage for minicomputers and larger mainframe computers. The earliest drives used 14-inch disks and could store only a few megabytes of data. They were housed in 6-foot-high cabinets. In addition to the 14-inch-diameter sizes, hard-disk media are now available in 9-inch, 8-inch, 5¼-inch, and 3½-inch diameters. The high capacity of hard disk drives is due, in part, to the stacking of as many as nine hard disks on a single spindle.

Multiple-disk, multiple read-write head disk drives permit simultaneous reading and writing of data on more than one disk surface. At the 5¼-inch size, most hard-disk drives usually have one less active disk surface (and read-write head) than twice the number of disks. For example, a drive with four disks on a spindle will have seven active surfaces and seven read-write heads. For the 8-inch size, there can be two unused disk surfaces.

Data is recorded on hard disks in 1) tracks (circular paths on the face of the disk) and 2) sectors (angular segments of the track). The address of any given byte must include reference to the cylinder, (the corresponding track on all disks), the actual track, (designated by the read-write head), and the sector of that track.

Winchester Drives The Winchester disk drive is form of rigid disk drive that uses a nonremovable rigid disk or stack of disks on a spindle. This design has been accepted for most hard-disk drives. Extremely high bit densities on disks require that the surfaces be protected from dust or airborne contaminants at all times. Such contaminants could cause data loss and damage the disk.

5¼-Inch Winchester Drives. Demand for smaller, high-density, rigid-disk drives led to the development of 5¼-inch Winchester disk drives with the same outline dimensions as the 5¼-inch floppy disk drive so the Winchester drives could replace the floppy drive in the same space in an equipment enclosure. Users wishing more data storage capacity can replace their floppy-disk drives with hard-disk drives to gain higher data storage capacity and faster access time. See Fig. 2.

The same evolutionary trends that were seen in the floppy-disk drives have taken place in Winchester hard-disk drives. After the acceptance of the 5¼-inch form factor, the demand increased for half-height models and then there was demand for the 3½-inch Winchester drive. The 5¼-inch Winchester style disk drive can have from two to eight disks on a single spindle. The number of disks per spindle will vary with the product and model, but the number of data surfaces is always one less than the number of disks. There is one read-write head for each active memory surface on the Winchester drive.

Capacities of low-end, full-height (approximately 3½-inch) mini Winchesters range from 5 Mbytes to nearly 150 Mbytes of formatted data, depending on the number of disks. Most of these machines use stepper motor band positioner technology. Access times are between 65 and 80 milliseconds.

Half-height mini Winchesters have memory storage capacities between 20 and 180 Mbytes. The average access time of popular 5¼-inch half-height drives with formatted capacity of about 26 Mbytes is 80 milliseconds.

Advanced full-height 5¼-inch Winchester disk drives have unformatted capacities of 182 to 442 Mbytes and

(Courtesy Imprimis Technology Inc.)

DISK DRIVE: Fig. 2. A 1.2-gigabyte (unformatted), 5¼-inch Winchester disk drive.

(Courtesy Imprimis Technology Inc.)

DISK DRIVE: Fig. 3. 1.3 gigabyte 8-inch Winchester disk drive for work stations and minicomputers.

have from six disks (11 data surfaces) to eight disks (15 data surfaces) per spindle. These machines use voice-coil head positioners and closed-loop control. Units with formatted capacities of 638 Mbyte are now available, and this number is increasing.

The most advanced commercial full-height 5¼-inch units can store 10,924 bits per inch. Average latency (the time required for the disk to make a single revolution) is 8 milliseconds and average seek or access time is 16 milliseconds. The interface standard is either ESDI or SCSI. Data transfer rates are from 9 to 15 megabits per second (Mbit/s). Advanced Winchester disk drives use thin-film metal alloy media.

3½-Inch Winchester Drives. The capacities of 3½-inch micro Winchesters have now reached 200 Mbytes unformatted or 172 Mbytes formatted with five disks and nine data surfaces. Average latency time is about 8 milliseconds, and average seek time is about 15 milliseconds. Transfer rate is 10 Mbits per second, and standard interfaces may be ESDI or SCSI. Some lower performance drives are mounted on add-in cards for mounting in slots on existing personal computers. Plated media is also employed on these drives.

8- to 14-Inch Winchester Drives. Winchester disk drives in 8- through 14-inch sizes are used individually and in serial arrays in minicomputers and larger computers. Low-end 8-inch Winchester drives now have capacities of 368 Mbytes, and at the high end they have 1.2 gigabytes. High-end units can have nine disks on a spindle but only 15 data surfaces. The spindle speed is typically 3600 revolutions per minute. Plated media is widely used on these drives. Average latency is about 8 milliseconds and average seek is about 16 milliseconds. These units can use SMD, SCSI, or IPI interfaces. Transfer rates may be 3 Mbytes per second. Disk drives in the 9-, 10-, and 14-inch sizes are in production, but the trend is toward the 8-inch drives. See Fig. 3.

Disk-Drive Interfaces. Eight-inch and larger disk drives have the following interfaces:

1. Storage-module drives (SMD) are high-performance disk interfaces developed by CDC in the 1970s. The E-SMD is the 3.0 Mbyte standard extension by CDC, and the H-SMD is a 2.4 Mbyte standard extension by Fujitsu.
2. Intelligent peripheral interface (IPI) is a new standard interface for high-performance peripherals.

Three different interfaces are available for 5¼-inch Winchester disk drives:

1. ST-506/512 was developed by Seagate for its low-performance drives.
2. Enhanced small device interface (ESDI) is a standard disk interface for high performance 5¼-inch drives.

3. Small computer systems interface (SCSI) is a bus-oriented interface that provides increased performance as well as compatibility with other computer peripherals.

The standard interface available for 5¼-inch floppy disk drives is the SA-400, an industry standard popularized by Shugart.

Disk Drive Controller The disk drive controller is a circuit that permits the host computer to read data from the disks or platters of a disk drive or write into them. The host computer gives commands to the controller and the circuit positions the multiple read-write heads over the track whose starting address was given. The controller, capable of reading the addresses of the sectors that pass under the head, seeks the designated address; it then either reads the contents of the sector located there into memory or writes data from memory into the sectors.

The disk-drive controller manages the exchange of data between the central processing unit (CPU) and storage devices at high speed. The common resource linking the high-performance peripheral controller and the CPU is memory. The controller includes a microprocessor to manage data transfers and has cache memory for temporary storage of data as it is moved between the host computer and the mass memory. *See* CACHE MEMORY, COMPUTER.

DISKETTE

See DISK DRIVE.

DISK RECORDING

Disk recording is the process of recording sound onto a phonograph disk. The disk recorder cuts a spiral groove into the disk, starting at the outer edge and moving inward toward the center. The standard speed for disk recording is 33⅓ revolutions per minute. Some records are made at a speed of 45 revolutions per minute. Early records were made at 78 revolutions per minute, but many modern phonographs do not have the ability to play these disks.

The sound vibrations in a disk recorder cause the cutter to vibrate back and forth. Most disks today are recorded in stereo, with two independent sound tracks. The cutter vibrates in two different planes at once. The planes are oriented at 90 degrees so that the vibrations in one can be completely independent of those in the other. The resulting groove contains two sound tracks on perpendicular faces (see drawing).

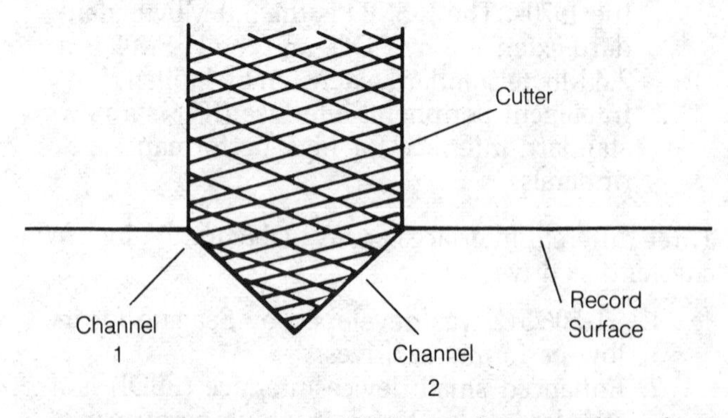

DISK RECORDING: In stereo disk recording, two channels are recorded at right angles to remain independent in a single groove.

Since the angular speed of the disk is constant, the cutter travels over the surface of the disk more rapidly at the outer edge than near the center. Theoretically, then, the high-frequency reproduction is best at the outer periphery, and worst at the center. Most disks are made of a polyester material that provides excellent audio reproduction even near the center of the disk. *See also* HIGH FIDELITY, STEREOPHONICS.

DISPERSION

In some materials, electromagnetic waves are propagated with a velocity that depends on their wavelength. Ordinary glass exhibits this property with visible light. Red light goes through the glass faster than violet light. This effect is known as dispersion.

The most familiar example of dispersion is the rainbow produced when white light is passed through a prism (see illustration). The index of refraction is different for the different light wavelengths. Thus, violet light is bent more than red light (*see* INDEX OF REFRACTION).

When microwave energy strikes an obstruction that is not completely transparent to it, dispersion occurs. Generally, the lower frequencies are passed more rapidly than the higher frequencies. Thus the velocity factor depends on the wavelength (*see* VELOCITY FACTOR). This is not true for all dielectric materials, however.

In acoustics, dispersion refers to the effectiveness with which a speaker propagates the different sound wavelengths widely and uniformly. The better the speaker dispersion, the more realistic the sound reproduction over a wider angle. *See also* ACOUSTICS, SPEAKER.

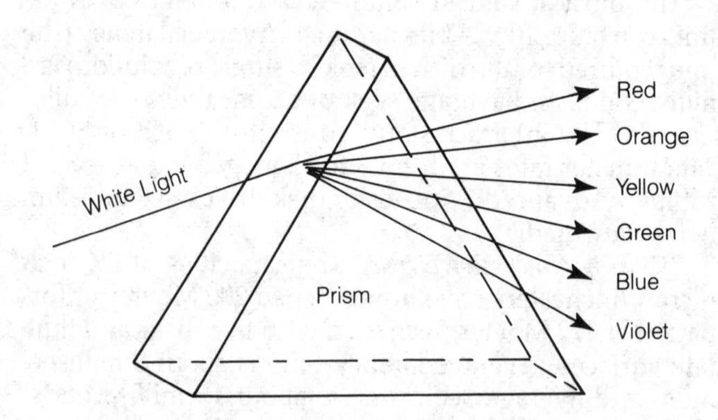

DISPERSION: White light is dispersed into its constituent colors by a glass prism.

DISPLACEMENT

Displacement is the movement of a particle or the change in position of a vector quantity (*see* VECTOR). Displacement is measured in linear units, such as centimeters or inches, or in angular units, such as degrees or radians.

The displacement of a particle or point in a multidimensional system can be defined in terms of the components parallel to each axis. In a polar coordinate system, the displacement may be defined in terms of a radial component and an angular component. Displacement might also be specified in terms of the overall distance the particle moves, and in a certain direction.

On the Cartesian (x,y) plane, a point may move from the origin (0,0) to the point (3,4). The x-axis component of the displacement is 3, and the y-axis component is 4. The overall distance traveled by the point is, in this case, 5 units, in a direction approximately 53 degrees counterclockwise from the x axis. In a polar system, this displacement would be given by $(\Theta,r) = 53$ degrees,5. In three dimensions, displacement becomes more difficult to define; in systems having many dimensions, the representations can get very complicated. *See also* CARTESIAN COORDINATES, POLAR COORDINATES.

DISPLACEMENT CURRENT

When a voltage is applied to a capacitor, the capacitor begins to charge. A current flows into the capacitor as soon as the voltage is applied. At first this displacement current is quite large. But as time passes, with the continued application of the voltage, it grows smaller and smaller, approaching zero in an exponential manner.

The rate of the decline in the displacement current depends on the amount of capacitance in the circuit, and also on the amount of resistance. The larger the product of the capacitance and resistance, the slower the rate of decline of the displacement current, and the longer the time necessary for the capacitor to become fully charged. The magnitude of the displacement current at the initial moment the voltage is applied depends on the quotient of the capacitance and resistance. The larger the capacitance for a given resistance, the greater the initial displacement current. The larger the resistance for a given capacitance, the smaller the initial displacement current. Of course, the larger the charging voltage, the greater the initial displacement current.

In electromagnetic propagation, a change in the electric flux causes an effective flow of current. This current is called displacement current. The more rapid the change in the intensity of the electric field, the greater the value of the displacement current. The displacement current is 90 degrees out of phase with the electric-field cycle. The displacement current is perpendicular to the direction of wave propagation. *See also* ELECTROMAGNETIC FIELD, ELECTROMAGNETIC RADIATION.

DISSECTOR TUBE

A dissector tube, also known as an image dissector, is a form of photomultiplier television camera tube (*see* PHOTOMULTIPLIER TUBE). Light is focused, by means of a lens, onto a translucent surface called a photocathode. This surface emits electrons in proportion to the light intensity. The electrons from the photocathode are directed to a barrier containing a small aperture. The vertical and horizontal deflection plates, supplied with synchronized

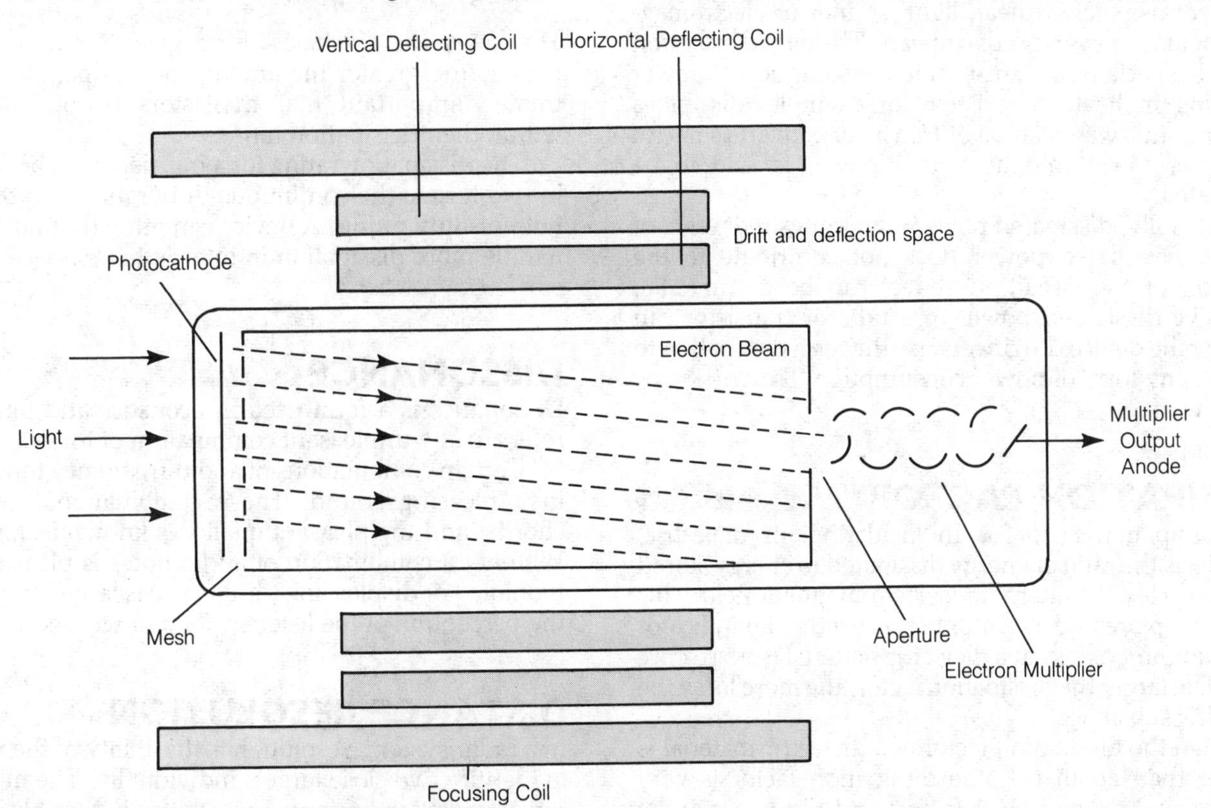

DISSECTOR TUBE: A dissector tube is a nonstorage TV camera tube with a light-sensitive photocathode. Electrons from the illuminated photocathode are focused onto an image plane and swept across an aperture by deflection coils. Those passing through the aperture (dissected) are multiplied to form the video output.

scanning voltages, move the beam from the photocathode across the aperture. Thus the aperture scans the entire image. The electron stream passing through the aperture is thus modulated depending on the light and dark nature of the image. A simplified diagram of a dissector tube is illustrated.

After the electrons have passed through the aperture, they strike a dynode or series of dynodes. Each dynode emits several secondary electrons for each electron that strikes it (*see* DYNODE). In this way, the electron stream is intensified. Several dynodes in cascade can provide an extremely large amount of gain.

The resolving power, or image sharpness, of the dissector tube depends on the size of the aperture. The smaller the aperture, to a point, the sharper the image. However, there is a limit to how small the aperture can be, while still allowing enough electrons to pass, and avoiding diffraction interference patterns.

The image dissector tube, unlike other types of television camera tubes, produces very little dark noise. That is, there is essentially no output when the image is dark. This results in an excellent signal-to-noise ratio. *See also* TELEVISION.

DISSIPATION

Dissipation is an expression of power consumption, and is measured in watts. Power is defined as the rate of expenditure of energy; dissipation always take place in a particular physical location.

When energy is dissipated, it can be converted into other forms, such as heat, light, sound, or electromagnetic fields. It never just disappears. The term dissipation is used especially with respect to consumption of power resulting in heat. A resistor, for example, dissipates power in this way. A tube or transistor converts some of its input power into heat; that power is said to be dissipated.

Generally, dissipated power is an undesired waste of power. Dissipated power does not contribute to the function of the circuit. Its effect can be detrimental; excessive dissipated power in a tube or transistor can destroy the device. Engineers use the term dissipation to refer to any form of power consumption. *See also* ENERGY, POWER, WATT.

DISSIPATION FACTOR

The dissipation factor of an insulating, or dielectric, material is the ratio of energy dissipated to energy stored in each cycle of an alternating electromagnetic field. This quantity, expressed as a number, is used as an indicator of the amount of loss in a dielectric material (*see* DIELECTRIC LOSS). The larger the dissipation factor, the more lossy the dielectric substance.

When the dissipation factor of a dielectric material is smaller than about 0.1, the dissipation factor is very nearly equal to the power factor, and the two may be considered the same for all practical purposes (*see* POWER FACTOR). The power factor is defined as the cosine of the angle by which the current leads the voltage. In general, the dissipation factor D is the tangent of the loss angle θ, which is the complement of the phase angle ϕ. Thus:

$$D = \tan \theta = \tan (90^\circ - \phi)$$

and, when $D < 0.1$:

$$D = \cos \phi$$

In a perfect, or lossless, dielectric material, the loss angle ϕ is zero. This means that the current leads the voltage by 90 degrees, as in a perfect capacitive reactance. In a conductor, the loss angle is 90 degrees. This means that the current and voltage are in phase. *See also* ANGLE OF LAG, ANGLE OF LEAD, LOSS ANGLE, PHASE ANGLE.

DISSIPATION RATING

The dissipation rating of a component is a specification of the amount of power it can safely consume, usually as heat loss. The dissipation rating is given in watts. Resistors are rated in this way; the most common values are 1/8, 1/4, 1/2, and 1 watt, although much larger ratings are available. Zener diodes are also rated in terms of their power-dissipation capacity.

Transistors are rated according to their maximum safe collector or drain dissipation. In an amplifier or oscillator, the dissipated power P_D is the difference between the input power P_I and the output power P_O. That is:

$$P_D = P_I - P_O$$

The lower the efficiency, for a given amount of input power, the greater the amount of dissipation. It is extremely important that transistors be operated well within their dissipation ratings.

The dissipation rating for a transistor can be specified in two forms: the continuous-duty rating and the intermittent-duty rating. A device can often (but not always) handle more dissipation in intermittent service. *See also* DUTY CYCLE, POWER.

DISSONANCE

Dissonance is a term used in acoustics and music that refers to any unpleasant combination of tones.

Certain combinations of audio-frequency tones result in a pleasing sound. These combinations are called chords, and the pleasant quality is known as harmony. Whether a combination of audio notes is pleasing (harmonious) or displeasing (dissonant) is largely a matter of the perception of the listener. *See also* ACOUSTICS.

DISTANCE RESOLUTION

In a radar system, resolution is the ability of the receiver to identify two close targets individually. The minimum radial target separation for which this is possible is called the distance resolution or range resolution.

When two targets are close to each other, they appear on the radar screen as a single echo, as shown at A in the illustration. But as they move farther apart, a radial separation is reached where the targets appear separated on the screen, as shown at B. This minimum distance is the distance resolution.

The distance resolution of a radar system depends on the precision with which the receiver can measure a short time interval. This, in turn, depends on the shape of the radiated pulse, and the frequency of pulse emission. Echoes from a more distant target arrive later than the echoes from a nearby target. *See also* AZIMUTH RESOLUTION, RADAR.

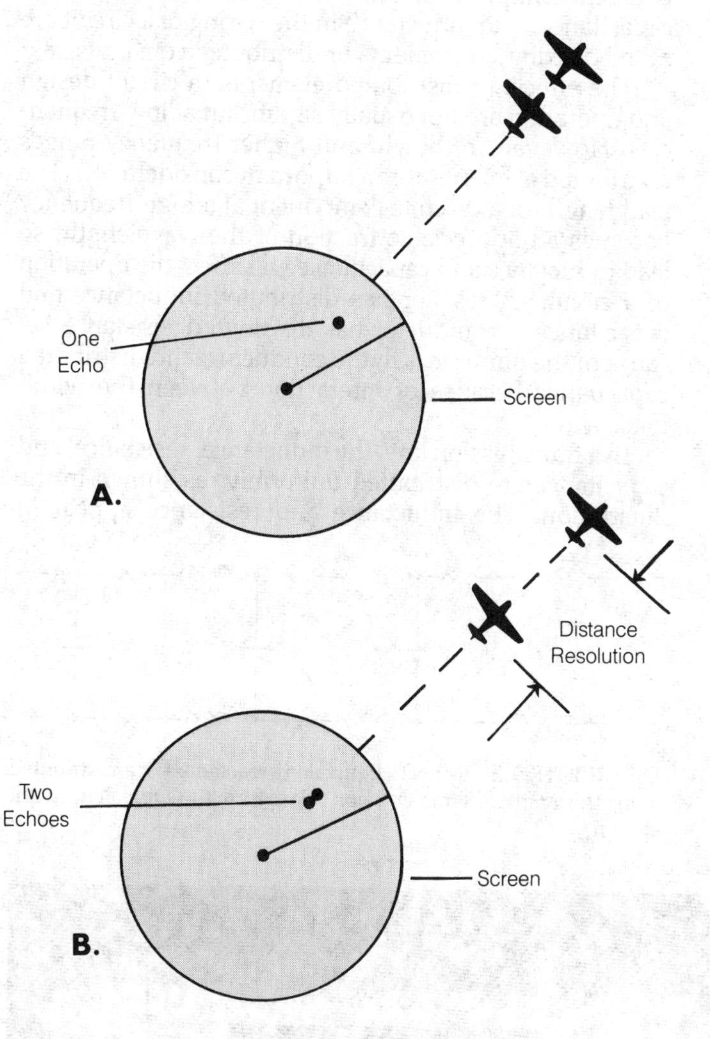

DISTANCE RESOLUTION: A radar display of two targets too close for distance resolution appears as a single dot (A), but with adequate radial resolution, they appear as two targets (B).

DISTORTION

Distortion is a change in the shape of a waveform. Generally, the term *distortion* is used to refer to an undesired change in a wave, but distortion is sometimes introduced into a circuit deliberately.

The figure illustrates a nearly perfect sine waveform, A, typical of radio-frequency transmitters. All of the energy is concentrated at a single wavelength. There is no harmonic energy in a perfect wave, but in practice there is always some distortion. Distortion results in the presence of the harmonics in various proportions (*see* HARMONIC), in addition to the fundamental frequency. Harmonics can be suppressed more than 160 decibels in radio-frequency practice. This means that a harmonic is just one part in 100 million in amplitude, compared with the fundamental frequency. The sine wave in this case is almost completely distortion-free.

A severely distorted waveform is shown at B. On a spectrum analyzer, distortion is visible at harmonic frequencies in addition to the fundamental frequency (*see* SPECTRUM ANALZYER). The harmonics and their amplitudes depend on the type and extent of the distortion.

In audio applications, distortion can occur with complex as well as simple waveforms. Sometimes the distortion can be so severe that a voice is unrecognizable. Sound quality is measured in *fidelity*, or accuracy of reproduction.

Distortion such as that illustrated is sometimes introduced into a circuit on purpose. Mixers and frequency multipliers require distortion to work properly. *See also* DISTORTION ANALYZER, FREQUENCY MULTIPLIER, MIXER.

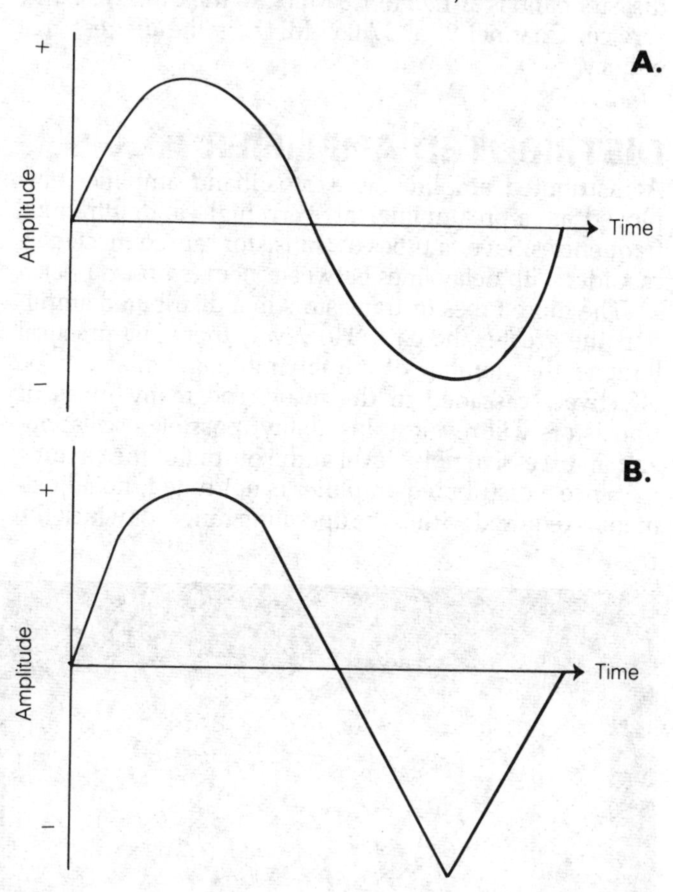

DISTORTION: An undistorted sine wave (A) and a distorted sine wave (B).

DISTORTION ANALYZER

A distortion analyzer is a meter for measuring total harmonic distortion in an audio circuit. This meter has a special notch filter set for 1000 Hz. The filter can remove the fundamental frequency of a standard audio test tone at the output of the amplifier. A voltmeter measures the audio signal remaining after notching out the fundamental frequency; this remaining energy consists entirely of harmonics, or distortion products. Total harmonic distortion is expressed as a percentage, measured with respect to the level of the output without the filter.

The figure shows a typical distortion analyzer. This particular instrument can also measure the SINAD, or signal-to-noise-and-distortion, sensitivity of a communications receiver. *See also* TOTAL HARMONIC DISTORTION.

DISTRESS FREQUENCY

In radio communication, a distress frequency is a frequency or channel specified for use in distress calling.

In the aeronautical mobile service, radiotelegraphy distress calls are made at 500 kHz; radiotelephone distress frequencies are 2.182 MHz and 156.8 MHz. Survival craft use 243 MHz. The distress frequencies are the same in the maritime mobile service.

In the fixed and land mobile radio services, the disaster band is at 1.75 to 1.8 MHz. In the Citizen's Band service, Channel 9, at 27.065 MHz, is the distress frequency.

DISTRIBUTED AMPLIFIER

A distributed amplifier is a broadband amplifier employed as a preamplifier at very high and ultrahigh frequencies. Several tubes or transistors are connected in cascade, with delay lines between them (*see* DELAY LINE).

The more tubes or transistors in a distributed amplifier, the greater the gain. However, there is a practical limit to the number of amplifying stages that can be effectively cascaded in this way. Too many tubes or transistors will result in instability, possible oscillation, and an excessive noise level at the output of the circuit.

Since a distributed amplifier is not tuned, no adjustment is required within the operating range for which the circuit is designed. Distributed amplifiers are useful as preamplifiers for television receivers and oscilloscopes.

DISTRIBUTED ELEMENT

A constant variable such as resistance, capacitance, or inductance is usually thought to exist in a discrete form in a component especially designed to have certain electrical properties (*see* DISCRETE COMPONENT). However, there is always some resistance, capacitance, and inductance in even the simplest circuits. The equivalent resistance, capacitance, and inductance in the wiring of a circuit are called distributed elements or distributed constants.

The effects of distributed elements in circuit design and operation are not usually significant at low frequencies. However, in the vhf and higher frequency ranges distributed elements are an important consideration. The lead length of a discrete component at a high frequency becomes an appreciable fraction of the wavelength, so lead inductance and capacitance will affect the operation of a circuit. A resistor has distributed inductance and capacitance. An inductor has distributed resistance because of the ohmic loss in the conductor, and distributed capacitance because of interaction between the windings.

In a transmission line, the inductance, resistance, and capacitance are distributed uniformly, as shown in the illustration. The inductance and resistance appear in

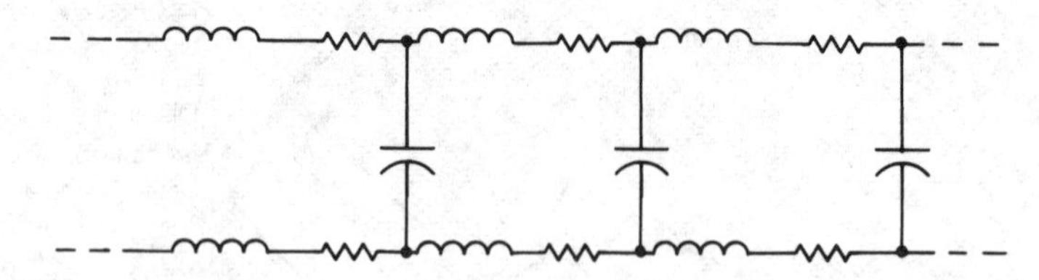

DISTRIBUTED ELEMENT: Equivalent reactances are distributed along the length of a transmission line to form the equivalent circuit shown.

DISTORTION ANALYZER: A distortion analyzer contains a meter, audio filter, and sensitivity controls.

series with the conductors. The capacitance appears across the conductors.

When discrete components are used in a circuit, or when the inductances, capacitances, or resistances are in defined locations, the elements are said to be lumped. *See also* LUMPED ELEMENT.

DISTRIBUTION

Distribution is a statistical property of a variable and is important in probability analysis. The most familiar distribution is the standard or normal distribution, represented by a bell-shaped curve (*see* NORMAL DISTRIBUTION). However, there are many different kinds of distribution functions.

An example of distribution is seen in the results of an examination. Most of the scores in any test tend to be lumped within a certain range, as shown in the graph. The average score is called the mean; half of the area under the curve is to the left of the mean, and half is to the right. The score at which half of the examinations are lower and half are higher is called the median. The degree to which the scores are spread out is expressed by the standard deviation. All distribution functions have a mean, a median, and a standard deviation. *See also* EXPONENTIAL DISTRIBUTION, MEAN, MEDIAN, STANDARD DEVIATION.

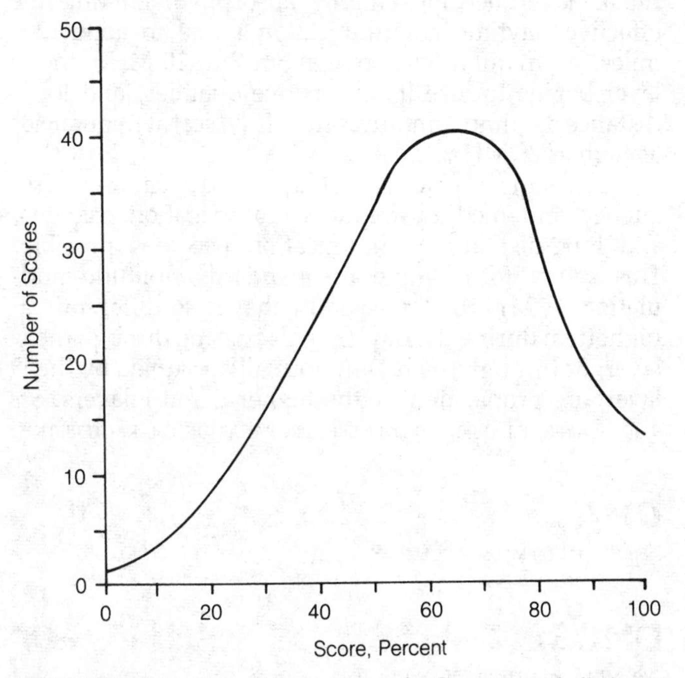

DISTRIBUTION: The approximation of a Gaussian or bell-shaped curve indicates a normal statistical distribution.

DISTRIBUTION FRAME

In a telephone switching system, a distribution frame is the point at which all of the lines come together for interconnection. The distribution frame is part of the central office (*see* CENTRAL-OFFICE SWITCHING SYSTEM).

At the distribution frame, any line may be connected to any other line. In modern telephone circuits, this is done by a computer, so no human operator is needed. Many cross-connections can be made at once, and the combination can be changed at any time. The distribution frame should be large enough to handle calls at peak periods. *See also* TELEPHONE SYSTEM.

DISTURBANCE

In electronics, the term *disturbance* generally refers to an unexpected or unwanted change in the propagation of radio signals. A disturbance in the geomagnetic field of the earth often causes a deterioration of ionospheric propagation at medium and high frequencies. This disturbance is caused by a solar flare and is accompanied by the aurora, or northern and southern lights. A severe geomagnetic disturbance may even affect telephone circuits.

An electrical storm is a form of disturbance, and the fields generated by these storms can cause severe local interference to radio circuits. Weather disturbances can also cause changes in the propagation at very high and ultrahigh frequencies. *See also* GEOMAGNETIC STORM, SOLAR FLARE.

DIVERGENCE LOSS

A radiated energy beam generally diverges from the point of transmission. Sound, visible light, and all electromagnetic fields diverge in this manner. The exception is the laser beam (*see* LASER), which forms a parallel beam of light.

Divergence results in a decrease in the intensity of the energy reaching a given amount of surface area, as a receiver is moved farther and farther from a transmitter. With most energy effects, the decrease follows the inverse-square law (*see* INVERSE-SQUARE LAW).

Given an energy source with a constant output, such as a speaker (see illustration), the amount of energy striking an area is inversely proportional to the square of the distance. If the distance doubles, the sound intensity is cut to one-fourth its previous level. If the distance becomes 10 times as great, the sound intensity drops to 1/100, or 0.01, of its previous value. Electromagnetic field strength, measured in microvolts per meter, decreases in proportion to the distance from the antenna. *See also* FIELD STRENGTH.

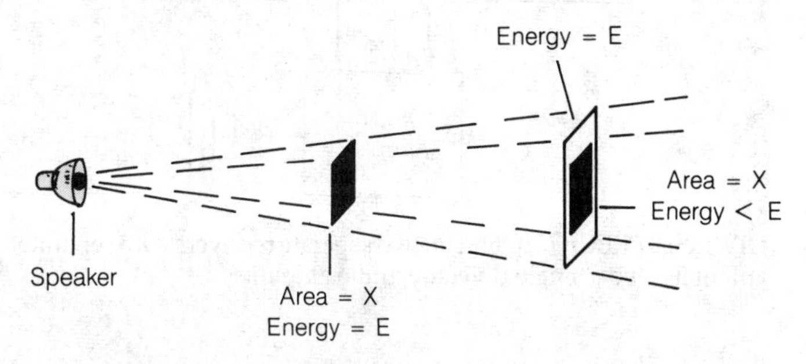

DIVERGENCE LOSS: Energy spreads out or diverges as a function of distance from the source, resulting in progressive loss.

DIVERSITY RECEPTION

Diversity reception is a technique of reception that reduces the effects of fading in ionospheric communication. Two receivers are used, and they are tuned to the same signal. Their antennas are spaced several wavelengths apart. The outputs of the receivers are fed into a common audio amplifier, as shown in the illustration.

Signal fading usually occurs over very small areas, and at different rates. Even at a separation of a few wavelengths in the receiving location, a signal may be very weak in one place and very strong in the other. Thus, when two antennas are used for receiving, and they are positioned sufficiently far apart, and they are connected to independent receivers, the chances are small that a fade will occur at both antenna locations simultaneously. At least one antenna almost always gets a good signal. The result of this, when the receiver outputs are combined, is the best of both situations. The fading is much less pronounced.

While diversity reception provides some measure of immunity to fading, the tuning procedure is critical and the equipment is expensive. More than two antennas and receivers can be used. To tune the receivers to exactly the same frequency, a common local oscillator should be used. Otherwise the audio outputs of the different receivers will not be in the proper phase, and the result may be a garbled signal, totally unreadable. *See also* FADING.

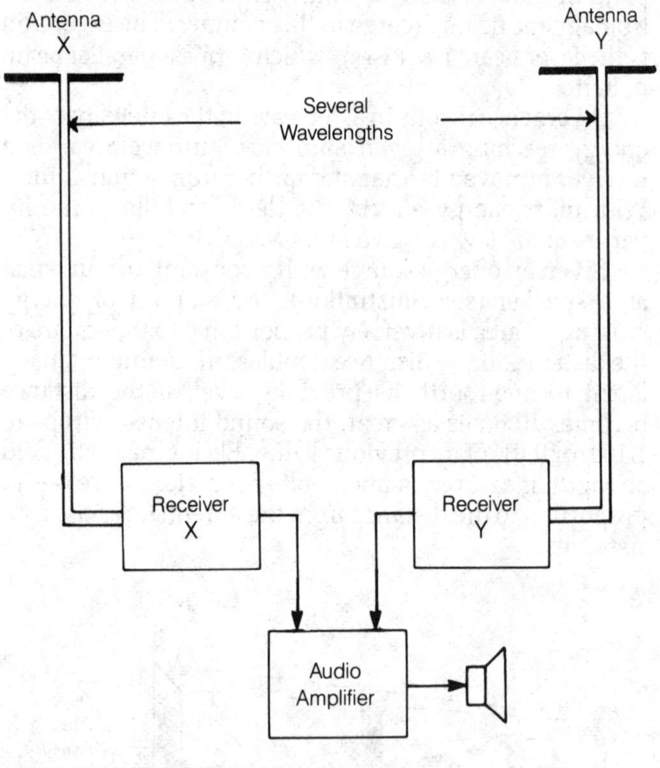

DIVERSITY RECEPTION: The two separate receivers with separate antennas feed a single diversity audio amplifier.

DIVIDER

A divider is a circuit used for splitting voltages or currents. A common example of a divider is the resistive network found in the biasing circuits of transistor oscillators and amplifiers.

In a calculator or computer, a divider is a circuit that performs mathematical division. A divider circuit is often used in a frequency counter to regulate the gate time. Divider circuits are used in digital systems to produce various pulse rates for frequency-control and measurement purposes. For example, a crystal calibrator having an oscillator frequency of 1 MHz may be used in conjunction with a divide-by-10 circuit, so that 100 kHz markers can be obtained. The divider can be switched in and out of the circuit so the operator can select the markers he wants. A divide-by-100 circuit will allow the generation of 10 kHz markers from a 1 MHz signal. *See also* FREQUENCY DIVIDER, VOLTAGE DIVIDER.

D LAYER

The D layer is the lowest region of ionization in the upper atmosphere of the earth. The D layer generally exists only while the sun is above the visual horizon. During unusual solar activity, such as a solar flare, the D layer may become ionized during the hours of darkness. The D layer is about 35 to 60 miles in altitude, or 55 to 95 kilometers.

At very low frequencies, the D layer and the ground combine to act as a huge waveguide, making worldwide communication possible with large antennas and high-power transmitters. At the low and medium frequencies, the D layer becomes highly absorptive, limiting the effective daytime communication range to about 200 miles. At frequencies above about 7 to 10 MHz, the D layer begins to lose its absorptive qualities, and long-distance daytime communication is typical at frequencies as high as 30 MHz.

After sunset, since the D layer disappears, low-frequency and medium-frequency propagation changes, and long-distance communication becomes possible. This is why, for example, the standard amplitude-modulation (A-M) broadcast band behaves so differently at night than during the day. Signals passing through the D layer, or through the region normally occupied by the D layer, are propagated by the higher E and F layers. *See also* E LAYER, F LAYER, IONOSPHERE, PROPAGATION CHARACTERISTICS.

DMA

See DIRECT MEMORY ACCESS.

DMOS

See METAL-OXIDE SEMICONDUCTOR.

DOHERTY AMPLIFIER

A Doherty amplifier is an amplitude-modulation amplifier consisting of two tubes or transistors used for different functions. One of the devices is biased to cutoff during unmodulated-signal periods, while the other acts as an ordinary amplifier. Enhancement of the modulation peaks is achieved by an impedance inverting line between the plate circuits of the two tubes, or between the collector circuits of the two transistors. The illustration shows a simplified schematic diagram of an FET (field-effect transistor) type Doherty amplifier. This circuit is also called a Terman-Woodyard modulated amplifier.

Under conditions of zero modulation, Q1, the carrier FET, acts as an ordinary radio-frequency amplifier. But Q2, the peak FET, contributes no power during unmodulated periods. During negative modulation, when the instantaneous input power is less than the carrier power under conditions of zero modulation, the output of the carrier FET Q1 drops in proportion to the change in the input amplitude. At the negative peak, where the instantaneous amplitude can be as small as zero, the output of the carrier tube is minimum.

When a positive modulation peak occurs, the peak FET contributes some of the power to the output of the amplifier. The carrier-FET output power doubles at a 100-percent peak, compared to the output under conditions of no modulation. The peak FET also provides twice the output power of the carrier FET with zero modulation. Consequently, the output power of the amplifier is quadrupled over the unmodulated-carrier level, when a 100-percent peak comes along. This is normal for amplitude modulation at 100 percent. *See also* AMPLITUDE MODULATION.

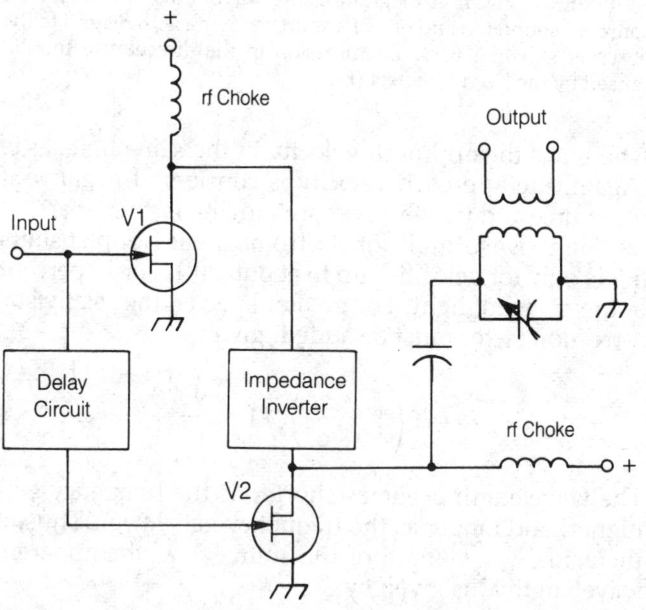

DOHERTY AMPLIFIER: One field-effect transistor amplifies the carrier signal, and the other amplifies the modulation peaks in this amplifier.

DOLBY

Dolby is the name for a method of enhancing the signal-to-noise ratio in a magnetic recording system. The technique is similar to compression in radio-frequency communications systems (*see* COMPRESSION). The low-amplitude sound components, most likely to be masked by background hiss in the recording process, are boosted in volume in the Dolby method of recording. During playback, the low-level components are attenuated, thus obtaining the original sound.

Dolby techniques are categorized as Dolby A, the original Dolby recording method, and Dolby B, a simpler system for use by nonprofessionals. The more sophisticated Dolby A system has four independent noise-reduction circuits, which operate at the bass, midrange, treble, and high-frequency parts of the audible sound spectrum. The Dolby B system has only one noise-reduction frequency band, effective primarily on those audio frequencies at which the most tape hiss occurs.

For best results, a Dolby tape should be played on a Dolby machine to prevent reduction of the dynamic range. The Dolby system is standardized to ensure that the fidelity is optimum. *See also* MAGNETIC RECORDING, MAGNETIC TAPE, TAPE RECORDER.

DOMAIN

The domain of a mathematical function is the set of values on which the function operates. In Cartesian coordinate graphs, the domain is a subset of the values on the horizontal axis (*see* CARTESIAN COORDINATES).

A function may not be defined for all possible values of the independent variable. The function $f(x) = x^2$ is defined for all real numbers x, and thus its domain is the entire set of real numbers. But the function $g(x) = \log_{10}(x)$ is defined only for the positive real numbers. If x is zero or negative, then $\log_{10}(x)$ is not defined.

In some cases, the domain of the function depends on the extent of the number system specified. The domain of the function $h(x) = \sqrt{x}$, for example, is restricted to the non-negative real numbers if h is allowed to vary only over the set of reals. But if the value of h is any complex number (*see* COMPLEX NUMBER), the domain of h becomes the entire set of real numbers. *See also* DEPENDENT VARIABLE, FUNCTION, INDEPENDENT VARIABLE, RANGE.

DOMAIN OF MAGNETISM

In any magnetic material, the domain is the region in which all of the magnetic dipoles are oriented in the same direction (*see* DIPOLE). When this occurs, the magnetic fields of the dipoles act in unison, and a large magnetic field is formed around the object, or in the vicinity of the domain.

An iron nail is normally not magnetized; its dipoles are oriented at random, and not all in the same direction, so the magnetic fields do not add together. However, a part of the nail may become magnetized under the influence of an electric current or external magnetic field. The part of the nail that becomes magnetized is called the magnetic domain. The domain may extend the whole length of the nail, or only through part of it. *See also* MAGNETIZATION.

DON'T-CARE STATE

In a binary logic function or operation, some states do not matter; they are unimportant in the outcome. This occurs when the function is defined for some, but not all, of the logical states. In this case, those states that do not affect conditions are called don't-care states. *See also* BINARY-CODED NUMBER.

DOPING

Doping is the addition of impurity materials to semiconductor substances. Doping alters the manner in which these substances conduct currents. This turns the semiconductor, such as silicon, into an N-type or P-type material.

When a doping impurity containing an excess of electrons is added to a semiconductor material, the impurity is called a donor impurity. This results in an N-type semiconductor (*see* N-TYPE SEMICONDUCTOR). The conducting particles in such a substance are the excess electrons passed along from atom to atom.

When an impurity containing a shortage of electrons is added to a semiconductor material, it is called an acceptor impurity. The addition of acceptor impurities results in a P-type semiconductor material (*see* P-TYPE SEMICONDUCTOR). The charge carriers in this kind of material are called holes, which are atoms lacking electrons. The flow of holes is from positive to negative in a P-type material (*see* HOLE).

Typical donor elements used in the manufacture of N-type semiconductors include antimony, arsenic, phosphorus, and bismuth. Acceptor elements, for P-type materials, include boron, aluminum, gallium, and indium. The N-type and P-type materials resulting from the addition of these elements to silicon and other semiconductor materials are the basis for all of solid-state electronics. *See also* SEMICONDUCTOR, SILICON.

DOPPLER EFFECT

An emitter of wave energy in the form of sound or electromagnetic fields shows a frequency that depends on the radial speed of the source with respect to an observer. If the source is moving closer to the observer, the apparent frequency increases. If the source moves farther from the observer with time, the apparent frequency decreases. For a given source of wave energy, different observers may measure different emission frequencies, depending on the motion of each observer with respect to the source. The illustration shows the Doppler effect for a moving wave source. Only the radial component of the motion with respect to the observer results in a Doppler shift.

For sound, where the speed of propagation in air at sea level is about 1100 feet per second, the apparent frequency f^* of a source having an actual frequency f is:

$$f^* = f\left(1 + \frac{v}{1100}\right)$$

where v is the approach velocity in feet per second. (The approach velocity is considered negative if the observer is moving away from the source.)

For electromagnetic radiation, where the speed of propagation is c, the apparent frequency f^* of the emission from a source having an actual frequency f is:

$$f^* = f\left(1 + \frac{v}{c}\right)$$

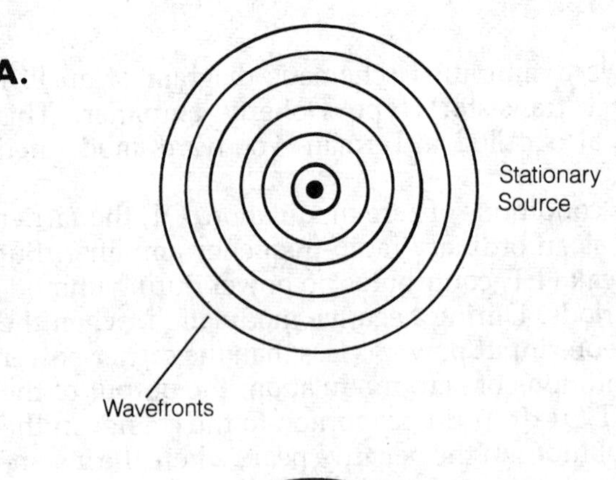

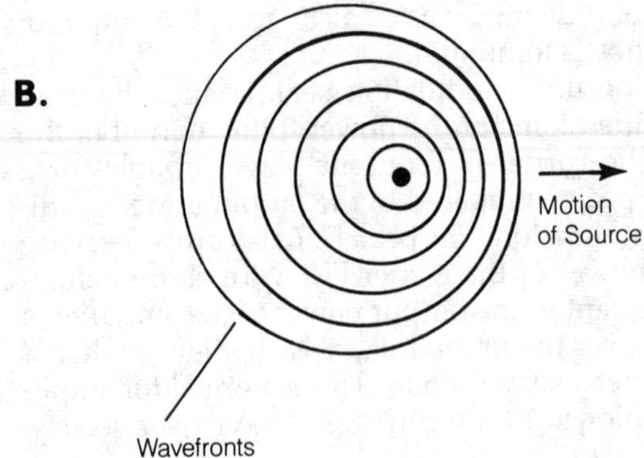

DOPPLER EFFECT: A diagram of the wavefronts of a stationary source of sound (A) and one of a source moving with respect to the observer showing wave compression in the direction of motion caused by the Doppler effect (B).

where v is the approach velocity in the same units as c. (Again, the approach velocity is considered negative if the source and the observer are moving farther apart.)

The above formula for electromagnetic Doppler shifts holds only for velocities up to about $\pm 0.1c$, or 10 percent of the speed of light. For greater speeds, the relativistic correction factor must be added, giving:

$$f^* = f\left(1 + \frac{v}{c}\right)\sqrt{1 - \frac{v^2}{c^2}}$$

The wavelength becomes shorter as the frequency gets higher, and longer as the frequency gets lower. Thus, if the actual wavelength of the source is λ, the apparent wavelength λ^* is given by:

$$\lambda^* = \frac{\lambda}{(1 + v/c)\sqrt{1 - v^2/c^2}}$$

If there is no radial change in the separation of two objects, there will be no Doppler effect observed between them, although, if the tangential velocity is great enough, there may be a lowering of the apparent frequency because of relativistic effects.

DOPPLER RADAR

A form of radar that allows measurement of the radial velocity of an object, by means of the Doppler effect, is called a Doppler radar. Electromagnetic bursts are transmitted in the direction of the moving object, and are received after they have been reflected from the object. An approaching object will cause an increase in the frequency of the signal, and the greater the speed, the greater the increase in the frequency. A retreating object will cause a decrease in the frequency of the signal, in direct proportion to the radial speed. *See also* DOPPLER EFFECT, RADAR.

DOSIMETRY

A dosimeter is an instrument for measuring the amount of atomic radiation to which a person or object has been exposed over a period of time. The standard unit of atomic radioactivity is the roentgen (*see* ROENTGEN). Most people get from 10 to 20 roentgens of atomic radiation within their lifetimes. Some of this comes from medical X rays; some comes from the environment in the form of natural and manmade background radiation.

A dosimeter consists of a Geiger counter or similar rate counter, in conjunction with an integrator circuit that adds up the number of counts in a certain period of time. Some dosimeters measure primarily alpha rays; some measure primarily the beta and gamma forms of radiation. It is also possible to measure the radiation resulting from high-speed bombardment by neutrons.

Exposure to about 100 roentgens of radiation, over a period of a few hours or days, will cause radiation sickness. Symptoms include fever, loss of appetite, possible loss of hair, and a sunburnlike redness of the skin. A dose of more than about 300 roentgens in a short time will usually cause death. *See also* ALPHA PARTICLE, BETA PARTICLE, COUNTER, GAMMA RAY, GEIGER COUNTER, INTEGRATION, X RAY.

DOT GENERATOR

In a color television receiver, it is important that the three color electron beams converge at all points on the phosphor screen. Red, green, and blue beams meet to form white. If the alignment is not correct, the picture will appear distorted in color. A dot generator is an instrument used by television technicians to aid in the adjustment of the beam convergence. The pattern produced by the dot generator contains a number of white circular regions distributed over the picture frame.

If the convergence alignment of a color television receiver is perfect, all of the dots will appear white and in good focus. But if the convergence alignment is faulty, the red, green, and blue electron beams will not merge at all points of the screen. Some of the white dots will then have colored borders. If the convergence alignment is very bad, the picture is hardly recognizable in color. *See also* CONVERGENCE, TELEVISION.

DOT-MATRIX PRINTER

See PRINTER.

DOUBLE BALANCED MIXER

A double balanced mixer is a circuit that operates in a manner similar to a balanced modulator (*see* BALANCED MODULATOR, MIXER, MODULATOR). The input energy in a double balanced mixer is suppressed from 40 to 60 dB, effectively isolating the input ports from the output ports, so the coupling between the input oscillators is negligible and the coupling between the input and output is also negligible, as in the figure. The double balanced mixer differs from the single balanced mixer; in the latter circuit, there is some degree of interaction between the input signal ports (*see* SINGLE BALANCED MIXER).

Hot-carrier diodes are generally used in modern balanced-mixer circuits. They can handle large signal amplitudes without distortion, and they generate very little noise. They are also effective at frequencies up to several gigahertz. Because these circuits, such as the one shown in the figure, are passive rather than active, they show some conversion loss. This loss is typically about 6 decibels. However, the loss is easily overcome by the use of amplifiers following the mixer.

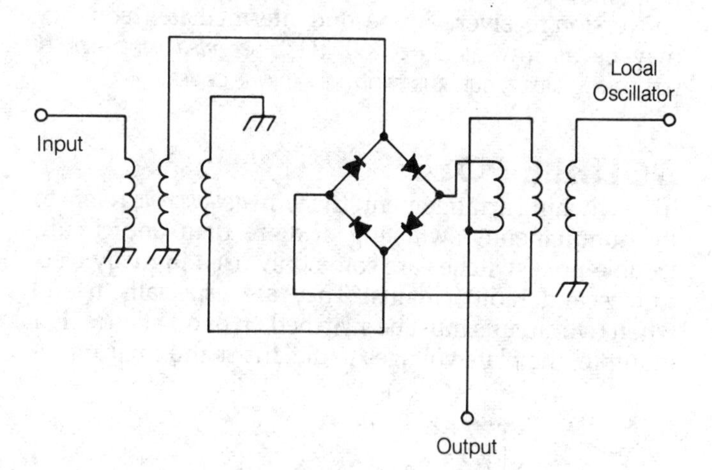

DOUBLE BALANCED MIXER: The ports of a double balanced mixer are electrically isolated from each other.

DOUBLE-BASE JUNCTION TRANSISTOR

See TRANSISTOR.

DOUBLE CIRCUIT TUNING

When the input circuit and the output circuit of an amplifier are independently tunable, the system is said to have double circuit tuning. In an oscillator, double tuning involves separate tank circuits in the base and collector portions for a bipolar transistor, the gate and plate portions for a vacuum tube. In a coupling transformer, double circuit tuning refers to the presence of a tuning capacitor across the secondary winding as well as the primary winding.

Double circuit tuning provides a greater degree of selectivity and harmonic suppression than the use of only one tuned circuit per stage. In a multistage amplifier, however, double circuit tuning increases the possibility of oscillation at or near the operating frequency. This tendency can be reduced by tuning each circuit to a slightly different frequency, thereby reducing the Q factor of the entire amplifier chain. In such a situation, the circuit is said to be stagger-tuned. *See also* STAGGER TUNING.

DOUBLE-CONVERSION RECEIVER

A double-conversion receiver, also known as a dual-conversion receiver, is a superheterodyne receiver that has two different intermediate frequencies. The incoming signal is first heterodyned to a fixed frequency, called the first intermediate frequency. Selective circuits are used at this point. In a single-conversion receiver (*see* SINGLE-CONVERSION RECEIVER) the first intermediate frequency signal is fed to the detector and subsequent audio amplifiers. However, in a double-conversion receiver, the first intermediate frequency is heterodyned to a second, much lower frequency, called the second intermediate frequency. The illustration is a block diagram of a double-conversion receiver. The second intermediate frequency may be as low as 50 to 60 kHz. *See also* INTERMEDIATE FREQUENCY, MIXER, SUPERHETERODYNE RADIO RECEIVER.

DOUBLE POLE

In a switching arrangement, the term *double pole* refers to the simultaneous switching of more than one circuit. Double-pole switches are commonly used in many electronics and radio circuits. They are especially useful when two circuits must be switched on or off at once. For example, the plate voltages to the driver and final amplifier stages of a vacuum-tube transmitter are generally of different values, and must be switched separately. A single switch having two poles is more practical than two single-pole switches. *See also* MULTIPLE POLE, SWITCH.

DOUBLER

A doubler is an amplifier circuit designed to produce an output at the second harmonic of the input frequency. This circuit is therefore a frequency multiplier (*see* FREQUENCY MULTIPLIER). The input circuit may be untuned, or it may be tuned to the fundamental frequency. The output circuit is tuned to the second-harmonic frequency. The active element in the circuit—that is, the transistor or tube—is deliberately biased to result in nonlinear operation. This means it must be biased for Class AB, B, or C. Alternatively, a nonlinear passive element, such as a diode, may be used. The nonlinearity of the device produces a signal rich in harmonic energy.

A circuit suitable for application as a doubler is the push-push circuit. In this arrangement, which resembles an active full-wave rectifier, two transistors or tubes are connected with their inputs in phase opposition, and their outputs in parallel. This tends to cancel the fundamental frequency and all odd harmonics in the output circuit. But the even harmonics are reinforced. *See also* PUSH-PUSH CONFIGURATION.

DOUBLE SIDEBAND

Whenever a carrier is amplitude-modulated, sidebands are produced above and below the carrier frequency.

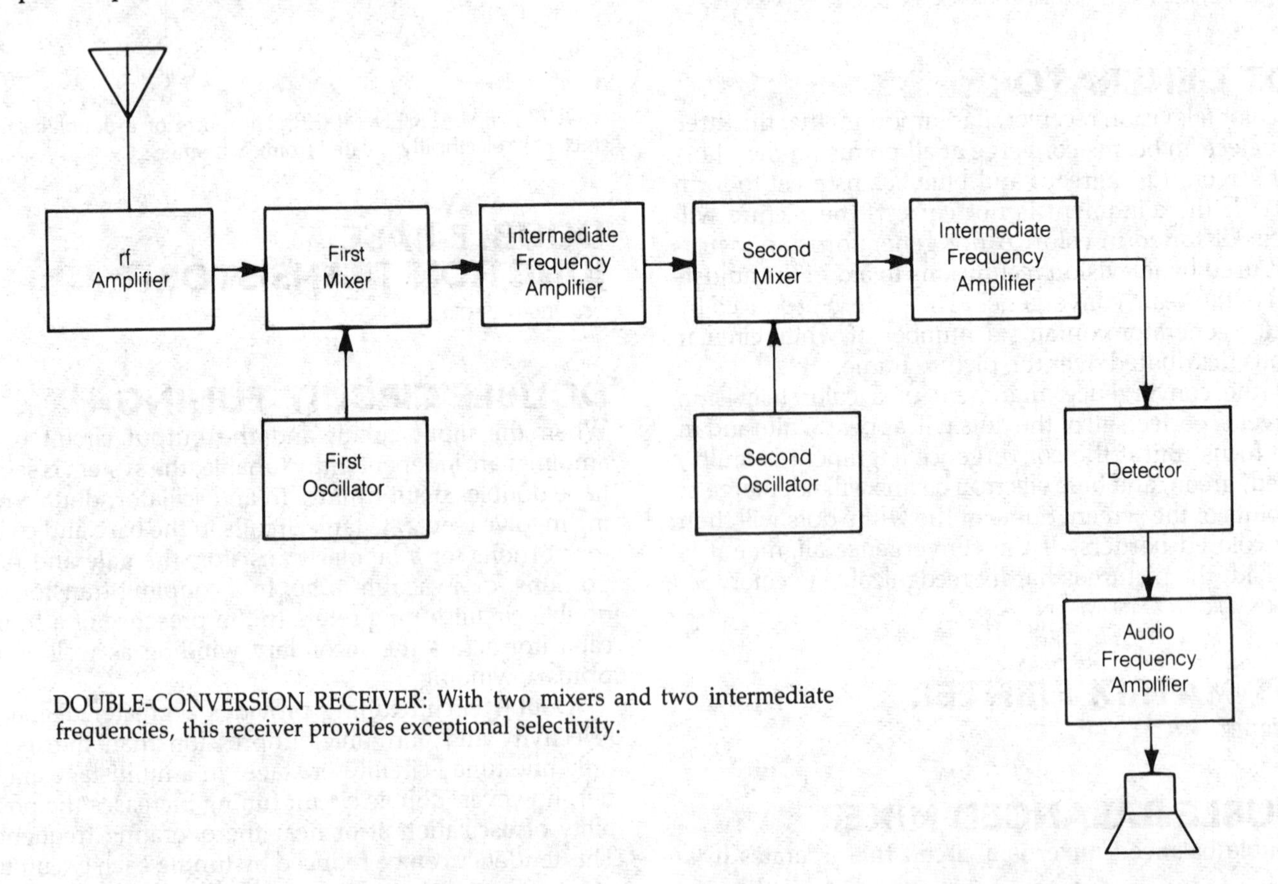

DOUBLE-CONVERSION RECEIVER: With two mixers and two intermediate frequencies, this receiver provides exceptional selectivity.

These sidebands represent the sum and difference frequencies of the carrier and the modulating signal (*see* SIDEBAND). A typical amplitude-modulated signal, as it might appear on a spectrum analyzer, is shown at A in the illustration. This is called a double-sideband, full-carrier signal; both sidebands and the carrier are plainly visible on the display.

If a balanced modulator (*see* BALANCED MODULATOR) is used to obtain the amplitude-modulated signal, the output lacks a carrier, as shown at B. The sidebands are left intact. Actually the carrier is not completely eliminated, but it is greatly attenuated, typically by 40 to 60 decibels or more. This signal is called a double-sideband, suppressed-carrier signal. To obtain a normal amplitude-modulated signal at the receiver, a local oscillator is used to re-insert the missing carrier. The double-sideband, suppressed-carrier signal makes more efficient use of the

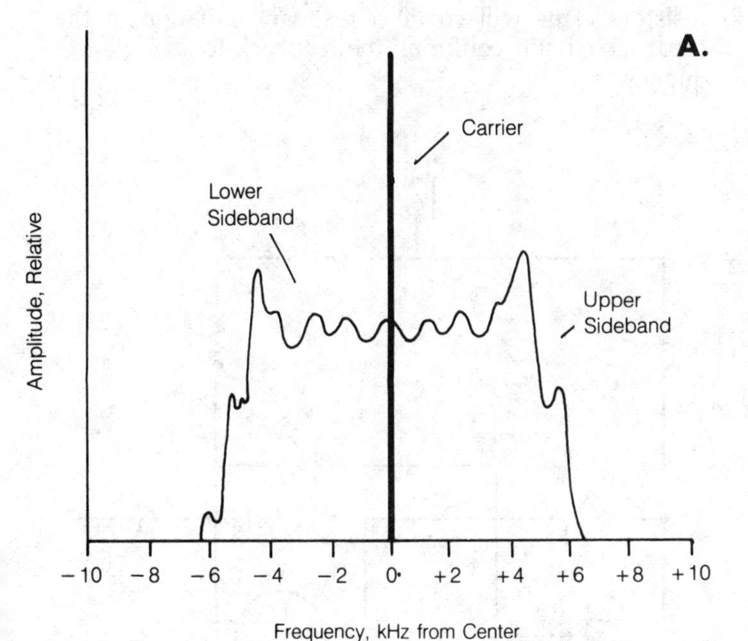

DOUBLE SIDEBAND: The waveform of an amplitude-modulated signal as viewed on a spectrum analyzer (A) compared with the waveform of a double-sideband, suppressed carrier signal (B).

audio energy than the full-carrier signal. In the suppressed-carrier signal, all of the transmitter output power goes into the sidebands, which carry the intelligence. In the full-carrier signal, no less than two-thirds of the transmitter power is used by the carrier, which alone carries no information at all.

A double-sideband, suppressed-carrier signal is difficult to receive, because the local oscillator must be on exactly the same frequency as the suppressed carrier. The slightest error will cause beating between the two sidebands, and the result will be a totally unreadable signal. Therefore, this kind of modulation is not often found in radiocommunication; single-side-band emission is much more common. *See also* AMPLITUDE MODULATION, SINGLE SIDEBAND.

DOUBLE-THROW SWITCH

A double-throw switch is used to connect one line to either of two other lines. For example, it might be desired to provide two different voltages to an amplifier for operation at two different power-input levels. In this case, the common point of a double-throw switch can be connected to the collector circuit, and the two terminals to the different power supplies. *See also* SWITCH.

DOWN CONVERSION

Down conversion refers to the heterodyning of an input signal with the output of a local oscillator, resulting in an intermediate frequency that is lower than the incoming signal frequency. Technically, any situation in which this occurs is down conversion. But usually, the term is used with reference to converters designed especially for reception of very high, ultrahigh, or microwave signals with communications equipment designed for much lower frequencies than that contemplated.

The available spectrum at ultrahigh and microwave frequencies is great, covering many hundreds of megahertz, and the operator of a receiver using a down converter may find the selectivity too high. For example, with a communications receiver designed to tune the high-frequency band in 1 MHz ranges, the 10,000 to 10,500 MHz band would appear, following down-conversion, to occupy 500 such ranges. Each of these 500 ranges would require the receiver to be tuned across its entire bandspread. *See also* MIXER.

DOWNLINK

The term *downlink* refers to the band on which an active communications satellite transmits its signal back to the earth. The downlink frequency differs from the uplink frequency on which the signal is sent up from the earth to the satellite. The different uplink and downlink frequencies allow the satellite to act as a repeater, retransmitting the signal immediately.

Satellite uplink and downlink frequencies are usually in the very high or ultrahigh frequency range. They may even be in the microwave range. The downlink transmitting antenna on a satellite must produce a fairly wide-angle beam so all of the desired receiving stations obtain some of the signal. *See also* SATELLITE COMMUNICATIONS, REPEATER, UPLINK.

DOWNTIME

Downtime is the time during which an electronic circuit or system is not operational. Downtime can be caused by routine servicing, power failure, or component failure.

DOWNWARD MODULATION

If the average power of an amplitude-modulated transmitter decreases when the operator speaks into the microphone, the modulation is said to be downward. Generally, in downward modulation, the instantaneous carrier amplitude is never greater than the amplitude under zero-modulation conditions. That is, the positive modulation peaks never exceed the level of the unmodulated carrier. Negative modulation peaks may go as low as the zero amplitude level, just as is the case with ordinary amplitude modulation, but clipping should not take place. The figure illustrates a typical amplitude-versus-time function for a downward-modulated signal.

In theory, downward modulation is less efficient than ordinary amplitude modulation because more transmitter power is required in proportion to the intelligence transmitted. That is, less effective use is made of the energy in the transmitter. Downward modulation is not often used in communications systems because the large amplitude of the unmodulated carrier is a waste of power. *See also* AMPLITUDE MODULATION.

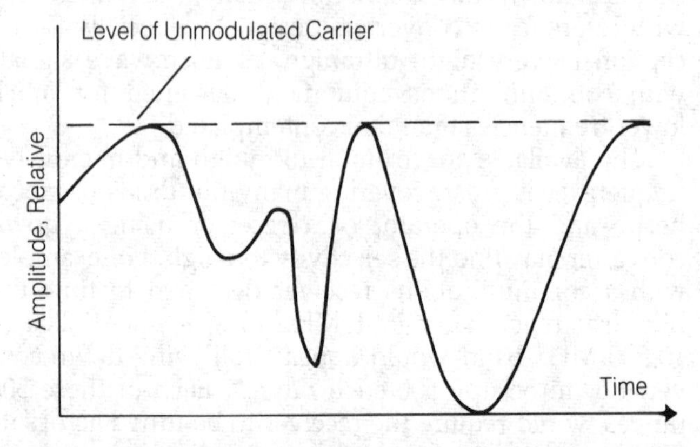

DOWNWARD MODULATION: The audio signal causes a decrease in the average power of a downward amplitude-modulated transmitter so the instantaneous carrier amplitude never exceeds the unmodulated carrier level.

DRAIN

The term *drain* refers to the current or power drain from a source. For example, the current drain of a 12-Ω resistor connected to a 12 V power supply is 1 ampere.

In a field-effect transistor, the drain is the electrode from which the output is generally taken. The drain of a field-effect transistor is at one end of the channel, and is the equivalent of the plate of a vacuum tube or the collector of a bipolar transistor. *See also* FIELD-EFFECT TRANSISTOR.

DRAIN-COUPLED MULTIVIBRATOR

A drain-coupled multivibrator is a form of oscillator that uses two field-effect transistors. The circuit is analogous to a tube type multivibrator (*see* MULTIVIBRATOR). The drain of each field-effect transistor is coupled to the gate of the other through a blocking capacitor, as shown in the diagram.

The drain-coupled multivibrator is, in effect, a pair of field-effect-transistor amplifiers, connected in cascade with positive feedback from the output to the input. Because each stage inverts the phase of the signal, the output of the two-stage amplifier is in phase with the input, and oscillation results. The values of the resistors and capacitors determine the oscillating frequency of the drain-coupled multivibrator. Alternatively, tuned circuits may be used in place of, or in series with, the drain resistors. This will create a resonant situation in the feedback circuit, confining the feedback to a single frequency.

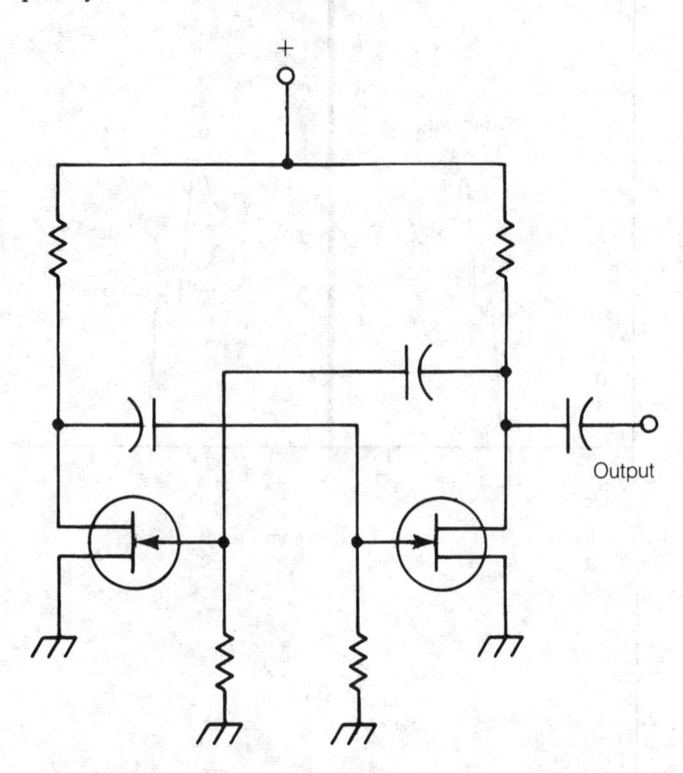

DRAIN-COUPLED MULTIVIBRATOR: This FET (field-effect transistor) circuit functions as an oscillator.

DRIFT

In a conductor or semiconductor, drift is the movement of charge carriers. The term is especially useful in conjunction with the flow of current in a semiconductor. In an N-type semiconductor material, the charge carriers are extra electrons among the atoms of the substance. In a P-type material, the charge carriers are atoms deficient in electrons; these are called holes (*see* ELECTRON, HOLE).

Drift velocity is often mentioned with respect to a semiconductor device. Drift velocity is given in centimeters per second or meters per second. In most materials, an increase in the applied voltage results in an increase in the drift velocity, but only up to a certain maximum. The greater the drift velocity for a given material, the greater the current per unit cross-sectional area.

The ease with which the charge carriers in a semiconductor are set in motion is called the drift mobility. Generally, the drift mobility for holes is less than that for electrons. The drift mobility determines, in a transistor or field-effect transistor, the maximum frequency at which the device will produce gain in an amplifier circuit. This frequency also depends on the thickness of the base or channel region.

The term *drift* is sometimes used in electronics to refer to an unwanted change in a variable. For example, the oscillator frequency of a radio transmitter may change in either an upward or a downward direction, because of temperature variations among the components. Frequency drift is not as serious a problem today as it was before solid-state technology. Transistors and integrated circuits generate almost no heat in an oscillator circuit. *See also* PHASE-LOCKED LOOP.

DRIVE

Drive is the application of power to a circuit for amplification, dissipation, or radiation. In particular, the term *drive* refers to the current, voltage, or power applied to the input of the final amplifier of a radio transmitter. The greater the drive, the greater the amplifier output. But excessive drive can cause undesirable effects such as harmonic generation and signal distortion.

Class A and Class AB amplifiers require very little driving power. The Class A amplifier theoretically needs no driving power at all; it runs from the voltage supplied to its input. Class C radio-frequency amplifiers must have large driving power to function properly. (*See* CLASS A AMPLIFIER, CLASS AB AMPLIFIER, CLASS B AMPLIFIER, CLASS C AMPLIFIER.)

In an antenna, the element that is connected directly to the feed line, and therefore receives power from the transmitter directly, is called the driven element. An antenna may have one or more elements (*see* DRIVEN ELEMENT).

The mechanism for handling magnetic floppy disks or hard disks of a computer is also called a drive. *See also* DIRECT-DRIVE TUNING, DISK DRIVE, VERNIER.

DRIVEN ELEMENT

In an antenna with parasitic elements (*see* PARASITIC ARRAY), those elements connected directly to the transmission line are called driven elements. In most parasitic arrays, there is one driven element, one reflector, and one or more directors (*see* DIRECTOR).

When several parasitic antennas are operated together, such as in a collinear or stacked array, there are several driven elements. Each driven element receives a portion of the output power of the transmitter. Generally, the power is divided equally among all of the driven elements. In some phased arrays, all of the elements are driven.

The driven element in a parasitic array is always resonant at the operating frequency. The parasitic elements are usually (but not always) slightly off resonance; the directors are generally tuned to a higher frequency than that of the driven element, and the reflector is generally set to a lower frequency. The impedance of the driven element, at the feed point, is a pure resistance when the antenna is operated at its resonant frequency. When parasitic elements are near the driven element, the impedance of the driven element is low compared to that of a dipole in free space.

For the purpose of providing an impedance match between a driven element and a transmission line, the driven element may be folded or bent into various configurations. Among the most common matching systems are the delta, gamma, and T networks (*see* DELTA MATCH, GAMMA MATCH). Sometimes the driven element is a folded dipole rather than a single conductor. *See also* FOLDED DIPOLE ANTENNA, QUAD ANTENNA, YAGI ANTENNA in the ANTENNA DIRECTORY.

DRIVER

A driver is an amplifier in a radio transmitter, designed for providing power to the final amplifier. The driver stage must be designed to provide the right amount of power into a load impedance. Too little drive will result in reduced output from the final amplifier, and reduced efficiency of the final amplifier. Excessive drive can cause harmonic radiation and distortion of the signal modulation envelope.

The driver stage receives a signal at the operating frequency and provides a regulated output with a minimum of tuning. Most or all of the transmitter tuning adjustments are performed in the final amplifier stage. *See also* DRIVE.

DROOP

Droop is a term used to define the shape of an electric pulse. Droop is generally specified as a percentage of the maximum amplitude of a pulse.

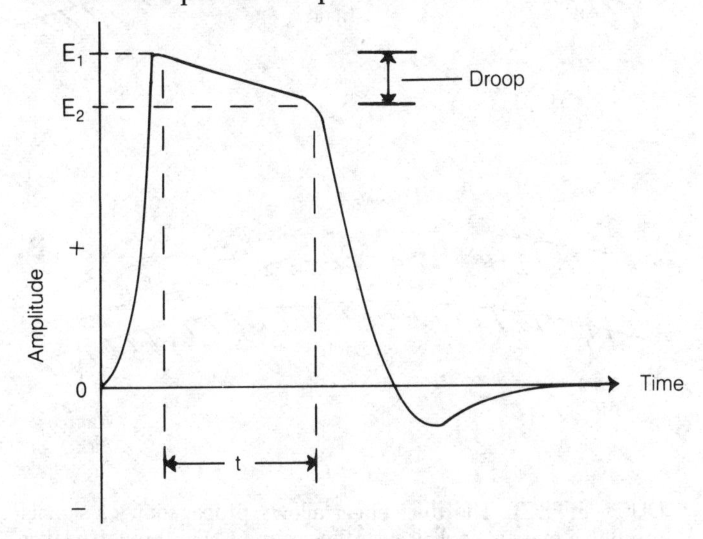

DROOP: Droop is a measure of the decay of pulse amplitude from the ideal flat top of a square pulse.

The figure illustrates a pulse as it might appear on an oscilloscope display screen. The maximum amplitude of the pulse is indicated by E_1. After the initial rise to maximum amplitude, the pulse should ideally remain at that amplitude for its duration, and then rapidly fall back to the zero or minimum level. In practice, this ideal is seldom attained.

DTL

See DIODE-TRANSISTOR LOGIC.

DUAL-BEAM OSCILLOSCOPE

See OSCILLOSCOPE.

DUAL IN-LINE PACKAGE

See SEMICONDUCTOR PACKAGE.

DUCT EFFECT

The duct effect is a form of propagation that takes place with very high frequency and ultrahigh-frequency electromagnetic waves. Also called ducting, this form of propagation is entirely confined to the lower atmosphere of earth.

A duct forms in the troposphere when a layer of cool air becomes trapped underneath a layer of warmer air, as shown at A in the illustration, or when a layer of cool air becomes sandwiched between two layers of warmer air, as shown at B. These kinds of atmospheric phenomena are fairly common along and near weather fronts in the temperate latitudes. They also take place frequently above water surfaces during the daylight hours, and over land surfaces at night. Cool air is denser than warmer air at the same humidity level, so it exhibits a higher index of refraction for radio waves of certain frequencies. Total internal reflection takes place inside the region of cooler air, in much the same way as light waves are trapped inside an optical fiber or under the surface of a body of water (*see* INDEX OF REFRACTION, TOTAL INTERNAL REFLECTION).

For the duct effect to provide communications, both the transmitting and receiving antennas must be located within the same duct, and this duct must be present continuously between the two locations. Sometimes, a duct is only a few feet wide. Over cool water, a duct may extend just a short distance above the surface, although it may persist for hundreds or even thousands of miles in a horizontal direction. Ducting allows over-the-horizon communication of exceptional quality on the very high and ultrahigh frequencies. Sometimes, ranges of over 1,000 miles can be obtained by this mode of propagation. *See also* TROPOSPHERIC PROPAGATION.

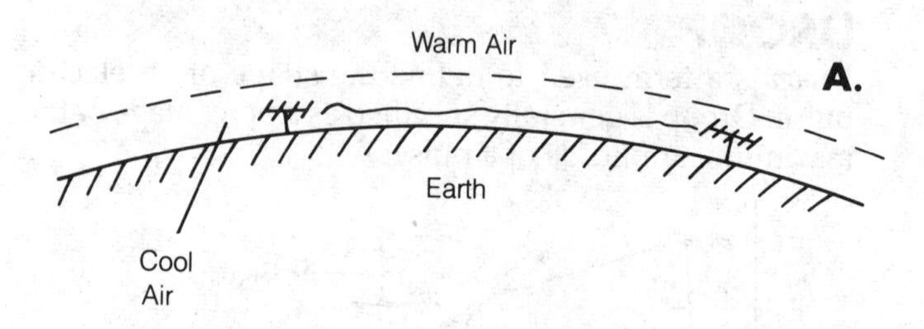

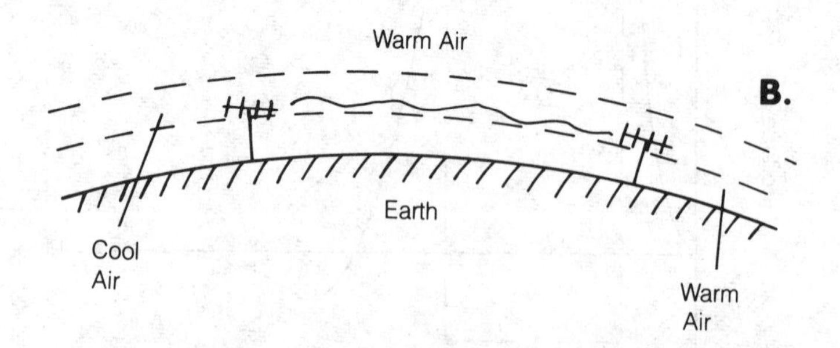

DUCT EFFECT: The duct effect allows propagation of signals through a mass of cool air near the ground (A) or through a layer of cool air sandwiched between layers of warmer air above the ground (B).

DUMMY ANTENNA

A dummy antenna is a resistor having a value equal to the characteristic impedance of the feed line and a power-dissipation rating at least as great as the transmitter output power. There is no inductance in the resistor. A dummy antenna is used for testing transmitters off the air so no interference will be caused on the frequencies at which testing is carried out. A dummy antenna can also be useful for testing communications receivers for internal noise generation when no outside noise sources can be allowed to enter the receiver antenna terminals.

The resistor in a dummy antenna is always noninductive. This ensures that the feed line can be terminated in a perfect impedance match. The sizes of dummy loads can range from very small to very large. The figure shows a dummy antenna for use with a Citizen's Band transceiver with 4 W of output power. Some dummy loads consist of large resistors immersed in oil-coolant containers. They can handle a kilowatt or more of power on a continuous basis.

Dummy antennas must be well shielded to prevent accidental signal radiation. This is especially important at the higher power levels.

DUMMY ANTENNA: A low-power coaxial dummy antenna.

DUMP

The term *dump* is used in computer science to refer to the transfer or loss of memory. Memory bits can be moved from one storage location to another, to make room in the first memory for more data. In the second memory, sometimes called a hard memory, the data is not affected by programming changes or power failures.

Some computer programs use dump points in their execution at periodic intervals. This transfers the data from the active, or soft, memory into a permanent, or hard, memory. During a pause in the execution of a program, the computer may backtrack to the most recent dump point, and use the data there to continue the program. This process is called dump-and-restart. *See also* SEMICONDUCTOR MEMORY.

DUPLEXER

A duplexer is a device in a communications system that allows duplex operation. Two operators can interrupt each other at any time, even while the other operator is actually transmitting. Duplex operation is usually carried out with two different frequencies. Notch filters between the transmitter and receiver at each station prevent overloading of the receiver front end. A repeater uses a duplexer to allow the simultaneous retransmission of received signals on a different frequency (*see* NOTCH FILTER, REPEATER). The illustration shows a block diagram of a duplex communications system, including the duplexer.

In a radar installation, a duplexer is a device that automatically switches the antenna from the receiver to the transmitter whenever the transmitter puts out a pulse. This prevents the transmitter from damaging or overloading the receiver. In this application, the duplexer acts as a high-speed, radio-frequency-actuated transmit-receive (TR) switch. The receiver is disabled while the transmitter sends out the pulse. *See also* TRANSMIT-RECEIVE SWITCH.

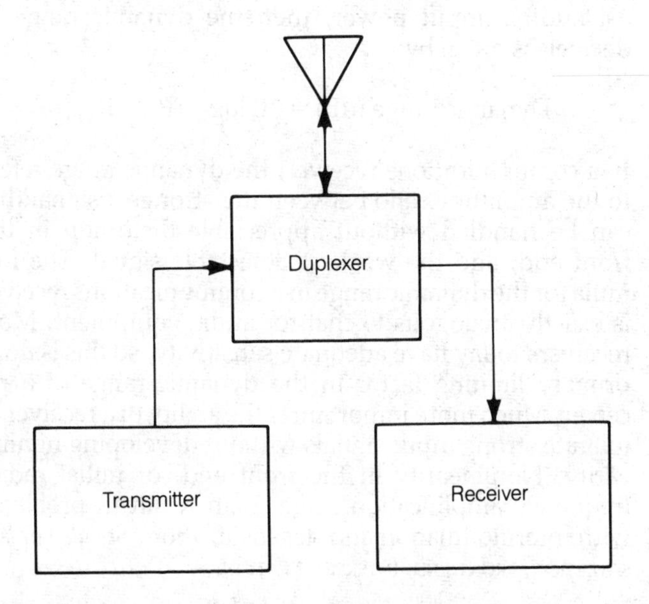

DUPLEXER: A duplexer permits simultaneous operation of a transmitter and receiver on different frequencies with a shared antenna.

DUPLEX OPERATION

Duplex operation is a form of radiocommunication in which the transmitters and receivers of both stations are operated simultaneously and continuously. Each station operator may interrupt the other at any time, even while the other is actually transmitting. Therefore, a normal conversation is possible, similar to that over a telephone.

To achieve full duplex operation, it is necessary that the transmitting and receiving frequencies differ. Two stations in duplex communication, say station X and station Y, must have opposite transmitting and receiving frequencies. For example, station X might transmit on 146.01 MHz and receiver on 146.61 MHz; then station Y must transmit on 146.61 MHz and receive on 146.01 MHz. A duplexer (*see* DUPLEXER) is usually required to prevent the station receiver from being desensitized or overloaded by the transmitter.

DURATION TIME

Duration time is the length of an electric pulse, measured from the turn-on time to the turn-off time. If the pulse is rectangular, then the duration time is determined as shown in the figure. But electric pulses are usually not perfectly rectangular. They often have rounded corners, droop, or other irregularities. Then the duration of the pulse may be expressed in several different ways. Generally, the duration time of a nonrectangular pulse is considered to be the theoretical duration time of an equivalent rectangular pulse. The equivalent pulse may be generated by area redistribution, or by amplitude averaging (*see* AREA REDISTRIBUTION, AVERAGE ABSOLUTE PULSE AMPLITUDE).

The average amplitude of a pulse train depends on the duration and the frequency of the pulses. *See also* DUTY FACTOR.

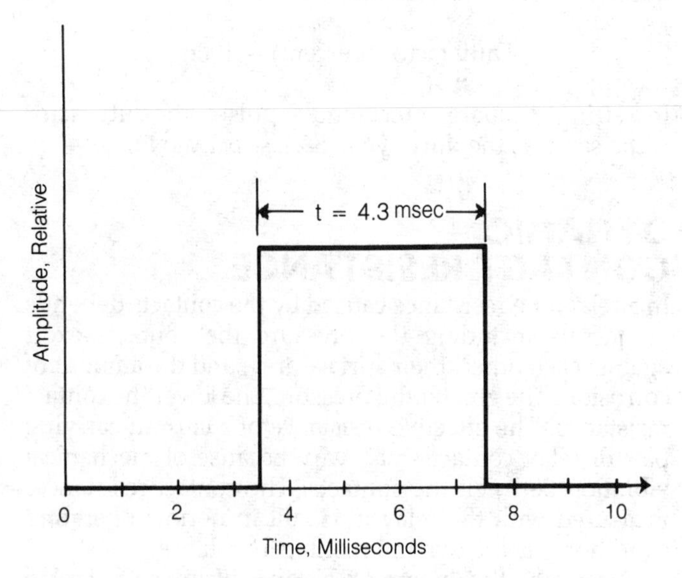

DURATION TIME: The duration time of a pulse is the time it remains in the high state.

DUTY CYCLE

The duty cycle is a measure of the proportion of time during which a circuit is operated. Suppose, for example, that during a given time period, t_0, a circuit is in operation for a total length of time t, measured in the same units as t_0. Of course, t cannot be larger than t_0; if $t = t_0$, then the duty cycle is 100 percent. In general, the duty cycle is given by:

$$\text{Duty cycle (percent)} = \left(\frac{100t}{t_0}\right)$$

When determining the duty cycle, it is important that the evaluation time t_0 be long enough so that an accurate representation is obtained. For example, a circuit may be on for one minute and then off for one minute, in a repeating cycle. In this case, an evaluation time of 10 seconds is obviously too short: t_0 must be at least several minutes for a reasonably accurate estimate of the duty cycle. *See also* DISSIPATION RATING.

DUTY FACTOR

The duty factor of a system is the ratio of average power to peak power, expressed as a percentage. The duty factor is similar to the duty cycle, except that the system does not have to shut down completely during the low period (*see* DUTY CYCLE). In a continuous-wave transmitter, the duty factor and the duty cycle are the same. Letting P_{max} be the peak power and P_{av} be the average power, then:

$$\text{Duty factor (percent)} = 100P_{av}/P_{max}$$

In a pulse train, or string of pulses, the duty factor is the product of the pulse duration and the frequency. As a percentage, letting the duration time be t and the frequency in Hertz be f, then:

$$\text{Duty factor (percent)} = 100tf$$

In a string of square or rectangular pulses, the duty factor is the same as the duty cycle. *See also* DURATION TIME.

DYNAMIC CONTACT RESISTANCE

In a relay, the resistance caused by the contacts depends on factors including the pressure the contacts exert against each other, their surface area, and the amount of corrosion. The greater the pressure, the lower the contact resistance. The effective resistance of a current-carrying pair of relay contacts may vary because of mechanical vibration between the contacts. The contact resistance, measured with the relay in circuit in normal operating conditions, is the dynamic contact resistance.

Normally, the dynamic resistance of any pair of relay contacts should be as low as possible. This can be ensured by keeping the contacts clean, and by making certain that the relay coil gets the right amount of current so that the contacts will open and close properly. *See also* RELAY.

DYNAMIC SPEAKER

See SPEAKER.

DYNAMIC MICROPHONE

See MICROPHONE.

DYNAMIC PICKUP

A dynamic pickup is a phonograph transducer that operates on the same principle as the dynamic microphone or loudspeaker (*see* SPEAKER, MICROPHONE). A stylus vibrates as it follows the groove of the phonograph disk. A small coil, attached to the stylus, moves along with it in the field of a permanent magnet. This results in audio-frequency currents in the coil.

Dynamic pickups are used in high-fidelity applications. The ceramic and crystal type pickups are also fairly common. The dynamic pickup has the advantage of being electrically and physically rugged. *See also* CERAMIC PICKUP, CRYSTAL TRANSDUCER.

DYNAMIC RANGE

In an audio sound system or a communications receiver, the dynamic range is often specified as an indicator of the signal variations that the system can accept and reproduce without objectionable distortion.

In high-fidelity sound systems, there is a weakest signal level that can be reproduced without being masked by the internal noise of the amplifiers. There is also a strongest signal level that can be reproduced without exceeding a certain amount of total harmonic distortion in the output. Usually, the maximum allowable level of total harmonic distortion is 10 percent (*see* TOTAL HARMONIC DISTORTION). If P_{max} represents the maximum level of input power that can be reproduced without appreciable distortion, and P_{min} represents the smallest audible input power, then the dynamic range in decibels is given by:

$$\text{Dynamic range (dB)} = 10 \log_{10} (P_{max}/P_{min})$$

In a communications receiver, the dynamic range refers to the amplitude ratio between the strongest signal that can be handled without appreciable distortion in the front end, and the weakest detectable signal. The formula for the dynamic range in a communications receiver is exactly analogous to that for audio equipment. Most receivers today have adequate sensitivity, so this is not a primary limiting factor in the dynamic range of a receiver. Much more important is the ability of a receiver to tolerate strong input signals without developing nonlinearity. Nonlinearity in the front end, or initial radio-frequency amplification stages, can result in problems with intermodulation and desensitization. *See also* DESENSITIZATION, FRONT END, INTERMODULATION.

DYNAMIC REGULATION

In a voltage-regulated power supply, dynamic regulation refers to the ability of the supply to handle large, sudden changes in the input voltage with a minimum of change in the output voltage or current. The most common kind of sudden change in the input voltage to a power supply is called a transient. A transient can be the result of a lightning stroke or arc in a power transmission line (*see* TRANSIENT).

Poor dynamic regulation in a power supply can result in sudden changes in the output voltage. This can cause damage to modern solid-state equipment. *See also* POWER SUPPLY.

DYNAMIC RESISTANCE

According to Ohm's law, the resistance between two points is the quotient of the voltage divided by the current (*see* OHM'S LAW). This quantity is the static resistance in a circuit. In practice, the effective resistance or impedance of a component may not be the same as the static resistance or impedance. The actual, or effective, operating resistance is called the dynamic resistance.

If a component is subjected to a changing voltage causing a changing current, the illustration shows two possible resulting curves. At A, the current increases linearly with the voltage. At B, the current does not change in direct proportion to the voltage.

A curve such as the one at B might result from a gradual change in the voltage across a component, with a resulting heat change that causes a fluctuation in the resistance. The dynamic resistance over a certain range is found by dividing the change in the voltage, ΔE (delta E), by the change in the current, ΔI, between two specific points along the curve. Thus, at A, the dynamic resistance is:

$$\frac{\Delta E}{\Delta I} = \frac{10}{0.5} = 20\ \Omega$$

and it does not matter what points on the curve we choose. The dynamic resistance is always the same in the illustration at A. But at B, the dynamic resistance changes depending on the points selected. For the two points shown, the dynamic resistance is:

$$\frac{\Delta E}{\Delta I} = \frac{10}{0.3} = 33\ \Omega$$

but this is the unique situation for these points. *See also* RESISTANCE.

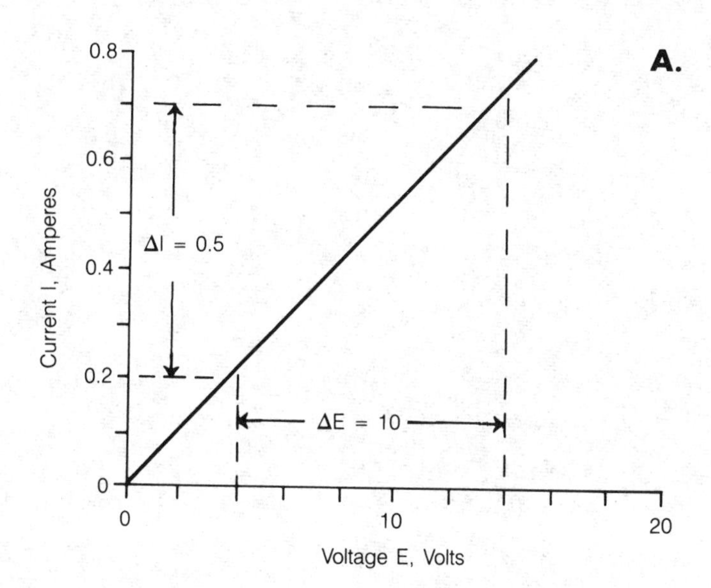

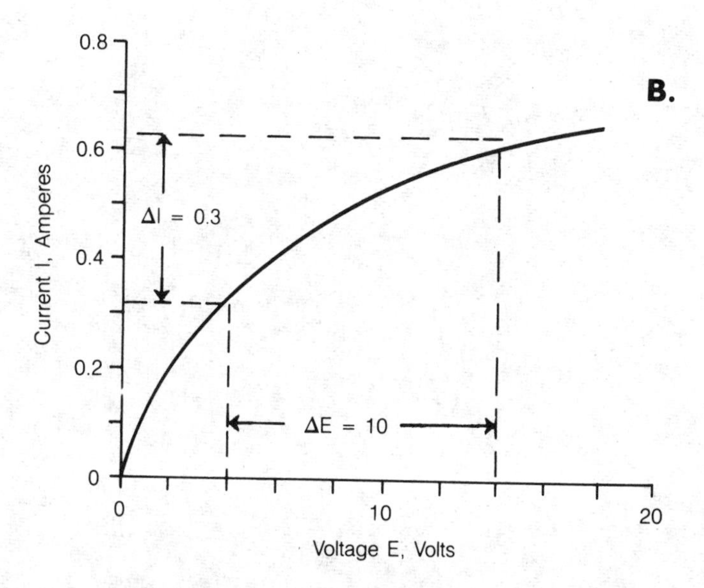

DYNAMIC RESISTANCE: Dynamic resistance is voltage divided by current. Where the function is linear, a range need not be specified (A). Beginning and end points must be defined for a curved function to give a meaningful value (B).

DYNAMO

See GENERATOR.

DYNODE

A dynode is an electrode used in some photomultiplier or dissector tubes to amplify a stream of electrons. When the dynode is hit by an electron, it emits approximately five to seven secondary electrons.

Dynodes in photomultiplier tubes are slanted in such a way that an electron must strike a series of the dynodes in succession. The electron hits first one dynode, where several secondary electrons are emitted along with it. The electrons then hit a second dynode, where one incident electron again produces secondary electrons. It does not take many dynodes to get very large gain figures in this way. For example, four dynodes, each of which emits five secondary electrons in addition to the one that strikes it, will result in a gain of 6^4 or 1296. This is 62 decibels in terms of current.

Dynodes in a photomultiplier or dissector tube are placed around the inner circumference and are oriented at the proper angles so that an impinging electron strikes them one after the other. With the electrons in the tube, as with light against a mirror, the angle of incidence is equal to the angle of reflection. *See also* DISSECTOR TUBE, PHOTOMULTIPLIER TUBE.

EARPHONE

An earphone is a small speaker designed to be worn in or over the ear. Earphones can be used in communications for privacy or for quiet reception. With an earphone worn on one ear, the other ear is free, allowing the operator to hear external sounds.

A pair of earphones is often mounted on a headband and worn over both ears. This combination, called a headphone or headset, obstructs the listener's sensitivity to external noises.

Earphones are available with impedances as low as 4 ohms and as high as several thousand ohms. An earphone consumes far less power than a speaker, so the earphone is useful when it is necessary to conserve power, as in battery operation. Some earphones operate on the dynamic principle; others are crystal or ceramic transducers. *See also* CRYSTAL TRANSDUCER, HEADPHONE, SPEAKER.

EAROM

See SEMICONDUCTOR MEMORY.

EARTH CONDUCTIVITY

The earth is a fair conductor of electric current. The conductivity of the soil is, however, highly variable, depending on the geographic location. In some places the soil might be an excellent conductor; in other locations it is a very poor conductor. In general, the better the soil for farming, the better it will be for electrical conductivity. Wet, black soil is a very good conductor, and rocky or sandy dry soil is poor. A brief rain shower can drastically alter the soil conductivity in some locations.

The conductivity of the earth is generally measured in millisiemens per meter or millimhos per meter. A mho (also known as a siemens) is the unit of conductance and is expressed as the reciprocal of the resistance. In the United States, the earth conductivity ranges from approximately 1 millisiemens per meter to more than 30 millisiemens per meter. All earth is, by comparison to sea water, a very poor conductor; salt water has an average conductivity of about 5,000 millisiemens per meter.

In radio communications earth conductivity is important for obtaining a good electrical ground. The earth conductivity is also important for surface-wave propagation at frequencies below 20 to 30 MHz. The better the conductivity of the earth, the better the radio-station ground system will be for a given installation, and the better the surface-wave propagation. Good ground conductivity is also advantageous in electrical installations, and for the grounding of equipment for protection against lightning. *See also* GROUND CONNECTION, SURFACE WAVE.

EARTH CURRENT

Electric wires under or near the ground, carrying alternating currents, cause currents to be induced in the ground. This occurs because the ground is a fairly good conductor, and the electromagnetic field around the wire causes movement of the charge carriers in any nearby conductor. High-voltage lines can induce considerable current in the ground near the wires. Radio broadcast stations cause radio-frequency currents to flow in the ground in the vicinity of their antennas.

Currents flow in the earth naturally. Two ground rods, driven several feet apart, might have a small potential difference, and this can be measured with an alternating-current voltmeter. The intensity of the ground current in the vicinity of a conductor carrying alternating current depends on the conductivity of the ground in the area, and on the current intensity in the wire and the distance of the wire above the ground. *See also* EARTH CONDUCTIVITY.

E BEND

In a waveguide, an E bend is a change in the direction of the waveguide, parallel with the plane of electric-field lines of flux. It is called an E bend because it follows the E-field component of the electromagnetic wave.

The bending of radio waves by the ionospheric E layer is sometimes called E bend of E skip, although both of these terms are technically inaccurate. Signal propagation through the E layer is fairly common at frequencies below about 50 MHz, and might occur well into the very high frequency spectrum. *See also* E LAYER, WAVEGUIDE.

ECCENTRICITY

When a dial or control of a unit of equipment is not properly centered, it is said to be eccentric. Eccentricity in a control results when the position of the knob with respect to the calibrated scale is not consistent as the setting is changed. Eccentricity is a mechanical, rather than an electrical, problem.

In mathematics, eccentricity is a measure of the degree to which an ellipse differs from a circle or an ellipsoid deviates from a sphere. A circle and sphere have zero eccentricity. If *c* represents the distance from the center of an ellipse or ellipsoid to either of the foci, and *a* is the larger (major) radius, then the eccentricity is given by *c*/*a*.

In a phonograph disk, eccentricity refers to a condition in which the spiral groove is not centered on the disk. This might happen because the hole is off center or because the grooves themselves are off center. Eccentricity in a phonograph disk results in warbling of the sound when it is played back; the stylus first speeds up and then slows down in its travel through the length of the groove. If the eccentricity is severe, this warbling will be objectionable. *See also* DISK RECORDING.

ECHO

Whenever energy transmitted from a source is reflected from a distant object and returns in detectable magnitude, that return signal is called an echo. Sound echoes are heard after a shout in the direction of a distant, acoustically reflective object. In radar, the term *echo* refers to any return of the transmitted pulse, creating a blip on the screen (*see* RADAR).

Echoes of radio signals in the high-frequency part of the electromagnetic spectrum can be heard where ionospheric propagation is prevalent. A few milliseconds after receiving a strong signal pulse, a weaker echo might be heard. This occurs because the signal has propagated along two paths: a short path and a long path. The echo delay time is the difference in the time required for the signal to propagate by means of the two paths. *See* LONG-PATH PROPAGTION.

Echoes are occasionally detected in wire transmission circuits. This can happen, for example, in a long-distance telephone connection. Electrical echoes result from impedance discontinuities along the line. These discontinuities cause some of the incident signal to be reflected back toward the transmitting station. A circuit called an echo eliminator in telephone circuits suppresses these reflected signals. Sometimes, however, the echo gets through and can be heard by the person talking on the telephone. *See also* TELEPHONE SYSTEM.

ECHO BOX

An echo box is a unit of equipment for testing and calibrating a radar set.

The transmitted pulse is fed into the echo box instead of the radar antenna. The echo box then produces a delayed return pulse, whose intensity decays at a certain rate. The decaying echo signal is fed to the radar receiver input. Eventually the pulse decays to the level where the radar receiver can no longer detect it.

If the amplitude-versus-time function of the output signal is known, the receiver sensitivity can be determined by measuring the length of time the receiver is able to detect the signal. When the echo can no longer be seen on the radar screen, the level of the signal is checked, and this represents the weakest signal that the radar can see. *See also* RADAR.

ECHO INTERFERENCE

When a high-frequency radio signal is propagated through the ionosphere, an echo can occur (*see* ECHO) as a result of long-path propagation in addition to the normal short-path propagation. Echo interference may also result from propagation of the signal over several more-or-less direct paths of variable distance.

When echo interference is severe, it can render a signal almost unintelligible because phase modulation is superimposed on the envelope of the signal. This phenomenon is especially noticeable with auroral propagation, when the signal path may change at a very rapid rate.

In a long-distance telephone circuit, the echo may have sufficient amplitude to create interference at the transmitting station. This can annoy the speaker because he hears his own words with a slight delay in the earphones. Echo eliminators are commonly used in telephone systems to prevent this kind of interference, which is caused by reflection of signals from impedance discontinuities along the line.

In a radar system, echo interference is caused by the presence of unwanted false targets such as buildings, birds, and thunderstorms. *See also* LONG-PATH PROPAGATION, RADAR, TELEPHONE SYSTEM.

ECLIPSE EFFECT

The conditions of the ionospheric D, E, and F layers vary greatly with the time of day. During a solar eclipse, when the moon passes between the earth and the sun, nighttime conditions can occur for a brief time within the shadow of the moon. This effect on the ionized layers of the upper atmosphere is called the eclipse effect.

The D layer, the lowest layer of the ionosphere, seems to be affected to a greater extent during a solar eclipse than the E or F layers, which are at greater altitudes and ionize more slowly. The ionization of the D layer depends on the immediate level of ultraviolet radiation from the sun, but the E and F layers have some lag time. An eclipse in the sun causes a dramatic decrease in the ionization level of the D layer, but the E and F layers are generally not affected to a great extent. The reduction in the ionization density of the D layer causes improved propagation at some frequencies because the absorption created by the D layer is not as great during an eclipse, as under normal circumstances during daylight hours. *See also* D LAYER, E LAYER, F LAYER, PROPAGATION CHARACTERISTICS.

E CORE

An E core is a form of transformer core made of ferromagnetic material that is shaped like a capital letter *E*. The figure illustrates a typical E core. Wire coils are wound on the horizontal bars of the letter E; up to three different windings are possible, so a transformer can have two secondary windings giving different voltages and impedances with respect to a single primary winding.

The interaction among the windings on an E core are enhanced by the action of the ferromagnetic material, which has a high permeability for magnetic fields. Most of the magnetic flux is contained within the E-shaped core, and very little exists in the surrounding air. *See also* TRANSFORMER.

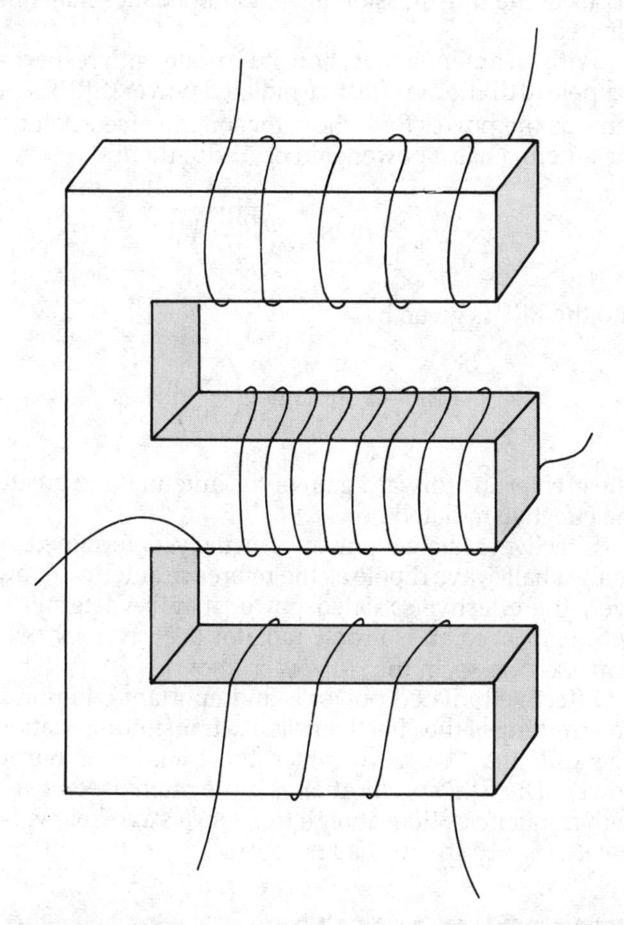

E CORE: An E-shaped transformer core can have primary and secondary windings on the central bar or windings on all three bars. This core can be closed with an I-shaped section to form a figure-8 core.

EDDY-CURRENT LOSS

In any ferromagnetic material, the presence of an alternating magnetic field produces currents that flow in circular or elliptical paths. These currents, because of their resemblance to vortex motions, are called eddy currents.

Eddy currents are responsible for some degradation in the efficiency of any transformer that uses an iron core. To minimize the effects of eddy currents, low-frequency transformers have laminated cores, consisting of many thin, flat pieces of ferromagnetic material insulated from each other. *See also* TRANSFORMER.

EDGE CONNECTOR

See CONNECTOR.

EDGE EFFECT

In a capacitor, most of the electric field is confined between the two plates. However, near the edges of the plates, some of the electric lines of flux bulge outward from the space between the plates. This is called the edge effect.

The edge effect is most pronounced with air-dielectric capacitors because the surrounding air has the same dielectric constant as the medium between the plates. However, the edge effect also occurs in other types of capacitors. It is least pronounced in electrolytic capacitors, which have shielded enclosures. The edge effect can result in unwanted capacitive coupling among components in a circuit. For this reason, the component layout of a circuit is critical at frequencies where the smallest intercomponent capacitance may cause feedback or other undesirable effects. *See also* CAPACITOR.

EDISON EFFECT

When the filament of an incandescent bulb is heated by passing an electric current through it, electrons are emitted. This is called the Edison effect, because Thomas Edison, in his early experiments with light bulbs, first discovered it. Edison used positively charged electrodes in the evacuated bulbs to attract these electrons, resulting in a flow of current.

EEPROM

See SEMICONDUCTOR MEMORY.

EFFECTIVE BANDWIDTH

All bandpass filters allow some signal to pass while they reject energy at frequencies outside the passband range (*see* BANDPASS RESPONSE). The bandwidth of this filter is determined at the 3-decibel power-attenuation points. However, there is another method of specifying the characteristic of a bandpass filter: with effective bandwidth.

Consider a hypothetical bandpass filter with a perfectly rectangular response (*see* FREQUENCY RESPONSE) centered at the same frequency f_0 as an actual bandpass filter. (This situation is illustrated in the figure.) Suppose that the bandwidth of the rectangular filter can be varied at will, all the way from zero to a value much greater than that of the real filter, but always remaining centered at f_0. At some value of bandwidth of the rectangular filter, the characteristics will be identical to that of the actual filter, with respect to the amount of energy transferred. This rectangular bandwidth is the effective bandwidth of the real filter.

The effective bandwidth of a filter depends on the steepness of the response and the nature of the response within the passband. *See also* BANDPASS FILTER.

EFFECTIVE GROUND

In radio communications below a frequency of about 100

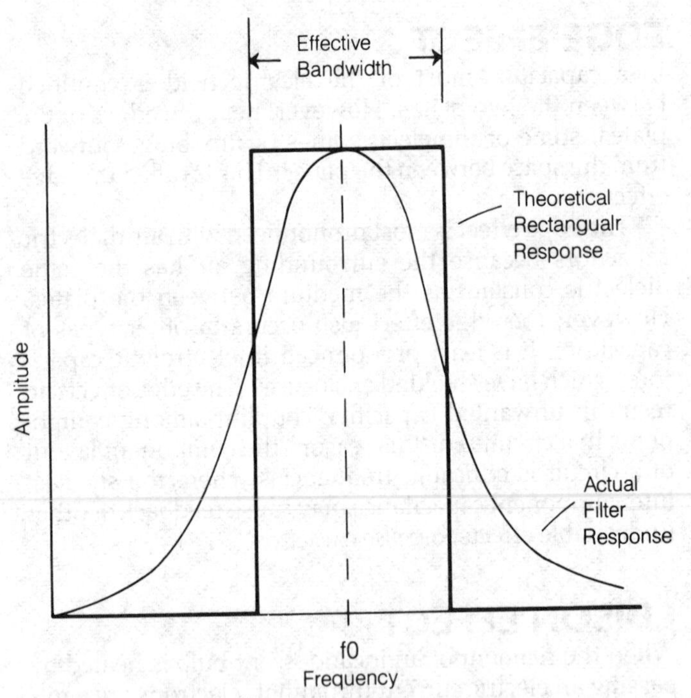

EFFECTIVE BANDWIDTH: The effective bandwidth of a bandpass filter is the bandwidth of a rectangular response with the same energy-transfer ratio.

MHz, the ground plays a significant role in the behavior of antenna systems, both receiving and transmitting. This is true whether or not an antenna is connected to an earth ground. The earth, a good conductor at most locations, absorbs some radio-frequency energy and reflects some. The lower the frequency, in general, the better the conductivity of the soil in a particular location (*see* EARTH CONDUCTIVITY). Some soils are good conductors at virtually all frequencies below 100 MHz; other soils are poor conductors even for direct current.

Because reflected electromagnetic waves combine with the direct wave for any antenna system, the height of the antenna above the ground is an important consideration; this height determines the elevation angles at which the direct and reflected waves will act together, and also the elevation angles at which they will oppose each other. The visible earth surface is not the surface from which the electromagnetic waves are reflected. The effective ground plane, in most cases, lies several feet below the actual earth surface. The poorer the earth conductivity, the farther below the surface lies the effective ground plane at any electromagnetic frequency.

Irregularity in the terrain near an antenna, whether natural or man-made, affects the level of the effective ground. In areas with many buildings, utility wires, fences, and other obstructions, the determination of the effective ground plane is practically impossible from theory. Antenna installations in these environments usually proceed on a trial-and-error basis. *See also* DIRECT WAVE, REFLECTED WAVE.

EFFECTIVE RADIATED POWER

When a transmitting antenna has power gain, the effect in the favored direction is equivalent to an increase in the transmitter power. To obtain a doubling of the effective radiated power in a certain direction, for example, either of two things may be done: The transmitter output power can actually be doubled, or the antenna power gain can be increased in that direction by 3 decibels. Effective radiated power takes into account both the antenna gain and the transmitter output power. It is also affected by losses in the transmission line and impedance matching devices.

With an antenna that shows zero gain with respect to a dipole (0 dBd), the effective radiated power (ERP) is the same as the power *P* reaching the antenna feed point. If the antenna has a power gain of A_p dB, then:

$$A_p = 10 \log_{10} \left(\frac{\text{ERP}}{\text{P}}\right)$$

and the ERP is given by:

$$\text{ERP} = \text{Pantilog}_{10} \left(\frac{A_p}{10}\right)$$

The greater the forward gain of the antenna, the greater the effective radiated power.

Effective radiated power is usually determined by using a half-wave dipole as the reference antenna. However, the effective radiated power may be determined with respect to an isotropic radiator (*see* DIPOLE ANTENNA, ISOTROPIC ANTENNA in the ANTENNA DIRECTORY).

Effective radiated power is an important criterion in determining of the effectiveness of a transmitting station. The ERP may be many times the transmitter output power. The ERP specification is more often used at very high frequencies than at high frequencies and below. *See also* ANTENNA POWER GAIN, dBd,dBi, DECIBEL.

EFFECTIVE VALUE

An electrical quantity can have a value that appears different from the actual value. This value is called the effective value of the quantity.

An example of the difference between an actual value and an effective value can be found when a wirewound resistor is used at different alternating-current frequencies. The actual value, at direct-current and very low alternating-current frequencies, might be 10 Ω. As the frequency is raised, the inductance of the windings introduces an additional resistance in the circuit. At a frequency of, say, 10 MHz, the effective resistance might be 20 Ω, consisting of 10 Ω pure resistance and 10 Ω of inductive reactance.

In alternating-current theory, the root-mean-square value of a waveform is sometimes called the effective value. *See also* ROOT MEAN SQUARE.

EFFICIENCY

Efficiency is the ratio, usually expressed as a percentage, between the output power from a device or system and the input power to that device or system. All kinds of electronic components from speakers, amplifiers, transformers, and antennas, to transistors, can be evaluated in terms of efficiency. Generally, only the output in the desired form is taken as the actual output power; dissipated power in unwanted forms is considered to have been wasted. The efficiency of all devices is less than 100 percent; while some circuits or devices approach 100-percent efficiency, this ideal is never actually achieved.

If the input power to a device is P_{IN} watts and the useful output power is P_{OUT} watts, then the efficiency in percent is given by:

$$\text{Efficiency (percent)} = \frac{100P_{OUT}}{P_{IN}}$$

EIA

See ELECTRONIC INDUSTRIES ASSOCIATION.

EIGHT-TRACK RECORDING

Eight-track recording is a common method of recording stereo audio signals on a magnetic tape. The cassette album uses eight-track tape, and it is only about one-quarter of an inch wide. In this cassette, an endless loop of tape, which can have a single-revolution playing time of from 8 to 30 minutes, contains four separate full-stereo recordings. Each recording contains two separate magnetic tracks on the tape, one for the left-hand channel and one for the right-hand channel.

At the end of each revolution of the tape, an electromechanical device causes the tape player to switch to the next pair of tracks. Thus the total playing time of the tape is four times the playing time of a single revolution.

Excellent channel separation is possible with modern eight-track tapes. The tapes can be stored almost indefinitely, provided the temperature and humidity are kept within reasonable limits. *See also* MAGNETIC RECORDING, MAGNETIC TAPE.

ELASTANCE

Elastance is a measure of the opposition a capacitor offers to being charged or discharged. The unit of elastance is called the daraf, which is *farad* spelled backwards. Elastance is the reciprocal of capacitance. That is, if C is the capacitance in farads, then:

$$\text{Elastance (darafs)} = 1/C$$

A value of capacitance of 1 picofarad is equivalent to an elastance of 1 teradaraf, or 10^{12} darafs. A capacitance of 1 nanofarad is an elastance of 1 gigadaraf (10^9 darafs); a value of capacitance of 1 microfarad is 1 megadaraf.

The daraf is an extremely small unit of elastance, seldom observed in capacitors. The elastance of 1 daraf is the value at which 1 volt of applied electric charge will produce 1 coulomb of displacement (*see* COULOMB). In most capacitors, 1 volt of charge produces no more than a few thousandths of a coulomb of charge displacement. *See also* DARAF.

E LAYER

The E layer of the atmosphere is an ionized region at an altitude of about 60 to 70 miles, or 100 to 115 kilometers, above sea level. The E layer permits medium-range communication on the low-frequency through very high frequency radio bands. The ionized atoms of the atmosphere at that altitude affect electromagnetic waves at wavelengths as short as about 1 meter.

The ionization density of the E layer is affected greatly by the height of the sun above the visual horizon. Ultraviolet radiation and X rays from the sun are primarily responsible for the existence of the E layer. The maximum ionization density generally occurs around midday. The minimum ionization density takes place during the early morning hours. The E layer almost disappears shortly after sundown under normal conditions.

Occasionally a solar flare will cause a geomagnetic disturbance, and the E layer will ionize at night in small areas or clouds. The resulting propagation is called sporadic E propagation. Often, the range of communication in sporadic-E mode can extend for over 1,000 miles. However, the range is not as great as with conventional F-layer propagation (*see* SPORADIC-E PROPAGATION).

At frequencies above about 150 MHz, or a wavelength of 2 meters, the E layer becomes essentially transparent to radio waves. Then, instead of being returned to earth, they continue on into space. *See also* D LAYER, F LAYER, IONOSPHERE, PROPAGATION CHARACTERISTICS.

ELECTRET

An electret is the electrostatic equivalent of a permanent magnet. While a magnet has permanently aligned magnetic dipoles, the electret has permanently aligned electric dipoles (*see* DIPOLE). An electret is generally made of an insulating material, or dielectric. Some varieties of wax, ceramics, and plastics are used to make electrets. The material is heated and then cooled while in a strong, constant electric field.

An electret element is used in the dynamic microphone. The impinging sound waves cause a diaphragm, attached to an electret, to vibrate. This produces a changing electric field, which produces an audio-frequency voltage at the output terminals. *See also* MICROPHONE.

ELECTRICAL ANGLE

A sine-wave alternating current or voltage can be represented by a counterclockwise rotating vector in a Cartesian coordinate plane. The origin of the vector is at the point (0,0), or the center of the plane, and the end of the vector is at a point having a constant distance from the origin.

In this representation, the amplitude of the current or voltage is indicated by the length of the vector. The greater the amplitude, the longer the vector, and the farther its end point is from the point (0,0). The frequency of the sine wave is represented by the rotational speed of the vector; one cycle is given by a full-circle, or 360-degree, rotation. This model is illustrated in the figure. The y axis, or ordinate, represents the amplitude scale, and the instantaneous amplitude of the wave is therefore given by the y value of the end point of the vector (*see* CARTESIAN COORDINATES, VECTOR).

The electrical angle θ is the angle, in degrees or radians, that the vector subtends relative to the positive x axis at any given instant. The wave cycle begins at 0 degrees, with the vector pointing along the x axis in the positive direction. The amplitude, or y component, at this time is zero. At an electrical angle of 90 degrees, the vector points along the y axis in a positive direction. This is the maximum positive point of the wave cycle. At 180 degrees, the vector points along the x axis in a negative direction, and the amplitude is zero once again. At 270 degrees, the vector points along the y axis in a negative direction, and this is the maximum negative amplitude of the wave.

While the y component of the electrical-angle vector gives the instantaneous amplitude, the x component gives the instantaneous rate of change of the signal amplitude. The speed of vector rotation, in radians per second, is called the angular frequency of the wave, and is equal to 2π (about 6.28) times the frequency in hertz. *See also* PHASE ANGLE, SINE WAVE.

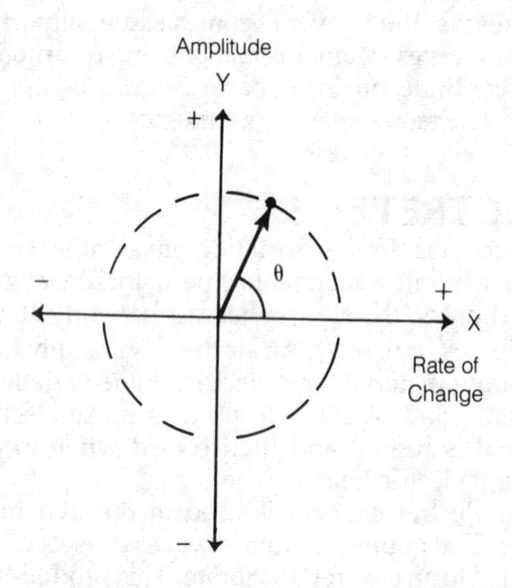

ELECTRICAL ANGLE: The electrical angle of a sine wave is shown as a vector in the Cartesian coordinate plane.

ELECTRICAL BANDSPREAD

In a radio receiver, electrical bandspread refers to bandspread obtained by adjustment of a separate inductor or capacitor, in parallel with the main-tuning inductor or capacitor. The main-tuning control in a shortwave receiver may cover several megahertz in one revolution. Bandspread is considered electrical if it is derived using a separate component; gearing down the main-tuning control constitutes mechanical bandspread.

An electrical-bandspread control can have any desired degree of resolution, consistent with the stability of the frequency of the receiver. Generally, one turn of a bandspread knob covers between 10 and 100 kHz of spectrum in a high-frequency receiver, with 25 kHz about average.

ELECTRICAL CONDUCTION

Electrical conduction is the movement of charge carriers through a material. Substances may be good conductors, semiconductors, or insulators. Even the best conductor offers some resistance, and even the best insulator conducts to a tiny extent. For all practical purposes, electrical conduction is of use only in good conductors and semiconductors.

Conduction normally occurs by the exchange of excess electrons between atoms, or by the exchange of electron deficiencies between atoms. The electron carries a negative charge, and thus the flow of electrons in a circuit is from negative to positive. An electron deficiency, called a hole, carries a positive charge, and the flow of holes is therefore from the positive to the negative.

Electrical conduction is measured in terms of current. The unit of current is the ampere (A); 1 ampere of current is the conduction of 1 coulomb of electrons per second past a given point in a circuit. *See also* AMPERE, COULOMB, CURRENT, ELECTRON, HOLE.

ELECTRICAL DISTANCE

Electrical distance is the distance, measured in wavelengths, required for an electromagnetic wave to travel between two points. The electrical distance depends on the frequency of the electromagnetic wave and on the velocity factor of the medium in which the wave travels.

In free space, an electrical wavelength at 1 MHz is about 300 meters. In general, in free space:

$$\lambda = 300/f$$

where λ is the length of a wavelength in meters and *f* is the frequency in megahertz. If *v* is the velocity factor, given as a fraction rather than a percent, then:

$$\lambda = 300\ v/f$$

Electrical distances are often specified for antenna radiators and transmission-line stubs. These distances are always given in wavelengths.

ELECTRICAL ENGINEERING

Electrical engineering is the study of the application of physical and mathematical skills to the solution of electrical problems. An electrical engineer is a trained professional in this field. College degrees in electrical engineering include the associate degree (ASEE), the bachelor's degree (BSEE), the master's degree (MSEE), and the doctor of engineering.

ELECTRICAL INTERLOCK

An electrical interlock is a switch designed to remove

power from a circuit when the enclosing cabinet is opened. Interlocks are especially important in high-voltage circuits because they protect service personnel from the possibility of electrical shock. (*See* ELECTRIC SHOCK).

An interlock switch is usually attached to the cabinet door or other opening in such a way that access cannot be gained without removing the high voltage from the interior circuits.

ELECTRICAL WAVELENGTH

In an antenna or transmission line carrying radio-frequency signals, the electrical wavelength is the distance between one part of the cycle and the next identical part. The electrical wavelength depends on the velocity factor of the antenna or transmission line, and also on the frequency of the signal.

In free space, where a signal has a frequency f in megahertz, the wavelength λ is given by the equations:

$$\lambda \text{ (meters)} = 300/f$$
$$\lambda \text{ (feet)} = 984/f$$

Along a thin conductor, the electrical wavelength at a given frequency is shorter than in free space. The velocity factor (*see* VELOCITY FACTOR) is about 95 percent, or 0.95, therefore:

$$\lambda \text{ (meters)} = 285/f$$
$$\lambda \text{ (feet)} = 935/f$$

along a wire conductor.

In general, in a conductor or transmission line with a velocity factor v, given as a fraction:

$$\lambda \text{ (meters)} = 300v/f$$
$$\lambda \text{ (feet)} = 984v/f$$

The electrical wavelength of a signal in a transmission line is always less than the wavelength in free space. *See also* ELECTRICAL DISTANCE.

ELECTRIC CHARGE

See CHARGE.

ELECTRIC CONSTANT

The electrical permittivity of free space is also called the electric constant, represented by ϵ_0. This value is about 8.854 picofarads per meter. Essentially all dielectric materials have an electrical permittivity greater than that of a vacuum.

The electric constant is the basis for the determination of dielectric constants for all insulating materials. *See also* DIELECTRIC CONSTANT, PERMITTIVITY.

ELECTRIC COUPLING

Electric coupling exists between, or among, any objects that show mutual capacitance. Electric coupling can be desirable, or it can be unwanted.

When two objects are electrically coupled, the charged particles on both objects exert a mutual attractive force or a mutual repulsive force. Like charges repel, and opposite charges attract. The plates of a capacitor provide a good illustration of the effects of electric coupling. A negative charge on one plate produces a positive charge on the other plate, by repelling the electrons in the other plate and literally pushing them from it. A positive charge on one plate produces a negative charge on the other, by attracting extra electrons to that plate. *See also* CHARGE, MAGNETIC COUPLING.

ELECTRIC FIELD

Any electrically charged body sets up an electric field in its vicinity. The electric field produces demonstrable effects on other charged objects. The electric field around a charged object is represented as lines of flux. A single charged object produces radial lines of flux as at A in the illustration. The direction of the field is considered to be from the positive pole outward, or inward toward the negative pole.

When two charged objects are brought near each other, their electric fields interact. If the charges are both positive or both negative, a repulsion occurs between their electric fields, as shown at B. If the charges are opposite, attraction takes place, as at C.

The intensity of an electric field in space is measured in volts per meter. Two opposite charges with a potential difference of 1 volt, separated by 1 meter, produce a field of 1 volt per meter. The greater the electric charge on an object, the greater the intensity of the electric field surrounding the object, and the greater the force that is exerted on other charged objects in the vicinity. *See also* CHARGE, DIPOLE, ELECTRIC COUPLING, ELECTRIC FLUX.

ELECTRIC FLUX

Electric flux is the presence of electric lines of flux in space or in a dielectric substance. The greater the intensity of the electric field, the more lines of flux present in a given space, and therefore the greater the electric flux density.

The lines of an electric field are imaginary with each line representing a certain amount of electric charge. Electric flux density is measured in coulombs per square meter.

The electric flux density decreases with increasing distance from a charged object, according to the law of inverse squares (*see* INVERSE-SQUARE LAW). If the distance from a charged object is doubled, the electric flux density is reduced to one-quarter its previous value. This relationship can be envisioned by imagining spheres centered on the charged particle shown at A in the illustration. No matter what the radius of the sphere, all of the electric flux around the charged particle must pass through the surface of the sphere. As the radius of the sphere is increased, the surface area of the sphere grows according to the square of the radius. Therefore, a region of the sphere with a certain area, such as 1 square meter,

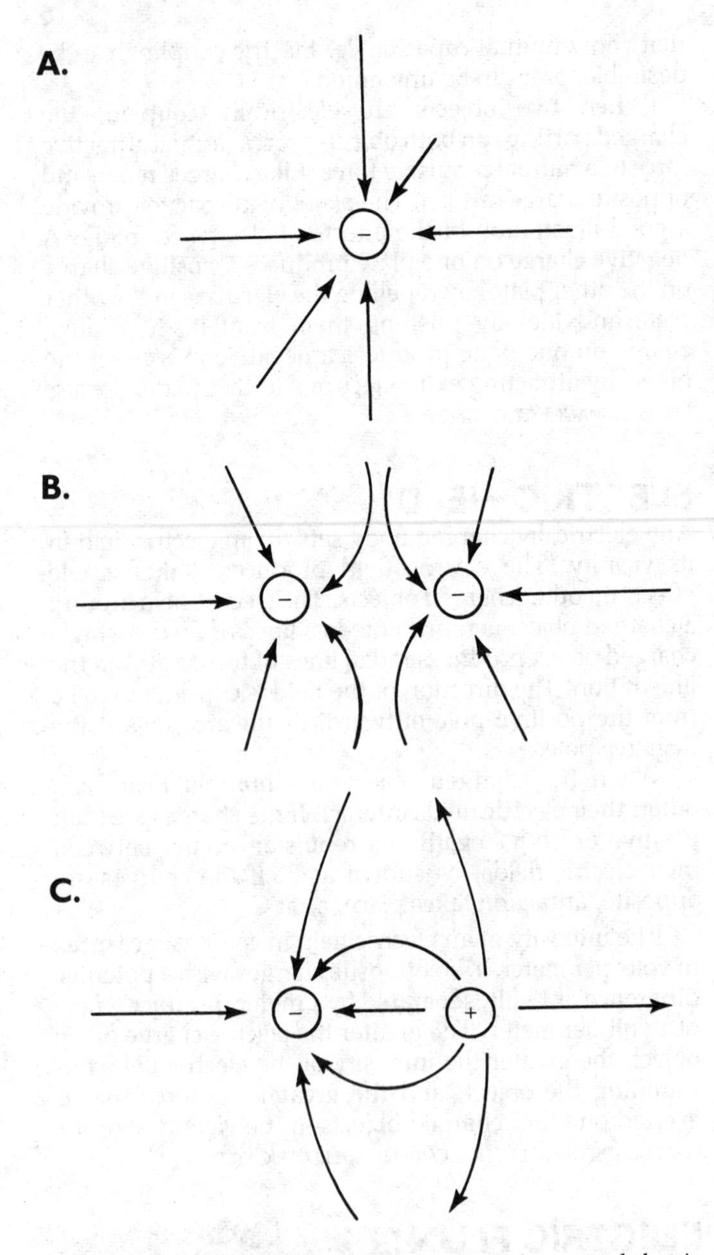

ELECTRIC FLUX: Examples of flux lines toward or around electric charges include: radial lines directed to a single charge (A), repulsive lines directed to two like charges (B), and attractive lines directed toward and away from two opposite electric charges (C).

becomes smaller in proportion to the square of the radius of the sphere; fewer electric lines of flux will cut through the region. *See also* CHARGE, ELECTRIC FIELD.

ELECTRICITY

Electricity is a discipline of science, and in particular, of physics. The study of electricity is concerned with the use and behavior of electric current, power, and voltage. This applies especially to power distribution and utilization.

Any electric current or voltage may be called electricity. For example, the 120 V electrical wires in homes, offices, or plants contain electricity. Electricity can be manifested as a potential difference between two objects, or as the flow of electrons (or other charged particles) past a certain point.

ELECTRIC SHOCK

Electric shock is the harmful flow of electrical current through living tissue. The effects of this current can be injurious, and may even cause death. All living cells contain some electrolytes and are capable of carrying electric current. Harmful currents are produced by varying amounts of voltage, depending on the resistance of the circuit through the living tissue.

In the human body, the most susceptible organ, as far as life-threatening effects are concerned, is the heart. A current of 100 to 300 mA for a short time can cause the heart muscle to stop its rhythmic beating and begin to twitch in an uncoordinated way. This is called heart fibrillation. An electric shock is much more dangerous when the heart is part of the circuit, as compared to when the heart is not in the circuit.

Under certain conditions, even a low voltage can produce a lethal electric shock. The 120 V utility lines produce an extreme hazard. However, precautions should be taken when working around or with all voltages greater than about 50 V, ac or dc.

ELECTROCHEMICAL EQUIVALENT

The electrochemical equivalent is a constant that expresses the rate at which a metal is deposited in the electroplating process (*see* ELECTROPLATING). The constant can be given in a variety of different ways. It is often specified in coulombs required to deposit one gram of the metal onto the electrode. The reciprocal of this quantity can also be specified.

The value of the electrochemical equivalent varies for different metals. The table lists the electrochemical equivalents in coulombs per gram for various metals. Electrochemical equivalents can be given for nonmetallic elements as well, although in electronics the value is of concern only for the metals used in electroplating. The larger the value of the electrochemical equivalent given in the table, the greater the amount of total electrical charge necessary to cause a given mass of the metal to be deposited.

ELECTROCHEMICAL EQUIVALENT: Electrochemical equivalents in coulombs per gram, for some common metals.

Element	Electrochemical Equivalent
Aluminum	1.1×10^4
Chromium	5.6×10^3
Copper	3.0×10^3
Gold	1.5×10^3
Iron	5.2×10^3
Lead	1.9×10^3
Magnesium	7.9×10^3
Nickel	3.3×10^3
Platinum	2.0×10^3
Silver	8.9×10^2
Tin	3.3×10^3
Zinc	3.0×10^3

ELECTRODE

An electrode is any object to which a voltage is deliberately applied. Electrodes are in vacuum tubes, electrolysis devices, cells, and batteries. An electrode is usually made of a conducting material such as metal or carbon.

In a device that has both a positive and a negative electrode, the positive electrode is called the anode, and the negative electrode is called the cathode. The cathode in a vacuum tube emits electrons, and the anode, sometimes called the plate, collects them.

All electrodes display properties of capacitance, conductance, and impedance. These characteristics depend on many factors, such as the size and shape of the electrodes, their mutual separation, and the medium between them. *See also* ELECTRODE CAPACITANCE, ELECTRODE CONDUCTANCE, ELECTRODE IMPEDANCE.

ELECTRODE CAPACITANCE

Electrode capacitance is the effective capacitance that an electrode shows relative to the ground, the surrounding environment, or other electrodes in a device. All electrodes have some capacitance, which depends on the surface area of the electrode, the distances and surface areas of nearby objects, and the medium separating the electrode from the nearby objects.

ELECTRODE CONDUCTANCE

The conductance of an electrode is the ease with which it emits or picks up electrons. Electrode conductance is measured in units called mhos or siemens, the standard units of conductance. These units are equivalent to the reciprocal of the ohmic resistance.

The conductance of an electrode or group of electrodes depends on many factors, including the physical size of the electrodes, the material from which they are made, the coating (if any) on electron-emitting electrodes, and the medium among the electrodes. In a vacuum tube, the interelectrode conductance is the ease with which electrons flow from the cathode to the plate. This depends on the grid bias and the characteristics of the tube itself.

ELECTRODE DARK CURRENT

In a photomultiplier or television camera tube, the electrode dark current is the level of current that flows in the absence of light input to the tube. Dark current is important as a measure of the noise generated by a television camera tube; the smaller the dark current, the better.

The electrode dark current in most photomultiplier tubes is very small. The anode dark current originates partially from the photocathode and partly from the other components within the tube. Some of the dark current also originates because of thermal disturbances. The electrode dark current results in what is called dark noise, as the electrons strike the anode. Dark noise limits the sensitivity of a television picture camera tube. *See also* PHOTOMULTIPLIER TUBE.

ELECTRODE IMPEDANCE

Electrode impedance is the resistance encountered by a current as it flows through an electrode. In the case of direct current, electrode impedance is simply the reciprocal of the electrode conductance (*see* ELECTRODE CONDUCTANCE). It is measured in ohms.

For alternating current, the electrode impedance consists of both resistive and reactive components. The electrode impedance is determined in the same manner as ordinary impedance, by adding, vectorially, the resistance and the reactance (*see* IMPEDANCE, REACTANCE, RESISTANCE).

Electrode impedance in a particular situation depends on several factors, including the surface area of the electrode, the material from which it is made, the surrounding medium, and, in a vacuum tube, the bias and the grid voltage. The frequency of the applied alternating-current signal also affects the electrode impedance because the capacitive and inductive reactances change as the frequency changes. *See also* ELECTRODE CAPACITANCE.

ELECTRODYNAMICS

Electrodynamics is a branch of physics and electricity concerned with the interaction between electrical and mechanical effects. The attraction of opposite charges and the repulsion of like charges are examples of electrodynamic phenomena. The forces and the electric charges are directly related. The forces of magnetism are also electrodynamic forces, since a magnetic field can be generated by a current in a wire.

Electric motors and generators are examples of electrodynamic devices. They operate with magnetic forces, generated by electric currents or resulting in electric currents.

ELECTRODYNAMOMETER

An electrodynamometer, also called a dynamometer, is an analog metering instrument. Its principle of operation is similar to that of the D'Arsonval meter movement, but the magnetic field is supplied by an electromagnet, rather than by a permanent magnet. The current to be measured supplies the field current for the electromagnet.

The drawing illustrates the operation of the electrodynamometer. Two stationary coils are connected in series with a moving coil. The moving coil is attached to the indicating needle and is mounted on bearings. The moving coil is held at the meter zero position, under conditions of no current, by a set of springs.

When a current is applied to the coils, magnetic fields are produced around both the stationary coils and the movable coil. The resulting magnetic forces cause the movable coil to rotate on its bearings. The larger the current, the more the coil is allowed to rotate by the springs in accordance with the motor rule..

Electrodynamometers can be used in any situation in which a D'Arsonval meter is used. The D'Arsonval meter

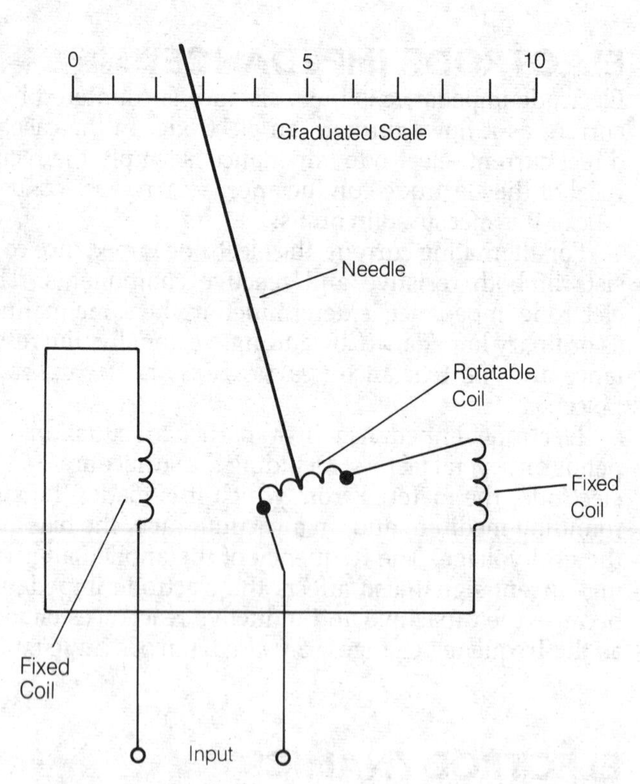

ELECTRODYNAMOMETER: Schematic showing the interaction between fixed and moving magnetic coils of an electrodynamometer to move the needle across the scale.

is generally more sensitive when it is necessary to measure very small currents. *See also* D'ARSONVAL MOVEMENT.

ELECTROKINETICS

Electrokinetics is concerned with the behavior of moving charged particles and the behavior of moving materials in electrical fields. Examples of electrokinetic effects are the electrolysis of salt water, electrophoresis, and the electrical operation of a battery. *See also* CAPACITOR, ELECTROLYSIS, ELECTROPHORESIS.

ELECTROLUMINESCENT DISPLAY

Electroluminescent (EL) displays generate light when an electric field is applied to an electroluminescent phosphor. Most electroluminescent materials require activators. Activators are impurities in the material that determine the characteristics of the emitted radiation. A typical electroluminescent phosphor is zinc sulfide doped with manganese. The illustration shows a cutaway view of an ac (alternating current) thin-film electroluminescent panel. The phosphor glows yellow orange.

The polycrystalline EL phosphor compound is sandwiched between two electrodes with dielectric layers between the electrodes and the phosphor. The dielectric layers might be silicon dioxide. The front electrodes deposited on the clear glass substrate are transparent conducting films, typically indium-tin-oxide; the back electrode is aluminum. If an ac voltage of about 290 V is applied between the two electrodes, light is emitted through the transparent electrodes and glass substrate.

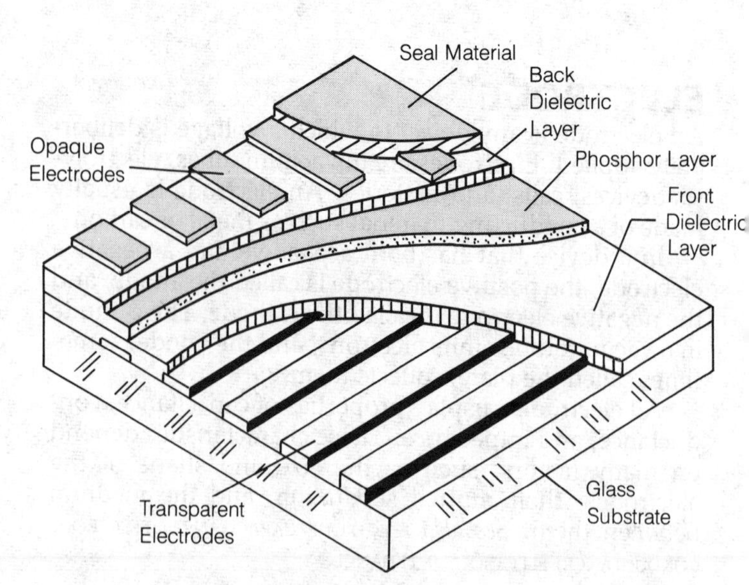

ELECTROLUMINESCENT DISPLAY: Cutaway view of a thin-film electroluminescent display panel with a manganese-doped zinc sulfide phosphor luminescent layer.

EL displays can be segmented numerical displays or organized as X-Y matrices to display dot-matrix characters, the principal use of the displays today. In dot-matrix displays the transparent electrode is the column electrode and the opaque electrode is the row electrode. A dot or pixel is the smallest independently addressable display area.

Electroluminescent panels are made by evaporating the transparent electrodes onto glass through screens to form columns. The first dielectric or insulator layer is deposited, and this layer is followed by the phosphor layer. Then a dielectric layer is deposited, and the opaque aluminum electrodes are deposited. The electroluminescent material can be a powder or a thin film (4000 angstroms thick).

The original EL displays were made with phosphor powder and operated in an ac mode. Work on dc (direct-current) powder EL displays has been pursued for years, and prototypes have been developed. Emphasis is on increasing the generally shorter life of these displays. Most applications use ac thin-film EL because of its demonstrated high luminescence and long life.

Because an insulating material is used between the transparent front column electrodes and the phosphor, EL display elements behave electrically like capacitors. As a result, the scanning rates and size of an EL panel are limited by the effective RC (resistance-capacitance) time constant set by the electrode resistance and the panel capacitance.

EL modules are available to replace cathode-ray tube displays in laptop portable personal computers. These displays provide up to 512 by 256 lines or (130 k pixels) and measure approximately 10 by 5 inches. EL modules are display panels with all the necessary drive and interface electronics on a backup PC (printed circuit) board. The electronics converts a typically serial image signal with synchronization signals to an image on the display panel. EL modules are now competitive with liquid crystal and ac plasma displays, but each display technology has its strengths and weaknesses. Users compare power consumption, contrast ratio, viewing angle, luminance, color, and price.

ELECTROLYSIS

Electrolysis is a process in which a compound is separated into its constituent elements through the application of an electric current. Electrolysis can occur in a solid, a liquid, or a gas. Most commonly, electrolysis occurs in liquids or semi-solid materials.

All compounds consist of elements, some of which have positive valence numbers, and some of which have negative valence numbers (*see* VALENCE NUMBER). Atoms with positive valence numbers are attracted to negatively charged electrodes, and atoms with negative valence numbers are attracted to positively charged electrodes.

Electrolysis can be used for many different purposes. Electroplating is an example of electrolysis (*see* ELECTROPLATING). A simple experiment can be performed to illustrate electrolysis in a liquid. Place two electrodes in a salt-water solution, and apply a potential difference of 12 V dc between the electrodes.

Bubbles will appear at the electrodes. Hydrogen gas will be formed at the negative electrode, and oxygen will be formed at the positive electrode. Sodium from the salt will accumulate at the negative electrode. It reacts with the water immediately to form hydrogen gas. Chlorine gas will be formed, to a limited extent, at the positive electrode, along with the oxygen from the water.

Electrolysis results not only in the deposition of elements because of an applied electric current; electrolysis can also result in the generation of electricity when the appropriate electrodes are placed into an electrolyte. *See* BATTERY, ELECTROLYTE.

ELECTROLYTE

An electrolyte is any substance which, in a solution carrying an electric current, will separate into its constituent elements. Common table salt (sodium chloride) is a good example of an electrolyte (*see* ELECTROLYSIS). So is common baking soda (sodium bicarbonate). Most acids and bases are electrolytes. Electrolyte solutions always conduct electricity.

Electrolyte substances are important in the manufacture of electric cells and batteries. The action of an electrolyte solution on a pair of electrodes results in the generation of an electric potential between the electrodes. Chemical energy is thus transformed into electrical energy. *See also* BATTERY.

ELECTROLYTIC CAPACITOR

See CAPACITOR.

ELECTROLYTIC CELL

An electrolytic cell is any device containing an electrolytic substance and at least two electrodes. All electrochemical cells fall into this category, as do electrolytic capacitors and electrolytic resistors.

Electrolytic cells may be used for either of two purposes: to generate direct currents, or to operate from or with direct currents. *See also* BATTERY, ELECTROLYTE, CAPACITOR.

ELECTROLYTIC CONDUCTION

Electrolytic conduction is the passage of current through an electrolytic solution. Any electrolytic solution tends to ionize when a voltage is applied to electrodes immersed in it. Negative ions are attracted to the positive electrode, and they give up their excess electrons there. Positive ions are attracted to the negative electrode, and at that point, they acquire electrons. If the ion is a gas, it can be seen bubbling out of the electrolyte solution. If the ion is a solid, a noticeable accumulation may eventually form on the electrode. The ionization will continue for as long as current is applied, or as long as the ions remain available in the solution.

The flow of current in an electrolyte is different from the movement of electrons in a metal wire. The current is, nevertheless, measurable in terms of unit charges per second, and the current can be evaluated by placing an ammeter in series with the electrolyte device. *See also* ELECTROLYSIS, ELECTROLYTE.

ELECTROMAGNET

An electromagnet is a temporary magnet that makes use of the effects of electric currents to produce a magnetic field. Electromagnets display the same properties, when in operation, as ordinary permanent magnets. Some components and instruments use electromagnets in

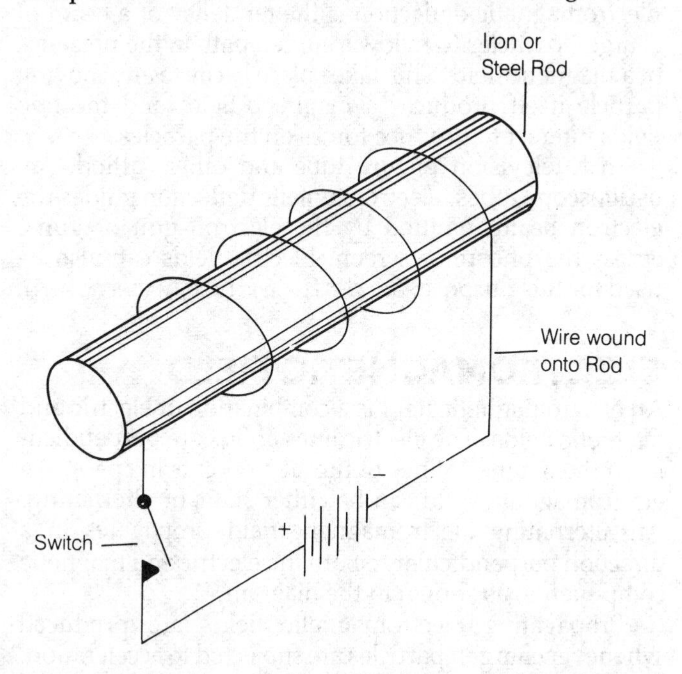

ELECTROMAGNET: A coil of wire wrapped around a magnetic core is an electromagnet. The core can be bent into a U shape.

place of permanent magnets (*see* ELECTRODYNAMOMETER, MICROPHONE, RELAY, SPEAKER).

An electromagnet is constructed by placing an iron or steel core inside a solenoidal coil, as shown in the figure. When a direct current is passed through the coil winding, a magnetic field is produced, and the rod behaves exactly like a permanent magnet until the current is shut off. If an alternating current is applied to the coil, an alternating magnetic field is produced. This will attract magnetic metals, but will not develop a constant north or south magnetic pole in the iron or steel core within the coil. It will produce a vibrating action.

The intensity of the magnetic field generated by an electromagnet depends on the number of turns in the coil, the metal in the rod, and the current through the coil windings. *See also* MAGNETIC FIELD.

ELECTROMAGNETIC CONSTANT

In a vacuum, the speed of propagation of electric, electromagnetic, and magnetic fields is a constant of 186,283 statute miles per second, or 299,793 kilometers per second, rounded off for most purposes to 186,000 miles per second or 3×10^8 meters per second. This constant is sometimes abbreviated by the lowercase letter *c*.

The electromagnetic constant is always the same in a vacuum, no matter how or where it is measured. A passenger on a rapidly moving space ship would observe the same value as a person standing on the earth. *See also* ELECTRIC FIELD, ELECTROMAGNETIC FIELD, MAGNETIC FIELD.

ELECTROMAGNETIC DEFLECTION

Electromagnetic deflection is the tendency of a beam of charged particles to follow a curved path in the presence of a magnetic field. This takes place because any moving particle itself produces a magnetic field, and the two fields interact to produce forces on the particles.

In a television picture tube and other cathode-ray oscilloscope tubes, electromagnetic deflection guides the electron beam, emitted by the electron gun or guns, across the phosphor screen. Electric fields can also be used for this purpose. *See also* ELECTRIC FIELD, MAGNETIC FIELD.

ELECTROMAGNETIC FIELD

An electromagnetic field is a combination of electric and magnetic fields. The electric lines of flux are perpendicular to the magnetic lines of flux at all points in space. An electromagnetic field can be either static or alternating. An alternating electromagnetic field propagates in a direction perpendicular to both the electric and magnetic components, as shown in the diagram.

Propagating electromagnetic fields are produced whenever charged particles are subjected to acceleration. The most common example of acceleration of charged particles is the movement of electrons in a conductor carrying an alternating current. The constantly changing

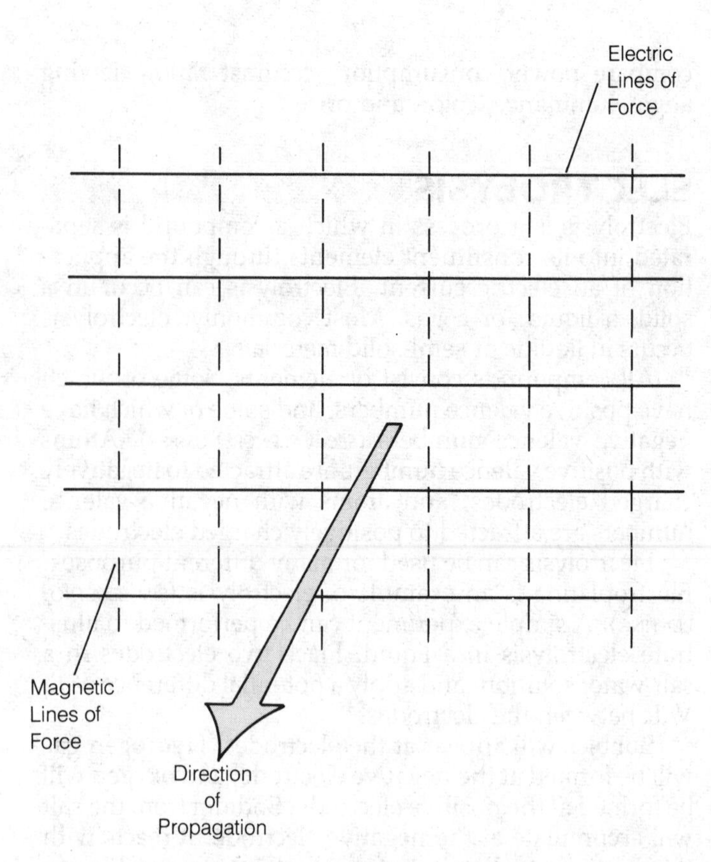

ELECTROMAGNETIC FIELD: The electric and magnetic lines of flux are perpendicular to each other in an electromagnetic field, and the direction of propagation is perpendicular to both fields.

velocity of the electrons causes the generation of a changing magnetic field, which results in an electromagnetic field that propagates outward from the conductor.

The frequency of an alternating electromagnetic field is the same as the frequency of the alternating current in the conductor. The frequency can be as low as a fraction of a cycle per second, or hertz; it can be as high as many trillions or even quadrillions of hertz. Visible light is a form of electromagnetic-field disturbance. So are infrared rays, ultraviolet rays, X rays, and gamma rays. Radio waves are an electromagnetic phenomenon.

All propagating electromagnetic fields display properties of wavelength λ as well as frequency f. In a vacuum, where the speed of electromagnetic propagation is about 3×10^8 meters per second, the wavelength in meters is given by:

$$\lambda = 300/f$$

where f is specifed in megahertz. The higher the frequency, the shorter the wavelength. *See also* ELECTROMAGNETIC RADIATION, ELECTROMAGNETIC SPECTRUM.

ELECTROMAGNETIC FOCUSING

Electromagnetic focusing is a form of electromagnetic deflection used in cathode-ray tubes (*see* ELECTROMAGNETIC DEFLECTION). A stream of electrons is brought to a sharp focus at the phosphor surface inside the tube with a set of coils that carry electric current. The coils produce magnetic fields, which direct the electrons so that they impact on the phosphor screen at the same point. *See also* CATHODE RAY TUBE.

ELECTROMAGNETIC INDUCTION

When an alternating current flows in a conductor, electron movement will occur at the same frequency in nearby conductors that are not connected to the current-carrying wire. This phenomenon is known as electromagnetic induction, and is a result of the effects of the electromagnetic field set up by a conductor having an alternating current. The varying electric and magnetic fields produced by acceleration of charged particles cause forces to be exerted on nearby charged particles.

All radio communication is possible because of electromagnetic induction acting over long distances. The effect, originally observed over small distances, actually occurs over an unlimited distance. The effect travels through space with the speed of light, or about 186,000 miles per second. This disturbance takes the form of mutually perpendicular electric and magnetic fields. *See also* ELECTROMAGNETIC FIELD.

ELECTROMAGNETIC INTERFERENCE

An electromagnetic field that causes interference is called electromagnetic interference (EMI). A very common form of electromagnetic interference is the alternating-current hum picked up by inadequately shielded microphone or interconnecting cables in audio-amplifier systems. The interfering field is the 60 Hz field produced by utility wires. The cure for this form of EMI is the proper shielding of the appropriate wires and, if necessary, bypassing the leads or the insertion of parallel choke coils.

Another common form of electromagnetic interference is seen in home entertainment equipment such as high-fidelity stereo systems. The electromagnetic fields from nearby radio broadcast stations, or other transmitting stations, are intercepted by dynamic pickups or wire leads, and are rectified by the amplifier circuits. If the nearby transmitter is amplitude-modulated, the modulating signal will actually be heard through the speakers of the system. If the signal is unmodulated, a buzz or change in volume will be observed. As with 60 Hz electromagnetic interference, the shielding of wire cables is helpful. Bypassing with small capacitors may be necessary as well; small radio-frequency chokes can be installed in series with audio leads to provide improvement.

Electromagnetic interference can be seen on the screens of home television sets in the form of moving bars or streaks as a result of interference from household appliances with motors, microwave ovens, and computers without line filters. *See also* LINE FILTER.

ELECTROMAGNETIC PULSE

An electromagnetic pulse (EMP) is a sudden burst of electromagnetic energy, caused by a single, abrupt change in the speed or position of a group of charged particles. An electromagnetic pulse does not generally have a well-defined frequency or wavelength, but can exist over the entire electromagnetic spectrum. This can include the radio wavelengths, infrared, visible light, ultraviolet, X rays, and even gamma rays. Electromagnetic pulses can be generated by arcing. On a radio receiver they sound like popping or static bursts. Lightning produces an electromagnetic pulse with a bandwidth extending from the very low frequency range to the ultraviolet range.

An electromagnetic pulse can contain a large amount of power for a short time. Lightning discharges have been known to induce current and voltage spikes in nearby electrical conductors of such magnitude that the equipment is destroyed and fires are started.

The detonation of an atomic bomb creates a strong electromagnetic pulse. The explosion of a multimegaton bomb at a very high altitude, while not creating a devastating shock wave or heat blast at the surface of the earth, can generate a damaging electromagnetic pulse over an area of many thousands of square miles (see illustration). The resulting electromagnetic pulses can induce damaging voltages and currents in radio antennas, telephone wires, and power transmission lines over a vast geographic area. *See also* ELECTROMAGNETIC FIELD.

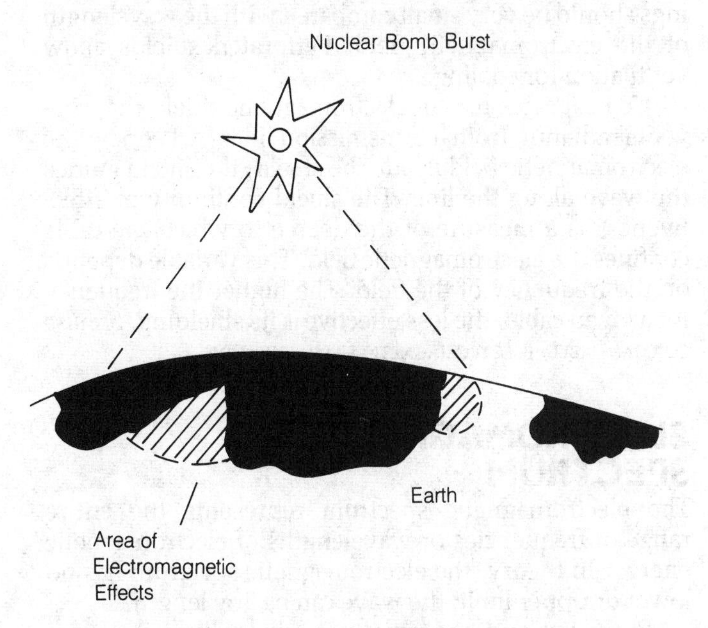

ELECTROMAGNETIC PULSE: A nuclear weapon exploding at high altitude generates electromagnetic pulses that have damaging effects over long distances.

ELECTROMAGNETIC RADIATION

Electromagnetic radiation is the propagation of an electromagnetic field through space (*see* ELECTROMAGNETIC

FIELD). All forms of radiant energy—such as infrared, visible light, ultraviolet, and X rays—are forms of electromagnetic radiation.

Radio waves may be as long as thousands of meters, or as short as a few millimeters. Other forms of electromagnetic radiation have very short wavelengths. The shortest known gamma rays measure less than 0.006 nm. This is so small that even the most powerful microscope could not resolve it. Theoretically, no limits exist for how long or how short an electromagnetic wave can be.

In a vacuum, electromagnetic radiation travels at the speed of light, or about 186,000 miles per second or 3×10^8 meters per second. In media other than a perfect vacuum, the speed of propagation is slower. The precise velocity depends on the substance and the wavelength of the electromagnetic radiation. *See also* ELECTROMAGNETIC CONSTANT, ELECTROMAGNETIC SPECTRUM.

ELECTROMAGNETIC SHIELDING

Electromagnetic shielding is a means of keeping an electromagnetic field from entering or leaving an area. The most common form of electromagnetic shielding consists of a grounded enclosure, made of sheet metal or perforated metal. A screen may also be used.

Electromagnetic shields are important in many different kinds of electronic systems. Shielding prevents unwanted electromagnetic coupling between circuits. If a shielded enclosure is not made of solid metal, the openings should be very small compared with the wavelength of the electromagnetic field. Perforated shields allow ventilation for cooling.

Coaxial cable has an electromagnetic shield that prevents radiation from a transmission line. By keeping the electromagnetic field inside the shield, the shield guides the wave along the line. The shield continuity or effectiveness is a measure of the degree to which the cable confines the electromagnetic field. This variable depends on the frequency of the field. The higher the frequency for a given cable, the less effective is its shielding. *See also* COAXIAL CABLE, ELECTROMAGNETIC FIELD, SHIELDING.

ELECTROMAGNETIC SPECTRUM

The electromagnetic spectrum represents the entire range of frequencies or wavelengths of electromagnetic energy. In theory, the electromagnetic spectrum has no lower or upper limit; the wave can be any length.

Physicists and engineers use a logarithmic scale, as shown in the figure, to illustrate the electromagnetic spectrum. The wavelength is shown in meters in this example. Some of the more familiar landmarks in the electromagnetic spectrum are indicated. At shorter wavelengths, more energy is contained in a single photon (*see* PHOTON).

Electromagnetic radiation has properties that vary with wavelength. Some radio waves are bent or reflected by the ionized layers in the upper atmosphere of the earth, while others are not affected. In general, radio waves shorter than about 2 meters will pass through the ionosphere into space. The air itself is practically opaque at some electromagnetic wavelengths. Some infrared cannot penetrate the atmosphere, and the short ultraviolet rays, X rays, and gamma rays are also blocked.

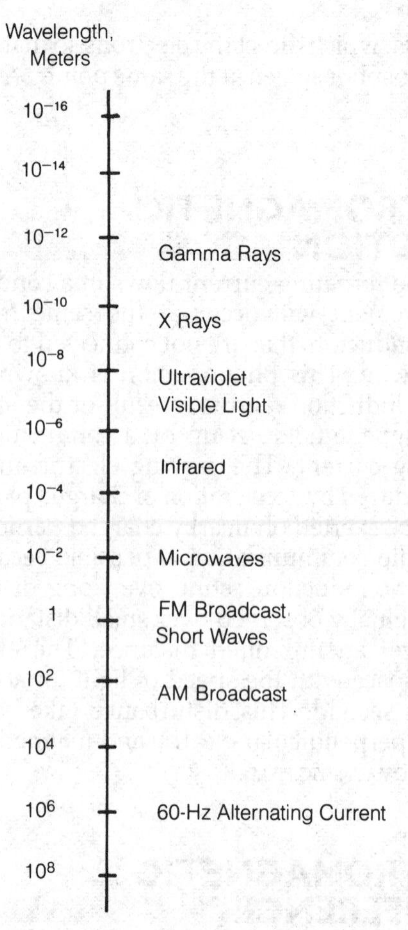

ELECTROMAGNETIC SPECTRUM: The wavelengths of the principal energy sources shown on a logarithmic scale.

The shortest known gamma rays have a wavelength shorter than 0.006 nanometers (nm). X rays are 0.006 to 5 nm; ultraviolet, 5 nm to 0.4 μm; visible light, 0.4 to 0.7 μm; infrared, 0.7 μm to 1 mm. The shortest radio-frequency waves are 1 mm. *See also* ELECTROMAGNETIC FIELD, ELECTROMAGNETIC RADIATION, GAMMA RAY, INFRARED, LIGHT, RADIO WAVE, ULTRAVIOLET RADIATION, X RAY.

ELECTROMAGNETISM

When a charged particle or a stream of charged particles is set in motion, a magnetic field is produced. The lines of magnetic flux occur in directions perpendicular to the motion of the charged particles. This effect is called electromagnetism.

Electromagnetism is well illustrated by the flow of current in a straight wire conductor. The magnetic lines of flux lie in a plane or planes orthogonal to the wire. They appear as concentric circles around the wire axis, as illustrated in the figure. The intensity of the magnetic field, or the number of flux lines in a given amount of area, is proportional to the intensity of the current in the wire.

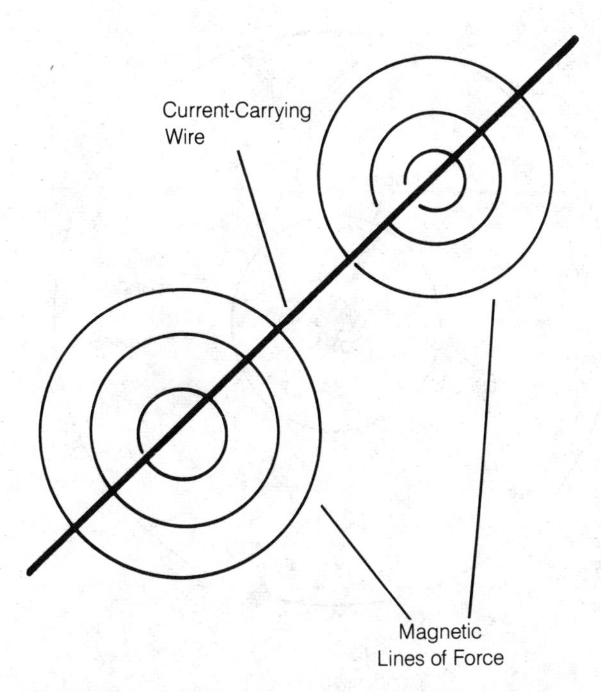

ELECTROMAGNETISM: A current-carrying wire is surrounded by magnetic flux.

If an alternating current is flowing in the wire, the intensity of the magnetic field varies, and these fluctuations result in a spatial charge differential or electric field. A changing, or oscillating, magnetic field therefore contains an electric-field component as well. The electric lines of flux occur parallel to the wire, or perpendicular to the magnetic lines of flux. The combination of the two fields has an ability to propagate long distances through space and is called an electromagnetic field. *See also* ELECTROMAGNET, ELECTROMAGNETIC FIELD, ELECTROMAGNETIC RADIATION.

ELECTROMECHANICAL TRANSDUCER

Any device that converts electrical energy into mechanical energy, or vice versa, is called an electromechanical transducer. Examples of transducers include motors and generators. The transformation is the result of electromagnetic effects: a moving charge creates magnetic forces, or a moving object with a magnetic field creates an electric current (*see* ELECTROMAGNETISM).

Devices that convert sound, or acoustic energy into electric energy are electromechanical transducers. Speakers, earphones, and microphones are examples. They may also be called electroacoustic transducers. The mechanical movement of sound can be converted into electrical energy, or vice versa, in a number of different ways, including electromagnetic, electrostatic, and piezoelectric.

ELECTROMETER

An electrometer is a highly sensitive instrument includinging a meter and an amplifier and used for measuring small voltages. The input resistance is extremely high, and may be as large as several quadrillion ohms (a quadrillion is 10^{15}, or a million billion). The electrometer, because of its high input resistance, draws essentially no current from the source being measured. This allows the electrometer to be used for measuring electrostatic charges.

An electrometer uses a special vacuum tube called an electrometer amplifier or electrometer tube. The interelectrode resistance is very high, and the amplifier circuit is designed in a manner similar to that of a Class A audio or radio-frequency amplifier. The noise generation is minimal; this is important for obtaining the maximum possible sensitivity. A small change in the input current, in some cases as small as 1 picoampere (a millionth of a millionth of an ampere), can be detected.

An electrometer can also be used to measure electrostatic voltages in which the amount of charge (quantity) is actually small.

ELECTROMOTIVE FORCE

Electromotive force is the force that causes movement of electrons in a conductor. The greater the electromotive force, the greater the tendency of electrons to move. Other charge carriers can also be moved by electromotive force; in some types of semiconductor, the deficiency of an electron in an atom can be responsible for conduction. This deficiency is called a hole; electromotive force can move holes as well as electrons. *See also* ELECTRON, HOLE, VOLTAGE.

ELECTRON

An electron is a subatomic particle that carries a unit negative electric charge. Electrons may be free in space, or they may be under the influence of the nuclei of atoms. The mass of an electron is 9.11×10^{-31} kilogram or 9.11×10^{-27} gram. A single electron carries a charge of 1.59×10^{-19} coulomb (*see* COULOMB).

Electrons are believed to move around the nuclei of atoms, in such a way that their average position is represented by a sphere at a discrete distance from the nucleus. The force of electrostatic repulsion prevents gravity from causing atoms to fall into one another. Electrons are responsible for this repulsion.

In an electric conductor, the flow of electrons occurs as the passage of particles from one atom to another. It does not take place as a simple flow, like water in a hose. Some atoms have a tendency to pick up extra electrons, and some have a tendency to lose electrons. Sometimes a flow of current occurs because of a deficiency of electrons, and sometimes it occurs because of an excess of electrons among the constituent atoms of a particular substance. The deficiency of an electron in an atom is called a hole. Hole conduction is important in the operation of semiconductor diodes and transistors. *See also* HOLE.

ELECTRON AVALANCHE

See AVALANCHE.

ELECTRON-COUPLED OSCILLATOR

See OSCILLATOR CIRCUITS.

ELECTRON EMISSION

Electron emission occurs whenever an object gives off electrons to the surrounding medium. This occurs in all vacuum tubes, and also in incandescent and fluorescent light bulbs. It also takes place in a variety of other devices. Some materials such as barium oxide or strontium oxide are good electron emitters. Electrodes are coated with these chemicals to enhance their ability to give off electrons when heated and subjected to a negative voltage.

Electron emission can occur because of thermionic effects, where the temperature of a negatively charged electrode is raised to the point that electrons break free from the forces that normally hold them to their constituent atoms. The cathodes of most vacuum tubes act in this way. They may be directly heated or indirectly heated (*See* CATHODE.)

Electron emission can be secondary in nature, resulting from the impact of high-speed electrons against a metal surface. This takes place at the plate of a vacuum tube, and it can be detrimental. (The suppressor grid of a pentode tube keeps the secondary electrons from escaping from the plate.) A dynode operates on the principle of secondary emission to amplify an electron beam (*see* DYNODE).

If an electric charge is applied to the surface of an object, and the voltage is great enough, electrons will be thrown off. Also, if a nearby positive voltage is sufficiently strong, electrons will be pulled from an object. This is called field emission. Cold-cathode tubes operate on this principle (*see* COLD CATHODE).

In a photoelectric tube, light striking a barrier called a photocathode results in the emission of electrons. This is called photoemission. The intensity of photoemission depends on the brightness of the light, and also on its wavelength (*see* PHOTOCATHODE, PHOTOMULTIPLIER TUBE).

Electrons emitted from an electrode are always the same elementary particles. They carry a unit charge and have the same mass, no matter what the cause of their emission. *See also* ELECTRON.

ELECTRON-HOLE PAIR

In an atom of a semiconductor material, the conduction band and the valence band are separated by an energy gap. An electron in the conduction band is free to move to another atom or escape entirely. But an electron in the valence band is held to the atom by the positive charge of the protons in the nucleus (*see* VALENCE BAND). An electron may move from the valence band to the conduction band of an atom if it received enough energy. This leaves an electron vacancy, called a hole, in the valence band of the atom, as shown in the illustration. The electron and the hole form what is called an electron-hole pair. *See also* ELECTRON, HOLE, N-TYPE SEMICONDUCTOR, P-TYPE SEMICONDUCTOR.

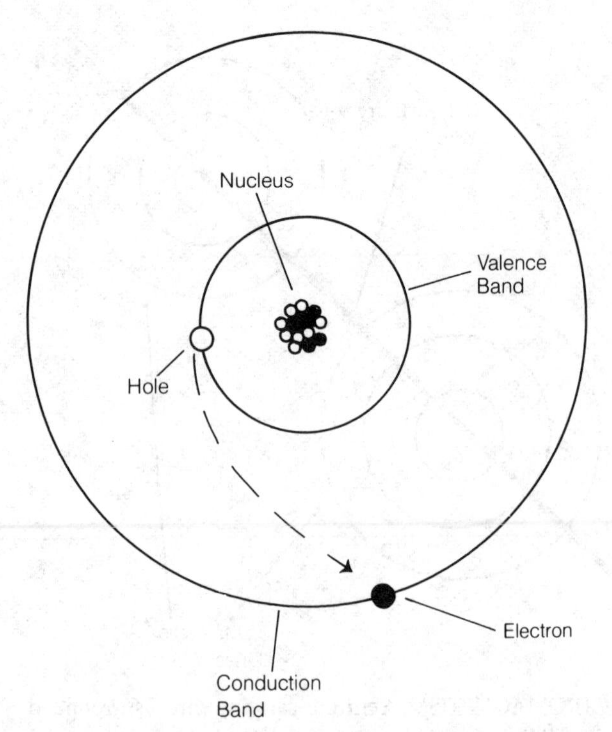

ELECTRON-HOLE PAIR: An electron vacating a valance band leaves a hole.

ELECTRONIC CALCULATOR

An electronic calculator is a machine with transistors and integrated circuits that performs arithmetic operations

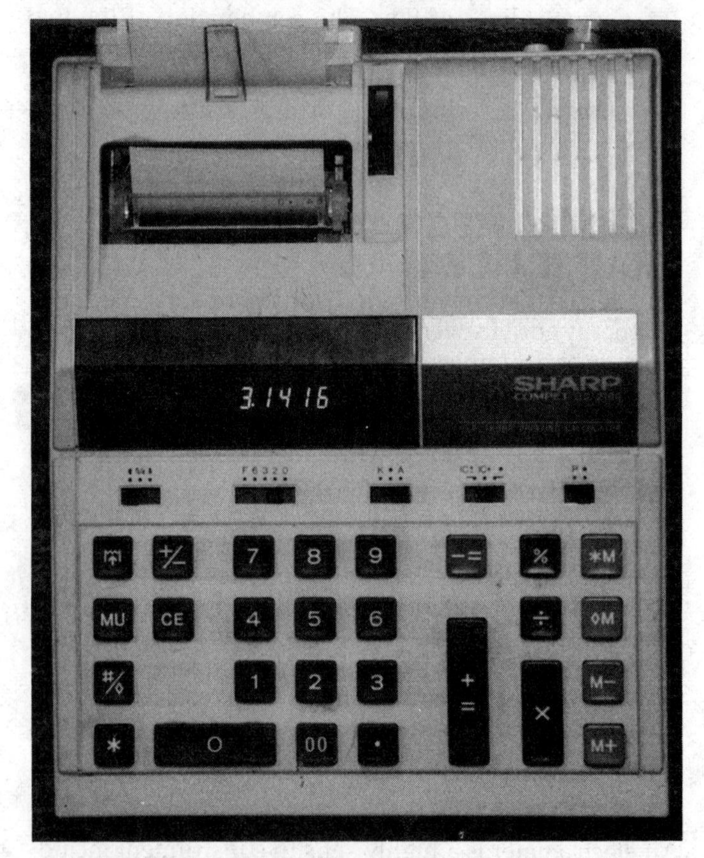

ELECTRONIC CALCULATOR: An electronic calculator that includes a printer for producing printed results.

such as addition, subtraction, multiplication, and division. Some electronic calculators can also perform more advanced operations such as extraction of square roots, logarithms, and trigonometric functions.

Electronic calculators are available for home use, business use, and scientific applications. The desktop calculator shown in the photograph is intended primarily for business and financial use. It contains a small printer that records, if desired, all of the operations on a sheet of narrow paper, in much the same way as a cash register.

ELECTRONIC CLOCK

An electronic clock is a timepiece of any size whose activity is governed by electronic circuitry. Electronic clocks have almost completely replaced the older mechanical and electrical clocks. They are more rugged and accurate than the clocks of the past. They can be made very small, and may be included in calculators or even in pens (see photograph).

A mechanical clock with a motor governed by an electronic oscillator is sometimes called an electronic clock. The oscillator uses a frequency standard such as a tuning fork or quartz crystal. Some electronic clocks have a built-in radio receiver tuned to one of the time-standard frequencies such as the radio station WWV/WWVH frequencies of 5, 10, 15, or 20 MHz and others. A tone, transmitted hourly by such stations, automatically resets these clocks (*see* WWV/WWVH).

A timing circuit, used in a digital system, is sometimes called an electronic clock. This circuit generates pulses at precise intervals to govern the operation of computers and other digital equipment.

ELECTRONIC CLOCK: An electronic clock with a liquid crystal digital display in a pen.

ELECTRONIC CONTROL

Electronic control is the operation or adjustment of a system by electronic, rather than mechanical, means. Electronic controls are now used widely in automobiles to control ignition, fuel consumption, exhaust emission, and braking. They are also used in electric garage door openers, washing machines, dishwashers, handheld and bench power tools, and microwave ovens, to mention but a few applications.

ELECTRONIC COUNTER

See COUNTER.

ELECTRONIC ENGINEERING

See ELECTRICAL ENGINEERING.

ELECTRONIC GAME

With the availability of low-cost computers and solid-state electronic circuits, electronic games have become commonplace in recent years. An electronic game uses electronic circuits and devices to create certain situations which may then be controlled by one or more operators. A video monitor, resembling a television receiver, may be used in an electronic game. Some electronic games can be connected directly to the antenna terminals of a home television receiver.

In the late 1970s and early 1980s, rapid advances in computer technology resulted in the proliferation of electronic games in amusement centers and bars, as well as in the home. Some electronic games have educational value and allow the operators to actively participate in a learning process. Other electronic games are intended primarily for entertainment. The variety of different situations obtainable with computer games is almost unlimited. Computer-game cassettes and diskettes are sold in electronics and department stores. *See also* COMPUTER.

ELECTRONIC INDUSTRIES ASSOCIATION

The Electronic Industries Association, or EIA, is an agency in the United States that sets standards for electronic components and test procedures. The EIA is also responsible for the setting of performance standards for various types of electronic equipment. Standardization is advantageous both to the manufacturers (because it broadens their markets) and to the users of electronic equipment (because it makes it easier for them to get replacement parts and peripheral devices).

The Electronic Industries Association works with other national and international standard agencies, including American National Standard Institute (ANSI). The International Electrotechnical Commission (IEC), headquartered in Geneva, Switzerland, sets electronic standards worldwide.

ELECTRONIC MUSIC

Audio-frequency electronic oscillators are capable of producing complex as well as simple waveforms. All musical instruments produce rather complex, but identifiable and reproducible, waveforms. Recent technology has made it possible to duplicate the sound of any musical instrument by means of audio-frequency oscillators and transducers. A system called a Moog synthesizer can produce many different kinds of sound and plays electronic music.

Electronically amplified or modified music is commonplace today. Even classical music is modified electronically when it is reproduced through a high-fidelity system. Popular music bands typically use some electronic means to create their sounds.

Very complicated musical themes can be created with a synthesizer and a computer. Many sound tracks can be recorded, one on top of the other, reproducing the sound of a whole band or orchestra. Some experimenters are working with computers to compose music. *See also* MOOG SYNTHESIZER.

ELECTRONIC TIMER

See ELECTRONIC CLOCK.

ELECTRONIC WARFARE

Electronic warfare uses electronic equipment and techniques to aid in the successful use of conventional weapons. Some examples include the use of jamming signals that render enemy radar ineffective or complex decoy signals that cause the enemy radar to indicate false targets or targets that are not where they are expected. Signals emitted by either a ground station or an aircraft can confuse an enemy missile, causing it to miss the intended target. Missiles can be programmed to follow an enemy radar beam to its source, thus destroying the radar site. Communications circuits can be disrupted by noise transmitters or penetrated by false signals, thus making information transfer more difficult.

Every transmitter, and every type of emission, has its own characteristics, called a signature. Identifying these sources and their role in any military situation is also part of electronic warfare. The technique of counteracting the enemy's use of electronic warfare is termed electronic countermeasures (ECM). Both fields are becoming increasingly important and complex, in today's military preparedness. The search for new and better methods of EW and ECM has resulted in discoveries useful to the electronics industry.

ELECTRONIC WATCH

See ELECTRONIC CLOCK.

ELECTRON MICROSCOPE

An electron microscope is a microscope that uses electron beams rather than visible light to obtain an image of a small object. The electron microscope allows extreme magnification, much greater than any optical microscope. The electron beam is focused and refracted by means of magnetic fields. The magnetic deflectors, sometimes called magnetic lenses, produce a magnified image of the object under observation when the object is placed at the proper point in the electron beam. The illustration shows a simplified diagram of an electron microscope.

With the electron microscope, the image of the object is viewed on a cathode-ray tube, which resembles a television picture screen. Alternatively, the image can be photographed on a special electron-sensitive film. The magnification factor can be as large as 300,000 to 400,000. Under ideal conditions, this means that a grain of sand could be magnified to have an effective diameter of over 100 feet.

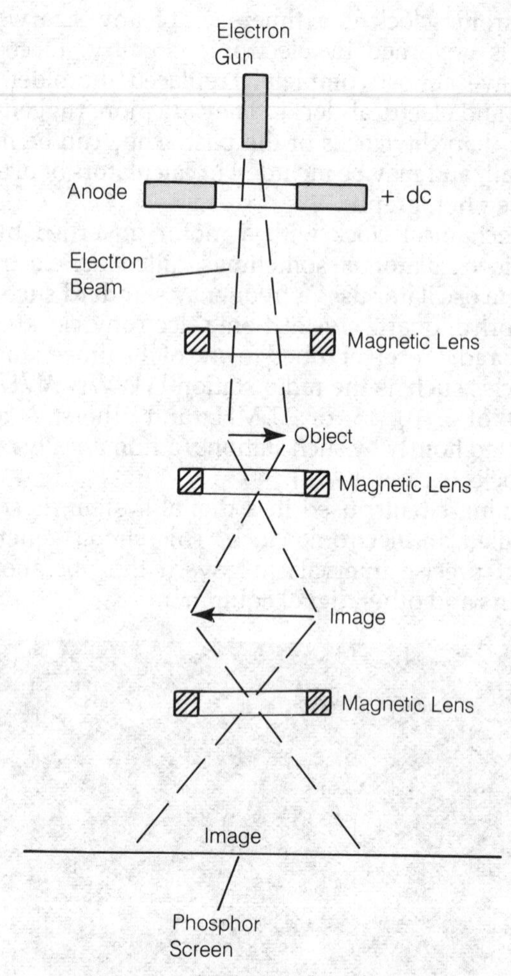

ELECTRON MICROSCOPE: A diagram showing the principal elements of an electron microscope in which electron beams replace visible light.

ELECTRON-MULTIPLIER TUBE

An electron-multiplier tube is a device that amplifies a beam of electrons. These tubes operate on the principle of secondary emission (*see* SECONDARY EMISSION). The electron beam is reflected off of a series of plates, each of which throws off electrons for each one that strikes it. With a series of plates, called dynodes, very large current gains can be realized. The gain may exceed 60 dB.

Electron-multiplier tubes are commonly used in an instrument called a photomultiplier. *See also* DYNODE, PHOTOMULTIPLIER TUBE.

ELECTRON ORBIT

In an atom, the electrons orbit the nucleus in defined ways. The positions of the electrons are sometimes called orbits or shells, but actually, the orbit or shell of an electron represents the average path that it follows around the nucleus. Electrons orbit only at precise distances from the nucleus, in terms of their average positions. The greater the radius of an orbit, the greater the energy contains in the individual electron.

The defined orbits or shells for the electrons in an atom are called, beginning with the innermost shell, the K, L, M, N, O, P, and Q shells. These are roughly diagrammed, although not to scale, in the illustration. The shells, which are actually spherical in shape, appear as circles around the nucleus. The K shell is the least energetic, and can contain two electrons. The L shell, slightly more energetic, can have up to eight electrons. The M shell may have as many as 18 electrons; the N shell, 32; the O shell, 50; the P shell, 72; and the Q shell, 98. The outermost shells of an atom are never filled. The inner ones tend to fill up first, and the number of electrons in an atom is limited.

An electron may change its orbit from one shell to another. If an electron absorbs enough extra energy, it will move into a more distant shell. By moving inward to a shell closer to the nucleus, an electron gives off energy. When an electron absorbs energy, it leaves a definite trace in a spectral dispersion, called an absorption line. When it gives off energy, it creates an emission line. Astronomers have used absorption and emission lines to detect the presence of various elements on planets, in the sun, and in other stars. *See also* ELECTRON.

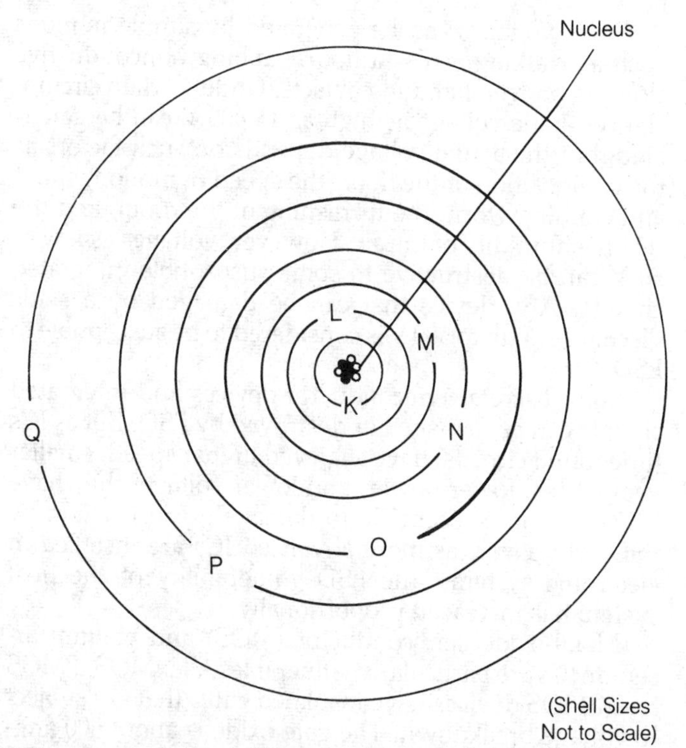

ELECTRON ORBIT: Electron orbits or shells exist only at discrete distances from the nucleus.

ELECTRON VOLT

The electron volt is a unit of electrical energy. An electron or electron charge moving through a potential difference of 1 volt has an energy of 1 electron volt (eV). An electron volt is a very small amount of energy, about 1.6×10^{-19} joule or 1.6×10^{-12} erg (*see* ENERGY, JOULE). Common units of energy in this system are the kilo-electron volt (keV), mega-electron volt (MeV), and giga-electron volt (GeV or BeV); these units represent, respectively, a thousand, a million, and a billion electron volts. *See also* ELECTRON, ENERGY.

ELECTROPHORESIS

Electrophoresis is a method of coating a metal with an insulating material. A suspension of dielectric particles in a liquid is prepared. Two electrodes are immersed into the suspension, and a voltage is applied between the electrodes.

When the current flows in the suspension, particles of the dielectric material are attracted to, and deposited on, the anode or positively charged electrode. The resulting coating of insulation can adhere very well. Electrophoresis is a superior means of applying insulation to certain metals.

ELECTROPLATING

Electroplating is a method of coating one material with another material by electrolysis. This process is done on metals to reduce oxidation or to improve appearance. A metal plating is applied to the surface of a metal that tends to corrode easily. Plating is especially important in marine environments because the chlorine ions from salt water can corrode metals.

ELECTROSCOPE

An electroscope is an instrument for indicating the presence of an electrical charge or a potential difference. There have been many different forms of electroscopes, but the gold-leaf version, invented about 200 years ago, is a landmark instrument in the history of science. This classical instrument is used to detect static charges. It consists of a conductive plate, rod, and leaf assembly mounted inside a grounded metal case with an insulating sleeve. The metal shaft has a circular metal plate on top and two thin gold strips called leaves fastened at its lower end. The electroscope has a glass window for observing the deflection of the leaves.

This form of electroscope is a true scientific instrument that permits a quantitative measure of electrostatic charge. Because the metal leaves are contained in the metal and glass case, they cannot be disturbed by air currents in the room. The modern version of this classical instrument shown in the figure usually has aluminum rather than gold leaves.

When the metal plate on top of the electroscope touches a positively charged body, the positively charged body will attract electrons from the leaves, rod,

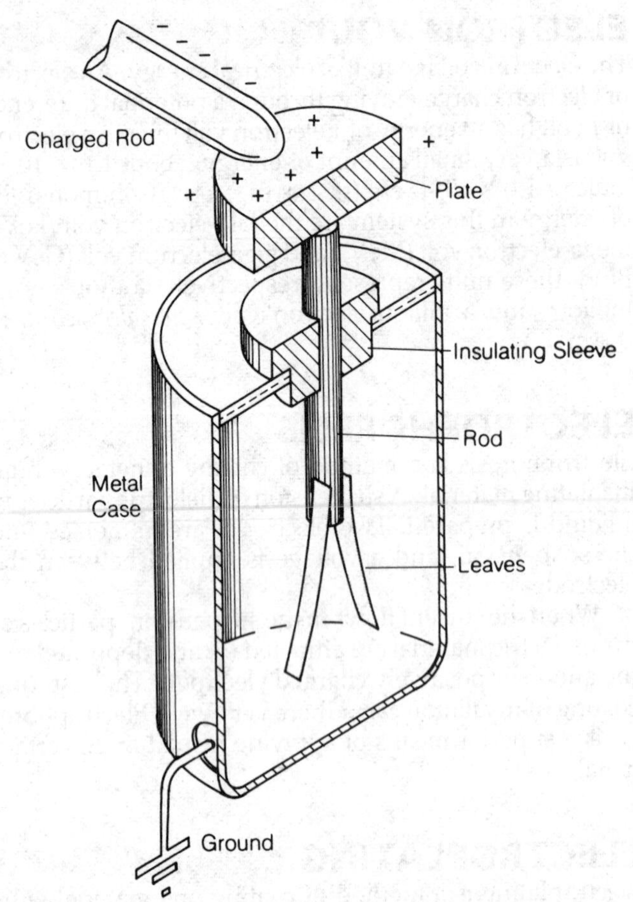

ELECTROSCOPE: An electroscope can determine charge polarity and magnitude.

and plate. This gives the leaves a positive charge. Because the leaves have like charges, they will fly apart from the force of repulsion. The greater the charge on these leaves, the farther the leaves will separate. This will cause a greater angle of divergence between the leaves.

The electroscope works equally well for negatively charged bodies. The negatively charged body forces electrons into the leaves which then separate because they also have like charges.

ELECTROSTATIC DEFLECTION

When a beam of electrons or other charged particles is passed through an electric field, the direction of the beam is altered. A beam of electrons is composed of particles with a negative charge, so the beam will be accelerated away from the negative pole and toward the positive pole of an electric field, as shown in the illustration.

Electrostatic deflection is used in oscilloscope cathode-ray tubes. A pair of deflecting plates causes the electron beam in these tubes to be deflected; if an alternating field is applied, the electron beam swings back and forth or up and down. This causes movement of the spot of the waveforms applied to the deflecting plates. Electrostatic deflection permits high sensitivity.

Electrostatic deflection is important in the operation of many different kinds of devices. However, televison tubes and CRTs in computer monitors typically use

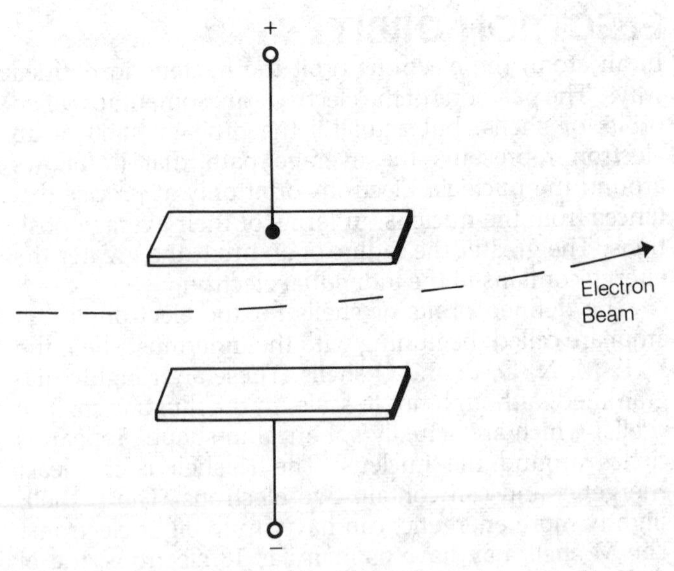

ELECTROSTATIC DEFLECTION: A positive charge on the plate of a cathode-ray tube causes an electron beam to deflect toward it.

electromagnetic deflection. *See also* CATHODE-RAY TUBE, ELECTROMAGNETIC DEFLECTION, OSCILLOSCOPE.

ELECTROSTATIC DISCHARGE

Electrostatic discharge (ESD) is the discharge of static electricity that can destroy or irreparably damage sensitive electronic devices and circuitry unless precautions are taken to prevent it. Electrostatic charges are generated and stored on nonconductive surfaces such as plastic, glass, paper, and natural fiber by friction or induction.

Static voltages can be generated by simple motions such as walking across a floor, rubbing nonconductive clothing or even handling objects. Under certain circumstances static voltage as high as 15,000 V can be generated, but the actual voltage depends on such factors as the composition of the floor, the speed of motion, shoes and clothing worn, the furnishing of the room, and the relative humidity of the air. However, voltages as low as 50 V can be destructive to some susceptible integrated circuits. Any device that can be damaged by a static discharge of up to 500 V is considered to be susceptible to ESD.

Both discrete semiconductor devices and integrated circuits can be damaged or destroyed by ESD. But as ICs (integrated circuits) have achieved higher speed, smaller geometries, lower power, and lower voltage, they have become more susceptible to damage or destruction by ESD. Moreover, as more advanced ICs are installed in electronic systems, the ESD vulnerability of the host system has increased proportionally.

Metal-oxide semiconductor (MOS) and gallium arsenide ICs are particularly vulnerable. All NMOS, PMOS and CMOS devices have insulated gates that are subject to voltage breakdown. The gate oxide is about 800 angstroms thick and breaks down at a gate source potential of about 100 V. Although the high-impedance gates on these devices can be protected by resistor-diode networks, the devices are not rendered immune to ESD. Laboratory tests have shown that devices might fail catastrophically after one high-voltage discharge, or they might fail from the cumulative effect of several lower voltage discharges.

ICs destroyed by ESD are easily detected because the input or output has been shorted or open-circuited. However, devices that have only been damaged are more difficult to detect because the damage might appear as intermittent failures, degraded performance, or increased leakage current. This means that damaged parts might pass tests and be assembled into a system with the defect showing up only in the field, making their detection and replacement far more costly and time consuming.

Some semiconductor manufacturers have been making linear CMOS ICs whose threshold of vulnerability to ESD have been raised with the addition of internal protective devices. Some of these ICs are said to be able to withstand ± 2000 V ESD, but they still require protection from ESD.

Protection against the destructive effects of ESD is best accomplished with a coordinated program covering four separate but related areas: 1) protection of devices and circuit boards during handling and shipment, 2) grounding of benchtop work surfaces, workplace equipment, tools, and furniture to a common point, 3) grounding of personnel with wrist straps and the use of conducive clothing, and 4) maintenance of a specified relative humidity and ionization balance within the work area. *See* ELECTROSTATIC DISCHARGE CONTROL.

ELECTROSTATIC DISCHARGE CONTROL

Control measures to prevent electrostatic discharge (ESD) damage or destruction of susceptible devices and circuits are directed in four related areas: 1) protection of the devices and circuits during handling and shipping, 2) electrical grounding of all work surfaces and equipment in the work area, 3) electrical grounding of all personnel, and 4) environmental conditioning. An effective program includes personnel training and enforcement or preventive measures.

The minimum ESD preventative measures in any program include: 1) use of suitable conductive tote boxes and packaging for susceptible devices and circuits in process or during shipment, 2) grounding of work surfaces, tools, equipment and sometimes even flooring at all workstations and, 3) mandatory use of grounded conductive wrist straps (or shoe ground straps) by all personnel within the work area. However, many programs go far beyond these minimum requirements to include the mandatory wearing of conductive outer clothing by all personnel in the work area, local control of relative humidity, and the use of appliances for ion neutralization of the work area, as illustrated in the figure.

Conductive and Shielded Containers. The use of special conductive tote trays or boxes and special conductive and shielded shipping bags is now accepted as standard procedure in handling and shipping most susceptible devices and circuits. The tote trays or boxes are molded or formed from plastics that contain antistatic conductive

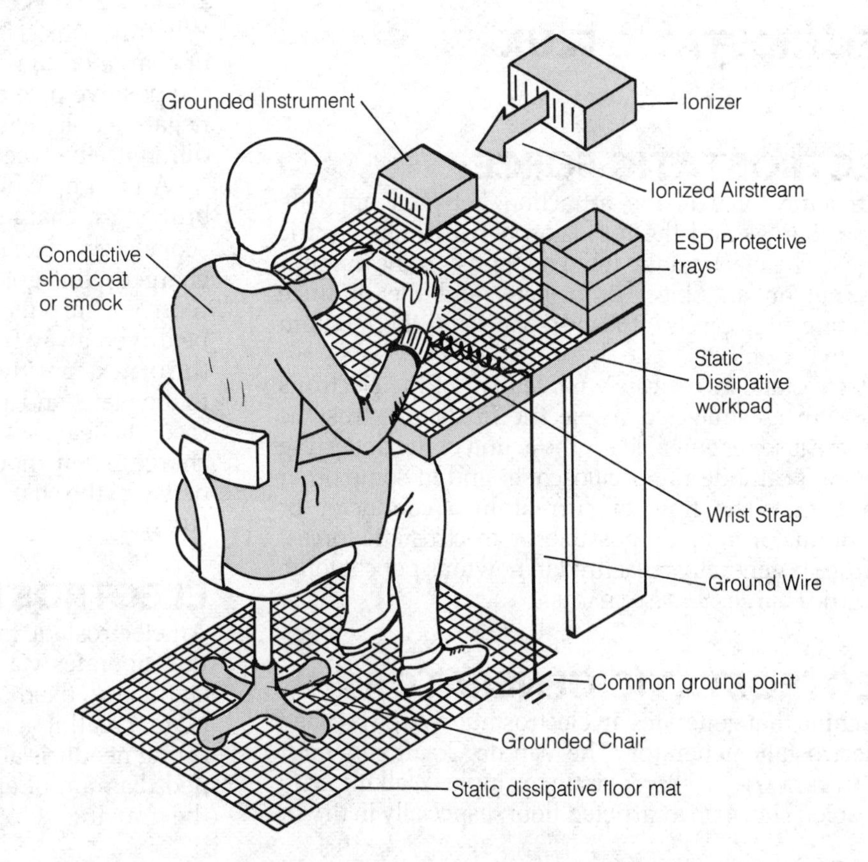

ELECTROSTATIC DISCHARGE CONTROL: Basic protective measures at an ESD-controlled workplace are illustrated.

materials like carbon. Special shipping tubes for ICs have conductive inner and outer surfaces. Susceptible semiconductors and circuit boards containing these devices are shipped in conductive bags. Some bags have metallized outer layers with high conductivity coatings and inner linings of low conductivity antistatic plastic.

Work and Floor Surfaces. Conductive benchtop and floor mats are grounded to a common point. All chairs, workstation furnishings, tools, and test instruments are also grounded by means of wires, conductive paints, or even coating with antistatic solution.

Personal Ground Straps and Clothing. If a grounded wrist strap cannot be worn, conductive shoe toe and heal grounders must be worn. Shopcoats or smocks made from fabrics with conductive threads woven in are worn over street clothing.

Air Ionizers. Air ionizers are appliances that emit free ions from electrostatic or radioactive ion sources. When combined with fans and mounted above the work surfaces, the ionizers inject free positive and negative ions into the room. The free ions neutralize or balance static charges of opposite polarity that have built up on nonconductive surfaces within the throw range of the fan.

Surface-Resistivity Meters. Surface-resistivity meters are available for measuring surface resistivity of materials in workstations. A typical instrument is a hand-held, battery-operated megohmeter. These instruments are designed to measure ohms per square in three ranges from 10^3 to 10^{10} Ω per square.

ELECTROSTATIC FIELD

See ELECTRIC FIELD.

ELECTROSTATIC FLUX

See ELECTRIC FLUX.

ELECTROSTATIC FORCE

Electrostatic force is the attraction between opposite electric charges and the repulsion between like electric charges. The electrostatic force depends on the amount of charge on an object or objects, and the distance separating the objects. It also depends on the medium between the objects.

Electrostatic forces cause movement of the electrons in substances subjected to electric fields. Electrostatic forces are also responsible for the action of the deflecting plates in a cathode-ray oscilloscope, and in some other instruments. The flow of current in a conductor or semiconductor is made possible by electrostatic forces. The force is generally measured in newtons per coulomb of electric charge. *See also* CHARGE.

ELECTROSTATIC GENERATOR

A machine that generates an electrostatic charge is called an electrostatic generator. The Van de Graaff generator (*see* VAN DE GRAAFF GENERATOR) is an example. Walking with hard-soled shoes on a carpeted floor, especially in dry or cool weather, makes the shoes an electrostatic generator. A charge is built up on the body with respect to the surrounding objects because of friction between the shoe soles and the carpet material. This potential difference can become quite large, sometimes as much as several thousand volts. All electrostatic generators operate on principles similar to the scuffing of shoes on a carpet. The largest Van de Graaff generators can generate millions of volts in this way, and the resulting spark discharge can jump across many feet of space.

The largest known electrostatic generators are the cumulonimbus clouds in thunderstorms. Charge differences in these clouds cause lightning.

ELECTROSTATIC HYSTERESIS

In a dielectric substance, the polarization of the electric field within the material does not always follow the polarization of the external field under alternating-current conditions. A small amount of delay, or lag, occurs, because any dielectric substance requires time to charge and discharge. This effect is called electrostatic hysteresis.

Electrostatic hysteresis is not usually significant in most dielectrics at the lower radio frequencies. As the frequency is increased into the very high and ultrahigh ranges, however, the effects of electrostatic hysteresis become noticeable. Different dielectric materials exhibit different electrostatic hysteresis properties. Ferroelectric substances are especially noted for their large amount of electrostatic hysteresis. *See also* DIELECTRIC ABSORPTION, FERROELECTRICITY, HYSTERESIS.

ELECTROSTATIC INDUCTION

When any object is placed in an electric field, a separation of charge occurs on that object. Electrons move toward the positive pole of the electric field, and away from the negative pole, within the object. This creates a potential difference between different regions in the object.

An example of electrostatic induction is provided by bringing a charged object near the plate of an electroscope (*see* ELECTROSCOPE). If the object is negatively charged, the gold-foil leaves of the electroscope will receive a surplus of electrons because the electrons will be driven away from the ball. If the object brought near the plate is positively charged, electrons will be attracted to the plate, and away from the gold-foil leaves. In either case, the leaves will separate indicating that they have a charge, even though no actual contact has been made between the charged object and the electroscope. *See also* CHARGE.

ELECTROSTATIC INSTRUMENT

An electrostatic instrument is a form of metering device that operates via electrostatic forces only. Generally, a stationary, fixed metal plate is mounted near a rotatable plate, creating a sort of air-variable capacitor. The indicating needle is attached to the rotatable plate. A simplified diagram of an electrostatic instrument is shown in the drawing.

When a voltage is applied between the two metal plates, they are attracted to each other because of their opposite polarity (*see* ELECTROSTATIC FORCE). A spring, attached to the rotatable plate, works against this force to regulate the amount of rotation. The greater the voltage between the plates, the greater the force of attraction, and the farther the rotatable plate will turn against the tension of the spring.

An electrostatic voltmeter draws very little current. Only the tiny leakage current through the air between the plates contributes to current drain. Thus, the electrostatic voltmeter has an extremely high impedance. When the appropriate peripheral circuitry is added, the electrostatic voltmeter can be used as an ammeter or wattmeter. *See also* CHARGE.

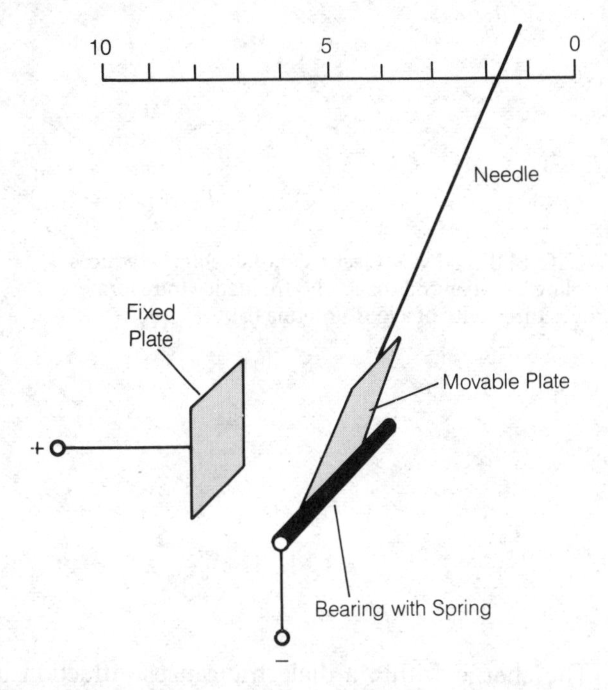

ELECTROSTATIC INSTRUMENT: The fixed plate attracts the movable plate of this electrostatic voltmeter in response to voltage across the plates. The movement can also measure current and power if the voltmeter has additional circuitry.

ELECTROSTATIC MICROPHONE

See MICROPHONE.

ELECTROSTATIC POTENTIAL

Any electric field produces a voltage gradient, measured in volts per meter of distance. Two points in space, separated by a given distance within the influence of an electric field, will have a voltage difference. This is called an electrostatic potential. The voltage between the two electrodes from which the field results, is the largest possible electrostatic potential within the field.

An electrostatic potential between two objects produces an electrostatic force, and this force is directly proportional to the potential difference for a given constant separation distance. For a given amount of electrostatic potential between two objects, the force drops off as the distance increases. Electrostatic potential is measured by an instrument that draws very little current because, although the voltages might be quite large, the total amount of electric charge may be rather small. *See also* ELECTROSTATIC FORCE, ELECTROSTATIC INSTRUMENT.

ELECTROSTATIC PRECIPITATION

Electrostatic precipitation is an effect produced on particles in a gas or liquid by an electric field. In a sufficiently intense electric field, particles become ionized, or charged. The charge causes the particles to be attracted toward one of the field-generating electrodes. Positive ions are attracted to the negative electrode, and negative ions are attracted to the positive electrode.

Electrostatic precipitation is used in the dust precipitator equipment for clearning the air. The effect of electrostatic precipitation is also useful in a process called electrophoresis in which metals can be coated with insulating material by applying a charge to a pair of electrodes immersed in a suspension. *See also* ELECTROPHORESIS.

ELECTROSTATICS

Electrostatics is the branch of physics that studies the behavior of electrical charges at rest, when no current is flowing. These charges produce forces among objects on which they exist. Other objects or particles in the vicinity are affected by these forces. Electrostatics differs from electrodynamics, which is concerned with the behavior of moving charges. *See also* CHARGE, ELECTRIC FIELD, ELECTRIC FLUX, ELECTRODYNAMICS, ELECTROSTATIC FORCE, ELECTROSTATIC DISCHARGE, ELECTROSTATIC DISCHARGE CONTROL.

ELECTROSTATIC SPEAKER

See SPEAKER.

ELECTROSTATIC SHIELDING

Electrostatic shielding, also commonly known as Faraday shielding, is a means of blocking the effects of an electric field, while allowing the passage of magnetic fields. Electrostatic shielding reduces the amount of capacitive coupling between two objects practically to zero. However, inductive coupling is not affected.

An electrostatic shield consists of a wire mesh, a screen, or sometimes a plate of nonmagnetic metal such as aluminum or copper, grounded at a single point. This causes the mesh, screen, or plate to acquire a constant electric charge, but little or no current can flow in it. Such a shield, placed between the primary and secondary windings of an air-core transformer as shown at A in the illustration, practically eliminates capacitive interaction between the windings. Electrostatic shields are often used in the tuned circuits of radio-frequency equipment, when transformer coupling is employed. The shield, by eliminating the electrostatic coupling between stages,

A.

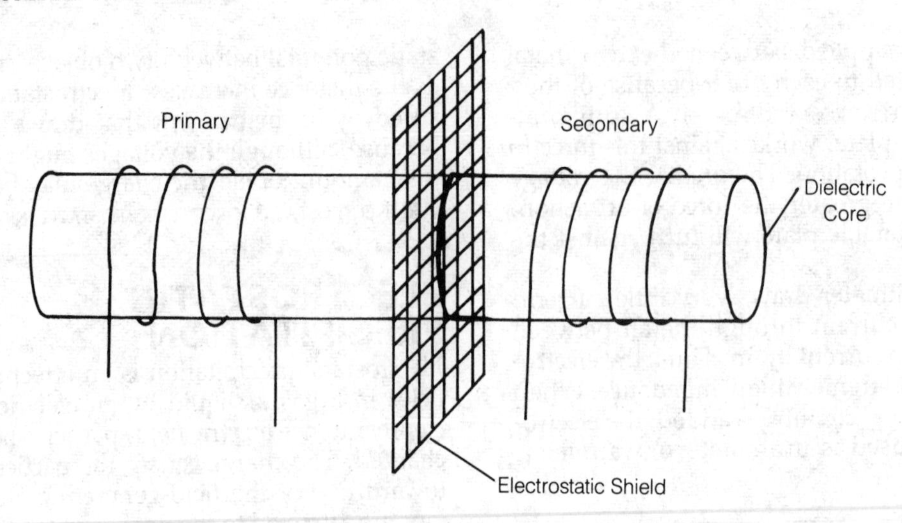

B.

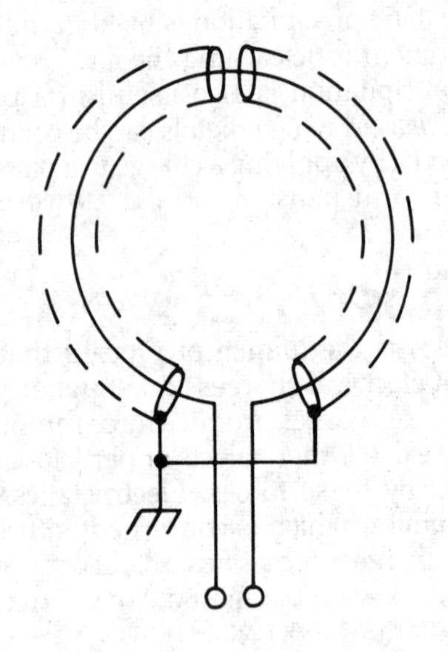

ELECTROSTATIC SHIELDING: Electrostatic shielding reduces capacitive coupling between coils of a radio-frequency transformer (A) and improves directivity of a loop antenna (B).

reduces the amount of unwanted harmonic and spurious-signal energy that is transferred from stage to stage (*see* TRANSFORMER COUPLING).

An electrostatic shield can be used in the construction of a direction-finding loop antenna, as shown at B. A section of nonmagnetic tubing or braid is placed around, and insulated from, the loop itself. The shield is grounded at the feed point, but is broken at the top, preventing the circulation of current but maintaining a fixed electric charge. This allows the magnetic component of an electromagnetic wave to get to the loop, but the electric component is kept out. This enhances the directional characteristics of the antenna. It also may improve the signal-to-noise ratio. *See also* LOOP ANTENNA in the ANTENNA DIRECTORY.

ELECTROSTATIC VOLTMETER

See ELECTROSTATIC INSTRUMENT.

ELECTROSTRICTION

When an electric field is applied to a dielectric material, the inward force created on the material is called electrostriction.

The charge within a dielectric causes attraction between one face of the dielectric and the other. The electrodes, with opposite polarity, are also attracted to each other. If the intensity of the electric field becomes too great, the dielectric might crack under the stress of electrostriction. This can happen to disk capacitors and other ceramic components when they are subjected to excessive voltages. *See also* ELECTROSTATIC FORCE.

ELEMENT

An element is one of the fundamental building blocks of matter. According to contemporary theories, all elements are made up of atoms. The atom consists of a nucleus with positively charged particles and neutral particles. Electrons, which are negatively charged, orbit around the nucleus. The positively charged particles in the nucleus are called protons. The number of protons in an atom is normally the same, or about the same, as the number of electrons, so the atom has no net charge. The neutral particles are called neutrons.

An atom might have only one proton and one electron as does the simplest atom, hydrogen. Hydrogen is the most abundant element in the universe. Some atoms have more than 100 protons and electrons. The elements with many protons and electrons are called heavy elements, and they tend to be unstable, breaking up into elements with fewer protons and electrons.

The table under the listing ATOMIC NUMBER is an alphabetical list of the known elements, according to their atomic number (number of protons) and atomic weight (relative mass). In general, the greater the atomic number, the greater the atomic weight, although there are occasional exceptions to this rule (*see* ATOM, ATOMIC WEIGHT).

Elements may exist in the form of a gas, a liquid, or a solid. Elements often combine to form compounds (*see* COMPOUND). Elements and compounds may exist in the same medium without being chemically attached; this combination is called a mixture. Elements, compounds, and mixtures make up all of the matter in the universe.

The term *element* is sometimes used to refer to parts of antennas. The antenna element that receives the energy directly from the feed line is called the driven element. Some antennas have elements that are not connected to the feed line directly and are called parasitic elements. *See also* DRIVEN ELEMENT, ELECTRON, NEUTRON, PARASITIC ELEMENT, PROTON.

ELEMENT SPACING

In an antenna with more than one element, such as a parasitic or phased array, the element spacing is the free-space distance, in wavelengths, between two specified antenna elements. This might be the director spacing with respect to the driven element or another director; it might be the reflector spacing with respect to the driven element; it might be the spacing between two driven elements. In an antenna with several elements, the spacing between adjacent elements can differ depending on which elements are specified.

For a given frequency in megahertz, the spacing in wavelengths is given by the simple formula:

$$s = \frac{df}{300}$$

if d is in meters. If d is in feet, then:

$$s = \frac{df}{984}$$

At shorter wavelengths, d can be given in centimeters or inches. If d is in centimeters, then:

$$s = \frac{df}{30{,}000}$$

and if d is in inches, then:

$$s = \frac{df}{11{,}800}$$

See also DIRECTOR, DRIVEN ELEMENT, and the articles QUAD ANTENNA, YAGI ANTENNA in the ANTENNA DIRECTORY.

ELEVATION

Elevation is the angle, in degrees, that an object subtends with respect to the horizon. It is also specified as the vertical deviation of the major lobe of an antenna from the horizontal. The smallest possible angle of elevation is 0 degrees, which represents the horizontal; the largest possible angle of elevation is 90 degrees, which represents the zenith.

Elevation can be called altitude. It is one of two coordinates needed for uniquely determining the direction of an object in the sky. The other coordinate, called the azimuth, represents the angle measured clockwise around the horizon from true north (*see* AZIMUTH).

ELLIPTICAL POLARIZATION

The polarization of an electromagnetic wave is the orientation of the electric lines of flux in the wave. This orientation usually remains constant; but sometimes it is deliberately made to rotate as the wave propagates through space. If the orientation of the electric lines of flux changes as the signal is propagated from the transmitting antenna, the signal is said to have elliptical

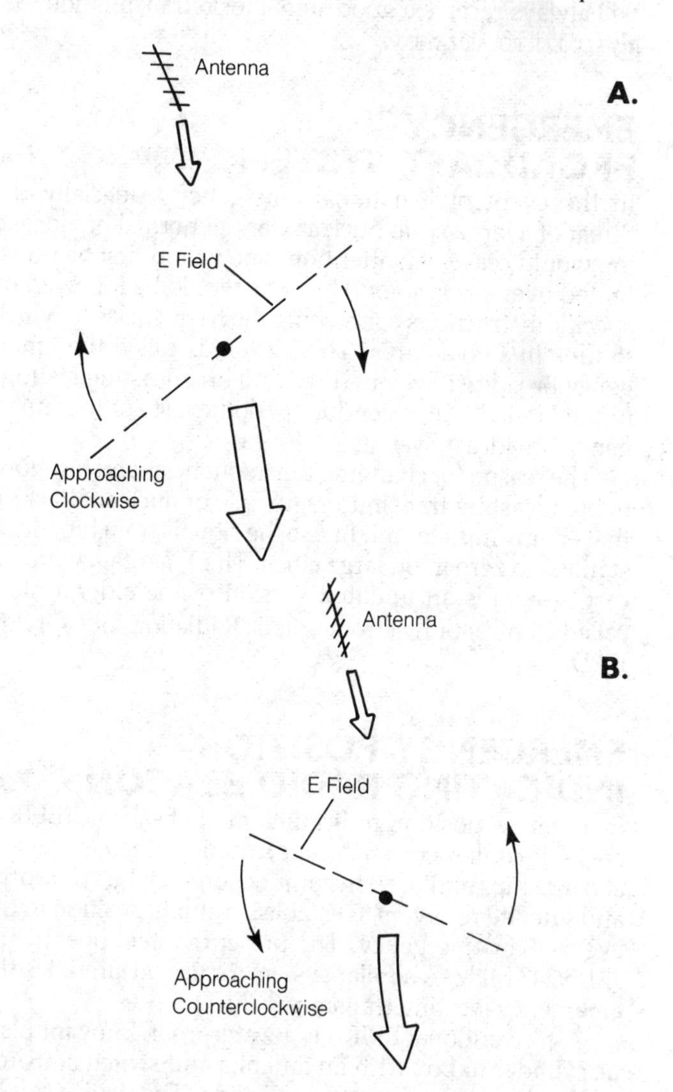

ELLIPTICAL POLARIZATION: Elliptical polarization is diagrammed as clockwise (A) and counterclockwise (B).

polarization. (*See* HORIZONTAL POLARIZATION, POLARIZATION, VERTICAL POLARIZATION.)

An elliptically polarized electromagnetic field may rotate either clockwise or counterclockwise as it moves through space, as shown in the illustration. The intensity of the signal might not remain constant as the wave rotates. If the intensity does remain constant as the wave rotates, the polarization is said to be circular (*see* CIRCULAR POLARIZATION).

Elliptical polarization is useful because it permits the reception of signals having unpredictable or changing polarization with a minimum of fading and loss. Ideally, the transmitting and receiving antennas should both have elliptical polarization, although signals with linear polarization can be received with an elliptically polarized antenna. If the transmitted signal has opposite elliptical polarization from the receiving antenna, however, there is substantial loss.

Elliptical polarization is generally used at ground stations for satellite communication. In receiving, the use of an elliptically polarized antenna reduces the fading caused by changing satellite orientation. In transmitting, the use of elliptical polarization ensures that the satellite will always receive a good signal for retransmission. *See also* LINEAR POLARIZATION.

EMERGENCY BROADCAST SYSTEM

In the event of a national emergency, especially the threat of a large-scale nuclear war, all normal broadcasting would cease. An attention tone would first be transmitted over all stations. This tone would be followed by specific instructions concerning the frequencies to which to tune in a given area. This system is called the Emergency Broadcast System (EBS). All broadcasting stations in the United States conduct periodic tests of the Emergency Broadcast System.

The reason for changing the frequencies and locations of broadcasting transmitters, in case of nuclear attack, is that enemy missiles might use the signals from broadcast stations to zero in on large cities. The Emergency Broadcast System is an updated version of the older system called Control of Electromagnetic Radiation, or CONELRAD.

EMERGENCY POSITION-INDICATING RADIO BEACON

Emergency position-indicating radio beacons (EPIRB) are self-contained emergency radio beacons carried aboard ships and used in conjunction with both airborne and satellite receivers to establish a ship's position in the event of an emergency. The present system operates at 121.5/243 MHz. A similar system for use on aircraft is the emergency locating transmitter (ELT).

A conventional EPIRB is a waterproof, buoyant plastic cylinder or box with an antenna and switch at its top end and a radio and batteries inside. The device can be stowed up to 10 years. Its radio signal is made up of two components—a distinctive audio signal (Whoop, Whoop, Whoop) on a carrier frequency of 121.5 MHz as well as on a military distress frequency of 243 MHz. Class A and B EPIRBs are restricted by law for use 20 miles or more offshore. Class A units have an automatic switch and can be mounted in a bracket so they will float freely when submerged. Class B EPIRBs must be removed from their mounts and turned on manually.

Class C units were developed for use within 20 miles of shore. They transmit on the marine frequency of 156.75/156.8 VHF (very high frequency), and they are monitored only by ships and shore stations.

Class A and B EPIRB signals can be received by aircraft or Soviet and U.S. satellites and ground stations that are part of the international search and rescue system COSPAS/SARSAT. Electronics installed on these polar orbiting satellites receive, compute, and transmit EPIRB signals from anywhere within their range.

The passing satellite computes the location of the beacon by measuring the inflection point—the point of closest approach. It marks the precise moment when the pitch of the received signal changes from increasing to decreasing (the Doppler effect). Decisions for search and rescue are made based on information relayed back to ground-control centers scattered throughout the Northern Hemisphere and in Brazil and Chile.

A newer system operating at 406 MHz is being installed. The EPIRBs broadcasting on this frequency will send a coded signal, identifying the craft from which it is transmitting—its name, size, registry, and in some units, last known position. *See also* BEACON, DOPPLER EFFECT.

EMERGENCY POWER SUPPLY

An emergency power supply is an independent, self-contained power source that can be used in the event of a failure in the normal utility power. An emergency supply is usually either a set of rechargeable batteries, or a fossil-fuel generator. The primary requirements for an emergency power supply are:

- Ability to supply the necessary voltages for the operation of the equipment to be used.
- Ability to deliver sufficient current to run the equipment to be used.
- Ability to operate for an extended period of time.
- Ability to be accessed immediately when needed.

The engine-powered gasoline or diesel generator is an excellent emergency power supply. These generators can supply 120 volts (V) at about the standard line frequency, up to power levels of several kilowatts.

A set of storage batteries, such as 12V automotive batteries, can also be used as a source of emergency power. Many radio receivers and transmitters, as well as some specialized appliances, will run directly from 12 V direct current. Power inverters can be used to get 120 V alternating current from such a source. A solar recharging system can be used to keep the batteries operational. *See also* GENERATOR, INVERTER, SOLAR ENERGY.

EMISSION CLASS

The emission class of a radio-frequency transmitter is a standard designation of the type of modulation used. These designators consist of a letter followed by a number; sometimes another letter comes after the number. The letter *A* stands for amplitude modulation; the letter *F* stands for frequency modulation; the letter *P* stands for pulse modulation. Emissions are designated, for each form of modulation, by numbers from 0 to 9. The table lists the standard emission classes, with short descriptions of the characteristics of each class.

The bandwidth of a signal varies greatly with the emission class used. A signal with no modulation, called A0 or F0, takes almost no spectrum space. A television signal such as A5 might take several megahertz of space. The bandwidth of a particular type of emission might be different in one communications service, as compared with another service. For example, the standard F3 signal in the frequency-modulation broadcast band is much wider than the same kind of signal in the two-way communications service. Some kinds of emissions are prohibited on certain frequencies. *See also* AMPLITUDE MODULATION, BANDWIDTH, FREQUENCY MODULATION, MODULATION, PULSE MODULATION.

EMISSION CLASS: Standard emission classifications.

Emission	Description
A0	Unmodulated, pure carrier
A1	Carrier telegraphy
A2	Amplitude-modulated telegraphy
A3	Amplitude-modulated telephony
A3A	Single-sideband reduced-carrier telephony
A3J	Single-sideband suppressed-carrier telephony
A4	Amplitude-modulated facsimile
A5	Amplitude-modulated television
A7	Multi-channel voice-frequency amplitude-modulated telegraphy
A9	Two independent sidebands
F0	Unmodulated, pure carrier
F1	Frequency-shift keying
F2	Frequency-modulated telegraphy
F3	Frequency-modulated telephony
F4	Frequency-modulated facsimile
F5	Frequency-modulated television
F6	Four-frequency diplex telegraphy
F9	Frequency-modulated emission other than above
P0	Unmodulated-pulse carrier
P1	Keyed pulsed carrier
P2	Pulse-modulated telegraphy
P3	Pulse-modulated telephony
P9	Pulse-modulated emission other than above

EMISSIVITY

Emissivity is the ease with which a substance emits or absorbs heat energy. This property is important in the choice of materials for heatsinks (*see* HEATSINK) because these radiators give up much of their heat by means of radiation. Emissivity is a measure of the brightness of an object in the infrared part of the electromagnetic spectrum. The darker the object appears in the infrared, the better the emissivity. Simply painting an object black does not guarantee that the emissivity will be improved. Black paint might not be as black in the infrared region.

The capability of a material to dissipate power depends on the temperature; the higher the temperature, in general, the more power can be dissipated by a given material. Emissivity depends on the shape of an object; irregular surfaces are better for heat radiation than smooth surfaces. A sphere is the worst possible choice for a heat emitter; in heatsink design, finned surfaces are employed to maximize the ratio of the surface area to the volume. Different substances have different emissivity characteristics under identical conditions. Graphite is one of the most heat-emissive substances known. *See also* INFRARED.

EMITTER

In a semiconductor bipolar transistor, the emitter is the region from which the current carriers are injected. The emitter of a transistor is analogous to the cathode of a vacuum tube, or the source of a field-effect transistor. But the operating principles of the bipolar-transistor emitter are different from those of the cathode of a tube or source of a field-effect device.

The emitter of a bipolar transistor may be made from either an N-type semiconductor wafer (in an NPN device) or a P-type semiconductor wafer (in a PNP device). The figure shows the schematic symbols for both of these bipolar transistors, illustrating the position of the emitter. The emitter lead is indicated by the presence of an arrowhead. *See* N-TYPE SEMICONDUCTOR, P-TYPE SEMICONDUCTOR.

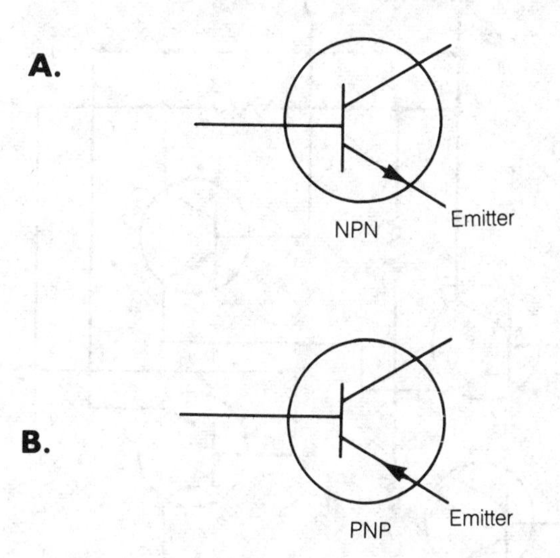

EMITTER: In the schematic of an NPN transistor, the emitter is shown as an arrowhead pointing away from the base (A), and in the PNP transistor it points toward the base (B).

EMITTER-BASE JUNCTION

The emitter-base junction of a bipolar transistor is the boundary between the emitter semiconductor and the base semiconductor. It is always a P-N junction (*see* P-N JUNCTION). In an NPN transistor, the emitter is made from N-type material and the base is made from P-type; in the PNP transistor the arrangement is reversed.

In most transistor oscillators and low-level amplifiers, the emitter-base junction is forward-biased. In the NPN device, therefore, the base is usually positive with respect to the emitter, and in the PNP device, negative. The emitter-base junction of a transistor will conduct only when it is forward-biased. In this respect, it behaves exactly like a diode. For the current to pass to the collector circuit, the emitter-base junction must be forward-biased. But a small change in this bias will result in a large change in the collector current. The input signal to a bipolar-transistor circuit is always applied at the emitter-base junction. *See also* TRANSISTOR.

EMITTER-COUPLED LOGIC

Emitter-coupled logic, abbreviated ECL, is a bipolar form of logic. Most digital switching circuits in the bipolar family use saturated transistors; they are either fully conductive or completely cut off. But in emitter-coupled logic, this is not the case. The figure illustrates a typical emitter-coupled logic gate. The emitter-coupled logic circuit acts as a comparator between two different current levels.

The input impedance of ECL is very high, its switching speed very rapid, and its output impedance is relatively low. The principal disadvantages of emitter-coupled logic are the large number of components required, and the susceptibility of the circuit to noise. The noise susceptibility of ECL arises from the fact that the transistors, not operated at saturation, tend to act as amplifiers of analog signals. *See also* DIGITAL-LOGIC TECHNOLOGY.

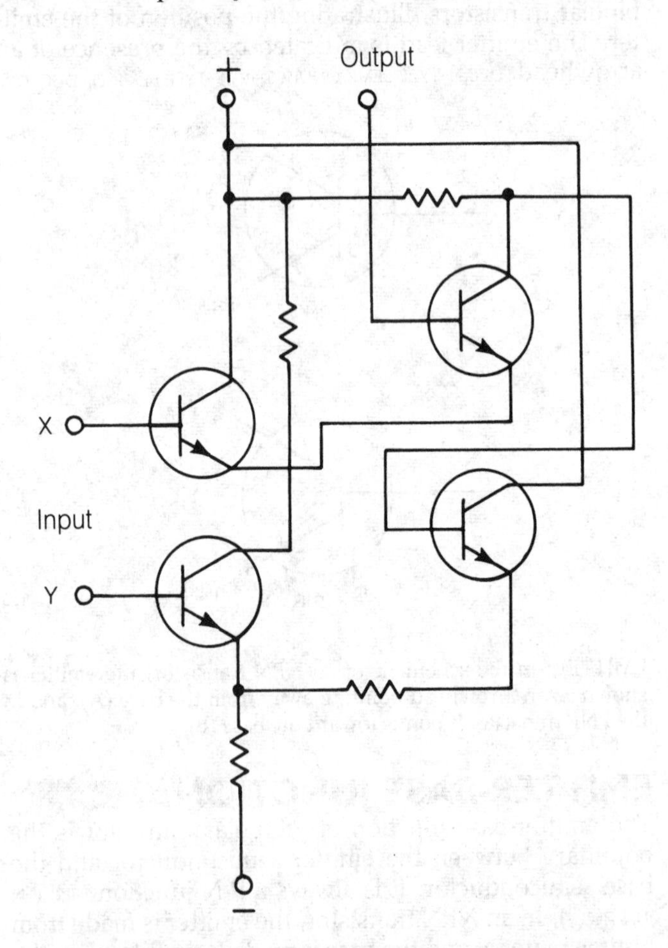

EMITTER-COUPLED LOGIC: Four bipolar transistors perform analog switching in emitter-coupled logic.

EMITTER COUPLING

When the output of a bipolar-transistor amplifier is taken from the emitter circuit, or when the input to a transistor amplifier is applied in series with the emitter, the device is said to have emitter coupling. Emitter coupling may be capacitive, or it may use transformers. The illustration shows an example of emitter coupling in a two-stage, NPN-transistor amplifier. Capacitive coupling is employed in this case. The output of the first stage is taken from the emitter.

Emitter coupling results in a low input or output impedance, depending on whether the coupling is in the input or output of the amplifier stage. Emitter input coupling is used with grounded-base amplifiers. These circuits are used as radio-frequency power amplifiers. Emitter output coupling is used when the following stage requires a low driving impedance. A circuit with emitter output coupling is called an emitter follower. *See also* EMITTER FOLLOWER.

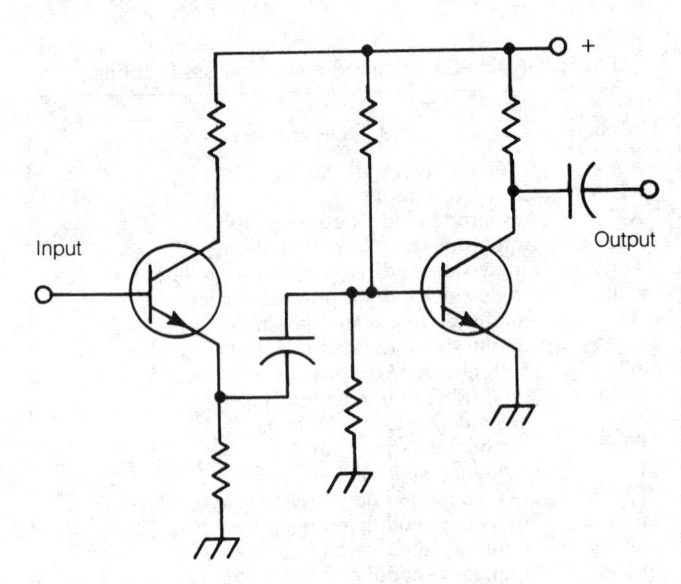

EMITTER COUPLING: The first transistor is emitter coupled to the second transistor through a capacitor in emitter coupling.

EMITTER CURRENT

In a bipolar-transistor circuit, emitter current is the rate of charge-carrier transfer through the emitter. Emitter current is generally measured with an ammeter connected in series with the emitter lead of a bipolar-transistor amplifier circuit.

Most of the emitter current flows through a transistor and appears at the collector circuit. A small amount of the emitter current flows in the base circuit. When the emitter-base junction is reverse-biased, no emitter current flows, and the transistor is cut off. As the emitter-base junction is forward-biased to higher voltages, the emitter current increases, as does the base current and the collector current. The emitter current varies along with the signal input at the emitter-base junction of a transistor amplifier. *See also* EMITTER-BASE JUNCTION.

EMITTER DEGENERATION

Emitter degeneration is a simple means of obtaining degenerative, or negative, feedback in a common-emitter transistor amplifier (*see* COMMON CATHODE/EMITTER/SOURCE). A series resistor is installed in the emitter lead of a transistor amplifier stage; no capacitor is connected across it.

When an alternating-current input signal is applied to the base of an amplifier with emitter degeneration, the emitter voltage changes along with the input signal. This happens so that the gain of the stage is reduced. The distortion level is also greatly reduced. The amplifier is able to tolerate larger swings in the amplitude of the input signal when emitter degeneration is used; the output is virtually distortion-free by comparison with an amplifier that does not use emitter degeneration.

Emitter degeneration is often used in high-fidelity audio amplifiers. Distortion must be kept to a minimum in these circuits. Emitter degeneration stabilizes an amplifier circuit, protecting the transistor against destruction by excess current or thermal runaway. *See also* EMITTER CURRENT, EMITTER STABILIZATION, EMITTER VOLTAGE.

EMITTER FOLLOWER

An emitter follower is an amplifier circuit in which the output is taken from the emitter circuit of a bipolar transistor. The output impedance of the emitter follower is low. The voltage gain is always less than 1; that is, the output signal voltage is smaller than the input signal voltage. The emitter-follower circuit is generally used for impedance matching. The amplifier components are often less expensive, and offer greater bandwidth, than transformers.

The illustration under EMITTER COUPLING shows a two-stage, NPN-transistor amplifier circuit in which the first stage is an emitter follower. The output of the emitter follower is in phase with the input. The impedance of the output circuit depends on the particular characteristics of the transistor used, and also on the value of the emitter resistor. Sometimes, two series-connected emitter resistors are employed. The output is taken, in these cases, from between the two resistors. Alternatively, a transformer output may be used. Emitter-follower circuits are useful because they offer wideband impedance matching. *See also* EMITTER COUPLING.

EMITTER KEYING

In an oscillator or amplifier emitter keying is the interruption of the emitter circuit of a bipolar-transistor stage for code transmission. Emitter keying completely shuts off an oscillator in the key-up condition. In an amplifier, only a negligible amount of signal leakage occurs when the key is up. The illustration shows emitter keying in a radio-frequency amplifier stage. Emitter keying is a satisfactory method of keying in bipolar circuits although, in high-power amplifiers, the switched current may be very large. Usually, emitter keying is done at low-level amplifier circuits. In a code transmitter, emitter keying is often performed on two or more amplifier stages simultaneously. A shaping circuit, illustrated as a series resistor and a parallel capacitor, provides reduction of key clicks at the instants of make and break in code transmission. The series resistor, in conjunction with the parallel capacitor, slows down the emitter voltage drop when the key is closed. This allows a controlled signal rise time. When the key is released, the capacitor discharges through the resistor and the emitter-base junction, slowing down the decay time of the signal. *See also* CODE TRANSMITTER, KEYING.

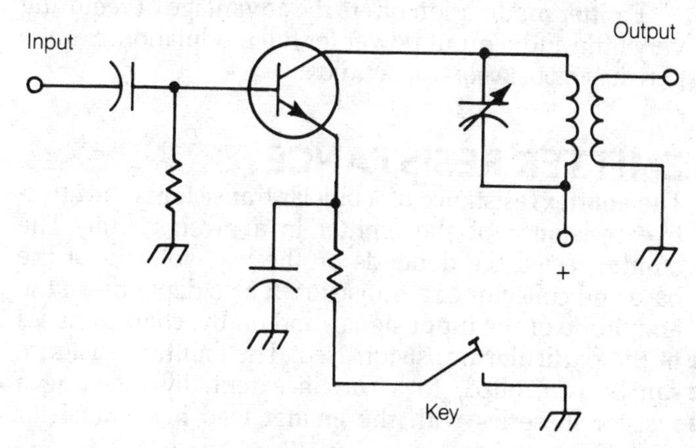

EMITTER KEYING: Emitter keying is a method of modulating the radio-frequency carrier in code transmission.

EMITTER MODULATION

Emitter modulation is a method of obtaining amplitude modulation in a radio-frequency amplifier. The radio-

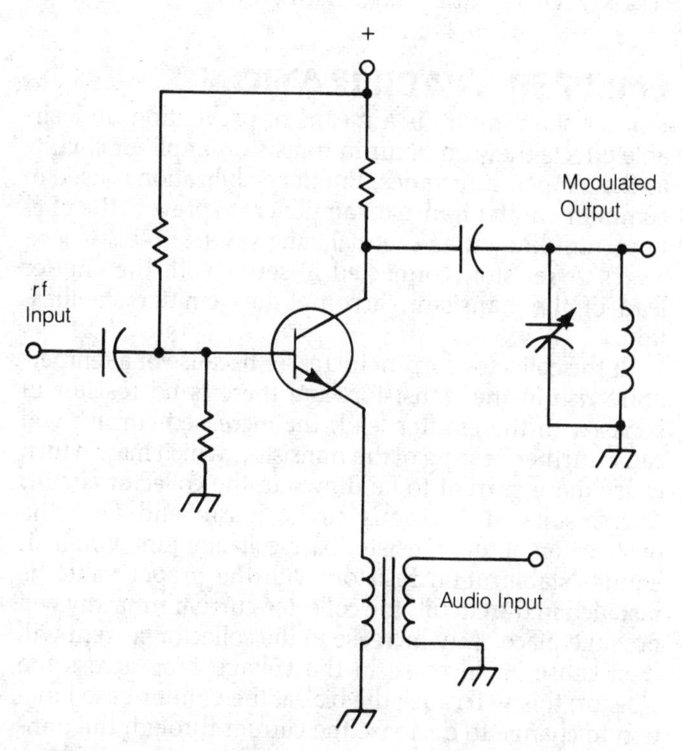

EMITTER MODULATION: In emitter modulation, the radio-frequency signal is applied to the base, and the audio-frequency signal is applied in series with the emitter.

frequency input signal is usually applied to the base of a bipolar transistor although it may be applied in series with the emitter. The audio signal is always applied in series with the emitter. The illustration shows a typical emitter-modulated radio-frequency amplifier circuit.

As the audio-frequency signal at the emitter swings negative in the NPN circuit, the current through the transistor increases. As the audio voltage swings positive, the current through the transistor decreases. When the audio voltage reaches its positive peak, the amplitude of the output signal drops to zero under conditions of 100-percent amplitude modulation.

Emitter modulation offers the advantage of requiring very little audio input power for full modulation. *See also* AMPLITUDE MODULATION, MODULATION.

EMITTER RESISTANCE

The emitter resistance of a bipolar transistor is the effective resistance of the emitter in a given circuit. The emitter resistance depends on the bias voltages at the base and collector of a transistor. It also depends on the amplitude of the input signal, and on the characteristics of the particular transistor used. The emitter resistance can be controlled, to a certain extent, by inserting a resistor in series with the emitter lead in a transistor circuit.

The external resistor in the emitter lead of a bipolar transistor oscillator or amplifier is, itself, sometimes called the emitter resistance. This resistor is used for impedance-matching purposes, for biasing, for current limiting, or for stabilization. The emitter resistance can have a capacitor connected in shunt. *See also* EMITTER CURRENT, EMITTER STABILIZATION, EMITTER VOLTAGE.

EMITTER STABILIZATION

Emitter stabilization is a means of preventing undesirable effects that can occur in transistor amplifier circuits as the temperature varies. Emitter stabilization is used in common-emitter transistor amplifiers to prevent the phenomenon known as thermal runaway (*see* THERMAL RUNAWAY). A resistor, connected in series with the emitter lead of the transistor, accomplishes emitter stabilization.

If the collector current increases because of a temperature rise in the transistor, and there is no resistor in series with the emitter lead, the increased current will cause further heating of the transistor, which may in turn cause more current to be drawn in the collector circuit. This results in a vicious circle. It can end with the destruction of the transistor base-collector junction. If an emitter stabilization resistor, with the proper value, is installed in the circuit, the collector-current runaway will not take place. Any increase in the collector current will then cause an increase in the voltage drop across the resistor; this will cause the bias at the emitter-base junction to change to decrease the current through the transistor. The emitter-stabilization resistor thus regulates the current in the collector circuit. A decrease in the current through the transistor will, conversely, cause a decrease in the voltage drop across the emitter resistor, and a tendency for the collector current to rise.

Emitter stabilization is especially important in transistor amplifiers using two or more bipolar transistors in a parallel or push-pull common-emitter configuration. *See also* COMMON CATHODE/EMITTER/SOURCE, EMITTER CURRENT, EMITTER VOLTAGE.

EMITTER VOLTAGE

In the bipolar transistor, the emitter voltage is the direct-current potential difference between the emitter and ground. If the emitter is connected directly to the chassis ground, then the emitter voltage is zero. But if a series resistor is used, then the emitter voltage will not be zero. In this case, if an NPN transistor is employed, the emitter voltage will be positive; if a PNP transistor is used, it will be negative. The emitter voltage increases as the current through the transistor increases, and also as the value of the series resistance is made larger. In the common-collector configuration, when the collector is directly connected to chassis ground, the emitter voltage is negative in the case of an NPN circuit, and positive in the case of a PNP circuit. Letting I_e be the emitter current in a transistor circuit, and R_e the value of the series resistor in the emitter circuit, specified in amperes and ohms respectively, then the emitter voltage V_e is given in volts by:

$$V_e = I_e R_e$$

A capacitor, placed across the emitter resistor, keeps the emitter voltage constant under conditions of variable input signal. If no such capacitor is employed, the emitter voltage will fluctuate along with the input cycle. *See also* EMITTER CURRENT, EMITTER DEGENERATION, EMITTER STABILIZATION.

ENABLE

The term *enable* means the initialization of a circuit. To enable a device is to switch it on, or make it ready for operation. This can be done by a simple switch or by a triggering pulse. The initializing command or pulse is itself called the enable command or pulse. The term is used mainly in computer and digital circuit applications.

When a magnetic core is induced to change polarity by an electronic signal, the signal is called an enable pulse. This signal allows a binary bit to be written onto, or removed from, the magnetic memory. *See also* MAGNETIC CORE, SEMICONDUCTOR MEMORY.

ENAMELED WIRE

When wire is insulated with a thin coating of enamel, it is called enamel wire. Enameled wire is available in sizes ranging from smaller than AWG40 to larger than AWG10 (*see* AMERICAN WIRE GAUGE). Enameled wire is often preferable to ordinary insulated wire because the enamel does not add significantly to the wire diameter.

Enameled wire is commonly used in coil windings when many turns must be put in a limited space. Examples include audio-frequency chokes, transformers, and coupling or tuning coils. Some radio-frequency chokes also use enameled wire. Because the enamel is thin, this type of wire is not suitable for high-voltage use. *See also* WIRE.

ENCAPSULATION

When a group of discrete components is enclosed or embedded in a rigid material such as wax or plastic resin, the circuit is said to be encapsulated, or potted. Encapsulation protects the components against physical damage and reduces the effects of mechanical vibration on the operation of a circuit. Room-temperature vulcanizing (RTV) resins are used for potting or encapsulation.

ENCODER

An encoder is an electromechanical or electro-optical transducer that monitors the motion of moving mechanisms and translates information about that motion into useful coded signals. Encoder output signals can be used to drive a speed or position indicator, or they can provide digital signals for indicating errors in closed-loop systems.

A class of analog-to-digital converter, encoders can be based on optical, magnetic, and conductive brush/contact technology. The optical encoder, more accurately an electro-optical or optoelectronic encoder, is favored for use in electronic systems because of its compact form, proven reliability, and a solid-state output circuitry that is compatible with standard transistor-transistor logic (TTL) and metal-oxide semiconductor (MOS) transistor logic levels.

There are two types of optical encoders in use: incremental and absolute. Incremental encoders have two output channels, and position is determined by counting pulses clockwise or counterclockwise. Absolute encoders have multiple output channels that provide absolute position information in a digital code. They do not have to be zeroed to determined position each time a system is restarted.

Incremental Rotary Encoder. Figure 1 is a cutaway drawing showing the principal components of a typical incremental optical encoder. The disk assembly consists of an encoding disk mounted perpendicular to its axial input shaft. The input shaft is coupled, directly or indirectly, to a rotating machine element. The transparent encoding disk has a uniform pattern of closely spaced opaque lines radiating from its center. It is positioned to intercept an internal optocoupled path between a LED light emitter and two photodetectors. *See* LIGHT-EMITTING DIODE, OPTOCOUPLER.

The illumination from a LED or IRED source is collimated with a lens and passed to the spinning disk and an additional element, the optical aperture, located behind the disk. Light reaches the two photodetectors only by passing through two closely spaced transparent sectors when they are aligned. As a result, the light is modulated by a shuttering or chopping effect.

A photodetector in each of the two channels converts the modulated light into sinusoidal signals that are 90 electrical degrees apart. These quadrature signals are amplified and conditioned to produce pulses in parallel channels that provide a directional output for counting or display. Position is determined by counting clockwise or counterclockwise pulses. Some incremental encoders have a reference marker for counting because all counts return to zero if the host machine is shut down. Incremental encoders can also be used as tachometers to measure shaft velocity.

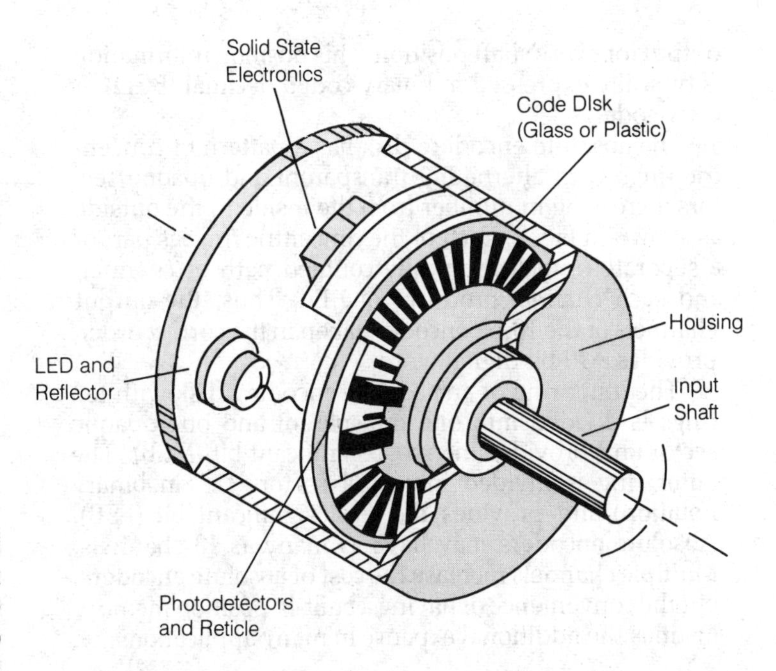

ENCODER: Fig. 1. Cutaway view of an incremental encoder showing moving disk and interruptor optoelectronics for counting turns.

Absolute Rotary Encoders. Absolute rotary encoders include the same components as incremental encoders, but their encoding disks generate a unique digital parallel

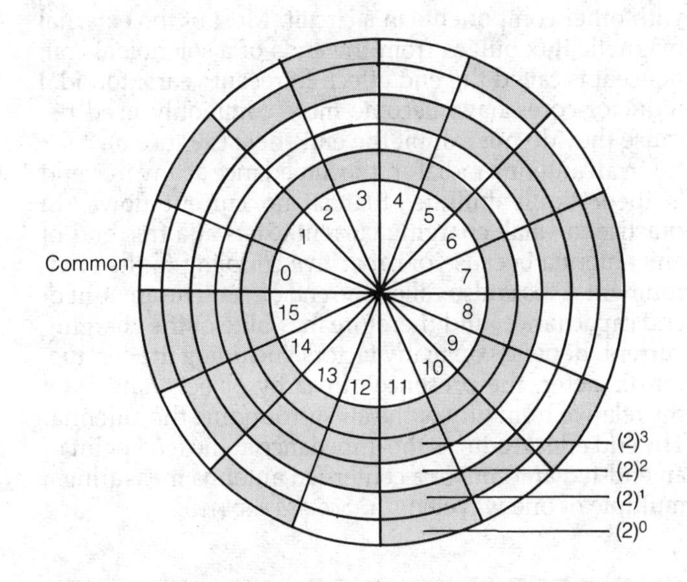

ENCODER: Fig. 2. Binary disk of an absolute optical encoder capable of producing the numbers 0 to 15. Shaded sectors indicate 1, and transparent sectors indicate 0.

output for every shaft position. This position information is typically expressed in binary-coded decimal (BCD) or Gray code.

The absolute encoding disk has a pattern of concentric rings with alternating transparent and opaque sectors increasing in number from the inside to the outside as shown in Fig. 2. Each of the concentric rings is part of a separate optoelectronically coupled path or channel, and each channel produces one bit. Thus, the output channels of the 10-bit encoder, taken in the correct order, provides a 10-bit digital word.

The inner ring of an absolute encoding disk with ten rings is divided into one transparent and one opaque sector and provides the most-significant bit (MSB). The outer ring is divided into 1024 sectors (2^{10} in binary notation) and provides the least significant bit (LSB). Absolute encoders may have as many as 13 channels. Multiple channels increase the cost of absolute encoders, but the convenience of having a built-in position memory justifies the additional expense in many applications.

ENCODING

Encoding is the process of translating a common spoken or written language, such as English or Japanese, into code. This process may be performed by a mechanical or electronic machine called an encoder (*see* ENCODER), or it can, in some cases, be done by a human operator. A Teletype machine is an electronic encoder.

END EFFECT

When a solenoidal inductor carries an alternating current, some of the magnetic lines of flux extend outside the immediate vicinity of the coil. Although the greatest concentration of magnetic flux lines is inside the coil itself, especially if the coil has a ferromagnetic core such as powdered iron, the external flux can result in coupling with other components in a circuit. Most of the external magnetic flux bulges from the ends of a solenoidal coil; hence it is called the end effect. In recent years, toroidal inductor cores have become more commonly used because they do not exhibit the end effect (*see* TOROID).

In an antenna radiator, the impedance at any free end is theoretically infinite; that is, no current flows. In practice, a small charging current exists at a free end of any antenna because of capacitive coupling to the environment. This is also called the end effect. The amount of end capacitance, and therefore the value of the charging current, depends on many factors, including the conductor diameter, the presence of nearby objects, and even the relative humidity of the air surrounding the antenna. The end effect reduces the impedance at the feed point of an end-fed antenna or a center-fed antenna measuring a multiple of one wavelength. *See also* END FEED.

END FEED

When the feed point of an antenna radiator is at one end of the radiator, the arrangement is called end feed. End feed always constitutes voltage feed because the end of an antenna radiating element is always at a current node (*see* VOLTAGE FEED). End feed is usually accomplished with a two-wire balanced transmission line. One of the feed-line conductors is connected to the radiator, and the other end is left free, as shown in the illustration.

In an end-fed antenna system, the radiator must have an electrical length that is an integral multiple of ½ wavelength. If *f* is the frequency in megahertz, then ½ wavelength, denoted by *L*, is given approximately by the formulas:

$$L \text{ (meters)} = 143/f$$
$$L \text{ (feet)} = 468/f$$

The length of the radiating element is critical when end feed is used. If the antenna is too long or too short, the current nodes in the transmission line will not occur in the same place. This will result in unbalance between the two conductors of the line, with consequent radiation from the line. Even if the antenna radiator is exactly a multiple of ½ electrical wavelength, the end effect will cause the terminated end of the line to have a different impedance than the unterminated end (*see* END EFFECT). This also results in unbalance between the currents in the two line conductors.

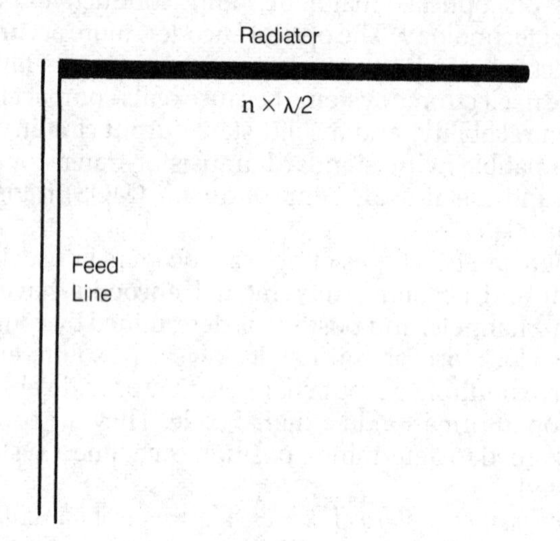

END FEED: An example of antenna end feed with the radiator that's a multiple of a half wavelength.

END-FIRE ANTENNA

See ANTENNA DIRECTORY.

ENERGY

Energy is the capacity for doing work. Energy is manifested in many forms, including chemical, electrical, thermal, electromagnetic, mechanical, and nuclear. Energy is expended at a rate that can be specified in various ways; the standard unit of energy is the joule (*see* JOULE). The rate of energy expenditure is known as power (*see* POWER); the unit of power is the watt, which is a rate of energy expenditure of 1 joule per second. The units specified for energy depend on the application. The British thermal unit is a measure of thermal energy, and the foot pound is a measure of mechanical energy. *See* BRITISH THERMAL UNIT, ELECTRON VOLT, ENERGY UNITS, ERG, FOOT-POUND, JOULE, KILOWATT HOUR, WATTHOUR.

ENERGY CONVERSION

Energy occurs in many different forms. Energy conversion is the process of changing energy from one form to another. For example, mechanical energy may be changed into electrical energy with a generator. Visible-light energy can be converted to electrical energy with a solar cell. Chemical energy can be converted to electrical energy with a battery.

Energy conversion always takes place with an efficiency of less than 100 percent. No device has ever been invented that will produce the same output energy, in desired form, as it receives in another form. Some devices are almost 100 percent efficient; others are much less than perfect. An incandescent light bulb, which converts electrical energy into visible-light energy, has a very poor efficiency; it is better for generating heat than light. The fluorescent light bulb is a much more efficient energy converter. *See also* BRITISH THERMAL UNIT, EFFICIENCY, ELECTRON VOLT, ENERGY UNITS, ERG, FOOT-POUND, JOULE, KILOWATT HOUR, WATTHOUR.

ENERGY DENSITY

In an energy-producing cell, the energy density is the ability of the cell to produce a given amount of power for a given length of time, for a given amount of cell mass. Theoretically, the energy density is expressed as the total energy in joules divided by the cell mass in kilograms. It is advantageous to have an energy density that is as high as possible.

Energy density in a cell may also be represented as the total energy in joules divided by the cell volume in cubic meters. Again, it is desirable to maximize the energy density as expressed in terms of cell volume. *See* BATTERY.

ENERGY LEVEL

In an atom of any element, the electrons orbit the nucleus in discrete average positions. These positions are called shells. The farther a shell is from the nucleus, or center, of the atom, the greater the energy contained in an electron orbiting in that shell. The innermost shell of an atom is called the K shell. Outside of the K shell are the L, M, N, O, P, and Q shells, with progressively increasing energy levels.

If an electron, orbiting in a given shell, absorbs sufficient additional energy, it will move to a more distant shell. Conversely, if an electron moves to a closer orbital shell, it will emit energy. If an electron absorbs enough energy, it will escape from the nucleus of the atom and become a free electron.

The effects of electron energy levels are responsible for the absorption and emission lines in the spectra of substances. The different elements have unique patterns of emission and absorption lines, and this enables astronomers to determine the composition of distant stars and planets. Electrons are believed to behave in the same manner everywhere in the universe. *See also* ELECTRON ORBIT.

ENERGY STORAGE

Energy can be stored in chemical form, as in the cell or battery; in electrical form, as in a capacitor; or in magnetic form, as in an inductor. Energy can also be stored as heat.

Whenever energy is stored for use at a later time, there is some loss. The amount of useful energy recovered is never as great as the amount of energy initially stored. However, in most practical energy-storage devices, the loss is quite small within a reasonable period of time. But the longer the elapsed time between the initial storage period and its use, the greater is the loss. *See also* BATTERY.

ENERGY UNITS

Energy is manifested in many different forms and used for different purposes. Energy may be measured in any of several different systems of units, depending on the application. The fundamental unit of energy is the joule, or watt second (*see* JOULE). The other units are used as a matter of convention, and each bears a relationship to the joule that can be given in terms of a numerical constant or conversion factor. The table shows the conversion factors for the most common energy units: the British thermal unit (Btu), the electron volt, the erg, the foot-pound, the kilowatt hour, and the watthour.

The unit chosen for a given application depends on the user. While joules represent a perfectly satisfactory way to express energy in any situation, values can be unwieldy. For example, an average residential dwelling might consume 1,000 kilowatt hours of electricity in a month; this is 3.6 billion joules. *See also* BRITISH THERMAL UNIT, ELECTRON VOLT, ENERGY, ERG, FOOT-POUND, JOULE, KILOWATT HOUR, WATTHOUR.

ENERGY UNITS: Conversion factors for various energy units.

Units	To convert to joules, multiply by	Conversely, multiply by
British thermal units (Btu)	1.055×10^3	9.480×10^{-4}
electron volt	1.602×10^{-19}	6.242×10^{18}
erg	10^{-7}	10^7
foot pound	1.356	7.376×10^{-1}
kilowatt hour	3.6×10^6	2.778×10^{-7}
watthour	3.6×10^3	2.778×10^{-4}

ENHANCEMENT MODE

The enhancement mode is an operating characteristic of metal-oxide-semiconductor field-effect transistor (MOSFET). Some MOSFETs have no channel under conditions of zero gate bias. In other words, they are normally pinched off. These MOSFETs are called enhancement-mode devices because a voltage must be applied for conduction to take place. The channel is enhanced, rather than depleted, with increasing bias. The other type of field-effect transistor is normally conducting, and the application of gate bias reduces the conduction. This is the depletion mode (*see* DEPLETION MODE).

In an N-channel enhancement-mode device, a positive voltage at the gate electrode causes drain current to flow. In the P-channel enhancement-mode device, a positive voltage at the gate is necessary to cut off drain current.

The larger the positive voltage, the wider the channel, and the better the conductivity from the source to the drain. This occurs, however, only up to a specific maximum. *See also* FIELD-EFFECT TRANSISTOR.

ENTROPY

Energy always moves from a region of greater concentration to areas of lesser concentration. For example, a hot object radiates its heat into the surrounding environment; a cold object absorbs heat from the environment. The tendency for energy to equalize its level everywhere in the universe is called entropy. Just as water seeks to find the lowest level of elevation, energy seeks to become uniformly distributed throughout the universe.

According to theoretical physics, the process of entropy is, in the general universe, irreversible.

ENVELOPE

The envelope of a modulated signal is the modulated waveform. Generally, the term *envelope* is used for amplitude-modulated or single-sideband signals. An imaginary line, connecting the peaks of the radio-frequency carrier wave, illustrates the envelope of a signal. A typical envelope display for a single-sideband signal is shown in the illustration.

The appearance of a signal envelope is an indicator of how well a transmitter is functioning. Different kinds of distortion appear as different abnormalities in the modulation envelope of a transmitted signal. The most serious kind of distortion in an amplitude-modulated wave is called peak clipping or flat topping. This condition causes the generation of sidebands at frequencies far removed from that of the carrier. This can result in interference to stations on frequencies outside the normal channel required by such a transmitter (*see* AMPLITUDE MODULATION, SINGLE SIDEBAND).

The glass enclosure of a vacuum tube is called the tube envelope.

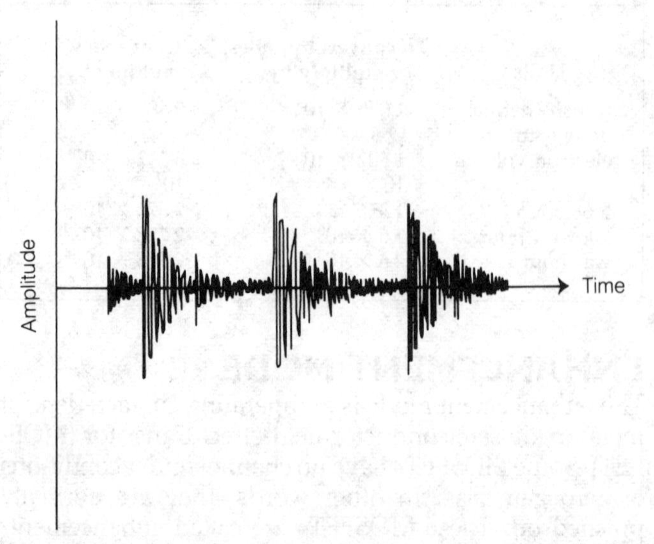

ENVELOPE: A single-sideband modulation envelope as viewed on an oscilloscope.

ENVELOPE DETECTOR

An envelope detector is a form of demodulator in which changes of signal amplitude are converted into audio-frequency impulses. One form of envelope detector is the half-wave rectifier circuit for radio frequencies. A semiconductor diode can function as an envelope detector.

Although an envelope detector is suitable for demodulating an amplitude-modulated signal having a full carrier, it is not satisfactory for the demodulation of a single-sideband, suppressed-carrier signal. When an envelope detector is used in an attempt to demodulate a suppressed-carrier signal, the resulting sound at the receiver is unintelligible.

To demodulate a single-sideband, suppressed-carrier signal properly, a beat-frequency oscillator is necessary. This replaces the missing carrier wave in the receiver.

An envelope detector can be used to demodulate a frequency-modulated signal with a technique called slope detection. The receiver is tuned slightly away from the carrier frequency of the signal, so that the frequency variations bring the signal in and out of the receiver passband. This produces amplitude changes in the receiver; when these are demodulated by the envelope detector, the result is amplitude variations identical to those of an amplitude-modulated signal. *See also* AMPLITUDE MODULATION, DETECTION, FREQUENCY MODULATION, SLOPE DETECTION.

EPIRB

See EMERGENCY POSITION-INDICATING RADIO BEACON.

EPITAXY

Epitaxy is an important process in the manufacture of discrete transistors and integrated circuits. It is the growth on a crystal substrate of a crystal layer that duplicates the substrate crystallographic structure (orientation of the substrate lattice). Epitaxial growth allows the formation of a crystal on the substrate that has fewer defects or different properties than the host substrate or seed. The term *epitaxy* is derived from the Greek, meaning arrangement upon.

In conventional silicon transistor and integrated-circuit fabrication technology, epitaxy provides a layer with doping that differs from that of the host substrate. Epitaxy is used in combination with the photomask process in device manufacture. A crystalline layer is grown on the substrate or wafer by exposing it to a molten or vaporized crystal material. The atomic orientation of the crystalline growth layer is controlled by the structure of the substrate.

Vapor-Phase Epitaxy (VPE). VPE is a process in which the wafers are placed in a furnace and exposed to a gas containing a compound of the wafer material. The resulting growth pattern follows the precise structure of the starting crystal. In the case of silicon wafers, silicon atoms freed by the breakdown of silicon tetrachloride (SiC_{14}) at a temperature of about 2,200°F grow on the surface of the silicon wafer.

Liquid-Phase Epitaxy (LPE). LPE is a process in which crystal growth takes place at the surface of a wafer that is slid along the surface of molten crystal containing a dopant within a furnace with a protective atmosphere.

By introducing small amounts of N- or P-type impurities or dopants into the gas or liquid, the epitaxial layer can be controlled to form either N- or P-type crystalline layers. Also, the distribution or concentration of impurities within the layer or thickness may be graduated to fabricate special devices. The thickness of the epitaxial layer is a function of exposure time. Epitaxy is used in the fabrication of gallium arsenide as well as silicon devices and is an alternative to diffusion and ion implantation as a means for introducing dopants. *See* INTEGRATED-CIRCUIT MANUFACTURE, ION IMPLANTATION, TRANSISTOR MANUFACTURE.

E PLANE

The E plane, or electric-field plane, is a theoretical plane in space surrounding an antenna. There are an infinite number of E planes surrounding any antenna radiator; they all contain the radiating element itself. The principal E plane is the plane containing both the radiating element and the axis of maximum radiation. Thus, for a Yagi antenna, the E plane is oriented horizontally; for a quarter-wave vertical element without parasitics, there are infinitely many E planes, all oriented vertically.

The magnetic lines of flux in an electromagnetic field cut through the E plane at right angles at every point in space. The direction of field propagation always lies within the E plane, and points directly away from the radiating element. In a parallel-wire transmission line, the E plane is that plane containing both of the feed-line conductors in a given location; the E plane follows all bends that are made in the line. In a coaxial transmission line, the E planes all contain the line of the center conductor. In waveguides, the orientation of the E plane depends on the manner in which the waveguide is fed. *See also* ELECTRIC FIELD, ELECTROMAGNETIC FIELD.

EPROM

See SEMICONDUCTOR MEMORY.

EQUALIZER

An equalizer is a high-fidelity circuit used to tailor the frequency response of an audio system. The simplest equalizers are the base-treble controls on all stereo amplifiers, phonographs, and tape players.

Different speakers generally produce different sounds when used with the same amplifier; also, a given set of speakers might have different sounds when used with different audio systems. By using a precision equalizer, these discrepancies can be avoided.

The more complex equalizers generally have several linear sliding-contact potentiometers located side by side, as illustrated. Each potentiometer controls the gain at a different range of frequencies. There may be eight or more of these potentiometers. The maximum gain for each frequency range is obtained with the contact at the top, and the low-to-high frequency setting is obtained from left to right. Thus the adjustment of the potentiometers gives the impression of a graphic display. These controls are, for this reason, called graphic equalizers. *See also* HIGH FIDELITY.

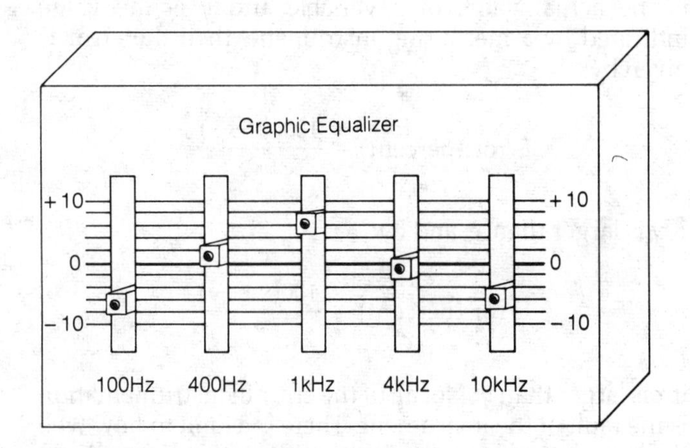

EQUALIZER: An equalizer for use in high-fidelity sound reproduction. Sliding vertical potentiometers control the gain in different frequency zones across the band.

EQUIVALENT CIRCUITS

When two different circuits behave the same way under the same conditions, the circuits are said to be equivalent. There are many examples of equivalent circuits.

Equivalent circuits are used in the test laboratory and on the technician's bench, for simulating actual operating conditions when designing and repairing electronic devices. The equivalent circuit is often simpler than the actual circuit. The equivalent must have the same current, voltage, impedance, and phase relationship as the actual circuit. It must also be capable of handling the same amount of power as the actual circuit.

ERASE

The erase process is the removal of information or data from a medium. Examples of erasable media include magnetic tape and disks, and various integrated-circuit memories (EPROM, EEPROM, EAROM, and flash).

In some media or devices, it is necessary to erase the stored data before new data can be recorded. In other instances, the new data is recorded over the old data, without the need to erase or re-initialize the medium.

ERG

The erg is a unit of energy generally abbreviated by the small letter e. A force of 1 dyne, acting through a distance of 1 centimeter, is an energy expenditure of 1 erg.

The erg is not often used as a unit of energy in electronics; more often, the watthour or kilowatt hour is specified. The universal unit of energy is the joule (*see* JOULE). *See also* ENERGY UNITS.

ERROR

Whenever a quantity is measured, there is always a difference between the instrument indication and the true value. This difference, expressed as a percentage, is called the error or instrument error. Mathematically, if x is the actual value of a variable and y is the value indicated by a measuring instrument, then the error is given by:

$$\text{Error (percent)} = \frac{100(y - x)}{x}$$

if y is larger than x, and by:

$$\text{Error (percent)} = \frac{100(x - y)}{x}$$

if x is larger than y. Not all of the error in instrumentation is the fault of the instrument. There is a limit to how well the person reading an instrument can interpolate values. For example, the needle of a meter might appear to be 3/10 of the way from one division to the next, but in reality it is 0.38 of the way from one division to the next. Digital meters eliminate this problem, but the resolution of a digital device is limited by the number of digits in the readout. The term error is also used in reference to an improper command to a computer. This will result in an undesired response from the computer, or will cause the computer to inform the operator of the error. *See also* INTERPOLATION.

ERROR ACCUMULATION

Errors in measurement (*see* ERROR) tend to add together when several measurements are made in succession. This is called error accumulation. If the maximum possible error in a given measurement is x units, for example, and the measurement is repeated n times, then the maximum possible error becomes nx units. The percentage of maximum error, however, does not change.

An example of error accumulation can be seen with the measurement of a long distance, such as the length of antenna wire, with a short ruler.

ERROR CORRECTION PROGRAM

An error correction program is a computer program for correcting certain types of errors automatically. An example of this routine is a program that maintains a large dictionary of common English words. The operator of a word processor may make mistakes in typing or spelling, but when the error-correction program is run, the computer automatically corrects all typographical errors in words contained in its dictionary.

ERROR-SENSING CIRCUIT

An error-sensing circuit is a circuit for regulating the output current, power, or voltage of a system. In a power supply, for example, an error-sensing circuit might be used to keep the output voltage at a constant value; such a circuit requires a standard reference with which to compare the output of the supply. If the power-supply voltage increases, the error-sensing circuit produces a signal to reduce the voltage. If the power-supply output voltage decreases, the error-sensing circuit produces a signal to raise it. An error-sensing circuit is therefore a form of amplitude comparator.

An error circuit does not necessarily have to provide regulation for the circuit to which it is connected; some error-sensing circuits simply indicate that something is wrong, by causing a bell to ring or a warning light to flash. *See also* ERROR SIGNAL.

ERROR SIGNAL

An error signal is the signal produced by an error-sensing circuit (*see* ERROR-SENSING CIRCUIT). The error signal is actuated whenever the output variable of a circuit or device differs from a standard reference value. An error signal can be used simply for indication.

An example of an error signal is the rectified voltage in an automatic-level-control circuit (*see* AUTOMATIC LEVEL CONTROL). The output of the amplifier chain in this circuit is held to a constant value. If the signal increases in amplitude, the error signal reduces the gain of the system.

Error signals are used to regulate the operation of servo systems. Each mechanical movement is monitored by an error-sensing circuit; error signals are generated whenever the system deviates from the prescribed operating conditions, and the signals cause corrective action. *See also* SERVO SYSTEM.

ESAKI DIODE

See TUNNEL DIODE.

ESNAULT-PELTERIE FORMULA

The Esnault-Pelterie formula is a formula for determining the inductance of a single-layer, solenoidal, air-core coil based on its physical dimensions. The inductance increases roughly in proportion to the coil radius for a given number of turns. The inductance increases with the square of the number of turns for a given coil radius. If the number of turns and the radius are held constant, the inductance decreases as the solenoid is made longer.

Letting r be the coil radius in inches, N the number of turns, and m the length of the coil in inches, then the inductance L in microhenrys is given by:

$$L = \frac{r^2N^2}{(9r + 10m)}$$

See also INDUCTANCE.

EVEN-ORDER HARMONIC

An even-order harmonic is any even multiple of the fundamental frequency of a signal. For example, if the fundamental frequency is 1 MHz, then the even-order harmonics occur at frequencies of 2 MHz, 4 MHz, 6 MHz, and so on.

Certain conditions favor the generation of even-order harmonics in a circuit, while other circumstances tend to cancel out such harmonics. The push-push circuit accentuates even-order harmonics (*see* PUSH-PUSH CONFIGURATION), and is often used as a frequency doubler. The push-pull circuit (*see* PUSH-PULL CONFIGURATION) tends to cancel out the even-order harmonics.

A half-wave dipole antenna generally discriminates against the even-order harmonics. The impedance at the feed point of a half-wave dipole is very high at the even-order harmonic frequencies; it may range from 50 to 150 Ω at the fundamental and odd harmonics, but it can be as large as several thousand ohms at the even harmonics. *See also* ODD-ORDER HARMONIC.

EXALTED-CARRIER RECEPTION

Selective fading causes severe distortion on amplitude-modulated signals, especially when the carrier frequency is attenuated significantly. When that happens, and the sidebands are much stronger than the carrier for a moment, the signal becomes nearly unintelligible (*see* SELECTIVE FADING). Exalted-carrier reception is a method of reducing the effects of selective distortion on the signal carrier.

In the exalted-carrier system, the signal is split into two branches in the intermediate-frequency section of the receiver. One branch amplifies the entire signal in the usual manner. The other branch contains a narrow-band filter that allows only the carrier signal to pass. The carrier signal is amplified greatly in the stages immediately following the narrow-band filter, so that when the carrier is recombined with the signal at a later stage, the amplitude of the carrier is much greater than that of the sidebands. The figure illustrates a block diagram of this principle.

If the carrier amplitude is exaggerated, or exalted, the signal readability is not adversely affected. Distortion occurs only when the amplitude of the carrier is not great enough. By overamplifying the carrier with the technique illustrated, it is always at a sufficient amplitude, even during a fade, to allow good signal readability. Thus the adverse effects of fading are reduced. An even better system more often used today is single sideband system, in which the entire carrier is supplied at the receiver. *See also* SINGLE SIDEBAND.

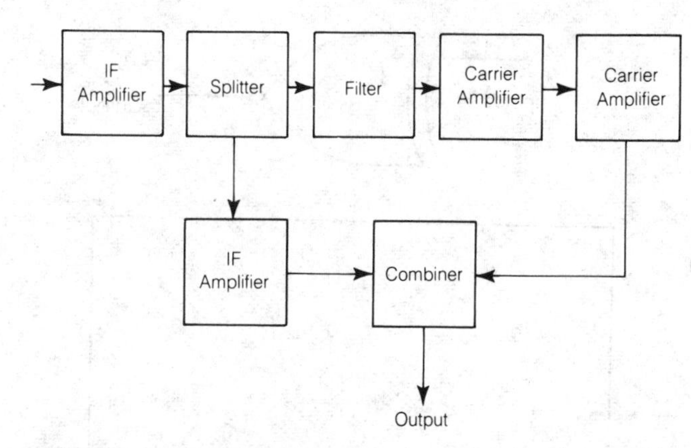

EXALTED-CARRIER RECEPTION: A block diagram of a receiver for exalted-carrier reception.

EXCITATION

Excitation is the driving power, current, or voltage to an amplifier circuit. The term is generally used in reference to radio-frequency power amplifiers. Excitation is sometimes called drive or driving power.

Class A amplifiers theoretically require only a driving voltage, but no excitation power. In practice, a small amount of excitation power is required in the Class A amplifier. In the Class AB and Class B amplifiers, some excitation is required to obtain proper operation. In the Class C amplifier, a large amount of excitation power is needed in order to obtain satisfactory operation.

The circuit that supplies the excitation to a power amplifier is called the driver or exciter. When a transmitter is used in conjunction with an external power amplifier, the transmitter itself is called an exciter. *See also* CLASS A AMPLIFIER, CLASS AB AMPLIFIER, CLASS B AMPLIFIER, CLASS C AMPLIFIER, DRIVE, DRIVER.

EXCITER

See EXCITATION.

EXCLUSIVE-OR GATE

The OR function may be either inclusive or exclusive. Unless otherwise specified, it is considered inclusive. The normal OR function is true if either or both of the inputs are true, and false only if both of the inputs are false. The exclusive OR function is true only if the inputs are opposite; if both inputs are true or both are false, the exclusive-OR function is false.

The illustration shows the schematic symbol of the exclusive-OR logic gate and a truth table of the function. The logic symbol 1, or high, indicates true, and 0, or low, indicates false. *See also* LOGIC GATE.

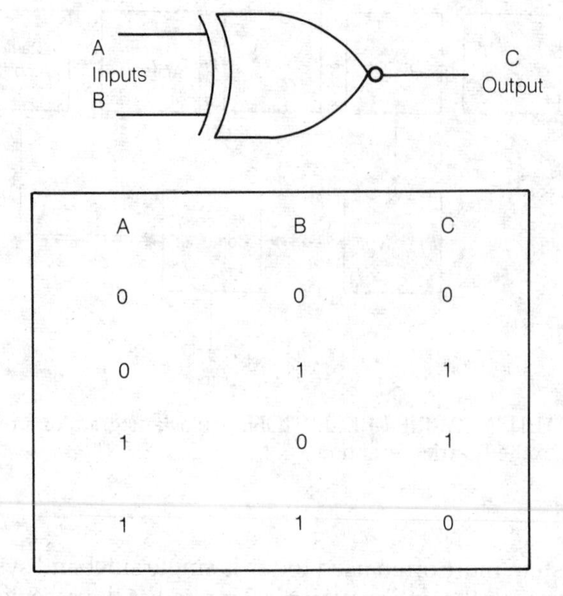

A	B	C
0	0	0
0	1	1
1	0	1
1	1	0

EXCLUSIVE-OR GATE: The schematic symbol and the truth table for an exclusive-OR gate.

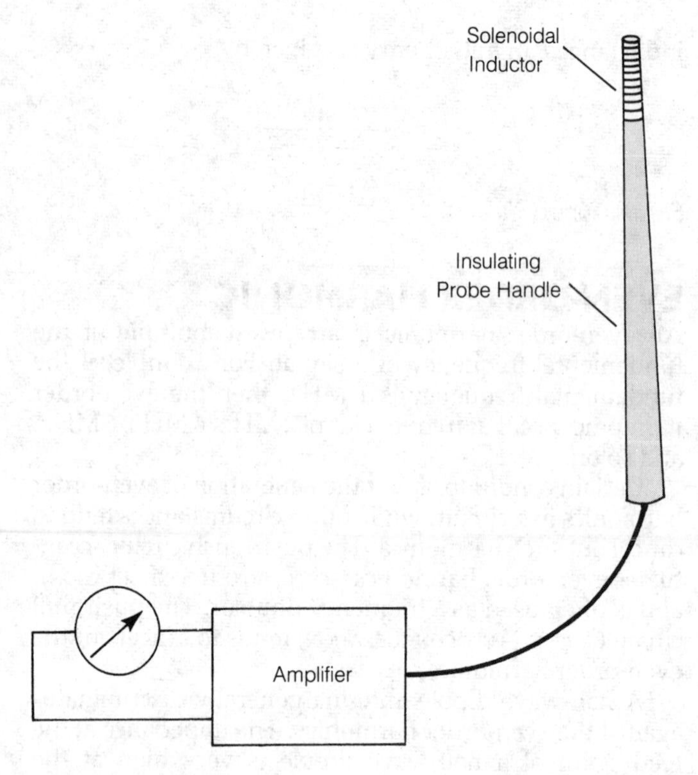

EXPLORING COIL: An exploring coil is an instrument for detecting magnetic fields.

EXPANDER

An expander is a circuit that increases the amplitude variations of a signal. Expansion is the opposite, electrically, of compression (*see* COMPRESSION, COMPRESSION CIRCUIT). At low signal levels, the expander has little effect; the amplification factor increases as the input-signal amplitude increases.

An amplitude expander is used at the receiving end of a circuit in which compression is employed at the transmitter. By compressing the amplitude variations at the transmitting end of a circuit, and expanding the amplitude at the receiver, the signal-to-noise ratio is improved. The weaker parts of the signal are boosted before propagation. When the weak components are attenuated at the receiver, the noise is attenuated also. A system that makes use of compression at the transmitter and expansion at the receiver is called an amplitude compandor. *See also* COMPANDOR.

EXPLORING COIL

An exploring coil, sometimes called a sniffer, is a form of radio-frequency or magnetic-field detector. The sensor consists of a small inductor at the end of a probe and a shielded or balanced cable connection to an amplifier. The amplifier output is connected to an indicating device such as a meter, light-emitting diode, or oscilloscope. The illustration is a simple diagram of an exploring-coil device.

When a magnetic field is present in the vicinity of the exploring coil, an indication is obtained whenever the coil is moved across the magnetic lines of flux. If a radio-frequency field is present, an indication will be obtained even if the coil is held still. An exploring coil can be used to determine if the shielding is adequate in a particular circuit. The exploring coil may also be used to obtain a small signal for monitoring purposes, without affecting the load impedance of the circuit under observation. *See also* EXPLORING ELECTRODE.

EXPLORING ELECTRODE

An exploring electrode is a sensor similar to an exploring coil. The sensor consists of a small pickup electrode connected to an amplifier circuit. The output of the amplifier circuit is connected to an indicating meter, light-emitting diode, or oscilloscope.

Either an exploring coil or an exploring electrode can be used to detect radio-frequency fields. However, for some applications, the electrode is preferable to the coil. An exploring electrode, like the exploring coil, causes little or no change in the load impedance of a circuit under test. The coupling mode of the exploring electrode is capacitive, or electrostatic; that of the exploring coil is inductive, or magnetic.

An exploring electrode generally presents an extremely high impedance to the input of the amplifier. Shielding of the cable from the probe to the amplifier is necessary, and the radio-frequency ground must be complete or stray electromagnetic fields will be picked up by the cable and amplified. *See also* EXPLORING COIL.

EXPONENTIAL DISTRIBUTION

The exponential distribution is a form of statistical distribution, used for determining the probability that an event will occur within a specified time interval. The longer the time interval, the greater the chances of an event occurring. In general, the exponential distribution function is given by:

$$P = fe^{-ft}$$

where P is the probability, f is the frequency of the occurrence, t is the length of the time interval, and e is approximately 2.718.

EXPONENTIAL FUNCTION

An exponential function is a function that describes the rise and decay curves of many different phenomena. The general form of the exponential function is:

$$f(x) = ke^{mx}$$

where k and m are constants and e is approximately 2.718.

If the value of the exponential constant, m, is positive, then the function is positive and increasing (if k is positive) or negative and decreasing (if k is negative). If m is negative, then the value of the function is positive and decreasing (if k is positive) or negative and increasing (if k is negative). These four situations are illustrated in the figure for the cases (k,m) = (1,1), (k,m) = (−1,1), (k,m) = (1,−1), and (k,m) = (−1,−1).

A charging or discharging capacitor behaves according to an exponential function. So does a damped oscillation. Radioactive substances decay according to an exponential function. *See also* EXPONENTIAL DISTRIBUTION, FUNCTION.

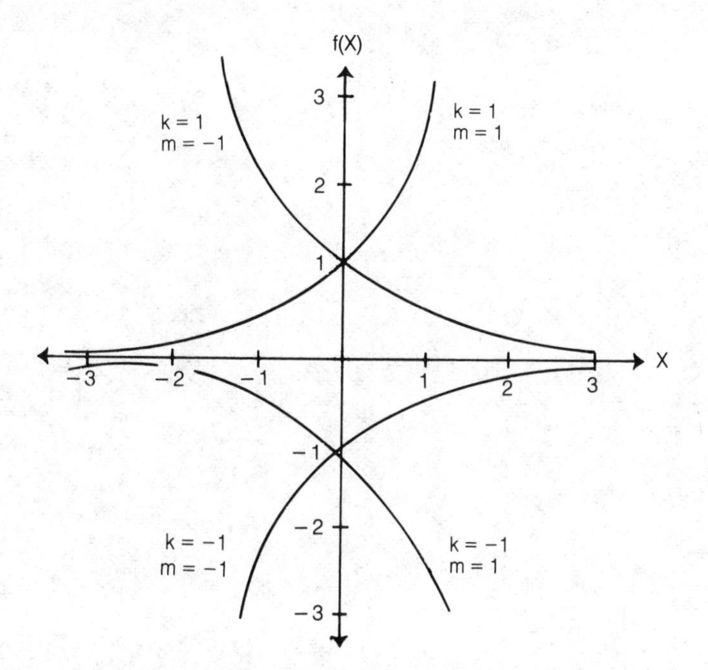

EXPONENTIAL FUNCTION: Four basic exponential functions.

EXTRAPOLATION

When data is available within a certain range, an estimate of values outside that range can be made by a technique called extrapolation.

An example of extrapolation is shown in the graph. An attenuation-versus-frequency curve is illustrated for a hypothetical bandpass filter. The center frequency is 3.600 MHz; data are provided for the range 3.595 to 3.605 MHz. To estimate the attenuation at a frequency of, say, 3.607 MHz, it is useful to extrapolate. This is done by extending the function over the range 3.605 to 3.607 MHz, based on the nature of the curve within the range given. Assuming the function contains no unnatural irregularities, a good estimate can be obtained for the attenuation at 3.607 MHz.

Random-number functions are impossible to extrapolate. Other very complicated functions can be difficult or impossible to extrapolate with reasonable accuracy. The accuracy of extrapolation diminishes as the independent variable gets farther from the domain of values given. *See also* FUNCTION, INTERPOLATION.

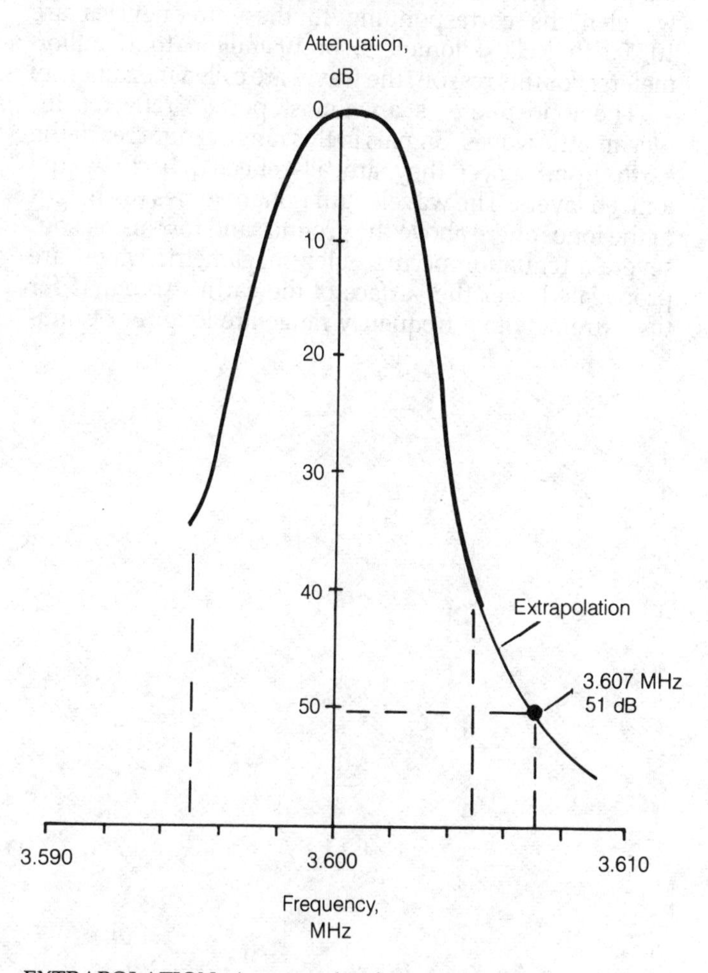

EXTRAPOLATION: An example of extrapolation of a curve. At 3.607 MHz, attenuation is approximately 51 dB.

EXTREMELY HIGH FREQUENCY

The extremely high frequency (EHF) band of the radio spectrum extends from 30 to 300 GHz. The wavelengths corresponding to these limit frequencies are 10 to 1 millimeters, and thus the extremely high frequency waves are called millimetric waves.

Electromagnetic waves in the extremely high frequency range are not affected by the ionosphere of the earth. Signals in this range pass through the ionosphere as if it were not there. However, they are affected by oxygen resonance in the atmosphere. Millimetric waves behave very much like infrared rays, visible light, and ultraviolet. They can be focused by parabolic reflectors of modest size. Because of their ability to propagate in straight lines unaffected by the atmosphere of the earth, millimetric waves are used in satellite communications. The frequencies allow modulation by wideband signals. *See also* ELECTROMAGNETIC SPECTRUM, FREQUENCY ALLOCATIONS.

EXTREMELY LOW FREQUENCY

The extremely low frequency (ELF) band of the electromagnetic spectrum extends from 30 to 300 Hz. The wavelengths corresponding to these frequencies are 10,000 to 1,000 kilometers, or 10 million to 1 million meters. For this reason, the waves are called megametric.

The ionosphere is an almost perfect reflector for megametric waves. Signals in this range cannot reach the earth from space; they are all reflected back by the ionized layers. The wavelength is many times the height of the ionosphere above the ground, and for this reason, severe attenuation occurs when megametric waves are propagated over the surface of the earth. Antennas for the extremely low frequency range are long. For example, at 30 Hz, a half-wave dipole must be approximately 3,100 miles long.

Megametric waves may be used to generate ground currents in which the entire planet may become resonant. Research is continuing in this area. Megametric waves penetrate the ocean, and therefore are useful in communicating with submarines. However, only very narrow band modulation is possible. *See also* ELECTROMAGNETIC SPECTRUM, FREQUENCY ALLOCATIONS.

EXTRINSIC SEMICONDUCTOR

An extrinsic semiconductor is a semiconductor to which an impurity has been deliberately added. The pure form of the semiconductor is called intrinsic. Impurities must generally be added to semiconductor materials to make them suitable for use in solid-state devices such as transistors and diodes. The process of adding impurities to semiconductors is called doping (*see* DOPING).

Generally, the greater the amount of impurity added to a material (such as silicon or gallium arsenide), the better the conductivity becomes. Certain impurities result in charge transfer by electrons; this is called an N-type semiconductor. Other doping elements form a material in which holes carry the charge; these materials are called P-type semiconductors. *See also* ELECTRON, HOLE, INTRINSIC SEMICONDUCTOR, N-TYPE SEMICONDUCTOR, P-TYPE SEMICONDUCTOR.

FACSIMILE

Facsimile (fax) is a process for scanning fixed graphic material (including written or printed text, photographs, and line drawing) and converting that information into signals. The signals are transmitted by radio or telephone lines, then reconstructed by the receiving station, and finally duplicated as hardcopy on paper.

Fax machines for use over dialed telephone lines are now available as consumer products. Individual units are capable of both the scanning and transmission and the receiving and reproduction processes.

FADING

Fading is a phenomenon that occurs with radio-frequency signals propagated through the ionosphere. Fading takes place because of changes in the conditions of the ionosphere between a transmitting and receiving station, at frequencies where ionospheric propagation is permitted.

The ionosphere is somewhat unstable, and as its height above the earth changes, the phase of a signal reflected from it also changes. Several signals might propagate over different paths simultaneously, and the resulting phase combination at a receiver can cause fading.

Fading can also occur because of conditions in the ionosphere between the transmitting and receiving station. This kind of fading is apt to take place near the maximum usable frequency for the ionosphere at a given time (*see* MAXIMUM USABLE FREQUENCY). A signal may be perfectly readable and then suddenly disappear.

Fading sometimes takes place because of multipath propagation. A signal is reflected from the ionosphere at several different points en route to the receiving station (*see* MULTIPATH FADING). When the ionosphere changes orientation rapidly, the fading may occur first at one frequency, and then at another, progressing across the passband of a signal from top to bottom or vice versa. This is called selective fading. The narrower the bandwidth of a signal, the less susceptible it is to selective fading. Since selective fading is common, narrowband signals are preferred over wideband signals in the frequency range where ionospheric propagation is predominant. *See also* SELECTIVE FADING.

FAHRENHEIT TEMPERATURE SCALE

The Fahrenheit temperature scale is used in the United States and other countries, mainly for weather reporting and forecasting. The scale is centered at the freezing point of a saturated saltwater solution; this temperature is assigned the value zero. Pure water freezes at 32 degrees above zero and boils at a temperature that depends on the atmospheric pressure. At sea level under normal (standard) pressure, water boils at a Fahrenheit temperature of 212°. Fahrenheit readings are indicated by the capital letter *F*. *See also* CELSIUS TEMPERATURE SCALE, KELVIN TEMPERATURE SCALE.

FAILSAFE

Failsafe is a term for a form of failure protection in electronic circuits and systems. A failsafe circuit is designed so that, in the event of malfunction of one or more functions, the circuit will shut down without catastrophic damage and no hazard will be presented to personnel using the equipment. A failsafe system can place a backup circuit in operation if a circuit fails. This allows continued operation. A warning signal alerts the personnel that a circuit has failed, and this expedites repair.

In a computer, failsafe is sometimes called failsoft. A malfunction in a failsoft computer will result in some degradation of efficiency, but will not cause a total shutdown. The computer alerts the user that a problem is present, giving its location in the system.

No system is 100 percent reliable, but failsafe circuits and systems reduce the amount of downtime in the event of component failures. *See also* DOWNTIME, FAILURE RATE.

FAILURE RATE

Failure rate is an expression of the frequency of component, circuit, or system failure. Generally, failure rate is specified in terms of the average number of failures per unit time. The longer the time interval chosen for the specification, the greater the probability that a given component, circuit, or system will malfunction.

An example of component failure rate is as follows: Suppose that out of 100 diodes three malfunction within a period of 1 year. Assume that all of the diodes are operational at the start of the 1-year period. Then the failure rate is 3 percent per year for that particular application of the diode. The probability is 3 percent that a given diode will fail within 1 year.

When determining the failure rate for a component, circuit, or system, it is customary to test many units at the same time. This quantity testing increases the number of observed failures and allows a better probability estimate than would be possible with just one test unit.

As the time interval approaches zero in the failure-rate specification, the failure rate itself does not approach zero. Some components, circuits, or systems fail the instant they are switched on. The probability that a component will fail at any given moment is called the hazard rate. *See also* QUALITY CONTROL, RELIABILITY.

FALL TIME

See DECAY TIME.

FAMILY

Any set of similar entities, such as components, devices, or mathematical functions, is called a family. For example, in digital logic, there are many families of bipolar logic including LS, ALS, FAST, ECL10K, and ECL10KH.

The illustration shows a family of curves that is a two-dimensional representation of a three-dimensional function. There are two independent variables: one is shown (in this example) along the horizontal axis, and the other is represented in incremental form by the different curves. The horizontal axis shows collector voltage, and the different curves are illustrated for various values of base voltage for a bipolar transistor. The dependent variable, shown on the vertical axis, is the collector current.

The constituents of a family are always related in some way. Totally different items cannot be members of the same family. However, there are many ways of specifying the relationship among members of a family.

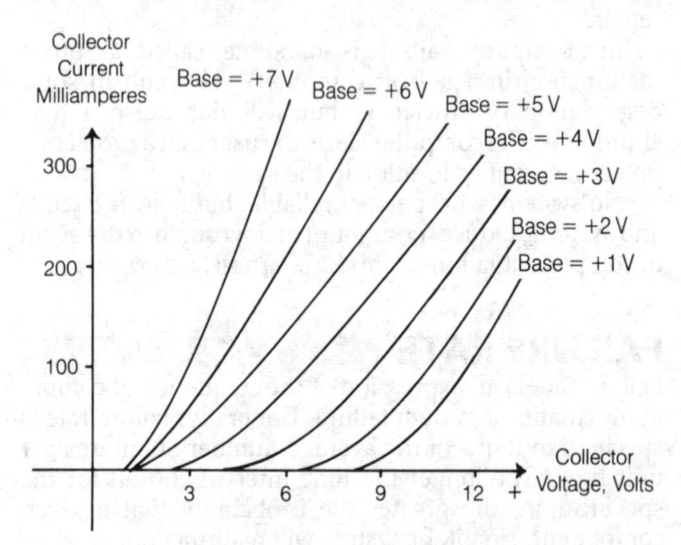

FAMILY: A family of curves for the operation of an NPN transistor.

FAN

A fan is an air-cooling machine with a rotating impeller. Fans are employed in electronic circuits that generate large amounts of heat, especially power supplies. Fans are sometimes also used for cooling solid-state equipment such as computers, because excessive temperatures can cause an increase in the frequency of errors in digital systems.

The illustration shows a small fan used in a computer system to keep the integrated circuits below their critical temperature. With the fan, the computer can be operated at a higher room temperature with a lower error rate. If the fan is not used, the system may malfunction at a lower ambient temperature. *See also* AIR COOLING.

FAN: A 4¹¹⁄₁₆-inch square, brushless direct-current fan for cooling electronics provides an air flow of 106 cubic feet per minute.

FAN-IN AND FAN-OUT

Fan-in and *fan-out* are terms used to describe the connection of several terminals to a common circuit. When a circuit has more than one input terminal, the inputs are said to be fanned in. When a circuit has more than one output terminal, the outputs are said to be fanned out.

In a digital computer, fan-in is the number of inputs that can be accommodated by a particular logic circuit, and fan-out is the number of outputs that can be fed by a particular logic circuit. For example, an AND gate might have many inputs; for the output to be high, all of the inputs must be high. There may be several outputs, as well, each feeding a subsequent circuit. Fan-out is a function of the transistors used in the logic. *See* DIGITAL-LOGIC TECHNOLOGY.

FARAD

A farad is the unit of capacitance. When the voltage across a capacitor changes at a rate of 1 volt per second, and a current flow of 1 ampere results, the capacitor has a value of 1 farad. A capacitance of 1 farad also results in 1 volt of potential difference for a charge of 1 coulomb.

The farad is, in practice, an extremely large unit of capacitance. Capacitance in electronics is generally measured in microfarads, or millionths of a farad (μF); for small capacitors, the picofarad, or trillionth of a farad (pF) is often specified. Sometimes the nanofarad, or billionth of a farad (nF) is specified. *See also* CAPACITANCE, CAPACITOR.

FARADAY CAGE

A Faraday cage is an enclosure that prevents electric fields from entering or leaving. However, magnetic fields can pass through. The Faraday cage consists of a wire screen or mesh, or a solid nonmagnetic metal, broken so that there is not a complete path for current flow. The Faraday cage concept is used to protect components and circuit boards from electrostatic discharge (ESD). *See* ELECTROSTATIC DISCHARGE.

Faraday cages are used in experimentation and testing, when the presence of electric fields is not desired. An entire room may be shielded. *See also* ELECTROMAGNETIC SHIELDING.

FARADAY EFFECT

When radio waves are propagated through the ionosphere, their polarization changes because of the effects of the magnetic field of the earth. This is called the Faraday effect or Faraday rotation. Because of the Faraday effect, sky-wave signals arrive with random, and fluctuating, polarization. It makes very little difference what the polarization of the receiving antenna may be; signals originating in a vertically polarized antenna may arrive with horizontal polarization some of the time, vertical polarization at other times, and slanted polarization at still other times.

The Faraday effect results in signal fading over ionospheric circuits. There are other causes of fading, as well. The fading caused by changes in signal polarization can be reduced by the use of a circularly polarized receiving antenna (*see* CIRCULAR POLARIZATION).

When light is passed through certain substances in the presence of a strong magnetic field, the plane of polarization of the light is made to rotate. This is a form of Faraday effect. For this to occur, the magnetic lines of flux must be parallel to the direction of propagation of the light. *See also* POLARIZATION.

FARADAY SHIELDING

See ELECTROSTATIC SHIELDING, FARADAY CAGE.

FARADAY'S LAWS

Faraday's laws are concerned with two different phenomena, and are therefore placed into two categories.

The law of electromagnetic induction is sometimes called Faraday's law. This law is the principle of the generation of a current in a wire that moves with respect to a magnetic field. The wire can be stationary and the field can be moving; but as long as the wire cuts across magnetic lines of flux, a current is induced. The current is directly proportional to the rate at which the wire crosses the lines of flux of the magnetic field. The faster the motion, the greater the current; the more intense the field, the greater the current. It is this effect, in part, that makes radio communication possible (*see* ELECTROMAGNETIC INDUCTION).

The other principle form of Faraday's law concerns electrolytic cells. In any electrolytic cell, the greater the amount of charge passed through, the greater the mass of substance deposited on the electrodes. The actual mass deposited for a given amount of electric charge depends on the electrochemical equivalent of the substance. *See also* ELECTROLYSIS.

FAR FIELD

The far field of an antenna is the electromagnetic field at a great distance from the antenna. The far field has essentially straight lines of electrical and magnetic flux, and the lines of electric flux are perpendicular to the lines of magnetic flux. The wavefronts are essentially flat planes.

The far field of an antenna has a polarization that depends on several factors. The polarization of the transmitting antenna dictates the polarization of the far field under conditions in which there is no Faraday effect (*see* FARADAY EFFECT). The power density of the far field diminishes with the square of the distance from the antenna. The field strength, in microvolts per meter, diminishes in direct proportion to the distance from the antenna.

The far field of an antenna is also called the Fraunhofer region. It begins at a distance that depends on many factors including the wavelength and the size of the antenna. The signal normally picked up by a receiving antenna is the far field except at extremely long wavelengths. *See also* ELECTROMAGNETIC FIELD, NEAR FIELD, TRANSITION ZONE.

FEDERAL COMMUNICATIONS COMMISSION

The Federal Communications Commission (FCC) is an agency of the United States Government. It is responsible for the allocation of frequencies for radiocommunications and broadcasting within the United States. The FCC is also responsible for the enforcement of the laws concerning telecommunications.

The FCC issues various kinds of licenses for communications and broadcasting personnel. Some licences can be obtained only by passing an examination. The FCC composes and administers these examinations. Other licenses require only the filing of an application.

When the FCC deems it necessary to create a new rule, it first publishes a notice of proposed rulemaking. Then, within a certain specified time period, interested or concerned parties are allowed to make comments on the subject. The FCC, after considering the comments, makes its decision. The power of the FCC is limited, however, by the U.S. Congress.

Each country determines its own frequency allocations and system of radiocommunications laws. The

International Telecommunication Union provides coherence among the many nations of the world, for the purpose of optimum utilization of the limited space in the spectrum. *See also* INTERNATIONAL TELECOMMUNICATION UNION.

FEED

Feed is the application of current, power, or voltage to a circuit. The term feed is used particularly with reference to antenna systems. There are basically three electrical methods of antenna feed: current feed, voltage feed, and reactive feed.

A current-fed antenna has its transmission line connected where the current in the radiating element is greatest. This occurs at odd multiples of ¼ wavelength from free ends of the radiator. A voltage-fed antenna has the transmission line connected at points where the current is minimum; such points occur at even multiples of ¼ wavelength from free ends, and also at the free ends. Both current feed and voltage feed are characterized by the absence of reactance; only resistance is present. In current feed, the resistance is relatively low; in voltage feed, it is high.

When an antenna is not fed at a current or voltage loop, the feed is considered reactive. This may be the case when an antenna is not resonant at the operating frequency, or when it is fed at a point not located at an integral multiple of ¼ wavelength from a free end.

Feed may be classified in other ways, such as according to the geometric position of the point where the transmission line joins the radiating element. In end feed, the line is connected to the end of the element; in center feed, to the center; in off-center feed, somewhat to either side of the center. A vertical radiator may be fed at the base, or part of the way up from the base to the top. A special method of feed for a vertical radiator is called shunt feed. *See also* CENTER FEED, CURRENT FEED, END FEED, OFF-CENTER FEED, TRANSMISSION LINE, VOLTAGE FEED.

FEEDBACK

When part of the output from a circuit is returned to the input, the situation is known as feedback. Sometimes feedback is deliberately introduced into a circuit; sometimes it is not wanted. Feedback is called positive when the signal arriving back at the input is in phase with the original input signal. Feedback is called negative (or inverse) when the signal arriving back at the input is 180 degrees out of phase with respect to the original input signal. Positive feedback often results in oscillation, although it can enhance the gain and selectivity of an amplifier if it is not excessive. Negative feedback reduces the gain of an amplifier stage, makes oscillation less likely, and enhances linearity.

Feedback can result from a deliberate arrangement of components. All oscillators use positive feedback. Unwanted feedback can result from coupling among external wires or equipment. This external coupling might be capacitive, inductive, or perhaps acoustic.

The number of amplifying stages that can be cascaded obtaining high gain is limited partly by the effects of feedback. The more stages connected in cascade, the higher the probability that some positive feedback will occur, with consequent oscillation. *See also* ACOUSTIC FEEDBACK, NEGATIVE FEEDBACK, OSCILLATION.

FEEDBACK AMPLIFIER

A feedback amplifier is a circuit placed in the feedback path of another circuit to increase the amplitude of the feedback signal. The phase can also be inverted. A feedback amplifier is used when the feedback signal would not otherwise be strong enough to obtain the desired operation.

An example of a feedback amplifier is shown in the illustration. In this circuit, the feedback signal is direct current obtained by rectifying the signal output. The greater the amplitude of the signal at the output, the larger the direct-current voltage applied to the input. The direct-current voltage reduces the gain of the amplifier. This example illustrates amplified automatic level control (*see* AUTOMATIC LEVEL CONTROL).

The term *feedback amplifier* is used to describe an amplifier that uses feedback to obtain a certain desired characteristic. Negative feedback, for example, is often used to reduce the possibility of unwanted oscillations in a radio-frequency amplifier. Combinations of negative and positive feedback can be used to modify the frequency response of an amplifier. This is especially true of operational-amplifier circuits. *See also* FEEDBACK, OPERATIONAL AMPLIFIER.

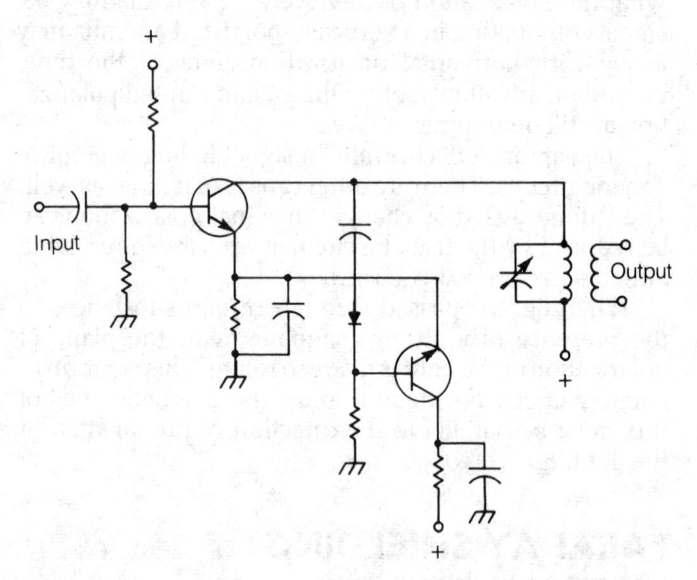

FEEDBACK AMPLIFIER: This amplifier is used for automatic level control.

FEEDBACK CONTROL

Whenever feedback is deliberately introduced into a circuit, some means must be provided to control the amount of feedback. Otherwise, the desired circuit operation may not be obtained. A feedback control may consist of a simple potentiometer to adjust the amount of feedback signal arriving at the input of the circuit. More sophisticated methods of feedback control consist of self-regulating circuits, such as the feedback amplifier employed in the amplified automatic level-control circuit (*see* AUTOMATIC LEVEL CONTROL, FEEDBACK AMPLIFIER).

The elimination of undesired feedback in a circuit is called feedback control. The output of a circuit can be partially coupled back to the input by the capacitance or inductance between the input and output peripheral wiring. In a public-address system, acoustic feedback can take place if the microphone picks up sound from the speakers. Coupling between the input and output should be minimized. This can be done with bypass capacitors, series chokes, or phase-inverting circuits. *See also* FEEDBACK.

FEEDBACK RATIO

The feedback ratio, usually represented by the Greek letter beta (β), is a measure of the amount of feedback in a feedback amplifier. If e_o is the output voltage of an amplifier with no load, and e_f is the feedback voltage, the feedback ratio is given simply as:

$$\beta = \frac{e_f}{e_o}$$

When the feedback ratio is to be expressed as a percentage, it is generally represented by *n*, and:

$$n = \frac{100e_f}{e_o}$$

This quantity is called the feedback percentage. The feedback ratio is never greater than 1, and the feedback percentage is never greater than 100.

If *A* is the open-loop gain of an amplifier and β is the feedback ratio, then the feedback factor *m* is the quantity given by:

$$m = 1 - \beta A$$

See also FEEDBACK.

FEEDER CABLE

Any cable that carries signal information from one place to another is a feeder cable. A coaxial transmission line for a radio-frequency antenna system is a feeder cable. Usually, the term *feeder cable* described a communication cable running from a central station to a distribution station. From the distribution station, secondary cables, called distribution cables, run to local stations and subscribers. The coaxial cable used to carry cable-television signals is called a feeder cable. *See also* CABLE TELEVISION.

FEED LINE

A feed line is a transmission line for transferring an electromagnetic field from a transmitter to an antenna. The point where the feed line joins the antenna is called the feed point.

Feed lines may take a variety of forms. The most common type of feed line is the coaxial line. Two-wire line, also called open-wire or parallel-wire line, is often used in the feed system of a television receiving antenna. *See also* COAXIAL CABLE, FOUR-WIRE TRANSMISSION LINE.

FEEDTHROUGH CAPACITOR

A feedthrough capacitor is a capacitor that permits passing a lead through a chassis while bypassing the lead to chassis ground. The center (axial) part of the feedthrough capacitor is connected to the lead; the outer part of the component is connected to chassis ground. The illustration shows a feedthrough capacitor.

Feedthrough capacitors are useful in power-supply leads when it is desirable to prevent radio-frequency energy from being transferred into or out of the equipment along these leads. Feedthrough capacitors are also used in the speaker or microphone leads of an audio amplifier circuit.

Feedthrough capacitors are available in many different physical and electrical sizes. *See also* BYPASS CAPACITOR, CAPACITOR.

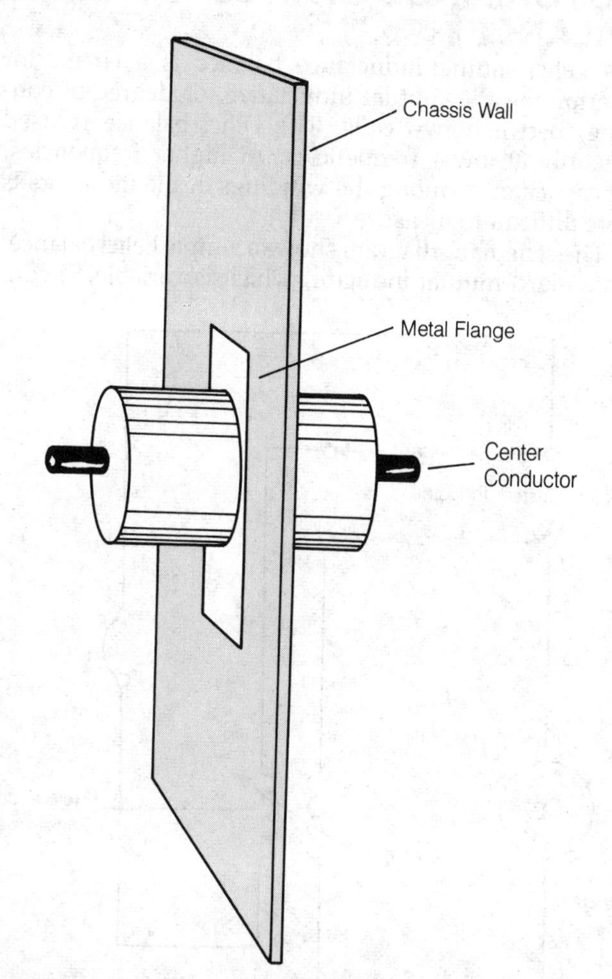

FEEDTHROUGH CAPACITOR: The metal flange is connected to the chassis ground in the installation of this capacitor.

FEEDTHROUGH INSULATOR

A feedthrough insulator is used for passing a lead

through a metal chassis while maintaining complete electrical insulation from the chassis. Feedthrough insulators are also used when it is necessary to pass leads through a partial conductor or nonconductor with a minimum amount of dielectric loss. Feedthrough insulators generally consist of a threaded shaft with two conical sections of porcelain or glass held in place by nuts on the shaft.

Feedthrough insulators are available in a variety of sizes. The voltage across a feedthrough insulator may, in some applications, reach substantial values. Therefore, a sufficiently large insulator should always be used. In an antenna-tuning network, the feed through insulators can be subjected to several thousand volts under certain conditions. The feedthrough insulators should have the lowest possible loss at the highest frequency of the equipment in which they are used. The insulators should also have the smallest possible amount of capacitance. *See also* INSULATOR.

FELICI MUTUAL-INDUCTANCE BALANCE

The Felici mutual-inductance balance is a circuit for determining the mutual inductance, or degree of coupling, between two coils. The Felici balance is used primarily at lower frequencies. At higher frequencies, the capacitance among the windings of the inductors is more difficult to measure.

The schematic diagram shows a simple Felici balance. A standard mutual inductor, which is variable, is connected together with the pair of inductors for which the mutual inductance is to be determined. The reference transformer is adjusted until a null is observed on the indicator. This indicator can be a meter or, if the signal generator produces audio-frequency energy, a headset or speaker. When the null is seen, it indicates that the reference inductor has been set to have the same mutual inductance as the unknown. The mutual inductance may then be read directly from a calibrated scale. *See also* MUTUAL INDUCTANCE.

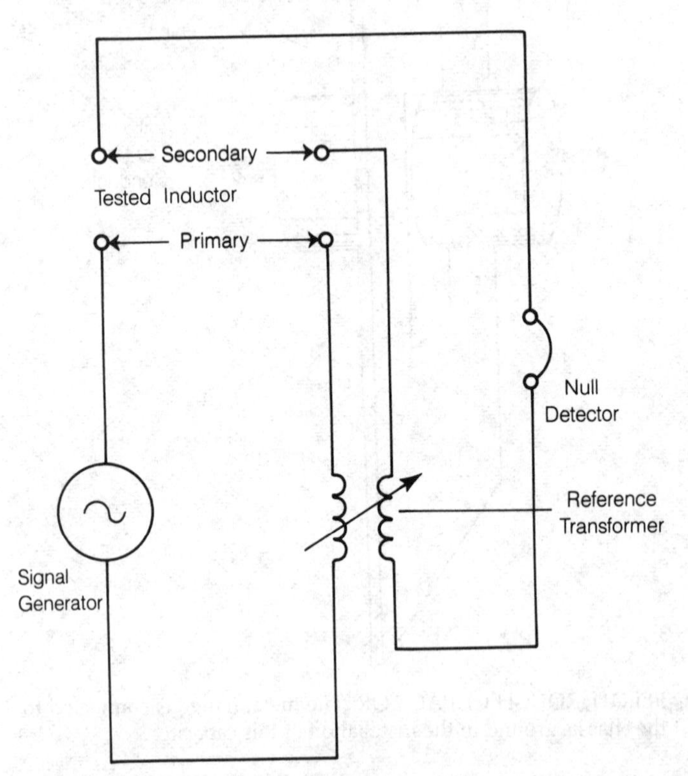

FELICI MUTUAL-INDUCTANCE BALANCE: This circuit determines the degree of coupling between the windings of a transformer.

FEMALE

Female is an electrical term used to describe a jack. A female plug is recessed, and the conductors are not exposed. An example of a female plug is the common utility wall outlet.

The female jack is usually, but not always, mounted in a fixed position on the panel of electronic equipment. The male plug is generally attached to the cord or cable and fits into the female plug. Connectors may have only one conductor, or they may have many conductors.

FEMTO

See PREFIX MULTIPLIERS.

FERRIC OXIDE

Ferric oxide is a form of iron oxide with a chemical formula of Fe_2O_3. Ferric oxide is reddish in color, and is commonly known as rust.

Ferric oxide has magnetic properties, and also is easily powdered and applied to surfaces. Ferric oxide is used in the manufacture of recording tapes. The material is ground into a very fine dust and is applied to the surface of the tape, usually made from Mylar or a similar flexible material. *See also* MAGNETIC RECORDING, MAGNETIC TAPE.

FERRITE

Ferrite is a substance capable of providing high magnetic permeability in an inductor core. Ferrite has a higher degree of permeability than ordinary powdered iron. Ferrite materials are classified as either soft or permanent. Ferrite acts as an electrical insulator, resembling a ceramic with respect to current conductivity. The magnetic conductivity, however, is excellent, and the eddy-current losses are very small.

Ferrite materials are available in a variety of shapes and permeability values. The small receiving antennas used in transistor radios consist of a coil wound on a solenoidal ferrite core. These antennas have excellent sensitivity at frequencies below the shortwave range. Ferrite is employed in toroidal inductor cores to obtain large values of inductance with a relatively small amount of wire. Ferrite is also used to make pot cores, which allows even larger inductances to be obtained. Typical permeability values for ferrites range from about 40 to more than 2,000. Ordinary powdered-iron cores generally have much smaller permeability values. Ferrite is well suited to low-frequency and medium-frequency applications in which loss must be kept to a minimum. *See also* FERRITE BEAD, FERRITE CORE, MAGNETIC CORE, PERMEABILITY, TOROID.

FERRITE BEAD

A ferrite bead is a small piece of ferrite toroid. Ferrite beads are used for choking off radio-frequency currents on wire leads and cables. The bead is slipped around the wire or cable, as illustrated. This introduces an inductance for high-frequency alternating currents without affecting the direct currents and low-frequency alternating currents.

Ferrite beads are especially useful for choking off antenna currents on coaxial transmission lines in the very high frequency range and higher. Antenna currents, or radio-frequency currents flowing on the outside of a coaxial feed line, can cause problems with transmitting equipment. One or more ferrite beads along the feed line are effective in choking off these unwanted currents. But the desired transmission of electromagnetic fields inside the cable is not affected.

Ferrite beads are used in ferrite core digital memory systems to store information with magnetic fields. The polarity of the magnetic field can be either clockwise or counterclockwise. The polarity will remain the same until sufficient current is sent through a conductor passing through the bead to reverse the polarity of the magnetic field stored in the ferrite. The polarity will remain reversed until another surge of current, in the opposite direction, changes the field orientation again. *See also* FERRITE, MAGNETIC CORE, TOROID.

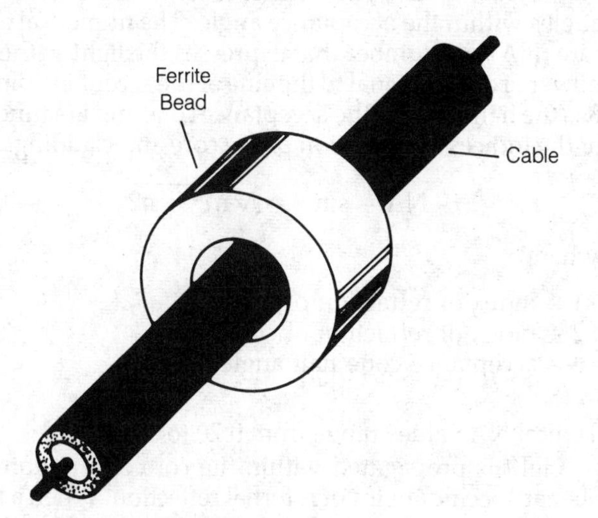

FERRITE BEAD: A ferrite bead blocks radio-frequency current on the outer conductor of a coaxial cable.

FERRITE CORE

A ferrite core is an inductor core made from ferrite material. Ferrite cores are extensively used in electronic applications at low, medium, and high frequencies for maximizing the inductance of a coil while minimizing the number of turns. This results in low-loss coils.

The most common ferrite-core configurations are the solenoidal type and the toroidal type. The entire magnetic flux is contained within the core material. This eliminates coupling to external components (*see* TOROID).

At audio frequencies and very low frequencies, the pot core is often used. This core actually surrounds the coil, whereas in other inductors the coil surrounds the core. Pot cores allow inductance values in excess of 1 henry to be realized with a moderate amount of wire. These coils, because of the low loss in the ferrite material, exhibit extremely high Q factors, and are useful in applications where extreme selectivity is needed.

At higher frequencies, ferrite becomes lossy, and powdered-iron cores are preferred. *See also* FERRITE, MAGNETIC CORE.

FERROELECTRICITY

Ferroelectricity is the polarization of electric dipoles in certain insulating materials (*see* DIPOLE). It is very similar to magnetism, in which magnetic dipoles become polarized. Ferroelectric substances are used to make certain kinds of microphones and speakers.

Examples of ferroelectric substances include barium titanate, barium strontium titanate, potassium dihydrogen phosphate, Rochelle salts, and triglycine sulfate. These materials are essentially ceramics. Certain waxes and plastics can be heated and then cooled in a strong electric field to become ferroelectric substances. *See also* DIELECTRIC, DIELECTRIC POLARIZATION.

FERROMAGNETIC MATERIAL

A ferromagnetic material is a material that has a high magnetic permeability. The most common ferromagnetic materials used in electronics are ferrite and powdered iron. These substances increase the inductance of a coil when used as the core.

Ferrite and powdered iron are highly resistive to the direct application of electric current, but will carry magnetic fields with very little loss. The permeability of a ferromagnetic material is a measure of the degree to which the magnetic lines of flux are concentrated within the substance (*see* PERMEABILITY).

Ferromagnetic materials are formed into different shapes for different applications. *See also* FERRITE, MAGNETIC CORE, TOROID.

FET

See FIELD-EFFECT TRANSISTOR.

FET VOLTMETER

An FET voltmeter is an instrument similar to a vacuum-tube voltmeter (*see* VACUUM-TUBE VOLTMETER). The input impedance is extremely high, so the instrument does not draw much current from the source. An FET voltmeter uses a field-effect-transistor amplifier circuit to achieve the high input impedance.

The illustration shows a simplified schematic diagram

of an FET voltmeter. The ranges are selected by varying the gain of the amplifying stage, or by connecting (through a switch) various resistances in parallel with the microammeter.

The FET voltmeter also can be used to measure current and resistance by using the meter and power supply in conjunction with various resistors. The field-effect transistor acts as an amplifier to increase the sensitivity. This combination meter is called an FET volt ohm-milliammeter, or FET VOM. *See also* FIELD-EFFECT TRANSISTOR.

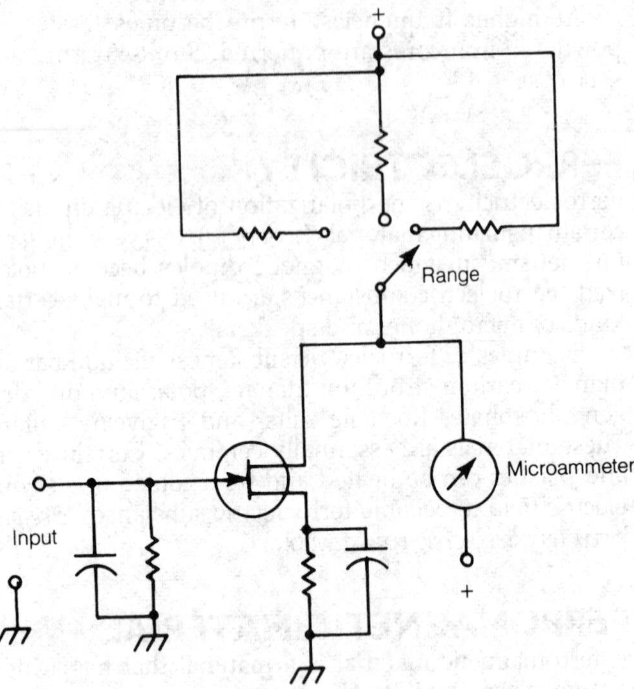

FET VOLTMETER: This sensitive FET (field-effect transistor) voltmeter has low current drain.

FIBEROPTIC CABLE AND CONNECTORS

Fiberoptic communication takes place over thin optical fiber, a transparent medium that connects a photoemitter to a photodetector in a transmission system. (*See* FIBEROPTIC COMMUNICATIONS.) Optical fiber consists of a cylinder of transparent dielectric material with a specific refractive index whose walls are in contact with a second dielectric material of a lower refractive index. Alternatively, it can be a cylinder whose core has a refractive index that gets progressively lower away from the center. The fiber relies on internal reflection to transmit light along its axial length. Light enters one end of the fiber and emerges from the opposite end with only minimal loss. It has also been called an optical waveguide, or light pipe.

Optical fiber is packaged in an assembly of materials called a fiberoptic cable. The materials provide tensile strength, external protection, and handling properties comparable to those of equivalent diameter coaxial cables.

Optical cable connectors are manufactured products that are fitted to the ends of the fiberoptic cable to permit coupling and decoupling with a minimal loss of light energy. They are similar in many respects to coaxial cable connectors.

Optic-Fiber Theory Light propagates along an optical fiber by internal reflection if the core of the fiber is surrounded by optical cladding with a lower refractive index. The light can be modulated to communicate intelligence. Transmission of light or optical energy is a form of electromagnetic wave propagation. The use of ray theory or geometric optics makes it easier to understand the concepts of light propagation.

The index of refraction of a material is the ratio of the velocity of light in a vacuum to the velocity of light in a given medium. (Light velocity is slower in all substances than it is in a vacuum.) The index of refraction is not constant; it varies with color or wavelength of light.

A light ray is refracted at a surface separating two regions with different indexes of refraction. (*See* SNELL'S LAW.) A ray of light also will be internally reflected at the boundary between two dielectric media when the ray is incident within the denser medium and the angle of incidence is greater than a critical angle θ defined by the refractive indexes of the media. (The critical angle for internal reflection can be calculated from Snell's law.)

A multimode fiber can accept and propagate nonaxial rays or modes that enter the fiber core at an angle (with respect to the core axis) that is less than the critical angle. The fiber consists of a core surrounded with cladding material of a lower index of refraction. Multimode fibers can be step index or graded index.

Light energy enters the end surface of a fiber at an infinite number of angles as shown in Fig. 1. It is accepted and transmitted down the core only for those entry angles within the acceptance angle. The numerical aperture (NA) is a number that expresses the light-gathering power of a fiber, equal to the sine of the acceptance angle. It is the half angle of the acceptance cone and is a function of the indices of refraction of the core and cladding:

$$\mathrm{NA} = \sin\theta = \sqrt{n1^2 - n2^2}$$

where

$n1$ = index of refraction of core.
$n2$ = index of refraction of cladding.
θ = acceptance cone half angle.

Typical NA values range from 0.20 to 0.27.

Light is propagated within the core of a multimode fiber at specific angles of internal reflection. When a light ray strikes the core-cladding interface, it is reflected and follows a zig-zag course down the core as shown in Fig. 1. If the light strikes the interface at an angle greater than the internal critical angle it will not be reflected.

Figure 2A illustrates propagation in a step index fiber, one of the two multimode fibers. The fiber has an abrupt change in refractive index because the core and cladding have different indices of refraction. The rays reflected at higher angles A and B must travel a greater distance than those such as C and D that enter at a shallower angle and

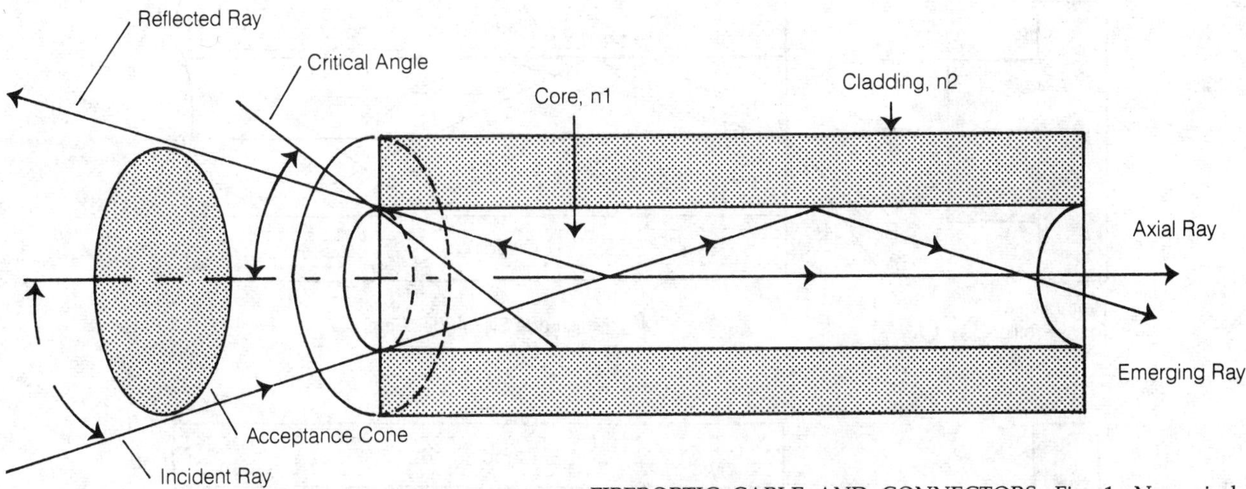

FIBEROPTIC CABLE AND CONNECTORS: Fig. 1. Numerical aperture or NA is the sine of the input half angle (θ) and the total acceptance angle is $2\theta_c$.

travel further before being reflected. Therefore, these higher order modes take more time to travel down the fiber than the lower order modes. This separation or spreading of input optical signals along the length of the optical fiber is called dispersion. It can cause distortion in communications.

A multimode fiber with a graded refractive index as shown Fig. 2B overcomes dispersion because it has a refractive index that gets progressively lower away from the axis. This causes the light rays to be continually refocused by refraction, rather than being guided by internal reflection. This causes the light to travel in smooth bending paths, more slowly as it approaches the core center and faster with increasing radial distance from the centerline. As shown in the figure, rays A, B, and C with different order modes all travel at approximately the same speed.

The single-mode fiber as shown in Fig. 2C is a low-loss optical waveguide with a very small core. It requires a laser source for the input signals because of the very small entrance aperture (acceptance cone). Special attention must be given to matching the laser source to the fiber and fiber to the detector. Single-mode core has a typical diameter of 1 to 8 microns to allow only a single, lower order mode of propagation.

Fiberoptic Cables The best fiber cores and cladding are made of silica glass doped with impurities to give them different refraction properties. The fibers are protected from damage by the same kinds of jacketing materials used to protect wire cable. The jacket might include a steel wire or plastic fibers to impart additional tensile strength. Special connectors are available for interconnecting fiberoptic cables and cables to active emitters and detectors. Cable is the costliest item in most medium- and long-haul fiberoptic systems.

Fiberoptic communications systems under one kilometer in length are considered to be short haul. A local area network (LAN) using fiberoptic cable in place of coaxial cable is an example of a short-haul system. (*See* COAXIAL CABLE, LOCAL AREA NETWORK.) They are typically used for the connection of computers and peripherals within buildings or clusters of building. Short-haul systems are usually privately owned and maintained. They are normally constructed from commercially available components.

However, medium- and long-haul systems connect computers and peripherals separated by tens to hundreds of kilometers. Most of these systems are owned by the government or public-service telecommunications utilities. Typically, the fiberoptic and electronic components as well as the cable itself in these longer links are custom-fabricated to meet customer requirements.

Fiber Selection. Step index fibers will usually transmit digital and analog data up to a limit of about 30 MHz. However, graded-index fibers have an upper limit of about 500 MHz. Single-mode fiber permits only very small numercal apertures.

The selection in commercial in optical fiber includes: 1) all-glass step- and graded-index fibers, 2) plastic and glass combined in step-index, 3) plastic-clad silica (PCS) fiber, and 4) all-plastic step-index fibers.

Three variables considered in addition to numerical aperture (NA) in selecting fiberoptic cable are: 1) attenuation, 2) bandwidth, and 3) core diameter.

Attenuation. Attenuation is the loss in optical signal power due to the absorption and scattering of optical radiation at a given wavelength in a length of fiber. It is expressed as a rate of loss in decibels of optical power per kilometer (dB/km). All glass fibers fabricated by depositing doped silica glass from high-purity gases have the lowest optical attenuation.

Bandwidth or Dispersion. Bandwidth is a measure of the highest sinusoidal modulation frequency that can be transmitted through a length of fiber at a specified optical wavelength without losing more than 50 percent of the signal power. It is expressed in megahertz per kilometer of length (MHz/km). Bandwidths of 200 MHz/km to 1000

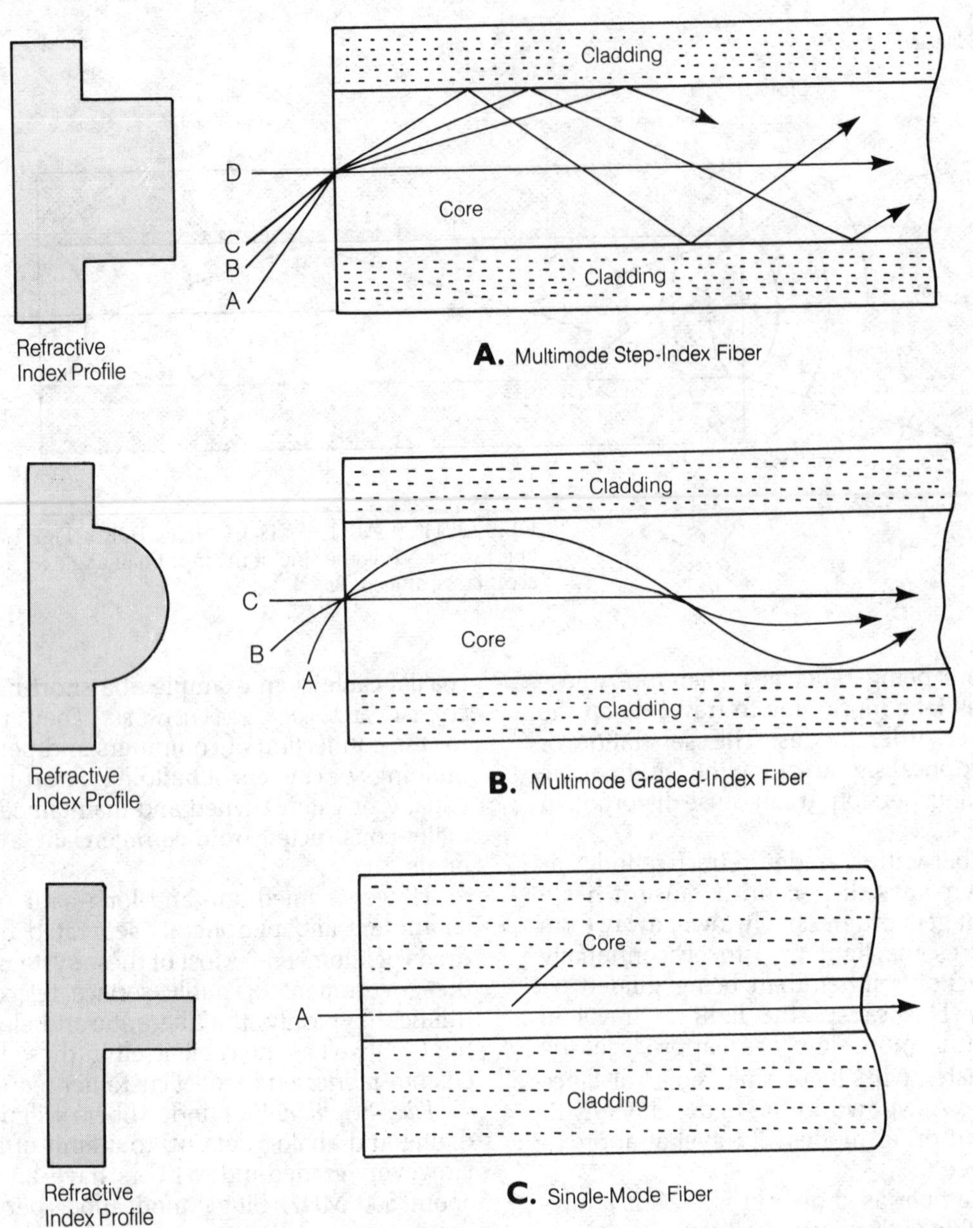

FIBEROPTIC CABLE AND CONNECTORS: Fig. 2. Principal optical fibers include multimode step index fiber (A), multimode graded-index fiber (B), and single-mode fiber (C).

MHz/km are obtained with graded 50-micron (μm) glass cores, but they can only be 20 MHz/km with 300 μm step-index glass cores.

Core Diameter. Core diameter is the diameter of the central region of an optic fiber. They range from 50 to 300 μm for commercial cable. The cladding diameter of 50 μm diameter glass core is typically 125 μm, while that of a 300 μm diameter glass core is typically 440 μm. Plastic-clad silica (PCS) fiber is available with core diameters from 125 μm to 1 mm.

Optical signals with only a few milliwatts of power can be detected after traveling dozens of kilometers in silica optical fiber. However the signal is attenuated as it travels through the fiber because of light scattering. This requires that repeaters be installed to boost and regenerate the signal to overcome losses over long distances.

At the shorter infrared wavelengths of about 850 nanometers (nm), the typical attenuation of a silica fiber is 2.5 to 3 decibels per kilometer. Attenuation declines to 0.3 to 1 dB/km at 1300 nm. However, it rises sharply again at about 1390 nm because of the presence of the oxygen-hydrogen (OH) impurity. The greatest transparency of silica is reached at around 1550 nanometer where attenuation is less than 0.15 dB/km. Thus light at that wavelength can travel 3 to 10 times as far as an 850 nanometer signal with the same power.

Chromatic dispersion also affects repeater spacing. Because the index of refraction varies for different wavelengths of light, chromatic dispersion introduced by the fibers causes signals with slightly different wavelengths to travel through through the fiber at different speeds. This becomes a problem in digital transmission. In extreme cases, each pulse becomes broad enough to interfere with neighboring pulses, increasing the bit-error rate. Chromatic dispersion forces a tradeoff between repeater spacing and bit rate.

Fiberoptic Cable Fabrication. A fiberoptic cable can contain one or many individual fibers. Typical standard commercial construction for glass core fibers contains 1, 2, 6, 12 or 18 fibers; however the actual construction can vary with application. The two choices that can be made in the selection of fiberoptic cable include the jacket compound and the strength member. The most commonly used jacket compounds to protect the fiber against impact and abrasion are polyvinyl chloride (PVC), polyethylene and polyurethane, generally in that order. (*See* CABLE.) Strength members are used inside the cable to support the optical fiber and relieve strain on it. Fiberglass rod, Kevlar yarn, and steel wire are commonly used strength members.

Fiberoptic Connectors. Connections must be carefully made between fiber ends in a cable and optoelectronic components. Special connectors have been designed for fiberoptic cable, many of them based on coaxial cable connectors. No standards have been universally accepted, but many proprietary cable connector styles have been introduced.

A fiberoptic connector must: 1) align mating fibers for efficient transfer of optical power, 2) couple fibers to optical devices, 3) protect the fiber from the environment and during handling, 4) terminate the cable strength member, and 5) provide cable strain relief. A major problem in connecting mating fibers is axial alignment.

Cable-to-cable connections may be accomplished by any of a number of different methods for aligning the axial fibers. These methods include clamping both fiber ends in vee-grooves or compressing them with multiple rods. The optical power loss of a connector-to-connector interface is usually between 0.5 and 2 decibels, depending on both the style of the connector and the quality of workmanship. Fiberoptic devices including IREDs, injection laser diodes and photodiodes are packaged in connector receptacles to simplify alignment to the fiber.

Splicing Optical Fiber. Many different methods have been developed for making low-loss splices in the field because it is impractical to removing long lengths of cable and return them to a repair shop. These include splice hardware employing the same techniques used in connectors for aligning the fibers.

FIBEROPTIC COMMUNICATION

Fiberoptic communication is the technique for communicating information in the form of light waves modulated by analog or digital signals and transmitted through glass or plastic fibers. Fiberoptic communication has many advantages over conventional coaxial cable and metallic-wire links for the transmission of voice and data. These advantages include increased bandwidth, smaller diameter, lower weight, lack of crosstalk, and complete immunity to inductive interference. Fiberoptic transmission is in use in telecommunications, computers, military systems, and many other applications.

A basic fiberoptic communications system is shown in the diagram. It consists of a light energy source within the transmitter module that includes a modulator. The input signal can be a serial digital bit stream or an analog signal, either audio or video. The source typically converts electrical energy directly to light energy while it is modulated. The light energy is closely coupled to the optical cable and a light energy detector at the receiving end. The receiver module must be capable of demodulating and amplifying the received signal.

For low data rates of less than 50 million bits per second (Mb/s) and for short-distance links (typically under one kilometer), an IRED is a suitable light source. It provides incoherent light. For data rates greater than 50 Mb/s and longer-distance links, a gallium arsenide laser diode is more suitable. It is a source of coherent light capable of transmitting light through small (10- to 12-micron) clad optical fibers. For telephone voice transmission, a data rate of 50 Mb/s allows 780 voice channels to be combined using time-division multiplexing (TDM). The 780 voice channels can be transmitted over a single transmission link. A data rate of 150 Mb/s allows 2340 voice channels.

At the receiving end of short-range links, a photodiode or phototransistor converts the light into electric signals while timing and decision circuits regenerate the information. For longer distances, avalanche photodiodes are used. Optical connectors couple the light from the source to the cable and from the cable to the detector. Long-haul links may include one or more repeaters, essentially combination receiver-transmitter modules. *See* FIBEROPTIC CABLE AND CONNECTORS, INFRARED EMITTING DIODE, LASER DIODE, PHOTODETECTOR.

Fiberoptic links couple computers and peripherals together in local area networks (LANs) confined to a single building or cluster of buildings. (*See* LOCAL AREA NETWORK.) The links can also replace copper cables in public and private dial-up telephone systems and in undersea cables. They are now competitive with microwave terrestrial links and satellite communications.

Fiberoptic links can be simplex, half duplex or full duplex. (*See* DATA COMMUNICATIONS.) Short, low-data rate fiberoptic link assemblies are available commercially for research and development or as a component in instruments and computer systems. More complex, higher data rate links may be assembled from off-the-shelf components, but high-speed, long-haul links are usually assembled from custom-made components.

The advantages of fiberoptic communications include:

- Increased bandwidth because they can handle wider bandwidths than copper wire or coaxial cable links.
- Smaller size and lower weight because a single optical fiber can replace a very large bundle of copper wires. For example, a typical telephone cable might contain 1000 pairs of copper wires and have a diameter of 100 millimeters (4 inches). By contrast, a single pair of glass optical fibers in a cable with a diameter of 5 millimeters (⅕ inch) is capable of handing the same traffic. Moreover, protective

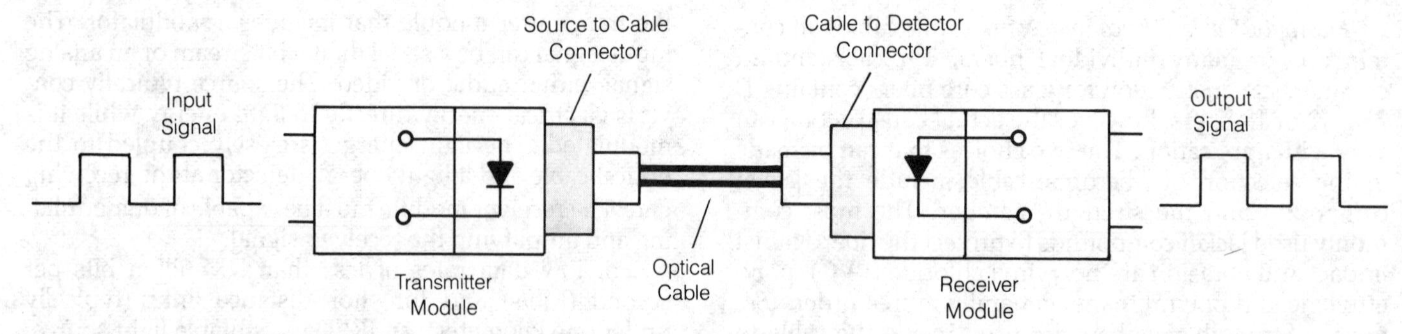

FIBEROPTIC COMMUNICATION: Basic components of a fiberoptic data transmission system. Long links include repeaters between the transmitter and receiver to boost signal level.

jacketing accounts for most of the diameter; the fibers have 125 micron (μm) diameters. The saving in weight makes fiberoptic cable easier and faster to install. It can be passed through ducts and conduits, placed under floors or put in locations that would not permit the passage of conventional wire or cable.

- Lower attenuation because losses in optical fiber are far less than in copper wire; a smaller number of repeaters is needed to amplify the signal over the same long distance.
- No electromagnetic interference because optical fibers are insulators and do not induce, generate or transmit electric current or magnetic fields. Fibers must be protected from crushing and abrasion, but they do not have to be armored or shielded.
- More rugged because optical fibers are inherently rugged and can actually be twisted and tied without breakage. They are also inert to the gases or acids that would corrode copper in above- or below-ground installations.
- Better safety because optical fibers are electrical insulators and conduct no current. Thus they present no short-circuit hazards for the equipment coupled to them or in their immediate vicinity. Moreover, they will not conduct voltage induced by nearby lightning strikes. Fiberoptic cables can pass through spaces where there are explosives, volatile gases, or air-gas mixtures without presenting a hazard. They can also pass through flammable materials or chemicals or be submerged under water or in some chemical solutions. These features by themselves often justify the installation of a fiberoptic link.
- Better security because it is usually necessary to break the optical fiber and install a coupler to tap information. This makes an intrusion easier to detect. Also, the fiber does not radiate energy that can be detected by remote sensors. The installation of an optical coupler is a difficult, time-consuming task even for an expert. The fiberoptic cable can be alarmed to send a warning signal automatically to the terminals if the fiber is broken. Alternatively, a section of the cable may be stripped so that it can be sharply bent to couple out light. But this is also time-consuming and a less reliable method for the intruder, and it also can be detected.

FIELD-EFFECT TRANSISTOR

A field-effect transistor (FET) is a voltage-operated transistor. Unlike a bipolar transistor (*see* TRANSISTOR), a FET requires very little input current, and it exhibits extremely high input resistance. There are two major classes of field-effect transistors: junction FETs (JFET) and metal-oxide semiconductor FETs (MOSFET), also known as insulated-gate FETs (IGFET). FETs are further subdivided into P-channel and N-channel devices. FETS are sometimes called unipolar transistors because, unlike the bipolar transistor, the drain current consists of only one type of charge carrier—electrons in an N-channel FET and holes in a P-channel FET.

JFETs and MOSFETs are both made as discrete transistors, but MOSFET technology has been adopted for the fabrication of power FET transistors (*see* POWER TRANSISTOR) and integrated circuits. (*see* INTEGRATED CIRCUIT). There are both NMOS and PMOS ICs. When both P-channel and N-channel transistors are integrated into the same gate circuit, it is known as complementary MOS or CMOS. (*See* COMPLEMENTARY METAL-OXIDE SEMICONDUCTOR.)

Junction Fets. The N-channel JFET, illustrated by the section view in Fig. 1A is made by a diffusion- or ion-implantation processes. An N-channel is diffused into a P-type substrate and then P-type impurities are diffused into the N-channel to form the gate. Metal terminals are formed directly on the N-type gate regions. All FETs have three terminals: source, gate, and drain. With the symmetrical construction shown, the drain and source are interchangeable.

If a positive voltage is applied at the drain and a negative voltage is applied at the source with the gate terminal open, a drain current flows. When the gate is biased negative with respect to the source, the PN junctions are reverse biased, and depletion regions are formed. The N-channel is more lightly doped than the P-type material, so the depletion region penetrates into the channel. This is a region depleted of charge carriers so it behaves like an insulator. The depletion region narrows the channel and increases its resistance. If the gate bias voltage is made even more negative, the drain current is cut off completely. The gate bias voltage that cuts off the drain current is called the pinch-off or gate-cutoff voltage. However, as the bias becomes positive, the depletion region recedes, the channel resistance is reduced, and drain current increases. Thus the FET gate controls the FET current.

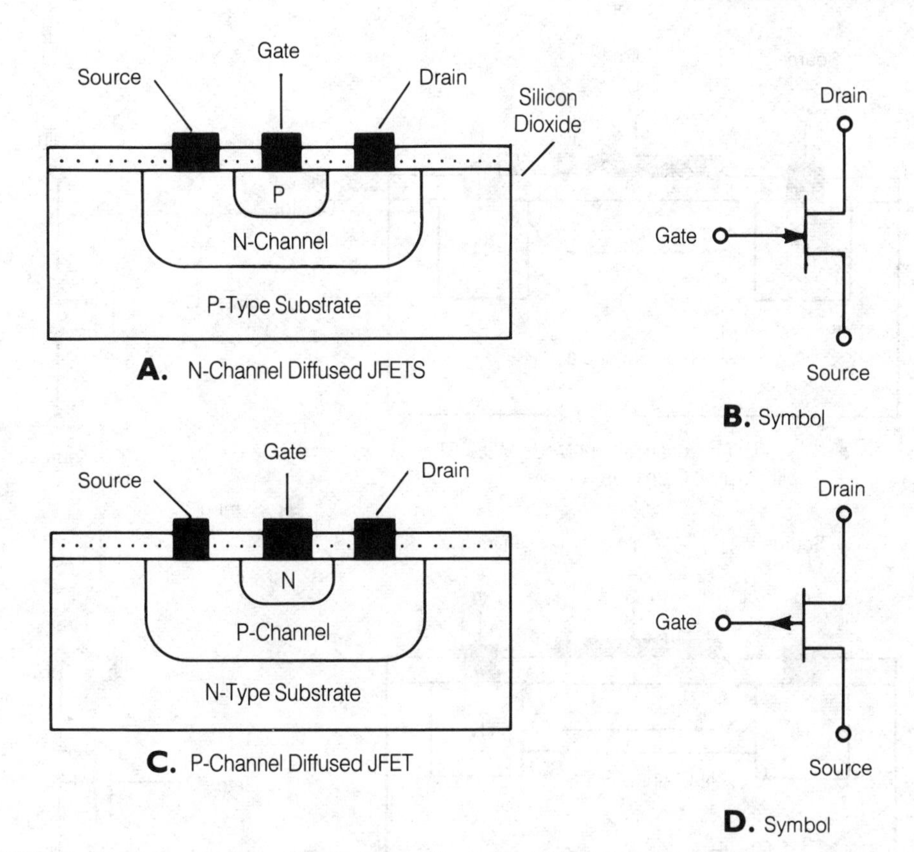

FIELD-EFFECT TRANSISTOR: Fig. 1. An N-channel JFET (A) and its symbol (B). A P-channel JFET (C) and its symbol (D).

The symbol for the N-channel JFET is shown in Fig. 1B. In accepted transistor symbols, the arrowhead points from P to N; for an N-channel FET, the arrowhead points from the P-type gate towards the N-type channel.

The P-channel JFET is illustrated in Fig. 1C. The channel in this device a P-type material, and the gate region is N-type. If a negative voltage is applied to the drain and a positive voltage is applied to the source, current flows (by convention) from the source to the drain. To reverse bias the junctions between the gate and the channel, the N-type gate must be made positive with respect to the P-type channel.

The symbol for the P-channel JFET is shown in Fig. 1D. The arrowhead also points from P-type material to N-type material. In this case, it points from the P-type channel to the N-type gate. The characteristics for the P-channel JFET are similar to those of the N-channel device, except that voltage and current polarities are reversed.

Metal-Oxide Semiconductor FET (MOSFET). The manufacture of an N-type MOSFET starts with a P-type wafer. Then N-type regions are diffused into the wafers as shown in Fig. 2A. The surface is coated with a layer of silicon oxide, and windows are etched through the oxide to contact the N-type regions. Metal is deposited on the oxide layer through the holes to form the drain and source terminals, and a metal gate is deposited on the oxide layer.

If the drain of an N-type MOSFET is made positive with respect to the source and no voltage is applied to the gate, the two N-type regions and the P-type substrate form back-to-back PN junctions. The junctions permit only minor leakage current flow. However, if the gate is made positive with respect to the source, negative charge carriers (electrons) are induced in the channel between the two N-type regions. As the gate voltage is increased, more electrons are induced in the channel. Because the electrons cannot flow across the oxide layer to the gate, they accumulate at the substrate surface below the gate oxide. The electrons form an N-type channel connecting the drain to the source. Thus, a drain current flows, and its magnitude depends on the channel resistance. Therefore, the MOSFET gate voltage controls the drain current.

Because the conductivity of the channel is enhanced by the positive bias on the gate, the device is known as an enhancement-mode MOSFET. The oxide insulation prevents leakage current and the MOSFET has a higher input impedance than the JFET.

The symbol for an N-channel enhancement-mode MOSFET is shown in Fig. 2B. In this symbol the gate does not make direct contact with the channel. The arrowhead points from the P-type substrate towards the (induced) N-type channel represented by a line broken into three sections. This indicates that the channel does not exist until a gate voltage is applied in the enhancement-mode MOSFET.

A P-channel enhancement-mode MOSFET is made the same way as the N-channel enhancement device except that P-type drain and source regions are diffused into an N-type substrate. The symbol for a P-type enhancement-mode MOSFET is the same as the one shown

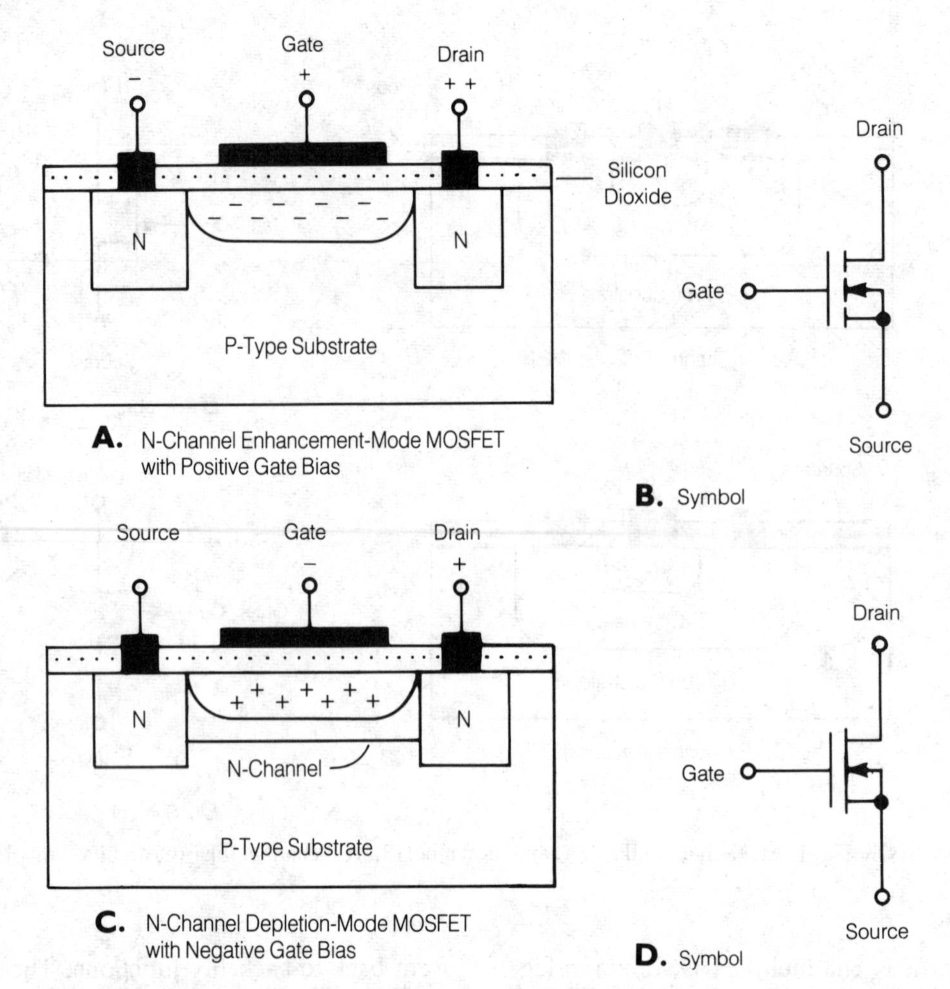

FIELD-EFFECT TRANSISTOR: Fig. 2. An N-channel enhancement mode MOSFET (A) and its symbol (B). An N-channel depletion mode MOSFET (C) and its symbol (D).

in Fig. 1B, except that the direction of the arrow is reversed.

Depletion-Mode MOSFET. If an N-type MOSFET is made with a lightly doped N-channel between the heavily doped source and drain regions as shown in Fig. 2C, a depletion mode MOSFET is formed. When the drain is made positive with respect to the source, a drain current will flow, even with zero gate voltage. However, if the gate is made negative with respect to the substrate, positive charge carriers (holes) induced in N-channel will combine with the electrons and cause channel resistance to increase. With increasing negative bias, the pinch-off voltage will be reached, and drain current will cease. However, if the gate is made positive with respect to the substrate, additional electrons are induced and channel current increases.

With increasing positive voltage, the N-type depletion-mode MOSFET functions as an enhancement-mode MOSFET. The symbol for the N-type depletion-mode MOSFET shown in Fig. 2B is similar to that of the N-type enhancement-mode MOSFET except that the line representing the channel is solid.

FIELD STRENGTH

Field strength is a measure of the intensity of an electric, electromagnetic, or magnetic field. The strength of an electric field is measured in volts per meter. The strength of a magnetic field is measured in gauss (*see* GAUSS). The intensity of an electromagnetic field is generally measured in volts per meter, as registered by a field-strength device. The intensity of an electromagnetic field can also be measured in watts per square meter.

The electromagnetic field strength from a transmitting antenna, as measured in volts, millivolts, or microvolts per meter, is proportional to the current in the antenna and to the effective length of the antenna. The field strength is inversely proportional to the wavelength and distance from the antenna. The field strength in watts, milliwatts, or microwatts per square meter varies with the power applied to the antenna and inversely with the square of the distance. The field strength always depends on the direction from the antenna as well, and is influenced by such factors as phasing systems and parasitic elements. *See also* ELECTROMAGNETIC FIELD.

FIELD-STRENGTH METER

A field-strength meter is an instrument designed for measuring the intensity of an electromagnetic field. This meter may be simple and broadbanded, or complex, incorporating amplifiers and tuned circuits. The field-strength meter usually provides an indication in volts, millivolts, or microvolts per meter.

The simplest type of field-strength meter consists of a microammeter, a semiconductor diode, and a short length of wire that serves as the pickup. This instrument is uncalibrated, but is useful for estimating the level of radio-frequency energy in a particular location.

A more sophisticated field-strength meter may use an amplification circuit to measure weak fields. Generally, the amplifier uses a tuned circuit to avoid confusion resulting from signals on frequencies other than those desired. The figure is a simplified schematic diagram of a field-strength meter.

The most complex field-strength meters are built into radio receivers. Accurately calibrated S meters can serve as field-strength meters in advanced receivers. Most S meters in common receivers are calibrated, but are not precise enough for actual measurements of electromagnetic field strength. The precision field-strength meter is employed in antenna testing and design, primarily for measurements of gain and efficiency. *See also* S METER.

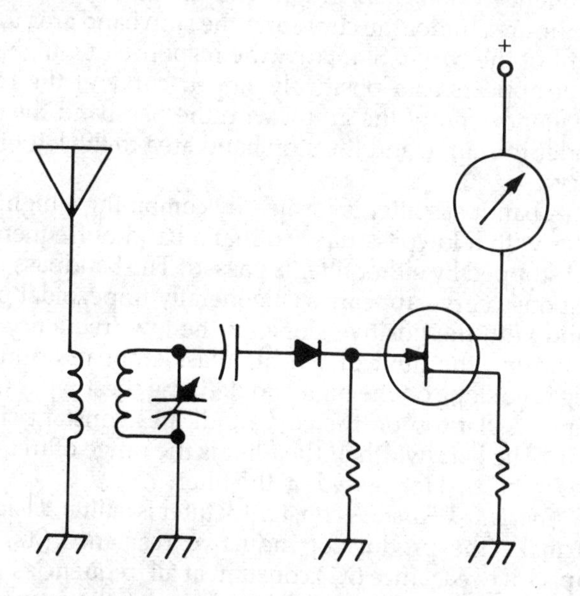

FIELD-STRENGTH METER: This meter has a FET (field-effect transistor) amplifier and a tuned circuit.

FIFO

See FIRST-IN/FIRST-OUT.

FIGURE OF MERIT

The figure of merit is a measure of the quality of a capacitor or inductor. It is specified as the ratio of the reactance to the resistance. The greater the resistance in proportion to the reactance, the larger the loss in the device, and the lower the figure of merit. In a tuned circuit, the figure of merit affects the selectivity, and is known as the Q factor.

Generally, the larger the value of an inductor or capacitor, the lower the figure of merit; large-value capacitors have more dielectric material to cause loss, and use lossier dielectrics. Larger value inductors require more wire and the core material tends to be lossier. *See also* Q FACTOR.

FILAMENT

A filament is a thin coil of wire that emits heat and light when an electric current is passed through it. The incandescent bulb filament glows white hot and produces light. A vacuum tube filament generates heat to drive electrons off the cathode.

Filaments in vacuum tubes take two basic forms. The directly heated cathode consists of a filament which also serves as the cathode. It is connected to a negative source of voltage for tube operation. The indirectly heated cathode has a thin, cylindrical cathode around the filament; the filament heats the cathode, but is not directly connected to it.

FILE

A file is a data store. All of the elements in a file are generally related according to format or application. Data files can be modified; it is easy to add, delete, or change a file.

An example of a file is a section of text in a word processor. The file is stored on a magnetic disk or tape. There is a limit to the size of the file. Any input data consisting of the standard alphabetic, numeric, and punctuation symbols, along with spaces in any arrangement, can be placed in the file by the operator of the system. The file can be deleted, changed, or appended at any time. The file is given a name and this name is used to retrieve the file and to distinguish it from others on the same disk or tape.

A file set is a collection of data consisting of smaller individual files. For example, a file may be kept of tax-deductible expenditures for a given year; the combination of all the annual files, over a long period of time, is the file set. *See also* DATA PROCESSING, WORD PROCESSING.

FILTER

A filter is a network able to discriminate between frequencies by passing components or signals in one frequency band while blocking components or signals outside that frequency band. Filters can be classified into five basic types. For example, a lowpass filter allows the lower frequency components or signal to pass while the higher frequencies are attenuated. A highpass filter does the opposite, allowing the high-frequency component or signal to pass. Filters are classed as passive or active.

Passive Filters. Passive filters are networks designed with resistors, capacitors, and inductors connected to be frequency selective. However, the components are reactive and dissipative so that gain is always less than one and output power is always less than input power.

Power supply filters are used in ac-to-dc (alternating current to direct current) and dc-to-dc power supplies to smooth ripples or pulsations in the raw dc output. Line filters are used to suppress radio-frequency interference (rfi) generated by the host system (emitted) or generated outside by other sources (received). Line filters are required in all systems powered by switching power supplies to comply with Federal Communications Commission regulations limiting EMI/RFI above 10 kHz.

Filter circuits generally combine inductive and capacitive components. Because inductive reactance increases with higher frequencies and capacitive reactance decreases, the two opposite effects improve the filtering action.

A capacitor has inherent filtering capability for alternating current because capacitive reactance, X_C, is inversely proportional to frequency. It blocks direct current entirely and opposes the passage of low-frequency signals but provides a progressively easier passage for signals as the frequency is increased (*see* CAPACITOR). By contrast, an inductor has filtering capability because inductive reactance, X_L, is directly proportional to frequency. There is no opposition to direct current, but an inductor provides progressively more opposition to alternating current as the frequency increases. (*See* INDUCTOR.) Filters exploit these opposing or complementary properties.

A resistor-capacitor (RC) coupling circuit is a highpass filter because the ac component of the input voltage is developed across the resistor and the dc voltage is blocked by the series capacitor. By contrast, a bypass capacitor is a lowpass filter because the high frequencies are bypassed. The lower the frequency, the less the bypassing action.

Filters with combinations of inductors (L) and capacitors (C) are named to correspond to the circuit configuration. The most common types are L, T, and pi. Any one of the three can function either as a lowpass or a highpass filter.

For either lowpass or highpass filters with L and C the reactance, X_L increases with higher frequencies and X_C decreases. The circuit connections are opposite to reverse the filtering action.

Highpass filters normally use:

1. Coupling capacitance C in series with the load. Then X_C can be low for high frequencies to be passed while low frequencies are blocked.
2. Choke inductance L in parallel across R_L. Then the shunt X_L can be high for high frequencies to prevent a short circuit across R_L, while low frequencies are bypassed.

The opposite characteristics for lowpass filters are:

1. Inductance L in series with the load. The high X_L for high frequencies can serve as a choke, and low frequencies can be bypassed to R_L.
2. Bypass capacitance C in parallel across R_L. Then high frequencies are bypassed by a small X_C, and low frequencies are not affected by the shunt path.

There are five basic types of filters:

1. Passband filters pass a specified frequency band while rejecting adjacent frequencies above and below that passband.
2. Band reject or notch filters have a function that is the inverse of the passband; they reject a specific band, the stopband.
3. Lowpass filters pass frequencies below a threshold or cutoff frequency and reject those above it.
4. Highpass filters pass frequencies above a specified cutoff and reject frequencies below that cutoff.
5. Allpass or phase-shift filters change the phase of the signal without affecting its amplitude.

The objective of practical filter design is to reach a compromise between the simplest filter that can accomplish the task and the ideal approximated by elaborate networks.

The response of a lowpass filter appears as a negatively sloped curve at the low-frequency end of the frequency versus gain graph with the passband defined by the area under the curve and the stop band area to the right of the curve. Similarly, the response of a highpass filter appears as a positively sloped curve at the high-frequency end of the graph with the passband the area under the curve and the stop band area to the left of the curve.

A bandpass filter is formed by combining a highpass filter with a lowpass filter so that a band of frequencies not stopped by either filter is passed. The bandpass filter response curve appears as a generally trapezoidal passband with the positive slope on the low-frequency end indicating the limit of the highpass stopband and the negative slope on the high end defining the lowpass stop band. A flat-top on the curve indicates constant signal gain. The bandwidth of the filter is the range of frequencies in hertz (Hz) passed by the filter.

Constant-k Filter. A constant-k filter is a filter designed to make the product of inductive reactance (X_L) and capacitive reactance (X_C) constant at all frequencies. This filter presents a constant impedance at the input and output terminals. A constant-k filter can be highpass, lowpass, bandpass, or other.

m-Derived Filter. An m-derived filter is a modified form of a constant-k filter whose design is based on the ratio of the filter cutoff frequency to the frequency of infinite attenuation. This ratio determines the m factor, which is generally between 0.8 and 1.25. The m-derived filter can also be highpass, lowpass, bandpass or other.

Resonant Filters. Tuned circuits provide a convenient means of filtering a band of radio frequencies because resonance can be achieved with relatively small values of L and C. A tuned circuit provides filtering with its maximum response at the resonant frequency. The width of the band of frequencies affected by resonance depends on the Q of the tuned circuit with a higher Q providing narrower bandwidth. Resonant filters are called band-stop or bandpass because resonance is effective for a band of frequencies above and below the resonant frequency. Series or parallel LC circuits can be used for either function, depending on the connection with respect to R_L.

Interference Filters. Voltage or current introduced into a circuit at other than the desired frequency can be considered interference that should be eliminated by a filter. Some typical examples are: 1) power-line filters that eliminate the passage of radio-frequency interference on the 60 Hz power line to or from the receiver, 2) highpass filter to eliminate radio-frequency interference from the signal received by a television receiving antenna, and 3) resonant filter to eliminate an interfering radio frequency signal from the desired radio-frequency signal. A resonant bandstop filter is called a wavetrap.

Power-line filters are manufactured as commercial components with different current ratings. A radio-frequency bypass capacitor across the line with two series radio-frequency chokes forms a lowpass, balanced, L-type filter. The choke in each side of the line balances the circuit with respect to ground.

Active Filters. Active filters are filters that include an operational amplifier with an external clock and other external networks. They can also be monolithic ICs (integrated circuits) for filtering requiring only an external clock and resistors. An example is the switched-capacitor filter. Active filters are suitable for use in solid-state circuitry because they eliminate the need for bulky inductors. Moreover, unlike passive filters, the output signal can exceed the input signal because of the gain through the operational amplifier. *See* OPERATIONAL AMPLIFIER, SWITCHED-CAPACITOR FILTER. *See also* ACTIVE ANALOG FILTER, BANDPASS FILTER, BAND-REJECTION FILTER, CERAMIC FILTER, CRYSTAL-LATTICE FILTER, FILTER ATTENUATION, FILTER CAPACITOR, FILTER CUTOFF, FILTER PASSBAND, FILTER STOPBAND, HIGHPASS FILTER, LOWPASS FILTER.

FILTER ATTENUATION

Filter attenuation is the loss caused by a selective filter at a certain frequency. If the specified frequency is within the passband of the filter, the attenuation is known as insertion loss.

Outside the passband of a selective filter, the attenuation depends on many factors including the distance of the frequency from the passband and the sharpness of the filter response. The ultimate attenuation of the filter is the greatest amount of attenuation obtained at any frequency. This normally occurs well outside the passband of the filter.

Filter attenuation is specified in decibels. If E is the root-mean-square signal voltage at a given frequency with the filter, and E_o is the root-mean-square signal voltage with the filter short-circuited, then the attenuation of the filter at that frequency is:

$$\text{Attenuation (dB)} = 20 \log_{10} (E_o/E)$$

When an active filter is used, the filter can produce gain. In that instance, the attenuation is negative. The filter gain is given by:

$$\text{Gain (dB)} = 20 \log_{10} (E/E_o)$$

See also FILTER.

FILTER CAPACITOR

A filter capacitor is used in a power supply to smooth out the ripples in the direct-current output of the rectifier circuit. These capacitors are usually quite large in value, ranging from a few microfarads in high-voltage, low-current power supplies to several thousand microfarads in low-voltage, high-current power supplies. Filter capacitors are often used in conjunction with other components such as inductors and resistors.

The filter capacitor holds the charge from the output-voltage peaks of the power supply. The smaller the load resistance, the greater the amount of capacitance that is required.

Filter capacitors in high-voltage power supplies can hold their charge even after the equipment has been shut off. Resistors of a fairly large value are placed in parallel with the filter capacitors in this supply, so that the shock hazard is reduced. *See also* POWER SUPPLY.

FILTER CUTOFF

In a selective filter, the cutoff frequency, or filter cutoff, is that frequency or frequencies at which the signal output voltage is 6 decibels below the level in the passband. A bandpass or band-rejection filter normally has two cutoff frequencies. A highpass or lowpass filter has just one cutoff point.

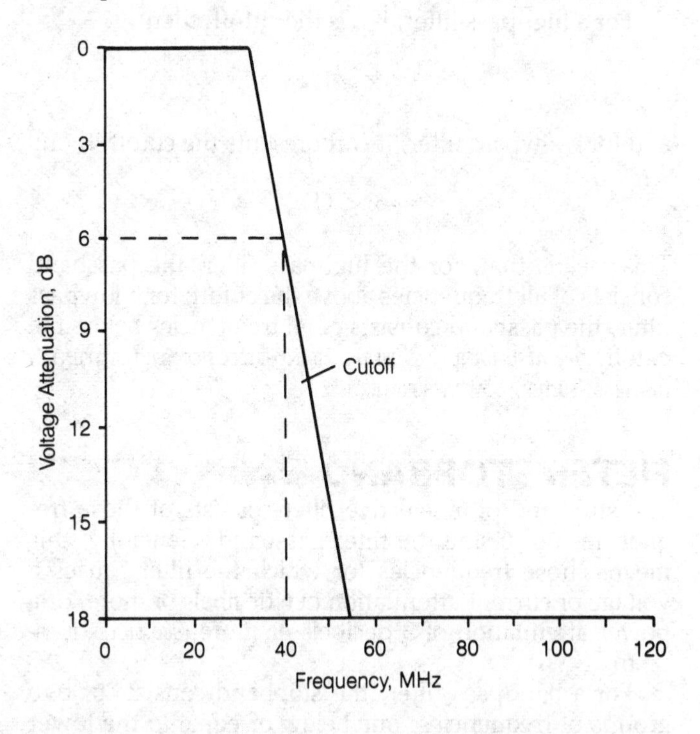

FILTER CUTOFF: The frequency of a filter at which the signal output voltage is 6 decibels below the attenuation in the passband.

The filter cutoff frequency is an important characteristic in choosing a filter for an application. The sharpness of the response is also important, as is the general shape of the response within the passband.

The drawing illustrates the response of a typical lowpass filter such as might be used in a high-frequency radio transmitter to reduce emissions at very high and ultra high frequencies. The cutoff frequency is clearly shown, marking the point at which the voltage drops by 6 decibels and the power drops by 3 decibels. The cutoff frequency depends on the impedance of the transmission line and antenna system. *See also* BANDPASS FILTER, BAND-REJECTION FILTER, FILTER, HIGHPASS FILTER, LOWPASS FILTER.

FILTER PASSBAND

The passband of a selective filter is the range of frequencies over which the attenuation is less than a certain value. Usually, this value is specified as 6 dB for voltage or current, and 3 decibels for power.

The passband of a filter depends on the kind of filter. If *L* represents the lower filter cutoff (*see* FILTER CUTOFF) and *U* represents the upper cutoff, then the passband of a bandpass filter is given according to:

$$L < x < U,$$

where *x* represents frequencies in the passband. For a band-rejection filter:

$$x < L \text{ or } x > U$$

That is, the passband consists of all frequencies outside the limit frequencies.

For a highpass filter, if *L* is the cutoff, then:

$$x > L$$

and for a lowpass filter, if *U* represents the cutoff, then:

$$x < U$$

This means that, for the highpass filter, the passband consists of all frequencies above the cutoff; for a lowpass filter, the passband consists of all frequencies below the cutoff. *See also* BANDPASS FILTER, BAND-REJECTION FILTER, FILTER, HIGHPASS FILTER, LOWPASS FILTER.

FILTER STOPBAND

The stopband of a selective filter consists of those frequencies not inside the filter passband. Generally, this means those frequencies for which the filter causes a voltage or current attenuation of 6 decibels or more, or a power attenuation of 3 decibels or more (*see* FILTER PASSBAND).

For a bandpass filter, the stopband consists of two groups of frequencies, one below or equal to the lower cutoff L, and the other above or equal to the upper cutoff *U*. For a band-rejection filter, the stopband consists of all frequencies between and including *L* and *U*. For a highpass filter, the stopband is that range of frequencies less than or equal to the cutoff. For a lowpass filter, the stop band is the range of frequencies higher than or equal to the cutoff. *See also* BANDPASS FILTER, BAND-REJECTION FILTER, FILTER, HIGHPASS FILTER, LOWPASS FILTER.

FINE TUNING

Fine tuning is the precise adjustment of the frequency of a radio transmitter or receiver. Fine tuning can be accomplished in many different ways, both electrical and mechanical. In a communications receiver, the bandspread control is used for fine tuning. In some communications transceivers, a clarifier control accomplishes fine tuning.

Mechanical fine-tuning controls are not often seen in modern equipment, but might be encountered in older receivers. Such a fine-tuning control uses knobs, connected to the main-tuning shaft through gears, to obtain a spread-out control.

Electrical fine tuning can be accomplished with small variable capacitors, inductors, or potentiometers. The fine-tuning control is normally connected in parallel with the main-tuning control, and has a much smaller minimum to maximum range.

FIRING ANGLE

The firing angle is an expression of the phase angle at which a thyratron or silicon-controlled rectifier fires (*see* SILICON-CONTROLLED RECTIFIER, THYRATRON). The firing angle is denoted by the lowercase Greek letter alpha (α). The firing angle is measured in degrees or radians; it represents the point on the control-voltage cycle at which the device is activated.

In a magnetic amplifier, the firing angle is denoted by the lowercase Greek letter phi (ϕ). As the input-voltage vector rotates, the core of the magnetic amplifier is driven into saturation at a certain point (*see* MAGNETIC AMPLIFIER). This point, measured as a phase angle in degrees or radians, is called the firing angle. *See also* PHASE ANGLE.

FIRMWARE

Firmware is a form of computer programming, or software (*see* SOFTWARE). Firmware is programmed into a circuit permanently. However, to change the programming, it is necessary to replace one or more memories in the system. The read-only memory (ROM) is an example of firmware.

FIRST HARMONIC

Whenever frequency multipliers are used in a circuit, harmonics are produced. The input signal to a frequency multiplier is called the first harmonic. Higher-order harmonics appear at integral multiples of the first harmonic.

A crystal calibrator, often used in communications receivers, provides a good example of deliberate harmonic generation. A 1 MHz crystal may be used in the oscillator circuit. A nonlinear device follows, resulting in harmonic output; the second harmonic is 2 MHz, the third harmonic is 3 MHz, and so on. The 1 MHz output signal is the first harmonic.

Some crystal calibrators use frequency dividers. For example, a divide-by-10 circuit may be switched in between the oscillator and the nonlinear device in the above example, producing markers at multiples of 100 kHz. The first harmonic of the divider then becomes 100 kHz, the second harmonic becomes 200 kHz, and so on. The first harmonic can be identified because all of the other signals are integral frequency multiples of it. This is true only of the first harmonic.

The first harmonic is sometimes called the fundamental frequency. In circuits where harmonic output is not desired, the term first harmonic is generally not used. *See also* FREQUENCY MULTIPLIER, FUNDAMENTAL FREQUENCY, HARMONIC.

FIRST-IN/FIRST-OUT

A first-in/first-out circuit (FIFO) is a form of read-write memory. The buffer circuits used in electronic typewriters, word processors, and computer terminals are examples of first-in/first-out memory stores.

The operation of a first-in/first-out buffer is evident from its name. The illustration shows a FIFO circuit with eight characters of storage. If certain characters are fed into the input at an irregular rate of speed, the buffer eliminates some of the irregularity without changing the order in which the characters are transmitted.

Not all buffers operate on the FIFO principle. Sometimes it is desirable to have a first-in/last-out buffer; as its name implies, this type of buffer inverts the order of the the characters it receives. *See also* BUFFER, SEMICONDUCTOR MEMORY.

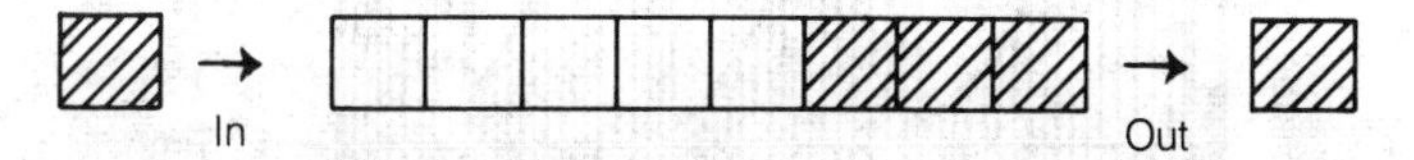

FIRST-IN/FIRST-OUT: A diagram of the operation of a FIFO buffer.

FIRST LAW OF THERMODYNAMICS

The First Law of Thermodynamics is an expression of the equivalence between different forms of energy: heat can be converted into mechanical work, and vice versa. For example, mechanical work of 4.183 joules results in the generation of 1 calorie of heat energy. *See also* ENERGY, SECOND LAW OF THERMODYNAMICS, THERMODYNAMICS, THIRD LAW OF THERMODYNAMICS.

FIXED BIAS

When the bias at the base, gate, or grid of an amplifying or oscillating transistor, field-effect transistor, or tube is unchanging with variations in the input signal, the bias is called fixed bias. Fixed bias can be supplied by resistive voltage-divider networks, or by an independent power supply. Fixed bias is often used in amplifying circuits for both audio and radio frequencies.

The schematics illustrate two ways of getting fixed bias with a bipolar-transistor amplifier. At A, a voltage divider is shown. The values of the resistors determine the voltage at the base of the transistor. At B, an independent power supply is used. The circuit at A is a Class A amplifier, because the base is biased to draw current even in the absence of signal; the resistors are chosen for operation in the middle of the linear part of the collector-current curve. At B, a Class C amplifier is shown; the independent power supply is used to bias the transistor beyond the cutoff point.

When the bias is not fixed, the characteristics of an amplifier will change along with any parameter responsible for the changes in bias. An automatic level-control circuit, for example, employs variable bias to change the gain of an amplifier. *See also* AUTOMATIC BIAS, AUTOMATIC LEVEL CONTROL.

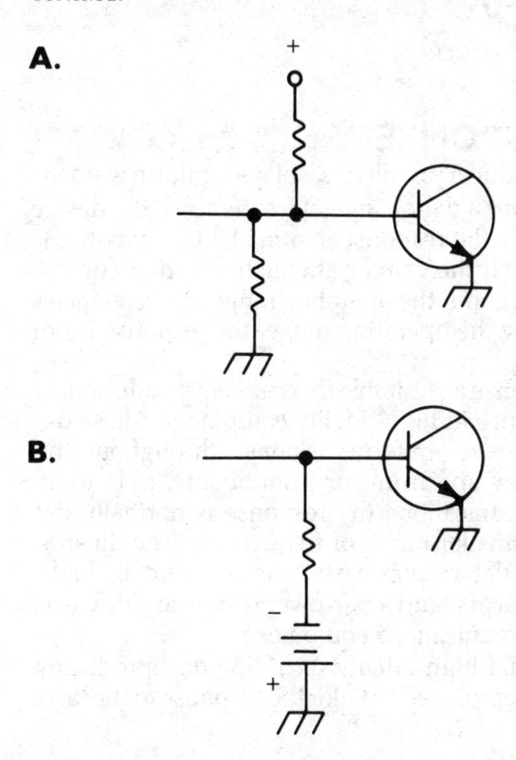

FIXED BIAS: Two methods of obtaining fixed bias are with a voltage divider (A) and with a separate power source (B).

FIXED DECIMAL

When the decimal point in a numeric display does not move, regardless of the operations performed, the display is said to have a fixed decimal. Fixed-decimal systems are commonly found in frequency-measuring instruments and some electronic calculators. The fixed-decimal display is always used in bookkeeping, for example; the decimal point is two places from the right in these applications.

For scientific calculations, a floating-decimal display is preferred. This allows the display of extremely small or large numbers without sacrificing accuracy. *See also* FLOATING DECIMAL.

FIXED FREQUENCY

A fixed-frequency device or circuit is an oscillator, receiver, or transmitter designed to operate on only one frequency.

Fixed-frequency communication offers the advantage of instant contact; no search is necessary at the receiving station to locate the frequency of the transmitter. However, if this advantage is to be realized, the frequencies must be accurately matched. To maintain the operating frequency, a standard source must be used, such as radio station WWV or WWVH. Phase-locked-loop circuits can, in conjunction with these frequency standards, keep the transmitter and receiver frequencies matched to a high degree of precision. *See also* FREQUENCY, PHASE-LOCKED LOOP, WWV/WWVH.

FLATPACK

See SEMICONDUCTOR PACKAGE.

FLAT RESPONSE

When a transducer or filter displays uniform gain or attenuation over a wide range of frequencies, the device is said to have a flat response. Normally, the response is considered flat if the gain or attenuation is more or less constant throughout the operating range. At frequencies above or below the operating range, the response is not important.

A flat response is desirable for speakers, headphones, and microphones in high-fidelity equipment. These devices should have uniform response throughout the audio-frequency spectrum, or from about 20 Hz to 20 kHz. In communications the response is normally flat over a much smaller range of frequencies. The illustration shows a flat response such as is found in high-fidelity transducers, and a narrower response, such as is typical of communications equipment.

Controls in a high-fidelity recording or reproducing system can be employed to tailor the response to the taste of a particular listener. In the recording process, the amplifier response is not always flat; the gain might be greater at some frequencies than at others. This is corrected in the playback process. *See also* FREQUENCY RESPONSE.

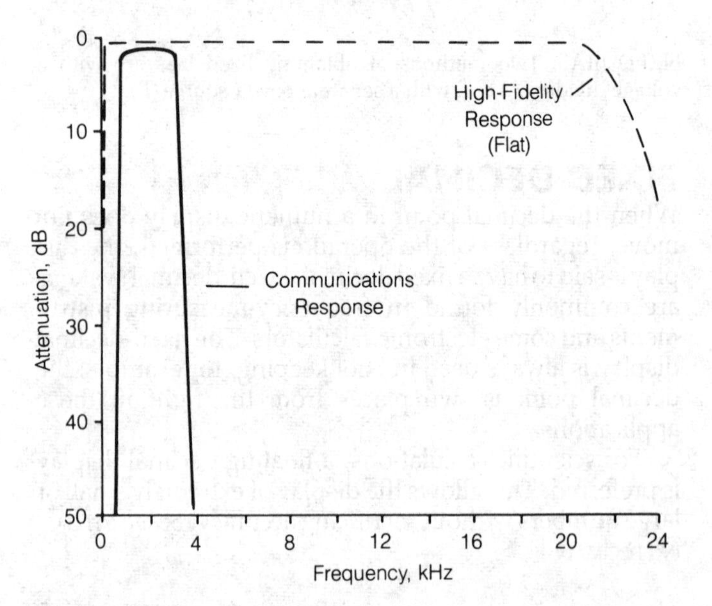

FLAT RESPONSE: The flat response for high-fidelity reception (dotted line) and the restricted response for communications reception (solid line) are shown.

FLAT TOPPING

Flat topping is a form of distortion that sometimes occurs on an audio-frequency waveform or a modulation envelope. Flat topping takes place because of a severe nonlinearity in an amplifying circuit. Flat topping is undesirable because it results in the generation of harmonic energy, and degrades the quality of a signal.

Flat topping in a modulation envelope of a single-sideband transmitter is shown in the illustration. The peaks of the signal are cut off, and appear flat. Flat topping on a single-sideband or amplitude-modulated signal results in excessive bandwidth because of the harmonic distortion. This can cause interference to stations on frequencies near the distorted signal. This type of distortion is sometimes called splatter.

Flat topping is usually caused by improper bias in an amplifying stage, or by excessive drive. If the bias is incorrect, a tube or transistor may saturate easily. Saturation can also be caused by too much driving voltage or power. *See also* DISTORTION.

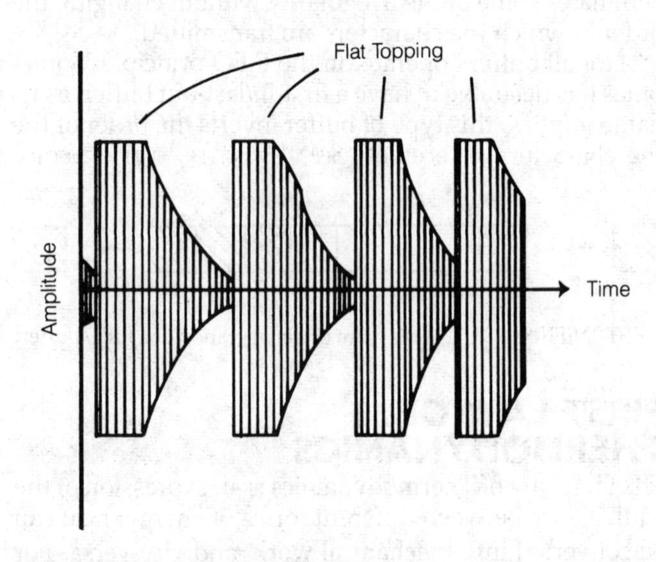

FLAT TOPPING: An oscilloscope display of flat-topping of a single-sideband envelope.

FLAT TRANSMISSION LINE

When a transmission line contains no reflected electromagnetic field, but only the forward field, the line is called flat. A flat line is terminated in an impedance with a resistive component equal to the characteristic impedance of the line; there is no reactance. A flat line has no standing waves, and the current and voltage remain in the same proportion everywhere along the length of the line.

A transmission line operates at maximum efficiency when it is flat; standing waves cause some additional loss. At very high frequencies and above, a flat line is desirable because the additional loss tends to be greater than at lower frequencies. Flatness is more important in coaxial lines than in two-wire transmission lines.

When a feed line is terminated in an impedance other than its characteristic impedance, the currents and voltages become nonuniform. The current and voltage change with the location in the line. *See also* CHARACTERISTIC IMPEDANCE, STANDING WAVE, STANDING WAVE RATIO.

F LAYER

The F layer is a region of ionization in the upper atmosphere of the earth. The altitude of the F layer ranges from about 100 to 260 miles, or 160 to 420 kilometers. The F layer of the ionosphere is responsible for most long-distance radio propagation at medium and high frequencies.

The F layer consists of two separate ionized regions during the daylight hours. The lower layer is called the F1 layer and the upper layer the F2 region. During the night, the F1 layer usually disappears. The F1 layer rarely affects radio signals, although under certain conditions it may intercept and return them.

The F layer attains its maximum ionization during the afternoon hours. But the effect of the daily cycle is not as pronounced in the F layer as in the D and E layers. Atoms in the F layer remain ionized for a longer time after sunset. During times of maximum sunspot activity, the F layer often remains ionized all night long.

Because the F layer is the highest of the ionized regions in the atmosphere, the propagation distance is longer via F-layer circuits than via E-layer circuits. The single-hop distance for signals returned by the F2 layer is about 500 miles for a radiation angle of 45 degrees. For a radiation angle of 10 degrees, the single-hop distance can be as great as 2,000 miles. For horizontal waves the single-hop F2-layer distance is from 2,500 to 3,000 miles. For signals to propagate over longer distances, two or more hops are necessary.

The maximum frequency at which the F layer will return signals depends on the level of sunspot activity. During sunspot maxima, the F layer may occasionally return signals at frequencies as high as about 80 to 100 MHz. During the sunspot minimum, the maximum usable frequency can drop to less than 10 MHz. *See also* D LAYER, E LAYER, IONOSPHERE, PROPAGATION CHARACTERISTICS.

FLEMING'S RULES

Fleming's rules are simple means of remembering the relationship among electric fields, magnetic fields, voltages, currents, and forces.

If the fingers of the right hand are curled and the thumb is pointed outward as shown in A, then a current flowing in the direction of the thumb will cause a magnetic field to flow in a circle, the sense of which is indicated by the fingers. This is called the right-hand rule for the magnetic flux generated by an electric current.

If the thumb, first finger, and second finger are

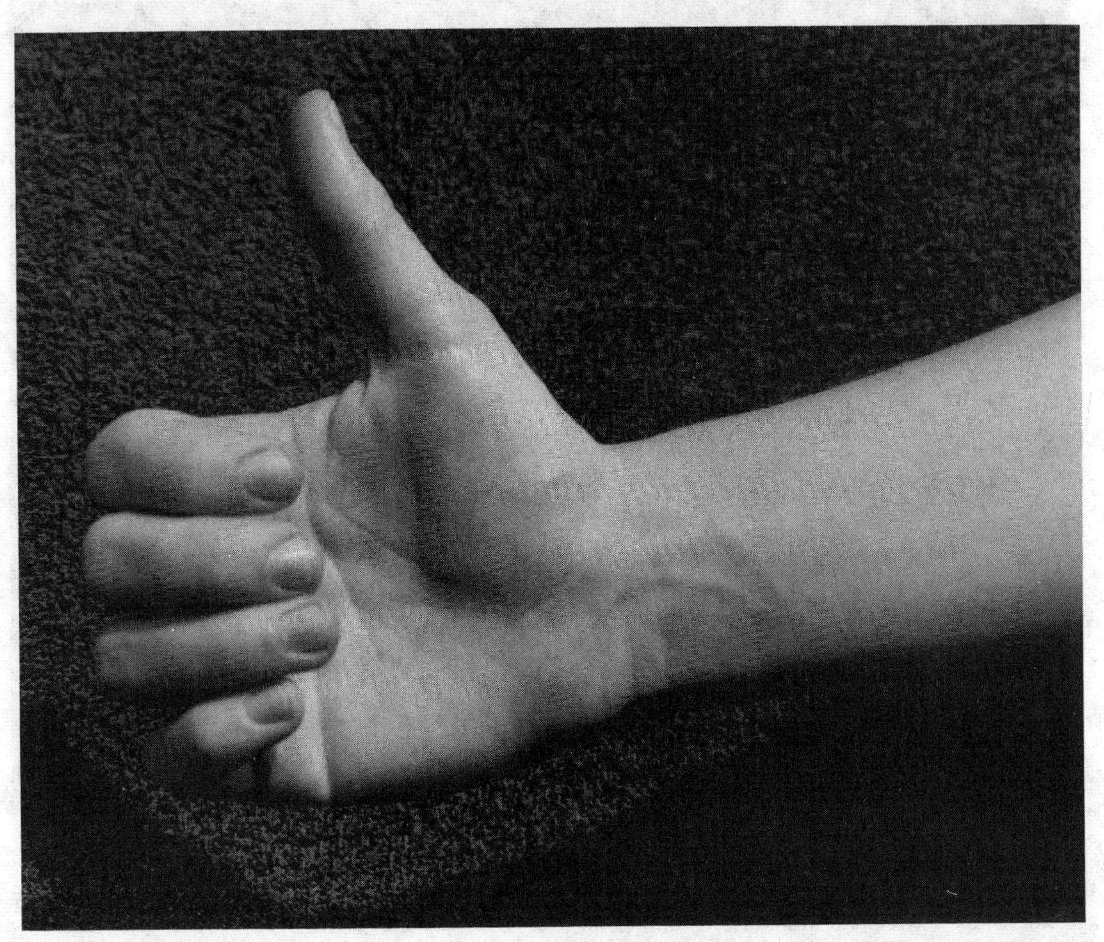

A.

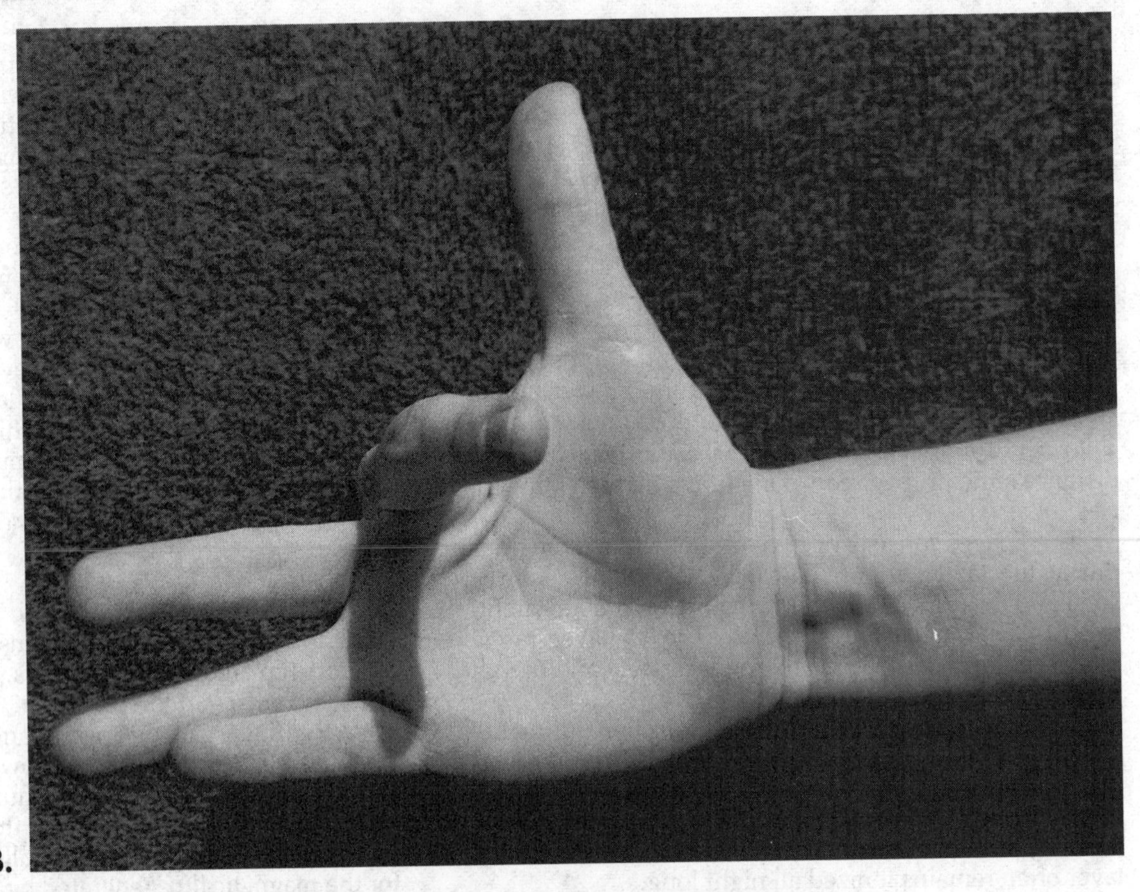

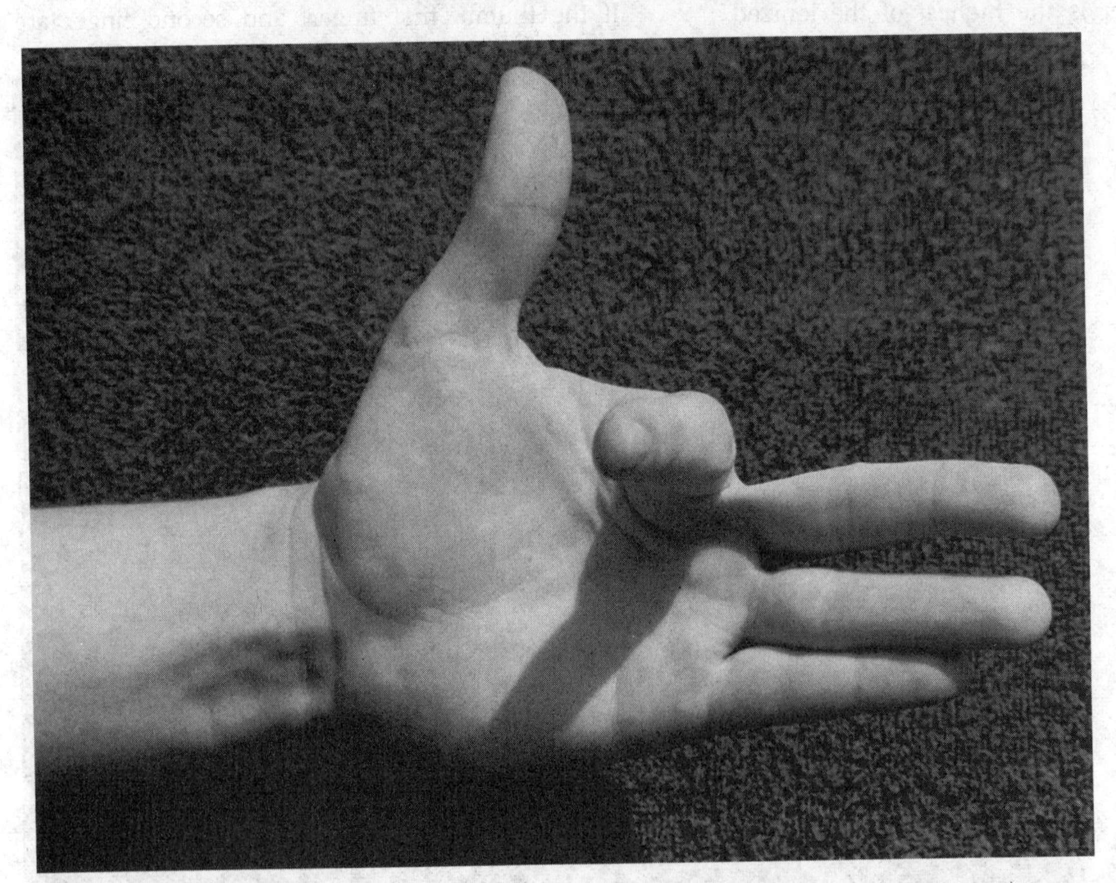

FLEMING'S RULES: The right-hand rules are shown for the direction of the magnetic field generated by electric current (A), and the relationships between the armature motion, the magnetic field, and the current in a generator (B). The left-hand rule is shown for the same relationships in a motor (C).

oriented at right angles to each other as illustrated in B and C, then the right hand will show the relationship among the direction of the wire motion, the magnetic field, and the current in an electric generator (B), and the left hand will indicate the relationship among the direction of wire motion, the magnetic field, and the current in an electric motor (C). The thumb shows the direction of wire motion. The index finger shows the direction of the magnetic field from north to south. The middle finger shows the direction of electric current from positive to negative. *See also* GENERATOR, MOTOR.

FLEXIBLE WAVEGUIDE

A flexible waveguide is a section of waveguide used to join rigid waveguides that are not in precise orientation with respect to each other. Flexible-waveguide sections make the installation of a waveguide transmission line much easier because exact positioning is not required. Flexible waveguides also allow for the expansion and contraction of rigid waveguides with changes in temperature.

Flexible-waveguide sections are made in a variety of ways: metal ribbons can be joined together and metal foil or tubing can be used. Some flexible waveguides can be twisted as well as bent or stretched. If a section of flexible waveguide is installed between two sections of rigid waveguide, the impedances should be matched. The flexible section should have the same characteristic impedance as the rigid waveguide. The joints themselves must be made properly, to prevent the formation of impedance bumps. *See also* WAVEGUIDE.

FLICKER FREQUENCY

The flicker frequency in a motion-picture projection system is the number of times the screen is illuminated each second. Normally, the flicker frequency is twice the number of frames per second. The screen is blanked as the frame is positioned, and again while the frame is projected. The standard frame rate is 24 Hz, and the flicker frequency is 48 Hz.

The human eye can detect a light-modulation frequency of only about 15 to 25 Hz. When the frequency is greater than this, a flashing light appears continuous. The flicker frequency is chosen so that the projected image appears constant. With a flicker frequency of twice the frame rate, motion is reproduced in a more realistic manner than would be the case if the flicker frequency were the same as the frame rate.

FLIP-FLOP

A flip-flop is a simple electronic circuit with two stable states. The circuit is changed from one state to the other by a pulse or other signal. The flip-flop maintains its state indefinitely unless a change signal is received. There are several different kinds of flip-flop circuits.

The D type flip-flop operates in a delayed manner, from the pulse immediately preceding the present pulse.

The J-K flip-flop has two inputs, commonly called the J and K inputs. If the J input receives a high pulse, the output is set to the high state; if the K input receives a high pulse, the output is set low. If both inputs receive high pulses, the output changes its state either from low to high or vice versa.

The R-S flip-flop has two inputs, called the R and S inputs. A high pulse at the R input sets the output low; a high pulse on the S input sets the output high. The circuit is not affected by high pulses at both inputs.

The R-S-T flip-flop has three inputs called R, S, and T. The R-S-T flip-flop operates exactly as the R-S flip-flop works, except that a high pulse at the T input causes the circuit to change states.

A T flip-flop has only one input. Each time a high pulse appears at the T input, the output state is reversed.

Flip-flop circuits are interconnected to form the familiar logic gates, which in turn comprise all digital apparatus. *See also* LOGIC GATE.

FLOAT CHARGE

See CHARGING.

FLOATING CONTROL

Most control components, such as switches, variable capacitors, and potentiometers, have grounded shafts. This is convenient from an installation standpoint, since the shafts of controls are normally fed through a metal front panel. However, there are certain instances in which it is not possible to ground the shaft. The control is then said to be floating, because it is not grounded.

The illustration shows an example of a grounded (nonfloating) control, at A, and a floating control, at B. The rotor plates of a variable capacitor are internally connected to the shaft; for this reason, it is customary to put the rotor plates at ground potential. This is the case at A. But at B, this is not practical, and the frame and shaft

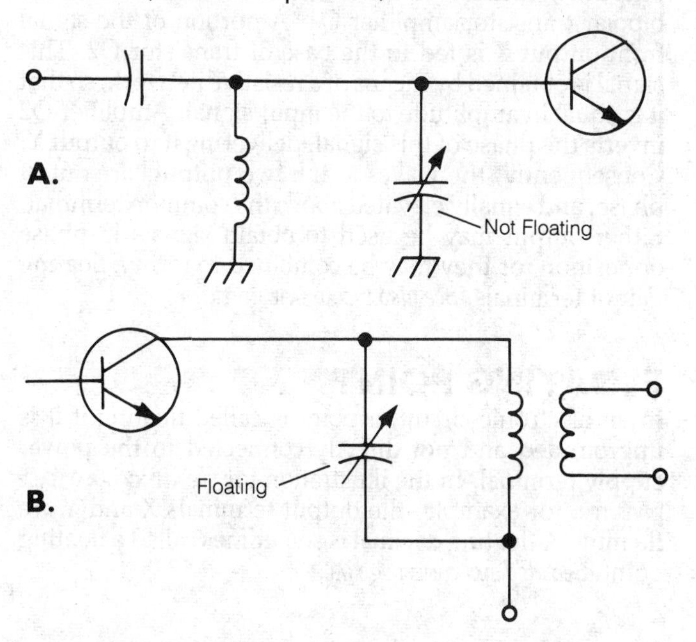

FLOATING CONTROL: The rotor plates of the capacitor are grounded (A), and they are floating (not grounded) (B).

of the variable capacitor must be floated. To accomplish this, the capacitor is mounted on insulators above the chassis, and the shaft must be insulated from the panel and knob or shaft.

Floating controls are more subject to the effects of body capacitance than grounded controls. Therefore, floating controls should be used only when there is no alternative. *See also* BODY CAPACITANCE.

FLOATING DECIMAL

A floating-decimal or floating-point display is a numeric display often found in calculators. All calculators have a fixed number of digits in the display, and accuracy is maximized by the use of a movable decimal point.

A ten-digit display with a fixed decimal point, two places from the far right, can accurately display quantities from 0.01 to 99,999,999.99. However, when the decimal point is movable, the display can accurately render quantities as small as 0.0000000001, or as large as 9,999,999,999. Because of the larger range using a floating-decimal display, this scheme is preferred in scientific work.

The range of a display can be further increased by using scientific notation. Some hand-held calculators have this feature, and it is actuated whenever the display can no longer render a quantity in the simple decimal form. *See also* FIXED DECIMAL, SCIENTIFIC NOTATION.

FLOATING PARAPHASE INVERTER

A floating paraphase inverter is a circuit designed for inverting the phase of a signal. Two transistors or vacuum tubes are used in the circuit. The schematic diagram shows a floating paraphase inverter.

A signal applied to the input is inverted in phase at output X, because of the phase-reversing effect of the bipolar transistor amplifier Q1. A portion of the signal from output X is fed to the base of transistor Q2. This signal is obtained by means of a resistive network, so that it is equal in amplitude to the input signal. Amplifier Q2 inverts the phase of this signal, delivering it to output Y. Consequently, the waves at the two outputs are out of phase, and equally elevated above the common terminal. Either output may be used to obtain signals in phase opposition, or they may be combined to form a floating pair of terminals. *See also* PARAPHASE INVERTER.

FLOATING POINT

In an electronic circuit, a point is called floating if it is ungrounded and not directly connected to the power supply terminal. In the illustration for FLOATING PARAPHASE INVERTER, for example, the output terminals X and Y are floating. A floating decimal is sometimes called a floating point. *See also* FLOATING DECIMAL.

FLOOD GUN

A flood gun is an electron gun employed in a storage cathode-ray tube. The flood gun operates in conjunction with another electron gun, called the writing gun, to store a display for a period of time. The writing gun emits a thin beam of electrons that strikes the phosphor screen and forms the image. The beam is modulated by means of deflecting plates, as in the conventional oscilloscope. A storage mesh holds the image. The flood gun is used to illuminate the phosphor screen to view the stored image. Low-energy electrons from the flood gun can pass through the mesh only where the image has been stored. When the image is to be erased, the flood gun is used to clear the mesh. *See also* OSCILLOSCOPE, STORAGE OSCILLOSCOPE, WRITING GUN.

FLOPPY DISK

See DISK DRIVE.

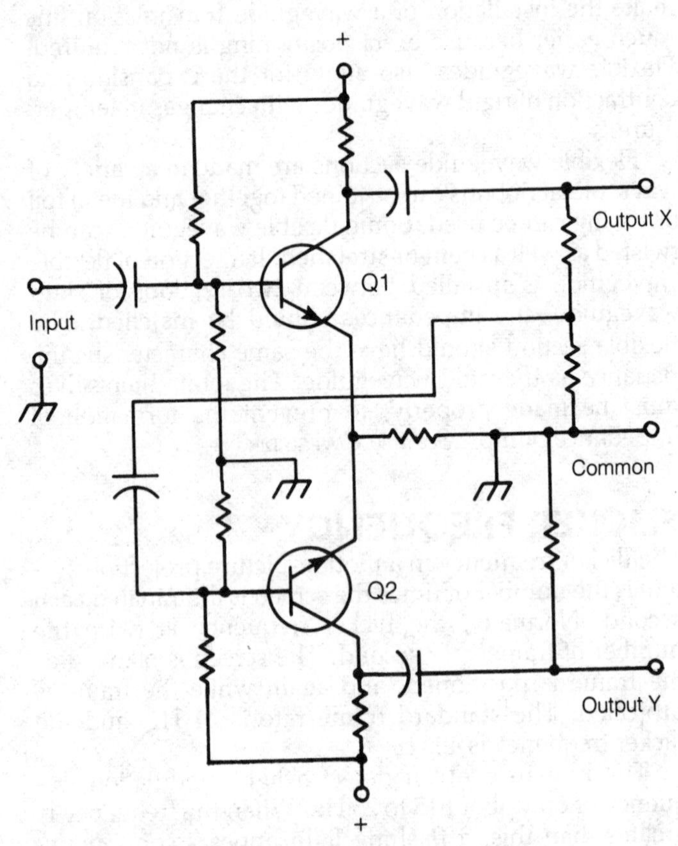

FLOATING PARAPHASE INVERTER: The Outputs X and Y are in phase opposition.

FLOWCHART

A flowchart is a diagram that depicts a logical algorithm or sequence of steps. The flowchart looks very much like a block diagram. Boxes indicate conditions, and arrows show procedural steps. Flowcharts are often used to develop computer programs. These charts can also document troubleshooting processes for all kinds of electronic equipment.

The symbology used in flowcharts is not well standardized. An example of a simple troubleshooting flowchart is shown in the illustration. Decision steps are

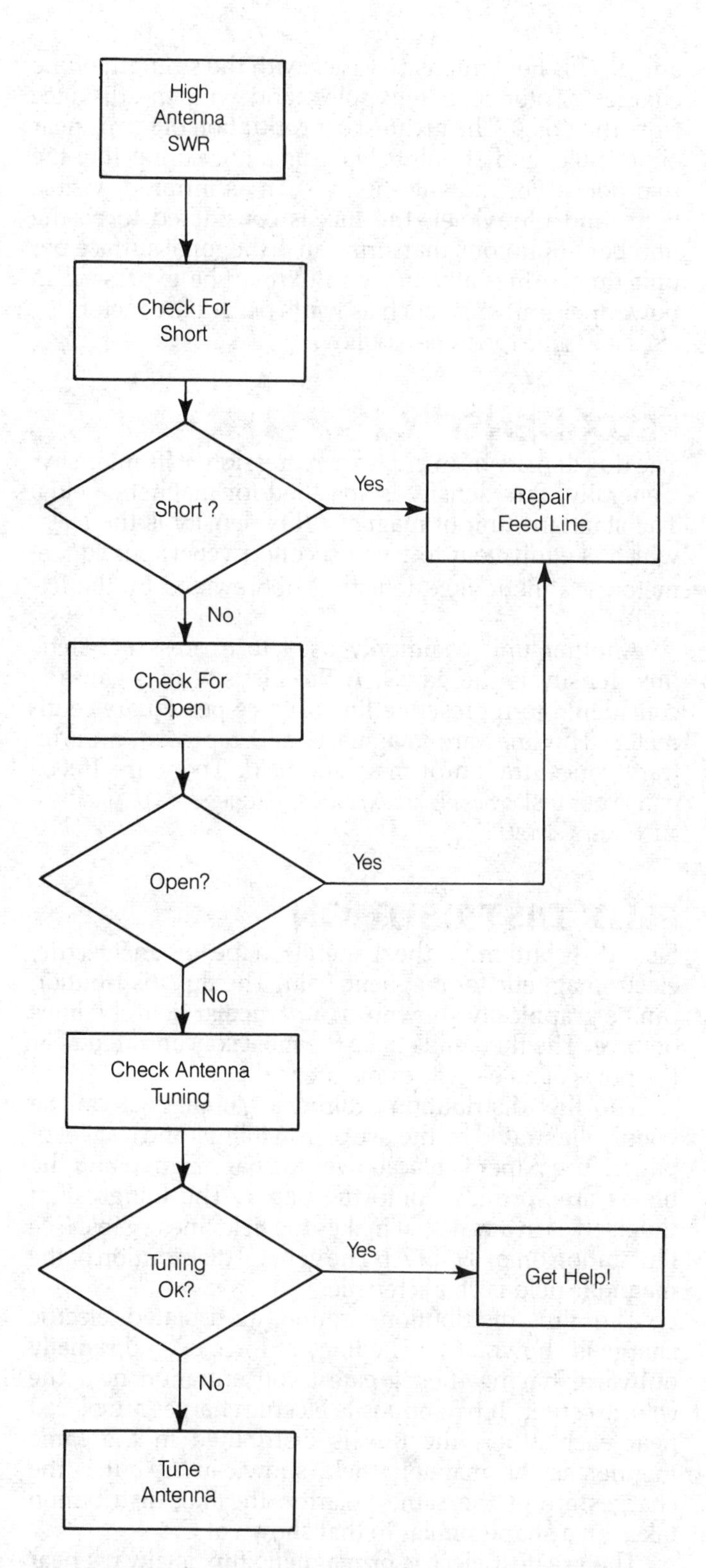

FLOWCHART: The procedure for repairing an antenna with a high standing-wave ratio (SWR) is illustrated in the flowchart.

indicated by diamonds. Various conditions are indicated by boxes.

A flowchart must represent a complete logical process. No matter what the combination of conditions and decisions, a conclusion must always be reached. Sometimes the conclusion consists simply of an instruction to return to an earlier stage in the algorithm, but infinite loops should not exist. *See also* ALGORITHM.

FLUID LOGIC

Fluid logic, also called fluid-flow or fluidic logic, is a technology for obtaining digital-logic circuits by means of fluid valves. There are no electronic elements or moving parts. The fluid can be either a liquid or a gas.

Fluid logic can operate in explosive environments—impossible with ordinary electronic logic circuits. Fluid logic is essentially immune to electromagnetic interference. However, fluid logic operates slowly, limiting the number of operations that can be performed per unit of time. Logic gates, flip-flops, inverters, and amplifiers can all be built with fluidic technology.

FLUORESCENCE

When a substance emits visible light because it has absorbed energy, the phenomenon is called fluorescence. Fluorescence can occur in gases, liquids, or solids. Most gases will exhibit this property when bombarded with, or forced to carry, an electron beam of sufficient intensity.

Fluorescent materials are found in many different electronic devices. The cathode-ray tube, for example, has a fluorescent coating on the inside surface of the screen to display images produced by electron beams. The phosphor coating on these screens is made up of fine crystals that emit visible light of various colors. Some examples of fluorescent substances employed in cathode-ray tubes are zinc sulfide, zinc fluoride, and zinc oxide.

Fluorescence provides an efficient means of producing light for household or business purposes. *See also* FLUORESCENT TUBE.

FLUORESCENT TUBE

The fluorescent tube or lamp is a source of illumination powered by electricity. An efficient light source, the fluorescent tube generates more visible light and less heat than incandescent bulbs of the same rating.

The illustration on p. 392 is a simple cross-section view of a fluorescent tube. A glass tube is evacuated and filled with a combination of argon gas and mercury. When an electric current flows through the gas, the mercury is vaporized. This results in ultraviolet radiation, which strikes the phosphor coating on the inside of the tube. The phosphor produces white light when excited by the radiation.

Fluorescent tubes are available in many different sizes and wattage ratings. The smallest commercially available fluorescent lamps are rated at 15 W at 12 V and measure just a few inches in length. The largest tubes are over 6 feet long and are rated at 100 W. *See also* FLUORESCENCE.

FLUORINE

Fluorine is an element, symbolized by the capital letter F. It has an atomic number of 9, and an atomic weight of 19. Fluorine is a member of the halogen family of elements (*see* HALOGEN).

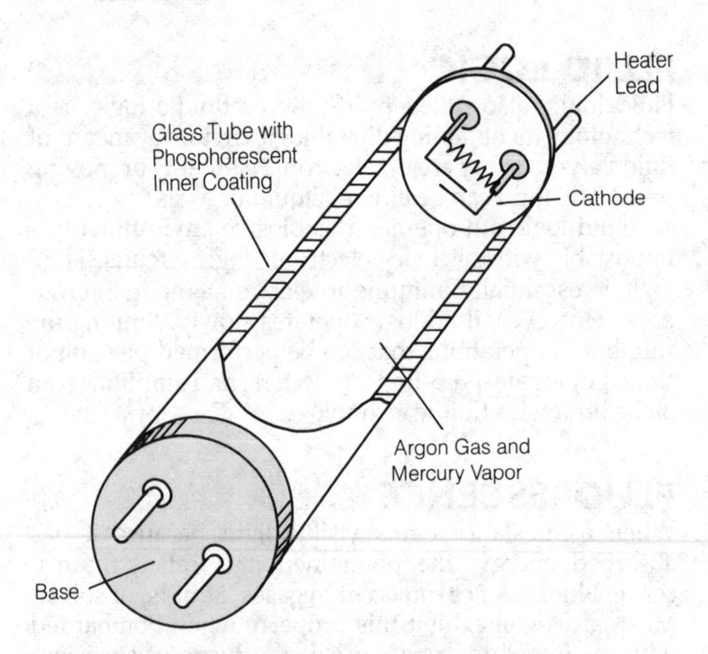

FLUORESCENT TUBE: Ionization in a fluorescent tube produces ultraviolet light, which strikes the phosphor coating on the inside of the tube to produce visible light.

FLUTTER

Flutter is the term used to describe a warbling, or change in pitch, of the sound in an audio recording/reproducing system. Slow flutter is sometimes called wow because of the way it sounds. Flutter can occur in any recording or reproducing process in which the vibrations are caused by a fixed medium.

Flutter may result from a periodic change in the speed of the recording medium relative to the transducer. An example of this can be seen by placing a record on a turntable off-center. (This can be done with standard 45 rpm records because of their large center hole.) The stylus speeds up and slows down as the disk rotates.

Flutter may also occur because of nonuniform motor speed in either a disk system or a magnetic system. This type of flutter is generally much more rapid than the periodic type. Random flutter may be so minor that it cannot be noticed on voice recordings, but it becomes evident when music is recorded and reproduced. Some tape recorders are intended for voice reproduction only. *See also* DISK RECORDING, MAGNETIC RECORDING, SUPERHETERODYNE RADIO RECEIVER, WOW.

FLUX

Flux is a measure of the intensity of an electric, electromagnetic, or magnetic field. Any field has an orientation, or direction, which can be denoted by imaginary lines of flux. In an electric field, the lines of flux extend from one charge center to the other in various paths through space. In a magnetic field, the lines extend between the two poles. In an electromagnetic field, the electric lines of flux are generally denoted. (*See* ELECTRIC FIELD, ELECTRIC FLUX, ELECTROMAGNETIC FIELD, MAGNETIC FIELD, MAGNETIC FLUX.)

The flux lines of any field have a certain concentration per unit of surface area through which they pass at a right angle. This field intensity varies with the strength of the charges and/or magnetic poles, and with the distance from the poles. The greatest concentration of flux is near either pole, and at points lying on a line connecting the two poles. For radiant energy such as infrared, visible light, and ultraviolet, the flux is considered to be the number of photons that strike an orthogonal surface per unit time. Alternatively, the flux may be expressed in power per unit area, such as watts per square meter. *See also* FLUX DENSITY, FLUX DISTRIBUTION.

FLUX DENSITY

The flux density of a field is an expression of its intensity. Generally, flux density is specified for magnetic fields. The standard unit of magnetic flux density is the tesla, which is equivalent to 1 volt second (weber) per square meter. Magnetic flux density is abbreviated by the letter B.

Another unit commonly used to express magnetic flux density is the gauss. A flux density of 1 gauss is considered to represent a line of force per square centimeter. The lines are imaginary, and represent an arbitrary concentration of magnetic field. There are 10,000 gauss per tesla. *See also* FLUX, GAUSS, MAGNETIC FIELD, MAGNETIC FLUX, TESLA, WEBER.

FLUX DISTRIBUTION

Flux distribution is the general shape of an electric, electromagnetic, or magnetic field. The flux distribution can be graphically shown, for a particular field, by lines of force. The flux tends to be the most concentrated near the poles of an electric or magnetic dipole.

The flux distribution around a bar magnet can be vividly illustrated by the use of iron fillings and a sheet of paper. The paper is placed over the bar magnet, and the filings are sprinkled onto the paper. The filings align themselves in a way that makes the field lines visible (see illustration on p. 393 at A). The general distribution of the magnetic field is characteristic.

The flux distribution around an isolated electric charge is shown at B. The lines of force extend radially outward, having their greatest concentration near the charge center. If two opposite electric charges are placed near each other, the flux is distributed in the same manner as the magnetic field shown at A; but if the charges are of the same polarity, the flux distribution takes on a shape similar to that shown at C.

The greatest electric or magnetic flux is always near the poles and along a line connecting two poles. *See also* DIPOLE, ELECTRIC FLUX, FLUX, MAGNETIC FLUX.

FLUXMETER

A fluxmeter is an instrument for measuring the intensity of a magnetic field. This meter is also called a gaussmeter.

The fluxmeter generally includes a wire coil that can be moved back and forth across the flux lines of the magnetic field. This causes alternating currents to flow in the coil, and the resulting current is indicated by a meter.

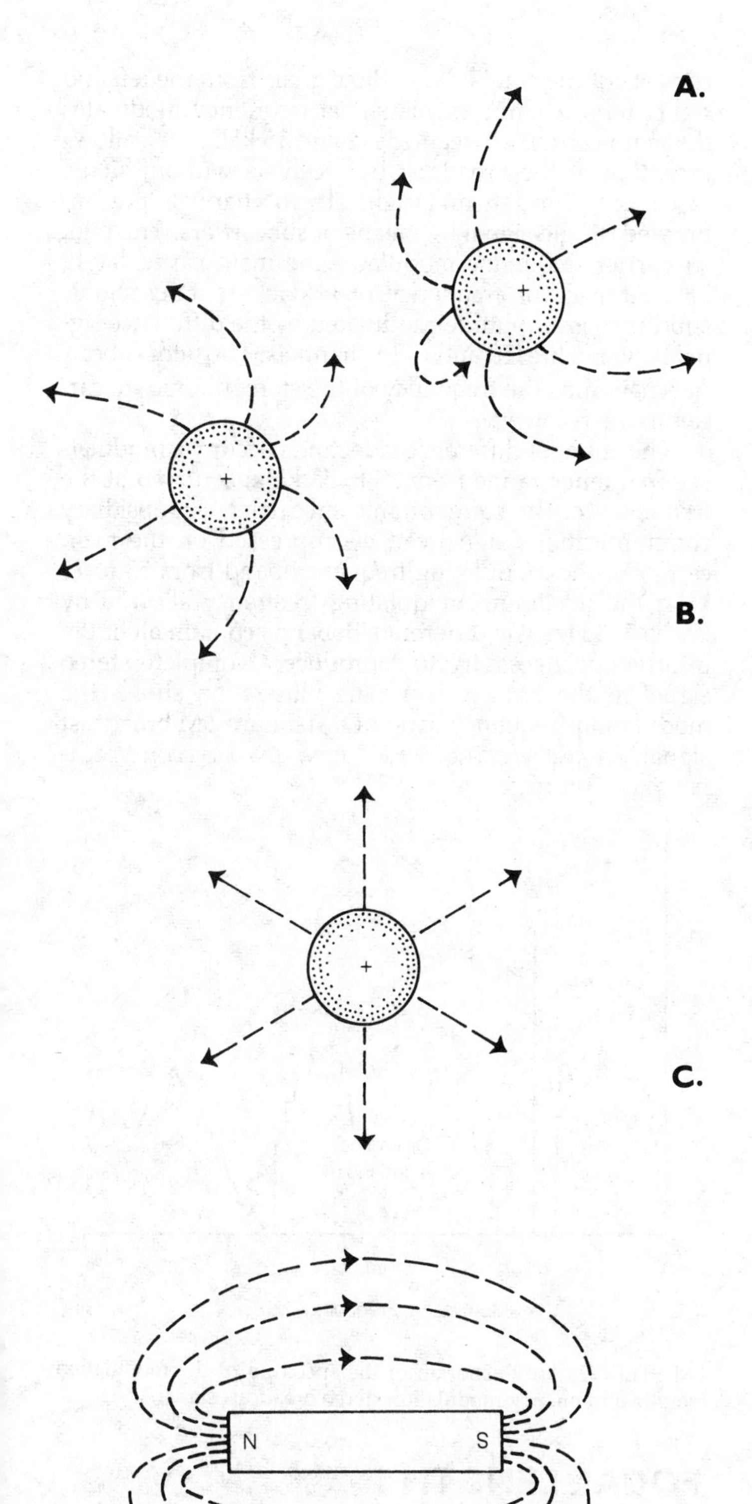

FLUX DISTRIBUTION: The flux distribution around a magnetic dipole is shown at (A), around an isolated electric charge (B), and around nearby like charges (C).

The greater the intensity of the magnetic field for a given coil speed, the greater the amplitude of the alternating currents in the coil. Knowing the rate of acceleration of the coil, and the amplitude of the currents induced in it, the flux density can be determined.

Fluxmeters are usually calibrated in gauss. A flux density of 1 gauss is the equivalent of a line of flux per square centimeter. *See also* FLUX, GAUSS, MAGNETIC FLUX.

FLYBACK

Flyback is a term that denotes the rapid fall of a current or voltage that has been increasing for a period of time. The flyback time is practically instantaneous under ideal conditions; in practice, however, it is finite. The illustration shows flyback in a sawtooth waveform.

Sawtooth waves are used in oscilloscopes, television receivers, and computer monitors to sweep the electron beam rapidly back after a line has been scanned. The flyback sweep is more rapid than the forward sweep; the display is blanked during the flyback to prevent any possible interference with the image created by the forward sweep. In an oscilloscope or television picture tube, the flyback is from right to left, while the visible sweep is from left to right.

Flyback also refers to a design of a switching power supply similar to the forward converter. These supplies require only a single switching transistor, and they have relatively simple magnetics. *See* POWER SUPPLY.

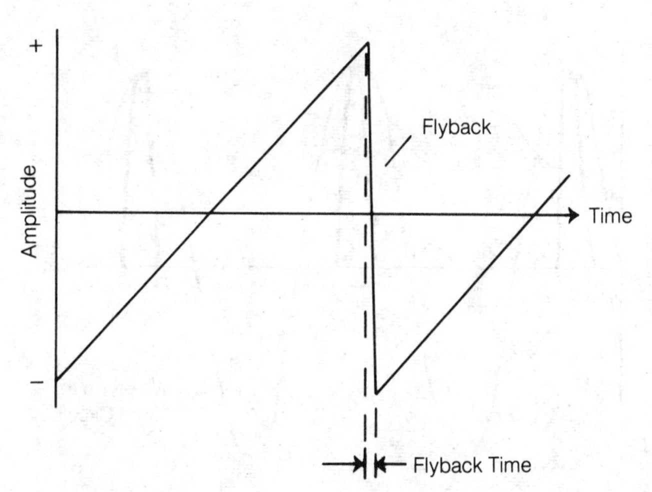

FLYBACK: Flyback is the return of a sawtooth pulse to its minimum, or starting, value.

FLYBACK CIRCUIT

In a television receiver, horizontal scanning is controlled by a flyback circuit. The flyback circuit provides a high-voltage sawtooth wave, which is fed to the horizontal deflecting coils in the picture tube.

The output waveform of the flyback circuit is formed so that the deflection of the electron beam occurs relatively slowly from left to right, but rapidly from right to left. The picture modulation is impressed on the electron beam during the left-to-right sweep. The beam is blanked out during the return sweep, to prevent interference with the picture. A similar circuit is employed in the cathode-ray oscilloscope, but the left-to-right (forward) scanning speed is variable rather than constant. *See also* FLYBACK, TELEVISION.

FLYWHEEL EFFECT

In any tuned circuit containing inductance and capacitance, oscillations continue at the resonant frequency of the circuit, after the energy has been removed. The higher the Q factor, or selectivity, of the circuit, the longer the decay period. The Q factor increases as the amount of resistance in the tuned circuit decreases. This phenomenon, which causes resonant circuits to ring, is called flywheel effect, and it occurs because the inductor and capacitor in a tuned circuit store energy.

The flywheel effect makes it possible to operate a Class AB, Class B, or Class C radio-frequency amplifier, with minimal distortion in the shape of the signal wave. The output waveforms of these amplifiers without tuned output circuits would be nonsinusoidal. The Class B and Class C output waveforms would be greatly distorted, and would contain considerable harmonic energy. But, because of the flywheel effect, the tuned output circuit requires only brief pulses, occurring at the resonant frequency, to produce a nearly pure sine-wave output (see illustration). *See also* CLASS AB AMPLIFIER, CLASS B AMPLIFIER, CLASS C AMPLIFIER, Q FACTOR, TUNED CIRCUIT.

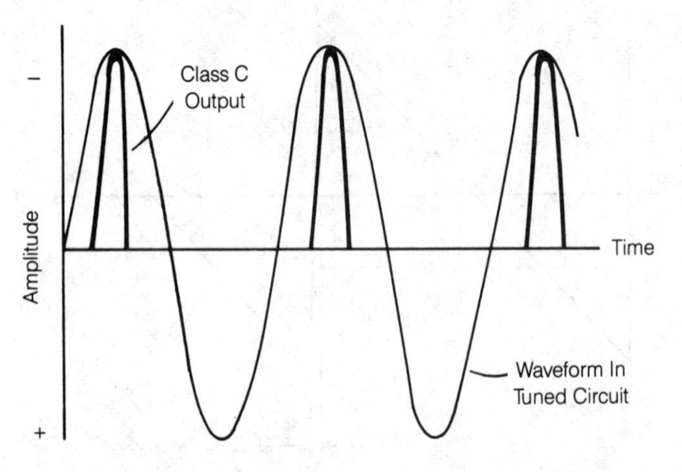

FLYWHEEL EFFECT: The flywheel effect completes the alternating current cycle in the output of Class B and Class C amplifiers.

FM

See FREQUENCY MODULATION.

FM STEREO

Most frequency-modulation (FM) broadcast stations use stereo transmission. Some FM broadcast receivers are capable of reproducing the stereo sound, and others are not. For this reason, FM stereo broadcasting must be accomplished in a special way, so it is compatible with both types of receivers. Modulated subcarriers (*see* SUBCARRIER) are used to obtain FM transmission of two independent channels.

An FM stereo signal consists of a main audio channel and two stereophonic channels. The main-channel audio consists of the combined audio signals from the left and right stereo channels; this signal frequency modulates the main carrier between 15 Hz and 15 kHz. This allows reception of the broadcast by receivers without stereo capability. The left and right stereo channels are impressed on the signal by means of subcarriers. The pilot subcarrier frequency modulates the main carrier at 19 kHz. At twice this frequency, or 38 kHz, the stereophonic subcarrier is amplitude-modulated by the difference signal between the left and right channels. The pilot subcarrier maintains the frequency of the stereophonic subcarrier in the receiver.

The channel-difference sidebands occupy a modulating frequency range from 23 to 53 kHz, centered at the frequency of the stereophonic subcarrier. A subsidiary communications signal can be impressed on the main carrier in the modulating-frequency band from 53 to 75 kHz; the maximum modulating frequency allowed by law is 75 kHz. The different sidebands contain all of the information necessary to reproduce a complete stereo signal in the FM receiver. The illustration shows the modulation-frequency band of a standard FM broadcast signal. *See also* FREQUENCY MODULATION, SUBSIDIARY COMMUNICATIONS AUTHORIZATION.

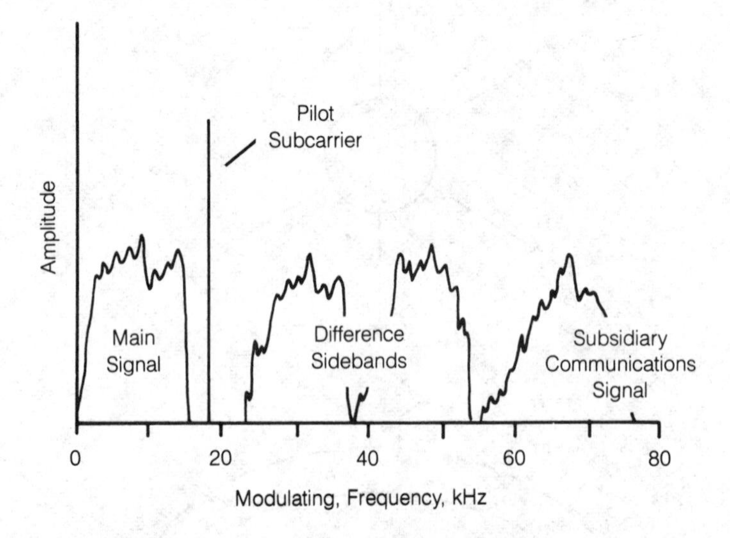

FM STEREO: An illustration of the spectrum of the modulation band in a frequency-modulation stereo broadcast signal.

FOCAL LENGTH

The focal length of a lens or reflector is the distance from its center to its focal point. Parallel rays of energy arriving at the lens or reflector from a great distance converge, and form an image, at the focal point. Conversely, a point source of energy at the focal point of a lens or reflector will produce parallel rays.

The drawing illustrates the focal lengths for typical reflectors and lenses. At A, the focal point of a spherical reflector is approximately halfway to its center; the focal length is thus about half the radius. For a parabolic reflector having a curve based on the function $y = x^2$, the focal point is at $(x,y) = (0,0.25)$; the focal length is 0.25 unit. At B, the focal point of a convex lens is illustrated.

The focal length of a lens or reflector determines the

size of the image resulting from impinging parallel rays. The greater the focal length, the larger the image. For a given source of energy, the greater the focal length, the more nearly parallel the transmitted rays. *See also* PARABOLOID ANTENNA in the ANTENNA DIRECTORY.

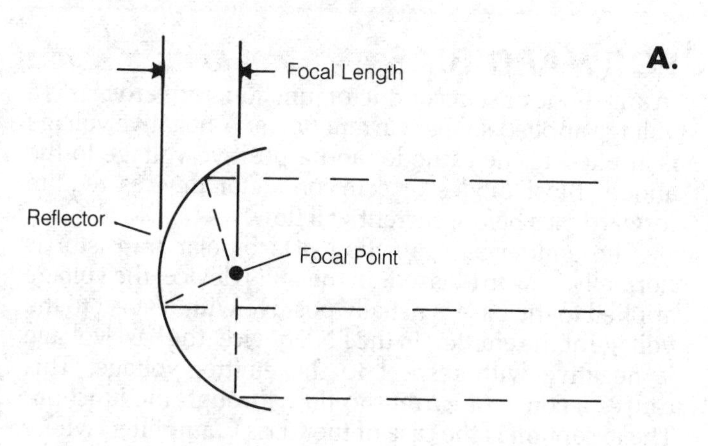

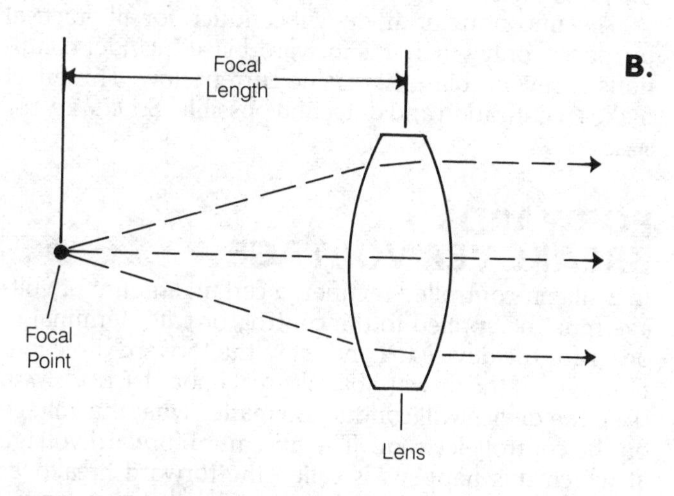

FOCAL LENGTH: The focal point and focal length of a reflector (A) and the focal point and focal length of a convex lens (B).

FOCUS CONTROL

In a television receiver or cathode-ray oscilloscope, it is desirable to keep the electron beam as narrow as possible. This is accomplished by means of focusing electrodes. The voltage applied to these electrodes controls the deflection of electrons so that they are brought to a point as they impact on the phosphor screen.

Most television receivers have internal focusing potentiometers, and the controls are preset. Once properly adjusted, they seldom require further attention. In the oscilloscope, the focus control is usually located on the front panel. Focusing can be accomplished by electromagnetic as well as electrostatic deflection. *See also* ELECTROMAGNETIC DEFLECTION, ELECTROMAGNETIC FOCUSING, ELECTROSTATIC DEFLECTION.

FOLDBACK

See CURRENT LIMITING.

FOLLOWER

A follower is a circuit in which the output signal is in phase with, or follows, the input signal. Follower circuits always have a voltage gain of less than 1. The output impedance is lower than the input impedance. Follower circuits are typically used as broadbanded impedance-matching circuits; they are often cheaper and more efficient than transformers.

The follower circuit is characterized by a grounded collector if a bipolar transistor is used, a grounded drain if a field-effect transistor is used, and a grounded plate if a vacuum tube is used. The vacuum-tube follower is called a cathode follower; the bipolar circuit is called an emitter follower; the field-effect-transistor circuit is called a source follower. *See also* EMITTER FOLLOWER, SOURCE FOLLOWER.

FOOTCANDLE

The footcandle (fc) is a unit of illuminance. One footcandle is equal to 1 lumen per square foot (lm/ft^2). *See also* CANDELA, ILLUMINANCE, LUMEN.

FOOT-LAMBERT

The foot-lambert (fl) is a unit of luminance. One foot-lambert is equal to $1/\pi$ candela per square foot. This term is obsolete. *See also* CANDELA, LUMINANCE.

FOOT-POUND

The foot-pound is a unit of energy or work. When a weight of 1 pound is lifted against gravity for a distance of 1 foot, the energy expended is equal to 1 foot-pound. In the metric system, the unit most commonly specified for work is the kilogram meter; 1 kilogram meter is equal to approximately 7.3 foot-pounds. The standard unit of energy is the joule; 1 foot-pound is equal to about 1.36 joules. *See also* ENERGY, WORK.

FORCE

Force is that effect which, when exerted against an object, induces motion or causes acceleration. Force is measured in different units. In general, a certain force F, exerted against a mass m, produces an acceleration a, such that $F = ma$. A force of 1 kilogram meter per second per second (1 kgm/s^2) is called 1 newton. A force of 1 gram centimeter per second per second (1 gmcm/s^2) is called 1 dyne. The newton represents 100,000 dynes.

When a force acts on an object, the resulting acceleration may consist of a change in the speed of the object, or a change in the direction of motion, or both. The acceleration is always in the same direction as the force. *See also* ACCELERATION.

FORCED-AIR COOLING

Forced-air cooling is a method for maintaining the proper temperature of electronic equipment by passing air through an enclosure. Forced-air cooling is accomplished with line- or direct-current powered fans or blowers. Fans are generally used to purge the air from chassis or individual electronic modules, and blowers are used to purge cabinets, enclosures, and consoles.

FORM FACTOR

The form factor is a function of the diameter-to-length ratio of a solenoidal coil. It is used for calculating the inductance. The value of the form factor, as a function of the ratio of diameter to length, is given by the graph. Given the value of the form factor *F*, the number of turns *N*, and the diameter of the coil *d* in inches, the coil inductance *L* in microhenrys is given by the formula:

$$L = FN^2d$$

In a tuned-circuit resonant response, the shape factor is sometimes called the form factor (*see* SHAPE FACTOR).

In an alternating-current wave, the ratio of the root-mean-square, or rms, value to the average value of a half cycle is sometimes called the form factor (*see* ROOT MEAN SQUARE).

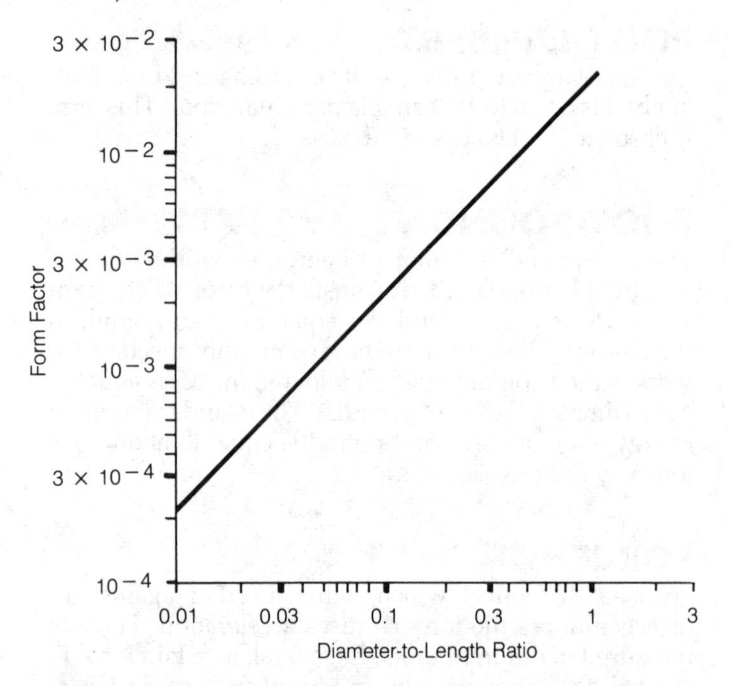

FORM FACTOR: The plot of form factor as a function of diameter-to-length ratio for determining the inductance of a solenoidal coil.

FORMULA

A formula is a mathematical expression that gives the value of a desired quantity as a function of other quantities. A formula contains variable quantities and constant quantities. The left side of a formula shows the desired quantity, and the right side is more or less complicated, depending on the nature of the function.

FORTRAN

FORTRAN is a form of higher-order computer language, designed for scientific and engineering applications. The word *FORTRAN* is a contraction of FORmula and TRANslation. The language was originally developed by International Business Machines (IBM). *See also* BASIC, COBOL, HIGHER-ORDER LANGUAGE.

FORWARD BIAS

In any diode or semiconductor junction, forward bias is a voltage applied so that current flows. A negative voltage is applied to the cathode, and a positive voltage to the anode. Most diodes or semiconductor devices require forward bias before current will flow.

The emitter-base junction of a bipolar transistor is normally forward-biased. In the NPN device, the voltage applied to the base is usually positive with respect to the voltage at the emitter; in the PNP device, the base voltage is negative with respect to the emitter voltage. This causes a constant current to flow through the junction. The exception is the case of the Class C amplifier, where the junction is reverse-biased to put it at cutoff.

A semiconductor diode will conduct, for all practical purposes, only when it is forward-biased. Under conditions of reverse bias, there is no current flow. This effect makes rectification and detection possible. *See also* REVERSE BIAS.

FORWARD BREAKOVER VOLTAGE

In a silicon-controlled rectifier, a certain amount of voltage must be applied to the control, or gate, terminal in order for the device to conduct in the forward direction. However, if the forward bias is great enough (*see* FORWARD BIAS), the device will conduct no matter what the voltage on the control electrode. The minimum forward voltage at which this happens is called the forward breakover voltage. This voltage is extremely high in the absence of a signal at the control electrode; the greater the signal voltage, the lower the forward breakover voltage.

After the control signal has been removed, the silicon-controlled rectifier continues to conduct. However, if forward bias is removed and then reapplied, the device will not conduct again until the control signal returns. *See also* SILICON-CONTROLLED RECTIFIER.

FORWARD CURRENT

In a semiconductor junction, the forward current is the value of the current that flows in the presence of a forward bias (*see* FORWARD BIAS). Generally, the greater the forward bias, the greater the forward current, although there are exceptions.

In any diode, a certain amount of forward bias is necessary to cause the flow of current in the forward direction. The function of forward current versus forward bias voltage is called the forward characteristic of a diode, transistor, or tube. *See also* REVERSE BIAS.

FORWARD RESISTANCE

The forward resistance of a diode or any device that

typically conducts in only one direction, is the value of the forward voltage divided by the forward current. If *E* is the forward voltage in volts and *I* is the forward current in amperes, then the forward resistance *R*, in ohms, is:

$$R = E/I$$

This is simply an expression of Ohm's law (*see* OHM'S LAW).

FORWARD SCATTER

When a radio wave strikes the ionospheric E and F layers, it may be returned to earth. This kind of propagation most often takes place at frequencies below about 30 MHz, although it may occasionally be observed at frequencies well into the very high part of the spectrum. When the electromagnetic field encounters the ionosphere, most of the energy is returned at an angle equal to the angle of incidence, much like the reflection of light from a mirror. However, some scattering does occur. Forward scatter is the scattering of electromagnetic waves in directions away from the transmitter.

Forward scatter results in the propagation of signals over several paths at once. This contributes to fading, because the signal reaches the receiving antenna in varying phase combinations. Some scattering occurs in the backward direction, toward the transmitter; this is called backscatter. *See also* BACKSCATTER, IONOSPHERE, PROPAGATION CHARACTERISTICS.

FORWARD VOLTAGE DROP

In a rectifying device such as a semiconductor diode or vacuum tube, the forward voltage drop is the voltage, under conditions of forward bias, that appears between the cathode and the anode. The forward voltage drop of most rectifying devices remains constant at all values of forward bias.

If the forward bias is less than the forward voltage drop for a particular device, the component will normally not conduct. When the voltage reaches or exceeds the forward voltage drop, the component abruptly begins to conduct. *See also* FORWARD BIAS, FORWARD CURRENT, FORWARD RESISTANCE.

FOSTER-SEELEY DISCRIMINATOR

The Foster-Seeley discriminator is a form of detector circuit for reception of frequency-modulated signals. A center-tapped transformer, with tuned primary and secondary windings, is connected to a pair of diodes. The center tap of the secondary winding is coupled to the primary. See the schematic diagram of a simple Foster-Seeley type discriminator. The circuit is recognized by the single, center-tapped secondary winding.

As the frequency of the input signal varies, the voltage across the secondary fluctuates in phase. The diodes produce audio-frequency variations from these changes in phase. The Foster-Seeley discriminator circuit is sensitive to changes in the signal amplitude as well as phase and frequency. Therefore, the circuit must have a limiter stage preceding it for best results. *See also* DISCRIMINATOR, FREQUENCY MODULATION, RATIO DETECTOR.

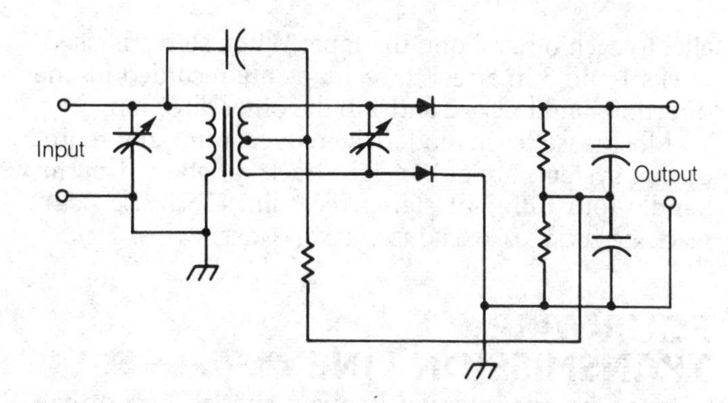

FOSTER-SEELEY DISCRIMINATOR: This discriminator is a common circuit for detecting frequency-modulated signals.

FOURIER SERIES

Any periodic function can be expressed in terms of an infinite series called a Fourier series. The Fourier series is a sum of sine and cosine functions. Any periodic wave, such as a sawtooth wave or square wave, consists of a certain combination of fundamental and harmonic frequencies. The Fourier series is a means of denoting this combination.

The general form of a Fourier series is given by:

$$f(x) = (A_o/2) + \sum_{n=1}^{\infty} (A_n\cos(nx) + B_n\sin(nx))$$

where A_o, A_n, and B_n are constants. The more terms are evaluated in the series, the more accurate the representation of the actual function. A Fourier series is an infinite series, however, so any finite sum represents an approximation of the actual function. *See also* FOURIER TRANSFORM.

FOURIER TRANSFORM

The Fourier transform is a mathematical method of obtaining the power spectrum, as a function of frequency, for a given signal waveform. This is done by determining the Fourier series representing the particular waveform, and is a mathematical approximation.

Fourier transformations allow the evaluation of nonperiodic as well as periodic waveforms. Any waveform can be mathematically approximated over a finite interval, provided there are a finite number of maxima and minima in that interval. *See also* FOURIER SERIES.

FOUR-TRACK RECORDING

Four-track recording is a method of recording stereo audio signals on magnetic tape. A reel-to-reel or cassette arrangement may be used. Each recording requires two tracks on the tape, one for the left channel and the other for the right channel. A four-track tape can therefore contain two separate stereo recordings or four separate monaural recordings. The four magnetic tracks run parallel to each other along the tape. When stereo is used, tracks 1 and 3 in a reel-to-reel tape are recorded in one direction, and tracks 2 and 4 in the other direction.

Most cassettes in modern stereo recording and reproducing systems are of the eight-track variety, which can handle four different stereo recordings. *See also* EIGHT-TRACK RECORDING, MAGNETIC RECORDING, MAGNETIC TAPE.

FOUR-WIRE TRANSMISSION LINE

A four-wire transmission line is a special form of balanced line used to carry radio-frequency energy from a transmitter to an antenna, or from an antenna to a receiver. Four conductors are run parallel to each other so that they form the edges of a long square prism (see illustration). Each pair of diagonally opposite conductors is connected together at either end of the line.

Four-wire transmission line may be used in place of parallel-wire line in almost any installation. For a given conductor size and spacing, four-wire line has a somewhat lower characteristic impedance because the four-wire line is essentially two two-wire lines in parallel. The four-wire line is not as susceptible to the influences of nearby objects as is the two-wire line; more of the electromagnetic-field energy is contained within the line. Thus, the four-wire line has inherently better balance, and is less likely to radiate or pick up unwanted noise. Four-wire line is more difficult to install, and is more expensive, than two-wire lines. Four-wire lines are not commonly used. *See also* BALANCED TRANSMISSION LINE, OPEN-WIRE LINE, TWIN-LEAD.

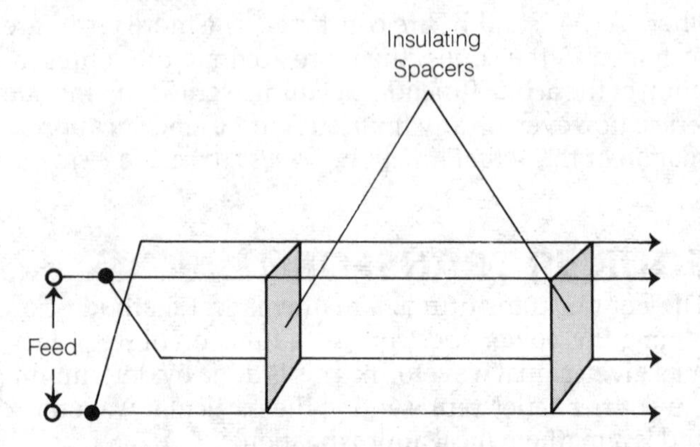

FOUR-WIRE TRANSMISSION LINE: This type of transmission line has excellent balance.

FRAME

A frame is a single, complete television picture. The picture is generally repeated 25 times per second. The electron beam scans 625 horizontal lines per frame, with a horizontal-to-vertical size ratio of 4:3. Scanning takes place from left to right along each horizontal line, and from top to bottom for the lines in a complete frame.

There are some variations, in different parts of the world, in the number of scanning lines in a frame and in the overall bandwidth of the picture signal. However, each frame is always a still picture. The human eye can perceive only about 15 to 25 still pictures per second; if the pictures are repeated more often than this, they combine to form an apparent moving picture. Television picture frames are repeated so often that they combine to form a moving image. *See also* TELEVISION.

FRANKLIN OSCILLATOR

See OSCILLATOR CIRCUITS.

FREE ELECTRON

When an electron is not held in an orbit around the nucleus of an atom, the electron is called a free electron. In order to become independent of the influence of atomic nuclei, electrons must receive or contain a minimum amount of energy. The lower the energy an electron possesses, the more tightly it is held in orbit around the nucleus of an atom because the positive charge of the nucleus attracts the negative charge of the electron. *See also* BOUND ELECTRON, ELECTRON, IONIZATION, PLASMA.

FREE-RUNNING MULTIVIBRATOR

See ASTABLE MULTIVIBRATOR.

FREE SPACE

Free space is a term used to denote an environment in which there are essentially no objects or substances that affect the propagation of an electromagnetic field. Perfect free space exists nowhere in the universe; the closest thing to true free space is probably found in interstellar space. Free space has a magnetic permeability of 1 and a dielectric constant of 1. Electromagnetic fields travel through free space at approximately 186,282 miles per second.

When designing and testing antenna systems, the free-space performance is evaluated for the purpose of determining the directional characteristics, impedance, and power gain. A special testing environment provides conditions approaching those of true free space. An antenna installation always deviates somewhat from the theoretical free-space ideal. However, antennas placed several wavelengths above the ground, and away from objects such as trees, utility wires, and buildings, operate as if they were in free space. *See also* FREE-SPACE LOSS, FREE-SPACE PATTERN.

FREE-SPACE LOSS

As an electromagnetic field is propagated away from its source, the field becomes less intense as the distance increases. In free space, this happens in a predictable and constant manner, regardless of the wavelength.

The field strength of an electromagnetic field, in volts per meter, varies inversely with distance. If F is the field strength expressed in volts per meter and d is the distance from the source, then:

$$F = k/d$$

where k is a constant that depends on the intensity of the source.

The field strength as measured in watts per square meter varies inversely with the square of the distance from the source. If *G* is the field strength in watts per square meter and *d* is the distance from the source, then:

$$G = m/d^2$$

where m is, again, a constant that depends on the source intensity (though not the same constant as k).

In free space, the field strength drops by 6 decibels when the distance from the source is doubled. This is true no matter how the field strength is measured in the far field. *See also* FAR FIELD, FIELD STRENGTH, FREE SPACE, INVERSE-SQUARE LAW.

FREE-SPACE PATTERN

The free-space pattern is the radiation pattern exhibited by an antenna in free space. Free-space patterns are usually specified when referring to the directional characteristics and power gain of an antenna. Under actual operating conditions, in which an antenna is surrounded by other objects, the free-space pattern is modified. If the antenna is close to the ground, or is operated in the midst of trees, buildings, and utility wires within a few wavelengths of the radiator, the antenna pattern may differ substantially from the free-space pattern.

An antenna that radiates equally well in all possible directions is called an isotropic antenna (*see* ISOTROPIC ANTENNA in the ANTENNA DIRECTORY). The free-space pattern of the isotropic antenna is a sphere, with the antenna at the center (see illustration at A). A dipole antenna (*see* HALF-WAVE ANTENNA in the ANTENNA DIRECTORY) radiates equally well in all directions perpendicular to the axis of the radiator, but the field strength drops as the direction approaches the axis of the wire. Along the axis of the wire, there is no radiation from the dipole. The free-space pattern of a dipole is torus-shaped, as at B. The dipole and the isotropic antenna are used as references for antenna power-gain measurements. *See also* ANTENNA PATTERN, ANTENNA POWER GAIN, dBd, dBi.

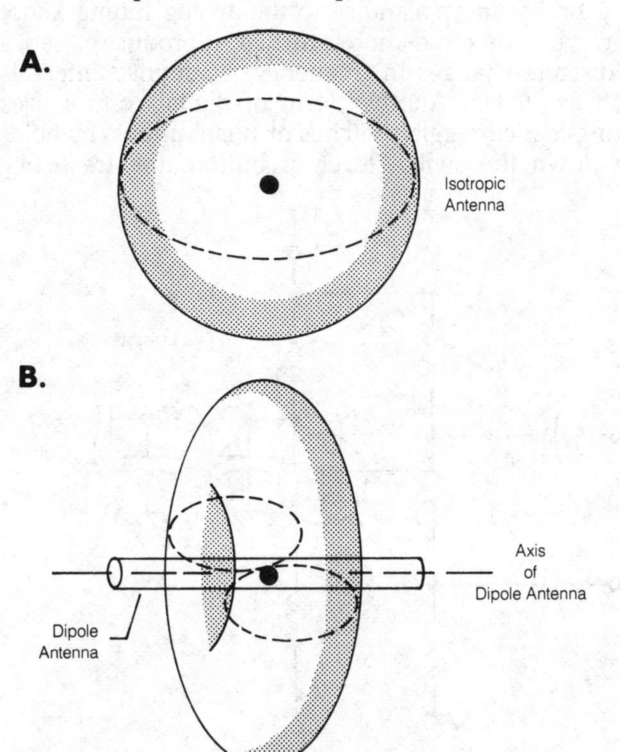

FREE-SPACE PATTERN: The free-space radiation pattern for an isotropic radiator (sphere), A, and for a dipole (toroid with point-sized hole), B.

FREON

Freon is a synthetic gas consisting of carbon, chlorine, and fluorine. Its chemical formula is CCl_2F_2. Freon is commonly used as a coolant in air conditioners and refrigerators. However, in gaseous form, freon can serve as an excellent dielectric.

Freon has about 2½ times the dielectric strength of air. This means that, at atmospheric pressure, freon can withstand about 2½ times as much voltage as dry, pure air. When large voltages must exist within a limited space, freon can be useful for insulation purposes. *See also* DIELECTRIC, DIELECTRIC BREAKDOWN.

FREQUENCY

For any periodic disturbance, the frequency is the rate at which the cycle repeats. Frequency is generally measured in cycles per second, or hertz, abbreviated Hz. Rapid oscillation frequencies are specified in kilohertz (kHz), megahertz (MHz), gigahertz (GHz), and terahertz (THz). A frequency of 1 kHz is 1,000 Hz; 1 MHz = 1,000 kHz; 1 GHz = 1,000 MHz; and 1 THz = 1,000 GHz. If the time required for 1 cycle is denoted by *p*, in seconds, then the frequency *f* is equal to $1/p$.

Wave disturbances can have frequencies ranging from less than 1 Hz to many trillions of teraherz. The audio frequency range, at which the human ear can detect acoustic energy, ranges from about 20 Hz to 20 kHz. The radio-frequency spectrum of electromagnetic energy ranges up to hundreds or thousands of gigahertz. The frequency of electromagnetic radiation is related to the wavelength according to the formula:

$$\lambda = 300v/f$$

where λ is the wavelength in meters, v is the velocity factor of the medium in which the wave travels, and f is the frequency in megahertz. *See also* ELECTROMAGNETIC SPECTRUM, VELOCITY FACTOR, WAVELENGTH.

FREQUENCY ALLOCATIONS

Frequency allocations are frequencies, or bands of frequencies in the electromagnetic spectrum assigned by a government or other authority for use in specific applications.

Frequency assignments are made on a worldwide basis by the International Telecommunication Union (ITU), headquartered in Geneva, Switzerland. The lowest allocated electromagnetic frequency is 10 kHz. The highest allocated frequency is 275 GHz. However, the maximum frequency can be expected to rise as technology improves at the short wavelengths. In the United States, frequency assignments are made by the Federal Communications Commission (FCC). A list of United States allocations can be obtained from the Superintendent of Documents, U.S. Government Printing Office, Washington, DC 20402.

FREQUENCY CALIBRATOR

A frequency calibrator is an instrument that generates unmodulated-carrier markers at precise, known frequencies. Frequency calibrators generally consist of crystal-controlled oscillators. The crystal may be housed in a special heat-controlled chamber for maximum stability; phase-locked-loop circuits may also be employed in conjunction with standard broadcast frequencies.

Frequency calibrators are a virtual necessity in all receivers. Without a properly adjusted calibrator, the operator of a receiver cannot be certain of the accuracy of the frequency dial or readout. Calibrators are especially important in transmitting stations because out-of-band transmission can cause annoying interference at other stations.

The frequency calibrator in a receiver should be checked periodically against a standard frequency source such as radio station WWV or WWVH. The calibrator should be adjusted, if it produces markers at even-megahertz points, to zero beat with the carrier frequency of the standard source. *See also* CRYSTAL CONTROL, CRYSTAL OVEN, PHASE-LOCKED LOOP, WWV/WWVH.

FREQUENCY COMPARATOR

A frequency comparator is a circuit that detects the difference between the frequencies of two input signals. The output of a frequency comparator may be in video or audio form. Basically, all frequency comparators operate on the principle of beating. The beat frequency, or difference between the two input frequencies, determines the output.

A simple frequency comparator circuit is illustrated in the schematic diagram. When the frequencies of the two input signals are identical, there is no output. When the frequencies are different, a beat note is produced, in proportion to the difference between the input-signal frequencies. This output signal is fed to an audio amplifier. The audio beat note appears at the output of the amplifier, where it can be fed to a speaker, oscilloscope, or frequency counter.

Frequency comparators are used by phase-locked loop circuits where two signals must be precisely matched in frequency. The output of the frequency comparator is used to keep the frequencies identical, by means of controlling circuits. *See also* PHASE-LOCKED LOOP.

FREQUENCY COMPENSATION

When the frequency response of a circuit must be tailored to have a certain characteristic, frequency-compensating circuits are used. The simplest form of frequency-compensating circuit is a capacitor or inductor, in conjunction with a resistor. More advanced frequency-compensating circuits use operational amplifiers or tuned circuits.

The bass and treble controls in a stereo hi-fi system are a form of frequency compensation. The desired response can be obtained by adjusting these controls. In frequency-modulation communications systems, a form of frequency compensation called preemphasis is used at the transmitter, with deemphasis at the receiver for improved intelligibility. Some communications receivers have an audio tone control, to compensate for the speaker response. *See also* DEEMPHASIS, PREEMPHASIS, TONE CONTROL.

FREQUENCY CONTROL

A frequency control is a knob, switch, screw, pushbutton, or other actuator that is used to set the frequency of a receiver, transmitter, transceiver, or oscillator.

The most common method of frequency control is the tuning knob; the frequency can be adjusted over a continuous range by rotating the knob. This is analog frequency control. The knob is connected either to a variable capacitor or a variable inductor, which in turn sets the frequency of an oscillator-tuned circuit.

In recent years, digital frequency control has become increasingly common. There are several different configurations of digital frequency control, but this method of frequency adjustment is always characterized by discrete frequency intervals instead of a continuous range. One method of digital frequency control uses a rotary knob, very similar in appearance to the analog tuning knob. However, when the knob is turned, it produces a series of discrete changes in frequency, at defined intervals such as 10 Hz. Another form of digital control uses spring-loaded toggle switches or pushbuttons; by holding down the switch lever or button, the frequency

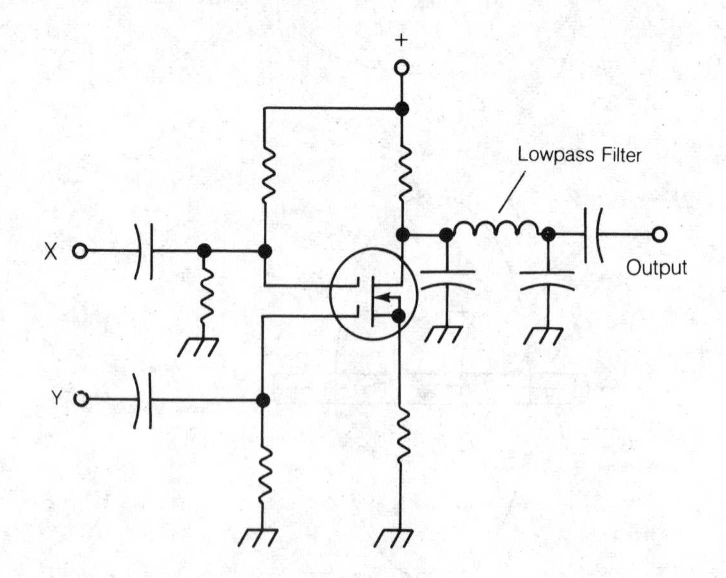

FREQUENCY COMPARATOR: The schematic of a frequency comparator that operates as a mixer, producing a difference signal in the output.

increments at a certain rate in discrete steps, such as 1 kHz per second in 10 Hz jumps. The third common form of digital frequency control uses several pushbuttons, each of which sets one digit of the desired frequency. Some communications equipment have more than one of these digital tuning methods so the operator can choose the particular frequency-control mode desired. The electrical control of the frequency is usually accomplished by means of varactor diodes in the oscillator tuned circuit, and/or synthesized oscillators.

Both analog and digital frequency-control methods have advantages and disadvantages. Often, the optimum method is determined largely by the preference of the individual operator. *See also* ANALOG, DIGITAL.

FREQUENCY CONVERSION

In some situations, it is necessary to change the carrier frequency of a signal without changing the modulation characteristics. When this is deliberately done, it is called frequency conversion. Frequency conversion is an integral part of the superheterodyne receiver, and it is found in many communications transmitters as well.

Frequency conversion is accomplished by means of heterodyning. The signal is combined with an unmodulated carrier in a circuit called a mixer (*see* MIXER). If the input signal has a frequency f, and the desired output frequency is h, then the frequency of the unmodulated oscillator g must be such that either $f - g = h$, $g - f = h$, or $f + g = h$. The output frequency must, in other words, be equal to either the sum or the difference of the input frequencies.

Frequency conversion is not the same thing as frequency multiplication. With certain types of modulation, the bandwidth changes when the frequency of the carrier is multiplied. But with true frequency conversion, the signal bandwidth is never affected. Frequency conversion is desirable in many applications because it simplifies the design of the selective circuits in a receiver or transmitter. *See also* FREQUENCY CONVERTER, FREQUENCY MULTIPLIER, SUPERHETERODYNE RADIO RECEIVER.

FREQUENCY CONVERTER

A frequency converter is a mixer designed for use with a communications receiver to allow operation on frequencies not within the normal range of the receiver. If the operating frequency is above the receiver range, the converter is called a down converter. If the operating frequency is below the receiver range, the converter is called an up converter. Converters are often used with high-frequency, general-coverage receivers to allow reception at very low, very high, and ultrahigh frequencies.

Converters are relatively simple circuits. With a good communications receiver, capable of operating in the range of 3 to 30 MHz, frequency converters can be used to obtain excellent sensitivity and selectivity throughout the radio spectrum. The use of a converter with an existing receiver is often not only superior to a separate receiver, but far less expensive.

Some converters, designed to operate with transceivers, allow two-way communication on frequencies outside the range of the transceiver. This kind of device has two separate converters, and the entire circuit is called a transverter. *See also* DOWN CONVERSION, FREQUENCY CONVERSION, MIXER.

FREQUENCY COUNTER

A frequency counter is an instrument or circuit that measures the frequency of a periodic wave by actually counting the pulses in a given interval of time.

The typical frequency-counter circuit includes a gate, which begins and ends each counting cycle at defined intervals. The accuracy of the frequency measurement is a direct function of the length of the gate time; the longer the time base, the better the accuracy. At higher frequencies, the accuracy is limited by the number of digits in the display. An accurate reference frequency is necessary to have a precision counter; crystals are used for this purpose. The reference oscillator frequency may be synchronized with a time standard by means of a phase-locked loop.

Frequency counters are available that allow accurate measurements of signal frequencies into the gigahertz range. Frequency counters are invaluable in the test laboratory, since they are easy to use and read. Typical frequency counters have readouts that show six to ten significant digits. *See also* FREQUENCY MEASUREMENT, PHASE-LOCKED LOOP.

FREQUENCY DEVIATION

See DEVIATION.

FREQUENCY DIVIDER

A frequency divider is a circuit whose output frequency is some fraction of the input frequency. All frequency dividers are digital circuits. While frequency multiplication can be accomplished by means of simple nonlinear analog devices, frequency division requires counting devices. Fractional components of a given frequency are sometimes called subharmonics; a wave does not naturally contain subharmonics.

Frequency dividers can be made from bistable multivibrators. Dividers, in conjunction with frequency multipliers, can be combined to produce almost any rational-number multiple of a given frequency. For example, four divide-by-two circuits can be combined with a tripler, resulting in multiplication of the input frequency by 3/16. *See also* FREQUENCY MULTIPLIER.

FREQUENCY-DIVISION MULTIPLEX

One method of sending several signals simultaneously over a single channel is called frequency-division multiplex. An available channel is subdivided into smaller segments, all of equal size, and these segments are called subchannels. Each subchannel carries a separate signal. Frequency-division multiplexing can be used in wire transmission circuits or radio-frequency links.

One method of obtaining frequency-division multiplexing is shown in the illustration. The available channel is the frequency band from 10,000 to 10.009 MHz, or a space of 9 kHz. This space is divided into nine subchannels, each 1 kHz wide. Nine different medium-speed carriers are impressed on the main carrier of 10.000 MHz. This is accomplished by using audio tones of 500, 1500, 2500, 3500, 4500, 5500, 6500, 7500, and 8500 Hz, all of which modulate the main carrier to produce a single-sideband signal. The main carrier is suppressed by means of a balanced modulator. Each audio tone may be keyed independently in a digital code such as Morse, Baudot, or ASCII. The tones must be fed into the modulator circuit using appropriate isolation methods to avoid the generation of mixing products. The resulting output signals appear at 1 kHz intervals from 10.0005 MHz to 10.0085 MHz.

The example is given only for the purpose of illustrating the concept of frequency-division multiplex. There are many different ways of obtaining this form of emission. Frequency-division multiplex necessitates a sacrifice of data-transmission speed in each subchannel, as compared with the speed obtainable if the entire channel were used. For example, each signal in the illustration must be no wider than 1 kHz; this limits the data-transmission speed to 1/9 the value possible if the whole 9 kHz channel were employed. Frequency-division multiplex is a form of parallel data transfer. *See also* MULTIPLEX, PARALLEL DATA TRANSFER.

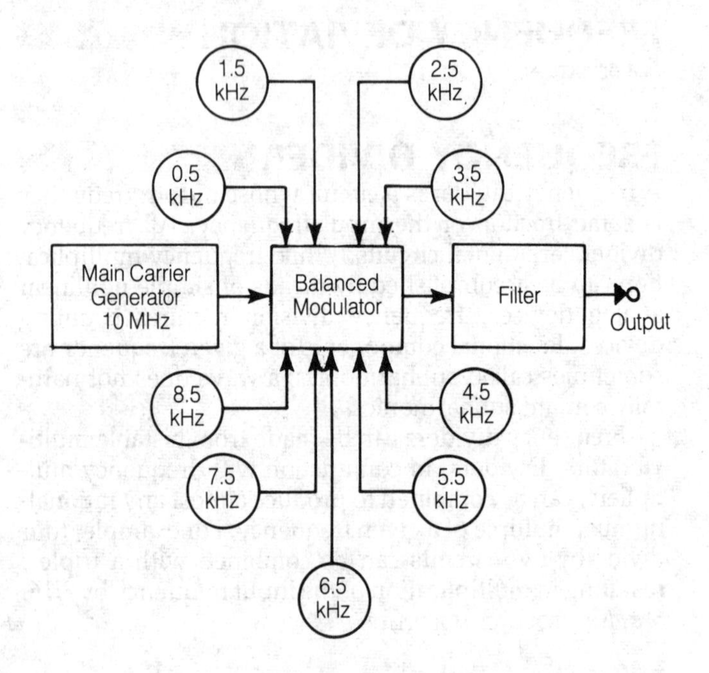

FREQUENCY-DIVISION MULTIPLEX: A schematic representation of the generation of a frequency-division multiplex signal.

FREQUENCY MEASUREMENT

There are many different ways of determining the frequency of a signal. Basically, these methods can be categorized in two ways: direct measurement and indirect measurement.

Direct frequency measurement is done by means of frequency counters. The number of alternations, or cycles, within a given time period is actually counted by means of digital circuits. The time interval need not necessarily be exactly 1 second. The longer the counting interval, the greater the accuracy of the measurement.

Indirect frequency measurement is done with frequency meters or calibrated receivers. The simplest frequency meter consists of an adjustable tuned circuit and an indicator that shows when the device is set for resonance at a particular frequency. The approximate frequency can then be read from a calibrated scale. If a receiver is available, the signal can be tuned in and the frequency read, with fair to good accuracy, from the tuning dial. If a calibrated signal generator is available, it may be set, by means of an indicating device, to zero beat with the signal under measurement. *See also* FREQUENCY, FREQUENCY METER.

FREQUENCY METER

A frequency meter is an instrument used for measuring the frequency of a periodic occurrence. There are several kinds of frequency meters. An increasingly common method of frequency measurement is the direct counting of the cycles within a given interval of time; this is done with a frequency counter. Frequency counters provide extremely accurate indications of the frequency of a periodic wave. In some cases the resolution is better than ten significant digits.

The absorption wavemeter uses a tuned circuit in conjunction with an indicating meter. The capacitor or inductor in the tuned circuit is adjustable, and has a calibrated scale showing the resonant frequency over the tuning range. When the device is placed near the signal source, the meter shows the energy transferred to the tuned circuit; the coupling is maximum, resulting in a peak reading of the meter, when the tuned circuit is set to the same frequency as the signal source.

The gate-dip meter operates on a principle similar to the absorption wavemeter, except that the gate-dip meter contains a variable-frequency oscillator. An indicating meter shows when the oscillator is set to the same frequency as a tuned circuit under test. Gate-dip meters are commonly used to determine unknown resonant frequencies in tuned amplifiers and antenna systems.

The heterodyne frequency meter contains a variable-frequency oscillator, a mixer, and an indicator such as a meter. The oscillator frequency is adjusted until zero beat is reached with the signal source. This zero beat is shown by a dip in the meter indication.

The cavity frequency meter is often used to determine resonant frequencies at the very high, ultra high, and microwave parts of the spectrum. The principle of operation is similar to that of the absorption wavemeter, except that an adjustable resonant cavity is used rather than a tuned circuit.

The lecher-wire system of frequency measurement employs a tunable section of transmission line to determine the wavelength of an electromagnetic signal in the very high, ultrahigh, and microwave regions. The system consists of two parallel wires with a movable shorting bar. The shorting bar is set until the system reaches resonance, as shown by an indicating meter.

Spectrum analyzers are sometimes used for the purpose of frequency measurement. General-coverage receivers can be used for determining the frequency of a signal. An oscilloscope can be used for very approximate measurements. *See also* ABSORPTION WAVEMETER, FREQUENCY COUNTER, HETERODYNE FREQUENCY METER, LECHER WIRES, OSCILLOSCOPE, SPECTRUM ANALYZER.

FREQUENCY MODULATION

Information may be transmitted over a carrier wave by frequency modulation (FM). The frequency of the carrier wave is made to vary in accordance with the modulating waveform.

Frequency modulation is generally accomplished by a circuit called a reactance modulator. The audio-frequency signal is applied to the tuned circuit of an oscillator across a varactor diode, which introduces a variable capacitance into the circuit (*see* REACTANCE MODULATOR, VARACTOR DIODE). The amount by which the signal frequency changes is called the deviation (*see* DEVIATION). In communications practice, the maximum deviation is typically plus or minus 5 kHz with respect to the unmodulated carrier frequency.

Frequency modulation results in the generation of sidebands, very similar to those in an amplitude-modulated system. However, as the deviation is increased, sidebands appear at greater and greater distances from the main carrier. The amplitude of the main carrier also depends on the amount of deviation. The amplitudes of the sidebands, which appear at integral multiples of the modulating-signal frequency above and below the carrier, as well as the amplitude of the carrier itself, are a function of the ratio of the deviation to the modulating frequency. The function is rather complicated, but in general, the greater the deviation, the greater the bandwidth of the signal. The ratio of the maximum frequency deviation to the highest modulating frequency is called the modulation index (*see* MODULATION INDEX).

When a frequency-modulated signal is put through a frequency multiplier, the deviation is multiplied by the same factor as the carrier frequency. This results in an increase in the bandwidth of the signal. This effect does not occur with mixing. These facts must be kept in mind when designing a frequency-modulation transmitter.

Reception of FM signals requires a special kind of detector called a discriminator. The discriminator, in conjunction with a limiter, is sensitive only to variations in the signal frequency, and is essentially immune to variations in amplitude. The ratio detector, another form of FM detector, needs no limiter circuit (*see* DISCRIMINATOR, RATIO DETECTOR).

A frequency-modulated signal should remain constant in amplitude, and should change only in frequency. The illustration shows the function of frequency versus time for a sine-wave frequency-modulated communications signal. Note that the function of frequency variation is a linear representation of the modulating waveform; the rate of frequency change is directly proportional to the rate of change in the voltage of the modulating signal.

The phase of a signal is inherently related to its frequency. With frequency modulation, changes in phase are introduced. With changes in phase, instantaneous variations also occur in the signal frequency. Phase modulation can be used in place of frequency modulation in a communications system, with only minor differences. *See also* PHASE MODULATION.

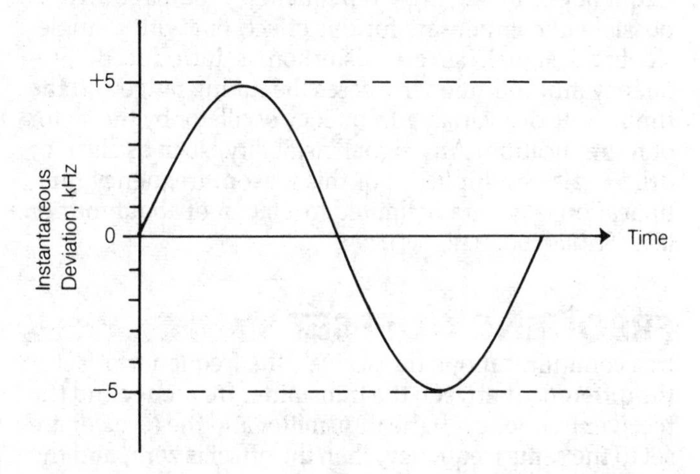

FREQUENCY MODULATION: The frequency versus time function for sine-wave modulation.

FREQUENCY MULTIPLIER

A frequency multiplier is a device that produces an integral multiple, or many integral multiples, of a given input signal. A frequency multiplier circuit may also be called a harmonic generator.

To generate harmonic energy, a nonlinear element must be placed in the path of the signal. A semiconductor diode is ideal for this purpose. (However, the type of diode must be chosen so that the capacitance is not too great.) Then, the signal waveform is distorted, and harmonics are produced. Following the nonlinear element, a tuned circuit may be used to select the particular harmonic desired. The tuned circuit is set to resonate at the

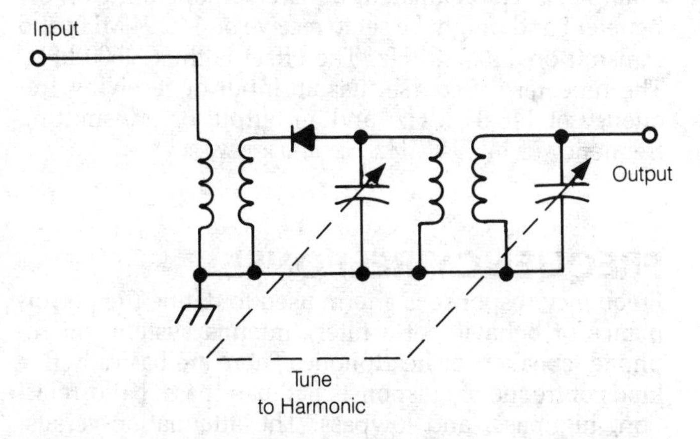

FREQUENCY MULTIPLIER: This frequency multiplier circuit uses a single diode.

harmonic frequency. The illustration is a simple schematic diagram of a frequency multiplier. There are other methods of obtaining frequency multiplication; odd multiples may be produced by means of a push-pull circuit, and even multiples by a push-push circuit (*see* PUSH-PULL CONFIGURATION, PUSH-PUSH CONFIGURATION).

Frequency multiplication does not change the characteristics of an amplitude-modulated or continuous-wave signal. However, the bandwidth of a single-sideband, frequency-shift-keyed, or frequency-modulated signal is multiplied by the same factor as the frequency. With frequency-shift keying or frequency modulation, it is possible to compensate for this effect, but with a single-sideband signal, severe distortion is introduced. Frequency multiplication increases the tuning range and the tuning rate of a variable-frequency oscillator by the factor of multiplication. Any signal instability, such as chirp or drift, is also multiplied. For this reason, frequency multiplication is generally limited to a factor of about four or less. *See also* DOUBLER, TRIPLER.

FREQUENCY OFFSET

In a communications transceiver, the frequency offset is the difference between the transmitter frequency and the receiver frequency. If the transmitter and the receiver are set to the same frequency, then the offset is zero, and the mode of operation is called simplex.

In continuous-wave (code) communications, the two stations usually set their transmitters to exactly the same frequency. This necessitates that their receivers be offset by several hundred hertz, so that audible tones appear at the speakers. The receiver offset is obtained by a control called a clarifier or a receiver-incremental-tuning (RIT) control.

In transceivers designed for operation with repeaters, the transmitter frequency is significantly different from the receiver frequency. The offset may be several megahertz at the ultrahigh-frequency range. This offset is needed because the repeater cannot transmit on the same frequency at which it receives. If the transmitter frequency of the transceiver is higher than the receiver frequency, the offset is called positive; if the transmitter frequency is lower, the offset is considered negative. For example, a typical amateur transceiver, operating on the 2-meter band, might be set to receive on 146.94 MHz and transmit on 146.34 MHz. The offset is thus −600 kHz. The repeater, of course, has an input or receiving frequency of 146.34 MHz, and an output or transmitting frequency of 146.94 MHz. *See also* REPEATER.

FREQUENCY RESPONSE

Frequency response is a term used to define the performance or behavior of a filter, antenna system, microphone, speaker, or headphone. There are basically five kinds of frequency response: flat, bandpass, band-rejection, highpass, and lowpass. The attenuation-versus-frequency functions, in simple generalized form, for each type of response are shown in the illustration.

The frequency responses of some devices are more complicated than the elementary functions illustrated. There may be several peaks and valleys in the attenuation-versus-frequency functions of some systems. An example of this is the audio high-fidelity system employing a graphic equalizer (*see* EQUALIZER). The equalizer allows precise adjustment of the response.

Devices with various kinds of frequency response are useful in particular electronics applications. Bandpass filters are employed in radio-frequency transmitters and receivers. Lowpass filters are used to minimize the harmonic output from a transmitter. Highpass filters can be used to reduce the susceptibility of television receivers to interference by lower-frequency signals. Band-rejection filters are often used to attenuate a strong, undesired signal in the front end of a receiver. *See also* BANDPASS RESPONSE, BAND-REJECTION RESPONSE, FLAT RESPONSE, HIGHPASS RESPONSE, LOWPASS RESPONSE.

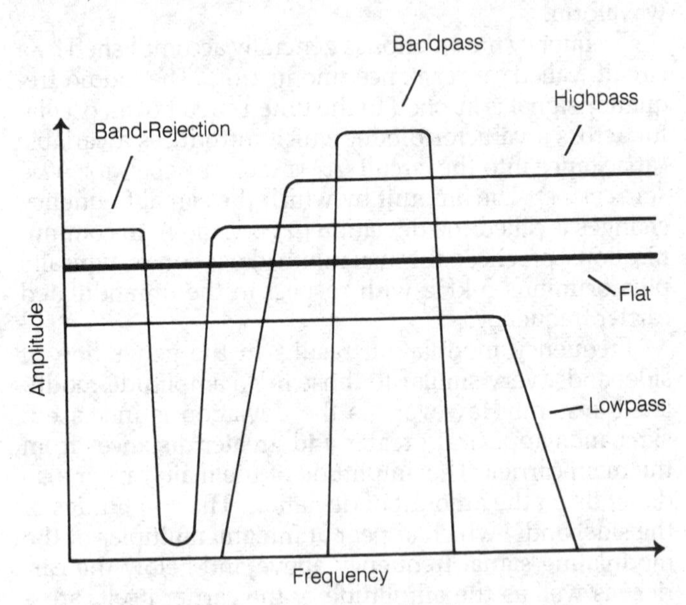

FREQUENCY RESPONSE: Five types of frequency response. (The zero-attenuation levels differ for clarity of illustration.)

FREQUENCY SHIFT

A frequency shift is a change in the frequency of a signal, either intentional or unintentional. The term *shift* implies a sudden, or rapid, change in frequency.

The frequency of a transmitter or receiver may suddenly shift because of a change in the capacitance or inductance of the oscillator tuned circuit. The amount of the shift is given by the difference between the signal frequencies before and after the change. An increase in frequency represents a positive shift; a decrease is a negative shift.

Frequency shift is deliberately introduced into a carrier wave in all forms of frequency-modulation transmission. *See also* FREQUENCY MODULATION, FREQUENCY-SHIFT KEYING.

FREQUENCY-SHIFT KEYING

Frequency-shift keying (FSK) is a digital mode of transmission, commonly used in radioteletype applications.

In a continuous-wave or code transmission, the carrier is present only while the key is down, and the transmitter is turned off while the key is up. In frequency-shift keying, however, the carrier is present all the time. When the key is down, the carrier frequency changes by a predetermined amount. The carrier frequency under key-up conditions is called the space frequency. The carrier frequency under key-down conditions is called the mark frequency (see illustration).

In high-frequency radioteletype communications, the space frequency is usually higher than the mark frequency by a value ranging from 100 to 900 Hz. The most common values of frequency shift are 170 to 425, and 850 Hz. In recent years, because of the improvement in selective-filter technology, the narrower shift values have become increasingly common.

Frequency-shift keying is considered a form of frequency modulation, and is designated type F1 emission. Frequency-shift keying can be obtained in two ways. The first, and obvious, method is the introduction of reactance into the tuned circuit of the oscillator in a transmitter. The other method uses an audio tone generator with variable frequency, the output of which is fed into the microphone input of a single-sideband transmitter. If this method is used, precautions must be taken to ensure that there is no noise in the injected signal, for such noise will appear on the transmitted signal.

In some radioteletype links, especially at the very high frequencies and above, audio frequency-shift keying is used. This is called F2 emission. An audio tone generator with variable frequency is coupled to the microphone input of an amplitude-modulated or frequency-modulated transmitter. The tone frequencies are between about 1000 and 3000 Hz. This mode of frequency-shift keying requires considerably greater bandwidth than F1 emission. Its primary advantage is that the receiver frequency setting is not critical, whereas with F1 emission, even a slight error in the receiver setting will seriously degrade the reception.

Frequency-shift keying provides a greater degree of transmission accuracy than is possible with an ordinary continuous-wave system. This is because the key-up condition, as well as the key-down condition, is positively indicated. The frequency-shift-keyed system is therefore less susceptible to interference from atmospheric or man-made sources. *See also* RADIOTELETYPE.

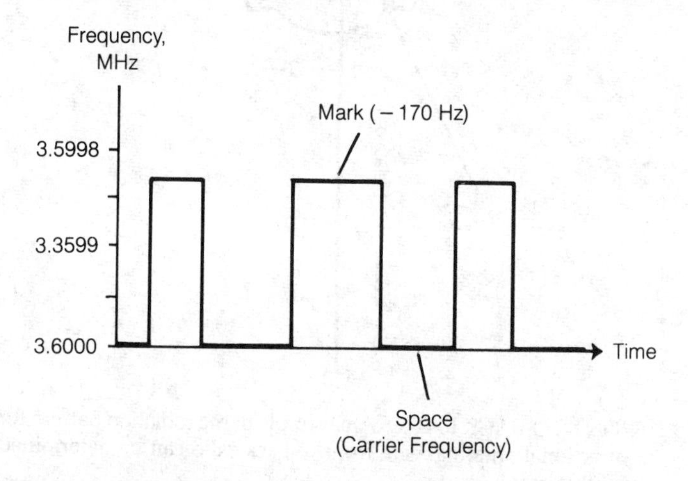

FREQUENCY-SHIFT KEYING: A frequency-versus-time waveform for an FSK signal.

FREQUENCY SWING

See CARRIER SWING.

FREQUENCY SYNTHESIZER

A frequency synthesizer is a circuit that generates precise frequency signals by means of a single crystal oscillator in conjunction with frequency dividers and multipliers. Frequency synthesizers are often used in modern communications equipment in place of the older variable frequency oscillators.

Frequency synthesizers are digital devices, and they are characterized by discrete steps in frequency rather than adjustability over a continuous range. The steps can be rather large or very small, depending upon the application in which the device is to be used. The crystal oscillator can be set to zero beat with a standard frequency source, greatly increasing the accuracy of the frequency output. Phase-locked-loop circuits are used for the purpose of maintaining the accuracy of the frequency output of a synthesizer. *See also* PHASE-LOCKED LOOP.

FREQUENCY TOLERANCE

Frequency tolerance is a means of expressing the accuracy of a signal generator. The more stable an oscillator, the narrower the frequency tolerance. Tolerances are always expressed as a percentage of the fundamental quantity under consideration; in this case, it is the frequency of the oscillator.

Suppose, for example, that a 1 MHz oscillator has a frequency tolerance of plus or minus 1 percent. Because 1 percent of 1 MHz is 10 kHz, this means that the oscillator can be expected to be within 10 kHz of its specified frequency of 1 MHz. This is a range of 0.990 to 1.010 MHz.

Frequency-tolerance specifications are often given for communications receivers and transmitters. The tolerances may be given as a percentage, or as an actual maximum expected frequency error. The frequency tolerance required for one communications service may be much different from that needed for another service. *See also* TOLERANCE.

FRONT END

Front end is another term for the first radio-frequency amplifier stage in a receiver. The front end is one of the most important parts of any receiver, because the sensitivity of the front end dictates the sensitivity of the entire receiver. A low-noise, high-gain front end provides a good signal for the rest of the receiver circuitry. A poor front end results in a high amount of noise or distortion.

The front end of a receiver becomes more important as the frequency gets higher. At low and medium frequencies, there is a great deal of atmospheric noise, and the design of a front-end circuit is rather simple. But at frequencies above 20 or 30 MHz, the amount of atmospheric noise becomes small, and the main factor that limits the sensitivity of a receiver is the noise generated within the receiver itself. Noise generated in the front end of a receiver is amplified by all of the succeeding stages, so it is most important that the front end generate very little noise.

In recent years, improved solid-state devices, especially field-effect transistors, have paved the way for excellent receiver front-end designs. As a result, receiver sensitivity is vastly superior to that of just a few years ago.

Another major consideration in receiver front-end design is low distortion. The front-end amplifier must be as linear as possible; the greater the degree of nonlinearity, the more susceptible the front end is to the generation of mixing products. Mixing products result from the heterodyning, or beating, of two or more signals in the circuits of a receiver; this causes false signals to appear in the output of the receiver.

The front end of a receiver should be capable of handling very strong as well as very weak signals, without nonlinear operation. The ability of a front end to maintain its linear characteristics in the presence of strong signals is called the dynamic range of the circuit. If a receiver has a front end that cannot handle strong signals, desensitization will occur when a strong signal is present at the antenna terminals. Mixing products may also appear because the front end is driven into nonlinear operation. *See also* DESENSITIZATION, INTERMODULATION, NOISE FIGURE, SENSITIVITY.

FRONT-TO-BACK RATIO

Front-to-back ratio is one expression of the directivity of an antenna system. The term applies only to antennas that are unidirectional, or directive in predominantly one direction. Such antennas include (among others) the Yagi, the quad, and the parabolic-dish antenna.

Mathematically, the front-to-back ratio is the ratio, in decibels, of the field strength in the favored direction to the field strength exactly opposite the favored direction, given an equal distance (see illustration). In terms of power, if P is the power gain of an antenna in the forward direction and Q is the power gain in the reverse direction, both expressed in decibels with respect to a dipole or isotropic radiator, then the front-to-back ratio is given simply by $P - Q$.

The other primary consideration in antenna directive behavior is the forward gain, or power gain. Power gain is more important in terms of transmitting; the front-to-back ratio is of more interest in terms of receiving directivity. *See also* ANTENNA POWER GAIN, dBd, dBi.

F SCAN

See RADAR.

FULL-SCALE ERROR

The full-scale error is a means of determining the accuracy of an analog metering device. Often, the accuracy specification of such a meter is given in terms of a percentage of the full-scale reading.

As an example, suppose that a milliammeter, with a range of 0 to 1 mA, is stated as having a maximum error of plus or minus 10 percent of full scale. This means that, if the meter reads 1 mA (or full scale), the actual current might be as small as 0.9 mA or as large as 1.1 mA. If the meter reads half scale, or 0.5 mA, the actual current might be as small as 0.4 mA or as large as 0.6 mA. In other words, the maximum possible error is always plus or minus 0.1 mA, regardless of the meter reading.

Meters are specified for full-scale readings because that is one point where all errors can conveniently be summed and compensated. Errors introduced by resistive components in the circuit are constant percentages, regardless of the current flowing through them. Errors introduced by the meter movement are not necessarily

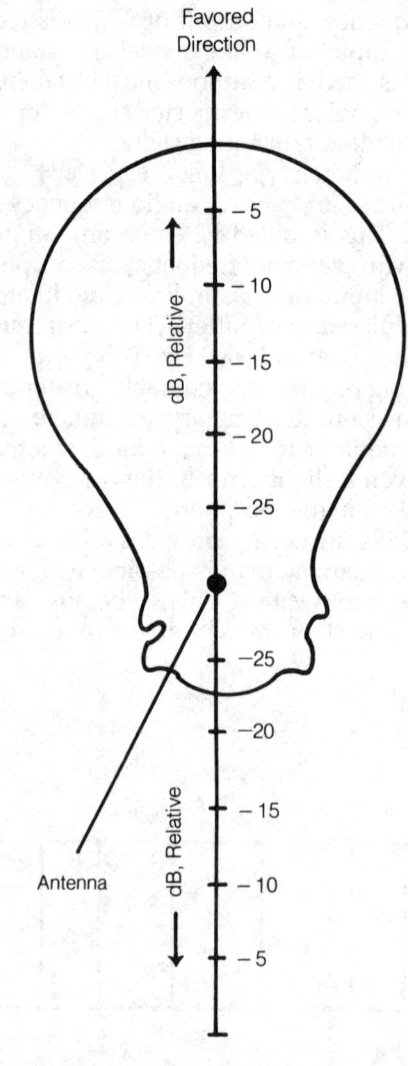

FRONT-TO-BACK RATIO: An example of the radiation pattern for a directional antenna. The front-to-back ratio can be determined from this graph.

linear; the springs that return the pointer to its zero position produce nonlinear forces which can cause an error at midscale to be significantly different from the error at full scale. Calibration of all the errors in a meter at many points on the scale would be costly and time-consuming. *See also* ANALOG PANEL METER.

FULL-WAVE RECTIFIER

A full-wave rectifier is a circuit that converts alternating current into pulsating direct current. The full-wave rectifier operates on both halves of the cycle. One half of the wave cycle is inverted, while the other is left as it is. The input and output waveforms of a full-wave rectifier in the case of a sine-wave alternating-current power source are illustrated at A.

A typical full-wave rectifier circuit is shown at B. The transformer secondary winding is center-tapped, and the center tap is connected to ground. The opposite sides of the secondary winding are out of phase. Two diodes, usually semiconductor types, allow the current to flow either into or out of either end of the winding, depending on whether the rectifier is designed to supply a positive or a negative voltage. A form of full-wave rectifier that does not require a center-tapped secondary winding, but necessitates the use of four diodes rather than two, is called a bridge rectifier (*see* BRIDGE RECTIFIER).

The primary advantage of a full-wave rectifier circuit over a half-wave rectifier circuit is that the output of a full-wave circuit is easier to filter. The voltage regulation also tends to be better. The ripple frequency at the output of a full-wave rectifier circuit is twice the frequency of the alternating-current input. *See also* HALF-WAVE RECTIFIER, RECTIFICATION.

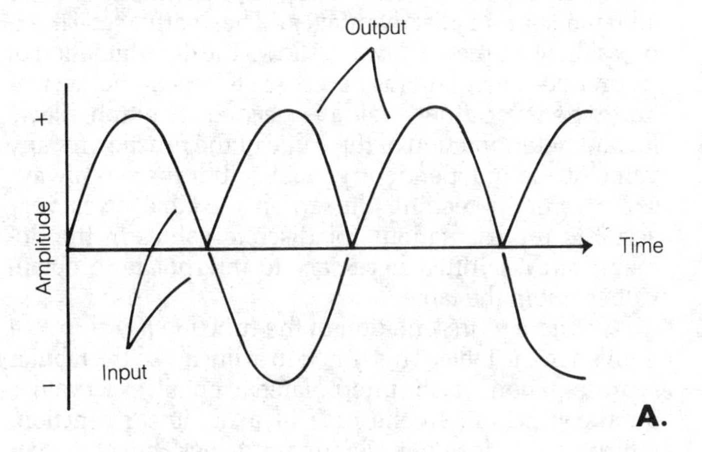

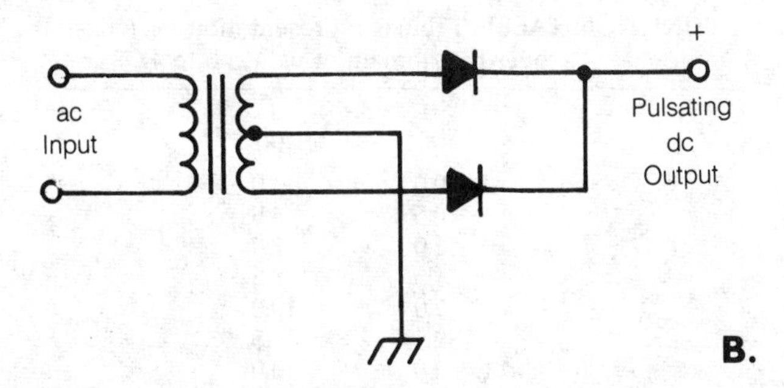

FULL-WAVE RECTIFIER: The alternating-current input and pulsating direct-current output of full-wave rectifier circuit (A), and a typical full-wave rectifier circuit (B).

FULL-WAVE VOLTAGE DOUBLER

A full-wave voltage doubler is a form of full-wave power supply. The direct-current output voltage is approximately twice the alternating-current input voltage. Voltage-doubler power supplies are useful because smaller transformers can be used to obtain a given voltage, as compared with ordinary half-wave of full-wave supplies. Medium-power and low-power vacuum-tube circuits sometimes have voltage-doubler supplies.

See the schematic diagram of a full-wave voltage doubler. During one-half of the cycle, the upper capacitor charges to the peak value of the alternating-current input voltage; during the other half of the cycle, the lower capacitor charges to the peak of the alternating-current input voltage, but in the opposite direction. Both capacitors hold their charge throughout the input cycle, resulting in the effective connection of two half-wave, direction-current power supplies in series, with twice the voltage of either supply.

Full-wave voltage doublers are not generally suitable for applications in which large amounts of current are drawn. The regulation and filtering tends to be rather poor in such cases. *See also* HALF-WAVE RECTIFIER, RECTIFICATION.

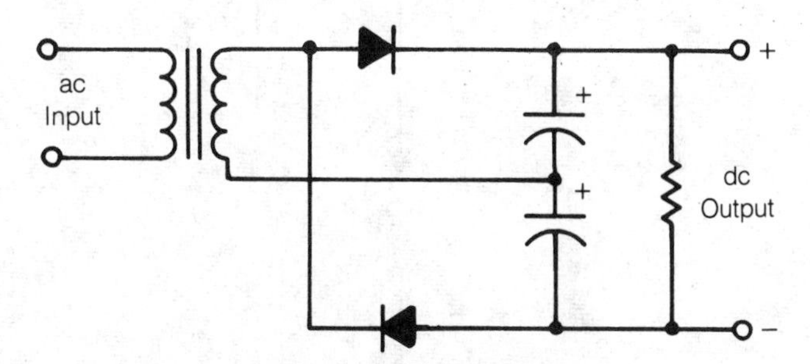

FULL-WAVE VOLTAGE DOUBLER: This power supply provides twice the direct-current voltage of a half-wave rectifier, but there is sacrifice in regulation.

FUNCTION

A function is a form of mathematical relation. The function assigns particular values to a variable or set of variables. The value of the function is called the dependent variable because it depends on the other variable or variables. The variables that affect the value of the function are called the independent variables. A function may be considered an operation, which assigns, or maps, the constituents of one set into those of another set.

For a relation to be a function, a special property must be satisfied: Each value of the independent variable or variables must correspond to at most one value of the dependent variable. In other words, each point in the domain must have no more than one constituent in the range. The illustration shows a function of one variable, with the independent variable on the horizontal or x axis, and the dependent variable on the vertical or y axis (A). Note that it is possible to draw a vertical line anywhere in this graph, and never does it intersect the function at more than one point. A relation that is not a function is shown at B; some of the values of the independent variable correspond to more than one value of the dependent variable.

Functions are generally written as a letter of the alphabet, followed by the independent variables in parentheses. For example, $f(x) = 2x + 3$, or $g(x,y,z) = x + y + z$. Sometimes, functions are written in simplified form, such as $y = 2x + 3$ or $x + y + z = w$. *See also* DEPENDENT VARIABLE, DOMAIN, INDEPENDENT VARIABLE, RANGE.

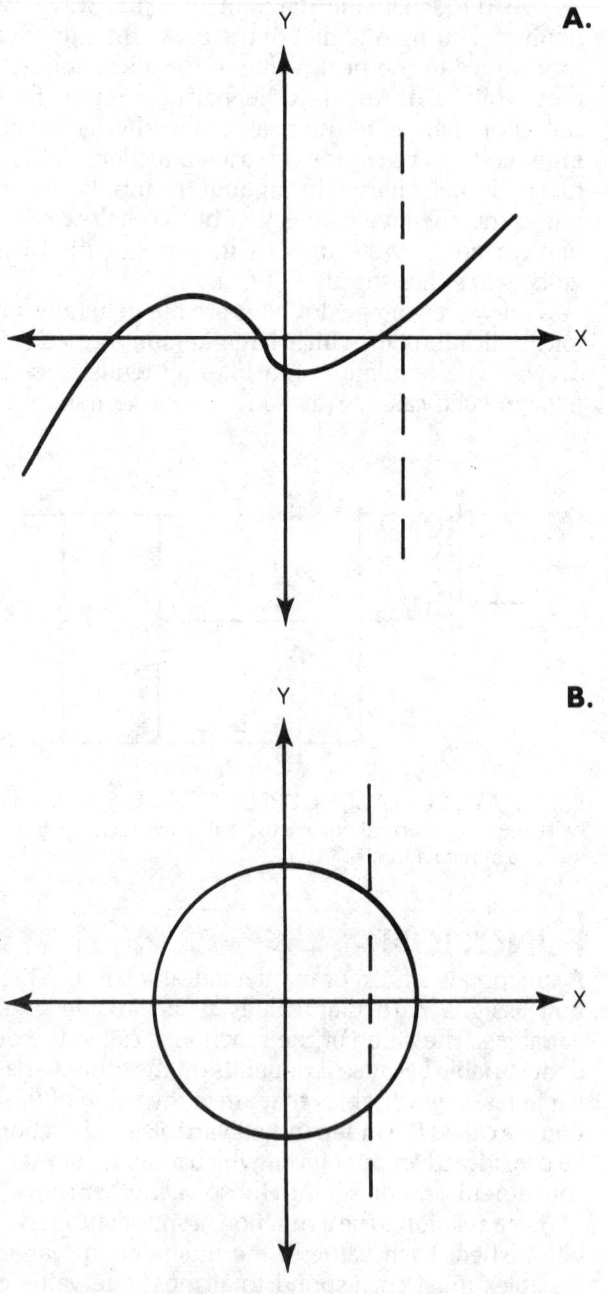

FUNCTION: A plot of a mathematical function (A) and a plot of a relation that is not a function (B).

FUNCTION GENERATOR

A function generator is a signal generator that can produce different waveforms. All electrical waveforms can be expressed as mathematical functions of time; for example, the instantaneous amplitude of a sine wave can be expressed in the form $f(t) = a \sin(bt)$, where a is a constant that determines the peak amplitude, and b is a constant that determines the frequency. Square waves, sawtooth waves, and all other periodic disturbances can be expressed as mathematical functions of time, although the functions are quite complicated in some cases (*see* FUNCTION).

Most function generators can produce sine waves, sawtooth waves, and square waves. Some can also produce sequences of pulses. More sophisticated function generators can create many different waveforms for testing the design, troubleshooting, and alignment of electronic apparatus such as audio amplifiers and digital circuits.

The Moog synthesizer is an example of a function generator for audio frequencies. The Moog can duplicate the sound of almost any musical instrument, as well as create unique sounds. *See also* MOOG SYNTHESIZER, SAWTOOTH WAVE, SINE WAVE, SQUARE WAVE.

FUNCTION TABLE

A function table is a list of the values of a mathematical function at discrete points within its domain. Functions can be represented by graphs, written in mathematical form, or listed as tables. Each representation has its advantages and disadvantages. The mathematical expression of a function always allows the determination of its precise value, but it is necessary to repeat the calculation whenever a new value is needed. A graph allows instant determination of the value of the function for any value of the independent variables, but there is always some error in reading the graph. A table gives very accurate representations for discrete values in the domain, but it is often necessary to interpolate to obtain values not in the table.

A simple representation of the function $f(x) = 3x + 4$ is given in the table. This function is linear, so the tabular representations can be interpolated to obtain exact values for the function. In the case of a nonlinear function, interpolation does not yield exact values, but the accu-

FUNCTION TABLE: Tabular representation for $f(x) = 3x + 4$, over the domain of values 0 to 4.

x	f(x)
0.0	4.0
0.5	5.5
1.0	7.0
1.5	8.5
2.0	10.0
2.5	11.5
3.0	13.0
3.5	14.5
4.0	16.0

racy is good if the representations are at closely spaced intervals.

Function tables are frequently seen in electronics because they provide convenient and accurate means for determining the values of functions at a glance. *See also* FUNCTION.

FUNDAMENTAL FREQUENCY

The fundamental frequency of an oscillator, antenna, or tuned circuit is the primary frequency at which the circuit is resonant. All waveforms or resonant circuits have a fundamental frequency, and integral multiples of the fundamental frequency, called harmonics. A theoretically pure sine wave contains energy at only its fundamental frequency. But in practice, sine waves are always slightly impure or distorted and contain some harmonic energy. Complex waves, such as the sawtooth or square wave, often contain large amounts of harmonic energy in addition to the fundamental.

The amplitude of the fundamental frequency is used as a reference standard for the determination of harmonic suppression. The degree of suppression of a particular harmonic is expressed in decibels with respect to the level of the fundamental. The fundamental frequency need not necessarily be the frequency at which the most energy is concentrated; a harmonic may have greater amplitude. The fundamental also might not be the desired frequency in a particular application; harmonics are often deliberately generated and amplified. *See also* HARMONIC, HARMONIC SUPPRESSION.

FUNDAMENTAL SUPPRESSION

In the measurement of the total harmonic distortion in an amplifier circuit, the fundamental frequency must be eliminated, so that only the harmonic energy remains. This is called fundamental suppression. Fundamental suppression is generally done with a trap circuit, or band-rejection filter, centered at the fundamental frequency (*see* TOTAL HARMONIC DISTORTION).

In a frequency multiplier, the fundamental frequency is suppressed at the output while one or more harmonics are amplified. In these circuits, the fundamental frequency is not wanted. Band-rejection filters may be used in the output of the frequency multiplier to attenuate the fundamental signal. Highpass filters may also be used. Bandpass filters, centered at the frequency of the desired harmonic, also provide fundamental suppression. *See also* FREQUENCY MULTIPLIER, HARMONIC.

FUSE

A fuse is a low-cost, slow-acting overcurrent protection device that causes a circuit to open when excess current burns out or *blows* a conductive metal wire or foil strip. A fuse is placed in series with the electric circuit it is to protect. When the current exceeds the rated value of the fuse, the conductive element in the fuse melts, opening the circuit.

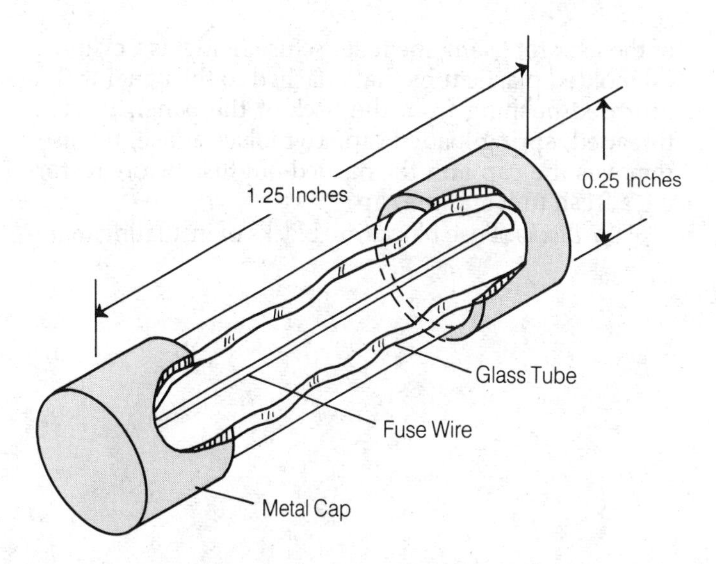

FUSE: Cartridge fuse for protecting electronics against overcurrent is 1¼-inches long and ¼ inch in diameter.

Fuses are manufactured in a wide range of electrical ratings to safeguard equipment from consumer products to industrial apparatus. However, for electronics applications, most fuses have glass tubular bodies and metal end caps. The most popular replaceable American styles have a diameter of ¼ inch and a length of 1¼ inches, as shown in the figure. Voltage ratings are 125 V ac and 250 V ac, and current ratings are 1⁄15 to 15 A. The European standard fuse has a diameter of 0.2 inch and is 0.8 inch long (5 by 20 millimeters).

Response time of fuses is measurable in milliseconds so fuses offer no protection against high-speed voltage transients. They are fail-safe, disposable components that respond to the same I^2t product as thermal protectors. Four classes of fuses are specified for electronics applications:

Very Fast Acting. Very fast acting fuses have the fastest response and are available with close time tolerances. These devices are widely used to protect semiconductor devices.

Non-Time Delay. Non-time-delay fuses offer a high-speed response and are typically used to protect power supplies. They have ratings of 250 V ac with minimum 1000 A interrupt capability under short-circuit conditions.

Time Delay. Time-delay fuses are slow-burning units with built-in time delays in the low range of overload. The time delay slows the blow time to prevent nuisance response caused by the routine surge or motor starting currents. They are widely specified where tungsten filament lamps are in the load.

Dual-Element Time Delay. Dual-element time delay fuses include two separate fusible elements in series within the fuse cartridge. This feature provides time delays in the overload region with only minimal reduction in the resistance value of the circuit. These fuses are specified for use in circuits with inductive loads such as motors, solenoids and transformers.

Fuses are mounted in fuse holders accessible from the front panel of the product if the manufacturer approves of the user replacing the fuse. A fuse holder is a cylindrical molded plastic tube that attached to the panel with a threaded bushing from the back of the panel. It has a threaded, spring-loaded cap. To replace a fuse, the user removes the cap and the burned-out fuse before restoring a fresh unit and the cap.

Fuse Blocks. Fuse blocks are blocks of insulating material with two metal spring clips attached at a spacing to mate with the metal end caps of the fuse. They permit the fuse to be snapped into position. Fuse blocks and clips are usually mounted within the equipment enclosure and may not be accessible to the end user. It may be necessary to open the enclosure to gain access to the fuse.

GAGE

See GAUGE.

GAIN

Gain is a term that describes the extent of an increase in current, voltage, or power. The gain of an amplifier circuit is an expression of the ratio between the amplitudes of the input signal and the output signal. In an antenna system, gain is a measure of the effective radiated power compared to the effective radiated power of some reference antenna (*see* ANTENNA POWER GAIN). Gain is almost always given in decibels.

In terms of voltage, if E_{IN} is the input voltage and E_{OUT} is the output voltage in an amplifier circuit, then the gain G, in decibels, is given by:

$$G = 20 \log_{10} \left(\frac{E_{OUT}}{E_{IN}} \right)$$

The current gain of an amplifier having an input current of I_{IN} and an output current of I_{OUT} is given in decibels by:

$$G = 20 \log_{10} \left(\frac{I_{OUT}}{I_{IN}} \right)$$

In a power amplifier, where P_{IN} is the input power and P_{OUT} is the output power, the power gain G, in decibels, is:

$$G = 10 \log_{10} \left(\frac{P_{OUT}}{P_{IN}} \right)$$

When the gain of a particular circuit is negative, there is insertion loss. Some circuits, such as the follower, exhibit insertion loss. For typical radio-frequency amplifiers, the gain ranges from about 15 to 25 decibels per stage of amplification, but it can reach 30 decibels per stage. *See also* AMPLIFICATION FACTOR, DECIBEL, INSERTION LOSS.

GAIN CONTROL

A gain control is an adjustable control such as a potentiometer that changes the gain of an amplifier circuit. The volume control in every radio receiver or audio circuit is a form of gain control. Communications equipment generally has an audio-frequency gain control and a radio-frequency gain control.

A potentiometer can be used to provide varying amounts of drive to an amplifier circuit; this should be done at a low power level. The illustration shows a typical means of obtaining variable gain in an audio amplifier. The position of the potentiometer determines the proportion of the input that is applied to the amplifier circuit.

A gain control should not affect the linearity of the amplifier circuit in which it is placed. Therefore, changing the bias on the transistor or tube in an amplifier is not an acceptable way of obtaining gain control. The gain control should be chosen so that succeeding stages cannot be overdriven if the gain is set at maximum. The gain control should have sufficient range so that, when the gain is set at minimum, the output of the amplifier chain is reduced essentially to zero. *See also* GAIN.

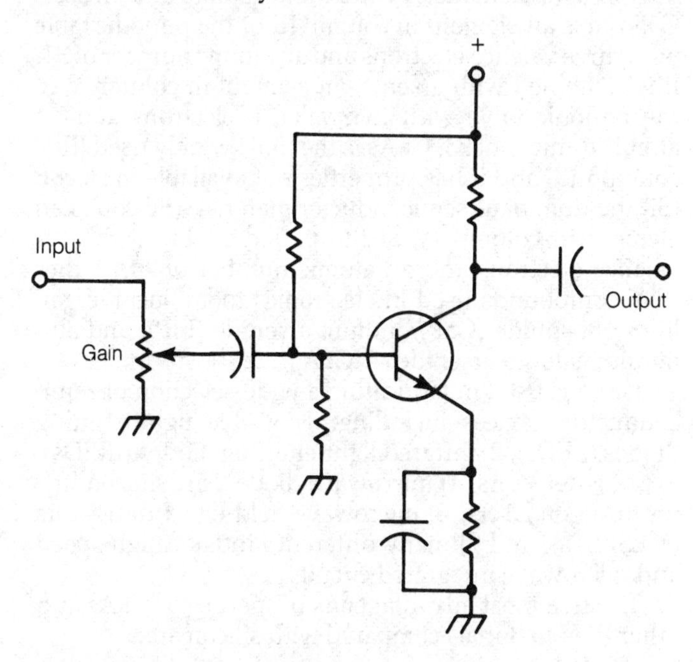

GAIN CONTROL: A gain control in the input of an audio amplifier.

GALACTIC NOISE

All celestial objects radiate energy at radio wavelengths, especially stars, galaxies, and the diffuse matter among the stars.

Galactic noise was first observed by Karl Jansky, a physicist working for the Bell Telephone Laboratories, in the 1930s. It was an accidental discovery. Jansky was investigating the source of atmospheric noise at a wavelength of about 15 m, or 20 MHz. Jansky's antenna consisted of a rotatable array.

Most of the noise from our galaxy comes from the

direction of the galactic center, where most of the stars are concentrated. Galactic noise contributes, along with noise from the sun, the planet Jupiter, and a few other celestial objects, to most of the cosmic radio noise arriving at the surface of the earth. Other galaxies radiate noise, but since the external galaxies are much farther away from earth than the center of this galaxy, sensitive equipment is needed to detect the noise from them. *See also* RADIO TELESCOPE.

GALENA

Galena is a compound of lead and sulfur with a chemical formula PbS. In nature, galena occurs in the form of silver-gray crystals with a roughly cubical shape.

Galena was used in the earliest semiconductor detectors. By placing a fine wire, called a cat's whisker, at the right point of the surface of a crystal of galena, rectification could be obtained at radio frequencies. The galena detector is an unpredictable rectifier and trial and error is usually necessary to obtain good performance.

GALLIUM ARSENIDE

Gallium arsenide (GaAs) is a III-V compound semiconductor material made by combining gallium and arsenic. Gallium is an element in column III of the periodic table with three valence electrons and an atomic number of 31. It is combined with arsenic, an element in column V of the periodic table with five valence electrons and an atomic number of 33. GaAs is the most widely used III-V compound, and it has properties not available in silicon (Si) the dominant semiconducter material. (Silicon is an element in column IV of the periodic table with four valence electrons and an atomic number of 14.) Other III-V compounds used in electronics today include gallium phosphide (GaP), indium arsenide (InP), and aluminum gallium arsenide (AIGaAs). *See* SILICON.

GaAs is used in fabricating a wide selection of semiconductor devices including: 1) visible light-emitting diodes (LEDs), 2) infrared light-emitting diodes (IREDs), 3) photodetectors, 4) microwave diodes, 5) semiconductor or diode lasers, 6) microwave field-effect transistors (MESFETs), and 7) many different kinds of high-speed and microwave integrated circuits.

The five most advantageous properties of GaAs and other III-V materials compared with silicon are:

1. Higher speed of electrons within the material.
2. Lower voltage requirements for devices.
3. Semi-insulating properties of the substrate.
4. Ability to merge optical and electronic functions on the same substrate.
5. Radiation hardness.

The electrical and optical properties of GaAs and other III-V compounds can be tailored for different applications and this is why they are such versatile compounds. Thin layers of these compounds can be deposited on a common substrate to achieve special results such as increased electron velocities.

When gallium and arsenic are combined, the compound has four valence electrons like silicon and germanium. GaAs can be doped with P-type and N-type materials to form PN junctions, the basis of many different kinds of diodes. The ability of GaAs to produce visible light and infrared emitting diodes is related to its energy band gap, which is higher than that of silicon. *See* BAND GAP.

The unusual properties of GaAs transistors and ICs depend on the first three properties listed above: speed, power, and dielectric characteristics. Electrons can move in a GaAs substrate at four or five times the speed of electrons in silicon because electrons are more mobile in GaAs. In addition, electrons can reach higher speeds in GaAs at lower voltages than in silicon.

Pure GaAs is a semi-insulator that provides natural isolation for individual transistors and eliminates the need for implanting resistive material around them to form isolation barriers. This quality saves space on the chip. The dielectric characteristics of silicon make it suitable as a substrate for monolithic microwave ICs.

The development of GaAs devices that merge optical and electronic functions on the same chip depend on the fourth important property of GaAs listed. In addition, the value of GaAs ICs in military and aerospace applications depends, in part, on their ability to withstand heavy doses of nuclear radiation without damage or failure.

GaAs is a costly and difficult crystal to grow and considerable effort and resources are being placed on the elimination of defects in the large wafers for the fabrication of large scale ICs. It is unlikely that GaAs will ever replace silicon as the predominant semiconductor material for low-frequency, general-purpose ICs because it is satisfactory and cost effective for the fabrication of those devices. GaAs is now being grown on silicon wafers as a way of reducing material costs and combining the best features of both crystals. GaAs on Si might become an important substrate for fabricating many different high-speed analog/interface, digital, and microwave ICs. *See* COMPOUND SEMICONDUCTOR, GALLIUM ARSENIDE TRANSISTOR, GALLIUM ARSENIDE INTEGRATED CIRCUIT, INFRARED EMITTING DIODE, LASER DIODE, LIGHT EMITTING DIODE, PHOTODETECTOR.

GALLIUM-ARSENIDE INTEGRATED CIRCUIT

Gallium arsenide (GaAs) integrated circuits (ICS) have been fabricated to perform a wide range of digital and analog functions at high speed and at frequencies up to 20 GHz. They are based on the integration of transistors on semi-insulating GaAs substrates generally following techniques developed for the fabrication of silicon ICs. However, many new and exotic processes have been developed for fabricating larger and faster GaAs ICs.

GaAs IC technology is emerging, and the list of standard commercial parts continues to grow even as scientists and researchers develop laboratory prototypes of higher performance devices with even higher levels of integration. However, the future of GaAs ICs is seen in custom and applications-specific devices rather than in

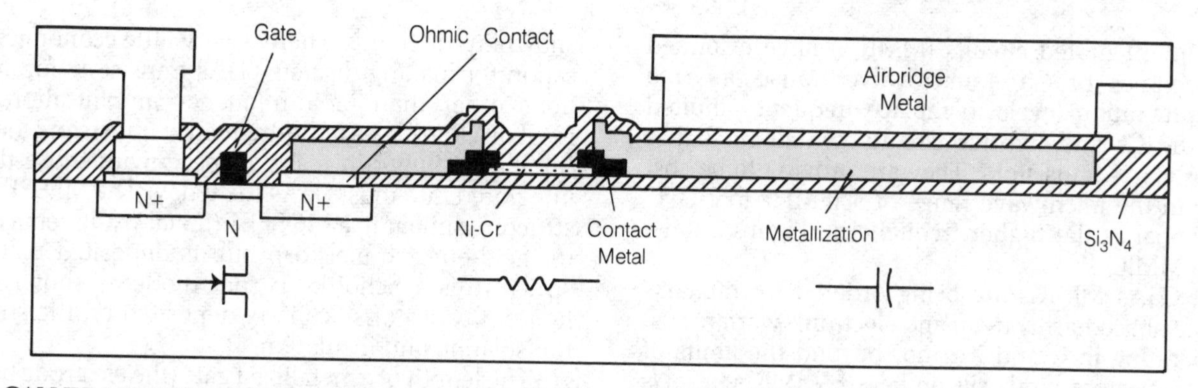

GALLIUM-ARSENIDE INTEGRATED CIRCUIT: Fig. 1. Cross section view of a gallium-arsenide (GaAs) integrated circuit.

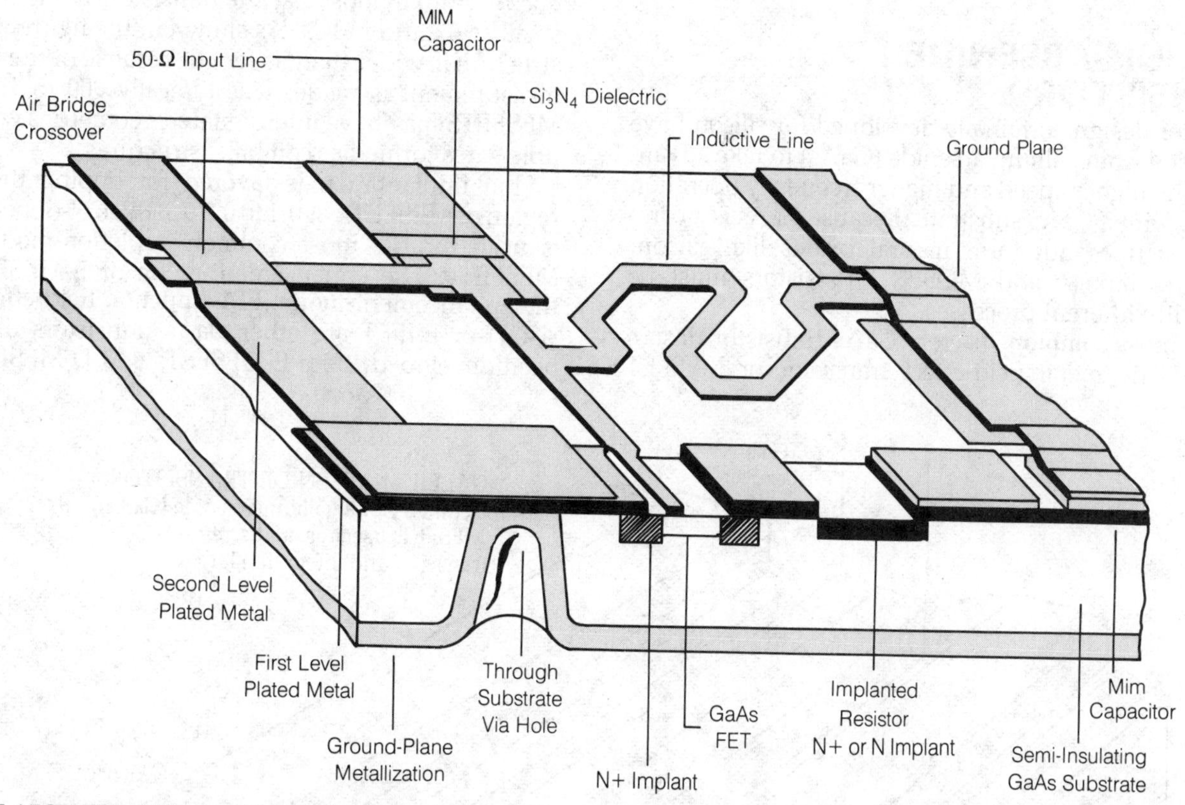

GALLIUM-ARSENIDE INTEGRATED CIRCUIT: Fig. 2. View and section of a GaAs microwave monolithic integrated circuit (MMIC).

the families of standard interchangeable devices so prevalent with silicon IC. Continuous improvements in silicon ICs have, however, eroded some of the performance margins that GaAs ICs held over silicon ICs. *See* APPLICATIONS-SPECIFIC DEVICE.

GaAs ICs are now generally classed as: 1) high-speed digital, 2) high-speed analog/interface, and 3) microwave. The digital and analog/interface devices are niche parts that supplement or complement existing silicon ICs. They are being used in applications requiring higher speed, lower power consumption (at high speed), and the ability to operate at higher temperatures and lower noise levels than comparable silicon ICs. However, their better radiation hardness might, by itself, justify their use. Figure 1 shows a crosssection of a GaAs IC.

Digital ICs. GaAs digital ICs are faster and consume less power than the nearest equivalent silicon digital bipolar ICs. They are made to be compatible with other logic families, particularly emitter-coupled logic (ECL). At the present time, GaAs ICs lag bipolar ICs in density of integration. SSI and MSI digital logic, memories, and gate arrays are now available in GaAs. *See* DIGITAL-LOGIC INTEGRATED CIRCUIT.

Analog/Interface ICs. GaAs analog and interface ICs supplement slower silicon bipolar and CMOS (complementary metal-oxide semiconductors) linear and interface ICs in systems requiring higher data rates or faster conversion. They also are used to interface microwave and digital systems. Operational amplifiers, comparators and analog-to-digital and digital-to-analog converters are available as GaAs ICs. *See* LINEAR INTEGRATED CIRCUIT.

Microwave ICs. Amplifiers and oscillators made from GaAs hold clear and undisputed advantages over those made from silicon, which are unable to amplify or oscillate efficiently much above 2 GHz. GaAs microwave monolithic integrated circuits (MMICs) have extended the capabilities of GaAs microwave transistors into higher integration levels to replace frequency limited silicon MMICs and hybrids circuits containing either silicon or GaAs transistors. They are proving to be cost effective in the microwave range of 500 MHz to 2 GHz and essential at the higher frequencies. Figure 2 illustrates an MMIC.

Most GaAs MMICs are being ordered for military/aerospace phased-array radar and electronic warfare systems operating in C and X band, beyond the limits of silicon transistors and silicon-based MMICs. GaAs MMICs now include amplifiers, oscillators, mixers, and switches. *See* GALLIUM ARSENIDE TRANSISTOR.

GALLIUM-ARSENIDE TRANSISTOR

Transistor designs originally developed for silicon have been made from gallium arsenide (GaAs) to take advantage of the higher speed and higher frequency operation possible with a GaAs substrate. Because GaAs is a compound, it does not form natural oxides like silicon. Therefore, bipolar and MOSFET transistors must be made with different processes.

The most common discrete GaAs transistor design today is the metal field-effect transistor or MESFET shown in the figure. There is very little economic justification for making discrete GaAs transistors for applications other than radio-frequency amplification. Thus most discrete GaAs radio-frequency transistors today are MESFETs similar in many respects with those that are integrated into most GaAs ICs today. The MESFET has a structure similar to a MOSFET (metal-oxide semiconductor field-effect transistor), but its deposited metal-gate structure is a Schottky barrier diode as shown in the figure. Oxides of silicon are deposited on the substrate for isolation and insulation.

The length of a metallized gate (the electrode between the source and collector) is critical in both discrete transistors and ICs. This measurement is typically 0.5 to 1.0 micron (μm) in most discrete transistors but it may be as small as 0.2 μm in ICs. As shown in the figure, the gate structure is wider than its length because of the way the measurement is made. It is typically 900 to 1200 μm. MESFETS may have interdigitated geometries with multiple gates formed as comblike structures.

Ion implantation is favored for doping the active region of MESFETs. A 0.1 to 0.2 μm thick N-doped region is used for the most common depletion-mode or D-MESFETs. The enhancement-mode or E-MESFET and the enhancement-mode JFET (junction field-effect transistor) or E-JFET are other GaAs transistors that have been developed. Both E-MESFETs and D-MESFETs are

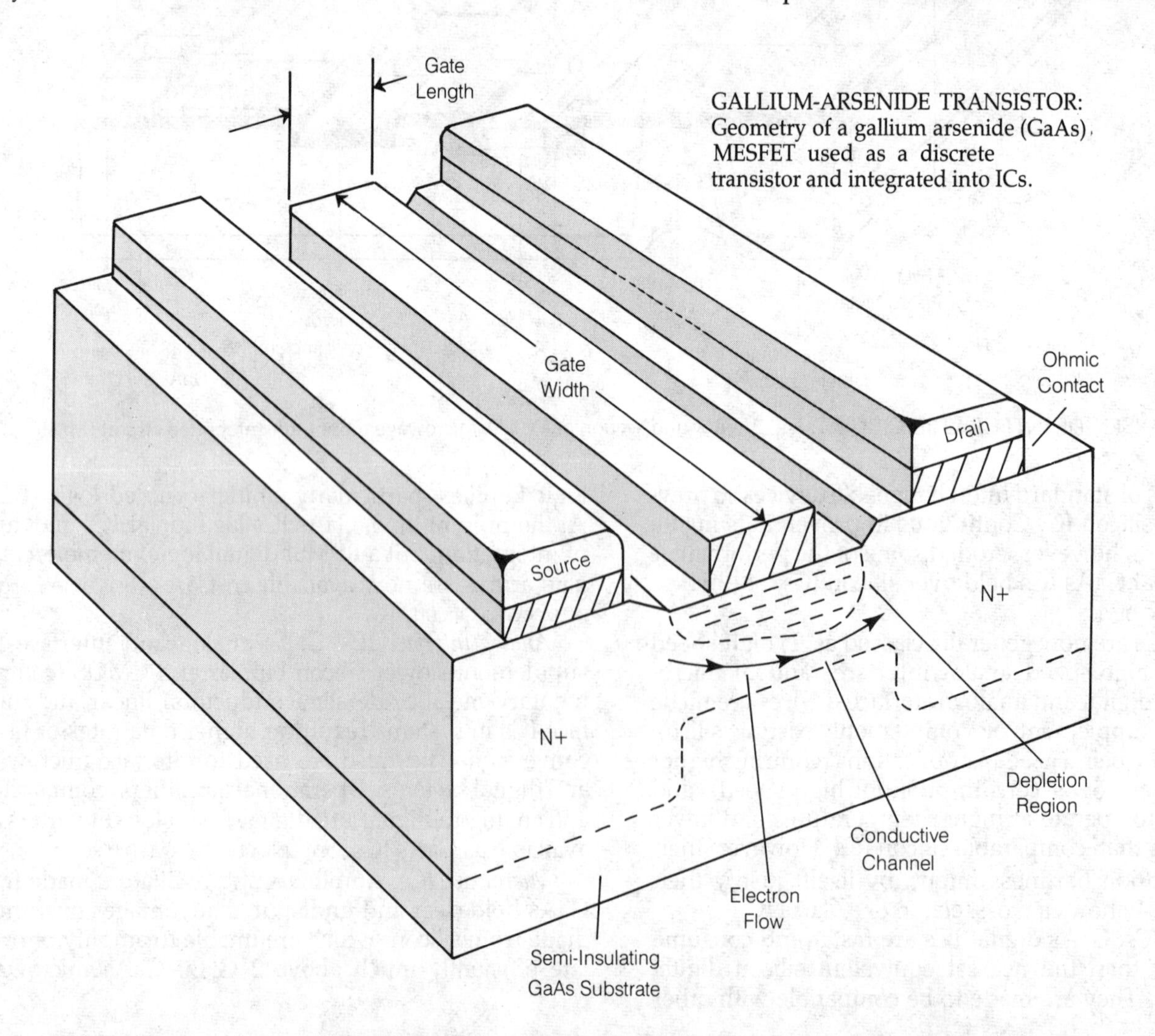

GALLIUM-ARSENIDE TRANSISTOR: Geometry of a gallium arsenide (GaAs) MESFET used as a discrete transistor and integrated into ICs.

combined in some ICs to form enhancement/depletion-mode (E/D) logic.

The high-electron mobility transistor or HEMT is a transistor designed primarily for ICs. It is made on a layer of aluminum gallium arsenide (AlGaAs) grown on a GaAs substrate, known as a heterojunction, to improve device performance and permit even higher levels of integration. Heterojunction E/D technology is seen as a method for achieving cost effective GaAs digital LSI and VLSI devices.

Another GaAs transistor developed on a heterojunction is the heterojunction bipolar transistor or HBT, a structure also designed to achieve higher levels of integration. Both HEMTs and HBTs require special processing to achieve precise, sharp heterojunctions.

GALVANISM

The production of electric current by chemical action is called galvanism. The word is derived from the name of the eighteenth-century scientist, Luigi Galvani. The principles of galvanism are used in cells and batteries. Galvanism also causes certain metals to corrode, or react with other substances to form compounds.

When two dissimilar metals are brought into contact with each other in the presence of an electrolyte, they can act like an energy cell, and a voltage difference will be produced between the metals. Salt in the air, or even slightly impure water vapor, can act as the electrolyte. The metals gradually corrode because of the chemical reaction; this is called galvanic corrosion.

Electrolytic action can be used to coat iron or steel with a thin layer of zinc. This is called galvanizing. Galvanized metal is more resistant to corrosion than bare metal, although in a marine or tropical environment, even galvanized iron or steel will eventually corrode. *See also* BATTERY, ELECTROLYSIS.

GALVANOMETER

A galvanometer is a sensitive device for detecting the presence of, and measuring, electric currents. The galvanometer is similar to an ammeter, but the needle of the galvanometer normally rests in the center position. Current flowing in one direction results in a deflection of the needle to the right; current in the other direction causes the needle to move to the left (see illustration).

The galvanometer is especially useful as a null indicator in the operation of various bridge circuits because currents in both directions can be detected. An ordinary meter in this situation will read zero when a reverse current is present, thus making an accurate null indication impossible to obtain. *See also* AMMETER.

GAMMA MATCH

A gamma match is a device for coupling a coaxial transmission line to a balanced antenna element, such as a half-wave dipole. The radiating element consists of a single conductor. The shield part of the coaxial cable is connected to the radiating element at the center. The

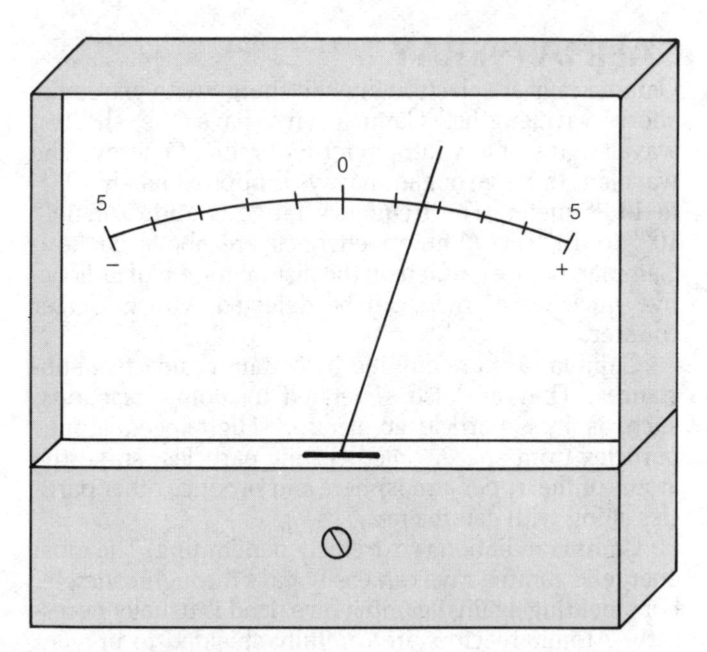

GALVANOMETER: Current intensity and direction are indicated on a galvanometer.

center conductor is attached to a rod or wire running parallel to the radiator, toward one end. The illustration shows the gamma match. The device is named because of its resemblance in shape to the upper-case Greek letter gamma.

The impedance-match ratio of the gamma match depends on the physical dimensions: the spacing between the matching rod and the radiator, the length of the matching rod, and the relative diameters of the radiator and the matching rod. The gamma match allows the low feed-point impedance of a Yagi antenna driven element to be precisely matched with the 50 or 75 Ω characteristic impedance of a coaxial feed line. The gamma match also acts as a balun, which makes the balanced radiator compatible with the unbalanced feed line. Gamma matching is often seen in multielement parasitic arrays, especially rotatable arrays. *See also* BALUN, IMPEDANCE MATCHING.

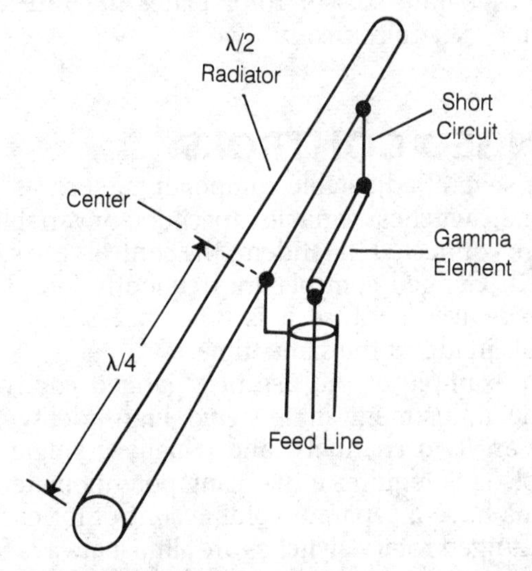

GAMMA MATCH: A gamma match provides an impedance match between a coaxial line and a half-wave radiator.

GAMMA (γ) RAY

Gamma rays are electromagnetic energy with extremely short wavelengths. Gamma rays have the shortest wavelengths of any known form of radiant energy. The wavelength range of gamma rays is approximately 10^{-13} to 10^{-10} meter. The frequency range is approximately 10^{20} to 10^{18} Hz. Photon energies are above 50 keV. Gamma rays are emitted on the disintegration of radioactive nuclei, and they can be detected with a Geiger counter.

Gamma rays are emitted by certain radioactive substances. They are also generated in atomic reactions, such as in a particle accelerator. High-speed atomic particles from space, called cosmic particles, strike the atoms of the upper atmosphere and produce other particles along with gamma rays.

Gamma radiation is extremely penetrating. The most energetic gamma rays can easily pass through concrete. For shielding against gamma rays, lead is usually necessary. Atomic reactors are carefully shielded to prevent the gamma radiation from escaping. Gamma rays damage living tissue. An overexposure to gamma radiation can result in radiation sickness and even death. Gamma radiation is measured in units called roentgens. *See also* ALPHA PARTICLE, BETA PARTICLE, DOSIMETRY, GEIGER COUNTER, RADIATION HARDNESS, ROENTGEN, X RAY.

GANG CAPACITOR

Variable capacitors can be connected together in parallel, on a common shaft. This combination capacitor is called a gang capacitor. The individual capacitors rotate together. The sections may or may not have the same values of capacitance.

Gang capacitors are used in circuits having more than one adjustable resonant stage, and a requirement for tuning each resonant stage along with the others. This is often the case in communications equipment.

The rotor plates of the gang capacitor are usually all connected to the frame of the unit, and the frame is in turn put at common ground. Gang capacitors with electrically separate sets of rotor plates are rare. *See also* CAPACITOR, GANGED CONTROLS.

GANGED CONTROLS

When several adjustable components, such as potentiometers, switches, variable capacitors, or variable inductors are connected in tandem, the controls are said to be ganged. Ganged controls are frequently found in electronic devices because it is often necessary to adjust several circuits at the same ttime.

An example of the use of a ganged control is the volume adjustment in a stereo high-fidelity system. There are two channels, and usually a single volume control. This requires a two-gang potentiometer. (Some systems have a separate volume control for each channel.) Ganged rotary switches are almost always found in multiband communications receivers and transmitters because of the need to switch several stages of amplification simultaneously. When several tuned circuits must be adjusted together, a gang of variable capacitors is employed. *See also* GANG CAPACITOR.

GAP

See AIR GAP.

GATE

The gate of a field-effect transistor is the electrode that allows control of the current through the channel of the device. The gate is analogous to the base of a bipolar transistor, or the grid of a vacuum tube. In an N-channel junction field-effect transistor (JFET), the gate voltage is usually negative with respect to the source voltage. In the P-channel device, the gate is biased positively.

In the JFET, the gate consists of two sections of P-type semiconductor material in the N-channel device, placed on either side of the channel. Alternatively, the gate can have the configuration of a cylinder of P-type material around the channel. In the P-channel device, of course, the gate is fabricated from N-type semiconductor material.

The greater the reverse bias applied to the gate, the less the channel of the JFET conducts, until, at a certain value of reverse bias, the channel does not conduct at all. This condition is called pinchoff. The pinchoff voltage depends on the voltage between the source and the drain of the field-effect transistor.

In the insulated-gate (metal-oxide semiconductor) field-effect transistor (MOSFET), the gate operates differently. There are two basic modes of operation, known as depletion mode and enhancement mode (*see* DEPLETION MODE, ENHANCEMENT MODE, METAL-OXIDE SEMICONDUCTOR FIELD-EFFECT TRANSISTOR). A MOSFET may have more than one gate.

Small variations in the gate voltage result in large changes in the current through the field-effect transistor. Thus amplification is possible with such devices. *See also* FIELD-EFFECT TRANSISTOR.

In digital logic, the term *gate* refers to a switching device with one or more inputs, and one output. The gate performs a logical function, depending on the states of the inputs. The three basic logic gates are the inverter, or NOT gate, and AND gate, and the OR gate. *See also* LOGIC GATE.

GATE ARRAY

A gate array is a semicustom semiconductor integrated circuit that is prefabricated as a matrix of uncommitted identical cells, each containing transistors and resistors. Gate arrays are also known as logic arrays, macrocell arrays and undefined logic arrays (ULAs). All gates, drains, sources, and channels are accessible, and all mask levels except the final one or two metal masks are predefined and fixed. The final metal masks uniquely define the interconnects for each application.

The gate array is an applications-specific integrated circuit (ASIC) along with custom-designed ICS, standard cells and programmable logic devices (PLDS). The dedication or personalizing of a gate array starts as a fully diffused or ion-implanted semiconductor wafer with a matrix of identical primary cells arranged in columns. There are routing channels between the cell columns in the X and Y directions and input/outout (I/O) devices around the periphery. Each prefinished array has an equivalent gate density, an indication of the number of functions that can be performed with the chip and the size of each individual device. *See* APPLICATIONS-SPECIFIC DEVICE, PROGRAMMABLE LOGIC DEVICE, STANDARD CELL.

Gate arrays can be fabricated in CMOS, ECL, and TTL (complementary metal-oxide semiconductor, emitter-coupled logic and transistor-transistor logic) bipoplar, gallium arsenide, silicon-on-sapphire, and various combinations of these technologies. Delivery time is reduced because gate array wafers are manufactured from wafers that are 70 to 80 percent complete in stock. Unit cost is lower than for other comparably sized custom and semicustom ICs because of the volume prefabrication of wafers. Nonrecurring engineering cost (NRE) is relatively low because only one or two masks are designed and made, and that work is done with the assistance of computer-aided design (CAD). The primary reason for using gate arrays is cost reduction that is possible due to:

1. Fewer components on the PC (printed circuit) card and smaller PC cards than are possible with standard digital logic devices.
2. Increased system reliability and performance.
3. Reduced interconnections and connectors.
4. Lower power supply requirements.

Intraconnection of the cells and interconnections between cells are customized with the aid of routing performed on a CAD workstation with appropriate software.

The gate array functional block might only permit the implementation of digital logic or they might include provisions for memory and analog circuitry. Digital gate arrays can include memory cells and analog elements such as operational amplifiers. There are also 100 percent analog gate arrays. The functional blocks, referred to as macrocells (or macros) and macrofunctions, permit a wide selection of precisely characterized logic or other functions.

Many different factors are considered by a user in selecting the most suitable gate array density and technology for a design. These include required gate complexity, bus structures, speed requirements, pin count and package type, input and output buffers, power pins, and ease in testing.

Gate array density is normally measured in equivalent gates, a generally accepted measure although not a standard definition in the industry. An equivalent gate is typically a two-input NAND gate. Macrocells, the basic gate array elements, include inverters, NANDs, NORs, latches, flip-flops, decoders, multiplexers, shift registers, and buffers. Macrocells have predefined metal interconections. Macrofunctions such as adders, arithmetic logic units (ALU), comparators, decoders, flip-flop registers, and counters are integrated from macrocells. In contrast with macrocells, macrofunctions do not have predefined metal interconnections; redundant parts of macrofunctions can be deleted if required.

Gate arrays are available commercially in both TTL and ECL bipolar technologies as well as CMOS and gallium arsenide. Many gate arrays are now made as mixtures of technologies with the I/O (input-output) circuits selected to provide a better match to other system circuits. For example, the array might be ECL and the I/O sections are TTL. Similarly, the arrays might be CMOS and the I/O sections are TTL.

ECL arrays provide subnanosecond (ns) delay times. CMOS and bipolar arrays with two-layer interconnections meet requirements in the 1 to 5 ns delay region. Geometries are generally in the 1.5- to 3-micron (μm) size range. Where delays in excess of 5 ns are acceptable, single-layer 5 μm CMOS arrays are used.

As progress continues in gate array technology, improvements are being made in speed and equivalent gates. Increases in equivalent gate density result in higher performance in all technologies because of the shorter time delays. Feature size in commercial gate arrays dropped from 5 μm to less than 1.0 μm in less than five years.

ECL arrays with more than 3,500 equivalent gates and CMOS arrays with more than 10,000 equivalent gates have been announced. However, volume purchasing is generally at a lower level than these upper limits because of the problems in finding CAD software capable of making optimum use of the large number of available equivalent gates.

A standard cell differs from full-custom ICs because circuit equivalent to the gate array, but the gate array is 70 to 80 percent prefabricated. By contrast, the standard cell is completely fabricated starting with a blank wafer. *See* STANDARD CELL. Both gate arrays and standard cells can use the same library of tested macrocells stored in CAD workstation memory, but in the standard cell process uses only those gates, memory bits, and I/O pads that are actually needed for the application; there are no redundant circuit elements. Thus a standard cell will have a smaller chip size than the equivalent gate array in the same technology. However, the standard cell typically will have higher NRE costs, longer lead times and, higher unit costs.

A standard cells differs from full-custom ICs because of more complete dependence on CAD/CAE (computer-aided design/computer-aided engineering) techniques and the comprehensive software library of predesigned cells stored in workstation memory. However, the full-custom designed IC makes more efficient use of the wafer than either a gate array or a standard cell because a skilled designer can manually modify the macrocells in workstation memory. If necessary, new cells can be introduced that are more appropriate to the application, further optimizing the design. This results in a more efficient circuit and a further reduction in chip size. However, custom NRE costs are higher and lead times are longer than for a standard cell.

Programmable logic devices (PLD) that permit the

user programming of logic arrays by various techniques up to densites of 1,000 equivalent gates are now available commercially. They are an alternative to the factory-programmed gate arrays for low-density semicustom logic arrays.

GATED-BEAM TUBE

A gated-beam tube is a vacuum tube used in a gated-beam or quadrature detector. The gated-beam has three grids, and is thus a pentode. As the electrons travel from the cathode to the plate, they first encounter the signal or limiter grid. The frequency-modulated signal is applied to this grid. The accelerator grid comes next; it imparts extra speed to the electrons. Finally, the electrons encounter the quadrature grid. Here, a signal is applied that is 90 degrees out of phase with the wave at the signal grid.

The drawing illustrates the internal details of the gated-beam tube. The tube can be used for detection of frequency-modulated or phase-modulated signals. *See also* FREQUENCY MODULATION, PHASE MODULATION.

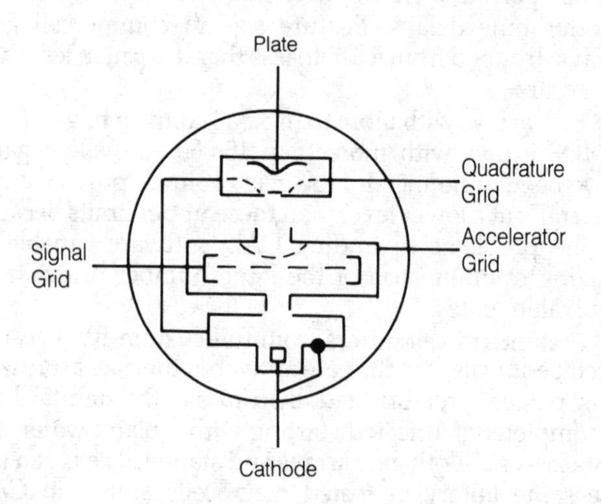

GATED-BEAM TUBE: The gated-beam tube is used in a frequency-modulation quadrature detector.

GATE-DIP METER

A gate-dip meter is a field-effect transistor-instrument that operates on the same principle as a grid-dip meter. The gate-dip and grid-dip meters are used for determining the resonant frequency of a tuned circuit or antenna.

The gate-dip meter has the obvious advantage of requiring less power than a grid-dip meter. There is no filament, and the necessary operating voltage is much smaller. The characteristics of the gate-dip meter are, however, identical to those of the grid-dip meter in practical applications. *See also* GRID-DIP METER.

GAUGE

The diameter of an electric wire is specified in terms of gauge, a number assigned to indicate the approximate size of a conductor. In the United States, the American Wire Gauge (AWG) designator is most commonly used for this purpose. There are other wire gauges. The term *gauge* is also used in reference to the thickness of a piece of sheet metal.

GAUSS

The gauss is a unit of magnetic flux density. The gauss is equal to 10^4 webers per square meter, or 1 maxwell per square centimeter. *See also* MAGNETIC FLUX, MAXWELL, WEBER.

GAUSSIAN DISTRIBUTION

See NORMAL DISTRIBUTION.

GAUSSIAN FUNCTION

A Gaussian function is a mathematical function used in the design of filters. A Gaussian filter passes a pulse with essentially no overshoot, but the rise time is rapid.

In order to pass a square pulse with maximum effectiveness, a filter must have a nearly flat response, with skirts of optimum steepness. *See also* BANDPASS FILTER, BANDPASS RESPONSE, BESSEL FUNCTION.

GAUSS'S THEOREM

Gauss's Theorem is an expression for determining the intensity of an electric field, depending on the quantity of charge.

For any closed surface in the presence of an electric field, the electric flux passing through that surface is directly proportional to the enclosed quantity of charge. *See also* ELECTRIC FIELD.

GEIGER COUNTER

The Geiger counter, also called a Geiger-Muller counter, is an instrument for measuring the intensity of high-energy radiation, such as alpha particles, beta particles, X rays, and gamma rays.

The heart of the Geiger counter is the counter tube as shown in the figure on p. 419. When a high-energy particle or photon enters the tube, the gas in the tube is ionized, causing a brief pulse of current to flow. This current pulse is amplified by means of a transistorized circuit. Then the pulse is fed to a digital counter, an audio amplifier, or a meter. A typical Geiger counter can register up to 2,000 pulses per second. *See also* ALPHA PARTICLE, BETA PARTICLE, GAMMA RAY, X RAY.

GENERATOR

A generator is a source of signal in an electronic circuit. It might be an oscillator, or it might be an electromechanical circuit. Signal generators are widely used in electronic design, testing, and troubleshooting (*see* SIGNAL GENERATOR).

A machine for producing alternating-current electricity, by means of a rotating coil and magnetic field, is called a generator. Whenever a conductor moves within a magnetic field so that the conductor cuts across mag-

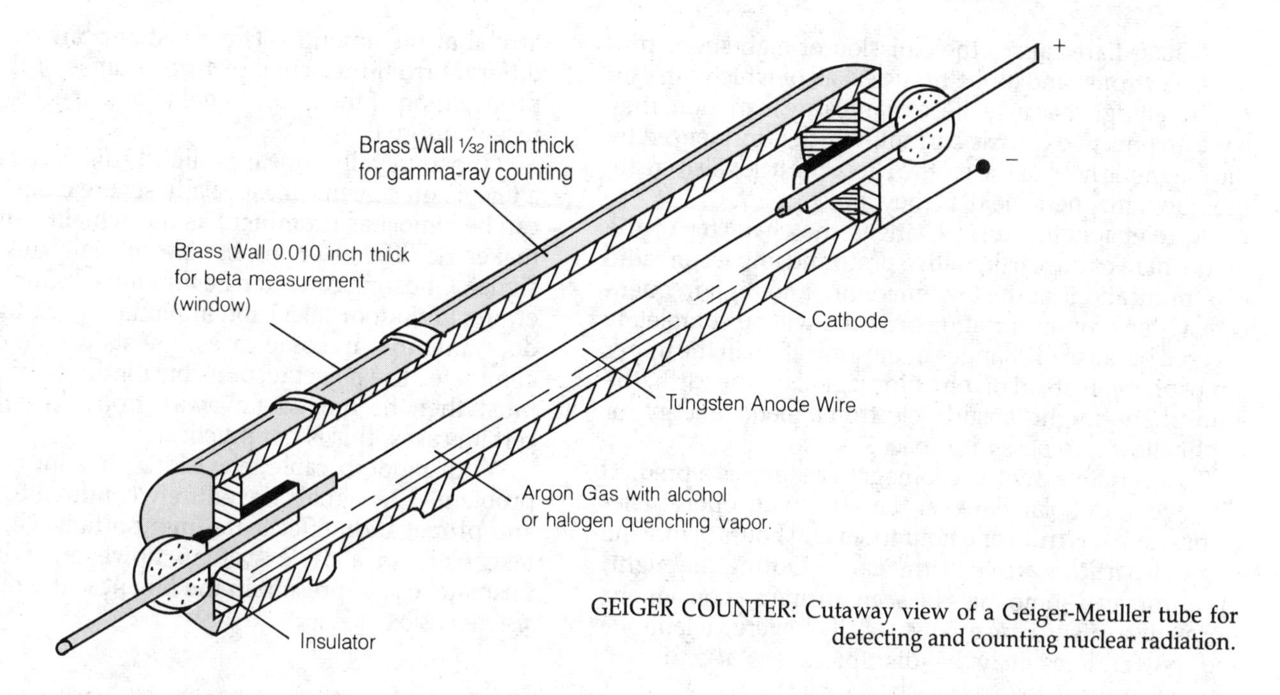

GEIGER COUNTER: Cutaway view of a Geiger-Meuller tube for detecting and counting nuclear radiation.

netic lines of flux, current is induced in the conductor. A generator can consist of either a rotating magnet inside a coil of wire, or a rotating coil of wire inside a magnet. In an alternating-current generator, the rotating portion of the generator is driven by an engine powered by oil, gasoline, or some other fossil fuel. Alternatively, steam turbines can be used.

A generator, like a motor, is an electromechanical transducer. The construction of an electric generator is, in fact, almost identical to that of a motor; some motors can operate as generators when their shafts are turned by an external force. *See also* ALTERNATING-CURRENT GENERATOR, MOTOR.

GEOMAGNETIC FIELD

The geomagnetic field is a magnetic field surrounding the earth. The earth is magnetized like a huge bar magnet. The north magnetic pole is located near the north geographic pole, and the south magnetic pole near the south geographic pole. The magnetic lines of flux extend far out into space (see illustration). The earth is not the only planet that has a magnetic field. Other planets, including Jupiter, as well as stars, such as the sun, are known to have magnetic fields.

The magnetic field of the earth is responsible for the behavior of the magnetic compass. The needle of a compass is actually a small bar magnet suspended on a bearing that allows free rotation. One pole of the bar magnet points south and the other pole points north in most locations on the surface of the earth. There is an error or deviation because of the difference in location between the magnetic and geographic poles.

The geomagnetic field attracts charged particles emitted by the sun, and causes them to be concentrated near the magnetic poles. When there is a solar flare, resulting in large amounts of charged-particle emission from the sun, the upper atmosphere glows in the vicinity of the magnetic poles. This is known as the aurora. The appearance of the aurora heralds a disturbance in the magnetic environment of the earth. This is called a geomagnetic storm. *See also* AURORA, GEOMAGNETIC STORM, MAGNETIC FIELD.

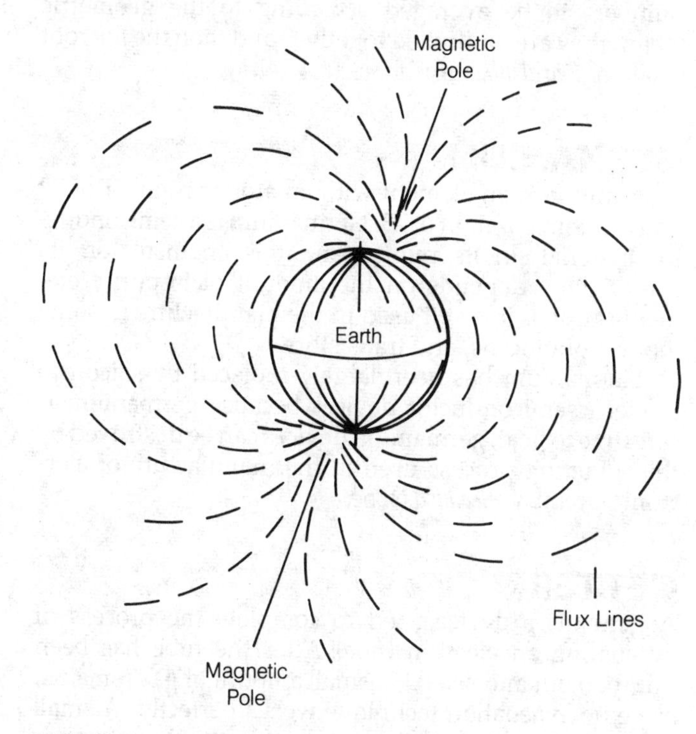

GEOMAGNETIC FIELD: Flux lines around the earth are similar to those around a bar magnet. Geomagnetic poles do not coincide with geographic poles.

GEOMAGNETIC STORM

A geomagnetic storm is a disturbance, or fluctuation, in the magnetic field of the earth. Such a storm is caused by charged particles from the sun.

A solar flare causes the emission of high-speed protons, electrons, and alpha particles, all of which carry an electric charge. Because these charges are in motion, they have an effective electric current, which is influenced by the magnetic field of the earth. These particles also create their own magnetic field as they move.

A geomagnetic storm changes the character of the ionosphere of the earth, with a profound impact on radio communications at the low, medium, and high frequencies. Often, communications circuits will be completely severed because of changes in the ionization in the upper atmosphere. Instead of returning signals to the earth, the ionosphere might absorb electromagnetic energy or might allow it to pass into space.

To a certain extent, a geomagnetic storm is a predictable event. A solar flare can be seen with optical telescopes, and it is from one hour to several hours before the charged particles arrive at the earth. During the night, the aurora can sometimes be seen during a geomagnetic storm. If a disturbance is especially severe, telephone and power lines may be disrupted. *See also* AURORA, GEOMAGNETIC FIELD, IONOSPHERE, SOLAR FLARE.

GEOMETRIC MEAN

The geometric mean is a form of mathematical average of two or more numbers. In general, if there are *n* different numbers to be averaged according to the geometric mean, they are multiplied together, and then the *nth* root is taken. *See also* ARITHMETIC MEAN.

GERMANIUM

Germanium is an element with an atomic number of 32 and an atomic weight of 73. Germanium is a semiconductor material. In its pure form, it is an insulator; its conductivity depends on the amount of impurity elements added. It is still used in the manufacture of some diodes, photocells, and transistors.

Germanium has been largely replaced by silicon in modern semiconductor devices because germanium is sensitive to heat; germanium devices can be destroyed by the soldering process used in the manufacture of electronic circuitry. *See also* DIODE.

GETTER

A getter is a device used to complete the process of evacuating an electron tube. After the tube has been pumped out and sealed, a small amount of gas remains, since no evacuation technique works perfectly. A small electrode, the getter, is activated with a radio-frequency voltage. This causes the remaining gas in the tube to react with the electrode, and remove additional contaminants from the tube. The getter serves no purpose once the evacuation is complete.

GHOST

A ghost is a false image in a television receiver caused by reflection of a signal from some object just prior to its arrival at the antenna. The ghost appears in a position different from the actual picture because of the delay in propagation of the ghost signal with respect to the actual received signal.

Ghosts usually appear as slightly displaced images on a television screen. In especially severe cases, the ghost can be almost as prominent as the actual picture, which makes it difficult or impossible to view the television signal. Ghosting can often be minimized simply by reorienting an indoor television antenna, or rotating an outdoor antenna. In some cases, ghosts can be difficult to eliminate. The object responsible for the signal reflection must then be moved far away from the antenna. In certain cases, this is not practical.

With modern cable television, ghosting is seldom a problem. The signals are entirely confined to the cable, and propagation effects are unimportant. Ghosting can take place in a cable system, however, if impedance mismatches are present in the line near the receiver. *See also* TELEVISION.

GIGA-

Giga- is a prefix meaning 1 billion (1,000,000,000). This prefix is attached to quantities to indicate the multiplication factor of 1 billion. For example, 1 gigaohm is 1 billion ohms; 1 gigaelectronvolt is 1 billion electron volts; 1 gigahertz is 1 billion hertz; and so on.

Giga- is one of many different prefix multipliers commonly used in electronics. *See also* PREFIX MULTIPLIERS.

GILBERT

The gilbert (Gb) is a unit of magnetomotive force named after the scientist William Gilbert. In the centimeter-gram-second (cgs) electromagnetic system, 1 gilbert is equal to 1 oersted-centimeter.

One gilbert is also equal 1.26 amperes. In a coil with *N* number of turns and *I* current in amperes that flows through the coil, the magnetomotive force in gilberts is equal to:

$$Gb = 1.26\ NI$$

See also MAGNETOMOTIVE FORCE.

GLASS FIBER

See FIBEROPTIC CABLE AND CONNECTORS, FIBEROPTIC COMMUNICATION.

GLASS INSULATOR

Glass is an excellent dielectric or insulating material. It has good mechanical strength as well. Thus, glass is used in the manufacture of insulators for electrical wiring and antenna installations.

Glass insulators are available in many different sizes and shapes. The surface of the insulator is generally ribbed to increase the amount of surface distance between the two ends. Glass insulators are less common than porcelain insulators. *See also* INSULATOR.

G-LINE

See SURFACE-WAVE TRANSMISSION LINE.

GLOBAL POSITIONING SYSTEM

The global positioning system (GPS) consists of a group of Navstar navigation satellites orbiting the earth in different planes for complete earth coverage and ground stations for control and synchronization of the satellites and receivers. The figure illustrates the general organization of GPS. Its receivers are capable of determining the latitude and longitude of their positions on the earth within 100 feet (30 meters) of their true locations. They can calculate the range and bearing of selected destinations from satellite signals. Advanced versions will also be able to determine altitude.

GPS is a ranging system, based on the precise knowledge of satellite position at any given time and the equally precise measurement of the time needed for a signal from the satellite to reach the receiver. GPS triangulates position almost instantly from several satellites within view at once.

Sponsored by the U.S. Department of Defense and managed by the U.S. Air Force, GPS is available for both military and nonmilitary applications. GPS provides navigational guidance for military aircraft and missiles, but it also provides navigational information for commercial and private aircraft and ships. The 1.575 MHz frequency is offered for civilian use. The GPS system can track the movement of vehicles on land, and it can be used to guide explorers in remote uncharted parts of the world. The system is expected to establish accurate terrestrial references for improved maps and charts.

The GPS system is now functioning with seven Navstar satellites but will, when completed in 1991, include 21 satellites. Four or five ground stations will continuously track the Navstar satellites and precisely calculate the position and orbital track of each satellite. This data is then transmitted back to the satellites for continuous tracking and synchronization. Each satellite broadcasts a radio-frequency signal giving its exact position over the earth as well as a time reference. The one-way transit time of this data from each satellite can be used to measure its distance from the receiver.

If the receiver clock is synchronized with the satellites clock and the time of satellite broadcast is known, the signal travel time (corrected for atmospheric conditions and other variables), yields a distance or range to the satellite. For a given range, the receiver could be anywhere on a sphere surrounding the satellite. When a second satellite is synchronized with the first, another range—and sphere of position (SOP)—is created. The interaction of the two spheres is a circle or line of position (LOP). With a third satellite range, the LOP becomes a point.

In practice, two satellites are needed for the first sphere. One acts as a time standard for the receiver and other satellites, and the second provides the range. As a result, three satellites are needed to get a two-dimensional fix (latitude and longitude).

Handheld portable receivers have been developed

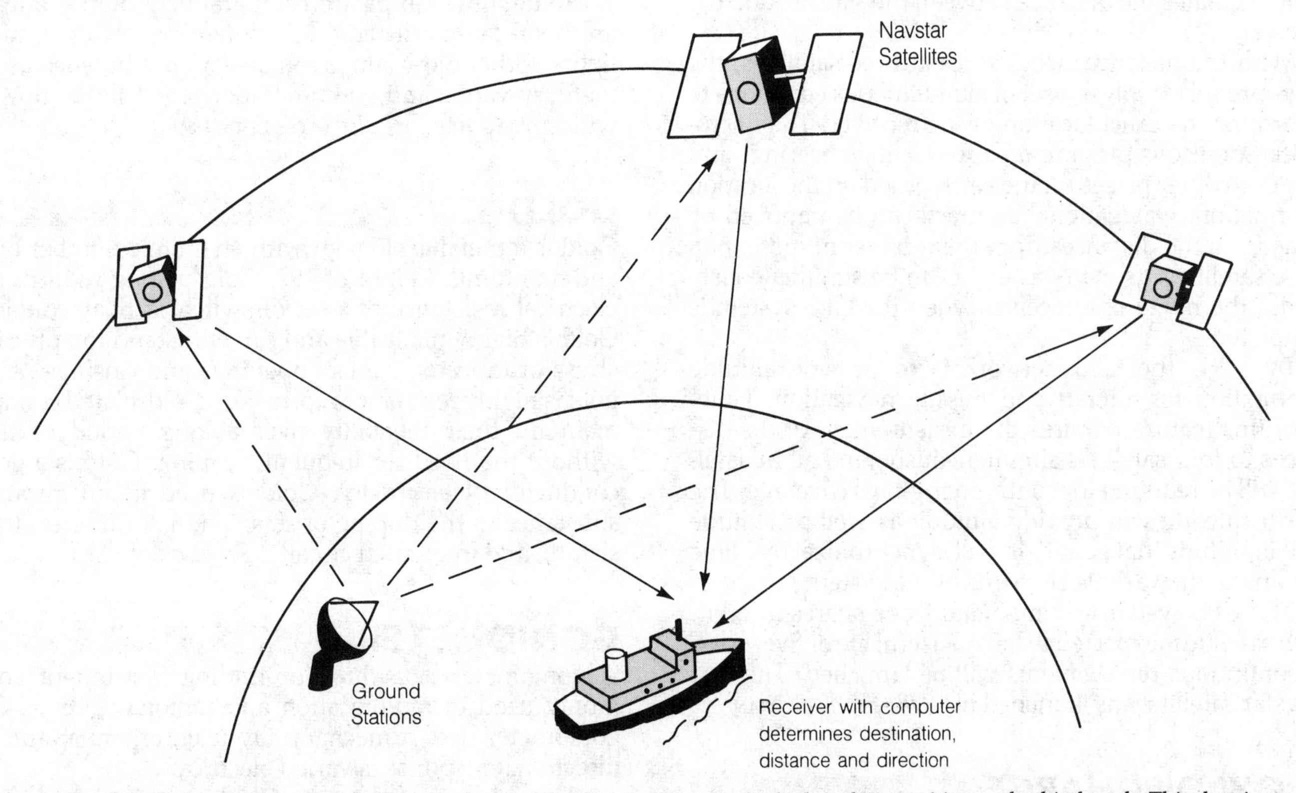

GLOBAL POSITIONING SYSTEM: Ground stations track Navstar satellites and calculate their position and orbital track. This data is sent to satellites for broadcast to provide their position and time checks. The receiver measures transit times to three different satellites, and its computer determines its exact position from these distances.

that will permit the user to determine the following information on a digital display:

1. Latitude and longitude of position within 100 feet (30 meters) of the exact position.
2. Speed and course over the ground.
3. Direction and exact distance along the great circle route of any of 50 destinations stored in memory.
4. Probable time of arrival at selected destination based on current course and speed.

Each Navstar satellite circles the earth twice a day in an orbit 10,900 miles high. The orbits are coordinated so the same number of satellites is above the horizon. Seven Navstar satellites were in orbit in 1988 in planes inclined 63 degrees from the equator. Six additional satellites are to be orbited each year until 21 are in position by 1991. These satellites will be assigned to six orbital planes inclined 55 degrees from the equator to provide global coverage from pole to pole, 24 hours a day.

Each Navstar satellite tracks its exact position over the earth with the help of signals broadcast from four or five fixed tracking stations on the ground. An atomic clock with an accuracy of one second in 300,000 years is on each of the satellites to keep time.

Each satellite broadcasts a time signal to the earth that permits the ground receiver to determine the exact time of the signal departure. The receiver registers the signal arrival time and compares it with the time the signal was sent from the satellite. The difference between these times is measured in microseconds. The microprocessor then calculates the distance between the satellite and the receiver.

With the distances from at least three satellites, the microprocessor solves a set of simultaneous equations to determine the exact location of the receiver. The microprocessor also is programmed to calculate bearings and ranges to other places on the earth based on the location information. Navigational accuracy can be impoved by measuring the distances from the receiver of more than three satellites: as many as eight can be simultaneously within the range of a receiver when the GPS system is completed.

By 1991, the GPS network is to provide altitude information for aircraft and missile navigation. However, this feature requires the measurement of the distances to four satellites simultaneously, and all 21 satellites will be required for continuous global coverage. The fourth satellite will provide altitude as well as latitude and longitude data, and it will synchronize the time keeping in the various elements of the system.

The GPS system will maintain three spares in orbit. Each satellite is expected to have a useful life of five years, so continuous replacements will be launched. The first Navstar satellite was launched in 1978. *See also* LORAN.

GLOW DISCHARGE

When an electric current is passed through a rarefied gas, the gas glows with a characteristic spectrum. Different gases appear different in color. This emission is called glow discharge.

When the glow-discharge spectrum is separated by means of a spectroscope, the energy can be observed to occur at discrete wavelengths, rather than over a continuous range. The discrete energy wavelengths are called emission lines. Each gas has its own set of emission lines, like a fingerprint that identifies it. Astronomers can recognize the emission lines of gases in stars, planets, and in interstellar space.

When a voltage is applied to a gas, causing the flow of current, electrons absorb energy from the voltage source. Some of the electrons fall back again to lower energy levels, and release discrete bursts of radiation in electromagnetic form. This can happen only in specific quantities, resulting in emission at defined wavelengths. Some emission lines are stronger than others; this is because some energy quantities are more common than others as the electrons fall back.

A mercury-vapor glow discharge appears bluish or bluish white. Sodium vapor has a candle-flame yellow color. Other gases have various colors from red to violet. Some of the emission lines are invisible, occurring at the infrared, ultraviolet, and even the X ray or radio wavelengths. *See also* ELECTROMAGNETIC SPECTRUM.

GLOW LAMP

Any lamp that makes use of glow discharge for generating visible light is called a glow lamp. They last longer than incandescent lamps, and produce more light and less heat, resulting in greater energy efficiency.

Some glow lamps are comparatively dim; examples are neon lamps in radio equipment panels or in night lights. Other glow lamps can be very bright, such as the mercury-vapor and sodium-vapor street lights now in widespread use. *See also* GLOW DISCHARGE.

GOLD

Gold is a metallic element with an atomic number of 79 and an atomic weight of 197. Gold plating reduces the electrical resistance of a set of switch or relay contacts. Gold is highly malleable and can withstand the physical stress of repeated contact openings and closings. Also, gold is highly resistant to corrosion. Gold-plated contacts maintain their reliability over a long period of time without the need for frequent cleaning. Gold is a good conductor of electricity. Gold is used as an impurity substance in the doping of semiconductor diodes, transistors, and integrated circuits. *See also* DOPING.

GONIOMETER

A goniometer is a direction-finding instrument commonly used in radiolocation and radionavigation. The goniometer uses a mechanically fixed antenna, and its directional response is varied electrically.

A simple pair of phased vertical antennas can serve as a goniometer. The two antennas can be spaced at any distance between ¼ and ½ wavelength. Depending on the phase in which the antennas are fed, the null can occur in any compass direction. The null can be unidirectional in some situations and bidirectional in other cases. The precise direction is determined by knowing the phase relationship of the two feed systems.

GRADED FILTER

A graded filter is a form of power-supply output filter which provides various degrees of ripple elimination. Some circuits can tolerate more ripple than others. Those circuits that can operate with higher ripple levels are connected to earlier points in the filter sequence (see illustration). Some circuits need less power-supply filtering because they draw very little current; these circuits are also connected to the earlier points in the graded filter.

The graded filter offers the advantage of superior ripple elimination and voltage regulation for those circuits that require the purest direct current. The current drain is reduced at the farthest point in the graded filter. *See also* POWER SUPPLY.

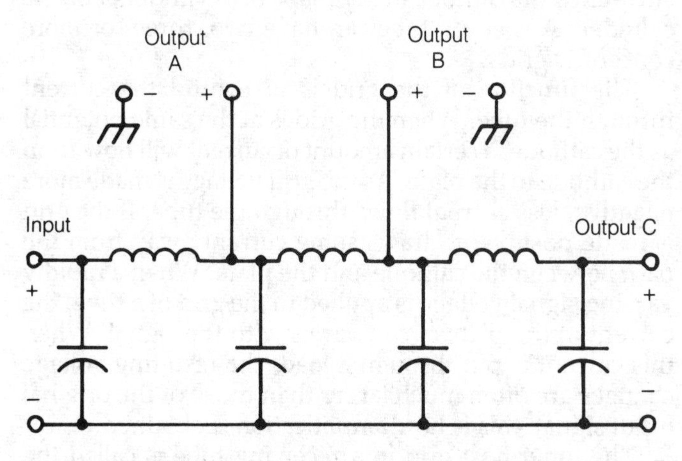

GRADED FILTER: Output A provides the least filtering; output B provides more filtering, and output C provides the most filtering.

GRADED JUNCTION

A graded semiconductor junction is a P-N junction that is grown by epitaxial methods in a carefully controlled manner.

Graded-junction devices have reverse-bias characteristics that differ from ordinary semiconductor devices. The capacitance across the junction of a reverse-biased, graded-junction diode decreases more rapidly, with increasing reverse voltage, than the capacitance across an ordinary P-N junction. *See also* P-N JUNCTION.

GRAPH

A graph is a means of illustrating a relation between two or more variable quantities. Graphs can be constructed in a variety of different ways, but the most common method is the Cartesian system (*see* CARTESIAN COORDINATES).

In the simple graph, the controlled or dependent variable is usually represented on the vertical axis. The axis may or may not be calibrated in linear form.

The illustration shows a simple graph, depicting the current-versus-voltage function for a typical semiconductor diode. The independent variable is the voltage across the P-N junction. The dependent variable is the current through the device. *See also* CURVILINEAR COORDINATES, DEPENDENT VARIABLE, FUNCTION, INDEPENDENT VARIABLE, P-N JUNCTION, POLAR COORDINATES.

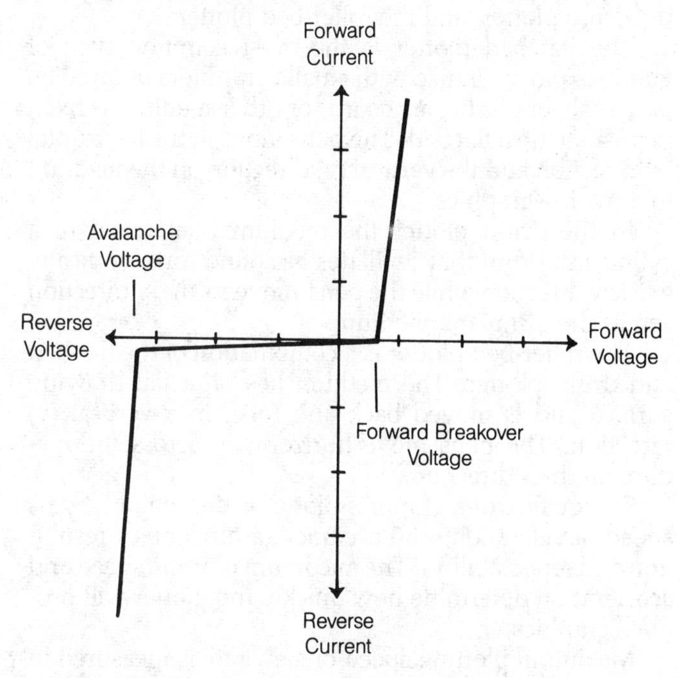

GRAPH: A graph showing the characteristics of a typical semiconductor diode.

GRAPHICAL ANALYSIS

Graphical analysis is a method of evaluating the interaction among different quantities. A graphical representation, such as that in the GRAPH illustration, is used as the basis for graphical analysis. The illustration is an example of the current-versus-voltage curve of a typical diode. The avalanche voltage can be seen as the reverse voltage at which the current abruptly rises. The forward breakover voltage can also be seen, as the forward voltage at which the current abruptly rises. The rate of current increase with increasing voltage, in the forward direction, can be determined by placing a straight edge tangent to the graph at the desired point. The slope of the line indicates the rate of change.

Graphical analysis is used in many different phases of electronic engineering. Antenna performance, transmission-line design, and the design of oscillators, amplifiers, and filters can all benefit from the use of graphical analysis. Modern computers can perform graphical analysis much more quickly and accurately than is possible with paper and pencil. *See also* GRAPH.

GRAPHIC EQUALIZER

See EQUALIZER.

GRAPHICS PLOTTER

A graphics plotter is a mechanical desktop computer

peripheral that, with appropriate software and pens, can plot graphs, histograms, pie charts, line drawings, maps and overhead projections in one or more colors. The color graphics can be drawn on paper, plastic, art boards or other media for reports, presentations and proposals.

There are three basic types of plotters, each with different mechanical characteristic: the flat-bed plotter, the drum plotter, and the roller-bed plotter.

The flat bed plotter is the most common type of graphics plotter in use with small computers today. The paper, sheet plastic, art board, or other medium is fixed in position on a flat bed. The pens move in the horizontal (x) direction and the vertical (y) direction on the medium to draw the graphics.

In the drum plotter the medium is attached to a cylindrical drum that oscillates back and forth vertically in the y direction while the pens move in the x direction across the drum and medium.

The roller-bed plotter is a combination of the flat-bed and drum plotter. The medium lies on a flat drawing surface and is moved back and forth in a vertical (y) direction. The pens move horizontally across the medium in the x direction.

The quality of a graphics plotter is determined by its speed, acceleration, and accuracy (a function of resolution and repeatability). The maximum plotting speed and acceleration determine how quickly the plotter will produce graphics.

Maximum plotting speed or slew rate is measured in inches per second. A typical range of plotter speeds when graphing is from 4 to 15 inches per second.

Acceleration is measured in g's—the force of gravity. The higher the g rating, the faster the acceleration of the pens across the medium. Typical ratings range from 1 to 3 g's.

Accuracy is measured in terms of resolution and repeatability. Resolution is the number of points a plotter can make in a given area. The higher the resolution, the greater the number of lines per square inch and the smaller the possibile pen motions. Ratings are given in thousandths of an inch and typical values are 0.05 to 0.001 inch (500 to 1000 steps per inch). The lower the decimal number, the better the resolution. High resolution graphics printers minimize the appearance of steps in drawing diagonal lines and curves.

Repeatability is a measure of how accurately the plotter pen can return precisely to any given point. It is measured in thousandths of an inch. A rating of 0.001 inch is excellent. In practice, high repeatability means circles being drawn will close and the sides of boxes being drawn will meet at the corners.

Graphics printers are available with multiple pens that can change color automatically. Available plotters offer a selection of fine, medium and wide pens. Some plotters offer automatic pen capping to prevent the ink in the pens from drying out too quickly.

GRATING

A grating, also called a diffraction grating, is a transparent piece of glass or plastic with many fine, straight, parallel dark lines. The diffraction grating acts like a prism for visible light. When a beam of light is passed through the grating, the beam is split into a spectrum of constituent wavelengths from red to violet. Some diffraction gratings allow the passage of infrared and ultraviolet wavelengths.

Gratings can be used at radio frequencies as well as at visible wavelengths. A set of parallel metal bars, or slots in a metal sheet, produce an interference pattern when radio waves pass through. If the bars or slots are spaced within certain limits, the radio waves are spread out into a spectrum. Frequency resolution can be obtained in this way; a movable detector can be set anywhere along the spectral distribution to choose a particular frequency.

Diffraction gratings are used in optical spectroscopes and other devices. *See also* DIFFRACTION, SPECTROSCOPE.

GRID

A grid is an element in a vacuum tube, placed between the cathode, which emits electrons, and the plate, which collects electrons. The grid resembles a screen, and surrounds the cathode in the shape of a cylinder or oblate cylinder. A vacuum tube can have two, three, or more concentric grids.

The function of the grid is to control the current through the tube. When the grid is at the same potential as the cathode, a certain amount of current will flow from the cathode to the plate. As the grid voltage is made more negative, less current flows through the tube. If the grid is made positive, it draws some current away from the path between the cathode and the plate. When a rapidly varying signal voltage is applied to the grid of a tube, the current in the plate circuit varies with the signal. When this current is put through a load, the resulting voltage changes are often much larger than those of the original input signal; this is how amplification is obtained.

The innermost grid in a receiving tube is called the control grid. The second grid is called the screen; the third grid is called the suppressor. All have different functions. The input signal is usually applied to the control grid.

GRID BIAS

In a vacuum tube, the grid bias is the direct-current voltage applied to the control grid. The grid bias is usually measured with respect to the cathode. The grid bias may be obtained in different ways. Three methods are shown in the illustration. At A, a resistor and capacitor are used to elevate the cathode circuit above direct-current ground, generating negative grid bias. At B, a separate power supply is used for obtaining the negative grid bias. At C, the grid bias is zero.

Most vacuum tubes require a negative grid bias, the exact value depending on the characteristics of the tube and the application for which the tube is to be used. Some tubes operate with zero bias.

As the grid bias is made progressively more negative, the plate current decreases until, at a certain point, the tube stops conducting altogether. This value of grid bias is called the cutoff bias, and it depends on the plate voltage as well as the particular type of tube used. If the grid bias is made positive, the grid will draw current away from the plate circuit. An input signal may drive the control grid positive for part of the cycle, but deliberate positive bias is almost never placed on the control grid of a tube.

The level of grid bias must be properly set for a tube to function according to its specifications in a given application. The required grid bias is different among Class A, Class AB, Class B, and Class C amplifiers. *See also* CLASS A AMPLIFIER, CLASS AB AMPLIFIER, CLASS B AMPLIFIER, CLASS C AMPLIFIER, GRID.

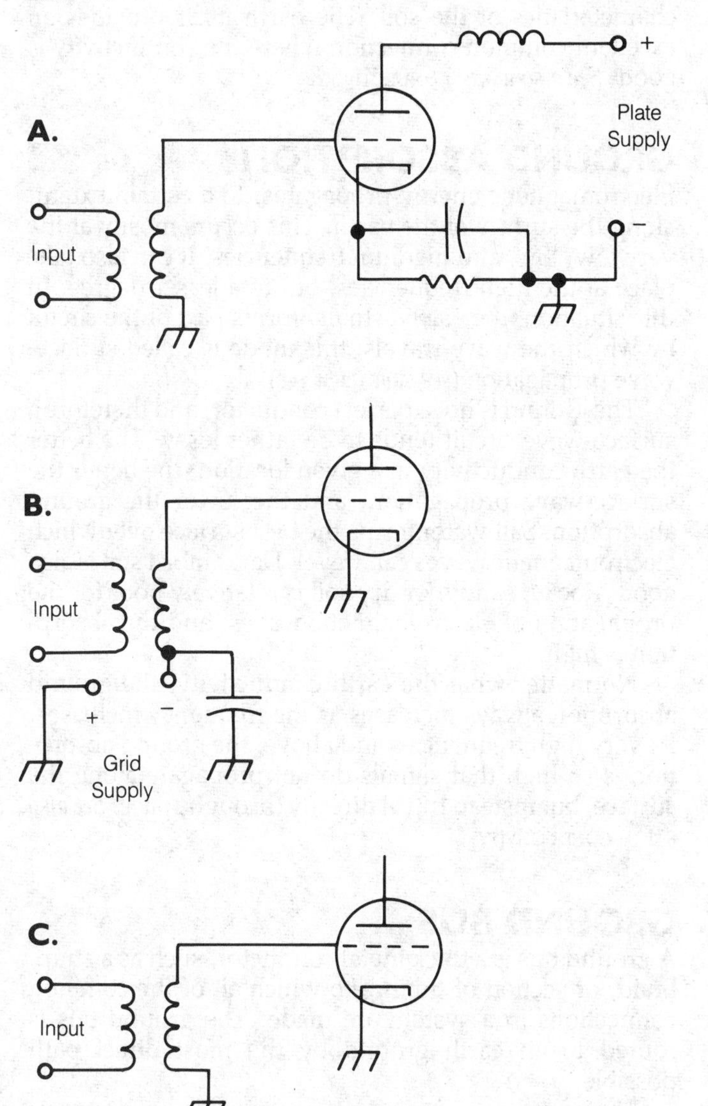

GRID BIAS: Three methods of obtaining grid bias are with an RC network in a cathode circuit (A), a separate power supply (B), and zero bias (C).

GRID-BLOCK KEYING

Grid-block keying, also called blocked-grid keying, is a form of amplifier keying. A large negative voltage is applied to the grid of one of the low-power amplifier stages of a transmitter during the key-up condition. This cuts the tube off, preventing the input signal from reaching the output. The illustration is a schematic diagram of a typical tube amplifier using grid-block keying.

When the key is pressed, the negative cutoff voltage is short-circuited, and the grid attains normal operating bias. Then, the tube amplifies normally, and the input signal is allowed to reach the following stages.

Grid-block keying offers the advantage of relatively low-current switching, and moderate voltages. Grid-block keying also allows the keying of a transmitter with minimal impedance change at the oscillator stage. This helps to prevent chirping. *See also* GRID BIAS.

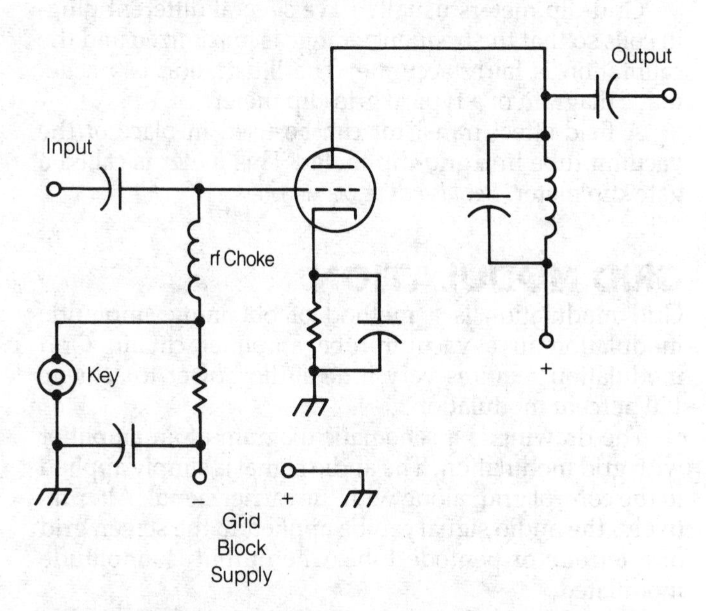

GRID-BLOCK KEYING: The key amplifier is cut off when the key is up.

GRID-DIP METER

A grid-dip meter is an instrument for determining the resonant frequency of a tuned circuit or antenna. The

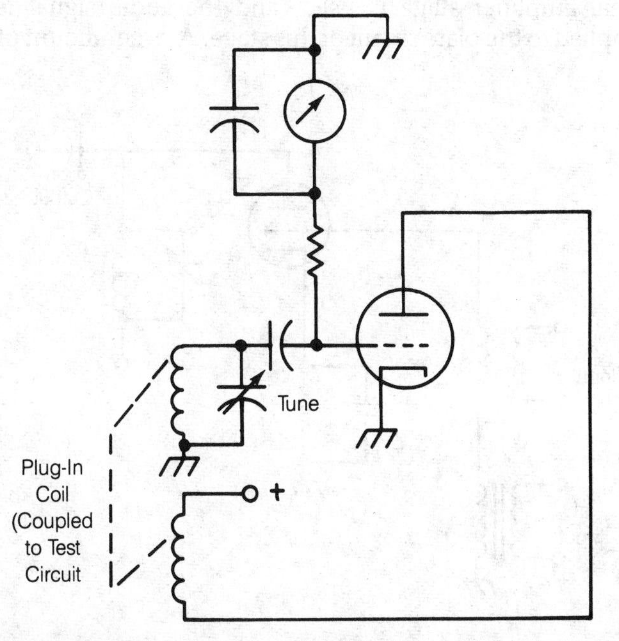

GRID-DIP METER: A simple grid-dip meter can determine the resonant frequency of tuned circuits.

heart of the grid-dip meter is an oscillator. The tuned circuit of the oscillator contains a large, solenoidal inductor, which allows coupling to external circuits.

As the frequency of the oscillator is varied with the coil near the resonant circuit to be tested, a sharp dip occurs in the grid current. This dip is indicated by the meter when the resonant frequency of the external circuit is reached. The external circuit may have several resonant frequencies; at each, the grid current dips sharply. The resonant frequency is read from a calibrated dial on the grid-dip oscillator.

Grid-dip meters usually have several different plug-in coils so that the frequency range is maximized and the calibration is fairly accurate. The illustration is a schematic diagram of a typical grid-dip meter.

A field-effect transistor can be used in place of the vacuum tube in a grid-dip meter. This meter is called a gate-dip meter. *See also* GATE-DIP METER.

GRID MODULATION

Grid modulation is a method of obtaining amplitude modulation in a vacuum-tube amplifier circuit. Grid modulation requires very little audio power to achieve 100 percent modulation.

The drawing is a schematic diagram of an amplifier with grid modulation. The audio signal is simply applied to the control grid, along with the carrier signal. Alternatively, the audio signal can be applied to the screen grid of a tetrode or pentode tube. The output is amplitude modulated.

The principal disadvantage of grid modulation is that all of the amplifying stages following the modulated stage must be linear. This means that they must be Class A, Class AB, or Class B amplifiers. The efficient Class C amplifier will cause distortion of an amplitude-modulated signal.

In some amplitude-modulated transmitters, a Class C final amplifier stage is used, and the audio signal is applied to the plate circuit of this stage. A large amount of audio signal power is necessary in order to achieve 100 percent modulation in this form of circuit. This method is called plate modulation. *See also* AMPLITUDE MODULATION.

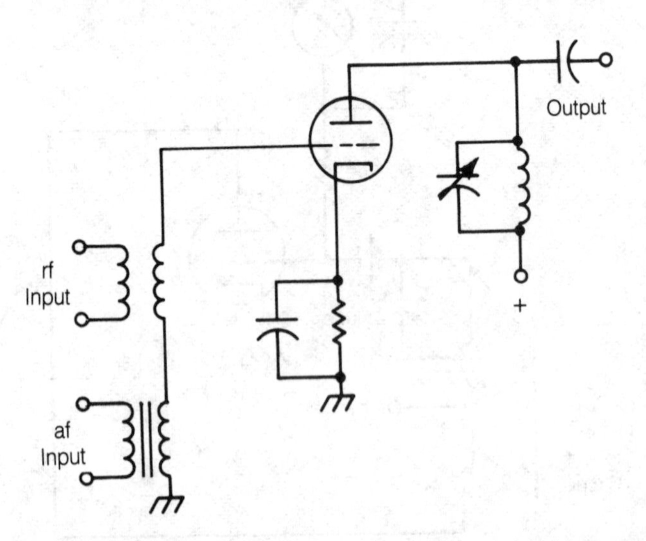

GRID MODULATION: The grid bias is varied by the modulating signal in grid modulation.

GROUND

Ground is the term used to describe the common connection in an electrical or electronic circuit. The common connection is usually at the same potential for all circuits in a system.

The common connection for electronic circuits is almost always ultimately routed to the earth. The earth is a fair to good conductor of electricity, depending on the characteristics of the soil. The earth itself provides an excellent common connection where the conductivity is good. *See also* EARTH CONDUCTIVITY.

GROUND ABSORPTION

Electromagnetic energy propagates, to a certain extent, along the surface of the earth. This occurs mostly at the very low, low, and medium frequencies. It can also take place at the high frequencies, but to a lesser degree. In this situation, the earth actually forms part of the circuit by which the wave travels; this mode is called surface-wave propagation (*see* SURFACE WAVE).

The ground is not a perfect conductor, and therefore a surface-wave circuit tends to be rather lossy. The better the earth conductivity in a given location, the better the surface-wave propagation, and the lower the ground absorption. Salt water forms the best surface over which electromagnetic waves can travel. Dark, moist soil is also good. Rocky, sandy, or dry soil is relatively poor for the propagation of electromagnetic waves, and the absorption is high.

No matter what the earth conductivity, the ground absorption always increases as the frequency increases. At very high frequencies and above, the ground absorption is so high that signals do not propagate along the surface, but instead travel directly through space. *See also* EARTH CONDUCTIVITY.

GROUND BUS

A ground bus is a thick metal conductor, such as a strap, braid, or section of tubing, to which all of the common connections in a system are made. The ground bus is routed to an earth ground by the most direct path possible.

The advantage of using a ground bus in an electronic system is the avoidance of ground loops (*see* GROUND LOOP). All of the individual components of the system are separately connected to the bus, as shown in the schematic diagram.

The size of the conductor used for a ground bus depends on the number of pieces of equipment in the system, and on the total current that the system carries. Typical ground-bus conductors are made from copper tubing having a diameter of ¼ to ½ inch. For radio-frequency applications, the larger conductor sizes are preferable; also, the distance to the ground connection

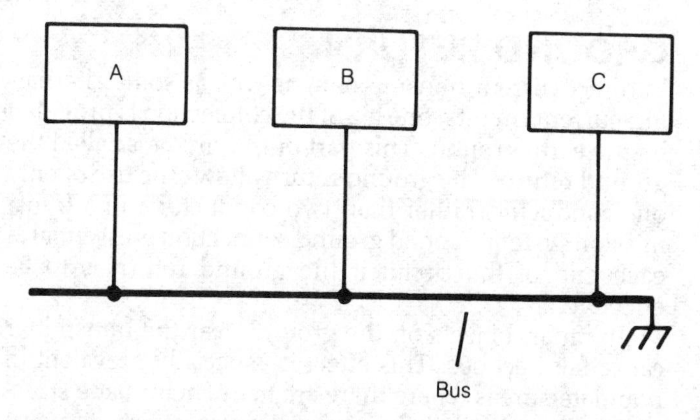

GROUND BUS: The ground bus is a convenient, effective grounding system.

should be made as short as possible in such cases. *See also* GROUND.

GROUND CLUTTER

See GROUND RETURN.

GROUND CONDUCTIVITY

See EARTH CONDUCTIVITY.

GROUND CONNECTION

A ground connection is the electrical contact between the common point of an electrical or electronic system and the earth. A ground connection should have the least possible resistance. The most effective ground connection consists of one or more ground rods driven into the soil to a depth of at least 6 to 8 feet. Alternatively, a ground connection can be made to a cold-water pipe.

The ground connection is of prime importance to the operation of unbalanced radio-frequency (rf) systems. In balanced rf systems, the ground connection is less critical, although from a safety standpoint, it should always be maintained. All components of a system should be connected together, preferably with a ground bus leading to an earth ground, to prevent possible potential differences that can cause electric shock. *See also* GROUND, GROUND BUS.

GROUND EFFECT

The directional pattern of an antenna system, especially at the very low, low, medium, and high frequencies, is modified by the presence of the surface of the earth beneath the antenna. This occurrence is called ground effect. It is more pronounced in the vertical, or elevation, plane than in the horizontal plane. The effective surface of the earth usually lies somewhat below the actual surface. The difference depends on the earth conductivity, and on the presence of conducting objects such as buildings, trees, and utility wires. The effective ground surface reflects radio waves to a certain extent. At a great distance, the reflected wave and the direct wave add together in variable phase.

The drawing illustrates an example of the effect of perfectly conducting ground and partially conducting ground on the vertical-plane radiation pattern of a half-wave dipole antenna. At A, the free-space pattern of the dipole is shown, in a situation where no ground surface is present. At B, the effect of a perfectly conducting ground surface ½ wavelength below the antenna is illustrated. At C, the effect of a partially conducting ground surface (typical of the actual earth), ½ wavelength below the antenna, is shown.

Ground effects have a significant influence on the takeoff angle of electromagnetic waves from an antenna. This, in turn, affects the optimum distance at which communications is realized. In general, the higher a horizontally polarized antenna is positioned above the ground, the better the long-distance communication will be. For a vertically polarized antenna with a good electrical ground system, the height of the radiator above the ground is of lesser importance.

Ground effect takes place in the same way for reception, with a given antenna system, as for transmission at the same frequency. *See also* ANGLE OF DEPARTURE, EARTH CONDUCTIVITY, EFFECTIVE GROUND, GROUND WAVE.

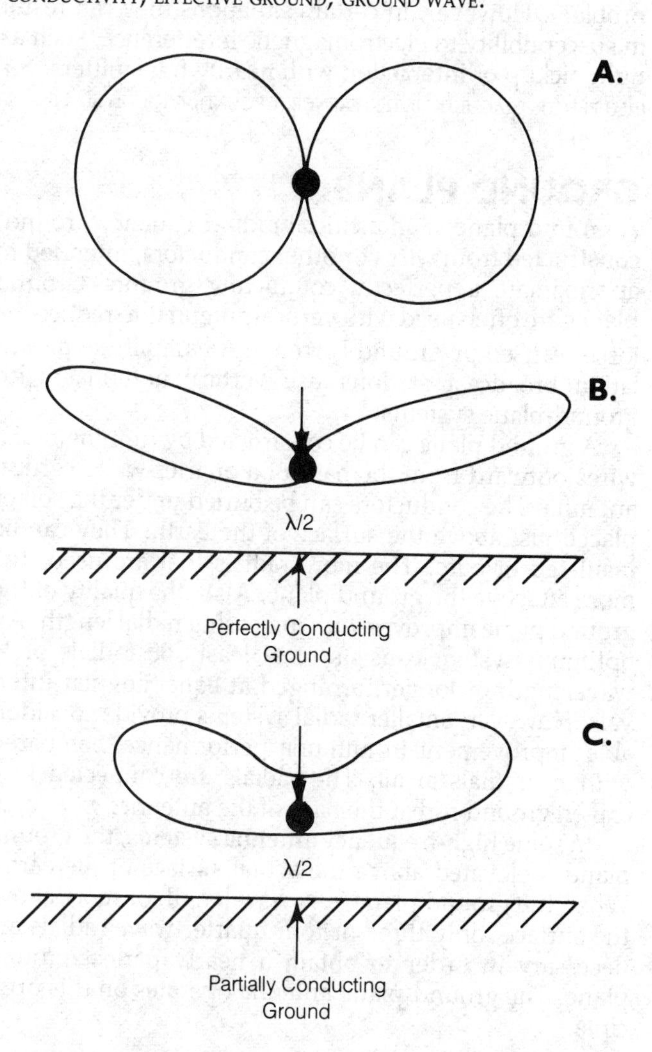

GROUND EFFECT: The radiation pattern of a vertical half-wave dipole is shown in the absence of ground (A), over perfectly conducting ground at a height of ½ wavelength (B), and over typical ground at a height of ½ wavelength (C).

GROUND FAULT

A ground fault is an unwanted interruption in the ground connection in an electrical system. This can result in loss of power to electrical equipment because there is no circuit in which the current can return. Excessive current can flow in the ground conductor of an ac electrical system, and dangerous voltages may be present at points that should be at ground potential. Ground faults can result in dangerous electrical shock.

Ground connections should, if possible, be soldered into place, so that corrosion at the connector joint will not result in ground faults. The number of joints in a ground circuit should be kept to a minimum. *See also* GROUND, GROUND CONNECTION.

GROUND LOOP

A ground loop occurs when two units of equipment are connected to a common ground bus, and are also connected together by separate wires or cables. Ground loops are quite common in all kinds of electrical and electronic apparatus; normally they do not present a problem. However, in certain situations they can result in susceptibility to electromagnetic interference, such as hum pickup or interaction with nearby transmitters. *See also* ELECTROMAGNETIC INTERFERENCE, GROUND, GROUND BUS.

GROUND PLANE

A ground plane is an artificial radio-frequency ground, constructed from wires or other conductors, intended to approximate a perfectly conducting ground. Ground planes are often used with vertical antennas to reduce the losses caused by ground currents. All amplitude-modulation broadcast stations use vertical antennas with ground-plane systems.

A ground plane can be constructed by running radial wires outward from the base of a quarter-wave vertical antenna. The conductors can be buried under the soil or placed just above the surface of the earth. They can be insulated or bare. The more radials that are used, the more effective the ground plane. Also, the quality of the ground plane improves with increasing radial length. An optimum system consists of at least 100 radials of ½ wavelength or longer, arranged at equal angular intervals. However, smaller radial systems provide considerable improvement in antenna performance, compared with no radials at all. The radials are connected to a driven ground rod at the base of the antenna.

In some high-frequency antenna systems, the ground plane is elevated above the actual surface of the earth. When the ground plane is ¼ wavelength or more above the surface, only three or four quarter-wave radials are necessary in order to obtain a nearly perfect ground plane. The ground-plane antenna operates on this principle.

GROUND RESISTANCE

See EARTH CONDUCTIVITY.

GROUND RETURN

In direct-current transmission, as well as some alternating-current circuits, one leg of the connection is provided by the earth ground. This part of the circuit is called the ground return. The ground return allows the use of only one conductor, rather than two conductors, in a transmission system. A good ground connection is essential at each end of the circuit if the ground return is to be effective.

In radar, objects on the ground near the transmitter cause false echoes. This effect is especially prevalent in populated areas where there are many man-made structures. This is called the ground-return effect; it is also sometimes called ground clutter. Ground-return effects make it difficult to track a target at a low altitude. *See also* RADAR.

GROUND ROD

A ground rod is a solid metal rod, usually made of copper-plated steel, used for the purpose of obtaining a good ground connection for electrical or electronic apparatus.

Ground rods are available in a wide variety of diameters and lengths. The most effective ground is obtained by driving one or more rods at least 8 feet into the soil, well away from the foundations of buildings. Smaller ground rods are sometimes adequate for very low current or low-power installations in locations where the soil conductivity is excellent. *See also* GROUND, GROUND CONNECTION.

GROUND WAVE

In radio communication, the ground wave is that part of the electromagnetic field that is propagated parallel to the surface of the earth. The ground wave actually consists of three distinct components: the direct or line-of-sight wave, the reflected wave, and the surface wave. Each of these three components contributes to the received ground-wave signal.

The direct wave travels in a straight line from the transmitting antenna to the receiving antenna. At most radio frequencies, the electromagnetic fields pass through objects such as trees and frame houses with little attenuation. Concrete-and-steel structures cause some loss in the direct wave at higher frequencies. Obstructions such as hills, mountains, or the curvature of the earth cut off the direct wave completely.

A radio signal can be reflected from the earth or from certain structures such as concrete-and-steel buildings. The reflected wave combines with the direct wave (if any) at the receiving antenna. Sometimes the two are exactly out of phase, in which case the received signal is extremely weak. This effect occurs mostly in the very high frequency range and above.

The surface wave travels along the earth, and occurs only with vertically polartized energy at the very low, low, medium, and high frequencies. Above 30 MHz, there is essentially no surface wave. At the very low and low frequencies, the surface wave propagates for hundreds or even thousands of miles. Sometimes the surface wave is called the ground wave, in ignorance of the fact that the direct and reflected waves can also contribute to the ground wave in very short-range communications. The actual ground-wave signal is the phase combination of all three components at the receiving antenna. *See also* DIRECT WAVE, REFLECTED WAVE, SURFACE WAVE.

GUARD BAND

In a channelized communications system, a guard band or guard zone is a small part of the spectrum allocated for minimizing interference between stations on adjacent channels. It is desirable to have some unused frequency space between channels so the sidebands of one signal will not cause interference in the passband of a receiver tuned to the next channel.

The width of the guard band should be sufficient to allow for the imperfections in the bandpass filters of the transmitter and receiver circuits. However, excessive guard-band space is wasteful of the spectrum.

In some bands, notably the international shortwave broadcast band, there is often no guard zone at all. Stations in this part of the spectrum usually have a bandwidth of 10 kHz but can be spaced only 5 kHz apart, resulting in severe adjacent-channel interference. *See also* ADJACENT-CHANNEL INTERFERENCE.

GUNN DIODE

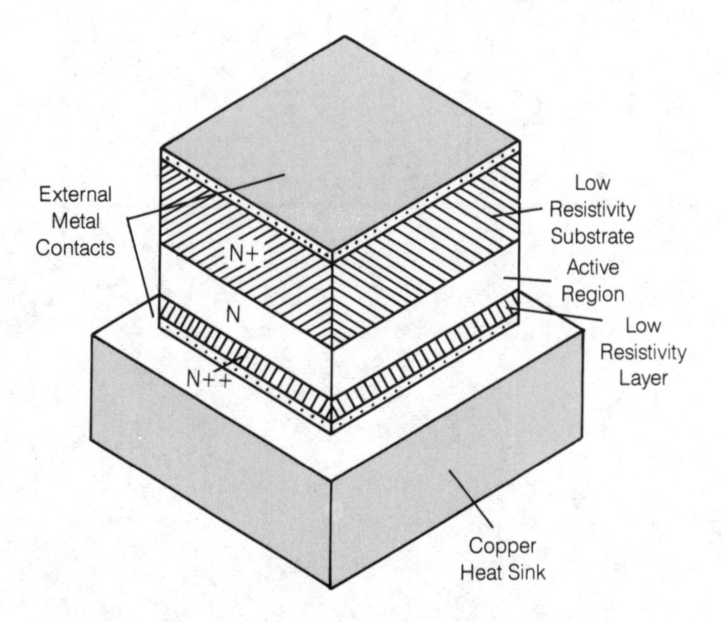

GUNN DIODE: A negative resistance device without a PN-junction, the Gunn diode self-oscillates in a tunable cavity to generate microwave signals.

A Gunn diode or transferred-electron oscillator (TEO) is a solid-state, negative-resistance microwave oscillator that can be used as a microwave local oscillator or as a radio-frequency generator in a transmitter. The diodes are designed for direct mounting in a waveguide cavity, and frequency can be adjusted with a mechanical tuning screw. Gunn diodes can produce from 15 mW to 1 W of output power. They exhibit low noise and good frequency stability under stable temperature and voltage conditions. Gunn diodes are widely used in microwave data links, low-power frequency-modulated and continuous-wave radars, intrusion detection alarm systems, and as parametric amplifier pump sources.

Most Gunn diodes are made from three-layer epitaxially grown gallium arsenide (GaAs) crystals. They also have been made from indium phosphide (InP). The figure shows a Gunn diode inverted from the position in which the layers were grown. The top N+ layer is acually the low-resistivity substrate that may be more than 50 microns (μm) thick. The N layer is the active region whose thickness determines the optimum center frequency of the device. This is about 18 μm at 6 GHz, 10 μm at 10 GHz and 6 μm at 18 GHz.

The third N++ layer has even lower resistivity than the substrate. Only about 1 to 2 μm thick, the layer improves device yield and life. It isolates the active layer from impurities that could enter during deposition of the gold contact. The contact is necessary to bond the diode to the copper heatsink. The substrate bottom (top layer) also is metallized to minimize thermal and electrical resistance.

Commercially available Gunn diodes are rated:

- 250 mW at 26 to 40 GHz
- 100 mW at 40 to 60 GHz
- 50 mW at 60 to 75 GHz
- 20 mW at 75 to 90 GHz

The diodes are mounted in ceramic packages 0.035 inch in diameter on copper heatsinks 3⁄16 inch in diameter with an overall height of about 3⁄16 inch.

Two other solid-state, negative-resistance microwave oscillators in use are the ESAKI or tunnel diode and the read or impatt diode. *See* IMPATT DIODE, TUNNEL DIODE.

HAIRPIN MATCH

A hairpin match is a means of matching a half-wave, center-fed radiator to a transmission line. The hairpin match is especially useful in the driven element of a Yagi antenna, where the feed-point impedance is lowered by the proximity of parasitic elements.

The hairpin match requires that the radiating element be split at the center. It also requires the use of a balanced feed system; if a coaxial line is used, a balun is needed at the feed point. The hairpin match consists of a section of parallel-wire transmission line, short circuited at the far end (see illustration). The length of the section is somewhat less than ¼ wavelength, and thus it appears as an inductance. *See* YAGI ANTENNA in the ANTENNA DIRECTORY.

A sliding bar can be used to vary the length of the section. The section itself should be perpendicular to the driven element; this generally necessitates mounting it along the boom of the Yagi antenna. The use of a hairpin match causes a slight lowering of the resonant frequency of a radiating element. Therefore, the radiator must be shortened by a few percent to maintain operation at the desired frequency. When the length of the hairpin section and the length of the radiating element are just right, a nearly perfect match can be obtained with either 50-ohm or 75-ohm coaxial feed lines. *See also* IMPEDANCE MATCHING.

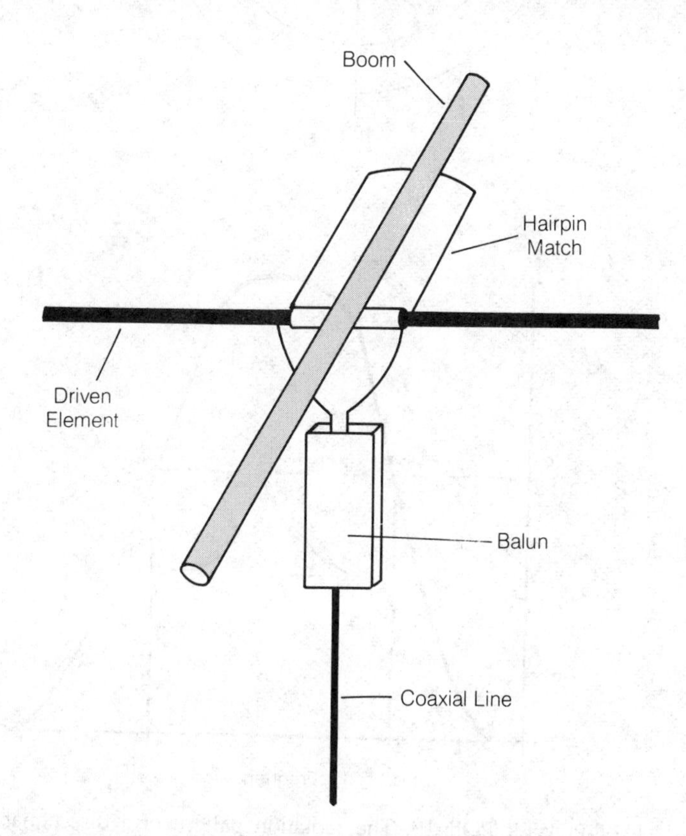

HAIRPIN MATCH: An illustration of a hairpin match installed on the driven element of a Yagi antenna.

HALF-ADDER

A half-adder is a digital logic circuit with two input terminals and two output terminals. The output terminals are called the sum and carry outputs.

The sum output of a half-adder circuit is the exclusive-OR function of the two inputs. That is, the sum output is 0 when the inputs are the same and I when they are different. The carry output is the AND function of the two inputs: It is 1 only when both inputs are 1. (See the table.)

The half-adder circuit differs from the adder in that the half-adder will not consider carry bits from previous stages. There are several different combinations of logic gates that can function as half adders. *See also* ADDER, BINARY-CODED NUMBER.

HALF-ADDER:
Sum and carry outputs as a function of logic input.

INPUTS X	INPUTS Y	SUM OUTPUT	CARRY OUTPUT
0	0	0	0
0	1	1	0
1	0	1	0
1	1	0	1

HALF BRIDGE

A half-bridge is a form of rectifier circuit, similar to the bridge rectifier except that two of the diodes are replaced by resistors (see illustration). The two resistors have equal values. The voltages at either end of the series combination are always equal and opposite. Thus, at the center point, the voltage is zero, and this point is generally grounded. In the schematic, a positive voltage appears at the output. By reversing the diodes, a negative-voltage supply is obtained.

The efficiency of the half-bridge circuit is less than

that of the conventional bridge rectifier because the resistors tend to dissipate some power as heat as the current flows through them. The half-bridge operates over the entire alternating-current input cycle, and is therefore a form of full-wave rectifier. *See also* BRIDGE RECTIFIER.

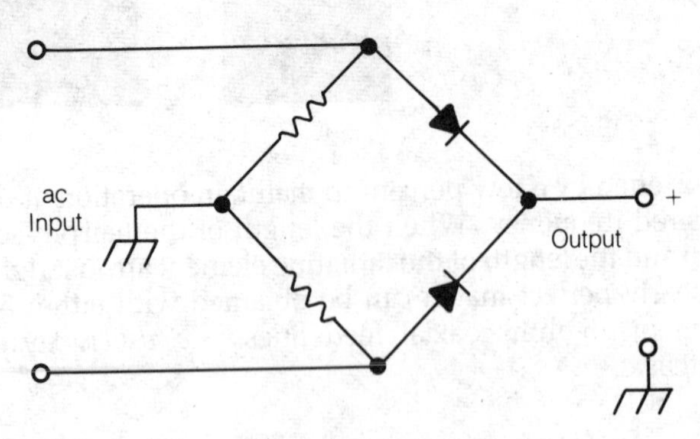

HALF BRIDGE: A half-bridge rectifier circuit.

HALF CYCLE

In any alternating-current system, a complete cycle occurs between any two identical points on the waveform. These points can be chosen arbitrarily; they can be positive peaks, zero-voltage points, negative peaks, or any other point. The complete cycle requires a certain length of time, *P*, which is called the period of the wave.

A half cycle is simply any part of the waveform that occurs during a time interval of *P*/2, if the period is *P*. Usually, the term *half-cycle* is used in reference to either the negative or positive portion of a sine-wave alternating-current waveform. A half-cycle represents 180 electrical degrees. *See also* ALTERNATING CURRENT, CYCLE, ELECTRICAL ANGLE.

HALF LIFE

The half life of a radioactive substance is the time it takes for the substance to decay to a point where its radiation intensity is 50 percent of its initial value. Half life is a convenient expression of the speed of radioactive material disintegration.

The half life of any particular substance is always the same. It does not depend on the intensity of the radiation. If a substance has a half life of 10 years, for example, its radioactivity will be ½ as great after 10 years as at the time of first measurement; after 20 years it will be ¼ as great; after 30 years, ⅛ as great; and so on. *See also* RADIOACTIVITY.

HALF-POWER POINTS

The sharpness of an antenna directive pattern, or of the selective response of a bandpass filter, can be specified in terms of the half-power points. In the case of a directive antenna, the variable parameter is compass direction (see A in the illustration). In the case of a bandpass filter, the variable parameter is frequency, as at B.

In antenna systems, the reference power level is the effective radiated power in the favored direction. This can occur in more than one direction. In most parasitic arrays, the favored direction is a single compass point; in the case of the half-wave dipole, the favored direction is two compass points; an unterminated longwire generally has favored directions at four compass points. The effective radiated power at the half-power points is 3 decibels below the level in the favored direction. The field strength, in volts per meter, is 0.707 times the field strength in the favored direction. The angle, in degrees, between the half-power points of a single lobe is called the beamwidth. *See also* BEAMWIDTH, FIELD STRENGTH.

For a bandpass filter, the half-power points are the frequencies at which the power from the filter drops 3 decibels below the power output at the center of the passband. The bandwidth of the filter can be specified in terms of the half-power points, but more often it is given in terms of the 6-decibel attenuation points and the 60-decibel attenuation points. This combined figure

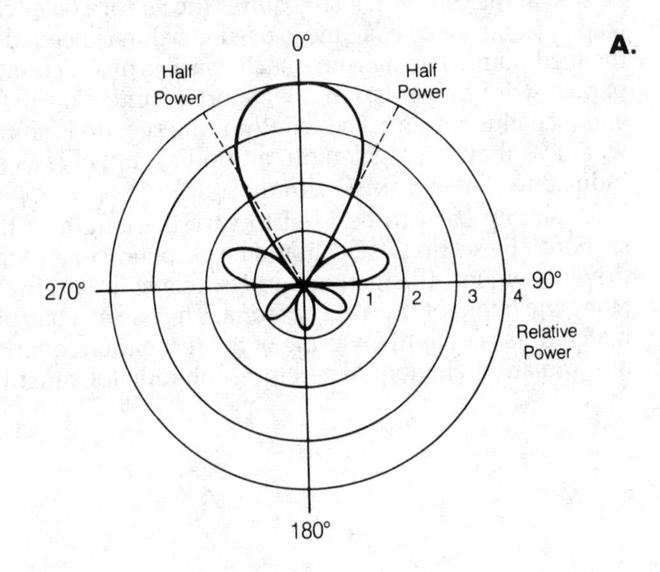

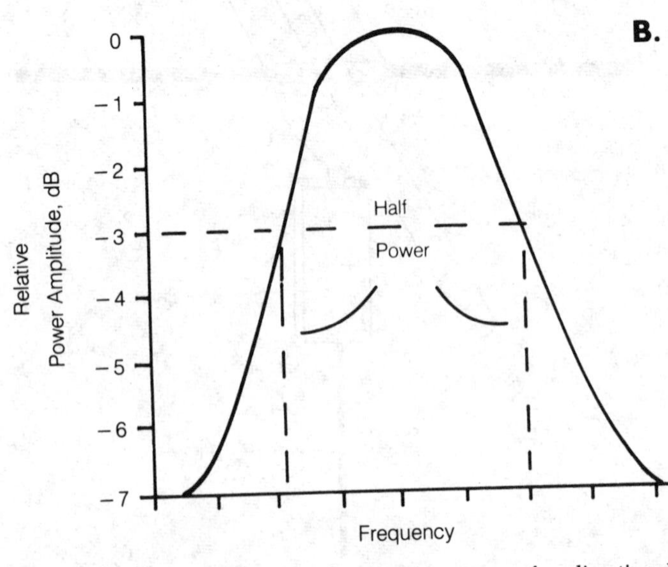

HALF-POWER POINTS: The radiation pattern of a directional antenna is shown at (A) and a bandpass frequency response at half-power points is shown at (B).

gives an indication of the skirt selectivity as well as the actual bandwidth. *See also* BANDPASS FILTER, BANDPASS RESPONSE, SHAPE FACTOR.

HALF STEP

A half step is a change in the frequency of an audio tone, or a difference in the frequencies of two audio tones, equivalent to 6 percent of the frequency of the lower-pitched tone. Any two immediately adjacent keys on a piano are a half step apart in frequency. In most musical arrangements, every tone frequency is at a defined half step with respect to the other tones. However, in some cultures, smaller intervals can be found in musical tones.

HALF-WAVE ANTENNA

See ANTENNA DIRECTORY.

HALF-WAVE DIPOLE

See HALF-WAVE ANTENNA in the ANTENNA DIRECTORY.

HALF-WAVE RECTIFIER

A half-wave rectifier is the simplest form of rectifier circuit. It consists of a diode in series with one line of an alternating-current (ac) power source, and a transformer (if necessary) to obtain the desired voltage. A is a schematic diagram of a half-wave rectifier circuit designed to supply a positive output voltage. The diode can be reversed to provide a negative output voltage.

The output of the half-wave rectifier contains only one half of the ac input cycle, as shown at B. The other half is simply blocked. The half-wave rectifier gets its name from the fact that it operates on only one half of the input cycle.

Half-wave rectifiers have the advantage of being an extremely simple design. An unbalanced ac input source can be used; in some instances, no transformer is needed and the direct-current (dc) voltage can be derived from a wall outlet. However, the half-wave rectifier has poor voltage-regulation characteristics. The pulsating output is more difficult to filter than that of the full-wave rectifier. Half-wave rectifier circuits are often used in situations where the current drain is low and the voltage regulation need not be especially precise. *See also* BRIDGE RECTIFIER, FULL-WAVE RECTIFIER.

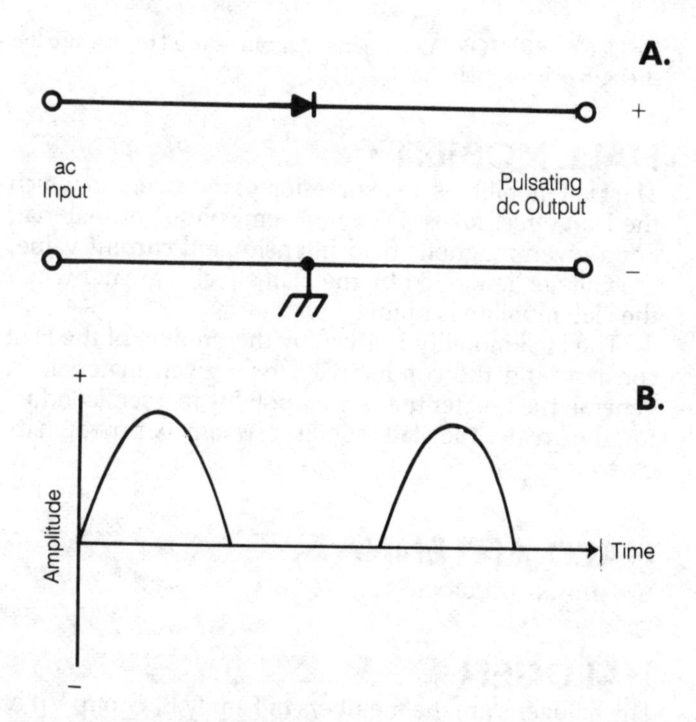

HALF-WAVE RECTIFIER: A simple half-wave rectifier circuit (A) and its output waveform (B).

HALF-WAVE TRANSMISSION LINE

A half-wave transmission line is a section of electromagnetic feed line that measures an electrical half wavelength. The physical length L of such a transmission-line section, for a frequency f in megahertz, is given in feet by:

$$L = \frac{492v}{f}$$

where v is the velocity factor of the line (*see* VELOCITY FACTOR). The length L in meters is:

$$L = \frac{150v}{f}$$

A half-wave section of transmission line has certain properties that make it useful as a tuned circuit. The impedance at one end of such a line is exactly the same, neglecting line loss, as the impedance at the other end. This is true, however, only within a narrow range of frequencies, centered at the half-wave resonant frequency. If the far end of a half-wave transmission line is an open circuit, the line behaves as a parallel-resonant tuned circuit. If the far end is short-circuited, the section behaves as a series-resonant tuned circuit.

At frequencies above approximately 20 MHz, where half-wave sections of transmission line have reasonable length, these sections are often used in place of coils and capacitors as resonant circuits. If the line loss is reasonably low, the Q factor of the half-wave transmission line is very high. The half-wave transmission line may be either balanced or unbalanced, depending on the circuit in which it is used. *See also* Q FACTOR.

HALL EFFECT

When a current-carrying electrical conductor is placed in a magnetic field, a voltage can develop between one side of the conductor and the other. For this to happen, the magnetic lies of force must be perpendicular, or nearly perpendicular, to the line containing the conductor. The voltage then appears at right angles to the magnetic lines of force. If the conductor is a strip of metal or semiconductor, and the magnetic lines of force are perpendicular to the strip, then the voltage will appear between opposite edges of the strip (see illustration). This is known as the Hall effect. The electric-field intensity E_{HV}, generated by the Hall effect, is given by the formula:

$$E_{HV} = \frac{B \times I_C \times K_H}{t}$$

where I_C is the current in the conductor, B is the magnetic-field strength, K_H is a constant called the Hall constant, and t is the thickness of the material.

In some metals, the voltage displays a polarity opposite to that in other metals under the same conditions. The polarity depends on the atomic structure of the metal. A device called a Hall generator makes use of the Hall effect in semiconductors to measure magnetic-field strength. The voltage is quite small unless the current and magnetic field are extremely intense. However, the voltage can produce current in an external circuit. *See also* HALL GENERATOR, HALL MOBILITY.

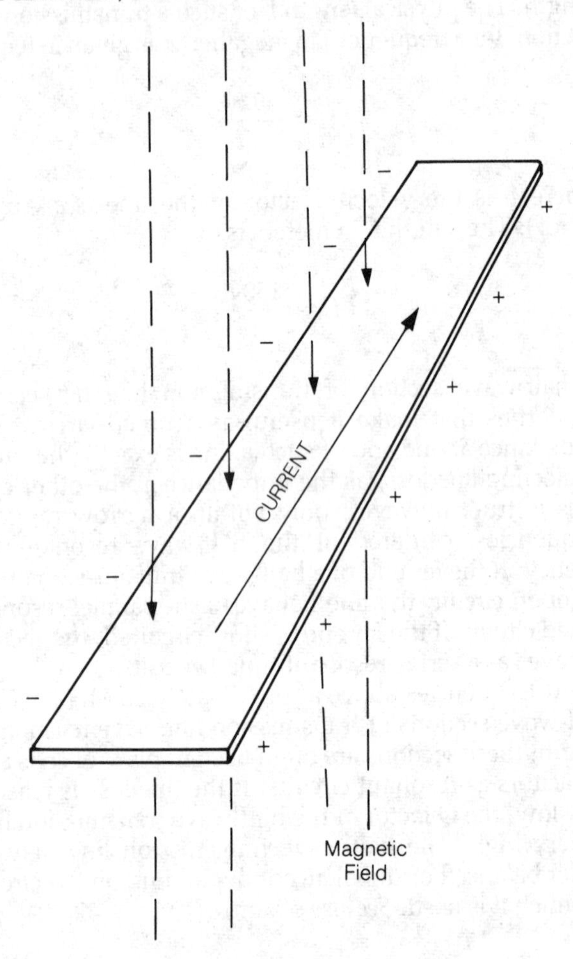

HALL EFFECT: A potential difference develops between opposite edges of the metal strip in a magnetic field demonstrating the Hall effect.

HALL GENERATOR

A Hall generator is a device that uses the Hall effect for generating a direct-current voltage in the presence of a magnetic field. A semiconductor chip is used. A battery, or other source of direct current, is connected to opposite ends of the wafer. A voltmeter is connected to the adjacent sides of the chip, as illustrated in the drawing. Some semiconductor materials are much better than others for use in the Hall generator. The greater the Hall mobility (*see* HALL MOBILITY), the better; indium antimonide is especially effective.

When a magnetic field exists in the vicinity of the wafer, so that the magnetic lines of force are perpendicular to the plane of the wafer, a voltage is induced and will register on the voltmeter. The value of the voltage is quite small, but is proportional to the current in the wafer and to the intensity of the magnetic field. The Hall generator is therefore useful for measuring the strength of a magnetic field. *See also* HALL EFFECT.

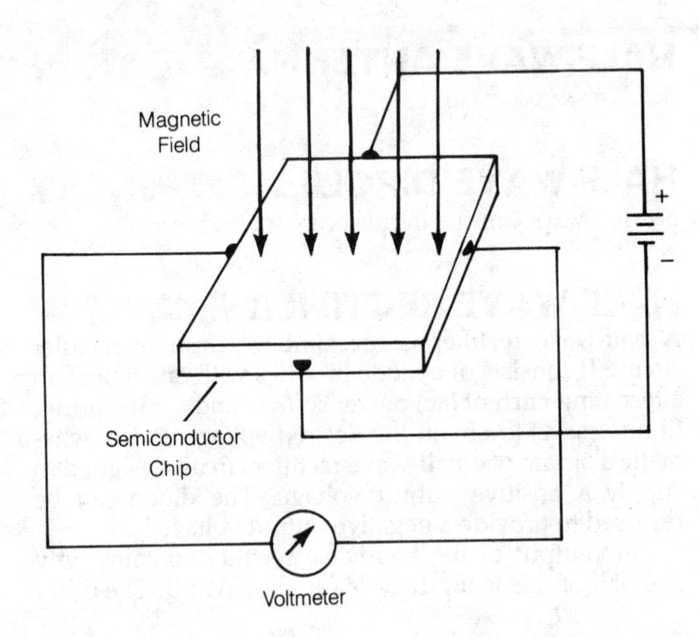

HALL GENERATOR: A Hall generator can be used to measure the strength of a magnetic field.

HALL MOBILITY

The Hall mobility is an expression of the extent to which the Hall effect takes place in a semiconductor material. For a given magnetic-field intensity and current value, the voltage generated by the Hall effect is greater when the Hall mobility is higher.

The Hall mobility is given by the product of the Hall constant and the conductivity for a given material. In general, the greater the carrier mobility in a semiconductor, the greater the Hall mobility. *See also* HALL EFFECT, HALL GENERATOR.

HALO ANTENNA

See ANTENNA DIRECTORY.

HALOGEN

The halogens are the members of Family B, Group VII of the Periodic Table. The halogens are: bromine, chlorine, fluorine, and iodine. These elements are highly reactive and have similar characteristics. Each of them contains seven electrons in its outer orbit. Each element possesses powerful electronegative or nonmetallic properties and tends to combine energetically with any electropositive substance.

HANDSET

A handset is a telephone type communications instrument, containing a microphone and an earphone in a configuration that can be easily held against the side of the head. Some portable radio transceivers use handsets, but the most familiar handset is that found in the common household telephone (see the photograph).

The telephone handset may have the dialing apparatus contained within itself, as shown, or the dialer may be in the base of the telephone. *See also* TELEPHONE SYSTEM.

HANDSET: A telephone handset.

HANDSHAKING

In a digital communications system, accuracy can be improved by synchronizing the transmitter and receiver precisely before the beginning of data transfer. This is called handshaking. The process may be repeated at intervals to maintain synchronization.

Handshaking is becoming more universal in electronic communications because it dramatically improves the signal-to-noise ratio. In general, the more frequently the handshaking operation is done in the process of signal transmission, the better the signal-to-noise ratio, although there is a point of diminishing returns.

In some digital systems, an independent reference standard, such as a time and frequency station, is used to synchronize the digital signals between the transmitter and receiver. This is called coherent or synchronized digital communications. *See also* SYNCHRONIZED COMMUNICATIONS.

HARDWARE

In computers, *hardware* is a term applied to the actual circuitry that makes up the system (wiring, circuit boards, integrated circuits, diodes, transistors, and resistors) as well as the peripheral devices (display terminals, printers, and keyboards). The computer programs, in contrast, are called software or firmware, depending on whether or not they are easily modified (*see* FIRMWARE, SOFTWARE).

HARMONIC

Any signal contains energy at multiples of its frequency, in addition to energy at the desired frequency. The lowest frequency component of a signal is called the fundamental frequency; all integral multiples are called harmonic frequencies, or simply harmonics.

In theory, a pure sine wave contains energy at only one frequency, and has no harmonic energy. In practice, this ideal is never achieved. All signals contain some energy at harmonic frequencies, in addition to the energy at the fundamental frequency. The signal having a frequency of twice the fundamental is called the second harmonic. The signal having a frequency of three times the fundamental is called the third harmonic, and so on.

Wave distortion always results in the generation of harmonic energy. While the nearly perfect sine wave has very little harmonic energy, the sawtooth wave, square wave, and other distorted periodic oscillations contain large amounts of energy at the harmonic frequencies. Whenever a sine wave is passed through a nonlinear circuit, harmonic energy is produced. A circuit designed to deliberately create harmonics is called a harmonic generator or frequency multiplier.

Harmonic output from radio transmitters is undesirable, and the designers and operators of such equipment often go to great lengths to minimize this energy as much as possible. *See also* FREQUENCY MULTIPLIER, FUNDAMENTAL FREQUENCY, HARMONIC SUPPRESSION.

HARMONIC GENERATOR

See FREQUENCY MULTIPLIER.

HARMONIC SUPPRESSION

Harmonic suppression is an expression of the degree to which harmonic energy is attenuated, with respect to the fundamental frequency, in the output of a radio transmitter. Harmonic suppression is also used to denote the process of minimizing harmonic energy in the output of a signal generator, especially a radio transmitter. Harmonics are undesirable in the output of this equipment because they cause interference to other services and operations.

If a given harmonic signal has a power level of *Q* watts in the output circuit of a transmitter, while the fundamental-frequency output is *P* watts, then the harmonic suppression *S* in decibels is given by:

$$S = 10 \log_{10} (P/Q)$$

Harmonic suppression may be accomplished in three ways. The most frequently used method is the insertion of one or more tuned bandpass filters in the output of the transmitter. The filter frequency is centered at the operating frequency of the transmitter. This provides additional harmonic attenuation in the output of the final amplifier.

The second method of obtaining harmonic suppression is the insertion of a lowpass filter in the transmitter output. The cutoff frequency of the filter should be the lowest that will result in negligible attenuation at the fundamental frequency.

The third method of obtaining harmonic suppression is the use of band-rejection filters, also sometimes called traps. The trap offers harmonic attenuation at only one frequency; the bandpass and lowpass filters provide rejection of all harmonics above the fundamental. However, the trap circuit often gives better results at the design frequency.

If an amplifier is intended to operate as a linear amplifier, the bias and drive should be maintained at the proper levels to ensure minimal generation of harmonic energy. The Class C amplifier causes more harmonics to be produced than other types of amplifiers. *See also* BANDPASS FILTER, BAND-REJECTION FILTER, HARMONIC, LOWPASS FILTER.

HARTLEY OSCILLATOR

See OSCILLATOR CIRCUITS.

HASH NOISE

Hash, or hash noise, is a form of electrical noise generated by gas and mercury-vapor tubes. It is generally broadband in nature, and may also be called white noise. Hash noise is also generated by semiconductor rectifiers.

Hash noise can be a problem with sensitive receiving circuits, since some of the noise occurs at radio frequencies. The hash noise can be suppressed by housing the power supply components in a shielded enclosure and placing radio-frequency chokes and bypass capacitors in the power-supply leads at the points where they leave the enclosure. *See also* WHITE NOISE.

HAY BRIDGE

The Hay bridge is a circuit designed for measuring the value of an unknown inductance. The Hay bridge also gives an indication of the Q factor of the inductor under test. A typical Hay bridge is illustrated in the schematic diagram.

Two balance controls, consisting of potentiometers, are used in the bridge. One control is calibrated in terms of inductance (millihenrys or microhenrys). The other is calibrated in terms of the Q factor, giving a measure of the reactance-to-resistance ratio of the tested coil. Current adjustment is indicated by a null meter or headset.

The Hay bridge is generally used for the measurement of relatively large inductances. The signal generator is thus an audio device in most cases. At very high frequencies, the inductance of the component leads affects the accuracy of the bridge. If the resistances (R_a R_b), as shown in the diagram, are given in ohms, and the capacitance (C_s) is specified in farads, then the inductance (L_x) of the coil in henrys is given by:

$$L_x = \frac{R_a R_b C_s}{(1 + \omega^2 C_s^2 R_s^2)}$$

where ω is the angular frequency in radians per second.

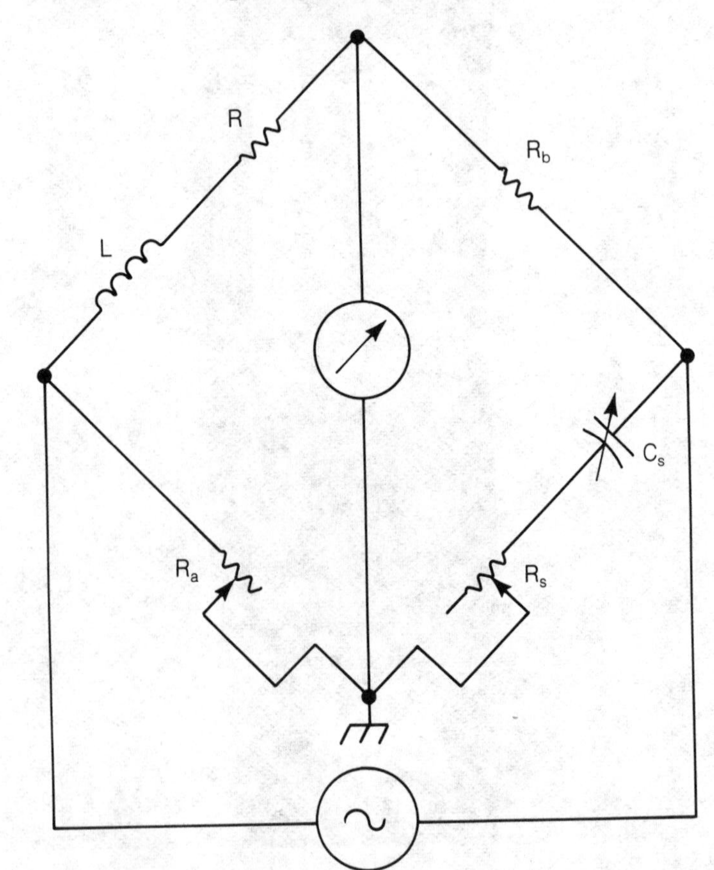

HAY BRIDGE: A circuit for measuring large inductances.

The angular frequency is approximately equal to 6.3 times the frequency in hertz.

The Q factor of the unknown combination *L* and *R* is:

$$Q_x = \frac{1}{\omega C_s R_s}$$

where ω is, again, the angular frequency in radians per second. *See also* INDUCTANCE, Q FACTOR.

HEADPHONE

A headphone is a pair of earphones, designed to be worn over the head so that both ears are covered. Headphones provide excellent intelligibility in communications because external sounds are minimized.

Some headphones operate with both earphones connected in parallel or in series. These devices are called monaural headphones. Other headphones have two independent earphones, and are used for listening to high-fidelity stereo (see photograph).

Headphones are available with impedances ranging from 4 ohms (Ω) to more than 2,000 Ω. Some headphones have a wide-range frequency response; these are ideal for high-fidelity applications. Other headphones have peaked responses, intended for communications. The proper impedance and frequency response should be chosen for the desired situation. Most headphones are of the dynamic type, and resemble pairs of miniature speakers. *See also* EARPHONE, SPEAKER.

HEADPHONE: A typical headphone for use with a portable stereo radio or tape player.

HEATSINK

A metal radiator that helps to remove excess heat from electronic devices by means of conduction, convection, and radiation is called a heatsink. The use of a heatsink increases the power-dissipating capability of a transistor, tube, integrated circuit, or other active device. Some heatsinks are small, and fit around or under a single miniature component. Other heatsinks are large, and can be used to mount several different components at once.

The heatsink is made of metal, usually aluminum, an excellent conductor of heat. The heat is carried away from the component by means of thermal conduction. The surface of the heatsink is deliberately designed to facilitate radiation and convection (see the figure on p. 438). This is accomplished with a finned surface. Heatsinks are often painted black to enhance the radiation.

When a heatsink is used, it is necessary that it be thermally bonded to the device. Otherwise, the benefits will be lost. Silicone compound is generally used for thermal bonding. If it is necessary to electrically insulate the component from the heatsink, a flat piece of mica may be placed between the component and the heatsink.

HEAVISIDE LAYER

See E LAYER, F LAYER, IONOSPHERE.

HEIGHT ABOVE AVERAGE TERRAIN

At very high frequencies and above, antenna performance is dependent on the height. The higher the antenna, the better the performance in most cases. In locations where the terrain is irregular, it is often difficult to determine the effective height of an antenna for communications purposes. Thus, the height-above-average-terrain (HAAT) figure has been devised.

To determine accurately the HAAT of an antenna, a topographical map is necessary. The location of the antenna must be found on the map. A circle is then drawn around the point corresponding to the location of the antenna. This circle should have a radius large enough to encompass at least several hills and valleys in the vicinity of the antenna. Within the circle, the average elevation of the terrain is calculated. This process involves choosing points within the circle in a grid pattern, adding up all of the elevation figures, and then dividing by the number of points in the pattern.

Once the average elevation, *A*, is found in the vicinity of the antenna, the elevation *B* of the antenna must be found by adding the height of the supporting structure to the ground-elevation figure at the antenna location. The HAAT is then equal to $B - A$.

The Federal Communications Commission imposes certain limitations, in some services, on antenna HAAT. This is especially true for repeater antennas. *See also* ANTENNA, REPEATER.

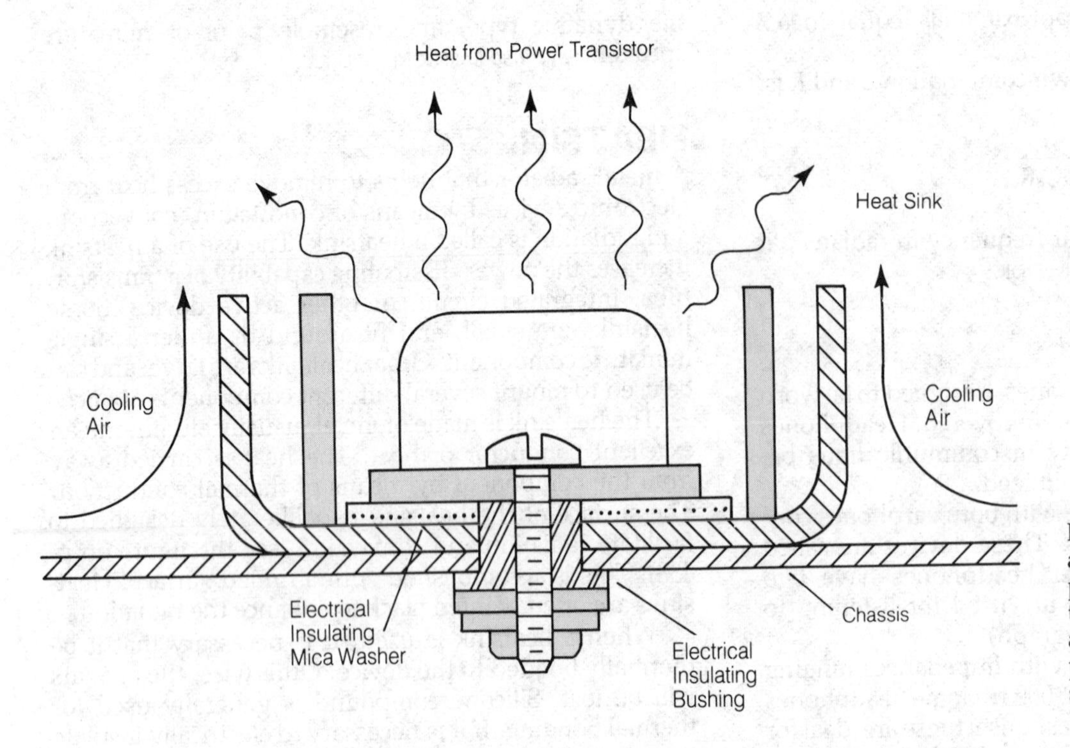

HEATSINK: Fins increase the radiating surface of a heatsink to improve heat exchange between a transistor case and surrounding air. Insulated washer and bushing provide electrical isolation of transistor from heatsink and chassis.

HELICAL ANTENNA

See ANTENNA DIRECTORY.

HELICAL FILTER

A helical filter, or helical resonator, is a quarter-wave filter used at very high and ultrahigh frequencies as a bandpass filter. A shielded enclosure contains a coil with one end connected to the enclosure and the other end either left free or connected to a variable capacitor.

The photograph on p. 439 shows a three-section helical filter designed for operation in the 142 to 150 MHz band. The three filter sections are tuned to slightly different frequencies, resulting in very little attenuation within the passband, but steep skirts and high attenuation outside the passband. Helical filters provide a high Q factor, which is important in the reduction of out-of-band interference. *See also* BANDPASS FILTER, Q FACTOR.

HELMHOLTZ COIL

A Helmholtz coil is an inductive coil that provides continuously variable phase shift for an alternating-current signal. Two primary windings are oriented at right angles and split into two sections, as shown in the illustration. The currents in the two primary coils differ by 90 degrees; this phase shift is provided by a resistor and capacitor. The secondary coil is mounted on rotatable bearings. As the secondary coil is turned through one complete rotation, the phase of the signal at the output terminals changes continuously from 0 to 360 degrees. Any desired signal phase may be chosen by setting the coil to the proper position.

The Helmholtz coil works because the fields from the primary windings add together vectorially. The magnitudes of the two component vectors change in the secondary coil as the secondary coil is turned. The Helmholtz coil is frequency sensitive and it will work at only one frequency. To change the operating frequency, the values of the resistor and capacitor must be changed to provide a 90-degree phase difference between the two primary coils. *See also* PHASE ANGLE.

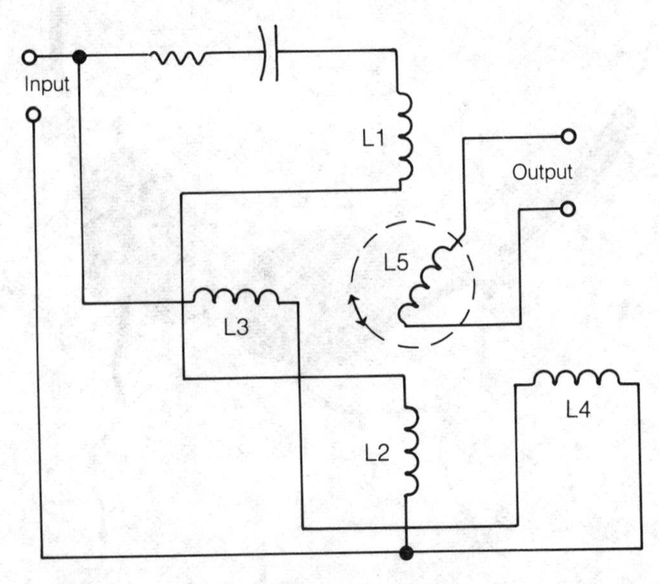

HELMHOLTZ COIL: The Helmholtz coil provides continuously variable adjustment of alternating current phase.

HELICAL FILTER: A three-section helical filter that will operate in the 142 to 150 MHz band range.

HENRY

The henry, abbreviated H, is the standard unit of inductance. In a circuit in which the current is changing at a constant rate of 1 ampere per second, an inductance of 1 henry results in the generation of 1 volt of potential difference across an inductor.

The henry is an extremely large unit of inductance. It is rare to find a coil with a value of 1 H. Therefore, inductance values are generally given in millihenrys (mH), microhenrys (μH), or nanohenrys (nH). An inductance of 1 mH is equal to 0.001 H; 1 μH is 0.001 mH; 1 nH is 0.001 μH. *See also* INDUCTANCE.

HEPTODE

A heptode is a vacuum tube with seven internal elements. In addition to the plate and cathode, there are usually five grids in the heptode. Alternatively, there may be one or more control electrodes for accelerating the electron beam.

HERMAPHRODITIC CONNECTOR

A hermaphroditic connector is an electrical plug that mates with another plug exactly like itself. This connector has an equal number of male and female contacts.

Hermaphroditic connectors can be put together in only one way. This makes them useful in polarized circuits such as direct-current power supplies. *See also* CONNECTOR.

HERMETIC SEAL

Some electronic components are susceptible to damage or malfunction from moisture in the air. As a result, these components—including relays, piezoelectric crystals, and certain semiconductor devices—are often enclosed in hermetically sealed cases. A hermetic seal is an airtight, durable seal. The common oscillator crystal, housed in a metal can, is hermetically sealed.

A hermetic seal must be long lasting and physically rugged. Hermetic sealing is achieved by welding or brazing metal cases, brazing or soldering metalized ceramic cases, and heat sealing glass envelopes. The interior of a hermetically sealed enclosure may be filled with an inert gas such as helium to retard further the deterioration of the component or components inside.

HERTZ

Hertz (Hz) is the standard unit of frequency. A frequency of 1 complete cycle per second is a frequency of 1 Hz. The term *hertz* is now used instead of the former term *cycles per second*.

In radio communication, signals are typically thousands, millions, or billions of hertz in frequency. A frequency of 1,000 Hz is called 1 kilohertz (kHz); a frequency of 1,000 kHz is 1 megahertz (MHz); a frequency of 1,000 MHz is gigahertz (GHz). Sometimes the terahertz (THz) is used as a measure of frequency; 1 THz = 1,000 GHz.

The angular frequency in radians per second is equal to approximately 6.3 times the frequency in hertz. *See also* FREQUENCY.

HETERODYNE

The term *heterodyne* can refer to either of two subjects. A heterodyne is a mixing product resulting from the combination of two signals in a nonlinear component or circuit. Mixing is sometimes called heterodyning.

When two waves having frequencies f and g are combined in a nonlinear component or circuit, the original frequencies appear at the output along with energy at two new frequencies. These frequencies are the sum and difference frequencies, $f + g$ and $f - g$. They are sometimes called heterodyne frequencies. A tuned circuit can be used to choose either the sum or the difference frequency. In the mixer, $f + g$ and $f - g$ are usually much different. In the heterodyne or product detector, they might be very close together.

Heterodyning can occur whenever two signals are present in the same medium. No circuit is perfectly linear, and some distortion always takes place. A diode, capable of effectively handling frequencies f and g, or a transistor biased to cutoff, is ideal for heterodyning purposes. *See also* FREQUENCY CONVERSION, FREQUENCY CONVERTER, HETERODYNE DETECTOR, MIXER.

HETERODYNE DETECTOR

A heterodyne detector is a detector that operates by beating the signal from a local oscillator against the received signal. This form of detector is also called a product detector. It is required for the reception of continuous-wave, frequency-shift-keyed, and single-sideband signals.

The incoming signal information is extracted by heterodyning, resulting in audible difference frequencies. In the case of a continuous-wave signal, the difference can be as small as about 100 Hz or as large as about 3 kHz; the same is true with frequency-shift keying. For single-sideband reception, the local oscillator frequency should correspond to the frequency of the suppressed carrier.

A heterodyne detector is used in many superheterodyne receivers and all direct-conversion receivers. *See also* DIRECT-CONVERSION RECEIVER, SUPERHETERODYNE RADIO RECEIVER.

HETERODYNE FREQUENCY METER

A heterodyne frequency meter is a circuit similar to a direct-conversion receiver used for measuring unknown frequencies. A calibrated local oscillator is used, along with a mixer and amplifier, as shown in the diagram.

The signal of unknown frequency is fed into the mixer along with the output of the local oscillator. It is helpful to have some idea of the frequency of the signal beforehand, because harmonics of the local oscillator will produce false readings. As the local oscillator is tuned, heterodynes occur in the output. These heterodynes occur at frequencies f, $f/2$, $f/3$, $f/4$, and so on, as read on the local-oscillator calibrated scale. The correct reading is f, the highest frequency. Readings can also be obtained at $2f$, $3f$, $4f$, and so on, corresponding to harmonics of the input signal; but these components are actually present, while the lower-frequency indications are false signals. (Some heterodyne frequency meters have tuned input circuits, which track along with the local oscillator fundamental frequency, reducing harmonic effects.) The correct reading usually gives the loudest heterodyne. The local oscillator should be adjusted for zero beat before the final reading is taken.

All receivers with product detectors can be used as heterodyne frequency meters within their operating ranges, provided the dial is calibrated with reasonable accuracy. It is helpful to have a crystal calibrator in such a receiver, and this calibrator should be adjusted against a frequency standard, such as that of radio station WWV or WWVH. *See also* FREQUENCY MEASUREMENT, HETERODYNE, WWV/WWVH.

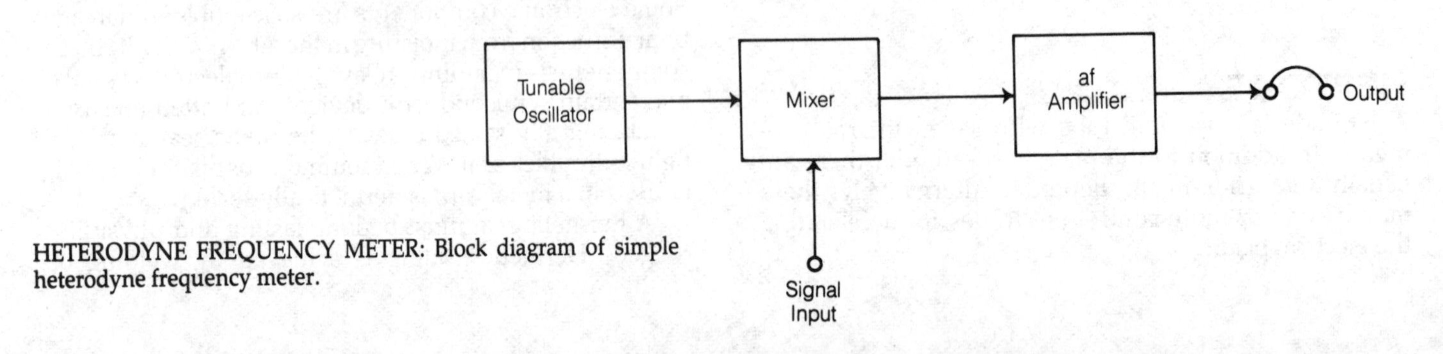

HETERODYNE FREQUENCY METER: Block diagram of simple heterodyne frequency meter.

HETERODYNE REPEATER

When a signal is amplified and retransmitted, its frequency usually must be changed, or feedback is likely to result in oscillation between the receiver and transmitter circuits. Virtually all repeaters use frequency converters, and they are therefore called heterodyne repeaters.

The input signal to the heterodyne repeater might have a frequency f. This signal is heterodyned with the output of a local oscillator at frequency g, producing mixed products at frequencies $f + g$ and $f - g$. Generally, the difference frequency is chosen. A tuned bandpass filter, centered at frequency $f - g$, serves to eliminate other signals. An amplifier tuned to frequency $f - g$ provides the output. At the input of the repeater, a notch or band-rejection filter may be used, centered at frequency $f - g$, to prevent desensitization of the receiver by the transmitter.

In general, the greater the difference between the input and output frequencies of a heterodyne repeater, the more easily the circuit is designed, and the less difficulty is encountered with receiver desensitization. *See also* FREQUENCY CONVERSION, REPEATER.

HEXODE

A hexode is a vacuum tube with six internal elements. In addition to the plate and the cathode, there can be four grids in the hexode. Alternatively, there may be one or more control electrodes for such purposes as accelerating the electron beam.

HEXADECIMAL NUMBER SYSTEM

The hexadecimal number system is a base-16 system. Hexadecimal notation is commonly used in computers because 16 is a power of 2. The numbers 0 through 9 are the same in hexadecimal notation as they are in the familiar decimal number system. However, the values 10 through 15 are represented by single digits, usually the letters A through F.

Addition, subtraction, multiplication, and division are somewhat different in hexadecimal notation. For example, in decimal language, we say that 5 + 7 = 12; but in hexadecimal, 5 + 7 = C. The carry operation does not occur until a sum is greater than F. Thus, 8 + 7 = F, and 8 + 8 = 10. *See also* NUMBER SYSTEM.

HIGHER-ORDER LANGUAGE

A higher-order or high-level computer language is the interface between the machine and the human operator. The computer itself thinks in binary or machine language, and the construction of programs in this language is tedious. The higher-order language is similar to ordinary language. A translation program, called an assembler, converts the machine language back into terms the operator can readily understand.

Examples of higher-order languages are BASIC (used for mathematical and scientific calculations), COBOL (used in business applications), and FORTRAN (used mostly by scientists and engineers). Other high-order, or high-level, languages include PL/1, Ada, APL, C, and Pascal. *See also* ASSEMBLER AND ASSEMBLY LANGUAGE, BASIC, COBOL, FORTRAN.

HIGH FIDELITY

High fidelity is a term used to describe audio systems engineered for highly accurate reproduction of the original sound. The typical high-fidelity recording and reproducing system is sensitive to sounds at all audio frequencies.

In the recording process, two channels, called the left and right channels, are used. The microphones have a wide frequency response. The recording tape must be capable of accurately storing impulses at frequencies of at least 20 kHz.

In the reproducing process, the audio amplifiers must be capable of linear operation over a range of frequencies from at least 20 Hz to 20 kHz. The speakers or headphones are an important part of the high-fidelity reproducing system. Communications type speakers or headphones are not satisfactory because of their peaked audio response.

In recent years, great progress has been made in the field of high-fidelity recording and reproducing. This has been largely due to solid-state technology. *See* MAGNETIC RECORDING, STEREO DISK RECORDING, STEREOPHONICS.

HIGH FREQUENCY

High frequency is the designation applied to the range of frequencies from 3 MHz to 30 MHz. This range corresponds to a wavelength of 100 meters to 10 meters. The high frequencies include all of the so-called shortwave bands. Ionospheric propagation is of great importance at the high frequencies. They are called decametric waves.

HIGHPASS FILTER

A highpass filter is a combination of capacitance, inductance, and/or resistance intended to produce large amounts of attenuation below a certain frequency and little or no attenuation above that frequency. The frequency at which the transition occurs is called the cutoff frequency (*see* CUTOFF FREQUENCY). At the cutoff frequency, the voltage attenuation is 3 decibels with respect to the minimum attenuation. Above the cutoff frequency, the voltage attenuation is less than 3 decibels. Below the cutoff, the voltage attenuation is more than 3 decibels.

The simplest highpass filters consist of a parallel inductor or a series capacitor. More sophisticated highpass filters have a combination of parallel inductors and series capacitors, such as the filters shown in the illustration. The filter at A is called an L-section filter; that at B is called a T-section filter. These names are derived from the geometric shapes of the filters as they appear in the schematic diagram.

Resistors are sometimes substituted for the inductors in a highpass filter. This is especially true if active devices are used, in which case many filter sections can be cascaded.

Highpass filters are used in a wide variety of situations in electronic apparatus. One common use for the highpass filter is at the input of a television receiver. The cutoff frequency of this filter is about 40 MHz. The installation of such a filter reduces the susceptibility of the television receiver to interference from sources at lower frequencies. *See also* FILTER, HIGHPASS RESPONSE.

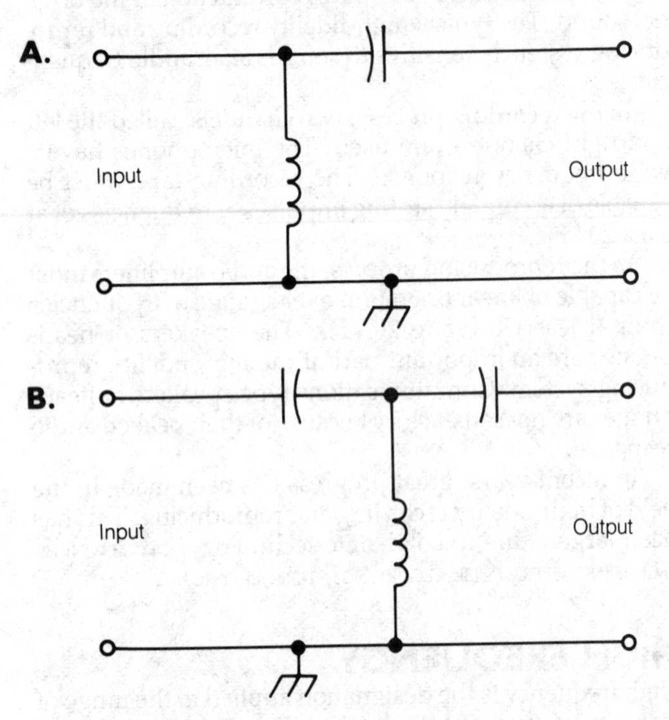

HIGHPASS FILTER: Two common forms of the unbalanced highpass filter are the L network (A) and T network (B).

HIGHPASS RESPONSE

A highpass response is an attenuation-versus-frequency curve that shows greater attenuation at lower frequencies than at higher frequencies. The sharpness of the response may vary considerably. Usually, a highpass response is characterized by a high degree of attenuation up to a certain frequency, where the attenuation rapidly decreases. Finally the attenuation levels off at near zero insertion loss. The highpass response is typical of highpass filters.

The cutoff frequency of a highpass response is that frequency at which the insertion loss is 3 decibels with respect to the minimum loss. The ultimate attenuation is the level of attenuation well below the cutoff frequency, where the signal is virtually blocked. The ideal highpass response should look like the attenuation-versus-frequency curve in the illustration. The curve is smooth, and the insertion loss is essentially zero everywhere well above the cutoff frequency. *See also* CUTOFF FREQUENCY, HIGHPASS FILTER, INSERTION LOSS.

HIGH Q

High Q is a term used to describe a filter circuit with a high selectivity. The term is usually applied to bandpass and band-rejection filters.

High-Q filters are desirable in situations where the response must be confined to a single signal, or to a very narrow range of frequencies. However, this characteristic is not always wanted. In general, high-Q circuits require low-loss inductors and capacitors, and the resistance should be as low as possible. *See also* Q FACTOR.

HISTOGRAM

A histogram is a form of graphical representation. The histogram shows occurrences of various events as a function of some variable. The histogram can be recognized by the presence of vertical rectangles of varying heights. These graphical illustrations are commonly used to show statistical distributions.

The histogram is an extremely simple and effective means of illustrating certain statistical functions. However, it is relatively imprecise. The example shown on p. 443 gives us an approximate idea of where to tune the receiver to find the most signals; but it does not show the exact frequencies of the signals. *See also* CARTESIAN COORDINATES, GRAPH.

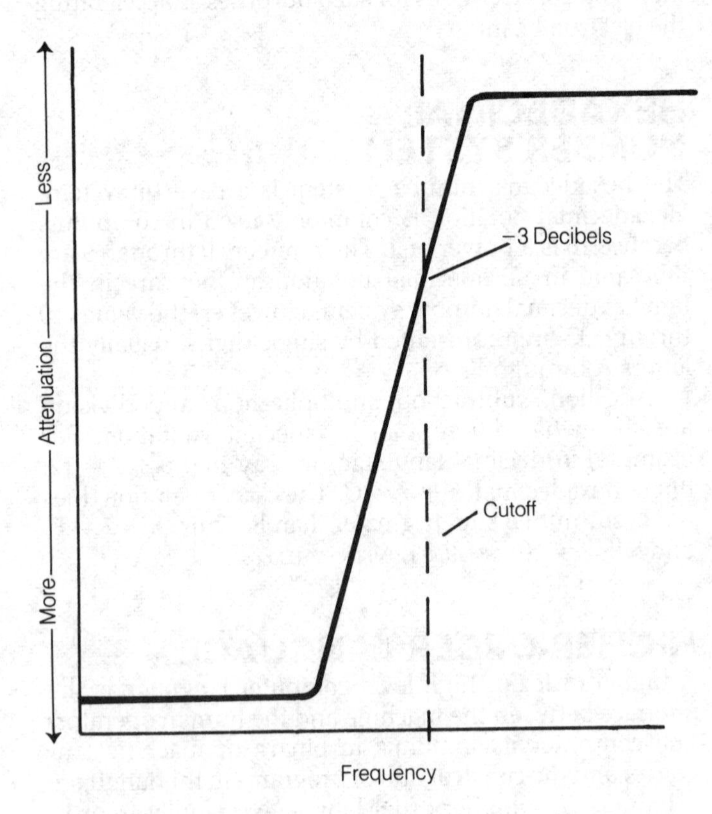

HIGHPASS RESPONSE: A filter characteristic in which attenuation is greater at the lower frequencies than at the higher frequencies.

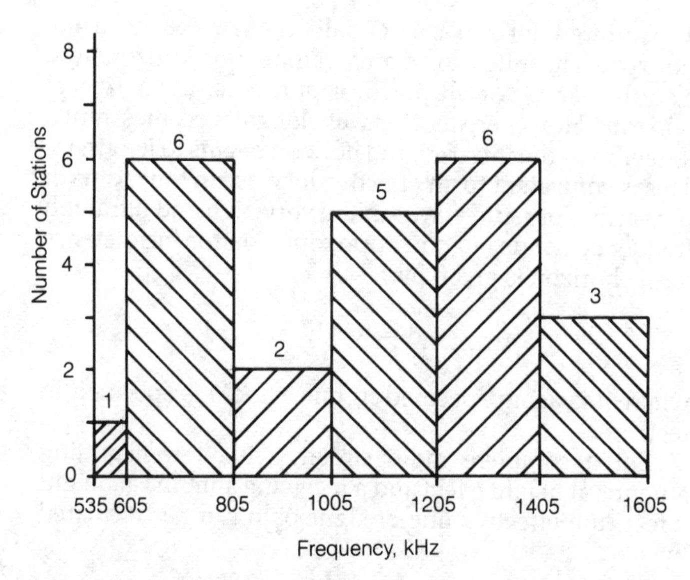

HISTOGRAM: A histogram showing the number of received broadcast stations in various segments of the amplitude-modulation broadcast band.

H NETWORK

An H network is a form of filter section sometimes seen in balanced circuits. The H network gets its name from the fact that the schematic-diagram component arrangement looks like the capital letter H turned sideways (see illustration).

The H configuration may be used in the construction of bandpass, band-rejection, highpass, and lowpass filters. The H configuration is a popular arrangement for attenuator design; noninductive resistors are employed. Several H networks may be cascaded to obtain better filter or attenuator performance.

The H network is not the only configuration for filters and attenuators. *See also* L NETWORK, PI NETWORK, T NETWORK.

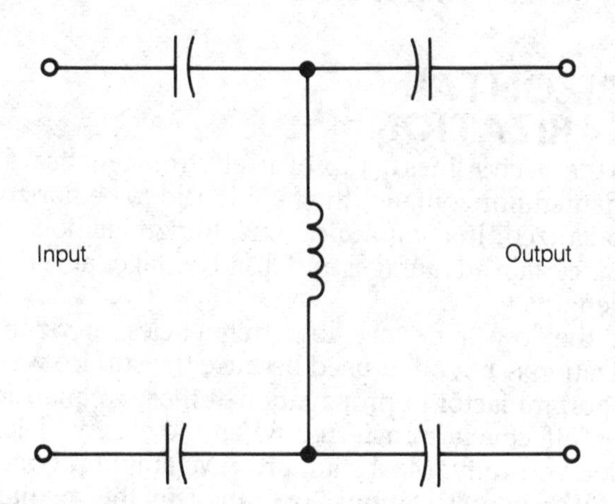

H NETWORK: An H network highpass filter.

HOLDING CURRENT

Holding current is the minimum amount of current that will keep a switching device actuated. The term applies to relays and also to various tube and solid-state devices, including the triac and the silicon-controlled rectifier.

In any switching device, a certain amount of current must flow before a change of state occurs. This turn-on current is always larger than the holding current. In relays, the difference between the turn-on and holding currents is usually rather small; the initial current must therefore be nearly maintained. But in the silicon-controlled rectifier and triac, the current may be reduced considerably following turn-on, and the device will remain actuated. Relays, silicon-controlled rectifiers, and triacs with various ratios of turn-on current on holding current are available. *See also* SILICON-CONTROLLED RECTIFIER, TRIAC.

HOLE

A hole is a carrier of electric charge. In certain semiconductor materials, the charge carriers are predominantly holes rather than electrons. A hole is an atom with one electron missing. Holes have positive charge, while electrons have negative charge.

An electric current in a P-type semiconductor material flows as the result of hole movement. The electron-deficient atoms do not themselves move; the electrons migrate from atom to atom, creating a chain of vacancies which moves from the positive to the negative. The holes in a P-type material behave in much the same way as the electrons in an N-type substance or ordinary wire conductor.

While an electron is an identifiable particle of matter, a hole is not. The concept of the hole makes it easier to explain the operation of a semiconductor device in some cases. *See also* ELECTRON, ELECTRON-HOLE PAIR, N-TYPE SEMICONDUCTOR, P-N JUNCTION, P-TYPE SEMICONDUCTOR, TRANSISTOR.

HOLLERITH CODE

Early digital computers used punched cards. The standard rectangular computer card measured 3 ¼ by 7 ⅜ inches, and was made of stiff paper. Holes in the card were arranged according to a code known as the Hollerith code.

There are 12 possible positions for data in a Hollerith character. Most punched cards had 12 rows and 80 columns. Each column contains one Hollerith character. The table on p. 444 indicates the Hollerith letters and digits, along with the more common punctuation marks. An *x* indicates that the hole position is punched through on the card.

The basic alphanumeric Hollerith code contains only the letters A through Z (uppercase) and the numerals 0 through 9. The extended Hollerith contains symbols and punctuation. The FORTRAN and commercial Hollerith codes contain additional symbols. Each Hollerith card, which normally had 80 characters, could contain one full line of data. *See also* COMPUTER.

HORIZON

The horizon is the apparent edge of the earth, as viewed from a specific height above the surface at a particular electromagnetic frequency. The radio horizon is farther

HOLLERITH CODE: Punched locations are marked by an X. In this table, rows are shown vertically and columns are shown horizontally.

Character	12	11	0	1	2	3	4	5	6	7	8	9
A	x			x								
B	x				x							
C	x					x						
D	x						x					
E	x							x				
F	x								x			
G	x									x		
H	x										x	
I	x											x
J		x		x								
K		x			x							
L		x				x						
M		x					x					
N		x						x				
O		x							x			
P		x								x		
Q		x									x	
R		x										x
S			x		x							
T			x			x						
U			x				x					
V			x					x				
W			x						x			
X			x							x		
Y			x								x	
Z			x									x
0			x									
1				x								
2					x							
3						x						
4							x					
5								x				
6									x			
7										x		
8											x	
9												x
.	x					x					x	
,			x			x					x	

away than the visual horizon. The radio horizon is of considerable importance in communication at very high frequencies because most communication in this part of the spectrum takes place by means of direct or line-of-sight propagation.

The distance to the horizon depends on several factors. In general, the higher the viewing point, the greater the distance to the horizon, regardless of frequency. The distance is also dependent on the frequency; at some wavelengths the atmosphere displays refractive effects to a greater extent than at other wavelengths. In calculating the effective distance to the radio horizon, it must be realized that the values obtained are theoretical, and are based on smooth terrain. In areas with many large hills or mountains, the actual distance to the radio horizon can vary greatly from the theoretical values.

Let h be the height of the viewpoint in feet over smooth earth, and let d be the distance to the horizon in miles. Then, at visual wavelengths:

$$d = \sqrt{1.53\ h}$$

The infrared horizon is essentially the same as the visual horizon. The ultraviolet and gamma-ray horizons are also the same, for all practical purposes, as the visual horizon. However, as the wavelength becomes much longer than the infrared, the horizon begins to lengthen. This is equivalent to an effective increase in the radius of the earth. For radio waves in the very high and ultrahigh frequency bands, a good approximation for the distance to the horizon is given by:

$$d = \sqrt{2h}$$

where d is again specified in miles and h is specified in feet.

For a complete radio circuit, with a transmitting antenna at height g feet and a receiving antenna at height h feet, the effective line-of-sight path can be calculated by:

$$d = \sqrt{2g} + \sqrt{2h}$$

See also DIRECT WAVE, LINE-OF-SIGHT COMMUNICATION.

HORIZONTAL LINEARITY

Horizontal linearity is one expression of the degree to which a television picture is an accurate reproduction of a scene. Most television receivers have an internal adjustment for calibrating the horizontal linearity. In a few receivers, the control is external.

Improper horizontal linearity results in a distorted picture. If the horizontal linearity is not properly set, objects that appear to have a certain width in one part of the screen will seem to be wider or narrower at another location on the screen.

All television picture signals are broadcast with a linear horizontal scan. Improper horizontal linearity is usually the fault of the receiver. *See also* TELEVISION.

HORIZONTAL POLARIZATION

When the electric lines of flux of an electromagnetic wave are oriented horizontally, the field is said to be horizontally polarized. In communications, horizontal polarization has certain advantages and disadvantages at various wavelengths.

At the low and very low frequencies, horizontal polarization is not often used because the surface wave, an important factor in propagation at these frequencies, is more effectively transferred when the electric field is oriented vertically. Most standard AM (amplitude-modulation) broadcast stations, operating in the medium-frequency range, also employ vertical rather than horizontal polarization.

In the high-frequency part of the electromagnetic spectrum, horizontal polarization becomes practical. The polarization is always parallel to the orientation of the radiating antenna element; horizontal wire antennas are simple to install above about 3 MHz. The surface wave is of lesser importance at high frequencies than at low and very low frequencies; the sky wave is the primary mode of propagation above 3 MHz. This becomes increasingly true as the wavelength gets shorter. Horizontal polarization is just as effective as vertical polarization in the sky-wave mode.

In the very high and ultrahigh frequency range, either vertical or horizontal polarization may be used. Horizontal polarization generally provides better noise immunity and less fading than vertical polarization in this part of the spectrum. *See also* CIRCULAR POLARIZATION, POLARIZATION, VERTICAL POLARIZATION.

HORIZONTAL SYNCHRONIZATION

In television broadcasting the picture signals must be synchronized at the transmitter and receiver. The electron beam in the television picture tube scans from left to right and top to bottom. At any given instant, in a properly operating television system, the electron beam in the receiver picture tube is in exactly the same relative position as the scanning beam in the camera tube. This requires synchronization of the horizontal as well as the vertical position of the beams.

When the horizontal synchronization in a television system is lost, the picture becomes totally unrecognizable. This is illustrated by misadjustment of the horizontal-hold control in any television set. Even a small synchronization error results in severe tearing of the picture. The transmitted television signal contains horizontal-synchronization pulses at the end of every line. This tells the receiver to move the electron beam from the end of one line to the beginning of the next line. Vertical synchronization pulses tell the receiver that one complete picture, called a frame, is complete, and that it is time to begin the next frame. *See also* TELEVISION, VERTICAL SYNCHRONIZATION.

HORN ANTENNA

See ANTENNA DIRECTORY.

HORSEPOWER

Horsepower (hp) is a unit of power, generally used in reference to mechanical devices such as motors and internal-combustion engines. One horsepower is equal to 746 watts. One watt is therefore equal to 0.00134 hp.

In electronic applications, horsepower is seldom used. The watt is the preferred unit. *See also* POWER.

HOT-WIRE METER

A hot-wire meter is an instrument that uses the thermal expansion characteristics of a metal wire for measuring current. When a current flows through the tightly stretched wire, the wire expands. This causes a pointer, attached to the wire, to move across a graduated scale. Hot-wire current-measuring devices can be used as ammeters, voltmeters, or wattmeters. Hot-wire ammeters can be attached to a variety of different devices to indicate such parameters as wind speed, motor speed, or rate of fluid flow.

The hot-wire meter is not particularly sensitive, but can register large amounts of current without damage. The damping, or rate at which the meter responds to change in the current passing through it, is slow. This is an advantage in situations where rapid fluctuations are of little interest, but is not desirable when precise indications are required for rapidly changing parameters. The hot-wire meter can measure alternating current just as well as it can measure direct current. For this reason, hot-wire ammeters are often used for the determination of radio-frequency current in antenna transmission lines. *See also* AMMETER.

HUE

Hue is a term that is essentially synonymous with color. The hue of an object or light beam is dependent on the wavelength of the light. Hue is determined by the wavelength at which the light intensity is the greatest.

Colors may have much different levels of saturation, even though they have the same hue. This can be illustrated with a color television receiver. The color-intensity knob controls the saturation. The tint knob controls the hue. *See also* CHROMA, SATURATION.

HUM

Hum is the presence of 60 or 120 Hz modulation in an electronic circuit. Hum can occur in the carrier of a radio transmitter, in a radio-frequency receiving system, or in an audio system.

When the filtering is inadequate in the output of a power supply, hum is often introduced into the circuits operating from the supply. With half-wave power supplies, the hum has a frequency of 60 Hz; with full-wave supplies, the frequency is 120 Hz. Both of these frequencies are within the range of human hearing, and produce objectionable modulation.

Hum can be picked up by means of inductive or capacitive coupling to nearby utility wires. This hum is always at a frequency of 60 Hz. Poorly shielded amplifier-input wiring is a major cause of hum in audio circuits. Hum may also be picked up by the magnetic heads of a tape recorder. Improper shielding or balance in the output leads of an audio amplifier can also cause hum modulation to be introduced into the circuit.

Power-supply hum can be remedied by the installation of additional filtering chokes and/or capacitors. Hum from utility wiring can be minimized by the use of shielding in the input leads to an amplifier and by proper shielding or balance in the output. A good ground system, without ground loops, can be helpful as well. *See also* ELECTROMAGNETIC SHIELDING, POWER SUPPLY.

HUMAN FACTORS ENGINEERING

The design of electronic equipment that considers the

physiological needs of the operator is called human factors engineering. Even if a product is well designed electrically, it will be difficult to use if proper attention has not been given to the mechanical and sensory capabilities and limitations of the human body. Human factors engineering involves judicious positioning of controls and indicators. Meters and displays should be easy to read, and controls should be convenient to adjust. The number of controls should be sufficient to accomplish the desired functions, but too many adjustments make it difficult to use the apparatus. The discipline is also called ergonomics.

HUMIDITY

Humidity is the presence of water vapor in the air. The humidity is usually specified in terms of a ratio, called the relative humidity. This is the amount of water vapor actually in the air, compared to the amount of water vapor the air is capable of holding without condensation. The higher the temperature, the more water vapor can be contained in the air.

Relative humidity is determined by the use of two thermometers, one with a dry bulb and the other with its bulb surrounded by a wet cloth or wick. Evaporation from the wick causes the wet-bulb reading to be lower than the dry-bulb reading. For a given temperature, the water from the wet bulb evaporates more rapidly as the humidity gets lower. The two readings are found on a table, and the relative humidity is indicated at the point corresponding to both readings.

HUNTING

Hunting is the result of overcompensation in a closed-loop system. Hunting is particularly common in direction-finding apparatus and improperly adjusted servosystems.

Any circuit that is designed to lock on some signal is subject to hunting if it is set so that overcompensation occurs. The circuit will then oscillate on either side of the desired direction, frequency, or other variable. The oscillation can be fast or slow. It might eventually stabilize, and the circuit could achieve the desired condition. But if the misadjustment is especially severe, the hunting may continue indefinitely.

Hunting can be eliminated by proper alignment of a system. Additional damping usually eliminates this problem. *See also* PHASE-LOCKED LOOP, RADIO DIRECTION FINDER, SERVO SYSTEM.

HYBRID CIRCUIT

A hybrid circuit is a microcircuit assembly made by bonding active and passive components to a dielectric substrate to perform a specific function. The active devices include diodes, transistors, integrated circuits (ICs), and passive devices such as resistors, capacitors, and inductors. The substrate has conductive paths or networks formed on its surface and may also have deposited resistors, capacitors, and inductors formed by thick- or thin-film technology. The term *hybrid circuit* is used interchangeably with hybrid integrated circuit, hybrid device and hybrid microcircuit.

Hybrid circuit manufacture is an alternative to IC fabrication where it might not be economical to build a circuit as a monolithic IC because of the low number of units required. However, a hybrid circuit can have some advantages over an equivalent IC where more precise resistors are required or there is a need for better thermal conductivity or dielectric isolation.

Hybrid fabrication technology is an alternative to modular circuit fabrication in which active and passive components are assembled and soldered to a printed circuit (PC) board or card and the entire assembly is potted or encapsulated in a protective case.

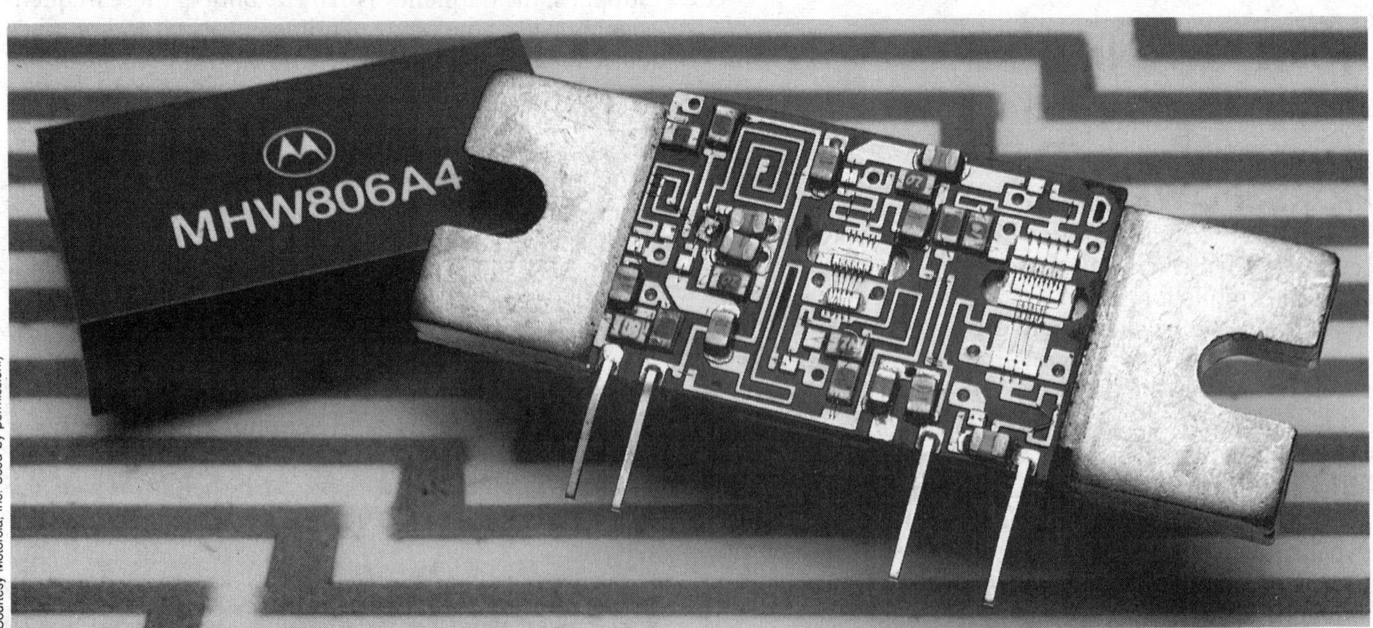

(Courtesy Motorola, Inc. Used by permission.)

HYBRID CIRCUIT: Hybrid circuit fabricated by chip-and-wire technology.

Hybrid circuits are widely used in military and aerospace applications for specialized analog and interface circuits and for microwave frequency circuits. They vary widely in: 1) choice of substrate material, 2) the use of components in packaged or bare chip form, 3) the use of chip resistors, capacitors and inductors as opposed to thick- or thin-film deposition of those functions, and 4) frequency of operation.

The most widely used substrate for hybrid circuits is aluminum oxide (Al_2O_3) or alumina ceramic. However, flexible substrates such as Kapton also are being used. The active semiconductor devices can be dies or chips, or they can be packaged in ceramic, metal, or plastic leadless packages.

Traditional hybrid circuits use semiconductors in chip form bonded to a ceramic substrate with precious metal eutectic solder rather than lead-tin eutectic solder. They are made by chip-and-wire techniques. In addition, most packaged hybrid circuits are hermetically sealed in metal and ceramic cases rather than being potted or encapsulated in plastic resin. Many similarities exist between hybrid circuit fabrication methods and surface-mount technology (SMT) although they are different processes. (*See* SURFACE MOUNT TECHNOLOGY.) The figure illustrates a conventional hybrid circuit.

The layout of the conductors and connecting pads on a hybrid circuits is made as in a conventional PC board and surface mount assembly. The pattern is drawn at an enlarged scale, often with a computer-aided design (CAD) workstation and is then photoreduced to form a mask. In thick-film technology, a 1:1 mask is then used to screen conductive metal inks on the substrate prior to furnace firing. This technology is similar to that used in manufacturing commercial resistor networks (*see* RESISTOR) and resistor-capacitor (RC) networks.

Alternatively, in thin-film technology the 1-to-1 mask is used in selective evaporation or sputtering of metal on the substrate in a vacuum chamber. The films are typically 5 microns (μm) thick or less.

Resistors and capacitors can be formed by successive screening of thick- or thin-film conductors and insulators. Inductors for microwave hybrid circuits can be screened on with these methods. Both thick- and thin-film resistors and capacitors can be trimmed to precise values by abrasive jet etching or laser-trimming techniques. The focused laser beam vaporizes the film material in a series of pulsed, overlapping spot holes along a preprogrammed pattern while the resistance is monitored to its desired value.

Automatic pick-and-place machines can be used to locate active and passive components on the substrate in the correct orientation. The previously tested devices can be in a feed tray or attached to plastic tape.

The term *attach* in hybrid technology means permanent joining with intermediary materials. Examples include attaching a device to a substrate or a substrate to a case. The selection of the process for attaching components to a hybrid substrate depends on the substrate size and hybrid application. For example, soldering is preferred for fastening active and passive components while conductive epoxy is the standard method for bonding face-up semiconductor dice and resistor chips; nonconductive epoxy is used to bond capacitors between end terminations.

Complex metals are used in attaching face-down active elements such as flip-chips and beam-lead devices; silver-glass is the choice for large-area chips (VLSI); and gold-eutectic preforms are options for power-device dice. Fine wire bonding between pads on the semiconductor chip or die and the substrate lands or headers is done by automatic wire bonding machines. The wire is made of gold or aluminum.

There are choices in fastening a substrate permanently inside a bottom-package case. An adhesive such as an epoxy or polyimide can be used to bond low-power hybrids, and a eutectic alloy preform is used for high-power hybrids.

Hybrid circuits can include the highest performance, highest density monolithic ICs as well as various discrete signal-level and power semiconductors. External trimming networks within the hybrid package can improve the performance of monolithic ICs such as operational amplifiers and digital-to-analog converters. The size and component density of a hybrid circuit are determined by end-use applications. The standard packages available for hybrid circuits include metal flatpacks and ceramic/metal dual-in-line (DIP) packages.

Hybrid circuits designed and built specifically for use in the microwave frequency range are called microwave integrated circuits (MICs) or, more precisely, hybrid microwave integrated circuits (HMICs). These circuits can include gallium arsenide (GaAs) as well as silicon amplifiers and oscillators. HMICs can be assembled in arrays within a single large hermetically sealed case to perform specific functions. An example is a radar transmit-receive (T/R) function for alternately connecting the receiver and transmitter to the antenna in phased arrays. (*See* RADAR.) HMICs are distinguished from microwave monolithic MMICs.

Circuits for fiberoptic and surface acoustic wave (SAW) functions may also be fabricated as hybrids. Hybrid technology can be considered as a transitional fabrication method. Hybrid technology will remain valid as long as it is able to incorporate the best available monolithic ICs and improve their performance with special attention to isolation, tuning, thermal stability, or other properties.

HYDROGEN

Hydrogen is the simplest element in the universe. It is also the most abundant; more than half of all matter is believed to be hydrogen. In its most common form, the hydrogen atom consists of one electron and one proton; its atomic number is 1, and its atomic weight is 1. Some hydrogen atoms have one neutron in the nucleus, resulting in an atomic weight of 2. This variant of hydrogen is called deuterium. It is possible for two neutrons to exist in the nucleus of a hydrogen atom; this results in tritium.

Hydrogen is a gas at room temperature. When combined with oxygen, hydrogen forms water. Hydrogen is extremely flammable. If a spark or flame ignites a large volume of hydrogen, it combines violently with oxygen in the atmosphere. The sole end product of this reaction is water.

HYPERBOLA

A hyperbola is a form of conic section, representing the intersection of a plane and a double cone. The plane must be parallel to the axis of the double cone (see illustration at A). The cone may have any apex angle of at least 0 degrees but less than 90 degrees with respect to the axis. The plane may be any finite, nonzero distance from the cone axis.

The general form of the equation for a hyperbola in the Cartesian coordinate system is:

$$\frac{(x - x_0)^2}{a^2} - \frac{(y - y_0)^2}{b^2} = 1$$

where a and b are constants, and (x_0, y_0) represents the center of the hyperbola, as shown at B.

The hyperbola with the simplest possible equation, $x^2 - y^2 = 1$, is sometimes called the unit hyperbola. It is centered at the origin, or the point (0,0). The unit hyperbola is the basis for a special form of trigonometric function, called the hyperbolic trigonometric function. This function is invaluable in the solution of differential equations. *See also* CARTESIAN COORDINATES, CONIC SECTION.

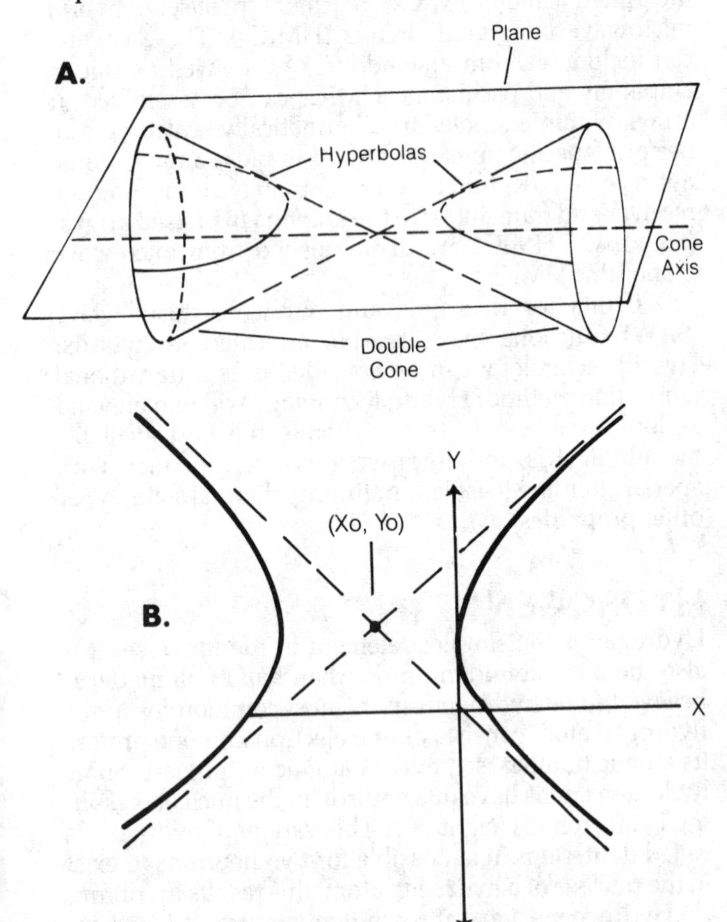

HYPERBOLA: An illustration of a conic section for deriving a hyperbola (A) and the graph of a hyperbola (B).

HYSTERESIS

Hysteresis is the tendency of electronic devices to act "sluggish." This effect is especially noticeable in the cores of transformers, because it limits the rate at which the core can be magnetized and demagnetized. This, in turn, limits the alternating-current frequency at which a transformer will operate efficiently.

Hysteresis effects occur in other components than transformer cores. For example, the thermostat in a heater or air conditioner must have a certain amount of hysteresis, or sluggishness. If there were no hysteresis in these devices, the heater or air conditioner would cycle on and off at a rapid rate, once the temperature reached the thermostat setting. With too much hysteresis, the temperature would fluctuate far above and below the thermostat setting. The correct amount of hysteresis in a thermostat results in a nearly constant temperature without excessive cycling of the heater or air conditioner. *See also* B-H CURVE, HYSTERESIS LOOP, HYSTERESIS LOSS.

HYSTERESIS LOOP

A hysteresis loop is a graphical representation of the effects of hysteresis. This graph is also called a hysteresis curve. The B-H curve is the graphical representation of magnetization versus magnetic force (*see* B-H CURVE), and thus is a special form of hysteresis loop.

The drawing illustrates a hysteresis loop for a typical thermostat used in conjunction with a heater or air conditioner. The temperature, in degrees Fahrenheit, is shown on the horizontal scale. The on and off conditions are shown on the vertical scale. For heating, the on condition is above the dotted line; for air conditioning,

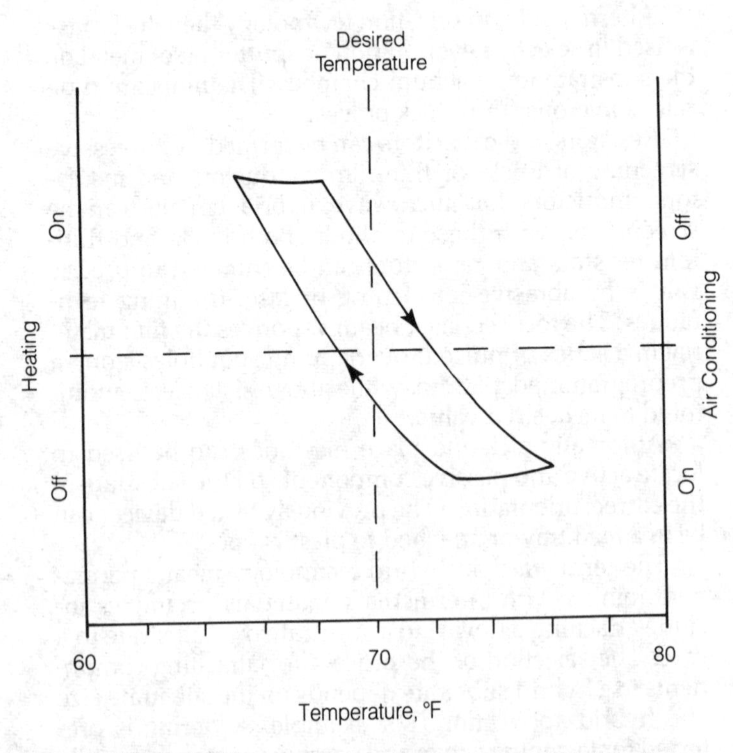

HYSTERESIS LOOP: A hysteresis loop for a hypothetical thermostat.

the on condition is below the dotted line. Time is depicted by tracing clockwise around the loop. *See also* HYSTERESIS, HYSTERESIS LOSS.

HYSTERESIS LOSS

The hysteretic properties of magnetic materials cause losses in transformers with ferromagnetic cores. Hysteresis loss occurs as a result of the tendency of ferromagnetic materials to resist rapid changes in magnetization. Hysteresis and eddy currents are the two major causes of loss in transformer core materials.

In general, the higher the alternating-current frequency of the magnetizing force, the greater the hysteresis loss for any given material. The way in which the material is manufactured affects the amount of hysteresis loss at a specific frequency.

Solid or laminated iron displays considerable hysteresis loss. For this reason, transformers with these cores are useful at frequencies up to only about 15 kHz. Ferrite has much less hysteresis loss, and can be used at frequencies as high as 20 to 40 MHz, depending on its composition. Powdered-iron cores have the least hysteresis loss of any ferromagnetic material, and are used well into the vhf range. Air has practically no hysteresis loss, and is preferred in many radio-frequency applications for this reason. *See also* EDDY-CURRENT LOSS, FERRITE, FERRITE CORE, FERROMAGNETIC MATERIAL, HYSTERESIS, LAMINATED CORE.

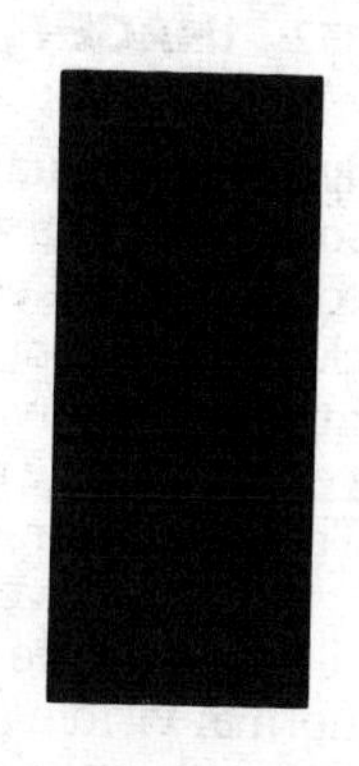

IC

See INTEGRATED CIRCUIT.

IEEE

See INSTITUTE OF ELECTRICAL AND ELECTRONIC ENGINEERS.

IF

See INTERMEDIATE FREQUENCY.

IGNITION NOISE

Ignition noise is a wideband form of impulse noise, generated by the electric arc in the spark plugs of an internal combustion engine. Ignition noise is radiated from many different sources, including automobiles and trucks, lawn mowers, and gasoline-engine-driven generators.

Ignition noise can usually be reduced or eliminated in a receiver by means of a noise blanker. The pulses of ignition noise are of very short duration, although their peak intensity can be very high. Ignition noise is a common problem for mobile radio operators, especially if communication in the high-frequency range is contemplated. Ignition noise can be intensified by radiation from the distributor to wiring in a truck or automobile. Special spark plugs, called resistance plugs, can be installed in place of ordinary spark plugs, and the ignition noise will be reduced. An excellent ground connection is imperative in the mobile installation. Mobile receivers for high-frequency work should have effective, built-in noise blankers.

Ignition noise is not the only source of trouble for the mobile radio operator. Noise can be generated by the friction of the tires against a pavement. High-voltage power lines can radiate noise at radio frequencies. *See also* IMPULSE NOISE, NOISE, NOISE LIMITER, POWER-LINE NOISE.

ILLUMINANCE

Illuminance is the density of the luminous flux incident at a point on a surface. Illuminance is measured in lumens per square meter (photometric system) and watts per square meter (radiometric system).

For a given light source, the illuminance impinging on a surface, oriented at a right angle to the incident radiation, is inversely proportional to the square of the distance of the object from the source. For example, if the distance is doubled, the illuminance becomes just ¼ of its previous value. This is the law of inverse squares. *See also* FOOTCANDLE, INVERSE-SQUARE LAW.

ILLUMINATION METER

An illumination meter is an instrument for measuring the intensity of light in a given place. Illumination meters are used by photographers and television camera operators. Illumination meters are calibrated in illuminance.

The schematic diagram shows a simple illumination meter. The instrument consists of a solar cell and a microammeter or milliammeter. The greater the illuminance on the solar cell, the more current it generates, and the higher the meter reading.

Illumination meters are wavelength-sensitive. That is, they respond more readily to light of certain wavelengths than to light at other wavelengths. For best results, the wavelength response of an illumination meter should be similar to that of the film used in photographic apparatus. Otherwise, overexposure may result in some situations and underexposure will occur in other cases. *See also* ILLUMINANCE.

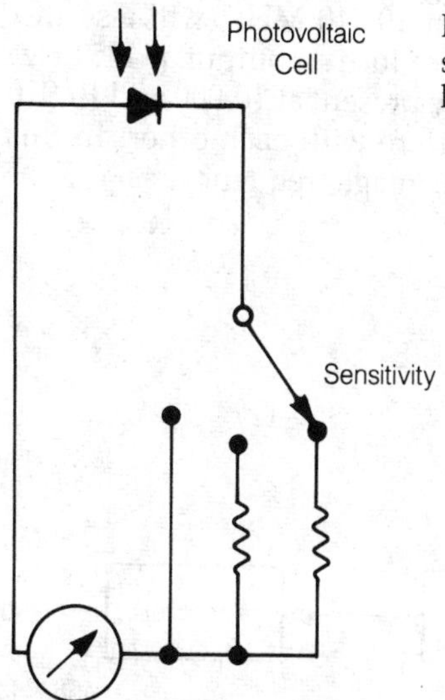

ILLUMINATION METER: A simple meter for measuring light intensity.

IMAGE CONVERTER

An image converter is a device that changes an electrical, acoustic, infrared, ultraviolet, X ray, or gamma-ray image into a visual image. Image converters include ultrasonic scanners and X ray machines. An infrared image converter makes it possible to see through some obstacles that are opaque to visible light. Astronomers use infrared, ultraviolet, X ray, and gamma-ray image converters to view celestial objects at invisible wavelengths. The electron microscope is a form of image converter.

Image converters may consist of special photographic films, sensitive to radiation at wavelengths outside the normal visible spectrum; these devices normally do not allow the observation of motion. More sophisticated image converters, such as the fluoroscope enable personnel to see motion. Image converters can be interconnected with computers for missile guidance and medical diagnosis. *See also* COMPUTER-AIDED MEDICAL SCANNING, ELECTRON MICROSCOPE, GAMMA RAY, INFRARED, ULTRAVIOLET RADIATION, X RAY.

IMAGE FREQUENCY

In a superheterodyne receiver, frequency conversion results in a constant intermediate frequency. This simplifies the problem of obtaining good selectivity and gain over a wide range of input-signal frequencies. However, it is possible for input signals on two different frequencies to result in output at the intermediate frequency. One of these signal frequencies is the desired one, and the other is undesired. The undesired response frequency is called the image frequency.

The illustration is a simple block diagram of a superheterodyne receiver with an intermediate frequency of 455 kHz. This is a common value for the intermediate frequency of a superheterodyne circuit. The local oscillator is tuned to a frequency 455 kHz higher than the desired signal; for example, if reception is wanted at 10.000 MHz, the local oscillator must be set to 10.455 MHz. However, an input signal 455 kHz higher than the local-oscillator frequency, at 10.910 MHz, will also mix with the local oscillator to produce an output at 455 kHz. If signals are simultaneously present at 10.000 and 10.910 MHz, then, they may interfere with each other. In this example, 10.910 MHz is the image frequency.

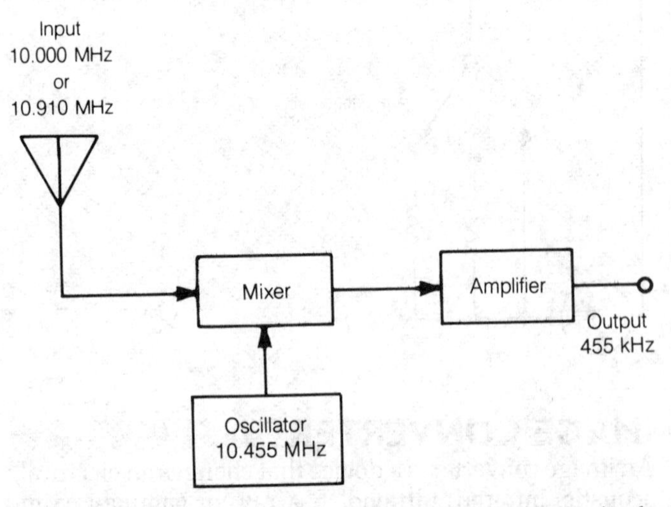

IMAGE FREQUENCY: In this superheterodyne receiver, the desired signal is at 10.0 MHz, and the image signal is at 10.910 MHz.

The image frequency in a superheterodyne receiver always differs from the signal frequency by twice the value of the intermediate frequency. A selective circuit in the front end of the receiver must be used to attenuate signals arriving at the image frequency, while passing those arriving at the desired frequency. The lower the intermediate frequency in a superheterodyne receiver, the closer the image frequency is to the desired frequency, and the more difficult it becomes to attenuate the image signal. For this reason, modern superheterodyne receivers use a high intermediate frequency, such as 9 MHz, so that the image and desired frequencies are far apart. The high intermediate frequency can later be heterodyned to a low frequency, making it easier to obtain sharp selectivity. *See also* DOUBLE-CONVERSION RECEIVER, IMAGE REJECTION, SINGLE-CONVERSION RECEIVER, SUPERHETERODYNE RADIO RECEIVER.

IMAGE IMPEDANCE

An image impedance is an impedance that is seen at one end of a network, when the other end is connected to a load or generator having a defined impedance. The image impedance can be the same as the terminating impedance at the other end of the network, but this is not always the case.

As an example, consider a pure resistive impedance Z_1 connected to one end of a quarter-wave transmission line having a characteristic impedance Z_0. The image impedance Z_2 at the opposite end of the line is related to the other impedances according to the formula:

$$Z_0 = \sqrt{Z_1 Z_2}$$

If the transmission line measures a half electrical wavelength, however, Z_1 and Z_2 will always be equal.

The relationships become more complicated when reactance is present in the terminating impedance. For a quarter-wave line, capacitive reactance at one end is transformed into an inductive reactance at the opposite end, and vice versa. For a half-wave line, the terminating and image impedances are equal even when reactance is present.

There are many different kinds of networks that produce various relationships between terminating and image impedances. *See also* IMPEDANCE, REACTANCE.

IMAGE ORTHICON

The image orthicon is a camera tube for television broadcasting. The image orthicon is a highly sensitive camera tube that responds rapidly to motion and changes in light intensity. A diagram of an image orthicon camera tube is shown.

Light is focused by means of a convex lens onto a translucent plate called the photocathode. The photocathode emits electrons according to the intensity of the light in various locations. A target electrode attracts these electrons, and accelerator grids give the electrons additional speed. When the electrons from the photocathode, called photoelectrons, strike this target electrode, sec-

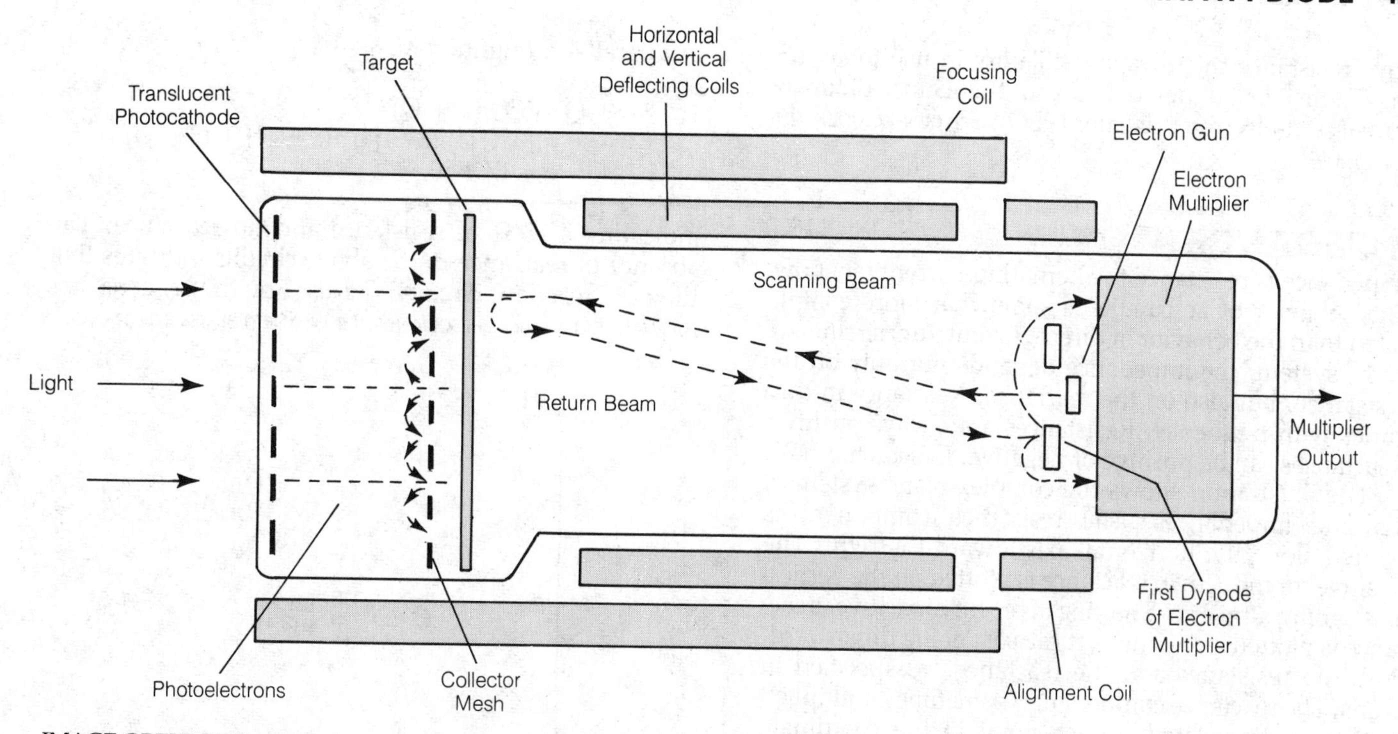

IMAGE ORTHICON: A light image incident on the photocathode emits electrons that are focused on the target. A scanning electron beam interacts with the target to form a modulated return beam that is multiplied to produce a video output.

ondary electrons are produced. Several secondary electrons are emitted from the target for each impinging photoelectron.

A fine electron beam, generated by an electron gun, scans the target electrode. (The scanning rate and pattern correspond with the scanning in television receivers.) Some of these electrons are reflected back toward the electron gun by the secondary emissions from the target. Areas of the target having greater secondary-electron emission result in an intense return beam; areas of the target that emit few secondary electrons result in lower returned-beam intensity. Therefore, the returned beam is modulated as it scans the target electrode. A receptor electrode picks up the returned beam.

The main disadvantage of the image orthicon is its rather high noise output. However, in situations where there is not much light, or where a fast response time is required, the image orthicon is superior to other types of camera tubes. *See also* CAMERA TUBE, VIDICON.

IMAGE PROCESSOR

An image processor is a specialized computer for changing, enhancing, and extracting information from images. Most image processors use digital computer technology and analog-to-digital converters to sample the analog image signal.

Image processors are used to analyze satellite images and in industrial inspection, among other tasks. An image processor often works in conjunction (as a coprocessor) with a standard, general-purpose computer. *See* COMPUTER VISION.

IMAGE REJECTION

In a superheterodyne receiver, image rejection is the amount by which the image signal is attenuated with respect to a signal on the desired frequency. Image rejection is always specified in decibels.

In general, the higher the intermediate frequency of a receiver, the better the image rejection. *See also* IMAGE FREQUENCY, SUPERHETERODYNE RADIO RECEIVER.

IMPATT DIODE

The IMPATT (impact avalanche transit time) diode is a multilayer, negative-resistance diode capable of generating microwave power by atomic oscillations in and around the avalanche condition. It is also known as the Read diode.

The most practical IMPATT diode geometry for millimeter wave applications is a double-drift silicon structure. The layers in a double-drift diode are a doped P region (P+), a moderately doped P region (P), a moderately doped N region (N), and a heavily doped N region (N+). The P+ and N+ regions allow ohmic contacts to be made to the external circuit.

Commercial IMPATT diodes are able to generate 500 milliwatts (mW) of continuous wave (CW) and 10 watts (W) of output power at 33 to 37 gigahertz (GHz). These values fall to 20 mW CW and 1 W pulse at 135 to 145 GHz. IMPATT diodes are packaged in hermetically sealed cases and supplied mounted to a copper heat sink. The packages measure about 3/16 inch (4 millimeters) in diameter and about 1/8 inch (3 millimeters) high.

IMPATT diodes are high-power members of the bulk-effect oscillator (BEO) family. Two other solid-state negative-resistance microwave oscillators in use today are the EASKI or tunnel diode and the Gunn diode or transferred-electron oscillator (TEO). *See* GUNN DIODE, TUNNEL DIODE.

IMPEDANCE

Impedance is resistance to alternating-current (ac) flow. The behavior of ac circuits is somewhat more complicated than the behavior of direct-current (dc) circuits. In an ac system, the impedance depends not only on the resistance, but also on the reactance. Reactance in turn varies with frequency. Resistances are always positive. Reactances can be positive or negative. (*See* REACTANCE.)

The illustration shows the complex-plane system of defining impedances. The resistance component is plotted along the horizontal axis toward the right. The positive, or inductive, reactance is plotted on the vertical axis, going upward. The negative, or capacitive, reactance is plotted along the vertical axis, going downward. Both the resistance and the reactance are specified in ohms. The reactance component is sometimes multiplied by $\sqrt{-1}$, abbreviated j. Each point in the coordinate plane therefore corresponds to exactly one impedance, and any impedance can be defined as a unique point on the plane.

If an inductor presents 3 ohms (Ω) of resistance and 4 Ω of reactance, then the impedance is defined on the plane by the point P = (3,4). Engineers write this as 3 + j4. Suppose a capacitor offers 3 Ω of resistance and 4 Ω of reactance. Then the impedance is defined by the point Q = (3,−4); engineers would write 3 − j4.

Sometimes the impedance is defined simply as the length of the line from the origin, or (0,0), to the point on the plane defining the impedance. This length is given by:

$$Z = \sqrt{R^2 + X^2}$$

where *R* is the resistance and *X* is the reactance, both given in ohms. This method of defining impedance does not tell the complete story, however, because Z does not contain information concerning the components. In the illustration, points *P* and *Q* both correspond to Z = 5 Ω, but the two impedances are vastly different.

Circuits have different impedance characteristics. Antennas, tuned circuits, and transmission lines show impedances at radio frequencies. A transmission line also displays a property called the characteristic impedance, which depends on its physical dimensions and construction.

Impedances in series add vectorially together. In the example, if the capacitor and inductor are connected in series, the impedance becomes:

$$P + Q = (3 + j4) + (3 - j4) = 6 + j0,$$

indicating 6 Ω of resistance and no reactance.

Impedances in parallel add like resistances in parallel, providing the complex representations are used. Thus, in a parallel configuration:

$$\begin{aligned} P + Q &= PQ / (P + Q) \\ &= (3 + j4)(3 - j4) / (3 + j4) + (3 - j4) \\ &= 4.17 + j0, \end{aligned}$$

indicating 4.17 Ω of resistance and no reactance. The absence of reactance in the above circuits indicates that they are resonant circuits. *See also* ADMITTANCE, CONJUGATE IMPEDANCE, PARALLEL RESONANCE, RESONANCE, SERIES RESONANCE.

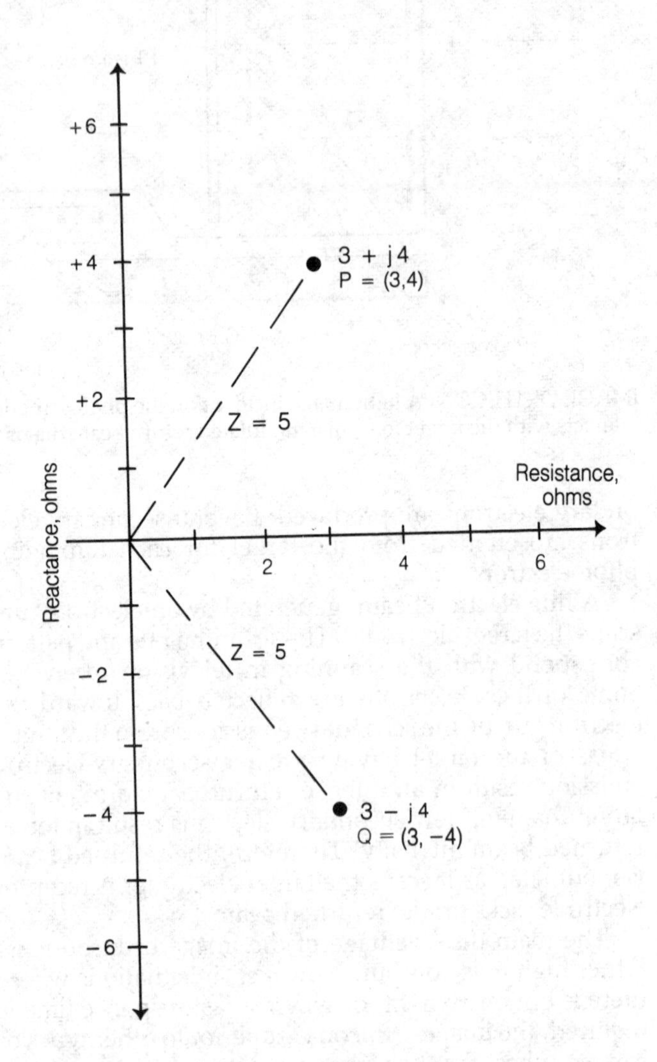

IMPEDANCE: The complex impedance plane showing a positive point (upper) and a negative point (lower) plotted.

IMPEDANCE BRIDGE

An impedance bridge is a circuit used for determining the resistive and reactive components of an unknown impedance. The impedance of any reactive circuit changes with the frequency, so the bridge can function only at a specific frequency. Impedance bridges are used to determine the impedances of tuned circuits and antenna systems.

The impedance bridge generally has a null indicator and two adjustments. One adjustment indicates the resistance, and the other indicates the reactance. *See also* IMPEDANCE.

IMPEDANCE MATCHING

An alternating-current circuit always functions best when the impedance of a power source is the same as the impedance of the load to which power is delivered. This does not happen by chance; often, the two impedances must be made the same by means of special transformers or networks. This process is called impedance matching. Impedance matching is important in audio-frequency as well as radio-frequency applications.

In a high-fidelity system, audio transformers ensure that the output impedance of an amplifier is the same as that of the speakers; this value is generally standardized at 8 ohms (Ω). Radio-frequency transmitting equipment is usually designed to operate into a 50-Ω, nonreactive load, although some systems have output tuning circuits that allow for small resistance fluctuations and/or small amounts of reactance in the load.

At very high, ultrahigh, and microwave frequencies, poorly matched impedances can result in large amounts of signal loss. If the generator and load impedances are not identical, some of the electromagnetic field is reflected from the load back toward the source. In any alternating-current system, the load will accept all of the power only when the impedances of the load and source are identical. *See also* IMPEDANCE, REACTANCE, RESISTANCE.

IMPEDANCE TRANSFORMER

An impedance transformer is used to change one pure resistive impedance to another. Impedance transformers can be used at audio frequencies to match the output impedance of an amplifier to the input impedance of a set of speakers. In radio-frequency work, impedance transformers comprise part of an impedance-matching system, generally used for optimizing the antenna-system for a transmitter.

The impedance of a purely resistive (nonreactive) load is Z_s, and this load is connected to the secondary winding of a transformer with a primary-to-secondary turns ratio of T. This situation is shown in the illustration. The impedance Z_P appearing across the primary winding, neglecting transformer losses, is given by:

$$Z_P = Z_S T^2$$

An impedance transformer must be designed for the proper range of frequencies. This generally means that the reactances of the windings must be comparable to the source and load reactances at the frequency in use. *See also* IMPEDANCE, IMPEDANCE MATCHING, TRANSFORMER.

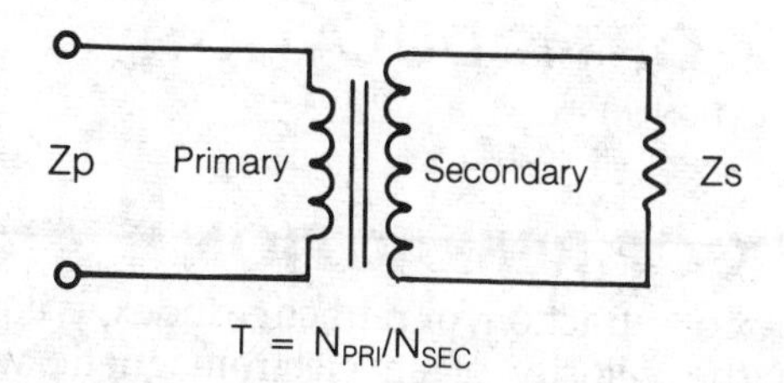

IMPEDANCE TRANSFORMER: The impedance-transfer ratio is equal to the square of the turns ratio.

IMPULSE

An impulse is a sudden surge in voltage or current. An impulse that appears at a utility outlet is sometimes called a transient or transient spike.

Impulses are of very short duration, so they contain high-frequency components. This can cause radio-frequency emissions from electrical wiring and appliances, resulting in interference for nearby radio receivers. *See also* IMPULSE NOISE.

IMPULSE NOISE

Any sudden, high-amplitude voltage pulse will cause radio-frequency energy to be generated. In electrical systems without shielded wiring, the result can be electromagnetic interference. Impulse noise is generated by internal combustion engines because of the spark-producing voltage pulses. This form of noise is called ignition noise (*see* IGNITION NOISE). Impulse noise can also be produced by household appliances such as vacuum cleaners, hair dryers, electric blankets, thermostat mechanisms, and fluorescent-light starters.

Impulse noise is usually the most severe at very low and low radio frequencies. However, serious interference can often occur in the medium- and high-frequency ranges. Impulse noise is seldom a problem above about 30 MHz. Impulse noise may be picked up by high-fidelity audio systems, resulting in interference with phonograph devices and tape-deck playback equipment.

Impulse noise in a radio receiver can be reduced by the use of a good ground system. Ground loops should be avoided. A noise blanker or noise limiter can be helpful. A receiver should be set for the narrowest response bandwidth consistent with the mode of reception. Impulse noise in a high-fidelity sound system can be more difficult to eliminate. An excellent ground connection, without ground loops, is imperative. It may be necessary to shield all speaker leads and interconnecting wiring. *See also* ELECTROMAGNETIC INTERFERENCE.

INCANDESCENT LAMP

An incandescent lamp is a device that produces visible light by means of a current in a resistive wire. The resistive wire is called the filament. With the application of sufficient voltage, the filament glows white hot. The filament is enclosed in an evacuated chamber to prevent oxidation. The photograph on p. 456 shows a typical incandescent lamp with a coiled tungsten filament. *See also* FLUORESCENT TUBE, GLOW LAMP.

INCLUSIVE-OR GATE

See OR GATE.

INCOHERENT RADIATION

Incoherent electromagnetic radation occurs when there are many different frequency and/or phase components in the wave. An example of incoherent radiation is the

INCANDESCENT LAMP: A small, high-intensity lamp rated for 120 V ac.

light from any ordinary incandescent lamp or fluorescent tube.

Most radiation, including radio-frequency noise, infrared, visible light, ultraviolet rays, X rays, and gamma rays, is incoherent because these emissions result from a random sort of disturbance. When radiation is concentrated at a single wavelength, and all of the wavefronts are lined up in phase, the radiation is said to be coherent. A radio transmitter generates coherent radiation, as does a maser or laser. *See also* COHERENT RADIATION.

INCREMENT

An increment is a change in the value of a variable, such as current, voltage, power, or frequency. The Greek uppercase letter delta (Δ) is often used to represent an incremental quantity; for example, Δf represents a defined change in frequency from one value to another.

Increments are important in the dynamic evaluation of some circuits. The slope of a line in the Cartesian (x,y) plane, for example, is given by $\Delta y/\Delta x$. This indicates the rate of change in y with respect to the rate of change in x. On a curve, the value of $\Delta y/\Delta x$ depends on the part of the graph in which the evaluation is made, and also on the size of the increments chosen. As the increments get smaller and smaller, the value of $\Delta y/\Delta x$ approaches a limiting value, known as the derivative of the function at a point. *See also* DERIVATIVE.

INDEFINITE INTEGRAL

The area under the curve of a function is found by a calculus technique called integration. The indefinite integral is an expression that can be evaluated for any two points in the domain of the function, so that any defined portion of the area under the curve may be calculated.

Given a function $f(x)$, the indefinite integral is written in the form:

$$\int f(x)\, dx$$

where dx is the differential of x, representing the limiting increment in the value of x. (For purposes of calculation, the differential is not of importance, except that is should always be written when specifying an indefinite integral.)

The table shows some common indefinite integrals. More comprehensive tables are available in a variety of mathematics data books. Note that all of the indefinite integral functions contain a constant, denoted by c. This constant is present because indefinite integration is just the opposite of differentiation. Any constant term, when differentiated, becomes zero. In the table, the functions after the integral signs are the derivates of their indefinite-integral functions. Definite integrals are found by first establishing the limits, x_1 and x_2, in the domain within which the area under the curve is to be evaluated. The value of the integral function is then calculated for x_1 and x_2. The definite integral is the difference between these two numerical values. *See also* CALCULUS, DEFINITE INTEGRAL, DERIVATIVE, FUNCTION.

INDEFINITE INTEGRAL: Indefinite integrals of some common mathematical functions. Constants are represented by the letters c, k, and n. The value of e is approximately 2.7183.

f(x)	∫ f(x) dx
$y = k$	$\int y\, dx = kx + c$
$y = kx$	$\int y\, dx = (k/2)x^2 + c$
$y = kx^2$	$\int y\, dx = (k/3)x^3 + c$
$y = kx^n$	$\int y\, dx = (\,(k/(n+1)\,)x^{n+1} + c$
$y = \log_e (x)$	$\int y\, dx = x \log_e (x) - x + c$
$y = e^x$	$\int y\, dx = e^x + c$
$y = \sin (x)$	$\int y\, dx = -\cos (x) + c$
$y = \cos (x)$	$\int y\, dx = \sin (x) + c$
$y = \tan (x)$	$\int y\, dx = -\log_e (\cos (x)) + c$

INDEPENDENT VARIABLE

In any relation or function, an independent variable is a value on which other variables depend. There may be one independent variable, or there may be several. In the Cartesian coordinate plane, wherein functions of one variable are represented, the independent variable is almost always plotted along the horizontal axis. In a polar coordinate system, the independent variable is usually plotted as an angle progressing counterclockwise (*see* CARTESIAN COORDINATES, DEPENDENT VARIABLE, FUNCTION, POLAR COORDINATES).

INDEX OF MODULATION

See MODULATION INDEX.

INDEX OF REFRACTION

The index of refraction, or refractive index, is equal to the ratio of the velocity of an electromagnetic wave in a vacuum to the velocity in a material. The velocity of an electromagnetic wave is approximately 300,000 kilometers per second in a vacuum or free space. However, in other dielectric media, the speed is slower. For a given material at a specified wavelength, the index of refraction, *n*, is defined as:

$$n = \frac{300{,}000}{v}$$

where *v* is the speed of propagation in the given medium, measured in kilometers per second.

For most substances, the index of refraction changes with the wavelength. For example, the index of refraction for glass is greater at the violet end of the visible spectrum than at the red end. This is called dispersion (*see* DISPERSION).

Refractive effects are evident at the visible wavelengths and radio frequencies. Radio waves travel more slowly in materials such as polyethlene (with a refractive index of 1.52) than in free space. The value 1/n in these situations is known as the velocity factor. *See also* REFRACTION, VELOCITY FACTOR.

INDIRECTLY HEATED CATHODE

In a vacuum tube, the cathode may consist of a tubular metal electrode, within which a heating filament is inserted. This type of cathode is called an indirectly heated cathode because the heating current does not actually pass through the cathode itself.

Indirectly heated cathodes have an advantage over the directly heated type. The cathode can be electrically separated from the filament. This makes it easy to control the bias between the grid and the cathode of the tube; a parallel resistor-capacitor combination, connected in series with the cathode lead, provides negative grid bias without the need for a separate power supply. The input signal can be easily applied to an indirectly heated cathode. The main disadvantage of the indirectly heated cathode is that the tube requires time to warm up. Cathode-ray tubes and microwave power tubes have indirectly heated cathodes. *See also* CATHODE.

INDIUM

Indium is an element with an atomic number of 49 and an atomic weight of 115. In the manufacture of semiconductor materials, indium is used as an impurity or dopant. Indium is an acceptor impurity. This means that it is electron-deficient. When indium is added to a gallium arsenide or silicon semiconductor, a P-type substance is the result. P-type semiconductors conduct via holes. *See also* DOPING, HOLE, P-TYPE SEMICONDUCTOR.

INDUCED EFFECT

Voltages and currents can be induced in a material in a variety of ways. When a conductor is moved through a magnetic field, current flows in the conductor. An object in an electric field exhibits a potential difference, in certain instances, between one region and another. Alternating currents in one conductor can induce similar currents in nearby conductors.

Induced effects are extremely important in many applications. The transformer works because a current in the primary winding induces a current in the secondary. Radio communication is possible because of the induced effects of alternating currents at high frequencies. *See also* ELECTRIC FIELD, ELECTROMAGNETIC FIELD, MAGNETIC FIELD.

INDUCTANCE

Inductance is the ability of a device to store energy in the form of a magnetic field. Inductance is represented by the capital letter *L* in mathematical equations. The unit of inductance is the henry. One henry is the amount of inductance necessary to generate 1 volt with a current that changes at the rate of 1 ampere per second. Mathematically:

$$L \text{ (henrys)} = \frac{E}{dI/dt}$$

where *E* is the induced voltage and *dI/dt* is the rate of current change in amperes per second.

In practice, the henry is an extremely large unit of inductance. Usually, inductance is specified in millihenrys (mH), microhenrys (μH), or nanohenrys (nH). These units are, respectively, a thousandth, a millionth, and a billionth of 1 henry:

$$1 \text{ mH} = 10^{-3} \text{ H}$$
$$1 \text{ μH} = 10^{-3} \text{ mH} = 10^{-6} \text{ H}$$
$$1 \text{ nH} = 10^{-3} \text{ μH} = 10^{-9} \text{ H}$$

Any length of electrical conductor displays a certain amount of inductance. When a length of conductor is deliberately coiled to produce inductance, the device is called an inductor (*See* INDUCTOR).

Inductances in series add together. For *n* inductances $L_1, L_2, \ldots, L_n$ in series, then, the total inductance *L* is:

$$L = L_1 + L_2 + \ldots + L_n$$

In parallel, inductances add according to the equation:

$$\frac{1}{L} = \frac{1}{L_1} + \frac{1}{L_2} + \ldots + \frac{1}{L_n}$$

See also INDUCTIVE REACTANCE.

INDUCTANCE MEASUREMENT

Inductance is usually measured with an instrument incorporating known capacitances that cause resonance effects at measurable frequencies. In a tuned circuit, an unknown inductance can be combined with a known capacitance, and the resonant frequency can be determined with a signal generator and an indicating device. The illustration shows this arrangement.

Once the resonant frequency *f* has been found for a known capacitance C and an unknown inductance *L*, the value of L can be determined according to the following equation:

$$L = \frac{1}{39.5f^2C}$$

where L is given in microhenrys, f in megahertz, and C in microfarads. Alternatively, L can be found in henrys if f is given in hertz and C is given in farads.

A more sophisticated circuit for determining unknown inductances is called the Hay bridge. This bridge permits both impedance and Q factor to be determined. *See also* HAY BRIDGE, INDUCTANCE, Q FACTOR, RESONANCE.

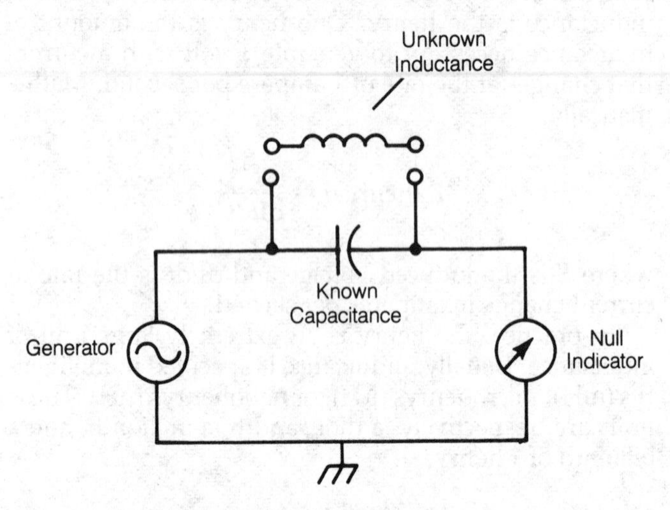

INDUCTANCE MEASUREMENT: The resonant-circuit method for determining inductance values.

INDUCTION

See INDUCED EFFECT.

INDUCTION COIL

An induction coil is a component that generates very high alternating-current (ac) voltages. The induction coil operates on the transformer principle (see illustration).

The primary winding of an ac step-up transformer is connected to a source of direct current through an interrupter, such as a vibrator or chopper. The interrupter produces a series of pulses in the primary winding. The secondary winding exhibits a large ac voltage. If the secondary-to-primary turns ratio of the transformer is large enough, this potential may reach several thousand volts. The maximum obtainable value is limited by the efficiency of the transformer. *See also* CHOPPER POWER SUPPLY, TRANSFORMER.

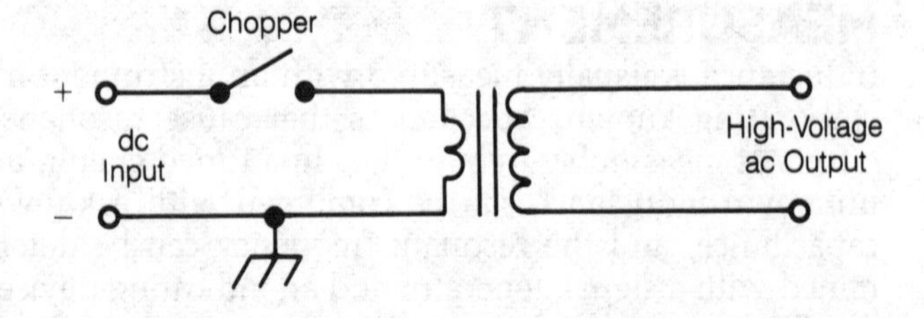

INDUCTION COIL: An induction coil generates very high alternating-current voltages.

INDUCTION HEATING

Induction heating is a method for heating a metal by causing induced current to flow. The induction heater consists of a large coil, known as a work coil, to which a high-power radio-frequency signal is applied. The metal object is placed inside this coil. If the radio-frequency signal is of the correct wavelength, and the power level is high enough, the metal object will become very hot.

Induction heating occurs because the oscillating magnetic field within the coil causes circulating currents in the metal; the ohmic losses in the metal result in the generation of heat. Induction heating makes use of magnetic effects in the same way that electric fields cause dielectrics to become hot. *See also* DIELECTRIC HEATING.

INDUCTION LOSS

When a conductor carries radio-frequency energy, it is important that the conductor either be shielded or kept away from metallic objects because the inductive coupling to nearby metallic objects can cause losses in the conductor.

Induction loss can occur when an antenna radiator is placed too close to utility wiring, a metal roof, or other conducting medium. Circulating currents occur, resulting in heating of the nearby material. A parallel-wire transmission line should be kept at least several inches away from metal objects such as wires, towers, or siding, so that the possibility of induction loss is minimized. Induction losses increase the overall loss in an antenna system.

INDUCTIVE COUPLING

See TRANSFORMER COUPLING.

INDUCTIVE FEEDBACK

Inductive feedback results from inductive coupling between the input and output circuits of an amplifier. The feedback may be intentional or unintentional; it can be positive (regenerative) or negative (degenerative), depending on the number of stages and the circuit configuration. Inductive feedback can take place at audio or radio frequencies.

Some oscillators use inductive feedback for signal generation. The Armstrong oscillator, for example, has a feedback coil from the output circuit deliberately placed near the tank coil.

Unwanted inductive feedback is generally more of a problem in the very high frequency range and above than at lower frequencies because the inductive reactance of intercomponent wiring becomes greater as the frequency increases. This in turn increases the inductive coupling among different parts of a circuit. Inductive feedback of a regenerative nature can result in spurious oscillations in a receiver or transmitter. The spurious signals in a transmitter may be radiated along with the desired signal, causing interference to other stations. In audio-frequency equipment, inductive feedback can occur if unshielded speaker leads are placed near the input cables of the amplifier. The result, if the feedback is regenerative, will be hum or an audio-frequency tone. *See also* FEEDBACK.

INDUCTIVE LOADING

In radio-frequency transmitting installations, it is often necessary to lengthen an antenna electrically, without making it longer physically. This is likely to be done at very low, low, and medium frequencies. Inductive loading is the most common method of accomplishing this.

In an unbalanced antenna, such as a quarter-wave resonant vertical working against ground, the inductor may be placed at any location less than approximately 90 percent of the way to the top of the radiator. The most common positions for the inductor are at the base and near the center. With a given physical radiator length, larger inductances result in lower resonant frequencies. There is no theoretical lower limit to the resonant frequency for an inductively loaded quarter-wave vertical. In practice, however, the losses become prohibitive when the physical length of the radiator is less than about 0.05 wavelength.

In a balanced antenna system, such as a half-wave resonant dipole, identical inductors are placed in each half of the antenna, in a symmetrical arrangement with respect to the feed line (see illustration). For a given antenna length, larger inductances result in lower resonant frequencies. As with the quarter-wave vertical antenna, there is no theoretical lower limit to the resonant frequency that can be realized with a given antenna length. However, the losses become prohibitive when the physical length becomes too short. If the dipole is placed well above the ground and away from obstructions, and if the conductors and inductors are made from material with the least possible ohmic loss, good performance can be obtained with antennas as short as 0.05 free-space wavelength.

Inductively loaded antennas always show a low radiation resistance because the free-space physical length of the radiator is so short. As the length of an antenna becomes shorter than 0.1 wavelength, the radiation resistance drops rapidly. This makes it difficult to achieve good transmitting efficiency with very short radiators. *See also* BASE LOADING, CENTER LOADING, RADIATION RESISTANCE.

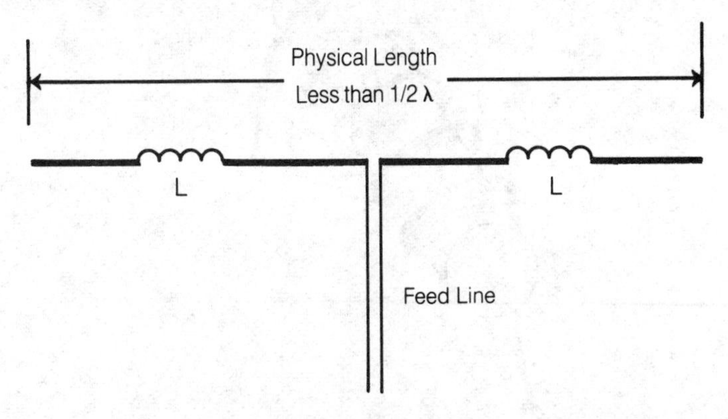

INDUCTIVE LOADING: Inductive loading provides for half-wave resonance in an antenna that is less than a half wavelength long.

INDUCTIVE REACTANCE

Reactance is the opposition that a component offers to alternating current. The reactance of a component can be either positive or negative, and it varies in magnitude depending on the frequency of the alternating current. Positive reactance is called inductive reactance; negative reactance is called capacitive reactance. These choices of positive and negative are purely arbitrary, and are used as a matter of mathematical convenience.

The reactance of an inductor of L henrys, at a frequency of f hertz, is given by:

$$X_L = 2\pi fL$$

where X_L is specified in ohms. If the frequency f is given in kilohertz and the inductance L in millihenrys, the formula also applies. The frequency also may be given in megahertz and the inductance in microhenrys.

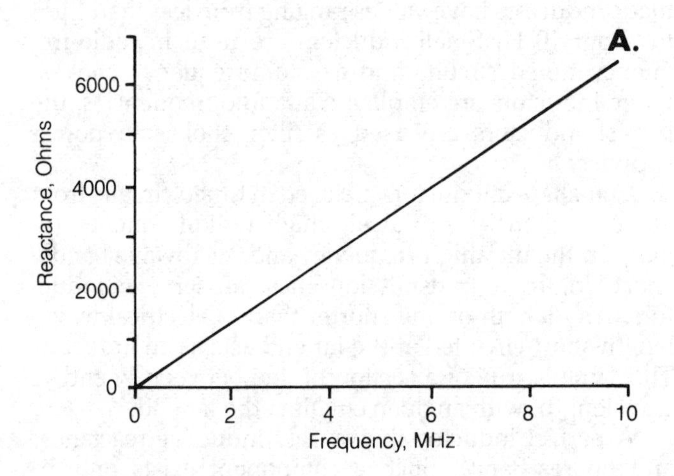

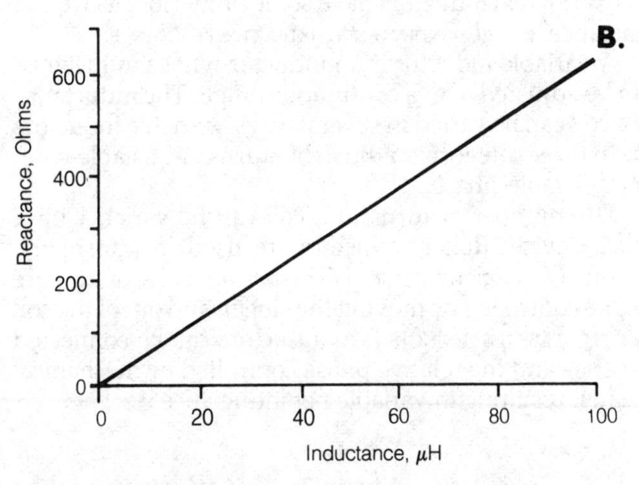

INDUCTIVE REACTANCE: The reactance of a 100 μH inductor as a function of frequency (A) and reactance versus inductance at a frequency of 1 MHz (B).

The magnitude of the inductive reactance, for a given component, approaches zero as the frequency is lowered. The inductive reactance gets larger without limit as the frequency is raised. This effect is illustrated in the graph. At A, the reactance of a 100 μH inductor is shown as a function of the frequency in megahertz. At B, the reactance of various values of inductance are shown for a constant frequency of 1 MHz.

Reactances, like direct-current resistances, are always expressed in ohms. In a complex circuit containing resistance and reactance, the reactance is multiplied by $\sqrt{-1}$, which is mathematically represented by the letter j in engineering notation. Thus an inductive reactance of +20 Ω is specified as +j20; a combination of 10 Ω resistance and +20 Ω inductive reactance is a complex impedance of 10 + j20. *See also* IMPEDANCE, J OPERATOR.

INDUCTOR

An inductor is an electronic component designed especially to provide a controlled amount of inductance. Inductors generally consist of a length of wire wound into a solenoidal or toroidal shape (see illustration). The inductance may be increased by placing a core with high magnetic permeability within the coil. Suitable materials include iron, powdered iron, and ferrite. Commercially made inductors have values ranging from less than 1 μH to about 10 H. Small inductors are used in radio-frequency tuned circuits and as radio-frequency chokes. Large inductors are employed at audio frequencies; the largest inductors are used as filter chokes in power supplies.

Coil-shaped inductors are used in tuned circuits from audio frequencies to the ultrahigh radio-frequency region. In the ultrahigh-frequency and microwave bands, short lengths of transmission lines can serve as inductors. Any length of line shorter than ¼ electrical wavelength short-circuited at the far end acts as an inductor. The same is true of a section of line between ¼ and ½ wavelength, with an open circuit at the far end.

A perfect inductor shows only inductive reactance, and no resistance. Such a component exists only in theory; all real inductors have some ohmic loss as well as reactance. *See also* INDUCTANCE, INDUCTIVE REACTANCE.

A variable inductor is an inductor whose inductance can be adjusted over a continuous range. The inductance of a coil can be varied in several ways. Variable inductors are in three categories: adjustable-turns, adjustable-core, or adjustable-phase.

The number of turns in a coil can be varied with a roller device. Roller inductors are used in transmatch circuits (*see* INDUCTOR TUNING). The permeability of the core can be controlled by moving the slug in and out of the coil (*see* PERMEABILITY TUNING). Two inductors can be connected in series and the relative phase controlled by mechanical means, resulting in variable net inductance.

INDUCTOR TUNING

Inductor tuning is a means of adjusting the resonant frequency of a tuned circuit by varying the inductance while leaving the capacitance constant. When this is done, the resonant frequency varies according to the inverse square root of the inductance. This means that if the inductance is doubled, the resonant frequency drops to 0.707 of its previous value. If the inductance is cut to one-quarter, the resonant frequency is doubled. Mathematically, the resonant frequency of a tuned inductance-capacitance circuit is given by the formula:

$$f = \frac{1}{2\pi\sqrt{LC}}$$

where f is the frequency in hertz, L is the inductance in henrys, and C is the capacitance in farads. Alternatively, the units can be given as megahertz, microhenrys, and microfarads.

Inductor tuning is more difficult to obtain than capacitor tuning. However, inductor tuning usually offers a more linear frequency readout than capacitor tuning. If the capacitance required for resonance is larger than the values normally provided for variable capacitors, inductor tuning is preferable. For inductor tuning, a rotary inductor or coil may be used. A movable core, consisting of powdered iron or ferrite, also can be used. The latter

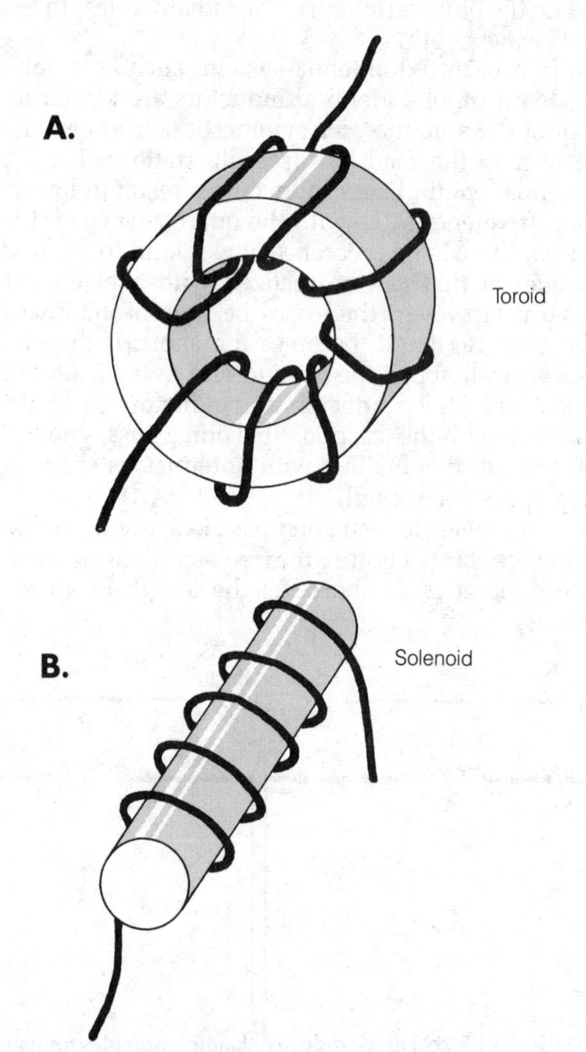

INDUCTOR: A toroid inductor (A) and a solenoid inductor (B).

form of inductor tuning is also called permeability tuning. *See also* CAPACITOR TUNING.

INFORMATION

Information is a term used to describe many kinds of data. In general, any message, whether analog or digital, contains information. The amount of information in an analog message is difficult to measure, but can be roughly estimated in terms of the size of an equivalent digital message. The size of a digital message is specified in bits, bytes, or words.

In general, for the transmission of a given amount of information, a minimum amount of bandwidth is required for a certain amount of time. The wider the bandwidth allotted, the more rapidly a given quantity of information can be conveyed. The longer the time allowed, the narrower the bandwidth can be. *See also* BIT, BYTE, DATA, WORD.

INFORMATION THEORY

Information theory is a form of statistical communication theory. The development of information theory is credited to C. E. Shannon who believed that a signal could be processed at the transmitting end of a circuit, as well as at the receiving end, for improved communications reliability under adverse conditions.

Information theory allows an engineer to calculate the highest possible reliability that can be expected for a communications circuit under specified conditions. For any given noise level, it is possible to transmit information with any desired degree of accuracy. The rate of data transfer is the primary variable. The slower the rate at which data is sent, the greater the accuracy.

Information theory has been refined to help engineers to determine the reliability of a particular system and to design the optimum system for a specified set of conditions. Information theory makes extensive use of probability and calculus.

INFRADYNE RECEIVER

An infradyne receiver is a form of superheterodyne receiver. Most superheterodyne receivers produce an intermediate-frequency signal that has a frequency equal to the difference between the input and local-oscillator frequencies. The infradyne receiver, however, produces an intermediate frequency that is equal to the sum of the input and local-oscillator frequencies. The term *infradyne* is derived from the fact that both the input and local oscillator signals are lower than the intermediate frequency.

If f is the input-signal frequency to an infradyne receiver and g is the frequency of the local oscillator, heterodynes are produced at frequencies corresponding to $f - g$ (if f is larger than g) or $g - f$ (if f is less than g); a signal is also produced at $f + g$. In the infradyne, $f + g$ is the intermediate frequency.

In the infradyne receiver, the signal always appears right side up. That is, the frequency components are not inverted. In a superheterodyne receiver using the difference frequencies, the signal sometimes appears upside down. The main disadvantage of the infradyne is that harmonics of the local-oscillator signal may fall within the input-tuning range of the receiver. *See also* HETERODYNE, SUPERHETERODYNE RADIO RECEIVER.

INFRARED

Infrared refers to a band of electromagnetic radiation with wavelengths that lie within the range of 770 nanometers (nm) or 0.77 micrometers (μm) to 1 millimeter (nm). The wavelength of infrared is longer than the wavelength of visible light. The shortest infrared wavelengths occur in the range just outside the visible spectrum. The longest infrared wavelengths border on the microwave part of the electromagnetic spectrum.

Infrared radiation is classified in four ways. The near infrared extends from a wavelength of 770 nm to 1,500 nm. The middle infrared extends from approximately 1,500 nm to 6,000 nm. The far infrared extends from 6,000 nm to 40,000 nm, and the far-far infrared encompasses wavelengths as great as 1 mm.

Infrared is considered to be heat radiation, although in a technical sense this is not correct. Infrared radiation feels warm as it strikes the skin because the radiant energy is transformed into heat as it is absorbed.

Infrared radiation is produced in large quantities by incandescent lamps. All stars produce some infrared radiation, especially the red stars. Infrared wavelengths can be used for communications purposes, although the atmosphere is opaque to the energy at certain wavelengths. Water in the air causes severe attenuation between about 4,500 and 8,000 nm; carbon dioxide interferes with infrared transmission between about 14,000 and 16,000 nm. There are various other infrared-opaque bands.

Special forms of light-emitting diode (IRED) produce emissions in the infrared region. Infrared photographs are used for evaluating the heat loss from a house or building. *See also* ELECTROMAGNETIC SPECTRUM.

INFRARED EMITTING DIODE

The infrared emitting diode (IRED) is a diode capable of converting electrical energy directly into infrared energy. In most applications it is matched to a silicon photodetector for signal transmission or object detection. Similar to the visible light-emitting diode, the IRED is fabricated as a P-N junction from gallium arsenide (GaAs) or aluminum gallium arsenide (AlGaAs).

An N-type epitaxial layer is grown on the GaAs or AlGaAs substrate, and a P-type diffusion forms the upper layer. IREDs also are made with alternate layers of GaAs and AlGaAs. The N-type substrate is metallized, and wire contacts are bonded to the upper P-type layer. *See* LIGHT-EMITTING DIODE.

When forward biased (positive lead connected to P-type, and negative lead connected to N-type material), charge-carrier recombination occurs. Electrons cross from the N side to recombine with holes on the P side of the junction. Because electrons are at a higher energy level than the holes, phonons (heat energy) and photons (infrared energy) are emitted. The output wavelength depends on the bandgap of the material. The peak GaAs energy is at about 900 nanometers (nm) and the peak AlGaAs energy is at about 830 nm.

GaAs and AlGaAs are transparent to infrared emission, so energy can be emitted from the upper P-type layer. With special epitaxial growth techniques, IREDs can be made to emit from both end surfaces. Instantaneous light-power output (in milliwatts) depends on drive current; typical forward voltage is about 1.2 V.

An IRED is usually matched to a silicon photodetector—photodiode, phototransistor, or photo Darlingtons. The peak of the silicon response curve is at approximately 850 nm, so silicon is a satisfactory material for IRED emission detectors. *See* PHOTODETECTOR.

IREDs are molded in the familiar bullet-shaped, radial-leaded LED T-1 and T-1 ¾ epoxy packages. The lens provides emission angles from 30 to 110 degrees depending on die position. They are also hermetically sealed in metal TO-5 cases with glass end lenses for use in extreme environments. Special packages adapt IREDs to fiberoptic systems; these can be TO-18 and TO-52 metal cases. Some packages include short lengths of optical fiber for better matching with fiberoptic cable. IRED dies are assembled into many optoelectronic devices including opto-isolators, optical interrupters, optical encoders, optical reflector modules and television and video cassette recorder (VCR) remote controls. *See* FIBEROPTIC COMMUNICATION, FIBEROPTIC CABLE AND CONNECTORS, SEMICONDUCTOR PACKAGE.

INFRASOUND

Infrasound is an acoustic disturbance that occurs at frequencies below the human hearing range. The lower limit of the audio-frequency range is about 16 to 20 Hz. The infrasonic range extends downward in frequency from this point.

Infrasound is characterized by long acoustic wavelengths. For this reason, an infrasonic disturbance can be propagated around many obstructions that would interfere with the transfer of ordinary sound. Infrasound can, if sufficiently intense, cause damage to physical objects because of resonance effects.

In air, sound waves travel at about 1100 feet per second or 335 meters per second. The same is true of infrasound. The wavelengths of infrasonic disturbances are generally greater than 55 feet or 17 meters. Some disturbances might contain components at wavelengths of over 100 meters.

INHIBIT

In a digital system such as a computer, an inhibit command is a signal or pulse that prevents or delays a certain operation. If the inhibit line is in the low or zero-voltage state, certain operations are forbidden, but they will occur if the inhibit line is high. In negative logic, the situation is reversed; a low inhibit line allows the operation to proceed, while the high state prevents it.

Inhibit gates are used in memory circuits to prevent a change of state. This keeps the contents of the memory intact and inalterable. Until the inhibit command is removed, the memory cannot be changed. *See also* SEMICONDUCTOR MEMORY.

INITIAL SURGE

When power is first applied to a device, circuit, or system that draws a large amount of current, an initial surge may occur in the circuit current. The surge is often considerably higher than the normal operating current load of the circuit. This effect is demonstrated by the momentary dimming of the lights in a house when the refrigerator, air conditioner, or electric heater first starts. The current surge is normally not dangerous, but it can occasionally result in a blown fuse or tripped circuit breaker. The same effect occurs when power is applied to radio equipment, although it is not as dramatic. Fuses are employed in products that can be damaged by an initial surge.

In the event of an electric power failure, an initial voltage surge can occur when electricity is restored. This surge can, in some instances, damage appliances.

INJECTION LASER

See LASER, LASER DIODE.

INPUT

Input is the signal applied to a circuit. An amplifier circuit, for example, receives its input in the form of a small signal; the output is a larger signal. A logic gate may have several signal inputs. The input terminals may be called inputs.

The input signal in a given situation must conform to certain requirements. The amplitude must be within certain limits, and the impedance of the signal source must match the impedance presented by the input terminals of the circuit. *See also* INPUT CAPACITANCE, INPUT IMPEDANCE, INPUT RESISTANCE.

INPUT CAPACITANCE

The input terminals of any amplifier, logic gate, or other circuit always contain a certain amount of parallel capacitance. The capacitance appears between the two terminals in a balanced circuit, and between the active terminal and ground in an unbalanced circuit.

The amount of input capacitance must, in most circuits, be smaller than a threshold value for proper circuit operation. In general, the higher the frequency, the smaller the maximum tolerable input capacitance. Also, the higher the input impedance of a circuit (*see* INPUT IMPEDANCE), the smaller the maximum tolerable input capacitance. Generally, the input capacitance should not exceed 10 percent of the resistive input impedance. When the input capacitance becomes too high, the input impedance changes so much that some of the incident signal is reflected back toward the source.

In some circuits, the input capacitance can be very

large. This is the case, for example, in the output filtering network of a power supply when a capacitor is used at the input. This condition is normal. *See also* INPUT RESISTANCE.

INPUT IMPEDANCE

The input impedance of a circuit is the complex impedance that appears at the input terminals. Input impedance normally contains resistance, but no reactance, under ideal operating conditions. However, in certain situations, large amounts of reactance may be present. This occurs either as an inductance or a capacitance.

Input impedances are often matched to the output impedance of the driving device. For example, an audio amplifier with a 600 ohm (Ω) resistive input impedance should be used with a microphone with an impedance of close to 600 Ω. A half-wave dipole antenna with a pure resistive input impedance of 73 Ω should be fed with a transmission line with a characteristic impedance of nearly this value; the transmitter output circuits should also have an impedance of nearly this value.

If reactance is present in the input impedance of a circuit, some of the electromagnetic energy is reflected back toward the source. This can happen in some circuits with little or no adverse effect. In other circuits, this condition must be avoided. *See also* IMPEDANCE, IMPEDANCE MATCHING, OUTPUT IMPEDANCE, REACTANCE.

INPUT/OUTPUT MODULE

An input/output (I/O) module is a standardized, factory-made solid-state device for interfacing computers with external sensors and actuators. I/O modules include both miniature alternating-current and direct-current (ac and dc) output solid state relays (SSRs) and input signal-conditioning circuits. The output SSR modules accept digital computer input signals to control the operation of motors, solenoids and other actuators and the input signal-conditioning modules accept signals from sensors or operators for conversion to computer-compatible digital input signals. Input signals confirm the completion of an output command and permit process monitoring and data recording.

Commercial industry-standard I/O modules are packaged in uniformly sized, rectangular cases that are color coded to identify them by function as ac or dc output or input modules. The cases measure 1.7 inches long, 1.3 inches high, and 0.6 inch thick. The have pins arranged to plug interchangeably into a master printed circuit board in various combinations to meet specific process control requirements. A circuit board with 16 I/O modules is shown in the illustration.

Output Modules. Both ac and dc output modules accept logic-level control signals to switch loads up to 3-½ A. They are available for 5, 15, or 24 V logic. Alternating-current output modules are solid-state relays with optoisolated input and output, zero-voltage switching, and

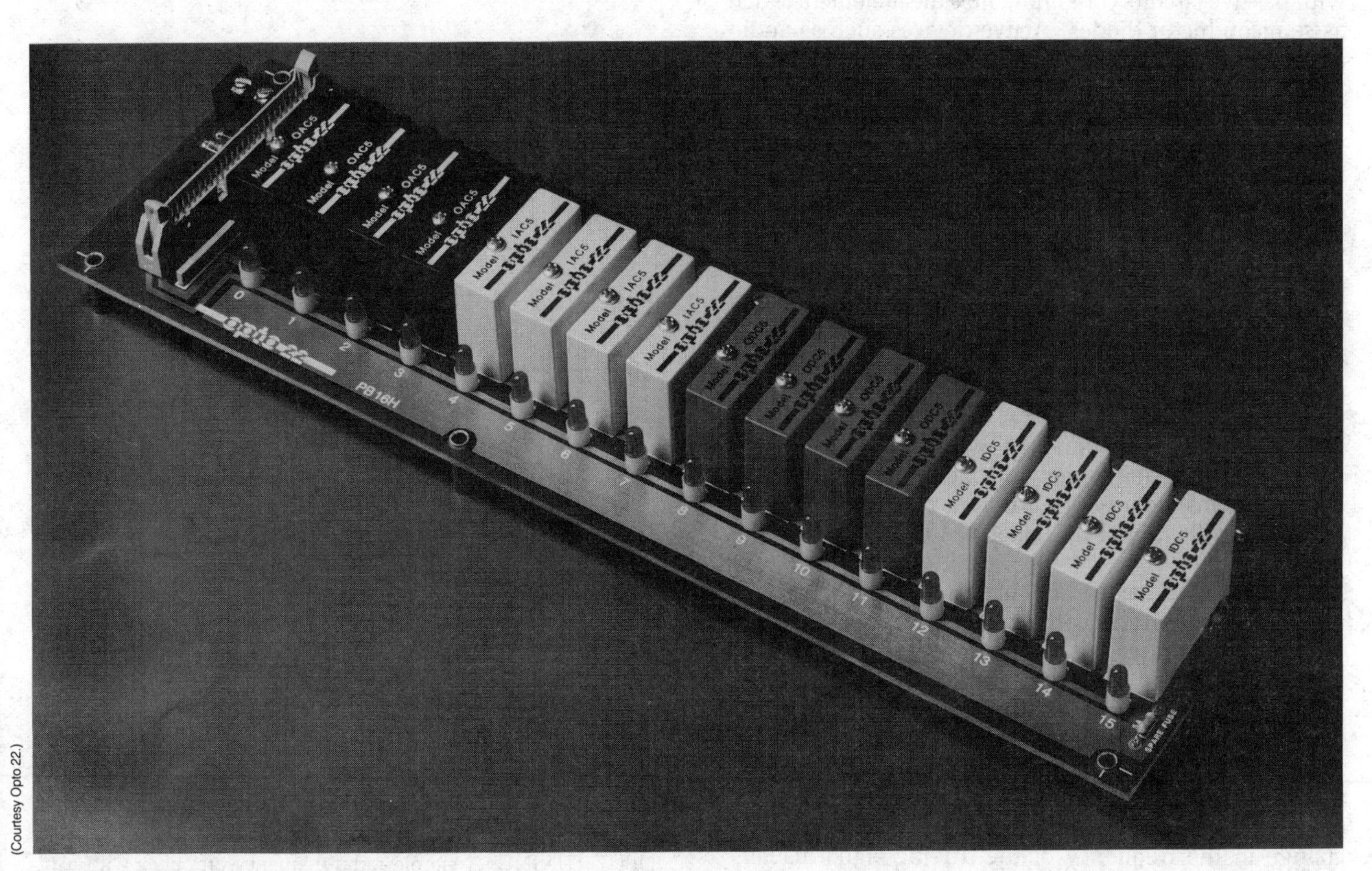

(Courtesy Opto 22.)

INPUT-OUTPUT MODULE: Alternating- and direct-current input and output modules in uniform color-coded cases are plugged into a mounting rack.

triac power switches. They are rated for 140 and 280 V ac. Direct-current output modules are dc solid state relays with optocoupler and transistor power switches. Load voltage ratings are up to 200 V dc. *See* SOLID STATE RELAY.

Input Modules. Input modules accept control signals from sensors, actuators, and transducers and convert them to an optoisolated logic level. Both ac and dc input modules have dual transistor output stages. Dc input modules are rated for 32 V dc.

Standard ac input modules are yellow, dc input modules are white and both are in 5-pin packages. Alternating-current output modules are black, dc output modules are red and both are in 4-pin packages. They can be mounted in standard circuit boards with barrier strips, fuses, and LED status lamps. There are 4-, 8-, 16-, or 24-module sizes. The modules are fastened to the board with captive screws.

INPUT RESISTANCE

In a direct-current (dc) circuit, a certain resistance is always present at the input terminals. This is called the input resistance, and is equivalent to the input voltage divided by the input current.

The input resistance of a dc circuit can remain constant, regardless of the applied voltage. This is generally the case with passive dc networks that contain only resistors. Often, however, the input resistance changes with the amount of applied voltage. This is generally true with passive circuits containing nonlinear elements such as semiconductor diodes. Active devices such as audio amplifiers and radio equipment usually have an input resistance that changes, not only with the level of applied voltage, but with the level of the input signal.

In an alternating-current (ac) circuit, the input resistance is the resistive component of the input impedance. *See also* INPUT IMPEDANCE.

INPUT TUNING

When the input circuit of an amplifier contains a tuned inductance-capacitance circuit, the amplifier is said to have input tuning. Not all amplifiers require input tuning. The illustration shows three radio-frequency amplifier circuits. The amplifier at A has a pi-network input-tuning circuit. The amplifier shown at B uses a resonant inductance capacitance network at its input. The two configurations are used for different purposes. At C, an amplifier is illustrated that does not make use of input tuning.

Input tuning has certain advantages and disadvantages. The primary advantage is realized in situations where the amplifier input impedance differs from the output impedance of the driving stage, or when additional selectivity is desired. The pi-network circuit (A) allows unequal impedances to be matched. The parallel-resonant network (B) provides high input selectivity. The main disadvantage of input tuning is that it increases the complexity of adjustment of the amplifier circuit. A change in the frequency of the driving signal usually necessitates retuning of the input network. *See also* IMPEDANCE MATCHING, PI NETWORK.

INSERTION GAIN

Insertion gain is an expression of signal amplitude with and without the use of some intermediate circuit. Insertion gain is expressed in decibels.

If the voltage amplitude at the output of a device is E_1 without the intermediary circuit and E_2 with the intermediary circuit, then the insertion gain G is given in decibels by:

$$G\ (\text{dB}) = 20 \log_{10}(E_2/E_1)$$

For current, the formula is similar; if I_1 represents the output current without the intermediary circuit and I_2 is the output current with the circuit in place, then the gain G is:

$$G\ (\text{dB}) = 20 \log_{10}(I_2/I_1)$$

For power, if P_1 is the power without the intermediary circuit and P_2 is the power with the circuit in place, then the insertion gain G is:

$$G\ (\text{dB}) = 10 \log_{10}(P_2/P_1)$$

It is evident from the above equations that if the insertion of the intermediary circuit causes a reduction in voltage, current, or power, then the insertion gain for that parameter is negative. In such cases, the intermediary circuit is

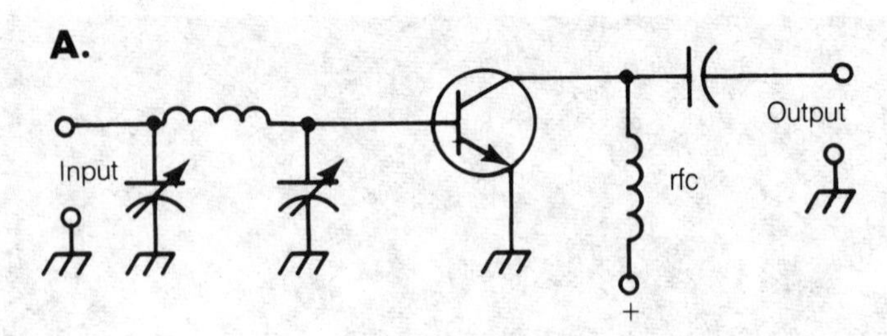

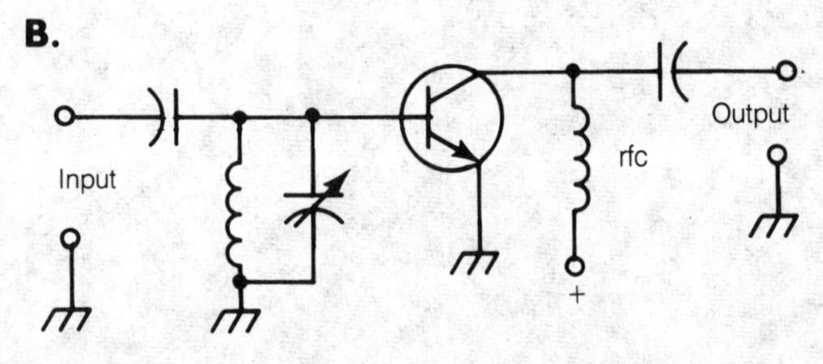

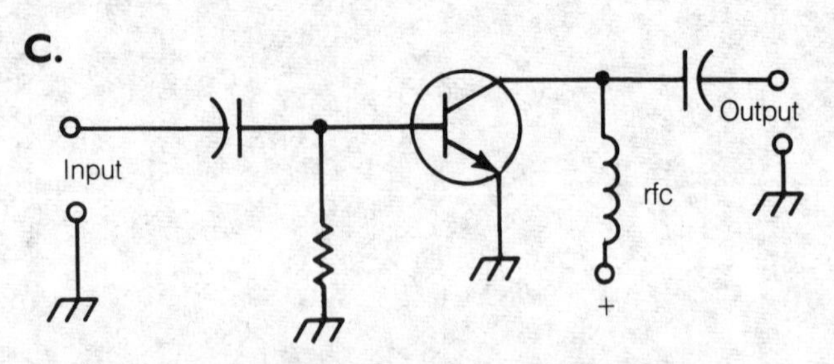

INPUT TUNING: A simple radio-frequency amplifier with a pi network input-tuning circuit (A), a parallel-resonant network in the amplifier input (B), and no input tuning (C).

said to have insertion loss (*see* INSERTION LOSS). Passive circuits always have negative insertion gain. Active circuits can display positive or negative insertion gain, depending on whether voltage, current, or power is evaluated.

Amplifier circuits may show insertion gain for some parameters, but a loss for others. It is necessary to define insertion gain in terms of current, voltage, or power, if the expression is to have meaning. *See also* DECIBEL, GAIN.

INSERTION LOSS

Insertion loss is a reduction in the output voltage, current or power of a system, resulting from the addition of an intermediary network such as a filter or attenuator. Insertion loss is specified in decibels.

For a selective filter, the insertion loss is normally specified for the frequency, or band of frequencies, at which the attenuation is the least. Most selective filters have low insertion loss at the frequencies to be passed, and high loss at other frequencies.

If the output voltage of a circuit is E_1 before the insertion of an intermediary network and E_2 after the insertion of the network, the insertion loss L is given by:

$$L\ (\text{dB}) = 20 \log_{10}(E_1/E_2)$$

For current, if I_1 is the output current before insertion of the device and I_2 is the current with the device installed, then the loss in decibels is given by:

$$L\ (\text{dB}) = 20 \log_{10}(I_1/I_2)$$

For power, if P_1 is the output wattage before the insertion of the device and P_2 is the wattage afterward, then:

$$L\ (\text{dB}) = 10 \log_{10}(P_1/P_2)$$

Passive networks always have a certain amount of insertion loss, but an efficient device normally exhibits less than 1 decibel of loss at the frequencies to be passed. Active networks can have negative insertion loss. In general, if the insertion loss in decibels is $-x$, then the insertion gain in decibels is x. That is, negative loss is the same as positive gain, and vice versa. *See also* DECIBEL, INSERTION GAIN, LOSS.

INSTANTANEOUS EFFECT

Whenever a parameter changes, rather than maintaining a constant magnitude or value, instantaneous effects become important. Amplitude and frequency are the most common parameters in electronic circuits that change value. However, resistance and impedance also change rapidly under some conditions.

The instantaneous magnitude of a parameter is the value at a given instant. This is a function of time. The sine-wave alternating-current waveform provides a good example of the difference between an instantaneous value and the average and effective values. The illustration shows the voltage-versus-time functions for the output of a typical household electrical outlet. The instantaneous voltage can attain any value between −165 and +165 V. The precise voltage changes with time, completing one cycle every 16.67 milliseconds. The average voltage is actually zero, but this is not the effective voltage. The root-mean-square (rms) voltage, which is usually considered the effective voltage, is approximately 117 V.

Instantaneous effects are generally of little concern, but in some cases they can be very important. A sudden surge in the voltage from a household utility line might last for only a few millionths of a second, and thus not affect the average or rms voltages significantly. However, the instantaneous voltage can reach several hundred volts during that brief time, and this can damage some electronic equipment. Instantaneous effects can be measured only with devices capable of displaying the details of a waveform. The oscilloscope is the most common instrument for this. *See also* AVERAGE VALUE, EFFECTIVE VALUE, OSCILLOSCOPE, PEAK VALUE, ROOT MEAN SQUARE.

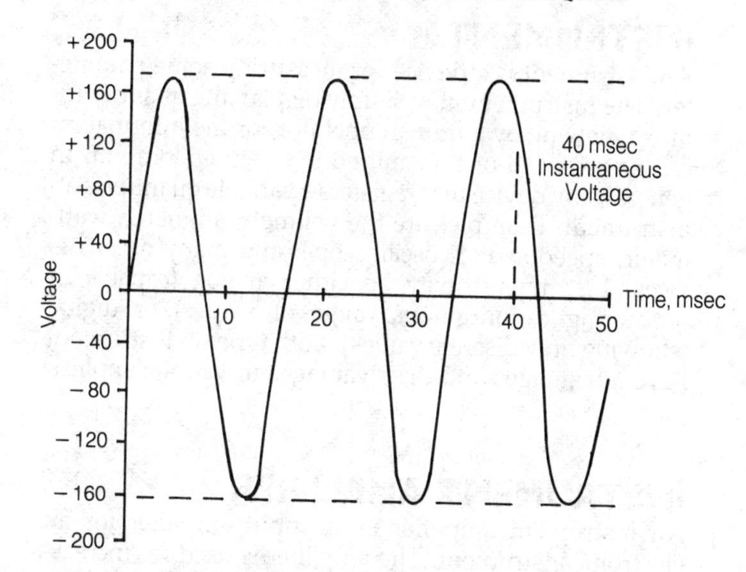

INSTANTANEOUS EFFECT: An oscilloscope display of the voltage at a utility outlet. All points on the curve show instantaneous voltage at a specified time.

INSTITUTE OF ELECTRICAL AND ELECTRONIC ENGINEERS

The Institute of Electrical and Electronic Engineers (IEEE) is a professional organization in the United States and some other countries. The purpose of the association is to advance the state of the art in electricity and electronics. Originally, the IEEE consisted of two separate groups called the American Institute of Electrical Engineers (AIEE) and the Institute of Radio Engineers (IRE).

The IEEE sets standards and specifications for design and performance of many kinds of electrical and electronic equipment. *See also* ELECTRONIC INDUSTRIES ASSOCIATION.

INSTRUCTION

An instruction is a set of data bits that directs a computer to follow certain procedures. A computer program, for example, is a string of instructions. The execution of the instructions results in the computer performing the complete operations desired. A program may contain in excess of 1,000,000 individual instructions.

A computer is capable of performing only a finite number of different kinds of instructions; the complete set of instructions that the machine can execute is called the instruction repertoire. The total number of individual instructions that a machine can handle, in a specific program, depends on the memory capacity.

An instruction normally consists of a code signal and a memory address. The code signal denotes the particular instruction, such as multiply or divide. The memory address tells the machine where to get and store the information. Instructions may be fed into a computer by means of a disk drive, a magnetic tape, a modem, or by a human operator with a keyboard and video display unit. *See also* COMPUTER, DATA.

INSTRUMENT

An instrument is a device for measuring some parameter. The instrument may simply display the quantity for direct viewing by a human operator, or the information can be recorded or transmitted to a remote location. In general, any device that registers a variable quantity is an instrument. Examples are the voltmetr, ammeter, wattmeter, speedometer, oscilloscope, and many other devices. Instruments may be either analog (capable of registering a continuous range of values) or digital (showing only discrete values). Both types of instrument have advantages and disadvantages in various applications. *See also* ANALOG PANEL METER, DIGITAL PANEL METER.

INSTRUMENT AMPLIFIER

An instrument amplifier is an input amplifier for an electronic instrument. The amplifier is used for increasing the sensitivity of the instrument. These amplifiers can be used with direct-current or alternating-current instruments. Some devices that use amplifiers include the active field-strength meter, the oscilloscope, and the FET (field-effect transistor) voltmeter.

There is a limit to the sensitivity that can be obtained with any instrument, no matter how high the gain of the amplifier. All amplifiers generate some noise, and a small amount of background noise occurs as a result of currents and thermal effects. Quantities of magnitude smaller than the level of the noise, therefore, cannot be reliably measured even when instrument amplifiers are used.

INSTRUMENT ERROR

All indicating instruments have some degree of inaccuracy; none display the actual value of a parameter. The difference between the actual value and the instrument reading is called the instrument error.

In an analog instrument, the amount of error depends on the electrical and mechanical calibration of the device, and also on the ease with which the operator can interpolate among the divisions on the scale. The error is usually specified, for a given instrument, as a maximum percentage at full scale (*see* FULL-SCALE ERROR).

In a digital instrument, operator error is not a factor because the parameter is displayed as a numeral. The limitations on the accuracy of a digital device are imposed by the electrical precision of the indicator circuit, and by the resolution of the display. A digital meter having a resolution of three decimal places for a given range (such as 0.00 V to 9.99 V) cannot provide as much accuracy as a meter with a resolution of four decimal places (such as 0.000 V to 9.999 V). The instrument error becomes just a tenth as great when one digit is added to the display, assuming the electrical precision is sufficient. *See also* ANALOG PANEL METER, DIGITAL PANEL METER.

INSTRUMENT TRANSFORMER

In an instrument intended for measuring alternating currents, an instrument transformer can be used to provide a variety of metering ranges. Transformers can be used for alternating-current ammeters of voltmeters. These transformers usually have several tap positions connected to a rotary switch for choosing the desired range (see illustration).

In a voltmeter, the range is inversely proportional to the number of turns in the secondary winding of the transformer. For example, if the secondary has n turns and the maximum voltage on the scale is x, then switching to $10n$ turns will change the full-scale reading to $0.1x$. In an ammeter, the situation is reversed; the full-scale range is directly proportional to the number of turns. Thus, if n turns provide a reading of x amperes at full scale, $10n$ turns will result in a full-scale reading of $10x$. These relationships are based on the assumptions that the input of the meter is provided at the primary winding of the transformer, and that the number of turns in the primary winding remains the same under all conditions. *See also* TRANSFORMER.

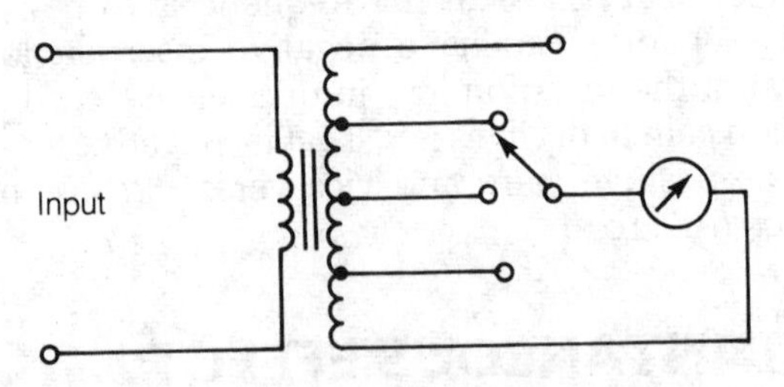

INSTRUMENT TRANSFORMER: A switchable instrument transformer allows selection of various full-scale ranges.

INSULATED CONDUCTOR

See CABLE.

INSULATING MATERIAL

Electrical insulating materials are used in situations where conduction must be prevented. Insulating materials are classified according to their ability to withstand heat. In order of ascending temperature rating, these categories are designed as follows: O (90°C), A (105°C), B (130°C), F (155°C), H (180°C), C (220°C), and over Class C (more than 220 °C).

Class O materials include paper and cloth. Class A insulating substances include impregnated paper and cloth. Class B, Class F, Class H, and Class C materials are substances such as glass fiber, asbestos, and mica. All insulators over Class C consist of inorganic materials such as specially treated glass, porcelain, and quartz. Teflon is able to withstand temperatures of more than 250° C, and it is therefore rated over Class C. Enamel is usually a Class O material. Vinyl, often used for insulating electrical wiring, is generally a Class A substance, although silicone enamel can rate as high as Class H.

The class of insulating material needed for a given application depends on the amount of current expected to flow in the conductors, and on the conductor size. If the class of insulation is too low, fire may occur. If the class if unnecessarily high, the expense is unwarranted.

Insulating materials are also rated according to dielectric constant, dielectric strength, tensile (breaking) strength, and leakage resistance. *See also* DIELECTRIC, DIELECTRIC CONSTANT, DIELECTRIC RATING, INSULATION RESISTANCE.

INSULATION RESISTANCE

The resistance of an insulating material is a measure of the ability of the substance to prevent the undesired flow of electric currents. In general, the higher the insulation resistance, the more effective the material for purposes of preventing current flow.

Insulation resistance is generally measured for direct currents. This value, given in ohm-centimeters (ohm-cm), may change for radio-frequency alternating currents. Among the substances with the highest known insulation resistance are polystyrene, at 10^{18} ohm-cm, fused quartz, at 10^{19} ohm-cm, and teflon and polyethylene, which have direct-current resistivity of 10^{17} ohm-cm. Vinyl substances have insulation resistances ranging from 10^{14} to 10^{16} ohm-cm. Rubbers rate from about 10^{12} to 10^{15} ohm-cm. *See also* DIELECTRIC, DIELECTRIC RATING.

INSULATOR

An insulating material is sometimes called an insulator (*see* INSULATING MATERIAL). However, the term *insulator* generally refers to a component specifically designed to prevent current from flowing between or among nearby electrical conductors. An insulator is a rigid object, usually made from porcelain or glass. Insulators are commonly used in radio-frequency antenna systems and in utility lines. Insulators are available in a wide range of sizes and shapes.

INTEGRATED CIRCUIT

An integrated circuit (IC) is a monolithic microcircuit consisting of interconnected active and passive electronic elements interconnected on or within a single semiconductor substrate. An IC is intended to perform an electronic circuit function. Figure 1 illustrates a bipolar integrated circuit showing how a transistor, resistor, and

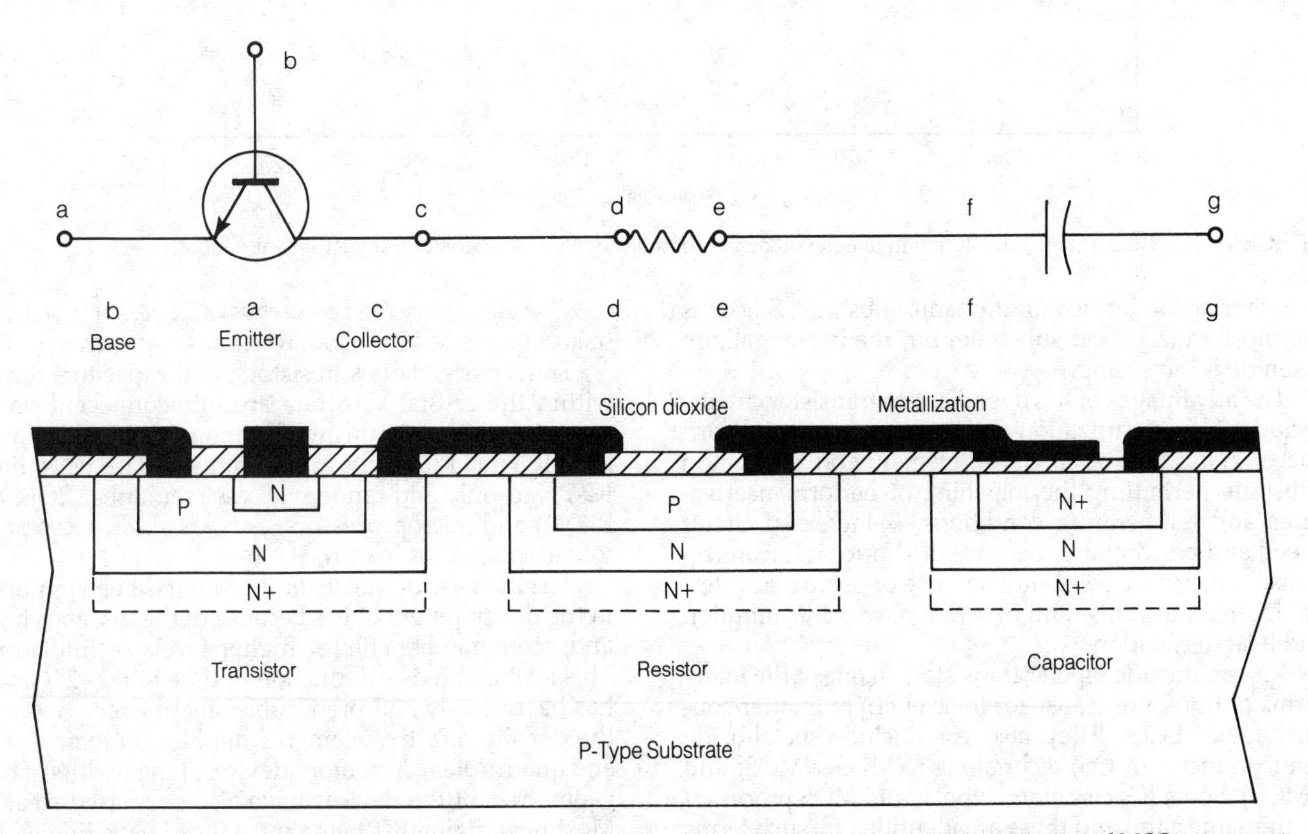

INTEGRATED CIRCUIT: Fig. 1. Section view of a bipolar integrated circuit with an NPN transistor, P-type resistor, and MOS-type capacitor.

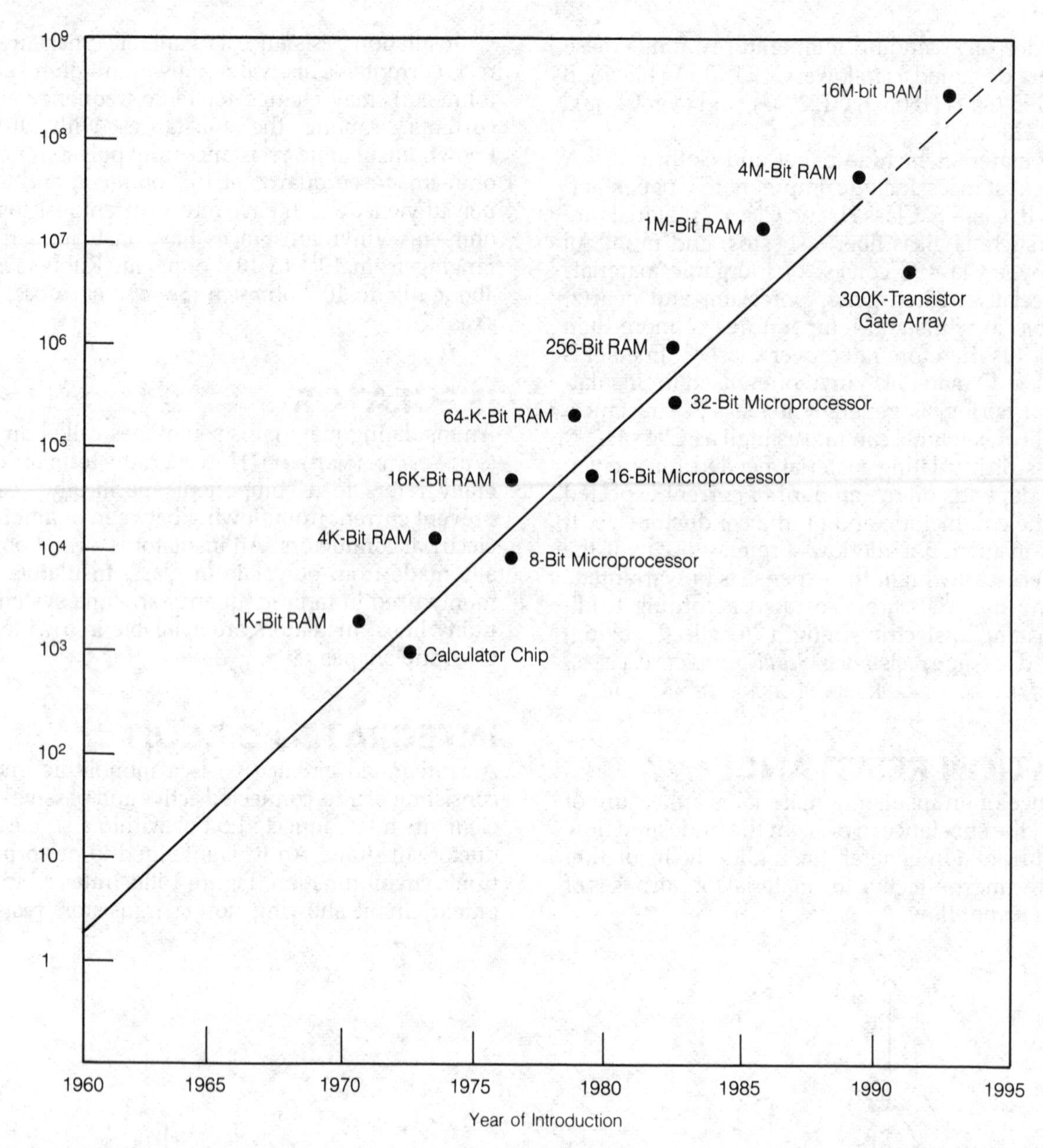

INTEGRATED CIRCUIT: Fig. 2. Evolution in integration density of milestone memory, logic, and microprocessor ICs.

capacitor are all formed on the same substrate. Silicon is the most widely used substrate, but the use of gallium arsenide is increasing.

The advantages of ICs over discrete transistor circuits include: 1) miniaturization with savings in circuit board space, 2) location of all circuit elements on a common substrate permitting the matching of performance features and temperature conditions, 3) increased circuit speed and performance because of shorter intraconnections, 4) increased reliability because of decreased external interconnections, and 5) lower power consumption and heat dissipation.

ICs can include bipolar transistors in one of the many forms of transistor-transistor logic (TTL) or emitter-coupled logic (ECL). They also can include metal-oxide semiconductor (MOS) technology (NMOS, PMOS, and CMOS). Some ICs combined bipolar and MOS processes on the same chip, and these are identified by such terms as BIFET, BIMOS, BIDFET and LinCMOS. *See* DIGITAL LOGIC INTEGRATED CIRCUIT, LINEAR INTEGRATED CIRCUIT, METAL OXIDE SEMICONDUCTOR LOGIC FAMILIES, POWER INTEGRATED CIRCUIT.

Transistors, diodes, resistors, and capacitors formed within the crystal substrate are intraconnected on the surface with aluminum metallization. ICs include digital-logic ICs including gate arrays and standard cells, linear ICs, read-only and random access memories (ROM and RAM), and microprocessors. *See* GATE ARRAY, MICROPROCESSOR, SEMICONDUCTOR MEMORY, STANDARD CELL.

The increase of transistor density has been important in the development of ICs because of the many technical and economic benefits of higher levels of integration. This is illustrated as a function of time in Fig. 2. Growth has been steady and predictable: about every two years the density of active elements doubles on logic devices and quadruples on memory devices. Line widths of basic geometries of the devices have also decreased steadily. Most new digital ICs being introduced have line widths of less than two microns. CMOS has become the preferred technology for the largest IC memories, microprocessors, and custom and semicustom digital ICs; CMOS is also gaining in linear IC technology.

The two-input NAND gate is the standard for measurement of IC integration density. An IC is said to be small-scale integration (SSI) if the complexity is equivalent to that of 2 to 10 of these NAND gates. Medium-scale integration (MSI) has the equivalent of 10 to 100 of these gates. Large-scale integration (LSI) has more than 100. Very large scale integrated (VLSI) ICs have 1000 or more equivalent gates.

INTEGRATED-CIRCUIT MANUFACTURE

Most IC design is done on a computer-aided design (CAD) workstation. The computer helps the designer maximize the number of elements in the critical chip area. (*See* COMPUTER-AIDED DESIGN AND COMPUTER-AIDED ENGINEERING). Important engineering decisions must be made in converting a discrete transistor circuit into an IC. For example, there must be adequate electrical isolation regardless of the density of elements and their geometrical size on the chip. Moreover, because there is no known way to integrate a low-frequency inductor on an IC, the circuit design must avoid the use of these components. Other constraints include the specification of low capacitance values and a narrow range of resistor values because of the limits imposed on these passive components by IC manufacturing process.

Capacitors and resistors are the most expensive devices in IC technology because they take up relatively large amounts of chip space compared with transistors. The design rules for IC circuits encourage the use of active elements—transistors and diodes—and discourage the use of resistive and capacitive elements.

The objective of layout is the preparation of a set of photomasks, each containing the pattern for a single layer. A photographically reduced image of the mask is next reproduced hundreds of times in a step-and-repeat process to form a set of master masks with a 1:1 size ratio. Many working plates are copied from the master masks. A working plate can be either a fixed image in an ordinary photographic emulsion or a pattern etched in a chromium film on a glass substrate.

Electron-beam technology is replacing optical methods for producing masks. Work is also being done on the use of X rays in mask production. With these techniques it is possible to write the pattern directly on the working mask from information stored in the computer workstation.

The silicon wafers that serve as the substrates for the ICs have diameters of 4 to 6 inches. They are obtained from crystals grown by the Czochralski process. (*See* CZOCHRALSKI CRYSTAL-GROWTH SYSTEM.) A flat is ground along the length of the long single crystal prior to slicing so that each wafer will have a reference edge parallel with a natural crystal plane. The wafers are obtained by sawing the crystal into slices about a half millimeter (0.020 inch) thick. These are polished, cleaned, and oxidized to prepare for the first patterning step.

In a typical integrated circuit, there might be 10 masking and etching operations, as well as dozens of other fabrication steps. A CMOS memory, for example, might have 14 masking levels and well over 100 processing steps. IC technology is an extension of the technology developed for fabricating small-signal and power transistors. *See* TRANSISTOR MANUFACTURE.

Basic Process Concepts More than a hundred production steps must be performed on a silicon wafer before the IC chip is finished and can become a functioning product. All processing includes a number of masking steps. The microelectronic circuits are built up layer by layer, each layer receiving a pattern from a mask designed specifically for it. The microlithography process involves etching patterns into an oxide as summarized in Fig. 1.

A wafer is oxidized as shown in Fig. 1A and is then coated with photoresist (Fig. 1B), a light-sensitive organic compound (*See* PHOTORESIST). Then a mask is placed over the photoresist-coated wafer and the wafer is exposed to ultraviolet (UV) light as shown in Fig. 1C and 1D. (Electron beams and soft X rays are alternatives to UV light for obtaining higher definition.) The photoresist is changed by exposure to the light so that exposed parts subsequently dissolve either faster or slower in a chemical solvent than unexposed parts. In Fig. 1E, the exposed photoresist dissolves faster and is washed away. The remaining resist is further hardened by baking (Fig. 1F).

Then the wafer with its hardened photoresist pattern is processed to remove the exposed oxide. This can be done by acid etching, but increasingly dry plasma etching is used. In dry etching, excited gas plasma removes the unprotected oxide layer defined by the mask, but it does not etch either the photoresist or the silicon wafer.

Silicon can be doped by diffusion, but ion implantation as shown in Fig. 1H is widely used in IC fabrication. This process permits selective doping at room temperature.

The uppermost layers of ICs are formed by depositing and patterning thin films. The two most important processes for the deposition of thin films are chemical vapor deposition (CVD) and evaporation. An important operation in silicon-gate MOS technology is the deposition of polycrystalline silicon. It is usually formed by chemical-vapor deposition of silane gas (SiH_4). The gas decomposes when it is heated, releasing silicon and hydrogen. If silicon wafers are heated in a silane atmosphere, a film of polycrystalline silicon forms on its surface. This silicon can also be doped, oxidized and patterned. The insulating films of silicon dioxide and silicon nitride may also be deposited by CVD.

Thin-film metallic conducting layers of aluminum can be deposited by evaporation techniques. The wafers are placed in a vacuum chamber and solid aluminum is heated by direct electron bombardment. This results in a pure aluminum film about a micro-meter thick, being vaporized uniformly on the wafers. The photoresist and

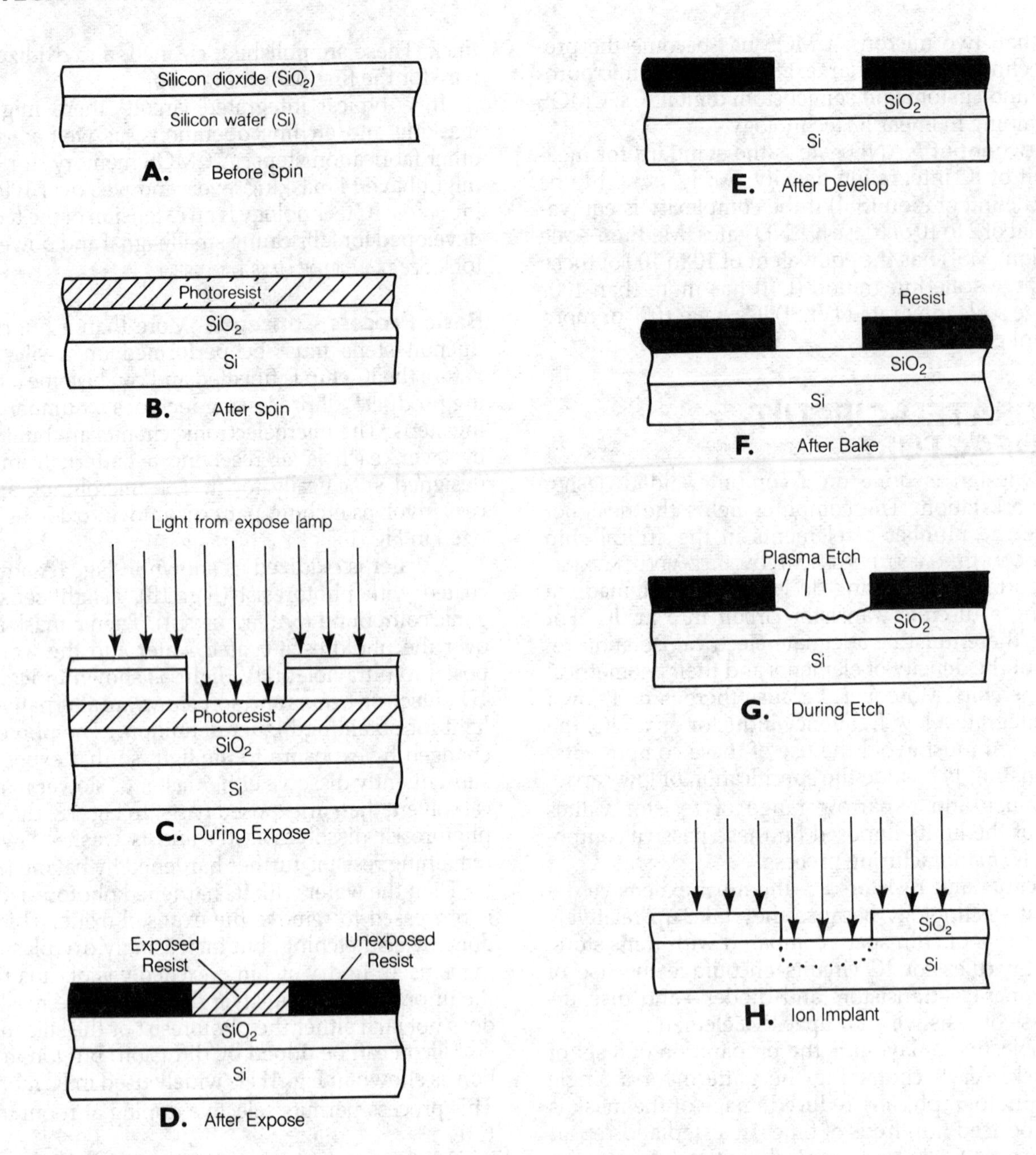

INTEGRATED-CIRCUIT MANUFACTURE: Fig. 1. Basic process steps in the manufacture of a silicon IC.

masking methods can be used to pattern the aluminum for selective etching of aluminum contacts.

There are usually more thin-film steps than diffusion steps in the fabrication of large-scale ICs. A layer of silicon nitride can be vapor-deposited, patterned and used as an oxidation mask. This process has been used in MOS technology for the formation of more precise definition of elements and improved isolation. It is also replacing conventional diffused isolation with oxide isolation in bipolar ICs permitting more transistor per chip area.

IC Component Formation

Integrated Transistors. Active circuit elements including bipolar and MOS transistors and diodes are formed in part within the silicon substrate. The most common integrated bipolar transistor is the npn configuration shown in Fig. 1 of INTEGRATED CIRCUIT. The techniques for manufacturing bipolar transistors for ICs are similar to those used for manufacturing planar transistors except that the collector contact is brought to the top surface with the emitter and base contacts.

All components are formed on a single wafer of conductive silicon so it is necessary to isolate each active component electrically from the others. The required intraconnections are then made on the substrate surface. Isolation is achieved in diffusing N-type regions into P-type silicon and by allowing adequate guard bands between the N-type regions. In the IC, each of these P-N junctions is biased in the reverse direction forming high resistance to isolate the N-type region and any component that is in it.

In the alternate epitaxial growth method, N-type silicon is grown on a P-type substrate as shown in Fig. 2. However, prior to the epitaxial growth, selected regions of low-resistance N-type material can be diffused into the P-type substrate to improve isolation. To complete the isolated regions, narrow P-type channels are diffused through the epitaxial layer to join up with the P-type substrate. This leaves separate islands of the N-type epitaxial layer for the formation of various components.

The isolated N layers become the collectors of the transistors formed in each IC as shown in Fig. 3. They are then connected from the top. This arrangement differs from that of the planar transistor where the substrate serves as the collector and is connected from the bottom. Boron is diffused in to form the P-type base regions, and finally phosphorous is diffused in to form the high concentration N-type (N+) emitter regions.

Integrated Diodes. IC diodes are prepared by forming P-N junctions at the same time the transistor regions are diffused. Both contacts of the diode are brought to the top surface, on contrast to discrete diodes in which one contact is made on each side of the wafer. The diodes can be formed either at the same time as the transistor collector-base junction or at the same time as the base-emitter junction.

Figure 4A shows a diode with collector-base diffusions. The P-type anode region of the diode is formed during the transistor-base diffusion. This is a general-purpose diode. If faster switching is required, emitter-base diodes as shown in Fig. 4B are used. The diode anode is formed during the base diffusion, and the cathode is formed during the emitter diffusion. To avoid unwanted transistor action with this diode, the anode contact short-circuits the P-type anode region to the N-type collector region. By using reverse voltage on the emitter-base junction a zener diode is formed. *See* DIODE.

Resistors. IC resistors are formed by diffusion in the IC chip simultaneously with the formation of the base P region of the npn transistors in the crosssectional view Fig. 5. The N-layer below the P-layer will be reverse biased for isolation from the substrate. Silicon dioxide provides insulation. External connections to the resistor are made by evaporating aluminum contacts. This type of monolithic resistor has a resistivity of about 200 ohms (Ω) for the entire chip. However, the ratio of length to width of the R strip can provide values of 100 to 25,000 Ω.

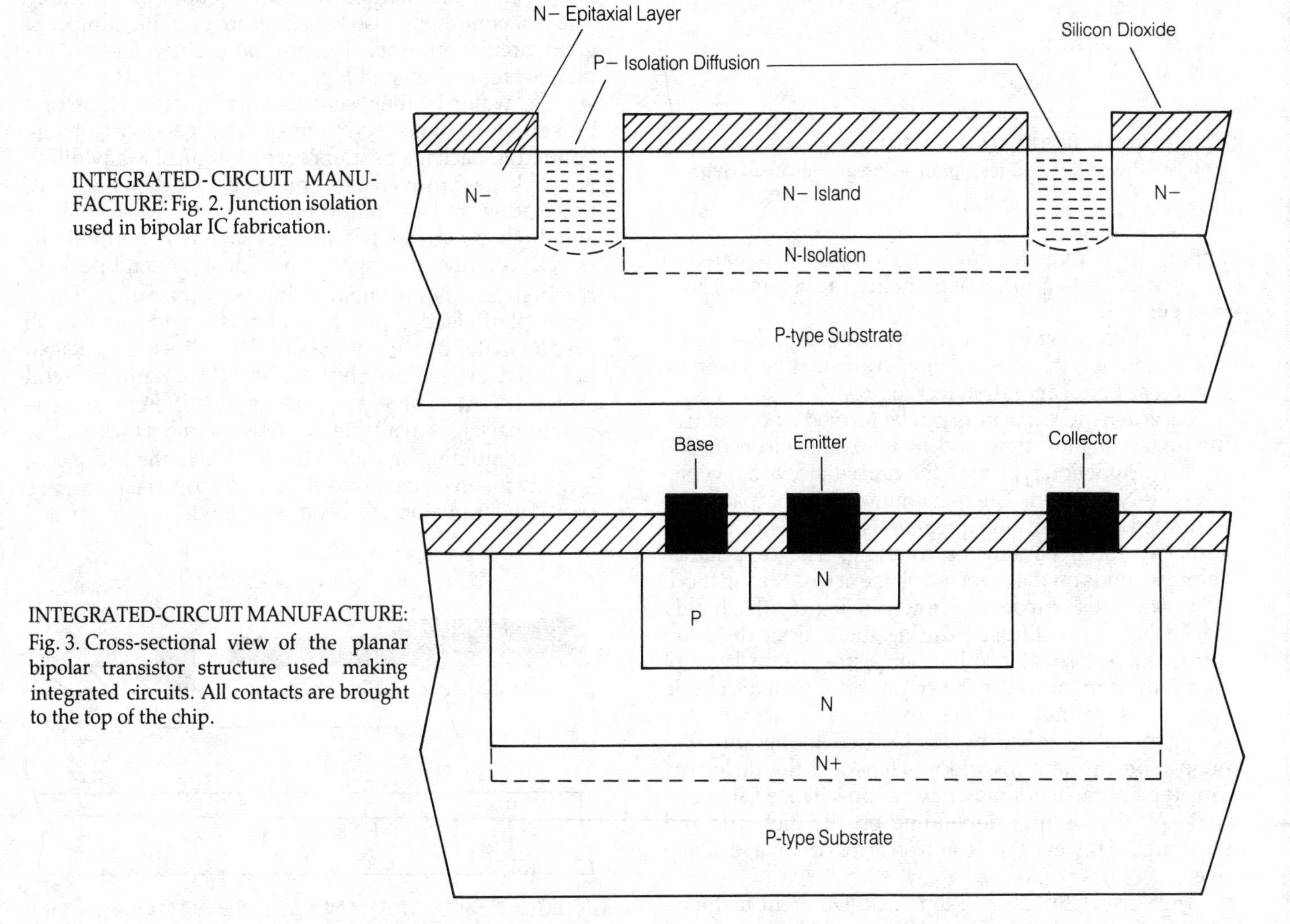

INTEGRATED-CIRCUIT MANUFACTURE: Fig. 2. Junction isolation used in bipolar IC fabrication.

INTEGRATED-CIRCUIT MANUFACTURE: Fig. 3. Cross-sectional view of the planar bipolar transistor structure used making integrated circuits. All contacts are brought to the top of the chip.

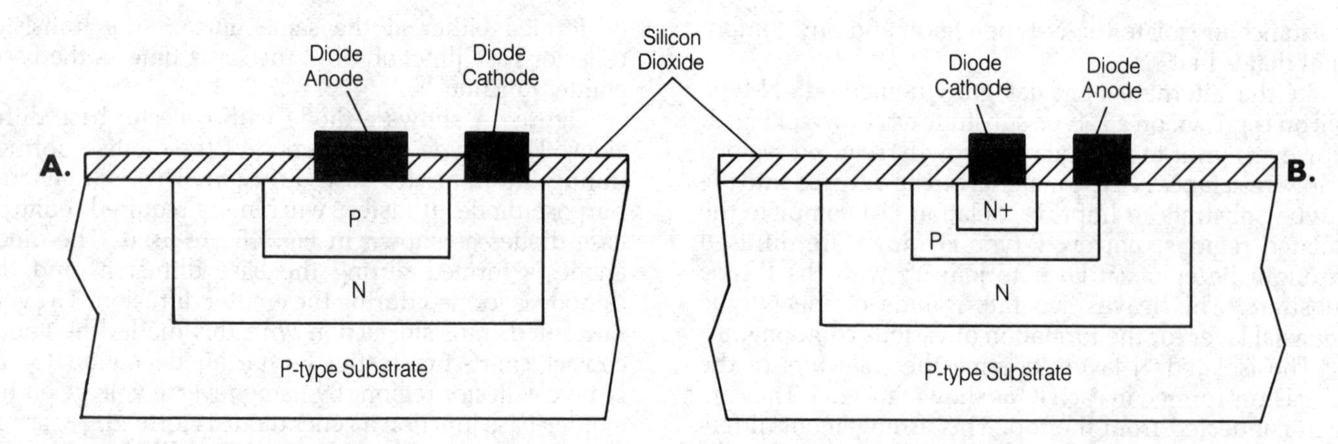

INTEGRATED-CIRCUIT MANUFACTURE: Fig. 4. Cross-sectional view of collector-base diode (A) and emitter-base diode (B) for integrated circuits.

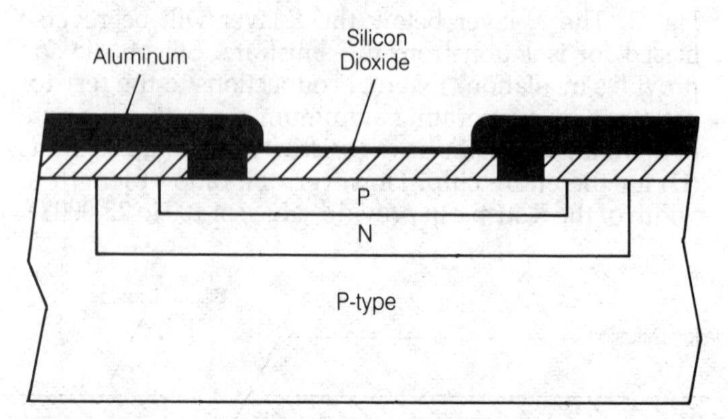

INTEGRATED-CIRCUIT MANUFACTURE: Fig. 5. Cross-sectional view of a P-type diffused resistor in an integrated-circuit chip.

Higher values of 10 to 50 Ω can be obtained by special doping techniques to reduce the crosssectional area of the R strip. The tolerance of monolithic resistors is ±30 to ±50 percent.

Thin-film resistors can also be deposited directly on the surface of the silicon dioxide insulating layer by photoresist or evaporation techniques.

Capacitors. IC capacitors can be formed by one of two methods: junction type and metal-oxide silicon (MOS) type. The junction type uses the capacitance of a reverse-biased P-N junction. The maximum value of capacitance is limited to about 100 picofarads (pF). This can be sufficient for bypassing or coupling. The capacitance value depends on the reverse voltage across the junction.

In the MOS capacitor shown in Fig. 6, the highly doped N+ layer diffused during the emitter diffusion forms the low-resistance bottom plate. A thin layer of silicon dioxide forms the dielectric and the top electrode is formed by aluminum metallization. Contact with the N+ material is made by an isolated aluminum film deposition through a window etched in the dielectric. This type of capacitor may have a capacitance value of 3 to 30 picofarads (pF) depending on the dielectric and plate area. However, it is not polarity or voltage sensitive.

MOS versus Bipolar IC Fabrication. Different methods are used in the fabrication of bipolar and MOS ICs based upon their different geometries and operating characteristics. CMOS technology uses simpler techniques that take advantage of less complex processing. P wells are diffused into an N substrate for N-channel transistors.

Wafer fabrication ends with the electrical probing of each chip to determine if it functions correctly. Automatic test equipment sequentially tests each circuit. Defective chips are marked with an ink spot for rejection. The test equipment also keeps records on the number of good circuits and their location on each wafer. A complete wafer is shown in Fig. 7.

The wafer is then scribed between the chips and broken along those scribe lines. The good circuits are bonded to package headers or lead frames. Many different packages are available, and the choice depends on IC application and operating environment. Most military and high reliability ICs are packaged in hermetically sealed ceramic packages, and most general-purpose commercial ICs are molded in epoxy packages. Dual-in-line (DIP) leaded packages of plastic and ceramic still predominate, but many LSI and VLSI ICs are packaged in leaded or leadless chip carriers (LCC) and pin-grid arrays (PGA). There are also many different surface-mount packages available. *See* SEMICONDUCTOR PACKAGE.

After molding or sealing the IC chip in the package, it is ready for an exhaustive series of electrical tests to make sure it will function reliably as specified.

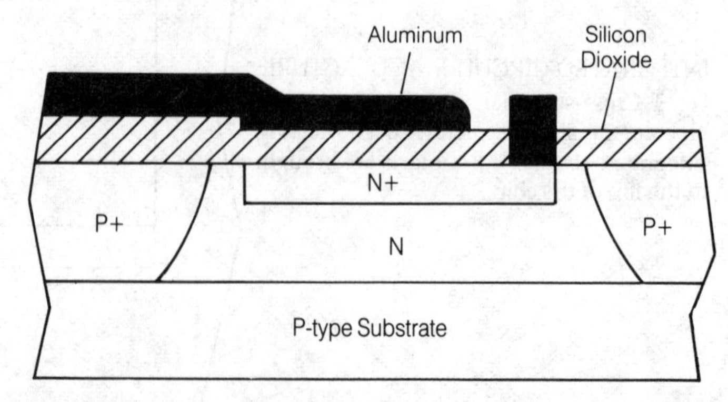

INTEGRATED-CIRCUIT MANUFACTURE: Fig. 6. Cross-sectional view of a monolithic capacitor in an integrated-circuit chip.

(Courtesy Motorola, Inc. Used by Permission.)

INTEGRATED-CIRCUIT MANUFACTURE: Fig. 7. Six-inch silicon wafer with completed integrated circuits compared in size with DIP- and PLCC-packaged LSI devices.

INTEGRATED INJECTION LOGIC

A bipolar form of logic, using both NPN and PNP transistors, is called integrated injection logic (I^2L). The illustration is a simple schematic diagram of an I^2L gate.

No resistors are needed; a single pair of transistors completes the circuit for one gate. The operating speed is high, approaching the speed of transistor-transistor logic. The current requirement of I^2L is quite small, resulting in low power consumption.

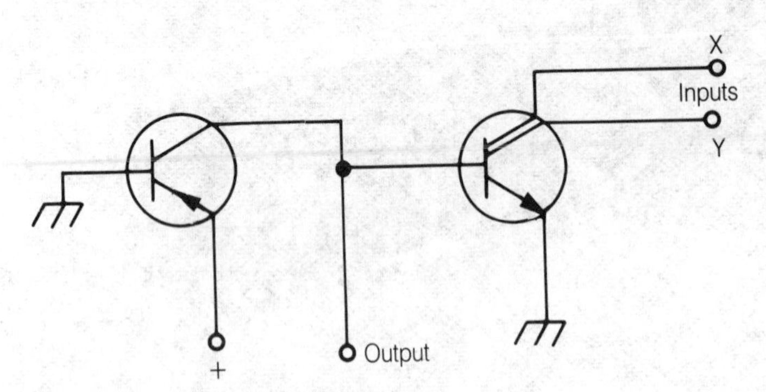

INTEGRATED INJECTION LOGIC: A schematic diagram of an integrated-injection logic (I^2L) gate.

INTEGRATED-SERVICES DIGITAL NETWORK

The integrated-services digital network (ISDN) is a planned, unified global telecommunications network. The eventual objective of ISDN is a network that will permit telephones, computer terminals, facsimile machines, video-display terminals, and alarm systems to be plugged into a single interface, for instant access to any form of information. For example, ISDN will permit a telephone and a computer terminal to be plugged into the same connector and carry on a dialog with another station halfway around the world. The ISDN will extend digital technology over the entire path between terminals, including the loop between the subscriber and the central office which is now analog. Eventually, it may be able to support video conferencing and computer-aided design.

The development of ISDN is being coordinated by the International Telegraph and Telephone Consultative Committee (CCITT), an arm of the International Telecommunication Union of the United Nations. It is expected to evolve from the digital telephone networks of various countries. CCITT is trying to standardize the interfaces and protocols before terminals are developed. ISDN will probably be most useful and cost effective for business, and eventually it will extend to residential customers.

INTEGRATION

Integration is a mathematical or electrical process by which the cumulative value of a function is constantly added. Integration is the opposite of differentation (*see* DIFFERENTIATION), in both the mathematical and the electrical sense. Graphically integration is represented by the area under a curve.

The graph shows the integration process beginning from a designated point. Time is depicted along the horizontal axis, and amplitude along the vertical axis. With each half cycle, the total area under the curve fluctuates, resulting in a 90-degree phase shift in this example. *See also* DEFINITE INTEGRAL, INDEFINITE INTEGRAL, INTEGRATOR CIRCUIT.

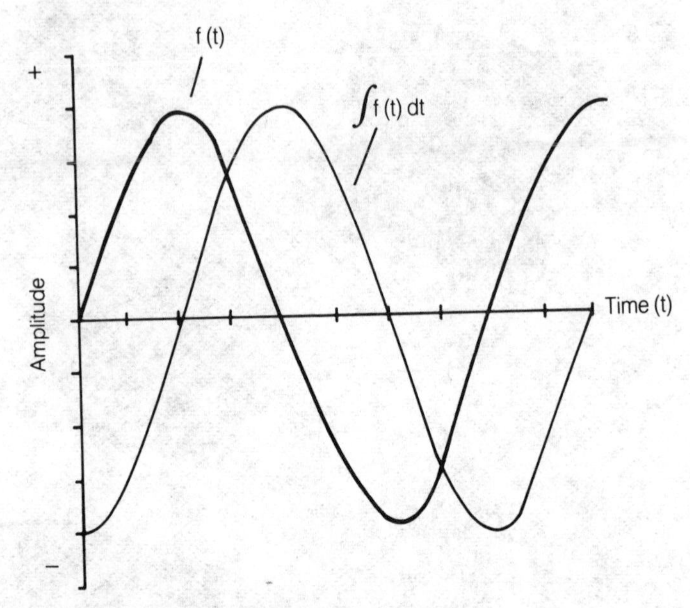

INTEGRATION: The integration of the sine wave f (t) (heavy) is shown as the second (lighter) waveform.

INTEGRATOR CIRCUIT

An electronic circuit that generates the integral, with respect to time, of a waveform is called an integrator circuit. When the input amplitude to an integrator is constant—that is, a direct current—the output constantly increases. If the input is zero, the output is zero. When the input amplitude changes alternately toward the positive and the negative, the output reflects the integral of the waveform (*see* illustration, for INTEGRATION).

The diagram on p. 475 shows two examples of circuits designed for electrical integration. At A, a resistance-capacitance network is shown. At B, an operational amplifier is used.

The integrator circuit has an action that is opposite to a differentiator. This does not mean, however, that cascading the two forms of circuit will always result in an output that is the same as the input. This can happen, but there are certain waveforms that are not duplicated when passed through an integrator and differentiator in cascade. *See also* DIFFERENTIATION, DIFFERENTIATOR CIRCUIT, INTEGRATION.

INTELLIGIBILITY

Intelligibility is a measure of the ease with which a signal is accurately received. It is measured as a percentage of correctly received single syllables in a plain-text transmission. The intelligibility can be affected by the context

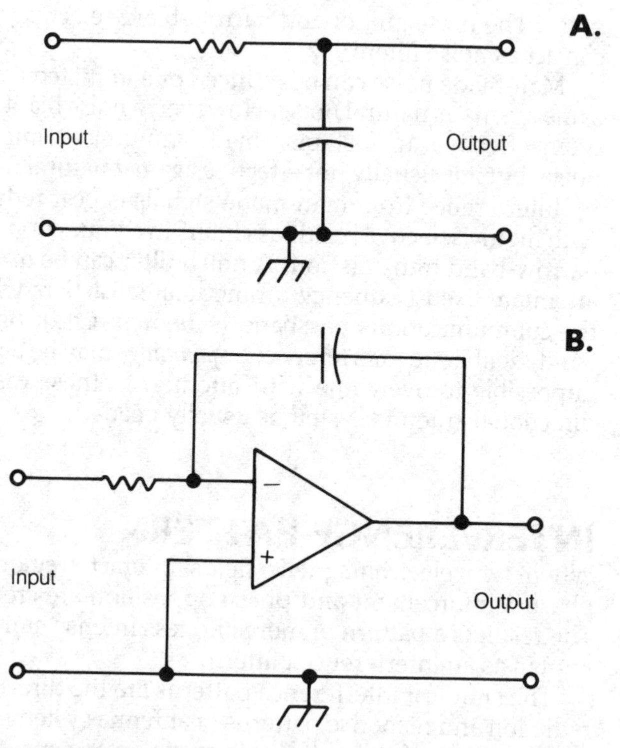

INTEGRATOR CIRCUIT: Integrator circuits include a simple passive resistance-capacitance network (A) and a circuit with an operational amplifier (B).

of the transmission, since the receiving operator can guess what is coming.

The percentage of correctly received syllables in a random transmission in which the receiving operator cannot guess what is coming is called articulation (*see* ARTICULATION).

Good intelligibility is imperative in a communications circuit. Normally, a band of audio frequencies ranging from about 300 to 3000 Hz is sufficient to convey a voice message with good intelligibility. Intelligibility is not highly correlated with fidelity; excellent fidelity requires a much wider band of audio frequencies, and this can actually reduce the intelligibility. Intelligibility is directly related to the signal-to-noise ratio in a voice circuit. *See also* SIGNAL-TO-NOISE RATIO.

INTENSITY

Intensity is a term that refers to the magnitude of a signal, sound, or beam of radiation. The magnitude of the electron beam in a cathode-ray tube is called the intensity. Most television receivers and oscilloscopes are equipped with intensity controls for varying the brightness of the picture.

The intensity of a signal may be rapidly modulated to convey information. Radio-frequency carriers can be amplitude or frequency modulated. If an electron beam is modulated, as in the television picture tube, the mode is called intensity modulation.

INTENSITY MEASUREMENT

Methods of measuring signal intensity depend on the kind of signal: acoustic, radio, infrared, visible, ultraviolet, X ray, gamma-ray, or electron-beam.

Acoustic disturbances, or sound waves, can be measured in terms of intensity with a microphone, audio amplifier, rectifier, and an indicating ammeter or voltmeter. Sound intensity is generally measured in decibels above the threshold of hearing (*see* DECIBEL).

The intensity of radio waves is generally measured by a field-strength meter. The most common units are microvolts per meter (*see* FIELD-STRENGTH METER).

Visible-light intensity is measured by means of a photocell or photovoltaic cell, a power supply if needed, and a meter. Light intensity can be specified in candela (*see* CANDELA).

The intensity of infrared or ultraviolet radiation can be measured in a manner similar to the method used for the measurement of visible-light intensity. A thermocouple can be used for measurement of infrared intensity, since radiation at these wavelengths causes heating of objects. For X rays and gamma rays, dosimeters are used (*see* DOSIMETRY).

The intensity of an electron beam is usually specified in terms of the equivalent current in amperes.

INTERCHANGEABILITY

Interchangeability is the degree of ease with which one component type can be replaced by another without compromising the acceptable operation of the system. Standardization of electrical and electronic components makes interchangeability feasible in many cases. The American National Standards Institute (ANSI) and the Electronic Industries Association (EIA) set standards for component interchangeability in commercial electronics in the United States. International standardization is carried out by a number of national organizations coordinated by the International Electrotechnical Commission (IEC).

INTERELECTRODE CAPACITANCE

When electrodes are placed in close proximity, they show a certain amount of mutual capacitance. For any two electrodes, this capacitance varies inversely with the separation, and directly according to the size of either electrode. Interelectrode capacitance is not very important at low and medium frequencies, but as the frequency increases, the interelectrode capacitance in a system becomes more significant. *See also* ELECTRODE CAPACITANCE.

INTERFACE

In an electronic system, an interface is a point at which data is transferred from one component to another. An interface generally involves conversion of data so that compatibility is achieved for both components. Thus there is a computer-to-operator interface (a terminal) or a radio-to-telephone interface (autopatch). The circuit that performs the data conversion is often an interface.

An interface might perform a relatively simple task

such as the connection of audio-frequency energy into a transmitter. The data conversion may be complicated, as in the translation of visual data and keyboard commands into machine language for the operation of a computer. In computer systems, there are many interface points; a personal computer, for example, requires a video monitor, a disk drive, and often a printer. It might also have a telephone modem. *See also* DATA, DATA CONVERSION, MODEM.

INTERFERENCE

Interference is the presence of unwanted signals or noise that increases the difficulty of radio reception. There are three kinds of interference: natural noise, man-made noise, and man-made signals.

Natural electromagnetic noise originates in outer space and in the atmosphere of the earth. There is nothing man can do to reduce the level of this noise. Atmospheric noise tends to be greatest at the lower frequencies and in the vicinity of thunderstorms. Noise from outer space is prevented from reaching the surface of the earth in the very low, low-, and medium-frequency ranges because of the shielding effect of the ionosphere. However, at higher frequencies, this noise can be significant in communications. In general, the narrower the signal bandwidth in a communications link, the greater the immunity to natural noise (*see* NOISE).

Man-made noise is caused mostly by electrical wiring, appliances, internal-combustion engines, and motors. Man-made noise is most intense at the very low frequencies. As the electromagnetic frequency increases, the level of man-made noise decreases. As with natural noise, the immunity of a system depends on the bandwidth. In some cases, special circuits can reduce the level of man-made noise.

Interference from man-made communications transmitters may be further categorized as either unintentional or intentional. The radio-frequency spectrum is widely used in modern industrialized nations, and a certain amount of accidental inteference is to be expected. The narrower the average signal bandwidth, in a given band of frequencies, the lower the probability of accidental interference. Intentional interference is called jamming, and is often used in wartime to interrupt enemy communications links. (*See* INTERFERENCE FILTER, INTERFERENCE REDUCTION.)

When two or more electromagnetic waves combine in variable phase, producing areas of stronger and weaker field intensity, the interaction is called interference. *See also* INTERFERENCE PATTERN, INTERFEROMETER.

INTERFERENCE FILTER

An interference filter is a circuit used in radio receivers to reduce or eliminate interference (*see* INTERFERENCE).

There are several different types of interference filters. For natural noise, such as that originating in the atmosphere, a noise limiter may provide some reduction in the level of interference (*see* NOISE LIMITER). Narrow-band transmission and reception is preferable to wide-band communications in the presence of high levels of natural noise. The noise limiter and narrow-band selective filter can act as noise filters.

Man-made noise can be reduced or eliminated by the same means as natural noise. However, a noise blanker is often effective against man-made ignition or impulse noise, but it is usually not effective against natural noise.

Interference from man-made signals is best reduced with highly selective bandpass filters in conjunction with narrow-band transmission. A notch filter can be used to attenuate fixed-frequency, unmodulated carriers within the communications passband (*see* NOTCH FILTER). Broadband, deliberate interference, or *jamming*, may be almost impossible to overcome with filtering. In these cases a directional antenna system is usually needed.

INTERFERENCE PATTERN

When two electromagnetic fields interact, regions of phase reinforcement and phase opposition are created. The result is a pattern of more and less intense emission known as an interference pattern.

The simplest interference patterns are the directional radiation and response patterns of antenna systems. The configuration of the pattern depends on the spacing of the antennas, measured in wavelengths, and on the relative phase of the signals at the different antennas.

When visible light is passed through a pair of narrow slits, an interference pattern occurs. This was noticed by physicists during the nineteenth century, and it led to the discovery of the wave theory of light (see illustration).

Interference patterns are often noticed with acoustic (sound) waves. These patterns must be considered when a high-fidelity system is designed. The interference pattern from a pair of stereo speakers, for example, causes "loud" and "dead" spots in certain locations unless the physical layout is carefully planned. *See also* PHASE ANGLE, PHASE OPPOSITION, PHASE REINFORCEMENT.

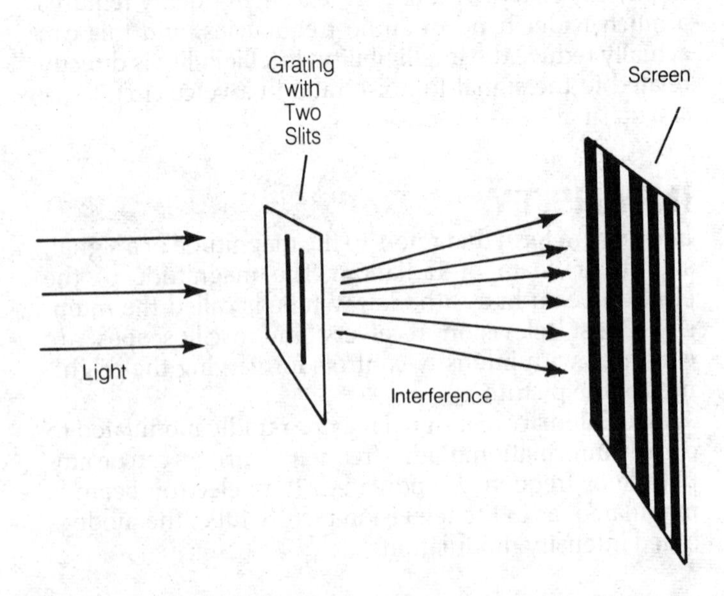

INTERFERENCE PATTERN: The pattern of light and dark bands caused by passing monochromatic light through a double-slit grating.

INTERFERENCE REDUCTION

The reduction of interference is important in a communications system. The lower the interference level, the better the signal-to-noise ratio in a communications link. Interference reduction can be effected at the transmitting station with the use of the narrowest possible emission bandwidth. With this single exception, however, all interference reduction must be accomplished at the receiving end of a circuit.

Interference can be reduced in a receiver with various filtering devices (*see* INTERFERENCE FILTER). Directional antenna systems (*see* DIRECTIONAL ANTENNA in the ANTENNA DIRECTORY) can also be used. If the interference originates at a single point in space and this point remains fixed or moves slowly, a loop antenna can be used to "null-out" the interference. If the interference comes from a broad, general direction different from the direction of the desired signal, a unidirectional antenna such as a quad, Yagi, or dish can be used.

In certain situations, interference reduction can be extremely difficult, especially in cases of deliberate interference, or jamming (*see* JAMMING). A new technique of communication called spread-spectrum operation can reduce interference from jamming. *See also* SPREAD-SPECTRUM TECHNIQUES.

INTERFEROMETER

The interferometer is a form of radio telescope in which two antennas are used in a phase configuration. The interferometer provides much greater resolving power, using two small antennas, than is possible with a single radio antenna. The technique was pioneered by two radio astronomers: Martin Ryle of England, and J. L. Pawsey of Australia.

A single dish antenna with a diameter of many wavelengths can yield good resolution in radio astronomy, but two of these antennas, separated by a great distance, allow far greater resolution. The sensitivity of the interferometer is not much higher than that of a single-antenna system, but in radio astronomy, resolving power is often more important than sensitivity.

The illustration shows, at A, a typical directional pattern for a single large dish antenna. At B is the response of a typical interferometer system, using two identical antennas spaced at a great distance. Each fine line in B represents a direction in which the incoming wavefronts add in phase at both antennas. The gaps between the lines represent directions in which the incoming wavefronts arrive in phase opposition. The interferometer is sensitive, therefore, in many directions. The radio astronomer, knowing the exact relationship between the two signals, can determine or select the exact direction of reception.

The greater the spacing between the antennas in the interferometer, the better is the resolution. If the spacing in meters between the aerials is L and the wavelength in meters is λ, then the angular separation a, in degrees, between the lobes of the directional pattern is given by:

$$a = 57.3\lambda/L$$

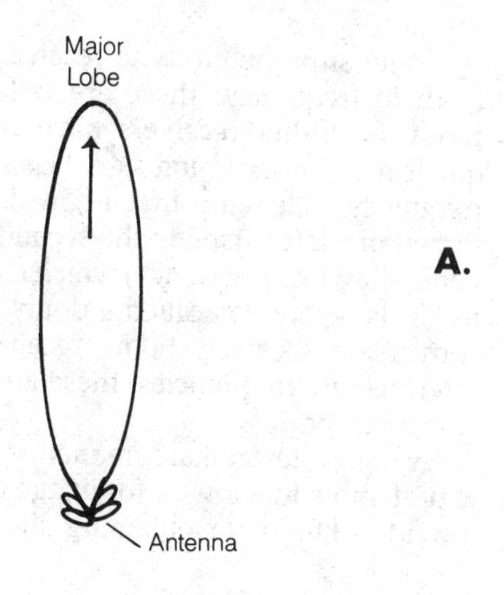

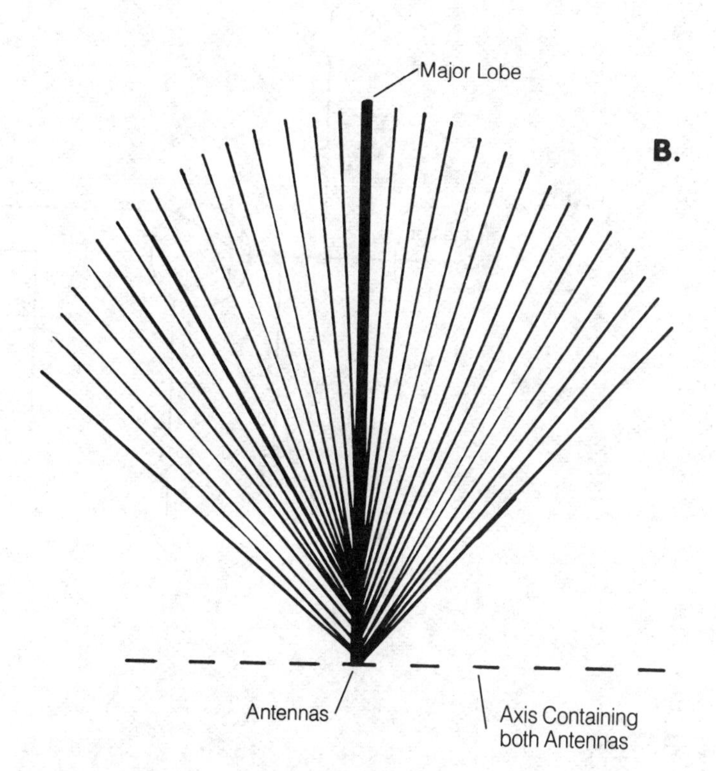

INTERFEROMETER: The directional response for a dish antenna (A) and the response pattern for an interferometer (B).

The shorter the wavelength, the greater the resolving power for a given antenna separation. Some interferometer systems have a resolution of a few seconds of arc at wavelengths of several centimeters. *See also* RADIO TELESCOPE.

INTERMEDIATE FREQUENCY

In a superheterodyne receiver the intermediate frequency (i-f) is the output frequency from the mixer stage (*see* MIXER, SUPERHETERODYNE RADIO RECEIVER). The intermediate frequency of the superheterodyne receiver is usually a fixed frequency. High gain and excellent selectivity can be obtained because the i-f amplifier stages can be tuned precisely for optimum performance at a single frequency (*see* INTERMEDIATE-FREQUENCY AMPLIFIER).

Some superheterodyne receivers employ one intermediate frequency; these are called single-conversion receivers. Other receivers have two intermediate frequencies. The incoming signal is heterodyned to a fixed frequency called the first i-f, which is in turn heterodyned in a later stage to the second i-f. The second i-f is generally a low frequency, which facilitates high selectivity. This receiver is called a double-conversion or dual-conversion receiver. Some receivers may have three intermediate frequencies; these are called triple-conversion receivers.

A high intermediate frequency of several megahertz is preferable to a low i-f for image rejection. However, a low i-f is better for obtaining sharp selectivity. This is why double-conversion receivers are common: They provide the advantages of both a high first i-f and a low second i-f. *See also* DOUBLE-CONVERSION RECEIVER, IMAGE REJECTION, SINGLE-CONVERSION RECEIVER.

INTERMEDIATE-FREQUENCY AMPLIFIER

An intermediate-frequency amplifier is a fixed radio-frequency amplifier commonly used in superheterodyne receivers. These amplifiers generally are cascaded two or more in a chain, with tuned-transformer coupling (see A in the figure). The intermediate-frequency (i-f) amplifiers follow the mixer stage, and precede the detector stage.

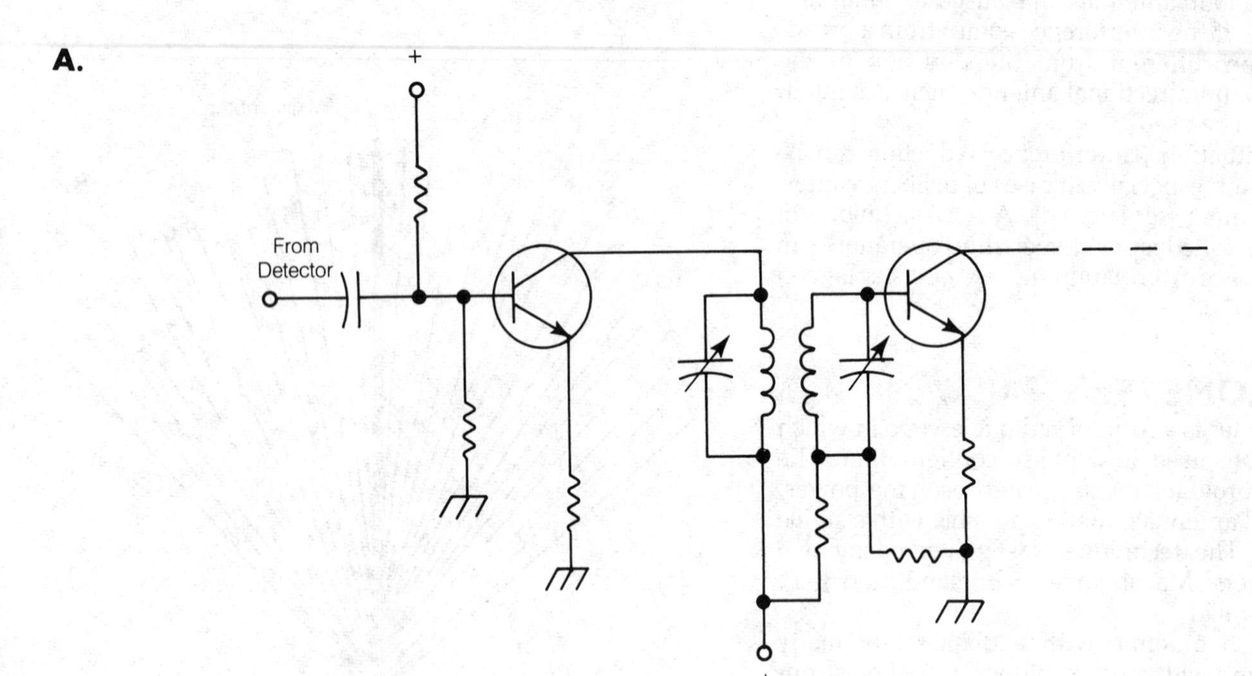

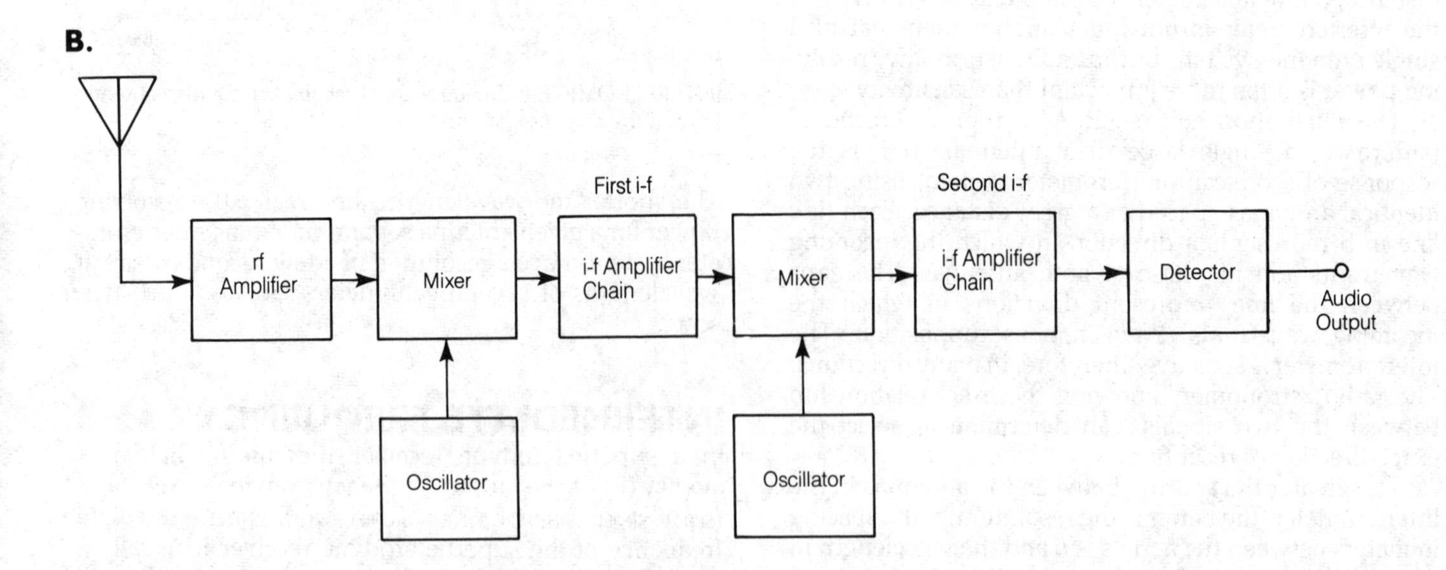

INTERMEDIATE-FREQUENCY AMPLIFIER: A typical superheterodyne circuit with transformer coupling (A) and a block diagram of a double-conversion receiver (B).

Double-conversion receivers (*see* DOUBLE-CONVERSION RECEIVER) have two sets of i-f amplifiers; the first set follows the first mixer and precedes the second mixer, and the second set follows the second mixer and precedes the detector, as at B in the figure.

The intermediate-frequency amplifier chain serves two main purposes: to provide high gain, and to provide excellent selectivity. Gain and selectivity are much easier to obtain with amplifiers that operate at a single frequency, as compared with tuned radio-frequency amplifiers. *See also* INTERMEDIATE FREQUENCY, SUPERHETERODYNE RADIO RECEIVER.

INTERMITTENT-DUTY RATING

The intermittent-duty rating of a component is its ability to handle current, voltage, or power with a duty cycle of less than 100 percent (*see* DUTY CYCLE). Usually intermittent duty is specified as a duty cycle of 50 percent. However, the length of a single active period must also be specified. The intermittent-duty rating of a device is almost always greater than the continuous-duty rating (*see* CONTINUOUS-DUTY RATING).

Some electronic components are intended primarily for intermittent service, while others are designed for continuous service. Sometimes, specifications will be given for both continuous and intermittent service.

INTERMITTENT FAILURE

When a system, circuit, or component malfunctions intermittently, the malfunction is called an intermittent failure, or simply an intermittent.

Troubleshooting is difficult if the problem is not always evident. In some cases, the failure does not show up under test conditions but only occurs in actual service, where repair is inconvenient or impossible. The failure might occur for only a fraction of 1 percent of the operating time.

Intermittent failures can be induced by mechanical vibration, thermal shock, excessive humidity, or operation at excessive voltage, current, or power levels. Sometimes there is no apparent cause for the failure. The repair process consists of locating and replacing the defective components, or improving the operating environment to conform with equipment specifications. It might be necessary to introduce cooling or localized heating devices. The troubleshooting process can be very time-consuming. When the problem appears to have been corrected, testing might be needed for verification.

INTERMODULATION

In a receiver, an undesired signal sometimes interacts with the desired signal. This often occurs in the first radio-frequency amplifier stages, or front end. The desired signal appears to be modulated by the undesired signal, as well as having its own modulation. This effect, known as intermodulation, is sometimes mistaken for interference from another signal and vice versa. Intermodulation can be comparatively mild, or it can be so severe that reception becomes impossible.

Intermodulation can take place for any or all of many reasons. Inadequate selectivity in the receiver input can result in unwanted signals passing into the front-end transistor. Nonlinear operation of the front-end transistor increases the chances of intermodulation. An excessively strong signal may drive the front end into nonlinear operation, although it is properly designed. A poor electrical connection in the antenna system might cause intermodulation in a receiver used with that antenna system. In the extreme, a poor electrical connection near the antenna, but physically external to it, can produce intermodulation effects. The poor connection effectively forms a diode mixer.

Intermodulation distortion (IMD) is never desirable, and modern receivers are designed to keep this form of distortion to a minimum. Manufacturers and designers refer to a specification called intermodulation spurious-response attenuation, usually expressed in decibels. Two signals are applied simultaneously to the antenna terminals of the receiver under test. The amplitude ratio is prescribed according to preset specifications. The strength of the unwanted signal required to produce a given amount of intermodulation is then compared with the strength of the desired signal, and the ratio is calculated in decibels.

Alternatively, the strength of the unwanted signal may be held constant, and the extent of the distortion measured as a modulation percentage in the desired signal. This percentage is called the intermodulation-distortion percentage.

Standards for intermodulation spurious-response attenuation are set in the United States by the Electronic Industries Association. When testing a receiver for intermodulation response, the procedures and standards of this association should be consulted. The susceptibility or immunity of a receiver to IMD is primarily dependent on the design of its front end. *See also* FRONT END.

INTERNAL IMPEDANCE

The internal impedance of a component or device is the impedance between its terminals at a given frequency, without external components. Internal impedance consists of direct-current or ohmic resistance and either positive (inductive) or negative (capacitive) reactance. Internal impedance is an important consideration when using test instruments such as meters and oscilloscopes.

The internal impedance of current-measuring devices should be as low as possible. When voltage is to be measured, the internal impedance should be high. In general, the internal impedance of a test instrument should be such that the performance of the circuit is not altered when the instrument is used. *See also* IMPEDANCE.

INTERNAL NOISE

In a radio receiver, noise is an important consideration because it limits the sensitivity of the circuit. Noise can originate outside the receiver in the form of atmospheric and cosmic noise (*see* COSMIC NOISE). It is also produced by the movement of atoms as a result of temperature (*see* THERMAL NOISE). Some noise is produced by the active components within the receiver itself and it is known as internal noise.

At very low, low, and medium frequencies, external noise is quite intense, and internal receiver noise normally does not play a significant part in limiting the sensitivity. In the high-frequency spectrum, external noise becomes less intense, and in the very high, ultra high, and microwave parts of the spectrum, internal noise is much more important than external noise. Low-noise receiver design is of the greatest concern, therefore, at frequencies above about 30 MHz.

Modern semiconductor technology permits the design and manufacture of low-noise receivers. In some scientific applications, noise is further reduced by supercooling, or operation of equipment at extremely low temperatures. *See also* NOISE, SENSITIVITY, SIGNAL-TO-NOISE RATIO, SUPERCONDUCTIVITY.

INTERNATIONAL MORSE CODE

The International Morse Code is a system of dot and dash symbols commonly used by radiotelegraph operators throughout the world. The International Morse Code differs from and is more often used than the American Morse Code.

The table shows the International Morse symbols. The dot represents one unit length, the dash represents three units, and the spaces between dots and dashes are one unit. The space between letters is three units, and between words is seven units. The International Morse Code is less confusing than the American Morse because it does not contain odd spaces and unit lengths. *See also* AMERICAN MORSE CODE.

INTERNATIONAL MORSE CODE: International Morse Code symbols.

Character	Symbol	Character	Symbol
A	•–	U	••–
B	–•••	V	•••–
C	–•–•	W	•––
D	–••	X	–••–
E	•	Y	–•––
F	••–•	Z	––••
G	––•	1	•––––
H	••••	2	••–––
I	••	3	•••––
J	•–––	4	••••–
K	–•–	5	•••••
L	•–••	6	–••••
M	––	7	––•••
N	–•	8	–––••
O	–––	9	––––•
P	•––•	0	–––––
Q	––•–	PERIOD	•–•–•–
R	•–•	COMMA	––••––
S	•••	QUESTION MARK	••––••
T	–		

INTERNATIONAL SYSTEM OF UNITS

The International System of Units (SI) is based on the meter, kilogram, second, ampere, degree Kelvin, candela, and mole (Avogadro constant). The system is sometimes called the meter-kilogram-second (MKS) or meter-kilogram-second-ampere (MKSA) system of units. All physical units can be derived in terms of these standard units. The SI, or Système International d'Unités, was agreed upon in the Treaty of the Meter in 1960.

Prior to 1948, the international system of units was based on the centimeter, gram, second, and degree Kelvin. The older international-unit system was discarded by international treaty on January 1, 1948. The most recent agreement was made in 1960. Sometimes a distinction is made between absolute and international electrical quantities. The difference relates to the standard quantities from which the units are derived. *See also* AMPERE, AVOGADRO CONSTANT, CANDELA, KELVIN TEMPERATURE SCALE, KILOGRAM, METER, SECOND.

INTERNATIONAL TELECOMMUNICATION UNION

The International Telecommunication Union (ITU) is an international organization that sets worldwide standards for electromagnetic communication. The ITU governs the frequency allocations and call-sign prefixes for all regions of the world. This helps maximize the efficiency of electromagnetic spectrum utilization. The ITU has established a set of Radio Regulations. These regulations may be obtained by contacting the Secretary General, International Telecommunication Union, Place des Nations, CH-1211, Geneva 20, Switzerland.

In the United States, telecommunications regulations are determined by the Federal Communications Commission, although their rules are based on those given by the ITU. *See also* FEDERAL COMMUNICATIONS COMMISSION.

INTERPOLATION

Interpolation is a mathematical process by which intermediate values are determined. Interpolation is a commonly used technique in all fields of engineering. It is especially useful in tabular function listings.

The table is an example of a listing of values for the sine function. Values are specified for each angular degree. To determine the sine of 35.5 degrees, (the value is not shown in the sine table), an approximate value can be obtained by averaging the values of sin(35 degrees) and sin(36 degrees):

$$\begin{aligned} \sin(35.5 \text{ degrees}) &= [\sin(35 \text{ degrees}) + \sin(36 \text{ degrees})]/2 \\ &= (0.574 + 0.588)/2 \\ &= 0.581 \end{aligned}$$

This value is not exact because the sine function is not a perfectly linear function, but the value can be considered accurate for practical purposes.

INTERPOLATION: Values of the sine function for angles between 30 and 40 degrees. Function values are split into tenths between 35 and 36 degrees for illustrative purposes (see text).

Angle, Degrees	sin θ
30	0.500
31	0.515
32	0.530
33	0.545
34	0.559
35	0.574
35.1	0.5754
35.2	0.5768
35.3	0.5782
35.4	0.5796
35.5	0.581
35.6	0.5824
35.7	0.5838
35.8	0.5852
35.9	0.5866
36	0.588
37	0.602
38	0.616
39	0.629
40	0.643

If the value of sin(35.7 degrees) is to be found according to the table, the interpolation process is more complicated. The interval between sin(35°) and sin(36°) must then be divided into ten equal parts. Thus, sin(35.7°) is the value that is 7/10 of the way from sin(35°) to sin(36°).

INTERPRETER

An interpreter is a complex computer program designed to translate ordinary-language commands into computer machine language.

The interpreter translates each command and executes it immediately. In this way, the interpreter differs from the compiler, which collects all the commands before translating and executing them.

In practice, an interpreter is somewhat more convenient to use than a compiler. *See also* COMPILER.

INTERRUPT

An interrupt is a computer instruction that tells the machine to stop a program and perform some other more important task. The interrupted program is called the background job. The background job has low priority, but may be very time-consuming. The more important task is called the foreground job; it is usually of short duration but high priority. A background job may be interrupted several times for different foreground jobs.

A computer program may be interrupted when it becomes evident that an error has been made. For example, a computer operator may notice that a program contains an infinite loop. Rather than let the computer run aimlessly for an indefinite period of time, the operator interrupts the execution of the program by issuing an interrupt command.

INTERRUPTER

An interrupter is a device that intermittently opens a circuit. The interrupter can be used for a variety of purposes, including generation of pulsating direct current from pure direct current, half-wave rectification of alternating current, or keying of a continuous-wave transmitter.

Interrupters generally fall into two categories: electrical and mechanical. Electrical interrupters include switching diodes and transistors and silicon-controlled rectifiers. Mechanical interrupters include switches and relays. In a direct-current to alternating-current power inverter, the interrupter is sometimes called a chopper. *See also* CHOPPER.

INTERSTAGE COUPLING

Interstage coupling is the means by which a signal is transferred from one circuit to another. The circuits may be oscillators, amplifiers, mixers, or detectors. There are several different methods of interstage coupling.

Capacitive coupling is commonly used at all frequencies from the low audio range into the ultra high frequency radio spectrum. A series capacitor passes the signal while allowing different direct-current voltages to be applied to the stages. (*See* CAPACITIVE COUPLING.)

Direct coupling consists of a short circuit between two stages. This method of coupling does not allow for different direct-current voltages between stages. Direct coupling exhibits a constant attenuation level over a wide range of frequencies. (*See* DIRECT COUPLING.)

Transformer coupling makes use of a two-winding audio-frequency or radio-frequency transformer for passing a signal. Either or both of the windings may contain a parallel capacitor for tuning purposes. An electrostatic shield may be placed between the windings to reduce the stray capacitance. Transformer coupling results in a narrow range of operating frequencies when tuning is used. Direct-current isolation is excellent. *See* TRANSFORMER COUPLING.

INTRINSIC SEMICONDUCTOR

A pure semiconductor material containing no added impurities is called an intrinsic semiconductor. These materials have nearly the resistive characteristics of insulators; they do not conduct electricity very well. Usually impurities must be added in manufacturing semiconductor devices. Then the semiconductor material becomes extrinsic. *See also* DOPING, ELECTRON, EXTRINSIC SEMICONDUCTOR, HOLE, N-TYPE SEMICONDUCTOR, P-TYPE SEMICONDUCTOR.

INVERSE FUNCTION

A mathematical function is an operation that assigns the members of one set to the members of another set according to certain rules. A relation is a function if, but only if, a given member from the domain has at most one constituent in the range (*see* DOMAIN, RANGE). Examples of functions are the sine, cosine, and tangent operations.

But there are infinitely many different mathematical functions.

An inverse relation or function is the operation that carries out exactly the reverse of the original function. If f is a given function, the inverse is written f^{-1}. The inverse of a function may not be a function unless its domain or range are restricted. This is the case with all of the trigonometric functions, as well as many others.

In general, if a function and its inverse are performed in succession, the original value is obtained. That is, for any x in the domain of a function f:

$$f^{-1}(f(x)) = x$$

and

$$f(f^{-1}(x)) = x$$

See also FUNCTION.

INVERSE-SQUARE LAW

The inverse-square law is a physical principle that defines the way three-dimensional radiation is propagated. The law applies to radio waves, infrared radiation, visible light, ultraviolet, X rays, and gamma rays. It also applies to barrages of particles—such as high-speed protons or neutrons—and to sound waves. The inverse-square law is used for determining or predicting the power intensity, per unit area, of a diverging effect at a given distance from a point source. The inverse-square law does not apply to parallel coherent energy beams from the maser or laser. The rule is also invalid when a parameter is measured in terms other than intensity per unit area.

To illustrate the derivation of the inverse-square principle, consider a light bulb that emits 100 watts (W) of power. If this bulb is surrounded by a sphere, 100 W of power is striking the inside surface of the sphere. This is true whatever the radius of the sphere, because any photon from the bulb must eventually encounter the sphere. Recall the formula for the surface area of the sphere:

$$A = 4\pi r^2$$

where A is the area, r is the radius, and π is approximately equal to 3.14. The energy intensity per unit area depends on the size of the sphere. If the radius of the sphere is doubled the surface area is quadrupled, and this results in just ¼ as much light for each square meter, square inch, or square foot of the sphere. By tripling the radius, the surface area increases by a factor of 9, resulting in 1/9 the energy density. In general, given a power level of P watts per square centimeter at a distance d from a source of energy, the power Q at distance x will be:

$$Q = P/x^2$$

where x can be any positive real number.

The inverse-square law only applies to quantities that are measured over a specific surface area. The rule does not apply to parameters that are measured along a line. Therefore, the field strength from an antenna in watts per square meter obeys the inverse-square law, but the field strength in microvolts per meter does not obey the principle. Effects such as magnetic-field intensity surrounding an inductor do not obey the inverse-square law, since flux density is determined partly in terms of three-dimensional volume and is not a simple radiant effect.

INVERSE VOLTAGE

The inverse voltage across a rectifier is the measure by which the cathode becomes positive with respect to the anode. Inverse voltage results in nonconduction of a rectifier device, unless the potential becomes so great that avalanche or flashover occurs. In a semiconductor device, inverse voltage is also called reverse voltage.

All rectifier devices are rated according to the maximum peak value of inverse voltage they can withstand. If this rating is exceeded, avalanche may take place. *See also* AVALANCHE, PEAK INVERSE VOLTAGE.

INVERTER

The term *inverter* denotes either of two different kinds of electronic circuits. A logical inverter is a NOT gate, while a power inverter converts low-voltage direct current (dc) into 120 V ac.

A logical inverter has one input and one output. If the input is high, the output is low; if the input is low, the output is high. The illustration shows the symbol for the logical inverter, along with a truth table showing its characteristics.

An example of a power inverter is a chopper power supply. An interrupting device produces a pulsating dc output from the power source, typically a 12 V battery. This pulsating dc is regulated in frequency so that it is as close as possible to 60 Hz. A step-up transformer then provides the 120 V output needed to operate low-power household appliances. *See also* CHOPPER POWER SUPPLY.

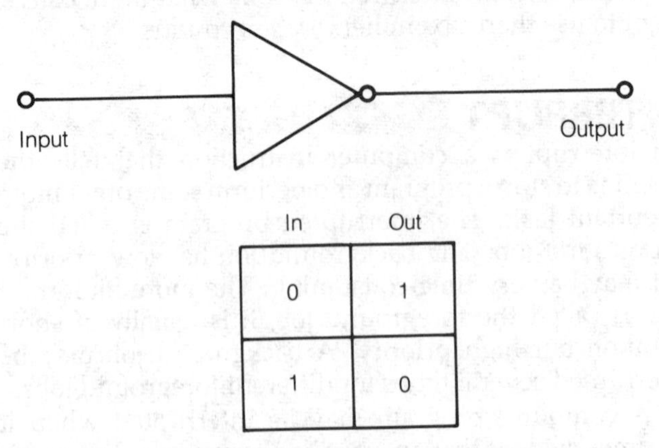

In	Out
0	1
1	0

INVERTER: The schematic symbol and truth table for an inverter.

INVERTING AMPLIFIER

An inverting amplifier is any amplifier that produces a 180-degree phase shift in the process of amplification. Most common-cathode, common-emitter, and common-source amplifiers are inverters (*see* COMMON CATHODE/EMITTER/SOURCE). In general, the grounded-grid, grounded-base, and grounded-gate configurations are not inverting amplifiers (*see* COMMON GRID/BASE/GATE).

A push-pull amplifier can be inverting or noninverting, depending on the transformer output arrangement.

Inverting amplifiers are less susceptible to oscillation than noninverting amplifiers because any stray coupling between input and output results in degenerative, rather than regenerative, feedback.

ION

See IONIZATION.

ION IMPLANTATION

Ion implantation is a selective doping process that embeds impurities into the surface regions of a semiconductor at room temperature. The dopant atoms are ionized (stripped of one or more of their electrons) and accelerated to a high speed by submitting them to a high electrical field of tens of thousands of volts. The charged atoms (ions) are then fired in a narrow beam into predetermined locations in the semiconductor wafer. Ion implantation is more precise than doping by diffusion. The depths of penetration of the ions depend on the mass and energy of the dopant atoms.

The wafer is selectively masked against the ions either by a patterned oxide layer, as in conventional diffusion, or by a photoresist pattern. As the accelerated ions penetrate the wafer, they damage the crystal lattice. This damage can be repaired by annealing the wafer in a furnace.

Very shallow, high-concentration doping can be obtained with ion implantation. Each of the ions bombarding the crystal has a charge, and the total charge that accumulates can be measured to determine the number of impurities precisely. Ion implantation is used whenever the doping level must be accurately controlled. It is also used to introduce impurities that are difficult to predeposit from a high-temperature vapor. Ion implantation can also drive impurities through a thin protective oxide.

A beam of ions is directed at a selected window in the masked surface of the wafer with enough energy to penetrate the surface and slow to rest within the solid. Ion implantation replaces diffusion steps in the fabrication of silicon MOSFET transistors, monolithic resistors, diodes, and integrated circuits as well as gallium arsenide MOSFET transistors and integrated circuits. *See* INTEGRATED CIRCUIT MANUFACTURE.

IONIZATION

An atom normally has the same number of negatively charged electrons and positively charged protons. Because both particles have equal charge quantities, most atoms are electrically neutral. Sometimes, however, electrons are lost or captured by an atom, so the charge is not neutral. The non-neutral atom is called an ion, and any process or event that creates ions is known as ionization.

Ionization occurs when substances are heated to high temperatures, or when large voltages are impressed across them. Lightning is the result of ionization of the air. An electric spark is caused by a large buildup of charges, resulting in forces on the electrons in the intervening medium. These forces pull the electrons away from individual atoms. Ionized atoms generally conduct electric currents with greater ease than electrically neutral atoms.

Ionized gases are prevalent in the upper atmosphere of the earth. These layers are known as the ionosphere. The ionosphere has refracting effects on electromagnetic waves at certain frequencies. This makes long-distance communication possible on the shortwave bands. *See also* IONIZATION VOLTAGE, IONOSPHERE.

IONIZATION VOLTAGE

In a gas-filled electron tube, a high voltage will cause ionization. As the voltage increases between the cathode and plate of the tube, a point is reached at which conduction begins because of ionization. This potential is called the ionization voltage of the tube. The ionization voltage of a gas tube depends on the kind of gas and its concentration. The neon gas discharge display and the neon glow lamp are examples of gas-filled electron tubes. Gas tubes were once used as voltage regulators and rectifiers. *See also* NEON GAS DISCHARGE DISPLAY.

IONOSPHERE

The atmosphere of the earth becomes less dense with increasing altitude. Because of this, the amount of ultraviolet and X ray energy received from the sun increases at higher altitude. At certain levels, the gases in the atmosphere become ionized by solar radiation. These regions comprise the ionosphere of the earth.

Ionization in the upper atmosphere occurs mainly at three levels, called layers. The lowest region, called the D layer, exists at altitudes ranging from 35 to 60 miles, and is ordinarily present only on the daylight side of the earth. The E layer, at about 60 to 70 miles above the surface, also exists mainly during the day, although night-time ionization is sometimes observed. The upper layer is called the F layer. This region sometimes splits into two regions, known as the F1 (lower) and F2 (upper) zones. The F layer may be found at altitudes as low as 100 miles and as high as 260 miles.

The ionosphere has a significant effect on the propagation of electromagnetic waves in the very low, low, medium, and high frequency bands. Some effect is observed in the very high frequency part of the spectrum. At frequencies above about 100 MHz, the ionosphere has essentially no effect on radio waves. Ionized layers cause absorption and refraction of the waves. This makes long-distance communication possible on some frequencies. At the longer wavelengths, energy from space cannot reach the surface of the earth because of the ionosphere. *See also* D LAYER, E LAYER, F LAYER.

IONOSPHERIC PROPAGATION

See PROPAGATION CHARACTERISTICS.

IRON

Iron is an element with an atomic number of 26 and an atomic weight of 56. Iron is best known for its magnetic properties and is found in great abundance in the earth. The core of the earth is believed to consist largely of iron. This may be the reason for the magnetic field surrounding the earth.

Iron is used in many applications in electricity and electronics. The atoms of iron tend to align themselves in the presence of a magnetic field. If the field is strong enough, the alignment persists even after the field is removed. In this case, the iron is said to be permanently magnetized. Permanent magnets are used in microphones, speakers, earphones, and other transducers. Permanent magnets are also used in moving-cell analog meters.

Because of its high magnetic permeability, iron makes an excellent core material for transformers. Various forms of iron are used at different frequencies. Laminated iron is used in transformers for frequencies in the audio range. Powdered iron is used at radio frequencies up to the very high region. Ferrite transformer cores are used at radio frequencies in the very low, low, medium, and high ranges. *See also* ALNICO, FERRITE, FERROMAGNETIC MATERIAL, LAMINATED CORE, PERMEABILITY, TRANSFORMER.

IRON CORE

See FERRITE CORE, LAMINATED CORE.

IRON-VANE METER

An iron-vane meter is an instrument for measuring alternating currents. The ordinary D'Arsonval moving-coil direct-current meter will not respond to alternating current because it produces directional readings, according to the current polarity.

The iron-vane instrument contains a movable iron vane attached to a pointer. The pointer is held at the zero position on a calibrated scale by a spring under conditions of zero current. A stationary iron vane is positioned next to the movable vane as shown in the illustration. Both vanes are placed inside a coil of wire, the ends of which form the meter terminals.

When current is supplied to the coil, a magnetic field is produced in the vicinity of the iron vanes. The field causes the vanes to become temporarily magnetized. The vanes are positioned so that like magnetic poles form near each other on each vane end; the strength of the magnetism is proportional to the current in the coil. Because like magnetic poles repel, the movable vane is deflected away from the stationary vane. The spring allows the pointer to move a specific distance, depending on the force of repulsion. This effect occurs for alternating or direct currents in the coil. The meter deflection is therefore proportional to the current.

The iron-vane instrument is, by itself, an ammeter. However, it can be used as a voltmeter or wattmeter with the addition of appropriate peripheral components. The scale of the iron-vane meter is usually nonlinear. The meter is accurate only at power-line frequencies because of the coil inductance and losses in the iron vanes. Thus, the iron-vane meter is used only for measurement of currents and voltages in ordinary utility devices. It is not suitable for use at audio or radio frequencies.

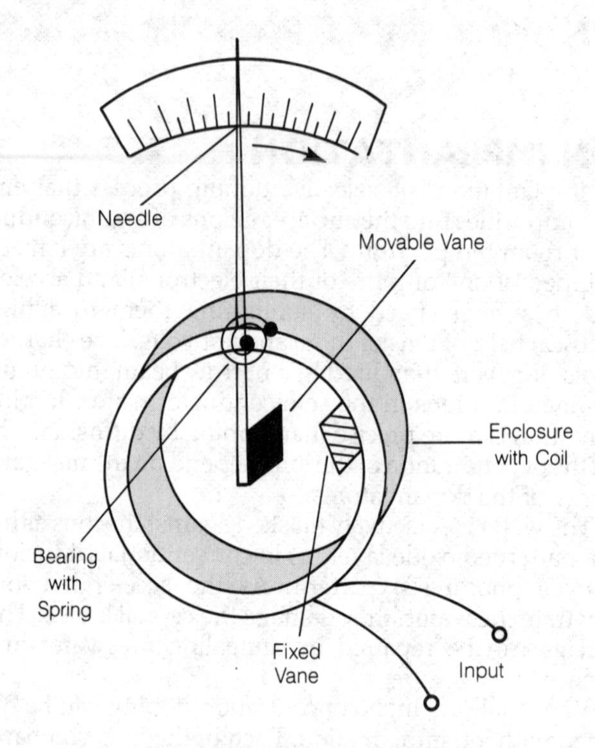

IRON-VANE METER: An illustration of an iron-vane meter for measuring either alternating or direct currents.

IRRADIANCE

Irradiance is the radiant power incident per unit area upon a surface, expressed in watts per square centimeter (W/cm^2) or watts per square meter (W/m^2). Irradiance can be measured at any electromagnetic wavelength.

In the visible-light range, irradiance is called illuminance. *See also* ILLUMINANCE.

IRRATIONAL NUMBER

An irrational number is a quantity that cannot be expressed as the quotient of two integers. In decimal form, an irrational number consists of an infinite sequence of digits, with no repeating block of digits.

Irrational numbers are much more common in mathematics than rational numbers. In fact, the number of irrational quantities is infinitely greater than the number of rational quantities. Irrational numbers include such factors as pi (approximately 3.14), e (approximately 2.72), and many others. *See also* RATIONAL NUMBER, REAL NUMBER.

I SCAN

See RADAR.

ISOTOPE

The nuclei of all atoms are identified according to the number of protons they contain. For example, a carbon atom is an atom with six protons; oxygen is an atom with eight protons. All atoms also contain neutrons in their nuclei, with the exception of simple hydrogen. The number of neutrons can vary, however, even while the number of protons remains the same.

The number of protons in an element, known as the atomic number, uniquely determines the identity of the element. The number of neutrons can vary, so that a given element can occur in more than one natural form. The different forms of an element, resulting from nuclei having different numbers of neutrons, are called isotopes of the element.

Hydrogen, while normally containing no neutrons, can have one or two neutrons in addition to the proton in its nucleus. If there is one neutron, the isotope is called deuterium. If there are two neutrons, the isotope is called tritium. Carbon, with atomic number 6, generally occurs with six neutrons, so that the atomic weight is 12. However, some carbon atoms have eight neutrons, and the atomic number of such atoms is 14. Carbon-14 is a radioactive material that decays rapidly. Radioactivity is a characteristic of many different isotopes.

In general, the larger the atomic number of an element, the greater the number of isotopes that can be formed. The most common naturally occurring isotopes are the most stable. Some isotopes are made by man for nuclear and medical research. *See also* ATOM, ATOMIC NUMBER, ATOMIC WEIGHT, COMPUTER-AIDED MEDICAL SCANNING.

ISOTROPIC ANTENNA

See ANTENNA DIRECTORY.

ITERATION

Iteration is a process of repeating a mathematical operation for making a calculation. Iteration is used for such purposes as determining square roots of numbers, and for addition of infinite series.

Some mathematical calculations involve approximations, the accuracy of which depend on the number of iterations. Long iterative processes are now carried out with computers. The more repetitions that are performed, the greater the accuracy of the final result.

ITU

See INTERNATIONAL TELECOMMUNICATION UNION.

JACKET

A jacket is the coating on a multiconductor cable, particularly coaxial cables. The sleeve on an electronic component, such as a capacitor, is also sometimes called a jacket.

The jacket is made of insulating material to isolate the component from surrounding objects. It also protects the component or cable against corrosion.

JACK PANEL

A jack panel is a flat, usually metallic, panel in which a number of jacks are mounted. Each jack is connected to a specific circuit in a system. An operator can insert one or more plugs into various jacks in the jack panel to connect various circuits.

Jack panels were common in the early days of wire telephone systems. Telephone operators made connections by inserting plugs into various jacks on a jack panel. Today, most telephone interconnections are performed by automatic switching devices controlled by computers. *See also* TELEPHONE SYSTEM.

JAMMING

Jamming is deliberate interference to radar or comunication signals. The simplest form of jamming is the intentional transmission of an unmodulated carrier on the frequency of a two-way radio link. This form of interference is prohibited by law in the United States.

Jamming is practiced by some countries that do not want their citizens to listen to certain international radio transmissions on the shortwave bands. A highly modulated or overmodulated, high-power signal is transmitted on the frequency to be jammed. A jamming signal can be recognized by its characteristic whining or buzzing sound as heard on an AM (amplitude-modulation) receiver.

In radar, jamming is used for masking a target. This can be done by the transmission of a synchronized set of pulses at or near the frequency of the radar transmitter. The radar receiver then sees false targets or large bright spots that cannot be identified. Jamming, when skillfully done, can be almost impossible to overcome. *See also* INTERFERENCE.

J DISPLAY

See RADAR.

JFET

See FIELD-EFFECT TRANSISTOR.

JITTER

A jitter is a rapid variation in the amplitude, frequency, or other characteristic of a signal. Usually the term *jitter* is used in video systems to refer to instability in the displayed picture on a cathode-ray tube.

Jitter in a cathode-ray-tube display can result from changes in the voltages on the deflecting plates. It can also occur because of variations in the current through the deflecting coils, or because of changes in the speed of the electrons passing through the tube. Mechanical vibration, an interfering signal, or internal noise can also cause jitter. Sometimes the synchronizing circuits fail to lock onto the signal in the display, causing vertical or horizontal jitter. *See also* CATHODE-RAY TUBE, OSCILLOSCOPE, TELEVISION.

J-K FLIP-FLOP

See FLIP-FLOP.

JOHNSON COUNTER

See RING COUNTER.

JOHNSON NOISE

See THERMAL NOISE.

J OPERATOR

The j operator is an imaginary number, mathematically defined as the square root of -1. Mathematicians denote this quantity by the lowercase letter i; however, electrical and electronic engineers prefer the lowercase letter j because i can be used to represent current.

The j operator is used to represent reactance. An inductive reactance of $X\ \Omega$, where X is any positive real number, is written $\mathrm{j}X$. Sometimes it may be written $X\mathrm{j}$. A capacitive reactance of $-X\ \Omega$ is written $-\mathrm{j}X$ or $-X\mathrm{j}$. The set of possible reactances is the set of imaginary numbers, or real-number multiples of j.

A complex impedance is represented by the sum of a pure resistance and an inductive (positive) or capacitive (negative) reactance. If the resistive component is R, then the impedance Z takes the form $R + \mathrm{j}X$ or $R - \mathrm{j}X$.

Complex impedance can be added together, or even multiplied, in this form. The important thing to remember is that when j is multiplied by itself, the result is −1. *See also* CAPACITIVE REACTANCE, IMPEDANCE, INDUCTIVE REACTANCE.

JOSEPHSON EFFECT

When two superconducting materials are brought close together, a current flows across the gap between the tips of the two objects. Superconductivity is characterized by extremely low ohmic resistance in a substance (*see* SUPERCONDUCTIVITY).

When the current crosses the gap between the superconducting objects, high-frequency electromagnetic energy is generated. This effect was noticed by Brian Josephson, and thus it is called the Josephson effect.

JOULE

The joule is the standard unit of energy or work. When a force of 1 newton is applied over a distance of 1 meter, the amount of work done is 1 joule. The joule is equivalent in energy to 0.24 calories; that is, 1 joule of energy will raise the temperature of 1 gram of water by 0.24°C. A kilowatt hour is equivalent to 3.6 million joules. A power level of 1 watt is a rate of expenditure of 1 joule per second. *See also* ENERGY.

JOULE'S LAW

When a current flows through a resistance, heat is produced. This heat is called joule heat or joule effect. The amount of heat produced is proportional to the power dissipated. The power P, in watts, dissipated in a circuit carrying I amperes, and having resistance of R Ω, is given by the formula:

$$P = I^2R$$

Joule's law recognizes this by stating that the amount of heat generated in a constant-resistance circuit is proportional to the square of the current.

JOYSTICK

A joystick is a control device that functions in two dimensions: the up-down dimension and the left-right dimension. A joystick consists of a movable lever or handle and a ball-bearing unit (see illustration). The joystick gets its name from its physical resemblance to the joystick of an airplane. Some joysticks have a rotational control dimension, in addition to the up-down and right-left functions. The lever is rotated either clockwise or counterclockwise. This makes it possible to control three different parameters.

The joystick is used extensively in modern electronic computer games to manipulate objects on the display screen. The joystick is also employed for entering coordinates in the x, y, and z input registers of a computer. The x and y coordinates are set by moving the joystick from side to side and up and down. The z coordinate is set by twisting. *See also* LIGHT PEN, MOUSE.

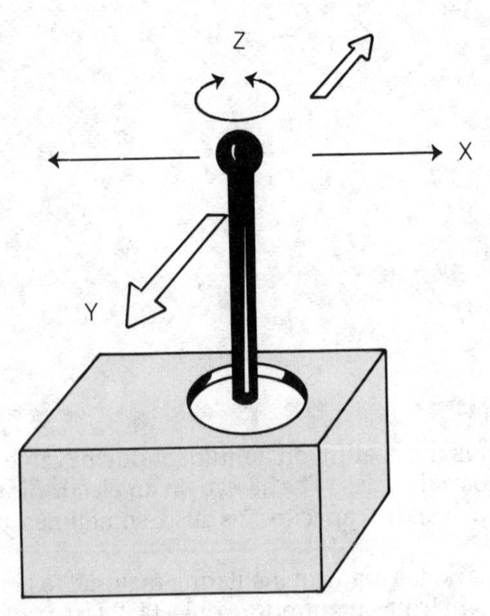

JOYSTICK: The joystick can be manipulated in two dimensions. Some have provision for a third dimension, accomplished by rotating the lever.

JUNCTION

A junction is the point, line, or plane at which two or more different components or substances are joined. A simple electrical connection may be called a junction. In a waveguide, a junction is a fitting used for the purpose of joining one section to another. In a transmission line, a junction is the splice between two sections of the feed system.

A semiconductor junction is a plane surface where P-type and N-type materials meet. This is called a P-N junction. This junction forms a one-way barrier for current; that is, it has diode action. *See also* P-N JUNCTION.

JUNCTION CAPACITANCE

When a semiconductor P-N junction is reverse-biased (*see* P-N JUNCTION), the resistance is high. However, there is some capacitance, known as junction capacitance, present.

The junction capacitance of a diode, under conditions of reverse bias, depends on several factors. The surface area of the junction is important; in general, the larger the area, the greater the junction capacitance. The value of the reverse voltage also affects the junction capacitance. This fact can be seen in the varactor diode, which acts as a voltage-controlled, variable capacitor (*see* VARACTOR DIODE).

The capacitance of a reverse-biased semiconductor junction determines the maximum frequency at which a device can be used. The smaller the junction capacitance, the higher the limiting frequency. At very high and ultrahigh frequencies, semiconductor devices must have small values of junction capacitance. Rectifier diodes have large junction capacitance. Gallium arsenide diodes usually have small values of junction capacitance, as do metal-oxide semiconductor devices. The point-contact junction exhibits a smaller capacitance than the P-N junction. *See also* DIODE CAPACITANCE, GALLIUM-ARSENIDE TRANSISTOR, METAL-OXIDE SEMICONDUCTOR, METAL-OXIDE SEMICONDUCTOR FIELD-EFFECT TRANSISTOR.

JUNCTION DIODE

A junction diode is a common type of semiconductor device manufactured by diffusing P-type material into N-type material, thereby obtaining a P-N junction (*see* P-N JUNCTION). Junction diodes are made from germanium, silicon, and gallium arsenide with added dopants.

The junction diode conducts when the N-type material is negatively charged with respect to the P-type material. If the potential difference is less than the threshold value, or if the P-type material is negative with respect to the N-type material, the junction diode will exhibit a very large direct-current resistance, infinite for all practical purposes.

Junction diodes are commonly used in rectifier and detector circuits. They can also be used as mixers, modulators, converters, and frequency-control devices. *See also* DIODE, DIODE CAPACITANCE, JUNCTION CAPACITANCE.

JUNCTION FIELD-EFFECT TRANSISTOR

See FIELD-EFFECT TRANSISTOR.

JUNCTION TRANSISTOR

See TRANSISTOR.

KARNAUGH MAP

A Karnaugh map is a truth table, arranged so that a logical expression is broken down into its constituents. Complicated logic functions can be obtained by locating logic gates in different arrangements. The Karnaugh map shows which of these arrangements is the simplest. This minimizes the number of individual gates to be used. *See also* BOOLEAN ALGEBRA, TRUTH TABLE.

KELVIN ABSOLUTE ELECTROMETER

The Kelvin absolute electrometer is an instrument for measuring electrostatic voltages. The meter consists of two stationary metal plates with a movable plate between them. The movable plate is attached to a pointer and is held at the zero position by a set of return springs. The illustration is a simplified diagram of the meter.

When an alternating-current or direct-current voltage is applied to the input terminals of the electrometer, the movable plate is deflected away from the stationary plates. The extent of the movement is determined by the voltage; the greater the input voltage, the farther the plate moves before equilibrium is reached with the return-spring tension. The pointer moves along a graduated scale.

The Kelvin absolute electrometer has extremely high input impedance and therefore draws essentially no current from the circuit under test. The meter does not use an amplifying device, which is why it is called an absolute electrometer. *See also* ELECTROMETER.

KELVIN ABSOLUTE ELECTROMETER: This instrument can measure either alternating or direct voltage.

KELVIN BALANCE

The Kelvin balance is an instrument for measuring alternating and direct electric currents. It uses a mechanical balance, similar to chemical balances. Coils attached to the arm of the balance are interconnected in series, so that a current at the input terminals will cause magnetic fields to form around the coils. The coils are arranged so that one pair attracts while the other repels (see illustration).

When there is no current at the input of the Kelvin balance, the indicator shows equilibrium. As a small alternating or direct current is applied to the coils, the needle is deflected. The greater the current, the more the needle is deflected. A sliding weight is moved along the balance arm until the needle returns to the center position. The balance arm has a calibrated scale, from which the value of the current is determined according to the position of the weight at equilibrium.

The Kelvin balance is an extremely sensitive device. However, it has a very slow response to changes in the current. If the current fluctuates, the weight must be repositioned. *See also* AMMETER.

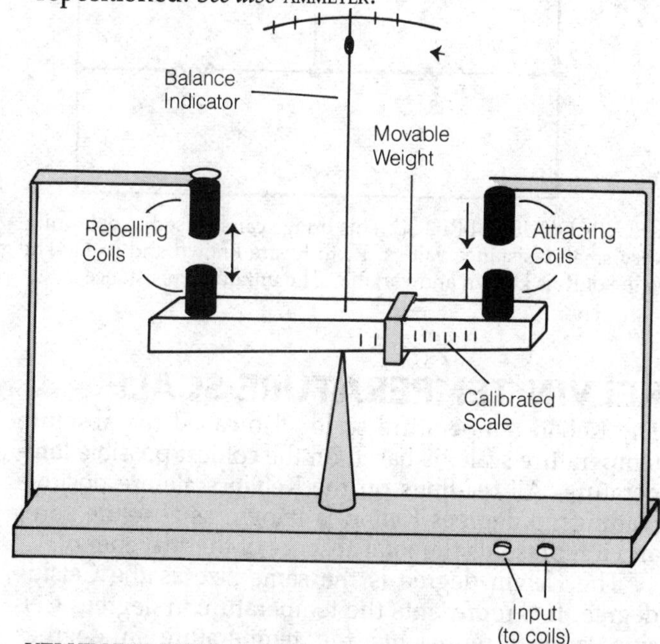

KELVIN BALANCE: An extremely sensitive current meter capable of measuring either alternating or direct current.

KELVIN DOUBLE BRIDGE

The Kelvin double bridge was designed for measuring low resistance (see illustration).

The unknown resistance, *R*, is connected as shown in the circuit with the other known resistances. The ratios R_1/R_2 and R_3/R_4 are known, and they are identical. The value of R_5, a precisely calibrated, low-resistance potentiometer, is adjusted for a zero reading on the meter. When balance is obtained, the ratio R/R_5 is equal to the ratios R_1/R_2 and R_3/R_4. The value R is calculated from the other values by the formulas:

$$R = \left(\frac{R_1}{R_2}\right) R_5$$

or:

$$R = \left(\frac{R_3}{R_4}\right) R_5$$

The Kelvin double bridge is sometimes called the Thomson bridge.

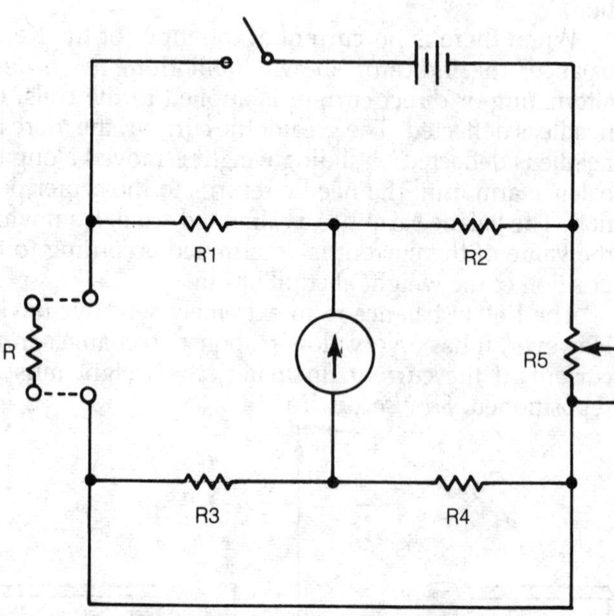

KELVIN DOUBLE BRIDGE: This bridge can be used to determine very small resistance values. R_1 to R_4 are known and fixed. The value of R_5 is known and variable. The unknown resistance is R.

KELVIN TEMPERATURE SCALE

The Kelvin temperature scale, also called the absolute temperature scale, is based on the coldest possible temperature. All readings on the Kelvin scale are positive readings; 0 degrees Kelvin is known as absolute zero, and it represents the total absence of thermal energy.

The Kelvin degree is the same size as the Celsius degree. If C represents the temperature in degrees Celsius, and *K* represents the temperature in degrees Kelvin, then the readings are approximately related according to the simple formula:

$$K = C + 273$$

For conversion from Fahrenheit (*F*) to Kelvin, the following formulas apply:

$$K = 0.555F + 255$$
$$F = 1.8K - 459$$

See also CELSIUS TEMPERATURE SCALE, FAHRENHEIT TEMPERATURE SCALE.

KEY

A key is a simple switch used for sending radiotelegraph code. It is sometimes called a straight key or brass pounder. The key consists of a spring-mounted lever with a knob, which is held by the fingers (see illustration). Characters of the Morse code are formed by pushing the lever down at intervals.

The radiotelegraph key was the earliest device for modulating a transmitter. Today, more sophisticated keying devices are used. The speed at which an operator can generate code with a simple key is limited. Some experts can send at speeds of 35 to 40 words per minute with a straight key, but most operators rarely exceed about 20 words per minute. *See also* CODE TRANSMITTER, KEYING.

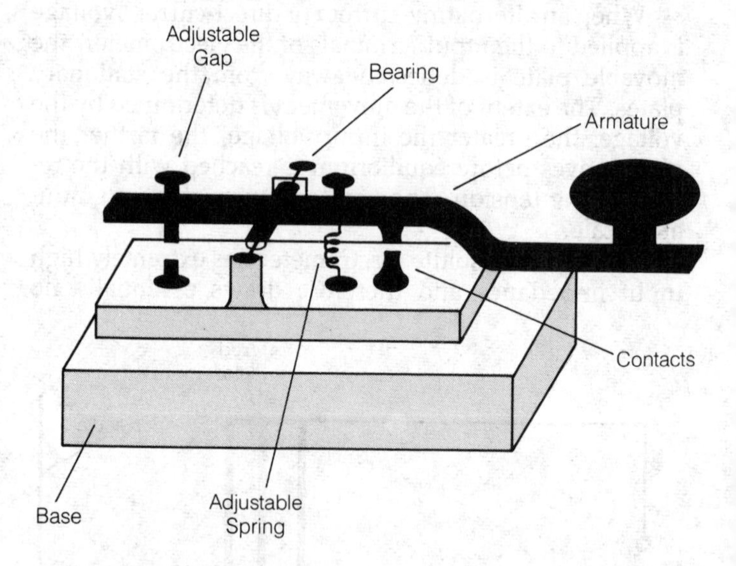

KEY: An illustration of a telegraph key.

KEYBOARD AND KEYPAD

Keyboards and keypads are electromechanical components for the entry of data into digital systems. Data can be entered as numerical codes, numbers, plaintext, or combinations of these. The full-travel, encoded ASCII keyboard, copied from the typewriter keyboard, is the most widely used computer data entry device. Fully assembled keyboards in desktop cases are available as interchangeable peripherals. They include all encoding and formatting electronics and interconnection cables that are plug compatible with computers. By contrast, keypads are typically matrices of keyswitches intended primarily to enter numerical codes and are sold without attached electronics.

Other devices can supplement the keyboard for computer data entry, but they are more limited in their applications. These include the graphics tablet, joystick, light pen, and mouse, a device that can enter data or commands as it is moved in two dimensions on a desktop. All of these devices are widely used in computer-generated graphics and in desktop publishing.

Keyboards. The conventional full-travel keyboard was accepted for use with computers because the first universally acceptable computer terminal was the commercial teletypewriter. Capable of both manual data entry and printing, the teleprinter keyboard was organized in the familiar QWERTY key format as shown in Fig. 1. It generated digital code in accordance with the American Standard Code for Information Exchange (ASCII). An ASCII keyboard can generate 128 numbers, letters, symbols, and special control codes with its double-function keys. As a result of the acceptance of the QWERTY key arrangement, persons skilled in typing are able to enter data in a computer with little additional training.

The first computer keyboards were assemblies of dumb electromechanical key switches mounted directly on the computer enclosure. It depended on the computer processor to carry out routine data formatting tasks. In the latest detachable units, an internal microcontroller performs key decoding and organizes the data for serial transmission over a cable to the host computer. Standard digital interfaces permit keyboard interchangeability.

Full-Travel Keyboard. The keycaps of a full-travel keyboard must be depressed from 0.120 to 0.150 inch for the key switch to make the positive closure for generating an electrical code. Many different technologies have been used in key switches: mechanical contact, capacitive, inductive, Hall effect, and reed switch. Human factors engineering or ergonomic studies have shown the benefits of adding the tactile responses or feel of an electric typewriter keyboard to the computer keyboard. The finger senses a change in pressure as the key cap is depressed. To obtain this nonlinear response, keyboard manufacturers include foam plastic blocks, rubber cones, or metal springs in the keyswitches. Ergonomic studies in the United States and Europe to reduce operator fatigue and errors have led to standard key cap size and shape, distance of key travel and outline dimensions of keyboards.

Short-Travel Keyboard. In certain slow-speed data entry applications not requiring the skills of a trained typist, the short-travel keyboard is acceptable. It can be organized in the ASCII format, or it can have keys with customized legends or icons. These keyboards are usually fabricated with a flexible membrane that permit actuation with micromotions. They are rugged and sealed against spilled liquids. These keyboards are found on point-of-sale terminals in retail stores and fast-food restaurants or on industrial robots or process-control systems. The keys are printed on the plastic membrane stretched over an array of micromotion switches; the membrane surface need be depressed only about 0.005 inch for key switch actuation.

Keypad. The term *keypad* applies to many different kinds of key switch assemblies primarily intended for numerical data entry into a digital system. The key panels of touchtone telephones and pocket calculators are examples. However, keypads available as standard or custom-made commercial products are usually supplied without electronics.

A keypad can be made as an assembly of separate full-travel or short-travel keyswitches, or it can be made as an assembled array of short-travel switch contacts covered with individual caps or a flexible membrane. The simplest mechanical keypad consists of an array of spring-loaded, short-travel keycaps (Fig. 2). When pressed, the keys short metal contacts on a supporting substrate. Keycap travel is typically 0.030 inch or less.

(Courtesy Micro Switch)

KEYBOARD AND KEYPAD: Fig. 1. IBM personal computer keyboard with capacitance key switches.

KEYBOARD AND KEYPAD: Fig. 2. Short-travel alphanumeric keypad for the entry of coded data into a digital system.

Membrane keypads offer unlimited opportunities for printing keys on the exterior membrane with symbols or legends in various colors. Internal-switch elements can be metallized rubber domes or plastic blisters that contact a matrix of conductors when pressed. They are made like the short-travel keyboards.

KEYING

Keying is the means for modulating a code transmitter. It is accomplished either by switching the carrier on and off, or by changing its frequency. The latter method is called frequency-shift keying (*see* FREQUENCY-SHIFT KEYING).

The code transmitter can be keyed at the oscillator or at any of the amplifiers following the oscillator. If a heterodyne circuit is used, any of the local oscillators can be keyed. Oscillator keying makes it possible for the operator to listen in between dots and dashes. This is called break-in operation. Amplifier keying usually results in the transmission of a local signal when the key is up; this disables the receiver unless excellent isolation is provided in the transmitter circuit.

The most common method of keying in a transmitter is called gate-block keying. In the bipolar circuit, the equivalent is called base-block keying. *See also* CODE TRANSMITTER.

KEYSTONING

Keystoning is a form of television picture distortion in which the horizontal gain is not uniform at various vertical picture levels. This form of distortion occurs when the picture tube circuitry is out of alignment (see illustration).

Keystoning can be detected by directing the camera toward a square object. The distortion will cause the object to appear trapezoidal; that is, narrower at the top than at the bottom, or vice versa. *See also* TELEVISION.

KICKBACK

When the current through an inductor is suddenly interrupted, a voltage surge, called kickback, occurs. The polarity of this voltage is opposite to the polarity of the original current. The surge may be quite large; the peak voltage depends on the coil inductance and the original current.

Kickback voltage can, in some instances, reach lethal values. Whenever large inductances carry large amounts of current, the components of the circuit must be handled carefully because of the kickback shock hazard.

KILO-

Kilo- is a prefix multiplier. The addition of kilo- before a designator denotes a quantity of 1,000 (one thousand) times as large as the designator without the prefix. *See also* KILOGRAM, KILOHERTZ, KILOWATT, KILOWATT HOUR, PREFIX MULTIPLIERS.

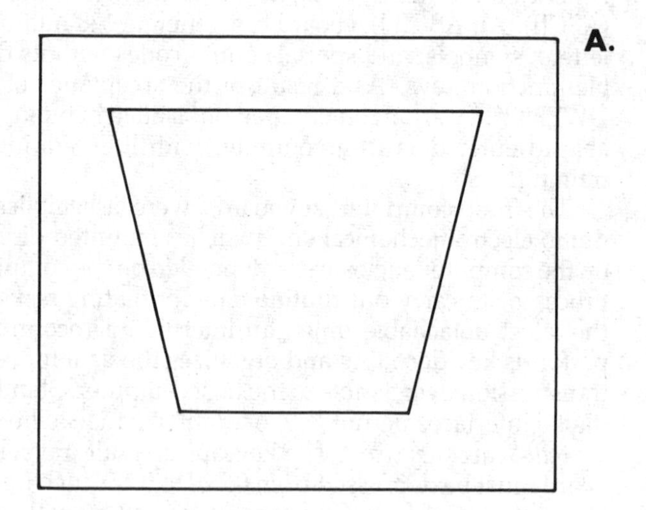

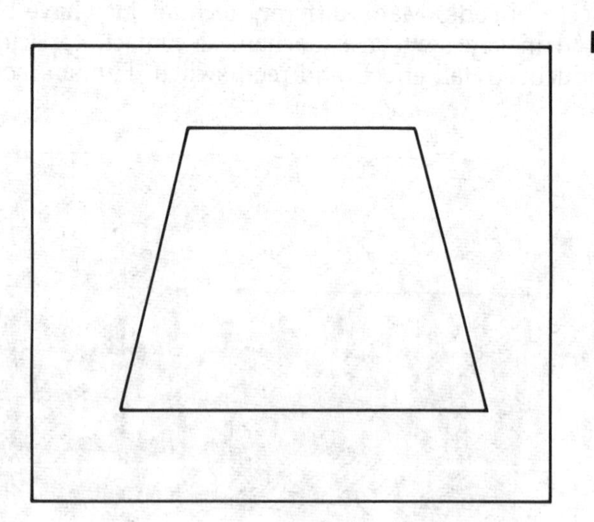

KEYSTONING: Television distortion can have greater gain at the top (A) or greater gain at the bottom (B).

KILOBYTE

The kilobyte is a commonly used unit by which information quantity is expressed. A kilobyte is equivalent to 1024, or 2^{10}, bytes.

Most personal computers have a random-access memory capacity of hundreds of kilobytes. The term *kilobyte* is abbreviated K, so a personal computer might have 640K or more of memory. *See also* BYTE, MEGABYTE.

KILOCYCLE

See KILOHERTZ.

KILOGRAM

The kilogram is the standard international unit of mass. At the surface of the earth, a mass of 1 kilogram weighs 2.205 pounds. The abbreviation for kilogram is kg. A kilogram is 1,000 grams. *See also* INTERNATIONAL SYSTEM OF UNITS.

KILOHERTZ

The kilohertz (kHz) is a unit of frequency, equal to 1,000 hertz, or 1,000 cycles per second.

Audio frequencies, and radio frequencies in the very low, low, and medium ranges, are often specified in kilohertz. The human ear can detect sounds at frequencies as high as 16 to 20 kHz. *See also* FREQUENCY.

KILOWATT

The kilowatt (kW) is a unit of power, or rate of energy expenditure. A power level of 1 kW represents 1,000 W. In terms of mechanical power, 1 kW is about 1.34 horsepower.

The kilowatt is a fairly large unit of power. An average incandescent bulb uses only about 0.1 kW. Radio broadcast transmitters produce radio-frequency signals ranging in power from less than 1 kW to over 100 kW. An average residence uses from about 1 kW to 10 kW. *See also* POWER, WATT.

KILOWATT HOUR

The kilowatt hour (kWh) is a unit of energy. An energy expenditure of 1 kWh represents an average power dissipation of P kilowatts, drawn for t hours, such that $Pt = 1$.

The kilowatt hour is the standard unit for measuring electrical energy consumption in industry and residences. An average household consumes about 500 to 1,500 kWh each month. *See also* ENERGY.

KINESCOPE

See TELEVISION.

KINETIC ENERGY

Kinetic energy is energy in the form of the motion of objects or particles. The most common form of kinetic energy is thermal energy, in which atoms and molecules constantly move about and collide with each other. Kinetic energy also takes place as organized motion, such as the movement of an automobile.

Kinetic energy may be measured in many different units. The standard unit for quantifying energy is the joule. *See also* ENERGY, JOULE.

KIRCHHOFF'S LAWS

The physicist Gustav Kirchhoff developed two fundamental rules for the behavior of currents in networks, called Kirchhoff's First and Second Laws.

According to Kirchhoff's First Law, the total current flowing into any point in a direct-current circuit is the same as the total current flowing out of that point. This is true regardless of the number of branches intersecting at the point (see A in the illustration).

Kirchhoff's Second Law states that the sum of all the voltage drops around a circuit is equal to zero (see B). This sum must include the voltage of the generator, and polarity must be taken into account: equal negative and positive voltages total zero.

Using Kirchhoff's Laws, the behavior of networks, from the simplest to the most complex, can be evaluated.

Lesser well known of Kirchhoff's discoveries is his work on black body radiation. From his principle, the Wien Displacement Law for black body radiation was derived. *See also* BLACK BODY.

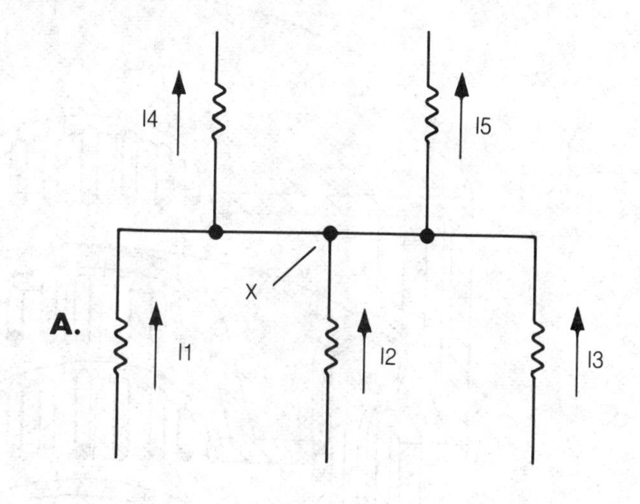

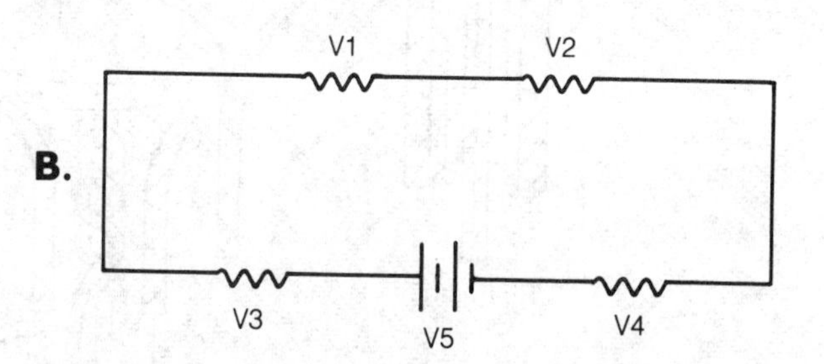

KIRCHHOFF'S LAWS: Current into point X is equal to current out of point X (A) and the sum of voltages V_1 through V_4 is equal but opposite in polarity to voltage V_5 (B).

KLYSTRON

A klystron is a velocity-modulated vacuum tube that includes one or more resonant cavities enclosing grid pairs. Each resonant cavity is controlled by the grids, and the cavity can receive or supply electromagnetic energy over a wide range of frequencies. A klystron can function as a microwave amplifier or oscillator.

A basic two-cavity klystron amplifier contains two toroidal resonant cavities aligned axially within the vacuum tube. An electron gun at one end generates an axial stream of electrons that is directed through the buncher grids of the first, or input, cavity toward a positively charged collector plate at the far end of the tube. A radio-frequency electromagnetic field applied to the first cavity modulates the speed of the electrons, forming them into groups or bunches like a highway traffic signal breaks streams of automobiles into bunches. The electron bunches coincide with the input-signal frequency.

The electron bunches pass through the grids and travel into the drift area between the two cavities where the faster-moving bunches overtake the slower ones and combine with them. This builds the concentration of electrons in the bunches. When the enhanced bunches pass the catcher grids of the second (or output) cavity, the power contained in the bunches is transferred to the cavity causing it to resonate. Amplification takes place between the input and output cavities, and the output of the amplifier is continuous. Some klystron amplifiers have a third or intermediate cavity for additional gain.

When the klystron is designed as an oscillator, some

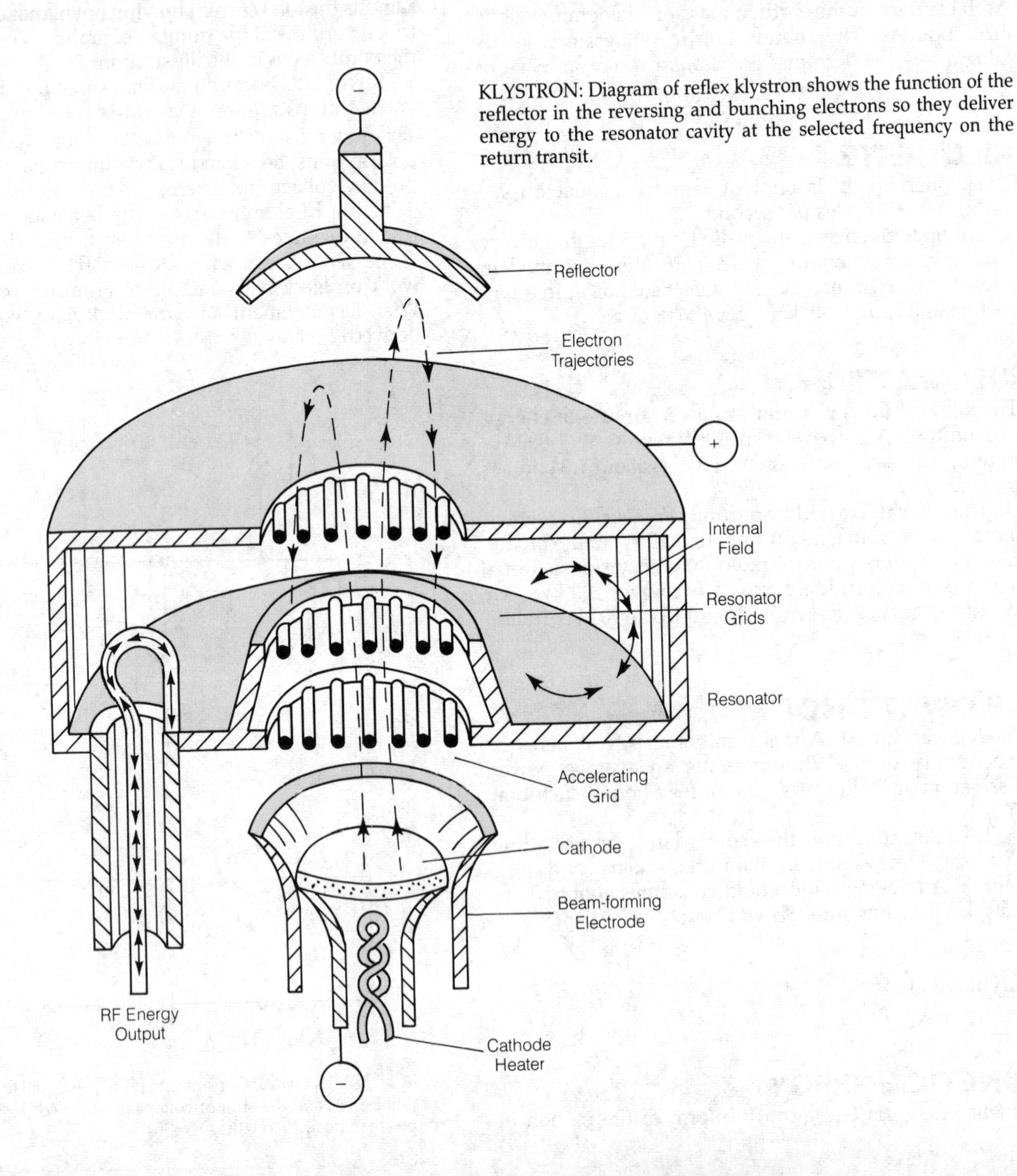

KLYSTRON: Diagram of reflex klystron shows the function of the reflector in the reversing and bunching electrons so they deliver energy to the resonator cavity at the selected frequency on the return transit.

of the output power is coupled back to the input cavity in the proper phase. In microwave oscillator applications the reflex klystron design has been in use. As shown in the figure, the reflex klystron has only cavity that performs both as input and output cavity. The electron gun emits a stream of electrons into the tube and these are bunched during their first transit through the grids. The electron bunches are repelled by a negatively charged reflector and are returned to the resonator grids where they are caught. They give up their energy to sustain continuous oscillation. The reflex klystron, once widely used as a local oscillator for radar systems, has largely been replaced by solid-state oscillator circuits. *See also* CAVITY RESONATOR, OSCILLATOR.

KNEE

A sharp bend in the response curve of an electronic device is called a knee. An example is the current-versus-voltage curve for a semiconductor P-N junction.

As the forward voltage across this junction increases, the current remains small until the voltage reaches 0.3 V to 0.6 V. At this point, the current abruptly rises; this is the knee (see illustration). Under conditions of reverse voltage, the response is similar, but the knee occurs at a much higher voltage than in the forward condition. *See also* DIODE, P-N JUNCTION.

KNIFE-EDGE DIFFRACTION

When an electromagnetic wave encounters a barrier with a sharp edge, some energy propagates around the edge. The sharper the edge with respect to the wavelength, the more pronounced is this effect, called knife-edge diffraction.

In general, the lower the electromagnetic frequency, the less the knife-edge diffraction loss in any given case. For this reason, knife-edge diffraction is common at the very low, low, and medium frequencies; it becomes less prevalent at high and very high frequencies. *See also* DIFFRACTION.

KOOMAN ANTENNA

See ANTENNA DIRECTORY.

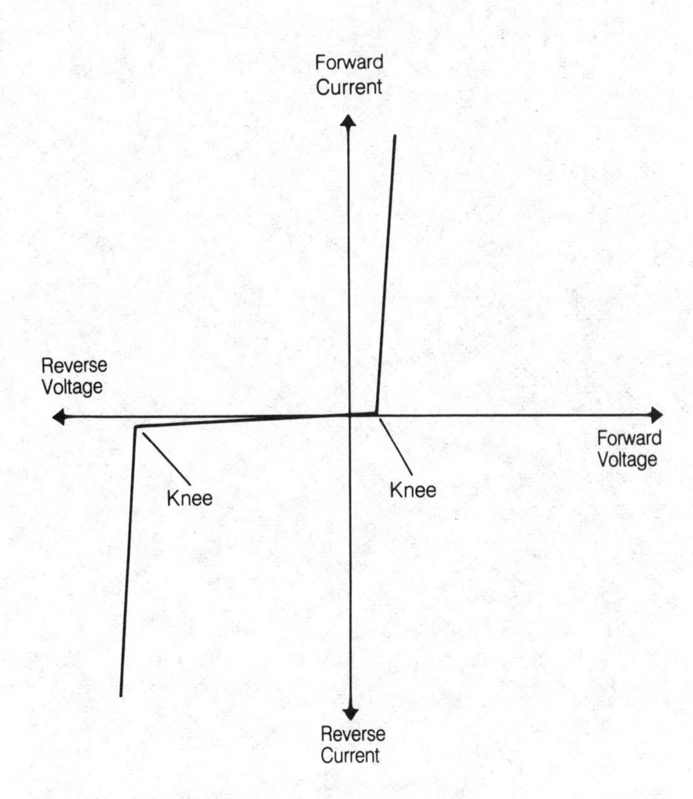

KNEE: Knees in current-versus-voltage curve for a semiconductor diode.

LABYRINTH SPEAKER

A labyrinth speaker is a special form of enclosed speaker containing a network of sound-absorbing chambers positioned behind the speaker cone. The chambers are oriented in different ways to reduce internal resonances. The inside walls of the speaker enclosure are lined with acoustically absorbent material, preventing reflection of the sound waves within the enclosure.

This speaker displays a flat response with very low acoustic standing waves over a wide range of audio frequencies, making it ideal for high-fidelity applications.

LADDER ATTENUATOR

A ladder attenuator is a network of resistors connected so that the input and output impedances remain constant as the attenuation is varied. The drawing illustrates a ladder attenuator equipped with a multiple-position switch for selecting various levels of attenuation. Some ladder attenuators have selector switches in banks, facilitating operation at any desired decibel value.

The ladder attenuator is often used in the test laboratory for comparing the amplitudes of different signals when calibrated oscilloscopes or spectrum analyzers are not sufficiently accurate. They are also used where the test signal is too strong for the input circuit of the test equipment. Ladder attenuators are characterized by negligible reactance at frequencies well into the ultrahigh range, provided the resistor leads are very short. *See also* ATTENUATION, ATTENUATOR.

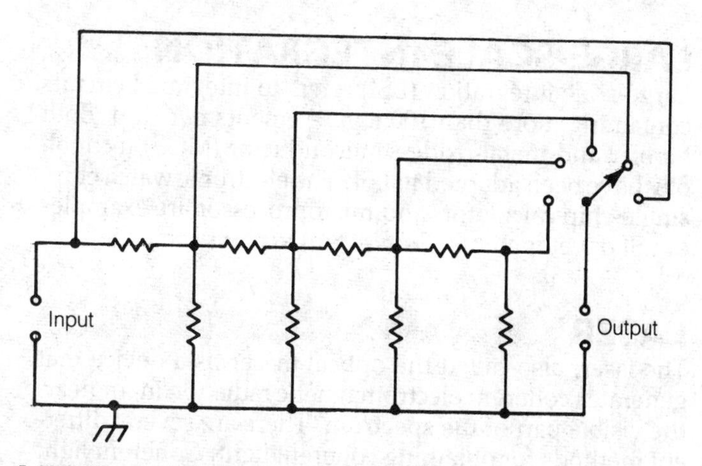

LADDER ATTENUATOR: A five-position ladder attenuator is illustrated.

LADDER NETWORK

A ladder network is, in general, any cascaded series of L networks or H networks (*see* H NETWORK, L NETWORK). The L-network sequence is used in unbalanced systems; a series of H networks is used in balanced systems. The ladder attenuator is a network of resistors (*see* LADDER ATTENUATOR). However, capacitors and inductors can also be used in ladder networks.

Ladder networks, consisting of inductances and capacitances, are used as lowpass and highpass filters in radio-frequency applications. By cascading several L-section or H-section filters, greater attenuation is obtained outside the passband, and a sharper cutoff frequency is achieved. Some power-supply filter sections consist of ladder networks, using series inductors and/or resistors, and parallel capacitors. *See also* HIGHPASS FILTER, LOWPASS FILTER.

LAGGING PHASE

When two alternating-current waveforms with identical frequencies do not precisely coincide, the waveforms are said to be out of phase. The difference in phase can be as great as ½ cycle.

In a display of amplitude versus time, two waveforms with different phase, but identical frequency, appear as shown in the graph. One wave occurs less than ½ cycle later than the other. The later wave is called the lagging

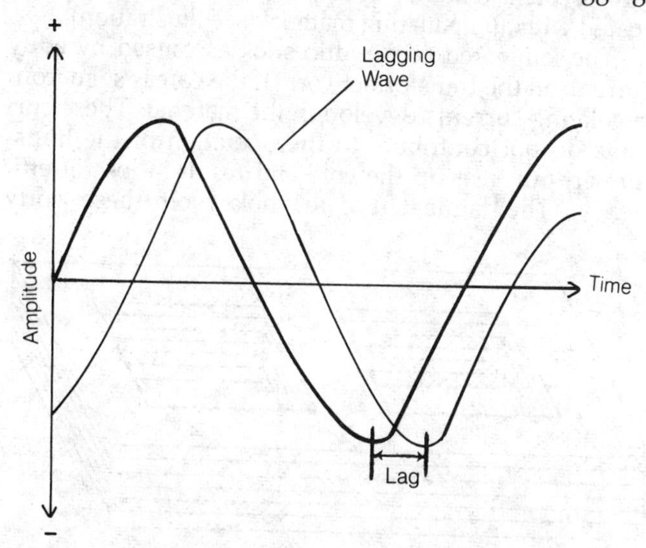

LAGGING PHASE: The lagging waveform has the same frequency as the fundamental waveform, but it occurs a fraction of a cycle later.

wave. If the waveforms are exactly ½ cycle apart in phase, the two signals are said to be in phase opposition. If they are precisely aligned in phase, they are reinforcing. *See also* ANGLE OF LAG, PHASE ANGLE.

LAMBERT

The lambert is a unit of luminance (photometric brightness) in the centimeter-gram-second (CGS) system. It is equal to $1/\pi$ candela per square centimeter and therefore equal to the uniform luminance of a perfectly diffusing surface emitting or reflecting light at the rate of 1 lumen per square centimeter. *See also* CANDELA, LUMINANCE.

LAMBERT'S LAW OF ILLUMINATION

The inverse-square law states that the illuminance of a surface is inversely proportional to the square of the distance between the surface and the light source. This is strictly true, however, only if the angle between the surface and the light beam is the same at all times. Lambert's law is a modification of the inverse-square law for visible light, incorporating an additional factor to compensate for possible changes in this angle.

If a light beam falls on a surface with illuminance of P watts per square meter at a 0 degree angle of incidence, and then the surface is deflected by a certain angle ϕ from that position, the illuminance will be multiplied by the cosine of that angle; that is, it will become equal to $P \cos \phi$. This multiplication factor is always less than 1, and thus the light intensity always decreases as a surface is deflected from the orthogonal position. *See also* ANGLE OF INCIDENCE, INVERSE-SQUARE LAW.

LAMINATED CORE

A laminated core is a form of transformer core generally used at power-line and audio frequencies. The transformer core is made up of stacks of thin metal stampings coated with an insulating material (see illustration).

The laminated core reduces losses caused by eddy currents in the transformer core. If the core is solid iron, circulating currents develop in the material. These currents do not contribute to the operation of the transformer, but heat up the core and result in loss of efficiency. The laminated core chokes off these eddy currents (*see* EDDY-CURRENT LOSS). At radio frequencies, laminated cores become too lossy for practical transformer use. For this reason, ferrite and powdered iron are preferred. *See also* FERRITE CORE, TRANSFORMER.

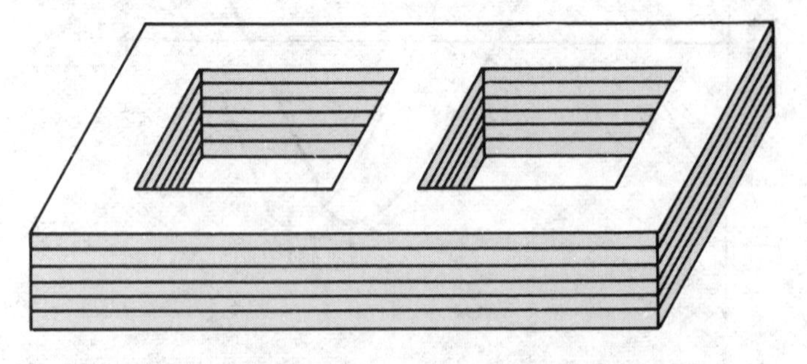

LAMINATED CORE: Many thin layers of iron are coated with an insulating varnish to form the core of a conventional transformer.

LAND-MOBILE RADIO SERVICE

A two-way radio station in a traveling land vehicle, such as a car or bus, engaged in nonamateur communication is called a land-mobile station. Any fixed station, used for the purpose of nonamateur communication with the operator of a land vehicle, is also a land-mobile station. An example is the two-way radio system for dispatching and directing a fleet of taxis.

The land-mobile radio service is allocated a number of frequency bands in the radio spectrum. Many of these bands are in the very high and ultrahigh ranges. The allocations vary somewhat from country to country.

LANGUAGE

In computer applications, a language is a system for representing data with digital bits, or by characters such as letters and numerals.

Languages are classified three basic categories. Machine language is the actual digital information that is processed by the computer. This language appears meaningless to the casual observer. The operator of the computer must converse in a higher-order language, such as BASIC, COBOL, or FORTRAN. These languages use words and phrases as well as standard numerals and symbols. The higher-order language is converted into the machine language by the compiler program. This program must be written in an intermediate language known as assembly language.

Different computer languages are best suited to different applications. For example, BASIC and FORTRAN are intended mainly for mathematical and scientific applications, while COBOL is used primarily by business. *See also* ASSEMBLER AND ASSEMBLY LANGUAGE, BASIC, COBOL, FORTRAN, HIGHER-ORDER LANGUAGE, MACHINE LANGUAGE.

LARGE-SCALE INTEGRATION

Large-scale integration (LSI) refers to integrated circuits containing more than 100 logic elements per chip. Both bipolar and metal-oxide-semiconductor (MOS) technology have been adapted to LSI. The electronic watch chip, single-chip calculator, and microprocessor are examples of LSI developments. *See also* INTEGRATED CIRCUIT.

LASER

The laser, also called the optical maser, is a device that generates coherent electromagnetic radiation in, or near, the visible part of the spectrum. There are several different methods for obtaining coherent light. Coherent light is characterized by a narrow beam and the alignment of all wavefronts in the disturbance (*see* COHERENT LIGHT, COHERENT RADIATION).

Laser action occurs in many different materials and the laser process can occur in many different ways. Lasers can be classified in the following categories: chemical, gas, liquid, metal vapor, semiconductor, and solid state.

A laser constructed from a ruby rod is called a ruby laser. The illustration is a simplified diagram of the ruby laser. A flashlamp surrounds the ruby rod which has mirrors at each end. The rear mirror is a total reflector, but the front mirror allows about 4 to 6 percent of the light to pass through (that is, it reflects 94 to 96 percent of the light). A resonant condition occurs within the ruby rod. If the gain of the system is greater than the loss, oscillation occurs, resulting in a coherent-light output at about 700 nanometers (7×10^{-7} meters). This falls in the red part of the visible spectrum.

A common form of commercially available laser uses helium and neon gas to produce a red visible beam. The laser diode, or injection laser, is a simple type of device that is universally available at a low cost (*see* LASER DIODE).

Lasers are used in many different applications. Low-power lasers can be employed for short-range visible-light communications, because laser light can be modulated, and it suffers far less angular divergence than ordinary light (*see* OPTICAL TRANSMISSION). Lasers have been used for such diverse purposes as measurement of distances, measurement of velocities, and medical surgery. High-power lasers can be used for heating and welding, and they can be deadly weapons. *See also* MASER.

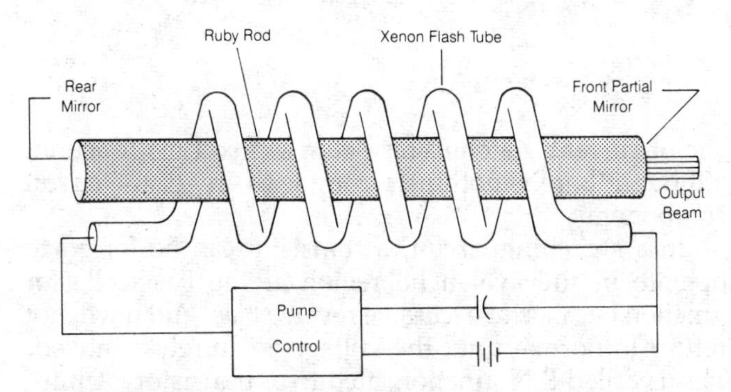

LASER: A simplified diagram of a ruby laser.

LASER DIODE

A laser diode is a semiconductor device capable of emitting coherent light with an internal reflection and reinforcement process similar to that of gas and crystal-rod lasers. The size of a crystal of table salt, a typical laser diode requires a 100 to 200 mW power supply. Also known as injection lasers, they were first operated in the pulse mode in 1962. They were then used in continuous wave (CW) operation in the 1970s.

Laser diodes have been widely used as emitters in short- and long-range fiberoptic communications and as sensors in compact disk (CD) players. Typical diodes produce from 3 to 5 mW of output power in a single stable beam. They are cheaper, more compact, and more reliable than gas lasers. Laser diodes are easily modulated by switching the input current on and off.

Single-mode laser diodes capable of emitting 20 to 50 mW are in demand for optical recording, high-speed printing, data distribution systems, analog signal transmission, long-distance optical communication at high data rates, and space communications between orbiting satellites.

The performance of the laser diode is determined by its chemical composition and its geometry. All diode lasers are essentially multilayered structures made up of several different types of semiconductor material, as shown in the illustration. The materials are chemically doped with impurities to give them either an excess of electrons (N-type) or an excess of electron vacancies (P-type).

Laser diodes that emit in the 0.78 to 0.9 micron region are made of layers of gallium arsenide (GaAs) and aluminum gallium arsenide (AlGaAs) grown on a GaAs substrate. The longer wavelength devices that emit at 1.3 to 1.67 microns are made as layers of indium gallium arsenide phosphide (InGaAsP) and indium phosphide (InP), grown on a substrate of InP.

The illustration shows the structural features common to all continuous wave (CW) laser diodes. The base of the diode is a substrate made of heavily doped N-type GaAs or InP. A lighter N-type doped planar layer of the same material is grown as a cladding on top of the substrate. An active layer of low-doped or undoped semiconductor (AlGaAs or InGaAsP) is grown on the N-type cladding layer. Then another lightly doped P-type layer of cladding is grown on the active layer followed by a heavily doped P-type capping layer.

When current passes through the metallic contacts, electrons injected from the N-type layer and holes injected from the P-type layer recombine in the thin active area, emitting light. The light travels back and forth between the partially reflective end facets of the diode. Lasing begins as the current is increased. The round trip optical gain must overcome losses due to absorption and scattering in the active layer to sustain lasing.

Many laser diodes have a thin layer of oxide deposited on top of the P-type capping layer. This oxide layer is etched so that a shallow recessed metal contact stripe can be formed longitudinally along the top surface of the diode. The index of refraction of the active layer is larger than the index of refraction of the P-type and N-type material (the cladding layers) above and below it. As a result, light is trapped in a dielectric waveguide formed by the two cladding layers and the active layer, propagating in both the active and cladding layers.

The beam of light emerging from the laser diode forms a vertical ellipse (in cross section), although the lasing spot is a horizontal ellipse. The light propagating inside the diode spreads out transversely (vertically) from the cladding layers above and below. When the diode is operating in the fundamental mode, the intensity profile of its emitted beam in the transverse plane is a bell-shaped Gaussian curve.

Light in the laser is amplified by traveling back and forth in the longitudinal direction between the crystal

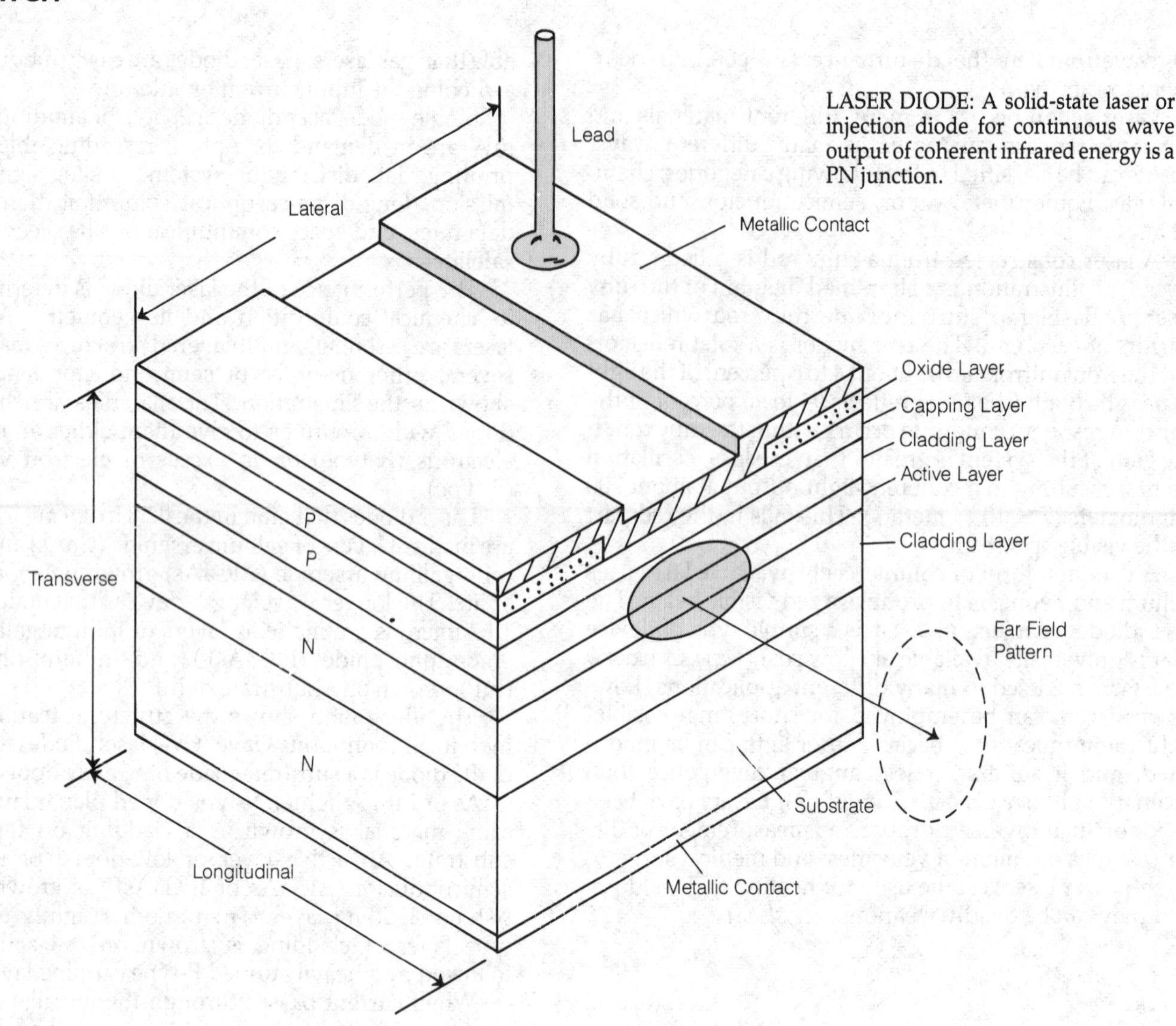

LASER DIODE: A solid-state laser or injection diode for continuous wave output of coherent infrared energy is a PN junction.

facets at each end of the diode. The resonant modes extending in a direction perpendicular to the P-N junction are called transverse modes. The injection of electrons and holes into the active layer directly below the narrow metallic contact stripe alters the refractive index of the active layer and confines the light laterally so it does not spread out on both sides of the center of the active layer. *See* FIBEROPTIC CABLE AND CONNECTORS, LIGHT-EMITTING DIODE, OPTICAL TRANSMISSION.

LATCH

A latch is a digital circuit for maintaining a particular logic condition. A latch consists of a feedback loop and it prevents changes from high to low, or vice versa, resulting from external causes.

A simple circuit for storing a logic element is called a latch. A flip-flop and a cross-coupled pair of logic gates can also be used. *See also* LOGIC GATE.

LATCHUP

Latchup is an undesirable and abnormal operating condition in which a transistor or logic circuit becomes disabled because of the application of an excessive voltage or current. Latchup results in a circuit malfunction, although there is nothing wrong with any of the circuit components.

In a switching circuit, a transistor can be forced to operate in the avalanche region at the base-collector junction because of excessive reverse bias, and it will not return to normal until the voltage is entirely removed. This is called P-N-junction latchup or transistor latchup. It can occur in regulated power supplies.

In a digital circuit, latchup can occur in many ways: the application of an excessive or improper voltage at some point in the circuit is usually responsible. Computer circuits can operate improperly because of a stray voltage. In these cases, reinitialization is usually needed to bring the circuit back to normal.

In an amplifier circuit, unwanted oscillation is sometimes called latchup because it disables the amplifier. This can occur in improperly shielded audio-frequency or radio-frequency systems. *See also* FEEDBACK, PARASITIC OSCILLATION.

LATENCY

In a digital computer system, the response to a command may be very rapid, but it is not instantaneous. The delay time is called the latent time, and the condition of the computer during this time is called latency. The latent time should be as short as possible when high operating speed is needed. In a serial storage system, latency is defined as the difference between access time and word time. *See also* ACCESS TIME.

LATITUDE

Latitude is one of two angular coordinates that uniquely define the location of a point on the surface of the earth.

The equator of the earth, or any sphere, is a great circle with latitude defined as 0 degrees. The north pole is assigned latitude +90, and the south pole is assigned latitude −90. Latitude is measured with respect to the plane containing the equator (see illustration). Therefore, intermediate-latitude lines are circles that become smaller as the latitude increases, either positively or negatively.

The latitude line +23.5 is known as the Tropic of Cancer; the latitude line −23.5 is called the Tropic of Capricorn. The arctic and antarctic circles are at +66.5 and −66.5 degrees, respectively.

For a point to be determined unambiguously, the coordinate longitude is needed. *See also* LONGITUDE.

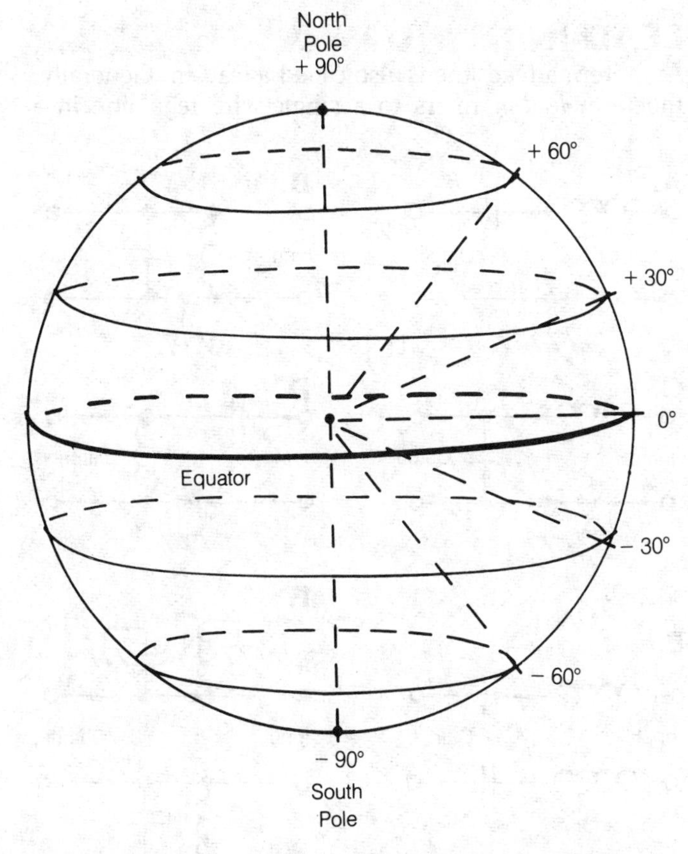

LATITUDE: Latitude lines are parallel planes passed through the earth.

LATITUDE EFFECT

The magnetic field surrounding the earth, known as the geomagnetic field, has an effect on charged atomic particles passing through the upper atmosphere. Charged particles, traveling at high speed through space, have an effective electric current, and consequently they produce magnetic fields. This is true of electrons, protons, alpha particles, and other atomic nuclei. When these particles come near the geomagnetic field, they are diverted toward the north and south magnetic poles by the interacting magnetic forces.

Because of this effect, more charged particles are observed near the poles than near the equator. This is called the latitude effect. The higher the latitude, either north or south, the more cosmic particles are observed at the surface of the earth at any given time. *See also* COSMIC RADIATION, GEOMAGNETIC FIELD.

LATTICE

In crystalline substances, the atoms are arranged in an orderly three-dimensional pattern. This pattern is the same wherever the substance is found. Silicon is an example of a material with a structure called a lattice structure. The term *lattice* is used to describe the arrangement of components in certain electronic circuits. *See also* CRYSTAL-LATTICE FILTER, LATTICE FILTER.

LATTICE FILTER

A lattice filter is a selective filter in which the components are arranged in a lattice configuration. The illustration at A shows one general arrangement of impedances in a lattice filter. (Variations are possible.) The details of the impedance can vary greatly; in the simplest form it might be an inductor or capacitor. However, the impedances in

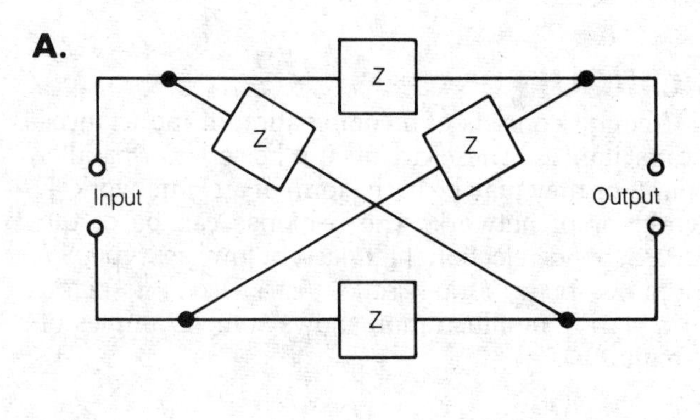

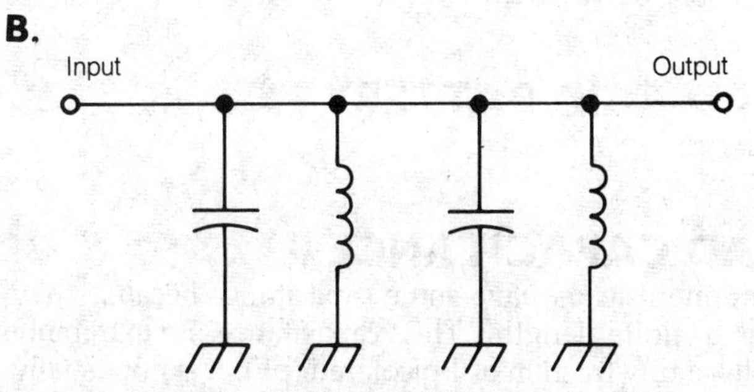

LATTICE FILTER: A lattice filter configuration (A) and a circuit configuration for the bandpass impedances in the filter (B).

a lattice filter are usually bandpass, or resonant, circuits, such as that shown at B. The impedances can also be piezoelectric crystals.

Lattice filters provide a better bandpass response than is obtainable with a single selective circuit. *See also* BANDPASS RESPONSE, CRYSTAL-LATTICE FILTER.

LAW OF AVERAGES

The law of averages is a theorem of probability and statistics. It states that, given a large sampling of events, the probability values obtained by experiment will closely approximate the theoretical values.

An example of the law of averages is illustrated by the repeated tossing of a coin. The probability of getting heads on any given toss is 0.5, or 50 percent. If the coin is tossed just eight times, it is possible that this theoretical value will not be realized. The coin might show heads all eight times, or not at all. However, if the coin is tossed 800 times, the theoretical value will be more closely approached. The greater the number of tosses, the more nearly the experimental and theoretical values will agree. As the number of tosses increases without limit, the proportion of heads will approach 0.5.

If a coin is tossed heads eight times in a row, the probability of heads on the ninth toss does not decrease; it is still 0.5 as for any toss. The idea that an event is due because of a disproportionate series of past events is, for a random sampling, a false notion. The situation might be altered if cause and effect is involved.

LAW OF INVERSE SQUARES

See INVERSE-SQUARE LAW.

LC CIRCUIT

An LC circuit consists of a combination of inductances and capacitances. These circuits can be series or parallel resonant, or they may be in the form of an H network, L network, or pi network. The response can be of the bandpass, band-rejection, highpass, or lowpass type. *See also* BANDPASS FILTER, BAND-REJECTION FILTER, HIGH-PASS FILTER, LOWPASS FILTER. The illustration shows some examples of LC circuits.

LCD

See LIQUID-CRYSTAL DISPLAY.

LEAD-ACID BATTERY

See BATTERY.

LEAD CAPACITANCE

Component leads have some capacitance because they have a finite length. This capacitance is extremely small—a tiny fraction of 1 picofarad (pF). It is not usually of concern at frequencies below the ultrahigh range. However, above about 300 MHz, even this small amount of capacitance can have a significant effect on the performance of a circuit.

In the design of ultrahigh frequency and microwave circuits, component leads should be as short as possible to minimize the capacitance. Lead inductance can also influence circuit operation at these frequencies.

Long leads, such as test leads for voltmeters and oscilloscopes, have much larger capacitance than component leads. Typical coaxial cable has a capacitance of 15 to 30 pF per foot. Parallel-wire leads exhibit somewhat lower lead capacitance than coaxial cable, although it can still be significant at high and very high frequencies. *See also* LEAD INDUCTANCE.

LEADER TAPE

A leader tape is a section of blank magnetic or paper tape that precedes the portion of the tape containing the recorded data. In a magnetic tape, the leader usually consists of clear plastic spliced to the main tape. In a paper tape, the leader is generally a blank section of tape without any punched holes.

The leader tape facilitates the loading or winding of the tape for recording or playback. *See also* MAGNETIC TAPE, PAPER TAPE.

LEAD-IN

An antenna feed line is also called a lead-in. Generally, the term *lead-in* refers to a single-wire feed line in a

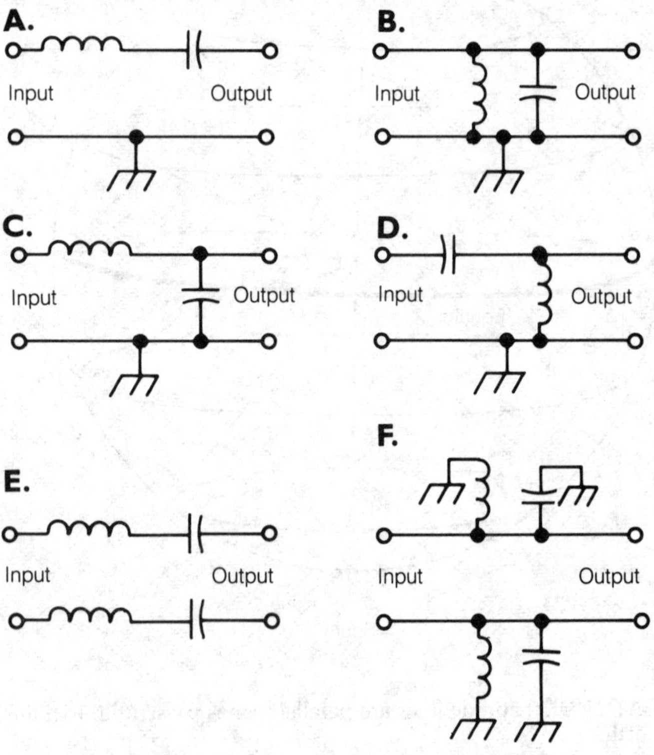

LC CIRCUIT: Examples of inductance-capacitance circuits include the unbalanced, series-resonant (A), the unbalanced, parallel-resonant (B), the unbalanced, lowpass L section (C), and the unbalanced highpass L section (D). The balanced, series-resonant (E) and balanced, parallel-resonant (F) circuits are also shown.

shortwave receiving antenna. A single-wire line displays a characteristic impedance of between 600 and 1,000 Ω under ordinary circumstances (*see* CHARACTERISTIC IMPEDANCE). Little or no attempt is made to match the characteristic impedance of a single-wire lead-in to the impedance of the antenna itself.

The lead-in, if it consists of a single wire, contributes to the received signal. However, most of the signal is picked up by the main antenna. If the lead-in consists of a coaxial cable or a parallel-wire line and the lead-in is properly terminated, the lead-in does not contribute to reception or transmission of signals. *See also* FEED LINE.

LEAD INDUCTANCE

Component leads of wire with measurable length display a certain amount of inductance. This inductance is normally very small—typically a few nanohenrys (billionths of a henry) for the average resistor or capacitor with 1-inch leads.

The lead inductance of electronic components is not generally of great concern at frequencies below the ultrahigh range. However, above 300 MHz the effects can become significant because the lead lengths become a substantial fraction of a wavelength. At ultrahigh and microwave frequencies, leads must be very short to minimize the effects of lead inductance.

Long leads, such as test leads for voltmeters and oscilloscopes, have much larger inductance than component leads. A typical cable or parallel-wire test lead has sufficient inductance to affect circuit operation at very high frequencies. In conjunction with the lead capacitance, resonance can occur at various frequencies in the very high or ultrahigh range. This will affect measurements at those frequencies. *See also* LEAD CAPACITANCE.

LEADING PHASE

When two alternating-current waveforms with the same frequency do not precisely coincide, the waveforms are said to be out of phase. The difference in phase can be as great as ½ cycle.

In a display of amplitude versus time, two waveforms with different phase but identical frequency appear as in the graph. One wave occurs less than ½ cycle earlier than the other. The earlier wave is called the leading wave. If the waveforms are exactly ½ cycle apart in phase, the two signals are said to be in phase opposition. If they are precisely aligned in phase, they are reinforcing. *See also* ANGLE OF LEAD, PHASE ANGLE.

LEAKAGE CURRENT

In a reverse-biased semiconductor junction, the current flow is ideally, or theoretically, zero. However, some current flows in practice, even when the reverse voltage is well below the avalanche value. This small current is called the leakage current. *See* P-N JUNCTION. Insulation resistance (IR) and leakage current are specified and tested in electrolytic capacitors. *See* CAPACITOR.

All dielectric materials conduct to a certain extent, although the resistance is high. Even a dry, inert gas allows a tiny current to flow when a potential difference exists between two points. This current is called the dielectric leakage current. *See also* DIELECTRIC CURRENT.

LEAKAGE FLUX

In a transformer, the coupling between the primary and the secondary windings is the result of a magnetic field that passes through both windings. In some transformers, not all of the magnetic lines of flux pass through both windings. Some of the flux generated by the current in the primary does not pass through the secondary. This is called the leakage flux.

In general, the closer the primary and secondary windings are located to each other, the smaller the leakage flux. An iron-core transformer has a smaller leakage flux than an air-core transformer of the same configuration. The toroidal transformer has the least leakage flux of any transformer. The leakage flux is inversely related to the coefficient of coupling between two inductive windings. *See also* LEAKAGE INDUCTANCE, MAGNETIC FLUX, TRANSFORMER.

LEAKAGE INDUCTANCE

In a transformer, the mutual inductance between the primary and secondary windings is often less than 1 because of the existence of leakage flux (*see* LEAKAGE FLUX). As a result, the primary and secondary windings contain inductance that does not contribute to the transformer action (see figure p. 506). This inductance, known as leakage inductance, is effectively in series with the actual primary and secondary windings. *See also* TRANSFORMER.

LEAKAGE REACTANCE

See LEAKAGE INDUCTANCE.

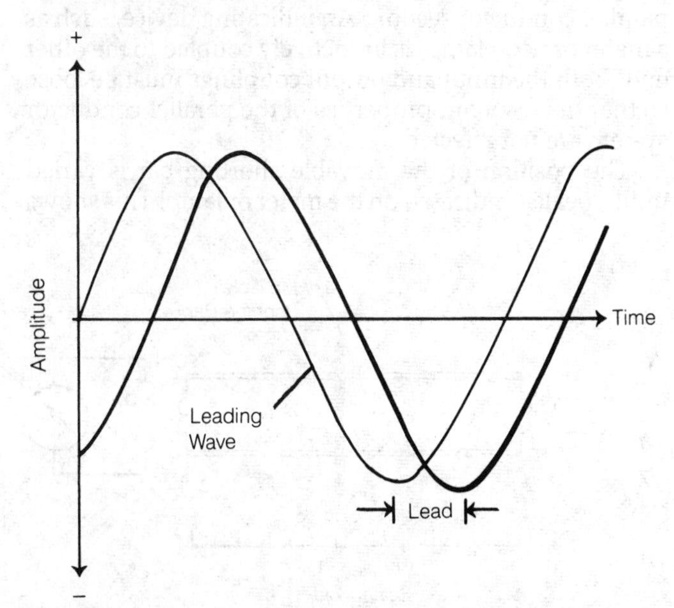

LEADING PHASE: The leading waveform has the same frequency as the fundamental waveform but occurs a fraction of a cycle earlier.

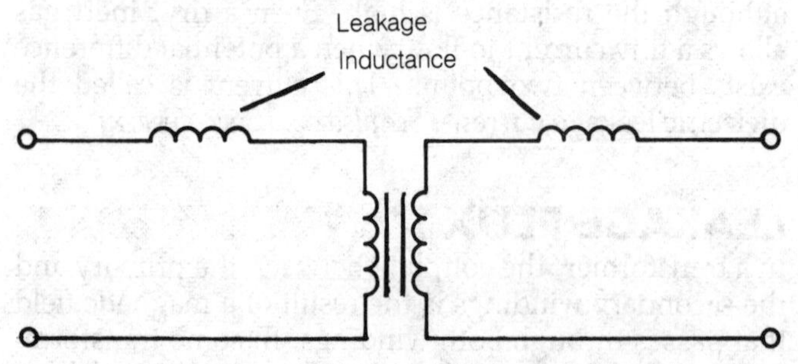

LEAKAGE INDUCTANCE: Leakage inductance appears in series with the transformer windings.

LEAKAGE RESISTANCE

In theory, a capacitor should not conduct any current when a direct-current voltage is placed across it except for the initial charging current. However, once the capacitor has become fully charged, a small current continues to flow. The leakage resistance of the capacitor is defined as the voltage divided by the current.

Air-dielectric capacitors generally have the highest leakage resistance. Ceramic and mylar capacitors also have very high leakage resistance. Electrolytic capacitors have somewhat lower leakage resistance. Typically, capacitors have leakage resistances of billions or trillions of ohms or more. *See also* DIELECTRIC CURRENT, LEAKAGE CURRENT.

LECHER WIRES

Lecher wires are a form of selective circuit used at ultrahigh frequencies for measuring the frequency of an electromagnetic wave. The assembly consists of a pair of parallel wires or rods mounted on an insulating framework. A movable bar allows adjustment of the length of the section of parallel conductors (see illustration).

The electromagnetic energy is coupled into the system by means of an inductor at the shorted end of the parallel-conductor section. An indicating device, such as a meter or neon lamp, is inductively coupled to the other end. Both the input and output couplings must be loose so that the resonant properties of the parallel-conductor system are not affected.

The position of the movable shorting bar is varied until a peak is indicated on the meter or lamp. This shows that the system of wires is resonant at the input frequency, or at some multiple of the input frequency. The shortest resonant length indicates the fundamental frequency of the input signal. A calibrated scale, adjacent to the movable bar, shows the approximate frequency or wavelength of the input signal. The fundamental frequency also corresponds to the distance between adjacent resonant positions of the bar.

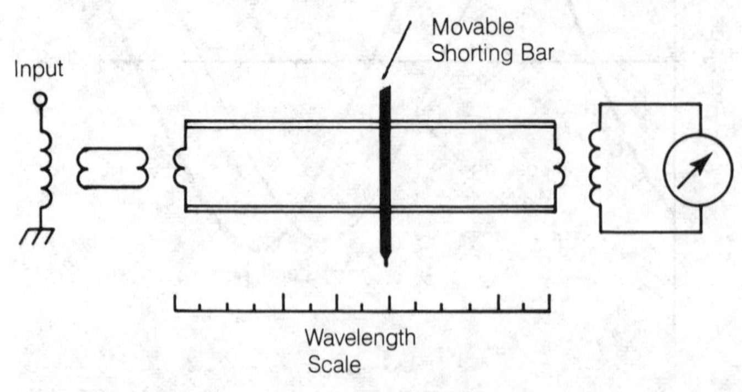

LECHER WIRES: Lecher wires are used for measuring frequency or wavelength. The movable bar is set for resonance, and the scale shows the wavelength.

LECLANCHE CELL

See BATTERY.

LED

See LIGHT-EMITTING DIODE.

LENZ'S LAW

When a current is induced by a changing magnetic field, or by the motion of a conductor across the lines of flux of a magnetic field, that current generates a magnetic field. According to Lenz's Law, the induced current causes a magnetic force that acts against the motion.

An example of Lenz's Law can be observed in the operation of an electric generator. When the output of the generator is connected to a load (so that current flows in the generator coils), considerable turning force (torque) is necessary to make the generator work. The lower the load resistance, the greater the current in the coils, and the greater the magnetic force that opposes the coil rotation. This is why a powerful engine or turbine is needed to operate a large generator. *See also* MAGNETIC FIELD.

LEYDEN JAR

A Leyden jar is a form of capacitor first used to demonstrate that electric charges can be stored. The Leyden jar

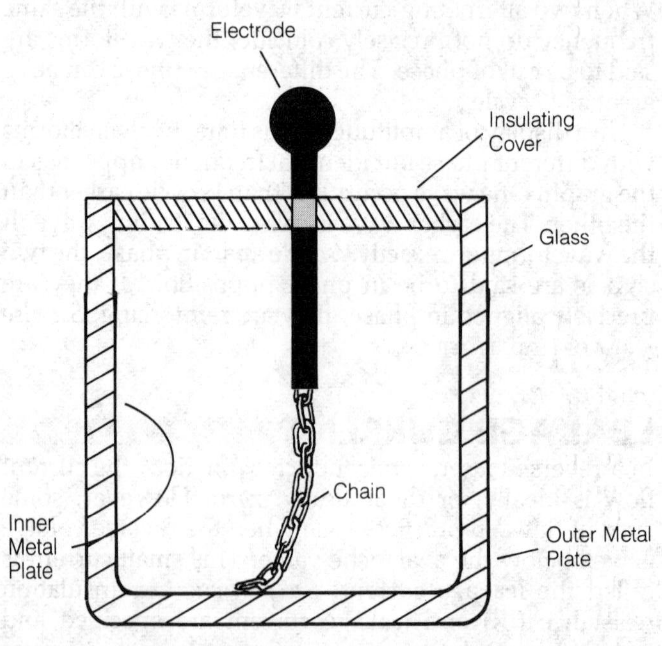

LEYDEN JAR: This early form of capacitor can store thousands of volts.

consists of an ordinary glass jar with aluminum foil or plating on the inside and outside surfaces (see illustration). A rod, generally made of brass or aluminum, extends through the top of the jar, and a chain provides an electrical connection between the rod and the inner metallic surface.

When a high voltage is placed on the rod, the Leyden jar becomes charged. It can retain a potential difference of more than 100,000 volts. The Leyden jar can be used in classroom demonstrations. When the Leyden jar is discharged, the spark may jump a gap of several inches.

LIGHT

Light is visible electromagnetic radiation in the wavelength range of 770 nanometers (nm) to 380 nm. (A nanometer is equal to a billionth of a meter.) The longest wavelengths appear red to the human eye, and the colors change as the wavelength gets shorter, progressing through orange, yellow, green, blue, indigo, and violet.

The earliest theory of light held that it is a barrage of particles. This is known as the corpuscular theory of light, and it is still accepted today, although it is known that light also has electromagnetic-wave properties. The light particle is called a photon. The shorter the wavelength of light, the more energy is contained in a single photon.

In a vacuum, light travels at a speed of about 186,282 miles per second (299,792 kilometers per second), the same speed at which all electromagnetic fields propagate. In some materials, light travels more slowly than this.

Light can be amplitude-modulated and polarization-modulated for the purpose of line-of-sight communications. This is commonly done in fiberoptic systems (*see* FIBEROPTICS COMMUNICATIONS OPTICAL TRANSMISSION).

Light can be generated by thermal reactions or by other means, such as ionization and lasing. Ordinary white or colored light is called incoherent light because the waves arrive in random phase. Laser light is called coherent light, because all of the waves are in phase. *See also* COHERENT LIGHT, INCOHERENT RADIATION, LIGHT INTENSITY, PHOTON.

LIGHT-ACTIVATED SILICON-CONTROLLED RECTIFIER

A light-activated silicon-controlled rectifier (LASCR) is a switching device activated by visible light. Incident light waves perform the same function as the gate current in the ordinary silicon-controlled rectifier (SCR).

Ordinarily, the LASCR does not conduct. However, when the incident light reaches a threshold intensity, the device conducts. The amount of light necessary for conduction is determined by the characteristics of the device and by the value of an externally applied bias. *See also* SILICON-CONTROLLED RECTIFIER.

LIGHT-EMITTING DIODE

The visible light-emitting diode (LED), also called the VLED, is the active diode in all visible light-emitting diode lamps, indicators, and displays. It is a pinhead-size PN junction made from such III-V compounds as gallium arsenide phosphide ($GaAs_{1-x}P_x$), gallium phosphide (GaP), and more recently aluminum gallium arsenide ($Al_xGa_{1-x}As$) and silicone carbide (SiC). LEDs emit light because of electron-hole recombination that takes place at the junction of the P-doped and N-doped regions. The wavelength of the light emission, also called electroluminescence or injection luminescence, is a function of the band gap of the materials from which the junction is made. *See* BAND GAP.

LEDs are grown as layered structures on an N-type substrate with an excess of freely traveling, negatively charged conduction electrons. The epitaxial layer is a P-type material with an excess of mobile positively charged electron vacancies, or holes. Metal contacts are attached to the N-type substrate and a P-type upper layer, and a forward bias (positive at the P contact, negative at the N contact) causes the electrons and holes to migrate into the active layer. As electrons are injected into the N region of the diode, they recombine with holes near the junction to create photons.

LEDs are widely used as lamps and indicators or as components of displays because of their high mechanical stability, low operating voltage, compatibility with digital logic drive circuits, ability to function at low ambient temperatures, and long service life.

Vapor-phase epitaxy (VPE) is a widely used process for manufacturing gallium arsenide phosphide (GaAsP) LEDs on a gallium arsenide (GaAs) substrate. These low-cost LEDs, mass-produced in volume, emit visible red light at a wavelength of about 648 nanometers (nm). Changing the ratio of arsenic to phosphorous and the introduction of nitrogen dopants cause the emission of yellow light at about 585 nm.

Gallium phosphide is another compound used to fabricate LEDs with liquid-phase epitaxy (LPE). Doping GaP wafers with the zinc-oxygen (ZnO) pair causes the GaP die to emit visible red light at approximately 720 nm; doping with nitrogen causes the LED to emit green light at approximately 569 nm. The electrical characteristics of GaP differs from those of GaAsP. GaP-ZnO will produce more red light at lower voltages, but it is unsuitable for multiplexing displays because of its low saturation current. Multiplexing is the sharing of switching circuitry among a string of characters in a display by pulsing the LED elements intermittently at a rate fast enough so there will be no apparent flicker.

High-efficiency LEDs that emit visible red light are being made with VPE growth of GaAsP-on-GaP substrates. The emission color is determined by the ratio of arsenic to phosphorous. The visible red emission of these LEDs is approximately 626 nm. The same technique is used to produce orange emission at 608 nm and yellow at 585 nm.

In another recent development, aluminum gallium arsenide (AlGaAs) is used to produce LEDs that emit red

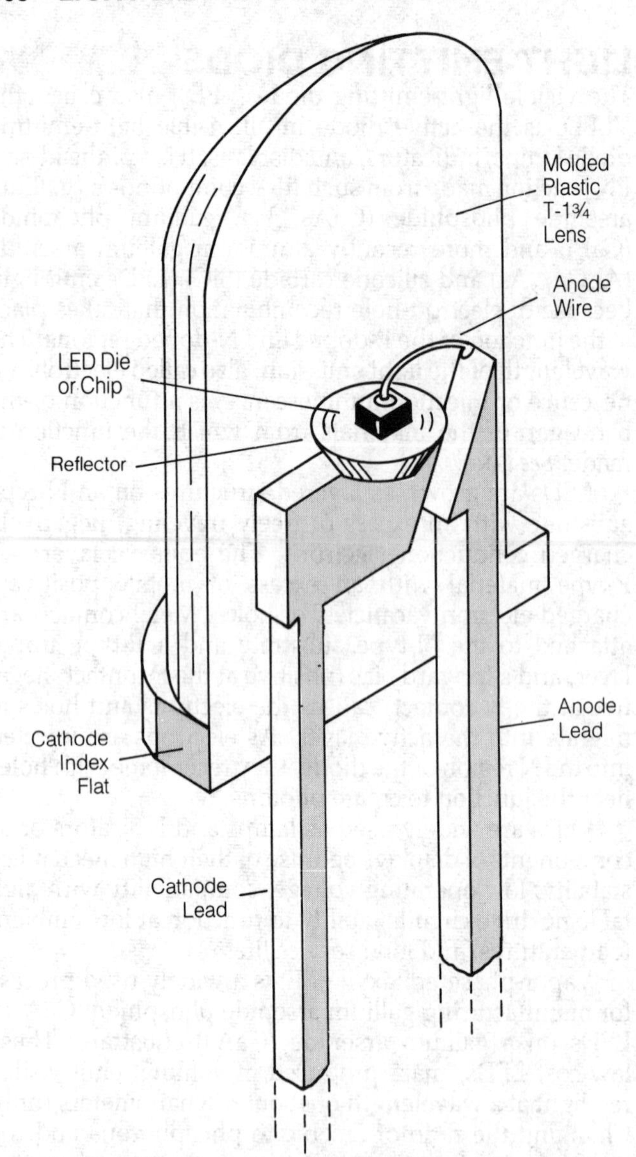

LIGHT-EMITTING DIODE: An LED lamp is a die attached to a reflector on a cathode lead and wire bonded to an anode lead. The molded protective plastic lens determines the viewing angle and lamp brightness.

light at about 646 nm. These LEDs are made by a double-heterojunction AlGaAs process. They can operate at lower currents than existing high-efficiency red materials. An N-type AlGaAs confining layer, a P-type AlGaAs active layer and a P-type AlGaAs confining layer is grown on an N + GaAs substrate. Both high-efficiency and AlGaAs LEDs are suitable for display multiplexing because they offer the higher brightness of GaP with the linear light-versus-current characteristic of GaAsP.

LEDs that emit visible blue light at 475 nm are being made with silicon carbide (SiC). This is the only material that allows reproducible P and N doping and has a band gap for emitting blue light. The blue wavelength of SiC is achieved by doping the diode with aluminum and nitrogen in an LPE process. These LEDs are used as color calibration sources for TV cameras and photographic equipment and as light sources for medical equipment.

Because the light output or luminous intensity of LEDs is generally proportional to current, alteration of the bias voltage can modulate light output. Changes of LED light output of about 2:1 are necessary for detection by the human eye so that slight variations in bias voltage will not be noticed. But, the human eye can easily detect minor light output differences between two adjacent LEDs.

Individual LEDs are packaged in a wide variety of cases with some form of optical lens. In the most common packages the dies are mounted on radial lead frames and molded in a bullet-shaped plastic form as shown in the figure. The lens may be diffused or nondiffused, and it can be tinted in specific color such as red, yellow or green.

The most widely purchased case styles are the T-1 and T-1 ¾ radial-leaded plastic packages. Other packages include flat top and surface-mount styles as well as rectangular molded cases, which produce rectangles of light from their end surfaces. For military and high-reliability applications, LEDs are usually packaged in hermetically sealed TO-5 style metal cases with a glass lens at the top.

The forward bias voltages for most LEDs are from 1.6 to 2.3 V, but the blue light emitting LED requires 5 V. Typical luminous intensity for LEDs is measured in millicandela (mcd) at currents of 2 to 20 mA. Viewing angles (angles within which the luminous intensity is at least half the axial value) vary from 18 to 150 degrees.

Viewing angle is important in LED applications because a lamp with a wide viewing angle can be seen when viewed from large offset angles. However, a LED with narrow viewing angle appears brighter because the light is concentrated in a narrower beam. *See* LIGHT-EMITTING DIODE DISPLAY.

LIGHT-EMITTING DIODE DISPLAY

Light-emitting diode (LED) displays are assemblies of individual LED diodes on a rigid substrate to form one or more alphanumeric characters. The displays can be segmented or dot-matrix displays. Segmented displays can form all numbers and some letters but dot matrix displays are capable of forming all of the ASCII characters. LED displays are used in a wide range of applications from readouts for consumer clocks and appliances (washing machine, microwave ovens, etc.) to readouts for aircraft instruments and industrial controls. When fitted with decoder and driver circuitry, LED dot-matrix displays have become peripheral equipment for computers capable of one or two lines of text.

Seven-segment (figure 8) display modules as shown in the illustration permit the formation of all numbers and uppercase letters A through E. Each of the bar elements is illuminated by an individual LED. Sixteen-segment LED displays are able to form 64 of the ASCII characters. A 5-by-7 dot matrix of individual LEDs permits the formation of all 128 ASCII characters including numbers, upper- and lowercase letters, and other symbols.

The LED dies are bonded to a common conductive surface. This could be a metal lead frame as illustrated, or

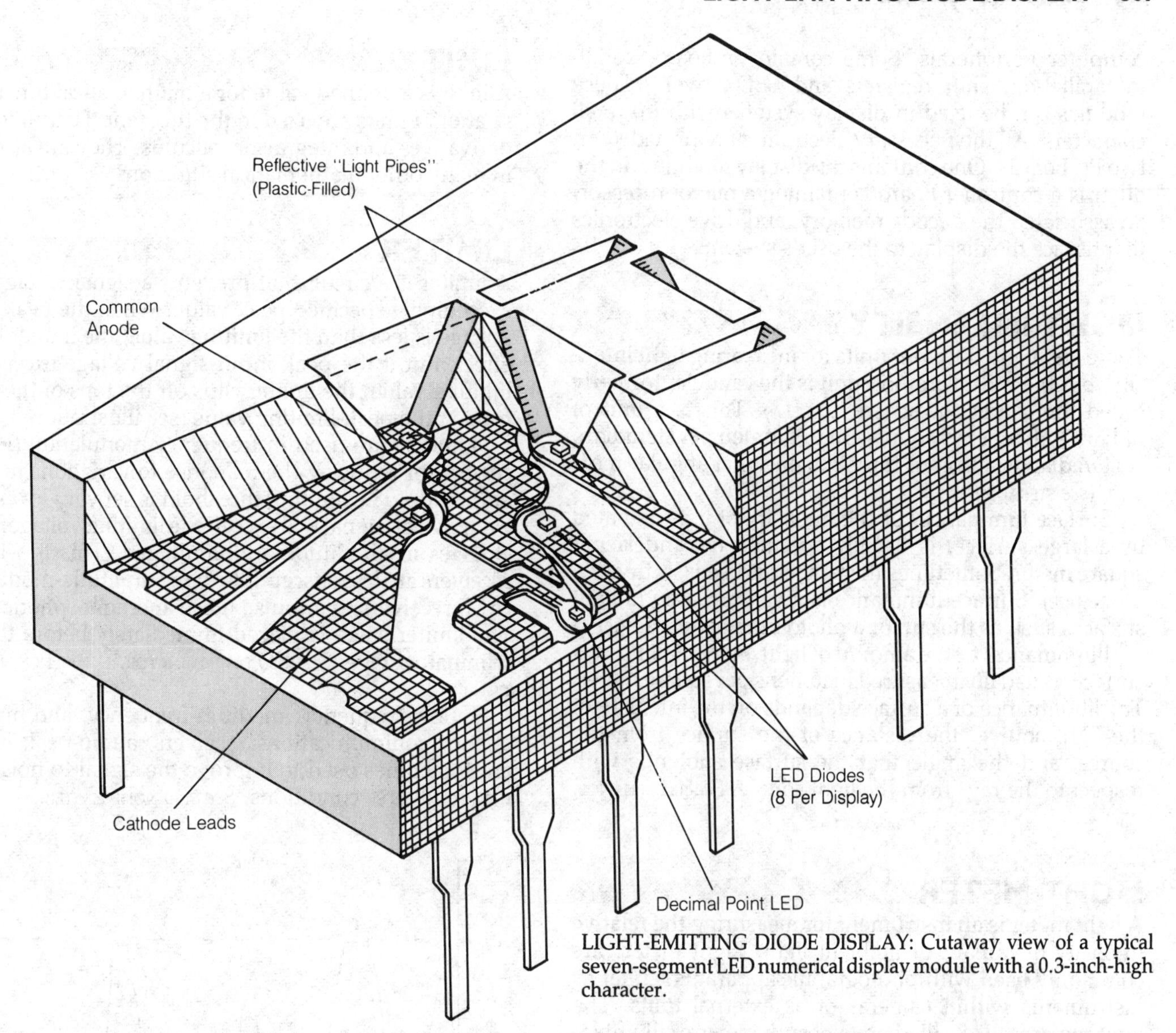

LIGHT-EMITTING DIODE DISPLAY: Cutaway view of a typical seven-segment LED numerical display module with a 0.3-inch-high character.

an insulating substrate with or without screened conductive paths. The substrate could be ceramic or an epoxy-filled glass fiber card. The second terminal on each die is wire bonded to an isolated, insulated conductor. When used to form segmented characters, shaped cavities are placed over each die and back filled with a translucent plastic to function as a lens and light diffuser. Displays are covered with tinted filter caps to give the characters a more uniform appearance. Dot-matrix displays are arrays of LEDs bonded in patterns that define character height. The color of the display is determined by the characteristics of the LED dies employed. *See* LIGHT-EMITTING DIODE.

Seven-segment display modules are available with character heights of 0.3, 0.43, 0.56, and 0.8 inch. These units can display numeric information on electronic instruments, point-of-sale terminals, appliances, and automobiles. They are packaged in dual-in-line (DIP) packages as shown for direct mounting on printed circuit (PC) boards. Dual-character versions of these modules are also available.

Monolithic seven-segment LED displays were developed for watches, and they now are used in instrument and control panels where space is restricted. Seven segments and a decimal point are etched on a single GaAsP chip measuring less than 1⁄10 inch square. Wires are bonded to each of the segments to complete the diode circuits for illuminated character formation. Usually groups of four to eight of these monolithic LED chips are end stacked and covered with bubble-shaped glass or plastic magnifying lenses to increase the apparent size of the display. Some of these are packaged in dual-in-line packages.

Multidigit smart monolithic sixteen-segment LED display modules are available with on-board CMOS integrated circuits (ICs) containing random access memory (RAM), ASCII decoders, multiplexing circuitry, and drivers. Also packaged in DIPs, they are used in portable data entry terminals, medical instruments, process control equipment, test instruments and computer peripherals.

Multicharacter 5-by-7 dot matrix alphanumeric display modules with character heights of 0.16 to 0.27 inch are also available for instruments, process controls, and computer peripherals. Some contain on-board, serial-in/parallel-out shift registers and LED drivers. These modules can be used in display systems with up to 40 characters. All the necessary electronics are provided on two PC boards. One contains the display module and the other is a controller board containing a microprocessor, an associated the decode memory, and drive electronics to interface the display to the user's system.

LIGHT INTENSITY

There are many different units for measuring light intensity, but the most common unit is the candela, formerly called the candlepower (*see* CANDELA). This is a unit of radiant-light energy equivalent to 1 lumen per steradian. A steradian is a unit of three-dimensional angular measure (*see* STERADIAN).

Surface luminance, or the intensity of light radiated by a large surface, is usually specified in candela per square meter. Sometimes the lambert is used (*see* LAMBERT, LUMINANCE). Surface luminance is given for large radiant surfaces such as the sun or a photoluminescent device.

Illuminance, or the amount of light energy falling on a surface, is usually measured in lumens per square meter. The illuminance of a surface depends on the intensity of the light source, the distance of the surface from the source, and the angle that the surface subtends with respect to the rays from the light source. *See* ILLUMINANCE, LUMEN.

LIGHT METER

A light meter is an instrument for measuring the relative intensity of radiant or ambient light. Light meters are commonly used with photographic apparatus, either as instruments within cameras or as external units. The light meter is also called an exposure meter or illumination meter. The most common form of light meter uses a light-sensitive semiconductor device, such as a phototransistor, photodiode, or photovoltaic cell, in conjunction with a milliammeter or microammeter. *See also* ILLUMINATION METER.

LIGHT PEN

A light pen is an input device for computers and video-display monitors that makes it possible to draw on or alter images that appear on the screen by moving the cursor.

The image on a CRT screen is created by an electron beam that scans rapidly from left to right, in horizontal lines beginning at the top of the screen and moving downward. When the light-sensitive tip of the light pen is placed over a certain spot on the screen, a signal is produced as the electron beam scans past that point. This signal is fed to the computer, which causes the spot to change state. The light pen can also be used to produce certain functions or effects, in video games. *See also* JOYSTICK, MOUSE.

LIMIT

A limit is a defined value for a mathematical function or relation. Limits are used in the theoretical definitions for derivatives and integrals in calculus. They are also used in many other mathematical situations.

LIMITER

A limiter is a circuit that prevents a signal voltage from exceeding a specified peak value. When the peak signal voltage is less than the limiting value, the limiter has no effect. But if the peak input-signal voltage exceeds the limiting value, the limiter clips off the tops of the waveform at the peak-limiting value (see illustration).

Limiters are used in frequency-modulation (FM) receivers for reducing the response to variations in signal amplitude. The limiting threshold is set very low so that even a weak signal will exceed the limiting voltage. Then, changes in amplitude are eliminated. This is why FM receivers are less susceptible than amplitude-modulation (AM) receivers to impulse noise and atmospheric static. The limiter stage is placed immediately before the discriminator stage. *See also* DISCRIMINATOR, FREQUENCY MODULATION.

In low-frequency, medium-frequency, and high-frequency communications receivers, audio-peak limiters are sometimes used to improve the signal-to-noise ratio under adverse conditions. *See also* NOISE LIMITER.

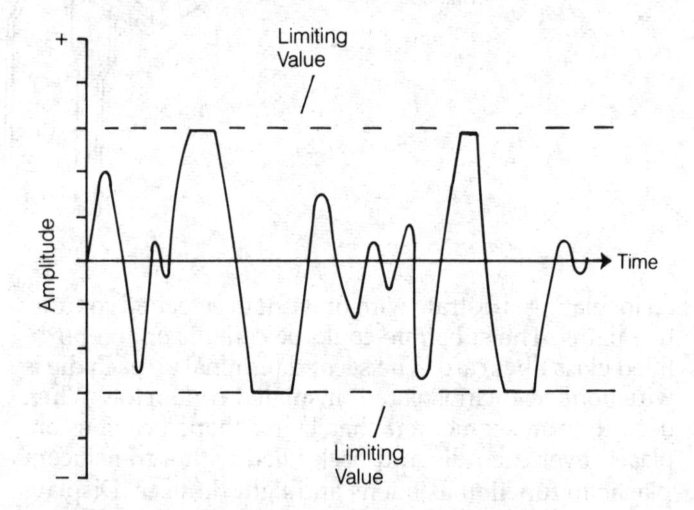

LIMITER: A signal processed through a limiter cannot exceed the specified peak negative and positive limiting values (dotted lines).

LIMIT SWITCH

A limit switch is an electromechanical component that opens or closes a circuit when a parameter, such as current, power, voltage, or illumination, reaches a specified value, either from below or above. A circuit breaker is a form of limit switch that is actuated by excessive current in a circuit. A voice-operated relay is another form of limit switch.

LINE

A line is a wire, or set of wires, over which currents or electromagnetic fields are propagated. A power line, telephone line, or radio transmission line may each be simply called a line. *See also* FEED LINE, POWER LINE, TRANSMISSION LINE.

Electric and magnetic fields are defined in terms of flux lines. The lines of flux represent the theoretical direction of the field. Each line represents a certain quantity of electric or magnetic flux. *See* ELECTRIC FLUX, FLUX, MAGNETIC FLUX.

LINEAR AMPLIFIER

A linear amplifier is a circuit that provides current gain, power gain, or voltage gain so that the instantaneous output amplitude is a constant multiple of the instantaneous input amplitude. The linear amplifier ideally produces no distortion in the waveform or envelope of a signal. All high-fidelity audio amplifiers are linear amplifiers. *See also* ENVELOPE.

LINEAR EQUATION

See LINEAR FUNCTION.

LINEAR FUNCTION

A linear function is a mathematical function of one or more variables such that no variable is raised to a power. That is, there are no exponents in the expression. A one-variable linear function is written in the form:

$$f(x) = mx + b$$

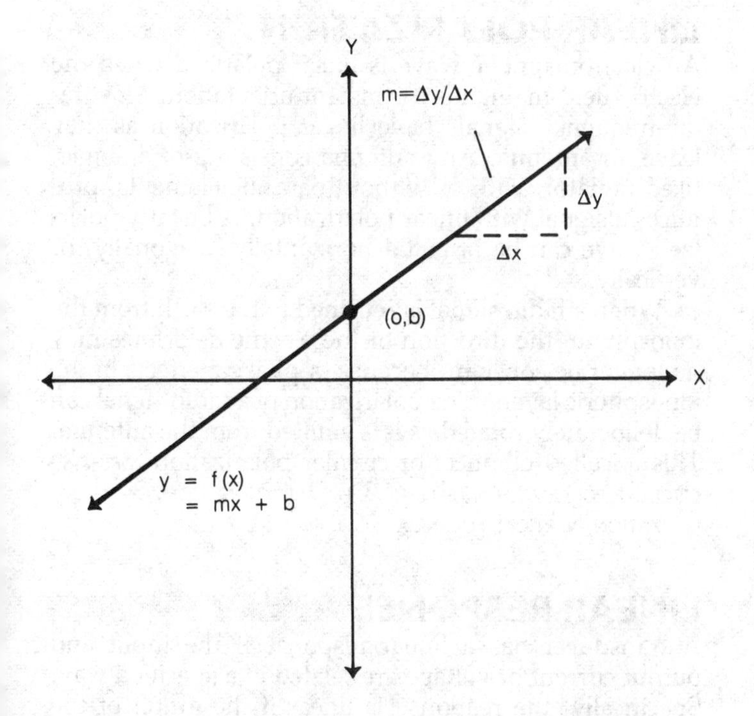

LINEAR FUNCTION: A graph of the line y = mx + b, where m is the slope and b is the y-intercept value.

where m and b are real-number constants. The value m is called the slope of the function, and b is the point at which the graph of the function crosses the dependent-variable axis, or ordinate (see illustration).

The linear function is so named because its plot is always free from curvature in the Cartesian coordinate system. A one-variable linear function appears as a straight line. If the linear function has two independent variables, the graph appears as a flat plane in space.

In electronic applications, a linear function is identical to a linear response (*see* LINEAR RESPONSE). A linear electrical function is a special form of linear mathematical function. If y represents the output current or voltage and x represents the input current or voltage, then a linear electrical function takes the form:

$$y = kx$$

where k is a real-number constant. *See also* FUNCTION, NONLINEAR FUNCTION.

LINEAR INTEGRATED CIRCUIT

A linear integrated circuit (LIC) is a monolithic integrated circuit capable of such analog functions as amplifying, modulating and detecting time variable signals. Also known as analog integrated circuits, they include updated versions of many analog circuits originally designed to include vacuum tubes and discrete transistors. These circuits had to be redesigned to eliminate transformers and inductors because of the impossibility of fabricating low-frequency inductors on a monolithic chip.

The most popular and widely used standard building block linear ICs include operational amplifiers, comparators, voltage regulators, timers, and sample-hold amplifiers. Other linear circuit available in monolithic IC form include phase-locked loops (PLLs) and sample-hold amplifiers (SHAs). Analog-to-digital, digital-to-analog, and voltage-to-frequency converters in monolithic IC form are classed either as linear or interface ICs. *See* ANALOG-TO-DIGITAL CONVERTER, DIGITAL-TO-ANALOG CONVERTER, PHASE-LOCKED LOOP, SAMPLE-HOLD AMPLIFIER.

Operational Amplifier. The key building block in all linear ICs is the operational amplifier (op amp) because of its ability to amplify without the need for inductors or transformers. They are used in many mathematical and signal processing functions: linear and nonlinear; static and dynamic. IC op amp circuits are available as monolithic ICs, and the circuitry is incorporated in many other monolithic linear or analog ICs. The op amp is a stable, high-gain amplifier whose operation depends on external feedback from the output to the input. The name operational amplifier is derived from its original use in analog computers for mathematical operations such as addition, subtraction, multiplication, division, integration, and differentiation. *See* ANALOG COMPUTER.

The function of the op amp is determined by the active and passive components in its external, closed-loop feedback circuit. The resistors, capacitors, or transistors (or combinations of them) dedicate the op amp to functions such as adding, subtracting, integrating, differentiating, or oscillating. Op amps also can be organized so the output is proportional to the difference between the signals at each of the two input terminals. Oscillators based on the op amp can be designed to meet the wide bandwidth, high gain requirements of TV video amplifiers as well as those for selective, narrow-band instrumentation amplifiers.

The performance of the op amp is closely related to the quality of the resistive elements in its feedback circuitry. Resistors must exhibit low temperature coefficients and high stability. Operational amplifiers are classified as low power, high performance, general purpose, low noise, low offset, and high gain. *See* OPERATIONAL AMPLIFIER.

Voltage Comparator IC. A voltage-comparator IC is a circuit that compares the signals at its two input terminals, A and B, to determine which has the greater magnitude. A comparator produces a logic-level output when a true difference occurs. The circuit can be organized so the output will switch from a low of 2 V to a high of 3 to 5 V if the input on terminal A is larger than the input on terminal B. However, if input at A is not greater than the input at B, the output level will remain low. The transition from one level to another occurs over a very narrow voltage range established by an internal reference voltage. A comparator IC can switch from one level to another in about 20 nanoseconds (ns).

The reference voltage establishes a threshold level. This eliminates false signals and the effects of noise in determining if an event has occurred. External resistors at both inputs determine the reference voltage level. When a true input occurs, the threshold will be exceeded and the output will be switched from low to high. In a typical application, a photodetector or other current source is connected to one input. The comparator output can be used to trigger a relay, power transistor or other active device. Voltage comparator ICs are classified as general purpose, high speed, video, and high-input impedance.

Voltage-Regulator IC. Voltage-regulator ICs control the voltage in power supplies within ±10 percent. Universal general-purpose linear ICs, they are widely used in digital circuits where a ±10 percent variation in voltage is acceptable. The most popular voltage regulator ICs are positive and negative three-terminal ICs. Current ratings are typically 0.1, 0.5, and 1.5 amperes. Output voltages are 5, 6, 8, 12 and 15 V. Both negative and positive adjustable voltage regulators are available.

Voltage Regulator ICs are available in two basic categories: linear and switchmode. The conventional or linear series regulator consists of an operational amplifier, a pass transistor, several sensing resistors and a voltage reference. There are four basic switchmode voltage regulators; step-up, step-down, inverter, and push-pull converter. Again all are based upon the operational amplifier and include a pulse width modulator, oscillator and voltage reference function in addition to capacitors, resistors and diodes. *See* POWER SUPPLY.

Timer IC. Timer ICs contain comparator and flip-flop circuits capable of producing accurate, repetitive oscillations that can be counted as time delays. When used to set time delays, an external resistor and capacitor establish the time interval. If the circuit is to be used as an oscillator, frequency and duty cycle are controlled by two external resistors and a capacitor. The timing function can be triggered and reset by the waveforms, and the output can drive logic circuits. Dual-precision timers in the same package are available.

Analog Multiplexer. The analog multiplexer is a circuit that serially switches a number of different analog input signals into a single line or channel. Any individual input channel is normally addressed with a digital address code applied to some digital switches. Time-division multiplexing can be viewed as a rotating commutator that momentarily and sequentially connects each of the several inputs to a common output. Multiplexers are also used in reverse as demultiplexers.

LINEARITY

Linearity is an expression of the resemblance between the input and output signals of a circuit. In general, the better the linearity, the less distortion is generated in a device.

In a strict mathematical sense, linearity is the condition in which the instantaneous input and output signal amplitudes are related by a constant factor. In audio-frequency applications, the input and output waveforms must be such that all points are related by this constant. In radio-frequency (rf) work, a circuit may be considered linear as long as the input and output envelope amplitudes are related by a constant. The actual rf waveform may be distorted. *See also* LINEAR AMPLIFIER.

LINEAR POLARIZATION

An electromagnetic wave is linear polarized when the electric field maintains a constant orientation. Most radio-frequency signals have linear polarization as they leave an antenna. Any antenna consisting of a single, fixed radiator, with or without parasitic elements, produces a signal with linear polarization. A linearly polarized wave can be oriented horizontally, diagonally, or vertically.

When a radio signal is returned to the earth from the ionosphere, the direction of the electric-field lines may no longer be constant, because of phasing effects in the ionospheric layers. The polarization of a radio signal can be deliberately rotated as it is emitted from the antenna. This is called elliptical or circular polarization. *See also* CIRCULAR POLARIZATION, ELLIPTICAL POLARIZATION, HORIZONTAL POLARIZATION, VERTICAL POLARIZATION.

LINEAR RESPONSE

A transducer has a linear response if the input and output current or voltage are related in a specified way. Specifically, the response is linear if the graph of the output versus input appears as a straight line passing through the origin point (0,0).

A linear response can be defined in algebraic terms. If an input current or voltage of x_1 results in an output current or voltage of y_1, and if an input current or voltage of x_2 results in an output current or voltage of y_2, then the response is linear if and only if $y_1 = kx_1$, $y_2 = kx_2$, and $y_1 + y_2 = k(x_1 + x_2)$.

A transducer can exhibit a linear response over a certain range of input current or voltage. A transducer can be linear only within a certain range; outside of that range it can become nonlinear. *See also* LINEARITY.

LINEAR TAPER

Some potentiometers have a resistance that varies linearly with respect to rotation. These potentiometers are said to have a linear taper. If a linear-taper potentiometer is rotated through a specified angle in any part of its range, the change in resistance is always the same and is repeatable.

Linear-taper panel potentiometers are used in many electronic circuits for alignment or adjustment purposes. For volume control, however, the linear-taper potentiometer is not generally used because sound is perceived in a logarithmic manner; audio-taper potentiometers are preferred. *See also* AUDIO TAPER, POTENTIOMETER.

LINEAR TRANSFORMER

A linear transformer is a radio-frequency (rf) transformer used at very high frequencies for obtaining an impedance match.

The linear transformer consists of a quarter-wave section of transmission line, short-circuited at one end and open at the other end (see illustration). The line can be a parallel-conductor or coaxial. Power is applied by a small link at the short-circuited end of the line. The output is taken from some point along the length of the transmission line.

The output impedance is near zero at the short-circuited end of the linear transformer, and is extremely large (theoretically infinite) at the open end. At intermediate points, the impedance is a pure resistance whose value depends on the distance from the short-circuited end.

The linear transformer is used, in various configurations, in many kinds of antenna systems for rf reception and transmission. *See also* IMPEDANCE MATCHING.

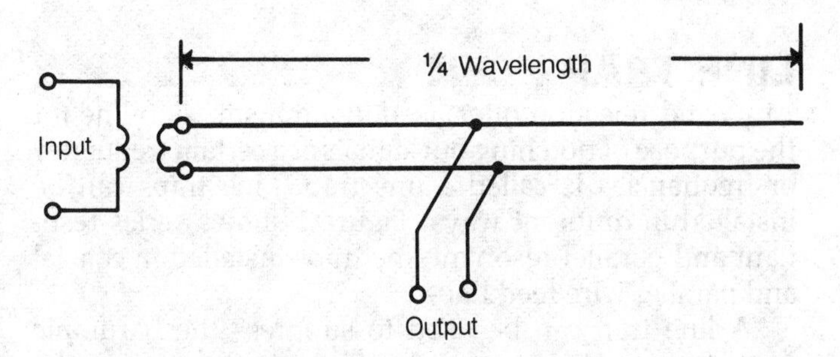

LINEAR TRANSFORMER: A quarter-wave section of transmission line forms a linear transformer.

LINE BALANCE

A parallel-conductor transmission line must be balanced to function properly. Line balance is achieved when the currents in the two conductors are equal in magnitude, but opposite in direction, at every point along the line.

In an antenna system, line balance might be difficult to obtain because of interaction between the line and the radiating part of the antenna. In a center-fed dipole antenna, the open-wire or twin-lead feed line should be oriented at a right angle to the radiating element. The two halves of the radiating element must be exactly the same electrical length, and they must present the same impedance at the feed point. In an antenna system, poor line balance results in radiation from the feed line. *See also* BALANCED TRANSMISSION LINE.

LINE FAULT

A line fault is an open or short circuit in a transmission line, resulting in partial or complete loss of signal or power at the output end of the line.

An open circuit in a line can be located by shorting the terminating, or output, end of the line, and measuring the circuit conductivity at various points with an ohmmeter or other device (see A in the illustration). Measurements should be taken first at the input end of the line; subsequent checks should be made at points closer and closer to the output end. The resistance will appear infinite, or very large, until the testing apparatus is moved past the line fault. Then the resistance will abruptly drop.

A short circuit in a line can be found without breaking the circuit if a high-frequency signal is applied at the input, as shown at B. An indicating device, which may consist of a meter with a radio-frequency diode in series, is moved back and forth along the line. The meter

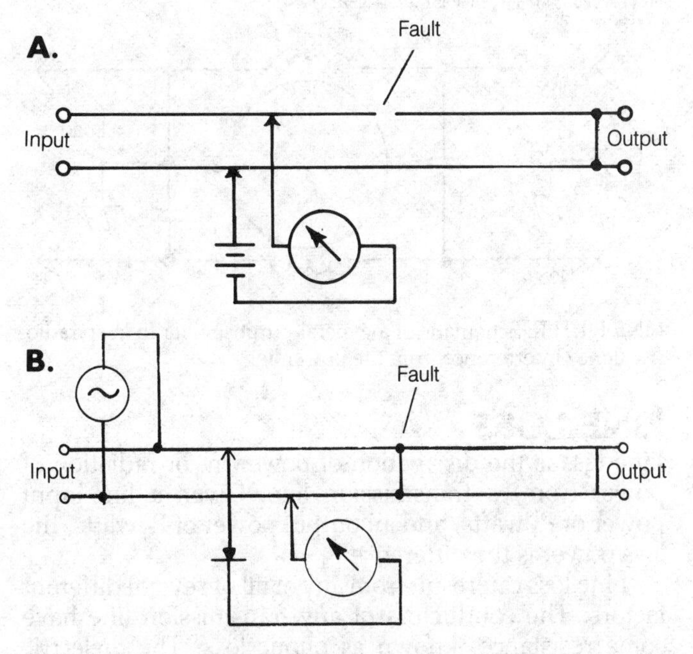

LINE FAULT: A method of checking for an open circuit is shown in (A), and a method for finding a short circuit is shown in (B).

produces readings that fluctuate, depending on the distance from the input, until the short is passed. Beyond the short, the meter reading drops to nearly zero and remains there as the meter is moved further.

LINE FILTER

A line filter is a packaged component that can be placed in series with the alternating-current (ac) power-line cord from electronic equipment to attenuate radio-frequency (rf) noise or interference (rfi) to an acceptable level, while permitting the 50 or 60 Hz current to pass with little or no attenuation. It functions essentially as a trap for rfi, preventing it from entering or leaving the equipment.

Line filters are made as dual, lowpass networks with inductors typically in series and capacitors typically in parallel (see the schematic diagram). Line filters are widely used in switching power supplies, computers, industrial controls, and medical equipment to suppress conducted rfi.

The radiation or conduction of rf energy from electrical and electronic sources that interferes with the operation of adjacent equipment is considered to be rfi. The conducted frequency range of most concern is 10 kHz to 30 MHz. Rfi can be conducted through a power line in two modes: asymmetric or common mode (measured between the line and ground), and symmetric or differential mode (measured from line-to-line). *See also* ELECTROMAGNETIC INTERFERENCE.

Line filters help U.S. equipment manufacturers meet the requirements of Federal Communications Commission (FCC) docket 20780. This docket requires the suppression of conducted rfi greater than 10 kHz on the power line, which can be caused by unfiltered clocks in computing equipment. Verband Deutscher Electrotechniker (VDE) 0871 imposes similar requirements on many European equipment manufacturers. *See also* FILTER, NOISE FILTER. POWER-LINE NOISE, TRANSIENT.

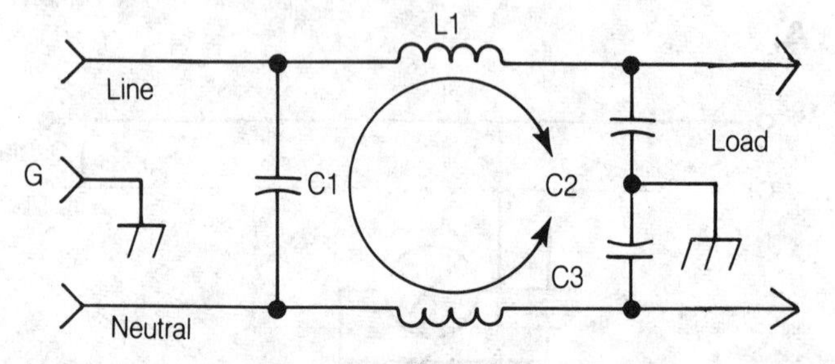

LINE FILTER: Schematic of a general-purpose filter to keep radio-frequency interference from the power line.

LINE LOSS

Line loss is the dissipation of power in, or radiation of power from, a transmission line. Given a line-input power of P_1 watts, and an output power of P_2 watts, the loss power is the difference $P_1 - P_2$.

Line loss can result from any or all of several different factors. The conductors of any transmission line have some resistance, known as ohmic loss. The dielectric material between the conductors of a transmission line produces some loss. Conductor and dielectric losses are converted to heat. If the line balance is poor, power is lost by radiation from the line.

Line loss is generally specified in decibels per unit length. For a particular transmission line, the loss usually increases as the frequency increases. The decibel value, or proportion, of line loss does not depend on the power level of the applied signal as long as the power does not exceed the maximum rating for the line. In a radio-frequency feed line, the loss increases as the standing-wave ratio increases. *See also* DECIBEL, STANDING-WAVE-RATIO LOSS.

LINE-OF-SIGHT COMMUNICATION

Radio communication by means of the direct wave is called line-of-sight communication. The range of line-of-sight communication depends on the height of the transmitting and receiving antennas above the ground, and on the nature of the terrain between the two antennas.

Line-of-sight communication is the primary mode at microwave frequencies. While the range is obviously limited in this mode, propagation is virtually unaffected by ionospheric or tropospheric disturbances. Line-of-sight communication range is limited to the radio horizon. *See also* DIRECT WAVE, HORIZON.

LINE PRINTER

See PRINTER.

LINE REGULATION

See POWER SUPPLY.

LINES OF FLUX

Electrical or magnetic fields can be described theoretically as consisting of lines of flux. The lines of flux in a field run in the general direction of the field effect. The field is considered to originate at one end of each line of flux and to terminate at the other end of the same line. The lines of flux converge at the electric charge centers or magnetic poles.

Lines of flux do not exist physically. They simply represent a certain quantity of electric or magnetic field flux. *See also* ELECTRIC FLUX, FLUX, MAGNETIC FLUX.

LINE TRAP

Any band-rejection filter used in a transmission line for the purpose of notching out signals at a certain frequency or frequencies is called a line trap. Line traps can be installed in different ways. Figure 1 shows series-resonant and parallel-resonant line traps installed in coaxial and parallel-wire feed lines.

A line trap can be used to suppress the harmonic output of a radio transmitter. The trap is tuned to the harmonic frequency. In a receiving antenna system, a line trap can be used to reduce the level of a strong local

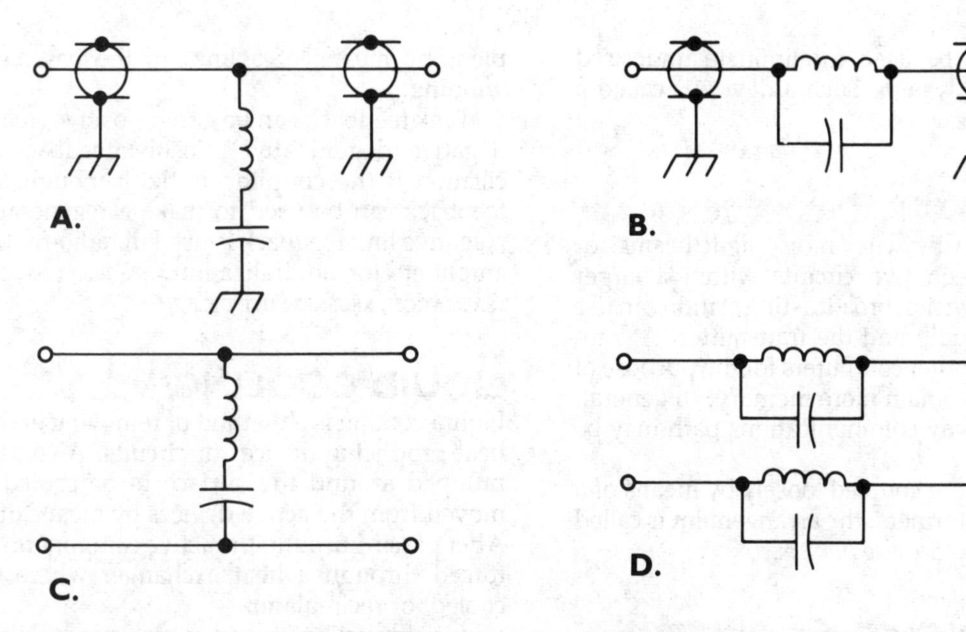

LINE TRAP: Fig. 1. Line traps include the series-resonant unbalanced (A), the parallel-resonant unbalanced (B), the series-resonant balanced (C), and the parallel-resonant balanced (D). These line traps are band-rejection filters.

signal that would otherwise overload the front end of the receiver.

A quarter-wavelength section of transmission line, either coaxial or parallel-wire, can be used as a series-resonant trap. One end of the quarter-wavelength section is connected to the receiver or transmitter antenna terminals along with the feed line, and the other end is simply left open (see Fig. 2). This trap can be called a line trap because it is constructed from transmission line. At the resonant frequency, it appears as a short circuit across the input terminals. *See also* TRAP.

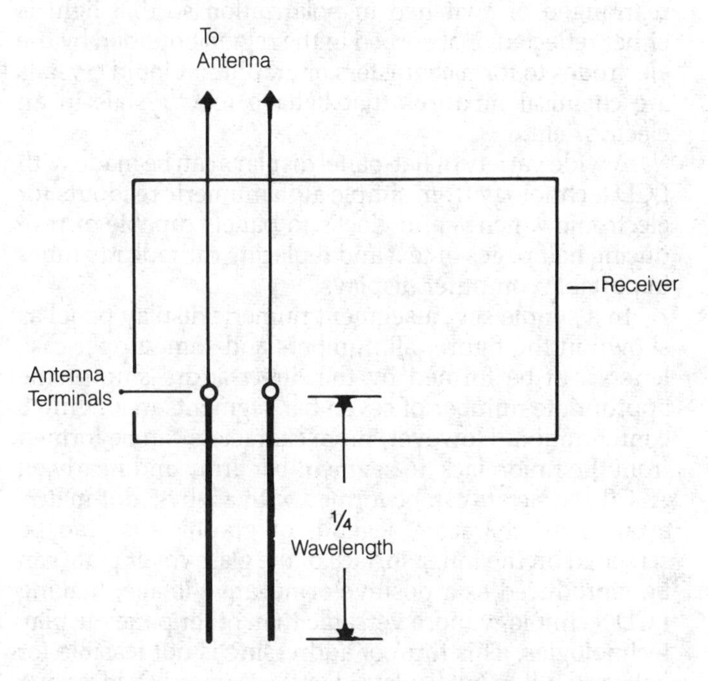

LINE TRAP: Fig. 2. A quarter-wave section of transmission line can be a balanced radio-frequency line trap.

LINE TUNING

Line tuning is a method of tank-circuit construction used at ultrahigh and microwave frequencies. A pair of parallel conductors, or a length of coaxial line, is used as a parallel-resonant circuit. This is accomplished by short circuiting one end of a quarter-wavelength section of line (see illustration). The other end of the line then attains the characteristics of a parallel-resonant, inductance-capacitance circuit.

The resonant line may be tuned by means of a movable shorting bar, or the frequency may be fixed. A tuned line has better frequency stability than an inductor-capacitor arrangement at ultrahigh and microwave frequencies. However, the quarter-wave tuned line is resonant at odd harmonic frequencies as well as the fundamental frequency.

A quarter-wavelength section of line can be left open at the far end, resulting in the equivalent of a series-resonant inductance-capacitance circuit. Half-wavelength sections of line are also used for tuning purposes. A shorted half-wavelength line acts as a series-resonant circuit; if the far end is open, the half-wavelength line acts as a parallel-resonant tank.

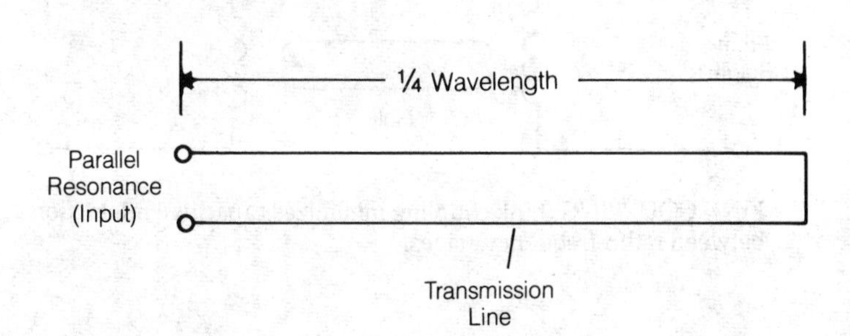

LINE TUNING: Line tuning is accomplished with a resonant length of transmission line.

A tuned line may be used to eliminate undesired signals in an antenna system. Such a device is called a line trap. *See also* LINE TRAP.

LINK

A link is a connection by wire, radio, light beams, or other medium, between two circuits within a larger system. A radio or television broadcasting station can use a link between the studio and the transmitter. A computer can be linked to other computers for the purpose of transferring data or to obtain more memory. In general, any one-way or two-way communications path may be called a link.

When two circuits are coupled loosely by means of a pair of inductive transformers, the arrangement is called a link, or link coupling. *See also* LINK COUPLING.

LINK COUPLING

Link coupling is a means of inductive coupling. The output of one circuit is connected to the primary winding of a step-down transformer. The secondary winding of the transformer is directly connected to the primary winding of a step-up transformer. The secondary of the step-up transformer provides the input signal for the next stage or circuit. The illustration is a schematic diagram illustrating link coupling.

The smaller, intermediate windings in a link-coupling arrangement usually have just one or two turns. As a result, there is very little capacitance between the two stages. This minimizes the transfer of unwanted harmonic energy. Link coupling also minimizes loading effects in the output of the first stage.

Test instruments can be connected to a circuit with link coupling. Because the capacitance is so small, the presence of the test instrument does not significantly affect the circuit under test. This is important in radio-frequency transmission, especially at the very high frequencies.

Link coupling can be used to generate feedback in an amplifier system. The feedback can be positive, resulting in oscillation; negative feedback can be employed for the purpose of neutralization. *See also* LINK FEEDBACK.

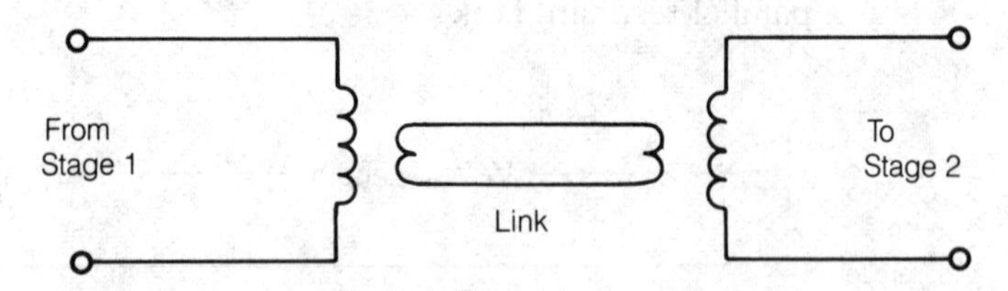

LINK COUPLING: Link coupling minimizes capacitive interaction between radio-frequency stages.

LINK FEEDBACK

Link feedback is the use of link coupling between the output and input of an ampliflier circuit. The collector, drain, or plate transformer winding is inductively coupled, by means of a link, to the base, gate, or grid winding.

Link feedback can be either positive (regenerative) or negative (degenerative). Positive feedback results in oscillation if the coupling is tight enough. Positive link feedback can be used to make a regenerative receiver. Negative link feedback is used in radio-frequency power amplifiers for neutralization. *See also* LINK COUPLING, NEUTRALIZATION, REGENERATIVE DETECTOR.

LIQUID COOLING

Liquid cooling is a method of removing excess heat from heat-producing devices or circuits. A confined liquid is pumped around the surface to be cooled. Heat is removed from the active devices by means of conduction. After passing around the active components, the liquid is forced through a heat exchanger where the liquid is cooled for recirculation.

Liquid cooling is used in high-power broadcast transmitters. It is also used in cooling power supplies and memory banks in supercomputers. *See also* AIR COOLING, CONDUCTION COOLING, CONVECTION COOLING, FORCED-AIR COOLING.

LIQUID-CRYSTAL DISPLAY

Liquid-crystal displays (LCDs) are electronically switched display panels that make use of changes in the reflective properties of liquid crystals in series with an electric field. A thin film of liquid crystals is sandwiched between glass plates imprinted with transparent electrodes. When a voltage is applied selectively across the electrodes, the liquid crystal molecules between them are rearranged or switched in polarization so that light is either reflected or absorbed in the region bounded by the electrodes to form characters or graphics. Liquid crystals are chemical mixtures that behave like crystals in an electric field.

A wide variety of flat-panel displays can be made with LCD technology from simple alphanumeric readouts for electronic watches and clocks to panels capable of producing half pages of text and replacing cathode ray tubes in portable computer displays.

In a simple seven-segment numeric display panel as shown in the figure, all numbers and some upper case letters can be formed by the direct addressing of the appropriate number of seven bar segments in a figure 8 configuration. However, more characters can be formed from the union jack 16-segment bar array and nearly all ASCII characters can be formed with a 5-by-7 dot-matrix array. Any character, legend, or graphic that can be screened on the inner surface of the glass cover plate can be reproduced as a positive or negative image, making LCD technology more versatile than other panel display technologies. This form of addressing is not feasible for half- and full-page displays. For these, matrix addressing is used.

The glass covers of display panels are assembled with spacers between them to keep a constant separation distance. They are then sealed around the edges to

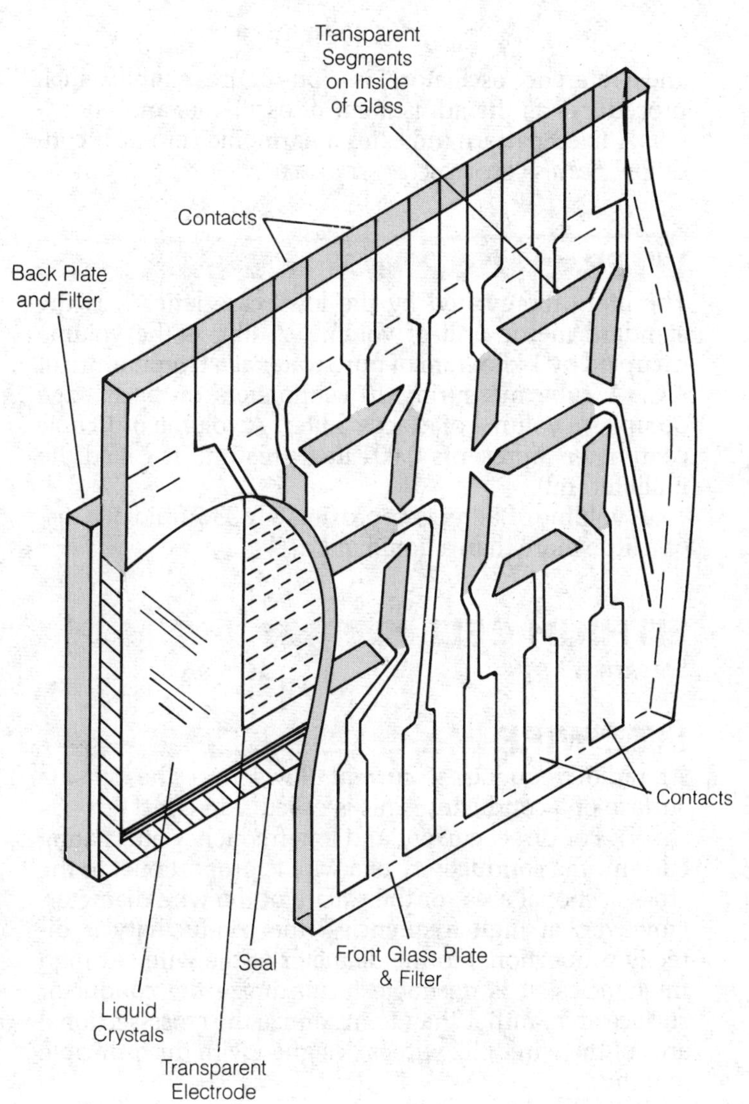

LIQUID-CRYSTAL DISPLAY: Liquid crystals are sandwiched between glass plates with transparent electrodes. Voltage across the electrode segments alters the optical properties of the crystals to form characters with reflected light.

contain the viscous liquid crystals. The panels are equipped with polarizing filters. The smaller LCD numeric displays can be packaged in dual-in-line (DIP) cases with elastomeric bonding to cushion them from shock. This packaging permits convenient socket mounting directly on printed circuit boards.

Because of their low voltage and power requirements, LCDs are widely specified as alphanumeric displays in battery-powered watches, clocks, pocket calculators, digital multimeters (DMMs), digital panel meters (DPMs), navigational instruments (loran and depth finders), games, toys, and other consumer products. Large LCDs capable of 640 by 200 pixels have been developed. This display size, which can show 25 lines of 80 characters is a page display. The large LCDs are intended for battery-powered laptop personal computers where they are competing with cathode ray tube, electroluminescent displays (ELD), alternating-current and gas-plasma displays.

Liquid-crystal displays are used in battery-operated portable television sets, low-cost radars, and unified displays for depth finders and fish finders, and loran for commercial vessels and small craft. They permit the display of sections of nautical charts stored in digital memory for use with loran readouts.

Most LCDs today employ twisted-nematic field-effect (TNFE) liquid crystals that provide either dark or black characters on a light (gray-white) field or the inverse. TNFE displays have threshold voltages of 1.6 V, but 3 to 15 V are required to obtain sufficient contrast for high readability. When subjected to an electric field, the TNFE molecules twist and act as small shutters to reflect light. When the voltage is removed, they recover their normal orientation and are transparent to light. The recovery time is called response time.

Alternating current or pulsed direct current is used to drive LCDs because direct current causes destructive electrolysis of the liquid crystals. Standard and custom MOS integrated circuits perform the decoding and provide the required drive pulses.

Dynamic scattering liquid crystals were introduced first and are still in limited use for large-area display panels. Guest host LCD materials, introduced after TNFE, are able to provide a range of colors in the display.

LCD quality has improved significantly since they were first introduced in the 1970s. Better liquid crystal mixtures permit them to function reliably over wider temperature ranges. In addition, problems with LCD seals have been solved and new assembly and mounting techniques have eliminated concern over accidental breakage of the displays.

The advantages of LCDs are:

- High visual contrast in direct sunlight eliminates the washout of most self-illuminating displays.
- Low voltage and power requirement eliminate shock hazard, permitting their use on medical instruments.
- Fine line and point electrodes permit high definition graphics and alphanumeric characters with minimal limits on electrode size and shape.

The disadvantages of LCDs are summarized as follows:

- Low contrast and legibility.
- Backlighting is required for night or dark viewing, which cancels savings in power consumption.
- Narrower operating temperature range than in self-illuminated displays because of blurring at high and low ambient temperatures.
- Limits on multiplexing because of slow response time.

In a large LCD matrix displays, each pixel is a twisted-nematic cell. Each cell is formed as an x-y coordinate matrix of fine width parallel electrodes formed on the inside surfaces of the opposing glass plates in the liquid crystal sandwich. These electrodes, like the segments and dots of the smaller displays, are formed by the deposition of a transparent conductive film on the inside of each plate and subsequent photolithographic and etching processes.

The glass plates are assembled so the addressable parallel electrodes on the top and bottom plates are at right angles to each other over the liquid crystal material. A pixel is obtained when the voltage pulses across the opposing electrodes and liquid crystal exceed a certain voltage threshold. The LCD is usually organized so that the rows are pulsed cyclically and the data is multiplexed into the columns synchronously.

Large LCD displays based on metal-insulator-metal (MIM) technology have been introduced. Another active matrix technology under investigation is thin-film, transistor-based (TFT-based) displays. These large-panel liquid-crystal displays use a transistor and capacitor at each pixel location to improve contrast and readability. Conventional LCD pixels are driven only by a short voltage pulse during the scanning cycle but active-matrix pixels are driven by a continuous voltage. The transistor and capacitor hold the pulsed-voltage level indefinitely.

LISSAJOUS FIGURE

A Lissajous figure is a pattern that appears on the screen of an oscilloscope when the horizontal and vertical signal frequencies are integral multiples of some base frequency.

When the horizontal and vertical signals are equal in amplitude and frequency and differ in phase by 90 degrees, a circle appears on the display. When the vertical signal has twice the frequency of the horizontal signal, a sideways figure-8 pattern is observed. If the vertical signal has half the frequency of the horizontal signal, an upright figure-8 pattern is traced. Different frequency combinations produce different patterns. The shapes of the ellipses vary depending on relative amplitude and/or phase angle. Some common Lissajous figures are shown in the illustration.

Lissajous figures are useful for precise adjustment of variable-frequency oscillators against a reference oscillator with a known audio frequency. Using an oscilloscope and a reference oscillator, Lissajous figures simplify such procedures as the adjustment of oscillators and encoders. A fixed pattern indicates a harmonic zero-beat condition. *See also* OSCILLOSCOPE, ZERO BEAT.

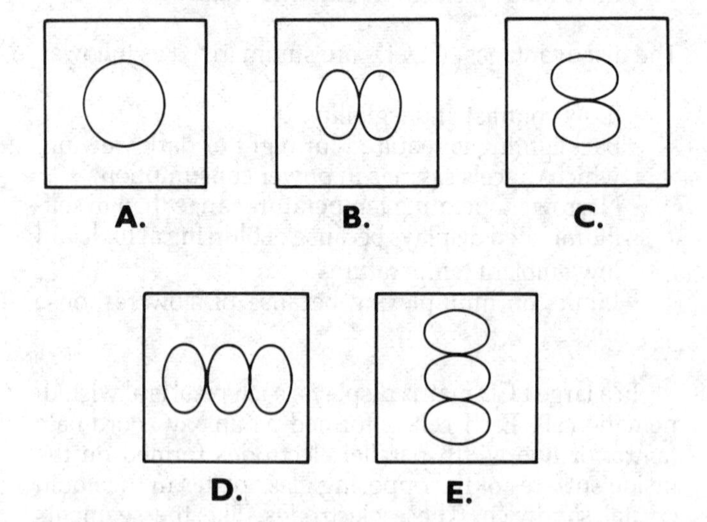

LISSAJOUS FIGURE: Simple Lissajous figures include vertical (v) = horizontal (h) (A), v = 2h (B) v = h/2 (C), v = 3h (D), and v = h/3 (E).

LITER

The liter, abbreviated by the lowercase letter *l*, is the standard metric unit of volume. A liter is the volume occupied by 1 kilogram of pure water at a temperature of 4°C. A cube measuring 10 centimeters on each edge occupies a volume of exactly 1 liter. A volume of 1 cubic centimeter represents 0.001 liter; this unit is called the milliliter (ml).

A volume of 1 liter is approximately 33.7 fluid ounces, or a little more than a liquid quart.

LITHIUM CELL

See BATTERY.

LITZ WIRE

At radio frequencies, current flows near the outside surface of a conductor. This is called skin effect (*see* SKIN EFFECT). For direct current and low-frequency alternating current, the conductivity of a wire is proportional to the cross-sectional area, or the square of the wire diameter. However, at high frequencies, the conductivity is directly proportional to the diameter of the wire. At high frequencies, it is desirable to maximize the conductor surface area, rather than to maximize the cross-sectional area of the wire. Litz wire is designed with this principle in mind.

Litz wire is a conductor composed of a number of fine, separately insulated strands interwoven in a special way. The conductor is fabricated so that all inner strands come to the outside at regular intervals, and all outer strands go to the center at equal intervals. Litz wire is basically a stranded wire in which the conductors are insulated from each other.

Litz wire exhibits low losses at radio frequencies because the conducting surface area is much greater than that of an ordinary solid wire of the same diameter.

L NETWORK

An L network is a form of filter section that is used in unbalanced circuits. The L network is so named because schematic-diagram component arrangement resembles the capital letter L (see illustration).

The L configuration can be used in the construction of highpass or lowpass filters. The series component may appear either before or after the parallel component. In the illustration, a typical highpass L network is shown at A; a lowpass network is shown at B. Several L networks can be cascaded to obtain a sharper cutoff response, as shown at C.

The L network is one of several different configurations for selective filters. *See also* H NETWORK, PI NETWORK, T NETWORK.

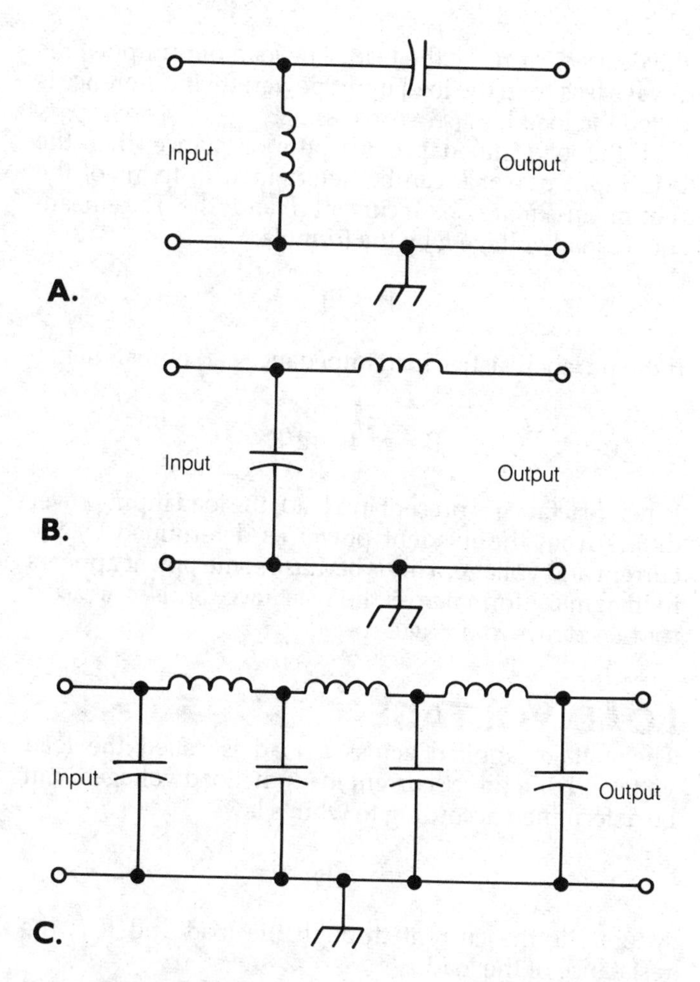

L NETWORK: Highpass network (A), lowpass network (B), and cascaded lowpass network (C).

LO

See LOCAL OSCILLATOR.

LOAD

A load is any circuit device or component that dissipates, radiates, or otherwise makes use of power. A direct-current load exhibits a definite resistance. This resistance may vary with the amount of power applied. An alternating-current load exhibits resistance, and may also show capacitive or inductive reactance. Examples of loads include audio speakers, electrical appliances, and radio-frequency transmitting antennas.

For optimum transfer of power from a circuit to its load, the load impedance should be the same as the output impedance of the circuit. *See also* IMPEDANCE MATCHING, LOAD IMPEDANCE, LOADING.

LOAD CURRENT

The amount of current flowing in a load is called the load current. For a direct-current load, the current I is given by Ohm's law:

$$I = E/R$$

where E is the voltage across the load, and R is the resistance of the load.

For an alternating-current load, the current depends on additional factors. In general, given a load impedance Z, and a root-mean-square voltage E across the load (*see* ROOT MEAN SQUARE), the current I is given by:

$$I = E/Z$$

The load current depends on the power supplied to the load and on the load impedance. *See also* LOAD, LOAD IMPEDANCE, OHM'S LAW.

LOADED ANTENNA

See ANTENNA MATCHING, ANTENNA TUNING, BASE LOADING, CENTER LOADING.

LOADED LINE

The characteristic impedance of a transmission line depends on its distributed capacitance and inductance. Any transmission line can be represented theoretically as a set of series inductors and parallel capacitors, although line reactances are not discrete components.

Sometimes it is necessary to change the characteristics of a transmission line at a certain location. This is done by the installation of capacitors and/or inductors in the line. The result of adding reactance to a transmission line is a change in the resonant frequency of a line. A transmission line in which reactance is deliberately added is called a loaded line. A line can be loaded for the purpose of matching the characteristic impedance to the output impedance of a radio transmitter. Line loading can also be used to match the characteristic impedance to the impedance of an antenna or other load. *See also* CHARACTERISTIC IMPEDANCE, FEED LINE, IMPEDANCE MATCHING, LOADING.

LOAD IMPEDANCE

Any power-dissipating or power-radiating load has an impedance. Part of the impedance is a pure resistance, and is given in ohms. Reactance may also be present; it is also specified in ohms. A direct-current load or a resonant alternating-current load has resistance but no reactance. A nonresonant alternating-current load may have capacitive or inductive reactance.

The load impedance should always be matched to the impedance of the power-delivering circuit. Normally, this means that the load contains no reactance (that is, the load is resonant), and the value of the load resistance is equal to the characteristic impedance of the feed line. If the load impedance is not the same as the characteristic impedance of the line, special circuits can be installed at the feed point to change the effective load impedance.

If the load impedance differs from the feed-system impedance, the load will not absorb all of the delivered power. Some of the electromagnetic field will be reflected back toward the source. This results in an increased amount of power loss in the system. The magnitude of this loss increase may or may not be significant. *See also* IMPEDANCE, IMPEDANCE MATCHING, REFLECTED POWER.

LOADING

The deliberate insertion of reactance into a circuit is called loading. Generally, this is done to obtain resonance. If a load contains capacitive reactance, then an equal amount of inductive reactance must be added to obtain resonance. Conversely, if a load contains inductive reactance, an equal amount of capacitive reactance must be added.

A common example of loading is the insertion of an inductor in series with a physically short antenna. A short antenna has capacitive reactance, and the loading inductor provides an equal and opposite reactance. The larger the value of the loading inductance, the lower the resonant frequency becomes. A capacitor may be inserted in series with a physically long antenna to obtain resonance. *See also* CAPACITIVE LOADING, INDUCTIVE LOADING, LOADED LINE, RESONANCE.

LOADING CAPACITOR

See CAPACITIVE LOADING.

LOADING COIL

See INDUCTIVE LOADING.

LOADING INDUCTOR

See INDUCTIVE LOADING.

LOAD LOSS

When power is applied to a load, the load ideally dissipates all of the power in the manner intended. For example, a light bulb should produce visible light but no heat; a radio antenna should radiate all of the energy it receives, and waste none as heat. However, in practice, no load is perfect. If P watts are delivered to a load, and the load dissipates Q watts in the manner intended, then the value $P-Q$ is called the load loss. Load loss may be expressed in decibels by the formula:

$$\text{Loss (dB)} = 10 \log_{10} \left(\frac{P}{Q}\right)$$

Some loads exhibit large losses. The incandescent lamp is an example of a lossy load. Other loads have very little loss; a half-wave dipole antenna in free space is an example of an efficient load. *See also* LOAD.

LOAD POWER

Load power can be defined in two ways: 1) load input power is the amount of power delivered to a load or 2) load output power is the amount of power dissipated, in the desired form, by the load. The load output power is always less than the load input power; their difference is called the load loss (*see* LOAD LOSS).

If the load impedance is a pure resistance, then the load input power P can be determined in terms of the root-mean-square load current I and the root-mean-square load voltage E by the formula:

$$P = EI$$

If the purely resistive load impedance is R ohms, then:

$$P = E^2/R = I^2R$$

When reactance is present in a load, the load input power differs from the incident power as determined by the current and voltage. This is because some power appears in imaginary form across the reactance. *See also* APPARENT POWER, REACTIVE POWER, TRUE POWER.

LOAD VOLTAGE

The voltage applied across a load is called the load voltage. For a direct-current load, the load voltage E can be determined according to Ohm's law:

$$E = IR$$

where I is the current through the load and R is the resistance of the load.

For an alternating-current load, the voltage depends on several different things. In general, given a load impedance Z and a root-mean-square current I through the load (*see* ROOT MEAN SQUARE), the root-mean-square voltage E is:

$$E = IZ$$

The load voltage depends on the power supplied to the load, and also on the load impedance. *See also* LOAD, LOAD IMPEDANCE.

LOBE

In the radiation pattern of an antenna, a lobe is a local angular maximum. An antenna pattern may have just one or several lobes. These lobes can have different magnitudes (see illustration). The strongest lobe is called the main or major lobe; the weaker lobes are called secondary or minor lobes. *See also* ANTENNA PATTERN.

Lobes occur in the directional responses of transducers such as microphones and speakers. *See also* DIRECTIONAL MICROPHONE.

LOCAL AREA NETWORK

A local area network (LAN) is a data communications network that allows many different computers and data processing devices to communicate with each other. It differs from other communications networks in its

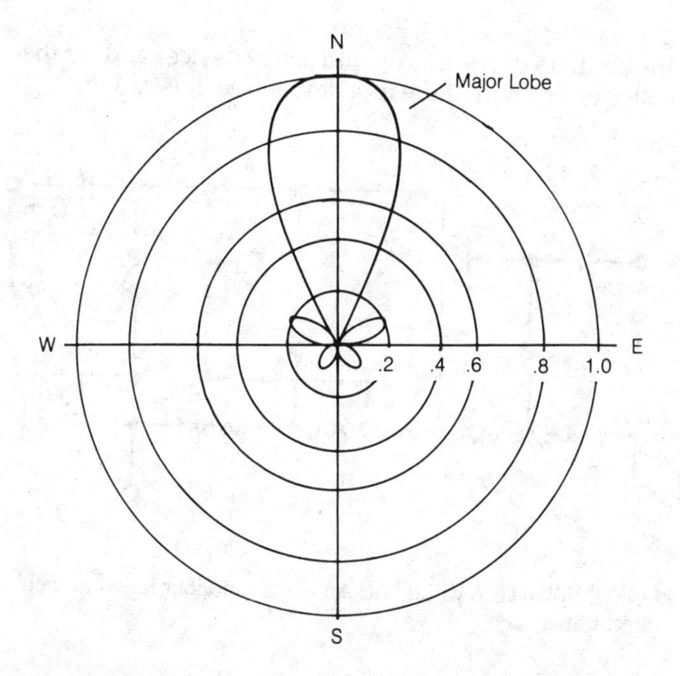

LOBE: The major lobe of the directional antenna pattern illustrated is pointed north. The side lobes are shown as the smaller loops with low field strength in all four quadrants.

smaller size and geographical coverage. A local area can be a single office building, a factory, a warehouse, or a college campus. LANs work well on communications channels with moderate to high data rates (1 to 20 Mbps—megabits per second), low delay and low error rates. The devices are directly interconnected (hard wired) with twisted wires, coaxial cables, or fiberoptic cables.

Most LANs are owned and operated by a single organization. This ownership contrasts to wide area networks (WANs) that interconnect facilities in different parts of the country and are owned and operated by national communications utilities for services such as packet-switched public data networks (PSPDN). LANs have a controlled structure because one organization determines the components to be used in the system. They will typically have a regular form and be connected with a bus or ring topology, as shown in the figure. This configuration contrasts with WANs that typically have hierarchical internal structures like the telephone network.

The principal components of LANs are personal computers (PCs). LANs can also include design workstations and video data terminals (VDTs) for limited data entry and retrieval. A LAN is likely to have a mass storage memory consisting of a cluster of Winchester disk drives with disk drive controllers acting as a file server on line to provide additional memory capability for the computers and terminals in the network.

Tape drives and write-once, read-many (WORM) optical memories also can be in the network for backup and archival storage. Conventional dot-matrix, daisy-wheel and laser printers can be connected in the LAN to provide printed copies and documentation services for users.

LANs are widely used in file transfer, word processing, electronic mail, database access, and office automation. It is expected that they will include graphics and digital voice.

A LAN in an industrial plant might have monitoring and process control equipment connected to the system to permit engineers and supervisors to monitor manufacturing processes from their desks. They might even be organized to permit changes in the manufacturing process to be made remotely from personal computers or workstations in offices. The scope of the LAN can be increased with interconnections to other networks which might or might not be under the same control.

In bus topology, stations are connected with baseband transceivers or broadband radio frequency modems. In ring topology, the stations are arranged in a single closed path and function as repeaters. Each topology has different advantages related to the transmission medium, routing requirements, and reliability.

An example of a local area network system is Ethernet from Xerox. Over 100 vendors offer Ethernet products. It operates at 10 Mbps over 500 meter baseband coaxial cable segments. IBM has three LANs: PC Network (a 2 Mbps broadband LAN), Token ring and Industrial LAN. AT&T has four different LANs.

A metropolitan area network (MAN) is an expanded LAN designed for a larger geographical area than a LAN. This area could be several blocks of buildings or an entire city. There can be shared ownership and responsibility for a MAN. These networks, like LANs, also depend on communications channels with moderate to high data rates.

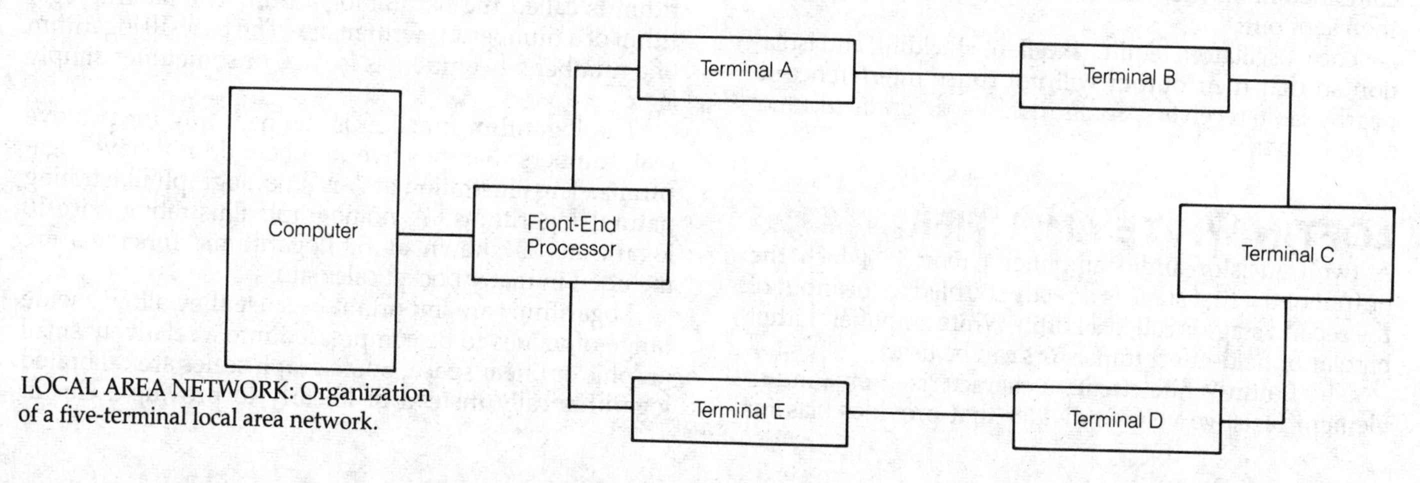

LOCAL AREA NETWORK: Organization of a five-terminal local area network.

IEEE 802 Standards define four types of medium-access technologies. These are specified for a variety of alternative physical media. A logical link control protocol standard is intended for use with the four medium-access standards. In addition an inter-networking standard is defined for use with the four medium-access standards and the logical link control standards. *See* DATA COMMUNICATIONS, DATA COMMUNICATIONS LINE SHARING EQUIPMENT, DATA SERIAL TRANSMISSION, MODEM.

LOCAL CHANNEL

In standard amplitude-modulation broadcasting, a local channel is a short-range channel. A station operating on a local channel is sometimes called a Class-IV station. The local-channel station has the least priority of any standard-broadcast station.

Local-channel stations are limited to a power level of 250 to 1,000 W during the daylight hours (sunrise to sunset). During the night, the maximum permissible power level is 250 W. These limitations apply in the United States and are set by the Federal Communications Commission. *See also* CLEAR CHANNEL.

LOCAL LOOP

In a teletype system, the local loop is an arrangement that allows paper tapes to be run through the machine for checking. In modern, computerized systems, the information is usually stored in a solid-state memory bank rather than on paper tape. By running the data through the machine locally, the station operator can correct errors in the text before they are transmitted. The local loop does not involve connection to an external line or circuit; it is contained entirely at one place. *See also* RADIOTELETYPE, TELETYPE.

LOCAL OSCILLATOR

A local oscillator is a radio-frequency oscillator for use within a receiver or transmitter to provide the low-power signal needed for the intermediate frequency.

All oscillators in a superheterodyne receiver are local oscillators. In transmitters, local-oscillator signals can be combined in mixers, amplified, perhaps multiplied, and then sent out.

Local oscillators require excellent shielding and isolation so that their output will not cause interference to nearby radio receivers. *See also* OSCILLATOR, SUPERHETERODYNE RADIO RECEIVER.

LOFTIN-WHITE AMPLIFIER

A two-transistor audio amplifier circuit in which the output of the first stage is directly coupled to the input of the second stage is called a Loftin-White amplifier. Either bipolar or field-effect transistors can be used.

The Loftin-White circuit is characterized by a multi-element resistive voltage divider that provides bias for the emitters or sources of both active devices and for the collector or drain of the first device (see illustration).

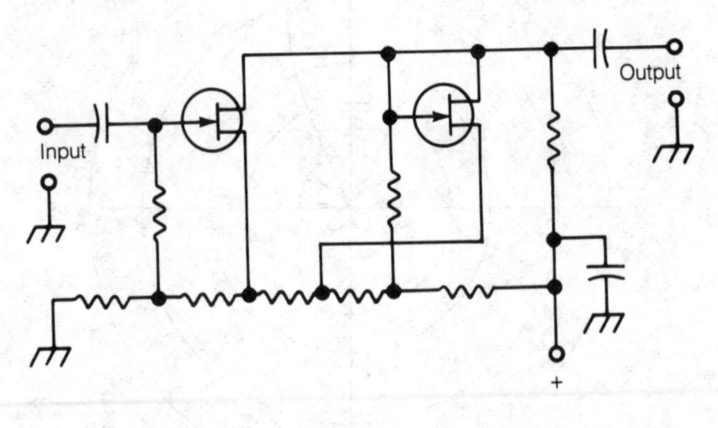

LOFTIN-WHITE AMPLIFIER: An audio amplifier based on field-effect transistors.

LOG

A log is a record of the operation of a radio station. Logs generally include such information as the date and time of a transmission, the nature of the transmission, the mode of emission, the power input or output of the transmitter, the frequency of transmission, and the message traffic handled. In some situations, logs must be kept by law. Logging requirements are determined, in the United States, by the Federal Communications Commission.

The mathematical logarithm operation is abbreviated as *log*. *See also* LOGARITHM.

LOGARITHM

A logarithm is a mathematical operation. The logarithm of a number x, in base b, is the power to which the number b must be raised to obtain the value x. If $y = \log_b x$, then:

$$b^y = x$$

Logarithms are usually specified either in base e, where e is about 2.718, or in base 10. The base-e logarithm is called the natural logarithm. The natural logarithm of a number x is written $\ln x$. The base-10 logarithm of a number x is written as $\log_{10} x$, or sometimes simply $\log x$.

The logarithm function is defined only for positive real numbers. Nonpositive numbers do not have logarithms. The illustration at A is a nomograph illustrating natural logarithms. A nomograph illustrating base-10 logarithms is shown at B. Logarithmic functions are included in many pocket calculators.

Logarithms are important because they allow a wide range of values to be compressed into a relatively small graphic or linear space. Some graph scales are calibrated logarithmically instead of linearly to provide a clearer

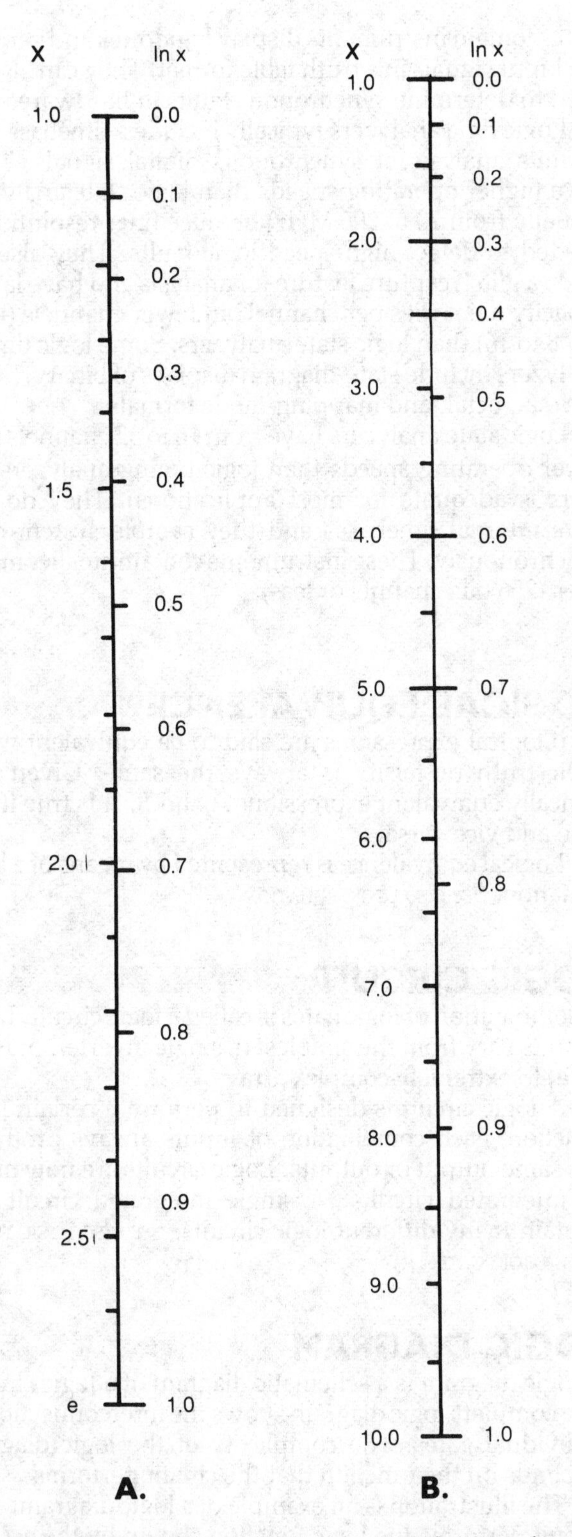

LOGARITHM: Natural logarithms for values between 1 and e, where e is approximately 2.718 (A) and base-10 logarithms for values between 1 and 10 (B).

rendition of certain mathematical functions. The amplitude scales of some test instruments are calibrated logarithmically. *See also* LOGARITHMIC SCALE.

LOGARITHMIC SCALE

When a coordinate scale is calibrated according to the logarithm of the actual scale displacement, it is called a logarithmic scale. This scaling is universally done according to the base-10 logarithm of the distance.

Logarithmic scales may be used in a Cartesian coordinate systen for one axis or both axes (see illustration). The coordinate system at A is called a semi-logarithmic graph; the coordinate system at B is known as a logarithmic or log-log graph (*see* CARTESIAN COORDINATES). In a polar graph, the radial coordinate axis may be calibrated in a logarithmic fashion (*see* POLAR COORDINATES). Logarithmic scales also are used in nomographs.

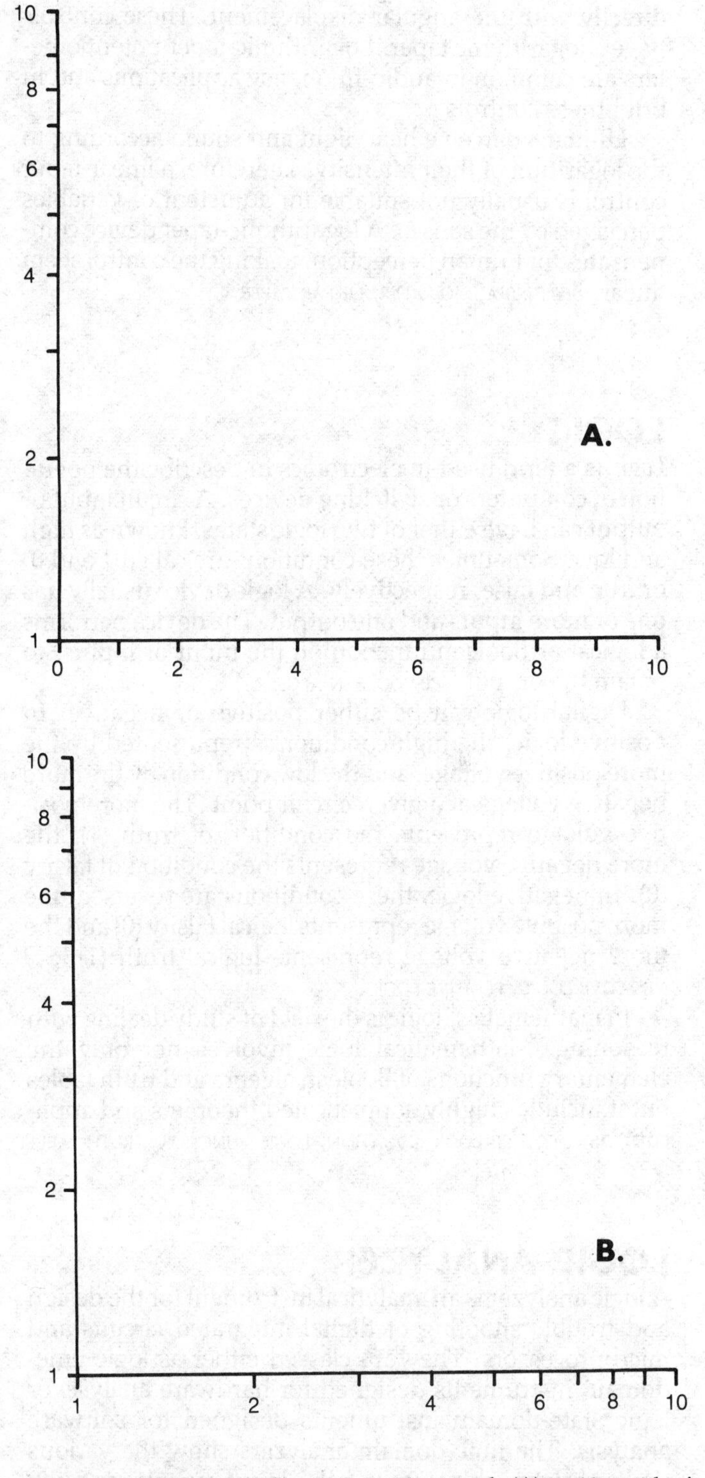

LOGARITHMIC SCALE: A semilogarithmic scale (A) and logarithmic or log-log scale (B).

Logarithmic scales are especially useful in plotting amplitudes when the range is expected to vary over many orders of magnitude. All decibel scales are logarithmic; some meters and oscilloscope displays are calibrated in this way. *See also* DECIBEL, LOGARITHM.

LOGARITHMIC TAPER

Some controls have a value or function that varies with the logarithm of the angular displacement rather than directly with the angular displacement. These controls have a logarithmic taper. Logarithmic-taper potentiometers are common in audio-frequency applications and in brightness controls.

Humans perceive heat, light and sound according to the logarithm of their intensity. Therefore, a linear-taper control is usually not suitable for adjustent of variables perceived by the senses. A logarithmic-taper device compensates for human perception, making the control seem linear. *See also* AUDIO TAPER, LINEAR TAPER.

LOGIC

Logic is a term used in electronics to describe the operation of computers or switching devices. A circuit input or output can have either of two logic states, known as high and low. Sometimes these conditions are called 1 and 0, or true and false, respectively. A logic device usually has one or more inputs and one output. The device performs a logical or Boolean function on the input or inputs, to obtain the output (*see* BOOLEAN ALGEBRA).

Digital logic can be either positive or negative. In positive logic, the high condition is represented by the more positive voltage, and the low condition by the more negative voltage at a given circuit point. The more positive voltage represents the condition of truth (1); the more negative voltage represents the condition of falsity (0). In negative logic, these conditions are reversed; the more positive voltage represents logical falsity (0) and the more negative voltage represents logical truth (1). *See* NEGATIVE LOGIC, POSITIVE LOGIC.

In mathematics, logic is the field of study dealing with reasoning. Mathematical logic involves not only the elementary functions of Boolean algebra and truth tables but it includes highly sophisticated theorems and applications. *See also* LOGIC EQUATION, LOGIC FUNCTION, TRUTH TABLE.

LOGIC ANALYZER

A logic analyzer is an analytical instrument for the design and trouble shooting of digital integrated circuits and microprocessors. They are classed either as logic time-domain instruments designed for hardware analysis or logic state-domain instruments designed for software analysis. The time-domain analyzers show the various timing relationships among the input signals, and the state-domain instruments display logic ones and zeros of the input signals in a truth table format. They can also be used to determine synchronous faults in hardware.

Logic time analyzers typically include a timebase that permits analysis of synchronous digital signals. They have higher operating speeds than logic state analyzers (usually from 20 to 200 MHz) because finer resolution is needed to detect high-speed logic faults. They also include a glitch capture feature for analysis and have larger capacity memories per channel but fewer channels (usually 8 to 16) than logic state analyzers. Some logic timing analyzers include state-diagram displays of binary, hexadecimal, octal, and mapping-mode formats.

Logic state analyzers have from 16 to 32 channels and lower operating speeds than logic timing analyzers (10 MHz is adequate for most applications). They do not have internal timebases and they sample system data synchronously. These instruments contain smaller memories (256 bits/channel or less).

LOGICAL EQUIVALENCE

Two logical expressions are said to be equivalent when their truth or falsity is always the same. Given two logically equivalent expressions *A* and *B*, *A* is true if *B* is true and vice versa.

Logical equivalence is represented by means of a logic equation. *See also* LOGIC EQUATION.

LOGIC CIRCUIT

A combination of logic gates is called a logic circuit. Logic circuits vary from the simplest (a single inverter, or NOT gate) to extremely complex arrays.

A logic circuit is designed to perform a certain logic function; each combination of inputs always produces the same output or outputs. Logic circuits are now made as integrated circuits. A single integrated circuit can contain many different logic circuits. *See also* LOGIC FUNCTION, LOGIC GATE.

LOGIC DIAGRAM

A logic diagram is a schematic diagram of a logic circuit. The complete logic diagram shows the interconnection of individual gates. The complexity of the logic diagram depends on the function that the circuit performs.

The illustration is an example of a logic diagram. This circuit performs the logic function shown by the accompanying truth table. There are other possible logic circuits that will accomplish this same logic function, and the diagrams of those circuits differ from the one illustrated. *See also* LOGIC CIRCUIT, LOGIC FUNCTION, LOGIC GATE.

LOGIC EQUATION

A logic equation is a symbolic statement showing two logically equivalent Boolean expressions on either side of an equal sign (*see* BOOLEAN ALGEBRA). The logic equation

$A = B$, where A and B are Boolean expressions, means literally, A is true if B is true, and B is true if A is true. Mathematicians would say: A if and only if B.

Logic equations are important in electronics. A particular logic circuit, intended to accomplish a given logic function, can often be assembled from logic gates in several different ways. One of these arrangements is usually simpler than all of the others, and is therefore most desirable from an engineering standpoint.

Many logic equations are known to be true, and are called theorems of Boolean algebra. By applying these theorems, the digital engineer can find the simplest possible logic circuit for a given function. *See also* LOGIC CIRCUIT, LOGIC FUNCTION.

LOGIC FUNCTION

A logic function is an operation, consisting of one or more input variables and one output variable. The logic function is a simple form of mathematical function (*see* FUNCTION), in which the input variables can achieve either of two states in various combinations. A logic function can be written in the form:

$$F(x_1, x_2, x_3, \ldots, x_n) = y$$

where x_1 through x_n represent the input variables and y represents the output. The values of the variables can be either 0 (representing falsity) or 1 (representing truth). If there are n input variables, then there are 2^n possible input combinations.

Every logic circuit performs a particular logic function. Logic functions may be written in Boolean form. Most logic functions can be represented in several different Boolean forms. Logic functions can also be written as a truth table, or a listing of all possible input combinations along with the corresponding output states. Logic functions can also be written in the form of a schematic diagram of the logic circuit, showing the interconnection of gates. *See also* BOOLEAN ALGEBRA, LOGIC CIRCUIT, LOGIC DIAGRAM, LOGIC EQUATION, LOGIC GATE, TRUTH TABLE.

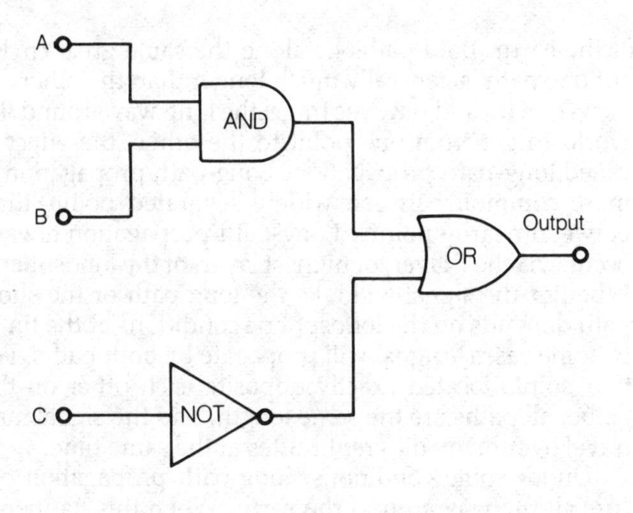

A	B	C	AB + C′
0	0	0	1
0	0	1	0
0	1	0	1
0	1	1	0
1	0	0	1
1	0	1	0
1	1	0	1
1	1	1	1

LOGIC DIAGRAM: A schematic diagram and truth table for a logic circuit containing an AND gate, an OR gate, and an inverter.

LOGIC GATE

A logic gate is a simple logic circuit. The gate performs a single AND, NOT, or OR operation, perhaps preceded by a NOT operation.

Logic gates are shown schematically in the illustration on p. 526. The AND gate, at B, produces a high output only if all of the inputs are high. The NOT gate or inverter (A) changes the state of the input. The OR gate, at C, produces a high output if at least one of the inputs is high. The AND gate may be followed by an inverter, resulting in a NAND gate (D). The OR gate can be followed by an inverter to form a NOR gate (E).

Logic gates are combined to perform complicated logical operations. This is the basis for all of digital technology. *See also* AND GATE, EXCLUSIVE-OR GATE, INVERTER, LOGIC CIRCUIT, LOGIC FUNCTION, NAND GATE, NOR GATE, OR GATE.

LOG-PERIODIC ANTENNA

See ANTENNA DIRECTORY.

LONGITUDE

Longitude is one of two coordinates used to locate a point on the surface of the earth. Longitude values range from −180 degrees to +180 degrees. The longitude line corresponding to 0 degrees runs through Greenwich, England, and is called the Greenwich Meridian. Points west of this line are assigned negative longitude values; points east of the line are given positive values.

All longitude lines on the globe are great circles intersecting at the north and south geographic poles, as shown in the illustration on p. 526. The two halves of each longitude line represent different longitude coordinates. If the eastern-hemisphere, half of a given circle represents x degrees east, then the western-hemisphere half of the same circle represents $180 - x$ degrees west.

The other coordinate commonly used for locating points on a sphere is called latitude. *See also* LATITUDE.

LONGITUDINAL WAVE

A wave disturbance in which the displacement is parallel to the direction of travel is called a longitudinal wave. The most common example of a longitudinal wave is the

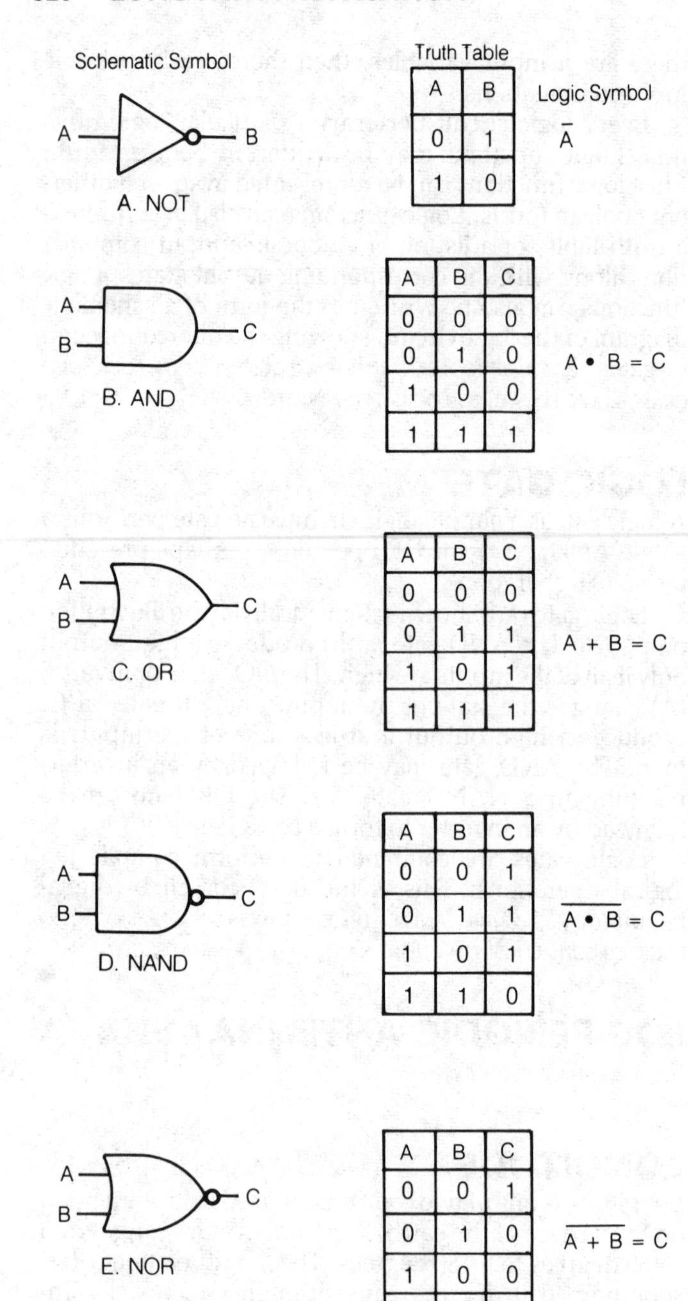

A	B
0	1
1	0

A	B	C
0	0	0
0	1	0
1	0	0
1	1	1

A	B	C
0	0	0
0	1	1
1	0	1
1	1	1

A	B	C
0	0	1
0	1	1
1	0	1
1	1	0

A	B	C
0	0	1
0	1	0
1	0	0
1	1	0

LOGIC GATE: The basic logic gate schematic symbols are shown with their truth tables and logic notations: NOT (A), AND (B), OR (C), NAND (D), and NOR (E).

sound wave in air.

All longitudinal waves are propagated by the movement of particles such as air molecules. Some waves are propagated by lateral, or transverse, motion of particles. Others are propagated by electromagnetic effects. *See also* ELECTROMAGNETIC FIELD, TRANSVERSE WAVE.

LONG-PATH PROPAGATION

At some radio frequencies, worldwide communication is possible because of the effects of the ionosphere. Radio waves travel over, or near, great-circle paths across the globe. There are two possible paths over which the radio waves may propagate between two points on the surface

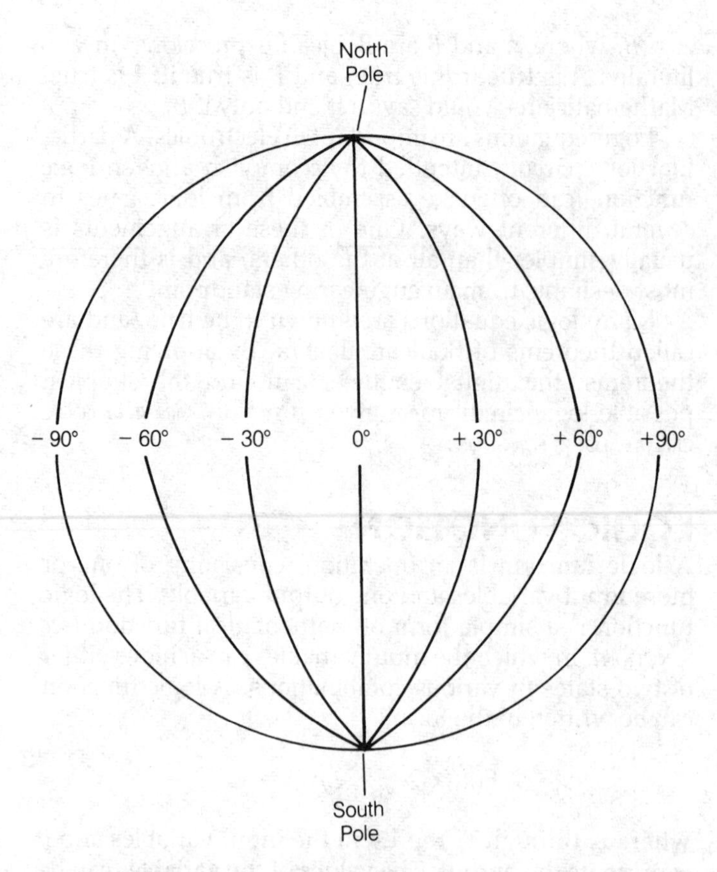

LONGITUDE: Longitude lines are great circles passing through the poles of the earth.

of the earth. Both paths lie along the same great circle, but one path is generally much longer than the other.

When the radio waves travel the long way around the world to get from one point to the other, the effect is called long-path propagation. Long-path propagation is more common between widely separated points than between nearby points. Long-path propagation always occurs via the F layer, or highest layer, of the ionosphere. Whether the signal will take the long path or the short path depends on the ionospheric conditions at the time. In some cases a signal will propagate by both paths. For two points located exactly opposite each other on the globe, all paths are the same length, and the signal may travel over many different routes at the same time.

Under some conditions, long-path propagation occurs all the way around the earth. When this happens, the signal from a nearby transmitter may seem to echo: The signal is heard first directly, and then a fraction of a second later as the long-path wave arrives. *See also* F LAYER, IONOSPHERE, PROPAGATION CHARACTERISTICS.

LONGWIRE ANTENNA

See ANTENNA DIRECTORY.

LOOP

In electronics, the term *loop* can have any of several different meanings.

A local current or voltage maximum on an antenna or transmission line is called a loop (*see* CURRENT LOOP, VOLTAGE LOOP).

A closed signal path in an amplifier, oscillator, or other circuit may be called a loop. Examples are the feedback loop in an oscillator, and the loop in an operational-amplifier circuit (*see* FEEDBACK).

In a computer program, a loop is a portion of the program through which the computer cycles many times. The loop ends when some condition is satisfied.

A single-turn coil of wire is called a loop. A large coil, having one or more turns and used for the purpose of receiving or transmitting radio signals, is called a loop or loop antenna. *See also* LOOP ANTENNA in the ANTENNA DIRECTORY.

LOOP ANTENNA

See ANTENNA DIRECTORY.

LOOSE COUPLING

When two inductors have a small amount of mutual inductance, they are said to be loosely coupled. Loose coupling provides relatively little signal transfer.

Loose coupling is desirable when it is necessary to pick off a small amount of signal energy for monitoring purposes, without disturbing the operation of the circuit under test. *See also* COEFFICIENT OF COUPLING, MUTUAL INDUCTANCE.

LORAN

Loran is an electronic navigation system for ships. The system is based on radio transmission. The name loran is derived from the descriptive phrase—*lo*ng *ra*nge *n*aviation. A passive system, loran does not require a transmitter on the receiving ship. It can be used to determine an accurate navigational fix (or the unique latitude and longitude coordinates) of the receiving ship. The original loran-A, also called standard loran, has been phased out and replaced with loran-C.

Pairs of shore-based transmitters send out high-powered pulses of low-frequency radio waves that are received by the ship's loran receiver. The stations are a known distance apart so there is a constant difference in

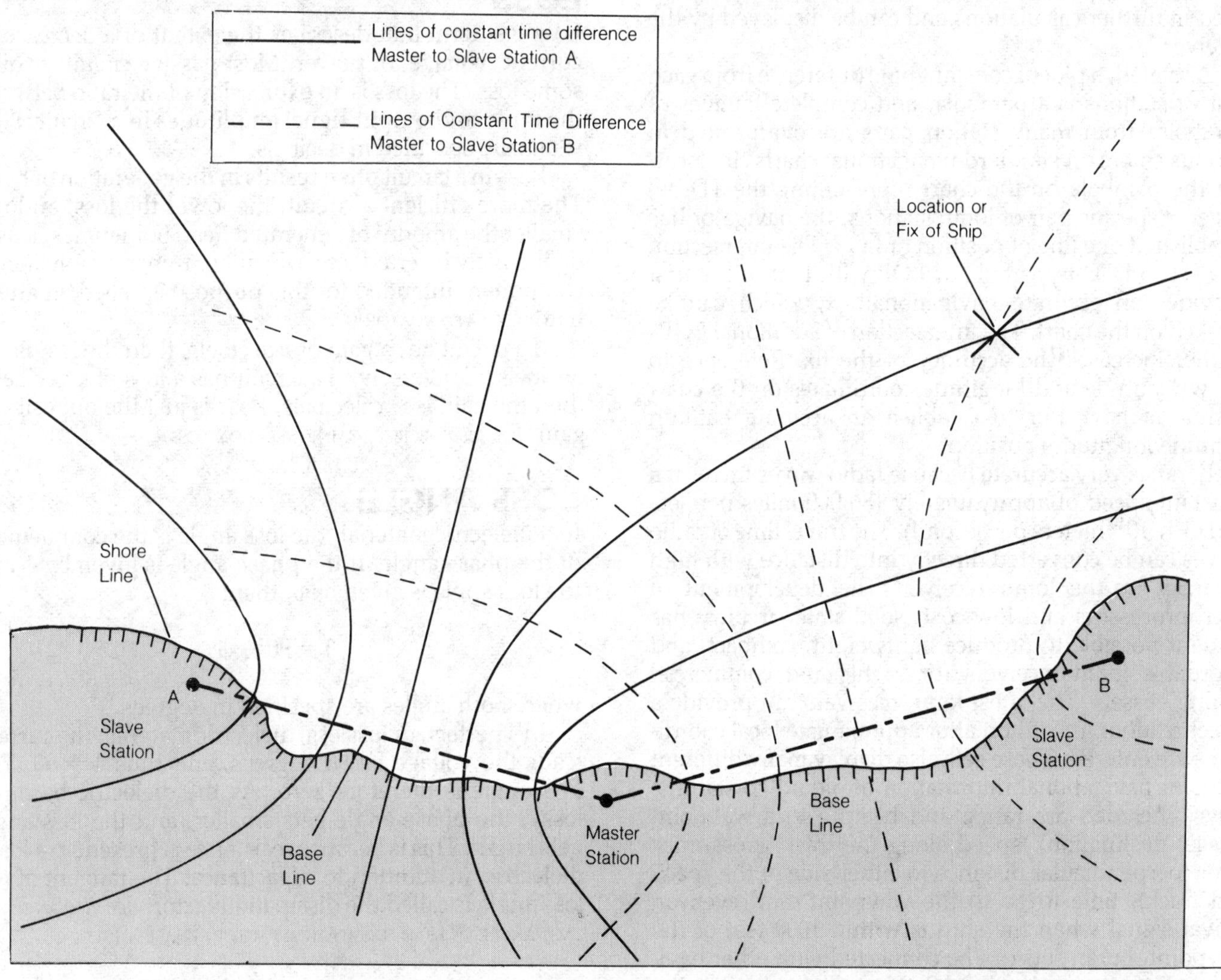

LORAN: Details of loran hyperbolic navigation system. Lines of constant time difference (TD) in microseconds are lines of position (LOPs).

time between the transmission of pulses from each of the stations. This difference results in a time difference (TD) in the arrival of pulses from each pair of stations (as measured at the receiver). The difference corresponds to the difference in distance of the receiver from the two transmitters. A specific difference in distance corresponds to a receiving point along a hyperbola. This relationship exists because a hyperbola is the locus of points that differ in distance from two fixed points by a constant amount. This distance is translated into microseconds.

The loran station common to each pair of stations is called the master and the other is called the slave. There can be two slave stations, as shown in the diagram, or as many as four. The master station transmits its pulses first. These pulses propagate outward over the region of the ocean covered by the system and travel toward the slave transmitter. When the master pulses are received at its slave station (at a known distance away along a baseline), the master pulses trigger the slave station, which then transmits its pulses. The pulses from the master station are first received by the ship's loran receiver, which then measures the time for the arrival of the slave pulses. This time difference in microseconds is used in further calculations and can be displayed by the receiver.

As stated, a plot of constant time difference from each pair of stations is a parabola, and complete families of parabolas from many station pairs are overprinted in various colors on standard navigational charts. By locating the parabola on the chart representing the TD between a specific pair of loran stations, the navigator has established one line of position or LOP. The intersection of a second TD hyperbola or LOP with that of the first provides an accurate navigational fix, which can be marked on the chart. The intersection of additional LOPs further increases the accuracy of the fix. Reference to known latitude and longitude coordinates on the chart permits a navigator to establish an accurate Lat/Lon (latitude/longitude) position.

Loran is very accurate because radio waves travel at a constant speed of approximately 186,000 miles per second (3×10^{10} meters per second). The travel time of radio waves can be converted directly into distance with high accuracy by the loran receiver. The development of microprocessors and low-cost, solid-state circuitry has made it possible to produce lightweight, compact, and affordable loran receivers for yachts and commercial fishing vessels. The latest loran receivers can provide a direct readout in Lat/Lon after approximate local coordinates are entered. These sets also display many different kinds of navigational information on liquid crystal displays. Included are range and bearing to a waypoint (preset destination), speed along the track, cross-track error (perpendicular distance to either side of the specified track), time to go to the waypoint, and even an arrival signal when the ship is within 1000 feet of the waypoint. Loran sets can be connected with other navigational instruments such as SatNav, Global Positioning System, and radar to share information. *See* GLOBAL POSITIONING SYSTEM, RADAR, SATELLITE NAVIGATION.

The original loran-A transmitted at frequencies between 1750 and 1950 kHz, a frequency region susceptible to changes in transmission conditions. Loran-A had different day and night ranges. Daytime reception using ground waves provided accurate fixes out to about 500 miles (925 kilometers). However, reception at night using sky waves could extend the range out to 1400 miles (2,600 km), but with less accuracy.

The transmitters of loran-C operate on 100 kHz in chains of a master and two to four subordinate stations. Actually each station transmits groups of pulses on the same frequency, but loran-A transmitted single pulses. Loran-C chains are identified by their pulse group repetition intervals. The transmitters are stabilized in time and frequency by atomic standards. Loran-C has a ground-wave range of up to 1200 miles (2200 km), with sky wave reception out as far as 3000 miles (5600 km). Microprocessor-based automatic receivers can establish the receiving ship's position within 50 yards (45 meters) of its true position at ranges within 500 miles (930 km).

Loran-D is a shorter-baseline and lower-power adaptation of loran-C for military tactical applications.

LOSS

Loss is a term that describes the extent of a decrease in current, voltage, or power. Most passive circuits exhibit some loss. The loss is an expression of the ratio between the input and output signal amplitudes in a circuit. Loss is usually specified in decibels.

Loss in a circuit often results in the generation of heat. The more efficient a circuit, the lower the loss, and the smaller the amount of generated heat. Sometimes, loss is deliberately inserted into a circuit or transmission line. A component intended for this purpose is called an attenuator (*see* ATTENUATOR).

If a circuit has a gain of x decibels, then the loss is $-x$ decibels. Conversely, if a circuit has a loss of x decibels, then the gain is $-x$ decibels. Loss is just the opposite of gain. *See also* DECIBEL, GAIN, INSERTION LOSS.

LOSS ANGLE

In a dielectric material, the loss angle is the complement of the phase angle. If the phase angle is given by ϕ and the loss angle is given by θ, then:

$$\theta = 90 - \phi$$

where both angles are specified in degrees.

In a perfect, or lossless, dielectric material, the current leads the voltage by 90 degrees, and thus $\phi = 90$. The loss angle is therefore zero. As the dielectric becomes lossy, the phase angle gets smaller, and the loss angle gets larger. This is because resistance is present in a lossy dielectric, in addition to capacitance. The tangent of the loss angle is called the dissipation factor. *See also* ANGLE OF LAG, ANGLE OF LEAD, DISSIPATION FACTOR, PHASE ANGLE.

LOSSLESS LINE

A theoretically perfect transmission line is called a lossless line. It does not exist in practice, but it is useful for mathematical calculations to assume that a transmission line has no loss.

In a lossless line, no power is dissipated as heat. The available power at the line output is the same as the input power. If the line is terminated in a pure resistance exactly equal to its characteristic impedance, the current and voltage along the lossless line are uniform at all points.

In a lossless line, an impedance mismatch at the terminating end would result in no additional power loss. But in a real feed line, a mismatch results in some additional loss because of the standing waves. *See also* FEED LINE, STANDING-WAVE-RATIO LOSS.

LOSS TANGENT

See DISSIPATION FACTOR, LOSS ANGLE.

LOUDNESS

Loudness is an expression of relative apparent intensity of sound. Loudness is sometimes called volume.

The human ear perceives loudness approximately in terms of the logarithm of the actual intensity of the disturbance. Loudness is usually expressed in decibels (*see* DECIBEL). An increase in loudness of 2 dB represents the smallest detectable change in sound level that a human listener can detect if he or she is expecting the change. The threshold of hearing is assigned the value 0 dB.

In an audio circuit, such as a radio receiver output or high-fidelity system, the control that affects the sound intensity is usually called the loudness control. *See also* SOUND.

LOWER SIDEBAND

An amplitude-modulated signal carries the information in the form of sidebands, or energy at frequencies just above and below the carrier frequency (*see* AMPLITUDE MODULATION, SIDEBAND). The lower sideband (LSB) is the group of frequencies immediately below the carrier frequency.

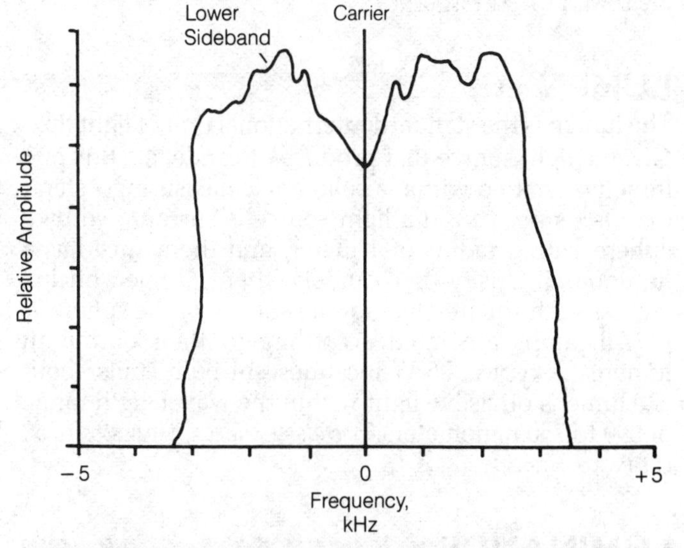

LOWER SIDEBAND: Signal information at a lower frequency than the carrier forms the lower sideband.

The LSB frequencies result from difference mixing between the carrier signal and the modulating signal. For a typical voice amplitude-modulated signal, the lower sideband occupies about 3 kHz of spectrum space (see illustration). For reproduction of television video information, the sideband may be several megahertz wide.

The lower sideband contains all of the modulating signal intelligence. For this reason, the rest of the signal (carrier and upper sideband) can be eliminated. This results in a single-sideband (SSB) signal on the lower sideband. *See also* SINGLE SIDEBAND, UPPER SIDEBAND.

LOWEST USABLE FREQUENCY

For any two points separated by more than a few miles, electromagnetic communication is possible only at certain frequencies. At very low frequencies, communication is almost always possible between any two locations in the world, provided the power level is high enough. Usually, however, communication is also possible at some higher band of frequencies, in the medium or high range.

Given some arbitrary medium or high frequency f, at which communication is possible between two specific points, assume this frequency is decreased until communication is no longer possible. This cutoff frequency could be called f_L. It is known as the lowest usable high frequency, or simply the lowest usable frequency (LUF). The LUF changes as the locations of the transmitting or receiving stations are changed. The LUF is a function of the ionospheric conditions, which fluctuate with the time of day, the season of the year, and the emissions from the sun. *See also* MAXIMUM USABLE FREQUENCY, MAXIMUM USABLE LOW FREQUENCY, PROPAGATION CHARACTERISTICS.

LOW FREQUENCY

The range of electromagnetic frequencies extending from 30 to 300 kilohertz is called the low-frequency band. The wavelengths are between 1 and 10 kilometers. For this reason, low-frequency waves are called kilometric waves.

The ionospheric E and F layers return all low-frequency signals to the earth, even if they are sent straight upward. Low-frequency signals from space cannot reach the surface of earth because the ionosphere blocks them.

Low-frequency waves, if vertically polarized, propagate well along the surface of the earth. Ease of propagation improves as the frequency decreases. *See also* SURFACE WAVE.

LOWPASS FILTER

A lowpass filter is a combination of capacitance, inductance, and/or resistance, intended to produce high attenuation above a specified frequency and little or no attenuation below that frequency. The frequency at which the transition occurs is called the cutoff frequency (*see* CUTOFF FREQUENCY). At the cutoff frequency, the attenuation is 3 decibels (dB) with respect to the minimum attenuation. Below the cutoff frequency, the attenuation is less than 3 dB. Above the cutoff, the attenuation is more than 3 dB.

The simplest lowpass filters consist of a series inductor or a parallel capacitor. More sophisticated lowpass filters have a combination of series inductors and parallel capacitors, as shown in the illustration. The filter at A is called an L-section lowpass filter; the filter at B is a pi-section lowpass filter. These names are derived from the geometric arrangement of the components as they appear in the diagram.

Resistors can be substituted for the inductors in a lowpass filter. When active devices are used, many filter stages can be cascaded.

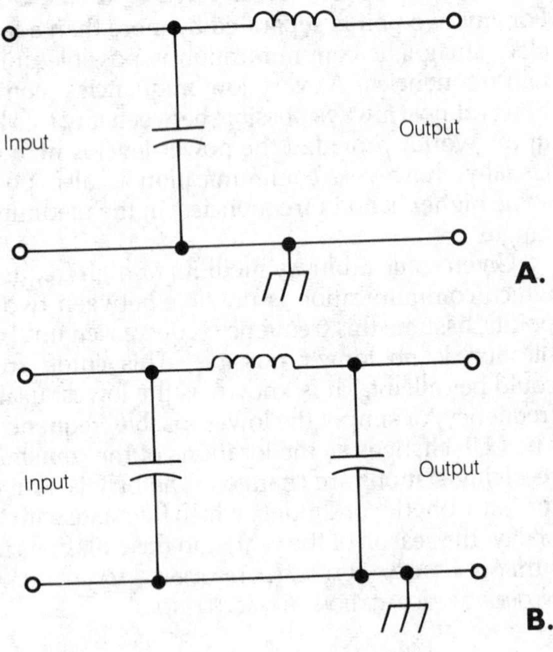

LOWPASS FILTER: A lowpass L-section filter (A) and a lowpass pi section filter (B). Both are for use in unbalanced circuits.

LOWPASS RESPONSE

A lowpass response is an attenuation-versus-frequency curve that shows greater attenuation at higher frequencies than at lower frequencies. The sharpness of the response may vary considerably. Usually, a lowpass response is characterized by a low degree of attenuation up to a specified frequency; above that frequency, the attenuation rapidly increases. Finally the attenuation levels off at a large value. Below the cutoff frequency, the attenuation is practically zero. The lowpass response is typical of lowpass filters.

The cutoff frequency of a lowpass response is that frequency at which the insertion loss is 3 decibels with respect to the minimum loss. The ultimate attenuation is the level of attenuation well above the cutoff frequency, where the signal is virtually blocked. The ideal lowpass response should look like the attenuation-versus-frequency curve in the illustration. The curve is smooth, and the insertion loss is essentially zero everywhere well below the cutoff frequency. *See also* CUTOFF FREQUENCY, INSERTION LOSS, LOWPASS FILTER.

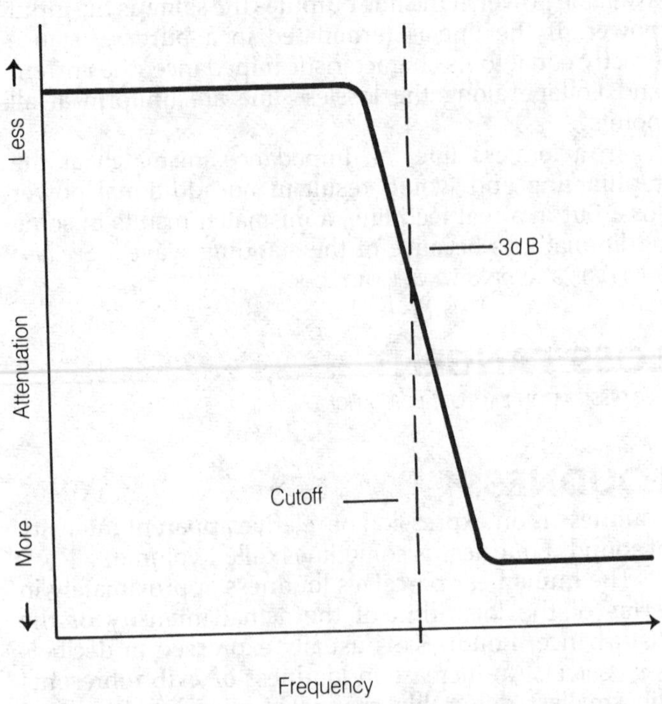

LOWPASS RESPONSE: A filter characteristic in which attenuation is greater at the higher frequencies than at the lower frequencies.

LSB

See LOWER SIDEBAND.

LSI

See LARGE-SCALE INTEGRATION.

LUF

See LOWEST USABLE FREQUENCY.

LUMEN

The lumen is the standard international unit of light flux. Given a light source that produces 1 candela, a flux of 1 lumen is emitted within a solid angle measuring 1 steradian (*see* STERADIAN). If a light source is surrounded by a sphere with a radius of 1 meter, and the source has a luminous intensity of 1 candela, then 1 lumen of flux crosses each square meter of the surface of the sphere.

All lamps are rated according to their output in lumens. A typical 25-W incandescent bulb emits about 220 lumens of visible light within the wavelength range of 390 to 750 nanometers. *See also* CANDELA, LUMINOUS FLUX, LUMINOUS INTENSITY.

LUMINANCE

Luminance is a measure of the luminous flux passing

through a given surface area. Luminance is also known as luminous sterance. In the standard international system of units, luminance is measured in candela per square meter. Luminance may also be measured in candela per square centimeter, candela per square foot, and other combinations.

Given a light source of fixed intensity, the luminance decreases as the distance from the source increases. If the distance is doubled, the luminance drops to one-fourth its previous value. In general, the luminance varies inversely with the square of the distance from the light source. *See* CANDELA, LUMEN, LUMINOUS FLUX, LUMINOUS INTENSITY.

LUMINESCENCE

Luminescence is any emission of light not attributable to an incandescent source. When a substance absorbs energy in a nonvisible form, and retransmits the energy as light, that material is said to be luminescent. Examples of luminescent substances include the phosphor on a cathode-ray-tube screen, certain semiconductors, and various other chemicals. Luminescence can result from bombardment by electrons or other subatomic particles, from the presence of an electric current, or from irradiation by electromagnetic fields such as infrared, ultraviolet, or X rays.

Luminescence always involves the conversion of energy from one form to another. Electrical energy is converted to light by means of an electroluminescent cell. Invisible radiation is converted into light energy by phosphorescent substances. *See also* ELECTROLUMINESCENT DISPLAY, PHOSPHORESCENCE.

LUMINOUS FLUX

Luminous flux is the rate at which light energy is emitted or transmitted. The unit of luminous flux is the lumen (*see* LUMEN).

In general, the greater the luminous flux produced by a light source, the brighter the source appears from a given distance. Luminous flux is determined within the optical range only. Optical wavelengths are considered to be between approximately 380 and 770 nanometers.

When luminous flux is measured within a given solid angle, it is called luminous intensity. *See also* LUMINOUS INTENSITY.

LUMINOUS INTENSITY

Luminous intensity is a measure of the amount of luminous flux emitted within a certain solid angle. Luminous intensity is generally measured in units called candela.

An intensity of 1 candela is equal to a flux of 1 lumen within a solid angle of 1 steradian. A solid angle of 1 steradian represents $1/(4\pi)$ of the total three-dimensional emission from an object. Therefore, a light source that produces 1 candela has a total luminous flux of about 4π lumens. *See also* CANDELA, LUMEN, LUMINOUS FLUX, STERADIAN.

LUMINOUS STERANCE

See LUMINANCE.

LUMPED ELEMENT

When a reactance, resistance, or voltage source appears in a discrete location, the component is called a lumped element. All ordinary capacitors, inductors, and resistors are lumped elements. However, a transmission line is not a lumped element because it exhibits inductance and capacitance that are distributed over a region. *See also* DISCRETE COMPONENT, DISTRIBUTED ELEMENT.

LUX

The lux is the SI unit of illuminance. It is the illuminance produced by a luminous flux of 1 lumen uniformly distributed over a surface of 1 square meter (lm/m^2). *See also* ILLUMINANCE.

MACHINE LANGUAGE

Digital computers operate in machine language. Small groups of binary digits, or bits, comprise instructions for the computer. These groups of bits are called microinstructions. One or more microinstructions together form a macroinstruction (*see* MACROINSTRUCTION, MICROINSTRUCTION).

Computers are usually not operated or programmed in machine language. Instead, an assembler or compiler determines the machine language from a higher-order language that is more easily understood by people. There are many different higher-order languages for computer applications.

When a new computer or a new higher-order language is developed, it might be necessary to work directly with machine language. In some program debugging situations, and in computer maintenance and repair work, machine language must be understood by the programmer or technician. *See also* ASSEMBLER AND ASSEMBLY LANGUAGE, HIGHER-ORDER LANGUAGE.

MACRO-

Macro- is a prefix that refers to a collection of things, or to an extremely large group. Macro- means large scale.

Sometimes a macroinstruction or macroprogram is called a macro for short. *See also* MACROINSTRUCTION, MACROPROGRAM.

MACROINSTRUCTION

A macroinstruction is a machine-language computer instruction made up of a group of microinstructions (*see* MICROINSTRUCTION). A macroinstruction corresponds to a specific, unique command in a higher-order language. The translation from command to macroinstruction is done by the compiler. *See also* MACHINE LANGUAGE.

MACROPROGRAM

A macroprogram is a computer program written in the language used by the operator. A typical computer program is actually a macroprogram, which takes the form of a series of statements in higher-order language.

A macroprogram may be written in assembly language for translating between machine language and the user language. This program is called a macroassembler or macroassembly program. *See also* ASSEMBLER AND ASSEMBLY LANGUAGE, HIGHER-ORDER LANGUAGE.

MAGNESIUM

Magnesium is an element with an atomic number of 12 and an atomic weight of 24. In its natural form, magnesium is a metal that looks like aluminum. It reacts readily with oxygen or chlorine, as well as with other elements.

Magnesium compounds are used as phosphor substances in cathode-ray tubes. Magnesium fluoride and magnesium silicate produce a red-orange glow when bombarded by electrons. Magnesium tungstate fluoresces with a bluish color when exposed to an electron beam.

MAGNET

A magnet is an object that produces a flux field because of the effects of molecular alignment or of electrical current. The molecules of iron or nickel, when aligned uniformly, produce a continuous magnetic field, and these materials have the properties of permanent magnets. A coil of wire with an iron core will produce a strong magnetic field when a current passes through the coil, but the field collapses when the current stops flowing. This is an electromagnet or temporary magnet.

Magnets are used for many purposes in electronics. The magnet of a dynamic microphone or speaker converts mechanical forces into electrical impulses, or vice-versa. Generators and motors use magnets and instruments such as meters use magnets to obtain needle deflection. *See also* ELECTROMAGNET, MAGNET COIL, MAGNETIC FIELD, MAGNETIC POLARIZATION, MAGNETIC POLE, MAGNETIZATION, PERMANENT MAGNET.

MAGNET COIL

An electric current always produces a magnetic field. The magnetic lines of flux occur in directions perpendicular to

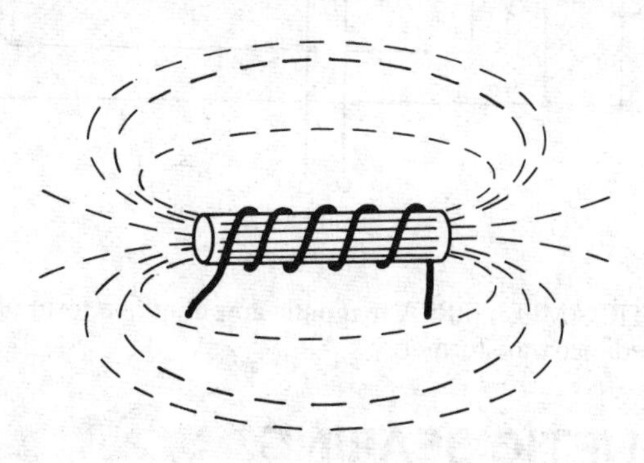

MAGNET COIL: A magnet coil produces a magnetic field with a flux identical to that of a bar magnet.

the flow of current. Thus, if a wire is wound into a helical coil, a magnetic field is produced similar to that of the magnetic field surrounding a bar magnet (see illustration). A coil wound around an iron rod to create a magnetic field is called a magnet coil or magnetic coil.

All electromagnets contain magnet coils. When a magnet coil carries a large amount of current, the magnetic field becomes strong enough to pick up large iron or steel objects. If the magnet coil is supplied with an alternating current, the magnetic field collapses and releases polarity in step with the change in the direction of the current.

Some meters, motors, generators, and other products use magnet coils to convert electrical energy to mechanical energy or vice versa. *See also* ELECTROMAGNET.

MAGNETIC AMPLIFIER

A magnetic amplifier is an amplifier that modulates or changes the voltage across the load in an alternating-current (ac) circuit. The magnetic amplifier consists of an iron-core transformer with an extra winding to which a control signal can be applied.

The schematic diagram illustrates a simple magnetic amplifier. Two windings appear in series with the ac power supply and the load. These windings are connected in opposing phase, and together they are called the output winding.

A third coil is wound around the center column of the transformer core. This coil is called the input coil. When a direct current is applied to the input coil, the impedance of the output coil changes because of the saturable-reactor principle (*see* SATURABLE REACTOR). A small change in the current through the input coil results in a large change in the output-coil impedance so amplification occurs. Magnetic amplifiers can, if properly designed, produce considerable power gain.

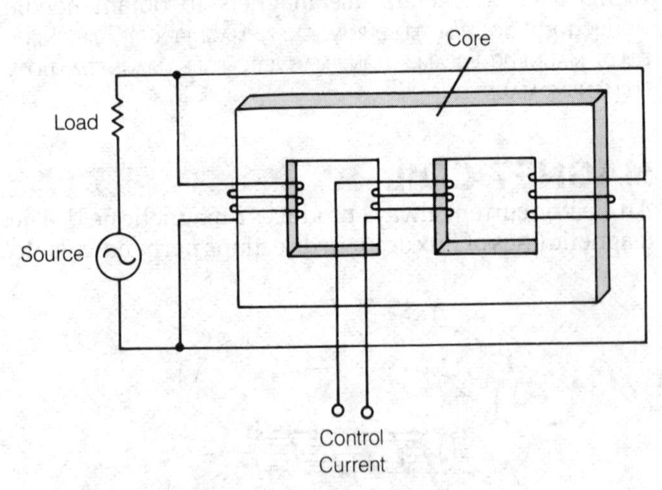

MAGNETIC AMPLIFIER: A magnetic amplifier is a form of variable-impedance transformer.

MAGNETIC BEARING

The magnetic bearing, or magnetic heading refers to the azimuth as determined by a magnetic compass. Magnetic bearings are measured in degrees clockwise from magnetic north.

The magnetic bearing usually differs from the true bearing because the geographic and geomagnetic poles are not in the same place. *See also* GEOMAGNETIC FIELD.

The shafts of some meters, gyroscopes, and other instruments may employ magnetic fields as bearings to minimize friction. These bearing are called magnetic bearings. The shaft is literally suspended in the air because of the repulsive magnetic force between like poles of permanent magnets or electromagnets. The friction of these bearings, when operated in a vacuum, is very nearly zero. *See also* MAGNETIC FORCE.

MAGNETIC CIRCUIT

A magnetic circuit is a complete path for magnetic flux lines (*see* MAGNETIC FLUX). The lines of flux are considered to originate at the north pole of a magnetic dipole and to terminate at the south pole. A magnetic circuit is analogous to an electric circuit.

The magnetic equivalent of voltage is magnetomotive force and is measured in gilberts. The magnetic equivalent of current is flux density and the magnetic equivalent of resistance is reluctance. These three parameters behave, in a magnetic circuit, in the same way as voltage, current, and resistance behave in an electric circuit. *See also* FLUX DENSITY, MAGNETOMOTIVE FORCE, RELUCTANCE.

MAGNETIC CORE

A magnetic core is a ferromagnetic ring or rod used for concentrating or storing a magnetic field. Magnetic cores are used in many different electrical and electronic products including electromagnets, transformers, and relays.

A special form of magnetic core stores digital information in some computers. A toroidal ferromagnetic bead surrounds a wire or set of wires, as shown in the illustration. When a current pulse is sent through the wire, a magnetic field is produced, and the toroid becomes magnetized in either the clockwise or counterclockwise sense (depending on the direction of the current pulse). The toroid remains magnetized in that sense indefinitely, until another current pulse is sent through

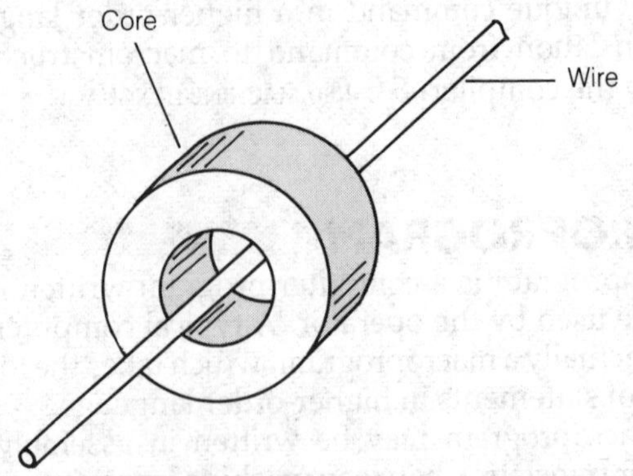

MAGNETIC CORE: A magnetic memory core is a toroidal, ferromagnetic ring. A pulse of current passed through the wire causes the magnetic state of the core to change to one of two states to perform a memory function.

the wire in the opposite direction. One sense represents the binary digit 0 (false); the other sense represents 1 (true). Magnetic-core storage is retained even in the total absence of power in a computer circuit. Several magnetic cores, when combined, allow the storage of large binary numbers.

MAGNETIC COUPLING

When a magnetic field influences a nearby object or objects, the effect is known as magnetic coupling. Examples of magnetic coupling include the deflection of a compass needle by the earth's magnetic field, the interaction among the windings of a motor or generator, and the operation of a transformer. Magnetic coupling can take place as a direct force, such as attraction or repulsion, between magnetic fields. Magnetic coupling can also occur because of motion among magnetic objects, or because of changes in the intensity of a magnetic field.

An alternating current produces a changing magnetic field which in turn produces changing currents in nearby objects. This is true when coils of wire are placed along a common axis. This form of magnetic coupling is called electromagnetic induction (*see* ELECTROMAGNETIC INDUCTION).

Magnetic coupling is enhanced by the presence of a ferromagnetic substance between two objects. At extremely low frequencies, solid iron is an effective magnetic medium. At very low frequencies, laminated iron or ferrite is preferable. At higher frequencies, ferrite and powdered iron are used.

The degree of magnetic coupling between two wires or coils is measured as mutual inductance. *See also* MUTUAL INDUCTANCE.

MAGNETIC DEFLECTION

Magnetic force, like any force, can cause objects to be accelerated. This acceleration may take place as a change in the speed of an object, a change in the direction of its motion, or both.

A moving charged object generates a magnetic field. The most common example of this is the magnetic field that surrounds a wire carrying an electric current. However, charged particles traveling through space also produce magnetic fields. When these moving charged particles encounter an external magnetic field, the particles are accelerated or deflected (see illustration).

MAGNETIC EQUATOR

The magnetic equator is an imaginary line around the earth, midway between the north and south magnetic poles. Every point on the magnetic equator is equidistant from either magnetic pole. At the magnetic equator, a compass needle experiences no dip; the geomagnetic lines of flux are parallel to the surface of the earth.

The magnetic equator does not coincide with the geographic equator because the magnetic poles do not lie exactly at the geographic poles. *See also* GEOMAGNETIC FIELD.

MAGNETIC FEEDBACK

See INDUCTIVE FEEDBACK.

MAGNETIC FIELD

A magnetic field is a region in which magnetic forces can occur. A magnetic field is produced by a magnetic dipole or by the motion of electrically charged particles.

Physicists consider that magnetic fields consist of flux lines. The intensity of the magnetic field is determined by the number of flux lines per unit surface area (*see* FLUX DENSITY). Although the magnetic lines of flux do not physically exist, they do have a direction that can be determined at any point in space. Magnetic fields can have various shapes, as shown in the illustration on p. 536.

A magnetic field is considered to originate at a north magnetic pole and to terminate at a south magnetic pole. A pair of magnetic poles, surrounded by a magnetic field, is called a magnetic dipole. *See also* MAGNETIC FLUX, MAGNETIC FORCE, MAGNETIC POLE.

MAGNETIC FIELD INTENSITY

See FLUX DENSITY.

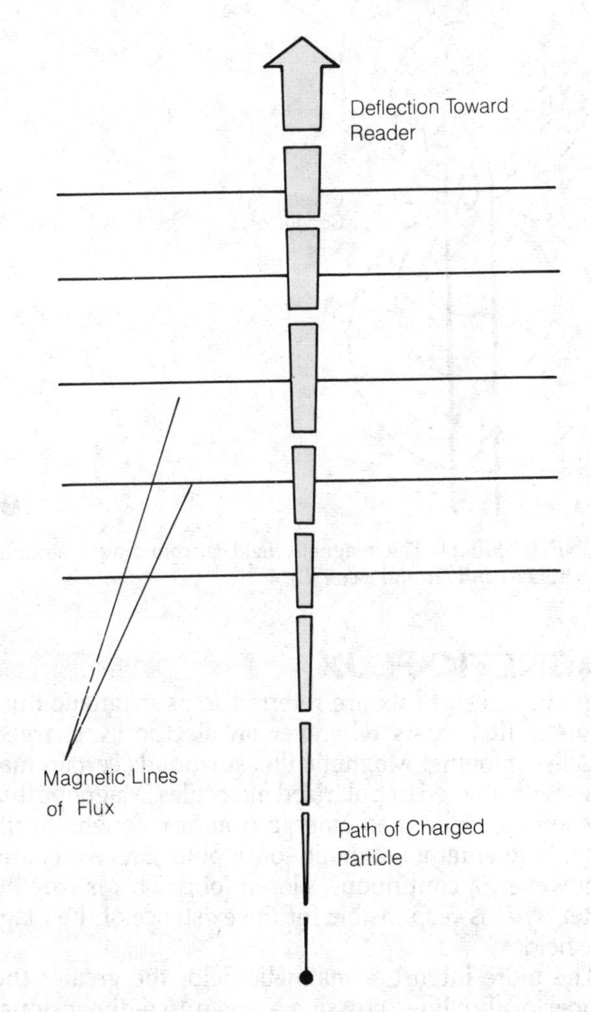

MAGNETIC DEFLECTION: A magnetic field exerts a lateral force on the moving charged particles.

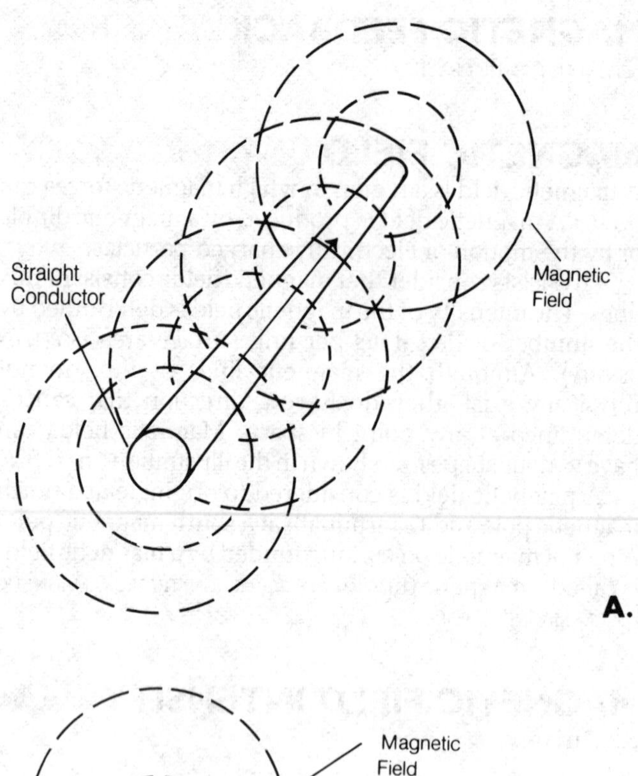

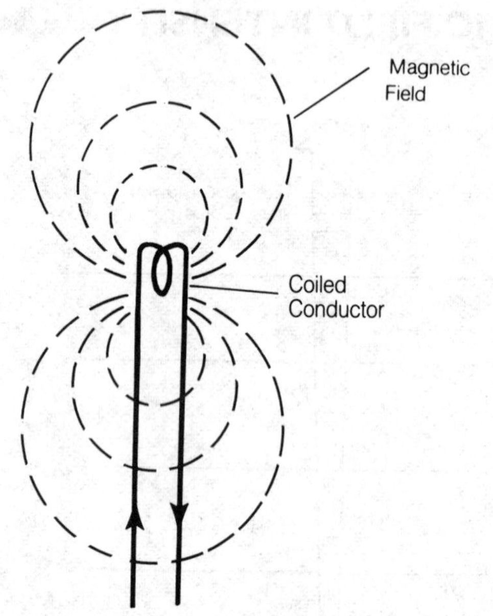

MAGNETIC FIELD: The magnetic field surrounding a straight conductor (A) and around a coil (B).

MAGNETIC FLUX

Magnetic lines of flux are referred to as magnetic flux. Magnetic flux exists wherever an electrically charged particle is moving. Magnetic flux surrounds certain materials with aligned or polarized molecules. Magnetic flux lines are considered to emerge from a magnetic north pole and to enter a magnetic south pole. Every flux line is, however, a continuous, closed loop. This is true no matter what is responsible for the existence of the magnetic field.

The more intense a magnetic field, the greater the number of flux lines crossing a given two-dimensional region in space. A line of flux is actually a quantity of magnetic flux, usually 1 weber or 1 maxwell. The strength of a field may be measured in webers per square meter. Another commonly used unit of flux density is the line of flux per square centimeter; this unit is known as the gauss. *See also* FLUX, FLUX DENSITY, FLUX DISTRIBUTION, GAUSS, MAGNETIC FIELD, MAXWELL, WEBER.

MAGNETIC FORCE

Magnetic fields produce forces that can be directly measured. Magnetic forces exist between magnetic poles and objects made with ferrous material. Forces also exist between any two magnetic poles, and between current-carrying wires.

When two like magnetic poles are brought near each other, repulsion occurs. When two unlike poles are brought near each other, attraction occurs. The magnitude of the force depends on the strengths of the magnetic fields surrounding the poles, and on the distance separating the poles. The force is proportional to the product of the flux densities and inversely proportional to the cube of the distance between the poles.

Magnetic force causes deflection in the paths of moving charged particles. The force occurs in a direction perpendicular to both the magnetic field and the path traveled by the particle (*see* MAGNETIC DEFLECTION).

MAGNETIC HEADING

See MAGNETIC BEARING.

MAGNETIC INDUCTION

See ELECTROMAGNETIC INDUCTION.

MAGNETIC MATERIAL

A magnetic material is any substance whose molecules become aligned, or polarized, in the presence of a magnetic field. These materials include iron and nickel, and various alloys derived from them.

Some magnetic materials can become permanently magnetized because their molecules remain aligned even after the external field has been removed. These objects are called permanent magnets (*see* MAGNET, PERMANENT MAGNET). Permanent magnets are used in transducers and meters.

MAGNETIC PERMEABILITY

See PERMEABILITY.

MAGNETIC POLARIZATION

In ferrous materials—iron, nickel, or steel—the molecules produce magnetic fields. Each molecule forms a magnetic dipole, containing a north and south magnetic pole (*see* MAGNETIC POLE).

The molecules in a ferrous material are normally aligned in random fashion so that the net effect of their magnetic fields is zero (A in the illustration). However, in the presence of an external magnetic field, the molecules become more nearly aligned. If the field is very strong, the molecules line up in nearly the same orientation. This greatly increases the strength of the magnetic field within and near the substance. When the field is removed, the molecules of some ferrous materials remain aligned, forming a permanent magnet (B).

Magnetic memory devices operate on the principle of magnetic polarization. A current through a wire produces a magnetic field. This field influences the properties of ferrous substances, such as powdered iron, causing the substance to become polarized. The magnetic memory retains its polarization until a current pulse is sent through the wire to reverse it (*see* MAGNETIC CORE, MAGNETIC RECORDING).

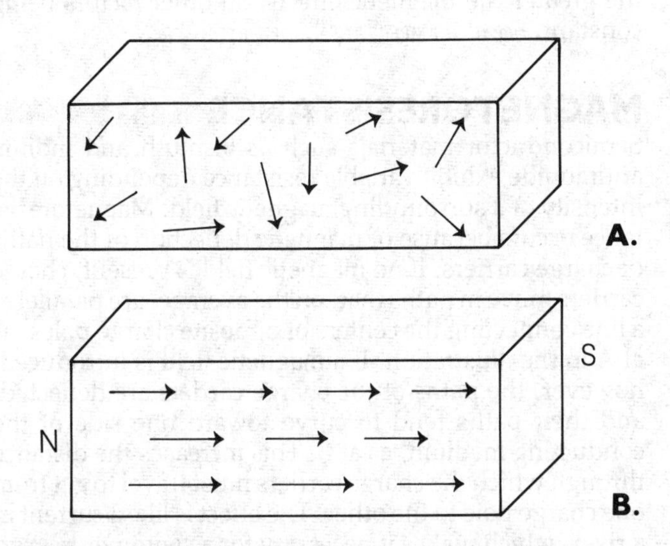

MAGNETIC POLARIZATION: The dipoles in unmagnetized material (A) and the dipoles in magnetized material (B).

MAGNETIC POLE

A magnetic pole is a point, or localized region, at which magnetic flux lines converge. Magnetic poles are called either north or south, a matter of convention. In the magnetic compass, one end of the magnetized needle points toward the north magnetic pole; this is called the north pole of the needle. Similarly, the end of a needle that orients itself toward the south is called the south pole. Magnetic flux lines, according to theory, leave north poles and enter south poles.

It is believed that magnetic poles must always exist in pairs. This conclusion is drawn from the fundamental laws of Maxwell. Magnetic flux lines, according to Maxwell, always take the form of closed loops. Thus the magnetic lines of flux must arise as a result of dipoles—pairs of north and south magnetic poles (*see* DIPOLE).

MAGNETIC RECORDING

Magnetic recording is a method of storing and retrieving data on magnetic tape or disks. The most commonly used magnetic medium is powdered iron oxide on a flexible plastic base. A magnetic transducer, called the head, generates a magnetic field that causes polarization of the molecules in the iron oxide. *See* DISK DRIVE, MAGNETIC TAPE.

Magnetic-tape recorders use reels or cassettes of magnetic tape, usually 0.5, 1.0, or 1.5 mil thick and ⅛, ¼, or 2 inches wide. In audio magnetic recording, the information is placed along straight paths called tracks. A tape can have one, two, four, or eight tracks, each with a different recording on it. For audio applications, the standard tape sizes are ⅛ inch or ¼ inch wide, and the most common speeds are 1 ⅞, 3 ¾, or 7 ½ inches per second. In video recording, the tape speed is 15 inches per second; the video information is recorded on oblique tracks with a frequency-modulated carrier, and the sound is recorded on one or two straight tracks. *See also* TAPE RECORDER, VIDEO TAPE RECORDING.

MAGNETIC TAPE

Magnetic tape is a form of information-storage medium commonly used for storing audio, video, and digital data. Magnetic tape is available on reels in several different thicknesses and widths for different applications.

Magnetic tape consists of millions of fine particles of iron oxide bonded to a plastic or mylar base. A magnetic field, produced by the recording head, causes polarization of the iron-oxide particles. As the field changes in intensity, the polarization of the iron-oxide particles also varies (see illustration). When the tape is played back, the magnetic field surrounding the individual iron-oxide particles produces current variations in the playback or pickup head.

For audio and computer-data recording, magnetic tape is available either in cassette form or wound on reels. The thickness may be 0.5, 1.0, or 1.5 mil. (A mil is 0.001 inch). The thicker tapes have better resistance to stretching, although the recording time, for a given length of tape, is proportionately shorter than with thin tape. In the cassette, the tape may be either ⅛ or ¼ inch wide. Reel-to-reel tape is generally ¼ inch wide. Wider tape is often used, however, in recording studios and in video applications. The tape may have up to 24 individual recording tracks. The most common speeds are 1 ⅞, 3 ¾, and 7 ½ inches per second. For voice recording, the slower speeds are adequate, but for music, higher speeds

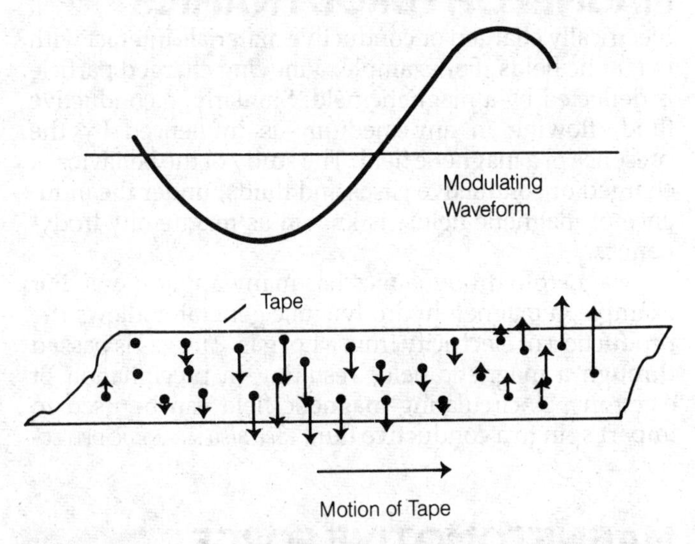

MAGNETIC TAPE: Magnetic dipoles in magnetic tape are aligned by the magnetizing force of the applied signal.

in excess of 15 inches per second are preferable for enhanced sensitivity to high-frequency sound.

Video tape ranges from ½ to 2 inches wide, and the standard recording speed is 15 inches per second. Even this speed is not high enough for the direct reproduction of the high frequencies encountered in video, so the recording is done in a slanted or oblique direction, resulting in much higher effective speeds.

MAGNETISM

Magnetism is a term used to describe any magnetic effects. Magnetic attraction and repulsion, interaction with moving charged particles, and polarization of molecules may all be categorized as magnetism. *See also* MAGNETIC DEFLECTION, MAGNETIC FIELD, MAGNETIC FORCE, MAGNETIC MATERIAL, MAGNETIC POLARIZATION.

MAGNETIZATION

When certain materials are subjected to magnetic fields, their molecules become aligned (*see* MAGNETIC POLARIZATION), resulting in intensification of the magnetic field within the substance. The substance is then said to be magnetized. Magnetization can be temporary, decaying as soon as the magnetizing force is removed; or it may be permanent, persisting indefinitely after removal of the external field. Transformer cores exhibit temporary magnetization. Magnetic tape is an example of a material that can be permanently magnetized.

Some materials can be magnetized and demagnetized rapidly, while other substances react sluggishly to a changing magnetic field. The rapidity with which a material can be magnetized, demagnetized, and remagnetized is called the degree of magnetic hysteresis for the material. (*See* HYSTERESIS.) Some materials are easily magnetized, while others are difficult to magnetize. The ease with which a material can be magnetized is called its permeability. (*See* PERMEABILITY.)

MAGNETOHYDRODYNAMICS

Electrically charged or conductive materials interact with magnetic fields. For example, a moving charged particle is deflected by a magnetic field. Similarly, a conductive fluid, flowing in any medium, is influenced by the presence of a magnetic field. The study of the behavior of charged or conductive gases and fluids, under the influence of magnetic fields, is known as magnetohydrodynamics.

Magnetohydrodynamics has many applications. For example, a magnetohydrodynamic generator allows the production of electricity from a hot gas. The gas is passed through a magnetic field, resulting in precipitation of electrons. A circulating magnetic field can be used to impart spin to a conductive fluid. *See also* MAGNETIC DEFLECTION.

MAGNETOMOTIVE FORCE

Magnetomotive force, also called magnetic potential or magnetic pressure, in a magnetic circuit is analogous to electromotive force in an electric circuit.

The most common unit of magnetomotive force is the gilbert. Another often-used unit is the ampere turn. A current of 1 ampere, passing through a one-turn coil of wire, sets up a magnetomotive force of 1 ampere turn or 1.26 gilberts. In general, the magnetomotive force M generated by a coil of N turns, carrying a current I of 1 A, is given in ampere turns by the formula:

$$M = NI$$

The larger the magnetomotive force in a region of space, the greater the magnetic effects, all other factors being constant. *See also* AMPERE TURN, GILBERT.

MAGNETORESISTANCE

Semiconductor materials such as bismuth and indium antimonide exhibit variable resistance depending on the intensity of a surrounding magnetic field. Magnetoresistance occurs because of magnetic deflection of the paths of charge carriers. If no magnetic field is present, charge carriers move in paths that, on the average, are parallel to a line connecting the centers of opposite charge poles, as at A in the illustration. If a magnetic field is introduced, however, the paths of the charge carriers are deflected, and their paths tend to curve toward one side of the conducting medium, as at B. This increases the distance through which the charge carriers must travel to get from one charge pole to the other. The effect is like a current in a river, which makes it necessary for a swimmer to exert more effort to reach a specified point on the opposite shore. *See also* MAGNETIC DEFLECTION.

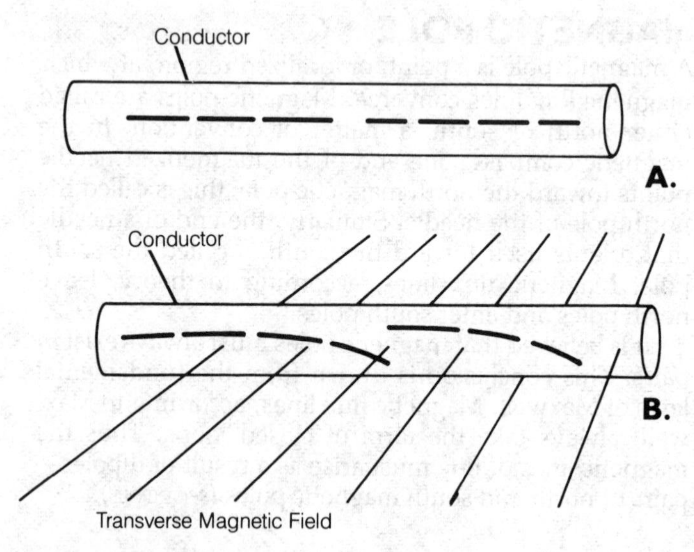

MAGNETORESISTANCE: Current in a conductor when no magnetic field is present (A) and current in the presence of a transverse magnetic field (B).

MAGNETOSPHERE

The magnetic field surrounding the earth produces dramatic effects on charged particles in space. The geomagnetic lines of flux extend far above the surface of the earth, and high-speed atomic particles constantly arrive from the sun and the stars. When these charged particles enter the earth's magnetic field, they are deflected toward the poles.

The magnetosphere is the boundary, roughly spherical in shape, surrounding the earth that represents the maximum extent of the magnetic-deflection effects of the geomagnetic field on charged particles from space. *See also* GEOMAGNETIC FIELD.

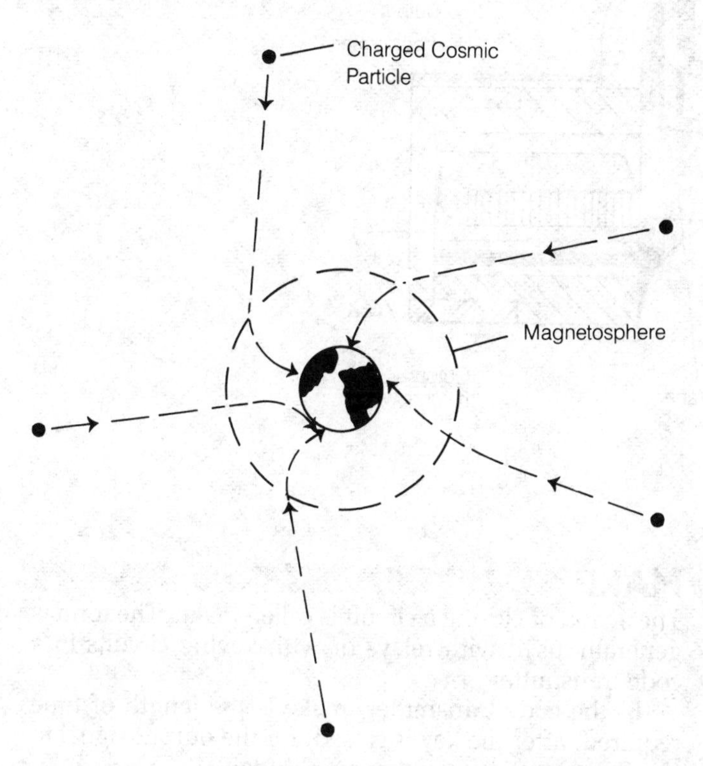

MAGNETOSPHERE: The magnetosphere of the earth is the outer boundary of the region within which charged particles are affected by the geomagnetic field.

MAGNETOSTRICTION

Magnetostriction is the characteristic of materials that causes them to expand or contract under the influence of a magnetic field. Magnetostriction in a magnetic field is analogous to electrostriction in an electric field (*see* ELECTROSTRICTION).

Materials that exhibit magnetostrictive properties include alloys of aluminum and iron as well as nickel and iron. Nickel exhibits negative magnetostriction; that is, it is compressed by a magnetic field. Permalloy, an alloy of iron and nickel, exhibits positive magnetostriction; it expands as the surrounding magnetic field becomes stronger.

If a magnetostrictive material is subjected to a sufficiently strong magnetic field, it may actually break under the strain. This has happened ferrite and powdered-iron inductor cores when the surrounding coils carry current far in excess of rated values.

Magnetostrictive materials can be employed in some of the same applications as piezoelectric crystals. For example, a magnetostriction oscillator works by resonance in a rod of magnetostrictive material. Sonar transducers can also be made with magnetostrictive materials. *See also* OSCILLATOR, PIEZOELECTRIC EFFECT, SONAR.

MAGNETOSTRICTION OSCILLATOR

See OSCILLATOR CIRCUITS.

MAGNETRON

A magnetron is a microwave power-tube oscillator capable of converting voltage at the cathode directly into microwave energy. Magnetrons are capable of generating microwave frequencies starting at 1 GHz and going into the millimeter wave region. The magnetron is a diode with a cylindrical anode structure lined with reentrant resonant cavities, as shown in the illustration on p. 540. A strong magnetic field parallel with the axial cathode causes an interaction between electrons from the cathode and the oscillating fields within the cavities. Magnetrons can be operated as pulsed or continuous wave oscillators or can be adapted for use as amplifiers with the addition of an input port.

The development of the multicavity magnetron permitted microwave frequency radar to be introduced before the end of World War II. Magnetrons remained the principal transmitter power tubes in radar systems for more than 30 years. They have been largely replaced by traveling-wave tubes (TWTs) in modern high-power radars. The TWT offers higher conversion efficiency and greater frequency agility.

High-voltage applied to the heated cathode of the magnetron drives out free electrons which are attracted to the surrounding positively charged anode structure. Instead of moving in a straight line from the cathode to the anode, their paths are curved by the right-angle force of the magnetic field. This causes electrons to bunch in pinwheel-shaped clouds that are swept around the cathode at high speed a result of their interaction with the anode cavities.

As the spokes of the electron pinwheel turn, they repel free electrons in the walls of the metal cavities, causing them to oscillate back and forth inside the cavities. The sides of the cavities alternate polarity at high frequency creating oscillating magnetic fields across the gaps. These fields, interacting with the bunched electron cloud, sustain its rotation.

The frequency of the electron oscillations within the cavities depends on cavity geometry and size. (Smaller cavities produce higher-frequency and shorter-wavelength microwaves.)

The high-frequency power generated by all the cavities is coupled out through one of the cavities by a loop to a coaxial cable or it may be coupled out directly through a dielectric window to a waveguide that connects to the antenna. Magnetrons may be pulsed at repetition rates as high as 1000 times per second to produce high-power bursts of microwave power.

Magnetrons rated for 300 to 800 watts of continuous wave output power are the heat sources of household microwave ovens. *See also* MICROWAVE, MICROWAVE OVEN.

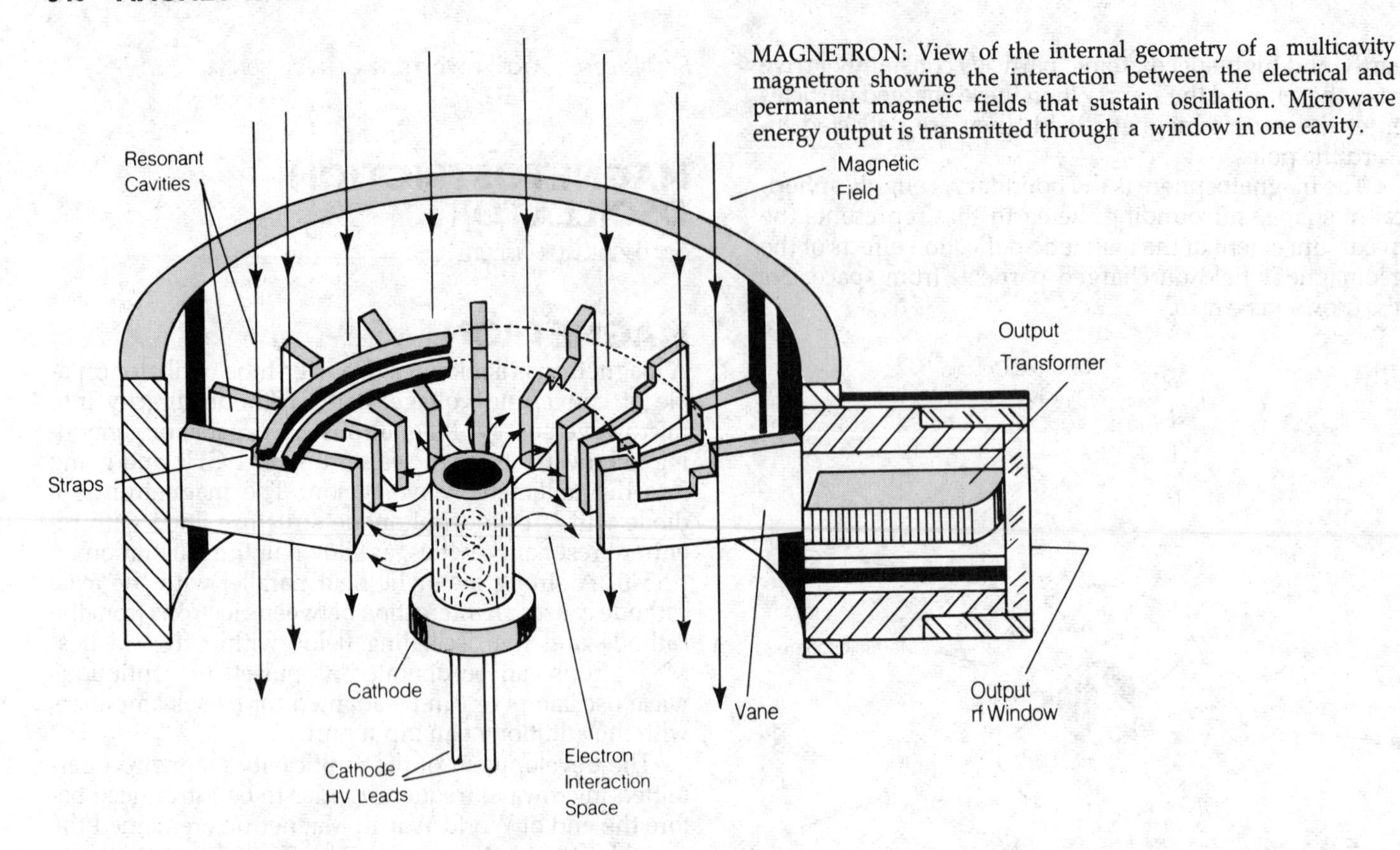

MAGNETRON: View of the internal geometry of a multicavity magnetron showing the interaction between the electrical and permanent magnetic fields that sustain oscillation. Microwave energy output is transmitted through a window in one cavity.

MAGNET WIRE

Magnet wire is a name applied to insulated solid-copper wire for low-current use. It is so named because it is popular for winding coils for electromagnets. Magnet wire may have plastic, rubber, or enamel insulation, and ranges in size from approximately AWG No. 14 to No. 40.

Magnet wire is rigid; when bent, it stays bent. This makes magnet wire especially convenient for coil winding. Magnet wire is sometimes used for shortwave receiving antennas and for transmitting antennas at low power levels. However, magnet wire tends to stretch when it is unsupported in a long span because the wire is made of soft copper.

MAINFRAME

See COMPUTER.

MAJORITY CARRIER

Semiconductors can carry current in two ways: electrons transfer negative charge from the negative pole to the positive pole; holes carry positive charge from the positive pole to the negative pole (*see* ELECTRON, HOLE). Electrons and holes are known as charge carriers.

In an N-type semiconductor material, electrons are the dominant charge carriers. In a P-type material, holes are the more dominant carriers. The dominant charge carrier is called the majority carrier. *See also* MINORITY CARRIER, N-TYPE SEMICONDUCTOR, P-TYPE SEMICONDUCTOR.

MAJOR LOBE

See LOBE.

MAKE

The action of closing a circuit is called make. The term is generally used with relays or with keying circuits in a code transmitter.

In the code transmitter, make is the length of time required, after the key is closed, for the output signal to rise from zero to maximum amplitude.

MAN-MADE NOISE

Man-made noise is any form of electromagnetic interference that can be traced to non-natural causes. In particular, man-made noise refers to such interference as ignition and impulse noise originating from internal-combustion engines and electrical appliances.

Man-made noise has become an increasing problem in radiocommunications at the very low, low, medium, and high frequencies because of the proliferation of household appliances that generate electromagnetic energy. At very high frequencies and above, man-made noise is not usually a serious problem, except when extreme receiving sensitivity is needed.

Man-made noise, unlike natural noise, is generally created by an identifiable source at a well-defined frequency. This makes it possible, in most cases, to cancel man-made noise with highly directional receiving antennas such as the ferrite rod or small loop. However, the apparent direction of the noise may change with frequency. Noise blankers and limiters can also help in the reduction of interference from man-made noise. *See also* ELECTROMAGNETIC INTERFERENCE, IGNITION NOISE, IMPULSE NOISE, NOISE, NOISE LIMITER.

MARCONI ANTENNA

See ANTENNA DIRECTORY.

MARCONI EFFECT

A center-fed or end-fed antenna can show resonance at a frequency determined by the length of the feed line and radiator combined. This resonance usually occurs at the frequency occurring when the combined length of the feed line and radiator is ¼ wavelength, or some odd multiple of it. Marconi effect can result in unwanted radiation from an antenna feed line because the whole system behaves like a Marconi antenna. This can occur with antenna systems using coaxial feeders, as well as with systems using open wire line. The currents flow on the outside of the coaxial outer conductor or in phase along both conductors of open-wire line. *See also* MARCONI ANTENNA in the ANTENNA DIRECTORY.

MARITIME-MOBILE RADIO SERVICE

A two-way radio station in any private boat or commercial vessel not engaged in amateur communication is called a maritime-mobile station. Any fixed station used for nonamateur communication with the operator of any private or commercial vessel is also part of the maritime-mobile service.

The maritime-mobile radio service is allocated a number of frequency bands in the radio spectrum. The bands range from the very low frequencies through the ultrahigh frequencies; precise allocations vary from country to country.

MARX GENERATOR

A Marx generator is a form of high-voltage, direct-current impulse generator. Its principle of operation is analogous to a voltage-multiplier power supply. A source of direct current is connected to a set of capacitors that are arranged in parallel with series resistances, as shown in the illustration. A set of spark gaps is connected to the network of capacitors so that the gaps are subjected to the series combination of the capacitor voltages. The voltage appearing across the spark gaps is therefore much greater than the voltage across the individual capacitors.

When the capacitors reach a certain critical charge value, they discharge to create a spark across all of the gaps at once. This process occurs repeatedly as the capacitors alternately charge and discharge. The sparks produce short pulses of extremely high voltage at the output terminals.

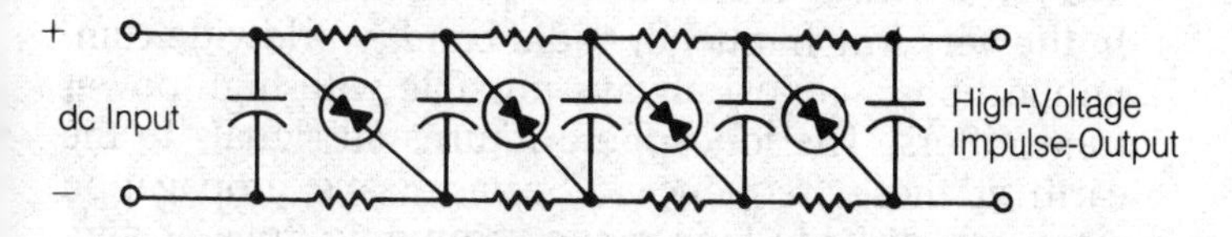

MARX GENERATOR: A high-voltage, pulsed output is produced from the low-voltage direct-current input to the Marx generator.

MASER

The maser is an amplifier for microwave energy. The term is an acronym for *m*icrowave *a*mplification by *s*timulated *e*mission of *r*adiation. Maser output is the result of quantum resonances in various substances.

When an electron moves from a high-energy orbit to an orbit with less energy, a photon is emitted (*see* ELECTRON ORBIT). For a particular electron transition or quantum jump, the emitted photon always has the same amount of energy, and therefore the same frequency. By stimulating a substance to produce many quantum jumps from a given high-energy level to a given low-energy level, an extremely stable signal is produced. This is the principle of the maser oscillator. Ammonia and hydrogen gas are used in maser oscillators as frequency standards. Rubidium gas can also be used in masers. The output-frequency accuracy is within a few billionths of 1 percent in the gas maser.

Solid materials such as ruby can be used to obtain maser resonances. This kind of maser is called a solid-state maser, and it can be used as an amplifier or oscillator. When an external signal is applied at one of the quantum resonance frequencies, amplification is produced. The solid-state maser must be cooled to very low temperatures for proper operation; the optimum temperature is absolute zero. While satisfactory maser operation can take place at temperatures as high as that of dry ice, the most common method of cooling employs liquid helium or liquid nitrogen.

The traveling-wave maser is one form of the device for use at ultrahigh and microwave frequencies. The cavity maser is an alternate configuration. Both the traveling-wave and cavity masers use the resonant properties of materials made to precise dimensions. The traveling-wave and cavity masers are solid-state devices, and the temperature must be reduced to a few degrees Kelvin for operation. The gain of a properly operating traveling-wave or cavity type maser amplifier may be as great as 30 to 40 decibels.

Maser amplifiers find applications in radio astronomy, communications, and radar. Some masers operate in the infrared, visible-light, and even ultraviolet parts of the electromagnetic spectrum. Visible-light masers are called optical masers or, more commonly, lasers. *See also* LASER.

MASS

Mass is a measure of the amount of matter present in an object. Mass is expressed in kilograms (kg) in the standard international system of units. The kilogram is the amount of mass which, when subjected to a force of 1 newton in free space, undergoes an acceleration of 1 meter per second per second.

Mass is often confused with weight. However, weight depends on the intensity of a gravitational field in which the mass exists; mass is independent of the environment in which an object is found. The pound (lb) is an expression of weight. On the surface of earth, a mass of 1 kg weighs about 2.2 lb, but on the moon, the same mass weighs just 0.36 lb, and on Jupiter, it weighs 5.5 lb.

MATCHED IMPEDANCE

See IMPEDANCE MATCHING.

MATCHED LINE

A matched line is a feed line or transmission line, terminated by a nonreactive load whose impedance is identical to the characteristic impedance of the line. In a matched line, there is no reflected power and there are no standing waves. A matched condition is desirable for any line, because the line loss is minimized.

Given a feed-line characteristic impedance of Z_o and a purely resistive load impedance R, where $Z_o = R$, the current and voltage appear in the same proportion at all points along the line. If E is the voltage at any point on the line, and I is the current at that same point, the following is true:

$$E/I = Z_o = R$$

This uniformity of voltage and current exists only in a matched line.

Because any feed line has some loss, the voltage and current are higher near the transmitter than near the load, as shown in the illustration. However, the ratio E/I is always the same. *See also* CHARACTERISTIC IMPEDANCE, IMPEDANCE MATCHING, MISMATCHED LINE, REFLECTED POWER, STANDING WAVE, STANDING-WAVE RATIO.

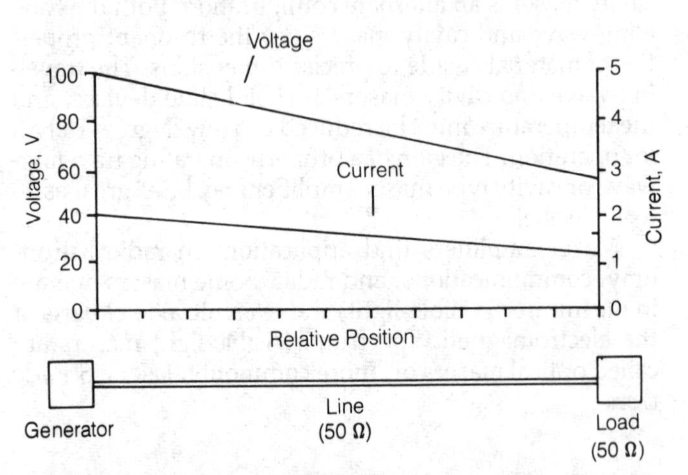

MATCHED LINE: Current and voltage along a matched transmission line are always in the same ratio, everywhere along the line.

MATCHING

When two or more active devices such as transistors or vacuum tubes are operated in parallel, push-pull, or push-push, it is desirable to ensure that they have operating characteristics that are identical. For a given type of transistor or tube, there can be considerable differences in operating characteristics. If the components are chosen at random, current hogging or other undesirable effects may occur during circuit operations. Component matching eliminates potential problems. *See also* CURRENT HOGGING, PUSH-PULL CONFIGURATION, PUSH-PUSH CONFIGURATION.

The term *matching* is also used to describe the process or condition of matched impedances. *See also* IMPEDANCE MATCHING.

MATRIX

A matrix is a high-speed switching array consisting of wires and diodes in a certain orderly arrangement. The matrix itself is an array of wires; some of the wires are interconnected with diodes (*see* DIODE MATRIX).

In linear algebra, a matrix is a rectangular array of numbers that can represent a set of simultaneous linear equations. The coefficients of the equations are represented in an orderly fashion, eliminating the variables. This facilitates algebraic operations with a minimum of confusion.

MAXIMUM USABLE FREQUENCY

For any two points separated by more than a few miles, electromagnetic communication is possible only at certain frequencies. At very low frequencies, using high transmitter power, worldwide communication is almost always possible. In general, there is a band of higher frequencies at which communication is also possible, and the power requirements are moderate.

Given some arbitrary medium or high frequency f, at which communication is possible between two specific points, the frequency can be raised until communication is no longer possible. The cutoff frequency can be called f_u. This cutoff point is called the maximum usable frequency, or MUF, for the two locations.

The MUF depends on the locations and separation distance of the transmitter and receiver. The MUF also varies with the time of day, the season of the year, and the sunspot activity. All of these factors affect the ionosphere. For a given signal path, the propagation generally improves as the frequency is increased toward the MUF; above the MUF the communication abruptly deteriorates. When a communications circuit is operated just below the MUF, and the MUF suddenly drops, a rapid and total fadeout occurs. *See also* LOWEST USABLE FREQUENCY, MAXIMUM USABLE LOW FREQUENCY, PROPAGATION CHARACTERISTICS.

MAXIMUM USABLE LOW FREQUENCY

In the very low frequency radio band, worldwide communication is almost always possible with high-power transmitters. The ionosphere returns all signals to the earth at these frequencies. Surface-wave propagation alone can provide long-range communications at very low frequencies.

Assume that a transmitter is operating at some very-low frequency f between two defined points on the earth and that the frequency is gradually raised. If the transmitter and receiver are more than a few miles apart, the path loss generally increases as the frequency is increased. This happens for two reasons: The ionospheric D layer becomes less reflective and more absorptive, and the surface-wave loss increases (*see* D LAYER, SURFACE WAVE). Under some conditions, a frequency f_u exists at which communication deteriorates to the point of cutoff. This frequency may be in the low or medium range, and is known as the maximum usable low frequency (MULF).

The MULF is not the same as the maximum usable frequency. If the frequency is raised above the MULF, communication will often become possible again at a frequency called the lowest usable high frequency (LUF). As the frequency is raised still higher propagation ultimately deteriorates at the maximum usable frequency (MUF). *See also* LOWEST USABLE FREQUENCY, MAXIMUM USABLE FREQUENCY, PROPAGATION CHARACTERISTICS.

MAXWELL

The maxwell is a unit of magnetic flux representing one line of flux in the centimeter-gram-second system of units. Magnetic flux density may be measured in maxwells per square centimeter.

The weber is the more common, standard international unit of magnetic flux. The weber is equivalent to 10^8 maxwells. *See also* MAGNETIC FLUX, WEBER.

MAXWELL BRIDGE

The Maxwell bridge is a circuit used for measuring unknown inductances. The internal resistance of a coil can also be determined. A typical Maxwell bridge circuit is illustrated by the schematic diagram.

The unknown inductance, *L*, and the unknown series resistance, *R*, are determined by manipulating the variable resistors R_1 and R_3. Balance is indicated by a null reading on the indicator meter. When balance has been reached, the value of *L* in henrys is found by:

$$L = C(R_1R_2)$$

where *C* is given in farads and R_1 and R_2 are given in ohms. The series resistance, *R*, is:

$$R = \frac{R_2^2}{R_3}$$

where all resistances are in ohms. *See also* INDUCTANCE.

MAXWELL'S EQUATIONS

Maxwell's equations are a set of four mathematical relations that describe the behavior of electromagnetic fields. A description of these equations is:

1. The amount of work necessary to move a unit magnetic pole completely around any closed path is equal to the total current linking the path. The total current consists of conduction and displacement currents.
2. The electromotive force induced in any non-moving, closed loop is proportional to the rate of change of magnetic flux within the loop.
3. For an electrically charged object, the total electric flux surrounding the object is precisely equal to the charge quantity.
4. Magnetic lines of flux are always closed loops. That is, magnetic lines of flux have no points of beginning or ending.

See also ELECTRIC FIELD, ELECTROMAGNETISM, MAGNETIC FIELD.

M-DERIVED FILTER

An M-derived filter is a variation of a constant-k inductance-capacitance filter (*see* CONSTANT-K FILTER). The m-derived filter is so named because the values of inductance and capacitance are multiplied by a common factor m. The illustration at A shows a typical T-section, constant-k, lowpass filter for unbalanced line. The illustration at B shows an m-derived filter with component values altered by the factor m. An additional inductor is placed in series with the capacitor in the m-derived filter.

The value of the factor m is always between 0 and 1. The optimum value for m in a given situation depends on the type of response and the cutoff frequency desired. A properly designed m-derived filter has a sharper cutoff than a constant-k filter for a given frequency. *See also*

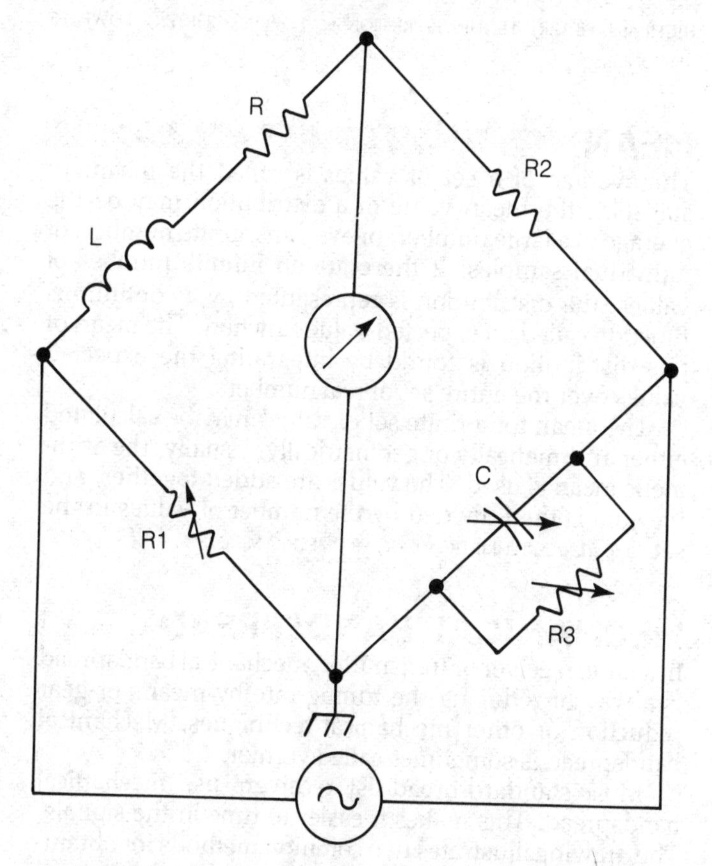

MAXWELL BRIDGE: A circuit for determining inductance.

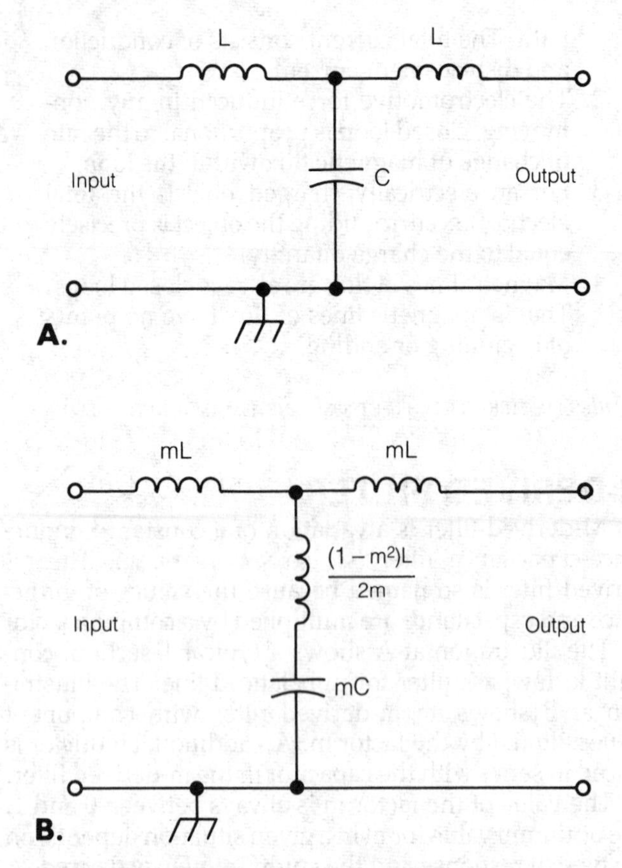

M-DERIVED FILTER: Examples of T-section lowpass filters include a typical constant-k (A) and m-derived (B).

HIGHPASS FILTER, HIGHPASS RESPONSE, LOWPASS FILTER, LOWPASS RESPONSE.

MEAN

The average of a set of values is called the mean. In statistics, the mean value of a distribution may be the average of a large number, or even an infinite number, of individual samples. If there are an infinite number of values, the distribution is represented by a continuous function called an expected-value function. The mean of this distribution is found by integrating the expected values over the entire set of real numbers.

The mean for a finite set of values may be calculated either arithmetically or geometrically. Usually, the arithmetic mean is used. The values are added together, and the sum is then divided by the number of values in the set. *See also* ARITHMETIC MEAN, GEOMETRIC MEAN.

MECHANICAL BANDSPREAD

In a radio receiver or transmitter, mechanical bandspread is a way of reducing the tuning rate by means of gear reduction or other mechanical techniques. Mechanical bandspread is sometimes called vernier.

Most standard-broadcast receivers use mechanical bandspread. This makes it easier to tune in the signals. The drawing illustrates two common methods for obtaining mechanical bandspread. At A, a pair of gears is used with the tuning knob connected to the smaller gear. At B, a rack-and-pinion arrangement is shown. The extent of the bandspreading is determined by the gear ratio.

In advanced communications receivers, electrical bandspread is generally preferred over mechanical bandspread because the precision obtainable by mechanical methods is limited. *See also* ELECTRICAL BANDSPREAD.

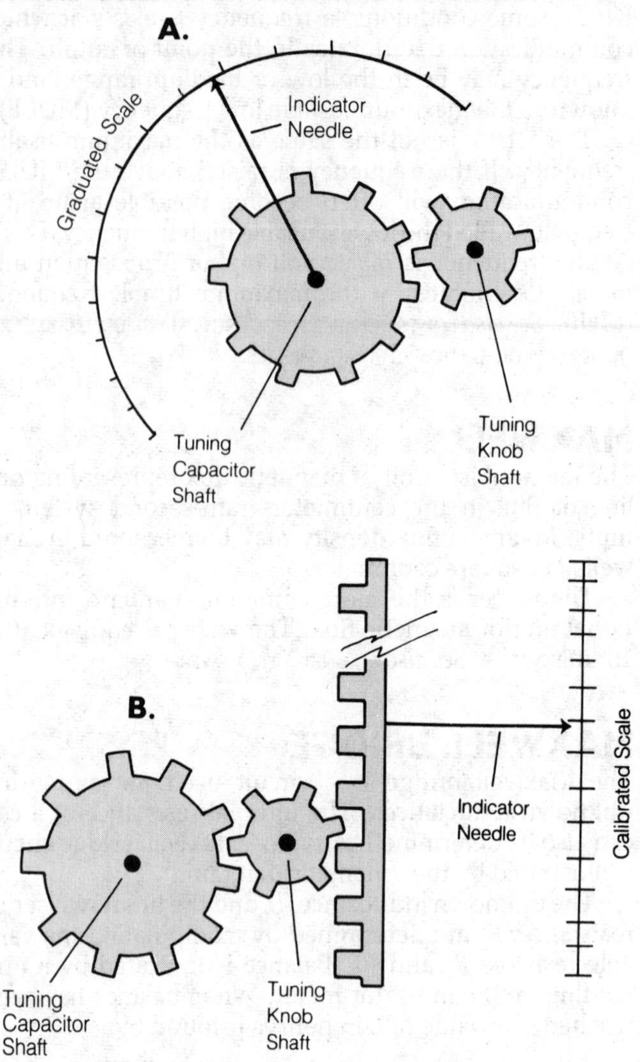

MECHANICAL BANDSPREAD: Two methods of obtaining mechanical bandspread are a rotary indicator (A) and rack-and-pinion (B).

MECHANICAL FILTER

A mechanical filter is a form of bandpass filter. It is sometimes called an ultrasonic filter.

The mechanical filter operates on a principle similar to the ceramic or crystal filter. Two transducers are employed, one for the input and the other for the output. The input signal is converted into electromechanical vibration by the input transducer. The resulting vibrations travel through a set of resonant disks to the output transducer. The output transducer converts the vibrations back into an electrical signal by means of magnetostriction.

The mechanical filter offers some advantages over electrical filters at low and medium frequencies. Mechanical elements are of reasonable size above about 75 kHz, allowing a filter to be enclosed in a small package. Above perhaps 750 kHz, mechanical filters become difficult to construct because the elements are too small. No adjustment is needed, since the resonant disks are solid and of fixed size. The resonant characteristics are extremely pronounced, often far superior to electrical filters of the same physical size. Mechanical filters can be designed to have a nearly flat response within the passband, and very steep skirts with excellent ultimate attenuation. The mechanical filter is a good choice for bandpass applications at intermediate frequencies from 75 to 750 kHz.

The elements of a mechanical filter have more than one resonant frequency. Therefore, it is necessary to provide some external means for attenuation af spurious frequencies. Mechanical filters are sensitive to physical shock, and care must be exercised to ensure that these filters are not subjected to excessive vibration.

The construction of a mechanical filter is very similar to the construction of a ceramic filter. Instead of ceramic disks, metal disks, usually made of nickel, are used. *See also* CERAMIC FILTER.

MEDIAN

In a given set of values or samples, the median value is that value for which an equal number of samples fall above and below.

If the number of values in a set is finite and odd, the median is the middle value. For example, if the values are 1, 3, 4, 5, 7, 8, and 9, the median is 5: There are three values above 5 and three below 5. If the number of values is even, then the median is generally the arithmetic mean of the two center values. For example, if the values are 2, 3, 7, and 8, then the median is equal to the arithmetic mean of 3 and 7. That value is 5. (Note that if the number of elements in a sample set is finite and even, then the median is not in the set.)

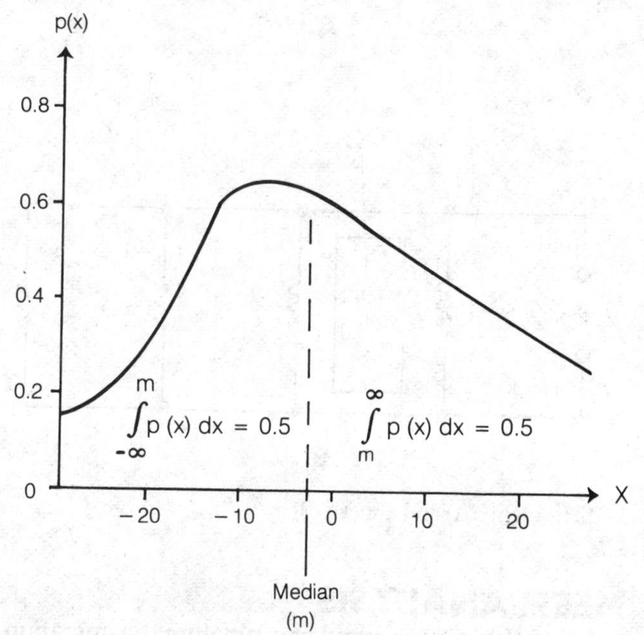

MEDIAN: The area under the curve is equal on both sides of the median.

In a statistical distribution with an infinite number of values, the median must be found by integration. The median is that value for which half of the area under the curve lies to the left, and half of the area under the curve lies to the right (see illustration). Thus, if the total area under the curve is 1, that is:

$$\int_{-\infty}^{\infty} p(x)\,dx = 1$$

where p is the continuous probability distribution, then the median is the value m so that:

$$\int_{-\infty}^{m} p(x)\,dx = \int_{m}^{\infty} p(x)\,dx = 0.5$$

The median can be the same, for a given set of values, as the arithmetic mean or average, but this relationship is not true in general. *See also* ARITHMETIC MEAN.

MEDIUM-SCALE INTEGRATION

Medium-scale integration refers to integrated circuits containing up to 100 individual gates per chip. Medium-scale integration, or MSI, allows considerable miniaturization of electronic circuits, but not to the extent of large-scale integration (LSI) or very large scale integration (VLSI).

Both bipolar and metal-oxide semiconductor (MOS) technology have been adapted to MSI. Various linear and digital logic circuits are made as MSI. *See also* INTEGRATED CIRCUIT.

MEGA

A prefix multiplier modifying a quantity to indicate 1 million (10^6), is mega. For example, a megahertz is 1 million hertz, and a megavolt is 1 million volts. *See also* MEGAHERTZ, MEGAWATT, MEGOHM, PREFIX MULTIPLIERS.

MEGABYTE

The megabyte (Mb) is a unit of information equal to 2^{20}, or 1,048,576 bytes. The megabyte is a common unit of memory in all digital computers. *See also* BYTE, KILOBYTE.

MEGACYCLE

See MEGAHERTZ.

MEGAHERTZ

The megahertz (MHz) is a unit of frequency, equal to 1 million hertz, or 1 million cycles per second.

Radio frequencies in the medium, high, very high, and ultrahigh ranges are specified in megahertz. A frequency of 1 MHz is approximately at the middle of the standard AM broadcast band in the United States. *See also* FREQUENCY.

MEGAWATT

The megawatt (MW) is a unit of power, or rate or energy expenditure. A power level of 1 MW represents 1 million watts. In terms of mechanical power, 1 MW is about 1,340 horsepower.

The megawatt is a very large unit of power. A small town of 5,000 residents uses approximately 1 MW of power on an average day. Some radio broadcast transmitters have an input power level approaching 1 MW. *See also* POWER, WATT.

MEGGER

A megger is an instrument used to measure extremely high resistances. The resistances of electrical insulators range from several megohms (millions of ohms) up to billions or trillions of ohms. For the measurement of these resistances, a high-voltage source is required and one is included in the megger.

The illustration is a schematic diagram of a typical megger. The high-voltage generator is turned by hand. The output voltage is dangerous, and presents an electric shock hazard.

The megger may be used for continuity testing and short-circuit location, as well as for the measurement of high direct-current resistances. *See also* OHMMETER.

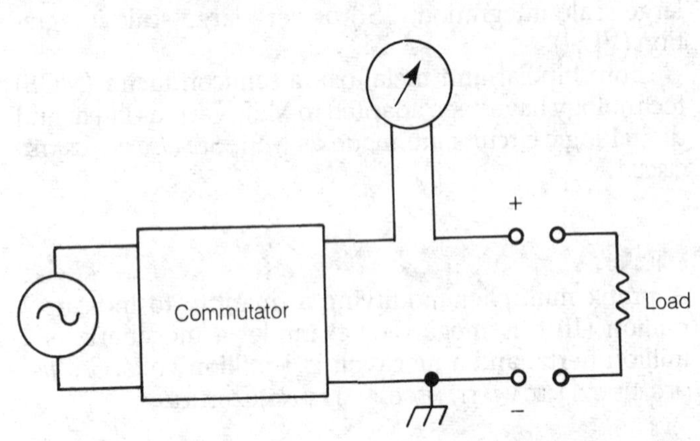

MEGGER: A megger measures extremely large resistances.

MEGOHM

A megohm is a unit of resistance equivalent to 1 million ohms. With a potential difference of 1 volt, a current of 1 μA (1 millionth of an ampere) flows through a resistance of 1 megohm.

Some common resistors have values that range as high as a few hundred megohms. Typical insulating materials have resistances of millions of megohms or more. *See also* OHM, RESISTANCE.

MEMORY

See SEMICONDUCTOR MEMORY.

MEMORY-BACKUP BATTERY

A memory backup battery is a power cell or battery used to retain data in a volatile semiconductor memory. The memory backup battery might be rechargeable and connected to the main power source, or it may be independent of the main power source.

Memory backup batteries are used in computer systems that are subject to power failure or are periodically shut down. *See* BATTERY, SEMICONDUCTOR MEMORY.

MERCURY CELL

See BATTERY.

MESH

A mesh is a collection of circuit branches with the following three properties: 1) The collection forms a closed circuit, 2) every branch point in the circuit is incident to exactly two branches, and 3) no other branches are enclosed by the collection. Although a circuit loop may contain smaller loops, as in the example at A in the illustration, a mesh is a smallest possible loop as at B.

The mesh provides a means for evaluating the currents and voltages in various parts of a complex circuit. Kirchhoff's laws can be applied to all of the individual meshes in a circuit. *See also* KIRCHHOFF'S LAWS, MESH ANALYSIS.

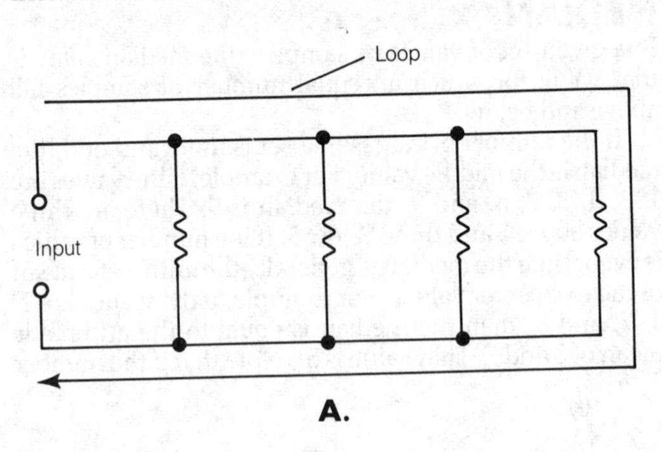

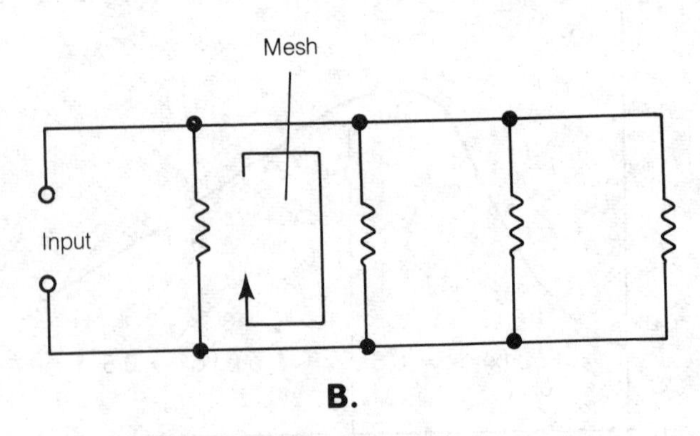

MESH: A loop (A) and a mesh (B).

MESH ANALYSIS

Mesh analysis is a method of evaluating the operation or characteristics of an electronic circuit containing one or more closed loops. The current, voltage, and resistance through or between various parts of the circuit can be determined according to Kirchhoff's laws. Kirchhoff's laws form the basis for mesh analysis.

Mesh analysis is relatively simple for direct-current circuits, although the equations become rather cumbersome for circuits with many meshes. Mesh analysis is more complex mathematically for alternating currents where reactance as well as resistance must be considered. *See* KIRCHHOFF'S LAWS, MESH.

METAL

Metal is a material characterized by good electrical conductivity because the electrons of a metal are easly stripped from their orbits around the atomic nuclei. Electrical conduction occurs as the electrons move from atom to atom. Metals also are generally good conductors of heat.

Common metals include aluminum, copper, and iron. All metals except mercury are solids at room temperature. Metals display different melting temperatures. Mercury melts at about −39°C, but tungsten remains in solid form at temperatures up to about 3,400°C.

METAL-OXIDE SEMICONDUCTOR

The oxides of certain metals exhibit insulating properties. In recent years, metal-oxide-semiconductor (MOS) devices have come into widespread use.

Metal oxides used in the manufacture of MOS devices include such compounds as aluminum oxide and silicon dioxide. Metal-oxide semiconductor devices are noted for their low power requirements. MOS integrated circuits have high component density and high operating speed. Metal oxides are also used in the manufacture of certain field-effect transistors.

All MOS devices are subject to damage by the discharge of static electricity. Therefore, care must be exercised when working with MOS components. *See also* ELECTRO-STATIC DISCHARGE CONTROL, METAL-OXIDE-SEMICONDUCTOR FIELD-EFFECT TRANSISTOR.

METAL-OXIDE-SEMICONDUCTOR FIELD-EFFECT TRANSISTOR

Most new integrated circuits and power transistors are based on the metal-oxide-semiconductor field-effect transistor (MOSFET).

The MOSFET can be recognized because the gate electrode is insulated from the channel by a thin layer of metal oxide. Because the gate is insulated from the channel, the MOSFET is sometimes called an insulated-gate field-effect transistor, or IGFET.

The MOSFET device has extremely high input impedance, typically billions or trillions of ohms. Thus, the MOSFET requires essentially no driving power. MOSFETs are commonly used in high-gain receiver amplifier circuits. A single-gate MOSFET amplifier stage can produce better than 15 decibels gain at a frequency of 100 MHz.

MOSFETs can have channels of either N-type material or P-type material. MOSFETs that require a gate bias for operation are called enhancement-mode MOSFETS. Others that are normally conducting, or saturated, under conditions of zero gate bias are called depletion-mode MOSFETs. *See also* FIELD-EFFECT TRANSISTOR, TRANSISTOR.

METAL-OXIDE-SEMICONDUCTOR LOGIC FAMILIES

Metal-oxide semiconductor technology permits the fabrication of both analog and digital integrated circuits. In recent years, several metal-oxide-semiconductor (MOS) logic families have been developed. Miniaturization is more easily obtained with MOS technology than with bipolar technology, and the power requirements are lower.

Metal-oxide-semiconductor logic families include complementary (CMOS), N-channel (NMOS), and silicon-on-sapphire (SOS).

CMOS uses both N-type and P-type materials on the same substrate. CMOS is characterized by extremely low power requirements and relative insensitivity to external noise (*see* COMPLEMENTARY METAL-OXIDE SEMICONDUCTOR). The operating speed is relatively high.

NMOS technology is still commonly used today. It permits high operating speed and has the advantage of simplicity. P-channel (PMOS) metal-oxide-semiconductor logic is occasionally used, although it operates at a slower speed than NMOS or bipolar logic devices.

Metal-oxide-semiconductor logic families are especially useful in high-density memory applications. Most microcomputer chips use MOS technology today. *See also* DIGITAL-LOGIC TECHNOLOGY, METAL-OXIDE SEMICONDUCTOR.

METAL-OXIDE VARISTOR

Metal-oxide varistors (MOVs) are variable resistors for protecting electronic circuits against ac (alternating-current) voltage transients. They behave like two reversed or back-to-back transient voltage suppression (TVS) diodes. Varistors are nonlinear resistors whose resistance value changes as a function of the applied voltage. Their symmetrical, bilateral characteristics provide clamping during both positive and negative swings of the ac waveform. When the voltage exceeds the varistor rating, internal resistance drops sharply, and the device becomes a short circuit. While the voltage transient is bypassed, the body of the device is able to absorb current flowing at the time of the transient without self-destruction. *See* CIRCUIT PROTECTION.

Metal-oxide varistors are made from powdered metal oxides, principally of zinc, mixed with a suitable binder and pressed into a suitable shape—disk, cylinder or block. The pressed slugs form a multiple-junction conductive matrix of metal-oxide grains suspended by highly resistive boundaries when fired at temperatures above 1000°C. Varistors are bulk protective devices whose energy absorption is determined by their volume: they can absorb energy uniformly throughout their volume rather than at a single boundry as is done by TVS diodes. As a result, metal-oxide varistors have a higher power handling ability per unit volume than comparably rated TVS diodes.

Metal-oxide varistors are now packaged for surface mounting. Like TVS diodes, they must be operated within their ratings, or they will be destroyed. In comparing the two devices with comparable ratings, TVS diodes have sharper voltage-current characteristics than MOVs. This gives them more precise breakdown values. In addition, TVS diodes are triggered at lower voltages than the metal-oxide varistors.

METER

A meter can be an instrument, either electrical or electromechanical, for measuring an electrical quantity. The most familiar kind of meter is the D'Arsonval moving-coil meter. However, digital meters are becoming increasingly common.

The basic unit of displacement in the metric system is the meter. Originally, the meter was defined as 1 ten-millionth of the distance from the north pole to the equator of the earth. That is, there were supposed to be 10,000,000 meters in that distance. Now, the meter is more precisely defined as the standard international unit of length. A distance of 1 meter represents 1,650,763.73 wavelengths, in a vacuum, of the radiation corresponding to the transition of electrons between the levels $2p_{10}$ (in the L shell) and $5d_5$ (in the 0 shell) of the atom of krypton 86. The meter is approximately 39.37 inches, or a little more than an English yard. *See also* METRIC SYSTEM.

METRIC SYSTEM

The metric system is a decimal system for measurement of length, area, and volume. Because it is a decimal system, the metric system has gained widespread use throughout the world. Scientists routinely use the metric system. The principal unit of length is the meter (*see* METER). The decimeter is 0.1 meter; the centimeter is 0.01 meter; the millimeter is 0.001 meter; the kilometer is 1,000 meters.

The decimal nature of the metric system makes it much easier to use than the older English system. However, the English system is still used for non-scientific applications in the United States. *See also* MKS SYSTEM, STANDARD INTERNATIONAL SYSTEM OF UNITS.

MHO

The mho is a unit of electrical conductance. The mho is now called the siemens in the standard international (SI) system of units. Electrical conductance is the mathematical reciprocal of resistance. Given a resistance of $R\ \Omega$, the conductance S in siemens in simply:

$$S = 1/R$$

A resistance of 1 Ω represents a conductance of 1 siemens. When the resistance is 1,000 ohms, the conductance is 0.001 siemens or 1 millisiemens. When the resistance is 1,000,000 ohms, the conductance is 1 microsiemens. The siemens is also used as the unit of admittance in alternating-current circuits. *See also* ADMITTANCE, CONDUCTANCE.

MICA

Mica is a silicate material that occurs naturally in the crust of the earth. Physically, mica appears as a transparent, sheet-like substance. Mica has excellent dielectric properties.

Mica conducts heat quite well, and it is therefore useful in transistor heat-sink applications when it is necessary to insulate electrically the case of the transistor from the heat sink. Mica is commonly used in the manufacture of low- to medium-value capacitors for use at moderately high voltages. *See also* CAPACITOR.

MICA CAPACITOR

See CAPACITOR.

MICRO-

Micro- is a prefix multiplier that means 0.000001, or 10^{-6}. For example, 1 microfarad is 10^{-6} farad, and 1 microvolt is 10^{-6} volt. *See also* PREFIX MULTIPLIERS.

The prefix micro- is often used to refer to something very small. For example, a microcircuit is a small, or miniaturized, electronic circuit; a microcomputer is a miniature computer.

MICROCIRCUIT

A microcircuit is a miniaturized electronic circuit. The term applies to monolithic integrated circuits and hybrid circuits, however it usually refers to an integrated circuit. *See also* INTEGRATED CIRCUIT.

MICROCODE

See MICROINSTRUCTION.

MICROCOMPUTER

See COMPUTER, MICROCONTROLLER, MICROPROCESSOR.

MICROCONTROLLER

A microcontroller (MCU) is a microcomputer-on-a-chip because it contains most of the functional elements of an operating computer including central processor, random access memory (RAM), read-only memory (ROM), and input/output (I/O) ports. Although microprocessors and microcontrollers have common origins, they are designed for different applications. Most microcontrollers are employed in real-time, hardware-intensive applications where both digital and analog signals are present.

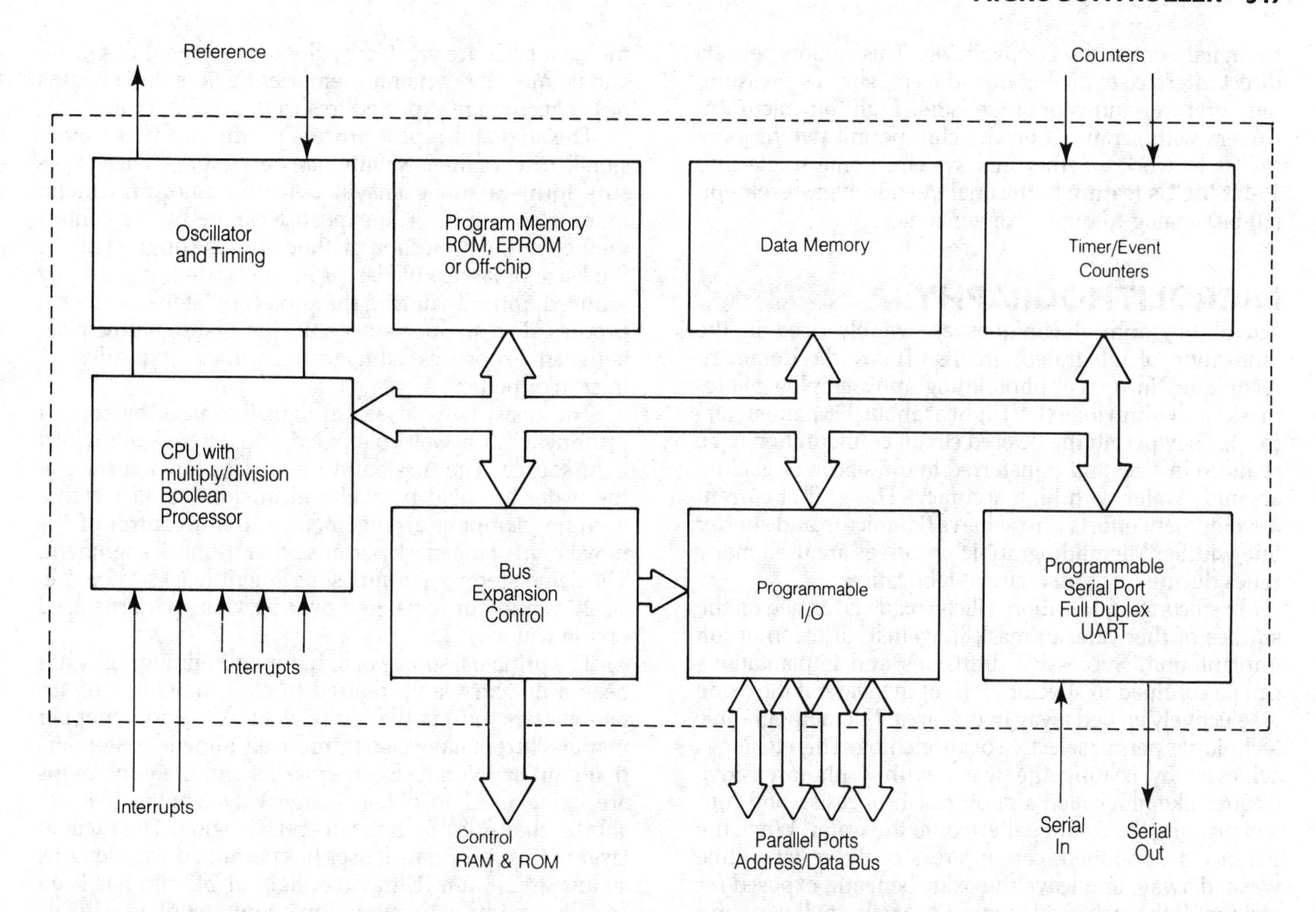

MICROCONTROLLER: Block diagram of an 8-bit, single-chip microcontroller with on-chip memory, input/output ports, and other support functions.

These range from the control of keyboard encoding to motors, industrial processes, and automobile engines. By contrast, the microprocessor (MPU), with continual improvements, is used primarily in the kinds of software-intensive applications encountered in personal computers, graphics workstations, and the new generation parallel computers.

The simplified block diagram illustrates an 8-bit, single-chip microcontroller containing a central processing unit (CPU) and RAM data memory. However, it also has functions not found on microprocessors such as program memory (ROM or EPROM), clock, timer/event counter, and programmable parallel and serial I/O ports. Advantages to the use of the microcontroller where appropriate include savings on external memory and support ICs integrated circuits (ICs), printed-circuit (PC) board space and power consumption.

A microcontroller occupies a silicon chip about the same size as a microprocessor, so compromises have to be made to accommodate the additional functions. Size limits are imposed on the on-chip RAM and ROM, for example. However, some advanced MCUs are able to perform duties that required an earlier generation microprocessor and five additional supportung devices.

Microcontrollers are available with 4-, 8-, and 16-bit architectures with performance generally correlated with bit length. They are usually offered in families to give the user performance and function options. For example, the MCUs can have different amounts of on-chip RAM and ROM to make it easier to select a cost effective match for an application requirement. The lead part in every family has a factory-programmed ROM memory. One 8-bit MCU might offer 1000 bytes of internal factory-programmed ROM and only 64 bytes of internal RAM, but another might offer 2000 bytes of internal ROM and 128 bytes of internal RAM.

The first MCUs were fabricated with NMOS technology. Some designs have converted to CMOS, and new devices are being introduced in that technology. They offer lower power consumption, important in control applications. MCUs have followed in the footsteps of MPU development, and they have benefited from technical advances in those devices.

ROM-less MCUs are available for product development. The designer might wish to do software development on an off-chip EPROM or EEPROM to permit faster and more economical program changes. When an error-free program is obtained, the user can order conventional MCUs in quantity with factory-programmed ROMs.

Alternatively, some MCU manufacturers are offering MCUs with EPROM and EEPROM on the chip because some customers might never be able to order in high enough volume to justify the purchase factory-programmed MCUs. Some 8-bit MCUs with on-chip analog-to-digital converters are available. This option permits direct interface to analog transducers, such as pressure, temperature, and motion sensors. Eight-bit microcontrollers with serial I/O on the chip permit two or more MCUs to work together in a system. Some single-chip 16-bit MCUs feature both serial I/O and high-resolution (10-bit) analog-to-digital conversion.

MICROLITHOGRAPHY

Microlithographic techniques are widely used in the fabrication of integrated circuits. Today the dominant techniques in optical photolithography employ photomasks and ultraviolet (UV) light at about 436 nanometers (nm). They permit the desired circuit configuration to be reduced in size, and transferred to the silicon or gallium arsenide wafer with high accuracy. The goal of current development efforts is to achieve 0.5-micron and shorter line widths. Microlithographic processes are used many times during integrated-circuit fabrication.

In silicon IC fabrication, silicon oxide is formed on the surface of the wafer to mask it against diffusion or ion implantation. Successive diffusions and implantations can be confined to specific areas of the wafer if the oxide is selectively etched away in that area. Microlithographic techniques permit selective oxide etching. This etching is achieved by coating the wafer with a photosensitive lacquer-like film called a photoresist, or resist, and contact printing the circuit pattern onto the wafer. When the pattern is developed, certain areas of the resist will be washed away, and leave the oxide beneath, exposed for etching. If the etch used is specific for silicon dioxide and does not attack silicon, then the oxide can be preferentially etched, to expose the silicon wafer beneath for diffusion.

Similarly, high-conductivity metallic interconnection patterns can be formed on the wafer by lithographic techniques. If a wafer is coated with high-conductivity metal such as aluminum, lithographic techniques permit exposing the excess aluminum, so that it can be etched away to leave an aluminum interconnection pattern on the wafer.

Three kinds of masks are used in IC fabrication: 1) a photographic mask, 2) a photoresist mask, and 3) an oxide mask. A photographic mask can be either a positive or negative image of a circuit pattern formed on a glass plate that has been photosensitized. This mask is used to form the resist mask on the wafer in the photoresist for selectively etching the oxide or aluminum it covers. Finally, the oxide mask is the etched pattern formed in the oxide when the resist is removed. It is used for selective diffusion or implantation. *See* INTEGRATED CIRCUIT MANUFACTURE, PHOTORESIST, TRANSISTOR MANUFACTURE.

Both discrete transistors and ICs are fabricated in batches, with each wafer carrying many identical individual devices or circuits. To process an IC, a set of precision photographic masks is required—one mask for each step in the process. Each mask must be precisely registered with all the others so that when the masks are superimposed, the registration of the masks is kept within narrow tolerances. The more demanding requirements for ICs are created by the development of 4-, 16-, and 64-megabit dynamic memories devices. Line widths of 1.5 micron are expected to shrink to 0.5 micron.

The step-and-repeat process produces a two-dimensional array of images with many exposures. Each exposure forms a single image, but the frame can contain more than one IC. The exposure can be by the contact method or the projection method, and the original can be final size or larger. If a larger image is used, it is usually reduced optically during the projection step-and-repeat process. The stepper must shift the photographic plate between exposures with great accuracy, typically ±1 micron or better.

The resist is exposed through the mask by contact printing with a well-collimated and filtered ultraviolet light source. The mask and the resist-coated surface of the wafer are held parallel and in close contact with a vacuum clamping arrangement. Careful control of the movement of the mask permits accurate mask alignment. The same stepper permits subsequent masks to be precisely aligned or registered over previous patterns used on the wafer.

The principal source of light for photolithography has been a mercury lamp filtered to emit the G-line of 436 nanometers (nm) in the ultraviolet (UV) region. Stepper manufacturers have also introduced I-line lens steppers that emit at 365 nm. Excimer lasers and mercury lamps are being used to obtain image wavelengths that are shorter than 308 nm in the deep-UV region. The excimer laser with appropriate lenses has permitted wavelengths as low as 248 nm. Ultraviolet light at 248 nm has been seen by many as a minimum requirement to achieve commercial half-micron devices. Some stepper systems have also been developed based on the use of electron or E-beam and X ray lithography. These systems are expected to make possible devices with line widths of 0.3 to 0.35 micron.

MICROELECTRONICS

Microelectronics is the application of miniaturization techniques to the design and construction of electronic equipment. *See* INTEGRATED CIRCUIT.

MICROINSTRUCTION

In computer machine language, a bit pattern comprising an elementary command is called a microinstruction. The microinstruction is the smallest form of machine-language instruction. Several microinstructions can be combined to form a macroinstruction, a common and identifiable computer instruction. *See also* MACHINE LANGUAGE, MACROINSTRUCTION.

MICROPHONE

A microphone is an electroacoustic transducer that produces alternating-current electrical impulses from sound waves. This can be done in many ways so there are many different types of microphones.

One of the earliest microphones was the carbon-

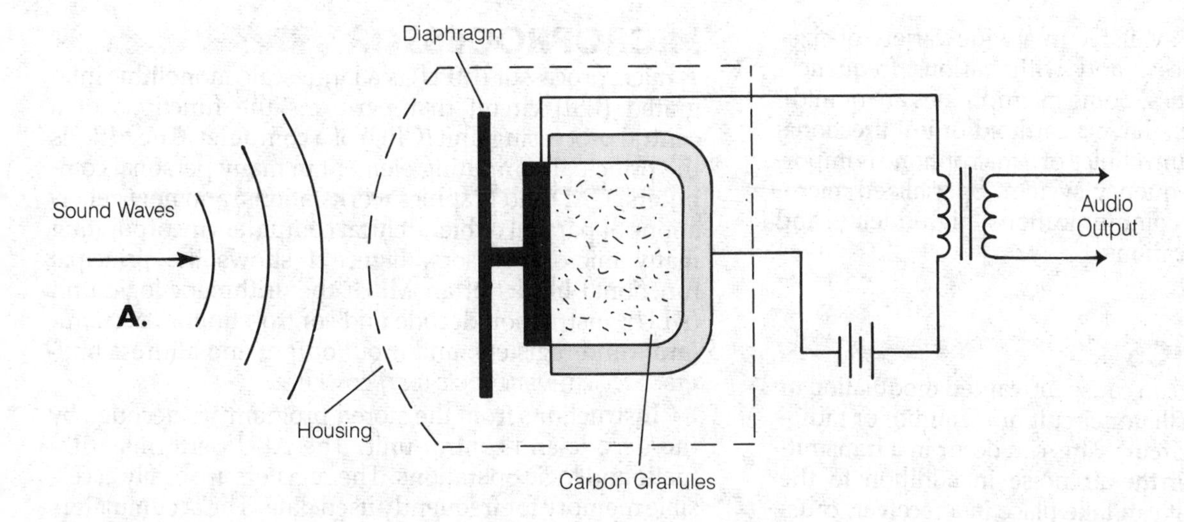

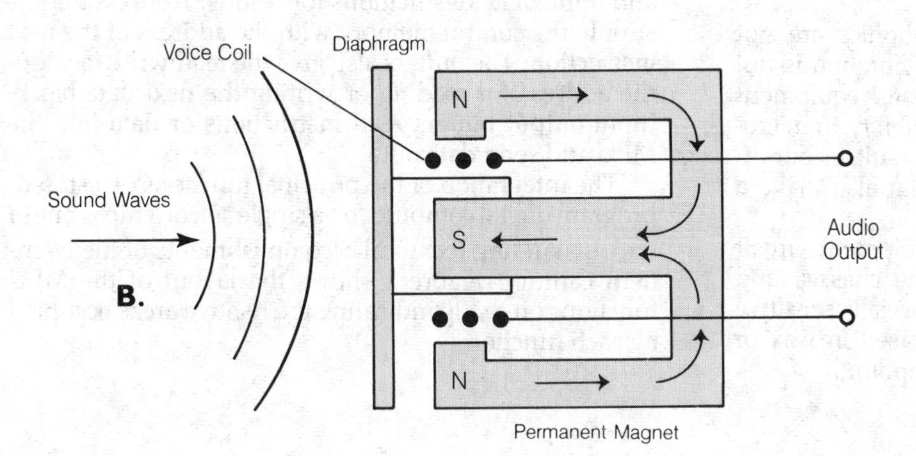

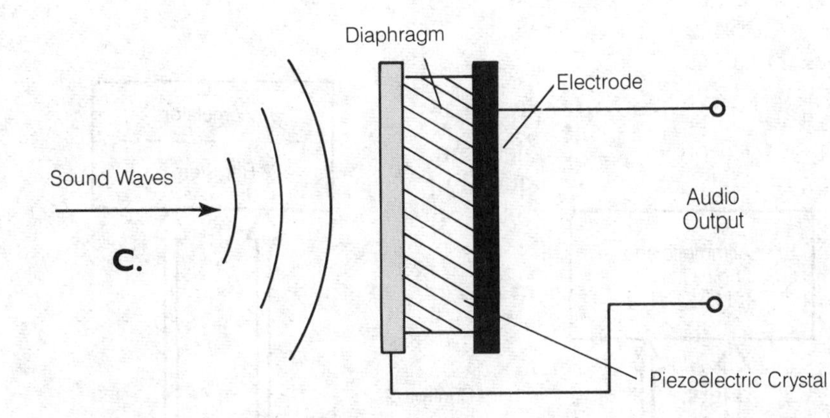

MICROPHONE: Carbon granules form a variable resistor in a carbon microphone (A), voice coil moves in permanent magnet field of dynamic microphone (B), and sound waves produce voltage by piezoelectric effect in crystal microphone (C).

granule microphone. This microphone required an external source of direct current. The sound waves caused changes in the resistance of a carbon-granule container, resulting in modulation of the current. This kind of microphone is shown in part A of the figure.

More recently, the piezoelectric effect has been used to generate electric currents from passing sound waves. Certain crystals, when subjected to vibration, produce weak electric impulses that can be amplified. The ceramic microphone operates on this principle as shown in part C of the figure.

A more rugged type of microphone operates by means of electromagnetic effects. A diaphragm, set in motion by passing sound waves, causes a coil and magnet to move with respect to each other. The changing magnetic field results in alternating currents through the coil. This product, called a dynamic microphone, is illustrated in part B of the figure.

There are other less common forms of microphones used for special purposes. Electrostatic and optical devices, for example, can be used as microphones in certain situations.

Microphones are available in a wide variety of sizes and input impedances, and with various frequency-response characteristics. Some microphones are omnidirectional, while others have a cardioid or unidirectional response. The optimum choice of a microphone is important in any audio-frequency system. Specialized microphones are made for communications, high-fidelity, and public-address applications.

MICROPHONICS

Mechanical vibration can cause unwanted modulation of a radio-frequency oscillator circuit or an audio- or radio-frequency amplifier circuit. This can occur in a transmitter, resulting in over-the-air noise in addition to the desired modulation; it can take place in a receiver, causing apparent noise on a signal. This unwanted modulation is known as microphonics.

In fixed-station equipment, microphonics are not usually a problem because mechanical vibration is not severe. However, in mobile applications equipment must be designed for minimum susceptibility to microphonics. Microphonics can, if severe, result in out-of-band modulation in a transmitter. It can also make a signal unintelligible.

For immunity to microphonics, equipment circuit boards must be firmly anchored to the chassis, and component leads must be kept short. Especially sensitive circuits, such as oscillators, can be encased in wax or some other shock-absorbing potting compound.

MICROPROCESSOR

A microprocessor (MPU) is a large-scale monolithic integrated (LSI) circuit that performs the functions of a central processing unit (CPU) of a computer. The MPU is the principal computing element in many personal computers, CAD and graphics workstations, and new generations of parallel or hierarchical computers incorporating many microprocessors. Figure 1 shows the principal functional blocks of an MPU: the arithmetic-logic unit (ALU), instruction decode and control, timing, accumulators and registers, and input/output and address buffers. *See* ARITHMETIC LOGIC UNIT, COMPUTER.

Instructions from the stored program are decoded by the decode-and-control unit. The ALU performs arithmetic and logic operations. The registers are easily accessible memory for frequently used data. The accumulators are special registers that serve the ALU as sources of data and immediate destinations for results. Address buffers supply the control memory with the address of the next instruction. The buffers also give the read/write memory the address for reading or writing the next data block. Input/output buffers read instructions or data into the MPU and send data out.

The integration of the principal functions of a stored-program digital computer on a single silicon chip is one of the outstanding technical accomplishments of the twentieth century. Figure 2 shows the layout of the MPU functions on a chip indicating the relative areas occupied by each function.

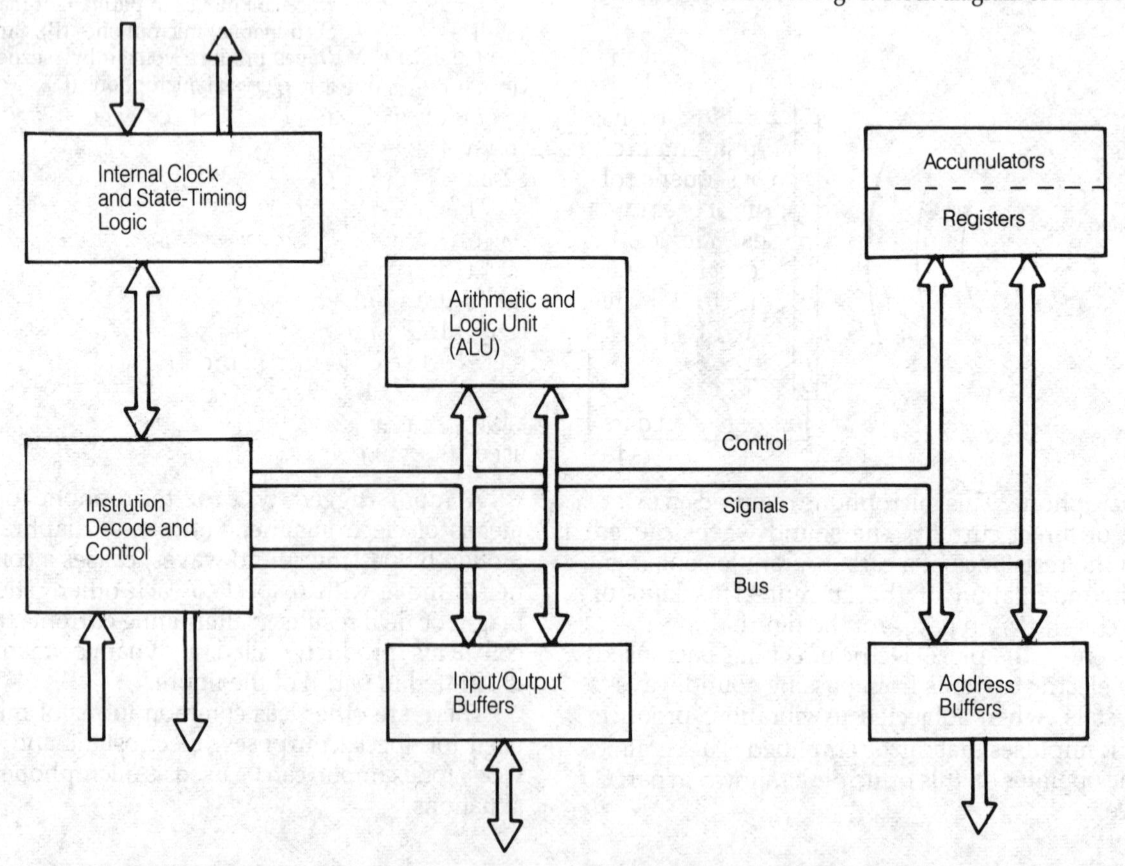

MICROPROCESSOR: Fig. 1. Block diagram of a microprocessor.

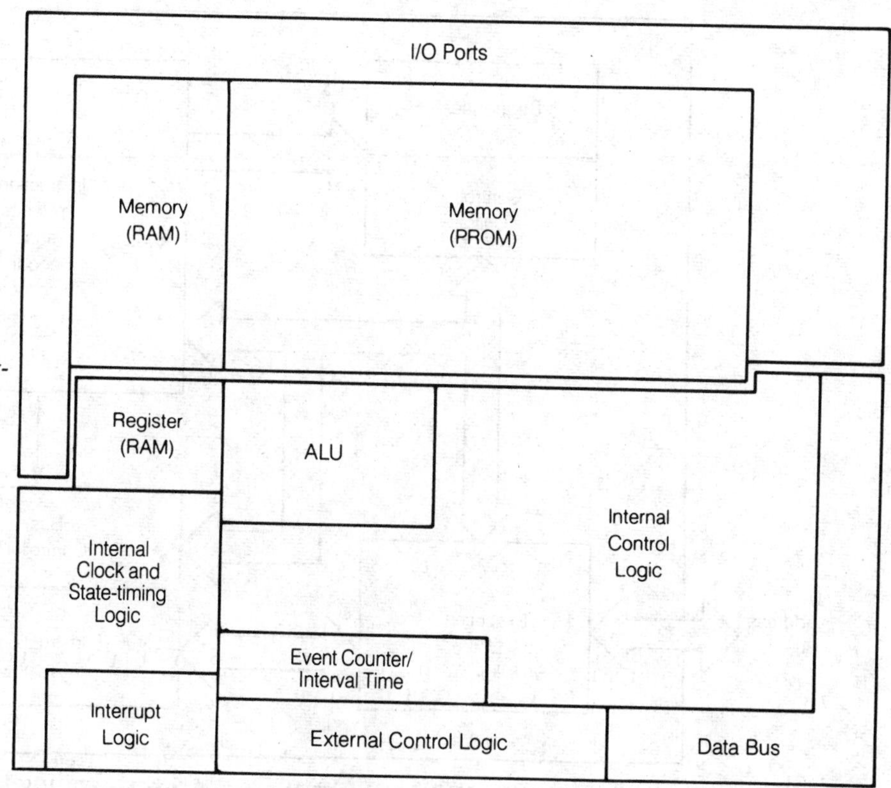

MICROPROCESSOR: Fig. 2. Functional layout of a single-chip 8-bit microprocessor.

The MPU is a universal LSI circuit capable of replacing many standard logic ICs (integrated circuits) and a smaller number of more costly custom LSI circuits. In other cases, MPUs are performing functions that could not have been built in the form of hard-wired logic. The MPU can perform a wider variety of different functions more reliably than arrays of logic devices.

The MPU is the computing component in a microcomputer and it can be coupled by suitable input/output (I/O) circuits to many different external devices which both provide its input signals and are controlled by its output. The microprocessor responds to inputs and produces outputs determined by the program or sequence of instructions which are stored in some form of memory connected to the MPU. In even the simplest microcomputer-based system, the MPU requires additional read-only memory (ROM) to handle instructions, read/write memory (RAM) to handle data, input/output (I/O) ports, and a clock oscillator and timer. *See* SEMICONDUCTOR MEMORY.

Microprocessors are optimized for software-intensive applications involving numerical computations and data processing. They are the principal computing devices in personal computer, graphics and CAD workstations, and the new-generation parallel computers. MPUs are available with 8-, 16-, and 32-bit architecture. They should be distinguished from microcontrollers (MCU) or single-chip microcomputers available with 4-, 8-, and 16-bit architectures. Both classes of device have similar origins, but different architecture and applications. The microcontroller has more of the functional elements of a complete computer on a single monolithic chip than does the microprocessor, but the microcontroller has lower performance. Microcontrollers are optimized for real-time, hardware-intensive applications such as those in automobile engine controls, motor controls, and keyboards. *See* MICROCONTROLLER.

The most advanced MPUs available have 32-bit internal architecture (32-bit internal address bus and 32-bit internal data bus) and 32-bit external data bus. However, some have 32-bit internal architecture, 16-bit external data buses and 24-bit address lines. Driven by intense competition, manufacturers are continually improving their existing MPUs because of the perceived customer demand for even higher speeds and more features for operation in a parallel-system bus, large-memory environment. Figure 3 is a block diagram of a single-chip 32-bit microprocessor.

The newer 32-bit microprocessors feature instruction and data caches, relatively high-speed, on-chip memory for holding the most recently used instructions and data for future re-use by the CPU. A memory management unit (MMU) provides protection by allowing programmers to use system resources without considering the actual size of the memory in megabytes. The MMU also allows multiple programs and operating systems to be used simultaneously.

The electronics industry has accepted 8-, 16-, and 32-bit microprocessors. The most concentrated efforts relate to finding new applications for the 32-bit devices. MPUs designed and built by Motorola Semiconductor and Intel Corporation dominate in the world marketplace. These are the 68000 family processors from Motorola (68020 and 68030) and the iAPX 286 and 386 from Intel (80286 and 80386).

The process of selecting a microprocessor for a new product is complicated by a lack of compatibility between the leading 32-bit products. Because of its compatibility with software developed for earlier and very popular

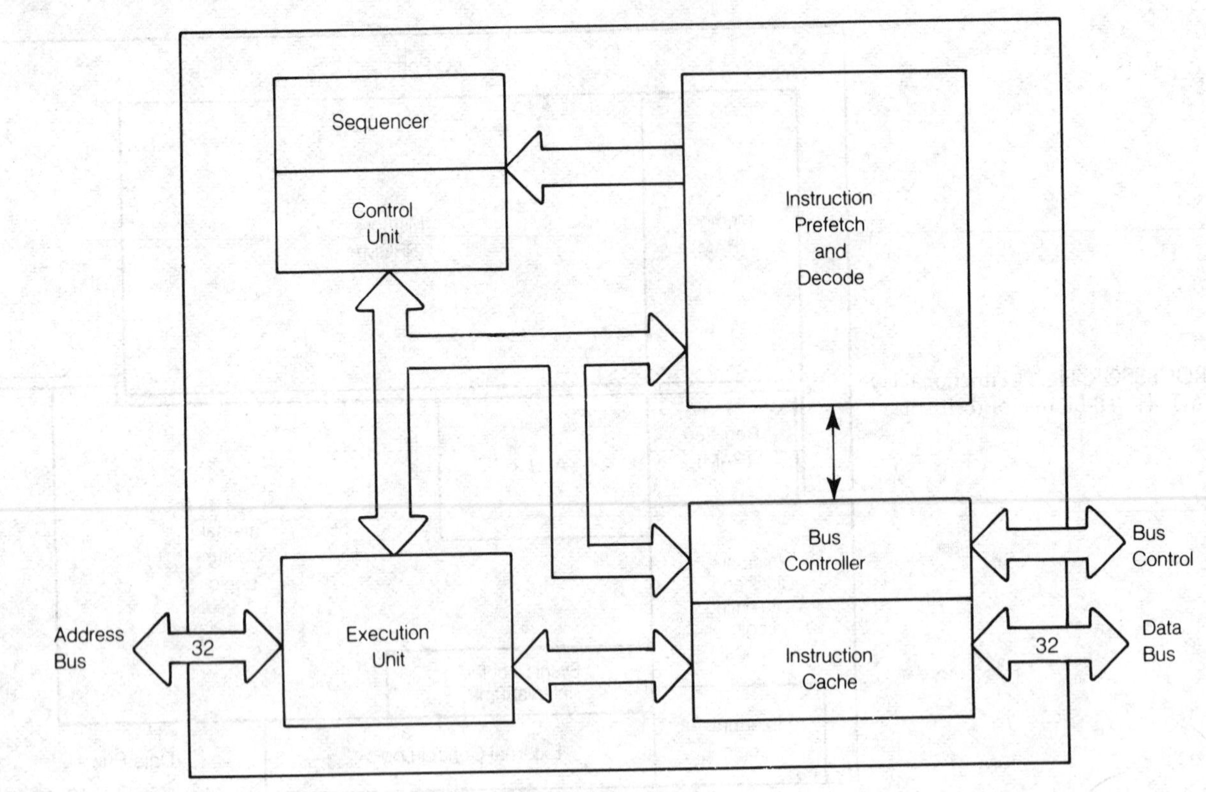

MICROPROCESSOR: Fig. 3. Block diagram of a single-chip 32-bit microprocessor.

Intel 8- and 16-bit microprocessors, many users are persuaded to stay with the Intel MPUs. However, the Motorola microprocessors have been successful in the important workstation marketplace and are compatible with the more versatile Unix operating system. They are attracting many users wishing to take advantage of this operating system that has benefits in intercomputer communications, graphics and multiprocessing (the sharing of a task by several machines). Efforts are underway to resolve differences between the processors and make them universally compatible. *See* OPERATING SYSTEM.

The first-generation microprocessors were able to manipulate 4-bit words and the next generation was able to handle 8-bit words. The ability of an MPU to handle longer 8-bit words or bytes made them more adaptable to communications applications where characters and symbols are encoded with 8-bits or bytes. The more advanced 8-bit MPUs require fewer supporting peripheral chips than the earlier models because more support functions have been integrated into the MPU. The early MPUs were fabricated with NMOS technology.

Sixteen-bit MPUs were introduced to handle two 8-bit words (bytes) at a time. The transition from NMOS technology to CMOS technology was made at the 16-bit level. In the next evolutionary step, 32-bit MPUs were introduced with 32-bit internal architecture. The first ones had a 16-bit external data bus, but they now have 32-bit external buses. The latest 32-bit MPUs have achieved the performance levels formerly attained only by CPUs in mainframe computers. Some 32-bit MPUs are capable of 25 MHz cycle times.

New MPUs under development are incorporating reduced instruction set (RISC) architecture. (*See* REDUCED INSTRUCTION SET COMPUTER). Multiple MPUs are being installed in new hierarchical computer designs for parallel operation to shorten computing time. *See* COMPUTER ARCHITECTURE.

Variable word length computers are being built by what is known as bit-slice architecture. In this design, the microprocessor consists of a common control section made up of read-only memory for decoding instructions and controlling the microprocessor and a functional unit called registers and an arithmetic and logic unit (RALU) consisting of separate monolithic chips called bit slices. The bit slices are identical 2-bit or 4-bit units connected in parallel so the processors may have different word lengths. The modular bit-slice approach permits the formation of processors with from 4 to 32 bits. Bit-slice architecture is used for bipolar microprocessors because of the improved heat dissipation for these high-density bipolar circuits.

MICROPROGRAM

A microprogram is a set of microinstructions at the machine-language level. It results in the execution of a specific function, independent of the functions of the program being run. The microprograms implement routine operating functions in a computer. Microprograms can be permanently placed in a computer or microcomputer in the form of firmware. *See also* FIRMWARE.

MICROSTRIP

Microstrip is a form of unbalanced transmission line. A flat conductor is bonded to a ground-plane strip by means of a dielectric material (see illustration). This results in an image conductor, parallel to the flat wire conductor, but on the other side of the ground-plane strip. The effective current in the image conductor is equal in magnitude to the current in the actual conductor, but flows in the opposite direction. Thus, very little radiation occurs (in theory) from a microstrip transmission line.

Microstrip lines are usually used at ultrahigh and microwave frequencies. Microstrip lines have somewhat lower loss than coaxial lines, but radiate less than most open-wire lines at the short wavelengths. The characteristic impedance of the microstrip line depends on the width of the flat wire conductor, the spacing between the flat wire conductor and the ground plane, and on the type of dielectric material used.

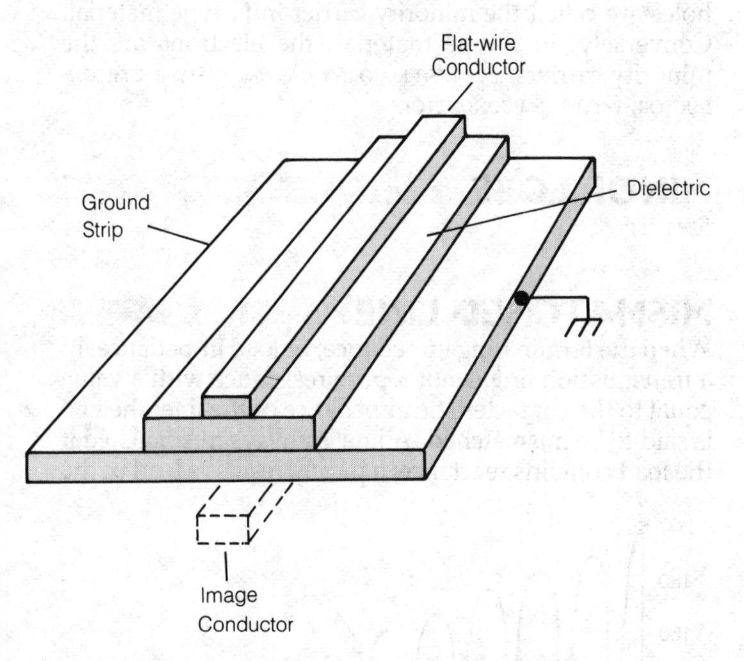

MICROSTRIP: A microstrip is a type of unbalanced transmission line for use at microwave frequencies.

MICROWAVE

Microwave refers to that part of the electromagnetic spectrum in which the wavelength is between about 1 millimeter and 30 centimeters. The microwave frequencies range from approximately 1 to 300 GHz.

Microwaves are very short electromagnetic radio waves, but they have longer wavelengths than infrared energy. Microwaves travel in essentially straight lines through the atmosphere, and are not affected by the ionized layers.

Microwave frequencies are used for military and commercial radar and television links. In a radio or television broadcasting system, the studio is usually at a different location than the transmitter; a microwave link connects them. Satellite communication and control is generally accomplished at microwave frequencies. The microwave region contains a vast amount of spectrum space, and can therefore hold many wideband signals.

Microwave radiation can cause heating of certain materials. This heating can be dangerous to human beings when the microwave radiation is intense. When working with microwave equipment, care must be exercised to avoid exposure to the rays. The heating of objects by microwaves has been put to practical use in industry and the home and restaurant microwave oven. *See also* MICROWAVE OVEN, RADAR.

MICROWAVE OVEN

A microwave oven is an appliance for cooking food by microwave heating. Organic molecules become agitated when subjected to microwave energy, and the material temperature rises. The more intense the microwave energy, the greater the rise in temperature for a given organic substance at a given frequency.

Microwave ovens make is possible to cook food in a fraction of the time required by a conventional oven. The microwave and heat source of microwave ovens is the magnetron. Because the microwave energy penetrates homogeneous food with virtually no attenuation, the food is uniformly heated, resulting in even cooking. Caution must be exercised when using a microwave oven because any metallic items inside the oven will result in microwave interference patterns that could cause arcing and fire. *See also* MAGNETRON, MICROWAVE.

MICROWAVE REPEATER

A microwave repeater is a receiver/transmitter combination for relaying signals at microwave frequencies. Basically, the microwave repeater works in the same way as a repeater at any lower frequency. The signal is intercepted by a horn or dish antenna, amplified, converted to another frequency, and retransmitted (see illustration on p. 556).

Microwave repeaters are used in long-distance overland communications links. With the aid of repeaters, microwave links supplant wire-transmission systems. *See also* REPEATER.

MIDRANGE

The middle portion of the audio-frequency spectrum is sometimes called the midrange. Although the lower and upper limits of the midrange are not specifically defined, midrange is considered to be the minimum band of frequencies necessary for the understanding of a human voice: about 300 Hz to 3 kHz.

In a high-fidelity system, a speaker must respond well to the midrange frequencies as well as lower and higher frequencies. However, the midrange should be in the proper proportion with respect to the lower and higher frequencies. Some high-fidelity speaker combinations have separate transducers for the low, midrange, and high frequencies. *See also* BASS, TREBLE.

MIL

A mil is a small unit of linear measure, equal to 0.001 inch or approximately 0.0254 millimeter.

Cross-sectional area is specified in circular mils, however. An area of 1 circular mil is equal to the area of a circle

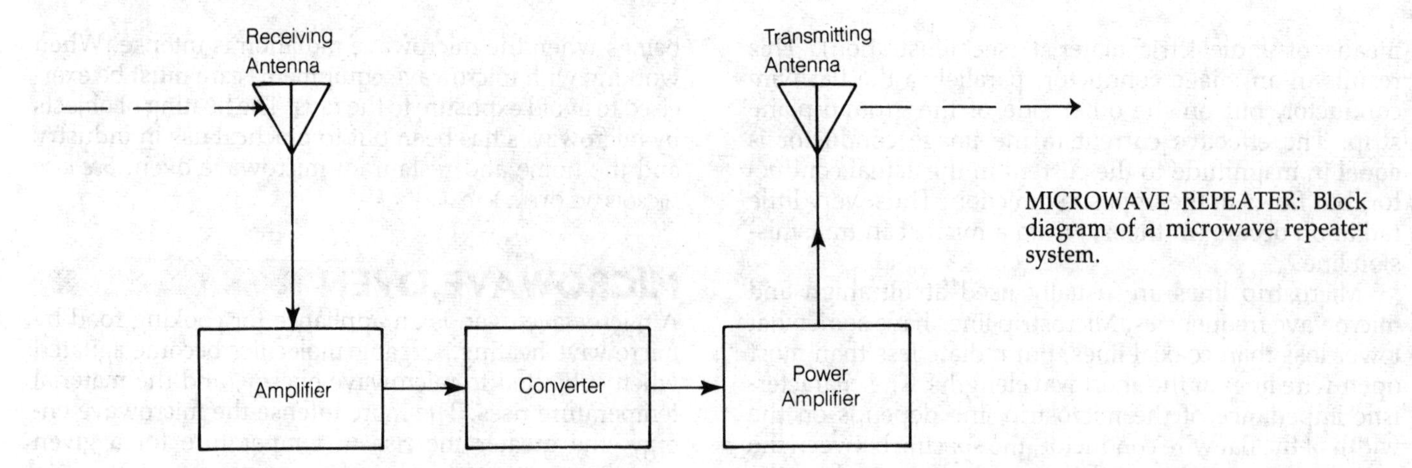

MICROWAVE REPEATER: Block diagram of a microwave repeater system.

whose diameter is 1 mil. Therefore, 1 circular mil is about 0.785 square mil, or 0.000507 square millimeter.

MILLER EFFECT

A transistor, field-effect transistor, or tube exhibits variable input capacitance under conditions of changing direct-current input bias. In general, an active amplifying device shows smaller input capacitance when it is biased at or beyond cutoff, as compared with bias below the cutoff point. This change in capacitance is called the Miller effect.

The Miller effect can result in rapidly alternating input impedance to a stage driven into the cutoff region during the alternating-current input cycle. For example, in a Class-B amplifier circuit, the input capacitance normally increases with increasing drive. This can result in nonlinearity unless the circuit is designed with the Miller effect in mind. *See also* INPUT IMPEDANCE.

MILLER OSCILLATOR

See OSCILLATOR CIRCUITS.

MILLI-

Milli- is a prefix multiplier meaning 0.001 or 1/1,000. When this prefix is added to a quantity, the magnitude is decreased by a factor of 1,000. Thus, for example, 1 millimeter is equal to 0.001 meter, and 1 milliampere is equal to 0.001 ampere. *See also* PREFIX MULTIPLIERS.

MINICOMPUTER

See COMPUTER.

MINORITY CARRIER

Semiconductor materials carry electric currents in two ways. Electrons transfer negative charge from the negative pole to the positive pole; holes carry positive charge from the positive pole to the negative pole (*see* ELECTRON, HOLE).

Electrons and holes are known as charge carriers. In an N-type semiconductor material, most of the charge is carried by electrons, and relatively little by holes. Thus, holes are called the minority carrier in N-type material. Conversely, in P-type material, the electrons are the minority carriers. *See also* MAJORITY CARRIER, N-TYPE SEMICONDUCTOR, P-TYPE SEMICONDUCTOR.

MINOR LOBE

See LOBE.

MISMATCHED LINE

When the terminating impedance, or load impedance, of a transmission line is not a pure resistance with a value equal to the characteristic impedance of the line, the line is said to be mismatched. A line is always mismatched if the load contains reactance; a purely resistive load of the

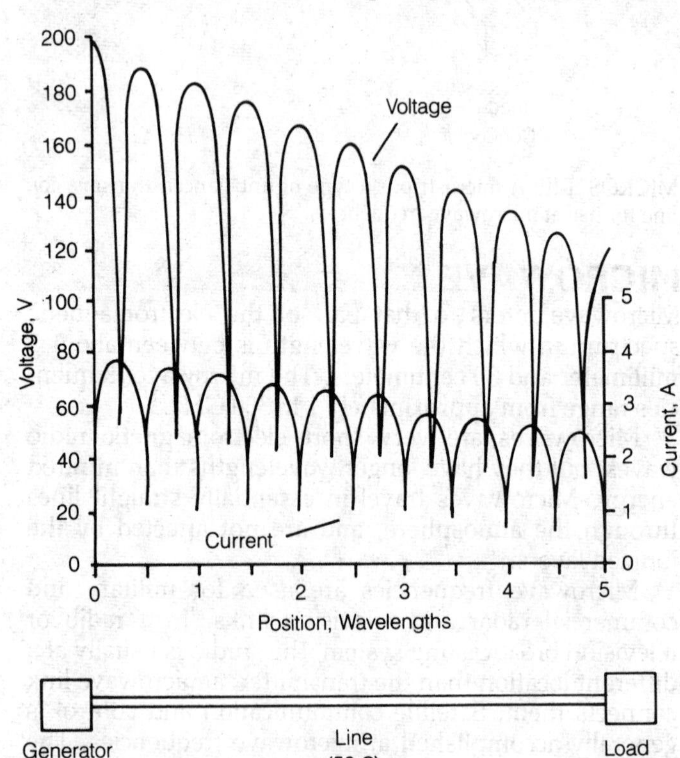

MISMATCHED LINE: In a mismatched transmission line voltage and current vary along its length, resulting in standing waves.

wrong value can also present a mismatch to a transmission line.

A mismatched line may degrade the performance of a system significantly. In a mismatched line, the voltage and current do not exist in uniform proportion. The current and voltage, instead, occur in a pattern of maxima and minima along the line (see illustration). Current minima occur at the same points as voltage maxima, and current maxima correspond to voltage minima. This pattern of current and voltage is known as a standing-wave pattern. Standing waves result in increased conductor and dielectric heating in a transmission line, causing power to be dissipated as heat. The efficiency of the system is then reduced.

Small mismatches in a transmission line can often be tolerated with little or no adverse effects. However, large mismatches can cause problems such as reduced generator efficiency, increased line loss, and overheating of the line. The degree of the mismatch is called the standing-wave ratio. *See also* STANDING-WAVE RATIO, STANDING-WAVE-RATIO LOSS.

MIXER

A mixer is a circuit that combines two signals of different frequencies to produce a third signal. The frequency of the third signal is either the sum or difference of the input frequencies. Usually, the difference frequency is used as the mixer output frequency. Mixers are widely used in superheterodyne receivers and in radio transmitters (*see* SUPERHETERODYNE RADIO RECEIVER).

A mixer requires some kind of nonlinear circuit element to function properly. The nonlinear element can be a diode or combination of diodes; this circuit is shown in the illustration at A. The diode mixer is a passive circuit because it does not require an external source of power. However, there is some insertion loss when a mixer is used.

Mixers can have gain; at B, a bipolar-transistor mixer is shown. At C, a dual-gate metal-oxide-semiconductor field-effect transistor (MOSFET) is shown. At D, a tetrode vacuum tube is shown.

The mixer operates on a principle similar to that of an amplitude modulator. In the mixer, both signals are usually in the radio-frequency spectrum, but in a modulator, one signal is much lower in frequency than the other (*see* MODULATOR). The output of the mixer circuit is tuned to the sum or difference frequency, as desired.

Mixing may occur in a circuit even when it is not desired. This is likely in high-gain receiver front-end amplifiers. Mixing can also occur between a desired signal and a parasitic oscillation in an amplifier or oscillator. Mixing sometimes takes place in semiconductor devices when two or more strong signals are present. This mixing is called intermodulation, and the resulting unwanted signals are called mixing products.

A device used for combining two or more audio signals in a broadcast or recording studio is known as a mixer. In this type of device, nonlinear operation, with the consequent generation of harmonics and distortion

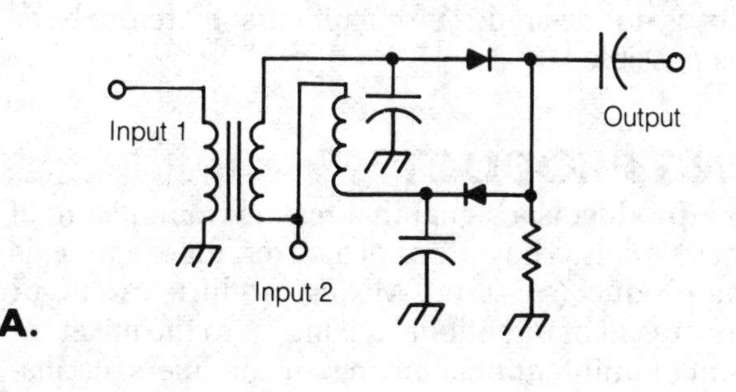

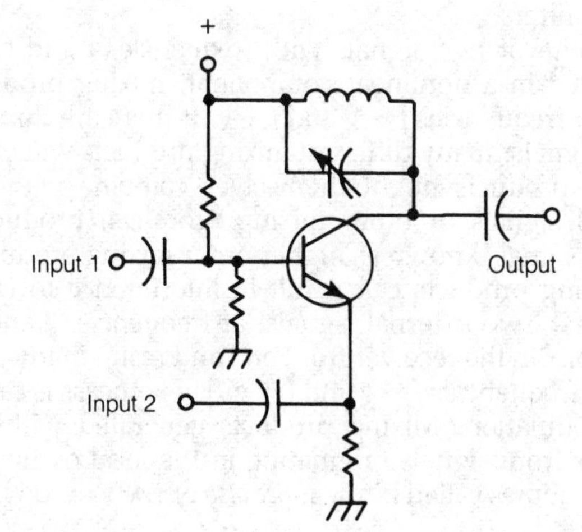

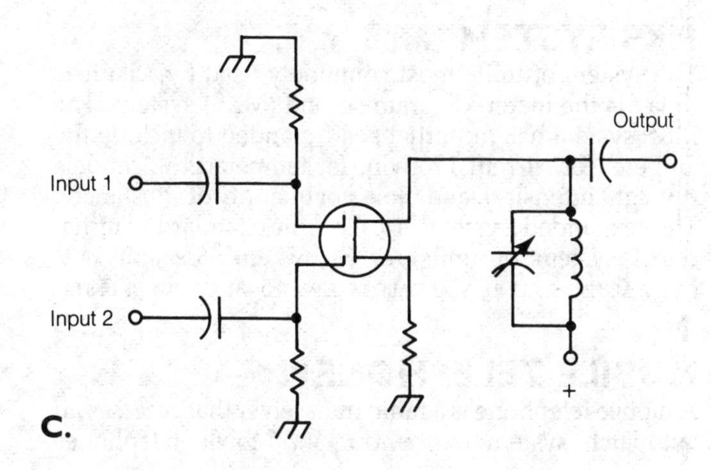

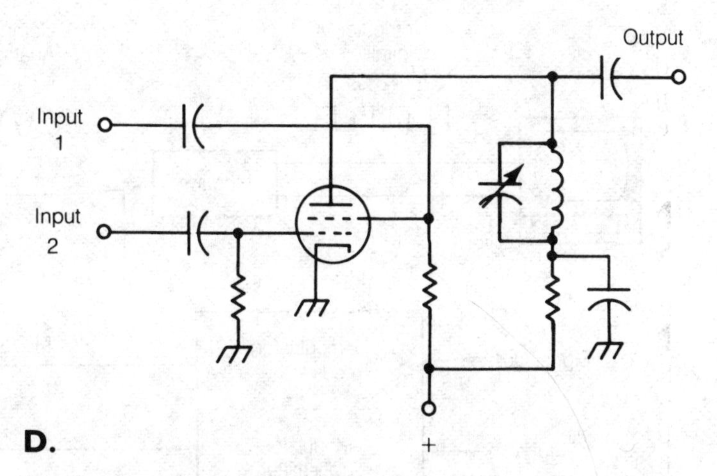

MIXER: A passive mixer circuit (A), a bipolar-transistor active mixer circuit (B), and a mixer employing a dual-gate MOSFET (C) and a mixer using a tetrode tube (D).

products, is not desired. The circuit must therefore be as linear as possible.

MIXING PRODUCT

A mixing product is a signal that results from mixing of two other signals. The output of a mixer, for example, is a mixing product (*see* MIXER). Mixing products can occur either as a result of intentional mixing, as in the mixer, or as a result of unintentional mixing, in amplifiers, oscillators, or filters.

Whenever two signals with frequencies *f* and *g* are combined in a nonlinear component, mixing products occur at frequencies $f - g$ and $f + g$. If there are three or more signals, many different mixing products will exist. Mixing products might themselves combine with the original signals or other mixing products, producing further signals known as higher-order mixing products.

Mixing products can result in interference to radio receivers. Two external signals, at frequencies *f* and *g*, might mix in the receiver front end and result in interference at frequencies $f - g$ and $f + g$. This process is called intermodulation. Mixing products generated within a receiver from signals originating in the local oscillators are sometimes called birdies. *See also* INTERMODULATION.

MKS SYSTEM

The system of units most commonly used by scientists today is the meter-kilogram-second (MKS) system. The MKS system has recently been expanded to include the ampere (for current), Kelvin (for temperature), candela (for light intensity), and mole (for quantity of substance). This expanded system is called the standard international system of units, or SI system. *See also* KILOGRAM, METER, SECOND, STANDARD INTERNATIONAL SYSTEM OF UNITS.

MOBILE TELEPHONE

A mobile telephone is a radio transceiver that accesses an autopatch system (*see* AUTOPATCH). Mobile telephones originally did not permit continuous two-way conversation; neither party could interrupt the other. However, mobile telephones now allow true duplex operation.

Mobile and portable telephones generally operate in the very high frequency and ultrahigh frequency radio bands, or between about 30 MHz and 3 GHz. This provides reliable operation within a radius of several miles from the home station or repeater system.

Mobile telephones are of obvious value to businesses, and can be useful in emergency situations as well.

A special form of portable telephone consists of a small handheld unit and a base station, with an operating range of several hundred feet. This telephone is called a cordless or wireless telephone. It operates as a short-range transceiver only between the handset and the base station. Messages received at and sent from the base station use the public dial-up telephone system to complete the call. The number of different frequencies for linking the handset to the base station is limited. The radio link can be overheard by other receivers, so cordless telephones do not provide secure communications links. *See also* CELLULAR MOBILE RADIO TELEPHONE, CORDLESS TELEPHONE.

MOBILITY

See CARRIER MOBILITY.

MODE OF EMISSION

See EMISSION CLASS.

MODEM

A modem is a circuit that uses digital data to modulate and demodulate a carrier wave so that the digital data can be transmitted over an analog communications line, typically between computers. Figure 1 is a block diagram of a modem. The modulation technique alters three properties of carrier waves with respect to time: frequency, amplitude, and phase. The receiving modem

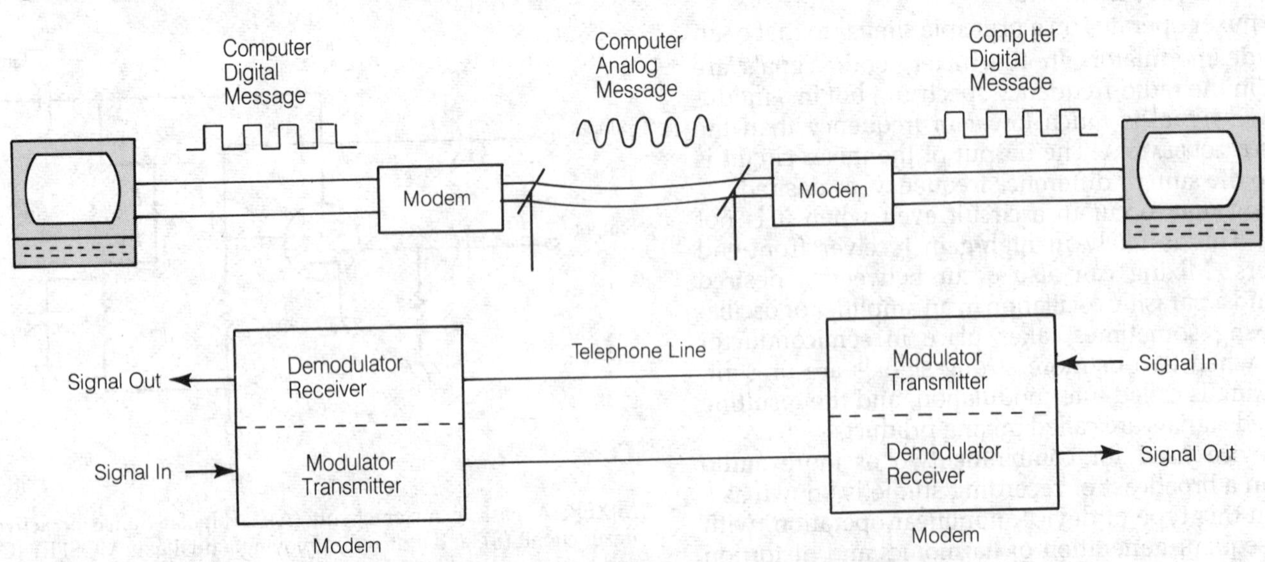

MODEM: Fig. 1. Modem conversion of digital to analog signals for transmission over the telephone line.

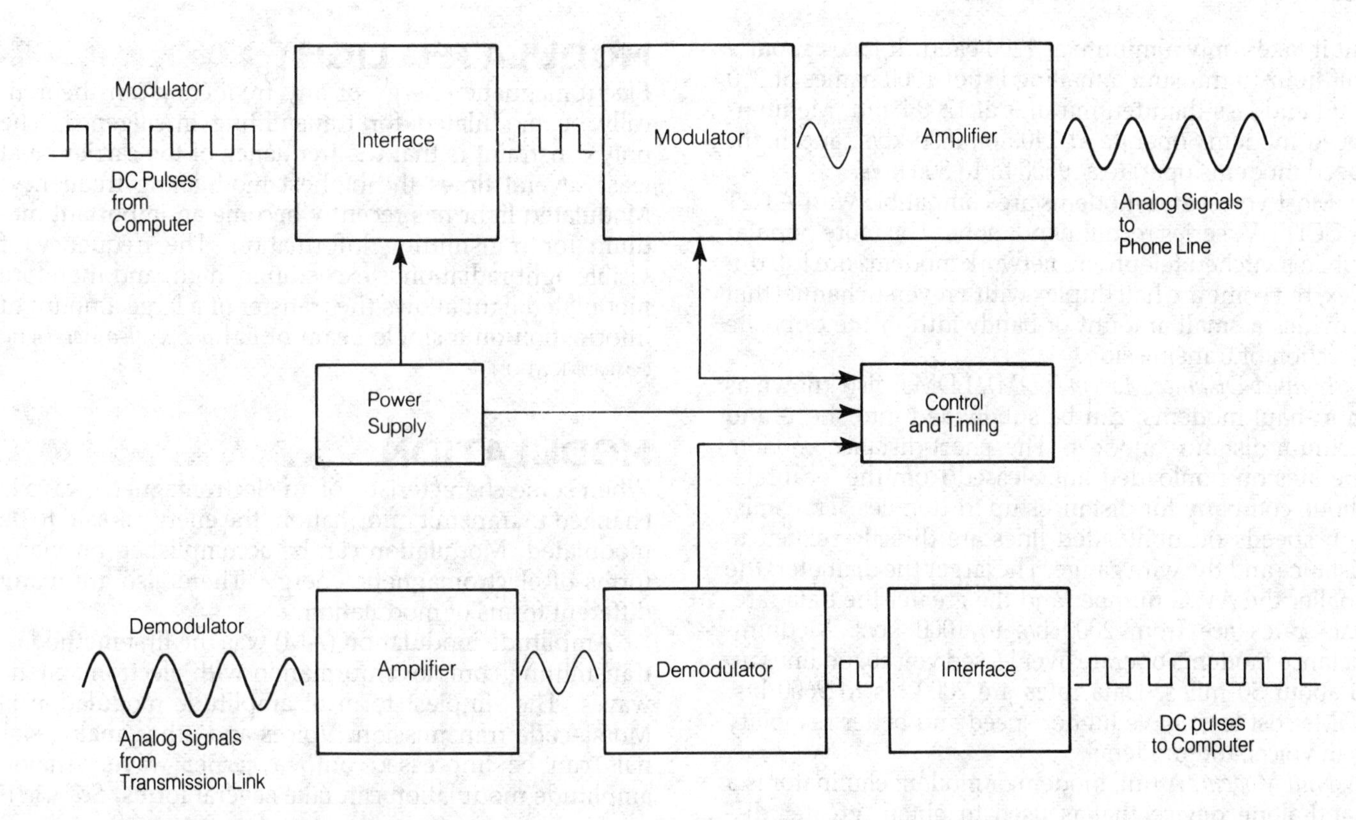

MODEM: Fig. 2. Basic functions of a computer modem.

reconverts the modulated carrier to digital data for use or storage by another computer, terminal or receiving device.

Modems accept a binary serial input from the transmitter and provide a binary serial output to the receiver. To control the communications, the modem and the data terminal or computer exchange selected digital signals across an interface. These signals are referred to as modem controls. The functions performed include line turnaround, control of a shared carrier, detection of a carrier, and timing-rate selection. The basic functions of a modem are shown in Fig. 2.

Modems are classified into categories that differ in operating variables and performance according to the communications channels used. Some of these categories are:

- Voiceband modems: public switched telephone networks and leased lines with 0 to 4 kHz bandwidth.
- Limited-distance modems (also called short-haul modems).
- Null modems (modem eliminators or line drivers).
- Wideband modems.

Voiceband Modem. Voiceband modems are the most popular class of modem. They are used on voice-grade lines with an operational frequency range of 300 to 3400 Hz (the actual frequency allocation is 0 to 4 kHz). These modems operate over public switched telephone lines (where they are called dial-up modems) or over 2- or 4-wire private leased lines.

Acoustic-Coupler Modem. Acoustic-coupler-modems are the simplest and lowest cost modems. They permit the user to dial a telephone number, listen for a signal from the computer on the other end of the telephone line, and then press the telephone receiver into two cushioned cups on the acoustic coupler.

Direct-Connect Modem. The acoustic coupler is rapidly being replaced by direct-connect modems, which reduce noise interference by plugging directly into the telephone wall outlet. Direct-connect modems provide automatic answer and/or automatic calling capability. Also, calls can be answered and placed manually through a telephone. The desired number is dialed by typing it on the computer keyboard. These modems take two forms: the free-standing modem and the internal or board modem.

The free-standing modem is a book-size box that hooks up to a serial port at the back of the computer. It has the advantage of being compatible with any computer. In addition, it can be reset or adjusted without opening the computer. Some early direct connect modems for personal computers were modules that plugged into the back of the computer without the need to open a card cage. But most are made as circuit boards designed to plug into a backplane within the computer housing. They are designed specifically for certain computer models. The most popular dial-up modem today is the AT&T 212A, which is also made by most independent modem manufacturers.

Most personal computer modems relay information at one of two rates: 300 or 1200 baud (signals per second) (*see* BAUD RATE). This is the equivalent of 0.3 and 1.2 kilobits per second (Kb/s) and they are termed low-speed voiceband modems. It takes slightly over four and one-half minutes to transmit the equivalent of five double-spaced typed pages from one computer to another at 300 baud, but it takes only a minute at 1200 baud. It takes about a half hour to transmit a mailing list of 1000 names at 300 baud and less than ten minutes at 1200 baud. Medium-speed modems operate at 2400 or 4800 kb/s, and high-speed modems operate at 9600 to 16,800 kb/s.

Most voiceband modems are compatible with AT&T or CCITT V. series recommendations. The more popular public-switched telephone network modems are full duplex, but some are half duplex with a reverse channel that provides a small amount of bandwidth in the opposite direction of transmission.

Limited-Distance Modem (LDM). LDMs, also known as short-haul modems, can be subdivided into short- and medium-distance models. The short-distance modem operates on nonloaded lines leased from the local telephone company for distances up to 10 miles. Transmission speeds on nonloaded lines are directly related to distance and the wire gauge. The larger the diameter, the smaller the AWG number and the greater the data rate. Data rates are from 2000 b/s to 1000 kb/s. Medium-distance modems operate over leased voiceband lines up to about 50 miles. Data rates are 2000 b/s to 9600 b/s. LDMs cost less, have higher speed and better reliability than voiceband modems.

Null Modem. A null modem or modem eliminator is a stand-alone device that is used to obtain greater distances between terminals and computers. The EIA RS-232C physical interface limits direct attachment to 15 meters. The null modem is placed between the terminal and computer to regenerate the digital signals in each direction. In effect, it is a digital repeater that presents a low-impedance output to prevent signal rise and fall times from degrading appreciably. In private-wire cable, data rates are 2 kb/s to 2 Mb/s.

Wideband Modem. Wideband modems operate over group facilities that bundle together several analog voice-grade circuits. A single group takes up to 12 voice-grade circuits to provide a 48 kHz bandwidth. Fiberoptic, radio-frequency, cable TV, and satellite modems operating up to 5 or 6 Mb/s are considered to be special wideband modems.

MODULAR CONSTRUCTION

In the modular method of construction, individual circuit boards are used. Each circuit board contains the components for a certain part or parts of the system. The circuit boards are entirely removable, usually with a simple tool resembling a pliers. Edge connectors facilitate easy replacement. The edge connectors are wired together for interconnection between circuit boards.

Modular construction has greatly simplified the maintenance and servicing of complicated apparatus. In-the-field repair consists of identification, removal, and replacement of the faulty module. The faulty circuit board is then sent to a central repair facility, where it can be fixed using highly sophisticated equipment. Once the faulty board has been repaired, it is ready to serve as a replacement module in another system.

MODULATED LIGHT

Electromagnetic energy of any frequency can theoretically be modulated for transmitting intelligence. The only constraint is that the frequency of the carrier be at least several times the highest modulating frequency. Modulated light has recently become an important medium for transmitting information. The frequency of visible light radiation is exceedingly high, and therefore modulated light allows the transfer of a large amount of information on a single beam of light. *See also* FIBEROPTIC COMMUNICATION.

MODULATION

When some characteristic of an electromagnetic wave is changed to transmit information, the energy is said to be modulated. Modulation can be accomplished on many forms of electromagnetic energy. There also are many different forms of modulation.

Amplitude modulation (AM) was the first method of transmitting complex information with electromagnetic waves. The simplest form of amplitude modulation is Morse-code transmission. Voices and other analog signals can be impressed onto a carrier wave. Analog amplitude modulation can take several forms. *See* AMPLITUDE MODULATION, DOUBLE SIDEBAND, SINGLE SIDEBAND.

Frequency modulation (FM) is another common method of conveying intelligence with electromagnetic waves. Phase modulation works in a similar way. *See* FREQUENCY MODULATION, PHASE MODULATION.

A more recent development has been the technique of pulse modulation. There are several ways of modulating signal pulses; the amplitude, frequency, duration, or position of a pulse may be modified to achieve modulation. *See* PULSE MODULATION.

To recover the intelligence from a modulated signal, some method is needed to separate the modulating signal from the carrier wave at the receiving end of a communications circuit. This process is called detection. Different forms of modulation require different processes for signal detection. *See* DETECTION, DISCRIMINATOR, ENVELOPE DETECTOR, HETERODYNE DETECTOR.

In general, if a characteristic of an electromagnetic wave can be made to change rapidly enough to convey intelligence at the desired rate, then that parameter offers a means of modulation. The more data that must be transmitted in a given amount of time, however, the more difficult it becomes to modulate efficiently an electromagnetic wave in a particular manner. *See also* EMISSION CLASS, MODULATOR.

MODULATION COEFFICIENT

The modulation coefficient is a specification of the extent of amplitude modulation of an electromagnetic wave. The modulation coefficient is abbreviated by the small letter m, and it can range from a value $m = 0$ (for an unmodulated carrier) to $m = 1$ (for 100-percent or maximum distortion-free modulation).

Let E_c be the peak-to-peak voltage of the unmodulated carrier wave. Let E_m be the maximum peak-to-peak voltage of the modulated carrier. Then:

$$m = \frac{(E_m - E_c)}{E_c}$$

Theoretically, m can be larger than 1, but this represents overmodulation, and distortion inevitably occurs. *See also* AMPLITUDE MODULATION, MODULATION PERCENTAGE.

MODULATION ENVELOPE

See ENVELOPE.

MODULATION INDEX

In a frequency-modulated transmitter, the modulation index is a specification that indicates the extent of modulation. Frequency modulation differs from all forms of amplitude modulation in the way it must be measured. There is a limit to the extent a signal can be usefully amplitude-modulated, but the only limitation on frequency deviation is imposed by the requirement that frequency cannot be lower than zero at any given instant. For all practical purposes, this limitation is meaningless. If the maximum instantaneous frequency deviation (*see* DEVIATION) of a carrier is f kHz, and the instantaneous audio modulation frequency is g kHz with the modulation index identified by the letter m, then:

$$m = f/g$$

For example, if the frequency deviation is plus or minus 5 kHz and the modulating frequency is 3 kHz, then $f = 5$ and $g = 3$; therefore:

$$m = 5/3 = 1.67$$

The modulation index is a useful means of measuring frequency modulation when a pure sine-wave tone and constant deviation are used. However, with a voice signal, this is not true and a more general form of the modulation index must be used. This specification is called the deviation ratio. The deviation ratio d is given by:

$$d = f/g$$

where f is the maximum instantaneous deviation and g is the highest audio modulating frequency, both specified in kilohertz. *See also* FREQUENCY MODULATION.

MODULATION PERCENTAGE

In an amplitude-modulated signal, the modulation percentage is a measure of the extent to which the carrier wave is modulated. A percentage of zero refers to the absence of amplitude modulation. A percentage of 100 represents the maximum modulating-signal amplitude that can be accommodated without envelope distortion.

Suppose the unmodulated-carrier voltage of a signal is E_c volts, and that the peak amplitude (with maximum modulation) is E_m volts. Then the percentage of modulation, m, is given by:

$$m = 100\,\frac{(E_m - E_c)}{E_c}$$

Under conditions of 100-percent amplitude modulation, the peak power is four times the unmodulated power, regardless of the waveform of the modulating signal. The extent to which the average power increases over the unmodulated-carrier power, for a given modulation percentage, depends on the waveform. The average-power increase with 100 percent sine-wave modulation is approximately 50 percent. For a human voice, the average power increases by only about half that amount. Generally, amplitude modulation levels of more than 100 percent are not used is because negative-peak clipping occurs under these conditions. This results in distortion of the signal and unnecessary bandwidth. *See also* AMPLITUDE MODULATION.

MODULATOR

A modulator is a circuit that combines information with a radio-frequency carrier for the purpose of transmission. Different forms of modulation require different kinds of modulator circuits.

The simplest modulator is the telegraph key for producing code transmissions. Code is a form of amplitude modulation. More complex amplitude modulation is produced by a circuit such as that shown in the illustration. This circuit is essentially an amplifier with variable gain. The gain is controlled by the input signal, in this case an audio signal from the microphone and audio amplifier.

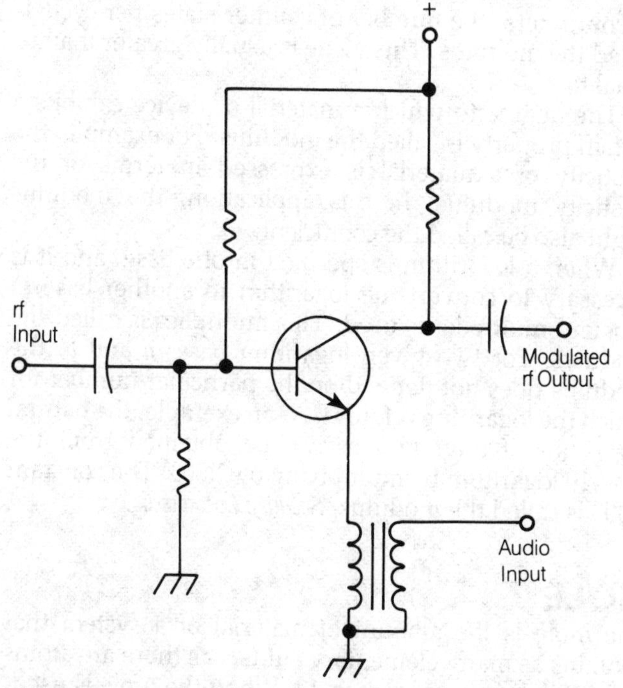

MODULATOR: A simple circuit for low-level amplitude modulation.

All amplitude modulators work according to this basic principle, although the details vary. *See* AMPLITUDE MODULATION.

A special type of amplitude modulator is designed to eliminate the carrier wave, leaving only the sideband energy. This circuit is called a balanced modulator. *See* BALANCED MODULATOR, DOUBLE SIDEBAND, SINGLE SIDEBAND.

Frequency or phase modulation requires the introduction of a variable reactance into an oscillator circuit. This can be done in various ways. The variable reactance is controlled by the modulating signal, and causes the phase and/or resonant frequency of the oscillator to fluctuate. *See* FREQUENCY MODULATION, PHASE MODULATION, REACTANCE MODULATOR.

Pulse modulation is obtained by circuits that cause changes in the amplitude, timing, or duration of high-powered radio-frequency pulses. The circuit details depend on the type of pulse modulation used. *See* PULSE MODULATION.

MODULATOR-DEMODULATOR

See MODEM.

MODULUS

The term *modulus* can have any of several different meanings in electricity and electronics.

The absolute value or magnitude of an impedance is sometimes called the modulus. An impedance generally consists of a resistive component and a reactive component (*see* IMPEDANCE), both specified in ohms. If the resistance is denoted by R and the reactance by X; then the modulus impedance Z is given by the equation:

$$Z = \sqrt{R^2 + X^2}$$

In computers, the number of counter states per cycle is called the modulus. This value is usually greater than or equal to 1.

The degree to which a material or device exhibits a certain property is called the modulus. For example, the elasticity of a material is expressed in terms of the elasticity modulus. In this application, the modulus might also be called the coefficient.

When a logarithm is specified in one base, and it is necessary to convert the logarithm to another base, a constant multiplier is used. This multiplier is called the modulus. For two given logarithm bases a and b, the modulus does not depend on the particular number for which the logarithm is found. As an example, the natural logarithm of any number can be obtained from the base-10 logarithm by multiplying by 2.303. The constant 2.303 is called the modulus. *See also* LOGARITHM.

MOLE

The mole is the amount of material of a system that contains as many elementary entities as there are atoms in 0.012 kilogram of carbon-12. When the mole is used, the elementary entities must be specified. They may be atoms, molecules, ions, electrons, or other particles. *See also* ATOMIC WEIGHT.

MOLECULE

A molecule is a group of atoms representing a substance as it occurs in nature. Every substance is made of molecules, and the molecules are always moving. Some molecules consist of just one atom. Others have thousands or even millions of atoms. Molecules are too small to see with ordinary optical microscopes although some molecules have been observed with the electron microscope.

An atom of oxygen consists of a nucleus with eight protons and eight neutrons. In the atmosphere, oxygen atoms tend to group themselves in pairs. Sometimes they group together in groups of three. A pair or triplet of oxygen atoms comprises an oxygen molecule in our air. Hydrogen atoms also tend to pair off; thus a normal hydrogen molecule consists of two hydrogen atoms. Oxygen and hydrogen atoms particularly like to group together; one oxygen atom joins with two hydrogen atoms (see illustration) to form the familiar water molecule. The foregoing examples are simple. Some molecules consist of complicated combinations of several different kinds of atoms. *See also* ATOM.

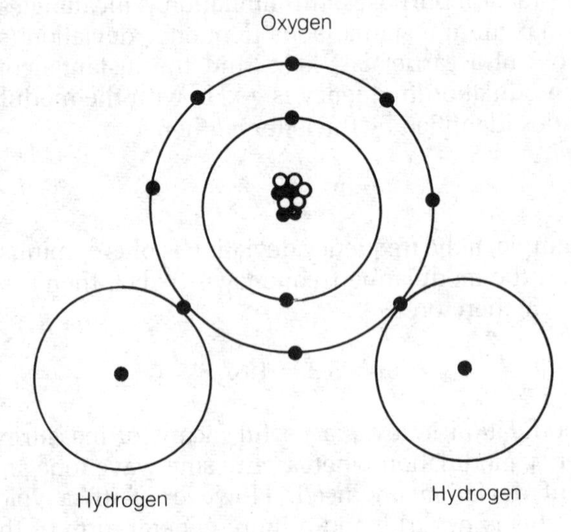

MOLECULE: A diagram of a water molecule.

MOLYBDENUM

Molybdenum is an element with an atomic number of 42 and an atomic weight of 96. Molybdenum occurs in nature as a metallic substance.

Molybdenum is used in the electrodes of certain vacuum tubes. Molybdenum alloys are used as ferromagnetic cores for audio coils. The permeability of molybdenum permalloy is comparable to that of powdered iron or ferrite.

MOMENT

Moment is a mathematical expression related to rotation around a point. Moment is determined by the product of a variable and a radial distance. For example, suppose a force F is applied to a lever of length d, as shown in the illustration. Then the product Fd, where F is given in newtons and d is given in meters, is called the moment of force.

In physics, moment is defined in many ways. For example, the moment of inertia is a measure of the inertia of a rotating or revolving object. For a magnet, the product of the pole strength and the distance between the poles is known as the magnetic moment.

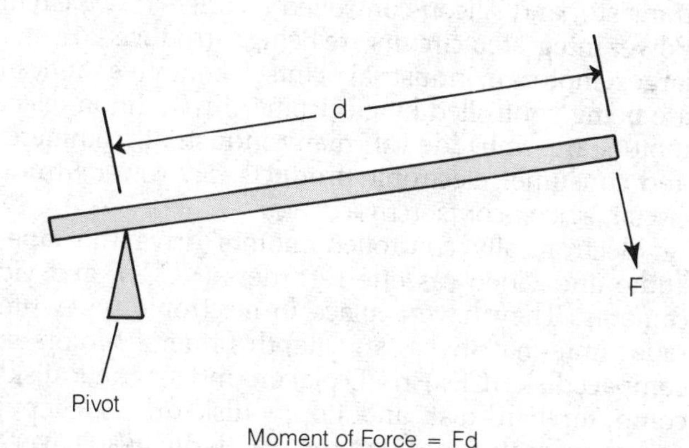

MOMENT: The moment of force is an angular quantity.

MONITOR

A monitor is an instrument that allows a signal to be analyzed. The oscilloscope is a common type of signal monitor (*see* OSCILLOSCOPE). However, more specialized types of monitors are used for analyzing the modulating waveform in an amplitude-modulated or frequency modulated signal or evaluating the spectral output of a radio transmitter.

In a personal computer a monitor is used to view the text or data in alphanumeric form. This monitor consists of a cathode-ray tube and associated circuitry, and looks like a television set. *See also* VIDEO DISPLAY TERMINAL.

MONOCHROMATICITY

Monochromaticity is the presence of light at only one wavelength, or within a single narrow band of wavelengths. A monochromatic light source has a definite color (hue), depending on the wavelength of the emission. The saturation may vary.

A laser emits highly monochromatic light because the output is essentially at only one wavelength. A color filter makes things appear monochromatic, although the transmitted light actually occupies a band of wavelengths. The greater the level of saturation of light, the more perfectly monochromatic it is. The precise level of saturation at which a light source is considered monochromatic is, however, not precisely defined. *See also* HUE, SATURATION.

MONOLITHIC CIRCUIT

See INTEGRATED CIRCUIT.

MONOPOLE

A single unit electric charge or group of electric charges is called a monopole if it is isolated in space and not accompanied by an equal and opposite charge center. Unlike a dipole, a monopole has a nonzero net charge.

The flux lines immediately surrounding an electric monopole are almost perfectly straight and they radiate outward in all directions from the charge center. The lines become curved near other electric charges. In theory, the electric lines of flux around a monopole extend indefinitely.

MONOPOLE ANTENNA

See CONICAL MONOPOLE ANTENNA, MARCONI ANTENNA in the ANTENNA DIRECTORY.

MONOSTABLE MULTIVIBRATOR

A monostable multivibrator is a circuit with only one stable condition. The circuit can be removed from this condition temporarily, but it always returns to that condition after a certain period of time. The monostable multivibrator is also called a one-shot multivibrator.

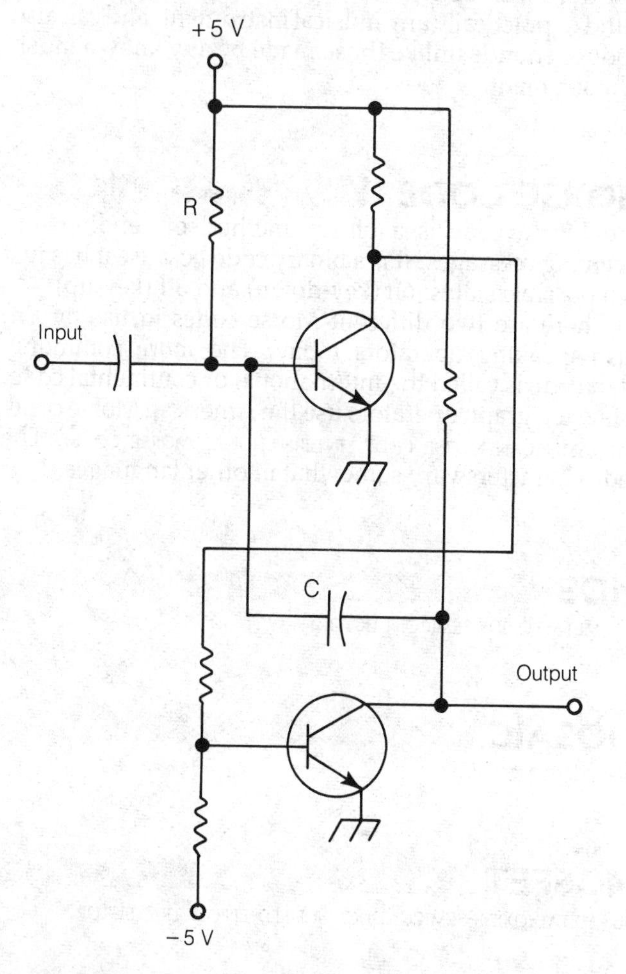

MONOSTABLE MULTIVIBRATOR: A monostable, or one-shot, multivibrator circuit.

The illustration is a simple schematic diagram of a monostable multivibrator. Normally, the output is high, at the level of the supply voltage (+5 V). When the positive triggering pulse is applied to the input, the output goes low (0 V) for a length of time that depends on the values of the timing resistor *R* and the timing capacitor *C*. If *R* is given in ohms and *C* is given in microfarads, then the pulse duration *T*, in microseconds, can be found by the equation:

$$T = 0.69RC$$

After the pulse duration time *T* has elapsed, the monostable multivibrator returns to the high state.

Monostable multivibrators are used as pulse generators, timing-wave generators, and sweep generators for cathode-ray tubes. *See also* MULTIVIBRATOR.

MOOG SYNTHESIZER

The Moog synthesizer is an electronic waveform generator that can produce almost any kind of periodic audio-frequency signal. A single audio-frequency tone can have thousands of different sound qualities, depending on the details of the waveform. A clarinet sounds different from a flute or a trumpet. The Moog can simulate the sound of practically any musical instrument, and can also produce sounds unlike those made by any known musical instrument.

MORSE CODE

The Morse code is a binary method of sending and receiving messages. It is a binary code because it has just two possible states: on (key-down) and off (key-up).

There are two different Morse codes in use by English-speaking operators today. The more commonly used code is called the international or continental code. A few telegraph operators use the American Morse code. (*See* AMERICAN MORSE CODE, INTERNATIONAL MORSE CODE.) The code characters vary somewhat in other languages.

MOS

See METAL-OXIDE SEMICONDUCTOR.

MOSAIC

See PHOTOMOSAIC.

MOSFET

See METAL-OXIDE-SEMICONDUCTOR FIELD-EFFECT TRANSISTOR.

MOTOR

An electric motor is a machine that converts electrical energy into rotational mechanical energy. The operation of a motor is based on the motor effect in which a force is experienced by a current-carrying conductor in a magnetic field. Figure 1 shows the principal parts of a simple direct-current (dc) motor with an eight-coil armature or rotor and a permanent magnet field. Motors are classified as alternating-current (ac) or dc powered.

Electronic circuits have been used to control electric motors for more than 50 years. First vacuum tube and thyratron circuits were used, and these were followed by transistor and silicon-controlled rectifiers (SCR) circuits. Power integrated circuits are being introduced. Even the largest motors in industrial plants, locomotives and ships are being controlled by electronics. In addition, electric motors are embedded in many industrial, commercial, and consumer electronic products. *See* POWER INTEGRATED CIRCUIT, SILICON-CONTROLLED RECTIFIER.

Electronically controlled motors drive the tape in audio and video cassette recorders (VCRs), and video cameras. They have a place in electronic typewriters, radar antenna drives, and depth finders. Motors spin compact disks (CDs) in CD players and magnetic disks of computer hard-disk and floppy-disk drives. Stepping and servo motors position the read write heads over the spinning disks. *See* DISK DRIVE.

Motors feed the paper in computer printers, X-Y plotters, and copying machines; they pump gas or fluids in scientific and medical instruments; they drive film in still and motion picture cameras, and they drive fans and blowers for the forced-air cooling of electronic circuitry.

The discussion in this section is limited to the fractional horsepower motors embedded in electronic equipment or motors that require electronics for operation and control.

DC Motor. Direct-current motors are widely used in portable electronics products because they run from either batteries or rectified ac. The dc motor is specified for many applications because direct current is the output of most power supplies for electronics. Motor speed can be controlled by adjusting the field or armature voltage.

Improvements in motor design, permanent magnets, and insulation materials have made dc motor size reduction possible, even in miniature frame sizes, without lowering the output torque. (Torque is the force that tends to produce rotation or twisting.) Direct-current motors with permanent magnets, as shown in Fig. 1, are popular in electronic application because of their simplicity. However, if a dc motor has a field winding instead of permanent magnets, the armature and field windings are connected in series, parallel (shunt) or a combination of these known as compound wound. The armature and winding are supplied with current from the same dc source.

Shunt-Wound DC Motor. A shunt-wound dc motor has its field and armature connected in shunt across the power supply, as shown in Fig. 2A. Speed is simply controlled by adjusting the armature voltage, and the field voltage remains constant. These motors can be reversed by reversing the armature voltage. Figure 2B is a typical performance curve for a shunt-wound motor. It

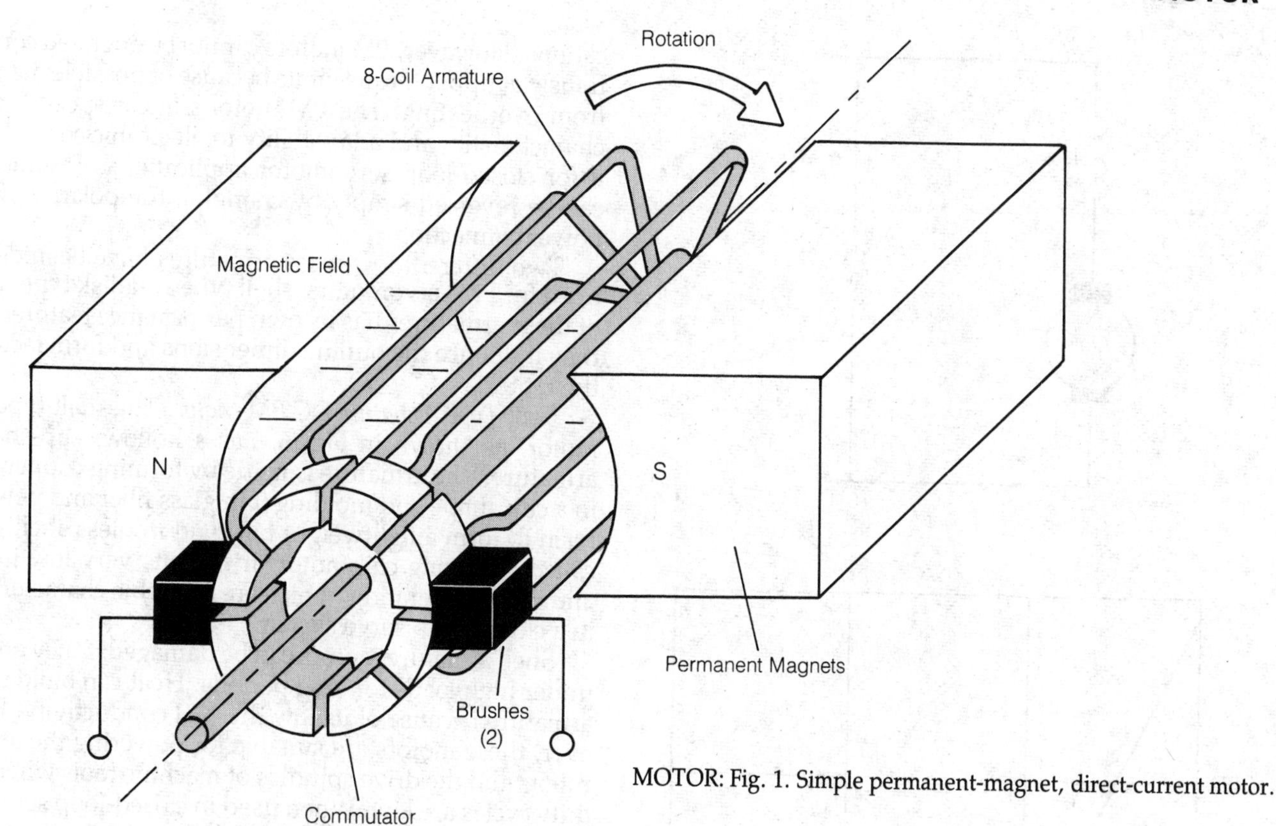

MOTOR: Fig. 1. Simple permanent-magnet, direct-current motor.

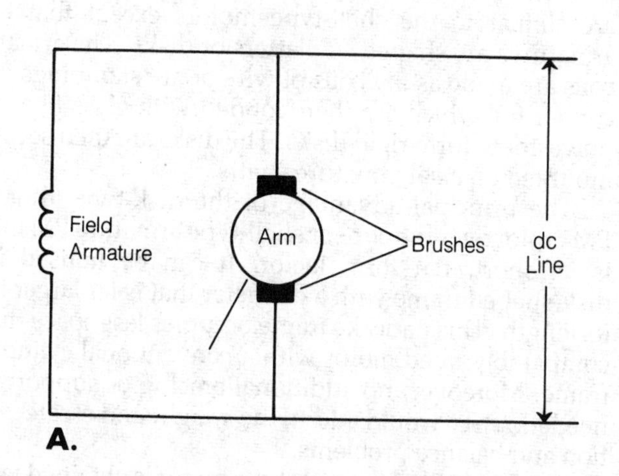

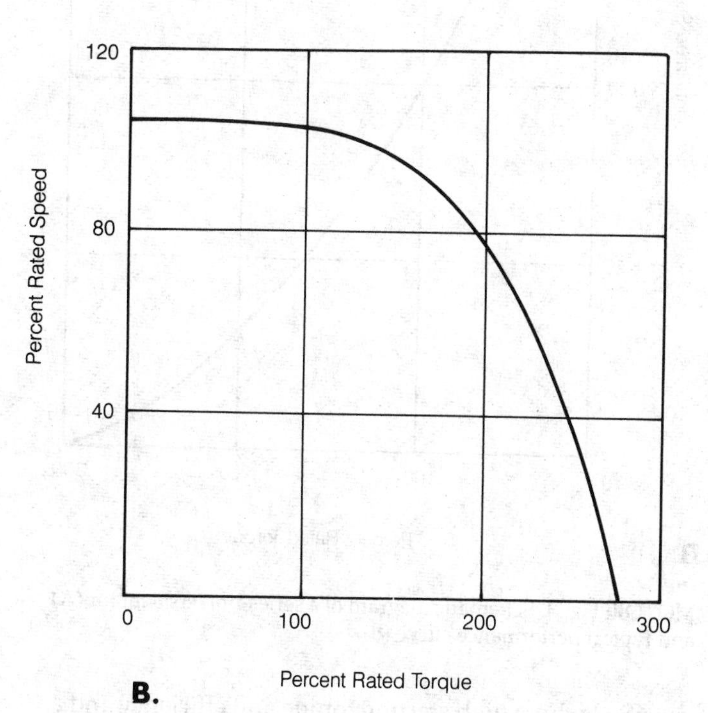

MOTOR: Fig. 2. Schematic diagram of a shunt-wound, direct-current motor (A) and typical performance curve (B).

shows constant speed until the torque exceeds 100 percent.

Series-Wound DC Motor. A series-wound shunt dc motor has its field and armature connected in series across the power supply as shown in Fig. 3A. This configuration offers high speed, high starting torque and wide drive capability. This motor is also called a universal motor because it runs with either ac or dc power, although it runs slower on ac power. Series-wound motors can run at more than 3,600 revolutions per minute (rpm) from a single-phase ac supply. The main drawback of the motor is shown in Fig. 3B: speed falls off sharply with increasing load.

Permanent-Magnet DC Motor. A permanent-magnet (PM) motor, as shown in Fig. 4, has no need for separate field excitation. The armature and commutator assemblies in the PM motors are similar to those in other dc motors. The PM motor is simple and reliable and is

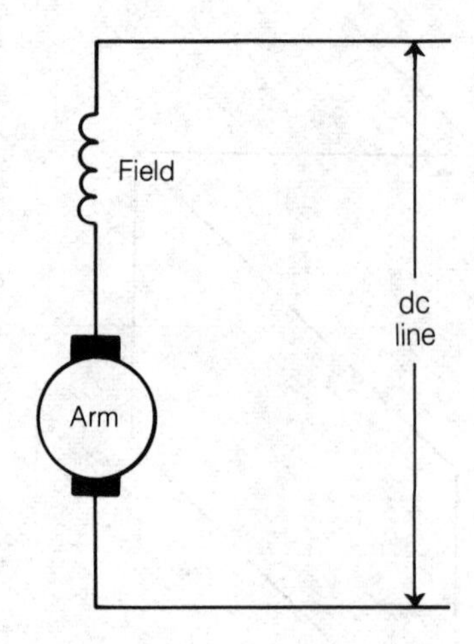

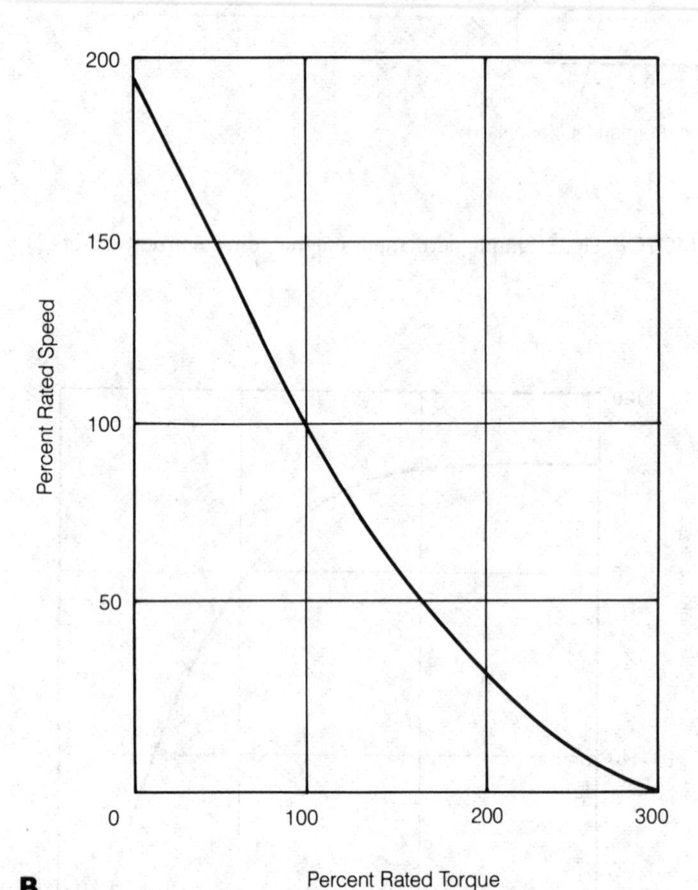

MOTOR: Fig. 3. Schematic diagram of a series-universal motor (A) and typical performance curve (B).

specified where high starting torque and efficiency and a linear speed-torque curve is desired. A family of PM dc motor curves is shown in Fig. 5.

The main advantage of a PM dc motor over a wound-field dc motor is smaller size for the same torque rating. This has been made possible by the use of Alnico V or samarium-cobalt permanent magnets. In addition to a smaller diameter and length, the weight is reduced. *See* ALNICO.

Because PM motors produce relatively high torques at low speed, they can replace gear motors in many applications. However, PM motors cannot be operated continuously at upper torque limits because of possible damage from overheating. The PM motor's linear speed/torque characteristics and adaptability to electronic control suit it for closed-loop servomotor applications. The motors can be reversed simply by changing the polarity of the power connection.

Two different low-mass armatures have been developed for PM servomotors: shell type and disk type. Each of these armatures has its own performance features and they influence the outline dimensions and form factor of the motor.

Shell-Type Armature DC PM Motor. The shell-type PM motor, as shown in Fig. 6, has a hollow, cup-shaped armature. The armature is made by forming copper coils in a cup shape and molding it in glass fiber and polymer resin to form a lightweight but rigid ironless shell. Also called a moving-coil motor, it exhibits very low inertia and high acceleration, which are desirable characteristics in a closed-loop servo system.

Shell armature motors can be damaged if they are run under high loads for long periods. Heat can build in the armature because of its low thermal conductivity. However, these motors are suitable for powering the axes of robots and the drive spindles of machine tools where the duty cycles are low. When used in closed-loop servosystems, they are coupled to on-shaft tachometers and/or resolvers for speed and position control. *See* SERVO SYSTEM.

Disk-Type Armature DC PM Motor. Disk-type meters are similar to the shell-type motors except that their armatures are shaped as platters or disks. The armature coils are made as flat coils of wire or are stampings from copper foil which are then bonded with glass fiber and polyester to form rigid disks. The disks are then securely mounted on axial armature shafts.

The principal advantage of the disk-type armature PM motor over the cup- or shell-type armature PM motor is its short, flat form factor. It can be housed in a disk-shaped frame with a diameter that is far larger than its length. This pancake frame occupies less space than a comparably rated motor with a conventional cylindrical frame. Moreover, no additional bracing or supports are needed. They would add to the weight and create vibration and balance problems.

The PM field of the disk-type motor is obtained with a ring of magnetic disks around the rim of the motor housing or a cast C-shaped magnet. The brushes of this motor ride directly on commutator bars in the thin laminated armature in a face commutation arrangement. This arrangement permits rapid acceleration and provides constant torque.

Disk-armature PM motors drive products from sewing machines to robots. In closed-loop controls they are also coupled to tachometers for speed control and brushless resolvers for position control. They are also not suited for long operation under heavy loads.

Brushless DC Motor. The brushless dc motor is an electronically commutated dc motor that operates on conventional dc-motor principles. Solid-state switching circuitry replaces brushes and a segmented commutator.

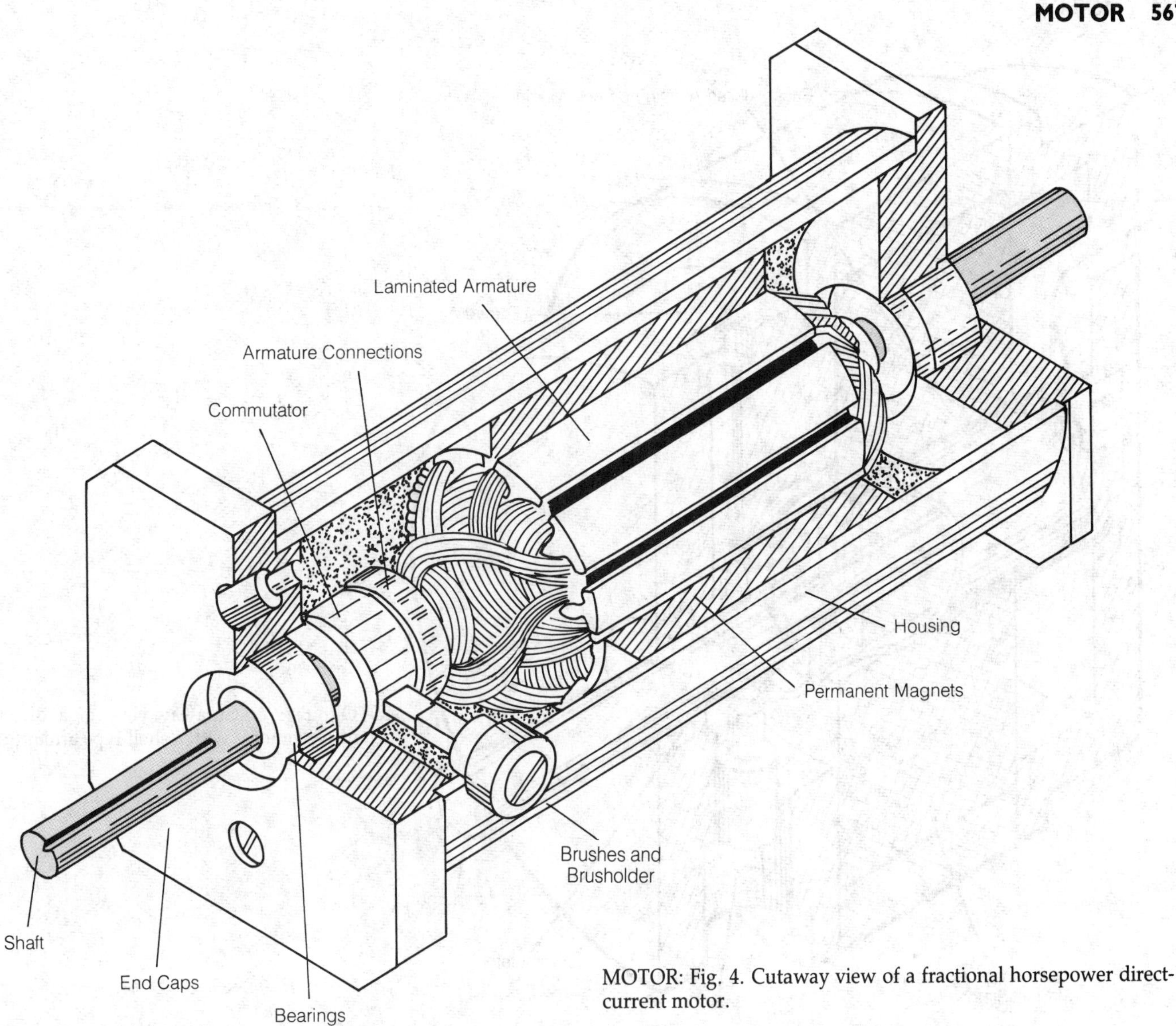

MOTOR: Fig. 4. Cutaway view of a fractional horsepower direct-current motor.

These motors are constructed inside out compared with other permanent magnet motors, as shown in Fig. 7. The rotor includes a permanent magnet and the stator has wound field coils. One or more transducers sense the position of the rotor magnetic field with respect to the field coils so that current flow in the coils can be switched in time to maintain continuous torque.

Sensor input to transistorized logic circuits triggers the power transistors controlling the flow of current to the coils. This switching action creates the revolving magnetic field in the stator that maintains the rotation of the rotor. Because no current is conducted by the rotor, there is no need for brushes, hence the name.

The elimination of brushes simplifies motor maintenance because there are no brushes to be serviced or replaced. Without brushes there is no arcing to create electromagnetic interference (EMI). This is an important feature for a motor running near sensitive circuitry. The elimination of arcing also eliminates any explosion hazard in the presence of flammable or explosive mixtures. Thus brushless motors can be used safely in hospitals, laboratories, and factories where these hazards are present.

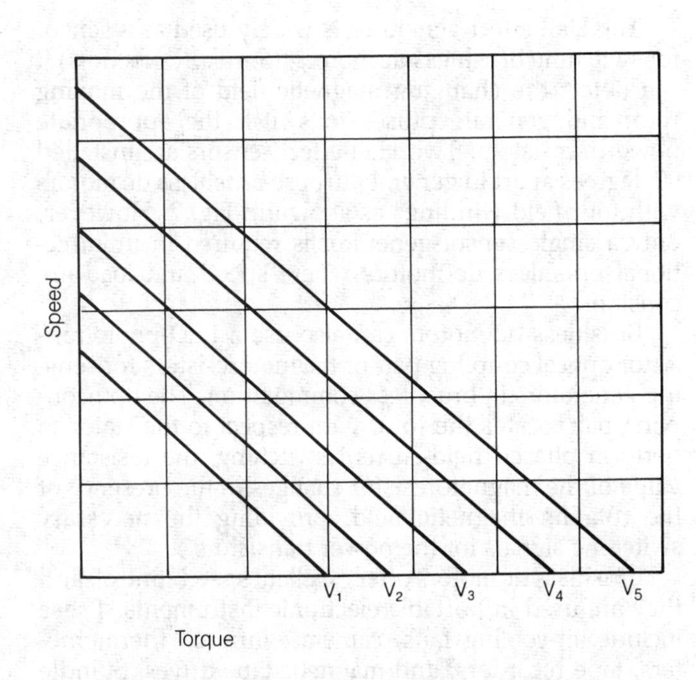

MOTOR: Fig. 5. Typical family of speed/torque curves for a permanent-magnet dc motor with a range of input voltages.

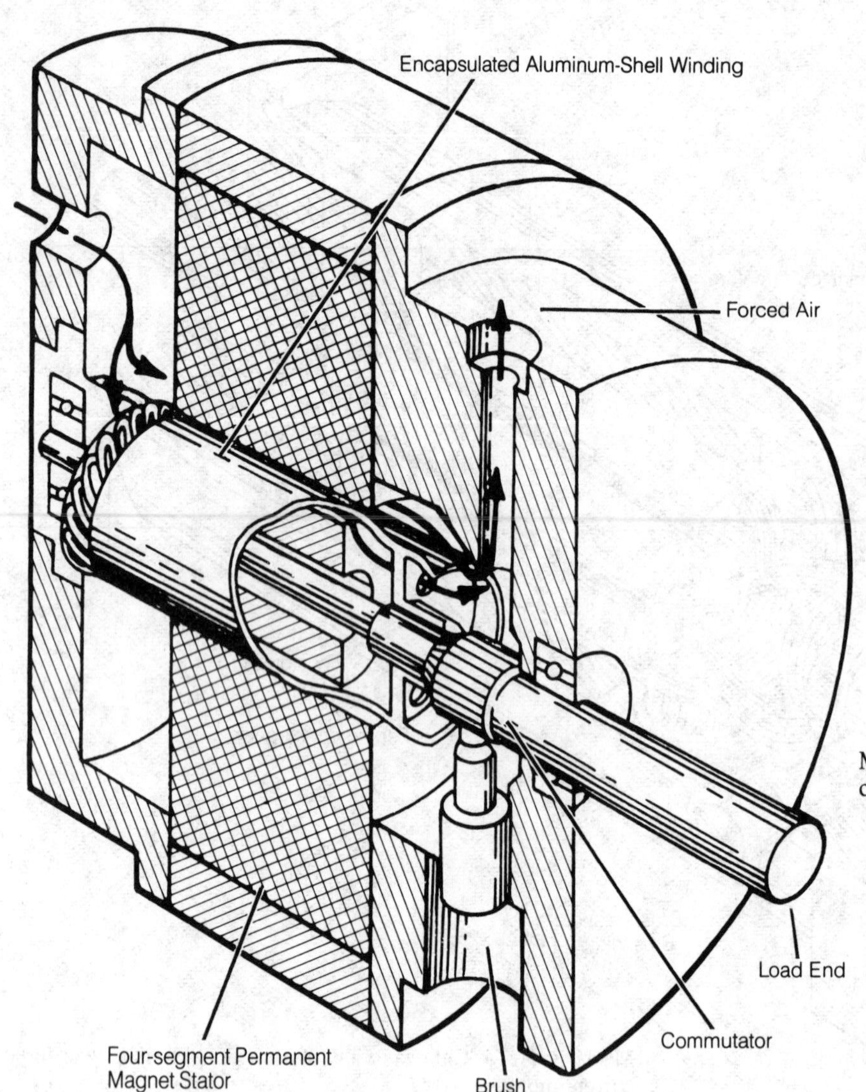

MOTOR: Fig. 6. Cutaway view of a direct-current servomotor with a shell-type armature.

The Hall-effect generator is widely used as a sensor for switching brushless dc motors. (*See* HALL GENERATOR.) It can detect the changing magnetic field of the moving rotor and generate pulses to switch the appropriate power transistors. Two Hall-effect sensors are installed 90 degrees apart in general purpose brushless dc motors with four field windings as shown in Fig. 7. However, only a single sensor-generator is required in unidirectional brushless dc motors where speed and load are constant.

Brushless dc motors can also use a LED-phototransistor optical coupling pair or magnetoresistors for sensing generators in brushless commutation. The optocoupling pair locates the rotor with respect to the stator to perform phased field-current switching; the resistance value of the magnetoresistor changes in the presence of the rotating magnetic field, providing the necessary switching signals for the power transistors.

Brushless dc motors offer excellent speed control, and they are used in portable electronic instruments. These include air-cooling fans, cameras, infrared thermometers, tape recorders, and magnetic tape drives. Spindle drive motors on hard-disk and floppy-disk drives are brushless dc motors. Complete electronic circuitry for switching the field coils may be included within the motor frame.

Stepping Motor. The stepping motor or stepper is classed as a dc motor although it is actually an ac motor operated by trains of pulses. These motors are designed so that their rotors move or are indexed a carefully controlled fraction of a revolution each time they receive an input step pulse. This permits shaft movement to be controlled with high precision which can be translated into precise rotational or linear movement. Input pulses are counted electronically to provide controlled motion without the need for a closed-loop feedback circuit to correct errors in commanded motion. Thus the stepper is usually an open-loop controller.

The three principal types of stepper motors are:

1. Variable reluctance (VR) motors with wire-wound stators and multipoled iron rotors can step through angles of 5 to 15 degrees. Because of low torque and load capacity, these motors are widely used in instruments.
2. Permanent magnet (PM) motors are capable of stepping angles from 5 degrees to 90 degrees. They have wound stators and permanent

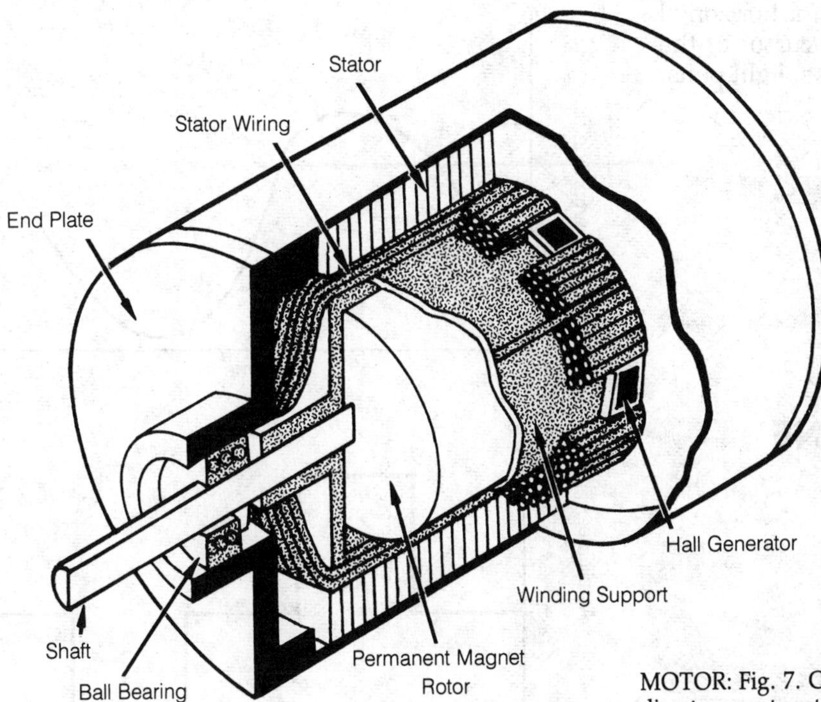

MOTOR: Fig. 7. Cutaway view of a permanent-magnet brushless direct-current motor.

magnet rotors. Torque output is low, but step accuracy is higher than for VR motors: ±10 percent.

3. PM/hybrid motors combine the construction and features of both PM and VR steppers. Coils are wound on toothed segments of the stator assembly. The rotor contains a permanent magnet and it has splined outer surfaces to aid in precise stepping. Torque capacities of 50 to 2000 ounce-inches are available and stepping is accurate to ±3 percent. Step angles vary from 0.5 to 15 degrees, with the 1.8-degree angle most often specified. Electronic switching causes the rotor to move in increments determined by the number of input pulses and it also locks the rotor in each incremental position when pulses are not introduced.

Steppers are specified in control systems where speed is not critical and the error tolerances introduced by discrete angular increments are acceptable. Stepping motor control systems save on the higher cost of closed-loop servo control by eliminating the tachometer, encoder or resolver and more complex circuitry. However, closed- loop controls have been used with steppers to improve the precision of motion.

PM/hybrid stepping motors are used in X-Y motion tables, machine tools, process controls, and dedicated robots—such as pick-and-place component insertion machines. They can be operated by numerical control, process controllers, or computers.

AC Motor. The ac motor can operate from single-phase or three-phase ac. The series-wound universal ac/dc motor is used in many household appliances and power tools. The majority of motors made and sold today are single-phase, fractional horsepower ac motors for consumer products. These motors can be split-phase, capacitor-start, shaded-pole or universal motors. Although less efficient than comparably rated dc and three-phase ac motors, they cost less. The speed of some ac motors can be controlled by electronic circuits. Both single-phase and three-phase ac motors are embedded in products and systems that are considered to be electronic.

MOTORBOATING

Excessive feedback in an audio amplifier can result in a fluttering or popping sound that resembles the sound of a motorboat. This oscillation is caused by undesirable coupling between the output and the input of an amplifier or chain of amplifiers. It can also occur as a result of capacitive or inductive coupling in the wiring or be caused by the power supply. Because of the characteristic sound, this low-frequency oscillation is called motorboating.

Motorboating can be eliminated by minimizing the coupling between the output and the input of an amplifier system. Interstage connecting wires should be balanced or shielded, and should be as short as possible. A filter may have to be placed in series with the leads to the power supply. This filter usually consists of a large-value choke and one or two capacitors. *See also* FEEDBACK.

MOUSE

In computer technology, a mouse is a small wheeled device with a cable (the mouse's tail), which is connected to the computer for control of the display. A mouse is used to overcome the time and effort required to move the screen's cursor through the control keys on the keyboard. The mouse is moved about on a horizontal surface and the motion is tracked by the cursor of the computer. Mice are alternatives to joysticks, light pens, and touch screens. *See* JOYSTICK, LIGHT PEN.

MOVING-COIL METER

See D'ARSONVAL MOVEMENT.

MOVING-COIL SPEAKER

See SPEAKER.

MOVING-COIL MICROPHONE

See MICROPHONE.

MOVING-COIL PICKUP

See DYNAMIC PICKUP.

MSI

See MEDIUM-SCALE INTEGRATION.

MU

Mu is a letter of the Greek alphabet. Its symbol is written like a small English letter *u* with a tail (μ). The English *u* is often used in place of the actual symbol μ.

Mu is used as a prefix multiplier meaning micro (*see* MICRO). The symbol μ is also used to indicate amplification factor, permeability, inductivity, magnetic moment, and molecular conductivity.

MUF

See MAXIMUM USABLE FREQUENCY.

MULTIBAND ANTENNA

See ANTENNA DIRECTORY.

MULTIELEMENT ANTENNA

See ANTENNA DIRECTORY.

MULTILEVEL TRANSMISSION

Multilevel transmission is a form of digital transmission in which some signal variable has three or more discrete values. The number of possible levels must, however, be finite. Therefore ordinary analog modulation is not considered multilevel transmission.

Multilevel transmission can be used to transmit digitally a complex waveform, provided the element (bit) duration is short enough, and provided the number of levels is large enough. The complex waveform shown at A in the illustration is converted to a coarse three-level amplitude-modulated signal at B. Finer multilevel conversion signals are illustrated at C and D.

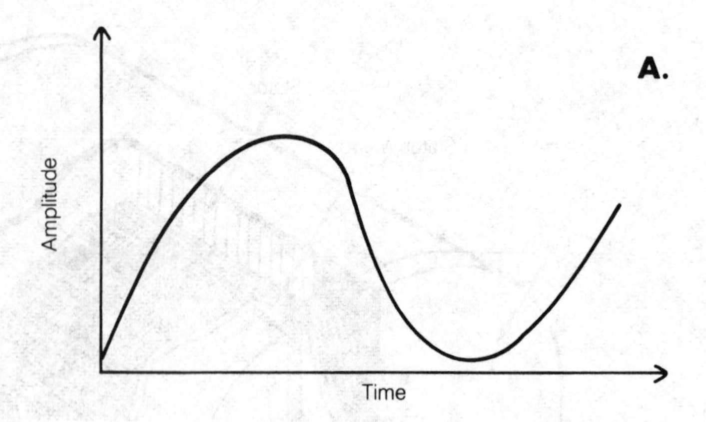

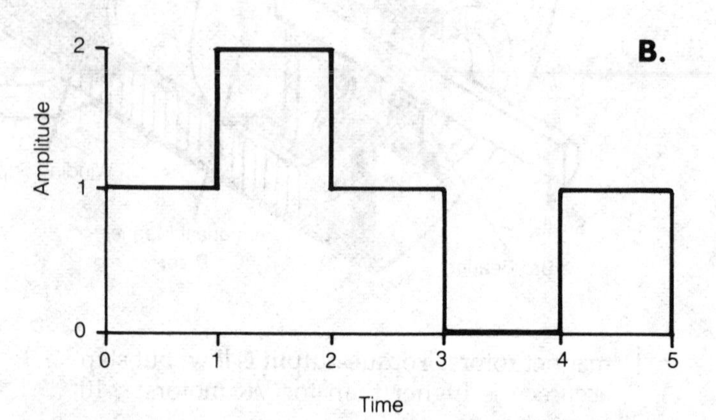

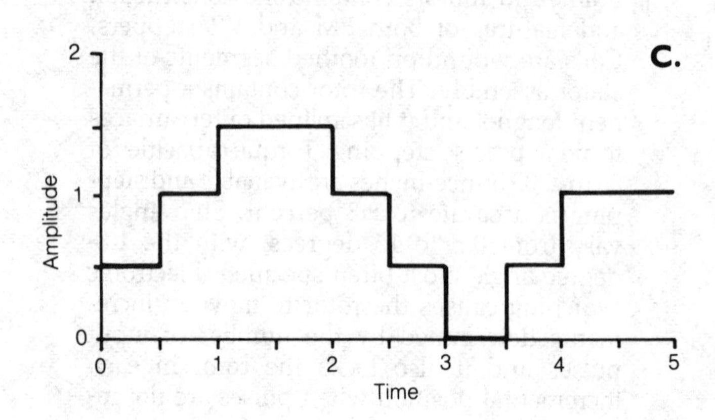

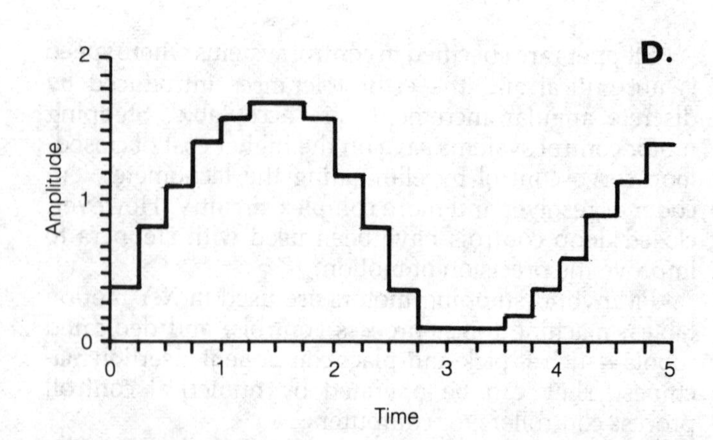

MULTILEVEL TRANSMISSION: An irregular analog waveform (A), a three-level digital representation (B), and finer representations of the waveform (C and D).

The amplitude is not the only parameter that can be varied in a multilevel signal.

Multilevel transmission can be used with any form of modulation. The primary advantage of multilevel transmission is its narrow bandwidth compared with ordinary analog modulation.

MULTIPATH FADING

Multipath fading is fading that occurs primarily at medium and high frequencies and results from the random phase combination of received ionospheric signals arriving along more than one path at the same time.

The ionosphere is not a smooth reflector of electromagnetic energy. Instead, the ionization occurs in irregular patches and with variable density. The result is that signals may be transmitted between two points by several different paths simultaneously (see illustration). These paths do not all have the same overall length and therefore the signals combine in random phase at the receiver. As the ionosphere undulates, the path lengths constantly vary and so does the overall phase combination at the receiver.

Multipath fading effects can be reduced by using two receivers for the same signal with antennas located at least several wavelengths apart. The probability that a severe fade will occur at both receivers, simultaneously, is less than the probability that a fade will take place at a single receiver. *See also* DIVERSITY RECEPTION, FADING.

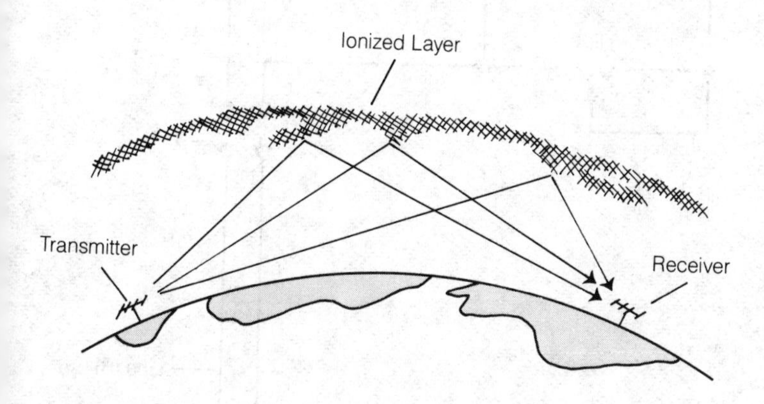

MULTIPATH FADING: Several signals combine at the receiver in varying phase in multipath fading.

MULTIPLE POLE

A multiple-pole relay or switch is designed for switching two or more circuits at once. Multiple-pole relays and switches may be of the single-throw or multiple-throw variety (*see* MULTIPLE THROW).

Schematic symbols for multiple-pole switches or relays generally show the same circuit designator for each pole, followed by A, B, C, and so on (see illustration). The various poles may be connected in the diagram by a dotted line. When one pole changes state, the others all change state. *See also* RELAY, SWITCH.

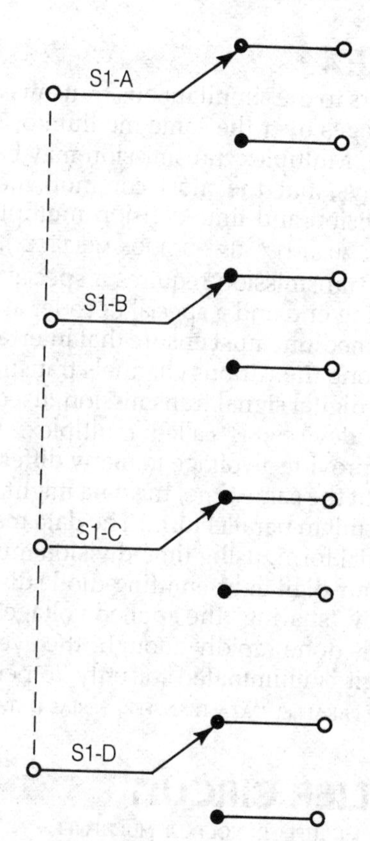

MULTIPLE POLE: A switch with four poles.

MULTIPLE THROW

A multiple-throw relay or switch can connect a given conductor to two or more other conductors. A relay normally has at most two throw positions because of mechanical constraints, but a switch may have several throw positions. Multiple-throw relays and switches can be ganged to form a multiple-pole, multiple-throw device (*see* MULTIPLE POLE).

The illustration shows the schematic symbol for a five-position multiple-throw switch. Generally, a switch with three or more throw positions is a wafer switch. *See also* SWITCH.

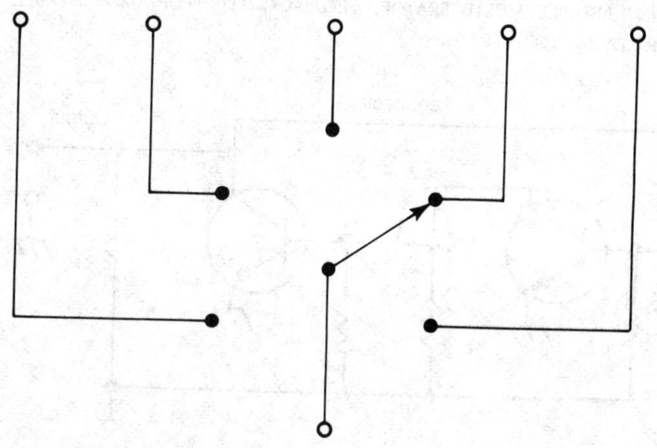

MULTIPLE THROW: A single-pole switch with five different positions.

MULTIPLEX

Multiplex refers to the simultaneous transmission of two or more messages over the same medium or channel at the same time. Multiplex transmission may be achieved in various ways, but the most common methods are frequency-division and time-division multiplex (*see* FREQUENCY-DIVISION MULTIPLEX, TIME-DIVISION MULTIPLEX).

Multiplex transmission requires a special encoder at the transmitting end and a special decoder at the receiving end. The medium must ensure that interference does not occur among the various channels transmitted.

A form of digital signal transmission, used especially with display devices, is called multiplex. When it is necessary to provide a voltage to many different display components at the same time, the data might be cumbersome to transmit in parallel form. The data may, instead, be sent in serial form, using time-division multiplex. For example, a four-digit light-emitting-diode display can be illuminated by "sharing" the applied voltage from left to right. If this is done rapidly enough, the eye cannot tell that each digit is illuminated for only 25 percent of the time. *See also* PARALLEL DATA TRANSFER, SERIAL DATA TRANSFER.

MULTIPLIER CIRCUIT

See FREQUENCY MULTIPLIER, VOLTAGE MULTIPLIER.

MULTIVIBRATOR

A multivibrator is a form of relaxation oscillator. A multivibrator circuit consists of a pair of inverting amplifiers connected in series. A direct feedback loop runs from the output of the second stage to the input of the first. A block diagram of the basic scheme is shown in the illustration. Multivibrators are extensively used in digital circuits.

There are three main types of multivibrator: astable, bistable, and monostable. The astable multivibrator is a free-running oscillator. The bistable multivibrator, often called a flip-flop, may attain either of two stable conditions. The monostable multivibrator maintains a single condition except when a triggering pulse is applied; then the state changes for a predetermined length of time. *See also* ASTABLE MULTIVIBRATOR, FEEDBACK, FLIP-FLOP, MONOSTABLE MULTIVIBRATOR.

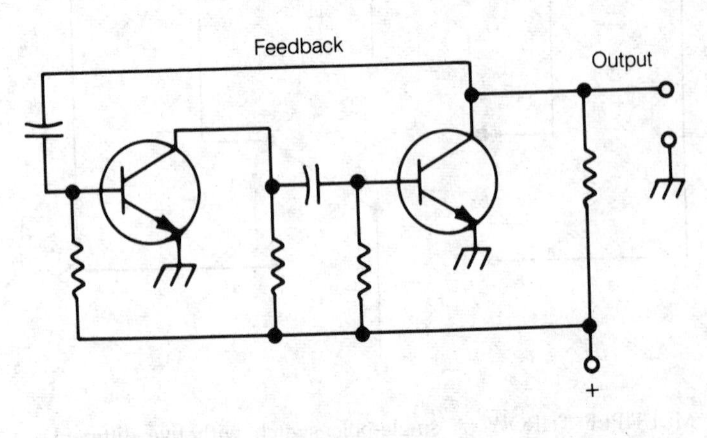

MULTIVIBRATOR: Two amplifiers in a multivibrator produce square or rectangular output pulses.

MUMETAL

Mumetal is a ferromagnetic material commonly used as a magnetic shield in cathode-ray tubes.

Mumetal consists of a mixture of iron, nickel, copper, and chromium. Its magnetic permeability is very high. The direct-current resistivity is moderately low, resulting in good magnetic shielding characteristics. *See also* FERROMAGNETIC MATERIAL.

MURRAY LOOP TEST

The Murray loop test is a method of determining the distance from a given point in a telephone line to a ground fault. An extremely accurate method of resistance measurement is needed. The instrument used for this purpose is called a Murray bridge, a special form of Wheatstone bridge.

The figure illustrates the connection of the Murray bridge to a telephone line. Resistor R_3 is adjusted until the indicator shows a null reading. The resistance along

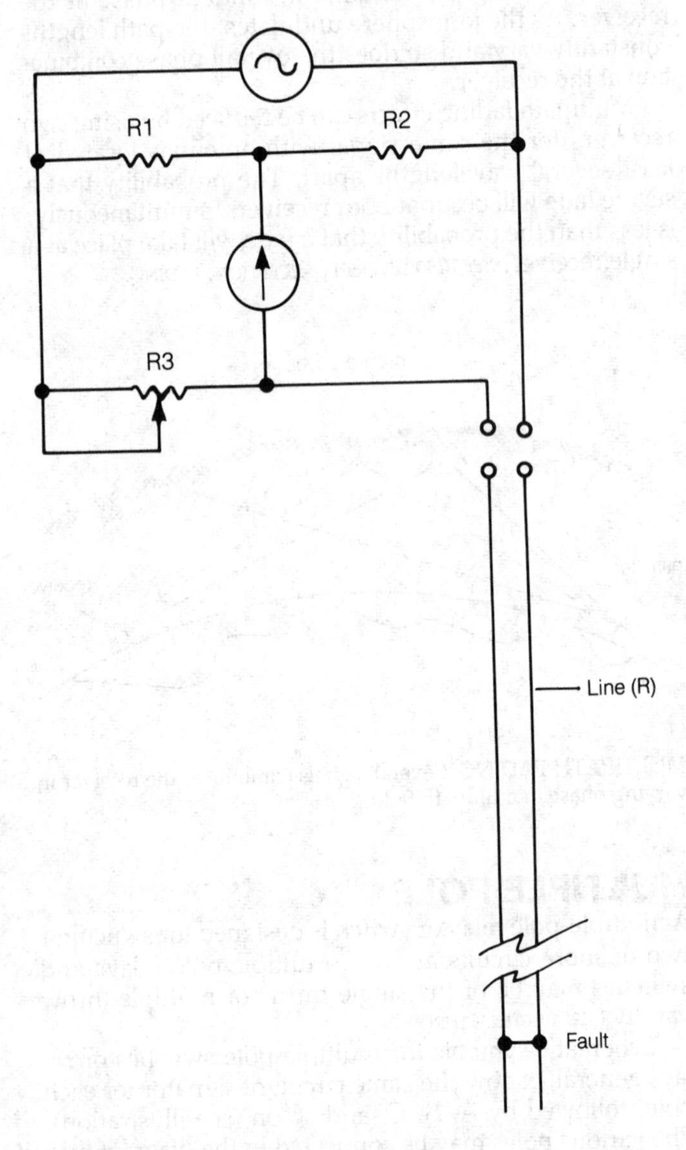

MURRAY LOOP TEST: A method for finding faults in telephone lines.

the line, R, is then given by:

$$R = \frac{R_2 - R_3}{R_1}$$

where R_1 and R_2 are fixed resistances.

Knowing the resistance of the line per unit length in either conductor (for example, x ohms per meter), the distance d, in meters, to the ground fault can be found by simple calculation:

$$d = R/(2x)$$

See also WHEATSTONE BRIDGE.

MUTING CIRCUIT

A muting circuit cuts off a television or radio receiver while the power is on. This prevents signals from being heard, but it allows instantaneous reactivation of the circuit when desired by the operator.

A muting circuit generally consists of a switching transistor or relay in series with one of the stages of the receiver. The muted stage is usually one of the radio-frequency or intermediate-frequency amplifiers in the receiver chain. However, an audio-frequency circuit may be muted if an extremely fast recovery is not needed. The schematic diagram shows a simple muting system using a transistor switch. A negative voltage at the muting terminal opens the emitter circuit of the amplifier. Other types of muting circuits also exist; some are actuated by a positive voltage, others by an open circuit at the input terminals, and still others by a short circuit.

Muting circuits are used for break-in operation for Morse-code communication. Muting circuits are also used to prevent receiver overload or acoustic feedback in two-way voice radio systems. They can also be found in television channel switching circuits.

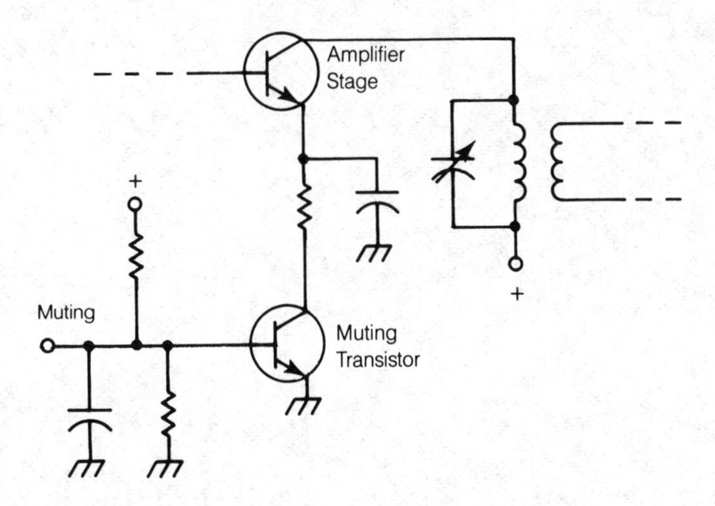

MUTING CIRCUIT: A circuit that can cut off a receiver temporarily.

MUTUAL CAPACITANCE

Any two conductors, no matter where they are located or how they are oriented, have some capacitance with respect to each other. With an instrument capable of measuring a sufficiently small value of capacitance, any set of two conductors can be shown to act as a capacitor. The existence of capacitance among electrical conductors is called mutual capacitance. The mutual capacitance between two conductors increases as the surface area of either conductor is made larger; the mutual capacitance decreases as the distance between conductors becomes greater.

Mutual capacitance is often too small to be measured and is of no consequence. However, in radio-frequency (rf) circuit wiring, mutual capacitance can be enough to cause feedback and other problems. This is especially true at very high and ultrahigh frequencies. Radio circuits at these frequencies are designed to minimize the mutual capacitance among wires and components.

Mutual capacitance exists among the elements within certain electronic components, particularly diodes, relays, switches, transistors, and vacuum tubes. In rf circuit design, this capacitance is often significant, and must be taken into account in the design process. *See also* ELECTRODE CAPACITANCE, INTERELECTRODE CAPACITANCE.

MUTUAL CONDUCTANCE

See TRANSCONDUCTANCE.

MUTUAL IMPEDANCE

Two adjacent electrical conductors display some mutual capacitance and mutual inductance (*see* MUTUAL CAPACITANCE, MUTUAL INDUCTANCE). The combination of mutual capacitance and inductance presents a complex impedance, known as mutual impedance, that varies with frequency. Because of this, a resonant circuit can be formed. The resonant frequency depends on the amount of mutual capacitance and inductance. Generally, the natural resonant frequency of two conductors or electrodes is very high—on the order of hundreds or even thousands of megahertz. *See also* PARASITIC OSCILLATION.

MUTUAL INDUCTANCE

Any two adjacent conductors have a certain amount of inductance with respect to each other. An alternating magnetic field around one conductor will invariably cause an alternating current in the other. The degree of inductive coupling between conductors or inductors is called mutual inductance.

Mutual inductance, represented by the letter M and expressed in henrys, is an expression of the degree of coupling between two coils. The mutual inductance is mathematically related to the coefficient of coupling, which may range from 0 (no coupling) to 1 (maximum possible coupling). If the coefficient of coupling is given by k and the inductances of the two coils, in henrys, are given by L_1 and L_2, then:

$$M = k\sqrt{L_1 L_2}$$

For example, suppose that $k = 0.5$, and the coils have inductances of 0.5 henry and 2 henrys. Then:

$$M = 0.5\sqrt{0.5 \times 2} = 0.5$$

In practice, because most coils have inductances that are a small fraction of 1 henry, the value of M is quite small.

For two inductors with values L_1 and L_2 (given in henrys) and connected in series with the flux linkages in the same direction, the total inductance L, in henrys, is:

$$L = L_1 + L_2 + 2M$$

If the flux linkages are in opposition, then:

$$L = L_1 + L_2 - 2M$$

See also COEFFICIENT OF COUPLING, INDUCTANCE.

MYLAR CAPACITOR

See CAPACITOR.

NAND GATE

A NAND gate is a logical AND gate, followed by an inverter (*see* AND GATE, INVERTER). The term *NAND* is a contraction of NOT AND. The illustration shows the schematic symbol for the NAND gate, along with a truth table indicating the output as a function of the input.

When both or all of the inputs of the NAND gate are high, the output is low. Otherwise, the output is high. *See also* LOGIC.

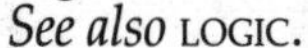

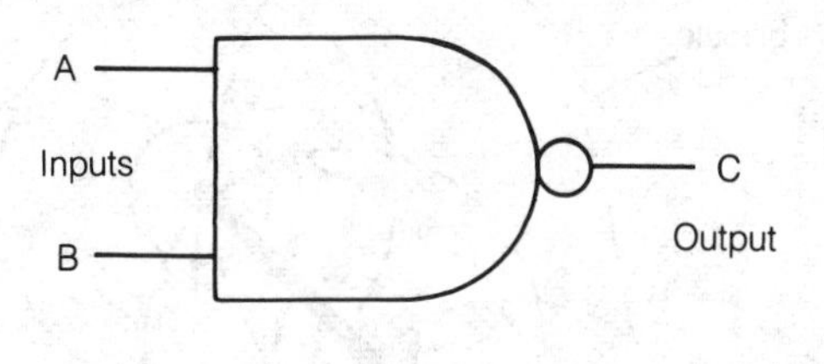

A	B	C
0	0	1
0	1	1
1	0	1
1	1	0

NAND GATE: Schematic symbol and truth table for a NAND gate.

NANO-

Nano- is a prefix multiplier representing 1 billionth (10^{-9}). For example, 1 nanofarad is equal to 10^{-9} farad, or 0.001 microfarad; 1 nanowatt is equal to 10^{-9} watt. The abbreviation for nano is the lowercase letter *n*. *See also* PREFIX MULTIPLIERS.

NAPIER

See NEPER.

NARROW-BAND VOICE MODULATION

Narrow-band voice modulation (NBVM) is a technique for reducing the bandwidth necessary for the transmission of a human voice. Normally, this bandwidth is about 2.5 to 3 kHz; a bandpass of approximately 300 Hz to 3 kHz is required. Using NBVM, the bandwidth can be substantially reduced.

The sounds of the human voice do not occupy the entire frequency range between 300 Hz and 3 kHz; rather, the sound tends to be concentrated within certain bands called formants. For a voice to be clear and understandable, three formants must be passed. The lower formant occurs at a range of about 300 to 600 Hz. The upper two formants exist within the range 1.5 to 3 kHz. Between 600 Hz and 1.5 kHz, very little sound occurs. In NBVM, this unused range is eliminated, reducing the overall signal bandwidth by about 900 Hz. Narrow-band voice modulation is thus a frequency-companding process.

Narrow-band voice modulation is achieved by mixing at the baseband, or audio-frequency, level. The resulting audio signal sounds scrambled. The audio output of the NBVM compressor can be fed to the microphone input of any voice transmitter. At the receiver, the signal is restored to its original form by frequency expansion. *See also* COMPANDOR.

NATIONAL BUREAU OF STANDARDS

The National Bureau of Standards (NBS) is an agency in the United States that maintains values for physical constants in the standard international system of units.

Among shortwave radio listeners and radio amateurs, the National Bureau of Standards is best known for its continuous standard time and frequency broadcasts. These signals are sent by radio station WWV in Fort Collins, Colorado, and WWVH on the island of Kauai, Hawaii. *See also* STANDARD INTERNATIONAL SYSTEM OF UNITS, WWV/WWVH.

NATIONAL ELECTRIC CODE

The National Electric Code (NEC) is a set of recommendations for safety in electric wiring. The NEC is prepared by the National Fire Protection Association and the Institute of Electrical and Electronic Engineers, and is purely advisory. However, the NEC is enforced by some local governments.

The NEC is primarily concerned with types of insulation, suitable wire sizes and conductor materials, and various levels of voltage and current. Maximum allowable current values are given for different kinds and sizes of wire and cable. Temperature derating curves are given.

The NEC also makes recommendations regarding the

wiring of electronic equipment, especially if high voltage and/or current levels are used. Detailed information can be obtained from the American National Standards Institute, New York, New York.

NATURAL FREQUENCY

The natural frequency of a circuit, antenna, or set of electrical conductors is the lowest frequency at which the system is resonant for electromagnetic energy. In acoustics, physical objects have a lowest resonant sound frequency; this is sometimes called the natural frequency.

Certain systems have a tendency to oscillate at the natural frequency. For example, a radio-frequency power amplifier, if not neutralized, may develop parasitic oscillations at the natural frequency of interelectrode impedance.

In an antenna system, the radiator and feed line together have a natural frequency that differs from the resonant frequency of the antenna by itself. When tuned feeders are used to obtain broadband operation, the feed line may radiate if the system is used at its natural frequency. *See also* PARASITIC OSCILLATION, RESONANCE, TUNED FEEDERS.

NATURAL LOGARITHM

See LOGARITHM.

NAUTICAL MILE

The nautical mile is a unit of distance used by mariners. It is slightly greater than the statute mile. A nautical mile is equal to 1.151 statute miles, or 1.85 kilometers.

Nautical speed is expressed in units called knots. A knot is one nautical mile per hour. Wind speed is sometimes given in knots; to obtain the speed in statute miles per hour, multiply by 1.151.

NEAR FIELD

The near field of radiation from an antenna is the electromagnetic field in the immediate vicinity of the antenna. The electric and magnetic lines of flux in the near field are curved because of the proximity of the radiating element (A in the illustration). The wavefronts are not flat, but instead are convex.

For a paraboloidal or dish antenna, the near field is within a cylindrical region having the same diameter as the dish, as at B. The near field of radiation from a dish antenna is sometimes called the Fresnel zone. The distance to which the near field extends depends on the diameter of the antenna, the antenna aperture, and the wavelength.

In radio communications, the near field is generally of little importance. The signal normally picked up by the receiving antenna results not from the near field, but from the far field or Fraunhofer region. *See also* FAR FIELD.

NECESSARY BANDWIDTH

The necessary bandwidth of a signal is the minimum bandwidth required for transmission of the data with reasonable accuracy. The degree of accuracy is somewhat arbitrary; it is impossible to obtain perfection. In general, the greater the signal bandwidth, the better the accuracy. As the bandwidth is increased beyond a certain point, however, the improvement is small and spectrum space is wasted. Typical values of necessary bandwidth for various emission types are:

- For a Morse-code signal, the necessary bandwidth in hertz, is generally considered to be about 2.4 times the speed in words per minute. Thus, for example, a Morse transmission at 20 words per minute (a typical speed) is 48 Hz. The receiver bandpass should be at least 48 Hz for accurate reception of a Morse signal at 20 words per minute.

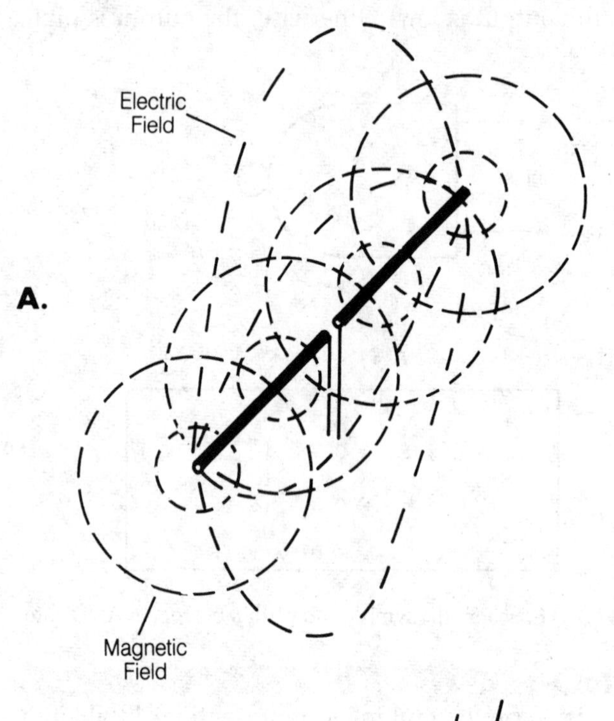

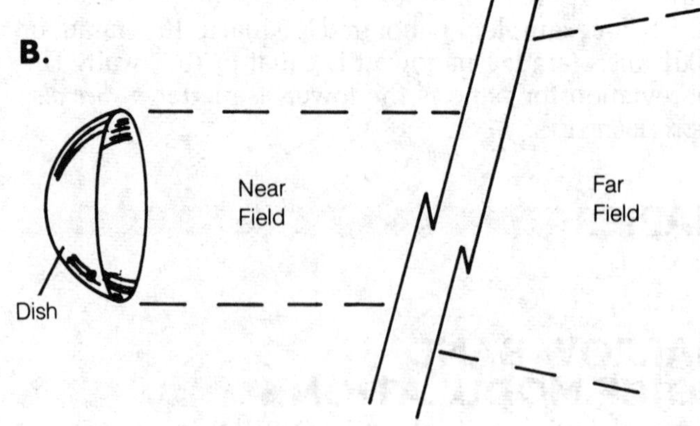

NEAR FIELD: The near field surrounds a half-wave dipole (A) but is adjacent to the antenna dish (B).

- For BAUDOT and ASCII codes, the necessary bandwidth also depends on the speed. Generally, the necessary bandwidth, in hertz, for these signals is about 2.4 times the speed in words per minute, or 3.2 times the speed in bauds. As with Morse transmission, this is a somewhat subjective value, and does not represent an absolute standard.
- For a single-sideband transmission, the necessary bandwidth is about 2.5 kHz. However, with narrow-band modulation techniques, this value can be reduced considerably (*see* NARROW-BAND VOICE MODULATION). A slow-scan television signal requires about 2.5 kHz.
- A normal amplitude-modulated or frequency-modulated voice signal requires approximately 5 to 6 kHz of spectrum space for reasonable intelligibility. Some frequency-modulated signals, however, are spread over a much larger space to improve the fidelity. The transmission of high-fidelity music requires a minimum of 40 kHz of spectrum space with amplitude or frequency modulation.
- Video signals and high-speed data transmissions have large necessary bandwidth. The typical television broadcast channel is 6 MHz wide.

See also BANDWIDTH, BAUD RATE.

NEGATION

Negation is a logical NOT operation. In Boolean algebra, negation is called complementation. Electronically, negation is performed by a NOT gate or inverter. *See also* BOOLEAN ALGEBRA, INVERTER.

NEGATIVE CHARGE

Negative charge, or negative electrification, is the result of an excess of electrons on a body. Friction between objects can result in an accumulation of electrons on one object (a negative charge) at the expense of electrons on the other object. When an atom has more electrons than protons, the atom is considered to be negatively charged.

The smallest unit of negative charge is carried by a single electron. Conversely, the smallest unit of positive charge is carried by the proton. The terms *negative* and *positive* are arbitrary. *See also* CHARGE, POSITIVE CHARGE.

NEGATIVE FEEDBACK

When the output of an amplifier circuit is fed back to the input in phase opposition, the feedback is said to be negative. Negative feedback is used for a variety of purposes in electronic circuits.

Some amplifiers break into oscillation easily. This can be prevented by a neutralizing circuit, a form of negative-feedback arrangement (*see* NEUTRALIZATION).

Negative feedback is used in some audio amplifiers to improve stability and fidelity. Generally, negative feedback can be provided by simply installing a resistor of the proper value in series with the cathode, emitter, or source of the active amplifying device. Negative feedback can also be obtained by directly applying some of the output signal to the input of the amplifier, 180 degrees out of phase, as shown in the illustration.

Negative feedback is routinely used in operational-amplifier circuits to control the gain and improve the bandwidth. When no negative feedback is used, the operational amplifier is said to be operating in the open-loop condition (*see* OPEN LOOP, OPERATIONAL AMPLIFIER). Negative feedback is provided by placing a resistor in the feedback path. The amount of negative feedback depends on the value of the resistor between the output and the input, denoted by R in the illustration. The smaller the value of R, the greater the negative feedback, and the smaller the gain of the circuit. *See also* FEEDBACK.

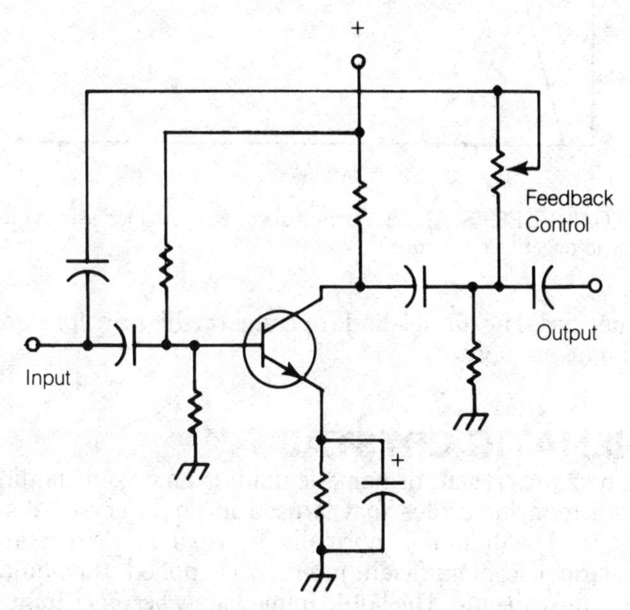

NEGATIVE FEEDBACK: An amplifier with negative feedback to control gain.

NEGATIVE LOGIC

Logical signals are defined by convention. Normally, the logic 1 is the more positive of the voltage levels, and the logic 0 is the more negative of the voltage levels. That is, logic 1 is high and logic 0 is low. When the voltages are reversed, the logic is said to be negative or inverted.

Either positive or negative logic will provide satisfactory operation of a digital device. *See also* LOGIC.

NEGATIVE RESISTANCE

Normally current through a device increases as the applied voltage increases. However, some components exhibit a negative characteristic within a limited range. The current decreases as the applied voltage is made larger. This is called negative resistance. The illustration shows a negative-resistance characteristic.

Negative resistance occurs in some diodes, transistors, and vacuum tubes. Negative resistance causes oscillation when the applied voltage is within its range. This oscillation usually occurs at ultrahigh or microwave fre-

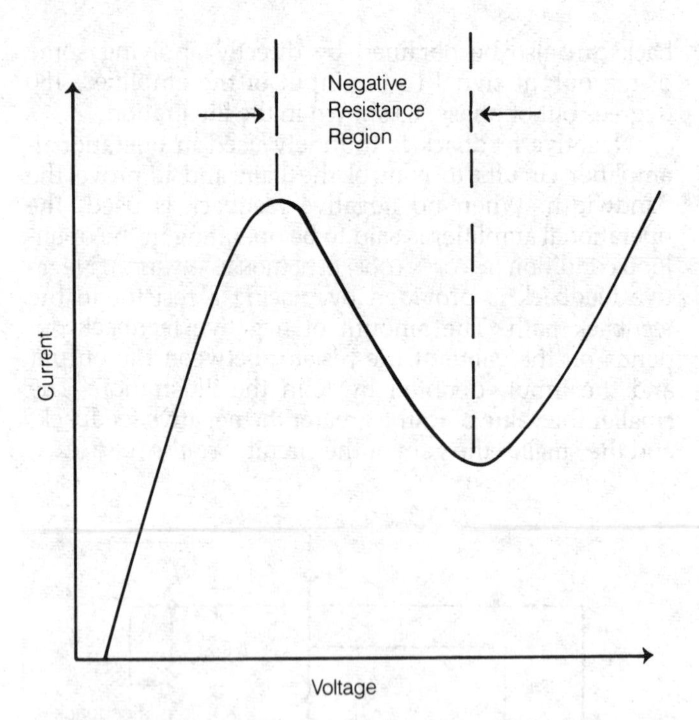

NEGATIVE RESISTANCE: Current decreases with increasing voltage in negative resistance.

quencies. The tunnel diode operates on this principle. *See also* TUNNEL DIODE.

NEMATIC CRYSTAL

A nematic crystal, or nematic fluid, is an organic liquid with long molecules that is used in liquid crystal displays. The liquid is normally transparent. When an electric field of sufficient intensity is applied, the liquid becomes opaque. The liquid immediately becomes transparent again when the field is removed.

Displays using nematic crystals have become increasingly popular in calculators, watches, and other electronic instruments. The nematic crystal draws very low current because the electric field alone causes the change of state. This is an advantage in battery-powered devices, where the current drain must be minimized. *See also* LIQUID-CRYSTAL DISPLAY.

NEON

Neon is an element with an atomic number of 10 and an atomic weight of 20. At room temperature, neon is a gas. Neon is nonreactive because the inner two electron shells are completely filled.

Neon is used in glow-discharge lamps and neon gas discharge lamps. When neon is ionized by a high voltage, some of the emission lines fall in the visible-light spectrum. *See also* NEON GAS DISCHARGE DISPLAY, NEON LAMP.

NEON GAS DISCHARGE DISPLAY

The neon gas discharge display produces illuminated alphanumeric characters and graphics by selective breakdown of pairs of electrodes in the sealed, neon-filled glass cavity. Approximately 150 V is required across opposing electrodes for initial ionization of a neon display device. However, once started, characters can be formed at lower voltages. The breakdown of the gas produces the high-visibility orange-red glow characteristic of neon. Neon displays are classified as 1) neon gas discharge displays which use a direct current power supply, and 2) plasma displays, which use an alternating current power supply.

Neon Gas Discharge. Neon gas alphanumeric displays for instrument readout are made as sealed and evacuated glass sandwiches enclosing neon gas between electrodes formed in the shape of bars or segments. The cathode electrodes on the inner surface of the cover plate are transparent as shown in Fig. 1. They can be formed as 7-segment (figure 8) bars or 16-segment (union jack) bars. All numbers and upper case letters from A through F can be formed with eight bars, but 16 segments permit the formation of the basic set of ASCII characters. *See* ASCII.

The Nixie tube from Burroughs Corporation was the first successful neon gas discharge numeric display tube introduced in the 1950s. It was constructed as a stack of shaped electrodes within a gas-filled tube. By selection of the appropriate electrode, the numbers from 1 to 9 and 0

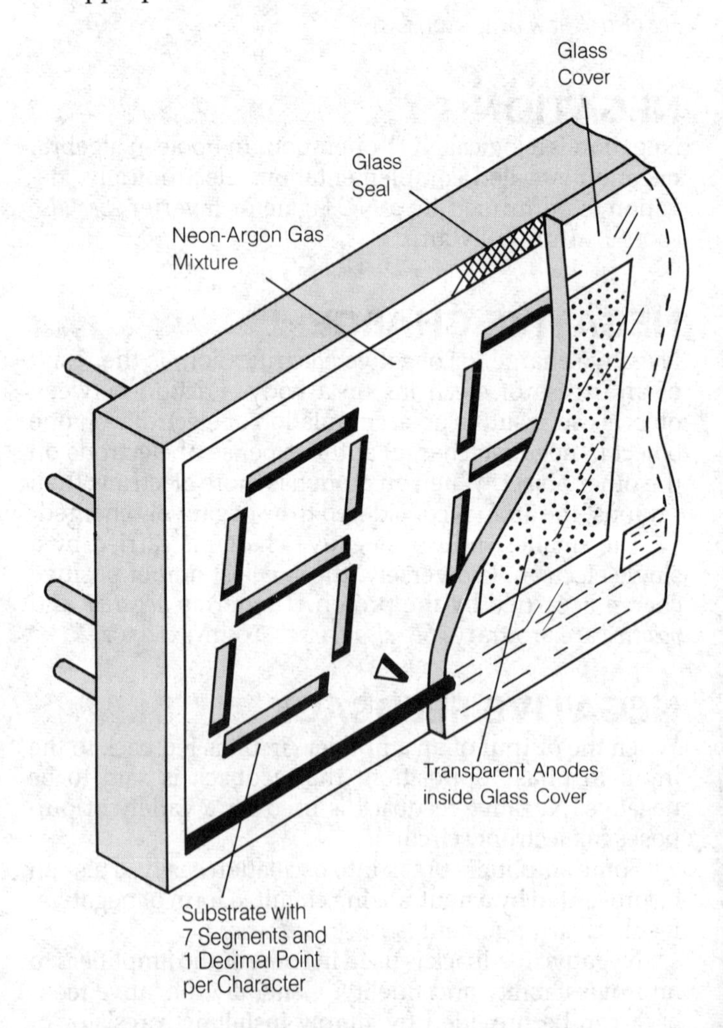

NEON GAS DISCHARGE DISPLAY: Fig. 1. Cutaway view of a seven-segment dc neon gas discharge display.

could be displayed. Nixie tube displays were handicapped by very narrow angles of visibility, high power requirements, and complexity. They also produced heat and electromagnetic interference (EMI).

Single and dual seven-segment modules in plug-in metal and glass cases were introduced in the 1970s. Display strings can be formed by end stacking the modules. They were produced in volume and sold as industrial instrument displays. High brightness and the ability to withstand elevated ambient temperatures in a factory setting offset their drawbacks as sources of heat and EMI. They were first introduced by Sperry Rand Corporation, but the manufacture was later taken over by Beckman Instruments.

Flat, multidigit display neon gas discharge display panels are available from many suppliers. The manufacturing technique of screening on transparent anodes and cathodes permitted both standard catalog products and custom units to meet user size and font requirements.

The panels are filled with a gas mixture of about 97 percent neon and about 3 percent argon to reduce the ionization voltage. However, these displays are handicapped by their high-voltage (150 to 200 V dc) power supply and high-voltage integrated-circuit (IC) driver requirements, although their current drain is low. Because they are difficult to start at low temperatures, minute amounts of radioactive gas (krypton 85 isotope or tritium) were added to assist in the ionization process. But the presence of radioactive gas barred them from many applications where they would be near people.

Neon discharge displays are in digital panel meters (DPMs), digital multimeters (DMMs), and other electronic test instruments. They were also in early desktop and handheld electronic calculators.

Neon Alternating-Current (ac) Plasma Display. Plasma displays are flat-panel displays capable of forming all the ASCII characters, text and graphics. They are alternatives to the CRT and the other flat-panel displays—liquid crystal and electroluminescent. The ac plasma panel consists of two glass plates with a conductor pattern on the inner surfaces of each plate, separated by a gas-filled gap as shown in Fig. 2. The conductors are configured in an X-Y matrix with horizontal row electrodes and vertical column transparent electrodes deposited at right angles to each other with thin-film techniques.

The electrodes of the ac plasma display, unlike those in the direct-current (dc) neon discharge display, are covered by a thin glass dielectric layer. The glass plates are put together to form a sandwich, with the distance between the two plates fixed by spacers. The edges of the plates are sealed and the cavity between the plates is evacuated and back-filled with a 99.9 percent neon and 0.1 percent argon gas mixture.

When the gas ionizes, the insulators charge like small capacitors so the sum of the drive voltage and the capacitive voltage is large enough to sustain the plasma. Alternating-current plasma displays are orange against a black background. Small, light-emitting pixels are formed as ac voltage is applied across the row and column electrodes.

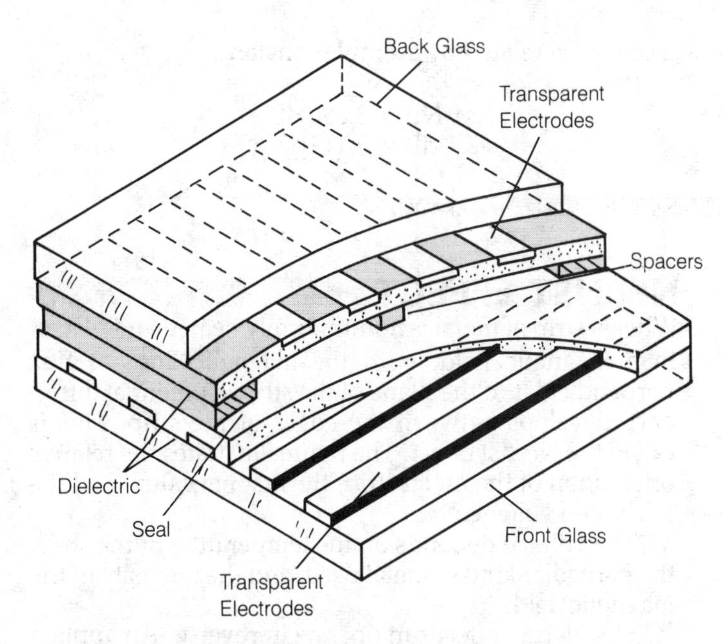

NEON GAS DISCHARGE DISPLAY: Fig. 2. Cutaway view of an ac neon plasma display for alphanumeric characters, text, and graphics to replace a cathode-ray tube.

An important feature of the ac plasma display is its data retention or memory; a pixel must be erased with a pulse of opposite polarity and magnitude. To write data into an ac plasma display, a wall charge must be created. This is an accumulated electrical charge across the dielectrics. These dielectric regions separate the electrodes from the ionized gas. When data are to be erased, some of the wall charge must be removed or subtracted. Because the ac plasma displays do not need to be refreshed to sustain an image, updating can be performed at a much lower rate than in electroluminescent or CRT displays.

Intended for computer control, ac plasma displays require high-voltage drivers and multiplexing circuitry. These displays are now available on portable personal computers with high-voltage (up to about 190 V) power supplies. Although ac plasma displays measuring up to 1 meter diagonally have been made, commercial 640 by 200 pixel (25-line-by-80-character) full-page displays are available.

NEON LAMP

A neon lamp is a device consisting of two electrodes in a sealed glass envelope containing neon gas. When a voltage is introduced between the electrodes, the neon gas becomes ionized, and the lamp glows. Neon lamps have a characteristic red-orange color.

Neon lamps are used in indicating devices, displays, oscillators, and voltage regulators. *See also* NEON GAS DISCHARGE DISPLAY.

NEPER

The neper (Np) is a unit for expressing a ratio of currents, voltages, or wattages. The neper also is called the napier. The neper is similar to the decibel. In fact, nepers and decibels are related by a simple constant:

$$1 \text{ Np} = 8.686 \text{ dB}$$
$$1 \text{ dB} = 0.1151 \text{ Np}$$

See also DECIBEL, LOGARITHM.

NERNST EFFECT

When a strip of metal is nonuniformly heated and placed in a magnetic field, with the magnetic lines of flux perpendicular to the plane of the strip, a potential difference develops between the edges of the strip. This is called the Nernst effect. The figure illustrates the relative orientation of the metal strip, the magnetic flux, and the generated voltage.

The voltage depends on the temperature of the strip, the particular kind of metal used, and the intensity of the magnetic field.

The Nernst effect can operate in reverse: An applied voltage can result in heating of a crystal strip in a magnetic field. *See* NERNST-ETTINGHAUSEN EFFECT.

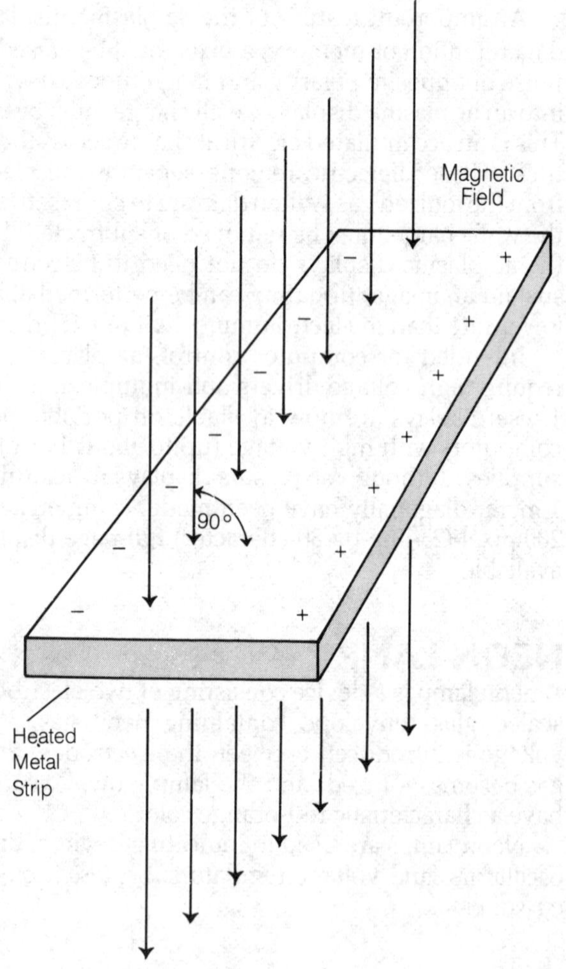

NERNST EFFECT: A voltage develops across a metal strip when it is heated in a magnetic field.

NERNST-ETTINGHAUSEN EFFECT

When certain substances are placed in a magnetic field and current is passed through them, heating occurs. The current flow must be perpendicular to the magnetic lines of flux. This thermomagnetic effect is known as the Nernst-Ettinghausen effect.

The extent of the temperature differential depends on the level of the current, the intensity of the magnetic field, and the particular crystal used. The temperature difference occurs along an axis that is perpendicular to the flow of the current, and is also perpendicular to the magnetic lines of flux.

The Nernst-Ettinghausen effect can operate in reverse: A heated metal strip in a magnetic field develops a potential difference between its edges. *See* NERNST EFFECT.

NET

A communications network, intended for the relaying and delivering of information or messages by radio, is called a net. Nets are especially popular among amateur radio operators. However, commercial nets also exist.

The term net is widely used as a synonym for the term network (*see* NETWORK).

A resultant or effective quantity is called a net quantity. For example, a circuit may have currents flowing in opposite directions at the same time; the net current is the effective current is one direction. A circuit can have two or more voltage sources, some of which buck each other; the net voltage is the sum of the individual voltages, taking polarity into account.

NETWORK

A network is a set of electronic components, interconnected for a specific purpose. In particular, a network is a circuit intended for insertion in a power-supply line or transmission line. Some networks are quite simple; some are exceedingly complicated.

Networks can be classified as either active or passive. An active network contains at least one component that requires an external source of power for operation. A passive network requires no external source of power; it is entirely made up of passive components.

Networks are often identified according to the way in which the components are interconnected. For example, a pi network consists of two parallel components, one on either side of a series-connected component. A T network consists of two series components, one on either side of a parallel-connected component. *See also* H NETWORK, L NETWORK, PI NETWORK, T NETWORK.

NEUTRAL CHARGE

When the total charge on an object is zero—the positive

balances the negative—the object is said to be electrically neutral. A neutral charge results in a zero net electric field.

Atoms normally have neutral charge, since the number of protons is the same as the number of orbiting electrons. However, atoms may gain or lose orbiting electrons, becoming charged. *See also* NEGATIVE CHARGE, POSITIVE CHARGE.

NEUTRALIZATION

Radio-frequency amplifiers, especially power amplifiers, have a tendency to oscillate. This unwanted oscillation is called parasitic oscillation (*see* PARASITIC OSCILLATION), and can occur at any frequency, regardless of the frequency at which the amplifier is operating. Neutralization is a method of reducing the likelihood of such oscillation.

The neutralization circuit is a negative-feedback circuit. A small amount of the output signal is fed back, in phase opposition, to the input. This may be done in a variety of different ways; the most common method is the insertion of a small, variable capacitor in the circuit. The figure illustrates two methods of neutralization in a simple bipolar-transistor power amplifier. The method at A is called collector neutralization, since the capacitor is connected to the bottom of the collector tank circuit. The method at B is called base neutralization, because the capacitor is connected to the bottom of the base tank circuit. The negative feedback occurs through the variable capacitor. The capacitance required for optimum neutralization depends on the operating frequency and impedance of the amplifier. Normally the neutralizing capacitance is a few picofarads.

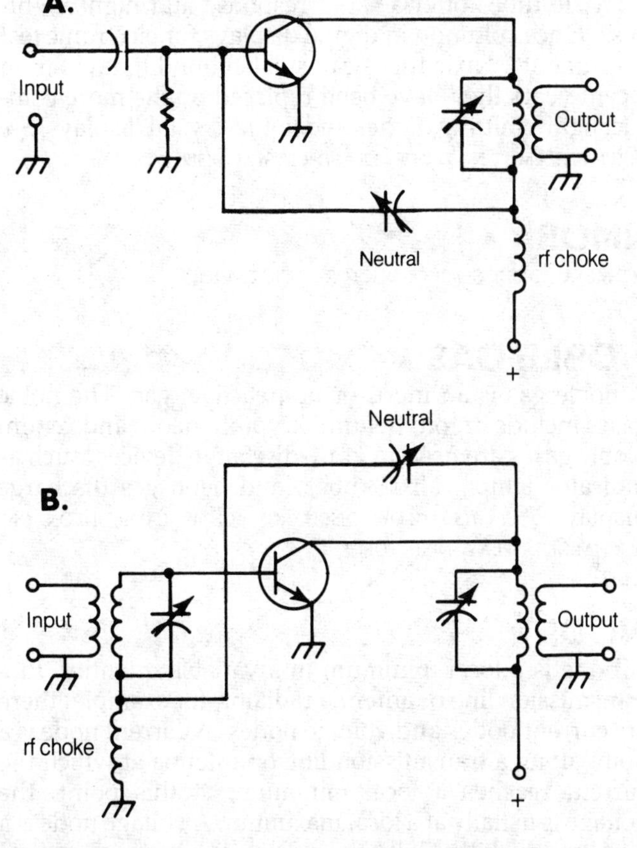

NEUTRALIZATION: Two methods of neutralizing a radio-frequency amplifier are: collector neutralization (A), and base neutralization (B).

The neutralizing capacitor is first adjusted by removing the driving power from the input of the amplifier. A sensitive broadband wattmeter, or other output-sensing indicator, is connected to the amplifier output. The capacitor is adjusted for minimum output indication. *See also* NEGATIVE FEEDBACK, NEUTRALIZING CAPACITOR, POWER AMPLIFIER.

NEUTRALIZING CAPACITOR

A neutralizing capacitor is a small-valued, variable capacitor used for providing negative feedback in a radio-frequency power amplifier. The negative feedback can be varied by adjusting the value of the capacitance.

The capacitor method is the most common means of neutralizing a power amplifier. However, other negative-feedback circuits may also be used. *See also* NEGATIVE FEEDBACK, NEUTRALIZATION, POWER AMPLIFIER.

NEUTRON

The neutron is a subatomic particle with zero electric charge. Neutrons are found in the centers (nuclei) of all atoms, with the exception of the hydrogen atom. The neutron has a slightly larger mass than the proton.

Neutrons are given off by certain radioactive materials as they decay. Neutron radiation, in large quantities, is dangerous to life. Some neutron radiation arrives at the earth from outer space.

Under certain conditions, a neutron will split into a proton and an electron. In other cases, a proton and electron can combine to form a neutron. *See also* ATOM, ELECTRON, PROTON.

NEWTON

The newton is the standard international (SI) unit of force. The unit is named for Sir Isaac Newton.

A force of 1 newton, applied to a mass of 1 kilogram, will result in an acceleration of 1 meter per second per second. Newtons are also called kilogram meters per second squared (kgm/s^2).

A force of 1 newton is the equivalent of 100,000 dynes. *See also* FORCE.

NEWTON'S LAWS OF MOTION

In the seventeenth century, Sir Isaac Newton changed the course of physics by formulating his three laws of motion. From experimentation, Newton deduced these facts concerning the behavior of objects and forces:

1. Any body at rest, or moving at a uniform velocity, tends to maintain that condition of motion until an external force is applied.
2. When a force is applied to a mass, the mass

accelerates. The acceleration occurs in direct proportion to the force, and in inverse proportion to the mass of the body. The acceleration takes place in the same direction in which the force is applied.
3. For every acting force, there is an equal reacting force that occurs in the opposite direction.

Newton's laws of motion are still the basis for classical mechanics today. These laws apply to charged particles moving at nonrelativistic speeds. As the speed of a particle becomes very high, however, Newton's laws no longer accurately explain its behavior.

NICHROME

Nichrome is an alloy of two common metals, nickel and chromium. Nichrome has high resistivity and a high melting point. It is used for electrical heating elements, and also in the fabrication of thin-film and wirewound resistors.

Nichrome is also called nickel-chromium. The word *Nichrome* was originally coined by Driver-Harris Company. *See also* RESISTOR.

NICKEL

Nickel is an element with an atomic number of 28 and an atomic weight of 59. Nickel is used for many different electrical and electronic purposes.

An alloy of nickel and chromium exhibits high resistance and a high melting temperature. This alloy, called Nichrome, is used in the manufacture of heating elements and resistors (*see* NICHROME).

Nickel is used as a plating material for steel. Nickel-plated steel has a dull, grayish-white appearance when unpolished; when polished it appears dark gray. Nickel plating retards the corrosion of steel, but does not entirely prevent it (*see* ELECTROPLATING).

Nickel is used in the manufacture of electrodes of some vacuum tubes. Nickel cathodes exhibit high electron emission at relatively low temperatures. However, the tungsten cathode can withstand higher temperature levels (*see* CATHODE).

Nickel is used in the manufacture of some ferromagnetic transformer-core and inductor-core materials. The characteristics of these cores vary greatly, depending on the method of manufacture (*see* FERROMAGNETIC MATERIAL).

A compound of nickel, hydrogen, and oxygen is used in the manufacture of a form of the rechargeable nickel-cadmium cell. Cells and batteries made from them are widely used in electronics (*see* BATTERY).

Nickel is a magnetostrictive material. In the presence of an increasing magnetic field, nickel contracts. For this reason, nickel is often used in the manufacture of magnetostrictive transducers (*see* MAGNETOSTRICTION).

NICKEL-CADMIUM BATTERY

See BATTERY.

NITROGEN

Nitrogen is an element with an atomic number of 7 and an atomic weight of 14. At room temperature, nitrogen is a gas. The atmosphere is 78 percent nitrogen.

Nitrogen is a relatively nonreactive gas, although it can form certain compounds under the right conditions. Because nitrogen is so abundant, it is relatively inexpensive to obtain in pure form. One method of getting a 99 percent pure sample of nitrogen is to remove the oxygen from the air in an enclosed chamber.

Nitrogen is used in the fabrication of sealed transmission lines because it is a noncorrosive gas. An air-dielectric coaxial cable, for example, may be evacuated and then refilled with pure, dry nitrogen gas. This improves the dielectric qualities, and prolongs the life of the transmission line.

NIXIE® TUBE

A Nixie tube, also called a readout tube or readout lamp contains many cathodes. The cathodes are arranged so that various alphanumeric characters can be displayed. Decimal points may also be included. The tube is filled with neon gas that glows when a voltage is applied between the anode and one or more of the cathodes.

Each of the cathodes is connected to a separate pin at the base of the tube. The anode has its own pin. When voltages are applied to various cathodes, the corresponding segments are illuminated with a reddish-orange color.

Nixie tubes offer a rapid response and high brightness. Once common in digital displays for electronic test instruments, Nixie tubes can still be found today. But in recent years they have been replaced by the more compact light-emitting diodes and liquid-crystal displays. *See also* LIGHT-EMITTING DIODE, LIQUID-CRYSTAL DISPLAY.

NMOS

See METAL-OXIDE SEMICONDUCTOR LOGIC FAMILIES.

NOBLE GAS

A noble gas is an inert, or nonreactive, gas. The noble gases include argon, helium, krypton, neon, and xenon. Noble gases are used in glow-discharge devices, such as indicator lamps, Nixie tubes, and neon gas discharge displays. *See also* GLOW DISCHARGE, GLOW LAMP, NEON GAS DISCHARGE DISPLAY, NIXIE TUBE.

NODE

A node is a local minimum in a variable quantity. In a transmission line or antenna radiator, for example, there are current nodes and voltage nodes. A current node is a point along a transmission line or antenna at which the current reaches a local minimum; at this point, the voltage is usually at a local maximum. A voltage node is a point at which the voltage reaches a local minimum, and the current is usually at a local maximum. Nodes are separated by electrical multiples of ½ wavelength. The opposite of a node is a loop (*see* CURRENT NODE, LOOP, VOLTAGE NODE).

In a circuit with two or more branches, a node is any circuit point that is common to at least two different branches. At a node, the current inflow is always the same as the current outflow, according to Kirchhoff's laws (*see* KIRCHHOFF'S LAWS).

NOISE

Noise is a broadbanded electromagnetic field generated by various environmental effects and man-made sources. Noise can be categorized as either natural or man-made.

Natural noise may be either thermal or electrical in origin. All objects radiate noise as a result of their thermal energy content. The higher the temperature, the shorter the average wavelength of the noise. This is known as black-body radiation (*see* BLACK BODY). Black-body radiation constantly bombards the earth from outer space. It also originates in all objects on the earth, and in the earth itself.

Cosmic disturbances, solar flares, and the movement of ions in the upper atmosphere all contribute to the natural electromagnetic noise present at the surface of the earth. Sferics, or noise generated by lightning, is a source of natural noise.

Electromagnetic noise is produced by many different man-made sources. In general, any circuit or appliance that produces electric arcing will produce noise. Such devices include fluorescent lights, heating devices, automobiles, electric motors, thermostats, and many appliances.

The level of electromagnetic noise affects the ease with which radio communications can be carried out. The higher the noise level, the stronger a signal must be if it is to be received. The signal-to-noise ratio (*see* SIGNAL-TO-NOISE RATIO) can be maximized in different ways. The passband of the receiver can be narrow if the bandwidth of the signal is narrow. This permits a better signal-to-noise ratio at a given frequency. This improvement occurs, however, at the expense of data-transmission speed capability. Circuits such as noise blankers and limiters are sometimes helpful in improving the signal-to-noise ratio. Noise-reducing antennas can also be used to advantage in some cases. But there is a limit to the noise-level reduction; a certain amount of noise will always exist. *See also* NOISE FILTER, NOISE FLOOR, NOISE LIMITER.

NOISE BANDWIDTH

See NOISE-EQUIVALENT BANDWIDTH.

NOISE-EQUIVALENT BANDWIDTH

Any noise source has a spectral distribution. Some noise is very broadbanded, and some noise occurs with a fairly well-defined peak in the spectral distribution (see illustration, A and B). The total noise can be expressed in terms of the area under the curve of amplitude versus frequency.

Let f be the frequency at which the power density of the noise is greatest in the spectral distribution in the graph at B. Let a represent the power density of the noise at the frequency f. A rectangle can be constructed with height a, centered at f, that has the same enclosed area P as the curve encloses. This is illustrated at C. The width of this rectangle, N, is a frequency span $f_2 - f_1$, so that $f_2 - f = f - f_1$. The value N is the noise-equivalent bandwidth at the frequency f.

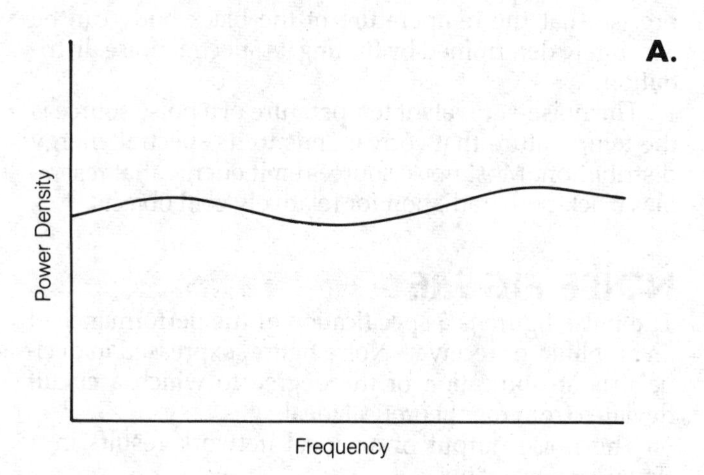

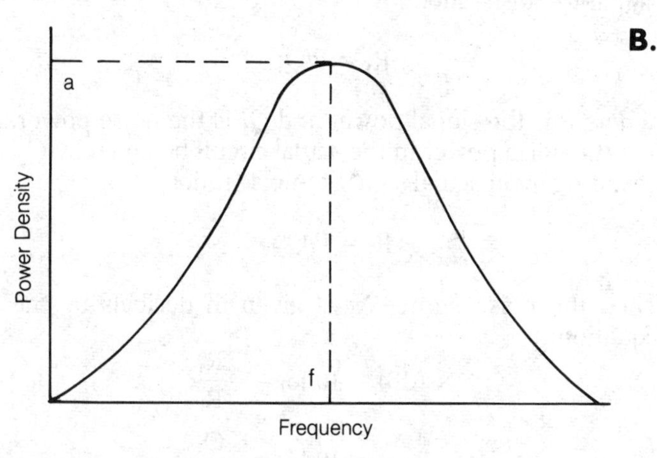

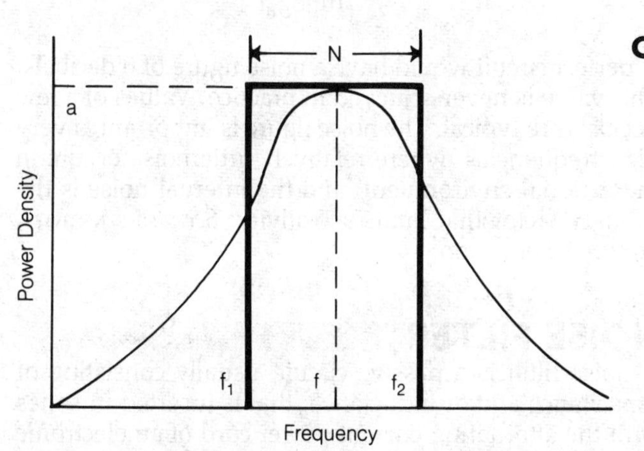

NOISE-EQUIVALENT BANDWIDTH: A typical noise spectral distribution (A), noise with peak in power density at a certain frequency (B), and the computation of noise equivalent bandwidth at frequency f (C).

Noise-equivalent bandwidth is often specified for devices such as filters, bolometers, and thermistors. The noise-equivalent bandwidth is an expression of the frequency characteristics of the devices. *See also* NOISE.

NOISE-EQUIVALENT TEMPERATURE

A black body produces a characteristic spectral noise distribution according to its absolute temperature. The higher the temperature, the higher the peak frequency of the noise output (*see* BLACK BODY). The correlation is so precise that the temperature of the black body can be accurately determined by finding its spectral noise distribution.

The noise-equivalent temperature of a noise source is the temperature that corresponds to its spectral energy distribution. Most noise sources emit energy that resembles black-body radiation for relatively cold objects.

NOISE FIGURE

The noise figure is a specification of the performance of an amplifier or receiver. Noise figure, expressed in decibels, is an indication of the degree to which a circuit deviates from the theoretical ideal.

The noise output of an ideal network results in a signal-to-noise ratio:

$$R_1 = P/Q_1$$

where P is the signal power and Q_1 is the noise power. Let the noise power in the actual circuit be given by Q_2, resulting in an actual signal-to-noise ratio:

$$R_2 = P/Q_2$$

Then the noise figure, N, is given in decibels by the equation:

$$N\text{ (dB)} = 10 \log_{10} \frac{R_1}{R_2} = 10 \log_{10} \frac{Q_1}{Q_2}$$

A perfect circuit would have a noise figure of 0 decibels. This value is never achieved in practice. Values of a few decibels are typical. The noise figure is important at very high frequencies, where relatively little noise occurs in the external environment, and the internal noise is the primary factor that limits sensitivity. *See also* SENSITIVITY, SIGNAL-TO-NOISE RATIO.

NOISE FILTER

A noise filter is a passive circuit, usually consisting of capacitance and/or inductance, that is inserted in series with the alternating-current power cord of an electronic device. The noise filter allows the 60-Hz alternating current to pass with essentially no attenuation, but higher-frequency noise components are suppressed.

A typical noise filter is simply a lowpass filter, consisting of a capacitor or capacitors in parallel with the power leads, and an inductor or inductors in series. *See also* LINE FILTER.

NOISE FLOOR

The level of background noise, relative to some reference signal, is called the noise floor. Signals normally can be detected if their levels are above the noise floor; signals below the noise floor cannot be detected. In a radio receiver, the level of the noise floor may be expressed as the noise figure (*see* NOISE FIGURE).

In a spectrum-analyzer display (see illustration on p. 585), the noise-floor level determines the sensitivity and the dynamic range of the instrument. The noise floor is generally specified in decibels (dB) with respect to the local-oscillator signal. A typical spectrum analyzer has a noise floor of −50 to −70 dB. *See also* SPECTRUM ANALYZER.

NOISE GENERATOR

Any electronic circuit that is designed to produce electromagnetic noise is called a noise generator. Noise generators are used in a variety of testing and alignment applications, especially with radio receivers.

Noise may be generated in many different ways. A diode tube, operated at saturation (full conduction), produces broadband noise. A semiconductor diode will also produce broadband noise when operated in the fully conducting condition. Some diodes generate noise when they are reverse-biased. A current-carrying resistor produces thermal noise. *See also* NOISE.

NOISE LIMITER

A noise limiter is a circuit, often used in radio receivers, that prevents externally generated noise from exceeding a certain amplitude. Noise limiters are also called noise clippers.

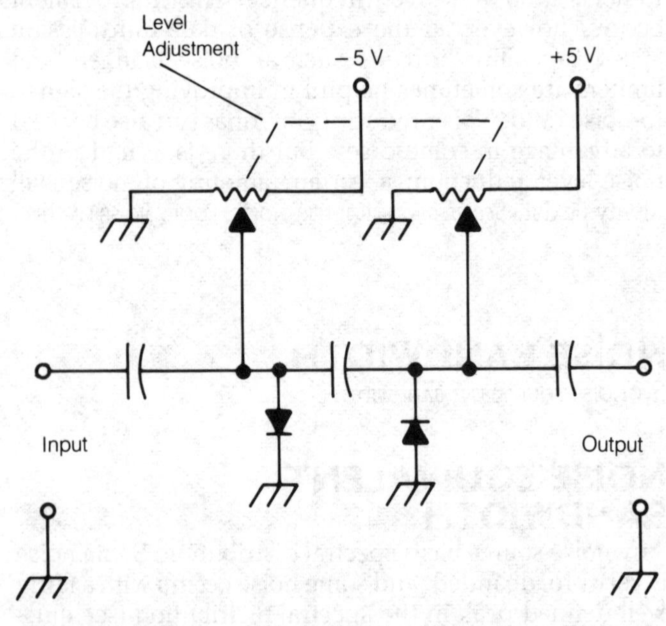

NOISE LIMITER: A variable-threshold noise-limiter circuit.

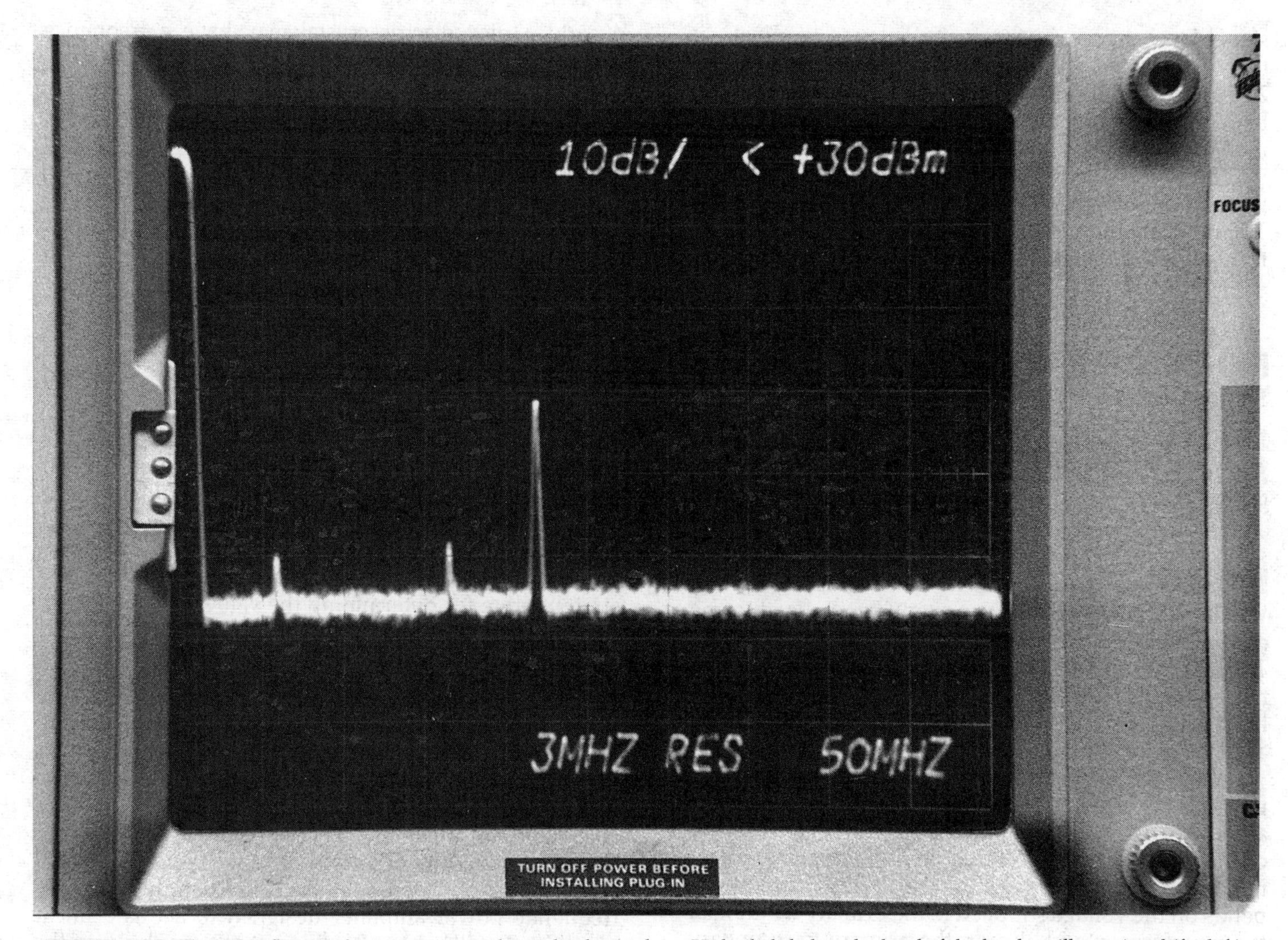

NOISE FLOOR: The noise floor in this spectrum-analyzer display is about 55 decibels below the level of the local-oscillator signal (far left).

A noise limiter may consist of a pair of clipping diodes with variable bias for control of the clipping level (see illustration). The bias is adjusted until clipping occurs at the signal amplitude. Noise pulses then cannot exceed the signal amplitude. This makes it possible to receive a signal that would otherwise be drowned out by the noise. The noise limiter is generally installed between two intermediate-frequency stages of a superheterodyne receiver. In a direct-conversion receiver, the best place for the noise limiter is just prior to the detector stage.

A noise limiter may use a circuit that sets the clipping level automatically, according to the strength of an incoming signal. This is a useful feature because it eliminates the need for continuous readjustment of the clipping level as the signal fades. This circuit is called an automatic noise limiter (*see* AUTOMATIC NOISE LIMITER).

Noise limiters are effective against all types of natural and manmade noise, including noise not affected by noise-blanking circuits. However, the noise limiter cannot totally eliminate the noise.

NOISE PULSE

A noise pulse is a short burst of electromagnetic energy. Often the instantaneous amplitude of a noise pulse rises to a much higher level than the amplitude of a received signal. However, the duration of a noise pulse is short, so the total electromagnetic energy is usually small.

Noise pulses are usually generated by electric arcing. The noise pulse may produce energy over a wide band of frequencies, from the very low to the ultrahigh, at wavelengths from kilometers to millimeters. *See also* IMPULSE NOISE, NOISE.

NOISE QUIETING

Noise quieting is a decrease in the level of internal noise in a frequency-modulation (FM) receiver, as a result of an incoming signal. With the squelch open (receiver unsquelched) and no signal, an FM receiver emits a loud hissing noise. This is internally generated noise. When a weak signal is received, the noise level decreases. As the signal gets stronger, the level of the noise continues to decrease until, when the signal is very strong, there is almost no hiss from the receiver.

The noise-quieting phenomenon provides a means of measuring the sensitivity of an FM receiver. The level of the noise at the speaker terminals is first measured under no-signal conditions. Then a signal is introduced, by means of a calibrated signal generator. The signal level is increased, without modulation, until the noise voltage drops by 20 decibels (dB) at the speaker terminals. The signal level, in microvolts at the antenna terminals, is then determined. Typical unmodulated signal levels for 20 dB noise quieting are in the range of 1 μV or less at very high and ultrahigh frequencies with receivers using modern solid-state amplifiers.

The noise-quieting method is one of two common ways of determining the sensitivity of an FM receiver. The other method is called the signal-to-noise-and-distortion (SINAD) method. *See also* SQUELCH.

NOISE SUPPRESSOR

See NOISE LIMITER.

NOISE TEMPERATURE

See NOISE-EQUIVALENT TEMPERATURE.

NO-LOAD CURRENT

When an amplifier is disconnected from its load, some current still flows in the collector, drain, or plate circuit. This current is usually smaller than the current under normal load conditions. The no-load current depends on the class of amplification and the full-load output impedance for which the circuit is designed.

When the second winding of an alternating-current power transformer is disconnected from its load, a small amount of current still flows. This current occurs because of the inductance of the secondary winding; it is called no-load current. The value of the no-load current depends on the voltage supplied to the primary winding of the transformer, and also on the primary-to-secondary turns ratio. *See also* TRANSFORMER.

NO-LOAD VOLTAGE

When the load is removed from a power supply, the voltage rises at the output terminals. The voltage at the output terminals of a power supply, with the load disconnected, is called the no-load voltage.

The difference between the no-load and full-load voltages of a power supply depends on the amount of current drawn by the load. The greater the current demanded by the load, the greater the difference between the no-load voltage and the full-load voltage.

The percentage difference between the no-load and full-load voltages of a power supply is called the regulation, or the regulation factor, of the supply. In some instances, it is important that the output voltage change very little with large fluctuations in the load current. In other cases, a large voltage change can be tolerated. *See also* POWER SUPPLY, REGULATION.

NOMINAL VALUE

All components have named, or specified, values. For example, a capacitor may have a specified value of 0.1 μF; a microphone may have a named impedance of 600 Ω; an amplifier may have a specified power-output rating of 25 W; and a piezoelectric crystal may have a frequency rating of 3.650 MHz. The named, or specified, value is called the nominal value.

The nominal value for a component or circuit is the average for a large sample of manufactured items, but individual units will vary somewhat either way. Nominal values are given without reference to the tolerance, or deviation that an individual sample may exhibit from the nominal value. *See also* TOLERANCE.

NOMOGRAPH

A nomograph is a graphic means of illustrating the relationship between the independent and dependent variables in a mathematical function. A nomograph can be called a nomogram.

The illustration on p. 587 is an example of a nomograph of the relationship between the frequency of an electromagnetic wave, in megahertz, and the free-space wavelength, in meters. This is a simple, linear nomograph. More complex nomographs require the use of a straight-edge for determining the values of functions of two variables. *See also* FUNCTION.

NONLINEAR CIRCUIT

A nonlinear circuit is a circuit with a nonlinear relationship between its input and output. That is, if the output is graphed against the input, the function is not a straight line.

Sometimes a circuit will behave in a linear manner within a certain range of input current or voltage, but will be nonlinear outside that range. An example is a radio-frequency power amplifier designed to act as a linear amplifier. It will behave in a linear manner only when the drive (input power) is not excessive. But if the amplifier is overdriven, it will no longer act as a linear amplifier. A similar effect is observed in other linear circuits.

Nonlinearity results in increased harmonic and intermodulation distortion in a circuit. In fact, harmonic generators and mixers make use of this characteristic.

Nonlinear circuits are extensively used as radio-frequency power amplifiers. Class AB, Class B, and Class C amplifiers all cause some distortion of the input waveform. Class AB and Class B amplifiers do not, however, cause envelope distortion of an amplitude-modulated signal. *See also* CLASS A AMPLIFIER, CLASS AB AMPLIFIER, CLASS B AMPLIFIER, CLASS C AMPLIFIER, LINEAR AMPLIFIER, LINEARITY.

NONLINEAR FUNCTION

A nonlinear function is a mathematical function that is not a linear function. A nonlinear function of one variable, for example, may have any appearance except a straight line in the Cartesian plane (see illustration on p. 587). A nonlinear function of two variables has any appearance except that of a flat plane in Cartesian three-space.

Any function in which an exponent appears is generally nonlinear. Logarithmic, trigonometric, and many other kinds of functions are nonlinear. *See also* CARTESIAN COORDINATES, FUNCTION, LINEAR FUNCTION.

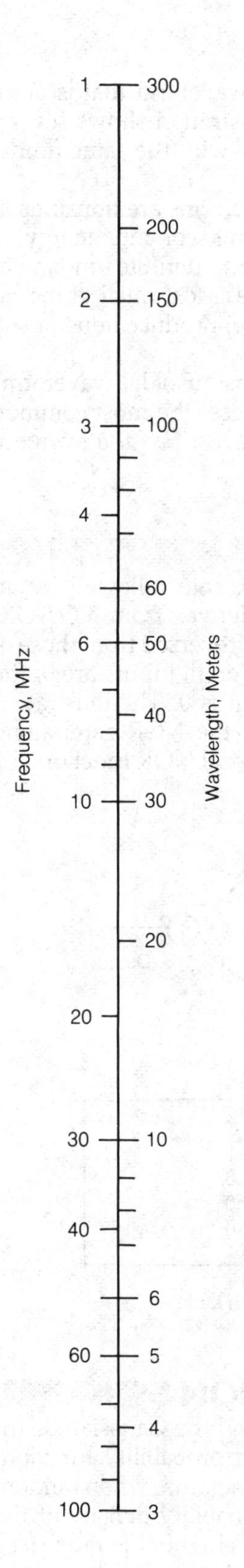

NOMOGRAPH: A nomograph showing the relation between frequency and wavelength.

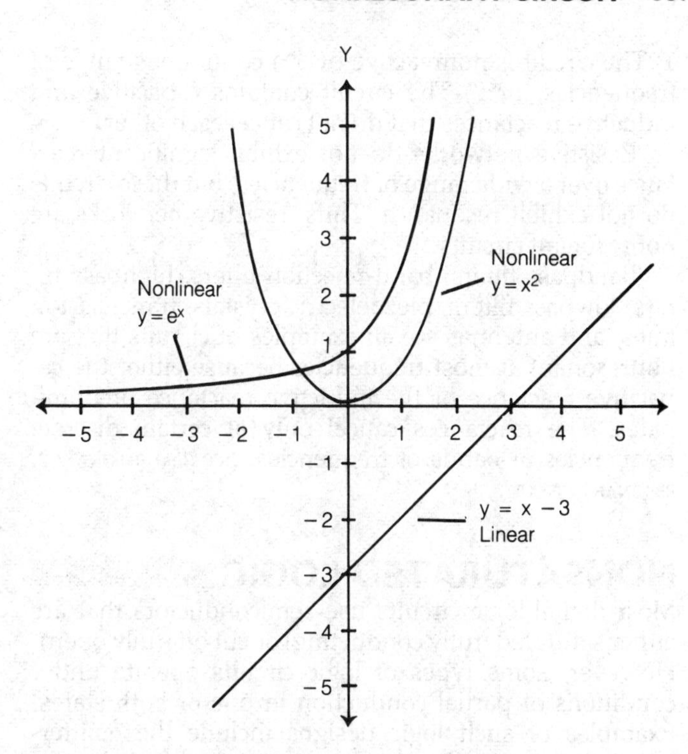

NONLINEAR FUNCTION: Two nonlinear functions are shown by curved lines. The straight line shows a linear function.

NONPOLARIZED COMPONENT

A component is considered nonpolarized if it can be inserted either way in a circuit with the same results. Examples of nonpolarized components are resistors, inductors, and many types of capacitors.

If the leads of a component cannot be interchanged or reversed without adversely affecting circuit performance, the component considered is polarized. *See also* POLARIZED COMPONENT.

NONPOLARIZED ELECTROLYTIC CAPACITOR

See CAPACITOR.

NONREACTIVE CIRCUIT

All circuits contain inductive or capacitive reactance because of effects among the various leads and electrodes. However, this reactance may be so small that it can be neglected in some instances. A circuit is nonreactive if it contains no effective inductive or capacitive reactance as seen from its input terminals.

A circuit may be nonreactive at just one frequency. An example is the common resonant circuit. At the resonant frequency, the inductive and capacitive reactances cancel, and only resistance is left (*see* RESONANCE).

A circuit may be nonreactive at one frequency and an infinite number of harmonic frequencies. For example, a quarter-wave section of transmission line with a fundamental frequency *f* is also resonant at frequencies 2*f*, 3*f*, 4*f* and so on. At all of these frequencies, the line measures an integral multiple of a quarter wavelength. A quarter-wave radiating element exhibits the same property.

Some circuits are essentially nonreactive over a wide range of frequencies. This type of circuit generally contains no reactive components, but only resistances. At all frequencies below a certain maximum frequency, the circuit can be considered nonreactive; but above this subjective maximum, component leads introduce significant reactance. *See also* REACTANCE, REACTIVE CIRCUIT.

NONRESONANT CIRCUIT

A circuit is nonresonant if either of the following is true:

1) The circuit is nonreactive over a continuous range of frequencies, or 2) The circuit contains capacitive and inductive reactances that do not cancel each other.

Resistive networks do not exhibit significant reactance over a wide range of frequencies, but these circuits do not exhibit resonance. Thus, resistive networks are nonresonant circuits.

Bandpass filters, band-rejection filters, highpass filters, lowpass filters, piezoelectric crystals, transmission lines, and antennas are all examples of circuits that are nonresonant at most frequencies because either the capacitive reactance or the inductive reactance predominates. The reactances cancel only at certain discrete frequencies or bands of frequencies. *See also* RESONANCE, RESONANT CIRCUIT.

NONSATURATED LOGIC

Most digital-logic circuits use semiconductors that are either saturated (fully conducting) or cut off (fully open). However, some types of logic circuits operate under conditions of partial conduction in one or both states. Examples of such logic designs include the emitter-coupled, integrated injection, and triple-diffused emitter-follower varieties (*see* EMITTER-COUPLED LOGIC, INTEGRATED INJECTION LOGIC).

Nonsaturated logic has certain advantages. The most notable advantage is its higher switching speed, compared with most saturated logic forms. However, a nonsaturated transistor is somewhat more susceptible to noise than is a saturated transistor. Recent technological improvements have reduced the noise susceptibility of nonsaturated bipolar logic devices.

NONSINUSOIDAL WAVEFORM

A waveform is nonsinusoidal if its shape cannot be precisely represented by the sine function. That is, a nonsinusoidal waveform is any waveform that is not a sine wave (*see* SINE WAVE). The illustration shows several periodic nonsinusoidal waves, all with the same fundamental period.

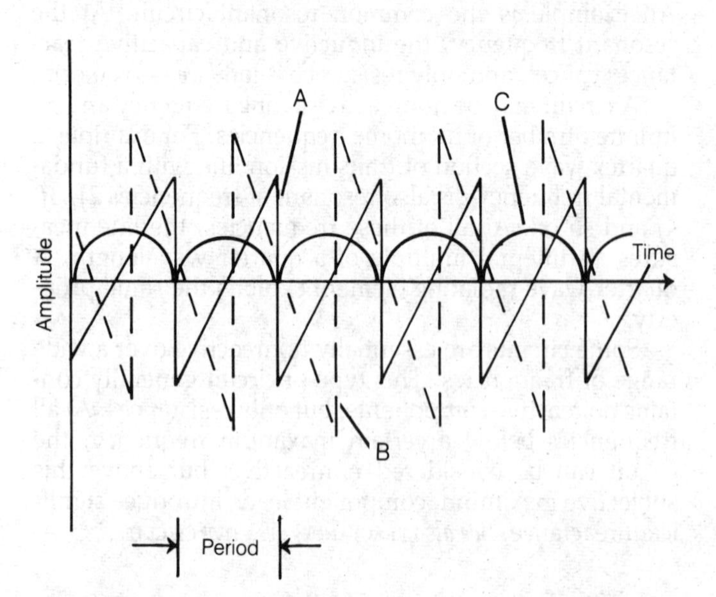

NONSINUSOIDAL WAVEFORM: A sawtooth wave (A), reversed sawtooth with some distortion (B), and a sine square root (C).

Most naturally occurring waveforms are nonsinusoidal. All nonsinusoidal waveforms contain energy at more than one frequency, although a definite fundamental period is usually ascertainable. Most musical instruments, as well as the human voice, produce nonsinusoidal waves.

Certain forms of periodic nonsinusoidal waveforms are used for test purposes. Of these, the most common are the sawtooth and square waves. *See also* SAWTOOTH WAVE, SQUARE WAVE.

NOR GATE

A NOR gate is an inclusive-OR gate followed by an inverter. The expression NOR derives from NOT-OR. The NOR gate outputs are exactly reversed from those of the OR gate. That is, when both or all inputs are 0, the output is 1; otherwise, the output is 0. The illustration shows the schematic symbol for the NOR gate, along with a truth table showing the logical NOR function. *See also* OR GATE.

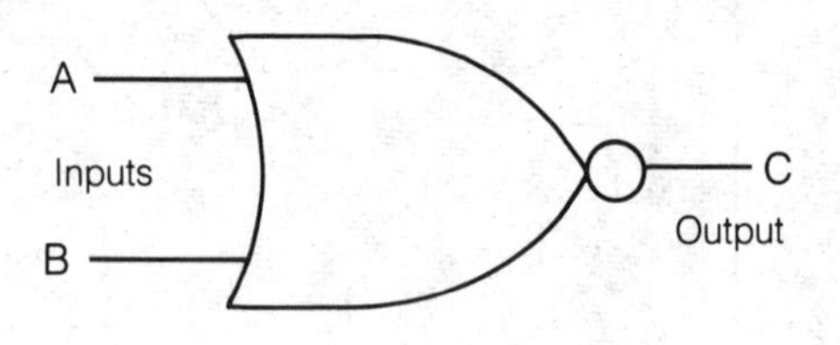

A	B	C
0	0	1
0	1	0
1	0	0
1	1	0

NOR GATE: Schematic symbol and truth table.

NORMAL DISTRIBUTION

The normal distribution, also called the Gaussian distribution or bell-shaped curve, is a probability function. The normal distribution is found in nature when random events occur. The maximum probability density of the normal distribution occurs in the center of the range (see illustration). The total area under the curve is 1.

There are infinitely many possible variations of the normal distribution, depending on the value of the standard deviation (s). The smaller the standard deviation, the more the distribution is concentrated toward the center. The larger the standard deviation, the flatter the function becomes. The functions illustrated show normal distributions for $s = 0.5$, $s = 1$, and $s = 2$. When $s = 0$, the function is entirely concentrated at the center of the range or x axis. As the value of s becomes very large, the

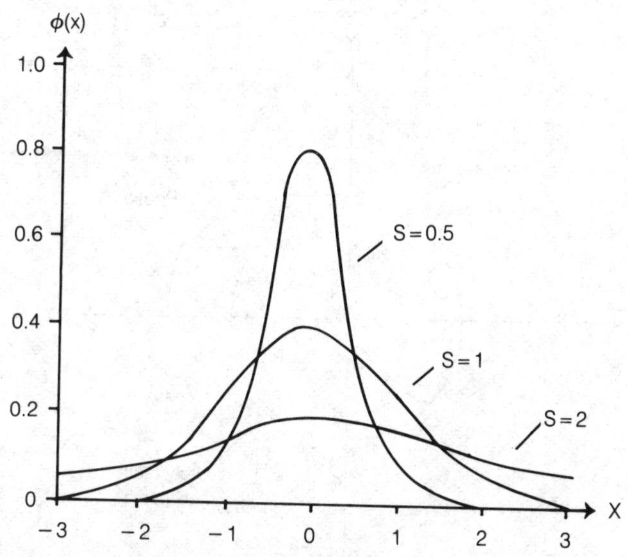

NORMAL DISTRIBUTION: Three examples of normal distributions with different values of standard deviation (s).

function approaches the x axis, and the central maximum becomes less and less defined. *See also* STANDARD DEVIATION.

NORTON'S THEOREM

Norton's theorem describes the properties of circuits with more than one generator and/or more than one constant impedance.

In a circuit with two or more constant-current, constant-frequency generators and/or two or more constant impedances, there is an equivalent circuit consisting of exactly one constant-current, constant-frequency generator and exactly one constant impedance. If the more complicated circuit is replaced by the simpler circuit, identical conditions will prevail. *See also* COMPENSATION THEOREM, RECIPROCITY THEOREM, SUPERPOSITION THEOREM, THEVENIN'S THEOREM.

NOTCH FILTER

A notch filter is a narrowband-rejection filter. Notch filters are found in many superheterodyne receivers. The notch filter is extremely convenient for reducing interference caused by strong, unmodulated carriers within the passband of a receiver.

Notch-filter circuits are generally inserted in one of the intermediate-frequency stages of a superheterodyne receiver, where the bandpass frequency is constant. There are several different kinds of notch-filter circuit. One of the simplest is a trap configuration, inserted in series with the signal path, see A in illustration. The notch frequency is adjustable, so that the deep null (B) can be tuned to any frequency within the receiver passband.

A properly designed notch filter can produce attenuation well in excess of 40 decibels. *See also* BAND-REJECTION FILTER.

NOT GATE

See INVERTER.

NPN TRANSISTOR

See TRANSISTOR.

NTSC COLOR TELEVISION

See COLOR TELEVISION, TELEVISION.

N-TYPE SEMICONDUCTOR

An N-type material is a semiconductor material that conducts mostly by electron transfer. N-type semiconductors are formed by adding certain impurity elements to the semiconductor. These impurity elements, called donors, contain an excess of electrons. Examples include antimony, phosphorus, and arsenic. The N-type semiconductor is so named because it conducts by negative charge carriers (electrons).

The carrier mobility in an N-type semiconductor is greater than the carrier mobility in a P-type material, because electrons are transferred more rapidly, from atom to atom, than are holes, for a given applied electric field.

N-type material combined with P-type material is the basis for manufacture of most semiconductor diodes, transistors, and integrated circuits. *See also* CARRIER MOBILITY, DOPING, ELECTRON, HOLE, P-N JUNCTION, P-TYPE SEMICONDUCTOR.

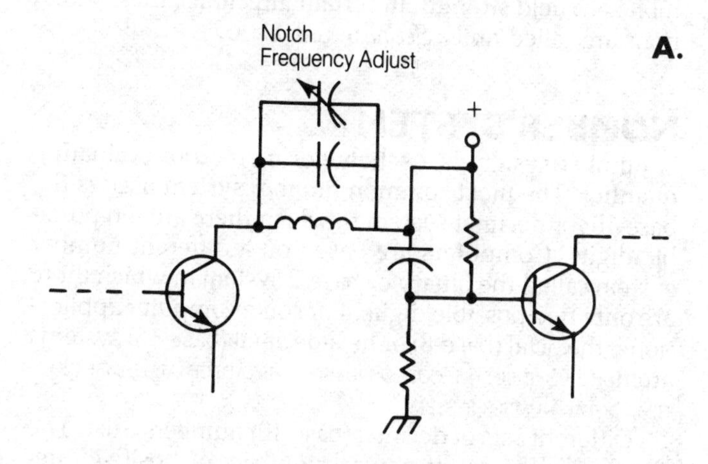

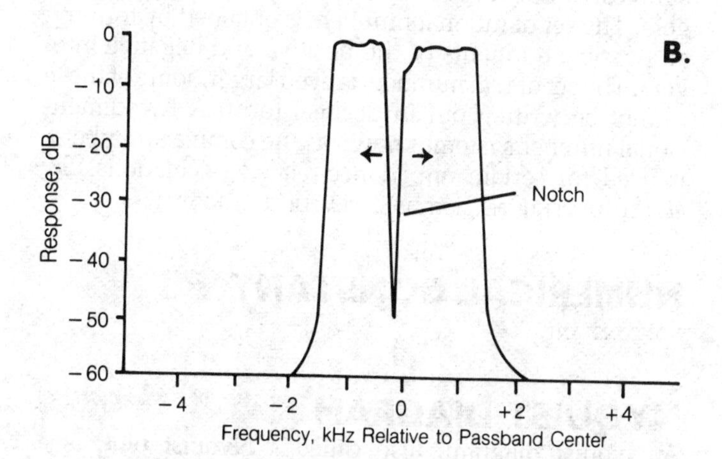

NOTCH FILTER: A simple notch-filter circuit (A) and a drawing of the spectral response with the notch in the passband (B).

NUCLEAR MAGNETIC RESONANCE

In the presence of a strong, alternating magnetic field, the nuclei of certain atoms will oscillate because of electromagnetic forces between the charged nucleus and the magnetic field. The phenomenon is known as nuclear magnetic resonance (NMR).

Nuclear magnetic resonance is used for medical diagnostic purposes. The patient's body is exposed to a strong, radio-frequency electromagnetic field, which causes the nuclei of the atoms in the body to oscillate and emit their own radio waves. These radio waves can be received and processed by a computer, producing detailed cross-sectional pictures of the interior of the body. The technique also allows physicians to evaluate the functioning of body organs. *See also* COMPUTER-AIDED MEDICAL IMAGING.

NULL

A null is a condition of zero output from an alternating-current circuit resulting from phase cancellation or signal balance.

An alternating-current bridge circuit is adjusted for a zero-output, or null, reading to determine the values of unknown capacitances, inductances, or impedances. The radiation pattern from an antenna system may exhibit zero field strength in certain directions; these directions are called nulls. *See also* PHASE BALANCE.

NUMBER SYSTEM

A number system is a scheme or method of evaluating quantity. The most common number system used is the base-10, or decimal system in which there are ten possible digits. Computers are based on a different number system called the binary or base-2 system in which there are only two possible digits. For some computer applications, the octal (base-8) or hexadecimal (base-16) systems are used. *See* BINARY-CODED NUMBER, HEXADECIMAL NUMBER SYSTEM, OCTAL NUMBER SYSTEM.

Different sets of decimal (base-10) numbers exist. The simplest is the set of natural numbers or positive integers. The set of rational numbers is obtained by forming all possible quotients of the positive and negative integers. The set of real numbers is even larger; some of these cannot be written out in decimal form. A two-dimensional number system, known as the complex numbers, is used for certain impedance-related calculations. *See also* COMPLEX NUMBER, RATIONAL NUMBER, REAL NUMBER.

NUMERICAL CONSTANT

See CONSTANT.

NYQUIST DIAGRAM

A Nyquist diagram, also called a Nyquist plot, is a complex-number graph for evaluating the performance of a feedback amplifier. The Nyquist-diagram plane is a

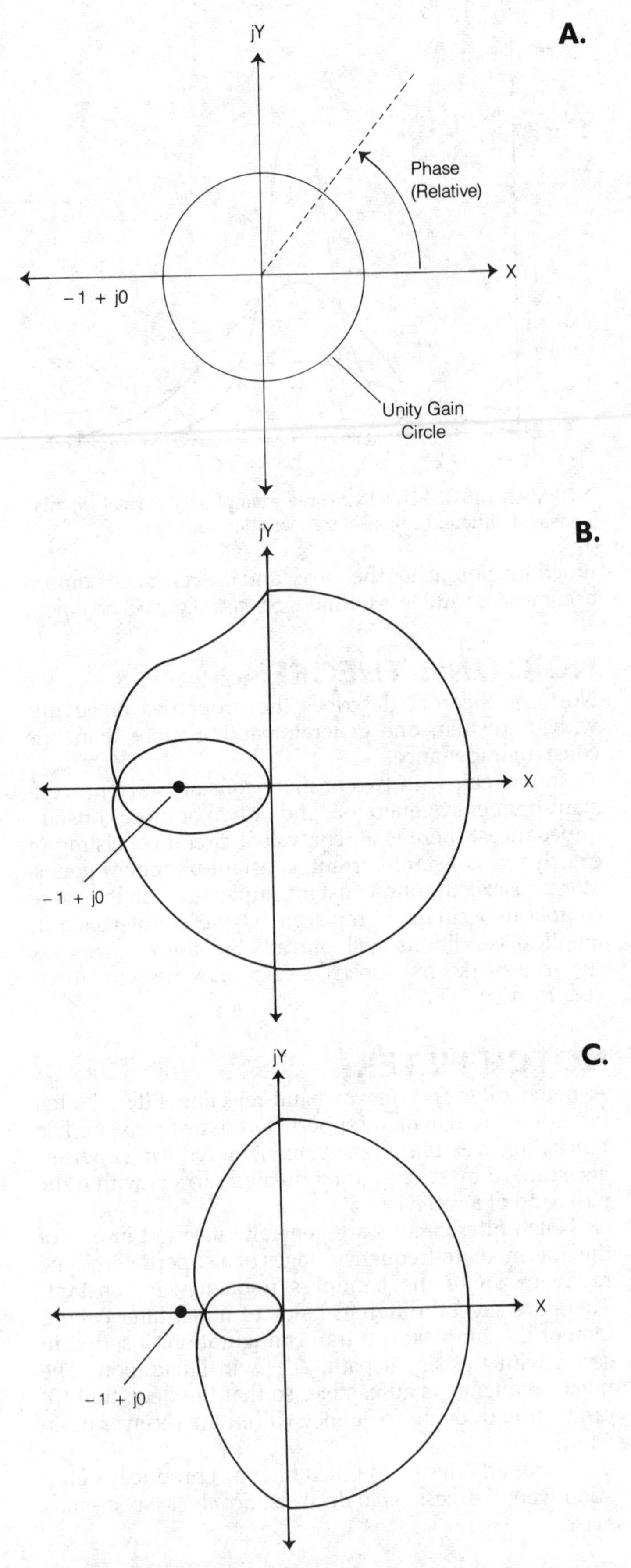

NYQUIST DIAGRAM AND NYQUIST STABILITY CRITERION: The Nyquist plane (A), the plot for unstable feedback amplifier (B), and a plot for a stable feedback amplifier (C).

Cartesian system. The real-number axis (or x axis) is horizontal, and the imaginary-number axis (or y axis) is vertical.

In the Nyquist diagram open-loop gain is represented by circles centered at the origin. The circle with radius 1 is called the unity-gain circle. Points inside this circle represent amplifier gain less than 1; points outside this circle correspond to gain figures greater than 1. The output phase is represented by radial lines passing through the origin. This is shown in A of the illustration. The Nyquist diagram is made by plotting the real and imaginary components of the amplifier output in the complex plane for all possible input frequencies. Two examples of these plots are shown at B and C.

The Nyquist diagram is used to determine if a feedback amplifier is stable. *See* NYQUIST STABILITY CRITERION.

NYQUIST FREQUENCY

See NYQUIST THEOREM.

NYQUIST RATE

The Nyquist rate is the rate at which an analog signal must be sampled (digitized) to ensure accurate representation and reconstruction. The Nyquist rate is twice the highest frequency in the signal. For example, a 100-Hz sine wave must be sampled 200 times per second or more to be accurately represented and reconstructed from the sampled data. Signals sampled below the Nyquist rate may exhibit aliasing. *See* ALIASING, NYQUIST THEOREM, SAMPLING THEOREM.

NYQUIST STABILITY CRITERION

To determine if a single-loop feedback amplifier is stable the Nyquist diagram can be used. This is known as the Nyquist stability criterion.

The Nyquist plot is obtained for all possible input frequencies. If the Nyquist plot contains a closed loop surrounding the point $-1 + jO$, then the system is unstable. If the plot contains no such loop and the open-loop configuration is stable, then the amplifier is stable.

The illustration under NYQUIST DIAGRAM shows the Nyquist plane at A. At B, an example of a Nyquist plot for an unstable single-loop feedback amplifier is shown. At C, a plot is illustrated for a stable single-loop feedback amplifier. *See also* NYQUIST DIAGRAM.

NYQUIST THEOREM

The Nyquist theorem is a rule applied in signal processing for sampling an analog signal frequency to minimize the effect of high-frequency noise on its digital conversion. It says that an analog waveform should be sampled at a rate that is at least twice the highest frequency found in the signal.

This sampling rate is called the Nyquist rate, while the signal frequency (at least half the Nyquist rate) is referred to as the Nyquist frequency. If unexpectedly high frequencies should be within the signal or accompanying noise on the input channel, they cause aliasing. This shows up as spurious low-frequency signals caused by an intermodulation of high high-frequency signal components (and input noise) with harmonics of the sampling frequency. In communications, aliasing is unwanted, but in CRT display technology intentional aliasing is used to blend the jagged edges between regions of contrasting color or tone. (*See* ALIASING.)

The desired frequencies are usually known in communications permitting the Nyquist rate to be set. But it may not be feasible to sample fast enough to prevent aliasing in all situations. In either case, a lowpass filter can be attenuate these unwanted high frequencies. (*See* FILTER, LOWPASS FILTER.) A filter that eliminates aliasing is called an anti-aliasing filter.

Lowpass anti-aliasing filters should have a flat passband response, steep rolloff, and good phase response. If the amplitude and phase response are critical, a Butterworth filter is recommended. But if superior amplitude characteristics are desired at the expense of the better low-frequency characteristics of the Butterworth filter, the Chebyshev filter is selected. *See* BUTTERWORTH FILTER, CHEBYCHEV FILTER.

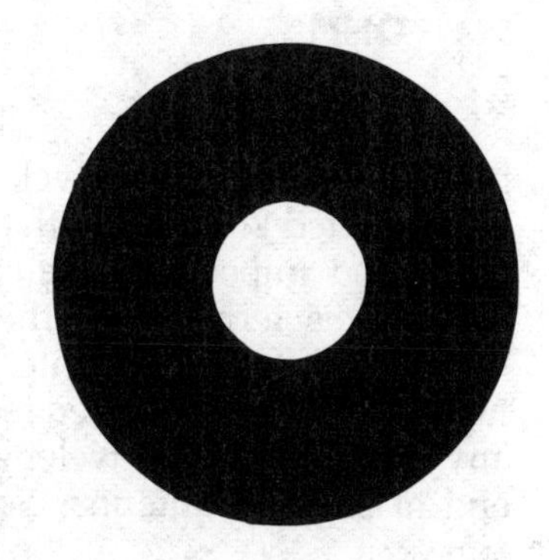

OCTAL NUMBER SYSTEM

The octal number system is a base-8 system. Only the digits 0 through 7 are used. When counting in base 8, the number following 7 is 10; the number following 77 is 100, and so on. Moving toward the left, the digits are multiplied by 1, then by 8, then by 64, and so on, upward in successively larger integral powers of 8.

The octal number system is used as shorthand for long binary numbers. The binary number is split into groups of three digits beginning at the extreme right, and then the octal representation is given for each three-digit binary number. For example, consider the binary number 11000111. This is an eight-digit binary number. Adding a zero to the left-hand end of this number, making it nine digits long (a multiple of three), yields 011000111. Breaking up this number gives 011 000 111. These three binary numbers correspond to 3, 0, and 7 respectively. The octal representation, then, of 11000111 is 307.

The octal number system is used by some computers because its base, 8, is a power of 2, making it easier to work with than the base-10 system. A base-16 number system is also sometimes used by computers. *See also* HEXADECIMAL NUMBER SYSTEM.

OCTAVE

An octave is a 2:1 range of frequencies. If the lower limit of an octave is represented in hertz by f, then the upper limit is represented by $2f$, or the second harmonic of f. If the upper limit of an octave is given in hertz by f, then the lower limit is $f/2$.

The term *octave* can be used when specifying the gain-versus-frequency characteristics of an operational amplifier. A circuit might, for example, have a rolloff characteristic (decrease in gain) of 6 decibels per octave. This means that if the gain is x dB at frequency f, then the gain is $x - 6$ dB at frequency $2f$, $x - 12$ decibels at frequency $4f$, $x - 18$ decibels at frequency $8f$, and so on. In audio frequencies, the octave can be recognized by ear. *See also* HARMONIC.

ODD-ORDER HARMONIC

An odd-order harmonic is any odd multiple of the fundamental frequency of a signal. For example, if the fundamental frequency is 1 MHz, then the odd-order harmonics occur at frequencies of 3 MHz, 5 MHz, 7 MHz, and so on.

Certain conditions favor the generation of odd-order harmonics in a circuit, while other circumstances tend to cancel out the odd harmonics. The push-pull circuit accentuates the odd harmonics (*see* PUSH-PULL CONFIGURATION). These circuits are often used as frequency triplers. The push-push circuit (*see* PUSH-PUSH CONFIGURATION) tends to cancel out the odd harmonics in the output.

A half-wave dipole antenna will operate at all odd-harmonic frequencies, assuming it is a full-size antenna (not inductively loaded). The impedance at the center of an antenna is purely resistive at all odd-harmonic frequencies. At the odd harmonics, the dipole antenna presents a slightly higher feed-point resistance than at the fundamental frequency, but the difference is not usually large enough to cause an appreciable mismatch when the antenna is operated at the odd harmonics. *See also* EVEN-ORDER HARMONIC.

OERSTED

The oersted is a unit of magnetic-field intensity in the centimeter-gram-second (CGS) system, expressed in terms of magnetomotive force per unit length. A field of 1 oersted is equivalent to a field of 1 gilbert, or 0.796 ampere turns per centimeter. *See also* AMPERE TURN, GILBERT, MAGNETOMOTIVE FORCE.

OFF-CENTER FEED

Off-center feed is a method of applying power to an antenna at a current or voltage loop not at the center or end of the radiating element. Off-center feed is sometimes called windom feed, but the windom is only one form of off-center-fed antenna (*see* ANTENNA DIRECTORY).

For off-center feed to be possible, an antenna must be at least 1 wavelength long. A 1-wavelength antenna has current loops at distances of ¼ wavelength from either end. A 1-wavelength radiator can thus be off-center fed with coaxial cable, as shown in the illustration.

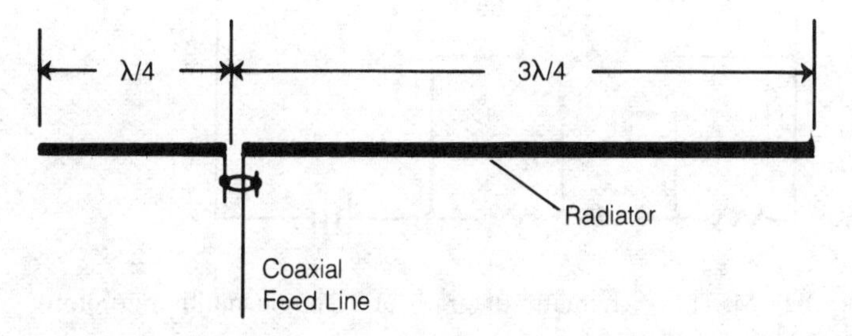

OFF-CENTER FEED: An off-center fed, full-wavelength radiator.

In general, for a longwire antenna measuring an integral multiple of ½ wavelength, a coaxial feed line can be connected ¼ wavelength from one end, and a reasonably good impedance match will be obtained. If the longwire is terminated, the feed point should be ¼ wavelength from the unterminated end. A low-loss, high-impedance, balanced feed line may be connected at any multiple of ¼ wavelength from the unterminated end of such an antenna. *See also* CENTER FEED, END FEED, ANTENNA.

OHM

The ohm is the standard unit of resistance, reactance, and impedance. The symbol is the Greek capital letter omega (Ω). A resistance of 1 ohm will conduct 1 A of current when a voltage of 1 V is placed across it. *See also* IMPEDANCE, REACTANCE, RESISTANCE.

OHMIC LOSS

Ohmic loss is the loss of power that occurs in a conductor because of resistance. Ohmic loss is determined by the type of material used to carry a current, and by the size of the conductor. In general, the larger the conductor diameter for a given material, the lower the ohmic loss. Ohmic loss results in heating of a conductor.

For alternating currents, the ohmic loss of a conductor generally increases as the frequency increases. This occurs because of a phenomenon known as skin effect. In high-frequency alternating currents tend to flow mostly near the surface of a conductor (*see* SKIN EFFECT).

In a transmission line, ohmic loss is just one form of loss. Additional losses can occur in the dielectric material. Loss can also be caused by unwanted radiation of electromagnetic energy, and by impedance mismatches. *See also* DIELECTRIC LOSS, RADIATION LOSS, STANDING-WAVE-RATIO LOSS.

OHMMETER

An ohmmeter is an instrument for measuring direct-current resistance. A source of known voltage, usually a battery, is connected in series with a switchable known resistance, a milliammeter, and a pair of test leads (see illustration).

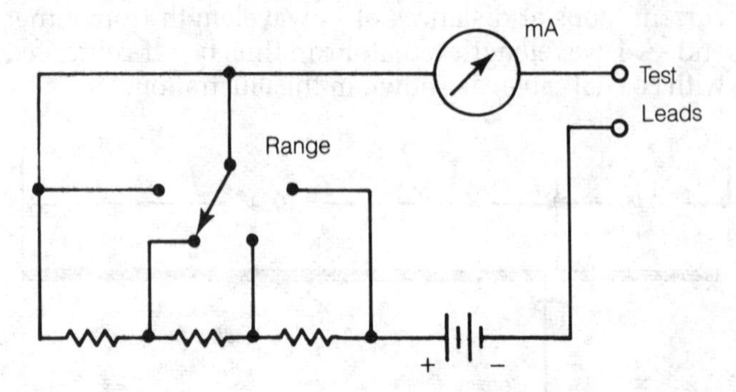

OHMMETER: Schematic diagram of a simple multirange ohmmeter.

Most ohmmeters have several ranges, labeled according to the magnitude of the resistances in terms of the scale indication. The scale is calibrated from 0 to infinity in a nonlinear manner. The range switch, by inserting various resistances, allows measurement of fairly large or fairly small resistances.

Most ohmmeters can accurately measure any resistance between about 1 ohm and several hundred megohms. Specialized ohmmeters are needed for measuring extremely small or large resistances. *See also* MEGGER, WHEATSONE BRIDGE.

OHM'S LAW

Ohm's law is a simple relation between the current, voltage, and resistance in a circuit. The current, voltage, and resistance in a direct-current situation are interdependent. If two of the quantities are known, the third can be found by a simple equation.

Letting I represent the current in amperes, E represent the voltage in volts, and R represent the resistance in ohms, the following relation holds:

$$I = E/R$$

The formula may also be restated as:

$$E = IR, \text{ or}$$
$$R = E/I$$

In an alternating-current circuit, pure reactance can be substituted for the resistance in Ohm's law. If both reactance and resistance are present, however, the situation becomes more complex. *See also* CURRENT, RESISTANCE, VOLTAGE.

OHMS PER VOLT

A voltmeter is rated for sensitivity according to a specification known as ohms per volt. The ohms per volt rating is an indication of the extent to which the voltmeter will affect the operation of a high-impedance circuit.

To determine the ohms-per-volt rating of a voltmeter, the resistance through the meter must be measured at a certain voltage. Normally this is the full-scale voltage. This resistance can be found by connecting the voltmeter in series with a milliammeter or microammeter, and then placing a voltage across the combination such that the meter reads full scale. The ohms-per-volt (R/E) rating is then equal to 1 divided by the milliammeter or microammeter reading expressed in amperes.

For example, suppose that a voltmeter with a full-scale reading of 100 V is connected in series with a microammeter. A source of voltage is connected across this combination, and the voltage adjusted until the voltmeter reads full scale. The meter reads 10μA. Then the ohms per volt rating of the meter is 1/0.00001, or 100,000, ohms per volt.

It is desirable that a voltmeter have a very high ohms per volt rating. The higher this figure, the smaller the current drain produced by the voltmeter when it is connected into a circuit.

OMEGA

The Omega electronic navigation system is a global network of very low frequency (vlf) radio stations broadcasting at frequencies from 10.2 to 13.6 kHz. Omega offers complete global coverage with only eight strategically located transmitters. This is possible because of the long range and day and night stability of vlf signals. Only six transmitters are actually required, but two additional transmitters provide backup for equipment failure or maintenance. The stations are located approximately 6,000 miles apart, and at any place on the earth at least four stations are usable.

The Omega system was developed by the U.S. Navy for communication with surfaced or submerged submarines as well as for ships and aircraft. Signals in the vlf range can be received under water. Omega receiving equipment is now available for commercial ships and aircraft. Omega stations transmit intermittent continuous-wave (CW) signals rather than pulses on each frequency used. As in Loran, signals from a single pair of stations on a single frequency can provide a hyperbolic line of position (LOP). However, for Omega to be useful, the ship or aircraft must know its approximate position so that the set of LOPs within which the receiver is located can be identified. *See* LORAN.

An Omega fix can be made with two or more LOPs. For the highest accuracy with station pairs, the LOPs selected should be as close as possible to 90 degrees apart. With three stations, the LOPs should be as close as possible to 60 degrees apart. Special charts overprinted with Omega LOPs. The latest Omega receivers include microcomputers, and they can display LOPs and compute and display positions directly in latitude and longitude.

The nominal all-weather accuracy of Omega is one mile in daytime and two miles at night due to the effects of sky-wave propagation conditions. Because Omega transmitter signals are controlled by atomic frequency standards, they can be used as accurate sources of time.

OMNIDIRECTIONAL ANTENNA

See ISOTROPIC ANTENNA in the ANTENNA DIRECTORY.

OMNIDIRECTIONAL MICROPHONE

An omnidirectional microphone is a microphone that responds equally well to sound from all directions. Omnidirectional microphones often have spherical heads.

Omnidirectional microphones are used when it is necessary to pick up background noises as well as the voice or music of the subject. Omnidirectional microphones are not generally desirable for communications applications. *See also* MICROPHONE.

ONE-SHOT MULTIVIBRATOR

See MONOSTABLE MULTIVIBRATOR.

OPEN CIRCUIT

A circuit is open when current cannot flow. An open circuit can occur as a result of a deliberate action, such as the opening of a switch or relay, or as a result of an accidental break in a line or disconnection of a soldered or connected termination.

In an open circuit, the current is zero, and the resistance is theoretically infinite. *See also* CLOSED CIRCUIT.

OPEN LOOP

In an amplifier, the condition of zero negative feedback is called the open-loop condition. With an open loop, an amplifier is at maximum gain.

In an operational-amplifier circuit, the open-loop gain is specified as a measure of the amplification factor of a particular integrated circuit. The input and output impedances are generally specified for the open-loop configuration. *See also* OPERATIONAL AMPLIFIER.

OPEN-WIRE LINE

Open-wire line is a form of radio-frequency feed line often called parallel-wire line. Open-wire feed lines are commercially manufactured with characteristic impedances of approximately 300 and 450 Ω. Open-wire lines can be made with characteristic impedances as high as 600 ohms.

The characteristic impedance of an air-dielectric balanced line depends on the size of the conductors and the spacing between the conductors. If the radius of the conductors is given by r and the center-to-center spacing by s, then the characteristic impedance Z_O is:

$$Z_O = 276 \log_{10} (s/r)$$

provided, of course, that r and s are specified in the same units. The presence of spacers tends to lower this value slightly. Also, if the line is placed near objects with a higher dielectric constant than air (that includes most materials), the characteristic impedance will be lowered slightly (*see* CHARACTERISTIC IMPEDANCE).

The illustration shows a typical open-wire line. The wires are spaced at a distance of from ½ inch to 6 inches in most cases. The spacers are placed at regular intervals to keep the wires at a roughly constant separation. Open-wire line is a balanced transmission line.

Open-wire feed line is characterized by low loss and high power-handling capacity, even in the presence of significant mismatches between the line and the antenna. Because open-wire line is an inherently balanced line, it should be used with a balanced load. A tuning network, having a balanced output, is needed at the transmitter or receiver (*see* BALANCED LOAD, BALANCED TRANSMISSION LINE).

Open-wire line is ideal for television reception in

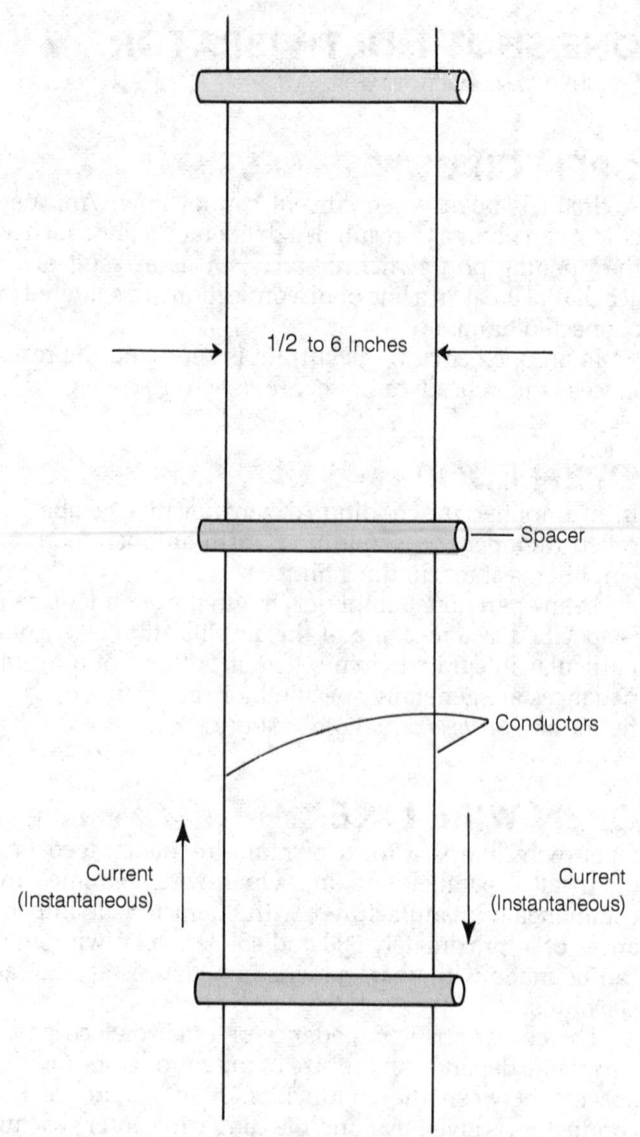

OPEN-WIRE LINE: Construction of an open-wire transmission line.

fringe areas, since it has lower loss, per unit length, than the standard twin-lead lines. *See also* TWIN-LEAD.

OPERATING ANGLE

In an amplifier, the operating angle is the number of degrees, for each cycle, during which current flows in the collector, drain, or plate circuit. The operating angle varies, depending on the class of amplifier operation.

In a Class A amplifier, the output current flows during the entire cycle. Therefore, in such an amplifier, the operating angle is 360 degrees. In a Class AB amplifier, the current flows for less than the entire cycle, but for more than half of the cycle; hence the operating angle is larger than 180 degrees but less than 360 degrees. In a single-ended Class B amplifier, the operating angle is about 180 degrees. In a Class C amplifier, the current flows for much less than half of the cycle, and the operating angle is thus smaller than 180 degrees.

The operating angle varies with the relative base/collector, gate/drain, or grid/plate bias of an amplifier. The operating angle is also affected by the driving voltage. *See also* CLASS A AMPLIFIER, CLASS AB AMPLIFIER, CLASS B AMPLIFIER, CLASS C AMPLIFIER, OPERATING POINT.

OPERATING BIAS

See OPERATING POINT.

OPERATING POINT

The operating point of an amplifier circuit is the point at which the direct-current bias is applied along the curve depicting the collector current versus base voltage (for a transistor), the drain current versus gate voltage (for a field-effect transistor), or the plate current versus grid voltage (for a tube). The choice of operating point determines the class of operation for the amplifier. *See* CLASS A AMPLIFIER, CLASS AB AMPLIFIER, CLASS B AMPLIFIER, CLASS C AMPLIFIER, FIELD-EFFECT TRANSISTOR, TRANSISTOR.

Certain transducers and semiconductor devices require direct-current bias for proper operation. When the signal is received or applied, the instantaneous current fluctuates above and below the no-signal current. The no-signal current is called the operating current or operating point. The correct choice of operating point ensures that the device will function according to its ratings.

OPERATING SYSTEM

An operating system is an integrated collection of supervisory programs and subroutines that controls the execution of computer programs and performs special system functions. This software organizes a central processor and peripheral devices into an active unit for the development and execution of programs. Some of the important functions performed by the operating system are:

1. Job scheduling selects programs waiting in an input queue and schedules them for processing.
2. Memory management assigns programs to specific locations in main memory and releases the main memory when the programs are completed.
3. Input/output (I/O) control directs I/O activities and handles interrupt conditions while optimizing channels and peripheral devices.
4. Multiprogramming schedules and controls the simultaneous execution of several programs.

The central component of the operating system is the supervisor or executive. The supervisor schedules and controls the operations of the computer. Supervisors perform the following duties:

1. Load processing programs into main memory from disk memory, as requested.
2. Schedule program sequence for maximum efficiency.

3. Schedule and control I/O operations and handle interrupts, or signals to the CPU.

Most operating systems have been developed as proprietary software by the computer manufacturer. Major computer companies such as International Business Machines Corporation (IBM) and Digital Equipment Corporation (DEC), still sell computers with proprietary operating systems. Other computer companies offering proprietary operating systems include Apple and Tandem Computer Corporation.

Because the operating system determines the applications programs that can be run on a computer, users do not want to be limited to software that is written specifically for a brand name computer. This has encouraged the development of standard operating systems that will permit wide software interchange. The best known and most popular standard operating system is MS-DOS (Microsoft Disk Operating System) from Microsoft developed for IBM personal computers. Others include CP/M and OASIS.

American Telephone and Telegraph Corp. (AT&T) developed a standard operating system for larger computers called UNIX. It is competitive with OS/2 (Operating System 2) developed by IBM for its more advanced personal computers. Many new large computer systems and CAD (computer-aided design) workstations have been designed to be compatible with UNIX. Large computer companies including DEC are now also offering UNIX options for some of their computer systems.

OPERATIONAL AMPLIFIER

An operational amplifier is an amplifier that exhibits high stability and linear characteristics. The ideal operational amplifier, under optimum conditions, would theoreti-

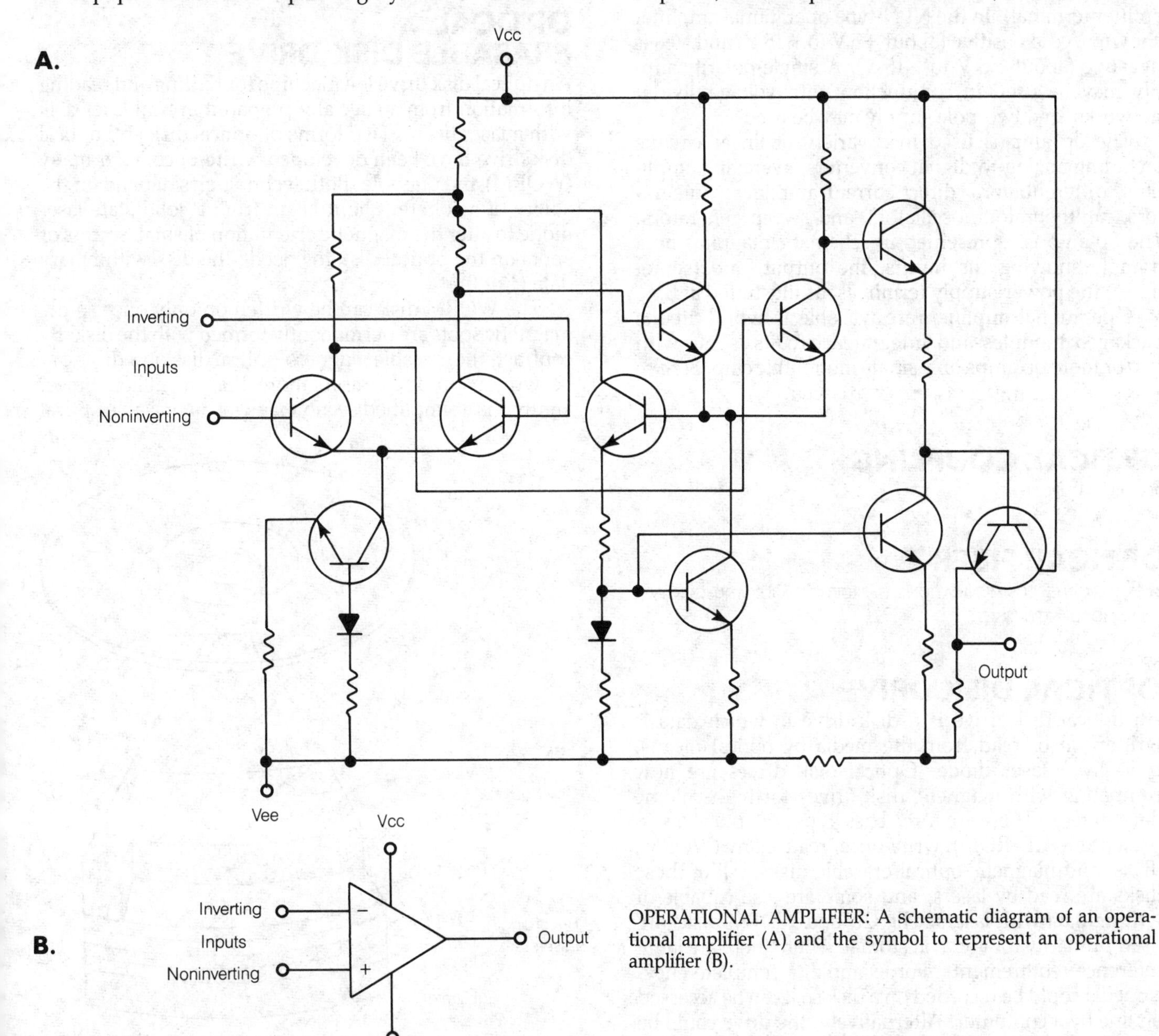

OPERATIONAL AMPLIFIER: A schematic diagram of an operational amplifier (A) and the symbol to represent an operational amplifier (B).

cally have infinite input impedance, zero output impedance, infinite gain, and infinite bandwidth. These ideals cannot be realized in practice, but many operational amplifiers (op amps) have very high input impedance (they draw almost no current and hence almost no power), very low output impedance, extremely high gain (more than 100 decibels in some cases), and large bandwidth (up to several megahertz).

A typical operational amplifier is shown in A of the schematic diagram. It has two inputs called the inverting and noninverting. The inverting input provides a 180-degree phase shift at the output. A negative-feedback loop, consisting of a resistor and/or capacitor between the inverting input and the output, controls the gain of the amplifier. The noninverting input is in phase with the output.

Two power-supply terminals are provided; they are usually called Vcc (the collector terminal) and Vee (the emitter terminal). In the NPN-type operational amplifier shown, Vcc is positive (about +5 V to +15 V) and Vee is negative (about −5 V to −15 V). A single-polarity supply may be used in conjunction with voltage-divider networks, or a two-pole supply may be used.

The op amp is used in a variety of linear circuits including analog-to-digital converters, averaging amplifiers, differentiators, direct-current amplifiers, integrators, multivibrators, oscillators, and sweep generators. The op amp is represented in schematic diagrams by a triangle showing the inputs, the output, and (sometimes) the power-supply terminals, as illustrated at B.

Operational amplifiers are available as hybrid circuits packaged modules and integrated circuits, some with two or more op amps on a single monolithic chip. *See also* INTEGRATED CIRCUIT, LINEAR INTEGRATED CIRCUIT.

OPTICAL COUPLING

See OPTOCOUPLER.

OPTICAL FIBER

See FIBEROPTIC COMMUNICATION, FIBEROPTIC CABLE AND CONNECTORS, MODULATED LIGHT.

OPTICAL DISK DRIVE

An optical disk drive is a disk drive in which data is written on or read from the media by optical means, typically a laser diode. Optical disk drives are now competing with magnetic disk drives for high-volume data storage. There are three basic types of optical disks: prewritten (CD-ROM); write-once, read-many (WORM) disks; and magnetic-optical erasable disks. All of these disks are read by lasers, and some are also capable of writing data with lasers. The CD and WORM disks are used primarily for archival storage where there are large reference requirements. For example, a complete encyclopedia could be recorded on a disk so it can be accessed on line by a computer. Alternatively, the drive could be used for writing a full year's business transactions so they can be accessed the following year.

CD-ROM Optical Disk Drive. Half-height 5¼-inch compact disk read-only memory (CD-ROM) drives can randomly access up to 600 Mbytes of data. Average access is 500 milliseconds, and transfer rate is 154 kbytes sustained, or 500 kbytes burst. One disk holds the equivalent of 200,000 pages. Disks are replicated from masters.

WORM Optical Disk Drive. Write-once, read-many (WORM) drives are available for 5¼-inch and 12-inch removable cartridge optical disks. The 5¼-inch disk cartridges have a capacity of 654 Mbytes and the drives offer 60 millisecond average seek. The 12-inch cartridges have capacities of 1 gigabyte per side or 2 gigabytes total.

Erasable Optical Disk Drive. Erasable optical disk technology uses a magneto-optical drive. A laser is used to heat a spot on the disk surface, and then a conventional magnetic read/write head is able to change the polarity of the heated spot. *See* OPTICAL ERASABLE DISK DRIVE.

OPTICAL ERASABLE DISK DRIVE

An optical disk drive is a machine for writing and reading information from a specially prepared optical laser disk with a laser beam. Two forms of optical disk and optical disk drive have been developed: write-once, read-many (WORM) and erasable. Both technologies depend on the ability of a coherent light beam from a solid state laser diode to alter the magnetic orientation of small sectors or spots on the optical disk to encode the disk with binary data (1 or 0).

The WORM disk can be written on only once so the magnetic spots are permanently formed with the disk. By contrast, the erasable magneto-optical disk on a drive can be written on and erased more than a million times. Figure 1 is a simplified section view of an erasable optical

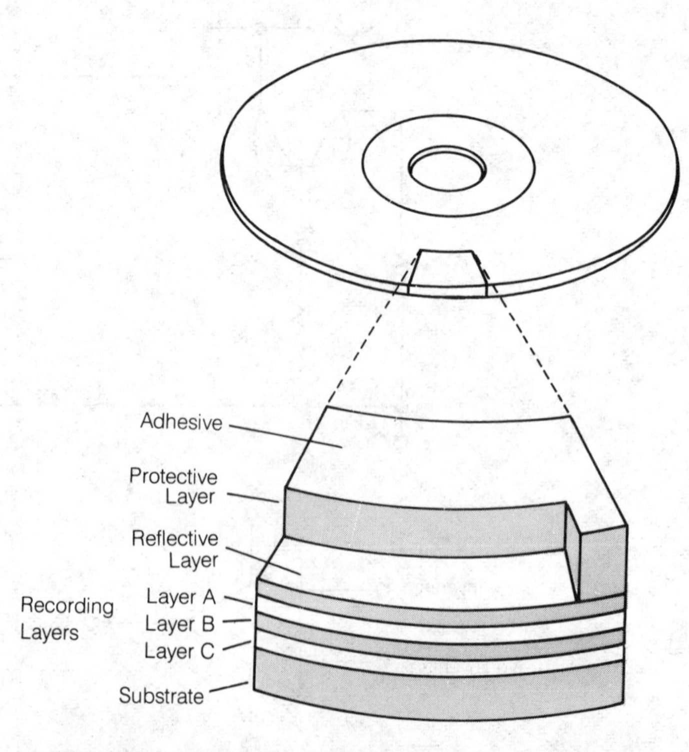

OPTICAL ERASABLE DISK DRIVE: Fig. 1. Sectional view of an erasable optical laser disk.

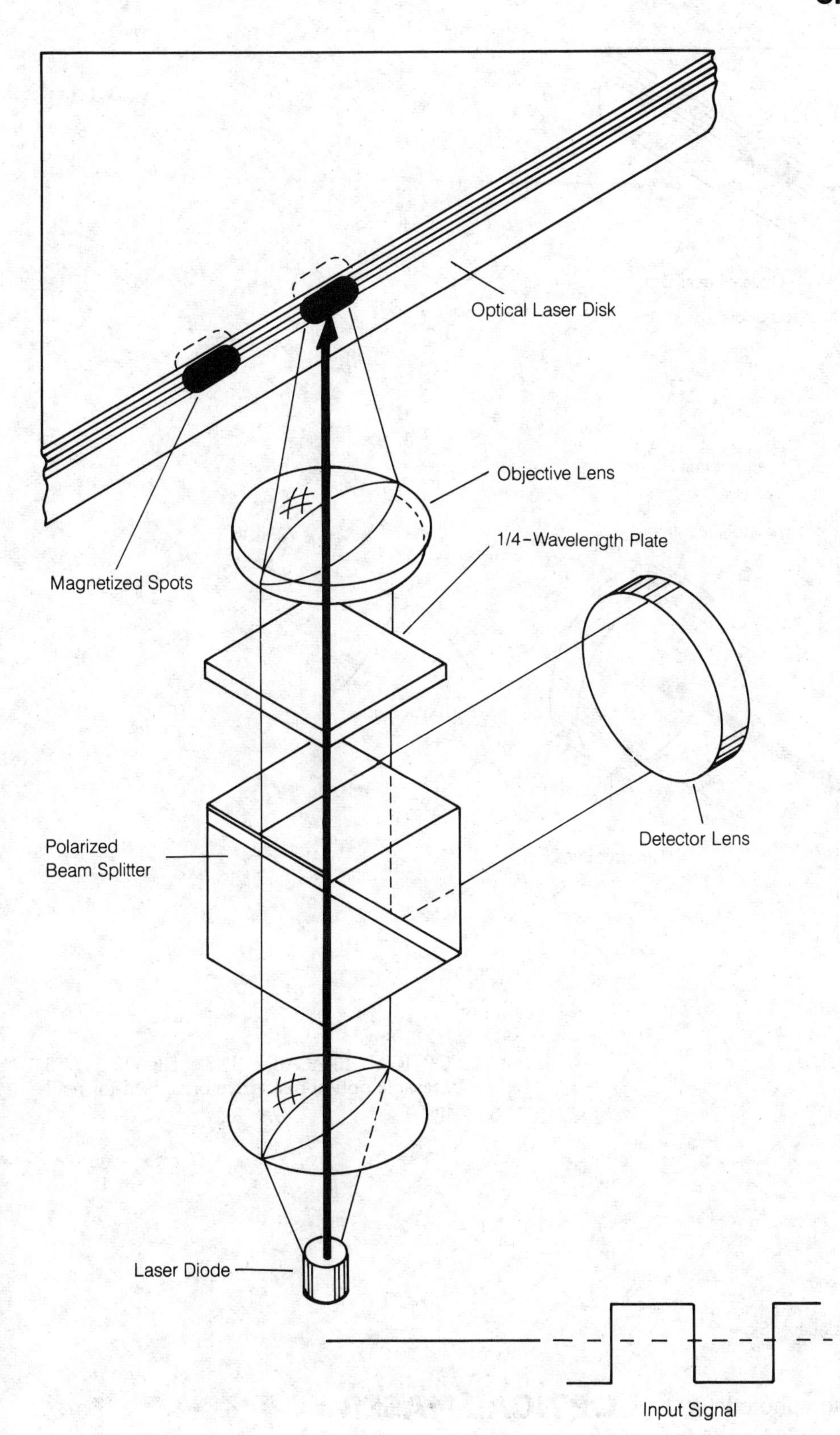

OPTICAL ERASABLE DISK DRIVE: Fig. 2. Erasable optical disk drive organized for writing.

laser disk. A four-layer magnetic film is deposited on the substrate of the disk, which rotates over the vertically oriented laser optical assembly. When a strong laser beam strikes the disk, it heats a microscopic spot and causes the atoms in the lower three layers to reform into a magnetized zone representing a bit of data.

Figure 2 illustrates the principles of writing digital data on an erasable optical laser disk. The disk is inverted so the laser beam strikes the disk from the under side. By switching the laser beam on and off as the disk rotates, the microscopic magnetic spots are formed along circular tracks. The size of each spot determines if a 1 or 0 is formed. The magnetic film can be written and erased more than 1 million times without a decline in accuracy.

Figure 3 illustrates the principles of reading digital data from an erasable optical laser disk. A weak polarized laser beam scans the magnetized spots on the recording layer. Recorded areas reflect back more light than unre-

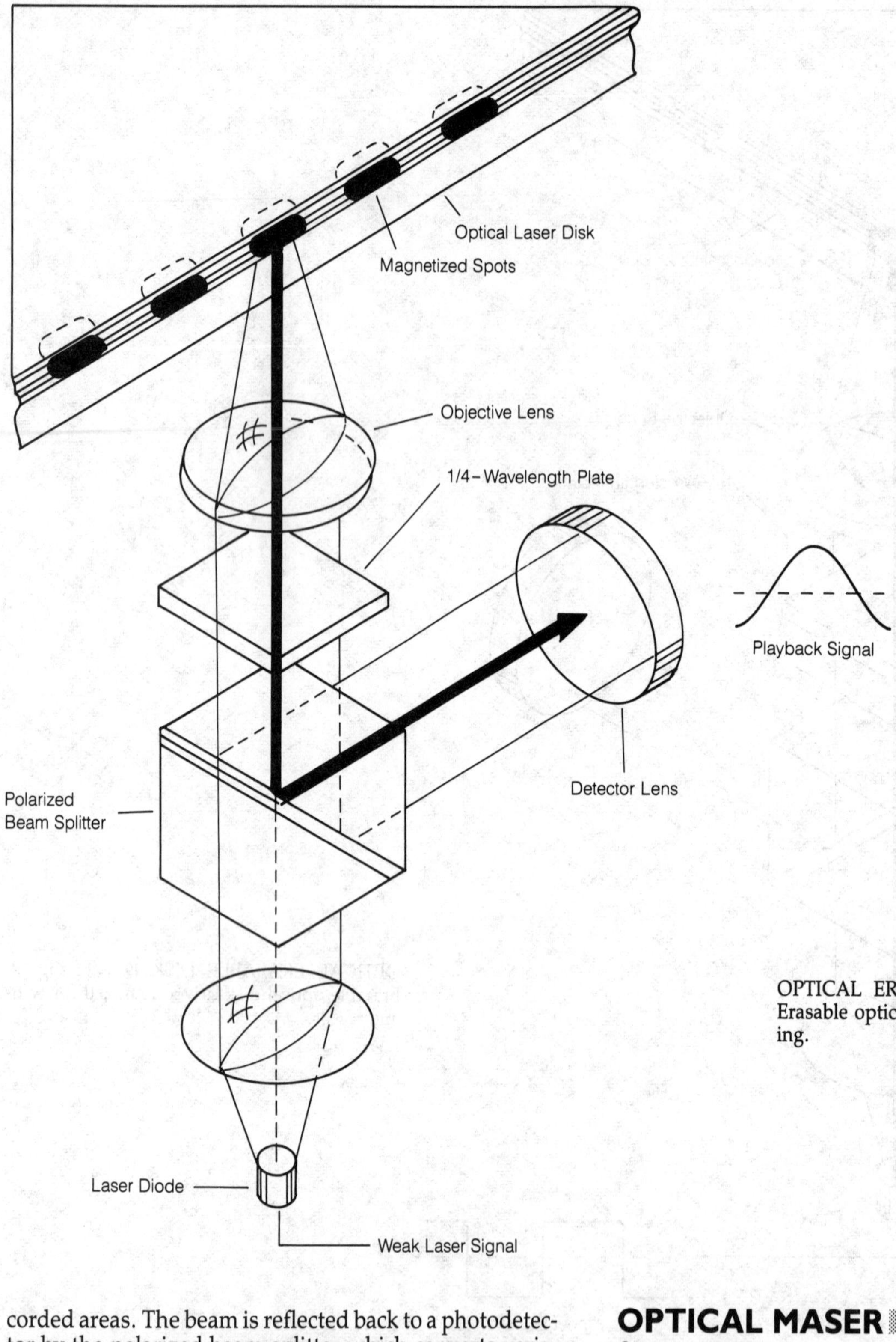

Optical Laser Disk
Magnetized Spots
Objective Lens
1/4-Wavelength Plate
Playback Signal
Detector Lens
Polarized Beam Splitter
Laser Diode
Weak Laser Signal

OPTICAL ERASABLE DISK DRIVE: Fig. 3. Erasable optical disk drive organized for reading.

corded areas. The beam is reflected back to a photodetector by the polarized beam splitter which converts variations in spot size into binary data.

High-capacity erasable magneto-optical drives store up to one billion bytes, or one gigabyte, of data. This amount is 12 to 50 times the amount of information that is now available on comparably sized magnetic hard disk drives. (The most advanced 5¼-inch Winchester hard disk drives store more than 700 megabytes of data.) The magneto-optical disk also can be used to store digitized audio data.

OPTICAL MASER

See LASER, MASER.

OPTICAL SOUND RECORDING AND REPRODUCTION

Movie-film sound tracks are recorded by means of a modulated-light source (*see* MODULATED LIGHT). As the film moves through the recorder, the modulated-light source causes the film to be exposed in a linear pattern that follows the sound vibrations. The sound track appears on the film as a band with variable density or width. The band is located to one side of the picture-frame sequence.

When the film is run through the projector, the sound is retrieved by a form of optical coupling (*see* OPTOCOUPLER). A constant-brightness light source is placed on one side of the film, and the beam of light is passed through the sound band. A photocell on the other side of the film picks up the modulated light and converts it into the original audio impulses.

OPTICAL TRANSMISSION

Optical transmission is the generation and propagation of a modulated-light signal for communications purposes. Optical transmission can be achieved in many different ways.

The most common method of optical transmission is the direct amplitude modulation of a light source. Light-emitting diodes and lasers are generally amplitude-modulated because these devices have a rapid response rate. However, even an incandescent lamp can be modulated at audio frequencies.

Other methods of optical transmission include polarization modulation, pulse modulation, and multiplexing. *See* MODULATED LIGHT.

The relative ease with which visible light propagates through a medium is called the optical transmission characteristic, or optical transmittivity, of the medium. The attenuation varies with the wavelength of the light.

OPTICS

Optics is a branch of physics concerned with the behavior of visible light. Optics is important in electronics because of its integration in the design and construction of optical communication and display systems. *See also* FIBEROPTIC COMMUNICATION, MODULATED LIGHT.

OPTIMIZATION

Optimization is a process or operation by which a set of variables is adjusted for the most efficient, the maximum, or the minimum result.

In electronics, the process of optimization is a part of all engineering practice. Once a circuit design works, it is usually possible to improve its performance. This might involve the changing of component values or the changing of component types. *See also* FUNCTION.

OPTIMUM ANGLE OF RADIATION

When a radio signal is returned to earth by the ionosphere, the angle of incidence is approximately equal to the angle of reflection. Although some irregularities exist in the ionosphere, this rule can be considered valid on the average. For a given distance between the transmitting and receiving station, and a given ionized-layer altitude, the optimum angle of radiation for single-hop propagation can be determined. The greater the distance

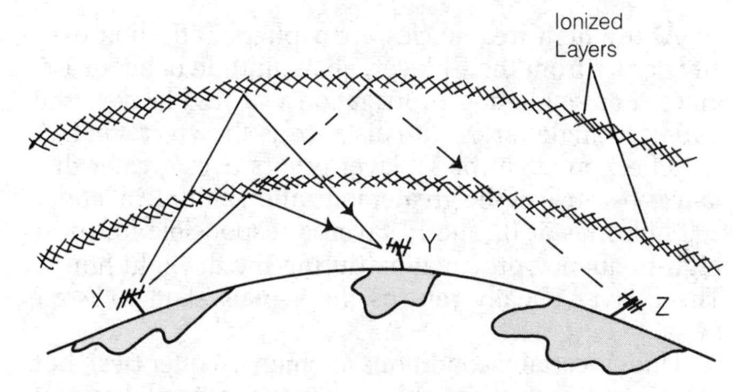

OPTIMUM ANGLE OF RADIATION: Fig. 1. This angle varies with the height of the ionized layer and with the distance between stations.

between the stations, the lower the optimum angle; the higher the ionized-layer altitude, the higher the optimum angle. An example is shown in Fig. 1.

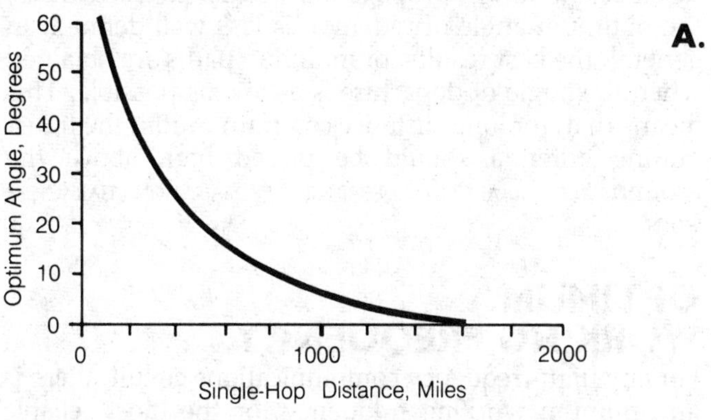

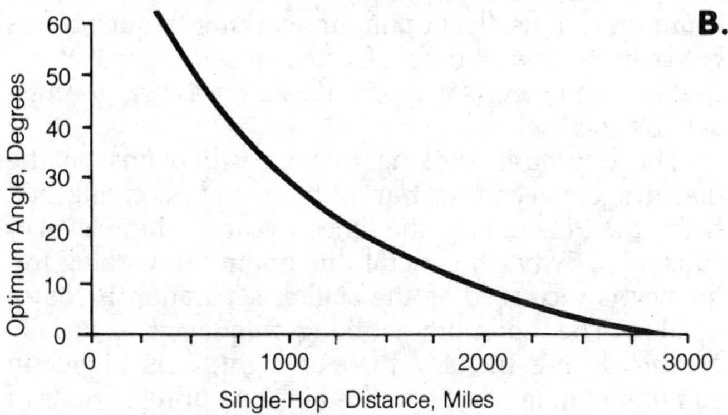

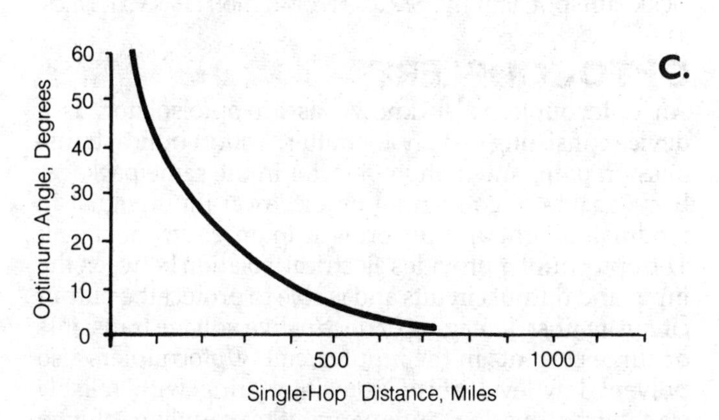

OPTIMUM ANGLE OF RADIATION: Fig. 2. Optimum angle as a function of station separation for the F1 layer (A), the F2 layer (B), and the E layer (C).

At the high frequencies, ionospheric reflection usually occurs from the F1 layer, at an altitude of about 100 miles. For single-hop propagation via the F1 layer, the optimum angle varies with distance as shown at A in Fig. 2. Reflection from the F2 layer occurs over greater distances because of the greater ionization altitude of about 200 miles, as at B. The F1 layer is responsible for most high-frequency propagation during the daylight hours. The F layer usually returns the signals at night (*see* F LAYER).

Under certain conditions at high frequencies, but more commonly at very high frequencies, the E layer, at an altitude of about 60 miles, returns signals to the earth. The resulting relation between the optimum angle and the station separation is shown at C (*see* E LAYER, SPORADIC-E PROPAGATION).

When the distance between the transmitting and receiving station is greater than the maximum single-hop distance, multihop propagation occurs. In such cases, the optimum angle of radiation is less well defined. In general, the best results for multihop paths are obtained when the angle of departure is as low as possible. This means that, for long-distance communication, the transmitting antenna should be placed high above the ground. *See also* ANGLE OF DEPARTURE, PROPAGATION CHARACTERISTICS.

OPTIMUM WORKING FREQUENCY

For any high-frequency communications circuit, there is an optimum working frequency for the most reliable communications. The optimum working frequency lies between the lowest usable frequency and the maximum usable frequency (*see* LOWEST USABLE FREQUENCY, MAXIMUM USABLE FREQUENCY).

The optimum working frequency depends on the distance between the transmitting and receiving stations, the time of day, the time of year, and the level of sunspot activity. In general, the optimum working frequency is increased as the station separation becomes greater. The optimum working frequency is usually greater during the day than at night; it is higher in summer than in winter, and is highest during periods of peak sunspot activity. *See also* PROPAGATION CHARACTERISTICS.

OPTOCOUPLER

An optocoupler, also known as an optoisolator, is a device consisting of a photoemitter, a short optical transmission path, and a photodetector in the same package. It is capable of converting an electrical input signal to modulated light and restoring it to an electrical signal. The optocoupler provides electrical isolation between the input and output circuits and is able to protect the output circuit against damaging or destructive voltage transients or surge currents in the input circuit. Optocouplers also prevent low-level noise from interfering with reliable signal transmission and permit the coupling of two circuits operating at different logic voltage levels with different ground points. Two related devices are the optointerrupter and the optoreflector.

The photoemitter in most optocouplers is an infrared-emitting diode (IRED), and the photodetector is typically a silicon photodiode, phototransistor, or other device whose sensitivity is matched to the output of the IRED. The detector produces an electrical output and might provide signal gain. Optocouplers are packaged in opaque cases which prevent the entry of extraneous light that could interfere with signal transmission. These can be conventional four-pin and six-pin dual-in-line (DIP) package or proprietary case styles. *See* INFRARED EMITTING DIODE, PHOTODETECTOR.

When not receiving an electrical input, the IRED and photodetector are normally off. The input signal to the IRED causes it to emit infrared energy that is transmitted through a short glass or plastic light guide to a photodetector. IR energy incident on the photodetector causes it to generate an electrical output that follows the input signal. These can be short pulses for switching relay contacts, digital, or analog signals. An IRED can be biased to establish a zero level for ac signals.

The current-transfer ratio (CTR) indicates the efficiency of the coupler. It depends on the radiative efficiency of the IRED, the separation between the IRED and the detector, and both the sensitivity and amplifying gain of the detector. Standoff protection, expressed as an isolation surge voltage, indicates the integrity of the coupler package as well as the dielectric strength of its insulation.

Optocouplers are classified by their photodetectors or output circuits. Schematics of these devices are shown in Fig. 1 as follows:

1. Photodiode output achieves the highest speeds but the output requires amplification to provide useful output levels (Fig. 1A).
2. Phototransistor output is adaptable to the widest range of applications because the collector-base junction can operate as a photodiode and the transistor provides output signal gain. This output is capable of switching at 300 kHz (Fig. 1B).
3. Photo-Darlington output provides up to ten times higher transfer ratios and higher output current than the phototransistor output. However, the tradeoff is in switching speed which is only about 10 percent of the phototransistor output with a typical value of about 30 kHz (Fig. 1C).
4. Photo-SCR (silicon-controlled rectifier) output provides the features of the silicon-controlled rectifier for switching ac or dc with a logic level input. The SCR can switch the 120/240 V, 50 or 60 Hz ac (alternating-current) line. A separate gate electrode controls SCR switching (Fig. 1D).
5. Triac output offers the benefit of the triac for switching the 120/240 V, 50/60 Hz ac line with a logic-level signal. A separate gate electrode also controls triac switching (Fig. 1E).
6. Integrated-circuit output provides logic-level control of a high-speed IC capable of interfac-

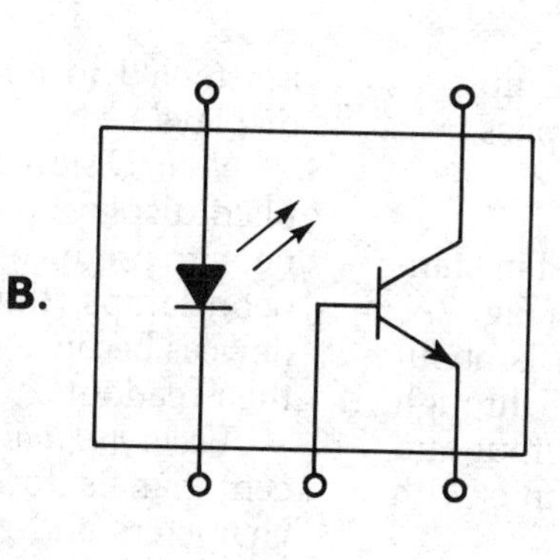

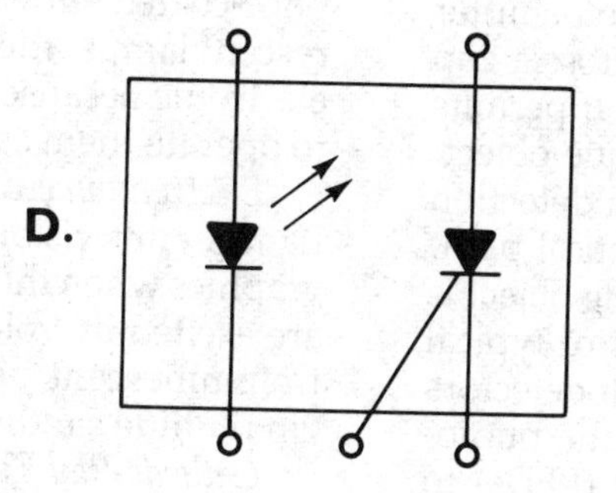

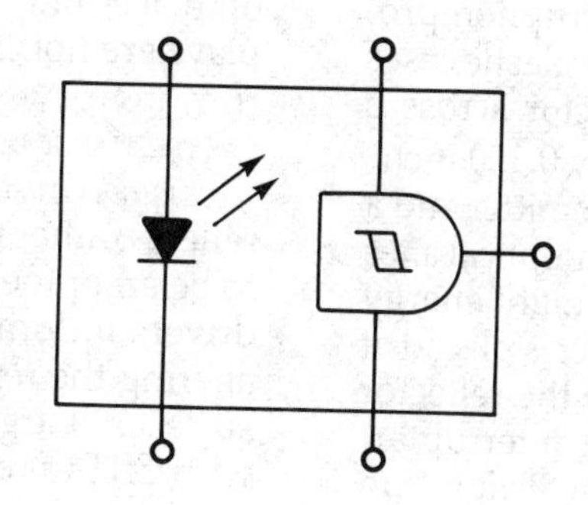

OPTOCOUPLER: Fig. 1. Optocouplers with photodiode output (A), phototransistor output (B), photo-Darlington output (C), photo-SCR output (D), phototriac output (E), and integrated circuit—Schmitt trigger—output (F).

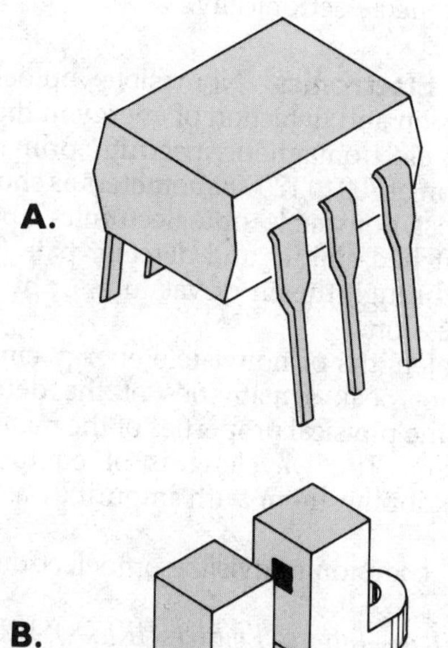

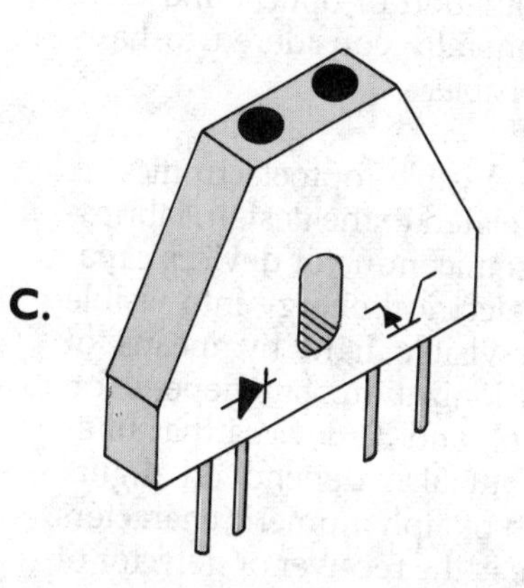

OPTOCOUPLER: Fig. 2. Typical packages for optocouplers are in a dual-in-line package (A), as a slotted optical switch (B), and as a focused reflective assembly (C).

ing computer terminals to peripheral equipment and digital control of power supplies, motors and other actuators (Fig. 1F).

Most commercial optocouplers are packaged in standard six-pin dual-in-line packages as shown in Fig. 2A. The accepted industry standard for isolation is about 5000 V ac peak. The industry designations 4N25 through 4N38 refer to popular units with transistor output and the designations 4N29 through 4N33 refer to units with photo-Darlington output.

An optointerrupter is a device with a photoemitter and photodetector coupled through the air across a gap in the modular package as shown in Fig. 2B. It permits the interruption of the light path by an opaque object. These normally on links permit object motion detection and rate counting as an object crosses the optical path. Useful in machine control systems, the moving object is typically a slotted wheel. The photoemitters of typical commercial products are IREDs and the photodetectors are photo-Darlington transistors. However, the photoemitters can be visible light-emitting diodes (LEDs) to permit direct visual observation of the interruption process. Interrupters are packaged in U-shaped plastic cases with the photoemitter facing a photodetector across a gap in the unified package of approximately 0.120 inch.

An optoreflector is a device with a photoemitter and a photodetector in the same module with their optical axes converging at a point outside the housing. Light energy is coupled between them only when an object is placed at or near the focal point. The components of the reflector module are similar to those of the optical interrupter. These normally on links permit motion detection and rate counting as an object moves through the focal point. Useful in machine control systems, the reflector is typically a moving machine part or element. The package is shown in Fig. 2C.

OPTOELECTRONICS

Optoelectronics is the combination of optical and electronic technologies. It is generally considered to have two parts—visible and nonvisible.

Visible Optoelectronics Visible optoelectronics is that part of optoelectronics related to the design, fabrication, and application of: 1) semiconductor devices capable of direct conversion of electrical energy into visible light; 2) devices that emit visible light by means of electroluminescence or gas ionization, but depend on digital electronics for control; and 3) devices that filter and reflect incident light but also depend on digital electronics for the formation of alphanumeric characters or graphics. The human eye is the receiver or detector of a visible optoelectronic device.

Visible light-emitting optoelectronic devices include:

Visible Light-Emitting Diode (VLED or LED), Lamp, and Display. LEDs are semiconductor diodes capable of converting electrical energy directly into visible light at many different lengths. The emitted color depends on the materials and doping used in the LED. The LEDs can be assembled to form visible light-emitting alphanumeric displays.

Neon Display Panel. Neon display panels are neon-filled displays that employ electrode pairs to ionize the neon to produce visible alphanumeric characters. Unlike neon lamps, they are considered to be optoelectronic devices because they require digital logic drivers to alter their readouts.

Vacuum-Fluorescent Display Panel. Vacuum-fluorescent panels form visible light-emitting alphanumeric characters and graphics by ion-bombardment of phosphor-coated surfaces with three electrodes. Unlike fluorescent lamps, these displays are considered to be optoelectronic because they require digital electronic drivers to operate them as alterable readouts.

Electroluminescent Display Panel. Electroluminescent panels emit visible light as alphanumeric characters or graphics when the manganese atoms in a phosphor layer are excited by voltage. The alternating-current (ac) electroluminescent display is a layer of manganese-doped zinc sulfide sandwiched between two electrodes.

Cathode-Ray-Tube (CRT) Display. CRTs are also visible optoelectronic devices, but incandescent lamps and displays are not. *See* CATHODE-RAY TUBE, ELECTROLUMINESCENT DISPLAY, LIGHT-EMITTING DIODE, LIGHT-EMITTING DIODE DISPLAY, NEON GAS DISCHARGE DISPLAY.

Liquid-Crystal Display. The liquid-crystal display (LCD) reflects rather than emits visible light, but it is considered to be an optoelectronic device because it requires digital drivers to form alphanumeric characters or graphics by altering the reflection of light from segmented or matrix surfaces. Large LCDs are considered to be competitive with CRTs as displays for portable battery-powered computers.

Gas and solid-state lasers that operate in the visible spectrum of approximately 500 to 700 nanometers (nm) are also considered to be optoelectronic devices because they are controlled electronically.

Nonvisible Electronics Nonvisible optoelectronics use the emission and detection of energy in the nonvisible part of the electromagnetic spectrum, primarily in the infrared region of 700 to 1200 nanometers as shown in the diagram. Most nonvisible optoelectronics applications require a matched emitter and detector pair. Transmission can be through the air or vacuum, or by fiberoptic cable transmission.

The wavelengths of nonvisible energy emitters and the regions of peak sensitivities of the detectors are functions of the physical properties of the materials used to make them. The wavelengths of emitters can be modified by doping them with impurities as in visible LEDs.

The most common nonvisible optoelectronic devices are:

Photoemitters—Infrared Emitters (IREDs). IREDs are infrared-emitting LEDs whose fundamental emission is determined by the bandgap of the materials from which they are fabricated. These are gallium arsenide (GaAs) or aluminum gallium arsenide (AlGaAs) diodes whose emission wavelength can be altered by impurity doping.

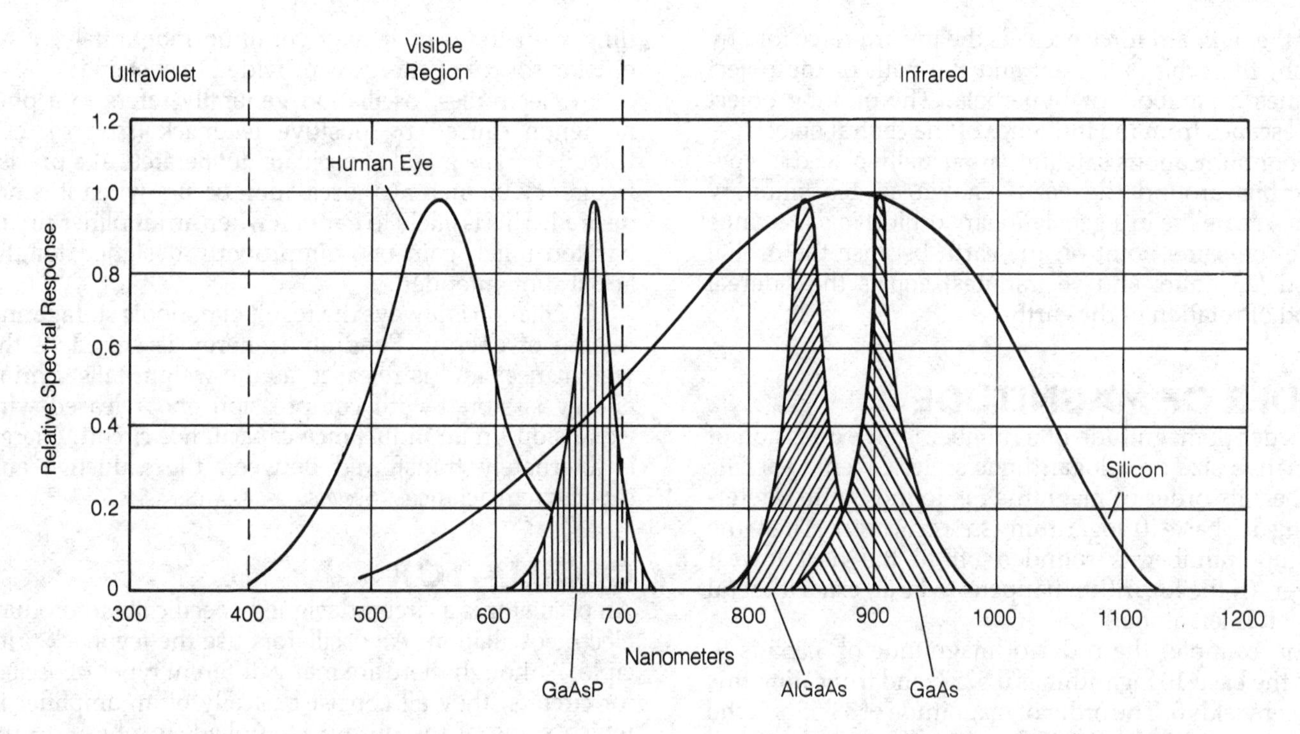

OPTOELECTRONICS: Response of emitters (GaAsP, AlGaAs, and GaAs) and receptors (human eye and silicon photoreceptors) plotted against wavelength (nanometers).

Photodetectors—Silicon Photodiodes, Phototransistors and Photo Darlingtons. Photodetectors are devices whose region of highest sensitivity is determined by the peak sensitivity of silicon.

Other photodetectors include photo SCRs (also called light-activated SCRs or LASCRS), phototriacs, P-I-N diodes, and avalanche photodiodes. Some integrated circuits such as Schmitt trigger circuits have been developed as individual photoresponsive devices. *See* AVALANCHE PHOTODIODE, PHOTODETECTOR.

Individually packaged and matched photoemitter and photodetector pairs separated by air can perform motion detection or modulated signal transmission. Applications include intrusion detection, remote television and VCR tuning, and optical encoding. When coupled with fiberoptic cable, they can transmit and receive digital and analog signals and are considered to be fiberoptic communications systems.

An optocoupler or optoisolator is a matched pair of photoemitter and photodetector within the same package able to transmit analog and digital signals without electromagnetic interference between two circuits operated at different voltage levels. Some optocouplers have Schmitt trigger integrated-circuit (IC) output devices that provide pulses. Matched pairs emitter-detector pairs are also used in optointerrupters and optoreflectors for detecting motion. (Some optointerrupters or optoreflectors can include visible light-emitting diodes rather than IREDs to permit operator verification of their function.) Optocouplers are also included in solid-state relays and input/output modules. More specialized photodetectors include P-I-N diodes and avalanche photodiodes. *See* ENCODER, INFRARED-EMITTING DIODE, INPUT/OUTPUT MODULE, OPTOCOUPLER, SOLID-STATE RELAY.

Other optoelectronic devices include solar cells, which, although sensitive to visible light, are primarily responsive to nonvisible radiation. Hence they are classified as nonvisible optoelectronic devices. They are matched with IREDs in certain types of solid-state relays. *See* SOLAR CELL.

Semiconductor injection laser diodes are nonvisible optoelectronic devices because their primary emissions are in the infrared region. Fabricated from aluminum gallium arsenide (AlGaAs), they are commercially available only as discrete packaged devices. These lasers are used as emitters in medium- and long-haul fiberoptic communications. *See* FIBEROPTIC COMMUNICATION, LASER DIODE.

OPTOISOLATOR

See OPTOCOUPLER.

ORBIT

An orbit is the path that one body follows around another body, or around the center of force between two bodies. Electrons follow orbits around the nuclei of atoms (*see* ELECTRON ORBIT). Satellites follow orbits around the earth, the earth orbits the sun, and the sun orbits the center of the Milky Way galaxy. In every case, the inward and outward forces exactly balance, resulting in a condition of equilibrium.

An orbit can be either circular or elliptical. If the orbit is circular, the center of force is at the center of the circle. If the orbit is elliptical, the center of force is at one focus of the ellipse.

If the outward force exceeds the inward force for any reason, the orbit is broken and the path of the object becomes a parabola or hyperbola. The orbiting object then escapes from the influence of the central object.

Communications satellites are usually placed in special orbits around the earth known as geostationary orbits. A satellite in a geostationary orbit always remains above the same point on the earth because the orbital period (23 hours and 56 minutes) equals the sidereal period of rotation of the earth.

ORDER OF MAGNITUDE

The order of magnitude of a number is an expression of its relative size on a logarithmic scale. Given a specific number, its order of magnitude is found by first determining its base-10 logarithm (*see* LOGARITHM). Then the resulting number is rounded off to the next lowest integer. (If the logarithm happens to be an exact integral value, it is left alone.)

For example, the order of magnitude of 3.355 is 0, since the base-10 logarithm is 0.5257, and truncating this number yields 0. The order of magnitude of 33.55 is 1; the order of magnitude of 335.5 is 2, and so on. Working down instead of up, the order of magnitude of 0.3355 is −1; the order of magnitude of 0.03355 is −2, and so on. *See also* SCIENTIFIC NOTATION.

OR GATE

An OR gate is a form of digital logic gate with two or more inputs and one output. The OR gate performs the Boolean inclusive-OR function. That is, if all inputs are 0 (false), the output is 0; if any or all inputs are 1 (true), the output is 1.

The illustration shows the schematic symbol for an OR gate, along with a truth table for a two-input device. *See* LOGIC.

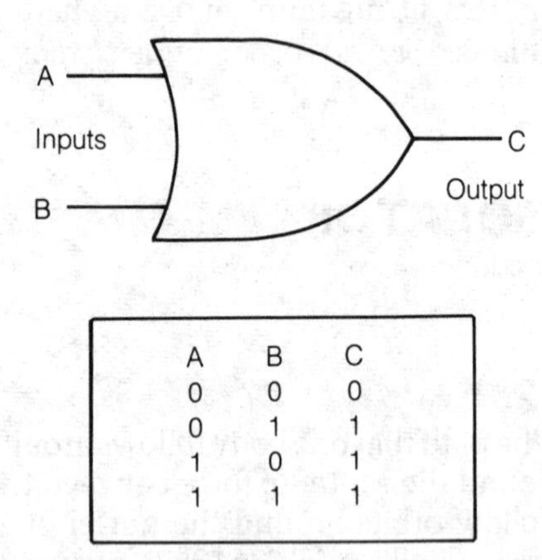

A	B	C
0	0	0
0	1	1
1	0	1
1	1	1

OR GATE: Schematic symbol and truth table.

OSCILLATION

Oscillation is any repetitive motion such as the swinging of a pendulum, the expansion and compression of a spring, or the continual movement of electrons in an electrical conductor. Oscillation can be short-lived, settling with time, or it may continue indefinitely, if an outside source of energy is provided to sustain it.

In electronics, oscillation generally refers to a phenomenon caused by positive feedback (*see* FEEDBACK). Circuits can be made to oscillate deliberately at a precise frequency. Sometimes, oscillation occurs when it is not desired. This is likely to happen when an amplifier circuit has too much gain or is improperly designed for the application intended.

Oscillation is always the result of periodic storage and release of energy. Pendulum energy is stored as the weight rises and is released as the weight falls. Spring energy is stored with compression and released with expansion. In an inductance-capacitance circuit, energy is alternately transferred between the inductive and capacitive reactances. *See also* OSCILLATOR.

OSCILLATOR

An oscillator is a circuit designed specifically to produce electric oscillation. All oscillators use the feedback principle. Although there are many different types of oscillator circuits, they all consist basically of an amplifier in which some of the output is applied, in phase, to the input.

An oscillator requires an active amplifying device, such as a transistor, field-effect transistor, or tube, and a source of power. If the oscillator is to produce alternating currents at a specific frequency, some form of resonant circuit, such as an inductance-capacitance network or a piezoelectric crystal, is necessary. Specialized devices, such as the Gunn diode and the klystron tube produce oscillation because of negative-resistance effects (*see* NEGATIVE RESISTANCE, OSCILLATOR CIRCUITS).

OSCILLATOR CIRCUITS

A directory of common oscillator circuits is presented here in alphabetical order.

Armstrong Oscillator An Armstrong oscillator is a circuit that produces oscillation by inductive feedback. See the simple circuit diagram on p. 607. A coil called the tickler is connected to the plate or collector, and is brought near the coil of the tuned circuit. The tickler is oriented to produce positive feedback. The amount of coupling between the tickler coil and the tuned circuit is adjusted so that stable oscillation takes place. The output is taken from the (tube) plate or (transistor) collector by means of a small capacitance, or by transformer coupling to the tuned circuit.

The frequency of the Armstrong oscillator is determined by the tuned-circuit resonant frequency. Usually, the capacitor is variable and the inductor is fixed so the amount of feedback remains relatively constant as the frequency is changed. Armstrong oscillators are generally used in regenerative receivers (*see* REGENERATIVE DETECTOR).

Balanced Oscillator Any oscillator with center-tapped, grounded tank-circuit inductors is called a balanced oscillator. A simple balanced oscillator consists of a

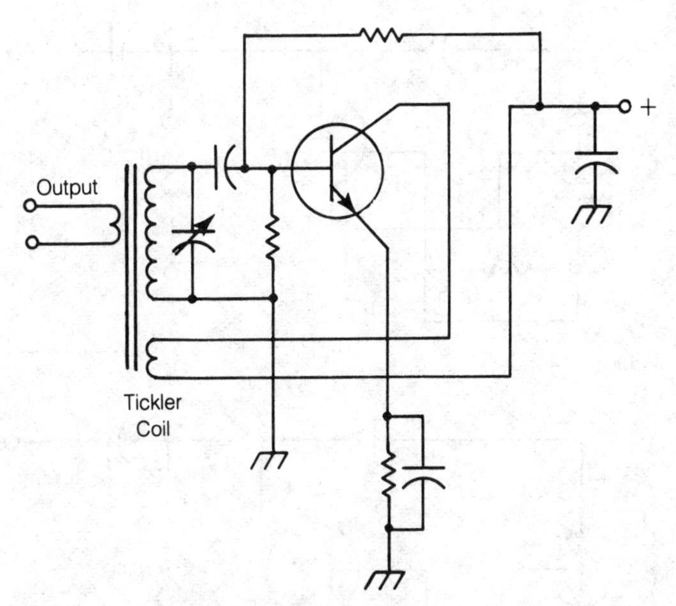

ARMSTRONG OSCILLATOR: The Armstrong oscillator is often used as a regenerative detector in a simple receiver.

tuned push-pull amplifier with inductive feedback, as in the circuit illustrated. The variable capacitors, C, determine the frequency of oscillation; the input circuit is kept in balance with a split-stator (ganged) capacitor, ensuring identical values on either side of the tank circuit. Transistors Q1 and Q2 must have the same gain characteristics for good balance to be maintained.

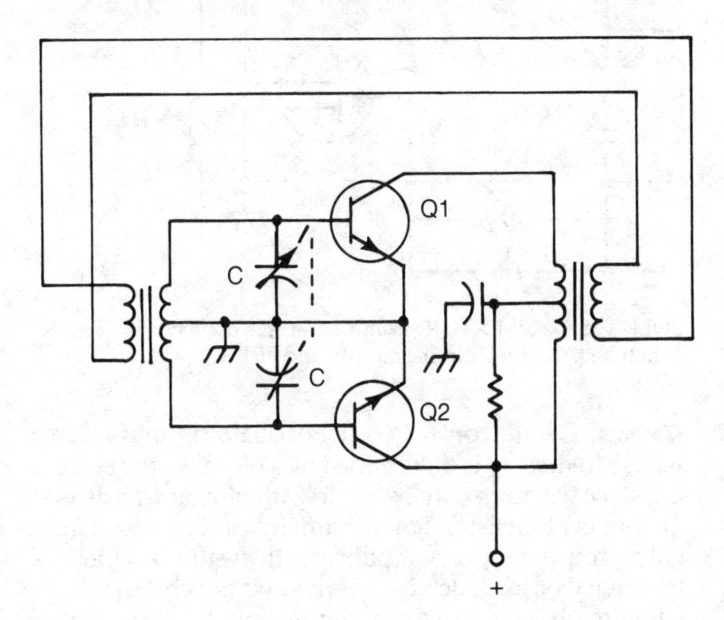

BALANCED OSCILLATOR: This oscillator uses inductive feedback.

Beat-Frequency Oscillator The beat-frequency oscillator is widely used in receivers for the detection of single-sideband (SSB), Morse code (CW), and radioteletype (RTTY) signals. The beat-frequency oscillator is generally located in the intermediate-frequency stages of the receiver. When tuned to the suppressed-carrier frequency of an SSB signal, or slightly away from the carrier frequencies of a CW or RTTY signal, the modulation information is retrieved. A detector that uses a beat-frequency oscillator is called a product detector. A block diagram of the beat-frequency oscillator is shown in the illustration.

Without the beat-frequency oscillator (abbreviated BFO), CW and RTTY signals would be inaudible except for the clicks and thumps of the make and break moments. An SSB signal would sound muffled and unintelligible without a product detector. Some receivers have the frequency of the BFO fixed at one side of the intermediate-frequency passband. The proper signal reproduction is obtained by adjusting the frequency of the receiver. In other receivers the frequency of the BFO is tunable from one end of the intermediate-frequency passband to the other. In still other receivers, the passband itself is tunable and the BFO frequency remains constant.

A direct-conversion receiver has a BFO that is tunable over the entire reception range. When the BFO frequency gets sufficiently close to a signal, the beat frequency is audible. *See also* DIRECT-CONVERSION RECEIVER.

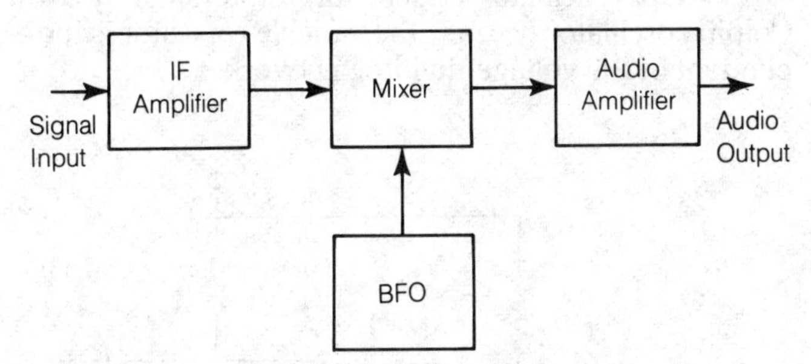

BEAT FREQUENCY OSCILLATOR: This oscillator produces audio output with radio-frequency signal input by mixing.

Blocking Oscillator A blocking oscillator is a form of relaxation oscillator that uses a feedback capacitor in conjunction with an inductance. The illustration shows a schematic diagram of a blocking oscillator. The feedback

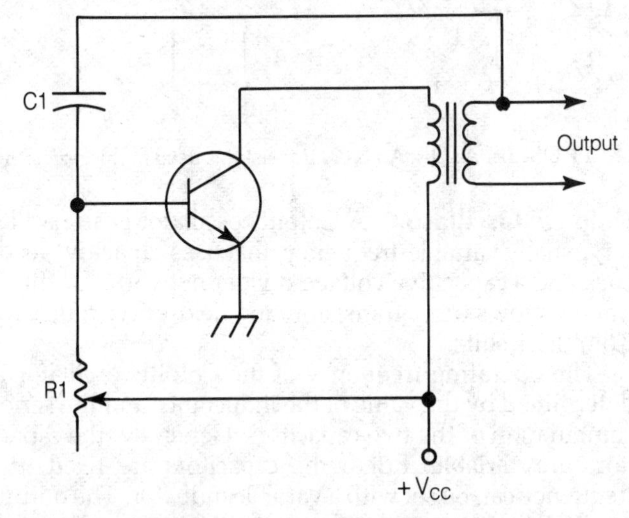

BLOCKING OSCILLATOR: Output voltage pulses are spaced by cutoff periods determined by the RC time constant of C1R1. Adjustment of R1 changes the frequency.

path is between the base of the transistor and the side of the output-transformer secondary that is in phase with the base.

The term *blocking oscillator* refers to an oscillator that is switched on and off periodically by an external source. A blocking oscillator is a form of relaxation oscillator that uses a feedback capacitor in conjunction with an inductance. The illustration shows a schematic diagram of a blocking oscillator. The feedback path is between the base of the transistor and the side of the output-transformer secondary that is in phase with the base.

Clapp Oscillator A specialized form of Colpitts oscillator with a series-tuned tank circuit is called a Clapp oscillator (*see* COLPITTS OSCILLATOR). The tank circuit is tapped with a capacitive voltage-divider network, but the voltage-divider capacitors are fixed, as shown in the figure. The frequency of the oscillator is adjusted either by varying the tank capacitance C or by varying the tank inductance *L*. The output can be taken from a secondary winding coupled to *L* or with a capacitive coupling, as shown.

A Clapp oscillator is generally more stable than a Colpitts oscillator because the variable capacitor is independent of the voltage-dividing network.

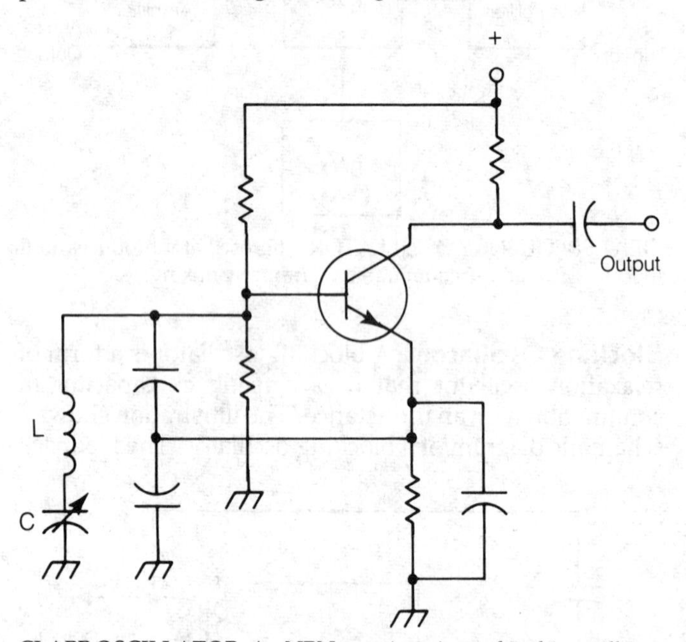

CLAPP OSCILLATOR: An NPN transistor is used in this oscillator.

Colpitts Oscillator A Colpitts oscillator is an oscillator, usually variable-frequency, that uses capacitive feedback and a capacitive voltage-divider network. The illustration shows tube, transistor, and field-effect-transistor Colpitts circuits.

The operating frequency of the Colpitts oscillator is determined by the value of the inductance and the series combination of the two capacitors. Generally, the capacitors are variable. But if the capacitors are fixed, the frequency can be set with a variable inductor. The output can be taken from the circuit by inductive coupling, but better stability is usually obtained by capacitive or transformer coupling from the plate, collector, or drain circuit.

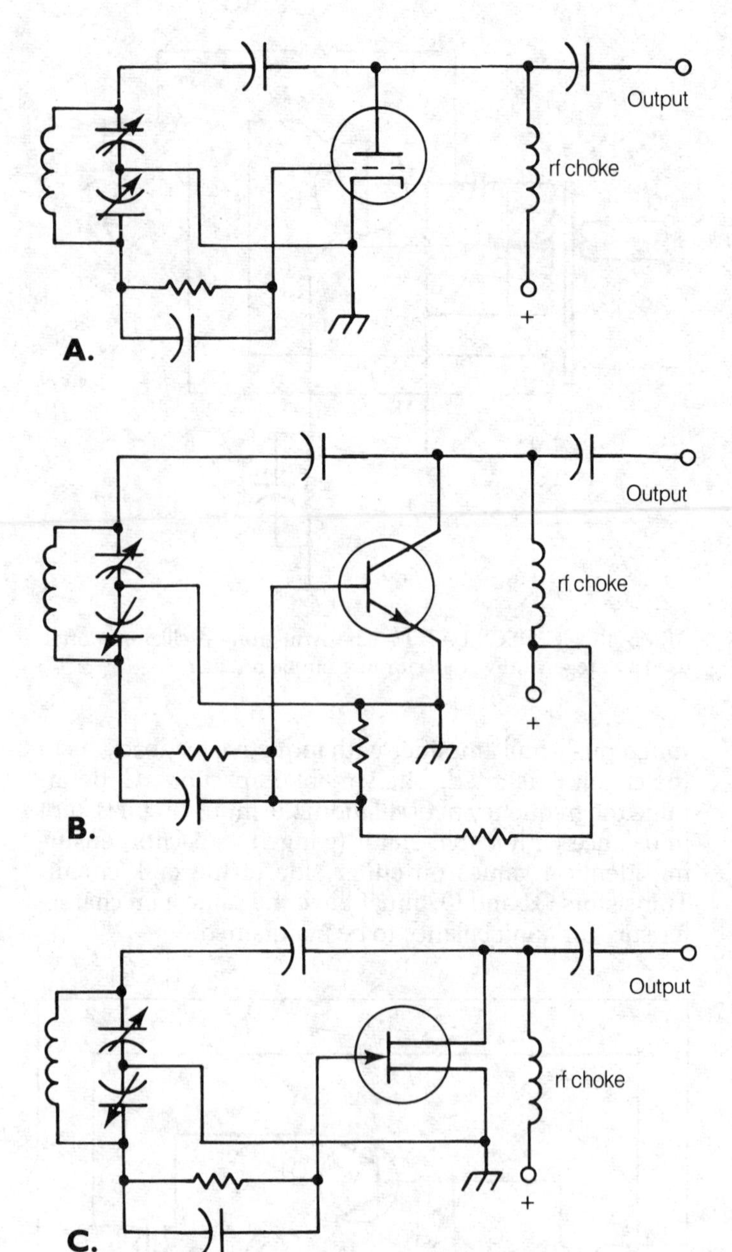

COLPITTS OSCILLATOR: Examples of this oscillator with a vacuum tube (A), bipolar transistor (B), and FET (C).

Crystal Oscillator A crystal oscillator is an oscillator whose frequency is determined by a piezoelectric crystal. Crystal oscillators may be made with bipolar transistors, field-effect transistors, or vacuum tubes. The circuit generally consists of an amplifier with feedback, and the frequency of feedback is governed by the crystal. Oscillation might take place at the fundamental frequency of the crystal or at one of the harmonic frequencies.

The schematic illustrates three common types of crystal-oscillator circuits. A tuned output circuit provides harmonic attenuation, or it may be used to select one of the harmonic frequencies. If the oscillator is used at the fundamental frequency of the crystal, and if the amount of feedback is properly regulated, a tuned output circuit is not usually needed. However, the oscillator output will contain more energy at unwanted frequencies when a tuned circuit is not used.

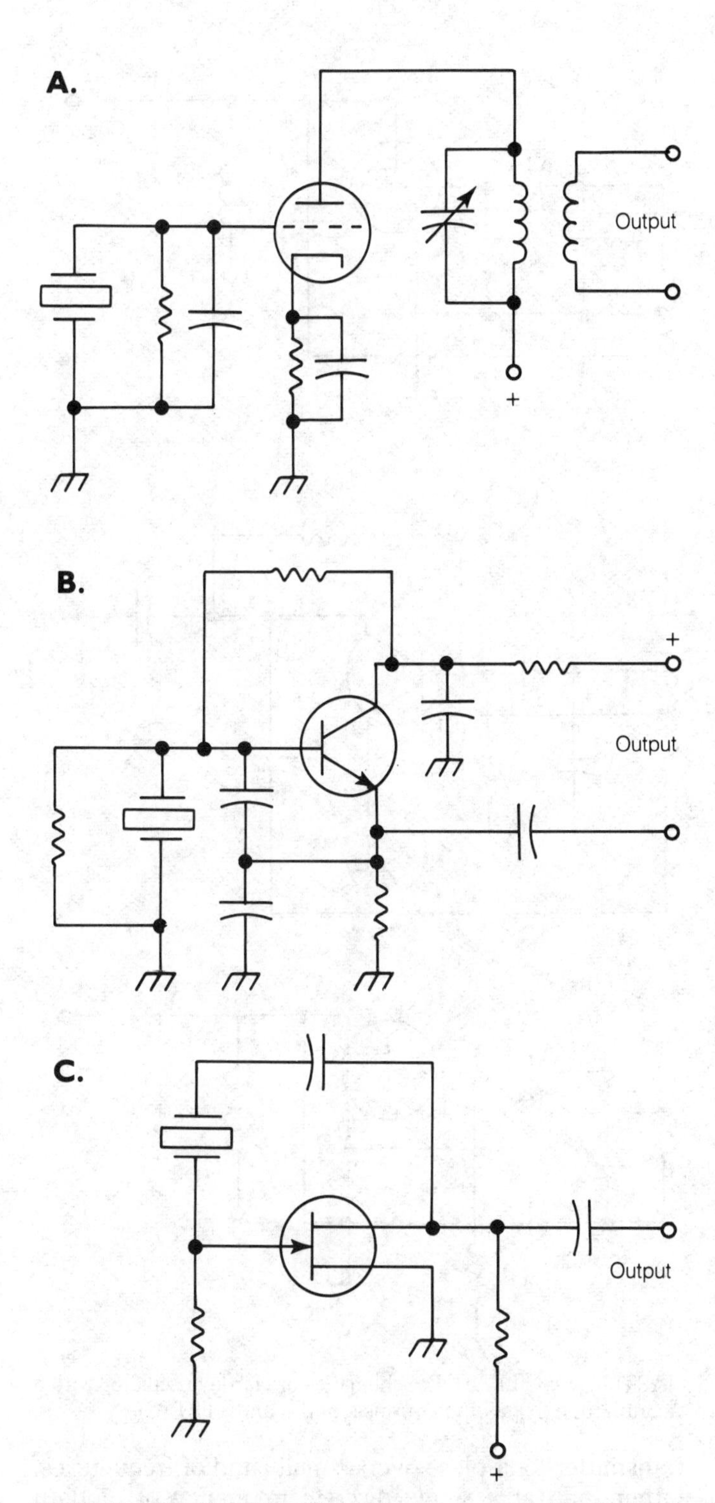

CRYSTAL OSCILLATOR: Schematics of crystal oscillator with a vacuum tube (A), a bipolar transistor (B), and a FET (C).

Electron-Coupled Oscillator An electron-coupled oscillator is a special form of vacuum-tube oscillator that is less susceptible to loading effects than is a conventional oscillator circuit. A change in the load impedance of an oscillator can cause frequency changes, and can sometimes even result in cessation of oscillation. The electron-coupled oscillator is designed to overcome these effects. A changing load impedance to an electron-coupled oscillator does not result in significant changes in the frequency, and almost never causes the oscillator to stop functioning.

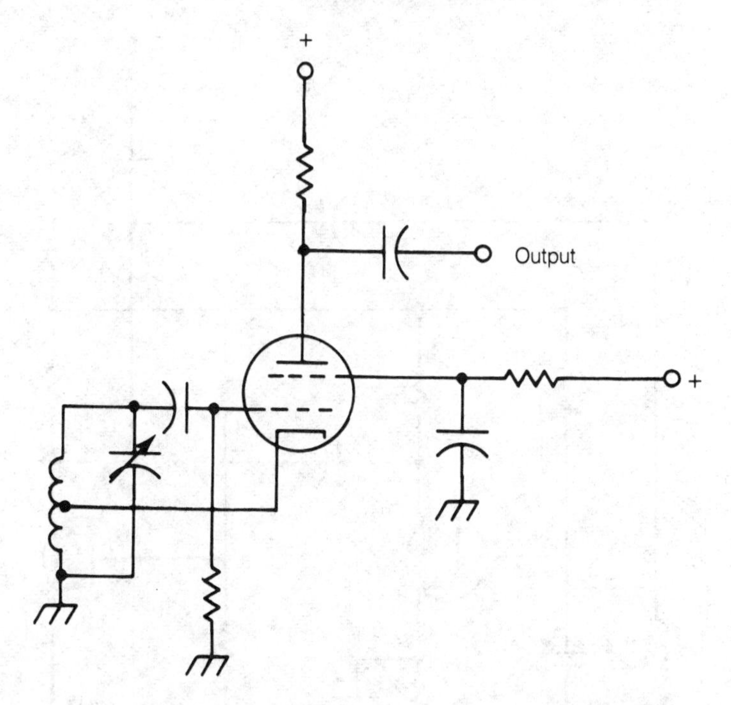

ELECTRON-COUPLED OSCILLATOR: In this oscillator, the screen grid acts as the plate for the oscillator, and the output is taken from the actual plate of the tube.

The illustration is a schematic diagram of a Hartley type electron-coupled oscillator. The screen grid acts as the plate for the oscillator circuit, while the output is taken from the actual plate of the tube. An intervening suppressor grid may or may not be used. The stream of electrons arriving at the plate of the tube is modulated by the oscillator. The coupling to the oscillator output is provided only by the electron beam as it traverses the inside of the tube from the control grid through the screen grid to the plate. The interelectrode capacitance is generally very small within a vacuum tube, and therefore a large change in the output impedance does not significantly load down the oscillator. The electron-coupled oscillator acts as its own buffer stage, and this often eliminates the need for such a stage following the oscillator. The disadvantage of the electron-coupled oscillator is that a vacuum tube must be used to obtain the necessary results.

Franklin Oscillator A Franklin oscillator is a form of variable-frequency oscillator circuit, using transistors, field-effect transistors, or tubes. A resonant circuit is connected in the base, gate, or grid circuit of one device. Feedback is provided by a second stage, which serves to amplify the signal and invert its phase. The output of the amplifying stage is coupled back to the base, gate, or grid of the first device.

In the common-emitter, common-source, or common-cathode configurations, active devices show a 180-degree phase difference between their input and output circuits. In the Franklin oscillator, since there are two devices, the output is in phase with the input. Coupling between the two devices is usually accomplished by means of capacitors. However, tuned circuits can be used. The output of the oscillator may be taken from any point in the circuit.

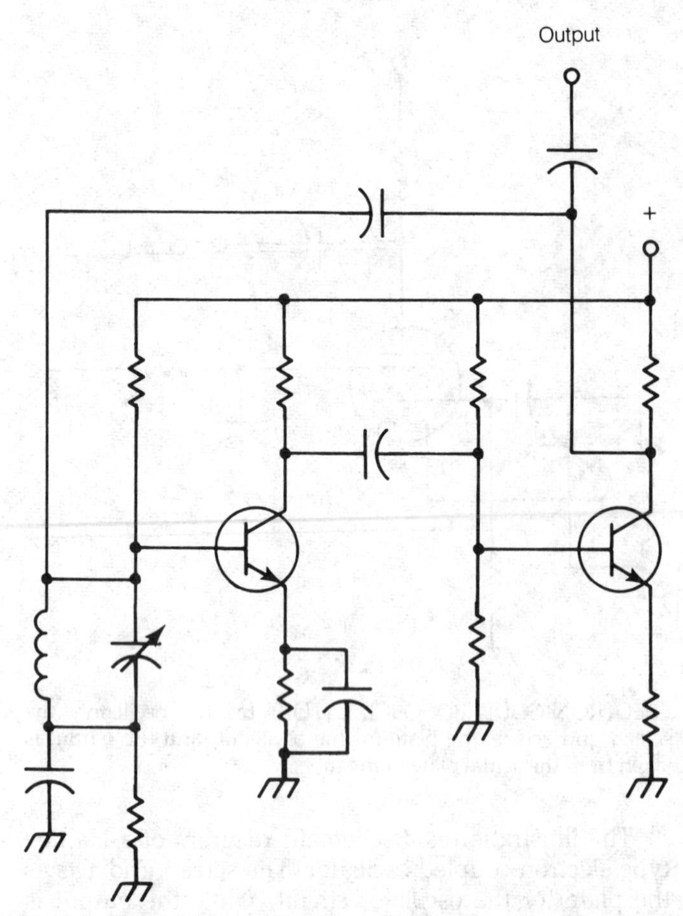

FRANKLIN OSCILLATOR: This oscillator uses a phase-inverting amplifier in the feedback circuit.

The illustration is a schematic diagram of a simple Franklin oscillator.

Hartley Oscillator The Hartley oscillator is a form of variable-frequency oscillator. The operating frequency is determined by a parallel combination of inductance and capacitance. The feedback system is provided by a tap in the coil of the tank circuit. The illustration shows schematic diagrams for Hartley oscillators using a vacuum tube, a bipolar transistor, and a field-effect transistor.

The cathode, emitter, or source of the amplifying device is always connected to the coil tap in the Hartley configuration. This is how it can be recognized. The output is usually taken from the plate, collector, or drain circuit of the oscillator. However, the output may also be obtained by means of a loosely coupled coil near the tank coil.

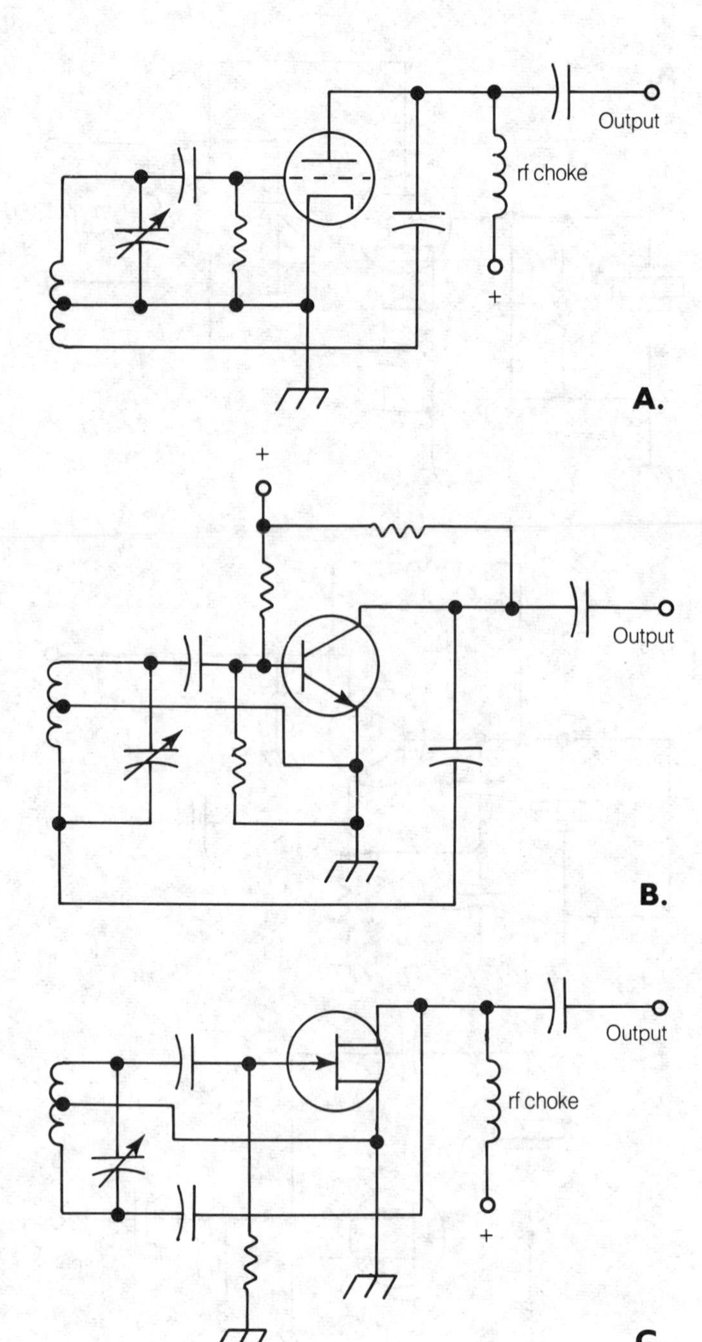

HARTLEY OSCILLATOR: Schematics of Hartley oscillator with a vacuum tube (A), a bipolar transistor (B), and a FET (C).

Hertz Oscillator The Hertz oscillator is a radio-frequency (rf) generating circuit developed by Hertz to demonstrate the action of electromagnetic induction.

The Hertz oscillator contains a spark gap operated from a high-voltage power source. The high voltage is derived from an induction coil. Resonant inductance-capacitance circuits, placed in series with the supply voltage, confine the noise from the spark to a narrow band of radio frequencies. Radiation takes place from two metal plates, which act as a large capacitor.

The earliest radio transmitters made use of the spark principle to generate their rf energy. The output of this transmitter took place over a small band of frequencies, rather than at a single discrete frequency of modern continuous-wave transmitters. In place of the metal plates, a tuned antenna system was connected to the output of the oscillator. The Hertz oscillator is not a true oscillator, but instead is a resonant device that concentrates the noise from a spark generator to a narrow band of frequencies.

Magnetostriction Oscillator A magnetostriction oscillator is similar to a crystal oscillator. Instead of a piezoelectric crystal, the magnetostriction oscillator makes use of a rod of magnetostrictive material. The rod serves two purposes: It vibrates at the frequency of oscillation, and it acts as the core for the feedback transformer (see illustration).

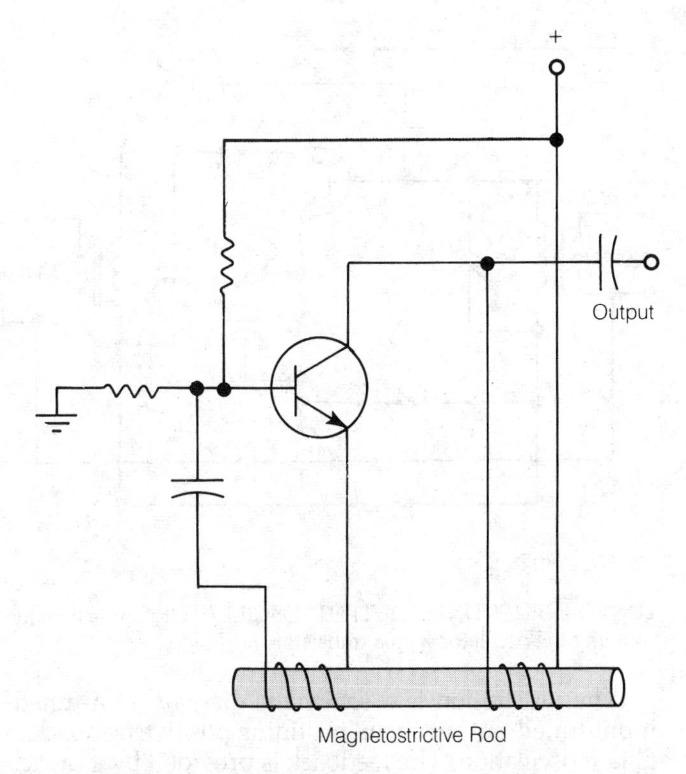

MAGNETOSTRICTION OSCILLATOR: This circuit uses a bipolar transistor.

The fundamental oscillating frequency is determined by the physical length and diameter of the rod, whether it exhibits positive or negative magnetostriction, and the extent of the magnetostrictive effects. *See also* MAGNETOSTRICTION.

Miller Oscillator A Miller oscillator is a special form of crystal oscillator. A transistor, field-effect transistor, or tube can be used as the active element in the circuit. The illustration shows Miller oscillators using an NPN bipolar transistor (A), an N-channel field-effect transistor (B), and a triode vacuum tube (C).

The Miller oscillator can be recognized by the presence of the crystal between the base, gate, or grid and ground. The output contains a tuned circuit. The internal capacitance of the active device provides the necessary feedback for oscillation. The Miller oscillator is sometimes called a conventional crystal oscillator.

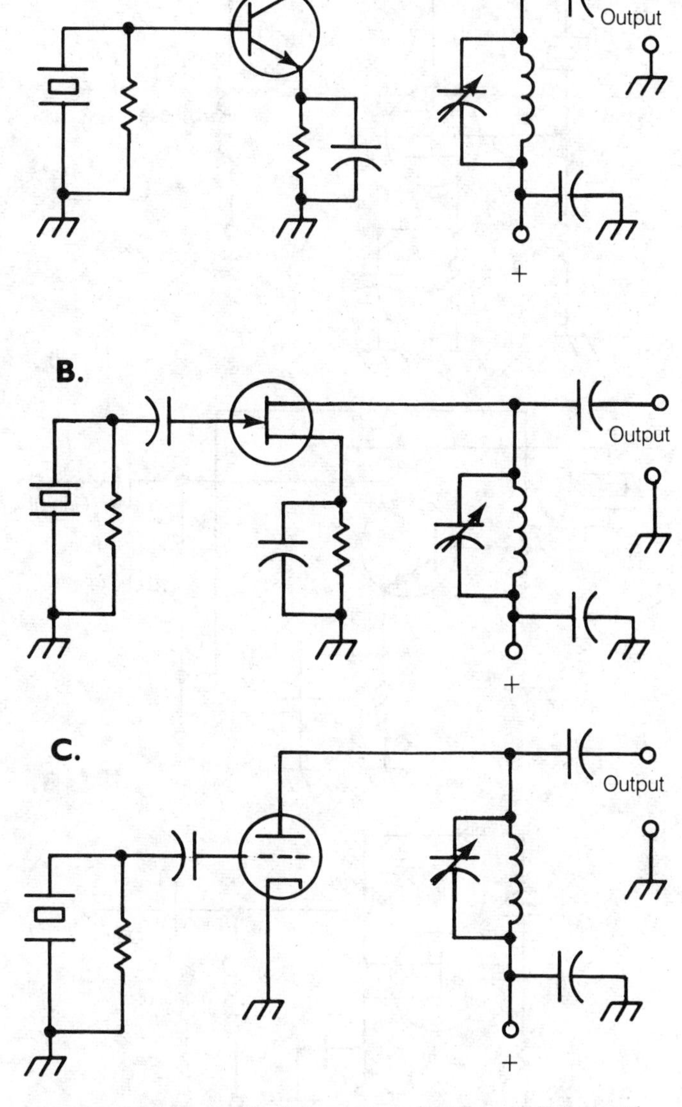

MILLER OSCILLATOR: Schematics of Miller oscillator with a bipolar transistor (A), a FET (B), and a vacuum tube (C).

Pierce Oscillator The Pierce oscillator is a form of crystal oscillator. The Pierce configuration may be recognized by the connection of the crystal between the grid and plate of a vacuum-tube circuit (see A in illustration on p. 612), between the base and collector of a bipolar-transistor circuit (B) or between the gate and drain of a field-effect-transistor circuit (C).

The Pierce circuit is used extensively in radio-frequency applications. The main advantage of the Pierce oscillator is that the crystal acts as its own tuned circuit. This eliminates the need for an adjustable inductance-capacitance tank circuit in the output.

Reinartz Crystal Oscillator Reinartz crystal oscillator is a special form of oscillator, characterized by high efficiency and little or no output at frequencies other than the fundamental. The Reinartz configuration can be used in oscillators with field-effect transistors, bipolar transistors, or vacuum tubes.

The illustration on p. 612 is a schematic diagram of a Reinartz oscillator that employs a dual-gate metal-oxide field-effect transistor (MOSFET). A tank circuit is inserted in the source line. This resonant circuit is tuned to approximately half the crystal frequency. The result is enhanced positive feedback, which allows the circuit to oscillate at a lower level of crystal current than would otherwise be possible.

Tuned-Input/Tuned-Output Oscillator A radio-frequency amplifier with tuned input and output circuits resonant at the same frequency, often breaks into oscillation. This can be avoided by neutralization (*see* NEUTRALIZATION). However, it can also be used to advantage if oscillation is wanted.

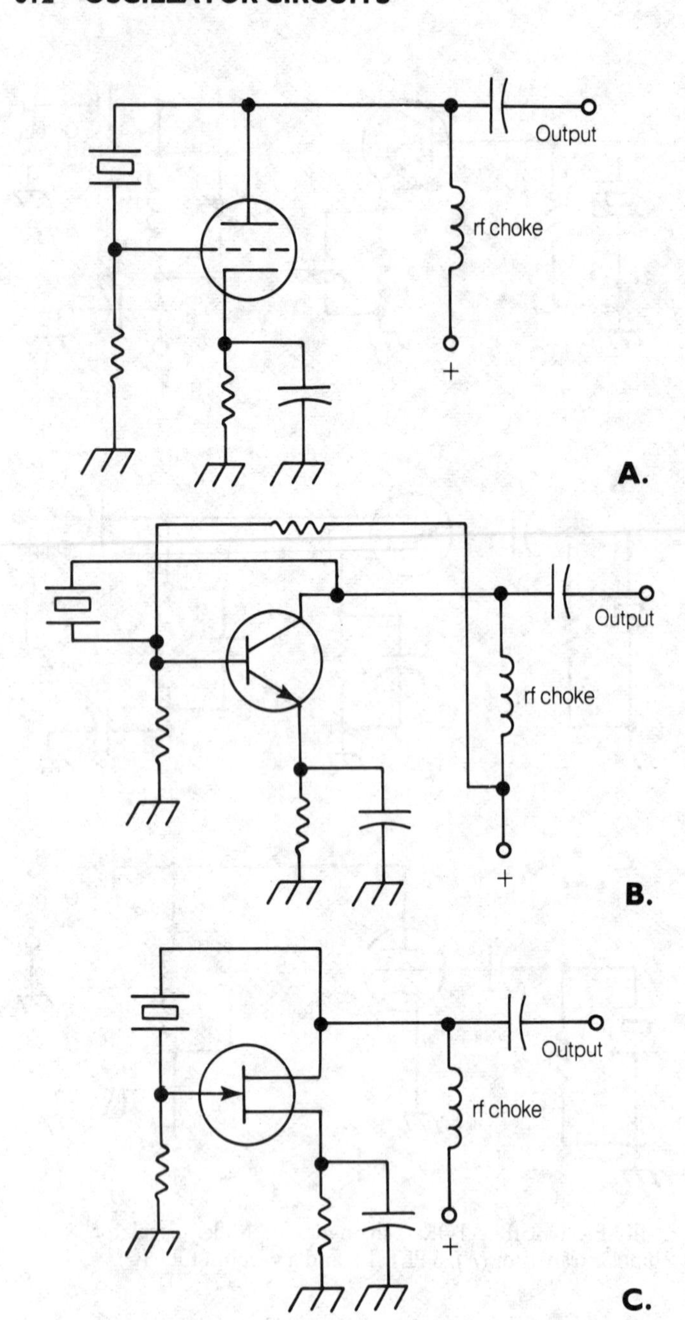

PIERCE OSCILLATOR: Schematics of Pierce oscillator with a vacuum tube (A), a bipolar transistor (B), and an FET (C).

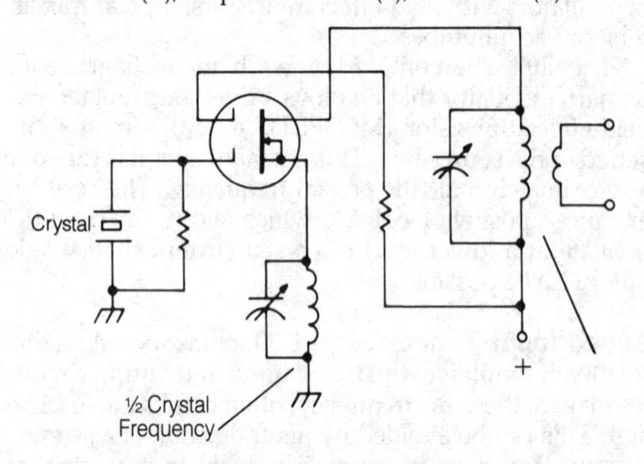

REINARTZ CRYSTAL OSCILLATOR: A Reinartz oscillator with a dual-gate metal-oxide field-effect transistor.

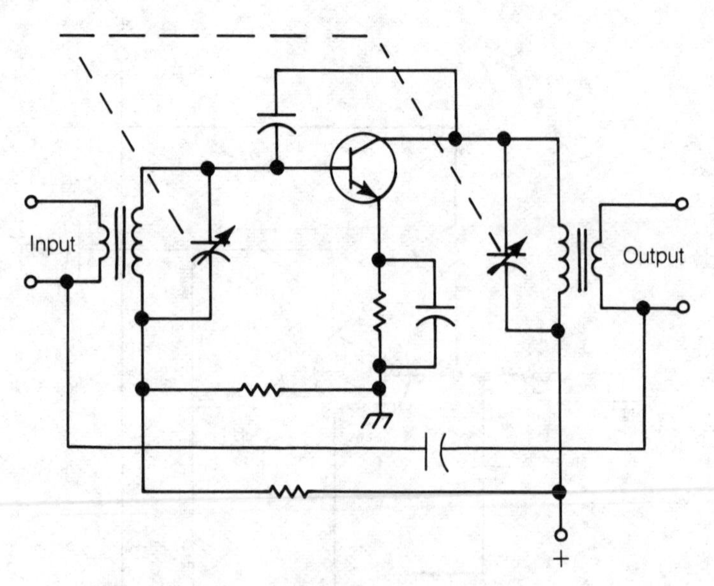

TUNED-INPUT/TUNED-OUTPUT OSCILLATOR: A schematic showing the oscillator with a transistor.

The illustration is a schematic diagram of a tuned-input/tuned-output circuit including positive feedback to obtain oscillation. The feedback is provided by a capacitor between the collector and the base. The frequency is adjusted with a dual-gang variable capacitor so the input and output circuits are always resonant at the same frequency.

Variable-Crystal Oscillator The oscillating frequency of a piezoelectric crystal is normally fixed. However, it can be varied slightly with the addition of a small capacitor or inductor. The components are connected in parallel or in series with the crystal. This is called pulling (*see* PULLING). A variable crystal oscillator (VXO) is a crystal

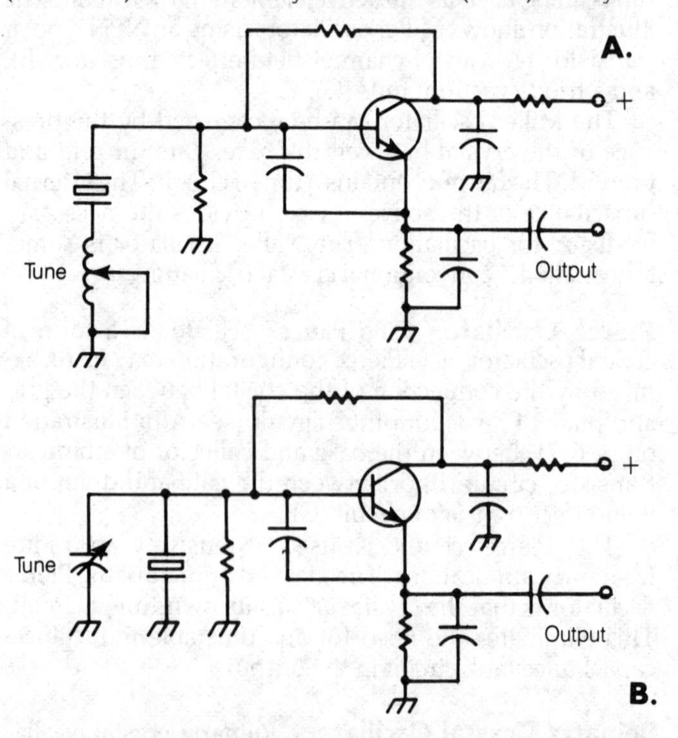

VARIABLE-CRYSTAL OSCILLATOR: Examples of variable crystal oscillator with a series inductor (A), and parallel capacitor (B).

oscillator in which the frequency can be adjusted over a small range by means of added reactances.

Variable crystal oscillators are used in radio transmitters or transceivers so they can operate over a small part of a band. Any crystal oscillator can be made into a VXO by adding the proper parallel or series reactances. Two examples of VXO circuits are shown in the illustration.

The main advantage of the VXO over an ordinary variable-frequency oscillator is excellent frequency stability. The main disadvantage of the VXO is its limited frequency coverage. Usually, the frequency of the VXO can be varied up or down by a maximum of approximately 0.1 percent of the operating frequency without crystal failure or loss of stability.

Variable-Frequency Oscillator Any oscillator in which the frequency can be adjusted, either in discrete channels or over a continuous range, is called a variable-frequency oscillator (VFO). Many different types of oscillators can be used for variable-frequency operation.

Some VFO circuits use inductance-capacitance tuning to obtain oscillation over a band of frequencies. The Colpitts and Hartley configurations are probably the most common. The tuned-input/tuned-output circuit is also sometimes used. The frequency of a crystal-controlled oscillator can be varied to a certain extent (*see* COLPITTS OSCILLATOR, HARTLEY OSCILLATOR, TUNED-INPUT/TUNED-OUTPUT OSCILLATOR, VARIABLE CRYSTAL OSCILLATOR).

In recent years, another type of VFO has evolved: the frequency synthesizer. This circuit contains a reference oscillator, usually crystal-controlled. Various multiplier and divider circuits, in conjunction with a phase-locked loop, allow operation at hundreds, thousands, or millions of discrete frequencies over a wide band. *See also* FREQUENCY SYNTHESIZER, PHASE-LOCKED LOOP.

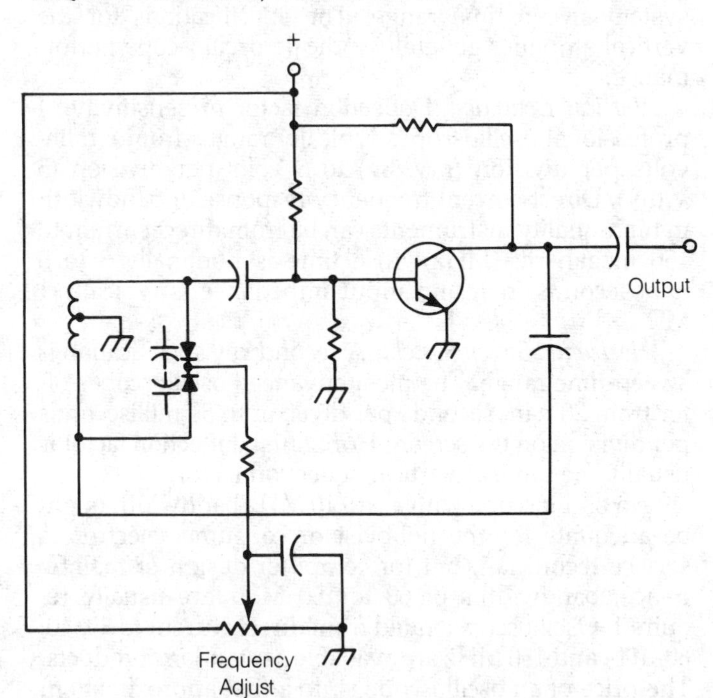

VOLTAGE-CONTROLLED OSCILLATOR: This voltage-controlled oscillator uses a Hartley configuration.

Voltage-Controlled Oscillator Varactor diodes can be used to adjust the frequency of an oscillator circuit. This is commonly done to provide frequency modulation (*see* FREQUENCY MODULATION, REACTANCE MODULATOR, VARACTOR DIODE). A direct-current voltage, supplied to the varactor diode or pair of diodes, facilitates variable-frequency operation in a circuit called a voltage-controlled oscillator (VCO).

In a VCO, one or two varactor diodes are connected in parallel with, or in place of, the tank capacitor of the oscillator. A pair of diodes can be connected in reverse series, replacing the tuning capacitor, as shown in the illustration. Frequency adjustment is accomplished by means of a potentiometer. The tuning range depends on the ratio of inductance to capacitance in the tank circuit. The maximum-to-minimum oscillator frequency ratio can be made as high as 2:1 without difficulty; considerably greater ratios are possible.

Wien-Bridge Oscillator A Wien bridge can be connected in the feedback circuit of an oscillator, resulting in a nearly pure sine wave at audio frequencies. This type of oscillator is a two-stage circuit, and is known as a Wien-bridge oscillator (see illustration).

In order for the Wien-bridge oscillator to produce a sine wave, the components must be selected so the Wien bridge is balanced. This requires that the values, as shown in the diagram, satisfy the equation:

$$C = C_1\left(\frac{R_2}{R_1} - \frac{R_3}{R}\right)$$

Under these conditions, the resonant frequency f, in hertz, is:

$$f = \frac{1}{2\pi C_1 R_3}$$

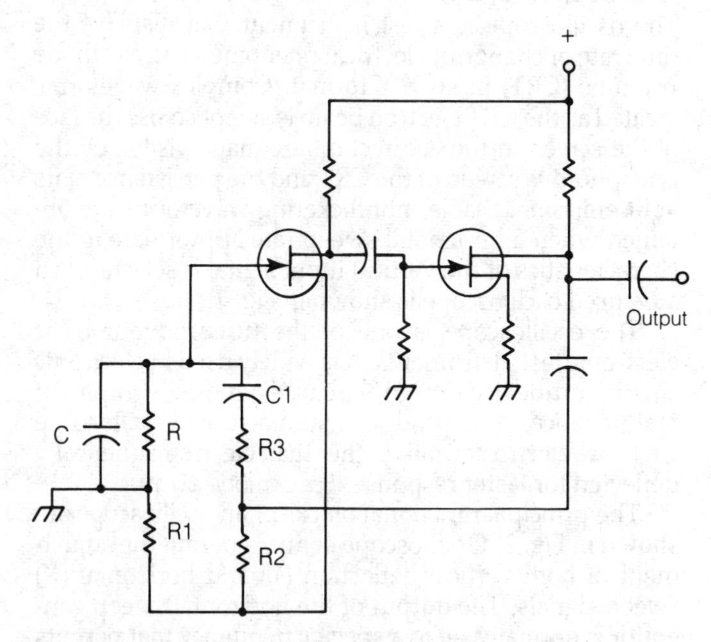

WIEN-BRIDGE OSCILLATOR: This Wien-bridge oscillator produces a sine wave at audio frequencies.

The primary advantage of the Wien-bridge circuit is its relatively pure output. The circuit is very stable. *See also* WIEN BRIDGE.

OSCILLATOR KEYING

Oscillator keying is a method of obtaining code transmission by interrupting the action of an oscillator circuit. Oscillator keying may be accomplished in various ways; the most common method is simply to open the cathode, emitter, or source circuit of the oscillator (see the schematic diagram).

Oscillator keying offers the advantage of completely shutting down the transmitter when the key is opened. This allows reception between code elements. However, oscillator keying results in chirp unless the circuit is carefully designed. *See also* CHIRP.

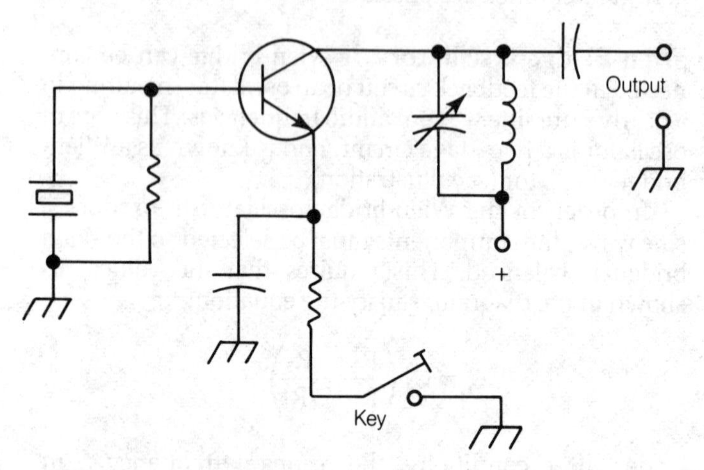

OSCILLATOR KEYING: The key opens the circuit between the emitter and ground.

OSCILLOSCOPE

The oscilloscope is a test instrument that displays the time rate of change of electrical phenomena on a cathode ray tube (CRT) in an X-Y format. Complex waveforms created as the CRT electron beam is swept across the face of the screen in the X direction are made visible by the phosphorous screen of the CRT and the persistence of its light emission. Stable, nonflickering waveforms are obtained when a horizontal sweep rate appropriate to the characteristics of the vertical input signal is selected. An advanced oscilloscope is shown in Fig. 1.

The oscilloscope is one of the three or four basic electronic test instruments. It is widely used in electronic circuit and device design, circuit diagnosis, equipment maintenance, and product manufacture. Oscilloscope CRTs are electrostatically rather than electromagnetically deflected for faster response. *See* CATHODE-RAY TUBE.

The principal functional blocks of an oscilloscope are shown in Fig. 2. Oscilloscope controls permit the adjustment of both vertical deflection (Y) and horizontal (X) sweep signals. The output of the horizontal sweep generator is normally set to a specific frequency that permits the most favorable viewing and analysis of the vertical deflection. The vertical signal is introduced to the vertical deflection plates of the CRT after amplification and conditioning. A horizontal sweep generator can be triggered internally from the vertical deflection section or by an external trigger.

The screen of the CRT is graduated into a uniform grid of lines called the graticule to permit visual estimation of the amplitude and time duration of displayed signals. An oscilloscope can function both as a voltmeter and a waveform indicator. It can display both negative and positive signals and both analog and digital or pulsed signals. Actual CRT waveforms can be compared with theoretical waveforms or those obtained from a correctly functioning circuit to pinpoint and isolate circuit problems. The oscilloscope can also give instant displays of rapidly dynamic changes in both active and passive components.

The principal functional blocks of an oscilloscope are: 1) CRT display, 2) beam control, 3) vertical amplifier, 4) probe and attenuator, 5) horizontal amplifier, 6) timebase generator, 7) synchronizing circuit, 8) intensity modulation circuit, 9) power supply, and 10) calibration circuit.

Professional-quality oscilloscopes are equipped with 6-inch (diagonal) rectangular CRT screens. The internal graticules are divided into 8-by-10 divisions with each division equal to 1 centimeter. Graticules are formed on the inside face of the CRT to eliminate angular distortion that would be visible if the waveforms were viewed at an angle through a grid outside of the CRT. The accelerating potential of a quality CRT typically ranges from 12 to 20 kV.

Some of the basic specifications considered in the selection of an oscilloscope include vertical amplifier deflection factor or sensitivity, rise time, and horizontal system sweep time range. The specifications for the vertical amplifier generally indicate oscilloscope performance.

Vertical Deflection. Deflection factor or sensitivity of professional oscilloscopes typically ranges from 2 millivolts per division (mV/div) to 5 volts per division (5 V/div). Direct-current frequency response or bandwidth in high-quality instruments can be from direct current to 150 megahertz (MHz). Rise time is nominally 3 to 5 nanoseconds (ns), and input impedance is typically 1 MΩ.

Horizontal Sweep System. A second key specification is sweep-time range. The most advanced oscilloscopes offer from 20 nanoseconds per division to 50 milliseconds per division on the screen. Horizontal deflection factor is usually the same as vertical deflection factor.

An oscilloscope with a 5 to 10 MHz bandwidth might be adequate for the hobbyist or consumer electronics service technician, but for computer design or maintenance, bandwidths of 60 to 150 MHz are usually required. Oscilloscopes rated from direct current to 20, 40, 60, 100, and 150 MHz are available as standard products. The price of an oscilloscope is, to a first approximation, directly related to bandwidth in megahertz. Oscilloscope pricing has dropped dramatically within recent years

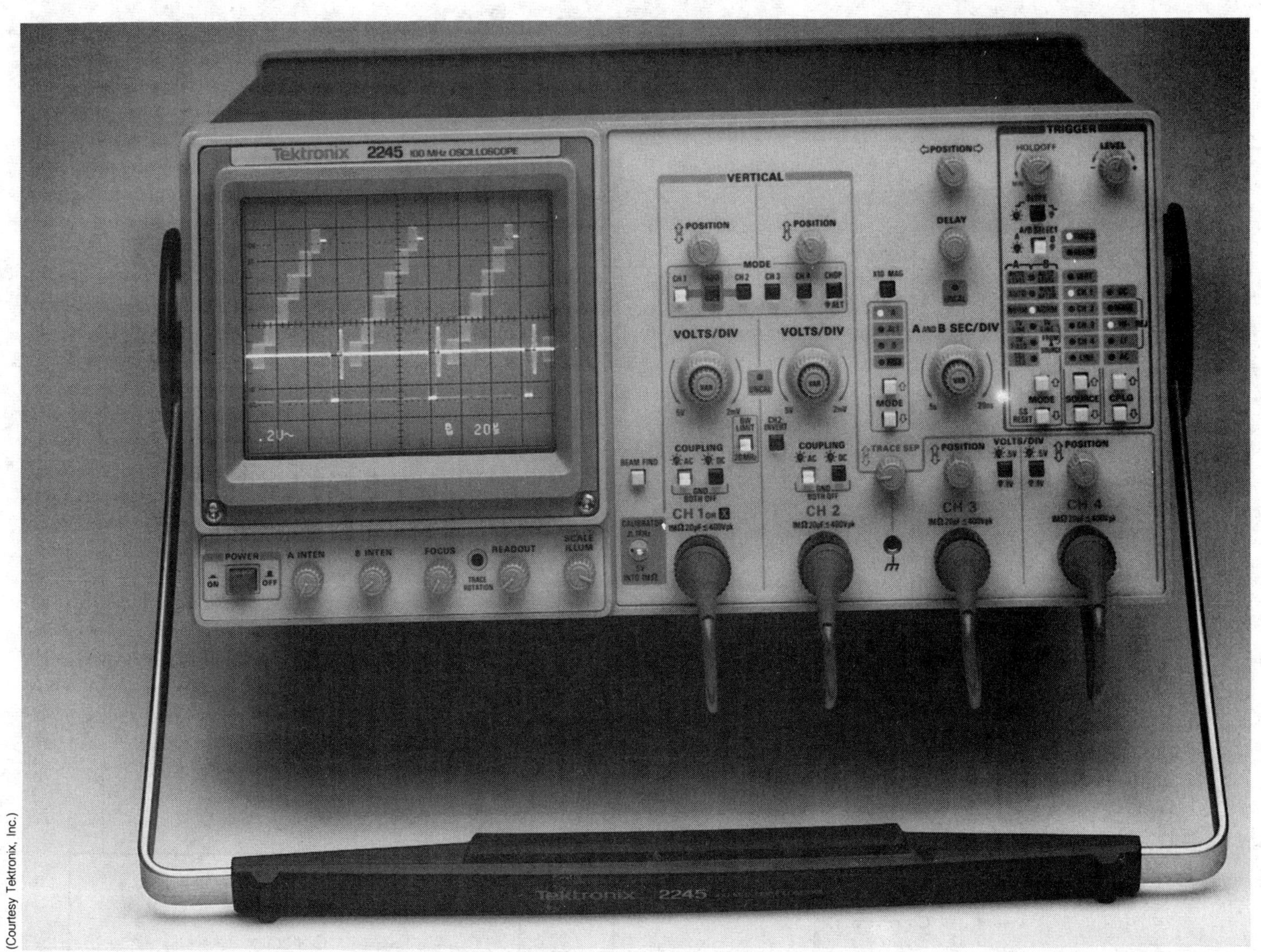

(Courtesy Tektronix, Inc.)

OSCILLOSCOPE: Fig. 1. Laboratory portable oscilloscope with a 100 MHz bandwidth.

while performance has improved significantly because of increased use of integrated circuits.

Dual and quad channels and delayed sweep are features on many new high-end oscilloscopes. A quad channel oscilloscope can have eight traces. A single CRT display can be split into as many as four separate displays for simultaneous viewing. This allows instant comparison of variables on the same or different time bases. Other features available include: 1) automatic setup and triggering, 2) options for external triggering, and 3) interfacing to an external computer with a general-purpose interface bus (GPIB).

Some oscilloscopes now have built-in digital voltmeters to measure alternating-current, direct-current, and peak-to-peak voltage. Built-in frequency meters with a frequency range of 10 Hz to 150 MHz may have a 5-digit readout. Events can be counted within delay time or delayed sweep time. Ground reference level can be displayed with dotted line waveforms. The user can prepare text for display on the CRT screen. In addition, vertical and horizontal deflection factors can be displayed digitally on the screen for quick reference.

Analog oscilloscopes equipped with special storage CRT tubes can store waveforms on the face of the CRT for indefinite periods to permit analysis and photography. However, analog storage oscilloscopes are significantly more expensive than comparable conventional instruments.

Digital-storage oscilloscopes transpose all input data to digital signal before it is used to write on the screen. Internal semiconductor memory preserves the data. This gives selective recall and storage in analog format on the screen. The digitized input data can also be sent to a computer over the GPIB (IEEE 488) interface bus for further processing. Stored analog output data can be used to drive a chart recorder or X-Y plotter. Most real-time digitizers for oscilloscopes are based on flash-analog converters (ADCs) that encode the outputs of as many as 64 comparators in parallel at the same time. (*See* ANALOG-TO-DIGITAL CONVERTER.) These oscilloscopes acquire an entire waveform for each trigger occurrence.

The digital-storage oscilloscope has advantages over the analog oscilloscope. The digital model will probably not replace the analog model because the analog model has unique advantages in circuit troubleshooting—spotting and identifying unusual or unpredictable signal characteristics. The analog oscilloscope circuitry closely tracks the instantaneous response of the CRT to permit

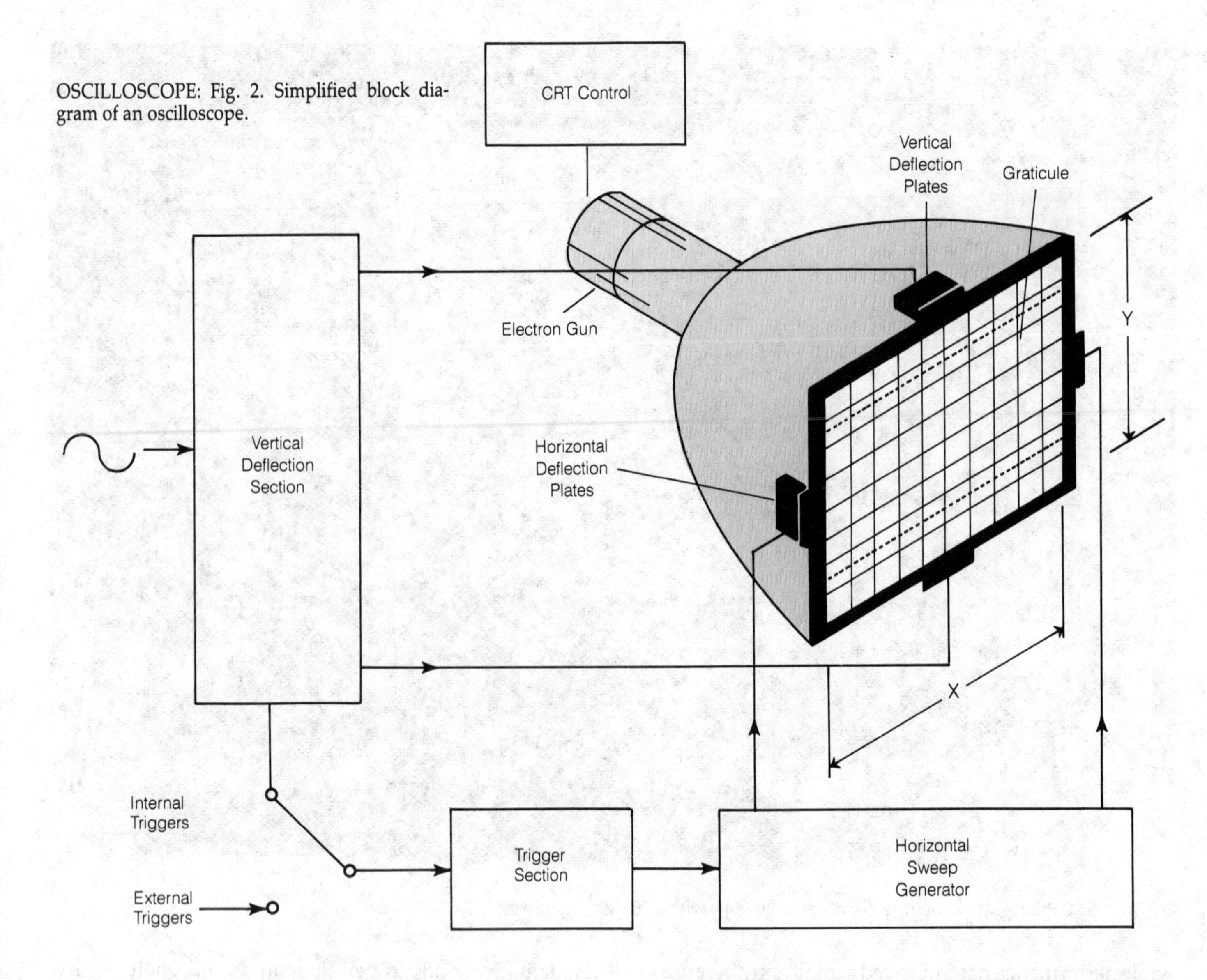

OSCILLOSCOPE: Fig. 2. Simplified block diagram of an oscilloscope.

fast signals to be captured in one pass.

Integrated circuitry has permitted dramatic weight and power reduction and shorter CRT length dimensions have permitted significant size reductions in oscilloscopes. Some of the latest and most advanced 100 to 150 MHz models are light enough to be considered portable, although they are line powered. However, there is still demand for true battery-portable oscilloscopes for use where stable 120/240 V, 50 to 60 Hz line power is not available.

Battery portability adds significantly to the price and weight of an oscilloscope. These instruments are designed to withstand rough handling and are provided with rechargeable nickel-cadmium battery packs and special carrying cases. But the selection and available bandwidth of these products is limited. As an alternative, some oscilloscope manufacturers offer add-on power modules for converting available on-site direct-current power to stable alternating current. If a direct-current source is not available, a separate rechargeable battery module can be assembled to form a self-contained power pack.

OSCILLOSCOPE CAMERA

An oscilloscope camera is a special camera designed for taking photographs of oscilloscope displays. This provides a permanent record of the display. Oscilloscope cameras are often used for comparing data obtained at different times in the design of electronic equipment.

Specially designed cameras are made for the purpose of photographing oscilloscope displays. These cameras are fitted with an opaque hood for keeping out stray light. The hood is attached to the screen frame of the oscilloscope. Many oscilloscopes are provided with mounting screws or flanges around the screen, for the attachment of an oscilloscope camera. *See also* OSCILLOSCOPE.

OUNCE

The ounce is a unit of weight. In the gravitational field of the earth, an ounce is the amount of weight corresponding to a mass of 28.35 grams. An ounce is 1/16 (0.0625) pound.

The fluid ounce is a unit of measurement of volume.

A fluid ounce is 1/16 (0.0625) pint, or 1/32 (0.03125) quart. A fluid ounce is equivalent to 0.02957 liter, or 29.57 cubic centimeters.

OUT OF PHASE

Two signals are out of phase when their phase difference is 180 degrees. That is, they must be ½ cycle different in phase.

For certain waveforms such as the sine wave and the square wave, two out-of-phase signals of equal amplitude totally cancel (see illustration). When this happens, the net signal is zero because the instantaneous signal amplitudes are equal and opposite. *See also* PHASE, PHASE ANGLE.

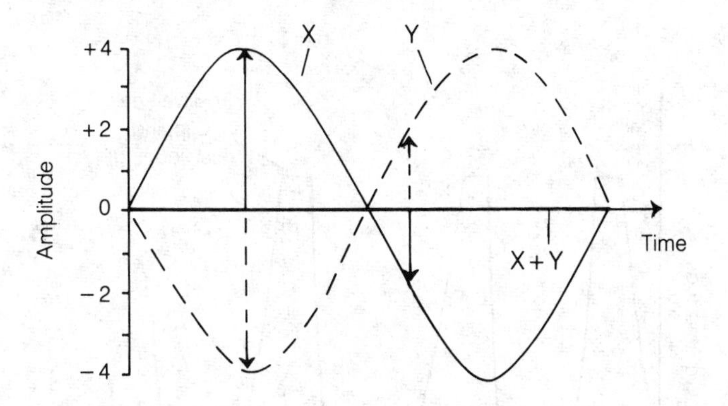

OUT OF PHASE: Sine waves of opposite phase and equal amplitude cancel each other, resulting in zero signal.

OUTPUT

Output is the signal produced or processed by a circuit. An oscillator generates its own output. An amplifier circuit produces its output from a signal of lesser intensity, applied to the input. A circuit might have just one set of output terminals, or it might have several. The output terminals themselves are sometimes called outputs. *See also* OUTPUT IMPEDANCE.

OUTPUT IMPEDANCE

The output impedance of a circuit is the load impedance which, when connected to the output terminals, results in optimum transfer of power. The output impedance of a circuit is normally a pure resistance. If a load contains reactance, that reactance must, for best operation, be tuned out by means of an equal and opposite reactance connected in series with the load.

The resistance of a load should always be matched to the output resistance of the driving circuit. Removing the reactance may not be sufficient. A transformer is required for resistance matching. For example, if an audio amplifier is designed to work into an 8-Ω load, a step-up transformer must be used for best results with 16-Ω speakers, and a step-down transformer should be used with 4-Ω speakers.

When the output impedance of a device differs from the load impedance, some of the electromagnetic field is reflected back from the load toward the source. This might or might not cause significant deterioration in the performance of the driving circuit. In the audio-amplifier example, the transformer can, in some cases, be removed with only a slight reduction in system efficiency. However, in many instances, the impedance match must be nearly perfect. *See also* IMPEDANCE, IMPEDANCE MATCHING, INPUT IMPEDANCE, REACTANCE.

OUTPUT RESISTANCE

See OUTPUT IMPEDANCE.

OVERCURRENT PROTECTION

Overcurrent protection is any means of preventing excessive current from flowing through a circuit. The most common forms of overcurrent-protection devices are the circuit breaker and the fuse (*see* CIRCUIT BREAKER, FUSE).

Some power supplies have built-in overcurrent protection. If the load resistance drops to a value that would cause excessive current to flow, the power supply automatically inserts an effective resistance in series with the load, preventing the current from rising above the maximum rated value. When the load resistance returns to normal, the power supply continues to deliver the rated current without the need for resetting a circuit breaker or replacing a blown fuse. *See also* CIRCUIT PROTECTION, CURRENT LIMITING, CURRENT REGULATION.

OVERDRIVE

Overdrive occurs when excessive input is applied to a power amplifier. Overdrive is an undesirable condition, because it results in distortion of the envelope of a modulated signal. It also causes the generation of excessive harmonics. In some cases, overdrive can damage a transistor or tube.

In a properly operating power amplifier, the output power increases, up to a certain point, as the driving

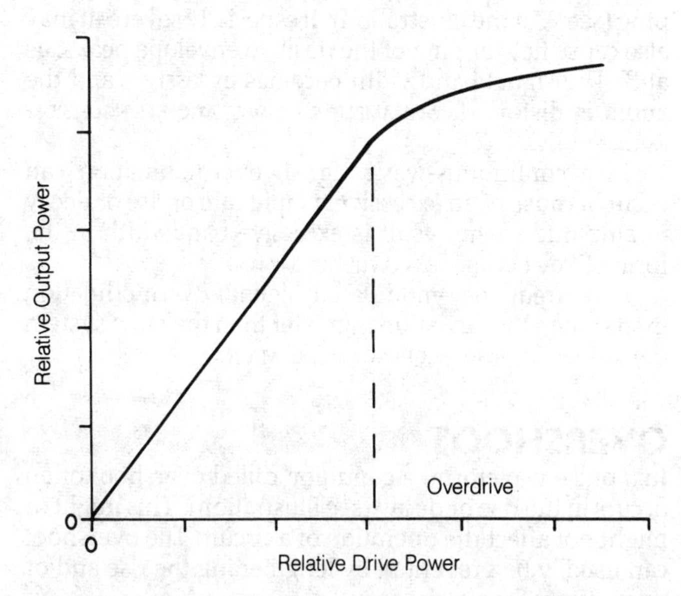

OVERDRIVE: When power output no longer increases at the normal rate, an amplifier is overdriven for linear operation.

power is increased. Beyond this point, the output power increases more gradually. It is in this range that overdrive occurs (see illustration). *See also* DRIVE.

OVERFLOW

The capacity of any memory device is limited. When the amount of data is too great for storage in a memory, an overflow condition occurs. This can be observed in computers.

Overflow can result in various responses from a calculator or computer. Generally, the excess data is ignored or compensated for in some way, or an error signal can be generated.

OVERLOADING

Overloading is the result of an attempt to draw too much current from a source of power, or to apply excessive voltage to a load.

An example of overloading is the improper adjustment of the output matching circuit in a radio transmitter, resulting in excessive collector or plate current in the final-amplifier stage, and reduced radio-frequency output power and efficiency.

In a radio receiver, an incoming signal may be so strong that the front end is driven into nonlinear operation or desensitization. This condition is sometimes called overloading. It results in false signals, intermodulation, or a general reduction in sensitivity.

OVERMODULATION

When a radio transmitter is modulated to an extent greater than that specified or required for normal operation, the transmitter is said to be overmodulated. Overmodulation can occur with any type of emission.

In an amplitude-modulated signal, overmodulation is generally considered to exist when the modulation is more than 100 percent. This causes negative-peak clipping (see A in the illustration). If especially severe, it may also cause flat-topping of the positive envelope peaks, as at B. The signal bandwidth becomes excessive, and the audio is distorted. *See* AMPLITUDE MODULATION, SINGLE SIDEBAND.

In a continuous-wave signal, overmodulation can occur because of an excessively rapid rate of rise or decay in amplitude. The result is excessive bandwidth in the form of key clicks. *See* CONTINUOUS WAVE.

In a frequency-modulated signal, overmodulation exists when the deviation is greater than the rated system deviation. *See also* FREQUENCY MODULATION.

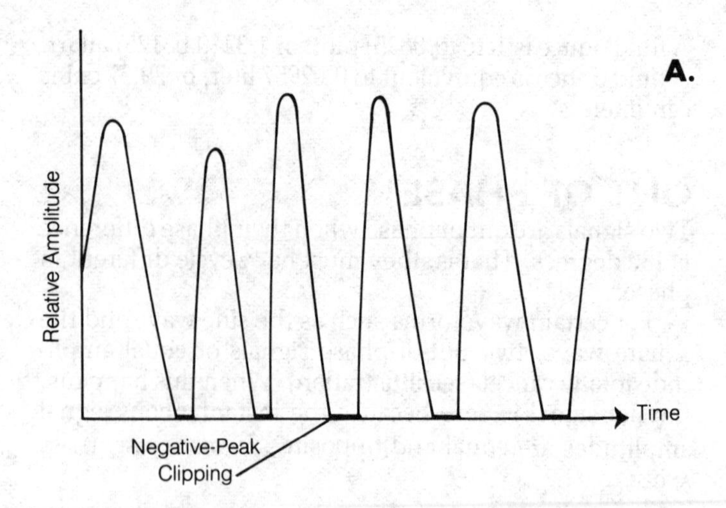

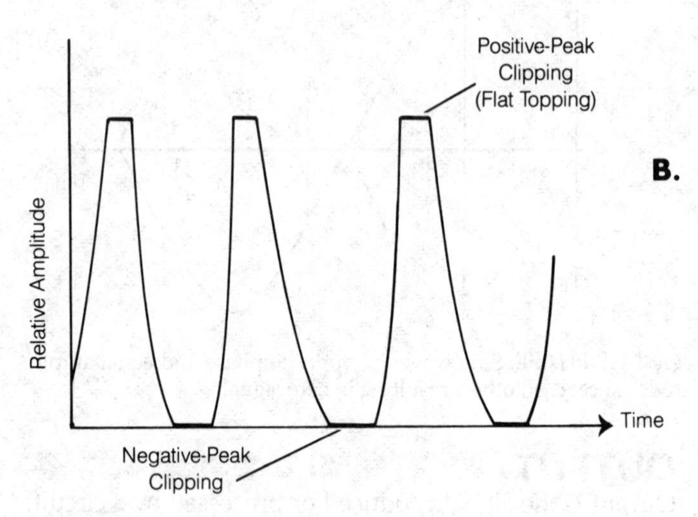

OVERMODULATION: Examples of overmodulation include negative-peak clipping (A) and positive-peak and negative-peak clipping (B).

OVERSHOOT

In a pulse waveform, a condition called overshoot often occurs in the rise or decay (see illustration). This might or might not affect the operation of a circuit. The overshoot can usually be prevented by lengthening the rise and/or decay time. *See also* DROOP.

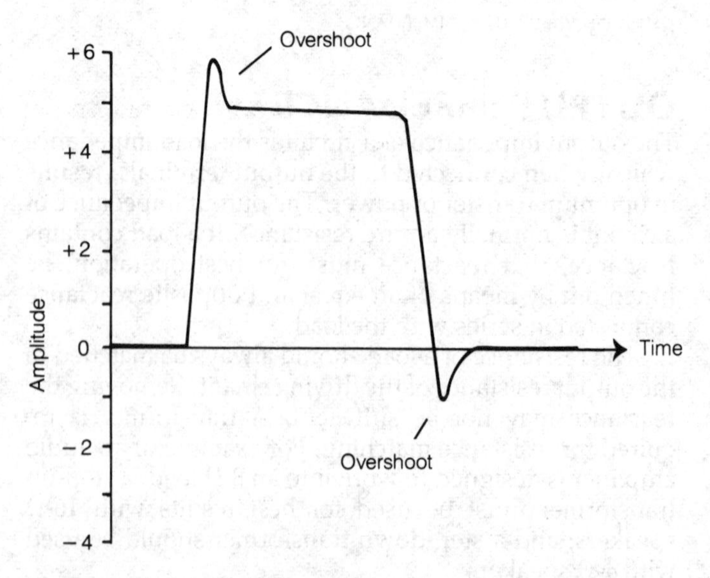

OVERSHOOT: Overshoot can occur on the rise and/or decay of a pulse.

OVERVOLTAGE PROTECTION

In a regulated power supply, the regulation system can fail. If this happens, and the power supply does not have an overvoltage-protection circuit, the voltage will rise dramatically. In some cases it may increase to more than twice the normal value. This can cause damage to equipment connected to the supply.

Most overvoltage-protection circuits consist of a zener diode for sensing the output voltage, and a silicon-controlled rectifier for short-circuiting the output terminals when overvoltage occurs. This causes the regulator circuit to switch off the supply, or the fuse can blow. A schematic diagram of this type of overvoltage-protection circuit is shown.

There are other methods of obtaining overvoltage protection besides that shown in the illustration. One common alternative method is called remote sensing. The remote-sensing circuit consists of a device that has a certain threshold voltage, shutting down the supply when the threshold is exceeded. A bipolar transistor can be used for this purpose. The threshold voltage depends on the base bias. Under normal conditions the transistor, which is connected in series with the supply output, conducts. If the voltage becomes excessive, the transistor opens the circuit. *See also* POWER SUPPLY, REGULATION.

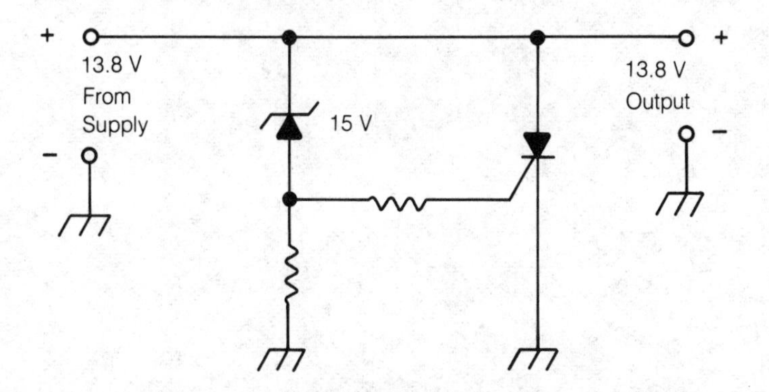

OVERVOLTAGE PROTECTION: A Zener diode causes the power supply to be shut down if the voltage exceeds 15 V.

OWEN BRIDGE

The Owen bridge is a circuit for measuring unknown values of inductance. The Owen bridge is noted for its extremely wide range—it can determine any inductance from less than 0.1 nH to more than 100 H. This covers essentially all possible inductance values.

The schematic diagram shows an Owen bridge. The unknown inductance, *L*, has an effective series resistance *R*. Potentiometers R1 and R2 are adjusted for a null reading on the indicator. The generator produces an audio-frequency signal. The indicator can be a speaker, a headset, or a meter with a rectifier diode connected in series. Capacitors C1 and C2, and resistor R3, are fixed in value.

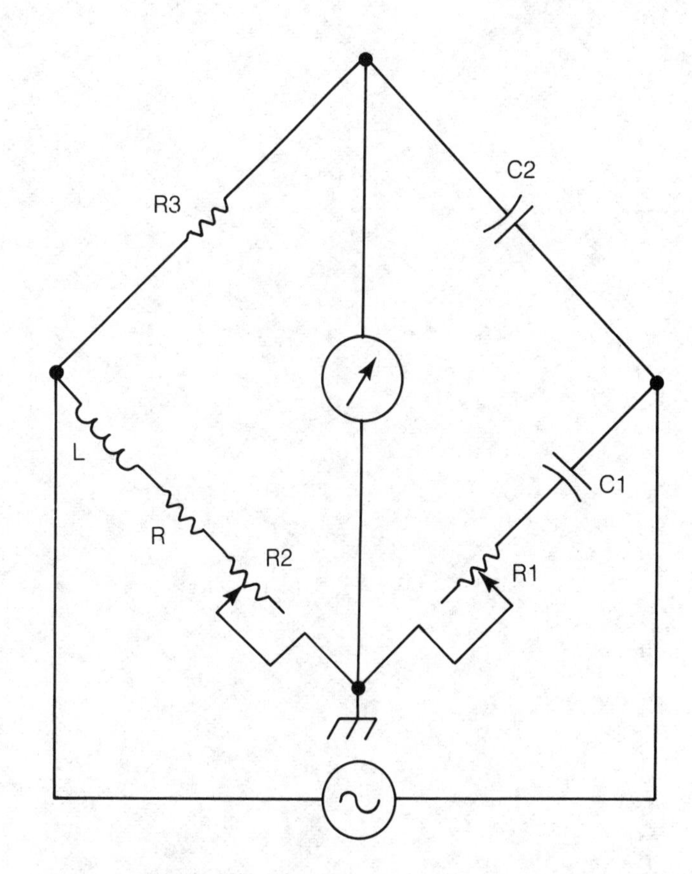

OWEN BRIDGE: The Owen bridge measures inductance over a wide range, and it also can determine the effective series resistance of an inductor.

When a balance indication is obtained, the unknown inductance can be determined according to the formula:

$$L = C_2R_1R_3$$

and the series resistance can be determined by the formula:

$$R = \frac{C_2R_3}{C_1} - R_2$$

See also INDUCTANCE.

OXIDATION

Oxidation is a form of corrosion. A material is oxidized when it combines with oxygen. Many different materials undergo oxidation when exposed to the air. The oxidation process is similar to combustion, but it occurs more slowly.

OXYGEN

Oxygen is an element with an atomic number of 8 and an atomic weight of 16. Oxygen is the second most abundant gas in the earth's atmosphere; the air is about 21 percent oxygen.

PACEMAKER

A pacemaker is an electronic circuit that regulates the heartbeat of a person whose heart cannot properly regulate itself. Pacemakers can be implanted in, or attached to, the body of a patient. Pacemakers are powered by lithium batteries.

Pacemakers generate electrical impulses at regular intervals. The impulses are transmitted to the nerves that control the heart muscles. Each impulse causes the heart to contract.

PAD

A pad is an attenuator network that displays no reactance and exhibits constant input and output impedances. Pads are used between radio transmitters and external linear amplifiers to regulate the amplifier drive. They are also used in some receiver front ends to reduce or prevent overloading in the presence of extremely strong signals. These pads can be switched in or out of the front end as desired. Pads are sometimes used for impedance matching.

A typical attenuator pad configuration is shown in the illustration. The resistors are noninductive and must be capable of dissipating the necessary amount of power. In a receiver front end, small resistors [¼ or ⅛ watts (W)] are satisfactory. For external power-amplifier applications, the resistors may have to dissipate as much as 100 W. *See also* ATTENUATOR.

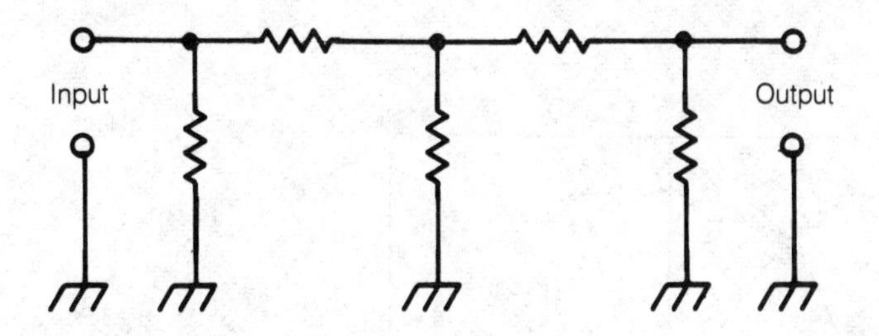

PAD: An attenuator pad for unbalanced lines.

PADDER CAPACITOR

A padder capacitor is a small, variable capacitor for adjusting the frequency of an oscillator circuit. The padder capacitor is placed in parallel with the tuning capacitor in the tank circuit of the oscillator (see the figure). The value of the padder capacitor is much smaller than that of the main tuning capacitor. Most padders have a maximum capacitance of only a few picofarads. The padder allows precise calibration of the tuning dial or frequency readout of a receiver or transmitter.

When a ganged variable capacitor is used to tune more than one resonant circuit at a time, padder capacitors can be placed across each section of the ganged capacitor. The padders are adjusted until proper tracking is obtained (*see* GANG CAPACITOR). Some gang capacitors have built-in padders. *See also* CAPACITOR.

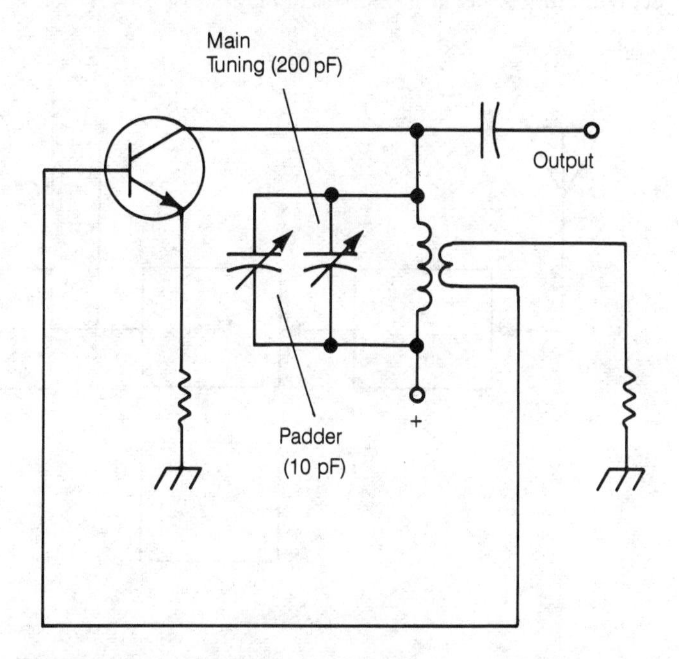

PADDER CAPACITOR: An oscillator with a padder capacitor in the tuning network.

PADDING

Padding is the process of precisely adjusting a circuit with small, series or parallel components. An example is the use of a padder capacitor for adjustment of oscillator frequency (*see* PADDER CAPACITOR).

In computer practice, padding is a process in which "blanks" or other meaningless data characters are added to a file to make the file a certain standard size. For example, the meaningful data in a computer file may comprise 12 kilobytes (K), but the standard file size might be 16 K. The file is padded by adding 4 K of blanks. The blanks can be added anywhere in the file; the most common location is at the beginning or end.

When an attenuator pad is inserted into a circuit, the process is sometimes called padding (*see* PAD).

PANORAMIC RECEIVER

A panoramic receiver is an instrument that allows continuous visual monitoring of a specified band of frequencies. Most receivers can be adapted for panoramic reception by connecting a specialized form of spectrum analyzer into the first intermediate-frequency chain, just after the first (see A in illustration).

A monitor screen displays the signals as vertical traces along a horizontal axis. The signal amplitude is indicated by the height of the trace. The position of the trace along the horizontal axis indicates its frequency. The frequency at which the receiver is tuned appears at the center of the horizontal scale. The number of kilohertz per horizontal division can be set for spectral analysis of a single signal or a narrow frequency band (B) or the scale can be set for observation of a wide range of frequencies (C). The maximum possible range is limited by the selectivity characteristics of the radio-frequency receiver stages. *See also* SPECTRUM ANALYZER.

PAPER CAPACITOR

See CAPACITOR.

PAPER TAPE

Paper tape is a permanent data-storage medium. It has been largely replaced by disk memory in modern computers.

Each character is punched into a paper tape as a set of small holes in a row perpendicular to the edges of the tape by a machine called a reperforator. The arrangement in which the holes are punched is shown at A in the illustration. The eight-channel hole code, according to Electronic Industries Association (EIA) Standard RS-244, is shown for the basic symbols A through Z, 0 through 9, and various punctuation and function characters. Holes correspond to the high (or 1) condition; blanks correspond to the low (or 0) condition. The eight-level code is now more commonly used than the old five-level code (see B).

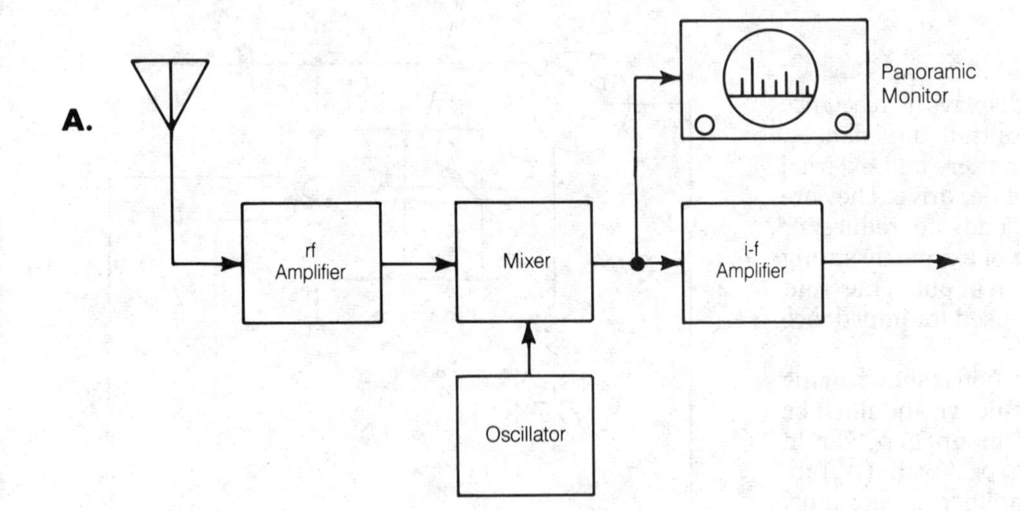

PANORAMIC RECEIVER: A block diagram showing a panoramic monitor in a receiver (A), a narrow band display (B), and a wide-band display (C).

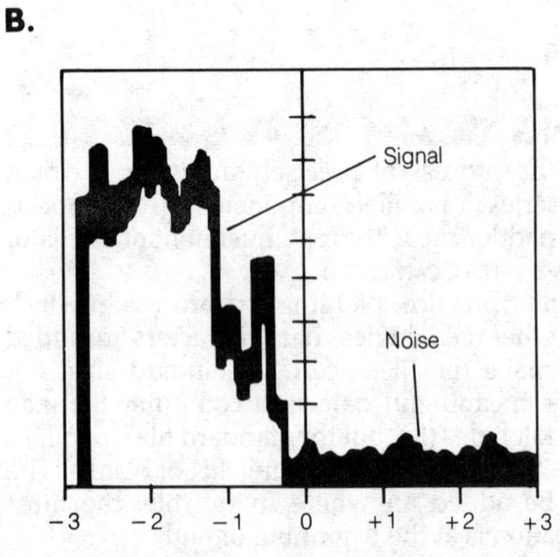

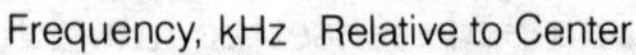

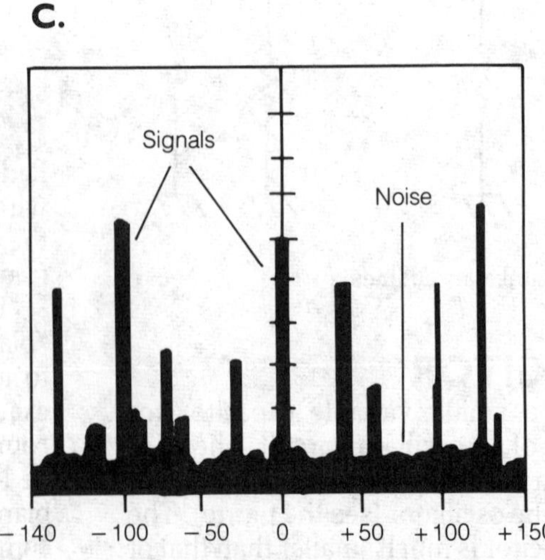

Paper tape can be read mechanically at a rate of approximately 60 characters per second. Electronic readers employing optical coupling can read the tape much faster. Paper tape has the advantage of being a permanent form of memory. It cannot be erased by accidental removal of power or exposure to magnetic fields. However, paper tape is bulky for storage of large amounts of data. A disk or magnetic-tape storage media are generally used for large files. *See also* DISK DRIVE, MAGNETIC TAPE.

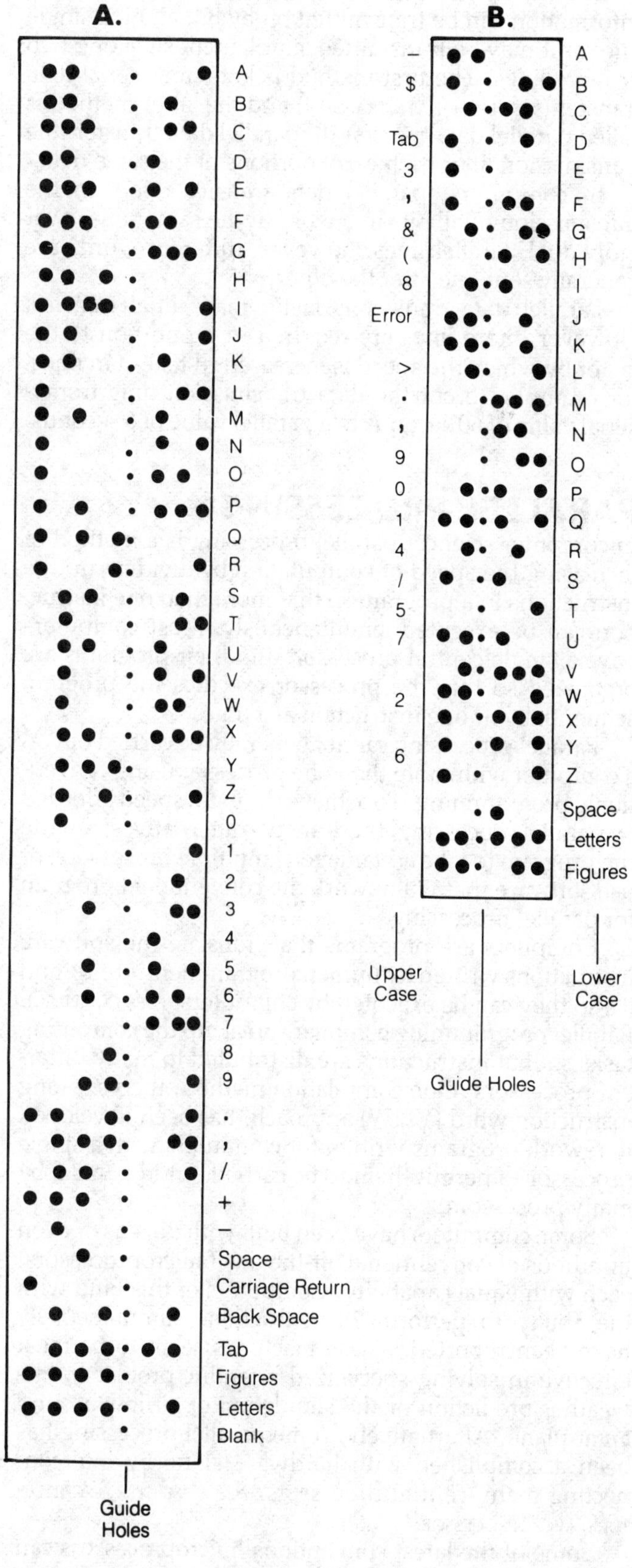

PAPER TAPE: A standard eight-level paper tape code (A) and a standard five-level code (B).

PARABOLA

A parabola is a conic section in two dimensions. In the Cartesian system of coordinates it is a function of the form:

$$y = ax^2 + bx + c$$

where *a*, *b*, and *c* are constants, *x* is the independent variable, and *y* is the dependent variable. Any parabolic shape, in terms of constant units, can be defined by selecting appropriate values for *a*, *b*, and *c*. The drawing illustrates the simple parabola $y = x^2$.

If a parabola is rotated about its axis the result, in three dimensions, is called a paraboloid. In the example shown the axis of the parabola is the y axis. Dish antennas are often of paraboloidal form. The paraboloid acts as a collimator for electromagnetic waves. *See also* PARABOLOID ANTENNA in the ANTENNA DIRECTORY.

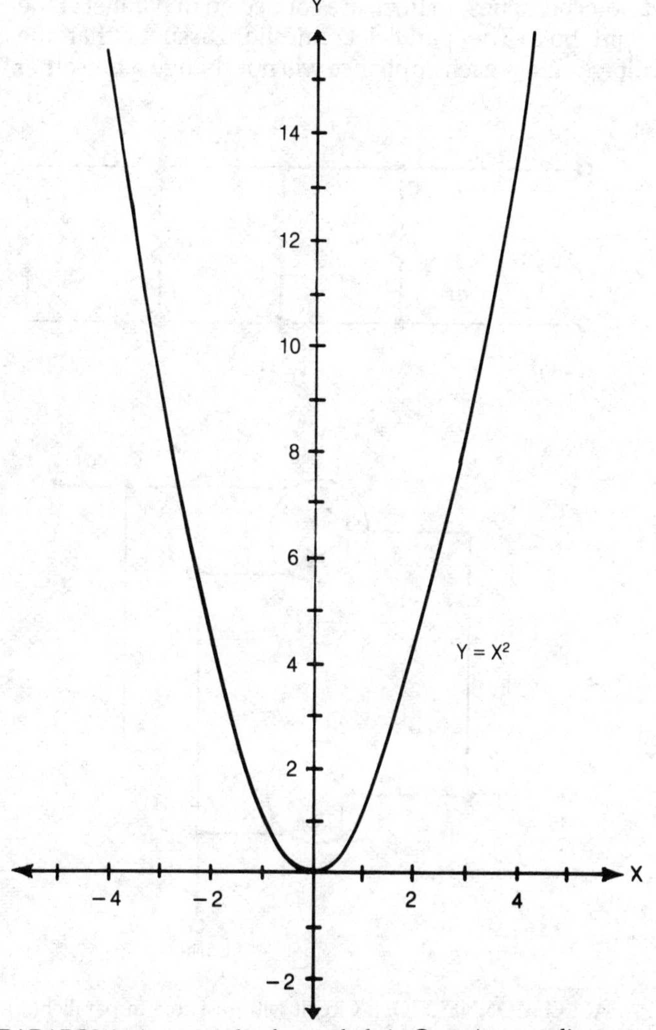

PARABOLA: An example of a parabola in Cartesian coordinates.

PARABOLOID

See PARABOLA.

PARABOLOID ANTENNA

See ANTENNA DIRECTORY.

PARALLEL CONNECTION

Components are connected in parallel when corresponding leads of each component are connected together. The drawing illustrates some examples of parallel connections; at A, resistors are connected in parallel, and at B, transistors are connected in parallel.

When components are connected in parallel, the voltage across all components is the same. The components share the load of current.

Parallel connections increase the current-handling or power-handling capability of a circuit. A parallel combination of ten 500 Ω, 1-watt (W) resistors can handle 10 W as a 50 Ω dummy load, for example. This is ten times the dissipation rating for a single 50 Ω, 1 W resistor. Transistors or tubes can be connected in parallel in an amplifier circuit to increase the obtainable power output.

In household electric circuits, all of the appliances are normally connected in parallel in each branch. The various branches, in turn, are connected in parallel at the circuit box. The parallel connection assures that the voltage across each appliance will not change when other appliances are used. *See also* PARALLEL TRANSISTORS, SERIES CONNECTION.

A.

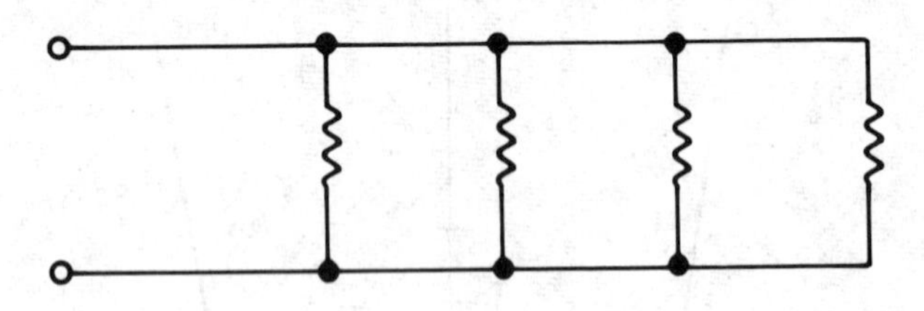

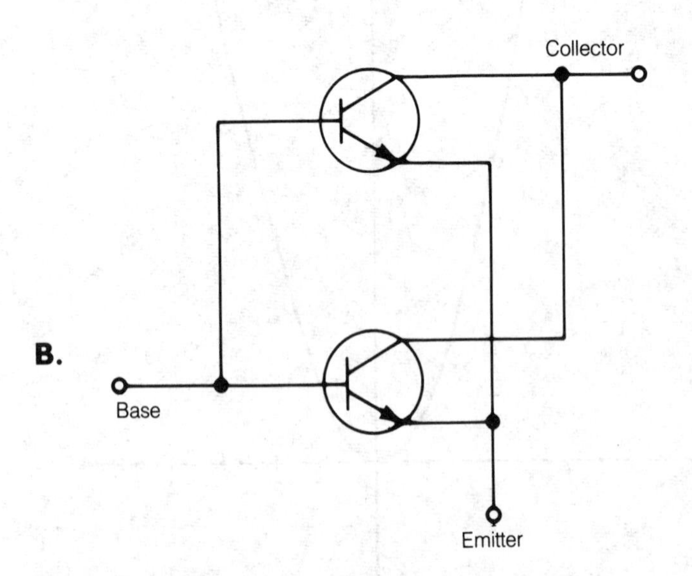

PARALLEL CONNECTION: Circuit with resistors in parallel (A) and transistors in parallel (B).

PARALLEL DATA TRANSFER

Information can be transmitted bit by bit, along a single line, or it may be transmitted simultaneously along two or more lines. The first method is known as serial data transfer (*see* SERIAL DATA TRANSFER) and the latter method is called parallel data transfer. In parallel data transfer, the transmission lines each carry portions of the same data.

In computers, parallel data transfer refers to the transmission of all bits in a word at the same time, over individual parallel lines; however, different words are generally sent one after the other.

Parallel data transfer is faster than serial transfer. However, more lines are required in proportion to the factor by which the speed is increased. It takes ten lines, for example, to cut the data transmission time from a serial value of 60 seconds to a parallel value of 6 seconds.

PARALLEL PROCESSING

In computer science, parallel processing is a method of increasing the speed of computation by dividing up the instructions in a program so that more than one instruction can be executed simultaneously. Most computers have a single central processor, and their programs are organized so that the processor executes the program sequentially or one instruction at a time.

Parallel processing cannot generally be carried out on a computer with more than one processor using conventional programming. To achieve the extra speed afforded by parallel processing, there are two alternatives: rewrite the program to take advantage of multiple processors, or use software that will rework the conventional program for parallel processing.

Compilers are programs that translate the software instructions written by human programmers into 1's and 0's so they can be executed by computers. (*See* COMPILER). Parallel programming compilers break up the computing tasks so that instructions are distributed to many different processors. One compilation method, the very long instruction word (VLIW) approach, has been developed to rework programs written for computers with a single processor to permit them to be carried out in parallel by many processors.

Some computers have been built with dozens or even hundreds of conventional off-the-shelf microprocessors, each with equal capability. Computers of this kind with the ability to perform 28 instructions simultaneously have been reported. These machines seem to be most effective in solving specialized scientific problems, like weather prediction or the simulation of airflow around an airplane. Alternatively, some parallel processing has been accomplished with hardware alone by interconnecting many central processors. *See* COMPUTER ARCHITECTURE, MICROPROCESSOR.

Some of the latest conventional microprocessors can carry out several instructions simultaneously. Future reduced instruction set computers (RISC) whose central processors are built as microprocessors also are expected to have the ability to execute more than one instruction at a time. *See* REDUCED INSTRUCTION SET COMPUTER.

PARALLEL RESONANCE

When an inductor and capacitor are connected in parallel, the combination will exhibit resonance at a specific frequency (*see* RESONANCE, RESONANT FREQUENCY). This condition occurs when the inductive and capacitive reactances are equal and opposite, thereby cancelling each other and leaving a pure resistance.

The resistive impedance that appears across a parallel-resonant circuit is extremely high. In theory, assuming zero loss in the inductor and capacitor, the impedance at resonance is infinite. In practice, since there is always some loss in the components, the impedance is finite, but it may be on the order of hundreds of kilohms. The exact value depends on the Q factor, which in turn depends on the resistive losses in the components and interconnecting wiring (*see* IMPEDANCE, Q FACTOR). Below the resonant frequency, the circuit appears capacitively reactive; above the resonant frequency it is inductively reactive (see graph). Resistance is also present under non resonant conditions, making the impedance complex.

Parallel-resonant circuits are commonly used in the output configurations of radio-frequency oscillators and amplifiers. They may also be seen in some input circuits and coupling-transformer arrangements.

A section of transmission line may be used as a parallel-resonant circuit. A quarter-wave section, short-circuited at the far end, behaves like a parallel, and resonant, inductance and capacitance. A half-wave section, open at the far end, also exhibits this property (*see* LINE TRAP). Such resonant circuits can be used in place of inductance-capacitance (LC) circuits. This is often done at very high and ultrahigh frequencies, where the lengths of quarter-wave or half-wave sections are reasonable.

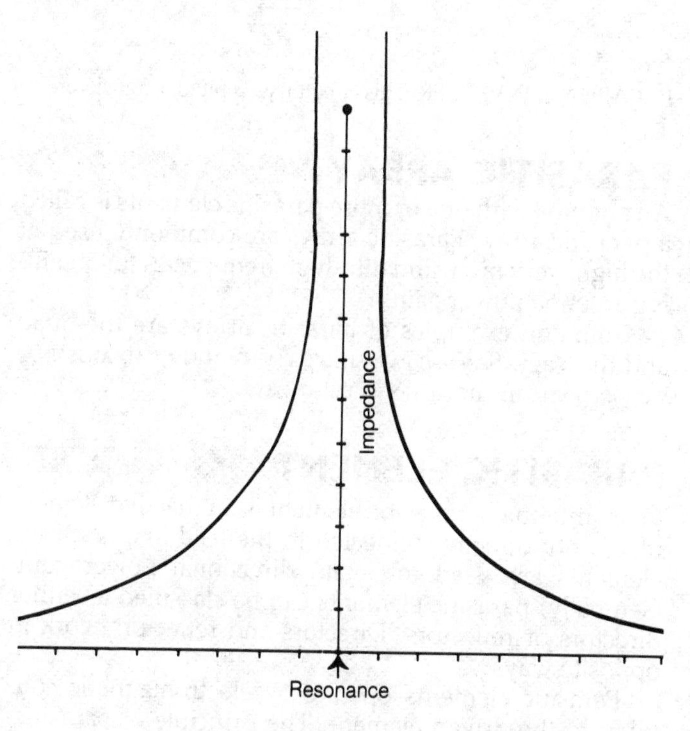

PARALLEL RESONANCE: A parallel-resonant circuit has a very high resistance at the resonant frequency.

PARALLEL TRANSISTORS

Two or more bipolar or field-effect transistors can be connected in parallel to increase power-handling capability. All of the emitters or sources, bases or gates, and collectors or drains, are connected together. All of the transistors should be identical and matched.

Ideally, the power-handling capability of a parallel combination of n transistors is n times that of a single transistor of the same kind. When two or three transistors are connected in parallel, this is essentially the case. However, in practice, certain limitations exist; it is not feasible to tie many transistors together, expecting an effective high-power amplifier to be the result. It is always better, if possible, to use one high-power transistor rather than several small ones. (See the figure.)

The main reason that massive parallel combinations of bipolar transistors do not work well is that the input impedance is reduced by this configuration, and most bipolar power transistors exhibit low input impedances to begin with. *See also* TRANSISTOR.

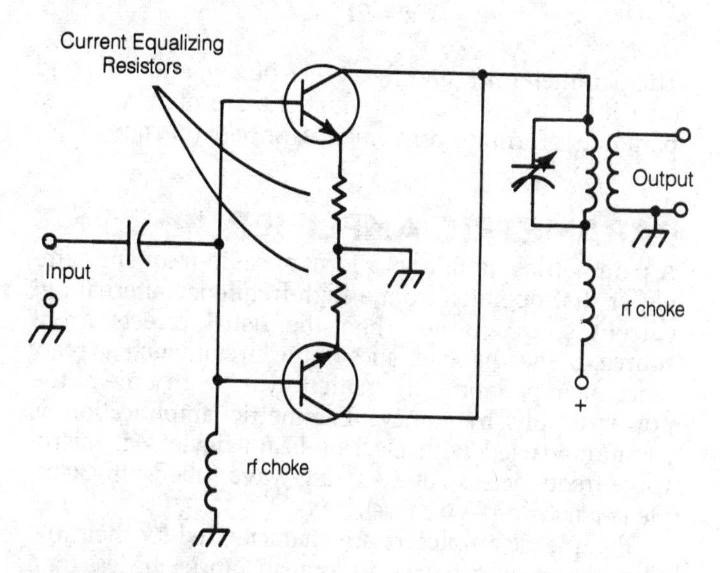

PARALLEL TRANSISTORS: Practical amplifier circuit using parallel transistors.

PARALLEL-WIRE LINE

See OPEN-WIRE LINE, TWIN-LEAD.

PARAMAGNETIC MATERIAL

A material with a magnetic permeability greater than 1 but less than that of a ferromagnetic material is called paramagnetic. Paramagnetic materials tend to maintain the same permeability even when the intensity of the magnetic field alternates. This is not characteristic of the ferromagnetics. Both ferromagnetic and paramagnetic materials exhibit changes in permeability with changes in temperature and frequency.

Some examples of paramagnetic materials are compounds of aluminum, beryllium, cobalt, magnesium, manganese, and nickel. *See also* DIAMAGNETIC MATERIAL, FERROMAGNETIC MATERIAL, PERMEABILITY.

PARAMETER

The term *parameter* is used to describe any factor that influences some other factor or factors. Parameter is frequently used as a synonym for the term *variable*. Some common examples of parameters used in electronics include such variables as radiation resistance, the value of the feedback resistor in an operational-amplifier circuit, and the value of the capacitor in a resonant circuit. Radiation resistance affects the operation of a radio antenna; the feedback resistance affects the gain of an operational amplifier; the capacitance in a resonant circuit affects the resonance frequency. These are just a few of many possible examples of parameters.

In mathematics, a parameter is a variable that affects or relates two or more other variables. Time is often specified as a parameter in physics or mathematics. Consider the following set of equations:

$$x = 3t + 4$$
$$y = -6t - 6$$
$$z = 2t - 2$$

The parameter is t, and the values of x, y, and z depend on t in this case. Sets of equations are often put into parametric form for convenience. *See also* FUNCTION.

PARAMETRIC AMPLIFIER

A parametric amplifier is a form of radio-frequency amplifier that operates from a high-frequency alternating-current source rather than the usual direct-current source. Some characteristics of the circuit, such as reactance or impedance, are made to vary with time at the power-supply frequency. Parametric amplification is commonly used with electron-beam devices at microwave frequencies. The traveling-wave tube is an example (*see* TRAVELING-WAVE TUBE).

Parametric amplifiers are characterized by their low noise figures which may, in some instances, be less than 3 decibels (dB). The gain is about the same as that of other types of amplifiers; typically it ranges from 15 to 20 dB.

PARAMETRIC EQUATION

See PARAMETER.

PARAPHASE INVERTER

A paraphase inverter is a circuit that provides two outputs, differing in phase by 180 degrees. A transistor or tube can be used.

The illustration shows an example of a paraphase inverter using a bipolar transistor. Two output connections are provided; one is at the collector and the other is at the emitter. The input is applied to the base.

The emitter output signal is in phase with the input signal, but the collector output signal is 180 degrees out of phase with the input signal. Because the collector output impedance is normally higher than the emitter output impedance, a voltage-dividing network can be used in the collector output if it is necessary for the two outputs to show equal impedances. If this is done, the circuit has gain of less than unity for both signals, and the output impedances are lower than the input impedance.

The paraphase inverter offers a simple and inexpensive way to obtain two signals in opposing phase.

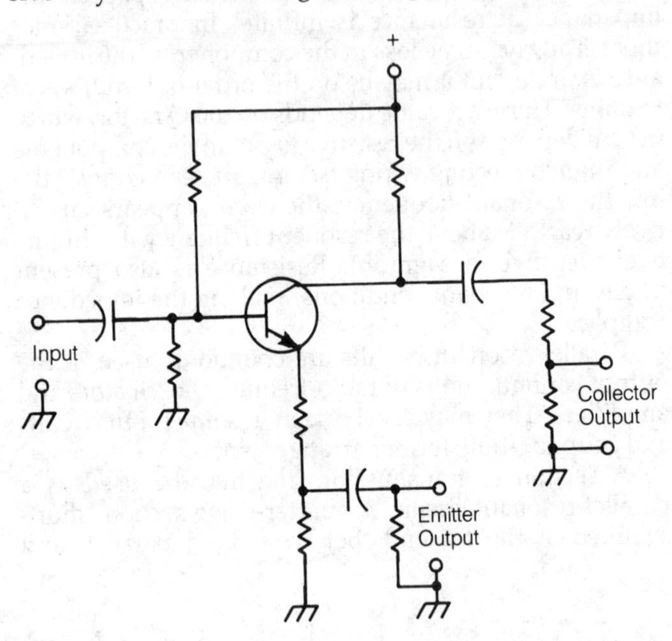

PARAPHASE INVERTER: This circuit uses a bipolar transistor.

PARASITIC ARRAY

An antenna with one or more parasitic elements is called a parasitic array. Parasitic arrays are commonly used at the high, very high, and ultrahigh frequencies for obtaining antenna power gain.

Common examples of parasitic arrays are the quad and the Yagi. *See also* PARASITIC ELEMENT and QUAD ANTENNA, YAGI ANTENNA in the ANTENNA DIRECTORY.

PARASITIC ELEMENT

In an antenna, a parasitic element is an inactive element that is not directly connected to the feed line. Parasitic elements are used to obtain directional power gain. Generally, parasitic elements can be classified as either directors or reflectors. Directors and reflectors work in opposite ways.

Parasitic elements operate by electromagnetic coupling to the driven element. The principle of parasitic-element operation was first discovered by a Japanese engineer named Yagi (the Yagi antenna is named after him). Yagi found that elements parallel to a radiating element, at a specific distance from it, and of a certain length, caused the radiation pattern to show gain in one direction and loss in the opposite direction.

At high frequencies, parasitic elements are often used in directional antennas. The most common of these, the quad and the Yagi antennas, are known as parasitic arrays. They exhibit a unidirectional pattern.

PARASITIC OSCILLATION

In a radio-frequency power amplifier, undesirable oscillation can occur at frequencies that differ from its operating frequency. These oscillations are called parasitic oscillations, or simply parasitics.

Parasitics can be eliminated by providing a certain amount of negative feedback in a power amplifier. This is accomplished by a neutralization circuit. A parasitic suppressor may also be used to choke off parasitics in some instances. *See also* NEUTRALIZATION, PARASITIC SUPPRESSOR.

PARASITIC SUPPRESSOR

A radio-frequency (rf) amplifier can oscillate at a frequency far removed from the operating frequency. This oscillation is called parasitic oscillation (*see* PARASITIC OSCILLATION).

When parasitic oscillations are at a frequency much higher or lower than the operating frequency, the unwanted oscillation can sometimes be choked off by means of parasitic suppressors. This method is especially common for eliminating parasitics in medium-frequency and high-frequency power amplifiers. The parasitic choke consists of a resistor and a small coil connected in parallel, and placed in series with the collector, drain, or plate lead of the amplifier (see illustration). The resistor is typically 50 to 150 Ω; it is a noninductive carbon type. The coil consists of three to five turns of wire, wound on the resistor. In an amplifier with two or more transistors or tubes in parallel suppressors are installed at each device individually.

Low-frequency parasitics are not affected by the type of suppressor shown. To eliminate low-frequency parasitics in an rf power amplifier, neutralization is usually necessary. *See also* NEUTRALIZATION.

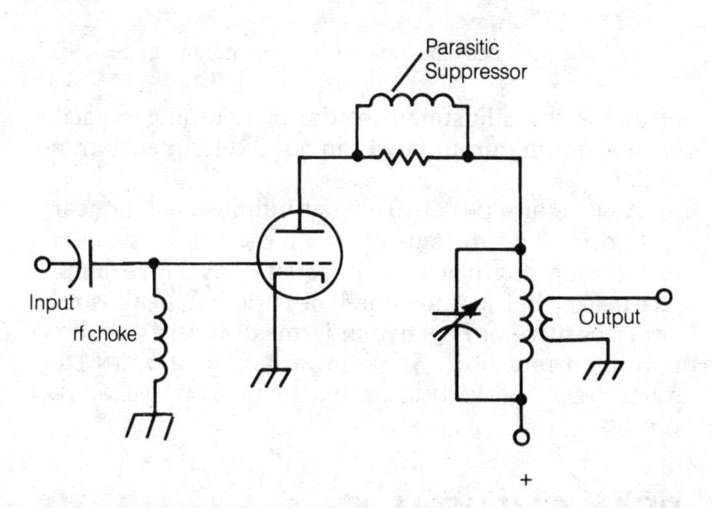

PARASITIC SUPPRESSOR: The RL circuit chokes off very high frequency parasitic oscillations.

PARITY

Parity is an expression that indicates whether the sum of the binary digits in a code word is even or odd. Accordingly, if the sum is even, the parity is called even; if the sum is odd, the parity is called odd. In certain situations, it is necessary to have the digits in all code words add up to an even number; in other cases the sum must always be odd.

For example, a code consists of five-digit words. Some examples might be 01001, 11000, and 11101 if the parity is even; in the case of odd parity, some code words might be 10000, 10101, and 00111.

If the sum of the digits in a word is incorrect (odd when it should be even, or vice versa), an extra bit can be added to the word to correct the discrepancy. Some codes have an extra bit that can be 0 (if the parity of the word is correct) or 1 (if not).

In the example given, a sixth digit could be added at the end of each word as a parity bit. Suppose that the parity is even. Then the first three words given would have a 0 added at the end and they would become, respectively, 010010, 110000, and 111010. The second three words, however, would require the addition of a 1, becoming 100001, 101011, and 001111. With the addition of the parity bit, the sum of the digits in all of the words is made to be even.

In a data-processing or transmission system, a parity check is sometimes made as a test for accuracy. If the parity is wrong, the state of the parity bit can be reversed, or the receiver might ask the transmitter to repeat the word.

PASSBAND

In a selective circuit, the passband is the band of frequencies at which the attenuation is less than a certain value with respect to the minimum attenuation. This value is usually specified as 3 decibels, representing half the power at the maximum-gain frequency.

In a superheterodyne receiver with a narrow bandpass filter, the passband may be as small as a few hundred hertz for continuous-wave signals, or as wide as 5 to 10 kHz for amplitude-modulated signals. In a receiver designed for reception of frequency-modulated signals, the passband may be as wide as 100 kHz or more. In a television receiver the passband is about 6 MHz in width.

If the passband is too wide for a given emission mode, excessive interference is received, reducing the sensitivity of the receiver, and the selectivity is degraded. There exists an optimum passband for every situation.

For continuous-wave signals of moderate speed, the passband can be as narrow as 50 to 100 Hz. For single-sideband and slow-scan television, 2 to 3 kHz is best. For amplitude-modulation, 6 kHz is standard for voice signals, about 10 kHz for music signals, and 6 MHz for fast-scan television. For frequency modulation, phase modulation, and pulse modulation, the optimum passband depends on the deviation, phase change, or pulse rate. *See also* BANDPASS FILTER, BANDPASS RESPONSE.

PASSIVE COMPONENT

In any electronic circuit, some components require no external source of power for their operation. Those components are called passive. Passive components include semiconductor diodes (in most cases), resistors, capacitors, and inductors. Components that require an external source of power, and/or produce gain in a circuit, are called active (*see* ACTIVE COMPONENT).

PASSIVE ELEMENT

See PARASITIC ELEMENT.

PASSIVE FILTER

A passive filter is a form of selective filter that does not require an external source of power for its operation. Passive filters may be of the bandpass, band-rejection, highpass, or lowpass variety (*see* BANDPASS FILTER, BAND-REJECTION FILTER, HIGHPASS FILTER, LOWPASS FILTER). Passive filters are commonly made using inductors, resistors, and/or capacitors. Crystal and mechanical filters are also passive (*see* CRYSTAL LATTICE FILTER, MECHANICAL FILTER).

Passive filters are commonly used in radio-frequency circuits. Active filters based on integrated-circuit operational amplifiers are widely used at audio frequencies. *See also* ACTIVE ANALOG FILTER.

PATENT

A patent is a legal guarantee of exclusive rights to manufacture a product. A patent for an original invention can be obtained in the United States by submitting the details to the Patent Office in Washington, DC.

To obtain a patent for an invention, such as an electronic circuit, the inventor must show novelty and originality. Improvements in previously existing devices can also be eligible for patent protection. For information concerning patents in the United States, the prospective inventor should contact a patent attorney or the U.S. Government Printing Office, Washington, DC 20402.

PEAK

In a waveform or other changing variable, a peak is an instantaneous or local maximum. The value of the variable falls off on either side of a maximum peak (see A in illustration), and rises on either side of a minimum peak (B).

The term peak is used in many different situations. A local maximum in a mathematical function may be called a peak. In the response curve of a bandpass filter, the frequency of least attenuation is often called the peak frequency or peak point. In the adjustment of a device, the peak setting is the control position that results in the maximum value for a certain parameter. An example

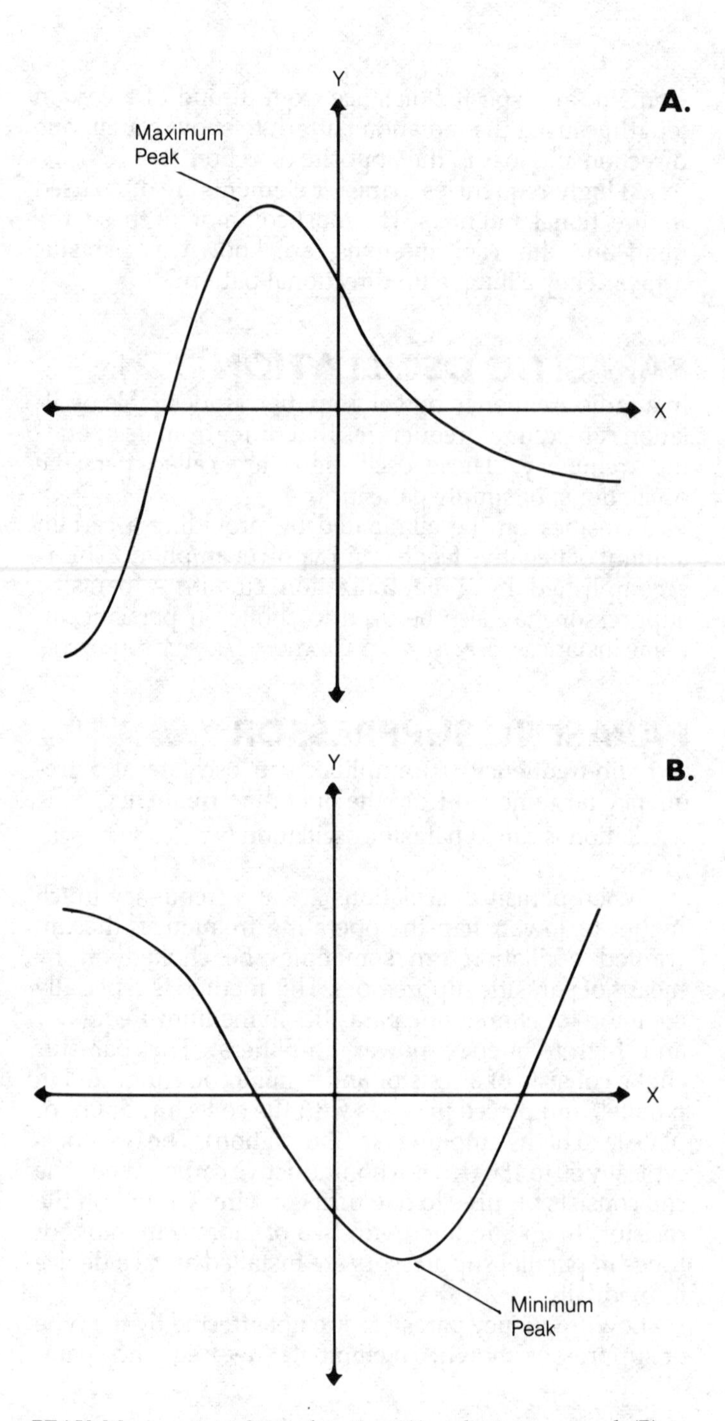

PEAK: Maximum peak in a function (A) and minimum peak (B).

of this is the adjustment of the plate-tuning capacitor for maximum output power in a radio-frequency transmitter.

A waveform peak usually has infinitesimal, or nearly zero, duration. (In some cases a peak may be sustained for a certain length of time; the square wave, for example, has peaks that last for one-half cycle.) A peak can be either positive or negative in terms of polarity or direction of current flow. *See also* PEAK CURRENT, PEAK ENVELOPE POWER, PEAK POWER, PEAK-READING METER, PEAK VALUE, PEAK VOLTAGE.

PEAK CURRENT

Peak current is the maximum instantaneous value

reached by the current in an alternating or pulsating waveform.

In a symmetrical alternating-current sine wave, the peak current is the same in either direction, and is reached once per cycle in either direction. In nonsymmetrical waveforms, the peak current may be greater in one direction than in the other. The peak current can be sustained for a definite length of time during each cycle, but more often the peak current is attained for only an infinitesimal time.

The peak current in a sinusoidal waveform is 1.414 times the root-mean-square (rms) current, assuming there is no direct-current component associated with the signal. For other waveforms, the peak-to-rms current ratio is different. *See also* ROOT MEAN SQUARE.

PEAK ENVELOPE POWER

In an amplitude-modulated or single-sideband signal, the radio-frequency carrier output power of the transmitter varies with time. The input power also varies. The maximum instantaneous value reached by the carrier power is called the peak envelope power. For any form of amplitude modulation, the peak envelope power is always greater than the average power. The peak envelope power differs from the actual peak waveform power. The peak waveform power is determined by the product of the instantaneous waveform voltage and current; the peak envelope power is the maximum root-mean-square waveform power (*see* PEAK POWER). Nevertheless, the terms *peak envelope power* and *peak power* are often used interchangeably.

In the case of frequency modulation or phase modulation, the carrier amplitude does not change, and the peak envelope power is the same as the average power. In a continuous-wave (CW) Morse-code signal, the peak envelope power is the key-down average power, which is approximately twice the long-term average power.

The peak envelope power is not shown by ordinary meters when signals are voice-modulated because the transmitter input and output power change too rapidly for any meter needle to follow. Special meters exist for measurement of peak envelope power in such cases (*see* PEAK-READING METER).

An oscilloscope can be used to show the modulation envelope of a transmitter; the peak envelope power can then be determined by comparison with a known, constant power level. This constant transmitter output can be obtained in a conventional amplitude-modulated transmitter by cutting off the modulating audio. In a single-sideband transmitter, a sine-wave tone can be supplied to the audio input, at a level sufficient to maximize the transmitter output without envelope distortion.

To determine the peak-envelope output power a wattmeter is first used to read the continuous output as obtained. The constant-amplitude signal is then displayed on an oscilloscope and the sensitivity of the scope is adjusted until the display indicates plus or minus 1 division (see A in illustration). Then normal modulation is applied. The peaks of the envelope can be seen,

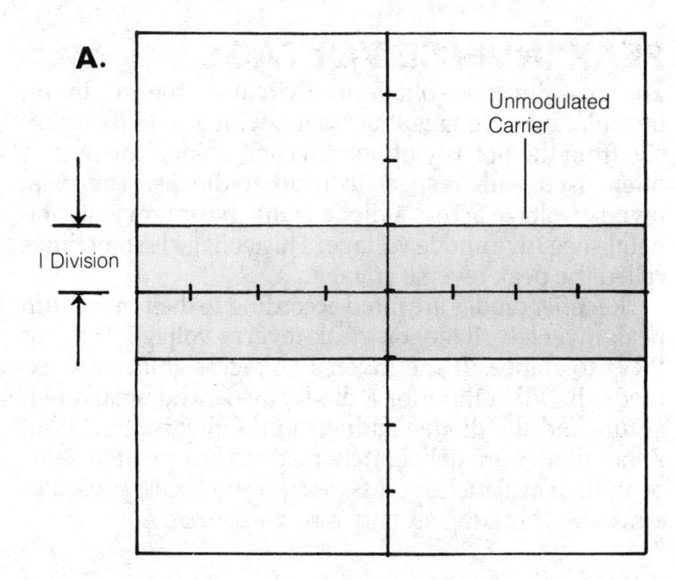

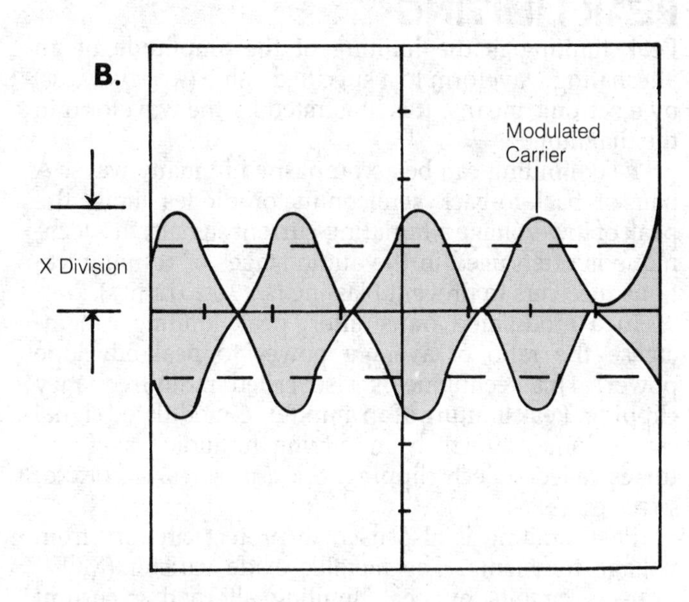

PEAK ENVELOPE POWER: An unmodulated carrier (A), and a fully modulated amplitude-modulated signal (B).

reaching a level of plus or minus x divisions, where x is greater than 1, as at B.

The oscilloscope indicates voltage, not power; thus, the ratio of the peak envelope power P to the constant-carrier power Q is:

$$P/Q = x^2$$

and the peak envelope power is given by:

$$P = x^2Q$$

Peak envelope power is often specified for single-sideband signals. The ratio of the peak envelope power to the average power in a single-sideband emission depends on the modulating waveform. In a single-sideband signal, the peak-envelope power for a human voice signal is approximately two to three times the average power. *See also* ENVELOPE, SINGLE SIDEBAND.

PEAK INVERSE VOLTAGE

The peak inverse voltage in a circuit is the maximum instantaneous voltage that occurs with a polarity opposite from the polarity of normal conduction. The term is often used with respect to rectifier diodes. The peak inverse voltage across a diode is the maximum instantaneous negative anode voltage. This voltage is sometimes called the peak reverse voltage.

Rectifier diodes are rated according to their maximum peak-inverse-voltage or peak-reverse-voltage (PIV or PRV) tolerance. If the inverse voltage significantly exceeds the PIV rating for a diode, avalanche breakdown occurs and the diode conducts in the reverse direction. Zener diodes are deliberately made to be operated near, or at, their avalanche points. *See also* AVALANCHE, AVALANCHE BREAKDOWN, RECTIFIER, RECTIFIER TUBE, ZENER DIODE.

PEAK LIMITING

Peak limiting is the limiting of the amplitude of an alternating waveform to a specified value (*see* PEAK VALUE) by electronic means. It is illustrated by the waveform in the diagram.

Peak limiting can be accomplished in many ways. A pair of back-to-back semiconductor diodes limits the peak of low-voltage alternating-current signals; this technique is often used in the audio stages of communications receivers to prevent blasting (*see* AUDIO LIMITER).

In a modulated transmitter, peak limiting can increase the ratio of average power to peak-envelope power. This technique is also called radio-frequency clipping. Peak limiting of an amplitude-modulated signal can be done indirectly by modifying the audio waveform; this is called speech clipping. *See also* CLIPPER, RF CLIPPING, SPEECH CLIPPING.

Peak limiting is also used to protect circuitry from voltage transients. The metallic oxide varistor (MOV) protects circuits by peak limiting alternating current circuits, and the transient voltage suppressor (TVS) protects circuits by peak limiting direct current. *See* CIRCUIT PROTECTION, METAL OXIDE VARISTOR.

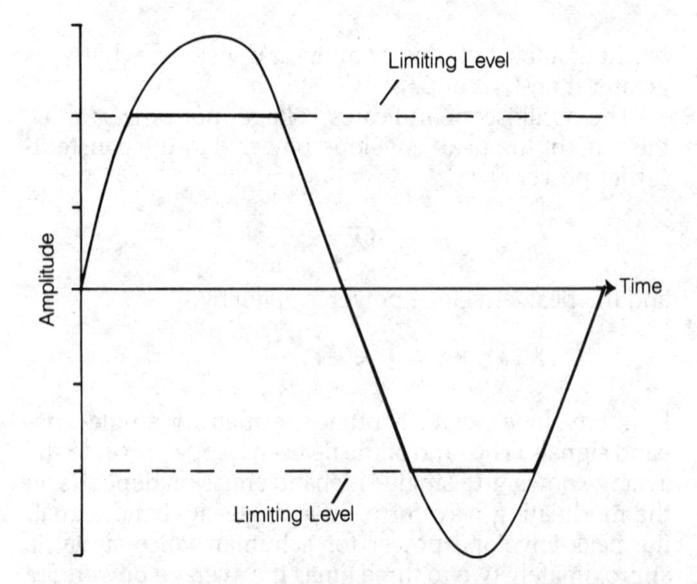

PEAK LIMITING: The amplitude cannot exceed the limiting level.

PEAK POWER

The peak power of an alternating-current signal is the maximum instantaneous power. The peak power is equivalent to the peak voltage multiplied by the peak current when the phase angle is zero (no reactance).

In a sinusoidal alternating-current signal, the peak power, P_p, can be determined from the root-mean-square (RMS) voltage, E_{RMS}, and the RMS current, I_{RMS}, according to the formula:

$$P_p = 2E_{RMS}I_{RMS}$$

where P_p is given in watts, E_{RMS} is given in volts, and I_{RMS} is given in amperes. The peak power, in terms of the peak voltage, E_p, and the peak current, I_p, is:

$$P_p = E_pI_p$$

In a modulated radio-frequency signal, the peak envelope power is sometimes called the peak power. *See also* PEAK ENVELOPE POWER.

PEAK-READING METER

A peak-reading meter is an instrument for measuring peak values. Most meters measure average values for signals that change rapidly because the meter needle or display time-constant cannot follow the variations.

Peak-reading meters can be designed for measurement of current, power, or voltage. These meters function in basically the same way, using voltmeters or ammeters with specialized peripheral circuitry.

A typical peak-reading meter operates as follows: (It is illustrated by the simplified schematic diagram.) A detecting circuit with an extremely fast time constant senses the peak voltage across a resistance. (A semiconductor diode can be used for this purpose.) A capacitor is charged by the rectified voltage. The circuit is wired so the capacitor charges rapidly but discharges slowly. The voltage across the capacitor is monitored by a voltmeter with an extremely high internal resistance. The peak

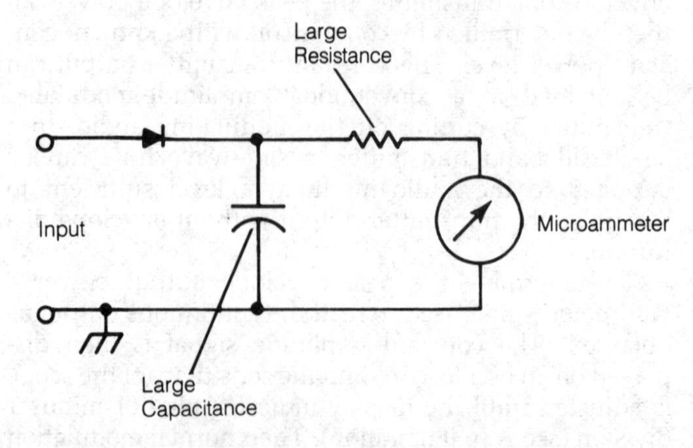

PEAK-READING METER: Simple peak-reading meter circuit.

current, power, or voltage can be determined from this voltmeter reading and the resistance in the circuit. The meter needle, upon application of signal, jumps up to the peak value and remains there for several seconds.

Peak-reading meters are employed for convenient reading of peak values without an oscilloscope. A peak-reading wattmeter may, for example, be used to measure the peak envelope power from a single-sideband transmitter. *See also* PEAK CURRENT, PEAK ENVELOPE POWER, PEAK POWER, PEAK VALUE, PEAK VOLTAGE.

PEAK-TO-PEAK VALUE

In an alternating waveform, the peak-to-peak value is the difference between the maximum positive instantaneous value and the maximum negative instantaneous value. Peak-to-peak is sometimes abbreviated as p-p or pk-pk.

The drawing illustrates peak-to-peak values for an alternating-current sine wave (A), a rectified sine wave (B), and a sawtooth wave (C). The peak-to-peak value of a waveform is a measure of signal intensity at a given frequency. Signal voltage is often given in terms of the peak-to-peak value. Note that changes in the direct-current component of a signal have no effect on the peak-to-peak value.

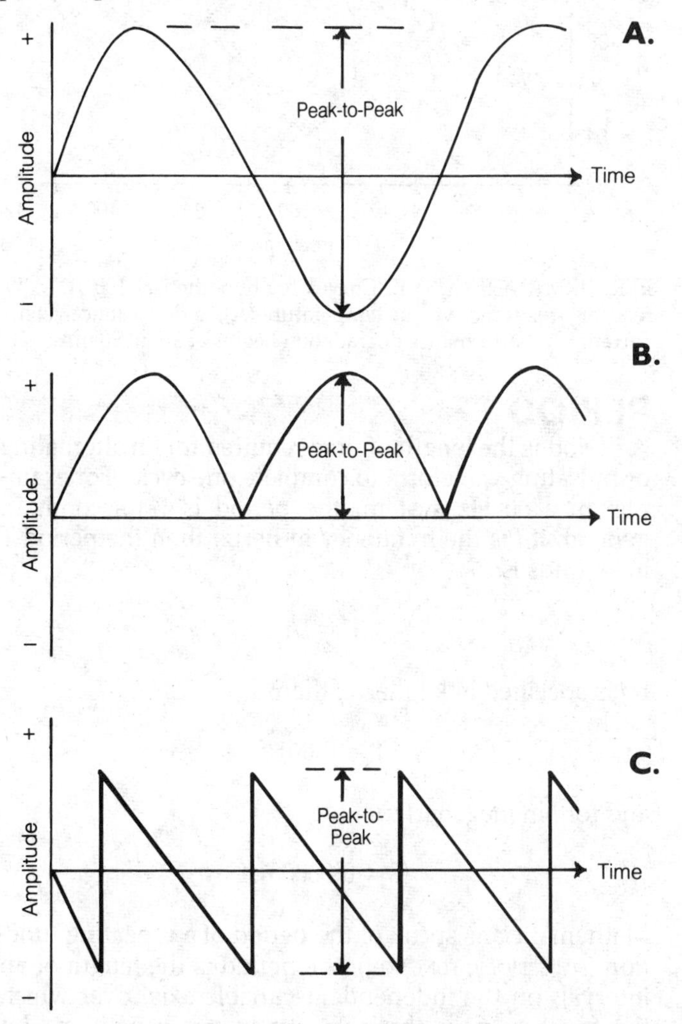

PEAK-TO-PEAK VALUE: Peak-to-peak amplitudes for a sine wave (A), a rectified sine wave (B), and a sawtooth wave (C).

Peak-to-peak values for any waveform are readily observed on an oscilloscope display. This is possible even if the frequency is too high to allow viewing of the waveform, or if waveform irregularity makes it difficult to determine the average value.

For an alternating-current sinusoidal waveform, the peak-to-peak value is equal to twice the peak value, and 2.828 times the root-mean-square value. *See also* PEAK VALUE, ROOT MEAN SQUARE.

PEAK VALUE

The peak value of an alternating waveform is the maximum instantaneous value. Some waveforms reach peaks in two directions of polarity or current flow (*see* PEAK CURRENT, PEAK VOLTAGE). These peaks might be equal in absolute magnitude. The peak value might last for a definite length of time; this is the case for a square wave. The peak value of a sine wave lasts for a short (essentially zero) time. The illustration at A shows peak values for a sine-wave alternating-current waveform. The peak value for the same waveform with the addition of a direct-current component is shown at B.

Peak values of power may be specified in either of two ways. The maximum instantaneous waveform power is the product of the peak voltage and the peak current (*see* PEAK POWER). The peak envelope power is the maximum carrier power in an amplitude-modulated or single-sideband signal (*see* PEAK ENVELOPE POWER).

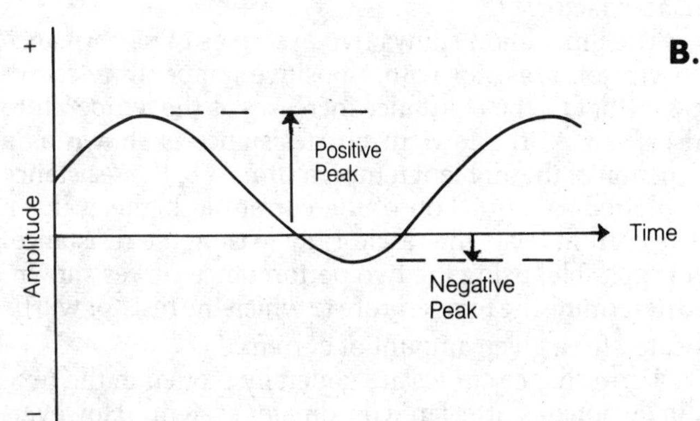

PEAK VALUE: Positive and negative peak values are equal (A), but they are not equal because of an added direct-current component (B).

PEAK VOLTAGE

The peak voltage of an alternating waveform is the instantaneous maximum positive or negative voltage. The peak positive voltage can be equal in magnitude to the peak negative voltage. The peak voltage may be sustained for a certain length of time, but in most cases it is sustained for near zero time.

A sinusoidal alternating-current wave with no direct-current component has a peak voltage equal to 1.414 times the root-mean-square voltage. For other types of waveforms, the relationship differs. *See also* ROOT MEAN SQUARE.

PELTIER EFFECT

When a current is passed through a junction of two disimilar metals, a change in temperature occurs. This change can either be an increase or decrease in the temperature, depending on the direction of the current through the junction. If a current in one direction causes heating, then a current in the other direction produces cooling. This effect is known as the Peltier effect. It is essentially the reverse thermocouple effect (*see* THERMOCOUPLE).

The amount of thermal energy transferred, for a given current, depends on the metals. The thermal energy that is emitted or absorbed with the passage of current through a junction of dissimilar metals is called the Peltier heat. The quotient of the Peltier heat and the current is called the Peltier coefficient.

PERFORMANCE CURVE

A graph, representing the characteristics of a component under variable conditions, is called a performance curve. A single component usually has several different performance curves, one for each independent variable for which the device is to be evaluated. Typical independent variables that affect component performance include the voltage across the component, the current through it, the frequency of the signal passing through or applied to the component, the temperature, and many other factors. The dependent variable may be the gain, the component value, the emitted power, or some other performance-related factor.

The illustration shows two examples of performance curves for a resistor with a positive temperature coefficient (that is, the resistance increases as the temperature rises). At A, the low-current resistance is shown as a function of the ambient temperature. At B, the resistance is plotted as a function of the current at higher values. The current heats the resistor, increasing the resistance. It is possible, using the two performance curves shown, to determine the temperature to which the resistor will be heated for a given amount of current.

Performance curves are generally plotted in the two-dimensional Cartesian coordinate system. However, other coordinate systems are used for special purposes. *See also* CARTESIAN COORDINATES, COORDINATE SYSTEM.

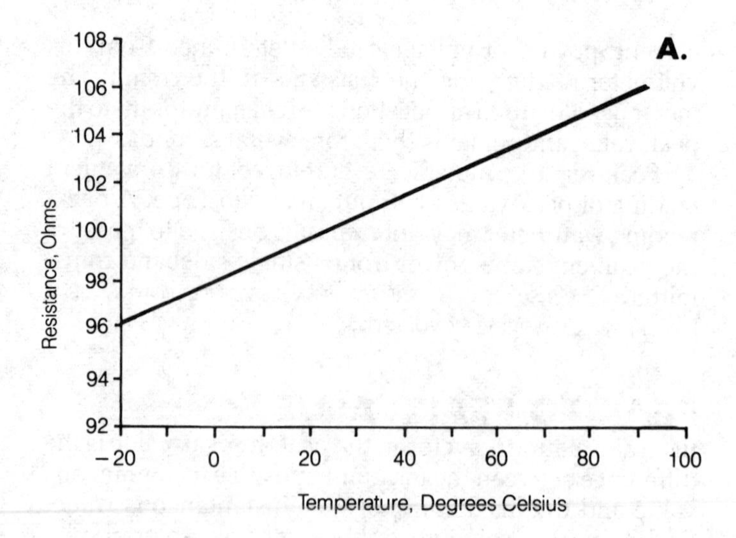

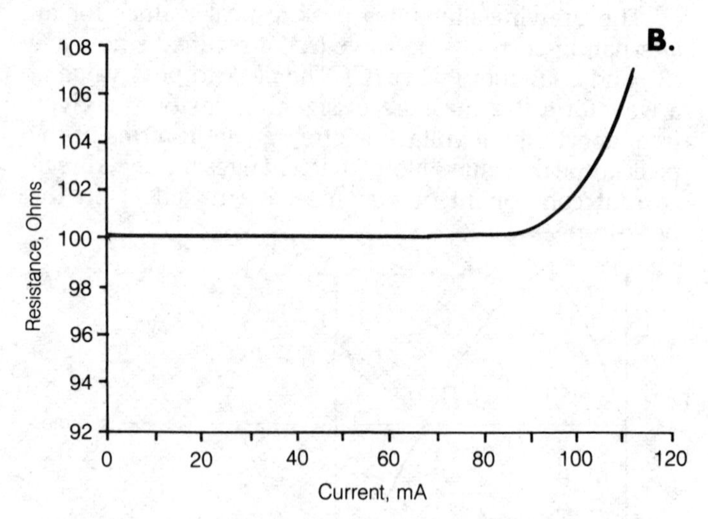

PERFORMANCE CURVE: Curves for hypothetical 100 Ω, ½ W resistor—resistance versus temperature (A), and resistance versus current (B). Performance degradation begins at about 90 mA.

PERIOD

A period is the length of time required for an alternating or pulsating waveform to complete one cycle. For example, in a 60 Hz system, the period is 1/60 second. In general, if f is the frequency in hertz, then the period T in seconds is:

$$T = 1/f$$

If f is specified in kilohertz, then:

$$T = 0.001/f$$

and for f in megahertz:

$$T = 0.000001/f$$

Mathematicians speak of the period of a repeating function (*see* PERIODIC FUNCTION); the period is the length of an interval, on the independent-variable axis, over which the function goes through exactly one repetition. An example of this is the sine function; its period is 2π, or about 6.28.

PERIODIC FUNCTION

A periodic function is a mathematical function (*see* FUNCTION) that repeats itself. This function attains all of the values in its range within certain defined intervals in the domain. The intervals are all of the same length. The interval length is called the period of the function. An arbitrary example of a periodic function is shown in the illustration.

In electronics, periodic functions are commonplace. The most familiar periodic function is the sine function and multiples of it. The cosecant, cosine, cotangent, secant, and tangent functions, and their multiples, are periodic. *See* COSECANT, COSINE, COTANGENT, SECANT, SINE, TANGENT.

Periodic functions can, in some instances, be very complicated. For example, several different multiples of the sine function may be superimposed on each other. This is the case in wave disturbances containing harmonic energy or having multiple frequency components. The period of this function is sometimes hard to determine. However, it can always be defined. *See also* PERIOD.

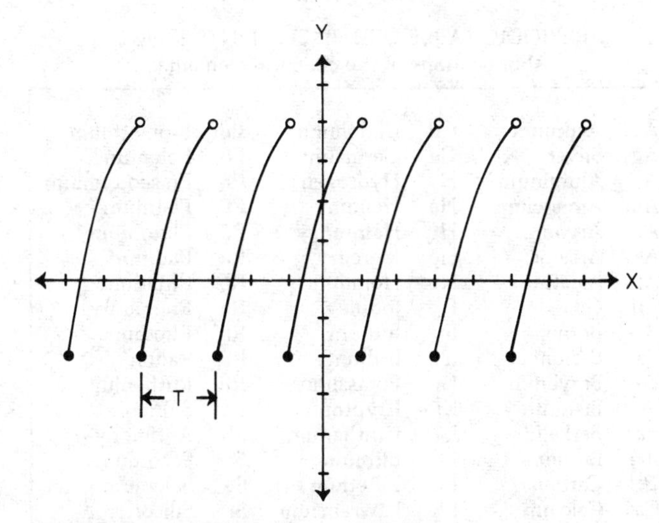

PERIODIC FUNCTION: The period is T.

PERIODIC TABLE OF THE ELEMENTS

The elements can be listed in a special arrangement known as the periodic table. Chemists and physicists use the periodic table because it groups elements with similar properties. There are seven different periods numbered 1 through 7.

The elements are also classified in 16 groups according to certain characteristics. The groups are designated by Roman numerals and English letters, as follows: I-A, II-A, III-B, IV-B, V-B, VI-B, VII-B, VIII, I-B, II-B, III-A, IV-A, V-A, VI-A, VII-A, and ZERO. This is the order of the groups as they appear across the periodic table from left to right.

Table 1 shows the most commonly reproduced long form of the periodic table. Atomic numbers are given with the element abbreviations. Table 2 on p. 634 lists the chemical abbreviations, in alphabetical order, followed by the full name of the element. *See also* ATOMIC NUMBER, ATOMIC WEIGHT.

PERMANENT MAGNET

Certain metals, notably iron and nickel, are affected by magnetic fields in a unique way. The magnetic dipoles within such substances can, in the presence of a sufficiently intense magnetizing force, become permanently aligned. This results in a permanent magnet—material that is constantly surrounded by a magnetic field.

PERIODIC TABLE OF THE ELEMENTS Table 1: Long form.

I A	II A	III B	IV B	V B	VI B	VII B	VIII	VIII	VIII	I B	II B	III A	IV A	V A	VI A	VII A	0	
H 1																H 1	He 2	1
Li 3	Be 4											B 5	C 6	N 7	O 8	F 9	Ne 10	2
Na 11	Mg 12											Al 13	Si 14	P 15	S 16	Cl 17	Ar 18	3
K 19	Ca 20	Sc 21	Ti 22	V 23	Cr 24	Mn 25	Fe 26	Co 27	Ni 28	Cu 29	Zn 30	Ga 31	Ge 32	As 33	Se 34	Br 35	Kr 36	4
Rb 37	Sr 38	Y 39	Zr 40	Nb 41	Mo 42	Tc 43	Ru 44	Rh 45	Pd 46	Ag 47	Cd 48	In 49	Sn 50	Sb 51	Te 52	I 53	Xe 54	5
Cs 55	Ba 56	La 57	Hf 72	Ta 73	W 74	Re 75	Os 76	Ir 77	Pt 78	Au 79	Hg 80	Tl 81	Pb 82	Bi 83	Po 84	At 85	Ru 86	6
Fr 87	Ra 88	Ac 89																7
			Ce 58	Pr 59	Nd 60	Pm 61	Sm 62	Eu 63	Gd 64	Tb 65	Dy 66	Ho 67	Er 68	Tm 69	Yb 70	Lu 71		**Lanthanide**
			Th 90	Pa 91	U 92	Np 93	Pu 94	Am 95	Cm 96	Bk 97	Cf 98	Es 99	Fm 100	Md 101	No 102	Lw 103		**Actinide**

PERIODIC TABLE OF THE ELEMENTS Table 2: Abbreviations of the chemical elements.

Ac	Actinium	Gd	Gadolinium	Pm	Promethium
Ag	Silver	Ge	Germanium	Po	Polonium
Al	Aluminum	H	Hydrogen	Pr	Praseodymium
Am	Americium	He	Helium	Pt	Platinum
Ar	Argon	Hf	Hafnium	Pu	Plutonium
As	Arsenic	Hg	Mercury	Ra	Radium
At	Astatine	Ho	Holmium	Rb	Rubidium
Au	Gold	I	Iodine	Re	Rhenium
B	Boron	In	Indium	Rh	Rhodium
Ba	Barium	Ir	Iridium	Rn	Radon
Be	Beryllium	K	Potassium	Ru	Ruthenium
Bi	Bismuth	Kr	Krypton	S	Sulfur
Bk	Berkelium	La	Lanthanum	Sb	Antimony
Br	Bromine	Li	Lithium	Sc	Scandium
C	Carbon	Lu	Lutetium	Se	Selenium
Ca	Calcium	Lw	Lawrencium	Si	Silicon
Cb	Columbium	Md	Mendelevium	Sm	Samarium
Cd	Cadmium	Mg	Magnesium	Sn	Tin
Ce	Cerium	Mn	Manganese	Sr	Strontium
Cf	Californium	Mo	Molybdenum	Ta	Tantalum
Cm	Curium	N	Nitrogen	Tb	Terbium
Co	Cobalt	Na	Sodium	Tc	Technetium
Cr	Chromium	Nb	Niobium	Te	Tellurium
Cs	Cesium	Nd	Neodymium	Th	Thorium
Cu	Copper	Ne	Neon	Ti	Titanium
Dy	Dysprosium	Ni	Nickel	Tl	Thallium
Er	Erbium	No	Nobelium	Tm	Thulium
Es	Einsteinium	Np	Neptunium	U	Uranium
Eu	Europium	O	Oxygen	V	Vanadium
F	Fluorine	Os	Osmium	W	Tungsten
Fe	Iron	P	Phosphorus	Xe	Xenon
Fm	Fermium	Pa	Proactinium	Y	Yttrium
Fr	Francium	Pb	Lead	yb	Ytterbium
Ga	Gallium	Pd	Palladium	Zn	Zinc
				Zr	Zirconium

Iron and steel can be permanently magnetized by stroking it with another permanent magnet. This action causes the magnetic dipoles, normally aligned at random, to be oriented more or less in a common direction. Their effects then average out to create a magnetic field around the object. (Normally, the effects of the dipoles average out to a zero magnetic field.)

Permanent magnets are used in speakers, microphones, meters, and certain types of transducers. In the laboratory or the factory, tools such as screwdrivers are weakly magnetized for convenience in dismantling or assembling electronic equipment. *See also* D'ARSONVAL MOVEMENT, DYNAMIC PICKUP, MAGNETIC FIELD, MAGNETIC RECORDING, MAGNETIZATION, SPEAKER.

PERMANENT-MAGNET METER

See D'ARSONVAL MOVEMENT.

PERMANENT-MAGNET MICROPHONE

See DYNAMIC PICKUP.

PERMANENT-MAGNET SPEAKER

See SPEAKER.

PERMANENT-MAGNET PICKUP

See DYNAMIC PICKUP.

PERMEABILITY

Certain materials affect the concentration of the lines of flux in a magnetic field (*see* MAGNETIC FIELD, MAGNETIC FLUX). Some substances cause the lines of flux to move farther apart, resulting in a decrease in the intensity of the field compared with its intensity in a vacuum. Other substances cause the density of the magnetic flux to increase.

The permeability of a material is the measure of how much better a given material is than air as a path for magnetic lines of flux. It is equal to the magnetic induction divided by the magnetizing force. This can be expressed as gausses (B) divided by oersteds (H). Free space (a vacuum) is assigned to permeability value 1. Materials that reduce the flux density, known as diamagnetic materials, have permeability less than 1. Substances that cause an increase in the flux density are known as ferromagnetic and paramagnetic materials. They have permeability factors greater than 1. The paramagnetic materials have values slightly larger than 1; the ferromagnetics have values much greater. (*See* DIAMAGNETIC MATERIAL, FERROMAGNETIC MATERIAL, PARAMAGNETIC MATERIAL.)

The table lists several common types of materials, along with their approximate permeability values at room temperature. The permeability of a given substance can change substantially with changes in the temperature. The permeability factors of some materials, notably the ferromagnetics, may vary depending on the intensity of the magnetic field. For these reasons, the values in the table are not precise. The table reflects the fact that no substances have permeability values much smaller than 1, although some have values far in excess of 1. Certain nickel-iron alloys exhibit permeability factors of more than 1,000,000.

Substances with high magnetic-permeability values are used primarily for increasing the inductances of coils. When a high-permeability core in inserted into a coil, the inductance is multiplied by the permeability factor. This effect is useful in the design of transformers and chokes

PERMEABILITY: Permeability factors of some common substances.

Substance	Permeability (Approx.)
Aluminum	Slightly more than 1
Bismuth	Slightly less than 1
Cobalt	60–70
Ferrite	100–3000
Free Space	1
Iron	60–100
Iron, refined	3000–8000
Nickel	50–60
Permalloy	3000–30,000
Silver	Slightly less than 1
Steel	300–600
Super-permalloys	100,000–1,000,000
Wax	Slightly less than 1
Wood, dry	Slightly less than 1

at all frequencies. *See also* CHOKE, FERRITE CORE, LAMINATED CORE, TRANSFORMER.

PERMEABILITY TUNING

In a resonant radio-frequency circuit, tuning may be accomplished by varying the value of either the inductor or the capacitor. The value of the inductor can be adjusted by changing the number of turns, or by changing the magnetic permeability of the core. The latter method of inductor tuning is known as permeability tuning.

Permeability tuning is accomplished by moving a powdered-iron core in and out of the coil (see illustration). A threaded shaft is attached to the core, and this shaft is rotated to allow precise positioning of the core. The farther into the coil the core is set, the larger the inductance, and the lower the resonant frequency becomes. Conversely, as the core moves farther outside the coil, the inductance becomes smaller, and the resonant frequency becomes higher.

The main advantage of permeability tuning, as compared with capacitor tuning, is that the control adjustment is more linear. This allows the use of a dial calibrated in linear increments. Permeability tuning also permits the capacitor in the tuned circuit to be fixed rather than variable. This reduces the effects of external capacitance and allows the use of a capacitor with an optimum temperature coefficient. *See also* CAPACITOR TUNING, INDUCTOR TUNING.

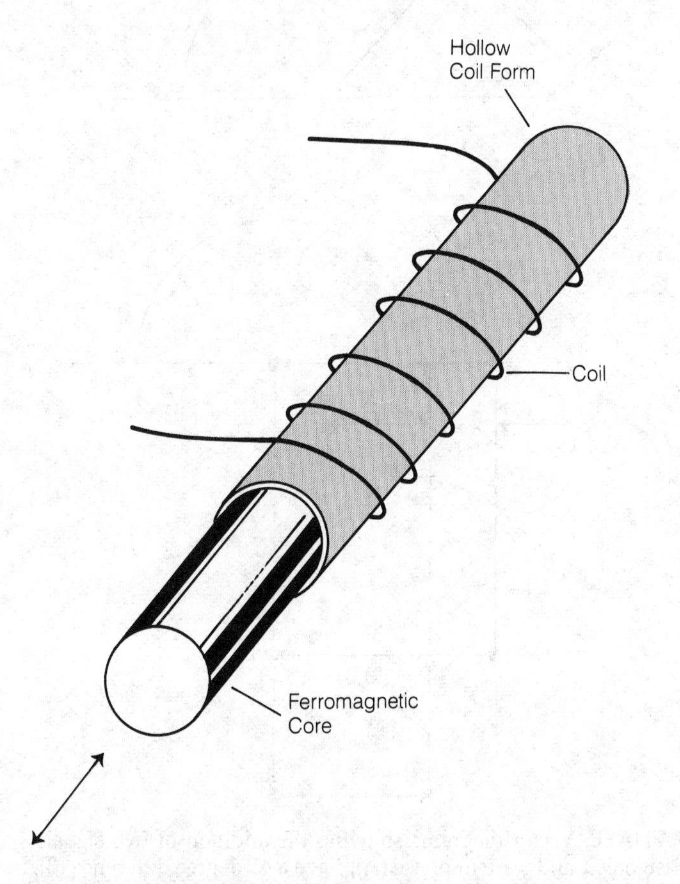

PERMEABILITY TUNING: A ferromagnetic core can be moved in and out of the coil for tuning.

PERMEANCE

In a magnetic circuit, permeance is an expression of the ease with which a magnetic field is conducted. Permeance is the reciprocal of reluctance. Permeance in a magnetic circuit is analogous to conductance in an electric circuit. Permeance is generally measured in webers per ampere. *See also* RELUCTANCE.

PERMITTIVITY

Permittivity is an expression of the absolute dielectric properties of a material or medium. The dielectric constant is determined from the permittivity.

Permittivity is represented by the lowercase Greek letter epsilon (ϵ). The permittivity of free space is represented by the symbol ϵ_o. The quantity is usually expressed in farads per meter; the permittivity of free space, in these units is:

$$\epsilon_o = 8.85 \times 10^{-12} \text{ farad per meter}$$

Given a substance with permittivity ϵ farad per meter, the dielectric constant, *k*, is determined according to the formula:

$$k = \frac{\epsilon}{\epsilon_o} = \frac{\epsilon}{8.85 \times 10^{-12}}$$

See also DIELECTRIC, DIELECTRIC CONSTANT.

PERMUTATION

A set of data characters or bits may be arranged in different ways. The possible arrangements, or ordering schemes, are called permutations.

Given n different objects or characters, the number of possible permutations, *P*, is equal to *n*! (*n* factorial), or:

$$P = n(n-1)(n-2)\ldots(1)$$

For example, the letters A, B, C, D, and E may be arranged in $P = 5 \times 4 \times 3 \times 2 \times 1 = 120$ different ways. The set of all letters in the alphabet may be arranged in 26! different ways; this is about 4.03×10^{26} permutations.

pH

The pH is an expression of the relative acidity or alkalinity of a liquid. The pH is represented by a value between 0 and 14. Numbers smaller than 7 represent acidity, and numbers greater than 7 represent alkalinity. A liquid is neutral when its pH value is exactly 7.

Theoretically, pH is the negative logarithm of the hydrogen-ion concentration, measured in gram equivalents per liter. The pH is significant as a measure of electrolyte concentration in electrochemical devices.

PHASE

Phase is a relative quantity, describing the time relationship between or among waves having identical frequency. The complete wave cycle is divided into 360 equal parts, called degrees of phase. The time difference between two waves can then be expressed in terms of these degrees. Phase is also expressed in radians. One radian corresponds to about 57.3 degrees of phase.

One wave may occur sooner than the other by as much as 180 degrees. The earlier wave is called the leading wave. A disturbance may occur as much as 180 degrees later than its counterpart; the later wave is called the lagging wave (*see* LAGGING PHASE, LEADING PHASE). When waves are exactly 180 degrees different in phase, they are said to be perfectly out of phase, or in phase opposition.

Two signals with equal frequency add together vectorially, depending on their phase difference. This is illustrated for two signals of equal frequency but different amplitude. At A, the phase difference is zero; that is, the signals are exactly in phase. The resulting amplitude is the sum of the amplitudes of the two signals. At B, the signals differ in phase by 45 degrees. The resulting amplitude is smaller than that at A, as can be seen by the parallelogram method for adding vectors. At C, the signals are 90 degrees out of phase, with the resulting amplitude still smaller. At D, the signals are 135 degrees out of phase, yielding an even smaller composite signal; and at E, they are in phase opposition, resulting in the minimum possible amplitude.

The length of time corresponding to one degree of phase depends on the frequency of the signal. If the frequency in hertz is given by f, then the time t, in seconds, corresponding to one degree of phase is:

$$t = \frac{1}{360f} = \frac{0.00278}{f}$$

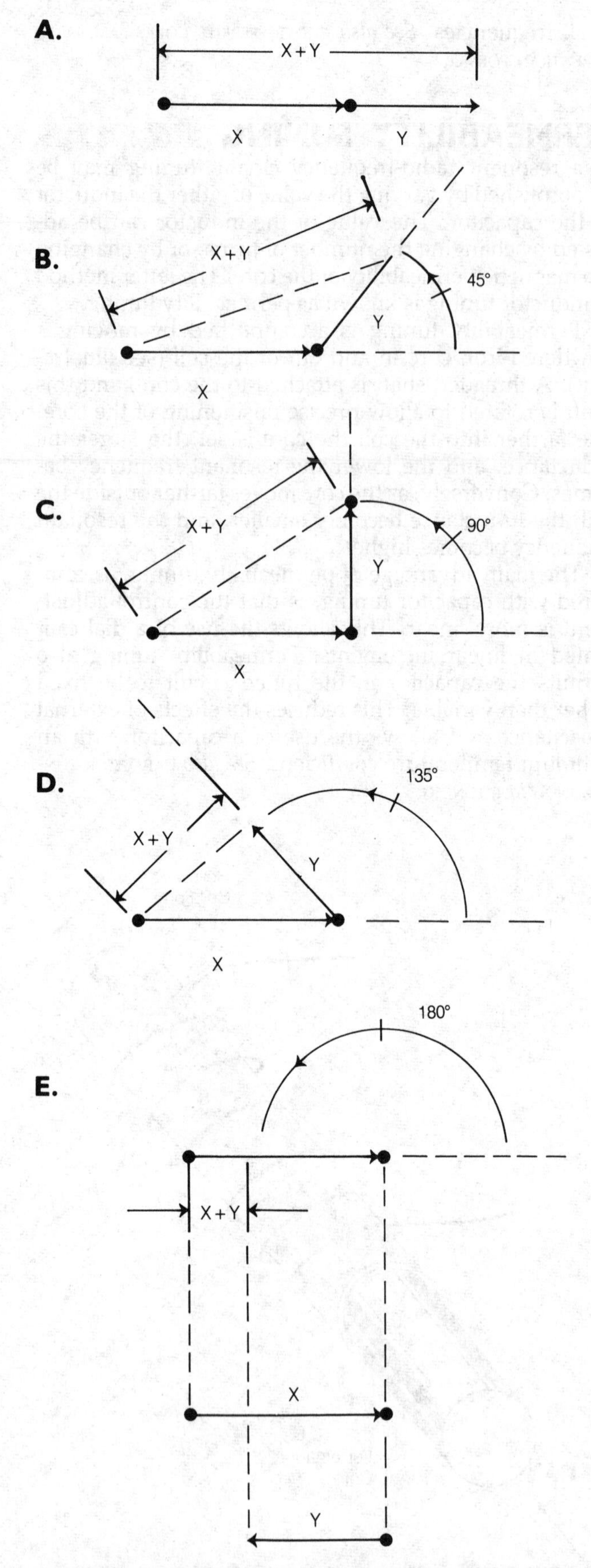

PHASE: Vector diagrams showing the addition of two signals—signals x and y are in phase (A), have a 45-degree difference (B), a 90-degree difference (C), a 135-degree difference (D), and a 180-degree difference (E).

PHASE ANGLE

Phase angle is an expression, given in degrees, for the relative difference in phase between two signals (*see* PHASE). Phase angle is often given to indicate the difference in phase between the current and the voltage in a circuit containing reactance (*see* LAGGING PHASE, LEADING PHASE).

In a dielectric material, the phase angle is the extent to which the current leads the voltage. The phase angle, indicated by ø, is equal to the complement of the loss angle θ. That is:

$$ø = 90 - \theta$$

The phase angle may vary from 0 to 90 degrees. The larger the phase angle, the lower the dielectric loss (*see* DISSIPATION FACTOR).

Phase angle also indicates the lossiness of an inductor. In a perfect inductive reactance, the current lags the voltage by 90 degrees. The phase angle ø might, again, vary from 0 to 90 degrees; the larger the phase angle, the lower the loss in the inductor.

A phase angle of 0 degrees indicates that the current

and voltage are in phase. This occurs only when there is no reactance in the circuit. *See also* CAPACITIVE REACTANCE, IMPEDANCE, INDUCTIVE REACTANCE.

PHASE BALANCE

Phase balance is an expression of the relative symmetry of a square wave. The phase-balance angle is defined as the difference between 180 degrees and the measured phase angle between the centers of the positive and negative pulses.

The drawing illustrates phase balance. The period, T, of the waveform, is the length of one complete cycle, from the midpoint of one positive pulse to the midpoint of the following positive pulse. This interval is divided into 360 degrees. The midpoint of the negative pulse is then located; ideally, it should be at the 180-degree point. If the midpoint of the negative pulse occurs at x degrees, then the phase-balance angle is $180 - x$.

In a square wave with zero transition time, the phase-balance angle is 0 degrees, since $x = 180$. This is not the case in the illustration; in this example, the phase-balance angle is about +10 degrees, since the midpoint of the negative pulse is at 170 degrees.

In a phased-array antenna, the condition of signal cancellation resulting in a directional null is called phase balance. As seen from a distant point along such a null, the currents in the various phased elements oppose each other in phase, resulting in a field strength of zero. *See* PHASED ARRAY ANTENNA in the ANTENNA DIRECTORY.

Phase opposition can be considered to be phase balance. This is the condition in which two signals of identical frequency differ in phase by 180 degrees. *See also* PHASE, PHASE OPPOSITION.

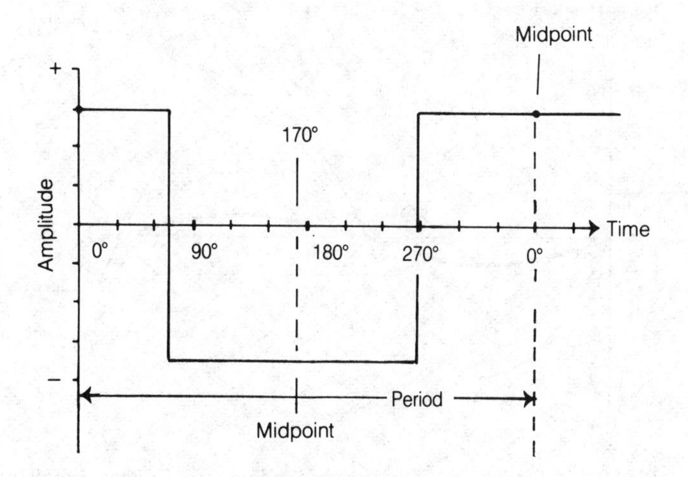

PHASE BALANCE: The phase balance is +10 degrees in this waveform.

PHASE CANCELLATION

See PHASE OPPOSITION.

PHASE COMPARATOR

A phase comparator is a part of a phase-locked-loop circuit (*see* PHASE-LOCKED LOOP). The phase comparator compares two signals in terms of their phase. The output voltage of the phase comparator depends on whether the signals are in phase or not. If the signals are not in phase, the voltage from the comparator causes the phase of an oscillator signal to be shifted back and forth until its signal is exactly in phase with the signal from a reference oscillator.

PHASED-ARRAY ANTENNA

See ANTENNA DIRECTORY.

PHASE DIFFERENCE

See PHASE, PHASE ANGLE.

PHASE DISTORTION

When the output phase of an amplifier alternates but the input does not, the amplifier is said to introduce phase distortion into the signal.

Phase distortion does not always cause problems in radio equipment. When phase distortion occurs, the effect is similar to phase modulation or frequency modulation of the signal. The signal might have a buzzing or fluttering sound as heard in a frequency-modulation receiver. In an amplitude-modulation or code receiver, phase distortion is usually not noticeable.

Phase distortion sometimes occurs when a phase-locked-loop circuit malfunctions. If the circuit is not able to lock properly, the frequency and/or phase of the signal will continually shift as the device attempts to stabilize. In a phase-modulation or frequency-modulation transmitter or receiver, phase distortion is extremely undesirable because it interferes with the transmitted information. *See also* PHASE, PHASE-LOCKED LOOP.

PHASE INVERTER

A phase inverter is a circuit that inverts the waveform. Phase inversion may be accomplished by most ordinary single-ended amplifier circuits or transformers. A phase inverter will operate effectively at all frequencies.

The paraphase inverter circuit with a single-ended input and a push-pull output is sometimes called a phase inverter. This circuit produces two output signals in phase opposition to each other. This circuit can be used to drive a push-pull or push-push amplifier without an input transformer. *See also* PARAPHASE INVERTER, PHASE.

PHASE LAG

See ANGLE OF LAG.

PHASE LEAD

See ANGLE OF LEAD.

PHASE-LOCKED LOOP

A phase-locked loop (PLL) is an electronic circuit for locking an oscillator in phase with an input signal. A PLL can act as a demodulator to demodulate a carrier frequency, or it can be used to track a carrier or synchronizing signal whose frequency varies with respect to time.

The basic circuit of a PLL, as shown in the figure, consists of a phase detector and lowpass filter with a feedback loop closed by a local voltage-controlled oscillator (VCO). The phase detector detects and tracks small differences in phase and frequency between the incoming signal and the VCO signal and provides output pulses that are proportional to the difference. The lowpass filter removes alternating current (ac) components to provide a direct-current (dc) voltage signal to drive the VCO. This input voltage will act to change the output frequency of the VCO to that of the input signal.

The phase detector and lowpass filter function as the mixer in a general feedback loop. As in the general loop, the output is driven in the direction that will minimize the error signal—in this case frequency. Thus the loop tends to drive the error signal back toward zero frequency. Once the two frequencies are made equal, the VCO will be locked to the input signal, and any phase difference between the two signals will be controlled.

The PLL was designed specifically for FM (frequency-modulation) demodulation. PLLs are now used to synchronize the horizontal and vertical scanning signals in television receivers, to remove the Doppler shift in satellite tracking, to stabilize the frequency of klystron oscillators, and to filter noise in communications circuits. They also are found in synchronous detection circuits, modems, tone decoders, and frequency shift keying (FSK) receivers.

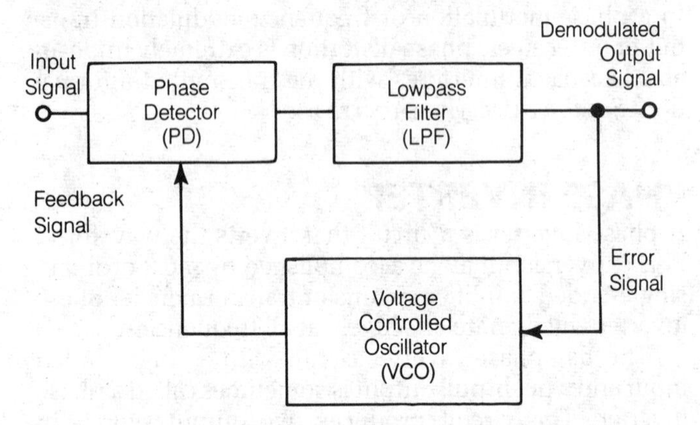

PHASE-LOCKED LOOP: Block diagram of phase-locked loop circuit.

PHASE MODULATION

Phase modulation is a method of conveying information with radio-frequency carrier waves. The instantaneous phase of the carrier is shifted in accordance with the modulating waveform.

Phase modulation is similar to frequency modulation because any change in the instantaneous phase of a carrier also results in an instantaneous change in the frequency, and vice versa.

In phase modulation, the extent of the phase shift is directly proportional to the amplitude of the modulating signal. The rapidity of the phase shift is directly proportional to both the amplitude and the frequency of the modulating signal. This differentiates phase modulation from frequency modulation; the result is a difference in the frequency-response characteristics.

Many frequency-modulated transmitters use phase modulation of one of the amplifier stages. When this is done, the high frequencies appear exaggerated at the receiver unless an audio-frequency lowpass filter is used at the transmitter. The output of this filter must decrease in direct proportion to the modulating frequency. When this modification has been made at the transmitter, it is impossible to distinguish between phase modulation and true frequency modulation at the receiver. *See also* FREQUENCY MODULATION, PHASE.

PHASE OPPOSITION

When two signals are exactly 180 degrees out of phase, they are said to be in phase opposition.

Phase opposition results in a net amplitude that is the absolute value of the difference in the amplitudes of two signals having identical frequency. The phase of the resultant signal is the same as the phase of the stronger of the two constituent signals (see illustration). If the two signals have the same amplitude, they combine in phase opposition to produce no signal. This effect is called phase cancellation.

Two signals can be in phase opposition for either of two reasons: One signal may be delayed with respect to the other by one-half cycle, or one signal may be inverted (upside down) with respect to the other signal. A phase delay of 180 degrees can be obtained with a half-wavelength delay line. Phase inversion can be accomplished with an amplifier circuit or a transformer. *See also* PHASE INVERTER.

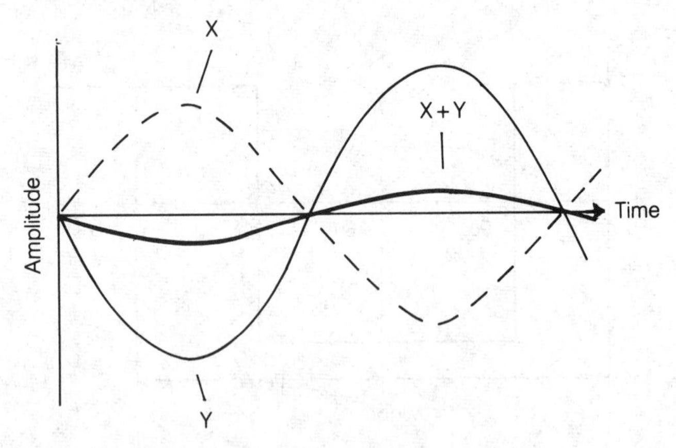

PHASE OPPOSITION: The sum of the two out-of-phase signals *x* and *y* is in phase with stronger signal, *y*.

PHASE REINFORCEMENT

Two signals of identical frequency are in phase reinforcement when they have the same phase.

Waves in phase reinforcement add in amplitude arithmetically. That is, the amplitude of the resultant signal is equal to the arithmetic sum of the amplitudes of the constituent signals (see illustration on p. 639). *See also* PHASE.

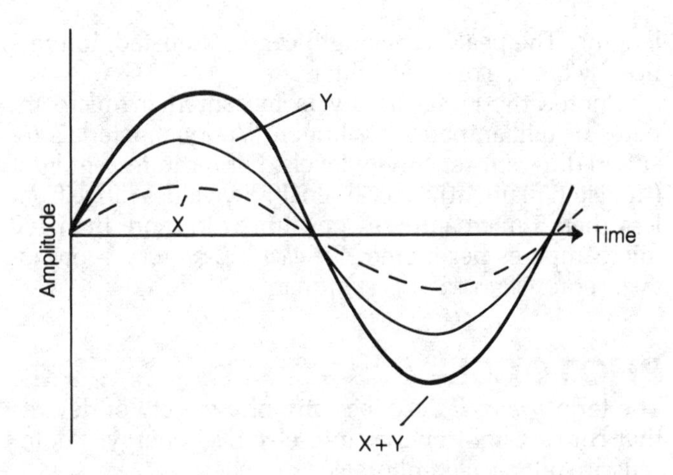

PHASE REINFORCEMENT: The two in-phase signals, x and y, add arithmetically.

PHASE SHIFT

A phase shift is a change in the phase of a signal. Phase shift can occur over a small fraction of one cycle, or it can occur over a span of many cycles. Phase shift is normally measured in degrees. One degree of phase shift corresponds to 1/360 (0.00278) cycle.

Certain electronic circuits shift the phase of an input signal. Phase shift is normally expressed as either positive or negative. A negative phase shift refers to a delay of up to one-half cycle (see A in the illustration). A positive

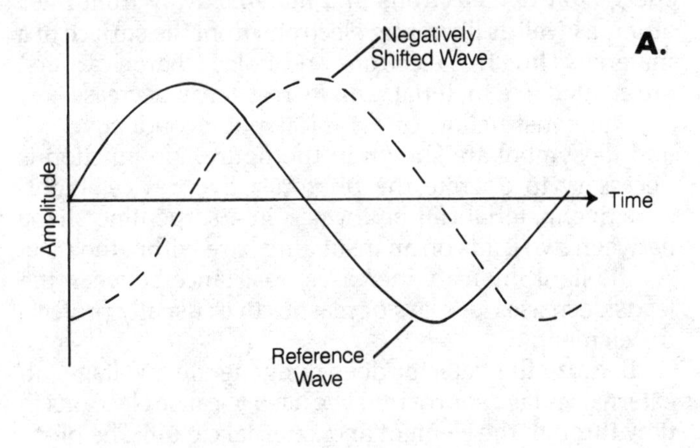

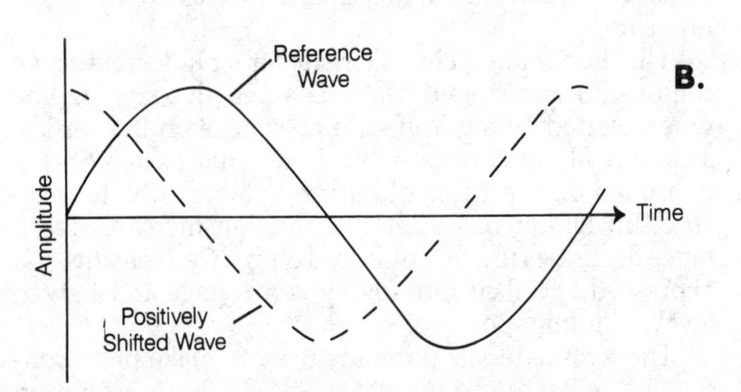

PHASE SHIFT: A, 90-degree negative phase shift (A) and a 90-degree positive phase shift (B).

phase shift is a delay of more than one-half cycle, and up to one full cycle, as at B. Generally, a phase delay of one full cycle is equivalent to zero phase shift.

A phase shift of more than 360 degrees can generally be expressed as equivalent to some value of shift smaller than 360 degrees. For example, if a signal is delayed by 500 degrees, the result is effectively equivalent to a delay of 500 − 360, or 140, degrees.

Phase shift can be used to convey information by radio-frequency carrier waves. In this system, the instantaneous phase of a carrier wave is made to follow the waveform of some input signal such as a human voice. This is called phase modulation. *See also* PHASE, PHASE ANGLE, PHASE MODULATION.

PHASE-SHIFT DISCRIMINATOR

See FOSTER-SEELEY DISCRIMINATOR.

PHASE-SHIFT KEYING

Phase-shift keying is a method of transmitting digital information. It is similar to frequency-shift keying (*see* FREQUENCY-SHIFT KEYING), except that the phase, not the frequency, is shifted.

In phase-shift keying, the carrier is transmitted at a constant amplitude. During the space, or key-up condition, the wave is at one phase; during the mark, or key-down, condition, the phase is altered by a predetermined amount.

Phase-shift keying may be used in place of frequency-shift keying in certain applications. The primary advantage of phase-shift keying is that it can be accomplished in an amplifier stage, whereas frequency-shift keying cannot. *See also* PHASE.

PHASING

Phasing is the technique for giving a phased-array antenna a directional characteristic (*see* PHASED-ARRAY ANTENNA in the ANTENNA DIRECTORY). A circuit that accomplishes this phasing is called a phasing harness.

Depending on the relative phase and spacing of the elements in a phased antenna, the radiation pattern may have one, two, three, or more major lobes. Phasing harnesses usually consist of simple delay lines. A transmission line, measuring an electrical quarter wavelength, produces a delay of 90 degrees. A half-wave section of line causes the signal to be shifted in phase by 180 degrees. *See also* PHASE.

PHASOR

A phasor is a complex number expressing the magnitude and phase of a time-varying quantity. It is used in reference to steady-state alternating linear systems.

PHOSPHOR

A phosphor material glows when bombarded by electrons, high-speed subatomic particles, or electromagnetic energy including ultraviolet rays, X rays, and infrared.

Phosphor materials are available that will glow in many different colors. A common color is yellow-green, used in oscilloscopes, radar, and video-display terminals. White phosphors are used in black-and-white television receivers. Color television receivers employ tricolor phosphors—red, blue, and green.

Different phosphors have different persistence characteristics. The persistence is a measure of how long the phosphor will continue to glow after the radiation has been removed. The optimum degree of persistence depends on the application. For example, in television reception, a short persistence is needed; in radar, a longer persistence is desirable.

Common phosphor materials include zinc, silicon, and potassium compounds. *See also* CATHODE-RAY TUBE.

PHOSPHORESCENCE

Phosphorescence is a property of substances that causes them to glow under bombardment by high-speed atomic particles or short-wave radiation such as ultraviolet, X rays, or gamma rays. Some phosphorescent materials stop glowing almost immediately after the energizing radiation is removed. Others continue to glow for seconds, minutes, or even hours.

Phosphorescence occurs because high-energy radiation imparts additional energy to the atoms in the substance. This causes the electrons to move into higher orbits (*see* ELECTRON ORBIT). When the energizing radiation is removed, the electrons fall back to their original lower-energy orbits more or less rapidly, giving off photons of visible light in the process. *See also* PHOSPHOR.

PHOT

The phot is the centimeter-gram-second unit of illuminance equal to 1 lumen per square centimeter. The more frequently used unit is the lux, equivalent to 1 lumen per square meter.

There are 10,000 square centimeters in one square meter, so 1 phot equals 10,000 lux. *See also* ILLUMINANCE, LUMEN.

PHOTOCATHODE

In a phototube, the photocathode is a light-sensitive electrode. The photocathode emits electrons when light of sufficient intensity falls on it. The electrons emitted by the photocathode, as a result of impinging light, are called photoelectrons. In the case of monochromatic light, the number of photoelectrons emitted is directly proportional to the energy in the arriving light. (For nonmonochromatic light, this is not true.)

The photocathode of a camera tube or phototube can respond to a wide spectrum of wavelengths. Typically, the response is maximum at one particular wavelength, and the sensitivity decreases at longer and shorter wavelengths. The peak wavelength can be adjusted, in practice, by the use of optical filters.

Photocathode sensitivity is measured in microamperes or milliamperes per lumen. Various materials result in different sensitivity levels. For incandescent light (tungsten filament), typical sensitivity values range from less than 5 microamperes per lumen to more than 100 microamperes per lumen. *See also* CAMERA TUBE, PHOTOMOSAIC, PHOTOMULTIPLIER TUBE, PHOTOTUBE.

PHOTOCELL

The term *photocell* describes any of a variety of devices that convert light energy into electrical energy. Photocells are often called photoelectric cells.

There are two kinds of photocells: those that generate a current by themselves in the presence of light and those that change in effective resistance when the intensity of incident light is varied. The first type of photocell is called a photovoltaic cell or solar cell. The second type is called a photoconductive cell. The latter type of photocell can be found in a variety of different forms. *See also* PHOTOCONDUCTIVE CELL, PHOTOCONDUCTIVITY, PHOTOTUBE, SOLAR CELL.

PHOTOCONDUCTIVE CELL

The photoconductive cell, also called a photoresistive detector, responds to incident illumination with a decrease in internal resistance. Light can provide sufficient energy to drive electrons in a material away from their atoms as well as liberating electrons from the surface of a material. Thus free electrons and holes (charge carriers) are created in a material, and its resistance decreases.

The construction of a typical photoconductive cell and its symbol are shown in the figure. No junction is necessary to operate the photoresistive device. Light-sensitive material can be formed in a serpentine shape between two leads on an insulating base within the case. As the light intensity increases, resistance between the leads decreases. A glass or plastic cover usually protects the element.

Because the detector does not generate a voltage, an external voltage source must be used to cause electrons to flow through the element and external circuit. The property of photoresistance decreases as light intensity increases, so current in the circuit increases with light intensity.

The illumination characteristic for a photoconductive cell or photoresistive detector is a negative sloping line when plotted on a graph with resistance on the vertical axis and illumination on the horizontal axis. Without illumination, the value of dark resistance is high (measured in kilohms). But as illumination increases resistance decreases to a few hundred ohms. Cell sensitivity is expressed as cell current for a given voltage and a given level of illumination.

The two materials normally used to make photoconductive cells are cadmium sulfate (CdS) and cadmium selenide (CdSe). Both respond slowly to changes in light intensity. They have different response times and tem-

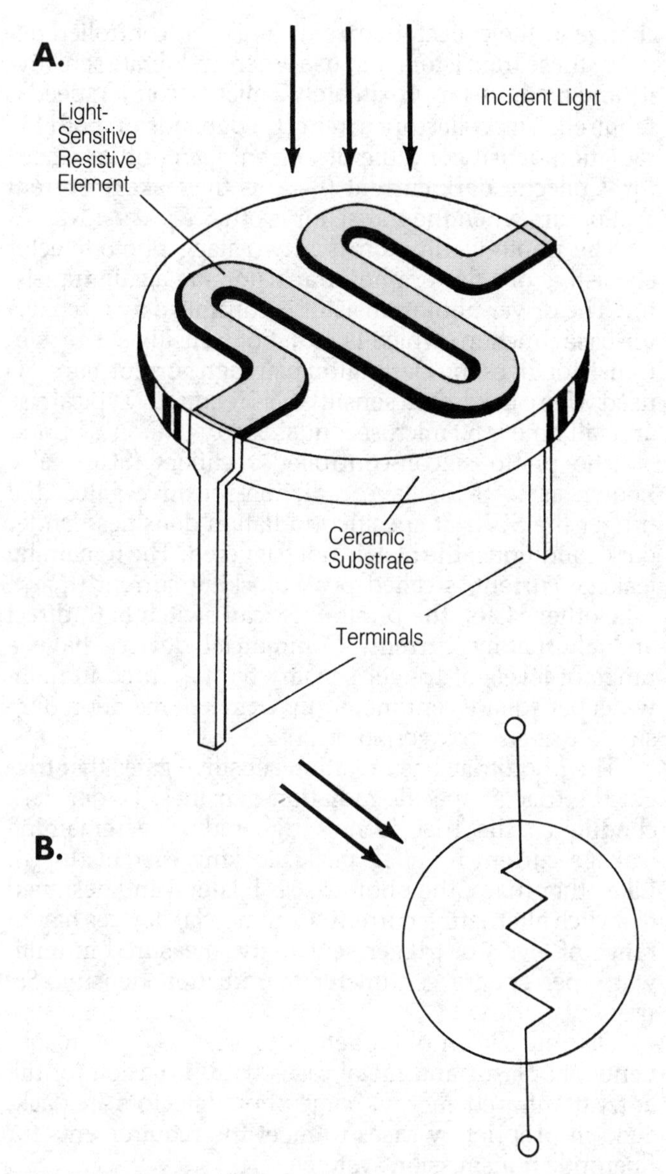

PHOTOCONDUCTIVE CELL: The resistance of the light-sensitive element decreases with increasing light intensity (A), and schematic symbol (B).

perature sensitivities. The spectral response of the CdS cell for visible light is similar to that of the human eye. By contrast, the spectral response for the CdSe cell starts at the long wavelength of the visible spectrum and extends into the infrared region.

PHOTOCONDUCTIVITY

Certain materials exhibit a resistance, or conductivity, that varies in the presence of visible, infrared, or ultraviolet light. These substances are called photoconductive materials. The property of changing resistance in accordance with impinging light intensity is called photoconductivity.

In general, a photoconductive substance has a finite resistance when no visible light is falling on it. As the intensity of the visible light increases, the resistance decreases. There is a limit, however, to the decrease of resistance with respect to brightness (see illustration).

Photoconductivity occurs in almost all materials, but it is much more pronounced in semiconductors. When light energy strikes a photoconductive material, the charge-carrier mobility increases. Thus, current can flow more easily when a voltage is applied. The more photons absorbed by the material for a given electromagnetic wavelength, the more easily the material conducts an electric current.

Photoconductive materials are used in the manufacture of photoelectric cells. Some examples of photoconductive substances are silicon and cadmium sulfate. *See also* PHOTOCONDUCTIVE CELL, PHOTODETECTOR, PHOTO- FET.

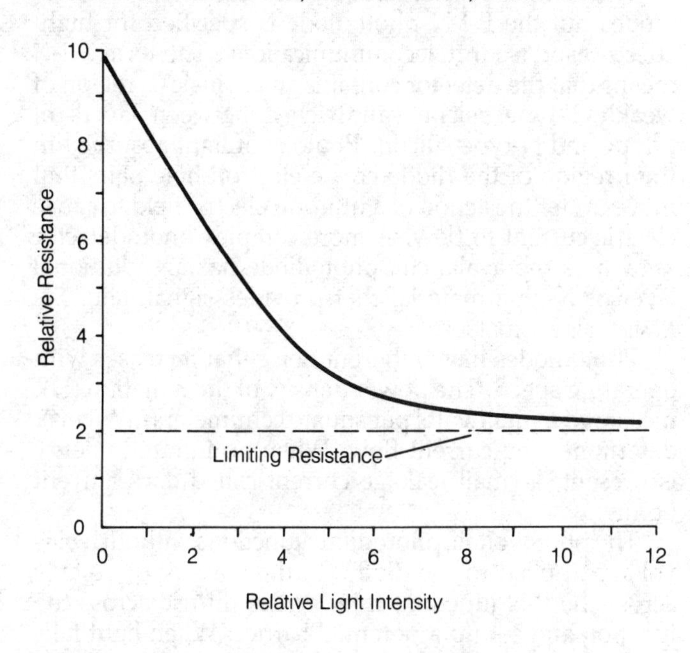

PHOTOCONDUCTIVITY: A typical photoconductivity curve.

PHOTODETECTOR

The term *photodetector* refers to any device capable of accepting light energy and converting it into an electrical signal. This conversion can be accompanied with signal gain in some devices. The most popular photodetectors today are solid state devices: silicon photodiodes, phototransistors, and photo-Darlingtons. They are packaged as discrete devices for through-air or fiberoptic communications systems; they are also components in optoelectronic devices such as optoisolators, optical interrupters, optical reflector modules, optical encoders, and solid-state relays. *See* ENCODER, OPTOCOUPLER, SOLID-STATE RELAY.

The radiation response of a silicon photodetector is a function of the silicon and the diffusion depth of the light-sensitive P-N junction. All silicon photosensors exhibit the same basic radiation frequency response, which peaks in the near infrared region at approximately 900 nanometers (nm). *See* OPTOELECTRONICS.

The photodiode is a P-N junction. When light energy

falls on the junction, it is absorbed into the crystal lattice and increases the energy of the electrons. Some electrons break away from their atoms and create free electrons and holes. The photodiode is both a photovoltaic and a photoconductive device. When the photodiode operates with a reverse voltage applied, it functions as a photoconductive device; without the reverse voltage, it functions as a photovoltaic diode. *See* DIODE.

The photoconductive photodiode functions when a reverse voltage is applied to the P-N junction, attracting the holes and electrons created near the junction across it in opposite directions. This causes electrons to flow out and through an external resistance to produce an output voltage that is generally proportional to light intensity.

The simplest photodiodes are silicon P-N photodiodes but the P-I-N photodiode is specified for high-speed response in telecommunications. The term *P-I-N* means that the detector contains an intrinsic (I) region of weakly P-type silicon sandwiched between layers of P-type and N-type silicon. Photons of light absorbed in the I region of the diode create electron-hole pairs that move under the action of a uniform electric field to cause electric current to flow. A more complex photodetector known as the avalanche photodiode has an additional layer of N-type material that provides signal gain. *See* AVALANCHE PHOTODIODE.

Photodiodes have inherent noise that increases with operating speed. The power density of the radiation (H), measured in milliwatts per square centimeter (mW/cm^2), determines the current flow. When no infrared energy is present, a small leakage current called dark current occurs.

The photovoltaic photodiode functions without a bias voltage applied to the diode. In the absence of voltage across the P-N junction, free carriers diffuse across the junction and set up a potential barrier. When light falls on the junction, the density of carriers diffusing across the junction will increase. If the region on one side of the junction is thin, it will become saturated with carriers. A differential voltage is set up across the junction, and electrons will flow through a load resistance connected across the device terminals.

The most common photovoltaic diode is the solar cell. It converts energy received from the sun directly into electric energy. Solar cells are optimized for production of voltage from incident solar radiation. They have a large detector area, low series resistance to provide maximum power transfer to the load, and very narrow depletion regions to provide a higher open-circuit output voltage. Cost effective in satellite applications, they also have limited specialized use in powering remote terrestrial and shipboard instruments. However, they are too expensive for general purpose use other than in special products such as pocket calculators. *See* SOLAR CELL.

The phototransistor is a junction transistor packaged so that light can be directed onto the collector junction where it increases the current through the collector junction. Because the base terminal of the transistor is left open circuit, current increase must flow through the emitter junction and it will be amplified to give a higher change in the collector current. Emission-controlled devices, these transistors find use where moderate sensitivity and medium (approximately 2 microseconds) speed is required. The collector current (I_C) depends on both the radiation density and the dc current gain of the transistor. Collector dark current (I_{CEO}) is the leakage current that occurs when the transistor is off. *See* TRANSISTOR.

The photo-Darlington is a two-stage photodetector consisting of a driver phototransistor and a gain transistor. The driver phototransistor is controlled by its collector-base junction, which is radiation sensitive. The gain transistor gives the Darlington pair high current gain; it is used where maximum sensitivity is required. Typical rise and fall time is 50 microseconds. *See* DARLINGTON AMPLIFIER.

The photo silicon-controlled rectifiers (SCR), also known as LASCR, have radiation-sensitive gates that trigger the SCRs at specified radiation densities. Under dark conditions, the SCR is not triggered. The remaining leakage current is called peak blocking current (I_{DRM}). Like other SCRs, the photo-SCR can switch both direct and alternating current. Commercial devices have a range of levels of trigger sensitivity, measured in milliwatts per square centimeter (mW/cm^2) to radiation density. *See* SILICON-CONTROLLED RECTIFIER.

The phototriac has radiation sensitive gates that trigger the triac at specific radiation densities. Under dark conditions, the triac is not triggered. The remaining leakage current is called peak blocking current (I_{DRM}). Like other triacs, the phototriac is bilateral and designed to switch alternating current. Commercial devices have a range of levels of trigger sensitivity, measured in milliwatts per square centimeter to radiation density. *See* TRIAC.

Discrete silicon photodetectors are packaged in conventional plastic and metal cases with provision for the entry of infrared energy. Some photodetectors are packaged in proprietary cases to meet the requirements for fiberoptic transmission systems.

PHOTODIODE

See PHOTODETECTOR.

PHOTOELECTRIC CELL

See PHOTOCELL.

PHOTOELECTRIC EFFECT

See PHOTOCELL, PHOTOCONDUCTIVE CELL, PHOTOCONDUCTIVITY, PHOTODETECTOR, PHOTO-FET, PHOTOMULTIPLIER TUBE, SOLAR CELL.

PHOTOELECTRON

See PHOTOCATHODE.

PHOTOEMISSION

See PHOTOCATHODE.

PHOTO-FET

A photo-FET is a field-effect transistor that exhibits photoconductive properties (*see* FIELD-EFFECT TRANSISTOR, PHOTOCONDUCTIVITY).

Most ordinary field-effect transistors would be suitable photo-FETs, except that their packages are opaque, and it is not possible for any light to reach the P-N junction.

The photoFET must be constructed so its P-N junction, forming the boundary between the gate and the channel, has the largest possible area. The P-N junction must be situated so that light can reach it. The package for the device must be transparent. Photo-FETs differ from other kinds of photocells primarily because of their higher impedance.

PHOTOMETRY

Photometry is the science of measurement of visible-light intensity. There are many ways to measure the brightness of a light source, but photometry is epecially concerned with the intensity of light as it relates to the human eye.

The human eye is more sensitive to light in the yellow green part of the spectrum than it is to other wavelengths. Most people can see electromagnetic radiation between about 380 and 770 nanometers wavelength (a nanometer is 10^{-9} meter), or about 3800 to 7700 Angstroms (an Angstrom unit corresponds to 0.1 nanometer or 10^{-10} meter). The longer wavelengths appear as a progressively deepening red; the shorter wavelengths as fading violet. Some people can see at wavelenths shorter than 390 nanometers, or longer than 750 nanometers. The human-eye response curve is approximately shown in the illustration.

Different photoconductive materials exhibit different sensitivity versus wavelength responses. In photometry, it is advantageous to use photoconductive substances with response closely resembling that of the human eye. A typical light-measuring instrument consists of a photocell, a source of voltage, perhaps a variable resistance, and a milliameter or microammeter, all connected in series. The meter scale should be calibrated before the instrument is used. *See also* LIGHT METER, PHOTOCELL, PHOTOCONDUCTIVITY.

PHOTOMETRY: The sensitivity curve for the human eye.

PHOTOMOSAIC

In a camera tube the photocathode is divided into a grid of small, photosensitive dots (*see* CAMERA TUBE, PHOTOCATHODE). This grid of dots is called the photomosaic.

Each small spot on the photomosaic receives a certain amount of visible-light energy, depending on the image

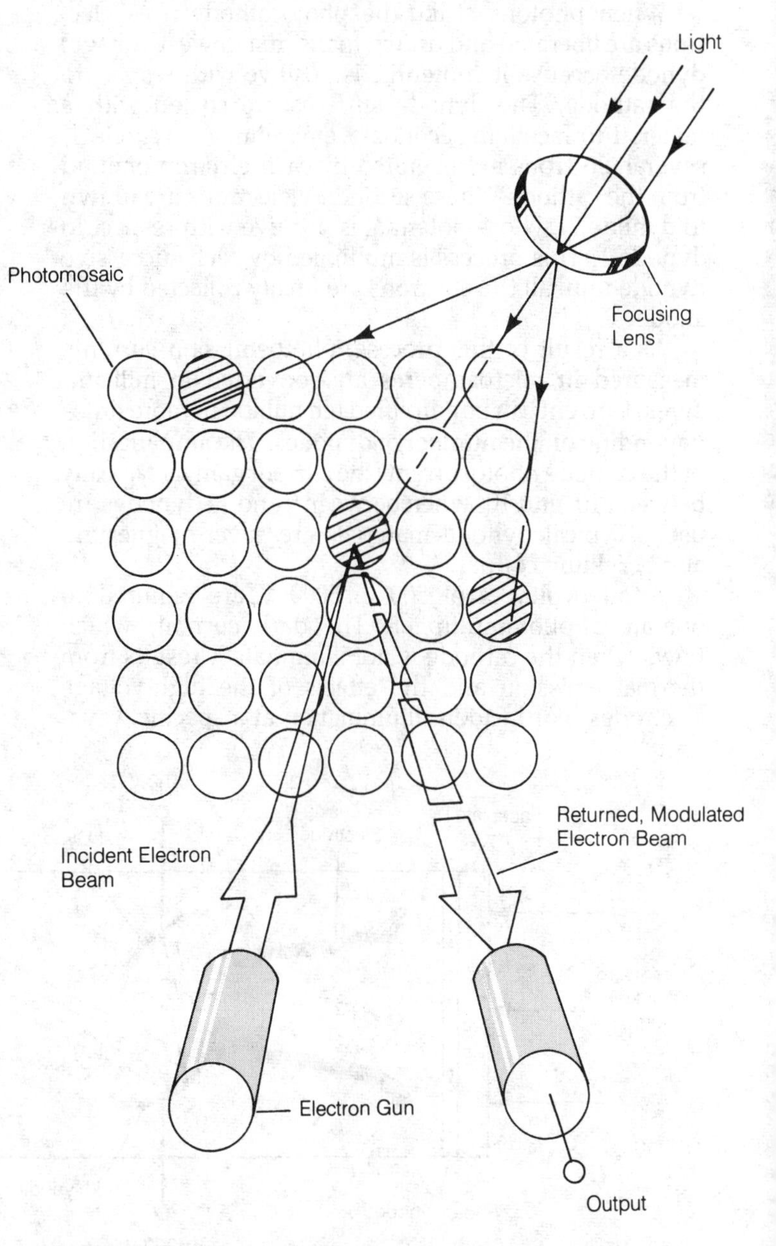

PHOTOMOSAIC: The electron beam is modulated according to the image.

received. A lens focuses the image onto the flat photomosaic. An electron beam then scans the photomosaic, and the beam is modulated according to the amount of light falling on each spot (see illustration). *See also* COMPOSITE VIDEO SIGNAL, IMAGE ORTHICON, VIDICON.

PHOTOMULTIPLIER TUBE

A photomultiplier tube is multi-electrode vacuum phototube containing a number of secondary emission dynodes between the photocathode and anode. By use of a voltage-divider network, successively larger voltages in steps of about 100 V are applied to the dynodes so the dynode nearest the anode has a potential close to the high anode voltage. The figure illustrates an internal configuration with curved plate-type dynodes facing each other. Another linear arrangement of grid-type venetian blind dynodes is frequently used.

When photons strike the photocathode, free electrons are liberated and drawn to the first (lowest voltage) dynode because its potential is positive with respect to the cathode. The dynode surfaces are coated with a material to facilitate secondary emission. At dynode 1, several electrons are liberated by each electron emitted from the cathode. These secondary electrons are drawn to dynode 2, whose potential is positive with respect to dynode 1. This process is multiplied by each successive dynode until all the electrons are finally collected by the anode.

As a result of this process, photoemission currents measured in microamperes are converted to milliamperes. Current can be amplified ten million or more times depending on the number of dynodes. The amplification of the cathode photocurrent (the current gain) is typically between 10^5 and 10^7 when between 9 and 14 dynodes are used. Typical dynode materials are silver-magnesium and beryllium copper.

Anode voltages of 500 to 5000 V are required to operate a photomultiplier. The dark current, which flows when the cathode is not illuminated, results from thermal emission and the effects of the high-voltage electrodes. For incident illumination at a specific wavelength, the number of emitted electrons is directly proportional to the intensity of the illumination. Thus anode current in a photomultiplier remains constant as the anode voltage is increased. However, dark current adds to the anode current produced by illumination, and secondary emission improves with applied voltage. As a consequence, the anode current increases with anode voltage.

Illumination levels in photomultiplier tubes are measured in microlumens. The tube is so sensitive that if it is exposed to ordinary daylight levels destructively large current could flow when voltage is applied to the electrodes.

PHOTON

A photon is a particle of electromagnetic radiation. Although a photon is a packet of visible-light energy, all electromagnetic radiation is made up of particles. Scientists can observe both particlelike and wavelike properties of electromagnetic radiation. The energy contained in a single photon depends on the wavelength.

If e represents the energy of one photon, in ergs, and the frequency in hertz is f, then:

$$e = hf$$

where h is Planck's constant, 6.626×10^{-34} joule seconds. If the wavelength is given by λ, in meters, then:

$$e = \frac{(3 \times 10^8)h}{\lambda}$$

Photons can be observed exerting pressure on objects, as a barrage of particles. *See also* ELECTROMAGNETIC RADIATION, LIGHT.

PHOTONICS

Photonics is the technology, analogous to electronics, for generating, transmitting, receiving and processing signals made up of light photons instead of electrons.

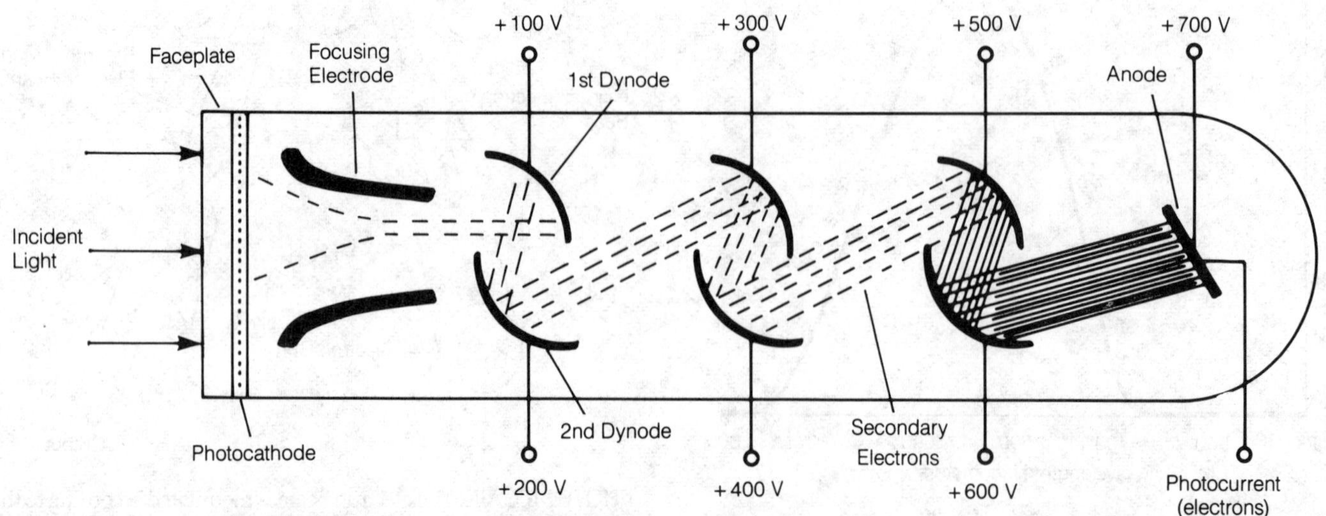

PHOTOMULTIPLIER TUBE: Schematic view of a photomultiplier vacuum tube.

Photonics is related to both optoelectronics and fiberoptics. Coherent monochromatic light emitted by the laser made it possible to modulate light fast enough to carry large amounts of information. Reductions in optical losses in glass over the past 25 years have also contributed to the advancements in photonic devices and systems.

Glass is the most familiar photonic material, but other complex materials have been developed for the transmission of information over glass-optic cable. A photonic transmission system not only guides a light signal over long distances, but it also converts electrical signals into light at the transmitting end and changes light back to electrical signals at the receiving end.

The new materials with the best prospects for present and future applications of photonics are the III-V materials, named after the columns of the periodic table. These materials have been used to make semiconductor injection or diode lasers, light-emitting diodes (LEDs), both visible and infrared, electro-optical repeaters that amplify the signal as it moves along a fiber, and detectors that convert the signal into an electronic pulse. (*See* FIBEROPTIC COMMUNICATION, LASER, LIGHT-EMITTING DIODE.) Pure photonic devices capable of assuming electronic functions other than information transmission have not yet proved to be economical.

A photonic device under development is the photonic transistor capable of amplifying a light signal without electronics. It switches light on and off or into alternate paths. Laboratory prototype photonic logic gates have been able to model the properties of the logical connectives AND, OR, and NOT. Photonic transistors and logic gates might some day be the basic components of a photonic computer.

A commercial light switch made from lithium niobate ($LiNbO_3$) changes the path of a light signal. The device depends on the changes made in the optical properties of lithium niobate in an applied electric field. The light is confined to waveguides in the crystal, defined by a surrounding material of lower refractive index.

PHOTORESIST

Photoresist is a light-sensitive polymeric material that is used in the photolithographic processing of circuit boards and semiconductor devices. It is applied to material substrates at various stages in the manufacturing process to define regions for selective etching. Photoresist is used for selective patterning of semiconductor masks and in the etching of various layers and the formation of metal contacts on the semiconductor wafer.

The most important property of photoresist is its change in solubility to selected chemical solvents caused by exposure to ultraviolet (UV) radiation. Photoresist coating on the wafer or substrate exposed to UV light might be more or less resistant to removal by a solvent than the unexposed part, depending on user selection. The solvent is applied to a wafer that has been oxidized or metallized. A drop of thin liquid applied to the spinning wafer spreads evenly over the wafer surface. The solvent then evaporates, leaving a polymeric film. Baking dries and improves the adhesion to the film.

The coated wafer is then exposed to UV light through a photographic mask. A negative photoresist cross links and polymerizes in a pattern wherever it is exposed. Washing in a selective solvent removes the unexposed film. The pattern is further hardened after development by heating. *See* INTEGRATED CIRCUIT MANUFACTURE, TRANSISTOR MANUFACTURE.

Photoresist liquids and film are also used to define conductive layers in additive and subtractive printed circuit board fabrication methods. It is most commonly used to mask the copper foil laminate for selective etching and plating. *See* PRINTED CIRCUIT BOARD.

PHOTORESISTOR

See PHOTOCONDUCTIVE CELL.

PHOTOTRANSISTOR

See PHOTODETECTOR.

PHOTOTUBE

A phototube is an electron tube that exhibits variable resistance, depending on the amount of light that strikes its cathode. Phototubes are generally sensitive to infrared, ultraviolet, X rays, and gamma rays as well as to visible light.

The greater the intensity of the radiation striking the cathode of the phototube, the greater the number of electrons that are emitted. Thus, the resistance of the tube goes down as the intensity of the light increases. The tube shows a certain finite resistance when there is no light; the value of the resistance eventually reaches a defined minimum as the light gets brighter. *See also* PHOTOMULTIPLIER TUBE.

PHOTOVOLTAIC CELL

See PHOTODETECTOR.

PI (π)

Pi is an irrational number, representing the ratio of the circumference of a circle to its diameter. Pi is symbolized by the Greek lowercase letter π. The value of π, to ten significant digits, is 3.141592654.

The number π occurs in many different mathematics, physics, and engineering equations. Some mathematicians have calculated the value of π, using computers, to hundreds of thousands of decimal places.

An unbalanced filter containing two parallel elements on either side of a single series element is called a pi network because of its resemblance (in schematic diagrams, at least) to the Greek letter π. *See also* PI NETWORK.

PICO-

Pico- is a prefix multiplier meaning one trillionth, or

10^{-12}. The abbreviation for pico- is the small letter p.

Some electrical units, such as the farad, are extremely large in practice. At the high, very high, and ultrahigh radio frequencies, capacitances are often specified in picofarads (pF). A capacitance of 1 pF is equivalent to a millionth of a microfarad (1 pF = 10^{-6} μF). *See also* PREFIX MULTIPLIERS.

PIERCE OSCILLATOR

See OSCILLATOR CIRCUITS.

PIEZOELECTRIC EFFECT

Certain crystalline or ceramic substances can act as transducers at audio and radio frequencies. When subjected to mechanical stress, these materials produce electric currents; when subjected to an electric voltage, the substances will vibrate. This effect is known as the piezoelectric effect. Piezoelectric substances include such materials as quartz, Rochelle salts, and various ceramics.

Piezoelectric devices are employed at audio frequencies as pickups, microphones, earphones, and buzzers. At radio frequencies, piezoelectric effect makes it possible to use crystals and ceramics as oscillators and tuned circuits. *See also* CERAMIC, CERAMIC FILTER, CERAMIC MICROPHONE, CERAMIC PICKUP, CRYSTAL, CRYSTAL CONTROL, CRYSTAL-LATTICE FILTER, CRYSTAL MICROPHONE, CRYSTAL TRANSDUCER, OSCILLATOR.

PIEZOELECTRIC FILTER

See CERAMIC FILTER, CRYSTAL-LATTICE FILTER.

PIEZOELECTRICITY

See PIEZOELECTRIC EFFECT.

PIEZOELECTRIC MICROPHONE

See MICROPHONE.

PIEZOELECTRIC PICKUP

See CERAMIC PICKUP, CRYSTAL TRANSDUCER.

PIEZOELECTRIC TRANSDUCER

See CERAMIC MICROPHONE, CERAMIC PICKUP, CRYSTAL MICROPHONE, CRYSTAL TRANSDUCER.

PINCHOFF

In a field-effect transistor, pinchoff is the condition in which the channel is completely blocked by the depletion region (*see* FIELD-EFFECT TRANSISTOR, METAL OXIDE-SEMICONDUCTOR FIELD-EFFECT TRANSISTOR). Pinchoff results in minimum conductivity from the source to the drain.

Pinchoff occurs with a large negative gate-to-source voltage in an N-channel junction field-effect transistor. In the P-channel device, pinchoff occurs with a large positive gate-to-source voltage. The minimum gate-to-source potential that results in pinchoff depends on the particular type of field-effect transistor and the voltage between the source and the drain.

In the N-channel device, as the gate voltage becomes more negative, the drain voltage required to cause pinchoff becomes less positive because the potential difference between the gate electrode and the channel is greater near the drain than near the source.

This effect is exaggerated as the drain-to-source voltage is increased. If the drain-to-source voltage is made sufficiently large, pinchoff will occur at zero gate-to-source bias. The graph shows the gate-to-source pinchoff voltages for a hypothetical N-channel-field-effect transistor. The different curves represent the pinchoff voltages for various values of the drain-to-source voltage.

In Class A and Class AB amplifiers, a field-effect transistor is normally biased at a lower value than that required to produce pinchoff. For operation in Class B, the device is biased at, or slightly beyond, the pinchoff voltage. In Class C operation, a field-effect transistor is biased considerably beyond pinchoff (*see* CLASS A AMPLIFIER, CLASS AB AMPLIFIER, CLASS B AMPLIFIER, CLASS C AMPLIFIER).

The field-effect transistor depicted is normally conductive through the channel; application of larger and larger bias voltages will eventually result in pinchoff. This operation is known as depletion-mode operation. Some types of field-effect transistors are normally nonconductive under conditions of zero bias. For conduction to occur in this kind of device, a gate-to-source bias must be applied. These field-effect transistors are called enhancement-mode transistors. *See also* ENHANCEMENT MODE.

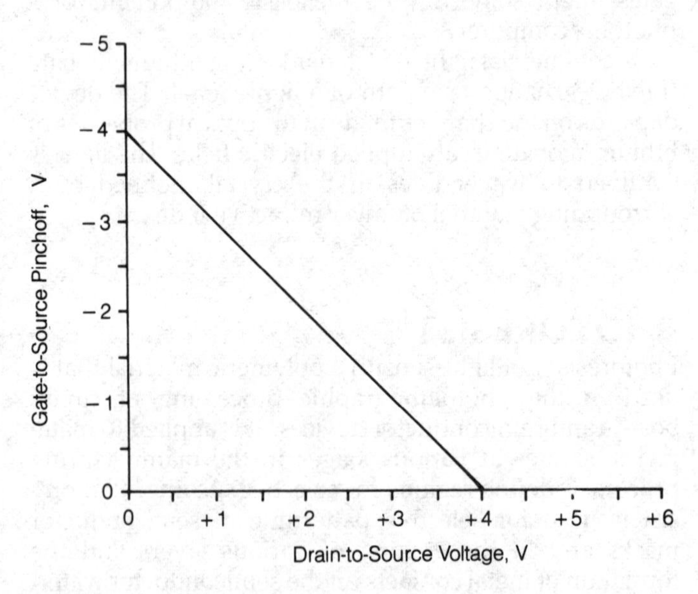

PINCHOFF: Gate-to-source pinchoff voltage is shown as a function of drain-to-source voltage for a field-effect transistor.

PINCHOFF VOLTAGE

In a depletion-mode field-effect transistor, the pinchoff voltage is the smallest gate-to-source bias voltage that results in complete blocking of the channel by the depletion region.

The pinchoff voltage depends on the kind of field-effect device and the voltage between the drain and the source. *See* PINCHOFF.

PINCUSHION DISTORTION

In a television system, pincushion distortion refers to distortion in which both sides, and the top and bottom, of the picture sag inward toward the center (see illustration). Pincushion distortion is the result of improper alignment of a television receiver. It often occurs because of mutual effects between the horizontal and vertical scanning circuits in the picture tube. *See also* TELEVISION.

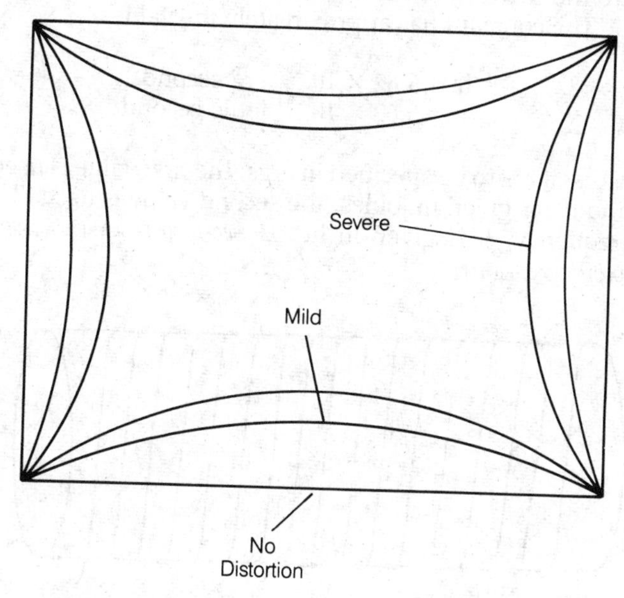

PINCUSHION DISTORTION: The borders of the picture are bowed inward.

PIN DIODE

A special form of diode with an intermediate layer of intrinsic semiconductor material between the P-type and N-type layers, is known as a PIN diode. The term *PIN* is derived from the terms P-type, intrinsic, and N-type.

The PIN diode has relatively low inherent capacitance. This makes it useful switching at high, very-high, and ultrahigh radio frequencies. The PIN diode can also be successfully used as a rectifier at these frequencies.

The PIN device is manufactured by a diffusion process; heavily doped P-type and N-type semiconductor materials are combined. The intrinsic layer may be quite thin or relatively thick; the thickness affects the performance of the diode. The I layer acts as a conductor when the diode is forward-biased. Under conditions of reverse bias, the I layer behaves like a dielectric with very low loss and low dielectric constant. *See also* DIODE CAPACITANCE, INTRINSIC SEMICONDUCTOR, PHOTODETECTOR.

PI NETWORK

A pi network is an unbalanced circuit widely used in the construction of attenuators, impedance-matching circuits, and various forms of filters. Two parallel components are connected on either side of a single series component. The drawing at A illustrates a pi-network circuit consisting of noninductive resistors. An inductance-capacitance impedance-matching circuit is shown at B. The pi network gets its name from its resemblance, at least in diagram form, to the Greek letter pi (π).

In a pi-network impedance-matching circuit, the capacitor nearer the generator or receiver is called the tuning capacitor, and the capacitor nearer the load or antenna is called the loading capacitor. By properly adjusting the values of the two capacitors and the inductor, matching between nonreactive impedances can be obtained over a fairly wide range in practice.

Pi-network impedance-matching circuits are extensively used in the output circuits of tube-type and transistor-type radio-frequency power amplifiers. In these circuits, the plate or collector impedance is often several times the actual load impedance. The loading-capacitor value is set so that when the tuning capacitor is adjusted, the dip in the plate or collector current results in the optimum operation of the final-amplifier tube or transistor.

In certain applications, a series inductor may be added to a pi-network impedance-matching circuit. This allows the cancellation of capacitive reactances in the load. This circuit (C) is called a pi-L network. Similarly, if inductive reactance exists in the load, a capacitor may be included in series with the network. This is known as a pi-C circuit (D). *See also* IMPEDANCE MATCHING.

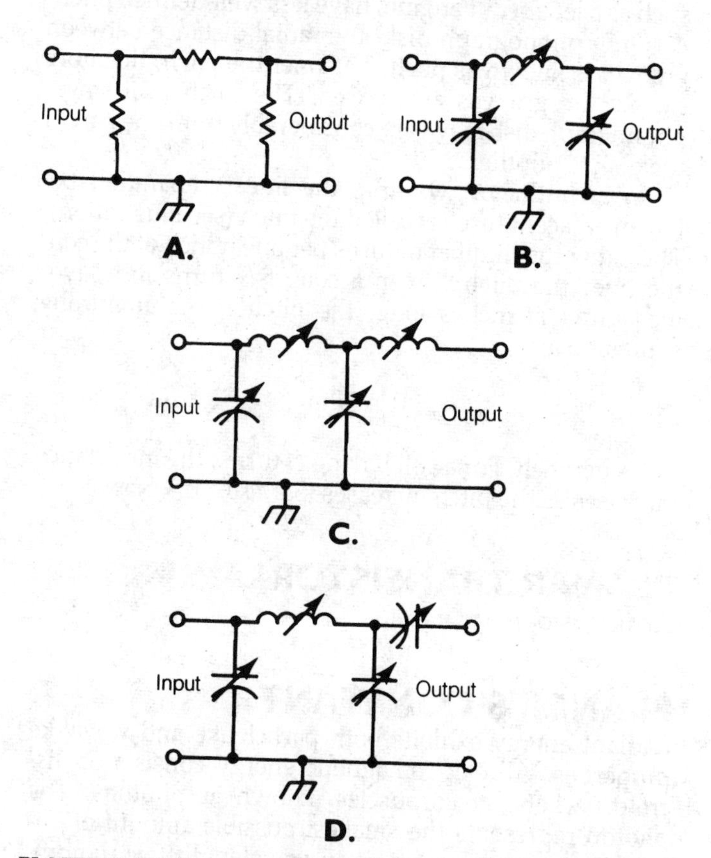

PI NETWORK: Forms of the pi network include an attenuator with resistors (A), an rf impedance-matching circuit (B), an rf pi-L network (C), and an rf pi-C network (D).

PIPELINE PROCESSOR

In computer technology, a pipeline processor is a computer system with a series of interconnected computing elements. Each element is specialized for a specific computing task. The work to be processed is passed along from one processor to the next, with each element performing a successive operation on the data. The pipeline is analogous to an assembly line in which all operations are conducted simultaneously but not on the same material.

Information flows along a fixed path and must move only a short distance between processing steps. The pipeline configuration is optimum as long as the same basic type of operation is to be performed. Functions are specialized and communications are minimized. Pipelining permits the arithmetic sections of high-speed computers to process sequences of numbers at high speed. Pipeline processors are less effective when the tasks to be performed are highly variable. Other parallel processors are the array processor and independent processors. *See* ARRAY PROCESSOR.

PITCH

Pitch is an expression for the perceived frequency of an acoustic disturbance. The shorter the sound wavelength or the greater the frequency, the higher the pitch. Some sounds have pitch that is easy to determine; an example is the note from a musical instrument. Other sounds, such as a jet aircraft engine, have less well-defined pitch.

On a phonograph disk, the radial distance between grooves is known as pitch. The finer the pitch, the more closely the grooves are spaced. The pitch in a single phonograph disk may vary considerably from one part of the disk to another.

In an inductor winding, the linear separation between adjacent turns is called the pitch (see illustration). The larger the number of turns per linear inch of the coil, the finer the pitch. Given a coil of N turns in a layer measuring m inches long, the pitch P is numerically expressed as:

$$P = N/m$$

turns per inch. For a coil having N turns, the inductance increases as the pitch increases. *See also* COIL WINDING.

PLANAR TRANSISTOR

See TRANSISTOR MANUFACTURE.

PLANCK'S CONSTANT

Radiant energy exhibits both particlelike and wavelike properties. All electromagnetic energy consists of discrete packets, or corpuscles, known as photons. The photon represents the smallest possible amount of energy that can exist at a given wavelength (*see* PHOTON). Electromagnetic energy cannot be divided into smaller components.

The shorter the wavelength (or the higher the frequency) of an electromagnetic disturbance, the more energy is contained in each photon. Conversely, the more energy in a photon, the shorter the wavelength and the higher the frequency. The energy and the wavelength are related according to a precise linear function:

$$e = hf$$

where e is the energy in the photon, f is the frequency of the electromagnetic wave, and h is a constant known as Planck's constant. The constant gets its name from the physicist Max Planck, one of several great scientists whose discoveries revolutionized particle physics around the turn of the century.

The constant has approximately the value:

$$h = 6.62 \times 10^{-27} \text{ erg second}$$
$$= 6.62 \times 10^{-34} \text{ joule second}$$

Thus, if e is to be specified in ergs, the first value is used, and if e is given in joules, the second value is used. The frequency, f, is given in hertz. *See also* ELECTROMAGNETIC RADIATION, LIGHT.

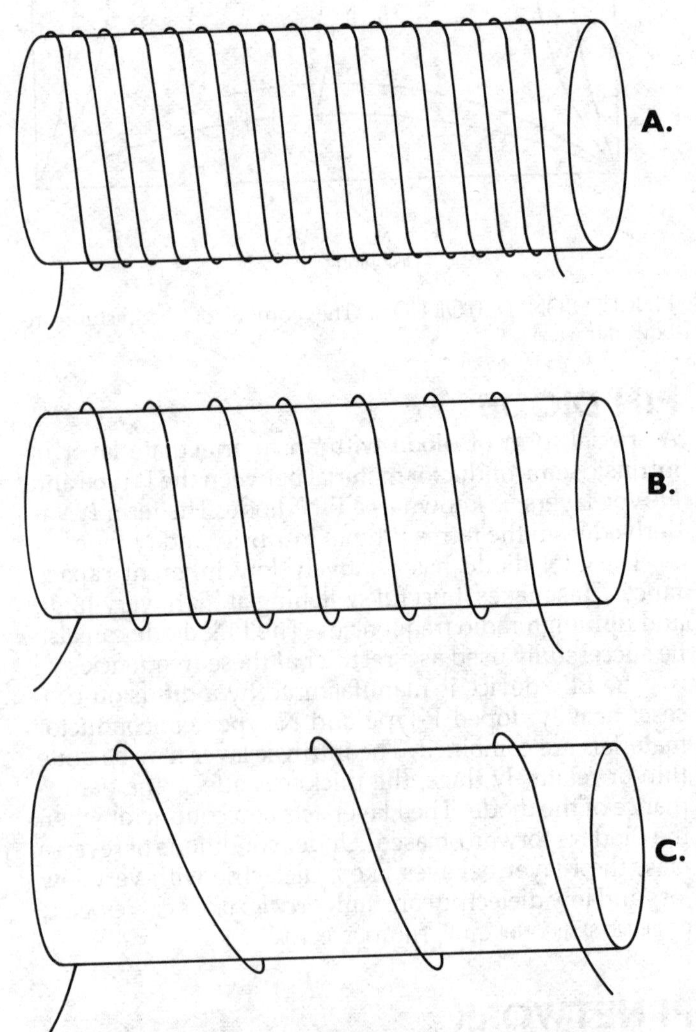

PITCH: Three examples of coil pitch: maximum (A), less pitch (B), and least pitch (C).

PLANE

A plane is a flat surface such as a floor or wall. In geometry, any three points in space lie in one plane. Any two intersecting lines determine a unique plane.

In three-dimensional Cartesian space, a plane is determined by a linear equation of the form:

$$ax + by + cz + d = 0$$

where a, b, c, and d are constants, and x, y, and z are the coordinates of the Cartesian three-space. *See also* CARTESIAN COORDINATES.

PLANE OF POLARIZATION

See POLARIZATION.

PLANE REFLECTOR

A plane reflector is a passive antenna reflector commonly used for transmitting and receiving at ultrahigh and microwave frequencies. The reflector is flat and electrically continuous. A plane reflector is positioned ¼ wavelength behind a set of dipole antennas, producing approximately 3 decibels of power gain.

The operation of the plane reflector is shown in the illustration. The direct, or forward, radiation from the antenna is not affected. The radiation in the opposite direction encounters the metal plane, where the wave is reversed in phase by 180 degrees and reflected back in the forward direction. The plane is ¼ wavelength away from the radiating element, so the total additional path distance for the reflected wave is ½ wavelength, or 180 degrees, with respect to the direct wave. The extra distance and the phase reversal together result in phase reinforcement between the direct wave and the reflected wave.

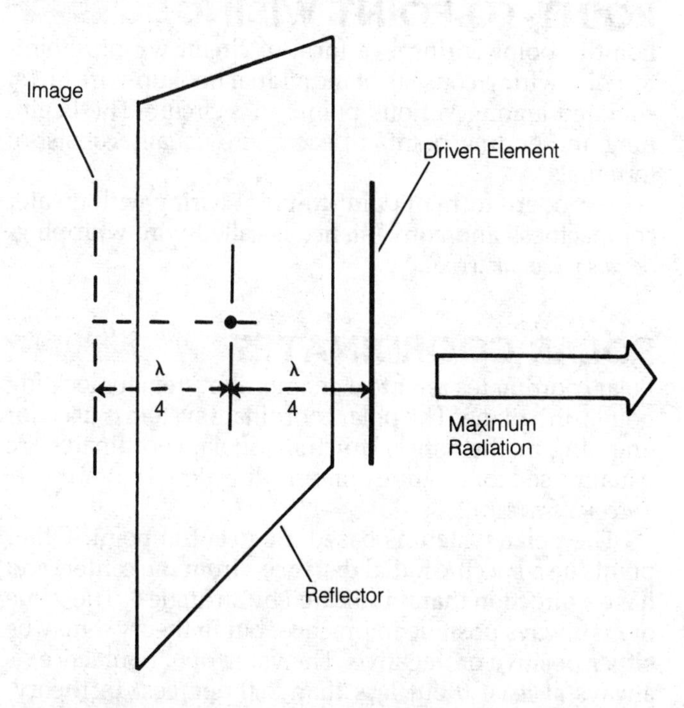

PLANE REFLECTOR: The gain of a plane reflector is about 3 decibels, doubling the effective radiated power.

Plane reflectors may be made from metal screen, parallel metal bars, or sheet metal. If wire screen or mesh is used, the spacing between the wires or bars should be less than about 0.05 wavelength for best results.

PLASMA

When a gas is heated to an extremely high temperature or is subjected to an intense electric field, the electrons are stripped away from their normal orbits around the atomic nuclei. Then the gas, which is ordinarily an excellent insulator, begins to conduct. The electrons are easily passed from nucleus to nucleus. This state of matter is called a plasma.

A plasma is affected by electric and magnetic fields in a manner different from a gas in its usual state. Since the plasma is an effective conductor, it can be confined or deflected by electric and magnetic fields. *See also* ELECTRON, ELECTRON ORBIT, NEON GAS DISCHARGE DISPLAY.

PLATE

The anode of a vacuum tube is also called the plate. The plate is supplied with a positive voltage so that it attracts electrons from the cathode. The plate of a tube is analogous to the collector of a bipolar transistor and to the drain of a field-effect transistor.

PLATINUM

Platinum is an element with an atomic number of 78 and an atomic weight of 195. In its pure form, platinum is a metallic substance, similar in appearance to silver.

Platinum is an excellent conductor of electricity, and is highly resistant to corrosion. It can be used as the electrode in chemical reactions.

PLL

See PHASE-LOCKED LOOP.

PLUG

Any male connector can be called a plug. Examples of plugs include the standard ¼-inch phone plug, lamp-core plugs, phono plugs, Jones plugs, and many others. A plug mates with a female jack of the same size and number of conductors. Sometimes the female socket or jack, together with the male connector, is called a plug.

PMOS

See METAL-OXIDE-SEMICONDUCTOR LOGIC FAMILIES.

P-N JUNCTION

A P-N junction is the boundary between layers of P-type and N-type semiconductor materials. A P-N junction conducts when the N-type material is negative with respect to the P-type material, but it exhibits high resistance when the polarity is reversed.

A P-N junction is said to be forward-biased when the N-type material is negative with respect to the P-type material (see A in illustration). The junction conducts well in this mode. The P-type material, with a deficiency of electrons, receives electrons from the N-type material (*see* N-TYPE SEMICONDUCTOR, P-TYPE SEMICONDUCTOR).

A P-N junction is said to be reverse-biased when the N-type material is positive with respect to the P-type material as at B. In this condition, electrons are pulled from the P-type material, which already has a deficiency of electrons. Electrons accumulate in the N-type material, which has an excess of electrons. The result is a region with extremely high resistance. This region occurs immediately on either side of the boundary between the two different semiconductor layers. Therefore, the P-N junction does not conduct well in the reverse direction; the resistance may be millions of megohms or more. If the reverse bias is increased to larger and larger values, however, the P-N junction suddenly begins to conduct at a certain bias level, known as the avalanche voltage (*see* AVALANCHE, AVALANCHE BREAKDOWN).

Because a P-N junction conducts well in one direction but not in the other, devices including P-N junctions are useful as detectors and rectifiers (*see* DETECTION, DIODE ACTION, RECTIFICATION). The P-N junction of a semiconductor device may be capable of accumulating and depleting

A.

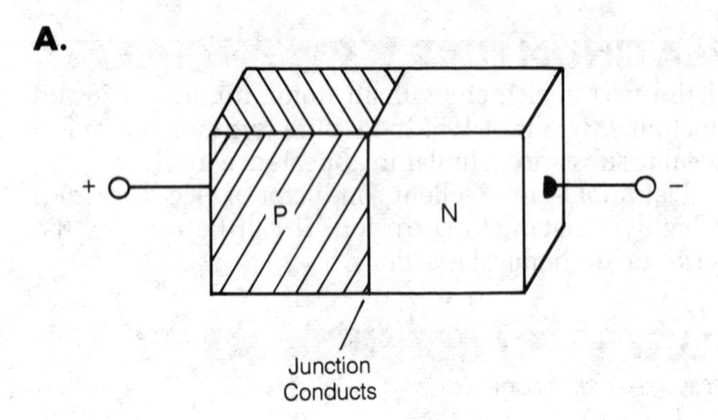

B.

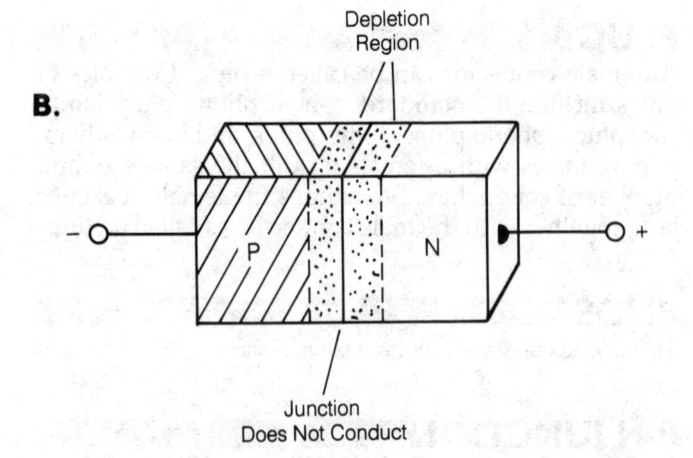

P-N JUNCTION: The P-N junction under conditions of forward bias (A) and reverse bias (B). With reverse bias, the depletion region is essentially an insulator.

the insulating layer very rapidly, perhaps on the order of millions or billions of times per second. In these instances, the resulting diode can be used for detection at megahertz and even gigahertz frequencies. Other P-N junction devices can follow alternating-current frequencies of only a few tens or hundreds of kilohertz. The ability of a P-N junction diode to handle high frequencies depends largely on the surface area of the junction. It also depends on the thickness of the depletion layer that results from a given amount of reverse voltage. *See also* DIODE CAPACITANCE.

PNP TRANSISTOR

See TRANSISTOR.

POINT SOURCE

A point source of radiation is, theoretically, an infinitely tiny spot in space from which the energy emanates. There is no such thing as a real point source of radiation, but for certain practical purposes it is useful to consider point sources. An energy source appears more like a point source as it is observed from greater distances.

The intensity of electromagnetic radiation from a point source, in terms of power per unit surface area, decreases with increasing distance, according to the inverse-square law (*see* INVERSE-SQUARE LAW). When the source is not a perfect point, the inverse-square law is not precisely correct. However, at long distances, most sources of electromagnetic radiation can be considered point sources. Thus, the inverse-square law is reasonably accurate at great distances from a source of radiation, regardless of the wavelength.

POINT-TO-POINT WIRING

Point-to-point wiring is a form of circuit wiring. Point-to-point wiring consists of individual hookup-wire links, soldered among various points in a circuit. The beginning and ending points of each wire usually consist of terminals.

A modern form of point-to-point wiring with greater compactness and convenience is called wire wrapping. *See also* WIRE WRAPPING.

POLAR COORDINATES

Polar coordinates refer to a geometric system for locating points on a plane. The polar coordinate system is used for graphing mathematical functions. Polar coordinates are usually used for plotting antenna directional patterns (*see* ANTENNA PATTERN).

The polar system is based on a central point. Other points lie a specific radial distance r from the center, and have a direction that is indicated by an angle θ. The value of r is always positive in practice, but in theory it may be either positive or negative. The value of θ, in practice, is always at least 0 but less than 360 degrees. In theory, however, θ can attain any value, positive or negative, representing multiple rotations about the center. The

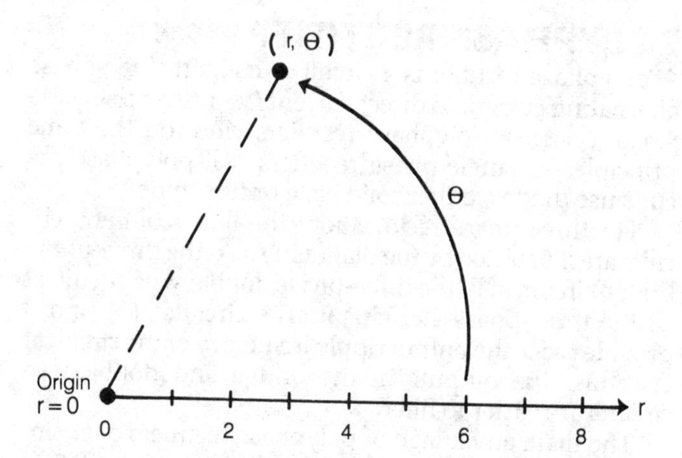

POLAR COORDINATES: A system for locating points in a plane.

angle θ is measured counterclockwise from a reference axis that coincides with the positive x axis of the Cartesian system (*see* CARTESIAN COORDINATES). The figure illustrates the polar coordinate system.

A given equation does not have the same graph in polar coordinates as in rectangular coordinates. The conversion from rectangular to polar coordinates is done according to the formulas:

$$r = \sqrt{x^2 + y^2}$$

$$\theta = \arctan (y/x)$$

Conversion from polar to rectangular coordinates is done according to the formulas:

$$x = r \cos \theta$$

$$y = r \sin \theta$$

Relations that appear simple when expressed in rectangular coordinates might become quite complicated when converted to polar coordinates. Similarly, simple polar equations are often complex in rectangular form. An illustrative example is the circle. In rectangular coordinates, the equation is difficult to work with. A circle of radius k, centered at the origin, appears in rectangular form as:

$$x^2 + y^2 = k^2$$

but in polar form the equation is simply:

$$r = k$$

See also COORDINATE SYSTEM.

POLARITY

Polarity refers to the relative voltage between two points in a circuit. One pole, or voltage point, is called positive, and the other pole is called negative. The polarity affects the direction in which the current flows in the circuit.

Physicists consider current to flow from the positive pole to the negative pole. But electron movement is from the negative pole to the positive pole, and electrons carry a negative charge.

In magnetism, polarity refers to the orientation of the north and south magnetic poles. *See also* DIPOLE, POLE.

POLARIZATION

Polarization is an expression of the orientation of the lines of flux in an electric, electromagnetic, or magnetic field. Polarization is of primary interest in electromagnetic effects. The polarization of an electromagnetic field is considered to be the orientation of the electric flux lines.

An electromagnetic field can be horizontally polarized, vertically polarized, or slantwise polarized (*see* HORIZONTAL POLARIZATION, VERTICAL POLARIZATION). The polarization is generally parallel with the active element of an antenna; thus, a vertical antenna radiates and receives fields with vertical polarization, and a horizontal antenna radiates and receives fields with horizontal polarization.

Electromagnetic fields may have polarization that continually changes. This kind of polarization can be produced in a variety of ways (*see* CIRCULAR POLARIZATION, ELLIPTICAL POLARIZATION).

Polarization effects occur at all wavelengths, from the very low frequencies to the gamma-ray spectrum. These effects are quite noticeable in the visible-light range. Light with horizontal polarization, for example, will reflect well from a horizontal surface (such as a pool of water), while vertically polarized light reflects poorly off the same surface. Simple experiments, conducted with a polarized lens, illustrate the effects of visible-light polarization. *See also* ELECTRIC FIELD, ELECTRIC FLUX, ELECTROMAGNETIC FIELD.

In an electrochemical cell or battery, the electrolysis process can result in the formation of an insulating layer of gas on one of the plates or sets of plates. This effect, called polarization, causes an increase in the internal resistance of the cell or battery. This, in turn, limits the amount of current that the battery can deliver. If the polarization is severe, the cell or battery might become useless. *See also* BATTERY.

POLARIZATION MODULATION

Information can be impressed on an electromagnetic wave by causing rapid changes in the polarization of the wave (*see* POLARIZATION). This method of modulation is known as polarization modulation. Polarization modulation can theoretically be achieved at any wavelength from the very low frequencies through the gamma-ray spectrum. However, polarization modulation is most often used at microwave and visible-light wavelengths.

POLARIZED COMPONENT

A polarized component is an electronic component that must be installed with its leads in a certain orientation. If a polarized component is installed incorrectly, the circuit will not work properly and the component may be damaged or destroyed.

Examples of polarized components include diodes, certain types of capacitors, and direct-current power sources. *See also* NONPOLARIZED COMPONENT.

POLARIZED LIGHT

Most visible light is randomly polarized. That is, the orientation of the electric-field component does not occur in any particular direction. When the electric lines of flux are all parallel, the light is polarized.

Polarized light can be produced by passing light through a polarized lens.

POLE

A pole is a point toward which the flux lines of an electric or magnetic field converge, or from which the flux lines diverge. Electric poles may be either positive, representing a deficiency of electrons, or negative, representing an excess of electrons. Magnetic poles can be either north or south.

Electric poles can exist alone—that is, a single positive or negative pole may be isolated—but magnetic poles are always paired. (*See* DIPOLE, MONOPOLE).

In a power supply the positive terminal can be called the positive pole and the negative terminal the negative pole. The way the supply is connected to its load or circuit is called its polarity. *See also* POLARITY.

POLYESTER

Polyester is a generic term for the plastic polyethylene glycol terephthalate. It is trade-named Mylar.® Polyester has high tensile strength, is extremely flexible, and is resistant to changes in temperature and humidity. Polyester film is the most popular general-purpose film dielectric for the manufacture of film capacitors. *See also* CAPACITOR.

POLYETHYLENE

Polyethylene is plastic resin with a chemical name of polymerized ethylene. This material is noted for its resistance to moisture, its low dielectric loss, and its high dielectric strength. Polyethylene is a tough, fairly flexible, semi-transparent plastic.

Polyethylene is widely used in electronics as an insulator, especially in coaxial cables and twin-lead transmission lines. It can be either in solid form or in foamed form. Solid polyethylene lasts longer than the foamed variety and is also less susceptible to contamination. Foamed polyethylene has somewhat lower dielectric loss and a lower dielectric constant. *See also* COAXIAL CABLE, TWIN-LEAD.

POLYPHASE CURRENT

See THREE-PHASE ALTERNATING CURRENT.

POLYPHASE RECTIFIER

A polyphase rectifier is a circuit to convert three-phase alternating current to direct current (*see* THREE-PHASE ALTERNATING CURRENT). Polyphase rectifiers work on the same principles as single-phase rectifiers. All polyphase circuits use diodes, either solid state or tube type.

For three-phase rectification, the most common circuits are illustrated by the diagram: At A, the three-phase bridge circuit; at B, the three-phase double-wye circuit; at C, the three-phase star circuit. The circuits at A and B provide twice the output ripple frequency of the circuit at C. Thus, the outputs of the bridge and double-wye circuits are easier to filter.

The main advantage of polyphase rectifiers over single-phase rectifiers is that the ripple frequency is higher, resulting in pure direct current after filtering.

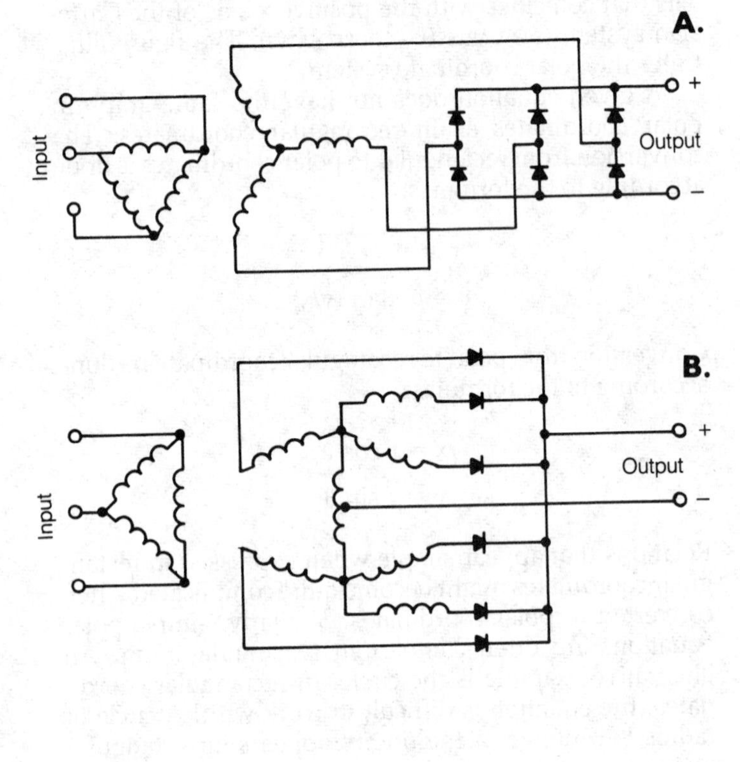

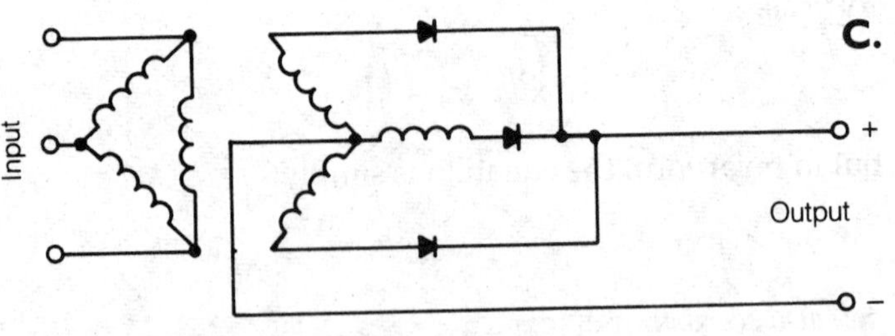

POLYPHASE RECTIFIER: Three common polyphase rectifier circuits are the three-phase bridge (A), the three-phase double wye (B), and the three-phase star (C).

POLYSTYRENE

Polystyrene is a plastic noted for its low dielectric loss at radio frequencies. The dielectric constant is approximately 2.6, regardless of the frequency. The direct-current resistivity is very high, on the order of 10^{18} ohm-centimeters. The power factor is very low at all frequencies. Polystyrene exhibits very little change in characteristics with large excursions in temperature and humidity. The material appears as a clear plastic in its pure form.

Polystyrene is widely used in the manufacture of film capacitors. *See also* CAPACITOR.

PORCELAIN

Porcelain is a ceramic material with excellent dielectric properties. It normally appears as a hard, white, dull substance when not glazed; it is shiny when glazed. Porcelain has high tensile strength, but it is brittle.

Porcelain is widely used in the manufacture of insulators for direct current as well as radio-frequency alternating current; it can withstand very high temperatures and voltages. The dielectric constant of porcelain decreases somewhat as the frequency increases; while it is about 5.4 at 1 kHz, it decreases to 5.1 at 1 MHz and 5.0 at 100 MHz.

The dielectric loss of dry porcelain is quite low. High-frequency and microwave fixed and trimmer capacitors have porcelain as the dielectric. Many air-variable capacitors, especially those designed for transmitters, incorporate porcelain for insulating purposes. Porcelain is also used as a coil-winding form, especially in power amplifiers. *See also* CERAMIC, CERAMIC FILTER, CERAMIC MICROPHONE, CERAMIC PICKUP.

POSITIVE CHARGE

Positive charge is the result of a deficiency of electrons in the atoms of an object. Friction between two objects can cause an electron imbalance, resulting in positive charge on one of the bodies. When an atom has fewer electrons than protons, that atom is said to be positively charged.

The smallest unit of positive charge is carried by a single proton. The smallest unit of negative charge is carried by the electron. The proton and electron have opposite charge polarity, but equal charge quantity. The terms positive and negative are chosen arbitrarily.

Electric lines of flux are considered to originate at positive-charge poles and terminate at negative-charge poles. An electric current is theoretically considered to flow from the positive pole of a circuit to the negative pole, but electrons flow from the negative pole to the positive pole. *See also* CHARGE, NEGATIVE CHARGE.

POSITIVE LOGIC

Logical signals are defined by convention. Normally, the logic 1 is the more positive of the voltage levels in a binary circuit, and the logic 0 is the more negative. Thus, the logic 1 is high and the logic 0 is low. This is called positive logic.

In some digital circuits, the logic 1 is the more negative (low) and the logic 0 is the more positive (high) of the voltage levels. This is known as negative logic (*see* NEGATIVE LOGIC).

From a practical standpoint, it makes no difference whether a circuit uses positive logic or negative logic, as long as the same form is consistently used throughout the circuit. *See also* LOGIC.

POSITRON

A positron is an atomic particle with a mass identical to the electron, or 9.11×10^{-31} kilogram, but carries a unit positive charge rather than a unit negative charge. If a positron happens to collide with an electron, the two particles annihilate each other and produce energy in the form of a photon. Because the electron (which is matter) and the positron disappear when they collide, the positron has been called a particle of antimatter.

Positrons are generated in atomic reactions. The hydrogen-fusion process causes a positron to be formed, as two protons combine to form a nucleus of deuterium.

POTENTIOMETER

A potentiometer is a variable resistor whose resistance value can be changed by moving a sliding contact or wiper along its resistive element to pick off the desired value. The potentiometer as shown in the basic schematic, Fig. 1, has two terminals; one is at the end of the fixed resistive element, and the other is coupled to the sliding wiper. If the wiper is moved toward the fixed contact, the resistance value decreases; if it is moved toward the other end, the resistive element the value will increase to its limit. Movement of the wiper is accomplished by turning a shaft or screw attached to the wiper or sliding a handle attached to the wiper.

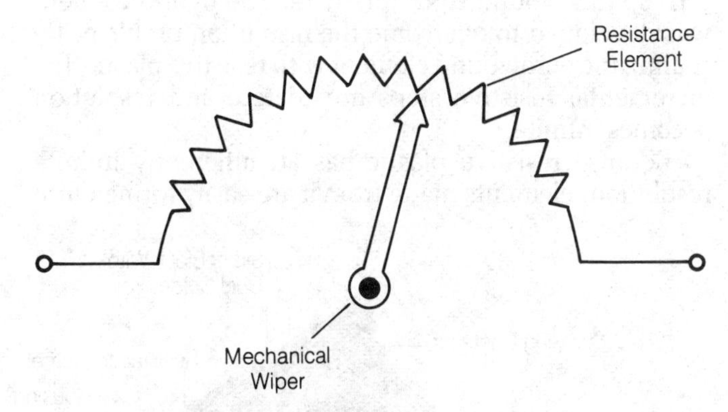

POTENTIOMETER: Fig. 1. Schematic for all variable resistors.

Commercial potentiometers for electronics are classed as: 1) precision potentiometers, 2) panel or volume-control potentiometers, and 3) trimmer potentiometers. In normal speech the term *potentiometer* is shortened to *pot* so the first two of these devices would be called precision pot and panel or control pot, respectively. However, the trimmer pot is customarily called a trimmer. Although the applications for these three classes of variable resistor vary widely, they have the same schematic symbol and many of the same resistive element technologies.

Precision Potentiometer A precision potentiometer

is an instrument-grade variable resistor with repeatable resistive accuracy of at least 1 percent. These pots were the basic components of analog computers and were widely specified for use in analog instrumentation and military and aerospace equipment. Precision pots are used primarily as sensors because, when a stable voltage source is applied, they are able to provide precise and resetable voltages corresponding to each setting of the control shaft. Vernier-type dials are widely used accessories to assure that the shaft can be returned to the same setting to obtain repeatable output voltage within a close tolerance.

Most precision pots have a cylindrical housing with an axial rotating shaft. The resistive elements of single-turn precision pots are formed in the shape of the letter C and rigidly fastened to the inside of the case. However, the multiturn precision pot, as shown in Fig. 2, has an element formed as a helix or spiral, also rigidly attached to the inside of the case. A sliding lead-screw assembly on the control shaft advances or retracts the wiper assembly with shaft motion and the wiper tracks around the inside of the helical element.

Resistance Element. Precision pots are classified by their resistance element. The most widely used elements are wound resistive wire (wirewound) and conductive plastic. The wirewound element is made by winding fine resistance wire on a heavy wire form or mandrel. Although wirewound elements have low temperature coefficients, they exhibit finite resolution. As the wiper slides along the resistive element, it spans increments of resistance equal to an individual turn of wire. As a result, accuracy has a tolerance of plus or minus one wire turn. The highest resolution wirewound elements are multiturn spirals wound from fine wire. The hybrid element was developed to overcome the resolution problem. By coating the wirewound element with resistive plastic, the incremental resistive steps are coated, and resolution becomes infinite.

Because resistive plastic has an inherently infinite resolution, elements made from it are easily formed into tapers, nonlinear elements contoured to produce an output voltage output with respect to shaft position that performs a mathematical function. For example, the voltage output can generate a sine or cosine function, a square law function, or a logarithmic function.

Ceramic-metal (cermet) elements, also capable of infinite resolution are specified where the precision pot will be used in high temperature environments. But these elements are abrasive and can cause wear on the wiper.

Turns. Precision pots can also be classified by their number of turns: single-turn or multiturn. Because of the nature of the resistive elements and the conventions of accepted in manufacturing them, wirewound and hybrid pots can be either single turn or multiturn. But conductive plastic and cermet precision pots are made only as single-turn.

Specifications. The principal specifications for precision pots are: 1) starting or running torque, 2) resistance range, 3) power rating, 4) ambient temperature range, and 5) rotational life or shaft revolutions before tolerance limits are exceeded. These factors determine the choice of number of turns and resistive element.

If a single-turn pot does not have a resistive element long enough to provide the desired accuracy, a multiturn pot is specified. The effective rotation of a single-turn pot is about 320 degrees, but the most common multiturn pots are three-turn (1080 degrees) and ten-turn (3600 degrees). Five-, 15-, 25-, and 40-turn units are available as standard.

Both single-turn and multiturn pots with linearities of 0.25 percent or better are available as stock items. The low resistance range for single-turn pots is about 10 to 150 Ω; the high-resistance range is about 200 kΩ to 1 MΩ. Similarly, the low resistance range of multiturn pots is about 3 Ω to 1 kΩ; the high resistance range is about 200 kΩ to more than 5 MΩ.

Precision pots are made as panel- or servo-mounted units. Cases of panel-mounted units are positioned behind a panel with threaded bushings projecting through a hole in the panel. They are fastened with ring nuts and

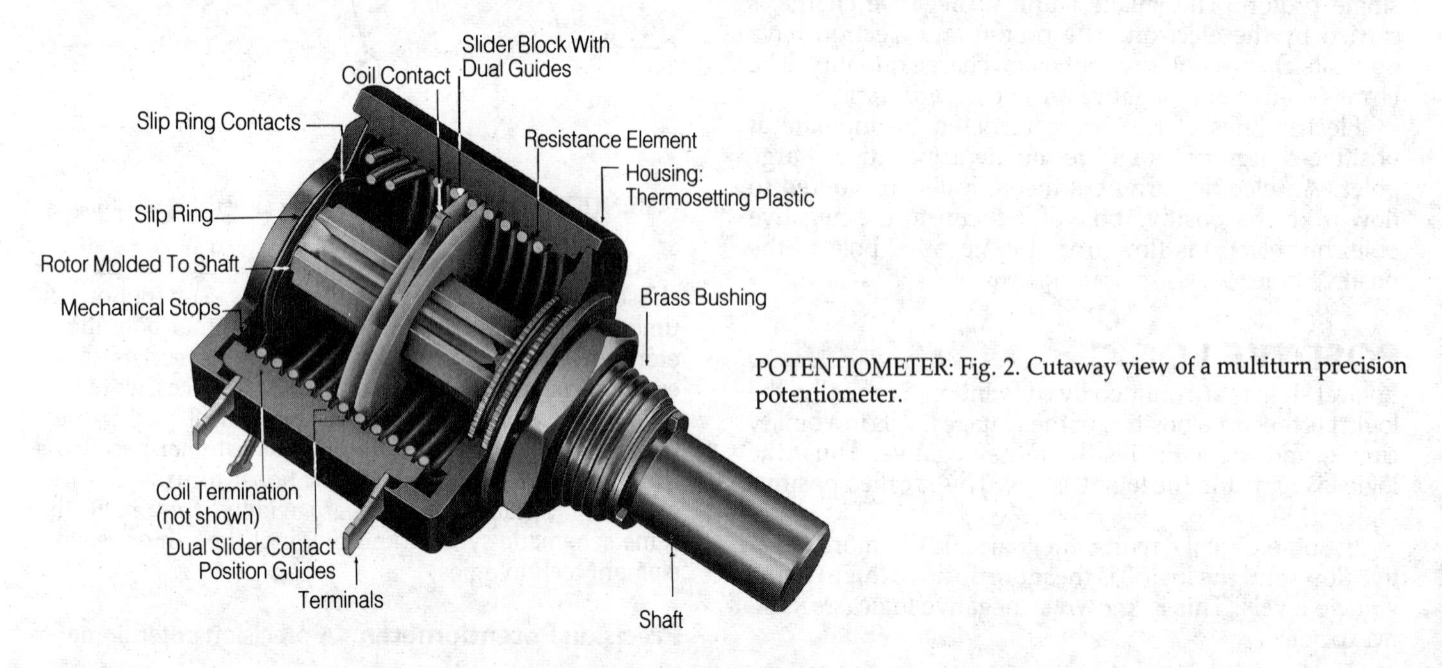

POTENTIOMETER: Fig. 2. Cutaway view of a multiturn precision potentiometer.

lockwashers. Servo-mounted pots are fastened face down on a metal baseplate and clamped with screw-type lugs secured in the clamping groove of the case. Precision pots are manufactured as both as standard catalog or custom products. However, even custom units may be packaged in standard case sizes.

Panel or Volume-Control Potentiometer Panel or volume-control pots are true variable resistors designed for long rotational life in applications such as tuning and adjustment of audio volume or tone and video chroma, intensity, and contrast. These pots are widely used on consumer entertainment products including radios, stereos, TV receivers, tape recorders, and telephone answering machines. They are also widely used on computer monitors, video display terminals (VDT), oscilloscopes, and other test instruments.

Unlike precision pots, panel pots are intended for subjective or personal adjustment of a physical variable. No attention is given to the precise relationship between the control shaft and resistive element. Panel pots look like precision pots and they also have cylindrical metal or plastic cases, axial control shafts, and "C"-shaped resistive single-turn elements.

Control pots are mounted behind the panel with a threaded bushing projecting through the front panel. They are fastened with a ring nut and lockwasher. However, some control pots with threadless bushings are designed for direct mounting on a PC card fastened behind the front panel. The bushing projects through a hole and the PC card provides the mechanical support. An on-off switch is often combined with the control pot in consumer entertainment products.

Resistive Elements. Resistive elements in panel pots can be hot-molded carbon, cermet or conductive plastic. Each has a different resistive range, tolerance and power rating. Tolerances are typically ±10 to 20 percent, and both carbon and conductive plastic elements can have resistive tapers. Cermet elements offer the highest power ratings.

Panel pots are made both as standard catalog and custom-made components. Some control pots are assembled from modular interchangeable parts permitting a wide choice of resistive elements. They are designed to be ganged with two or more resistive modules elements on the same coaxial shaft to save front panel space. Control pots are manufactured to both commercial and military specifications.

Trimmer Potentiometer Trimmer potentiometers, or trimmers, are small set-and-forget variable resistors for infrequent post-manufacturing adjustment, usually in linear circuits. Trimmers are normally adjusted during final test and checkout or during routine calibration. They are used in consumer radios and television sets, audio equipment, video data terminals, computer monitors and different kinds of test instruments and communications products. There would be no need for trimmers if resistive and capacitive components were precisely made and not subject to change due to temperature or aging. Trimmers are usually inaccessible to the user within the enclosure of the product.

There are many variations in trimmer design, style, size, and resistive element. They are manufactured both to commercial and military specifications. Two general trimmer classifications are rotary and linear.

Rotary Trimmers. Rotary trimers are single-turn rotary units set directly by screwdriver in their slotted shaft ends. The popular packages are round, open, PC (printed-circuit) board-mounting cases with ¼- and ⅜-inch diameters. But larger ½-inch-diameter units are available. Multiturn rotary units are set by turning slotted top-, side-, or end-mounted leadscrews. Popular sizes are open or sealed ¼- and ⅜-inch square packages with pins for PC-board mounting. Rotating mechanisms permit the wiper to move around the element in up to 20 turns. Surface-mount (SMT) versions of popular trimmers are now available.

Rectangular Trimmers. Rectangular trimmers are traditional multiturn trimmers in a ¾-inch-long rectangular case. Turning an internal leadscrew drives the wiper across the resistive element in up to 20 turns. The packages have pins for PC board mounting. Some styles of these trimmers have manually set sliding wipers. These trimmers have also been adapted for SMT.

Trimmers can have any of the commercial resistive elements: carbon film, bulk carbon, wirewound, cermet, conductive plastic and bulk metal. Most trimmers can dissipate ½ W, but some of the larger 1¼-inch multiturn units are able to dissipate 1 W. Power rating is largely determined by size although the resistive element is a factor.

Trimmers are manufactured both to military and commercial specifications. Individual military specifications cover cermet, wirewound, and high-reliability wirewound trimmers.

POT

See POTENTIOMETER.

POTENTIAL

See VOLTAGE.

POTENTIAL DIFFERENCE

Two points in a circuit are said to have a difference of potential when the electric charge at one point is not the same as the electric charge at the other point. Potential difference is measured in volts (*see* VOLTAGE).

When two points with a potential difference are connected by a conducting or semiconducting medium, current flows. Physicists consider the current to flow from the more positive point to the more negative point. If the charge carriers are electrons, they actually move from the more negative point to the more positive point. If the charge carriers are holes, they move from positive to negative. *See also* CHARGE, ELECTRON, HOLE.

POTENTIAL ENERGY

Potential energy is an expression of the capability of a

body to produce useful energy. For example, if a weight is lifted up several feet, it gains potential energy; when the weight is dropped, the energy output is realized.

POUND

The pound is the standard English unit of weight. In the gravitational field of the earth, a mass of 0.4536 kilogram weighs 1 pound. The abbreviation of pound is lb.

A mass that weighs 1 pound on the earth will have different weights in gravitational fields of different intensity. For example, on Mars, a mass of 0.4536 kilogram would weigh only about 6 ounces (6/16 pound). On Jupiter, the same mass would weigh approximately 2.5 pounds. *See also* KILOGRAM, MASS.

POWER

Power is the rate at which energy is expended or dissipated. Power is expressed in joules per second, more often called watts.

In a direct-current (dc) circuit, the power is the product of the voltage and the current. A source of E volts, delivering I amperes to a circuit, produces P watts, as follows:

$$P = EI$$

From Ohm's law, you can find the power in terms of the current and the resistance, R, in ohms:

$$P = I^2R$$

Similarly, in terms of the voltage and the resistance:

$$P = E^2/R$$

In an alternating current (ac) circuit with no reactance, the power is determined in the same manner with root-mean-square values for the current and the voltage (*see* ROOT MEAN SQUARE).

If reactance is present in an ac circuit, some of the power appears across the reactance, and some appears across the resistance. The power that appears across the reactance is not usually dissipated in real form. For this reason, it has been called imaginary or reactive power. The power appearing across the resistance is actually dissipated in real form, and it is thus called true power. The combination of real and reactive power is known as the apparent power (*see* APPARENT POWER, REACTIVE POWER).

The proportion of the real power to apparent power in an ac circuit depends on the proportion of the resistance to the total or net impedance. The smaller this ratio, the lower the real power. In the case of a pure reactance, none of the power is real. *See also* POWER FACTOR.

POWER AMPLIFIER

A power amplifier is an amplifier that delivers alternating-current power to a load. Power amplifiers are used in audio-frequency and radio-frequency applications.

The illustration is a schematic diagram of a typical audio-frequency power amplifier using bipolar transistors in a push-pull circuit. This amplifier might be used in a radio receiver, tape recorder, or record player. The input requirements are modest; the circuit needs almost no drive power to produce an output of several watts. A transformer is used at the output of an audio-frequency power amplifier to ensure that the transfer of power is optimized. Audio-frequency power amplifiers are usually operated in Class A, or in a push-pull Class B configuration, so distortion is kept to a minimum.

Audio-frequency power is measured as the product of the root-mean-square (rms) current and voltage delivered to a nonreactive load (*see* ROOT MEAN SQUARE). Audio-frequency power amplifiers are rated according to the rms power they can deliver without producing more than a specified amount of distortion, such as 3 percent to 10 percent. Some audio-frequency power amplifiers can produce several kilowatts of output power.

Radio-frequency power amplifiers are usually operated in Class AB, Class B, or Class C, depending on the intended application. *See also* CLASS A AMPLIFIER, CLASS AB AMPLIFIER, CLASS B AMPLIFIER, CLASS C AMPLIFIER, POWER.

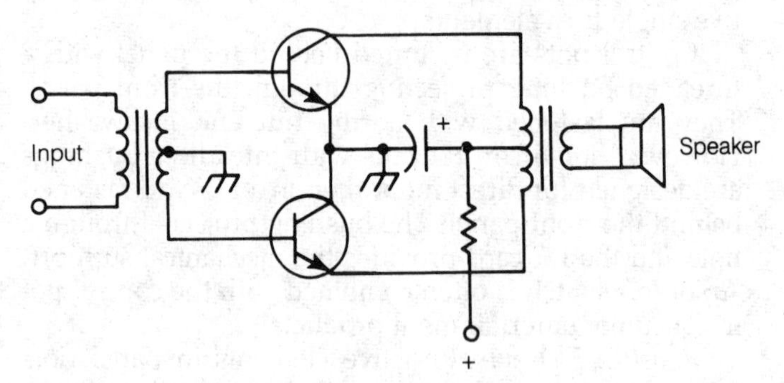

POWER AMPLIFIER: A push-pull audio power amplifier circuit using bipolar transistors.

POWER DENSITY

Power density is a term for expressing the strength of an electromagnetic field. Power density is expressed in watts per square meter or watts per square centimeter (W/m^2 or W/cm^2), as determined in a plane oriented parallel to both the electric and magnetic lines of flux.

In the vicinity of a source of electromagnetic radiation, the power density decreases according to the law of inverse squares. *See also* FIELD STRENGTH, INVERSE-SQUARE LAW.

POWER FACTOR

In an alternating-current (ac) circuit containing reactance and resistance, not all of the apparent power is true power. Some of the power occurs in reactive form (*see* APPARENT POWER, POWER, REACTIVE POWER, TRUE POWER). The ratio of the true power to the apparent power is known as the power factor. Its value may range from 0 to 1.

Power factor may be determined according to the phase angle in a circuit. The phase angle is the difference, in degrees, between the current and the voltage in an alternating-current circuit. If the phase angle is represented by ϕ, then the power factor, PF, is:

$$PF = \cos \phi$$

The power factor, as a percentage, is:

$$PF = 100 \cos \phi$$

The power factor in an ac circuit can also be determined by plotting the impedance on the complex plane, as shown in the illustration. If the resistance is R ohms and the reactance is X ohms, the phase angle is:

$$\phi = \arctan \left(\frac{X}{R} \right)$$

and thus the power factor is:

$$PF = \cos \left(\arctan \frac{X}{R} \right)$$

See also IMPEDANCE, PHASE ANGLE.

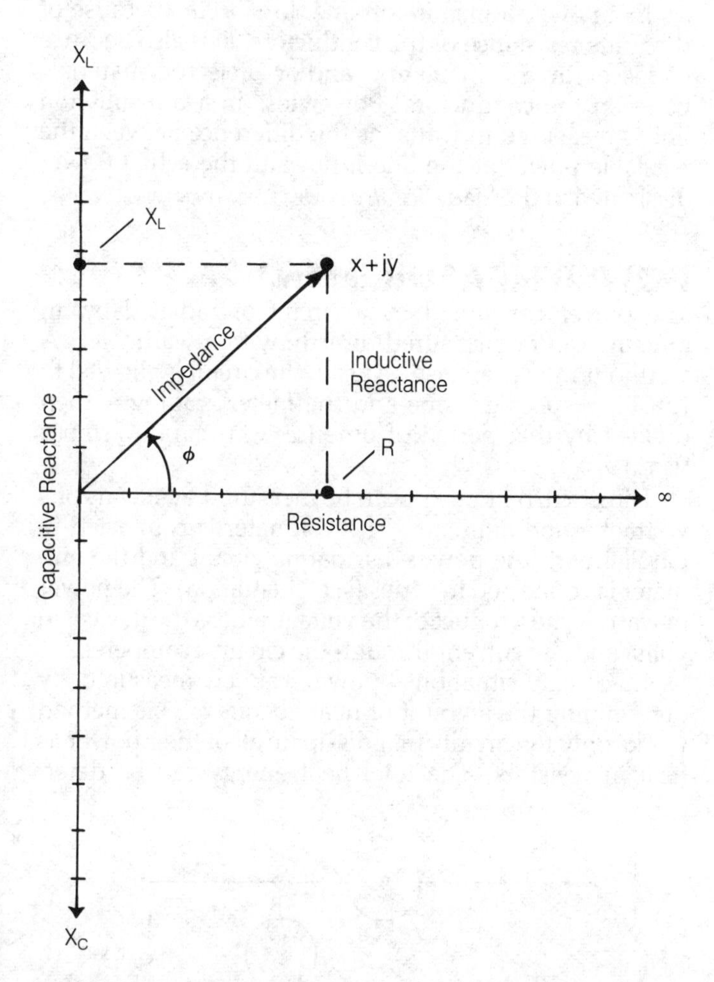

POWER FACTOR: Once the phase angle ϕ is known, the power factor can be determined from the impedance-plane diagram.

POWER GAIN

Power gain is an increase in signal power between one point and another. Power gain is used as a specification for power amplifiers.

If the input to a power amplifier, in watts, is given by P_{IN} and the output, in watts, is given by P_{OUT}, then the power gain in decibels is:

$$G = 10 \log_{10} \left(\frac{P_{OUT}}{P_{IN}} \right)$$

See also DECIBEL, GAIN.

POWER INTEGRATED CIRCUIT

A power integrated circuit is a monolithic circuit that combines signal-level analog circuitry or digital logic on the same chip with one or more power transistors capable of handling at least 2 A or 2 W. Power ICs save printed-circuit (PC) board space by placing the power switch and driver electronics on the same IC. They were made possible by special fabrication techniques that permit a power device to coexist on the same chip with small-signal transistors without destroying them or interfering with their operation.

The first power ICs were power drivers for high-voltage, neon seven-segment displays. They combined bipolar digital logic with a bipolar power transistor on the same chip. Later power ICs mixed bipolar and metal-oxide semiconductor (MOS) power devices with analog circuitry. CMOS logic was combined with bipolar transistors in BiMOS technology. Then CMOS logic was combined with DMOS (MOSFET) transistors in CMOS/DMOS. If lateral layout or topology is employed, the IC may have two or more on-chip power devices. But with vertical topology, the IC is limited to only one power device.

BiMOS technology is suitable for medium voltage and currents; it has been used to make motor controllers, solenoid switchers, and printhead-switching ICs. By contrast, CMOS/DMOS technology is suitable either for low-voltage, high-current, fast-switching ICs or for high-voltage, low-current, fast-switching ICs. The first option has been used to fabricate power supply pulse-width modulator, multiplexer, and voltage regulators ICs; the second option has been used to fabricate ac plasma and electroluminescent display drivers.

The three different techniques are used to isolate control circuitry from the power device to prevent interference and breakdown on the monolithic IC are:

1. Self isolation, an extension of CMOS technology in which a reverse-biased junction is located between the source and the drain region. This technique is usually limited to devices drawing less than 2 A, but voltage can be as high as 500 V.
2. Dielectric isolation (DI) uses single-crystal islands or tubs grown on a polycrystalline silicon substrate for the IC functions. Current must be brought out of the top of the chip within the tub, so the voltage levels are limited. DI is said to produce the lowest parasitic

capacitance and permit full isolation on the chip.
3. Junction isolation (JI) permits both lateral and vertical ICs. An epitaxial layer is formed on the substrate and deep junctions are diffused to obtain isolated areas. Current flow is similar to that in discrete power devices.

Power ICs are packaged in the same conventional IC packages except more attention is given to keeping junction temperatures below set limits. Some manufacturers use copper rather than kovar lead frames. As in other power devices, IC power handling ability is improved by mounting them on heat sinks and cooling them with forced air. Power ICs are in dual-in-line (DIP) packages with 8 to 28 pins and small outline transistor (SOT) cases where dissipated power is less than 2 W. More complex parts are in single-in-line (SIP) cases with from 11 to 23 leads. Plastic TO-220 style packages are used for power ICs that dissipate from 5 to 10 W. Some power ICs are in plastic leaded chip carriers (PLCCs). *see* SEMICONDUCTOR PACKAGE.

POWER INVERTER

See INVERTER.

POWER LINE

A utility line carrying electricity for household use is called a power line or power-transmission line. The wire, usually carrying 234 V, 60 Hz alternating current, that runs from a house to a nearby utility pole, is a power line. However, the higher voltage distribution wires are also called power lines.

The term *power line* is technically imprecise because power does not actually travel from place to place. Electrons in the conductors and the electromagnetic fields travel. Power is dissipated, or used, at the terminating points of the power line (*see* POWER). Some power is dissipated in the line because of losses in the conductors and insulating materials. This dissipation is undesirable because power lost in the line is unavailable at the terminating points. *See also* TRANSMISSION LINE.

POWER-LINE NOISE

Power lines for utility 60 Hz alternating current also act as conduits for unwanted currents. These currents cover a wide spectrum and result in an effect called power-line noise.

The currents are usually generated by electric arcing at some point in the circuit. The arcing may originate in appliances connected to the terminating points or in faulty or obstructed transformers; it can even occur in high-voltage lines as a corona discharge in humid air. The broadband currents cause an electromagnetic field to be radiated from the power line because they flow in the same direction along the conductors.

Power-line noise causes a hum or hiss when picked up by a radio receiver. Some types of power-line noise can be greatly attenuated with a line filter. For other kinds of noise, a limiter can be used to improve the reception. Phasing may be used in an antenna system to reduce the level of received power-line noise. *See also* LINE FILTER, NOISE LIMITER.

POWER LOSS

Power loss occurs as dissipation, usually resulting in the generation of heat. Power loss occurs in all electrical conductors and dielectric materials. Power loss can also take place in the cores of transformers and inductors.

The power loss in a circuit is usually specified in decibels. If the input power is represented by P_{IN} and the output power by P_{OUT}, then the loss L, in decibels, is:

$$L = 10 \log_{10} \left(\frac{P_{IN}}{P_{OUT}} \right)$$

A power loss is equivalent to a negative power gain. That is, if the loss (according to the formula), is x dB, then the gain is $-x$ dB. *See* DECIBEL, LOSS.

In a power-transmission line, loss occurs because of the finite resistance of the conductors, and also because of losses in the insulating and/or dielectric materials between the conductors. Power loss in a transmission line is measured in watts, as the difference between the available power at the line input and the actual power dissipated in the load. *See also* POWER LINE, TRANSMISSION LINE.

POWER MEASUREMENT

The power consumed by a circuit or radiated by an antenna can be measured directly with a wattmeter. A watthour meter can also be used; the circuit is allowed to run for a specified time and the meter reading is then divided by that period of time (*see* WATTMETER, WATTHOUR METER).

Direct-current power can be measured by means of a voltmeter and ammeter. The voltmeter is connected in parallel with the power-dissipating circuit and the ammeter is connected in series (see illustration). The power in watts is the product of the voltage across the device, in volts, and the current through the circuit in amperes.

In certain situations, power can be measured by determining the amount of heat liberated. This method works only for circuits that dissipate all of their power as heat in resistors. The total heat energy can be deter-

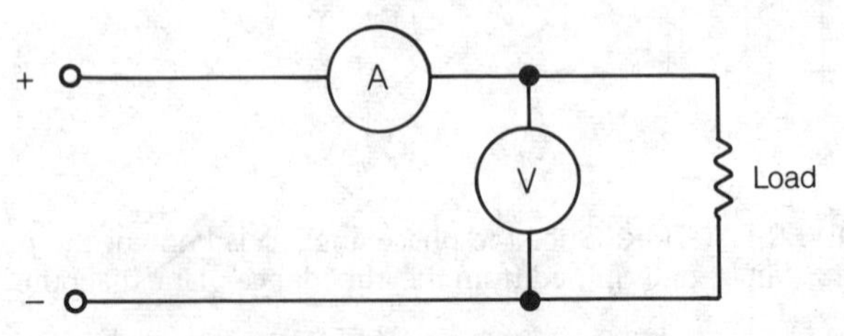

POWER MEASUREMENT: DC power is measured by multiplying the voltmeter (V) reading by the ammeter (A) reading.

mined, over a period of time, and the dissipated power can then be obtained by dividing the total heat energy by the time period.

POWER SENSITIVITY

Some Class A amplifiers based on field-effect transistors consume essentially no power while producing measurable output power (*see* CLASS A AMPLIFIER). These amplifiers theoretically draw no current and need only an alternating voltage at the gate to function. The performance of this amplifier can be expressed by a parameter called power sensitivity.

If the root-mean-square (rms) input-signal voltage is E, and the rms output power is P, then the power sensitivity S is given by the ratio:

$$S = P/E$$

Power sensitivity is expressed in watts per volt.

The power sensitivity of a Class A amplifier does not vary with the driving voltage as long as the amplifier is correctly biased and its not overdriven. If the amplifier is driven into the nonlinear portion of the characteristic curve, the power sensitivity decreases.

POWER SUPPLY

A power supply is a circuit for converting alternating current or unregulated direct current to regulated direct current. In electronics, the term usually refers to a separate packaged alternating-current to direct-current (ac-to-dc) conversion product—circuit board, potted module, or enclosed and shielded module. It does not normally refer to a dc-to-dc converter, uninterruptable power source (UPS), battery, or power-conversion circuitry integrated in and inseparable from entertainment electronics or test instruments. The two principal categories of ac-to-dc power supply are linear regulated and switching regulated.

A dc power supply converts power from the ac line to direct current and steady voltage of a desired value. The ac input voltage is first rectified to provide a pulsating dc which is filtered to produce a smooth voltage. Then the voltage may be regulated to assure that a constant output level is maintained despite variations in the power line voltage or circuit loading.

Linear Regulated dc Supply. There are three principal functional blocks in a voltage-regulated linear power supply as shown in the block diagram Fig. 1A. A linear power supply can have either series or shunt voltage regulator as determined by the arrangement of the pass element with respect to the load. However, the series-pass regulator with a transistor pass element as shown in Fig. 1B is most commonly used.

The rectifier, typically a bridge, delivers unregulated dc to the filter. Isolation from the ac line is provided by the transformer. The rectifier output is a steady dc voltage with an alternating voltage superimposed on it. The filter network removes the alternating voltage (ripple) and delivers it to the pass transistor in series with the load (*See* FILTER, RECTIFIER.) Regulation is accomplished by variation of the current through the series-pass transistor in response to a change in line voltage or circuit loading. Thus, the voltage drop across the pass transistor is varied, and voltage delivered to the load (RL) remains essentially constant. A comparator, voltage or current reference, and amplifier form a closed feedback loop in the regulator as shown in Fig. 1A.

Commercial linear power supplies are rated from 3 to about 1000 W with single or multiple outputs. These power supplies provide close regulation, are highly reliable, and are insensitive to minor shifts in frequency. Initial cost is low, particularly in the lower power ratings where low-cost transformers are used. The drawback of these supplies is inefficiency caused by continuous current drain in the pass transistor. It acts as a variable resistor and dissipates power continuously, keeping efficiencies of 15 to 35 percent. Another drawback is the use of large, relatively heavy 50 or 60 Hz transformers that account for most of the size and weight of the supply.

Ferroresonant Power Supply. The ferroresonant power supply (also known as a constant-voltage supply), is basically a tuned transformer that also functions as a filter. Both reliable and efficient, this supply is, however, sensitive to changes in line frequency. Moreover, at 50 or 60 Hz, a ferroresonant transformer is large, heavy, and expensive. It is about three times as heavy and occupies about twice the space of a 50 or 60-Hz transformer for a comparable linear supply. Although its regulation is excellent and its efficiency can reach 80 percent, the ferroresonant transformer depends on stable line frequency for efficient operation. As an additional drawback, it hums, making it annoying and unsuitable for use in occupied rooms.

Switching-Regulated Power Supply. The switching-regulated power supply (also known as the switch-mode power supply), offers the advantages of smaller size, lighter weight, and higher efficiency than comparably rated series-regulated or ferroresonant power supplies. A block diagram of switching regulator is shown in Fig. 2. It includes a reference, a comparator, a pulse-width modulator (PWM) and an amplifier. The comparator compares the output voltage to a reference voltage and the difference determines the on time of the switching transistor or pulse width.

If the output voltage falls below its reference value because of increasing load, the width (duration) of the drive pulses to the switching transistor will be increased by the PWM circuit. This increases the on time of the transistor to offset the drop in voltage. Similarly, if the output voltage increases over its reference value due to decreasing load, the width of the pulses will be decreased. This reduces the transistor on time to offset the increase in output voltage.

The secret to the higher efficiency of the switching regulated power supply lies in the intermittent operation of its regulator circuit. Power is drawn only when the transistors are switching at frequencies from 20,000 to 100,000 times per second (20 to 100 kHz) or more. By varying the length of time the switch is on during each cycle, the amount of energy delivered to the filter can be

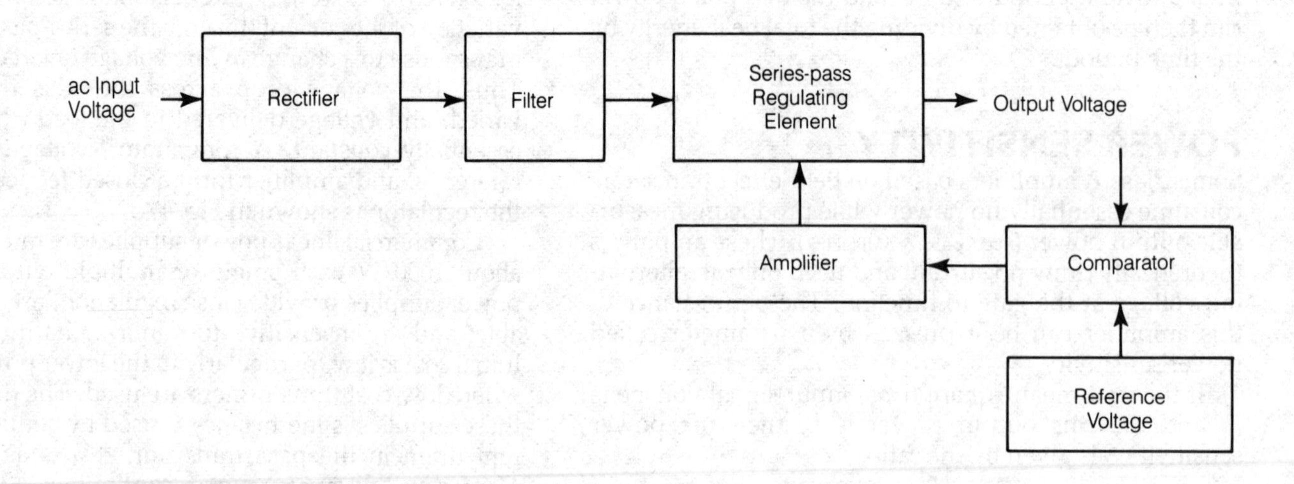

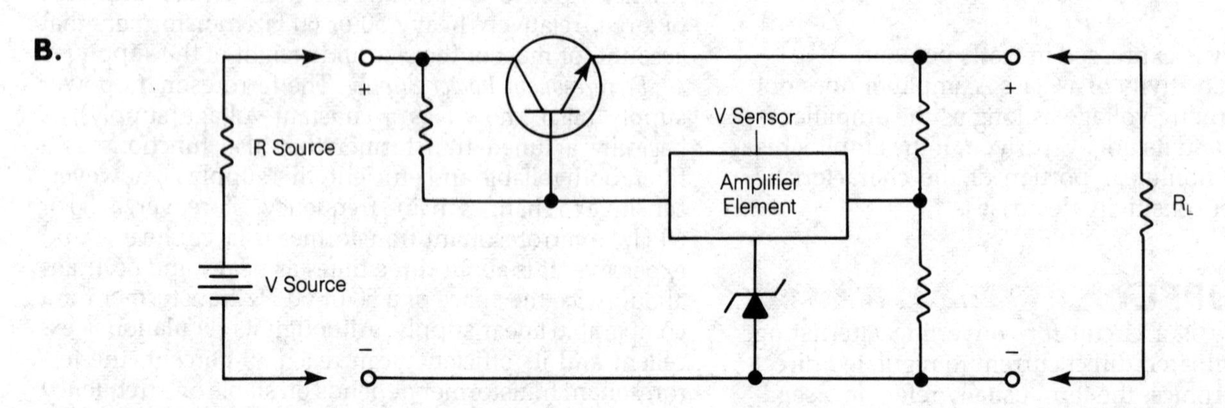

POWER SUPPLY: Fig. 1. Block diagram of a series voltage-regulating dc (direct-current) power supply (A) and a schematic for a basic series voltage regulator (B).

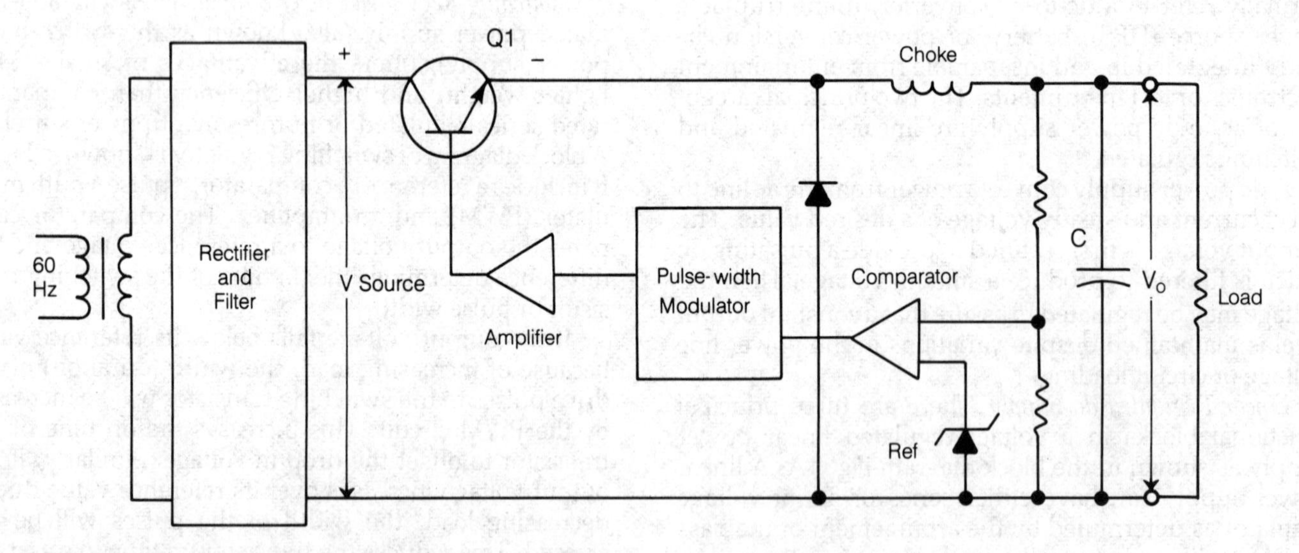

POWER SUPPLY: Fig. 2. Schematic for a basic switching-regulator power supply.

controlled. The switching regulator can have an efficiency as high as 85 percent in supplies with 12 to 15 V dc output—more than twice that of the most efficient linear supplies.

Because switching is done at a high frequency, the input transformer can be far smaller and lighter than the 50 or 60 Hz transformers used in series-regulated and ferroresonant power supplies, and heat dissipation is lower. This transformer size and weight reduction permits smaller and lighter switching-regulated supplies.

The flyback configuration of an off-line switching

power supply as shown in Fig. 3 is suitable for low-power, low-cost switching power supplies, generally rated under 100 W. An alternative off-line configuration is the single-transistor forward converter configuration shown in Fig. 4. Both require only a single switching transistor, and they have relatively simple, inexpensive magnetics. These supplies are usually manufactured as open-frame units.

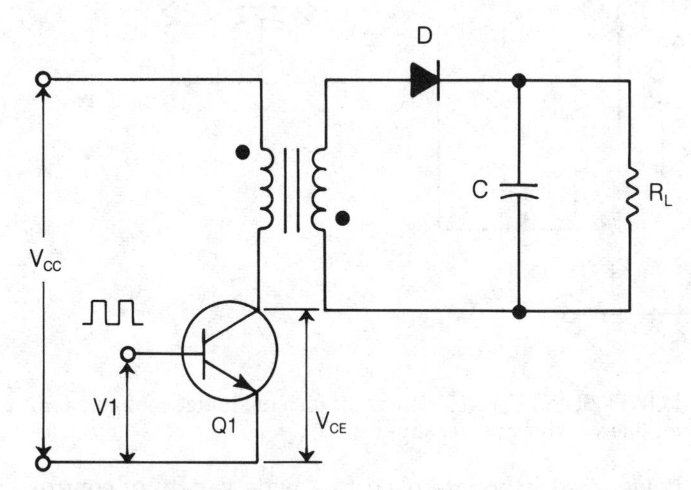

POWER SUPPLY: Fig. 3. Schematic for a flyback configuration, off-line switching power supply.

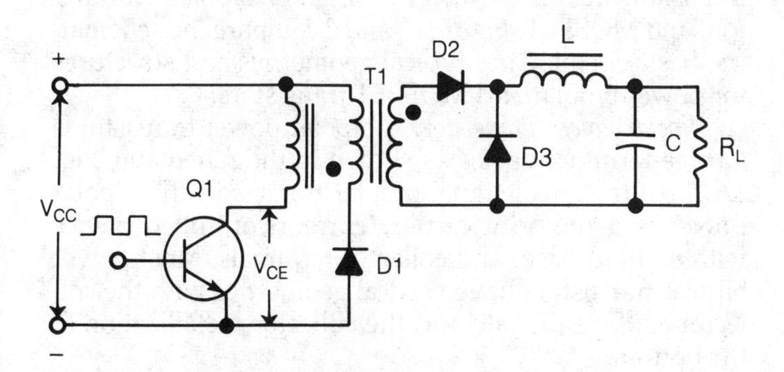

POWER SUPPLY: Fig. 4. Schematic for a single-transistor, forward-configuration, off-line switching power supply.

The push-pull (Fig. 5) and half-bridge (Fig. 6) configurations are widely used in the 100 to 500 W power range. With two switching transistors, they are made either as open-frame units or enclosed modules. For switching supplies rated at 500 W or higher, the full bridge configuration shown in Fig. 7 is most widely used.

The switching power supply has become widely accepted for powering computers and computer peripherals. This popularity is based on its smaller size, lighter weight, and higher efficiency. The regulation of most switching power supplies is typically ±5 percent, lower than that of most linear power supplies. However, this is acceptable for powering most digital circuitry.

These supplies gained popularity for commercial applications only after many years of improving on the original military designs. Many components such as high-frequency bipolar and MOSFET power transistors, ferrite-core transformers, fast-recovery and Schottky rec-

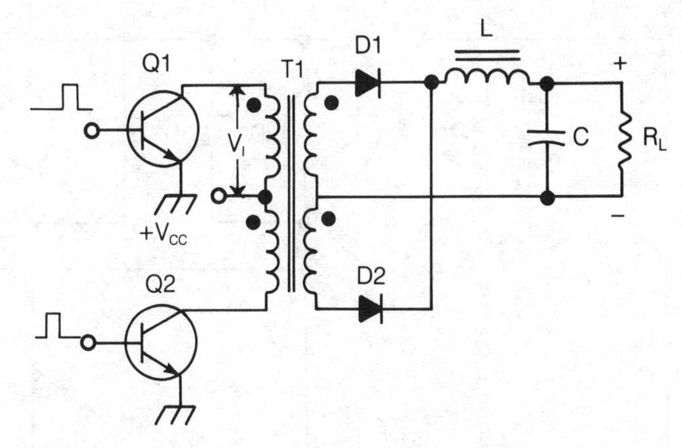

POWER SUPPLY: Fig. 5. Schematic for a push-pull configuration, off-line switching power supply.

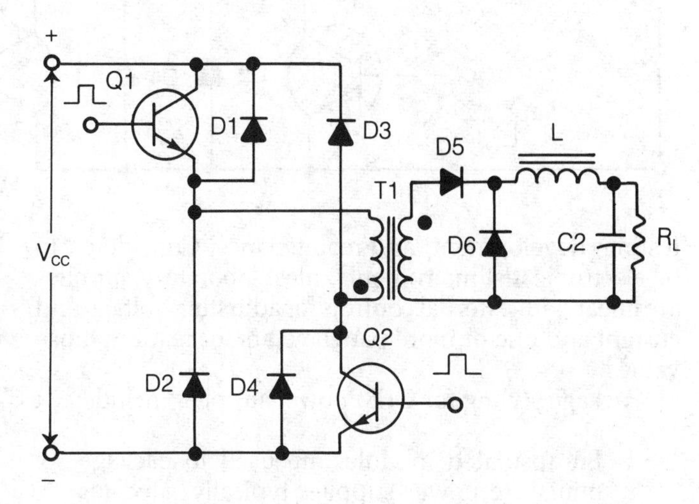

POWER SUPPLY: Fig. 6. Schematic for a half-bridge configuration, off-line switching power supply.

tifiers, and aluminum electrolytic capacitors are designed and made specifically for switchmode applications. *See* CAPACITOR, POWER TRANSISTOR, RECTIFIER, RECTIFIER BRIDGE.

Switching power supplies generate electromagnetic interference (EMI) that can adversely affect other electronic circuits in the vicinity. This problem is only partially solved with filters installed by the manufacturer. Many applications require additional filtering and shielding tailored to the host system.

Power supplies are classified by end user or application and packaging style. For example, a power supply can be commercial or military, standard or custom, or any combination of these. Commercial power supplies are made for the original equipment manufacturer (OEM) for resale as part of a host system. They can be custom designed or built as standard or catalog products. Military power supplies are generally custom designed and built, but standard, off-the-shelf units are available. The outputs of commercial supplies are set at the factory. The supplies may be completely or partially packaged at the factory.

Laboratory power supplies are fully enclosed bench-type or rack-mounted units capable of providing a range of output voltages and currents for circuit and product

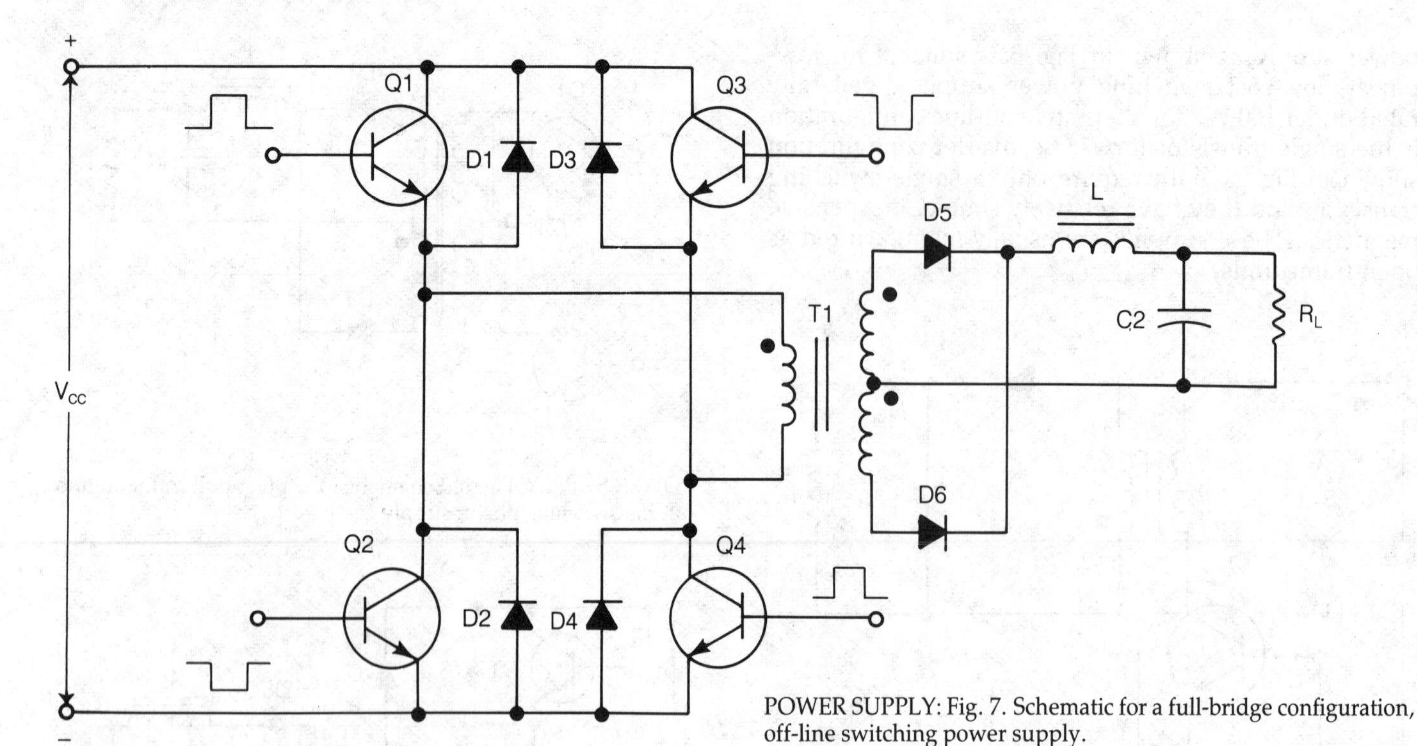

POWER SUPPLY: Fig. 7. Schematic for a full-bridge configuration, off-line switching power supply.

testing, development, and maintenance. Considered to be electronic test instruments, most laboratory supplies are linear with external controls for adjusting voltage and current and one or more panel meters for reading those values.

Package styling for OEM power supplies include:

1. Encapsulated modules are used to package miniature power supplies typically rated for 10 W output or less. The circuitry is potted or encapsulated within rigid rectangular plastic boxes.
2. Open frame packaging is used for supplies rated from 100 to 500 W. Components are assembled on an open circuit board for card cage installation or mounting in open L-shaped metal chassis for installation within an enclosure.
3. Enclosed modules are widely used to package power supplies rated for 500 W or higher. The supply is complete within a covered metal enclosure. The cover is usually pierced or has vents for ventilation while providing radio frequency shielding. Modular switching power supplies are equipped with emi/rfi filters, and many have fans or blowers for forced-air cooling.

POWER TRANSFORMER

See TRANSFORMER.

POWER TRANSISTOR

A power transistor is a transistor capable of handling 1 W or more of power or drawing 1 A or more of current during normal operation without damage to the device. Power transistors are used in a wide variety of control functions, including amplification, oscillation, switching, and frequency conversion. Three principal types of transistors qualify as power transistors: bipolar, Darlington, and MOSFET. Figures 1 and 2 compare the schematics, basic circuits, and typical geometries and structures for power bipolar and MOSFET transistors.

Bipolar Power Transistor. A bipolar power transistor is a three-terminal device as shown in the schematic, Fig. 1A. Like the small-signal bipolar transistor, the bipolar power is a minority carrier, current-controlled device with emitter, base, and collector terminals. Most power bipolar transistors have vertical geometries with the collector as the substrate and the collector metallization at the bottom.

In normal operation, the emitter-to-base junction is biased in the forward direction, while the collector-to-base junction is reverse biased in the NPN transistor shown in Fig. 1B. Conventional current must flow between the base and emitter terminals as shown by the direction of the arrow in Fig. 1A to produce a flow of current in the collector. (Electrons flow between the emitter and base terminals to produce a flow of electrons to the collector.) The amount of drive required to produce a given output also depends on the transistor's gain. Bipolar power transistors can be either NPN or PNP. Electrons flow from the emitter to the collector in npn bipolar transistors, but the direction is reversed in pnp transistors.

Bipolar power transistors are made as two parallel P-N junctions as shown in Fig. 1B and Fig. 1C. There is controlled spacing between the junctions and there are controlled impurity levels on both sides of each junction. Many different structures and geometries have been developed during power transistor evolution. Structure refers to the junction depth, the concentration, and profile of the impurities (doping) and to the spacing of the

A.

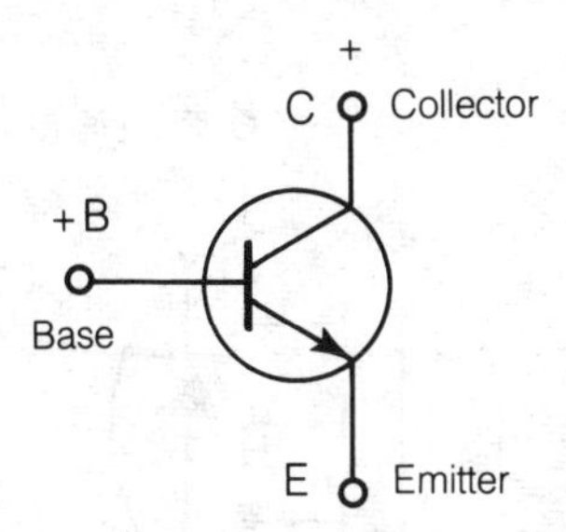

B.

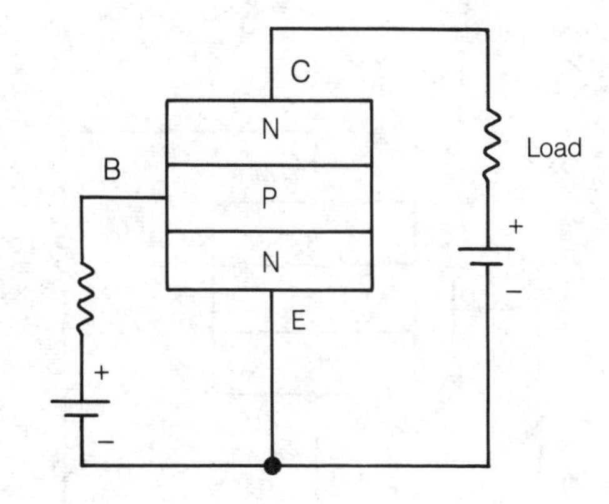

C.

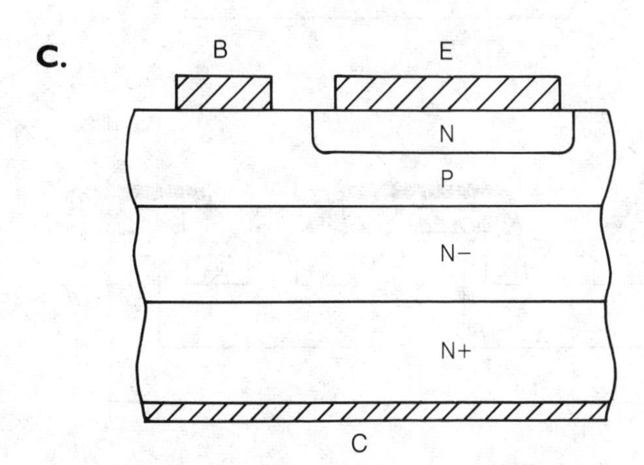

POWER TRANSISTOR: Fig. 1. Symbol for an NPN bipolar power transistor (A), basic circuit (B), and structure (C).

various levels of the device. Geometry refers to the topography of the transistor. The applications for power transistors are influenced by transistor structure, geometry and packaging, and the choices selected determine the gain, frequency, voltage, current, and dissipation of the power transistors.

Many different bipolar power transistor structures have been developed to provide a range of electrical and thermal characteristics at different price levels; each has advantages and disadvantages or compromises. Structures can generally be classed by the number of diffused layers, the use of an epitaxial base, or combinations of these. Bipolar power transistors may have a mesa or a planar structure. Some of the bipolar transistor structures are as follows:

1. Single-diffused (hometaxial).
2. Double-diffused (mesa, planar, epitaxial mesa, planar mesa, and multiple epitaxial mesa).
3. Triple-diffused (mesa and planar).
4. Epitaxial-base (mesa).
5. Multiple-epitaxial base (mesa).

A mesa is a raised portion of the chip with the emitter and base geometry in relief above the level of the silicon collector substrate. The mesa is formed by selectively etching away, by chemical means, all but the corners of a completed, double-diffused chip. A planar transistor is made in basically the same as the mesa version, but the collector-base junction terminates under a protective oxide layer at the surface. Power transistors made with these geometries have different voltage ratings, switching speed, saturation resistance, and leakage current. The most advanced bipolar switching transistors have multiple epitaxial, double-diffused structures.

Power bipolar transistors are specified with reference to the following parameters:

1. Voltage rating, collector to emitter.
2. Current rating of the collector.
3. Power rating.
4. Switching speed.
5. Direct-current gain.
6. Gain-bandwidth product.
7. Rise and fall times.
8. Safe operating area (SOA).
9. Thermal properties.

The popularity of the switching regulated power supply has created a heavy demand for power bipolar transistors capable of switching at high frequencies higher than 10 kHz. To be acceptable for this application, the transistors must be able to withstand voltage that is typically twice the input voltage. The transistor must also have collector current ratings and SOAs that are high enough for the intended application.

Safe operating area (SOA) defines the ability of the power transistor to sustain simultaneous high currents and voltages. It is shown by a SOA plotted on the collector current versus collector-to-emitter voltage axes. The curve defines, for both steady-state and pulsed operation, the voltage-current boundaries that result from the combined limitations imposed by voltage and current ratings, the maximum allowable dissipation, and the second breakdown capabilities of the transistor.

A bipolar transistor operated at high power densities is subject to a failure termed second breakdown, which occurs when a thermal hot spot forms within the transistor chip and the emitter-collector voltage drops 10 to 25 V. Unless power is quickly removed, current concentrates in a small region and temperatures rise until the transistor is degraded or destroyed.

Darlington Power Transistor. A Darlington power transistor consists of two bipolar power transistors on a single chip dc coupled internally as emitter followers. The device is packaged in a single case with three external leads. A Darlington configuration provides higher input resistance and more current gain than a single transistor.

Power MOSFET. A power MOSFET is a high-input impedance, voltage-controlled device because of its electrically isolated gate. Like the small-signal MOSFET, it is a majority-carrier device that stores no charge, so it can switch faster than a bipolar transistor. It has the same three electrodes: source, gate, and drain. The schematic for an N-channel, enhancement-mode MOSFET power transistor is shown in Fig. 2A. Unlike the small-signal transistor, most power MOSFETs are fabricated in a vertical geometry with the substrate as the drain and the gate and source formed on top.

Figure 2B illustrates the basic N-channel enhancement mode MOSFET circuit. Application of voltage between the gate and the source terminals enhances the conductivity and produces a flow of current in the drain. The gate is isolated from the source by a layer of silicon oxide. With no voltage applied between the gate and source electrodes, the impedance between the drain and source terminals is very high.

When a voltage is applied between the gate and source terminals an electric field is set up within the MOSFET. This field alters the resistance between the drain and the source terminals and permits conventional current to flow in the drain in response to the applied drain circuit voltage. (Electron flow is from the source to the drain.) There are also P-channel enhancement mode power MOSFETs, in which conventional current flows in the opposite direction.

Most power MOSFETs are made with the vertical double-diffused (DMOS) process as shown in Fig. 2C. This geometry and structure has largely replaced the V-groove or VMOS process widely used in the 1970s. Channels are formed by double diffusion at the periphery of each source cell as shown in the figure. An insulating gate oxide layer covers the channels, and polysilicon gates covers both the insulating oxide and channel. Each silicon gate, in turn, is insulated from the source by an additional oxide layer. All of the source cells are then parallel connected by a continuous deposition of aluminum metallization which forms the source terminal.

Multiple silicon gates of power MOSFETs are formed in basket-weave or hexagonal patterns on the top surface of the DMOS transistor chip, as shown in Fig. 3. The source cells consist of closed rectangular or hexagonal channels, which separate a source region from the substrate drain body. They are formed by an integration process and density may be more than a half million cells per square inch.

The widely used vertical DMOS process permits saving of as much as 60 percent of the silicon substrate over the requirements of the earlier planar MOSFET fabrication processes. The DMOS MOSFET contains an inherent P-N junction diode, and its equivalent circuit can be considered as a diode in parallel with an ideal MOSFET as shown in the schematic Fig. 2A. International Rectifier Corporation (IRC), uses hexagonal cells and calls its process and products HEXFET. Motorola Semiconductor, by contrast, uses rectangular cells, but has named its products TMOS because of the T-shaped flow of current in the device.

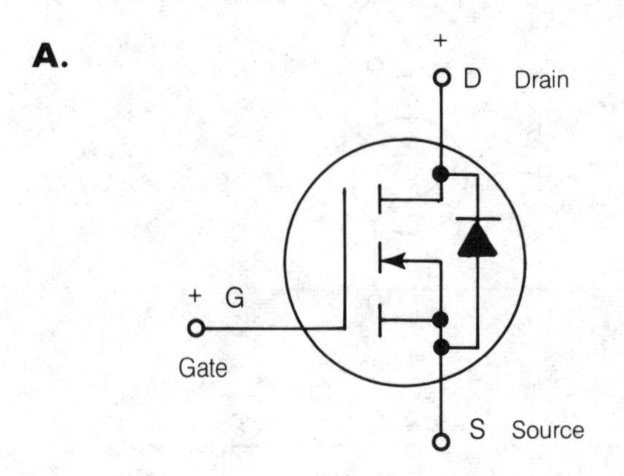

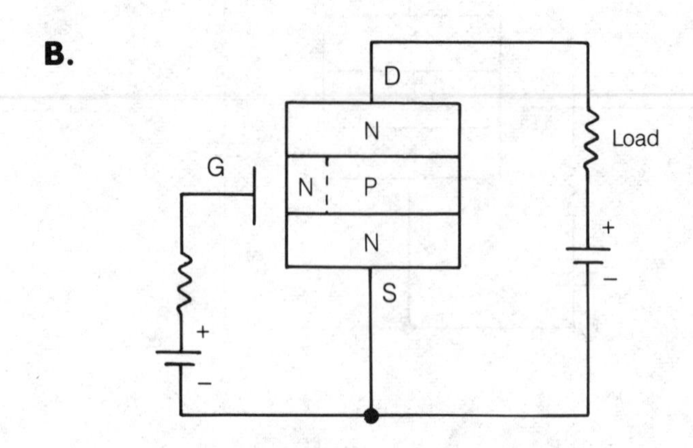

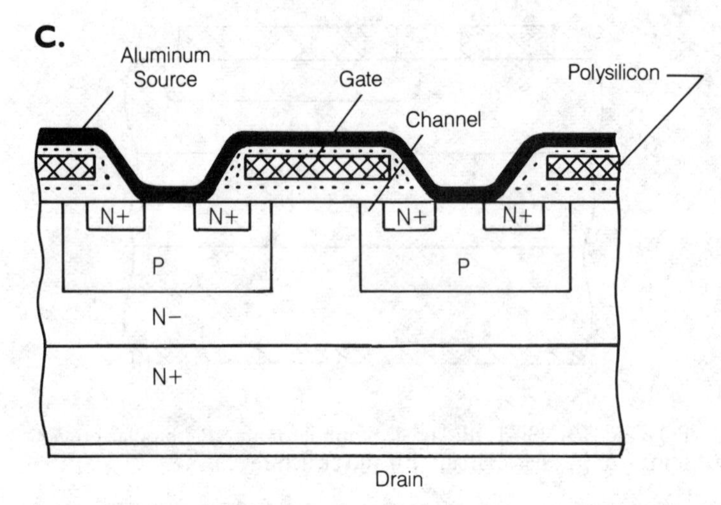

POWER TRANSISTOR: Fig. 2. Symbol for an N-channel power metal-oxide MOSFET (silicon field-effect transistor) (A), basic circuit (B), and structure (C).

Power MOSFETs are widely specified for high-frequency switching power supplies, chopper and inverter systems for dc and ac motor speed control, high-frequency generators for induction heating, ultrasonic generators, audio amplifiers and AM transmitters. Power MOSFETs have the following advantages over bipolar transistors:

1. Faster switching speeds and low switching losses.
2. Absence of second breakdown.
3. Wider safe operating area.
4. Higher input impedance.

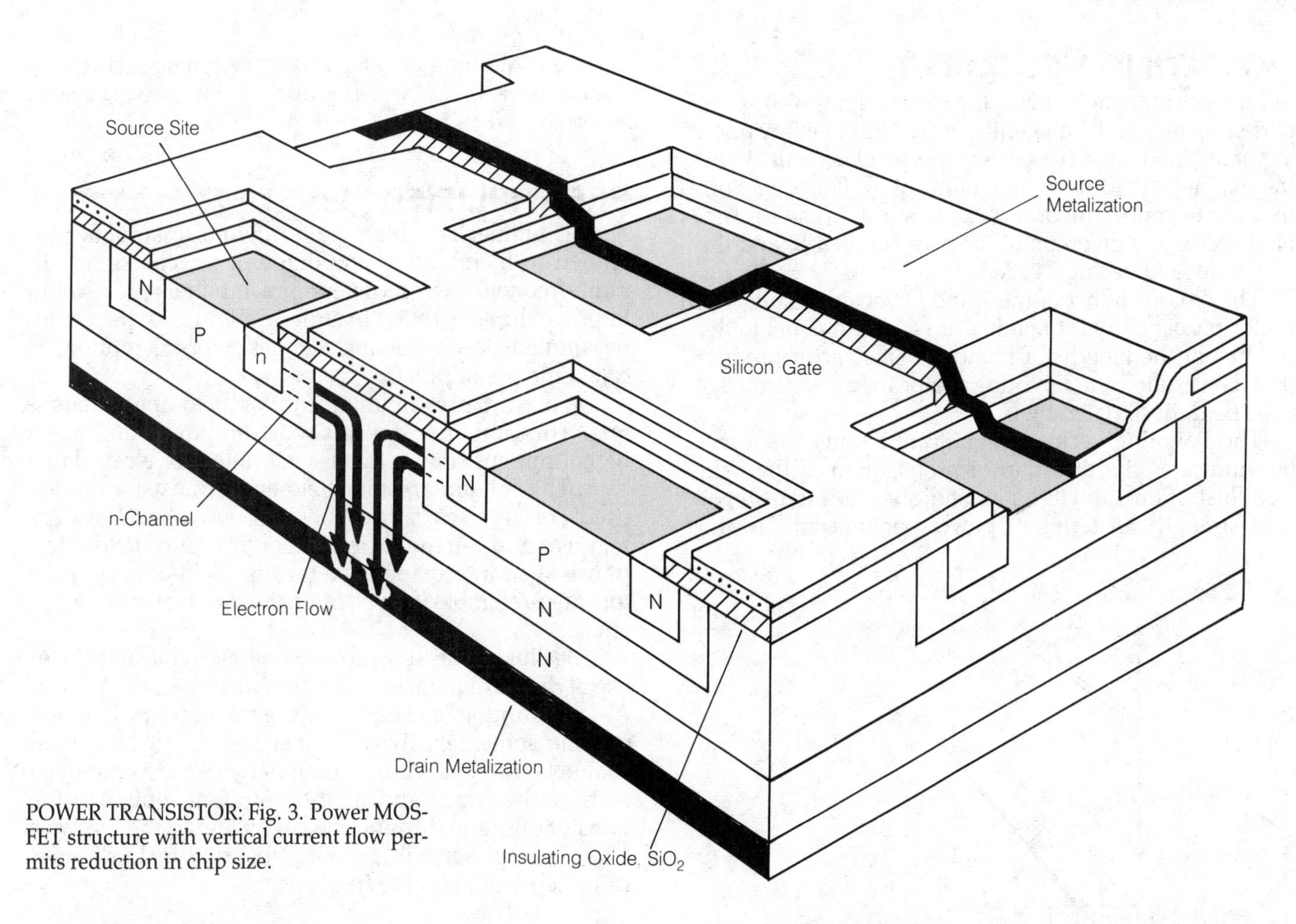

POWER TRANSISTOR: Fig. 3. Power MOSFET structure with vertical current flow permits reduction in chip size.

5. High if not higher gain.
6. Faster rise and fall times.
7. Simple drive circuitry.

The principal disadvantages of power MOSFETs have been higher cost and a higher static drain-to-source on-state resistance (typically 0.1 to 2.5 Ω) that caused power losses. Manufacturers have been trying to reduce this loss.

In evaluating a power MOSFET for applications the designer is most likely to consider: 1) speed, 2) SOA, 3) blocking voltage, 4) on-voltage, 5) thermal stability, and 6) paralleling.

A conductivity-modulated field-effect transistor (COMFET) is a variation on the basic MOSFET in which conductivity of the N-type epitaxial drain region is increased by the use of a P-type substrate on the drain side of a conventional N-channel power MOSFET. It is intended to overcome on-resistance increased with increasing drain-source voltage encountered in the standard MOSFET. The COMFET operates basically the same as the MOSFET and combines the characteristics of a power MOS transistor, a bipolar transistor, and a thyristor in the same device.

Power Transistor Packaging. Case selection for power transistors is strongly influenced by the device current ratings. The most popular cases for all power transistors rated 15 A or less are the plastic TO-218AC, TO-220AB, and TO-220AC packages. Metal TO-204AA and TO4AE (TO-3), TO-205AF, and TO-61 cases are also widely used. Power transistors are also sold as chips. Some manufacturers use their own proprietary variations of the industry-standard metal and plastic packages. *See* FIELD-EFFECT TRANSISTOR, METAL-OXIDE SEMICONDUCTOR FIELD-EFFECT TRANSISTOR, SEMICONDUCTOR, TRANSISTOR.

POWER TRANSMISSION

See POWER LINE, TRANSMISSION LINE.

POWER TUBE

A power tube is a vacuum tube designed for use in power amplifiers. A power tube can be a triode, tetrode, or pentode, or it can be a microwave tube.

Power tubes have large plates because the plate is where most of the dissipation occurs. The plate of a power tube may be finned, thus maximizing the heat-radiating surface area. A power tube with a plate-dissipation rating of more than a few watts is usually cooled by means of a fan or heat sink (*see* AIR COOLING, HEATSINK).

Power tubes can be used at frequencies from direct current well into the microwave range. At the low, medium, and high frequencies, single tubes can deliver several thousand watts of useful radio-frequency output power. Large power tubes are housed in room-sized enclosures, and are cooled by forced-air or liquid circulation systems.

Power tubes typically require high plate voltages.

POYNTING VECTOR

In an electromagnetic field, the electric lines of flux are perpendicular to the magentic lines of flux at every point in space (*see* ELECTROMAGNETIC FIELD). The electric field and the magnetic field have magnitude as well as direction; they can be represented as vector quantities. The electric-field vector at a given point in space is called E, and the magnetic-field vector is called H.

The cross product of the E and H vectors results in a vector perpendicular to both, with a length equal to the product of the lengths of E and H. This vector product (E × H) is called the Poynting vector for an electromagnetic field, designated by U.

The Poynting vector is significant because its direction indicates the direction of propagation of the field (see illustration), and its length indicates the intensity, or field strength, in terms of power per unit area in the surface containing the electric and magnetic vectors. *See also* ELECTRIC FIELD, ELECTRIC FLUX, FIELD STRENGTH, MAGNETIC FIELD, MAGNETIC FLUX.

POYNTING VECTOR: The Poynting vector is the cross product of electric-field vectors (E) and magnetic-field vectors (H).

PREAMPLIFIER

A preamplifier is a high-grain, low-noise amplifier intended to increase the amplitude of a weak signal. In radio-frequency (rf) receivers, preamplifiers are used to improve the sensitivity. In audio applications, the amplifier immediately following the pickup or microphone is called the preamplifier.

Modern rf preamplifiers employ field-effect transistors. This amplifier draws almost no power and has a high input impedance that is well suited for weak-signal gain. The gallium-arsenide field-effect transistor is often used (*see* FIELD-EFFECT TRANSISTOR, GALLIUM-ARSENIDE TRANSISTOR). For audio-frequencies, either bipolar or field-effect transistors can provide good results. Field-effect transistors are preferable if noise reduction is a major consideration.

The illustration at A shows a typical circuit suitable for use at radio frequencies. The amplifier operates in Class A. Input tuning is used to reduce the noise pickup and provide some selectivity against signals on unwanted channels. Impedance-matching networks are employed both at the input and at the output to optimize the transfer of signal through the circuit and to the receiver. This preamplifier will produce about 10 to 15 decibels gain, depending on the frequency.

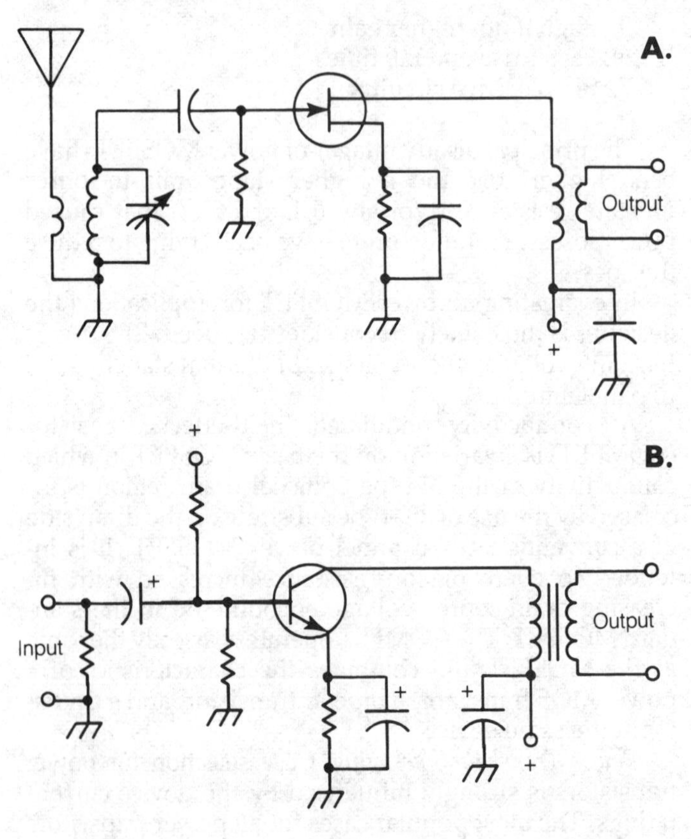

PREAMPLIFIER: Two examples of preamplifiers include an rf preamplifier using a FET (A) and an audio preamplifier using a bipolar transistor (B).

A preamplifier suitable for use at audio frequencies is shown at B. This circuit has a broadband response, since it is intended for high-fidelity receivers. The gain is essentially constant, at about 15 to 20 decibels, up to frequencies well beyond the range of human hearing.

In the design of any preamplifier, regardless of its intended frequencies of operation, linearity is important. Nonlinearity in a preamplifier results in intermodulation distortion at radio frequencies and harmonic distortion at audio frequencies. *See also* CLASS A AMPLIFIER, INTERMODULATION, LINEARITY.

PRECIPITATION STATIC

Precipitation static is a form of radio interference caused by electrically charged water droplets or ice crystals as they strike objects. The resulting discharge produces wideband noise that sounds like noise generated by electric motors, fluorescent lights, or other appliances.

Precipitation static is often observed in aircraft flying through clouds containing rain, snow, or sleet. But occasionally, precipitation static occurs in mobile radio installations. This is most likely to happen when it is snowing; then the noise is called snow static. Dust storms can also cause precipitation static. Precipitation static can be quite severe at times, making radio reception difficult, especially at the very low, low, and medium frequencies.

A noise blanker or limiter is effective in reducing interference caused by precipitation static. A means of facilitating discharge, such as an inductor between the antenna and ground, might be helpful. Improvement may also be obtained by blunting any sharp points in antenna elements. *See also* NOISE LIMITER.

PREEMPHASIS

Preemphasis is the deliberate accentuation of the higher audio frequencies in a frequency-modulated (FM) transmitter. Preemphasis is obtained in a transmitter by inserting a highpass filter in the audio stages. This highpass filter is shown in the illustration at A.

In a phase-modulated transmitter, the preemphasis occurs automatically because of the nature of phase modulation as compared with frequency modulation (*see* FREQUENCY MODULATION, PHASE MODULATION). The graph at B shows the attenuation-versus-frequency characteristic of a typical preemphasis network.

Preemphasis in an FM transmitter, in conjunction with deemphasis at the receiver, improves the signal-to-noise ratio at the upper end of the audio range. *See also* DEEMPHASIS.

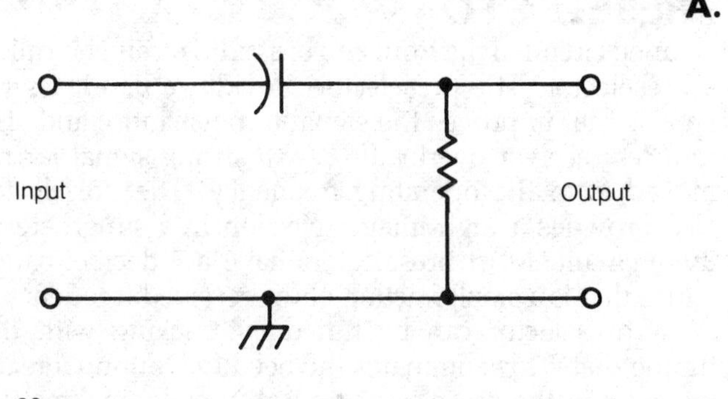

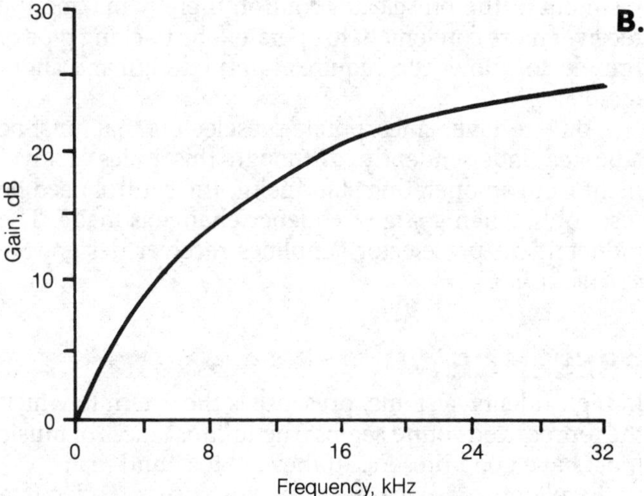

PREEMPHASIS: A simple network for introducing preemphasis is shown at (A) and a typical preemphasis curve is shown at (B).

PREFIX MULTIPLIERS

Physical units are often subdivided or multiplied for mathematical convenience. Special prefixes, known as prefix multipliers, indicate the constant factor by which the unit is changed.

Generally, prefix multipliers are given in increments of three, such as 10^3, 10^6, and so on for large units, and 10^{-3}, 10^{-6}, and so on for small quantities. However, intermediate values are sometimes given. The table shows the prefix multipliers used for factors ranging from 10^{-18} to 10^{18}.

As exponents become more negative, the unit size diminishes. As the exponents become more positive, the unit size increases. Each exponential increment represents one order of magnitude, or a factor of 10. *See also* ORDER OF MAGNITUDE, SCIENTIFIC NOTATION.

PREFIX MULTIPLIERS:
The most commonly used prefix multipliers.

Prefix	Abbreviation	Multiple
atto-	a	10^{-18}
femto-	f	10^{-15}
pico-	p	10^{-12}
nano-	n	10^{-9}
micro-	μ or u	10^{-6}
milli-	m	10^{-3}
centi-	c	10^{-2}
deci-	d	10^{-1}
deca-	da	10
hecto-	h	10^2
kilo-	k	10^3
mega-	M	10^6
giga-	G	10^9
tera-	T	10^{12}
peta-	P	10^{15}
exa-	E	10^{18}

PRESELECTOR

A tuned circuit in the front end of a radio receiver is called a preselector. The preselector provides a bandpass response that improves the signal-to-noise ratio, and also reduces receiver overloading by a strong signal far removed from the operating frequency. The preselector also provides a high image rejection in a superheterodyne circuit. Most preselectors have a 3 decibel bandwidth that is a small fraction of the received frequency.

A preselector can be tuned by tracking with the tuning dial. This eliminates the need for continual readjustment of the preselector control, thereby making the receiver more convenient to operate. The tracking type of preselector, however, requires careful design and alignment.

Some receivers incorporate preselectors that must be adjusted independently. Although this is less convenient from an operating standpoint, the control need be reset only when a large frequency change is made. The independent preselector simplifies receiver design. *See also* FRONT END.

PRESENCE

In high-fidelity systems, presence is the extent to which the reproduced sound seems true-to-life. Voices or music must have good presence if they are to sound real.

Excellent presence requires the use of adequate speakers or headphones and good amplifier design. The speakers or headphones must have a flat frequency response over the entire human hearing range; the same must be true of the amplifiers. The linearity and dynamic range of the system must also be excellent.

To a certain extent, a speaker/headphone frequency-response deficiency can be offset by adjustments in the amplifier circuit. An equalizer (also known as a graphic equalizer) can improve the presence of a marginal system. An equalizer may be valuable in any audio system. *See also* DYNAMIC RANGE, EQUALIZER, FLAT RESPONSE, HIGH FIDELITY, LINEARITY.

PRIMARY

See TRANSFORMER.

PRIMARY COLORS

The primary colors are the three colors that, when combined with equal intensity, result in white. These colors are red, blue, and green for radiant light. By mixing the three primary colors in various brightness combinations, all possible hues and saturation values can be obtained (*see* CHROMA, HUE, SATURATION).

For reflecting surfaces, colors are technically called pigments. The three primary pigments, when combined in equal concentration, result in no reflected light (black). The primary pigments are red, yellow, and blue.

In color television, primary-color combinations result in the various shades and hues. Many clusters or triads of phosphor dots on the picture-tube screen are activated to form all colors. One set of dots is red, one is blue, and one is green. Each set or triad contains red, green, and blue phosphor dots. *See also* CATHODE RAY TUBE, COLOR PICTURE SIGNAL, COLOR TELEVISION.

PRIMARY EMISSION

When electrons emanate directly from an object such as an electron gun or the cathode of a vacuum tube, the emission is said to be primary. Primary emission results from the application of a negative charge to an electrode. The greater the applied charge, assuming all other variables remain constant, the greater is the primary emission.

When an electrode is struck by high-speed particles, electrons may be knocked from it. This occurs, for example, in the plate of a vacuum tube or the dynode of a photomultiplier. This emission is called secondary emission. *See also* DYNODE, PHOTOMULTIPLIER TUBE, SECONDARY EMISSION.

PRINTED CIRCUIT BOARD

A printed circuit (PC) board or, more accurately, a printed wiring (PW) board, is a component manufactured from rigid base material upon which completely processed printed wiring has been formed. The terms *PC board* (PCB) and *PW board* (PWB) are used interchangeably to refer to rigid or flexible, single, double, and multilayer boards. A printed circuit or printed wiring assembly is a printed circuit or wiring board on which separately manufactured components and parts have been added. It may also be known as a board-level assembly.

PC boards are classified by the number of layers, as implied by the terms *single-sided*, *double-sided*, and *multilayer*. Multilayer boards may have internal conductors and internal ground or power planes, as well as conductors on both sides. The standard thickness for commercial, industrial and military PCBs, regardless of the number of layers, is about 0.0625 inch.

Most industrial and commercial PCBs today are manufactured by the subtractive process from rigid copper-clad epoxy-impregnated glass fiber laminate. The copper foil is removed selectively leaving the desired conductive paths and pads on the insulated base material. However, PC boards are also fabricated by the additive process, in which the desired conductive paths are selectively added by copper plating on a chemically prepared bare base material. Both processes usually include tin, lead-tin or gold plating steps. Many consumer product PCBs are fabricated from rigid phenolic-impregnated paper. The process may include screening resist patterns on the substrates rather than using photolithography.

The pattern or mask for the conductors and pads on the PCB, regardless of process, is produced by photographic reduction of a larger scale (perhaps 10 times) drawing (see the figure). Simple masks may be drawn or prepared manually, but complex masks are usually drawn on a CAD workstation and reproduced with an X-Y plotter. The master artwork is a 1:1 reduction used to

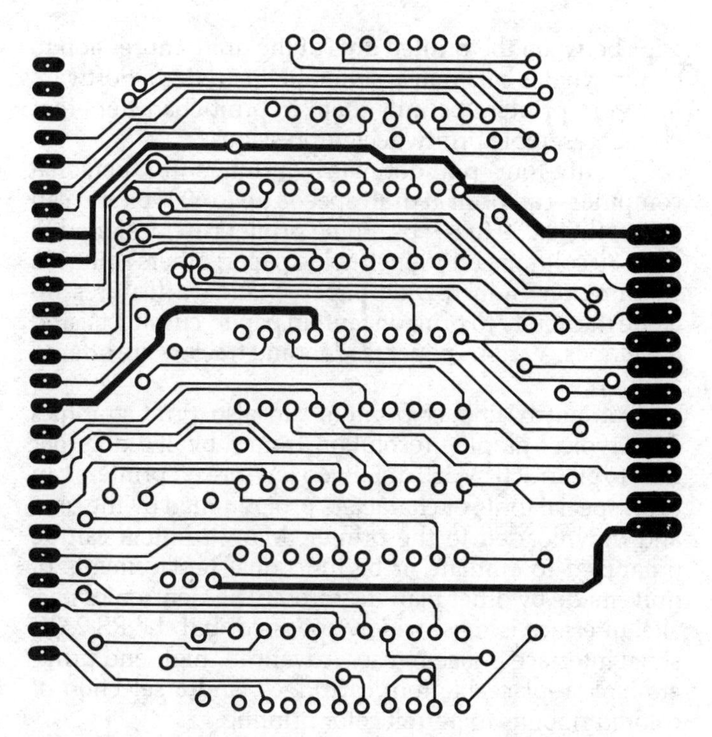

PRINTED CIRCUIT BOARD: The artwork for one conductive layer of a PC board showing the pin locations for components.

produce the production master or working mask. The mask, like a photographic negative, has opaque and transparent areas outlining the desired conductor pattern, and it can be positive or negative.

The intended conductive pattern on a positive mask is opaque to light, and the areas from which conductive material are to be removed are transparent. A negative mask has its conductive pattern transparent to light.

The production master can be used to prepare the fine-wire screens for the direct screening of resistive paints, a process that can be used on relatively simple PCBs with wide line widths. However, for fine-line definition, the production master is placed directly over the PCB blank coated with an ultraviolet (UV) light-sensitive resist. The resist is first applied as a liquid or dry film over the blank and covered with the master. Then both are exposed together. UV light passing through the transparent regions of the master hardens the resist so that the unexposed resist can easily be removed with a chemical solvent.

There are many possible variations in the process. In the positive process, the resist over the desired conductor pattern is exposed to UV light and hardened; the unexposed resist is first removed and the unwanted copper is removed by acid etching. After removal of the hardened resist, the bare copper conductors remain. In the negative process, the resist over the areas of unwanted copper is exposed to UV light and hardened. Then the unexposed resist over the desired conductor pattern is removed. The bare copper is lead-tin plated, and the hardened resist is removed. The plating can then be used as a mask for removing the unwanted bare copper.

Multilayer boards are made by forming conductive paths on partially cured board material. The individual layers of the multilayer board are then stacked and bonded under heat and pressure to complete the cure of the laminate and form a monolithic board.

In the additive process, the conductive lines and pads are first deposited by electroless deposition on the bare surface of the board. A plating solution that does not need the application of electric current is applied selectively. Copper adheres to the coating. In later steps, the thickness of the initial base can be increased with copper plating. As in the subtractive process, the conductors can be lead-tin plated or solder dipped to improve solderability.

Plated-through hole fabrication was an important advancement in the fabrication of double-sided and multilayer PCBs. Internal conductors can be interconnected with external conductors in any order with these plated vias or sleeves. Holes are formed or drilled at various points through the PCB where component leads are to be inserted. The inside walls of the holes are coated with copper by electroless deposition. The thin initial copper lining is reinforced with copper plating and may be lead-tin plated. Plated-through holes act as sockets for component leads. During wave soldering, molten solder is drawn up around the leads by capillary action to form a secure bond.

Plated-through holes eliminate the need to insert metal eyelets for layer interconnection, a costly, time consuming and unreliable process. However, in surface mounting the need for plated-through holes is eliminated because leadless components are soldered to the surface conductors and to the pads of the holeless, surface-mount PCB.

Edge-board contacts are formed at the same time as other conductors and component mounting pads if edge-board connectors are to be used. They are frequently gold plated to permit mating and removal of the connectors without damage to the contacts. These contacts are not needed with two-piece printed-circuit board connectors. *See* BACKPLANE, CIRCUIT BOARD OR CARD, SURFACE-MOUNT TECHNOLOGY.

PRINTER

In computer systems, the printer is a machine that prints output (data, calculations, text, and graphics) on paper. It is required for word processing, desktop publishing, and many kinds of computer-aided design tasks. The first computer printers printed out programs, data, and the contents of memory for computer operators and programmers. Recently there has been a shift away from this housekeeping and data-logging function to the use of computers for the final preparation of letters, reports, and reproducible copy. More emphasis has been placed on printing that matches that of the best electric typewriters or even the printing press.

The speed and print quality of even the lowest priced computer printers have been improved dramatically, and machines are available that meet wide user needs at all levels of machine performance and price. In addition to impact printing with black ribbon, multicolor ribbons are available. The alternatives to impact printing include thermal, ink-jet and laser printing. Printing is no longer confined to conventional ASCII characters because dot matrix printers can draw graphics under software control. *See* ASCII.

The first computer printers were commercial teleprinters adapted for use as computer terminals. Later, electric typewriters were adapted as computer printers, and finally new generations of printers were designed specifically for computers. They can be classed as: 1) formed character or printwheel, 2) dot matrix, and 3) line printers.

Printwheel. A printwheel printer has a printwheel with as many as 130 ASCII characters formed as type faces on the ends of the spokes of the wheel. The daisy wheel design is a widely used flat, spoked printwheel; however, other printwheels are cup or thimble shaped. Printwheels are positioned electrically for the serial printing of each individual character. The wheel is hammered against an inked ribbon to form each character on paper in the same way it is done by an electric typewriter. Printwheel printers suitable for use with personal computers can print 16 to 55 characters per second (cps), translating into a word rate of 160 to 550 words per minute (wpm). They are available with either parallel Centronics or serial RS-232C interfaces.

Dot Matrix. The dot-matrix printer is the most popular low-cost computer peripheral printer today because of its versatility and ability to print at different speeds and levels of quality. One model is shown in Fig. 1. These printers are suitable for use with both microcomputers and minicomputers. A dot-matrix printer forms approximations of characters with its computer-controlled cluster or matrix of printhead pins.

Dot-matrix printheads with from 9 to 24 pins print the ASCII characters as the printhead traverses the width of the paper and cogged wheels or friction rollers scroll the paper. Characters are printed as solenoids advance the appropriate pins in unison and hammer them against an inked ribbon onto the paper. Draft characters are formed in a single pass, but a second pass is needed to fill in the

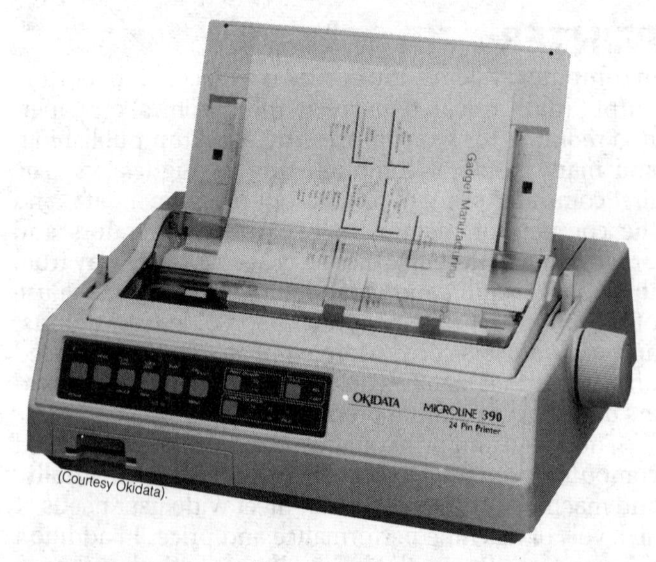

PRINTER: Fig. 1. Dot-matrix computer peripheral printer with 24-pin matrix printhead.

gaps between the printed dots of the draft impression to obtain what is called near-letter quality (NLQ). Software in 24-pin printers permits them to print characters that closely resemble printwheel impressions.

Twenty-four pin dot-matrix printers for personal computers can print at draft speeds up to 400 cps and can print NLQ at 30 to 80 cps. Some printers are designed to print 80 columns on 10-inch-wide paper; others print 136 columns on 16-inch-wide paper. Standard features include the ability to print on fanfold paper, cut sheets, and envelopes. Various paper feeders and tractors are offered as options.

Dot-matrix impact printers can also print graphics with typical graphic resolution from 8 by 480 dots per inch (dpi) to 360 by 360 dpi. Even low-priced printers can print special fonts or characters programmed by the user and downloaded to the printer. Many printers can be organized to emulate or be functional replacements of units made by other manufacturers. The Centronics parallel interface is most widely specified, but the RS-232C serial interface is also in use. Advanced, high-end printers have replaceable font cartridges, and a selection of colored ribbons to permit color printing.

Ink-Jet. The ink-jet printer draws ink from a reservoir and sprays it through a fine nozzle in the printhead to produce one or more columns of dots; these form the characters as the printhead traverses across the paper in a procedure similar to that of the dot-matrix printer. Ink-jet printheads can have a single column of nine nozzles or as many as three columns of eight nozzles. Less noisy than dot-matrix impact printers, they offer draft printing speeds from 160 to 176 cps and NLQ speeds from 32 to 106 cps.

Thermal. The thermal printer has a printhead made as a semiconductor chip etched with a matrix of points that can be selectively heated under computer control. Early models printed only on special chemically treated paper that discolored when heated. The hot points on the printhead burned dots on the paper to form characters as the printhead traversed over it. Later models did away with special paper by using a heat-sensitive ribbon. Points on the printhead melt ink on a special ribbon, forming characters from the dots as the printhead traverses plain paper. Alternatively, heat from the dots prepares the paper to accept ink from another type of ribbon. All thermal printers are dot-matrix printers and they make less noise than impact printers.

Laser Beam. The laser-beam printer is based on a dry electrostatic printing technique originally developed by Xerox Corp. for its copying machines. Figure 2 is a diagram of a laser-beam printer. The light beam source shown is a semiconductor laser now used in most of these machines. The laser is directly modulated or the light beam is modulated by an electro-optical modulator. The modulated beam is reflected from a facet of the polygon mirror rotated by a scanner motor and then reflected through a lens so it scans horizontally across the rotating photosensitive drum.

The drum is coated with selenium or other suitable photosensitive material so that it holds successive lines of tiny electrostatically charged dots corresponding to hor-

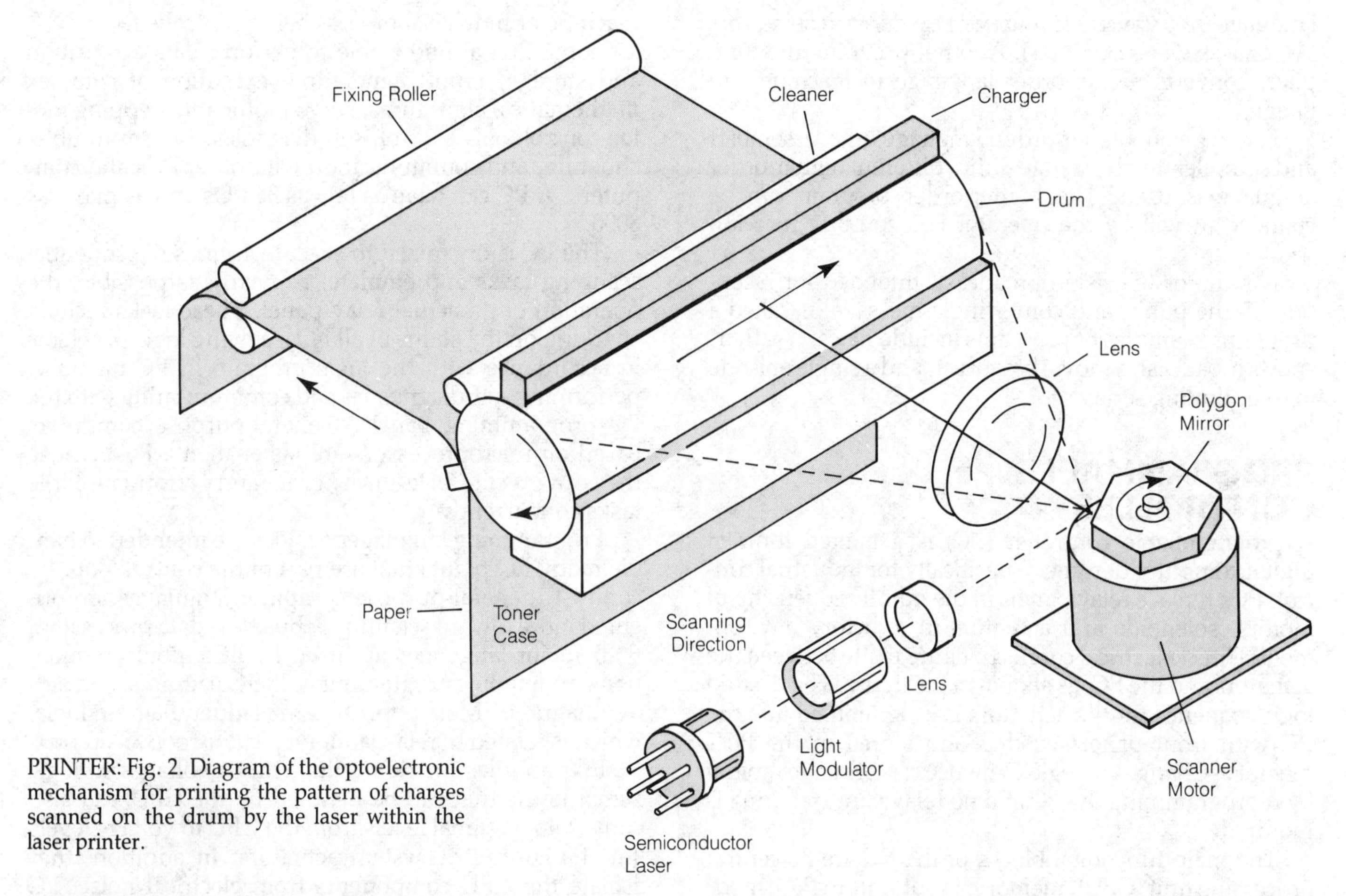

PRINTER: Fig. 2. Diagram of the optoelectronic mechanism for printing the pattern of charges scanned on the drum by the laser within the laser printer.

izontal slices of the text or graphics to be printed. (Eight to 10 horizontal scans can be required to reconstruct a legible line of text.) The drum passes continuously through the toner tray, and the opaque toner is attracted to the charged spots on the drum. Plain paper is then brought into contact with the drum, and the toner powder is transferred and heat fused to the paper to complete the printing of a page.

A wide range of laser printers is available in various price and performance ranges. Some models suitable for use with personal computers print at speeds up to 10 pages per minute. The internal assembly of electronics and optics is termed the engine. Laser-beam printers are rated in page-per-month capacities with claims being made of 20,000 pages per month. Laser-beam printers are offered with RS-232C serial and Centronics parallel interfaces, and some can be organized to emulate popular impact printers.

PROBABILITY

Probability is an expression of the mathematical likelihood that a given event will occur within a certain amount of time, or in a certain number of trials or tests. It is treated in the field of statistics.

PROBABLE ERROR

In any measurement, errors are likely to occur. The extent of the error depends on the accuracy of the test equipment, and on the precision of the calibration standard. If a measurement is repeated several times, the error will generally differ each time. The probable error is defined as the median error that is expected to occur.

PROBE

A probe is a device coupled or connected to a test instrument for picking a signal from a specific point in a circuit for test and measurement. The probe normally does not disturb the operation of the circuit.

Probes can be used to measure both alternating- and direct-current voltages and currents and they are used with ohmmeters to determine the resistance between two points in a circuit. Probes can also be used with oscilloscopes to determine the voltage, waveform characteristics, frequency, and phase of an alternating-current signal. Oscilloscope probes are also used in testing and troubleshooting digital equipment.

Most probes have extremely high impedance. An insulated handle prevents shock to personnel and reduces the effects of hand capacitance.

PROGRAM

A program is a set of instructions that the computer is directed to carry out in a specified sequence. The program sets down the procedure the computer is to follow. One program, called the assembler, is used to derive the machine language of the computer from the assembly language (*see* ASSEMBLER AND ASSEMBLY LANGUAGE, HIGHER-ORDER LANGUAGE, MACHINE LANGUAGE). Another program, the compiler, converts higher-order language to machine language.

For a given higher-order language, the assembler and compiler do not change until a different higher-order language is used. The higher-order program can be changed at will by the operator (*see* COMPUTER PROGRAMMING).

A radio or television broadcast, intended for reception by the public and conveying a message, is called a program. Examples of programs include news, weather, sports broadcast, a movie broadcast, and a single episode in a continuing series.

PROGRAMMABLE CONTROLLER

A programmable controller (PC) is a limited form of digital computer designed specifically for industrial control. PCs replace relay panels in the on-off sequencing of motors, solenoids and actuators in a factory environment. Direct electrical connections are made between the actuators and the PC, as shown in the figure. The desired logic sequence for the actuators is programmed into the PC with front panel switches and stored in the PC's memory. Changes in logic sequence can be made rapidly by reprogramming the PC and no relay panel rewiring is required.

The main functional blocks of the PC are its central processing unit (CPU), memory, input/output (I/O) modules, and power supply. The control logic program stored in the memory directs the CPU selection of events and their logical sequence. For example, if switch A is closed, then solenoid B will be energized. The CPU scans all of the I/Os in a fraction of a second, and records the status of each input and output in memory.

Necessary control actions are initiated after the instructions are executed and the results are scanned. The logic determining new output states is solved after the I/Os are scanned. Scanning is repeated frequently so that any changes in inputs and outputs can be detected by the CPU. Inputs are signals from sensors and switches in the control loop, and outputs are signals that actuate the machine or industrial process being controlled.

Some PCs are now able to perform data acquisition and storage, report generation, execution of complex mathematical algorithms, servo motor and stepping motor control, axis control, self diagnosis, system troubleshooting, and communication with other PCs and computers. A PC can have as few as 32 I/Os and as many as 8000.

The PC is organized to execute a series of sequential scanning tasks and emulate, as nearly as possible, the operation of an actual relay panel. These tasks include checking on the status of all I/Os, solving logic problems in accordance with the logic program in PC memory, performing self-diagnostics and communicating with the PC programming panels. General-purpose computers based on microprocessors are faster than a PC because they can execute tasks in any order and perform multiple tasks concurrently.

Programming languages for PCs are intended to handle many I/O points that are part of the control loop. By contrast, general-purpose computer languages are oriented more toward scientific or business data processing. Four major languages are used by PCs: Boolean equations, mnemonic programming, logic diagrams, and ladder diagrams. Most popular is the ladder diagram logic, which is related to relay ladder logic. *See* BOOLEAN ALGEBRA.

I/O modules for PCs reduce input voltages to logic signal levels that can be handled by the CPU and also convert logic signal levels from the CPU to voltage levels for the control of system actuators. In addition, they isolate the CPU components from electrical noise. I/O modules accept inputs from limit switches, pushbuttons, strain gauges, thermocouples, and other sensors. They also provide ASCII serial interfacing. Some I/O modules now include microcontrollers to preprocess information before it reaches the PC's CPU. This increases the speed of signal processing over that which could be accomplished by the PC's CPU.

PCs ordinarily give only on-off commands that cannot directly control machine speed or position. However, if the scan rate of the PC is fast enough for motion control, the machine can be equipped with digital position sensors, actuators, and control panels. With this equipment, PCs can perform motion control in either

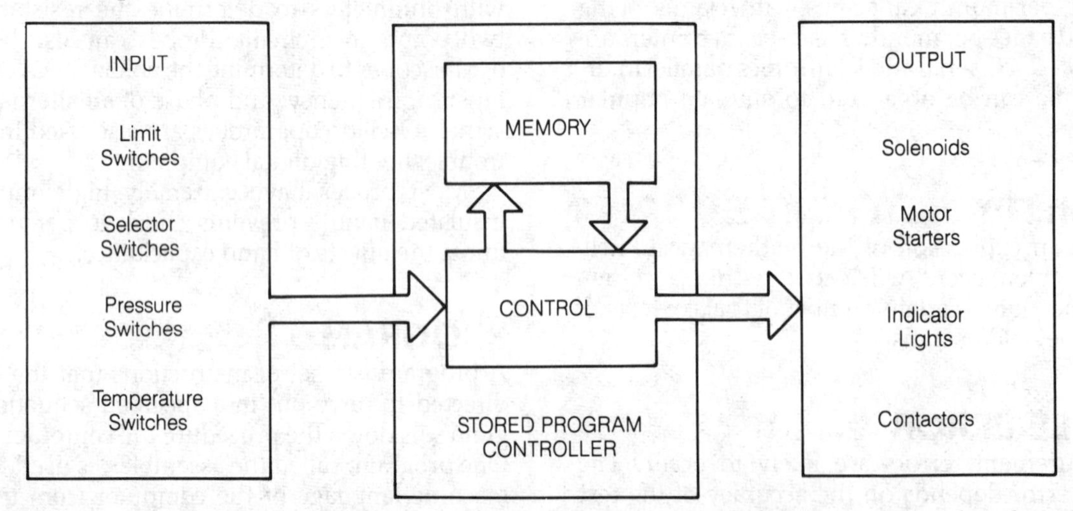

PROGRAMMABLE CONTROLLER: Input data is processed by the stored program controller to actuate selected output devices.

closed-loop feedback systems or open-loop positioning systems. In a closed-loop feedback system, the PC can control the movement of a sliding part of a machine. For example, command signals to direct the part to move a specified distance may be entered digitally into the PC by a keyboard or preset program. A digital-to-analog (D/A) converter then sends an analog signal to an electric gearmotor drive. The gear motor then turns a leadscrew to move the machine element. When the part reaches the desired distance, a resolver, also geared to the leadscrew, sends a feedback signal to stop the motor at the position specified.

Similarly, open-loop control can be achieved with a PC and a stepping-motor module that can control the motion of a stepping motor. The module issues commands in the form of pulses that actuate the stepping motor. The PC sends data and motion commands to the module and compares actual and programmed positions. If the positions are not the same, the PC issues a new command to the motor to adjust the actual position of the part.

The increase of automation in industry has created a need for machines that communicate with each other by means of plantwide communications networks. Available networks link PCs with each other and with minicomputers, mainframe computers, graphics-display terminals, line printers and video display terminals (VDTs).

PROGRAMMING LANGUAGE

Programming languages for computers are easier to use and are more versatile than machine language. Use of these languages require additional software, either a compiler or an interpreter. A compiler translates the entire program into a machine language version to be executed at a later time. An interpreter translates and executes the program line by line.

The most popular programming language for personal computers is BASIC (Beginner's All-purpose Symbolic Instruction Code). Other high-level programming languages are ADA, APL, C, COBOL, FORTRAN, and PASCAL.

PROGRAMMABLE LOGIC DEVICE

A programmable logic device (PLD) is an unstructured array of AND and OR logic gates that can be organized to perform dedicated logic functions by selectively opening or otherwise altering the interconnections between the gates. Alteration's can be accomplished by "burning out" fusible links or reducing the conductivity of their interconnections by selectively applying overvoltage. PLDs can meet the requirements of a custom-designed logic device with the off-the-shelf availability of standard logic integrated circuits (ICs).

Popular PLDs are available from many sources and are relatively low in cost because they are made in volume. PLD technology is an extension of concepts originally developed for field-programmable read-only memory (PROM). A PROM may be programmed to perform some logic functions. *See* SEMICONDUCTOR MEMORY.

PLDs are now manufactured with bipolar transistor-transistor logic (TTL) and emitter-coupled logic (ECL), as well as three different CMOS technologies: fuse programmable, ultraviolet erasable, and electrically erasable. *See* COMPLEMENTARY METAL-OXIDE SEMICONDUCTOR, DIGITAL-LOGIC INTEGRATED CIRCUIT.

PLDs are best known by their proprietary names, PAL and FPLA. The PAL (programmable logic array) was developed by Monolithic Memories Inc. (MMI, now a subsidiary of Advanced Micro Devices, AMD). The FPLA (field-programmable array) was developed by Signetics Inc. Altera Corporation refers to its CMOS PLDs as ELPDs (electrically programmable devices). The use of the generic term *PLD* avoids reference to these proprietary products in general discussions. Other terms such as *integrated fuse logic* (IFL) and *fuse programmable logic* (FPL) are used interchangeably with PLD.

The first PLDs (PALs and FPLAs) were introduced primarily as replacements for small-scale integration (SSI) logic devices and had from 4 to 100 gates. The density of both these classes of PLD has been increased to 100 to 500 gates, or medium-scale integration (MSI). These products replace four or more conventional logic ICs, thus conserving circuit board space and reducing power consumption.

The latest CMOS large-scale PLDs now have 1200 to 2000 logic gates, more than some low-density factory-programmed gate arrays, another form of semi-custom logic device. (*See* GATE ARRAY.) The choice of a PLD vs. gate array is determined by many considerations, including the quantity of devices to be ordered, critical delivery requirements, performance, and price.

PLDs are designed to be electrically programmed with commercial benchtop equipment originally developed to program PROMs. This equipment can be modified with plug-in "personality cards" or circuit boards with circuits capable of providing the voltages specified for blowing the fuses or altering the interconnection conductivity of each of the commercial PLDs.

However, before the PLD can be programmed, the logic must be designed. Simple PLD programming can be performed with Boolean algebra, but this method becomes increasingly difficult as the number of gates increases. It is now usually accomplished with appropriate software running in a personal computer or workstation. The availability of this computer software as a design aid has been important in the commercial success of PLDs, as it has been for other programmable devices. It will be crucial to the success of all large-scale PLDs in the future. The convenience of user-programmed PLDs makes them competitive with gate arrays.

PLAs and PALs have internal structures similar to those of PROMs. All three devices have the same basic internal AND/OR structures, but there are differences in the flexibility and programmability of these arrays. The AND/OR structure consists of the first-level AND structure that accepts the inputs, performs the desired AND functions on the inputs, and then outputs these functions to the second level, the OR array. The OR array combines various AND functions, producing the desired (AND/OR) outputs. This structure permits PLDs to carry out Boolean sum-of-product logic.

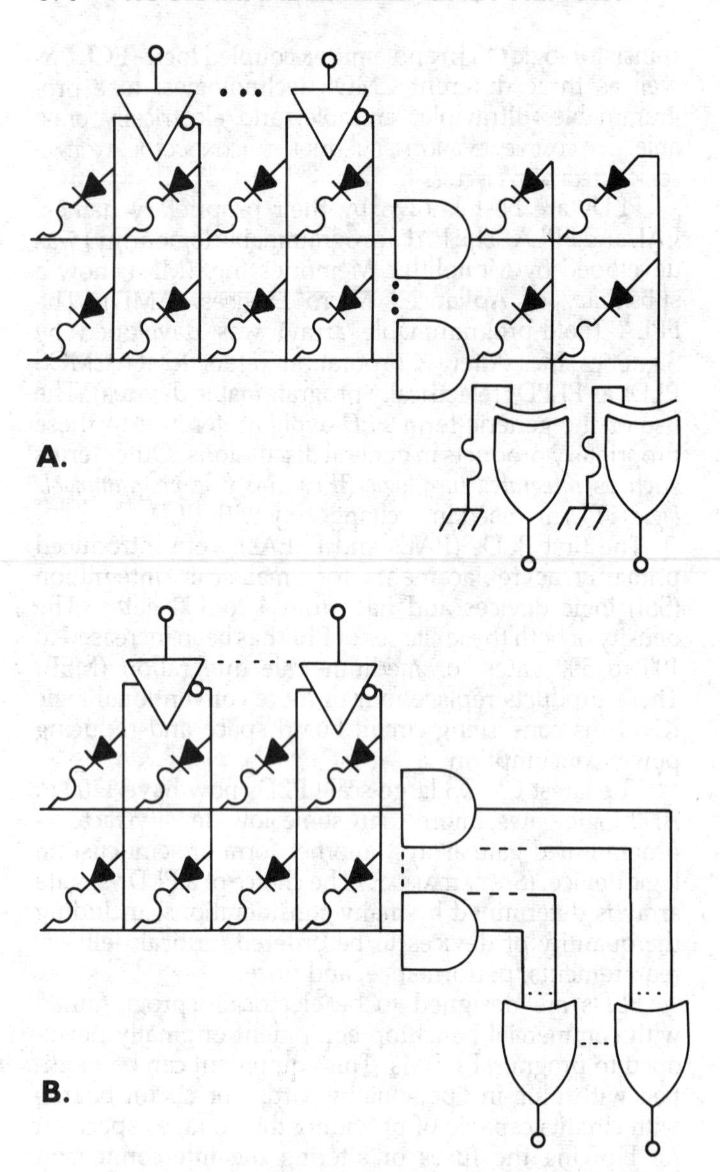

PROGRAMMABLE LOGIC DEVICE: Programmable logic array (PLA), also called integrated fuse logic (IFL), has a programmable AND array followed by programmable OR array (A). Programmable array logic (PAL) has a programmable AND array followed by a fixed OR array (B).

The PROM has a fixed AND array and a programmable OR array; the PAL has a programmable AND array and a fixed OR array; and the PLA has both programmable AND and OR arrays. The programmable AND array allows the PAL to solve equations with many inputs and architectural features—programmable input/output, internally registered feedback, and choice of output polarity. By contrast, FPLAs with both programmable AND and OR arrays have more flexibility than PALs. However, the programmable AND arrays in both PAL and FPLAs architectures permit them to overcome the limitations of PROMs in programming logic.

FPLAs were developed before PALs. However, because PALs have simpler structures and are easier to program (their OR arrays are fixed), they are accepted before PALs. The availability of computer software for computer-aided design was also a factor in the PAL's acceptance. The popularity of PAL architecture encouraged independent software designers to provide better support for PALs, which further increased their popularity and encouraged alternate sources for them.

The original TTL FPLAs had 60- to 70-nanosecond (ns) input-to-output propagation delays (time for the input to alter the output), or speed. In contrast, PALs had propagation delays of only 30 to 40 ns. However, as result of improvements, typical propagation delays for TTL versions of both devices are now in the range of 25 to 35 ns.

A popular PAL, the 16L8, has 300 logic gates and is packaged in a 20-pin dual-in-line (DIP) package. Because its OR array is not programmable, the gates are organized so that eight gates are connected to each output. By contrast, the pin-for-pin interchangeable FPLA, the 82S153, has fewer gates but a more flexible programmable OR array, which permits any gate to be connected to any output.

In this architectural tradeoff, programmable logic gates require more chip area than nonprogrammable logic gates. As a result, an FPLA chip of equivalent gate density is larger than a PAL chip. If the designer is able to accept the simpler structure, the PAL is favored; however, if greater flexibility is needed, the FPLA is favored.

The fabrication processing for all PLDs involves some tradeoff between speed (propagation delay) and power (supply current). Both PALs and FPLAs initially were made in bipolar TTL technology, and TTL PLDs are still sold in volume today. PLDs subsequently were made in faster bipolar ECL. ECL PLDs have shorter propagation delays—typically less than 10 ns—but they require a higher supply current than their TTL equivalents.

CMOS PLDs draw less current than do TTL PLDs, but they are slower than the fastest bipolar products. The principal advantage of CMOS in applications where speed is not crucial is their higher gate density and lower power requirement. At present, CMOS PLDs with up to 2000 usable gates are available. This gate count is larger than that of many factory-programmable gate arrays. Some CMOS PLDs are programmed by blowing fuses (as in bipolar PROMs) and others are programmed electrically.

Electrically programmable CMOS devices can be programmed and erased, permitting 100 percent functional testing. Erasable PLDs can be reprogrammed without the need to replace each device. Some PLDs can be erased with ultraviolet (UV) light, like EPROMs through quartz windows in their ceramic packages. Alternate one-time programmable (OTP) versions in plastic packages are also available for use where the design is fixed. Other CMOS PLDs that can be erased electrically, such as EEPROMs, are also packaged in plastic cases.

PLDs may have a variety of output structures: combinatorial, registered, and programmable macrocells. *See also* BOOLEAN ALGEBRA.

PROGRAMMABLE READ-ONLY MEMORY

See SEMICONDUCTOR MEMORY.

PROGRAMMABLE UNIJUNCTION TRANSISTOR

A programmable unijunction transistor (PUT) is a silicon-controlled rectifier (SCR) type device used in a particular way to simulate a unijunction transistor (UJT). A PNPN device, the PUT has a four-layer construction. The gate is one of the four layers. An external pair of series resistors can be used to set the device firing voltage. The device will trigger on when the anode becomes positive with respect to the gate. *See also* SILICON-CONTROLLED RECTIFIER, UNIJUNCTION TRANSISTOR.

PROGRAMMING

See COMPUTER PROGRAMMING, PROGRAM.

PROM

See SEMICONDUCTOR MEMORY.

PROPAGATION

Propagation is the transfer of energy through a medium or through space. Certain disturbances, such as sound waves or electric currents, can propagate only through a material. Electromagnetic fields can propagate through empty space (*see* ELECTRIC FIELD, ELECTROMAGNETIC FIELD, MAGNETIC FIELD).

Electromagnetic-wave propagation occurs in straight lines through a perfect vacuum in the absence of intervening forces or effects. In and around the atmosphere of the earth, however, electromagnetic waves are often propagated in bent paths. *See* PROPAGATION CHARACTERISTICS.

PROPAGATION CHARACTERISTICS

Radio waves are affected in various ways as they travel through the atmosphere of the earth. The effects vary with the wavelength. The troposphere, or lowest part of the atmosphere, cause electromagnetic waves to be bent or scattered at some frequencies. The ionized layers (*see* IONOSPHERE) affect radio waves at other frequencies.

A general description of propagation characteristics for various electromagnetic frequencies follows.

Very Low Frequencies (Below 30 kHz). At very low frequencies (VLF), propagation takes place mainly by waveguide effect between the earth and the ionosphere. Surface-wave propagation also occurs for considerable distances. With high-power transmitters and large antennas, communication can take place over distances of several thousand miles at frequencies between 10 and 30 kHz. The earth ionosphere waveguide has a low cutoff frequency—slightly below 10 kHz. For this reason, signals much below 10 kHz suffer severe attenuation and propagate over very short distances.

Antennas for very low frequency transmission must be vertically polarized. Otherwise, surface-wave propagation will not occur and the proximity of the ground tends to short-circuit the electromagnetic field.

Propagation at very low frequencies is remarkably stable; there is very little fading. Solar flares occasionally disrupt communication in this frequency range by making the ionosphere absorptive and by raising the cutoff frequency of the earth-ionosphere waveguide. *See* SOLAR FLARE, SURFACE WAVE.

Low Frequencies (30 kHz to 300 kHz). In the low-frequency (LF) range, propagation takes place in the surface-wave mode, and also as a result of ionospheric effects.

At the lower end of this band, wave propagation is similar to that in the very low frequency range. As the frequency is increased, surface-wave attenuation increases. While a surface-wave range of more than 3,000 miles is common at 30 kHz, it is unusual for the range to be greater than a few hundred miles at 300 kHz. Surface-wave propagation requires that the electromagnetic field be vertically polarized.

Ionospheric propagation at low frequencies usually occurs through the E layer. This increases the useful range during the nighttime hours, especially at the upper end of the band. Intercontinental communication is possible with high-power transmitters.

Solar flares can disrupt communication at low frequencies. After a flare, the D layer becomes highly absorptive, preventing ionospheric communication. *See* D LAYER, E LAYER, SOLAR FLARE, SURFACE WAVE.

Medium Frequencies (300 kHz to 3 MHz). Propagation at medium frequencies (MF) occurs by means of the surface wave, and by E-layer and F-layer ionospheric modes.

Near the lower end of the band, surface-wave communications path lengths can be up to several hundred miles. As the frequency is increased, the surface-wave attenuation increases. At 3 MHz, the range of the surface wave is limited to about 150 or 200 miles.

Ionospheric propagation at medium frequencies is usually not possible during the daylight hours because the D layer prevents electromagnetic waves from reaching the higher E and F layers. During the night, ionospheric propagation takes place mostly through the E layer in the lower portion of the band, and primarily through the F layer in the upper part of the band. The communications range increases as the frequency increases. At 3 MHz, worldwide communication is possible.

As at other frequencies, medium-frequency propagation is severely affected by solar flares. The 11-year sunspot cycle and the season of the year also affect propagation at medium frequencies. Propagation is usually better in the winter than in the summer. *See* D LAYER, E LAYER, F LAYER, SOLAR FLARE, SUNSPOT CYCLE, SURFACE WAVE.

High Frequencies (3 MHz to 30 MHz). Propagation at the high frequencies (HF) exhibits widely variable characteristics. Effects are much different in the lower part of this band than in the upper part. The lower portion is considered to be the range of 3 to 10 MHz, and the upper portion is the range of 10 to 30 MHz.

Some surface-wave propagation occurs in the lower part of the high-frequency band. At 3 MHz, the maximum range is 150 to 200 miles; at 10 MHz it decreases to about the radio-horizon distance, perhaps 15 miles.

Above 10 MHz, surface-wave propagation is essentially nonexistent.

Ionospheric communication occurs mainly through the F layer. In the lower part of the band, there is very little daytime ionospheric propagation because of D-layer absorption; at night, worldwide communication is possible. In the upper part of the band, this situation is reversed; ionospheric communication can occur on a worldwide scale during the daylight hours, but the maximum usable frequency often drops below 10 MHz at night. The transition is gradual, and the range in which it occurs is variable.

Communication in the lower part of the high-frequency band are generally better during the winter months than during the summer months. In the upper part of the band, this situation is reversed.

Some E-layer propagation is occasionally observed in the upper part of the high-frequency band. This is usually of the sporadic-E type. This may occur even when the F-layer maximum usable frequency is below the communication frequency.

Near 30 MHz, there is often no ionospheric communication. This is especially true when the sunspot cycle is at or near a minimum. The extreme upper portion of the high-frequency band then behaves similarly to the very high frequencies.

Solar flares cause dramatic changes in conditions in the high-frequency band. Sometimes ionospheric communications are almost totally lost within a matter of minutes by the effects of a solar flare. *See* D LAYER, E LAYER, F LAYER, MAXIMUM USABLE FREQUENCY, SOLAR FLARE, SPORADIC-E PROPAGATION, SUNSPOT CYCLE, SURFACE WAVE.

Very High Frequencies (30 MHz to 300 MHz). Propagation in the very high frequency (VHF) band occurs along the line-of-sight path, or via tropospheric modes. Ionospheric F-layer propagation is rarely observed, although it can occur at times of sunspot maxima at frequencies as high as about 70 MHz. Sporadic-E propagation is fairly common, and can occur up to approximately 200 MHz.

Tropospheric propagation can occur in three different ways: bending, ducting, and scattering. Communications range varies, but can often be had at distances of several hundred miles.

Meteor-scatter and auroral propagation can sometimes be observed in the very high frequency band. The range for meteor scatter is typically a few hundred miles; auroral propagation can sometimes produce communications over distances of up to about 1,500 miles. Code transmission must usually be employed for auroral communications; the moving auroral curtains introduce phase modulation that makes voices unintelligible.

Repeaters are used extensively at very high frequencies to extend the range of mobile communications equipment. A repeater at a high elevation can provide coverage over an area of thousands of square miles.

At the very high frequencies, satellites are widely used to provide worldwide communication on a reliable basis. The satellites are actually orbiting repeaters. *See* AURORAL PROPAGATION, E LAYER, F LAYER, REPEATER, SATELLITE COMMUNICATIONS, TROPOSPHERIC PROPAGATION, TROPOSPHERIC-SCATTER PROPAGATION.

Ultrahigh Frequencies (300 MHz to 3 GHz). Propagation in the ultrahigh frequency (UHF) band occurs almost exclusively over the line-of-sight distances and by satellites and repeaters. No ionospheric effects are ever observed. Auroral, meteor-scatter, and tropospheric propagation are occasionally observed in the lower portion of this band. Ducting can result in propagation over distances of several hundred miles.

The main feature of the ultrahigh-frequency band is its width—2,700 megahertz of spectrum. Another advantage of this band is its relative immunity to the effects of solar flares. Because ultrahigh frequency energy does not interact with the ionosphere, disruptions in the ionosphere have no effect. However, an intense solar flare can interfere with ultrahigh frequency circuits because of electromagnetic effects caused by fluctuations in the magnetic field of the earth. *See* AURORAL PROPAGATION, LINE-OF-SIGHT COMMUNICATION, REPEATER, SATELLITE COMMUNICATIONS, TROPOSPHERIC PROPAGATION, TROPOSPHERIC-SCATTER PROPAGATION.

Super High Frequency Microwaves (Above 3 GHz). In these bands, energy travels essentially in straight lines through the atmosphere. Microwaves are affected very little by temperature inversions and scattering, and are unaffected by the ionosphere.

The primary mode of propagation in the microwave range is the line-of-sight mode. This facilitates repeater and satellite communications.

The super high frequency (SHF) and higher extremely high frequency (EHF) bands are much wider than the ultrahigh frequency band. The upper limit of the microwave range, in theory, is the infrared spectrum. Transmitters capable of operation at frequencies higher than 300 GHz have been developed.

At some microwave frequencies, atmospheric attenuation becomes significant. Rain, fog, and other weather effects cause changes in the path attenuation as the wavelength approaches the diameter of water droplets. *See* LINE-OF-SIGHT COMMUNICATION, REPEATER, SATELLITE COMMUNICATIONS.

Infrared, Visible Light, Ultraviolet, X Rays, and Gamma Rays. Propagation at infrared and shorter wavelengths occurs in straight lines. The only factor that varies is the atmospheric attenuation.

Water in the atmosphere causes severe attenuation of infrared radiation between the wavelengths of approximately 4,500 and 8,000 nanometers (nm). Carbon dioxide interferes with the transmission of infrared radiation at wavelengths ranging from about 14,000 nm and 16,000 nm. Rain, snow, fog, and dust interfere with the transmission of infrared radiation.

Light is transmitted fairly well through the atmosphere at all visible wavelengths. But rain, snow, fog, and dust interfere with the transmission of visible light through the air.

Ultraviolet light at the longer wavelengths can penetrate the air with comparative ease. At shorter wavelengths, attenuation increases. Rain, snow, fog, and dust interfere.

X rays and gamma rays do not propagate well for long distances through the air because air molecules over long propagation paths are sufficient to block this radiation. *See* GAMMA RAY, INFRARED, OPTICAL TRANSMISSION, ULTRAVIOLET RADIATION, X RAY.

PROPAGATION DELAY

See PROPAGATION SPEED.

PROPAGATION SPEED

Electromagnetic waves in free space travel about 186,280 miles per second (299,800 kilometers per second). In media other than free space, this speed can be reduced considerably.

In air, electromagnetic waves travel at essentially the same speed as in free space. However, in a transmission line, the speed of electromagnetic field propagation is much slower. *See also* VELOCITY FACTOR.

PROTECTIVE GROUNDING

In an antenna system, a substantial voltage can develop between the active elements and the ground. This frequently happens in the proximity of thunderstorms. It can also result from electromagnetic interaction with nearby power lines. This voltage is often large enough to present a shock hazard.

To keep an antenna system at direct-current (dc) or 60 Hz ground potential, protective grounding is used. The most common method of protective grounding is the installation of radio-frequency chokes between the antenna feed line and ground.

In an unbalanced system, one choke is connected as shown in A in the illustration. In a balanced system, two chokes are needed (B). The chokes should have about 10 times the impedance of the antenna system at the lowest operating frequency. For example, in a 50 Ω unbalanced system, the choke should present 500 Ω of inductive reactance at the lowest frequency on which operation is contemplated. The inductor windings must be made of wire that is large enough to handle several amperes of current.

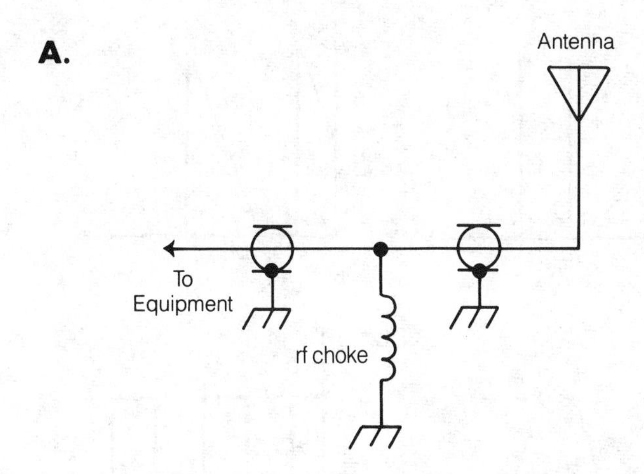

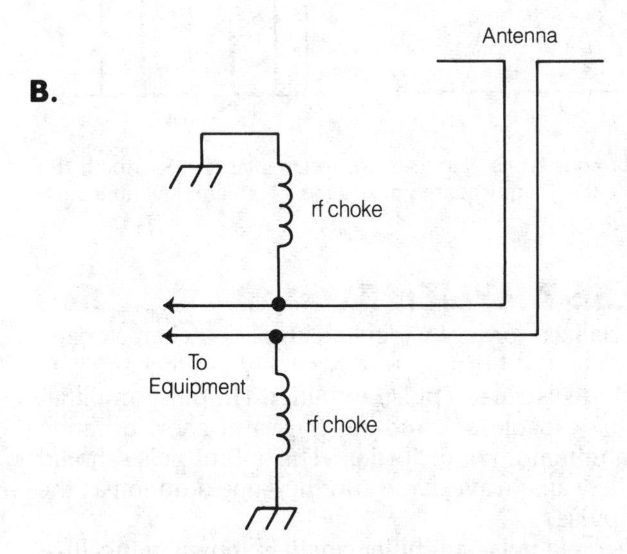

PROTECTIVE GROUNDING: Protective grounding is shown for an unbalanced antenna system (A) and for a balanced antenna system (B).

PROTON

The proton is a charged particle in the nucleus of an atom and the nucleus of the most common form of hydrogen atom. A proton at rest has a mass of about 1.67×10^{-27} kilograms. The proton carries the same charge quantity as the electron, but the opposite polarity—positive rather than negative (*see* ELECTRON).

The elements are classified according to the number of protons in their nuclei. This number is called the atomic number (*see* ATOMIC NUMBER). A given element always has the same number of protons in each of its nuclei, although the number of neutrons or electrons in the atom can vary (*see* IONIZATION, ISOTOPE).

P-TYPE SEMICONDUCTOR

A P-type semiconductor is a semiconductor that conducts by the transfer of electron deficiencies among atoms. The electron deficiencies are called holes and carry a positive charge. The P-type semiconductor is so named because it conducts mainly positive carriers.

P-type material is manufactured by adding certain impurities to the semiconductor base. Examples include boron, gallium, and indium. The impurities are called acceptors, because they are electron-deficient.

The carrier mobility in a P-type material is generally less than that of N-type material.

P-type material is used with N-type material in most semiconductor devices, including diodes, transistors, and integrated circuits. *See also* DOPING, ELECTRON, HOLE, N-TYPE SEMICONDUCTOR, P-N JUNCTION.

PULLING

When a capacitor or inductor is placed in parallel with a piezoelectric crystal, the natural frequency of the crystal is lowered. This effect is called pulling.

Pulling is routinely used in crystal oscillators and filters for adjusting the crystal frequency to exactly the desired value. The frequency of the average crystal can be pulled down by less than 0.1 percent of its natural frequency, or approximately 1 kHz/MHz.

When two variable-frequency oscillators are connected to a common circuit, such as a mixer, one of the oscillators may drift into tune with the other. This effect is also called pulling. It is most likely to occur when tuned circuits are too tightly coupled. *See also* CRYSTAL, CRYSTAL-LATTICE FILTER, OSCILLATOR CIRCUITS.

PULSATING DIRECT CURRENT

In a direct-current power supply with the filter disconnected, the output voltage pulsates. If the alternating-current frequency is 60 Hz at the supply input, then the pulsating direct current has a frequency of either 60 Hz or 120 Hz. The half-wave rectifier circuit produces pulsating direct current with a frequency of 60 Hz; the bridge and full-wave circuits produce 120 Hz pulsating direct current. The pulse waveshape for a sinusoidal, alternating-current input is shown at A in the illustration for a half-wave rectifier and at B for the bridge and full-wave circuits.

In general, 120 Hz pulsating direct current is easier to filter than 60 Hz pulsating direct current. The filter capacitors and inductors need their charge for only half as long when the frequency is 120 Hz. *See also* BRIDGE RECTIFIER, FULL-WAVE RECTIFIER, HALF-WAVE RECTIFIER, RECTIFIER.

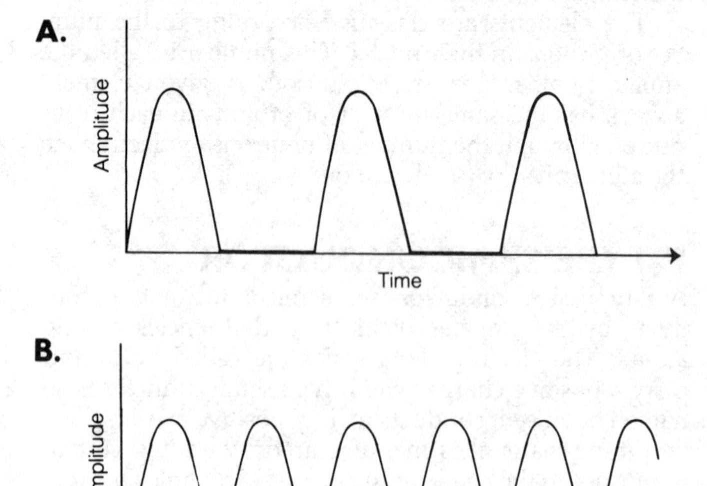

PULSATING DIRECT CURRENT: Pulsating direct-current from a half-wave rectifier (A) and from a full-wave rectifier (B).

PULSE

A pulse is a burst of current, voltage, power, or electromagnetic energy. A pulse can have an extremely short duration (as little as a fraction of a nanosecond), or it can have a long duration (thousands or millions of centuries).

The amplitude of a pulse is expressed either in terms of the maximum instantaneous (peak) value, or in terms of the average value (*see* AVERAGE VALUE, PEAK VALUE). The average and peak amplitudes are sometimes, but not always, the same.

The duration of a pulse can be expressed either in actual terms or in effective terms. The actual duration of a pulse is the length of time between its beginning and ending. The effective duration may be shorter than the actual duration (*see* AREA REDISTRIBUTION). Pulse duration is also called pulse width.

The interval, or length of time between pulses, has meaning if the pulses are recurrent. The pulse interval is the length of time from the end of one pulse to the beginning of another. The pulse interval may be zero in some cases. An example is the output of a full-wave rectifier.

In electronics, a pulse generally has a well-defined waveshape, such as rectangular, sawtooth, sinusoidal, or square. These pulse shapes, shown in the illustration, are the most common. However, a pulse may have a highly irregular shape, and the number of possible configurations is infinite.

A pulse of current or voltage normally maintains the same polarity from beginning to end, unless droop or overshoot occur. *See also* DROOP, OVERSHOOT.

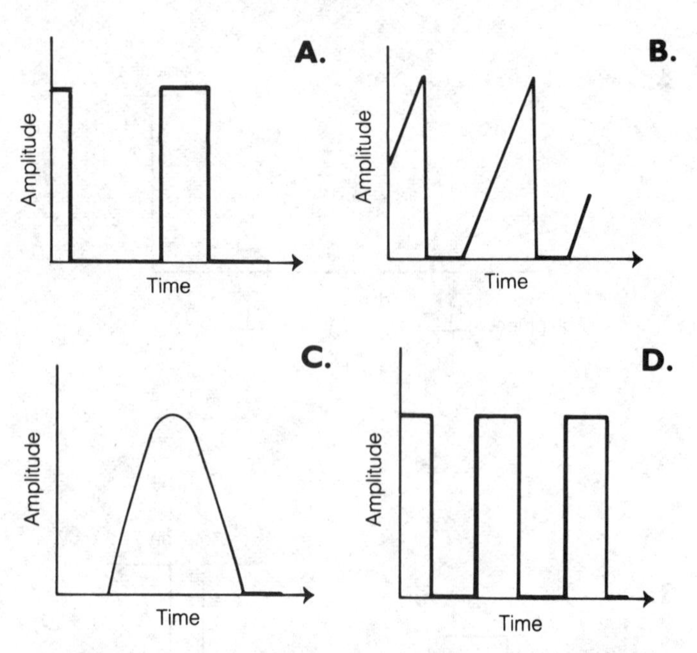

PULSE: Four types of pulses are rectangular (A), sawtooth (B), sinusoidal (C), and square pulse—equal high and low time duration (D).

PULSE AMPLIFIER

A specialized form of wideband amplifier, characterized by the ability to respond to extremely rapid changes in amplitude, is called a pulse amplifier. The pulse amplifier must be capable of handling pulses of short duration with a minimum of distortion. The output pulse should have the same waveshape and the same duration as the input pulse.

A simple pulse-amplifier circuit is shown in the illustration on p. 679. Pulse amplifiers are used in such applications as digital circuits, pulse-modulated transmitters, and radar transmitters. *See also* PULSE.

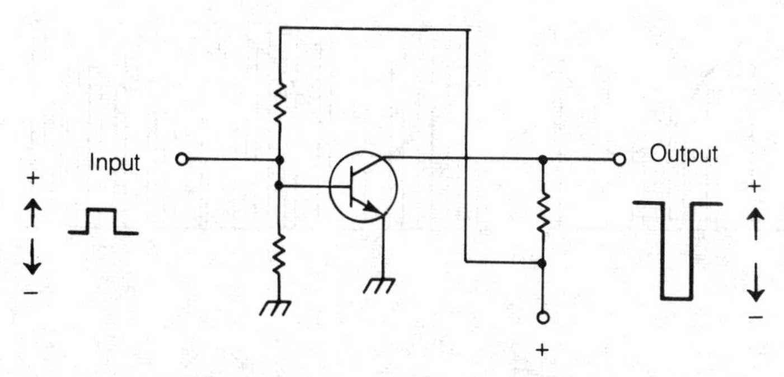

PULSE AMPLIFIER: A simple wideband pulse amplifier.

PULSE AMPLITUDE

See PULSE.

PULSE-AMPLITUDE MODULATION

See PULSE MODULATION.

PULSE-CODE MODULATION

See PULSE MODULATION.

PULSE COUNTER

See COUNTER.

PULSE DURATION

See PULSE.

PULSE-DURATION MODULATION

See PULSE MODULATION.

PULSE FREQUENCY

See PULSE.

PULSE-FREQUENCY MODULATION

See PULSE MODULATION.

PULSE GENERATOR

A pulse generator is a circuit specifically designed for producing electronic pulses. Commercially manufactured pulse generators are available for a variety of laboratory test purposes.

A typical pulse generator can produce rectangular, sawtooth, sinusoidal, and square pulses. The pulse amplitude, duration, and frequency are independently adjustable. Some advanced generators can reproduce pulses of any desired waveshape with the aid of a computer. The operator draws the desired pulse function with a light pen, on a video-monitor screen. The computer then produces a pulse or pulse train with that waveshape. *See also* PULSE.

PULSE INTERVAL

See PULSE.

PULSE-INTERVAL MODULATION

See PULSE MODULATION.

PULSE-LENGTH MODULATION

See PULSE MODULATION.

PULSE MODULATION

Pulse modulation is the transmission of intelligence by varying the characteristics of a series of electromagnetic pulses. Pulse modulation can be accomplished by varying the amplitude, the duration, the frequency, or the position of the pulses. Pulse modulation can also be obtained by means of coding. A brief description of each of these pulse-modulation methods follows:

Pulse-Amplitude Modulation. A complex waveform can be transmitted by varying the amplitude of a series of pulses. Generally, the greater the modulating-signal amplitude, the greater the pulse amplitude at the output of the transmitter. However, this can be reversed. The drawing illustrates pulse-amplitude modulation at A.

Pulse-Duration Modulation. The effective energy contained in a pulse depends not only on its amplitude, but also on its duration. In pulse-duration modulation, the width of a given pulse depends on the instantaneous amplitude of the modulating waveform. Usually, the greater the amplitude of the modulating waveform, the longer the transmitted pulse, but this can be reversed. Pulse-duration modulation may also be called pulse-length or pulse-width modulation. The principle is illustrated at B.

Pulse-Frequency Modulation. The number of pulses per second can be varied in accordance with the modulating-waveform amplitude. The duration and intensity of the individual pulses remains constant in pulse-frequency modulation. However, the effective signal power is increased when the pulse frequency is increased.

Usually, the pulse frequency increases as the modulating-signal amplitude increases; but this can be reversed. Pulse-frequency modulation is illustrated at C.

Pulse-Position Modulation. Pulse modulation can be obtained even without varying the frequency, amplitude, or duration of the pulses. The actual timing of the pulses can be varied, as shown at D. This is known as pulse-position modulation. It may also be called pulse-interval or pulse-time modulation.

Pulse-Code Modulation. All of the schemes for pulse modulation described earlier are analog methods. That is, the amplitude, duration, frequency, or position of the pulses can be varied in a continuous manner. A digital

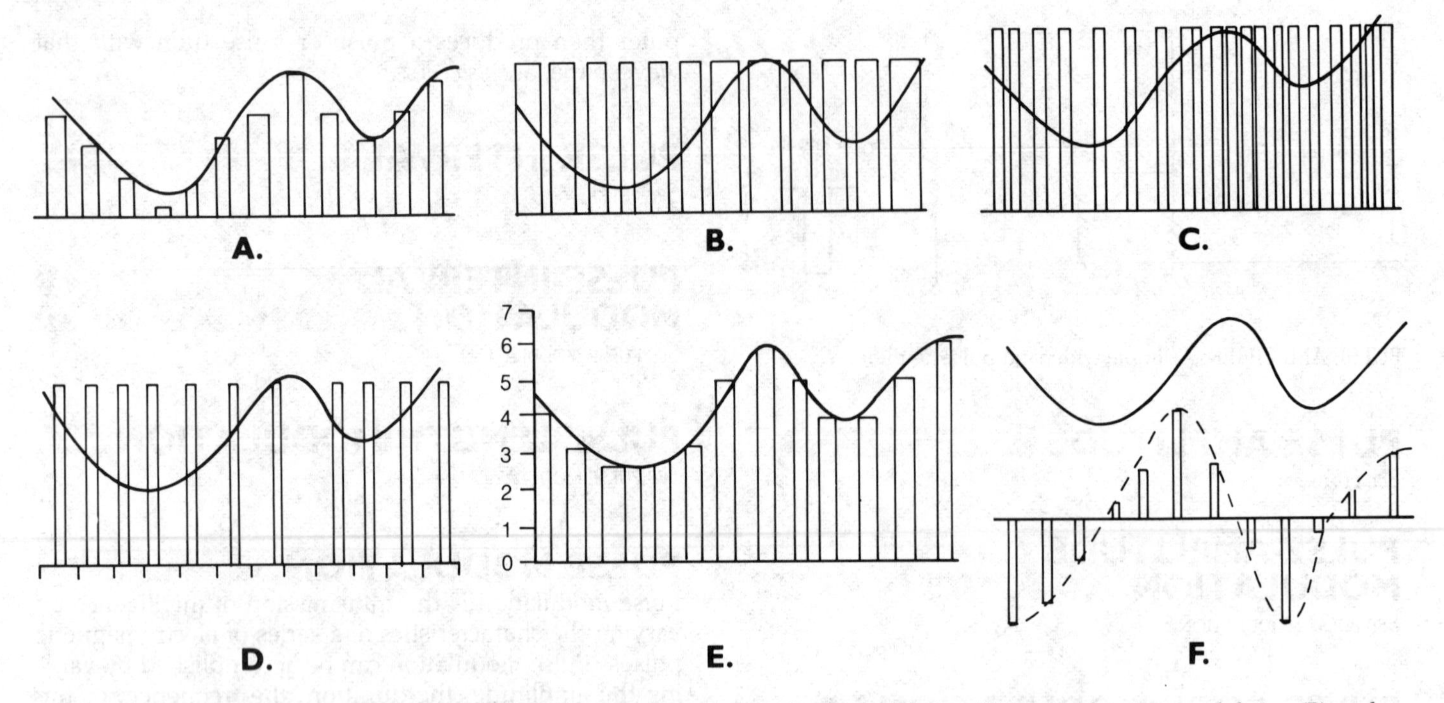

PULSE MODULATION: Examples of pulse modulation include pulse-amplitude modulation (A), pulse-duration modulation (B), pulse-frequency modulation (C), pulse-position modulation (D), pulse-code modulation (E), and delta modulation (F).

method for pulse modulation is called pulse-code modulation. In pulse-code modulation, some pulse variable (usually amplitude) of the pulses may reach only discrete values. The drawing at E illustrates the concept of pulse-code modulation in which the pulse amplitude can attain any of eight discrete levels.

A special form of pulse-code modulation makes use of the derivative of the modulating-signal waveform. This is known as delta modulation. The pulse amplitude (or other variable) may attain only discrete values according to the derivative of the modulating-waveform amplitude function, as at F.

Pulse modulation is used for a variety of communications purposes. It is especially well suited for use with communications systems incorporating time-devision multiplexing. *See also* TIME-DIVISION MULTIPLEX.

PULSE OVERSHOOT

See OVERSHOOT.

PULSE-POSITION MODULATION

See PULSE MODULATION.

PULSE RISE AND DECAY

An electric or electromagnetic pulse has a specific, measurable rise time and a measurable decay time.

The rise time of a pulse is defined as the time required for the pulse to reach its maximum, beginning at zero amplitude. The decay time is the length of time needed for the pulse to reach zero amplitude from its maximum value. These definitions, illustrated at A, apply to pulses with identifiable geometric shapes.

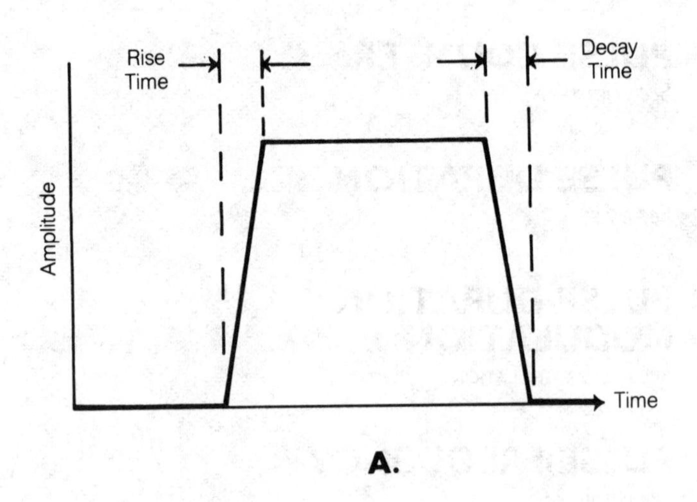

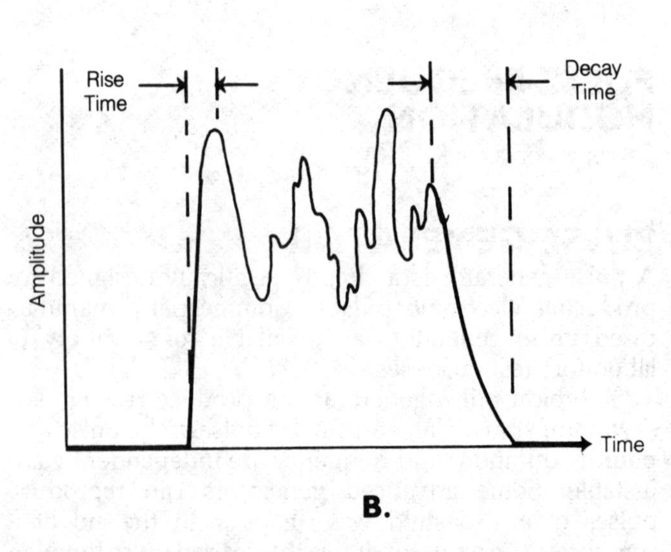

PULSE RISE AND DECAY: A well-defined, regular pulse (A) and an irregular pulse (B).

An irregular pulse can have rise or decay times that are difficult to define. One method of defining the rise and decay times of an irregular pulse is to consider the time required for the amplitude to reach the first local peak, beginning at zero amplitude, and the time required for the amplitude to fall to zero following the last local peak (B). *See also* PULSE.

PULSE TRANSFORMER

A pulse transformer is a special transformer designed to accommodate rapid rise and decay times with a minimum of distortion. A normal transformer may introduce distortion into a square pulse by slowing down the rise and decay times (*see* PULSE RISE AND DECAY).

Pulse transformers generally have windings with lower inductance than the windings of a transformer for sine waves. The core material of a pulse transformer usually has lower permeability than the core material in a typical transformer. *See also* TRANSFORMER.

PULSE WIDTH

See PULSE.

PULSE-WIDTH MODULATION

See PULSE MODULATION.

PUSHDOWN STACK

A first-in/last-out memory is called a pushdown stack. It is a read-write memory; information can be both stored and retrieved. The pushdown stack differs from the first-in/first-out memory (*see* FIRST-IN/FIRST-OUT) in that the data bits inserted first must be retrieved last; the bits inserted last are recalled first.

The illustration shows the principle of the pushdown stack. The term describes the operation—data bits are stored and retrieved as if they are stacked in a confined column.

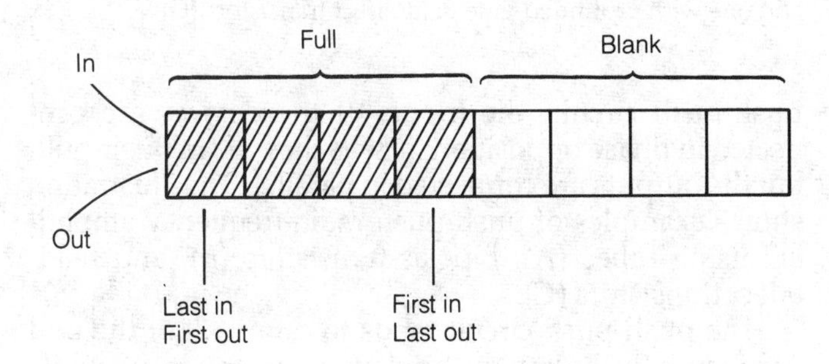

PUSHDOWN STACK: Principle of the pushdown stack.

PUSH-PULL AMPLIFIER

A push-pull amplifier is a specialized, low-distortion amplifier used at audio frequencies and at radio frequencies. The drawing illustrates push-pull amplifiers incorporating tubes (A), bipolar transistors (B), and field-effects transistors (C). These are audio-frequency circuits.

Any push-pull amplifier requires a pair of transistors or tubes. The cathodes, emitters, or sources are grounded through small resistors. The grids, bases, or gates receive the input signal in phase opposition; they are connected to opposite ends of a transformer secondary. The center tap at the transformer helps to balance the system, and also provides a point for bias application. The plates, collectors, or drains are connected to opposite ends of the primary winding of the output transformer. As with the input, the center tap helps to balance the system and also provides the point at which voltage is applied.

The input signal may be applied at the cathodes, emitters, or sources of the active devices, leaving the grids, bases, or gates at ground potential. This configuration also provides push-pull operation (*see* PUSH-PULL, GROUNDED-GRID/BASE/GATE AMPLIFIER).

Push-pull amplifiers are operated either in Class AB or in Class B, and are always used as power amplifiers (*see* CLASS AB AMPLIFIER, CLASS B AMPLIFIER, POWER AMPLIFIER). The main advantage of a push-pull amplifier is its ability to cancel all of the even harmonics in the output circuit. This reduces the distortion in audio-frequency applications and enhances the even-harmonic attenuation in a radio-frequency circuit.

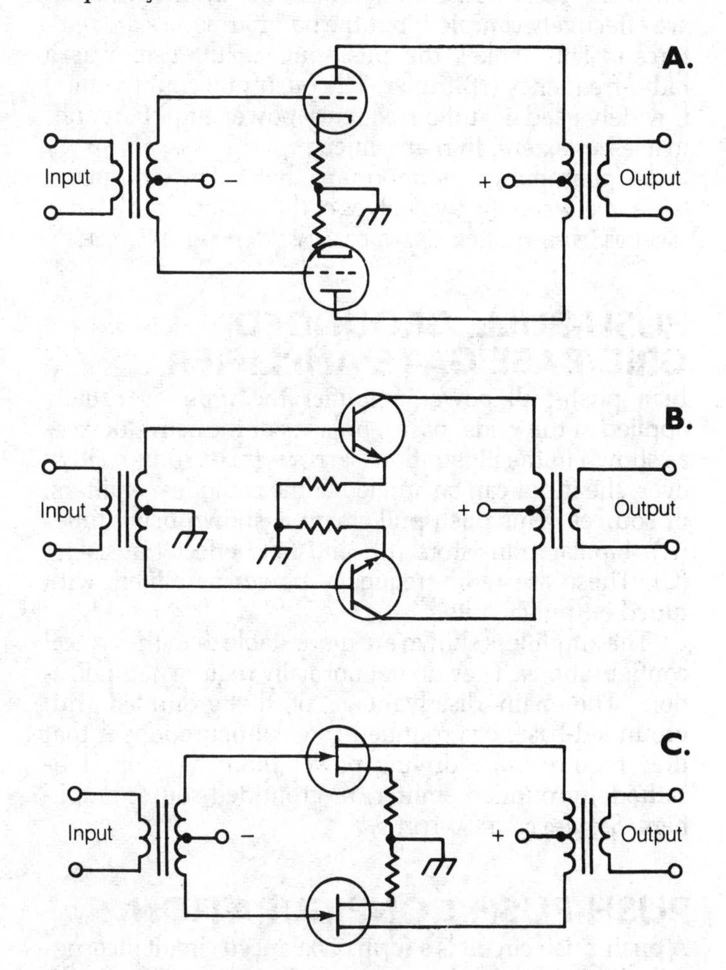

PUSH-PULL AMPLIFIER: Examples of push-pull amplifiers based on triode tubes (A), bipolar transistors (B), and FETs (C).

A push-pull radio-frequency amplifier may be used to multiply the frequency of a signal by a factor of 3. This circuit is called a tripler. The odd harmonics are not attenuated in the push-pull configuration; if the output circuit is tuned to an odd-harmonic frequency, the push-pull circuit will favor the harmonic over the fundamental frequency (*see* TRIPLER).

Push-pull circuits can be used in oscillators, modulators, and power amplifiers. *See also* PUSH-PULL CONFIGURATION.

PUSH-PULL CONFIGURATION

Any circuit in which both the input and the output are divided between two identical devices, each operating in phase opposition with respect to the other, is called a push-pull circuit. The push-pull configuration requires two transformers, one at the input and one at the output. The secondary of the input transformer, and the primary output transformer, are usually center-tapped. The illustration for PUSH-PULL AMPLIFIER shows three examples of push-pull configurations for audio-frequency amplifiers.

The term *push-pull* describes the operation of the circuit. While one side of the circuit carries current in one direction, the other side carries current in the opposite direction.

In the push-pull configuration, the even harmonics are effectively canceled, but the odd harmonics are reinforced. This makes the push-pull circuit useful as a radio-frequency (rf) tripler. The push-pull configuration is widely used in audio-frequency power amplifiers and, to a lesser extent, in rf amplifiers.

In push-pull, it is important that balance be maintained between the two halves of the circuit. *See* PUSH-PULL AMPLIFIER, PUSH-PULL, GROUNDED-GRID/BASE/GATE AMPLIFIER, TRIPLER.

PUSH-PULL, GROUNDED-GRID/BASE/GATE AMPLIFIER

In a push-pull power amplifier the input is usually applied at the grids, bases, or gates of the active devices as shown in the illustration (*see* PUSH-PULL AMPLIFIER). However, the input can be applied at the cathodes, emitters, or sources. This push-pull circuit is shown using tubes (A), bipolar transistors (B), and field-effect transistors (C). These are radio-frequency power amplifiers with tuned output circuits.

The amplifiers shown are more stable than the typical configurations. They do not normally require neutralization. The main disadvantage of the grounded-grid, grounded-base, or grounded-gate configurations is that they require more driving power than the grounded-cathode, grounded-emitter, or grounded-source amplifiers. *See also* POWER AMPLIFIER.

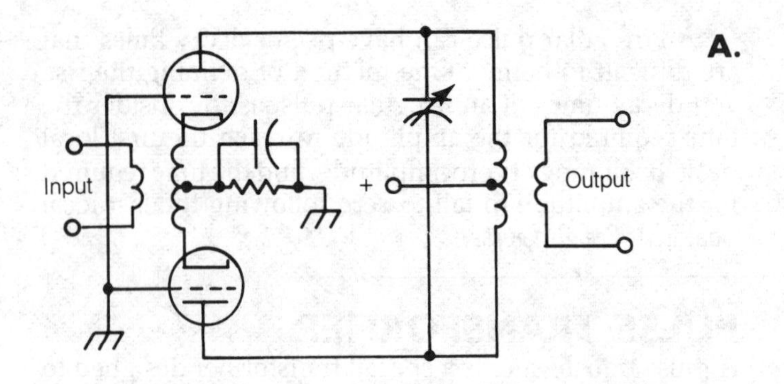

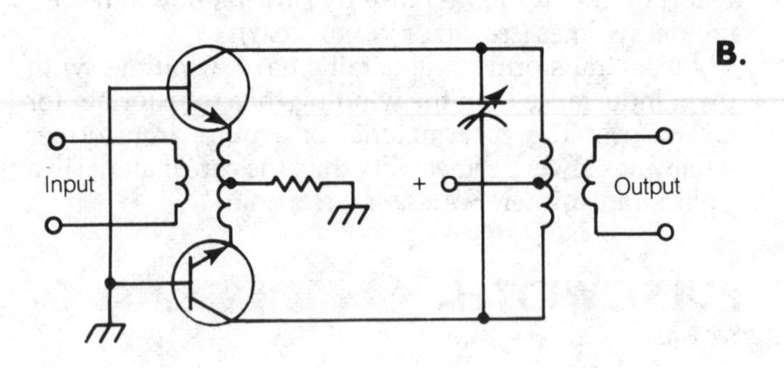

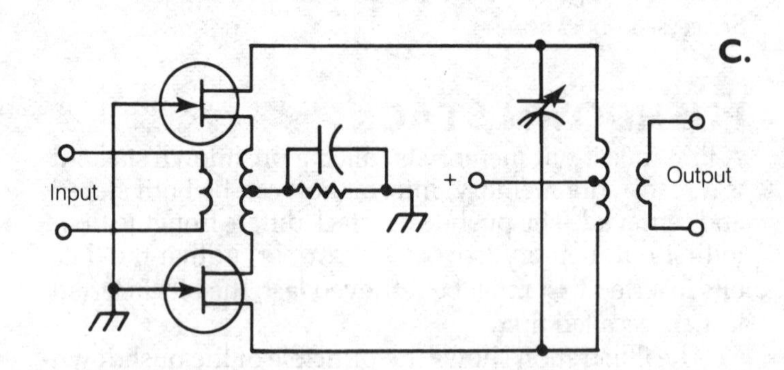

PUSH-PULL, GROUNDED-GRID/BASE/GATE AMPLIFIER: Examples of push-pull amplifiers include one with grounded-grid vacuum tubes (A), one with grounded-base bipolar transistors (B), and one with grounded-gate field-effect transistors (C).

PUSH-PUSH CONFIGURATION

A push-push circuit is a form of balanced circuit incorporating two identical passive or active halves. The push-push configuration is similar to push-pull (*see* PUSH-PULL CONFIGURATION), but there is one important distinction. In a push-push circuit, the inputs of the devices are connected in phase opposition, just as they are in push-pull, but the outputs are connected in parallel. The illustration shows examples of push-push radio-frequency amplifiers using tubes (A), bipolar transistors (B), and field-effect transistors (C).

The push-push circuit tends to cancel all of the odd harmonics (including the fundamental frequency), while reinforcing the even harmonics. This is just the opposite of the push-pull configuration. The push-push circuit is not often used at audio frequencies, but it can be used as a radio-frequency doubler or quadrupler.

The simplest form of push-push circuit is the full-wave rectifier. This is, in effect, a frequency doubler. With 60 Hz alternating-current input, the full-wave rectifier produces a 120 Hz pulsating direct current output. An identical circuit can be used as a passive radio-fre-

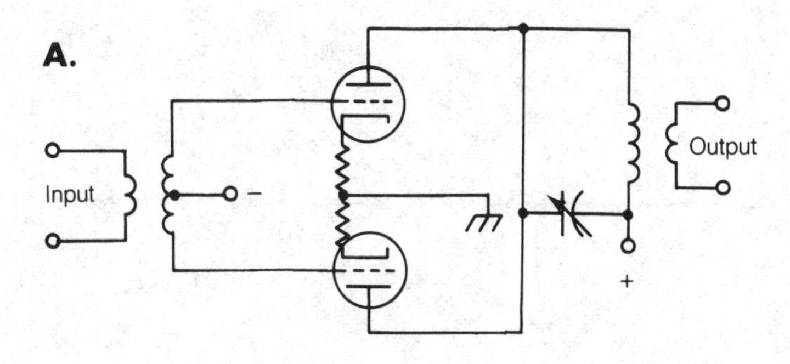

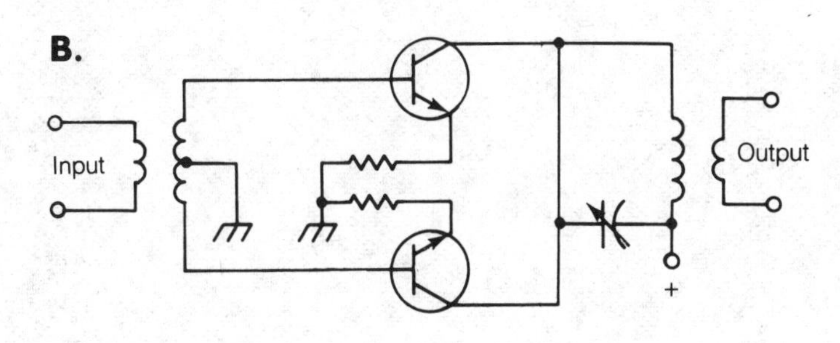

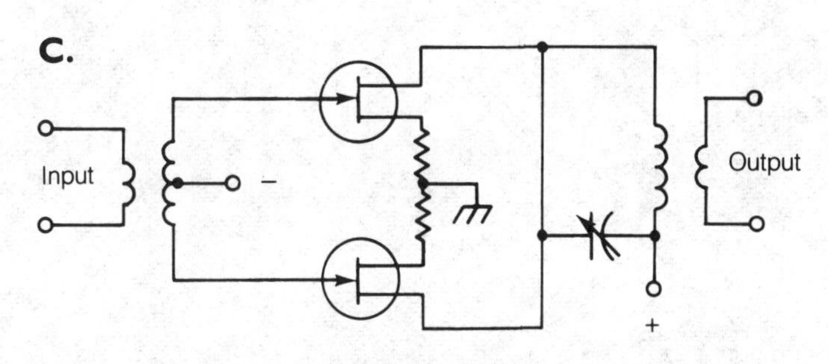

PUSH-PUSH CONFIGURATION: Examples of push-push amplifiers with vacuum tubes (A), bipolar transistors (B), and field-effect transistors (C).

quency doubler. *See also* DOUBLER, FULL-WAVE RECTIFIER, QUADRUPLER.

PYRAMID-HORN ANTENNA

See HORN ANTENNA in the ANTENNA DIRECTORY.

PYTHAGOREAN THEOREM

The Pythagorean Theorem is a simple rule of plane geometry discovered by the mathematician Pythagoras in ancient times. In a right triangle, the square of the length of the hypotenuse (the side opposite the right angle) is always equal to the sum of the squares of the lengths of the other two sides. This principle applies regardless of the shape of the triangle. The figure illustrates this principle. *See also* CARTESIAN COORDINATES.

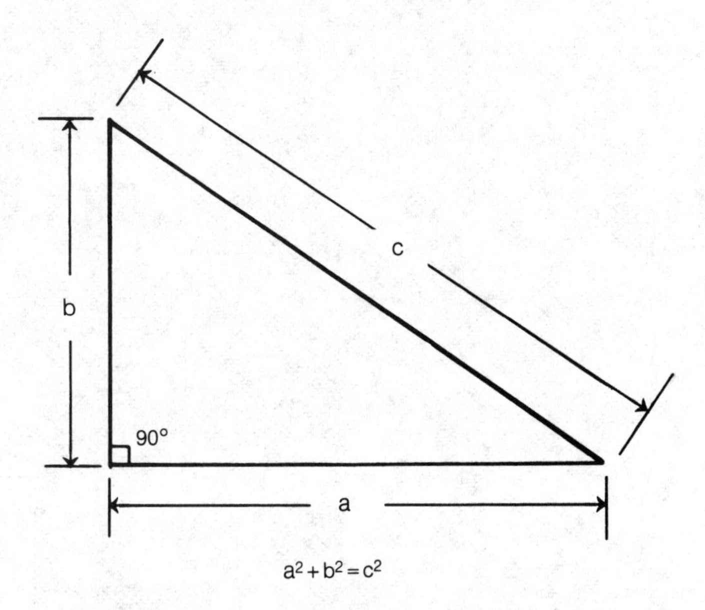

PYTHAGOREAN THEOREM: This theorum by Pythagoras relates the lengths of the sides of a right triangle.

Q FACTOR

For a capacitor, inductor, or tuned circuit, the Q factor (also known simply as the Q) is a figure of merit. The higher the Q factor, the lower the loss and the more efficient the component or tuned circuit becomes. In tuned circuits, the Q factor is directly related to the selectivity.

In the case of a capacitor or inductor, the Q factor is given in terms of the reactance and the resistance. If X_L represents the reactance and R represents the resistance, in a series circuit, both given in ohms, then Q for an inductor is:

$$Q = X_L/R$$

If X_c is the reactance, Q for a capacitor is:

$$Q = -X_c/R$$

(Capacitive reactance is considered negative.)

Because both inductive and capacitive reactance vary with the frequency, the Q factor of an inductor or capacitor depends on the frequency (*see* CAPACITIVE REACTANCE, INDUCTIVE REACTANCE).

In a series-resonant inductance-capacitance (LC) circuit, the Q factor is given by the same formula, where X represents the absolute value of the reactance of either the coil or the capacitor (the absolute values are equal at resonance) and R represents the resistive impedance of the circuit at resonance.

In a parallel-resonant LC circuit, the Q factor is given by:

$$Q = R/X$$

where R represents the resistive impedance of the circuit at resonance and X represents the absolute value of the reactance of either the inductor or the capacitor (again, the absolute values are identical at resonance). *See also* BANDWIDTH, RESONANCE, TRAP.

QUADRATIC FORMULA

The quadratic formula is a formula for solving quadratic equations. A quadratic equation is an equation of the form:

$$ax^2 + bx + c = 0$$

where a, b, and c are constants.

Every quadratic equation has two solutions. The solutions may be real numbers or complex numbers. In some instances, both solutions are the same; then they are said to be coincident or redundant. *See also* COMPLEX NUMBER, REAL NUMBER.

QUADRIFILAR WINDING

A set of four coil windings, usually oriented in the same sense and having the same number of turns on a common core, is called a quadrifilar winding. Quadrifilar windings are used in the manufacture of broadband balun transformers.

Quadrifilar windings on a toroidal core can be connected to obtain many different impedance-transfer ratios. *See also* BALUN.

QUADROPHONICS

Quadrophonics is another name for four-channel high-fidelity recording and reproduction. The spelling of the term varies considerably; it might be called quadraphonics, quadriphonics, or simply quad.

An ordinary stereophonic system has two channels, commonly called the left channel and the right channel. In a quadrophonic system there are four channels, usually designated left front, right front, left rear, and right rear.

A true quadrophonic system has four entirely independent channels providing true 360-degree reproduction of sound for a listener. Most quadrophonic systems in use today, however, have only two recording tracks. The two channels are combined according to some predetermined scheme, obtaining four different sound tracks in the reproduction.

QUADRUPLER

A quadrupler is a circuit that multiplies a radio-frequency signal by a factor of 4. The quadrupler generally consists of a push-push amplifier with tuned input and output circuits. The input circuit is tuned to the frequency of the incoming signal, while the output circuit is tuned to the fourth harmonic of the incoming-signal frequency (*see* PUSH-PUSH CONFIGURATION).

A power-supply circuit that multiplies the incoming voltage by 4 is also called a quadrupler. This power supply is used when the secondary of the power transformer will not provide adequate voltage and when the regulation need not be precise. *See also* VOLTAGE MULTIPLIER.

QUALITY CONTROL

Quality control, also known as quality assurance, is the control of variation of workmanship, processes, and materials in a planned, systematic manner to produce a consistent, uniform product. It is necessary in order to ensure that electronic (or other) equipment performs according to the specifications.

Common methods of quality control are: the complete testing of each unit at various stages during, and after, manufacture; the complete testing of each unit only after manufacture is completed; the complete testing of a certain proportion of units at various stages during, and after, manufacture; and the complete testing of a certain proportion of units only after manufacture is completed.

The first or second method is preferable, but when it is not economically feasible to test every unit, a less rigorous quality-control procedure may be used.

QUANTIZATION

Many natural phenomena occur in the form of discrete packets or bits, and not as a continuous, smooth process. This is known as quantization—the occurrence of discrete, although perhaps minute, steps or intervals in nature.

The most familiar example of quantization is seen in the particle nature of electromagnetic energy. All electromagnetic radiation consists of discrete packets of energy, called photons. For radiation of any fixed wavelength, the total energy is always an integral multiple of the energy contained in a single photon. There can be no intermediate values. (*See* ELECTROMAGNETIC RADIATION, PHOTON.)

Another example of quantization is the flow of electricity. Electric current is an expression of the number of unit charge carriers (generally either electrons or holes) passing a given point in a certain interval of time. If 1 coulomb of electrons passes a point in one second, that represents 1 ampere of electric current. Although 1 coulomb is a very large number, it is still an integer, and represents a discrete, finite number of charge carriers. Electric charge can exist only in integral multiples of a unit charge (*see* CHARGE, COULOMB, ELECTRON, HOLE).

Matter is quantized because it is made up of discrete particles. No one knows what the smallest particle actually is, but the discrete nature of molecules, atoms, electrons, neutrons, and protons has been demonstrated (*see* ATOM, MOLECULE).

QUANTUM MECHANICS

The study of phenomena related to the quantum theory, both theoretical and practical, is known as quantum mechanics. The quantum theory of matter and energy holds that the emission or absorption of energy by matter always occurs in discrete packets called photons (*see* PHOTON).

When an atom absorbs a photon of a certain wavelength, an electron moves to a higher orbit in the atom. This requires a photon with the right energy content to give up its energy to the electron, allowing it to gain speed and move farther from the nucleus of its parent atom.

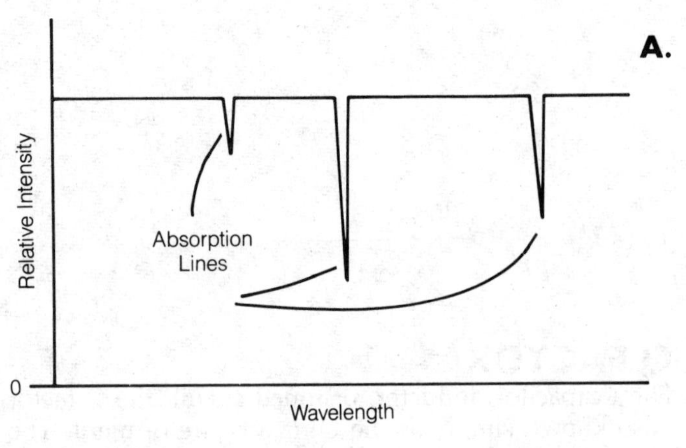

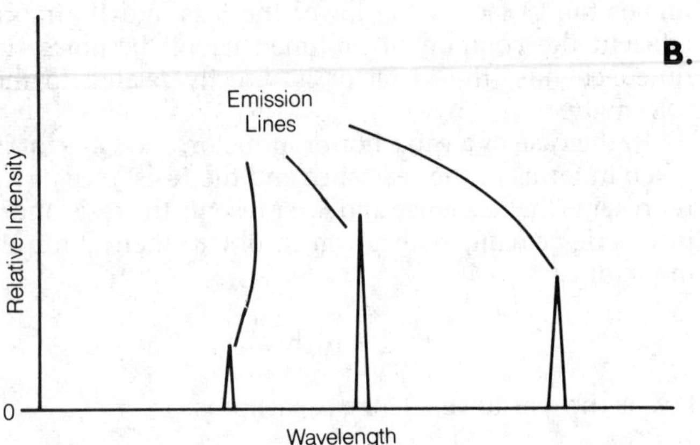

QUANTUM MECHANICS: Absorption lines (A) and emission lines (B) in a spectrum.

Absorption of energy by any atom can occur at one or more discrete wavelengths. As radiant energy passes through a material, the identity of the material can be determined because of an absorption-line signature unique to that material (see A in illustration). Astronomers use spectroscopes and radio telescopes to determine the makeup of interstellar gas and dust clouds. *See* ELECTRON ORBIT.

When an electron in an atom moves to a lower energy level, a photon with a precise wavelength is given off. This happens when substances are heated to high temperatures. Under these conditions, the electrons gain energy because of heating, and move into higher orbits. The electrons are unstable in the higher orbits, however, and tend to fall back into lower orbits, emitting photons as they do so. The continued presence of heat energy causes electrons to move alternately into higher and lower orbits. The result is continuous electromagnetic radiation at discrete wavelengths, as at B.

The possible number of electron-orbit changes and the number of emitted wavelengths are finite for any element. Each element has its own unique emission signature. Astronomers use radio telescopes and spectroscopes to observe the emission wavelengths of stars. This makes it possible to determine the material composition of distant stars, and even of whole galaxies.

QUARTER-WAVE ANTENNA

See MARCONI ANTENNA in the ANTENNA DIRECTORY.

QUARTER WAVELENGTH

A quarter wavelength is the distance corresponding to 90 degrees of phase as an electromagnetic disturbance is propagated (see illustration). In free space, it is related to the frequency by a simple equation:

$$\frac{\lambda}{4} = \frac{246}{f}$$

where $\lambda/4$ represents a quarter wavelength in feet, and f represents the frequency in megahertz. If $\lambda/4$ is expressed in meters, then the formula is:

$$\frac{\lambda}{4} = \frac{75}{f}$$

for the frequency f given in megahertz.

In media other than free space, electromagnetic waves propagate at speeds less than their speed in free space. The wavelength of a disturbance with a given frequency depends on the speed of propagation. In general, if v is the velocity factor in a given medium (*see* VELOCITY FACTOR), then:

$$\frac{\lambda}{4} = \frac{246v}{f}$$

in feet, and:

$$\frac{\lambda}{4} = \frac{75v}{f}$$

in meters.

Conductors and transmission lines have special properties when they have a length of $\lambda/4$. *See also* CYCLE, QUARTER-WAVE TRANSMISSION LINE, WAVELENGTH.

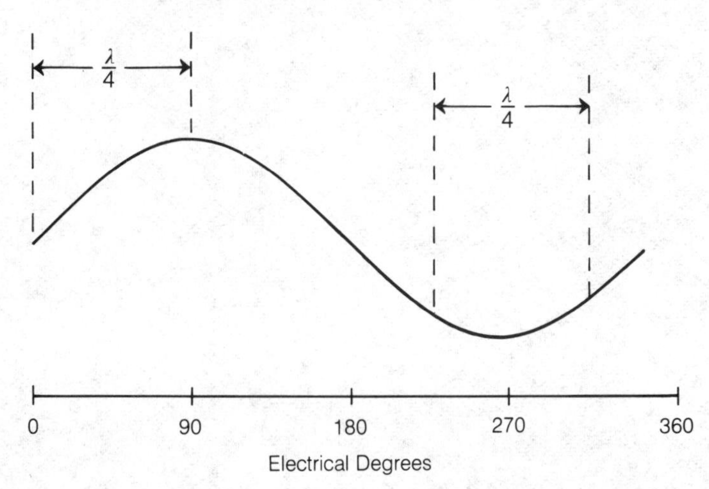

QUARTER WAVELENGTH: This wavelength corresponds to 90 electrical degrees.

QUARTER-WAVE MATCHING SECTION

See QUARTER-WAVE TRANSMISSION LINE.

QUARTER-WAVE TRANSMISSION LINE

A quarter-wave transmission line is any section of transmission line measuring an electrical quarter wavelength. If the velocity factor of a line is given by v (*see* VELOCITY FACTOR), then a quarter-wave transmission line measures:

$$L = \frac{246v}{f}$$

in feet, where f is the frequency in megahertz. The length in meters is:

$$L = \frac{75v}{f}$$

A quarter-wave section of transmission line can be used for impedance transformation. If a quarter-wave line has a characteristic impedance of Z_o ohms and a purely resistive impedance of Z_1 ohms is connected to one end of this section, then at the other end the impedance Z_2 will also be a pure resistance with a value of:

$$Z_2 = \frac{Z_o^2}{Z_1}$$

Quarter-wave sections are used in a variety of antenna systems, especially phased arrays, for impedance matching. Quarter-wave sections can also act as series-resonant or parallel-resonant tuned circuits. *See also* CHARACTERISTIC IMPEDANCE, COAXIAL TANK CIRCUIT, QUARTER WAVELENGTH.

QUARTZ

Quartz is a natural clear mineral with a higher index of refraction than glass and is harder than glass. It has a characteristic sparkle like that of diamond. Quartz can be grown artificially. Chemically, it is composed of silicon dioxide.

Quartz is used in electronics because of its piezoelectric properties. When quartz is subjected to electric currents, it will produce vibrations. The vibration frequency is determined by the size and shape of the crystal. The frequency can range from audio to several megahertz.

Probably the most important use of quartz in electronics is in the manufacture of oscillator crystals. Quartz crystals can be made to oscillate at frequencies of several megahertz with high stability. *See* CRYSTAL.

RAD

A rad is an absorbed radiation unit equivalent to 100 ergs per gram of an absorbing material. The material must be specified because this energy will differ with each material. For example:

1 rad (silicon) = 100 ergs/gram (silicon)

RADAR

Radar (for radio detection and ranging) is an active system for detecting and determining the range and direction (bearing) of distant objects such as ships and aircraft. Radar does this by illuminating them with radio frequency (rf) energy and then receiving, analyzing, and displaying the reflected energy. Radar systems have been built for use at ground stations, in aircraft, on ships, and in vehicles.

Figure 1 is a simplified block diagram of a typical radar system. The transmitter emits high-power radio waves in short pulses through a directive antenna to illuminate an object or target. The returned echo is received, usually by the same antenna, passed by a transmit-receive switch and amplified by a high-gain, wideband receiver. The output of the receiver is usually displayed on a cathode-ray tube (CRT) indicator display. Object direction (bearing) is determined by reference to the focal direction of the narrow-beam antenna at the time the echo is received. Range is measured in units of time for direct conversion into distance (nautical miles, yards, or feet) because radio energy travels at the speed of light, 300 meters one way per microsecond. It takes about 10 microseconds for radar energy to make a one-mile round trip.

High-frequency radiation, typically in the microwave region, is used because shorter wavelengths permit higher antenna directivity and more accurate range readings. Analysis of the display provides information about the course, speed, and closest point-of-approach (CPA) of the target.

The format of the radiated signal can differ, depending on the radar system application. Some systems emit pulses of longer duration and radiation duty cycles up to 100 percent, cw (continuous wave). By measuring the rf Doppler velocity in the receiver circuits, a target velocity can be computed. The radiation can also be frequency modulated so that object range is determined by comparing phase or modulation characteristics.

Radar has been adapted for many different applications, power ratings, and capabilities. They range from massive ground-based systems capable of searching out targets over the horizon or identifying spacecraft to low-cost systems for private aircraft and boats. Specialized commercial radars are designed specifically for air navigation, and long- and short-range ship navigation. Government radars are designed for air-traffic control, ship-traffic control, meteorology, and geological surveys. Military radars can be designed to perform all of these applications in addition to the more specialized military tasks—including surveillance and tracking of space vehicles, long-range early warning of missiles or aircraft, long- and short-range weapons fire control, and missile guidance.

The largest radars are the massive phased-array systems for tracking space vehicles. Other very large ground or shipboard radars with phased array or rotating antennas are used for air-traffic control or long-range air or missile search. Most airborne radars are small, compact short-range systems regardless of application. However, the smallest, lowest-cost packaged systems are used for small boat navigation.

Radars have four principal functional groups as described below: transmitter, antenna, receiver, and display.

Transmitters. The transmitter provides an rf signal with sufficient power for the intended application. The important transmitter factors are: pulse length in microseconds, pulse rate in hertz, duty cycle, average power in watts, peak power in watts, and carrier wavelength in centimeters.

Pulse lengths are typically measured in microseconds. Longer pulses can be used for greater range if the transmitter has the capability. However, for range resolution in feet, pulses must be relatively short. The pulse repetition frequency (prf) must be low enough to permit the pulse to reach the outer limit of the intended range and return before the next pulse. Most pulsed radars are designed so the prf, antenna beamwidth, and rotation rate permit 20 to 40 pulses to be transmitted while the antenna remains directed on the target, allowing a repetition of returns.

Peak power and average power are determined by the intended application. To improve range resolution when performance is limited by transmitter peak power, pulse compression techniques can be used. These waveforms permit range resolutions that are shorter than those corresponding to the radiated pulse width. The choice of carrier frequency is also determined by the application

RADAR: Fig. 1. Block diagram of a typical radar system showing timing data originating from the modulator.

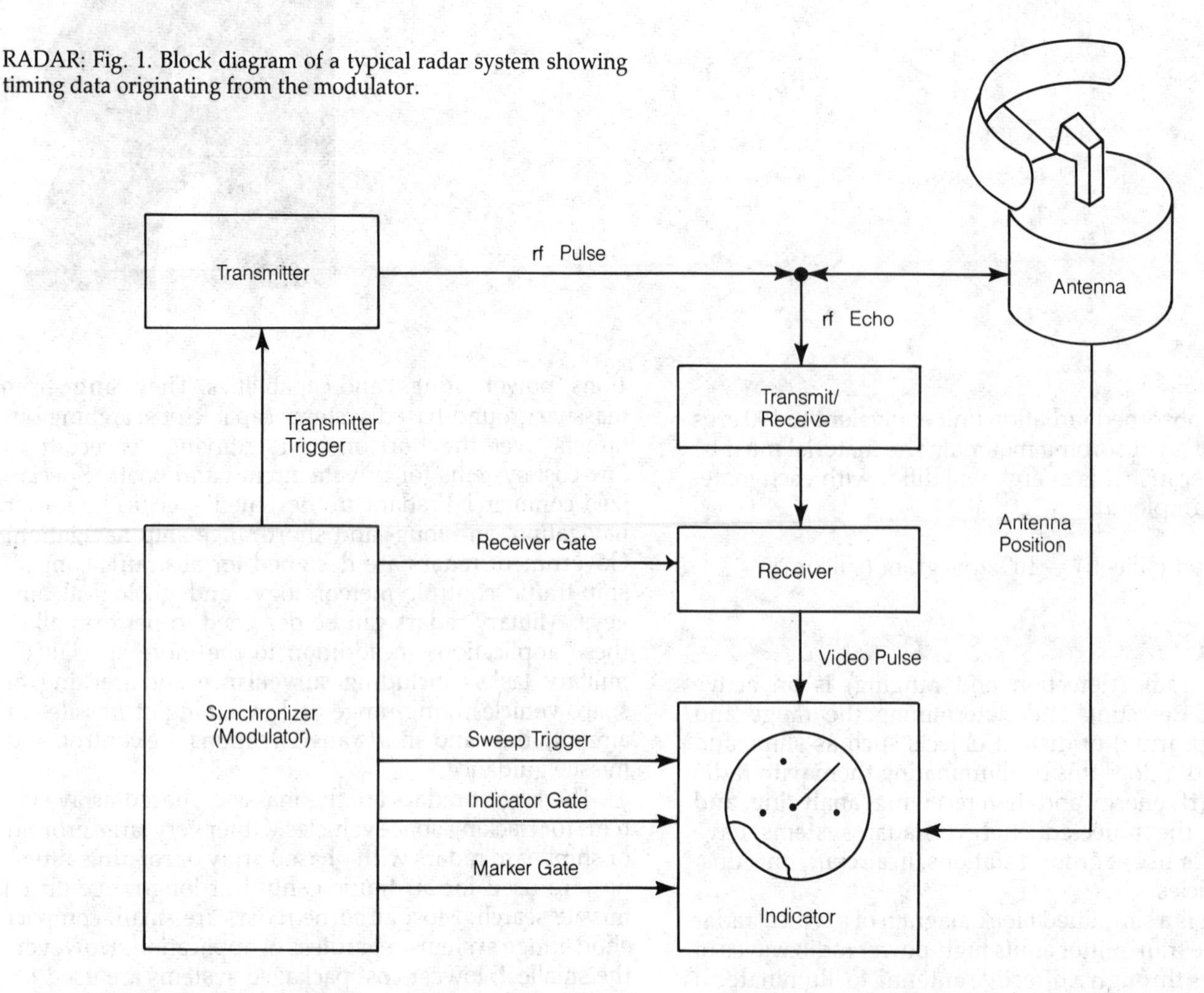

and is influenced by the available power supply, permitted or practical antenna size, and space constraints on or in the platform.

The rf power source in radar transmitters can be a microwave oscillator power tube such as a magnetron (*see* MAGNETRON) or a master oscillator-power amplifier chain. The master oscillator and driver power amplifier can be solid state devices, but the final amplifier for high power systems is typically a traveling-wave tube (TWT). *See* TRAVELING-WAVE TUBE.

The power tube oscillator or final power amplifier stage converts most of the electrical power to rf pulse power under the control of the pulse-forming modulator. The modulator supplies a faithful replica of the pulse to the modulating grid of the amplifier or the cathode of the magnetron. The modulator generally includes either solid state or vacuum tube switches.

The duplexer or transmit-receive switch permits one antenna to be used for both transmission and reception. It is equipped with a protective device that shorts out and blocks the strong transmission signal from the sensitive receiver, thus preventing the signal from damaging the receiver. *See* TRANSMIT-RECEIVE SWITCH.

Antennas. Radar antennas are made in many different shapes and sizes, but the most common today are parabolic dishes or paraboloidal sections. These can be sheet metal barrel staves or open-frame, wire-mesh covered structures on a motor-driven mount that is capable of turning through 360 degrees.

A shipboard antenna for surface search or navigation is typically a section of a parabola with its long axis horizontal. It produces a vertical fan-shaped beam that permits target detection even if the ship is pitching or rolling. For aircraft-height finding, the long axis of the parabolic section is vertical, and the antenna forms a horizontal fan beam. This beam can be moved through a 90-degree angle. Both of these antennas can be rotated through 360 degrees.

The size and shape of the antenna are determined by the transmitted frequency. The antennas of radars operating at the low end of the microwave band can be large, but those operating at the higher frequency ranges are smaller.

Receivers. The received signals from the transmit-receive switch are mixed with a local oscillator signal to produce an intermediate frequency (i-f) signal that is easier to amplify and process. Typical values are 30 or 60 MHz.

Most local oscillators are solid state devices, which have replaced reflex klystrons. Mixing occurs in a crystal cavity and the resulting i-f signal is amplified by the i-f amplifier and then fed to the detector, which produces a video signal. There are several parallel receiver channels in radar that process more than one return simultaneously.

The voltage of the video signal is proportional to the strength of the received signal. This video is amplified for use in the CRT of the radar display.

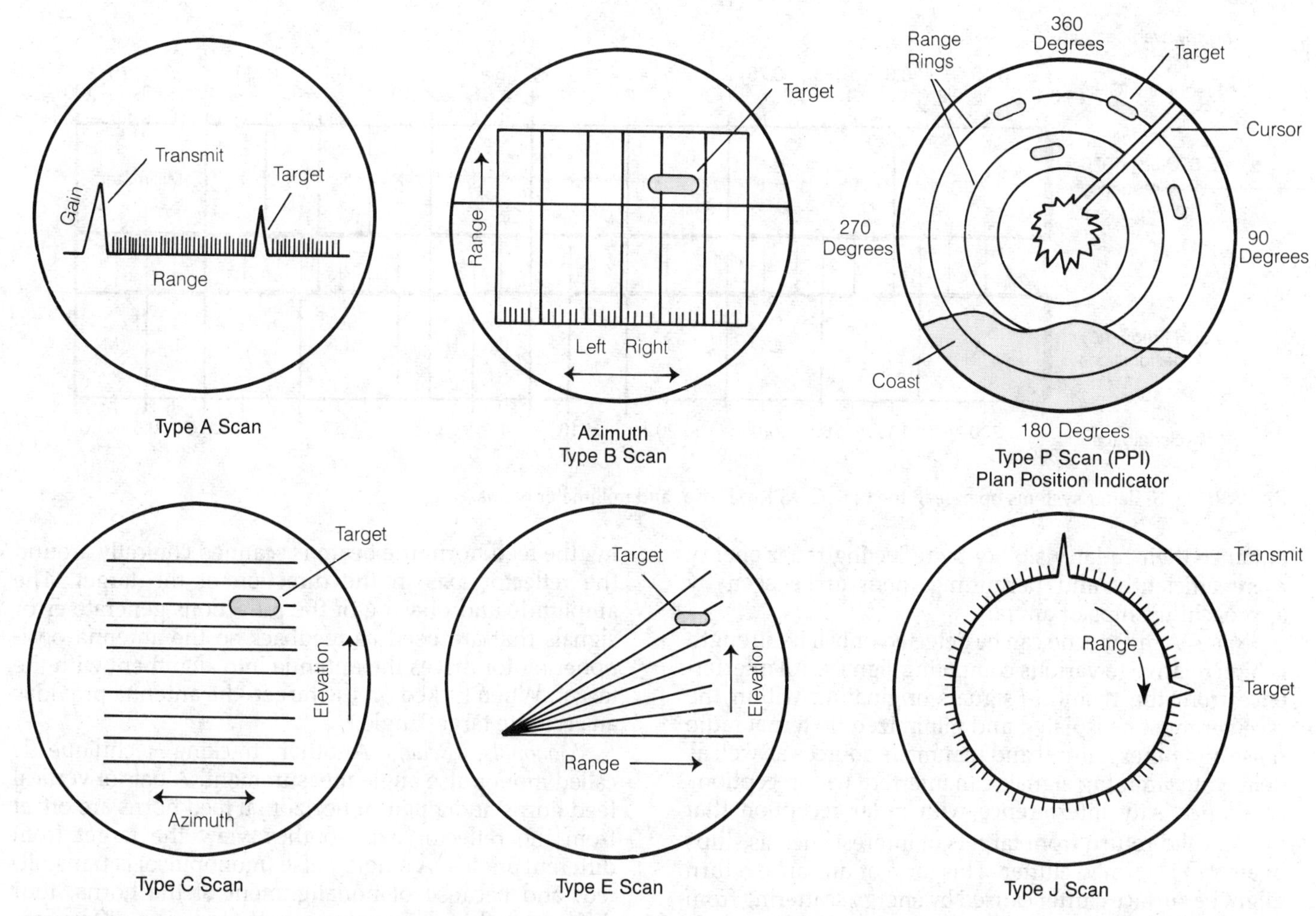

RADAR: Fig. 2. Oscilloscope displays offer many different presentations for target information, but the plan position indicator (PPI) is the one most widely used.

Radar Displays. The presence of the target can be indicated in many different CRT presentations as shown in Fig. 2. Of these, the plan position indicator (PPI) display is the most commonly used intensity-modulated display. It is also the easiest to interpret. In this polar coordinate presentation, the radar is represented at the center of the screen. Objects (targets) show up as illuminated spots (pips) with their positions correctly oriented with respect to true compass direction and other objects. Their movement on the screen is relative to that of the platform (ship or aircraft). Range and azimuth information can be obtained directly from the display. Land masses appear as streaks of light with contours or prominent features such as harbors, cliffs, or piers recognizable for navigation purposes.

Many displays other than the PPI are in use, but these are used primarily for fire control and weapons aiming. The type A scan was the first presentation used in early radars, but range determination had to be made manually. It is the best example of a deflection-modulated display.

PPI format and other displays are being produced as computer-generated synthetic video conveying the same information more clearly. Raster scanning replaces the polar display of the PPI. These displays are now able to show targets as symbols and provide alphanumeric notation about the targets. Moving objects can be labeled for special attention, and stationary objects can be deleted.

PPI Radar-Frequency Bands. Radar carrier frequencies are presented in Fig. 3. The spectrum is divided into bands. Most long-range air-search radars operate in the L band of 1 to 2 GHz. Long-range maritime navigation radars operate in S band (2 to 4 GHz) known as the 10 centimeter band. Weather radars are in C band (4 to 8 GHz). Short-range maritime navigation and higher definition weather radars operate in X band (8 to 12 GHz) known as the 3 centimeter band. Radars operating in the millimeter wave length K bands have the highest definition, but their ranges are limited for terrestrial applications because of atmospheric attenuation.

Target Characteristics. The amount of signal reflected by a target or object is a function of its size, shape, and composition. The return will change with shifts in target position or radar carrier frequency. Some targets made of plastic, wood, or other nonmetallic substances can be virtually invisible to radar. Some materials have been designed to absorb radar energy. Stealth aircraft with low vertical profiles exploit these materials. Small boats and aircraft are often equipped with metal corner reflectors to

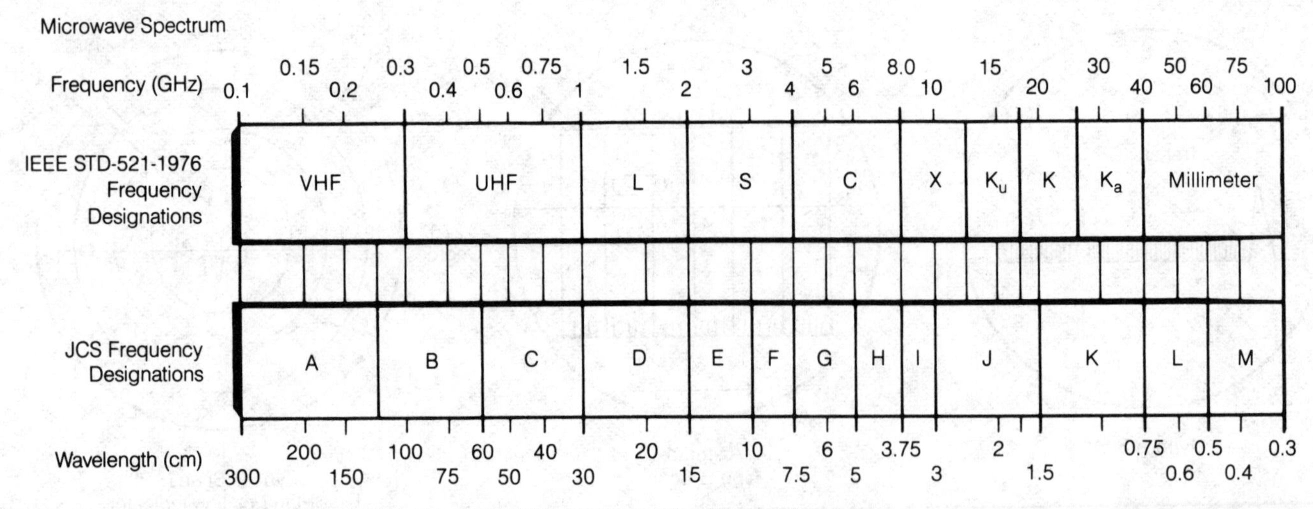

RADAR: Fig. 3. Radar systems operate in the L, S, C, X, Ku, K, Ka, and millimeter bands.

enhance their radar visibility by reflecting radar energy more efficiently and permitting them to be seen by approaching ships or aircraft.

Noise. A radar echo can be detected only if its strength is high relative to various competing signals. All interference from the rf and i-f signals originating within the receiver must be isolated and minimized. External radio noise and other natural and manmade sources as well as deliberate jamming signals can interfere with reception.

Clutter. Any interference with radar reception that masks valid return from targets of interest such as ships or aircraft is termed clutter. This form of nuisance return might be surface clutter caused by energy scattering from the sea near the transmitter; it increases with waveheight or sea-state conditions. Volume clutter includes fog, clouds, rain, snow, and birds. Deliberate clutter to disable a radar can be introduced with chaff (strips of metal foil) scattered in the vicinity of the target. The strips must be cut to fractional wavelengths of the transmitter frequency to be most effective. Widely used as a form of radar countermeasure or jamming, chaff is also considered volume clutter. The effects of clutter backscatter are more pronounced at the higher frequencies. For example, it is generally worse for an X-band than an L-band system.

Height-Finding Radars. For the rapid determination of the height of multiple targets, three-dimensional antennas can be used. One type has a vertical parabolic cylindrical reflector with multiple feed horns. The reflected energy from the horns produces a stack of horizontal fan beams. Another type of 3-D antenna has a motor-driven elevation scanning feed on a parabolic dish. This antenna provides a rapidly oscillating horizontal fan beam. Both antennas are rotated through 360 degrees to obtain the third dimension.

Tracking Radars. Radars have been built specifically for tracking one target very accurately. The Cassegrain dish antenna of these radars is steered in the direction of a target that has been identified. After the tracking radar acquires the target, it receives a continuous stream of data. The feed horn, mounted at the focal point of parabolic dish, produces a narrow pencil beam. By nutating the feed horn, the beam is scanned conically around the reflector axis in the direction of the target. The amplitude and phasing of the reflections generate error signals that are used as feedback so the antenna positioner motor drives the antenna into alignment with the target. When locked on the target, the antenna provides an accurate target angle.

Monopulse Radars. Another tracking technique is called monopulse angle measurement. A pair of vertical feed horns and a pair of horizontal feed horns are offset from the reflector axis so they view the target from different angles. A single pulse (monopulse) is transmitted, and because of nonalignment of the horns, four different reflected signals are returned. The differences between the signals create error signals to position the antenna axis directly on the target. Monopulse antennas are used for weapon fire control and missile guidance. They are difficult radars to jam because of their low repetition rate.

Phased-Array Antennas. Massive ground-based, phased-array antennas are used to detect and track multiple satellites, missiles, and other space vehicles. Large shipboard phased-array antennas are used for early warning and tracking multiple missile or aircraft threats. A phased-array antenna system is shown in Fig. 4. The array consists of hundreds of individually fed dipole elements. Under computer control, the rf output of the elements is formed into beams that can be electronically scanned at high speed in two dimensions. This permits more rapid sampling of multiple targets than would be possible with a rotating antenna. Three arrays provide hemispheric coverage. Although complex and expensive, the phased-array radar has a gradual failure mode and can continue to function even if many individual elements fail. *See* PHASED-ARRAY ANTENNA in the ANTENNA DIRECTORY.

Continuous-Wave Doppler Radar. Discrimination between fixed and moving targets is possible with Doppler radar. A continuous-wave transmitter is used, and the return energy is detected by mixing it with some of the transmitter power. Fixed targets produce a constant voltage, but moving targets produce an alternating volt-

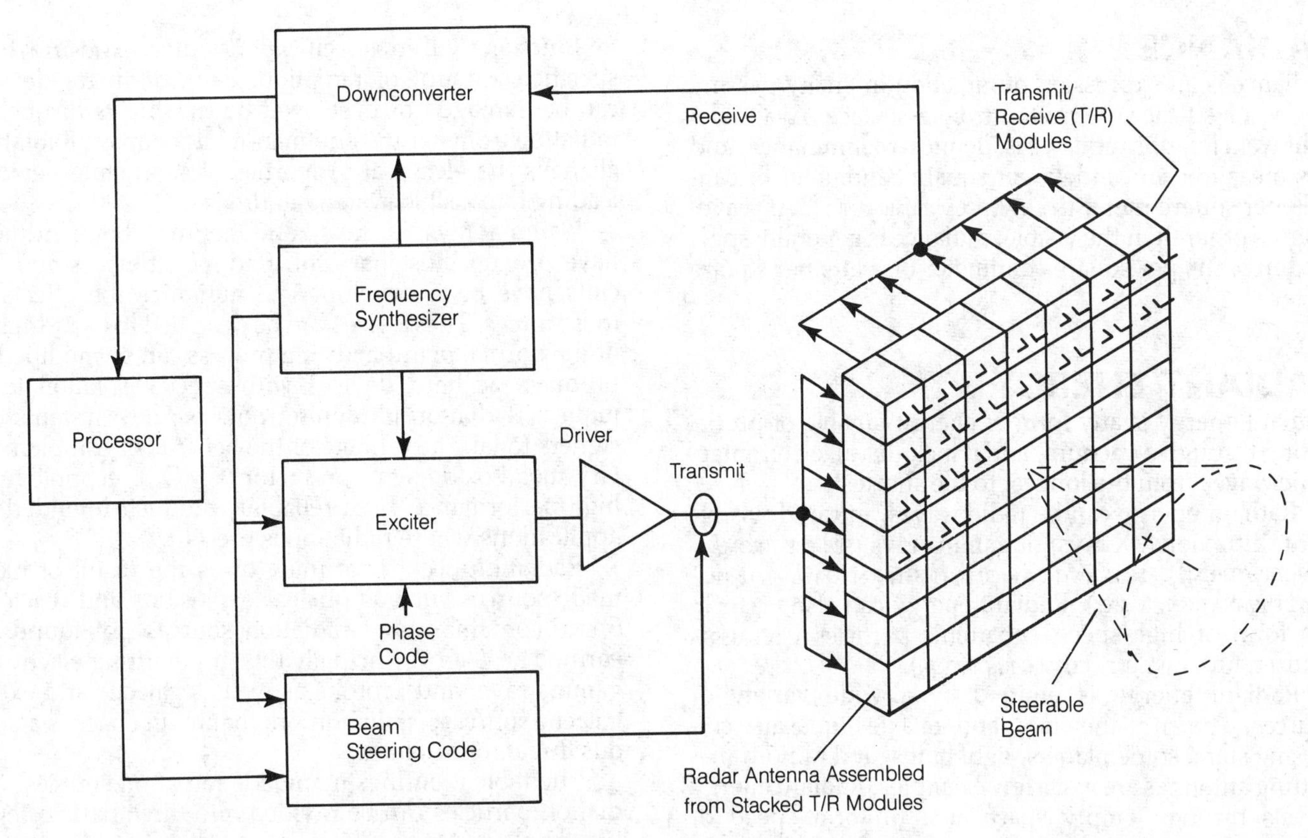

RADAR: Fig. 4. A phased-array antenna permits the radar to be scanned electronically in three dimensions with no movement of the antenna. The output from the active transmit-receive modules is formed into a steerable beam.

age at the Doppler frequency difference between the transmitted and received signals. This type of radar is suitable only for measuring radial velocities of targets and for the detection of presence rather than accurate position of moving targets. However, the cw source can be pulsed to obtain accurate range information.

Moving Target Indication (MTI) Radars. To obtain range information, the transmitted cw carrier is pulse modulated. The received pulses will be small segments of the cw return. A fixed target produces uniform pulses, but pulses from a moving target vary in amplitude periodically. Moving Target Indicator or MTI radars differentiate between the stationary and moving targets by subtracting echo pulses from exact replicas of their corresponding transmitted pulses. This subtraction process produces constant amplitude pulses for stationary targets and varying amplitude pulses for moving targets. It is possible to display only the moving targets by cancelling out the constant-amplitude returns.

RADIAN

The radian is the angle measure subtended around the perimeter of a circle over a length equal to the radius of the circle. The circumference of a circle is 2π times the radius, so there are 2π radians in a full circle of 360 degrees.

The radian is used as an angle of measure by mathematicians and physicists. To 10 significant digits, a radian is equivalent to 57.29577951 degrees.

For most practical purposes, degrees can be obtained from radians by multiplying by 57.3; an angle measure in radians can be obtained from degrees by multiplying by 0.0175. Use the nomograph in the illustration for reference. *See also* DEGREE.

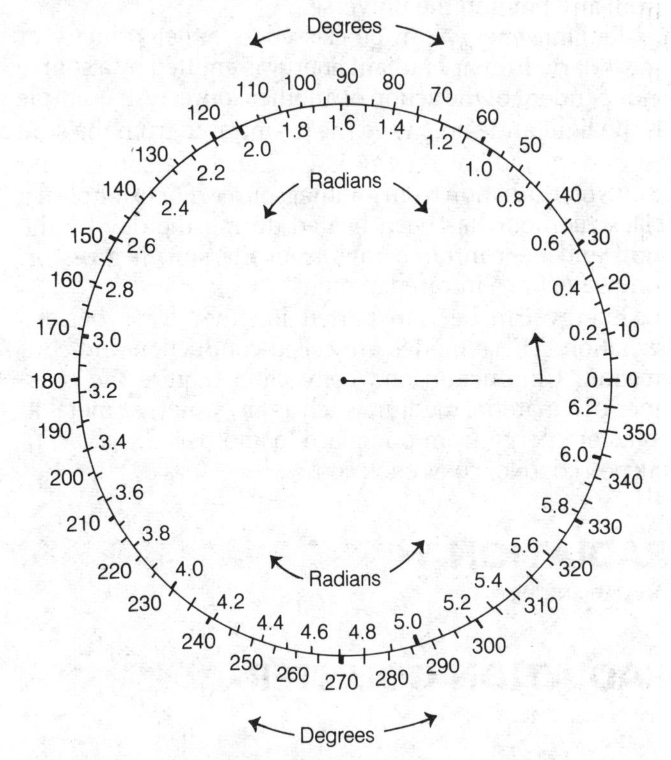

RADIAN: A nomograph relating radians and degrees of angular measure.

RADIANCE

Radiance is an expression of radiation intensity passing through a surface or emitted from a surface. At visible-light wavelengths radiance is identical to luminance, and it is measured in candela per square centimeter or candela per square meter (*see* CANDELA, LUMINANCE). At wavelengths other than the visible, radiance is generally specified in watts per square centimeter or watts per square meter.

RADIANT ENERGY

Radiant energy is any form of energy capable of propagating through a vacuum, and it includes all electromagnetic waves from the longest to the shortest.

Radiant energy can be radio waves, infrared, visible light, ultraviolet, X rays, or gamma rays (*see* ELECTROMAGNETIC SPECTRUM, GAMMA RAY, INFRARED, LIGHT, RADIO WAVE, ULTRAVIOLET RADIATION, X RAY). Radiant energy can also occur in the form of high-speed subatomic particles (*see* ALPHA PARTICLE, BETA PARTICLE, COSMIC RADIATION).

Radiant energy is emitted by a wide variety of sources. The sun, the stars, and galaxies, quasars, collapsing stars, some planets, light bulbs, and radio-transmitting antennas are just a few examples. Radiant energy travels through empty space at a uniform speed of approximately 186,280 miles per second (299,800 kilometers per second). This speed is the same when measured from any point in the universe.

Radiant energy can be classed as either primary or secondary. Primary radiant energy is emitted by a source independent of the action of another source. An example is the light and shortwave (near-) infrared from the sun. Secondary radiant energy is emitted by a source as a result of irradiation from another source. For example if a black tile floor has been heated during the day by the light and near-infrared rays from the sun, it gives off longwave (far-) infrared at night.

Energy can be transported in other ways besides radiation. These modes are called conduction and convection. Conduction and convection require the presence of a material medium such as air, water, or metal to transfer energy from one place to another. *See also* CONDUCTION COOLING, CONVECTION COOLING.

RADIATION

See RADIANT ENERGY.

RADIATION COUNTER

See GEIGER COUNTER.

RADIATION HARDNESS

Radiation hardness refers to the ability of a semiconductor device to withstand absorbed nuclear radiation without alteration of its electrical characteristics. A semiconductor device is said to be radiation hardened (rad-hard), radiation tolerant, or radiation resistant if it can continue to function within specifications after exposure to a specified amount of radiation. Semiconductor devices can be damaged or destroyed by the effects of nuclear radiation from natural and man-made sources. Radiation changes the electrical properties of solid state devices, leading to possible system failure.

Gamma rays, X rays, and neutron bombardment have proven most harmful. Rad-hard devices and circuits have been developed to minimize the effects of these forces. The devices can be designed to be rad-hard, or the normal manufacturing process can be modified to produce rad-hard devices with special isolation techniques. Radiation hardening now permits systems designers to take advantage of the benefits of complementary metal-oxide semiconductor (CMOS) technology in high-performance, high-reliability products intended for applications where radiation is present.

Radiation occurs naturally or as the result of man-made sources such as nuclear explosions and reactors. Space contains many radiation sources not found on earth. These are principally the sun emitting electrons, gamma rays, and protons as well as galactic and extra-galactic sources. Radiation originating in space is a serious threat for satellites.

The table identifies manmade radiation sources. Radiation particles can be divided into three basic categories: photons, charged particles, and neutrons.

Charged Particles. An alpha particle is a helium nucleus or a helium atom without its electrons. It also travels at speeds close to the speed of light, but it can be stopped by a sheet of paper.

A beta particle is an electron traveling at nearly the speed of light. The same as an electron orbiting a nucleus, it differs only in speed. The beta particle can travel about 20 feet in air and can be stopped by 1/16-inch-thick aluminum barriers.

Ions are charged particles formed when one or more electrons are removed or added to a previously neutral atom or molecule. There are, for example, silicon ions.

Neutrons. A neutron is a particle without an electric charge. It has a mass that is approximately the same as a proton. Neutrons are naturally bound to the nucleus of

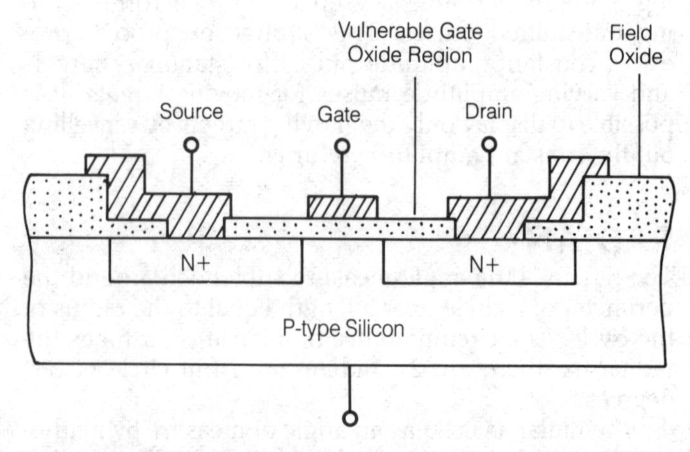

RADIATION HARDNESS: A sectional view of an N-channel metal-oxide semiconductor transistor showing where gate oxide is vulnerable to radiation damage.

RADIATION HARDNESS: Summary of irradiating particles.

Source	Charged Particles			Photons	Neutrons
Nuclear Explosion or Reactor	α (Alpha) Particles	β (Beta) Particles	Ions	γ (Gamma) Rays	Neutrons
	Protons	Electrons		X Rays	

an atom, but they can be displaced in various types of nuclear interactions.

Photons. Gamma rays are photons or quanta of energy with characteristics that are identical to those of X rays. The difference between gamma rays and X rays is their source. Gamma rays come from the atomic nucleus, but X rays are generated by processes outside the nucleus. Gamma rays and X rays have no electrical charge, travel at the speed of light, and can be stopped by a concrete or lead shield.

The interactions of a particle with a material will depend on the properties of each. The properties of particles are: 1) mass, 2) charge, and 3) kinetic energy. The properties of materials are: 1) mass, 2) charge, and 3) density. The interaction of particles and energies can be classified into two main mechanisms. These dominate the effect of radiation in materials of concern in electronics and both can cause temporary (transient) or permanent damage to semiconductors:

1. Displacement of atoms from their lattice structure (displacement damage).
2. Generation of electron-hole pairs (ionization). Both effects can cause temporary (transient) or permanent damage.

The energy transferred to a material by ionizing radiation is measured in rads (radiation absorbed dose). One rad is equal to the energy of 100 ergs per gram of material. The material must be specified because this energy will differ with each material:

One rad (Si) = 100 ergs per gram (Si)

One gray (gy) = 100 rads

The ionizing dose rate is referenced in rad (Si)/seconds. Particles are referred to in terms of concentration as well as the time integral of concentration:

Flux = Particles per square centimeters $\times$ seconds
Fluence = Particles per square centimeters

Displacement damage, one of the two mechanisms that affect semiconductor devices, is caused by heavy charged particles and neutrons. Neutron radiation presents more problems in bipolar devices than CMOS ICs (integrated circuits) until a fluence of 10^{15} neutrons per square centimeters or greater is reached. Heavily charged particles can cause single event upset.

Ionization is the principal agent that damages or destroys CMOS devices. It is caused by photon (gamma ray) interactions, fast neutron ($E > 1$ MeV) interactions, and charged (alpha and beta) particles. However, of these, gamma radiation is the primary source of this ionization.

Radiation exists naturally and can also be generated by man at various levels. These levels range from less than one rad to over a megarad (10^6). Human beings will become sick if they absorb radiation doses of 200 rads or more. Exposure to 500 rads will prove fatal to about 90 percent of the population in one to two weeks. Exposure to a dose of 10,000 rads will cause rapid death. Against this background, a conventional CMOS IC will absorb about 3000 rads before failure, but a radiation-hardened CMOS will withstand doses of up to a megarad.

As the dose increases, the number of carriers generated in silicon will increase. Out in space it might take many years for a device to absorb high levels of radiation. For example, it could take 20 years for an IC to absorb a total dose of 100,000 rads (Si). However, in the presence of a nuclear explosion, a device might reach this total dose within hundreds of nanoseconds. This type of photon exposure is referred to as transient radiation.

A typical MOS transistor, as shown in the figure, can be viewed as a capacitor with the metal gate and P-type silicon as the capacitor plates and the gate oxide (silicon dioxide SiO_2) as the dielectric. Ionizing radiation affects the gate oxide and the field oxide regions. It causes threshold voltage shifts and degrades channel mobility. Increased radiation is believed to add to the number of irregularities that already exist at the interface between the silicon and silicon dioxide.

Semiconductor manufacturers have developed various methods for radiation hardening semiconductor devices. The device can be initially designed to be rad-hard, or it can be processed to be rad-hard. It has been found recently that standard CMOS devices can be hardened without the need for special protective guard bands.

RADIATION LOSS

In a radio-frequency feed line there is always some loss because no feed line is 100-percent efficient. The conductor resistance and the dielectric loss contribute to the dissipation of power in the form of heat. Some energy can also be radiated from the line, and this energy, because it cannot reach the antenna, is considered loss.

In an unbalanced feed line such as a coaxial cable, radiation loss can occur because of inadequate shield continuity (*see* COAXIAL CABLE). Radiation loss can also occur because of currents induced on the outer shield by the field surrounding the antenna. Radiation loss in a coaxial feed line can be minimized by using cable with the highest degree of shielding continuity, and also by ensuring that the ground system is adequate at radio frequencies. A balun, placed at the feed point where a coaxial cable joins a balanced antenna system, might also be helpful.

In a balanced feed line such as open wire or twin-lead, radiation loss results from unbalance between the currents in the line conductors. Ideally, the currents should be equal in magnitude and opposite in phase everywhere along a balanced line. If this is not the case, radiation loss will occur. Imbalance in a parallel-conductor line can result from a physically or electrically asymmetrical antenna, lack of symmetry in the antenna/feed configuration, or poor feed-line installation. *See also* BALANCED TRANSMISSION LINE, OPEN-WIRE LINE, TWIN-LEAD.

RADIATION RESISTANCE

When radio-frequency energy is fed into an electrical conductor some of the power is radiated. If a nonreactive resistor is substituted for the antenna, in combination with a capacitive or inductive reactance equivalent to the reactance of the antenna, the transmitter will behave in precisely the same manner as it would when connected to the actual antenna. The resistor will dissipate the same amount of power as the antenna would radiate. For any antenna there is an equivalent resistance in ohms. The value of this theoretical resistor is called the radiation resistance of the antenna.

Radiation resistance depends on several factors including the length of the antenna, as measured in free-space wavelengths. The presence of objects near the antenna such as trees, buildings, and utility wires can also affect the radiation resistance.

If an infinitely thin, perfectly straight, lossless vertical monopole antenna is placed over perfectly conducting ground and there are no objects in the vicinity to affect the radiation resistance, its value, as a function of the vertical-antenna height in wavelengths, is as shown at A in the illustration.

For a quarter-wavelength vertical antenna, the radiation resistance is approximately 37 Ω. As the conductor length decreases, the radiation resistance also decreases, becoming zero when the conductor vanishes. As the conductor becomes longer than a quarter wavelength, the radiation resistance increases, and becomes larger without limit as the height approaches a half wavelength. These are theoretical values. In practice, the radiation resistance is somewhat lower than the figures given in A. However, the graph is a good approximation for most practical purposes. If a perfectly straight, infinitely thin, lossless conductor is located in free space and fed at the exact center, the value of the radiation resistance, as a function of the antenna length in wavelengths, is given by the graph at B.

When the conductor measures a half wavelength, the radiation resistance is approximately 73 Ω. As the conductor becomes shorter, the radiation resistance becomes smaller, approaching zero. As the conductor is lengthened, the radiation resistance increases without limit as the full-wavelength value is approached. As with

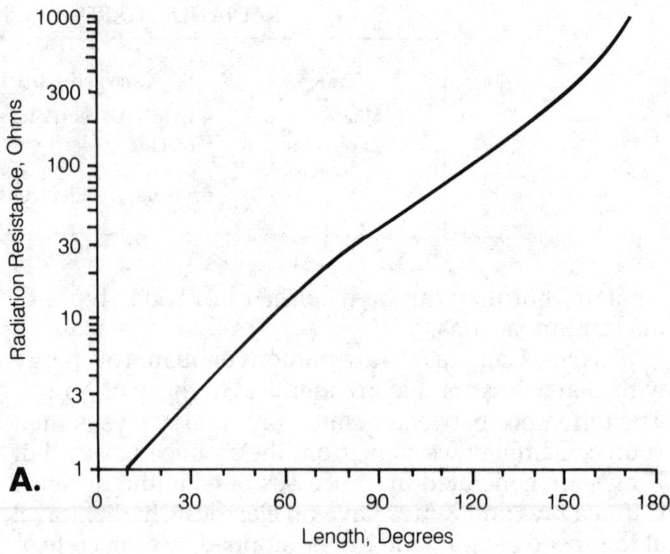

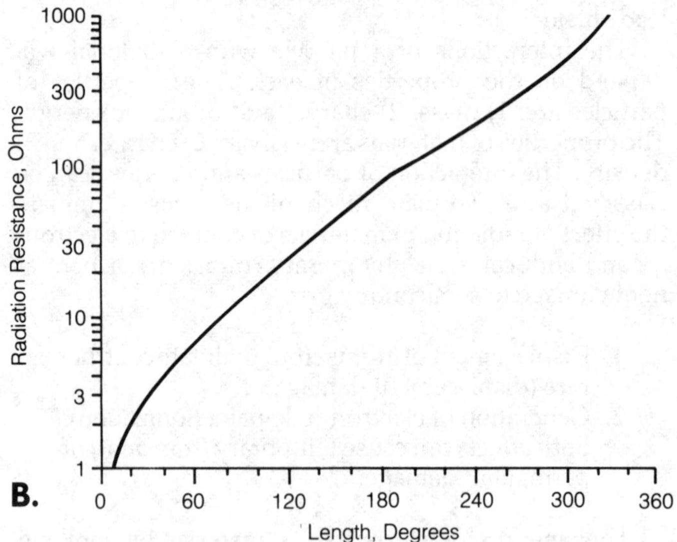

RADIATION RESISTANCE: Radiation resistance plotted for a vertical monopole (A) and a center-fed radiator (B) in free space.

the vertical system, the values shown are theoretical, but they are approximate for most center-fed vertical or horizontal antenna systems.

In practice, it is desirable to have a very large radiation resistance because the efficiency of an antenna depends on the ratio of the radiation resistance to the total system resistance. If the radiation resistance is R and the loss resistance is S, then the total system resistance, Z, is given by:

$$Z = R + S$$

and the efficiency, in percent, is:

$$\text{Eff}\ (\%) = \frac{100R}{Z}$$

The loss resistance in an average antenna system is a few ohms, but may be as high as 30 to 50 Ω or more. If the radiation resistance is very low, an antenna is inefficient. *See also* ANTENNA EFFICIENCY.

RADIOACTIVITY

Radioactivity is the emission of electromagnetic energy of extremely short wavelength. In particular, radioactivity refers to X ray, gamma-ray emission, and also to high-speed atomic particles (*see* ALPHA PARTICLE, BETA PARTICLE, GAMMA RAY, X RAY). Intense or excessive radioactivity is hazardous to life because its quanta contain large amounts of energy. High-energy photons or other particles can alter the nuclei of atoms upon impact. Radioactivity can cause genetic damage in humans, animals, and plants, and it can damage semiconductor devices. (*See* RADIATION HARDNESS).

Radioactivity can be measured in terms of relative intensity, or it can be measured as a total exposure (dose) over a given period of time. The unit of X-ray and gamma-ray dosage is the roentgen. The unit of alpha-particle and beta-particle dosage is the rem. *See also* DOSIMETRY, RADIANT ENERGY, REM, ROENTGEN.

RADIO DIRECTION FINDER

A radio direction finder (RDF) is a radio receiver with a directional antenna and a visual null indicator. When visibility is poor, an RDF can help to fix position, and the navigator can home in on a transmitter near his or her destination. The directional antenna on most RDFs can be rotated so the receiver is fastened to its support. An RDF is basically an improved version of a portable receiver with a rotating loop antenna on top. The antenna can either be a loop or a bar. The loop is about a foot in diameter, and the plastic-covered bar measures about an inch square by about six inches long.

As the antenna is rotated through 360 degrees, the output signal strength will pass through two positions of maximum signal strength and through two positions of minimum sensitivity called nulls. In theory, the peaks will be 180 degrees apart, and the nulls are located 90 degrees on either side. Because the nulls are sharp and precise and the peaks are broad and poorly defined, the nulls are used for direction finding. On most sets, the antenna is turned by mechanical gearing to keep the user's hand from interfering with the reception.

The operator can judge by ear the position of the antenna at a null or peak, but most RDFs have an analog meter to give a visual indication. RDFs normally cover three frequency bands: a low-frequency (lf) beacon band, the standard amplitude modulation (AM) radio broadcast band, and the 2 to 3 MHz communications band. The marine radio beacons along or just off the U.S. coast and near the Great Lakes are operated by the Coast Guard on frequencies between 285 and 325 kHz. They provide the most accurate bearings for shipping. However, aeronautical beacons that operate at lower and higher frequencies can also be used by ships for direction finding.

RADIO FREQUENCY

An electromagnetic disturbance is called a radio-frequency (rf) wave if its wavelength falls within the range of 33 km to 1 mm, the frequency range of 9 kHz to 3000 GHz.

The rf spectrum is divided into eight bands, each representing one order of magnitude in frequency and wavelength. These bands are called the very low, low, medium, high, very high, ultrahigh, super high, and extremely high frequencies. They are abbreviated, respectively, as vlf, lf, mf, hf, vhf, uhf, shf, and ehf. Super high frequency and extremely high frequency rf waves are also called microwaves (*see* BAND).

Radio-frequency waves propagate in different ways, depending on the wavelength. Some waves are affected by the ionosphere, troposphere, or other environmental factors (*see* PROPAGATION CHARACTERISTICS).

Radio frequencies represent a sizable part of the electromagnetic spectrum. Wavelengths shorter than 1 millimeter are (in descending order) infrared, visible light, ultraviolet, X rays, and gamma rays. *See also* ELECTROMAGNETIC SPECTRUM.

RADIO-FREQUENCY INTERFERENCE

See ELECTROMAGNETIC INTERFERENCE, LINE FILTER.

RADIO-FREQUENCY TRANSFORMER

See TRANSFORMER.

RADIO INTERFERENCE

See INTERFERENCE.

RADIOISOTOPE

An element always has the same number of protons in its nucleus. The number is called the atomic number. The number of neutrons, however, can vary, resulting in changes in the atomic weight. The elements with different possible nuclei are called isotopes. Normally, one isotope of a given element is the most stable and is therefore the most common (*see* ATOMIC NUMBER, ATOMIC WEIGHT, ISOTOPE). Some isotopes are unstable and tend to decay, producing radioactivity. These isotopes exist for most elements and are known as radioisotopes.

Carbon, with an atomic number of 6, normally has six neutrons in its nucleus, and its atomic weight is 12. However, some atoms of carbon have eight neutrons, resulting in an atomic weight of 14. This isotope, called carbon-14, is a radioisotope.

Carbon-14 is a relatively stable radioisotope; it takes a long time to decay. Archeologists and other earth scientists can determine the age of objects from the amount of carbon-14 still remaining, as determined with a radiation counter.

Heavy elements generally have more radioisotopes than lighter elements because there are a larger number of different isotope possibilities as the atomic number increases. Some radioisotopes, such as carbon-14, occur naturally. Others are man-made. *See also* ELEMENT.

RADIOMETRY

The science of measuring the heat resulting from optical energy is known as radiometry. Instruments used to measure this heat are called radiometers, and there are several types of radiometers.

The bolometer and thermocouple are two sensors that can measure the heating caused by radiant energy (*see* BOLOMETER, THERMOCOUPLE). These sensors, in conjunction with peripheral measuring equipment, facilitate quantitative measurement of the heat power in microwatts, milliwatts, or watts.

A radiometer is shown in the illustration. An assembly of three or four flat vanes is mounted on a bearing. Each vane is painted black on one side, (the clockwise-facing side), and white on the other side. When light strikes the black side of a vane, heat is produced. This agitates the air molecules adjacent to the vane. When light strikes the white side of a vane, no heat is produced, and the air molecules next to the vane are not affected. Because there is more molecular movement on the black side of each vane, greater pressure is exerted, and the vane assembly moves in the direction of the white faces. The whole assembly rotates with a speed that depends on the intensity of the light. Quantitative measurements can be obtained by counting the number of revolutions of the assembly per unit time.

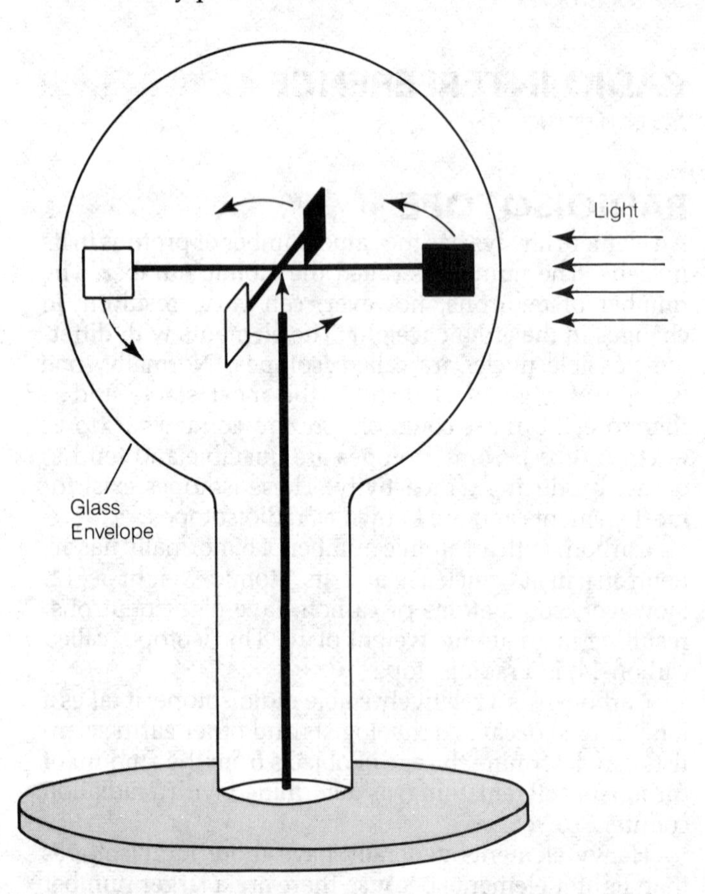

RADIOMETRY: A radiometer measures the ability of light to generate heat energy.

RADIOTELEGRAPH

The term radiotelegraph is used to describe any form of radio emission in which Morse code signals are sent. This may include A1, A2, F1, and F2 emission (*see* EMISSION CLASS).

A transmitter intended for sending messages by Morse code is called a radiotelegraph (*see* CODE TRANSMITTER).

RADIOTELEGRAPH CODE

See INTERNATIONAL MORSE CODE.

RADIOTELEPHONE

Radiotelephone refers to any form of two-way communication by voice. The earliest form of radiotelephone was amplitude modulation or emission-type A3. Today, radiotelephones can use A3 emission and any of the A3A, A3J, A9, F3, F9, P3, or P9 types of modulation. (*See* AMPLITUDE MODULATION, EMISSION CLASS, FREQUENCY MODULATION, PULSE MODULATION, SINGLE SIDEBAND.)

A transmitter or transceiver, designed for the purpose of sending or communicating by voice, can also be called a radiotelephone. *See also* MOBILE TELEPHONE, VOICE TRANSMITTER.

RADIO TELESCOPE

A radio telescope is a sensitive, highly directional radio receiver, intended for intercepting and analyzing radio-frequency noise from space. The first radio telescopes were built by Karl Jansky and Grote Reber in the 1930s. Modern radio telescopes use the most advanced antennas, receiver preamplifiers, and signal processing techniques. Radio astronomy is generally carried out at wavelengths shorter than about 10 meters, usually at ultrahigh and microwave frequencies.

A radio telescope system includes an antenna, a feed line, a preamplifier, the main receiver, a signal processor, and a recorder. (See the illustration.)

A radio-telescope antenna can consist of a large dish or an extensive phased array. The dish antenna can be fully steerable so that it can be pointed in any direction without changing the resolution or sensitivity. The phased array, which may consist of a set of Yagis, dipoles, or dish antennas, provides better resolution because it can be made very large. However, the steerability is more limited; phased arrays can usually be steered only along the meridian (north and south). The rotation of the earth must then provide full coverage of the sky.

Two or more large antennas, spaced a long distance apart, can obtain extreme resolution for a radio telescope. This technique is called interferometry (*see* INTERFEROMETER). The interferometer allows the radio astronomer to probe into the details of celestial radio sources. In some cases, what was originally thought to be a single source has been found to be multiple sources.

The antenna feed line is usually a coaxial cable for very high and ultrahigh frequencies. A waveguide is

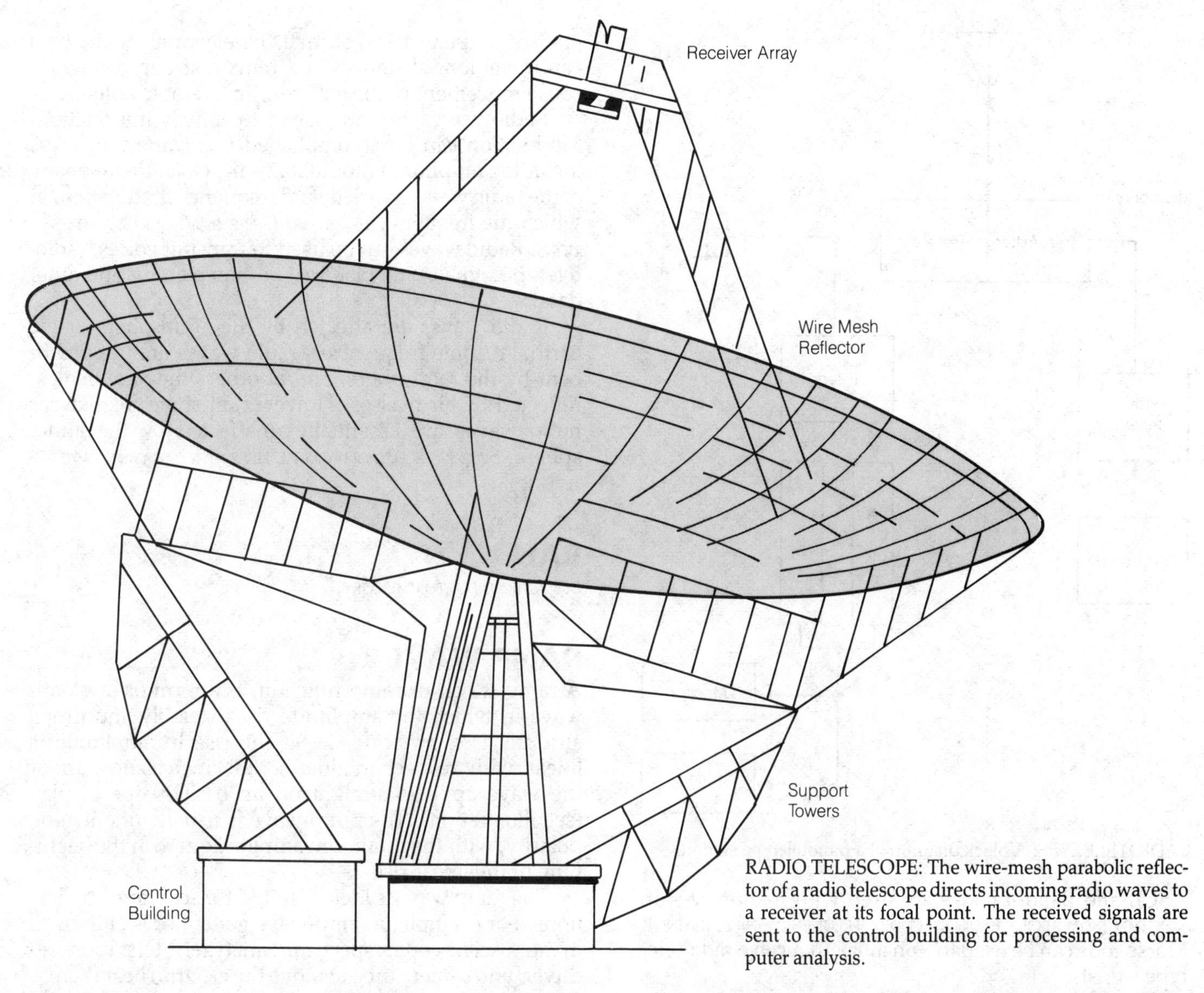

RADIO TELESCOPE: The wire-mesh parabolic reflector of a radio telescope directs incoming radio waves to a receiver at its focal point. The received signals are sent to the control building for processing and computer analysis.

used for microwave reception. The lowest possible loss, and the highest possible shielding continuity, must be maintained. This minimizes interference from earth-based sources, and ensures that the antenna directivity and sensitivity are optimum.

The radio-telescope receiver differs from an ordinary communications receiver. Sensitivity is most important. A special preamplifier is needed (*see* PREAMPLIFIER). The components may be cooled down to just a few degrees Kelvin to reduce the noise in the circuit (*see* CRYOGENICS, SUPERCONDUCTIVITY).

The main receiver of a radio telescope has a relatively large response bandwidth, in contrast to a communications receiver, which usually exhibits a very narrow bandwidth. The wavelength must be adjustable over a wide and continuous range.

A pen recorder and/or oscilloscope, and perhaps a spectrum analyzer, are used to evaluate the characteristics of the cosmic noise (*see* OSCILLOSCOPE, PANORAMIC RECEIVER, SPECTRUM ANALYZER). Computer control has now been integrated into most radio telescopes.

Radio telescopes can see things that cannot be detected with optical telescopes. For example, the hydrogen line at the 21-centimeter wavelength can be easily observed with a radio telescope, although it is invisible with optical apparatus.

The radio telescope, with its high-gain, high-resolution antennas, can be used to transmit signals as well as receive them. This has allowed astronomers to ascertain the distances to, the motions of, and the surface characteristics of the Moon, Venus, and Mercury. When a radio telescope is used in this way, it is called a radar telescope.

RADIOTELETYPE

Radioteletype is a communications method for sending and receiving printed messages by radio. The transmitter must be modulated so that the printed characters are conveyed. A demodulator is needed to convert the audio output of the receiver into the electrical impulses that drive the teleprinter or cathode-ray tube display. Both the modulator and demodulator are usually combined into a single circuit called a terminal. The terminal is a form of modem (*see* MODEM, TERMINAL UNIT).

The most common method of transmitting radioteletype signals is with frequency-shift keying. The two different carrier frequencies used are called the mark (key-down) and the space (key-up) conditions. The

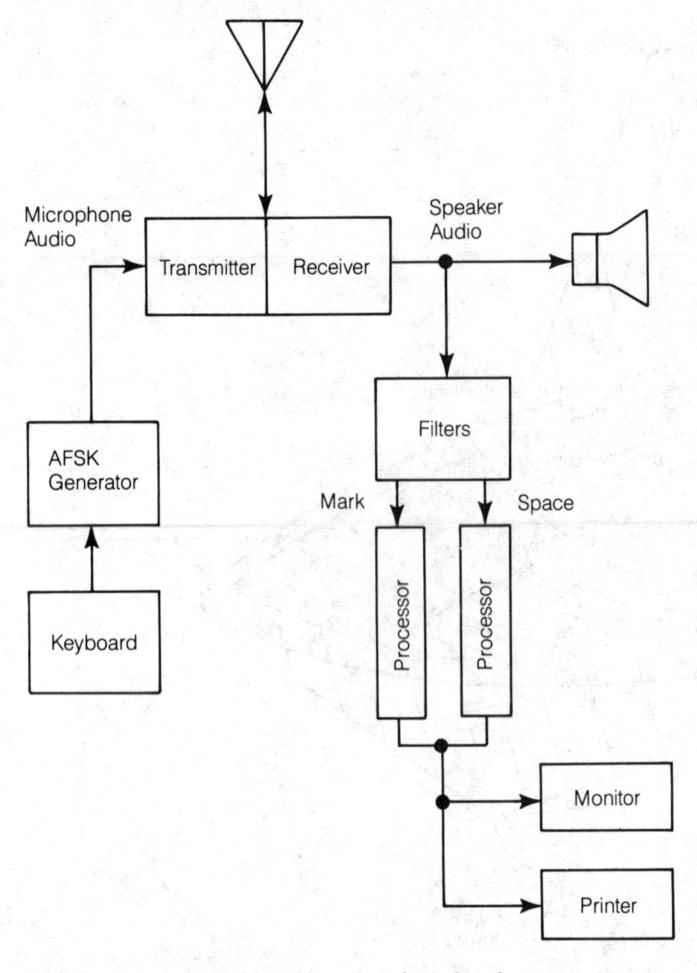

RADIOTELETYPE: A block diagram of a radioteletype station.

ASCII and Baudot codes are used with frequency-shift keying (*see* ASCII, BAUDOT CODE, FREQUENCY-SHIFT KEYING). Morse code can be used to transmit and receive radioteletype signals.

The illustration is a simple block diagram of a radioteletype station with a single-sideband transceiver. The received signals are fed into the terminal-unit demodulator from the speaker terminals of the transceiver. The modulator generates a pair of audio tones which are fed into the microphone terminals of the transceiver. This arrangement can provide communication by ASCII or Baudot at speeds up to approximately 300 baud.

RADIO WAVE

A radio wave is an electromagnetic field with a wavelength between 30 kilometers (km) to 1 millimeter (mm). Radio waves propagate through free space approximately at 186,280 miles per second or 299,800 kilometers per second, the same as the speed of light and all other electromagnetic radiation in free space. (*See* RADIO FREQUENCY.)

Radio waves are generated when electrons or other charged particles are induced to oscillate back and forth at a rate between 10 kHz and 300 GHz. For communications, electrons are made to accelerate in a circuit called an oscillator (*see* OSCILLATOR, OSCILLATOR CIRCUITS). The magnitude of this acceleration is increased by power amplifiers (*see* POWER AMPLIFIER) so that the electromagnetic field can travel long distances and remain strong enough to cause movement of the electrons in a remote antenna.

Radio waves are modulated to convey information. Modulation can be accomplished in a variety of ways including amplitude modulation, in which the intensity of the radio wave is varied, and frequency modulation, in which the frequency is varied (*see* EMISSION CLASS, MODULATION). Radio waves can be used to transmit voices, radioteletype signals, still pictures, motion pictures and other data.

Radio waves are affected by the atmosphere of the earth. At some frequencies, radio waves are trapped or bent by the ionosphere, and at others they are bent or reflected by air masses. However, at some frequencies radio waves are essentially unaffected by the atmosphere. *See also* ELECTROMAGNETIC FIELD, PROPAGATION CHARACTERISTICS.

RAM

See SEMICONDUCTOR MEMORY.

RAMP WAVE

A ramp wave, or ramp function, is a form of sawtooth wave in which the amplitude rises steadily and drops abruptly (*see* SAWTOOTH WAVE). The rise in amplitude is linear with respect to time so the increasing part of the wave appears straight on an oscilloscope display (see illustration). The ramp wave is usually of only one polarity, with the minimum amplitude zero at the beginning of the linear rise.

The ramp wave is ideally suited for scanning applications. For example, a ramp-wave generator is employed in an oscilloscope, spectrum analyzer, television receiver, and camera tube to move the electron beam across the screen at a constant speed. A ramp wave is also used in a circuit called a sweep generator (*see* SWEEP GENERATOR).

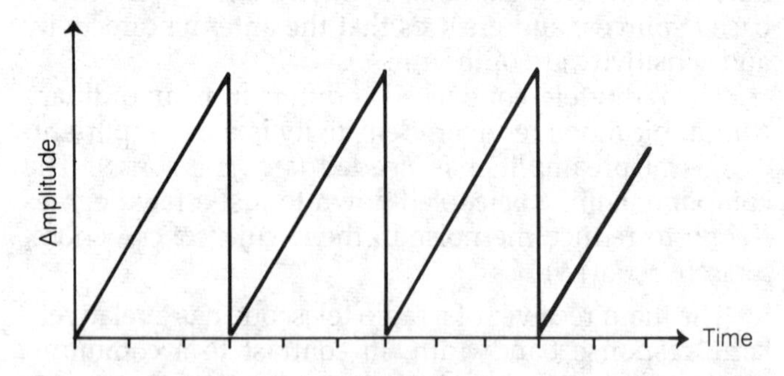

RAMP WAVE: A ramp wave has a linear rise and an abrupt decay.

RANDOM-ACCESS MEMORY

See SEMICONDUCTOR MEMORY.

RANDOM NOISE

Certain forms of electromagnetic noise, especially man-made noise, exhibit a definite pattern of amplitude variation versus time. Other noise shows no apparent relationship between amplitude and time and is called random noise. An example of nonrandom noise is the impulse type. Examples of random noise are thermal noise and shot noise.

Random noise is more difficult to suppress than noise with an identifiable waveform. If noise shows discernible amplitude-versus-time patterns, a circuit can be designed to recognize and eliminate it. No single circuit can eliminate random noise because there are no repetitive patterns.

A limiting circuit that allows the desired signal to compete with the noise is a good defense against strong random noise. The use of narrowband emission in conjunction with a narrow receiver bandpass is also helpful in dealing with strong random noise. The use of frequency modulation may give better results than amplitude modulation if the receiver is equipped with an effective limiter or ratio detector. Directional or noise-cancelling antenna systems can sometimes improve communications in the presence of random noise. *See also* NOISE, NOISE LIMITER.

RANDOM NUMBER

A set of digits is called random if the order exhibits absolutely no pattern. Truly random numbers are difficult, if not impossible, to generate. It is essentially impossible for a person to choose sequences of digits 0 through 9 at random. Algorithms can be used to generate numbers that are almost random, but the existence of the algorithm suggests that such numbers are not truly random.

Irrational numbers are believed to contain digits in random sequence. An example of an irrational number is the natural-logarithm base, e. This number, and all irrational numbers, are nonterminating (infinitely long) and nonrepeating (the digits exhibit no apparent pattern).

Random numbers are used for statistical calculations. A large sample of random digits 0 through 9 has certain properties. For example, as larger and larger samples are taken, the proportion of times each digit occurs, as a fraction of the total number of digits, approaches 0.1. The probability that any given digit will occur in a particular place is 0.1; the probability that any given digit will occur twice in sequence is 0.1×0.1 or 0.01; the probability that any given digit will occur n times in sequence is 0.1^n. There is, however, absolutely no pattern in a sequence of random numbers, so it is impossible to predict which digit will occur in a given place in the sequence.

Statisticians use random-number tables when they need truly random digits. (Whether these digits are truly random, or merely pseudorandom, is an open question.) Pseudorandom digits can be obtained by such exercises as extracting the square root of 2, or by using a computer algorithm designed for pseudorandom-number generation.

RANGE

The range of a mathematical function is the set of values that the dependent variable attains. In the Cartesian coordinate system, the range of a function is usually a subset of the values on the vertical axis (*see* CARTESIAN COORDINATES).

The range of a function may be the entire set of real numbers, or it may consist of a subset of the real numbers. Sometimes the range of a function is an interval. Some functions have ranges consisting of discrete numbers. The range of a function may even be the empty set.

As an example, consider the function $f(x) = x^2$, for the domain consisting of all real numbers. Then the range of f is the set of nonnegative real numbers. The range of the function $g(x) = 1$ is simply the set containing the number 1.

The range of a function often depends on the extent of the domain of values that the independent variable is allowed to attain. For example, the range of $f(x) = x^2$ is reduced to the set of all real numbers greater than 100, if the domain must contain only real numbers greater than 10. *See also* DEPENDENT VARIABLE, DOMAIN, FUNCTION, INDEPENDENT VARIABLE.

In a radio broadcasting or communications system, the distance over which messages can be sent and received is called the range. The communications range depends on the transmitter power, the type and location of the antennas, the receiver sensitivity, and various propagation factors. *See also* PROPAGATION CHARACTERISTICS.

In radar, the distance from the station to a target is called the range. Most radar systems display the range radially from the center of a circular screen. *See also* RADAR.

The term *range* is used in a variety of other situations in electronics. For example, the range of a meter refers to the measurement limits on a specific scale setting. A voltmeter may be capable of measuring from zero to 100 volts on a specific scale. In a power supply, the range is considered to be the upper and lower limits of output voltage that can be obtained with manual adjustment. A power supply may have a range of +5 volts, plus or minus 1 volt. A site for making antenna measurements is also called a range.

RANKINE TEMPERATURE SCALE

An absolute temperature scale, called the Rankine scale, can be used to express temperature relative to absolute zero. The coldest possible temperature on the scale is 0 degrees; above this, temperature is specified in increments that are the same size as the degrees on the Fahrenheit temperature scale.

Since absolute zero is −459.69° F, the freezing point of pure water is 491.69 Rankine. Water boils at 671.69 Rankine.

Temperatures on the Rankine scale can be converted to degrees Kelvin by multiplying by 5/9. To convert degrees Rankine to degrees Celsius, first multiply by 5/9, and then subtract 273. *See also* CELSIUS TEMPERATURE SCALE, FAHRENHEIT TEMPERATURE SCALE, KELVIN TEMPERATURE SCALE.

RASTER

A raster is a uniform rectangular pattern of light produced on the screen of a cathode-ray tube with no signal present. It is formed by deflecting the electron beam rapidly from left to right and from top to bottom in a predetermined pattern of scanning lines..

In the United States, the television-broadcast raster is standardized at 525 lines per frame. The aspect ratio, or ratio of width to height, is 4:3. In some other countries, the number of lines per frame is 625, rather than 525; however, the aspect ratio is almost universally 4:3. *See also* MONITOR, TELEVISION.

RATIO-ARM BRIDGE

A ratio-arm bridge is a circuit for measuring resistance. The bridge is so named because it compares the value of a standard resistance with that of the unknown resistance by electrically determining their ratio.

A schematic diagram of a simple ratio bridge is shown. A potentiometer is adjusted until a null reading is obtained on the galvanometer. If the standard resistance is designated by R_1, and the partial values of the potentiometer are given by R_2 and R_3, then the unknown is:

$$R = \frac{R_1 R_2}{R_3}$$

since the ratio $R{:}R_1$ is equal to the ratio $R_2{:}R_3$.

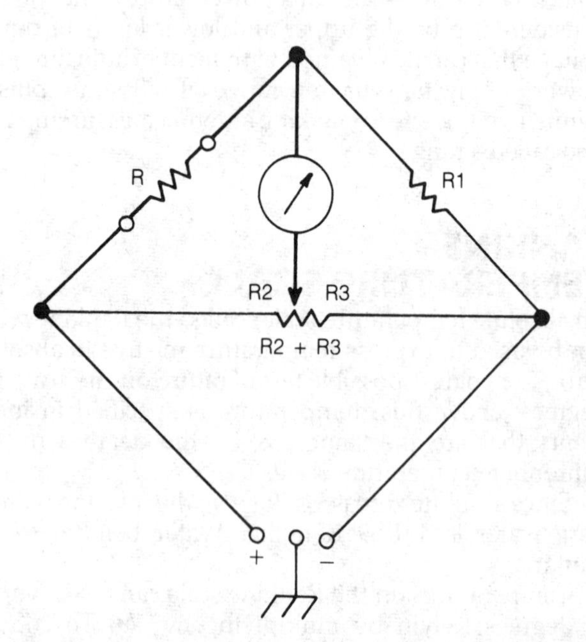

RATIO-ARM BRIDGE: This bridge is used for measuring high resistance values.

RATIO DETECTOR

The ratio detector is a special form of detector for frequency-modulated or phase-modulated signals. The ratio detector, unlike other detectors for these two modes, is insensitive to changes in amplitude. This eliminates the need for a preceding limiter stage. The ratio detector was developed by RCA. Used in most modern high-fidelity and television receivers, the ratio detector is not as commonly used in communications equipment.

A typical passive ratio detector is shown schematically in the illustration. A transformer with tuned primary and secondary windings splits the signal into two components. A change in amplitude will result in equal and opposite changes in both halves of the circuit; thus they cancel out. But if the frequency changes, an instantaneous phase shift is introduced, unbalancing the circuit and producing output.

The ratio detector is not as sensitive as the discriminator, but additional amplification can be included to correct for this slight deficiency, which amounts to approximately 3 decibels. *See also* DISCRIMINATOR, FREQUENCY MODULATION, PHASE-LOCKED LOOP.

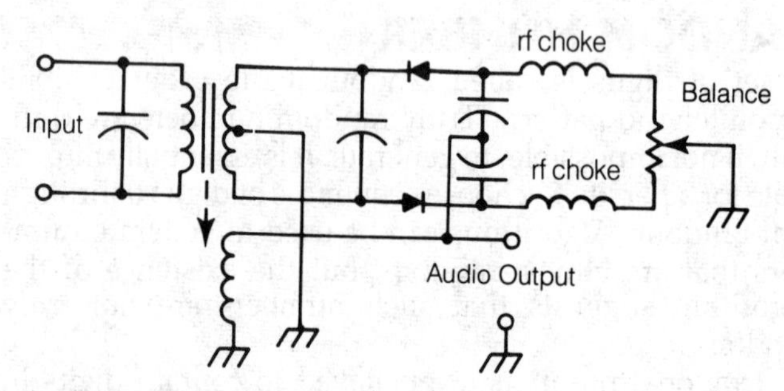

RATIO DETECTOR: A passive ratio detector is insensitive to amplitude changes.

RATIONAL NUMBER

Any number that can be expressed as a ratio between two integers is called a rational number. If x is a rational number, then x can be written in the form:

$$x = m/n$$

where m and n are integers. The set R of all rational numbers is an infinite set. The size of R is the same as the size of the set of integers.

Examples of rational numbers include 4, − 7, 3⅜, and −5.232323. Note that the last of these is nonterminating in decimal form. This is called a repeating decimal. All nonterminating-decimal rational numbers are of the repeating type.

Certain fundamental observations can be made about rational numbers. A sum, difference, product, or quotient of two rational numbers is always rational. However, some functions produce irrational or complex numbers when performed on rational numbers. Examples of such functions are square roots, logarithms, and trigonometric functions.

Some numbers are nonterminating in decimal form, and show no apparent pattern of repeating digits. These are the irrational numbers. The rational and irrational numbers, together, form the set of real numbers. *See also* IRRATIONAL NUMBER, REAL NUMBER.

RAY

In mathematics, a ray, also known as a half line, is a set of points that propagates straight from a single end point. In the Cartesian set of coordinates (*see* CARTESIAN COORDINATES), linear equations can be restricted to obtain rays.

In physics and electronics, a ray is a beam of radiant energy. The amount of energy contained in a ray, as well as the diameter of the ray, is not defined. One way to imagine a ray is to consider that it is the path that a single photon follows (*see* PHOTON). Rays always travel in straight lines through free space in the absence of electric, magnetic, electromagnetic, or gravitational disturbances.

A barrage of high-energy electrons, protons, neutrons, atomic nuclei, or other particles is called a ray. For example, the electron stream in a television picture tube can be called a ray. A beam of alpha particles can be called alpha rays.

High-energy electromagnetic waves are often called rays. For example, infrared, visible, ultraviolet, X rays, and gamma rays are generally thought of as rays, while radio signals (having a wavelength greater than 1 millimeter) are thought of as waves. *See also* ELECTROMAGNETIC FIELD, ELECTROMAGNETIC SPECTRUM.

RC CIRCUIT

See RESISTANCE-CAPACITANCE CIRCUIT.

RC CONSTANT

See RESISTANCE-CAPACITANCE TIME CONSTANT.

RC COUPLING

See RESISTANCE-CAPACITANCE COUPLING.

REACTANCE

Reactance is the opposition, independent of resistance, that a circuit or component offers to alternating current. Reactance is exhibited by inductors, capacitors, antennas, transmission lines, waveguides, and cavities. The unit of reactance is the ohm, and the symbol for reactance is *X*.

Reactances do not dissipate power. Instead, they affect the phase relationship between the current and the voltage in an alternating-current circuit. In a circuit with inductive reactance, the current lags behind the voltage from 0 to 90 degrees. In a capacitive reactance, the current leads the voltage from 0 to 90 degrees (*see* ANGLE OF LAG, ANGLE OF LEAD). The exact phase angle depends on the amount of resistance, as well as reactance, in a circuit or component (*see* PHASE ANGLE).

Mathematically, reactance can be either positive or negative. An inductive reactance is considered positive, and a capacitive reactance is considered negative. This is a matter of convention (*see* CAPACITIVE REACTANCE, INDUCTIVE REACTANCE). In engineering mathematics, reactance takes imaginary values. The ohmic quantities are multiplied by the j factor (*see* J OPERATOR). This results in a model that accurately reflects the way in which reactances behave.

When resistance and reactance occur together, the result is a complex impedance. The resistance forms the real part of a complex impedance, and the reactance forms the imaginary part. A complex impedance can be represented in the form:

$$Z = R + jX \text{ or } Z = R - jX$$

where *R* and *X* represent resistance and reactance, respectively, and $j = \sqrt{-1}$. *See also* COMPLEX NUMBER, IMPEDANCE.

REACTANCE GROUNDING

When a point in a circuit is grounded through a capacitor or inductor, that point is said to be reactively grounded. Reactive grounding is used in radio-frequency (rf) circuits.

A capacitor connected between a circuit point and ground allows some signals to pass while others are shorted to ground. This capacitor is called a bypass (*see* BYPASS CAPACITOR).

An inductor can be connected between a circuit point and chassis ground when it is desired to keep the point at a stable direct-current (dc) voltage while allowing rf signals to exist there. Transistors and tubes, especially in power amplifiers, are often biased in this manner. An antenna can be kept at dc ground while still accepting rf energy by means of reactance grounding (*see* PROTECTIVE GROUNDING).

REACTANCE MODULATOR

A reactance modulator is a circuit that produces a variable reactance from an audio-frequency input signal. The variable reactance is coupled to the crystal or tuned circuit of an oscillator, resulting in frequency modulation or phase modulation.

A reactance modulator can be made with a bipolar transistor, a field-effect transistor or a vacuum tube. However, the most common reactance-modulator circuit today uses a device called a varactor diode. The varactor exhibits a variable capacitance, depending on the amount of reverse-bias applied to it (*see* VARACTOR DIODE).

The illustration on p. 704 is a schematic diagram of a crystal oscillator incorporating a varactor diode to obtain frequency modulation. The capacitance of the diode pulls the crystal frequency by an amount that varies with the instantaneous audio-frequency amplitude. The deviation produced by this method is fairly small, but frequency multipliers can be used to increase it. *See also* FREQUENCY MODULATION, PHASE MODULATION.

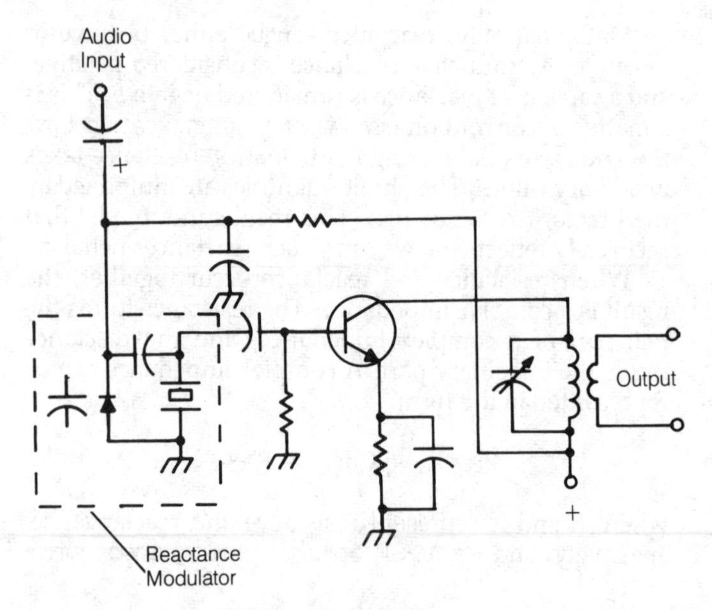

REACTANCE MODULATOR: This oscillator generates frequency-modulated signals.

REACTIVE CIRCUIT

An alternating-current (ac) circuit is called reactive if it contains inductance and/or capacitance, in addition to direct-current (dc) resistance.

In a reactive circuit, the resistance dissipates power, but the reactance does not. Examples of reactive circuits include antennas, transmission lines, waveguides, and tuned circuits. At radio frequencies, most circuits contain some reactance as well as resistance, and thus they must be considered reactive circuits.

A reactive circuit may become nonreactive at one or more frequencies. This occurs when there are equal amounts of inductive and capacitive reactance. Such a condition is known as resonance. *See also* CAPACITIVE REACTANCE, IMPEDANCE, INDUCTIVE REACTANCE, REACTIVE POWER, RESONANCE.

REACTIVE POWER

In an alternating-current (ac) circuit, power can be dissipated in a resistance but not in a reactance. Nevertheless, voltages and currents do occur in reactances. The product of the voltage and the current in a reactance is called reactive power. Reactive power is the difference between the apparent power and true power in a complex impedance (*see* APPARENT POWER, TRUE POWER). Reactive power is sometimes called reactive volt-amperes, or imaginary power.

The proportion of apparent power that is reactive depends on the phase angle in the circuit. The phase angle, in turn, depends on the ratio of reactance to resistance. The larger the phase angle, the more of the apparent power is reactive. In a pure resistance, all of the apparent power is true power. In a pure capacitance or inductance, none of the apparent power is true power (*see* POWER FACTOR).

Reactive power can cause confusing measurements. A radio-frequency wattmeter installed in an antenna system containing reactance will exhibit an exaggerated reading. On a directional wattmeter (reflectometer), the reactive power appears as reflected power and the apparent power appears as forward power. The true power—that which can actually be radiated by the antenna—is the difference between the forward and reflected readings. *See also* REFLECTOMETER.

READ-ONLY MEMORY

See SEMICONDUCTOR MEMORY.

READ-WRITE MEMORY

See SEMICONDUCTOR MEMORY.

REAL NUMBER

A real number is any number that is either rational or irrational. This includes integers such as 3 or −7 and numbers that can be written as a fraction, such as 4/5; it also includes nonterminating, nonrepeating decimals such as the natural-logarithm base e or the value of $\sqrt{2}$. Imaginary and complex numbers, however, are not real numbers (*see* COMPLEX NUMBER).

There are an infinite number of real numbers. A real number is an ordinary number, the size of which can be readily imagined. Numbers can be represented geometrically for example, by an infinitely long line known as the number line. On this line, once the unit size has been decided, every point corresponds to exactly one real number. Furthermore, every real number can be paired up with exactly one point on the line. Mathematicians say that the set of real numbers and the set of points on a line have a one-to-one correspondence or homomorphism.

REAL-TIME OPERATION

In any broadcast, communications, or data-processing system, operation done live is called real-time operation. The term *real time* also applies to computers. Real-time data exchange allows a computer and the operator to converse. This is not true of prerecorded modes.

Real-time operation is a great convenience when it is necessary to store and verify data in a short time. This is the case, for example, when making airline reservations, checking a credit card, or making a bank transaction.

RECEIVER

A receiver is a circuit that intercepts a signal, processes the signal, and converts it to a useful form. The signal can be in any form, from electric currents in a wire, radio waves, modulated light, or ultrasound. A receiver converts signals into audio information, video information, or both.

All closed-circuit or wire-operated receivers have at least three basic components: a signal filter, an amplifier, and a transducer (see A in the illustration). The signal filter eliminates all incoming signals except the desired one. The amplifier boosts the strength of the signal so that it can be detected when applied to the transducer. The transducer converts the signal into a form such as sound, a still picture, or a motion picture. The telephone is a common example of a closed-circuit receiver.

All radio-frequency (rf), visible-light, or other non-wire receivers include five or six fundamental functions: an antenna or receptor (except in a closed-circuit wire system), a front end/wideband filter, an amplification chain/narrowband filter, a detector, an audio and/or video amplifier, and a transducer (B). Electromagnetic waves strike the antenna or other receptor, setting up alternating currents in the conductors.

The front end/wideband filter provides amplification and some selectivity. The amplification chain boosts the weak signal to a level suitable for operating the detector, which extracts the modulation information from the rf energy. The audio or video amplifier gives the detached signal sufficient amplitude to drive the transducer. The transducer converts the detected signal to a form suitable for listening, viewing, driving a set of recording instruments, or some combination of these responses.

The most common type of rf receiver in use today is the superheterodyne circuit. *See also* SUPERHETERODYNE RADIO RECEIVER.

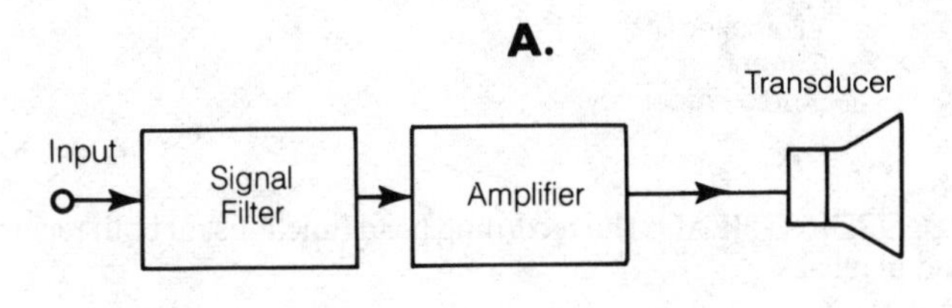

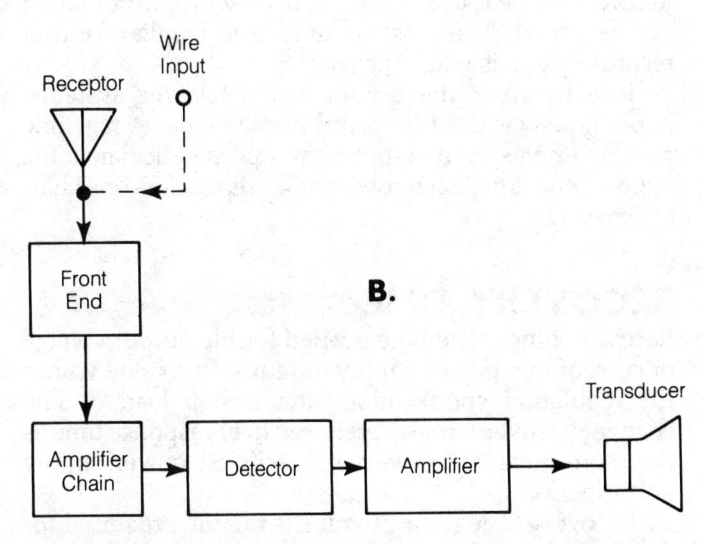

RECEIVER: A block diagram of a basic receiver (A) and a radio-frequency receiver (B).

RECHARGEABLE CELL

See BATTERY.

RECIPROCAL

In mathematics, the reciprocal of a number is the number 1 (unity) divided by that number. If the number is x, its reciprocal is $1/x$. A unique reciprocal exists for every integer, rational number, or real number except 0. Every complex number also has a unique reciprocal.

- The reciprocal of an integer is always a rational number between, and including, −1 and 1.
- The reciprocal of a rational number is always another rational number.
- The reciprocal of an irrational number is always irrational.
- The reciprocal of a real number is always real; if the number is between −1 and 1, the reciprocal is outside that range, and vice versa.
- The reciprocal of a pure imaginary number is always another pure imaginary number.
- The reciprocal of a nonreal complex number is always nonreal and complex.

In electronics, the term *reciprocal* is sometimes used with reference to bilateralism or interchangeability. *See, for example,* RECIPROCITY THEOREM.

RECIPROCITY THEOREM

The Reciprocity Theorem is a rule that describes the behavior of currents and voltages in certain multi-branch circuits.

In a passive, linear, bilateral circuit with two or more separate, current-carrying branches, such as that shown in the illustration, a power supply delivering E_x volts is placed across branch X of this circuit, resulting in a

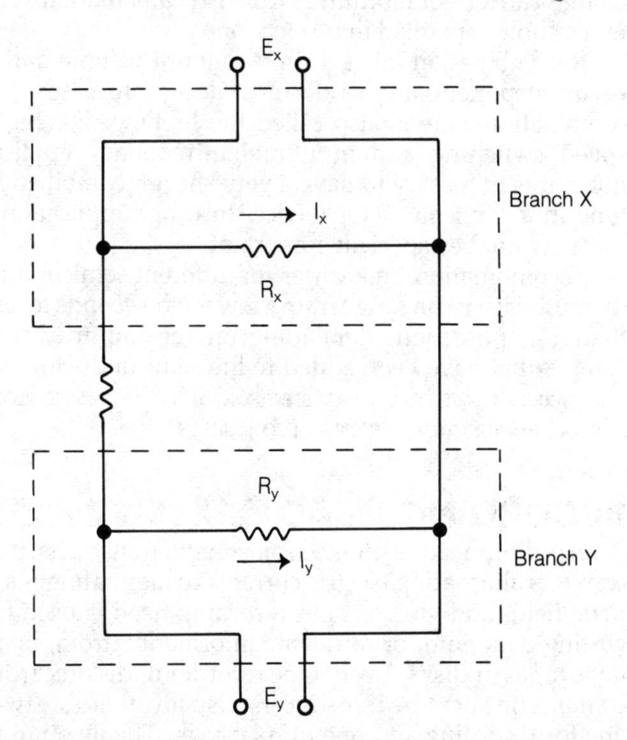

RECIPROCITY THEOREM: This theorem is a direct result of Kirchhoff's and Ohm's Laws.

current I_x in branch X, and a current I_y in branch Y. If the power supply has zero internal resistance, the voltage E_y across branch Y is found from the current I_y and the resistance R_y according to Ohm's Law:

$$E_y = I_y R_y$$

Assume the following circuit changes are made: The power supply with voltage E_x is disconnected from the branch X of the circuit shown and a different supply with voltage E_y and zero internal resistance is connected across branch Y and the same current as before, I_y, continues to flow through resistance R_y.

According to the Reciprocity Theorem, under these circumstances, the current through the resistance R_x will be the same as before, or I_x. Therefore, the voltage across branch X will be:

$$E_x = I_x R_x$$

This is precisely the same voltage E_x of the original supply.

The Reciprocity Theorem is a direct result of Kirchhoff's laws and Ohm's law (see the figure). This theorem can be used to determine currents and voltages in complicated networks. *See also* KIRCHHOFF'S LAWS, OHM'S LAW.

RECOMBINATION

When a charge is applied to a semiconductor material, excess holes and/or electrons are present. When the charge is removed, equilibrium is restored, assuming there is no other disturbance present that might produce excess holes and/or electrons. The process of restoring charge-carrier equilibrium is called recombination. The excess holes are filled in by electrons.

Recombination takes a finite amount of time and it occurs exponentially, as do most decay processes. Recombination time is also called the lifetime. For high-speed switching and in ultrahigh-frequency applications, it is necessary to have a very short recombination time in a semiconductor device. In other applications, this may not be especially important.

Recombination time varies for different semiconductor materials; it can range from a few microseconds to less than 1 nanosecond, depending on the impurity substances that have been added to the semiconductor. *See also* CARRIER MOBILITY, DOPING, ELECTRON, HOLE, N-TYPE SEMICONDUCTOR, P-N JUNCTION, P-TYPE SEMICONDUCTOR.

RECORDING HEAD

A recording head is an electromagnetic transducer that converts alternating electric currents to alternating magnetic fields and vice versa. A recording head is used for writing data onto, or retrieving information from, magnetic tapes or disks. Every tape recorder or disk recorder contains one or more transducers. Usually, there are two: one for recording and one for playback. The illustration shows the typical recording-head configuration in a tape recorder.

The recording head is placed adjacent to a tape or disk. The tape or disk is drawn past the head at a uniform, precise speed. If the head is to be used for recording, an electric current is supplied to it. Alternating magnetic fields, which occur in synchronization with the alternating currents, cause polarization of the fine magnetic particles in the tape or disk. If the head is to be used for playback or information retrieval, the polarized particles cause electric currents to flow in the recording head. *See also* DISK DRIVE, MAGNETIC RECORDING.

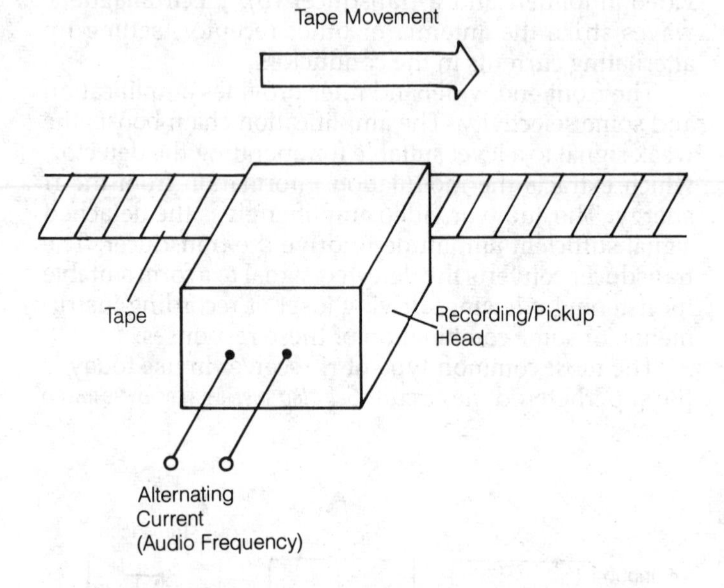

RECORDING HEAD: The recording head functions for both record and playback.

RECORDING TAPE

Recording tape is a linear medium on which information can be stored. The most common and familiar form of recording tape is magnetic tape.

In early computer terminals and teletype systems, paper tape was used to record printed data. A machine punched holes in the tape in rows perpendicular to the edges of the tape. Each row of holes represents one character (*see* PAPER TAPE).

RECOVERY TIME

Recovery time is the time needed for the output voltage or current in a power supply to return to a value within the regulation specification after a step load or line change. Recovery time is preferred over response time as the more meaningful way of specifying power supply performance.

Recovery time for a gas tube is the time required for the electrode to regain control after anode current interruption. *See* RECOMBINATION, REVERSE RECOVERY TIME.

RECTANGULAR COORDINATES

See CARTESIAN COORDINATES.

RECTANGULAR WAVEGUIDE

See WAVEGUIDE.

RECTIFICATION

The conversion of alternating current to direct current is called rectification. Rectification can be accomplished in different ways, but most commonly depends on the usage of diodes or commutators. Diodes are generally used for electronic-circuit rectification; commutators are used in motors and generators when direct current is needed (*see* COMMUTATOR, DIODE, DIODE ACTION, MOTOR).

The diode conducts current in only one direction. Therefore, when a diode is placed in a circuit where alternating current (ac) flows, the output is pulsating direct current (dc). Rectification is used in all dc power supplies designed for operation from utility lines. *See also* RECTIFIER, RECTIFIER CIRCUIT, RECTIFIER TUBE.

RECTIFIER

A rectifier is a two-terminal junction diode that converts alternating current (ac) into direct current (dc). As a class, rectifiers are considered to be capable of conducting 1 A or more of current or dissipating 1 W or more of power. Most semiconductor rectifiers today are made from silicon. *See* DIODE.

Most silicon rectifiers are PN junctions with an interface between one region containing N-type impurities and another region containing P-type impurities. The N-doped silicon has a surplus of electrons, and the P-doped silicon has a shortage of electrons or a surplus of holes, vacant sites that can be filled by electrons. Holes, acting like charged particles, conduct current.

If a positive voltage is applied to the P-type anode and a negative voltage is applied to the N-type cathode, the rectifier is forward biased. Electrons flow from the cathode across the junction to the anode. Conventional current flows from the anode to the cathode across the junction.

If the polarity of the input signal is reversed, the rectifier is back or reverse biased. In this condition, neither electrons nor conventional current will flow across the junction. A reverse-biased rectifier effectively becomes an insulator with resistance that can be measured in megohms due to the expansion of the depletion region that forms around the PN junction.

Rectifier chips are made with large PN junctions to prevent their self-destruction due to overheating. They can be connected in parallel to increase their power-handling ability. Rectifiers are typically packaged as discrete devices. Low-power rectifiers with current ratings of less than 6 A are packaged in axial-leaded glass or plastic cases. However, rectifiers with current ratings of 8 to 20 A are being packaged in flat plastic cases with internal metal heat sinks while those with ratings from about 12 to 75 A are usually packaged in all-metal cases. These cases have some provision for direct case-to-external heat sink mounting to aid in heat dissipation.

The most important electrical ratings of rectifiers are: peak repetitive reverse voltage—V_{RRM}, average rectified forward current—I_O, and peak repetitive forward surge current—I_{FSM}.

Rectifiers are widely used components in power supplies for electronics. Conventional junction rectifiers are acceptable for linear power supplies operating from input frequencies up to 300 Hz, but are not efficient in switching power supplies, typically operating at frequencies of 10 kHz or higher because of their slow recovery time. A finite amount of time is required for the recombination of minority and majority carriers—electrons and holes—after a polarity change in the input source. The minority carriers must be removed before full blocking voltage is obtained. *See* POWER SUPPLY, REVERSE RECOVERY TIME.

Despite slow recovery time, standard PN junction rectifiers have lower reverse currents, higher operating junction temperatures, and higher inverse-voltage capability than rectifiers designed to overcome the speed limitation. Three faster types of silicon rectifiers have been developed to overcome the losses that occur in high-frequency switching of conventional PN junctions: 1) fast recovery, 2) ultrafast or super-fast recovery, and 3) Schottky. There is no industry standard definitions for these devices.

Fast Recovery. A fast-recovery rectifier is made by diffusing gold atoms into a silicon substrate to accelerate the recombination of minority carriers, thereby reducing the reverse recovery time. Rectifiers in this class can be switched in 200 to 750 nanoseconds (ns). They have current ratings of 1 to 50 A and voltage ratings to 1200 V. Forward voltage drop is typically 1.4 V, higher than the 1.1 to 1.3 V of the PN junction. The maximum allowable junction temperature is about 25°C lower than a conventional PN junction, and maximum reverse voltage is limited to about 600 V.

Ultrafast or Super-Fast. Ultrafast or super-fast recovery rectifiers are PN junction rectifiers with reverse recovery times between 25 and 100 ns. These rectifiers also are made by diffusing gold or platinum into the silicon to speed up minority carrier recombination. They are widely specified for power supplies with output voltages of 12, 24, and 48 V.

Schottky. A Schottky rectifier does not have a PN junction so they have no minority charge carriers. The semiconductor material is in direct contact with one metal electrode. This device has no delays due to recombination, and recovery current is due principally to junction capacitance. Recovery time, although not specified, is typically less than 10 ns. Schottky rectifiers provide lower forward voltages (V F) than the PN rectifiers (0.4 to 0.8 V vs. 1.1 to 1.3 V). As a result, power dissipation is lower and efficiency is higher.

One drawback of the Schottky rectifier is its low blocking voltage, typically 35 to 50 V. However, some Schottky rectifiers with maximum blocking voltages of 200 V are available. These rectifiers require transient protection and they have inherently higher leakage current (I_{RRM}), than any of the PN junction rectifiers. Thus Schottky rectifiers are more susceptible to destruction by overheating (thermal runaway). Schottky rectifiers can be connected in parallel for use in the output stages of switching power supplies where they are usually used with output terminals of 5 V or less.

RECTIFIER BRIDGE

A rectifier bridge is an interconnected assembly of four rectifier diodes for full-wave, single-phase rectification or six rectifiers for three-phase rectification of alternating current. The term usually refers to a factory-made assembly.

Factory-made bridges save the user's assembly labor and time, conserve PC (printed-circuit) board space, and improve heat dissipation of the rectifiers. Low-power, low-cost bridges are made by bonding passivated or glass-encapsulated rectifier chips to metal lead frames. The frames are then epoxy molded in flatpacks or dual-in-line (DIP) packages. Bridges also are made by connecting leaded glass-sealed rectifiers to form the bridge circuits. The assemblies are then mounted in copper boxes that act as heat sinks and are potted for protection.

Standard commercial bridge rectifiers have ratings of 1 to 40 A. Power bridges with rating of 20 A or higher might have fast-connect, solder, or wirewrap terminals for external connections. Half bridges composed of pairs of Schottky or fast-recovery rectifiers for switching power are packaged in standard metal and plastic power semiconductor cases. (*See* RECTIFIER CIRCUIT.) Rectifier bridges are tested to U.S. and international specifications, and they can be recognized by the Underwriters Laboratories (UL) or certified by the Canadian Standards Association (CSA).

RECTIFIER CIRCUIT

A rectifier circuit is an arrangement of rectifiers intended to accomplish a specific application. The figure illustrates the widely used single-phase rectifier circuits: half wave, full wave with center-tapped transformer, and full-wave bridge. The optimum type of rectifier circuit for a specific application depends upon the direct-current (dc) voltage and current requirements, the maximum amount of ripple (undesirable variation in the dc output caused by an alternating-current component) that can be tolerated in the output voltage, and the type of power available. Single-phase circuits provide the relatively low dc power required for television receivers, VCRs, personal computers, and other consumer electronics.

Polyphase rectifier circuits usually provide the dc power for mainframe computers, industrial robots, and machine tools. These circuits make more effective use of the capabilities of rectifiers and power transformers. Moreover, they provide a dc output with very low ripple. As a result, polyphase circuits require less dc output voltage filtering than is required from single-phase rectifier circuits.

RECTIFIER TUBE

A rectifier tube is a two-electrode vacuum tube for use in high-voltage, direct-current power supplies. Rectifier tubes are not generally used in modern circuits, because semiconductor diodes can handle several thousand volts or hundreds of amperes. However, rectifier tubes can still be seen in high-voltage power supplies, especially for high-power transmitters.

There are two basic kinds of rectifier tubes. The high-vacuum rectifier consists of a cathode, a plate, and a filament. Electrons are driven from the cathode by the heat of the filament. Electron flow occurs easily from the cathode to the plate, but not from the plate to the cathode. The high-vacuum rectifier can be thought of as an ordinary tube with no grids. The mercury-vapor rectifier is not a vacuum tube; it contains gaseous mercury. Because of the characteristics of mercury vapor, this rectifier tube conducts well in only one direction. The mercury-vapor rectifier also has a constant forward voltage drop.

Tubes are more forgiving than semiconductor devices in the event of a temporary current or voltage overload. This is the main advantage of tubes; a semiconductor diode can be destroyed instantly by voltage spikes or current surges. Tubes are, however, more bulky than semiconductor diodes and require a source of power for the filament. *See also* DIODE, POWER SUPPLY, RECTIFIER CIRCUIT.

REDUCED INSTRUCTION SET COMPUTER

A reduced instruction set computer (RISC) has a smaller and simpler instruction set than the conventional, complex instruction set computer (CISC). An analysis of typical software programs for CISCs showed that 20 percent of the instruction set was used 80 percent of the time. An RISC attempts to improve program execution speed by making these frequently used instructions execute very fast—in one or a few Central Processing Unit (CPU) clock cycles.

To gain additional execution speed, an RISC may also use large register sets, as well as parallel and pipelined processing. RISCs depend upon sophisticated compilers for efficient use of the limited instruction set and hardware features such as pipelining. RISCs are faster than CISCs in many applications. *See* COMPILER.

REED RELAY

A reed relay is an electromechanical relay that includes a sealed reed switch capsule to provide isolated contacts. A cutaway view of a reed relay is shown in the diagram. The reed switch capsule is coaxially mounted within the coil. A magnetic field is induced when direct current (dc) is applied to the coil terminals. The magnetic field opens or closes the reed contacts within the capsule, making or breaking the load, depending on the form of the reed contacts. *See* REED SWITCH.

Reed relays are used in telecommunications, medical instrumentation, and automated test equipment (ATE). The reed capsule and coil are selected for the application. Some require that the relay switch have very low level inputs such as those from a thermocouple. These re-

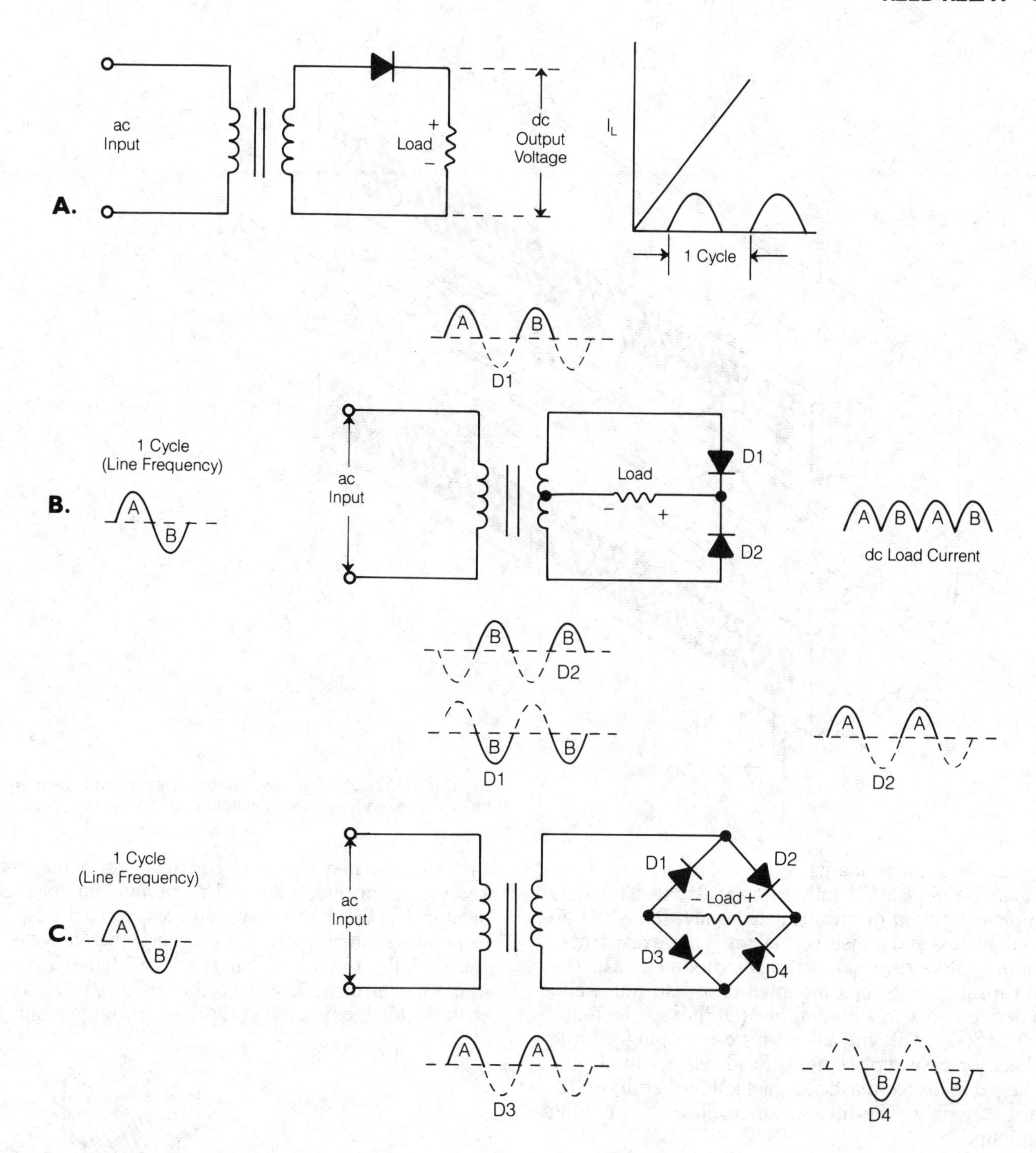

RECTIFIER CIRCUIT: Single-phase rectifier circuits are half-wave (A), full-wave with center-tapped transformer (B), and full-wave bridge (C). Waveforms show the phase and current in each rectifier.

quirements differ from those for telephone switching or sensitive logic circuit isolation from high voltage in test equipment.

The most popular reed relays are Form A (single-pole, single throw; normally open) and Form C (single-pole, double-throw). A single reed capsule within a coil is a single-pole reed relay; two reed capsules within the coil become a two-pole relay; two reed capsules within the coil become a two-pole relay, and three or more become multipole relays. Control voltages are typically 5 to 24 V dc. Most reed relays are packaged for printed circuit board mounting.

Open Style. An open-style reed relay is a simple assembly consisting of a reed capsule within a coil mounted on an insulating base with leads. These relays are subject to damage and failure because of accumulated dust,

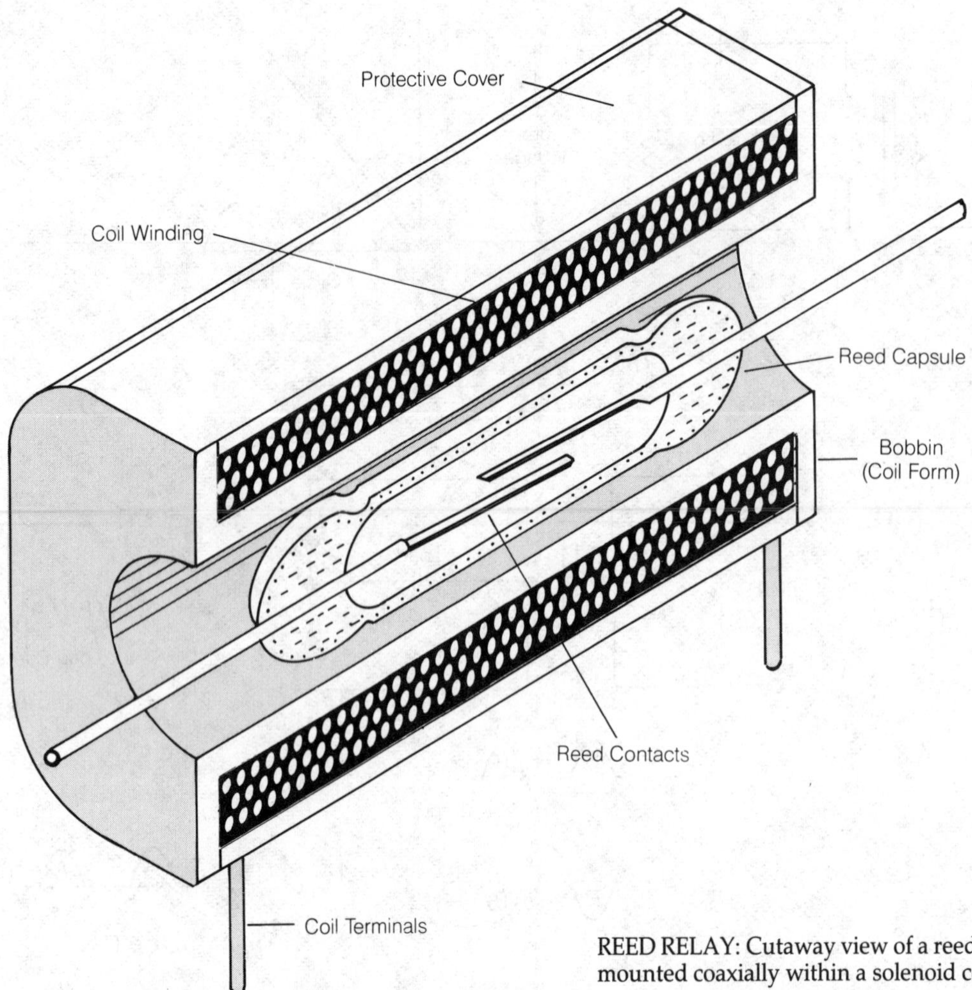

REED RELAY: Cutaway view of a reed relay showing reed capsule mounted coaxially within a solenoid coil.

moisture and contaminants.

Fully Encapsulated. A fully encapsulated reed relay is a completely potted or encapsulated relay that is not affected by dust and moisture, making it more resistant to handling and a severe operating environment. The coil and capsule are usually mounted on a lead frame and molded in epoxy in a dual-in-line (DIP) or single-in-line (SIP) package. DIP and SIP relays can be plugged into sockets for easy replacement. Reed relays can also be packaged in leaded plastic or shielded metal tubs for PC board assembly. The tubs are then filled with potting compound.

Mercury Wetted. A mercury-wetted reed relay has a mercury pool within the reed capsule to lubricate the moving reeds. The mercury film on the reeds also keeps the contact interface free of erosion. Under low-level loads, mercury-wetted contacts provide consistent and predictable contact resistance over wide ranges of temperature and contract load current. The mercury eliminates contact chatter or bounce. However, the mercury slows the speed of contact operation to about 2 milliseconds. The mercury-wetted reed relay, like the mercury-wetted switch, is position sensitive.

REED SWITCH

A reed switch is an assembly of a reed switch capsule that provides isolated electrical contacts and a magnet as shown in the illustration. The permanent magnet is arranged so that it can move with respect to the capsule, as shown in the figure. The magnet, when it is close to the capsule, provides a magnetic field that closes or opens the ferromagnetic reeds that make or break contact with the load, depending on the reed arrangement. The

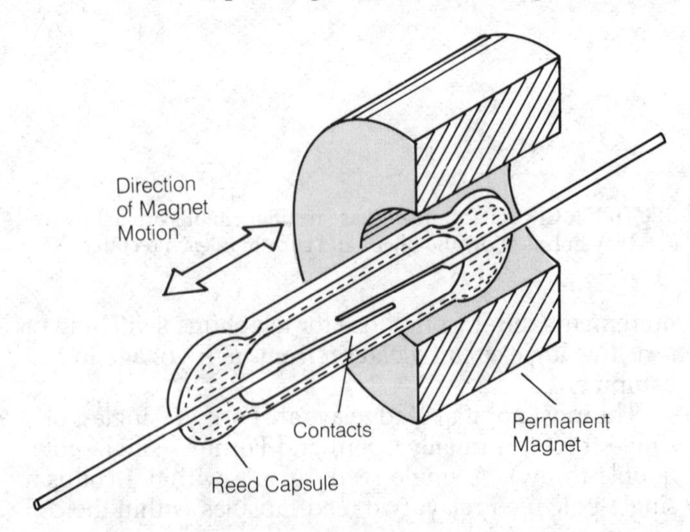

REED SWITCH: Reed switch capsule movement into the proximity of a permanent magnetic field opens or closes the reed contacts, depending on contact arrangement.

reeds are supported by the hermetically sealed glass capsule in which they are mounted. The difference between a reed switch and a reed relay is that a switch has a permanent magnet actuator and a relay is actuated by an electromagnetic coil. *See* REED RELAY.

The reed capsule has very high isolation (10^{12} Ω) when off and very low resistance (0.75 Ω) when on. Because the reeds are hermetically sealed in a glass capsule, the contact area remains clean and free of oxidation. The reeds will open and close reliably for millions of operations.

A practical reed switch is made by assembling a reed switch capsule with a spring-loaded magnet in the same housing. A magnet pushbutton keyboard switch is an example. The reed contacts will close (or open) when the magnet is manually pushed into the presence of the reeds (or the capsule is pushed into the presence of the magnet). When the button is released, the spring retracts the magnet, allowing the reed contacts to return to their normal state.

Reed-switch capsules are available with single-pole, single-throw, normally open (SPST-NO) contacts (Class A) and single-pole, double-throw (SPDT) contacts (Class C).

Mercury added to a reed capsule acts as a conductive medium, providing a low resistance path. The mercury-wetted switch damps out contact bounce or chatter, a response to a step input signal that causes the contacts to close and rebound one or more times before final closing. Bounce or chatter introduces ambiguity and delay to the output signal. The addition of mercury adds to the cost of the reed capsule and also makes it position sensitive. The mercury-wetted reed relay switch must be mounted so the capsule in an upright or nearly upright position to keep the reed ends immersed in mercury.

REFERENCE FREQUENCY

In the operation and calibration of radio receivers, a reference frequency is often used. A standard signal generator, such as a crystal oscillator or broadcast station whose frequency is precisely known, can be used for this purpose.

A reference-frequency source is considered primary if it is received directly from a standard-frequency broadcast station such as CHU or WWV/WWVH. If a commercial broadcast station or crystal calibrator is used, the reference frequency is considered secondary. *See also* FREQUENCY CALIBRATOR, FREQUENCY MEASUREMENT, WWV/WWVH.

REFERENCE LEVEL

When signal strength is measured in relative terms, a reference level must be specified. A reference level can be established for current, field strength, flux density, power, voltage, or any other quantity that varies. The reference level is assigned a gain factor of 1, corresponding to 0 decibels.

Certain standard reference levels are used. A common reference level for current is the milliampere; for power, it is the milliwatt; and for voltage, it is the millivolt. Other standard values, however, can be specified. If the variable exceeds the reference level, the gain factor is greater than 1 and the gain figure in decibels is positive. If the variable is smaller than the reference level, the gain factor is less than 1 and the gain figure in decibels is negative. *See also* DECIBEL, GAIN.

REFLECTANCE

When an electromagnetic field strikes a surface or boundary, some of the field may be reflected. The ratio of the reflected energy or field intensity to the incident intensity is known as reflectance. If the incident field intensity (known as incident or forward power) is P_1 and the reflected field intensity (called the reflected power) is P_2, then the reflectance factor R is:

$$R = \frac{P_2}{P_1}$$

As an example, consider a mirror. When visible light strikes a mirror, almost all of the energy is reflected. Therefore, the reflectance factor is almost 1, or 100 percent. A pane of glass reflects some light, but most of it is transmitted. Thus, the reflectance of glass is just a few percent. The reflectance of a surface varies depending on the angle at which the light strikes.

In a power-transmission line and load system, some of the electromagnetic field traveling from the generator to the load may be reflected upon reaching the load. This results in less than optimum current in the load. The ratio of actual current to optimum current is called reflectance or, more commonly, the reflection factor. *See also* REFLECTED POWER, REFLECTION COEFFICIENT, REFLECTION FACTOR.

REFLECTED IMPEDANCE

In an impedance transformer with a specific load connected to the secondary winding, the impedance across the primary is called reflected impedance (*see* IMPEDANCE TRANSFORMER). If the secondary, or load, impedance does not contain reactance, the reflected impedance will contain no reactance. This is generally true of impedance transformers. However, if the load does contain reactance, then the reflected impedance will also contain reactance (*see* CAPACITIVE REACTANCE, IMPEDANCE, INDUCTIVE REACTANCE).

In a transmission line operating at radio frequencies, the reflected impedance is the impedance at the transmitter end of the line. In the ideal case, where the antenna exhibits a pure resistance equal to the characteristic impedance of the line, the reflected impedance is the same as the load impedance or any line length. If the load contains reactance, or if the resistance of the load differs from the characteristic impedance of the line, the reflected impedance will vary according to the line length.

Certain transmission lines have special properties. Neglecting line loss, a transmission line measuring a multiple of a half wavelength will always cause reflected impedance identical to the load impedance; a line measuring an odd multiple of a quarter wavelength will invert the resistance and reactance. *See also* CHARACTERISTIC IMPEDANCE, HALF-WAVE TRANSMISSION LINE, MATCHED LINE, MISMATCHED LINE, QUARTER-WAVE TRANSMISSION LINE.

REFLECTED POWER

Ideally, the impedance of the load in a power-transmission system should be a pure resistance, and should have the same ohmic value as the characteristic impedance of the line. However, when the load impedance differs from the characteristic impedance of the line, the line is said to be mismatched.

The electromagnetic field, traveling along a mismatched line from the generator to the load, is not completely absorbed by the load. Some of the electromagnetic field is reflected at the feed point, and travels back toward the generator. Because the electromagnetic field causes power to be dissipated in a resistance, the forward-moving field is called forward power and the backward-moving field is called reflected power.

The amount, or proportion, of incident power that is reflected depends on the severity of the mismatch between the line and the load. The greater the mismatch, the more of the incident power is reflected at the feed point. In the extreme, when the load is a pure reactance, or if it is a short circuit or open circuit, all of the incident power is reflected. The degree of mismatch is expressed by a figure known as the standing-wave ratio (*see* STANDING-WAVE RATIO).

If the standing-wave ratio is represented by S, the forward power by P_1, and the reflected power by P_2, then:

$$\frac{P_2}{P_1} = \left(\frac{S-1}{S+1}\right)^2 = \frac{S^2 - 2S + 1}{S^2 + 2S + 1}$$

The proportion of the reflected power to forward power is equal to the square of the reflection coefficient (*see* REFLECTION COEFFICIENT).

Reflected power can be directly read in a radio-frequency transmission line with a calibrated reflectometer (*see* REFLECTOMETER). If the reflectometer is not calibrated in forward and reflected watts, the reflected power can be determined from the standing-wave ratio observed on the meter.

When reflected power reaches the transmitter end of a radio-frequency transmission line, all of the electromagnetic field is reflected back again toward the load, assuming the transmitter is adjusted for optimum output. If the transmitter output circuit is not adjusted to compensate for the reactance at the input end of the line, some of the reflected power is absorbed by the final amplifier. This is observed as a reduction in power output, although the final-amplifier direct-current input is just as great, or greater, than it would be if the output circuit were adjusted properly. *See also* MATCHED LINE, MISMATCHED LINE, REFLECTION FACTOR.

REFLECTED WAVE

In line-of-sight communication, the signal at the receiving antenna consists of two components. One component, the direct wave, follows the path through the air between the two antennas (*see* DIRECT WAVE). The other component consists of one or more waves that are reflected from the ground and/or other objects such as buildings (see illustration). They are called the reflected wave or waves.

The direct wave and the reflected wave(s) add together in varying phase. The reflected-wave signal is usually weaker than the direct wave. In certain situations, the direct-wave and reflected-wave signals are equal in amplitude and opposite in phase at the receiving antenna. This renders communication difficult or impossible. The remedy is to move either the transmitting antenna or the receiving antenna a fraction of a wavelength so that the coincidence is eliminated. This phase problem is observed primarily at frequencies above about 10 MHz and is especially acute at microwave frequencies.

The effects of reflected waves can be noticed in a mobile frequency-modulation receiver such as a car stereo. As the received station weakens, the sound begins to flutter. This is the result of alternate phase reinforcement and phase cancellation between the direct-wave and reflected-wave components. The faster the car moves, the more rapid the flutter. If the car comes to a stop in a dead zone where total phase cancellation occurs, the signal will disappear until the car is moved again. *See also* LINE-OF-SIGHT COMMUNICATION.

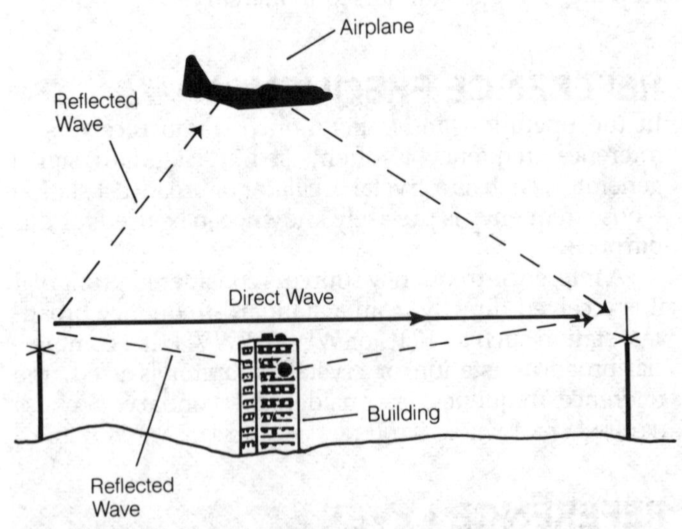

REFLECTED WAVE: The reflected wave combines with the direct wave at the receiving antenna in short-range communications.

REFLECTION COEFFICIENT

In a power transmission line, not all of the incident power is absorbed unless a perfect match exists (*see* MATCHED LINE, MISMATCHED LINE). When there is a mismatch, not all of the available or forward current is absorbed by the load; some current is reflected. The same thing happens with the voltage. The ratio of reflected current to forward current, or reflected voltage to forward voltage, is called the reflection coefficient. The reflection coefficient is represented by the lowercase Greek rho (ρ).

Let Z_o be the characteristic impedance of a transmission line and let R be the ohmic value of a purely resistive load. Then the reflection coefficient is:

$$\rho = \frac{R - Z_o}{R + Z_o}$$

if R is larger than Z_o, and:

$$\rho = \frac{Z_o - R}{R + Z_o}$$

if R is smaller than Z_o. The above formulas are accurate only for loads that contain no reactance.

The reflection coefficient can also be derived from the standing-wave ratio, S:

$$\rho = \frac{S - 1}{S + 1}$$

This second formula is accurate whether or not the load contains reactance. *See also* REFLECTED POWER, STANDING-WAVE RATIO.

REFLECTION FACTOR

In a matched transmission line, all of the incident current is absorbed by the load, but this is not true in a mismatched line (*see* MATCHED LINE, MISMATCHED LINE). The reflection factor is the ratio of the current actually absorbed by a load to the current that would be absorbed under perfectly matched conditions. The reflection factor is also called reflectance.

Let R be the ohmic value of a purely resistive load and let Z_o be the characteristic impedance of the transmission line. Then the reflection factor, F, is given by:

$$F = \frac{\sqrt{4RZ_o}}{R + Z_o}$$

This formula holds only if the load impedance is a pure resistance. *See also* REFLECTED POWER, REFLECTION COEFFICIENT.

REFLECTION LAW

If a particle or wave disturbance encounters a flat (plane) reflective surface, the angle of incidence is equal to the angle of reflection, measured with respect to the normal to the surface. If the surface is not flat, then the angles of incidence and reflection are identical as measured with respect to the normal to a plane tangent to the surface at the point of reflection (see illustration). This rule is sometimes called the reflection law or the law of reflection.

The reflection law holds for radio waves striking the ground, light striking a mirror, sound waves bouncing off a wall or baffle, and all other situations in which reflection takes place. *See also* ANGLE OF INCIDENCE, ANGLE OF REFLECTION, SNELL'S LAW.

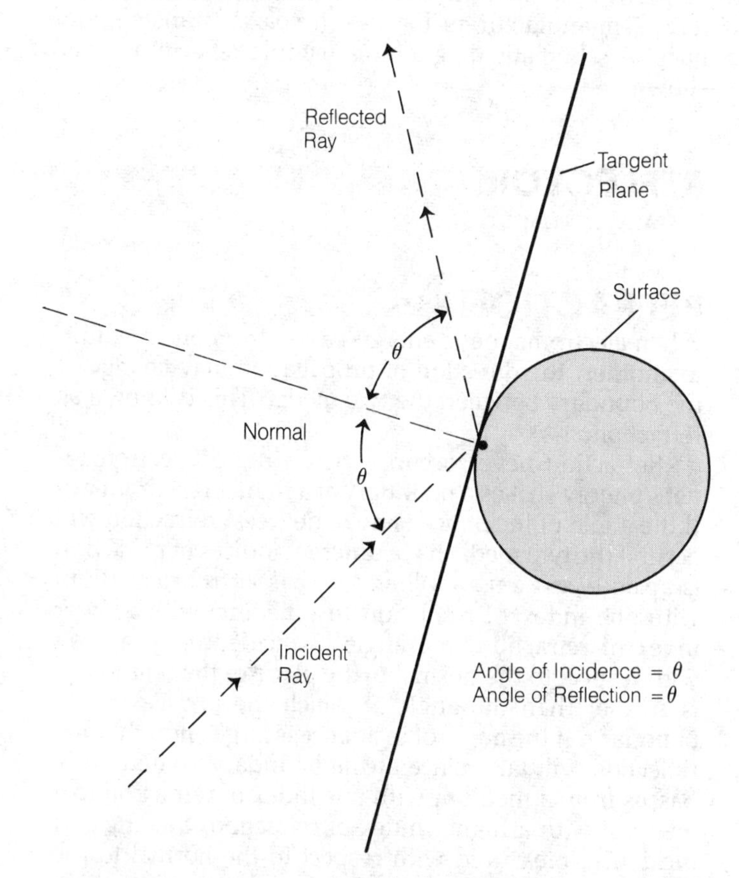

REFLECTION LAW: The angle of incidence is equal to the angle of reflection according to the reflection law.

REFLECTOMETER

A reflectometer is an instrument installed in a transmission line for measuring the standing-wave ratio (*see* STANDING-WAVE RATIO). Some reflectometers are calibrated in forward watts and reflected watts. A reflectometer may

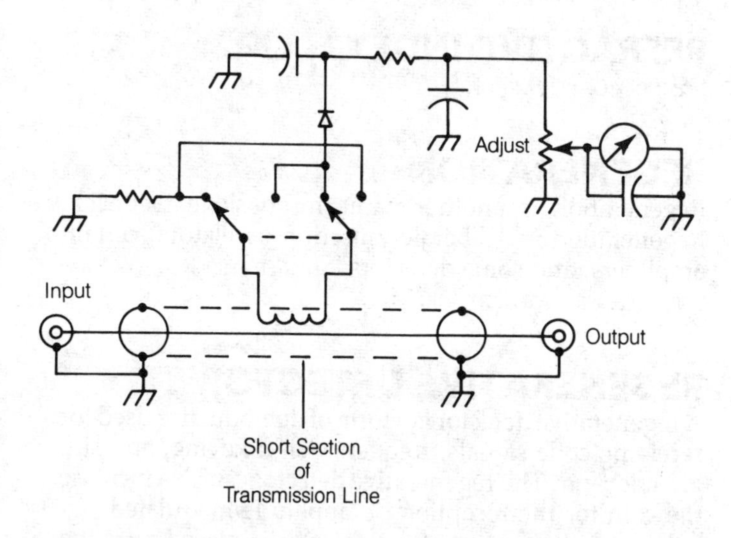

REFLECTOMETER: A schematic diagram of an uncalibrated reflectometer.

also be called a directional coupler, directional power meter, or monimatch.

Many different reflectometers are available from commercial manufacturers for use in coaxial transmission lines. A schematic diagram of a simple reflectometer is shown.

REFLECTOR

See PARASITIC ELEMENT.

REFRACTION

When electromagnetic energy passes from one medium to another, the direction of propagation may change at the boundary between the two media. This is known as refraction.

Refraction never occurs when a ray of electromagnetic energy strikes a boundary at a right angle (normal). If the angle of incidence is not 0 degrees, refraction will occur if the two media have different indices of refraction (*see* INDEX OF REFRACTION). When a ray passes from a medium with one index of refraction to a medium with a lower index of refraction, the angle of incidence, measured with respect to the normal to the plane of the boundary, is smaller than the angle at which the ray leaves the boundary. If the angle of incidence is large enough, total reflection will take place at the boundary. When a ray passes from a medium with one index of refraction to a medium with a higher index of refraction, the angle of incidence, measured with respect to the normal to the plane of the boundary, is larger than the angle at which the ray leaves the boundary. *See* ANGLE OF INCIDENCE, ANGLE OF REFRACTION, TOTAL INTERNAL REFLECTION.

Radio waves are refracted when they pass from one air mass into another air mass with a different density. Total internal reflection can occur if the angle of incidence is very large. These effects are called tropospheric bending and ducting, respectively. *See also* TROPOSPHERIC PROPAGATION.

REFRACTIVE INDEX

See INDEX OF REFRACTION.

REGENERATION

Regeneration is another name for positive feedback. Regeneration is deliberately used in oscillators, certain amplifiers, and some detectors. *See also* FEEDBACK, OSCILLATOR, REGENERATIVE DETECTOR.

REGENERATIVE DETECTOR

A regenerative detector is a form of demodulator used for receiving code signals, frequency-shift keying, and single sideband. The regenerative detector can also increase the gain for the reception of amplitude-modulated signals, provided an envelope detector is placed after the circuit. Regnerative detectors are not often used today, although they can still be found in simple direct-conversion receivers.

The regenerative detector consists of an amplifier circuit with some of the output applied in phase to the input. This increases the gain of the amplifier. If the positive feedback is made large enough, the circuit will oscillate. Incoming signals then mix with the resulting carrier, producing beat notes that can be heard in the speaker or headset. This makes code, frequency-shift-keyed, or single-sideband signals discernible. The schematic diagram shows a simple regenerative detector.

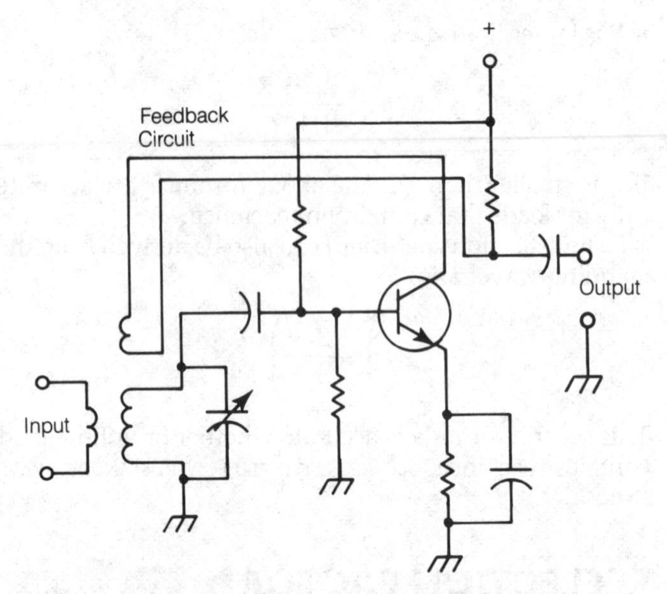

REGENERATIVE DETECTOR: This is an amplifier with inductive feedback that produces self-oscillation.

REGENERATIVE RECEIVER

See REGENERATIVE DETECTOR, SUPERREGENERATIVE RECEIVER.

REGENERATIVE REPEATER

A regenerative repeater is a circuit that reshapes pulses. In teletype or code transmission, the pulses often become distorted in transit. This is typically the case for signals that have been propagated through the ionosphere. The regenerative repeater restores the pulses to their original shape and duration.

A regenerative repeater responds to the appropriate length, spacing, and shape for a pulse. For example, the duration of a Morse-code dot at a given speed might be 0.1 second. At this same speed, the space between code elements in a character is also 0.1 second; the length of a dash is 0.3 second; the space between characters is 0.3 second; the space between words is 0.7 second. If a code signal with some distortion but the appropriate speed is applied to the input of the regenerative repeater, perfect code will appear at the output.

Regenerative repeaters can reduce the error rate in digital communications, especially when a machine is used to receive the signals. The higher the data speed, the greater is the chance for distortion in the signal.

Therefore, regenerative repeaters become more useful as the data speed is increased.

REGISTER

A short-term memory-storage circuit is called a register. A register may store any kind of information and usually has a small capacity. In a computer, registers store data for the central processing unit (CPU). A register configuration can be first-in/first-out (FIFO), pushdown stack, or random access (*see* FIRST-IN/FIRST-OUT, PUSHDOWN STACK, SEMICONDUCTOR MEMORY).

REGULATED POWER SUPPLY

See POWER SUPPLY.

REGULATING TRANSFORMER

There are two types of regulating transformers, both designed to maintain a constant output voltage under varying load conditions.

Some transformers use resonance and core saturation to maintain a constant output voltage. This type of transformer actually varies in efficiency as the load changes. As the load resistance decreases, the transformer operates more efficiently, compensating for the drop in voltage that would otherwise occur.

A small transformer may be connected in series with the secondary winding of the main transformer in a power supply to adjust the voltage. This arrangement is shown in the schematic. The small transformer is called a regulating transformer. When the current from the regulating transformer is in phase with the current from the main transformer, the output voltage is increased. When the phase of the current from the regulating transformer is opposite to that of the main transformer, the output voltage is decreased. By applying a small voltage of the appropriate phase to the primary of the regulating transformer, the output voltage can be held constant under alternating load conditions. *See also* REGULATION, TRANSFORMER.

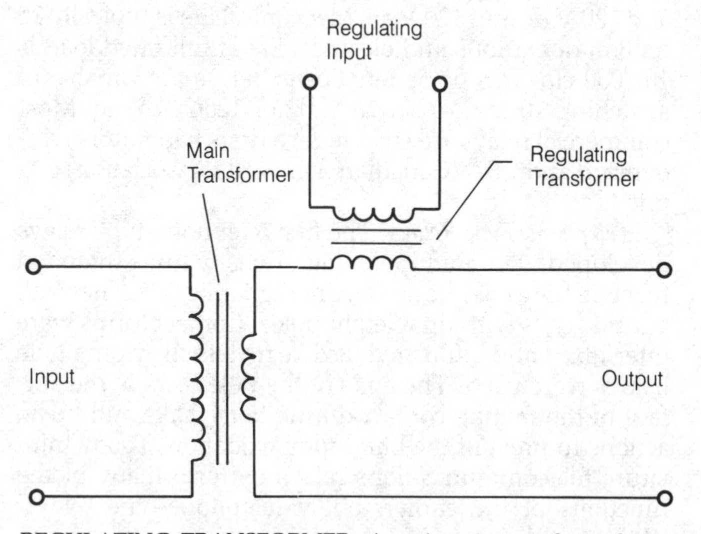

REGULATING TRANSFORMER: A series-connected regulating transformer.

REGULATION

Regulation refers to the maintenance of a constant condition in a circuit. Any variable, such as current, voltage, resistance, or power, can be regulated.

The extent of variation can be expressed mathematically as a percentage. This figure is called the regulation. Let the optimum, or rated, value of a parameter be represented by *P*, and consider that the actual value of the variable alternates between the minimum value P_1 and the maximum value P_2, so that $P_1 < P < P_2$. Then the regulation *R*, in percent, is given by:

$$R = \frac{100(P_2 - P_1)}{P}$$

Regulation can also be expressed as a plus-or-minus number. In the example, the regulation can be specified as a percentage if the alternation is the same above and below the ideal; that is, if:

$$P_2 - P = P - P_1$$

RELATIVE MEASUREMENT

The magnitude of a quantity can be measured by comparing it with a known standard. This is called a relative measurement. Relative measurements are often made for current, power, radiant energy, sound, and voltage.

A relative measurement can be expressed as a numerical ratio between the unknown and the standard. But the most commonly used method of relative measurement is the decibel system. *See* dBa, dBd, dBi, dBm, DECIBEL.

RELATIVE-POWER METER

An uncalibrated voltmeter, installed at the output of a transmitter or in a transmission line can be used for relative measurement of power. This instrument is called a relative-power meter.

A simple relative-power meter is shown in the schematic diagram. The sensitivity of a relative-power meter

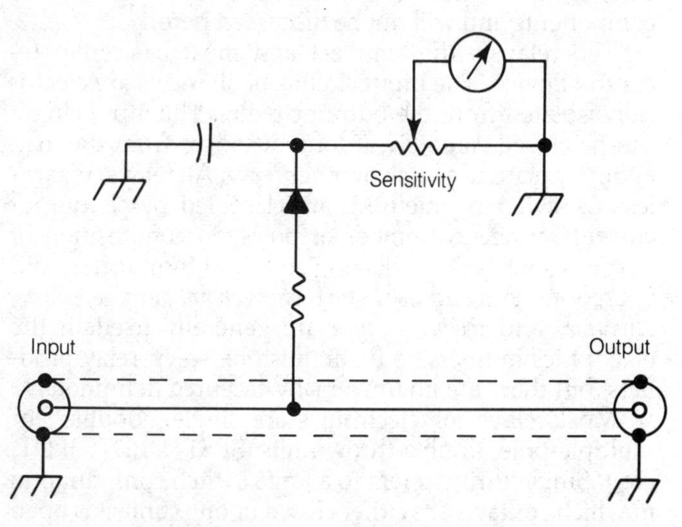

RELATIVE-POWER METER: A diagram of a relative-power meter.

usually varies with frequency. The presence of standing waves on a transmission line will also affect the sensitivity of a meter as shown in the illustration. The power at a given frequency is proportional to the square of the meter reading. Thus, if the power is doubled, the meter reading increases by a factor of 1.414; if the power is quadrupled, the meter reading increases by a factor of 2, and so on.

Relative-power meters are extensively used in low-power and medium-power transmitters, especially those intended for amateur radio or CB operation. A relative-power meter makes transmitter adjustment easy, and it also serves to indicate if the unit is functioning. Many transceivers use the receiver S meter as a relative power indicator in the transmit mode.

RELAY

A relay is a remotely controlled electrical switch that controls power to a load by energizing an isolated input circuit. The traditional relay as shown in the figure is an electromechanical (EM) coil and contact device consisting of electrically isolated input and output circuits. An input signal energizes an electromagnet that attracts a hinged and spring-loaded element called an armature. Output contacts, attached to but insulated from the armature, are opened or closed by the movement of the armature. In the closed position, the contacts apply power to the load; in the open position they remove power from the load.

Early EM relays were components in telegraph systems that amplified the coded signal being transmitted making possible long-distance communication. Electromagnetic relays also have been developed specifically to switch analog and digital telephone communications signals. These relays are still widely used in the control of electrical power, motors and other actuators. An EM relay based on the reed switch, the reed relay, has been developed. (*See* REED RELAY.) Electronic analogs of the EM relay, the solid-state relay, are suited for many of the same applications as EM relays. (*See* SOLID-STATE RELAY.) Heavy-duty EM relays called contactors are used to switch higher power levels than conventional EM relays. They are considered to be electrical rather than electronic components and will not be discussed here.

The relay is the simplest and most basic remote-control device. The input circuits of all relays are electrically isolated from the output circuits. The input circuit can be closed remotely at long distances from the load either by manual switch or other relay. All relays, regardless of switching method, are classified by: 1) output current rating, 2) number of poles, 3) composition of output contacts, 4) packaging style and form factor, and 5) predominant end use. The terms *general purpose, power, sensitive,* and *telephone type* are generally used in the electronics industry to distinguish between relay products, but there are no universally accepted definitions.

Most relays for electronics are single-, double-, or multiple-pole, double-throw units (SPST, DPDT, 3PDT, etc.). Single throw refers to a knife switch configuration in which contacts are either closed or one contact is open (suspended in the air like an open drawbridge). Double throw refers to a configuration in which contacts are closed in one of two possible positions. The terms *normally closed* (NC) and *normally open* (NO) refer to contact positions when the relay coil is de-energized.

When the relay coil is energized, the electromagnet attracts the armature against the tension of the return spring, causing the upper NC contact to break (open) and the lower NO contacts to make (close) in a break-before-make sequence. When the electromagnet is de-energized, the return spring pulls the armature open and the upper NC contacts are closed.

Relay contacts are made from low-resistance metal, typically silver alloy, bonded to the ends of the poles and base terminals. Fine silver is used to make contacts rated to 5 A and nickel-cadmium-silver is used for those rated to 10 A. The fine silver contacts of low-level or dry circuit relays can be lightly gold plated or flashed to prevent oxidation that would increase contact resistance during storage or periods of inactivity.

General-purpose EM relays are considered to be those rated to handle up to 10 A, although ratings of 2 to 3 A are more typical of relays mounted on of printed circuit boards. Power relays, similar in design to general-purpose relays, are typically rated 10 to 30 A. They are specified for large power-consuming systems like computers, radars, and radio transmitters.

The illustration of a general-purpose EM relay identifies its components. The electromagnet is a toroidal coil with an iron core whose coil winding is terminated at exterior terminals as shown. The armature and its return spring are shown at right. The contacts are at the ends of both leaf-spring poles attached to, but insulated from, the armature. Alternate parts of each contact are also terminated at the insulator block. A steel frame supports the relay and completes the magnetic circuit through the steel armature.

There are many different styles, sizes and ratings of general-purpose relays. Variables include: 1) the number of poles (up to eight), 2) terminal forms (solder, plug in, quick connect and printed circuit) and, 3) packaging (open or covered). Popular voltage ratings are 6, 12, and 24 V, alternating current or direct current (ac or dc), 48 and 120 V dc and 120 V ac. Mechanical life is typically 20 million operations and electrical life at full rated load is 100,000 closures or better. Power relays are capable of switching 10 to 50 A at 28 V dc or 120/240 V ac. Most commercial relays are UL (Underwriters Laboratory) recognized and CSA (Canadian Standards Association) certified.

Telephone-Type Relays. The first telephone-type relays developed were multipole, high-density units intended for switching telephone voice or digital signals. They had ac or dc coils with up to eight poles. Contact forms were intermixed and bifurcated flexible metal self-wiping pole blades were used. The ends of the poles scrape the surface of the mating contact during both make and break actions to prevent the buildup of oxidation. Today miniature telecommunications relays perform many of the functions of the earlier, bulky telephone-type relays. With typical ratings of 1 A or less, they are packaged in printed circuit (PC) board mounting flatpacks and sugar

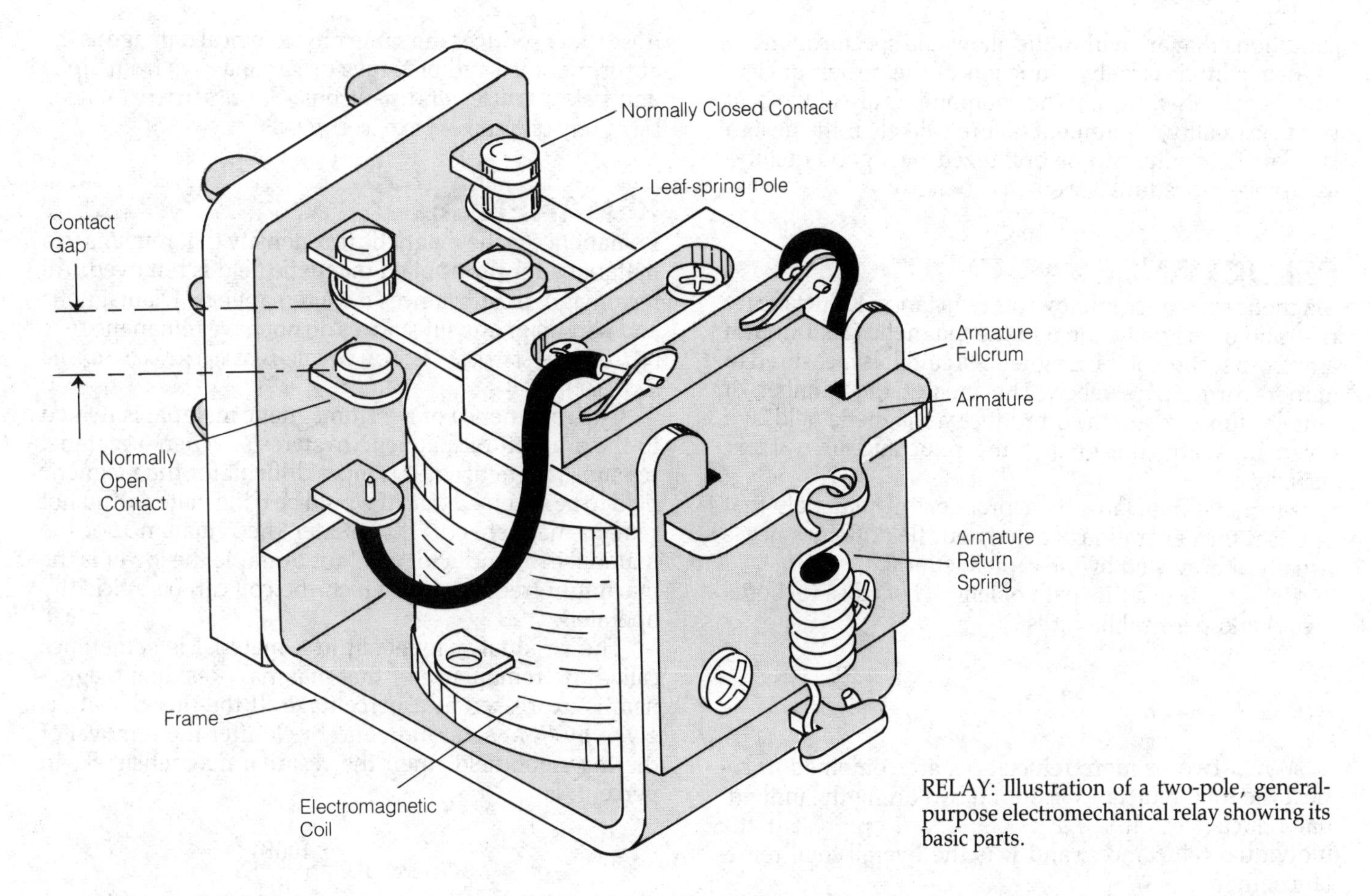

RELAY: Illustration of a two-pole, general-purpose electromechanical relay showing its basic parts.

cube-sized rectangular cases.

Sensitive Relays. Sensitive relays are general purpose or telephone-type EM relays designed to switch with pickup power that is compatible with solid-state logic. Any EM relay able to switch with an input of less than 10 mW is considered to be a sensitive relay.

Time-Delay Relays. Time delay relays are general purpose ac and dc EM relays with provision for delaying contact closure after actuation. Used primarily used in industrial controls, they are available as interval timers, flashers, one-shot and slow-operate, slow-release units. Time ranges are from 0.2 seconds to 120 minutes or longer. Timing can be derived by setting the resistance value in an internal resistive-capacitive (RC) circuit, or it can be counted down from the fixed-frequency 50/60 Hz power line.

Polarized Relays. Polarized relays combine a movable magnet and an electromagnet to concentrate the magnetic field of an EM relay and improve volumetric and electrical efficiency. There are two types: pivoting core and magnetically biased. The magnetically biased relay has a samarium-cobalt permanent magnet that is about one-fourth the size of ferrite magnets used in other relays. It occupies about half the volume of a comparably rated miniature PC board-mounted EM relay. Polarized relays are used in computers, copiers, industrial controls, home video systems, private branch telephone exchanges (PBX), and other telephone equipment.

RELAY LOGIC

A digital logic circuit can be made with electromechanical relays rather than semiconductor devices. This logic system is called relay logic (RL).

Relay logic is considerably slower than solid-state logic because the relay armatures require time to change state. Relay logic is a fully saturated form of logic; the resistance in the open state is practically infinite, and the resistance in the closed state is practically zero.

The main advantages of relay logic are its ability to handle rather large currents, and its high saturation level.

RELIABILITY

Reliability is a measure of the probability that a circuit or system will operate according to specifications for a specified period of time. Reliability is related, inversely, to the failure rate and the hazard rate of the components in a circuit (*see* FAILURE RATE).

Reliability can be expressed as a percentage, based on the proportion of units that remain in proper operation after a given length of time. For example, if 1,000,000 units are placed in operation on January 1, 1990 and 900,000 units are operating properly on January 1, 1991, the reliability is 90 percent per year. It is assumed that the units are operated under electrical and environmental conditions that are within the allowable specifications.

Reliability is largely a function of the design of electronic apparatus. Even if the components are of the best possible quality, equipment failure is likely if the design is poor. Reliability can be optimized by a good quality-assurance procedure (*see* QUALITY CONTROL).

RELUCTANCE

Magnetic resistance is known as reluctance. Reluctance is the ratio of magnetic force to the magnetic flux through any cross section of the magnetic circuit. It is measured in ampere turns per weber. The greater the number of ampere turns required to produce a magnetic field of a given intensity, the greater the reluctance (*see* AMPERE TURN, WEBER).

Magnetic reluctance is a property of materials that opposes the generation of a magnetic field. Reluctance is usually abbreviated by the capital letter *R*.

Reluctance of a flux path of length *l* of cross-sectional area *A* and permeability μ is:

$$R = \frac{l}{\mu A}$$

When two or more reluctances are combined in series, the total reluctance is found by adding the individual reluctances. If R_1, R_2, R_3, . . . , R_n represent the individual reluctances, and R is the overall total reluctance, then:

$$R = R_1 + R_2 + R_3 + \ldots + R_n$$

When combined in parallel, the same reluctances total:

$$\frac{1}{R} = \frac{1}{R_1} + \frac{1}{R_2} + \frac{1}{R_3} + \ldots + \frac{1}{R_n}$$

Reluctances add the same way as resistances in series and in parallel.

If *F* is the magnetomotive force in a magnetic circuit, Φ is the magnetic flux, and *R* is the reluctance, the three variables are related according to the equations:

$$\Phi = F/R, \text{ or}$$

$$F = \Phi R, \text{ or}$$

$$R = F/\Phi$$

This formula is known as Ohm's law for magnetic circuits. *See also* MAGNETIC FLUX, MAGNETOMOTIVE FORCE, OHM'S LAW, PERMEABILITY.

RELUCTIVITY

Reluctivity is a measure of the reluctance of a material per unit volume. Reluctivity is the reciprocal of magnetic permeability. *See also* PERMEABILITY, RELUCTANCE.

REM

The rem (for roentgen equivalent man) is the radiation dose that produces the same physiological damage as the absorption of 1 rad of X-rays or gamma rays from alpha and beta particles and neutrons. *See also* ALPHA PARTICLE, BETA PARTICLE, NEUTRON, RAD, ROENTGEN.

REMANENCE

Remanence is the magnetic flux density that remains in a material after the applied magnetic field is removed. All ferromagnetic substances have remanence. Diamagnetic and paramagnetic substances do not have remanence (*see* DIAMAGNETIC MATERIAL, FERROMAGNETIC MATERIAL, PARAMAGNETIC MATERIAL).

The remanence of a ferromagnetic material is related to the amount of magnetic hysteresis. When a material remains magnetized, it is more difficult for the magnetic field to be reversed than it would be if the material did not remain magnetized. The greater the remanance of the material in an inductor core, for example the lower is the maximum frequency at which the coil can be efficiently operated.

The residual magnetism in a material is sometimes called the remanence of that material. Residual magnetism is expressed as a percentage. If the flux density is given by B_1 at saturation and by B_2 after the removal of the magnetic field, then the residual magnetism B_r, in percent, is:

$$B_r \text{ (percent)} = \frac{100B_2}{B_1}$$

Residual magnetism is dependent on the type of material, and on the shape and size of the sample.

REMOTE CONTROL

Electronic equipment can be operated from controls located at a distance from the equipment. Called remote control, it is useful and convenient in many situations.

Remote control can be accomplished either by wire or electromagnetic energy. For example, a telephone answering system can receive messages and record them; it is then possible to call the machine and instruct it to play back the messages over the telephone. This is remote control by wire.

Model airplanes, boats, and cars can be controlled by hand-held radio transmitters. Television sets, video cassette recorders, and stereo systems can be controlled (turned on and off and made to perform specialized functions) with remote hand-held infrared (IR) transmitters. These units are called remote commanders. *See* OPTOELECTRONICS.

REPEATER

A repeater is a circuit that intercepts and retransmits a signal to provide long-distance communications. Repeaters are generally used at the very high, ultrahigh, and microwave frequencies. Repeaters are especially useful for mobile operation. The effective range of a mobile station is greatly enhanced by a repeater.

A radio repeater consists of an antenna, a receiver, a transmitter, and an isolator. The transmitter and receiver are operated at slightly different frequencies. The separation is approximately 0.3 percent to 1 percent of the transmitter frequency. This separation of the receiver and transmitter frequencies allows the isolator to work at maximum efficiency, preventing undesirable feedback.

Repeaters are placed aboard satellites. All active communications satellites use repeaters. A satellite in a synchronous orbit can provide coverage over approximately 30 percent of the globe. *See also* CELLULAR MOBILE RADIO TELEPHONE, MICROWAVE REPEATER, SATELLITE COMMUNICATIONS.

RESIDUAL MAGNETISM

See REMANENCE.

RESISTANCE

Resistance is the opposition that a material offers to the flow of electric current. It is the bulk property of a material that depends on the material's dimensions, electrical resistivity, temperature, and also voltage in non-ohmic materials. The resistance of the material determines the current (electron flow) produced by a given voltage. *See* OHM'S LAW.

The standard unit of resistance is the ohm, abbreviated by the uppercase Greek letter omega (Ω). In equations, the symbol for resistance is R. The range of possible resistance values, in ohms, is represented by the set of positive real numbers. In theory, it is possible to have resistances that are zero or infinite as well. Resistance is the mathematical reciprocal of conductance. The higher the conductance, the lower the resistance, and vice versa (*see* CONDUCTANCE).

The resistance of wire is usually measured in ohms per unit length. Silver wire has the lowest resistance of any metal wires. Copper and aluminum also have very low resistance per unit length. The direct-current (dc) resistance of a wire depends on its diameter. The table shows the dc resistance of copper wire at 20°C, for various sizes in the American Wire Gauge (*see* AMERICAN WIRE GAUGE).

RESISTANCE: Resistance of solid copper wire for American Wire Gauge (AWG) 1 through 40, in Ω per kilometer.

AWG	Ω/km	AWG	Ω/km
1	0.42	21	43
2	0.52	22	54
3	0.66	23	68
4	0.83	24	86
5	1.0	25	110
6	1.3	26	140
7	1.7	27	170
8	2.1	28	220
9	2.7	29	270
10	3.3	30	350
11	4.2	31	440
12	5.3	32	550
13	6.7	33	690
14	8.4	34	870
15	11	35	1100
16	13	36	1400
17	17	37	1700
18	21	38	2200
19	27	39	2800
20	34	40	3500

Resistance is often introduced into a circuit deliberately to limit the current or to provide various levels of voltage. This is done with components called resistors (*see* RESISTOR).

An inductor has extremely low resistance for dc, but it may exhibit a considerable amount of resistance for ac (alternating current). This effective resistance is known as inductive reactance; it increases as the ac frequency becomes higher. A capacitor has practically infinite resistance for dc, but it will exhibit a finite effective resistance for ac with a sufficiently high frequency. This effective resistance is called capacitive reactance, and it decreases as the ac frequency increases. An ac circuit may contain both resistance and reactance. The combination of resistance and reactance is represented by a complex number, and is known as impedance (*see* CAPACITIVE REACTANCE, IMPEDANCE, INDUCTIVE REACTANCE, REACTANCE).

The resistance of a material is, to some extent, affected by temperature. For most materials, resistance increases with temperature. A few materials exhibit very little change in resistance with temperature changes, while other materials become better conductors as the temperature increases. The behavior of a resistance under varying temperature conditions is known as the temperature coefficient (*see* TEMPERATURE COEFFICIENT). Certain materials, notably the highly conductive metals, attain extremely low resistance values at temperatures near absolute zero (*see* SUPERCONDUCTIVITY).

A radio transmitting antenna offers inherent opposition to the radiation of electromagnetic energy. This property is called radiation resistance, and is a function of the physical size of the antenna in wavelengths (*see* RADIATION RESISTANCE).

Resistance can be defined for phonomena other than electric currents. For example, a mechanical device may show resistance to changes in position; a pipe resists the flow of water to a greater or lesser degree. The opposition that a substance offers to the flow of heat is called thermal resistance; its reciprocal is thermal conductivity (*see* THERMAL CONDUCTIVITY).

RESISTANCE BRIDGE

See RESISTANCE MEASUREMENT.

RESISTANCE-CAPACITANCE CIRCUIT

A circuit that contains only resistors and capacitors, or only resistance and capacitive reactance, is called a resistance-capacitance (RC) circuit. This circuit can consist of just one resistor and one capacitor, or it can be a complicated network of components.

An RC circuit can be used for highpass or lowpass filtering. An example of an RC highpass filter is shown at

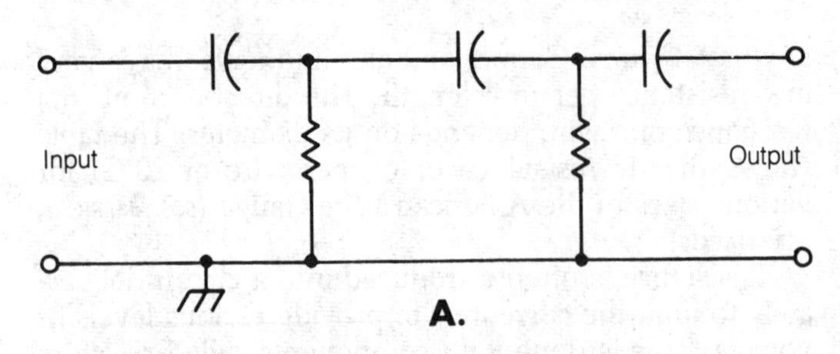

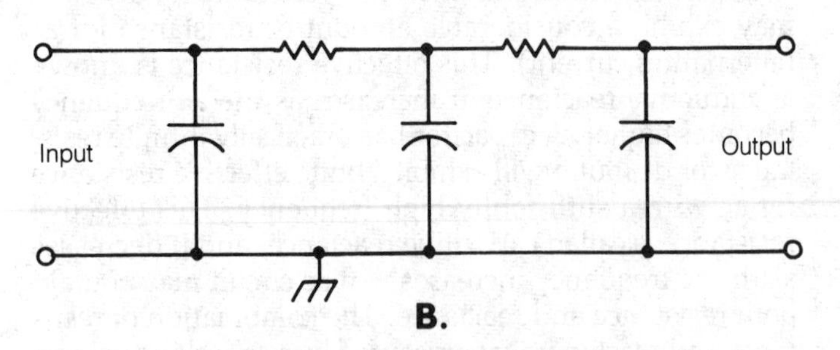

RESISTANCE-CAPACITANCE CIRCUIT: Examples of RC circuits are a highpass filter (A) and a lowpass filter (B).

A in the illustration. A lowpass network is shown at B. RC filters have a less well-defined cutoff characteristic than inductance-capacitance (LC) circuits (*see* HIGHPASS FILTER, LOWPASS FILTER). The RC-type highpass or lowpass filter is generally used at audio frequencies, and the LC type is more often used at radio frequencies.

An RC filter is commonly used in a code transmitter to obtain shaping of the keying envelope. Simple RC circuits provide filtering in low-current power supplies. More complicated RC circuits can be used to measure unknown capacitances.

Every RC circuit displays a certain charging and discharging characteristic. This property depends on the circuit configuration, and also on the values of the resistor(s) and capacitor(s). *See also* RESISTANCE-CAPACITANCE TIME CONSTANT.

RESISTANCE-CAPACITANCE COUPLING

Resistance-capacitance coupling is a form of capacitive coupling that is used in audio-frequency and radio-frequency circuits. Biasing resistors provide the proper voltages and currents at the output of the first stage and at the input of the second stage. The signal is transferred by a series blocking capacitor.

Resistance-capacitance coupling is simple and inexpensive. However, if a high degree of interstage isolation is required, or if selectivity is needed, transformer coupling is preferable. *See also* CAPACITIVE COUPLING, TRANSFORMER COUPLING.

RESISTANCE-CAPACITANCE TIME CONSTANT

The time required for a resistance-capacitance (RC) circuit to charge and discharge depends on the values of its resistance and capacitance. The larger the resistance or capacitance, the longer it takes for a circuit to charge and discharge.

Charging and discharging are actually exponential processes. The full-charge condition is theoretically never reached from a zero-voltage start; the full-discharge condition is theoretically never reached from a full-charge start.

The period required for an RC circuit to reach 63 percent of the full-charge condition, starting at zero charge, is called the RC time constant. The RC time constant can also be defined as the length of time required for an RC circuit to discharge to 37 percent of the full-charge condition, starting at full charge.

The time constant t, in seconds, of an RC circuit with an effective resistance of R ohms and an effective capacitance of C farads is:

$$t = RC$$

The RC time constant is important in the design of power-supply filtering circuits, transmitter circuits, and RC filtering devices. *See also* RESISTANCE-CAPACITANCE CIRCUIT.

RESISTANCE-INDUCTANCE CIRCUIT

A circuit that contains only resistors and inductors, or only resistance and inductive reactance, is called a resistance-inductance (RL) circuit. This circuit can be extremely simple, consisting of just one resistor and one inductor, or it can be a complicated network of components.

An RL circuit can be used for highpass or lowpass filtering. An example of an RL highpass filter is shown at A in the illustration. A lowpass network is shown at B.

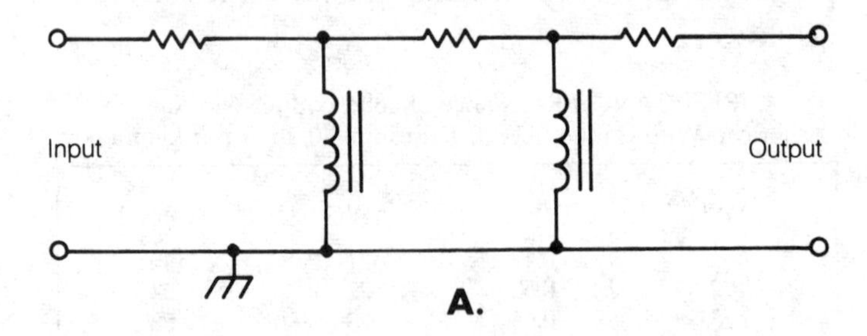

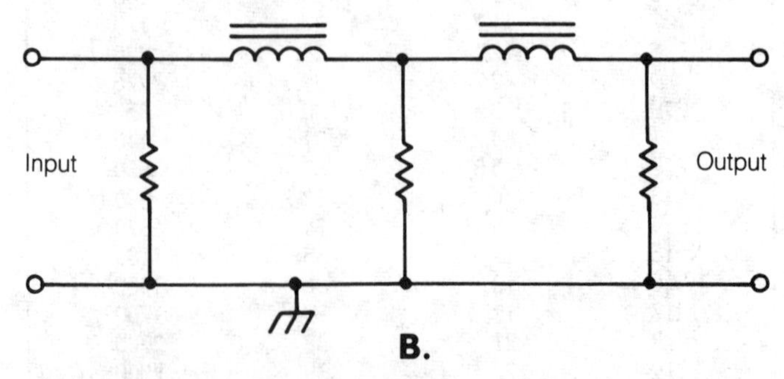

RESISTANCE-INDUCTANCE CIRCUIT: Examples of RL circuits are a highpass filter (A) and a lowpass filter (B).

The RL filters have a less well-defined cutoff characteristic than inductance-capacitance (LC) circuits (*see* HIGHPASS FILTER, LOWPASS FILTER). The RL highpass or lowpass filter is generally used at audio frequencies, while the LC type is more often used at radio frequencies. Resistance-inductance circuits can be used to measure unknown values of inductance; there are different types of RL circuit designed for this purpose.

Every RL circuit displays a charging and discharging characteristic. This depends on the circuit configuration, and on the values of the resistor(s) and capacitor(s). *See also* RESISTANCE-INDUCTANCE TIME CONSTANT.

RESISTANCE-INDUCTANCE TIME CONSTANT

The time required for a resistance-inductance (RL) circuit to charge and discharge depends on the values of its resistances and inductances. The larger the resistance and inductance, the longer it takes for a circuit to charge and discharge.

Charging and discharging are actually exponential processes. The full-charge condition is theoretically never reached for a zero-voltage start; the full-discharge condition is theoretically never reached from a full-charge start.

The period required for an RL circuit to reach 63 percent of the full-charge condition, starting at zero charge, is called the RL time constant. The RL time constant can also be defined as the length of time required for an RL circuit to discharge to 37 percent of the full-charge condition, starting at full charge.

Any resistance-inductance circuit can, ideally, be reduced to a simple equivalent combination of one resistor and one inductor. The simpler equivalent circuit has the same charging and discharging times as the more complicated circuit. The time constant t, in seconds, of an RL circuit with an effective resistance of R ohms and effective inductance of L henrys is:

$$t = RL$$

See also RESISTANCE-INDUCTANCE CIRCUIT.

RESISTANCE LOSS

In any electrical or electronic circuit some power is lost in the wiring because of resistance in the conductors. In an antenna system some power is lost in the earth because of ground resistance. The amount of power lost because of resistance is called resistance loss, I^2R loss, or ohmic loss.

Resistance loss affects the efficiency of a circuit. The efficiency is 100 percent only if the resistance loss is zero. As the resistance loss increases in proportion to the load resistance, efficiency goes down. If the loss resistance is R and the load resistance is S, the efficiency in percent is:

$$\text{Efficiency} = \frac{100S}{R + S}$$

The resistance loss in any circuit is directly proportional to the resistance external to the load. The loss is also proportional to the square of the current that flows. Mathematically, if R is the resistance in ohms and I is the current in amperes, the resistance power loss P, in watts, is:

$$P = I^2R$$

Resistance loss, unless deliberately introduced for a specific purpose, should be kept to a minimum. This is especially true in alternating-current (ac) power transmission and in radio-frequency transmitting antennas.

In a power-transmission system, resistance loss is minimized by using large-diameter conductors and by ensuring that splices have excellent electrical conductivity. The loss is also minimized by using the highest possible voltage, thereby reducing the current for delivery of a given amount of power.

In a transmitting antenna system, the resistance loss is reduced by using large-diameter wire or metal tubing for the radiating conductors, by using a heavy-duty tuning network and feed system, by providing excellent electrical connections where splices are necessary, and by ensuring that the ground conductivity is as high as possible. The resistance loss in a transmitting antenna can also be reduced by deliberately maximizing the radiation (load) resistance. *See also* ANTENNA EFFICIENCY, RADIATION RESISTANCE.

RESISTANCE MEASUREMENT

The value of a resistance can be measured in various ways. All methods of resistance measurement involve the application of voltage across the resistive element and the direct or indirect measurement of the current that flows as a result of the applied voltage.

The most common method of resistance measurement is with an instrument called an ohmmeter. The voltage source is usually a cell or battery which provides from 1.1 to 12 V. The ohmmeter is used for measurement of moderate values of resistance. Digital multimeters with resistance scales simplify resistance measurement and provide more accurate readings than classical ohmmeters. For determination of extremely high values of resistance, a special form of ohmmeter, known as a megger, is used. The megger includes a high-voltage generator instead of the low-voltage cell or battery (*see* MEGGER, OHMMETER).

An alternative means of resistance measurement is the comparison, or balance, method. The balance method gives more accurate results than the ohmmeter when the resistance is extremely low or high. The resistance-balance circuit is known as a ratio-arm bridge. The unknown resistance is compared with a known, variable, resistance, resulting in a balanced condition when a specific ratio is achieved (*see* RATIO-ARM BRIDGE).

RESISTANCE POWER LOSS

See RESISTANCE LOSS.

RESISTANCE STANDARD

See STANDARD RESISTOR.

RESISTANCE TEMPERATURE COEFFICIENT

See TEMPERATURE COEFFICIENT.

RESISTANCE THERMOMETER

The resistance of most substances changes with variations in temperature. This makes it possible to measure the temperature with an instrument similar to an ohmmeter. A length of wire, whose resistance-versus-temperature function is accurately known, is connected in series with a low-voltage power source, a current-limiting resistor, and a microammeter (see illustration). This apparatus is called a resistance thermometer.

As the temperature rises, the resistance of the wire increases, and the current reading falls. Conversely, when the temperature falls, the resistance of the wire drops, and thus more current is indicated. It is important that the temperature variations not affect the current-limiting resistor; the use of a standard resistor with a zero temperature coefficient is imperative (*see* TEMPERATURE COEFFICIENT).

A resistance thermometer, as shown in the diagram, can measure very high or low temperatures. The instrument is not accurate for determining moderate temperatures. However, greater resolution can be obtained by substituting a thermistor in place of the resistance wire. *See also* THERMISTOR.

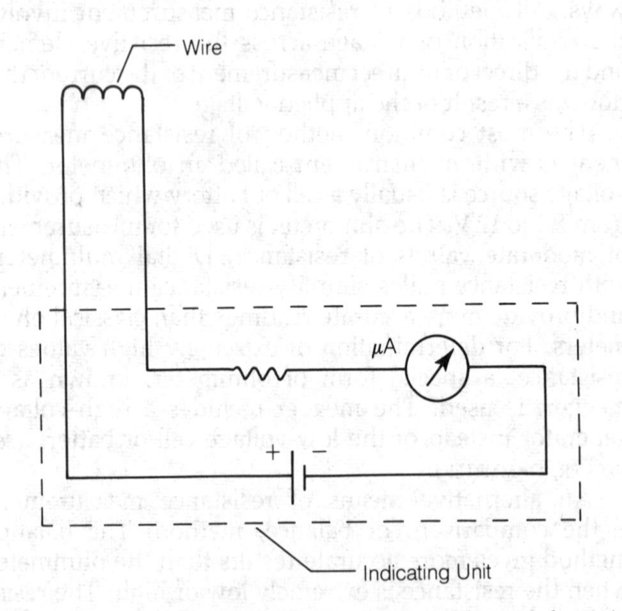

RESISTANCE THERMOMETER: Temperature changes the resistance of the wire element of a resistance thermometer.

RESISTIVE LOAD

A load is said to be resistive, or purely resistive, when it contains no reactance for alternating current at a specific frequency. A load may be resistive for either of two quite different reasons: It may contain no reactive components, or it may consist of equal but opposite reactances.

A simple, noninductive resistor presents a load with essentially zero reactance. This is true at all frequencies until the resistor lead lengths become a substantial fraction of a wavelength. Noninductive resistors are used for testing radio transmitters off the air; this load is called a dummy load or dummy antenna (*see* DUMMY ANTENNA).

A tuned circuit, consisting of a capacitor and an inductor, is resistive at one specific frequency, known as the resonant frequency. The same is true of an antenna system. This resistive load differs from the noninductive resistor. When the capacitive and inductive reactances cancel each other, a tuned circuit or antenna is said to be resonant. The remaining resistance is the result of losses in a tank circuit, or the combination of radiation resistance and resistance loss in an antenna.

A resistive load is desirable in any power-transfer circuit. If the load contains reactance, some of the incident power is reflected upon reaching the load. Ideally, the load resistance should be identical to the characteristic impedance of the feed line. *See also* CAPACITIVE REACTANCE, IMPEDANCE, INDUCTIVE REACTANCE, RADIATION RESISTANCE, REFLECTED POWER, RESISTANCE LOSS, RESONANCE.

RESISTIVE LOSS

See RESISTANCE LOSS.

RESISTIVITY

Resistivity is a measure of the intrinsic ability of a material to conduct current. Its value is independent of the dimensions of the material. Both conductors and nonconductors have resistivity. The unit of volume resistivity is the ohm-centimeter (ohm-cm). The unit of surface resistivity is ohms per square.

Surface resistivity and volume resistivity depend on the material and the temperature. Metals such as aluminum, copper, and silver have very low resistivity. Alloys, compounds, and mixtures can be prepared to have different amounts of resistivity for different applications. Dielectric substances have high resistivity. *See also* RESISTANCE.

RESISTOR

A resistor is a circuit component that provides a fixed value of resistance in ohms to oppose the flow of electrical current. A resistor can control or limit the amount of current flowing in a circuit, provide a voltage drop in accordance with Ohm's law, or dissipate energy as heat.

Fixed resistors are discrete devices classed by construction as cylindrical or planar. The most common cylindrical style is the axial-leaded part shown in Fig. 1. The resistive element is wound or deposited on a cylindrical ceramic core with a cap and lead assembly at each end. These elements include wire, metal film, carbon film, cermet, and metal oxide. Resistor networks and chip resistors are examples of planar construction as shown in Fig. 2.

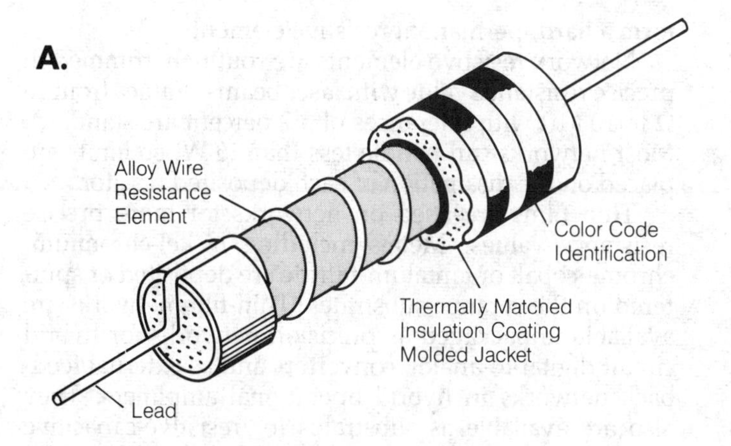

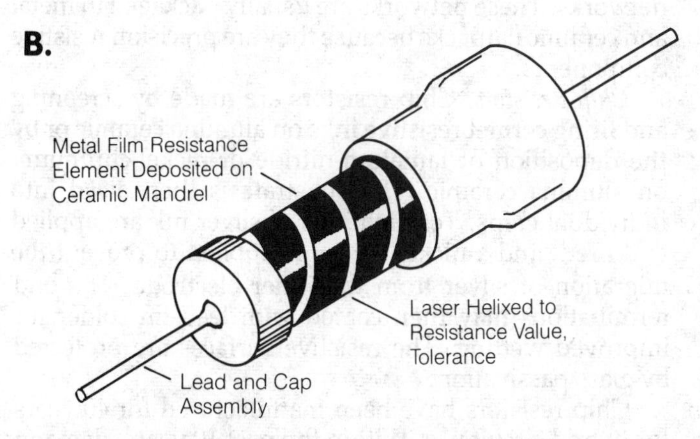

RESISTOR: Fig. 1. Cylindrical resistors include axial-leaded wirewound unit (A) and axial-leaded metal-film unit (B).

Specifications All fixed resistors are rated for a nominal resistance value in ohms covering fractions of an ohm, thousands of ohms (kilohms) or millions of ohms (megohms). Other electrical ratings include: 1) resistive tolerance as a percentage of nominal value in ohms, 2) power dissipation in watts (W), 3) temperature coefficient (tempco) in parts per million per degree Celsius temperature change (ppm/deg C), and 4) maximum working voltage in volts (V). Some resistors also have electrical noise and effective inductance and capacitance ratings.

Power dissipation in a resistor is directly related to its size. With the exception of power supply applications, most resistors for electronics are rated under 5 W. (The majority are rated for 1 W or less.) A 5 W cylindrical resistor is about an inch long with a diameter of ¼ inch. The ½, ¼, and ⅛ W units are correspondingly smaller. Resistors exhibit the unwanted parasitics of inductance and capacitance due to their construction. These residual properties must be considered in selecting resistors for applications.

Cylindrical Construction

Wirewound Resistors. Wirewound resistors are favored for their high temperature stability and are in two general classes: power and general purpose. General-purpose wirewound resistors have resistive values of 10 Ω to 1 MΩ, tolerances of ±2 percent, and tempcos of ±100 ppm/deg C (parts per million per degree Celsius). Power wirewounds are rated for more than 5 W and tolerances can exceed ±10 percent. These resistors are made by winding fine resistive wire on plastic or ceramic insulating mandrels. The most common resistance wire is nickel-chromium (nichrome). Lead and cap assemblies are placed over each end of the mandrel to complete the electrical connection.

Wire wound on a mandrel is an inductor in an alternating-current (ac) circuit. The inductive reactance that results adds to its resistance value, often ruling out wirewound resistors for use at high frequency. Bifilar winding cancels some of this reactance at low frequencies. (*See* INDUCTIVE REACTANCE.) These resistors are made with both axial and radial leads. Insulation of epoxy or silicone is applied to some low-power wirewounds.

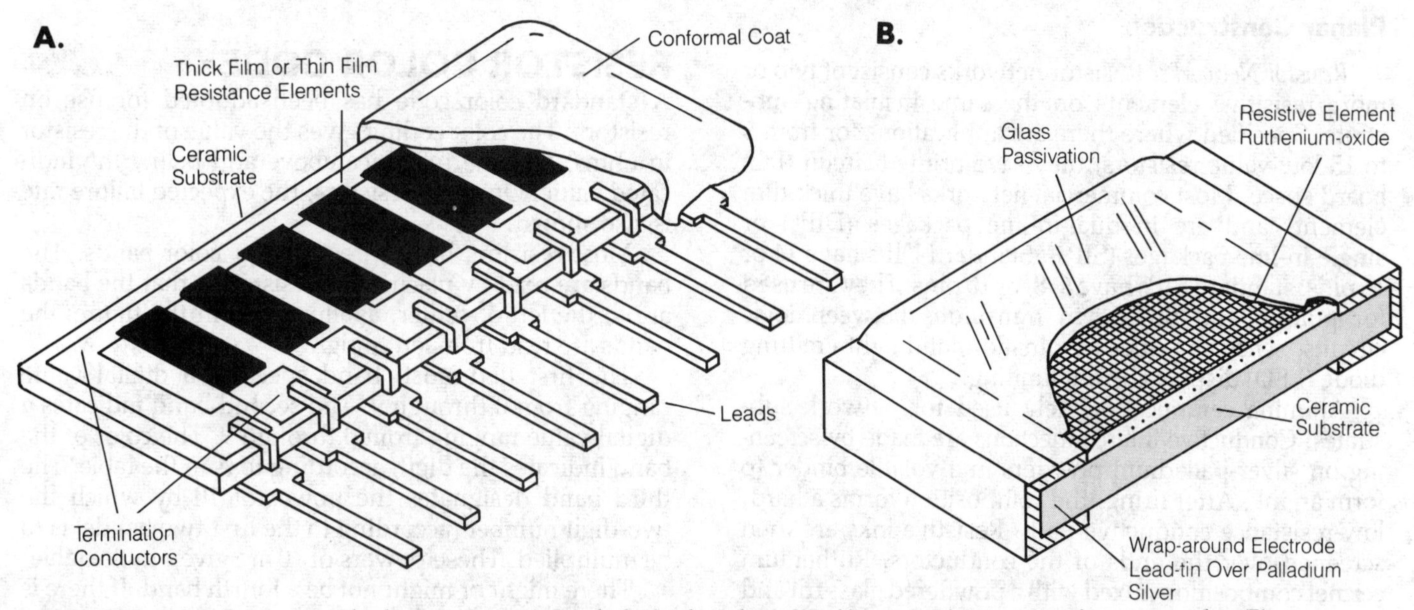

RESISTOR: Fig. 2. Resistors with planar construction include single-in-line (SIP) film network (A) and surface-mount chip (B).

High-power units are encased in ceramics or placed in aluminum cases with heat dissipating fins.

Film Resistors. Film resistors are available with metal and carbon films. Films may be applied to the mandrels as thick or thin films. Carbon-film resistors are made by screening carbon-based resistive inks on a long ceramic mandrel and firing it before it is cut into individual resistors. After attaching leaded end caps, the resistor is trimmed to a precise value and coated with an insulating material. Carbon-file resistive tolerances are typically ±5 percent. Largely used in consumer products, they have ratings of ½, ¼, and ⅛ W.

Metal-Film Resistors. Metal-film resistors are made by the same general methods as the carbon-film resistors. Thin metal films are sputtered or vacuum deposited on glass or alumina (aluminum oxide) ceramic mandrels in a vacuum chamber or thick films are applied to ceramic mandrels in air. Nickel-chromium and tin oxide are widely used thin films. Cermet ink, made from precious metal and glass powder in a volatile binder, is the most common thick film. Resistive values are set after firing by laser trimming on a closed-loop control system.

Commercial metal-film resistors are offered in two grades: those with tolerances of ±1 percent and tempcos of 25 to 100 ppm/deg C and those with tolerances of ±5 percent and tempcos of 200 ppm/deg C. Demand is highest for ¼ and ⅛ W, but 1/20 W units are available. Resistive values can be up to 100 M Ω, but are generally less than 10 kilohms.

Carbon-Composition Resistors. Carbon-composition resistors are made from carbon powder mixed with a phenolic binder that is hot molded to form a homogeneous resistive body. The molded resistive elements are leaded and fired to form rugged units capable of withstanding both temperature and transient electrical shock. They have resistive ratings of 1 Ω to 100 MΩ, but demand is concentrated in the 10 to 100 Ω range. Power ratings are typically ⅛ to 2 W, and tolerances are typically ±5 and ±10 percent.

Planar Construction

Resistor Networks. Resistor networks consist of two or more resisitive elements on the same insulating substrate. Specified where there are applications for from 6 to 15 low-value resistors, they save printed circuit (PC) board space. Most commercial networks have thick-film elements and are in dual-in-line packages (DIPs) or single-in-line packages (SIPs). Standard DIPs have 14 or 16 pins; standard SIPs have 6, 8, or 10 pins. They are used for pull-up and pull-down transitions between logic circuits, sense amplifier termination, and light-emitting diode (LED) display current limiting.

Alumina ceramic is widely used for network substrates. Conductive interconnections are made by screening on silver-palladium powders in a volatile binder to form an ink. After firing, the composition forms a hard, low-resistance conductive path. Resistive inks are then screened over the ends of the conductors. Ruthenium cermet composition mixed with a powdered glass frit and a volatile binder is most widely used. This ink is fired to form a hard, permanent resistive element.

Network resistive elements are routinely trimmed to precise resistance value with laser beams. Values from 10 Ω to 10 MΩ with tolerances of ±2 percent are standard. Most networks can handle less than ½ W, so limits are placed on the dissipation of each deposited resistor.

Thin films are used on networks for more precise resistance values. Metals including nickel-chromium, chrome-cobalt or tantalum-nitride are deposited or sputtered on the ceramic substrates. Thin-film networks are available unpackaged as precision networks for hybrid circuit digital-to-analog converters and as external feedback networks in hybrid operational amplifiers. They also are available as substrates for resistive-capacitive networks. These networks are usually packaged in metal and ceramic flatpacks because they are precision resistive components.

Chip Resistors. Chip resistors are made by screening and firing cermet resistive inks on alumina ceramic or by the deposition of tantalum-nitride or nickel-chromium on alumina ceramic. The substrate is then diced into individual chips. Terminations of silver ink are applied and fired, and a nickel barrier is applied to prevent the migration of silver from the inner electrode. The end terminations may then coated with lead-tin solder for improved wetting. The resistive surfaces are protected by glass passivation.

Chip resistors have been manufactured for 40 years for hybrid circuit use. Within the past 10 years, demand for them has increased for surface-mount assembly. Typical power ratings are ⅛ W or less. Resistor chips for surface mounting have been standardized to a 1.6- by 3.2-millimeter dimension for handling by automatic component pick-and-place machines. This is the same size as the 1206 chip capacitor that measures 0.125 by 0.063 inch. *See* SURFACE-MOUNT TECHNOLOGY.

An alternate resistor for surface mounting is the leadless MELF chip resistor. Metallized bands around each end of the cylindrical body form the end terminations.

RESISTOR COLOR CODE

A standard color code has been adopted for use on resistors. The color coding gives the value of the resistor in ohms, and the tolerance above and below the indicated value. On some resistors, the expected failure rate is also shown.

Most resistors have three or four color bands. The bands are read by placing the resistor so that the bands are to the left of center, as illustrated in the figure; the bands are read from left to right.

The first (left-most) band specifies a digital value ranging from 0 through 9. The second band indicates a digital value ranging from 0 through 9. The color of the band indicates the digit, according to A in the table. The third band designates the power of 10 by which the two-digit number (according to the first two bands) is to be multiplied. These powers of 10 are given in the table.

There might or might not be a fourth band. If there is no fourth band, then the resistor value may be as much as

RESISTOR COLOR CODE: At A, digital color bands. At B, multiplier bands. At C, failure-rate bands.

Color (A)	Digit	Color (B)	Multiplier
Black	0	Black	1
Brown	1	Brown	10
Red	2	Red	10^2
Orange	3	Orange	10^3
Yellow	4	Yellow	10^4
Green	5	Green	10^5
Blue	6	Blue	10^6
Violet	7		
Gray	8		
White	9		

Color (C)	Failure Rate Percent/1000 hours
Brown	1
Red	0.1
Orange	0.01
Yellow	0.001

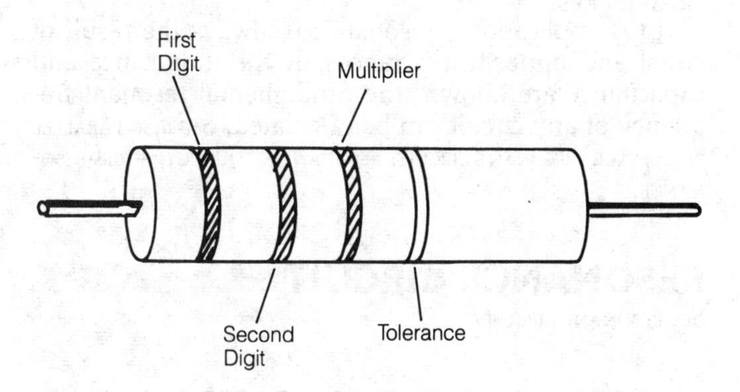

RESISTOR COLOR CODE: Each band corresponds to a digit of the resistance value.

20 percent greater than or less than the indicated value. A silver band indicates that the tolerance is plus or minus 10 percent. A gold band indicates that the tolerance is plus or minus 5 percent.

A fifth band might exist; this indicates the failure rate. The failure rate is expressed as a percentage of units expected to malfunction within 1,000 hours of operation at the maximum rated power dissipation. Failure rate is shown in C in the table.

As an example, consider that a resistor has the following sequence of color bands, from left to right: yellow, violet, red, silver, red. The first two bands indicate the digits 47. The third band designates that the number 47 is to be multipled by 100; thus the rated resistance is 4700 ohms. The fourth band designates a tolerance of plus or minus 10 percent (470 ohms above or below the rated value). The fifth band indicates an expected failure rate of 0.1 percent per 1,000 hours; in other words, one out of 1,000 resistors can be expected to fail within this length of time.

Some precision resistors have 5 color bands. The first three bands are significant figures, the fourth is the multiplier, and the fifth is the tolerance. Example: A resistor colored green yellow red red silver is 54200 Ω ±5 percent. *See also* RESISTOR.

RESISTOR-TRANSISTOR LOGIC

Resistor-transistor logic (RTL) is a form of bipolar digital logic circuit that uses, as its name implies, resistors and transistors. Resistor-transistor logic was among the first bipolar digital logic families to be manufactured as integrated circuits. The operating speed is moderate, but it can be increased by placing capacitors in shunt across the resistors. A schematic diagram of a RTL gate is shown.

The power-dissipation rate of RTL is fairly high. The design is quite simple and economical. However, RTL is susceptible to noise impulses. *See also* DIGITAL-LOGIC TECHNOLOGY.

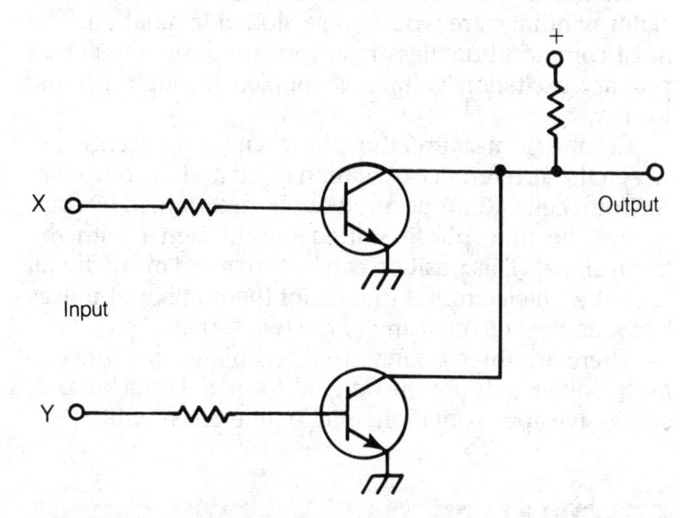

RESISTOR-TRANSISTOR LOGIC: A schematic diagram of an RTL logic gate.

RESOLUTION

Resolution is an expression of the ability of a radar, radio telescope, or optical telescope to distinguish between objects or targets that are close together. Resolution can be expressed in angular form, in radial (distance) form, or in absolute form.

Angular resolution is defined as the minimum angle, with respect to the measuring station, that two objects or targets may subtend while still appearing separate. The two objects or targets may be at much different distances from the measuring station, but almost lined up. Angular resolution is the only expression of resolution in optical or radio telescopes. Angular resolution may be given in degrees, minutes of arc (abbreviated by an apostrophe: 1′ = 1/60 degree), or in seconds of arc (abbreviated by a quotation mark: 1″ = 1/3600 degree).

Radial resolution is defined as the minimum difference in range for which a radar can distinguish between two targets. Radial resolution is usually specified in feet, meters, miles, or kilometers. Radial resolution is of importance in radar, but not in astronomy.

The absolute resolution is defined as the minimum actual separation, in feet, meters, miles, or kilometers, for which a radar can distinguish between two targets. *See also* DIGITAL PANEL METER, RADAR.

RESOLVER

A resolver is a rotary electromechanical transformer that can sense position in servo control systems. Related to a synchro unit, the resolver contains a rotor and a stator. It accepts an alternating-current (ac) reference excitation at a pair of rotor input terminals and produces voltages 90 electrical degrees apart at two pairs of stator output terminals. The output signals from the stator are at carrier frequency and their amplitudes are proportional to the sine and cosine, respectively, of the angular position of the rotor shaft.

Similar to an electric motor in construction, the two stator windings are wound on a slotted lamination. The most common brushless resolvers are designed so that primary excitation voltage is coupled through a transformer.

Resolvers measure the phase angle difference between the ac reference excitation input and the output of the rotor coils. Rotor position can be determined by comparing the time phase shifted output signal with the input signal. This position can be converted into a digital format with electronics that count the number of pulses between the zero crossing of the two signals.

There are three widely used techniques for converting resolver outputs into digital format: 1) tracking, 2) successive approximation, and 3) time phase shift.

RESONANCE

Resonance is a condition in which the frequency of an applied signal coincides with a natural response frequency of a circuit or object. In radio-frequency (rf) applications, resonance is a circuit condition in which there are equal amounts of capacitive reactance and inductive reactance. There must be reactances in a circuit for resonance to be possible; a purely resistive circuit cannot have resonance. Resonance occurs at a specific, discrete frequency or frequencies.

In a parallel-tuned or series-tuned inductance-capacitance (LC) circuit, the reactances balance at just one frequency. A parallel-tuned circuit exhibits maximum impedance at the resonant frequency; neglecting conductor losses, the resonant impedance of such a circuit is infinite, and decreases as the frequency departs from resonance. In a series-tuned circuit, neglecting conductor losses, the impedance at resonance is theoretically zero (*see* CAPACITIVE REACTANCE, IMPEDANCE, INDUCTIVE REACTANCE). The impedance of a series-tuned circuit rises as the frequency departs from resonance (*see* RESONANCE CURVE).

In an antenna radiator, resonance occurs at an infinite number of frequencies; the lowest of these is the fundamental frequency, and the integral-multiple frequencies are the harmonics (*see* ANTENNA RESONANT FREQUENCY, FUNDAMENTAL FREQUENCY, HARMONIC). The impedance of an antenna at resonance consists of radiation resistance and loss resistance. This is a finite value that depends on many factors (*see* RADIATION RESISTANCE).

Radio-frequency response can occur within a metal enclosure known as a cavity. Cavities are used as tuned circuits at ultrahigh and microwave frequencies. A cavity can exhibit resonance at an infinite number of frequencies (*see* CAVITY RESONATOR). A length of transmission line exhibits resonance at the frequency for which it measures ¼ electrical wavelength, and also on all integral multiples of this frequency (*see* HALF-WAVE TRANSMISSION LINE, QUARTER-WAVE TRANSMISSION LINE).

Cavities and rigid objects have acoustic resonant frequencies (*see* REVERBERATION). A tuning fork is an example of an acoustically resonant object. Most musical instruments exhibit acoustic resonance; a piano has resonant wires, for example, while horns and woodwinds have resonant air cavities. Mechanical resonance is a form of low-frquency acoustic resonance that occurs in rigid objects. Civil and mechanical engineers must take mechanical resonance into account when designing structures so the structures will not vibrate when exposed to external forces such as wind, wave motion, and earthquakes.

In rf applications, resonance is always the result of equal and opposite reactances. If the inductance and capacitance are known, the fundamental resonant frequency of any circuit can be calculated. *See also* PARALLEL RESONANCE, RESONANT CIRCUIT, RESONANT FREQUENCY, SERIES RESONANCE.

RESONANCE CIRCUIT

See RESONANT CIRCUIT.

RESONANCE FREQUENCY

See RESONANT FREQUENCY.

RESONANCE CURVE

A resonance curve is any graphic representation of the resonant properties of a circuit or object. Resonant curves are almost always plotted on the Cartesian plane, with the frequency as the independent variable. The dependent variable may be any characteristic that displays a peak or dip at resonance. In radio-frequency circuits, these parameters include the attenuation, current, absolute-value impedance, and voltage. Two examples of resonant curves are shown in the illustration.

At A, the absolute-value impedance of a lossless parallel-resonant inductance-capacitance (LC) circuit is plotted against the frequency. The absolute-value impedance of a lossless parallel-resonant circuit is given by:

$$Z = \frac{X_L X_C}{X_L + X_C}$$

where X_L and X_C represent the inductive and capacitive reactances, respectively. The curve at A is based on an inductance of 1 μH and a capacitance of 1 μF, resulting in a resonant frequency of about 159 kHz. The absolute-value impedance is theoretically infinite at resonance. (In

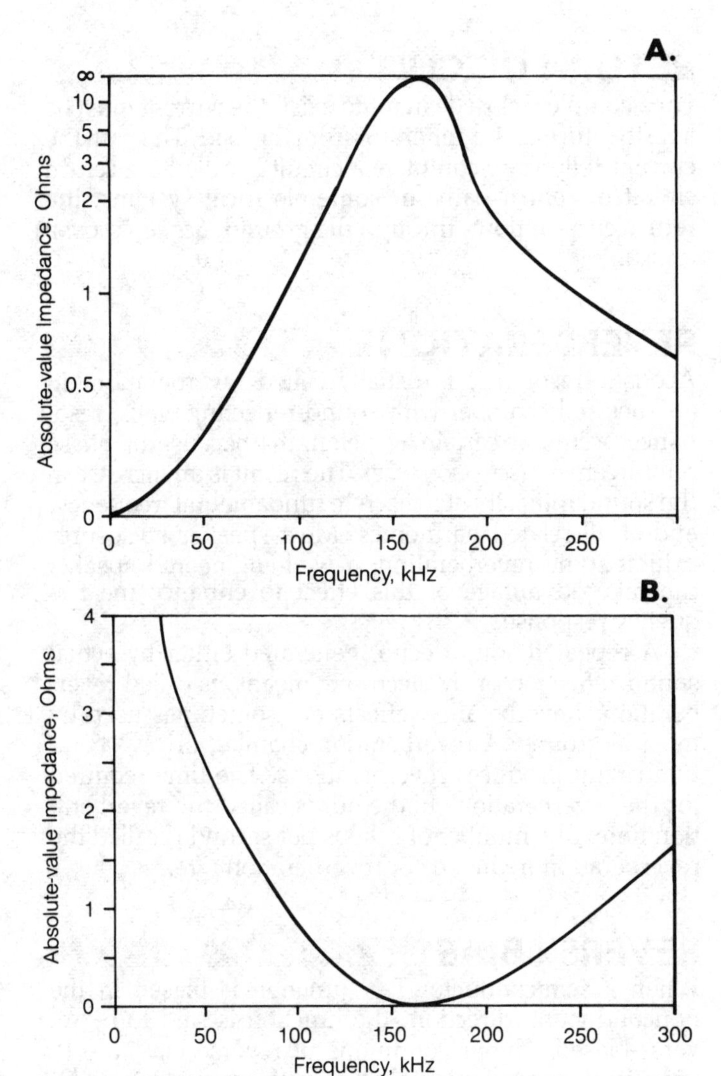

RESONANCE CURVE: Examples of resonance curves are those for a parallel-resonant circuit (A), and for a series-resonant circuit (B).

a practical circuit, losses in the components and wiring result in a finite, but very large, absolute-value impedance at resonance.)

At B, the absolute-value impedance is plotted as a function of frequency for a lossless series-resonant circuit with an inductance of 1 μH and a capacitance of 1 μF. This yields the same resonant frequency as the parallel configuration, but the absolute-value impedance shows a minimum rather than a maximum. In a lossless series-resonant circuit, the absolute-value impedance is:

$$Z = X_L + X_C$$

At the resonant frequency, the absolute-value is theoretically zero. (In a practical circuit, losses in the components and wiring result in a finite, but very small, absolute-value impedance.) *See also* CAPACITIVE REACTANCE, IMPEDANCE, INDUCTIVE REACTANCE, PARALLEL RESONANCE, RESONANCE, RESONANT CIRCUIT, RESONANT FREQUENCY, SERIES RESONANCE.

RESONANT CAVITY

See CAVITY RESONATOR.

RESONANT CIRCUIT

A circuit is considered resonant if it contains finite reactances which cancel each other. A parallel or series inductance-capacitance (LC) circuit is an example of a resonant circuit at a specific frequency. Antennas, lengths of transmission line, and cavities are resonant circuits at many different frequencies. Piezoelectric crystals have resonant properties similar to those of parallel LC circuits. Some ceramic and metal objects act as resonant circuits under certain conditions.

All resonant circuits exhibit variable attenuation, depending on frequency. Resonant circuits are extensively used in audio-frequency and radio-frequency design, for obtaining selectivity, impedance matching, and notching. *See also* ANTENNA RESONANT FREQUENCY, CAPACITIVE REACTANCE, CAVITY RESONATOR, HALF-WAVE TRANSMISSION LINE, IMPEDANCE, INDUCTIVE REACTANCE, NOTCH FILTER, PARALLEL RESONANCE, Q FACTOR, QUARTER-WAVE TRANSMISSION LINE, RESONANCE, RESONANCE CURVE, RESONANT FREQUENCY, SELECTIVITY, SERIES RESONANCE, TANK CIRCUIT, TRAP, TUNED CIRCUIT.

RESONANT FREQUENCY

The resonant frequency, or resonance frequency, is any frequency at which resonance occurs.

The resonant frequency of a tuned inductance-capacitance circuit depends on the values of the components. The larger the product of the inductance and capacitance, the lower the resonant frequency will be. If a series-tuned or parallel-tuned circuit has an inductance of *L* henrys and a capacitance of *C* farads, the resonant frequency *f*, in hertz, is given by the formula:

$$f = \frac{1}{2\pi\sqrt{LC}}$$

This formula applies for values of *L* in microhenrys, *C* in microfarads, and *f* in megahertz.

The resonant frequency of a piezoelectric crystal depends on the thickness of the crystal, the manner in which it is cut, and, in some cases, the temperature. The presence of a coil or capacitor in series or parallel with a crystal also affects the resonant frequency (*see* CRYSTAL).

The resonant frequencies of an antenna radiator, assuming that no loading reactances are present, can be determined from the length of the radiator. The fundamental frequency *f*, in magahertz, for a free-space radiator can be determined from the length *s* in feet according to the formula:

$$f = 468/s$$

For *s* in meters:

$$f = 143/s$$

The radiator is resonant at all positive integral multiples of the fundamental frequency *f*. For a radiator operating against a ground plane, the fundamental resonant frequency *f*, in megahertz, is given according to the length *s*, in feet, by:

$$f = 234/s$$

For *s* in meters:

$$f = 71/s$$

The radiator is resonant at all positive integral multiples of the fundamental frequency *f*.

The resonant frequencies of any antenna radiator are affected by the presence of reactive elements. An inductor in series with an antenna radiator lowers the resonant frequency below the value that would be given by the formula. A capacitor in series will result in a higher resonant frequency than the formulas would yield (*see* CAPACITIVE LOADING, INDUCTIVE LOADING).

A cavity exhibits resonance at frequencies that depend on its length. The same is true for half-wave and quarter-wave section of transmission line, when the velocity factor is taken into account. *See also* CAVITY RESONATOR, HALF-WAVE TRANSMISSION LINE, QUARTER-WAVE TRANSMISSION LINE, RESONANCE, RESONANT CIRCUIT.

RESONANT RESPONSE

See RESONANCE CURVE.

RESPONSE TIME

Response time is the elapsed time between the occurrence of an event and the response of an instrument or circuit to that event. Response time is important in switching circuits and measuring instruments.

If the response time of a switching circuit is not fast enough, the switch will fail to actuate properly in accordance with an applied signal. Similarly, if the response time of a measuring instrument is too slow, the instrument will not give accurate readings.

RETENTIVITY

See REMANENCE.

RETRACE

In a cathode-ray tube, the electron beam normally scans from left to right at a predetermined, and sometimes adjustable, rate of speed. At the end of a scanned line, when the beam reaches the right-hand side of the screen, the beam moves rapidly back to the left-hand side to begin the next line or trace. This rapid right-to-left movement is called the retrace or return trace.

The forward trace creates the image that is displayed on the screen. The retrace is blanked out so that it will not interfere with the viewing of the image. This is called retrace blanking. The ratio of the retrace time to the trace time is called the retrace ratio. *See also* CATHODE-RAY TUBE, OSCILLOSCOPE, TELEVISION.

RETURN CIRCUIT

For a complete closed circuit to exist, the current must be able to return to the generator from the load. The path the current follows to complete a circuit is called the return circuit or return path. In some electrical systems, the return current flows through the ground. *See also* GROUND RETURN.

REVERBERATION

Acoustic resonance is usually called reverberation. In an enclosed chamber with sound-reflecting walls, resonance occurs at certain wavelengths because of phase reinforcement (*see* RESONANCE). The result is an increase in the sound intensity at a discrete fundamental frequency, and at all related harmonics. Most speaker enclosures exhibit some reverberation; a well-engineered speaker can take advantage of this effect to enhance tne frequency response.

A repeated sound echo, generated either by actual sound reflection or by electronic means, is called reverberation. Reverberation effects are sometimes used by musical groups. A reverberation chamber or reverberation circuit produces reverberations. The time required for the reverberations to die out is called the reverberation time. The number of echoes per second is called the reverberation frequency or reverberation rate.

REVERSE BIAS

When a semiconductor P-N junction is biased in the nonconducting direction, the junction is said to be reverse-biased. Under conditions of reverse bias, the P-type semiconductor is negative with respect to the N-type semiconductor. This creates an ion-depletion region at the junction (*see* P-N JUNCTION).

If the reverse bias at a P-N junction is made large enough, the junction begins to conduct because of avalanche breakdown (*see* AVALANCHE BREAKDOWN). Some semiconductor diodes are designed to take advantage of this effect; they are called Zener diodes. The minimum reverse voltage at which avalanche breakdown takes place is called the avalanche voltage or peak inverse voltage.

A vacuum tube is reverse-biased when the anode (plate) is negative with respect to the cathode. The resistance of a tube under reverse-bias conditions is extremely high. However, if the voltage becomes too great, conduction will occur. The minimum voltage at which reverse-bias conduction takes place is called the reverse-breakdown voltage or peak inverse voltage.

Reverse bias may occur during half of an applied alternating-current cycle, as in detection and rectification. Reverse bias is applied deliberately by an external voltage source to varactor diodes and Zener diodes. *See also* DIODE, DIODE ACTION, FORWARD BIAS, PEAK INVERSE VOLTAGE, VARACTOR DIODE, ZENER DIODE.

REVERSE POWER

See REFLECTED POWER.

REVERSE RECOVERY TIME

Reverse recovery time in thyristors and semiconductor diodes is the time required for the principal current or voltage to recover to a specified value after switching from an on state to a reverse voltage or current.

REVERSE VOLTAGE

See REVERSE BIAS.

RF

The abbreviation rf stands for radio frequency. In practice, the term *rf* may serve as a descriptive indicator by itself. For example, radio-frequency current is rf current, or simply rf. The same holds true for power and voltage. *See also* RADIO FREQUENCY, RF CURRENT, RF POWER, RF VOLTAGE.

RF AMMETER

An rf ammeter is an instrument for measuring radio-frequency (rf) current. An rf ammeter can be used for the indirect determination of the rf power output from a transmitter by making measurements at the antenna.

When a transmission line is perfectly matched (*see* MATCHED LINE), and the antenna impedance is a pure resistance of Z ohms, the rf power P, in watts, can be determined from the current I according to the formula:

$$P = I^2Z$$

This formula holds only with a perfect match. *See also* RF CURRENT, RF POWER.

RF AMPLIFIER

The two principal types of radio-frequency amplifiers can be broadly classified as signal amplifiers and power amplifiers.

Signal amplifiers are intended for receiving. A signal amplifier draws essentially no power from the source, and its input impedance is extremely high. Signal amplifiers are used in the front ends of all radio-frequency (rf) receivers (*see* FRONT END, PREAMPLIFIER).

Radio-frequency power amplifiers are used in the driver and final stages of transmitters. Power amplifiers are usually Class AB, Class B, or Class C type (*see* CLASS AB AMPLIFIER, CLASS B AMPLIFIER, CLASS C AMPLIFIER, POWER AMPLIFIER).

Radio-frequency amplifiers can have narrow bandwidth, or they may be broadband. Narrow-band rf amplifiers offer superior unwanted-signal rejection in receivers and provide attenuation of harmonics and spurious signals in transmitters. Broadband amplifiers are easier to operate because no tuning is required, but undesired signals can be amplified along with the primary signal. Highpass and/or lowpass filters are often used in conjunction with broadband rf amplifiers to solve this problem. *See also* HIGHPASS FILTER, LOWPASS FILTER.

RF CHOKE

A radio-frequency (rf) choke is an inductor for blocking rf signals while allowing lower-frequency and direct-current (dc) signals to pass. These chokes are used in electronic circuits when it is necessary to apply an audio-frequency or dc bias to a component without allowing rf to enter or leave.

Radio-frequency chokes typically have inductance values in the range of about 100 μH. The exact value depends on the impedance of the circuit and the frequency of the signals to be choked. If the impedance of a circuit is Z ohms at the frequency to be choked, an rf choke is usually selected to have an impedance of approximately $10Z$ ohms.

Radio-frequency chokes are commercially manufactured in many configurations. Most have powdered-iron or ferrite cores and solenoidal windings, but some have air cores or toroidal windings. *See also* CHOKE, INDUCTIVE REACTANCE.

RF CLIPPING

The intelligibility of an amplitude-modulated or single-sideband speech signal can be increased by either of two basic methods: audio compression or radio-frequency-envelope compression. Radio-frequency envelope compression is called rf clipping or rf speech processing.

In rf speech clipping, the signal envelope is increased in average amplitude by a combination of an amplifier, clipper, and filter. The amplifier boosts the signal voltage so that the minima are quite strong; the clipper cuts off the maxima, resulting in a signal with much greater average amplitude than the original. The filter has a bandpass response with a bandwidth corresponding to the minimum needed for transfer of the modulation information.

The illustration is a block diagram of a simple rf clipping circuit that can be installed in the intermediate frequency chain of a single-sideband transmitter.

If rf clipping is used, the readability of a signal can be improved when conditions are marginal, as compared with the readability when no clipping is employed. However, if the clipping is excessive, or if the filtering is inadequate, the signal may be distorted so that the intelligibility is made worse. *See also* SPEECH CLIPPING, SPEECH COMPRESSION.

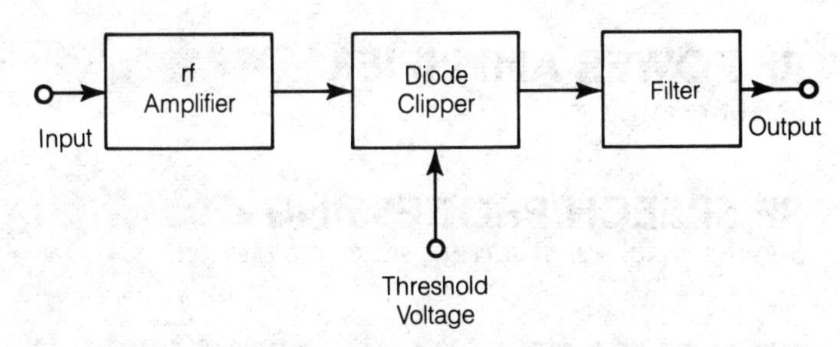

RF CLIPPING: Block diagram of an rf clipping circuit.

RF CURRENT

An alternating current with a frequency of at least 10 kHz is known as a radio-frequency (rf) current. The alternating component of a direct current that pulsates in magnitude at a rate of at least 10 kHz is also be called an rf current.

The intensity of an rf current is measured in root-mean-square amperes (*see* ROOT MEAN SQUARE). In a circuit with a nonreactive impedance of *Z* ohms and an rf power level of *P* watts, the rf current *I* is given by:

$$I = \sqrt{\frac{P}{Z}}$$

Radio-frequency current is measured with an rf ammeter. *See also* RF AMMETER.

RF FIELD STRENGTH

See FIELD STRENGTH.

RF FIELD-STRENGTH METER

See FIELD-STRENGTH METER.

RFI

See ELECTROMAGNETIC INTERFERENCE.

RF POWER

At radio frequencies, power can be consumed in three ways: it can be radiated, it can be dissipated as heat or light, and it can appear in reactive form.

Power can be radiated or dissipated only across a resistance; this is called true power (*see* TRUE POWER). A current and voltage can exist across a reactance, but no radiation or dissipation can take place in a reactance. This form of power is called reactive power (*see* REACTIVE POWER). The sum of true power and reactive power is known as apparent power (*see* APPARENT POWER).

Power at radio frequencies is usually measured with a wattmeter in a transmission line. The wattmeter indicates the apparent power. If reactive power is present in a transmission line and antenna system, it causes an exaggerated reading on the wattmeter. *See also* REFLECTED POWER, REFLECTOMETER, WATTMETER.

RF POWER AMPLIFIER

See POWER AMPLIFIER.

RF SPEECH PROCESSING

See RF CLIPPING, SPEECH CLIPPING, SPEECH COMPRESSION.

RF VOLTAGE

An rf voltage is an alternating potential difference between two points, with a frequency of 10 kHz or more. If a voltage switches at a rate of 10 kHz or more, but maintains the same polarity at all times, the alternating component constitutes an rf voltage.

The intensity of rf voltage is measured in root-mean-square volts (*see* ROOT MEAN SQUARE). In a circuit with a nonreactive impedance of *Z* ohms and an rf power level of *P* watts, the rf voltage *E* is given by:

$$E = \sqrt{PZ}$$

A radio-frequency voltage is measured with an rf voltmeter. An oscilloscope can be used to measure the peak or peak-to-peak rf voltage; the root-mean-square value is obtained by multiplying the peak value by 0.707 or the peak-to-peak value by 0.354. *See also* OSCILLOSCOPE, RF VOLTMETER.

RF VOLTMETER

At radio frequencies, the measurement of voltage is more difficult than at the 60 Hz ac (alternating-current) line frequency. Radio-frequency voltage can be measured directly with an oscilloscope, but meters can only make indirect measurements.

Most rf voltmeters rectify and filter the signal, obtaining a direct-current potential that is proportional to the radio-frequency voltage. An uncalibrated reflectometer uses this type of rf voltmeter (*see* REFLECTOMETER).

An rf voltmeter is usually frequency sensitive because the rectifier diode exhibits a capacitance that depends on the frequency of the applied voltage.

An rf voltmeter can be used to determine rf power indirectly. If the load contains no reactance and has an impedance of *Z* ohms, the rf power *P*, in watts, dissipated or radiated by the load is:

$$P = E^2/Z$$

where *E* is the voltmeter reading in root-mean-square volts. *See also* RF VOLTAGE.

RF WATTMETER

See REFLECTOMETER, WATTMETER.

RHEOSTAT

A rheostat is a variable resistor that includes a solenoidal or toroidal winding of resistance wire, two fixed end contacts, and a sliding or rotary contact. A solenoidal rheostat is shown in the illustration at A; a rotary or toroidal type is shown at B.

Rheostats have inductance as well as resistance, so they are not suitable for radio-frequency use. The rheostat is not continuously adjustable; its resistance is determined by a whole number of wire turns, presenting a finite number of discrete values.

Rheostats can dissipate large amounts of power, so they are useful for voltage dropping in high-current circuits. *See also* POTENTIOMETER.

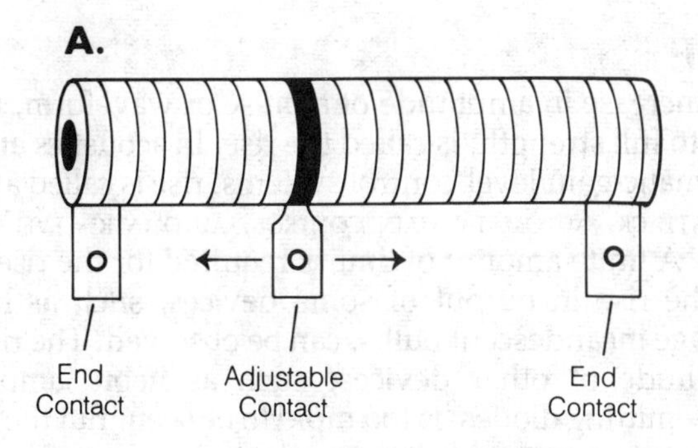

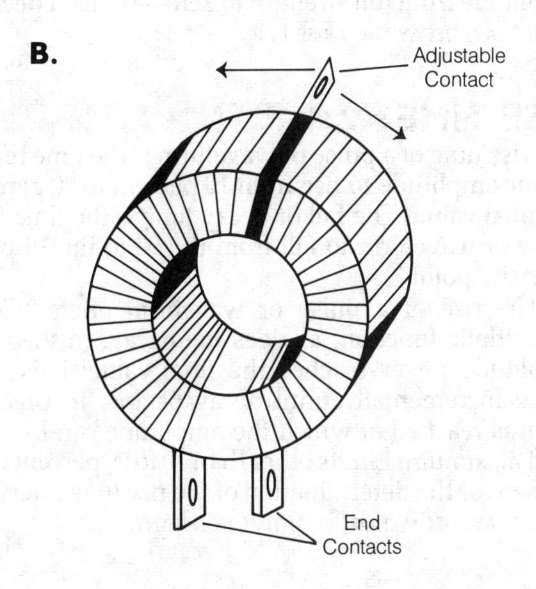

RHEOSTAT: A solenoidal rheostat (A) and a toroidal rheostat (B).

RHOMBIC ANTENNA

See ANTENNA DIRECTORY.

RIBBON CABLE

See CABLE.

RIGHT-HAND RULE

See FLEMING'S RULES.

RING COUNTER

A ring counter is a shift register in which data bits move in circular or loop fashion. If a data bit reaches the end of the ring counter, the next pulse will cause the data bit to revert to the beginning of the loop. This circular data movement can occur in either a left-to-right (clockwise) or right-to-left (counterclockwise) direction.

If a ring counter has *n* data positions, the information pattern is repeated every *n* pulses, as each data bit completes the circle exactly once. *See also* SHIFT REGISTER.

RINGING

When an alternating-current signal is applied to a tuned circuit at the resonant frequency, circulating currents are set up in the inductance and capacitance. These circulating currents, called ringing, continue for a short time after the signal is removed.

The total ringing time depends on the Q factor of the tuned circuit; the higher the Q, the longer it takes for the ringing to decay (*see* Q FACTOR). Ringing can be either desirable or undesirable.

In a Class B or Class C radio-frequency power amplifier, the tuned output circuit stores energy during the conduction period and releases it during the nonconduction period of the transistor or tube. This is a form of ringing called the flywheel effect (*see* CLASS B AMPLIFIER, CLASS C AMPLIFIER, FLYWHEEL EFFECT). It results in a nearly pure sine-wave output signal from the amplifier, although the amplifier does not conduct for the full cycle. The flywheel effect also enhances the operation of most tuned oscillators.

In an audio filter with extremely narrow bandwidth, ringing limits the maximum data speed that can be received. If the bandpass is too small, the maximum speed will be too slow, and reception will be impaired. Less frequently, the problem can be encountered in intermediate-frequency bandpass filters. Ringing in a bandpass filter can be reduced by designing the circuit so the response is rectangular. *See also* BANDPASS FILTER, BANDPASS RESPONSE.

RINGING SIGNAL

In a telephone system, alternating-current or direct-current (nonvoice) signals are sent along the line between two stations for two purposes: to inform the receiving operator of an incoming call, and to inform the sending operator that the distant telephone has been reached. These are called ringing signals.

The current that actuates the bell, tone oscillator, buzzer, or other device at the receiving station is called ringdown. The signal is transmitted at a rate of about 20 Hz, and is interrupted at regular intervals. Ringdown consists of a fairly high level of current because it must actuate a loud transducer. The return signal is called ringback; it is an audio-frequency signal of the same amplitude as the voice signals that are sent and received in a telephone conversation. The ringback signal is usually sent at a frequency of 500 Hz, modulated at 20 Hz to create a rippling sound. *See also* TELEPHONE SYSTEM.

RING MODULATOR

A ring modulator is a circuit for mixing or modulating radio-frequency signals. The ring modulator consists of four semiconductor diodes connected in a loop, allowing current to circulate. The signal inputs and outputs are taken from the apex points of the loop.

The ring modulator is a passive circuit, and its main advantage is simplicity. The ring modulator does not produce gain, but this is not a handicap in most applications because the ring modulator can be followed by a simple amplifier circuit.

The ring modulator can be used as a balanced modulator for generating single-sideband signals; it can be employed as a mixer in superheterodyne circuits; it can be used as a product detector. *See also* BALANCED MODULATOR, DOUBLE BALANCED MIXER, MIXER, MODULATOR.

RING WINDING

See TOROID.

RIPPLE

Ripple is the presence of an alternating-current (ac) component on a direct-current signal. Usually, the term refers to the residual 60 Hz or 120 Hz ac component in the output of a direct-current (dc) power supply. The ripple in the output of a power supply can modulate a radio-frequency transmitter and cause apparent modulation of received signals. This modulation is also called ripple.

The ripple in the output of a power supply can be expressed quantitatively in two ways. The ripple voltage is determined by eliminating the dc component, and then calculating the root-mean-square value of the remaining alternating current (*see* ROOT MEAN SQUARE). The base dc is designated E_r, and the steady direct-current voltage is E_d. Ripple can be expressed as:

$$R_p = E_r/E_d$$

These expressions of ripple are shown in the illustration.

Ripple can be reduced with large-value capacitors and chokes in the filter of a power supply. The amount of filtering depends on whether the ripple frequency is 60 or 120 Hz; less filtering is needed at 120 Hz. The amount of filtering also depends on the load resistance; as the load resistance in a power supply declines, more filtering is required. *See also* FILTER, FILTER CAPACITOR, POWER SUPPLY.

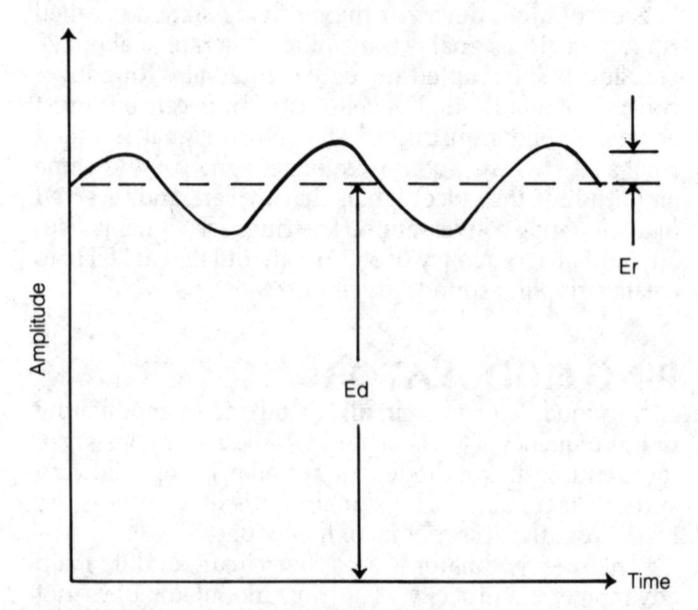

RIPPLE: Ripple voltage is expressed as the root-mean-square ripple divided by the average direct-current voltage.

RISE

The increase in amplitude of a pulse or waveform, from zero to full strength, is called the rise. In acoustics and in automatic-gain/level-control systems, rise is called attack (*see* ATTACK, AUTOMATIC GAIN CONTROL, AUTOMATIC LEVEL CONTROL). A finite amount of time is required for the rise.

The rise in output of some devices, such as high-wattage incandescent bulbs, can be observed. The rise in amplitude of other devices, such as neon lamps or light-emitting diodes, is too rapid to be seen. But the rise, no matter how rapid, is never instantaneous. The rise curve is logarithmic. The opposite of rise—the drop in amplitude from full strength to zero—is called decay. *See also* DECAY, DECAY TIME, RISE TIME.

RISE TIME

The rise time of a pulse or waveform is the time required for the amplitude to rise from 10 percent to 90 percent of the final value. The fall or decay time is the time for the pulse or waveform to fall from the 90 to the 10 percent intensity point.

The rise of a pulse or waveform often follows a logarithmic function, as does the decay. In theory, the amplitude never reaches the final value; it is always rising incrementally until decay begins. In practice, a point is reached at which the amplitude can be considered maximum. Limits other than 10 to 90 percent may be chosen for the determination of the rise time interval. *See also* DECAY, DECAY TIME, RISE, TIME CONSTANT.

RL CIRCUIT

See RESISTANCE-INDUCTANCE CIRCUIT.

RL TIME CONSTANT

See RESISTANCE-INDUCTANCE TIME CONSTANT.

RMS

see ROOT MEAN SQUARE.

ROBOT

According to the Robot Institute of America (RIA), a robot is a reprogrammable, multifunction manipulator designed to move material, parts, tools, or specialized devices, through variable programmed motions for the performance of a variety of tasks. This definition differs from those found in *Webster's New World Dictionary* because of its use of the term "reprogrammable, multifunction manipulator."

The dictionary defines a robot as: "(1) Any manlike mechanical being as those in Czech playwright Karel Capek's play R.U.R. (Rossum's Universal Robots), built to do routine manual work for human beings. (2) Any remote device operated automatically, especially by remote control, to perform in a seemingly human way. (3) An automaton." The dictionary goes on to note that the word *robot* is derived either from the Czech word *robota*, meaning forced labor, or the Bulgarian word *rabata*, meaning menial labor.

There is no universal agreement on any of the preceding definitions. Some believe the word *manipulator* confuses robots and another class of robotlike but human-operated machines called telecherics. A true robot operates independently and automatically from a self-contained program built into the machine. It is characterized by such features as an articulated and powered arm free to move in three or more degrees of freedom; it might also have a mechanical hand or gripper and even wheels or treads for mobility. But it is operated directly by humans only when it is being set up to perform a task and it is being taught a program. Most cost-effective, practical robots today have programs stored in a computer memory or on magnetic disks or tapes.

A telecheric machine can have many of the robot features but it is under human direction by cable or radio link at all times. Telecheric machines extend human manipulative capabilities in hostile situations. For example, they can handle radioactive objects or explosive materials, or retrieve bombs or dangerous objects while the operator remains at a safe distance behind protective barriers. Telecherics can also sample specimens on the sea floor while the operator remains in the protected environment of a ship on the surface, a submarine or diving bell.

Most true robots are stationary industrial robots in factories. Their assignments include heavy-duty materi-

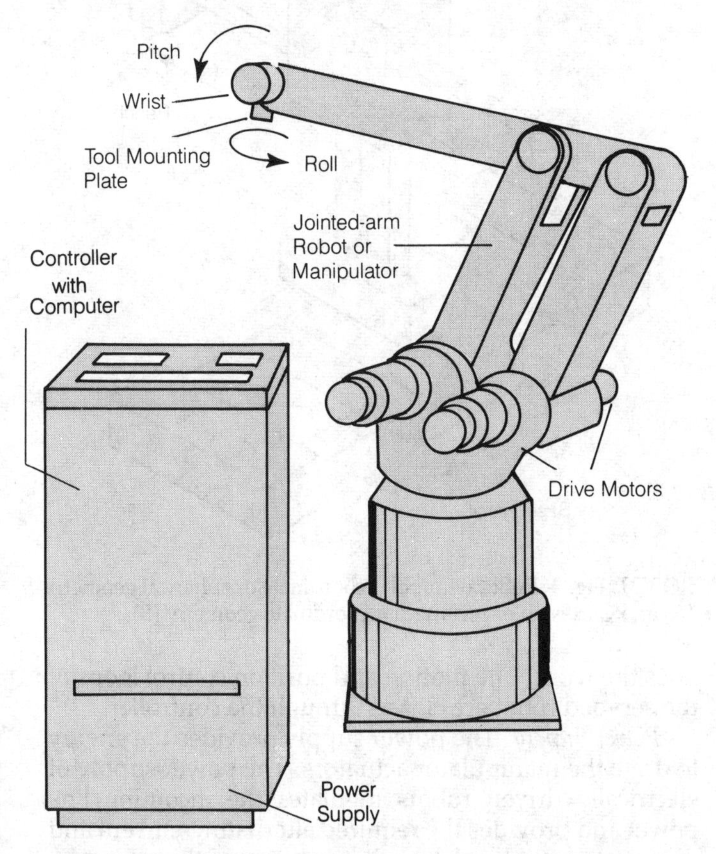

ROBOT: Fig. 1. Principal components of a robot are the arm or manipulator, the controller, and the power supply.

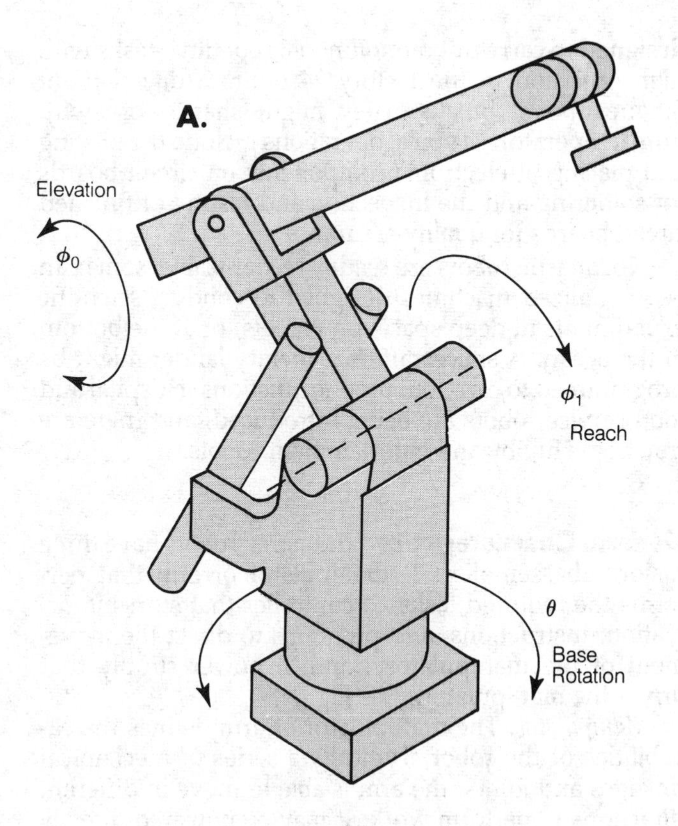

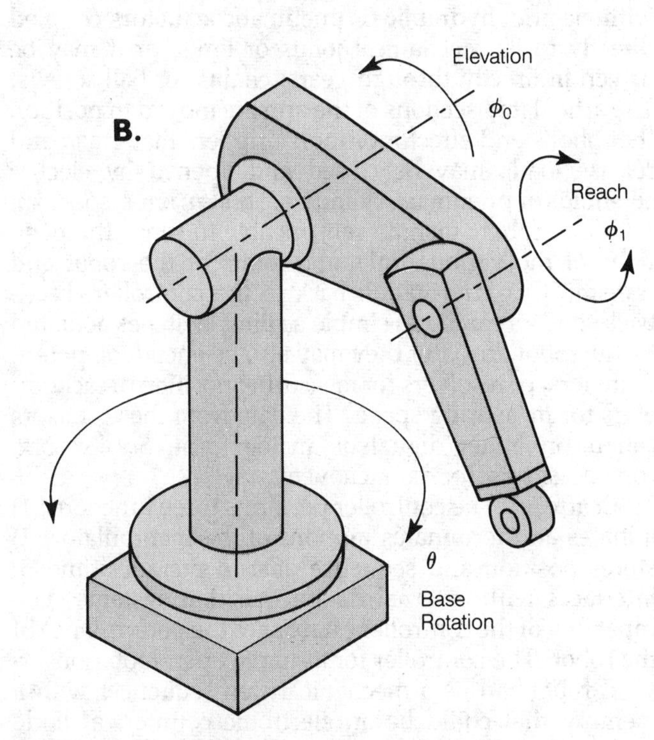

ROBOT: Fig. 2. Articulated, revolute, or jointed robot geometry with low shoulder (A) and high shoulder (B).

als handling, welding, or painting assignments. This work is often done in locations that present constant hazards for human operators or have stressful or extreme environments due to noise, heat, welding arcs, molten metal, noxious vapors, or toxic chemicals.

However, the newer, light-duty assembly and inspection robots address different problems. They are designed to carry out monotonous, repetitive tasks with high precision in the factory 24 hours a day without fatigue or rest breaks. They might share work with human operators. Typical operations include the picking and placing of electronic components on circuit boards for soldering and the inspection and testing of finished circuit boards for quality assurance.

Not all true robots are readily recognizable; some can be specialized machines designed to conduct scientific experiments in deep space, on planets, or at the bottom of the ocean. A spacecraft or planetary lander might be programmed to carry out robotic functions. Hospital and food service robots are being introduced, and there are true home hobby and entertainment robots.

General Characteristics Industrial robots have three major subassemblies: 1) manipulator or arm that performs the required tasks; 2) controller that stores information, instructions, and programs to direct the movement of the manipulatory; and 3) power supply that drives the manipulator. (See Fig. 1.)

Manipulator. The manipulator or arm defines the capabilities of the robot. Typically a series of mechanical linkages and joints, the arm is able to move in different directions to perform work. It may be powered directly with electric, hydraulic or pneumatic actuators coupled directly to its mechanical joints or links, or it may be driven indirectly through gears, chains, or ball screws. The articulated sections of the arm are moved to position the robot's end effector or tool. Grippers that grasp and release loads may be closed and opened by electric solenoids or pneumatic cylinders. *See also* MOTOR, SOLENOID.

Many robots include sensors able to sense the positions of the various links and joints of the robot and transmit this information back to the controller. Feedback sensors can be as simple as limit switches actuated by the robot arm. But they may also be encoders, potentiometers, or resolvers for measuring position or tachometers for measuring speed. The data from these sensors can be in either digital or analog form. *See* ENCODER, POTENTIOMETER, RESOLVER, TACHOMETER.

Controller. The controller performs three functions: 1) initiates and terminates motions of the manipulator, 2) stores position and sequence data in memory, and 3) interfaces with external data acquisition systems. The capability of the controller establishes the performance of the robot. The controller for a simple open-loop nonservoed robot can be a mechanical step sequencer with a memory that could be an electronic counter, a diode matrix, or a series of potentiometers.

Controllers for closed-loop servoed robots are typically microcomputers or minicomputers with disk or tape drives. In some robots the program might be completely stored in read-only memory (ROM). The controller can be part of the manipulator or it can be housed in a separate cabinet. The controller initiates and terminates the motions of the manipulator through interfaces with the manipulator control valves or electric feedback sensors. It can also perform arithmetic operations to control the path, speed, and position of the end effector, which

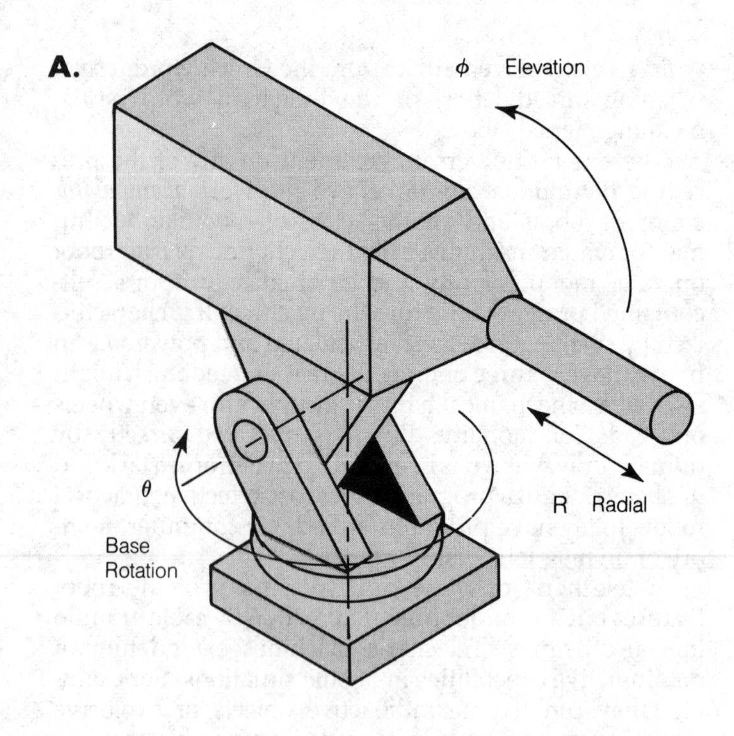

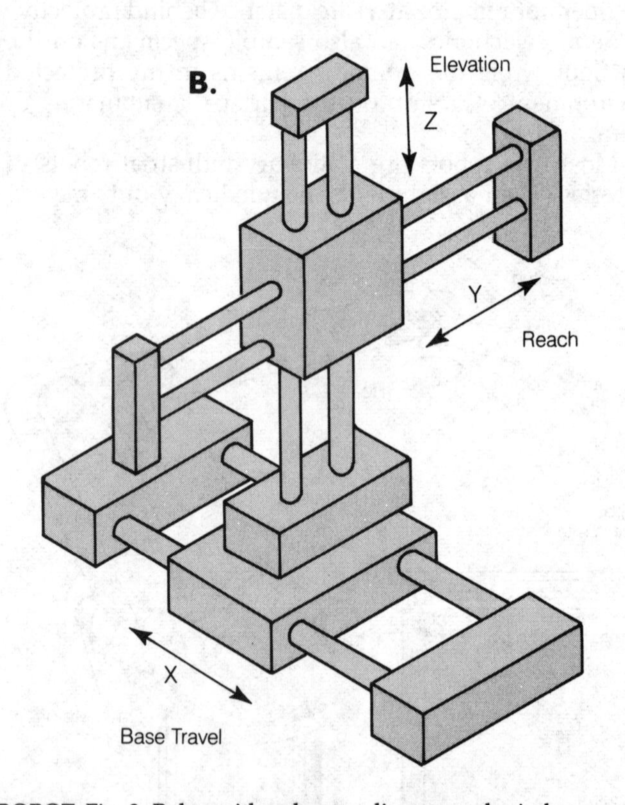

ROBOT: Fig. 3. Robot with polar coordinate or spherical geometry (A) and Cartesian or rectangular coordinate geometry (B).

does the work. The motion and position control loops of the servoed robot are closed through the controller.

Power Supply. The power supply provides the energy to drive the manipulator actuators. The power supply of electrically driven robots regulates the incoming line power and provides the required alternating-current and direct-current (ac and dc) voltages to power the actuators. The power supply of hydraulically driven robots includes a hydraulic reservoir and pump. However, com-

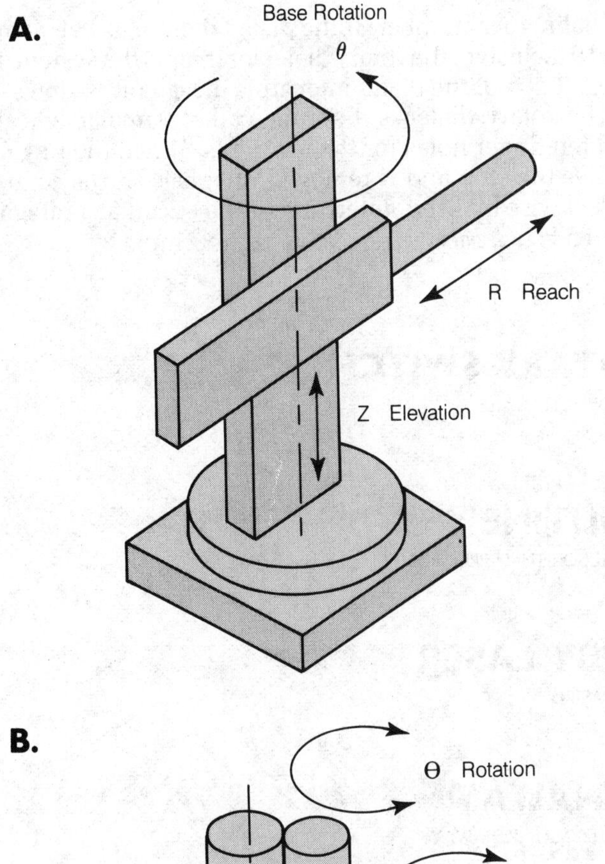

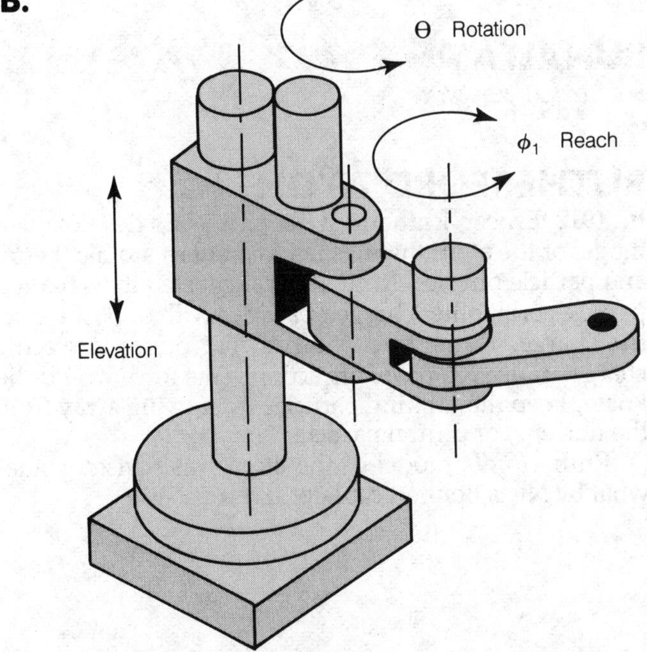

ROBOT: Fig. 4. Robot with cylindrical coordinate geometry (A) and vertically jointed articulated or jointed geometry (B).

pressed air for a pneumatic robot is typically supplied from a shop air compressor.

Robot Classifications Robots can be: 1) nonservoed or open loop or 2) servoed or closed loop. Nonservoed robots depend on the accuracy of the actuators and the precision of the gears and links to position the end effector with respect to the work (which is usually prepositioned). Most nonservoed robots are limited-sequence robots. By contrast, servoed robots include appropriate sensors in the closed loops to permit the robot to know where the end effector is at all times.

A second classification relates to motion of the end effector or tool in carrying out a programmed task: 1) point-to-point or 2) continuous. Nonservoed or limited-sequence robots are usually limited to point-to-point motion of their end effectors. They can be powered by hydraulic or pneumatic cylinders and vane motors or electric stepping motors. By contrast, servoed robots are usually capable of continuous-path or controlled-path motion. They move their end effectors in continuous sweeping motions and they may be programmed to control both speed and path contour.

Degrees of Freedom. The number of degrees of freedom of a robot determine its applications and flexibility. Most limited-sequence robots have two or three primary degrees of freedom, and their tasks are usually limited to the opening and closing of a gripper. Servo-controlled robots have at least three primary degrees of freedom for the placement of the wrist or the end of the forearm in the most favorable position to perform useful work. The wrist can have two or three additional degrees of freedom so that a tool or gripper can be positioned at the optimum angle of attack to perform the required task.

Geometrical Configuration. There are four principal industrial robot geometrical configurations: 1) articulated, revolute, or jointed arm—Figs. 2A and 2B; 2) polar coordinate or spherical—Fig. 3A; 3) Cartesian or rectangular—Fig. 3B; and 4) cylindrical—Fig. 4A. There are also a number of variations on these schemes including vertically oriented (Fig. 4B); vertically jointed, horizontally mounted; and gantry or overhead mounted.

Envelope. The envelope of a robot is the three-dimensional contour formed by the motion of the end effector or wrist moved completely through its outer limits of motion.

ROLLOFF

When a circuit exhibits a gradual increase in attenuation as the input frequency changes, the characteristic is known as rolloff. Rolloff can result from an increase in attenuation as the frequency increases, or it can result from an increase in attenuation as the frequency decreases.

Rolloff is usually applied in reference to audio-frequency characteristics. If the attenuation increases as the audio frequency rises, a bass or lowpass response is indicated. If the attenuation increases as the frequency drops, a treble or highpass response is indicated (*see* BASS RESPONSE, HIGHPASS RESPONSE, LOWPASS RESPONSE, TREBLE RESPONSE).

Rolloff is usually expressed quantitatively in decibels per octave. If the amplitude of a signal drops 6 dB per octave (as measured in volts), it means that the amplitude is E volts at $2f$, for a given frequency f, if the amplitude is $2E$ volts at frequency f. *See also* DECIBEL, OCTAVE.

ROENTGEN

The roentgen is a unit of radioactive flux used to express the total exposure to X rays and gamma rays. One roentgen is equivalent to the absorption in air of 1.065×10^{-8} joule per cubic centimeter.

A person in an average environment can expect to receive between 10 and 50 roentgens of radiation in a lifetime. An exposure of more than 100 roentgens received within a few hours or days can cause radiation sickness. Higher levels can result in violent sickness and death. *See also* GAMMA RAY, RAD, RADIATION HARDNESS, REM, X RAY.

ROM

See SEMICONDUCTOR MEMORY.

ROOT MEAN SQUARE

The current, power, or voltage in an alternating-current (ac) signal can be determined in various ways. The most common method of expressing the effective value of an ac waveform is the root-mean-square (rms) method. The rms current, power, or voltage is an expression of the effective value of a signal.

The rms current, power, or voltage is determined with the following procedure. First, the amplitude is squared so that the negative and positive halves of a waveform are made identical. Then the value is averaged over time. Finally, the square root of the average square value is determined. Mathematically, the rms value is an expression of the dc effective magnitude of an ac or pulsating-direct-current waveform.

For a sine wave, the rms value is 0.707 times the peak value, or 0.354 times the peak-to-peak value. For a square wave, the rms value is the same as the peak value, or half the peak-to-peak value. For other waveforms, the ratio varies. *See also* PEAK-TO-PEAK VALUE, PEAK VALUE.

ROTARY DIALER

In older telephone systems, a rotary dialer is used instead of the more modern tone dialer. The rotary dialer interrupts the telephone circuit a specific number of times for the dialing of a number. If the dialed digit, n, is between 1 and 9 inclusive, the rotary dialer interrupts the system n times. If $n = 0$, the dialer interrupts the circuit 10 times.

The rotary dialer is a spring-loaded, circular wheel with ten finger holes for the ten digits. When the finger reaches the stop and is removed, the dialer turns counterclockwise by itself, interrupting the circuit at a rate of about 5 Hz. *See also* TELEPHONE SYSTEM, TOUCHTONE®.

ROTARY SWITCH

See SWITCH.

ROUTINE

See ALGORITHM, PROGRAM.

RUBY LASER

See LASER.

RUNAWAY

See THERMAL RUNAWAY.

RUTHERFORD ATOM

In 1912 Ernest Rutherford, a physicist, developed a model of the atom that has led to modern atomic theory and particle physics. Rutherford suggested that charged particles orbit other charged particles with opposite electric charge. Rutherford realized that opposite electric charges cause a force of attraction. This force would, he knew, keep the orbiting particles from flying away from the nucleus, or central particle.

Rutherford's model of the atom was revised somewhat by Niels Bohr a year later. *See* BOHR ATOM.

SAFETY FACTOR

Many electrical and electronic components are rated according to the maximum current, voltage, or power they can handle. For example, a resistor may be rated at ½ W; a transistor may be rated at 10 W of continuous collector dissipation; a diode might be rated at 500 peak inverse volts. These ratings incorporate a factor called the safety factor.

The safety factor allows for the possibility that a component or device might be operated at, or near, the maximum limit of its rated capability.

SAFETY GROUNDING

When different units of electrical or electronic apparatus are used together, a potential difference can develop between them. This can present dangerous potential differences among various constituents of a system that can be eliminated by bonding all chassis together electrically, and connecting them to a common ground. This is called safety grounding. The ground-bus method is the most often-used means of accomplishing safety grounding. Safety grounding should always be used when dangerous voltages are present. Ground loops, which can cause problems in electronic equipment, are avoided by the bus system. *See also* ELECTROSTATIC DISCHARGE CONTROL, GROUND BUS.

SAMPLE-HOLD AMPLIFIER

A sample-hold amplifier (SHA) is a circuit capable of sampling a data stream and storing an analog signal level. It has an analog signal input, a control input, and an analog signal output. It always operates in one of two modes: sample or hold. In the sample mode, the SHA acquires the input signal as rapidly as possible and tracks it faithfully until commanded to hold. In the hold mode, the SHA circuit retains the last value of the input signal when the control signal ordered a mode change. The SHA is, in effect, an analog switch that, on command, samples the instantaneous level of an analog signal and retains that signal as a direct-current level. The figure shows a basic diagram of an SHA.

Sample-hold amplifiers consist of a pair of operational amplifiers connected in a voltage follower configuration. The control input is usually compatible with standard transistor-transistor logic (TTL) or metal-oxide semiconductor (MOS) logic levels. The SHA function can be constructed from discrete components: an analog switch, a holding capacitor, and an operational amplifier. It is also available in module, hybrid circuit, or integrated circuit (IC) form. The IC contains an input operational amplifier (op amp), a transistor switch, and an output (op amp). An external holding capacitor is then added to the IC. *See* OPERATIONAL AMPLIFIER.

The SHA is sometimes referred to as a track-hold amplifier (THA). The difference between the two depends on the time of switch closure. In track-hold operation, the switch is closed for a relatively long period, permitting the output to change significantly. However, the output holds the level present at the instant the switch is opened. By contrast, in SHA operation, the switch is closed only long enough to charge the holding capacitor completely. There is no difference between the

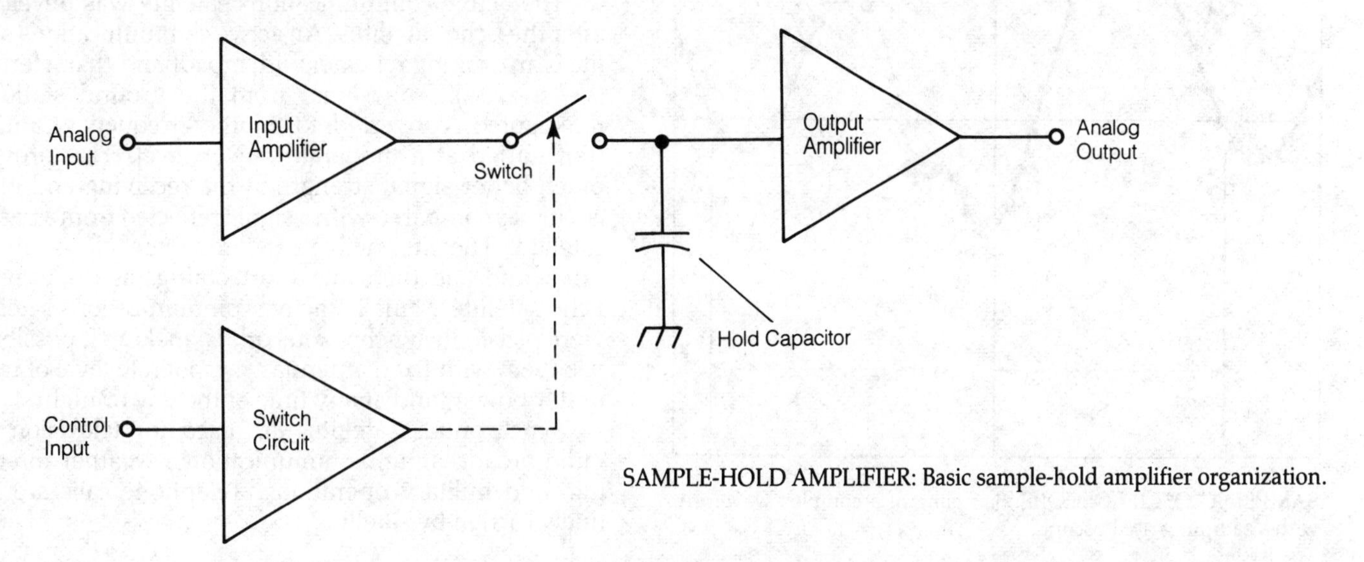

SAMPLE-HOLD AMPLIFIER: Basic sample-hold amplifier organization.

two circuits in data acquisition systems operating at update rates greater than 1 MHz.

In data-acquisition systems, the SHA is frequently assigned to freeze fast-moving signals prior to their processing by the system. Accurately holding the amplitude of the signal for the appropriate length of time is critical in measurement systems involving analog-digital conversion, peak detection, multiplexing, and other functions where precise timing is essential.

SAMPLING OSCILLOSCOPE

An oscilloscope can be used to obtain a prolonged view of a regular, repeating waveform by adjusting the sweep rate to correspond with the period of the wave. Most oscilloscopes can perform this function automatically with a triggering circuit (*see* TRIGGERING).If the waveform is very irregular, however, this cannot be done. A different method must be used for the analysis of irregular, repeating waveforms or waveforms of extreme frequency. A sampling oscilloscope provides one means for evaluating these waveforms.

The sampling oscilloscope measures the instantaneous voltage of a waveform at various moments. The samples are taken at the rate of one per cycle until the entire waveform has been sampled at intervals and the signals are amplified and stored. Then they are displayed simultaneously on the cathode-ray screen (see illustration). The display appears segmented or digitized, because a finite number of samples are used. The result is an approximation of the waveform. The accuracy of this approximation depends on the number of samples taken in a given amount of time, and also on the complexity of the waveform itself. *See also* OSCILLOSCOPE.

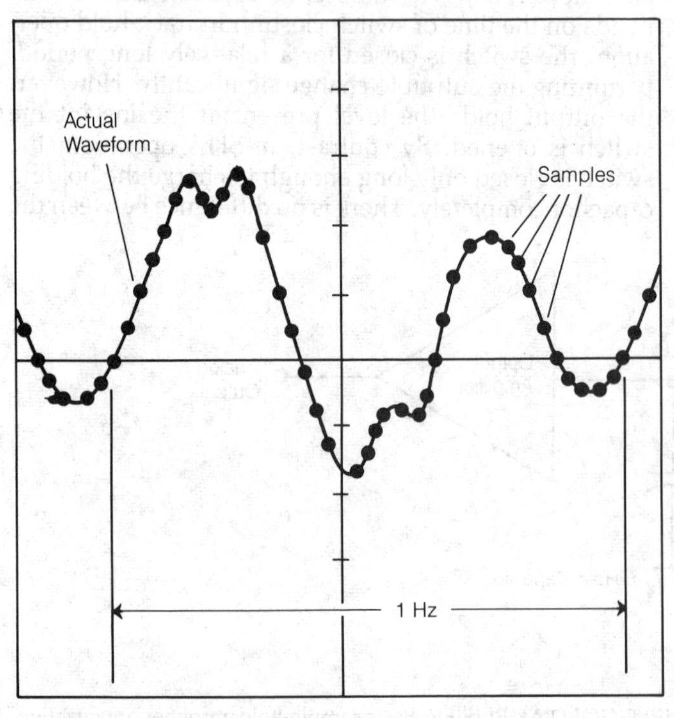

SAMPLING OSCILLOSCOPE: Sampling of a complex waveform with a sampling oscilloscope.

SAMPLING THEOREM

When a signal is evaluated by sampling, it is necessary to obtain a minimum number of samples per cycle, or the sample may not be representative. For example, in the illustration for SAMPLING OSCILLOSCOPE, it is apparent that the sampling oscilloscope must obtain a minimum number of samples per cycle to show accurately what the waveform looks like. The minimum sampling rate for accurate representation of a waveform is defined by a rule known as the sampling theorem. The sampling theorem was developed by the engineer Nyquist.

If the highest-frequency component of a signal has a frequency of f Hz. According to the sampling theorem, at least two samples per cycle are needed for this component. That is, the sampling rate must be at least $2f$. The maximum allowable interval between samples is therefore $1/2f$ seconds.

It is a good practice to obtain more than the minimum number of samples as defined by this theorem. The greater the number of samples, the more accurate the representation will be. It is rarely necessary, however, to obtain more than 10 to 15 samples for each cycle of the highest-frequency component. *See also* NYQUIST RATE, SAMPLING OSCILLOSCOPE.

SATELLITE COMMUNICATIONS

The earth is surrounded by artificial satellites today. Many of these satellites carry repeaters and are used for communications. In recent years satellites have been placed in synchronous orbits (*see* SYNCHRONOUS ORBIT), providing continuous communications capability among almost all possible parts of the globe.

In the 1960s, a series of passive satellites was launched into orbit around the earth. These were large metallized balloons that reflected radio waves sent up to them. The Echo satellites were placed in low orbits because the existing boosters could not loft a satellite in a synchronous orbit. The area of coverage for each Echo satellite was limited by the low orbit, and access time was brief.

The active communications satellite was developed after the Echo satellites. An active communications satellite is an orbiting repeater with broadband characteristics (*see* REPEATER). The signal from the ground station is intercepted, converted to another frequency, and retransmitted at a moderate power level. This provides much better signal strength at the receiving end of the circuit, as compared with a signal reflected from a passive satellite. The first active satellites were placed in low orbits and had the same shortcomings as the previous Echo satellites. Finally, active communications satellites were placed in synchronous orbits, making it possible to use them with fixed antennas, a moderate level of transmitter power, and at any time of the day or night.

Synchronous satellites are used for television and radio broadcasting, communications, weather forecasting, and military operations. Telephone calls are routinely carried by satellite.

SATELLITE NAVIGATION

Satellite navigation or SAT-NAV, a navigation system available to commercial shipping and private boat owners, can provide a position that is accurate within 100 feet of true position. SAT-NAV receivers are radio navigational aids for ships that receive signals from transit satellites as they pass overhead. The satellites are able to update the ship's position about every 90 minutes.

SAT-NAV receivers compute latitude and longitude and provide data for a range and bearing display to specific locations known as waypoints. Some receivers store the ship's position and the time and positions of up to 20 previous fixes. Some receivers can store predictions of the favorable positions of as many as 50 upcoming satellites for navigation and provide alarms to indicate their availability.

With SAT-NAV it might be necessary to dead reckon or plot the ship's position with known information on speed, heading, and course to obtain a current position between satellite passes. Some SAT-NAV receivers are equipped with a computer that accepts these manual inputs and computes an updated position between fixes.

The internal computer can provide ship's speed made good (SMG) and course made good (CMG), the course to steer (true or magnetic) and distance to waypoints. Some also offer waypoint approach alarms. When used in conjunction with loran, current position can be determined accurately. *See also* GLOBAL POSITIONING SYSTEM, LORAN.

SATELLITE SYSTEM

In electronics the term *satellite* has two meanings: 1) an unmanned spacecraft put either into an orbit around the earth or a space trajectory, or 2) an electronics system such as a transmitter or power station that is subordinate to or dependent on a larger or more complete electronic system.

Spacecraft. Satellites are lofted into orbit by rocket propulsion to perform one or more functions. Most are specialized for: 1) relay of terrestrial communications; 2) active or passive electromagnetic surveillance, photographic reconnaissance, or television reconnaissance; 3) navigation aids; 4) unmanned exploration of space by instrumentation; 5) monitoring of physical phenomena; and 6) scientific experimentation. A satellite system includes both the spacecraft and ground-based support systems for control and communications, usually by radio link. However, the signals from many broadcast and navigation satellites can be received by large numbers of suitable directional or omnidirectional antennas distributed over a wide geographical region.

Some satellites are put in low orbit to travel around the earth on a predetermined trajectory at scheduled times. Other satellites can be placed in high or geostationary orbits so they remain over fixed points on the Earth's surface. Space probes, however, are usually lofted out of a low orbit into an interplanetary trajectory.

Military reconnaissance or surveillance satellites are

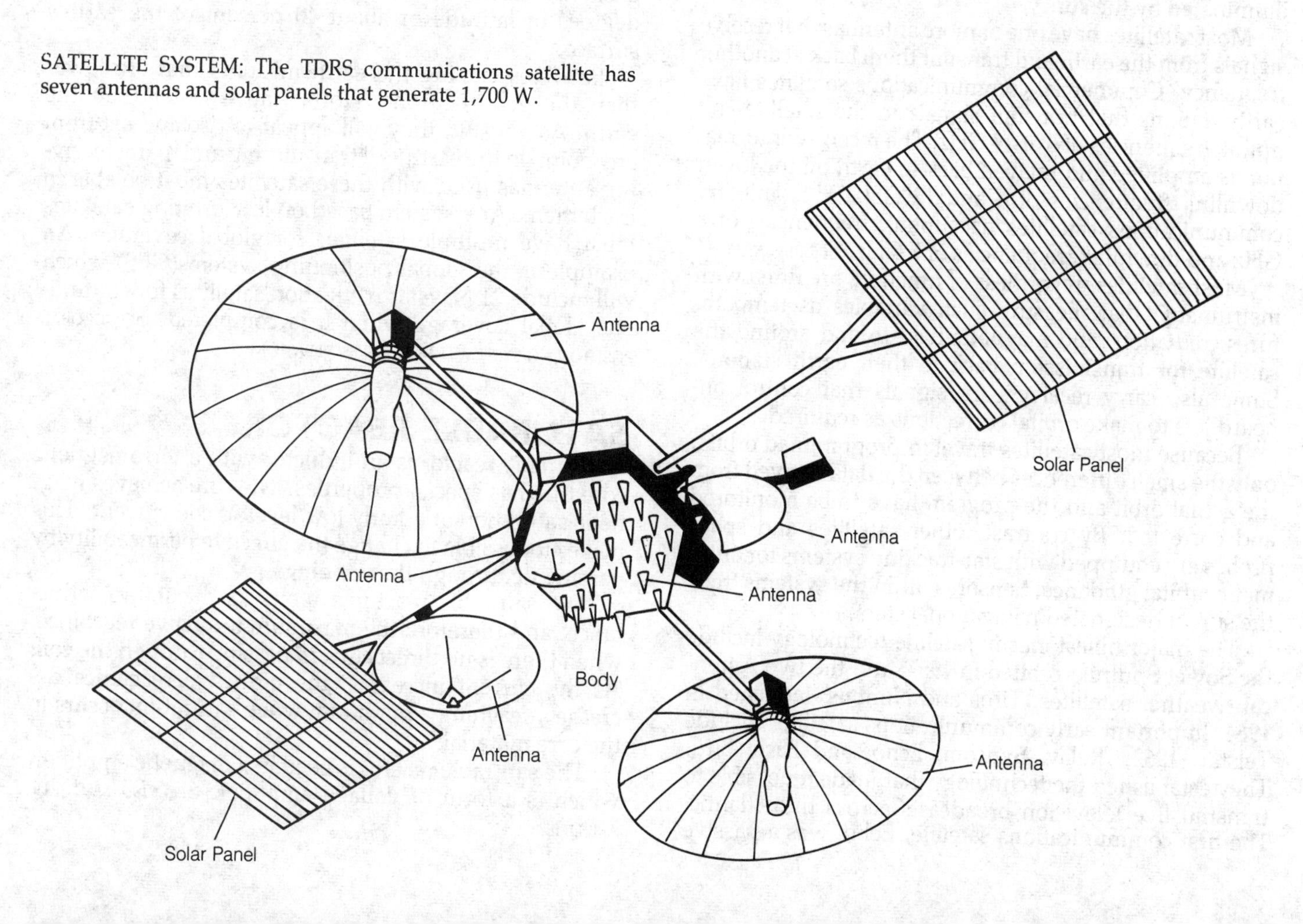

SATELLITE SYSTEM: The TDRS communications satellite has seven antennas and solar panels that generate 1,700 W.

usually placed in low orbit so they pass repeatedly over regions of the earth of interest to the governments that launched them. They might be equipped with radar, infrared sensors, broadband radio receivers, photographic cameras, television cameras, and various combinations of these.

Commercial satellites in low orbit can use infrared sensors and photographic or television cameras. They can monitor and map changes on the earth, in the sea, and in the atmosphere. Scientific satellites monitor weather patterns, magnetic fields, and changes in radiation strength. Navigation satellites make regular transits of the earth while transmitting coded signals that permit aircraft and ships to establish their positions with respect to latitude and longitude. *See* LATITUDE, LONGITUDE, SATELLITE NAVIGATION.

Most active satellites are powered by solar cells that convert the Sun's radiation directly into electric power. (Some Soviet satellites have been nuclear powered.) *See* SOLAR CELL. The solar cells can be mounted on the outside of the cylindrical satellite body that is continuously rotated so the solar cells always face the sun while the antenna is kept pointed at the earth. The spinning body acts as a gyroscope to stabilize the satellite in space.

Other satellites are stabilized with built-in gyroscopes in both the body and large solar panel arrays. This satellite configuration can provide more solar power than the spin-stabilized units because larger solar panels with a larger area can be positioned so they are continuously illuminated by the sun.

Most satellites have one or more antennas that receive signals from the earth and transmit them back at another frequency. Commercial communications satellites have earth stations that transmit signals to the satellites on uplink frequencies of 6 or 14 GHz. The received information is amplified and relayed back to earth on the lower downlink frequencies of 4 and 12 GHz. U.S. military communications satellites use uplink frequencies of 8 GHz and downlink frequencies of 7 GHz. *See* REPEATER.

Most satellites, regardless of function, are fitted with instruments that measure such variables as temperatures, radiation, and magnetic field in and around the satellite for transmission back to their earth stations. Some also carry receivers for signals that control on-board jets to make orbital corrections as required.

Because most satellites travel in programmed orbits, only the small differences between the data derived from the actual orbit and the program have to be monitored and corrected. By contrast, other satellites and space probes are equipped with star-tracking systems for automatic orbital guidance. Sensors within the systems track the sun or designated navigational stars.

The major milestones in satellite technology include the Soviet Sputnik, orbited in 1957, and the two American weather satellites, Tiros and Nimbus, launched in 1964. Important early communications satellites include Telstar (1962), Relay, Syncom, Echo, and Early Bird. They established the technology that made it possible to transmit live television broadcasts across the Atlantic. The first communications satellite, Echo, was a passive reflector of radio signals, but Telstar and all communications satellites that followed have been active repeaters. Scientific exploration of interplanetary space started with the American Ranger and Mariner and Russian Lunik space probes.

The ground tracking stations for space probes and communications satellites have highly sensitive antennas with typical gains of 60 decibels (1 million times) for receiving weak signals. The tracking antenna may be rotated with a directional accuracy of about one-thousandth of a degree (3.6 seconds of arc).

The communications satellite illustrated in the figure is the TDRS (tracking and data relay satellite). It has a hexagonal body that is 9 feet across; it weighs 4,668 pounds. The satellite is equipped with seven antennas, the largest of which are parabolic antennas with a 16-foot diameter. Two solar panels produce 1,700 W of power.

A module in the lower half of the hexagonal body houses the communications equipment and the payload module in the upper half contains the electronic equipment for communications with other spacecraft. An attitude control subsystem stabilizes the TDRS and keeps its antennas and solar panels properly oriented.

These communications satellites are in circular geosynchronous orbits at a radius distance of about 22,250 miles (35,800 kilometers) from the center of the earth. Also the planes of their orbits must contain the equator. Because the periodic times of these orbits are 24 hours, they are synchronous with the earth's rotation. At the geosynchronous altitude a satellite can see about 120 degrees of latitude or about 40 percent of the earth's surface.

Satellites circling the earth in orbits that are lower than the geosynchronous orbit move faster than the earth. As a result, they will appear to rise and set from any point on the earth. All transmitting and many receiving antennas used with these satellites must be able to track them. Any system based on low-orbiting satellites must have multiple satellites for global coverage. An example is the global positioning system (GPS) which will include 21 Navstar navigation satellites for continuous global coverage when it is completed. *See* GLOBAL POSITIONING SYSTEM, SATELLITE NAVIGATION.

SATURABLE REACTOR

A saturable reactor is an inductor with a ferromagnetic core that has special properties. The core achieves magnetic saturation at a fairly low level of coil current. This makes it possible to change the effective permeability by passing dc through the windings.

The saturable reactor exhibits its maximum inductance, and therefore the maximum inductive reactance, when there is no direct current passing through the coil. As the current increases, the inductive reactance decreases, reaching a minimum when saturation occurs in the core material.

The saturable reactor is used in a magnetic amplifier, which is a form of voltage amplifier. *See also* MAGNETIC AMPLIFIER.

SATURATION

In a switching or amplifying device, the fully conducting state is called saturation. The term is widely used in bipolar-transistor and field-effect-transistor (FET) circuitry.

An NPN bipolar transistor becomes saturated when the base voltage is sufficiently positive relative to the emitter; a PNP transistor becomes saturated when the base voltage is sufficiently negative relative to the emitter. An N-channel junction FET becomes saturated when the gate voltage is sufficiently positive relative to the source; a P-channel junction FET becomes saturated at high negative gate voltages relative to the source.

The voltage at which saturation occurs is called the saturation point or saturation voltage. The current that flows under these conditions is called the saturation current (*see* SATURATION CURRENT, SATURATION VOLTAGE). In the characteristic curve of an amplifying or switching device, saturation is indicated by a leveling off of the collector or drain current as the base or gate voltage changes. The saturation voltage and current are affected by the voltage at the collector or drain of the amplifying or switching device. The illustration at A is an example of saturation in

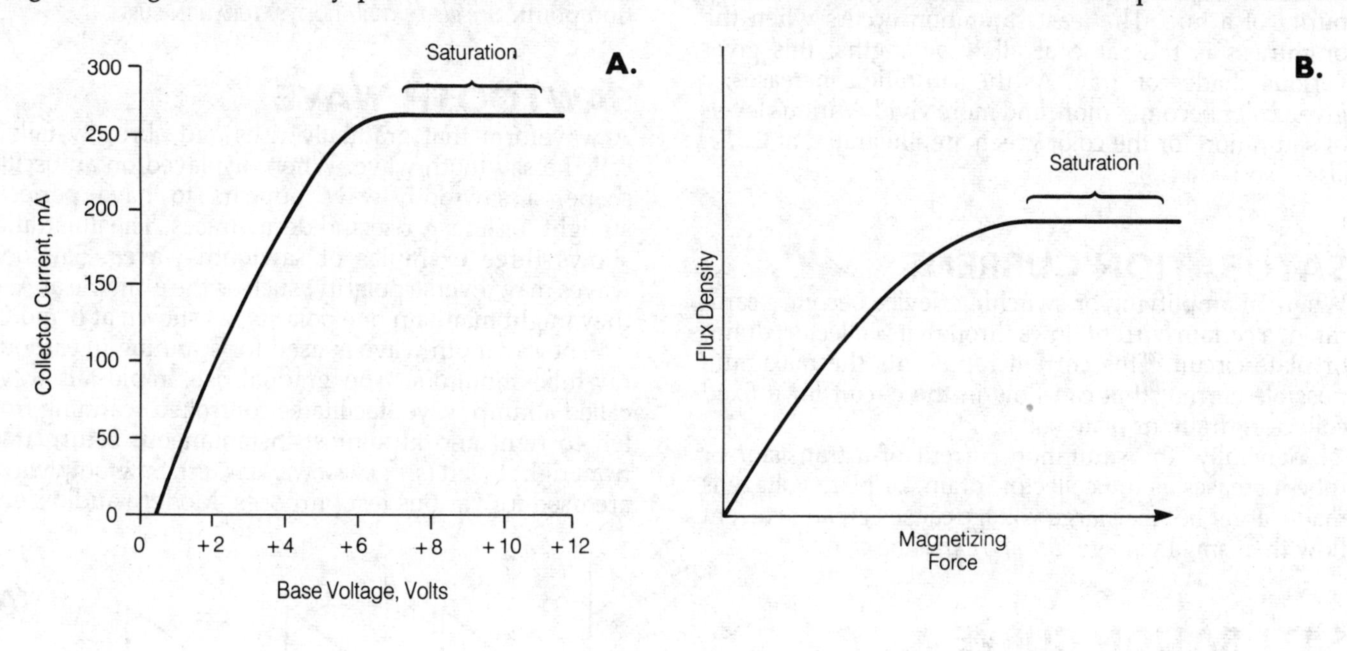

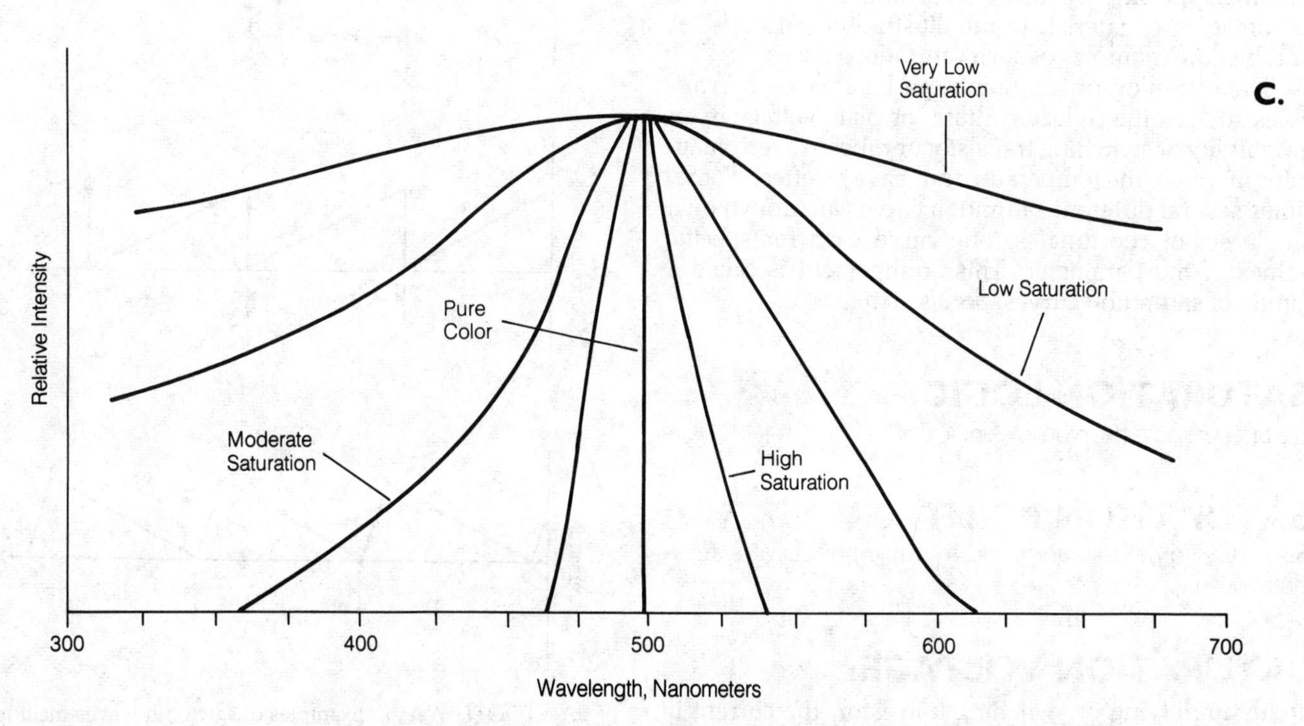

SATURATION: A saturation curve for a bipolar transistor (A), a saturation curve for ferromagnetic material (B), and varying saturation levels for green light (C).

an NPN bipolar transistor. *See also* TRANSISTOR.

In a FET, the condition of pinchoff is sometimes called saturation (*see* FIELD-EFFECT TRANSISTOR, PINCHOFF).

In a ferromagnetic substance, the flux density generally increases as the magnetizing force increases. However, there is a limit to the flux density that can be obtained in a ferromagnetic material. When this limit is reached, further increases in the magnetizing force do not produce an increase in the flux density (B). This condition is called saturation. *See also* FERROMAGNETIC MATERIAL.

In the expression of color, saturation is the relative purity of a hue. The least saturation occurs when the brightness is the same at all wavelengths; this gives various shades of gray. As the saturation increases, a given color becomes more and more vivid. Various levels of saturation for the color green are illustrated at C. *See also* CHROMA, HUE.

SATURATION CURRENT

When an amplifying or switching device becomes saturated, a certain current flows through its collector, drain, or plate circuit. This current represents the maximum possible current that can flow in the circuit for a fixed collector, drain, or plate voltage.

Generally, the saturation current of a transistor or tube increases as the collector, drain, or plate voltage is made larger because large voltage causes greater current flow than small voltage. *See also* SATURATION.

SATURATION CURVE

A saturation curve is a graphical representation of saturation in an amplifying or switching device, or in a ferromagnetic material. In the illustration SATURATION, A and B show examples of such saturation curves.

The condition of saturation can be affected by variables such as the collector, drain, or plate voltage in an amplifying or switching transistor or tube. In a ferromagnetic material, the temperature can have an effect. Sometimes several different saturation curves are drawn on a single set of coordinates, one curve each for specific values of the parameter. This arrangement is called a family of saturation curves. *See also* SATURATION.

SATURATION LOGIC

See DIGITAL-LOGIC TECHNOLOGY, LOGIC.

SATURATION POINT

See SATURATION, SATURATION CURRENT, SATURATION CURVE, SATURATION VOLTAGE.

SATURATION VOLTAGE

In an amplifying or switching transistor, the current in the collector or drain circuit changes with changing voltage at the control electrode. In an NPN transistor or N-channel field-effect transistor (FET), the current rises with increasing positive voltage at the base or gate. In a PNP transistor or P-channel FET, the current rises with increasing negative voltage at the base or gate.

As the voltage is changed so that the collector or drain current rises, a point is reached at which no further increase occurs. The voltage at the base or gate, measured with respect to the potential at the emitter or source, is called the saturation voltage. The saturation voltage depends on the type of device, and also on the collector-emitter or drain-source voltage. The saturation voltage is sometimes called the saturation bias or saturation point. *See also* SATURATION, SATURATION CURRENT.

SAWTOOTH WAVE

A waveform that gradually rises and abruptly falls is called a sawtooth wave. When displayed on an oscilloscope, a sawtooth wave appears to have perfectly straight, or linear, rise and decay traces. The illustration shows three examples of sawtooth waves. Sawtooth waves may reverse polarity, such as the example at A, or they might maintain one polarity, as shown at B and C.

The sawtooth wave is used for scanning in cathode-ray-tube monitors. The gradual-rise, rapid-fall wave, called a ramp wave, facilitates controlled scanning from left to right and an almost instantaneous return trace from right to left (*see* RAMP WAVE, RETRACE). Sawtooth waves are used for various test purposes. Modern audio signal

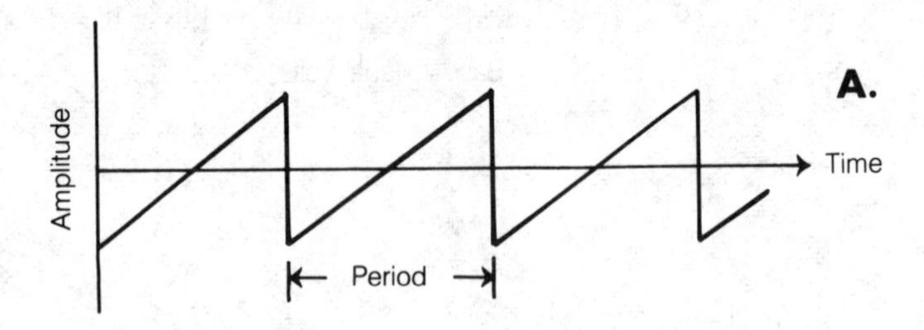

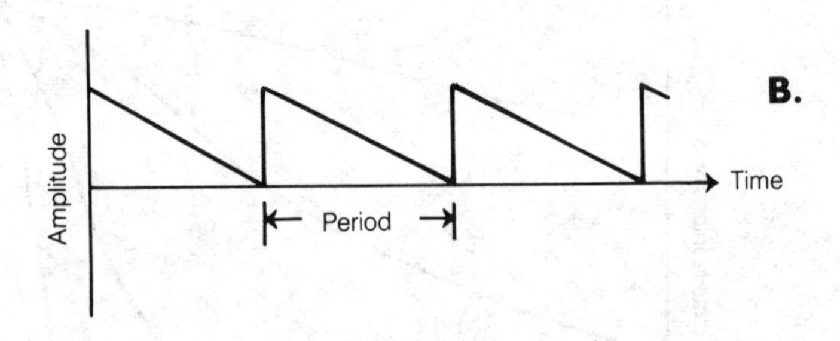

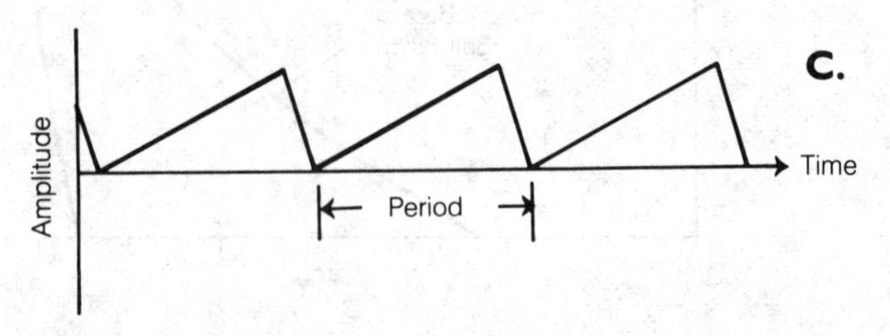

SAWTOOTH WAVE: Examples of sawtooth waves include those with slow rise and rapid decays (A), rapid rise and slow decay (B), and slow rise and moderate decay (C).

generators can produce sawtooth waves as well as square waves and sine waves.

Some sawtooth waves have nonzero rise and fall times. An example of this wave is shown at C. If the rise and fall times are identical, the wave is called a triangular wave.

The period of a sawtooth wave is the length of time from any point on the waveform to the same point on the next pulse in the train. This period is divided into 360 degrees of phase, with the zero point usually corresponding to the moment of instantaneous change, as at A and B, or the end of the more rapid change (C). Alternatively, the zero-degree point can be set at the time of zero polarity and increasing amplitude. The frequency of a sawtooth wave, in hertz, is the reciprocal of the period in seconds.

SCALAR

An ordinary, nondirected number is called a scalar quantity, or simply a scalar. Examples of scalar quantities are speed, current, and voltage, in which polarity or direction are not expressed. If direction is implied, the quantity is called a vector. Examples of vector quantities are velocity, left-to-right current, and positive voltage. *See also* VECTOR.

SCAN FUNCTION

A scan function is a mode of scanning in a cathode-ray-tube system. Different scan functions are used for different purposes. The waveform used to actuate scanning is often referred to as the scan function. The scan waveform in an oscilloscope or television system, for example, is a ramp wave (*see* RAMP WAVE).

In radar, several different scan functions are used. The most common is the circular scan. In television, a left-to-right, top-to-bottom scan function is used. In most oscilloscopes, a left-to-right, single-line scan function is used. *See also* OSCILLOSCOPE, RADAR, RASTER, TELEVISION.

An automatic scanning receiver can have different scan functions or scan modes. The scanner can search for an occupied channel among empty channels, stopping when a signal is encountered and remaining on that channel until the signal disappears. This is called the busy scan mode. The scanner can stop at an occupied channel and remain there only for a predetermined length of time, such as 5 seconds; this is called free scanning. The scanner may search for an empty channel among busy ones; this is called the vacant scan mode. *See also* AUTOMATIC SCANNING RECEIVER.

SCANNER

See AUTOMATIC SCANNING RECEIVER.

SCANNING

Scanning is a term that describes two quite different electronic processes.

In a television or facsimile system, the picture is scanned both in the camera tube and in the receiver. The picture is generally scanned from left to right and top to bottom, in the same way one reads lines of a book page. Scanning is controlled by a circuit with a ramp waveform (*see* RAMP WAVE). In an oscilloscope, the electron beam in the cathode-ray tube scans from left to right along a single line. The scanning rate can be controlled, and is measured in terms of the time per graticule division (about 1 centimeter). Scanning in an oscilloscope is sometimes called sweeping. As in television and facsimile, a ramp wave is used to produce the scanning effect in an oscilloscope. *See also* CATHODE-RAY TUBE, FACSIMILE, OSCILLOSCOPE, RASTER, TELEVISION.

In computers or microcomputer-controlled devices, scanning refers to the sampling of data in a continuous manner. If there are n data channels, having addresses 1, 2, 3, . . . , n, then scanning can proceed from channel 1 upward to channel n, or from channel n downward to channel 1. The process of scanning might occur just once across the range of channels, or it may be repeated over and over. A good example of this kind of scanning is illustrated by the operation of an automatic scanning receiver, which constantly checks communications channels for signals. The scanning rate is designated in channels per second. *See also* AUTOMATIC SCANNING RECEIVER.

SCATTERING

When any disturbance such as sound waves, electromagnetic waves, or particle radiation passes through a material, scattering can occur. The effects of scattering may go unnoticed, or they may be evident. The atmosphere, for example, scatters blue light. This is why the sky looks blue. Water has the same effect; this is why water in a swimming pool has a blue appearance.

Radio waves at very high and ultrahigh frequencies are scattered by molecules in the atmosphere. This is called tropospheric-scatter propagation. The ionosphere scatters radio waves at low, medium, high, and sometimes very high frequencies. *See also* PROPAGATION CHARACTERISTICS, TROPOSPHERIC-SCATTER PROPAGATION.

SCATTER PROPAGATION

See BACKSCATTER, TROPOSPHERIC-SCATTER PROPAGATION.

SCHEMATIC DIAGRAM

A schematic diagram is a technical illustration of the interconnection of components in a circuit. Most schematic diagrams include component values, and perhaps tolerances. Standard symbols are used. A schematic diagram does not indicate the physical arrangement of the components on the chassis or circuit board; it shows only how the components are interconnected.

SCHERING BRIDGE

The Schering bridge is a circuit for measuring capacitance. The Schering bridge operates by comparing an unknown capacitance with a standard, known capacitance. Balance is indicated either by a meter, as shown in the illustration, or by a set of headphones. The Schering bridge is employed for the determination of relatively large capacitances.

If the unknown capacitance, in farads, is C, and the other component values are as labeled (capacitances in farads, resistances in ohms), then:

$$C = \frac{C_2 R_1}{R_2}$$

and the Q factor of the unknown capacitor, assuming an effective series resistance of R ohms and a frequency at balance of f hertz, is:

$$Q = \frac{1}{2\pi f C R}$$

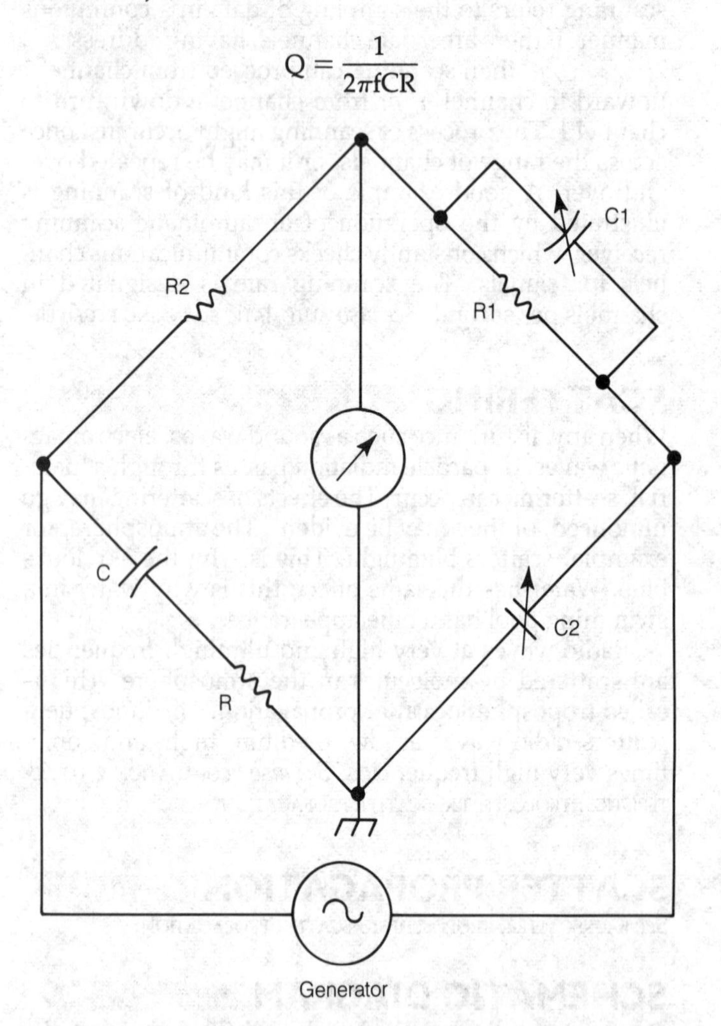

SCHERING BRIDGE: A bridge circuit for determining capacitance.

SCHMITT TRIGGER

A Schmitt trigger is a form of multivibrator circuit that is actuated by an input signal with a constant frequency. The Schmitt trigger produces rectangular waves, regardless of the input waveform. The schematic illustrates a typical Schmitt trigger circuit.

The circuit remains off until a specified rise threshold voltage is crossed; then it is actuated, and the output voltage abruptly rises. When the input voltage falls back below the fall triggering level, the output voltage drops to zero almost instantly. The rise and fall thresholds are generally different.

The Schmitt trigger is used in applications where square waves with a constant amplitude are needed. A Schmitt trigger may also be used to convert sine waves to square waves. The Schmitt trigger is a form of bistable multivibrator or flip-flop. *See also* FLIP-FLOP.

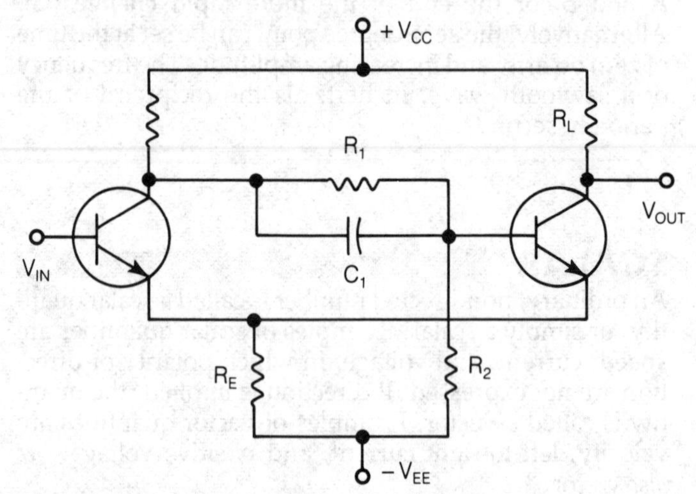

SCHMITT TRIGGER: A form of multivibrator, the Schmitt trigger is actuated by an external signal.

SCHOTTKY BARRIER DIODE

A Schottky barrier diode, also known as a Schottky diode or hot-electron diode, is a solid-state diode junction formed by metal-semiconductor contact. The Schottky barrier diode is usually fabricated from lightly doped N-type material and aluminum. The metal is evaporated or sputtered onto the semiconductor. The hot-carrier diode is a form of Schottky barrier diode.

Schottky barrier diodes are characterized by extremely rapid switching capability. The reverse-bias capacitance is very low. Schottky barrier diodes are useful at very high and ultrahigh frequencies because of their high-speed properties. They can be used as mixers, harmonic generators, detectors, and in other applications requiring diodes. The Schottky diode is used in the manufacture of transistor-transistor logic (TTL).

SCHOTTKY LOGIC

Schottky logic is any form of logic that incorporates Schottky diodes or transistors. Schottky diodes are used in the manufacture of integrated-injection-logic devices and Schottky transistor-transistor logic.

Schottky logic is characterized by high switching speed. *See also* INTEGRATED INJECTION LOGIC.

SCIENTIFIC NOTATION

Very large or small numbers are cumbersome to write in

the conventional fashion. Scientific notation provides a shortcut for expressing extreme numerical values.

A number in scientific notation consists of a decimal number with a value of at least 1 but less than 10, followed by a power of 10. The decimal number indicates the first few digits of the actual value as it would be written in longhand, or conventional, form. The power of 10 gives the factor by which the decimal number is to be multiplied. The exponent is always an integer; it may be positive or negative.

The decimal part of a scientific-notation expression may have any number of significant figures, depending on the accuracy needed. For example, 2.3×10^6 is an approximation accurate to two significant figures of 2,345,678.

When two numbers are multiplied or divided in scientific notation, the decimal numbers are first multiplied or divided by each other. Then the powers of 10 are added (for multiplication) or subtracted (for division). Finally, the product or quotient is reduced to standard form. That is, the decimal part of the expression should be at least 1 but less than 10. For example,

$$3 \times 10^2 \times 7 \times 10^3 = 21 \times 10^5 = 2.1 \times 10^6$$

When working with scientific notation, it is important to be aware of the number of significant figures in the expressions. Values are often expressed approximately, not precisely. *See also* SIGNIFICANT FIGURES.

SCINTILLATION

Scintillation is the variation in the intensity of electromagnetic signals received, caused by changes in the transmission medium and transmission path with time. It is perceived as a rapid fluctuation of the amplitude and phase of the wave.

Scintillation of radio waves is also called fading. It is a result of changes in the atmosphere, which cause irregular changes in the transmission path.

In radar, variation in the signal received from a complex target such as a ship or airplane is called scintillation. It is caused by rapid changes in the position of the target that alter the reflective qualities (aspect ratio) of the target.

When a high-speed atomic particle strikes a phosphorescent substance such as sodium iodide, there is a brief flash of visible light. This effect is called scintillation. A special type of radiation counter, known as a scintillation counter, uses phosphor crystals and light-detection apparatus to measure the intensity of radioactivity. *See also* SCINTILLATION COUNTER.

SCINTILLATION COUNTER

A scintillation counter is an instrument for measuring the intensity of atomic (nuclear) radiation. This counter is sensitive to all forms of nuclear radiation, including alpha, beta, gamma, and X rays.

A scintillation counter includes a phosphorescent material such as a sodium-iodide crystal and a photomultiplier tube. When a photon or subatomic particle strikes the phosphor, a momentary flash of visible light is produced. This flash is converted into a weak electrical impulse by the photocathode, and the photomultiplier tube amplifies the impulse. The impulse is then transmitted to a counting device or to a speaker, oscilloscope, or other indicator.

Scintillation counters are capable of detecting more than 1 billion particles per second, compared with only about 2,000 particles per second for a Geiger counter. *See also* ALPHA PARTICLE, BETA PARTICLE, GAMMA RAY, GEIGER COUNTER, PHOTON, X RAY.

SCR

See SILICON-CONTROLLED RECTIFIER.

SCRAMBLER

A scrambler is a circuit used to encode signals making it difficult for unauthorized persons to intercept them. Scramblers can be used for digital or analog signals of any kind, and at any speed.

A common type of scrambler used with voice signals inverts the frequency characteristics of the voice. The human voice can be transmitted within a passband of 2.7 kHz with frequency components between 300 Hz and 3 kHz. The scrambler turns the voice frequencies "upside down." A given impulse with a frequency f, in hertz, is converted to a frequency g, in hertz, so that:

$$g = 3300 - f$$

When a human voice signal is applied to the input of this circuit, the resulting output signal is unintelligible. However, by passing the unreadable chatter through a second, identical scrambler circuit, the voice is restored to its original form.

Many different scrambling methods are used by commercial and government communications services. Most scramblers are more complex than the type described, and may incorporate such devices as digital-to-analog and analog-to-digital converters, multiple-frequency operation, and other sophisticated schemes, making it nearly impossible for unauthorized persons to decode the signals without sophisticated and powerful computer-based decoding systems. Scrambling schemes are now designed into specialized coding and decoding integrated circuit chips.

SCRATCH-PAD MEMORY

A low-capacity, random-access memory, used in computers and some calculators for temporary data storage, is called a scratch-pad memory. Scratch-pad memory is used in complex arithmetic computations involving more than one operation. *See* SEMICONDUCTOR MEMORY.

SECANT

The secant is a trigonometric function equal to the reciprocal of the cosine function (*see* COSINE). In a right triangle, as shown in the drawing at A, the secant of an angle θ between 0 and 90 degrees is equal to the length of the hypotenuse divided by the length of the side adjacent to the angle. The larger the angle, the larger the secant becomes for values between 0 and 90 degrees. The values of the secant for angles between 0 and 360 degrees are shown at B. The secant is undefined for θ = 90 degrees and θ = 270 degrees.

In mathematical calculations, the secant function is abbreviated sec, and is given by the formula:

$$\sec \theta = \frac{1}{\cos \theta}$$

where cos represents the cosine function. *See also* TRIGONOMETRIC FUNCTION.

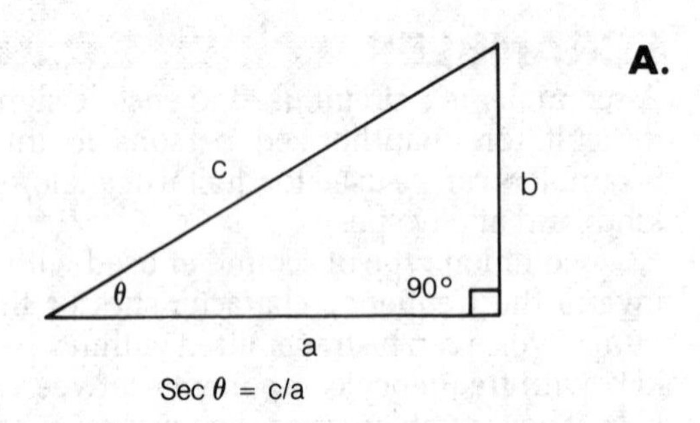

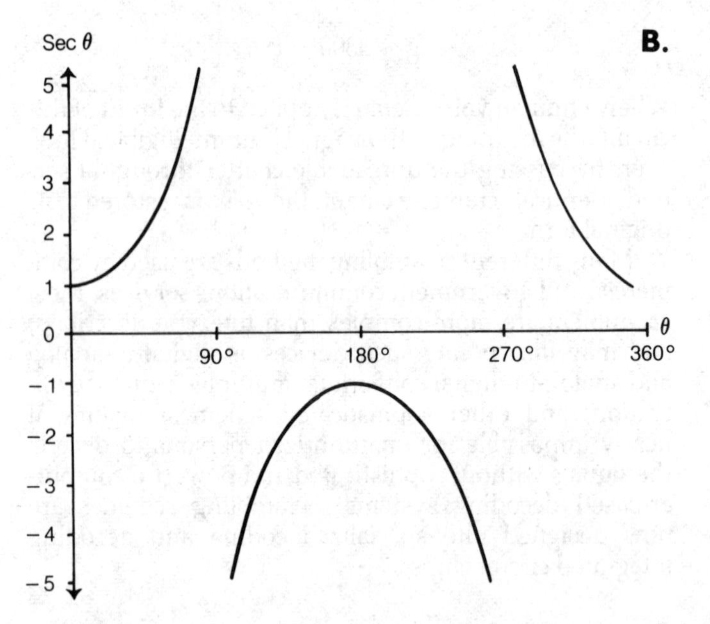

SECANT: The secant is explained with the triangle (A) and a graph of the secant function (B).

SECOND

The second is the standard international unit of time. A second corresponds to 1/86,400, or 1.1574×10^{-5}, mean solar day. With modern atomic clocks, a more precise definition has been formulated: the second is the duration of 9,192,631,770 periods of the radiation corresponding to the transition between the two hyperfine levels of the ground state of the cessium-133 atom. Seconds are transmitted over time-and-frequency standard stations such as CHU and WWV/WWVH.

In angular measure, the second is 1/3600, or 2.778×10^{-4}, angular degree. This unit is known as the arc second or the second of arc. It is abbreviated by a quotation mark or a double apostrophe ("). The arc second is used by astronomers as an indicator of the resolving power of an optical or radio telescope. The angular size of a celestial object may be expressed in seconds of arc.

In the expression of right ascension, one second corresponds to 1/3600 of an hour circle, or 15 angular seconds of arc.

SECONDARY

See TRANSFORMER.

SECONDARY BATTERY

See BATTERY.

SECONDARY COLOR

A secondary color is any color that is not a primary color. Any color can be obtained by combining primary colors in certain proportions, with certain amounts of saturation. *See also* HUE, PRIMARY COLORS, SATURATION.

SECONDARY ELECTRON

See SECONDARY EMISSION.

SECONDARY EMISSION

When an electrode is bombarded by high-speed electrons, other electrons are knocked from the atoms of the electrode. Electron emission is known as secondary emission.

The photomultiplier tube employs secondary emission. A single electron can knock many electrons free from an electrode. This makes it possible to magnify the intensity of an electron beam by a large factor (*see* DYNODE, PHOTOMULTIPLIER TUBE).

SECONDARY WINDING

See TRANSFORMER.

SECOND BREAKDOWN

Second breakdown is a phenomenon that occurs in some bipolar transistors under certain adverse conditions. Second breakdown can be the result of flaws in manufacture. It can also take place if the device is subjected to excessive voltage or current.

Second breakdown causes a dramatic fall in the output impedance of a transistor. The emitter-base voltage, and any input signal, no longer affects the current through the device, and the output therefore drops nearly to zero. The collector becomes effectively short-circuited to the emitter. A transistor may be permanently damaged by second breakdown. The probability of this happening can be reduced by ensuring that transistors are operated well within their ratings. *See also* POWER TRANSISTOR, TRANSISTOR.

SECOND LAW OF THERMODYNAMICS

The Second Law of Thermodynamics, also known as Carnot's Principle, states that no physical system or heat source can continuously transfer heat from a lower temperature to a higher temperature without continuous work input from an external source.

The second law says that heat always tends to flow from a hot body to a cooler body, so it includes a statement about the direction in which thermal processes go. *See also* ENERGY, FIRST LAW OF THERMODYNAMICS, THIRD LAW OF THERMODYNAMICS.

SEEBECK COEFFICIENT

The Seebeck coefficient is an expression of the amount of potential difference developed by a thermocouple. The Seebeck coefficient is expressed in volts per degree Celsius.

If the temperature difference between the two conductors of a thermocouple is t degrees Celsius, and the voltage between the conductor is E volts, then the Seebeck coefficient S is expressed as a limit:

$$S = \lim_{t \to 0} \frac{E}{t}$$

The Seebeck coefficient may be either negative or positive, depending on which conductor is named first. *See also* SEEBECK EFFECT, THERMOCOUPLE.

SEEBECK EFFECT

When two dissimilar metals are joined, a potential difference develops between them. This is the basis of a thermocouple. (*See* THERMOCOUPLE.) The development of a voltage occurs when the temperature of the junction differs from that of the rest of the metal, or when the two metals have different temperatures. This phenomenon is called the Seebeck effect or the thermoelectric effect. The resulting voltage is known as the Seebeck EMF, Seebeck potential, or Seebeck voltage.

In general, the greater the temperature gradient in a thermocouple junction, the greater the voltage. Different combinations of metals produce different voltages for a given temperature gradient. The voltage arising from a given temperature differential is expressed as the Seebeck coefficient.

SELECTANCE

Selectance is an expression of the selectivity of a receiver or a resonant circuit (*see* SELECTIVITY).

In a receiver, selectance is given as a sensitivity ratio between two channels, one desired and the other undesired. If the receiver is tuned to a channel at $f1$ MHz and the sensitivity is E_1 microvolts for a 10-decibel signal-to-noise ratio, and another signal is applied at some frequency $f2$ MHz, different from $f1$, the sensitivity will be less; the voltage E_2, needed to produce a 10-decibel signal-to-noise ratio will be larger than E_1. The selectance S can be expressed in different ways. The general form is:

$$S = \frac{E_2 - E_1}{f2 - f1}$$

if $f2 > f1$, and:

$$S = \frac{E_2 - E_1}{f1 - f2}$$

if $f2 < f1$. These expression are given in microvolts per megahertz. Of course, other units of voltage and frequency might be used.

A specific expression of selectance is obtained by determining the frequency $f2$ at which $E_2 = 2E_1$. Then:

$$S = \frac{2E_1 - E_1}{f2 - f1} = \frac{E_1}{f2 - f1}$$

if $f2 > f1$, and:

$$S = \frac{2E_1 - E_1}{f1 - f2} = \frac{E_1}{f1 - f2}$$

if $f2 < f1$.

In the determination of selectance, there will be two frequencies $f2$; one larger than $f1$ and the other smaller than $f1$. The selectance for $f2 > f1$ might differ from that in the case $f2 < f1$.

Selectance can be expressed for resonant circuits. If the voltage is E_1 at the resonant frequency $f1$, and E_2 at some nonresonant frequency $f2$, then selectance is specified in the same ways as outlined above. There are, again, two selectance figures: one for $f2 > f1$ and one for $f2 < f1$. These two figures may differ for a fixed ratio E_2/E_1.

The selectance of a radio receiver determines the degree of adjacent-channel rejection. The greater the selectance, the less the possibility of adjacent-channel interference. The selectance of a resonant circuit depends on the Q factor. The higher the Q, the greater the selectance. *See also* ADJACENT-CHANNEL INTERFERENCE, Q FACTOR.

SELECTIVE FADING

When a signal is propagated through the ionosphere, fading is commonly experienced at the receiving station (*see* FADING, PROPAGATION CHARACTERISTICS). This fading usually does not occur at the same time for all frequencies. For a given communications system, fading will not normally be observed at 5 and 7 MHz at the same time, for example. Because fading is a phase-related phenomenon, frequency and phase changes are interdependent.

Normally, the effects of frequency are not noticeable for channels separated by just a few hundred hertz. However, under certain conditions, fading may take place several seconds apart at two frequencies that are very close together. If this occurs, a signal can become distorted.

The narrower the bandwidth of a signal, the less likely it is to be affected by selective fading. Code (continuous wave type A1 emission) signals are essentially unaffected by selective fading, since the bandwidth is very small. But radioteletype (F1 emission) might be affected; single-sideband transmissions are often distorted; amplitude-modulated signals can be severely affected; and wideband frequency-modulated signals can be distorted beyond recognition. The effects of selective fading can be reduced by using the minimum bandwidth needed to carry on communications. Selective fading can also be alleviated, to some extent, by the use of diversity receiving systems. *See also* DIVERSITY RECEPTION.

SELECTIVITY

Selectivity is the ability of a radio receiver to distinguish between a signal at the desired frequency and signals at adjacent frequencies.

In the front end of a receiver, selectivity is important because it reduces the chances of overloading or image reception (*see* FRONT END, PRESELECTOR). In the intermediate-frequency stages of a superheterodyne receiver, selectivity is important because it reduces the probability of adjacent-channel interference (*see* ADJACENT-CHANNEL INTERFERENCE). Selectivity is obtained with bandpass filters (*see* BANDPASS FILTER, BANDPASS RESPONSE).

The selectivity of the intermediate-frequency chain in a superheterodyne receiver can be expressed mathematically. The bandwidths are compared for two voltage-attenuation values, usually 6 and 60 decibels. For example, specifications for a new receiver might include:

Selectivity: ± 8 kHz or more at −6 dB
± 16 kHz or less at −60 dB

A single receiver might have several different bandpass filters, each with a different degree of selectivity. The ideal amount of selectivity depends on the mode of communication that is contemplated. The figures are typical of a very high frequency receiver for frequency-modulation communications. For amplitude-modulated signals, the plus-or-minus bandwidth at the 6-decibel points is usually 3 to 5 kHz. For single-sideband, it might be roughly 1 to 1.5 kHz, representing an overall 6-decibel bandwidth of 2 to 3 kHz. For code reception, the overall bandwidth figure can be as small as 30 to 50 Hz at lower speeds, and perhaps 100 Hz at higher speeds. Examples of receiver selectivity curves for various emission modes are shown in the illustration. (These examples are representative, and are not intended as absolute standards.)

The ratio of the 60-decibel selectivity to the 6-decibel selectivity is called the shape factor. It is of interest, because it denotes the degree to which the response is rectangular. A rectangular response is the most desirable response in most receiving applications. The smaller the shape factor, the more rectangular the response (*see* SHAPE FACTOR). Filters such as the ceramic, crystal-lattice, and mechanical types can provide a good rectangular response in the intermediate-frequency chain of a receiver (*see* CERAMIC FILTER, CRYSTAL-LATTICE FILTER, MECHANICAL FILTER).

Selectivity is usually provided in the radio-frequency and intermediate-frequency stages of a receiver, but additional selectivity is sometimes added in the audio-frequency amplifier chain. This is especially useful for reception of code and radioteletype signals. Many different commercially manufactured audio filters are available for enhancing the selectivity of a receiver.

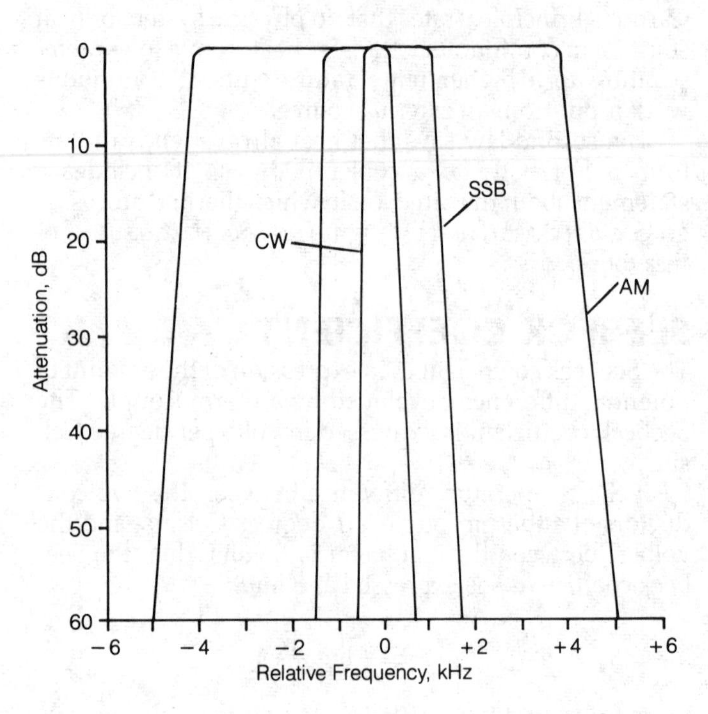

SELECTIVITY: Examples of selectivity curves for various reception modes.

SELENIUM

Selenium is an element with an atomic number of 34 and an atomic weight of 79. It is a semiconductor material that exhibits variable resistance, depending on the amount of ambient light. Selenium is used in the manufacture of photocells and rectifier diodes. *See also* PHOTOCELL, SELENIUM RECTIFIER.

SELENIUM RECTIFIER

A selenium rectifier is a diode made by joining selenium and another metal, usually aluminum. Electrons flow easily from the metal to the selenium, but the resistance in the opposite direction is high. The metal forms the cathode, and the selenium forms the anode of the rectifier. A selenium rectifier can be recognized by its heat sink, which makes it look similar to a stack of cards.

Selenium rectifiers are useful only at low frequencies, such as the 60 Hz line frequency. Although they are light in weight, they tend to be physically large because of the heat sink that is needed. The forward drop is about 1 V in normal operation. Selenium rectifiers are especially resistant to transient spikes on the power line. This is their main advantage over other types of rectifier diodes. *See also* DIODE, RECTIFIER.

SELF-BIAS

Self-bias is a means of providing effective negative grid or gate bias in a vacuum tube or field-effect transistor. Although negative grid bias can be obtained from a special power supply, it can also be provided by elevating the cathode above direct-current ground potential. A noninductive resistor and a capacitor are used for this purpose (see illustration).

There is a limit to the effect of negative gate bias that can be obtained by the method shown. The bias voltage is the product of the current (which is essentially constant) and the ohmic value of the resistor. The greater the bias voltage that is needed, the larger the resistor must be, and thus the more power it must dissipate. Most noninductive resistors have rather low dissipation ratings.

Self-bias is often used in low-power tube amplifiers in Class A, Class AB, or Class B. Self-bias cannot normally be used to obtain sufficient bias for Class C operation. Self-bias can be used in audio-frequency or radio-frequency circuits.

Biasing can be obtained in circuits using bipolar transistors, by means of the same technique described here. In solid-state circuits, the method is also called automatic bias. *See also* AUTOMATIC BIAS.

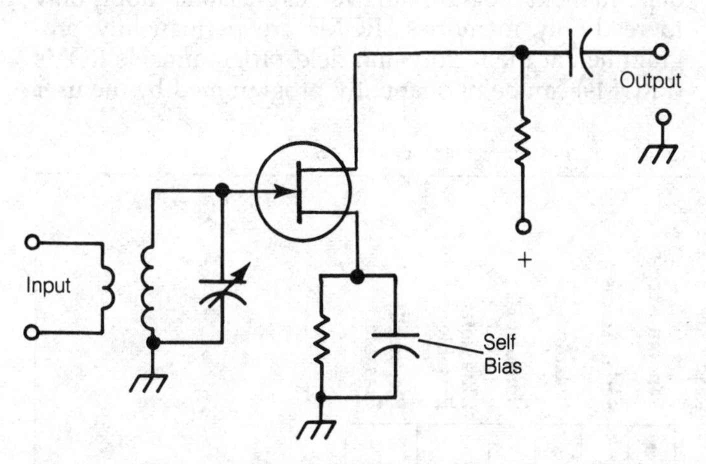

SELF-BIAS: A diagram of a self-biased field-effect transistor amplifier.

SELF OSCILLATION

An amplifier circuit might oscillate if positive feedback occurs. This type of oscillation degrades the performance of the amplifier. In audio-frequency circuits, self oscillation causes a fluttering or popping sound known as motorboating. In radio-frequency circuits, it can result in the emission of signals on undesired frequencies, and is called parasitic oscillation. *See also* FEEDBACK, MOTORBOATING, PARASITIC OSCILLATION.

SELF RESONANCE

Most electronic components exhibit a certain amount of capacitive reactance and inductive reactance. A component can be deliberately designed to show reactance. But the physical dimensions of a component, especially the lead lengths, result in some reactance of both types, even if the component is not a capacitor or an inductor.

Self resonance usually occurs at ultrahigh or microwave frequencies. In some larger capacitors and inductors, self resonance might exist at the high or very high frequencies. Self resonance is of primary concern in the design of circuits for operation above approximately 300 MHz.

Inductors, because of their inherent inter-winding capacitance, are the most likely components to have significant self-resonant effects. Some inductors are deliberately manufactured to be self-resonant.

SEMICONDUCTOR

A semiconductor is an element or compound that exhibits a moderate or large resistance. The resistance of a semiconductor is not as low as that of an electrical conductor, but it is much lower than that of an insulator or dielectric. The resistance of a semiconductor depends on the particular impurities, or dopants, added to it (*see* DOPING). Semiconductors conduct electricity with two forms of charge carrier: the electron and the hole (*see* ELECTRON, HOLE).

The most commonly used semiconductor materials are silicon and gallium arsenide. Others include germanium, selenium, and indium antimonide.

SEMICONDUCTOR JUNCTION

See P-N JUNCTION.

SEMICONDUCTOR MEMORY

A semiconductor memory is an integrated circuit that retains information in a digital format and can be used to store programs and data in a digital computer. Initially developed to replace magnetic core memory as the main memory in digital computers, semiconductor memories have been commercial devices since the early 1970s. The use of core memories is now limited to certain military computers because cores are less susceptible to data loss caused by nuclear radiation than are semiconductor memories. The bit storage density of semiconductor memories has increased dramatically since their introduction, and their prices have declined sharply.

The simplest semiconductor memory is a flip-flop, a two-transistor circuit whose output can be switched with the introduction of an appropriate signal at the input terminal. A flip-flop, like a toggle switch, is a 1-bit memory. Eight flip-flops in series become an 8-bit memory able to store an 8-bit word or byte of information. *See* FLIP-FLOP.

Semiconductor memories are classified by type, characteristics and fabrication technology as summarized in the table. All semiconductor memory devices can be randomly accessed or serially accessed. Data can be written so that it can be altered easily (read/write), altered by following an extraordinary procedure (erasable read-only) or permanently written (read-only). Stored data can be altered by writing over existing data with a low voltage, or erasing stored data partially or completely. The special procedures for erasing memories are exposure of the device chip to ultraviolet light or altering the charge of the memory cells by the application of an overvoltage.

Some semiconductor memory devices will retain data only as long as power is applied (volatile memory) or at all times (nonvolatile memory). However, certain volatile memories retain data without loss as long as they are powered (static), but others must be refreshed to retain readable data even if the power is on (dynamic).

Semiconductor memories can also be grouped by fabrication process: bipolar or metal-oxide semiconductor (MOS). The bipolar technologies used in the fabrication of memory devices include transistor-transistor logic (TTL) and emitter-coupled logic (ECL). Three different MOS technologies have been employed; P-channel MOS (PMOS); N-channel MOS (NMOS); and complementary MOS (CMOS).

Read/Write vs. Read-Only. The term *read/write* applies only to a memory in which data can be written and erased with logic-level voltages. This property is useful when the data and instructions change frequently and is available only with so-called (but inaccurately named) random-access memories (RAMs). All of these memories are volatile. A read-only memory (ROM) either stores data permanently or requires one of the special erasure procedures. However, all data is retained when power is removed. These techniques are described in detail further on in this section.

Random Access vs. Serial Access. Only random-access memories are read/write memories although most semiconductor memories can be randomly accessed. (The data can be written or read without searching the entire memory as must be done with a serial or sequentially accessed memory.) Read-only memories are randomly accessible. Shift registers (digital logic) and charge-coupled devices (CCDs) are serially accessed semiconductor memories. However, the non-semiconductor memories, magnetic bubble memory (MBM), and magnetic tapes are serially accessed, but magnetic floppy and rigid disks and optical disks are randomly accessible. *See* BUBBLE MEMORY, CHARGE-COUPLED DEVICE, MAGNETIC TAPE, SHIFT REGISTER.

Volatile vs. Nonvolatile. The two semiconductor random access read/write memories are volatile: dynamic RAM (DRAM) and static RAM (SRAM). Data stored in these devices will be lost on power shutdown unless they have backup battery power. (Computers with RAM have provision for transferring the data to a more permanent memory (magnetic disk or tape) prior to planned shutdown. All read-only memories are nonvolatile.

Static vs. Dynamic. The terms *static* and *dynamic* apply only to random-access memory (DRAM or SRAM). Static RAM memory cells retain all written data as long as power is maintained. However, the simpler DRAM memory cells must be refreshed by frequently rewriting the data.

Permanently Programmed vs. Erasable. The alternatives of permanent programming versus erasability apply only to read-only memories. ROMs are permanently programmed at the factory and field-programmable ROMs (PROMs) can be permanently programmed by the user

SEMICONDUCTOR MEMORY: Characteristics of semiconductor memory devices.

Type	Static	Dynamic	Read/Write	Read-Only	Random Access	Serial Access	Volatile	Nonvolatile	Ultraviolet Erasable	Electrically Erasable
Dynamic RAM		•	•		•		•			
Static RAM	•		•		•		•			
ROM				•	•			•		
PROM				•	•			•		
EPROM				•	•			•	•	
EEPROM				•	•			•		•
NVRAM	•		•		•			•		
CCD						•		•		
Magnetic Bubble[1]			•			•		•		

1. Thin-film device.

with special equipment outside the factory. Data once entered in either device cannot be altered. Erasable ROMs (EPROMs, EEPROMs, and flash memories) may be programmed, erased, and reprogrammed outside the factory. EPROMs need special ultraviolet erasure equipment.

Memory Technology The ideal semiconductor memory device would have access time less than 1 nanosecond (ns), high cell density per chip, low power dissipation, and a low cost. It would be randomly accessible, nonvolatile, easy to test, highly reliable, and standardized throughout the industry. Because this ideal has yet to be realized, memories are still made in alternate technologies.

CMOS is now the preferred semiconductor memory technology, having taken over from NMOS. Aside from speed, CMOS has come closer to the goal of the ideal memory than the alternative technologies. CMOS is almost exclusively used in the popular DRAMs.

Bipolar TTL and ECL memories account for only a small percentage of semiconductor memories in two general categories: SRAM and fuse-link PROM. These devices offer high speed and can be used in cache, buffer, and scratchpad memories where short cycle times and fast access are critical. These memories are also used in small computer systems with auxiliary add-on memory boards, military computers, and high-speed data-processing systems with bit-slice microprocessors. While the bipolar memory demand is specialized and small, many manufacturers still make them. Although advances continue to be made, progress is slower than in MOS memories. They still lag MOS memories in size, and density. There are 256-by-1, 256-by-4, and 1,024-by-1 TTL SRAMs with access times of 35 to 70 nanoseconds.

ECL memories have the advantage of very high speed, but power consumption is also high. Available ECL PROMs have 32-by-8, 256-by-4, and 1024-by-1 organization with access times between 10 and 35 nanoseconds.

Random-Access Memories Random access memories are organized in rectangular arrays of rows and columns as shown in Fig. 1. The diagram shows an eight-by-eight array for the storage of 64 bits; one bit is stored in the cell at each intersection. A specific memory location can be specified with three binary digits as row and column addresses. Row address 100 (binary for 4) and column address 101 (binary for 5) give the memory location of 4, 5.

Dynamic RAM. The DRAM is an array of simple data storage cells consisting of a select transistor and storage capacitor as shown in Fig. 2. Digital data is stored as a charge on the capacitor in each memory cell. When one of the selection lines, or rows, in an array is actuated, it turns on all transistor switches connected to it. The transistor functions as an on-off switch to connect the capacitor to its particular data line, a column in the array. The simultaneous activation of a row and a column picks the cell for reading or writing (here cell 4, 5).

The term *dynamic* refers to the periodic refresh discussed under the heading *Static vs. Dynamic.* Refreshing regenerates the charge that is lost both by reading and leakage. The charge is usually regenerated about once every two milliseconds. The charge is supplied by the refresh amplifier when the switch in the data line is closed. The charge must, at all times, be large enough so the memory state can be read unambiguously. The charge on a continually refreshed DRAM will remain indefinitely unless new data is written over it or the power to the memory is shut off. The latest DRAMs have on-chip refresh circuitry which has reduced the cost and design effort of including DRAMs in computers.

The simple memory cell of the DRAM permits rapid access, draws less power than a SRAM cell, and permits large dense arrays on a single chip. The SRAM cell, as shown in Fig. 3A, may consist of as many as six transistors—two select and four flip-flop. In addition to its larger size, it consumes power continuously. Because cost of memory relates to silicon chip size, large cell arrays on a single chip reduce memory cost per bit. This reduction has been the incentive for all efforts to increase DRAM density that now exhibits the lowest cost per bit of any semiconductor memory.

The first DRAMs had a capacity of 1024 bits (1k) per chip and were fabricated with the PMOS process. After the NMOS process was introduced, improvements in processing and circuit design permitted rapid increases in density. These took place in multiples of four: 4096 (4k); 16,384 (16k); 65,536 (64k); 262,144 (256k); and then 1,048,576 bits (1 Mbit—megabit) where most commercial activity is now centered. Both 4 Mbit and 16 Mbit DRAMs have been developed and are at the prototype stage. The CMOS process became dominant at the 256k level. Intense competition between American and Japanese manufacturers continues over these large-scale integrated DRAMs.

Until 1983 all DRAMs were organized with 1-bit word widths and depths of 1024, 4096, 16,384, and 65,536 words. These were satisfactory for large mainframe computers that use memories in multiples of eight. The 4-bit and 8-bit (byte-wide) organizations permit the economical use of high-density memories in small computers. Specialized DRAMs were introduced for special memory-access modes, multiple ports and high-speed video displays. For example, 8k by 9 and 32k by 9 DRAMs permit on-chip parity generation.

Static RAM. In contrast with DRAMs, SRAMs do not need refreshing. Their major advantages are avoidance of refresh circuitry and ease of application. However, the SRAM cell as shown in Fig. 3A, is larger and more complex than the DRAM cell shown in Fig. 2. Basically a transistor flip-flop, it does not depend on a capacitor to store data. In alternative designs, load resistors might replace select transistors. Once data is written into the flip-flop cell, it remains there indefinitely unless it is changed or the power is turned off.

SRAMs have lagged DRAMs in density by a factor of four. When 1024-bit DRAMs were introduced, SRAM densities were 256 bits per chip. Similarly, when 64k DRAMs were introduced, SRAMs were just reaching the 16k density level. This pattern continued through the

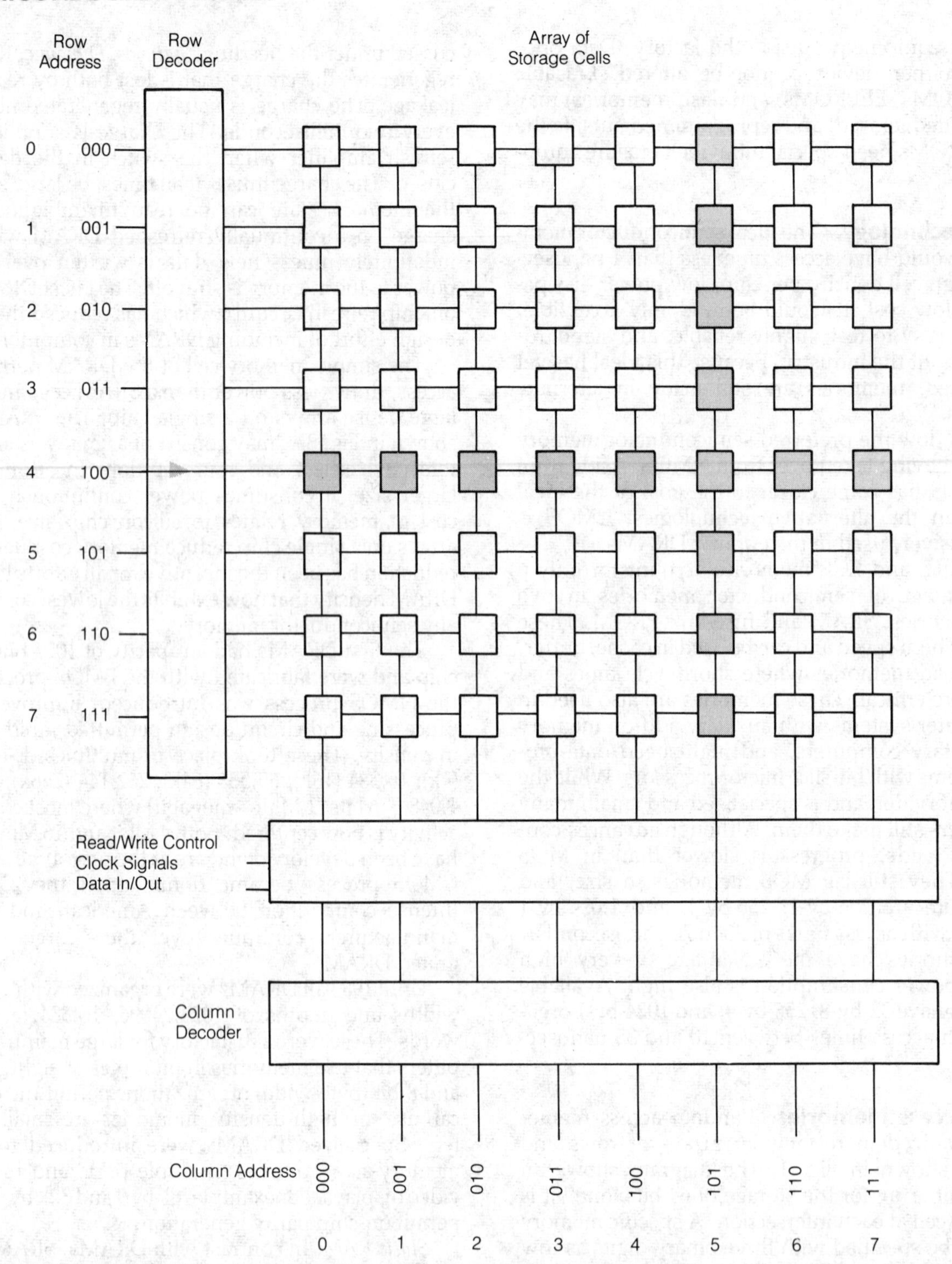

SEMICONDUCTOR MEMORY: Fig. 1. Random access memory is arranged in rectangular arrays of rows and columns. An 8-by-8 array for the storage of 64 bits is shown.

development of 64k and 256k DRAMs. CMOS 256k SRAMs are commercially available. Most SRAMs with densities above 16 k are made with a CMOS process, but below that density, NMOS was widely used. CMOS 1k SRAMs have 15 ns access times, and 256k CMOS SRAMs have attained 50 ns access times. These memories are manufactured in two speed ranges; fast SRAMs have faster access times than other SRAMs at the same density level and are premium priced. The fastest SRAMs are made from gallium arsenide (GaAs) and a 1k GaAs SRAM with a 2-ns access time has been reported. Very fast SRAMs are also made with bipolar Schottky TTL and ECL technology. However, bipolar SRAMs consume two to three times the power of NMOS SRAMS.

Some SRAMS offer faster access times than DRAMs. Even where access time is not a factor in the decision, SRAMS are more suitable for small and personal computers because they reduce the design complexity. SRAMs have been organized in 4- and 8-bit-wide architectures. These organizations require larger and more expensive packages than 1-bit wide devices because extra pins are needed.

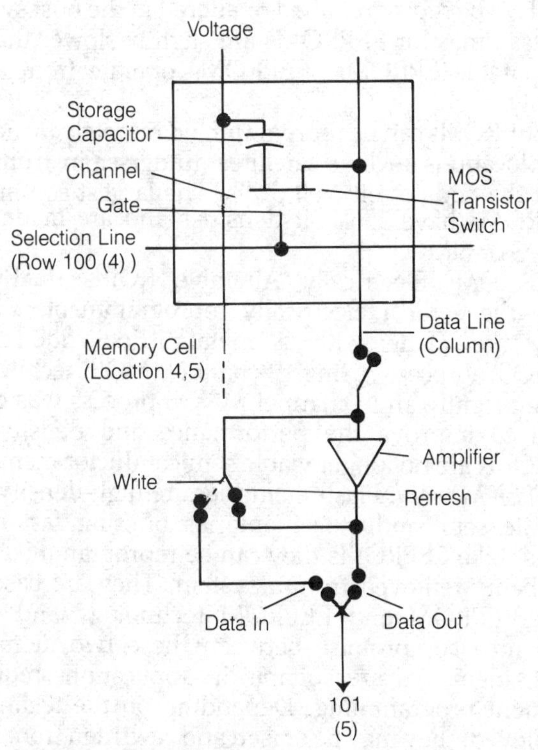

SEMICONDUCTOR MEMORY: Fig. 2. Schematic of a single-transistor dynamic RAM cell with refresh circuitry.

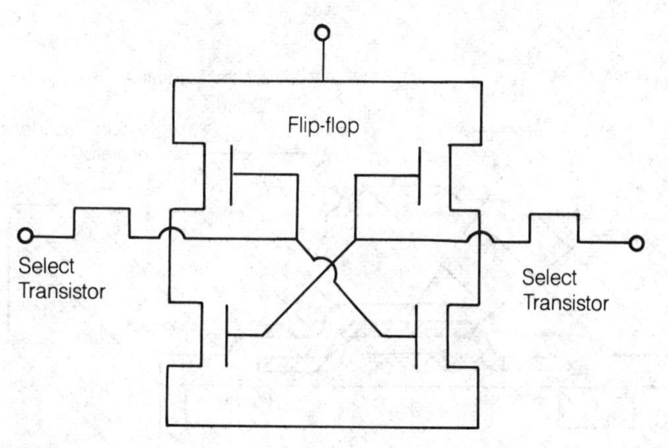

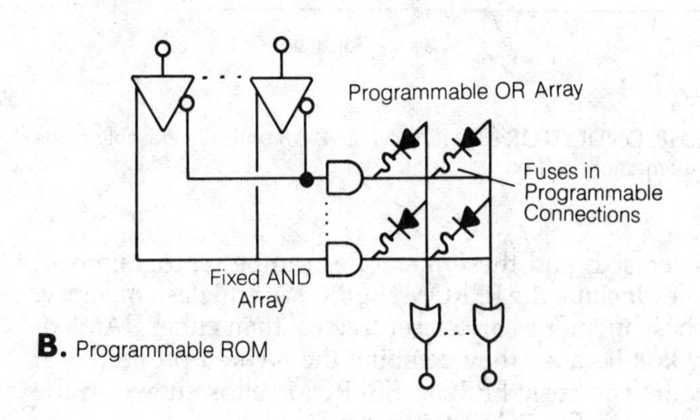

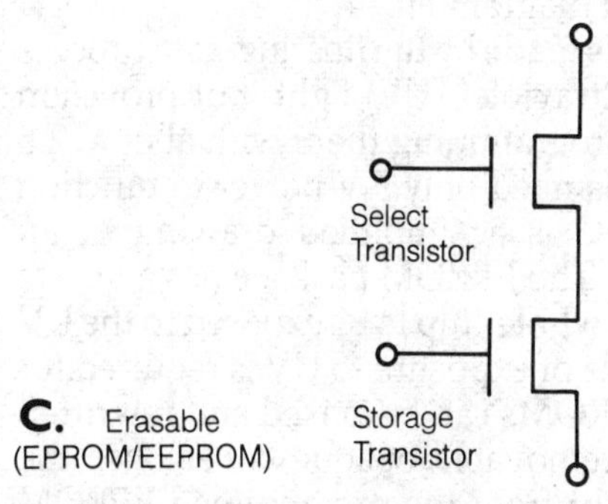

SEMICONDUCTOR MEMORY: Fig. 3. Organization of a MOS static random access memory cell (A), a programmable read-only memory (B), and a MOS EPROM/EEPROM memory storage cell (C).

Permanently Programmed Read-Only Memories The ROM is permanently programmed at the factory during the metal masking step while the device is still on the wafer. There are two types of ROM: the semicustom or final-mask programmable ROM, and the full-custom ROM. The final mask of the semicustom ROM makes the connections to the internal transistors, establishing the contents that will be read out when the ROM is accessed. The full-custom ROM is completely fabricated as a dedicated device and all masks used in the multiple mask process are custom generated. This ROM makes more efficient use of the silicon than the semi-custom ROM so chip size may be smaller. In large-volume purchases, this usually results in lower unit prices. However, the turnaround time for the fabrication of last-mask ROMs is shorter. The user must commit to a code pattern and pay an up-front nonrecurring engineering (NRE) charge for ROM fabrication.

ROMs can have capacitances of 1 megabit and most of these are available with 8-bit word (byte) widths. However, some ROMs are organized with 16-bit word widths. Large-scale ROMs can include built in detection and correction circuits. Most are fabricated in NMOS and CMOS technology.

PROM. The fuse-link PROM was the first user-programmable semiconductor memory introduced. These are now standardized products available with from a few hundred to more than 64k bits. PROMs are one-time programmable because none of the data programmed in can be altered. Two different technologies are used in PROM fabrication: fuse link and alterable transistor. In fuse-link technology, selected fuse links are burned out or blown by overcurrent to define a logic 1 or 0 for each cell in the memory. In alterable transistor PROMs, base-emitter junctions in an array of transistors can be shorted out to program the 1's or 0's in the memory.

The fastest PROMs have 4k density and are fabricated in bipolar ECL technology. The largest PROMs have 64k density level and are fabricated with bipolar TTL, but medium-speed PROMs are now being fabricated from CMOS.

PROMs can perform logic functions because they contain a fixed AND array followed by programmable OR array, as shown in Fig. 3B. This allows them to solve logic equations. They are the antecedents of the programmable logic device (PLD). *See* PROGRAMMABLE LOGIC DEVICE.

Reprogrammable Read-Only Memories Information stored in alterable, reprogrammable memories can

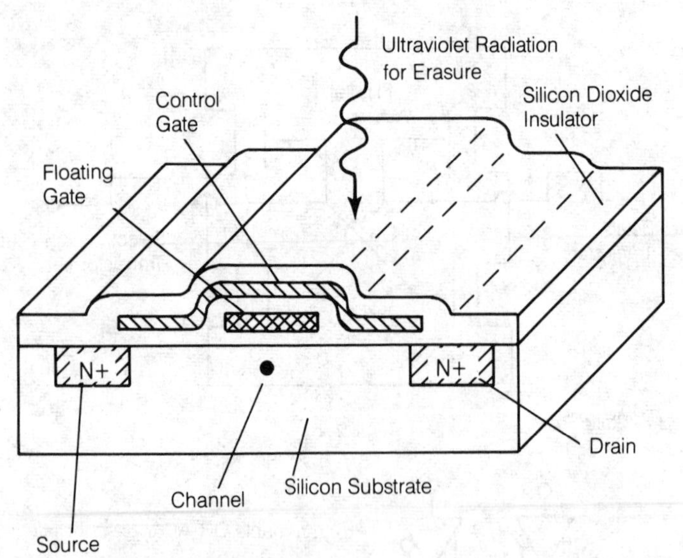

SEMICONDUCTOR MEMORY: Fig. 4. An optically erasable read-only memory cell.

be erased, and these memories can be reprogrammed. They include the EPROM, EEPROM, and flash memory. These memories are closer to ideal than either RAMs or ROMs because they combine the favored properties of both. The basic EPROM/EEPROM cell is shown in the schematic Fig. 3C and diagram Fig. 4.

EPROMs. EPROMs made by the floating-gate process can be erased with ultraviolet (UV) light, but provision must be made for reprogramming them with about 12.5 V. However, EPROMs need only 5 V power to function as a ROM. Figure 3C is a simplified drawing of an EPROM memory cell. The EPROM package has a quartz window to permit the whole chip to be exposed to the UV light. UP to 20 minutes of exposure to UV is required for complete erasure. EPROMs can be erased and rewritten about 100 times. A removable opaque cover over the window blocks all light from the programmed EPROM chip. Only a ceramic case is suitable for mounting the quartz window needed and this adds to the cost of the EPROM. The largest commercial UV EPROMs have 1-Mbit densities, and they are fabricated in CMOS. Other EPROMs are fabricated with NMOS technology.

A one-time only (OTO) EPROM is a windowless EPROM that permits data to be written but not erased because UV light cannot penetrate the opaque plastic case. Although the OTO functions like a PROM, it has the same electrical characteristics as the UV erasable EPROM. OTOs are specified when a program has been accepted and large quantities of identical ROMs are required. They are competitive with PROMs.

*EEPROMs.*The electrically erasable ROM or EEPROM can be selectively or completely erased depending on how the external erase voltage is applied. Unlike the EPROM, EEPROMs do not have to be removed from the host circuit to be erased; they may be reprogrammed in place on the circuit board. Although packaged in inexpensive plastic cases, provision must be made for the additional erasure circuitry and package pins if the device is to be reprogrammed or altered in the host system. Access times for EEPROMs are slightly slower than for comparable EPROMs. EEPROMs operate from a 5 V supply.

EEPROMs can be reprogrammed remotely in inaccessible locations such as satellites, undersea instruments, or at the ends of oil well drills. The largest commercial EEPROMs have 256k bit densities and are made with CMOS or NMOS.

EAROMs. Electrically Alterable ROMs (EAROMs) were the earliest electrically reprogrammable ROMs. They are made with a metal-nitride-oxide silicon (MNOS) process using P-channel MOS technology. More recently an N-channel MNOS process was developed to improve the performance and density, but EAROMs are no longer viable semiconductor memories.

Flash Memory. Flash memories are high-density nonvolatile semiconductor memories offering fast access times. Like EEPROMs they can be reprogrammed without being removed from a system. They are based on either EPROM and EEPROM technology and today present a compromise between these two memories. Flash memories are suitable in applications requiring frequent programming. Depending on the technology employed, they may be erased and rewritten from 100 to 10,000 or more times.

Flash memories are erased electrically, but they do not permit the selective erasure EEPROMs do. However, they can be bulk-erased in seconds, far faster than EPROMs. They consume less power than either EPROMs or EEPROMs and some exhibit access times that are faster than those of EEPROMs. These memories are seen as competititve with battery-backed SRAMs.

Nonvolatile Random-Access Memory

NV RAMs. Nonvolatile semiconductor RAMS or NV RAMs combine the properties of the static RAM with a backup, nonvolatile memory array. During normal system operation, the volatile RAM section is used as a RAM. But, when it receives a special store signal (indicating a power shutdown), RAM data is transferred to the nonvolatile section for storage. Thus the RAM section provides an unlimited read and write function while the nonvolatile section provides backup when power is removed. These devices are available with densities of 4 k bits and they are fabricated in NMOS.

A SRAM can be made into a nonvolatile memory by using a backup battery. The SRAM operates as a static RAM when it is normally powered, but when the power is shut off, the battery preserves the memory contents. CMOS SRAMs are best suited for this function.

Semiconductor Memory Packaging Most semiconductor memories are packaged in ceramic or plastic dual-in-line packages (DIP) as determined by the end-use application. Package width depends on the size of the chip and the number of pins depends on the input-output organization. There is also increasing demand for memory devices in the surface-mount packages. *See* SEMICONDUCTOR PACKAGE.

SEMICONDUCTOR PACKAGE

A semiconductor package is a protective case for a discrete semiconductor device or integrated circuit. It must be strong enough to withstand the stresses of testing, handling, and end use. It must not add unwanted resistive, conductive and inductive effects to the enclosed device. It also must insulate the die or chip from the effects of temperature extremes, electrical fields and humidity, and the package must be thermally stable. The semiconductor packages widely used include sealed metal cans, plastic and ceramic dual-in-line packages (DIPs), small-outline packages (SOTs), flat packs, ceramic and plastic leaded and leadless chip carriers, and pin-grid arrays. Many of these packages are registered with the Joint Electron Device Engineering Council (JEDEC). There are also many specialized and proprietary packages in use.

Glass/Plastic Packages. Small-signal and low-power diode chips including rectifiers and Zener diodes may be sealed in glass or plastic axial-leaded packages. The plastic packages may be molded or conformally coated. The DO-41 is an example of a glass axial-leaded package and the DO-15 is an example of a plastic axial-leaded case. Transistors and thyristors are packaged in popular three-lead plastic TO-226 type packages (formerly TO-92).

Metal Packages. Metal packages are widely used for small-signal, power, and radio-frequency (rf) transistors as well as thyristors (SCR and triac). These cases are made in two parts: a header with pins located in a circle for mounting the chip or die and a cap. The cap and header are bonded to form a hermetic seal. Some of the popular case styles are the TO-206 series (formerly TO-18, TO-39, TO-46, TO-52 and TO-72). See Fig. 1.

Metal Power Device Cases. Power transistors, rectifiers and thyristors rated up to 15 amperes are widely packaged in TO-204 type metal cases (formerly TO-3). Some packages have threaded studs in place of pins to act as heat sinks. Examples are the DO-4 and DO-5 metal cases.

Plastic Flat Pack Power-Device Cases. Plastic flat pack cases have been developed for discrete power devices including rf power transistors capable of handling 15 A. This flat package includes a metal substrate used as a heatsink. The three-terminal devices are transfer molded. The TO-218 and TO-220 packages are examples. There are many variations of this package. They are used to package military/aerospace ICs.

The standard DIP package has two parallel rows of leads spaced 0.1 inch on centers. The rows are spaced 0.3, 0.4, 0.6, and 0.9 inch apart, depending on lead count. Each additional two pins adds an additional 0.100 inch length to the DIP. The leads are short and stiff with shoulders so they stand off from the printed circuit (PC) board. These packages permit easy insertion into the PC boards or sockets, either manually or with automatic equipment. Analog, logic, memory, and optoelectronic devices as well as microprocessors and microcontrollers are packaged in DIPs. The packages are available with from 4 to 64 pins. A 64-pin DIP is 3.2 inches long by 0.9 inch wide and occupies about 3 square inches of PC board space. See Fig. 2.

The plastic or epoxy DIP has been the dominant package for integrated circuits over the past 20 years. Compact DIPs are DIPs with modified standard dimensions. Skinny DIPs with 20 to 28 pins have pin spacing of 0.1 inch, and row spacing of 0.3 in. Shrink DIPs have pin spacing of 0.07 inch and row spacing of 0.4, 0.6, and 0.75 inch for 28, 42 and 64 pins, respectively. Other DIPs have been introduced with 0.5-inch pin spacing.

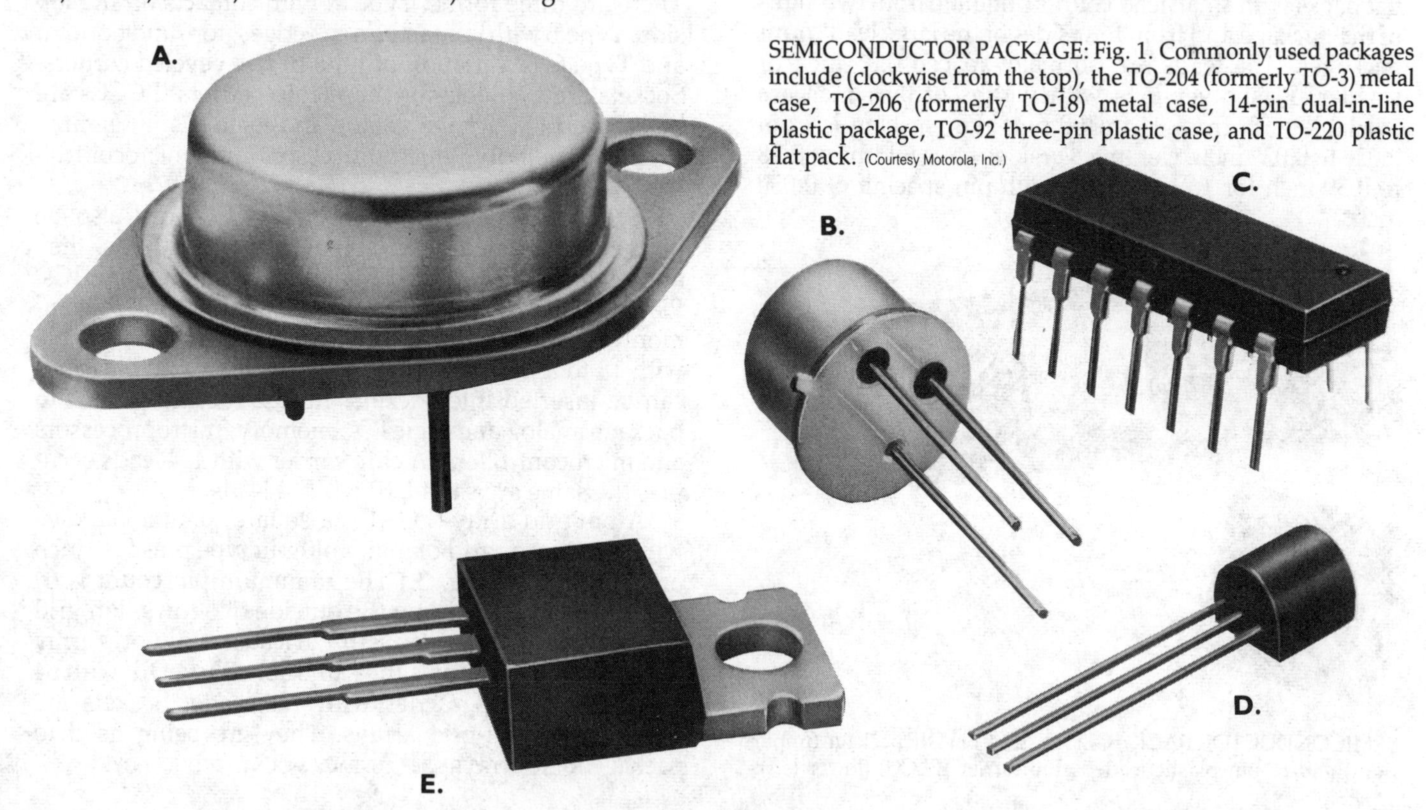

SEMICONDUCTOR PACKAGE: Fig. 1. Commonly used packages include (clockwise from the top), the TO-204 (formerly TO-3) metal case, TO-206 (formerly TO-18) metal case, 14-pin dual-in-line plastic package, TO-92 three-pin plastic case, and TO-220 plastic flat pack. (Courtesy Motorola, Inc.)

The die or chip in a plastic DIP is bonded to a metal lead frame with a eutectic solder, and gold wires are thermocompression bonded from the die to the lead frame pins as necessary. The leadframe/die assembly is molded in epoxy or other suitable plastic resin.

Two different types of ceramic hermetic DIP package are available: CERDIP (ceramic DIP) and solder-sealed (side brazed). These cases are used in military and high-reliability applications where the devices will be exposed to temperature extremes. The CERDIP is made of two alumina-ceramic parts: the base and the lid. Both have recesses to provide a cavity for the die. The lead frame is embedded in the base. The die is bonded to the base with a eutectic solder and aluminum wires are ultrasonically bonded to the lead frame. The lid is sealed over the base and included die with glass frit.

The side-brazed ceramic DIP is made of three layers of alumina ceramic with the lead frame brazed to the sides of the package. The die is bonded to the case, but aluminum wires are ultrasonically bonded to the lead frame. A metal or ceramic lid hermetically seals the cavity containing the die.

The SO packages (SO for small-outline) are rectangular plastic packages developed in Europe to replace DIPs for surface mounting. The leads can be J, gull wing, butt, or flat. The small-outline transistor (SOT) package has three leads for transistors. An example is the TO-236 (formerly TO-23). Small-outline integrated circuit (SOIC) packages have from 8 to 28 pins. The SOJ packages are J-bend SOIC cases with their leads bent under. They are available with 20 to 28 pins with package widths from 0.3 to 0.4 inch.

The flat pack is a small, lightweight package whose leads project in a plane parallel to the case. It is designed to be attached parallel to the seating plane. The leads of flat packs with small lead counts originate from two sides of the package and from four sides of the large lead count packages. The leads are normally spaced 0.05 inch or smaller. Plastic versions with widths of 0.55 inch are available with from 44 to 100 pins. Pin spacing is from 0.026 to 0.039 inch. Ceramic versions have widths of 0.28 to 0.39 inch for 16 to 24 pins with pin spacing of 0.050 inch.

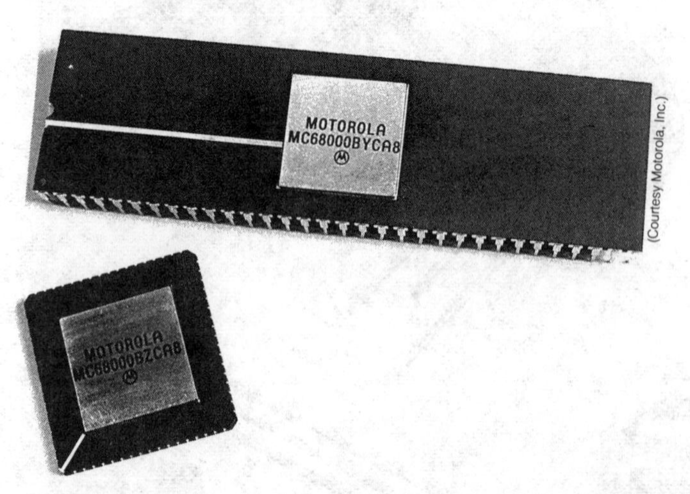

SEMICONDUCTOR PACKAGE: Fig. 2. 64-pin DIP package (upper right) and 68-pin plastic leaded chip carrier (PLCC) (lower left).

SEMICONDUCTOR PACKAGE: Fig. 3. A 149-pin grid array (PGA) package.

The chip carrier is a low-profile package, usually square, whose chip cavity or mounting area is a large fraction of the package size. Intended for surface mounting, metal terminations are arranged along all four edges. They are either metal pad surfaces on leadless versions, or leads formed around and under the sides on leaded versions. Chip carriers permit size reduction over DIPs with comparable pin numbers. Pin spacing is normally either 0.040 or 0.050 inch on centerlines.

The leadless chip carrier (LCC) is small, lightweight, and provides shorter connection paths from the die to the terminals. It is built as a ceramic single or multilayer package and can be hermetically sealed. Ceramic leadless chip carriers (LCCC) are available with 18 to 124 conductors running from the edges of the carrier to the chip. There are three forms: Type A with contacts on the top side, Type B with contacts on the edges, top, and bottom, and Type C (a variation of type B) has beveled corners. Sockets are available for these chip carriers. LCCCs are being used to package analog and logic ICs (integrated circuits), memory, microprocessors and microcontrollers.

Ceramic leaded chip carriers (CLCC) can also be hermetically sealed. Available with up to 84 pins, they are being used to package analog, memory and logic ICs. Plastic-leaded chip carriers (PLCC) are the most commonly used chip carrier packages. They are available with 18 to as many as 124 J-type leads. These packages can be inserted into sockets. PLCCs are being used to package analog and logic ICs, memory, microprocessors and microcontrollers. A chip carrier with 124 leads occupies the same area as a DIP with 64 leads.

A pin-grid array (PGA) is a square ceramic package with an array or grid of many plug-in type pins 0.10 inch on centers. (See Fig. 3.) The minimum pin count is 64 pins (10 by 10 grid) using the outside two rows. Pin-grid arrays are available with 68 to 128 leads. A pin-grid array with 256 leads occupies the same space as a DIP with 64 leads and a chip carrier with 124 leads. Sockets are available for pin-grid arrays. They are being used to package logic arrays. *See* SURFACE MOUNT TECHNOLOGY.

SEMICONDUCTOR RECTIFIER

See P-N JUNCTION, RECTIFIER, SELENIUM RECTIFIER.

SEMILOGARITHMIC GRAPH

See LOGARITHMIC SCALE.

SENSE AMPLIFIER

Many memory circuits retain logic states at very low voltage levels. A logic circuit requires a difference of several volts—normally 1 or 3 V—between the low and high conditions. A sense amplifier is a circuit that magnifies the relatively small voltages of the memory, so that they differ by the required amount for use by the logic circuit.

A sense amplifier is a straightforward, low-level, direct-current amplifier. It is advantageous for the sense amplifier to draw as little current as possible from the memory. A sense amplifier must therefore have a high input impedance. Sense amplifiers are usually low-noise circuits; this reduces the chances of interference from stray signal sources. *See also* DC AMPLIFIER.

SENSITIVITY

In electronics, sensitivity is an expression of the change in input necessary to cause a certain change in the output of a device or current. Usually, sensitivity is measured in terms of voltage.

In an amplifier or a radio receiver, sensitivity is a measure of the ability of the circuit to distinguish between a signal and noise at the output. Sensitivity also can be given for microphones and other transducers, and for electromechanical devices like relays and meters.

In an amplifier, sensitivity is expressed in terms of the input-signal voltage needed to produce a certain signal-to-noise ratio, in decibels, at the output. The sensitivity is indirectly related to the gain of the amplifier circuit or amplifier chain. But the noise figure is just as important (*see* GAIN, NOISE FIGURE, SIGNAL-TO-NOISE RATIO). Sensitivity is limited absolutely by the level of the noise that is generated in the amplifier, especially the first amplifier in a chain.

In a microphone or other transducer, sensitivity is expressed as the acoustic pressure, light intensity, or other flux required to produce a certain amount of signal at the output. Alternatively, the sensitivity can be expressed as the minimum actuating flux that results in satisfactory circuit operation. The sensitivity of a relay is the smallest coil current that will reliably close the armature contacts.

In a radio receiver, sensitivity is expressed as the number of microvolts at the antenna terminals that is required to produce a certain signal-to-noise ratio or level of noise quieting at the speaker. In amplitude-modulation, continuous-wave, frequency-shift-keying, and single sideband applications, the sensitivity is usually given as the number of microvolts needed to produce a 10-decibel signal-to-noise ratio. In frequency-modulation equipment, the sensitivity is specified either as the number of microvolts needed to cause 20-decibel noise quieting, or as the number of microvolts that results in a 12-decibel signal-to-noise-and-distortion ratio.

SENSOR

See TRANSDUCER.

SEQUENCE

A sequence is an ordered set of numbers, objects, or pieces of data. The specific order in which data bits, characters, or words appear is known as the sequence. An algorithm may also be called a sequence (*see* ALGORITHM).

In mathematics, a sequence of numbers is an ordered finite set:

$$S = \{x1, x2, x3, \ldots, xn\}$$

or an ordered infinite set:

$$S = \{x1, x2, x3, \ldots\}$$

such that for every term xk, the following term $x(k+1)$ is related to xk by a specific function f. The sum of a sequence is known as a series. *See also* FUNCTION.

SEQUENTIAL-ACCESS MEMORY

Sequential-access memory, or serial-access memory, is a form of memory in which data can be recalled only in a certain predetermined order. Any form of memory may have sequential access. The most common forms of sequential-access memory are the first-in/first-out (FIFO) memory and the pushdown stack (*see* FIRST-IN/FIRST-OUT, PUSHDOWN STACK).

If the addresses in a memory are numbered, such as $m1, m2, m3, \ldots, mn$, then the most common sequence for recall is upward from $m1$ to mn. A downward sequence, from mn to $m1$, is less common. Still less common are odd sequences in which the addresses do not ascend or descend in numerical order.

The data from a sequential-access memory may sometimes be recalled in two or more specific sequences. If the data can be recalled in any sequence possible, then the memory is called a random-access memory. *See also* SEMICONDUCTOR MEMORY.

SERIAL DATA TRANSFER

Information can be transmitted simultaneously along two or more lines, or it can be sent bit by bit along a single line. The first method is known as parallel data transfer (*see* PARALLEL DATA TRANSFER), and the latter method is called serial data transfer. The term *serial* is used because the data is sent in order, according to some predetermined sequence.

Serial data transfer requires only one transmission line, while parallel data transfer requires several lines.

However, serial data transfer is usually slower than parallel transfer.

SERIES CONNECTION

Components are said to be connected in series when the current follows a single path through them all. The diagram illustrates series connections of resistors (at A), inductors (at B), and capacitors (at C).

When components are connected in series, the current through each is the same. The voltage across each component depends on its resistance, reactance, or absolute-value impedance. (The absolute-value impedance is the distance from the origin to the impedance point in the complex plane.) The voltage across a given component is equal to the product of the current and the absolute-value impedance, according to Ohm's law (*see* IMPEDANCE, OHM'S LAW).

Series connections can be used for increasing voltage-handling capability. A series connection of five rectifier diodes, each with a peak-inverse-voltage (PIV) rating of 100 V, will yield a PIV rating of 500 V.

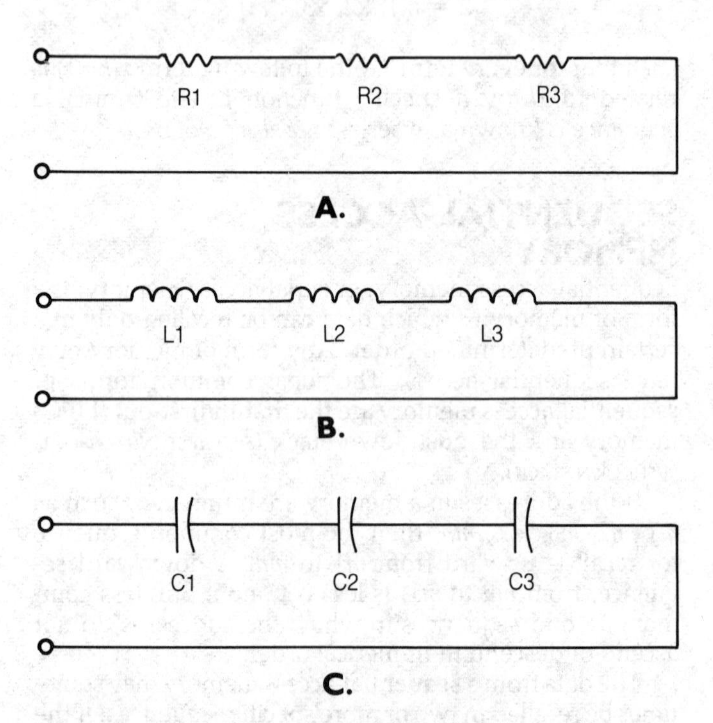

SERIES CONNECTION: Example of series connections of resistors (A), inductors (B), and capacitors (C).

SERIES MODULATION

Amplitude modulation can be obtained in a radio-frequency (rf) amplifier by the insertion of a variable-resistance component in the power-supply lead. This method of modulation is known as series modulation. Series modulation can be used with bipolar-transistor circuits and field-effect-transistor circuits.

The figure illustrates two methods of series modulation in a bipolar-transistor amplifier circuit. At A, the modulating component is in series with the emitter of the amplifying transistor. At B, the modulating component is in series with the collector lead. Either method will provide satisfactory results.

Series modulation can be used in the final amplifier circuit of a transmitter to obtain amplitude modulation. However, this requires that the modulating component handle a large amount of current or voltage. It also requires a high level of audio power. These disadvantages are offset by the fact that the amplifier itself can be operated in Class C, yielding higher rf efficiency than would be possible with a Class AB or Class B configuration. Also, no attention must be paid to the linearity of the rf amplifier circuits when series modulation is used at the final amplifier.

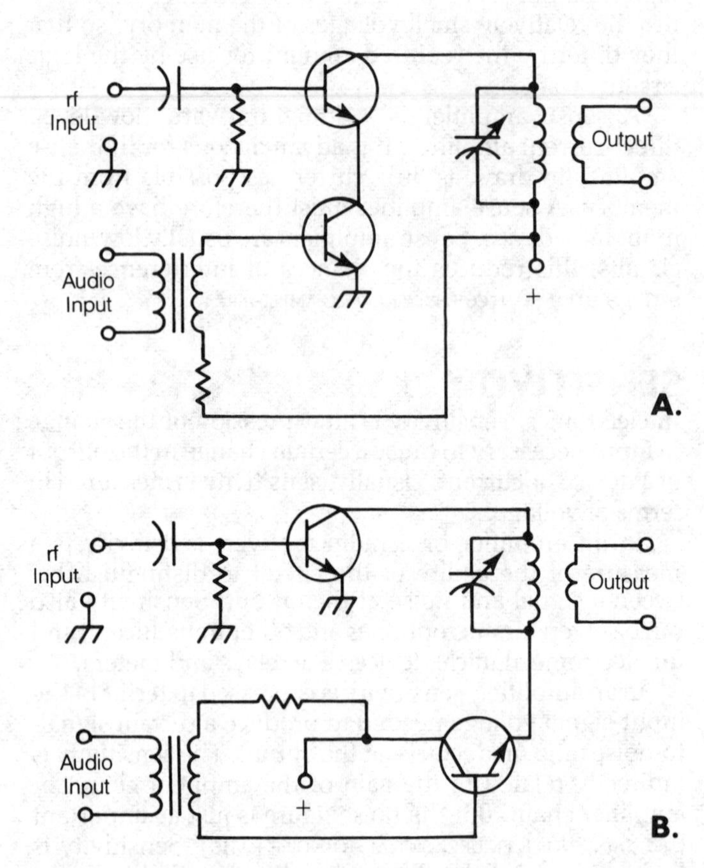

SERIES MODULATION: Series modulation in the emitter circuit of a bipolar transistor stage (A), and in the collector circuit (B).

SERIES-PARALLEL CONNECTION

A series-parallel connection is a method of increasing the power-handling capacity of a component. Series-parallel connections are used with resistors to obtain a higher power rating. Series-parallel connections also can be used with other components, such as inductors, capacitors, and diodes.

Series-parallel connections can be made in two basic configurations. One method is shown at A in the illustration, and a second method is shown at B. In both of these arrangements, if the value of each resistor is *R* ohms, then the value of the composite is *R* ohms. If the power-dissipation rating of each resistor is *P* watts, then the composite will be capable of dissipating $9P$ watts. In general, the power-dissipation rating is increased by a factor of n^2.

Series-parallel configurations can be assembled so the number of parallel elements differs from the number of series elements. This causes a change in the composite value, depending on the particular arrangement used. *See also* PARALLEL CONNECTION, SERIES CONNECTION.

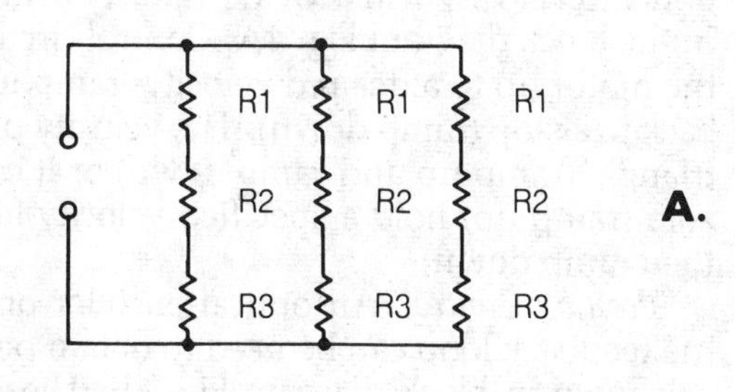

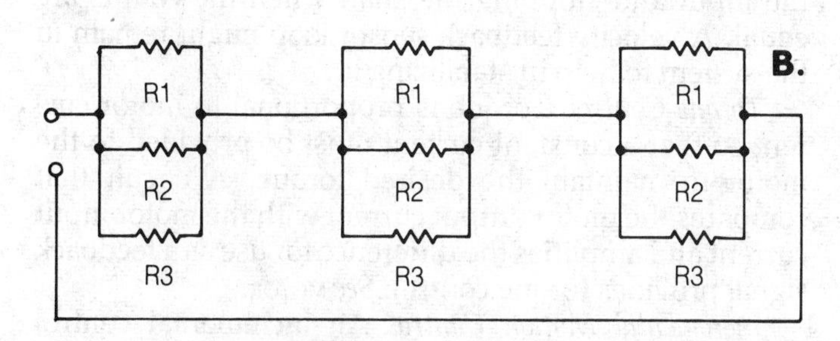

SERIES-PARALLEL CONNECTION: Examples of series-parallel connections (A and B) that provide the same overall resistance value if individual resistors have the same rating.

SERIES REGULATOR

See POWER SUPPLY.

SERIES RESONANCE

When an inductor and capacitor are connected in series, the combination will exhibit resonance at a certain frequency (*see* RESONANCE, RESONANT FREQUENCY). This condition occurs when the inductive and capacitive reactances are equal and opposite, thereby cancelling each other and leaving a pure resistance.

The resistive impedance that appears across a series-resonant circuit is very low. In theory, assuming that the inductor and capacitor are lossless components, the impedance at resonance is zero (see illustration). In practice there is always some loss in the components, so the impedance is greater than zero, but may be less than 1 ohm. The exact value depends on the Q factor; the higher the Q, the lower the impedance at resonance. The Q factor is a function of the losses in the components (*see* IMPEDANCE, Q FACTOR).

Below the resonant frequency, a series-resonant circuit shows capacitive reactance. Above the resonant frequency, it shows inductive reactance. Resistance is also present under nonresonant conditions because of the losses in the components; therefore, the impedance is complex at frequencies other than the resonant frequency.

Series-resonant circuits are used as traps (*see* TRAP). They are also used in some oscillators and amplifiers for tuning.

A section of transmission line can be used as a series-resonant circuit. A quarter-wave section, open at the far end, behaves like a series-resonant circuit. A half-wave section, short-circuited at the far end, also exhibits this property. These resonant circuits can be used in place of coil-capacitor (LC) combinations.

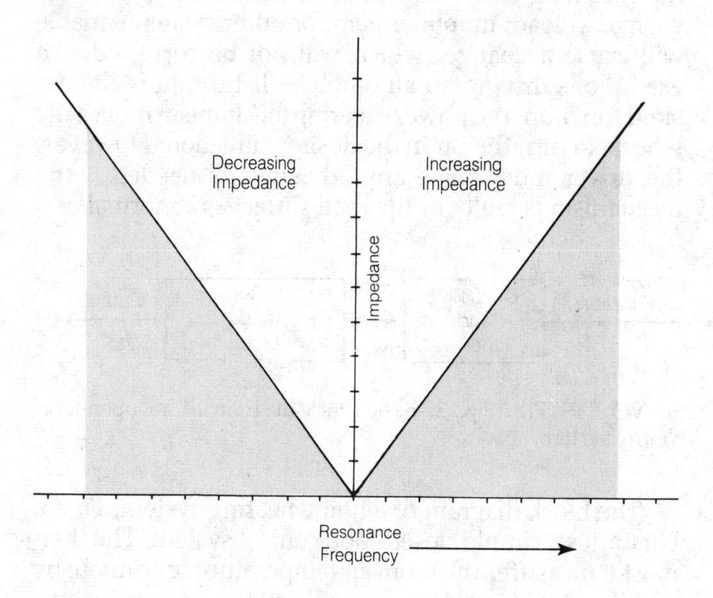

SERIES RESONANCE: Theoretical impedance curve for a series-resonant circuit.

SERVO SYSTEM

A servo system is a closed-loop control system that produces an error signal used to cancel any differences between the output and the input command. The error signal drives an actuator that corrects the difference so the output will agree continuously with the input. The basic closed-loop servo system shown in block diagram Fig. 1 includes a controller and amplifier, actuator (motor), load, and a feedback-sensing device.

In a closed-loop system, the output variable is measured, fed back, and compared to the desired input

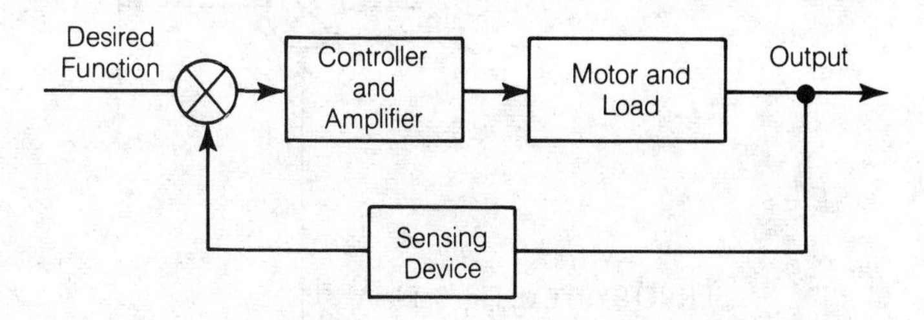

SERVO SYSTEM: Fig. 1. Basic block diagram of a closed-loop servo control system.

function at the summing point (symbolized as an X in a circle). Any difference between the two is a deviation or error, which is amplified as part of the correction process. The response of the system depends on how the loop is closed. The closed-loop transfer function is the relationship between the output and the input.

A closed-loop system is essentially insensitive to changing conditions in the system and therefore will continue to function correctly despite changes in load conditions, amplifier gain, wear on the mechanical components and even changes in the ambient temperature.

This contrasts with the open-loop system illustrated by the block diagram in Fig. 2. The output follows the input as long as system variables remain constant. Any change in load, amplifier gain, or other system variable will cause a change, which will not be corrected. An example is driving an automobile. If turning is the desired function, the power-steering mechanism moves the wheels to turn the car in the desired direction. However, the driver must make any adjustments needed if the mechanism is faulty or the road surface is abnormal.

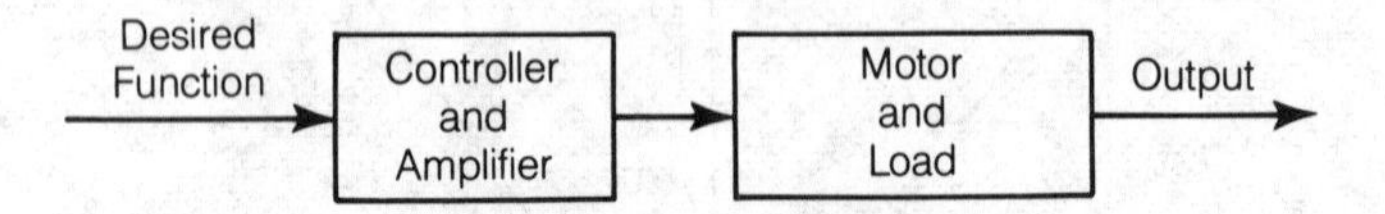

SERVO SYSTEM: Fig. 2. Basic block diagram of an open-loop control system.

The block diagram of a home heating system, Fig. 3, illustrates a simple closed-loop control system. The thermostat measures the room air temperature continuously, and if it falls below the manual setting, an error signal is developed. This signal switches the controller (a relay which acts as an amplifier) to turn on the heating plant (furnace, fuel pump motor, and blower) to heat the room. When the temperature rises high enough to cancel the error signal, the heating plant is shut down.

A system containing an amplifier, actuator, and feedback element is inherently unstable, and care must be taken to assure that the system is not subjected to too great an error or it may go into uncontrolled oscillations. In the case of the heating plant, the components must be selected so the system remain sensitive but stable so the heat does not cycle on and off too rapidly. In servo system design, considerable attention is given to system stability and response time.

Closed-Loop Systems Closed-loop servo systems are classified according to the variable being controlled. The most common forms of control are: velocity, position, torque, and combinations of these.

Velocity Control. A feedback sensor such as a tachometer in the feedback loop with an output that is proportional to velocity will provide velocity control, as shown in the block diagram Fig. 4A. Control circuitry can bring the motor up to a desired velocity (ramp up) and ease it back to a stop (ramp down). The velocity profile can be a triangle (ramp up and ramp down) or it can be a trapezoid (ramp up, hold a specified velocity for a time, and then ramp down).

Position Control. An optical encoder or a resolver in the feedback loop can be used to obtain position control as shown in block diagram Fig. 4B. These sensors can determine that the shaft has arrived at the desired angular position by counting pulses and comparing them with the input and stopping the shaft when the counts are equal. A velocity feedback sensor loop might remain in the system to help in stabilizing it.

Torque Control. Torque is proportional to motor current so that a constant current must be provided to the motor to maintain the desired torque. A circuit that compares the motor output current with the motor input current and amplifies the difference for use as a feedback signal provides torque control. *See* MOTOR.

Incremental Motion Control. An incremental control system includes a means for switching from one control mode to another. A velocity/position control, for example, can control motion with velocity in accordance with a desired velocity profile but be switched to position control to stop the shaft at a desired position with greater accuracy.

Servo System Characteristics The objective of all servo systems is to maintain zero error with a response that is as rapid as possible. Closed-loop feedback control provides accurate positioning because it continually tries to correct any error that exists. However, if there is a delay in error correction due to poor system response, the error will increase until the system becomes unstable. An unstable control system is unable to reach zero error

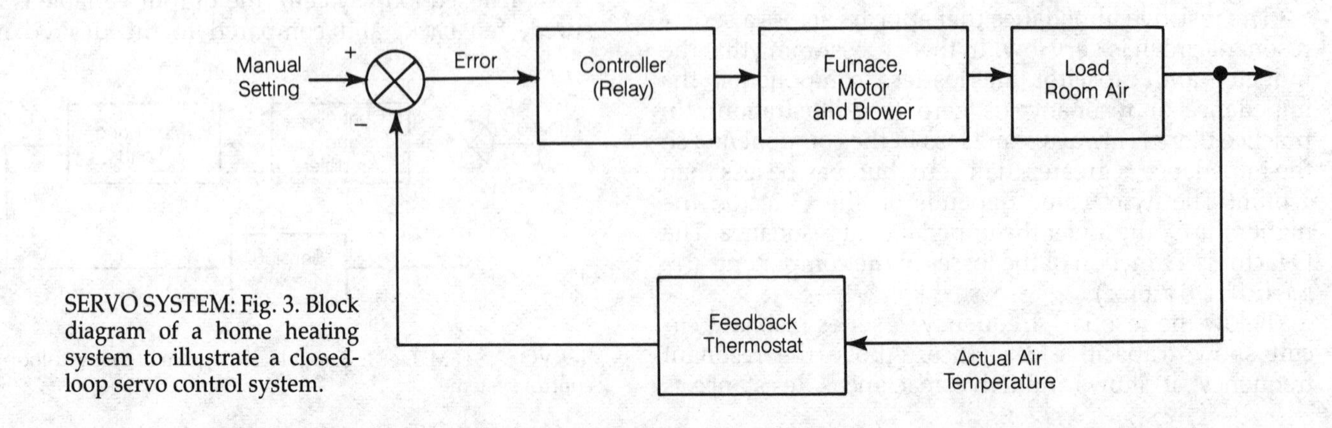

SERVO SYSTEM: Fig. 3. Block diagram of a home heating system to illustrate a closed-loop servo control system.

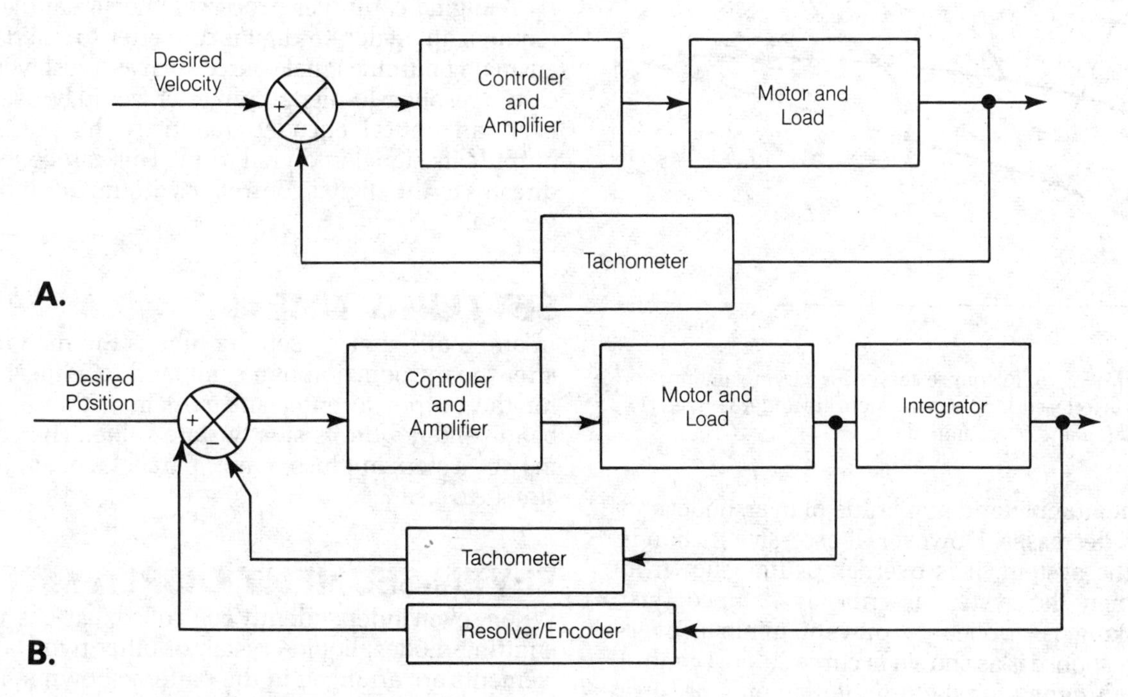

SERVO SYSTEM: Fig. 4. Block diagram of a closed-loop servo system with velocity control (A) and position control (B) with the velocity control loop retained.

and may wander or even go into sustained oscillation that can damage or destroy the system.

A servo system with high amplification must have rapid error correction, but this is less critical in a low-gain system. After corrective action has been initiated, it is important that overshooting the desired position be anticipated. There must be sufficient compensation for the inherent time delays of the system. The objective is to achieve a balance between positioning accuracy and stability.

The stability of a servo system may be determined by subjecting it to a large input error called a step function or step command and observing its response. A stable servo system will always return to a stable operating state unless there is a component failure. However, a frictionless system when subjected to a step function will go into oscillation, known as hunting, which will continue unless a means is found to remove energy from the load.

Figure 5B illustrates the events that occur when a large input voltage (step command), as shown in Fig. 5A, is introduced into a frictionless system. The large positive error is amplified to start the motor and drive the shaft and load in the positive direction (indicated by the positive slope of the curve). But with no friction in the system, the load continues on past the input-output alignment or stopping point because of energy stored in the load or inertia. This is called overshoot.

When overshoot occurs, the error signal will be reversed. But it takes time for the reverse torque of the motor to build to the value that will stop the shaft and load. Since the amplified negative error persists, the motor torque now accelerates the shaft and load from zero in the reverse direction (indicated by the negative slope of the curve). Again, because of its inertia, the shaft continues on past the stopping position. This is termed undershoot. A cycle of events called hunting will continue until the energy is removed from the load. An effective method for damping out or suppressing hunting is friction braking.

There are three possible responses to a step command in a servo system with friction or provision for braking as shown in Fig. 6: underdamped (curve 1); critically damped (curve 2), and overdamped (curve 3). As friction

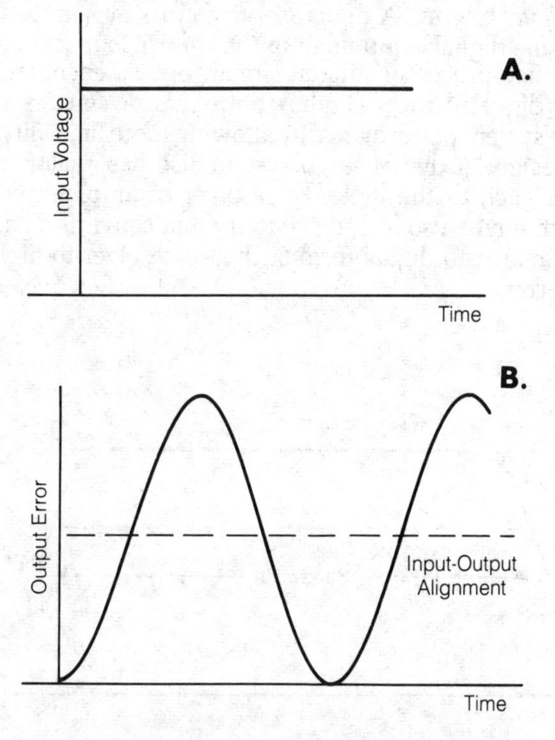

SERVO SYSTEM: Fig. 5. Step command (A) and the effect of step command on a frictionless closed-loop servo system (B).

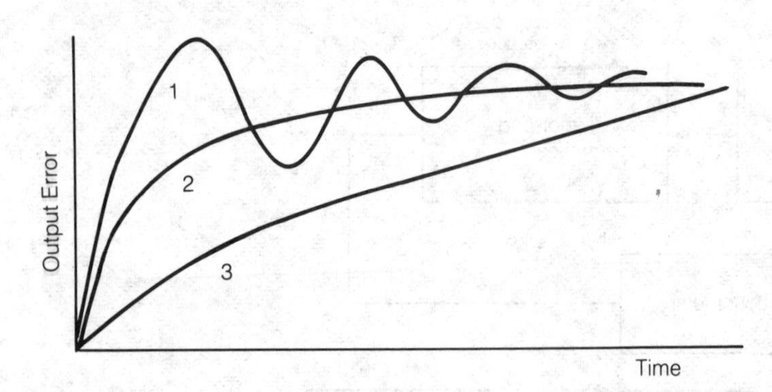

SERVO SYSTEM: Fig. 6. Response curves for varying amounts of system friction in a closed-loop servo system are underdamped (1), critical damped (2), and overdamped (3).

increases, the number and amplitude of overshoots and undershoots decreases. However, if excessive friction is applied to the system, it is overdamped as shown in curve 3. When the system is critically damped (just enough braking is applied to prevent minimal overshoot), the response is as shown in curve 2. Most control systems are designed for slight underdamping because the system is more responsive than if it is either critically damped or overdamped.

System Bandwidth. The bandwidth of a servo system is the frequency band over which amplifier gain is substantially constant. This is defined as the difference in frequency between the half-power points, as shown in Fig. 7 plotting output voltage, versus frequency. The half-power points are determined from the frequencies at which the output voltage is approximately 70 percent of the mid-band voltage. If the load resistance is constant, the mid-band output voltage is the amplitude of the constant output voltage between the half power points.

Digital Control. A digital servo control system has at least one digital component in the control loop. It might be a microprocessor, digital circuit, optical encoder, or other digital device. Digital controllers or sensors improve system performance by allowing more flexibility in the design. A digital servo system also has analog elements such as the motor, a resolver or amplifier. The system might also include circuitry that converts signals from analog-to-digital form such as a resolver-to-digital converter.

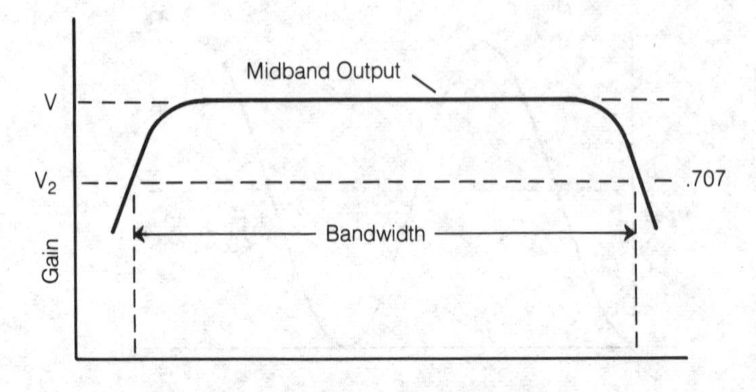

SERVO SYSTEM: Fig. 7. Diagram defining bandwidth for a closed-loop servo system.

A digital controller processes the signal digitally. It requires an analog-to-digital converter to interface with an analog output signal source such as a resolver. In this case, a resolver-to-digital converter would be used. However, an optical encoder quantizes the position and outputs its signal in digital form. This can be interfaced directly to the digital controller with no additional conversion.

SETTLING TIME

In automatic control, settling time is the time required, after the introduction of a command to a linear system, for the output to enter and remain within a specified band centered on its steady-state value. The command may be a step, impluse, ramp, parabola, or sinusoid. *See also* SERVO SYSTEM.

SEVEN-SEGMENT DISPLAY

When seven independently controlled, bar-shaped light-emitting diodes, liquid-crystal, or other two-state visual elements are arranged in the pattern shown in the illustration, the combination is known as a seven-segment display. The seven-segment display is used in many different digital displays because all numerals and some letters can be represented by some combination of states.

Although the seven-segment display is commonly used in digital clocks, calculators, meters, and other indicators, there are other methods of obtaining an alphanumeric display. The dot-matrix method, for example, is often used, especially for languages characters that cannot be represented on a seven-segment display. *See also* LIGHT-EMITTING DIODE DISPLAY, LIQUID-CRYSTAL DISPLAY, NEON GAS DISCHARGE DISPLAY.

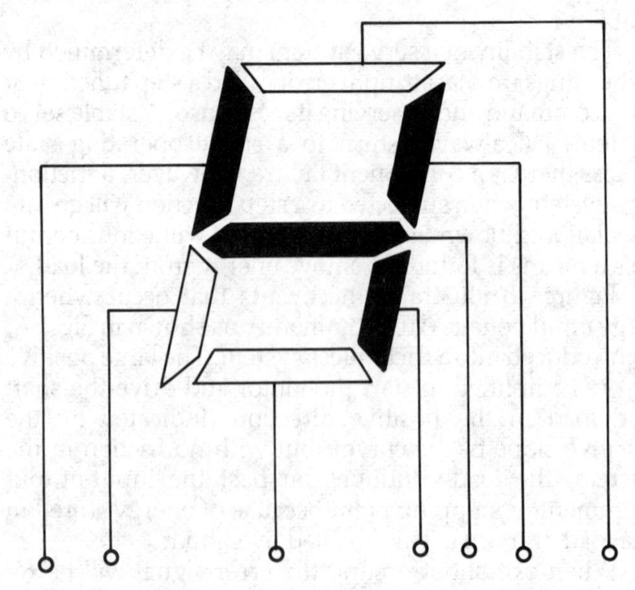

SEVEN-SEGMENT DISPLAY: All numbers and capital letters A through F can be displayed with seven segments.

SEVEN-UNIT TELEPRINTER CODE

The seven-unit code is identical to the Baudot teleprinter

code (*see* BAUDOT CODE) except that each character is preceded by a start signal and followed by a stop signal. Thus, instead of five bits or units per character, there are seven. The start unit is a space, and the stop unit is a mark.

The advantage of the seven-unit teleprinter code is that it provides synchronization between the transmitter and receiver. If some data bits are missed, a single character may be lost, but the error will not continue for several characters. All modern Baudot systems use the seven-unit code.

SFERICS

See ELECTROMAGNETIC INTERFERENCE.

SHANNON'S THEOREM

The amount of noise present in a communications channel affects the ease with which information can be transmitted and received accurately. The speed of transmission also affects the accuracy. These two factors are interdependent. The probability of obtaining error-free data transfer, based on the channel noise and the speed of transmission, is specified by Shannon's Theorem. Shannon's Theorem is also known as the noisy-channel coding theorem.

Shannon's Theorem is based on the concept of entropy. In information theory, entropy is an expression of the amount of intelligence contained in each communications symbol.

Shannon's Theorem states that if the entropy is no greater than the channel capacity, then the probability of error in data transmission is negligible. If the entropy is greater than the channel capacity, the probability of error is large.

No matter how large the entropy, the error rate can be made very small, as long as the data speed is slow enough. *See also* INFORMATION THEORY.

SHAPE FACTOR

The attenuation-versus-frequency response of a bandpass filter, especially in a communications receiver, is often evaluated according to an expression known as the shape factor. The shape factor determines the extent to which the response is rectangular.

The shape factor is generally given as the ratio of the bandwidth of a filter at the −60-decibel (dB) point to the bandwidth at the −6 dB point. The signal levels are measured in terms of voltage attenuation in the intermediate-frequency chain.

Various shape factors are shown in the illustration. In general, the smaller the shape factor, the more rectangular the response. (A shape factor of 1:1 indicates a perfectly rectangular response—a theoretical ideal, but not obtainable in practice.)

A rectangular response is desirable in a bandpass filter because a signal generally has a well-defined band over which the information is carried. It is essential that all of the signal components pass through the filter, but it

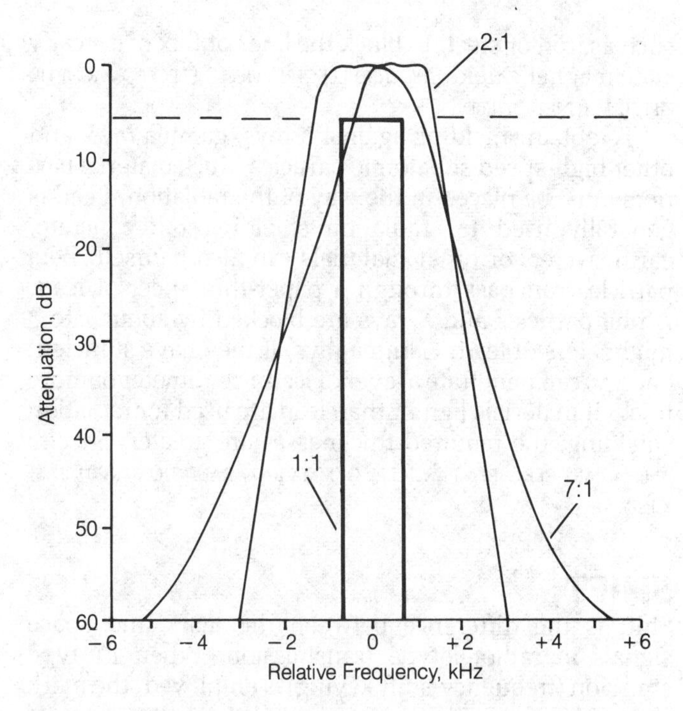

SHAPE FACTOR: A rectangular response is indicated by a 1:1 selectivity shape factor.

is undesirable for frequencies outside the signal band to be passed. *See also* BANDPASS RESPONSE, BANDWIDTH, SELECTIVITY.

SHELF LIFE

The length of time during which a cell or battery is usable is called the shelf life. Ordinary dry cells, and other types of cells and batteries, have shelf lives of 2 or 3 years. However, lithium batteries and cells have shelf lives of five to 10 years.

The shelf life can be specified for any type of component. Some devices, such as resistors, diodes, and transistors, will last for decades in storage under favorable conditions. Other components such as electrolytic capacitors, oil-filled capacitors, potentiometers, switches, and some tubes have more measurable shelf lives.

SHELL

See ELECTRON ORBIT.

SHIELDING

The intentional blocking of an electric, electromagnetic, or magnetic field is known as shielding. The deliberate blocking of high-energy radiation, such as X rays, gamma rays, or subatomic particles, is also called shielding.

Electric shielding is also called electrostatic shielding or Faraday shielding. A grounded metal screen or plate will block the lines of flux of an electric field. Electromagnetic shielding requires a completely enclosed cage or box made of a conducting material (*see* ELECTROMAGNETIC SHIELDING, ELECTROSTATIC SHIELDING). Magnetic shielding calls for an enclosure made from a ferromagnetic sheet metal, such as iron or steel, to block the lines of flux of a steady state magnetic field. *See also* ELECTRIC FIELD, ELECTROMAGNETIC FIELD, MAGNETIC FIELD.

To obtain shielding against X rays, gamma rays, and other high-speed subatomic particles, thick material barriers must be placed in the way of the radiation. Lead is generally used for radiation shielding but concrete, earth, water, or other materials can also be used. Beta particles can pass through a paper-thin sheet of lead. Alpha particles and X rays are blocked by about 1 to 3 millimeters of lead. Gamma rays, if they have sufficient energy, can penetrate a layer of lead 1 centimeter or more thick. If materials lighter than lead are used for radiation shielding, the required thicknesses are greater. *See also* ALPHA PARTICLE, BETA PARTICLE, GAMMA RAY, RADIATION HARDNESS, X RAY.

SHIFT

Shift is the difference between the mark and space signals in radioteletype transmission. When F1 type emission (frequency-shift keying) is employed, the mark and space frequencies may both be on the order of hundreds of megahertz, but the difference is usually less than 900 Hz. The most common values are 170 Hz, 425 Hz, and 850 Hz (*see* FREQUENCY-SHIFT KEYING). When F2 type emission (audio frequency-shift keying) is used, the audio frequencies are normally 2125 Hz and 2295 Hz, with a shift of 170 Hz. Occasionally, tones of 1275 Hz and 1445 Hz are used. Sometimes, audio tone pairs of 2125 Hz and 2975 Hz, or 1275 and 2125 Hz, are used, giving 850 Hz shift. In two-way modem systems, standard frequencies are 1270 Hz (mark) and 1070 Hz (space) in one direction, and 2225 Hz (mark) and 2025 Hz (space) in the other. Other shift values are less common.

In a memory or other circuit containing data in specific order, a shift is the displacement of each data bit one place to the right or one place to the left. In the case of a shift to the right, the far right-hand bit moves to the extreme left; in the case of a shift to the left, the far left-hand bit moves to the extreme right. This movement of data is also known as a cyclic shift. *See* BUBBLE MEMORY, CYCLIC SHIFT, SHIFT REGISTER.

SHIFT REGISTER

A shift register is a digital memory circuit used in calculators and computers. Information is fed into the shift register at one end and emerges from the other end. These ends are usually called the left and the right ends.

A clock provides timing pulses for the shift register. Each time a clock pulse is received, every data bit moves one place to the right or left. As data moves to the right, the right-hand bits leave the shift register one by one (see illustration). The opposite happens with the left-hand movement of data. Some shift registers can operate in only one direction, and are therefore called unidirectional. Other shift registers can operate in either direction, and are called bidirectional.

There are two basic categories of unidirectional or bidirectional shift register. These are known as the dynamic type and the static type. The dynamic shift register uses temporary storage methods. Data is lost if the clock operates at an insufficient rate of speed. In the static shift register, flip-flops are used to store the information (*see* FLIP-FLOP). The clock rate can be as slow as desired, because the data remains as long as power is supplied to the flip-flops. Shift registers are available in integrated-circuit form.

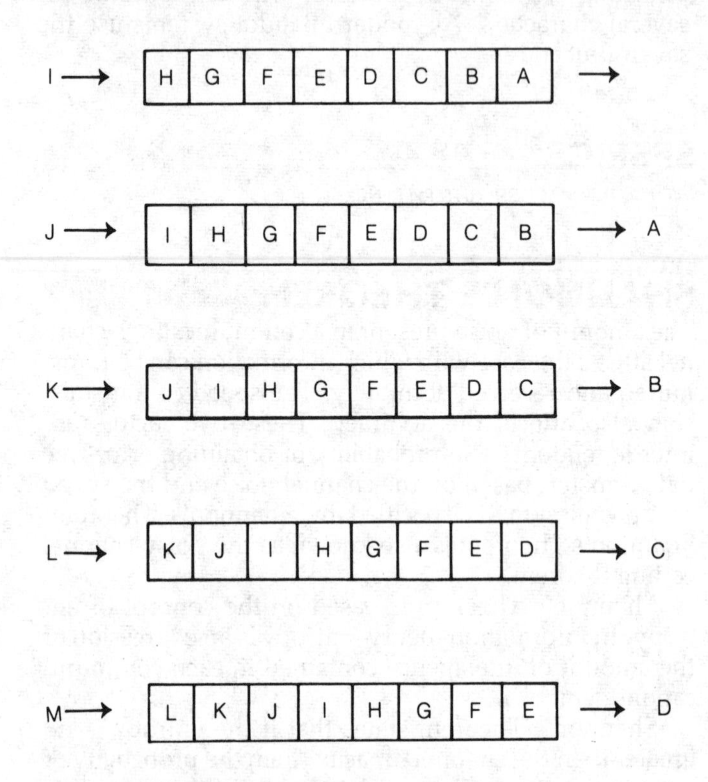

SHIFT REGISTER: In this diagram shift is from left to right. Consecutive data bits are inserted at left.

SHOCK

See ELECTRIC SHOCK.

SHOCK HAZARD

A shock hazard exists whenever there is a sufficient potential difference between two exposed conductors to cause harmful or lethal currents to pass through a person who touches both objects at the same time. Standard 117 V line voltage can, under some circumstances, be lethal. A hazard can exist at voltages of 50 V or more.

A current of just a few milliamperes causes pain. If a current of approximately 100 to 300 mA flows through the heart, there is a high probability that fibrillation will occur. Fibrillation is possible even with currents as low as 30 mA and can cause electrocution. High current can cause severe burns.

Generally, a shock hazard is considered to exist if there is sufficient voltage to cause a current of 5 mA to flow through a resistance of 500 Ω. The resistance of the human body may be much less than 500 Ω, especially if water is present. Damp hands, for example, greatly lower the effective resistance of the body.

SHORT CIRCUIT

A short circuit is a condition that occurs when current is partially or wholly diverted from its normal or proper path. It is called a short circuit because the path of the current is usually shorter than it otherwise would be.

In a 117 V electric system or power supply, a short circuit causes the tripping of a circuit breaker or the blowing of a fuse installed for short-circuit protection (*see* CIRCUIT BREAKER, FUSE, SHORT-CIRCUIT PROTECTION). In an audio-frequency or radio-frequency system, a short circuit causes a reduction in the level of the signal at the output.

SHORT-CIRCUIT PROTECTION

When a short circuit occurs in the utility lines or in the output of a power supply, overheating will occur unless the circuit is protected. This overheating can damage wiring, transformers, and series-connected resistors and chokes. If the current flow is large enough, fire can result.

The most widely used short-circuit protectors are the circuit breaker and the fuse. These devices open the circuit when the current exceeds a specified level (*see* CIRCUIT, PROTECTION, FUSE).

Current limiting, also called foldback, is provided in some power supplies to protect the components against a short circuit. The current-limiting circuit inserts an effective resistance in series with the output of the supply. *See* CIRCUIT BREAKER, CURRENT LIMITING.

SHORTWAVES

The high-frequency band, from 3 to 30 MHz (100 to 10 meters), is called the shortwave band. The wavelengths are not really very short; compared to microwaves, for example, the short waves are extremely long.

The term *shortwave* originated in the early days of radio when practically all communication and broadcasting was done at frequencies below 1.5 MHz (200 meters).

A high-frequency, general-purpose, communications receiver is still called a shortwave receiver. Most shortwave receivers cover the range from 1.5 MHz through 30 MHz. Some also operate in the standard broadcast band at 535 kHz to 1.605 MHz. Shortwave receivers can operate below 535 kHz into the so-called longwave band.

SHOTEFFECT

In a transistor, tube, or any current-carrying medium, the individual charge carriers cause noise impulses as they move from atom to atom. The individual impulses are extremely faint, but the large number of moving carriers results in a hiss that can be heard in any amplifier or radio receiver. This is called shot effect. The noise of shot effect is known as shot-effect noise or shot noise.

Shot-effect noise limits the ultimate sensitivity that can be obtained in a radio receiver. A certain amount of noise is always produced in the front end, or first amplifying stage (*see* FRONT END). This noise is amplified by succeeding stages along with the desired signals. The amount of shot-effect noise that a receiver produces is roughly proportional to the current it carries. In recent years, low-current solid-state devices such as the gallium-arsenide field-effect transistor (GaAsFET) have been developed for optimizing the sensitivity of a receiver front end. *See also* NOISE FIGURE, SENSITIVITY, SIGNAL-TO-NOISE RATIO, THERMAL NOISE.

SHOT-EFFECT NOISE

See SHOT EFFECT.

SHOT NOISE

Shot noise is a term used to describe three different forms of noise.

Whenever current flows, the movement of the electrons or holes causes electrical noise. This is called shot effect, and the noise is known as shot-effect noise or shot noise (*see* SHOT EFFECT). All semiconductor devices produce this kind of noise; the amount of noise depends primarily on the current at the area of the P-N junction(s) through which the current flows.

In a vacuum tube, noise is produced when the electrons bombard the plate. A smaller amount of noise also results from electrons striking the screen grid in a pentode or tetrode tube. This noise is called shot noise.

Whenever an electric spark occurs, an electromagnetic field is produced. Some appliances and all internal-combustion engines with spark plugs produce noise, called impulse noise or sometimes referred to as shot noise. *See also* IMPULSE NOISE.

SHUNT

The term *shunt* refers to a parallel connection of one component across another component or group of components. When one component is placed across another for a specific purpose, the inserted component is called a shunt.

A shunt can be any type of component. Examples of shunts include bypass capacitors, bleeder resistors, diode limiters, and various types of voltage regulators. Shunting resistors or coils are often used in ammeters to increase the indicated current range. *See also* PARALLEL CONNECTION, SHUNT RESISTOR.

SHUNT REGULATOR

A shunt regulator is a voltage-regulation circuit connected in parallel with the output of a power supply. Various types of transistors and tubes can be used as shunt regulators. In low-current power supplies, Zener diodes provide shunt regulation. *See also* OVERVOLTAGE PROTECTION, POWER SUPPLY, REGULATION, ZENER DIODE.

SHUNT RESISTOR

The current-indicating range of an ammeter can be increased by placing small-value resistors across the coil of the ammeter. These resistors can be noninductive, or they can consist of coils of resistance wire.

A general example of the use of a shunt resistor is shown in the illustration. Let the resistance of the meter coil be R_c (in ohms) and the full-scale range of the ammeter be A (in amperes). If a shunt resistor with an ohmic value R_s is placed across the meter coil as shown, then the new full-scale range, $A1$, will be:

$$A1 = \frac{A(R_c + R_s)}{R_s}$$

See also AMMETER.

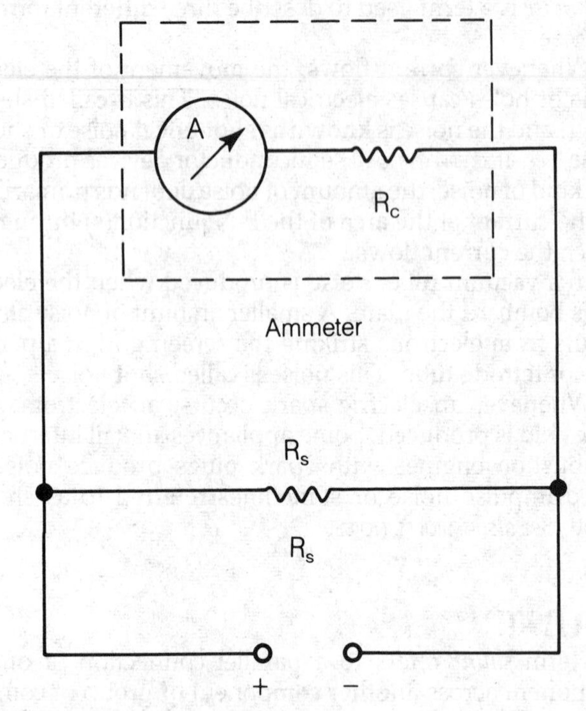

SHUNT RESISTOR: The addition of the shunt resistor R_s increases the range of an ammeter from A to A1.

SHUNT TUNING

See PARALLEL RESONANCE.

SIBILANT

A sibilant is a high-frequency component of speech. Sibilants result from the pronunciation of certain consonants. These consonant sounds are the soft *C*, the soft *G*, and the sounds of the letters *J*, *S*, *X*, and *Z*. Sibilants contain frequency components up to 10 kHz or higher.

In high-fidelity audio applications, it is important that all of the sibilants, as well as the other components of a voice, be passed with little or no attenuation. In communications, however, it has been found that a cutoff frequency of approximately 3 kHz will allow a voice to be understood well, including the sibilant sounds, even though some of the frequencies are not passed. *See also* VOICE FREQUENCY CHARACTERISTICS.

SIDEBAND

When a carrier is modulated for conveying information at any speed, sidebands are produced. Sideband signals occur immediately above and below the carrier frequency.

Sideband signals are the result of mixing between the carrier and the modulating signal. The greater the data speed, the higher the frequency components of the modulation signal, and the farther from the carrier the sidebands will appear (see illustration on p. 767).

Other influences on bandwidth include the type of modulation, the percentage or index of the modulation, and the efficiency of the data-transmission method. *See also* AMPLITUDE MODULATION, BANDWIDTH, EMISSION CLASS, FREQUENCY MODULATION, MODULATION.

SIDETONE

In a code transmitter or transceiver in which receiver muting is used, the actual transmitted signal cannot be heard. Even if receiver muting is not used, the transmitter and receiver are often operated on frequencies separated by an amount that results in an inaudible tone, or a tone with an objectionably high or low pitch. In these situations, a sidetone is employed.

A sidetone is generated by a simple audio oscillator such as a multivibrator or relaxation circuit. Most keyers have built-in sidetone oscillators, some with adjustable pitch. Many code transmitters and transceivers also have built-in sidetone oscillators, which are keyed along with the transmitter.

SIEMENS

The siemens is the unit of electrical conductance. It was formerly called the mho. Given a resistance of *R* ohms, the conductance *S* in siemens is $1/R$. *See also* MHO.

SIGNAL AMPLITUDE

See AMPLITUDE.

SIGNAL CONDITIONING

Signal conditioning is the process of modifying or changing a signal for a specific purpose. Signal conditioning is also called signal processing. It is used extensively in audio-frequency and radio-frequency equipment.

An example of signal conditioning is the detection and filtering of an alternating-current wave for indirect voltage measurement using a direct-current meter. Another example is the use of an equalizer in a high-fidelity system. *See* EQUALIZER.

SIGNAL ENHANCEMENT

Any method of improving the quality of intelligibility of a signal is called signal enhancement.

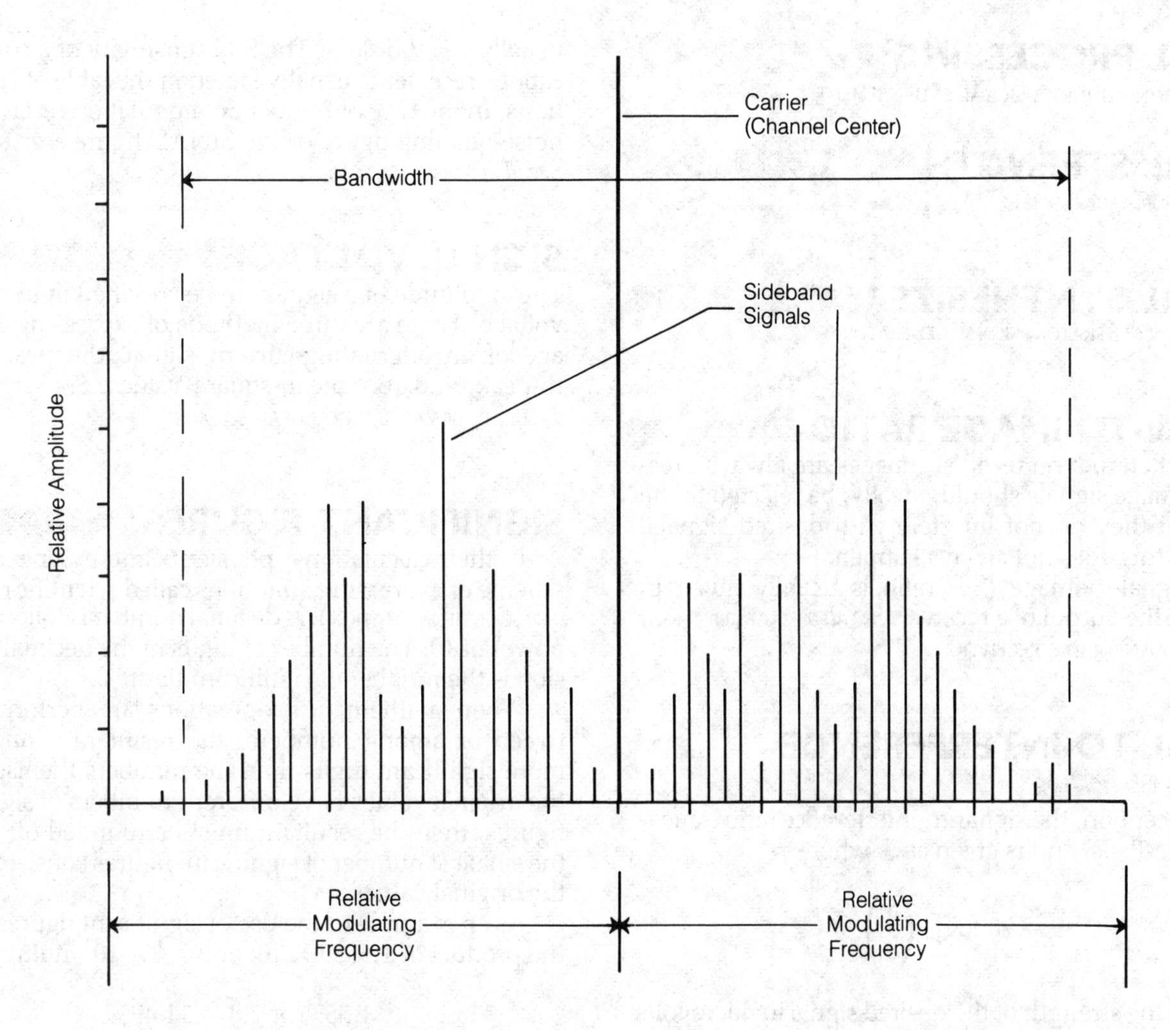

SIDEBAND: Sidebands result from the mixing of a carrier wave with a modulating signal.

In radio-frequency receiving apparatus, signal enhancement generally refers to the optimization of the signal-to-noise ratio or the signal-to-interference ratio. In high-fidelity audio receivers, signal enhancement is the reduction of noise and the tailoring of the sound to suit the taste of the listener. *See also* EQUALIZER, HIGH FIDELITY, SELECTIVITY, SIGNAL-TO-INTERFERENCE RATIO, SIGNAL-TO-NOISE RATIO.

SIGNAL ENVELOPE

See ENVELOPE.

SIGNAL GAIN

See AMPLIFICATION, GAIN.

SIGNAL GENERATOR

A signal generator is an oscillator, often equipped with a modulator, used for testing audio-frequency (af) or radio-frequency (rf) equipment. Most signal generators are intended for either af or rf applications, but not both.

In its simplest form, a signal generator consists of an oscillator that produces a sine wave of a specified amplitude measured in microvolts or millivolts, and a frequency measured in hertz, kilohertz, or megahertz. Some af signal generators can produce several different types of waveforms. The more sophisticated signal generators for rf testing have amplitude modulators or frequency modulators.

For the testing, adjustment, and servicing of radio-frequency transmitters and receivers, a combined signal generator and monitor is often used. *See also* FUNCTION GENERATOR, SPECTRUM ANALYZER.

SIGNAL MONITOR

A signal monitor is a test instrument used for analyzing radio-frequency signals.

An ordinary radio receiver can be used as a signal monitor in many situations. For a more detailed view of the signal than can be obtained by ear, a panoramic receiver is used.

An oscilloscope is often used for signal monitoring. In the test laboratory, a service monitor/generator is used. A spectrum analyzer is a form of signal monitor. *See* OSCILLOSCOPE, PANORAMIC RECEIVER, SPECTRUM ANALYZER.

SIGNAL PROCESSING

See SIGNAL CONDITIONING, SIGNAL ENHANCEMENT.

SIGNAL STRENGTH

See AMPLITUDE.

SIGNAL SYNTHESIZER

See FUNCTION GENERATOR, SIGNAL GENERATOR.

SIGNAL-TO-IMAGE RATIO

In a superheterodyne receiver, images are always present. The image signals should, ideally, be rejected to the extent that they do not interfere with desired signals. However, this does not always happen.

The signal-to-image (S/I) ratio is usually given in decibels at the output of a receiver. *See also* IMAGE REJECTION, SUPERHETERODYNE RADIO RECEIVER.

SIGNAL-TO-INTERFERENCE RATIO

In radio reception, the signal-to-interference ratio is measured in decibels, and is given as *S*, where:

$$S = 20 \log_{10} \left(\frac{a}{b} \right)$$

where *a* is the strength of the desired signal in microvolts at the antenna terminals, and *b* is the sum of the strengths of all of the interfering signals in microvolts at the antenna terminals.

The signal-to-interference ratio is important as a specification of sensitivity and selectivity in a radio receiver. *See also* SELECTIVITY, SENSITIVITY, SIGNAL-TO-NOISE RATIO.

SIGNAL-TO-NOISE RATIO

The sensitivity of a communications receiver is specified in terms of the audio signal-to-noise ratio that results from an input signal of a certain number of microvolts. This ratio is abbreviated S/N or S:N.

If the root-mean-square (rms) signal strength at the antenna terminals of a receiver is E_s, given in microvolts, and the rms noise level is E_n, also in microvolts, then the ratio S/N, in decibels, is:

$$S/N = 20 \log_{10} \left(\frac{E_s}{E_n} \right)$$

Usually, the sensitivity is specified as the signal strength in microvolts that is necessary to cause a S/N ratio of 10 decibels.

Modern communications receivers generally require about 0.5 μV, or less, to produce a S/N ratio of 10 decibels at the high frequencies in the continuous-wave or single-sideband modes. For amplitude modulation, the rating is usually 1 μ V or less. The S/N sensitivity of a communications receiver is usually stated in the table of specifications. In the case of frequency-modulation receivers, the noise-quieting figure or the SINAD figure are standard. *See also* NOISE QUIETING.

SIGNAL VOLTAGE

The amplitude of a signal can be specified in terms of its voltage. There are three methods of expressing the voltage of an alternating-current signal: the peak, peak-to-peak, and root-mean-square values. *See* PEAK-TO-PEAK VALUE, PEAK VALUE, ROOT MEAN SQUARE.

SIGNIFICANT FIGURES

In their calculations, physicists and engineers use a scheme of expressing quantities called scientific notation (*see* SCIENTIFIC NOTATION). A decimal number is followed by a power of 10. The number of digits in the decimal expression is the number of significant figures.

When mathematical operations are performed between or among numbers, the resultant cannot have more significant digits than the numbers themselves. If the original values have different numbers of significant figures, then the resultant must be rounded off to have the smallest number of significant figures consistent with the original values.

As an example of the use of significant figures to find the product of 1.5535×10^2 and 7.4×10^3 multiply:

$$1.5535 \times 7.4 = 11.4959$$

The powers of 10 multiply as follows:

$$10^2 \times 10^3 = 10^5$$

The complete product is thus:

$$\begin{aligned} 1.5535 \times 10^2 \times 7.4 \times 10^3 &= 11.4959 \times 10^5 \\ &= 1.14959 \times 10^6 \end{aligned}$$

The answer does not have six significant figures. The original decimal expressions, 1.5535 and 7.4, have only five and two significant figures, respectively. In rounding off the answer to two significant figures, 1.1×10^6 is obtained.

SILICON

Silicon is a Group IV element in the periodic table. Silicon has the ability to form natural oxides; that characteristic is fundamental to all silicon semiconductor processes. Bipolar transistors, metal-oxide silicon field-effect transistors (MOSFETs) and integrated circuits (ICs) are made from silicon. Because pure silicon is a semiconductor, its ability to conduct electrons and holes must be improved by doping, or injecting, it with impurity elements.

The silicon atom has four valance, or outermost, electrons available for bonding and other interactions. In a silicon crystal, each atom is surrounded by four other atoms arranged in the shape of a tetrahedron; the atoms are bound to one another by shared pairs of valance electrons. As a result, few electrons are available to move about because most are holding the crystal together. The added electrons from the impurities improve the conductivity of silicon.

The performance of bipolar transistors, MOSFETs and ICs formed with these transistors depends on the quality of the silicon wafer or substrate on which they are formed. High-quality, relatively uniform silicon crystals that are free of defects can be grown with a crystal puller in a technique called the liquid-encapsulated Czochralski (LEC) process. *See* CZOCHRALSKI CRYSTAL-GROWTH SYSTEM.

A crucible within a high pressure chamber is filled with silicon and heated to a temperature about 100 degrees higher than the melting point of silicon and held there (1,958°C). A small seed crystal of silicon is then dipped into the molten silicon and pulled slowly up from the melt while it is simultaneously rotated. As it is raised, silicon atoms in the melt adhere to the crystal. The crystal grows until it becomes a cylinder with a diameter from about 100 to 200 millimeters (4 to 8 inches).

The diameter of silicon ingots is limited by unwanted fluid currents in the molten crystal. These currents, caused by temperature and density differences in the melt, cause irregularities to form in the silicon as it solidifies. Research on methods to slow fluid motion are in progress to improve the quality of large-diameter crystals. Some of the techniques being applied for improving the uniformity of bulk LEC grown silicon crystals are: 1) reducing the thermal gradient from the center to the outside of the crystal, 2) doping the melt with indium, 3) growing the crystal in a magnetic field, and 4) automatically controlling crystal diameter with computer control.

The silicon ingot is sliced and polished to make wafers. Finished wafers have mirrorlike smoothness with surface variations of less than 10 angstrom units, or 1 nanometer from the highest point to the lowest point. The wafer is then ready to be processed. The wafer is first heated to approximately 1,073°C, causing a thin layer of silicon dioxide (SiO_2) to grow on the surface of the wafer. *See* INTEGRATED-CIRCUIT MANUFACTURE, TRANSISTOR MANUFACTURE.

SILICON BILATERAL SWITCH

A silicon bilateral switch (SBS) is a four-layer semiconductor device that has characteristics similar to those of the silicon unilateral switch (SUS), except that it exhibits those characteristics in both directions. *See* SILICON UNILATERAL SWITCH.

SILICON COMPILER

A silicon compiler is a computer program for designing integrated circuits based on a description of circuit function and providing an output suitable for further graphic processing. In practice, silicon compilers usually produce a layout in terms of placement and connection of predefined cells rather than a direct geometric layout.

The input data to a silicon compiler program includes a description of the end circuit's functional requirements: number of bits, arithmetic, logic, and control operations—often combined with structural information such as the number of registers and buses. The compiler can choose a top-level architecture—the pattern of how the functional blocks will be arranged for speed, area, and other variables. It can also verify that specified inputs will give the correct outputs and assure that logic errors are caught by checking the logic against the expected output.

The silicon compiler is similar to a software compiler that accepts algorithmic descriptions of what a computer program is supposed to do and produces machine-language instructions. This is analogous to the specification of individual transistor-level cells by the silicon compiler. However, the silicon compiler is limited to the instruction set of the computer in which it is resident.

Software compilers trade off large memory size and slower speed for ease of programming. Similarly, silicon compilers trade off design efficiency, in terms of device size and speed, to save human design effort. Developed as a byproduct of attempts to automate circuit design methods, these compilers emphasize regular structures, hierarchical design, and correctness of construction. They can encode rules for going from one stage of design to the next and for making sure that only correct layouts are generated.

There is no general agreement on a precise definition of a silicon compiler. Differences of opinion exist on the level of complexity of the description a program must accept, the kind of layout it must generate, and the degree of human intervention. Some believe that the term should be applied only to systems that include testing, simulation, and other verification tools, in addition to synthesis programs.

Some have considered gate-array design programs to be silicon compilers. (A gate array is an integrated circuit consisting of a large number of prefabricated, identical, regularly spaced cells containing transistors that can be interconnected by final masking steps to perform functions. *See* GATE ARRAY.) A program can transform a circuit description directly into a layout of devices on a premanufactured wafer. However, this kind of program is not considered by others to qualify as a true silicon compiler because they say the circuit designs produced are inefficient and waste silicon. Gate arrays use more silicon than hand-crafted custom designs and standard cells that are completely designed on the computer workstation. Moreover, gate arrays usually include some leftover gates that are nonfunctioning.

Standard cell design programs have also been considered to be silicon compilers. They permit the design of an integrated circuit (IC) from scratch based on gate and cell data stored in computer workstation memory. (Standard cells are IC building blocks that include such functions as adders or multiplexers that can be slightly modified to meet specific requirements for speed, power consumption, or bit length.) Although this approach allows the designer to specify cell placement in architectural and functional description terms, many do not consider these programs to be silicon compilers.

Programmed logic arrays (PLAs) and programmable read-only memory (ROM) generators are considered by some to be silicon compilers because they transform an algorithmic description into a layout. However, most design professionals consider that a true silicon compiler should apply only to high-level ICs although PLA and ROM programs can produce layouts as densely packed as handcrafted ones.

Silicon compilers available today are best suited for designing well-understood devices such as signal processors and microprocessors. Some compilers are optimized for the design of devices like microprocessors. However, results indicate that ICs designed with silicon compilers use more silicon and have fewer LSI functional blocks than comparable ICs that have been handcrafted for optimum performance. The most serious deficiencies in these compilers are seen as their inability to make good decisions about the layout and positioning of functional blocks for maximum speed and efficiency and the determination of the degree of pipelining required for parallel operations.

SILICON-CONTROLLED RECTIFIER

A silicon-controlled rectifier (SCR) is a four-layer P-N-P-N unidirectional device for bistable switching. The SCR has three junctions and three terminals: anode, cathode, and gate. The figure shows the schematic of an SCR and a sectional view of an SCR chip showing its layers and junctions.

The SCR is an ordinary rectifier diode with a control element, the gate. Current to the gate determines the anode-to-cathode voltage at which the device begins to conduct. The gate bias can keep the device off, or it can permit conduction to begin at any desired point in the forward half cycle of an alternating current across the anode and cathode terminals. The SCR is widely used to control alternating-current (ac) power.

For conventional forward-bias operation, the SCR anode potential must be positive. SCRs are turned on by placing a positive voltage on the gate electrode. Once turned on the SCR remains on, even if the gate voltage is cut off or made negative. The anode-to-cathode voltage must be reduced to the threshold level or forward current must be reversed. Also, if it is in an ac circuit, the ac must cross the zero level.

The SCR is most often used to switch currents in one direction only, making it useful for direct current (dc) and half-wave ac. SCRs can function as controlled rectifiers in bridges. They can handle hundreds of amperes and peak voltages to 1500 V with triggering currents of less than a few milliamperes. However, most SCRs used in electronics-related applications are rated 40 A or less. The triac (essentially two back-to-back SCRs) switches full-wave alternating current. *See* TRIAC.

The gate turn-off silicon-controlled rectifier (GTO) is a variation of the SCR designed to permit it to be turned off with a negative bias on the gate terminal.

SCRs are packaged in the same style cases as other power semiconductor devices. TO-206 style metal cases (formerly TO-18) and TO-226 style metal cases (formerly TO-92) are widely used for sensitive-gate SCRs rated for on-state root means square (rms) current under 1 A. TO-220 plastic cases are used to package SCRs for ratings from 1 to 16 A, and a wide selection of metal cases is used for SCRs rated 10 A or higher. These include TO-204 metal cases (formerly TO-3) and those with stud mounting for better heat dissipation.

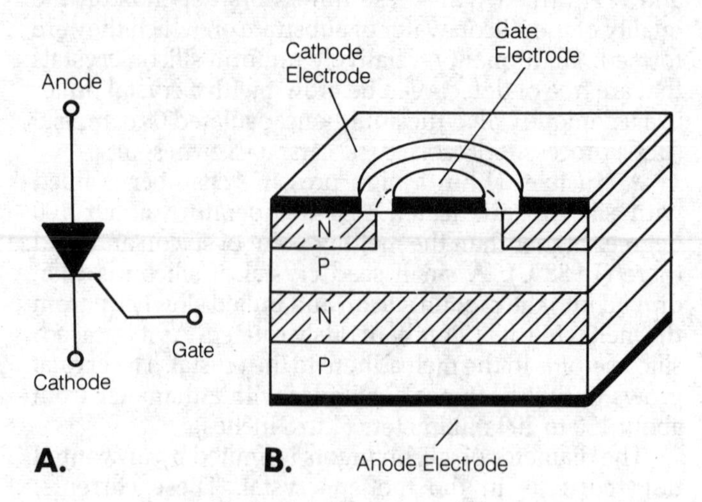

SILICON-CONTROLLED RECTIFIER: Symbol for an SCR (A) and a sectional view of an SCR chip (B).

SILICON-CONTROLLED SWITCH

The silicon-controlled switch (SCS) is a low-current silicon-controlled rectifier (SCR) with two gate terminals: anode and cathode. A negative pulse on the anode gate turns the SCS on; a positive pulse turns the SCS off. A positive pulse at the cathode gate can switch the device on, but a negative pulse is required to switch it off. *See* SILICON-CONTROLLED RECTIFIER.

SILICON DIOXIDE

Silicon dioxide is a common compound of silicon and oxygen. Much of the earth's silicon is found in this form. The chemical formula for silicon dioxide is SiO_2.

Silicon dioxide permits the fabrication of integrated circuits with extremely high density and low current consumption. Silicon dioxide is used in conjunction with other metal oxides and silicon nitride. *See also* INTEGRATED CIRCUIT MANUFACTURE, METAL-OXIDE SEMICONDUCTOR, METAL-OXIDE-SEMICONDUCTOR LOGIC FAMILIES, SILICON NITRIDE, TRANSISTOR MANUFACTURE.

SILICONE

Silicone is a polymerized material consisting of silicon and oxygen atoms. It is an excellent electrical insulating material, but is a good conductor of heat. Silicone can withstand very high temperatures.

Silicone is commonly used as a heat-transfer agent for semiconductor power transistors and diodes. The silicone is applied between the heat-conducting metal base of the device and a metal heat sink. This ensures efficient transfer of heat away from the semiconductor device.

Silicone is commercially available in paste form, packaged in squeeze tubes.

SILICON MONOXIDE

Silicon monoxide is a compound consisting of silicon and oxygen; its chemical formula is SiO. Some silicon is found in this form in the earth.

Silicon monoxide is an excellent insulating material and is useful as a capacitor dielectric in the manufacture of integrated circuits.

SILICON NITRIDE

In the manufacture of integrated circuits, silicon nitride is widely used in conjunction with silicon dioxide to facilitate the etching of components. The main advantage of a silicon-nitride/silicon-dioxide combination is that it prevents migration of impurities. *See also* SILICON DIOXIDE.

SILICON RECTIFIER

See P-N JUNCTION, RECTIFIER, SILICON-CONTROLLED RECTIFIER.

SILICON SOLAR CELL

See SOLAR CELL.

SILICON STEEL

A small amount of silicon can be added to steel to form an alloy called silicon steel. Silicon steel normally contains about 4 percent silicon and 96 percent steel.

Silicon steel is used as a core material for some inductors, chokes, and transformers at alternating-current frequencies. The core sections are laminated to minimize eddy-current loss (*see* EDDY-CURRENT LOSS, LAMINATED CORE). Silicon-steel cores exhibit very high efficiency at frequencies up to several kilohertz. At much higher frequencies, silicon steel becomes lossy because of hysteresis. *See also* TRANSFORMER.

SILICON TRANSISTOR

See TRANSISTOR.

SILICON UNILATERAL SWITCH

A silicon unilateral switch (SUS) is a four-layer semiconductor device, identical to the silicon-controlled rectifier (SCR) except that a Zener diode is placed in the gate lead. The Zener diode reduces the sensitivity of the switch, by blocking trigger pulses with voltages smaller than a predetermined level of approximately 8 V.

The SUS is less likely than an ordinary SCR to be triggered accidentally by stray noise pulses. *See also* SILICON BILATERAL SWITCH, SILICON-CONTROLLED RECTIFIER, ZENER DIODE.

SILVER

Silver is an element with an atomic number of 47 and an atomic weight 108. In its pure form silver is a shiny almost white metal. Silver is considered to be a precious metal.

Silver is an excellent conductor of electricity and is highly resistant to corrosion. Silver has a relatively high melting point. For these reasons, switch and relay contacts are plated with or made from silver. Silver is a constituent metal in a high-quality hard solder called silver solder. Silver wire is used in the manufacture of high-Q coils. Silver is also used in the manufacture of low-loss capacitors. *See also* CAPACITOR, SILVER SOLDER.

SILVER-OXIDE CELL

See BATTERY.

SILVER SOLDER

Silver solder is a hard solder used in the manufacture of electronic devices. Silver solder consists of copper, zinc, and silver.

Silver solder melts at a higher temperature (600°F or 320°C) than ordinary solder composed of tin and lead. Silver solder maintains a better electrical bond than tin-lead solder, primarily because the metals used are better conductors. *See also* SOLDER.

SILVER-ZINC CELL

See BATTERY.

SIMPLEX

Two-way telephone and radio system transmitters and receivers can be operated on a single frequency. This is known as simplex or simplex operation. The two stations communicate directly with each other; no repeater or other intermediary is used.

In simplex operation, it is possible for only one station to transmit at a time because neither station can receive and transmit signals simultaneously on the same frequency. If it is necessary to send and receive data simultaneously, two different frequencies must be used. This is called duplex operation (*see* DUPLEX OPERATION).

It is not normally possible for one station to interrupt the other station in simplex operation. However, it can be accomplished if continuous-wave (type A1) or single-sideband, suppressed-carrier (type A3J) emissions are used. This is called break-in operation. It requires the use of a special switch and muting system. The receiver is activated during brief pauses in a transmission.

SIMULCASTING

Simulcasting is the transmission of a single program with two or more different channels or modes at the same time.

A play-by-play broadcast of a football game, for example, may be transmitted over an amplitude-modulated

(AM) broadcast station at 830 kHz and a frequency-modulated (FM) station at 90.5 MHz.

In television broadcasting, an enhanced (stereo) sound track can be broadcast on the standard FM band. This form of simulcasting is often done for such events as orchestra concerts. The viewer can turn to the appropriate television channel, turn the voltage control of the television receiver down, and tune in to the appropriate station on an FM stereo receiver to hear the sound.

SINE

The sine function is a trigonometric function. In a right triangle, the sine is equal to the length of the far or opposite side, divided by the length of the hypotenuse, as shown in the illustration at A. In the unit circle $x^2 + y^2 = 1$, plotted on the Cartesian (x,y) plane, the sine of the angle θ, measured counterclockwise from the x axis, is equal to *y*. This is shown at B. The sine function is periodic, and begins with a value of 0 at θ =0. The shape of the sine function is identical to that of the cosine function (*see* COSINE), except that the sine function is displaced to the right by 90 degrees.

A.

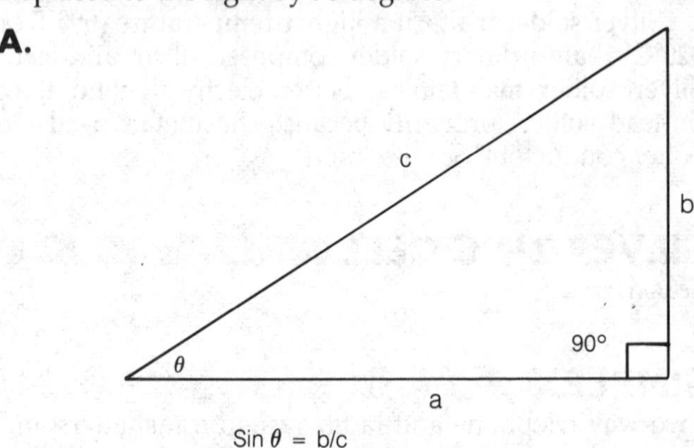

B.

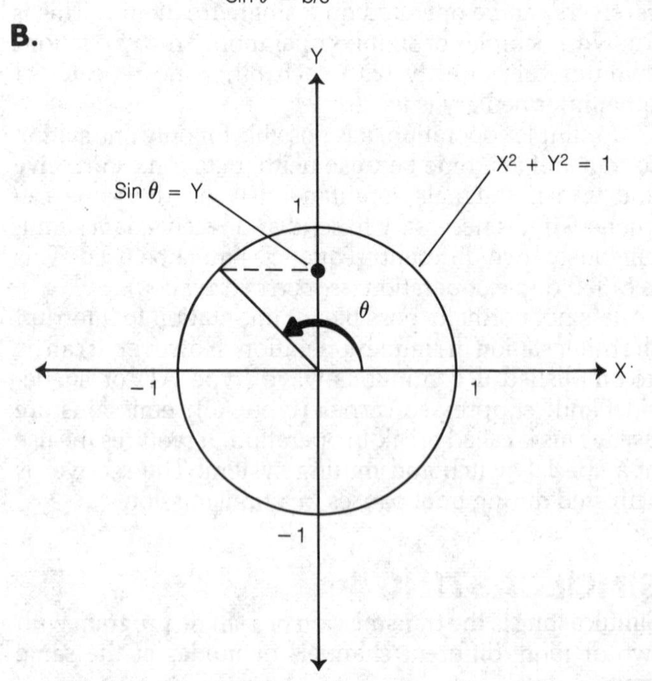

SINE: A triangle to explain the sine function (A) and a graph of the function (unit circle model) (B).

The sine function represents the waveform of a pure, harmonic-free, alternating-current disturbance. In mathematical calculations, the sine function is abbreviated sin. Values of sin θ for various angles of θ are given in the table on p. 773. *See also* SINE WAVE, TRIGONOMETRIC FUNCTION.

SINE LAW

See LAW OF SINES.

SINE WAVE

A sine wave, also called a sinusoidal waveform, is an alternating-current disturbance with only one frequency. The harmonic content and the bandwidth are theoretically zero (*see* BANDWIDTH, HARMONIC). The sine wave is so named because the amplitude-versus-time function is identical to the trigonometric sine function. The cosine function also is a perfect representation of the shape of a sine wave (*see* COSINE).

When displayed on an oscilloscope, the sine wave has a characteristic shape as shown in the illustration. Each cycle is represented by one complete alternation as shown. The cycle of a sine wave can be divided into 360 electrical degrees. *See also* SINE.

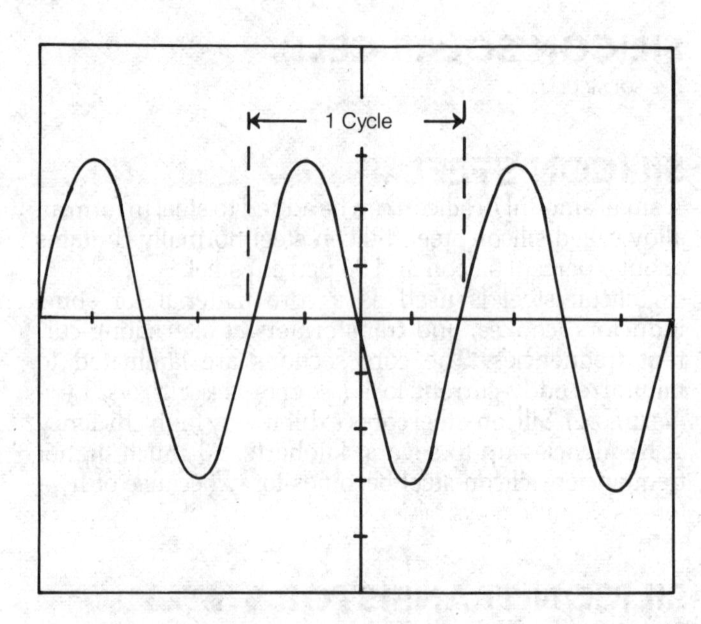

SINE WAVE: A sine wave as it would appear on an oscilloscope.

SINGLE BALANCED MIXER

A single balanced mixer is a mixer circuit that operates in a manner similar to a balanced modulator (*see* BALANCED MODULATOR, MIXER, MODULATOR). The input and output ports are not completely isolated in the single balanced mixer; some of either input signal leaks through to the output. If isolation is needed, the double balanced mixer is preferable (*see* DOUBLE BALANCED MIXER).

A typical single-balanced-mixer circuit is shown in the illustration. This circuit will work at frequencies up to several gigahertz. The circuit shown is a passive circuit,

SINE: Values of sin θ for values of θ between 0 and 90 degrees.
For values between 90 and 180 degrees, calculate 180 degrees − θ and read from this table.
For values between 180 and 270 degrees, calculate θ − 180 degrees, read from this table, and multiply by −1.
For values between 270 and 360 degrees, calculate 360 degrees − θ, read from this table, and multiply by −1.

θ, degrees	sin θ	θ, degrees	sin θ	θ, degrees	sin θ
0	0.000				
1	0.017	31	0.515	61	0.875
2	0.035	32	0.530	62	0.883
3	0.052	33	0.545	63	0.891
4	0.070	34	0.559	64	0.899
5	0.087	35	0.574	65	0.906
6	0.105	36	0.588	66	0.914
7	0.122	37	0.602	67	0.921
8	0.139	38	0.616	68	0.927
9	0.156	39	0.629	69	0.934
10	0.174	40	0.643	70	0.940
11	0.191	41	0.656	71	0.946
12	0.208	42	0.669	72	0.951
13	0.225	43	0.682	73	0.956
14	0.242	44	0.695	74	0.961
15	0.259	45	0.707	75	0.966
16	0.276	46	0.719	76	0.970
17	0.292	47	0.731	77	0.974
18	0.309	48	0.743	78	0.978
19	0.326	49	0.755	79	0.982
20	0.342	50	0.766	80	0.985
21	0.358	51	0.777	81	0.988
22	0.375	52	0.788	82	0.990
23	0.391	53	0.799	83	0.993
24	0.407	54	0.809	84	0.995
25	0.423	55	0.819	85	0.996
26	0.438	56	0.829	86	0.998
27	0.454	57	0.839	87	0.999
28	0.469	58	0.848	88	0.999
29	0.485	59	0.857	89	0.999
30	0.500	60	0.866	90	1.000

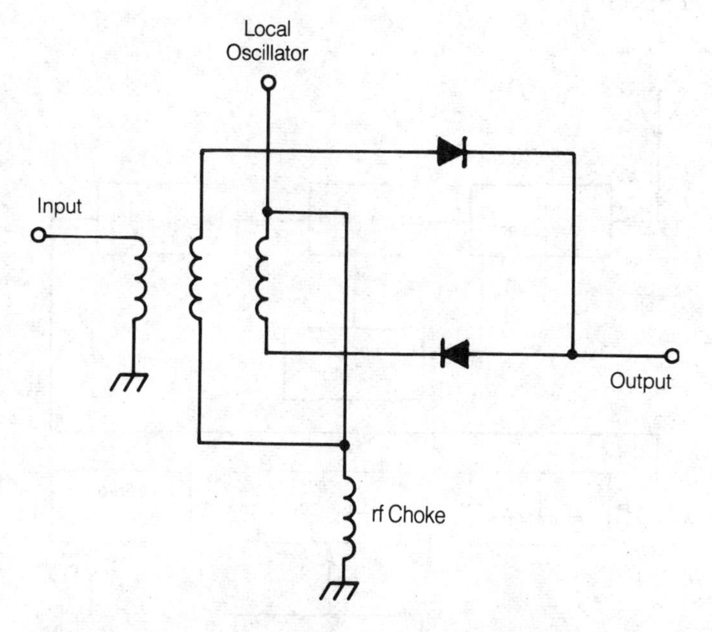

SINGLE BALANCED MIXER: A single balanced mixer using diodes.

and therefore some loss will occur. However, the loss can be overcome by means of an amplifier following the mixer.

SINGLE-CONVERSION RECEIVER

A single-conversion receiver is a form of superheterodyne receiver that has one intermediate frequency. The incoming signal is heterodyned to a fixed frequency. Selective circuits are used to provide discrimination against unwanted signals. The output from the mixer is amplified and fed directly to the detector.

The main advantage of the single-conversion receiver is its simplicity. However, the intermediate frequency must be rather high—on the order of several megahertz for a typical high-frequency communications receiver—and this limits the degree of selectivity that can be obtained. In recent years, excellent ceramic and crystal-lattice filters have become available for improving the performance of single-conversion receivers. *See also* DOUBLE-CONVERSION RECEIVER, INTERMEDIATE FREQUENCY, MIXER, SUPERHETERODYNE RADIO RECEIVER.

SINGLE-ENDED CIRCUIT

See UNBALANCED SYSTEM.

SINGLE IN-LINE PACKAGE

The single in-line package (SIP) is a form of packaging for

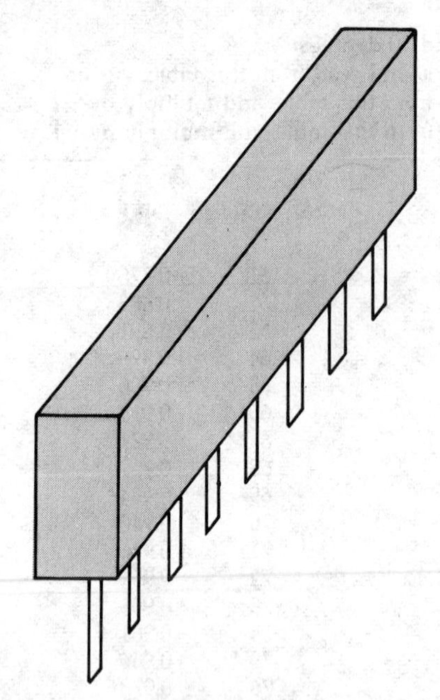

SINGLE IN-LINE PACKAGE: All of the pins are along one edge.

hybrid integrated circuits, resistor networks, resistor-capacitor networks, liquid-crystal displays and many other active and passive components. The advantage of the package is conservation of circuit board space. Leads project from only one edge (see illustration). SIPs can be protected by molding or dipping them in a resin. *See also* SEMICONDUCTOR PACKAGE.

SINGLE-PHASE ALTERNATING CURRENT

In the United States, electric utility line power is single-phase 120-V alternating current. The voltage has a root-mean-square (rms) amplitude of about 117 V.

SINGLE SIDEBAND

Single sideband is a form of amplitude modulation. An ordinary amplitude-modulated (AM) signal consists of a carrier and two sidebands, one above the carrier frequency and one below the carrier frquency (*see* AMPLITUDE MODULATION). A single-sideband signal results from the removal of the carrier and one of the sidebands.

Single-sideband, suppressed-carrier emission, also called A3J emission, provides greater communications efficiency than ordinary amplitude modulation, or type A3 emission. This is because ⅔ (67 percent) of the power in an AM signal is taken up by the carrier wave, which conveys no intelligence. In a single-sideband signal, all of the power is concentrated into one sideband, and this yields an improvement of about 8 decibels over A3 emission. A single-sideband signal has a bandwidth of approximately half that required for amplitude modulation. Therefore, it is possible to get twice as many A3J signals as A3 signals into a given amount of spectrum space.

To obtain A3J emission, a balanced modulator followed by a filter or phasing circuit must be used. The balanced modulator produces a double-sideband signal with a suppressed carrier (*see* BALANCED MODULATOR, DOUBLE SIDEBAND). The filter or phasing network then removes either the lower sideband or the upper sideband. If the lower sideband is removed, the resulting A3J signal is called an upper-sideband (USB) signal. If the upper sideband is removed, a lower-sideband (LSB) signal results (*see* LOWER SIDEBAND, UPPER SIDEBAND).

Most voice communication at the low, medium, and high frequencies is carried out as single sideband. Generally, the lower sideband is used at frequencies below about 10 MHz; the upper sideband is preferred at frequencies higher than 10 MHz. This is convention; either sideband will provide equally good communication at a particular frequency. *See* SINGLE SIDEBAND RECEIVER.

SINGLE-SIDEBAND RECEIVER

A single-sideband receiver is a form of superheterodyne receiver containing a carrier reinsertion oscillator that is connected to the amplitude-modulation (AM) detector to restore the original modulating signal. The conventional superheterodyne receiver will not detect single sideband transmission because the signal lacks a carrier. The carrier reinsertion oscillator compensates for the loss of the radio-frequency (rf) carrier that was suppressed at the transmitter after it produced the sideband signal.

The figure is a block diagram of a single-sideband receiver. The input sideband signal is amplified by an rf amplifier, and converted to an intermediate-frequency (i-f) signal by a mixer and the local oscillator as in a conventional superheterodyne receiver. Then the i-f output of the mixer is passed through a selective i-f filter to a series of i-f amplifiers. This i-f filter sharply blocks the passage of adjacent rf signals that are close to the i-f

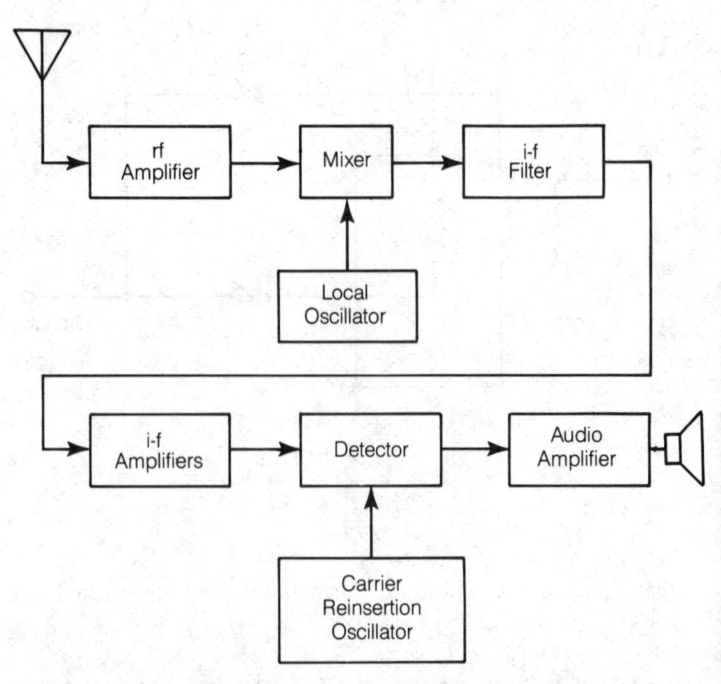

SINGLE-SIDEBAND RECEIVER: The reinserted carrier is heterodyned with the intermediate-frequency sideband signal to restore the original modulated AM signal at the transmitter.

signal from reaching the detector. After the signal is amplified by the i-f amplifiers, it is sent to the sideband detector.

The detector also receives the carrier signal from the carrier reinsertion oscillator for reinsertion. The output of this oscillator is at a frequency that produces an AM signal with the same envelope variations as the modulating signal at the transmitter when it is heterodyned with the i-f sideband signal. In effect, the detector heterodynes the sideband signal with the oscillator output, then detects the envelope variations of the resulting signal. The output of the detector is an audio signal that corresponds to the intelligence carried by the sideband signal. The detected signal is amplified and sent to the loudspeaker. *See* SINGLE SIDEBAND, SUPERHETERODYNE RADIO RECEIVER.

SKIN EFFECT

In a solid wire conductor, direct current flows uniformly along the length of the wire. The electrons pass uniformly through a given cross section of the wire. The same holds for alternating currents at relatively low frequencies. The conductivity of the wire is proportional to the cross-sectional area, which is in turn proportional to the square of the diameter.

At radio frequencies, the conduction in a solid wire becomes nonuniform. Most of the current tends to flow near the outer surface of the wire. This is called skin effect. It increases the effective ohmic resistance of a wire at radio frequencies. The higher the frequency becomes, the more pronounced the skin effect will be. At high and very high frequencies, the conductivity of a wire is more nearly proportional to the diameter than to the crosssectional area. *See also* LITZ WIRE.

SKIP

Skip is the tendency for signals to pass over a certain geographical region. At high frequencies, skip is sometimes observed. This effect is shown in the skip-zone illustration. A transmitting station *X* is heard by station *Z*, located thousands of miles distant, but not by station *Y*. The ionization of the F layer is insufficient to bend the signals back to earth at the sharper angle necessary to allow reception by station *Y*. *See also* PROPAGATION CHARACTERISTICS, SKIP ZONE.

SKIP ZONE

When skip occurs in ionospheric F-layer communication, the signals from a transmitting station cannot be received by other stations located within a specific geographical area. This dead area is called the skip zone.

Under most conditions, the skip zone begins at a distance of about 10 to 15 miles from the transmitting station, or the limit of the range provided by the direct wave, the reflected waves, and the surface wave (*see* DIRECT WAVE, REFLECTED WAVE, SURFACE WAVE). The outer limit of the skip zone varies considerably, depending on the operating frequency, the time of day, the season of the year, the level of sunspot activity, and the direction in which transmission is attempted. An example is shown in the illustration.

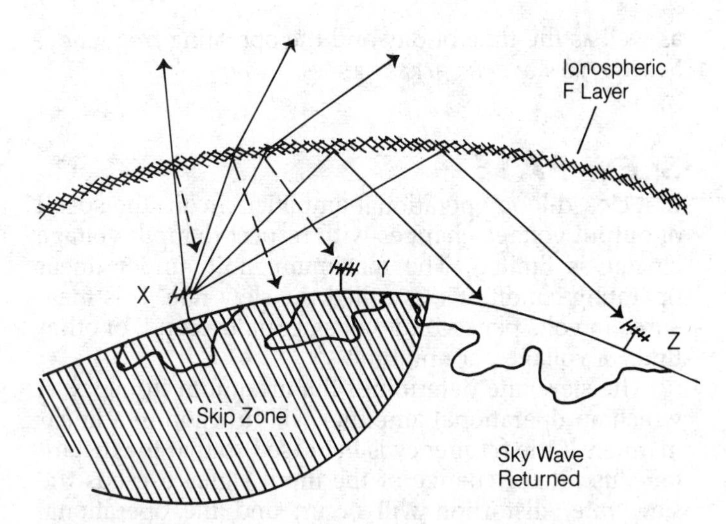

SKIP ZONE: Ionospheric signals are not returned to earth within the skip zone.

At the very low, low, and medium frequencies, a skip zone is never observed. In the high-frequency spectrum, however, a skip zone is often present. In the upper part of the high-frequency band, a skip zone is almost always observed.

At times, certain conditions arise that allow signals to be heard from points within the skip zone. Densely ionized areas may form in the E layer, causing propagation over shorter paths than would normally be expected (*see* E LAYER, SPORADIC-E PROPAGATION). Auroral propagation and backscatter sometimes allow communication between stations that would otherwise be isolated by the skip zone (*see* AURORAL PROPAGATION, BACKSCATTER). Above about 20 MHz, tropospheric effects can partially fill in the skip zone (*see* TROPOSPHERIC PROPAGATION).

If the frequency of operation is increased, the skip zone widens. The outer limit of the skip zone might be several thousand miles away. The same widening effect takes place at a constant frequency as darkness falls. At frequencies above a certain maximum, the outer limit of the skip zone disappears entirely, and no F-layer propagation is observed. *See also* F LAYER, IONOSPHERE, PROPAGATION CHARACTERISTICS, SKIP.

SKY WAVE

A radio frequency signal is called a sky wave if it has been returned to the earth by the ionosphere. Sky-wave propagation can be caused by either the E layer or the F layer of the ionosphere. Sky-wave propagation can also result from reflection from auroral curtains. (*See* AURORAL PROPAGATION, E LAYER, F LAYER, SPORADIC-E PROPAGATION.)

Sky waves are observed at various times of day at different frequencies. This form of propagation is responsible for most of the long-distance communication that takes place on the shortwave bands. This propagation occurs almost every night on the standard amplitude-modulation broadcast band. Sky waves are affected by the season of the year and the level of sunspot activity, as well as the time of day and the operating frequency. *See also* PROPAGATION CHARACTERISTICS.

SLEW RATE

In a closed-loop operational-amplifier circuit, the speed of output voltage changes with respect to input voltage change is limited. The maximum limit, under linear operating conditions, is called the slew rate. It is measured in volts per second, volts per millisecond, or other units of voltage per unit time.

The slew rate determines the maximum frquency at which an operational amplifier will function in a linear manner. If the frequency is increased so that the instantaneous rate of change of the input signal exceeds the slew rate, distortion will occur, and the operational amplifier will not perform according to the specifications. *See also* OPERATIONAL AMPLIFIER.

SLOPE DETECTION

A narrow-band frequency-modulated (FM) or phase-modulated (PM) signal can be demodulated with a receiver designed for amplitude modulation (AM). This is done with a technique called slope detection.

A standard FM and PM communications signal has a deviation of plus-or-minus 5 kHz. If an AM receiver with its typical passband of plus-or-minus 3 to 5 kHz is tuned slightly above or below the carrier frequency of the FM or PM signal, the signal will move in and out of the receiver passband with each cycle of modulation. The receiver should be tuned approximately 5 kHz above or below the FM or PM carrier for optimum results (see illustration).

Slope detection provides excellent sensitivity for reception of FM or PM signals. However, some distortion is likely to occur because the selectivity curves of most AM receivers do not have the gradual, uniform rolloff that would be necessary for distortion-free slope detection.

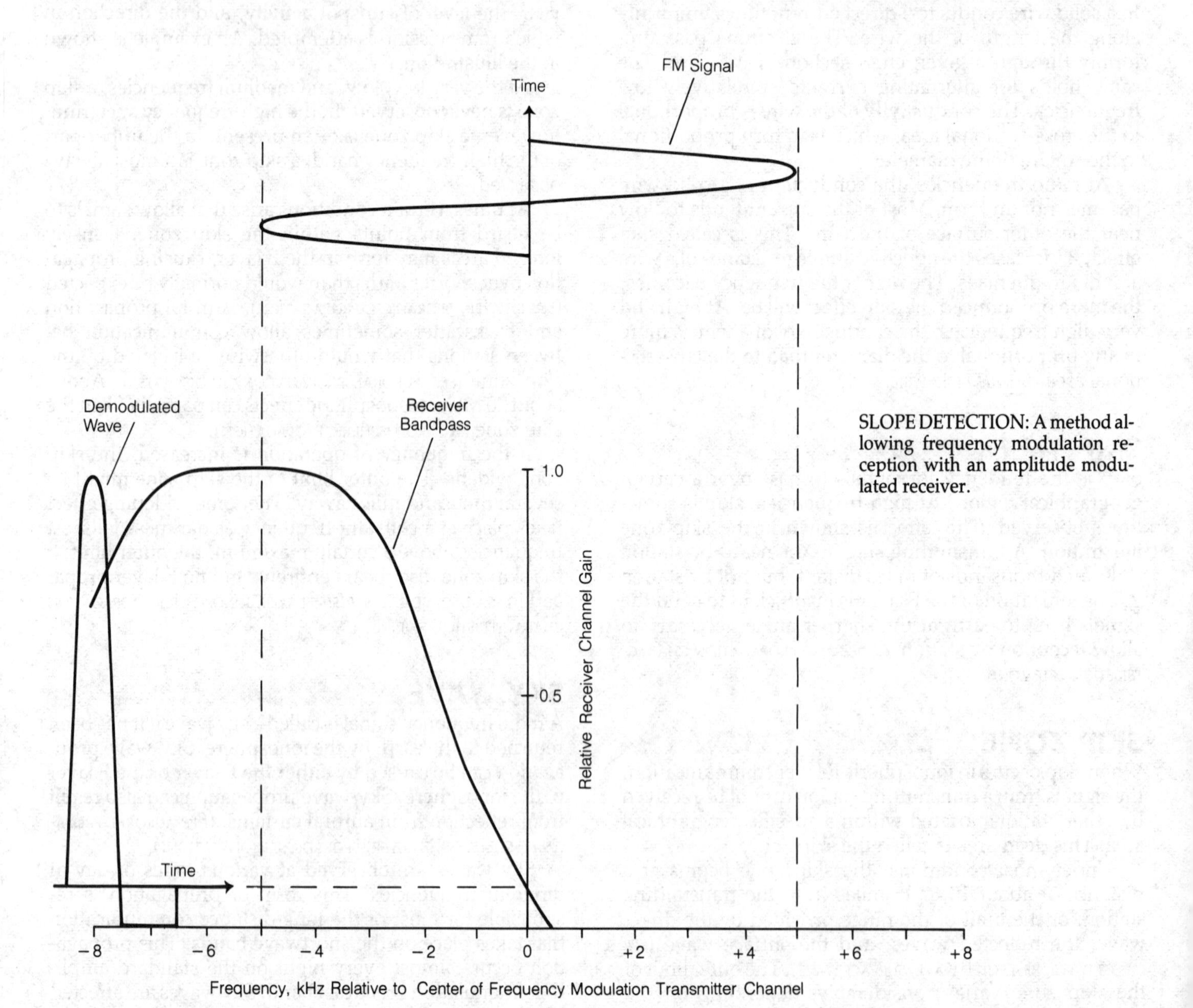

SLOPE DETECTION: A method allowing frequency modulation reception with an amplitude modulated receiver.

Also, the signal will have an exaggerated treble sound unless a deemphasis circuit is used in the audio stages of the receiver—a feature that AM equipment does not generally have. *See also* DEEMPHASIS, DETECTION, FREQUENCY MODULATION, PHASE MODULATION, PREEMPHASIS.

SMITH CHART

The Smith chart is a circular coordinate system used for plotting complex impedances (*see* IMPEDANCE). Smith charts are especially useful for determining the resistance and reactance at the input end of a transmission line when the resistance and reactance at the antenna feed point are known. The Smith chart is named after the engineer P.H. Smith, who developed it.

The resistance coordinates on the Smith chart are eccentric circles, mutually tangent at the bottom of the circular graph. The reactance coordinates are partial circles with variable diameter and centering. Resistance and reactance values can be assigned to the circles in any desired magnitude, depending on the characteristic impedance of the transmission line. The illustration is an example of a Smith chart intended for analysis of feed systems in which the characteristic impedance of the line is 50 Ω ($Z_o = 50$).

Complex impedances appear as points on the Smith chart. Pure resistances (impedances having the form $R + j0$, where R is a nonnegative real number) lie along the resistance line; the top of the line represents a short circuit and the bottom represents an open circuit. Pure reactances (having the form $0+jX$, where X is any real number except 0) lie on the perimeter of the circle, with inductance on the right and capacitance on the left. Impedances of the form $R + jX$, containing finite, nonzero resistances and reactances, correspond to points within the circle. Several different complex impedance points are illustrated.

The Smith chart can be used to determine the standing-wave ratio (SWR) on a transmission line if the characteristic impedance of the line and the complex antenna impedance are known (*see* STANDING-WAVE RATIO). For determination of SWR, a set of concentric circles is added to the Smith chart. These circles are called SWR circles.

The center point of the chart corresponds to an SWR of 1:1. Higher values of SWR are represented by progressively larger circles. The radii of the SWR circles on a

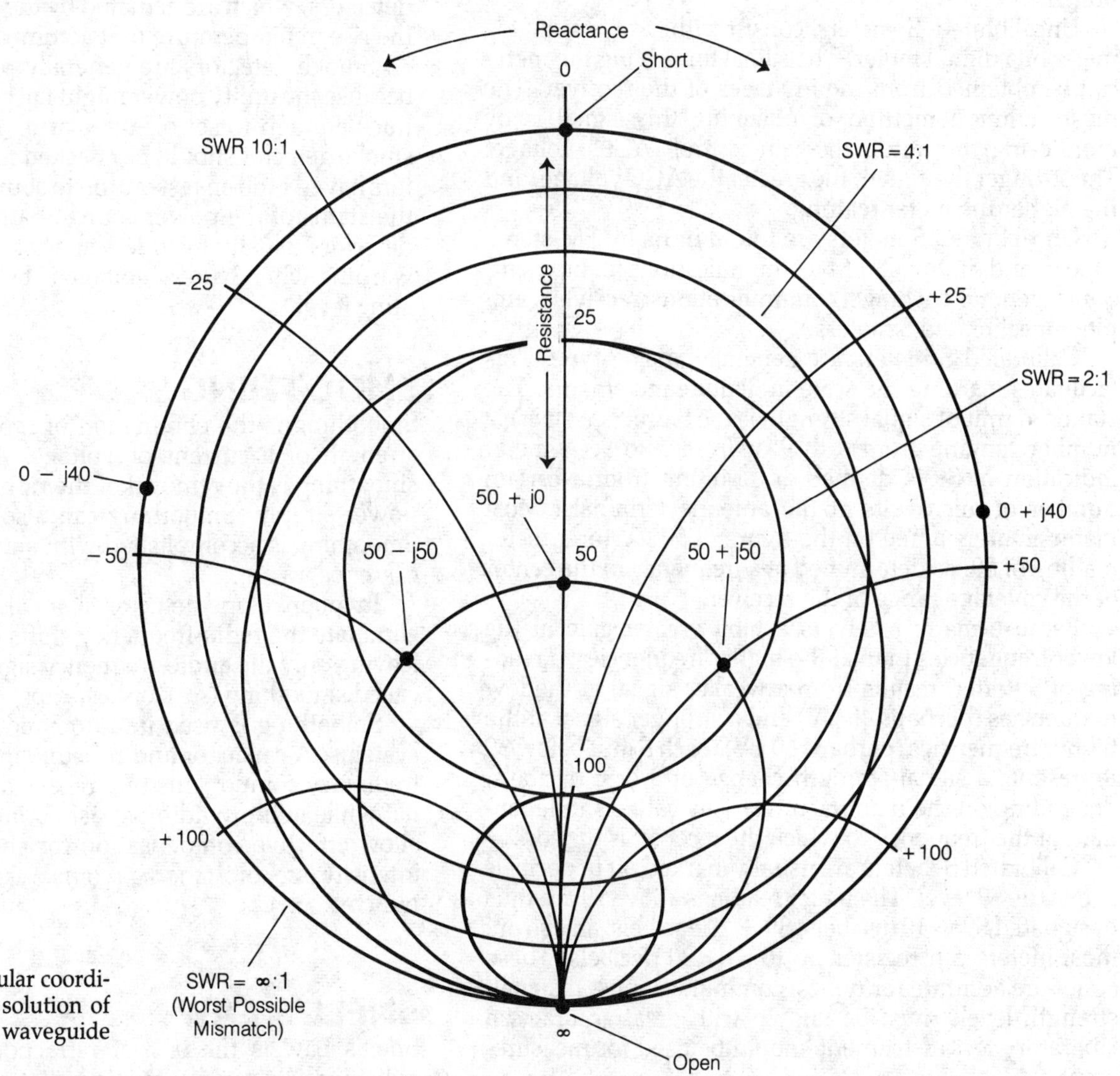

SMITH CHART: A circular coordinate chart used in the solution of transmission-line and waveguide problems.

given Smith chart can be determined according to points on the resistance line. In the example, the 2:1 SWR circle passes through the 100 Ω point on the resistance line; the 4:1 SWR circle passes through the 200 Ω point on the resistance line, and so on. In general, for any SWR value x:1, the x:1 SWR circle passes through the point on the resistance circle corresponding to 50x ohms.

The Smith chart illustrates why an SWR of 1:1 can be obtained only if the impedance of the load is a pure resistance equal to the characteristic impedance of the feed line. The SWR cannot be 1:1 if reactance exists in the load. A given SWR greater than 1:1 can occur in infinitely many ways, corresponding to the infinite number of points on a circle. An SWR of infinity exists when the load is a short circuit, an open circuit, a pure capacitive reactance, or a pure inductive reactance.

S METER

An S meter is an instrument in a radio receiver that indicates the relative or absolute amplitude of an incoming signal. Many receivers are equipped with these meters. There are many different styles of S meter, but they can be categorized as either uncalibrated or calibrated.

Uncalibrated S meters consist either of an analog meter or a digital meter. The signal for driving the meter can be obtained from the i-f stages of the receiver. The most common method of obtaining this signal is by monitoring the automatic-gain-control (AGC) voltage. The stronger the signal, the greater the AGC voltage, and the higher the meter reading.

Uncalibrated S meters are found in many FM stereo tuners and in most FM communications equipment. Some general-coverage communications receivers employ uncalibrated S meters.

Calibrated S meters are generally analog type. This facilitates marking the scale in definite increments. The standard unit of signal strength is the S unit (*see* S UNIT), a number ranging from 0 to 9 or from 1 to 9. A meter indication of S9 is defined as resulting from a certain number of microvolts at the antenna terminals. Most manufacturers agree on the figure of 50 μV for a meter reading of S9, as determined at a frequency in the center of the coverage range of the receiver.

Because many receivers exhibit greater gain at the lower frequencies than at the higher frequencies, a reading of S9 often results from a weaker signal at the low frequencies (perhaps 40 μV) and a stronger signal at the higher frequencies (perhaps 60 μV). Each S unit below S9 represents a signal-strength change of 3 or 6 decibels, depending on the manufacturer. This value is independent of the frequency to which the receiver is tuned.

Calibrated S meters are usually marked off in decibels above the S9 level. The meter readings above S9 are thus designated S9 + 10 decibels, S9 + 20 decibels, and so on; most meters can register up to S9 + 30 decibels. These scales are accurate for typical communications. If signal-strength levels must be known with great accuracy, a laboratory test instrument should be used for measurement.

A receiver S meter provides an indication of how strong an incoming signal is. All signal-strength values are subjective, however, because band conditions might vary. A 50 μV signal might be masked by static at 1.8 MHz, while the same signal stands out as that of a local station at 28 MHz. While many radio operators define signal strength solely on the basis of the S-meter indication, others use their own judgment, taking the meter reading into account as a subjective quantity. Relative signal-strength information is exchanged among radio operators by means of an S (strength) number ranging from 1 to 9.

SMOKE DETECTOR

A smoke detector is an electronic instrument that produces a loud buzz or tone when it is exposed to smoke. This system protects people against fires and asphyxiation by smoke in homes and other enclosed spaces.

Smoke detectors are based on several different principles. Some are sensitive to the dielectric properties of the air which change when smoke is present. Others use optical devices, sensitive to the light transmittivity of the air. Smoke detectors should not be confused with heat detectors, which are actuated by the infrared radiation or the rise in temperature that accompanies a fire.

Smoke detectors are generally powered by dry cells because the utility power might fail before the smoke gets thick enough to set off the alarm. The cell or cells in a smoke detector should be checked regularly; most detectors have a built-in test button to sound the alarm and test the status of the power source. Battery-powered smoke detectors usually include a special circuit that sounds a warning when battery voltage is below its normal operating level.

SMOOTHING

Smoothing is the elimination of rapid variations in the strength of a current or voltage. A good example of smoothing is the removal of the ripple in the output of a power supply. Smoothing can also be called filtering. Smoothing is accomplished with a capacitor (*see* FILTER CAPACITOR, RIPPLE).

In an envelope detector, a small capacitor is used to eliminate the radio-frequency shifts of the carrier wave, leaving only the audio-frequency signals. This process is called smoothing (*see* DIODE DETECTOR).

Smoothing is used in automatic-gain-control (AGC) systems. A capacitor and resistor smooth out the audio-frequency components of a received or transmitted signal while still providing a fast enough time constant to allow effective compensation for changes in the signal intensity (*see* AUTOMATIC GAIN CONTROL, RESISTANCE-CAPACITANCE TIME CONSTANT).

SNELL'S LAW

Snell's Law is the law of refraction that predicts the behavior of electromagnetic radiation as it passes from

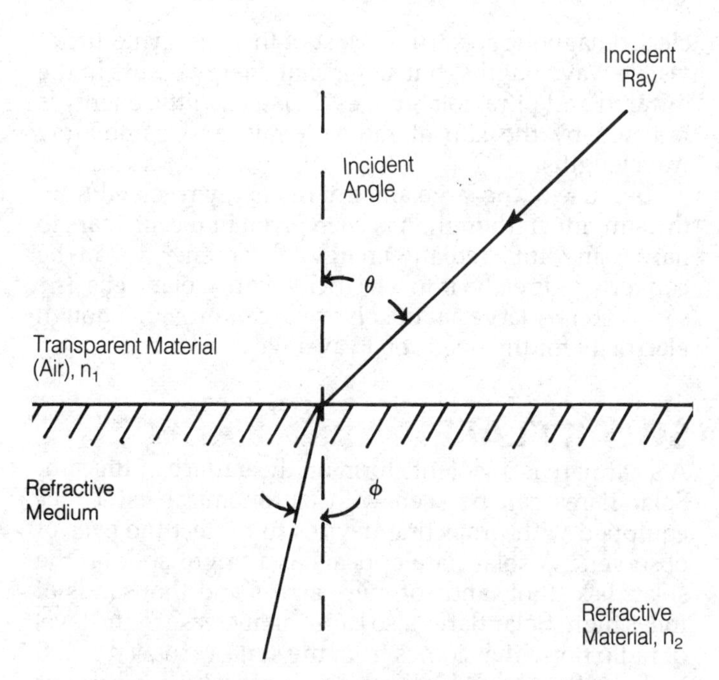

SNELL'S LAW: A light beam is refracted at the boundary between two media with different indices of refraction.

one homogeneous media to another, expressed as:

$$n_1 \sin\theta = n_2 \sin\phi$$

when $n_2 > n_1$,

$$\sin \phi = n_1 \sin\theta / n_2$$

where n_1 and n_2 are refractive indexes and θ and ϕ refer to the angles between the rays and the normal to the interface (see illustration). *See also* FIBEROPTIC CABLE AND CONNECTORS, INDEX OF REFRACTION.

SOCKET

A socket is an interconnection component that permits the easy insertion and removal of a leaded device from a circuit board without unsoldering connections. Sockets are made for semiconductor integrated circuits and discrete devices as well as passive devices such as relays, crystals and displays. Sockets are used where there is a high probability that the device will be replaced during the life of the host equipment or to avoid the thermal and mechanical stress in soldering a component directly to the circuit board. This is particularly applicable to devices with 40 or more pins. Sockets are inserted and soldered permanently in the host circuit board.

Sockets are generally made as insulated blocks or bodies with conductive contacts molded inside. The contacts interface between the input/output (I/O) pins of the device to be mounted and the circuit board. Socket contacts are designed to provide a secure electrical connection between the device and the circuit board. A clamping action on the inserted I/O pins is usually achieved with individual contacts stamped from metal spring stock. Alternatively, clamping can be achieved

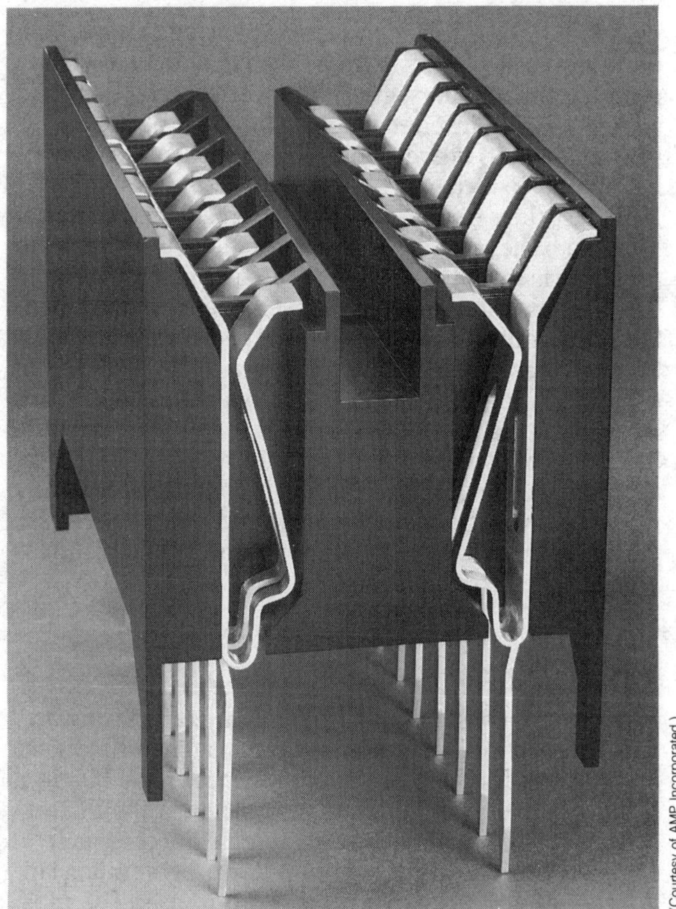

(Courtesy of AMP Incorporated.)

SOCKET: Sectional view of a 24-pin dual-in-line socket showing pin and contact arrangement.

with so-called zero-insertion force (ZIF) sockets. A lever-rows of contacts for easy insertion of the leaded part without external force. When the lever is switched, the rows of pins are securely clamped.

Sockets are made as commercial products for dual-in-line (DIP), plastic leaded chip carrier (PLCC), and pin-grid array (PGA) packaged devices. Although most sockets are made as commodity components for permanent use in an end product, some sockets are specifically made for test, burn-in and programming equipment. Many are made to withstand above-normal temperatures. Test sockets are used in equipment for programming permanently programmed as well as erasable read-only memories and logic devices. *See* PROGRAMMABLE LOGIC DEVICE, SEMICONDUCTOR MEMORY.

Dip sockets are made in the largest quantities and are used to mount DIP packages with from six to 64 I/O pins. The figure illustrates an example. (*See* SEMICONDUCTOR PACKAGE.) Socket pin and row spacing matches that of the device to be mounted: standard center-to-center spacing of pins is 0.10 inch, but distance between parallel rows of pins varies from 0.3 to 0.6 inch.

Most production DIP sockets are molded from glass-filled polyester, glass-filled nylon, and polyphenelene sulfide. If these thermoplastic resins are compounded to meet the Underwriters Laboratories standard 94V-O for flame retardance, the sockets will meet safety standards.

They will also be easier and less expensive to mold than the thermoset plastics such as diallyl phthalate widely used to make military-grade and test sockets. Standard sockets have an inserted height of 0.25 inch above the PC (printed-circuit) board, while low-profile sockets have a height of 0.16 inch.

DIP sockets can have stamped or machined contacts. The two basic styles of stamped contact are single beam (or single leaf) and dual beam (or dual leaf), a reference to the way the internal contacts are bent and formed. Stamped contacts are usually stamped from phosphor-bronze, brass, or beryllium copper. A wide choice of contact plating and thickness is available: nickel, tin, lead-tin, and gold. Two-piece machined socket contacts have internal springs that grip the inserted DIP pin in as many as four places. The inner contacts are usually machined from beryllium copper, and outer sleeves are usually machined from phosphor bronze. Gold plating is widely specified as a finish.

SOFTWARE

In a computer system, the programs are called software. Software can exist in written form, as magnetic impulses on tapes or disks, or as electrical or magnetic bits in a computer memory. Software also includes the instructions that tell personnel how to operate the computer.

There are many types of computer-programming languages, each with its own special purpose. The most basic form of software language is called machine language. It consists of the actual binary information used by the electronic components of the computer.

Software can be programmed temporarily into a memory, or it can be programmed permanently by various means. When software is not alterable (that is, it is programmed permanently), it is called firmware. *See also* ASSEMBLER AND ASSEMBLY LANGUAGE, COMPUTER, COMPUTER PROGRAMMING, FIRMWARE, HIGHER-ORDER LANGUAGE, MACHINE LANGUAGE, MICROPROCESSOR, SEMICONDUCTOR MEMORY.

SOLAR CELL

A solar cell or solar energy converter is a large photovoltaic photodiode designed to function as a power source. Low-current solar cells used as sensors or to power light loads such as pocket calculators can be packaged in metal cases with glass windows or in transparent plastic packages. However, those cells intended as solar energy converters require larger surface areas to provide maximum current capacity. The figure shows the construction of a typical power solar cell used as an energy converter.

The surface layer of P-type material is thin so that light can penetrate to the junction. Solar power cells are also available in flat strip form for efficient coverage of available surface areas. The circuit symbol for the photovoltaic cell is also shown in the figure. *See* PHOTOCELL.

SOLAR ENERGY

The sun radiates a large amount of energy across the electromagnetic spectrum. Most of this energy occurs at visible wavelengths, but significant energy occurs in the infrared and ultraviolet ranges. Relatively little energy is radiated by the sun at radio, X ray, and gamma-ray wavelengths.

Because of the large amount of energy received from the sun, much thought has been given in recent years to harnessing this radiant energy. Solar energy can be converted directly into electricity with solar cells (*see* SOLAR CELL). A large set of solar cells can provide enough electricity for the needs of an average residence.

SOLAR FLARE

A solar flare is a violent storm on the surface of the sun. Solar flares can be seen with astronomical telescopes equipped with projecting devices to protect the eyes of observers. A solar flare appears as a bright spot on the solar disk, thousands of miles across and thousands of miles high. Solar flares also cause an increase in the level of radio noise that comes from the sun (*see* SOLAR FLUX).

Solar flares emit large quantities of high-speed atomic particles. These particles travel through space and arrive at the earth a few hours after the occurrence of the flare. Since the particles are charged, they are attracted toward the geomagnetic poles. Sometimes a geomagnetic disturbance results (*see* GEOMAGNETIC FIELD, GEOMAGNETIC STORM). An aurora can be seen at night and there is a sudden, dramatic deterioration of ionospheric radiopropagation conditions. At some frequencies, communications can be completely cut off. Wire communications circuits and power lines can be affected.

Solar flares can occur at any time, but they take place most often near the peak of the 11-year sunspot cycle.

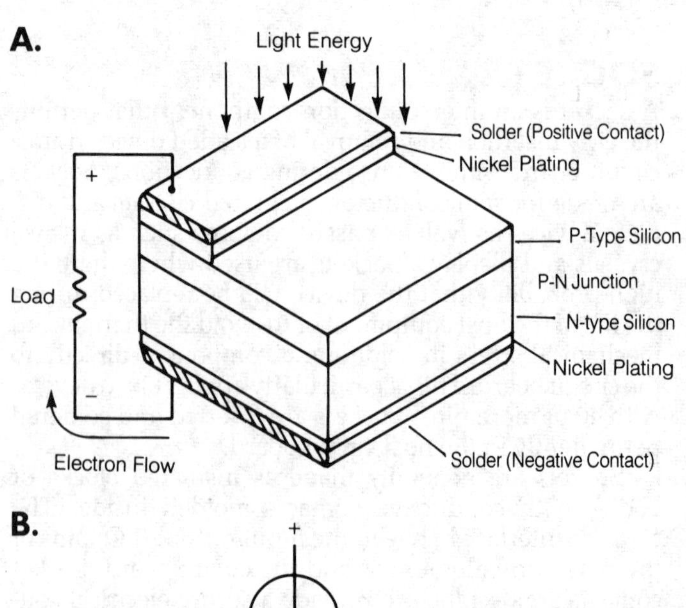

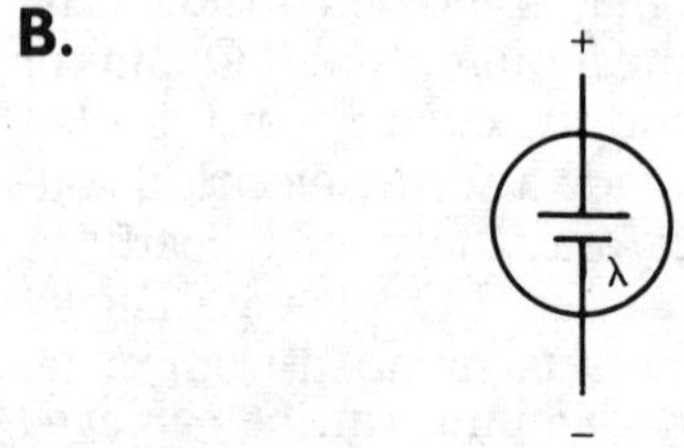

SOLAR CELL: Sectional view of a solar cell (A) and its symbol (B).

Scientists do not know exactly what causes solar flares, but they are evidently associated with sunspots, another type of solar storm. *See also* PROPAGATION CHARACTERISTICS, SUNSPOT, SUNSPOT CYCLE.

SOLAR FLUX

The amount of radio noise emitted by the sun is called the solar radio-noise flux, or simply the solar flux. The solar flux varies with frequency. However, at any frequency, the level of solar flux increases abruptly when a solar flare occurs (*see* SOLAR FLARE). This makes the solar flux useful for propagation forecasting: A sudden increase in the solar flux indicates that ionospheric propagation conditions will deteriorate within a few hours.

The solar flux is most often monitored at a wavelength of 10.7 centimeters, or a frequency of 2800 MHz. At this frequency, the troposphere and ionosphere have no effect on radio waves, making observation easy.

The 2800-MHz solar flux is correlated with the 11-year sunspot cycle. On the average, the solar flux is higher near the peak of the sunspot cycle, and lower near a sunspot minimum. *See also* SUNSPOT, SUNSPOT CYCLE.

SOLDER

Solder is a metal alloy used for making electrical connections between conductors in electronic applications. Lead-tin alloys are the most widely used solders for the manufacture and repair of electronics. The solder is typically in the form of wire with a central core of rosin flux.

The most common variety of solder consists of tin and lead, with a rosin core. Some types of solder have an acid core. In electronics rosin-core solder should be used; acid-core solder will result in rapid corrosion. Acid-core solder is more commonly used for bonding sheet metal.

The ratio of tin to lead in a rosin-core solder determines the temperature at which the solder will melt. In general, the higher the ratio of tin to lead, the lower the melting temperature. For general soldering purposes, the 60:40 solder is used. For heat-sensitive components, 63:37 solder is better because it melts at a lower temperature. An ordinary soldering gun or iron can be used for applying and removing all types of tin-lead solder. *See* DESOLDERING, SOLDERING GUN, SOLDERING IRON.

SOLDER: Characteristics of various types of solder

Solder Type	Melting Point °F/°C	Principal Uses
Tin-lead 60:40	370/190	Electronic Circuits
Tin-lead 63:37	361/183	Electronic Circuits Low-heat
Tin-lead 50:50	430/220	Non-electronic metal bonding
Silver	600/320	High-current, High-heat

Tin-lead solders are suitable for use with most metals except aluminum. For soldering to aluminum, a special form of solder, called aluminum solder, is available. It melts at a much higher temperature than tin-lead solder, and requires the use of a gas torch.

In high-current applications, a special form of solder, silver solder, is used. It can withstand higher temperatures than tin-lead solder. A gas torch is usually necessary for applying silver solder.

The table lists the most common types of solder, and their principal characteristics and applications. *See also* SURFACE-MOUNT TECHNOLOGY.

SOLDERING GUN

A soldering gun is a quick-heating soldering tool. It is called a gun because of its shape, which resembles a handgun (see photograph on p. 782). A trigger-operated switch is pressed with the finger, allowing the element to heat up within a few seconds.

Soldering guns are convenient in the assembly and repair of electronic equipment. Soldering guns are available in various wattage ratings for different electronic applications. The unit shown has two power levels: 100 W, for electronic connections and components; and 140 W, for such purposes as wire splicing.

Some electronic engineers and technicians prefer soldering irons to soldering guns. Soldering irons are generally easier to use when soldering miniature components such as in handheld and compact radio transceivers. Soldering guns are most often used in the assembly and repair of large, point-to-point wired apparatus such as power amplifiers. *See also* SOLDERING IRON.

SOLDERING IRON

A soldering iron is a soldering tool used by engineers and technicians in the assembly and maintenance of electronic apparatus. The heating element is now copper.

A typical soldering iron consists of a heating element and a handle. The soldering iron requires from about 1 minute to 5 minutes to heat up after it is plugged in; the larger the iron, the longer the warm-up time. Soldering irons are available in a wide range of watt ratings. The smallest irons for electronics are rated at only a few watts, and are used to solder miniature and delicate components.

Some electronic engineers and technicians prefer soldering guns to soldering irons. Soldering guns are often more convenient in situations where moderately high heat is needed, and where point-to-point wiring is used. The soldering gun heats up and cools down very quickly. *See also* SOLDERING GUN.

SOLENOID

A solenoid is an electromagnetic actuator that translates electrical energy applied to a coil into linear mechanical motion from an armature or plunger. The basic solenoid consists of a helical coil of copper wire wound on a sleeve or bobbin with a ferrous metal plunger, a fixed pole

SOLDERING GUN: This tool can be used for point-to-point wiring and splicing.

piece, or backstop, and a plunger free to slide axially within the bobbin.

A magnetic field is created around the coil when it is energized by current passing through it. The magnetic field is concentrated within the core, and this field acts to attract the plunger to the backstop. A spring is located on a reduced outside diameter of the plunger within the coil bobbin. It compresses between the plunger shoulder and the backstop when the solenoid is energized and the plunger is pulled in. When the coil is de-energized, the spring expands to move the armature back to its nonenergized position.

There are two fundamental solenoid designs: pull-type and push-type. Both are shown in the figure, but a typical solenoid would have only one action. The push-type solenoid is so named because the force of the plunger pushes against the external load (end A) as it moves inward toward the backstop against spring pressure. The plunger has a smaller diameter push rod on end A that projects through an axial hole in the backstop. The pull-type plunger (end B) moves in the same way, but it pulls against the load as the plunger is drawn in against spring pressure. The pull-type end of the plunger might have provision for mechanical linkage with the load as shown by the clevis in the figure.

The force/stroke characteristic is an important sole-

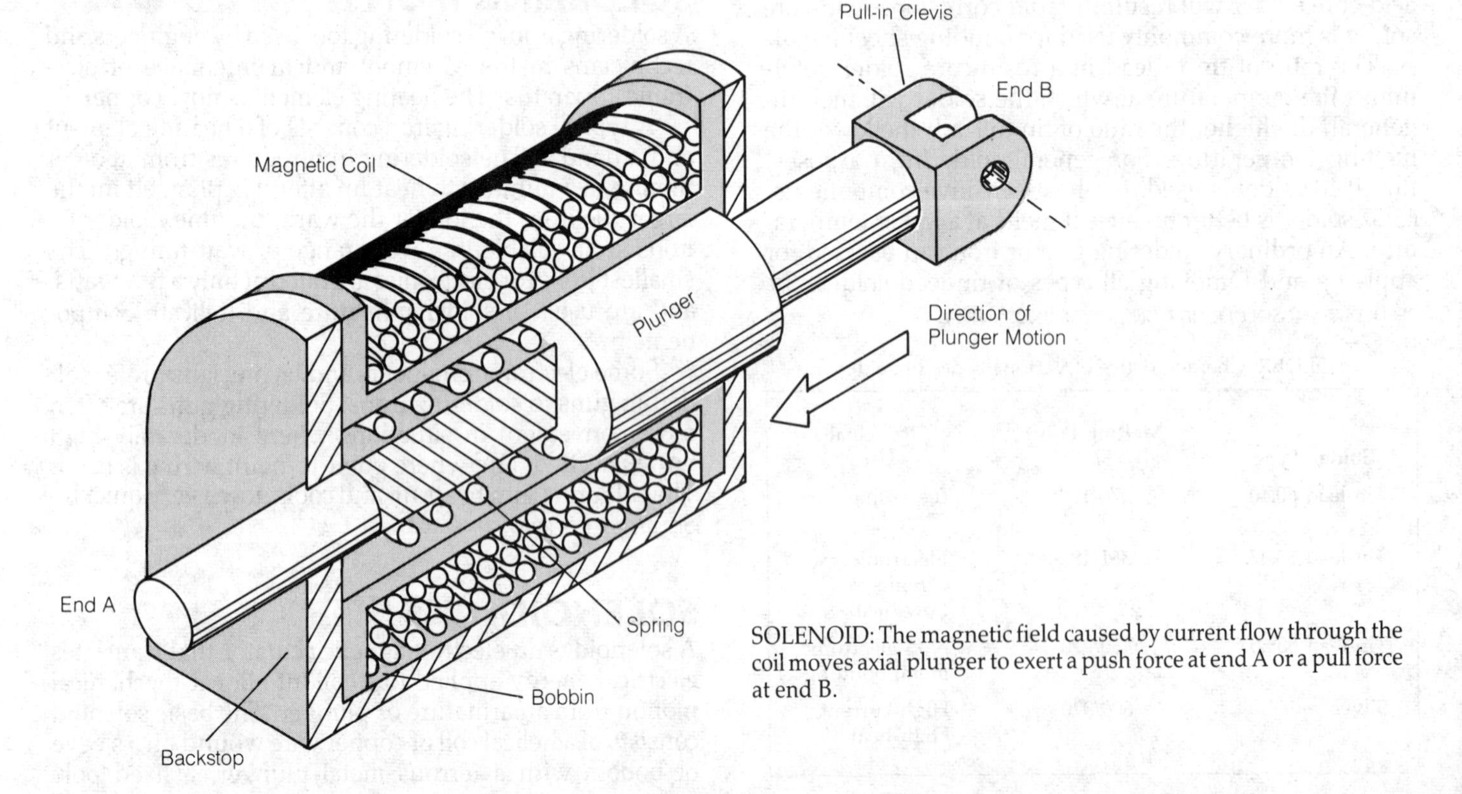

SOLENOID: The magnetic field caused by current flow through the coil moves axial plunger to exert a push force at end A or a pull force at end B.

noid specification. In most applications, the force can be minimum at the start of the plunger stroke, but must increase at a rapid rate to reach its maximum value before the plunger reaches the backstop. The magnetizing force of a solenoid is proportional to the current through the coil and the number of turns in the coil. The pull force required by the load must not be greater than the force developed by the solenoid during any portion of its required stroke. If this condition is not met, the solenoid will not pull in completely and the load will not be moved the required distance.

The current in an alternating-current (ac) powered solenoid, however, depends on the plunger position. At the starting point of the stroke, the force is greater than it is at the seated position. Although the force/stroke characteristic of the ac solenoid differs from that of the direct-current (dc) solenoid, the force developed by the solenoid must also exceed the force required by the load during the entire stroke or the ac coil will overheat and burn out.

Duty Cycle. The duty cycle of a solenoid is the proportion of time that the solenoid is energized as a percentage of its overall operating time. If the temperature of the coil is permitted to rise above the temperature rating of its insulation, performance will be degraded, and it might fail prematurely. Ambient temperature, power input, and duty cycle determine the operating temperature of a solenoid in its application.

The amount of work performed by a solenoid is directly related to its size. A large solenoid can develop more force at a given stroke than a small one with the same coil current because it has more turns of wire in its coil.

Open-Frame Solenoid. Open-frame or C-frame solenoids are widely used in appliances, in vending machines, and as a valve operators because they are made in volume at low cost. In spite of their low cost, they might have operational lives of millions of cycles. The coils, usually potted, are located in a steel frame formed in the shape of a C to complete the magnetic circuit through the core. Standard commercial open-frame units are capable of strokes up to ½ inch.

Box-Frame Solenoid. Box-frame or D-frame solenoids (so named because of their wrap-around frames) are specified for high-end applications such as tape-drive decks, industrial controls, tape recorders, and printers. They offer mechanical and electrical performance that is superior to those of open-frame solenoids. The frames of some open- and box-type solenoids are made from stacks of thin insulated sheets of iron to control eddy currents or to keep stray circulating currents confined in ac solenoids. Standard commercial box-frame units have strokes in excess of ½ inch.

Tubular Solenoid. Tubular solenoids are completely enclosed in cylindrical metal cases for improved magnetic circuit return and protection. These dc solenoids offer the highest volumetric efficiency of any design; they are specified for industrial and military/aerospace equipment where space is at a premium. They are used in printers, computer disk and tape drives, and military weapons systems. Both pull and push styles are available. Some commercial tubular solenoids may have strokes up to 1½ inches.

Rotary Solenoid. Rotary solenoids function the same way as linear solenoids except that the input electrical energy is converted to rotary motion. There are two basic rotary solenoid designs. The first has helical lands on the plunger and grooves on the inside walls of the bobbin. The plunger rotates as it is drawn in because the helical lands on the plunger twist in the grooves on the bobbin. Alternately, there can be ball bearing races on the plunger. In the second basic design, the armature is suspended inside the coil and rotated by the magnetic field. Rotary solenoids can act as limited-motion motors.

Specialized solenoids are used in the print heads of computer-peripheral dot-matrix printers to form alphanumeric characters. As many as 24 separate wires are controlled by solenoids to impact on the printer ribbon to print the characters. (*See* PRINTER.) Solenoids are also used in hydraulic circuit breakers to introduce time delay that prevents false tripping due to harmless transients. *See* CIRCUIT BREAKER.

SOLID-STATE RELAY

A solid-state relay (SSR) is an electronic circuit consisting of a signal-level trigger circuit coupled to a power semiconductor switch, either a transistor or thyristor. An SSR is widely understood to mean a factory-made and tested product rather than a solid state relay function built from separate components on a printed circuit (PC) board by the user.

The SSR differs significantly in both structure and operation from the coil-and-contact electromagnetic relay (EMR). However, both provide power gain. The input circuit of an SSR can be an optocoupler, reed relay or transformer. It is analogous to the coil of the EMR, and it is electrically isolated from the power semiconductor switch that acts as the contact. SSRs require only relatively low energy control circuits to switch the output power. *See* OPTOCOUPLER, REED RELAY, RELAY, TRANSFORMER.

SSRs for switching alternating current (ac) require either two inverse-parallel (back-to-back) silicon-controlled rectifiers (SCRs) or an electrically equivalent triac. However, if direct current (dc) is to be switched, a power bipolar transistor or MOSFET (metal-oxide semiconductor field-effect transistor) is used. *See* SILICON-CONTROLLED RECTIFIER, THYRISTOR, TRIAC.

The classification of an SSR is based on its input circuit or method of achieving input-output (I/O) isolation. True SSRs achieve electrical isolation between the input and output circuits with optocouplers. However, hybrid solid-state relays use reed relays or transformers for isolation. Figure 1 is a block diagram of a true solid-state relay.

Optocoupled ac SSRs. SSRs can be controlled by either ac or dc input signals applied to the terminals of an infrared emitting diode (IRED) matched to a photodetector to provide I/O isolation. The IR output of the IRED is detected by a matched photodetector (phototransistor, photodiode or photocell). The signal output from the photodetector triggers the output device to switch the

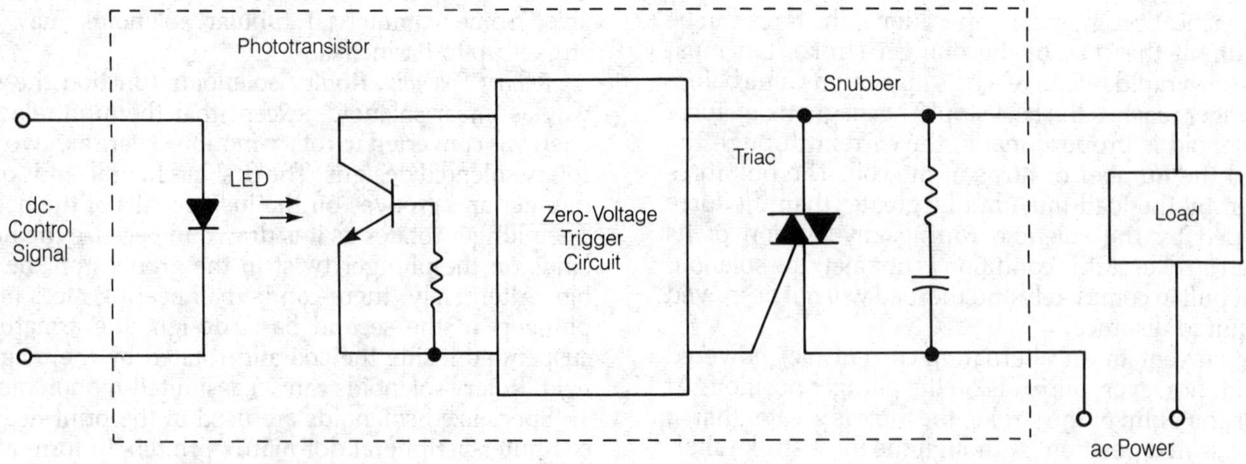

SOLID-STATE RELAY: Fig. 1. Diagram of an optically coupled solid-state relay providing zero-voltage triggering of a triac to switch the load. The snubber provides overload protection.

load current. *See* INFRARED-EMITTING DIODE, PHOTODETECTOR.

Reed-Relay Coupled ac SSRs. Reed-relay coupled SSRs are hybrid SSRs. The input control signal is applied to the coil of a reed switch which provides I/O isolation. The closure of the reed contacts in the switch actuates circuitry which, in turn, triggers the output device to switch the load current.

Transformer-Coupled ac SSRs. Transformer-coupled SSRs are also hybrid SSRs. The input control signal is applied to the primary winding of a small transformer, which provides isolation. The voltage from the transformer secondary winding triggers the output device to switch the load current.

Solid-state relays have the following advantages over electromechanical relays:

1. Longer life and higher reliability.
2. Better compatibility with logic-level circuits.
3. Higher-speed switching.
4. High resistance to shock and vibration.
5. Absence of mechanical contacts.

The absence of mechanical contacts eliminates: contact bounce, arcing due to contact opening, electromagnetic interference (EMI), and hazards due to the presence of explosive or flammable gases, liquids or solids.

Solid-state relays can be classified in five general groups follows:

1. Alternating-current power relays, capable of switching 24 to 530 V ac at 2 to 75 A with dc input (typically 3 to 32 V dc) or ac input (typically 90 to 280 V ac) using triacs or dual SCRs.
2. Direct-current power relays, capable of switching 100 to 500 V dc at 7 to 40 A under dc control with power transistors.
3. Alternating current low-power relays, for PC (printed-circuit) board mounting, capable of switching 60 to 240 V ac at 0.3 to 4 A with triacs.
4. Direct-current low-power relays, for PC-board mounting, capable of switching up to 60 V dc at 3 A with power transistors.
5. Input-output modules, specialized miniature low-power ac and dc relays for PC-board mounting, for use in interfacing computer systems with external sensors and actuators. *See* INPUT/OUTPUT MODULE.

The most popular ac power relays with ratings of 2 to 75 A are packaged in four-terminal flatpacks suitable for panel or heat-sink mounting. They have three functional sections as shown in Fig. 1: an optoisolator, a zero-voltage detector, and an output thyristor (back-to-back dual SCRs or triac).

The zero-voltage detector assures that the thyristor will be triggered only when the ac voltage crosses the zero reference (in either the negative or positive direction) to minimize the effect of surge currents at the time the thyristor is switched. Surge currents can result from switching tungsten-filament incandescent lamps and capacitive loads. For example, the cold resistance of a tungsten lamp is less than 10 percent of its illuminated resistance. If the SSR is turned on when the voltage is not at a zero crossing, the high instantaneous load current drawn by the lamp could destroy the SSR.

The thyristor, once triggered, will not stop conducting until the load current it is conducting falls to zero. A resistor and capacitor in series, called a snubber, bypasses voltage transients that occur with inductive loads when current and voltage are out of phase. Triacs are the thyristor of choice for general-purpose ac relays with ratings up to 10 A at 120 to 240 V. Dual SCRs are capable of switching ac power loads in excess of 40 kW.

Important specifications for ac SSRs include: 1) isolation voltage, 2) operating temperature range, 3) control-signal range, 4) must-operate voltage, 5) must-release voltage, and 6) input current. UL (Underwriters Laboratories) recognition and CSA (Canadian Standards Association) approval are commonly obtained for all power SSRs.

A standard package has been accepted by industry for ac SSRs rated 2 to 40 A. A rectangular flatpack for panel mounting, it is 2¼ inches long by 1¾ inches wide by 0.9 inch high. Four screw terminals permit easy field attachment of wires with formed loops or lugs as shown in Fig. 2.

SOLID STATE RELAY: Fig. 2. A solid-state relay in a four-terminal chassis-mount package with transparent safety cover.

Direct-Current SSRs. Optically coupled dc SSRs with power MOSFET output transistors are available as replacements for reed relays in low-level analog telecommunications switching. Some of these relays include optoisolators containing IREDs matched with photovoltaic cell arrays. Voltage from the array turns on a bidirectional output switch MOSFET permitting it to control ac as well as dc of either polarity. (*See* PHOTODETECTOR.) Direct-current SSRs are available in a wide range of package styles including dual-in-line (DIP) and single-in-line (SIP), depending on ratings and application. There is virtually no package standardization in dc SSRs.

SONAR

A sonar (from the words *so*und *na*vigation *r*anging) system detects and locates objects under water with acoustic energy. There are two types of sonar: active and passive. An active sonar system can determine the range and bearing of an object or target under the sea by measuring the time taken for acoustic energy projected toward the target to be reflected back as an echo. An active sonar can also determine the course and speed of ships on the surface or submarines under the water. By contrast, a passive sonar can only determine the presence and bearing of a target on or under the sea if it generates detectable noise.

An active sonar system is analogous to an active radar system except that its medium is high-frequency sound rather than radio frequency. (*See* RADAR.) A sonar transducer (analogous to an antenna) can be trained in azimuth through up to 360 degrees to determine the range and bearing of the target. This information can be used to determine the target course and speed. Sonar can also determine the depth of a target (submarine, for example) under the sea.

Active sonar has a principle of operation similar to a depth finder or fathometer. However, a depth finder is dedicated to the task of finding the depth of water under a ship's keel (distance to the sea bottom), and it does not

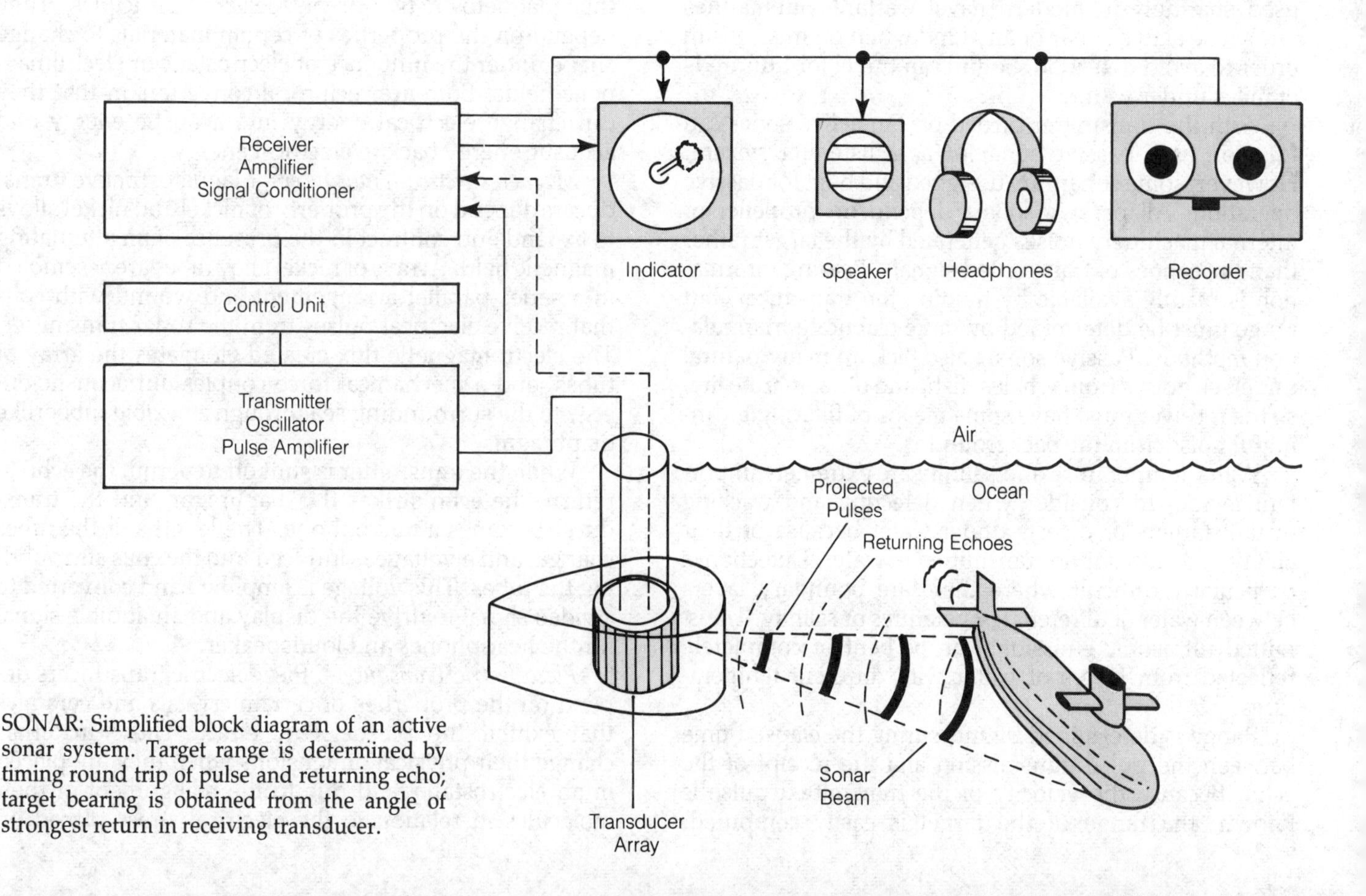

SONAR: Simplified block diagram of an active sonar system. Target range is determined by timing round trip of pulse and returning echo; target bearing is obtained from the angle of strongest return in receiving transducer.

provide bearing information. (*See* DEPTH SOUNDER.)

Sonar systems are used almost exclusively in Navy ships and submarines for antisubmarine warfare (ASW) or submarines in attacking surface vessels and submarines. Both surface ships and submarines also use sonar to detect the presence of mines and other underwater hazards (nets or manmade obstructions to navigation, for example). Some sonar systems are designed for use with helicopters.

The figure is a simplified block diagram of an active sonar system. It includes a transducer or array of transducers, a hydrophone, a transmitter and a receiver. The transducer projects the sound, and in most modern sonar systems the transducer serves double duty as projector and hydrophone to receive the returning echoes. The receiver section contains the signal conditioning circuitry needed to detect and amplify the output of the hydrophone. The audio equipment includes headphones and speaker and the visual display, typically a cathode-ray tube plan-position indicator (PPI), provides a visual indication of range and bearing of the target. Tracking over time permits the calculation of the target range and bearing. The sonar system can include a computer to analyze the returns and a tape recorder to record them for further study.

The transmission of sound pulses or pings of an active sonar needed to obtain clear echoes of the target can easily be detected by the target under surveillance. This permits the target ship to determine the bearing of the sonar and this noise source can itself become a target for a torpedo or missile capable of homing on an active sound source. As a result, active transmission of sound is used sparingly in modern naval warfare. Submarines rarely use active sonar of any kind when on missions in order to avoid detection. Sound can travel for hundreds of miles under water.

With the transmitter turned off, an active sonar can function as a passive sonar or as a listening system. However, some sonars are designed and built for passive operation. All passive sonars depend on propeller or internal machinery noises generated by the target rather than the echoes of transmitted signals. Bearing information is readily available by training the transducer, but range must be determined by more elaborate triangulation methods. Passive sonars also pick up many natural undersea noises from whales, fish, and other marine life, so the receiver must have some means of filtering meaningful noise from the background.

Water temperature and salinity are extremely important factors to consider when detecting and tracking distant targets or objects under water because of their ability to distort sound transmitted in water. Detection is particularly difficult where there are boundary layers between water of different temperatures or salinity. Transmitted ultrasonic emissions can be bent or completely reflected from layers of water with different temperatures.

Range is determined by measuring the elapsed time between the pulse transmission and the receipt of the echo. Because the velocity of the transmitted pulse is known, the range of the target is easily computed. Bearing is determined from the direction of the center line of the transducer or array as it is trained.

The sound waves from a sonar transducer are compression waves propagated through the water in a manner that is similar to the compression waves propagated through the air by a loudspeaker. A transducer transfers acoustic energy into the water with a flexible diaphram or other vibrating surface. The energy is transmitted as a series of alternate compressions and rarefactions. Wavelength of a sound wave is the distance between two successive rarefactions or compressions. The number of wavelengths (cycles) per second is the frequency measured in hertz (Hz). Frequencies of 10 to 50 kHz are commonly used in active sonar.

The velocity of sound in salt water increases from about 4700 feet per second (ft/s) at 30°F to 5300 ft/s at 85°F. (This is more than four times the speed of sound in air.) A 20,000 Hz sound in sea water has a wavelength of about 3 inches. *See* SOUND.

Sonar beams entering warm water are bent downwards by a refractive effect. Conversely, they are bent upwards if surface water is colder than subsurface water. In passages over long distances, the beam might be bent many times in a serpentine pattern, causing it to miss some targets (such as submarines) completely. Reverberations are another form of interference to sonar reception. These are the sum of multiple sound reflections from the ocean floor and the underside of the surface of the sea. Storms at sea increase surface reverberations and operation in shallow water increases bottom reverberations. *See* REVERBERATION.

Sonar Transducers. Sonar transducers operate on either magnetostrictive or piezoelectric principles. They depend on the properties of certain materials to change shape under the influence of electrostatic or electromagnetic fields. Both are reciprocal converters in that they can change electrical energy into acoustic energy and acoustic energy back to electrical energy.

Magnetostrictive Transducers. Magnetostrictive transducers depend on the property of nickel and nickel alloys to expand and contract in the presence of an alternating magnetic field. Arrays of nickel alloy tubes are assembled in a series-parallel arrangement and wound with coils that receive electrical pulses from the sonar transmitter. The electromagnetic flux created elongates the array of tubes, and a mechanical force couples out acoustic energy to the surrounding sea through a flexible rubberlike diaphragm.

When the transmitter is shut off to permit the echo to return, the echo strikes the diaphragm, and the transducer becomes a microphone. The lengths of the tubes change, and a voltage is induced into the coils surrounding the tubes. This voltage is amplified and converted to a video signal to drive the display and an audible signal for the headphones and loudspeaker.

Piezoelectric Transducers. Piezoelectric transducers depend on the properties of certain crystals and ceramics that exhibit the piezoelectric effect. These materials change their physical dimensions when they are placed in an electrostatic field due to the realignment of their molecules in relation to the electrical stress placed on them. The material shortens along one axis and lengthens along another axis. When subjected to mechanical pressure, a voltage will be produced across two of its faces. This property of crystals and ceramics is used in microphones and hydrophones and is the basis for surface acoustic wave (SAW) devices. (*See* SURFACE ACOUSTIC WAVE DEVICE.) Lead zirconate titanate and barium titanate are widely used in sonar transmit-receive transducers. However, a crystal, dihydrogen phosphate or ADP is widely used in hydrophones.

Unlike magnetostrictive transducers, piezoelectric transducers need no intermediate transistion of electric energy to magnetic flux, so they are more efficient and easier to build. These transducers also produce sound energy at the same frequency as the voltage used to excite them.

Sonar transmitters are set to transmit at specific frequencies. If the distance between the transmitter and target (perhaps a submarine) is decreasing, the echo frequency will be higher than the transmitted frequency due to the Doppler effect. This is referred to as up Doppler. However, if the target is moving away, the frequency of the echo will be lower than the transmitted frequency; this is known as down Doppler. But if the relative motion between the transmitter and target ceases, the Doppler effect is no longer present. The Doppler effect is useful in determining relative motion between two ships with either active or passive sonar.

Antisubmarine sonars towed on long cables by surface ships allow the receiving transducers to sink deep enough to avoid the warm surface layers of water that attenuate and distort sound. The long cable also keeps the transducer away from any noise interference caused by machinery on the towing ship. It also keeps the active transducer far enough away from the host ship so that only the towed sonar would be destroyed by a homing mine or acoustic torpedo.

Passive sonars with computer-assisted signal analysis equipment can detect and identify surface ships from their machinery and propeller noises. Ship identification is made possible by comparing the noise profiles or signatures received from the sonar with known noise profiles stored in computer memory.

Some sonar domes on naval ships are built permanently into the ship's bow below the water line as blisters protruding from the stem. Streamlining reduces noise generated by the host ship's motion through the water. The dome is filled with seawater to provide a better acoustic match to the ocean. Other sonar domes are designed to be retracted into the ship's hull when not in use or when the ship is moving in shallow water.

Submarines have spherical sonar transducer arrays in their bows that can project and receive narrow beams 1 to 2 degrees wide. The projection and listening beams can be moved through wide bearing and elevation angles. Passive sonar phased arrays permit submarines to determine the number of ships on the surface or other submarines underwater. Their bearings, courses, and speeds can be calculated from their propeller and machinery noises.

Specialized sonars have been developed for scientific exploration and mapping of the sea floor. Short-range, continuous-wave sonars are used on scientific minisubmarines for underwater navigation. Side-looking sonar is used for archaeological and geophysical surveys. These sonars plot strip contour maps on both sides of the ship as it moves along a prescribed course. Underwater geological features and sunken ships have been located with this equipment.

Acoustic transducers can be modulated by the human voice to permit the transmission of voice messages through the water. This equipment can be used to communicate with submerged submarines or divers.

SOUND

Sound is an acoustic disturbance that can be heard by the average person. The frequency of sound waves ranges from about 16 or 20 Hz to 20 kHz; this corresponds to a wavelength range of 55 feet to 5/8 inch, or 17 meters to 1.7 centimeters. Acoustic disturbances at frequencies below 20 Hz are called infrasound. Disturbances at frequencies above 20 kHz are known as ultrasound (*see* INFRASOUND, ULTRASOUND).

Sound travels through the air by compression of the molecules. The molecules themselves move back and forth parallel to the direction of sound propagation. Thus, sound waves in air are longitudinal disturbances (*see* LONGITUDINAL WAVE). Sound waves expand outward from a source in spherical wavefronts.

Sound can consist of a single wave disturbance at a single frequency. However, sound usually consists of simultaneous disturbances at many frequencies. The waveforms of most sounds are extremely complex.

Sound propagates faster in liquids and solids than in air. Water cannot be compressed, so sound waves propagate by means of lateral motion of the molecules, in a manner similar to wind waves on the surface of a lake. In water, sound is a transverse wave. Sound is also a transverse wave in most solid substances (*see* SONAR, TRANSVERSE WAVE).

The speed of sound in air depends on the temperature and pressure. In dry air at room temperature at a

SOUND: Speed of sound through various substances.

Substance	Speed of sound, Feet per second	Speed of sound, Meters per second
Air[1]	1100	335
Fresh water	4600	1400
Salt water	4900	1500
Petroleum	4200	1300
Turpentine	4600	1400
Aluminum	17000	5200
Copper	12000	3600
Silver	12000	3600
Gold	6600	2000
Lead	4100	1290
Tin	8900	2700
Wood	3300 - 16000	1000 - 5000
Glass	9800 - 20000	3000 - 6000

1. Room temperature at a pressure of one atmosphere.

pressure of 1 atmosphere (30 inches of mercury), sound travels about 1,100 feet, or 335 meters, per second. In other mixtures of gases, even at the same pressure, the speed of sound differs from what it is in air. (Air is a mixture of about 78 percent nitrogen, 21 percent oxygen, and 1 percent other gases.)

The table gives the speed of sound in various common liquids and solids.

The wavelength λ of a sound wave is a function of the frequency *f* and the velocity *v*. The general relation is:

$$\lambda = v/f$$

The frequency *f* is always specified in hertz. If λ is given in feet, then *v* must be given in feet per second; if λ is to be specified in meters, then *v* must be given in meters per second.

The intensity of sound is usually specified in decibels above the threshold of hearing. Sound loudness may also be expressed in units called sones, or in terms of the pressure it exerts (*see* DECIBEL, LOUDNESS).

SOURCE

The term *source* refers to the emitting electrode of a field-effect transistor (FET). The output of a FET amplifier can be taken from the source or the input signal may be applied to the source. The source of a FET corresponds to the cathode of a tube or the emitter of a transistor (*see* FIELD-EFFECT TRANSISTOR).

The originating generator of a signal is called the source of the signal. For example, there can be a radio-frequency source, a sound source, or a light source.

SOURCE COUPLING

When the output of a field-effect-transistor (FET) amplifier or oscillator is taken from the source circuit, or when the input to a FET amplifier is applied in series with the source, the amplifier or oscillator is said to employ source coupling. Source coupling can be capacitive, or it might use transformers. The illustration shows an example of source coupling in a two-stage amplifier. Capacitive coupling is used here. The output of the first stage is taken from the source.

Source coupling generally results in a low input or output impedance, depending on whether the coupling is in the input or output of the amplifier stage. Source coupling is often used with grounded-gate amplifiers. If the following stage needs a low driving impedance, source coupling can be used; this is called a source follower. *See also* SOURCE FOLLOWER.

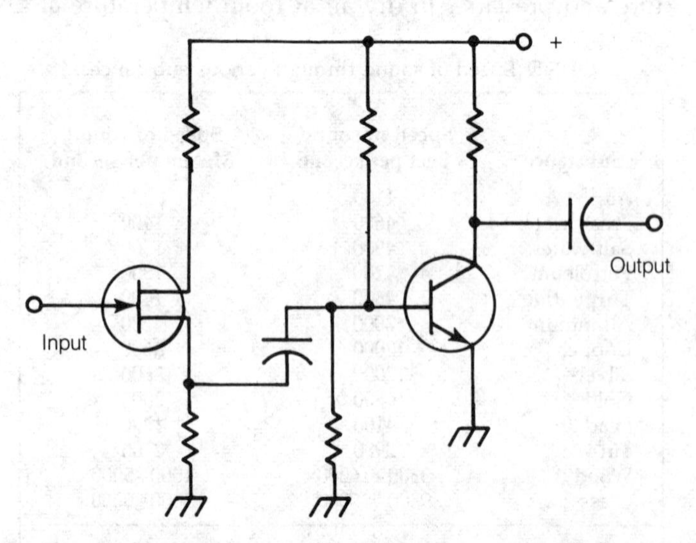

SOURCE COUPLING: A two-stage amplifier with source coupling between the stages.

SOURCE FOLLOWER

A source follower is an amplifier circuit in which the output is taken from the source circuit of a field-effect transistor (FET). The output impedance of the source follower is low and the voltage gain is always less than 1. Thus, the output signal voltage is smaller than the input signal voltage. The source-follower circuit is generally used for impedance matching. The amplifier components are often less expensive and offer greater bandwidth than transformers.

The diagram illustrates a two-stage FET amplifier circuit in which the first stage is a source follower. The output of the source follower is in phase with the input. The impedance of the output circuit depends on the particular characteristics of the FET used, and on the value of the source resistor. Sometimes two series-connected source resistors are used. In this case, the output is taken from between the two resistors. A transformer output may also be employed. Source-follower circuits are useful because they offer wideband impedance matching at low cost. *See also* SOURCE COUPLING.

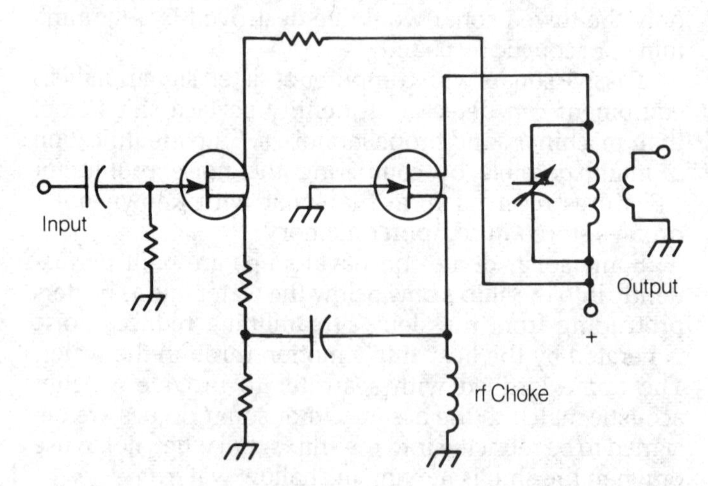

SOURCE FOLLOWER: This circuit provides an impedance match for driving low-impedance circuits or loads.

SOURCE MODULATION

Source modulation is a method of obtaining amplitude modulation in a radio-frequency (rf) amplifier. The carrier is applied to the gate of the device in most cases. Sometimes the carrier is applied to the source. The audio signal is applied in series with the source across a resistor, or by means of an audio-frequency transformer (see illustration). Source modulation is the counterpart of cathode modulation in a vacuum-tube circuit, or emitter modulation in a bipolar-transistor amplifier (*see* EMITTER MODULATION).

As the audio-frequency signal at the source swings negative, the instantaneous rf output voltage rises if an N-channel field-effect transistor is used. If a P-channel device is used, the instantaneous signal output voltage increases when the audio signal swings positive. Under conditions of 100-percent modulation, the instantaneous amplitude just drops to zero at the negative rf signal peaks.

Source modulation requires very little audio power for 100-percent amplitude modulation provided that the modulation is done at a low-level stage and not in the final amplifier circuit of the transmitter. *See also* AMPLITUDE MODULATION, FIELD-EFFECT TRANSISTOR, MODULATION.

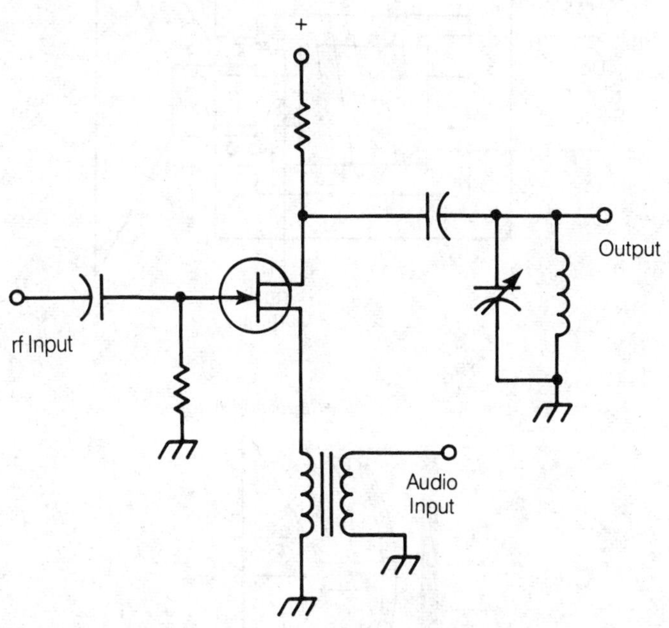

SOURCE MODULATION: The modulating signal is applied in the source circuit.

SOURCE RESISTANCE

The source resistance of a field-effect transistor (FET) is the effective resistance of the source in a given circuit. The source resistance depends on the bias voltages at the gate and the drain. It also depends on the input-signal level and on the characteristics of the particular FET used. The source resistance can be controlled, to some extent, by inserting a resistor in series with the source lead.

The external resistor in the source circuit of a FET oscillator or amplifier is usually called the source resistor, although its value is sometimes called the source resistance. A source resistor can be used for impedance-matching, biasing, current limiting, or stabilization. The source resistor can have a capacitor connected across it to bypass radio-frequency energy to ground. *See also* FIELD-EFFECT TRANSISTOR, SOURCE FOLLOWER, SOURCE STABILIZATION, SOURCE VOLTAGE.

SOURCE STABILIZATION

Source stabilization is a method of providing bias control in a power-FET (field-effect transistor) circuit using two or more FETs in parallel.

A resistor with a value that depends on the circuit application and input impedance is connected in series with the source of each FET to reduce the effects of minor differences in the characteristics of the FETs. Bypass capacitors can be connected across each resistor if it is necessary to keep the sources at signal ground.

If the channel current increases because of a temperature rise in a FET and there is no resistor in series with the source lead, the change in current will reduce the gain of the amplifier, and thereby reduce its efficiency. If a source-stabilization resistor is connected in series with the current path, however, any increase in the current will cause an increase in the voltage drop across the resistor. This will change the bias to stabilize the current through the channel.

Stabilization is more important in bipolar-transistor circuits than in FET circuits. In fact, stabilization is used almost universally in bipolar-transistor amplifiers and oscillators. By contrast, FET stabilization is usually necessary only for power amplifiers. *See also* COMMON CATHODE/EMITTER/SOURCE, EMITTER STABILIZATION.

SOURCE VOLTAGE

In a field-effect-transistor circuit, the source voltage is the direct-current potential difference between the source and ground. If the source is connected directly to chassis ground, or is grounded through an inductor, then the source voltage is zero. But if a series resistor is used for stabilization, as is the case in some common-source power amplifiers, the source voltage is positive in an N-channel device and negative in a P-channel device.

If I is the channel current under no-signal conditions and R is the value of the source resistor (assuming there is a source resistor), the source voltage V is obtained by Ohm's law as:

$$V = IR$$

A capacitor across the source resistor keeps the source voltage constant under conditions of variable input signal. If no capacitor is used, the instantaneous source voltage will vary along with the input signal because of changes in the instantaneous current through the channel. *See* SOURCE STABILIZATION.

In common-gate and common-drain circuits, it is usually necessary to provide a voltage at the source. In a common-gate amplifier, the source voltage is normally positive in the N-channel case and negative in the P-channel case. In a common-drain configuration, the source voltage must be negative if an N-channel device is used, and positive if a P-channel device is employed.

In these situations, the source voltage is provided directly by the power supply. *See also* COMMON CATHODE/EMITTER/SOURCE, COMMON GRID/BASE/GATE, COMMON PLATE/COLLECTOR/DRAIN, FIELD-EFFECT TRANSISTOR.

SPACE CHARGE

In a vacuum tube, the cathode initially emits more electrons than are drawn off to other electrodes. This results in a negatively charged cloud of electrons in the space around the cathode. The total negative charge of these electrons is called the space charge.

The space charge in a vacuum tube depends on the voltages at the cathode, grids, and plate. In general, the greater the potential difference between the cathode and the plate, the greater the space charge will be. The grid voltages have a lesser effect on the space charge. *See also* ELECTRON.

In a solid-state device, the depletion region near the P-N junction is often called the space-charge region (*see* P-N JUNCTION).

When electrically charged particles or objects are scattered in space, an electric charge per unit volume exists. This charge concentration is also called space charge. It is expressed in coulombs per cubic meter. *See also* CHARGE, COULOMB.

SPEAKER

A speaker, also known as a loudspeaker, is an electroacoustic transducer. The speaker converts alternating electric currents into sound waves with the same frequency characteristics and waveforms within the range of human hearing (approximately 16 Hz to 20 kHz).

Different types of speakers are intended for different applications. The woofer is designed to operate most efficiently at the lower audio frequencies; the midrange speaker is designed to be used at frequencies near the middle of the human hearing range; and the tweeter is intended for reproduction of sound at the upper end of the range. Some tweeters will operate at ultrasonic frequencies (*see* MIDRANGE, TWEETER, WOOFER). Many high-fidelity speaker enclosures consist of one of each of these three transducers. Communications speakers usually consist of a simple midrange speaker.

As shown in A of the diagram, a coil of wire, called the voice coil, is held in the field of a permanent magnet. When alternating currents flow in the coil, a changing magnetic field is produced around the coil. This field interacts with the field from the permanent magnet to produce reciprocal forces on the coil. The coil is mounted so that it can move as these forces occur. The coil is physically attached to a cone shaped diaphragm, which vibrates along with the coil. The moving diaphragm produces sound waves in the air.

As shown in B of the figure, some speakers use electromagnets, rather than permanent magnets, to produce the stationary magnetic field. In public-address systems and some high-fidelity midrange and high-frequency applications, horn-shaped speakers are used.

Another common speaker configuration is shown in C. Electrostatic speakers operate with electric fields, rather than magnetic fields. An electrostatic speaker converts audio-frequency electrical currents into sound waves by means of electrostatic forces. The speaker consists of fixed metal plate and a thin, flexible plate. The

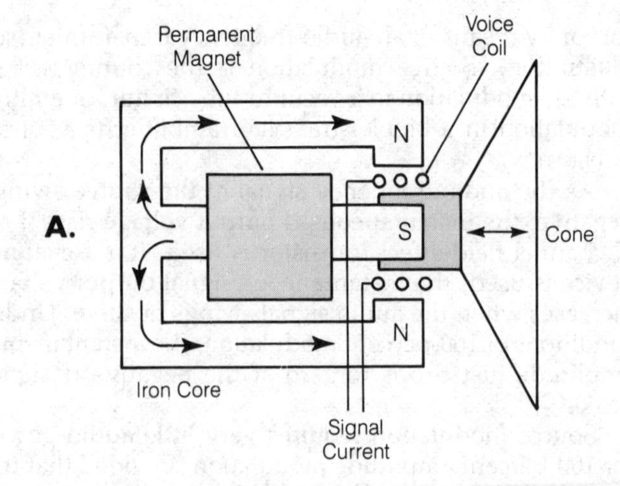

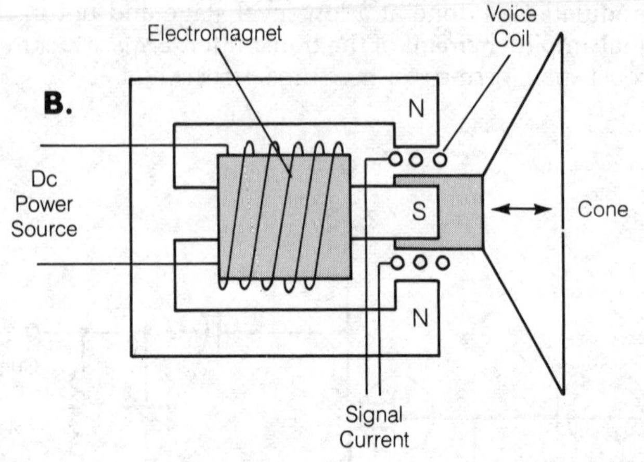

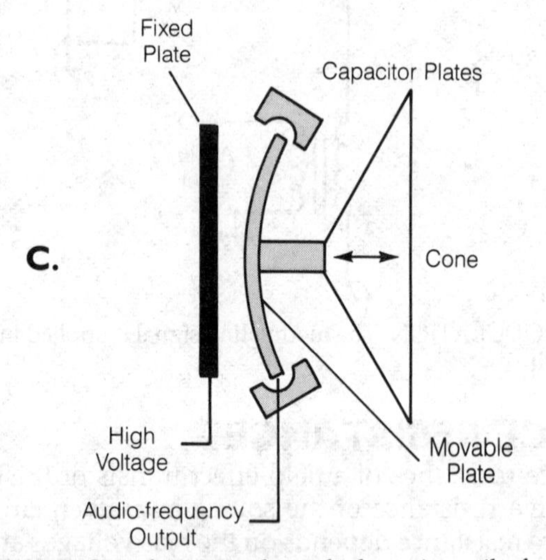

SPEAKER: Signal current through the voice coil of a permanent magnet speaker produces sound from the cone (A). The electrodynamic speaker has an electromagnet (B), and electrostatic effect causes sound output from electrostatic speaker (C).

illustration is a simplified diagram of the construction of an electrostatic speaker.

SPECIFICATIONS

The specifications of an electronic product or system are the operating characteristics, expressed in table form. Specifications give information of importance to the users of the equipment.

SPECIFICATIONS: Specifications for a typical VHF (very high frequency) FM (frequency modulation) transceiver.

GENERAL	
Frequency Range	144–148 MHz
Display Type	LED
Frequency Control	Microcomputer PLL VCO
Emission Type	F3
Memory Channels	8
Temperature Tolerance Range	−20 to +60 degrees C
Power-Supply Requirements	12 to 15 V dc, 5 A
Semiconductors	17 IC, 20 FET, 29 Tr, 59 Di
Dimensions	HWD 2.5 × 6 × 9 in (64 × 152 × 229 mm)
Weight	3 lbs (1.4 kg)
RECEIVER SECTION	
Intermediate Frequencies	17.0 MHz, 455 kHz
Sensitivity	Better than 0.35 μV for 20− dB noise quieting
Selectivity	Plus/minus 5 kHz at −6 dB Plus/minus 15 kHz at −60 dB
Audio Output	2 watts or more
Speaker Requirements	Impedance 4–8 ohms
TRANSMITTER SECTION	
RF Output	15 watts
Frequency Deviation	Plus/minus 5 kHz
Spurious Radiation	Less than −60 dB
Antenna Requirements	50 ohms resistive, SWR < 2:1
Microphone Requirements	Impedance 300–600 ohms

In a radio receiver, important specifications include the sensitivity, selectivity, tuning range, emission modes that can be received, and frequency stability. In a transmitter, the specifications include the frequency range, emission types, frequency stability, and radio-frequency power output. (The table is a list of specifications for a commercially manufactured frequency modulation (FM) transceiver.) For test instruments, accuracy and repeatability are the most important specifications. For power supplies, the voltage output, regulation, and current-delivering capability are important specifications.

SPECIFIC GRAVITY

Specific gravity is an expression of the relative density of a material.

If a given volume of material has a mass of *x* grams, while an equal volume of water has a mass of *y* grams, the specific gravity, *S*, of the unknown material is given by:

$$S = x/y$$

provided the two materials are at the same temperature.

SPECIFIC HEAT

The specific heat of a substance is an expression of the capacity of that material to be raised in temperature by the application of heat energy. Specific heat is expressed in calories per degree Celsius for 1 gram of the material.

Water has a specific heat of 1; this means that the application of 1 calorie of heat energy will raise 1 gram of water by 1 degree Celsius. Most materials have a specific heat values of less than 1.

SPECIFIC RESISTANCE

The resistance of an electrical conductor can be expressed as specific resistance. Specific resistance is given in ohms per unit length per unit area, usually circular mil feet.

The specific resistance of a conductor depends on the material and the temperature. The specific resistance of a material is generally expressed for direct current. *See also* RESISTANCE, RESISTIVITY.

SPECTRAL DENSITY

A radio-frequency signal or other electromagnetic emission usually contains many wavelength components. For example, an amplitude-modulated signal contains not only the carrier frequency, but sideband frequencies as well. Spectral density is an expression of the concentration of energy in terms of frequency.

In general, for a given signal power, the spectral density increases as the signal bandwidth decreases. The maximum possible spectral density exists when a signal is a steady carrier (emission type A0 or F0). The faster the rate of data transmission, the lower the spectral density. Also, the spectral density decreases as the efficiency of data transmission decreases.

Spectral density is stated as the average number of watts per kilohertz of spectrum space at the fundamental frequency of the signal. Spectral density can be expressed in relative terms, as the percentage of the total signal power per kilohertz of spectrum space at the fundamental frequency.

SPECTRAL ENERGY DISTRIBUTION

The electromagnetic output of an energy source can be

expressed graphically as an energy-versus-frequency function called the spectral energy distribution.

Most natural phenomena exhibit a characteristic spectral energy distribution with a peak at a specific wavelength. This is called black-body radiation (*see* BLACK BODY). Some man-made sources such as incandescent lamps have a similar spectral energy distribution. Radio transmitters have a spectral energy distribution that differs from natural electromagnetic-energy sources because the energy is concentrated within a very small band of frequencies.

Spectral energy distribution can be given as power rather than energy, called the spectral power distribution. *See also* ELECTROMAGNETIC SPECTRUM.

SPECTRAL RESPONSE

The spectral response, also called spectral sensitivity characteristic, is the relative effect produced by energy at any one wavelength on a sensor. Spectral response is often expressed for the human eye (for visible-light wavelengths) or for the human ear (for sound frequencies).

The human eye is most sensitive in the yellow and green parts of the visible spectrum, near the center of the visible-light range. The human ear is most sensitive at a frequency near 1 kHz.

Most radio receivers exhibit variable sensitivity throughout the range of coverage. The function of sensitivity versus frequency for a radio receiver is sometimes called the spectral response of the receiver. Spectral response may also be given for transducers. *See also* SENSITIVITY, TRANSDUCER.

SPECTROSCOPE

A spectroscope is an instrument that displays the spectral energy distribution of a visible-light source.

The spectroscope splits visible light into its constituent wavelengths by means of a slit and a prism or diffraction grating, as shown in the illustration. The resulting spectrum can be photographed or viewed directly.

The spectroscope is used by astronomers to evaluate the spectral energy distribution of light from distant stars and planets. Certain absorption lines and emission lines are characteristic of various elements and compounds. The patterns formed by these lines allow astronomers to determine the material composition of planetary atmospheres, stellar surfaces, and diffuse nebulae in space.

Doppler shifts in the wavelengths of spectral absorption lines make it possible to determine whether an object is moving toward or away from the viewer, and if it is moving, to determine its radial speed.

SPECTRUM

When visible light is passed through a prism or diffraction grating, the light is bent at different angles depending on wavelength. This creates a rainbow-like pattern called a spectrum. The spectral colors are generally recognized as red, orange, yellow, green, blue, indigo, and violet, with intervening shades (*see* VISIBLE SPECTRUM).

Any range or band of frequencies can be called a spectrum. The whole range of electromagnetic wavelengths, from the longest to the shortest, is called the electromagnetic spectrum. *See also* ELECTROMAGNETIC SPECTRUM.

SPECTRUM ANALYSIS

Spectrum analysis is the process of evaluating the spectral energy distribution from a source of electromagnetic waves (*see* SPECTRAL ENERGY DISTRIBUTION). Astronomers analyze the spectra of planetary atmospheres, stars, galax-

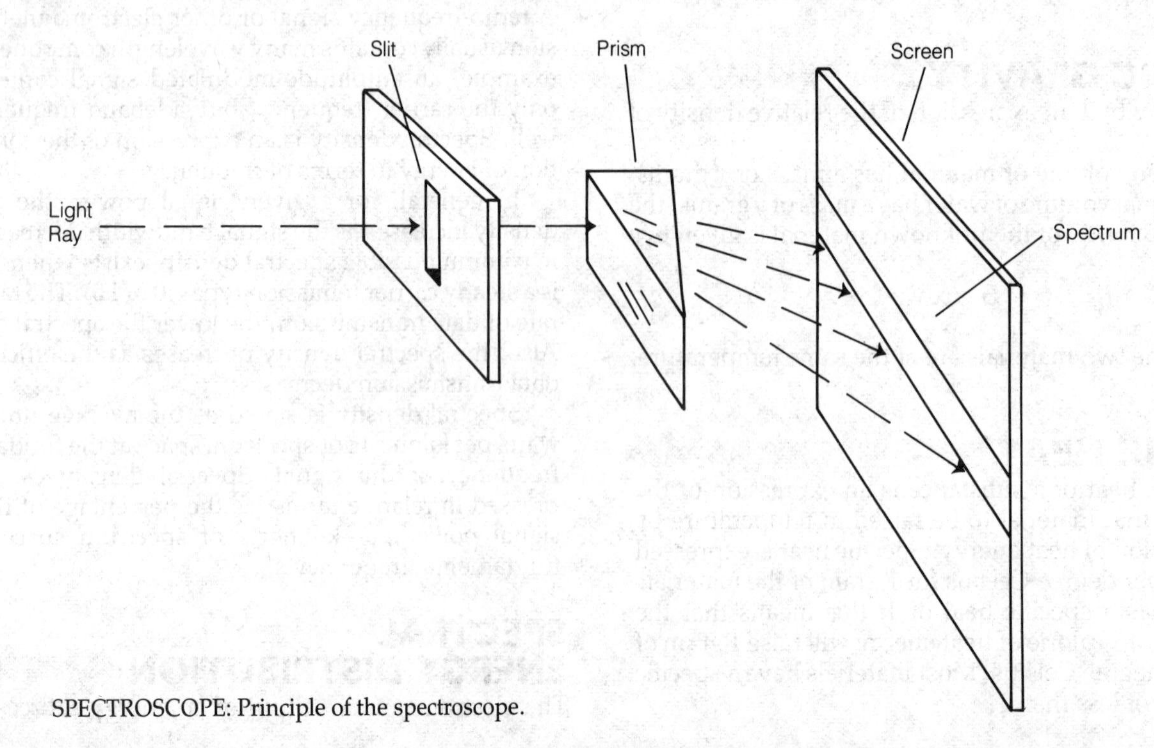

SPECTROSCOPE: Principle of the spectroscope.

ies, and other celestial objects to determine their material composition (*see* SPECTROSCOPE). Electronic engineers evaluate the spectral energy distribution of radio waves with an instrument known as a spectrum analyzer (*see* SPECTRUM ANALYZER).

SPECTRUM ANALYZER

A spectrum analyzer is a scanning receiver used as a test instrument that automatically scans through a frequency spectrum and displays a plot of amplitude versus frequency of the signals present at its input. A spectrum analyzer is able to display Fourier analysis on the screen of a cathode-ray tube.

At microwave frequencies, a spectrum analyzer is able to show the frequency distribution of the energy emitted by the transmitter tube or power train. It can also measure the Q of resonant cavities.

The spectrum analyzer is useful for determining the spurious-signal and harmonic content of the output of a radio transmitter. In the United States, radio equipment must meet government-imposed standards of spectral purity. The spectrum analyzer gives an immediate indication of proper transmitter functioning. It also can be used to observe the bandwidth of a modulated signal. Such improper operating conditions as splatter or overmodulation can be easily seen as excessive bandwidth.

Spectrum analyzers are often used in conjunction with sweep generators, for evaluating the characteristics of bandpass, band-rejection, highpass, or lowpass filters.

Some radio receivers are equipped with narrow-band spectrum analyzers for monitoring an entire communications band at once. These receivers are called panoramic receivers. *See also* PANORAMIC RECEIVER.

SPECTRUM MONITOR

A spectrum monitor is a narrow-band spectrum analyzer that can be connected in the intermediate-frequency chain of a receiver to observe signals at or near the operating frequency. The spectrum monitor converts any ordinary superheterodyne receiver into a panoramic receiver.

The spectrum-monitor display is centered at the intermediate frequency of the receiver. Signals appear as pips to the left or right of the center of a cathode-ray-tube screen. The bandwidth is adjustable. This makes it possible to observe an entire communications band, or the operator can choose one signal and observe its modulation characteristics. *See also* PANORAMIC RECEIVER, SPECTRUM ANALYZER.

SPEECH CLIPPING

Speech clipping is a method of increasing the average power of a voice signal without increasing the peak power. This increase can be done either in the audio-frequency microphone-amplifier stages of a transmitter, or in the intermediate-frequency or radio-frequency (rf) stages. The latter method is usually called rf clipping (*see* RF CLIPPING).

To accomplish audio-frequency speech clipping, a voice signal is first amplified. Then the signal is passed through a limiter that has a cutoff voltage substantially below the peak voltage of the amplified voice signal. The output of the limiter is then amplified so that the clipped peak voltage is the same as the unclipped peak voltage ahead of the limiter. A simplified circuit for audio-frequency speech clipping is shown in the illustration.

Speech clipping, if done properly, can improve the intelligibility of an amplitude-modulated or single-sideband signal. If done improperly, however, speech clipping can cause severe distortion and actually reduce the intelligibility as well as cause objectionable splatter. *See also* SPEECH COMPRESSION.

SPEECH COMPRESSION

Speech compression is a method of increasing the average power in a voice signal without increasing the peak

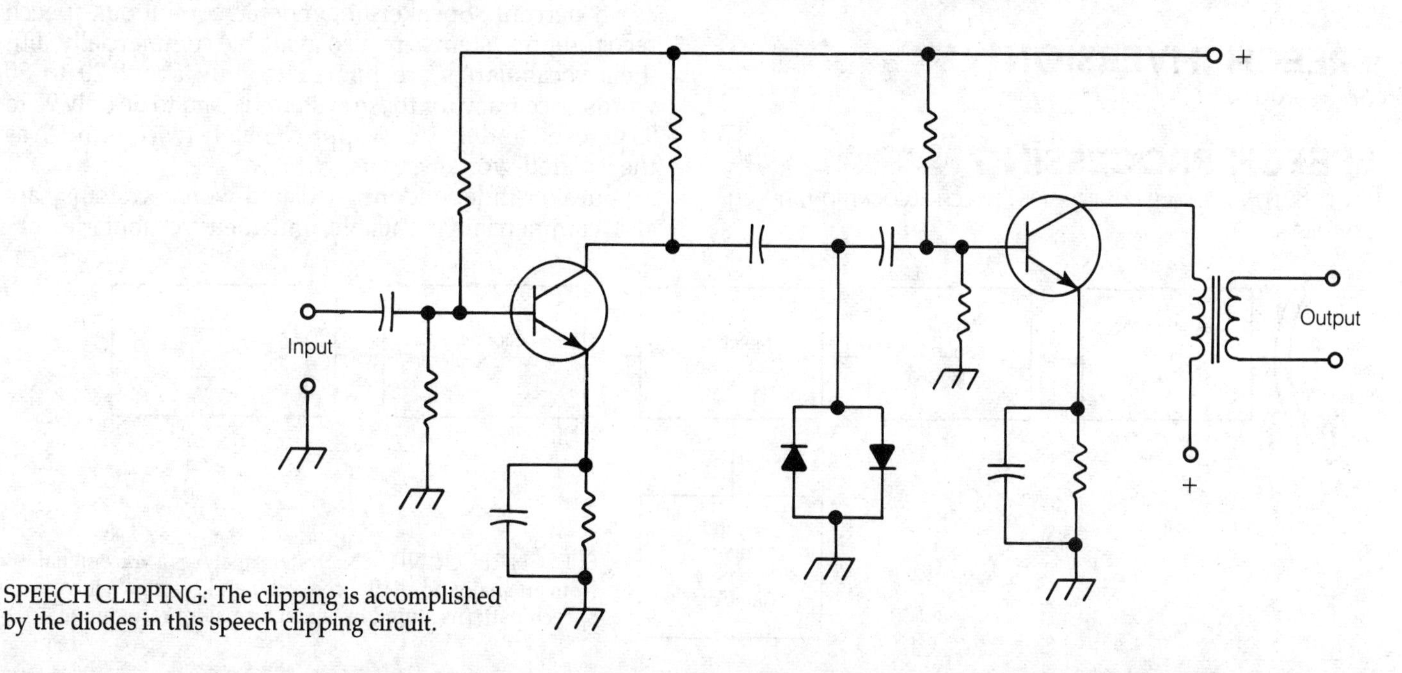

SPEECH CLIPPING: The clipping is accomplished by the diodes in this speech clipping circuit.

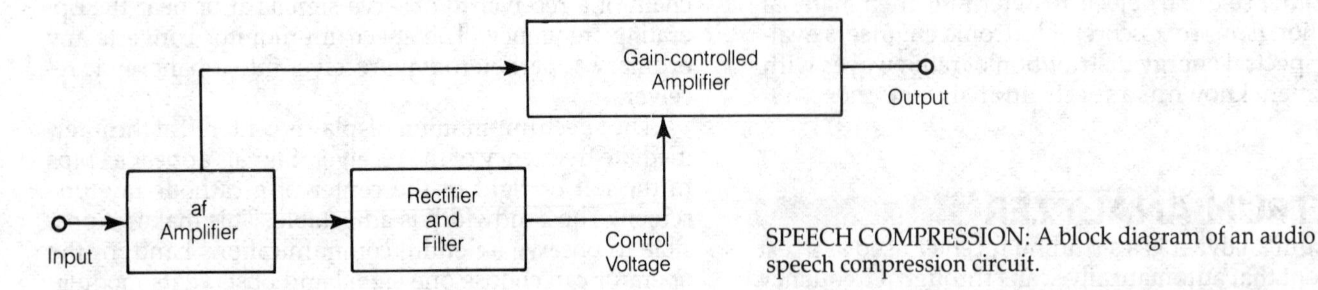

SPEECH COMPRESSION: A block diagram of an audio speech compression circuit.

power. Speech compression is used in many amplitude-modulated and single-sideband communications transmitters.

The speech compression circuit operates in the same way as an automatic-level-control (ALC) circuit (*see* AUTOMATIC LEVEL CONTROL). But speech compression carries the process farther than ordinary ALC. Although ALC is typically used only to prevent overmodulation, speech compression employs additional amplification of the low-level components of a voice. The result is greatly reduced dynamic range, but the intelligibility of the signal is often considerably improved because more effective use is made of the modulating-voice signal. A speech-compression circuit is also called an amplified-ALC (AALC) circuit. The illustration is a block diagram of an audio speech-compression circuit.

Speech compression is usually done in the audio (microphone-amplifier) circuits of a transmitter. But it can be done in the intermediate-frequency or radio-frequency (rf) stages. There are both audio speech compression and rf speech compression. Radio-frequency speech compression is also called envelope compression.

Speech compression, like speech clipping, can cause problems if it is not done correctly. There is a maximum increase in the ratio of average power to peak power that can be realized without objectionable envelope distortion. Too much speech compression will actually degrade the intelligibility of a voice signal. *See also* SPEECH CLIPPING.

SPEECH INVERSION

See SCRAMBLER.

SPEECH PROCESSING

See RF CLIPPING, SPEECH COMPRESSION, SPEECH RECOGNITION, SPEECH SYNTHESIS.

SPEECH RECOGNITION

Electronic speech recognition is the recognition of human speech with electronic circuitry, usually computer based, for practical control applications. These applications range from simply switching circuits on or off to the entry of data into computer memory in hands-free recording. In a speech-to-text application, spoken words and phrases are converted directly into printed text with a computer printer.

Speech-recognition systems convert utterances into word sequences stored in a computer. Sound waves travel from the speaker's mouth through a microphone to an analog-to-digital converter, where they are digitized. The digitized signal is filtered and compared with stored speech modules to determine the most likely choice for the word spoken.

Speech recognition is a continually developing field of electronics research because of its importance in direct interaction with computers. Many practical and effective speech recognition systems have been developed so far, but no system is capable of understanding complete sentences in natural language spoken without pauses by many different persons.

Speech recognition systems can be classified as speaker-dependent or speaker-independent, isolated word, or continuous speech. At the present time speaker-dependent, isolated-word systems are widely available with vocabularies of 500 to 1,000 words. Their manufacturers claim word-recognition accuracy as high as 98 percent. Speaker-dependent, continuous-speech recognition systems are also available commercially, but their vocabularies are limited to only about 20 to 50 words. Accuracy for these systems is said to be only 92 to 95 percent, and they cost approximately twice as much as the isolated-word systems.

Speaker-independent, isolated-word systems are also commercially available, but their vocabularies are

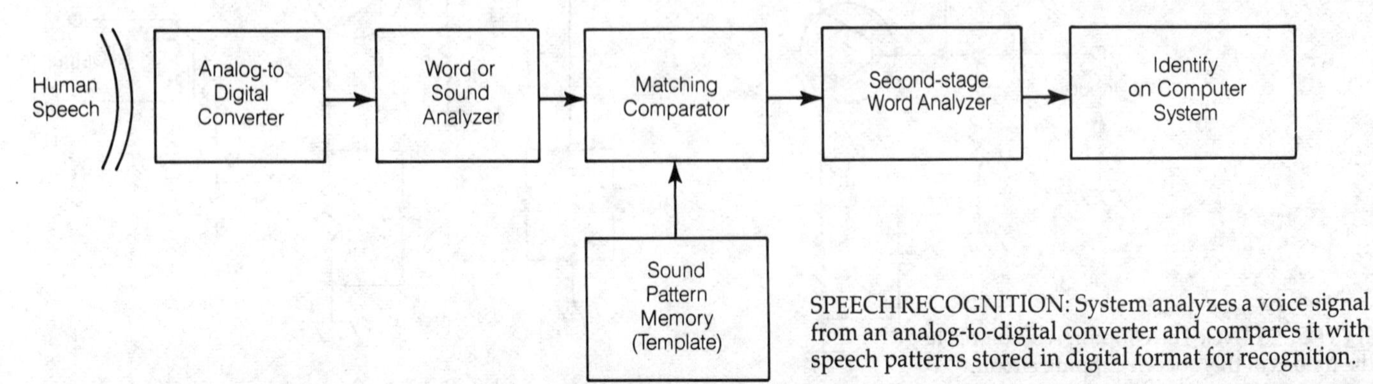

SPEECH RECOGNITION: System analyzes a voice signal from an analog-to-digital converter and compares it with speech patterns stored in digital format for recognition.

typically 20 words. However, accuracy of 95 percent is claimed. Speaker-independent, continuous-speech systems are still under development, and prototype models recognize only about 10 words. Speech recognition accuracy claimed is only about 90 percent.

Most of the available speaker-independent, isolated-word commercial speech-recognition products are packaged as single boards that plug into host computers. They are used for command and control, dictation, inspection, and voice data entry. However, some speaker-dependent, continuous-speech systems for voice data entry and dictation and some speaker-independent, isolated-word systems for command and control are packaged as subsystems.

The ideal speech-recognition system would have a vocabulary of as many as 20,000 words and be able to process instructions at the rate of 100,000 megainstructions per second. It would contain a knowledge base to evaluate the context of speech to assist in recognition and would be programed to learn from errors.

Human speech can be considered to be encoded data because each voice has distinctive characteristics. When combined with a codified language, unique speech patterns are produced. The objective of much of the speech recognition research is to eliminate individual acoustic characteristics to make the system independent of the speaker.

To understand speech, a human must evaluate the specific information conveyed to the ear and consider the context in which the information is presented. Humans have the ability to understand speech even when it is muffled or distorted, but this understanding requires a knowledge of language and experience in the interpretation of speech.

Speech recognition is limited by the ability of the computer program to recognize isolated words or continuous speech with a precisely structured vocabulary. The speaker-dependent, isolated-word and connected or continuous-speech systems must be trained to recognize one voice. There must be clear pauses between words in both speaker-dependent and speaker-independent, isolated-word recognition (IWR) systems.

By contrast, the most sophisticated speaker-independent, continuous-speech recognition (CSR) systems now in the research and development stage are capable of recognizing continuous speech from more than one speaker. Some systems use statistical methods or algorithms to identify words on a basis of preceding words, to supplement phonetics.

Speech is a sequence of discrete sound segments called phonemes linked in time. English, as spoken in the United States, for example, contains 16 vowel and 24 consonant sounds. Each phoneme has distinguishable acoustic characteristics, and together they are the building blocks of syllables and words. The human listener or electronic recognition system must be able to distinguish different words by sound units such as van from ban. *See* SPEECH SYNTHESIS.

In systems where each word is treated as a unit, recognition is achieved by matching variables of the input word with stored word patterns in computer memory. A word is recognized when the system achieves the best match between the input and the stored-word pattern. Time-alignment procedures are widely used because they allow for the inherent variability of the speech signal.

New users of most IWR systems must receive training, but the need to train a system for each user limits the size of the vocabulary that it can accommodate. For vocabularies with more than 5,000, the training procedure becomes more time consuming. For this reason, emphasis on the more advanced systems is placed on speaker independence. If a speech recognition system is to be used for dictation (speech-to-text), a vocabulary of several thousand words is considered to be a minimum requirement. It would be more practical if it were speaker independent. Even with an input of isolated words, speaker-independent recognition from a large vocabulary becomes very complex.

Commercial IWR systems requiring pauses between spoken words avoid the alteration of acoustic input at word boundaries that make it difficult to tell where one word ends and the other begins. An example is the ending of one word and the beginning of the next word as in *bus station* which merge into a single sound. Speaking with pauses between words is unnatural, and this drawback limits the acceptances of certain systems to specific applications.

One system that has been developed employs connected-word recognition achieved by applying the recognition algorithm twice, once for single-word matches and once for groups of connected words. Commercial speaker-dependent connected-speech systems are limited to short, simple sentences, typically derived from a vocabulary of less than 200 words. The system can be recognizing the simple sentence as an entity like a single word.

The development of CSR systems is handicapped by changes that occur in the acoustic properties of words due to usage. These changes depend on the position of the word in a sentence and also can be modified by adjacent words. Moreover, semantics, syntax, and knowledge of the situation in which the speech occurs can also directly modify the speech.

If the speech-recognition system is to have practical benefits beyond toys or games, the recognition rate must be 95 percent or better. The lowest error rates are achieved with speech inputs in the form of words with pauses between them.

Researchers are investigating voicegrams, which plot frequency on a vertical axis against time on the horizontal axis as a means of improving speech-recognition systems, as shown in Fig. 2 for SPEECH SYNTHESIS. (Amplitude at a given frequency is indicated by the darkness of lines.) People have been trained to identify spoken words by examining the spectrograms of the speech signal. The spectrogram contains all of the information content of the acoustic signal. It is believed that a set of rules can be formulated for the identification of phonemes (independent of the speaker) and these rules can be programmed into a speech-recognition system.

Artificial intelligence techniques developed for expert systems can be used to analyze the context of the subject under discussion and interpret what is actually being said. These techniques can be applied to the back end of the speech-recognition system.

Speech-recognition systems capable of recognizing isolated words from large vocabularies—independent of the speaker—are likely to be available as commercial products in the 1990s. However, dictation machines capable of continuous-speech recognition—also independent of the speaker—could be introduced during this same period. More research must be done to solve the remaining problems, and this will call for substantial resources. But the availability of lower-cost computers and improved programming techniques could shorten the development time.

SPEECH SYNTHESIS

Electronic speech synthesis (ESS) is the technology of reproducing understandable human speech by means of electronic components from stored digital data. Speech synthesis, which includes speech analysis, depends on the operating rules and techniques for reproducing human speech. The motivation for the original work on electronic speech synthesis was a need to improve the efficiency of digital communications.

The human voice and the vocal tract that produce it have been studied for centuries. Many scientists have tried to reproduce the human vocal tract artificially with various mechanical and electrical analogs. However, despite some very ingenious models, satisfactory reproduction of the vocal system was not achieved until it was done electronically in the 1980s with digital conversion techniques and computers.

The ability to reproduce the human voice electronically from small, low-cost, low-power circuits has found many practical applications. As an alternative to the use of cumbersome and unreliable tape recorders, ESS circuits are now found in toys, electronic games, and educational learning aids. They have also been put to use as warning devices in automobiles, trains, aircraft, and power plants, as well as on machine tools and process apparatus.

ESS systems are being used to announce the location of store merchandise in elevators and to give directions, and guidance in operating automatic bank teller machines. They are also providing instructions for industrial assembly and wiring procedures. Automatic verbal announcemenets from ESS systems give directions in the event of fire or other emergencies in public buildings.

ESS can take advantage of the predictability and redundancy of speech and provide an output code that is interpreted as speech by the human ear. Speech also can be reproduced by reconstructing the original waveform from a knowledge of changes in pitch, energy, and other features.

Three different approaches to electronic speech synthesis are in use today: analysis synthesis, constructive synthesis, and hybrid synthesis. The principal manufacturers of ESS integrated circuits have selected specific approaches and made extensive investments in their development. This has narrowed the concentration of effort to only a few of the many possible approaches.

The integrated-circuit (IC) manufacturers participating in ESS include certain proprietary variations on the recognized generic approaches in their products. There are genuine differences in the analysis synthesis techniques in use, but the average listener is unable to discriminate between them when the synthesized speech is limited to simple words and phrases, or even sentences.

Analysis Synthesis. Analysis synthesis is the name used to cover methods that derive synthetic speech from the recording and compression of the actual human voice as shown in Fig. 1. The two general approaches in use are: 1) time-domain synthesis, which produces a synthetic waveform representation of the original speech and 2) frequency-domain synthesis, which produces a parametric representation of the frequency spectrum of the original speech.

Time-Domain Synthesis. TDS constructs a synthetic speech waveform that sounds like the original waveform but has a very different voicegram as shown in Fig. 2, a plot of voice frequency versus time. About half of a synthetic waveform is silence. The waveform consists of many symmetric segments which range over a very restricted set of amplitude values.

This construction permits the TDS waveform to be described and stored with about 1 percent of the number of memory bits that are required to reproduce the original speech. The TDS waveform is constructed by adjusting the phases of all the frequencies in the original signal while keeping the amplitudes (the power spectrum) as close as possible to the signal.

Frequency-Domain Synthesis. FDS is applied in two general approaches that use parametric encoding of the (frequency) power spectrum of the speech waveform: formant synthesis and linear-predictive coding (LPC). In parametric encoding schemes, characteristics other than those in the original waveform are used in the analysis and synthesis of speech. These characteristics are used to control a mathematical synthesis model to create an output speech signal that is similar to the original. The speech output waveform of an FDS signal typically does not resemble the original speech introduced into the output analyzer. However, the objective of FDS is to duplicate the spectral shape of the speech signal as it is formed by the human vocal tract. It is not necessary that the speech output waveform be a reasonable match for the input waveform.

Formant Synthesis. Formant synthesis is an FDS technique that reproduces speech by recreating the spectral shape of the waveform. The formant center frequencies, their bandwidth, and the pitch periods are used as inputs. (A formant is a frequency region where the energy of the vowel sound is concentrated.) Formants are seen as frequency peaks in the voice spectrum.

Linear Predictive Coding. LPC, the second important frequency domain synthesis encoding technique, is based on a mathematical model of the human voice tract. Pitch and energy information as well as speech variables used to model the vocal tract are derived from recorded speech. The speech data are analyzed and encoded to produce input data for the digital model. The speaker's

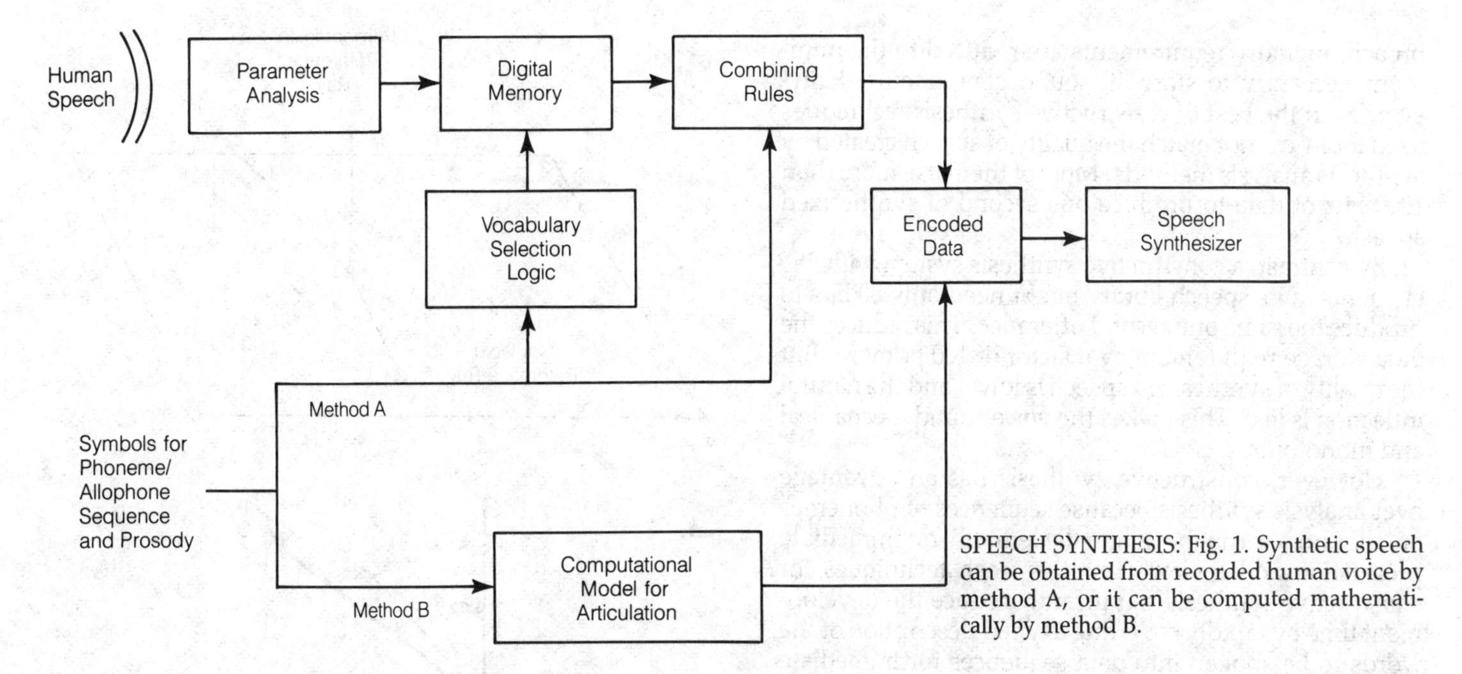

SPEECH SYNTHESIS: Fig. 1. Synthetic speech can be obtained from recorded human voice by method A, or it can be computed mathematically by method B.

voice, including intonation, accent, dialect, and pitch are accurately preserved by the LPC method of speech analysis in any language.

Pulse-Code Modulation. PCM is the easiest and lowest cost method for digitizing human speech and it is widely used in telecommunications. However, if used as a method for ESS, it would call for the largest amount of memory, making it a costly and inefficient method. *See* PULSE MODULATION.

The circuitry required for speech compression in TDS is more costly than the circuitry needed for PCM. But it results in a reduction in data rate and a savings in the amount of memory needed to store a given amount of speech.

The most widely used time domain synthesis encoding technique occurs at approximately 1800 bits per second (b/s). However, frequency domain techniques require even more sophisticated circuitry. For example, the circuitry needed for formant synthesis is typically able to encode at 1200 b/s, while the equipment used for the LPC techniques is capable of handling 1200 to 2400 b/s data rates.

The memory devices available from a number of semiconductor manufacturers are appropriate for storing words or phrases that can be played back through their integrated-circuit synthesizer chips on command. The naturalness and prosody (combination of pitch, duration, and intensity of speech) are preserved. In the simplest systems, every phrase that the system is able to reproduce must be stored in memory, and the phrases cannot be rearranged.

By contrast, the more complex systems offer constructive synthesis based on analysis synthesis approaches. These overcome the twin drawbacks of inflexibility and a limited amount of semiconductor memory. Based on speech building-block approaches, these produce individual sounds derived from analysis synthesis techniques that are joined together to create words and phrases. Individual sounds are based on speech elements called phonemes or extensions of these sounds called allophones.

A directory of individual words in one system can be stored in memory and then connected together to form phrases. This is a process called word concatenation. The fact that each word is stored only once and can be used repeatedly to form many different phases makes this approach flexible. However, despite this advantage, the technique is not popular because of the unnatural and artificial quality of phrases formed from words strung together with inadequate coarticulation.

The maximum flexibility of synthesized speech is achieved with the largest number of elemental sounds. There are approximately 490 phonemes in the English language—14 to 156 vowel sounds and 24 consonant sounds. By contrast, there are 128 allophones based on phonemes and they include the coarticulation variants of phonemes.

Almost any word or phrase can be created with constructive synthesis techniques today. With this ap-

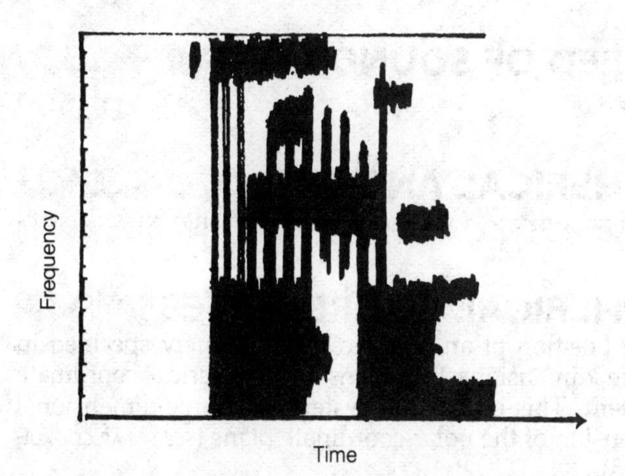

SPEECH SYNTHESIS: Fig. 2. A voicegram is a plot of frequency versus time showing the energy distribution in the spoken words.

proach, memory requirements are reduced to the minimum necessary to store all sound components. However, even the best of constructive synthesis techniques used today cannot match the quality of speech created by synthesis analysis methods. Most of them use more than 1000 bits of data to produce one second of synthesized speech.

By contrast, a constructive synthesis system with 256 elements in its speech library might need only 80 bits to produce the same one second utterance. This reduces the data storage requirement by a factor of 100 or more. But the quality of synthesized speech is lower and the natural inflection is lost. This makes the voice sound mechanical and monotonous.

However, constructive synthesis has an advantage over analysis synthesis because sequences of phonemes or allophones can be manually edited for immediate production and review. Text-to-speech techniques for constructive synthesis significantly reduce the development time by rapidly coverting a typed description of the words to be spoken into data sequences for immediate audition. Analysis synthesis, by contrast, requires sophisticated data collection and analysis before synthetic speech is produced.

As various approaches to speech synthesis are developed or improved, it is expected that some of their technical differences will disappear. Competitively priced encoding systems for LPC are expected to reduce the limitations on vocabulary development for analysis synthesis systems. Text-to-speech systems using constructive synthesis are also expected to become available as low-cost integrated circuits.

The hybrid approach to speech synthesis uses ICs to combine analysis synthesis and constructive synthesis methods in one system. The hybrid system can provide a flexible output including high quality speech with a large data storage requirement as well as low-quality speech with unlimited vocabulary.

SPEED OF LIGHT

See LIGHT.

SPEED OF SOUND

See SOUND.

SPHERICAL ANTENNA

See DISH ANTENNA, PARABOLOID ANTENNA in the ANTENNA DIRECTORY.

SPHERICAL COORDINATES

The position of an object can be uniquely specified in three dimensions by means of a spherical coordinate system. This coordinate system is a three-dimensional extension of the polar-coordinate plane (*see* POLAR COORDINATES).

The spherical coordinate scheme requires a reference point, a geometric plane passing through that point, a reference ray that lies in the plane and originates at the

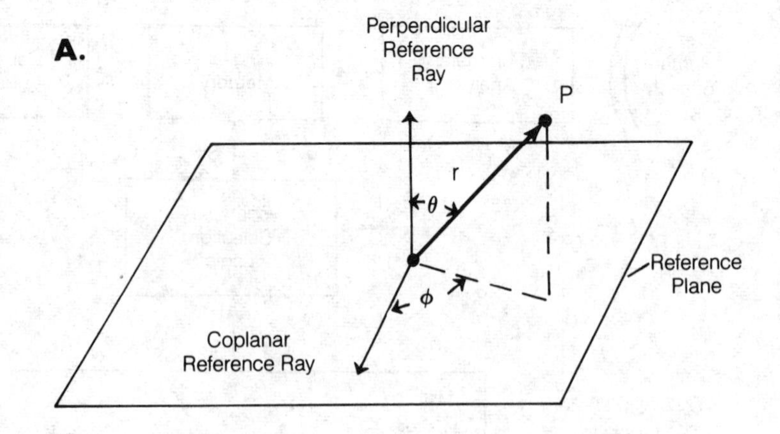

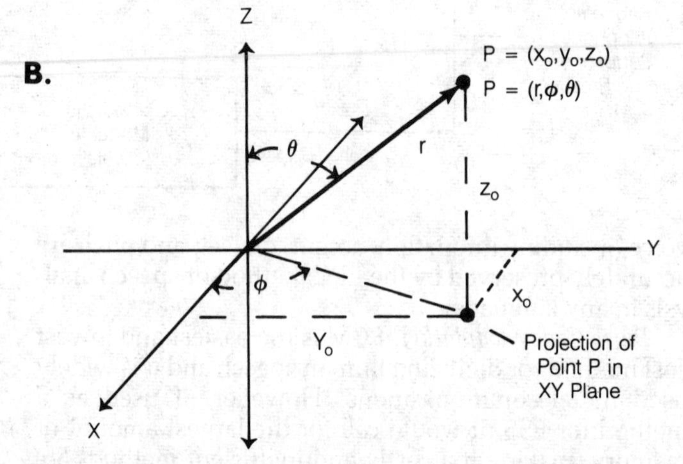

SPHERICAL COORDINATES: The scheme for spherical coordinates (A) and the coordinate orientation for transformation between the spherical and Cartesian coordinates (B).

point, and a reference ray that lies perpendicular to the plane and originates at the reference point. The position of a point in space is determined according to its distance *r* from the central point, and also according to two angles θ and ϕ, measured with respect to the two rays as shown in the illustration at A.

The distance *r* may be any nonnegative real number. The angle ϕ is measured counterclockwise from the ray in the reference plane. Accordingly, ϕ may range between 0 degrees and 360 degrees. The angle θ is measured between the perpendicular reference ray and a line connecting the origin point with the point in question. The value of θ can range from 0 to 180 degrees. The constraints, expressed mathematically, are:

$$r \geqslant 0$$
$$0 \leqslant \phi < 360$$
$$0 \leqslant \theta \leqslant 180$$

Using this scheme, each point in space corresponds to exactly one set of values (*r*, ϕ,θ), and each set of values (*r*, ϕ,θ) corresponds to exactly one point in space.

It is possible to convert three-dimensional Cartesian coordinates (x, y, z) to spherical coordinates (*r*, ϕ,θ), according to the illustration at B, by means of the following set of formulas:

$$r = \sqrt{x^2 + y^2 + z^2}$$

$$\phi = \arctan(y/x)$$
$$\theta = \arccos(z / \sqrt{x^2 + y^2 + z^2})$$

The inverse of this conversion is given by the formulas:

$$x = r \sin\theta \cos\phi$$
$$y = r \sin\theta \sin\phi$$
$$z = r \cos\theta$$

Spherical coordinates are used in various forms by astronomers. This coordinate scheme is also used in satellite communications and tracking. *See also* CARTESIAN COORDINATES, LATITUDE, LONGITUDE.

SPLITTER

A splitter is an impedance-matching device that allows more than one receiver to be used with a single antenna system. Splitters are commonly employed in television antenna systems and in cable-television installations where multiple receivers are desired.

Twin-lead television antenna systems generally have 300 ohms impedance. If two receivers with input impedances of 300 Ω are connected in parallel to a single 300 Ω line, a 2:1 impedance mismatch occurs. This will result in degraded reception at both receivers. A two-way splitter will eliminate this mismatch and improve the reception, especially if signals are weak.

Cable television systems have a typical characteristic impedance of 75 Ω. (A 1:4 impedance step-up transformer is used at television receivers to provide a match to the 300-ohm television input impedance.) Cable installations often have many receivers connected to a single incoming feed line. Each receiver has its own branch cable of 75 Ω impedance. A splitter provides a 1:1 match to all of the receivers, and to the incoming signal at the main line. Without a splitter, the mismatch at the junction might be 10:1, 20:1, or greater, resulting in very poor reception at each receiver.

Two-way splitters are available commercially with either 300 or 75 Ω systems. A four-way splitter can be made by combining three two-way splitters as shown in the illustration. A three-way splitter can be obtained by terminating one of the four-way splitter outputs with either a 300 or 75 Ω noninductive resistor, as appropriate.

Splitters are used in audio circuits, especially public-address systems with many speakers, or in multiple-listener systems with several sets of headphones connected to the output of a single audio amplifier. *See also* CHARACTERISTIC IMPEDANCE, IMPEDANCE MATCHING.

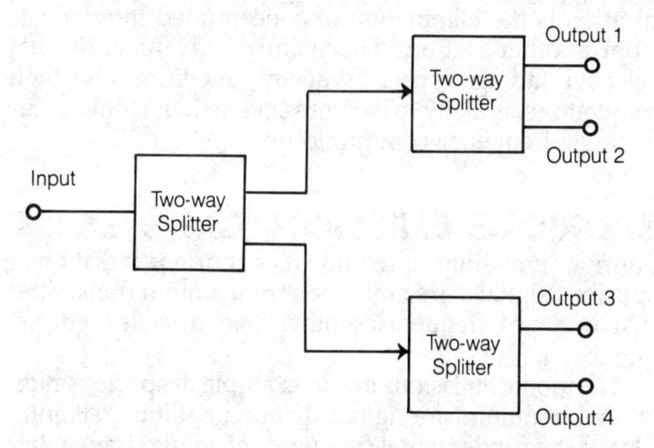

SPLITTER: Three two-way splitters can be combined to form a four-way splitter.

SPONTANEOUS EMISSION AND ABSORPTION

When an electron in an atom falls from a higher-energy shell to a lower-energy shell, a photon is produced. The wavelength of this photon depends on the energy transition through which the electron passes. This emission is called spontaneous emission.

If a photon with the right amount of energy is absorbed by an electron, that electron will jump to a higher-energy shell in the atom. This is known as spontaneous absorption (*see* ELECTRON ORBIT, PHOTON).

All energized matter, especially gases, produces emissions at specific wavelengths. Cooler gases exhibit absorption at well-defined electromagnetic wavelengths. These wavelengths show up as bright lines or dark lines in the spectrum of the gas, and they are called emission lines or absorption lines.

SPORADIC-E PROPAGATION

At certain radio frequencies the ionospheric E layer returns signals to earth. This propagation tends to be intermittent and conditions can change rapidly. For this reason it is known as sporadic-E propagation.

Sporadic-E propagation is most likely to occur at frequencies between 20 and 150 MHz. It is rarely observed at frequencies as high as 200 MHz. The propagation range is about several hundred miles, but occasionally communication is received over distances of 1,000 to 1,500 miles.

The standard frequency-modulation broadcast band can be affected by sporadic-E propagation. The same is true of the lower television channels, especially channel 2. Sporadic-E propagation often occurs on the amateur bands at 21 MHz through 148 MHz. *See also* E LAYER, IONOSPHERE, PROPAGATION CHARACTERISTICS.

SPREAD-SPECTRUM TECHNIQUES

Communications engineers have tried to minimize the bandwidth of transmitted signals to fit more signals into a given frequency band. Also, with narrow bandwidth less noise is received in proportion to the signal.

However, developments in communications have led to spread-spectrum communications techniques, in which the bandwidth is made very large deliberately, rather than very small.

Spread-spectrum transmission and reception are achieved by frequency modulation of a transmitter and receiver in exact synchronization. The variety of possible

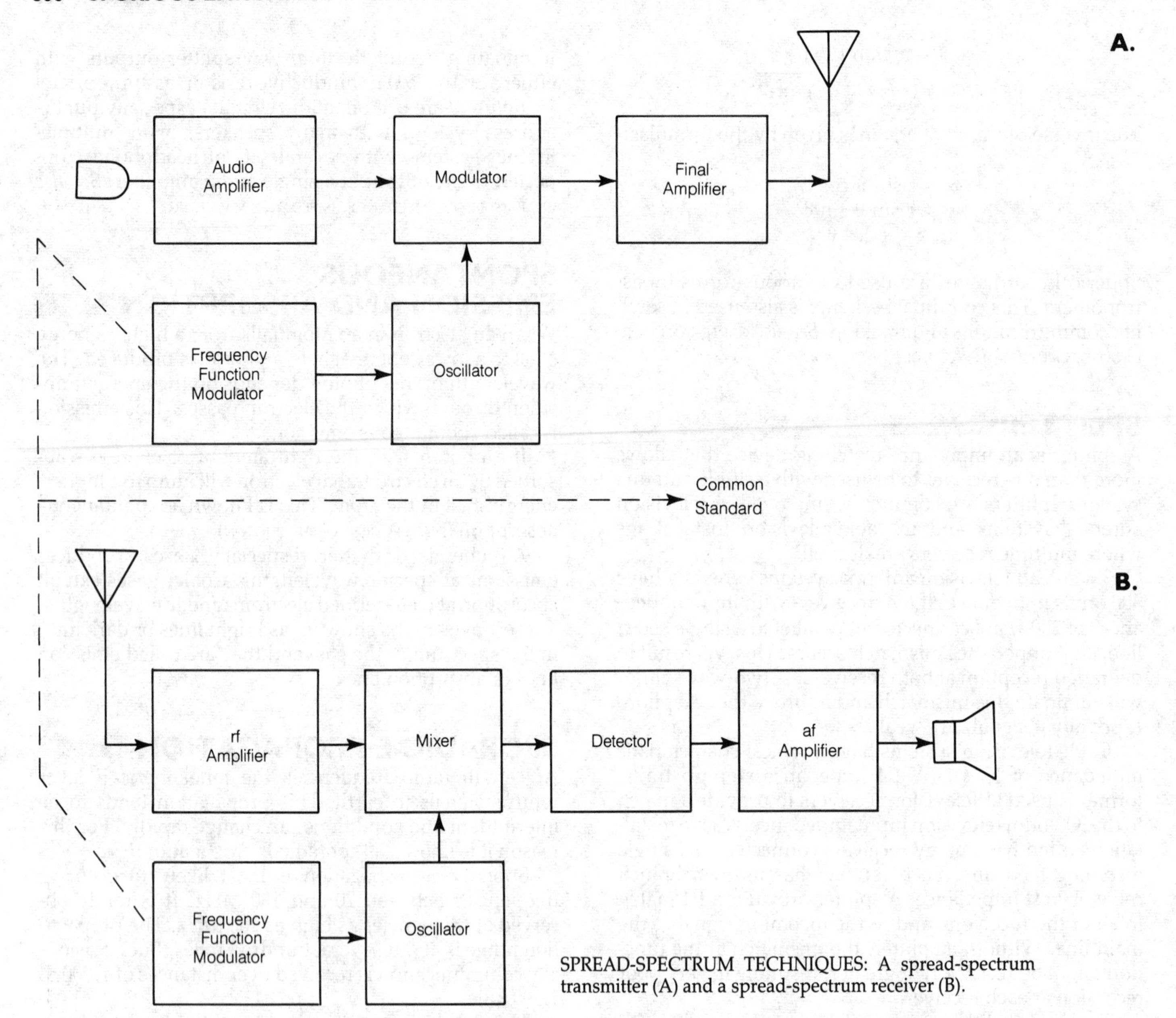

SPREAD-SPECTRUM TECHNIQUES: A spread-spectrum transmitter (A) and a spread-spectrum receiver (B).

deviation values and modulating waveforms is theoretically infinite. A signal with any type of emission—single sideband, frequency-shift keying, or any other mode—can be frequency modulated according to a specific scheme or function. The only requirement is that the receiver frequency follow exactly the changes in the transmitter frequency.

Different communications circuits can use different spread-spectrum functions so that no two will ever match. As more signals are placed in a given band, the apparent noise level rises, but the concentrated interference common to fixed-channel systems will not take place.

The illustration is a block diagram of a simple spread-spectrum transmitting and receiving setup. The local oscillators should, ideally, have the same nominal frequency. Both local oscillators are frequency-modulated by a function generator and reactance modulator. The function generators are synchronized so that the transmitter and the receiver are always on the same frequency. The deviation can be any value, although there is a practical limit to the size it can be without exceeding the transmitter-tuning or receiver-front-end passbands.

The primary advantage of spread-spectrum communication is the elimination of concentrated interference from another station. The main disadvantage is that selective fading, especially at low, medium, and high frequencies, is likely to be more severe than would occur in single-frequency communications.

SPURIOUS EMISSION

Spurious emissions, also known as spurious radiation or spurious signals, are emissions from a radio transmitter that occur at frequencies other than the desired frequency.

Harmonic emissions are an example of spurious radiation. Harmonics are signals that occur at integral multiples of the fundamental frequency of a radio transmitter (*see* HARMONIC, HARMONIC SUPPRESSION).

Spurious emissions can occur as a result of parasitic oscillation (*see* PARASITIC OSCILLATION, PARASITIC SUPPRESSOR). Spurious emissions can also result from inadequate selectivity in the output stages of a transmitter.

In the United States, the maximum allowable level of spurious emissions is dictated by the Federal Communications Commission (FCC). This protects the various radio services from interference that could be caused by improperly operating radio transmitters.

A transmitter can be checked for spurious emissions with an instrument called a spectrum analyzer. *See also* SPECTRUM ANALYZER.

SPURIOUS RESPONSE

A radio receiver may pick up signals that exist at frequencies other than the frequency to which the receiver is tuned. These signals are known as spurious responses.

Spurious responses in a receiver can take place as a result of mixing between two or more external signals. This mixing can occur in the front end of a receiver, especially if one signal is strong enough to cause nonlinear operation of this stage. Spurious signals can result from mixing in nonlinear junctions external to the receiver (*see* FRONT END, INTERMODULATION).

Image signals in a superheterodyne receiver are an example of spurious responses. These signals can cause severe interference in an improperly designed receiver (*see* IMAGE FREQUENCY, IMAGE REJECTION).

A receiver can be checked for image responses by using a signal generator and for intermodulation distortion by using two signal generators.

SQUARE-LAW DETECTOR

A square-law detector is a form of envelope detector (*see* ENVELOPE DETECTOR). The square-law detector is so named because the root-mean-square output-signal current is proportional to the square of the root-mean-square input-signal voltage.

The square-law detector operates because of the nonlinear response of the circuit. This detector also is called a weak-signal detector.

SQUARE-LAW METER

Some analog meters respond to an applied quantity according to the square of that quantity. This is true, for example, in wattmeters that actually measure the voltage across a particular resistance. This meter is called a square-law meter.

The square-law meter has a characteristic nonlinear scale. The graduations are progressively closer together toward the right-hand end of the scale, and are widely spaced near the left-hand end (see illustration). *See also* ANALOG PANEL METER.

SQUARE-LAW RESPONSE

A circuit or device has a square-law response when the output is proportional to the square of the input, or when the deflection of a meter is proportional to the square root of the input signal magnitude. A square-law curve is parabolic (*see* PARABOLA).

In general, if x represents the input quantity for a square-law circuit and y represents the output magnitude, then:

$$y = kx^2 + c$$

where k and c are constants. In the case of a square-law meter, if x represents the input-signal magnitude and y represents the meter-needle deflection, then:

$$x = ky^2 + c$$

where k and c, again, are constants. In both cases, the value of k depends on the units specified for input and output or scale deflection, and the value of c depends on the bias of the circuit or the starting point of the meter scale. *See also* SQUARE-LAW DETECTOR, SQUARE-LAW METER.

SQUARE WAVE

A square wave is a special form of alternating-current (ac) or pulsating direct-current (dc) waveform. The ideal amplitude transitions, both rise and decay, take place instantaneously. When displayed on an oscilloscope, a square wave appears as two parallel, dotted lines, sometimes with faint vertical traces connecting the ends of the line segments (see illustration at A).

Square waves can have equal positive and negative peaks, or the peaks might be unequal. Unequal peaks result from the combination of an alternating square-wave current and a direct current. If the magnitude of the dc component is exactly equal to the peak magnitude of the square-wave current, a train of pulses results, as

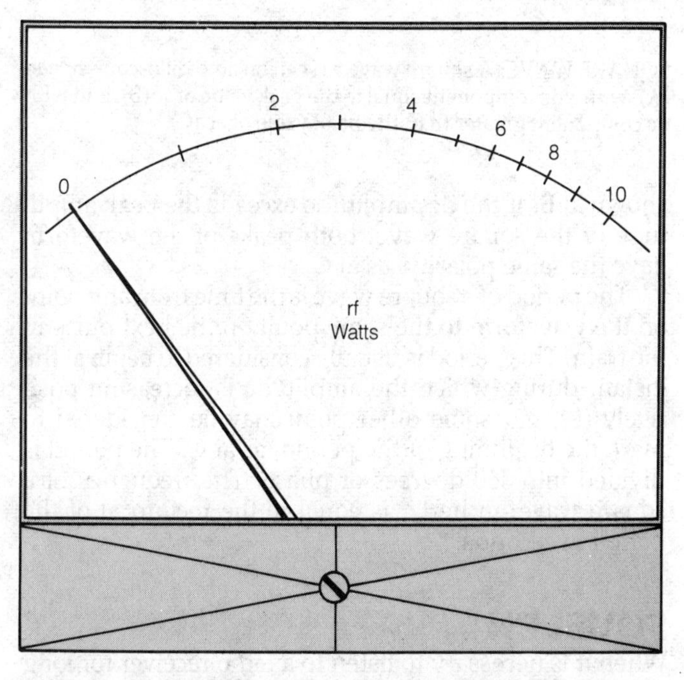

SQUARE-LAW METER: The scale of a square-law meter is nonlinear.

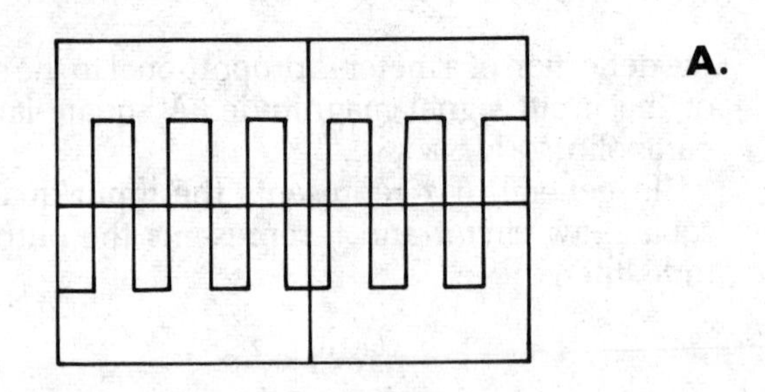

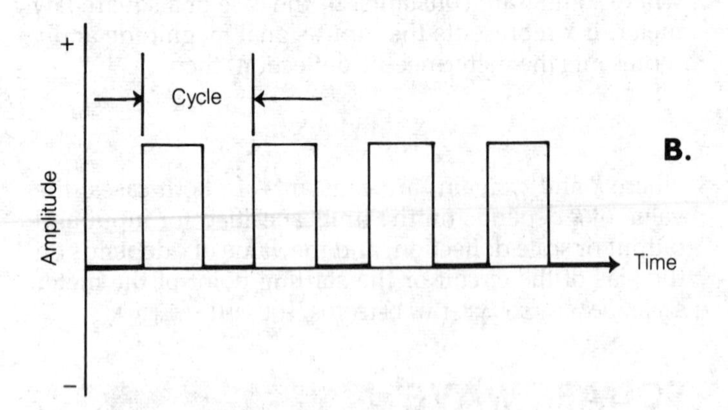

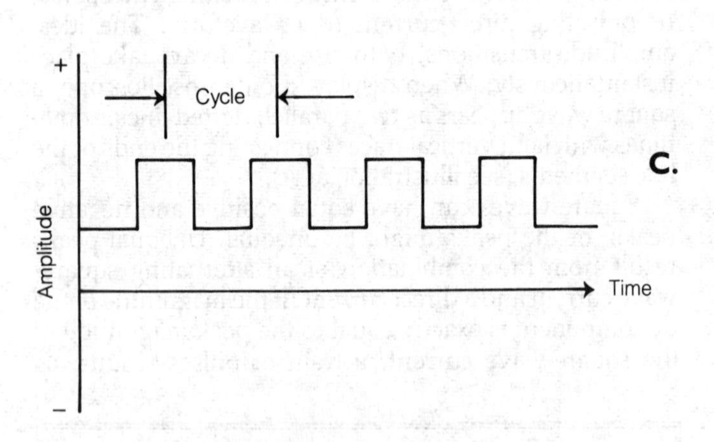

SQUARE WAVE: A square wave as seen on an oscilloscope screen (A), with a dc component equal to the peak value of ac (B), and with dc component greater than the peak value of ac (C).

shown at B. If the dc amplitude exceeds the peak amplitude of the square wave, both peaks of the waveform have the same polarity, as at C.

The period of a square wave is the time from any point on the waveform to the same point on the next pulse in the train. This period is usually considered to begin at the instant during which the amplitude is increasing positively (B), but some other point may be considered to mark the beginning of the period, as at C. The period is divided into 360 degrees of phase. The frequency of a square wave, in hertz, is equal to the reciprocal of the period in seconds.

SQUELCH

When it is necessary to listen to a radio receiver for long periods and useful signals are present infrequently, the constant hiss or roar can become annoying. A squelch circuit silences a receiver when no signal is present while allowing reception of useful signals when they arrive.

Squelch circuits are used in channelized Citizen's-Band and very high frequency communications transceivers. Squelch circuits are less common in continuous-tuning radio receivers. Most frequency-modulation (FM) receivers use squelching systems.

A typical squelching circuit is shown in the diagram. This squelch circuit operates in the audio-frequency stages of an FM receiver. The squelch is actuated by the incoming audio signal. When no signal is present, the rectified hiss produces a negative direct-current voltage, cutting off the field-effect transistor (FET) and keeping the noise from reaching the output. When a signal appears, the hiss level is greatly reduced, and the FET conducts, allowing the audio to reach the output. The cut-off squelch circuit is said to be closed; the conducting squelch is said to be open.

In most receivers the squelch is normally closed when no signal is present. A potentiometer permits adjustment of the squelch so that it can be opened, if desired, allowing the receiver hiss to reach the speaker. This control also provides for adjustment of the squelch sensitivity. Most squelch controls are open when the knob is turned fully counterclockwise. As the knob is rotated clockwise, the receiver hiss abruptly disappears; this point is called the squelch threshold. At this point, even the weakest signals will open the squelch. As the squelch knob is turned farther clockwise, stronger and stronger signals are required to open the squelch.

In some receivers the squelch will not open unless the signal has certain characteristics. This is called selective squelching. It is used in some repeaters and receivers to prevent undesired signals from being heard. The most common methods of selective squelching use subaudible-tone generators or tone-burst generators. Selective squelching is also known as private-line (PL) operation. *See also* SQUELCH SENSITIVITY.

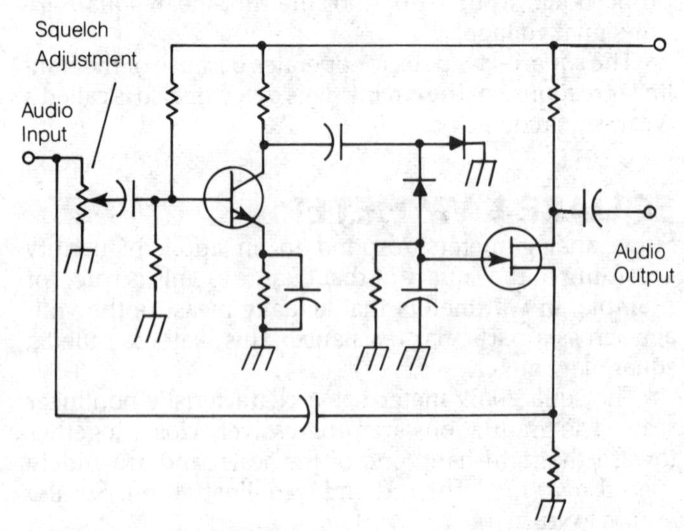

SQUELCH: An audio squelch circuit for use with a frequency-modulated receiver.

SQUELCH SENSITIVITY

The squelch sensitivity of a squelched receiver is the

signal level, in microvolts at the antenna terminals, that is required to keep the squelch open (*see* SQUELCH). The squelch sensitivity of a receiver depends on several factors.

Virtually all squelch controls allow for some adjustment of the squelch sensitivity with a control knob. The squelch sensitivity is greatest (the least signal is required to open it) at the threshold setting of the control.

An unmodulated carrier will usually open a squelch at a lower level than will a modulated carrier. If a frequency-modulated (FM) signal has excessive deviation, the squelch might cut off on modulation peaks. This effect is known as squelch blocking.

The squelch sensitivity of an FM receiver is related to its noise-quieting sensitivity. The better the quieting sensitivity, the less signal is required to actuate the squelch at the threshold.

Squelch sensitivity is measured with a calibrated signal generator. The signal is modulated at a frequency of 1 kHz with a deviation of plus or minus 3 kHz. The generator output is increased until the squelch remains continuously open at the threshold setting. *See also* NOISE QUIETING.

STABILITY

Stability is a measure of how well a component, circuit, or system maintains constant operating conditions over a period of time. For example, there is frequency stability of a radio receiver, stability of the output power of a radio transmitter, or the voltage-output stability of a power supply. The stability of an electronic component is usually called tolerance, and the stability of a power supply is known as the regulation (*see* REGULATION, TOLERANCE).

STAGGER TUNING

Stagger tuning is a method of aligning a bandpass filter or the intermediate-frequency chain of a receiver or transmitter. In a stagger-tuned filter, there are several tuned circuits, each set for a slightly different frequency. In the stagger-tuned amplifier chain, each stage is tuned to a higher or lower frequency than the stage immediately before or after it.

In a bandpass filter, stagger tuning results in a more nearly rectangular response than would be obtained if all of the resonant circuits or devices were tuned to the same frequency. Stagger tuning broadens the bandpass response and provides steep skirts (*see* BANDPASS RESPONSE). In a multistage, radio-frequency amplifier, stagger tuning reduces the possibility of interstage oscillation. If all of the tuned circuits were set for exactly the same resonant frequency, the positive feedback among the various amplifiers might be enough to cause instability (*see* FEEDBACK).

STANDARD BROADCAST BAND

See BROADCAST BAND.

STANDARD CAPACITOR

A standard capacitor is a capacitor manufactured to have a known, stable value. Standard capacitors are used in precision test equipment, particularly frequency meters and impedance bridges, to ensure the highest possible degree of accuracy. The tolerance of a standard capacitor is very low, and the temperature coefficient is essentially zero.

STANDARD CELL

A standard cell is an electrochemical cell used to provide a constant, known voltage. There are different standard cells used for making electronic measurements. The most common is the Weston cell. *See* WESTON STANDARD CELL.

STANDARD-CELL INTEGRATED CIRCUIT

A standard cell is a semicustom integrated circuit (IC) that is completely designed with the aid of a computer workstation and software that includes an extensive library of characterized and pretested macrocells and macrofunctions. The standard cell is considered to be an option to the custom-designed IC and the gate array. It is an applications specific IC as are the gate array and programmable logic device (PLD). *See* APPLICATIONS-SPECIFIC DEVICE, GATE ARRAY, PROGRAMMABLE LOGIC DEVICE.

In contrast to the gate array, which is 70 to 80 percent complete prior to the final one or two masking steps that dedicate the device, the standard cell is manufactured completely from a blank semiconductor wafer. The computer-aided design (CAD) procedure permits the selection of only those gates, memory bits, and I/O pads that will actually be used in the final IC. This makes it likely that the chip size will be smaller than that of an equivalent gate array in the same technology.

Standard cells have all the advantages of gate arrays, and they are selected over gate arrays where the chip size reduction is significant for economic and technical reasons. They are also selected if the purchase quantity is larger and more predictable. Other benefits of standard cells include:

1. Fewer IC packages on a printed-circuit (PC) card and smaller PC cards than are possible with standard digital logic devices.
2. Increased system reliability and performance.
3. Reduced interconnections and connectors.
4. Lower power supply requirements.

The standard cell can be completely digital logic (including memory and microprocessors), completely analog circuitry, or combinations of them. Standard cell libraries include memory cells and analog elements such as operational amplifiers. The functional blocks, referred to as macrocells (or macros) and macrofunctions, permit a wide selection of precisely characterized logic or other circuitry. Because they are completely designed on a CAD workstation, the selection of macrocells and macrofunctions may be optimized. This eliminates the redundant elements that occur on gate arrays. However, complete computer-aided design results in higher nonrecurring engineering (NRE) charges, generally longer delivery time and higher unit cost than are incurred with gate arrays.

Standard cells differ from full hand-crafted custom ICs primarily in their more complete dependence on CAD and the library of predesigned macrocells and macrofunctions. However, full custom design might make more efficient use of the wafer than a standard cell because of the intercession of a skilled designer. Although many of the same workstation and software tools will be used, the designer can modify the macrocells or macrofunctions in the software library or even introduce new designs more appropriate to the application.

This approach permits a higher degree of optimizing in the custom design because designer experience can provide further reduction in the amount of wafer area required for the IC. However, NRE charges are higher for a custom IC, and delivery time is longer. Unit costs will be lower only in high-volume purchases.

Many different factors must be considered by a user in determining the most suitable standard cell technology for a design. These include gate complexity, bus structures, speed requirements, pin count, package style, input and output buffers, power pins, and testability.

Standard cell density is normally measured in equivalent gates, a generally accepted measure in the industry, although not a fixed standard. An equivalent gate is typically a two-input NAND gate. Macrocells, the basic logic elements, include inverters, NAND logic, NOR logic, latches, flip-flops, decoders, multiplexers, shift registers, and buffers. Macrofunctions include adders, arithmetic logic units, comparators, decoders, flip-flop registers, and counters.

Standard cells are available in TTL (transistor-transistor logic) and ECL (emitter-coupled logic) bipolar and CMOS (complementary metal-oxide semiconductor) technology, gallium arsenide, and silicon-on-sapphire. They are also fabricated in mixed technologies with the I/O circuits selected to provide a better match to the host system. For example, the array might be ECL while the I/O sections are TTL. Also, the array might be CMOS while the I/O sections are TTL.

ECL standard cells permit subnanosecond (ns) delay times. CMOS and bipolar standard cells meet requirements in the 1 to 5 ns delay region. Geometries are in the 1.5- to 3-micron size range. Where delays in excess of 5 ns are acceptable, single-layer 5-micron CMOS devices are available. *See* LOGIC.

Commercial manufacturers are continually increasing the number of equivalent gates in standard cells as in gate arrays and other ICs. In a short span of five years, feature size or line width has been reduced from 5 microns to 1.0 micron.

Distinctions between standard cells and full custom ICs are being erased as the technologies converge. Expert system software in CAD workstations called silicon compilers includes the accumulated experience of skilled designers and permits the workstation operator to make an optimum selection of macrocells and macrofunctions for any given application. *See* SILICON.

STANDARD DEVIATION

Standard deviation is a measure of the variation of a set of data values varies above and below the mean value. The standard deviation is usually designated by the Greek lowercase letter sigma (σ) although other symbols, such as the capital S or the small s, are also used.

The standard deviation of a distribution with a finite number of values is defined as the root mean square of the deviations of all the individual values. For a large number of samples, standard deviation is best calculated with a computer. In the case of a continuous distribution, the standard deviation is determined by integration. *See also* NORMAL DISTRIBUTION.

STANDARD FREQUENCY

See REFERENCE FREQUENCY.

STANDARD INDUCTOR

A standard inductor is an inductor manufactured to have a precise, known, and essentially unchanging value. Standard inductors are used in precision test equipment, especially frequency-measuring apparatus or impedance bridges. The tolerance is very low, and the temperature coefficient is essentially zero.

STANDARD INTERNATIONAL SYSTEM OF UNITS

In 1960, the scientists of the world agreed on a universal scheme of measurement units known as the Standard International (SI) System of Units. This system is summarized in the table.

The meter-kilogram-second (MKS) measurement scheme is considered a subsystem of the SI System. Some scientists use a nonstandard scheme in which the basic units are the centimeter, the gram, and the second. This is called the CGS system. *See also* AMPERE, CANDELA, KELVIN TEMPERATURE SCALE, KILOGRAM, METER, MOLE, SECOND.

STANDARD INTERNATIONAL SYSTEM OF UNITS: Quantities, units, and symbols.

Quantity	Unit	Most Common Symbol
Displacement	Meter	m
Mass	Kilogram	kg
Time	Second	s
Current	Ampere	A
Temperature	Degree Kelvin	K
Luminous Intensity	Candela	cd
Quantity of Atoms	Mole	mol

STANDARD RESISTOR

A standard resistor is a noninductive, fixed-value resistor with a precise value that does not change significantly in time. Standard resistors are commonly used in precision test instruments, especially resistance-measuring bridges. The standard resistor has a low tolerance and a temperature coefficient near zero.

STANDARD TEST MODULATION

Modulated signals are used to evaluate the distortion characteristics of radio receivers. A signal generator that produces modulation at a frequency and percent of modulation required for test purposes is called standard test modulation.

For amplitude-modulation receivers, a test signal is 100-percent modulated by a pure sine-wave audio tone with a frequency of 1 kHz. For frequency-modulation and phase-modulation communications receivers the carrier is modulated at a deviation of plus or minus 3 kHz by a pure sine-wave tone of 1 kHz. For single-sideband equipment, two sine-wave tones of equal amplitude, both within the passband (between 300 Hz and 3 kHz), and spaced 1 kHz apart, are used. (A common combination is 1 kHz and 2 kHz.) The audio output of the receiver is then analyzed with a distortion analyzer and/or oscilloscope. *See also* DISTORTION ANALYZER.

STANDARD WORLD TIME

The common worldwide time standard is based on the solar time at the Greenwich Meridian. This meridian is assigned 0 degrees of longitude and passes through Greenwich, England. This common time is known as Coordinated Universal Time, and is abbreviated UTC (*see* COORDINATED UNIVERSAL TIME). Other countries of the world have their own local standard times, which can differ from UTC.

The continental United States is divided into four time zones: Eastern, Central, Mountain, and Pacific. During the part of the year in which standard time is in effect, these zones are behind UTC by 5, 6, 7, and 8 hours, respectively. During the part of the year in which daylight-savings time is in effect these zones are behind UTC by 4, 5, 6, and 7 hours, respectively. (Some states and counties do not use daylight time, so their time lags behind UTC remain constant all year.)

Locations in the western hemisphere of the world run from 0 to 12 hours behind UTC, and places in the eastern hemisphere run from 0 to 12 hours ahead of UTC. An approximate idea of the local standard time at any given location can be determined from its longitude (*see* LONGITUDE). Consider west longitude values to be negative, ranging from 0 to −180 degrees and east longitude values to be positive, from 0 to +180 degrees.

For the western hemisphere, divide the west longitude in degrees by 15 to get a number between 0 and −12. Round this number off to the nearest whole integer and add the result to the time in UTC. (Use 24-hour time: 0000 for 12:00 midnight, 0100 for 1:00 A.M., and so on up to 2400 for 12:00 midnight the following night.) If the final value is less than 0000, add 2400 and subtract one day.

For the eastern hemisphere, divide the east longitude in degrees by 15 to get a number between 0 and +12. Round this number off to the nearest whole integer and add the result to the time in UTC. If the final value is greater than 2400, subtract 2400 and add one day. This will give the approximate local time in the part of the world of interest.

This scheme is not completely reliable because time zones are not strictly based on longitude. Some countries, such as China, have chosen to use the same standard time everywhere in their country, although the land spans much more than 15 degrees of longitude. Other countries run a fractional number of hours ahead of UTC or behind UTC. (Tonga, for example, runs 12 hours and 20 minutes ahead of UTC.)

STANDING WAVE

A standing wave is an electromagnetic wave that occurs on a transmission line or antenna radiator when the terminating impedance differs from the characteristic impedance. The transmission line can be a simple pair of wires, a coaxial cable, or a waveguide for microwave frequencies.

Standing waves are generally present in resonant antenna radiators. The terminating impedance at the end of a typical radiator, which has a finite physical length, is a practically infinite, pure resistance. This means that essentially no current flows there. The standing-wave pattern set up is shown in the illustration on p. 806.

The pattern of standing waves on a half-wave resonant radiator fed at the center is shown at A. The pattern of standing waves on a full-wave resonant radiator fed at the center is shown at B. The pattern of standing waves on a full-wave resonant radiator fed at one end is shown at C. Notice the difference between the pattern for current and the pattern for voltage.

Some antenna radiators do not exhibit standing waves; instead, the current and voltage are uniform along the length of the radiating element. This happens in antennas that have a terminating resistor with a value equal to the characteristic impedance of the radiator (about 600 ohms). Some longwire and rhombic antennas use resistors to obtain a unidirectional pattern (*see* ANTENNA PATTERN).

In a transmission line that is terminated in a pure resistance with a value equal to the characteristic impedance of the line, no standing waves occur. This is a desirable condition because standing waves on a transmission line contribute to power loss. *See also* CHARACTERISTIC IMPEDANCE, STANDING-WAVE RATIO.

STANDING-WAVE RATIO

A transmission line terminated in an impedance that differs from the characteristic impedance (*see* CHARACTERISTIC IMPEDANCE) exhibits a nonuniform distribution of current and voltage along the length of the line. The greater

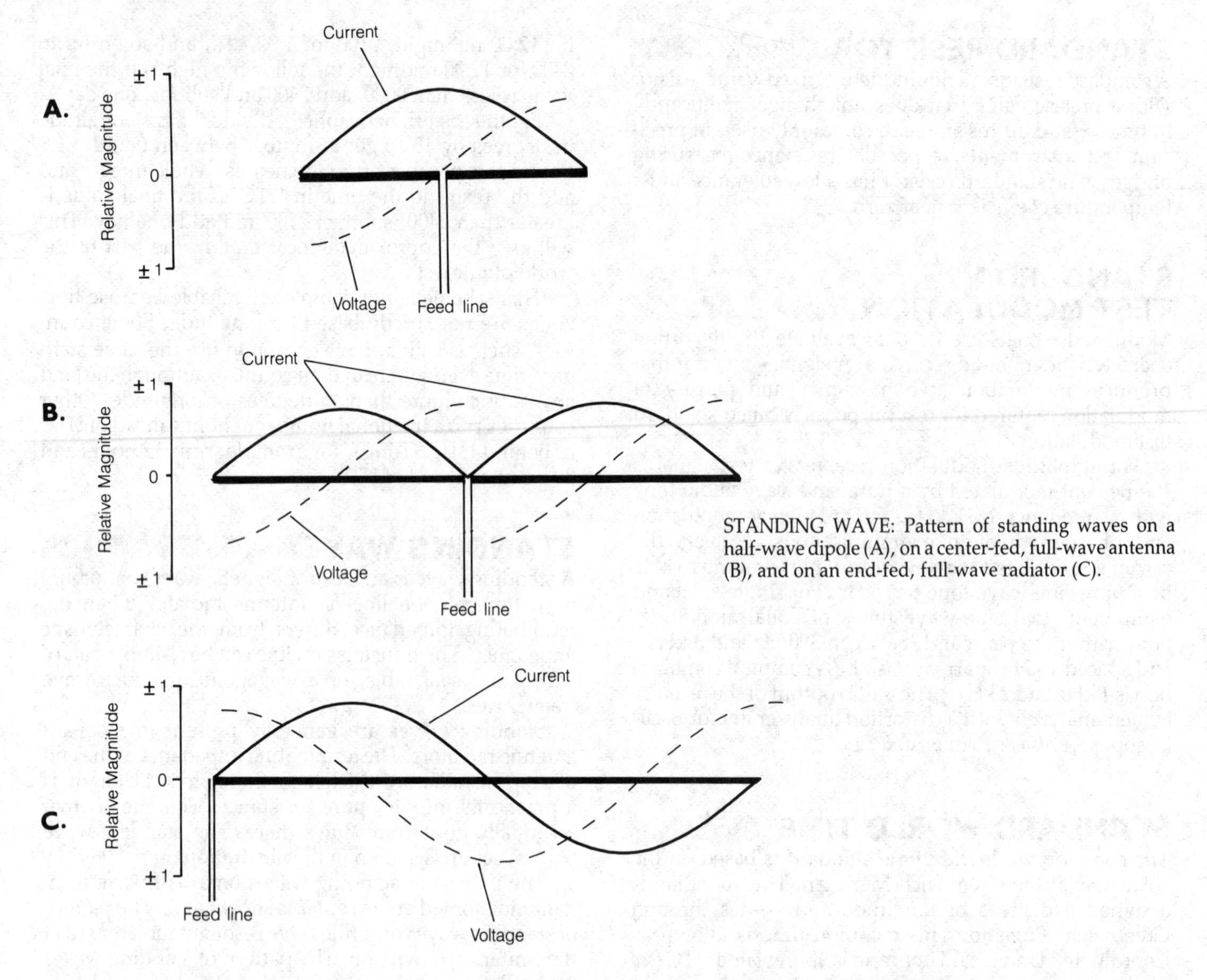

STANDING WAVE: Pattern of standing waves on a half-wave dipole (A), on a center-fed, full-wave antenna (B), and on an end-fed, full-wave radiator (C).

the mismatch, the more nonuniform this distribution becomes. The transmission line can be two-wire, coaxial cable, or waveguide.

The ratio of the maximum voltage to the minimum voltage, or the maximum current to the minimum current, is called the standing-wave ratio (SWR) on the transmission line. The standing-wave ratio is 1:1 if the current and voltage are in the same proportions everywhere along the line.

The SWR can be 1:1 only when a transmission line is terminated in a pure resistance with the same ohmic value as the characteristic impedance of the line. If the load contains reactance, or if the resistive component of the load impedance is not the same as the characteristic impedance of the line, the SWR cannot be 1:1. Then, standing waves will exist along the length of the line (*see* STANDING WAVE). In theory, there is no limit to the magnitude of SWR. In the extreme cases of a short circuit, open circuit, or pure reactance at the load end of the line, the SWR is theoretically infinite. In practice, line losses and loading effects prevent the SWR from becoming infinite, but it may be as great as 100:1 or more.

The SWR is important as an indicator of the performance of an antenna system because a high SWR indicates a severe mismatch between the antenna and the transmission line. This can have an adverse effect on the performance of a transmitter or receiver connected to the system. An extremely large SWR can result in significant signal loss in the transmission line. If a high-power transmitter is used, the large currents and voltages caused by a high SWR can actually damage the line. *See also* IMPEDANCE, REFLECTED POWER, REFLECTOMETER, STANDING-WAVE-RATIO LOSS.

STANDING-WAVE-RATIO LOSS

No transmission line is lossless. Even the most carefully designed line has some loss because of the ohmic resistance of the conductors and the imperfections of the dielectric material. In any transmission line, the loss is smallest when the line is terminated in a pure resistance equal to the characteristic impedance of the line: that is, when the standing-wave ratio (SWR) is 1:1.

If the SWR is not 1:1, the line loss increases over its base value at an SWR of 1:1. This additional loss is called

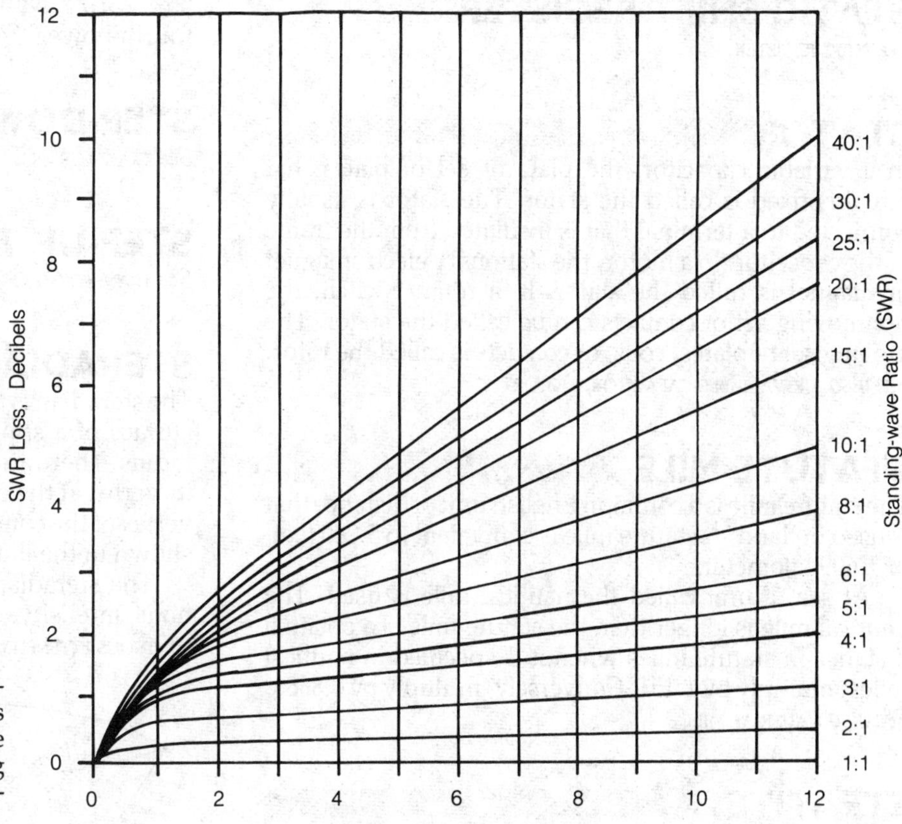

STANDING-WAVE-RATIO LOSS: Standing-wave ratio loss can be determined with this chart. The matched-line loss is found on the horizontal axis and the SWR is found among the family of curves. Additional loss is indicated on the vertical scale.

standing-wave-ratio loss, or SWR loss. The illustration shows the SWR loss that occurs for various values of matched-line loss and SWR. The SWR loss is generally not serious until the SWR becomes greater than 2:1. *See also* STANDING-WAVE RATIO.

STANDING-WAVE-RATIO METER

See REFLECTOMETER.

STATIC

See STATIC CHARACTERISTIC, STATIC ELECTRICITY.

STATIC CHARACTERISTIC

The steady-state behavior of a component, circuit, or system under fixed conditions (as contrasted to dynamic behavior under changing conditions) is known as the static characteristic. The static characteristic of a component is determined with direct currents that do not change rapidly with time. For example, the characteristic curve of a transistor is determined by making measurements as various currents or voltages are applied. The static characteristic of a circuit or system is an expression of its behavior with no signal input or with a direct-current signal input.

When a component is tested with an alternating current or a fluctuating direct current, its characteristics usually differ from those in the static condition. A diode, for example, does not function the same way at 1 GHz as it does with the application of direct current. Even simple components such as resistors and capacitors behave differently with alternating currents than direct currents. The behavior of a component under rapidly changing conditions is called the dynamic characteristic.

STATIC ELECTRICITY

Static electricity is stationary electricity. It is characterized by the existence of a potential difference between two objects without the flow of current, or the existence of an excess or deficiency of electrons on an object. The modifier *static* applies because no charge transfer takes place.

A static electric charge generates an electric field, the intensity of which depends on the potential difference and distance between two objects, or the quantity of charge and its concentration on a single object (*see* CHARGE, ELECTRIC FIELD).

If a static electric potential between two objects becomes large enough, a discharge will occur. The greater the separation distance between the objects, the larger the potential difference must be in order to cause the discharge. *See also* ELECTROSTATIC DISCHARGE, ELECTROSTATIC DISCHARGE CONTROL.

STATIC MEMORY

See SEMICONDUCTOR MEMORY.

STATIC SHIFT REGISTER

See SHIFT REGISTER.

STATOR

In a variable capacitor, the plate or set of plates that remains fixed is called the stator. The stator is usually connected to a terminal that is insulated from the frame of the capacitor. In a motor, the stationary electromagnet or magnet is called the stator. In a rotary switch, the nonmoving set of contacts can be called the stator. The moving set of plates, coils, or contacts is called the rotor. *See also* AIR-VARIABLE CAPACITOR, MOTOR.

STATUTE MILE

The statute mile is a common English unit of distance that is used on land. A statute mile is equivalent to 5,280 feet, or 1.609 kilometers.

At sea, a unit called the nautical mile is used. The nautical mile is longer than the statute mile. To obtain a distance in statute miles when it is specified in nautical miles, multiply by 1.151. Conversely, multiply by 0.8688. *See also* NAUTICAL MILE.

STEATITE

Steatite is a ceramic material made principally from magnesium silicate. Steatite has excellent dielectric properties and is used to make insulators for radio-frequency (rf) equipment. The dielectric constant of steatite is approximately 5.8 throughout the rf spectrum. Steatite looks something like porcelain. *See also* DIELECTRIC.

STEEL WIRE

Steel is extensively used in the manufacture of wire requiring high tensile (breaking) strength. In antenna systems, copper-clad steel wire is used (*see* COPPER-CLAD WIRE). Steel is also commonly employed in the manufacture of guy wire.

Steel wire is available in solid or stranded form in the American Wire Gauge (AWG) sizes. It has a maximum breaking strength of approximately twice that of annealed copper wire, or 1.5 times that of hard-drawn copper wire. *See also* AMERICAN WIRE GAUGE.

STEFAN-BOLTZMANN LAW

The Stefan-Boltzmann law is a function that relates the temperature of a black body, or perfectly radiating object, with the amount of power it radiates per square meter of surface area. If M represents the radiant emittance in watts per square meter, and T represents the absolute temperature in degrees Kelvin, then:

$$M = \sigma T^4$$

where σ is a constant called the Stefan-Boltzmann constant. The value of this constant is approximately 5.67×10^{-8} watts per square meter per degrees K to the fourth power. *See also* BLACK BODY.

STEP-DOWN TRANSFORMER

See TRANSFORMER.

STEP-UP TRANSFORMER

See TRANSFORMER.

STERADIAN

The steradian is the solid angle subtending an area on the surface of a sphere equal to the square of the sphere's radius. There are 4 π steradians in a sphere. The cone has its vertex at the center of the sphere, and the angle at the vertex of the cone has a solid measure of one steradian, as shown in the illustration.

The steradian is used in the determination of luminous intensity. *See also* CANDELA, LUMEN, LUMINOUS FLUX, LUMINOUS INTENSITY.

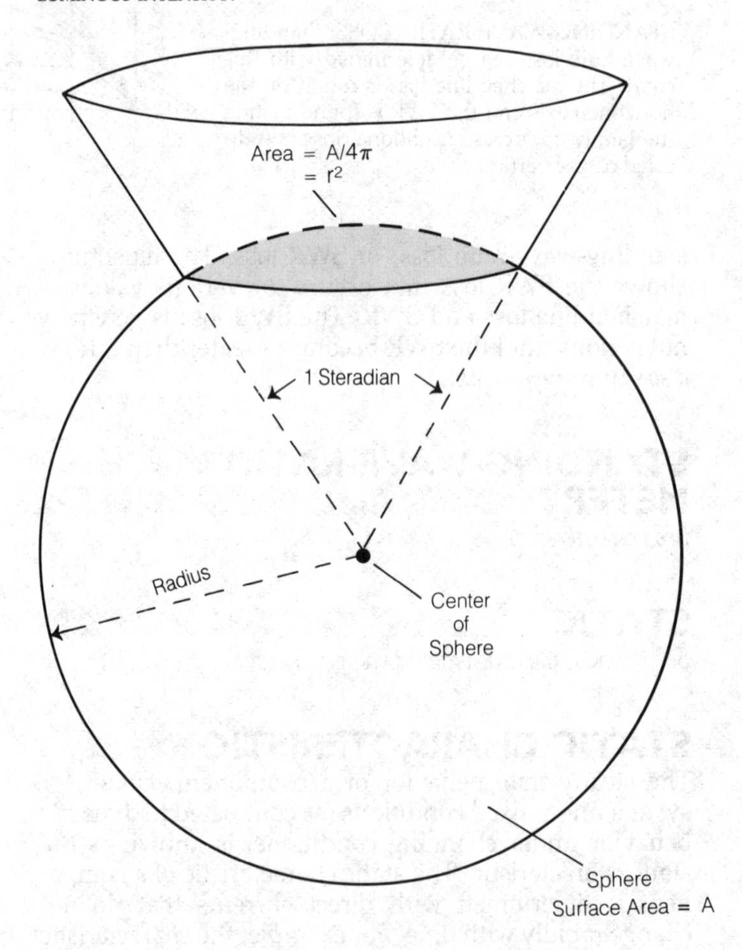

STERADIAN: The steradian is a unit of solid angular measure.

STEREO

See STEREOPHONICS.

STEREO BROADCASTING

See FM STEREO, STEREO MULTIPLEX.

STEREO DISK RECORDING

Two independent sound channels can be impressed onto a single groove in a disk, thereby obtaining stereo sound. This is done by a stereo disk recorder.

The stereo disk recorder cuts a groove by vibrating in the plane perpendicular to the groove. The left-hand-channel information is impressed in the groove by vibrations that occur at a 45-degree angle with respect to the surface of the disk. The right-hand-channel information is impressed in the groove by vibrations at a right angle to those of the left-hand channel (see illustration).

Once the disk recording process is complete, a mold is made from the disk. They are mass produced by pressing this metal mold into disks of vinyl plastic.

When a stereo disk is played back the stylus follows in the groove, vibrating in the same way as did the original cutting stylus. Since the left-hand and right-hand channel vibrations occur at right angles to each other, their instantaneous amplitudes are entirely independent. *See also* STEREOPHONICS.

STEREO MULTIPLEX

Stereo multiplex is a method of broadcasting two independent channels of sound over a single radio-frequency carrier to obtain stereo sound in a specially designed receiver.

To accomplish stereo multiplex, the left-hand and right-hand audio signals modulate low-frequency carriers (subcarriers). The two subcarriers have different frequencies. The main carrier is modulated by both subcarrier signals at the same time. Linearity in the circuits is required if separation between channels is to be maintained. A block diagram of a stereo-multiplex transmitter is shown at A in the illustration on p. 810.

The stereo-multiplex receiver uses a demodulator to obtain the two subcarrier signals from the main carrier. A pair of selective circuits separates the subcarrier channels. A demodulator is then used in each channel to obtain the original left-hand and right-hand audio signals (B). *See also* FREQUENCY-DIVISION MULTIPLEX.

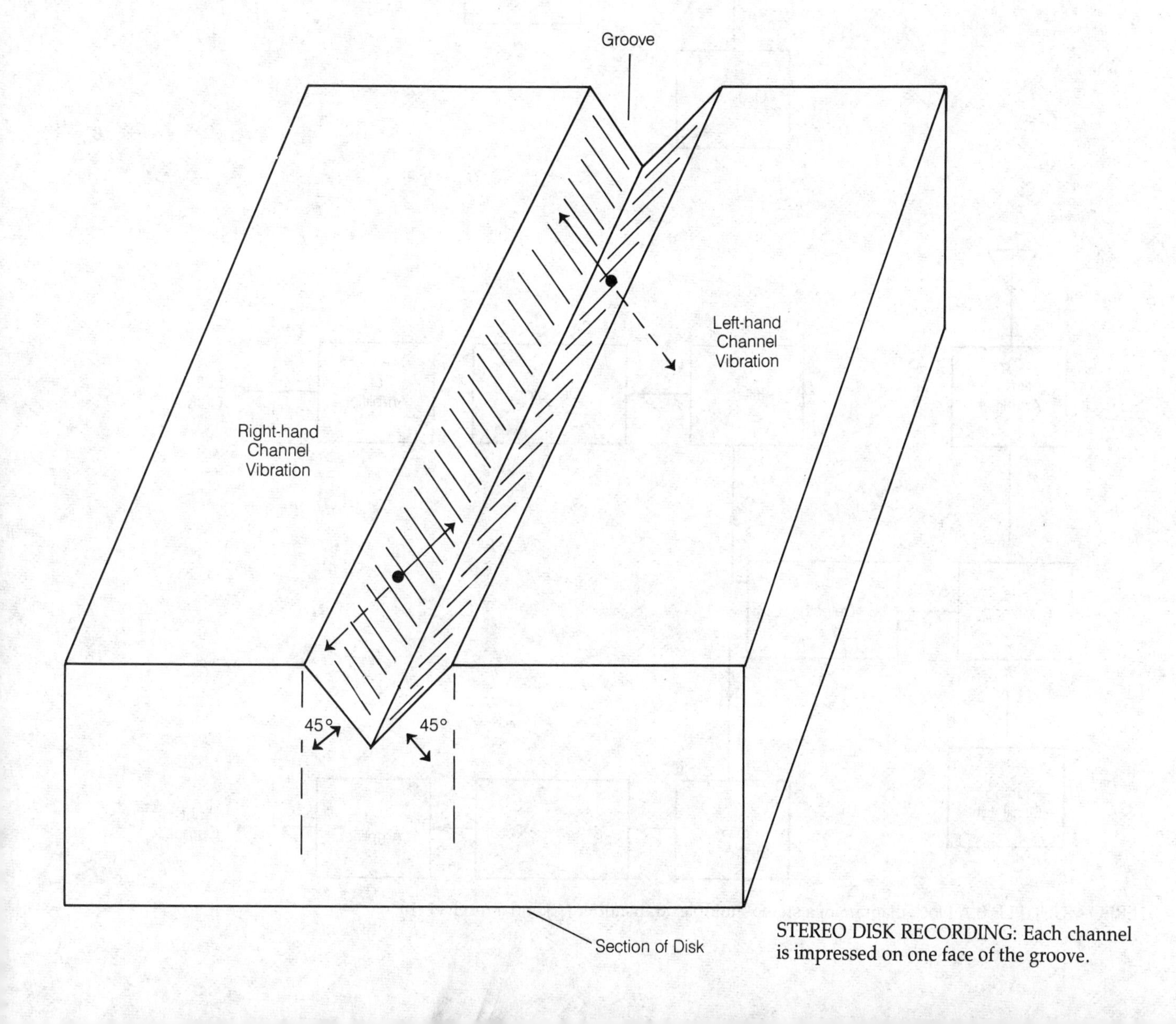

STEREO DISK RECORDING: Each channel is impressed on one face of the groove.

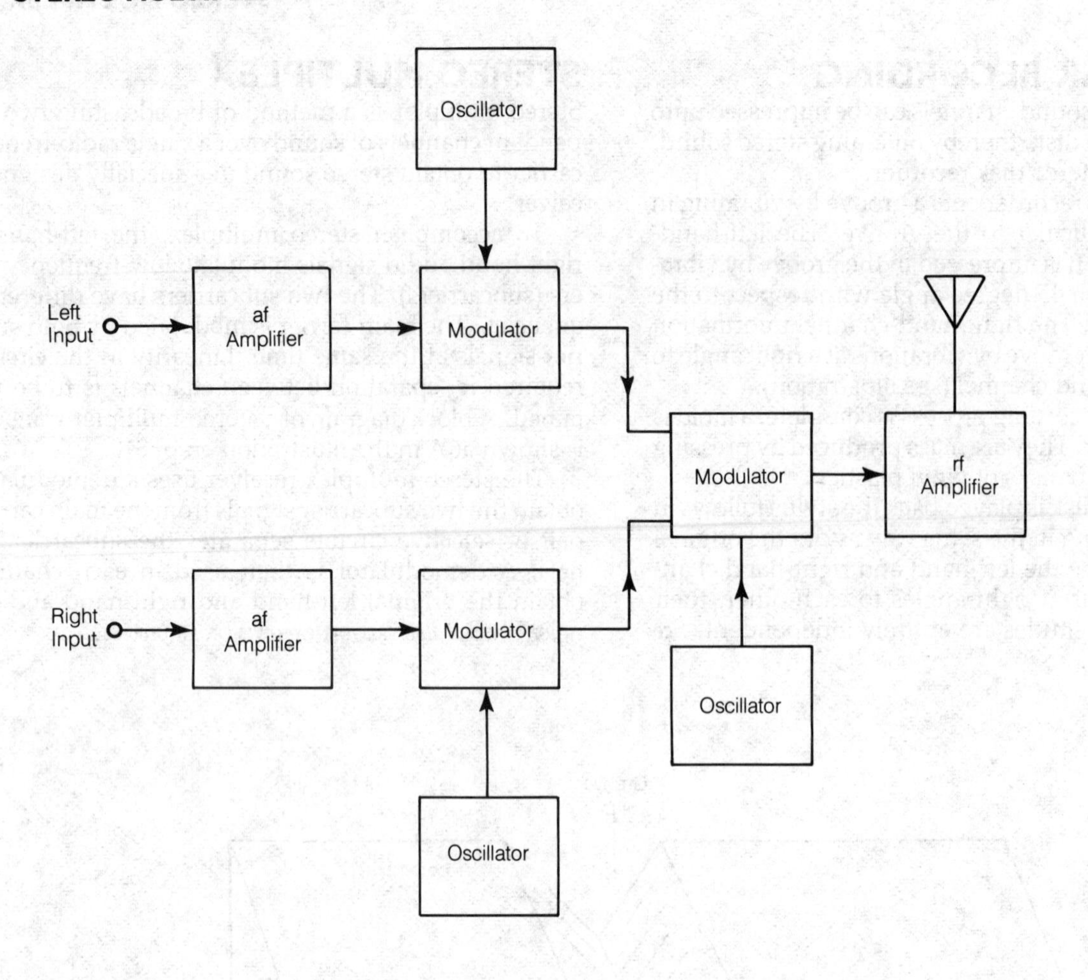

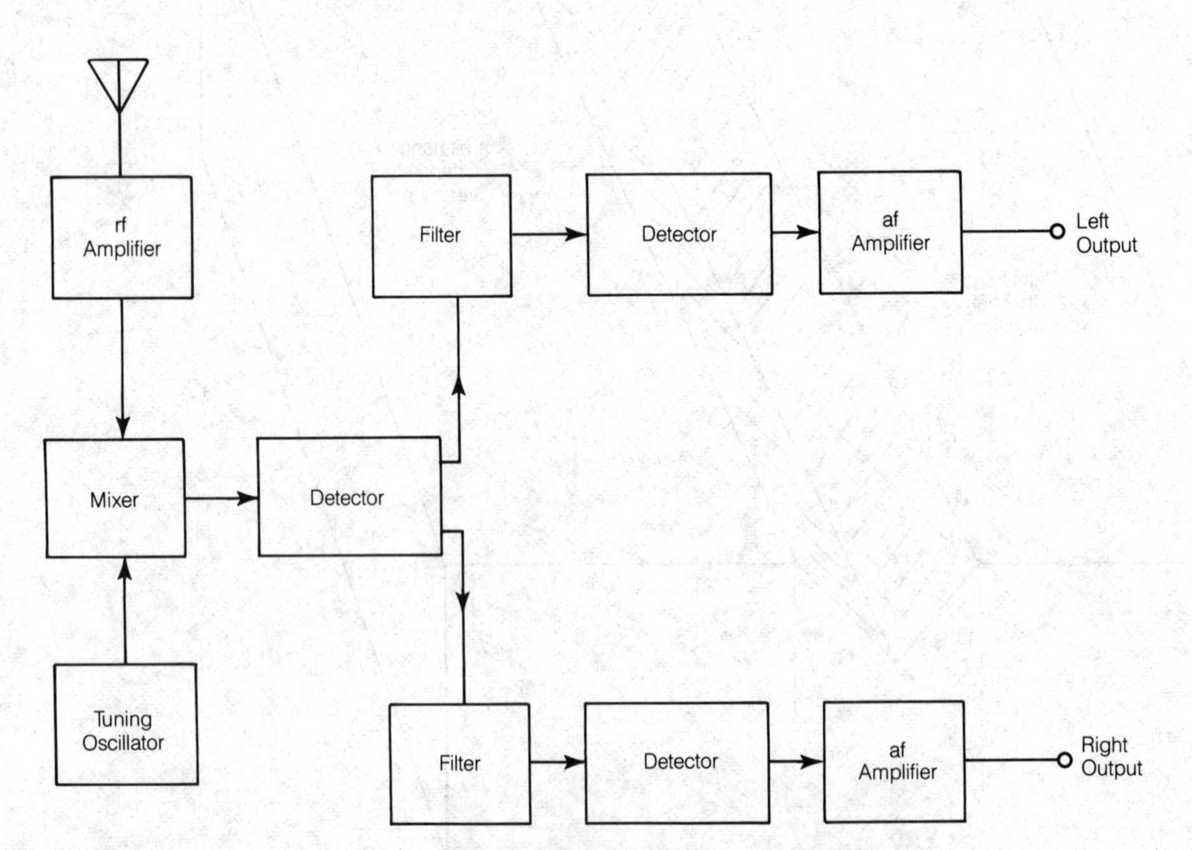

STEREO MULTIPLEX: A block diagram of a stereo-multiplex transmitter (A) and a receiver (B).

STEREOPHONICS

Stereophonics, also known as stereo, is the technology of two-channel sound reproduction, especially of high-fidelity music. Stereo music sounds more realistic than monaural music because the listener gets a sense of sounds originating at different sources.

A stereo recording has two channels, generally called the left-hand and right-hand channels. There may also be a center channel, obtained by combining the left and right channels. The two channels are independent. They are recorded with separate microphones and played back through separate speakers or earphones. This requires two sets of audio recording and reproduction systems.

Stereo is almost universally used in the recording of commercial music (*see* STEREO DISK RECORDING, STEREO TAPE RECORDING). Some stereo systems employ four channels for even more enhanced effects. A wide variety of stereo systems and accessories is manufactured commercially. *See also* HIGH FIDELITY, QUADROPHONICS.

STEREO TAPE RECORDING

Most stereophonic recording is done, at least initially, on magnetic tape. The left-hand and right-hand channels are recorded on parallel paths or tracks simultaneously.

A stereo tape recorder has two separate and independent sets of microphones, audio amplifiers, and recording heads. Some stereo tape systems have room for two or four simultaneous stereophonic recordings. In some stereo tape recorders, the input (microphone) gain can be controlled separately for each channel. All stereo tape recorders have high-fidelity audio circuits for reproduction of music.

In stereo recording, it is important that the sound be synchronized for each channel. Otherwise, an echoing or false-motion effect takes place. The heads must be precisely aligned, both while recording and while playing back.

The tape speed for good stereophonic recording must be fast enough to allow excellent reproduction of high-frequency sounds. The most commonly used speeds for commercial recording are 7.5 and 15 inches per second. Some home systems use a speed of 3.75 inches per second. It is extremely important that the tape speed be constant. The slightest variation will result in noticeable and objectionable flutter.

Stereo tape must be thick enough so that stretching does not occur while recording, playing back, or rewinding the tape. Even a tiny bit of tape stretching will seriously degrade the quality of sound reproduction. The shorter-playing tapes are generally thicker, for a given cassette or reel size, than longer-playing tapes. *See also* EIGHT-TRACK RECORDING, FOUR-TRACK RECORDING, MAGNETIC TAPE, STEREOPHONICS, TAPE RECORDER, TWO-TRACK RECORDING.

STIMULATED EMISSION

When electrons move from a higher energy level to a lower energy level, photons are emitted at discrete wavelengths. The energy contained in the photon, and consequently the wavelength of the photon, depends on the difference between the electron energy levels before and after the transition. The farther an electron falls, the more energy it gives up, and the shorter will be the wavelength of the emitted photon.

Stimulated emission is the basis for the operation of the mercury-vapor, neon, and sodium-vapor lamps, and for light-emitting diodes (LEDs). *See also* ELECTRON ORBIT, PHOTON, PLANCK'S CONSTANT, SPECTROSCOPE.

STOCHASTIC PROCESS

A stochastic process is any process that has been deliberately randomized. Stochastic processes are applied in many ways in science, industry, and electronics. These processes are used in quality control and in evaluating the behavior of molecules, atoms, and subatomic particles.

Consider the problem of determining the number of faulty diodes in a large lot. Suppose that 1,000,000 diodes have just been made. To avoid testing all of them, 1000 are selected at random—by stochastic process—and it is found that three of them do not function. Thus, the hazard rate, or proportion of diodes that are bad, is 0.3 percent. It can be expected that there will be about 3,000 bad diodes in the lot.

The application of a stochastic process requires the use of random numbers. Truly random numbers are not easily obtained. Usually, pseudorandom numbers, generated by a computer from a special algorithm, are used. *See also* RANDOM NUMBERS.

STOPBAND

A selective filter causes attenuation of energy at some band of frequencies. The band of frequencies at which the filter has high attenuation is called the stopband. Generally, the stopband is that band in which the filter blocks at least half of the applied energy. In other words, the stopband is that part of the spectrum for which the power attenuation is 3 decibels or 6 decibels.

There may be a stopband for any kind of filter, but this term is most often used with band-rejection filters. *See also* BANDPASS FILTER, BANDPASS RESPONSE, BAND-REJECTION FILTER, BAND-REJECTION RESPONSE, FILTER, HIGHPASS FILTER, HIGHPASS RESPONSE, LOWPASS FILTER, LOWPASS RESPONSE.

STORAGE BATTERY

See BATTERY.

STORAGE OSCILLOSCOPE

A storage oscilloscope is an oscilloscope able to freeze and store waveforms on the face of its cathode-ray tube (CRT). This oscilloscope can be operated in the same way as a conventional oscilloscope if desired.

The heart of the storage oscilloscope is a special cathode-ray tube that can hold an electric charge in a pattern corresponding to the waveform displayed. The

CRT has two electron guns, called the writing or main gun and the flooding gun.

A fine grid of electrodes, just inside the phosphor screen, is saturated with negative charge prior to storing a waveform. Then the oscilloscope is adjusted until the waveform appears as desired. The saturating charge is removed from the grid, and it acquires localized positive charges wherever it is struck by the electron beam from the main gun. This positive charge decays very slowly unless it is deliberately removed.

When a stored waveform is to be observed, the grid of electrodes is irradiated by electrons from the flooding gun. The grid prevents these electrons from reaching the phosphor, except in the areas of localized positive charge. Thus the original waveform is reproduced on the phosphor screen. The waveform can be viewed for periods as long as several minutes. The stored waveform is erased by saturating the grid with negative charge.

The storage oscilloscope is being superseded by the digital oscilloscope, which also has the ability to store waveforms. However, the digital oscilloscope stores the waveform in its internal semiconductor memory. The digital oscilloscope can preserve the desired waveform on its screen for an indefinite period. *See also* OSCILLOSCOPE.

STORAGE TIME

When the voltage at the base of a bipolar transistor is beyond the saturation level, an excess current flows in the emitter-base circuit. This current is greater than the amount required to keep current flowing in the collector circuit. Because of this excess current, an electrical charge accumulates in the base of the transistor.

If the bias is suddenly removed from the base of the transistor, the accumulated charge must dissipate before the transistor will come out of saturation. This requires a finite amount of time, known as the storage time, abbreviated t.

Storage time depends on many factors, including the size of the base electrode, the type and concentration of impurity in the semiconductor material, and the extent to which the transistor is driven beyond saturation. The storage time limits the speed at which switching can be accomplished by means of saturated logic. *See also* LOGIC, SATURATION, SATURATION VOLTAGE.

STORE

A memory file is sometimes called a store. A store contains an amount of information usually specified in bytes, kilobytes, or megabytes.

The act of putting information into computer memory is called a store operation. *See also* BYTE, FILE, SEMICONDUCTOR MEMORY, STORING.

STORING

Storing is the process of putting information into a computer's memory (*see* SEMICONDUCTOR MEMORY). Storing is usually accomplished with a simple sequence of operations, or even a single operation, such as pushing a button.

STRIPLINE

Stripline is a generic term referring to transmission line for use at the microwave frequencies and consisting of metal foil strips that act as conductors above or between extended parallel conducting surfaces. The conductors are typically separated by a suitable dielectric such as ceramic. Stripline permits the miniaturization of microwave circuitry.

Microstrip is a form of stripline, consisting of a foil strip above a metallic ground plane separated by a layer of dielectric. It is analogous to a two-wire line in which one of the wires is represented by the image in the ground plane of the wire that is physically present. Strip transmission line differs from microstrip in that a second ground plane is placed above the conductor strip.

STRIP TRANSMISSION LINE

See STRIPLINE.

STROBOSCOPE

A stroboscope is an instrument used to measure the rate of rotation or revolution of a motor or object, attached to the shaft of a motor.

A stroboscope emits bright, extremely brief flashes of light at regular intervals. The flash rate can be adjusted. The stroboscope flash rate is adjusted until the turning object appears to stand still. This represents one flash per rotation or revolution. The angular speed can then be directly read from a calibrated scale.

STUB

A stub is a section of transmission line, usually either a half-wave or quarter-wave section, connected in parallel or in series with the feed system of an antenna. Stubs can be used as impedance-matching transformers, as notch filters, or as bandpass filters.

For impedance matching between a transmission line and an antenna having purely resistive impedances of different values, a quarter-wave stub can be inserted in series with the feed system. The section is placed between the antenna and the feed point (see A in the illustration). The characteristic impedance of the matching stub is equal to the geometric mean of the two impedances to be matched (*see* GEOMETRIC MEAN).

A trap circuit can be constructed from a quarter-wavelength stub open at the far end (B) or from a half-wavelength stub shorted at the far end (C). This circuit is placed in parallel with a transmission line to eliminate signals at an unwanted frequency.

A narrowband resonant stub can be connected in parallel with a feed line to provide additional selectivity. A short-circuited, quarter-wave stub (D) or an open half-wave stub (E) will facilitate this. *See also* HALF-WAVE TRANSMISSION LINE, QUARTER-WAVE TRANSMISSION LINE.

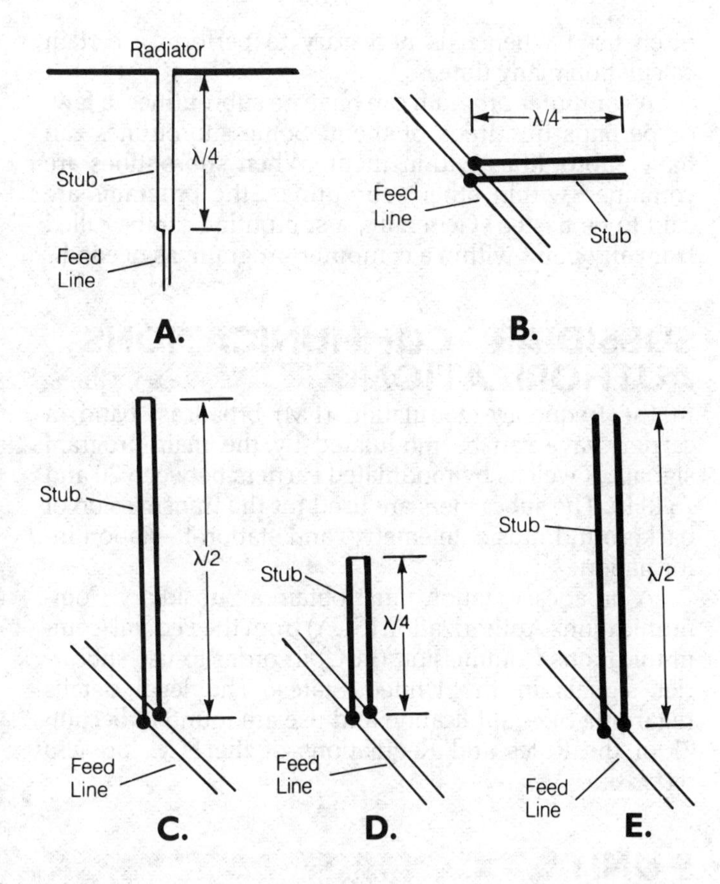

STUB: Examples of stubs include a quarter-wave matching section (A); a quarter-wave, series-resonant (B); a half-wave, series-resonant (C); a quarter-wave, parallel-resonant (D); and a half-wave, parallel-resonant (E).

STYLUS

A stylus is a tool used in disk recording and reproduction as an electromechanical transducer.

In disk recording, the stylus is fed with electrical impulses at audio frequencies. The stylus vibrates as the disk rotates underneath it. As a result, the stylus cuts a serpentine groove in the disk with a shape that corresponds to the audio-frequency vibrations. In monaural disk recording, the stylus moves back and forth, parallel to the surface of the disk and perpendicular to the direction of movement, as at A in the illustration. In stereo disk recording, the stylus moves along two independent axes, as shown at B (*see* STEREO DISK RECORDING).

When a disk is played back for reproduction, the stylus vibrates as it moves along the groove in the disk. These movements are translated into electrical impulses, usually by means of a dynamic or piezoelectric transducer (*see* CERAMIC PICKUP, CRYSTAL TRANSDUCER, DYNAMIC PICKUP).

A stylus, also known as a needle, is fabricated from a hard substance such as quartz, diamond, or carborundum. This maximizes its life because these materials are much harder than the vinyl plastic from which disks are made.

SUBAUDIBLE TONE

A subaudible tone is a tone used to modulate a communications transmitter to obtain private-line or tone-squelch operation. Subaudible tones have frequencies below 300 Hz, the approximate lower audio cutoff for communications.

SUBBAND

A given frequency allocation can be subdivided for various types of radio transmission. A subband is a band within a band.

Users of a subband can agree about how its use is shared, but this agreement is not enforceable by law. In the nonvoice subbands of the high-frequency amateur bands, for example, the users of A1 and F1 emissions stay out of each other's way by operating within different frequency ranges. *See also* BAND.

SUBCARRIER

A radio-frequency carrier can be modulated by another radio-frequency signal. This modulating carrier signal has a frequency much lower than that of the main carrier. The lower-frequency carrier is called a subcarrier.

Subcarriers are used in some multiplex transmissions (*see* FREQUENCY-DIVISION MULTIPLEX, MULTIPLEX, STEREO MULTIPLEX). A main carrier can be modulated by more than one subcarrier. Subcarriers generally have frequencies between 20 and 75 kHz. This allows the subcarrier to be modulated at audio frequencies, and prevents it from interfering with the main program (because listeners cannot hear modulation above 20 kHz).

In a receiver, the subcarriers are retrieved from the main carrier by a detector. A tuned circuit filters out all

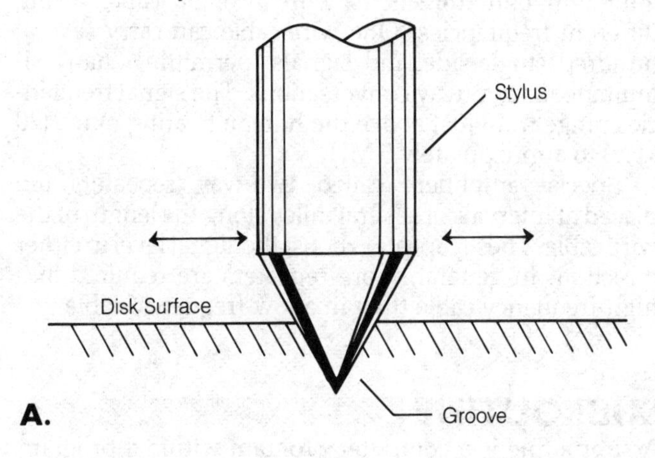

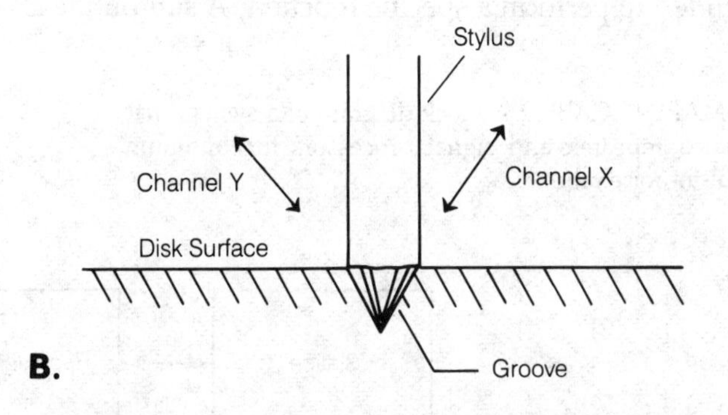

STYLUS: A monaural stylus (A) and a stereophonic stylus (B).

subcarrier signals except the desired one. A second detector retrieves the information that modulates the selected subcarrier.

In frequency-modulation (FM) broadcasting, subcarriers are used or leased for various purposes. Many of the stations heard on a stereo receiver are modulated by inaudible subcarrier signals, as well as by the voices and music. *See also* SUBSIDIARY COMMUNICATIONS AUTHORIZATION.

SUBHARMONIC

A subharmonic is a signal with a frequency that is an integral fraction of the frequency of a specified signal. A signal can have an infinite number of subharmonics. If the main-signal frequency is f, then the subharmonic frequencies are $f/2$, $f/3$, $f/4$, and so on. Subharmonic signals occur at wavelengths that are integral multiples of the main-signal wavelength.

While most alternating-current signals inherently contain some energy at harmonic frequencies (*see* HARMONIC), signals do not naturally contain energy at subharmonic frequencies. Subharmonics must be generated by means of a frequency divider (*see* FREQUENCY DIVIDER).

SUBMARINE CABLE

A submarine cable is a cable used to transmit signals by fiber or wire for long distances under water. Fiberoptic submarine cables are regaining traffic lost to satellites. These cables carry signals under the oceans of the world.

The illustration is a simplified diagram of a submarine-cable communications system. The signals are transmitted in both directions with a single cable, using different frequencies. One wire cable can carry several hundred single-sideband signals, permitting many simultaneous two-way conversations. The signal frequencies range from just above the human hearing range (20 kHz) to approximately 5 MHz.

Special amplifiers called two-way repeaters are placed at intervals of several miles along the length of the wire cable. These repeaters boost the signal level in either direction. In general, more repeaters are required in a high-frequency cable than in a low-frequency cable.

SUBROUTINE

A subroutine is a computer program within a program intended to perform a specific function. A subroutine is often used when it is necessary to perform a certain calculation many times.

A computer program can have no subroutines, a few, or perhaps hundreds of them. Some subroutines can have subroutines within them. When subroutines are contained within other subroutines, the programs are said to be nested. Generally, a subroutine can be called from any point within a computer program, as needed.

SUBSIDIARY COMMUNICATIONS AUTHORIZATION

In the frequency-modulation (FM) broadcast band, a carrier wave can be modulated by the main-program signal, as well as by modulated carriers between 20 and 75 kHz. The subcarriers are used for the transmission of background music, telemetry, and station-to-station information.

A broadcast station must obtain a Subsidiary Communications Authorization (SCA) from the Federal Communications Commission (FCC) in order to use subcarrier signals in the United States. The legal details regarding SCA application and use are found in Section 73 of the Rules and Regulations of the FCC. *See also* SUBCARRIER.

S UNIT

The S unit is a unit of signal strength. Generally, 1 S unit represents a change in signal strength amounting to 6 decibels of voltage, or a ratio of 2 to 1. A root-mean-square (rms) signal level of 50 μV is considered to represent 9 S units. Some engineers use other values to represent S9, or other voltage ratios to represent 1 S unit.

The S-unit scale ranges from 1 to 9. A signal strength of S8 means that the signal voltage at the antenna terminals is 50/2 or 25 μV; a signal strength of S7 means that the voltage is 25/2 or 12.5 μV, and so on. The signal strength is sometimes expressed in fractions of an S unit, such as S5.5. The illustration on p. 815 is a graph showing the rms signal voltages for S units ranging from 1 to 9, according to the most common definition of the scale.

SUNSPOT

Sunspots are dark regions on the incandescent surface of the sun or photosphere. Some have been large enough to contain 70 earth-size planets. Sunspots are believed to be created by the interaction of intense magnetic fields of

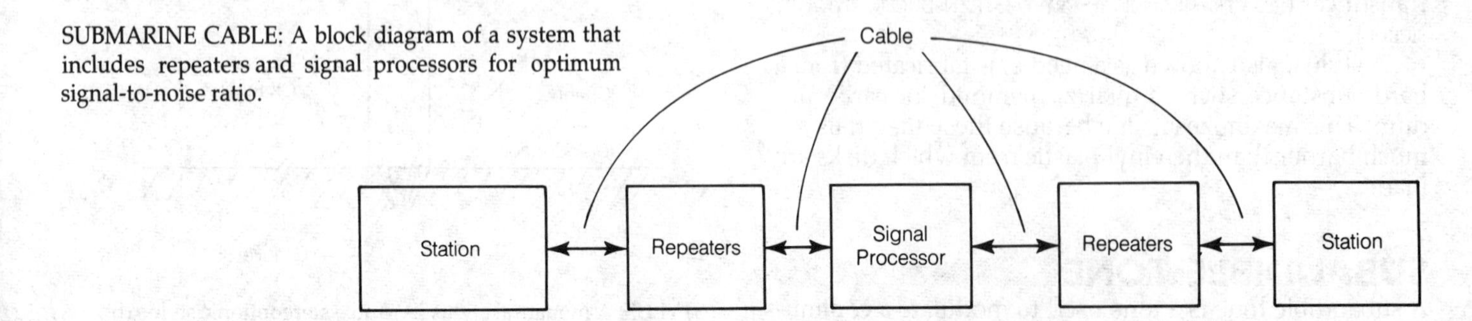

SUBMARINE CABLE: A block diagram of a system that includes repeaters and signal processors for optimum signal-to-noise ratio.

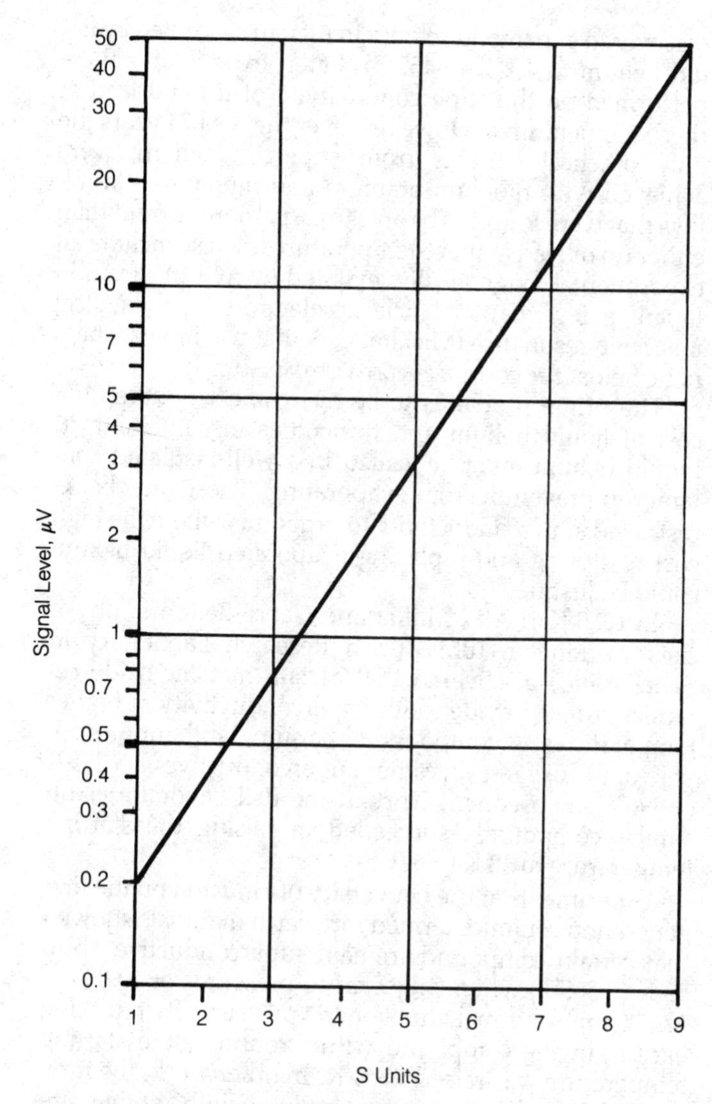

S UNIT: Graph showing the relation between S units and signal strength at antenna terminals.

the sun which generally extend from a north to a south pole. The spots have cooler, darker inner regions about 3,600°F cooler than the surrounding photosphere. Sunspots are a subject of concern in electronics because at periods of high sunspot activity they can cause disruptions in both radio and wire communications, and even power failures. (*See* SUNSPOT CYCLE.) They can be safely viewed with telescope projectors.

The sun consists of hot gases, predominantly hydrogen (72 percent) and helium (27 percent). It is now believed that energy is produced in the sun's core, whose volume occupies 2 percent of the sun's volume but contains half the mass. Hydrogen is fused into helium at 27 million °F. Energy from the core radiates slowly through an intermediate zone and a convective zone of circling currents of hot gas in the outer one-third of the sun carries energy to the surface. The photosphere is the narrow 300-mile thick zone at the surface of the sun where the sun's atmosphere changes from opaque to incandescent, and visible light escapes. The photosphere has an average temperature of 10,000°F.

Huge bursts of energy called solar flares are released in sunspot areas, and they can extend more than 200,000 miles into space and reach temperatures of up to 36 million °F. These flares eject streams of charged particles into the solar winds through the corona or sparse outer atmosphere of the sun. (*See* SOLAR FLARE.) The corona itself reaches temperatures up to 2 million °F and is a strong source of X rays. The sun also radiates ultraviolet as well as visible light. Enough of this energy penetrates the atmosphere to deliver some 100 trillion kilowatts of power to the earth. This amounts to 1.35 kilowatts falling on every square meter of earth, a number called the solar constant.

Solar wind, made up mostly of protons and electrons, radiates in all directions from the sun. Continuously flowing past the earth at 200 to 500 miles per hour, the wind feeds particles into the earth's Van Allen radiation belt and distorts the earth's magnetic field. It also sets off the frequent minor auroral displays visible at higher latitudes. *See* SOLAR FLUX, VAN ALLEN RADIATION BELTS.

SUNSPOT CYCLE

In the mid-nineteenth century it was discovered that sunspot activity followed approximately an 11-year cycle. The number of sunspots each month has varied as shown in the figure. Each cycle begins when spots show up in both the northern and southern hemispheres about 35 degrees away from the solar equator. (The spots can last hours, days, or even months.) As the cycle matures and the older spots fade away, new and more numerous spots begin to appear at lower latitudes. Near the end of

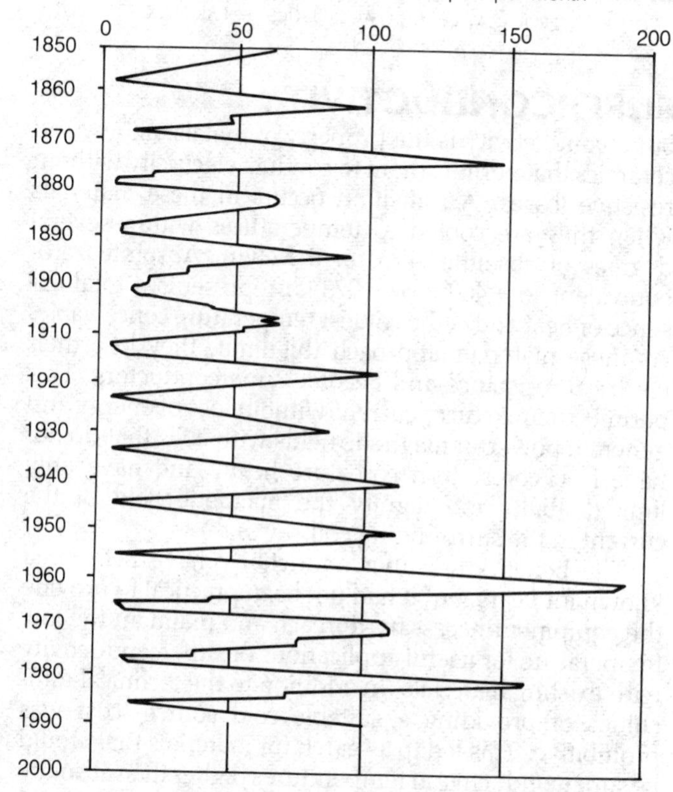

SUNSPOT CYCLE: Annual average number of sunspots per month, as a function of the year.

the cycle the numbers diminish, but some are only 5 degrees from the equator.

Sunspots usually travel in pairs or groups of opposite polarity formed by the twisted magnetic field lines emerging from the sun. The spots traverse the face of the sun in an east-west direction. The leading spots of each group in the northern hemisphere typically have positive polarity, the trailing spots negative. In the southern hemisphere, the leading spots are negative. During the next 11-year cycle, the hemisphere polarities reverse.

The surges of protons and electrons carried on the solar winds to the earth from sunspot flares have interrupted shortwave transmissions, satellite communications and loran navigation systems. Powerful transient magnetic fields generated in the upper atmosphere by the flares have induced excessive electrical currents in transmission lines, causing power disruptions over wide regions.

The darker central portions of sunspots, or umbras, have the strongest magnetic fields; the lighter exteriors, or penumbras, the weaker fields. Occasionally the penumbras of two sunspots of opposite polarity merge as they move past each other to set off a flare.

The sunspot cycle affects propagation conditions in the very high frequency spectrum up to about 70 MHz for F-layer propagation and 150 to 200 MHz for sporadic-E propagation (*see* E LAYER, F LAYER, SPORADIC-E PROPAGATION). At or near a sunspot maximum, the maximum usable frequency is high because the upper atmosphere is densely ionized. When there are few sunspots, the maximum usable frequency is low. The solar flux directly affects ionization in the upper atmosphere, and it correlates directly with the number of sunspots. *See* MAXIMUM USABLE FREQUENCY, PROPAGATION CHARACTERISTICS, SOLAR FLUX, SUNSPOT.

SUPERCONDUCTIVITY

Superconductivity is the property of metals, alloys, and ceramics that permits them to conduct electricity without resistive losses. A transition occurs in these materials when they are cooled to temperatures within several degrees of absolute zero, or 0 Kelvin. Absolute zero, equivalent to −460°F or −273°C, represents a total absence of heat and is the coldest temperature conceivable. As these materials approach this limit, they lose their electrical resistance and become superconductors. This permits them to carry current without loss of energy and generate powerful magnetic fields with coils that do not have iron cores. Iron cores are heavy and have only limited ability to magnify the magnetic field of the current in the surrounding coil.

The benefits of superconductivity have been well known for years, but it had not been practical to provide the equipment necessary to reach and maintain the low temperature for useful applications of superconductivity with existing materials. In addition to the technical difficulties of providing a suitable cold source, cost was prohibitive. This led to a search for materials that would be superconductive at temperatures higher than absolute zero.

In 1911, Dutch physicist Heike Kamerlingh Onnes discovered superconductivity in mercury—cooled by liquid helium to 4.2 K (−452°F). Later, experiments were performed on the superconductivity of tin and lead at these temperatures. However, over the next 74 years the only practical benefits from superconductivity were achieved with niobium-titanium and niobium-3-tin, alloys that were found to be able to carry more current than either tin or lead at these temperatures. These include an experimental magnetically levitated (MAGLEV) train in Japan, a few giant particle accelerators, and medical magnetic-resonance (MR) imagers that use intense magnetic fields. *See* COMPUTER-AIDED MEDICAL IMAGING.

The prime problems to be overcome were the high cost of liquid helium and difficulties in containing it. Liquid helium must be sealed in a well-insulated container to prevent it from evaporating. These drawbacks restricted superconductivity to programs where the high cost of storing and replacing evaporated liquid helium could be justified.

In 1985 Karl Alex Muller and Georg Bednorz, physicists working at IBM Zurich Research Laboratory in Switzerland, discovered that certain metallic oxide ceramics showed evidence of superconductivity at higher temperatures. A compound of barium, lanthanum, copper, and oxygen became superconductive at 35 K (−396°F). Subsequent work at the Bell Laboratories on similar compounds succeeded in raising the starting temperature to 38 K (−391°F).

Experiments at the University of Houston on the first generation ceramics termed rare earth materials showed that certain compounds remain superconductive up to 52 K (−365°F) when subjected to pressures of 10,000 to 12,000 times normal atmospheric pressure. By replacing barium in the compound with strontium, the starting temperature was raised to 54 K. In other work, the rare-earth element lanthanum was replaced with another rare earth element, yttrium. Resistance dropped sharply at 98 K when the compound was immersed in liquid nitrogen.

At this time it was discovered that less expensive liquid nitrogen at a higher temperature 77 K (−320°F) could replace costly liquid helium in research and development of superconductivity. The resistance of the ceramic compound containing yttrium dropped sharply at 98 K when it was immersed in liquid nitrogen.

In 1988, two additional superconducting materials were announced. The National Research Institute for Metals in Japan announced a compound of bismuth, strontium, calcium, copper, and oxygen. The bismuth material exhibits one transition temperature at about 80 to 90 K and the other at about 120 K (−243°F). A fourth material containing thallium, barium, calcium, copper, and oxygen was discovered by a group at the University of Arkansas. Tests at IBM confirmed that the effective temperature for superconductivity was 125 K (−234°F).

One of the earliest effects of superconductivity was the levitation of a small magnet in midair above a cooled superconductive material. Recently scientists have discovered the suspension effect. Under proper conditions, a chip of superconducting material will hang beneath a magnet as if suspended by an invisible thread.

The suspension effect was observed with a ceramic called yttrium-barium-copper oxide. Silver oxide was introduced into the ceramic to permit the material to handle the large currents of electricity that would be required for use in extremely powerful magnetics. The suspension phenomenon is believed to occur because the magnetic field induced in the superconducting material remains in a state of equilibrium with the magnetic field of the permanent magnet.

High-temperature superconducting magnets in particle accelerators and atomic fusion equipment could yield significant savings in the cost of electrical power by the substitution of liquid nitrogen for liquid helium. These magnets could also reduce the size and cost of the insulation required to maintain the liquid helium coolant in magnetic resonance imagers (MRIs) as well as the cost of liquid helium.

Thin-film, high-temperature superconducting materials are expected to find applications in computer circuitry and magnetic field strength sensors. Flexible high-temperature superconducting ceramics have been developed, and they are expected to be used in magnetic coils, transmission lines, and even conventional rotating electric-power generators.

SUPERHETERODYNE RADIO RECEIVER

The superheterodyne radio receiver was designed to overcome the basic disadvantages of the tuned radio frequency (trf) receiver by converting all incoming frequencies to a single intermediate frequency. As a result, this receiver has tuning flexibility and all radio-frequency (rf) amplification takes place at the same frequency.

Figure 1 is a block diagram of a superheterodyne radio receiver. The rf amplifier before the mixer boosts the amplitude of the incoming signals to improve selectivity and signal-to-noise ratio. The input from the rf amplifier is combined in the mixer with the output from the local oscillator (LO). The mixer is frequency selective; it accepts only the signal frequency to which it is tuned, and it rejects all other frequencies. Because the LO and mixer are tuned together, the mixer heterodynes these two signals and the resulting output is a difference signal or intermediate frequency.

Typically, the LO operates at a higher frequency than the signal frequency. For example, if the receiver has an intermediate frequency of 456 kHz and the mixer is tuned to a 1200 kHz signal, then the LO is simultaneously tuned to 1656 kHz, the sum of 1200 and 456 kHz.

The intermediate-frequency (i-f) from the mixer is amplified by a series of i-f amplifiers and then applied to the detector for demodulation. The i-f amplifiers are similar to rf amplifiers except that they always operate at the intermediate frequency. After the audio intelligence is removed from the i-f signal in the detector, it is amplified by one or more audio amplifiers before being reproduced by the loudspeaker.

For more complex long-range communications receivers, additional circuits are added to the superheterodyne receiver to improve its performance: an automatic volume control (AVC) circuit between the amplitude modulation (AM) detector and the rf amplifier, and a noise limiter between the AM detector and the audio amplifier.

The AVC circuit maintains constant audio output from the receiver despite changes in rf signal strength and produces a direct-current (dc) voltage that is proportional to the audio signal strength of the detector. If the rf signal strength decreases, the AVC circuit senses this as a decrease in the audio level of the detector and feeds back voltage to the rf amplifier to increase its gain. Conversely, if the rf signal strength increases, the AVC feedback to the rf amplifier decreases its gain. The noise limiter prevents excess noise riding on the signal from reaching the receiver output by preventing the signal amplitude from exceeding a preset level.

Conventional superheterodyne receivers recover the audio component from the i-f signal so they are unable to receive interrupted continuous wave (CW) signals with no audio components. To adapt the superheterodyne receiver to CW reception, a beat-frequency oscillator (BFO) is added to provide an audio output as shown in Fig. 2. The output of the BFO is connected to the AM detector by the BFO switch. When the switch is open, the receiver operates as a standard AM receiver. But, when the switch is closed, the BFO output is heterodyned with the i-f signal in the AM detector.

The output frequency of the BFO differs from the i-f by a selected audio frequency. This could be 500 or 1000 Hz. A difference or beat frequency in the audio range is produced by the heterodyning action whenever the CW is present. The CW may take the form of Morse code, a series of dots and dashes. These audio signals can then be heard on the speaker.

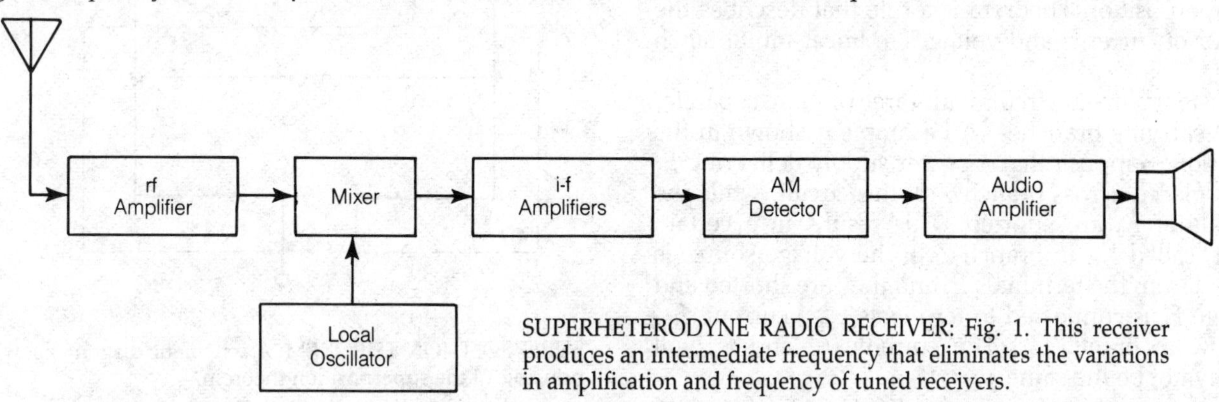

SUPERHETERODYNE RADIO RECEIVER: Fig. 1. This receiver produces an intermediate frequency that eliminates the variations in amplification and frequency of tuned receivers.

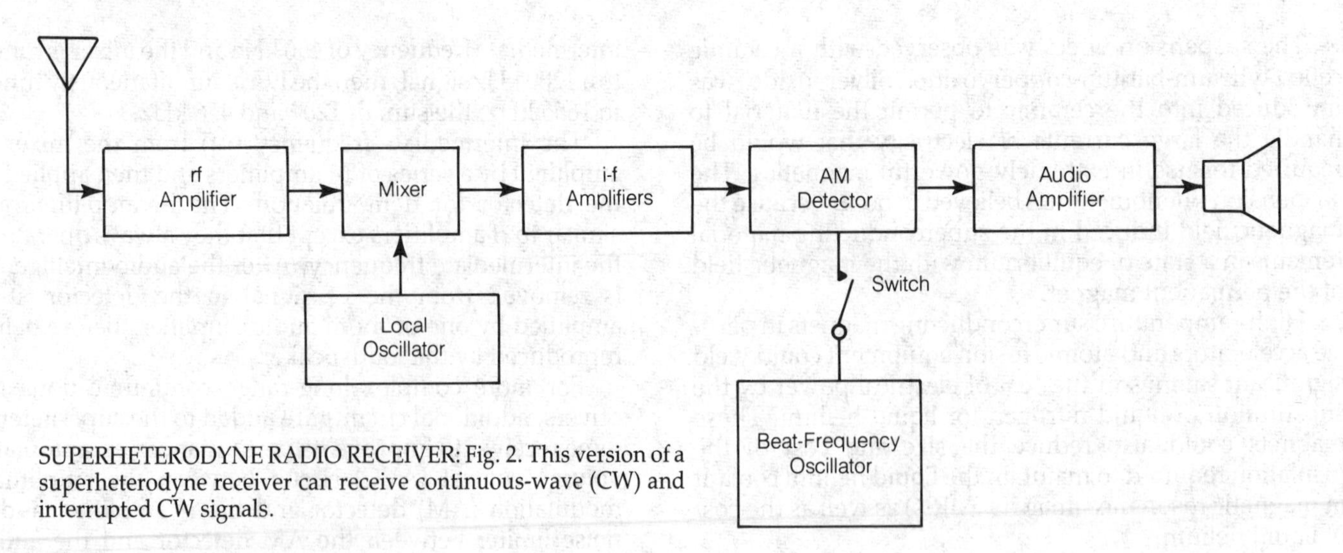

SUPERHETERODYNE RADIO RECEIVER: Fig. 2. This version of a superheterodyne receiver can receive continuous-wave (CW) and interrupted CW signals.

SUPERHETERODYNE RECEIVER

See SUPERHETERODYNE RADIO RECEIVER.

SUPERHIGH FREQUENCY

The superhigh-frequency (shf) range of the radio spectrum is the range from 3 to 30 GHz. The wavelengths corresponding to these limit frequencies are 10 and 1 centimeters, and thus the shf waves are sometimes centimetric waves. Superhigh-frequency waves are also microwaves. (*See* MICROWAVE.)

Superhigh-frequency waves are not affected by the ionosphere of the earth. Signals in this range pass through the ionosphere as if it were not there. Centimetric waves behave like waves at higher frequencies. They can be focused by medium-sized parabolic or spherical reflectors. Because of their tendency to propagate in straight lines unaffected by the atmosphere, centimetric waves are extensively used in satellite communications. Centimetric waves are also used for radar, radar beacons, electronic warfare jamming, telephone bypass, and broadcasting studio-to-transmitter links. Because the frequency of a centimetric signal is so high, wideband modulation is practical. *See also* ELECTROMAGNETIC SPECTRUM, FREQUENCY ALLOCATIONS.

SUPERPOSITION THEOREM

The Superposition Theorem is a rule that describes the behavior of currents and voltages in linear multibranch circuits.

Consider a linear circuit with three or more separate, current-carrying branches. An example is shown in the illustration. Suppose that a power supply delivering E_x volts is placed across branch X of this circuit, while the terminals at E_y are shorted. This results in a certain current, called I_{zx}, in branch Z. If the voltage source is removed from the terminals E_x and they are shorted and a voltage E_y is connected to terminals E_y, a current, I_{zy}, will flow in branch Z. (The currents I_{zx} and I_{zy} will probably not be the same.)

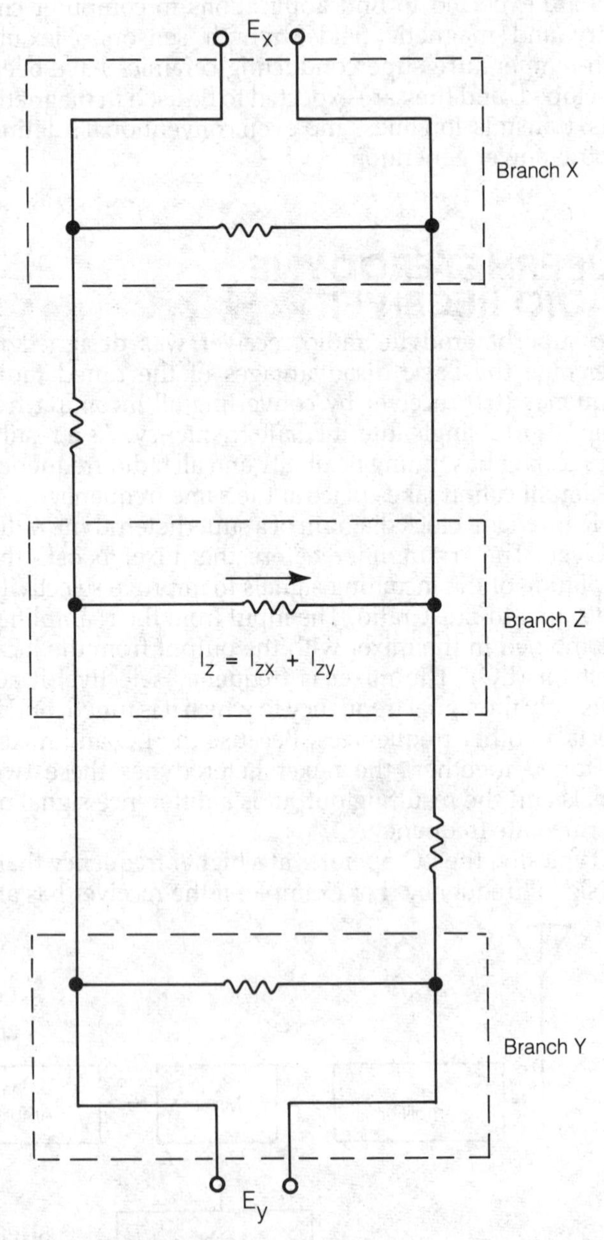

SUPERPOSITION THEOREM: A schematic diagram showing the principle of the superposition theorem.

The superposition theorem states what current will flow in branch Z of the circuit when the supplies E_x and E_y are connected at the same time. It is the sum of the currents I_{zx} and I_{zy}. That is, if the current in branch Z is I_z with both E_x and E_y connected, then:

$$I_z = I_{zx} + I_{zy}$$

This theorem can be used to determine currents and voltages in complicated circuits. *See also* KIRCHHOFF'S LAWS.

SUPERREGENERATIVE RECEIVER

A special form of regenerative detector is used in a circuit called a superregenerative receiver. This circuit provides greater gain and better sensitivity than an ordinary regenerative circuit. A superregenerative receiver is a form of direct-conversion circuit (*see* DIRECT-CONVERSION RECEIVER).

In the superregenerative detector, the positive feedback is periodically increased and decreased. This makes it possible to increase the feedback without causing unwanted oscillation of the detector. The changes in feedback cause alternations in the circuit gain, and this modulates incoming signals. However, if the feedback is varied at a rate of more than 20 kHz, the modulation cannot be heard, and it does not affect the sound of an incoming signal.

An example of a superregenerative detector is shown in the illustration. The amount of feedback is controlled by a varactor diode. The diode capacitance changes with the applied ultrasonic signal. The changes in capacitance cause the amount of feedback to vary. The detector is adjusted by manipulating the series variable capacitor and the amplitude of the applied ultrasonic signal. *See also* REGENERATIVE DETECTOR.

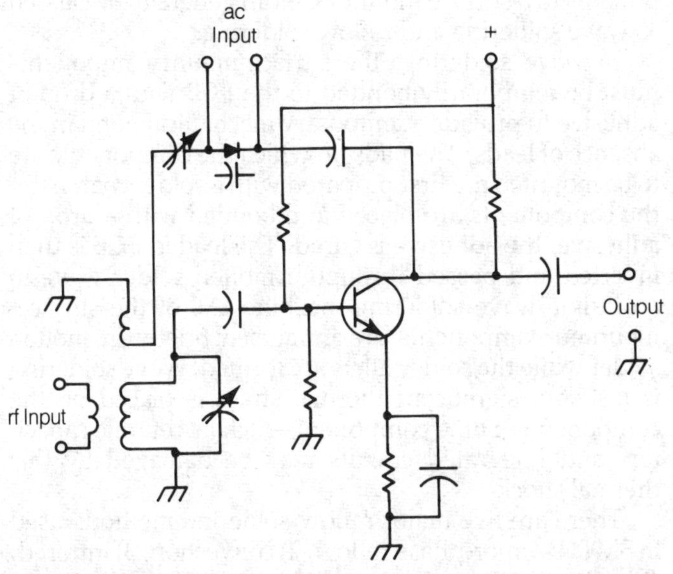

SUPERREGENERATIVE RECEIVER: An external oscillator is used to maximize gain.

SUPPRESSED CARRIER

A suppressed carrier in single-sideband transmission is the carrier wave that is greatly reduced in amplitude. This reduction also can occur in double-sideband transmission.

The advantage of suppressed-carrier transmission is that all of the signal power goes into the sidebands, resulting in optimal efficiency. Suppressed-carrier signals are generated with a balanced modulator. *See also* BALANCED MODULATOR, SINGLE SIDEBAND.

SURFACE ACOUSTIC WAVE DEVICE

A surface acoustic wave (SAW) device takes advantage of the reduction in signal propagation speed achieved by converting an electromagnetic (radio frequency or light signal) into an acoustic wave for propagation in rigid solid. This transition permits the SAW device to function as a delay line, filter, or signal generators. In a typical application, the input radio-frequency (rf) signal is converted to an acoustic or Rayleigh wave that is confined to the surface of the substrate. Acoustic energy is transmitted as a composite of longitudinal and transverse waves. A simple SAW filter is illustrated in the diagram.

The substrate for a SAW device must be a suitable elastic solid such as quartz or the single-crystal, piezoelectric ceramic material lithium niobate ($LiNbO_3$). These materials provide for the efficient exchange of energy between mechanical and electrical forms. Moreover, the periodicity of the crystal lattice supports the propagation of wave motion throughout the crystal in a variety of modes. One of these, the surface mode, propagates along the polished surface like ripples in a pond that has been disturbed by the dropping of a stone. The mass and spacing of the crystal ions in quartz and lithium niobate results in a popagation velicity of about 3 millimeters per microsecond.

Efficient coupling is achieved by periodic quarter wavelength surface features such as metal electrodes (forming a transducer). Electrodes for the necessary electromechanical coupling are formed as interdigital fingers on the surface of the substrate with photolithographic and thin-film deposition methods.

Acoustic waves created by electromagnetic frequencies travel five orders of magnitude slower in the solid

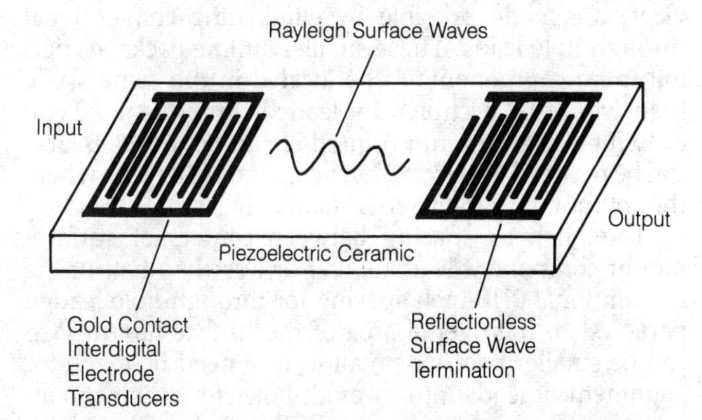

SURFACE ACOUSTIC WAVE DEVICE: A SAW delay line converts rf input into an elastic surface wave that travels at far slower speed and then reconverts it to a delayed rf signal at the other end.

ceramic than they do in air (a ratio of 1 to 10,000). This ratio permits a SAW delay line to be shorter by that ratio than a cable or waveguide used for signal delay. Acoustic energy injected at one point on the surface of the substrate can be collected at another point on the same structure. This permits addition, subtraction or division of input frequencies.

The fine line width and spacing of the interdigital thin-film transducers determine the frequency response of the SAW device and its efficiency as a signal processor. Unwanted signals are removed with filters. SAWs have been applied as delay lines, bandpass and dispersive filters, resonators, and radar pulse compressors and expanders. They also serve as pulse compression filters for chirped or linear frequency-modulated radars making them useful in discriminating radar patterns in electronic warfare systems and air-traffic control.

Small, lightweight hybrid oscillators stabilized by high-Q SAW resonators and/or delay lines provide very low phase noise performance. They can operate at frequencies as high as several GHz without the need for external frequency multiplication. These properties make them an attractive alternative to the quartz crystal oscillator as a source of pure microwave energy for electronic warfare, as well as radar and communications applications.

Other SAW device applications include their use as pulse sequence or coded digital signal generators and decoders. They are also being developed for use in electronic interaction and optical scanning. SAW devices can also deflect and modulate light beams.

SURFACE-MOUNT TECHNOLOGY

Surface-mount technology (SMT) is the technology of assembling electronic circuit boards by attaching leadless components to one or both sides of a printed-circuit board (PCB) that does not have plated-through holes. The benefits of SMT include increased density of components on a circuit board (compared with conventional assembly of leaded components) and reduction in assembly costs.

Packaging density is increased because smaller, lower component packages (particularly semiconductor devices) are made possible by eliminating conventional through-hole leads. These smaller outline packages permit more components to be located in the same space than would be occupied by leaded components. Typically the surface-mount printed-circuit-board (PCB) area can be reduced 40 percent over conventional PCBs where the components are inserted manually.

The pitch or spacing between centers of surface-mount component leads is half or less than half of the conventional 0.10-inch spacing for through-hole leaded parts. Also, the overall area of the surface-mount PCB can be smaller because no allowance need be made for alignment of leads into through-holes for automatic insertion. The surface-mount PCB can be 50 percent smaller than a conventional PCB designed for automated component insertion. Lower package height permits closer stacking of PCBs in card cages. *See* SEMICONDUCTOR PACKAGE.

Increased circuitry density shortens circuit paths between components and smaller packages reduce circuit paths within the package. This permits the components to operate at higher speeds while reducing radio-frequency interference (RFI) and electromagnetic interference (EMI).

Significant cost and time savings are gained with automated robotic component pick-and-place machines. Automatic grippers pick the components (transistors, diodes, resistors, capacitors, etc.) from a tray or tape, and place them—under computer control—at the location specified in a placement program. There are also cost savings due to the smaller size of the PCB required and the reduction or elimination of hole-forming and through-hole plating operations.

SMT is similar to hybrid circuit assembly in that the leadless components are bonded to surface conductors. Popular substrate materials for SMT PCBs include epoxy fiberglass (G-10) and polyimide fiberglass. The most widely used substrate for hybrid circuits is alumina ceramic. (*See* HYBRID CIRCUIT.) Hybrid circuit components are usually bonded with precious metal solders while SMT components are bonded with lead-tin solder—63 percent tin (Sn) and 37 percent lead (Pb) or 60 percent Sn and 40 percent Pb. (*See* SOLDER.) Other substrates include polyimide fiberglass.

The coefficient of thermal expansion of the surface mount PCB must be carefully matched to that of the component or the solder bonds may fail. The soldering of ceramic components with coefficients of thermal expansion of 5 to 7 ppm/°C (parts per million per degree Celsius) to an epoxy fiberglass PCB with a coefficient of 14 to 16 ppm/deg C can lead to solder joint cracking and failures.

Soldering. The different methods for soldering components to a surface mount PCB can generally be classed as wave soldering and reflow soldering.

In wave soldering, the surface-mount components must be temporarily bonded to the PCB with a drop of adhesive to provide a temporary mechanical bond in the absence of leads. The pads on which the components are to be mounted are first prepared with a solder coat. After the components are placed and bonded with a drop of adhesive, the adhesive is cured. The loaded PCB is then inverted and passed through a molten solder wave in an in-line wave soldering machine. All of the surface-mounted components are immersed briefly in molten solder while the solder fillets are formed. Wave soldering is fast, but significant thermal stress is placed on the components. Some components such as ceramic capacitors and integrated circuits may be damaged by this thermal shock.

There are five major reflow soldering methods used in SMT: 1) vapor-phase reflow, 2) convection, 3) infrared, 4) hot plate conduction, and 5) laser. The only difference in these methods is the process of applying heat to melt the solder. Each method uses one or more of the three heat transfer mechanisms: condution, convection and radiation.

The solder is usually applied to SMT boards for reflow soldering by screening or stenciling solder paste on the metallized contact pads. The solder paste may be cured after the components have been placed as an additional step.

Wave soldering can be readily adapted to PC boards with a combination of surface-mount and through-hole components. Leads of through-hole components are inserted into the reverse side of the circuit board (opposite the side on which the surface-mounted components are located), and soldered simultaneously.

In vapor-phase reflow, the PCB is prepared with solder and flux and the components are placed on it. The assembly enters into a chamber, where it is exposed to the superheated vapor of perfluorotrianylamine; it has a constant vapor-phase temperature of 215°C (419°F). This is above the melting point of electronic-grade lead-tin solders in paste form (183° to 190°C). The vapor condenses onto the entire assembly and heats it uniformly for 20 to 60 seconds. At about 188°C (370°F) the solder paste under the component pads liquefier, and solder fillets form to bond the components to the PCB pads.

The controlled vapor temperature prevents charring or degradation of the components. Some machines employ a secondary vapor that forms a vapor blanket to prevent the escape of the primary vapor into the air. The lower temperature secondary vapors also minimize thermal shock as the PCB is placed in or removed from the primary saturated vapor. However, even with a secondary vapor, the thermal shock of a vapor-phase process is more severe than other reflow methods. For this reason, many vapor-phase machines also include infrared (IR) heaters.

The convection process consists of passing the surface mount board assembly through an oven on a conveyor belt. The temperature of the atmosphere in the oven and the speed of the belt establish the temperature profile. It can take 3 minutes for the temperature to reach a sharp peak before another 2- to 3-minute falloff. The process is so named because the primary mode of heat transfer is with heated air. Secondary heating also occurs as a result of radiation and conduction from hot surfaces within the furnace. An inert gas such as nitrogen can be used instead of air to reduce oxidation.

In infrared/laser heating solder in paste form can also be reflowed. There is a choice of unfocused or focused infrared or focused laser light. The primary advantage of this method is its ability to heat very local areas. Heat might build for 3 minutes before reaching a sharp peak of less than a second. This is followed by a cooldown of several minutes. Inert gases also can be used.

Surface-Mount Component Package. Component packages suitable for surface mount soldering are made from appropriate materials. Leads on SMC packages may be cut to stub length, bent in a gull-wing form or rolled as J-leads. (gull-wing leads extends out from the sides of the package and J-leads are rolled under the package.) Higher temperature polyester resins replace epoxy as a suitable molding material. These higher temperature materials include polyester, polyamide and polyphenylene.

Standard SMC packages have been developed for resistors, capacitors, diodes, transistors, integrated circuits, and other components to facilitate automatic, robotic pick-and-place positioning. Standard length, width, and height dimensions have been established for leadless chip SMCs such as capacitors, resistors, and diodes. Resistors, capacitors and diodes are also packaged in metal electrode face-bonding (MELF) cylindrical leadless packages. (*See* CAPACITOR, RESISTOR.) Similar small-outline diode (SOD) packages are used for diodes.

Semiconductor devices such as transistors and thyristors are now packaged in small-outline transistor (SOT) packages and integrated circuits are in small-outline IC (SOIC) packages. ICs with more than 64 input/output pins can be in leaded or leadless chip carriers. *See* SEMICONDUCTOR PACKAGE.

Many other electronic components including inductors, surface acoustic wave (SAW) devices, trimmers, quartz crystals, switches, and relays are also packaged for surface mounting.

SURFACE WAVE

At some radio frequencies, the electromagnetic field is propagated over the horizon because of the effects of the ground. This is observed at very low, low, and medium frequencies, and to a limited extent at high frequencies. Surface-wave propagation does not take place at very high frequencies because the ground becomes progressively more lossy as the frequency increases.

At very low frequencies, surface-wave propagation occurs in conjunction with waveguide propagation (*see* PROPAGATION CHARACTERISTICS).

SURFACE-WAVE TRANSMISSION LINE

Surface-wave propagation can be deliberately introduced along wire conductors at very high, ultrahigh, and microwave frequencies. Surface-wave propagation does not occur along the ground at these wavelengths because the ground is not a very good conductor (*see* SURFACE-WAVE PROPAGATION), but it can occur when a good conductor is used. A single-wire line, designed to transfer surface-wave signals along its length, is called a surface-wave transmission line.

A surface-wave line consists of a wire conductor surrounded by a thick layer of low-loss dielectric material. The signals are coupled into and out of the line by horns called launchers. The surface-wave line does not radiate significantly because the electromagnetic field travels in contact with the conductor. The line is unbalanced since there is only one conductor. This type of line is also called G-line, or Goubau line, after its inventor, Dr. George Goubau.

The surface-wave line is more efficient than coaxial cable or a parallel-wire line with two or four conductors. However, the wavelength must be short or the electromagnetic field will tend to be radiated rather than propagated along the line. Surface-wave transmission lines are not effective at the very low, low, medium, and high frequencies. Shielded or balanced feed lines are used at those wavelengths.

SURGE

When power is first applied to a circuit or device that draws considerable current, a surge can occur. For a moment, excessive current flows. This surge can result in damage to the circuit or appliance.

Surges occur on a small scale in electronic circuits, especially those that draw significant current and operate from utility mains.

SURGE IMPEDANCE

See CHARACTERISTIC IMPEDANCE.

SURGE PROTECTION

See CIRCUIT PROTECTION.

SURGE SUPPRESSOR

See CIRCUIT PROTECTION.

SUSCEPTANCE

Susceptance is the alternating-current equivalent of conductance, and the reciprocal of reactance. Susceptance is a mathematically imaginary quantity.

Given an inductive reactance $X_L = jX$ the susceptance Y_L is given by:

$$Y_L = 1/X_L$$

where X is a positive real number.

Given a capacitive reactance X_C, the susceptance Y_C is given by:

$$Y_C = 1/X_C$$

where X is, again, a positive real number. *See also* REACTANCE.

SWEEP

Sweep is a continuous, usually linear change of the value of a variable from one defined limit to another. The movement of an electron beam across the face of a television picture tube is an example of sweep. An oscillator or radio transmitter can be made to scan in frequency; this is another example of sweep.

If the limits of a variable are defined as A and B, sweep is characterized by a controlled change from limit A to limit B, and an almost instantaneous return from limit B to limit A. Sweep is also called scanning. *See also* SCANNING.

SWEEP-FREQUENCY FILTER OSCILLATOR

A sweep-frequency filter oscillator is a specialized form of sweep generator (*see* SWEEP GENERATOR). It produces a signal that varies rapidly in frequency between two defined and adjustable limits. The oscillator is used in conjunction with a spectrum analyzer for testing and alignment of selective filters (*see* SPECTRUM ANALYZER).

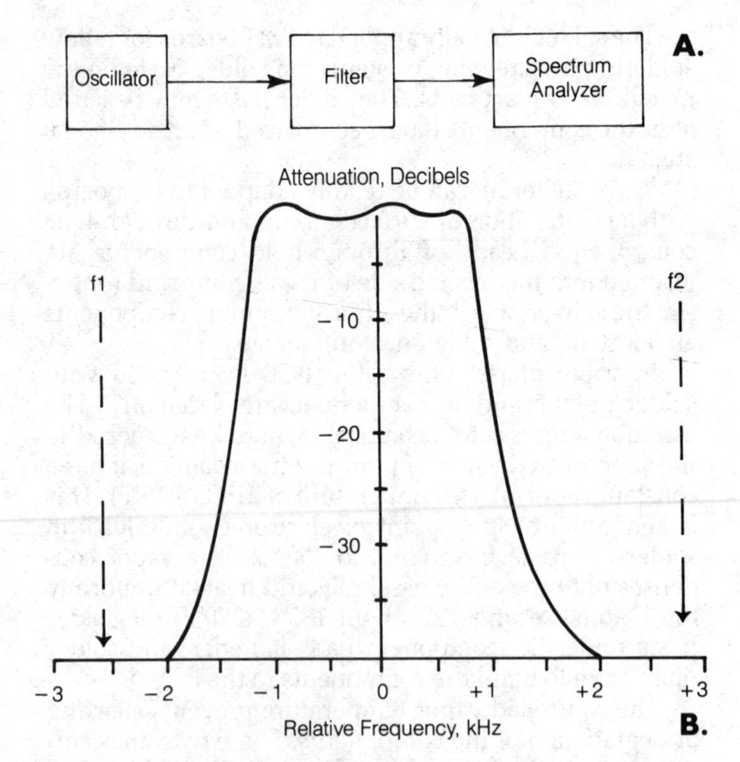

SWEEP-FREQUENCY FILTER OSCILLATOR: The connection of a sweep-frequency filter analyzer (A) and a typical display on a spectrum analyzer connected to a sweep-frequency filter analyzer (B).

The interconnection scheme for a sweep-frequency filter oscillator, spectrum analyzer, and test filter is shown at A in the illustration. The oscillator frequency sweeps between two limits, $f1$ and $f2$, where $f1 < f2$. The frequency $f1$ is well below the cutoff or passband; the frequency $f2$ is well above the cutoff or passband. The sweep rate is very rapid, so that the spectrum-analyzer display appears continuous.

An example is shown at B of a typical spectrum-analyzer display that might be obtained for a 3-kHz single-sideband filter in the intermediate-frequency chain of a superheterodyne receiver. This test arrangement gives a graphic illustration of the performance of the filter. The filter can, if necessary, be adjusted for optimum rectangular response while the oscillator and spectrum analyzer are connected.

The sweep-frequency filter oscillator can be used with all types of filters for checking the response. *See also* BANDPASS RESPONSE, BAND-REJECTION RESPONSE, HIGHPASS RESPONSE, LOWPASS RESPONSE.

SWEEP GENERATOR

A sweep generator is a circuit that generates the alternating voltages or currents that operate the deflection system in a cathode-ray tube. These voltages or currents move the electron beam from left to right, causing a trace

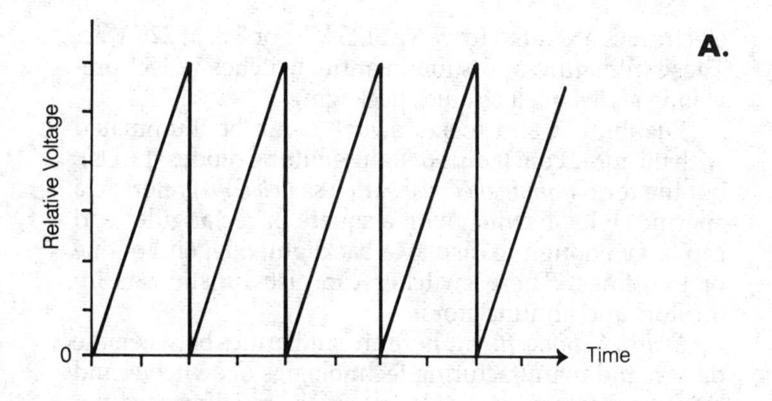

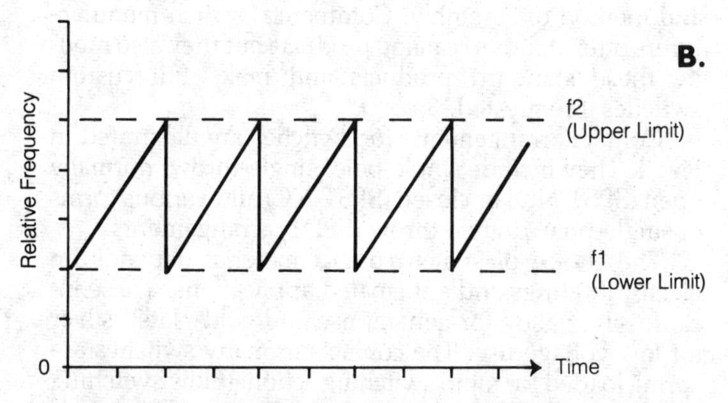

SWEEP GENERATOR: A waveform for a generator used with a cathode-ray tube (A) and the frequency versus time waveform for a sweep-frequency oscillator (B).

to appear on the screen. The voltage or current generally varies with time as shown in the graph at A. *See also* CATHODE-RAY TUBE.

A wideband oscillator with a frequency output that scans continuously between two limits in a sweeping action is also called a sweep generator. Sweep generators are used in conjunction with spectrum analyzers to evaluate the frequency-response characteristics of receivers, tuned circuits, and selective filters of various types. The frequency varies with time in a manner similar to that illustrated by the function at B. *See also* SPECTRUM ANALYZER, SWEEP, SWEEP-FREQUENCY FILTER OSCILLATOR.

SWINGING CHOKE

A swinging choke is a special form of power-supply filter choke designed to exhibit an inductance that varies inversely with the flow of current. When the current level is low, the swinging choke has maximum inductance, but the core saturates easily as the level of current rises.

In the average swinging choke, the inductance is halved when the current is tripled and quartered when the current increases by a factor of 9. The low-current inductance of a swinging choke depends on its application, but it is generally in the range of 1 to 15 henrys.

SWITCH

A switch is an electromechanical device with metallic contacts for manually opening and closing an external electrical circuit. Commercial switches suitable for electronics are packaged in cases designed to be securely mounted to a panel or printed-circuit board (PCB). They include provisions for making external connections to conductor terminations, movable metal contacts insulated and isolated from a manual actuator, and a spring mechanism to accelerate the on-off transition.

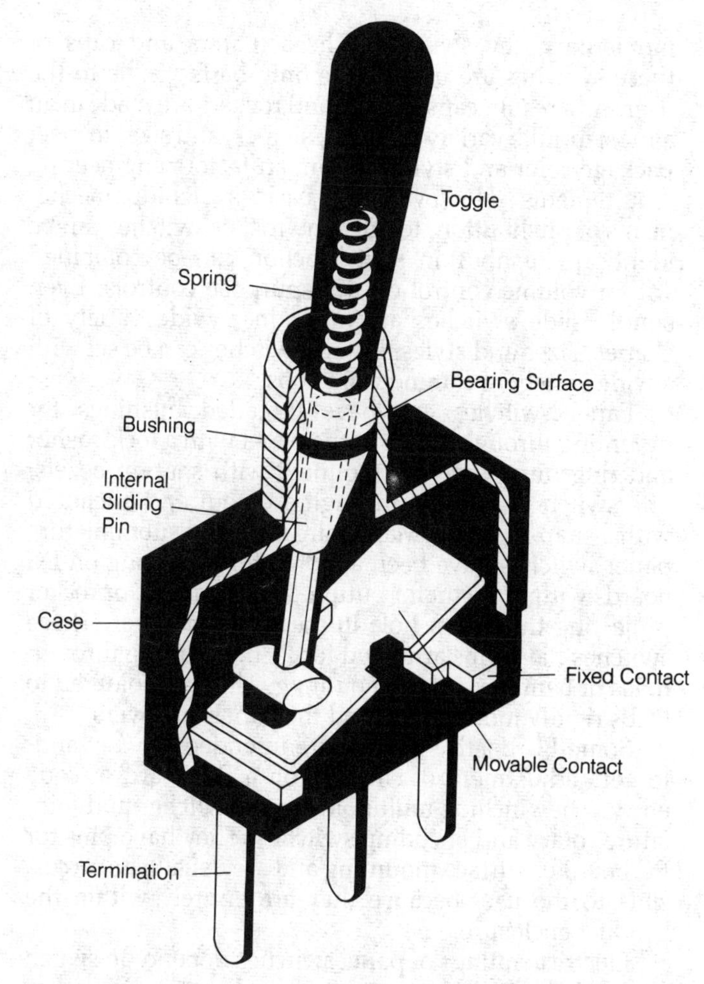

SWITCH: Fig. 1. Cutaway view of a PC board/panel-mounted toggle switch with a teeter-totter, snap-action contact switching mechanism.

Figure 1 shows a miniature switch suitable for use in electronics. The simplest and most basic control device, the switch can take on many forms and styles. Many switches for electronic applications have been scaled down in size and electrical rating from conventional electrical power switches. Miniature or subminiature panel switches suitable for electronics are generally classed by switch action as:

1. Pushbutton (latching or momentary action).
2. Toggle (including paddle).
3. Rocker.
4. Slide.
5. Rotary.
6. Thumbwheel encoding.

Pushbutton, toggle, and rocker switches can be lighted or unlighted, and pushbutton switches can be momentary key switches. The actuators and caps of these switches are usually the only parts visible to the user. As a result, caps, levers, and rockers are made in an almost infinite variety of colors, shapes, and sizes to meet package color and style and human-factors engineering requirements. However, the basic switching mechanisms of pushbutton, toggle, and rocker switches can be identical. Pushbutton switch action can be combined with a volume control on dual-purpose controls. Even simple slide switches are made in a wide variety of shapes, sizes and styles. Rotary switches can be set with a wide variety of commercial knobs.

Panel switches can have threaded bushings for mounting through a hole in the panel with a lockwasher and ring nut. Some are mounted with snap-in bezels. The switch is inserted through a cutout and anchored with a snap-in frame. Many miniature and subminiature panel switches have been adapted for mounting on PC boards with the actuator button, knob, handle, or rocker projecting through a hole in the panel or cover. These switches have pins arranged for PCB mounting through holes or bent for surface mounting. They are soldered to PCBs rigidly mounted behind the panels or covers.

Some PC board switches are intended for "set-and-forget" adjustment of circuitry on a PC board or code entry. They include multipole toggle switches and miniature rotary and encoding switches. They have pins for PC board or surface mounting and are usually not accessible to the user because they are entirely within the product enclosure.

Electrical ratings of panel switches depend on switch size and construction materials. The dielectric properties of the switch case insulation material the spacing between conductors and contacts influence rating. International safety standards apply to switches rated 6 V dc or more but are usually mandated for those rated 115 V ac or more. Most commercially available panel switches for electronics are rated for 5 A at 125 V ac or 3 A at 220 V ac. These miniature and subminiature switches reflect prevailing styles in electronics packaging.

Pushbutton and rocker switches can be illuminated with incandescent lamps or light-emitting diodes (LEDs), but the term *illuminated pushbutton switch* (IPB) refers to a specific style of switch with a square or rectangular end cap large enough to display a backlighted legend of one or two lines. These switches can also function as indicators and annunciators.

Switch Manufacture. There are similarities between the design and manufacturing technologies of switches and relays including selection of materials, molding, plating, and method of assembly. Commercial switch manufacturers offer standard catalog products but they also modify these standard products and make full custom switches if requested. *See* RELAY.

Contact arrangements for switches are illustrated in Fig. 2. They include single-pole, single-throw, normally open (SPST-NO) or closed (SPST-NC) and various forms of single-pole, double-throw (SPDT) arrangements.

The choice of switch contact material depends on electrical ratings and anticipated applications. These include silver alloy for general use and gold-plated silver for low-voltage use. The contacts in many switches are spring loaded for rapid switching action. (Slow switching permits contact burning, sticking and arcing.) The spring also provides a tactile and sometimes an audible response to indicate that the switch has been actuated.

Very low voltage circuits are called dry circuits. Microampere currents and microvoltages are blocked by the resistance of layers of oxidation or contamination on the contacts. The contacts on some switches are lightly gold plated to prevent this resistive buildup, particularly when the switch is not in active use.

All switches are rated for contact resistance, working current and voltage, dielectric strength, insulation resis-

Design	Sequence	Symbol	Form
SPST-NO	Make (1)	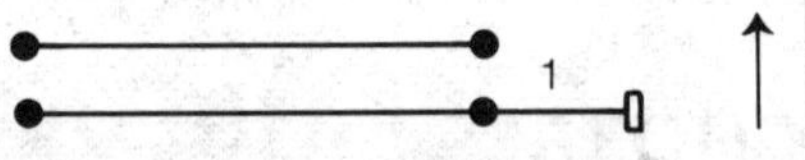 1	A
SPST-NC	Break (1)	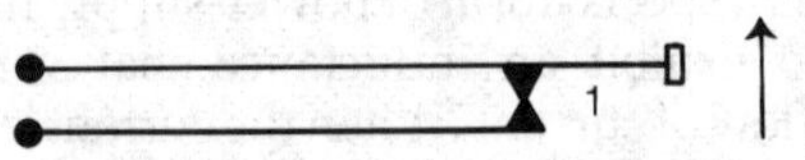 1	B
SPDT	Break (1)—Make (2)	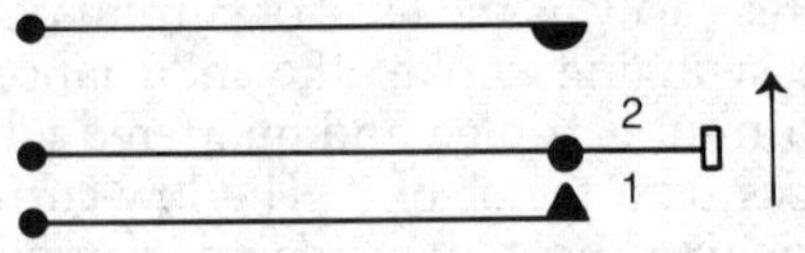 2 1	C
SPDT	Make (1) before Break (2)	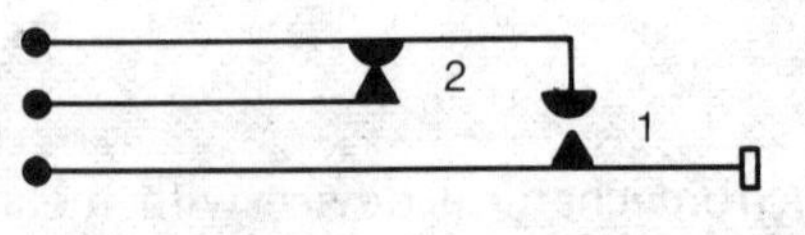 2 1	D

SWITCH: Fig. 2. The most commonly used switch contact arrangements.

tance, life, and ambient temperatures. Switch contacts are derated for high ambient temperature regardless of their electrical ratings. Contacts can be inadvertently welded together in direct-current (dc) circuits, but alternating-current (ac) reversals minimize this possibility. As a result, dc switch ratings are lower than ac switch ratings.

Switch cases can be molded from various thermosetting or thermoplastic resins or from ceramic materials. Flame-retardant thermoplastics cost less than the thermosetting plastics such as diallyl phthalate or phenolic. The metal parts of panel switches are typically formed or stamped from sheet steel, stainless steel or copper alloys, such as phosphor-bronze or beryllium copper.

Panel-Switch Specifications

Slide. Slide switches permit making and breaking a circuit as an insulated handle attached to a conductive element is moved in a channel. Typical ratings for miniature and subminiature versions are 1 A or less at 125 V ac. Standard-size units may be rated for 6 A at 125 V ac. Typical electrical life is 10,000 cycles and mechanical life is 100,000 cycles.

Toggle/Rocker. Toggle and rocker switches use similar contact mechanisms. Available in miniature, subminiature and standard sizes, a typical contact rating for a miniature toggle switch is 3 A for resistive loads at 120 V ac or 28 V dc. Electrical life is typically 50,000 make-and-break cycles.

Pushbutton. Pushbutton switches are available in standard, miniature, subminiature, and ultraminiature sizes. Typical electrical ratings are 1 A or less at 125 V ac or 28 V dc (miniature) to 3 A at 250 V ac (standard). Most are SPST, but there are momentary and push-on, push-off actions. Typical electrical ratings are 50,000 to 60,000 make-and-break cycles.

Lighted Pushbutton. Lighted-pushbutton switches (LPBs) with one or more lamps in a rectangular housing are available in two size ranges for snap-in mounting. Snap-on translucent plastic end caps are available in a range of colors and combinations. Caps are large enough (1 by ¾ inch) to permit the engraving or printing of legends. LPBs are designed for front-panel relamping; illumination may be with LED, neon or incandescent lamps. Typical electrical ratings are 1 A at 125 V ac.

Rotary. Rotary switches switch alternate circuits in a sequence dictated by a coaxial shaft with a rotary-detent action. A pointer knob selects the options. Open-deck rotary switches can have up to 6 poles or sections, each with up to 24 positions. There are choices of detent style, rotational stops, shaft length, bushing, and locating key. Most of these switches are panel mounted. Typical contact ratings are 0.5 A or less at 125 V ac. There are also miniature enclosed rotary switches used on instruments and military equipment.

Encoding Switch Specifications

Thumbwheel/Lever Wheel. Thumbwheel and lever-wheel panel switches convert numbers selected by individual sections or modules into digital codes. They are assembled from stacks of four to eight modular rotary switches, clamped together and snapped into position in panels through cutouts. Each module has a round printed-circuit encoder with a brush and rotary detent assembly that outputs a coded digital equivalent to any number selected with the rotary module actuator or lever. These switches permit nonelectronic encoding and they also provide memory of the last digits entered. From 8 to 16 dial positions per module permit the entry of decimal or hexadecimal codes. Typical electrical ratings are 125 mA at 28 V dc or ac. Some commercial units include LED or incandescent lamps.

PCB Encoding. PC-board data encoding switches do encoding on PC boards to personalize them for specific applications. These switches are usually set at the factory and are not accessible to the user. The two general classes are: rotary PCB-mounting and dual-in-line packaging (DIP).

Dual-In-Line Package. DIP switches are miniature rectangular assemblies of 4 to 12 isolated rocker or toggle switches that are set with a fine-point tool. Switch arrangements include SPST-NO and SPST-NC. Intended for switching logic-level signals (5 to 25 V ac, at 25 to 100 mA), many of these plastic switches have gold-plated contacts.

Rotary PCB. Rotary PC-board switches are miniature single-decade thumbwheel switches made for PC board mounting.

Underwriters' Laboratories (UL) listings and Canadian Standards Association (CSA) certification are typical requirements on all power switches today. These assure the user that the manufacturer has met safety standards by using adequate insulation and spacing between conductive elements to minimize shock hazard and make combustion or electrical failure unlikely.

SWITCHED-CAPACITOR FILTER

A switched capacitor filter (SCF) is an active integrated circuit filter that does not require discrete capacitors or inductors and is able to replace passive inductance-capacitance (LC) filters. The figure shows a simplified diagram of a switched capacitor filter consisting of a switching section and operational amplifier (op amp) integrator, all on the same chip. The actual switches are MOS transistors (shown in the diagram as equivalent switch elements S1 and S2) and a capacitor C1, which can simulate a resistor. The output of this section is fed to the op amp integrator with capacitor C2 in the feedback circuit. A simple clock switches the transistor on and off and determines the filter's cutoff or center frequency.

The transistors are alternately switched on and off in a "break-before-make" fashion by the clock. When S1 is open and S2 is closed, the charge on C1 flows to ground. With a fixed-input voltage, the faster the switching rate, the greater the flow of charge per unit time. Since current is the rate of charge flow, an average current flow can be determined. Thus an equivalent resistance can then be determined by dividing the known input voltage by the average current. Because the equivalent value of resistance also equals the inverse of the product of the clock

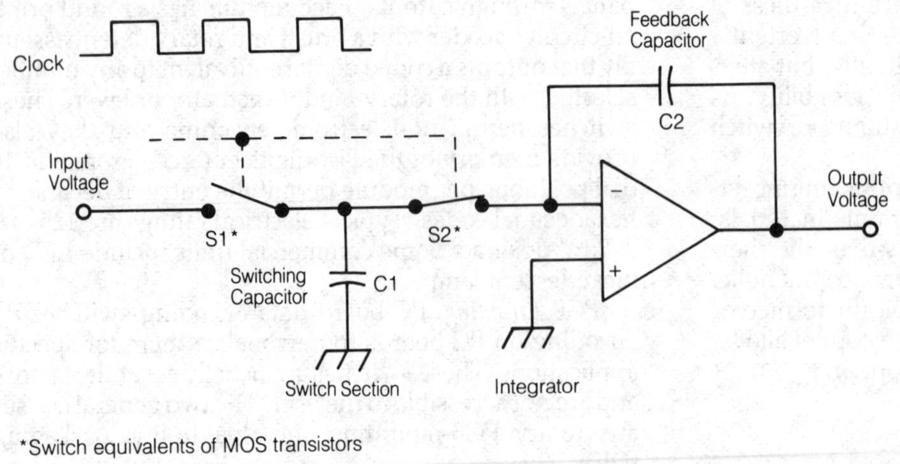

SWITCHED-CAPACITOR FILTER: Basic diagram of a switched-capacitor filter showing switch and op amp integrator sections. The switches are MOS transistors triggered by clock pulses.

frequency and switching capacitor C1, it can be seen that the equivalent input resistance to the integrator can be changed by adjusting clock frequency.

The cutoff frequency of an op amp integrator can be determined from the value of the equivalent input resistance and capacitor C2 in the integrator. The equivalent value of the cutoff frequency of the switched capacitor filter is directly proportional to the product of the clock frequency and C1, and is inversely proportional to the value of C2.

The designer can set the frequencies passed and rejected by the filter (its response) and the shape of the gain versus frequency graph by selecting the clock switching frequency, the capacitor ratio, and in some cases the value of external resistors. A single SCF chip can be adapted to function as a bandpass, highpass, lowpass, band-reject (notch), or all-pass filter in Butterworth, Chebyshev, Bessel, or Cauer (elliptic) response formats. The analog frequency range is typically 1 Hz to 20 kHz or more. SCFs are now used in audio systems, electronic musical instruments, speech synthesis and recognition equipment, and test instruments.

Neither discrete capacitors nor inductors are required for the operation of the switched capacitor filter. This eliminates the bulk, inaccuracy, and temperature sensitivity associated with those components. However, discrete resistors may be used to determine filter response and Q. Tunable or swept-frequency operation can be achieved with a variable-frequency clock.

Switched capacitor filters may contain more than one section. The most popular SCF configuration is the industry standard MF-10, which includes two second-order (two-pole) sections fabricated with the CMOS process. It can be configured for any common filter type. The two second-order sections on the IC can be cascaded to form fourth-order (four-pole) filters. Higher orders can be obtained by cascading the dual chips.

MOS switching transistors contribute to the high level of integration of the SCF because they are only about one-sixth the size of bipolar transistors. In addition, MOS op amps are also compact and can be made with precisely controlled characteristics. MOS capacitors can be formed in precise ratios and can store a charge for several milliseconds. The equivalent resistance created by the switched capacitors saves chip space and avoids dependence on integrated resistors that are inherently imprecise. *See* FILTER.

SWITCHING AMPLIFIER

See DC AMPLIFIER.

SWITCHING DIODE

See DIODE, P-N JUNCTION.

SWITCHING POWER SUPPLY

See POWER SUPPLY.

SWITCHING TRANSISTOR

See POWER TRANSISTOR.

SWR

See STANDING-WAVE RATIO.

SWR LOSS

See STANDING-WAVE-RATIO LOSS.

SWR METER

See REFLECTOMETER.

SYMBOLIC LOGIC

See BOOLEAN ALGEBRA.

SYMMETRICAL RESPONSE

A bandpass filter has a symmetrical response if the attenuation-versus frequency curve is balanced. Mathematically, a symmetrical response can be defined in terms of the center frequency.

Let f be the center frequency of a bandpass filter. Let $f1$ be some frequency below the center frequency, and let

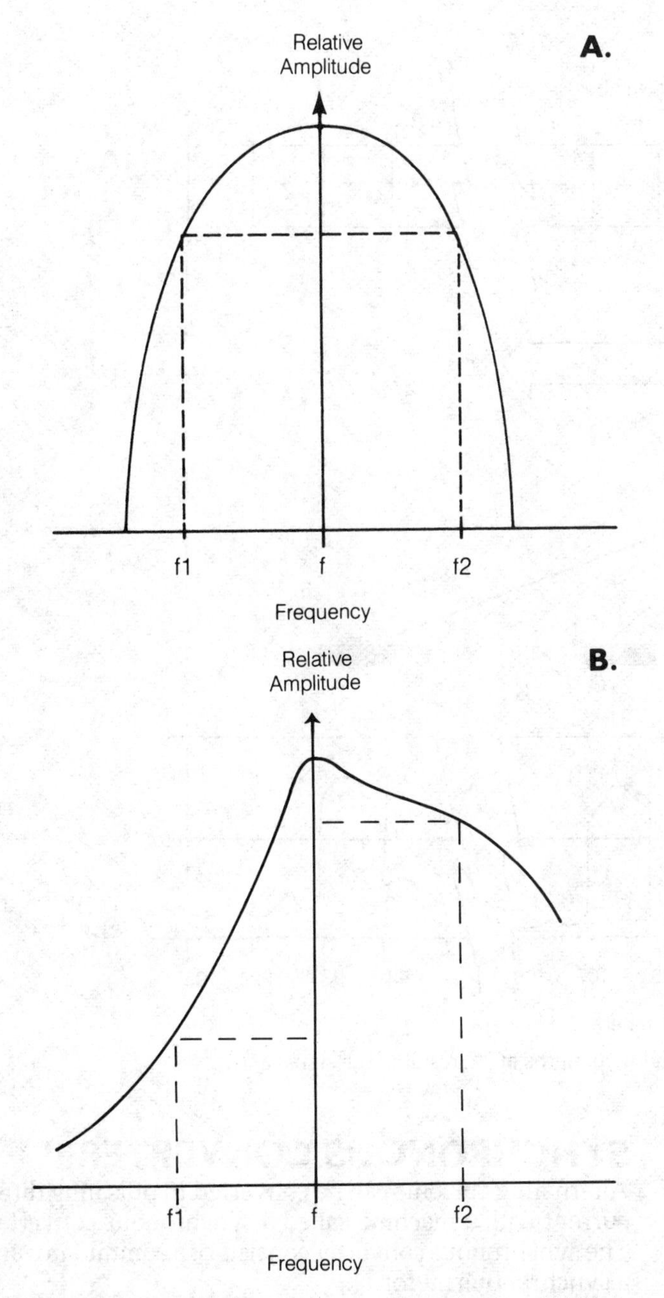

SYMMETRICAL RESPONSE: A symmetrical response (A) and a nonsymmetrical response (B).

$f2$ be some frequency above the center of the passband, so that $f2 - f = f - f1$. If the response is symmetrical, the filter attenuation at frequency $f1$ will be the same as the filter attenuation at $f2$, no matter what values are given for $f1$ and $f2$ within the stated constraints.

A symmetrical filter response is illustrated at A. A nonsymmetrical response is shown at B. *See also* BANDPASS FILTER, BANDPASS RESPONSE.

SYNC

See SYNCHRONIZATION.

SYNCHRONIZATION

When two signals or processes are exactly aligned, they are said to be in synchronization. Two identical waveforms, for example, are synchronized if they are in phase. Two clocks are synchronized if they agree.

In a television transmitting and receiving system, the electron beam in the picture tube must move in synchronization with the beam in the camera tube. Otherwise, the picture will appear to split, roll, or tear (*see* HORIZONTAL SYNCHRONIZATION, VERTICAL SYNCHRONIZATION). Pulses are sent by the transmitting station to keep the receiver in synchronization so that viewers see a clear picture.

In an oscilloscope, the sweep rate can be synchronized with the frequency of the applied signal to obtain a motion-free display. This is called triggering (*see* TRIGGERING).

In some communications systems, the transmitter and receiver are synchronized against an external, common time standard. This is called synchronous or coherent communications (*see* SYNCHRONIZED COMMUNICATIONS).

A synchro system connects two shafts electrically so the angular position or rotation of one shaft will always be in synchronism with the angular position of the other one. *See* SYNCHRO SYSTEM.

SYNCHRONIZED COMMUNICATIONS

It is important in communications that the maximum number of signals be accommodated within a given band of frequencies. This has traditionally been done by attempting to minimize the bandwidth of a signal. There is a limit, however, to how small the bandwidth can be if the information is to be effectively received (*see* BANDWIDTH).

Digital signals, such as Morse code, occupy less bandwidth than analog signals, such as voice. Perfectly timed Morse code consists of regularly spaced bits, each bit with the duration of one dot. The length of a dash is three bits; the space between dots and/or dashes in a single character is one bit; the space between characters in a word is three bits; the space between words and sentences is seven bits. This makes it possible to identify every single bit by number, even in a long message (see A in illustration). Therefore, the receiver and transmitter can be synchronized, so that the receiver knows which bit of the message is being sent at a given moment.

In synchronized or coherent Morse code, the receiver and transmitter are synchronized so the receiver hears and evaluates each bit individually. This makes it possible to use a receiving filter with an extremely narrow bandwidth. The synchronization requires an external, common frequency or time standard. The broadcasts of stations WWV/WWVH are used for this purpose (*see* WWV/WWVH). Frequency dividers provide the necessary synchronizing frequencies. A tone is generated in the receiver output for a particular bit, if, and only if, the average signal voltage exceeds a certain value over the duration of that bit, as at B. False signals, such as might be caused by filter ringing, sferics, or other noise, are generally ignored because they do not cause sufficient average voltage.

Synchronization can be used with codes other than Morse, such as Baudot or ASCII. Experiments with

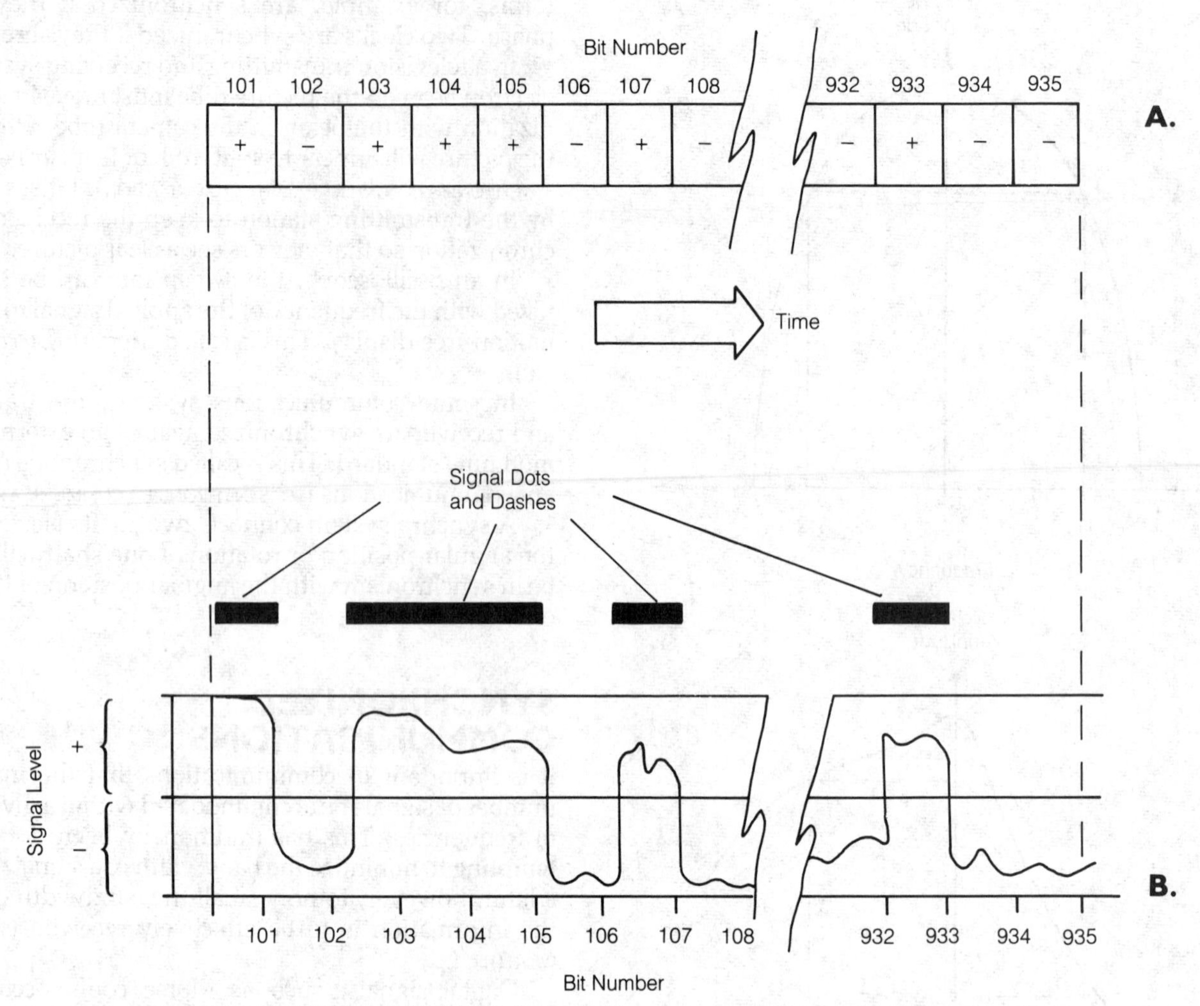

SYNCHRONIZED COMMUNICATIONS: High/low patterns at (A) result in the signal at (B).

synchronized communications have shown that the improvement in signal-to-noise ratio, compared with nonsynchronized systems, is nearly 20 decibels at low speeds (on the order of 15–20 words per minute). The reduced bandwidth of coherent communications allows proportionately more signals to be placed in any given band of frequencies.

SYNCHRONIZED VIBRATOR POWER SUPPLY

A mechanical vibrator can generate high-voltage alternating current (ac) from low-voltage direct current (dc). (*See* VIBRATOR POWER SUPPLY.) Generally, if a dc output is desired, the resulting alternating current is rectified with semiconductor diodes; however, rectification can be done with a vibrator. Then the device is called a synchronized or synchronous vibrator.

The synchronized vibrator has two sets of contacts: One set performs the chopping (interrupting) function and the other reverses the polarity of the alternating current in the transformer secondary, inverting either the negative or positive pulses to obtain pulsating direct current. This pulsating direct current is then filtered in the conventional manner.

SYNCHRONOUS CONVERTER

Alternating currents can be converted to pulsating direct current with a machine called a synchronous converter. The synchronous converter consists of a commutator and a synchronous motor.

For mechanical rectification of 60 Hz alternating current, the synchronous motor turns at a rate of 60 revolutions per second or 3,600 revolutions per minute. The commutator reverses the polarity of the output twice for each cycle. This results in inversion of either the negative or positive parts of the waveform in the output.

SYNCHRONOUS DATA

Synchronous data is any data, usually digital, that is transmitted according to a precise time function. Baudot and ASCII are examples of codes that can be sent as synchronous data. Each bit has the same predetermined duration. Mechanically sent Morse code can also be synchronous.

In some synchronous-data communications circuits, the receiver is kept in step with the transmitter by means of pulses sent at the beginning and/or end of each character, or at regular intervals. In radioteletype, some codes use these pulses to indicate the start and finish of each character. Television transmitters send out synchronizing pulses to keep the receiver scanning properly.

A special form of synchronous-data communication uses an independent reference standard to lock the receiver and transmitter precisely. This results in great improvement in the signal-to-noise ratio. *See also* SYNCHRONIZATION, SYNCHRONIZED COMMUNICATIONS.

SYNCHRONOUS DETECTOR

A synchronous detector is a circuit that is similar to a product detector. A local-oscillator signal is mixed with the incoming signal to recover the sidebands.

The synchronous detector uses a phase-locking system to keep the local-oscillator output in exact synchronization with the carrier wave of the incoming signal. This ensures that the output of the detector will be identical with the original modulating signals.

Synchronous detectors are used in color television receivers to recover the chrominance sidebands with optimum precision. If the local-oscillator signal were not locked with the carrier or subcarrier, distortion would result from even the smallest frequency difference.

SYNCHRONOUS MOTOR

See MOTOR.

SYNCHRONOUS MULTIPLEX

Synchronous multiplex is a form of time-division multiplex in which two or more signals are transmitted over a single circuit at the same time. The signals are split into discrete intervals with equal duration, and interwoven at a higher speed. Synchronous multiplexing is a form of serial data transfer (*see* SERIAL DATA TRANSFER, TIME-DIVISION MULTIPLEX).

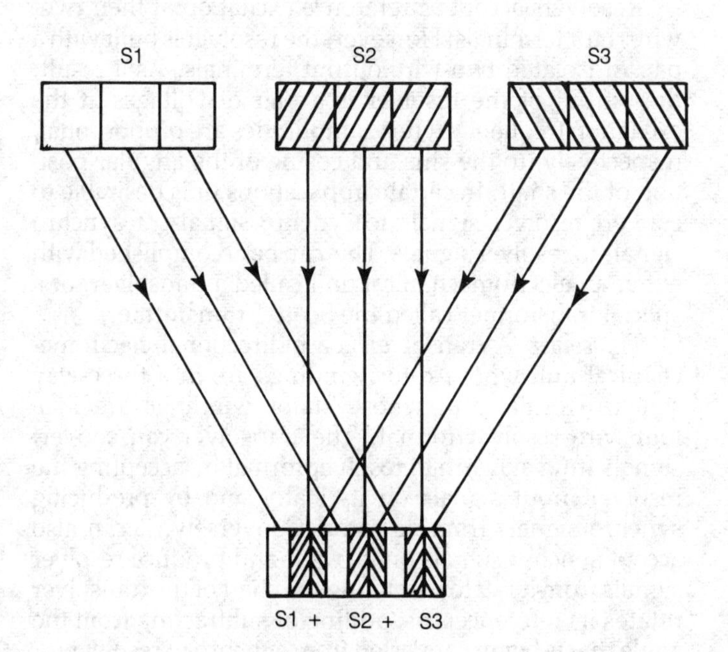

SYNCHRONOUS MULTIPLEX: The principle of synchronous multiplex diagrammed.

If there are n different signals, designated $s1$, $s2$, $s3$, . . ., sn, that are to be combined in series with synchronous multiplexing, let each signal be divided into intervals of time duration t. When the signals are recombined, they are sent one interval at a time in sequence $s1$, $s2$, $s3$, . . ., sn. To maintain synchronization of the composite signal with each of the original signals, the information in each interval must be speeded up by a factor of n, resulting in interval durations of t/n. This situation is illustrated for $n = 3$.

SYNCHRONOUS ORBIT

The period of revolution of a satellite around the earth depends on the altitude of the satellite. Objects in low orbits go around the earth very rapidly; the Mercury and Gemini spacecraft, for example, orbiting at altitudes of a few hundred miles, had periods of only about 90 minutes. A satellite at an altitude of 22,500 miles takes 24 hours to complete one orbit. This type of orbit is called synchronous, since the period is the same as the length of the earth day.

If a satellite is placed in a synchronous orbit above the equator so it revolves in the same direction the earth rotates, that satellite will remain above a fixed point on the surface of the earth. Many communications satellites have been placed in these orbits.

SYNCHROSCOPE

A synchroscope is an oscilloscope that shows a waveform or train of pulses in a motionless display, as a result of triggering. *See* OSCILLOSCOPE, TRIGGERING.

SYNCHRO SYSTEM

A synchro system connects two shafts electrically so the angular position or rotation of one shaft will always be in synchronism with the angular position of the other one. Synchros, sometimes called selsyns, are small self-synchronous alternating-current (ac) machines. A synchro system acts as if two trains of gears were coupled together with a flexible shaft. However, electrical transmission permits the transmitter and receiver to be much further apart than would be practical with mechanical coupling. Synchros are classified as angle-sensing transducers. Shaft angle is used in the measurement and control of position, velocity, and acceleration.

Control Transmitter and Receiver. The simplest synchro system consists of two units, a transmitter and a receiver, each with electrically and mechanically identical rotor and stator. These are shown schematically in the figure. Both transmitter and receiver rotors have single windings which are connected across the single-phase ac excitation source. Both stators have three wye-connected field windings spaced 120 degrees apart. The ends of each winding on the transmitter stator are connected to the ends of the corresponding windings on the receiver stator to form a three-wire circuit.

When the ac exciting circuit is energized, each rotor winding acts as the primary of a transformer, and a

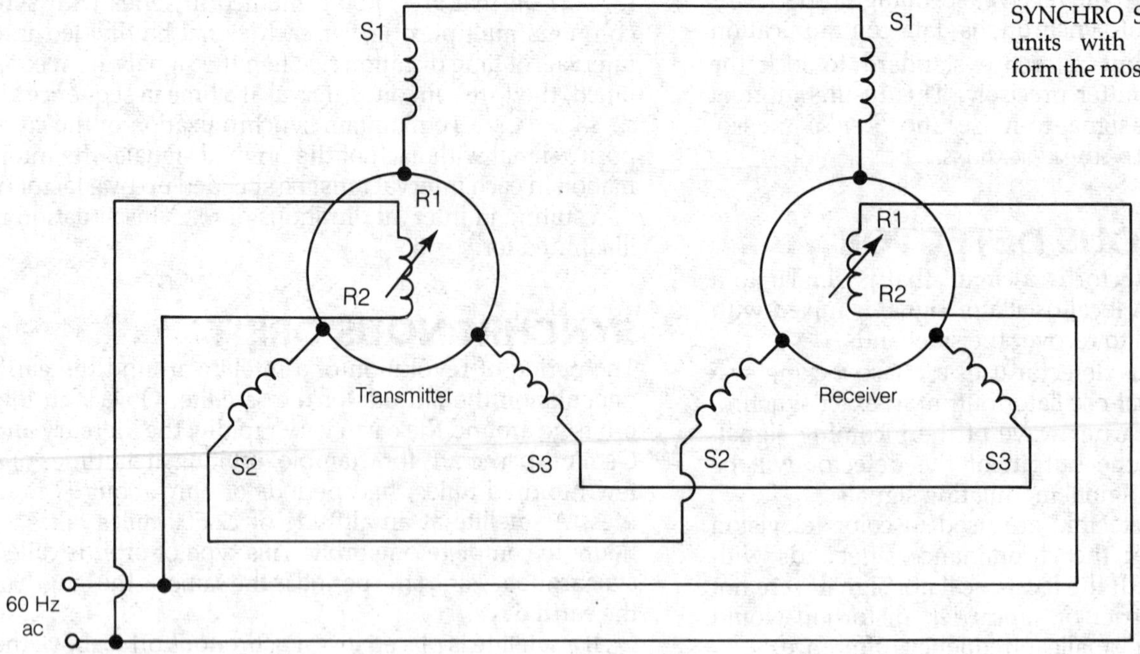

SYNCHRO SYSTEM: Two synchro units with electrical connections form the most basic synchro system.

voltage will be induced in the three windings of its stator. The magnitude of the voltage in each stator winding depends on the position of the rotor. When the transmitter and receiver rotors are in corresponding positions, the stator voltages across corresponding stator terminals in the two units are equal in magnitude and opposite in phase. Thus there is no current flowing in the connecting wires between the stators of the two units.

If the rotors are not in correspondence, the voltages produced by transformer action in the transmitter stator windings differ from those similarly induced in the receiver stator windings. The resulting currents that flow between the two stators establish a torque that turns the receiver rotor to a position that corresponds with the transmitter rotor. Signals representing an angular difference are transmitted by turning the transmitter rotor through an angle. The receiver rotor responds to these signals and immediately moves through the same angle.

Control Transformer. A control transformer is similar in construction to the control transmitter. However, in acting as a transformer, it accepts, at its three-wire stator terminals, a set of signals of the kind produced by a synchro control transmitter corresponding electrically to some shaft angle. It produces at its rotor terminals, a signal with the carrier frequency proportional to the sine of the angular difference between the electrical input angle and the mechanical angular position of its shaft.

Control Differential Transformer. A control differential transformer is a synchro component similar in construction to a control transmitter except that its rotor has a three-wire ac output. It accepts a set of signals at the carrier frequency (the kind produced by a control transmitter) at its three-wire stator terminals. The line-to-line amplitude ratios of this signal correspond to a remote shaft angle. It produces, at its three-wire rotor terminals, a three-wire set of signals at the carrier frequency whose line-to-line amplitude ratios are proportional to the difference between the input angle, and the mechanical angular position of its shaft.

In addition to synchros, there are other angle-sensing transducers. These include the encoder, the potentiometer, and the resolver. Resolvers are related to synchros in that they also present information about the angular position of a shaft in the form of relative amplitudes of the excitation frequency. All synchro and resolver signals, rotor and stator, input and output, are sine waves at the same frequency and in time phase synchronization.

Resolvers accept ac reference excitation at their two-wire rotor terminals. However, the resolver is built with a pair of isolated two-wire output terminals. As a result, the output of the resolver is a pair of voltages at the excitation frequency whose amplitudes are proportional, respectively, to the sine and cosine of the angular position of the shaft. In certain applications, it is desirable to convert resolver signals to synchro signals or synchro signals to resolver signals. This can be accomplished with either an electromechanical unit called a transolver, or a special transformer called the Scott-T transformer.

Transolver. A transolver is a bidirectional electromechanical unit whose rotor windings are in a three-way synchro format and whose stator windings are in a four-wire resolver format. The transolver can convert signals from synchro to resolver format by accepting the resolver input signals on its stator and by producing synchro signals from its rotor. Conversely, it can also accept synchro signals on its rotor and produce resolver signals from its stator. Rotating the shaft of the transolver rotates its reference axis, adding or subtracting from the angle that is being converted from synchro to resolver (or resolver to synchro) format.

Scott-T Transformer. The Scott-T transformer is a specialized transformer able to transform signals from synchro to resolver format or resolver to synchro format. It performs the same function as the transolver with its rotor set at the zero shaft position. It has a tapped winding on the synchro side providing three terminals, and two pairs of windings on the resolver side providing four terminals. *See also* ENCODER, POTENTIOMETER, RESOLVER, SERVO SYSTEM.

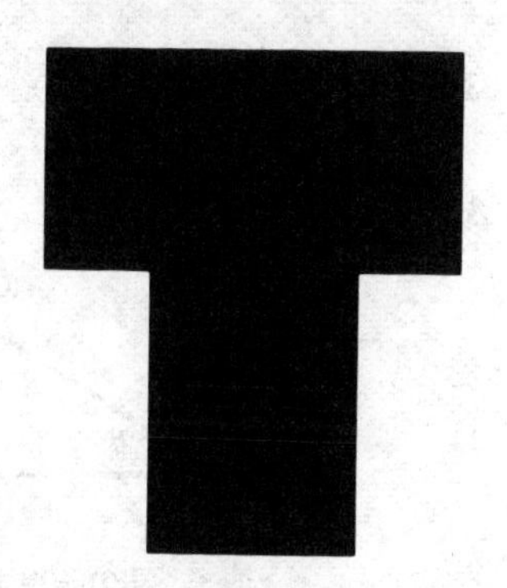

T-1 CARRIER

The T-1 Carrier is a telephone company service introduced in 1962 that employs digital transmission at 1.5444 megabits per second (Mb/s) to carry 24 voice channels with time-division multiplexing (TDM). Pulse-code modulation (PCM) is used to code 8 bits per signal element for each voice channel sampled in 256 discrete amplitudes. The 4 kHz voice channel is sampled at the Nyquist rate (8 kb/s) to produce a channel data rate of 64 kb/s. The pulses are transmitted over copper wire pairs or microwave radio. Repeaters are placed at 6,000-foot intervals to regenerate and retime the digital waveform. *See* PULSE MODULATION, TIME-DIVISION MULTIPLEX.

The T-1 Carrier was designed to transmit voice signals between central offices that are less than 50 miles apart. AT&T Communications will locate T-1 circuits on a customer's premises for Accunet T1.5 Service. M-44 Multiplexing Service can be used with Accunet T1.5 to carry up to 44 simultaneous voice conversations.

The T-1 Carrier System is the multiplexing structure for PCM systems in North America and is the first level and basic building block of this hierarchy.

T-1 CARRIER: Types and characteristics.

Line Type	Characteristics
T1	1.544 Mb/s[1]; 24 or 44 TDM voice channels
T1C	3.152 Mb/s; 2 TDM T1 lines
T2	6.312 Mb/s; 4 TDM T1 lines
T3	44.736 Mb/s; 28 TDM T1 lines
T4	274.176 Mb/s; 168 TDM T1 lines

1. Megabits per second.

TACHOMETER

A tachometer is an electromechanical generator that is capable of counting revolutions. The tachometer can be operated open loop to provide a readout in revolutions per minute (rpm), or it can be used in a closed-loop feedback system for velocity control. *See* SERVO SYSTEM.

TACTICAL AIR NAVIGATION

Tactical Air Navigation (TACAN) is a form of radionavigation used by commercial and military aircraft. The TACAN system operates at ultrahigh or microwave frequencies where the atmosphere has a minimal effect on electromagnetic propagation.

The TACAN system provides constant information on the distance and bearing to one or more ground stations with respect to an aircraft. The distance (range) and bearing (azimuth) can be directly read out on a display in the aircraft. Altitude is not indicated.

TANGENT

The tangent function is a trigonometric function. In a right triangle, the tangent is equal to the length of the opposite side divided by the length of the adjacent side (see illustration on p. 834). In the unit circle $x^2 + y^2 = 1$, plotted on the Cartesian (x,y) plane, the tangent of the angle θ, measured counterclockwise from the positive x axis, is equal to y/x. This is illustrated at B. For values of θ that are an odd multiple of 90 degrees, the tangent is not defined, because for those angles, $x = 0$.

The tangent function is periodic and discontinuous. The discontinuities appear at odd multiples of 90 degrees. The tangent function ranges through the entire set of real numbers shown at C.

In mathematical calculations, the tangent function is abbreviated tan. Mathematically, the tangent function is always equal to the value of the sine divided by the value of the cosine:

$$\tan \theta = \sin \theta / \cos \theta$$

Values of tan θ for various angles θ are given in the table on p. 834. *See also* TRIGONOMETRIC FUNCTION.

TANGENTIAL SENSITIVITY

The sensitivity of a receiver is usually expressed in terms of the signal voltage that results in a certain signal-to-noise ratio or amount of noise quieting in the audio output (*see* NOISE QUIETING, SENSITIVITY, SIGNAL-TO-NOISE RATIO). The most frequently used figures are a 10-decibel (dB) signal-to-noise ratio or 20 dB of noise quieting. These specifications are not always a precise indication of how well a receiver will respond to weak signals. A more accurate indication is given by the tangential-sensitivity figure. Tangential sensitivity is sometimes called threshold sensitivity or weak-signal sensitivity.

Tangential sensitivity is defined as the signal level, in microvolts at the antenna terminals, that results in a barely discernible signal at the output under conditions of minimum external noise. The tangential sensitivity is also defined as the signal level that produces a 3 dB

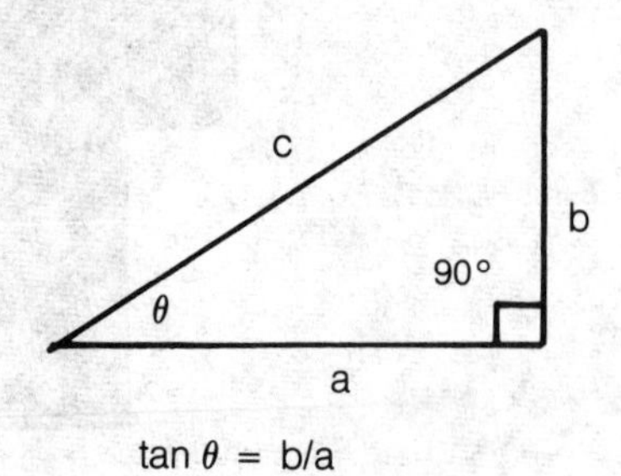

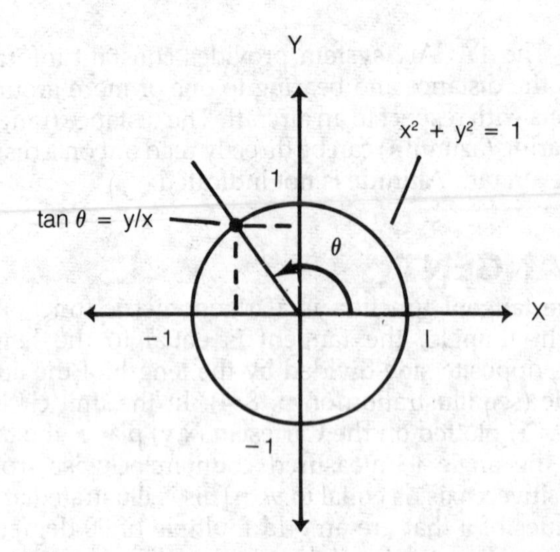

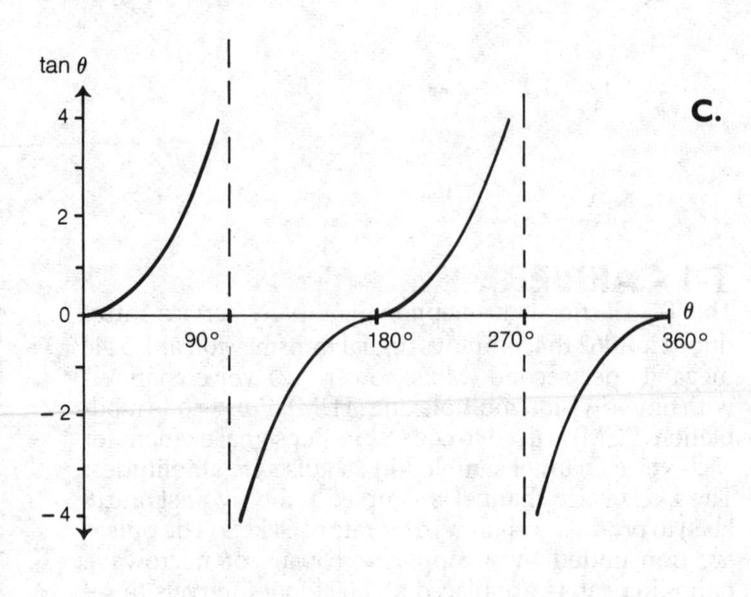

TANGENT: A triangle for defining the tangent function (A), the unit circle model for the tangent function (B), and a graph of the tangent function between 0 and 360 degrees (C).

TANGENT: Values of tan θ for values of θ between 0 and 90 degrees.
For values between 90 and 180 degrees, calculate $180 - \theta$, read from this table and multiply by -1.
For values between 180 and 270 degrees, calculate $\theta - 180$ and read from this table.
For values between 270 and 360 degrees, calculate $360 - \theta$, read from this table, and multiply by -1.

θ, degrees	tan θ	θ, degrees	tan θ	θ, degrees	tan θ
0	0.000				
1	0.017	31	0.601	61	1.80
2	0.349	32	0.625	62	1.88
3	0.052	33	0.649	63	1.96
4	0.070	34	0.675	64	2.05
5	0.087	35	0.700	65	2.15
6	0.105	36	0.727	66	2.25
7	0.123	37	0.754	67	2.36
8	0.141	38	0.781	68	2.48
9	0.158	39	0.810	69	2.61
10	0.176	40	0.839	70	2.75
11	0.194	41	0.869	71	2.90
12	0.213	42	0.900	72	3.08
13	0.231	43	0.933	73	3.27
14	0.249	44	0.966	74	3.49
15	0.268	45	1.00	75	3.73
16	0.287	46	1.04	76	4.01
17	0.306	47	1.07	77	4.33
18	0.325	48	1.11	78	4.70
19	0.344	49	1.15	79	5.14
20	0.364	50	1.19	80	5.67
21	0.384	51	1.23	81	6.31
22	0.404	52	1.28	82	7.12
23	0.424	53	1.33	83	8.14
24	0.445	54	1.38	84	9.51
25	0.466	55	1.43	85	11.4
26	0.488	56	1.48	86	14.3
27	0.510	57	1.54	87	19.1
28	0.532	58	1.60	88	28.6
29	0.554	59	1.66	89	57.3
30	0.577	60	1.73	90	—

signal-to-noise ratio, or 3 dB of noise quieting, at the output. These voltages are considerably smaller than the voltages required to produce a 10 dB signal-to-noise ratio or 20 dB of noise quieting. Two different receivers, both with the same sensitivity in terms of the 10 dB signal-to-noise or 20 dB noise-quieting figures, can differ in tangential sensitivity.

TANK CIRCUIT

A tank circuit is an electrical circuit that stores energy by exchanging it alternately between two reactances. A tank circuit operates at one frequency or integral multiples of that frequency. All parallel-resonant circuits are tank circuits. A resonant antenna, cavity, or length of transmission line can also act as a tank circuit if it is fed at a voltage loop. At resonance, the impedance of a tank circuit is theoretically infinite. In practice, despite some component losses, tank circuit impedance is still extremely high.

Many radio-frequency power amplifiers use parallel-resonant tuned circuits in their output. This circuit is called the tank circuit or simply the tank. The inductor, capacitor, or both components, are adjustable so the tank circuit can be made resonant at the desired frequency. In the tuned power amplifier, tank-circuit resonance is indicated by a dip in the collector, drain, or plate current. *See also* PARALLEL RESONANCE, RESONANCE, TUNED CIRCUIT.

TANTALUM CAPACITOR

See CAPACITOR.

TAPE

See MAGNETIC TAPE.

TAPE DRIVE

A tape drive is a computer peripheral with magnetic tape storage media for backing up the storage of digital data from disk drives and for long-term data storage or archival use. A tape drive is a tape recorder adapted for computers with controller and interface electronics. The principal advantage of magnetic tape data storage is high reliability and low media cost. The drawback is long access time because the tape drive is a serial data storage system.

Portable audio tape recorders with audiocassettes were adapted as mass data storage units for low-end personal computers. They stored programs and data but have now been replaced by floppy-disk drives in later-model PCs.

There are two principal types of tape drives in use: cartridge and streaming. Their design is determined by the choice of magnetic tape media.

Cartridge Tape Drive. Cartridge tape drives accommodate the ½-inch-wide tape cartridges able to provide 240 megabytes of disk backup in the space occupied by a 5¼-inch floppy-disk drive or Winchester. The accepted standard cartridge was developed by IBM for the IBM 3480. Compatible cartridges contains 600 feet of ½-inch plastic tape with a chromium-dioxide media.

A drive for this cartridge provides 24 tracks in a two-track, serial serpentine recording mode. Recording density is 12.7 kilobits per inch (kbits/in.), and tape speed is 79 inches per second (in./s). With a small computer systems interface (SCSI), the average data transfer rate is 250 kbits per second. The drive allows backup of 240 megabytes in 17 minutes. The complete drive with cassette measures approximately 3½ by 6 by 8 inches.

Streaming Tape Drive. A streaming tape drive writes data on 2400 feet of reel-to-reel, ½-inch-wide magnetic tape. Streaming tape drives provide start/stop tape speed of 12.5 to 25.0 in./s, low-speed streaming speed of 12.5 to 25 in./s, and high-speed streaming speed of 50 to 100 in./s.

There is a choice of interfaces on streaming tape drives: Pertec, STC, and SCSI. Recording density can be from 1600 to 6250 b/in., depending on the formatting technique. A streaming tape drive typically measures about 24 by 20 by 14 inches.

TAPE RECORDER

A tape recorder is a system designed for recording audio or video signals on, and recovering them from magnetic tape. Many types of tape recorders are available for applications such as voice reproduction, music reproduction, video reproduction, and the storage of data for computers and word processors.

Tape recorders can be classified as either cassette type or reel-to-reel type. Cassettes are made in several sizes, ranging from the small cartridges used in dictating machines to the large units used for recording video signals. Reel-to-reel tape is also available in various sizes.

All tape recorders operate according to the same principle: They convert electrical impulses to alternating magnetic fields and vice versa. In the record mode, a tape recorder produces an alternating magnetic field that causes polarization of fine particles in the tape. In the playback mode, the tape is pulled at constant speed through a unit that converts the fields around the particles into electrical impulses. A block diagram of a tape recorder is shown on p. 836. *See also* MAGNETIC TAPE.

TELECOMMUNICATION

Any form of information transmission or communication by radio, optical fibers, or wires or cables between stations is known as telecommunication. Telecommunication can be carried out at any wavelength from the audio frequencies to ultraviolet and above. Telecommunication covers all forms of electrical and electronic communications.

TELEGRAPH

A telegraph is a system for sending Morse code signals over wire. The telegraph was the earliest method of electrical communication over long distances. It has been

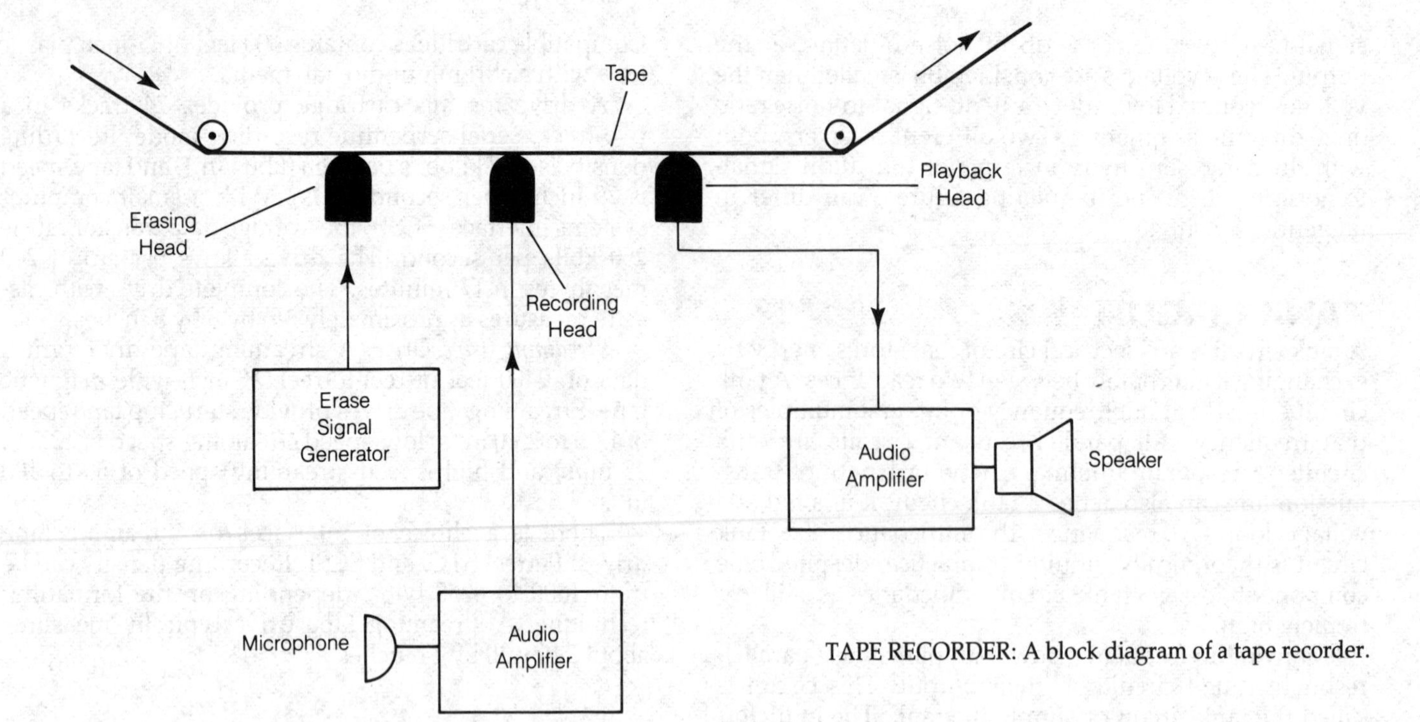

TAPE RECORDER: A block diagram of a tape recorder.

in use since the 1800s. A few telegraph systems still exist.

The simplest form of telegraph consists of a power supply, a key, a relay or other indicating device, and a long cable or two-wire line (see illustration). The range of the telegraph is limited by losses in the line. The earliest telegraph systems used direct current. If a system did not have enough range, it was necessary to relay the message with additional relay equipment.

Modern telegraph systems use amplifiers along the line, greatly increasing the range. Although the telegraph is still in limited use today in underdeveloped countries, it has largely been replaced by telephone and telex systems in the technologically advanced countries. *See also* TELEPHONE SYSTEM, TELEX.

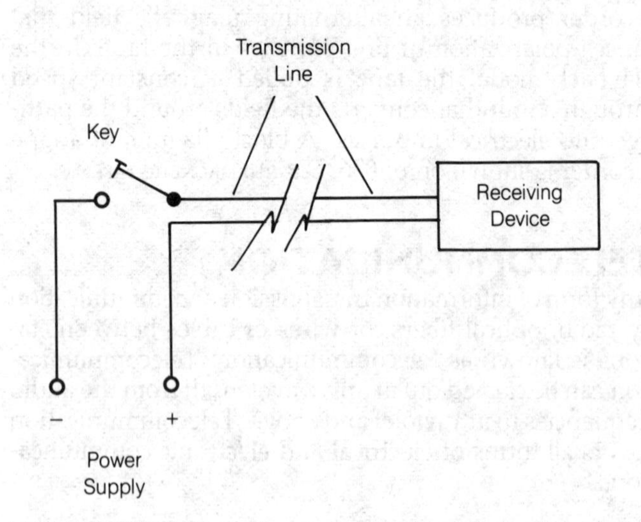

TELEGRAPH: A simple direct-current telegraph circuit.

TELEGRAPH CODES

See AMERICAN MORSE CODE, INTERNATIONAL MORSE CODE.

TELEGRAPH KEY

See KEY.

TELEMETERING

See TELEMETRY.

TELEMETRY

Telemetry is the practice of sensing, measuring, transmitting, and receiving the measurement of variables for the purpose of using the information at some remote location over radio-frequency channels. In typical systems, the measurement data is obtained at its source with sensors or transducers and is conditioned locally for transmission in a coded format to the receiving station. Data gathering at the source is usually done by automatic means from inaccessible locations, such as aircraft or missles in flight, spacecraft or satellites in space, oceanographic buoys, submersibles, and exploratory drilling rigs.

The kinds of information telemetered include engineering data (velocity, pressure, temperature, flow rates, stress, etc.), scientific data (radiation counts, salinity, magnetic field strength, etc.), and biological data from human or animal subjects (blood pressure, pulse rate, body temperature, respiration, blood chemistry, etc.)

The receiving terminals for telemetry are fixed or mobile stations that include receivers, recorders, and suitable equipment for analyzing and perhaps displaying the data. Each variable is typically assigned to a single channel. Data may be transmitted in either analog or digital format.

A telemetry transmitter consists of a measuring sensor, an encoder that translates the sensor readings into electrical impulses, and a modulated radio transmitter

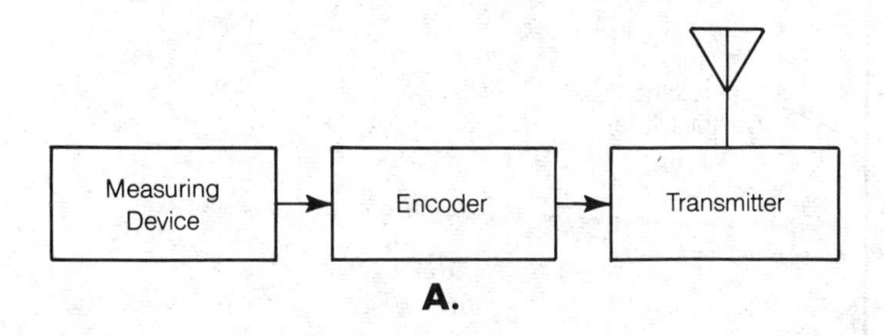

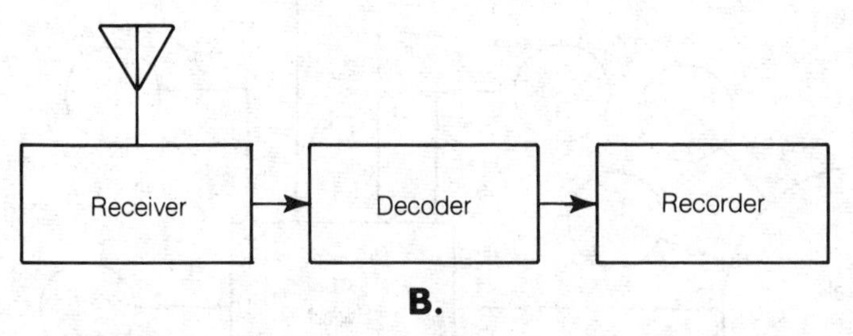

TELEMETRY: A block diagram of a telemetry transmitter (A) and a block diagram of a telemetry receiver (B).

with an antenna (see illustration at A). A telemetry receiver consists of a radio receiver with an antenna, a demodulator, and a recorder (B). Computers are widely used to process the data received.

TELEPHONE DIALER

See ROTARY DIALER, TOUCHTONE®.

TELEPHONE MODEM

See MODEM.

TELEPHONE SYSTEM

A telephone system is a combination of apparatus used for converting speech energy into electrical signals and transmitting those signals over a wire to a distant point, where they are reconverted to audible sound.

The most basic telephone system consists of a pair of handsets connected by a length of wire with a direct-current source. Each handset consists of a micrphone for converting sound into electrical signals and an earphone for converting the electrical signals back to sounds in the audio range. The direct current on the wire is modulated by the microphones in the headsets. Provisions for switching or interconnecting various lines permit one person to talk to all other persons in the network.

Modern telephones use single-sideband-modulated carriers, digital multiplexing, and other sophisticated techniques for transmission. A large system is usually operated by a computer located at a master control center. The computer performs all switching functions and it has a recorder-announcer that informs subscribers of disconnected or changed numbers, system overload, and other problems. Operators must still handle special problems.

Operating Features of Telephone Systems. One of the most important considerations in designing and building a telephone system is switching. The complexity of the switching network depends on the number of users or subscribers. A telephone system normally has the following two characteristics:

1. Accessibility: Any user can call any other user most of the time.
2. No redundancy: Whenever a subscriber X calls a subscriber Y, only Y will receive the call.

Some telephone systems have other features such as:

- Conference calling: Three or more subscribers can communicate on a single line.
- Call forwarding: A subscriber X can arrange to have incoming calls directed to any other subscriber Y.
- Call waiting: If subscriber X calls subscriber Y and the line is busy, subscriber Y is informed that someone is trying to call. Subscriber Y can then communicate with X at any time.
- Radio links: Subscribers may have mobile or cordless telephone sets, linked by radio into the system. Examples of this are the cellular radio network for automobiles and other mobile vehicles and cordless phones for household and office use.
- Local switching networks: A subscriber can have a single line split up into local lines (extensions). This arrangement is used by businesses and in homes.

Interconnection. A telephone system consists of one or more main lines, called trunks, one or more central-office switches, numerous branch lines, and the telephone sets themselves. There may also be operators, radio links, satellite links, and local switching systems. Provision is made for connecting with other systems. This is illustrated in the block diagram on p. 838.

When a subscriber X wishes to contact a subscriber Y, X dials a number consisting of multiple digits. The number of digits depends on the complexity of the switch. (In the United States telephone numbers have 11 digits, consisting of a 1 (in many areas) followed by a three-digit area code and a seven digit subscriber number.) If the subscriber Y has a switching system, an operator answers the call and directs it to the appropriate extension.

Telephone systems cannot operate properly if every subscriber attempts to use it at the same time. The system can handle a limited number of calls, based on the assumption that only a percentage of users will try to make calls at a given time. In areas of rapid population growth, or during holidays, a master control center can become overloaded. Callers are then informed that all lines are busy.

TELEPHONY

See TELEPHONE SYSTEM.

TELETYPE®

Teletype is a tradename of the Teletype Corporation. A

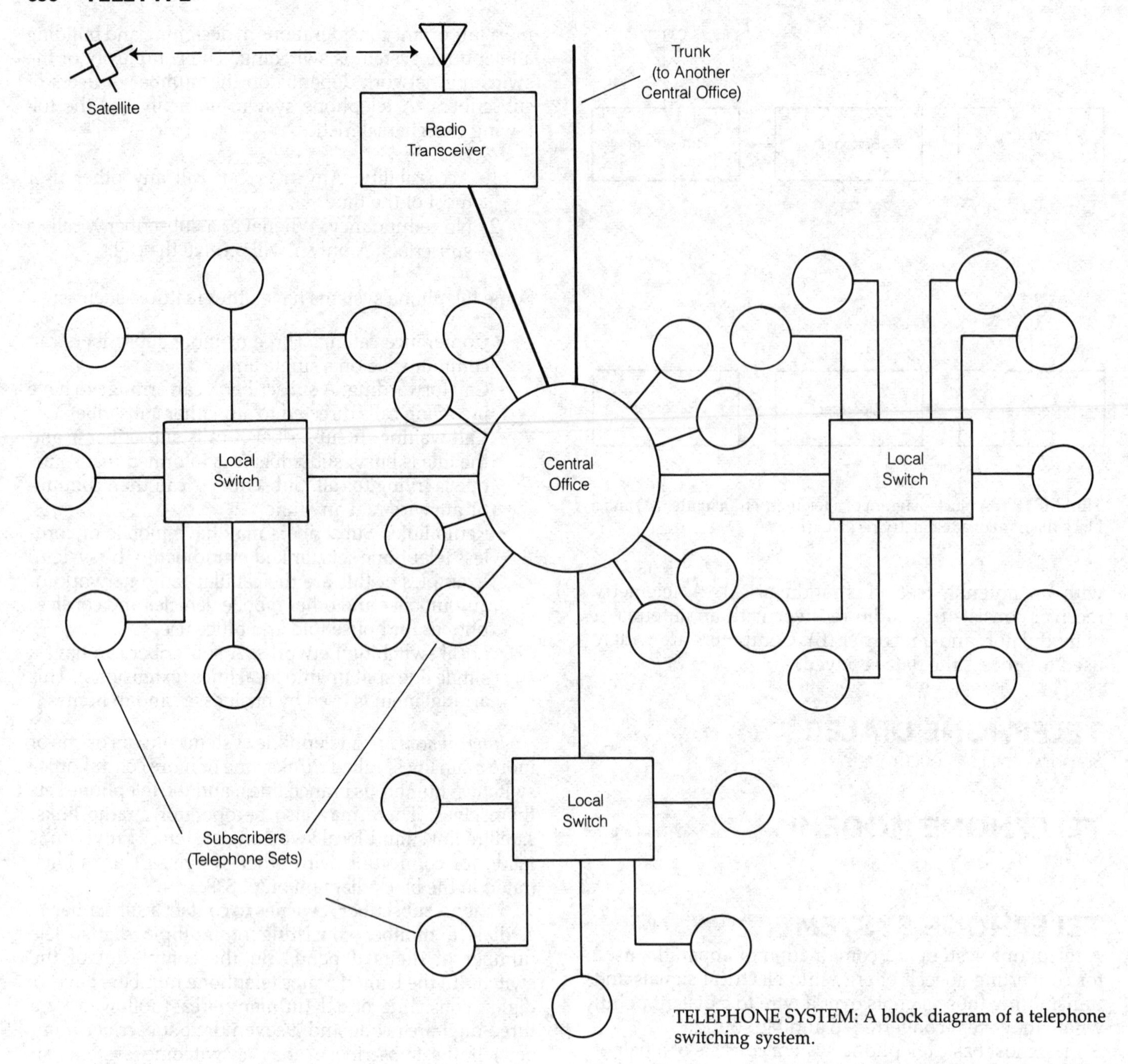

TELEPHONE SYSTEM: A block diagram of a telephone switching system.

teletype system may use wire transmission only, a combination of wire and radio circuits, or radio circuits exclusively. The transmission of printed material by radio-only circuits is called radioteletype and sometimes abbreviated RTTY. (*See* RADIOTELETYPE.)

Teletype signals are digital, consisting of two levels or tones corresponding to on (high) and off (low) conditions. In wire systems, these conditions are represented by direct currents. In radio systems, the high and low conditions are represented by different carrier frequencies. The on state is called mark, and the off state is called space (*see* FREQUENCY-SHIFT KEYING).

Teletype systems normally use either the Baudot code or the ASCII code. The speed may vary, although the most often used speeds are 45.45 baud for Baudot and 110 baud for ASCII (*see* ASCII, BAUDOT CODE, BAUD RATE, WORDS PER MINUTE).

The heart of a teletype installation is the teleprinter, which resembles a typewriter. Messages are sent by typing on the keyboard. The message is printed out simultaneously in the transmitting station and at all receiving stations.

Many teleprinters have a means of preloading and storing a message prior to transmitting it. In modern systems, this is done with electronic memory. In older teleprinters, messages are encoded on paper tape by a device called a reperforator. The reperforator punches holes in the tape according to a five-level or eight-level code (*see* PAPER TAPE). When the paper tape is run through the sending apparatus, the message goes out at the normal system speed.

Some teletype systems operate over telephone lines, making possible the quick sending of messages at moderate cost. Two-way communications can also be accomplished with a telex system. It is extensively used by businesses throughout the world. *See also* TELEX.

TELEVISION

Television is the transfer of moving visual images from one location to another by modulating a radio-frequency (rf) signal. Video signals are combined with audio signals.

Television can be used for broadcasting or for two-way communications. The video signals of television are normally sent and received along with audio signals.

Television systems can be categorized as either fast-scan or slow-scan. In broadcasting, fast-scan television is always used.

The Television Picture and Signal. To obtain a realistic impression of motion, it is necessary to transmit at least 20 still pictures per second, and the detail must be adequate. A fast-scan television system provides 30 images, or frames, each second. There are 525 lines in each frame, running horizontally across the picture, which is 1.33 times as wide as it is high. Each line contains shades of gray in a black-and-white system, and shades of brightness and color in a color system. The image is sent as an amplitude-modulated (AM) signal, and the sound is sent as a frequency-modulated (FM) signal.

A standard television-broadcast channel in the North American system takes up 6 MHz of spectrum space. All television broadcasting is done at very high and ultra-high frequencies for this reason.

The Transmitter and Receiver. A television transmitter consists of a camera tube, an oscillator, an amplitude modulator, and a series of amplifiers for the video signal. The audio system consists of a microphone, an oscillator, a frequency modulator, and a feed system that couples the rf output into the video amplifier chain. There is also an antenna or cable output. A simplified block diagram of a television transmitter is illustrated at A.

A television receiver is shown in simplified block form at B. The receiver contains an antenna, or an input with an impedance of either 75 or 300 Ω, a tunable front end, an oscillator and mixer, a set of intermediate-frequency amplifiers, a video demodulator, an audio demodulator and amplifier chain, a picture tube with associated peripheral circuitry, and a speaker.

For a television picture to appear normal, the transmitter and the receiver must be exactly synchronized. The studio equipment generates pulses at the end of each line and at the end of each complete frame. These pulses are transmitted with the video signal. In the receiver, the demodulator recovers the synchronizing pulses and sends them to the picture tube. The electron beam in the

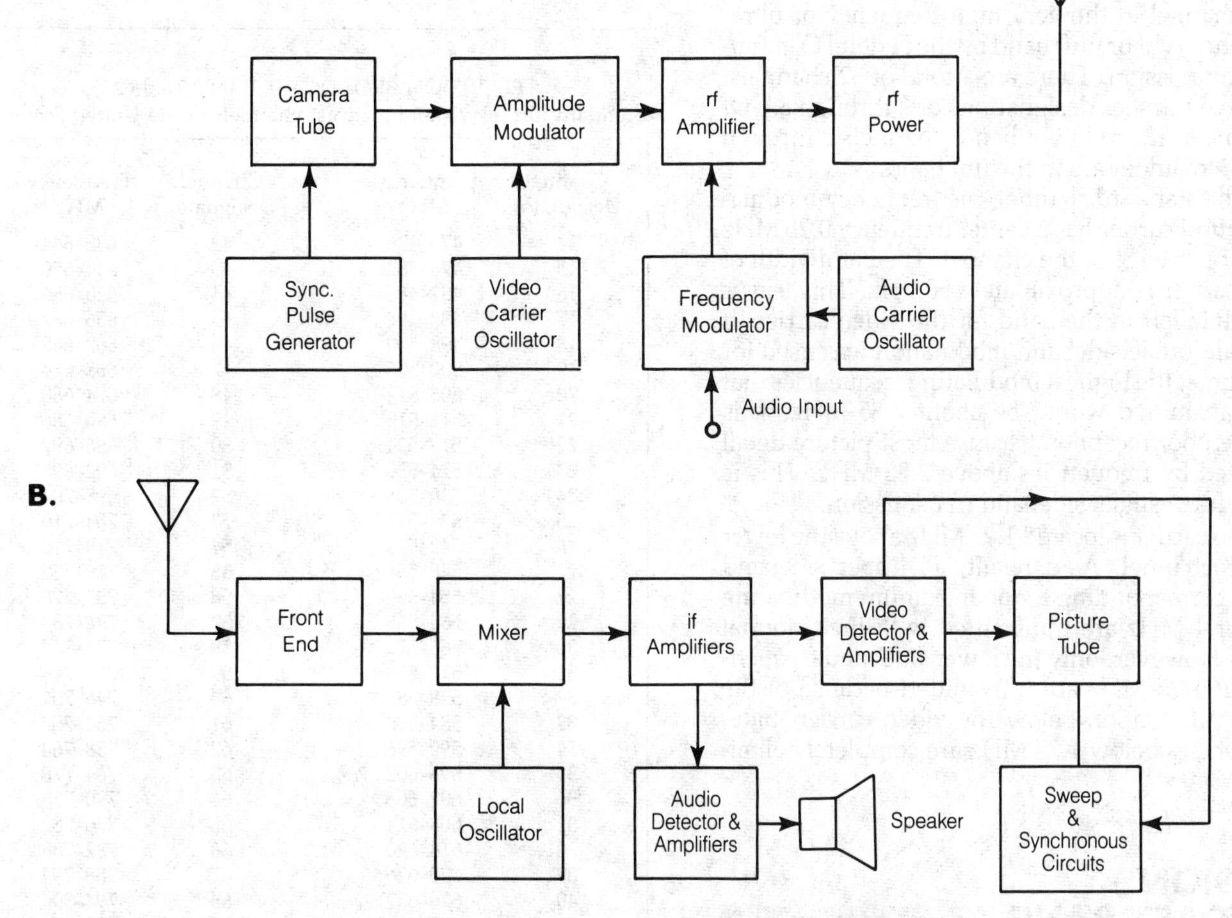

TELEVISION: Block diagrams of a television transmitting system (A) and receiver (B).

picture tube moves in exact synchronization with the scanning beam in the camera tube. If the synchronization is upset, the picture appears to roll or tear.

New Trends in Television. In recent years, more television transmission has been done over cable. In most major metropolitan areas, cable television is available to the general public. An increasing number of television stations also broadcast through geostationary satellites. Subscribers can select from dozens, or even hundreds, of different channels.

Because of the use of cable and satellite systems, television picture quality is much better than ever before. Some thought has been given to increasing the number of lines per frame in the television picture to get a sharper image. This is the objective of worldwide development of high-definition television (HDTV). There are now American, Japanese, and European proposals for HDTV.

Television can be combined with a computer. Services are being offered in some areas allowing viewers to do video shopping or respond to opinion polls.

See also ASPECT RATIO, CABLE TELEVISION, CAMERA TUBE, COLOR PICTURE SIGNAL, COLOR TELEVISION, COMPOSITE VIDEO SIGNAL, HORIZONTAL SYNCHRONIZATION, IMAGE ORTHICON, RASTER, TELEVISION BROADCAST BAND, VERTICAL SYNCHRONIZATION, VIDICON.

TELEVISION BANDWIDTH

All television signals are assigned a specific 6 MHz wide frequency channel in the very high frequency or ultra-high-frequency (vhf or uhf) band by the Federal Communications Commission. There are a total of 82 channels, and they have number designations of 2 through 83. Of the 82 channels, 12 are in vhf band (channels 2 through 13) and the remainder are in the uhf band.

Within the standard channel, the frequency-modulation (FM) sound carrier has a center frequency 0.25 MHz below the upper edge of the channel. The bandwidth of the sound carrier is approximately 50 kHz. This leaves about 5.7 MHz left in the band for the video carrier. If conventional double sideband modulation was used for the video signal, the highest modulating frequencies that could be transmitted would be about 2.85 mHz. This would cause poor reception because most picture detail is represented by frequencies above 2.85 MHz. This is overcome with vestigial sideband transmission.

The video carrier is located 1.25 MHz above the lower edge of the channel. As a result, all upper sideband frequencies corresponding to the maximum modulating frequency of 4 MHz are transmitted with their normal amplitudes. However, only the lower sideband frequencies to about 1.25 MHz are transmitted because of the progressive attenuation below the video carrier. Sideband frequencies below 1.25 MHz are completely eliminated.

TELEVISION BROADCAST BAND

In the United States television broadcasts are made on 68 different channels in the very high frequency and ultra-high-frequency (vhf and uhf) ranges. Each channel is 6 MHz wide, including both the video and audio information.

There is no channel 1. Channels 2 through 13 are the vhf television channels (see Table 1). Channels 14 through 69 are the uhf channels (Table 2).

Ionospheric propagation occasionally affects channels 2 through 6. Long-distance propagation can take place because of dense ionization in the E layer (*see* SPORADIC-E PROPAGATION) on these channels. Tropospheric propagation is occasionally observed on all of the vhf channels (*see* TROPOSPHERIC PROPAGATION). The uhf channels are unaffected by the ionosphere, although tropospheric effects can occur to some extent.

Cable television signals are transmitted at many different wavelengths. Because cable prevents signals from getting in or out, the entire electromagnetic spectrum can

TELEVISION BROADCAST BAND Table 1:
Standard vhf television broadcast channels in the United States.

Channel Designator	Frequency, MHz	Channel Designator	Frequency MHz
2	54.0–60.0	8	180–186
3	60.0–66.0	9	186–192
4	66.0–72.0	10	192–198
5	76.0–82.0	11	198–204
6	82.0–88.0	12	204–210
7	174–180	13	210–216

TELEVISION BROADCAST BAND Table 2:
Standard uhf television broadcast channels in the United States.

Channel Designator	Frequency, MHz	Channel Designator	Frequency MHz
14	470–476	42	638–644
15	476–482	43	644–650
16	482–488	44	650–656
17	488–494	45	656–662
18	494–500	46	662–668
19	500–506	47	668–674
20	506–512	48	674–680
21	512–518	49	680–686
22	518–524	50	686–692
23	524–530	51	692–698
24	530–536	52	698–704
25	536–542	53	704–710
26	542–548	54	710–716
27	548–554	55	716–722
28	554–560	56	722–728
29	560–566	57	728–734
30	566–572	58	734–740
31	572–578	59	740–746
32	578–584	60	746–752
33	584–590	61	752–758
34	590–596	62	758–764
35	596–602	63	764–770
36	602–608	64	770–776
37	608–614	65	776–782
38	614–620	66	782–788
39	620–626	67	788–794
40	626–632	68	794–800
41	632–638	69	800–806

be used. Satellite television systems also use frequencies other than those listed in the tables. *See also* CABLE TELEVISION, TELEVISION.

TELEX™

Teletype® signals can be sent over the telephone line with a frequency-shift-keyed audio tone. This is known as teletype exchange or Telex. A Telex system consists of a conventional teleprinter and a modem (*see* FREQUENCY-SHIFT KEYING, MODEM) connected into a telephone line leased specifically for the purpose of telex operation.

Most telex systems operate at 100 words per minute. Older systems may use a speed of 60 words per minute. Either speed facilitates the sending and receiving of short messages in just a few seconds.

A modern telex machine has an electronic memory for storing text before it is sent. The operator loads the text into the memory, dials the desired telex number, and sends the message at full system speed by pressing a single memory-recall button. If necessary, text can be sent manually. The machine at the receiving end is switched on automatically, no matter whether it is day or night, and the message is printed. Telex systems are used by businesses throughout the world.

TEMPERATURE

The atoms of all substances are in constant motion and the rate at which the atoms and molecules in a material move is proportional to the amount of energy present. This movement can be measured directly, giving a quantity called molecular temperature. All materials radiate some energy; the wavelength of this radiant energy can be measured, yielding a quantity known as spectral temperature.

Molecular Temperature. There is a limit to the degree of cold. Absolute zero is the complete lack of all movement among the atoms or molecules of a material. No place in the known universe is this cold, although in intergalactic space the temperature is very nearly absolute zero. At the other extreme, there is no theoretical limit to heat.

The highest known temperatures occur at the centers of large stars where the atoms are in such violent motion that elemental changes take place.

Molecular temperature can be measured directly with a thermometer. Temperature is expressed as a number, based on a scale at which zero represents some specific phenomenon or condition. In the United States the Fahrenheit scale is most often used. In other countries and among scientists, the Celsius scale is used. The Fahrenheit and Celsius scales are based on specific properties of water, and are therefore most often used for expressing molecular temperature.

Spectral Temperature. In deep space there are practically no molecules, and molecular temperature therefore becomes meaningless. Moreover, it is impossible to measure the temperature of a distant celestial object using a thermometer. Astronomers use a radio telescope or spectroscope to determine the wavelength distribution of the radiation from a planet, star, or gas cloud in outer space. From this, the temperature can be determined according to Wien's Displacement Law.

Astronomers have determined that the surfaces of the sun and distant stars have temperatures of thousands of degrees Fahrenheit. These are spectral temperatures, because a thermometer could not be used in these places.

The Kelvin scale is most often used for expressing spectral temperature. This scale is based on absolute zero, which is the complete lack of radiant energy. Occasionally, spectral temperature is expressed according to another absolute scale called the Rankine scale. *See also* BLACK BODY, CELSIUS TEMPERATURE SCALE, ENERGY, FAHRENHEIT TEMPERATURE SCALE, KELVIN TEMPERATURE SCALE, RADIO TELESCOPE, RANKINE TEMPERATURE SCALE, SPECTROSCOPE, SUPERCONDUCTIVITY, SUNSPOT.

TEMPERATURE COEFFICIENT

Most electronic components are affected by changes in temperature. Resistors and capacitors, for example, change value when the temperature varies over a wide range. The tendency of a component to change in value with temperature variations is known as temperature coefficient.

If the value of a component decreases as the temperature rises, that component is said to have a negative temperature coefficient. If the value increases as the temperature rises, a component has a positive temperature coefficient. A few components exhibit relatively constant value regardless of the temperature; these devices are said to have a zero temperature coefficient. The temperature coefficient (TC or tempco) is usually expressed in parts per million per degree Celsius (ppm/degC). For piezoelectric crystals, the temperature coefficient is often expressed in hertz or kilohertz per degree Celsius. *See also* CAPACITOR, RESISTOR.

TEMPERATURE COMPENSATION

Changes in temperature can cause instability in electronic circuits, especially oscillators. While crystal oscillators are generally more stable than variable-frequency oscillators, either type of circuit will exhibit some drift as the ambient temperature rises or falls. To compensate for changes, special components can be added to an oscillator circuit.

Suppose that a crystal has a negative temperature coefficient of 10 Hz/°C. This means that, for a rise in temperature of 1°C, the frequency will fall by 10 Hz; for a rise of 10°C, the frequency will drop by 100 Hz. Conversely, as the temperature falls, the frequency increases.

To compensate for this, a small capacitor, with a known negative temperature coefficient, can be installed in parallel with the crystal. As the temperature rises, the value of the capacitor decreases, effectively raising the frequency of the crystal-capacitor combination. As the temperature falls, the value of the capacitor increases, pulling the frequency down. If the capacitor is carefully selected, the temperature-coefficient effects of the capacitor will just offset those of the crystal, and the frequency will not change as the temperature fluctuates.

In any oscillator circuit, it is desirable to use components with temperature coefficients as close to zero as possible. This makes temperature compensation simpler than when components with large positive or negative temperature coefficients are used. The temperature should be kept as constant as possible. *See also* TEMPERATURE COEFFICIENT.

TEMPERATURE DERATING

Some electronic equipment, especially power supplies, generate significant heat. If this heat becomes excessive, components may be damaged.

The component's temperature is more likely to exceed the maximum limit in a high ambient temperature than in a cold ambient temperature. A component can dissipate more power when the ambient temperature is low, and less power when it is hot. For this reason, some electronic devices must be operated at a reduced power level when the ambient temperature is elevated. The higher the temperature the more the power must be reduced. This deliberate reduction of power is called temperature derating.

Temperature derating is specified according to a graph of power level versus temperature. A radio transmitter, for example, might normally be run at 100 W output. However, if the temperature exceeds a specified limit, the power input is reduced according to a derating curve as shown in the illustration. Note that, above 250°F, the curve drops to zero, indicating that the equipment should not be operated at all when the temperature reaches that level.

Individual components such as solid state relays have temperature derating curves.

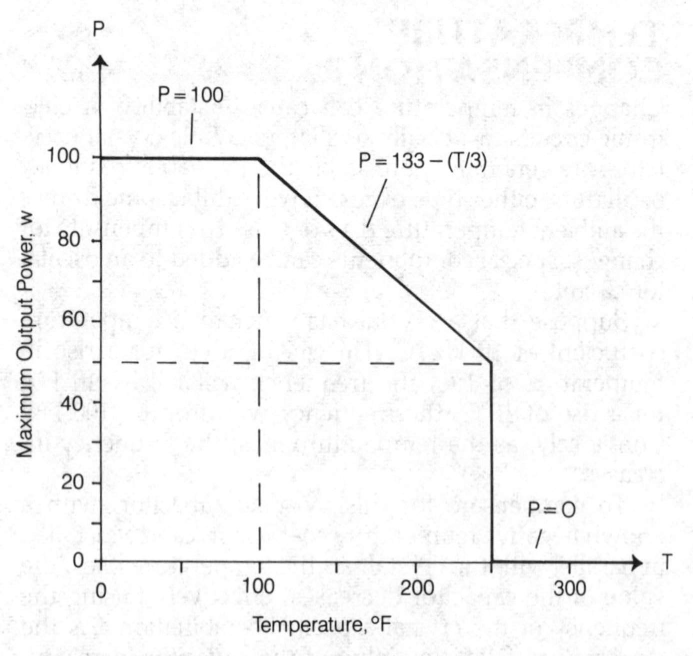

TEMPERATURE DERATING: A temperature derating curve for an electronic component.

TEMPERATURE MEASUREMENT

See CELSIUS TEMPERATURE SCALE, FAHRENHEIT TEMPERATURE SCALE, KELVIN TEMPERATURE SCALE, RANKINE TEMPERATURE SCALE, TEMPERATURE.

TEMPERATURE PROTECTION

See CIRCUIT PROTECTION.

TEMPERATURE RUNAWAY

See THERMAL RUNAWAY.

TEMPERATURE SCALE

See CELSIUS TEMPERATURE SCALE, FAHRENHEIT TEMPERATURE SCALE, KELVIN TEMPERATURE SCALE, RANKINE TEMPERATURE SCALE.

TENSILE STRENGTH

The amount of tension that a wire can withstand without breaking is known as the tensile strength of the wire. It is sometimes called the breaking load. Tensile strength is usually indicated in pounds per square inch, or in kilograms per square centimeter, of cross-sectional area. It can also be expressed in terms of the breaking load in pounds or kilograms. The larger the size of the wire for a given material, the greater the tensile strength.

Steel has the highest tensile strength of common wire materials. The breaking load varies from about 75,000 to 150,000 pounds per square inch of crosssectional area. Copper and aluminum wire have lower tensile strength; typically it ranges from 30,000 to 60,000 pounds per square inch for copper and 20,000 to 40,000 pounds per square inch for aluminum.

The average breaking loads for various gauges of steel, copper, and aluminum wire are given in the table on p. 843. *See also* WIRE.

TERA-

Tera- is a prefix multiplier that means 1,000,000,000,000 (10^{12}). For example, 1 terahertz is 10^{12} Hz or 1,000,000 MHz. (The wavelength of a signal at this frequency is 0.3 mm.) The abbreviation for tera- is the capital letter T. *See also* PREFIX MULTIPLIERS.

TERMINAL

A terminal is a point at which two or more wires are connected together, or where voltage or power is applied or taken from a circuit. There are the antenna terminals of a receiver or transmitter or the speaker terminals of a stereo high-fidelity amplifier. Input and output terminals are generally equipped with binding posts or connectors.

In a large computer, the operating console or consoles may be located separately from the computer mainframe. An assembly of a keyboard and cathode-ray-tube display, is called a terminal. Some computer terminals can

TENSILE STRENGTH: Tensile strength for various American Wire Gauge (AWG) sizes of steel, copper, and aluminum wire. All figures are given in pounds and are approximate.

AWG No.	Steel Wire	Copper Wire	Aluminum Wire
1	5,000–10,000	2,000–4,000	1,300–2,600
2	3,900–7,800	1,600–3,200	1,000–2,000
3	3,100–6,200	1,200–2,400	830–1,700
4	2,500–4,900	980–2,000	650–1,300
5	2,000–3,900	780–1,600	520–1,000
6	1,600–3,200	630–1,300	420–840
7	1,200–2,500	490–980	330–650
8	980–2,000	390–780	260–520
9	770–1,500	310–620	210–410
10	610–1,200	240–490	160–330
11	490–970	190–390	130–260
12	380–770	150–310	100–200
13	305–610	120–240	80–160
14	240–480	97–190	65–130
15	190–380	77–150	51–100
16	150–300	61–120	41–81
17	120–240	48–97	32–64
18	96–190	38–77	26–51
19	76–150	30–61	20–41
20	60–120	24–48	16–32

be connected to a telephone line and used with a distant computer. *See* VIDEO DISPLAY TERMINAL.

In a Teletype or radioteletype station, the keyboard, modem, and printer or cathode-ray-tube display are called the terminal. *See also* MODEM, TERMINAL UNIT, VIDEO DISPLAY TERMINAL.

TERMINAL UNIT

In a Teletype® system, a modem is used to encode signals for transmission and to decode signals for reception (*see* MODEM). This equipment is called a terminal unit.

Teletype systems today normally use either the ASCII or Baudot codes at various speeds (*See* ASCII, BAUDOT CODE, BAUD RATE). Sometimes, the International Morse code is used. A terminal unit consists of a receiving section and a transmitting section. A terminal unit may have a built-in cathode-ray-tube monitor and keyboard, forming a complete communications terminal.

TERMINATION

The point where a transmission line is connected to a load is called the termination. The load itself also can be called the termination.

A termination can consist of any type of load, such as an antenna, a dummy antenna, a telephone set, or an electrical appliance. Sometimes a transmission line is terminated in a short or open circuit. In a radio-transmitting antenna system, the termination is usually called the feed point.

Ideally, all of the available power at a termination is radiated, absorbed, or dissipated by the load in a radio-transmitting antenna, a cable-television system, a telephone line, and an audio-frequency amplifier/speaker system. The load impedance should be purely resistive with an ohmic value equal to the characteristic impedance of the transmission line. *See also* CHARACTERISTIC IMPEDANCE, IMPEDANCE, IMPEDANCE MATCHING.

TERTIARY COIL

Some audio-frequency or radio-frequency (rf) transformers have a small coil coupled to the main windings that can be used for feedback purposes (see illustration). This coil is called a tertiary coil. In an rf transformer, a tertiary coil is called a tickler coil.

A tertiary winding can be used to provide either positive or negative feedback. The Armstrong oscillator uses a tertiary coil to obtain the positive feedback needed to maintain oscillation. Some regenerative detectors use tertiary windings to obtain positive feedback. A negative-feedback tertiary winding is sometimes used to keep audio-frequency or rf power amplifiers from oscillating. *See also* FEEDBACK.

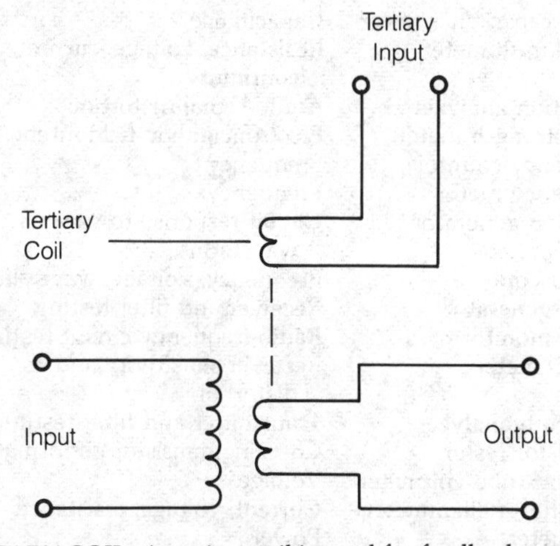

TERTIARY COIL: A tertiary coil is used for feedback purposes.

TESLA

The tesla is the standard international (SI) unit of magnetic flux, equal to 1 weber per square meter. Other units express magnetic flux density. The most common of these is the gauss. A flux density of 1 tesla is equal to 1×10^4 or 10,000 gauss. *See also* GAUSS, MAGNETIC FLUX, WEBER.

TESLA COIL

A Tesla coil is a special form of transformer and spark gap used to generate high voltages at high frequencies. A typical Tesla coil has an air-core transformer with a small primary winding and a large secondary winding. It operates at radio frequencies.

A Tesla coil can produce sparks several inches long. Although the voltage is high, the current flow is moderate to low.

TEST INSTRUMENT

In the design, alignment, and troubleshooting of electronic equipment, test instruments are used extensively. Test instruments vary in complexity from the simple ammeter or voltmeter to highly sophisticated instruments such as digital oscilloscopes and spectrum analyzers.

The most commonly used test instrument is a digital multimeter (DMM). This meter permits measurement of moderate to large voltages, currents, and resistances in typical electronic apparatus. Digital multimeters are relatively inexpensive, and are usually powered by batteries. They can be used in the field as well as in the laboratory.

Other common test instruments found in most electronic test laboratories are listed in the table.

TEST INSTRUMENT: Commonly used test instruments and functions or quantities measured.

Instrument	Function or Quantities Measured
Digital capacitance meter	Capacitance
Digital multimeter	Resistance, voltage, current, continuity
Distortion analyzer	Audio-signal distortion
Field-strength meter	Electromagnetic-field intensity
Frequency counter	Frequency
Frequency meter	Frequency
Function generator	Circuit response to various waveforms
Oscilloscope	Frequency, voltage, wave shape
Signal generator	Receiver and filter testing
Signal monitor	Radio-frequency circuit testing
SINAD meter	Receiver sensitivity and distortion
Spectrum analyzer	Transmitter and filter testing
Transistor tester	Checking transistor performance
Vacuum-tube voltmeter	Voltage
Volt-ohm-milliammeter	Current, voltage, resistance
Wattmeter	Power

TEST PATTERN

In television, special patterns are used to check the alignment of the transmitter and/or receiver. Various patterns facilitate the quantitative evaluation of such operating characteristics as resolution, horizontal and vertical linearity, brightness, contrast, and color reproduction. Test patterns are also used to help the receiving operator tune in the signal in a two-way television communications system.

Each television station has its own distinctive video test pattern. The audio channel might have a steady tone, music, or other test signals. The test pattern is used by the station technicians for regular checking of the transmitter alignment. Test patterns can be picked up by a television camera, or they may be electronically generated.

Test patterns are used by technicians who repair television receivers. The most common of these patterns is the bar pattern. *See also* BAR GENERATOR.

TEST PROBE

See PROBE.

TEST RANGE

A test range is a large area, usually located outdoors, in which transmitting and receiving antennas are tested for such properties as power gain and front-to-back ratio.

Testing of Transmitting Antennas. The illustration at A shows a typical test range for evaluating the performance of a transmitting antenna. A field-strength meter is placed at a distance of several wavelengths from the antenna under test. A reference antenna, such as a dipole or isotropic radiator, is placed near the antenna under test. The distance between the test antenna and the field-strength meter is identical to the distance between the reference antenna and the field-strength meter. The test and reference antennas are resonant at the same frequency.

For testing of antenna power gain, the reference antenna is supplied with a specified amount of radio-frequency power at its resonant frequency. The field-strength-meter sensitivity is adjusted until the scale reads 0 decibels (dB). Then the same amount of power, at the same frequency, is supplied to the test antenna. The reading on the scale of the field-strength meter, in decibels, indicates the power gain of the test antenna in dBd (if a dipole is used as the reference) or dBi (if an isotropic antenna is used as the reference). *See* dBd, dBi.

For the testing of front-to-back ratio, if applicable, the test antenna is turned so that it points away from the field-strength meter. The scale of the field-strength meter is set to read 0 decibels when a specified amount of power is applied to the test antenna. Then the test antenna is rotated so that it points directly at the field-strength meter. The reading on the meter scale, in decibels, indicates the front-to-back ratio. The antenna can be continuously turned, and meter readings taken at intervals of a few degrees, to get a complete directional pattern.

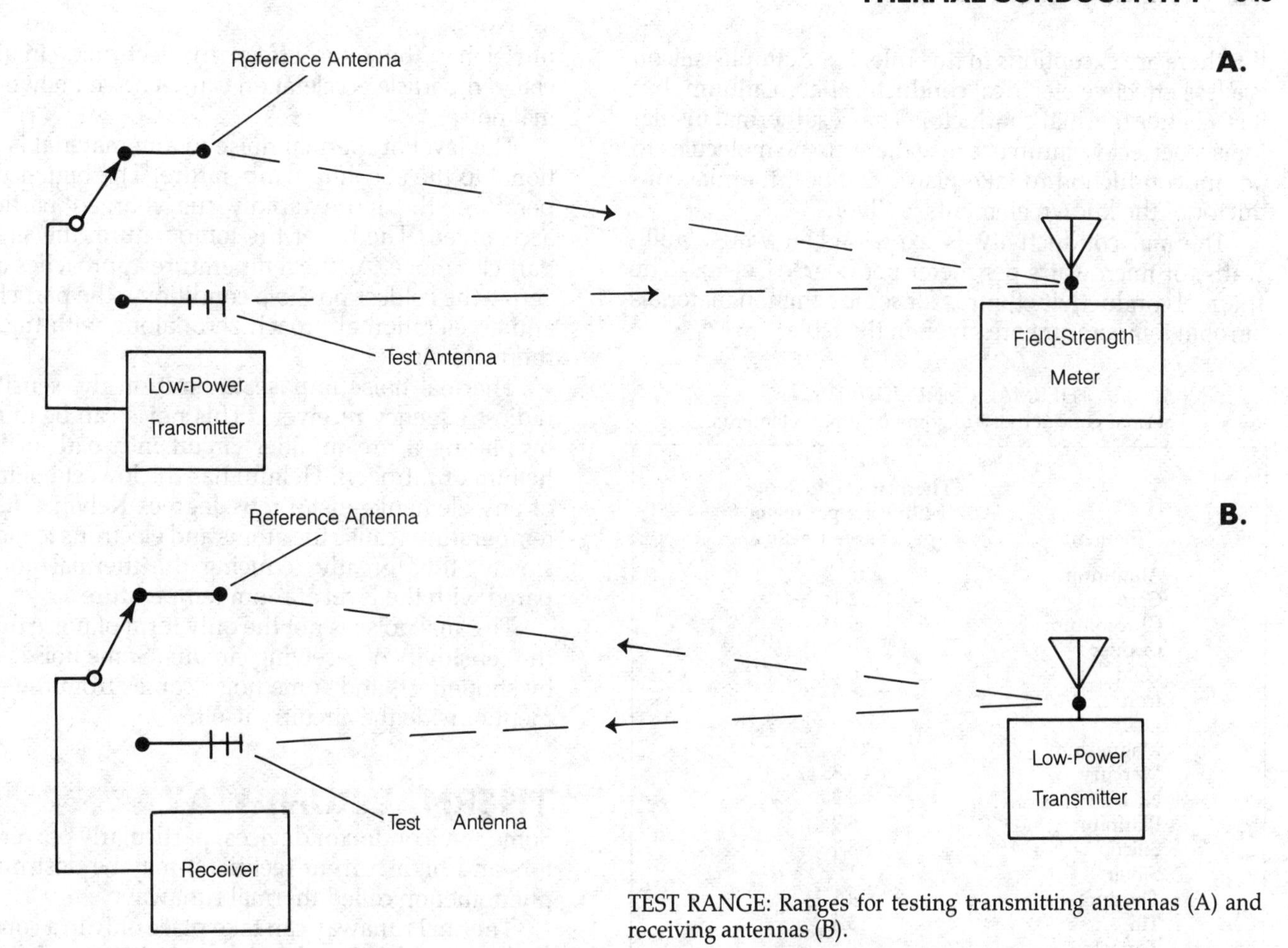

TEST RANGE: Ranges for testing transmitting antennas (A) and receiving antennas (B).

Testing of Receiving Antennas. The illustration at B shows a test-range layout for the evaluation of receiving-antenna performance. A low-power transmitter is located several wavelengths from the test and reference antennas. The test and reference antennas are both resonant at the same frequency, and are both located at the same distance from the transmitter.

For the testing of effective receiving gain, the test antenna is pointed directly at the transmitter, and the transmitter delivers a constant amount of power to its antenna. The receiver is switched between the reference antenna and the test antenna. The receiver has a calibrated S meter that allows precise determination of relative signal levels in decibels. The effective receiving gain of the test antenna is the increase, in decibels, that occurs in the signal when the receiver is switched from the reference antenna to the test antenna. (If the signal is weaker with the test antenna, the gain is negative.)

For measurement of front-to-back ratio, the receiver is connected to the test antenna, and signal levels are compared with the antenna pointed directly toward and directly away from the transmitter. A plot of the directional response can be obtained by checking the received signal level at many compass points.

THERMAL BREAKDOWN

In an electric conductor or semiconductor, the flow of current always generates some heat. Because no conductor is completely lossless, resistance causes dissipation of power in the form of heat. If the flow of current is excessive, the dissipated power may cause the temperature to rise so high that the material deteriorates. This is called thermal breakdown. It can occur in any device that carries current. *See also* THERMAL RUNAWAY.

An electric field in a dielectric material, or a magnetic field in a ferromagnetic material, causes a rise in the temperature of that substance because no dielectric or ferromagnetic material is completely without loss. If the field strength becomes great enough, the dielectric or ferromagnetic material may get so hot that it is physically damaged. This is another form of thermal breakdown. *See also* DIELECTRIC, DIELECTRIC LOSS, EDDY-CURRENT LOSS, FERROMAGNETIC MATERIAL, HYSTERESIS LOSS.

THERMAL COEFFICIENT

See TEMPERATURE COEFFICIENT.

THERMAL CONDUCTIVITY

All substances conduct some heat but some conduct heat much better than others. The extent to which a substance can transfer heat efficiently from one place to another can be expressed quantitatively and is known as the thermal conductivity of the material.

In general, materials are good thermal conductors if they are good electrical conductors, and vice versa. Electrical insulators tend to be poor thermal conductors.

But there are exceptions to this rule. For example, silicon is a less effective electrical conductor than platinum, but it is a better thermal conductor. The best thermal insulator is a perfect vacuum because there are no molecules to permit conduction to take place. The best thermal conductor of the known elements is silver.

Thermal conductivity is expressed in watts, milliwatts, or microwatts per meter per degree Celsius. The thermal conductivity figures for some common materials at room temperature are given in the table.

THERMAL CONDUCTIVITY:
Thermal conductivity of some common elements.

Element	Thermal conductivity, Milliwatts per meter per degree Celsius
Aluminum	22
Carbon	2.4
Chrominum	6.9
Copper	39
Gold	30
Iron	7.9
Lead	3.5
Magnesium	16
Mercury	0.85
Nickel	8.9
Platinum	6.9
Silicon	8.4
Silver	41
Thorium	4.1
Tin	6.4
Tungsten	20
Zinc	11

THERMAL ENERGY

Thermal energy, also called heat energy, is the kinetic energy of the moving atoms and molecules in matter. Thermal energy is proportional to the molecular temperature of a substance (*see* TEMPERATURE). The higher the temperature, the more rapidly the particles move, and the greater is the thermal energy. Thermal energy can be converted into other forms of energy by man-made devices. *See also* ENERGY, SOLAR ENERGY, THERMODYNAMICS.

THERMAL INSTABILITY

See THERMAL STABILITY.

THERMAL METER

See HOT-WIRE METER.

THERMAL NOISE

The electrons in all materials are in constant motion because they move in curved paths; they are always accelerating, even if they stay in the same orbital shell of a single atom (*see* ELECTRON, ELECTRON ORBIT). This acceleration of charged particles produces electromagnetic fields over a wide spectrum of wavelengths. To some extent, the random movement of the positively charged atomic nuclei has the same effect. In electronic circuits, this charged-particle acceleration causes noise known as thermal noise.

The level of thermal noise in any material is proportional to the absolute temperature. The higher the temperature, the more rapidly the charged particles are accelerated. The lower the temperature, the slower the particles move. As the temperature approaches absolute zero—the coldest possible condition—the particle speed and acceleration approach zero, along with the level of thermal noise.

Thermal noise imposes a limit on the sensitivity of radio-frequency receivers. This noise can be minimized by placing a preamplifier circuit in a bath of liquified helium or nitrogen. Helium has the lowest boiling point of any element—just a few degrees Kelvin. These cold temperatures cause the atoms and electrons to move very slowly, thus greatly reducing the thermal noise compared with the level at room temperature.

Thermal noise is not the only form of noise that limits the sensitivity of receiving circuits. Some noise is caused by shot effect, and some noise comes from the environment outside the circuitry itself.

THERMAL RUNAWAY

Some semiconductor devices, particularly power transistors and high-current rectifiers, may be destroyed by a phenomenon called thermal runaway.

Thermal runaway can take place only in a component that exhibits increased resistance with a rise in temperature. The problem begins as the current heats up the device, raising the resistance slightly. This increase in resistance may or may not produce a significant decrease in the flow of current, depending on the external circuitry. If the current does not decrease, the power dissipated in the component will increase in direct proportion to the resistance. This increased dissipation heats the device even faster. If allowed to go on unchecked, the end result will be thermal breakdown.

THERMAL STABILITY

Thermal stability is the capability of a circuit to function properly under conditions of variable temperature. The thermal stability of component values is expressed in terms of the temperature coefficient (*see* TEMPERATURE COEFFICIENT). The thermal stability of a circuit depends, to a large extent, on the temperature coefficients of the individual components.

Thermal stability can be quantitatively described in two ways. The simpler way is to state the temperature range in which the equipment will operate normally. The specifications for a small computer might state an ambient-temperature range of 10° to 50°C for example. A more comprehensive way of expressing thermal stability is to indicate how the various operating characteristics (such as frequency, power output, or sensitivity) change as a function of temperature.

Thermal stability is especially important in equipment that must be operated in diverse environments, where thermal shock is likely to be encountered. Thermal stability is not as important a specification for equipment that is always used in a controlled environment. *See also* TEMPERATURE COMPENSATION.

THERMAL TIME CONSTANT

When the ambient temperature around an object changes suddenly, the object gradually gets hotter or cooler until the object and the environment are at the same temperature. The rapidity with which this occurs depends on the material and mass of the object. The heating or cooling process is exponential, but the relative speed of change can be expressed in terms of the thermal time constant.

Suppose the ambient temperature around an object suddenly changes from T_1 to T_2 (in the Kelvin scale). The thermal time constant T_c is the time required for the object to reach a temperature of:

$$T_c = T_1 + 0.63\ (T_2 - T_1)$$

if $T_2 > T_1$, that is, heating; or:

$$T_c = T_1 - 0.63\ (T_1 - T_2)$$

if $T_2 < T_1$, that is, cooling.

The thermal time constant can be expressed for electronic components such as resistors and transistors. If the current suddenly changes, the temperature of the component will change exponentially, starting at T_1 and approaching a final value of T_2. The thermal time constant for an electronic component is determined with the formula.

THERMIONIC EMISSION

When an electrical conductor is heated to a high temperature in a vacuum, the electrons move so fast that they are easily stripped from the material. This is called thermionic emission. It is the principle of operation of a vacuum-tube cathode.

Thermionic emission from a surface is expressed in watts per square meter or watts per square centimeter. The thermionic emission from a cathode depends on the temperature, the voltage supplied to the cathode, and the type of material from which the cathode is made. Oxide-coated cathodes generally have the greatest emission at a given absolute temperature and voltage. Specially pressed and fired mixtures of nickel and tungsten also have high thermionic emissivity. *See* CATHODE.

THERMISTOR

A thermistor is a thermal resistor with definite thermal characteristic. Most thermistors have a negative temperature coefficient (NTC), but units with positive temperature coefficients (PTCs) are also available. They are used for the measurement and control of temperature, liquid level, and gas flow. Most thermistors are made from metal oxides including various mixtures of manganese, nickel, cobalt, copper, iron, and uranium. These are pressed into the desired shapes and fired at high temperature to form the thermistor. Electrical connections are made by including fine wires during the molding process, or by silvering the surfaces after firing.

Thermistors are made in the shape of beads, probes, disks, or washers. They can also be made by sintering metal-oxide coating on ceramic or foil substrates. Beads can be glass passivated or sealed in vacuum- or gas-filled capsules for protection against corrosion, as shown in the section view.

The typical thermistor resistance/temperature characteristic is a negative sloping curve when plotted on a resistance (y axis) versus temperature (x axis) graphic. Thermistor resistance decreases rapidly when it is heated. Current through a thermistor causes power dissipation which raises the device temperature. Thus, thermistor temperature is dependent upon ambient temperature and self heating. For a fixed ambient temperature, the thermistor resistance is dependent on its own power dissipation.

The static voltage-versus-current characteristic of the thermistor is generally in the shape of a bell. With voltage plotted on the y axis and current plotted on the x axis, the thermistor characteristic behaves initially like a constant resistance. However, when a peak voltage is reached, the heating effect of the current changes the thermistor resistance significantly and further increases in current cause a progressive reduction in resistance; this results in a voltage reduction across the device.

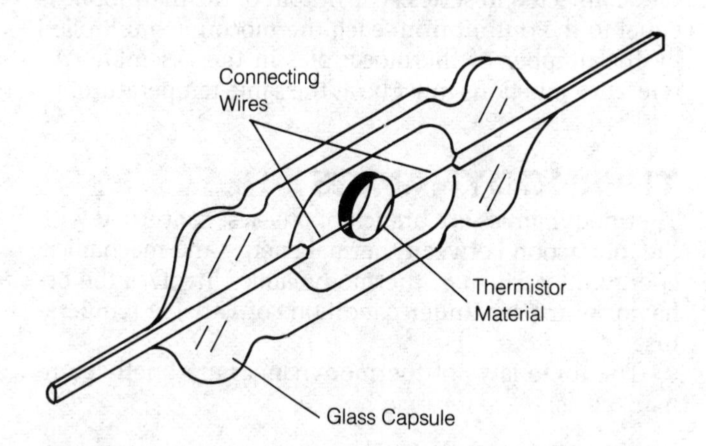

THERMISTOR: Sectional view of a simple bead-type thermistor capable of converting radio frequency to heat.

THERMOCOUPLE

A thermocouple is a temperature sensor made from a pair of wires of different metals joined together at one end to form a junction. The basic thermocouple circuit shown in the diagram includes a thermocouple whose wire ends are terminated by a reference junction maintained at a reference temperature. When there is a temperature difference between the sensing junction and the reference junction, an electromotive force is produced. Called the Seebeck effect, this electromotive force causes current to flow through the circuit. (*See* SEEBECK COEFFICIENT, SEEBECK EFFECT.)

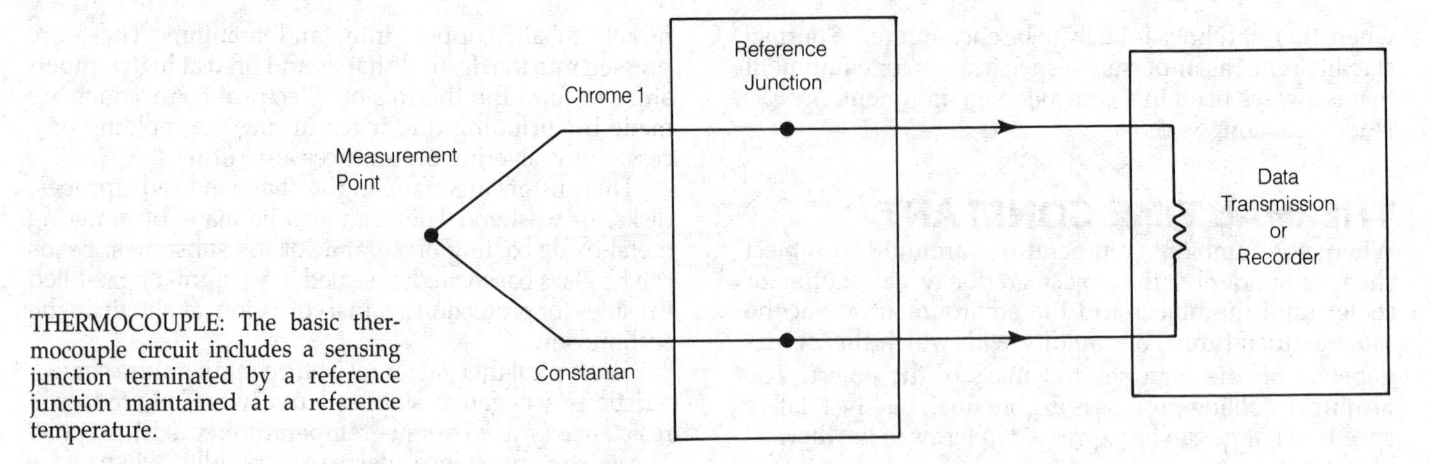

THERMOCOUPLE: The basic thermocouple circuit includes a sensing junction terminated by a reference junction maintained at a reference temperature.

The magnitude of the thermoelectric potential produced depends on the wire materials selected and the temperature difference between the two junctions. Some common examples of thermocouples include copper-Constantan, iron-Constantan, Chromel-Alumel, and Chromel-Constantan. These cover the temperature range of +371°C (700°F) to +1260°C (2300°F). A Chromel-Constantan junction, for example, will generate 70 millivolts at 1000°C.

Chromel is an alloy of 90 percent nickel, 10 percent chromium; Alumel is 95 percent nickel with traces of other metals, and Constantan is 55 percent copper and 45 percent nickel. Other thermocouples are selected for higher and lower temperatures.

A thermopile is a combination of several thermocouples connected in series. The output of the thermopile is equal to the output from each thermocouple multiplied by the number of thermocouples in the assembly. All reference junctions must be at the same temperature.

THERMODYNAMICS

Thermodynamics is a branch of physics, concerned with the interaction between thermal energy and mechanical energy. In particular, thermodynamics involves the behavior of matter under conditions of variable temperature.

The three laws of thermodynamics are briefly summarized as:

1. Thermal energy can be converted to mechanical energy, and vice versa.
2. In a reversible system, not all of the available thermal energy is converted to mechanical energy.
3. As the temperature of an isothermal process approaches absolute zero, the entropy approaches zero. At absolute zero, the entropy is zero.

Thermodynamics has application in electricity and electronics. Thermal energy is extensively used for generating electricity. Steam turbines convert the thermal energy into mechanical energy, and this energy is used to drive electric generators. Thermal energy may come from atomic reactors, geothermal springs or from the burning of fossil fuels. It may also be obtained from solar radiation. *See also* ENERGY, FIRST LAW OF THERMODYNAMICS, SECOND LAW OF THERMODYNAMICS, THERMOELECTRIC CONVERSION, THIRD LAW OF THERMODYNAMICS.

THERMOELECTRIC CONVERSION

Thermal energy can be converted into electricity in a process known as thermoelectric conversion.

Many power-generating plants use thermoelectric conversion to generate electricity. The heat from atomic reactions, the burning of fossil fuels gas, oil, or coal, the interior of the earth, or the radiation from the sun can be harnessed to drive a steam turbine. The turbine provides the mechanical energy to turn the electric generators. This is an example of indirect thermoelectric conversion because thermal energy is converted first to mechanical energy, then to electrical energy.

Direct thermoelectric conversion can be achieved with semiconductor materials. A junction between dissimilar metals acquires a potential difference at high temperatures. *See also* SEEBECK COEFFICIENT, SEEBECK EFFECT, THERMOCOUPLE.

THERMOELECTRIC EFFECT

See SEEBECK COEFFICIENT, SEEBECK EFFECT, THERMOCOUPLE.

THERMOSTAT

A thermostat is a temperature-sensor used to open or close electrical contacts at specified temperatures. The sensing element is a bimetal strip of dissimilar metals. A bimetal-strip thermostat opens and closes because of the curvature of the strip caused by different rates of expansion of the dissimilar metals. In a heating system, the thermostat switches the circuit on when the temperature falls below a certain point; in a cooling system, the system is switched on when the temperature rises above a certain point.

A typical bimetal-strip thermostat operates as shown in the illustration. There are two modes: heating and cooling. In the heating mode, the furnace is switched on when the temperature falls below a preset value. In the cooling mode, an air conditioner is switched on when the temperature rises above a preset value. Hysteresis, or sluggishness, in the thermostat prevents it from cycling on and off rapidly and continuously.

A thermostat can be used for thermal-protection purposes. If a device gets too hot, the thermostat actuates a switch that removes power. *See also* CIRCUIT PROTECTION.

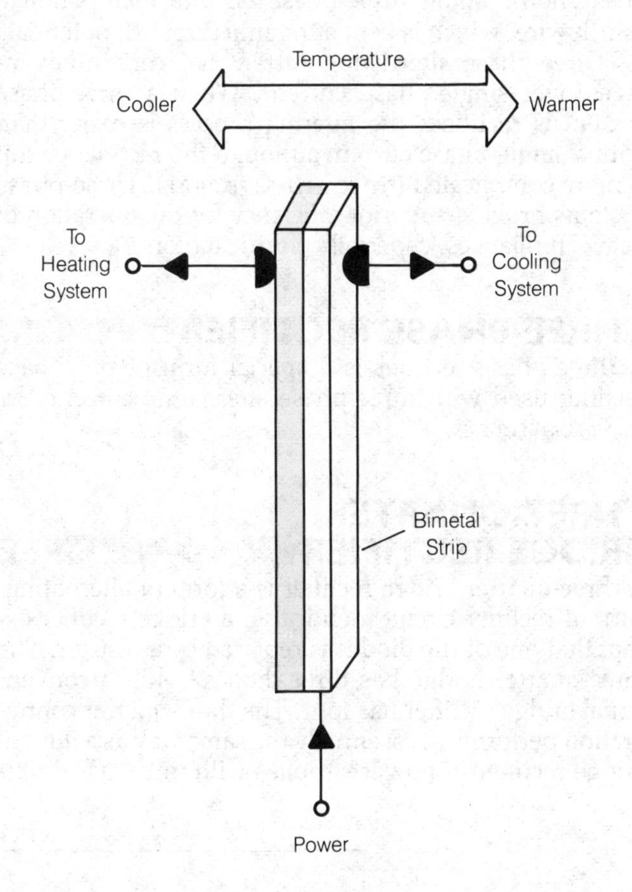

THERMOSTAT: A bimetal-strip thermostat.

THEVENIN'S THEOREM

In any linear circuit with more than one source or more than one load, simplification is possible with a rule known as Thevenin's theorem.

If there is a circuit, no matter how complex, consisting of power sources and linear loads, it will operate in the same manner as a hypothetical circuit consisting of one source and one load of the appropriate values. This hypothetical circuit is called the Thevenin equivalent circuit.

Thevenin's theorem can be stated in another way. If an impedance Z_x is connected into a linear system in which the open-circuit voltage is E and the existing impedance is Z, then the current I_x in the circuit will be:

$$I_X = \frac{E}{Z + Z_X}$$

Thevenin's theorem is used in network analysis for determination of current and voltage. *See also* KIRCHHOFF'S LAWS.

THICK-FILM AND THIN-FILM INTEGRATED CIRCUITS

See HYBRID CIRCUIT.

THIRD LAW OF THERMODYNAMICS

The Third Law of Thermodynamics is a theorem explaining the behavior of substances at low temperatures. The principle involves entropy, an expression of the distribution of thermal energy among substances or objects.

In any isothermal process involving a solid or liquid, the entropy approaches zero as the temperature approaches absolute zero. Also, the entropy is zero at absolute zero. This means simply that the temperature distribution becomes more uniform as the temperature gets lower.

In practice, absolute zero is never actually reached anywhere in the universe. If this temperature, about −459°F or −273°C were attained, all molecular and atomic motion would cease. *See also* FIRST LAW OF THERMODYNAMICS, SECOND LAW OF THERMODYNAMICS.

THOMSON BRIDGE

See KELVIN DOUBLE BRIDGE.

THOMSON EFFECT

When a current flows in a conductor with an uneven temperature distribution, heat is either produced or absorbed. This is called the Thomson effect. The extent to which heat is produced or absorbed depends on the material and also on the direction of current flow. If heat is liberated when current flows in a given direction, heat will be absorbed when the current flows in the opposite direction, and vice versa.

The tendency for the Thomson effect to occur is expressed as a quantity called the Thomson coefficient. Let E_1 be the voltage at some point P_1 along a metal conductor, and let E_2 be the voltage at some other point P_2. Let point T_1 be the absolute temperature (in degrees Kelvin) at point P_1, and let T_2 be the absolute temperature at point P_2. This is illustrated in the figure. The Thomson

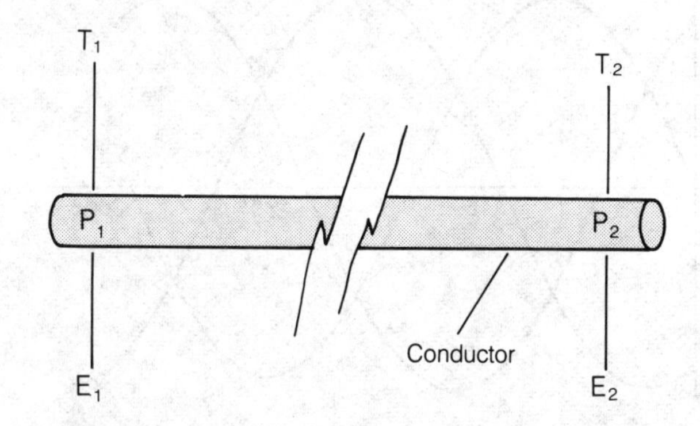

THOMSON EFFECT: A voltage $E_2 - E_1$ between two points P_1 and P_2, differing in temperature by $T_2 - T_1$, causes heating or cooling.

coefficient is given approximately by:

$$C = \frac{E_2 - E_1}{T_2 - T_1}$$

in volts per degree Kelvin.

THORIATED CATHODE/FILAMENT

See THORIUM.

THORIUM

Thorium is an element with an atomic number of 90. The most common isotope has an atomic weight of 232. Thorium is a heavy metallic substance in its pure form.

Thorium compounds, especially thorium oxides, are added to the cathodes of some vacuum tubes to improve the thermionic emission. Under conditions of high temperature, thorium and its compounds are excellent emitters of electrons. A filament or cathode is called thoriated when a small amount of this element is present. *See also* THERMIONIC EMISSION.

THREE-PHASE ALTERNATING CURRENT

Three-phase alternating current is current delivered through three wires. Each wire serves as a return for the other two and the three currents differ in phase. The 230-V utility power lines generally carry three waves simultaneously. The waves are identical except that they are separated by 120 degrees of phase (⅓ cycle). The current is supplied through three wires, with each wire carrying a single sine wave differing in phase by ⅓ cycle with respect to the currents in the other two wires (see illustration). Some three-phase circuits incorporate a fourth wire, which is kept at neutral (ground) potential.

Three-phase alternating current has certain advantages over single-phase current. When a three-phase current is rectified, the filtering process is easier than with a single-phase current although the rectifier circuit is more complicated (*see* POLYPHASE RECTIFIER). Three-phase systems provide superior efficiency for the operation of heavy appliances, especially electric motors.

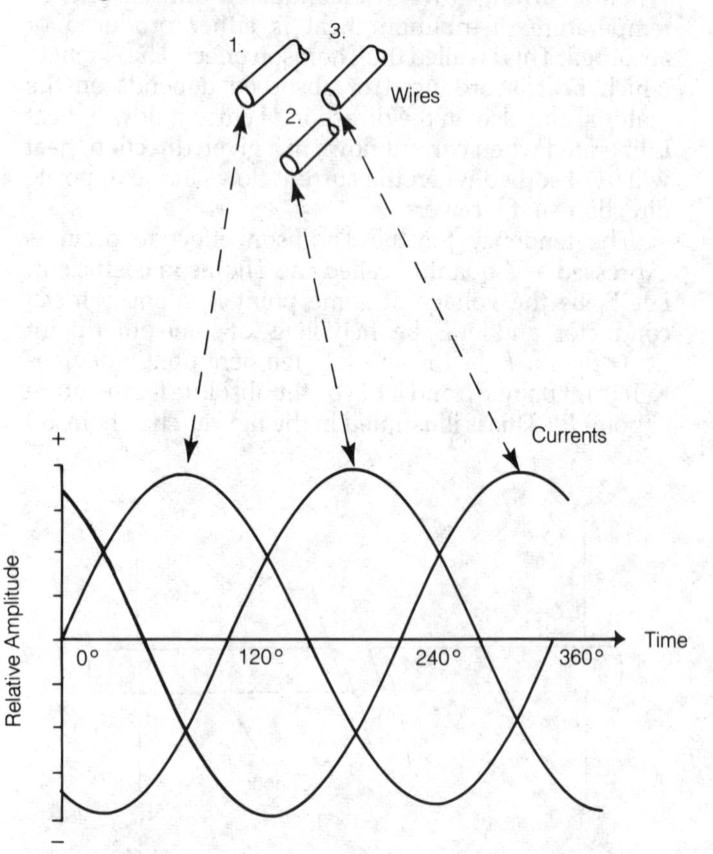

THREE-PHASE ALTERNATING CURRENT: The waves differ by 120 degrees of phase.

THREE-PHASE RECTIFIER

A three-phase rectifier is a special form of polyphase rectifier used with three-phase alternating current. *See* POLYPHASE RECTIFIER.

THREE-QUARTER BRIDGE RECTIFIER

A three-quarter bridge rectifier is a form of alternating-current rectifier circuit identical to a bridge rectifier except that one of the diodes is replaced by a resistor. The three-quarter bridge has three diodes, while a conventional bridge rectifier has four. The three-quarter configuration performs in essentially the same way as a normal bridge rectifier in power supplies with unbalanced out-

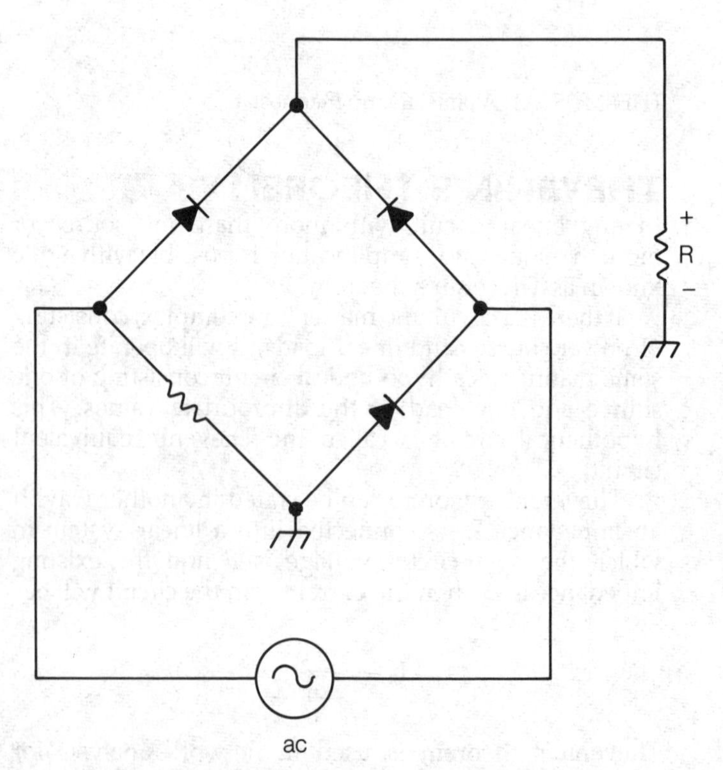

THREE-QUARTER BRIDGE RECTIFIER: This bridge can be used in unbalanced power supplies.

puts. The illustration is a schematic diagram of a three-quarter bridge. *See also* BRIDGE RECTIFIER.

THREE-WIRE SYSTEM

A three-wire system is a scheme for transmitting electric power, especially alternating current (ac). Three-wire systems are used in utility wiring. In recent years, the three-wire system has become increasingly popular because it affords better shock-hazard protection than the older two-wire system.

A three-wire system consists of one grounded (neutral) wire and two live wires. The live wires carry equal and opposite currents; in most three-wire systems, one of the live wires is kept at neutral potential, the same as the grounded wire, while the other carries 120 V of ac potential.

Three-wire electrical systems facilitate grounding of the chassis of radio equipment and appliances. This keeps lethal voltages from appearing at places where there might be a shock hazard.

THRESHOLD DETECTOR

A threshold detector is a circuit that allows strong signals to pass, but blocks weaker signals. The transition, or threshold, is sharply defined. A squelch circuit is a form of threshold detector (*see* SQUELCH).

The simplest form of threshold detector consists of two diodes, connected in reverse parallel with respect to each other and in series with the signal path. Signals will not be passed until the peak amplitude exceeds the forward breakover voltage of either diode. If silicon diodes are used, the peak voltage must be 0.6 V or greater.

A threshold detector can be used to obtain enhanced selectivity in an audio filter, as shown in the illustration. The first tuned circuit provides a sharp resonant audio response. The threshold detector cuts off the skirts at levels below 0.6 V if silicon diodes are used. The second tuned circuit, with values *L* and *C* identical to those of the first, reduces the waveform distortion caused by the threshold detector. The transistor circuit provides amplification if needed.

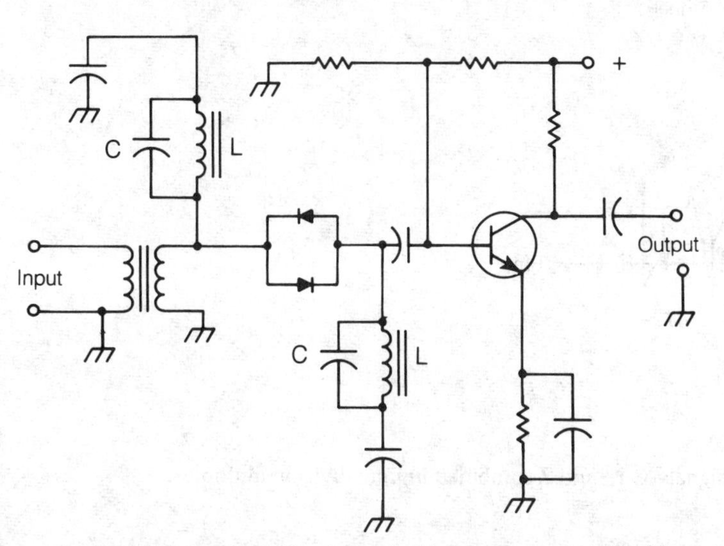

THRESHOLD DETECTOR: A threshold detector in an audio filter.

THROUGHPUT

In data transmission and computers, throughput is the total capability of equipment to process or transmit data during a specified period of time.

THYRATRON

A thyratron is an electron tube for power switching. It operates in a manner similar to the silicon-controlled rectifier. The envelope of the thyratron is filled with a gas such as argon, neon, or xenon for low-voltage applications. Mercury vapor is generally used for high-voltage applications.

Initially, the thyratron does not conduct. When a positive pulse is applied to the grid, the tube conducts because the gas becomes ionized. The tube continues to conduct until the plate voltage is removed or made negative. The voltage required to trigger the thyratron depends on the plate voltage.

The main advantage of the thyratron over semiconductor devices is its higher tolerance for momentary excessive current or voltage. *See also* SILICON-CONTROLLED RECTIFIER, THYRISTOR.

THYRISTOR

Thyristors are a class of gate-controlled semiconductor power rectifiers formed from four or more alternate layers of P- and N-doped silicon. They are three-terminal devices with one terminal that acts to trigger conduction. The most familiar thyristors are the silicon-controlled rectifier (SCR), and the bidirectional triode thyristor or triac. These devices behave like conventional P-N rectifiers in the presence of current passing in the reverse direction and as combination switch/rectifiers in the presence of current in the forward direction. *See* SILICON-CONTROLLED RECTIFIER, TRIAC.

TIME CONSTANT

Certain circuits change state exponentially as a function of time. Combinations of inductances, capacitances, and/or resistance exhibit this property. Materials heat up and cool down exponentially. The rate at which a change of state occurs, in a given situation, is known as the time constant.

The time constant is generally defined as the time, in seconds, required for a change of state to reach 63 percent of completion. This is illustrated graphically in the drawing on p. 852. *See also* RESISTANCE-CAPACITANCE TIME CONSTANT, RESISTANCE-INDUCTANCE TIME CONSTANT, THERMAL TIME CONSTANT.

TIME-DELAY RELAY

See RELAY.

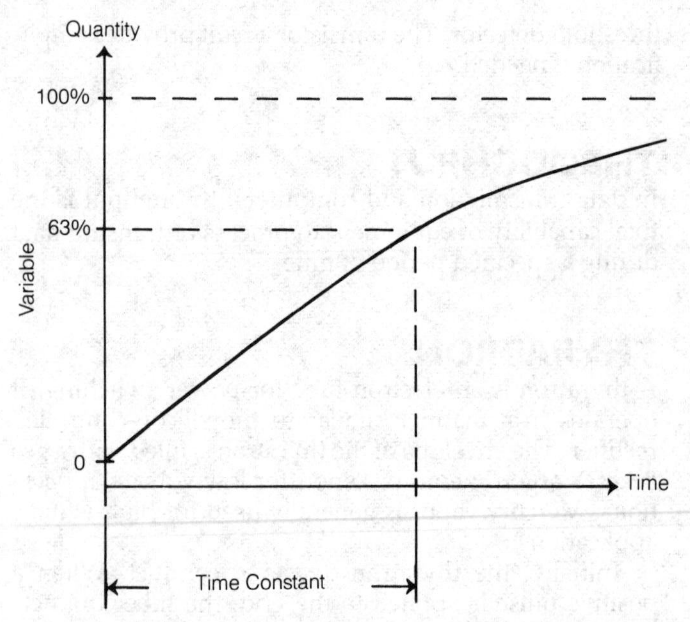

TIME CONSTANT: An illustration of a time constant for a variable quantity.

TIME-DIVISION MULTIPLEX

Time-division multiplex is a method of combining analog signals for serial transfer (*see* SERIAL DATA TRANSFER). The signals are sampled at intervals, and interwoven for transmission. At the receiving end the process is reversed; the signals are separated again (see illustration).

Time-division multiplex is used for transmitting two or more channels of information over a single carrier. The speed of the multiplexed signal is increased with respect to the individual channel speed by a factor equal to the number of signals combined. For example, if 10 signals are to be multiplexed by time-division process, the data speed of each signal must be multiplied by 10 to maintain synchronization.

Time-division multiplexing results in an increase in the bandwidth of a signal. The increase takes place by a factor equal to the number of signals combined. This is because of the increased data speed.

Time-division multiplex can be used with digital signals as well as analog signals. Digital time-division multiplex is usually called synchronous multiplex, since the

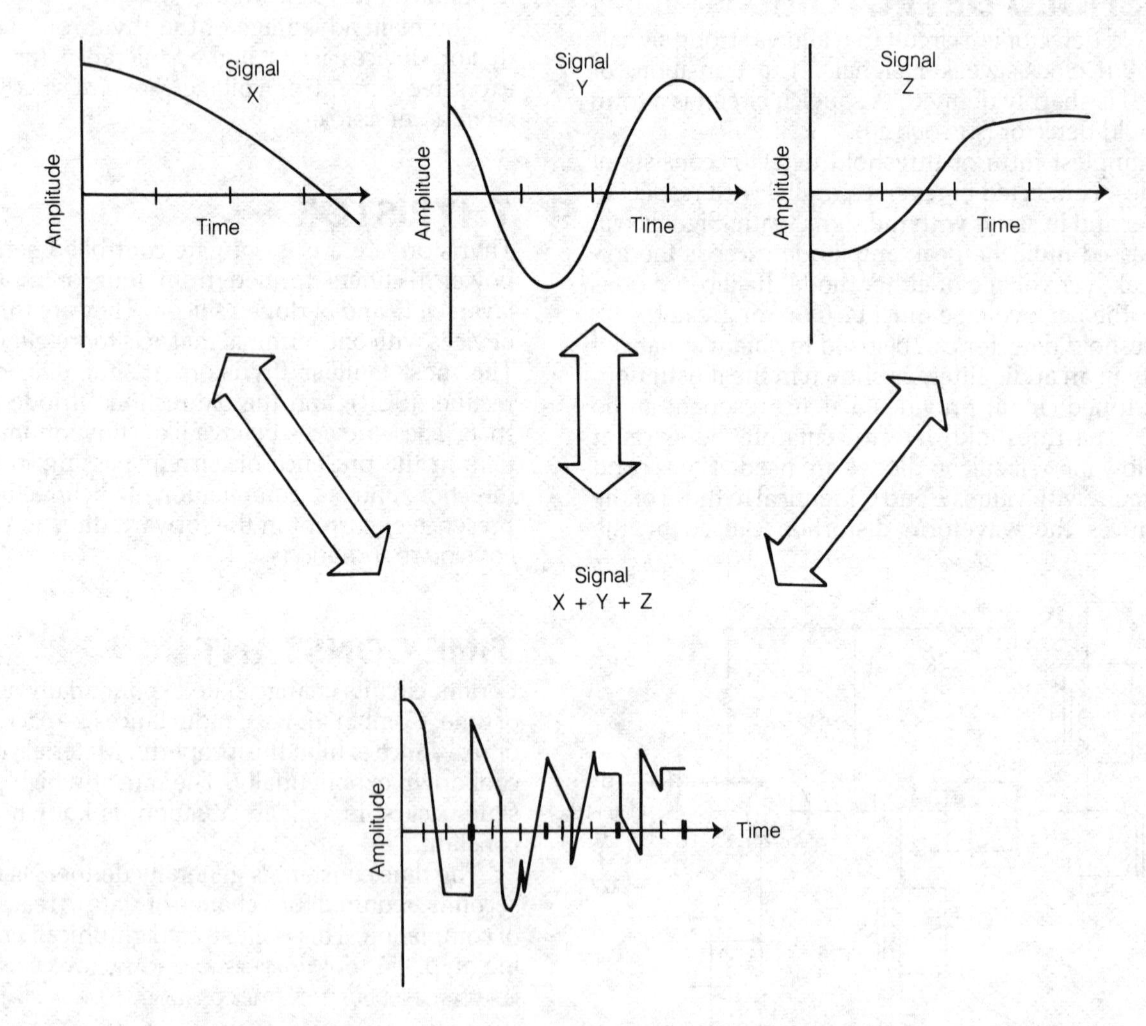

TIME-DIVISION MULTIPLEX: An illustration of three signals X, Y, and Z combined in time-division multiplex.

data from each channel is split bit by bit, in accordance with the speed of transmission. *See also* MULTIPLEX, SYNCHRONOUS MULTIPLEX.

TIME SHARING

Large computers can be used by many different operators simultaneously at independent terminals with a technique called time sharing. The principle of time sharing is similar to that of time-division multiplex (*see* TIME-DIVISION MULTIPLEX).

In time sharing, the computer connects each terminal for a short period, then moves to the next terminal, and so on in a repeating sequence. If there is just one terminal, the computer spends all of its time communicating with that terminal; if there are n terminals, the computer works with each terminal for 1/*n*th of the time.

The sequencing process is very rapid, so it seems to each operator that the computer is communicating only with his or her terminal. The computer appears to operate more slowly, however, when many terminals are connected, as compared to when only a few terminals are connected.

Time-sharing systems are operated over telephone lines. A subscriber can work with a computer hundreds of miles away by calling a number and using a telephone modem in conjunction with a terminal. *See also* MODEM, TERMINAL.

TIME SIGNALS

See WWV/WWVH.

TIME ZONE

The earth rotates at an angular rate of 15 degrees per hour. To match this rate, the time zones of the world are about 15 degrees wide so there is one zone for each hour of the day. Time zoning is complicated by the preferences of individual states and countries that elect to set their clocks an hour earlier or later than the natural time zone.

In the continental United States, there are four time zones: the Eastern, Central, Mountain, and Pacific. Eastern Standard Time (EST) is 5 hours behind Coordinated Universal Time (UTC). Eastern Daylight Time (EDT) is 4 hours behind UTC. Central, Mountain, and Pacific time are 1, 2, and 3 hours behind Eastern time, respectively. *See also* COORDINATED UNIVERSAL TIME.

TIN

Tin is an element with an atomic number of 50 and an atomic weight of 119. In its pure form, tin is a soft, malleable, silver-colored metal.

Tin is in the soft solder widely used in the electronic industry. The content of tin in lead-tin solder varies from 63 percent. *See also* SOLDER.

TIN-LEAD SOLDER

See SOLDER.

T MATCH

A T match is a method for coupling a balanced transmission line to a balanced antenna element such as a half-wave, center-fed radiator. The radiating element consists of a single conductor. A rod or wire is run parallel to the radiator, and the feed line is connected to the center of the rod or wire. The ends of the rod or wire are connected to the radiator at equal distances from the ends of the radiator. The figure illustrates a T match.

The impedance matching ratio of the T match depends on physical dimensions such as the spacing between the radiator and the rod or wire, the length of the matching element, and the relative diameters of the radiator and the matching element. The T match allows the low feed-point impedance of a Yagi antenna driven element to be matched to a 75-ohm balanced line. If a balun is used at the feed point, the antenna can be fed with coaxial line. Sometimes a half-T, or gamma match, is used with a coaxial line (*see* GAMMA MATCH).

There are several variations of the T match that can be used when large impedance-step-up-ratios are needed. If the rod or wire of a T match has the same length and diameter as the radiating element, the result is a folded dipole antenna. This configuration produces a step-up impedance ratio of 4:1, relative to a single element alone. Three conductors can be connected in parallel to obtain a step-up ratio of 9:1, or four conductors can be used to obtain a factor of 16:1. *See also* IMPEDANCE MATCHING.

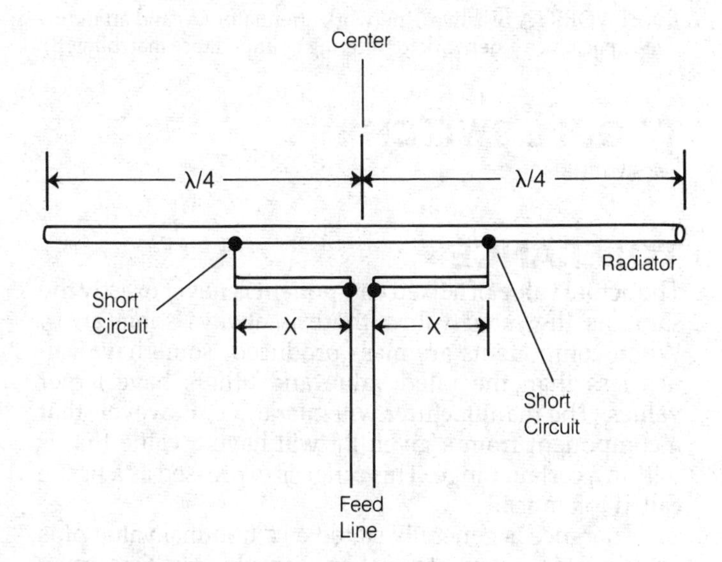

T MATCH: A T match is used between a radiator and a balanced transmission line for impedance matching.

T NETWORK

A T network is an unbalanced circuit used in the construction of filters, attenuators, and impedance-matching devices. Two series components are connected on either side of a single parallel component.

The diagram at A illustrates a T-network circuit consisting of noninductive resistors. This circuit can be used as an attenuator. B shows a T-network circuit that can be used for impedance matching. The T network is named from its shape in the schematic diagrams.

In a T-network impedance-matching circuit, the capacitor and/or inductors may be adjustable to vary the impedance-transfer ratio. A fairly wide range of load impedances can be matched to a fixed source impedance. The T-network circuit at B might also be used as a lowpass filter. *See also* IMPEDANCE MATCHING.

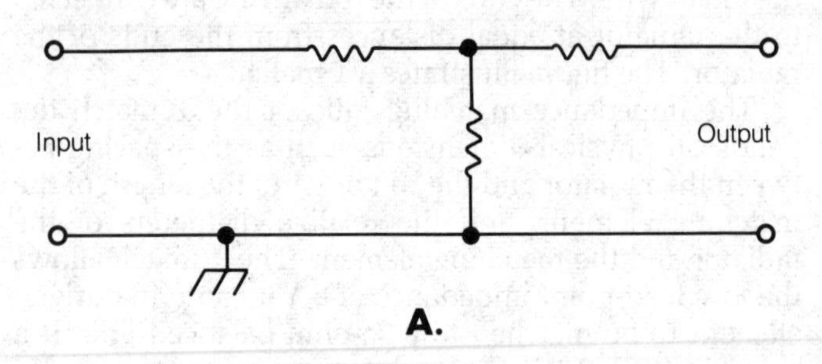

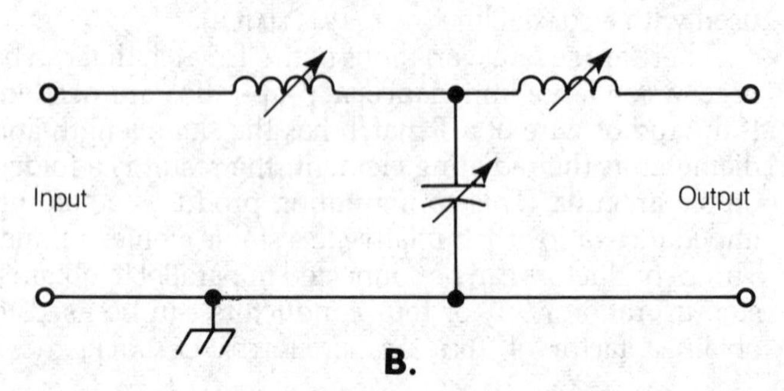

T NETWORK: A resistive T network attenuator (A) and an inductive-capacitance T network for filtering or impedance matching (B).

TOGGLE SWITCH

See SWITCH.

TOLERANCE

The actual value of a fixed component is never exactly the same as the rated value; there is always some error. When components are mass produced, some have values less than the rated value and others have larger values. The manufacturer will guarantee, however, that a component from a given lot will have a value that is within a certain range. This range is expressed as a figure called tolerance.

Tolerance is generally stated as a nominal value plus or minus (±) percentage. For example, a resistor may have a value of 1000 ohms ± 5 pecent. However, the nominal value may be a limit so that the tolerance is the nominal value plus a percentage or minus a percentage.

TONE CONTROL

The amplitude-versus-frequency characteristics of a high-fidelity sound system are adjusted by means of a tone control. Tone controls vary greatly in complexity.

In its simplest form, a tone control consists of a single rotary or slide potentiometer in series with a capacitor. The counterclockwise or lower position boosts the bass and attenuates the treble; the clockwise or upper position boosts the treble and attenuates the bass (*see* BASS, TREBLE). This is shown in the schematic diagram at A.

A more sophisticated tone control has two potentiometers and two capacitors, one combination in series and the other in parallel. One potentiometer controls the treble and the other controls the bass, as at B.

The most sophisticated tone controls have several filters that allow the listener to adjust the circuit response precisely. This system is called an equalizer or graphic equalizer (*see* EQUALIZER).

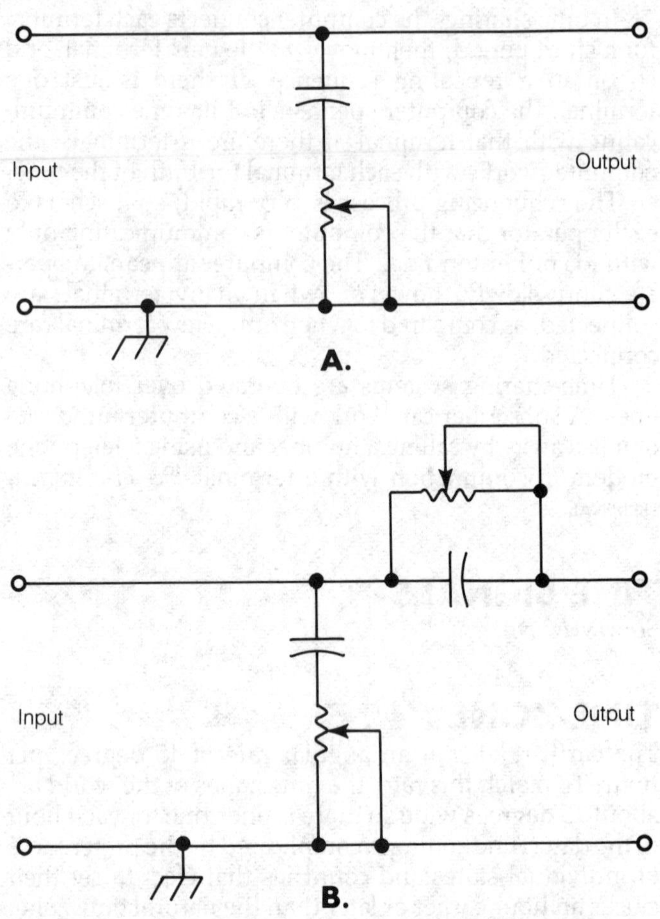

TONE CONTROL: Single tone control (A) and dual tone control (B).

TOROID

A toroid is a coil wound on a doughnut-shaped ferromagnetic core (see illustration). An advantage of the toroidal coil is that all of the magnetic flux is contained within the core material. Therefore, a toroid is not affected by surrounding components or objects. Also, nearby components are unaffected by the coil field. Another advantage of the toroid is its smaller physical size, for a given amount of inductance, than a solenoidal coil. Fewer coil turns are required to obtain a given inductance, compared with a solenoidal coil. This enhances the Q factor.

Toroidal cores can be used for winding simple inductors, chokes, and transformers. They are used at audio frequencies, and at radio frequencies up to several hundred megahertz. Ferromagnetic toroidal cores are available in a variety of sizes and permeability ratings for

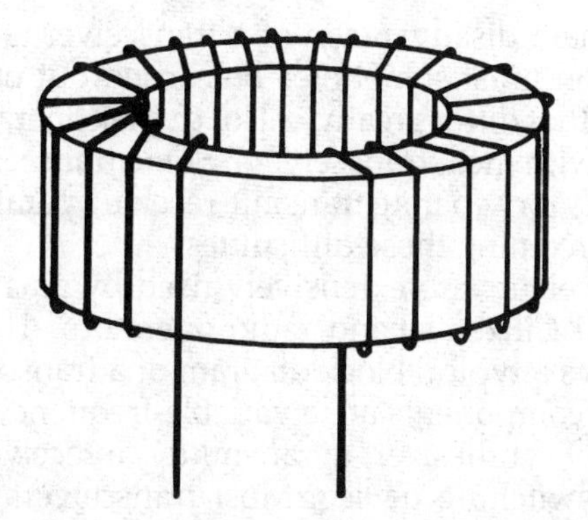

TOROID: A toroidal coil is wound on a ring-shaped form.

various frequencies and power levels. *See also* INDUCTOR, TRANSFORMER.

TORQUE

Torque is a measure of twisting or angular force. The turning force produced by a motor is a form of torque. Torque occurs around a defined axis, in either a clockwise or a counterclockwise direction, depending on the point of view.

TOTAL HARMONIC DISTORTION

The total root-mean-square (rms) harmonic voltage in a signal, as a percentage of the voltage at the fundamental frequency, is called total harmonic distortion (THD). Total harmonic distortion may vary from zero percent (no harmonic energy) to theoretically infinite (no fundamental-frequency energy).

The THD figure is used as an expression of the performance of an audio-frequency amplifier. The amplifier is provided with a pure sine-wave input, and the harmonic content is measured using a distortion analyzer or SINAD voltmeter (*see* DISTORTION ANALYZER).

Ideally, an audio-amplifier circuit should have as little THD as possible at the rated audio-output level. Most amplifiers are rated according to the amount of root-mean-square power they can deliver to a speaker of a certain impedance (usually 8 Ω) with less than a specified THD (usually 10 percent).

TOTAL INTERNAL REFLECTION

When a ray of light passes from one medium to another, and the index of refraction of the first substance is different from that of the second, refraction occurs (*see* INDEX OF REFRACTION, REFRACTION). Under certain circumstances, however, reflection takes place instead. This phenomenon, known as total internal reflection, occurs only when the following two conditions exist:

- The first medium must have a higher index of refraction than the second
- The angle of incidence must measure more than a certain limiting value

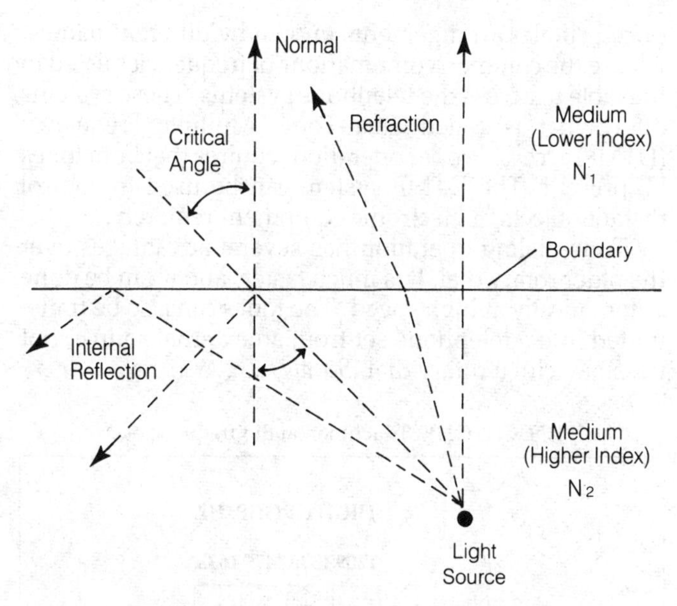

TOTAL INTERNAL REFLECTION: Total internal reflection occurs at the boundary between media with different indices of refraction if the angle of incidence exceeds the critical angle.

This principle is illustrated in the accompanying illustration. (*See* SNELL'S LAW.)

As the incidence angle increases, the angle at which total internal reflection begins depends on the ratio of the indices of refraction of the two substances. If the indices are nearly the same, reflection will occur only if the incidence angle is very large. If the first medium has a much greater index of refraction than the second, total internal reflection will be observed for fairly small incidence angles.

Optical fibers transmit light because of total internal reflection. Glass or plastic has a higher index of refraction than air, so a light beam will stay inside the fiber because the light will be reflected from the inside surfaces (*see* FIBEROPTIC COMMUNICATION).

Under some circumstances, total internal reflection affects radio signals. The boundary between two air masses, one much denser than the other, can produce reflection of electromagnetic fields. This is called the duct effect (*see* DUCT EFFECT). It is observed primarily at very high frequencies. Total internal reflection can also affect sound waves in a similar way.

TOUCHTONE®

Most telephones use pushbutton dialing instead of the rotary dialing. The buttons actuate tone pairs that cause automatic dialing of the numbers. This is called Touchtone® dialing (the term is a trademark of American Telephone and Telegraph Company).

The original Touchtone dial, still found on most telephone sets, has 12 buttons corresponding to digits 0 through 9, the star symbol (*) and the pound symbol (#). Some dialers have four additional keys, designated A, B, C, and D. The tone-pair frequencies for the 16 designators are listed in the table.

Although the original push-button keypads were called Touchtone pads, other manufacturers have produced similar arrangements known by different names. All use the common combinations of frequencies listed in the table to access the telephone systems. These systems offer what is called Dual-Tone, Multiple Frequency (DTMF) access. Proper operation requires that both tones be present. The DTMF system can be used to control distant objects or electronic equipment remotely.

Tone-dialing operation has several advantages over the older rotary dial. It is much faster, and it can be done automatically at high speed. The tones can also be transmitted into a telephone set from an external source, not possible with a rotary dial. *See also* ROTARY DIALER.

TOUCHTONE: Touchtone audio frequencies.

LOW GROUP, HZ	HIGH GROUP, HZ 1209	1336	1477	1633
697	1	2	3	A
770	4	5	6	B
852	7	8	9	C
941	*	0	#	D

TRANSCEIVER

A transceiver is a combination transmitter and receiver with a common frequency control and usually enclosed in a single case. Transceivers are extensively used in two-way radio communication at all frequencies, and in all modes.

The main advantage of a transceiver over a separate transmitter and receiver is economic. Many of the components can be used in both the transmit and receive modes. Another advantage is that most two-way operation is carried out at a single frequency, and transceivers are more easily tuned than separate units.

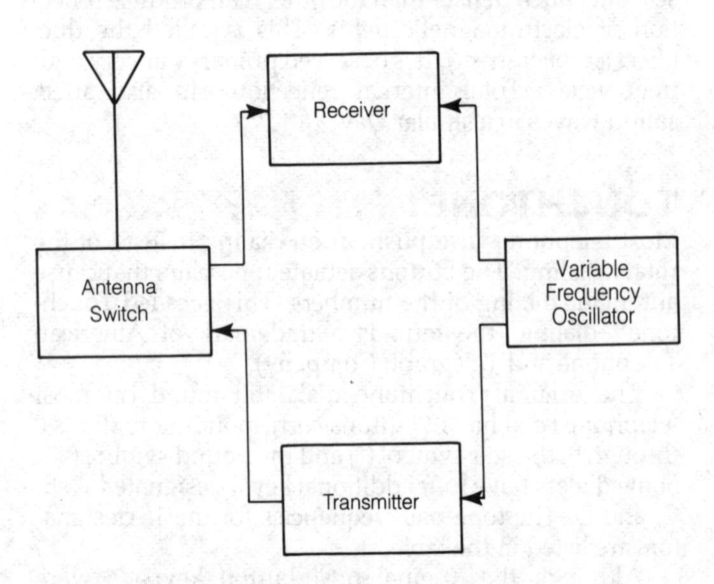

TRANSCEIVER: A block diagram of a basic transceiver.

The main disadvantage of a transceiver is that communication must sometimes be carried out on two frequencies that differ greatly. Also, duplex operation is not possible with most transceivers. Some transceivers have provisions for separate transmit-receive operation, however, overcoming these difficulties.

Transceivers are extensively used by amateur-, marine-, and Citizen's-band radio operators. The illustration shows a typical block diagram of a transceiver. The principal components are a variable-frequency oscillator or channel synthesizer, a transmitter, a receiver, and an antenna-switching device. Most transceivers use a superheterodyne design. *See also* RECEIVER, TRANSMITTER.

TRANSCONDUCTANCE

Transconductance is a measure of the performance of a vacuum tube or transistor. The symbol for transconductance is g_m, and the unit is the mho or siemens (*see* CONDUCTANCE, MHO, SIEMENS).

Transconductancce is defined as the ratio of the change in plate, collector or drain current to the change in grid, base or gate voltage over a defined, arbitrarily small interval on the plate-current-versus-grid-voltage, collector-current-versus-base-voltage, or drain-current-versus-gate-voltage characteristic curve. If ΔI represents a change in plate, collector or drain current caused by a change in the grid, base or gate voltage ΔE, then the transconductance is approximately:

$$g_m = \Delta I / \Delta E$$

and corresponds to the slope of the tangent line to the curve at that point (see illustration). *See also* CHARACTERISTIC CURVE.

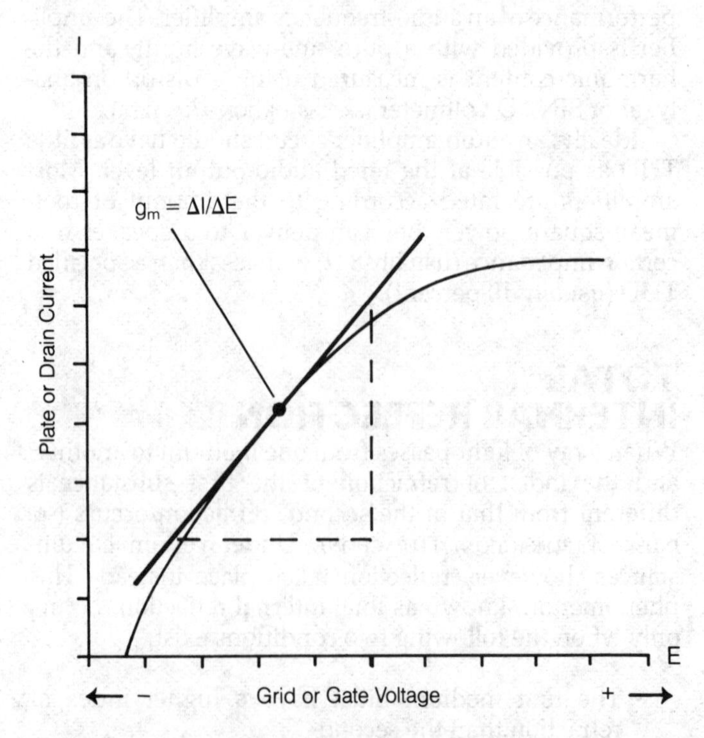

TRANSCONDUCTANCE: A plot of transconductance at a point on the characteristic curve of an amplifier.

TRANSDUCER

A transducer is a device that converts one form of energy into another. In electronics, transducers convert alternating or direct electric current into sound, light, heat, radio waves, or other forms. Transducers also convert sound, light, heat, radio waves, or other energy forms into alternating or direct electric current.

Common examples of electrical and electronic transducers include buzzers, speakers, microphones, piezoelectric crystals, light-emitting and infrared-emitting diodes, photocells, and radio antennas. *See* MICROPHONE, SPEAKER, SONAR.

TRANSFER FUNCTION

See FREQUENCY RESPONSE.

TRANSFER IMPEDANCE

In a direct-current or alternating-current circuit, a voltage applied at one point may result in a current at another point. The ratio E/I of the applied voltage to the resultant current is known as the transfer impedance.

Consider the example of a transformer with a step-down ratio of 5:1, an input voltage E of 120 V, and a load impedance Z of 12 Ω, as shown in the illustration. The output voltage, E_o, is:

$$E_o = 120/5 = 24 \text{ V}$$

and the load current I is therefore:

$$I = E_o/Z = 24/12 = 2 \text{ A}$$

therefore, the transfer impedance, Z_t, is given by:

$$Z_t = E/I = 120/2 = 60\Omega$$

This example assumes that the efficiency of the transformer is 100 percent. *See also* IMPEDANCE.

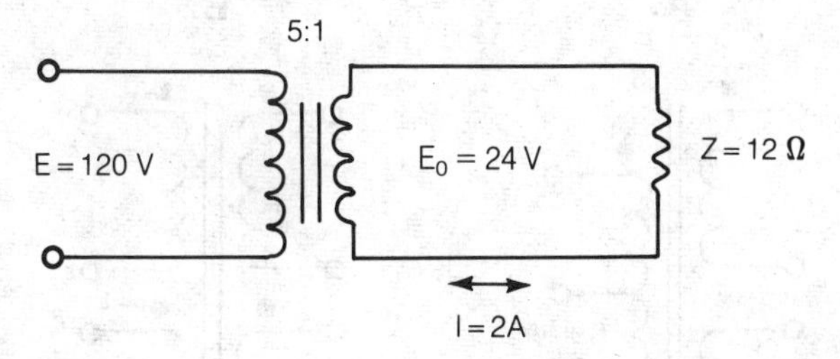

TRANSFER IMPEDANCE: A schematic to explain transfer impedance.

TRANSFORMER

A transformer is an electrical component that transfers electrical energy from one or more primary circuits to one or more secondary circuits by electromagnetic induction. It consists of at least one primary winding and one secondary winding of insulated wire on a common core. No electrical connection exists between any primary or input circuit and any secondary or output circuit, and no change in frequency occurs between the two circuits. *See* ELECTROMAGNETIC INDUCTION.

Most transformers step up (increase) or step down (decrease) voltage or current. The number of turns on the primary winding usually differs from the number of turns on the secondary winding. However, an isolation transformer provides secondary voltage and current that is essentially the same as the primary voltage and current (except for resistive losses) because both windings have the same number of turns. These transformers prevent the transfer of unwanted electrical noise from the primary to the secondary winding.

The primary and secondary windings of conventional transformers for electronic applications are wound on tubular bobbins made of plastic or other insulating materials. The wound bobbins are then enclosed by iron or steel cores formed in the shape of a figure 8. The cores for low-frequency (50 to 400 Hz) transformers are made as stacks of E- and I-shaped laminations assembled through and around the wound bobbins. The laminations are then clamped to form a rigid assembly as shown in Fig 1. Some transformers have plastic shrouds to insulate the windings from the core. Transformers may be placed in a U-shaped metal channels with padeyes at the ends to permit them to be fastened to a metal panel or chassis.

Both primary and secondary windings can be wound on the same bobbin, but it is now common practice to wind the primary and secondary separately on a split bobbin for improved electrical isolation. Primary and secondary terminals may be connected to rigid pins on the bobbin that also function as printed circuit board mounting pins.

If an alternating voltage is applied to the primary winding of a transformer, an electromagnetic field forms around the core and expands and contracts at the input frequency. This changing field cuts the wires in the secondary winding and induces a voltage within it. The voltage at the secondary winding depends on the voltage at the primary winding and the ratio of turns in the primary and secondary windings.

If the secondary winding has twice as many turns as the primary winding, the voltage at the secondary will be twice that of the primary; similarly, if there are half as many turns in the secondary as in the primary, the secondary voltage will be half that of the primary voltage. In accordance with the law of conservation of energy, the product of voltage and current remains the same in both primary and secondary windings (except for losses). These topics are discussed later.

In theory, if the voltage at the secondary is double the voltage at the primary, the current at the secondary side will be half of that at the primary to keep the volt-ampere product constant. Because the product of voltage and current is power, the power output would equal the power input. However, this occurs only if the transformer is 100 percent efficient. (Most transformers are only about 90 percent efficient). Losses are due to ohmic resistance (copper losses), eddy current induction, and hysteresis (molecular friction), caused by the changing

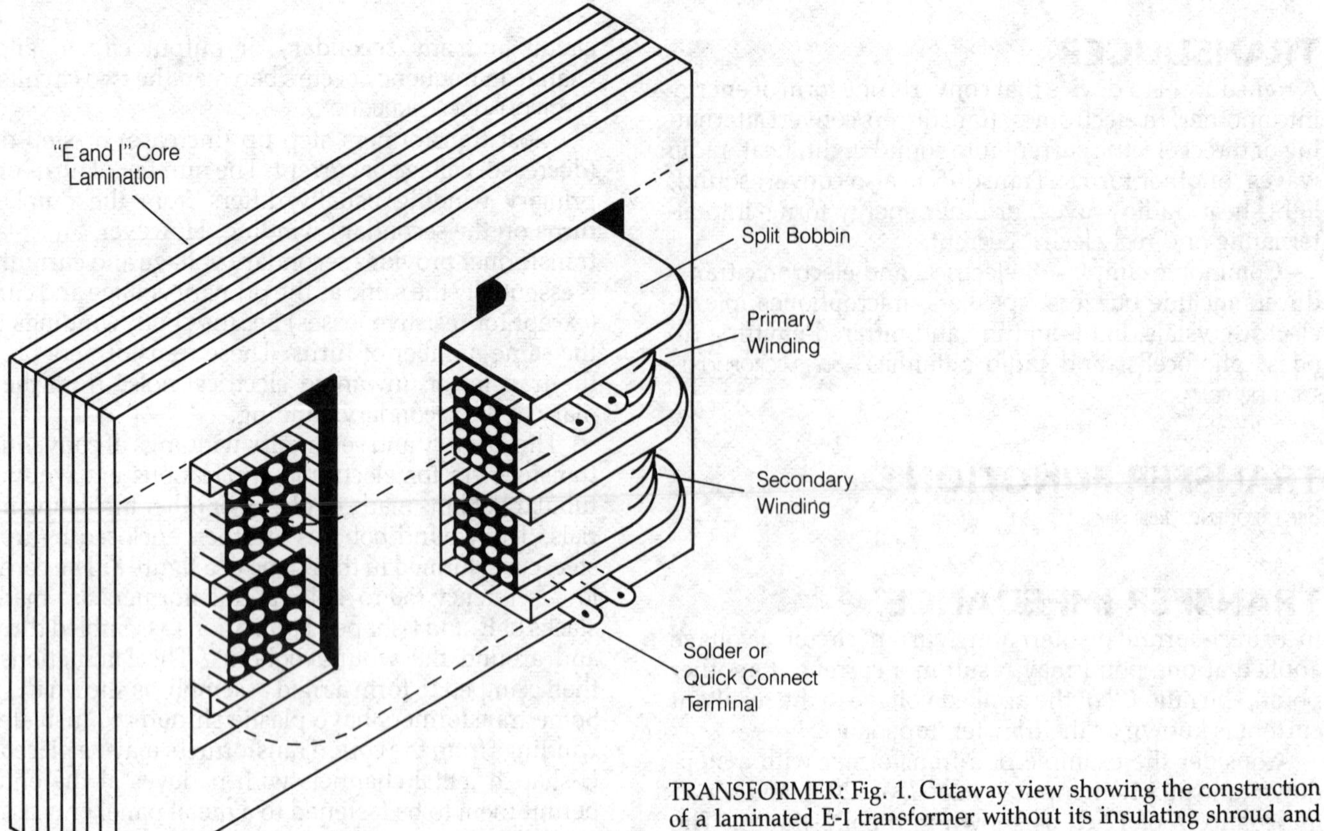

TRANSFORMER: Fig. 1. Cutaway view showing the construction of a laminated E-I transformer without its insulating shroud and channel frame.

polarity of the applied current. *See* EDDY-CURRENT LOSS, HYSTERESIS LOSS.

In electronics, transformers are classified by application. The most common examples are: 1) power, 2) audio or communications, and 3) pulse. These transformers are rated according to the product of their secondary voltage and current in volt-amperes (VA) or watts (W). Transformers for most electronic applications are rated for less than 100 VA or 100 W, but some large switching power supplies have transformers rated at 1 kW. Figure 2 illustrates schematically some common winding arrangements.

Power Transformers. Power transformers change the input voltage and current levels to meet circuit power requirements. Designed for operation from the alternating current line at 50 to 60 Hz, these transformers are made as standard products in high volume. Inductive reactance in transformers increases as a function of frequency and the amount of iron or steel in the core. The losses due to reactance limit efficient operation of iron- or steel-core transformers to about 400 Hz. At frequencies of 400 Hz to 50 kHz or higher, ferrite cores permit higher efficiency with lower core losses.

Audio or Voice Transformers. Audio or voice transformers are similar to power transformers, but they are designed to operate over a wider frequency range. They can carry direct current in one or more windings, transform voltage and current levels, act as impedance matching and coupling devices, or perform filtering. In addition, they may pass a limited range of voice frequencies—20 to 20,000 Hz.

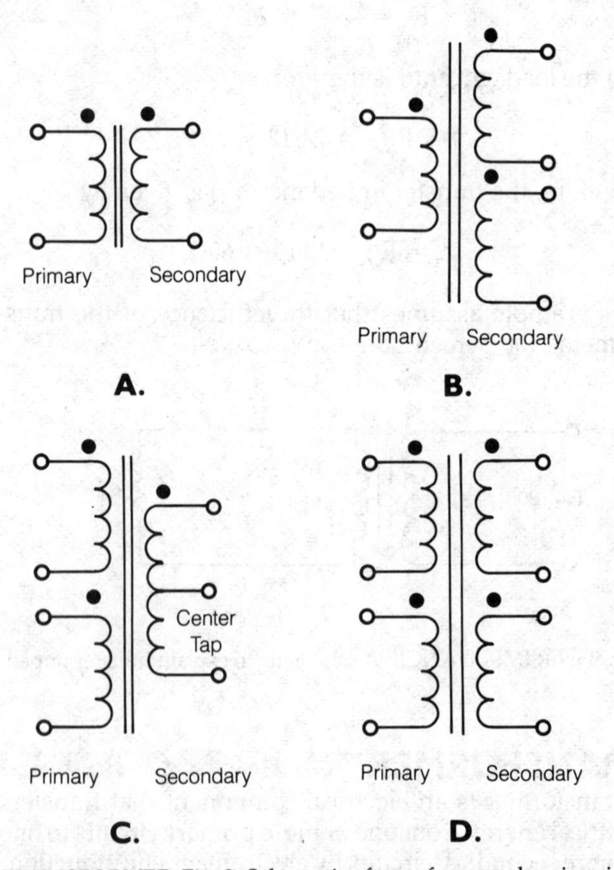

TRANSFORMER: Fig. 2. Schematic of transformers showing single primary, single secondary (A), single primary, dual secondary (B), dual primary, center-tapped secondary (C), and dual-primary, dual secondary (D). Dot indicates polarity.

Pulse Transformers. Pulse transformers are specialized miniature transformers that generate fast-rising output pulses for timing, counting, and triggering such electronics devices as SCRs (silicon-controlled rectifiers), triacs, radio transmitter tubes, and photoflash lamps. These transformers are usually made for printed-circuit-board (PCB) mounting.

Printed Circuit Board Transformers. Many miniature power, audio and pulse transformers are designed for PCB mounting. Some have low profiles to permit closer stacking of PC boards in card cages. These miniature transformers can be dipped in epoxy to seal them from dirt and moisture. The windings are terminated by pins on the bobbins for through-hole PCB mounting.

Radio-Frequency Transformers. Radio-frequency transformers are wound on bobbins because at high frequencies air is used as the core rather than steel laminations or ferrite.

Toroidal Transformers. Toroidal transformers are wound on a closed ring-shaped metal cores made by rolling up a long thin continuous strip of sheet steel. The process of winding a toroidal transformer is more complex and expensive than winding a conventional transformer. Both the primary and secondary windings of the toroid are wound on special machines through the closed core. Losses in toroidal transformers are lower than those of conventional transformers so they are more efficient. In addition they are also smaller and lighter than conventional transformers with the same power rating.

Specifications. Military standard MIL-T-27 is the mandatory specification for workmanship on military specification transformers, but it is also widely used as a guide in commercial manufacture. Most commercial transformers have Underwriters' Laboratories (UL) recognition and Canadian Standards Association (CSA) certification. Some also conform to the requirements of the Verband Deutscher Electrotechniker e.v. (VDE) and the International Electrotechnical Commission (IEC).

TRANSFORMER COUPLING

In an audio-frequency or radio-frequency system, the signal can be transferred from stage to stage by transformer coupling.

The illustration at A shows transformer coupling between two audio-frequency amplifier stages. The transformer is chosen so that the output impedance of the first stage is matched to the input impedance of the second stage. B shows transformer coupling in the intermediate-frequency chain of a radio receiver. In this arrangement, the windings are tapped for optimum impedance matching. Capacitors are connected across the transformer windings to provide resonance at the intermediate frequency.

An important advantage of transformer coupling is that it minimizes the capacitance between stages. This is especially desirable in radio-frequency circuits because stray capacitance reduces the selectivity. In transmitter power amplifiers, the output is almost always coupled to the antenna by means of a tuned transformer. This minimizes unwanted harmonic radiation. An electrostatic shield provides additional attenuation of harmonics by eliminating nearly all stray capacitance between the primary and secondary windings (*see* ELECTROSTATIC SHIELDING).

The main disadvantage of transformer coupling is cost: It is more expensive than simpler coupling methods. Transformer coupling is not widely used in modern circuitry. *See also* CAPACITIVE COUPLING.

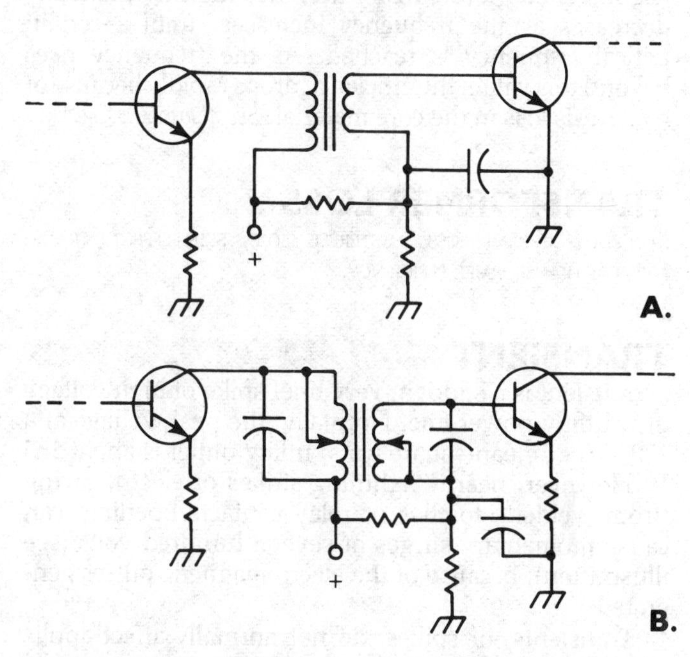

TRANSFORMER COUPLING: Transformer coupling in an audio-frequency circuit (A) and in an intermediate-frequency circuit of a radio receiver (B).

TRANSFORMER EFFICIENCY

Some power is lost in the coil windings of all transformers. If the transformer has a ferromagnetic core, some power is lost in the core. Conductor losses occur because of ohmic resistance and skin effect (*see* SKIN EFFECT). Core losses occur because of eddy currents and hysteresis (*see* EDDY-CURRENT LOSS, HYSTERESIS LOSS). The efficiency of a transformer is therefore always less than 100 percent.

Let E_p and I_p represent the primary-winding voltage and current in a hypothetical transformer, and let E_s and I_s be the secondary-winding voltage and current. In a perfect transformer, the product E_pI_p equals the product E_sI_s. However, in a real transformer, E_pI_p is always greater than E_sI_s. The efficiency, in percent, of a transformer is:

$$\text{Eff} = \frac{100E_sI_s}{E_pI_p}$$

The power P lost in the transformer windings and core is equal to the difference between E_pI_p and E_sI_s:

$$P = E_pI_p - E_sI_s$$

The efficiency of a transformer depends on the load connected to the secondary winding. If the current drain is excessive, the efficiency is reduced. Transformers are generally rated according to the maximum amount of power they can deliver without serious degradation in efficiency.

In audio and radio circuits, the frequency affects the efficiency of a transformer. In transformers with air cores, the efficiency gradually decreases as the frequency increases because of skin effect in the windings. In a ferromagnetic-core transformer, the efficiency gradually decreases as the frequency increases, until a certain critical frequency is reached. As the frequency rises beyond this value, the efficiency drops rapidly because of hysteresis loss in the core material. *See also* TRANSFORMER.

TRANSFORMER LOSS

See EDDY-CURRENT LOSS, HYSTERESIS LOSS, SKIN EFFECT, TRANSFORMER, TRANSFORMER EFFICIENCY.

TRANSIENT

A transient is a sudden, very brief spike of high voltage on a utility power line. Normally, the peak voltage at a 120-V root-mean-square (rms) utility outlet is about 165 V. However, nearby lightning strikes or electric arcing (from welder's torches or relay contacts opening) can cause momentary surges of several hundred volts (see illustration), because of the electromagnetic pulses generated.

Transients, or spikes, do not normally affect appliances, but they can cause damage to some types of electronic products that are not properly protected with devices capable of clipping or suppressing transients.

Transients are quite common, and all solid-state electronic equipment should have some protection to reduce the chances of damage. *See also* CIRCUIT PROTECTION, ELECTROMAGNETIC PULSE, SURGE.

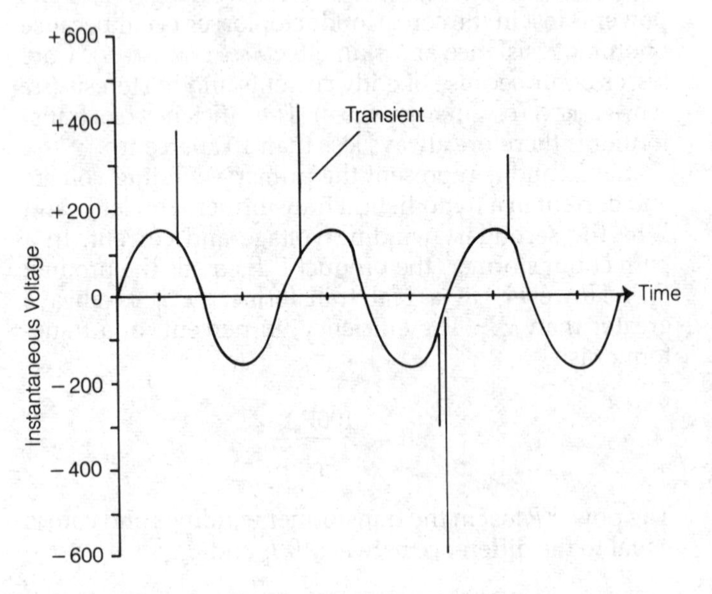

TRANSIENT: Transients of an alternating-current wave.

TRANSIENT PROTECTION

See CIRCUIT PROTECTION.

TRANSISTOR

A transistor is a three-terminal solid-state electronic device capable of amplification and switching. It is the solid-state analogy of the triode vacuum tube. Two different transistor designs predominate today: bipolar and metal-oxide semiconductor field effect (MOSFET). They are made as discrete small-signal and power devices and are integrated into digital and analog or linear silicon integrated circuits. Bipolar transistors have dominated where high-speed switching has been stressed. For example, they have been used in the central processing units (CPUs) of most computers. By contrast, MOSFETs have, until recently, been used principally in semiconductor memory devices.

Improvements in the switching speed of complementary metal-oxide semiconductor (CMOS) FETs, combined with their low power consumption, have qualified them for use in microcontrollers and the microprocessors now used as CPUs for microcomputers, minicomputers and computer-aided design (CAD) workstations. CMOS digital logic families based on MOSFETs are being used more extensively for general purpose digital logic and semicustom gate arrays and standard cells. The growing acceptance of commercially available CMOS 16-bit and 32-bit microprocessors is shifting the balance of high-speed processing toward CMOS.

There are differences in the construction of bipolar transistors and MOSFETs, but their principles of operation are similar. All transistors are switches. By applying electric charge to a specific location on a transistor, the transistor can be turned on or off. When a transistor is on, electric current flows from one region to another; when the transistor is off, the current flow stops.

Bipolar Transistor. Unless otherwise stated, the term *transistor* implies a silicon bipolar transistor. There are both NPN and PNP silicon bipolar transistors. Figure 1 is a section view of a silicon bipolar NPN transistor. A voltage applied to the base of the NPN transistor shown in the diagram causes electrons to flow between the

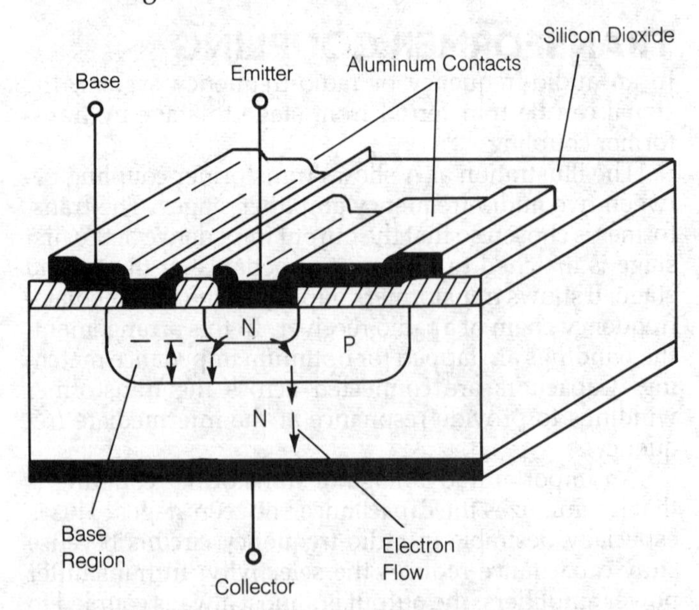

TRANSISTOR: Fig. 1. Structure of an NPN bipolar transistor.

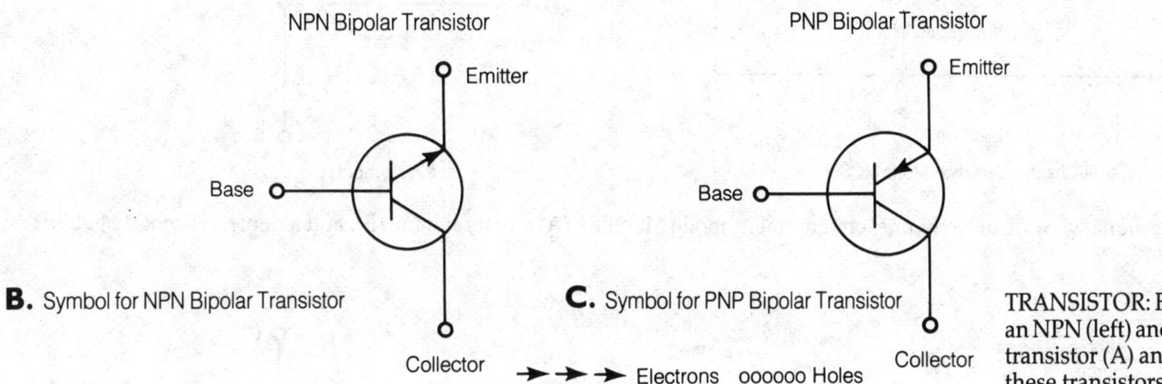

TRANSISTOR: Fig. 2. Section views of an NPN (left) and a PNP (right) bipolar transistor (A) and graphic symbols for these transistors (B and C).

emitter and collector. (Conventional current flows from the collector to the emitter.) In this vertical transistor, the collector terminal is the metallized base of the N-type substrate. The N-doped silicon contains excess electrons to carry current, but P-doped silicon has holes or vacant sites that can be occupied by electrons. However, holes act as positively charged particles to carry current.

Current in an NPN transistor cannot normally flow from the emitter to the collector because electrons cannot enter the P-doped base region. When a positive bias voltage is applied to the base and the collector, however, some holes, repelled by a positive voltage, flow from the base region into the emitter region. Electrons also flow from the emitter into the base. Most electrons flow through the base region into the lightly N-doped collector region and then to the collector terminal.

Figure 2 shows alternate geometries for bipolar transistors with all terminals brought to the surface. (This geometry used in bipolar integrated circuits.) A sectional view of an NPN bipolar transistor is shown on the left side of part (A) and a PNP bipolar transistor is shown on the right side of (A). The arrows represent electron flow and the dots represent hole flow. The polarities and materials of the NPN and PNP transistors are reversed.

The schematic symbol for an NPN bipolar transistor is shown at (B) and the schematic symbol for a PNP bipolar

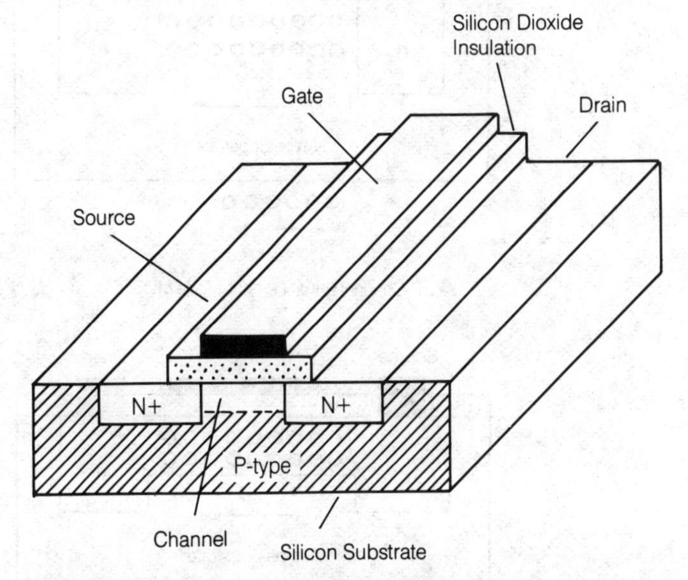

TRANSISTOR: Fig. 3. Structure of an N-channel metal-oxide semiconductor field-effect transistor (MOSFET).

transistor is shown at (C). The arrowhead points from the P-type material to the N-type material.

Metal-Oxide Semiconductor Field-Effect Transistor. Figures 3, 4, and 5 show silicon metal-oxide semiconductor

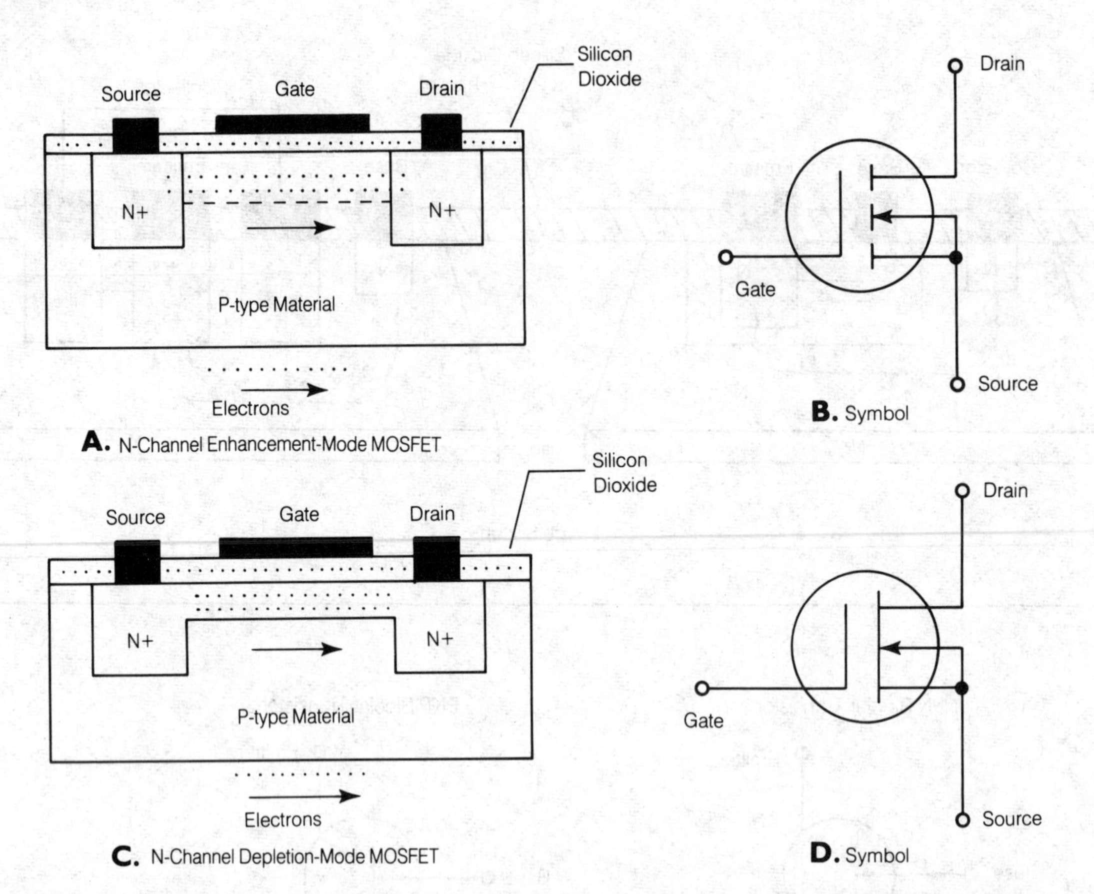

TRANSISTOR: Fig. 4. Sectional view of an N-channel enhancement-mode MOSFET (A) and its symbol (B), and a depletion-mode MOSFET (C) and its symbol (D).

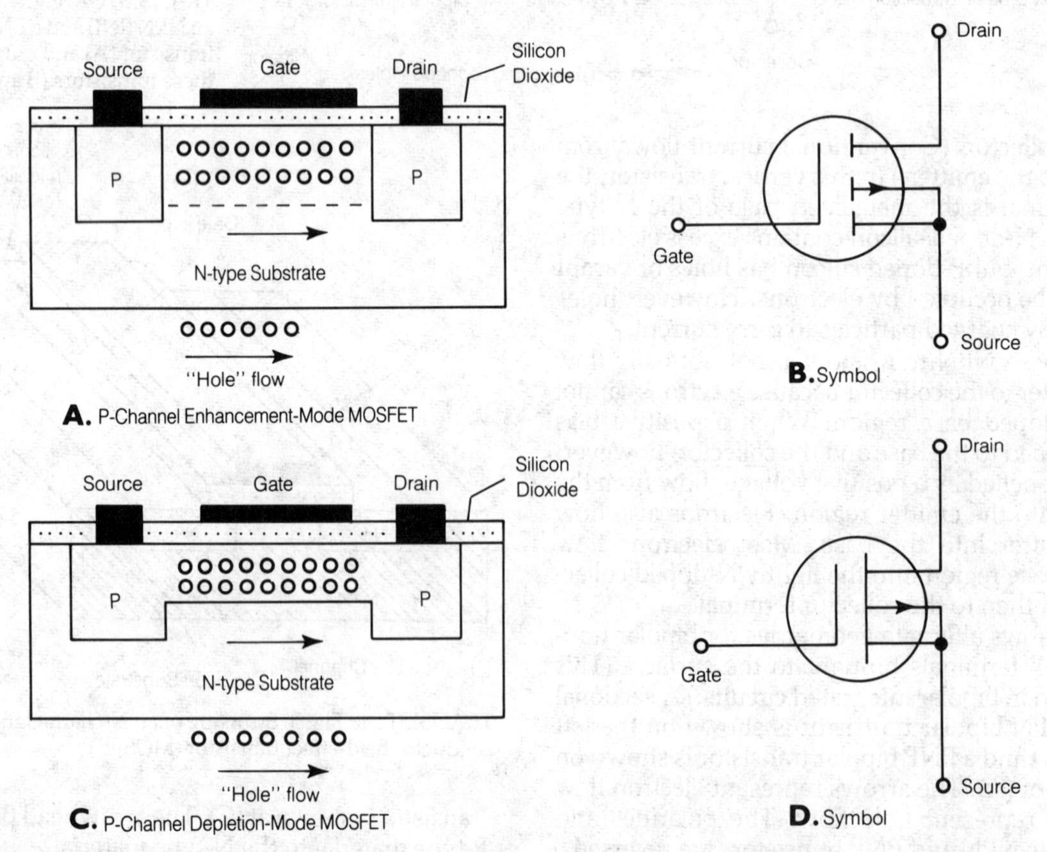

TRANSISTOR: Fig. 5. Sectional view of P-channel enhancement-mode MOSFET (A) and its symbol (B), and a depletion-mode MOSFET (C) and its symbol (D).

field-effect transistors (MOSFETs). In the N-channel NMOS version shown in Fig. 3, current cannot flow through the P-doped region between the source and the drain unless a positive voltage is applied to the gate, which is insulated by a thin layer of silicon dioxide. The voltage attracts a layer of electrons, called an inversion layer, forming a conductive channel permitting current to flow. The electrons flow from source to drain. *See* FIELD-EFFECT TRANSISTOR, METAL-OXIDE SEMICONDUCTOR FIELD-EFFECT TRANSISTOR.

TRANSISTOR MANUFACTURE

The manufacture of a discrete transistor requires a complex series of steps starting with a blank wafer of silicon. The process is similar to that used in the manufacture of an integrated circuit except that the end product is a discrete device. There are many similarities in the fabrication of all discrete silicon devices including both small-signal and power transistors and thyristors. The figure is a brief summary of some of the important steps in the fabrication of a simple diffused NPN bipolar transistor.

The process starts with a thin wafer of N-type silicon, perhaps four to six inches in diameter and 0.020 inch thick. The large single crystal from which these wafers are sliced is typically grown by the Czochralski process. (*See* CZOCHRALSKI CRYSTAL-GROWTH SYSTEM.) The long crystal is ground to remove uneven regions and a flat is ground along the length of the crystal prior to slicing so that each wafer will have a reference edge parallel to a natural plane. Each wafer is then ground and polished smooth and flat to precision tolerances.

Oxidation. A layer of silicon dioxide grows on the silicon substrate as hot gas at from 1000° to 1200°C flows over it. The desired thickness of silicon dioxide can be achieved by control of temperature, time, and gas flow rate. This silicon dioxide layer will shield selected areas of the silicon surface from penetration by dopants.

Lithography. Open areas are established in the silicon dioxide so that dopants can enter the silicon substrate. The silicon dioxide layer is coated with a photoresist—a material that resists chemical attack after it has been exposed to light. (*See* PHOTORESIST).

Next, a photographic mask or photomask is placed in contact with the photoresist the wafer is exposed to ultraviolet light. The photomask is a photographic reduction of large-scale artwork for a single layer of the transistor, accurate to millionths of an inch. Because the transistor chip is very small, hundreds can be fabricated simultaneously in an array on a single wafer of silicon. Each working photomask, typically a glass plate, has a pattern for a single layer.

The transparent or open areas defined by the photomask allow ultraviolet light to alter the chemical composition of the photoresist under the photomask to form an etch-resistant layer on the substrate. This makes it easier to remove the unexposed areas with solvents.

Etching. Following exposure to ultraviolet light, the wafer is immersed in a chemical solvent to remove the unexposed photoresist, leaving a pattern of photoresist on the oxide. The wafer, with its photoresist pattern, is

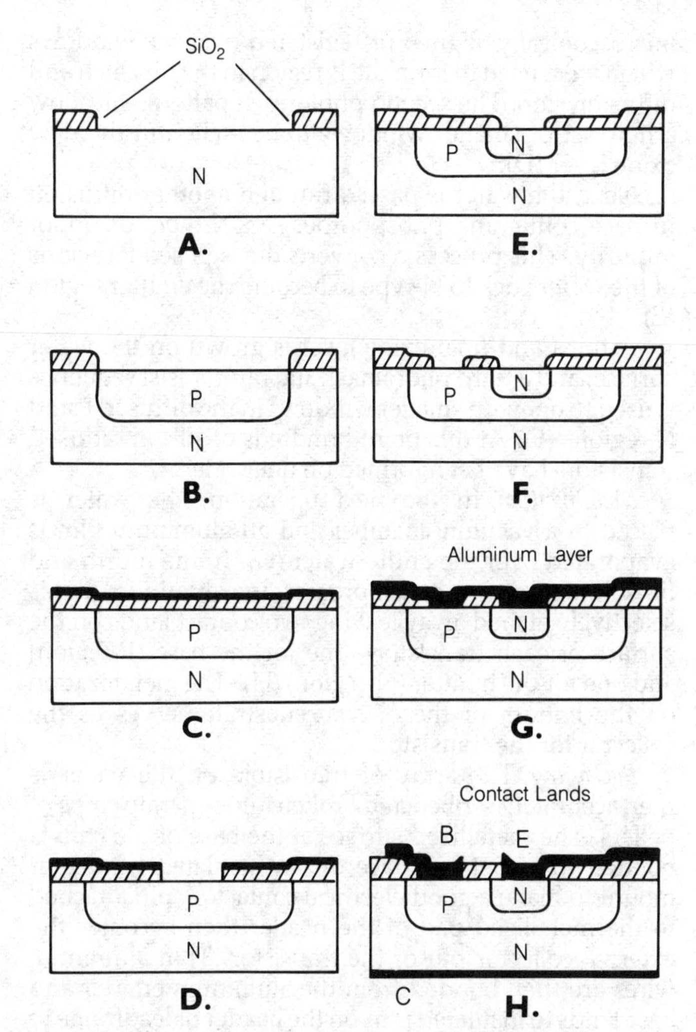

TRANSISTOR MANUFACTURE: Basic process steps in the manufacture of a simple NPN bipolar planar transistor.

then placed in a solution of hydrofluoric acid to dissolve the unprotected oxide layer, but the acid does not remove either the remaining photoresist or the silicon. Alternatively the oxide layer can be removed by dry etching with a plasma of reactive gas. Finally the hardened photoresist is removed. A crosssection of the result is shown in (A) of the figure.

Diffusion. With selected areas of the N-type substrate exposed, the wafer is placed in a furnace and heated in the presence of acceptor (P-type) impurity vapors such as boron. This is called the diffusion process. Minute amounts of atomic boron penetrate into the exposed silicon wafer (B in the figure). The amount is sufficient to cancel and overcompensate for the donor impurities present. This process converts the exposed area from N-type to P-type silicon. The penetration depth of the impurity is carefully controlled by the exposure time of the vaporized acceptor impurity. Diffusion depths may be over a thousandth of an inch.

At this point, a PN junction has been formed. If no further diffusion processing were performed, this device could be used as a diode. After the first diffusion step, another layer of silicon dioxide is grown over the entire wafer surface (C). Then a second photoresist operation is performed, using a second photomask. This mask is microscopically aligned or registered over the windows which were used to form the P-region in the first etch and diffusion step. The second photoresist pattern will allow a new set of smaller windows to be etched in the new oxide layer (D).

Next, the wafer is passed through another diffusion furnace containing phosphorous, an N-type, or donor impurity. This process reconverts the exposed P regions of the wafer back to N-type to become the emitter region (E).

A third and final oxide layer is grown on the wafer surface, and a third photomask and photoresist sequence is used to open up smaller windows in the diffused P and N regions (F). At this point hundreds of silicon diffused transistors have been formed on one wafer.

Metallization. In the next operation, the wafer is placed in a vacuum chamber and an aluminum film is evaporated over the entire wafer (G). In the fourth and final lithography and etch process, the aluminum film is selectively etched away leaving two contact lands on the surface of each transistor—one for the base (P-region) and one for the emitter (N-region) (H). The metallization on the bottom of the N-type substrate serves as the collector for the transistor.

Packaging. The array of transistors on the wafer is then accurately scribed and broken into separate chips or pellets. The metallized surface at the base of the chip is bonded to a metal surface on an insulated header or mount to make a good electrical contact. A pin attached to the metallized pad of the header then becomes the external collector pin of the transistor. Thin aluminum wires are then bonded from the aluminum emitter and base lands to insulated pins on the header or leadframe to become the external emitter of the transistor and base leads.

Transistor packaging can be completed with the bonding of a metal can to the metal rim of the header to form a hermetically sealed metal case. Alternatively, the transistor can be immersed in plastic resin which is then molded into a plastic case. *See* SEMICONDUCTOR PACKAGE.

TRANSISTOR BATTERY

See BATTERY.

TRANSISTOR TESTER

A transistor tester is an instrument used to determine if a transistor is functioning properly, and if it is, to give a quantitative indication of the operating characteristics.

Sophisticated transistor testers are commercially available for checking the performance of bipolar and field-effect transistors. These instruments measure the alpha and beta of a bipolar device and the transconductance of a field-effect device (*see* ALPHA, BETA, TRANSCONDUCTANCE, TRANSISTOR).

If it is suspected that a bipolar transistor has been destroyed, an ohmmeter can be used for a quick check. All ohmmeters produce some voltage at the test terminals. The ohmmeter should be switched to the lowest resistance ($R \times 1$) scale, and the polarity of the test leads determined by means of an external voltmeter. Designate P+ as the lead at which a positive voltage appears, and P− the lead at which negative voltage appears. (Note: In a volt-ohm-milliammeter, the ohmmeter-mode polarity may not correspond to the red/black lead colors used for measurement of voltage and current!) To test an NPN transistor:

- Connect P+ to the collector and P− to the emitter. The resistance should appear infinite.
- Connect P+ to the emitter and P− to the collector. The resistance should again appear infinite.
- Connect P+ to the emitter and P− to the base. The resistance should appear infinite.
- Connect P+ to the base and P− to the emitter. The resistance should appear finite and measurable.
- Connect P+ to the collector and P− to the base. The resistance should appear infinite.
- Connect P− to the collector and P+ to the base. The resistance should appear finite and measurable.

If any of the above conditions is not met, the transistor may be faulty.

To check a PNP transistor the procedure is just the same as described, except that every P+ should be changed to P−, and every P− should be changed to P+. Again, if any of the described conditions is not met, the transistor may be faulty.

TRANSISTOR-TRANSISTOR LOGIC

See DIGITAL-LOGIC INTEGRATED CIRCUIT.

TRANSITION ZONE

The electromagnetic field surrounding a transmitting antenna can be divided into three distinct zones: the far field, the near field, and the transition zone. The far field exists at a great distance from the antenna, and is called the Fraunhofer region. The near field is very close to the transmitting antenna and is called the Fresnel zone. The region between the far field and the near field is known as the transition zone.

The far field, near field, and transition zone are not precisely defined. In general, the transition zone occurs at distances where the electric and magnetic lines are somewhat curved. The lines of flux are decidedly curved in the near field, but they are practically straight in the far field. In communications practice, the far field is the only important consideration; the near field and transition zone fall within a few wavelengths of the radiating element. *See also* FAR FIELD, NEAR FIELD.

TRANSIT TIME

Transit time is the time required for a charge carrier to move from one place to another. Transit time is significant in the operation of vacuum tubes and semiconductor devices.

In a tube, the electrons travel from the cathode to the plate through a vacuum or near vacuum. Although the electrons move very fast, they require a finite amount of transit time to cross the gap. The transit time depends on the spacing of the electrodes, the plate voltage, and the current drawn. Transit time limits the frequency of microwave oscillator tubes.

In a semiconductor diode or bipolar transistor, electrons and holes move and diffuse through the chip. The transit time is the effective time required for a charge carrier to move to the anode from the cathode in a diode or from the emitter-base junction to the collector-base junction in a bipolar transistor. In a field-effect transistor, the transit time is the time required for the electrons or holes to pass through the channel. Transit time in a semiconductor device is related to the carrier mobility, which depends on the type of semiconductor material and the density of doping. *See also* CARRIER MOBILITY, FIELD-EFFECT TRANSISTOR, TRANSISTOR.

TRANSMISSION LINE

A transmission line is a conduit for the transmission of power to distant loads. A transmission line makes power available at distant points by means of electromagnetic-field propagation (*see* ELECTROMAGNETIC FIELD).

Transmission lines generally consist of one or two conductors in a balanced or unbalanced configuration. The most common type of balanced transmission line has two conductors parallel to each other. The most common type of unbalanced line consists of a wire surrounded by a cylindrical shield (*see* BALANCED TRANSMISSION LINE, UNBALANCED TRANSMISSION LINE). In both the balanced and unbalanced configurations, the electric and magnetic components of the propagated field are always perpendicular to the line conductors. The electromagnetic field follows the transmission line.

In an alternating current (ac) power system, a transmission line is balanced. In a radio-frequency (rf) system, the transmission line is either balanced or unbalanced, depending on the load. An rf transmission line is often called a feed line, especially if the load is an antenna (*see* FEED LINE).

For a transmission line to provide the highest possible available power to the load, two conditions must be met: First, the load must have no reactance, only resistance. Second, the characteristic impedance of the line must have the same ohmic value as the resistive impedance of the load. If these conditions are not met, the load will not convert all of the electromagnetic field into power. Some of the field will be reflected at the load, and sent back toward the generator. *See also* IMPEDANCE MATCHING, REFLECTED POWER.

TRANSMIT-RECEIVE SWITCH

When a transmitter and a receiver share a common antenna, a switch is used between the two functions. It is important that the receiver be disconnected while the transmitter is operating. A transmit-receive (T-R) switch accomplishes this. Powerful microwave radar transmitters use gas-discharge tubes, called T-R tubes.

The simplest form of transmit-receive switch is a relay. When the transmitter is keyed, the relay connects the antenna to the transmitter. When the transmitter is unkeyed, the relay connects the antenna to the receiver. This arrangement is used in many transceivers.

A more advanced T-R switch consists of a radio-frequency-actuated electronic circuit. The transmitter is always connected to the antenna, even while receiving. But when the transmitter is keyed, the receiver is disconnected. The schematic diagram shows this kind of T-R switch.

The T-R tube used to protect the receiver in a powerful pulsed microwave radar system is a gas-filled waveguide cavity. It ionizes and acts as a short circuit in the presence of a transmitted pulse but is transparent to low-power energy for the receiver when it is un-ionized. *See also* RADAR.

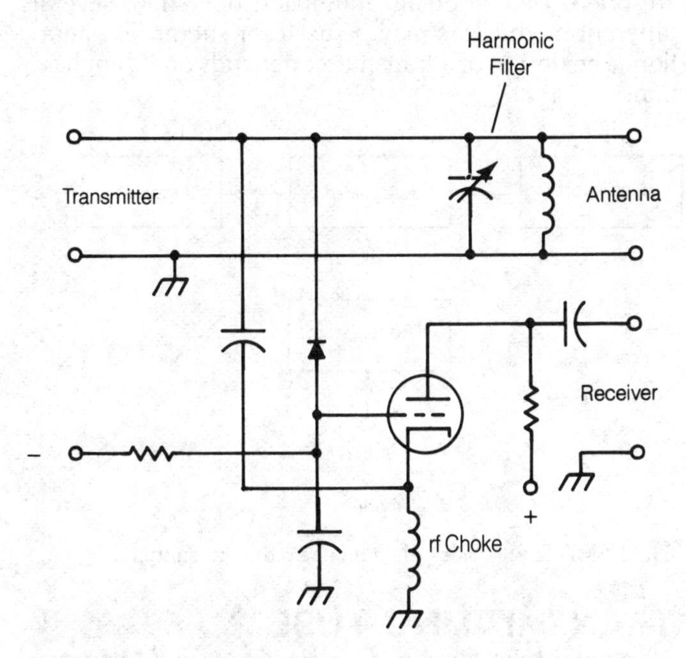

TRANSMIT-RECEIVE SWITCH: An example of a vacuum-tube T-R switch used to protect the receiver from strong rf voltage.

TRANSMIT-RECEIVE TUBE

See TRANSMIT-RECEIVE SWITCH.

TRANSMITTANCE

Transmittance is the ratio of transmitted power to incident power. In optics, transmittance is frequently expressed as optical density or percent.

A transmittance value of 0 indicates that a material is perfectly opaque, or that all of the available power is absorbed by a circuit. A transmittance figure of 100 percent indicates that a substance is perfectly transparent, or that none of the power is absorbed by a circuit.

In communications applications, transmittance is generally expressed in decibels (dB).

TRANSMITTER

A transmitter is a circuit that produces a signal for broadcasting or communications purposes. The signal might consist of an electric current, radio waves, light, ultrasound, or any other form of energy. A transmitter converts audio and/or video information into a signal that can be sent to a distant receiver (*see* RECEIVER).

A basic transmitter consists of an oscillator, a transducer, a modulator, and a signal amplifier (see illustration). The amplifier output is connected to a wire transmission line or an antenna system. The oscillator provides the carrier wave. The transducer converts audio and/or video information into electrical signals. The modulator impresses the output of the transducer onto the carrier wave. The amplifier boosts the signal level to provide sufficient power for transmission over the required distance.

Most transmitters incorporate more advanced designs than the basic scheme shown. For example, mixers are often used to obtain multiband operation; several different modulators may be used for subcarrier operation. The design of a transmitter depends on its application.

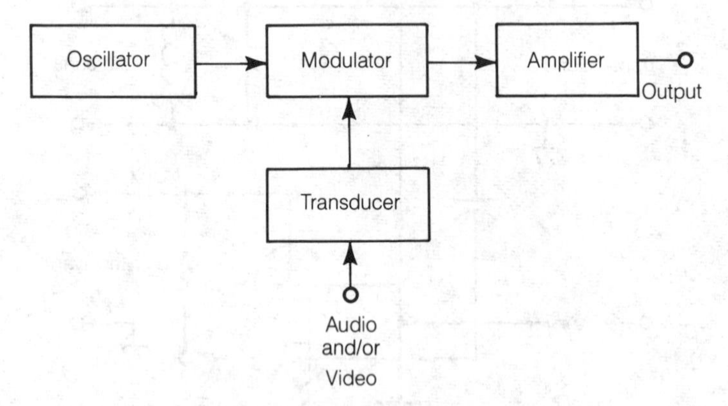

TRANSMITTER: A block diagram of a radio transmitter.

TRANSMITTING TUBE

A vacuum tube, designed for use in a radio-frequency power amplifier, is called a transmitting tube. Transmitting tubes are characterized by their ability to dissipate large amounts of power.

Transmitting tubes are physically larger than receiving tubes because the high power level dictates that the plate and cathode be large. Some transmitting tubes have finned plates to maximize the amount of power they can dissipate. Other have graphite plates for the same reason.

Although receiving tubes have been largely replaced by transistors in recent years, tubes are still used in some transmitters. This is especially true of high-power transmitters operating at microwave frequencies. Tubes are more tolerant of momentary excessive power dissipation than transistors.

TRANSMITTIVITY

See TRANSMITTANCE.

TRANSPONDER

A transponder is a receiver-transmitter that sends a radio signal whenever it receives a radio command from a distant station. A transponder is both a radio beacon and a relay containing receiving as well as transmitting circuits in a single package. A command signal, called an interrogation, is necessary to elicit a transponder return signal, called a response.

Transponders are used in aircraft for air traffic control and identification. A transponder aboard an aircraft allows its position to be continuously monitored by ground-based air traffic control personnel. A ground-based traffic control radar transmits pulses at a specified rate, and the transponder aboard the aircraft replies at the same rate.

The interrogating radar determines the position of each aircraft, just as does a conventional radar. However, the transponder reply contains information about the identity, altitude, course, and speed of each aircraft. This data is presented on the air traffic control radar screen in place of the anonymous blip that is received by conventional radar. These messages simplify the task of air traffic control in the air space around major airports with heavy air traffic. The message box provides a plan position indication of the aircraft relative to geographic coordinates and other aircraft in the vicinity. *See also* RADAR.

If the aircraft should be forced down over land or sea away from witnesses, the transponder can be interrogated by the radars on the search and rescue aircraft. The response can be used to speed up search efforts, and timely location of the downed aircraft will ensure faster rescue and medical aid for the survivors.

The transponder is the basis for military Identification, Friend or Foe (IFF) systems to prevent "friendly" aircraft from being accidentally attacked and shot down. The interrogation codes of military aircraft transponders are changed frequently to prevent enemy aircraft from posing as friendly aircraft to evade defensive weapons.

A marine variation of the aircraft transponder is used as an underwater marker or reference point for exploration and salvage. Marine transponders are self-contained receiver-transmitter packages that can be dropped to the seafloor in specified locations. They receive sonic or ultrasonic interrogation signals from surface vessels or submersibles and emit coded response signals only in response to interrogation, permitting them to conserve their power sources. Marine transponders permit ships to return to precise locations over the seafloor after weeks or months of absence. *See also* SONAR.

TRANSVERSE WAVE

A transverse wave is a disturbance in which the displacement is perpendicular to the direction of travel. All electromagnetic waves are transverse waves because the electric and magnetic lines of flux are perpendicular to the direction of propagation. In some materials sound propagates as transverse waves.

The waves on a lake or pond illustrate the principle of transverse disturbances. When a pebble is dropped into a mirror-smooth pond, waves form and are propagated outward. The effects of the disturbance travel at a defined and constant speed. The waves have measurable length and amplitude. The actual water molecules, however, move up and down, at right angles to the wave motion. *See also* SURFACE ACOUSTIC WAVE DEVICE.

Some waves are propagated by movement of particles in directions parallel to the wave motion. This kind of disturbance is called a longitudinal wave. *See also* LONGITUDINAL WAVE.

TRANSVERTER

A transverter is a circuit that allows operation of a transceiver on a frequency that differs from the design frequency. A transverter consists of a transmitting converter and a receiving converter in a single package.

The illustration is a block diagram of a hypothetical transverter. The transceiver operates in the band 28.0 to 30.0 MHz. The actual operating frequency band is 144.0 to 146.0 MHz. A common local oscillator at 116.0 MHz provides the conversion.

The receiving converter heterodynes the incoming frequency fr_1 MHz to a new frequency fr_2 MHz according to the relationship:

$$fr_2 = fr_1 - 116.0$$

The transmitting converter heterodynes the transceiver output frequency ft_1 to a new frequency ft_2 according to:

$$ft_2 = 116.0 + ft_1$$

Transverters are commercially manufactured for use with most kinds of transceivers. *See also* MIXER, RECEIVER, TRANSCEIVER, TRANSMITTER.

TRAP

A trap is a form of band-rejection filter designed for blocking energy at one frequency while allowing energy to pass at all other frequencies. A trap consists of a parallel-resonant circuit in series with the signal path (see illustration at A) or a series-resonant circuit in parallel with the signal path, as at B. The configuration at A is the more common.

Traps can be made from discrete inductors and capacitors, as shown, or they can be fabricated from sections of transmission line. In the latter case, the trap is usually called a stub (*see* STUB).

Traps are used extensively in the design and construction of multiband antenna systems.

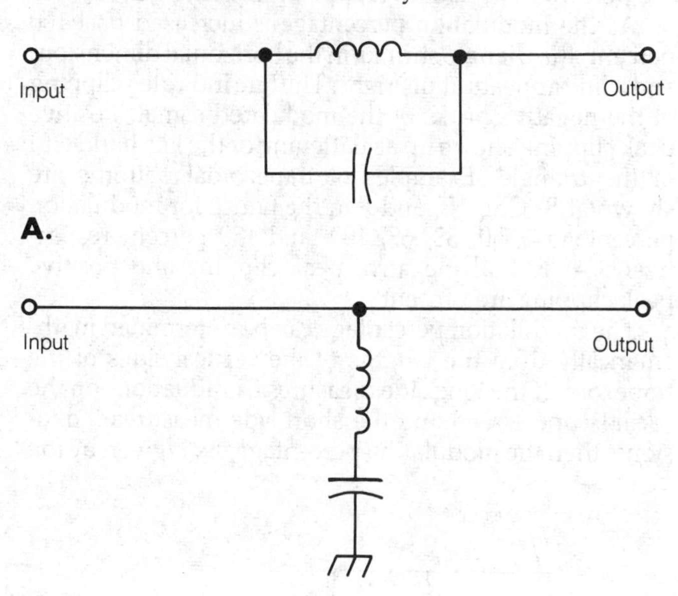

TRAP: Examples of trap circuits are parallel resonant (A) and series resonant (B).

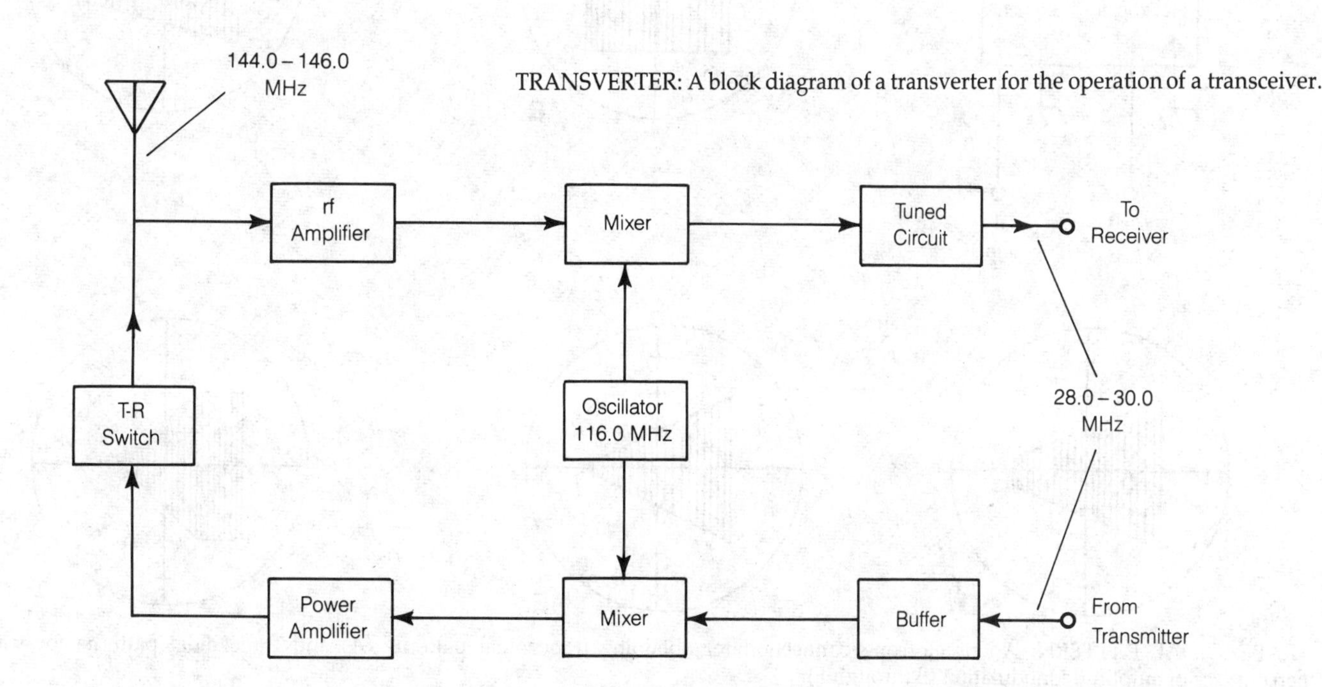

TRANSVERTER: A block diagram of a transverter for the operation of a transceiver.

TRAPEZOIDAL PATTERN

The percentage of modulation of an amplitude-modulated signal can be determined with an oscilloscope connected so that a trapezoid-shaped pattern appears on the screen. The radio-frequency signal is applied to the vertical deflection plates of the oscilloscope, and the modulating audio signal is applied to the horizontal deflection plates (see illustration at A).

The modulation percentage is determined by the variation in shape of the trapezoid from a perfect rectangle. If there is no modulation, the pattern appears as a perfect rectangle. As the modulation percentage increases, the pattern becomes more distorted until, with 100 percent modulation, it appears as a triangle.

As the modulation percentage is increased past 100 percent, the triangle shrinks in the horizontal dimension, and a line appears at the right. The line indicates clipping of the negative peaks of the modulated signal. Positive peak clipping shows up as flattening of the left-hand end of the triangle. Examples of trapezoidal patterns are shown at B, C, D, E, and F in the figure for modulation percentages of 0, 33, 67, 100, and 150 percent respectively. At F, both negative peak clipping and positive peak clipping are present.

The modulation percentage can be determined mathematically from the lengths of the vertical sides of the trapezoid. If the long side measures *L* graduations on the oscilloscope screen and the short side measures *S* divisions, then the modulation percentage, *m*, is given by the formula:

$$m = \frac{100(L - S)}{L + S}$$

within the limits of 0 to 100 percent. The formula does not hold for modulation percentages over 100. *See also* AMPLITUDE MODULATION, MODULATION PERCENTAGE.

TRAPEZOIDAL WAVE

A trapezoidal wave results from the combination of a square wave and a sawtooth wave, with identical frequencies but perhaps different amplitudes. The trapezoidal wave is so named because its shape resembles a trapezoid.

The period of a trapezoidal wave is the length of time from any point on the waveform to the same point on the next pulse in the train. This period is divided into 360 degrees of phase, with the zero point usually corresponding to the moment at which the amplitude is zero and rapidly changing (see illustration on p. 869). The frequency of the trapezoidal wave, in hertz, is the reciprocal of the period in seconds.

In a cathode-ray tube, the voltage applied to the deflecting coils always has a trapezoidal waveshape. This provides the necessary sawtooth-wave variation in the currents through the coils, and thus in the magnetic fields, ensuring a linear forward sweep and a fast return sweep. *See also* CATHODE-RAY TUBE, SAWTOOTH WAVE.

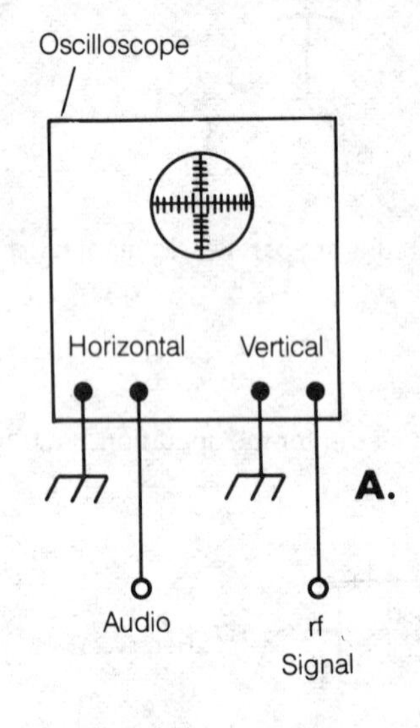

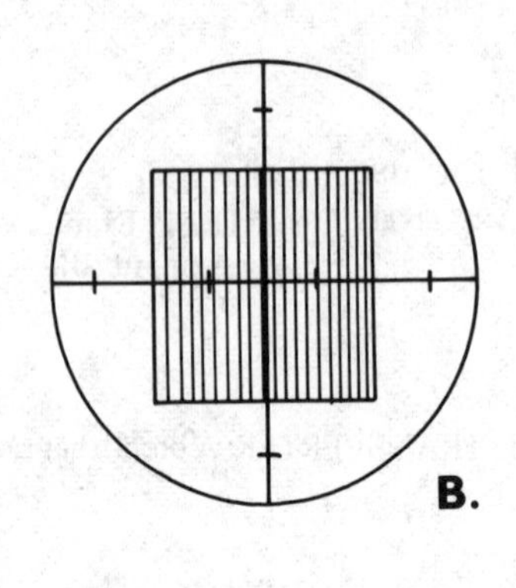

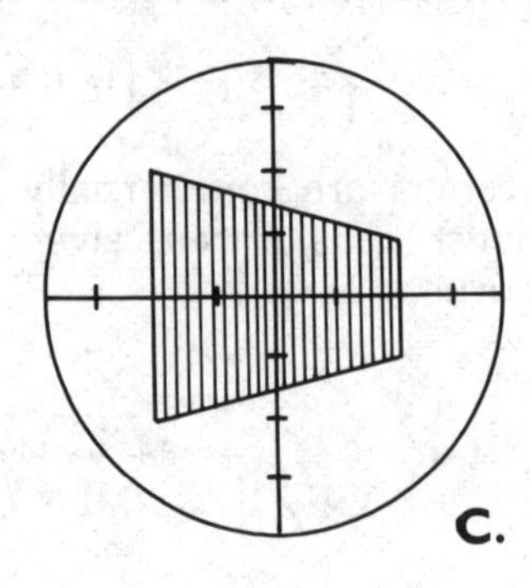

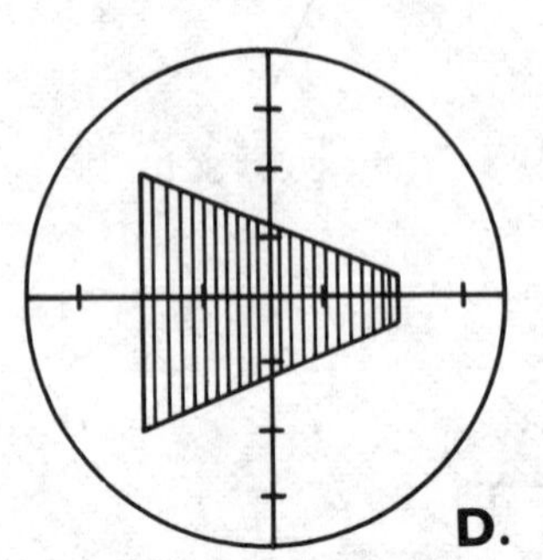

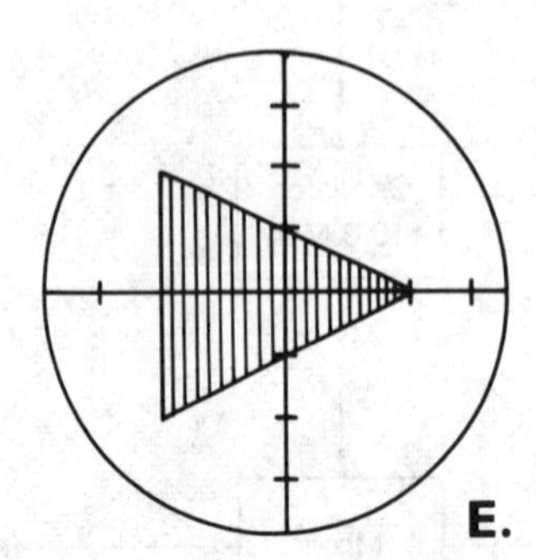

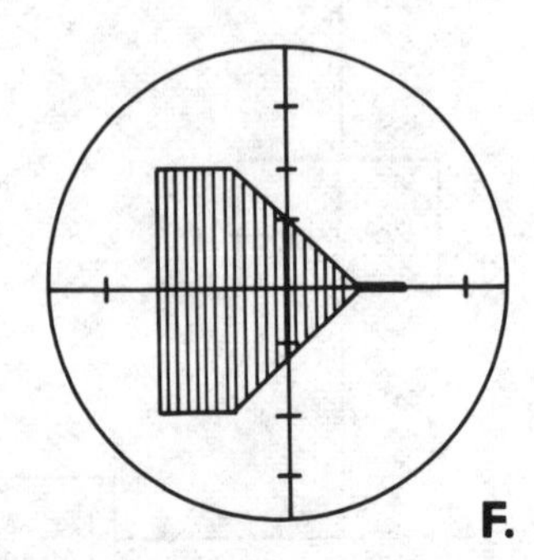

TRAPEZOIDAL PATTERN: An oscilloscope connection for obtaining trapezoidal patterns (A) and trapezoidal patterns for various percentages of amplitude modulation (B through F).

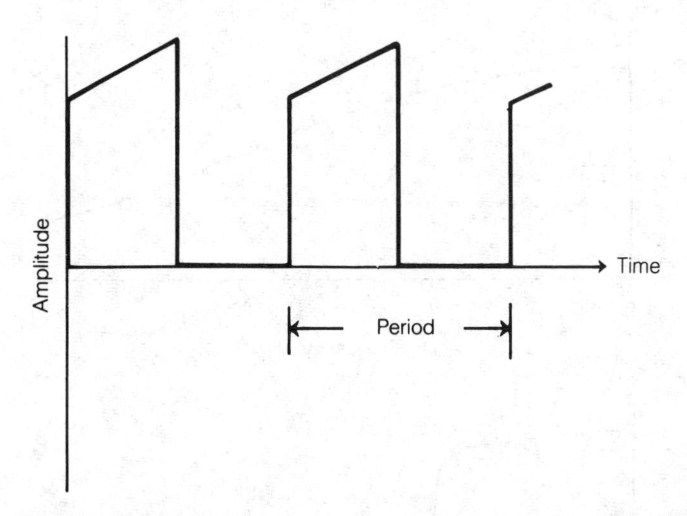

TRAPEZOIDAL WAVE: This wave is a combination of square and sawtooth waves.

TRAVELING-WAVE TUBE

A traveling-wave tube (TWT) is a specialized vacuum tube that is used in oscillators and amplifiers at ultrahigh and microwave frequencies. There are many different types of traveling-wave tubes. The most common configuration is shown in the illustration.

The electron gun produces a high-intensity beam of electrons that travels in a straight line to the anode. A helical conductor is wound around the tube. The distributed inductance and capacitance of this winding result in a very low velocity factor (*see* VELOCITY FACTOR), typically from 10 to 20 percent, approximately matching the speed of the electrons in the beam inside the tube. A signal is applied at one end of the helix and is taken from the other end.

When the helical winding is energized, the electron beam inside the tube is phase-modulated. Some of the electrons travel in synchronization with the wave in the helix because of the low velocity factor of the winding. This produces waves in the electron beam. The waves can travel in the same direction as the electrons (forward-wave mode) or in the opposite direction (backward-wave mode). In either case, energy from the electrons is transferred to the signal in the winding, producing gain when the beam voltage is within a certain range. Gain figures of 15 to 20 dB are common and some traveling-wave tubes can produce more than 50 decibels of gain.

A traveling-wave tube can produce energy at ultrahigh and microwave frequencies by coupling some of the output back into the input. This type of oscillator is called a backward-wave oscillator because the feedback is applied opposite to the direction of movement of the electrons inside the tube. The backward-wave oscillator can produce up to 100 mW of radio-frequency power at frequencies up to several gigahertz. The backward-wave arrangement can also be used for amplification.

TREBLE

Treble refers to high-frequency sound energy. On a piano or musical staff, any note at or above middle C (261.6 Hz) is considered to be in the treble clef. Any note below middle C is in the bass clef (*see* BASS).

A tone control in a high-fidelity sound system can have separate bass and treble adjustments, or it can consist of a single knob or slide type potentiometer. The amount of treble, especially at frequencies above 1000 Hz, affects the crispness of musical sound. Too little treble results in a muffled quality, and too much treble makes music sound tinny. *See also* AUDIO RESPONSE, TONE CONTROL, TREBLE RESPONSE.

TREBLE RESPONSE

Treble response is the ability of a sound amplification or reproduction system to respond to treble audio-frequency energy. The treble response can be defined for microphones, speakers, radio receivers, record and compact disk players, and recording tape—anything that

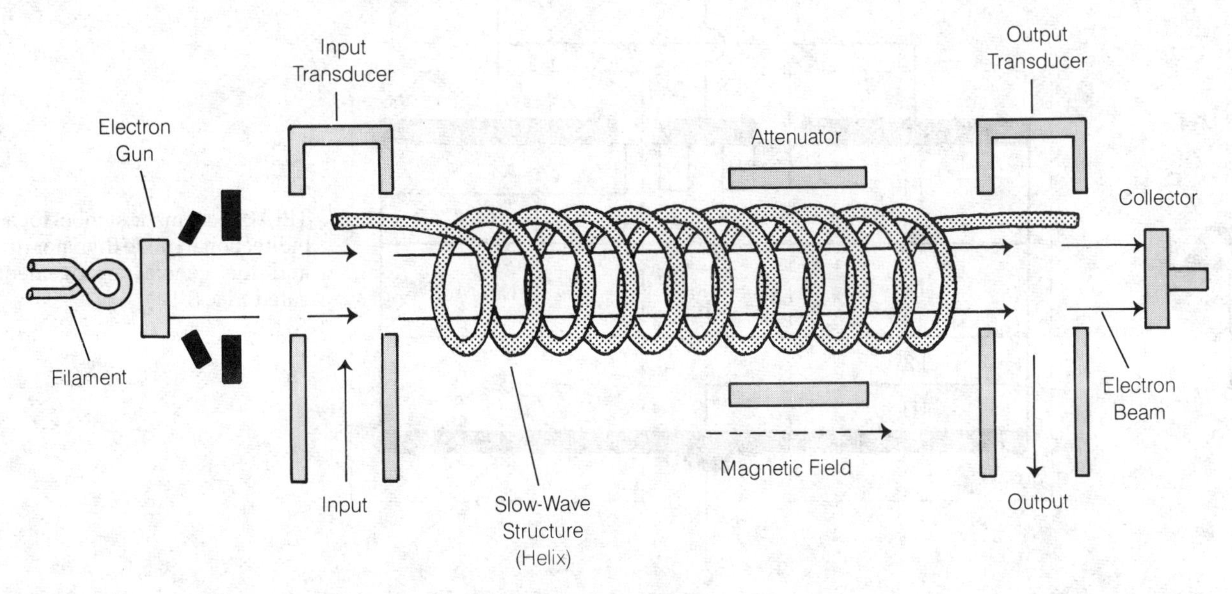

TRAVELING-WAVE TUBE: Schematic of a forward-wave TWT amplifier.

involves the transmission of audio energy. In most high-fidelity sound systems, the treble response is adjustable by means of a tone control or a treble gain control (*see* TONE CONTROL).

A good treble response involves more than the simple ability of a sound system to reproduce high-frequency audio. The sound will not be accurately reproduced unless the audio response is optimized (*see* AUDIO RESPONSE). Many systems have a gradual roll-off starting at 3–5 kHz (see illustration at A). Some high-fidelity amplifiers have a flat response at frequencies as high as 40 kHz or more. If the gain drops off anywhere within the human hearing range, the quality of the reproduction will be degraded (B).

A circuit called an equalizer (*see* EQUALIZER) facilitates compensation for differences in sound-system response. The bass and treble response can be tailored for the best possible sound reproduction with this circuit.

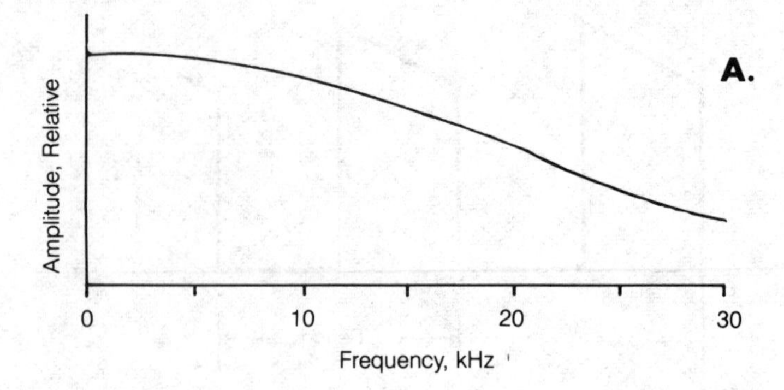

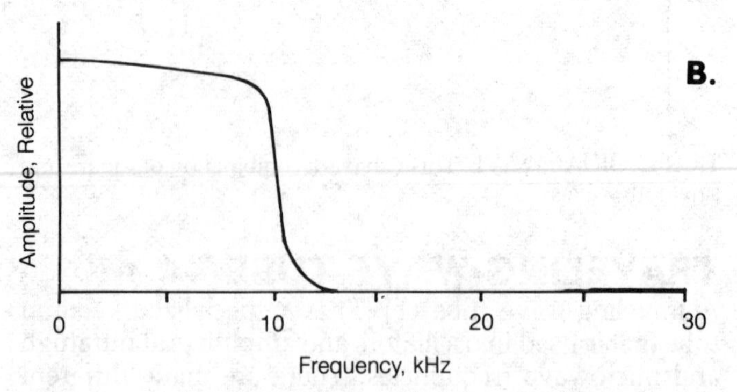

TREBLE RESPONSE: The plot of a good treble response for music (A) and a poor treble response for music (B).

TRIAC

A triac is a silicon bidirectional triode thyristor that behaves like two inverse parallel connected silicon-controlled rectifiers (SCRs). A single gate electrode turns the device on for current in both directions. It is sometimes called a gated-symmetrical switch. The figure shows the schematic symbol for a triac and its construction as an N-P-N-P-N structure. Because they are bidirectional, triacs are widely used in lower current (under 40 A) alternating current power applications.

The electrodes of the triac are the main terminals 1 and 2 and the gate. When a positive voltage is applied across the main terminals, a positive pulse at the gate will trigger the device into forward conduction; when a negative voltage is applied, a negative gate pulse will trigger the device into reverse conduction. Once a triac is turned on, the gate loses control and the device remains in the on state until the voltage across the terminals is reduced to the sustaining value or the ac passes through the zero value.

Some triacs can handle up to 1500 V, and others are able to handle currents up to about 40 A. The restriction on current is imposed because excessive current will

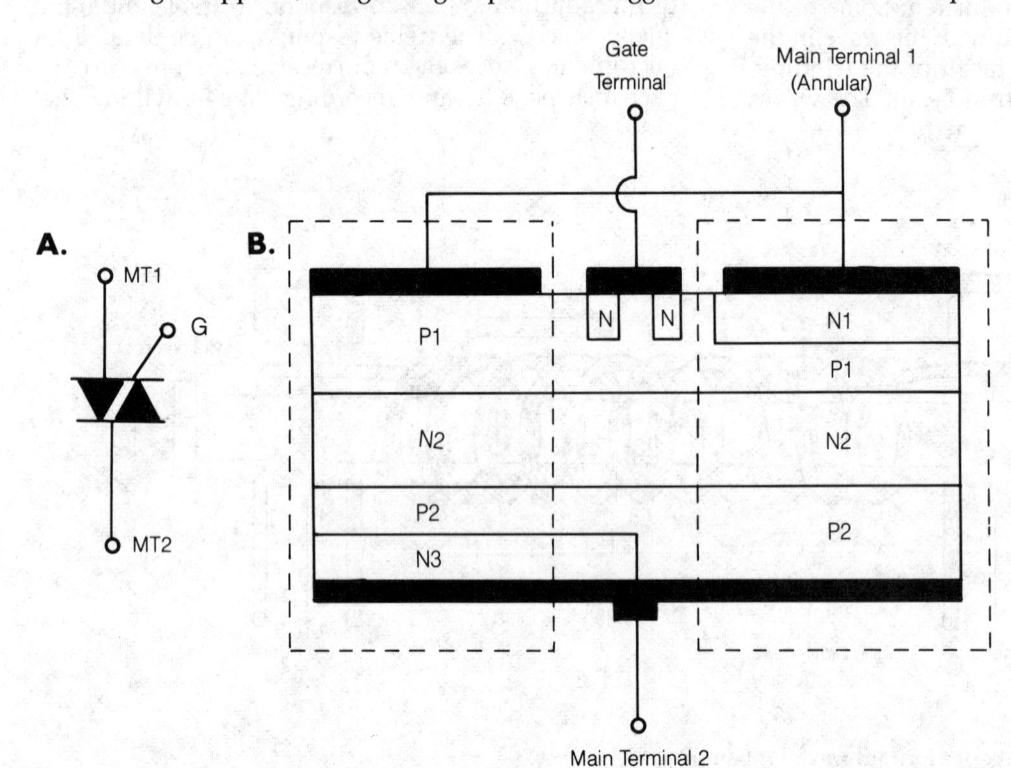

TRIAC: Schematic symbol for a silicon bidirectional triode thyristor (triac) (A) and the general arrangement of a gated triac (B).

cause an internal breakdown of the single chip carrying currents in opposite directions.

Triacs are used in consumer products including power switches, lamp dimmers, appliance motor speed controls, microwave ovens, and space heaters. In addition, they are widely used as power switches in industrial and commercial solid-state relays and alternating-current (ac) output modules.

The packaging of triacs follows the same basic guide lines as the packaging of other power semiconductor devices and thyristors. TO-226 (formerly TO-92) plastic packages are used to package triacs rated for less than 1 A, TO-220 plastic cases are used to package triacs rated to 25 A, and different metal cases are used for triacs rated 15 to 40 A. *See* SEMICONDUCTOR PACKAGE.

A diac is a triac without a gate terminal. The device is switched on by raising the applied voltage to the breakover voltage.

TRIANGULAR WAVE

A triangular wave is a form of sawtooth wave with rise and decay times that are identical. The triangle wave is therefore symmetrical with respect to its amplitude peak. *See* SAWTOOTH WAVE.

TRIBOELECTRIC EFFECT

When certain materials are rubbed together, an electrostatic charge develops on each object. One object becomes positive because electrons are transferred to the other object, which becomes negative. This is called the triboelectric effect.

The most common example of the triboelectric effect is the transfer of charge from a carpeted floor to the shoes of a person walking across the floor. This action charges the person. If that person then touches a metal doorknob or other metal object, he or she will be discharged, often with a painful electrostatic discharge in the form of an electric arc or spark. *See also* ELECTROSTATIC DISCHARGE, STATIC ELECTRICITY, TRIBOELECTRIC SERIES.

TRIBOELECTRIC SERIES

A triboelectric series is a list of materials in order of their tendency to become positively or negatively charged as a result of triboelectric effect (*see* TRIBOELECTRIC EFFECT).

The table in the illustration is an example of a triboelectric series. An object made of a given material becomes negatively charged when rubbed against an object made from some material higher on the list. The farther apart two substances appear on the list, the greater the tendency for transfer of charge.

Consider the following two examples, based on the table. When silk is rubbed against glass, the glass attains a positive charge and the silk attains a negative charge. The friction of rubber-soled shoes against a wool carpet results in a negative charge on the person wearing the shoes. *See also* ELECTROSTATIC DISCHARGE, STATIC ELECTRICITY.

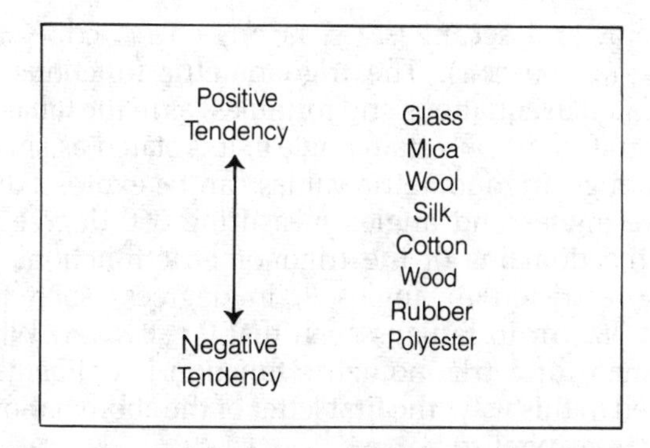

TRIBOELECTRIC SERIES: Materials near the top of the list tend to acquire positive charge; those near the bottom of the list tend to acquire negative charge.

TRIFILAR WINDING

A set of three coil windings oriented in the same way with the same number of turns on a common core is called a trifilar winding. Trifilar windings are used in transformers at audio or radio frequencies.

Trifilar windings on a toroidal core can be connected to obtain different impedance-transfer ratios. This is useful in radio-frequency impedance-matching applications. *See also* TRANSFORMER.

TRIGGER

In electronics, a trigger refers to the initiation of a circuit by the action of another (trigger) circuit. A trigger signal is typically a pulse. A trigger pulse can be applied to a silicon-controlled rectifier, relay, flip-flop, or other circuit. The device or circuit triggered then functions for a certain length of time under its own control. *See also* TRIGGERING.

TRIGGERING

Triggering is a means of synchronizing the sweep rate of an oscilloscope with the frequency of the applied signal. Most oscilloscopes have triggering capability. Triggering keeps the display steady on the screen so that the waveform can be easily analyzed. Some oscilloscopes can be triggered with a signal that differs from the incoming signal, or from the 60 Hz power-line voltage.

When a triggered oscilloscope is used, the ratio of the sweep frequency to the signal frequency can be varied by adjusting the sweep rate. In this way, a single cycle can be viewed for "closeup" analysis, or many cycles can be viewed for determination of peak voltage, frequency, and general wave shape. *See also* OSCILLOSCOPE.

TRIGONOMETRIC FUNCTION

A trigonometric function, also called a circular function, is a mathematical function that relates angular measure to the set of real numbers. There are six trigonometric functions. The three most commonly used are the cosine, sine, and tangent. Less commonly used are the cosecant, cotangent, and secant (*see* COSECANT, COSINE, COTANGENT, SECANT, SINE, TANGENT). The trigonometric functions are abbreviated in equations and formulas, as in the table. To specify the sine of a certain angle θ, it is stated as sin θ.

Although trigonometric values can be expressed for negative angles and angles measuring 360 degrees or more, the domains of the trigonometric functions are usually restricted to angles θ, in degrees, such that $0 \leq \theta < 360$, or in radians such that $0 \leq \theta < 2\pi$. When the domain of a trigonometric function is deliberately restricted in this way, the first letter of the abbreviation is sometimes capitalized.

All of the trigonometric functions can be expressed in either of two ways: the unit-circle model and the triangle model.

The Unit-Circle Model. Mathematicians usually express the trigonometric functions in terms of values on a circle in the Cartesian plane. The circle has a radius of 1 unit and is centered at the origin (see illustration at A). The equation of the circle is:

$$x^2 + y^2 = 1$$

If θ represents an angle, measured counterclockwise from the positive x axis, then the trigonometric-function values of this angle can be determined from the values of *x* and *y* at the point (*x*,*y*) on the circle that corresponds to the angle θ, as shown. The functions are:

$$\cos \theta = x$$
$$\sin \theta = y$$
$$\tan \theta = y/x$$
$$\csc \theta = 1/y$$
$$\cot \theta = x/y$$
$$\sec \theta = 1/x$$

A special category of trigonometric functions exists, based on the unit hyperbola with the equation:

$$x^2 - y^2 = 1$$

These are known as hyperbolic trigonometric functions.

The Triangle Model. A right triangle (B) can be used to express trigonometric-function values. The angle θ corresponds to one of the angles not measuring 90 degrees. If the sides of the triangle have lengths *a*, *b*, and *c*, as shown in the figure, and the angle θ is opposite the side of length *b*, then:

$$\cos \theta = a/c$$
$$\sin \theta = b/c$$
$$\tan \theta = b/a$$
$$\csc \theta = c/b$$
$$\cot \theta = a/b$$
$$\sec \theta = c/a$$

These formulas are valid no matter how large or small the right triangle is, in terms of the magnitudes of the numbers *a*, *b*, and *c*.

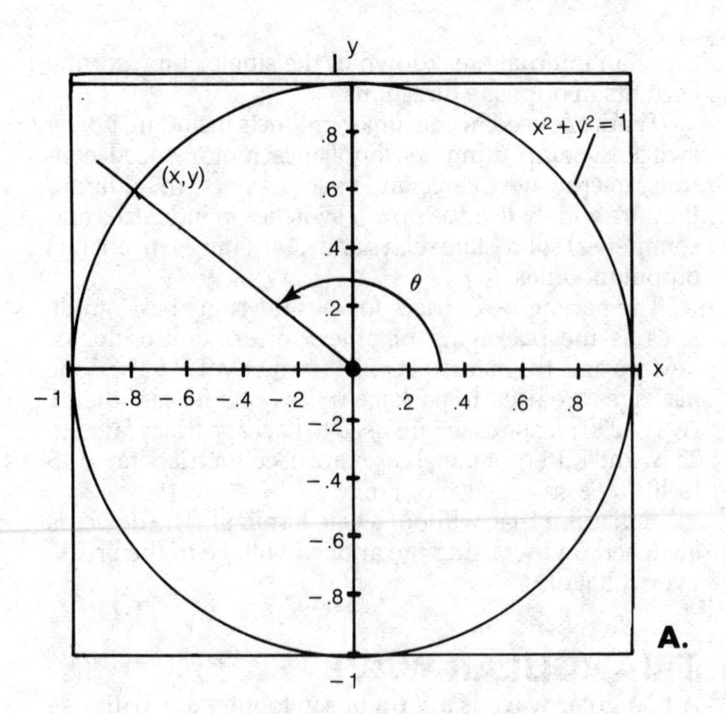

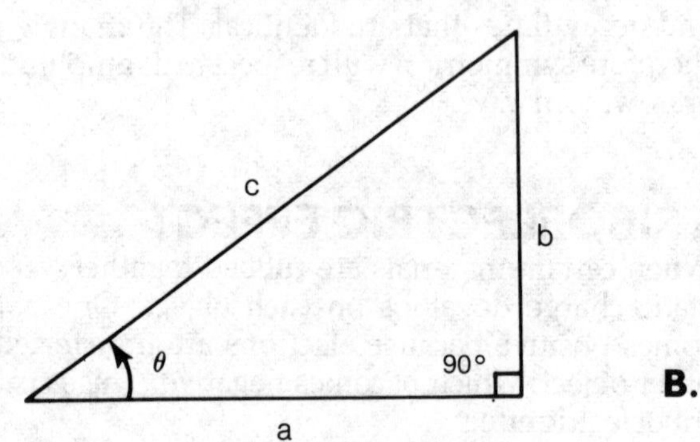

TRIGONOMETRIC FUNCTION: The trigonometric functions are defined according to a unit circle definition (A) and right-triangle definition (B).

The triangle model is normally used only for angles θ measuring between 0 and 90 degrees.

Inverse Trigonometric Functions. When the domain of a trigonometric function is properly restricted, an inverse function exists that maps a real number into a unique angular value. This function is called an inverse trigonometric function. Each of the six trigonometric functions has an inverse. The inverse of a given function can be denoted by the prefix *arc−* or by an exponent −1. The inverse trigonometric functions are defined only when the angles are restricted to certain values, some of which are negative:

If $\cos \theta = z$, then $\arccos z = \theta$ for $0 \leq \theta \leq 180$
If $\sin \theta = z$, then $\arcsin z = \theta$ for $-90 \leq \theta \leq 90$
If $\tan \theta = z$, then $\arctan z = \theta$ for $-90 \leq \theta \leq 90$
If $\csc \theta = z$, then $\text{arccsc } z = \theta$ for $-90 \leq \theta < 0$ and $0 < \theta \leq 90$
If $\cot \theta = z$, then $\text{arccot } z = \theta$ for $0 < \theta < 180$
If $\sec \theta = z$, then $\text{arcsec } z = \theta$ for $0 \leq \theta < 90$ and $90 < \theta \leq 180$

TRIGONOMETRIC FUNCTION:
Abbreviations for common trigonometric functions.

cosine: cos	hyperbolic cosine: cosh
sine: sin	hyperbolic sine: sinh
tangent: tan	hyperbolic tangent: tanh
cosecant: csc	hyperbolic cosecant: csch
cotangent: cot	hyperbolic cotangent: coth
secant: sec	hyperbolic secant: sech

Significance and Uses of Trigonometric Functions. Trigonometric functions, especially the sine and cosine functions, are observable in nature. These functions are representations of circular motion which occurs frequently in the physical universe.

The most common example of a trigonometric function in electricity and electronics is the alternating-current wave. A wave containing energy at only one frequency can be perfectly represented by a sine or cosine function. The wave cycle is divided into 360 degrees of phase, beginning at the point where the amplitude is zero and increasing in a positive direction (*see* PHASE ANGLE, SINE WAVE). Trigonometric functions arise in certain antenna and feed-line calculations, frequency and phase modulation, and many other instances. Trigonometric functions are important in polar and spherical coordinate systems. *See also* COORDINATE SYSTEM, FUNCTION, INVERSE FUNCTION, POLAR COORDINATES, SPHERICAL COORDINATES, TRIGONOMETRIC IDENTITY.

TRIGONOMETRIC IDENTITY

In calculations involving trigonometric functions, it may be necessary to simplify an expression or to substitute one expression for another. Trigonometric identities serve this purpose. A trigonometric identity is an equation that holds for all possible angles, or for angles within a specified range.

Table 1 lists common trigonometric identities for circular functions. Table 2 lists some identities that hold for the inverse of circular functions. *See also* TRIGONOMETRIC FUNCTION.

TRIGONOMETRIC IDENTITY Table 1:
Some common trigonometric identities for circular functions.

$$\csc \theta = 1 / \sin \theta$$
$$\sec \theta = 1 / \cos \theta$$
$$\tan \theta = \sin \theta / \cos \theta$$
$$\cot \theta = \cos \theta / \sin \theta$$
$$\sin^2 \theta + \cos^2 \theta = 1$$
$$\sec^2 \theta - \tan^2 \theta = 1$$
$$\csc^2 \theta - \cot^2 \theta = 1$$
$$\sin (\theta + \phi) = \sin \theta \cos \phi + \cos \theta \sin \phi$$
$$\sin (\theta - \phi) = \sin \theta \cos \phi - \cos \theta \sin \phi$$
$$\cos (\theta + \phi) = \cos \theta \cos \phi - \sin \theta \sin \phi$$
$$\cos (\theta - \phi) = \cos \theta \cos \phi + \sin \theta \sin \phi$$
$$\tan (\theta + \phi) = (\tan \theta + \tan \phi) / (1 - \tan \theta \tan \phi)$$
$$\tan (\theta - \phi) = (\tan \theta - \tan \phi) / (1 + \tan \theta \tan \phi)$$
$$\sin -\theta = -\sin \theta$$
$$\cos -\theta = \cos \theta$$
$$\tan -\theta = -\tan \theta$$
$$\csc -\theta = -\csc \theta$$
$$\sec -\theta = \sec \theta$$
$$\cot -\theta = -\cot \theta$$
$$\sin 2\theta = 2 \sin \theta \cos \theta$$
$$\cos 2\theta = \cos^2 \theta - \sin^2 \theta$$
$$\tan 2\theta = 2 \tan \theta / (1 - \tan^2 \theta)$$
$$\sin \theta + \sin \phi = 2 \sin (\theta/2 + \phi/2) \cos (\theta/2 - \phi/2)$$
$$\sin \theta - \sin \phi = 2 \cos (\theta/2 + \phi/2) \sin (\theta/2 - \phi/2)$$
$$\cos \theta + \cos \phi = 2 \cos (\theta/2 + \phi/2) \cos (\theta/2 - \phi/2)$$
$$\cos \theta - \cos \phi = 2 \sin (\theta/2 + \phi/2) \sin (\theta/2 - \phi/2)$$

TRIGONOMETRIC IDENTITY Table 2:
Some common trigonometric identities for inverses of circular functions.

$$\arcsin z + \arccos z = 90°$$
$$\arctan z + \text{arccot } z = 90°$$
$$\text{arcsec } z + \text{arccsc } z = 90°$$
$$\arcsin z = \text{arccsc } (1/z)$$
$$\arccos z = \text{arcsec } (1/z)$$
$$\arctan z = \text{arccot } (1/z)$$
$$\arcsin (-z) = -\arcsin z$$
$$\arccos (-z) = 180° - \arccos z$$
$$\arctan (-z) = -\arctan z$$
$$\text{arccot } (-z) = 180° - \text{arccot } z$$
$$\text{arccsc } (-z) = -\text{arccsc } z$$
$$\text{arcsec } (-z) = 180° - \text{arcsec } z$$

TRIGONOMETRY

Trigonometry is a branch of mathematics concerned with angles and straight lines in a plane or in space. *See* COSECANT, COSINE, COTANGENT, SECANT, SINE, TANGENT, TRIGONOMETRIC FUNCTION, TRIGONOMETRIC IDENTITY.

TRIMMER CAPACITOR

See CAPACITOR.

TRIODE TUBE

A triode tube is a vacuum tube with three electrodes. Early in the twentieth century, engineer and inventor Lee deForest discovered that the conductance of a diode tube could be controlled by putting an electrode, called the grid, between the cathode and the plate. A small signal applied to the grid caused large changes in the plate current, resulting in amplification; if some of the output was coupled back to the input and shifted 180 degrees in phase, oscillation occurred.

The control grid in a triode is usually biased at a voltage negative with respect to the cathode. The bias can be supplied by a resistance-capacitance network (*see* SELF BIAS), or by a separate power supply. The illustration is the schematic symbol for a triode tube showing a typical biasing arrangement.

As the instantaneous grid voltage becomes more negative, the instantaneous plate current decreases. As the instantaneous grid voltage becomes less negative, the instantaneous plate current rises. The change in the plate current, passing through a suitable load, results in an output-signal voltage that is many times larger than the input-signal voltage.

Some tubes have extra grids added to enhance the gain and improve the stability. A tube with two grids is called a tetrode because it has four elements altogether. A tube with three grids is called a pentode.

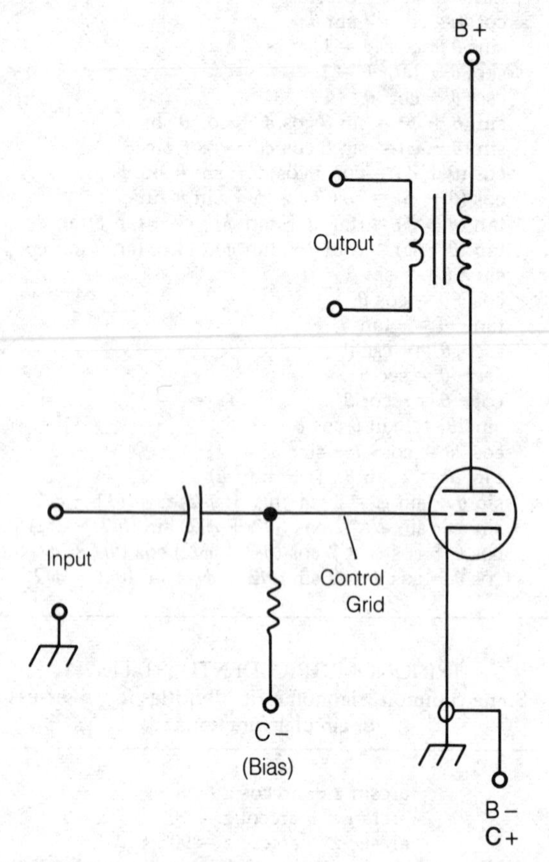

TRIODE TUBE: Typical biasing arrangement for a triode tube.

TRIPLER

A tripler is a radio-frequency circuit that produces an output signal at a frequency three times that of the input signal. There are different methods to obtain frequency multiplication by 3; two common methods are shown in the illustration.

The circuit at A uses a push-pull configuration. This causes reinforcement of the odd harmonics of the input signal, and suppresses the even harmonics. The output circuit is tuned to the third harmonic of the fundamental input signal. This circuit, if tuned properly, provides some amplification as well as frequency multiplication.

The circuit at B makes use of the nonlinear characteristics of semiconductor diodes to obtain harmonic energy. The output tank is tuned to the third harmonic of the fundamental input signal. This circuit has a small amount of insertion loss since only passive components are used.

Triplers are commonly used with radio transmitters to obtain operation at a higher frequency. Two triplers can be connected in cascade to obtain multiplication by 9. *See also* FREQUENCY MULTIPLIER, HARMONIC, PUSH-PULL CONFIGURATION.

TROPOSPHERE

The troposphere is the lower layer of the earth's atmosphere, extending to about 60,000 feet at the equator and 30,000 feet at the poles. The temperature generally decreases with altitude in this region, clouds form, and there is active convection. *See also* TROPOSPHERIC PROPAGATION, TROPOSPHERIC-SCATTER PROPAGATION.

TROPOSPHERIC PROPAGATION

The lower part of the earth's atmosphere has an effect on electromagnetic-field propagation at certain frequencies. At wavelengths shorter than about 15 meters (or a frequency of 20 MHz), refraction and reflection take place within and between air masses of different density. The air also produces some scattering of electromagnetic energy at wavelengths shorter than 3 meters (or a frequency of 100 MHz). All of these effects, known as tropospheric scatter, affect tropospheric propagation.

Tropospheric propagation can result in communication over distances of hundreds of miles. The most common type of tropospheric propagation occurs when radio waves are refracted in the lower atmosphere. This takes place most of the time, but is most noticeable in the vicinity of a weather front, where warm, relatively light air lies above cool, denser air. The cooler air has a higher index of refraction than the warm air, causing the electro-

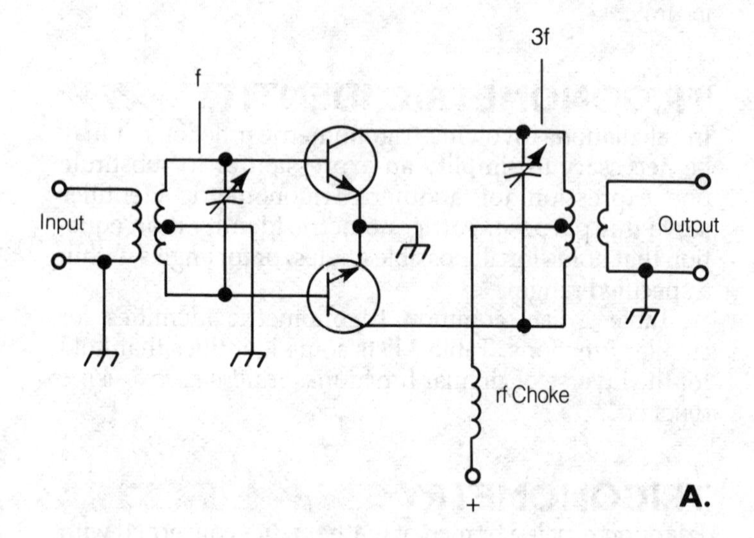

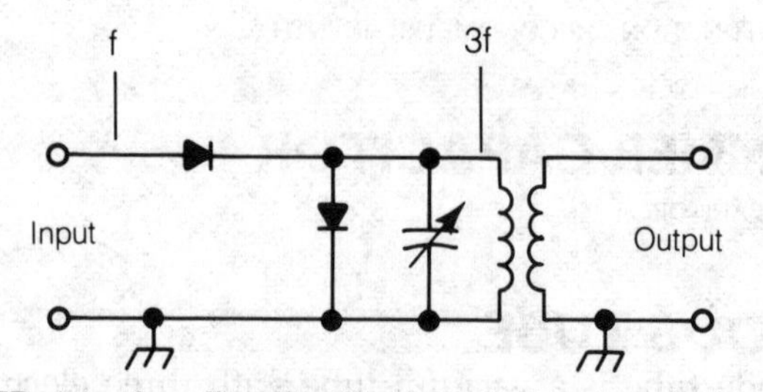

TRIPLER: A push-pull tripler (A) and a diode tripler (B).

magnetic fields to be bent downward at a considerable distance from the transmitter (see illustration at A). This phenomenon is called tropospheric bending.

If the boundary between a cold air mass and a warm air mass is extremely well-defined, and an electromagnetic field strikes the boundary at a near-grazing angle of incidence from within the cold air mass, total internal reflection occurs. This is known as tropospheric reflection, as at B. If a cold air mass is sandwiched between warm air masses, the energy may be propagated for long distances because of repeated total internal reflection. This is called the tropospheric ducting or duct effect (*see* DUCT EFFECT, TOTAL INTERNAL REFLECTION).

Tropospheric propagation is often responsible for anomalies in the reception of television and frequency-modulation broadcast signals. A television station hundreds of miles away may suddenly appear on a channel that is normally dead. Unfamiliar stations may be received on a hi-fi radio tuner. Sometimes two or more distant stations come in on a single channel, causing interference. Sporadic-E propagation has similar effects, and sometimes it is difficult to tell whether tropospheric or sporadic-E propagation is responsible for long-distance reception or communication (*see* SPORADIC-E PROPAGATION). Usually, sporadic-E events result in propagation over longer distances than tropospheric effects.

At very high, ultrahigh, and microwave frequencies, the atmosphere scatters electromagnetic energy in much the same way as it scatters visible light. This can result in propagation over the horizon. It is called tropospheric scatter or troposcatter. *See also* PROPAGATION CHARACTERISTICS, TROPOSPHERIC-SCATTER PROPAGATION.

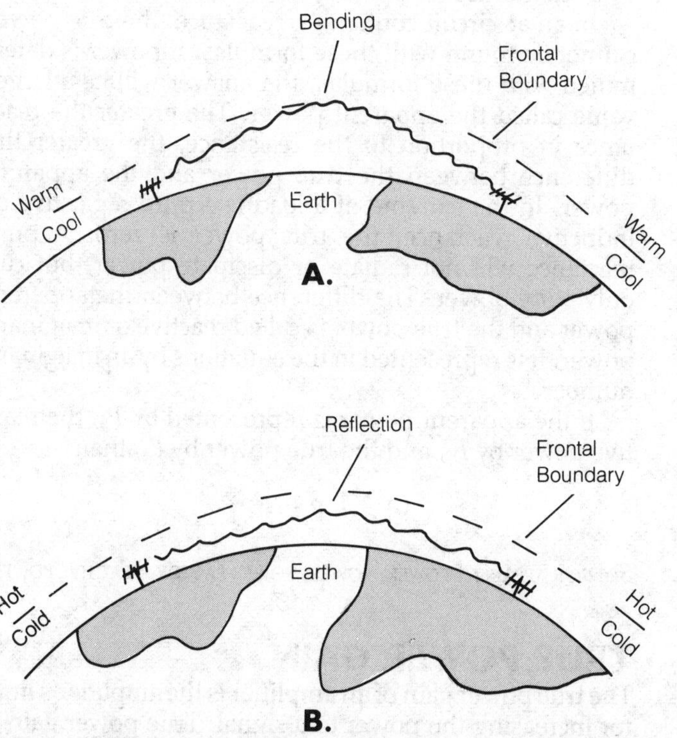

TROPOSPHERIC PROPAGATION: Tropospheric bending (A) and tropospheric reflection (B).

TROPOSPHERIC-SCATTER PROPAGATION

At frequencies above about 150 MHz, the atmosphere has a scattering effect on electromagnetic fields. The scattering allows over-the-horizon communication at very high, ultrahigh, and microwave frequencies. This mode of propagation is called tropospheric scatter, or troposcatter for short.

Dust and clouds in the air increase the scattering effect, but some troposcatter occurs regardless of the weather. Troposcatter takes place mostly at low altitudes, but some effects occur at altitudes up to about 10 miles. Troposcatter propagation can provide reliable communication over distances of several hundred miles when the appropriate equipment is used.

Communication by troposcatter requires the use of high-gain antennas. Fortunately, the size of a high-gain antenna is manageable at ultrahigh and microwave frequencies. The transmitting and receiving antennas are aimed at a common region, ideally located midway between the two stations and at as low an altitude as possible (see illustration). The maximum obtainable range depends not only on the gain of the antennas used for transmitting and receiving, but on their height above the ground: the higher the antennas, the greater the range. The terrain also affects the range; flat terrain is best, while mountains seriously impede troposcatter propagation.

To obtain troposcatter communication over long distances, sensitive receivers and high-power transmitters must be used, because the path loss is high. *See also* DUCT EFFECT, PROPAGATION CHARACTERISTICS, TROPOSPHERIC PROPAGATION.

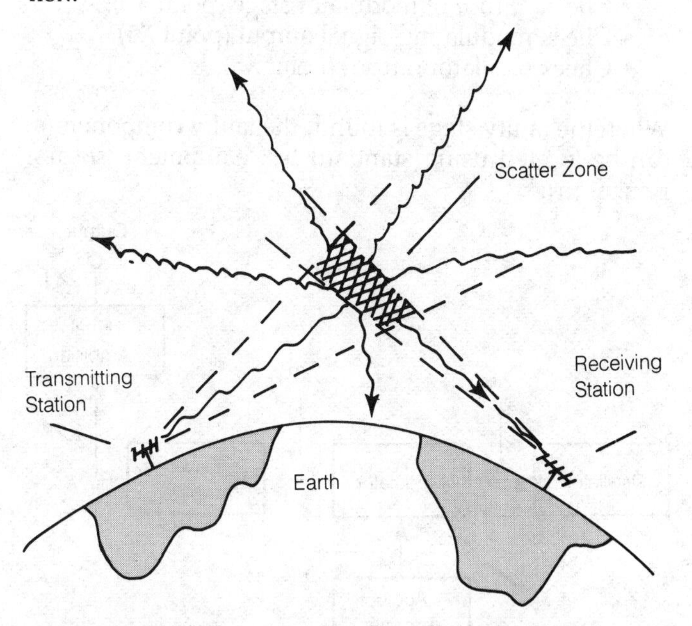

TROPOSPHERIC-SCATTER PROPAGATION: Random scattering allows over-the-horizon communication.

TROUBLESHOOTING

Troubleshooting is the process of finding the faulty component or components in a malfunctioning unit of equipment. In electronics, troubleshooting saves time and expense in the repair of equipment. Faulty components can be located without extensive removal and replacement when the proper troubleshooting procedures are followed.

The Basic Concept. Most complicated electronic products are provided with service manuals containing instructions on how to troubleshoot for specific problems. The most effective troubleshooting instructions are given in the form of flowcharts for potential malfunctions (*see* FLOWCHART). Troubleshooting procedures are also written in tabular form.

Digital computers have built-in self diagnostics to find certain kinds of problems. The two most common methods are the diagnostic program and the service read-only memory (ROM). These methods are used in conjunction with a comprehensive service manual. Troubleshooting of faulty computer programs and equipment prototypes is called debugging (*see* DEBUGGING).

In any troubleshooting situation, the most effective procedure is to work from output to input, from general systems to specific circuits and components, or from the most obvious to the least obvious potential problems. For example, if a radio transmitter does not work, the troubleshooting procedure should be as follows (assuming a transmitter design as shown in the illustration):

- Check to be sure equipment is operated within the recommended ambient temperature limits from the proper power source.
- Check power supply.
- Check output at antenna terminals (point X1).
- Check output of driver stage (point X2).
- Check output of modulator stage (point X3).
- Check modulating-signal output (point X4).
- Check oscillator output (point X5).

When the faulty stage is found, the faulty component(s) can be located using standard test equipment (*see* TEST INSTRUMENT).

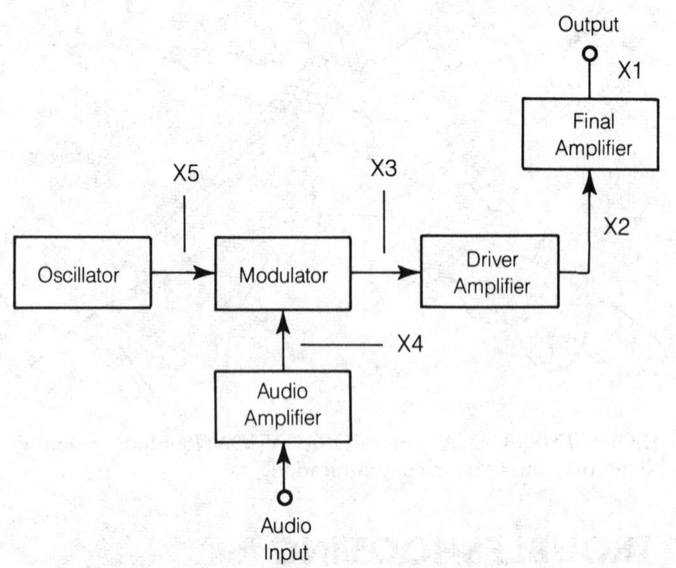

TROUBLESHOOTING: Sequence of points to be checked in troubleshooting follow the sequence X1 through X5.

Intermittent Problems. Electronic equipment does not always fail catastrophically; sometimes the problem occurs intermittently. To find the source of an intermittent problem, the equipment must be evaluated while the malfunction can be observed. The process can be time-consuming, tedious, and frustrating.

Intermittents usually occur either because of a faulty connection or as a result of temperature sensitivity. Some intermittents, however, occur for no apparent reason. A bad connection can often be located visually, or by moving the leads on a suspected component. Thermal intermittents require manual temperature control. Small heaters alternated with an aerosol coolant can subject individual components to thermal shock until the source of a thermal intermittent is found.

T-R SWITCH

See TRANSMIT-RECEIVE SWITCH.

TRUE POWER

In a direct-current circuit, or in an alternating-current circuit containing resistance but not reactance, true power can be determined from a simple formula. Given a root-mean-square (rms) voltage E across a load of resistance R, and an rms current I through the load, the dissipated or radiated power P_t is given by the formulas:

$$P_t = EI = E^2/R = I^2R$$

In this case, the value P_t is called the true power because it represents power that is dissipated or radiated. True power is often called real power because it is represented by a real number.

In an ac circuit containing reactance, the true power cannot be found with these formulas. If power is determined with these formulas, the answer will be a larger value called the apparent power. The greater the reactance in proportion to the resistance, the greater the difference between the true power and the apparent power. In the extreme, if a load is a pure capacitive or inductive reactance, the true power is zero. A pure reactance will not radiate or dissipate power, but can only store power. The difference between the apparent power and the true power is called reactive or imaginary power. It is represented in the equations by an imaginary number.

If the apparent power is represented by P_a, the reactive power by P_r, and the true power by P_t, then:

$$P_t^2 + P_r^2 = P_a^2$$

See also APPARENT POWER, POWER, POWER FACTOR, REACTIVE POWER.

TRUE POWER GAIN

The true power gain of an amplifier is the amplifier's limit for increasing the power of a signal. True power gain is generally given in decibels.

Consider an amplifier with a true power input of P_{ti}

watts. If the amplifier input circuit is replaced by a dissipating or radiating load of the same impedance, the load will consume P_{ti} watts. Let the true power output of the amplifier, in watts, be denoted P_{to}. Then the true power gain, G, in decibels, is:

$$G = 10 \log_{10} \left(\frac{P_{to}}{P_{ti}} \right)$$

The true power gain of an amplifier is always maximized when the input and output impedances are perfectly matched. This means there must be no reactance, and the resistance must be of the proper value. Any impedance mismatch results in a reduction in the true power output, and therefore in the true power gain. *See also* DECIBEL, GAIN, TRUE POWER.

TRUNK LINE

A trunk line is a transmission line that interconnects two central offices in a telephone system. Subscribers communicate between their respective central offices, and through a trunk line for long distances (*see* CENTRAL OFFICE SWITCHING SYSTEM).

A trunk line carries many conversations at once making it necessary to multiplex signals. The most common multiplex method is frequency-division multiplex with a carrier-current system (*see* FREQUENCY-DIVISION MULTIPLEX). In some systems, time-division multiplex may be used (*see* TIME-DIVISION MULTIPLEX).

A trunk line does not have to be a wire or cable. Microwave links and communications satellites have become alternate telephone trunk lines. *See also* TELEPHONE SYSTEM.

TRUTH TABLE

A truth table is an expression of a logical, or Boolean, function. Truth tables are useful for showing logical equivalences and for analyzing complicated logical expressions. *See also* BOOLEAN ALGEBRA, LOGIC.

TTL

See LOGIC.

TUNABLE CAVITY RESONATOR

See CAVITY RESONATOR.

TUNED CIRCUIT

Any circuit that displays resonance at one or more frequencies is called a tuned circuit. A tuned circuit can be either fixed or adjustable.

Tuned circuits are available in many different forms including inductor-capacitor combinations, sections of transmission lines, or metal enclosures. The most familiar tuned circuit is a combination of a discrete inductor and capacitor. A resonant section of transmission line is called a stub (*see* STUB). A resonant metal enclosure is called a cavity resonator (*see* CAVITY RESONATOR). An antenna can be a form of tuned circuit.

There are four types of tuned circuit. Some act as bandpass or band-rejection filters, and are classified as either series-resonant or parallel-resonant. Other tuned circuits may exhibit highpass or lowpass properties (*see* BANDPASS FILTER, BAND-REJECTION FILTER, HIGHPASS FILTER, LOWPASS FILTER).

A series-resonant tuned circuit acts as a short circuit at the resonant frequency or frequencies. A parallel-resonant tuned circuit acts as an open circuit at resonance. At nonresonant frequencies, a series- or parallel-resonant tuned circuit behaves as a pure reactance, whose value depends on the frequency. A parallel-resonant tuned circuit is also called a tank circuit (*see* PARALLEL RESONANCE, SERIES RESONANCE, TANK CIRCUIT).

TUNED FEEDER

An antenna feed line is usually operated with a standing-wave ratio that is as low as possible. This calls for operating the antenna at resonance, and matching the characteristic impedance of the feeder to the radiation resistance of the antenna (*see* CHARACTERISTIC IMPEDANCE, RADIATION RESISTANCE, RESONANCE, STANDING-WAVE RATIO). This feed system is called untuned because the length of the feed line does not affect the antenna-system impedance at the transmitter/receiver.

An antenna system with a low-loss feed line can be operated with a high standing-wave ratio. This can be done with an open-wire line (*see* OPEN-WIRE LINE). The electrical feed-line length is important, and therefore the line is said to be tuned. This feed system was popular in the early days of radio, when open-wire line was in universal use.

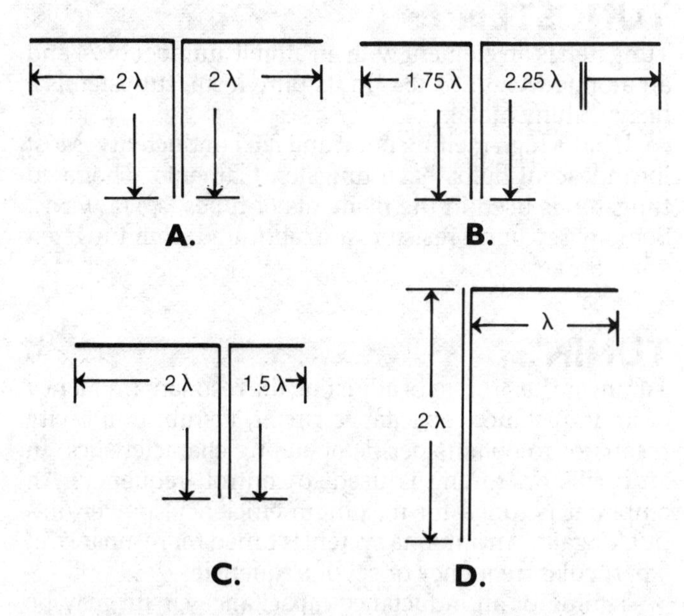

TUNED FEEDER: Antennas with tuned feeders include a symmetrical antenna (A), off-center tuned-feed antennas (B and C), and a tuned end-feed antenna (D).

A tuned feeder is trimmed to a length that makes the input impedance a pure resistance. This requires that the system have the following properties:

- Each side of the whole system must measure an integral multiple of ¼ electrical wavelength.
- The lengths of the two sides of the system must either be identical, or differ by an integral multiple of ½ electrical wavelength.

Ideally, both halves of the system should measure the same.

The illustration shows four examples of antennas with tuned feeders. The system at A is symmetrical, with each element measuring 2 wavelengths. The system at B incorporates off-center feed, with one element measuring 1¾ wavelength and the other element measuring 2¼ wavelength. The system at C uses off-center feed, with elements measuring 1½ and 2 wavelengths. The antenna system at D uses end feed, in which the feed line measures 2 wavelengths and the radiating element measures 1 wavelength.

The impedance at the input of a tuned feed system can range from a few ohms to several hundred ohms in practice. It is, therefore, usually necessary to use an impedance-matching transformer between a transmitter and a tuned feeder system (*see* IMPEDANCE TRANSFORMER).

If a transmatch capable of tuning out reactance is used at the input of a tuned-feeder system, the feed line length can be varied at will. This simplifies the installation procedure by eliminating the need for trimming the feed line. The two halves of the antenna must, however, still differ in length by some integral multiple of ½ wavelength. A center-fed radiator, fed with open-wire line and tuned with a transmatch, can be operated over a wide range of frequencies.

TUNGSTEN

Tungsten is an element with an atomic number of 74 and an atomic weight of 184. In its pure form, tungsten is a heavy, shiny metal.

It has a high melting point and fair conductivity. Most incandescent bulbs have tungsten filaments. Thoriated tungsten is used in the filaments of tubes (*see* FILAMENT). Some wirewound resistors are fabricated from tungsten wire. *See* RESISTOR.

TUNING

Tuning is the process of adjusting the resonant frequency of an inductance-capacitance circuit, a stub, or a cavity resonator to obtain specific operating characteristics. In an oscillator, tuning is used to control frequency. An amplifier is tuned for maximum efficiency, power output, or gain. An antenna system is tuned for resonance at a particular frequency or set of frequencies.

Tuning of an inductance-capacitance circuit may be accomplished with variable inductors or capacitors. Inductors can be tapped in various places so different parts are switched in; some inductors have continuously variable taps. Inductors with ferromagnetic cores can be permeability-tuned (*see* PERMEABILITY TUNING). If the capacitance in a tuned circuit is less than 0.001 μF (1000 pF), air-variable capacitors can obtain continuous tuning (*see* AIR-VARIABLE CAPACITOR).

Tuning of a stub involves the use of a movable shorting bar. The longer the stub, the lower the resonant frequency or frequencies (*see* STUB). Tuning of a cavity resonator is accomplished by a movable piston inside the enclosure (*see* CAVITY RESONATOR). An antenna is tuned either by cutting the radiator to a specified length, or by adding reactances in series. Antenna tuning is called loading (*see* CAPACITIVE LOADING, INDUCTIVE LOADING).

TUNNEL DIODE

The tunnel diode is a two-terminal, solid-state negative-resistance device also known as the Esaki diode. It can function as an amplifier, an oscillator, or a switch. Because of its very fast response to inputs, it is almost exclusively used as a microwave component. It offers low-noise amplification at microwave frequencies.

A tunnel diode is a P-N junction with very heavily doped P and N regions. This produces a very abrupt junction with a very thin junction barrier. The diode exhibits a very important negative resistance characteristic. As a forward voltage is increased from zero, majority carriers tunnel through the very thin barrier and very high values of electric field are created with small voltages because the barrier is so thin. As the majority carriers cross the junction, the effective width of the junction becomes greater. With further increases in voltage, the tunneling current ceases and the normal forward current due to minority carrier injection builds up.

Between the tunneling current and the minority-carrier current there is a negative resistance region. The current decreases from a peak to a valley before rising again as voltage is increased. This negative resistance characteristic provides very fast switching.

Gallium arsenide tunnel diodes have been used in the first radio-frequency amplifier stages of microwave receivers for microwave relay links, Doppler navigation radar, and weather radar. Because of their very low power capability, they are used only as receiver local oscillators and not as power amplifiers.

Two other solid-state, negative-resistance microwave oscillators in use today are the Gunn diode or transferred electron oscillator (TEO) and the IMPATT or impact-avalanche transit time diode. *See* GUNN DIODE, IMPATT DIODE.

TURNS RATIO

In a transformer, the turns ratio is defined as the number of turns in the primary winding divided by the number of turns in the secondary winding. In a step-up transformer, the turns ratio is between 0 and 1; in a step-down transformer, the turns ratio is greater than 1.

The primary-to-secondary turns ratio of a transformer is usually denoted by two integers, separated by a colon to indicate the ratio. The ratio is reduced to lowest terms.

A transformer with 500 turns in the primary and 200 turns in the secondary, for example, has a turns ratio of 5:2. *See also* TRANSFORMER.

TURNTABLE

A turntable is a mechanical component for the reproduction of disk recording consisting of a motor, a rotating, round table, a spindle (perhaps equipped with a disk changer), a speed control, and a tone arm with a stylus.

The standard turntable speed for recordings in the United States is 33⅓ revolutions per minute (rpm). This is 0.555 revolutions per second (rps). Most turntables can also operate at 45 rpm or 0.750 rps; a few can run at 78 rpm or 1.30 RPS. The standard disk size for 33⅓ rpm recording is 12 inches in diameter, and the playing time per side is 15 to 20 minutes. The standard size for a 45 rpm disk is 7 inches in diameter and the playing time per side is 3 to 5 minutes. The 78 rpm disk is the same size as the 33⅓ rpm disk, and the playing time is approximately 5 minutes per side. (The 78 rpm speed is not commonly used in disk recording, although many older disks were recorded at that speed.)

In high-fidelity reproduction, it is important that the turntable maintain a constant speed. This requires a precision motor and drive system. Most turntables are equipped with shock-absorbing devices to minimize the effects of mechanical vibration.

TWEETER

A tweeter is a speaker designed to radiate sound at high audio frequencies. A typical tweeter has a nearly flat response from about 3 kHz to well above the human hearing range. Tweeters are used in high-fidelity systems in conjunction with midrange speakers and woofers to obtain a flat audio response (*see* MIDRANGE, SPEAKER, WOOFER).

The most common tweeters are either dynamic or electrostatic units. Some tweeters can be used as ultrasonic transducers.

TWIN-LEAD

Twin-lead is a prefabricated, parallel-wire, balanced transmission line, molded in polyethylene and extensively used for television reception. Most twin-lead lines have a characteristic impedance of 300 Ω. A few have a characteristic impedance of 75 Ω. Twin-lead is sometimes called ribbon line.

Twin-lead is available in a wide variety of forms. The most common form has conductors of stranded copper, size American Wire Gauge (AWG) No. 18 or No. 20. One form of twin-lead has a foamed-polyethylene dielectric for reduced loss.

Twin-lead can be used in low-power transmitting applications in place of open-wire line. However, the loss is somewhat greater in twin-lead than in a conventional open-wire line. This is attributable to the higher dielectric loss of polyethylene as compared with air. *See also* BALANCED TRANSMISSION LINE, CABLE, OPEN-WIRE LINE.

TWO-TONE TEST

The distortion in a single-sideband transmitter can be evaluated using a procedure called a two-tone test. The intermodulation distortion of a receiver can be evaluated with a similar procedure, also called a two-tone test.

Testing a Single-Sideband Transmitter. The single-sideband (SSB) transmitter two-tone test consists of the application of a pair of audio tones, both within the audio passband, to the microphone terminals. Any two audio frequencies between about 300 Hz and 3 kHz can be used; the most common are 1 kHz and 2 kHz. The output of the transmitter is analyzed with an oscilloscope. The ideal pattern is shown in the illustration. Any deviation from this pattern indicates distortion, which may result in unnecessarily large signal bandwidth. *See also* SINGLE SIDEBAND.

Testing a Receiver. For testing the intermodulation distortion of a receiver, two unmodulated signals are applied at the antenna terminals. These signals are usually close in frequency; the spacing may vary between a few kilohertz and about 100 kHz.

If the two applied frequencies are $f1$ and $f2$, then the intermodulation products will be most noticeable at frequencies $f3$ and $f4$, so that:

$$f3 = 2f1 - f2$$
$$f4 = 2f2 - f1$$

The lower the levels of signals at frequencies $f3$ and $f4$, the better the intermodulation-distortion performance of the receiver. *See also* INTERMODULATION.

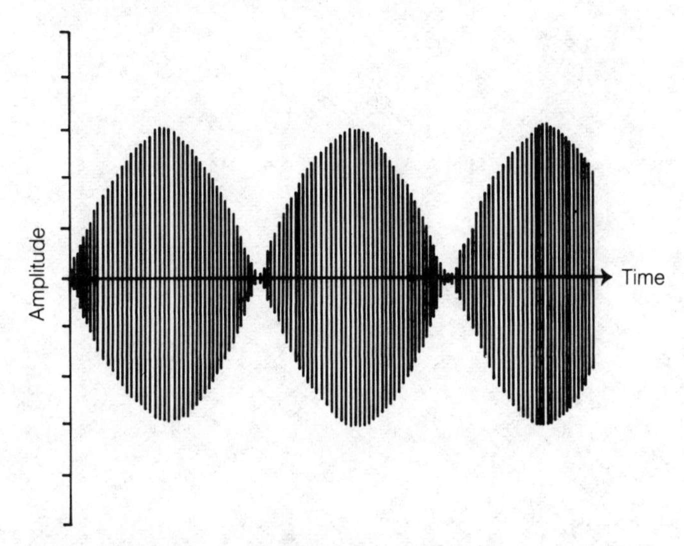

TWO-TONE TEST: An oscilloscope display showing a two-tone pattern for a properly operating single-sideband transmitter.

TWO-TRACK RECORDING

Two-track recording is a common method of recording stereo audio signals on magnetic tape. Some reel-to-reel recorders use two-track recording. Smaller cassette recorders also use this method. In stereo recording one track is used for the left-hand channel and the other is for the right-hand channel. The illustration shows the principle of two-track recording.

Excellent stereo channel separation is obtainable with modern two-track tapes. If desired, the recording time can be doubled by using each track separately in monaural form. Two-track magnetic tapes can be stored almost indefinitely, provided the temperature and humidity are kept within specified limits. *See also* MAGNETIC RECORDING, MAGNETIC TAPE.

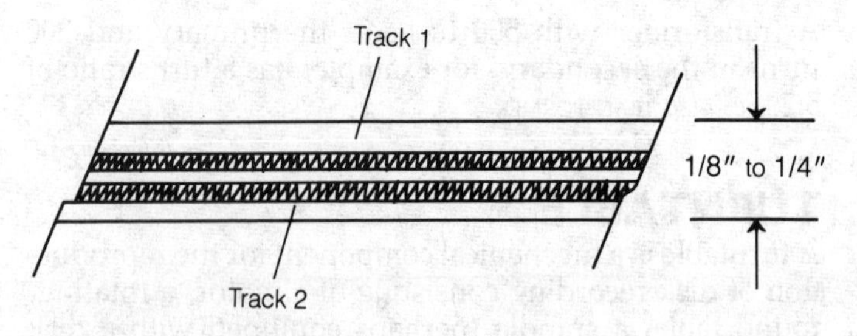

TWO-TRACK RECORDING: Two magnetic paths can be used simultaneously for stereo or separately (in opposite directions) for monaural recording.

UHF

See ULTRAHIGH FREQUENCY.

UHF CONNECTOR

See CONNECTOR.

UHF TELEVISION

See TELEVISION BROADCAST BAND.

UL

See UNDERWRITERS' LABORATORIES, INC.

ULTRAHIGH FREQUENCY

The ultrahigh-frequency (uhf) range of the radio spectrum is the band extending from 300 MHz to 3 GHz. The wavelengths corresponding to these limit frequencies are 1 meter and 10 centimeters. Ultrahigh-frequency waves are sometimes called decimetric waves because the wavelength is on the order of tenths of a meter. Channels and bands at ultrahigh frequencies are allocated by the International Telecommunication Union, headquartered in Geneva, Switzerland.

At uhf, electromagnetic fields are unaffected by the ionosphere of the earth. Signals in this range pass through the ionized layers without being bent or reflected in any way. The uhf waves behave very much like waves at higher frequencies. In the upper portion of the uhf band, waves can be focused by moderate-sized dish antennas for high gain and directivity. Because of their tendency to propagate through the ionosphere, uhf waves are extensively used in satellite communications. Because the frequency of a decimetric signal is so high, wideband modulation is practical.

Signals in the lower portion of the uhf band are sometimes bent or reflected within or between air masses of different temperatures. A certain amount of scattering also takes place in the atmosphere. *See also* DUCT EFFECT, ELECTROMAGNETIC SPECTRUM, FREQUENCY ALLOCATIONS, TROPOSPHERIC PROPAGATION, TROPOSPHERIC-SCATTER PROPAGATION.

UART/USART/USRT

See UNIVERSAL RECEIVER-TRANSMITTER.

ULTRASONIC TRANSDUCER

An ultrasonic transducer is a device that converts electrical impulses into ultrasound, or ultrasound into electrical impulses. Generally, the electrical waves have the same frequency as the acoustic waves that are produced or picked up by the transducer. An ultrasonic transducer resembles a speaker or microphone, depending on whether it is used to generate ultrasound or receive ultrasound.

Most ultrasonic transducers operate by one of three principles: dynamic interaction, magnetostriction, or piezoelectric effect (*see* CRYSTAL TRANSDUCER, ELECTROSTRICTION, MAGNETOSTRICTION, MICROPHONE, SONAR, SPEAKER).

ULTRASOUND

Ultrasound is an acoustic disturbance that occurs at frequencies too high to be heard by humans. Ultrasound ranges in frequency from near 20 kHz to several hundred kilohertz, or even 1 MHz. (There is no well-defined upper limit.) In air, ultrasonic waves measure less than ⅝ inch, or 1.7 centimeters, in length. Ultrasound propagates through materials in the same way, and at the same speed, as audible sound (*see* SOUND).

Ultrasound is produced along with audible sound by many natural phenomena. Ultrasound can be heard by certain animals, notably dogs, who can detect frequencies up to 30 or 40 kHz. Ultrasound can be generated and detected by means of ordinary oscillators, amplifiers, and special transducers for various purposes.

ULTRAVIOLET RADIATION

Ultraviolet radiation refers to electromagnetic radiation at wavelengths somewhat shorter than visible light. Ultraviolet waves measure between approximately 390 nanometers (nm) and 4 nm. The ultraviolet spectrum is sometimes divided into two categories, known as long-wave ultraviolet, abbreviated UV (390 nm to 50 nm) and short-wave or extreme ultraviolet, abbreviated XUV (50 nm to 4 nm). Ultraviolet light propagates through a vacuum in the same way, and at the same speed, as visible light. The photons of ultraviolet contain more energy than those of visible light, but less energy than photons of X rays and gamma rays (*see* LIGHT, PHOTON).

Although humans cannot see ultraviolet radiation, it can have pronounced effects. Intense ultraviolet can be damaging to the eyes; for this reason, you should never look directly at a source of ultraviolet radiation. A familiar effect of ultraviolet is the tan or burn that it produces on skin. Ultraviolet from the sun is responsible for the ionization of the upper atmosphere. This makes long-distance radio communication possible at some frequencies (*see* D LAYER, E LAYER, F LAYER, IONOSPHERE, PROPAGATION CHARACTERISTICS).

Excessive exposure to ultraviolet will kill most forms of life. The rarefied ozone in the earth's atmosphere blocks out much of the ultraviolet radiation from the sun. This minimizes damaging effects to life at the surface.

Ultraviolet can be generated by some gas-filled tubes. Specialized arc lamps and mercury-vapor lamps are the most common devices for generating ultraviolet. Ultraviolet is generated or absorbed when the electrons of certain atoms change orbital shells. Electric arcs produce energy at ultraviolet wavelengths.

Ultraviolet exposes most types of camera film; specially designed pinhole cameras are used by astronomers to photograph the sun and other celestial objects in ultraviolet. Some camera tubes are also sensitive to ultraviolet light.

Ultraviolet sources can be modulated and thus used for communications purposes.

UNBALANCED CIRCUIT

See UNBALANCED SYSTEM.

UNBALANCED LINE

Any single-conductor electrical line with a defined potential (either alternating current or direct current) with respect to ground, is called an unbalanced line. A shield may or may not be present.

The advantage of an unbalanced line is simplicity. The main disadvantage of unbalanced line is that it must be shielded if leakage is to be prevented. The loss in a shielded, balanced line is usually greater, per unit length, than the loss in a balanced line. *See also* BALANCED TRANSMISSION LINE, TRANSMISSION LINE, UNBALANCED TRANSMISSION LINE.

UNBALANCED LOAD

An unbalanced load is a load with one side or terminal at ground potential and the other side above ground potential. An unbalanced load is required at the termination of an unbalanced transmission line, such as coaxial cable, to ensure that unwanted currents do not flow on the ground side of the transmission line (*see* UNBALANCED TRANSMISSION LINE).

A good example of an unbalanced load is an end-fed or asymmetrical antenna. Ground-plane and vertical antennas are the most common types of unbalanced antennas (*see* GROUND-PLANE ANTENNA, VERTICAL ANTENNA in the ANTENNA DIRECTORY).

UNBALANCED OUTPUT

An unbalanced-output circuit is designed for use with an unbalanced load and an unbalanced line. There is only one output terminal in an unbalanced circuit. The output is usually shielded.

A balanced output can be used as an unbalanced output by simply grounding one of the terminals. However, if an unbalanced output is to be used with a balanced load or line, a special transformer, called a balun, is required. *See also* BALANCED OUTPUT, BALUN, UNBALANCED TRANSMISSION LINE, UNBALANCED LOAD.

UNBALANCED SYSTEM

An unbalanced system is any circuit in which one side is at ground potential and the other side is at some alternating-current or direct-current (ac or dc) voltage different from ground potential.

Unbalanced systems are common in electronics. Many types of audio-frequency and radio-frequency oscillators, amplifiers, and transmission lines are unbalanced. Unbalanced systems are somewhat more stable than balanced systems, since a condition of near-perfect balance is difficult to maintain. *See also* BALANCED CIRCUIT, BALANCED LOAD, BALANCED OUTPUT, BALANCED TRANSMISSION LINE.

UNBALANCED TRANSMISSION LINE

An unbalanced transmission line is a form of unbalanced line used in radio-frequency antenna transmitting and receiving applications. Such a line is usually a coaxial cable (*see* COAXIAL CABLE). In some instances, an unbalanced transmission line consists of a single wire, or a parallel-

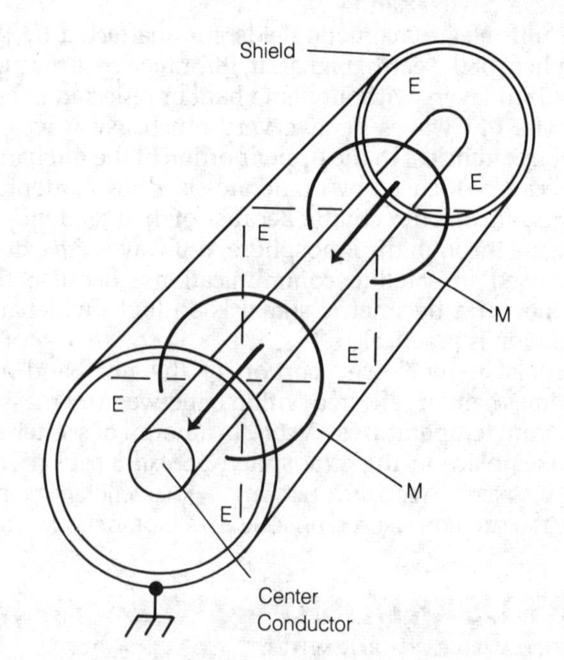

UNBALANCED TRANSMISSION LINE: The electric field (E) is represented by dotted lines and the magnetic field (M) is represented by solid lines. Current in the center conductor is shown by arrows.

wire line in which one conductor is at ground potential. The illustration shows the principle of operation of a coaxial unbalanced transmission line.

The current in the center conductor flows back and forth at a specific frequency (heavy arrow) setting up an electric, or E, field, shown by the dotted lines and a magnetic, or M, field, shown by the solid circles. The E and M fields are mutually orthogonal. They are prevented from leaking outside the line by the shield.

The E and M fields are perpendicular within the coaxial line so an electromagnetic field exists there. This field propagates in a direction perpendicular to both the E and M lines of flux along the length of the transmission line. The speed of propagation depends on the dielectric material between the center conductor and the shield. If the dielectric is air, the speed of propagation is about 95 percent of the speed of light. If the dielectric is polyethylene, the speed varies between 66 and 80 percent of the speed of light (*see* VELOCITY FACTOR).

With an unbalanced transmission line as shown, it is important that the shielding be as complete as possible. This keeps electromagnetic fields from being transmitted or received by the center conductor. In the case of an unshielded, unbalanced transmission line, radiation is inevitable, and is accepted as a characteristic of the line. Radiation from, or reception of signals by, a transmission line degrades the directional characteristics of the antenna with which the line is used. This may or may not be important, depending on the type of antenna.

Unbalanced transmission lines usually have more loss than balanced transmission lines because the electromagnetic field in an unbalanced transmission line must pass through more dielectric material than it does in a balanced transmission line. Shielded, unbalanced lines are easier to install, however, than are balanced transmission lines, and this is their main advantage.

UNDERWRITERS' LABORATORIES, INC.

In the United States, electrical equipment and appliances are tested for safety by an independent organization called Underwriters' Laboratories, Inc. (UL). The products checked by UL include wires, motors, fuses, circuit breakers, outlets, and components.

The UL organization is primarily concerned with the safety of users of electrical equipment. Fire and shock hazards must be minimized. An appliance or product carries a UL approval tag or sticker if it meets UL standards. A component receives a UL recognized label.

UNIJUNCTION TRANSISTOR

A unijunction transistor (UJT) is a three-terminal transistor with a single PN junction whose operation differs from that of bipolar and field effect transistors. It offers a negative-resistance characteristic, which makes the UJT useful in timing and oscillator circuits. Also called a double-base diode, the UJT has an emitter and two base terminals.

Part A of the figure shows the symbol for the UJT and part B is a diagram of an early UJT, useful in explaining its operation. The UJT was made from a bar of lightly doped N-type silicon, with a small piece of heavily doped P-type material joined along one side. The end terminals of the bar are designated base 1 (B1) and base 2 (B2) as shown, and the P-type region is termed the emitter.

When a positive voltage is applied across the base terminals B1 and B2, the current that flows is determined by the high resistance value of the N-type silicon bar. The P-type emitter forms a P-N junction with the N-type silicon bar.

If a positive bias is then applied between the emitter terminal E and the base terminal B1, the PN junction becomes forward biased. Holes are injected into the N-type region and flow toward B1. The resistance of this region decreases rapidly because of the presence of the additional carriers. As a result, the B1-B2 voltage drop across the E-B1 region will decrease although current through the region increases. This creates a negative resistance region that can be controlled by the B1-B2 voltage.

Modern unijunction transistors are made by conventional silicon oxide masking and etching processes. A cross section of a diffused UJT is shown in part C of the figure. The emitter is formed by diffusing a P-type boron impurity into the high-resistivity N-type silicon wafer. The B1 ohmic contact and an annular ring are diffused

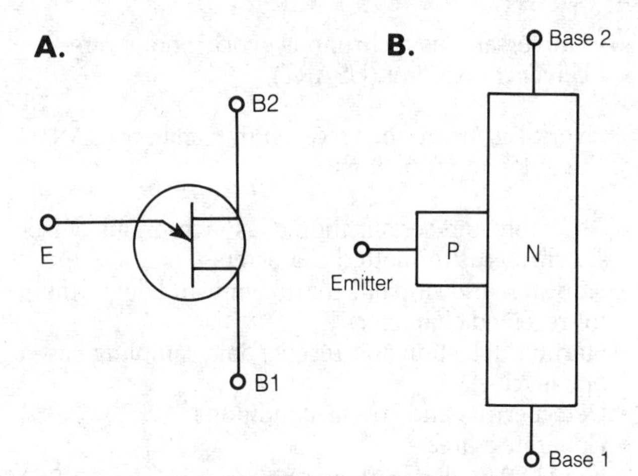

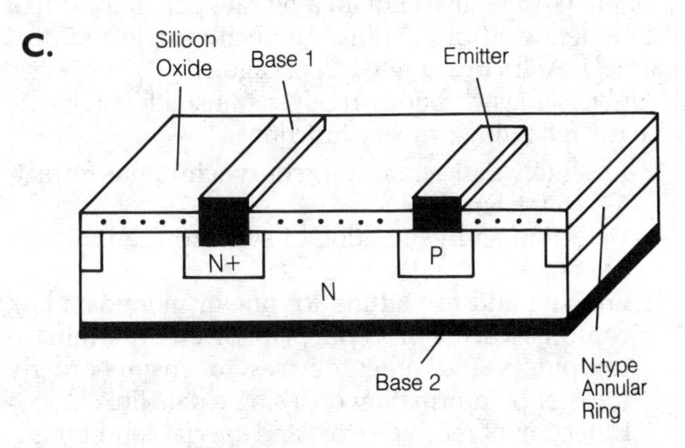

UNIJUNCTION TRANSISTOR: Symbol for a UJT (A), basic structure of a UJT transistor (B), and a modern UJT diffused structure (C).

into the silicon using N-type phosphorous. The annular ring provides junction protection. Aluminum is evaporated on the surface of the wafer for the B1 and the emitter contacts, and gold is evaporated onto the back of the structure for the B2. The wafer is then diced into many individual transistors.

A simple relaxation oscillator can be made with a UJT, a capacitor, and a resistor. Additional resistors, capacitors and a potentiometer can be used to make a variable frequency UJT relaxation oscillator. Unijunction transistors are frequently used to control silicon-controlled rectifiers. *See* SILICON-CONTROLLED RECTIFIER.

UNIVERSAL RECEIVER-TRANSMITTER

Universal receiver-transmitters are LSI (large-scale integrated) receiver-transmitter circuits that have helped in the design, manufacture, and growth of data communications equipment and have helped to reduce their cost. These LSI circuits (which differ in timing mode) carry out framing, formatting, modem control, and microprocessor bus interface functions on a single chip. The three receiver-transmitter LSI circuits are:

1. Universal asynchronous receiver transmitter (UART).
2. Universal synchronous receiver transmitter (USRT).
3. Universal asynchronous/synchronous receiver transmitter (USART).

Universal asynchronous receiver transmitters (UARTs) have the following functions:

- Assembly and serialization of asynchronous or isosynchronous formatted characters.
- Add start and stop bits for transmit and delete them for received characters.
- Start bit detection and receive data sampling based on fast clocks.
- Detect errors and special conditions.
- Generate a break.

Some UARTs also contain a bit rate generator (BRG) and modem control I/O pins. There are single and dual channel UARTs in a single IC package.

Universal synchronous receiver transmitters (USRTs) have the following primary functions:

- Assembly and serialization of synchronous formatted characters.
- Detection and generation of synchronization characters.
- Framing and formatting for one or more data link controls, a set of procedures followed by terminals, computers and other devices to ensure orderly transfer of information on a single data link.
- Detection of receive errors and special conditions.

Universal synchronous or asynchronous receiver transmitters (USARTs) are receiver-transmitters that can support either asynchronous or synchronous transmission by the software programming of internal mode registers. In asynchronous mode, a USART functions like a UART. In the synchronous mode, a UART can do the framing and formatting for character or bit-oriented data link controls or it might just handle the character synchronization of character-oriented data link controls.

These integrated circuits (ICs) are available with one or two serial communication channels per package. Some UARTs and USARTs include bit-rate generators, and many USRTs and USARTs support character and bit-oriented data link controls. UARTs, USRTs, and USARTs are specified for terminals and printers, personal computers, intelligent modems, cluster controllers, front-end processors, multiplexers, remote-data concentrators, data PBXs, and communications test equipment.

UPLINK

An uplink is a ground-based link for an active communications satellite. The uplink frequency differs from the downlink frequency on which the satellite transmits signals back to the earth. The different uplink and downlink frequencies allow the satellite to receive and retransmit signals at the same time, so it can function as a repeater.

At a ground-based satellite-communications station, the uplink antenna can be relatively nondirectional if high transmitter power is used. But for more distant satellites, and especially for satellites in geostationary orbits, directional atennas are preferable. *See also* DOWNLINK, REPEATER.

UPPER SIDEBAND

An amplitude-modulated signal carries the information in the form of sidebands, or energy at frequencies just

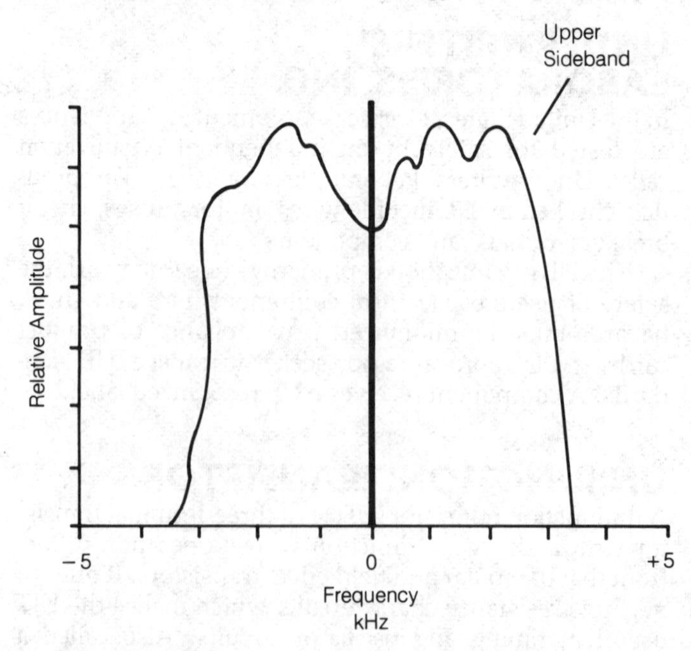

UPPER SIDEBAND: An AM signal spectrally displayed, showing the upper sideband.

above and below the carrier frequency (*see* AMPLITUDE MODULATION, SIDEBAND). The upper sideband (USB) is the group of frequencies immediately above the carrier frequency.

The USB frequencies result from additive mixing between the carrier signal and the modulating signal. For a typical voice amplitude-modulated signal, the upper sideband occupies approximately 3 kHz of spectrum space (see illustration). For reproduction of television video information, the sideband can be several megahertz wide.

The upper sideband contains all of the modulating-signal intelligence. For this reason, the rest of the signal (carrier and lower sideband) can be eliminated. This results in a single-sideband (SSB) signal on the upper sideband. *See also* LOWER SIDEBAND, SINGLE SIDEBAND.

USART

See UNIVERSAL RECEIVER-TRANSMITTER.

USASCII

See ASCII.

USA STANDARD CODE

See ASCII.

USB

See UPPER SIDEBAND.

USRT

See UNIVERSAL RECEIVER-TRANSMITTER.

VACUUM-TUBE VOLTMETER

A vacuum-tube voltmeter (VTVM) is an instrument used to measure voltages in electronic circuits without drawing significant current. The input impedance is theoretically infinite, although in practice it is finite but very large. The vacuum-tube voltmeter is so named because it uses a tube in a Class A configuration to obtain the high input impedance.

The illustration is a simplified schematic diagram of a vacuum-tube voltmeter. The ranges are selected by varying the gain of the tube, or by switching various resistances in parallel with the microammeter.

Some vacuum-tube voltmeters have provision for measuring current and resistance. The tube is not required for these modes; the operation is identical to that of the volt-ohm-milliammeter (*see* VOLT-OHM-MILLIAMMETER). A VTVM with the capability to measure current and resistance is sometimes called a vacuum-tube volt-ohm-milliammeter (VTVOM).

Field-effect transistors have replaced vacuum tubes for measurement of low and medium voltage when it is essential that the current drain be small. The VTVM is still sometimes used, however, for measuring high voltages. *See also* FET VOLTMETER.

VACUUM-TUBE VOLTMETER: A schematic diagram of an instrument for measuring extremely high resistance values.

VACUUM-TUBE VOLT-OHM-MILLIAMMETER

See VACUUM-TUBE VOLTMETER.

VACUUM-VARIABLE CAPACITOR

A vacuum-variable capacitor is a variable capacitor that is enclosed in an evacuated chamber. The construction of the vacuum-variable capacitor is similar to that of an air-variable capacitor.

Vacuum-variable capacitors have extremely low losses and high voltage ratings because a vacuum cannot become ionized. This makes the vacuum-variable capacitor ideal for high-power radio-frequency transmitting applications. *See also* AIR-VARIABLE CAPACITOR.

VALENCE BAND

The electrons in an atom orbit the nucleus at defined distances following average paths that lie in discrete spheres called shells. The more energy an electron possesses, the farther its orbit will be from the nucleus. If an electron gets enough energy, it will escape the nucleus altogether and move to another atom.

An electron is most likely to escape from a nucleus when the electron is in the outermost shell, or conduction band, of an atom. Electrons are stripped easily from nuclei in good conductors such as copper and silver, and less readily in semiconductors such as germanium and silicon. In insulators, the nuclei hold onto the electrons very tightly. The willingness of a nucleus to give up its outermost electrons depends on how full the outer shell is. The outer shell of an atom is called the valence shell. The range of energy states of electrons in this shell is known as the valence band. *See also* BAND GAP, ELECTRON ORBIT, VALENCE ELECTRON, VALENCE NUMBER.

VALENCE ELECTRON

In an atom, any electron that orbits in the partially filled outermost shell, without escaping from the nucleus, is called a valence electron. In an atom with a full outer shell (*see* ELECTRON ORBIT), there are, by definition, no valence electrons.

Some atoms have valence shells that are occupied by just one or two electrons, and thus the electrons are given up easily. Other atoms have nearly full valence shells; such elements tend to attract electrons. The various elements have properties that depend, to some extent, on the state of the outer shell. *See also* VALENCE NUMBER.

VALENCE NUMBER

The valence number of an atom is an expression of the conditions in the outer (valence) shell of that element in its natural state (*see* VALENCE ELECTRON). The valence numbers of various elements indicate how their atoms conduct an electric current, and how they behave in chemical reactions and mixtures. In some cases, relative valence numbers are important for elements combined as mixtures.

Definition. An element with a filled valence shell has a valence number of 0. If the outer shell contains either an excess or a shortage of electrons, the valence number is the number of electrons that must be added or taken away to result in an outer shell that is completely filled.

Valence numbers are sometimes expressed as positive or negative quantities, indicating an excess or deficiency of electrons compared to the most stable possible state. These numbers are called oxidation numbers. If the outer shell is less than half full, containing n electrons, the oxidation number is positive, and is defined as $+n$. If the outer shell is more than half full, containing n electrons but having a capacity to hold k electrons, then the oxidation number is negative, and is defined as $-(k - n)$. Substances with positive oxidation numbers tend to give up electrons; substances with negative oxidation numbers tend to accept electrons from outside the atom.

Sometimes both the oxidation values $+n$ and $-(k - n)$ are given, especially if the shell is approximately half filled. Such an element has two effective valence numbers. A few elements behave in such a way that they can be considered to have several different oxidation numbers.

Conduction and Chemical Reactions. The oxidation number of an element determines how readily it will conduct an electric current. Metallic substances have low positive oxidation numbers, indicating that the outer shell has just a few electrons. Semiconductors have outer shells that are approximately half full. Poor conductors have valence shells that are almost completely filled, and thus their oxidation numbers are small and negative. Those elements that have totally filled outer shells are the worst conductors, but the best dielectric substances.

The oxidation number of an element also indicates how readily it will react with certain other elements. When the nonzero oxidation numbers of two or more atoms add up to 0, the elements will react very easily when brought together. For example, two atoms of hydrogen (each with oxidation number $+1$) will combine readily with one atom of oxygen (oxidation number -2), since the three values add up to 0.

When an element has two or more oxidation numbers, it may react with other atoms according to any of the values. An example of such a substance is hydrogen, which can be considered to have an outer shell that is either half full or half empty, having oxidation value -1 in some cases and $+1$ in other cases. Hydrogen thus reacts with many other elements; it can either accept or give up an electron.

Relative Valence in Semiconductors. In a semiconductor material, impurity substances are added to modify the conducting characteristics. The valence of the impurity, relative to that of the original semiconductor, determines the major and minor charge carriers.

If an impurity has more electrons than the semiconductor, it is called a donor. If the impurity has fewer electrons than the semiconductor, it is called an acceptor. The addition of a donor impurity causes conduction mostly via electrons, and the resulting material is called N-type. The addition of an acceptor impurity causes conduction mostly in the form of electron-deficient atoms called holes, and the material is called P-type. *See also* ELECTRON, N-TYPE SEMICONDUCTOR, P-TYPE SEMICONDUCTOR.

VAN ALLEN RADIATION BELTS

The earth is surrounded by a magnetic field, known as the geomagnetic field (*see* GEOMAGNETIC FIELD). This field causes charged particles to be accelerated and deflected toward the poles. The particles are mostly protons and electrons, although there are some alpha particles and a heavier nuclei. The charged particles come primarily from the sun, and to a lesser extent from other stars (*see* ALPHA PARTICLE, COSMIC RADIATION, ELECTRON, PROTON).

The deflection of charged particles around the earth results in zones of radiation in outer space. High-energy protons and alpha particles circulate at an altitude of approximately 2,000 miles. High-energy electrons are accelerated at an altitude of about 10,000 miles. These particles, especially the protons and alpha particles, produce gamma rays if they strike heavy materials such as rock or metal. The regions in which the particles move are known as the Van Allen radiation belts, after one of the scientists who first theorized their existence. The gamma radiation might present a hazard to space travelers if they spend a long time at altitudes of 2,000 or 10,000 miles.

VAN DE GRAAFF GENERATOR

The Van De Graaff generator is a machine for generating high-voltage electrostatic voltages up to 6 million volts. The electric charge is stored at a high electric potential on a hollow metal sphere supported by an insulating column. The moving endless belt of silk or rubber is also housed within the column as shown in the figure. The belt, passing over an idler pulley at the top, is moved up

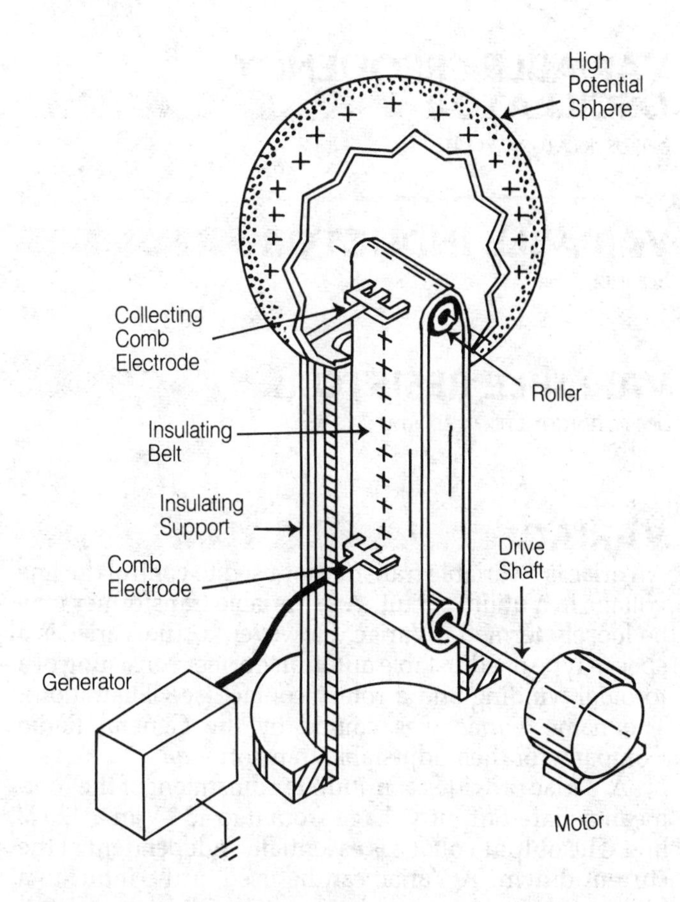

VAN DE GRAAFF GENERATOR: A Van de Graaff generator operates by means of triboelectric effect. The charge is stored in the metal sphere.

the column by a motor-driven sheave. An electrode in the shape of a comb with pointed teeth is positioned near the bottom of the moving belt. It is connected by cable to a source of high voltage with respect to ground. The high-intensity electric field on the points produces inductive charge separation in the belt, removing electrons to the points by corona discharge and leaving the belt (+) charged.

A second comb electrode, attached to the inside of the hollow sphere with a conductive rod is positioned close to the top of the belt to collect the charge. Inductive charge separation in the metal sphere provides a high electric field strength near the points of the comb, and electrons pass to the belt by corona action. This discharges the belt and leaves the outside of the sphere positively charged because electrostatic charges always reside on the exterior of a conductor.

The charge energy gradually built up on the sphere comes from the work done by the motor in driving the charged belt upwards against the repulsive electric forces exerted by the similarly charged sphere. If the charge on the sphere reaches a saturation level with respect to the size of sphere, it will repel additional charges, and the outside charge will leak off into the air.

A large-size Van De Graaff generator can have a sphere with a five-foot diameter mounted on a 50-foot insulating column. A machine of this size is likely to be equipped with a generator capable of 10-kV output. These large machines are used to simulate lightning, operate high-voltage X ray machines and high-energy particle accelerators in nuclear fission experiments. Smaller benchtop units several feet tall are used in classroom demonstrations of electrostatics.

VANE ATTENUATOR

A piece of resistive material placed in a waveguide to absorb some of the electromagnetic field is called a vane attenuator. The device can be moved back and forth to provide varying degrees of attenuation, from zero to maximum (see illustration). The resistivity of the vane is selected so that it does not upset the characteristic impedance of the waveguide.

Vane attenuators can provide exact resettability by means of a calibrated scale marked on the sliding element. Vane attenuators are extensively used at ultrahigh and microwave frequencies in various applications. *See also* ATTENUATOR, WAVEGUIDE.

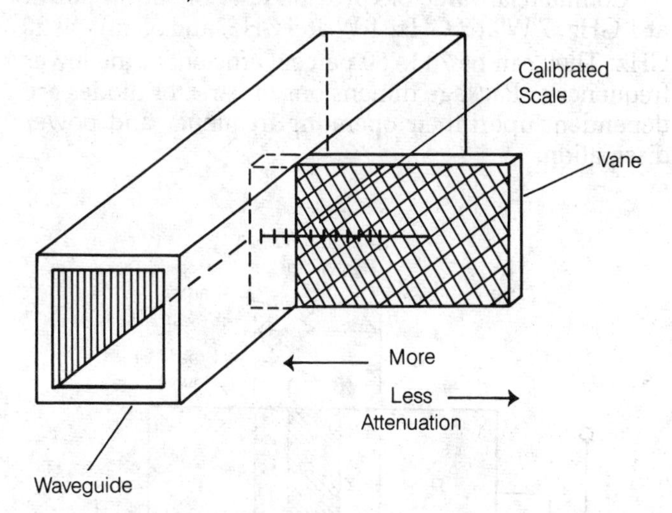

VANE ATTENUATOR: A vane attenuator controls propagation of the electromagnetic field through a waveguide.

VAR

The unit of reactive power is called the reactive volt-ampere, abbreviated VAR. Reactive power does not represent power actually dissipated, although it has an effect on power measurement. *See* REACTIVE POWER.

VARACTOR DIODE

The varactor diode, also known as a voltage-variable capacitor diode or varicap, is a back-biased P-N junction whose operation depends on its variation of junction capacitance with reverse bias. Special impurity profiles are grown in the depletion layer to enhance the capacitance variation and minimize series resistance losses. *See* DIODE.

The figure shows the schematic symbol for the varactor diode and its principle of operation. The varactor is made of junction materials whose doping concentration is graded throughout the device. The heaviest concentration of doping is in the regions adjacent to the junction, thus forming a graded junction. The junction area is made small to take advantage of the basic variation of junction capacitance with reverse voltage. Varactor diodes are designed to keep internal resistance as low as possible so that the P-N junction, when reverse biased, is purely capacitive. Because the junction is abrupt, junction capacitance varies inversely as the square root of the reverse voltage.

Most varactor diodes are made from silicon, but gallium arsenide varactors have a higher frequency response. Low-power varactors are used as voltage-variable capacitors for electronic tuning. In nonlinear operation, they also perform phase shifting and switching in the very high frequency and microwave regions. The diodes also can be used as very low loss frequency multipliers in solid-state transmitters. In addition they are used for limiting, pulse shaping, and parametric amplification.

Commercial varactors provide 12 W of output power at 1 GHz, 7 W at 2 GHz, 1 W at 5 GHz, and 50 mW at 20 GHz. They can be 70 to 80 percent efficient at the lower frequencies. Package dimensions of varactor diodes are dependent upon their operating frequency and power dissipation.

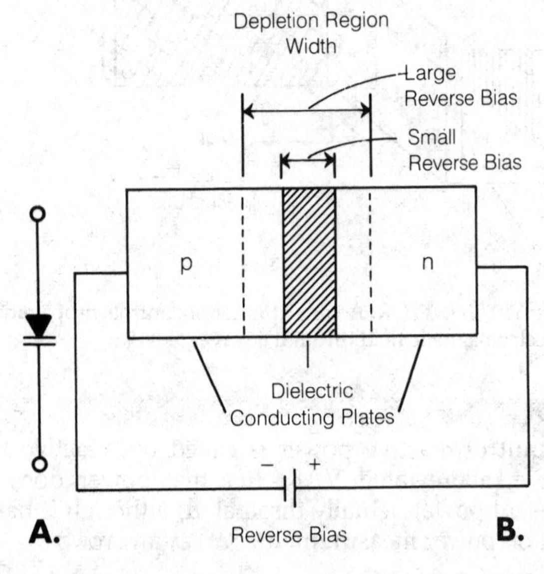

VARACTOR DIODE: Symbol of varactor diode (A) and a diagram showing is operation regions (B).

VARIABLE CAPACITOR

See CAPACITOR.

VARIABLE CRYSTAL OSCILLATOR

See OSCILLATOR CIRCUITS.

VARIABLE-FREQUENCY OSCILLATOR

See OSCILLATOR CIRCUITS.

VARIABLE INDUCTOR

See INDUCTOR.

VARIABLE RESISTOR

See POTENTIOMETER, RHEOSTAT.

VARIAC

A Variac is a variable transformer used to control the line voltage in a utility circuit. Any variable transformer may be loosely termed a Variac. However, a true Variac is a special type of adjustable autotransformer consisting of a toroidal winding and a rotary contact (see illustration). The name Variac was coined by the General Radio Company for their adjustable transformers.

A Variac provides continuous adjustment of the root-mean-square output voltage from 0 to 134 V in a 120 V line. The output voltage is essentially independent of the current drawn. A Variac can be used at the input to a high-voltage power supply to obtain adjustable direct-current output. This scheme is sometimes used in tube type radio-frequency power amplifiers.

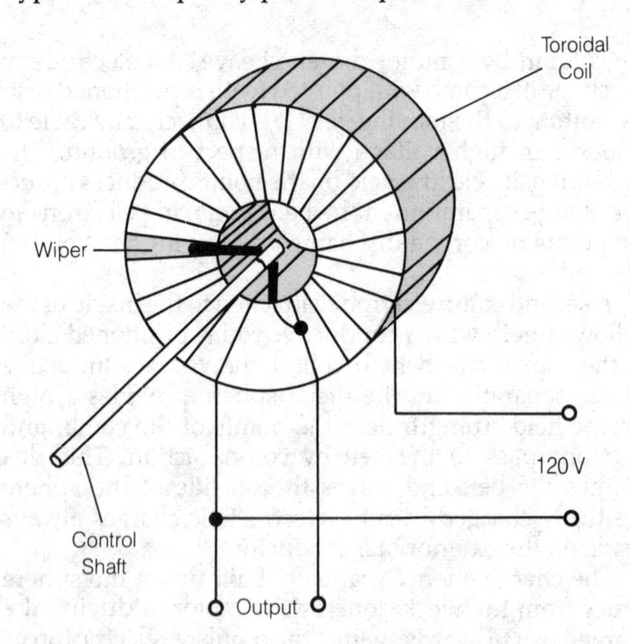

VARIAC: Voltage is controlled by turning the shaft of the Variac, moving the wiper along the wire turns.

VARICAP

See VARACTOR DIODE.

VAR METER

The reactive power in a radio-frequency (rf) circuit can be determined by a VAR meter. A simple VAR meter consists of an rf ammeter in series and a radio-frequency voltmeter in parallel with a circuit (see illustration). If the true power P_t is known, the reactive power P_r, in VAR, is given by:

$$P_r^2 = E^2I^2 - P_t^2$$

where E and I are the rf voltage and current as indicated by the meters, in volts and amperes respectively.

A reflectometer can be used as a VAR meter if the scales are calibrated in watts. The reactive power in VAR is simply the wattmeter reading on the reflected-power scale, or with the function switch set for reflected power. *See also* POWER, REACTIVE POWER, REFLECTED POWER, REFLECTOMETER, TRUE POWER, VAR.

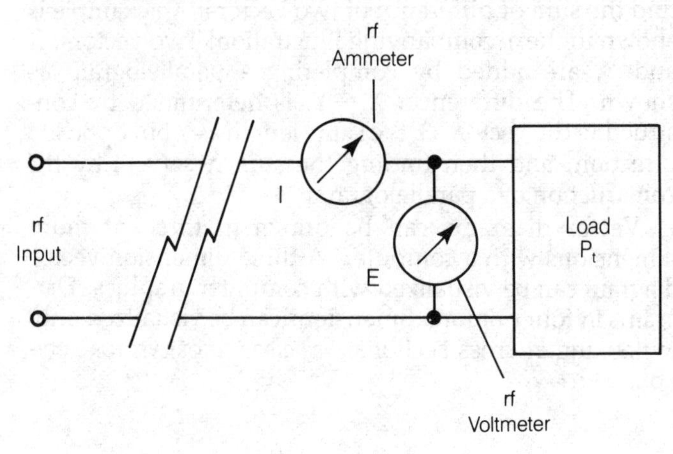

VAR METER: An ammeter and a voltmeter are connected at the same point on the line—preferably at the load—to measure the reactive power in a radio-frequency circuit.

V BEAM ANTENNA

See ANTENNA DIRECTORY.

VCR

See VIDEO-CASSETTE RECORDER.

VARISTOR

A varistor is a voltage-sensitive polycrystalline semiconductor device that provides a constant voltage drop of about 2.5 V useful in circuit protection. It has a characteristic that is similar to two back-to-back zener diodes. *See* ZENER DIODE. The most widely used varistors are metal-oxide varistors (MOVs) formed from zinc oxide and small amounts of selected metal oxides pressed and fired at high temperature. The polycrystalline ceramic body produced is a matrix of conductive zinc oxide grains separated by a resistive intergranular boundary.

Varistor voltage ratings are proportional to body thickness because more grain boundaries are connected in series. Similarly, current ratings are proportional to body area because more grain boundaries are connected in parallel. As a result, the energy handling capability of a varistor is proportional to volume. *See* CIRCUIT PROTECTION.

Metal-oxide varistors are packaged in a wide range of sizes and styles. The most common type of varistor is the radial-leaded dipped disk. Varistors are also packaged as encapsulated blocks where high energy dissipation is a requirement and in chip form for surface mounting.

VDT

See VIDEO DISPLAY TERMINAL.

VECTOR

A vector is a quantity that has magnitude and direction. Examples of vectors include the force of gravity, the movement of electrons in a wire, and the propagation of an electromagnetic field through space. If direction is not specified, a quantity is called a scalar (*see* SCALAR). A vector for which just two directions can exist is called a directed number.

A vector is usually represented in equations by a capital letter, such as X. Vectors may also be represented in equations by capital letters in boldface or italics, or both.

Vectors are geometrically represented by arrows whose length indicates magnitude and whose orientation indicates direction. Some examples of vectors are shown in the illustration. These are velocity vectors, corresponding to various speeds in miles per hour (mph) and azimuth bearings in degrees (°) as follows:

X = 30 mph at azimuth 45°
Y = 60 mph at azimuth 180°
Z = 20 mph at azimuth 280°

The examples shown are vectors in polar coordinates. Vectors can also be expressed in rectangular coordinates.

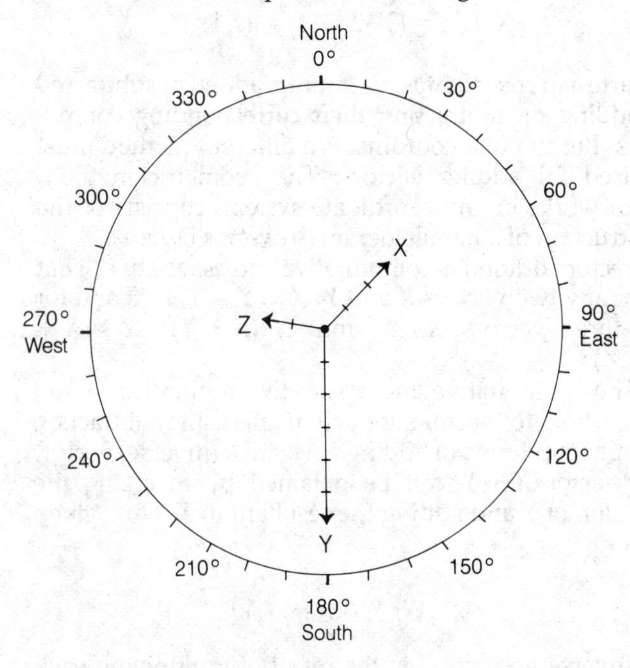

VECTOR: Examples of velocity vectors in polar coordinates.

Complex numbers are sometimes denoted as ordered pairs in a Cartesian plane (*see* CARTESIAN COORDINATES, COMPLEX NUMBER, POLAR COORDINATES).

Sometimes it is necessary to consider only the length of a vector. The magnitude alone is denoted by placing vertical lines on either side of the designator, such as |X|, |Y|, and |Z|. Alternatively, the length of a vector can be denoted by removing the arrow above the capital letter, or by eliminating the boldface or italic notation.

Vectors are important in applied mathematics. In electronics, vectors simplify the definitions and concepts of such abstractions as impedance and the behavior of fields. Vector quantities can be added, subtracted, and multiplied, although the procedures differ from the familiar arithmetic operations used with scalars. *See also* VECTOR ADDITION, VECTOR DIAGRAM.

VECTOR ADDITION

It is often necessary to add two or more vectors together in order to determine certain net effects. For example, there might be two magnetic fields acting in different directions with different intensities; their net effect would be found by means of a vector sum.

When vectors are expressed in Cartesian coordinates, sums are simple to determine. For two vectors *X* and *Y* in *n* dimensions, represented by ordered *n*-tuples:

$$X = (x_1, x_2, x_3, \ldots, x_n)$$
$$Y = (y_1, y_2, y_3, \ldots, y_n)$$

the sum is:

$$X + Y = (x_1 + y_1, x_2 + y_2, x_3 + y_3, \ldots, x_n + y_n)$$

For example, if $X = (3,5)$ and $Y = (-2,-4)$, then:

$$X + Y = (3-2, 5-4) = (1,1)$$

In Cartesian coordinates, vectors are added or subtracted by adding or subtracting their corresponding components. But in polar coordinates a different method must be used for adding vectors. The geometric method, which works in any coordinate system, consists of the construction of a parallelogram (*see* VECTOR DIAGRAM).

Vector addition is commutative and associative. That is, for any two vectors *X* and *Y*, $X + Y = Y + X$; and for any three vectors *X*, *Y*, and *Z*, $(X + Y) + Z = X + (Y + Z)$.

The commutative and associative properties do not always hold for vector subtraction unless the subtraction is defined in terms of adding a negative (inverse) vector. The vector $X - Y$ can be obtained by reversing the direction of *Y* and adding the resultant to *X*. For a given vector:

$$Y = (y_1, y_2, y_3, \ldots, y_n)$$

the inverse vector $-Y$ is the result of multiplying each component by -1:

$$-Y = (-y_1, -y_2, -y_3, \ldots, -y_n)$$

The difference $X - Y$ is equal to the sum $X + (-Y)$. This definition of vector subtraction allows us to use the commutative and associative properties in expressions containing both vector addition and subtraction. *See also* VECTOR.

VECTOR DIAGRAM

Vector quantities are geometrically represented with a vector diagram. In the vector diagram, the magnitude is denoted as the length of a line segment, and the direction is indicated by the orientation of the line segment and an arrowhead.

Vector diagrams can be drawn in two dimensions to find the sum or difference of two vectors. An example is shown in the accompanying illustration. Two vectors, *X* and *Y*, are added by completing a parallelogram as shown. The difference, $X - Y$, is determined by constructing the vector $-Y$ the same length as *Y* but opposite direction, and then finding the sum $X + (-Y)$ by the construction of a parallelogram.

Vector diagrams can be drawn in three or more dimensions with a computer. A three-dimension vector diagram can be visualized with computer graphics. Diagrams in four or more dimensions can be visualized only by looking at cross sections. *See also* VECTOR, VECTOR ADDITION.

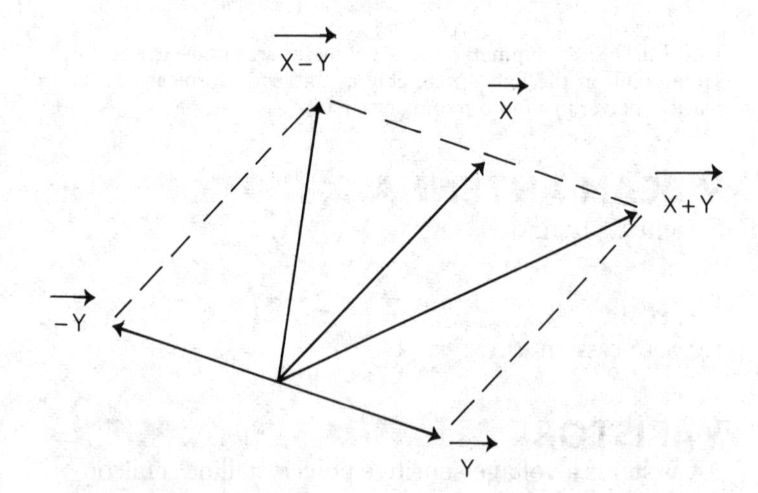

VECTOR DIAGRAM: The addition and subtraction of vectors by use of the parallelogram method.

VECTOR SUBTRACTION

See VECTOR ADDITION.

VELOCITY FACTOR

The speed of electromagnetic propagation through a vacuum is approximately 186,282 miles (299,792 kilometers) per second. In materials, however, the speed is reduced. The velocity factor, *v*, is expressed as a percentage by:

$$v = 100c'/c$$

where *c'* represents the speed of propagation in a particular substance, and *c* is the speed of light in a perfect vacuum. (Both *c'* and *c* must be specified in the same units.)

The velocity factor affects the wavelength of an electromagnetic field at a fixed frequency. For a disturbance with a frequency *f* in megahertz, the wavelength λ in a substance with velocity factor *v* is given in feet by:

$$\lambda = 984v/f$$

and in meters by:

$$\lambda = 300v/f$$

Velocity factor is an important consideration in the construction of antennas. An electromagnetic field travels along a single wire in air at a speed of about 0.95*c*. An electrical wavelength is therefore only 95 percent as long as in free space. As the conductor diameter is made larger, the velocity factor decreases a bit, perhaps to 92 or 93 percent for large-diameter metal tubing at very high frequencies. The velocity factor must be taken into account when sizing an antenna to a resonant length for a particular frequency. The radiating and parasitic elements must be cut to length *vm*, if *m* represents the length determined in terms of a free-space wavelength.

Velocity factor is also important when it is necessary to size a transmission line to a certain length. To make a half-wavelength stub, for example, from a transmission line with velocity factor v, multiply the free-space half wavelength by v to get the proper length for the line. This must also be done if tuned feeders are used with an antenna (*see* STUB, TUNED FEEDER).

In an open-wire transmission line with few spacers, the velocity factor is near 95 percent. But when dielectric material is present, for example in a twin-lead line, the velocity factor is reduced. In a coaxial line, the velocity factor is still smaller. The velocity factor of a transmission line depends on:

- The type of dielectric material used in construction.
- The amount of the electromagnetic field that is contained within the dielectric material.
- The presence of objects near a parallel-wire line.
- The extent of contamination of the dielectric because of aging in a coaxial cable.

The most common dielectric materials used in fabrication of feed lines are solid polyethylene, with a velocity factor of 66 percent, and foamed polyethylene, with a velocity factor of 75 to 85 percent.

The table lists the velocity factors of some common types of transmission lines, assuming ideal operating conditions. These values are approximate. For precise velocity-factor ratings in critical transmission-line applications, the manufacturer should be consulted.

VELOCITY FACTOR:
Velocity factors of various types of transmission lines.

Line Type or Manufacturer's No.	Velocity Factor, Percent
Coaxial cable, RG-58/U, solid dielectric	66
Coaxial cable, RG-59/U, solid dielectric	66
Coaxial cable, RG-8/U, solid dielectric	66
Coaxial cable, RG-58/U, foam dielectric	75–85
Coaxial cable, RG-59/U, foam dielectric	75–85
Coaxial cable, RG-8/U, foam dielectric	75–85
Twin-lead, 75-ohm, solid dielectric	70–75
Twin-lead, 300-ohm, solid dielectric	80–85
Twin-lead, 300-ohm, foam dielectric	85–90
Open-wire with plastic spacers, 300-ohm	90–95
Open-wire with plastic spacers, 450-ohm	90–95
Open-wire, homemade, 600-ohm	95

VELOCITY-MODULATED OSCILLATOR

See TRAVELING-WAVE TUBE, VELOCITY MODULATION.

VELOCITY MODULATION

A beam of electrons can be modulated by varying the speed at which the electrons move. This is the principle of operation of certain types of vacuum tubes. When the electrons are accelerated, their energy is increased, and when they are decelerated, their energy is reduced. Although the number of electrons per unit volume changes, the actual current remains constant or nearly constant.

When an electron beam is velocity-modulated a pattern of waves is produced. The density of the beam is inversely proportional to the acceleration of the electrons and 90 degrees out of phase with their instantaneous velocity (see illustration on p. 824). The pattern of waves may stand still along the length of the beam, move toward the anode, or move toward the cathode. If the pattern of waves moves toward the anode, the beam is forward-modulated; if it moves toward the cathode, the beam is backward-modulated.

Velocity modulation is used in traveling-wave tubes to obtain amplification and oscillation at ultrahigh and microwave frequencies. *See also* KLYSTRON, PARAMETRIC AMPLIFIER, TRAVELING-WAVE TUBE.

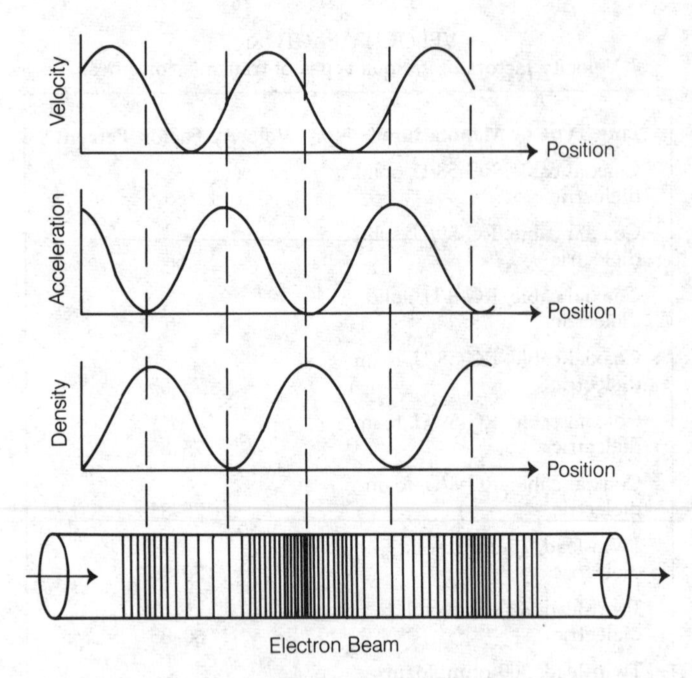

VELOCITY MODULATION: A diagram showing the relationship between electron density, acceleration and velocity.

VELOCITY OF LIGHT

See LIGHT.

VELOCITY OF SOUND

See SOUND.

VENN DIAGRAM

A Venn diagram is a graphic method for illustrating the relationships among sets. Venn diagrams can be used to find the unions and intersections of various sets and their complements.

The union of two sets is the set of elements in one set or the other, or both. There is a clear resemblance between set union and the Boolean OR operation. The intersection of two sets is the set of elements in both sets; this resembles the logic AND. The complement of a set is the set of all elements outside the set; there is a resemblance to the logic NOT.

The illustration demonstrates a Venn diagram showing three sets X, Y, and Z, with elements as follows:

X = {x1, x2, x3, x4, x5}
Y = {y1, y2, y3, y4, y5, y6}
Z = {x4, x5, y1, y2, z1, z2}

There are also three elements, $w1$, $w2$, and $w3$, that do not belong to any of the sets X, Y, or Z. The set of all elements in the diagram is called the universal set.

The Venn diagram makes it easy to find union and intersection sets. Complements can also be observed. If union is represented by addition, intersection by multi-

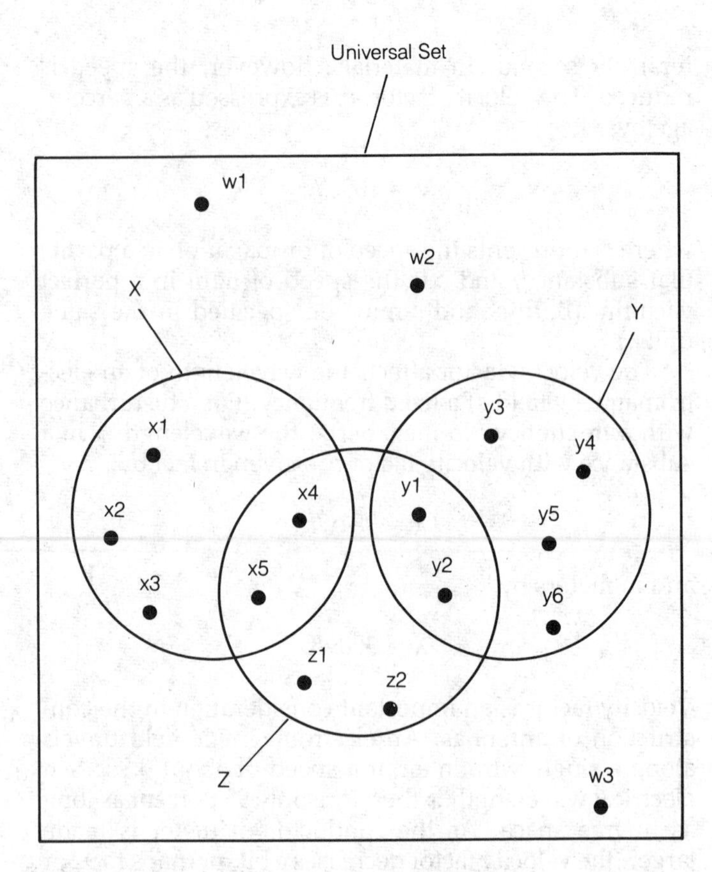

VENN DIAGRAM: A Venn diagram showing the relationship between elements of various sets.

plication, and complementation by an apostrophe (′), then:

X′ = {w1, w2, w3, y1, y2, y3, y4, y5, y6, z1, z2}
Y′ = {w1, w2, w3, x1, x2, x3, x4, x5, z1, z2}
Z′ = {w1, w2, w3, x1, x2, x3, y3, y4, y5, y6}
XY = ϕ (the empty set)
XZ = {x4, x5}
YZ = {y1, y2}
X + Y = {x1, x2, x3, x4, x5, y1, y2, y3, y4, y5, y6}
X + Z = {x1, x2, x3, x4, x5, y1, y2, z1, z2}
Y + Z = {x4, x5, y1, y2, y3, y4, y5, y6, z1, z2}

Venn diagrams are applied to Boolean algebra because of the similarities between set union and logic OR, set intersection and logic AND, and set complementation and logic NOT. *See also* AND GATE, BOOLEAN ALGEBRA, INVERTER, OR GATE.

VERNIER

A vernier is a special form of gear-drive and scale used for precision adjustment of variable capacitors, inductors, potentiometers, and other controls.

The vernier drive is a simple gear mechanism. The control knob must be rotated several times for the shaft to turn through its complete range. The gear ratio determines the ratio of the knob and shaft rotation rates.

The vernier scale is designed to aid an operator in interpolating readings between divisions. An auxiliary

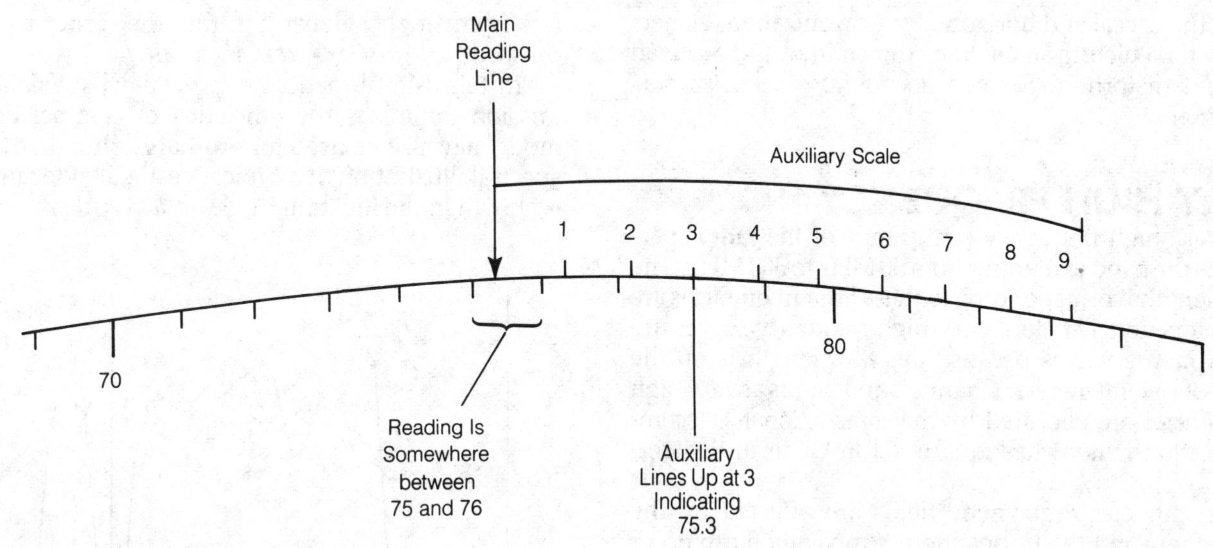

VERNIER: The Vernier scale allows precise interpolation of dial readings to the tenth of a division.

scale, located above the main scale, is marked in divisions that are 90 percent of the size of the main-scale divisions (see illustration). The auxiliary-scale lines are numbered 0 through 9, representing tenths of a division on the main scale.

When the dial is set to a point between two divisions on the main scale, one mark on the auxiliary scale will line up with some mark on the main scale. The operator looks for this coincidence, and notes the number of the auxiliary-scale mark at which the alignment occurs. This is the number of tenths of the way that the control is set between the two main-scale divisions. In the illustration, for example, the control is set to 75.3 units.

VERTICAL ANTENNA

See ANTENNA DIRECTORY.

VERTICAL LINEARITY

In a television video system, vertical linearity refers to the faithful reproduction of the image along the vertical axis. Ideally, all of the scanning lines in the raster are equally spaced. If this is not the case, the image will appear distorted. Most television receivers have a vertical-linearity control.

VERTICAL POLARIZATION

Vertical polarization occurs when the electric lines of flux of an electromagnetic wave are vertical or perpendicular to the surface of the earth. In communications, vertical polarization has certain advantages and disadvantages, depending on the application and the wavelength.

At low and very low frequencies vertical polarization is ideal because surface-wave propagation, the major mode of propagation at these wavelengths, requires a vertically polarized field. Surface-wave propagation is effective in the standard amplitude-modulation (AM) broadcast band as well; most AM broadcast antennas are vertical.

Vertical polarization is often used at high frequencies because vertical antennas can be erected in a very small space. At very high and ultrahigh frequencies, vertical polarization is used for mobile communications, and also in repeater communications.

The main disadvantage of vertical polarization is that most man-made noise tends to be vertically polarized. Thus, a vertical antenna picks up more of this interference than a horizontal antenna. At very high and ultrahigh frequencies, vertical polarization results in more flutter in mobile communications, as compared with horizontal polarization. *See also* CIRCULAR POLARIZATION, ELLIPTICAL POLARIZATION, HORIZONTAL POLARIZATION, POLARIZATION.

VERTICAL SYNCHRONIZATION

In television communications, the picture signals must be synchronized at the transmitter and receiver. The electron beam in the television picture tube scans from left to right and top to bottom, just as pages of a book are read. At any given instant of time, in a properly operating television system, the electron beam in the receiver picture tube is in exactly the same relative position as the scanning beam in the camera tube. This requires synchronization of the vertical and the horizontal positions of the beams.

When the vertical synchronization of a television system is lost, the picture appears split along a horizontal line. The picture may appear to roll upward or downward continually. Even a small error in synchronization results in a severely disturbed picture at the receiver.

The transmitted television picture signal contains vertical-synchronization pulses at the end of every frame (complete picture). These tell the receiver to move the electron beam from the end of one frame to the beginning of the next.

Both vertical and horizontal synchronization are necessary for a picture signal to be transmitted and received without distortion. *See also* HORIZONTAL SYNCHRONIZATION, TELEVISION.

VERY HIGH FREQUENCY

The very high frequency (vhf) range of the radio spectrum is the band extending from 30 MHz to 300 MHz. The wavelengths corresponding to these limit frequencies are 10 meters and 1 meter. Very high frequency waves are called metric waves because the wavelength is on the order of several meters. Channels and bands at very high frequencies are allocated by the International Telecommunication Union, headquartered in Geneva, Switzerland.

At vhf, electromagnetic fields are affected by the ionosphere and the troposphere. Propagation can occur in the E and F layers of the ionosphere (*see* E LAYER, F LAYER, PROPAGATION CHARACTERISTICS) in the lower part of the vhf spectrum. Tropospheric bending, ducting, reflection, and scattering take place throughout the vhf band (*see* DUCT EFFECT, TROPOSPHERIC PROPAGATION, TROPOSPHERIC-SCATTER PROPAGATION).

The vhf band is very popular for mobile communications and for repeater operation. Some satellite communication is also done at vhf. Wide band modulation is used to some extent; the most common example is television broadcasting. *See also* ELECTROMAGNETIC SPECTRUM, FREQUENCY ALLOCATIONS.

VERY LARGE SCALE INTEGRATION

An integrated-circuit chip that incorporates from 100 to 1,000 individual logic gates is called a very large scale integration (VLSI) chip. VLSI technology is widely used in the manufacture of microprocessor and peripheral circuits.

VERY LOW FREQUENCY

The range of electromagnetic frequencies extending from 9 kHz to 30 kHz is known as the very low frequency (vlf) band. The wavelengths corresponding to these frequencies are 33 and 10 kilometers. Some engineers consider the lower limit of the vlf band to be 3 kHz, or a wavelength of 100 kilometers.

The ionosphere returns all vlf signals to the earth, even if they are sent straight upward. If a vlf signal arrives from space, it cannot be heard on earth because the ionosphere will completely block it.

Electromagnetic fields at vlf propagate very well along the surface of the earth if they are vertically polarized.

VESTIGIAL SIDEBAND

Vestigial-sideband transmission is a form of amplitude modulation (AM) in which one of the sidebands has been largely eliminated. The carrier wave and the other sideband are unaffected. Vestigial-sideband transmission differs from single sideband in that the carrier is not suppressed (*see* SIDEBAND, SINGLE SIDEBAND).

In television broadcasting, vestigial-sideband transmission optimizes the efficiency of channel use. This mode may also be used for ordinary communications. A spectral illustration of a typical vestigial-sideband signal is shown in the illustration. *See also* AMPLITUDE MODULATION.

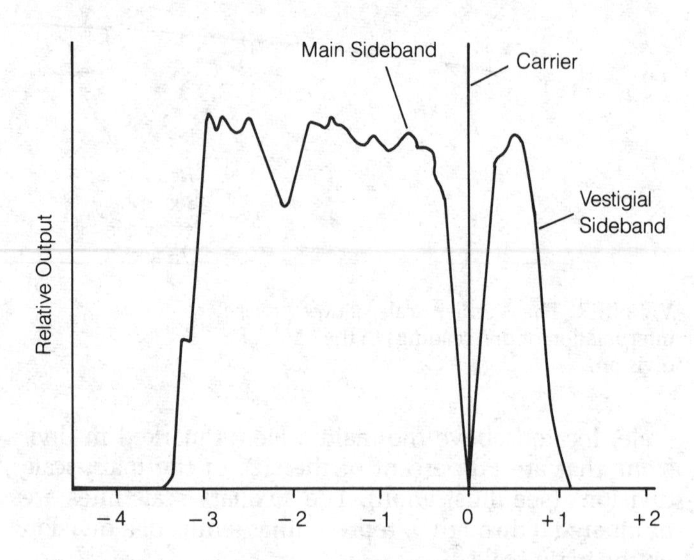

VESTIGIAL SIDEBAND: A spectral illustration of a vestigial-sideband signal.

VFO

See OSCILLATOR.

VHF

See VERY HIGH FREQUENCY.

VIBRATOR

See CHOPPER, CHOPPER POWER SUPPLY, VIBRATOR POWER SUPPLY.

VIBRATOR POWER SUPPLY

A vibrator power supply is a circuit that produces a high direct-current (dc) voltage from a lower dc voltage. These supplies have been used to power mobile or portable equipment containing vacuum tubes. Some vibrator power supplies can provide 120 V of alternating current, allowing certain appliances to be operated from low-voltage dc power.

A block diagram of a vibrator power supply is shown in the illustration. The principle of operation is the same as that of the chopper supply (*see* CHOPPER POWER SUPPLY). The vibrator consists of a relay connected in a relaxation-oscillator configuration. The relay contacts chatter, interrupting the current from the source. The resulting dc pulses are converted into true alternating current, as well as boosted in voltage, by a transformer. If the output voltage is 120 V root-mean-square (rms) with a frequency

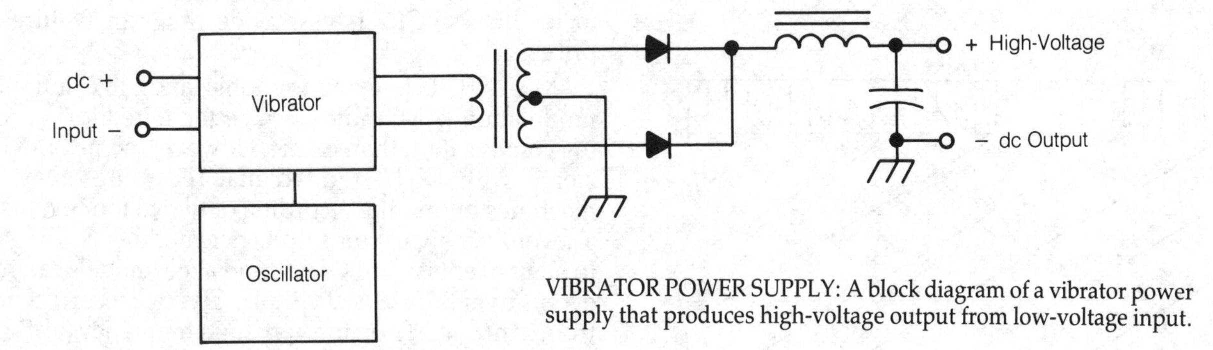

VIBRATOR POWER SUPPLY: A block diagram of a vibrator power supply that produces high-voltage output from low-voltage input.

near 60 Hz, the vibrator supply will provide satisfactory of 60 Hz, the power supply will provide power for simple appliances. Products that require a good sine wave, or a frequency of precisely 60 Hz cannot generally be run from vibrator supplies.

The supply shown incorporates a rectifier and filter after the transformer. Thus, the supply operates as a dc-to-dc converter (*see* DC-TO-DC CONVERTER).

VIDEOCASSETTE RECORDER

A videocassette recorder (VCR) is a video recorder that employs videotape cassettes to record and play back television broadcasts. Modern VCRs are equipped with automatic features that permit recording at times and for durations specified by the user without user supervision. The VCR can also play back prerecorded commercial videotapes in the compatible format.

Intended for home entertainment and education, VCRs available are designed to accept videocassettes in one of two different formats: Betamax and VHS. Videocassettes in both formats have built in supply and take-up reels and are similar in many respects to the popular audio cassettes. Videocassettes are made in internationally standardized case sizes. The most popular format is VHS, and the flat rectangular cassette for this format measures 7⅜ by 4 by 1 inches.

Sony Corporation of Tokyo developed the Betamax format and The Victor Company of Japan, known as JVC Limited, of Yokohama, developed the VHS format. The VHS carrier signals are between 3.4 and 4.4 megahertz, and the Betamax signals are between 4.4 and 5.6 MHz. Helical scanning and videocassettes with ½-inch-wide tape is used in both techniques. *See also* MAGNETIC RECORDING.

Video signals can have frequencies up to 4.2 megahertz, and they contain far more information than audio signals that are limited to 20 kilohertz. Videocassette recording uses helical scanning in contrast to audiocassette recording, in which the audio tape is simply pulled past an immobile recording head. Videocassette tape recording is done on tracks that run diagonally across the tape, and the tape is spiraled around a rotating drum with two, three, or four recording heads on it. (See Fig. 1.)

To accomplish this method of recording, videotape is pulled out of the cassette by the VCR mechanism and wrapped around the drum so that it will not slip out of position. This differs from audiocassette recording, in which the tape is left in the cassette and simply moved past the head.

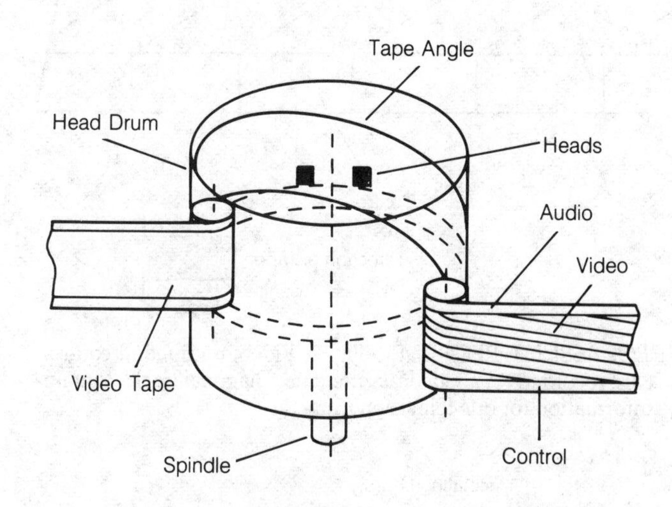

VIDEOCASSETTE RECORDER: Fig. 1. Magnetic tape from cassette is wrapped around a rotating drum at an angle in the helical video recording. The heads record the video, audio and control data diagonally on the tape.

Helical scanning, which wraps the tape around the drum at an angle, permits quad recording; each diagonal track contains information for one full television frame. (See Fig. 2). Video head gap size for both technologies is 0.3 micrometer, and both use cobalt alloy magnetic coating on the tape. Color signals in adjacent tracks are recorded 90 degrees out of phase.

The Betamax design uses a U-loading mechanism in which a single arm reaches into the cassette, pulls out the tape and wraps it around the head. The VHS M-loading mechanism has two arms on either side of the recording head that grab the tape and pull it against the head. (See Fig. 3.) The JVC VHS cassette is 30 percent larger in volume than the Sony Betamax cassette, but it has a recording speed of 5.80 meters per second (m/s), slightly slower than the 7.00 m/s of Betamax.

Automated VHS video recording and playback machines are packaged in cases that typically measure less than 14 by 14 by 4 inches. The front panel of the typical machine contains a digital clock and a function display indicating record, rewind and playback. The panel also contains rows of switches for preprogramming selected channels and setting the times, dates, and durations of programs to be recorded. It also contains controls for rewinding and rapidly advancing the videotape.

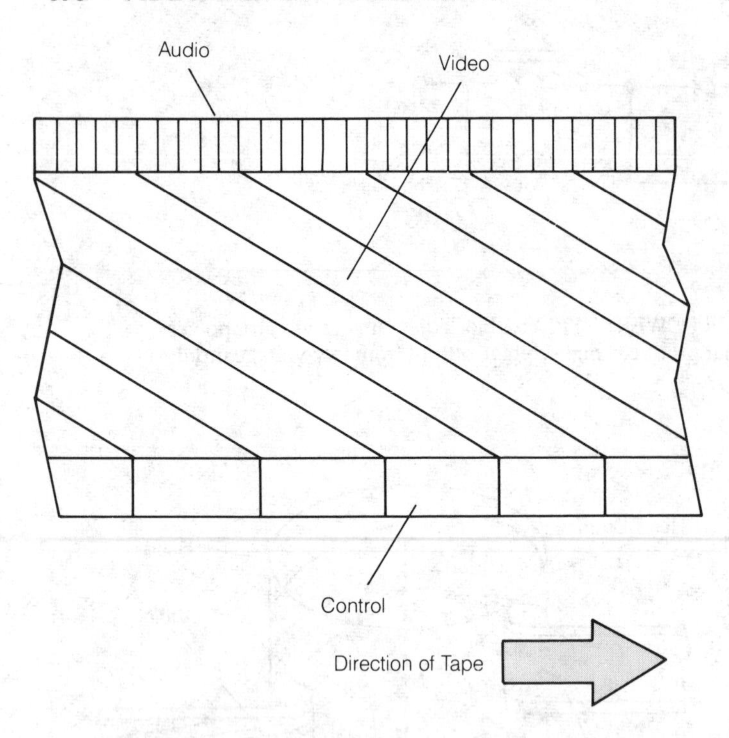

VIDEOCASSETTE RECORDER: Fig. 2. Video, audio, and control data are recorded on a videocassette tape. Diagonal tracks contain the information for one television frame.

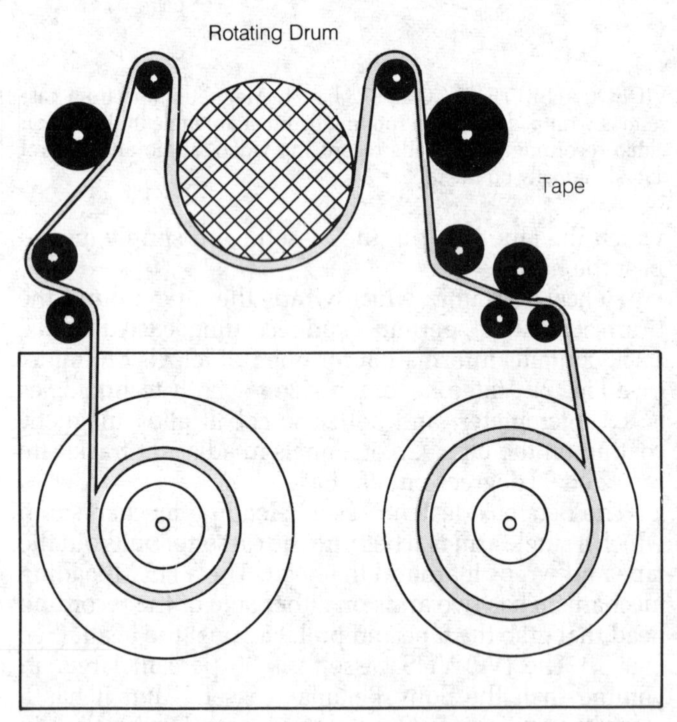

VIDEOCASSETTE RECORDER: Fig. 3. Tape from the VHS videocassette is pulled out and passed around the rotating drum in an M-shaped loading pattern by the VCR mechanism.

Most VCRs are equipped with remote, handheld, battery-powered control units that duplicate many of the front-panel controls and permit recording with one-touch of a button. Typical VCRs record about 240 TV raster lines. VCRs have a wide range of features and prices.

Most VCRs are monophonic and, like amplitude-modulation (AM) radios, are better suited for reproducing conversation than music. However, some offer linear stereo with Dolby, and digital recording. Most VHS machines permit the recording of two to four events over a seven-day span, some units permit the recording of up to eight events over 14 days, and a few models can record events over 365 days. Typical VCRs can record channels from 2 through 13 on the very high frequency (vhf) band, 14 through 83 on ultrahigh frequency (uhf), and 28 CATV (cable) channels. Some models permit simplified on-screen programming with a time and date table display.

Some VCRs are able to slow down and stop action in a live telecast or prerecorded cassette. VHS videocassettes can record up to two hours of telecasts with high resolution, but as many as six hours with low resolution. High-end innovations in both VHS and Beta include horizontal resolutions of 400 to 500 lines, bar-code scanners, and light pens to aid in programming.

VIDEO DISPLAY TERMINAL

The video display terminal (VDT) is a computer input/output I/O terminal with a CRT display and keyboard that interacts with a remote host computer. The VDT has replaced the teleprinter, the first computer terminal in wide use. The early VDTs were called glass Teletypes because the CRT replaced the printer of the Teletype machines used for both input and output. These VDTs without microprocessors were also called dumb terminals because they performed little more than keyboard and display functions and depended on the host computer for all data and text formatting.

A VDT, as shown in the illustration, includes a video monitor, detached keyboard and monitor base. Some VDTs include a flexible monitor mounting that can be swiveled and locked in the most favorable viewing position. The keyboard is full-travel, fully encoded, separately packaged unit with a plug-in cable connection for serial data transmission. VDTs are designed for 24-hour per day online computer connection at such places as banks, hotels, airline ticket counters, car rental stations, and travel agencies.

The CRT monitor enclosure also contains a power supply and card cage for the electronics used in managing the display, encoding, and formatting data, and performing routine communications functions. The circuit cards can contain software- or firmware-programmed microprocessors and memory devices to provide visual attributes, the storage of data pages, and editing. The extensive logic circuitry of earlier VDTs has been replaced, relieving the host computer of many routine tasks. *See* CATHODE-RAY TUBE, MICROPROCESSOR.

The internal microprocessors also make it possible to organize the VDT to emulate the characteristics of other brand-name VDTs and their proprietary commands and protocols. This feature is important for selling the units into the aftermarket to replace older VDTs or for adding new terminals to an existing system.

(Courtesy Wyse Technology.)

VIDEO DISPLAY TERMINAL: Video display terminal with a detached keyboard and a monitor that swivels for optimum viewing.

Monocolor CRT displays are standard, but color displays are available. VDTs can communicate directly with computers in the same building or over great distances with various private and public networks including local area networks (LANs). They can also communicate over dialed-up telephone lines with modems. *See* LOCAL AREA NETWORK, MODEM.

The design of VDTs is strongly influenced by international specifications for human factors engineering or ergonomics. These relate to keyboard keycap and keyboard arrangement and dimensions as well as the characteristics of the CRT display. VDT specifications standards have evolved for the display, keyboard and communications protocol. VDTs are either compatible with ASCII (*see* ASCII) or ANSI (American National Standards Institute X3.64) standards covering protocol, commands and encoding. *See* KEYBOARD AND KEYPAD.

Key specifications for a VDT terminal include: 1) format of the CRT screen—lines and columns or characters per line, 2) screen size—measured diagonally in inches, 3) display color, 4) video attributes, and 5) number of pages of copy that can be stored in memory. Most VDTs can display 24 lines of 80 characters, key selectable to 132 characters per line. Characters are formed by 7 by 9 or 7 by 12 dot matrixes. Screen size is typically 14 inches (diagonal) and displays can be green, amber, black-on-white, or white-on-black. Video attributes include: 1) underlining, 2) blinking, 3) blanking, 4) character boldness, 5) doubling character size, and 6) reverse video. Some VDTs can store up to four pages in memory. Other features include bidirectional scrolling, protected fields for menus or formats, and split-screen presentation.

The most popular telecommunications interface for VDTs is RS232C, but EIA RS422 is becoming more popular. The standard mode data rates are 300, 1200 and 2400 baud. *See* DATA SERIAL TRANSMISSION.

VIDEOTAPE RECORDER

A videotape recorder (VTR) operates in a manner similar to an audio tape recorder, but it has a higher tape speed, which is necessary to allow storage and reproduction of the high-frequency components of a video signal. The VTR uses more advanced circuitry and a different tape track configuration than an audio tape recorder.

The principle of videotape recording is shown in the illustration. The sound track runs along one edge of a tape 2 in (51 mm) wide. The opposite edge contains a control track. The video information is recorded at an angle with respect to the direction of the tape, thus greatly increasing the effective speed of the tape. The actual forward speed is 15 inches per second (IPS). Some VTR units run at 7.5 IPS.

The composite video signal from the camera or television receiver frequency modulates a carrier for recording on the tape. In the playback mode, a demodulator recovers the composite signal. This optimizes the signal-to-noise ratio in the recording and playback processes.

In the past few years, magnetic disks have begun to replace magnetic tape in some video recording applications. The VTR is still widely used, however, especially in television broadcasting. *See also* COMPOSITE VIDEO SIGNAL, MAGNETIC RECORDING, TAPE RECORDER.

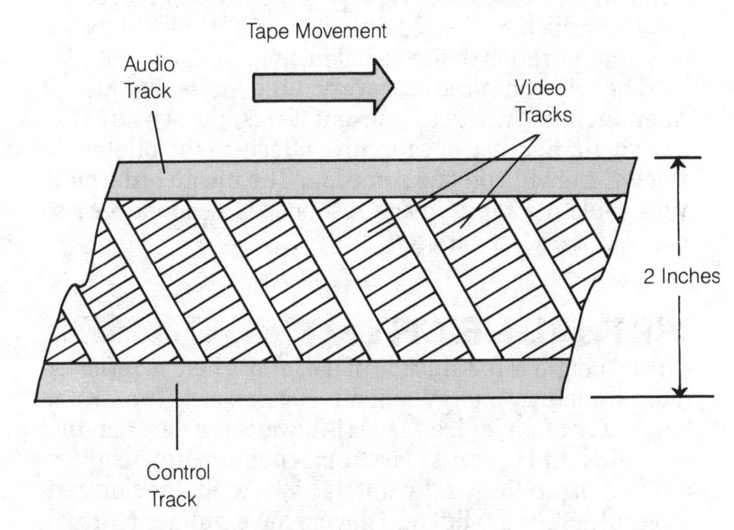

VIDEOTAPE RECORDER: A videotape contains separate tracks for video, audio, and control information.

VIDICON

A vidicon is a small, relatively simple television camera tube. It is preferred over other camera tubes in portable and mobile applications because it is less bulky and heavy. The vidicon is extensively used in closed-circuit television surveillance systems in business and industry.

The operation of a vidicon is illustrated by the diagram. The image is focused by a lens onto a layer of

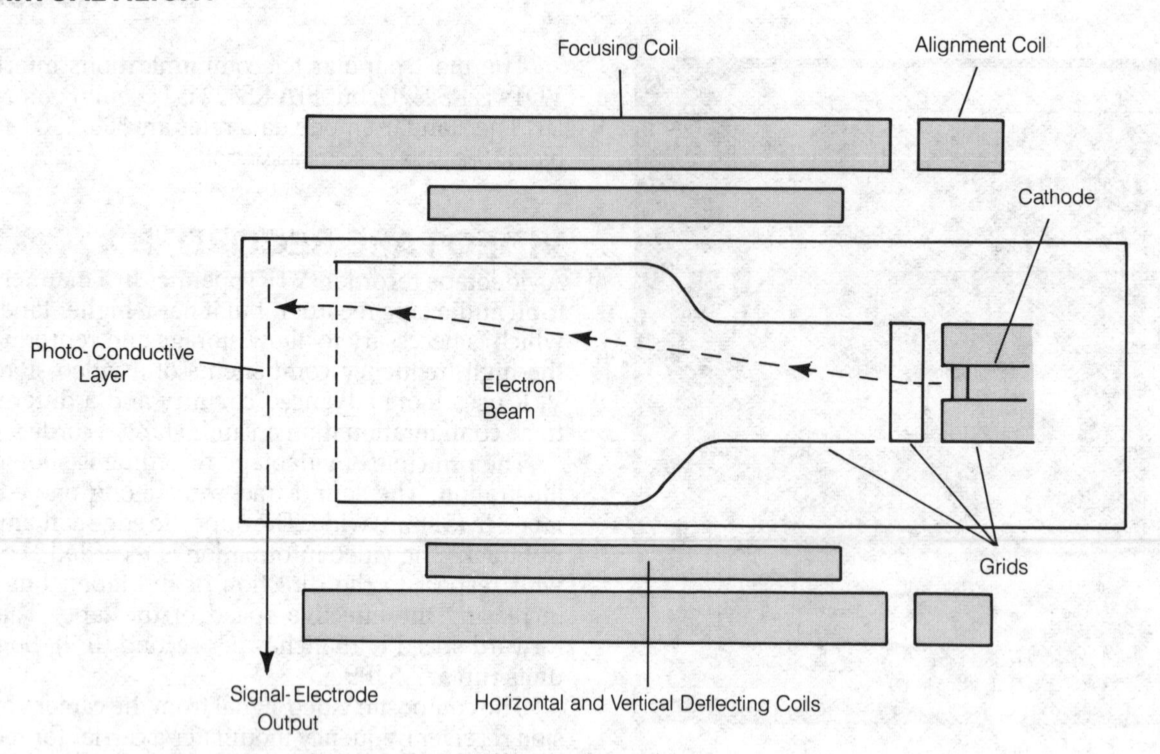

VIDICON: A positive potential pattern on the photoconductive layer corresponds to the pattern of light imaged on the layer. A scanning electron beam reads the layer by depositing electrons and provides a video output signal.

photoconductive material. The electron gun and grids generate an electron beam that scans the photoconductor via the action of external deflecting coils. This beam causes the photoconductive material to become electrically charged. The rate of discharge is roughly proportional to the intensity of the light in a particular part of the photoconductor. The result is different conductivity for different portions of the visual image.

The vidicon produces very little noise because it operates at a low level of current. Thus, the sensitivity is very high. Because of capacitive effects in the photoconductor, the vidicon has some lag. The image orthicon is preferable if a rapid image response is needed. *See also* CAMERA TUBE, IMAGE ORTHICON.

VIRTUAL HEIGHT

Virtual height is the altitude of the ionosphere in terms of radio-frequency energy sent directly upward. The virtual height Hv of an ionized layer is always greater than the actual height H because electromagnetic energy requires time to be reflected by interaction with the ionized molecules. The D, E, and F layers have different virtual heights as well as different actual altitudes (*see* D LAYER, E LAYER, F LAYER).

The virtual height of the ionosphere is determined from the delay in the return of a pulse sent straight upward (see illustration). If the pulse returns after t milliseconds, then the virtual height H_v is given in miles by:

$$H_v = 186t$$

and in kilometers by:

$$H_v = 300t$$

The virtual height of an ionized layer depends to some extent on the frequency. As the frequency rises, the virtual height increases very slightly up to a limit. As the frequency increases further, the virtual height rapidly grows greater. At a certain frequency called the critical

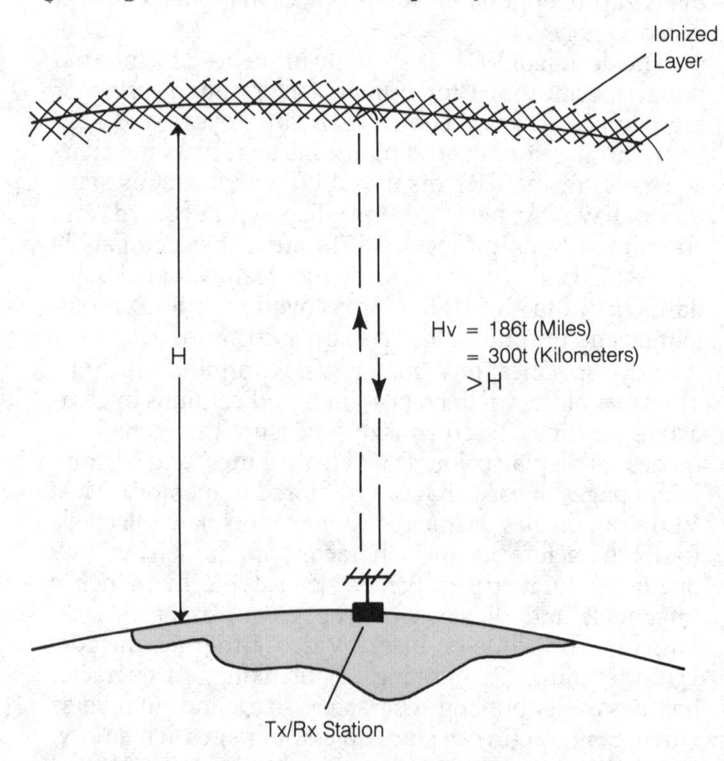

VIRTUAL HEIGHT: Virtual height is determined by the time required for a radio signal to return from the ionosphere.

frequency, the ionosphere no longer returns signals sent directly upward. *See also* CRITICAL FREQUENCY, IONOSPHERE.

VISIBLE LIGHT

See LIGHT, VISIBLE SPECTRUM.

VISIBLE SPECTRUM

The visible spectrum covers the range of electromagnetic wavelengths from 750 nanometers (nm) to 390 nm. This corresponds to frequencies of 400 and 770 THz. These wavelength and frequency values are approximate, since some people can visually detect energy at wavelengths slightly outside this range. The visible spectrum is bordered by the infrared and ultraviolet (*see* INFRARED, ULTRAVIOLET RADIATION).

The wavelength of visible light is a function of the energy contained in each photon. The photons contain the least energy at the longest wavelengths of lowest frequencies, and the most energy at the shortest wavelengths or highest frequencies (*see* PHOTON, PLANCK'S CONSTANT).

The wavelength of light is correlated with the colors seen. For monochromatic light, the longest wavelengths appear deep red. As the wavelength is shortened, the colors progress from red to red-orange, orange, yellow-orange, yellow, yellow-green, green, blue-green, blue, indigo (blue-violet), and violet. There is an infinite range of colors possible in the visible spectrum. The eyes of the average person are most sensitive in the green part of the spectrum; the sensitivity decreases toward the red and violet.

Different beams of monochromatic light can be combined to obtain colors that do not appear in the spectrum. By combining red, green, and blue light in the proper proportions, any hue can be obtained. *See also* LIGHT, PRIMARY COLORS.

VLSI

See VERY LARGE SCALE INTEGRATION.

VOCODER

A vocoder is a circuit that greatly reduces the bandwidth occupied by a voice signal. The unprocessed voice signal needs 2 to 3 kHz of spectrum space for intelligible transmission. A vocoder can compress a voice into a much smaller band with some sacrifice in inflection.

The basic principle of the vocoder is the replacement of certain voice signals with electronically synthesized impulses. A typical vocoder reduces the signal bandwidth to approximately 1 kHz with essentially no degradation of intelligibility. Greater compression ratios are possible with some sacrifice in signal quality. *See also* VODER, VOICE FREQUENCY CHARACTERISTICS.

VODER

A voder is a speech synthesizer for the transmission of recognizable voice signals over a channel with a narrow bandwidth. The voder is similar to the vocoder, except that the voder produces entirely synthesized signals.

A voder generates the various voice sounds with a keyboard. In theory, the voder could reduce the occupied bandwidth of a voice signal by a factor of several hundred. This might result in a bandwidth as narrow as a few hertz. But in practice, the quality of a synthesized voice, compressed to such a tiny part of the spectrum, is degraded. Virtually all inflection, and therefore the emotional content of the voice, is lost in the process. Therefore, the realizable compression factor is smaller than that predicted by theory. *See also* VOCODER, VOICE FREQUENCY CHARACTERISTICS.

VOICE FREQUENCY CHARACTERISTICS

Humans recognize and mentally decode the sounds of speech, but the electrical characteristics of a voice wave are complicated. A voice signal transmits information not only as characters and words, but emotional content as a result of subtle inflections.

A voice signal contains energy in three distinct bands of frequencies called formants. The lowest frequency band is called the first formant, and it occupies the range from a few hertz to near 800 Hz. The second formant ranges in frequency from approximately 1.5 to 2.0 kHz. The third formant ranges from near 2.2 to 3.0 kHz and above. The exact frequencies differ slightly from person to person.

It has been discovered that speech information can be adequately conveyed by restricting the audio response to two ranges, approximately 300 to 600 Hz and 1.5 to 3.0 kHz.

Five basic types of sounds are made by humans. These include the vowels, the semivowels, the plosives, the fricatives, and the nasal consonants. The lowest audio frequencies are generated by the utterance of vowels and semivowels: A, E, I, L, O, R, U, V, W, and Y. Medium frequencies are generated by nasal sounds: M, N, and NG. The highest frequencies are generated by the utterance of fricatives and plosives: B, C, D, F, G, H, J, K, P, PL, Q, S, SH, T, TH, X, and Z. (There are a few irregularities because of language anomalies.)

The analysis of speech characteristics is important in communications, especially for minimizing the occupied bandwidth of voice signals. Devices such as the vocoder and the voder have been used to synthesize some voice sounds, resulting in improved communications efficiency. When speech synthesis is used, however, some degradation occurs in voice inflection. Thus, while the bandwidth may be reduced, the subtle meanings, conveyed by emotion, are sacrificed or even lost. This may or may not be important, depending on the intended communications applications. *See also* SPEECH SYNTHESIS, VOCODER, VODER.

VOICE TRANSMITTER

A voice transmitter is any radio transmitter intended for

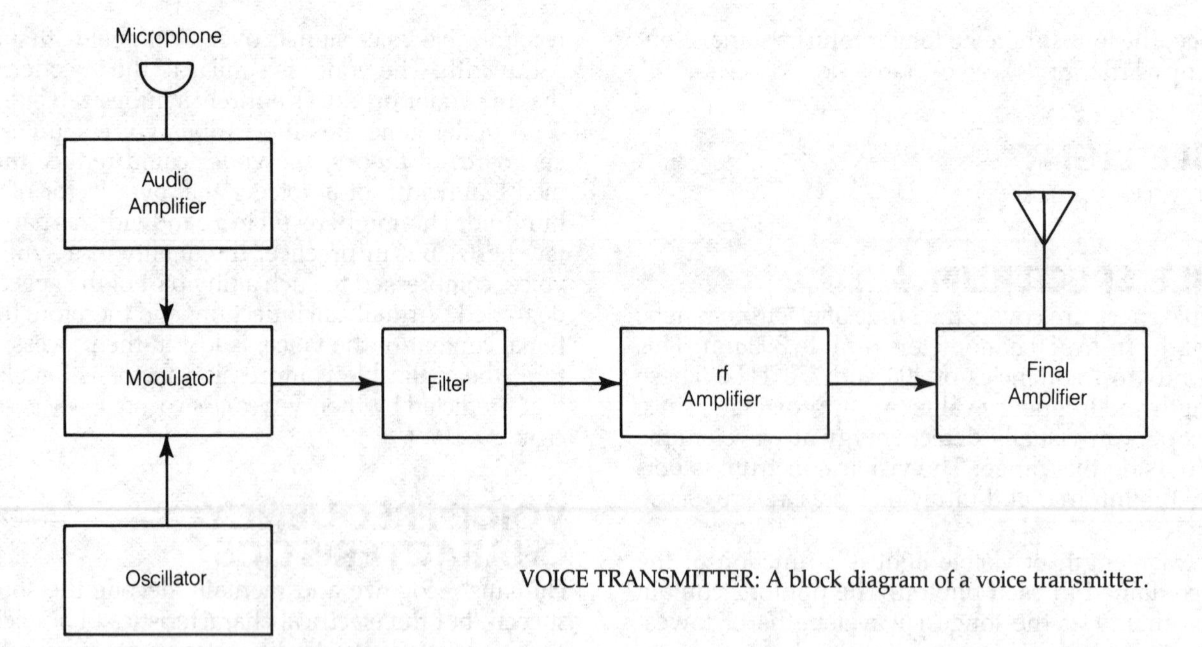

VOICE TRANSMITTER: A block diagram of a voice transmitter.

conveying information with the human voice. The most common methods of voice transmission are amplitude modulation, frequency modulation, and single sideband. Less often used are phase modulation, polarization modulation, pulse modulation, and other methods.

A voice transmitter consists of an oscillator, a modulator, a filter (if needed), and a chain of amplifiers (see illustration). In variable-frequency voice transmitters, mixing circuits are often used for multiband operation. *See also* AMPLITUDE MODULATION, FREQUENCY MODULATION, PHASE MODULATION, POLARIZATION MODULATION, PULSE MODULATION, SINGLE SIDEBAND.

VOLT

The volt is the unit of electric potential. A potential difference of one volt across a resistance of one ohm will result in a current of one ampere, or one coulomb of electrons per second (*see* AMPERE, COULOMB, OHM, OHM'S LAW).

Various units smaller than the volt are used to measure electric potential, especially weak radio signals. These can have root-mean-square magnitudes of less than a millionth of a volt (*see* ROOT MEAN SQUARE). A millivolt (mV) is one thousandth of a volt. A microvolt (μV) is one millionth of a volt. A nanovolt (nV) is a billionth of a volt. It is not likely that a voltage smaller than 1 nV can be detected with conventional instruments.

VOLTAGE

Voltage is the existence of a potential (charge) difference between two objects or points in a circuit. Normally, a potential difference is manifested as an excess of electrons at one point, and/or a deficiency of electrons at another point. Sometimes, other charge carriers, such as holes or protons, can be responsible for a voltage between two objects or points in a circuit (*see* ELECTRON, HOLE, PROTON).

Electric voltage is measured in units called volts. A potential difference of one volt causes a current of one ampere to flow through a resistance of one ohm (*see* AMPERE, COULOMB, OHM, OHM'S LAW). Voltage can be either alternating or direct. In the case of alternating voltage, the value can be expressed in any of three ways: peak, peak-to-peak, or root-mean-square (*see* PEAK-TO-PEAK VALUE, PEAK VALUE, PEAK VOLTAGE, ROOT MEAN SQUARE). Voltage is symbolized by the letter E or V in most electrical equations.

If a given point P has more electrons than another point Q, the point P is said to be negative with respect to Q, and Q is said to be positive with respect to P. This is relative voltage. Relative voltage is not the same as absolute voltage, which is determined with respect to a point at ground (neutral) potential.

When two points with different voltage are connected together, current is considered to flow away from the point having the more positive voltage, although the movement of electrons is actually toward the positive pole. *See also* CURRENT.

VOLTAGE AMPLIFICATION

Voltage amplification is the increase in the magnitude of a voltage between the input and the output of a circuit. It is also sometimes called voltage gain.

Some circuits are designed specifically for amplifying a direct or alternating voltage. Other circuits are intended for amplification of current, while still others are designed to amplify power.

In theory, a voltage amplifier can operate with zero driving power and an infinite input impedance. In practice, the input impedance can be made extremely high, and the driving power is therefore very small, but not zero. A Class A amplifier using a field-effect transistor is the most common circuit used for voltage amplification (*see* CLASS A AMPLIFIER). Bipolar transistors can also be used for this purpose.

Voltage amplification is measured in decibels. Mathematically, if E_{in} is the input voltage and E_{out} is the output voltage, then:

$$\text{Voltage gain (dB)} = 20 \log_{10} (E_{out}/E_{in})$$

See also DECIBEL, GAIN.

VOLTAGE-CONTROLLED OSCILLATOR

See OSCILLATOR CIRCUITS.

VOLTAGE DIVIDER

A voltage divider is a network of passive resistors, inductors, or capacitors, for obtaining different voltages for various purposes.

A resistive voltage divider is commonly employed to provide the direct-current bias in an amplifier or oscillator circuit (see illustration at A). The desired bias is obtained by selecting resistors with the proper ratio of values. If the supply voltage is E and the resistors have values R_1 and R_2 as shown, then the bias voltage is:

$$E_b = \frac{ER_1}{R_1 + R_2}$$

Although there are an infinite number of values of R_1 and R_2 that will provide a given bias voltage E_b, the actual values are chosen on the basis of the circuit impedance.

A capacitive voltage divider is shown at B, and an inductive divider is shown at C. These dividers are used for ac voltages. The ratio of reactances determines the voltage at the tap point, in exactly the same manner as with resistances (see above equation). An example of a capacitive voltage divider is the Colpitts oscillator, and an example of inductive voltage division is found in the Hartley oscillator. *See* OSCILLATOR CIRCUITS.

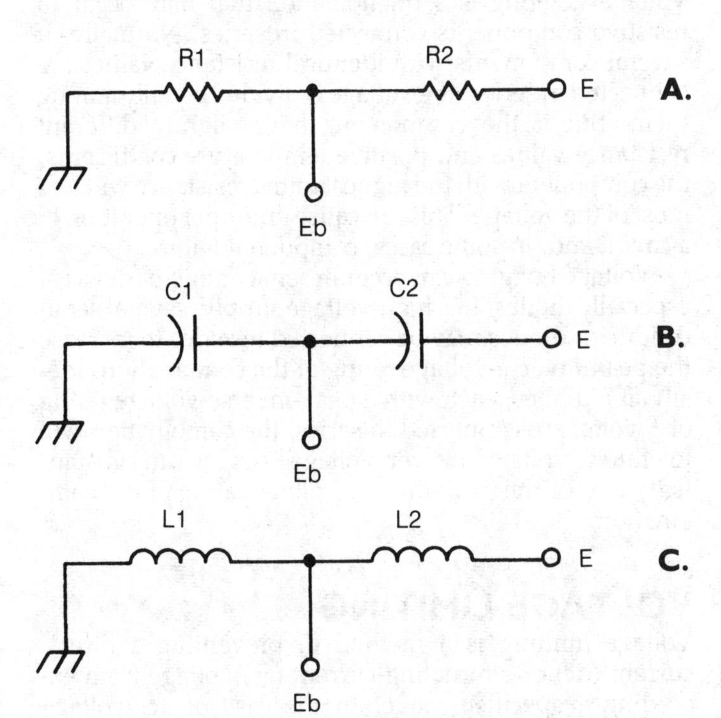

VOLTAGE DIVIDER: A resistive voltage divider circuit (A) and reactive voltage divider circuits (B and C).

VOLTAGE DOUBLER

To obtain high voltages needed for many circuits, a voltage-doubler power supply is often used. The voltage doubler makes it possible to use a transformer with a lower step-up ratio than would be needed if an ordinary full-wave supply were used. Voltage doublers are sometimes used in radio-frequency-actuated circuits to obtain the control voltage. Voltage-doubler circuits are not generally used when excellent regulation is needed, or when the current drain is high.

The illustration shows two types of voltage-doubler supplies. The circuit at A is called a conventional doubler; the circuit at B is known as a cascade doubler. In both circuits, the direct-current (dc) output voltage is twice the peak alternating-current (ac) input voltage, or 2.8 times the root-mean-square input voltage. The conventional doubler provides superior regulation and less output ripple, but the cascade circuit can be used without a transformer. Two or more cascade circuits can be connected in series to obtain voltage multiplication by any power of 2. *See also* VOLTAGE MULTIPLIER.

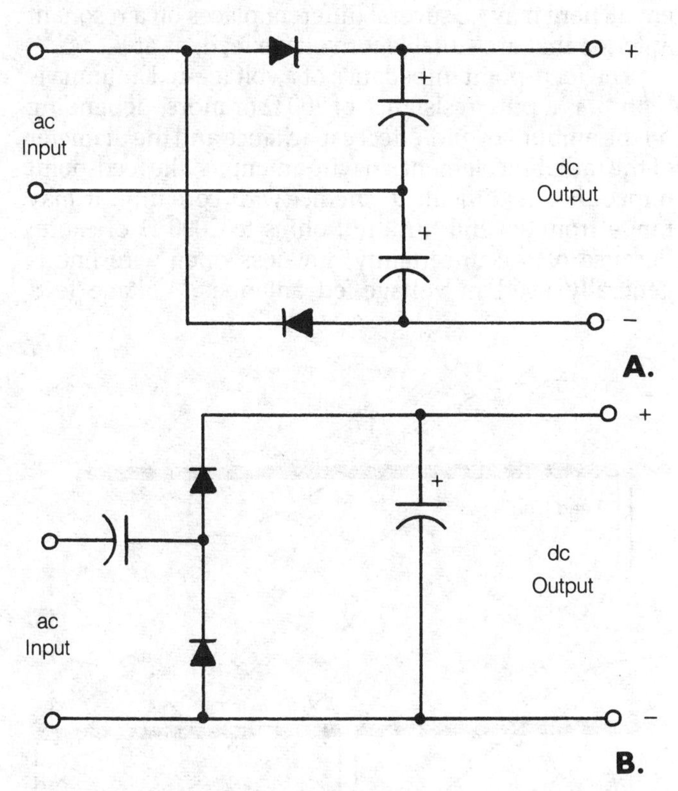

VOLTAGE DOUBLER: A conventional voltage-doubler supply (A) and a cascade voltage doubler (B).

VOLTAGE DROP

Voltage drop is the potential difference that appears across a current-carrying impedance.

For direct currents and resistances, the voltage drop E

across a resistance of R ohms that carries a current of I amperes is given by Ohm's law as:

$$E = IR$$

For alternating currents and reactances or complex impedances, the root-mean-square voltage drop E is determined from the root-mean-square current I and the absolute-value impedance $R + jX$ as:

$$E = I\sqrt{R^2 + X^2}$$

See also CURRENT, IMPEDANCE, OHM'S LAW, REACTANCE, RESISTANCE, VOLTAGE.

VOLTAGE FEED

Voltage feed is a method of connecting a radio-frequency feed line to an antenna at a point on the radiator where the voltage is maximum. This point is called a voltage loop or a current node (*see* CURRENT NODE, VOLTAGE LOOP). In a half-wavelength radiator, the voltage maxima occur at the ends. Therefore, voltage feed for a half-wavelength antenna can only be accomplished by connecting the line to either end of the radiating element (see illustration at A). In an antenna longer than ½ wavelength, voltage maxima exist at multiples of ½ wavelength from either end. There may be several different places on a resonant antenna that are suitable for voltage feed, as at B.

The feed-point impedance of a voltage-fed antenna is high. It is a pure resistance of 200 Ω or more, depending on the amount of end-effect capacitance and the diameter of the radiating element. In wire antennas, the feed-point impedance is difficult to predict with certainty; it may range from several hundred ohms to 2000 Ω or more. Because of the uncertainty, low-loss open-wire line is generally used in voltage-fed antennas. Voltage feed

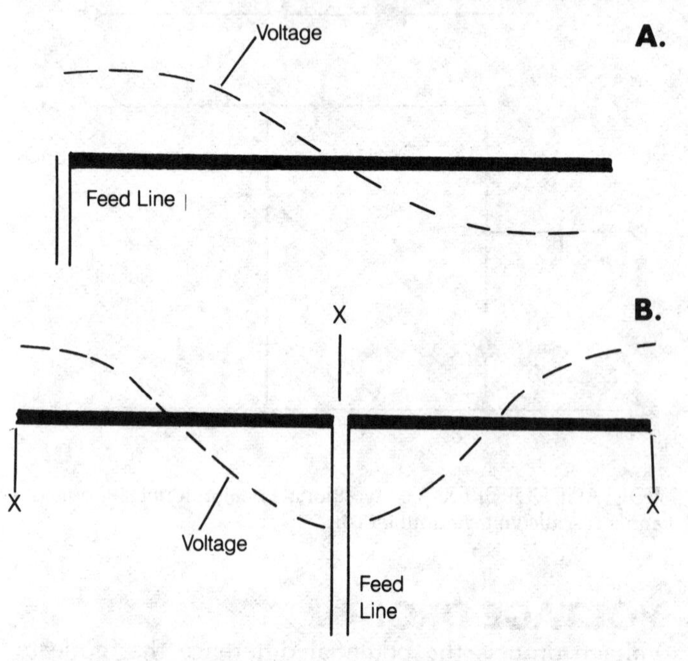

VOLTAGE FEED: End-fed antenna is always voltage fed (A), and points on a one-wavelength antenna for voltage feed are marked X (B).

results in good electrical balance for a two-wire line, provided the current and voltage distribution are reasonably symmetrical in the antenna. *See also* CURRENT FEED.

VOLTAGE GAIN

See VOLTAGE AMPLIFICATION.

VOLTAGE GRADIENT

Whenever a potential difference exists between two points (such as at either end of a long wire or resistor), there are points where the voltage is at some intermediate value. The rate of voltage change per unit length is called the voltage gradient.

The voltage gradient in a given circuit or conductor depends on the potential difference between the two end points, and on the distance between these points. In general, if the end points have voltages E_1 and E_2, and the length of the conductor or component is d meters, then the voltage gradient, in volts per meter, is:

$$G = \frac{E_2 - E_1}{d}$$

Voltage gradient also can be specified in volts per foot, centimeter, millimeter, or micron.

The voltage gradient is sometimes specified for semiconductor materials. This is especially true for field-effect transistors, in which a voltage gradient exists along the channel (*see* FIELD-EFFECT TRANSISTOR).

VOLTAGE HOGGING

Voltage hogging is a phenomenon that may occur in resistive components connected in series. Normally, if several components with identical resistance values are connected in series, the voltage is divided equally among them. But if the components have slightly different resistance values and positive temperature coefficients, the component with the highest initial resistance will hog most of the voltage. This can cause improper operation of a circuit and, in some cases, component failure.

Voltage hogging can occur in semiconductor devices, especially diodes. In a high-voltage supply, several semiconductor diodes may be connected in series to increase the peak-inverse-voltage rating of the combination. Ideally, if n diodes, each with a peak-inverse-voltage rating of E volts, are connected in series, the combination will tolerate E_n volts. However, voltage hogging can substantially reduce the peak-inverse-voltage rating of the combination.

VOLTAGE LIMITING

Voltage limiting is a method of preventing a direct-current (dc) or alternating-current (ac) voltage from exceeding a specified value. In the case of ac, voltage limiting is usually called clipping (*see* CLIPPER).

Voltage limiting is used for regulation in a dc power supply. A Zener diode connected across the source of voltage serves this purpose. *See also* POWER SUPPLY, ZENER DIODE.

VOLTAGE LOOP

In an antenna radiating element the voltage in the conductor depends on the location. At any free end the voltage is maximum. At a distance of ½ wavelength from a free end, or any multiple of ½ wavelength from a free end, the voltage is also at a maximum. The points of maximum voltage are called voltage loops.

A ½-wavelength radiator has two voltage loops, one at each end. A full-wavelength radiator has three voltage loops: one at each end and one at the center. In general, the number of voltage loops in an antenna radiator is 1 plus the number of half wavelengths. The illustration shows the voltage distribution in an antenna radiator of 3⁄2 wavelength, showing the locations of the voltage loops.

Voltage loops will occur along the length of a feed line not terminated in an impedance identical to its characteristic impedance. These loops occur at multiples of ½ wavelength from the resonant antenna feed point when the antenna impedance is greater than the characteristic impedance of the line. The loops exist at odd multiples of ¼ wavelength from the feed point when the resonant antenna impedance is less than the characteristic impedance of the line. Ideally, the voltage on a transmission line should be the same everywhere and equal to the product of the characteristic impedance and the line current. *See also* VOLTAGE NODE, STANDING WAVE.

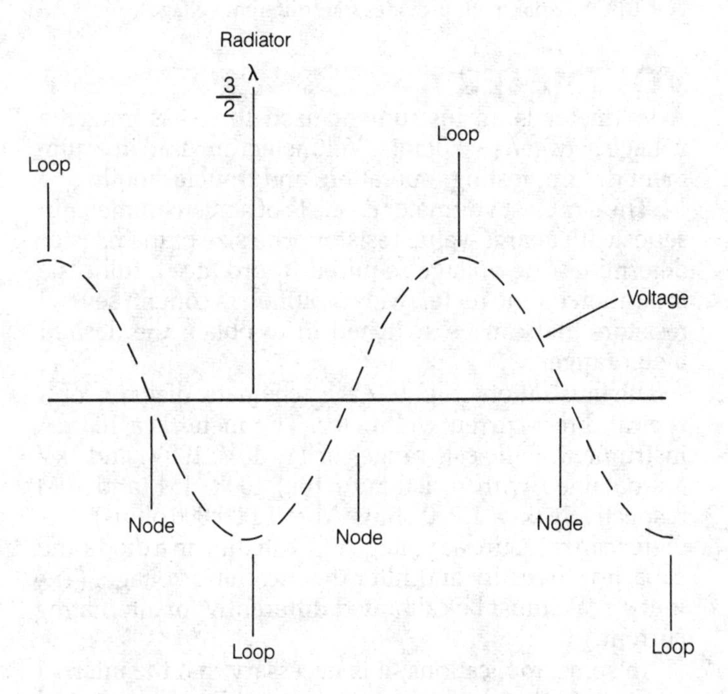

VOLTAGE LOOP: Locations of voltage loops and nodes on 3⁄2-wave radiator.

VOLTAGE MULTIPLIER

A voltage multiplier is a form of power supply designed to deliver a very high voltage from a transformer with a secondary that supplies only a moderate voltage. The most common type of voltage multiplier is the voltage doubler (*see* VOLTAGE DOUBLER). However, some supplies have multiplication factors of 3, 4, or more.

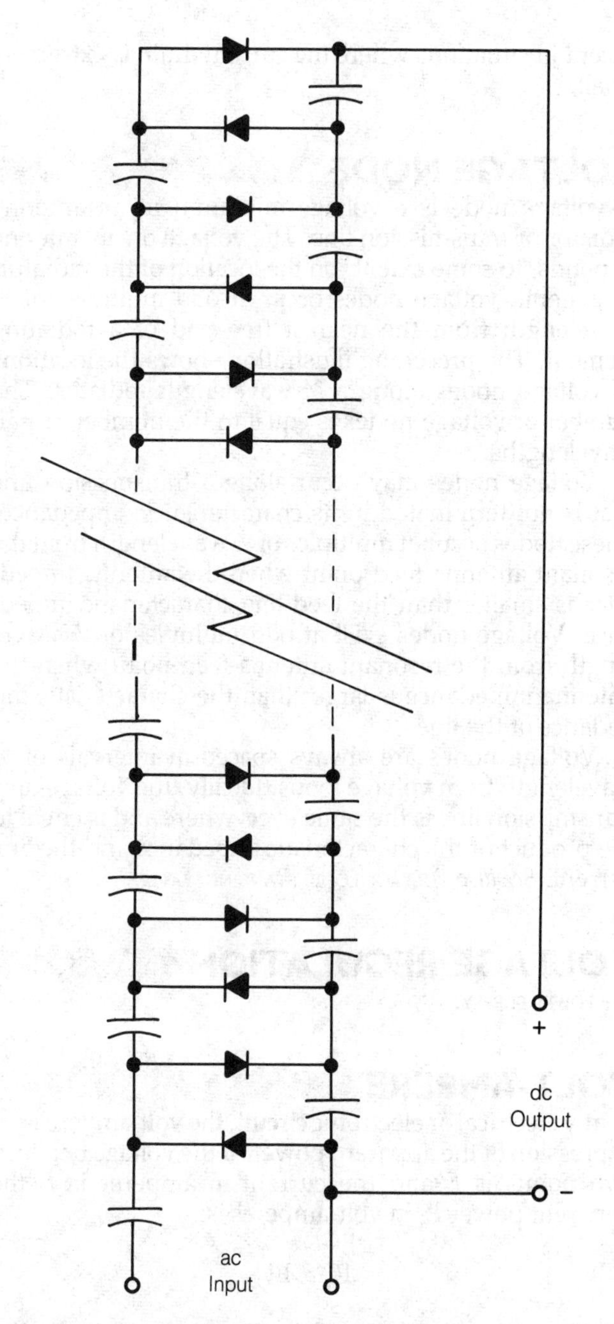

VOLTAGE MULTIPLIER: General configuration of voltage-multiplier power supply.

The illustration shows the general principle of operation for a voltage multiplier. Capacitors and diodes are connected in a lattice configuration. For a multiplication factor of n, the circuit requires $2n$ capacitors and $2n$ diodes. The direct-current output voltage is n times the peak input voltage, or about $1.4n$ times the root-mean-square input voltage.

In practice, there is a limit to voltage multiplication in the circuit shown. The larger the multiplication factor n, the lower the voltage regulation becomes. For this reason, high-order voltage multipliers are not often used except in situations where the current drain is extremely small.

VOLTAGE NODE

A voltage node is a voltage minimum in an antenna radiator or transmission line. The voltage on an antenna depends, to some extent, on the location of the radiator. In general, voltage nodes occur at odd multiples of ¼ wavelength from the nearest free end of a radiating element. The preceding illustration shows the locations of voltage nodes along a 3⁄2-wavelength radiator. The number of voltage nodes is equal to the number of half wavelengths.

Voltage nodes may occur along a transmission line that is not terminated in its characteristicic impedance. These nodes occur at multiples of ½ wavelength from the resonant antenna feed point when the antenna impedance is smaller than the feed-line characteristic impedance. Voltage nodes exist at odd multiples of ¼ wavelength from the resonant antenna feed point when the antenna impedance is larger than the characteristic impedance of the line.

Voltage nodes are always spaced at intervals of ¼ wavelength from voltage loops. Ideally, the voltage on a transmission line is the same everywhere and is equal to the product of the characteristic impedance and the line current. *See also* VOLTAGE LOOP, STANDING WAVE.

VOLTAGE REGULATION

See POWER SUPPLY.

VOLT-AMPERE

In any electrical or electronic circuit, the volt-ampere is an expression of the apparent power. If the voltage between two points is *E* and the current in amperes is *I*, the apparent power P_a in volt-amperes is:

$$P_a = EI$$

In direct-current circuits, the volt-ampere is identical to the watt, because the apparent power is the same as the true power. However, in alternating-current circuits, the volt-ampere may not be a true indication of the dissipated power. *See also* APPARENT POWER, POWER, REACTIVE POWER, TRUE POWER, VAR, WATT.

VOLT-AMPERE-HOUR METER

A volt-ampere-hour meter is an instrument that determines the product of the voltage, current, and time in an electrical circuit. The common utility meter operates as a volt-ampere-hour meter. This meter integrates the current in amperes over a period of time, obtaining the reading in watthours or kilowatt hours, under the assumption that the voltage is constant (234 V in most cases). *See* WATTHOUR METER.

VOLT BOX

A volt box is a precision voltage-divider circuit, used in the calibration of meters and other test instruments. A typical volt box is made with series-connected, fixed resistors, as shown in the diagram. There is one pair of input terminals and several output terminals. The various voltages are obtained from the appropriate output terminals, provided the load resistance is significantly higher than that of the series combination in the box. This is always the case for voltmeters and oscilloscopes. *See also* VOLTAGE DIVIDER.

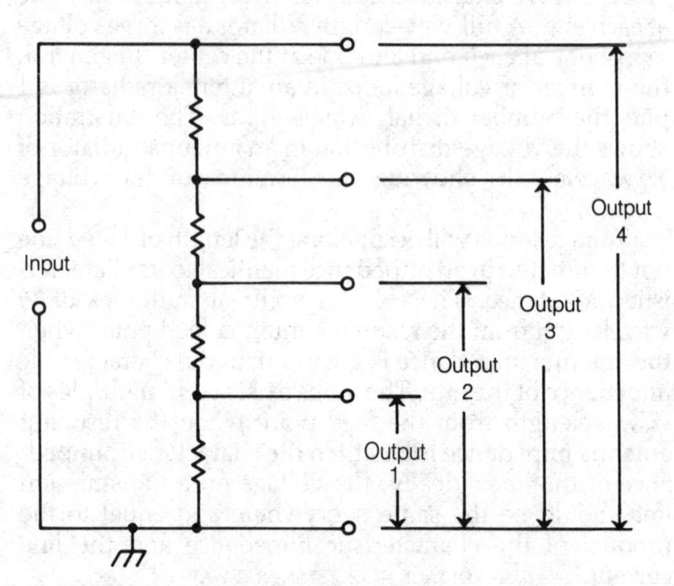

VOLT BOX: This circuit provides four different voltages.

VOLTMETER

A voltmeter is an instrument used for measuring the voltage between two points. Voltmeters are used in equipment design, testing, operation, and troubleshooting.

The simplest voltmeter consists of a microammeter in series with a large-value resistor. The size of the resistor determines the voltage required to produce a full scale indication on the meter. Most voltmeters contain several resistors that can be switched in to obtain the desired meter range.

The illustration on p. 907 is a schematic diagram of a typical direct-current voltmeter. The meter is a 100 μA instrument. Full-scale ranges of 1V, 10V, 100V, and 1kV are obtained with resistors of 10k, 100k, 1M, and 10M respectively (k = 1,000 ohms; M = 1,000,000 ohms). The meter can measure ac voltage by switching in a diode and capacitor to rectify and filter the incoming voltage. (The meter scale must be calibrated differently for alternating current.)

In some applications, it is necessary that the internal resistance of a voltmeter be extremely high. The internal resistance of a voltmeter is expressed in ohms per volt. The meter illustrated has an internal resistance of 10K ohms per volt—a value too low for voltage measurement in high-impedance circuits. *See also* DIGITAL MULTIMETER, FET VOLTMETER, VACUUM-TUBE VOLTMETER.

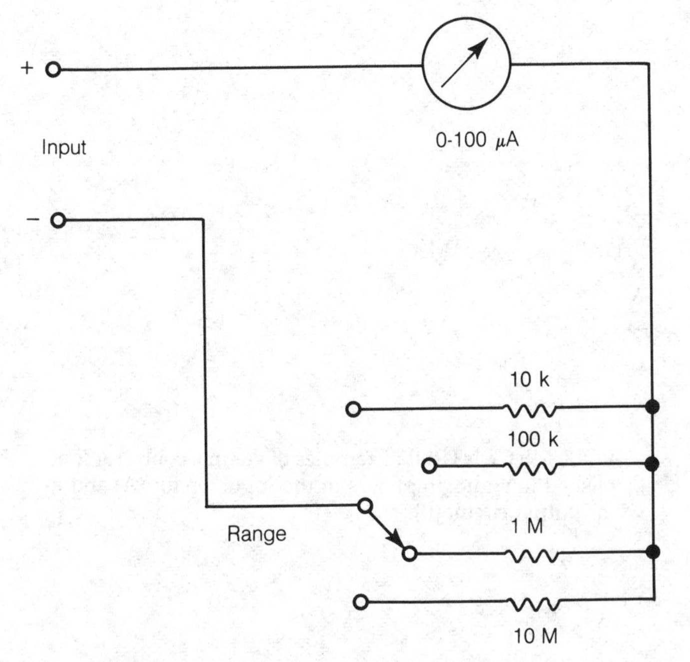

VOLTMETER: A diagram of a typical voltmeter with four range settings of 0-1, 0-10, 0-100 and 0-1000 volts.

VOLT-OHM-MILLIAMMETER

A versatile test meter consisting of a combination voltmeter, ohmmeter, and milliammeter is known as a volt-ohm-milliammeter (VOM) or multimeter.

A multimeter contains a microammeter, a source of direct current, and a switchable network of resistors. In the voltmeter mode, series resistances are selected to determine the meter range. In the ohmmeter mode, the meter is connected in series with the test leads and the power source. In the milliammeter mode, shunt resistances are placed across the meter to obtain the desired full-scale range.

The photograph shows a typical VOM. This instrument can measure voltages up to 1 kV (either alternating or direct current), direct currents to 10 A, and resistances to several megohms. A range-doubler switch allows for optimal meter reading.

While the instrument shown is adequate for most testing and service purposes, the ohms-per-volt rating might not be high enough for precise measurements. *See also* FET VOLTMETER, VACUUM-TUBE VOLTMETER.

VOLUME

Volume is an expression of the loudness of the sound produced by the speakers of a radio, tape recorder, or record player. This is a function of the audio-frequency current and voltage supplied to the speakers.

Volume can be expressed in decibels relative to the threshold of hearing. It is also specified in terms of electrical decibels relative to a power level of 2.51 mW in a load of 600 ohms resistive impedance. These units are called volume units. (*See* DECIBEL, LOUDNESS, SOUND, VOLUME UNIT, VOLUME-UNIT METER.)

VOLUME COMPRESSION

Volume compression is a technique for improving the signal-to-noise ratio in a communications, high-fidelity, or public-address system. The basic principle is to enhance low-level sounds at some or all frequencies in the transmitting or recording process. There are several volume-compression schemes used for different purposes. *See* AUTOMATIC GAIN CONTROL, AUTOMATIC LEVEL CONTROL, COMPRESSION, COMPRESSION CIRCUIT, DOLBY, SPEECH COMPRESSION.

VOLUME CONTROL

A volume control is a component, usually a potentiometer or set of potentiometers, used for adjusting the volume in an audio circuit.

There are two basic types of volume control: input adjustment and output adjustment. These two methods of volume control are shown in the illustration on p. 908. Both methods will provide satisfactory results. In high-power audio circuits, input adjustment is more commonly used.

VOLUME UNIT

In high-fidelity applications, the amplitude of a music signal is generally expressed in terms of volume units

VOLT-OHM-MILLIAMMETER: A widely used portable test instrument.

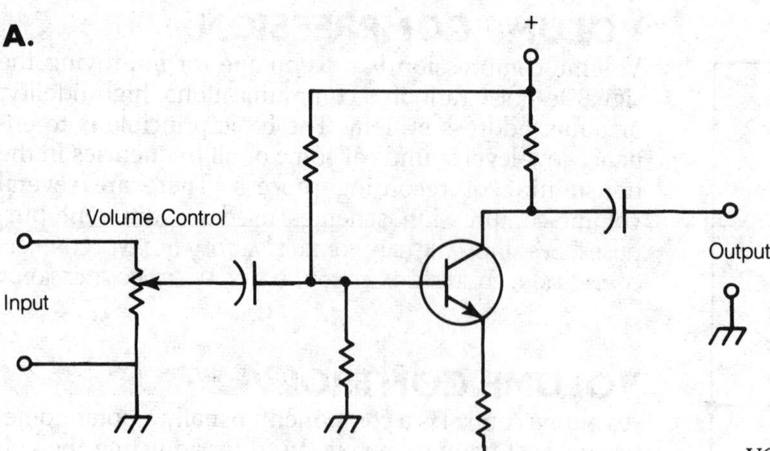

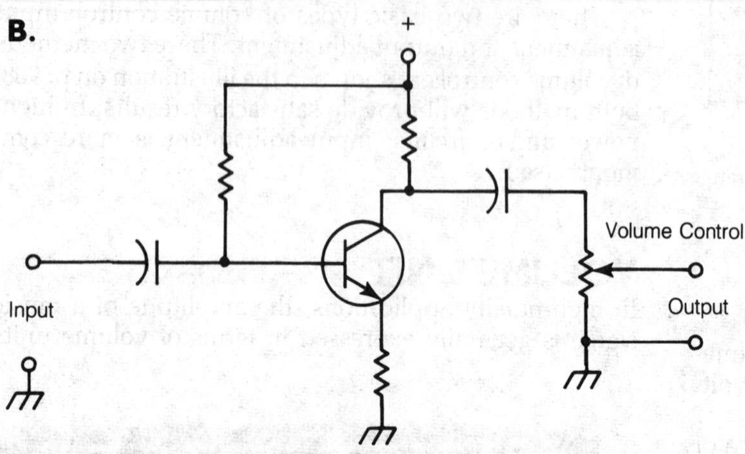

VOLUME CONTROL: Examples of volume control accomplished in audio amplifiers in the input circuit (A) and in the output circuit (B).

(VU). The VU is a relative indicator of the root-mean-square (rms) power output from an audio amplifier.

A level of 0 VU corresponds to +4 dBm, or 2.51 mW, across a purely resistive load of 600 Ω impedance. This is a rms voltage of 1.23 V and a rms current of 2.05 mA. In general, a level of x VU corresponds to a signal x dB louder than the reference level. Thus, for a power level of *P* mW, the VU level is:

$$x = 10 \log_{10} \frac{P}{2.51}$$

Volume units are measured by means of a special meter, called a volume-unit meter, or VU meter, at the output of an audio amplifier. *See also* DECIBEL, VOLUME, VOLUME-UNIT METER.

VOLUME-UNIT METER

A volume-unit (VU) meter is an instrument for measuring the root-mean-square volume level in an audio amplifier. The VU meter is calibrated in decibels relative to +4 dBm (*see* VOLUME UNIT).

Most VU meters are fast-acting instruments with just enough damping to allow easy reading. In a sophisticated stereo high-fidelity amplifier, each channel has a VU meter. The scale is marked off in black and red numerals with a black and red reference line (see illustration). The amplifier gain should normally be set so that the meter needle never enters the red range indicating that distortion is likely to occur on audio peaks.

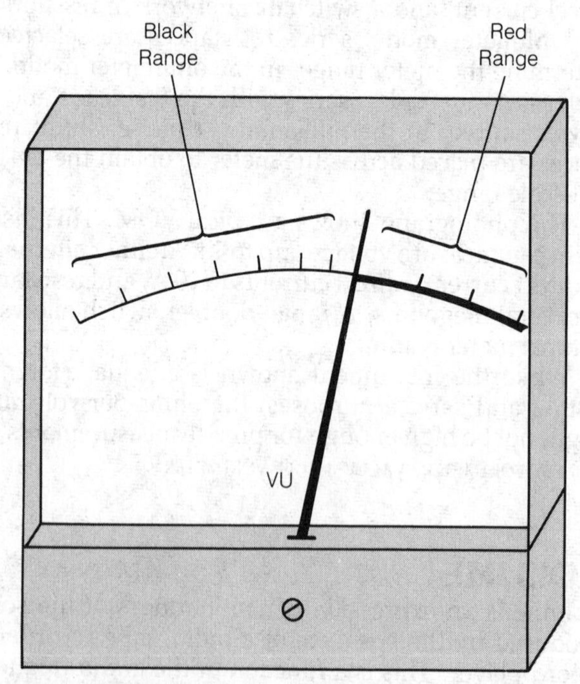

VOLUME-UNIT METER: If the needle of the volume-unit meter is in the black range, distortion is low. The red range indicates increasing distortion in an audio amplifier.

VOM

See VOLT-OHM-MILLIAMMETER.

VOX

VOX is a method for actuating a radio transmitter with voice signals from a microphone or audio amplifier through a relay or electronic switch.

VOX circuits are commonly used in communications transceivers, usually single-sideband units. VOX eliminates the need for pressing a push-button to change from the receive mode to the transmit mode. This feature is especially useful in mobile communications when both hands are needed for other tasks or taking notes. The VOX circuit can be disabled and push-to-talk switching can be used if the operator desires.

In VOX operation, the transmitter is actuated within a few milliseconds after the operator speaks into the microphone. The transmitter remains actuated for a short time after the operator stops speaking; this prevents unwanted tripping out during short pauses.

The schematic diagram shows a VOX circuit. The sensitivity and delay are adjustable. An anti-VOX circuit prevents the received signals from actuating the transmitter if a speaker is used for listening.

VOX is a form of semi break-in operation. If full break-in operation is required with single-sideband communications, a more advanced circuit is needed.

VSWR

See STANDING-WAVE RATIO.

VTVM

See VACUUM-TUBE VOLTMETER.

VTVOM

See VACUUM-TUBE VOLTMETER, VOLT-OHM-MILLIAMMETER.

VU METER

See VOLUME-UNIT METER.

VXO

See OSCILLATOR CIRCUITS.

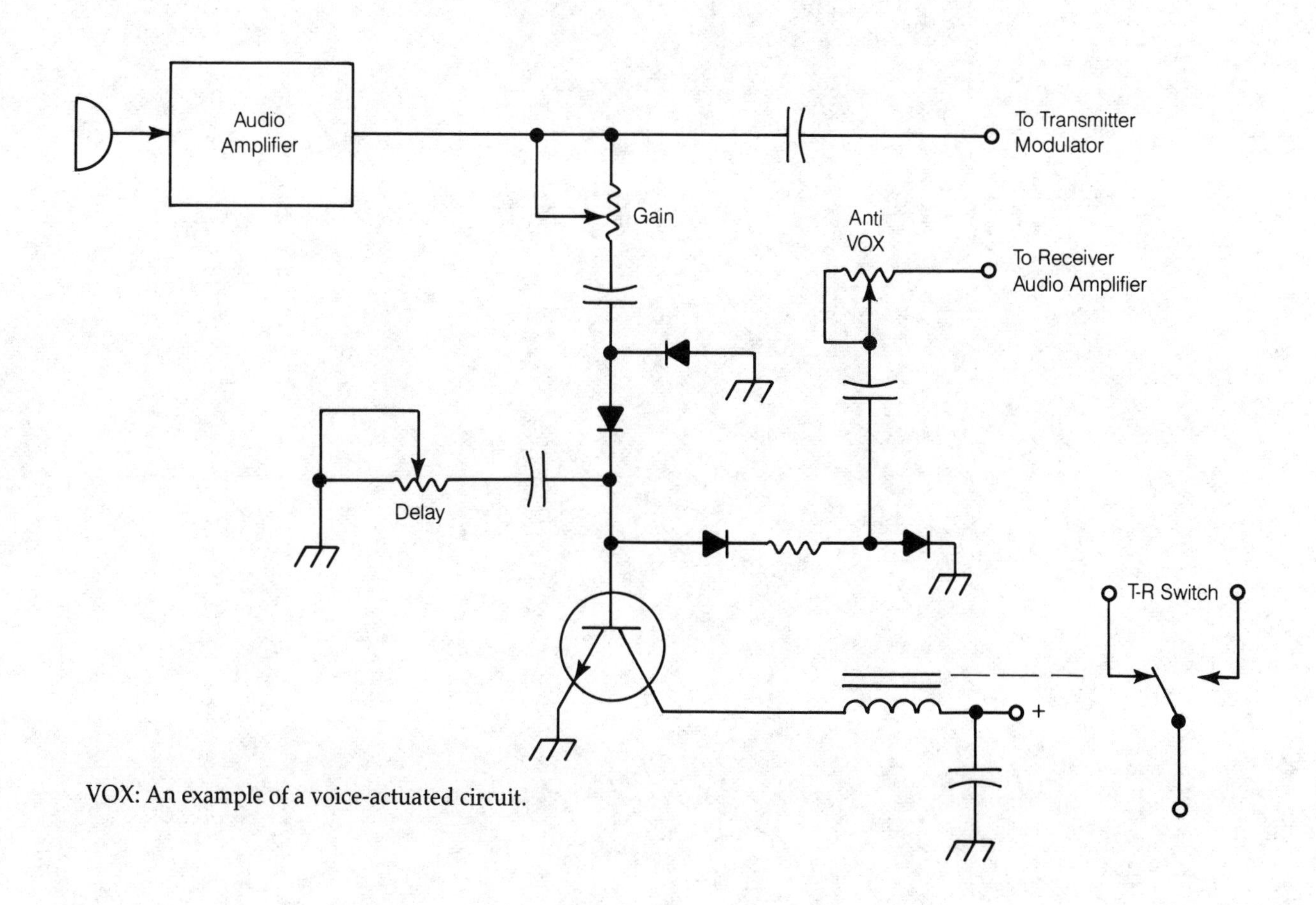

VOX: An example of a voice-actuated circuit.

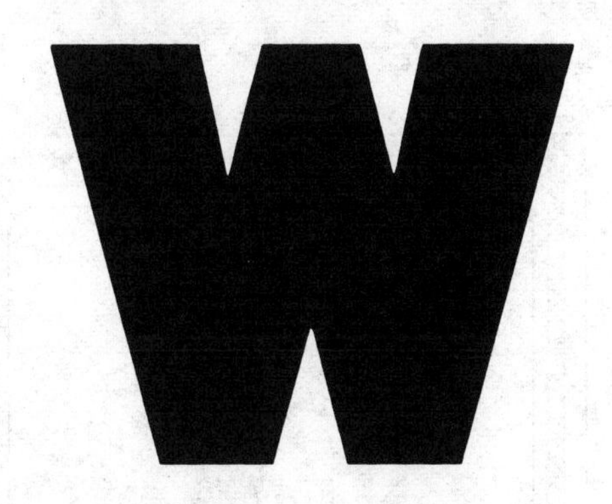

WAFER

See INTEGRATED CIRCUIT MANUFACTURE.

WAFER SWITCH

See SWITCH.

WATT

The watt is the unit of true, or real, power, abbreviated as W. In a direct-current (dc) circuit, or in an alternating-current (ac) circuit containing no reactance, the power *P*, in watts, can be found from any of the formulas:

$$P = EI = I^2R = E^2/R$$

where *E* is the root-mean-square voltage in volts, I is the rms current in amperes, and *R* is the resistance in ohms. In an ac circuit containing reactance as well as resistance, the true power in watts must be determined in a different way (*see* POWER, TRUE POWER).

A power level of 1 W represents the expenditure of 1 joule of energy per second. A flashlight or small lantern bulb dissipates approximately 1 W of power. The audio output from a typical small portable radio is about 1 W when the volume is turned to maximum. An average fluorescent light bulb consumes 15 to 50 W; an incandescent bulb, as much as 500 W; a heavy appliance, as much as 2,000 W.

Fractions of the watt are used to express low power. The milliwatt (mW) is a thousandth (0.001) watt; the microwatt (uW) is a millionth (0.000001) watt. For larger power levels, the kilowatt (kW) or megawatt (MW) are used: 1 kW = 1,000 W and 1 MW = 1,000,000 W.

The radio-frequency output of a transmitter can be expressed in watts. This is the power dissipated in a pure resistor with a value equal to the matched output impedance of the transmitter—usually 50 ohms. *See also* APPARENT POWER, JOULE, REACTIVE POWER, TRUE POWER, VAR, VOLT-AMPERE, WATTMETER.

WATTHOUR

The watthour (wh) is a unit of energy equivalent to 3,600 joules (watt seconds). The watthour is used to express the energy consumed by an electrical product over a period of time. It can also be used to express the storage capacity of a battery. Energy in watthours is the product of the average power drawn, in watts, and the time, in hours.

If a 60 W bulb burns for one hour, the energy consumed by the bulb is 60 Wh. In two hours the same bulb consumes 120 wh; in three hours it consumes 180 Wh; and so on. A 12 V battery with 50 ampere hours of storage capacity can deliver 12 × 50, or 600, watthours of energy. This might constitute 1 watt for 600 hours, 2 watts for 300 hours, or any of a number of other combinations. In electrical utility readings, the kilowatt hour is more often specified than the watthour. This is equivalent to 1,000 watt hours. The abbreviation for kilowatt hour is kWh. *See also* ENERGY, JOULE, KILOWATT HOUR, WATTHOUR METER.

WATTHOUR METER

A watthour meter is an instrument used to measure consumed energy. An electric meter is a form of watthour meter although it actually registers kilowatt hours (*see* KILOWATT HOUR, WATTHOUR). The abbreviation for watthour is Wh, and the abbreviation for kilowatt hour is kWh.

Most watthour meters actually register the integral of the root-mean-square current in a circuit over a period of time. Thus, a meter is suitable for operation only at a single rated voltage such as 234 V. A motor in the meter runs at a speed that is directly proportional to the current drawn at any given time. The motor is connected to a set of gears attached to pointers or drum indicators that numerically indicate the energy used to a given time.

A typical electric meter is shown in the figure. The points turn in opposite directions. The meter illustrated indicates 1604 kWh. Some electric meters have rotating drums with the numerals painted directly on their surfaces allowing direct and simple reading.

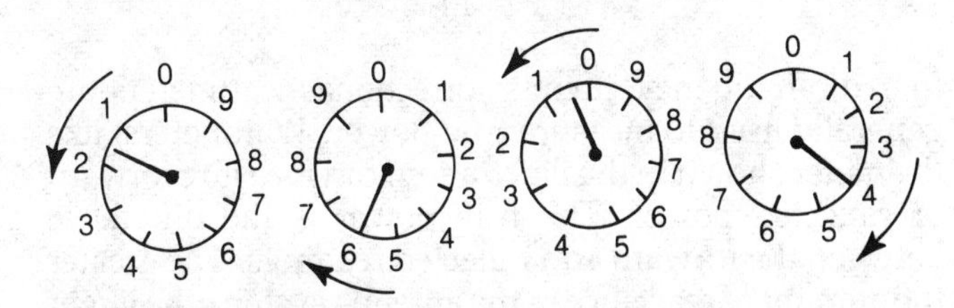

WATTHOUR METER: This meter reads out in kilowatt hours and shows 1604 kWh consumed.

WATTMETER

A wattmeter is an instrument that measures power.

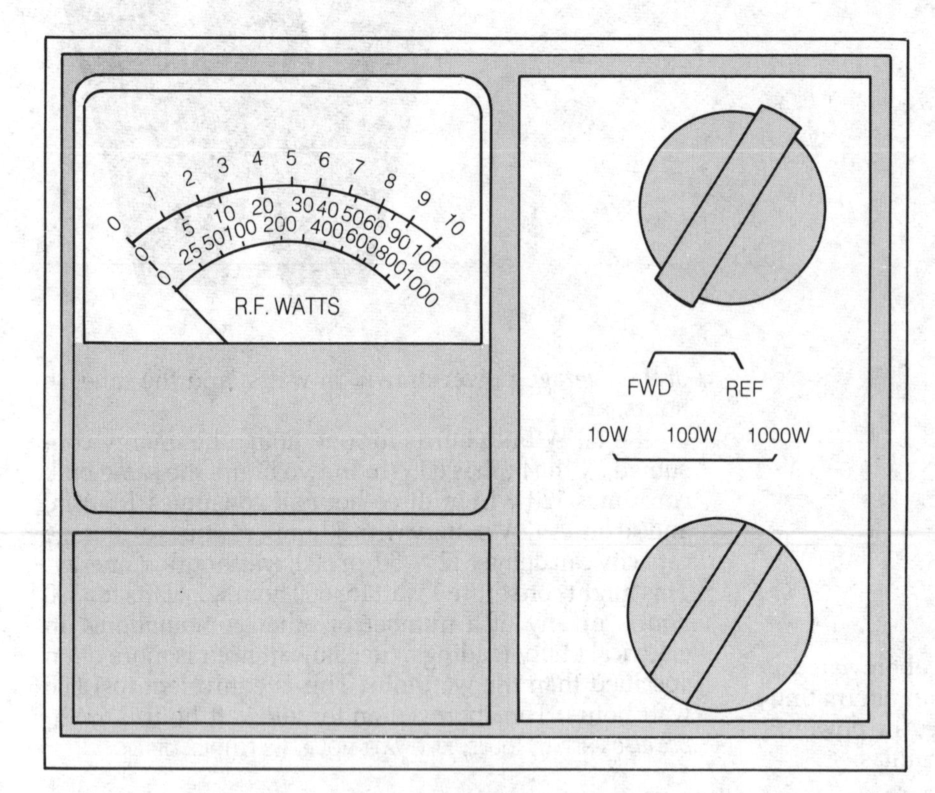

WATTMETER: A typical radio-frequency wattmeter measures both forward and reflected power.

There are many types of wattmeters; some measure true power, while others measure only the apparent power (*see* APPARENT POWER, POWER, REACTIVE POWER, TRUE POWER).

The simplest wattmeter consists of an ammeter in series with a circuit, and a voltmeter in parallel with the component or set of components for which the power consumption is to be determined. The power (in watts) is given by the product of the voltage (in volts) and the current (in amperes). This method is satisfactory for direct-current circuits and for alternating-current circuits in which no reactance is present. However, if reactance exists, the product of current and voltage is artificially large, and does not represent true power.

In a typical house circuit there is usually no reactance. Therefore, a root-mean-square ammeter can be used to measure the power consumed by a device at a particular voltage. The ammeter is connected in series with the appliance. The consumed power, P, is given (in watts) in terms of the current reading, I (in amperes) and the line voltage, E (in volts) as:

$$P = EI$$

In radio-frequency circuits, directional wattmeters are generally used to measure true power. Wattmeters, like the meter shown in the illustration, can measure forward or reflected power. The true power is the difference between the forward and reflected readings. The greater the amount of reactance in the antenna system, the larger the reflected reading will be, in proportion to the forward reading. *See also* REFLECTED POWER, REFLECTOMETER.

WAVE

A wave is any form of disturbance that exhibits a periodic pattern. Examples of waves include acoustic disturbances, pulsating or alternating currents or voltages, and electromagnetic fields.

All waves have a definable period T (usually given in seconds), a specific frequency f, a propagation speed c (usually expressed in units per second), and a wavelength λ. These quantities are related according to the formulas:

$$c = f\lambda = \lambda/T$$

Some disturbances exhibit changes in amplitude that have no identifiable period. These disturbances are not true waves although they may produce noticeable effects such as acoustic or electromagnetic noise over a wide band of frequencies. *See also* ALTERNATING CURRENT, ELECTROMAGNETIC FIELD, LONGITUDINAL WAVE, SOUND, TRANSVERSE WAVE, WAVEFORM.

WAVEFORM

Waveform is an expression used to describe the shape of a wave disturbance, either as seen directly or as observed on a display instrument such as an oscilloscope. A wave disturbance can have any of an infinite number of forms.

The simplest waveform occurs when a disturbance has only one frequency. The resulting waveform is sinusoidal and is known as a sine wave because its shape is identical to the graph of the sine function (*see* SINE, SINE WAVE). More complex waveforms result when energy is concentrated at frequencies that are integral multiples of a lowest, fundamental frequency (*see* FUNDAMENTAL FREQUENCY, HARMONIC). Examples include the sawtooth, square, trapezoidal, and triangular waves (*see* SAWTOOTH WAVE, SQUARE WAVE, TRAPEZOIDAL WAVE, TRIANGULAR WAVE).

When energy is concentrated at many different frequencies, waveforms become exceedingly complicated.

WAVEGUIDE

A waveguide is a hollow metal tube that directs energy from one point to another. Two types of waveguides are the rectangular type shown in A of Fig. 1 and the cylindrical type shown in B of Fig. 1. The term *waveguide* has come to mean a hollow metal tube through usage, but it could, logically, apply to other transmission lines including solid or stranded wire conductors, parallel metal plates, coaxial line or cable, dielectric rod, or even optical fiber.

The transmission of electromagnetic waves along a waveguide is similar to their transmission through space. The energy transmitted in a waveguide is contained in the electromagnetic fields that travel down the waveguide, and the current flow in the guide walls provides a boundary for these electric and magnetic fields.

The rectangular hollow-metal waveguide is more widely used than the circular waveguide because it is difficult to control the plane of polarization and mode of operation in a circular waveguide. Moreover, it is difficult to join curved surfaces when a junction is required. However, circular waveguide is used in rotating joints because of its physical and electrical symmetry.

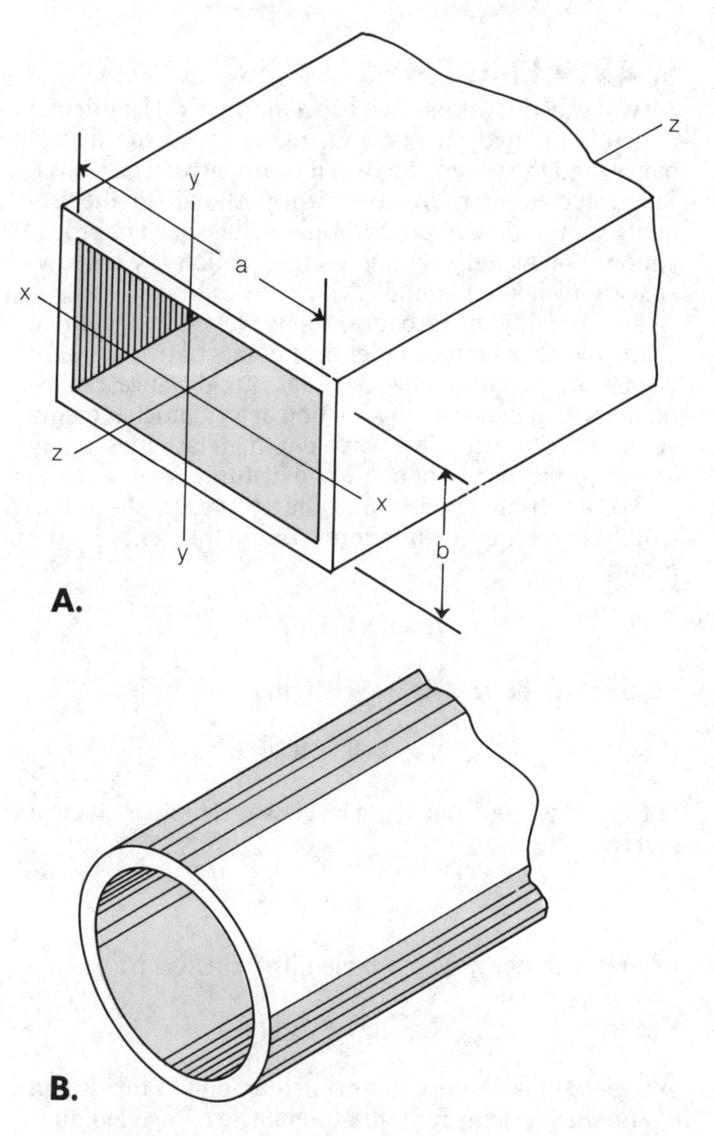

WAVEGUIDE: Fig. 1. A waveguide with commonly used rectangular crosssection (A), and a waveguide with less widely used circular crosssection (B).

A hollow waveguide has lower losses than either an open-wire line or a coaxial line in the frequency range where all are practical. An open-wire line has three kinds of loss: 1) radiation loss, 2) dielectric loss, and 3) copper loss. There is no radiation loss from either the waveguide or the coaxial line because both are perfectly shielded lines. The waveguide is a metal tube and the coaxial line has an outer braided conductor that also confines the magnetic and electric fields within them (*see* COAXIAL CABLE). However, a coaxial cable has significant dielectric and copper losses.

Because the waveguide is hollow and normally filled with air, it has no solid or beaded dielectric to cause dielectric losses. The dielectric loss of air is negligible at any frequency. Also, because the waveguide has no inner conductor to cause high copper losses like the coaxial line, its copper losses are far lower. Thus the total losses of a waveguide above the cutoff frequency are less than those of a coaxial cable of the same size operating at the same frequency.

The waveguide is simple in construction and more rugged than a coaxial cable. It has neither an inner conductor that could be displaced or broken nor the dielectric supports. The construction of a waveguide system is more like plumbing than wiring.

The minimum size of the waveguide that will transmit a specified frequency is proportional to the wavelength at that frequency. This proportionality depends on the shape of the waveguide and the way in which the electromagnetic fields are organized within the guide. In all cases there is a minimum or cutoff frequency that can be transmitted. The lowest cutoff frequency is determined by the inside dimensions shown in A of Fig. 1. The wavelength, λ, corresponding to the cutoff frequency is equal to twice the inside width of the guide or:

$$\lambda = 2a$$

However, higher frequencies can be transmitted. The width of the guide for these frequencies is greater than their corresponding free-space half wavelengths.

The dimension b is not critical for frequency, but it determines the arc-over voltage level of the waveguide. Therefore, the dimension *b* is larger for the transmission of high power than low power. In practice *b* is from 0.2 to 0.5 times the wavelength in air and *a* is about 0.7 times the wavelength in air.

Waveguides are rarely used at frequencies below about 2 GHz (15 centimeters) because the dimensions become excessively large. The cutoff frequency corresponds to a wavelength that is equal to twice the inside width (a) of the guide.

The radius bends in the guide must be greater than two wavelengths to avoid excessive attenuation and the cross section of the guide must be uniform around the bend. Dents or solder permitted to run inside the joints increase line attenuation, reduce the breakdown voltage and may cause standing waves.

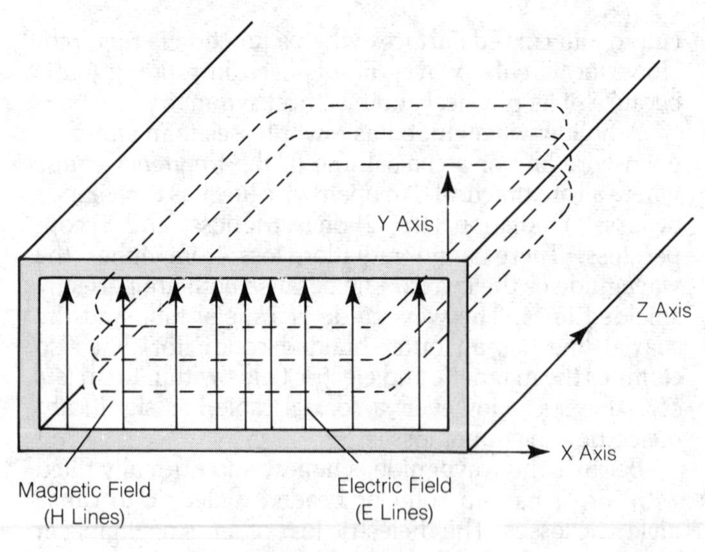

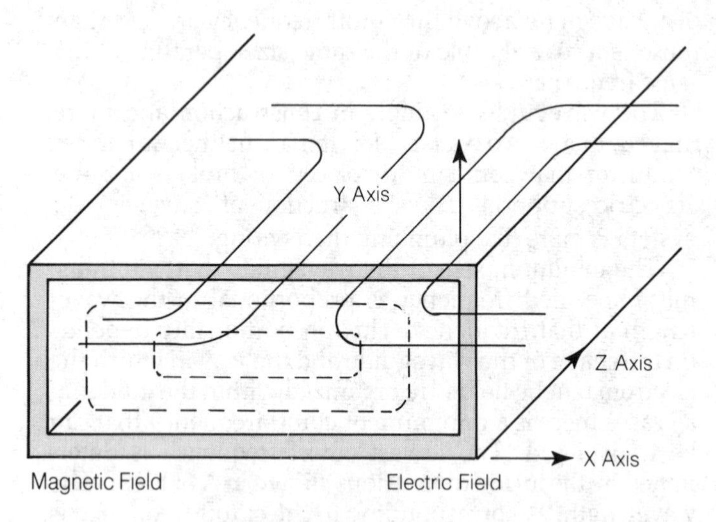

WAVEGUIDE: Fig. 2. The transverse electric (TE) mode of waveguide operation shown at (A) and the transverse magnetic (TM) mode of operation shown at (B).

Modes of Transmission. A system of letters and subscript numbers is used to describe waveguide nodes. Figure 2A shows the TE mode of operation of a rectangular waveguide. The letters TE indicate that the electric field (shown as parallel E lines) is in a transverse plane that contains the x and y axes. The E lines are parallel to the y axis and are perpendicular to the longitudinal Z axis. Similarly, the letters TM (Fig. 2B) indicates that the magnetic field (composed of closed loops) lies in transverse planes that contain the x and y axes and are wholly transverse to the guide axis.

In rectangular waveguides, a subscript system is used in which the first subscript number states the number of half-wave variations of the wide (or a) dimension of the guide (see Fig. 1), and the second number states the number of half wave variations of the same field in the or narrow (or b) dimension. The TE_{10} mode, for example, means that the electric field has a half-wave variation along the x axis, and none along the y axis. The TM_{11} mode means that the magnetic field has a one half-wave variation in both the wide and narrow dimensions. The mode with the lowest cutoff frequency for a given waveguide size is called the dominant mode for that waveguide. The dominant mode, TE_{10}, is most commonly used for rectangular waveguides. This mode is easily excited, is plane-polarized, and is easily matched to a radiator. Moreover, its cutoff frequency depends only on the waveguide width dimension, making system design easier.

Waveguide Coupling. The three principal ways to couple energy into or out of a waveguide are with the use of:

1. A loop of wire to cut or couple the H lines of the magnetic field, as in a transformer.
2. A probe placed parallel to the E lines of the electric field, to act as an antenna.
3. Slots or holes in the walls of the guide to link the external fields with the fields inside the guide.

WAVELENGTH

All wave disturbances have a specific physical length that depends on two factors: the frequency of the disturbance, and the speed at which it is propagated (*see* WAVE). The wavelength is inversely proportional to the frequency, and directly proportional to the speed of propagation. Wavelength is denoted in equations by the lowercase Greek letter lambda (λ).

In a periodic disturbance, the wavelength is defined as the distance between identical points of two adjacent waves. In electromagnetic fields, this distance can be millions of meters or a tiny fraction of a millimeter, or any value in between. The wavelength determines many aspects of the behavior of a wave disturbance.

For electromagnetic waves in free space, the wavelength, in feet, is given in terms of the frequency by the formula:

$$\lambda = 9.84 \times 10^8/f$$

where f is in hertz. The wavelength in meters is:

$$\lambda = 3.00 \times 10^8/f$$

For sound waves in air at sea level, wavelength is given in feet by:

$$\lambda = 1.10 \times 10^3/f$$

where f is in hertz. The wavelength in meters is:

$$\lambda = 3.35 \times 10^2/f$$

Wavelength is an important consideration in the design of antenna systems for radio frequencies. Wavelength is also important to the designers of optical apparatus and various acoustic devices. *See also* ELECTROMAGENTIC FIELD, ELECTROMAGNETIC SPECTRUM, FREQUENCY, SOUND, VELOCITY FACTOR.

WAVEMETER

See ABSORPTION WAVEMETER, COAXIAL WAVEMETER.

WAVE POLARIZATION

See POLARIZATION.

WAVESHAPE

See WAVEFORM.

WAVE THEORY OF LIGHT

See LIGHT.

WAVE TRAP

See TRAP, WAVE-TRAP FREQUENCY METER.

WAVE-TRAP FREQUENCY METER

A trap can be used to measure the frequency of a radio signal with a circuit called a wave trap or wave-trap frequency meter. The meter consists of a tunable trap (parallel-resonant circuit) in series with the signal path, and an indicating device such as a meter (see illustration).

The tunable trap is adjusted until a dip is observed in the meter reading. This condition occurs when the trap is set for resonance at the frequency of the applied signal. The frequency is then read from a calibrated scale.

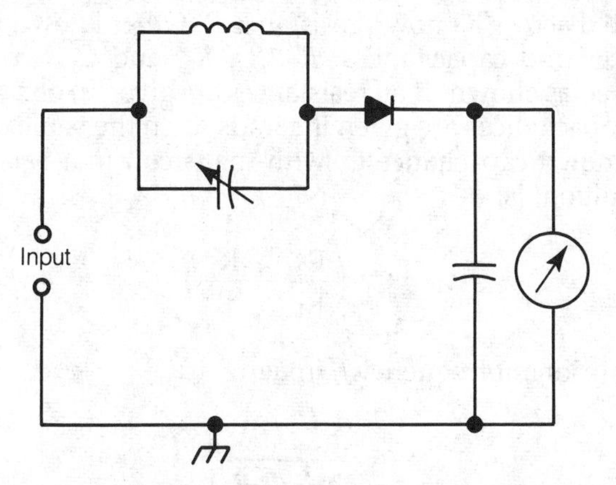

WAVE-TRAP FREQUENCY METER: Resonant frequency is adjusted in this meter by varying the capacitor.

WEBER

The weber is a unit of magnetic flux, representing one line of flux in the meter-kilogram-second (MKS) or standard international (SI) system of units. The weber is equivalent to a volt-second.

Magnetic flux is sometimes expressed in units called maxwells. The maxwell is equal to 10^{-8} weber. *See also* MAGNETIC FIELD, MAXWELL.

WESTON STANDARD CELL

A Weston standard cell is an electrochemical cell that provides a constant 1.0183 V at 20°C. The electrolyte is a solution of cadmium sulfate liquid. The positive electrode is a paste of mercurous sulfate and the negative electrode is fabricated from a combination of mercury and cadmium.

The Weston standard cell is generally housed in a pair of glass enclosures connected by a tube. The cell must be kept upright for proper operation. This type of cell is most often used as a voltage standard. The illustration is a cross-sectional diagram of a Weston standard cell. *See also* BATTERY.

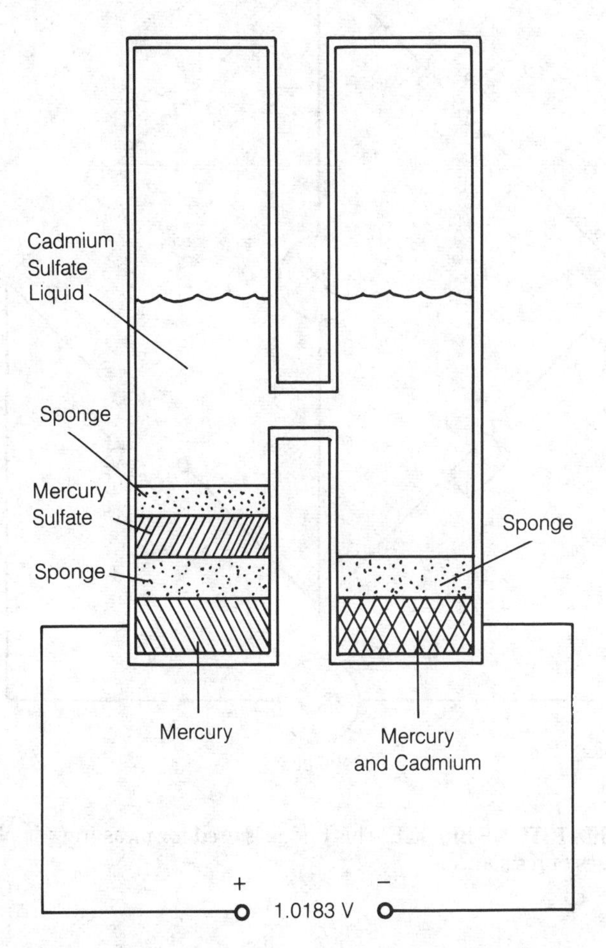

WESTON STANDARD CELL: A diagram of a Weston standard cell.

WHEATSTONE BRIDGE

A Wheatstone bridge is a circuit for measuring unknown resistances. The Wheatstone bridge operates by comparing the ratio of the unknown resistance to three other known resistances.

The illustration is a schematic diagram of a Wheatstone bridge. The indicator may be a galvanometer if a direct-current supply is used. If all the resistances are nonreactive, an audio-frequency generator can be used in conjunction with a headphone. The unknown resistance is *R*. Two fixed resistors, R1 and R2, and an adjustable resistor, R3, are used for comparison.

The value of R3 is adjusted until the galvanometer reads zero, or until a null is heard in the headset. This indicates that the ratio $R:R_3$ is equal to the ratio $R_1:R_2$. The unknown resistance R is given by the formula:

$$R = \frac{R_3R_1}{R_2}$$

The Wheatstone bridge is similar to the ratio-arm bridge. *See also* RATIO-ARM BRIDGE, RESISTANCE MEASUREMENT.

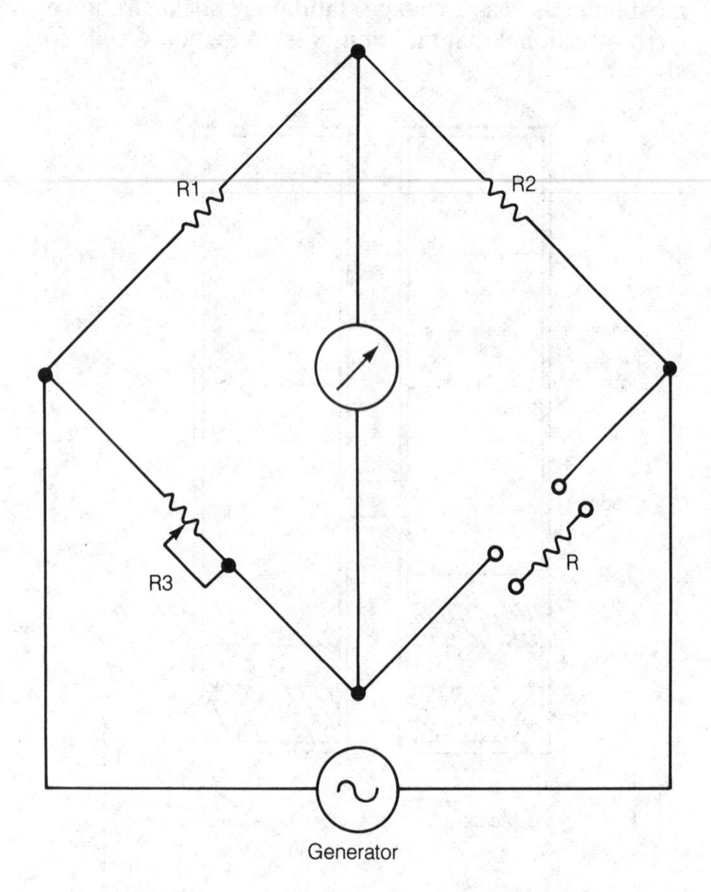

WHEATSTONE BRIDGE: This bridge is used for measuring small values of resistance.

WHITE NOISE

Broadband noise, also known as white noise, is so named because white light contains energy at all visible wavelengths. By analogy, wideband electromagnetic energy (wideband noise) is therefore, in a sense, white.

White noise at audio frequencies consists of a more or less uniform distribution of energy at wavelengths from 20 Hz to 20 kHz. At radio frequencies, the term white noise is used if a band of frequencies contains noise of essentially equal amplitude at all points. *See also* NOISE.

WHITE TRANSMISSION

White transmission is a form of amplitude-modulated facsimile signal (*see* FACSIMILE) in which the greatest copy density, or darkest shade, corresponds to the minimum amplitude of the signal. White transmission is the opposite of black transmission, in which the darkest shade corresponds to the maximum signal amplitude (*see* BLACK TRANSMISSION). White transmission may be considered, in a sense, right-side up.

In a frequency-modulated facsimile system, white transmission means that the darkest copy corresponds to the highest transmitted frequency.

WIDE-AREA TELEPHONE SERVICE

Wide-area telephone service (WATS) is a special form of long-distance telephone service. Most long-distance calls are charged on a timed basis, such as 35 cents per minute. A WATS line allows a subscriber to obtain long-distance service for a flat monthly charge. In some cases, surcharges are applied for use of the line for more than a specified amount of time in a month.

A WATS line may be either incoming or outgoing. The most familiar example of an incoming WATS line is an 800 line, so named because the area code is 800. Outgoing WATS lines are used by many businesses that engage in extensive long-distance telephone calling.

WIDEBAND MODULATION

See BROADBAND MODULATION, SPREAD-SPECTRUM TECHNIQUES.

WIEN BRIDGE

A Wien bridge is a resistance-capacitance circuit used for measuring unknown values of capacitance. The Wien bridge can also be used as a resonant circuit.

The configuration of a Wien bridge is shown in the diagram on p. 917. The unknown capacitance C is connected across a known resistance R. Other known resistances and capacitances, R_1, R_2, R_3, and C_1, are connected as shown. If all resistances are given in ohms and all capacitances are given in farads, then the value of the unknown capacitance C, with the circuit in a balanced condition, is:

$$C = C_1 \left(\frac{R_2}{R_1} - \frac{R_3}{R} \right)$$

The resonant frequency f, in hertz, is:

$$f = \frac{1}{2\pi \sqrt{C_1R_3}}$$

The Wien bridge requires an audio-frequency generator, and an indicating device such as a meter or headset. *See also* OSCILLATOR CIRCUITS.

WIEN-BRIDGE OSCILLATOR

See OSCILLATOR CIRCUITS.

WIRE

See CONDUCTOR.

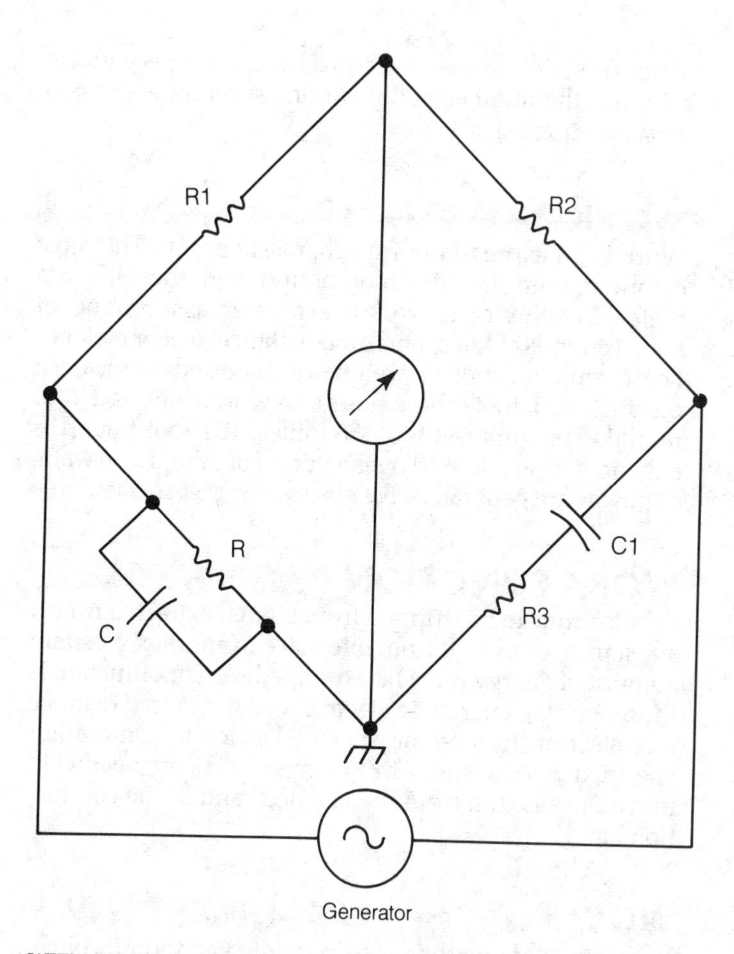

WIEN BRIDGE: This bridge is used for measuring capacitance.

WIRE GAUGE

See AMERICAN WIRE GAUGE, BIRMINGHAM WIRE GAUGE, BRITISH STANDARD WIRE GAUGE.

WIREWOUND RESISTOR

See RESISTOR.

WIRE WRAPPING

Wire wrapping is a technique for terminating fine insulated wire in point-to-point connections without the need for soldering. The bare ends of the wire are stretch-wrapped around square posts (typically 25 mils on a side) to form strong mechanical and electrical connections. The wire is wrapped under tension as high as 130,000 pounds per square inch, and up to three connections can be made per post.

The square posts for wire wrapping are securely driven into circuit boards or backplanes for connection with components. Wire wrapping can be done with special hand tools or by semiautomatic or automatic methods. The product usually does not have to be modified to accommodate method changes. *See also* BACKPLANE, CIRCUIT BOARD OR CARD.

Wire wrapping is used for custom-wiring equipment, such as that used in telecommunications. It also permits the rapid prototyping of new equipment designs for evaluation or shipment by saving the time required to fabricate new printed circuit boards.

Wire wrapping is also used where wiring is subject to change or update. It permits quick field changes; existing wiring can be unwrapped and the posts can be reused for rewiring in only minutes.

WIRING DIAGRAM

A wiring diagram is a drawing that shows all of the electrical wiring or cabling and point-to-point connections between components in electrical and electronic products and systems. A form of block diagram, the wiring diagram does not use schematic symbols.

WOOFER

A woofer is a speaker designed to respond to low audio and subaudible frequencies. A woofer resembles an ordinary speaker (*see* SPEAKER), except that it is larger. The large size is necessary because of the length of sound waves at low frequencies.

Woofers are used in high-fidelity speaker systems to enhance the reproduction of bass sound (*see* BASS RESPONSE). Many high-fidelity speakers contain a woofer, a midrange speaker, and a tweeter in a single enclosure. *See also* MIDRANGE, TWEETER.

WORD

In a digital computer, a word is a group of bits or characters with a specified length or number of units. A word conveys a specific amount of information, and occupies a specific amount of space in memory. Words can be used as data, instructions, designators, or numerical values. *See also* BIT, BYTE, CHARACTER.

In teletype and telegraph communications, a word consists of five characters plus one space. To obtain the total word count in a message, the number of characters, symbols, and spaces is counted. (In teletype, symbols and punctuation marks count as single characters, but in Morse code, they count as two characters each.) The result is divided by 6. Speed of transmission may then be expressed in terms of the number of words sent per minute. *See also* WORDS PER MINUTE.

WORD PROCESSING

Computers and personal computers are used in writing and editing letters and manuscripts. This is known as word processing. Some computers are designed specifically for this purpose. These computers are called word processors. They can be completely self-contained units, with a video terminal, keyboard, disk drive, and printer.

Some Common Word-Processing Capabilities. Different writing styles require different word-processing functions. For example, letters require a different set of functions than magazine articles. Tables and charts might call for another set of functions. Any of many specialized software packages can be used with a single word processor for various kinds of writing. A few of the most common word-processing functions are listed below.

- Left-margin adjustment and justification: all rows are aligned along the left side of the page with a margin width the operator can select.
- Right-margin adjustment and justification: If desired, all rows can be aligned along the right side of the page. The margin width is adjustable.
- Insertion and deletion of characters, words, lines, and paragraphs: Text can be added or removed and the continuity of the manuscript will be readjusted.
- Exchange of characters, words, lines, and paragraphs: Text can be altered as necessary.
- Row and column alignment: Used for making tables and charts.
- Word wrap around: Automatically moves a word to next line in the text if the word runs past the right-hand margin limit.
- Double spacing: Inserts an extra space between each line of text.

Data Storage and Printing. Most word processors allow storage of information on magnetic disks. The floppy disk is by far the most commonly used (*see* DISK DRIVE). Information is easily erased or rewritten onto the disks.

Every word processor has a printer. There are many printers available for use with word processors. The dot-matrix printer operates at high speed, and is useful when a lot of text must be printed in a short time. Daisy-wheel printers provide letter-press quality but run at a somewhat slower speed than dot-matrix printers (*see* PRINTER).

An important feature of a word processor is the memory capacity for each file of text. All computers have a capacity for each file of text, and a certain maximum memory limit. In general, each page of double-spaced text, containing 220 words, has 1400 to 1500 bytes. The number of pages that can be stored in a file is about 70 percent of the number of kilobytes of memory.

WORDS PER MINUTE

The speed of transmission of a digital code is measured in words per minute (WPM). A word generally consists of five characters plus one space (*see* WORD).

For teleprinter codes, speeds range from 60 WPM to hundreds or even thousands of words per minute. The original Baudot speed was 60 WPM, and this speed is still in widespread use (*see* BAUDOT CODE). The standard ASCII speed is 110 WPM, and the highest commonly used ASCII speed is 19,200 WPM (*see* ASCII). Sometimes the speed is given in bauds rather than words per minute. For Baudot transmissions, the speeds in words per minute is about 33 percent greater than the baud rate. For ASCII transmissions, the speed in words per minute is the same as the baud rate (*see* BAUD RATE).

The speed of an International Morse Code transmission, in words per minute, is about 1.2 times the baud rate, where one baud is the length of a single dot or the space between dots and dashes in a character. If a string of dots is sent, the speed in words per minute is equal to 2.4 times the number of dots in one second (*see* INTERNATIONAL MORSE CODE).

WORK

Work is an expression of mechanical energy. The most common units are the foot pound and the kilogram meter. In either case, work is expressed as a product of force (pounds or kilograms) and distance (feet or meters). For example, when a weight of 1 pound is raised a distance of 1 foot, the amount of work done is 1 foot pound. This represents 1.356 joules. If 1 foot pound is expended every second over a period of time, 1.356 watts of power are generated. *See also* ENERGY, ENERGY UNITS.

WORK FUNCTION

For electrons to be stripped from atoms, extracted from a substance, or moved from one place to another, a certain amount of energy must be expended. A work function is the amount of energy, in electron volts, required to move one electron from some specified place to some other specified place. Usually this means the energy needed to move the electron from the valence band to the conduction band.

WOW

Wow is a slow warbling, or periodic variation in pitch, that may occur in audio recording or reproduction. Wow may also be called flutter, although the term *flutter* is more often used to describe a rapid change in the pitch of reproduced sound (*see* FLUTTER).

WRITING GUN

A writing gun is an electron gun in a storage cathode-ray tube. The writing gun is basically an ordinary electron gun which emits a narrow beam of electrons that strikes the phosphor surface at a certain point, resulting in a bright spot (*see* CATHODE-RAY TUBE).

In a storage oscilloscope the writing-gun beam has constant intensity, and scans from left to right at a constant speed. The vertical position is varied by the incoming signal (*see* OSCILLOSCOPE, STORAGE OSCILLOSCOPE). The intensity of the writing-gun beam may be modulated, and the beam moved through a standard raster, for storage of television pictures. The display is observed using another electron gun, known as the flood gun (*see* FLOOD GUN).

WWV/WWVH

The National Bureau of Standards maintains two shortwave time-and-frequency broadcasting stations. Station WWV is located at Fort Collins, Colorado, and WWVH is located on the island of Kauai, Hawaii.

Both stations transmit signals on frequencies of 2.5, 5, 10, 15, and 20 MHz. Station WWV also broadcasts on 25 MHz. The frequencies are accurate to a tiny fraction of 1

Hz. Standard amplitude modulation is used.

Time broadcasts are sent continuously, 24 hours a day, with brief exceptions. Voice announcements indicate the hour and minute. Audible clicks indicate seconds. Standard audio-frequency tones are transmitted at 440, 500, and 600 Hz according to a prescribed format.

The time and frequency broadcasts of WWV and WWVH are generally accurate to within 1 part in 10^{12}, and never vary up or down by more than 1 part in 10^{11}.

In addition to time-and-frequency information, WWV and WWVH transmit data about weather and radio-propagation conditions. Propagation conditions are expressed according to a two-character code consisting of a phonetic designator and a number. Geophysical information, if of great significance, is also sent.

In addition to WWV and WWVH, a low-frequency station, WWVB, is operated at 60 kHz from Fort Collins, Colorado. Time information is sent using a carrier-shift code. The signals from WWVB can be used directly with a phase-locked loop to obtain an extremely accurate frequency standard.

WYE CONNECTION

In a three-phase alternating-current system, a wye connection is a method of interconnecting the windings of a transformer to a common point. The configuration is so named because, in a schematic diagram, it appears like a capital letter Y.

The diagram illustrates a wye transformer connection. The phases at points X1, X2, and X3 are 120 electrical degrees (⅓ cycle) out of phase with respect to each other (*see* THREE-PHASE ALTERNATING CURRENT). The center point is neutral. Wye configurations are often used in polyphase rectifiers (*see* POLYPHASE RECTIFIER).

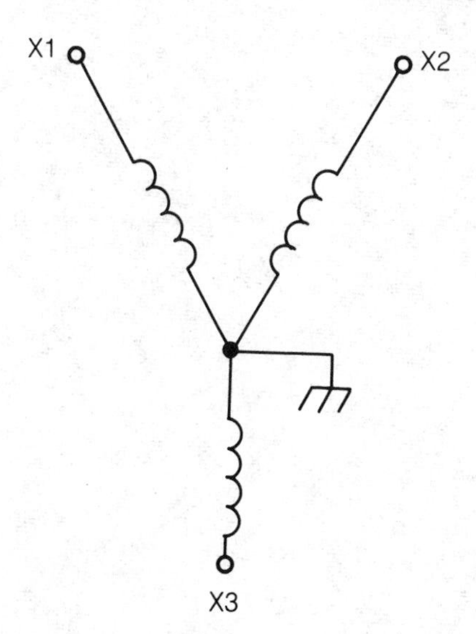

WYE CONNECTION: In the wye transformer currents at points X1, X2 and X3 differ by ⅓ cycle.

WYE MATCH

See DELTA MATCH.

X AXIS

In a two-dimensional Cartesian plane, the variables are often labeled *x* and *y*. The x axis is usually the horizontal axis, corresponding to the independent variable. The independent-variable axis is also called the abscissa. *See* CARTESIAN COORDINATES, INDEPENDENT VARIABLE, Y AXIS.

X BAND

The range of frequencies from 8.0 to 12 GHz, corresponding to wavelengths of 2.5 and 3.8 cm, is known as the X band. This band is used for radar, line-of-sight communications, and satellite communications. The X band falls in the superhigh range, which is considered to be in the microwave part of the electromagnetic spectrum (*see* SUPER HIGH FREQUENCY).

XENON

Xenon is an element with an atomic number of 54 and an atomic weight of 131. Xenon is a gas at room temperature. Because its outer electron shell is completely filled, xenon is an inert, or noble, gas.

Xenon is used in certain types of flash tubes, and in some voltage-regulator tubes. *See also* XENON FLASHTUBE.

XENON FLASHTUBE

A xenon flashtube is a device that generates an extremely brilliant, white, incoherent light. Two electrodes are enclosed in a glass envelope that has been evacuated and then filled with xenon gas (*see* XENON). Xenon flashtubes are used to energize certain types of lasers (*see* LASER).

The light from a xenon tube originates from two sources: the hot electrodes, and the ionized gas. Although the xenon tube emits energy at all visible wavelengths, the greatest amount of energy occurs at about 500 nanometers (nm) and 800 nm. These wavelengths are approximately in the blue and red ranges. The two wavelengths combine to produce a whitish light.

Xenon can be used as a continuous light source, in the same way as neon gas (*see* NEON LAMP).

X RAY

X rays are electromagnetic radiation at wavelengths shorter than ultraviolet, but longer than gamma rays. The precise cutoff between ultraviolet and X rays is not universally agreed upon, but it can be considered to be about 4 nanometers (nm). The lower wavelength limit is around 0.01 nm. X rays propagate through a vacuum in the same way, and at the same speed, as other forms of electromagnetic radiation, including visible light. The photons of X rays contain considerable energy, however, which gives X rays their penetrating power.

X rays are invisible but they cause some phosphor materials to glow. X rays also expose camera film. This, in conjunction with the ability of X rays to penetrate soft

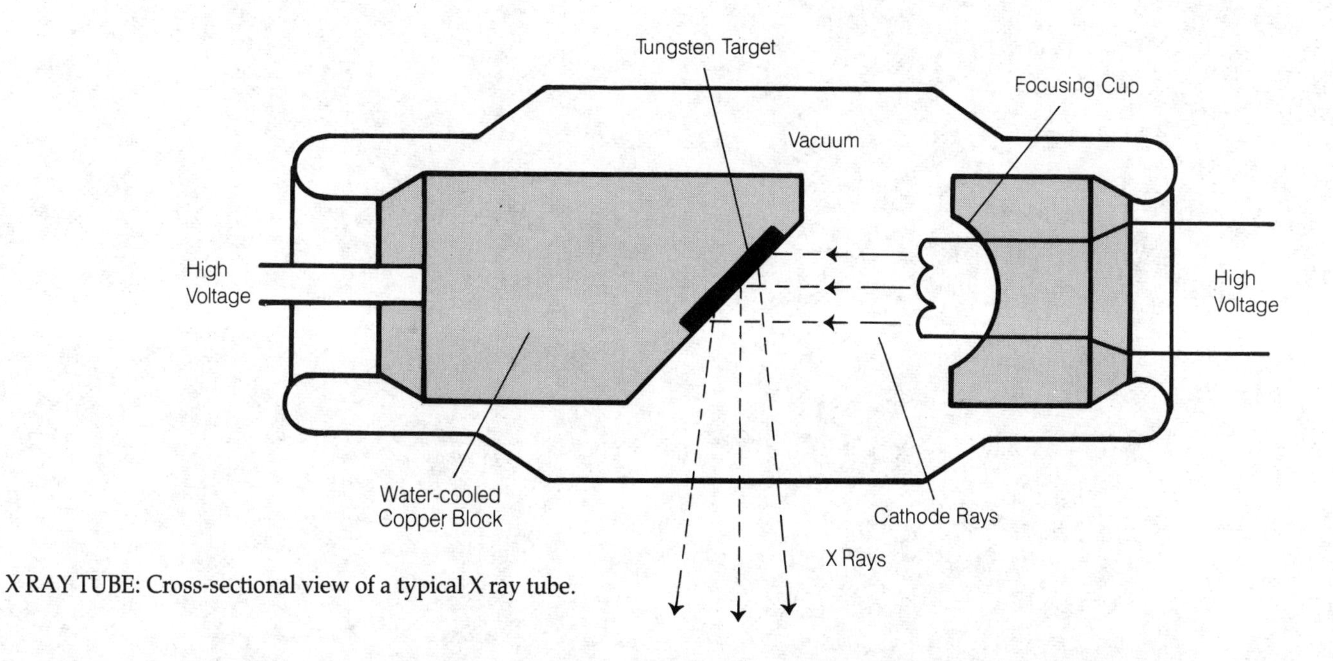

X RAY TUBE: Cross-sectional view of a typical X ray tube.

tissues of the body, makes them so useful in medicine. *See* X RAY TUBE.

Overexposure to X rays can be dangerous, and if too much radiation is received in a short time, illness can occur (*see* DOSIMETRY, RADIOACTIVITY). Excessive exposure to X rays will kill most forms of life. The atmosphere blocks almost all of the X radiation from the sun and other stars.

X RAY TUBE

When high-speed electrons strike a massive barrier such as lead or tungsten, X rays are generated (*see* X RAY) with an X ray tube.

The illustration on p. 921 is a simplified drawing of an X ray tube. A glass envelope is evacuated, and contains a heated cathode and an anode. The anode is usually made from tungsten so that it will resist destruction by heating as energetic electrons bombard it. The anode surface is turned at an angle. The gap between the cathode and the anode is a few inches.

When a positive voltage is supplied to the anode, electrons are emitted by the hot cathode, just as in an ordinary vacuum tube. A high-voltage power supply causes the electrons to accelerate to high speed. The electrons strike the anode at an angle of approximately 45 degrees. The X rays are thrown off in various directions, but mostly from the side of the tube, at approximately 90 degrees with respect to the electron beam.

YAGI ANTENNA

See ANTENNA DIRECTORY.

Y AXIS

In a two-dimensional Cartesian plane, the variables are often labeled x and y. The y axis is usually the vertical axis, corresponding to the dependent variable. The dependent-variable axis is also called the ordinate. *See* CARTESIAN COORDINATES, DEPENDENT VARIABLE, X AXIS.

Y CONNECTION

See WYE CONNECTION.

Y MATCH

See DELTA MATCH.

YOKE

In a cathode-ray tube, the electron beam may be deflected by charged plates, or coils. A complete set of magnetic deflection coils is called the yoke of the tube (*see* CATHODE-RAY TUBE, TELEVISION).

In an electric motor or generator, a ring-shaped piece of ferromagnetic material is used to hold the pole pieces together, and also to provide magnetic coupling between the poles. This assembly is known as a yoke (*see* GENERATOR, MOTOR).

ZENER DIODE

The Zener diode or reference diode is a silicon PN junction that provides a specified reverse reference voltage when it is operated into the reverse bias avalanche breakdown region. The diode exhibits a well-defined sharp reverse breakdown at less than about 6 V. A large number of electrons within the depletion region break the bonds with their atoms causing a large reverse current to flow. This is ionization by an electric field.

The figure shows the symbol for a Zener diode and a typical characteristic curve. The forward characteristic is similar to that of the normal high-voltage rectifier. In the reverse direction, the leakage current is very low, but when the breakdown point is reached, the reverse current increases very rapidly for only a slight increase of voltage. The curve shows a pronounced knee or sharp bend.

By biasing the diode permanently in the breakdown region, the Zener diode can be used as a stable voltage reference because the voltage across the device will remain essentially constant for quite large variations of current through the device. Zener diodes are widely used as general-purpose regulators and for clipping or bypassing voltages above a specified level. Specialized Zener diodes are used for transient suppression.

Commercial Zener diodes have nominal Zener voltages of 1.8 to 200 V and power ratings from 250 milliwatt (mW) to 50 watt (W). They are packaged in a variety of glass, metal, and plastic cases. Included are hermetically sealed glass axial-leaded DO-35 and DO-41 cases and leadless packages. The diodes are also in surface mount small outline transistor SOT-236 (formerly SOT-23) plastic cases. Metal packages include the leaded DO-13 and stud-mount DO-203 and DO-5 cases. Zener overvoltage transient suppressors have ratings from 5 to 200 V and can handle up to 5 W steady state (1500 W peak) power. *See* CIRCUIT PROTECTION.

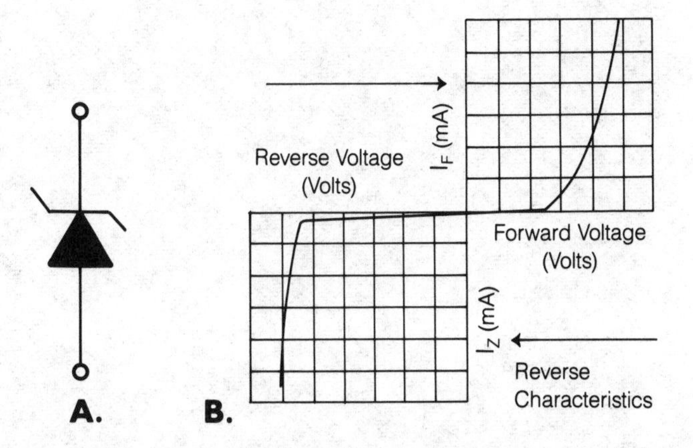

ZENER DIODE: Symbol for Zener diode (A) and the characteristic Zener curve with a sharp knee (B).

ZERO BEAT

When two signals are mixed, beat notes, or heterodynes, are generated at the sum and difference frequencies (*see* BEAT, HETERODYNE). As the frequencies of two signals are brought closer and closer together, the difference frequency decreases. When the two signals have the same frequency, the difference signal disappears altogether. This condition is called zero beat.

In a receiver product detector, a carrier produces an audible beat note. As the receiver is tuned, the frequency of the audio note varies. When the receiver is tuned closer to the carrier frequency, the beat note falls in pitch and then disappears, indicating zero beat between the carrier and the local oscillator.

In communications practice, two stations often zero beat each other. This means that they both transmit on exactly the same frequency, reducing the amount of spectrum space used.

ZERO BIAS

When the control grid of a vacuum tube is at the same potential as the cathode, the tube is said to be operating at zero bias. In a field-effect transistor (FET), zero bias occurs when the gate and the source are provided with the same voltage. Similarly, in a bipolar transistor, zero bias means that the base is at the same potential as the emitter.

In vacuum tubes and depletion-mode FETs, zero bias usually results in fairly large plate or drain current. In enhancement-mode FETs and in bipolar transistors, little or no drain or collector current normally flows with zero bias.

Index

This index contains article titles, cross references, and illustration and table titles. It also contains many terms that are not article titles. **Boldface** page numbers indicate that a term is represented by an article on that page. *Italicized* page numbers indicate that the term is represented by an illustration and/or a table on that page. For terms that are not represented as articles, page numbers are listed for occurrences of that term that are of special significance.

A

C

E

I

J

K

M

N

O

P

Q

R